RSMeans
Quarterly Updates
700 Longwater Drive
Norwell, MA 02061
USA

Tips for Customizing Your RSMeans Cost Data By Location

The cost data found throughout the book reflects national averages.

Customize your cost data for over 930 locations throughout the U.S. and Canada.

☑ To adjust costs to your region, use the *Location Factors* found in the back of the book.

☑ To compare costs from region-to-region, use the *City Cost Indexes* found in the back of the book.

☑ For additional guidance, refer to the *How RSMeans Data Works* section towards the front of the book.

RSMeans
Reed Construction Data®

Metric Conversion Formulas

Length

cm = 0.3937 in.	in. = 2.5400 cm
meter = 3.2808 ft.	ft. = 0.3048 m
meter = 1.0936 yd.	yd. = 0.9144 m
km = 0.6214 mile	mile = 1.6093 km

Weight

gram = 15.4324 grains	grain = 0.0648 g
gram = 0.0353 oz.	oz. = 28.3495 g
kg = 2.2046 lbs.	lb. = 0.4536 kg

Area

sq. cm = 0.1550 sq. in.	sq.in. = 6.4516 sq. cm
sq. m = 10.7639 sq. ft.	sq. ft. = 0.0929 sq. m
sq. m = 1.1960 sq. yd.	sq. yd. = 0.8361 sq. m
hectare = 2.4710 acres	acre = 0.4047 hectare
sq. km = 0.3861 sq. mile	sq. mile = 2.5900 sq. km

Weight, cont.

kg = 0.0011 ton (short)	ton (short) = 907.1848 kg
ton (met.) = 1.1023 ton (short)	ton (short) = 0.9072 ton (met.)
ton (met.) = 0.9842 ton (large)	ton (long) = 1.0160 ton (met.)

Common Units of Measure

Abbreviation	Measurement	Formula
SF	Square feet	Length (in feet) x Width (in feet)
SY	Square yards	Square feet / 9
CF	Cubic feet	Length (in feet) x Width (in feet) x Depth (in feet)
CY	Cubic yards	Cubic feet / 27
BF	Board foot	Length (in inches) x width (in inches) x thickness (in inches)/144
MBF	Thousand board foot	Board foot / 1,000
LF	Linear feet	12 inches of the item
SFCA	Square foot of contact area	A square foot of concrete formwork
SQ	Square	100 square feet
Ton	Ton	2,000 pounds

$177.95 per copy (in United States)
Price is subject to change without prior notice.

0012

The *2012 Building Construction Cost Data*, 70th Annual Edition, and all of the RSMeans annual cost data books are dedicated to the memory of our respected friend and colleague, Hsiao-Cheng (John) Chiang, PE.

RSMeans
Building
Construction
Cost Data

70TH ANNUAL EDITION

2012

RSMeans
A division of Reed Construction Data
Construction Publishers & Consultants
700 Longwater Drive
Norwell, MA 02061
USA
1-800-334-3509
www.rsmeans.com

Copyright 2011 by RSMeans
A division of Reed Construction Data
All rights reserved.

Printed in the United States of America
ISSN 0068-3531
ISBN 978-1-936335-29-9

Senior Editor
Phillip R. Waier, PE

Contributing Editors
Christopher Babbitt
Ted Baker
Barbara Balboni
Adrian C. Charest, PE
Gary W. Christensen
David G. Drain, PE
Cheryl Elsmore
Robert J. Kuchta
Robert C. McNichols
Melville J. Mossman, PE
Jeannene D. Murphy
Marilyn Phelan, AIA
Stephen C. Plotner
Eugene R. Spencer

Cover Design
Wendy McBay
Paul Robertson
Wayne Engebretson

President & CEO
Iain Melville

Vice President, Product & Development
Richard Remington

Vice President, Finance & Operations
David C. Walsh

Vice President, Data Aquisition & Fulfillment
Marco Piovesan

Chief Customer Officer
Lisa Fiondella

Senior Director, Product Management
Daimon Bridge

Vice President, Marketing & Business
Steven Ritchie

Engineering Director
Bob Mewis, CCC

Product Manager
Andrea Sillah

Product Marketing Manager
Ryan Cummings

Manager, Product Management
Jennifer Johnson

Production Manager
Michael Kokernak

Production Coordinator
Jill Goodman

Production Supervisor
Paula Reale-Camelio

Production
Jonathan Forgit
Mary Lou Geary
Debbie Panarelli
Sheryl A. Rose
Caroline Sullivan

Technical Support
Christopher Anderson
Tom Dion
Gary L. Hoitt
Genevieve Medeiros
Sharon Proulx
Kathryn S. Rodriguez

This book is printed using soy-based printing ink.
This book is recyclable.

Related RSMeans Products and Services

This book is aimed primarily at commercial and industrial projects or large multi-family housing projects costing $3,500,000 and up. For civil engineering structures such as bridges, dams, highways, or the like, please refer to *RSMeans Heavy Construction Cost Data*.

The engineers at RSMeans suggest the following products and services as companion information resources to *RSMeans Building Construction Cost Data*:

Construction Cost Data Books
Assemblies Cost Data 2012
Square Foot Costs 2012

Reference Books
Estimating Building Costs
RSMeans Estimating Handbook
Green Building: Project Planning & Estimating
How to Estimate with RSMeans Data
Plan Reading & Material Takeoff
Project Scheduling & Management for Construction

Seminars and In-House Training
Unit Price Estimating
Means CostWorks® Training
RSMeans Data for Job Order Contracting (JOC)
Practical Project Management for Construction Professionals
Scheduling with MSProject for Construction Professionals
Mechanical & Electrical Estimating

RSMeans on the Internet
Visit RSMeans at **www.rsmeans.com** for a one-stop portal for the most reliable and current resources available on the market. Here you can learn about our more than 20 annual cost data books, also available on CD and online versions, as well as about RSMeans seminars.

RSMeans Electronic Data
Get the information found in RSMeans cost books electronically on CD or on the Web at **MeansCostWorks.com**.

RSMeans Business Solutions
Cost engineers and analysts conduct facility life cycle and benchmark research studies and predictive cost modeling, as well as offer consultation services for conceptual estimating, real property management, and job order contracting. Research studies are designed with the application of proprietary cost and project data from Reed Construction Data and RSMeans' extensive North American databases. Analysts offer building product manufacturers qualitative and quantitative market research, as well as time/motion studies for new products, and Web-based dashboards for market opportunity analysis. Clients are from both the public and private sectors, including federal, state, and municipal agencies; corporations; institutions; construction management firms; hospitals; and associations.

RSMeans for Job Order Contracting (JOC)
Best practice job order contracting (JOC) services support the owner in the development of their program. RSMeans develops all required contracting documents, works with stakeholders, and develops qualified and experienced contractor lists. RSMeans JOCWorks™ software is the best of class in the functional tools that meet JOC specific requirements. RSMeans data is organized to meet JOC cost estimating requirements.

Construction Costs for Software Applications
Over 25 unit price and assemblies cost databases are available through a number of leading estimating and facilities management software providers (listed below). For more information see the "Other RSMeans Products" pages at the back of this publication.

RSMeansData™ is also available to federal, state, and local government agencies as multi-year, multi-seat licenses.

- 4Clicks-Solutions, LLC
- ArenaSoft Estimating
- Assetworks
- Beck Technology
- BSD – Building Systems Design, Inc.
- CMS – Construction Management Software
- Corecon Technologies, Inc.
- CorVet Systems
- Hard Dollar Corporation
- Maxwell Systems
- MC^2
- Parsons Corporation
- ProEst
- Sage Timberline Office
- UDA Technologies, Inc.
- US Cost, Inc.
- VFA – Vanderweil Facility Advisers
- WinEstimator, Inc.
- R & K Solutions
- Tririga

Table of Contents

Foreword

Our Mission

Since 1942, RSMeans has been actively engaged in construction cost publishing and consulting throughout North America.

Today, more than 65 years after RSMeans began, our primary objective remains the same: to provide you, the construction and facilities professional, with the most current and comprehensive construction cost data possible.

Whether you are a contractor, owner, architect, engineer, facilities manager, or anyone else who needs a reliable construction cost estimate, you'll find this publication to be a highly useful and necessary tool.

With the constant flow of new construction methods and materials today, it's difficult to find the time to look at and evaluate all the different construction cost possibilities. In addition, because labor and material costs keep changing, last year's cost information is not a reliable basis for today's estimate or budget.

That's why so many construction professionals turn to RSMeans. We keep track of the costs for you, along with a wide range of other key information, from city cost indexes . . . to productivity rates . . . to crew composition . . . to contractor's overhead and profit rates.

RSMeans performs these functions by collecting data from all facets of the industry and organizing it in a format that is instantly accessible to you. From the preliminary budget to the detailed unit price estimate, you'll find the data in this book useful for all phases of construction cost determination.

The Staff, the Organization, and Our Services

When you purchase one of RSMeans' publications, you are, in effect, hiring the services of a full-time staff of construction and engineering professionals.

Our thoroughly experienced and highly qualified staff works daily at collecting, analyzing, and disseminating comprehensive cost information for your needs. These staff members have years of practical construction experience and engineering training prior to joining the firm. As a result, you can count on them not only for accurate cost figures, but also for additional background reference information that will help you create a realistic estimate.

The RSMeans organization is always prepared to help you solve construction problems through its variety of data solutions, including online, CD, and print book formats, as well as cost estimating expertise available via our business solutions, training, and seminars.

Besides a full array of construction cost estimating books, RSMeans also publishes a number of other reference works for the construction industry. Subjects include construction estimating and project and business management, special topics such as green building and job order contracting, and a library of facility management references.

In addition, you can access all of our construction cost data electronically in convenient CD format or on the Web. Visit **MeansCostWorks.com** for more information on our 24/7 online cost data.

What's more, you can increase your knowledge and improve your construction estimating and management performance with an RSMeans construction seminar or in-house training program. These two-day seminar programs offer unparalleled opportunities for everyone in your organization to become updated on a wide variety of construction-related issues.

RSMeans is also a worldwide provider of construction cost management and analysis services for commercial and government owners.

In short, RSMeans can provide you with the tools and expertise for constructing accurate and dependable construction estimates and budgets in a variety of ways.

Robert Snow Means Established a Tradition of Quality That Continues Today

Robert Snow Means spent years building RSMeans, making certain he always delivered a quality product.

Today, at RSMeans, we do more than talk about the quality of our data and the usefulness of our books. We stand behind all of our data, from historical cost indexes to construction materials and techniques to current costs.

If you have any questions about our products or services, please call us toll-free at 1-800-334-3509. Our customer service representatives will be happy to assist you. You can also visit our Web site at **www.rsmeans.com**

How the Book Is Built: An Overview

The Construction Specifications Institute (CSI) and Construction Specifications Canada (CSC) have produced the 2010 edition of MasterFormat™, a system of titles and numbers used extensively to organize construction information.

All unit price data in the RSMeans cost data books is now arranged in the 50-division MasterFormat™ 2010 system.

A Powerful Construction Tool

You have in your hands one of the most powerful construction tools available today. A successful project is built on the foundation of an accurate and dependable estimate. This book will enable you to construct just such an estimate.

For the casual user the book is designed to be:

- quickly and easily understood so you can get right to your estimate.
- filled with valuable information so you can understand the necessary factors that go into the cost estimate.

For the regular user, the book is designed to be:

- a handy desk reference that can be quickly referred to for key costs.
- a comprehensive, fully reliable source of current construction costs and productivity rates so you'll be prepared to estimate any project.
- a source book for preliminary project cost, product selections, and alternate materials and methods.

To meet all of these requirements, we have organized the book into the following clearly defined sections.

New: Quick Start

See our new "Quick Start" instructions on the following page to get started right away.

Estimating with RSMeans Unit Price Cost Data

Please refer to these steps for guidance on completing an estimate using RSMeans unit price cost data.

How to Use the Book: The Details

This section contains an in-depth explanation of how the book is arranged . . . and how you can use it to determine a reliable construction cost estimate. It includes information about how we develop our cost figures and how to completely prepare your estimate.

Unit Price Section

All cost data has been divided into the 50 divisions according to the MasterFormat system of classification and numbering. For a listing of these divisions and an outline of their subdivisions, see the Unit Price Section Table of Contents.

Estimating tips are included at the beginning of each division.

Reference Section

This section includes information on Equipment Rental Costs, Crew Listings, Historical Cost Indexes, City Cost Indexes, Location Factors, Reference Tables, Change Orders, Square Foot Costs, and a listing of Abbreviations.

Equipment Rental Costs: This section contains the average costs to rent and operate hundreds of pieces of construction equipment.

Crew Listings: This section lists all of the crews referenced in the book. For the purposes of this book, a crew is composed of more than one trade classification and/or the addition of power equipment to any trade classification. Power equipment is included in the cost of the crew. Costs are shown both with bare labor rates and with the installing contractor's overhead and profit added. For each, the total crew cost per eight-hour day and the composite cost per labor-hour are listed.

Historical Cost Indexes: These indexes provide you with data to adjust construction costs over time.

City Cost Indexes: All costs in this book are U.S. national averages. Costs vary because of the regional economy. You can adjust costs by CSI Division to over 700 locations throughout the U.S. and Canada by using the data in this section.

Location Factors: You can adjust total project costs to over 900 locations throughout the U.S. and Canada by using the data in this section.

Reference Tables: At the beginning of selected major classifications in the Unit Price section are reference numbers shown in a shaded box. These numbers refer you to related information in the Reference Section. In this section, you'll find reference tables, explanations, estimating information that support how we develop the unit price data, technical data, and estimating procedures.

Change Orders: This section includes information on the factors that influence the pricing of change orders.

Square Foot Costs: This section contains costs for 59 different building types that allow you to make a rough estimate for the overall cost of a project or its major components.

Abbreviations: A listing of abbreviations used throughout this book, along with the terms they represent, is included in this section.

Index

A comprehensive listing of all terms and subjects in this book will help you quickly find what you need when you are not sure where it occurs in MasterFormat.

The Scope of This Book

This book is designed to be as comprehensive and as easy to use as possible. To that end we have made certain assumptions and limited its scope in two key ways:

1. We have established material prices based on a national average.

2. We have computed labor costs based on a 30-city national average of union wage rates.

For a more detailed explanation of how the cost data is developed, see "How To Use the Book: The Details."

Project Size/Type

The material prices in RSMeans data cost books are "contractor's prices." They are the prices that contractors can expect to pay at the lumberyards, suppliers/distributers warehouses, etc. Small orders of speciality items would be higher than the costs shown, while very large orders, such as truckload lots, would be less. The variation would depend on the size, timing, and negotiating power of the contractor. The labor costs are primarily for new construction or major renovation rather than repairs or minor alterations.

With reasonable exercise of judgment, the figures can be used for any building work.

Absolute Essentials for a Quick Start

If you feel you are ready to use this book and don't think you will need the detailed instructions that begin on the following page, this Absolute Essentials for a Quick Start page is for you. These steps will allow you to get started estimating in a matter of minutes.

1 Scope

Think through the project that you will be estimating, and identify the many individual work tasks that will need to be covered in your estimate.

2 Quantify

Determine the number of units that will be required for each work task that you identified.

3 Pricing

Locate individual Unit Price line items that match the work tasks you identified. The Unit Price Section Table of Contents that begins on page 1 and the Index in the back of the book will help you find these line items.

4 Multiply

Multiply the Total Incl O&P cost for a Unit Price line item in the book by your quantity for that item. The price you calculate will be an estimate for a completed item of work performed by a subcontractor. Keep adding line items in this manner to build your estimate.

5 Project Overhead

Include project overhead items in your estimate. These items are needed to make the job run and are typically, but not always, provided by the General Contractor. They can be found in Division 1. An alternate method of estimating project overhead costs is to apply a percentage of the total project cost.

Include rented tools not included in crews, waste, rubbish handling, and cleanup.

6 Estimate Summary

Include General Contractor's markup on subcontractors, General Contractor's office overhead and profit, and sales tax on materials and equipment.

Adjust your estimate to the project's location by using the City Cost Indexes or Location Factors found in the Reference Section.

Editors' Note: We urge you to spend time reading and understanding the supporting material in the front of this book. An accurate estimate requires experience, knowledge, and careful calculation. The more you know about how we at RSMeans developed the data, the more accurate your estimate will be. In addition, it is important to take into consideration the reference material in the back of the book such as Equipment Listings, Crew Listings, City Cost Indexes, Location Factors, and Reference Numbers.

Estimating with RSMeans Unit Price Cost Data

Following these steps will allow you to complete an accurate estimate using RSMeans Unit Price cost data.

1 Scope Out the Project

- Identify the individual work tasks that will need to be covered in your estimate.
- The Unit Price data in this book has been divided into 50 Divisions according to the CSI MasterFormat 2010—their titles are listed on the back cover of your book.
- Think through the project and identify those CSI Divisions needed in your estimate.
- The Unit Price Section Table of Contents on page 1 may also be helpful when scoping out your project.
- Experienced estimators find it helpful to begin with Division 2 and continue through completion. Division 1 can be estimated after the full project scope is known.

2 Quantify

- Determine the number of units required for each work task that you identified.
- Experienced estimators include an allowance for waste in their quantities. (Waste is not included in RSMeans Unit Price line items unless so stated.)

3 Price the Quantities

- Use the Unit Price Table of Contents, and the Index, to locate individual Unit Price line items for your estimate.
- Reference Numbers indicated within a Unit Price section refer to additional information that you may find useful.
- The crew indicates who is performing the work for that task. Crew codes are expanded in the Crew Listings in the Reference Section to include all trades and equipment that comprise the crew.
- The Daily Output is the amount of work the crew is expected to do in one day.
- The Labor-Hours value is the amount of time it will take for the crew to install one unit of work.

- The abbreviated Unit designation indicates the unit of measure upon which the crew, productivity, and prices are based.
- Bare Costs are shown for materials, labor, and equipment needed to complete the Unit Price line item. Bare costs do not include waste, project overhead, payroll insurance, payroll taxes, main office overhead, or profit.
- The Total Incl O&P cost is the billing rate or invoice amount of the installing contractor or subcontractor who performs the work for the Unit Price line item.

4 Multiply

- Multiply the total number of units needed for your project by the Total Incl O&P cost for each Unit Price line item.
- Be careful that your take off unit of measure matches the unit of measure in the Unit column.
- The price you calculate is an estimate for a completed item of work.
- Keep scoping individual tasks, determining the number of units required for those tasks, matching each task with individual Unit Price line items in the book, and multiply quantities by Total Incl O&P costs.
- An estimate completed in this manner is priced as if a subcontractor, or set of subcontractors, is performing the work. The estimate does not yet include Project Overhead or Estimate Summary components such as General Contractor markups on subcontracted work, General Contractor office overhead and profit, contingency, and location factor.

5 Project Overhead

- Include project overhead items from Division 1 – General Requirements.
- These items are needed to make the job run. They are typically, but not always, provided by the General Contractor. Items include, but are not limited to, field personnel, insurance, performance bond, permits, testing, temporary utilities, field office and storage facilities, temporary scaffolding and platforms, equipment mobilization and demobilization, temporary roads and sidewalks, winter protection, temporary barricades and fencing, temporary security, temporary signs, field engineering and layout, final cleaning and commissioning.
- Each item should be quantified, and matched to individual Unit Price line items in Division 1, and then priced and added to your estimate.
- An alternate method of estimating project overhead costs is to apply a percentage of the total project cost, usually 5% to 15% with an average of 10% (see General Conditions, page ix).
- Include other project related expenses in your estimate such as:
  - Rented equipment not itemized in the Crew Listings
  - Rubbish handling throughout the project (see 02 41 19.19)

6 Estimate Summary

- Include sales tax as required by laws of your state or country.

- Include the General Contractor's markup on self-performed work, usually 5% to 15% with an average of 10%.

- Include the General Contractor's markup on subcontracted work, usually 5% to 15% with an average of 10%.

- Include General Contractor's main office overhead and profit.

- RSMeans gives general guidelines on the General Contractor's main office overhead (see section 01 31 13.50 and Reference Number R013113-50).

- RSMeans gives no guidance on the General Contractor's profit.

- Markups will depend on the size of the General Contractor's operations, his projected annual revenue, the level of risk he is taking on, and on the level of competition in the local area and for this project in particular.

- Include a contingency, usually 3% to 5%, if appropriate.

- Adjust your estimate to the project's location by using the City Cost Indexes or the Location Factors in the Reference Section.

- Look at the rules on the pages for How to Use the City Cost Indexes to see how to apply the Indexes for your location.

- When the proper Index or Factor has been identified for the project's location, convert it to a multiplier by dividing it by 100, and then multiply that multiplier by your estimated total cost. The original estimated total cost will now be adjusted up or down from the national average to a total that is appropriate for your location.

Editors' Notes:

1) *We urge you to spend time reading and understanding the supporting material in the front of this book. An accurate estimate requires experience, knowledge, and careful calculation. The more you know about how we at RSMeans developed the data, the more accurate your estimate will be. In addition, it is important to take into consideration the reference material in the back of the book such as Equipment Listings, Crew Listings, City Cost Indexes, Location Factors, and Reference Numbers.*

2) *Contractors who are bidding or are involved in JOC, DOC, SABER, or IDIQ type contracts are cautioned that workers' compensation insurance, federal and state payroll taxes, waste, project supervision, project overhead, main office overhead, and profit are not included in bare costs. The coefficient or multiplier must cover these costs.*

How to Use the Book: The Details

What's Behind the Numbers? The Development of Cost Data

The staff at RSMeans continually monitors developments in the construction industry in order to ensure reliable, thorough, and up-to-date cost information. While overall construction costs may vary relative to general economic conditions, price fluctuations within the industry are dependent upon many factors. Individual price variations may, in fact, be opposite to overall economic trends. Therefore, costs are constantly tracked and complete updates are published yearly. Also, new items are frequently added in response to changes in materials and methods.

Costs—$ (U.S.)

All costs represent U.S. national averages and are given in U.S. dollars. The RSMeans City Cost Indexes can be used to adjust costs to a particular location. The City Cost Indexes for Canada can be used to adjust U.S. national averages to local costs in Canadian dollars. No exchange rate conversion is necessary.

[G] The processes or products identified by the green symbol in our publications have been determined to be environmentally responsible and/or resource-efficient solely by the RSMeans engineering staff. The inclusion of the green symbol does not represent compliance with any specific industry association or standard.

Material Costs

The RSMeans staff contacts manufacturers, dealers, distributors, and contractors all across the U.S. and Canada to determine national average material costs. If you have access to current material costs for your specific location, you may wish to make adjustments to reflect differences from the national average. Included within material costs are fasteners for a normal installation. RSMeans engineers use manufacturers' recommendations, written specifications, and/or standard construction practice for size and spacing of fasteners. Adjustments to material costs may be required for your specific application or location. Material costs do not include sales tax.

Labor Costs

Labor costs are based on the average of wage rates from 30 major U.S. cities. Rates are determined from labor union agreements or prevailing wages for construction trades for the current year. Rates, along with overhead and profit markups, are listed on the inside back cover of this book.

- If wage rates in your area vary from those used in this book, or if rate increases are expected within a given year, labor costs should be adjusted accordingly.

Labor costs reflect productivity based on actual working conditions. In addition to actual installation, these figures include time spent during a normal weekday on tasks such as, material receiving and handling, mobilization at site, site movement, breaks, and cleanup. Productivity data is developed over an extended period so as not to be influenced by abnormal variations and reflects a typical average.

Equipment Costs

Equipment costs include not only rental, but also operating costs for equipment under normal use. The operating costs include parts and labor for routine servicing such as repair and replacement of pumps, filters, and worn lines. Normal operating expendables, such as fuel, lubricants, tires, and electricity (where applicable), are also included. Extraordinary operating expendables with highly variable wear patterns, such as diamond bits and blades, are excluded. These costs are included under materials. Equipment rental rates are obtained from industry sources throughout North America—contractors, suppliers, dealers, manufacturers, and distributors.

Rental rates can also be treated as reimbursement costs for contractor-owned equipment. Owned equipment costs include depreciation, loan payments, interest, taxes, insurance, storage and major repairs.

Equipment costs do not include operators' wages; nor do they include the cost to move equipment to a job site (mobilization) or from a job site (demobilization).

Equipment Cost/Day—The cost of power equipment required for each crew is included in the Crew Listings in the Reference Section (small tools that are considered as essential everyday tools are not listed out separately). The Crew Listings itemize specialized tools and heavy equipment along with labor trades. The daily cost of itemized equipment included in a crew is based on dividing the weekly bare rental rate by 5 (number of working days per week) and then adding the hourly operating cost times 8 (the number of hours per day). This Equipment Cost/Day is shown in the last column of the Equipment Rental Cost pages in the Reference Section.

Mobilization/Demobilization—The cost to move construction equipment from an equipment yard or rental company to the job site and back again is not included in equipment costs. Mobilization (to the site) and demobilization (from the site) costs can be found in the Unit Price Section. If a piece of equipment is already at the job site, it is not appropriate to utilize mob./demob. costs again in an estimate.

Overhead and Profit

Total Cost including O&P for the *Installing Contractor* is shown in the last column on the Unit Price and/or the Assemblies pages of this book. This figure is the sum of the bare material cost plus 10% for profit, the bare labor cost plus total overhead and profit, and the bare equipment cost plus 10% for profit. Details for the calculation of Overhead and Profit on labor are shown on the inside back cover and in the Reference Section of this book. (See the "How to Use the Unit Price Pages" for an example of this calculation.)

General Conditions

Cost data in this book is presented in two ways: Bare Costs and Total Cost including O&P (Overhead and Profit). General Conditions, or General Requirements, of the contract should also be added to the Total Cost including O&P when applicable. Costs for General Conditions are listed in Division 1 of the Unit Price Section and the Reference

Section of this book. General Conditions for the *Installing Contractor* may range from 0% to 10% of the Total Cost including O&P. For the *General* or *Prime Contractor*, costs for General Conditions may range from 5% to 15% of the Total Cost including O&P, with a figure of 10% as the most typical allowance. If applicable, the Assemblies and Models sections of this book use costs that include the installing contractor's overhead and profit (O&P).

Factors Affecting Costs

Costs can vary depending upon a number of variables. Here's how we have handled the main factors affecting costs.

Quality——The prices for materials and the workmanship upon which productivity is based represent sound construction work. They are also in line with U.S. government specifications.

Overtime——We have made no allowance for overtime. If you anticipate premium time or work beyond normal working hours, be sure to make an appropriate adjustment to your labor costs.

Productivity——The productivity, daily output, and labor-hour figures for each line item are based on working an eight-hour day in daylight hours in moderate temperatures. For work that extends beyond normal work hours or is performed under adverse conditions, productivity may decrease. (See the section in "How To Use the Unit Price Pages" for more on productivity.)

Size of Project——The size, scope of work, and type of construction project will have a significant impact on cost. Economies of scale can reduce costs for large projects. Unit costs can often run higher for small projects.

Location——Material prices in this book are for metropolitan areas. However, in dense urban areas, traffic and site storage limitations may increase costs. Beyond a 20-mile radius of large cities, extra trucking or transportation charges may also increase the material costs slightly. On the other hand, lower wage rates may be in effect. Be sure to consider both of these factors when preparing an estimate, particularly if the job site is located in a central city or remote rural location. In addition, highly specialized subcontract items may require travel and per-diem expenses for mechanics.

Other Factors—
- season of year
- contractor management
- weather conditions
- local union restrictions
- building code requirements
- availability of:
  - adequate energy
  - skilled labor
  - building materials
- owner's special requirements/restrictions
- safety requirements
- environmental considerations

Unpredictable Factors——General business conditions influence "in-place" costs of all items. Substitute materials and construction methods may have to be employed. These may affect the installed cost and/or life cycle costs. Such factors may be difficult to evaluate and cannot necessarily be predicted on the basis of the job's location in a particular section of the country. Thus, where these factors apply, you may find significant but unavoidable cost variations for which you will have to apply a measure of judgment to your estimate.

Rounding of Costs

In general, all unit prices in excess of $5.00 have been rounded to make them easier to use and still maintain adequate precision of the results. The rounding rules we have chosen are in the following table.

Prices from . . .	Rounded to the nearest . . .
$.01 to $5.00	$.01
$5.01 to $20.00	$.05
$20.01 to $100.00	$.50
$100.01 to $300.00	$1.00
$300.01 to $1,000.00	$5.00
$1,000.01 to $10,000.00	$25.00
$10,000.01 to $50,000.00	$100.00
$50,000.01 and above	$500.00

How Subcontracted Items Affect Costs

A considerable portion of all large construction jobs is usually subcontracted. In fact, the percentage done by subcontractors is constantly increasing and may run over 90%. Since the workers employed by these companies do nothing else but install their particular product, they soon become expert in that line. The result is, installation by these firms is accomplished so efficiently that the total in-place cost, even adding the general contractor's overhead and profit, is no more, and often less, than if the principal contractor had handled the installation himself/herself. Companies that deal with construction specialties are anxious to have their product perform well and, consequently, the installation will be the best possible.

Contingencies

The allowance for contingencies generally provides for unforeseen construction difficulties. On alterations or repair jobs, 20% is not too much. If drawings are final and only field contingencies are being considered, 2% or 3% is probably sufficient, and often nothing need be added. Contractually, changes in plans will be covered by extras. The contractor should consider inflationary price trends and possible material shortages during the course of the job. These escalation factors are dependent upon both economic conditions and the anticipated time between the estimate and actual construction. If drawings are not complete or approved, or a budget cost is wanted, it is wise to add 5% to 10%. Contingencies, then, are a matter of judgment. Additional allowances for contingencies are shown in Division 1.

Important Estimating Considerations

The "productivity," or daily output, of each craftsman includes mobilization and cleanup time, break time and plan layout time, as well as an allowance to carry stock from the storage trailer or location on the job site up to 200' into the building and to the first or second floor. If material has to be transported over greater distances or to higher floors, an additional allowance should be considered by the estimator. An allowance has also been included in the piping and fittings installation time for leak checking and minor tightening. Equipment installation time includes the following applicable items: positioning, leveling and securing the unit in place, connecting all associated piping, ducts, vents, etc., which shall have been estimated separately, connecting to an adjacent power source, filling/bleeding, startup, adjusting the controls up and down to ensure proper response, setting the integral controls/valves/regulators/thermostats for proper operation (does not include external building type control systems, DDC systems, etc.), explaining/training owner's operator, and

warranty. A reasonable breakdown of the labor costs is as follows:

1.	Movement into building, installation/setting of equipment	35%
2.	Connection to piping/duct/power, etc.	25%
3.	Filling/flushing/cleaning/touchup, etc.	15%
4.	Startup/running adjustments	5%
5.	Training owner's representative	5%
6.	Warranty/call back/service	15%

Note that cranes or other lifting equipment are not included on any lines in the Mechanical divisions. For example, if a crane is required to lift a heavy piece of pipe into place high above a gym floor, or to put a rooftop unit on the roof of a four-story building, etc., it must be added. Due to the potential for extreme variation—from nothing additional required to a major crane or helicopter—we feel that including a nominal amount for "lifting contingency" would be useless and detract from the accuracy of the estimate. When using equipment rental from RSMeans, remember to include the cost of the operator(s).

Estimating Labor-Hours

The labor-hours expressed in this publication are based on Average Installation time, using an efficiency level of approximately 60%–65% (see item 7), which has been found reasonable and acceptable by many contractors. The book uses this national efficiency average to establish a consistent benchmark. For bid situations, adjustments to this efficiency level should be the responsibility of the contractor bidding the project. The unit labor-hour is divided in the following manner. A typical day for a journeyman might be:

1.	Study Plans	3%	14.4 min.
2.	Material Procurement	3%	14.4 min.
3.	Receiving and Storing	3%	14.4 min
4.	Mobilization	5%	24.0 min.
5.	Site Movement	5%	24.0 min.
6.	Layout and Marking	8%	38.4 min.
7.	Actual Installation	64%	307.2 min.
8.	Cleanup	3%	14.4 min.
9.	Breaks, Nonproductive	6%	28.8 min.
		100%	480.0 min.

If any of the percentages expressed in this breakdown do not apply to the particular work or project situation, then that percentage or a portion of it may be deducted from or added to labor-hours.

Final Checklist

Estimating can be a straightforward process provided you remember the basics. Here's a checklist of some of the steps you should remember to complete before finalizing your estimate. Did you remember to . . .

- factor in the City Cost Index for your locale?
- take into consideration which items have been marked up and by how much?
- mark up the entire estimate sufficiently for your purposes?
- read the background information on techniques and technical matters that could impact your project time span and cost?
- include all components of your project in the final estimate?
- double check your figures for accuracy?
- call RSMeans if you have any questions about your estimate or the data you've found in our publications? Remember, RSMeans stands behind its publications. If you have any questions about your estimate . . . about the costs you've used from our books . . . or even about the technical aspects of the job that may affect your estimate, feel free to call the RSMeans editors at 1-800-334-3509.

Unit Price Section

Table of Contents

1

2

Table of Contents (cont.)

How RSMeans Data Works

All RSMeans unit price data is organized in the same way.

It is important to understand the structure, so that you can find information easily and use it correctly.

RSMeans **Line Numbers** consist of 12 characters, which identify a unique location in the database for each task. The first 6 or 8 digits conform to the Construction Specifications Institute MasterFormat 2010. The remainder of the digits are a further breakdown by RSMeans in order to arrange items in understandable groups of similar tasks. Line numbers are consistent across all RSMeans publications, so a line number in any RSMeans product will always refer to the same unit of work.

RSMeans engineers have created **reference** information to assist you in your estimate. If there is information that applies to a section, it will be indicated at the start of the section. In this case, R033105-10 provides information on the proportionate quantities of formwork, reinforcing, and concrete used in cast-in-place concrete items such as footings, slabs, beams, and columns. The reference section is located in the back of the book on the pages with a gray edge.

RSMeans **Descriptions** are shown in a hierarchical structure to make them readable. In order to read a complete description, read up through the indents to the top of the section. Include everything that is above and to the left that is not contradicted by information below. For instance, the complete description for line 03 30 53.40 3550 is "Concrete in place, including forms (4 uses), reinforcing steel, concrete, placement, and finishing unless otherwise indicated; Equipment pad (3000 psi), 4' x 4' x 6" thick."

When using **RSMeans data**, it is important to read through an entire section to ensure that you use the data that most closely matches your work. Note that sometimes there is additional information shown in the section that may improve your price. There are frequently lines that further describe, add to, or adjust data for specific situations.

03 30 Cast-In-Place Concrete

03 30 53 – Miscellaneous Cast-In-Place Concrete

03 30 53.40 Concrete In Place

0010	**CONCRETE IN PLACE**	R033105-10
0020	Including forms (4 uses), reinforcing steel, concrete, placement,	R033105-20
0050	and finishing unless otherwise indicated	R033105-50
0300	Beams (3500 psi), 5 kip per L.F., 10' span	R033105-65
0350	25' span	R033105-70
0500	Chimney foundations (5000 psi), industrial, minimum	R033105-85
0510	Maximum	
0700	Columns, square (4000 psi), 12" x 12", minimum reinforcing	
3500	Add per floor for 3 to 6 stories high	
3520	For 7 to 20 stories high	
3540	Equipment pad (3000 psi), 3' x 3' x 6" thick	
3550	4' x 4' x 6" thick	
3560	5' x 5' x 8" thick	
3570	6' x 6' x 8" thick	

The data published in RSMeans print books represents a "national average" cost. This data should be modified to the project location using the **City Cost Indexes** or **Location Factors** tables found in the reference section (see pages 713–716). Use the location factors to adjust estimate totals if the project covers multiple trades. Use the city cost indexes (CCI) for single trade projects or projects where a more detailed analysis is required. All figures in the two tables are derived from the same research. The last row of data in the CCI, the weighted average, is the same as the numbers reported for each location in the location factor table.

Crews include labor or labor and equipment necessary to accomplish each task. In this case, Crew C-14H is used. RSMeans selects a crew to represent the workers and equipment that are typically used for that task. In this case, Crew C-14H consists of one carpenter foreman (outside), two carpenters, one rodman, one laborer, one cement finisher, and one gas engine vibrator. Details of all crews can be found in the reference section.

Crews

Crew No.	Bare Costs		Incl. Subs O&P		Cost Per Labor-Hour	
Crew C-14H	Hr.	Daily	Hr.	Daily	Bare Costs	Incl. O&P
1 Carpenter Foreman (outside)	$46.10	$368.80	$70.95	$567.60	$43.48	$66.89
2 Carpenters	44.10	705.60	67.85	1085.60		
1 Rodman (reinf.)	49.10	392.80	78.40	627.20		
1 Laborer	35.10	280.80	54.00	432.00		
1 Cement Finisher	42.35	338.80	62.30	498.40		
1 Gas Engine Vibrator		27.40		30.14	.57	.63
48 L.H., Daily Totals		$2114.20		$3240.94	$44.05	$67.52

The **Daily Output** is the amount of work that the crew can do in a normal 8-hour workday, including mobilization, layout, movement of materials, and cleanup. In this case, crew C-14H can install thirty 4' x 4' x 6" thick concrete pads in a day. Daily output is variable, based on many factors, including the size of the job, location, and environmental conditions. RSMeans data represents work done in daylight (or adequate lighting) and temperate conditions.

Bare Costs are the costs of materials, labor, and equipment that the installing contractor pays. They represent the cost, in U.S. dollars, for one unit of work. They do not include any markups for profit or labor burden.

Crew	Daily Output	Labor-Hours	Unit	Material	2012 Bare Costs Labor	Equipment	Total	Total Incl O&P
C-14A	15.62	12.804	C.Y.	305	565	47.50	917.50	1,275
"	18.55	10.782		320	480	40	840	1,125
C-14C	32.22	3.476		144	147	.83	291.83	385
"	23.71	4.724		168	200	1.13	369.13	495
C-14A	11.96	16.722		350	740	62	1,152	1,600
	31800	.002			.07	.02	.09	.12
	21200	.003			.10	.03	.13	.20
C-14H	45	1.067	Ea.	44	46.50	.61	91.11	121
	30	1.600		66	69.50	.91	136.41	181
	18	2.667		118	116	1.52	235.52	310
	14	3.429		159	149	1.95	309.95	405

The **Total Incl O&P column** is the total cost, including overhead and profit, that the installing contractor will charge the customer. This represents the cost of materials plus 10% profit, the cost of labor plus labor burden and 10% profit, and the cost of equipment plus 10% profit. It does not include the general contractor's overhead and profit. Note: See the inside back cover for details of how RSMeans calculates labor burden.

The **Total column** represents the total bare cost for the installing contractor, in U.S. dollars. In this case, the sum of $66 for material + $69.50 for labor + $.91 for equipment is $136.41.

The figure in the **Labor Hours** column is the amount of labor required to perform one unit of work—in this case the amount of labor required to construct one 4' x 4' equipment pad. This figure is calculated by dividing the number of hours of labor in the crew by the daily output (48 labor hours divided by 30 pads = 1.6 hours of labor per pad). Multiply 1.600 times 60 to see the value in minutes: 60 x 1.6 = 96 minutes. Note: the labor hour figure is not dependent on the crew size. A change in crew size will result in a corresponding change in daily output, but the labor hours per unit of work will not change.

All RSMeans unit cost data includes the typical **Unit of Measure** used for estimating that item. For concrete-in-place the typical unit is cubic yards (C.Y.) or each (Ea.). For installing broadloom carpet it is square yard, and for gypsum board it is square foot. The estimator needs to take special care that the unit in the data matches the unit in the take-off. Unit conversions may be found in the Reference Section.

How RSMeans Data Works (continued)

Sample Estimate

This sample demonstrates the elements of an estimate, including a tally of the RSMeans data lines, and a summary of the markups on a contractor's work to arrive at a total cost to the owner. The RSMeans Location Factor is added at the bottom of the estimate to adjust the cost of the work to a specific location.

Work Performed: The body of the estimate shows the RSMeans data selected, including line number, a brief description of each item, its take-off unit and quantity, and the bare costs of materials, labor, and equipment. This estimate also includes a column titled "SubContract." This data is taken from the RSMeans column "Total Incl O&P," and represents the total that a subcontractor would charge a general contractor for the work, including the sub's markup for overhead and profit.

General Requirements: This item covers project-wide needs provided by the general contractor. These items vary by project, but may include temporary facilities and utilities, security, testing, project cleanup, etc. For small projects a percentage can be used, typically between 5% and 15% of project cost. For large projects the costs may be itemized and priced individually.

Bonds: Bond costs should be added to the estimate. The figures here represent a typical performance bond, ensuring the owner that if the general contractor does not complete the obligations in the construction contract the bonding company will pay the cost for completion of the work.

Location Adjustment: RSMeans published data is based on national average costs. If necessary, adjust the total cost of the project using a location factor from the "Location Factor" table or the "City Cost Index" table. Use location factors if the work is general, covering multiple trades. If the work is by a single trade (e.g., masonry) use the more specific data found in the "City Cost Indexes."

Project Name:	Pre-Engineered Steel Building	Architect: As Shown		
Location:	**Anywhere, USA**			
Line Number	**Description**	**Qty**	**Unit**	**Material**
03 30 53.40 3950	Strip footing, 12" x 36"	33.33	C.Y.	$4,699.53
03 30 53.40 3940	Strip footing, 12" x24"	14.8	C.Y.	$2,175.60
03 31 05.35 0300	Concrete ready mix, 4000 psi, for slab	185.2	C.Y.	$19,631.20
03 31 05.70 4300	Placing concrete, slab on grade	185.2	C.Y.	$0.00
03 11 13.65 3000	Concrete forms	500	L.F.	$190.00
03 22 05.50 0200	WWF	150	C.S.F.	$3,975.00
03 39 23.13 0300	Curing	150	C.S.F.	$787.50
03 35 29.30 0250	Concrete finishing	15000	S.F.	$0.00
03 35 29.35 0160	Control joints	650	L.F.	$84.50
Division 03	**Subtotal**			**$31,543.33**
08 33 23.10 0100	10' x 10' OH Door	8	Ea.	$14,000.00
08 33 23.10 3300	Add for enamel finish	800	S.F.	$1,216.00
Division 8	**Subtotal**			**$15,216.00**
13 34 19.50 1000	Pre-Engineered Steel Building	15,000	SF Flr.	
13 34 19.50 6050	Framing for doors, 3' x 7'	4	Opng.	$0.00
13 34 19.50 6100	Framing for doors, 10' x 10'	8	Opng.	$0.00
13 34 19.50 6200	Framing for windows, 3' x 4'	6	Opng.	$0.00
13 34 19.50 5750	PESB doors	4	Opng.	$1,740.00
13 34 19.50 7750	PESB windows	6	Opng.	$2,100.00
13 34 19.50 6550	Gutters	380	L.F.	$2,679.00
13 34 19.50 8650	Roof vent	15	Ea.	$382.50
13 34 19.50 6900	Insulation	25,000	S.F.	$12,750.00
Division 13	**Subtotal**			**$19,651.50**
Estimate Subtotal				**$66,410.83**
		Gen. Requirements		4,648.76
		Sales Tax		3,320.54
		Subtotal		74,380.13
		GC O & P		7,438.01
		Subtotal		81,818.14
		Contingency		@5%
		Subtotal		
		Bond	$12/1000 +10% O&P	
		Subtotal		
		Location Adjustment		Factor
		Grand Total		

This estimate is based on an interactive spreadsheet.
A copy of this spreadsheet is located on the RSMeans website at
http://www.reedconstructiondata.com/rsmeans/extras/546011.
You are free to download it and adjust it to your methodology.

This example shows the cost to construct a pre-engineered steel building. The foundation, doors, windows, and insulation will be installed by the general contractor. A subcontractor will install the structural steel, roofing, and siding.

	01/01/12			
Labor	Equipment	SubContract	EstimateTotal	
$2,383.10	$14.33	$0.00		
$1,324.60	$7.99	$0.00		
$0.00	$0.00	$0.00		
$2,666.88	$87.04	$0.00		
$1,010.00	$0.00	$0.00		
$3,450.00	$0.00	$0.00		
$795.00	$0.00	$0.00		
$7,500.00	$300.00	$0.00		
$247.00	$58.50	$0.00		
$19,376.58	$467.86	$0.00	$51,387.77	Division 03
$4,080.00	$0.00	$0.00		
$0.00	$0.00	$0.00		
$4,080.00	$0.00	$0.00	$19,296.00	Division 0_
		$287,250.00		
$0.00	$0.00	$1,900.00		
$0.00	$0.00	$8,400.00		
$0.00	$0.00	$3,450.00		
$572.00	$0.00	$0.00		
$483.00	$61.80	$0.00		
$851.20	$0.00	$0.00		
$2,685.00	$0.00	$0.00		
$7,000.00	$0.00	$0.00		
$11,591.20	$61.80	$301,000.00	$3_2,304.50	Division 13
$35,047.78	$529.66	$301,000.00	$402,988.27	
2,453.34	37.08	21,070.00		Gen. Requirements
	26.48	7,525.00		Sales Tax
37,501.12	593.22	329,595.00		Subtotal
3,750.11	59.32	32,959.50		GC O & P
41,251.24	652.54	362,554.50	$486,276.42	Subtotal
			24,313.82	Contingency
			$510,590.24	Subtotal
			6,739.79	Bond
			$517,330.03	Subtotal
102.30			11,898.59	Location Adjustment
			$529,228.62	Grand Total

Sales Tax: If the work is subject to state or local sales taxes, the amount must be added to the estimate. Sales tax may be added to material costs, equipment costs, and subcontracted work. In this case, sales tax was added in all three categories. It was assumed that approximately half the subcontracted work would be material cost, so the tax was applied to 50% of the subcontract total.

GC O&P: This entry represents the general contractor's markup on material, labor, equipment, and subcontractor costs. RSMeans' standard markup on materials, equipment, and subcontracted work is 10%. In this estimate, the markup on the labor performed by the GC's workers uses "Skilled Workers Average" shown in Column F on the table "Installing Contractor's Overhead & Profit," which can be found on the inside-back cover of the book.

Contingency: A factor for contingency may be added to any estimate to represent the cost of unknowns that may occur between the time that the estimate is performed and the time the project is constructed. The amount of the allowance will depend on the stage of design at which the estimate is done, and the contractor's assessment of the risk involved. Refer to section 01 21 16.50 for contingency allowances.

Estimating Tips

01 20 00 Price and Payment Procedures

- Allowances that should be added to estimates to cover contingencies and job conditions that are not included in the national average material and labor costs are shown in section 01 21.

- When estimating historic preservation projects (depending on the condition of the existing structure and the owner's requirements), a 15%–20% contingency or allowance is recommended, regardless of the stage of the drawings.

01 30 00 Administrative Requirements

- Before determining a final cost estimate, it is a good practice to review all the items listed in Subdivisions 01 31 and 01 32 to make final adjustments for items that may need customizing to specific job conditions.

- Requirements for initial and periodic submittals can represent a significant cost to the General Requirements of a job. Thoroughly check the submittal specifications when estimating a project to determine any costs that should be included.

01 40 00 Quality Requirements

- All projects will require some degree of Quality Control. This cost is not included in the unit cost of construction listed in each division. Depending upon the terms of the contract, the various costs of inspection and testing can be the responsibility of either the owner or the contractor. Be sure to include the required costs in your estimate.

01 50 00 Temporary Facilities and Controls

- Barricades, access roads, safety nets, scaffolding, security, and many more requirements for the execution of a safe project are elements of direct cost. These costs can easily be overlooked when preparing an estimate. When looking through the major classifications of this subdivision, determine which items apply to each division in your estimate.

- Construction Equipment Rental Costs can be found in the Reference Section in section 01 54 33. Operators' wages are not included in equipment rental costs.

- Equipment mobilization and demobilization costs are not included in equipment rental costs and must be considered separately in section 01 54 36.50.

- The cost of small tools provided by the installing contractor for his workers is covered in the "Overhead" column on the "Installing Contractor's Overhead and Profit" table that lists labor trades, base rates and markups and, therefore, is included in the "Total Incl. O&P" cost of any Unit Price line item. For those users who are constrained by contract terms to use only bare costs, there are two line items in section 01 54 39.70 for small tools as a percentage of bare labor cost. If some of those users are further constrained by contract terms to refrain from using line items from Division 1, you are advised to cover the cost of small tools within your coefficient or multiplier.

01 70 00 Execution and Closeout Requirements

- When preparing an estimate, thoroughly read the specifications to determine the requirements for Contract Closeout. Final cleaning, record documentation, operation and maintenance data, warranties and bonds, and spare parts and maintenance materials can all be elements of cost for the completion of a contract. Do not overlook these in your estimate.

Reference Numbers

Reference numbers are shown in shaded boxes at the beginning of some major classifications. These numbers refer to related items in the Reference Section. The reference information may be an estimating procedure, an alternate pricing method, or technical information.

Note: Not all subdivisions listed here necessarily appear in this publication.

01 11 31 – Professional Consultants

01 11 31.10 Architectural Fees

		Crew	Daily Output	Labor-Hours	Unit	Material	2012 Bare Costs Labor	Equipment	Total	Total Incl O&P
0010	**ARCHITECTURAL FEES** R011110-10									
0020	For new construction									
0060	Minimum				Project					4.90%
0090	Maximum									16%
0100	For alteration work, to $500,000, add to new construction fee									50%
0150	Over $500,000, add to new construction fee									25%
2000	For "Greening" of building G									3%

01 11 31.20 Construction Management Fees

		Crew	Daily Output	Labor-Hours	Unit	Material	2012 Bare Costs Labor	Equipment	Total	Total Incl O&P
0010	**CONSTRUCTION MANAGEMENT FEES**									
0020	$1,000,000 job, minimum				Project					4.50%
0050	Maximum									7.50%
0300	$50,000,000 job, minimum									2.50%
0350	Maximum									4%

01 11 31.30 Engineering Fees

		Crew	Daily Output	Labor-Hours	Unit	Material	2012 Bare Costs Labor	Equipment	Total	Total Incl O&P
0010	**ENGINEERING FEES** R011110-30									
0020	Educational planning consultant, minimum				Project					.50%
0100	Maximum				"					2.50%
0200	Electrical, minimum				Contrct					4.10%
0300	Maximum									10.10%
0400	Elevator & conveying systems, minimum									2.50%
0500	Maximum									5%
0600	Food service & kitchen equipment, minimum									8%
0700	Maximum									12%
0800	Landscaping & site development, minimum									2.50%
0900	Maximum									6%
1000	Mechanical (plumbing & HVAC), minimum									4.10%
1100	Maximum									10.10%
1200	Structural, minimum				Project					1%
1300	Maximum				"					2.50%

01 11 31.50 Models

		Crew	Daily Output	Labor-Hours	Unit	Material	2012 Bare Costs Labor	Equipment	Total	Total Incl O&P
0010	**MODELS**									
0020	Cardboard & paper, 1 building, minimum				Ea.	720			720	790
0050	Maximum					1,650			1,650	1,800
0100	2 buildings, minimum					965			965	1,050
0150	Maximum					2,175			2,175	2,400
0200	Plexiglass and metal, basic layout				SF Flr.	.07			.07	.08
0210	Including equipment and personnel				"	.32			.32	.35
0300	Site plan layout, minimum				Ea.	1,375			1,375	1,525
0350	Maximum				"	2,300			2,300	2,550

01 11 31.75 Renderings

		Crew	Daily Output	Labor-Hours	Unit	Material	2012 Bare Costs Labor	Equipment	Total	Total Incl O&P
0010	**RENDERINGS** Color, matted, 20" x 30", eye level,									
0020	1 building, minimum				Ea.	2,000			2,000	2,200
0050	Average					2,800			2,800	3,075
0100	Maximum					4,500			4,500	4,950
1000	5 buildings, minimum					4,000			4,000	4,400
1100	Maximum					8,000			8,000	8,800
2000	Aerial perspective, color, 1 building, minimum					2,900			2,900	3,200
2100	Maximum					7,900			7,900	8,700
3000	5 buildings, minimum					5,700			5,700	6,275
3100	Maximum					11,400			11,400	12,500

01 21 Allowances

01 21 16 – Contingency Allowances

01 21 16.50 Contingencies		Crew	Daily Output	Labor-Hours	Unit	Material	2012 Bare Costs Labor	Equipment	Total	Total Incl O&P
0010	**CONTINGENCIES**, Add to estimate									
0020	Conceptual stage				Project					20%
0050	Schematic stage									15%
0100	Preliminary working drawing stage (Design Dev.)									10%
0150	Final working drawing stage									3%

01 21 55 – Job Conditions Allowance

01 21 55.50 Job Conditions

		Crew	Daily Output	Labor-Hours	Unit	Material	2012 Bare Costs Labor	Equipment	Total	Total Incl O&P
0010	**JOB CONDITIONS** Modifications to applicable									
0020	cost summaries									
0100	Economic conditions, favorable, deduct				Project				2%	2%
0200	Unfavorable, add								5%	5%
0300	Hoisting conditions, favorable, deduct								2%	2%
0400	Unfavorable, add								5%	5%
0500	General Contractor management, experienced, deduct								2%	2%
0600	Inexperienced, add								10%	10%
0700	Labor availability, surplus, deduct								1%	1%
0800	Shortage, add								10%	10%
0900	Material storage area, available, deduct								1%	1%
1000	Not available, add								2%	2%
1100	Subcontractor availability, surplus, deduct								5%	5%
1200	Shortage, add								12%	12%
1300	Work space, available, deduct								2%	2%
1400	Not available, add								5%	5%

01 21 57 – Overtime Allowance

01 21 57.50 Overtime

		Crew	Daily Output	Labor-Hours	Unit	Material	2012 Bare Costs Labor	Equipment	Total	Total Incl O&P
0010	**OVERTIME** for early completion of projects or where R012909-90									
0020	labor shortages exist, add to usual labor, up to				Costs		100%			

01 21 61 – Cost Indexes

01 21 61.10 Construction Cost Index

		Crew	Daily Output	Labor-Hours	Unit	Material	2012 Bare Costs Labor	Equipment	Total	Total Incl O&P
0010	**CONSTRUCTION COST INDEX** (Reference) over 930 zip code locations in									
0020	the U.S. and Canada, total bldg. cost, min. (Guymon, OK)				%					67.50%
0050	Average									100%
0100	Maximum (New York, NY)									132.20%

01 21 61.20 Historical Cost Indexes

0010	**HISTORICAL COST INDEXES** (See Reference Section)									

01 21 61.30 Labor Index

		Crew	Daily Output	Labor-Hours	Unit	Material	2012 Bare Costs Labor	Equipment	Total	Total Incl O&P
0010	**LABOR INDEX** (Reference) For over 930 zip code locations in									
0020	the U.S. and Canada, minimum (Guymon, OK)				%		31%			
0050	Average						100%			
0100	Maximum (New York, NY)						166.90%			

01 21 61.50 Material Index

		Crew	Daily Output	Labor-Hours	Unit	Material	2012 Bare Costs Labor	Equipment	Total	Total Incl O&P
0010	**MATERIAL INDEX** (Reference) For over 930 zip code locations in									
0020	the U.S. and Canada, minimum (Elizabethtown, KY)				%	91.40%				
0040	Average					100%				
0060	Maximum (Ketchikan, AK)					141.30%				

01 21 63 – Taxes

01 21 63.10 Taxes

		Crew	Daily Output	Labor-Hours	Unit	Material	2012 Bare Costs Labor	Equipment	Total	Total Incl O&P
0010	**TAXES** R012909-80									
0020	Sales tax, State, average				%	5.07%				
0050	Maximum R012909-85					9.25%				

11

01 21 Allowances

01 21 63 – Taxes

01 21 63.10 Taxes		Crew	Daily Output	Labor-Hours	Unit	Material	2012 Bare Costs Labor	2012 Bare Costs Equipment	Total	Total Incl O&P
0200	Social Security, on first $110,100 of wages				%		7.65%			
0300	Unemployment, combined Federal and State, minimum						.80%			
0350	Average						7.80%			
0400	Maximum						14.36%			

01 31 Project Management and Coordination

01 31 13 – Project Coordination

01 31 13.20 Field Personnel

01 31 13.20	Field Personnel	Crew	Daily Output	Labor-Hours	Unit	Material	Labor	Equipment	Total	Total Incl O&P
0010	**FIELD PERSONNEL**									
0020	Clerk, average				Week		420		420	650
0100	Field engineer, minimum						995		995	1,550
0120	Average						1,300		1,300	2,000
0140	Maximum						1,475		1,475	2,275
0160	General purpose laborer, average						1,375		1,375	2,125
0180	Project manager, minimum						1,850		1,850	2,850
0200	Average						2,125		2,125	3,275
0220	Maximum						2,425		2,425	3,750
0240	Superintendent, minimum						1,800		1,800	2,775
0260	Average						1,975		1,975	3,050
0280	Maximum						2,250		2,250	3,475
0290	Timekeeper, average						1,150		1,150	1,775

01 31 13.30 Insurance

01 31 13.30	Insurance		Crew	Daily Output	Labor-Hours	Unit	Material	Labor	Equipment	Total	Total Incl O&P
0010	**INSURANCE**	R013113-40									
0020	Builders risk, standard, minimum					Job					.24%
0050	Maximum	R013113-50									.64%
0200	All-risk type, minimum										.25%
0250	Maximum	R013113-60									.62%
0400	Contractor's equipment floater, minimum					Value					.50%
0450	Maximum					"					1.50%
0600	Public liability, average					Job					2.02%
0800	Workers' compensation & employer's liability, average										
0850	by trade, carpentry, general					Payroll		14.96%			
0900	Clerical							.49%			
0950	Concrete							12.70%			
1000	Electrical							5.58%			
1050	Excavation							9.01%			
1100	Glazing							12.57%			
1150	Insulation							11.85%			
1200	Lathing							7.82%			
1250	Masonry							12.10%			
1300	Painting & decorating							10.70%			
1350	Pile driving							16.76%			
1400	Plastering							10.78%			
1450	Plumbing							6.91%			
1500	Roofing							28.83%			
1550	Sheet metal work (HVAC)							8.47%			
1600	Steel erection, structural							36.86%			
1650	Tile work, interior ceramic							8.01%			
1700	Waterproofing, brush or hand caulking							6.41%			
1800	Wrecking							30.43%			
2000	Range of 35 trades in 50 states, excl. wrecking, min.							1.80%			

01 31 Project Management and Coordination

01 31 13 – Project Coordination

01 31 13.30 Insurance		Crew	Daily Output	Labor-Hours	Unit	Material	2012 Bare Costs Labor	Equipment	Total	Total Incl O&P
2100	Average				Payroll		13.70%			
2200	Maximum				↓		124.10%			

01 31 13.40 Main Office Expense

0010	**MAIN OFFICE EXPENSE** Average for General Contractors R013113-50									
0020	As a percentage of their annual volume									
0125	Annual volume under 1 million dollars				% Vol.				17.50%	
0145	Up to 2.5 million dollars								8%	
0150	Up to 4.0 million dollars								6.80%	
0200	Up to 7.0 million dollars								5.60%	
0250	Up to 10 million dollars								5.10%	
0300	Over 10 million dollars				↓				3.90%	

01 31 13.50 General Contractor's Mark-Up

0010	**GENERAL CONTRACTOR'S MARK-UP** on Change Orders									
0200	Extra work, by subcontractors, add				%					10%
0250	By General Contractor, add									15%
0400	Omitted work, by subcontractors, deduct all but									5%
0450	By General Contractor, deduct all but									7.50%
0600	Overtime work, by subcontractors, add									15%
0650	By General Contractor, add				↓					10%

01 31 13.60 Installing Contractor's Main Office Overhead

0010	**INSTALLING CONTRACTOR'S MAIN OFFICE OVERHEAD** R013113-50									
0020	As percent of direct costs, minimum				%				5%	
0050	Average								13%	
0100	Maximum				↓				30%	

01 31 13.80 Overhead and Profit

0010	**OVERHEAD & PROFIT** Allowance to add to items in this									
0020	book that do not include Subs O&P, average				%				25%	
0100	Allowance to add to items in this book that									
0110	do include Subs O&P, minimum				%					5%
0150	Average									10%
0200	Maximum									15%
0300	Typical, by size of project, under $100,000								30%	
0350	$500,000 project								25%	
0400	$2,000,000 project								20%	
0450	Over $10,000,000 project				↓				15%	

01 31 13.90 Performance Bond

0010	**PERFORMANCE BOND** R013113-80									
0020	For buildings, minimum				Job					.60%
0100	Maximum				"					2.50%

01 32 Construction Progress Documentation

01 32 13 – Scheduling of work

01 32 13.50 Scheduling

		Crew	Daily Output	Labor-Hours	Unit	Material	2012 Bare Costs Labor	Equipment	Total	Total Incl O&P
0010	**SCHEDULING**									
0020	Critical path, as % of architectural fee, minimum				%					.50%
0100	Maximum				"					1%
0300	Computer-update, micro, no plots, minimum				Ea.				455	500
0400	Including plots, maximum				"				1,450	1,600
0600	Rule of thumb, CPM scheduling, small job ($10 Million)				Job					.05%
0650	Large job ($50 Million +)									.03%
0700	Including cost control, small job									.08%
0750	Large job									.04%

01 32 33 – Photographic Documentation

01 32 33.50 Photographs

		Crew	Daily Output	Labor-Hours	Unit	Material	2012 Bare Costs Labor	Equipment	Total	Total Incl O&P
0010	**PHOTOGRAPHS**									
0020	8" x 10", 4 shots, 2 prints ea., std. mounting				Set	475			475	520
0100	Hinged linen mounts					530			530	580
0200	8" x 10", 4 shots, 2 prints each, in color					415			415	460
0300	For I.D. slugs, add to all above					5.30			5.30	5.85
0500	Aerial photos, initial fly-over, 6 shots, 1 print ea., 8" x 10"					845			845	925
0550	11" x 14" prints					1,025			1,025	1,125
0600	16" x 20" prints					1,200			1,200	1,325
0700	For full color prints, add					40%				40%
0750	Add for traffic control area					305			305	335
0900	For over 30 miles from airport, add per				Mile	5.45			5.45	6
1000	Vertical photography, 4 to 6 shots with									
1010	different scales, 1 print each				Set	1,125			1,125	1,225
1500	Time lapse equipment, camera and projector, buy					845			845	930
1550	Rent per month					305			305	335
1700	Cameraman and film, including processing, B.&W.				Day	1,375			1,375	1,525
1720	Color				"	1,375			1,375	1,525

01 41 Regulatory Requirements

01 41 26 – Permit Requirements

01 41 26.50 Permits

		Crew	Daily Output	Labor-Hours	Unit	Material	2012 Bare Costs Labor	Equipment	Total	Total Incl O&P
0010	**PERMITS**									
0020	Rule of thumb, most cities, minimum				Job					.50%
0100	Maximum				"					2%

01 45 Quality Control

01 45 23 – Testing and Inspecting Services

01 45 23.50 Testing

		Crew	Daily Output	Labor-Hours	Unit	Material	2012 Bare Costs Labor	Equipment	Total	Total Incl O&P
0010	**TESTING** and Inspecting Services									
0015	For concrete building costing $1,000,000, minimum				Project				4,725	5,200
0020	Maximum								38,000	41,800
0050	Steel building, minimum								4,725	5,200
0070	Maximum								14,800	16,300
0100	For building costing, $10,000,000, minimum								30,100	33,100
0150	Maximum								48,200	53,000
0200	Asphalt testing, compressive strength Marshall stability, set of 3				Ea.				145	165
0220	Density, set of 3								86	95
0250	Extraction, individual tests on sample								136	150

01 45 23.50 Testing		Crew	Daily Output	Labor-Hours	Unit	Material	2012 Bare Costs Labor	Equipment	Total	Total Incl O&P
0300	Penetration				Ea.				41	45
0350	Mix design, 5 specimens								182	200
0360	Additional specimen								36	40
0400	Specific gravity								41	45
0420	Swell test								64	70
0450	Water effect and cohesion, set of 6								182	200
0470	Water effect and plastic flow								64	70
0600	Concrete testing, aggregates, abrasion, ASTM C 131								136	150
0650	Absorption, ASTM C 127								42	46
0800	Petrographic analysis, ASTM C 295								775	850
0900	Specific gravity, ASTM C 127								50	55
1000	Sieve analysis, washed, ASTM C 136								59	65
1050	Unwashed								59	65
1200	Sulfate soundness								114	125
1300	Weight per cubic foot								36	40
1500	Cement, physical tests, ASTM C 150								320	350
1600	Chemical tests, ASTM C 150								245	270
1800	Compressive test, cylinder, delivered to lab, ASTM C 39								12	13
1900	Picked up by lab, minimum								14	15
1950	Average								18	20
2000	Maximum								27	30
2200	Compressive strength, cores (not incl. drilling), ASTM C 42								36	40
2250	Core drilling, 4" diameter (plus technician)				Inch				23	25
2260	Technician for core drilling				Hr.				45	50
2300	Patching core holes				Ea.				22	24
2400	Drying shrinkage at 28 days								236	260
2500	Flexural test beams, ASTM C 78								59	65
2600	Mix design, one batch mix								259	285
2650	Added trial batches								120	132
2800	Modulus of elasticity, ASTM C 469								164	180
2900	Tensile test, cylinders, ASTM C 496								45	50
3000	Water-Cement ratio curve, 3 batches								141	155
3100	4 batches								186	205
3300	Masonry testing, absorption, per 5 brick, ASTM C 67								45	50
3350	Chemical resistance, per 2 brick								50	55
3400	Compressive strength, per 5 brick, ASTM C 67								68	75
3420	Efflorescence, per 5 brick, ASTM C 67								68	75
3440	Imperviousness, per 5 brick								87	96
3470	Modulus of rupture, per 5 brick								86	95
3500	Moisture, block only								32	35
3550	Mortar, compressive strength, set of 3								23	25
4100	Reinforcing steel, bend test								55	61
4200	Tensile test, up to #8 bar								36	40
4220	#9 to #11 bar								41	45
4240	#14 bar and larger								64	70
4400	Soil testing, Atterberg limits, liquid and plastic limits								59	65
4510	Hydrometer analysis								109	120
4530	Specific gravity, ASTM D 354								44	48
4600	Sieve analysis, washed, ASTM D 422								55	60
4700	Unwashed, ASTM D 422								59	65
4710	Consolidation test (ASTM D2435), minimum								250	275
4715	Maximum								430	475
4720	Density and classification of undisturbed sample								73	80

01 45 Quality Control

01 45 23 – Testing and Inspecting Services

01 45 23.50 Testing

		Crew	Daily Output	Labor-Hours	Unit	Material	2012 Bare Costs Labor	Equipment	Total	Total Incl O&P
4735	Soil density, nuclear method, ASTM D2922				Ea.				35	38.50
4740	Sand cone method ASTM D1556								27	30
4750	Moisture content, ASTM D 2216								9	10
4780	Permeability test, double ring infiltrometer								500	550
4800	Permeability, var. or constant head, undist., ASTM D 2434								227	250
4850	Recompacted								250	275
4900	Proctor compaction, 4" standard mold, ASTM D 698								123	135
4950	6" modified mold								68	75
5100	Shear tests, triaxial, minimum								410	450
5150	Maximum								545	600
5300	Direct shear, minimum, ASTM D 3080								320	350
5350	Maximum								410	450
5550	Technician for inspection, per day, earthwork								320	350
5650	Bolting								400	440
5750	Roofing								480	530
5790	Welding								480	530
5820	Non-destructive metal testing, dye penetrant				Day				310	340
5840	Magnetic particle								310	340
5860	Radiography								450	495
5880	Ultrasonic								310	340
6000	Welding certification, minimum				Ea.				91	100
6100	Maximum				"				250	275
7000	Underground storage tank									
7500	Volumetric tightness test ,<=12,000 gal.				Ea.				435	480
7510	<=30,000 gal.				"				615	675
7600	Vadose zone (soil gas) sampling, 10-40 samples, min.				Day				1,375	1,500
7610	Maximum				"				2,275	2,500
7700	Ground water monitoring incl. drilling 3 wells, min.				Total				4,550	5,000
7710	Maximum				"				6,375	7,000
8000	X-ray concrete slabs				Ea.				182	200
9000	Thermographic testing, for bldg envelope heat loss, average 2,000 S.F. [G]				"					500

01 51 Temporary Utilities

01 51 13 – Temporary Electricity

01 51 13.80 Temporary Utilities

		Crew	Daily Output	Labor-Hours	Unit	Material	2012 Bare Costs Labor	Equipment	Total	Total Incl O&P
0010	**TEMPORARY UTILITIES**									
0100	Heat, incl. fuel and operation, per week, 12 hrs. per day	1 Skwk	100	.080	CSF Flr	36.50	3.63		40.13	45.50
0200	24 hrs. per day	"	60	.133		69.50	6.05		75.55	86
0350	Lighting, incl. service lamps, wiring & outlets, minimum	1 Elec	34	.235		2.80	12.15		14.95	21
0360	Maximum	"	17	.471		5.85	24.50		30.35	43
0400	Power for temp lighting only, 6.6 KWH, per month								.92	1.01
0430	11.8 KWH, per month								1.65	1.82
0450	23.6 KWH, per month								3.30	3.63
0600	Power for job duration incl. elevator, etc., minimum								47	51.50
0650	Maximum								110	121
1000	Toilet, portable, see Equip. Rental 01 54 33 in Reference Section									

01 52 Construction Facilities

01 52 13 – Field Offices and Sheds

01 52 13.20 Office and Storage Space	Crew	Daily Output	Labor-Hours	Unit	Material	2012 Bare Costs Labor	Equipment	Total	Total Incl O&P
0010	**OFFICE AND STORAGE SPACE**									
0020	Office trailer, furnished, no hookups, 20' x 8', buy	2 Skwk	1	16	Ea.	8,550	725		9,275	10,500
0250	Rent per month					144			144	158
0300	32' x 8', buy	2 Skwk	.70	22.857		13,600	1,050		14,650	16,600
0350	Rent per month					185			185	203
0400	50' x 10', buy	2 Skwk	.60	26.667		22,300	1,200		23,500	26,500
0450	Rent per month					279			279	305
0500	50' x 12', buy	2 Skwk	.50	32		27,000	1,450		28,450	32,000
0550	Rent per month					365			365	400
0700	For air conditioning, rent per month, add					41.50			41.50	45.50
0800	For delivery, add per mile				Mile	4.60			4.60	5.05
0890	Delivery each way				Ea.	200			200	220
0900	Bunk house trailer, 8' x 40' duplex dorm with kitchen, no hookups, buy	2 Carp	1	16		37,300	705		38,005	42,100
0910	9 man with kitchen and bath, no hookups, buy		1	16		38,000	705		38,705	42,900
0920	18 man sleeper with bath, no hookups, buy		1	16		49,000	705		49,705	55,000
1000	Portable buildings, prefab, on skids, economy, 8' x 8'		265	.060	S.F.	86	2.66		88.66	98.50
1100	Deluxe, 8' x 12'		150	.107	"	96	4.70		100.70	113
1200	Storage boxes, 20' x 8', buy	2 Skwk	1.80	8.889	Ea.	2,625	405		3,030	3,500
1250	Rent per month					72			72	79
1300	40' x 8', buy	2 Skwk	1.40	11.429		3,325	520		3,845	4,450
1350	Rent per month					94			94	103
5000	Air supported structures, see Section 13 31 13.13									

01 52 13.40 Field Office Expense

		Crew	Daily Output	Labor-Hours	Unit	Material	Labor	Equipment	Total	Total Incl O&P
0010	**FIELD OFFICE EXPENSE**									
0100	Office equipment rental average				Month	200			200	220
0120	Office supplies, average				"	75			75	82.50
0125	Office trailer rental, see Section 01 52 13.20									
0140	Telephone bill; avg. bill/month incl. long dist.				Month	81			81	89
0160	Lights & HVAC				"	152			152	167

01 54 Construction Aids

01 54 09 – Protection Equipment

01 54 09.50 Personnel Protective Equipment

		Crew	Daily Output	Labor-Hours	Unit	Material	Labor	Equipment	Total	Total Incl O&P
0010	**PERSONNEL PROTECTIVE EQUIPMENT**									
0015	Hazardous waste protection									
0020	Respirator mask only, full face, silicone				Ea.	270			270	297
0030	Half face, silicone					42.50			42.50	46.50
0040	Respirator cartridges, 2 req'd/mask, dust or asbestos					2.64			2.64	2.90
0050	Chemical vapor					3.74			3.74	4.11
0060	Combination vapor and dust					6.45			6.45	7.10
0100	Emergency escape breathing apparatus, 5 min					465			465	510
0110	10 min					500			500	550
0150	Self contained breathing apparatus with full face piece, 30 min					1,750			1,750	1,925
0160	60 min					2,925			2,925	3,225
0200	Encapsulating suits, limited use, level A					1,200			1,200	1,325
0210	Level B					335			335	370
0300	Over boots, latex				Pr.	6.35			6.35	7
0310	PVC					23			23	25.50
0320	Neoprene					44			44	48.50
0400	Gloves, nitrile/PVC					21			21	23.50
0410	Neoprene coated					27.50			27.50	30.50

01 54 Construction Aids

01 54 09 – Protection Equipment

01 54 09.60 Safety Nets

		Crew	Daily Output	Labor-Hours	Unit	Material	2012 Bare Costs Labor	Equipment	Total	Total Incl O&P
0010	**SAFETY NETS**									
0020	No supports, stock sizes, nylon, 4" mesh				S.F.	1.10			1.10	1.21
0100	Polypropylene, 6" mesh					1.59			1.59	1.75
0200	Small mesh debris nets, 1/4" & 3/4" mesh, stock sizes					.74			.74	.81
0220	Combined 4" mesh and 1/4" mesh, stock sizes					2.05			2.05	2.26
0300	Monthly rental, 4" mesh, stock sizes, 1st month					.50			.50	.55
0320	2nd month rental					.25			.25	.28
0340	Maximum rental/year					1.15			1.15	1.27

01 54 16 – Temporary Hoists

01 54 16.50 Weekly Forklift Crew

		Crew	Daily Output	Labor-Hours	Unit	Material	2012 Bare Costs Labor	Equipment	Total	Total Incl O&P
0010	**WEEKLY FORKLIFT CREW**									
0100	All-terrain forklift, 45' lift, 35' reach, 9000 lb. capacity	A-3P	.20	40	Week		1,800	2,400	4,200	5,350

01 54 19 – Temporary Cranes

01 54 19.50 Daily Crane Crews

		Crew	Daily Output	Labor-Hours	Unit	Material	2012 Bare Costs Labor	Equipment	Total	Total Incl O&P
0010	**DAILY CRANE CREWS** for small jobs, portal to portal									
0100	12-ton truck-mounted hydraulic crane	A-3H	1	8	Day		380	825	1,205	1,475
0200	25-ton	A-3I	1	8			380	955	1,335	1,625
0300	40-ton	A-3J	1	8			380	1,175	1,555	1,875
0400	55-ton	A-3K	1	16			710	1,625	2,335	2,850
0500	80-ton	A-3L	1	16			710	2,275	2,985	3,575
0600	100-ton	A-3M	1	16			710	2,275	2,985	3,575

01 54 19.60 Monthly Tower Crane Crew

		Crew	Daily Output	Labor-Hours	Unit	Material	2012 Bare Costs Labor	Equipment	Total	Total Incl O&P
0010	**MONTHLY TOWER CRANE CREW**, excludes concrete footing									
0100	Static tower crane, 130' high, 106' jib, 6200 lb. capacity	A-3N	.05	176	Month		8,425	22,100	30,525	37,000

01 54 23 – Temporary Scaffolding and Platforms

01 54 23.60 Pump Staging

		Crew	Daily Output	Labor-Hours	Unit	Material	2012 Bare Costs Labor	Equipment	Total	Total Incl O&P
0010	**PUMP STAGING**, Aluminum [R015423-20]									
0200	24' long pole section, buy				Ea.	330			330	365
0300	18' long pole section, buy					257			257	282
0400	12' long pole section, buy					173			173	190
0500	6' long pole section, buy					91.50			91.50	100
0600	6' long splice joint section, buy					67.50			67.50	74.50
0700	Pump jack, buy					136			136	149
0900	Foldable brace, buy					55			55	60.50
1000	Workbench/back safety rail support, buy					71.50			71.50	78.50
1100	Scaffolding planks/workbench, 14" wide x 24' long, buy					665			665	735
1200	Plank end safety rail, buy					288			288	315
1250	Safety net, 22' long, buy					320			320	355
1300	System in place, 50' working height, per use based on 50 uses	2 Carp	84.80	.189	C.S.F.	5.70	8.30		14	19.10
1400	100 uses		84.80	.189		2.86	8.30		11.16	15.95
1500	150 uses		84.80	.189		1.92	8.30		10.22	14.90

01 54 23.70 Scaffolding

		Crew	Daily Output	Labor-Hours	Unit	Material	2012 Bare Costs Labor	Equipment	Total	Total Incl O&P
0010	**SCAFFOLDING** [R015423-10]									
0015	Steel tube, regular, no plank, labor only to erect & dismantle									
0090	Building exterior, wall face, 1 to 5 stories, 6'-4" x 5' frames	3 Carp	8	3	C.S.F.		132		132	204
0200	6 to 12 stories	4 Carp	8	4			176		176	271
0301	13 to 20 stories	5 Clab	8	5			176		176	270
0460	Building interior, wall face area, up to 16' high	3 Carp	12	2			88		88	136
0560	16' to 40' high		10	2.400			106		106	163
0800	Building interior floor area, up to 30' high		150	.160	C.C.F.		7.05		7.05	10.85

01 54 Construction Aids

01 54 23 – Temporary Scaffolding and Platforms

01 54 23.70 Scaffolding	Crew	Daily Output	Labor-Hours	Unit	Material	2012 Bare Costs Labor	2012 Bare Costs Equipment	Total	Total Incl O&P
0900	Over 30' high	4 Carp	160	.200	C.C.F.		8.80		8.80	13.55
0906	Complete system for face of walls, no plank, material only rent/mo				C.S.F.	36			36	39.50
0908	Interior spaces, no plank, material only rent/mo				C.C.F.	3.42			3.42	3.76
0910	Steel tubular, heavy duty shoring, buy									
0920	Frames 5' high 2' wide				Ea.	81.50			81.50	90
0925	5' high 4' wide					93			93	102
0930	6' high 2' wide					93.50			93.50	103
0935	6' high 4' wide					109			109	120
0940	Accessories									
0945	Cross braces				Ea.	15.85			15.85	17.40
0950	U-head, 8" x 8"					19.10			19.10	21
0955	J-head, 4" x 8"					13.90			13.90	15.30
0960	Base plate, 8" x 8"					15.50			15.50	17.05
0965	Leveling jack					33.50			33.50	36.50
1000	Steel tubular, regular, buy									
1100	Frames 3' high 5' wide				Ea.	75			75	82.50
1150	5' high 5' wide					88			88	96.50
1200	6'-4" high 5' wide					118			118	129
1350	7'-6" high 6' wide					158			158	173
1500	Accessories cross braces					15.50			15.50	17.05
1550	Guardrail post					16.40			16.40	18.05
1600	Guardrail 7' section					6.30			6.30	6.95
1650	Screw jacks & plates					21.50			21.50	24
1700	Sidearm brackets					25.50			25.50	28.50
1750	8" casters					31			31	34
1800	Plank 2" x 10" x 16'-0"					24.50			24.50	27
1900	Stairway section					275			275	305
1910	Stairway starter bar					32			32	35
1920	Stairway inside handrail					53			53	58
1930	Stairway outside handrail					81.50			81.50	90
1940	Walk-thru frame guardrail					41.50			41.50	46
2000	Steel tubular, regular, rent/mo.									
2100	Frames 3' high 5' wide				Ea.	5			5	5.50
2150	5' high 5' wide					5			5	5.50
2200	6'-4" high 5' wide					5.15			5.15	5.65
2250	7'-6" high 6' wide					7			7	7.70
2500	Accessories, cross braces					1			1	1.10
2550	Guardrail post					1			1	1.10
2600	Guardrail 7' section					1			1	1.10
2650	Screw jacks & plates					2			2	2.20
2700	Sidearm brackets					2			2	2.20
2750	8" casters					8			8	8.80
2800	Outrigger for rolling tower					3			3	3.30
2850	Plank 2" x 10" x 16'-0"					6			6	6.60
2900	Stairway section					40			40	44
2940	Walk-thru frame guardrail					2.50			2.50	2.75
3000	Steel tubular, heavy duty shoring, rent/mo.									
3250	5' high 2' & 4' wide				Ea.	5			5	5.50
3300	6' high 2' & 4' wide					5			5	5.50
3500	Accessories, cross braces					1			1	1.10
3600	U - head, 8" x 8"					1			1	1.10
3650	J - head, 4" x 8"					1			1	1.10
3700	Base plate, 8" x 8"					1			1	1.10

01 54 Construction Aids

01 54 23 – Temporary Scaffolding and Platforms

01 54 23.70 Scaffolding

		Crew	Daily Output	Labor-Hours	Unit	Material	2012 Bare Costs Labor	Equipment	Total	Total Incl O&P
3750	Leveling jack				Ea.	2			2	2.20
5700	Planks, 2" x 10" x 16'-0", labor only to erect & remove to 50' H	3 Carp	72	.333	↓		14.70		14.70	22.50
5800	Over 50' high	4 Carp	80	.400	↓		17.65		17.65	27
6000	Heavy duty shoring for elevated slab forms to 8'-2" high, floor area									
6100	Labor only to erect & dismantle	4 Carp	16	2	C.S.F.		88		88	136
6110	Materials only, rent.mo				"	29.50			29.50	32.50
6500	To 14'-8" high									
6600	Labor only to erect & dismantle	4 Carp	10	3.200	C.S.F.		141		141	217
6610	Materials only, rent/mo				"	43			43	47.50

01 54 23.75 Scaffolding Specialties

		Crew	Daily Output	Labor-Hours	Unit	Material	2012 Bare Costs Labor	Equipment	Total	Total Incl O&P
0010	**SCAFFOLDING SPECIALTIES**									
1200	Sidewalk bridge, heavy duty steel posts & beams, including									
1210	parapet protection & waterproofing									
1220	8' to 10' wide, 2 posts	3 Carp	15	1.600	L.F.	32	70.50		102.50	144
1230	3 posts	"	10	2.400	"	48.50	106		154.50	217
1500	Sidewalk bridge using tubular steel									
1510	scaffold frames, including planking	3 Carp	45	.533	L.F.	5.60	23.50		29.10	42
1600	For 2 uses per month, deduct from all above					50%				
1700	For 1 use every 2 months, add to all above					100%				
1900	Catwalks, 20" wide, no guardrails, 7' span, buy				Ea.	133			133	146
2000	10' span, buy					186			186	205
3720	Putlog, standard, 8' span, with hangers, buy					66.50			66.50	73
3730	Rent per month					10			10	11
3750	12' span, buy					99			99	109
3755	Rent per month					15			15	16.50
3760	Trussed type, 16' span, buy					228			228	251
3770	Rent per month					20			20	22
3790	22' span, buy					274			274	300
3795	Rent per month					30			30	33
3800	Rolling ladders with handrails, 30" wide, buy, 2 step					229			229	252
4000	7 step					710			710	780
4050	10 step					985			985	1,075
4100	Rolling towers, buy, 5' wide, 7' long, 10' high					1,200			1,200	1,325
4200	For 5' high added sections, to buy, add				↓	207			207	228
4300	Complete incl. wheels, railings, outriggers,									
4350	21' high, to buy				Ea.	2,025			2,025	2,225
4400	Rent/month = 5% of purchase cost				"	113			113	124

01 54 23.80 Staging Aids

		Crew	Daily Output	Labor-Hours	Unit	Material	2012 Bare Costs Labor	Equipment	Total	Total Incl O&P
0010	**STAGING AIDS** and fall protection equipment									
0100	Sidewall staging bracket, tubular, buy				Ea.	45			45	49.50
0110	Cost each per day, based on 250 days use				Day	.18			.18	.20
0200	Guard post, buy				Ea.	19.85			19.85	22
0210	Cost each per day, based on 250 days use				Day	.08			.08	.09
0300	End guard chains, buy per pair				Pair	27			27	30
0310	Cost per set per day, based on 250 days use				Day	.18			.18	.20
1000	Roof shingling bracket, steel, buy				Ea.	8.90			8.90	9.80
1010	Cost each per day, based on 250 days use				Day	.04			.04	.04
1100	Wood bracket, buy				Ea.	16.80			16.80	18.50
1110	Cost each per day, based on 250 days use				Day	.07			.07	.07
2000	Ladder jack, aluminum, buy per pair				Pair	97			97	107
2010	Cost per pair per day, based on 250 days use				Day	.39			.39	.43
2100	Steel siderail jack, buy per pair				Pair	97.50			97.50	107

01 54 Construction Aids

01 54 23 – Temporary Scaffolding and Platforms

01 54 23.80 Staging Aids	Crew	Daily Output	Labor-Hours	Unit	Material	2012 Bare Costs Labor	2012 Bare Costs Equipment	Total	Total Incl O&P
2110 Cost per pair per day, based on 250 days use				Day	.39			.39	.43
3000 Laminated wood plank, 2" x 10" x 16', buy				Ea.	49.50			49.50	54.50
3010 Cost each per day, based on 250 days use				Day	.20			.20	.22
3100 Aluminum scaffolding plank, 20" wide x 24' long, buy				Ea.	785			785	865
3110 Cost each per day, based on 250 days use				Day	3.15			3.15	3.46
4000 Nylon full body harness, lanyard and rope grab				Ea.	172			172	189
4010 Cost each per day, based on 250 days use				Day	.69			.69	.76
4100 Rope for safety line, 5/8" x 100' nylon, buy				Ea.	56			56	61.50
4110 Cost each per day, based on 250 days use				Day	.22			.22	.25
4200 Permanent U-Bolt roof anchor, buy				Ea.	31			31	34
4300 Temporary (one use) roof ridge anchor, buy				"	32			32	35.50
5000 Installation (setup and removal) of staging aids									
5010 Sidewall staging bracket	2 Carp	64	.250	Ea.		11.05		11.05	16.95
5020 Guard post with 2 wood rails	"	64	.250			11.05		11.05	16.95
5030 End guard chains, set	1 Carp	64	.125			5.50		5.50	8.50
5100 Roof shingling bracket		96	.083			3.67		3.67	5.65
5200 Ladder jack		64	.125			5.50		5.50	8.50
5300 Wood plank, 2" x 10" x 16'	2 Carp	80	.200			8.80		8.80	13.55
5310 Aluminum scaffold plank, 20" x 24'	"	40	.400			17.65		17.65	27
5410 Safety rope	1 Carp	40	.200			8.80		8.80	13.55
5420 Permanent U-Bolt roof anchor (install only)	2 Carp	40	.400			17.65		17.65	27
5430 Temporary roof ridge anchor (install only)	1 Carp	64	.125			5.50		5.50	8.50

01 54 26 – Temporary Swing Staging

01 54 26.50 Swing Staging	Crew	Daily Output	Labor-Hours	Unit	Material	2012 Bare Costs Labor	2012 Bare Costs Equipment	Total	Total Incl O&P
0010 **SWING STAGING**, 500 lb. cap., 2' wide to 24' long, hand operated									
0020 steel cable type, with 60' cables, buy				Ea.	4,900			4,900	5,375
0030 Rent per month				"	490			490	540
0600 Lightweight (not for masons) 24' long for 150' height,									
0610 manual type, buy				Ea.	10,200			10,200	11,200
0620 Rent per month					1,025			1,025	1,125
0700 Powered, electric or air, to 150' high, buy					25,600			25,600	28,200
0710 Rent per month					1,800			1,800	1,975
0780 To 300' high, buy					26,000			26,000	28,600
0800 Rent per month					1,825			1,825	2,000
1000 Bosun's chair or work basket 3' x 3.5', to 300' high, electric, buy					10,400			10,400	11,400
1010 Rent per month					730			730	800
2200 Move swing staging (setup and remove)	E-4	2	16	Move		800	61	861	1,500

01 54 36 – Equipment Mobilization

01 54 36.50 Mobilization	Crew	Daily Output	Labor-Hours	Unit	Material	2012 Bare Costs Labor	2012 Bare Costs Equipment	Total	Total Incl O&P
0010 **MOBILIZATION** (Use line item again for demobilization) R015433-10									
0015 Up to 25 mi haul dist (50 mi RT for mob/demob crew)									
0020 Dozer, loader, backhoe, excav., grader, paver, roller, 70 to 150 H.P.	B-34N	4	2	Ea.		70.50	126	196.50	246
0100 Above 150 HP	B-34K	3	2.667			94	283	377	455
0300 Scraper, towed type (incl. tractor), 6 C.Y. capacity		3	2.667			94	283	377	455
0400 10 C.Y.		2.50	3.200			113	340	453	545
0600 Self-propelled scraper, 15 C.Y.		2.50	3.200			113	340	453	545
0700 24 C.Y.		2	4			141	425	566	680
0900 Shovel or dragline, 3/4 C.Y.		3.60	2.222			78.50	236	314.50	380
1000 1-1/2 C.Y.		3	2.667			94	283	377	455
1100 Small equipment, placed in rear of, or towed by pickup truck	A-3A	8	1			34.50	17.40	51.90	71
1150 Equip up to 70 HP, on flatbed trailer behind pickup truck	A-3D	4	2			68.50	62	130.50	172
2000 Crane, truck-mounted, up to 75 ton, (driver only, one-way)	1 Eqhv	7.20	1.111			53		53	80

01 54 Construction Aids

01 54 36 – Equipment Mobilization

01 54 36.50 Mobilization

		Crew	Daily Output	Labor-Hours	Unit	Material	2012 Bare Costs Labor	Equipment	Total	Total Incl O&P
2100	Crane, truck-mounted, over 75 ton	A-3E	2.50	6.400	Ea.		266	55.50	321.50	460
2200	Crawler-mounted, up to 75 ton	A-3F	2	8			330	435	765	985
2300	Over 75 ton	A-3G	1.50	10.667			445	660	1,105	1,400
2500	For each additional 5 miles haul distance, add						10%	10%		
3000	For large pieces of equipment, allow for assembly/knockdown									
3001	For mob/demob of vibrofloatation equip, see Section 31 45 13.10									
3100	For mob/demob of micro-tunneling equip, see Section 33 05 23.19									
3200	For mob/demob of pile driving equip, see Section 31 62 19.10									
3300	For mob/demob of caisson drilling equip, see Section 31 63 26.13									

01 54 39 – Construction Equipment

01 54 39.70 Small Tools

			Crew	Daily Output	Labor-Hours	Unit	Material	Labor	Equipment	Total	Total Incl O&P
0010	**SMALL TOOLS**	R013113-50									
0020	As % of contractor's bare labor cost for project, minimum					Total		.50%			
0100	Maximum					"		2%			

01 55 Vehicular Access and Parking

01 55 23 – Temporary Roads

01 55 23.50 Roads and Sidewalks

		Crew	Daily Output	Labor-Hours	Unit	Material	2012 Bare Costs Labor	Equipment	Total	Total Incl O&P
0010	**ROADS AND SIDEWALKS** Temporary									
0050	Roads, gravel fill, no surfacing, 4" gravel depth	B-14	715	.067	S.Y.	4.15	2.49	.47	7.11	8.90
0100	8" gravel depth	"	615	.078	"	8.30	2.89	.54	11.73	14.20
1000	Ramp, 3/4" plywood on 2" x 6" joists, 16" O.C.	2 Carp	300	.053	S.F.	1.23	2.35		3.58	4.98
1100	On 2" x 10" joists, 16" O.C.	"	275	.058	"	1.70	2.57		4.27	5.80

01 56 Temporary Barriers and Enclosures

01 56 13 – Temporary Air Barriers

01 56 13.60 Tarpaulins

		Crew	Daily Output	Labor-Hours	Unit	Material	2012 Bare Costs Labor	Equipment	Total	Total Incl O&P
0010	**TARPAULINS**									
0020	Cotton duck, 10 oz. to 13.13 oz. per S.Y., minimum				S.F.	.78			.78	.86
0050	Maximum					.80			.80	.88
0100	Polyvinyl coated nylon, 14 oz. to 18 oz., minimum					.48			.48	.53
0150	Maximum					.68			.68	.75
0200	Reinforced polyethylene 3 mils thick, white					.15			.15	.17
0300	4 mils thick, white, clear or black					.20			.20	.22
0400	5.5 mils thick, clear					.23			.23	.25
0500	White, fire retardant					.35			.35	.39
0600	12 mils, oil resistant, fire retardant					.28			.28	.31
0700	8.5 mils, black					.57			.57	.63
0710	Woven polyethylene, 6 mils thick					.35			.35	.39
0730	Polyester reinforced w/integral fastening system 11 mils thick					1.07			1.07	1.18
0740	Mylar polyester, non-reinforced, 7 mils thick					1.17			1.17	1.29

01 56 13.90 Winter Protection

		Crew	Daily Output	Labor-Hours	Unit	Material	2012 Bare Costs Labor	Equipment	Total	Total Incl O&P
0010	**WINTER PROTECTION**									
0100	Framing to close openings	2 Clab	750	.021	S.F.	.39	.75		1.14	1.58
0200	Tarpaulins hung over scaffolding, 8 uses, not incl. scaffolding		1500	.011		.25	.37		.62	.86
0250	Tarpaulin polyester reinf. w/integral fastening system 11 mils thick		1600	.010		.80	.35		1.15	1.42
0300	Prefab fiberglass panels, steel frame, 8 uses		1200	.013		.85	.47		1.32	1.66

01 56 Temporary Barriers and Enclosures

01 56 16 – Temporary Dust Barriers

01 56 16.10 Dust Barriers, Temporary	Crew	Daily Output	Labor-Hours	Unit	Material	2012 Bare Costs Labor	Equipment	Total	Total Incl O&P
0010 **DUST BARRIERS, TEMPORARY**									
0020 Spring loaded telescoping pole & head, to 12', erect and dismantle	1 Clab	240	.033	Ea.		1.17		1.17	1.80
0025 Cost per day (based upon 250 days)				Day	.24			.24	.26
0030 To 21', erect and dismantle	1 Clab	240	.033	Ea.		1.17		1.17	1.80
0035 Cost per day (based upon 250 days)				Day	.39			.39	.42
0040 Accessories, caution tape reel, erect and dismantle	1 Clab	480	.017	Ea.		.59		.59	.90
0045 Cost per day (based upon 250 days)				Day	.05			.05	.05
0060 Foam rail and connector, erect and dismantle	1 Clab	240	.033	Ea.		1.17		1.17	1.80
0065 Cost per day (based upon 250 days)				Day	.10			.10	.11
0070 Caution tape	1 Clab	384	.021	C.L.F.	2.65	.73		3.38	4.03
0080 Zipper, standard duty		60	.133	Ea.	8	4.68		12.68	16
0090 Heavy duty		48	.167	"	9.50	5.85		15.35	19.45
0100 Polyethylene sheet, 4 mil		37	.216	Sq.	3.23	7.60		10.83	15.25
0110 6 mil		37	.216	"	4.54	7.60		12.14	16.70
1000 Dust partition, 6 mil polyethylene, 1" x 3" frame	2 Carp	2000	.008	S.F.	.27	.35		.62	.84
1080 2" x 4" frame	"	2000	.008	"	.29	.35		.64	.86

01 56 23 – Temporary Barricades

01 56 23.10 Barricades	Crew	Daily Output	Labor-Hours	Unit	Material	2012 Bare Costs Labor	Equipment	Total	Total Incl O&P
0010 **BARRICADES**									
0020 5' high, 3 rail @ 2" x 8", fixed	2 Carp	20	.800	L.F.	5.30	35.50		40.80	60.50
0150 Movable	"	30	.533	"	4.35	23.50		27.85	41
0300 Stock units, 6' high, 8' wide, plain, buy				Ea.	475			475	520
0350 With reflective tape, buy				"	540			540	595
0400 Break-a-way 3" PVC pipe barricade									
0410 with 3 ea. 1' x 4' reflectorized panels, buy				Ea.	160			160	176
0500 Plywood with steel legs, 32" wide					99.50			99.50	109
0600 Warning signal flag tree, 11' high, 2 flags, buy					330			330	365
0800 Traffic cones, PVC, 18" high					10			10	11
0850 28" high					11.60			11.60	12.80
1000 Guardrail, wooden, 3' high, 1" x 6", on 2" x 4" posts	2 Carp	200	.080	L.F.	1.15	3.53		4.68	6.70
1100 2" x 6", on 4" x 4" posts	"	165	.097		2.26	4.28		6.54	9.10
1200 Portable metal with base pads, buy					15.25			15.25	16.75
1250 Typical installation, assume 10 reuses	2 Carp	600	.027		2.50	1.18		3.68	4.56
1300 Barricade tape, polyethylene, 7 mil, 3" wide x 500' long roll				Ea.	25			25	27.50
5000 Barricades, see Section 01 54 33.40									

01 56 26 – Temporary Fencing

01 56 26.50 Temporary Fencing	Crew	Daily Output	Labor-Hours	Unit	Material	2012 Bare Costs Labor	Equipment	Total	Total Incl O&P
0010 **TEMPORARY FENCING**									
0020 Chain link, 11 ga, 5' high	2 Clab	400	.040	L.F.	4.50	1.40		5.90	7.10
0100 6' high		300	.053		5.30	1.87		7.17	8.75
0200 Rented chain link, 6' high, to 1000' (up to 12 mo.)		400	.040		6.30	1.40		7.70	9.10
0250 Over 1000' (up to 12 mo.)		300	.053		3.29	1.87		5.16	6.50
0350 Plywood, painted, 2" x 4" frame, 4' high	A-4	135	.178		5.45	7.50		12.95	17.45
0400 4" x 4" frame, 8' high	"	110	.218		10.20	9.20		19.40	25.50
0500 Wire mesh on 4" x 4" posts, 4' high	2 Carp	100	.160		9.25	7.05		16.30	21
0550 8' high	"	80	.200		13.90	8.80		22.70	29

01 56 29 – Temporary Protective Walkways

01 56 29.50 Protection	Crew	Daily Output	Labor-Hours	Unit	Material	2012 Bare Costs Labor	Equipment	Total	Total Incl O&P
0010 **PROTECTION**									
0020 Stair tread, 2" x 12" planks, 1 use	1 Carp	75	.107	Tread	3.83	4.70		8.53	11.45

01 56 Temporary Barriers and Enclosures

01 56 29 – Temporary Protective Walkways

01 56 29.50 Protection

		Crew	Daily Output	Labor-Hours	Unit	Material	2012 Bare Costs Labor	Equipment	Total	Total Incl O&P
0100	Exterior plywood, 1/2" thick, 1 use	1 Carp	65	.123	Tread	1.41	5.45		6.86	9.90
0200	3/4" thick, 1 use		60	.133	↓	2.10	5.90		8	11.35
2200	Sidewalks, 2" x 12" planks, 2 uses		350	.023	S.F.	.64	1.01		1.65	2.25
2300	Exterior plywood, 2 uses, 1/2" thick		750	.011		.24	.47		.71	.98
2400	5/8" thick		650	.012		.31	.54		.85	1.18
2500	3/4" thick		600	.013	↓	.35	.59		.94	1.29

01 56 32 – Temporary Security

01 56 32.50 Watchman

		Crew	Daily Output	Labor-Hours	Unit	Material	2012 Bare Costs Labor	Equipment	Total	Total Incl O&P
0010	**WATCHMAN**									
0020	Service, monthly basis, uniformed person, minimum				Hr.				25	27.50
0100	Maximum								45.50	50
0200	Person and command dog, minimum								31	34
0300	Maximum				↓				54.50	60
0500	Sentry dog, leased, with job patrol (yard dog), 1 dog				Week				290	320
0600	2 dogs				"				390	430
0800	Purchase, trained sentry dog, minimum				Ea.				1,375	1,500
0900	Maximum				"				2,725	3,000

01 58 Project Identification

01 58 13 – Temporary Project Signage

01 58 13.50 Signs

		Crew	Daily Output	Labor-Hours	Unit	Material	2012 Bare Costs Labor	Equipment	Total	Total Incl O&P
0010	**SIGNS**									
0020	High intensity reflectorized, no posts, buy				S.F.	34			34	37.50

01 71 Examination and Preparation

01 71 23 – Field Engineering

01 71 23.13 Construction Layout

		Crew	Daily Output	Labor-Hours	Unit	Material	2012 Bare Costs Labor	Equipment	Total	Total Incl O&P
0010	**CONSTRUCTION LAYOUT**									
1100	Crew for layout of building, trenching or pipe laying, 2 person crew	A-6	1	16	Day		705	78.50	783.50	1,150
1200	3 person crew	A-7	1	24			1,150	78.50	1,228.50	1,850
1400	Crew for roadway layout, 4 person crew	A-8	1	32	↓		1,500	78.50	1,578.50	2,375

01 71 23.19 Surveyor Stakes

		Crew	Daily Output	Labor-Hours	Unit	Material	2012 Bare Costs Labor	Equipment	Total	Total Incl O&P
0010	**SURVEYOR STAKES**									
0020	Hardwood, 1" x 1" x 48" long				C	66			66	72.50
0100	2" x 2" x 18" long					74			74	81.50
0150	2" x 2" x 24" long				↓	130			130	143

01 74 Cleaning and Waste Management

01 74 13 - Progress Cleaning

01 74 13.20 Cleaning Up	Crew	Daily Output	Labor-Hours	Unit	Material	2012 Bare Costs Labor	Equipment	Total	Total Incl O&P
0010 **CLEANING UP**									
0020 After job completion, allow, minimum				Job					.30%
0040 Maximum				"					1%
0050 Cleanup of floor area, continuous, per day, during const.	A-5	24	.750	M.S.F.	1.75	26.50	2.20	30.45	45
0100 Final by GC at end of job	"	11.50	1.565	"	2.78	55	4.59	62.37	92
0200 Rubbish removal, see Section 02 41 19.23									

01 76 Protecting Installed Construction

01 76 13 - Temporary Protection of Installed Construction

01 76 13.20 Temporary Protection	Crew	Daily Output	Labor-Hours	Unit	Material	2012 Bare Costs Labor	Equipment	Total	Total Incl O&P
0010 **TEMPORARY PROTECTION**									
0020 Flooring, 1/8" tempered hardboard, taped seams	2 Carp	1500	.011	S.F.	.38	.47		.85	1.14
0030 Peel away carpet protection	1 Clab	3200	.003	"	.11	.09		.20	.26

01 91 Commissioning

01 91 13 - General Commissioning Requirements

01 91 13.50 Building Commissioning	Crew	Daily Output	Labor-Hours	Unit	Material	2012 Bare Costs Labor	Equipment	Total	Total Incl O&P
0010 **BUILDING COMMISSIONING**									
0100 Basic building commissioning, minimum				%					.25%
0150 Maximum									.50%
0200 Enhanced building commissioning, minimum									.50%
0250 Maximum				↓					1%

Division Notes

	CREW	DAILY OUTPUT	LABOR-HOURS	UNIT	2012 BARE COSTS				TOTAL INCL O&P
					MAT.	LABOR	EQUIP.	TOTAL	

Estimating Tips

02 30 00 Subsurface Investigation

In preparing estimates on structures involving earthwork or foundations, all information concerning soil characteristics should be obtained. Look particularly for hazardous waste, evidence of prior dumping of debris, and previous stream beds.

02 41 00 Demolition and Structure Moving

The costs shown for selective demolition do not include rubbish handling or disposal. These items should be estimated separately using RSMeans data or other sources.

- Historic preservation often requires that the contractor remove materials from the existing structure, rehab them, and replace them. The estimator must be aware of any related measures and precautions that must be taken when doing selective demolition and cutting and patching. Requirements may include special handling and storage, as well as security.

- In addition to Subdivision 02 41 00, you can find selective demolition items in each division. Example: Roofing demolition is in Division 7.

02 42 10 Building Deconstruction

This section provides costs for the careful dismantling and recycling of most of low-rise building materials.

02 56 13 Containment of Hazardous Waste

This section addresses on-site hazardous waste disposal costs.

02 80 00 Hazardous Material Disposal/ Remediation

This subdivision includes information on hazardous waste handling, asbestos remediation, lead remediation, and mold remediation. See reference R028213-20 and R028319-60 for further guidance in using these unit price lines.

02 91 10 Monitoring Chemical Sampling, Testing Analysis

This section provides costs for on-site sampling and testing hazardous waste.

Reference Numbers

Reference numbers are shown in shaded boxes at the beginning of some major classifications. These numbers refer to related items in the Reference Section. The reference information may be an estimating procedure, an alternate pricing method, or technical information.

Note: Not all subdivisions listed here necessarily appear in this publication.

02 21 13 – Site Surveys

02 21 13.09 Topographical Surveys

02 21 13.09 Topographical Surveys	Crew	Daily Output	Labor-Hours	Unit	Material	2012 Bare Costs Labor	Equipment	Total	Total Incl O&P
0010 **TOPOGRAPHICAL SURVEYS**									
0020 Topographical surveying, conventional, minimum	A-7	3.30	7.273	Acre	18.90	350	24	392.90	590
0100 Maximum	A-8	.60	53.333	"	58	2,500	131	2,689	4,025

02 21 13.13 Boundary and Survey Markers

	Crew	Daily Output	Labor-Hours	Unit	Material	Labor	Equipment	Total	Total Incl O&P
0010 **BOUNDARY AND SURVEY MARKERS**									
0300 Lot location and lines, large quantities, minimum	A-7	2	12	Acre	33.50	580	39.50	653	970
0320 Average	"	1.25	19.200		53.50	925	63	1,041.50	1,550
0400 Small quantities, maximum	A-8	1	32	↓	71.50	1,500	78.50	1,650	2,475
0600 Monuments, 3' long	A-7	10	2.400	Ea.	31.50	116	7.85	155.35	221
0800 Property lines, perimeter, cleared land	"	1000	.024	L.F.	.04	1.16	.08	1.28	1.91
0900 Wooded land	A-8	875	.037	"	.06	1.71	.09	1.86	2.79

02 21 13.16 Aerial Surveys

	Crew	Daily Output	Labor-Hours	Unit	Material	Labor	Equipment	Total	Total Incl O&P
0010 **AERIAL SURVEYS**									
1500 Aerial surveying, including ground control, minimum fee, 10 acres				Total					4,400
1510 100 acres									8,800
1550 From existing photography, deduct				↓					1,500
1600 2' contours, 10 acres				Acre					160
1650 20 acres									80
1800 50 acres									35
1850 100 acres									28
2000 1000 acres									28
2050 10,000 acres				↓					28
2150 For 1' contours and									
2160 dense urban areas, add to above				Acre					20%
3000 Inertial guidance system for									
3010 locating coordinates, rent per day				Ea.					4,400

02 32 Geotechnical Investigations

02 32 13 – Subsurface Drilling and Sampling

02 32 13.10 Boring and Exploratory Drilling

	Crew	Daily Output	Labor-Hours	Unit	Material	Labor	Equipment	Total	Total Incl O&P
0010 **BORING AND EXPLORATORY DRILLING**									
0020 Borings, initial field stake out & determination of elevations	A-6	1	16	Day		705	78.50	783.50	1,150
0100 Drawings showing boring details				Total		310		310	390
0200 Report and recommendations from P.E.				↓		720		720	900
0300 Mobilization and demobilization	B-55	4	6	↓		209	246	455	590
0350 For over 100 miles, per added mile		450	.053	Mile		1.86	2.18	4.04	5.25
0600 Auger holes in earth, no samples, 2-1/2" diameter		78.60	.305	L.F.		10.65	12.50	23.15	30
0650 4" diameter		67.50	.356			12.40	14.55	26.95	35
0800 Cased borings in earth, with samples, 2-1/2" diameter		55.50	.432		22	15.05	17.70	54.75	66.50
0850 4" diameter	↓	32.60	.736		34.50	25.50	30	90	111
1000 Drilling in rock, "BX" core, no sampling	B-56	34.90	.458			18.30	40.50	58.80	72.50
1050 With casing & sampling		31.70	.505		22	20	45	87	104
1200 "NX" core, no sampling		25.92	.617			24.50	55	79.50	98
1250 With casing and sampling	↓	25	.640	↓	26.50	25.50	57	109	131
1400 Borings, earth, drill rig and crew with truck mounted auger	B-55	1	24	Day		835	985	1,820	2,350
1450 Rock using crawler type drill	B-56	1	16	"		640	1,425	2,065	2,525
1500 For inner city borings add, minimum									10%
1510 Maximum									20%

02 32 Geotechnical Investigations

02 32 19 – Exploratory Excavations

02 32 19.10 Test Pits		Crew	Daily Output	Labor-Hours	Unit	Material	2012 Bare Costs Labor	2012 Bare Costs Equipment	Total	Total Incl O&P
0010	**TEST PITS**									
0020	Hand digging, light soil	1 Clab	4.50	1.778	C.Y.		62.50		62.50	96
0100	Heavy soil	"	2.50	3.200			112		112	173
0120	Loader-backhoe, light soil	B-11M	28	.571			23.50	13	36.50	50
0130	Heavy soil	"	20	.800	▼		32.50	18.20	50.70	69.50
1000	Subsurface exploration, mobilization				Mile				6.25	7.85
1010	Difficult access for rig, add				Hr.				240	300
1020	Auger borings, drill rig, incl. samples				L.F.				23	28
1030	Hand auger								29	36
1050	Drill and sample every 5', split spoon				▼				29	36
1060	Extra samples				Ea.				35	44

02 41 Demolition

02 41 13 – Selective Site Demolition

02 41 13.15 Hydrodemolition

02 41 13.15		Crew	Daily Output	Labor-Hours	Unit	Material	Labor	Equipment	Total	Total Incl O&P
0010	**HYDRODEMOLITION** R024119-10									
0015	Hydrodemolition, concrete pavement, 4000 PSI, 2" depth	B-5	500	.112	S.F.		4.33	2.63	6.96	9.50
0120	4" depth		450	.124			4.81	2.92	7.73	10.55
0130	6" depth		400	.140			5.40	3.29	8.69	11.85
0410	6000 PSI, 2" depth		410	.137			5.30	3.21	8.51	11.60
0420	4" depth		350	.160			6.20	3.76	9.96	13.60
0430	6" depth		300	.187			7.20	4.38	11.58	15.85
0510	8000 PSI, 2" depth		330	.170			6.55	3.99	10.54	14.45
0520	4" depth		280	.200			7.75	4.70	12.45	16.95
0530	6" depth	▼	240	.233	▼		9	5.50	14.50	19.85

02 41 13.17 Demolish, Remove Pavement and Curb

02 41 13.17		Crew	Daily Output	Labor-Hours	Unit	Material	Labor	Equipment	Total	Total Incl O&P
0010	**DEMOLISH, REMOVE PAVEMENT AND CURB** R024119-10									
5010	Pavement removal, bituminous roads, 3" thick	B-38	690	.058	S.Y.		2.30	1.71	4.01	5.40
5050	4" to 6" thick		420	.095			3.78	2.82	6.60	8.85
5100	Bituminous driveways		640	.063			2.48	1.85	4.33	5.80
5200	Concrete to 6" thick, hydraulic hammer, mesh reinforced		255	.157			6.25	4.64	10.89	14.60
5300	Rod reinforced		200	.200	▼		7.95	5.90	13.85	18.60
5400	Concrete, 7" to 24" thick, plain		33	1.212	C.Y.		48	36	84	113
5500	Reinforced	▼	24	1.667	"		66	49.50	115.50	155
5600	With hand held air equipment, bituminous, to 6" thick	B-39	1900	.025	S.F.		.94	.11	1.05	1.55
5700	Concrete to 6" thick, no reinforcing		1600	.030			1.11	.13	1.24	1.84
5800	Mesh reinforced		1400	.034			1.27	.15	1.42	2.11
5900	Rod reinforced	▼	765	.063	▼		2.32	.27	2.59	3.86
6000	Curbs, concrete, plain	B-6	360	.067	L.F.		2.55	.93	3.48	4.92
6100	Reinforced		275	.087			3.34	1.21	4.55	6.45
6200	Granite		360	.067			2.55	.93	3.48	4.92
6300	Bituminous	▼	528	.045	▼		1.74	.63	2.37	3.36

02 41 13.33 Minor Site Demolition

02 41 13.33		Crew	Daily Output	Labor-Hours	Unit	Material	Labor	Equipment	Total	Total Incl O&P
0010	**MINOR SITE DEMOLITION** R024119-10									
0015	No hauling, abandon catch basin or manhole	B-6	7	3.429	Ea.		131	47.50	178.50	254
0020	Remove existing catch basin or manhole, masonry		4	6			230	83.50	313.50	440
0030	Catch basin or manhole frames and covers, stored		13	1.846			70.50	25.50	96	137
0040	Remove and reset	▼	7	3.429			131	47.50	178.50	254
0100	Roadside delineators, remove only	B-80	175	.183			6.90	3.82	10.72	14.75
0110	Remove and reset	"	100	.320	▼		12.10	6.70	18.80	26

02 41 Demolition

02 41 13 – Selective Site Demolition

02 41 13.33 Minor Site Demolition

		Crew	Daily Output	Labor-Hours	Unit	Material	2012 Bare Costs Labor	2012 Bare Costs Equipment	Total	Total Incl O&P
0800	Guiderail, corrugated steel, remove only	B-80A	100	.240	L.F.		8.40	2.64	11.04	15.85
0850	Remove and reset	"	40	.600	"		21	6.60	27.60	40
0860	Guide posts, remove only	B-80B	120	.267	Ea.		10	1.94	11.94	17.45
0870	Remove and reset	B-55	50	.480			16.75	19.65	36.40	47
0900	Hydrants, fire, remove only	B-21A	5	8			350	89.50	439.50	630
0950	Remove and reset	"	2	20	↓		875	224	1,099	1,575
1000	Masonry walls, block, solid	B-5	1800	.031	C.F.		1.20	.73	1.93	2.64
1200	Brick, solid		900	.062			2.41	1.46	3.87	5.30
1400	Stone, with mortar		900	.062			2.41	1.46	3.87	5.30
1500	Dry set	↓	1500	.037	↓		1.44	.88	2.32	3.17
1600	Median barrier, precast concrete, remove and store	B-3	430	.112	L.F.		4.18	5.30	9.48	12.15
1610	Remove and reset	"	390	.123			4.60	5.85	10.45	13.45
2900	Pipe removal, sewer/water, no excavation, 12" diameter	B-6	175	.137			5.25	1.91	7.16	10.15
2930	15"-18" diameter	B-12Z	150	.160			6.30	11	17.30	21.50
2960	21"-24" diameter		120	.200			7.85	13.75	21.60	27
3000	27"-36" diameter	↓	90	.267			10.50	18.35	28.85	36
3200	Steel, welded connections, 4" diameter	B-6	160	.150			5.75	2.09	7.84	11.10
3300	10" diameter	"	80	.300			11.50	4.18	15.68	22
3500	Railroad track removal, ties and track	B-13	330	.170	↓		6.45	2.11	8.56	12.20
3600	Ballast	B-14	500	.096	C.Y.		3.56	.67	4.23	6.20
3700	Remove and re-install, ties & track using new bolts & spikes		50	.960	L.F.		35.50	6.70	42.20	62
3800	Turnouts using new bolts and spikes	↓	1	48	Ea.		1,775	335	2,110	3,100
4000	Sidewalk removal, bituminous, 2-1/2" thick	B-6	325	.074	S.Y.		2.83	1.03	3.86	5.45
4050	Brick, set in mortar		185	.130			4.97	1.81	6.78	9.60
4100	Concrete, plain, 4"		160	.150			5.75	2.09	7.84	11.10
4200	Mesh reinforced	↓	150	.160	↓		6.15	2.23	8.38	11.80
4300	Slab on grade removal, plain	B-5	45	1.244	C.Y.		48	29	77	106
4310	Mesh reinforced		33	1.697			65.50	40	105.50	144
4320	Rod reinforced	↓	25	2.240			86.50	52.50	139	190
4400	For congested sites or small quantities, add up to								200%	200%
4450	For disposal on site, add	B-11A	232	.069			2.82	5.35	8.17	10.20
4500	To 5 miles, add	B-34D	76	.105	↓		3.72	8.70	12.42	15.20

02 41 13.60 Selective Demolition Fencing

		Crew	Daily Output	Labor-Hours	Unit	Material	2012 Bare Costs Labor	2012 Bare Costs Equipment	Total	Total Incl O&P
0010	**SELECTIVE DEMOLITION FENCING** R024119-10									
1600	Fencing, barbed wire, 3 strand	2 Clab	430	.037	L.F.		1.31		1.31	2.01
1650	5 strand	"	280	.057			2.01		2.01	3.09
1700	Chain link, posts & fabric, 8' to 10' high, remove only	B-6	445	.054			2.07	.75	2.82	3.99
1750	Remove and reset	"	70	.343	↓		13.15	4.77	17.92	25.50

02 41 16 – Structure Demolition

02 41 16.13 Building Demolition

		Crew	Daily Output	Labor-Hours	Unit	Material	2012 Bare Costs Labor	2012 Bare Costs Equipment	Total	Total Incl O&P
0010	**BUILDING DEMOLITION** Large urban projects, incl. 20 mi. haul R024119-10									
0011	No foundation or dump fees, C.F. is vol. of building standing									
0020	Steel	B-8	21500	.003	C.F.		.12	.14	.26	.33
0050	Concrete		15300	.004			.16	.19	.35	.46
0080	Masonry		20100	.003			.12	.15	.27	.35
0100	Mixture of types	↓	20100	.003			.12	.15	.27	.35
0500	Small bldgs, or single bldgs, no salvage included, steel	B-3	14800	.003			.12	.15	.27	.35
0600	Concrete		11300	.004			.16	.20	.36	.46
0650	Masonry		14800	.003			.12	.15	.27	.35
0700	Wood	↓	14800	.003			.12	.15	.27	.35
0750	For buildings with no interior walls, deduct				↓				50%	
1000	Demoliton single family house, one story, wood 1600 S.F.	B-3	1	48	Ea.		1,800	2,275	4,075	5,250

02 41 Demolition

02 41 16 – Structure Demolition

02 41 16.13 Building Demolition

02 41 16.13 Building Demolition		Crew	Daily Output	Labor-Hours	Unit	Material	2012 Bare Costs Labor	2012 Bare Costs Equipment	Total	Total Incl O&P
1020	3200 S.F.	B-3	.50	96	Ea.		3,600	4,550	8,150	10,500
1200	Demoliton two family house, two story, wood 2400 S.F.		.67	71.964			2,700	3,400	6,100	7,850
1220	4200 S.F.		.38	128			4,800	6,075	10,875	14,000
1300	Demoliton three family house, three story, wood 3200 S.F.		.50	96			3,600	4,550	8,150	10,500
1320	5400 S.F.	↓	.30	160			5,975	7,575	13,550	17,500
5000	For buildings with no interior walls, deduct				↓			50%		

02 41 16.15 Explosive/Implosive Demolition

		Crew	Daily Output	Labor-Hours	Unit	Material	Labor	Equipment	Total	Total Incl O&P
0010	**EXPLOSIVE/IMPLOSIVE DEMOLITION** R024119-10									
0011	Large projects,									
0020	No disposal fee based on building volume, steel building	B-5B	16900	.003	C.F.		.12	.15	.27	.34
0100	Concrete building		16900	.003			.12	.15	.27	.34
0200	Masonry building	↓	16900	.003	↓		.12	.15	.27	.34
0400	Disposal of material, minimum	B-3	445	.108	C.Y.		4.04	5.10	9.14	11.75
0500	Maximum	"	365	.132	"		4.92	6.25	11.17	14.35

02 41 16.17 Building Demolition Footings and Foundations

		Crew	Daily Output	Labor-Hours	Unit	Material	Labor	Equipment	Total	Total Incl O&P
0010	**BUILDING DEMOLITION FOOTINGS AND FOUNDATIONS** R024119-10									
0200	Floors, concrete slab on grade,									
0240	4" thick, plain concrete	B-9	500	.080	S.F.		2.84	.42	3.26	4.83
0280	Reinforced, wire mesh		470	.085			3.02	.45	3.47	5.15
0300	Rods		400	.100			3.55	.52	4.07	6.05
0400	6" thick, plain concrete		375	.107			3.79	.56	4.35	6.45
0420	Reinforced, wire mesh		340	.118			4.18	.62	4.80	7.15
0440	Rods		300	.133			4.73	.70	5.43	8.05
1000	Footings, concrete, 1' thick, 2' wide	B-5	300	.187	L.F.		7.20	4.38	11.58	15.85
1080	1'-6" thick, 2' wide		250	.224			8.65	5.25	13.90	19.05
1120	3' wide		200	.280			10.80	6.60	17.40	24
1140	2' thick, 3' wide	↓	175	.320			12.35	7.50	19.85	27
1200	Average reinforcing, add								10%	10%
1220	Heavy reinforcing, add				↓				20%	20%
2000	Walls, block, 4" thick	1 Clab	180	.044	S.F.		1.56		1.56	2.40
2040	6" thick		170	.047			1.65		1.65	2.54
2080	8" thick		150	.053			1.87		1.87	2.88
2100	12" thick	↓	150	.053			1.87		1.87	2.88
2200	For horizontal reinforcing, add								10%	10%
2220	For vertical reinforcing, add								20%	20%
2400	Concrete, plain concrete, 6" thick	B-9	160	.250			8.90	1.31	10.21	15.10
2420	8" thick		140	.286			10.15	1.49	11.64	17.25
2440	10" thick		120	.333			11.85	1.74	13.59	20
2500	12" thick	↓	100	.400			14.20	2.09	16.29	24.50
2600	For average reinforcing, add								10%	10%
2620	For heavy reinforcing, add								20%	20%
4000	For congested sites or small quantities, add up to				↓				200%	200%
4200	Add for disposal, on site	B-11A	232	.069	C.Y.		2.82	5.35	8.17	10.20
4250	To five miles	B-30	220	.109	"		4.26	9.70	13.96	17.10

02 41 19 – Selective Demolition

02 41 19.13 Selective Building Demolition

0010	**SELECTIVE BUILDING DEMOLITION**									
0020	Costs related to selective demolition of specific building components									
0025	are included under Common Work Results (XX 05)									
0030	in the component's appropriate division.									

02 41 19 – Selective Demolition

02 41 19.16 Selective Demolition, Cutout		Crew	Daily Output	Labor-Hours	Unit	Material	2012 Bare Costs Labor	2012 Bare Costs Equipment	Total	Total Incl O&P
0010	**SELECTIVE DEMOLITION, CUTOUT**	R024119-10								
0020	Concrete, elev. slab, light reinforcement, under 6 C.F.	B-9	65	.615	C.F.		22	3.22	25.22	37
0050	Light reinforcing, over 6 C.F.		75	.533	"		18.95	2.79	21.74	32
0200	Slab on grade to 6" thick, not reinforced, under 8 S.F.		85	.471	S.F.		16.70	2.46	19.16	28
0250	8 – 16 S.F.		175	.229	"		8.10	1.20	9.30	13.80
0255	For over 16 S.F. see Line 02 41 16.17 0400									
0600	Walls, not reinforced, under 6 C.F.	B-9	60	.667	C.F.		23.50	3.49	26.99	40.50
0650	6 – 12 C.F.	"	80	.500	"		17.75	2.62	20.37	30.50
0655	For over 12 C.F. see Line 02 41 16.17 2500									
1000	Concrete, elevated slab, bar reinforced, under 6 C.F.	B-9	45	.889	C.F.		31.50	4.65	36.15	53.50
1050	Bar reinforced, over 6 C.F.		50	.800	"		28.50	4.18	32.68	48
1200	Slab on grade to 6" thick, bar reinforced, under 8 S.F.		75	.533	S.F.		18.95	2.79	21.74	32
1250	8 – 16 S.F.		150	.267	"		9.45	1.39	10.84	16.10
1255	For over 16 S.F. see Line 02 41 16.17 0440									
1400	Walls, bar reinforced, under 6 C.F.	B-9	50	.800	C.F.		28.50	4.18	32.68	48
1450	6 – 12 C.F.	"	70	.571	"		20.50	2.99	23.49	34.50
1455	For over 12 C.F. see Lines 02 41 16.17 2500 and 2600									
2000	Brick, to 4 S.F. opening, not including toothing									
2040	4" thick	B-9	30	1.333	Ea.		47.50	6.95	54.45	80.50
2060	8" thick		18	2.222			79	11.60	90.60	134
2080	12" thick		10	4			142	21	163	241
2400	Concrete block, to 4 S.F. opening, 2" thick		35	1.143			40.50	6	46.50	69
2420	4" thick		30	1.333			47.50	6.95	54.45	80.50
2440	8" thick		27	1.481			52.50	7.75	60.25	89.50
2460	12" thick		24	1.667			59	8.70	67.70	101
2600	Gypsum block, to 4 S.F. opening, 2" thick		80	.500			17.75	2.62	20.37	30.50
2620	4" thick		70	.571			20.50	2.99	23.49	34.50
2640	8" thick		55	.727			26	3.80	29.80	43.50
2800	Terra cotta, to 4 S.F. opening, 4" thick		70	.571			20.50	2.99	23.49	34.50
2840	8" thick		65	.615			22	3.22	25.22	37
2880	12" thick		50	.800			28.50	4.18	32.68	48
3000	Toothing masonry cutouts, brick, soft old mortar	1 Brhe	40	.200	V.L.F.		7.10		7.10	10.75
3100	Hard mortar		30	.267			9.50		9.50	14.35
3200	Block, soft old mortar		70	.114			4.07		4.07	6.15
3400	Hard mortar		50	.160			5.70		5.70	8.60
6000	Walls, interior, not including re-framing,									
6010	openings to 5 S.F.									
6100	Drywall to 5/8" thick	1 Clab	24	.333	Ea.		11.70		11.70	18
6200	Paneling to 3/4" thick		20	.400			14.05		14.05	21.50
6300	Plaster, on gypsum lath		20	.400			14.05		14.05	21.50
6340	On wire lath		14	.571			20		20	31
7000	Wood frame, not including re-framing, openings to 5 S.F.									
7200	Floors, sheathing and flooring to 2" thick	1 Clab	5	1.600	Ea.		56		56	86.50
7310	Roofs, sheathing to 1" thick, not including roofing		6	1.333			47		47	72
7410	Walls, sheathing to 1" thick, not including siding		7	1.143			40		40	61.50

02 41 19.18 Selective Demolition, Disposal Only

02 41 19.18 Selective Demolition, Disposal Only		Crew	Daily Output	Labor-Hours	Unit	Material	2012 Bare Costs Labor	2012 Bare Costs Equipment	Total	Total Incl O&P
0010	**SELECTIVE DEMOLITION, DISPOSAL ONLY**	R024119-10								
0015	Urban bldg w/salvage value allowed									
0020	Including loading and 5 mile haul to dump									
0200	Steel frame	B-3	430	.112	C.Y.		4.18	5.30	9.48	12.15
0300	Concrete frame		365	.132			4.92	6.25	11.17	14.35
0400	Masonry construction		445	.108			4.04	5.10	9.14	11.75

02 41 Demolition

02 41 19 – Selective Demolition

02 41 19.18 Selective Demolition, Disposal Only

		Crew	Daily Output	Labor-Hours	Unit	Material	2012 Bare Costs		Total	Total Incl O&P
							Labor	Equipment		
0500	Wood frame	B-3	247	.194	C.Y.		7.25	9.20	16.45	21.50

02 41 19.19 Selective Facility Services Demolition

		Crew	Daily Output	Labor-Hours	Unit	Material	Labor	Equipment	Total	Total Incl O&P
0010	**SELECTIVE FACILITY SERVICES DEMOLITION**, Rubbish Handling R024119-10									
0020	The following are to be added to the demolition prices									
0400	Chute, circular, prefabricated steel, 18" diameter	B-1	40	.600	L.F.	52	21.50		73.50	90.50
0440	30" diameter	"	30	.800	"	46.50	28.50		75	95
0600	Dumpster, weekly rental, 1 dump/week, 6 C.Y. capacity (2 Tons)				Week	460			460	505
0700	10 C.Y. capacity (3 Tons)					535			535	590
0725	20 C.Y. capacity (8 Tons) R024119-20					630			630	695
0800	30 C.Y. capacity (7 Tons)					810			810	890
0840	40 C.Y. capacity (10 Tons)					860			860	945
2000	Load, haul, dump and return, 50' haul, hand carried	2 Clab	24	.667	C.Y.		23.50		23.50	36
2005	Wheeled		37	.432			15.20		15.20	23.50
2040	51' to 100' haul, hand carried		16.50	.970			34		34	52.50
2045	Wheeled		25	.640			22.50		22.50	34.50
2080	Over 100' haul, add per 100 L.F., hand carried		35.50	.451			15.80		15.80	24.50
2085	Wheeled		54	.296			10.40		10.40	16
2120	In elevators, per 10 floors, add		140	.114			4.01		4.01	6.15
2130	Load, haul, dump and return, up to 50' haul, incl. up to 5 rsr stair, hand		23	.696			24.50		24.50	37.50
2135	Wheeled		35	.457			16.05		16.05	24.50
2140	6 – 10 riser stairs, hand carried		22	.727			25.50		25.50	39.50
2145	Wheeled		34	.471			16.50		16.50	25.50
2150	11 – 20 riser stairs, hand carried		20	.800			28		28	43
2155	Wheeled		31	.516			18.10		18.10	28
2160	21 – 40 riser stairs, hand carried		16	1			35		35	54
2165	Wheeled		24	.667			23.50		23.50	36
2170	100' haul, incl. 5 riser stair, hand carried		15	1.067			37.50		37.50	57.50
2175	Wheeled		23	.696			24.50		24.50	37.50
2180	6 – 10 riser stair, hand carried		14	1.143			40		40	61.50
2185	Wheeled		21	.762			26.50		26.50	41
2190	11 – 20 riser stair, hand carried		12	1.333			47		47	72
2195	Wheeled		18	.889			31		31	48
2200	21 – 40 riser stair, hand carried		8	2			70		70	108
2205	Wheeled		12	1.333			47		47	72
2210	Over 100' haul, add per 100 L.F., hand carried		35.50	.451			15.80		15.80	24.50
2215	Wheeled		54	.296			10.40		10.40	16
2220	For each additional flight of stairs, up to 5 risers, add		550	.029	Flight		1.02		1.02	1.57
2225	6 – 10 risers, add		275	.058			2.04		2.04	3.14
2230	11 – 20 risers, add		138	.116			4.07		4.07	6.25
2235	21 – 40 risers, add		69	.232			8.15		8.15	12.50
3000	Loading & trucking, including 2 mile haul, chute loaded	B-16	45	.711	C.Y.		25.50	13	38.50	53.50
3040	Hand loading truck, 50' haul	"	48	.667			24	12.20	36.20	50
3080	Machine loading truck	B-17	120	.267			10	5.75	15.75	21.50
5000	Haul, per mile, up to 8 C.Y. truck	B-34B	1165	.007			.24	.50	.74	.92
5100	Over 8 C.Y. truck	"	1550	.005			.18	.38	.56	.69

02 41 19.20 Selective Demolition, Dump Charges

		Crew	Daily Output	Labor-Hours	Unit	Material	Labor	Equipment	Total	Total Incl O&P
0010	**SELECTIVE DEMOLITION, DUMP CHARGES** R024119-10									
0020	Dump charges, typical urban city, tipping fees only									
0100	Building construction materials				Ton	82			82	90
0200	Trees, brush, lumber					70			70	77
0300	Rubbish only					70			70	77
0500	Reclamation station, usual charge					82			82	90

02 41 Demolition

02 41 19 – Selective Demolition

02 41 19.21 Selective Demolition, Gutting		Crew	Daily Output	Labor-Hours	Unit	Material	2012 Bare Costs Labor	Equipment	Total	Total Incl O&P
0010	**SELECTIVE DEMOLITION, GUTTING**	R024119-10								
0020	Building interior, including disposal, dumpster fees not included									
0500	Residential building									
0560	Minimum	B-16	400	.080	SF Flr.		2.85	1.46	4.31	6
0580	Maximum	"	360	.089	"		3.17	1.62	4.79	6.65
0900	Commercial building									
1000	Minimum	B-16	350	.091	SF Flr.		3.26	1.67	4.93	6.85
1020	Maximum	"	250	.128	"		4.56	2.34	6.90	9.55

02 41 19.25 Selective Demolition, Saw Cutting		Crew	Daily Output	Labor-Hours	Unit	Material	Labor	Equipment	Total	Total Incl O&P
0010	**SELECTIVE DEMOLITION, SAW CUTTING**	R024119-10								
0015	Asphalt, up to 3" deep	B-89	1050	.015	L.F.	.34	.60	.41	1.35	1.73
0020	Each additional inch of depth	"	1800	.009		.11	.35	.24	.70	.91
1200	Masonry walls, hydraulic saw, brick, per inch of depth	B-89B	300	.053		.42	2.11	2.46	4.99	6.35
1220	Block walls, solid, per inch of depth	"	250	.064		.42	2.53	2.95	5.90	7.55
2000	Brick or masonry w/hand held saw, per inch of depth	A-1	125	.064		.34	2.25	.53	3.12	4.41
5000	Wood sheathing to 1" thick, on walls	1 Carp	200	.040			1.76		1.76	2.71
5020	On roof	"	250	.032	↓		1.41		1.41	2.17

02 41 19.27 Selective Demolition, Torch Cutting		Crew	Daily Output	Labor-Hours	Unit	Material	Labor	Equipment	Total	Total Incl O&P
0010	**SELECTIVE DEMOLITION, TORCH CUTTING**	R024119-10								
0020	Steel, 1" thick plate	1 Clab	360	.022	L.F.	.37	.78		1.15	1.61
0040	1" diameter bar	"	210	.038	Ea.		1.34		1.34	2.06
1000	Oxygen lance cutting, reinforced concrete walls									
1040	12" to 16" thick walls	1 Clab	10	.800	L.F.		28		28	43
1080	24" thick walls	"	6	1.333	"		47		47	72

02 42 Removal and Salvage of Construction Materials

02 42 10 – Building Deconstruction

02 42 10.10 Estimated Salvage Value or Savings

	02 42 10.10 Estimated Salvage Value or Savings		Crew	Daily Output	Labor-Hours	Unit	Material	Labor	Equipment	Total	Total Incl O&P
0010	**ESTIMATED SALVAGE VALUE OR SAVINGS**										
0015	Excludes material handling, packaging, container costs and										
0020	transportation for salvage or disposal										
0050	All Items in Section 02 42 10.10 are credit deducts and not costs										
0100	Copper Wire Salvage Value	G				Lb.				1.30	1.30
0110	Disposal Savings	G								.03	.03
0200	Copper Pipe Salvage Value	G								1.34	1.34
0210	Disposal Savings	G								.03	.03
0300	Steel Pipe Salvage Value	G								.06	.06
0310	Disposal Savings	G								.03	.03
0400	Cast Iron Pipe Salvage Value	G								.06	.06
0410	Disposal Savings	G								.03	.03
0500	Steel Doors or Windows Salvage Value	G								.06	.06
0510	Aluminum	G								.66	.66
0520	Disposal Savings	G								.03	.03
0600	Aluminum Siding Salvage Value	G								.49	.49
0630	Disposal Savings	G								.03	.03
0640	Wood Siding (no lead or asbestos)	G				C.Y.				10.90	10.90
0800	Clean Concrete Disposal Savings	G				Ton				60	60
0850	Asphalt Shingles Disposal Savings	G				"				60	60
1000	Wood wall framing clean salvage value	G				M.B.F.				50	50
1010	Painted	G				↓				40	40

02 42 Removal and Salvage of Construction Materials

02 42 10 – Building Deconstruction

02 42 10.10 Estimated Salvage Value or Savings

			Crew	Daily Output	Labor-Hours	Unit	Material	2012 Bare Costs Labor	Equipment	Total	Total Incl O&P
1020	Floor framing	G				M.B.F.				50	50
1030	Painted	G								40	40
1050	Roof framing	G								50	50
1060	Painted	G								40	40
1100	Wood beams salvage value	G				↓				50	50
1200	Wood framing and beams disposal savings	G				Ton				60	60
1220	Wood sheating and sub-base flooring	G								60	60
1230	Wood wall paneling (1/4 inch thick)	G				↓				60	60
1300	Wood panel 3/4-1 inch thick low salvage value	G				S.F.				.50	.50
1350	high salvage value	G				"				2	2
1400	Disposal savings	G				Ton				60	60
1500	Flooring tongue and groove 25/32 inch thick low salvage value	G				S.F.				.50	.50
1530	High salvage value	G				"				1	1
1560	Disposal savings	G				Ton				60	60
1600	Drywall or sheet rock salvage value	G				↓				20	20
1650	Disposal savings	G				↓				60	60

02 42 10.20 Deconstruction of Building Components

			Crew	Daily Output	Labor-Hours	Unit	Material	2012 Bare Costs Labor	Equipment	Total	Total Incl O&P
0010	**DECONSTRUCTION OF BUILDING COMPONENTS**										
0012	Buildings one or two stories only										
0015	Excludes material handling, packaging, container costs and										
0020	transportation for salvage or disposal										
0050	Deconstruction of Plumbing Fixtures										
0100	Wall hung or countertop lavatory	G	2 Clab	16	1	Ea.		35		35	54
0110	Single or double compartment kitchen sink	G		14	1.143			40		40	61.50
0120	Wall hung urinal	G		14	1.143			40		40	61.50
0130	Floor mounted	G		8	2			70		70	108
0140	Floor mounted water closet	G		16	1			35		35	54
0150	Wall hung	G		14	1.143			40		40	61.50
0160	Water fountain, free standing	G		16	1			35		35	54
0170	Wall hung or deck mounted	G		12	1.333			47		47	72
0180	Bathtub, steel or fiberglass	G		10	1.600			56		56	86.50
0190	Cast iron	G		8	2			70		70	108
0200	Shower, single	G		6	2.667			93.50		93.50	144
0210	Group	G	↓	7	2.286	↓		80		80	123
0300	Deconstruction of Electrical Fixtures										
0310	Surface mount incandescent fixtures	G	2 Clab	48	.333	Ea.		11.70		11.70	18
0320	Fluorescent, 2 lamp	G		32	.500			17.55		17.55	27
0330	4 lamp	G		24	.667			23.50		23.50	36
0340	Strip Fluorescent, 1 lamp	G		40	.400			14.05		14.05	21.50
0350	2 lamp	G		32	.500			17.55		17.55	27
0400	Recessed drop-in fluorescent fixture, 2 lamp	G		27	.593			21		21	32
0410	4 lamp	G	↓	18	.889	↓		31		31	48
0500	Deconstruction of appliances										
0510	Cooking stoves	G	2 Clab	26	.615	Ea.		21.50		21.50	33
0520	Dishwashers	G	"	26	.615	"		21.50		21.50	33
0600	Deconstruction of millwork and trim										
0610	Cabinets, wood	G	2 Carp	40	.400	L.F.		17.65		17.65	27
0620	Countertops	G		100	.160	"		7.05		7.05	10.85
0630	Wall paneling, 1 inch thick	G		500	.032	S.F.		1.41		1.41	2.17
0640	Ceiling trim	G		500	.032	L.F.		1.41		1.41	2.17
0650	Wainscoting	G		500	.032	S.F.		1.41		1.41	2.17
0660	Base, 3/4" to 1" thick	G	↓	600	.027	L.F.		1.18		1.18	1.81

02 42 Removal and Salvage of Construction Materials

02 42 10 – Building Deconstruction

02 42 10.20 Deconstruction of Building Components		Crew	Daily Output	Labor-Hours	Unit	Material	2012 Bare Costs Labor	Equipment	Total	Total Incl O&P
0700	Deconstruction of doors and windows									
0710	Doors, wrap, interior, wood, single, no closers	G	2 Carp	21	.762	Ea.	5.75	33.50		39.25	58
0720	Double	G		13	1.231		11.50	54.50		66	96
0730	Solid core, single, exterior or interior	G		10	1.600		5.75	70.50		76.25	115
0740	Double	G		8	2		11.50	88		99.50	149
0810	Windows, wrap, wood, single	G		21	.762		5.75	33.50		39.25	58
0812	with no casement or cladding	G									
0820	with casement and/or cladding	G	2 Carp	18	.889	Ea.	5.75	39		44.75	67
0900	Deconstruction of interior finishes									
0910	Drywall for recycling	G	2 Clab	1775	.009	S.F.		.32		.32	.49
0920	Plaster wall, first floor	G		1775	.009			.32		.32	.49
0930	Second floor	G		1330	.012			.42		.42	.65
1000	Deconstruction of roofing and accessories									
1010	Built-up roofs	G	2 Clab	570	.028	S.F.		.99		.99	1.52
1020	Gutters, facia and rakes	G	"	1140	.014	L.F.		.49		.49	.76
2000	Deconstruction of wood components									
2010	Roof sheeting	G	2 Clab	570	.028	S.F.		.99		.99	1.52
2020	Main roof framing	G		760	.021	L.F.		.74		.74	1.14
2030	Porch roof framing	G		445	.036			1.26		1.26	1.94
2040	Beams 4" x 8"	G	B-1	375	.064			2.29		2.29	3.52
2050	4" x 10"	G		300	.080			2.86		2.86	4.40
2055	4" x 12"	G		250	.096			3.43		3.43	5.30
2060	6" x 8"	G		250	.096			3.43		3.43	5.30
2065	6" x 10"	G		200	.120			4.29		4.29	6.60
2070	6" x 12"	G		170	.141			5.05		5.05	7.75
2075	8" x 12"	G		126	.190			6.80		6.80	10.50
2080	10" x 12"	G		100	.240			8.60		8.60	13.20
2100	Ceiling joists	G	2 Clab	800	.020			.70		.70	1.08
2150	Wall framing, interior	G		1230	.013			.46		.46	.70
2160	Sub-floor	G		2000	.008	S.F.		.28		.28	.43
2170	Floor joists	G		2000	.008	L.F.		.28		.28	.43
2200	Wood siding (no lead or asbestos)	G		1300	.012	S.F.		.43		.43	.66
2300	Wall framing, exterior	G		1600	.010	L.F.		.35		.35	.54
2400	Stair risers	G		53	.302	Ea.		10.60		10.60	16.30
2500	Posts	G		800	.020	L.F.		.70		.70	1.08
3000	Deconstruction of exterior brick walls									
3010	Exterior brick walls, first floor	G	2 Clab	200	.080	S.F.		2.81		2.81	4.32
3020	Second floor	G		64	.250	"		8.80		8.80	13.50
3030	Brick chiminey	G		100	.160	C.F.		5.60		5.60	8.65
4000	Deconstruction of concrete									
4010	Slab on grade, 4" thick, plain concrete	G	B-9	500	.080	S.F.		2.84	.42	3.26	4.83
4020	Wire mesh reinforced	G		470	.085			3.02	.45	3.47	5.15
4030	Rod reinforced	G		400	.100			3.55	.52	4.07	6.05
4110	Foundation wall, 6" thick, plain concrete	G		160	.250			8.90	1.31	10.21	15.10
4120	8" thick	G		140	.286			10.15	1.49	11.64	17.25
4130	10" thick	G		120	.333			11.85	1.74	13.59	20
9000	Deconstruction process, support equipment as needed									
9010	Daily use, portal to portal, 12-ton truck-mounted hydraulic crane crew	G	A-3H	1	8	Day		380	825	1,205	1,475
9020	Daily use, skid steer and operator	G	A-3C	1	8			360	281	641	850
9030	Daily use, backhoe 48 H. P., operator and labor	G	"	1	8			360	281	641	850

02 42 Removal and Salvage of Construction Materials

02 42 10 – Building Deconstruction

02 42 10.30 Deconstruction Material Handling

		Crew	Daily Output	Labor-Hours	Unit	Material	2012 Bare Costs Labor	2012 Bare Costs Equipment	Total	Total Incl O&P
0010	**DECONSTRUCTION MATERIAL HANDLING**									
0012	Buildings one or two stories only									
0100	Clean and stack brick on pallet	G 2 Clab	1200	.013	Ea.		.47		.47	.72
0200	Haul 50' and load rough lumber up to 2" x 8" size	G	2000	.008	"		.28		.28	.43
0210	Lumber larger than 2" x 8"	G	3200	.005	B.F.		.18		.18	.27
0300	Finish wood for recycling stack and wrap per pallet	G	8	2	Ea.	46	70		116	159
0350	Light fixtures		6	2.667		83	93.50		176.50	235
0375	Windows		6	2.667		78	93.50		171.50	230
0400	Miscellaneous materials		8	2		23	70		93	134
1000	See Section 02 41 19.19 for bulk material handling									

02 43 Structure Moving

02 43 13 – Structure Relocation

02 43 13.13 Building Relocation

		Crew	Daily Output	Labor-Hours	Unit	Material	2012 Bare Costs Labor	2012 Bare Costs Equipment	Total	Total Incl O&P
0010	**BUILDING RELOCATION**									
0011	One day move, up to 24' wide									
0020	Reset on new foundation, patch & hook-up, average move				Total					11,200
0040	Wood or steel frame bldg., based on ground floor area	G B-4	185	.259	S.F.		9.20	2.50	11.70	16.90
0060	Masonry bldg., based on ground floor area	G "	137	.350			12.45	3.37	15.82	23
0200	For 24' to 42' wide, add									15%
0220	For each additional day on road, add	G B-4	1	48	Day		1,700	460	2,160	3,125
0240	Construct new basement, move building, 1 day									
0300	move, patch & hook-up, based on ground floor area	G B-3	155	.310	S.F.	9.20	11.60	14.70	35.50	44

02 56 Site Containment

02 56 13 – Waste Containment

02 56 13.10 Containment of Hazardous Waste

		Crew	Daily Output	Labor-Hours	Unit	Material	2012 Bare Costs Labor	2012 Bare Costs Equipment	Total	Total Incl O&P
0010	**CONTAINMENT OF HAZARDOUS WASTE**									
0020	OSHA hazard level C									
0030	OSHA Hazard level D decrease labor and equipment, deduct						-45%	-45%		
0035	OSHA Hazard level B increase labor and equipment, add						22%	22%		
0040	OSHA Hazard level A increase labor and equipment, add						71%	71%		
0100	Excavation of contaminated soil & waste									
0105	Includes one respirator filter and two disposable suits per work day									
0110	3/4 C.Y. excavator to 10 feet deep	B-12F	51	.314	B.C.Y.	1.67	13	12.70	27.37	35.50
0120	Labor crew to 6' deep	B-2	19	2.105		11.20	74.50		85.70	127
0130	6' - 12' deep	"	12	3.333		17.75	118		135.75	202
0200	Move contaminated soil/waste upto 150' on-site with 2.5 C.Y. loader	B-10T	300	.040	L.C.Y.	.28	1.71	1.57	3.56	4.63
0210	300'	"	186	.065	"	.46	2.76	2.54	5.76	7.45
0300	Secure burial cell construction									
0310	Various liner and cover materials									
0400	Very low density polyethelene (VLDPE)									
0410	50 mil top cover	B-47H	4000	.008	S.F.	.39	.37	.07	.83	1.07
0420	80 mil liner	"	4000	.008	"	.39	.37	.07	.83	1.07
0500	Chlorosulfunated polyethylene									
0510	36 mil hypalon top cover	B-47H	4000	.008	S.F.	1.21	.37	.07	1.65	1.97
0520	45 mil hypalon liner	"	4000	.008	"	1.27	.37	.07	1.71	2.04
0600	Polyvinyl chloride (PVC)									
0610	60 mil top cover	B-47H	4000	.008	S.F.	.67	.37	.07	1.11	1.38

02 56 Site Containment

02 56 13 – Waste Containment

02 56 13.10 Containment of Hazardous Waste

02 56 13.10 Containment of Hazardous Waste		Crew	Daily Output	Labor-Hours	Unit	Material	2012 Bare Costs Labor	Equipment	Total	Total Incl O&P
0620	80 mil liner	B-47H	4000	.008	S.F.	.79	.37	.07	1.23	1.51
0700	Rough textured H.D. polyethylene (HDPE)									
0710	40 mil top cover	B-47H	4000	.008	S.F.	.39	.37	.07	.83	1.07
0720	60 mil top cover		4000	.008		.41	.37	.07	.85	1.09
0722	60 mil liner		4000	.008		.37	.37	.07	.81	1.05
0730	80 mil liner		3800	.008		.44	.39	.07	.90	1.16
1000	3/4" crushed stone, 6" deep ballast around liner	B-6	30	.800	L.C.Y.	31.50	30.50	11.15	73.15	94
1100	Hazardous waste, ballast cover with common borrow material	B-63	56	.714		14.50	26.50	2.85	43.85	59.50
1110	Mixture of common borrow & topsoil		56	.714		18.75	26.50	2.85	48.10	64
1120	Bank sand		56	.714		17	26.50	2.85	46.35	62.50
1130	Medium priced clay		44	.909		23	33.50	3.63	60.13	81
1140	Mixture of common borrow & medium priced clay		56	.714		18.75	26.50	2.85	48.10	64

02 58 Snow Control

02 58 13 – Snow Fencing

02 58 13.10 Snow Fencing System

		Crew	Daily Output	Labor-Hours	Unit	Material	Labor	Equipment	Total	Total Incl O&P
0010	**SNOW FENCING SYSTEM**									
7001	Snow fence on steel posts 10' O.C., 4' high	B-1	500	.048	L.F.	1.09	1.72		2.81	3.84

02 65 Underground Storage Tank Removal

02 65 10 – Underground Tank and Contaminated Soil Removal

02 65 10.30 Removal of Underground Storage Tanks

			Crew	Daily Output	Labor-Hours	Unit	Material	Labor	Equipment	Total	Total Incl O&P
0010	**REMOVAL OF UNDERGROUND STORAGE TANKS**	R026510-20									
0011	Petroleum storage tanks, non-leaking										
0100	Excavate & load onto trailer										
0110	3000 gal. to 5000 gal. tank	G	B-14	4	12	Ea.		445	83.50	528.50	770
0120	6000 gal. to 8000 gal. tank	G	B-3A	3	13.333			500	320	820	1,125
0130	9000 gal. to 12000 gal. tank	G	"	2	20			750	480	1,230	1,675
0190	Known leaking tank, add					%				100%	100%
0200	Remove sludge, water and remaining product from tank bottom										
0201	of tank with vacuum truck										
0300	3000 gal. to 5000 gal. tank	G	A-13	5	1.600	Ea.		71.50	147	218.50	270
0310	6000 gal. to 8000 gal. tank	G		4	2			89.50	184	273.50	335
0320	9000 gal. to 12000 gal. tank	G		3	2.667			119	245	364	450
0390	Dispose of sludge off-site, average					Gal.				6	6.60
0400	Insert inert solid CO_2 "dry ice" into tank										
0401	For cleaning/transporting tanks (1.5 lb./100 gal. cap)	G	1 Clab	500	.016	Lb.	.94	.56		1.50	1.89
0403	Insert solid carbon dioxide, 1.5 lb./100 gal.	G	"	400	.020	"	.94	.70		1.64	2.11
0503	Disconnect and remove piping	G	1 Plum	160	.050	L.F.		2.73		2.73	4.12
0603	Transfer liquids, 10% of volume	G	"	1600	.005	Gal.		.27		.27	.41
0703	Cut accessway into underground storage tank	G	1 Clab	5.33	1.501	Ea.		52.50		52.50	81
0813	Remove sludge, wash and wipe tank, 500 gal.	G	1 Plum	8	1			54.50		54.50	82.50
0823	3,000 gal.	G		6.67	1.199			65.50		65.50	99
0833	5,000 gal.	G		6.15	1.301			71		71	107
0843	8,000 gal.	G		5.33	1.501			82		82	124
0853	10,000 gal.	G		4.57	1.751			95.50		95.50	144
0863	12,000 gal.	G		4.21	1.900			104		104	156
1020	Haul tank to certified salvage dump, 100 miles round trip										
1023	3000 gal. to 5000 gal. tank					Ea.				745	820

02 65 Underground Storage Tank Removal

02 65 10 – Underground Tank and Contaminated Soil Removal

02 65 10.30 Removal of Underground Storage Tanks		Crew	Daily Output	Labor-Hours	Unit	Material	2012 Bare Costs Labor	Equipment	Total	Total Incl O&P
1026	6000 gal. to 8000 gal. tank				Ea.				850	9,350
1029	9,000 gal. to 12,000 gal. tank				↓				1,000	1,100
1100	Disposal of contaminated soil to landfill									
1110	Minimum				C.Y.				125	140
1111	Maximum				"				400	440
1120	Disposal of contaminated soil to									
1121	bituminous concrete batch plant									
1130	Minimum				C.Y.				70	77
1131	Maximum				"				110	120
1203	Excavate, pull, & load tank, backfill hole, 8,000 gal. +	G	B-12C	.50	32	Ea.		1,325	2,500	3,825	4,775
1213	Haul tank to certified dump, 100 miles rt, 8,000 gal. +	G	B-34K	1	8			282	850	1,132	1,375
1223	Excavate, pull, & load tank, backfill hole, 500 gal.	G	B-11C	1	16			655	335	990	1,350
1233	Excavate, pull, & load tank, backfill hole, 3,000 – 5,000 gal.	G	B-11M	.50	32			1,300	730	2,030	2,800
1243	Haul tank to certified dump, 100 miles rt, 500 gal.	G	B-34L	1	8	↓		360	211	571	770
2010	Decontamination of soil on site incl poly tarp on top/bottom										
2011	Soil containment berm, and chemical treatment										
2020	Minimum	G	B-11C	100	.160	C.Y.	7.10	6.55	3.34	16.99	21.50
2021	Maximum	G	"	100	.160		9.20	6.55	3.34	19.09	23.50
2050	Disposal of decontaminated soil, minimum									125	140
2055	Maximum					↓				400	440

02 81 Transportation and Disposal of Hazardous Materials

02 81 20 – Hazardous Waste Handling

02 81 20.10 Hazardous Waste Cleanup/Pickup/Disposal		Crew	Daily Output	Labor-Hours	Unit	Material	2012 Bare Costs Labor	Equipment	Total	Total Incl O&P
0010	**HAZARDOUS WASTE CLEANUP/PICKUP/DISPOSAL**									
0100	For contractor rental equipment, i.e., Dozer,									
0110	Front end loader, Dump truck, etc., see 01 54 33 Reference Section									
1000	Solid pickup									
1100	55 gal. drums				Ea.				225	250
1120	Bulk material, minimum				Ton				170	190
1130	Maximum				"				575	635
1200	Transportation to disposal site									
1220	Truckload = 80 drums or 25 C.Y. or 18 tons									
1260	Minimum				Mile				3.85	4.25
1270	Maximum				"				7	7.10
3000	Liquid pickup, vacuum truck, stainless steel tank									
3100	Minimum charge, 4 hours									
3110	1 compartment, 2200 gallon				Hr.				125	140
3120	2 compartment, 5000 gallon				"				180	200
3400	Transportation in 6900 gallon bulk truck				Mile				7.70	8.50
3410	In teflon lined truck				"				9.90	10.90
5000	Heavy sludge or dry vacuumable material				Hr.				125	140
6000	Dumpsite disposal charge, minimum				Ton				125	140
6020	Maximum				"				400	440

02 82 13.39 Asbestos Remediation Plans and Methods	Crew	Daily Output	Labor-Hours	Unit	Material	2012 Bare Costs Labor	Equipment	Total	Total Incl O&P
0010 **ASBESTOS REMEDIATION PLANS AND METHODS**									
0100 Building Survey-Commercial Building				Ea.				2,000	2,200
0200 Asbestos Abatement Remediation Plan				"				1,200	1,325

02 82 13.41 Asbestos Abatement Equip.

	Crew	Daily Output	Labor-Hours	Unit	Material	2012 Bare Costs Labor	Equipment	Total	Total Incl O&P
0010 **ASBESTOS ABATEMENT EQUIP.** R028213-20									
0011 Equipment and supplies, buy									
0200 Air filtration device, 2000 CFM				Ea.	885			885	975
0250 Large volume air sampling pump, minimum					365			365	400
0260 Maximum					530			530	585
0300 Airless sprayer unit, 2 gun					4,900			4,900	5,400
0350 Light stand, 500 watt					46.50			46.50	51.50
0400 Personal respirators									
0410 Negative pressure, 1/2 face, dual operation, min.				Ea.	26			26	28.50
0420 Maximum					32.50			32.50	35.50
0450 P.A.P.R., full face, minimum					109			109	120
0460 Maximum					199			199	219
0470 Supplied air, full face, incl. air line, minimum					191			191	210
0480 Maximum					395			395	435
0500 Personnel sampling pump					226			226	248
1500 Power panel, 20 unit, incl. G.F.I.					565			565	620
1600 Shower unit, including pump and filters					1,200			1,200	1,300
1700 Supplied air system (type C)					5,425			5,425	5,975
1750 Vacuum cleaner, HEPA, 16 gal., stainless steel, wet/dry					1,200			1,200	1,325
1760 55 gallon					1,950			1,950	2,150
1800 Vacuum loader, 9 – 18 ton/hr.					92,000			92,000	101,000
1900 Water atomizer unit, including 55 gal. drum					235			235	258
2000 Worker protection, whole body, foot, head cover & gloves, plastic					8.90			8.90	9.80
2500 Respirator, single use					29.50			29.50	32.50
2550 Cartridge for respirator					7			7	7.70
2570 Glove bag, 7 mil, 50" x 64"					17.60			17.60	19.35
2580 10 mil, 44" x 60"					5.50			5.50	6.05
2590 6 mil, 44" x 60"					5.50			5.50	6.05
3000 HEPA vacuum for work area, minimum					1,400			1,400	1,525
3050 Maximum					1,475			1,475	1,600
6000 Disposable polyethylene bags, 6 mil, 3 C.F.					2.35			2.35	2.59
6300 Disposable fiber drums, 3 C.F.					17.25			17.25	19
6400 Pressure sensitive caution labels, 3" x 5"					3.22			3.22	3.54
6450 11" x 17"					7.15			7.15	7.90
6500 Negative air machine, 1800 CFM					820			820	905

02 82 13.42 Preparation of Asbestos Containment Area

	Crew	Daily Output	Labor-Hours	Unit	Material	2012 Bare Costs Labor	Equipment	Total	Total Incl O&P
0010 **PREPARATION OF ASBESTOS CONTAINMENT AREA**									
0100 Pre-cleaning, HEPA vacuum and wet wipe, flat surfaces	A-9	12000	.005	S.F.	.02	.26		.28	.43
0200 Protect carpeted area, 2 layers 6 mil poly on 3/4" plywood	"	1000	.064		2	3.13		5.13	7.10
0300 Separation barrier, 2" x 4" @ 16", 1/2" plywood ea. side, 8' high	2 Carp	400	.040		3	1.76		4.76	6
0310 12' high		320	.050		3	2.21		5.21	6.70
0320 16' high		200	.080		2	3.53		5.53	7.65
0400 Personnel decontam. chamber, 2" x 4" @ 16", 3/4" ply ea. side		280	.057		4	2.52		6.52	8.30
0450 Waste decontam. chamber, 2" x 4" studs @ 16", 3/4" ply ea. side		360	.044		4	1.96		5.96	7.40
0500 Cover surfaces with polyethylene sheeting									
0501 Including glue and tape									
0550 Floors, each layer, 6 mil	A-9	8000	.008	S.F.	.05	.39		.44	.66
0551 4 mil		9000	.007		.03	.35		.38	.58

02 82 Asbestos Remediation

02 82 13 – Asbestos Abatement

02 82 13.42 Preparation of Asbestos Containment Area		Crew	Daily Output	Labor-Hours	Unit	Material	2012 Bare Costs Labor	Equipment	Total	Total Incl O&P
0560	Walls, each layer, 6 mil	A-9	6000	.011	S.F.	.05	.52		.57	.86
0561	4 mil	↓	7000	.009	↓	.03	.45		.48	.74
0570	For heights above 12', add						20%			
0575	For heights above 20', add						30%			
0580	For fire retardant poly, add					100%				
0590	For large open areas, deduct					10%	20%			
0600	Seal floor penetrations with foam firestop to 36 sq. in.	2 Carp	200	.080	Ea.	7.60	3.53		11.13	13.85
0610	36 sq. in. to 72 sq. in.		125	.128		15.25	5.65		20.90	25.50
0615	72 sq. in. to 144 sq. in.		80	.200		30.50	8.80		39.30	47
0620	Wall penetrations, to 36 square inches		180	.089		7.60	3.92		11.52	14.45
0630	36 sq. in. to 72 sq. in.		100	.160		15.25	7.05		22.30	27.50
0640	72 sq. in. to 144 sq. in.	↓	60	.267	↓	30.50	11.75		42.25	51.50
0800	Caulk seams with latex	1 Carp	230	.035	L.F.	.15	1.53		1.68	2.53
0900	Set up neg. air machine, 1-2k CFM/25 M.C.F. volume	1 Asbe	4.30	1.860	Ea.		91		91	142

02 82 13.43 Bulk Asbestos Removal

		Crew	Daily Output	Labor-Hours	Unit	Material	2012 Bare Costs Labor	Equipment	Total	Total Incl O&P
0010	**BULK ASBESTOS REMOVAL**									
0020	Includes disposable tools and 2 suits and 1 respirator filter/day/worker									
0100	Beams, W 10 x 19	A-9	235	.272	L.F.	.84	13.30		14.14	22
0110	W 12 x 22		210	.305		.94	14.90		15.84	24
0120	W 14 x 26		180	.356		1.10	17.40		18.50	28
0130	W 16 x 31		160	.400		1.24	19.55		20.79	32
0140	W 18 x 40		140	.457		1.42	22.50		23.92	36.50
0150	W 24 x 55		110	.582		1.80	28.50		30.30	46.50
0160	W 30 x 108		85	.753		2.33	37		39.33	60
0170	W 36 x 150		72	.889	↓	2.75	43.50		46.25	70.50
0200	Boiler insulation	↓	480	.133	S.F.	.48	6.50		6.98	10.70
0210	With metal lath, add				%				50%	
0300	Boiler breeching or flue insulation	A-9	520	.123	S.F.	.38	6		6.38	9.80
0310	For active boiler, add				%				100%	
0400	Duct or AHU insulation	A-10B	440	.073	S.F.	.23	3.56		3.79	5.80
0500	Duct vibration isolation joints, up to 24 sq. in. duct	A-9	56	1.143	Ea.	3.54	56		59.54	91
0520	25 sq. in. to 48 sq. in. duct		48	1.333		4.13	65		69.13	107
0530	49 sq. in. to 76 sq. in. duct		40	1.600	↓	4.96	78.50		83.46	127
0600	Pipe insulation, air cell type, up to 4" diameter pipe		900	.071	L.F.	.22	3.48		3.70	5.65
0610	4" to 8" diameter pipe		800	.080		.25	3.91		4.16	6.35
0620	10" to 12" diameter pipe		700	.091		.28	4.47		4.75	7.25
0630	14" to 16" diameter pipe		550	.116	↓	.36	5.70		6.06	9.25
0650	Over 16" diameter pipe		650	.098	S.F.	.31	4.82		5.13	7.85
0700	With glove bag up to 3" diameter pipe		200	.320	L.F.	6.05	15.65		21.70	31
1000	Pipe fitting insulation up to 4" diameter pipe		320	.200	Ea.	.62	9.80		10.42	15.95
1100	6" to 8" diameter pipe		304	.211		.65	10.30		10.95	16.75
1110	10" to 12" diameter pipe		192	.333		1.03	16.30		17.33	26.50
1120	14" to 16" diameter pipe		128	.500	↓	1.55	24.50		26.05	39.50
1130	Over 16" diameter pipe		176	.364	S.F.	1.13	17.80		18.93	28.50
1200	With glove bag, up to 8" diameter pipe		75	.853	L.F.	6.55	41.50		48.05	72
2000	Scrape foam fireproofing from flat surface		2400	.027	S.F.	.08	1.30		1.38	2.12
2100	Irregular surfaces		1200	.053		.17	2.61		2.78	4.24
3000	Remove cementitious material from flat surface		1800	.036		.11	1.74		1.85	2.83
3100	Irregular surface		1000	.064		.14	3.13		3.27	5.05
4000	Scrape acoustical coating/fireproofing, from ceiling		3200	.020		.06	.98		1.04	1.59
5000	Remove VAT and mastic from floor by hand	↓	2400	.027		.08	1.30		1.38	2.12
5100	By machine	A-11	4800	.013	↓	.04	.65	.01	.70	1.08

02 82 13 – Asbestos Abatement

02 82 13.43 Bulk Asbestos Removal

		Crew	Daily Output	Labor-Hours	Unit	Material	2012 Bare Costs Labor	2012 Bare Costs Equipment	Total	Total Incl O&P
5150	For 2 layers, add				%				50%	
6000	Remove contaminated soil from crawl space by hand	A-9	400	.160	C.F.	.50	7.85		8.35	12.75
6100	With large production vacuum loader	A-12	700	.091	"	.28	4.47	1.05	5.80	8.40
7000	Radiator backing, not including radiator removal	A-9	1200	.053	S.F.	.17	2.61		2.78	4.24
8000	Cement-asbestos transite board	2 Asbe	1000	.016		.14	.78		.92	1.37
8100	Transite shingle siding	A-10B	750	.043		.21	2.09		2.30	3.49
8200	Shingle roofing	"	2000	.016		.08	.78		.86	1.31
8250	Built-up, no gravel, non-friable	B-2	1400	.029		.08	1.01		1.09	1.65
8260	Bituminous flashing	1 Rofc	300	.027		.08	1		1.08	1.76
8300	Asbestos millboard	2 Asbe	1000	.016	↓	.08	.78		.86	1.31
9000	For type B (supplied air) respirator equipment, add				%					10%

02 82 13.44 Demolition In Asbestos Contaminated Area

		Crew	Daily Output	Labor-Hours	Unit	Material	2012 Bare Costs Labor	2012 Bare Costs Equipment	Total	Total Incl O&P
0010	**DEMOLITION IN ASBESTOS CONTAMINATED AREA**									
0200	Ceiling, including suspension system, plaster and lath	A-9	2100	.030	S.F.	.09	1.49		1.58	2.42
0210	Finished plaster, leaving wire lath		585	.109		.34	5.35		5.69	8.70
0220	Suspended acoustical tile		3500	.018		.06	.89		.95	1.45
0230	Concealed tile grid system		3000	.021		.07	1.04		1.11	1.70
0240	Metal pan grid system		1500	.043		.13	2.09		2.22	3.40
0250	Gypsum board		2500	.026	↓	.08	1.25		1.33	2.04
0260	Lighting fixtures up to 2' x 4'	↓	72	.889	Ea.	2.75	43.50		46.25	70.50
0400	Partitions, non load bearing									
0410	Plaster, lath, and studs	A-9	690	.093	S.F.	.86	4.54		5.40	8
0450	Gypsum board and studs	"	1390	.046	"	.14	2.25		2.39	3.67
9000	For type B (supplied air) respirator equipment, add				%					10%

02 82 13.45 OSHA Testing

		Crew	Daily Output	Labor-Hours	Unit	Material	2012 Bare Costs Labor	2012 Bare Costs Equipment	Total	Total Incl O&P
0010	**OSHA TESTING**									
0100	Certified technician, minimum				Day				200	220
0110	Maximum								300	330
0120	Industrial hygienist, minimum								300	330
0130	Maximum				↓				400	440
0200	Asbestos sampling and PCM analysis, NIOSH 7400, minimum	1 Asbe	8	1	Ea.	2.75	49		51.75	79
0210	Maximum		4	2		3	97.50		100.50	155
1000	Cleaned area samples		8	1		2.63	49		51.63	79
1100	PCM air sample analysis, NIOSH 7400, minimum		8	1		30.50	49		79.50	110
1110	Maximum	↓	4	2		3.20	97.50		100.70	156
1200	TEM air sample analysis, NIOSH 7402, minimum								90	120
1210	Maximum				↓				360	450

02 82 13.46 Decontamination of Asbestos Containment Area

		Crew	Daily Output	Labor-Hours	Unit	Material	2012 Bare Costs Labor	2012 Bare Costs Equipment	Total	Total Incl O&P
0010	**DECONTAMINATION OF ASBESTOS CONTAINMENT AREA**									
0100	Spray exposed substrate with surfactant (bridging)									
0200	Flat surfaces	A-9	6000	.011	S.F.	.36	.52		.88	1.21
0250	Irregular surfaces		4000	.016	"	.31	.78		1.09	1.56
0300	Pipes, beams, and columns		2000	.032	L.F.	.56	1.57		2.13	3.06
1000	Spray encapsulate polyethylene sheeting		8000	.008	S.F.	.31	.39		.70	.95
1100	Roll down polyethylene sheeting		8000	.008	"		.39		.39	.61
1500	Bag polyethylene sheeting		400	.160	Ea.	.77	7.85		8.62	13.05
2000	Fine clean exposed substrate, with nylon brush		2400	.027	S.F.		1.30		1.30	2.03
2500	Wet wipe substrate		4800	.013			.65		.65	1.02
2600	Vacuum surfaces, fine brush		6400	.010	↓		.49		.49	.76
3000	Structural demolition									
3100	Wood stud walls	A-9	2800	.023	S.F.		1.12		1.12	1.74
3500	Window manifolds, not incl. window replacement	↓	4200	.015	↓		.75		.75	1.16

02 82 Asbestos Remediation

02 82 13 – Asbestos Abatement

02 82 13.46 Decontamination of Asbestos Containment Area	Crew	Daily Output	Labor-Hours	Unit	Material	2012 Bare Costs Labor	Equipment	Total	Total Incl O&P
3600	Plywood carpet protection	A-9	2000	.032	S.F.		1.57		1.57	2.44
4000	Remove custom decontamination facility	A-10A	8	3	Ea.	15	147		162	246
4100	Remove portable decontamination facility	3 Asbe	12	2	"	12.75	97.50		110.25	166
5000	HEPA vacuum, shampoo carpeting	A-9	4800	.013	S.F.	.05	.65		.70	1.08
9000	Final cleaning of protected surfaces	A-10A	8000	.003	"		.15		.15	.23

02 82 13.47 Asbestos Waste Pkg., Handling, and Disp.

		Crew	Daily Output	Labor-Hours	Unit	Material	Labor	Equipment	Total	Total Incl O&P
0010	**ASBESTOS WASTE PACKAGING, HANDLING, AND DISPOSAL**									
0100	Collect and bag bulk material, 3 C.F. bags, by hand	A-9	400	.160	Ea.	2.35	7.85		10.20	14.80
0200	Large production vacuum loader	A-12	880	.073		.80	3.56	.83	5.19	7.35
1000	Double bag and decontaminate	A-9	960	.067		2.35	3.26		5.61	7.70
2000	Containerize bagged material in drums, per 3 C.F. drum	"	800	.080		17.25	3.91		21.16	25
3000	Cart bags 50' to dumpster	2 Asbe	400	.040			1.95		1.95	3.04
5000	Disposal charges, not including haul, minimum				C.Y.				60	66
5020	Maximum				"				350	385
5100	Remove refrigerant from system	1 Stpi	40	.200	Lb.		11.15		11.15	16.80
9000	For type B (supplied air) respirator equipment, add				%					10%

02 82 13.48 Asbestos Encapsulation With Sealants

		Crew	Daily Output	Labor-Hours	Unit	Material	Labor	Equipment	Total	Total Incl O&P
0010	**ASBESTOS ENCAPSULATION WITH SEALANTS**									
0100	Ceilings and walls, minimum	A-9	21000	.003	S.F.	.27	.15		.42	.53
0110	Maximum		10600	.006		.42	.30		.72	.92
0200	Columns and beams, minimum		13300	.005		.27	.24		.51	.67
0210	Maximum		5325	.012		.47	.59		1.06	1.44
0300	Pipes to 12" diameter including minor repairs, minimum		800	.080	L.F.	.37	3.91		4.28	6.50
0310	Maximum		400	.160	"	1.04	7.85		8.89	13.35

02 83 Lead Remediation

02 83 19 – Lead-Based Paint Remediation

02 83 19.21 Lead Paint Remediation Plans and Methods

		Crew	Daily Output	Labor-Hours	Unit	Material	Labor	Equipment	Total	Total Incl O&P
0010	**LEAD PAINT REMEDIATION PLANS AND METHODS**									
0100	Building Survey-Commercial Building				Ea.				2,000	2,200
0200	Lead Abatement Remediation Plan								1,200	1,325
0300	Lead Paint Testing, AAS Analysis								50	55
0400	Lead Paint Testing, X-Ray Fluorescence								50	55

02 83 19.23 Encapsulation of Lead-Based Paint

		Crew	Daily Output	Labor-Hours	Unit	Material	Labor	Equipment	Total	Total Incl O&P
0010	**ENCAPSULATION OF LEAD-BASED PAINT**									
0020	Interior, brushwork, trim, under 6"	1 Pord	240	.033	L.F.	2.46	1.27		3.73	4.62
0030	6" to 12" wide		180	.044		3.28	1.70		4.98	6.15
0040	Balustrades		300	.027		1.98	1.02		3	3.71
0050	Pipe to 4" diameter		500	.016		1.19	.61		1.80	2.23
0060	To 8" diameter		375	.021		1.57	.82		2.39	2.95
0070	To 12" diameter		250	.032		2.36	1.22		3.58	4.43
0080	To 16" diameter		170	.047		3.47	1.80		5.27	6.50
0090	Cabinets, ornate design		200	.040	S.F.	2.97	1.53		4.50	5.55
0100	Simple design		250	.032	"	2.36	1.22		3.58	4.43
0110	Doors, 3' x 7', both sides, incl. frame & trim									
0120	Flush	1 Pord	6	1.333	Ea.	30.50	51		81.50	110
0130	French, 10 – 15 lite		3	2.667		6.05	102		108.05	160
0140	Panel		4	2		36.50	76.50		113	154
0150	Louvered		2.75	2.909		33.50	111		144.50	203
0160	Windows, per interior side, per 15 S.F.									

43

02 83 19 – Lead-Based Paint Remediation

02 83 19.23 Encapsulation of Lead-Based Paint	Crew	Daily Output	Labor-Hours	Unit	Material	2012 Bare Costs Labor	Equipment	Total	Total Incl O&P
0170	1 to 6 lite	1 Pord	14	.571	Ea.	21	22		43	55.50
0180	7 to 10 lite		7.50	1.067		23	41		64	86.50
0190	12 lite		5.75	1.391		31	53		84	114
0200	Radiators		8	1		73.50	38.50		112	138
0210	Grilles, vents		275	.029	S.F.	2.15	1.11		3.26	4.03
0220	Walls, roller, drywall or plaster		1000	.008		.59	.31		.90	1.11
0230	With spunbonded reinforcing fabric		720	.011		.68	.43		1.11	1.39
0240	Wood		800	.010		.74	.38		1.12	1.38
0250	Ceilings, roller, drywall or plaster		900	.009		.68	.34		1.02	1.26
0260	Wood		700	.011		.84	.44		1.28	1.57
0270	Exterior, brushwork, gutters and downspouts		300	.027	L.F.	1.98	1.02		3	3.71
0280	Columns		400	.020	S.F.	1.47	.77		2.24	2.76
0290	Spray, siding		600	.013	"	.99	.51		1.50	1.85
0300	Miscellaneous									
0310	Electrical conduit, brushwork, to 2" diameter	1 Pord	500	.016	L.F.	1.19	.61		1.80	2.23
0320	Brick, block or concrete, spray		500	.016	S.F.	1.19	.61		1.80	2.23
0330	Steel, flat surfaces and tanks to 12"		500	.016		1.19	.61		1.80	2.23
0340	Beams, brushwork		400	.020		1.47	.77		2.24	2.76
0350	Trusses		400	.020		1.47	.77		2.24	2.76

02 83 19.26 Removal of Lead-Based Paint

		Crew	Daily Output	Labor-Hours	Unit	Material	2012 Bare Costs Labor	Equipment	Total	Total Incl O&P
0010	**REMOVAL OF LEAD-BASED PAINT** R028319-60									
0011	By chemicals, per application									
0050	Baseboard, to 6" wide	1 Pord	64	.125	L.F.	.66	4.78		5.44	7.90
0070	To 12" wide		32	.250	"	1.32	9.55		10.87	15.75
0200	Balustrades, one side		28	.286	S.F.	1.32	10.95		12.27	17.80
1400	Cabinets, simple design		32	.250		1.32	9.55		10.87	15.75
1420	Ornate design		25	.320		1.32	12.25		13.57	19.75
1600	Cornice, simple design		60	.133		1.32	5.10		6.42	9.10
1620	Ornate design		20	.400		5.05	15.30		20.35	28.50
2800	Doors, one side, flush		84	.095		1	3.64		4.64	6.55
2820	Two panel		80	.100		1.32	3.83		5.15	7.15
2840	Four panel		45	.178		1.32	6.80		8.12	11.60
2880	For trim, one side, add		64	.125	L.F.	.66	4.78		5.44	7.90
3000	Fence, picket, one side		30	.267	S.F.	1.32	10.20		11.52	16.70
3200	Grilles, one side, simple design		30	.267		1.32	10.20		11.52	16.70
3220	Ornate design		25	.320		1.32	12.25		13.57	19.75
4400	Pipes, to 4" diameter		90	.089	L.F.	1	3.40		4.40	6.20
4420	To 8" diameter		50	.160		2	6.10		8.10	11.35
4440	To 12" diameter		36	.222		3	8.50		11.50	16
4460	To 16" diameter		20	.400		4	15.30		19.30	27.50
4500	For hangers, add		40	.200	Ea.	2.54	7.65		10.19	14.25
4800	Siding		90	.089	S.F.	1.17	3.40		4.57	6.40
5000	Trusses, open		55	.145	SF Face	1.85	5.55		7.40	10.35
6200	Windows, one side only, double hung, 1/1 light, 24" x 48" high		4	2	Ea.	25.50	76.50		102	142
6220	30" x 60" high		3	2.667		34	102		136	191
6240	36" x 72" high		2.50	3.200		41	122		163	228
6280	40" x 80" high		2	4		51	153		204	285
6400	Colonial window, 6/6 light, 24" x 48" high		2	4		51	153		204	286
6420	30" x 60" high		1.50	5.333		68	204		272	380
6440	36" x 72" high		1	8		102	305		407	570
6480	40" x 80" high		1	8		102	305		407	570
6600	8/8 light, 24" x 48" high		2	4		51	153		204	286

02 83 Lead Remediation

02 83 19 – Lead-Based Paint Remediation

02 83 19.26 Removal of Lead-Based Paint	Crew	Daily Output	Labor-Hours	Unit	Material	2012 Bare Costs Labor	Equipment	Total	Total Incl O&P
6620	40" x 80" high	1 Pord	1	8	Ea.	102	305		407	570
6800	12/12 light, 24" x 48" high		1	8		102	305		407	570
6820	40" x 80" high	↓	.75	10.667	↓	136	410		546	760
6840	Window frame & trim items, included in pricing above									

02 85 Mold Remediation

02 85 16 – Mold Remediation Preparation and Containment

02 85 16.40 Mold Remediation Plans and Methods

		Crew	Daily Output	Labor-Hours	Unit	Material	Labor	Equipment	Total	Total Incl O&P
0010	**MOLD REMEDIATION PLANS AND METHODS**									
0020	Initial inspection, average 3 bedroom home				Total				250	275
0030	Average 5 bedroom home								400	440
0040	Testing, air sample each								125	138
0050	Swab sample								125	138
0060	Tape sample								125	140
0070	Post remediation air test								125	140
0080	Mold abatement plan, average 3 bedroom home								1,200	1,300
0090	Average 5 bedroom home								1,600	1,750
0100	Packup & removal of contents, average 3 bedroom home, excl storage								8,000	8,800
0110	Average 5 bedroom home, excl storage				↓				15,000	16,500
0600	For demolition in mold contaminated areas, see Section 02 85 33.50									
0610	For personal protection equipment, see Section 02 82 13.41									

02 85 16.50 Preparation of Mold Containment Area

		Crew	Daily Output	Labor-Hours	Unit	Material	Labor	Equipment	Total	Total Incl O&P
0010	**PREPARATION OF MOLD CONTAINMENT AREA**									
0100	Pre-cleaning, HEPA vacuum and wet wipe, flat surfaces	A-9	12000	.005	S.F.	.02	.26		.28	.43
0300	Separation barrier, 2" x 4" @ 16", 1/2" plywood ea. side, 8' high	2 Carp	400	.040		3	1.76		4.76	6
0310	12' high		320	.050		3	2.21		5.21	6.70
0320	16' high		200	.080		2	3.53		5.53	7.65
0400	Personnel decontam. chamber, 2" x 4" @ 16", 3/4 ply ea. side		280	.057		4	2.52		6.52	8.30
0450	Waste decontam. chamber, 2" x 4" studs @ 16", 3/4 ply each side	↓	360	.044	↓	4	1.96		5.96	7.40
0500	Cover surfaces with polyethylene sheeting									
0501	Including glue and tape									
0550	Floors, each layer, 6 mil	A-9	8000	.008	S.F.	.06	.39		.45	.68
0551	4 mil		9000	.007		.04	.35		.39	.59
0560	Walls, each layer, 6 mil		6000	.011		.09	.52		.61	.91
0561	4 mil	↓	7000	.009	↓	.07	.45		.52	.78
0570	For heights above 12', add						20%			
0575	For heights above 20', add						30%			
0580	For fire retardant poly, add					100%				
0590	For large open areas, deduct					10%	20%			
0600	Seal floor penetrations with foam firestop to 36 sq. in.	2 Carp	200	.080	Ea.	7.60	3.53		11.13	13.85
0610	36 sq. in. to 72 sq. in.		125	.128		15.25	5.65		20.90	25.50
0615	72 sq. in. to 144 sq. in.		80	.200		30.50	8.80		39.30	47
0620	Wall penetrations, to 36 square inches		180	.089		7.60	3.92		11.52	14.45
0630	36 sq. in. to 72 sq. in.		100	.160		15.25	7.05		22.30	27.50
0640	72 sq. in. to 144 sq. in.	↓	60	.267	↓	30.50	11.75		42.25	51.50
0800	Caulk seams with latex caulk	1 Carp	230	.035	L.F.	.15	1.53		1.68	2.53
0900	Set up neg. air machine, 1-2k CFM/25 M.C.F. volume	1 Asbe	4.30	1.860	Ea.		91		91	142

02 85 Mold Remediation

02 85 33 – Removal and Disposal of Materials with Mold

02 85 33.50 Demolition in Mold Contaminated Area	Crew	Daily Output	Labor-Hours	Unit	Material	2012 Bare Costs Labor	Equipment	Total	Total Incl O&P
0010 **DEMOLITION IN MOLD CONTAMINATED AREA**									
0200 Ceiling, including suspension system, plaster and lath	A-9	2100	.030	S.F.	.09	1.49		1.58	2.42
0210 Finished plaster, leaving wire lath		585	.109		.34	5.35		5.69	8.70
0220 Suspended acoustical tile		3500	.018		.06	.89		.95	1.45
0230 Concealed tile grid system		3000	.021		.07	1.04		1.11	1.70
0240 Metal pan grid system		1500	.043		.13	2.09		2.22	3.40
0250 Gypsum board		2500	.026		.08	1.25		1.33	2.04
0255 Plywood		2500	.026		.08	1.25		1.33	2.04
0260 Lighting fixtures up to 2' x 4'		72	.889	Ea.	2.75	43.50		46.25	70.50
0400 Partitions, non load bearing									
0410 Plaster, lath, and studs	A-9	690	.093	S.F.	.86	4.54		5.40	8
0450 Gypsum board and studs		1390	.046		.14	2.25		2.39	3.67
0465 Carpet & pad		1390	.046		.14	2.25		2.39	3.67
0600 Pipe insulation, air cell type, up to 4" diameter pipe		900	.071	L.F.	.22	3.48		3.70	5.65
0610 4" to 8" diameter pipe		800	.080		.25	3.91		4.16	6.35
0620 10" to 12" diameter pipe		700	.091		.28	4.47		4.75	7.25
0630 14" to 16" diameter pipe		550	.116		.36	5.70		6.06	9.25
0650 Over 16" diameter pipe		650	.098	S.F.	.31	4.82		5.13	7.85
9000 For type B (supplied air) respirator equipment, add				%					10%

02 91 Chemical Sampling, Testing and Analysis

02 91 10 – Monitoring, Sampling, Testing and Analysis

02 91 10.10 Monitoring, Chemical Sampling, Testing and Analysis	Crew	Daily Output	Labor-Hours	Unit	Material	2012 Bare Costs Labor	Equipment	Total	Total Incl O&P
0010 **MONITORING, CHEMICAL SAMPLING, TESTING AND ANALYSIS**									
0015 Field Sampling of waste									
0100 Field samples, sample collection, sludge	1 Skwk	32	.250	Ea.		11.35		11.35	17.55
0110 Contaminated soils	"	32	.250	"		11.35		11.35	17.55
0200 Vials and bottles									
0210 32 oz. clear wide mouth jar (case of 12)				Ea.	43.50			43.50	48
0220 32 oz. Boston round bottle (case of 12)					35			35	38.50
0230 32 oz. HDPE bottle (case of 12)					32			32	35
0300 Laboratory analytical services									
0310 Laboratory testing 13 metals				Ea.	187			187	205
0312 13 metals + mercury					230			230	253
0314 8 metals					160			160	176
0316 Mercury only					40			40	44
0318 Single metal (only Cs, Li, Sr, Ta)					35			35	38.50
0320 Single metal (excludes Hg, Cs, Li, Sr, Ta)					35			35	38.50
0400 Hydrocarbons standard					65			65	71.50
0410 Hydrocarbons fingerprint					143			143	157
0500 Radioactivity gross alpha					169			169	186
0510 Gross alpha & beta					170			170	187
0520 Radium 226					84			84	92.50
0530 Radium 228					65			65	71.50
0540 Radon					140			140	154
0550 Uranium					80			80	88
0600 Volatile organics without GC/MS					170			170	187
0610 Volatile organics including GC/MS					170			170	187
0630 Synthetic organic compounds					1,000			1,000	1,100
0640 Herbicides					180			180	198

02 91 Chemical Sampling, Testing and Analysis

02 91 10 – Monitoring, Sampling, Testing and Analysis

02 91 10.10 Monitoring, Chemical Sampling, Testing and Analysis	Crew	Daily Output	Labor-Hours	Unit	Material	2012 Bare Costs Labor	Equipment	Total	Total Incl O&P
0650	Pesticides				Ea.	160			160	176
0660	PCB's				↓	160			160	176

Division Notes

	CREW	DAILY OUTPUT	LABOR-HOURS	UNIT	2012 BARE COSTS				TOTAL INCL O&P
					MAT.	LABOR	EQUIP.	TOTAL	

Estimating Tips

General

- Carefully check all the plans and specifications. Concrete often appears on drawings other than structural drawings, including mechanical and electrical drawings for equipment pads. The cost of cutting and patching is often difficult to estimate. See Subdivision 03 81 for Concrete Cutting, Subdivision 02 41 19.16 for Cutout Demolition, Subdivision 03 05 05.10 for Concrete Demolition, and Subdivision 02 41 19.23 for Rubbish Handling (handling, loading and hauling of debris).

- Always obtain concrete prices from suppliers near the job site. A volume discount can often be negotiated, depending upon competition in the area. Remember to add for waste, particularly for slabs and footings on grade.

03 10 00 Concrete Forming and Accessories

- A primary cost for concrete construction is forming. Most jobs today are constructed with prefabricated forms. The selection of the forms best suited for the job and the total square feet of forms required for efficient concrete forming and placing are key elements in estimating concrete construction. Enough forms must be available for erection to make efficient use of the concrete placing equipment and crew.

- Concrete accessories for forming and placing depend upon the systems used. Study the plans and specifications to ensure that all special accessory requirements have been included in the cost estimate, such as anchor bolts, inserts, and hangers.

- Included within costs for forms-in-place are all necessary bracing and shoring.

03 20 00 Concrete Reinforcing

- Ascertain that the reinforcing steel supplier has included all accessories, cutting, bending, and an allowance for lapping, splicing, and waste. A good rule of thumb is 10% for lapping, splicing, and waste. Also, 10% waste should be allowed for welded wire fabric.

- The unit price items in the subdivisions for Reinforcing In Place, Glass Fiber Reinforcing, and Welded Wire Fabric include the labor to install accessories such as beam and slab bolsters, high chairs, and bar ties and tie wire. The material cost for these accessories is not included; they may be obtained from the Accessories Division.

03 30 00 Cast-In-Place Concrete

- When estimating structural concrete, pay particular attention to requirements for concrete additives, curing methods, and surface treatments. Special consideration for climate, hot or cold, must be included in your estimate. Be sure to include requirements for concrete placing equipment, and concrete finishing.

- For accurate concrete estimating, the estimator must consider each of the following major components individually: forms, reinforcing steel, ready-mix concrete, placement of the concrete, and finishing of the top surface. For faster estimating, Subdivision 03 30 53.40 for Concrete-In-Place can be used; here, various items of concrete work are presented that include the costs of all five major components (unless specifically stated otherwise).

03 40 00 Precast Concrete
03 50 00 Cast Decks and Underlayment

- The cost of hauling precast concrete structural members is often an important factor. For this reason, it is important to get a quote from the nearest supplier. It may become economically feasible to set up precasting beds on the site if the hauling costs are prohibitive.

Reference Numbers

Reference numbers are shown in shaded boxes at the beginning of some major classifications. These numbers refer to related items in the Reference Section. The reference information may be an estimating procedure, an alternate pricing method, or technical information.

Note: Not all subdivisions listed here necessarily appear in this publication.

03 01 Maintenance of Concrete

03 01 30 – Maintenance of Cast-In-Place Concrete

03 01 30.62 Concrete Patching

		Crew	Daily Output	Labor-Hours	Unit	Material	2012 Bare Costs Labor	Equipment	Total	Total Incl O&P
0010	**CONCRETE PATCHING**									
0100	Floors, 1/4" thick, small areas, regular grout	1 Cefi	170	.047	S.F.	1.23	1.99		3.22	4.28
0150	Epoxy grout	"	100	.080	"	7.30	3.39		10.69	13
2000	Walls, including chipping, cleaning and epoxy grout									
2100	1/4" deep	1 Cefi	65	.123	S.F.	7.30	5.20		12.50	15.65
2150	1/2" deep		50	.160		14.55	6.80		21.35	26
2200	3/4" deep	↓	40	.200	↓	22	8.45		30.45	36.50

03 05 Common Work Results for Concrete

03 05 05 – Selective Concrete Demolition

03 05 05.10 Selective Demolition, Concrete

		Crew	Daily Output	Labor-Hours	Unit	Material	2012 Bare Costs Labor	Equipment	Total	Total Incl O&P
0010	**SELECTIVE DEMOLITION, CONCRETE** R024119-10									
0012	Excludes saw cutting, torch cutting, loading or hauling									
0050	Break up into small pieces, minimum reinforcing	B-9	24	1.667	C.Y.		59	8.70	67.70	101
0060	Average reinforcing		16	2.500			89	13.10	102.10	151
0070	Maximum reinforcing	↓	8	5	↓		178	26	204	300
0150	Remove whole pieces, up to 2 tons per piece	E-18	36	1.111	Ea.		54.50	26.50	81	124
0160	2 – 5 tons per piece		30	1.333			65.50	31.50	97	149
0170	5 – 10 tons per piece		24	1.667			82	39.50	121.50	186
0180	10 – 15 tons per piece	↓	18	2.222			109	53	162	248
0250	Precast unit embedded in masonry, up to 1 C.F.	D-1	16	1			40		40	60
0260	1 – 2 C.F.		12	1.333			53		53	80
0270	2 – 5 C.F.		10	1.600			64		64	96.50
0280	5 – 10 C.F.	↓	8	2	↓		79.50		79.50	120
0990	For hydrodemolition see Section 02 41 13.15									

03 05 13 – Basic Concrete Materials

03 05 13.20 Concrete Admixtures and Surface Treatments

		Crew	Daily Output	Labor-Hours	Unit	Material	2012 Bare Costs Labor	Equipment	Total	Total Incl O&P
0010	**CONCRETE ADMIXTURES AND SURFACE TREATMENTS**									
0040	Abrasives, aluminum oxide, over 20 tons				Lb.	1.57			1.57	1.73
0050	1 to 20 tons					1.73			1.73	1.90
0070	Under 1 ton					1.80			1.80	1.98
0100	Silicon carbide, black, over 20 tons					2.34			2.34	2.57
0110	1 to 20 tons					2.69			2.69	2.96
0120	Under 1 ton				↓	2.80			2.80	3.08
0200	Air entraining agent, .7 to 1.5 oz. per bag, 55 gallon drum				Gal.	11.60			11.60	12.75
0220	5 gallon pail					15.95			15.95	17.55
0300	Bonding agent, acrylic latex, 250 S.F. per gallon, 5 gallon pail					21			21	23
0320	Epoxy resin, 80 S.F. per gallon, 4 gallon case				↓	59.50			59.50	65.50
0400	Calcium chloride, 50 lb. bags, TL lots				Ton	670			670	735
0420	Less than truckload lots				Bag	20			20	22
0500	Carbon black, liquid, 2 to 8 lb. per bag of cement				Lb.	7.55			7.55	8.30
0600	Colors, integral, 2 to 10 lb. per bag of cement, minimum					2.43			2.43	2.67
0610	Average					3.13			3.13	3.44
0620	Maximum				↓	4.92			4.92	5.40
0920	Dustproofing compound, 250 S.F./gal., 5 gallon pail				Gal.	7.75			7.75	8.55
1010	Epoxy based, 125 S.F./gal., 5 gallon pail				"	58.50			58.50	64
1100	Hardeners, metallic, 55 lb. bags, natural (grey)				Lb.	.90			.90	.99
1200	Colors					1.48			1.48	1.63
1300	Non-metallic, 55 lb. bags, natural grey					.52			.52	.57
1320	Colors				↓	.79			.79	.87

03 05 13 – Basic Concrete Materials

03 05 13.20 Concrete Admixtures and Surface Treatments	Crew	Daily Output	Labor-Hours	Unit	Material	2012 Bare Costs Labor	Equipment	Total	Total Incl O&P
1550	Release agent, for tilt slabs, 5 gallon pail				Gal.	15.60			15.60	17.15
1570	For forms, 5 gallon pail					12.35			12.35	13.55
1590	Concrete release agent for forms, 100% biodegradeable, zero VOC, 5 gal pail **G**					18.60			18.60	20.50
1595	55 gallon drum **G**					17.65			17.65	19.45
1600	Sealer, hardener and dustproofer, epoxy-based, 125 S.F./gal., 5 gallon, min					58.50			58.50	64
1620	5 gallon pail, maximum					64			64	70.50
1630	Sealer, solvent-based, 250 S.F./gal., 55 gallon drum					20			20	22
1640	5 gallon pail					27			27	29.50
1650	Sealer, water based, 350 S.F., 55 gallon drum					19.80			19.80	22
1660	5 gallon pail					23			23	25.50
1900	Set retarder, 100 S.F./gal., 1 gallon pail				▼	19.10			19.10	21
2000	Waterproofing, integral 1 lb. per bag of cement				Lb.	1.76			1.76	1.94
2100	Powdered metallic, 40 lb. per 100 S.F., minimum					2.09			2.09	2.30
2120	Maximum				▼	2.92			2.92	3.21
3000	For integral colors, 2500 psi (5 bag mix)									
3100	Red, yellow or brown, 1.8 lb. per bag, add				C.Y.	22			22	24
3200	9.4 lb. per bag, add					114			114	126
3400	Black, 1.8 lb. per bag, add					28			28	31
3500	7.5 lb. per bag, add					117			117	129
3700	Green, 1.8 lb. per bag, add					44.50			44.50	48.50
3800	7.5 lb. per bag, add				▼	185			185	203
6000	Concrete ready-mix additives, recycled coal fly ash, mixed at plant **G**				Ton	56			56	61.50
6010	Recycled blast furnace slag, mixed at plant **G**				"	87			87	95.50

03 05 13.25 Aggregate

		Crew	Daily Output	Labor-Hours	Unit	Material	2012 Bare Costs Labor	Equipment	Total	Total Incl O&P
0010	**AGGREGATE** R033105-20									
0100	Lightweight vermiculite or perlite, 4 C.F. bag, C.L. lots **G**				Bag	21.50			21.50	23.50
0150	L.C.L. lots R033105-40 **G**				"	23.50			23.50	26
0250	Sand & stone, loaded at pit, crushed bank gravel				Ton	17.30			17.30	19.05
0350	Sand, washed, for concrete R033105-50					16.60			16.60	18.25
0400	For plaster or brick					16.60			16.60	18.25
0450	Stone, 3/4" to 1-1/2"					19.25			19.25	21
0470	Round, river stone					30			30	33
0500	3/8" roofing stone & 1/2" pea stone					24			24	26.50
0550	For trucking 10 miles (5 mile R/T), add to the above	B-34B	117	.068			2.41	5	7.41	9.15
0600	30 miles (15 mi R/T), add to the above	"	72	.111	▼		3.92	8.10	12.02	14.90
0850	Sand & stone, loaded at pit, crushed bank gravel				C.Y.	24			24	26.50
0950	Sand, washed, for concrete					23			23	25.50
1000	For plaster or brick					23			23	25.50
1050	Stone, 3/4" to 1-1/2"					37			37	40.50
1055	Round, river stone					33.50			33.50	37
1100	3/8" roofing stone & 1/2" pea stone					41			41	45
1150	For trucking 10 miles (5 mi R/T), add to the above	B-34B	78	.103			3.62	7.50	11.12	13.75
1200	30 miles (15 mi R/T), add to the above	"	48	.167	▼		5.90	12.20	18.10	22.50
1310	Quartz chips, 50 lb. bags				Cwt.	19.45			19.45	21.50
1330	Silica chips, 50 lb. bags					10.65			10.65	11.75
1410	White marble, 3/8" to 1/2", 50 lb. bags				▼	15.85			15.85	17.40
1430	3/4"				Ton	88			88	97

03 05 13.30 Cement

		Crew	Daily Output	Labor-Hours	Unit	Material	2012 Bare Costs Labor	Equipment	Total	Total Incl O&P
0010	**CEMENT** R033105-20									
0240	Portland, Type I/II, TL lots, 94 lb. bags				Bag	11.50			11.50	12.65
0250	LTL/LCL lots R033105-30				"	13.50			13.50	14.85
0300	Trucked in bulk, per cwt				Cwt.	7.80			7.80	8.60

03 05 Common Work Results for Concrete

03 05 13 – Basic Concrete Materials

03 05 13.30 Cement

		Crew	Daily Output	Labor-Hours	Unit	Material	2012 Bare Costs Labor	Equipment	Total	Total Incl O&P
0400	Type III, high early strength, TL lots, 94 lb. bags	R033105-40			Bag	12.60			12.60	13.85
0420	L.T.L. or L.C.L. lots					14.10			14.10	15.50
0500	White, type III, high early strength, T.L. or C.L. lots, bags	R033105-50				25			25	27.50
0520	L.T.L. or L.C.L. lots					28			28	31
0600	White, type I, T.L. or C.L. lots, bags					24.50			24.50	27
0620	L.T.L. or L.C.L. lots					26			26	28.50

03 05 13.80 Waterproofing and Dampproofing

		Crew	Daily Output	Labor-Hours	Unit	Material	2012 Bare Costs Labor	Equipment	Total	Total Incl O&P
0010	**WATERPROOFING AND DAMPPROOFING**									
0050	Integral waterproofing, add to cost of regular concrete				C.Y.	10.55			10.55	11.60

03 05 13.85 Winter Protection

		Crew	Daily Output	Labor-Hours	Unit	Material	2012 Bare Costs Labor	Equipment	Total	Total Incl O&P
0010	**WINTER PROTECTION**									
0012	For heated ready mix, add, minimum				C.Y.	4.25			4.25	4.68
0050	Maximum				"	5.30			5.30	5.85
0100	Temporary heat to protect concrete, 24 hours, minimum	2 Clab	50	.320	M.S.F.	695	11.25		706.25	780
0150	Maximum	"	25	.640	"	905	22.50		927.50	1,025
0200	Temporary shelter for slab on grade, wood frame/polyethylene sheeting									
0201	Build or remove, minimum	2 Carp	10	1.600	M.S.F.	259	70.50		329.50	395
0210	Maximum	"	3	5.333	"	325	235		560	720
0500	Electrically heated pads, 110 volts, 15 watts per S.F., buy				S.F.	6.55			6.55	7.20
0600	20 watts per S.F., buy					8.70			8.70	9.55
0710	Electrically, heated pads, 15 watts/S.F., 20 uses, minimum					.23			.23	.26
0800	Maximum					.39			.39	.43

03 11 Concrete Forming

03 11 13 – Structural Cast-In-Place Concrete Forming

03 11 13.20 Forms In Place, Beams and Girders

		Crew	Daily Output	Labor-Hours	Unit	Material	2012 Bare Costs Labor	Equipment	Total	Total Incl O&P
0010	**FORMS IN PLACE, BEAMS AND GIRDERS**	R031113-40								
0500	Exterior spandrel, job-built plywood, 12" wide, 1 use	C-2	225	.213	SFCA	2.51	9.15		11.66	16.85
0550	2 use	R031113-60	275	.175		1.32	7.50		8.82	13
0600	3 use		295	.163		1.01	7		8.01	11.85
0650	4 use		310	.155		.82	6.65		7.47	11.15
1000	18" wide, 1 use		250	.192		2.20	8.25		10.45	15.10
1050	2 use		275	.175		1.21	7.50		8.71	12.90
1100	3 use		305	.157		.88	6.75		7.63	11.35
1150	4 use		315	.152		.72	6.55		7.27	10.85
1500	24" wide, 1 use		265	.181		2	7.80		9.80	14.15
1550	2 use		290	.166		1.13	7.10		8.23	12.20
1600	3 use		315	.152		.80	6.55		7.35	10.95
1650	4 use		325	.148		.65	6.35		7	10.45
2000	Interior beam, job-built plywood, 12" wide, 1 use		300	.160		3.03	6.85		9.88	13.90
2050	2 use		340	.141		1.46	6.05		7.51	10.95
2100	3 use		364	.132		1.21	5.65		6.86	10.05
2150	4 use		377	.127		.99	5.45		6.44	9.50
2500	24" wide, 1 use		320	.150		2.05	6.45		8.50	12.15
2550	2 use		365	.132		1.15	5.65		6.80	9.95
2600	3 use		385	.125		.81	5.35		6.16	9.15
2650	4 use		395	.122		.66	5.20		5.86	8.80
3000	Encasing steel beam, hung, job-built plywood, 1 use		325	.148		2.57	6.35		8.92	12.55
3050	2 use		390	.123		1.41	5.30		6.71	9.70
3100	3 use		415	.116		1.03	4.97		6	8.80

03 11 Concrete Forming

03 11 13 – Structural Cast-In-Place Concrete Forming

03 11 13.20 Forms In Place, Beams and Girders

		Crew	Daily Output	Labor-Hours	Unit	Material	2012 Bare Costs Labor	Equipment	Total	Total Incl O&P
3150	4 use	C-2	430	.112	SFCA	.83	4.79		5.62	8.25
3500	Bottoms only, to 30" wide, job-built plywood, 1 use		230	.209		3.46	8.95		12.41	17.60
3550	2 use		265	.181		1.94	7.80		9.74	14.10
3600	3 use		280	.171		1.38	7.35		8.73	12.80
3650	4 use		290	.166		1.12	7.10		8.22	12.20
4000	Sides only, vertical, 36" high, job-built plywood, 1 use		335	.143		3.72	6.15		9.87	13.55
4050	2 use		405	.119		2.04	5.10		7.14	10.10
4100	3 use		430	.112		1.49	4.79		6.28	9
4150	4 use		445	.108		1.21	4.63		5.84	8.50
4500	Sloped sides, 36" high, 1 use		305	.157		3.62	6.75		10.37	14.40
4550	2 use		370	.130		2.02	5.55		7.57	10.75
4600	3 use		405	.119		1.45	5.10		6.55	9.45
4650	4 use		425	.113		1.18	4.85		6.03	8.75
5000	Upstanding beams, 36" high, 1 use		225	.213		4.53	9.15		13.68	19.10
5050	2 use		255	.188		2.50	8.10		10.60	15.20
5100	3 use		275	.175		1.83	7.50		9.33	13.55
5150	4 use		280	.171		1.48	7.35		8.83	12.95

03 11 13.25 Forms In Place, Columns

			Crew	Daily Output	Labor-Hours	Unit	Material	2012 Bare Costs Labor	Equipment	Total	Total Incl O&P
0010	**FORMS IN PLACE, COLUMNS**	R031113-40									
0500	Round fiberglass, 4 use per mo., rent, 12" diameter		C-1	160	.200	L.F.	8.15	8.35		16.50	22
0550	16" diameter	R031113-60		150	.213		9.70	8.95		18.65	24.50
0600	18" diameter			140	.229		10.85	9.55		20.40	26.50
0650	24" diameter			135	.237		13.50	9.90		23.40	30
0700	28" diameter			130	.246		15.10	10.30		25.40	32.50
0800	30" diameter			125	.256		15.75	10.70		26.45	34
0850	36" diameter			120	.267		21	11.15		32.15	40
1500	Round fiber tube, recycled paper, 1 use, 8" diameter	G		155	.206		1.51	8.65		10.16	14.95
1550	10" diameter	G		155	.206		2.10	8.65		10.75	15.60
1600	12" diameter	G		150	.213		2.44	8.95		11.39	16.45
1650	14" diameter	G		145	.221		3.54	9.25		12.79	18.10
1700	16" diameter	G		140	.229		4.29	9.55		13.84	19.40
1720	18" diameter	G		140	.229		5	9.55		14.55	20
1750	20" diameter	G		135	.237		6.85	9.90		16.75	23
1800	24" diameter	G		130	.246		8.75	10.30		19.05	25.50
1850	30" diameter	G		125	.256		12.90	10.70		23.60	30.50
1900	36" diameter	G		115	.278		16.15	11.65		27.80	35.50
1950	42" diameter	G		100	.320		40	13.40		53.40	64.50
2000	48" diameter	G		85	.376		46	15.75		61.75	75
2200	For seamless type, add						15%				
3000	Round, steel, 4 use per mo., rent, regular duty, 14" diameter	G	C-1	145	.221	L.F.	15.45	9.25		24.70	31
3050	16" diameter	G		125	.256		15.75	10.70		26.45	34
3100	Heavy duty, 20" diameter	G		105	.305		17.40	12.75		30.15	39
3150	24" diameter	G		85	.376		18.90	15.75		34.65	45
3200	30" diameter	G		70	.457		22	19.15		41.15	53.50
3250	36" diameter	G		60	.533		23.50	22.50		46	60
3300	48" diameter	G		50	.640		69.50	27		96.50	118
3350	60" diameter	G		45	.711		42.50	30		72.50	93
4500	For second and succeeding months, deduct						50%				
5000	Job-built plywood, 8" x 8" columns, 1 use		C-1	165	.194	SFCA	2.62	8.10		10.72	15.40
5050	2 use			195	.164		1.49	6.85		8.34	12.20
5100	3 use			210	.152		1.05	6.40		7.45	10.95
5150	4 use			215	.149		.87	6.25		7.12	10.55

03 11 Concrete Forming

03 11 13 – Structural Cast-In-Place Concrete Forming

03 11 13.25 Forms In Place, Columns

		Crew	Daily Output	Labor-Hours	Unit	Material	2012 Bare Costs Labor	Equipment	Total	Total Incl O&P
5500	12" x 12" columns, 1 use	C-1	180	.178	SFCA	2.38	7.45		9.83	14.05
5550	2 use		210	.152		1.31	6.40		7.71	11.25
5600	3 use		220	.145		.95	6.10		7.05	10.40
5650	4 use		225	.142		.77	5.95		6.72	10
6000	16" x 16" columns, 1 use		185	.173		2.31	7.25		9.56	13.70
6050	2 use		215	.149		1.26	6.25		7.51	11
6100	3 use		230	.139		.93	5.80		6.73	10
6150	4 use		235	.136		.76	5.70		6.46	9.60
6500	24" x 24" columns, 1 use		190	.168		2.57	7.05		9.62	13.65
6550	2 use		216	.148		1.41	6.20		7.61	11.10
6600	3 use		230	.139		1.03	5.80		6.83	10.10
6650	4 use		238	.134		.83	5.65		6.48	9.55
7000	36" x 36" columns, 1 use		200	.160		1.84	6.70		8.54	12.35
7050	2 use		230	.139		1.06	5.80		6.86	10.10
7100	3 use		245	.131		.74	5.45		6.19	9.20
7150	4 use	▼	250	.128	▼	.60	5.35		5.95	8.90
7400	Steel framed plywood, based on 50 uses of purchased									
7420	forms, and 4 uses of bracing lumber									
7500	8" x 8" column	C-1	340	.094	SFCA	1.82	3.94		5.76	8.05
7550	10" x 10"		350	.091		1.59	3.83		5.42	7.65
7600	12" x 12"		370	.086		1.35	3.62		4.97	7.05
7650	16" x 16"		400	.080		1.05	3.35		4.40	6.30
7700	20" x 20"		420	.076		.93	3.19		4.12	5.95
7750	24" x 24"		440	.073		.67	3.04		3.71	5.40
7755	30" x 30"		440	.073		.86	3.04		3.90	5.65
7760	36" x 36"	▼	460	.070	▼	.75	2.91		3.66	5.30

03 11 13.30 Forms In Place, Culvert

			Crew	Daily Output	Labor-Hours	Unit	Material	2012 Bare Costs Labor	Equipment	Total	Total Incl O&P
0010	**FORMS IN PLACE, CULVERT**	R031113-40									
0015	5' to 8' square or rectangular, 1 use		C-1	170	.188	SFCA	3.13	7.90		11.03	15.55
0050	2 use	R031113-60		180	.178		1.86	7.45		9.31	13.50
0100	3 use			190	.168		1.44	7.05		8.49	12.45
0150	4 use		▼	200	.160	▼	1.23	6.70		7.93	11.65

03 11 13.35 Forms In Place, Elevated Slabs

			Crew	Daily Output	Labor-Hours	Unit	Material	2012 Bare Costs Labor	Equipment	Total	Total Incl O&P
0010	**FORMS IN PLACE, ELEVATED SLABS**	R031113-40									
1000	Flat plate, job-built plywood, to 15' high, 1 use	R031113-60	C-2	470	.102	S.F.	3.18	4.38		7.56	10.25
1050	2 use			520	.092		1.75	3.96		5.71	8
1100	3 use			545	.088		1.27	3.78		5.05	7.20
1150	4 use			560	.086		1.03	3.68		4.71	6.80
1500	15' to 20' high ceilings, 4 use			495	.097		1.03	4.16		5.19	7.55
1600	21' to 35' high ceilings, 4 use			450	.107		1.30	4.58		5.88	8.50
2000	Flat slab, drop panels, job-built plywood, to 15' high, 1 use			449	.107		3.57	4.59		8.16	11
2050	2 use			509	.094		1.96	4.05		6.01	8.40
2100	3 use			532	.090		1.43	3.87		5.30	7.50
2150	4 use			544	.088		1.16	3.79		4.95	7.15
2250	15' to 20' high ceilings, 4 use			480	.100		2.08	4.29		6.37	8.90
2350	20' to 35' high ceilings, 4 use			435	.110		2.34	4.74		7.08	9.85
3000	Floor slab hung from steel beams, 1 use			485	.099		2.14	4.25		6.39	8.90
3050	2 use			535	.090		1.69	3.85		5.54	7.80
3100	3 use			550	.087		1.54	3.75		5.29	7.45
3150	4 use			565	.085		1.47	3.65		5.12	7.20
3500	Floor slab, with 1-way joist pans, 1 use			415	.116		5.35	4.97		10.32	13.50
3550	2 use			445	.108		3.72	4.63		8.35	11.25

03 11 Concrete Forming

03 11 13 – Structural Cast-In-Place Concrete Forming

03 11 13.35 Forms In Place, Elevated Slabs

		Crew	Daily Output	Labor-Hours	Unit	Material	2012 Bare Costs Labor	Equipment	Total	Total Incl O&P
3600	3 use	C-2	475	.101	S.F.	3.18	4.34		7.52	10.20
3650	4 use		500	.096		2.92	4.12		7.04	9.55
4500	With 2-way waffle domes, 1 use		405	.119		5.45	5.10		10.55	13.85
4520	2 use		450	.107		3.85	4.58		8.43	11.30
4530	3 use		460	.104		3.31	4.48		7.79	10.55
4550	4 use		470	.102		3.04	4.38		7.42	10.10
5000	Box out for slab openings, over 16" deep, 1 use		190	.253	SFCA	3.93	10.85		14.78	21
5050	2 use		240	.200	"	2.16	8.60		10.76	15.55
5500	Shallow slab box outs, to 10 S.F.		42	1.143	Ea.	9.75	49		58.75	86.50
5550	Over 10 S.F. (use perimeter)		600	.080	L.F.	1.31	3.43		4.74	6.75
6000	Bulkhead forms for slab, with keyway, 1 use, 2 piece		500	.096		1.85	4.12		5.97	8.40
6100	3 piece (see also edge forms)		460	.104		1.97	4.48		6.45	9.05
6200	Slab bulkhead form, 4-1/2" high, exp metal, w/keyway & stakes G	C-1	1200	.027		1.13	1.12		2.25	2.96
6210	5-1/2" high G		1100	.029		1.25	1.22		2.47	3.25
6215	7-1/2" high G		960	.033		1.47	1.39		2.86	3.77
6220	9-1/2" high G		840	.038		1.67	1.59		3.26	4.29
6500	Curb forms, wood, 6" to 12" high, on elevated slabs, 1 use		180	.178	SFCA	1.29	7.45		8.74	12.85
6550	2 use		205	.156		.71	6.55		7.26	10.85
6600	3 use		220	.145		.52	6.10		6.62	9.90
6650	4 use		225	.142		.42	5.95		6.37	9.60
7000	Edge forms to 6" high, on elevated slab, 4 use		500	.064	L.F.	.17	2.68		2.85	4.31
7500	Depressed area forms to 12" high, 4 use		300	.107		.88	4.46		5.34	7.80
7550	12" to 24" high, 4 use		175	.183		1.19	7.65		8.84	13.05
8000	Perimeter deck and rail for elevated slabs, straight		90	.356		9.75	14.90		24.65	33.50
8050	Curved		65	.492		13.35	20.50		33.85	46
8500	Void forms, round plastic, 8" high x 3" diameter G		450	.071	Ea.	1.89	2.98		4.87	6.65
8550	4" diameter G		425	.075		2.66	3.15		5.81	7.80
8600	6" diameter G		400	.080		3.97	3.35		7.32	9.50
8650	8" diameter G		375	.085		6	3.57		9.57	12.10

03 11 13.40 Forms In Place, Equipment Foundations

		Crew	Daily Output	Labor-Hours	Unit	Material	2012 Bare Costs Labor	Equipment	Total	Total Incl O&P
0010	**FORMS IN PLACE, EQUIPMENT FOUNDATIONS** R031113-40									
0020	1 use	C-2	160	.300	SFCA	2.93	12.90		15.83	23
0050	2 use R031113-60		190	.253		1.61	10.85		12.46	18.45
0100	3 use		200	.240		1.17	10.30		11.47	17.15
0150	4 use		205	.234		.96	10.05		11.01	16.50

03 11 13.45 Forms In Place, Footings

		Crew	Daily Output	Labor-Hours	Unit	Material	2012 Bare Costs Labor	Equipment	Total	Total Incl O&P
0010	**FORMS IN PLACE, FOOTINGS** R031113-40									
0020	Continuous wall, plywood, 1 use	C-1	375	.085	SFCA	5.95	3.57		9.52	12
0050	2 use R031113-60		440	.073		3.26	3.04		6.30	8.25
0100	3 use		470	.068		2.37	2.85		5.22	7
0150	4 use		485	.066		1.93	2.76		4.69	6.35
0500	Dowel supports for footings or beams, 1 use		500	.064	L.F.	.74	2.68		3.42	4.93
1000	Integral starter wall, to 4" high, 1 use		400	.080		.80	3.35		4.15	6.05
1500	Keyway, 4 use, tapered wood, 2" x 4"	1 Carp	530	.015		.18	.67		.85	1.21
1550	2" x 6"		500	.016		.27	.71		.98	1.38
2000	Tapered plastic		530	.015		.60	.67		1.27	1.68
2250	For keyway hung from supports, add		150	.053		.74	2.35		3.09	4.43
3000	Pile cap, square or rectangular, job-built plywood, 1 use	C-1	290	.110	SFCA	2.36	4.62		6.98	9.70
3050	2 use		346	.092		1.30	3.87		5.17	7.40
3100	3 use		371	.086		.94	3.61		4.55	6.60
3150	4 use		383	.084		.77	3.50		4.27	6.25
4000	Triangular or hexagonal, 1 use		225	.142		2.75	5.95		8.70	12.20

03 11 Concrete Forming

03 11 13 – Structural Cast-In-Place Concrete Forming

03 11 13.45 Forms In Place, Footings

		Crew	Daily Output	Labor-Hours	Unit	Material	2012 Bare Costs Labor	Equipment	Total	Total Incl O&P
4050	2 use	C-1	280	.114	SFCA	1.51	4.78		6.29	9
4100	3 use		305	.105		1.10	4.39		5.49	7.95
4150	4 use		315	.102		.89	4.25		5.14	7.55
5000	Spread footings, job-built lumber, 1 use		305	.105		1.90	4.39		6.29	8.85
5050	2 use		371	.086		1.06	3.61		4.67	6.70
5100	3 use		401	.080		.76	3.34		4.10	6
5150	4 use		414	.077	▼	.62	3.23		3.85	5.65
6000	Supports for dowels, plinths or templates, 2' x 2' footing		25	1.280	Ea.	4.92	53.50		58.42	88
6050	4' x 4' footing		22	1.455		9.85	61		70.85	104
6100	8' x 8' footing		20	1.600		19.70	67		86.70	125
6150	12' x 12' footing		17	1.882	▼	24	79		103	147
7000	Plinths, job-built plywood, 1 use		250	.128	SFCA	2.49	5.35		7.84	11
7100	4 use	▼	270	.119	"	.82	4.96		5.78	8.55

03 11 13.47 Forms In Place, Gas Station Forms

				Crew	Daily Output	Labor-Hours	Unit	Material	2012 Bare Costs Labor	Equipment	Total	Total Incl O&P
0010	**FORMS IN PLACE, GAS STATION FORMS**											
0050	Curb fascia, with template, 12 ga. steel, left in place, 9" high		G	1 Carp	50	.160	L.F.	13.10	7.05		20.15	25.50
1000	Sign or light bases, 18" diameter, 9" high		G		9	.889	Ea.	82.50	39		121.50	152
1050	30" diameter, 13" high		G		8	1		131	44		175	212
2000	Island forms, 10' long, 9" high, 3'- 6" wide		G	C-1	10	3.200		365	134		499	610
2050	4' wide		G		9	3.556		380	149		529	645
2500	20' long, 9" high, 4' wide		G		6	5.333		610	223		833	1,025
2550	5' wide		G	▼	5	6.400	▼	635	268		903	1,100

03 11 13.50 Forms In Place, Grade Beam

			Crew	Daily Output	Labor-Hours	Unit	Material	2012 Bare Costs Labor	Equipment	Total	Total Incl O&P
0010	**FORMS IN PLACE, GRADE BEAM**	R031113-40									
0020	Job-built plywood, 1 use		C-2	530	.091	SFCA	2.68	3.89		6.57	8.95
0050	2 use	R031113-60		580	.083		1.48	3.55		5.03	7.05
0100	3 use			600	.080		1.07	3.43		4.50	6.50
0150	4 use		▼	605	.079	▼	.87	3.41		4.28	6.20

03 11 13.55 Forms In Place, Mat Foundation

			Crew	Daily Output	Labor-Hours	Unit	Material	2012 Bare Costs Labor	Equipment	Total	Total Incl O&P
0010	**FORMS IN PLACE, MAT FOUNDATION**	R031113-40									
0020	Job-built plywood, 1 use		C-2	290	.166	SFCA	2.55	7.10		9.65	13.75
0050	2 use	R031113-60		310	.155		1.05	6.65		7.70	11.40
0100	3 use			330	.145		.67	6.25		6.92	10.35
0120	4 use		▼	350	.137	▼	.62	5.90		6.52	9.75

03 11 13.65 Forms In Place, Slab On Grade

				Crew	Daily Output	Labor-Hours	Unit	Material	2012 Bare Costs Labor	Equipment	Total	Total Incl O&P
0010	**FORMS IN PLACE, SLAB ON GRADE**		R031113-40									
1000	Bulkhead forms w/keyway, wood, 6" high, 1 use			C-1	510	.063	L.F.	.87	2.63		3.50	4.99
1050	2 uses		R031113-60		400	.080		.48	3.35		3.83	5.65
1100	4 uses				350	.091		.28	3.83		4.11	6.20
1400	Bulkhead form for slab, 4-1/2" high, exp metal, incl keyway & stakes		G		1200	.027		1.13	1.12		2.25	2.96
1410	5-1/2" high		G		1100	.029		1.25	1.22		2.47	3.25
1420	7-1/2" high		G		960	.033		1.47	1.39		2.86	3.77
1430	9-1/2" high		G		840	.038	▼	1.67	1.59		3.26	4.29
2000	Curb forms, wood, 6" to 12" high, on grade, 1 use				215	.149	SFCA	1.97	6.25		8.22	11.75
2050	2 use				250	.128		1.09	5.35		6.44	9.45
2100	3 use				265	.121		.79	5.05		5.84	8.65
2150	4 use				275	.116	▼	.64	4.87		5.51	8.20
3000	Edge forms, wood, 4 use, on grade, to 6" high				600	.053	L.F.	.29	2.23		2.52	3.75
3050	7" to 12" high				435	.074	SFCA	.62	3.08		3.70	5.40
3500	For depressed slabs, 4 use, to 12" high				300	.107	L.F.	.58	4.46		5.04	7.50
3550	To 24" high				175	.183		.75	7.65		8.40	12.60
4000	For slab blockouts, to 12" high, 1 use			▼	200	.160	▼	.60	6.70		7.30	10.95

03 11 Concrete Forming

03 11 13 – Structural Cast-In-Place Concrete Forming

03 11 13.65 Forms In Place, Slab On Grade

		Crew	Daily Output	Labor-Hours	Unit	Material	2012 Bare Costs Labor	2012 Bare Costs Equipment	Total	Total Incl O&P
4050	To 24" high, 1 use	C-1	120	.267	L.F.	.76	11.15		11.91	18
4100	Plastic (extruded), to 6" high, multiple use, on grade	↓	800	.040	↓	7.90	1.67		9.57	11.30
5000	Screed, 24 ga. metal key joint, see Section 03 15 05.25									
5020	Wood, incl. wood stakes, 1" x 3"	C-1	900	.036	L.F.	.68	1.49		2.17	3.04
5050	2" x 4"		900	.036	"	.67	1.49		2.16	3.03
6000	Trench forms in floor, wood, 1 use		160	.200	SFCA	1.56	8.35		9.91	14.60
6050	2 use		175	.183		.86	7.65		8.51	12.70
6100	3 use		180	.178		.62	7.45		8.07	12.15
6150	4 use		185	.173	↓	.51	7.25		7.76	11.70
8760	Void form, corrugated fiberboard, 4" x 12", 4' long [G]		3000	.011	S.F.	2.64	.45		3.09	3.59
8770	6" x 12", 4' long		3000	.011		3.10	.45		3.55	4.10
8780	1/4" thick hardboard protective cover for void form	2 Carp	1500	.011	↓	.57	.47		1.04	1.35

03 11 13.85 Forms In Place, Walls

		Crew	Daily Output	Labor-Hours	Unit	Material	2012 Bare Costs Labor	2012 Bare Costs Equipment	Total	Total Incl O&P
0010	**FORMS IN PLACE, WALLS** R031113-10									
0100	Box out for wall openings, to 16" thick, to 10 S.F.	C-2	24	2	Ea.	21	86		107	156
0150	Over 10 S.F. (use perimeter) R031113-40	"	280	.171	L.F.	1.83	7.35		9.18	13.30
0250	Brick shelf, 4" w, add to wall forms, use wall area abv shelf									
0260	1 use R031113-60	C-2	240	.200	SFCA	1.97	8.60		10.57	15.35
0300	2 use		275	.175		1.08	7.50		8.58	12.75
0350	4 use		300	.160	↓	.79	6.85		7.64	11.40
0500	Bulkhead, wood with keyway, 1 use, 2 piece	↓	265	.181	L.F.	1.60	7.80		9.40	13.70
0600	Bulkhead forms with keyway, 1 piece expanded metal, 8" wall [G]	C-1	1000	.032		1.47	1.34		2.81	3.68
0610	10" wall [G]		800	.040		1.67	1.67		3.34	4.42
0620	12" wall [G]	↓	525	.061	↓	2	2.55		4.55	6.10
0700	Buttress, to 8' high, 1 use	C-2	350	.137	SFCA	3.08	5.90		8.98	12.45
0750	2 use		430	.112		1.69	4.79		6.48	9.20
0800	3 use		460	.104		1.23	4.48		5.71	8.25
0850	4 use		480	.100	↓	1.02	4.29		5.31	7.70
1000	Corbel or haunch, to 12" wide, add to wall forms, 1 use		150	.320	L.F.	1.91	13.75		15.66	23
1050	2 use		170	.282		1.05	12.10		13.15	19.80
1100	3 use		175	.274		.76	11.80		12.56	18.95
1150	4 use		180	.267	↓	.62	11.45		12.07	18.30
2000	Wall, job-built plywood, to 8' high, 1 use		370	.130	SFCA	2.18	5.55		7.73	10.95
2050	2 use		435	.110		1.39	4.74		6.13	8.85
2100	3 use		495	.097		1.01	4.16		5.17	7.50
2150	4 use		505	.095		.82	4.08		4.90	7.20
2400	Over 8' to 16' high, 1 use		280	.171		2.40	7.35		9.75	13.95
2450	2 use		345	.139		1.02	5.95		6.97	10.30
2500	3 use		375	.128		.73	5.50		6.23	9.25
2550	4 use		395	.122		.60	5.20		5.80	8.70
2700	Over 16' high, 1 use		235	.204		2.07	8.75		10.82	15.80
2750	2 use		290	.166		1.14	7.10		8.24	12.20
2800	3 use		315	.152		.83	6.55		7.38	10.95
2850	4 use		330	.145		.67	6.25		6.92	10.35
4000	Radial, smooth curved, job-built plywood, 1 use		245	.196		1.92	8.40		10.32	15.05
4050	2 use		300	.160		1.06	6.85		7.91	11.70
4100	3 use		325	.148		.77	6.35		7.12	10.60
4150	4 use		335	.143		.63	6.15		6.78	10.15
4200	Below grade, job-built plywood, 1 use		225	.213		2.17	9.15		11.32	16.50
4210	2 use		225	.213		1.20	9.15		10.35	15.40
4220	3 use		225	.213		1	9.15		10.15	15.20
4230	4 use	↓	225	.213	↓	.71	9.15		9.86	14.90

03 11 Concrete Forming

03 11 13 – Structural Cast-In-Place Concrete Forming

03 11 13.85 Forms In Place, Walls	Crew	Daily Output	Labor-Hours	Unit	Material	2012 Bare Costs Labor	Equipment	Total	Total Incl O&P
4300	Curved, 2' chords, job-built plywood, 1 use	C-2	290	.166	SFCA	1.64	7.10		8.74	12.75
4350	2 use		355	.135		.90	5.80		6.70	9.95
4400	3 use		385	.125		.66	5.35		6.01	8.95
4450	4 use		400	.120		.53	5.15		5.68	8.55
4500	Over 8' high, 1 use		290	.166		.70	7.10		7.80	11.70
4525	2 use		355	.135		.39	5.80		6.19	9.35
4550	3 use		385	.125		.28	5.35		5.63	8.55
4575	4 use		400	.120		.23	5.15		5.38	8.20
4600	Retaining wall, battered, job-built plyw'd, to 8' high, 1 use		300	.160		1.55	6.85		8.40	12.25
4650	2 use		355	.135		.85	5.80		6.65	9.90
4700	3 use		375	.128		.62	5.50		6.12	9.15
4750	4 use		390	.123		.50	5.30		5.80	8.70
4900	Over 8' to 16' high, 1 use		240	.200		1.69	8.60		10.29	15.05
4950	2 use		295	.163		.93	7		7.93	11.75
5000	3 use		305	.157		.68	6.75		7.43	11.15
5050	4 use		320	.150		.55	6.45		7	10.50
5500	For gang wall forming, 192 S.F. sections, deduct					10%	10%			
5550	384 S.F. sections, deduct					20%	20%			
5750	Liners for forms (add to wall forms), A.B.S. plastic									
5800	Aged wood, 4" wide, 1 use	1 Carp	256	.031	SFCA	3.12	1.38		4.50	5.55
5820	2 use		256	.031		1.72	1.38		3.10	4.01
5830	3 use		256	.031		1.25	1.38		2.63	3.49
5840	4 use		256	.031		1.01	1.38		2.39	3.24
5900	Fractured rope rib, 1 use		192	.042		4.52	1.84		6.36	7.80
5925	2 use		192	.042		2.49	1.84		4.33	5.55
5950	3 use		192	.042		1.81	1.84		3.65	4.82
6000	4 use		192	.042		1.47	1.84		3.31	4.45
6100	Ribbed, 3/4" deep x 1-1/2" O.C., 1 use		224	.036		4.52	1.57		6.09	7.40
6125	2 use		224	.036		2.49	1.57		4.06	5.15
6150	3 use		224	.036		1.81	1.57		3.38	4.41
6200	4 use		224	.036		1.47	1.57		3.04	4.04
6300	Rustic brick pattern, 1 use		224	.036		3.12	1.57		4.69	5.85
6325	2 use		224	.036		1.72	1.57		3.29	4.31
6350	3 use		224	.036		1.25	1.57		2.82	3.79
6400	4 use		224	.036		1.01	1.57		2.58	3.54
6500	3/8" striated, random, 1 use		224	.036		3.12	1.57		4.69	5.85
6525	2 use		224	.036		1.72	1.57		3.29	4.31
6550	3 use		224	.036		1.25	1.57		2.82	3.79
6600	4 use		224	.036		1.01	1.57		2.58	3.54
6850	Random vertical rustication, 1 use		384	.021		5.90	.92		6.82	7.90
6900	2 use		384	.021		3.26	.92		4.18	4.99
6925	3 use		384	.021		2.37	.92		3.29	4.01
6950	4 use		384	.021		1.92	.92		2.84	3.53
7050	Wood, beveled edge, 3/4" deep, 1 use		384	.021	L.F.	.24	.92		1.16	1.67
7100	1" deep, 1 use		384	.021	"	.42	.92		1.34	1.87
7200	4" wide aged cedar, 1 use		256	.031	SFCA	3.12	1.38		4.50	5.55
7300	4" variable depth rough cedar		224	.036		4.52	1.57		6.09	7.40
7500	Lintel or sill forms, 1 use		30	.267		2.62	11.75		14.37	21
7520	2 use		34	.235		1.44	10.40		11.84	17.55
7540	3 use		36	.222		1.05	9.80		10.85	16.25
7560	4 use		37	.216		.85	9.55		10.40	15.60
7800	Modular prefabricated plywood, based on 20 uses of purchased									
7820	forms, and 4 uses of bracing lumber									

03 11 Concrete Forming

03 11 13 – Structural Cast-In-Place Concrete Forming

03 11 13.85 Forms In Place, Walls		Crew	Daily Output	Labor-Hours	Unit	Material	2012 Bare Costs Labor	Equipment	Total	Total Incl O&P
7860	To 8' high	C-2	800	.060	SFCA	.90	2.58		3.48	4.95
8060	Over 8' to 16' high		600	.080		.94	3.43		4.37	6.35
8600	Pilasters, 1 use		270	.178		2.52	7.65		10.17	14.55
8620	2 use		330	.145		1.39	6.25		7.64	11.15
8640	3 use		370	.130		1.01	5.55		6.56	9.65
8660	4 use		385	.125		.82	5.35		6.17	9.15
9010	Steel framed plywood, based on 50 uses of purchased									
9020	forms, and 4 uses of bracing lumber									
9060	To 8' high	C-2	600	.080	SFCA	.74	3.43		4.17	6.10
9260	Over 8' to 16' high		450	.107		.74	4.58		5.32	7.85
9460	Over 16' to 20' high		400	.120		.74	5.15		5.89	8.75
9475	For elevated walls, add						10%			
9480	For battered walls, 1 side battered, add					10%	10%			
9485	For battered walls, 2 sides battered, add					15%	15%			

03 11 19 – Insulating Concrete Forming

03 11 19.10 Insulating Forms, Left In Place

			Crew	Daily Output	Labor-Hours	Unit	Material	Labor	Equipment	Total	Total Incl O&P
0010	**INSULATING FORMS, LEFT IN PLACE**										
0020	S.F. is for exterior face, but includes forms for both faces (total R22)										
2000	4" wall, straight block, 16" x 48" (5.33 S.F.)	G	2 Carp	90	.178	Ea.	17.60	7.85		25.45	31.50
2010	90 corner block, exterior 16" x 38" x 22" (6.67 S.F.)	G		75	.213		19.95	9.40		29.35	36.50
2020	45 corner block, exterior 16" x 34" x 18" (5.78 S.F.)	G		75	.213		20	9.40		29.40	36.50
2100	6" wall, straight block, 16" x 48" (5.33 S.F.)	G		90	.178		17.85	7.85		25.70	31.50
2110	90 corner block, exterior 16" x 32" x 24" (6.22 S.F.)	G		75	.213		21.50	9.40		30.90	38.50
2120	45 corner block, exterior 16" x 26" x 18" (4.89 S.F.)	G		75	.213		18.95	9.40		28.35	35.50
2130	Brick ledge block, 16" x 48" (5.33 S.F.)	G		80	.200		22.50	8.80		31.30	38
2140	Taper top block, 16" x 48" (5.33 S.F.)	G		80	.200		21	8.80		29.80	36.50
2200	8" wall, straight block, 16" x 48" (5.33 S.F.)	G		90	.178		18.40	7.85		26.25	32
2210	90 corner block, exterior 16" x 34" x 26" (6.67 S.F.)	G		75	.213		23.50	9.40		32.90	40.50
2220	45 corner block, exterior 16" x 28" x 20" (5.33 S.F.)	G		75	.213		20.50	9.40		29.90	37
2230	Brick ledge block, 16" x 48" (5.33 S.F.)	G		80	.200		23	8.80		31.80	39
2240	Taper top block, 16" x 48" (5.33 S.F.)	G		80	.200		21.50	8.80		30.30	37

03 11 19.60 Roof Deck Form Boards

			Crew	Daily Output	Labor-Hours	Unit	Material	Labor	Equipment	Total	Total Incl O&P
0010	**ROOF DECK FORM BOARDS**	R051223-50									
0050	Includes bulb tee sub-purlins @ 32-5/8" O.C.										
0070	Non-asbestos fiber cement, 5/16" thick		C-13	2950	.008	S.F.	2.66	.39	.04	3.09	3.64
0100	Fiberglass, 1" thick			2700	.009		3.64	.42	.05	4.11	4.78
0500	Wood fiber, 1" thick	G		2700	.009		2.20	.42	.05	2.67	3.19

03 11 23 – Permanent Stair Forming

03 11 23.75 Forms In Place, Stairs

			Crew	Daily Output	Labor-Hours	Unit	Material	Labor	Equipment	Total	Total Incl O&P
0010	**FORMS IN PLACE, STAIRS**	R031113-40									
0015	(Slant length x width), 1 use		C-2	165	.291	S.F.	4.66	12.50		17.16	24.50
0050	2 use	R031113-60		170	.282		2.65	12.10		14.75	21.50
0100	3 use			180	.267		1.99	11.45		13.44	19.80
0150	4 use			190	.253		1.65	10.85		12.50	18.50
1000	Alternate pricing method (1.0 L.F./S.F.), 1 use			100	.480	LF Rsr	4.66	20.50		25.16	36.50
1050	2 use			105	.457		2.65	19.65		22.30	33
1100	3 use			110	.436		1.99	18.75		20.74	31
1150	4 use			115	.417		1.65	17.90		19.55	29.50
2000	Stairs, cast on sloping ground (length x width), 1 use			220	.218	S.F.	1.86	9.35		11.21	16.45
2025	2 use			232	.207		1.03	8.90		9.93	14.80
2050	3 use			244	.197		.75	8.45		9.20	13.80

03 11 Concrete Forming

03 11 23 – Permanent Stair Forming

03 11 23.75 Forms In Place, Stairs	Crew	Daily Output	Labor-Hours	Unit	Material	2012 Bare Costs Labor	Equipment	Total	Total Incl O&P
2100	4 use	C-2	256	.188	S.F.	.61	8.05		8.66	13.05

03 15 Concrete Accessories

03 15 05 – Concrete Forming Accessories

03 15 05.02 Anchor Bolt Accessories

		Crew	Daily Output	Labor-Hours	Unit	Material	2012 Bare Costs Labor	Equipment	Total	Total Incl O&P
0010	**ANCHOR BOLT ACCESSORIES**									
0015	For anchor bolts set in fresh concrete, see Section 05 05 23.05									
8150	Anchor bolt sleeve, plastic, 1" diam. bolts	1 Carp	60	.133	Ea.	7.90	5.90		13.80	17.75
8500	1-1/2" diameter		28	.286		14	12.60		26.60	35
8600	2" diameter		24	.333		17.55	14.70		32.25	42
8650	3" diameter		20	.400		32	17.65		49.65	62.50

03 15 05.12 Chamfer Strips

		Crew	Daily Output	Labor-Hours	Unit	Material	2012 Bare Costs Labor	Equipment	Total	Total Incl O&P
0010	**CHAMFER STRIPS**									
2000	Polyvinyl chloride, 1/2" wide with leg	1 Carp	535	.015	L.F.	.40	.66		1.06	1.45
2200	3/4" wide with leg		525	.015		.72	.67		1.39	1.82
2400	1" radius with leg		515	.016		.33	.68		1.01	1.41
2800	2" radius with leg		500	.016		.57	.71		1.28	1.72
5000	Wood, 1/2" wide		535	.015		.16	.66		.82	1.19
5200	3/4" wide		525	.015		.24	.67		.91	1.29
5400	1" wide		515	.016		.42	.68		1.10	1.51

03 15 05.15 Column Form Accessories

			Crew	Daily Output	Labor-Hours	Unit	Material	2012 Bare Costs Labor	Equipment	Total	Total Incl O&P
0010	**COLUMN FORM ACCESSORIES**										
1000	Column clamps, adjustable to 24" x 24", buy	G				Set	136			136	150
1100	Rent per month	G					17.85			17.85	19.65
1300	For sizes to 30" x 30", buy	G					147			147	162
1400	Rent per month	G					18.90			18.90	21
1600	For sizes to 36" x 36", buy	G					170			170	187
1700	Rent per month	G					21			21	23
2000	Bull winch (band iron) 36" x 36", buy	G					60			60	66
2100	Rent per month	G					6.10			6.10	6.70
2300	48" x 48", buy	G					61			61	67
2400	Rent per month	G					7.30			7.30	8.05
3000	Chain & wedge type 36" x 36", buy	G					83			83	91.50
3100	Rent per month	G					9.30			9.30	10.25
3300	60" x 60", buy	G					113			113	124
3400	Rent per month	G					13.40			13.40	14.75
4000	Friction collars 2'-6" diam., buy	G					845			845	930
4100	Rent per month	G					83			83	91.50
4300	4'-0" diam., buy	G					970			970	1,075
4400	Rent per month	G					107			107	118

03 15 05.20 Dovetail Anchor System

			Crew	Daily Output	Labor-Hours	Unit	Material	2012 Bare Costs Labor	Equipment	Total	Total Incl O&P
0010	**DOVETAIL ANCHOR SYSTEM**										
0500	Anchor slot, galv., filled, 24 ga.	G	1 Carp	425	.019	L.F.	.72	.83		1.55	2.07
0600	20 ga.	G		400	.020		1.57	.88		2.45	3.09
0625	12 ga.	G		400	.020		1.83	.88		2.71	3.37
0900	26 ga. stainless steel, foam filled	G		375	.021		1.15	.94		2.09	2.72
1200	Brick anchor, corr., galv., 3-1/2" long, 16 ga.	G	1 Bric	10.50	.762	C	28	33.50		61.50	81.50
1300	12 ga.	G		10.50	.762		39.50	33.50		73	94
1500	Flat, galv., 3-1/2" long, 16 ga.	G		10.50	.762		43.50	33.50		77	98.50
1600	12 ga.	G		10.50	.762		61	33.50		94.50	118

03 15 Concrete Accessories

03 15 05 – Concrete Forming Accessories

03 15 05.20 Dovetail Anchor System

		Crew	Daily Output	Labor-Hours	Unit	Material	2012 Bare Costs Labor	Equipment	Total	Total Incl O&P
2000	Cavity wall, corr., galv., 5" long, 16 ga.	1 Bric	10.50	.762	C	38	33.50		71.50	92.50
2100	12 ga.		10.50	.762		54.50	33.50		88	110
3000	Furring anchors, corr., galv., 1-1/2" long, 16 ga.		10.50	.762		14.70	33.50		48.20	66.50
3100	12 ga.		10.50	.762		25.50	33.50		59	78.50
6000	Stone anchors, 3-1/2" long, galv., 1/8" x 1" wide		10.50	.762		94.50	33.50		128	155
6100	1/4" x 1" wide		10.50	.762		108	33.50		141.50	170

03 15 05.25 Expansion Joints

		Crew	Daily Output	Labor-Hours	Unit	Material	2012 Bare Costs Labor	Equipment	Total	Total Incl O&P
0010	**EXPANSION JOINTS**									
0020	Keyed, cold, 24 ga., incl. stakes, 3-1/2" high	1 Carp	200	.040	L.F.	.96	1.76		2.72	3.77
0050	4-1/2" high		200	.040		1.13	1.76		2.89	3.95
0100	5-1/2" high		195	.041		1.25	1.81		3.06	4.16
0150	7-1/2" high		190	.042		1.47	1.86		3.33	4.48
0160	9-1/2" high		185	.043		1.67	1.91		3.58	4.77
0300	Poured asphalt, plain, 1/2" x 1"	1 Clab	450	.018		.65	.62		1.27	1.68
0350	1" x 2"		400	.020		2.60	.70		3.30	3.94
0500	Neoprene, liquid, cold applied, 1/2" x 1"		450	.018		3	.62		3.62	4.26
0550	1" x 2"		400	.020		12	.70		12.70	14.30
0700	Polyurethane, poured, 2 part, 1/2" x 1"		400	.020		1.33	.70		2.03	2.54
0750	1" x 2"		350	.023		5.30	.80		6.10	7.10
0900	Rubberized asphalt, hot or cold applied, 1/2" x 1"		450	.018		.36	.62		.98	1.36
0950	1" x 2"		400	.020		1.45	.70		2.15	2.68
1100	Hot applied, fuel resistant, 1/2" x 1"		450	.018		.54	.62		1.16	1.55
1150	1" x 2"		400	.020		2.18	.70		2.88	3.47
2000	Premolded, bituminous fiber, 1/2" x 6"	1 Carp	375	.021		.44	.94		1.38	1.93
2050	1" x 12"		300	.027		2.01	1.18		3.19	4.02
2140	Concrete expansion joint, recycled paper and fiber, 1/2" x 6"		390	.021		.39	.90		1.29	1.82
2150	1/2" x 12"		360	.022		.78	.98		1.76	2.37
2250	Cork with resin binder, 1/2" x 6"		375	.021		.83	.94		1.77	2.36
2300	1" x 12"		300	.027		2.32	1.18		3.50	4.36
2500	Neoprene sponge, closed cell, 1/2" x 6"		375	.021		1.93	.94		2.87	3.57
2550	1" x 12"		300	.027		7.25	1.18		8.43	9.80
2750	Polyethylene foam, 1/2" x 6"		375	.021		.62	.94		1.56	2.13
2800	1" x 12"		300	.027		1.99	1.18		3.17	4
3000	Polyethylene backer rod, 3/8" diameter		460	.017		.05	.77		.82	1.23
3050	3/4" diameter		460	.017		.07	.77		.84	1.26
3100	1" diameter		460	.017		.14	.77		.91	1.33
3500	Polyurethane foam, with polybutylene, 1/2" x 1/2"		475	.017		.95	.74		1.69	2.19
3550	1" x 1"		450	.018		2.40	.78		3.18	3.85
3750	Polyurethane foam, regular, closed cell, 1/2" x 6"		375	.021		.70	.94		1.64	2.22
3800	1" x 12"		300	.027		2.50	1.18		3.68	4.56
4000	Polyvinyl chloride foam, closed cell, 1/2" x 6"		375	.021		1.90	.94		2.84	3.54
4050	1" x 12"		300	.027		6.55	1.18		7.73	9
4250	Rubber, gray sponge, 1/2" x 6"		375	.021		1.60	.94		2.54	3.21
4300	1" x 12"		300	.027		5.75	1.18		6.93	8.15
4400	Redwood heartwood, 1" x 4"		400	.020		1.16	.88		2.04	2.64
4450	1" x 6"		375	.021		1.74	.94		2.68	3.36
5000	For installation in walls, add						75%			
5250	For installation in boxouts, add						25%			

03 15 05.30 Hangers

0010	**HANGERS**									
0020	Slab and beam form									
0500	Banding iron, 3/4" x 22 ga, 14 L.F. per lb. or									

03 15 Concrete Accessories

03 15 05 – Concrete Forming Accessories

03 15 05.30 Hangers

		Crew	Daily Output	Labor-Hours	Unit	Material	2012 Bare Costs Labor	Equipment	Total	Total Incl O&P
0550	1/2" x 14 ga, 7 L.F. per lb.	G			Lb.	1.30			1.30	1.43
1000	Fascia ties, coil type, to 24" long	G			C	370			370	405
1500	Frame ties to 8-1/8"	G				435			435	480
1550	8-1/8" to 10-1/8"	G				455			455	505
5000	Snap tie hanger, to 30" overall length, 3000#	G				370			370	405
5050	To 36" overall length	G				405			405	445
5100	To 48" overall length	G				480			480	530
5500	Steel beam hanger									
5600	Flange to 8-1/8"	G			C	435			435	480
5650	8-1/8" to 10-1/8"					455			455	505
6000	Tie hangers to 30" overall length, 4000#	G				420			420	460
6100	To 36" overall length	G				455			455	500
6500	Tie back hanger, up to 12-1/8" flange	G				1,175			1,175	1,275
8500	Wire, black annealed, 9 ga	G			Cwt.	142			142	156
8600	16 ga				"	171			171	189

03 15 05.35 Inserts

		Crew	Daily Output	Labor-Hours	Unit	Material	2012 Bare Costs Labor	Equipment	Total	Total Incl O&P
0010	**INSERTS**									
1000	Inserts, slotted nut type for 3/4" bolts, 4" long	G	1 Carp	84	.095	Ea.	17.10	4.20		21.30	25.50
2100	6" long	G		84	.095		19.25	4.20		23.45	27.50
2150	8" long	G		84	.095		25.50	4.20		29.70	35
2200	Slotted, strap type, 4" long	G		84	.095		18.50	4.20		22.70	27
2250	6" long	G		84	.095		21	4.20		25.20	29.50
2300	8" long	G		84	.095		27.50	4.20		31.70	37
2350	Strap for slotted insert, 4" long	G		84	.095		10.15	4.20		14.35	17.60
4100	6" long	G		84	.095		10.95	4.20		15.15	18.50
4150	8" long	G		84	.095		13.65	4.20		17.85	21.50
4200	10" long	G		84	.095		16.60	4.20		20.80	24.50
7000	Loop ferrule type									
7100	1/4" diameter bolt	G	1 Carp	84	.095	Ea.	1.96	4.20		6.16	8.60
7350	7/8" diameter bolt	G	"	84	.095	"	9.15	4.20		13.35	16.50
9000	Wedge type									
9100	For 3/4" diameter bolt	G	1 Carp	60	.133	Ea.	8.10	5.90		14	18
9800	Cut washers, black									
9900	3/4" bolt	G				Ea.	1.02			1.02	1.12
9950	For galvanized inserts, add						30%				

03 15 05.70 Shores

		Crew	Daily Output	Labor-Hours	Unit	Material	2012 Bare Costs Labor	Equipment	Total	Total Incl O&P
0010	**SHORES**									
0020	Erect and strip, by hand, horizontal members									
0500	Aluminum joists and stringers	G	2 Carp	60	.267	Ea.		11.75		11.75	18.10
0600	Steel, adjustable beams	G		45	.356			15.70		15.70	24
0700	Wood joists			50	.320			14.10		14.10	21.50
0800	Wood stringers			30	.533			23.50		23.50	36
1000	Vertical members to 10' high	G		55	.291			12.85		12.85	19.75
1050	To 13' high	G		50	.320			14.10		14.10	21.50
1100	To 16' high	G		45	.356			15.70		15.70	24
1500	Reshoring	G		1400	.011	S.F.	.50	.50		1	1.33
1600	Flying truss system	G	C-17D	9600	.009	SFCA		.40	.08	.48	.71
1760	Horizontal, aluminum joists, 6-1/4" high x 5' to 21' span, buy	G				L.F.	13.40			13.40	14.70
1770	Beams, 7-1/4" high x 4' to 30' span	G				"	15.65			15.65	17.25
1810	Horizontal, steel beam, W8x10, 7' span, buy	G				Ea.	47.50			47.50	52
1830	10' span	G					56			56	61.50
1920	15' span	G					96			96	106

03 15 Concrete Accessories

03 15 05 – Concrete Forming Accessories

03 15 05.70 Shores

		Crew	Daily Output	Labor-Hours	Unit	Material	2012 Bare Costs Labor	Equipment	Total	Total Incl O&P
1940	20' span	G			Ea.	135			135	149
1970	Steel stringer, W8x10, 4' to 16' span, buy	G			L.F.	5.60			5.60	6.15
3000	Rent for job duration, aluminum joist @ 2' O.C., per mo	G			SF Flr.	.33			.33	.37
3050	Steel W8x10	G				.14			.14	.15
3060	Steel adjustable	G			↓	.14			.14	.15
3500	#1 post shore, steel, 5'-7" to 9'-6" high, 10000# cap., buy	G			Ea.	128			128	141
3550	#2 post shore, 7'-3" to 12'-10" high, 7800# capacity	G				148			148	163
3600	#3 post shore, 8'-10" to 16'-1" high, 3800# capacity	G			↓	162			162	178
5010	Frame shoring systems, steel, 12000#/leg, buy									
5040	Frame, 2' wide x 6' high	G			Ea.	93.50			93.50	103
5250	X-brace	G				15.85			15.85	17.40
5550	Base plate	G				15.50			15.50	17.05
5600	Screw jack	G				33.50			33.50	36.50
5650	U-head, 8" x 8"	G			↓	19.10			19.10	21

03 15 05.75 Sleeves and Chases

		Crew	Daily Output	Labor-Hours	Unit	Material	2012 Bare Costs Labor	Equipment	Total	Total Incl O&P
0010	**SLEEVES AND CHASES**									
0100	Plastic, 1 use, 12" long, 2" diameter	1 Carp	100	.080	Ea.	2.19	3.53		5.72	7.85
0150	4" diameter		90	.089		6.45	3.92		10.37	13.15
0200	6" diameter		75	.107		11.35	4.70		16.05	19.70
0250	12" diameter	↓	60	.133	↓	32	5.90		37.90	44

03 15 05.80 Snap Ties

		Crew	Daily Output	Labor-Hours	Unit	Material	2012 Bare Costs Labor	Equipment	Total	Total Incl O&P
0010	**SNAP TIES**, 8-1/4" L&W (Lumber and wedge)									
0100	2250 lb., w/flat washer, 8" wall	G			C	78			78	86
0200	12" wall	G				91			91	100
0250	16" wall	G				105			105	116
0300	18" wall	G				111			111	122
0500	With plastic cone, 8" wall	G				71			71	78
0600	12" wall	G				78			78	86
0650	16" wall	G				86			86	94.50
0700	18" wall	G				89			89	98
1000	3350 lb., w/flat washer, 8" wall	G				111			111	122
1150	12" wall	G				125			125	138
1200	16" wall	G				143			143	157
1250	18" wall	G				149			149	164
1500	With plastic cone, 8" wall	G				111			111	122
1600	12" wall	G				125			125	138
1650	16" wall	G				143			143	157
1700	18" wall	G			↓	149			149	164

03 15 05.85 Stair Tread Inserts

		Crew	Daily Output	Labor-Hours	Unit	Material	2012 Bare Costs Labor	Equipment	Total	Total Incl O&P
0010	**STAIR TREAD INSERTS**									
0015	Cast iron, abrasive, 3" wide	G	1 Carp	90	.089	L.F.	10	3.92	13.92	17.05
0020	4" wide	G		80	.100		13.30	4.41	17.71	21.50
0040	6" wide	G		75	.107		20	4.70	24.70	29.50
0050	9" wide	G		70	.114		30	5.05	35.05	41
0100	12" wide	G	↓	65	.123		40	5.45	45.45	52.50
0300	Cast aluminum, compared to cast iron, deduct						10%			
0500	Extruded aluminum safety tread, 3" wide	G	1 Carp	75	.107		11.45	4.70	16.15	19.85
0550	4" wide	G		75	.107		15.30	4.70	20	24
0600	6" wide	G		75	.107		23	4.70	27.70	32.50
0650	9" wide to resurface stairs	G	↓	70	.114	↓	34.50	5.05	39.55	46
1700	Cement fill for pan-type metal treads, plain		1 Cefi	115	.070	S.F.	1.89	2.95	4.84	6.40
1750	Non-slip		"	100	.080	"	2.08	3.39	5.47	7.25

03 15 Concrete Accessories

03 15 05 – Concrete Forming Accessories

03 15 05.95 Wall and Foundation Form Accessories	Crew	Daily Output	Labor-Hours	Unit	Material	2012 Bare Costs Labor	Equipment	Total	Total Incl O&P
0010 **WALL AND FOUNDATION FORM ACCESSORIES**									
2000 Footings, turnbuckle form aligner G				Ea.	15.90			15.90	17.50
2050 Spreaders for footer, adjustable G				"	22			22	24
3000 Form oil, coverage varies greatly, minimum				Gal.	7.30			7.30	8.05
3050 Maximum				"	19.05			19.05	21
3500 Form patches, 1-3/4" diameter				C	45			45	49.50
3550 2-3/4" diameter				"	79			79	87
4000 Nail stakes, 3/4" diameter, 18" long G				Ea.	2.96			2.96	3.26
4050 24" long G					3.82			3.82	4.20
4200 30" long G					4.91			4.91	5.40
4250 36" long G					5.90			5.90	6.50

03 15 13 – Waterstops

03 15 13.50 Waterstops	Crew	Daily Output	Labor-Hours	Unit	Material	2012 Bare Costs Labor	Equipment	Total	Total Incl O&P
0010 **WATERSTOPS**, PVC and Rubber									
0020 PVC, ribbed 3/16" thick, 4" wide	1 Carp	155	.052	L.F.	1.11	2.28		3.39	4.72
0050 6" wide		145	.055		2.05	2.43		4.48	6
0500 With center bulb, 6" wide, 3/16" thick		135	.059		1.75	2.61		4.36	5.95
0550 3/8" thick		130	.062		3.43	2.71		6.14	7.95
0600 9" wide x 3/8" thick		125	.064		5.05	2.82		7.87	9.90
0800 Dumbbell type, 6" wide, 3/16" thick		150	.053		1.83	2.35		4.18	5.65
0850 3/8" thick		145	.055		3.41	2.43		5.84	7.50
1000 9" wide, 3/8" thick, plain		130	.062		5.50	2.71		8.21	10.25
1050 Center bulb		130	.062		7.70	2.71		10.41	12.65
1250 Ribbed type, split, 3/16" thick, 6" wide		145	.055		1.79	2.43		4.22	5.70
1300 3/8" thick		130	.062		3.75	2.71		6.46	8.30
2000 Rubber, flat dumbbell, 3/8" thick, 6" wide		145	.055		8.60	2.43		11.03	13.20
2050 9" wide		135	.059		13.05	2.61		15.66	18.35
2500 Flat dumbbell split, 3/8" thick, 6" wide		145	.055		1.79	2.43		4.22	5.70
2550 9" wide		135	.059		3.75	2.61		6.36	8.15
3000 Center bulb, 1/4" thick, 6" wide		145	.055		7.50	2.43		9.93	12
3050 9" wide		135	.059		16.85	2.61		19.46	22.50
3500 Center bulb split, 3/8" thick, 6" wide		145	.055		9.10	2.43		11.53	13.80
3550 9" wide		135	.059		15.75	2.61		18.36	21.50
5000 Waterstop fittings, rubber, flat									
5010 Dumbbell or center bulb, 3/8" thick,									
5200 Field union, 6" wide	1 Carp	50	.160	Ea.	23	7.05		30.05	36
5250 9" wide		50	.160		29.50	7.05		36.55	43.50
5500 Flat cross, 6" wide		30	.267		37.50	11.75		49.25	59.50
5550 9" wide		30	.267		54	11.75		65.75	77.50
6000 Flat tee, 6" wide		30	.267		38.50	11.75		50.25	60
6050 9" wide		30	.267		52	11.75		63.75	75
6500 Flat ell, 6" wide		40	.200		37.50	8.80		46.30	54.50
6550 9" wide		40	.200		48.50	8.80		57.30	67
7000 Vertical tee, 6" wide		25	.320		29.50	14.10		43.60	54
7050 9" wide		25	.320		42	14.10		56.10	67.50
7500 Vertical ell, 6" wide		35	.229		28	10.10		38.10	46.50
7550 9" wide		35	.229		37.50	10.10		47.60	56.50

03 21 Reinforcement Bars

03 21 05 – Reinforcing Steel Accessories

03 21 05.10 Rebar Accessories		Crew	Daily Output	Labor-Hours	Unit	Material	2012 Bare Costs Labor	Equipment	Total	Total Incl O&P
0010	**REBAR ACCESSORIES**									
0030	Steel & plastic made from recycled materials	G								
0100	Beam bolsters (BB), lower, 1-1/2" high, plain steel	G				C.L.F.	51			51	56
0102	Galvanized	G					61			61	67.50
0104	Stainless tipped legs	G					480			480	530
0106	Plastic tipped legs	G					55			55	60.50
0108	Epoxy dipped	G					104			104	114
0110	2" high, plain	G					55			55	60.50
0120	Galvanized	G					66			66	72.50
0140	Stainless tipped legs	G					485			485	535
0160	Plastic tipped legs	G					56			56	61.50
0162	Epoxy dipped	G					115			115	127
0200	Upper (BBU), 1-1/2" high, plain steel	G					97			97	107
0210	3" high	G					112			112	123
0300	Beam bolster with plate (BBP), 1-1/2" high, plain steel	G					192			192	211
0310	3" high	G					305			305	340
0500	Slab bolsters, continuous (SB), 1" high, plain steel	G					27			27	29.50
0502	Galvanized	G					47			47	51.50
0504	Stainless tipped legs	G					455			455	505
0506	Plastic tipped legs	G					36			36	39.50
0510	2" high, plain steel	G					29			29	32
0515	Galvanized	G					59			59	65
0520	Stainless tipped legs	G					460			460	505
0525	Plastic tipped legs	G					41			41	45
0530	For bolsters with wire runners (SBR), add	G					37			37	40.50
0540	For bolsters with plates (SBP), add	G					85			85	93.50
0700	Bag ties, 16 ga., plain, 4" long	G				C	3			3	3.30
0710	5" long	G					4			4	4.40
0720	6" long	G					3			3	3.30
0730	7" long	G					4			4	4.40
1200	High chairs, individual (HC), 3" high, plain steel	G					40			40	44
1202	Galvanized	G					48			48	53
1204	Stainless tipped legs	G					470			470	515
1206	Plastic tipped legs	G					54			54	59.50
1210	5" high, plain	G					58			58	64
1212	Galvanized	G					69.50			69.50	76.50
1214	Stainless tipped legs	G					490			490	535
1216	Plastic tipped legs	G					73			73	80.50
1220	8" high, plain	G					118			118	130
1222	Galvanized	G					142			142	156
1224	Stainless tipped legs	G					550			550	605
1226	Plastic tipped legs	G					151			151	166
1230	12" high, plain	G					273			273	300
1232	Galvanized	G					325			325	360
1234	Stainless tipped legs	G					705			705	775
1236	Plastic tipped legs	G					282			282	310
1400	Individual high chairs, with plate (HCP), 5" high	G					170			170	187
1410	8" high	G					235			235	259
1500	Bar chair (BC), 1-1/2" high, plain steel	G					26			26	28.50
1520	Galvanized	G					37			37	40.50
1530	Stainless tipped legs	G					455			455	500
1540	Plastic tipped legs	G					39			39	43
1550	Joist chair (JC), 3/4" x 5", plain steel	G					33			33	36.50

03 21 05 – Reinforcing Steel Accessories

03 21 05.10 Rebar Accessories			Crew	Daily Output	Labor-Hours	Unit	Material	2012 Bare Costs Labor	Equipment	Total	Total Incl O&P
1580	Galvanized	G				C	39.50			39.50	43.50
1600	Stainless tipped legs	G					465			465	510
1620	Plastic tipped legs	G					55			55	60.50
1630	Epoxy dipped	G					121			121	133
1700	Continuous high chairs (CHC), legs 8" O.C., 4" high, plain steel	G				C.L.F.	48			48	53
1705	Galvanized	G					57.50			57.50	63.50
1710	Stainless tipped legs	G					480			480	525
1715	Plastic tipped legs	G					95			95	105
1718	Epoxy dipped	G					87			87	95.50
1720	6" high, plain	G					69			69	76
1725	Galvanized	G					83			83	91
1730	Stainless tipped legs	G					500			500	550
1735	Plastic tipped legs	G					95			95	105
1738	Epoxy dipped	G					114			114	125
1740	8" high, plain	G					101			101	111
1745	Galvanized	G					121			121	133
1750	Stainless tipped legs	G					530			530	585
1755	Plastic tipped legs	G					151			151	166
1758	Epoxy dipped	G					173			173	190
1900	For continuous bottom wire runners, add	G					39			39	43
1940	For continuous bottom plate, add	G					172			172	189
2100	Paper tubing, 4' lengths, for #4 bar	G					69			69	76
2120	For #6 bar	G					81			81	89
2200	Screed chair base, 1/2" coil thread diam., 2-1/2" high, plain steel	G				C	281			281	310
2210	Galvanized	G					295			295	325
2220	5-1/2" high, plain	G					340			340	375
2250	Galvanized	G					355			355	390
2300	3/4" coil thread diam., 2-1/2" high, plain steel	G					355			355	390
2310	Galvanized	G					475			475	520
2320	5-1/2" high, plain steel	G					435			435	480
2350	Galvanized	G					610			610	670
2400	Screed holder, 1/2" coil thread diam. for pipe screed, plain steel, 6" long	G					340			340	375
2420	12" long	G					525			525	575
2500	3/4" coil thread diam. for pipe screed, plain steel, 6" long	G					490			490	540
2520	12" long	G					790			790	870
2700	Screw anchor for bolts, plain steel, 3/4" diameter x 4" long	G					465			465	510
2720	1" diameter x 6" long	G					760			760	835
2740	1-1/2" diameter x 8" long	G					955			955	1,050
2800	Screw anchor eye bolts, 3/4" x 7" long	G					3,975			3,975	4,375
2820	1" x 9" long	G					7,850			7,850	8,650
2840	1-1/2" x 14" long	G					19,900			19,900	21,900
2900	Screw anchor bolts, 3/4" x 9" long	G					1,375			1,375	1,525
2920	1" x 12" long	G					2,675			2,675	2,925
3001	Slab lifting inserts, single pickup, galv, 3/4" diam., 5" high	G					1,300			1,300	1,450
3010	6" high	G					1,325			1,325	1,475
3030	7" high	G					1,350			1,350	1,500
3100	1" diameter, 5-1/2" high	G					1,375			1,375	1,525
3120	7" high	G					1,425			1,425	1,575
3200	Double pickup lifting inserts, 1" diameter, 5-1/2" high	G					2,675			2,675	2,950
3220	7" high	G					2,950			2,950	3,250
3330	1-1/2" diameter, 5-1/2" high	G					3,250			3,250	3,575
3800	Subgrade chairs, #4 bar head, 3-1/2" high	G					30			30	33
3850	12" high	G					43			43	47.50

03 21 Reinforcement Bars

03 21 05 – Reinforcing Steel Accessories

03 21 05.10 Rebar Accessories

		Crew	Daily Output	Labor-Hours	Unit	Material	2012 Bare Costs Labor	Equipment	Total	Total Incl O&P
3900	#6 bar head, 3-1/2" high	G			C	30			30	33
3950	12" high	G				43			43	47.50
4200	Subgrade stakes, no nail holes, 3/4" diameter, 12" long	G				264			264	290
4250	24" long	G				380			380	420
4300	7/8" diameter, 12" long	G				335			335	370
4350	24" long	G				560			560	620
4500	Tie wire, 16 ga. annealed steel	G			Cwt.	171			171	189
4550	Tie wire holder, plastic case	G			Ea.	45.50			45.50	50
4600	Aluminum case	G			"	60.50			60.50	66.50

03 21 05.75 Splicing Reinforcing Bars

		Crew	Daily Output	Labor-Hours	Unit	Material	2012 Bare Costs Labor	Equipment	Total	Total Incl O&P
0010	**SPLICING REINFORCING BARS** R032110-70									
0020	Including holding bars in place while splicing									
0100	Standard, self-aligning type, taper threaded, #4 bars	G	C-25	190	.168	Ea.	5.95	6.50		12.45	17.10
0105	#5 bars	G		170	.188		7.25	7.25		14.50	19.75
0110	#6 bars	G		150	.213		8.40	8.20		16.60	22.50
0120	#7 bars	G		130	.246		9.75	9.45		19.20	26
0300	#8 bars	G		115	.278		16.60	10.70		27.30	35.50
0305	#9 bars	G	C-5	105	.533		18.15	25.50	6.65	50.30	68
0310	#10 bars	G		95	.589		20	28.50	7.35	55.85	74.50
0320	#11 bars	G		85	.659		21.50	31.50	8.20	61.20	82.50
0330	#14 bars	G		65	.862		32	41.50	10.75	84.25	112
0340	#18 bars	G		45	1.244		49	60	15.50	124.50	165
0500	Transition self-aligning, taper threaded, #18-14	G		45	1.244		51	60	15.50	126.50	167
0510	#18-11	G		45	1.244		52	60	15.50	127.50	168
0520	#14-11	G		65	.862		34	41.50	10.75	86.25	114
0540	#11-10	G		85	.659		23.50	31.50	8.20	63.20	85
0550	#10-9	G		95	.589		22	28.50	7.35	57.85	77
0560	#9-8	G	C-25	105	.305		20	11.75		31.75	41
0580	#8-7	G		115	.278		18.60	10.70		29.30	38
0590	#7-6	G		130	.246		11.75	9.45		21.20	28.50
0600	Position coupler for curved bars, taper threaded, #4 bars	G		160	.200		27	7.70		34.70	42.50
0610	#5 bars	G		145	.221		28.50	8.50		37	45.50
0620	#6 bars	G		130	.246		34.50	9.45		43.95	53.50
0630	#7 bars	G		110	.291		36.50	11.20		47.70	58.50
0640	#8 bars	G		100	.320		38	12.30		50.30	61.50
0650	#9 bars	G	C-5	90	.622		41.50	30	7.75	79.25	101
0660	#10 bars	G		80	.700		44.50	33.50	8.70	86.70	112
0670	#11 bars	G		70	.800		46.50	38.50	9.95	94.95	122
0680	#14 bars	G		55	1.018		58	49	12.70	119.70	155
0690	#18 bars	G		40	1.400		83.50	67.50	17.45	168.45	217
0700	Transition position coupler for curved bars, taper threaded, #18-14	G		40	1.400		85.50	67.50	17.45	170.45	219
0710	#18-11	G		40	1.400		86.50	67.50	17.45	171.45	220
0720	#14-11	G		55	1.018		60	49	12.70	121.70	157
0730	#11-10	G		70	.800		48.50	38.50	9.95	96.95	125
0740	#10-9	G		80	.700		46.50	33.50	8.70	88.70	114
0750	#9-8	G	C-25	90	.356		43.50	13.70		57.20	70
0760	#8-7	G		100	.320		40	12.30		52.30	64
0770	#7-6	G		110	.291		38.50	11.20		49.70	60.50
0800	Sleeve type w/grout filler, for precast concrete, #6 bars	G		72	.444		21.50	17.10		38.60	52
0802	#7 bars	G		64	.500		26	19.25		45.25	60
0805	#8 bars	G		56	.571		30.50	22		52.50	69.50
0807	#9 bars	G		48	.667		35.50	25.50		61	80.50

03 21 Reinforcement Bars

03 21 05 – Reinforcing Steel Accessories

03 21 05.75 Splicing Reinforcing Bars

			Crew	Daily Output	Labor-Hours	Unit	Material	2012 Bare Costs Labor	Equipment	Total	Total Incl O&P
0810	#10 bars	G	C-5	40	1.400	Ea.	43	67.50	17.45	127.95	173
0900	#11 bars	G		32	1.750		48	84	22	154	209
0920	#14 bars	G	↓	24	2.333		74	112	29	215	291
1000	Sleeve type w/ferrous filler, for critical structures, #6 bars	G	C-25	72	.444		59	17.10		76.10	93
1210	#7 bars	G		64	.500		59.50	19.25		78.75	97
1220	#8 bars	G	↓	56	.571		63	22		85	105
1230	#9 bars	G	C-5	48	1.167		64.50	56	14.55	135.05	176
1240	#10 bars	G		40	1.400		69	67.50	17.45	153.95	201
1250	#11 bars	G		32	1.750		83	84	22	189	248
1260	#14 bars	G		24	2.333		104	112	29	245	325
1270	#18 bars	G	↓	16	3.500		106	168	43.50	317.50	430
2000	Weldable half coupler, taper threaded, #4 bars	G	E-16	120	.133		8.90	6.70	1.01	16.61	23
2100	#5 bars	G		112	.143		10.45	7.20	1.09	18.74	25.50
2200	#6 bars	G		104	.154		16.65	7.75	1.17	25.57	33.50
2300	#7 bars	G		96	.167		19.35	8.40	1.27	29.02	38
2400	#8 bars	G		88	.182		20	9.15	1.38	30.53	40
2500	#9 bars	G		80	.200		22	10.05	1.52	33.57	44
2600	#10 bars	G		72	.222		22.50	11.20	1.69	35.39	47
2700	#11 bars	G		64	.250		24.50	12.60	1.90	39	51
2800	#14 bars	G		56	.286		28	14.40	2.17	44.57	59
2900	#18 bars	G	↓	48	.333	↓	45.50	16.80	2.53	64.83	83

03 21 10 – Uncoated Reinforcing Steel

03 21 10.60 Reinforcing In Place

			Crew	Daily Output	Labor-Hours	Unit	Material	2012 Bare Costs Labor	Equipment	Total	Total Incl O&P
0015	**REINFORCING IN PLACE**, 50-60 ton lots, A615 Grade 60	R032110-10									
0020	Includes labor, but not material cost, to install accessories										
0030	Made from recycled materials	G									
0100	Beams & Girders, #3 to #7	G	4 Rodm	1.60	20	Ton	980	980		1,960	2,650
0150	#8 to #18	G R032110-20		2.70	11.852		980	580		1,560	2,000
0200	Columns, #3 to #7	G		1.50	21.333		980	1,050		2,030	2,750
0250	#8 to #18	G		2.30	13.913		980	685		1,665	2,175
0300	Spirals, hot rolled, 8" to 15" diameter	G		2.20	14.545		1,525	715		2,240	2,850
0320	15" to 24" diameter	G R032110-40		2.20	14.545		1,475	715		2,190	2,750
0330	24" to 36" diameter	G		2.30	13.913		1,400	685		2,085	2,625
0340	36" to 48" diameter	G R032110-50		2.40	13.333		1,325	655		1,980	2,500
0360	48" to 64" diameter	G		2.50	12.800		1,475	630		2,105	2,600
0380	64" to 84" diameter	G R032110-70		2.60	12.308		1,525	605		2,130	2,675
0390	84" to 96" diameter	G		2.70	11.852		1,600	580		2,180	2,700
0400	Elevated slabs, #4 to #7	G R032110-80		2.90	11.034		1,050	540		1,590	2,025
0500	Footings, #4 to #7	G		2.10	15.238		930	750		1,680	2,225
0550	#8 to #18	G		3.60	8.889		880	435		1,315	1,675
0600	Slab on grade, #3 to #7	G		2.30	13.913		930	685		1,615	2,125
0700	Walls, #3 to #7	G		3	10.667		930	525		1,455	1,850
0750	#8 to #18	G	↓	4	8	↓	930	395		1,325	1,650
0900	For other than 50 – 60 ton lots										
1000	Under 10 ton job, #3 to #7, add						25%	10%			
1010	#8 to #18, add						20%	10%			
1050	10 – 50 ton job, #3 to #7, add						10%				
1060	#8 to #18, add						5%				
1100	60 – 100 ton job, #3 to #7, deduct						5%				
1110	#8 to #18, deduct						10%				
1150	Over 100 ton job, #3 to #7, deduct						10%				
1160	#8 to #18, deduct						15%				

03 21 Reinforcement Bars

03 21 10 – Uncoated Reinforcing Steel

03 21 10.60 Reinforcing In Place

		Crew	Daily Output	Labor-Hours	Unit	Material	2012 Bare Costs Labor	Equipment	Total	Total Incl O&P
1200	Reinforcing in place, A615 Grade 75, add G				Ton	90			90	99
1220	Grade 90, add					120			120	132
2000	Unloading & sorting, add to above	C-5	100	.560			27	6.95	33.95	50
2200	Crane cost for handling, add to above, minimum		135	.415			19.95	5.15	25.10	37
2210	Average		92	.609			29.50	7.60	37.10	54.50
2220	Maximum		35	1.600			77	19.90	96.90	143
2400	Dowels, 2 feet long, deformed, #3 G	2 Rodm	520	.031	Ea.	.41	1.51		1.92	2.86
2410	#4 G		480	.033		.72	1.64		2.36	3.40
2420	#5 G		435	.037		1.12	1.81		2.93	4.12
2430	#6 G		360	.044		1.62	2.18		3.80	5.25
2450	Longer and heavier dowels, add G		725	.022	Lb.	.54	1.08		1.62	2.32
2500	Smooth dowels, 12" long, 1/4" or 3/8" diameter G		140	.114	Ea.	.70	5.60		6.30	9.70
2520	5/8" diameter G		125	.128		1.23	6.30		7.53	11.40
2530	3/4" diameter G		110	.145		1.52	7.15		8.67	13.05
2600	Dowel sleeves for CIP concrete, 2-part system									
2610	Sleeve base, plastic, for 5/8" smooth dowel sleeve, fasten to edge form	1 Rodm	200	.040	Ea.	.67	1.96		2.63	3.88
2615	Sleeve, plastic, 12" long, for 5/8" smooth dowel, snap onto base		400	.020		1.56	.98		2.54	3.29
2620	Sleeve base, for 3/4" smooth dowel sleeve		175	.046		.56	2.24		2.80	4.20
2625	Sleeve, 12" long, for 3/4" smooth dowel		350	.023		1.23	1.12		2.35	3.14
2630	Sleeve base, for 1" smooth dowel sleeve		150	.053		.86	2.62		3.48	5.15
2635	Sleeve, 12" long, for 1" smooth dowel		300	.027		1.28	1.31		2.59	3.50
2700	Dowel caps, visual warning only, plastic, #3 to #8	2 Rodm	800	.020		.29	.98		1.27	1.89
2720	#8 to #18		750	.021		.64	1.05		1.69	2.38
2750	Impalement protective, plastic, #4 to #9		800	.020		1.79	.98		2.77	3.54

03 21 10.70 Glass Fiber Reinforced Polymer Bars

		Crew	Daily Output	Labor-Hours	Unit	Material	2012 Bare Costs Labor	Equipment	Total	Total Incl O&P
0010	**GLASS FIBER REINFORCED POLYMER BARS**									
0020	Includes labor, but not material cost, to install accessories									
0050	#2 bar, .043 lb./L.F.	4 Rodm	9500	.003	L.F.	.35	.17		.52	.65
0100	#3 bar, .092 lb./L.F.		9300	.003		.48	.17		.65	.80
0150	#4 bar, .160 lb./L.F.		9100	.004		.70	.17		.87	1.05
0200	#5 bar, .258 lb./L.F.		8700	.004		1.05	.18		1.23	1.45
0250	#6 bar, .372 lb./L.F.		8300	.004		1.38	.19		1.57	1.82
0300	#7 bar, .497 lb./L.F.		7900	.004		1.73	.20		1.93	2.22
0350	#8 bar, .620 lb./L.F.		7400	.004		2.32	.21		2.53	2.89
0400	#9 bar, .800 lb./L.F.		6800	.005		2.92	.23		3.15	3.58
0450	#10 bar, 1.08 lb./L.F.		5800	.006		3.50	.27		3.77	4.28
0500	For Bends, add per bend				Ea.	1.25			1.25	1.38

03 21 13 – Galvanized Reinforcement Steel Bars

03 21 13.10 Galvanized Reinforcing

		Crew	Daily Output	Labor-Hours	Unit	Material	2012 Bare Costs Labor	Equipment	Total	Total Incl O&P
0010	**GALVANIZED REINFORCING**									
0150	Add to uncoated reinforcing price for galvanizing				Ton	450			450	495

03 21 16 – Epoxy-Coated Reinforcement Steel Bars

03 21 16.10 Epoxy-Coated Reinforcing

		Crew	Daily Output	Labor-Hours	Unit	Material	2012 Bare Costs Labor	Equipment	Total	Total Incl O&P
0010	**EPOXY-COATED REINFORCING**									
0100	Add to uncoated reinforcing price for coating with epoxy				Ton	470			470	515

03 22 Fabric and Grid Reinforcing

03 22 05 – Uncoated Welded Wire Fabric

03 22 05.50 Welded Wire Fabric

		Crew	Daily Output	Labor-Hours	Unit	Material	2012 Bare Costs Labor	Equipment	Total	Total Incl O&P
0010	**WELDED WIRE FABRIC** ASTM A185	R032205-30								
0020	Includes labor, but not material cost, to install accessories									
0030	Made from recycled materials	G								
0050	Sheets									
0100	6 x 6 - W1.4 x W1.4 (10 x 10) 21 lb. per C.S.F.	G	2 Rodm	35	.457	C.S.F.	13.80	22.50		36.30	51
0200	6 x 6 - W2.1 x W2.1 (8 x 8) 30 lb. per C.S.F.	G		31	.516		21.50	25.50		47	64
0300	6 x 6 - W2.9 x W2.9 (6 x 6) 42 lb. per C.S.F.	G		29	.552		24	27		51	70
0400	6 x 6 - W4 x W4 (4 x 4) 58 lb. per C.S.F.	G		27	.593		33	29		62	83
0500	4 x 4 - W1.4 x W1.4 (10 x 10) 31 lb. per C.S.F.	G		31	.516		20.50	25.50		46	63
0600	4 x 4 - W2.1 x W2.1 (8 x 8) 44 lb. per C.S.F.	G		29	.552		26	27		53	72
0650	4 x 4 - W2.9 x W2.9 (6 x 6) 61 lb. per C.S.F.	G		27	.593		41.50	29		70.50	92
0700	4 x 4 - W4 x W4 (4 x 4) 85 lb. per C.S.F.	G		25	.640		51.50	31.50		83	107
0750	Rolls									
0800	2 x 2 - #14 galv., 21 lb./C.S.F., beam & column wrap	G	2 Rodm	6.50	2.462	C.S.F.	44	121		165	242
0900	2 x 2 - #12 galv. for gunite reinforcing	G	"	6.50	2.462	"	65	121		186	265

03 23 Stressed Tendon Reinforcing

03 23 05 – Prestressing Tendons

03 23 05.50 Prestressing Steel

		Crew	Daily Output	Labor-Hours	Unit	Material	2012 Bare Costs Labor	Equipment	Total	Total Incl O&P
0010	**PRESTRESSING STEEL**	R034136-90								
0100	Grouted strand, post-tensioned in field, 50' span, 100 kip	G	C-3	1200	.053	Lb.	3.86	2.42	.09	6.37	8.15
0150	300 kip	G		2700	.024		1.45	1.07	.04	2.56	3.32
0300	100' span, 100 kip	G		1700	.038		3.86	1.71	.06	5.63	7
0350	300 kip	G		3200	.020		3.27	.91	.03	4.21	5.05
0500	200' span, 100 kip	G		2700	.024		3.86	1.07	.04	4.97	5.95
0550	300 kip	G		3500	.018		3.26	.83	.03	4.12	4.92
0800	Grouted bars, 50' span, 42 kip	G		2600	.025		1.39	1.12	.04	2.55	3.34
0850	143 kip	G		3200	.020		1.32	.91	.03	2.26	2.92
1000	75' span, 42 kip	G		3200	.020		1.42	.91	.03	2.36	3.03
1050	143 kip	G		4200	.015		1.14	.69	.03	1.86	2.37
1200	Ungrouted strand, 50' span, 100 kip	G	C-4	1275	.025		.63	1.25	.02	1.90	2.71
1250	300 kip	G		1475	.022		.63	1.08	.02	1.73	2.43
1400	100' span, 100 kip	G		1500	.021		.63	1.06	.02	1.71	2.40
1450	300 kip	G		1650	.019		.63	.96	.02	1.61	2.25
1600	200' span, 100 kip	G		1500	.021		.63	1.06	.02	1.71	2.40
1650	300 kip	G		1700	.019		.63	.93	.02	1.58	2.20
1800	Ungrouted bars, 50' span, 42 kip	G		1400	.023		.49	1.13	.02	1.64	2.37
1850	143 kip	G		1700	.019		.49	.93	.02	1.44	2.05
2000	75' span, 42 kip	G		1800	.018		.49	.88	.02	1.39	1.97
2050	143 kip	G		2200	.015		.49	.72	.01	1.22	1.71
2220	Ungrouted single strand, 100' slab, 25 kip	G		1200	.027		.63	1.32	.03	1.98	2.83
2250	35 kip	G		1475	.022		.63	1.08	.02	1.73	2.43
3000	Slabs on grade, 0.5-inch diam. non-bonded strands, HDPE sheathed,									
3050	attached dead-end anchors, loose stressing-end anchors									
3100	25' x 30' slab, strands @ 36" O.C., placing		2 Rodm	2940	.005	S.F.	.59	.27		.86	1.08
3105	Stressing		C-4A	3750	.004			.21	.01	.22	.34
3110	42" O.C., placing		2 Rodm	3200	.005		.52	.25		.77	.96
3115	Stressing		C-4A	4040	.004			.19	.01	.20	.32
3120	48" O.C., placing		2 Rodm	3510	.005		.46	.22		.68	.86
3125	Stressing		C-4A	4390	.004			.18	.01	.19	.30
3150	25' x 40' slab, strands @ 36" O.C., placing		2 Rodm	3370	.005		.57	.23		.80	1

03 23 Stressed Tendon Reinforcing

03 23 05 – Prestressing Tendons

03 23 05.50 Prestressing Steel

		Crew	Daily Output	Labor-Hours	Unit	Material	2012 Bare Costs Labor	Equipment	Total	Total Incl O&P
3155	Stressing	C-4A	4360	.004	S.F.		.18	.01	.19	.30
3160	42" O.C., placing	2 Rodm	3760	.004		.49	.21		.70	.87
3165	Stressing	C-4A	4820	.003			.16	.01	.17	.27
3170	48" O.C., placing	2 Rodm	4090	.004		.44	.19		.63	.79
3175	Stressing	C-4A	5190	.003			.15	.01	.16	.25
3200	30' x 30' slab, strands @ 36" O.C., placing	2 Rodm	3260	.005		.57	.24		.81	1.01
3205	Stressing	C-4A	4190	.004			.19	.01	.20	.31
3210	42" O.C., placing	2 Rodm	3530	.005		.51	.22		.73	.93
3215	Stressing	C-4A	4500	.004			.17	.01	.18	.29
3220	48" O.C., placing	2 Rodm	3840	.004		.46	.20		.66	.83
3225	Stressing	C-4A	4850	.003			.16	.01	.17	.27
3230	30' x 40' slab, strands @ 36" O.C., placing	2 Rodm	3780	.004		.55	.21		.76	.94
3235	Stressing	C-4A	4920	.003			.16	.01	.17	.26
3240	42" O.C., placing	2 Rodm	4190	.004		.48	.19		.67	.83
3245	Stressing	C-4A	5410	.003			.15	.01	.16	.24
3250	48" O.C., placing	2 Rodm	4520	.004		.44	.17		.61	.76
3255	Stressing	C-4A	5790	.003			.14	.01	.15	.23
3260	30' x 50' slab, strands @ 36" O.C., placing	2 Rodm	4300	.004		.52	.18		.70	.86
3265	Stressing	C-4A	5650	.003			.14	.01	.15	.23
3270	42" O.C., placing	2 Rodm	4720	.003		.46	.17		.63	.78
3275	Stressing	C-4A	6150	.003			.13	.01	.14	.21
3280	48" O.C., placing	2 Rodm	5240	.003		.41	.15		.56	.69
3285	Stressing	C-4A	6760	.002			.12	.01	.13	.20

03 24 Fibrous Reinforcing

03 24 05 – Reinforcing Fibers

03 24 05.30 Synthetic Fibers

					Unit	Material	Labor	Equipment	Total	Total Incl O&P
0010	**SYNTHETIC FIBERS**									
0100	Synthetic fibers, add to concrete				Lb.	5.20			5.20	5.75
0110	1-1/2 lb. per C.Y.				C.Y.	8.05			8.05	8.85

03 24 05.70 Steel Fibers

					Unit	Material	Labor	Equipment	Total	Total Incl O&P
0010	**STEEL FIBERS**									
0150	Steel fibers, add to concrete	G			Lb.	.90			.90	.99
0155	25 lb. per C.Y.	G			C.Y.	22.50			22.50	25
0160	50 lb. per C.Y.	G				45			45	49.50
0170	75 lb. per C.Y.	G				69.50			69.50	76
0180	100 lb. per C.Y.	G				90			90	99

03 30 Cast-In-Place Concrete

03 30 53 – Miscellaneous Cast-In-Place Concrete

03 30 53.40 Concrete In Place

			Crew	Daily Output	Labor-Hours	Unit	Material	Labor	Equipment	Total	Total Incl O&P
0010	**CONCRETE IN PLACE**	R033105-10									
0020	Including forms (4 uses), reinforcing steel, concrete, placement,	R033105-20									
0050	and finishing unless otherwise indicated	R033105-50									
0300	Beams (3500 psi), 5 kip per L.F., 10' span	R033105-65	C-14A	15.62	12.804	C.Y.	305	565	47.50	917.50	1,275
0350	25' span	R033105-70	"	18.55	10.782		320	480	40	840	1,125
0500	Chimney foundations (5000 psi), industrial, minimum	R033105-85	C-14C	32.22	3.476		144	147	.83	291.83	385
0510	Maximum		"	23.71	4.724		168	200	1.13	369.13	495
0700	Columns, square (4000 psi), 12" x 12", minimum reinforcing		C-14A	11.96	16.722		350	740	62	1,152	1,600

03 30 Cast-In-Place Concrete

03 30 53 – Miscellaneous Cast-In-Place Concrete

03 30 53.40 Concrete In Place	Crew	Daily Output	Labor-Hours	Unit	Material	2012 Bare Costs Labor	2012 Bare Costs Equipment	Total	Total Incl O&P
0720	Average reinforcing	C-14A	10.13	19.743	C.Y.	550	875	73.50	1,498.50	2,025
0740	Maximum reinforcing		9.03	22.148		830	980	82.50	1,892.50	2,525
0800	16" x 16", minimum reinforcing		16.22	12.330		274	545	46	865	1,200
0820	Average reinforcing		12.57	15.911		470	705	59	1,234	1,675
0840	Maximum reinforcing		10.25	19.512		730	865	72.50	1,667.50	2,200
0900	24" x 24", minimum reinforcing		23.66	8.453		231	375	31.50	637.50	870
0920	Average reinforcing		17.71	11.293		415	500	42	957	1,275
0940	Maximum reinforcing		14.15	14.134		665	625	52.50	1,342.50	1,750
1000	36" x 36", minimum reinforcing		33.69	5.936		205	263	22	490	655
1020	Average reinforcing		23.32	8.576		365	380	32	777	1,025
1040	Maximum reinforcing		17.82	11.223		620	495	42	1,157	1,500
1100	Columns, round (4000 psi), tied, 12" diameter, minimum reinforcing		20.97	9.537		297	420	35.50	752.50	1,025
1120	Average reinforcing		15.27	13.098		500	580	48.50	1,128.50	1,500
1140	Maximum reinforcing		12.11	16.515		765	730	61.50	1,556.50	2,025
1200	16" diameter, minimum reinforcing		31.49	6.351		272	281	23.50	576.50	760
1220	Average reinforcing		19.12	10.460		475	465	39	979	1,275
1240	Maximum reinforcing		13.77	14.524		720	645	54	1,419	1,850
1300	20" diameter, minimum reinforcing		41.04	4.873		274	216	18.15	508.15	655
1320	Average reinforcing		24.05	8.316		460	370	31	861	1,100
1340	Maximum reinforcing		17.01	11.758		725	520	43.50	1,288.50	1,650
1400	24" diameter, minimum reinforcing		51.85	3.857		257	171	14.35	442.35	560
1420	Average reinforcing		27.06	7.391		460	325	27.50	812.50	1,050
1440	Maximum reinforcing		18.29	10.935		710	485	40.50	1,235.50	1,575
1500	36" diameter, minimum reinforcing		75.04	2.665		258	118	9.90	385.90	475
1520	Average reinforcing		37.49	5.335		440	236	19.85	695.85	870
1540	Maximum reinforcing		22.84	8.757		690	390	32.50	1,112.50	1,400
1900	Elevated slab (4000 psi), flat slab with drops, 125 psf Sup. Load, 20' span	C-14B	38.45	5.410		239	239	19.30	497.30	655
1950	30' span		50.99	4.079		261	180	14.55	455.55	580
2100	Flat plate, 125 psf Sup. Load, 15' span		30.24	6.878		221	305	24.50	550.50	740
2150	25' span		49.60	4.194		234	185	14.95	433.95	560
2300	Waffle const., 30" domes, 125 psf Sup. Load, 20' span		37.07	5.611		239	248	20	507	670
2350	30' span		44.07	4.720		224	209	16.85	449.85	585
2500	One way joists, 30" pans, 125 psf Sup. Load, 15' span		27.38	7.597		281	335	27	643	860
2550	25' span		31.15	6.677		267	295	24	586	775
2700	One way beam & slab, 125 psf Sup. Load, 15' span		20.59	10.102		237	445	36	718	990
2750	25' span		28.36	7.334		229	325	26	580	780
2900	Two way beam & slab, 125 psf Sup. Load, 15' span		24.04	8.652		231	385	31	647	880
2950	25' span		35.87	5.799		202	256	20.50	478.50	640
3100	Elevated slabs, flat plate, including finish, not									
3110	including forms or reinforcing									
3150	Regular concrete (4000 psi), 4" slab	C-8	2613	.021	S.F.	1.34	.84	.27	2.45	3.05
3200	6" slab		2585	.022		1.97	.85	.28	3.10	3.75
3250	2-1/2" thick floor fill		2685	.021		.87	.82	.27	1.96	2.47
3300	Lightweight, 110# per C.F., 2-1/2" thick floor fill		2585	.022		1.16	.85	.28	2.29	2.86
3400	Cellular concrete, 1-5/8" fill, under 5000 S.F.		2000	.028		.78	1.09	.36	2.23	2.91
3450	Over 10,000 S.F.		2200	.025		.75	.99	.33	2.07	2.69
3500	Add per floor for 3 to 6 stories high		31800	.002			.07	.02	.09	.12
3520	For 7 to 20 stories high		21200	.003			.10	.03	.13	.20
3540	Equipment pad (3000 psi), 3' x 3' x 6" thick	C-14H	45	1.067	Ea.	44	46.50	.61	91.11	121
3550	4' x 4' x 6" thick		30	1.600		66	69.50	.91	136.41	181
3560	5' x 5' x 8" thick		18	2.667		118	116	1.52	235.52	310
3570	6' x 6' x 8" thick		14	3.429		159	149	1.95	309.95	405
3580	8' x 8' x 10" thick		8	6		335	261	3.42	599.42	775

03 30 53 – Miscellaneous Cast-In-Place Concrete

03 30 53.40 Concrete In Place	Crew	Daily Output	Labor-Hours	Unit	Material	2012 Bare Costs Labor	Equipment	Total	Total Incl O&P
3590	10' x 10' x 12" thick	C-14H	5	9.600	Ea.	575	415	5.45	995.45	1,275
3800	Footings (3000 psi), spread under 1 C.Y.	C-14C	28	4	C.Y.	166	169	.96	335.96	445
3825	1 C.Y. to 5 C.Y.		43	2.605		192	110	.63	302.63	380
3850	Over 5 C.Y.	↓	75	1.493		177	63	.36	240.36	293
3900	Footings, strip (3000 psi), 18" x 9", unreinforced	C-14L	40	2.400		121	98.50	.70	220.20	285
3920	18" x 9", reinforced	C-14C	35	3.200		145	135	.77	280.77	370
3925	20" x 10", unreinforced	C-14L	45	2.133		118	87.50	.62	206.12	265
3930	20" x 10", reinforced	C-14C	40	2.800		137	118	.67	255.67	335
3935	24" x 12", unreinforced	C-14L	55	1.745		117	72	.51	189.51	240
3940	24" x 12", reinforced	C-14C	48	2.333		136	98.50	.56	235.06	305
3945	36" x 12", unreinforced	C-14L	70	1.371		114	56.50	.40	170.90	212
3950	36" x 12", reinforced	C-14C	60	1.867		131	79	.45	210.45	267
4000	Foundation mat (3000 psi), under 10 C.Y.		38.67	2.896		202	122	.70	324.70	415
4050	Over 20 C.Y.	↓	56.40	1.986		178	84	.48	262.48	325
4200	Wall, free-standing (3000 psi), 8" thick, 8' high	C-14D	45.83	4.364		165	192	16.25	373.25	495
4250	14' high		27.26	7.337		200	320	27.50	547.50	745
4260	12" thick, 8' high		64.32	3.109		151	136	11.55	298.55	390
4270	14' high		40.01	4.999		161	219	18.60	398.60	540
4300	15" thick, 8' high		80.02	2.499		144	110	9.30	263.30	340
4350	12' high		51.26	3.902		144	171	14.50	329.50	440
4500	18' high	↓	48.85	4.094	↓	162	180	15.25	357.25	470
4520	Handicap access ramp (4000 psi), railing both sides, 3' wide	C-14H	14.58	3.292	L.F.	305	143	1.88	449.88	555
4525	5' wide		12.22	3.928		315	171	2.24	488.24	610
4530	With 6" curb and rails both sides, 3' wide		8.55	5.614		315	244	3.20	562.20	725
4535	5' wide	↓	7.31	6.566	↓	320	286	3.74	609.74	795
4650	Slab on grade (3500 psi), not including finish, 4" thick	C-14E	60.75	1.449	C.Y.	118	63	.45	181.45	228
4700	6" thick	"	92	.957	"	114	41.50	.30	155.80	190
4701	Thickened slab edge (3500 psi), for slab on grade poured									
4702	monolithically with slab; depth is in addition to slab thickness;									
4703	formed vertical outside edge, earthen bottom and inside slope									
4705	8" deep x 8" wide bottom, unreinforced	C-14L	2190	.044	L.F.	3.18	1.80	.01	4.99	6.25
4710	8" x 8", reinforced	C-14C	1670	.067		5.50	2.83	.02	8.35	10.45
4715	12" deep x 12" wide bottom, unreinforced	C-14L	1800	.053		6.55	2.19	.02	8.76	10.60
4720	12" x 12", reinforced	C-14C	1310	.086		10.75	3.61	.02	14.38	17.40
4725	16" deep x 16" wide bottom, unreinforced	C-14L	1440	.067		11.05	2.74	.02	13.81	16.40
4730	16" x 16", reinforced	C-14C	1120	.100		16.10	4.23	.02	20.35	24
4735	20" deep x 20" wide bottom, unreinforced	C-14L	1150	.083		16.85	3.43	.02	20.30	24
4740	20" x 20", reinforced	C-14C	920	.122		23	5.15	.03	28.18	33.50
4745	24" deep x 24" wide bottom, unreinforced	C-14L	930	.103		24	4.24	.03	28.27	32.50
4750	24" x 24", reinforced	C-14C	740	.151	↓	32	6.40	.04	38.44	45.50
4751	Slab on grade (3500 psi), incl. troweled finish, not incl. forms									
4760	or reinforcing, over 10,000 S.F., 4" thick	C-14F	3425	.021	S.F.	1.29	.84	.01	2.14	2.69
4820	6" thick		3350	.021		1.89	.86	.01	2.76	3.38
4840	8" thick		3184	.023		2.59	.91	.01	3.51	4.21
4900	12" thick		2734	.026		3.88	1.06	.01	4.95	5.85
4950	15" thick	↓	2505	.029	↓	4.88	1.15	.01	6.04	7.10
5000	Slab on grade (3000 psi), incl. textured finish, not incl. forms									
5001	or reinforcing, 4" thick	C-14G	2873	.019	S.F.	1.32	.77	.01	2.10	2.62
5010	6" thick		2590	.022		2.07	.85	.01	2.93	3.57
5020	8" thick	↓	2320	.024		2.70	.95	.01	3.66	4.41
5200	Lift slab in place above the foundation, incl. forms,									
5210	reinforcing, concrete (4000 psi) and columns, minimum	C-14B	2113	.098	S.F.	6.55	4.35	.35	11.25	14.30
5250	Average	↓	1650	.126	↓	7.15	5.55	.45	13.15	17

03 30 53 – Miscellaneous Cast-In-Place Concrete

03 30 53.40 Concrete In Place

		Crew	Daily Output	Labor-Hours	Unit	Material	2012 Bare Costs Labor	Equipment	Total	Total Incl O&P
5300	Maximum	C-14B	1500	.139	S.F.	7.80	6.15	.50	14.45	18.55
5500	Lightweight, ready mix, including screed finish only,									
5510	not including forms or reinforcing									
5550	1:4 (2500 psi) for structural roof decks	C-14B	260	.800	C.Y.	136	35.50	2.86	174.36	207
5600	1:6 (3000 psi) for ground slab with radiant heat	C-14F	92	.783		133	31.50	.30	164.80	194
5650	1:3:2 (2000 psi) with sand aggregate, roof deck	C-14B	260	.800		133	35.50	2.86	171.36	204
5700	Ground slab (2000 psi)	C-14F	107	.673		133	27	.26	160.26	187
5900	Pile caps (3000 psi), incl. forms and reinf., sq. or rect., under 10 C.Y.	C-14C	54.14	2.069		159	87.50	.50	247	310
5950	Over 10 C.Y.		75	1.493		152	63	.36	215.36	265
6000	Triangular or hexagonal, under 10 C.Y.		53	2.113		122	89.50	.51	212.01	273
6050	Over 10 C.Y.		85	1.318		138	55.50	.32	193.82	237
6200	Retaining walls (3000 psi), gravity, 4' high see Section 32 32	C-14D	66.20	3.021		140	133	11.25	284.25	370
6250	10' high		125	1.600		134	70	5.95	209.95	262
6300	Cantilever, level backfill loading, 8' high		70	2.857		154	125	10.65	289.65	375
6350	16' high		91	2.198		147	96.50	8.20	251.70	320
6800	Stairs (3500 psi), not including safety treads, free standing, 3'-6" wide	C-14H	83	.578	LF Nose	5.10	25	.33	30.43	44.50
6850	Cast on ground		125	.384	"	4.35	16.70	.22	21.27	30.50
7000	Stair landings, free standing		200	.240	S.F.	4.12	10.45	.14	14.71	20.50
7050	Cast on ground		475	.101	"	3.37	4.39	.06	7.82	10.50

03 31 Structural Concrete

03 31 05 – Normal Weight Structural Concrete

03 31 05.25 Concrete, Hand Mix

		Crew	Daily Output	Labor-Hours	Unit	Material	2012 Bare Costs Labor	Equipment	Total	Total Incl O&P
0010	**CONCRETE, HAND MIX** for small quantities or remote areas									
0050	Includes bulk local aggregate, bulk sand, bagged Portland cement,									
0060	and water, using gas powered cement mixer									
0125	2500 psi	C-30	135	.059	C.F.	3.60	2.08	1.19	6.87	8.45
0130	3000 psi		135	.059		3.88	2.08	1.19	7.15	8.80
0135	3500 psi		135	.059		4.03	2.08	1.19	7.30	8.95
0140	4000 psi		135	.059		4.21	2.08	1.19	7.48	9.15
0145	4500 psi		135	.059		4.41	2.08	1.19	7.68	9.35
0150	5000 psi		135	.059		4.74	2.08	1.19	8.01	9.70
0300	Using pre-bagged dry mix and wheelbarrow (80-lb. bag = 0.6 C.F.)									
0340	4000 psi	1 Clab	48	.167	C.F.	5.80	5.85		11.65	15.35

03 31 05.35 Normal Weight Concrete, Ready Mix

			Crew	Daily Output	Labor-Hours	Unit	Material	2012 Bare Costs Labor	Equipment	Total	Total Incl O&P
0010	**NORMAL WEIGHT CONCRETE, READY MIX**, delivered	R033105-10									
0012	Includes local aggregate, sand, Portland cement, and water										
0015	Excludes all additives and treatments	R033105-20									
0020	2000 psi					C.Y.	91.50			91.50	101
0100	2500 psi	R033105-30					94			94	103
0150	3000 psi						102			102	112
0200	3500 psi	R033105-40					99.50			99.50	110
0300	4000 psi						103			103	113
0350	4500 psi	R033105-50					106			106	116
0400	5000 psi						109			109	120
0411	6000 psi						124			124	136
0412	8000 psi						202			202	223
0413	10,000 psi						287			287	315
0414	12,000 psi						345			345	380
1000	For high early strength cement, add						10%				
1010	For structural lightweight with regular sand, add						25%				

03 31 Structural Concrete

03 31 05 – Normal Weight Structural Concrete

03 31 05.35 Normal Weight Concrete, Ready Mix	Crew	Daily Output	Labor-Hours	Unit	Material	2012 Bare Costs Labor	Equipment	Total	Total Incl O&P
1300	For winter concrete (hot water), add				C.Y.	4.25			4.25	4.68
1400	For hot weather concrete (ice), add					9.35			9.35	10.25
1410	For mid-range water reducer, add					4.13			4.13	4.54
1420	For high-range water reducer/superplasticizer, add					6.35			6.35	6.95
1430	For retarder, add					2.71			2.71	2.98
1440	For non-Chloride accelerator, add					4.83			4.83	5.30
1450	For Chloride accelerator, per 1%, add					3.28			3.28	3.61
1460	For fiber reinforcing, synthetic (1 lb./C.Y.), add					6.65			6.65	7.30
1500	For Saturday delivery, add					8.85			8.85	9.70
1510	For truck holding/waiting time past 1st hour per load, add				Hr.	87.50			87.50	96.50
1520	For short load (less than 4 C.Y.), add per load				Ea.	112			112	124
2000	For all lightweight aggregate, add				C.Y.	45%				

03 31 05.70 Placing Concrete

	03 31 05.70 Placing Concrete	Crew	Daily Output	Labor-Hours	Unit	Material	2012 Bare Costs Labor	Equipment	Total	Total Incl O&P
0010	**PLACING CONCRETE** R033105-70									
0020	Includes labor and equipment to place, strike off and consolidate									
0050	Beams, elevated, small beams, pumped	C-20	60	1.067	C.Y.		40	12.85	52.85	75.50
0100	With crane and bucket	C-7	45	1.600			61	26	87	122
0200	Large beams, pumped	C-20	90	.711			27	8.55	35.55	50.50
0250	With crane and bucket	C-7	65	1.108			42	18.10	60.10	84
0400	Columns, square or round, 12" thick, pumped	C-20	60	1.067			40	12.85	52.85	75.50
0450	With crane and bucket	C-7	40	1.800			68.50	29.50	98	137
0600	18" thick, pumped	C-20	90	.711			27	8.55	35.55	50.50
0650	With crane and bucket	C-7	55	1.309			50	21.50	71.50	99.50
0800	24" thick, pumped	C-20	92	.696			26	8.40	34.40	49
0850	With crane and bucket	C-7	70	1.029			39	16.80	55.80	78
1000	36" thick, pumped	C-20	140	.457			17.25	5.50	22.75	32.50
1050	With crane and bucket	C-7	100	.720			27.50	11.75	39.25	55
1400	Elevated slabs, less than 6" thick, pumped	C-20	140	.457			17.25	5.50	22.75	32.50
1450	With crane and bucket	C-7	95	.758			29	12.40	41.40	57.50
1500	6" to 10" thick, pumped	C-20	160	.400			15.10	4.82	19.92	28.50
1550	With crane and bucket	C-7	110	.655			25	10.70	35.70	50
1600	Slabs over 10" thick, pumped	C-20	180	.356			13.40	4.28	17.68	25
1650	With crane and bucket	C-7	130	.554			21	9.05	30.05	42
1900	Footings, continuous, shallow, direct chute	C-6	120	.400			14.65	.46	15.11	23
1950	Pumped	C-20	150	.427			16.10	5.15	21.25	30
2000	With crane and bucket	C-7	90	.800			30.50	13.10	43.60	61
2100	Footings, continuous, deep, direct chute	C-6	140	.343			12.55	.39	12.94	19.60
2150	Pumped	C-20	160	.400			15.10	4.82	19.92	28.50
2200	With crane and bucket	C-7	110	.655			25	10.70	35.70	50
2400	Footings, spread, under 1 C.Y., direct chute	C-6	55	.873			32	.99	32.99	50
2450	Pumped	C-20	65	.985			37	11.85	48.85	69.50
2500	With crane and bucket	C-7	45	1.600			61	26	87	122
2600	Over 5 C.Y., direct chute	C-6	120	.400			14.65	.46	15.11	23
2650	Pumped	C-20	150	.427			16.10	5.15	21.25	30
2700	With crane and bucket	C-7	100	.720			27.50	11.75	39.25	55
2900	Foundation mats, over 20 C.Y., direct chute	C-6	350	.137			5	.16	5.16	7.80
2950	Pumped	C-20	400	.160			6.05	1.93	7.98	11.30
3000	With crane and bucket	C-7	300	.240			9.15	3.92	13.07	18.20
3200	Grade beams, direct chute	C-6	150	.320			11.70	.36	12.06	18.30
3250	Pumped	C-20	180	.356			13.40	4.28	17.68	25
3300	With crane and bucket	C-7	120	.600			23	9.80	32.80	46
3500	High rise, for more than 5 stories, pumped, add per story	C-20	2100	.030			1.15	.37	1.52	2.15

03 31 05 – Normal Weight Structural Concrete

03 31 05.70 Placing Concrete		Crew	Daily Output	Labor-Hours	Unit	Material	2012 Bare Costs Labor	Equipment	Total	Total Incl O&P
3510	With crane and bucket, add per story	C-7	2100	.034	C.Y.		1.31	.56	1.87	2.61
3700	Pile caps, under 5 C.Y., direct chute	C-6	90	.533			19.55	.61	20.16	30.50
3750	Pumped	C-20	110	.582			22	7	29	41
3800	With crane and bucket	C-7	80	.900			34.50	14.70	49.20	68
3850	Pile cap, 5 C.Y. to 10 C.Y., direct chute	C-6	175	.274			10.05	.31	10.36	15.70
3900	Pumped	C-20	200	.320			12.05	3.85	15.90	22.50
3950	With crane and bucket	C-7	150	.480			18.30	7.85	26.15	36.50
4000	Over 10 C.Y., direct chute	C-6	215	.223			8.20	.25	8.45	12.80
4050	Pumped	C-20	240	.267			10.05	3.21	13.26	18.85
4100	With crane and bucket	C-7	185	.389			14.80	6.35	21.15	29.50
4300	Slab on grade, up to 6" thick, direct chute	C-6	110	.436			16	.50	16.50	25
4350	Pumped	C-20	130	.492			18.55	5.95	24.50	35
4400	With crane and bucket	C-7	110	.655			25	10.70	35.70	50
4600	Over 6" thick, direct chute	C-6	165	.291			10.65	.33	10.98	16.60
4650	Pumped	C-20	185	.346			13.05	4.17	17.22	24.50
4700	With crane and bucket	C-7	145	.497			18.90	8.10	27	38
4900	Walls, 8" thick, direct chute	C-6	90	.533			19.55	.61	20.16	30.50
4950	Pumped	C-20	100	.640			24	7.70	31.70	45.50
5000	With crane and bucket	C-7	80	.900			34.50	14.70	49.20	68
5050	12" thick, direct chute	C-6	100	.480			17.60	.55	18.15	27.50
5100	Pumped	C-20	110	.582			22	7	29	41
5200	With crane and bucket	C-7	90	.800			30.50	13.10	43.60	61
5300	15" thick, direct chute	C-6	105	.457			16.75	.52	17.27	26
5350	Pumped	C-20	120	.533			20	6.40	26.40	37.50
5400	With crane and bucket	C-7	95	.758			29	12.40	41.40	57.50
5600	Wheeled concrete dumping, add to placing costs above									
5610	Walking cart, 50' haul, add	C-18	32	.281	C.Y.		9.95	1.77	11.72	17.25
5620	150' haul, add		24	.375			13.25	2.37	15.62	23
5700	250' haul, add		18	.500			17.65	3.16	20.81	30.50
5800	Riding cart, 50' haul, add	C-19	80	.113			3.97	1.18	5.15	7.40
5810	150' haul, add		60	.150			5.30	1.57	6.87	9.90
5900	250' haul, add		45	.200			7.05	2.10	9.15	13.15

03 35 29 – Tooled Concrete Finishing

03 35 29.30 Finishing Floors

		Crew	Daily Output	Labor-Hours	Unit	Material	2012 Bare Costs Labor	Equipment	Total	Total Incl O&P
0010	**FINISHING FLOORS**									
0012	Finishing requires that concrete first be placed, struck off & consolidated									
0015	Basic finishing for various unspecified flatwork									
0100	Bull float only	C-10	4000	.006	S.F.		.24		.24	.36
0125	Bull float & manual float		2000	.012			.48		.48	.71
0150	Bull float, manual float, & broom finish, w/edging & joints		1850	.013			.52		.52	.77
0200	Bull float, manual float & manual steel trowel		1265	.019			.76		.76	1.13
0210	For specified Random Access Floors in ACI Classes 1, 2, 3 and 4 to achieve									
0215	Composite Overall Floor Flatness and Levelness values up to F35/F25									
0250	Bull float, machine float & machine trowel (walk-behind)	C-10C	1715	.014	S.F.		.56	.03	.59	.86
0300	Power screed, bull float, machine float & trowel (walk-behind)	C-10D	2400	.010			.40	.05	.45	.65
0350	Power screed, bull float, machine float & trowel (ride-on)	C-10E	4000	.006			.24	.06	.30	.43
0352	For specified Random Access Floors in ACI Classes 5, 6, 7 and 8 to achieve									
0354	Composite Overall Floor Flatness and Levelness values up to F50/F50									
0356	Add for two-dimensional restraightening after power float	C-10	6000	.004	S.F.		.16		.16	.24

03 35 Concrete Finishing

03 35 29 – Tooled Concrete Finishing

03 35 29.30 Finishing Floors

		Crew	Daily Output	Labor-Hours	Unit	Material	2012 Bare Costs Labor	Equipment	Total	Total Incl O&P
0358	For specified Random or Defined Access Floors in ACI Class 9 to achieve									
0360	Composite Overall Floor Flatness and Levelness values up to F100/F100									
0362	Add for two-dimensional restraightening after bull float & power float	C-10	3000	.008	S.F.		.32		.32	.48
0364	For specified Superflat Defined Access Floors in ACI Class 9 to achieve									
0366	Minimum Floor Flatness and Levelness values of F100/F100									
0368	Add for 2-dim'l restraightening after bull float, power float, power trowel	C-10	2000	.012	S.F.		.48		.48	.71
0400	Integral topping and finish, using 1:1:2 mix, 3/16" thick	C-10B	1000	.040		.11	1.52	.25	1.88	2.68
0450	1/2" thick		950	.042		.29	1.60	.26	2.15	3.02
0500	3/4" thick		850	.047		.44	1.79	.29	2.52	3.50
0600	1" thick		750	.053		.59	2.03	.33	2.95	4.07
0800	Granolithic topping, laid after, 1:1:1-1/2 mix, 1/2" thick		590	.068		.33	2.58	.42	3.33	4.72
0820	3/4" thick		580	.069		.49	2.62	.43	3.54	4.96
0850	1" thick		575	.070		.65	2.64	.43	3.72	5.20
0950	2" thick		500	.080		1.30	3.04	.50	4.84	6.55
1200	Heavy duty, 1:1:2, 3/4" thick, preshrunk, gray, 20 M.S.F.		320	.125		.74	4.75	.78	6.27	8.85
1300	100 M.S.F.		380	.105		.44	4	.66	5.10	7.25
1600	Exposed local aggregate finish, minimum	1 Cefi	625	.013		.23	.54		.77	1.05
1650	Maximum		465	.017		.68	.73		1.41	1.82
1800	Floor abrasives, .25 psf, aluminum oxide		850	.009		.45	.40		.85	1.09
1850	Silicon carbide		850	.009		.70	.40		1.10	1.36
2000	Floor hardeners, metallic, light service, .50 psf, add		850	.009		.65	.40		1.05	1.31
2050	Medium service, .75 psf		750	.011		.98	.45		1.43	1.74
2100	Heavy service, 1.0 psf		650	.012		1.31	.52		1.83	2.21
2150	Extra heavy, 1.5 psf		575	.014		1.96	.59		2.55	3.03
2300	Non-metallic, light service, .50 psf		850	.009		.26	.40		.66	.87
2350	Medium service, .75 psf		750	.011		.39	.45		.84	1.08
2400	Heavy service, 1.00 psf		650	.012		.52	.52		1.04	1.34
2450	Extra heavy, 1.50 psf		575	.014		.77	.59		1.36	1.72
2800	Trap rock wearing surface for monolithic floors									
2810	2.0 psf	C-10B	1250	.032	S.F.	.02	1.22	.20	1.44	2.08
3000	Floor coloring, dusted on (0.6 psf), add to above	1 Cefi	1300	.006		.67	.26		.93	1.11
3050	(1.0 psf), add to above	"	625	.013		1.11	.54		1.65	2.02
3100	Colored powder only				Lb.	1.11			1.11	1.22
3600	1/2" topping using 0.6 psf powdered color	C-10B	590	.068	S.F.	4.98	2.58	.42	7.98	9.85
3650	1/2" topping using 1.0 psf powdered color	"	590	.068		5.45	2.58	.42	8.45	10.30
3800	Dustproofing, solvent-based, 1 coat	1 Cefi	1900	.004		.27	.18		.45	.55
3850	2 coats		1300	.006		.96	.26		1.22	1.44
4000	Epoxy-based, 1 coat		1500	.005		.13	.23		.36	.48
4050	2 coats		1500	.005		.27	.23		.50	.63
4400	Stair finish, float		275	.029			1.23		1.23	1.81
4500	Steel trowel finish		200	.040			1.69		1.69	2.49
4600	Silicon carbide finish, .25 psf		150	.053		.45	2.26		2.71	3.82

03 35 29.35 Control Joints, Saw Cut

		Crew	Daily Output	Labor-Hours	Unit	Material	2012 Bare Costs Labor	Equipment	Total	Total Incl O&P
0010	**CONTROL JOINTS, SAW CUT**									
0100	Sawcut control joints in green concrete									
0120	1" depth	C-27	2000	.008	L.F.	.03	.34	.07	.44	.62
0140	1-1/2" depth		1800	.009		.05	.38	.08	.51	.70
0160	2" depth		1600	.010		.07	.42	.09	.58	.79
0180	Sawcut joint reservoir in cured concrete									
0182	3/8" wide x 3/4" deep, with single saw blade	C-27	1000	.016	L.F.	.05	.68	.15	.88	1.22
0184	1/2" wide x 1" deep, with double saw blades		900	.018		.10	.75	.16	1.01	1.40
0186	3/4" wide x 1-1/2" deep, with double saw blades		800	.020		.20	.85	.19	1.24	1.67

03 35 Concrete Finishing

03 35 29 – Tooled Concrete Finishing

03 35 29.35 Control Joints, Saw Cut

		Crew	Daily Output	Labor-Hours	Unit	Material	2012 Bare Costs Labor	Equipment	Total	Total Incl O&P
0190	Water blast joint to wash away laitance, 2 passes	C-29	2500	.003	L.F.		.11	.03	.14	.20
0200	Air blast joint to blow out debris and air dry, 2 passes	C-28	2000	.004	↓		.17	.01	.18	.26
0300	For backer rod, see Section 07 91 23.10									
0340	For joint sealant, see Sections 03 15 05.25 or 07 92 13.20									
0900	For replacement of joint sealant, see Section 07 01 90.81									

03 35 29.60 Finishing Walls

		Crew	Daily Output	Labor-Hours	Unit	Material	2012 Bare Costs Labor	Equipment	Total	Total Incl O&P
0010	**FINISHING WALLS**									
0020	Break ties and patch voids	1 Cefi	540	.015	S.F.	.03	.63		.66	.96
0050	Burlap rub with grout		450	.018		.03	.75		.78	1.15
0100	Carborundum rub, dry		270	.030			1.25		1.25	1.85
0150	Wet rub	↓	175	.046			1.94		1.94	2.85
0300	Bush hammer, green concrete	B-39	1000	.048			1.78	.21	1.99	2.96
0350	Cured concrete	"	650	.074			2.74	.32	3.06	4.54
0600	Float finish, 1/16" thick	1 Cefi	300	.027		.31	1.13		1.44	2
0700	Sandblast, light penetration	E-11	1100	.029		.74	1.15	.19	2.08	2.95
0750	Heavy penetration	"	375	.085	↓	1.49	3.38	.56	5.43	7.90
0850	Grind form fins flush	1 Clab	700	.011	L.F.		.40		.40	.62

03 35 33 – Stamped Concrete Finishing

03 35 33.50 Slab Texture Stamping

		Crew	Daily Output	Labor-Hours	Unit	Material	2012 Bare Costs Labor	Equipment	Total	Total Incl O&P
0010	**SLAB TEXTURE STAMPING**									
0050	Stamping requires that concrete first be placed, struck off, consolidated,									
0060	bull floated and free of bleed water. Decorative stamping tasks include:									
0100	Step 1 - first application of dry shake colored hardener	1 Cefi	6400	.001	S.F.	.39	.05		.44	.51
0110	Step 2 - bull float		6400	.001			.05		.05	.08
0130	Step 3 - second application of dry shake colored hardener	↓	6400	.001		.19	.05		.24	.29
0140	Step 4 - bull float, manual float & steel trowel	3 Cefi	1280	.019			.79		.79	1.17
0150	Step 5 - application of dry shake colored release agent	1 Cefi	6400	.001		.09	.05		.14	.18
0160	Step 6 - place, tamp & remove mats	3 Cefi	2400	.010		1.41	.42		1.83	2.17
0170	Step 7 - touch up edges, mat joints & simulated grout lines	1 Cefi	1280	.006			.26		.26	.39
0300	Alternate stamping estimating method includes all tasks above	4 Cefi	800	.040		2.09	1.69		3.78	4.78
0400	Step 8 - pressure wash @ 3000 psi after 24 hours	1 Cefi	1600	.005			.21		.21	.31
0500	Step 9 - roll 2 coats cure/seal compound when dry	"	800	.010	↓	.53	.42		.95	1.20

03 35 43 – Polished Concrete Finishing

03 35 43.10 Polished Concrete Floors

		Crew	Daily Output	Labor-Hours	Unit	Material	2012 Bare Costs Labor	Equipment	Total	Total Incl O&P
0010	**POLISHED CONCRETE FLOORS**									
0015	Processing of cured concrete to include grinding, honing,									
0020	and polishing of interior floors with 22" segmented diamond									
0025	planetary floor grinder (2 passes in different directions per grit)									
0100	Removal of pre-existing coatings, dry, with carbide discs using									
0105	dry vacuum pick-up system, final hand sweeping									
0110	Glue, adhesive or tar	J-4	1.60	15	M.S.F.	19.20	600	114	733.20	1,050
0120	Paint, epoxy, 1 coat		3.60	6.667		19.20	266	50.50	335.70	470
0130	2 coats	↓	1.80	13.333	↓	19.20	530	101	650.20	930
0200	Grinding and edging, wet, including wet vac pick-up and auto									
0205	scrubbing between grit changes									
0210	40-grit diamond/metal matrix	J-4A	1.60	20	M.S.F.	31	775	226	1,032	1,450
0220	80-grit diamond/metal matrix		2	16		31	620	180	831	1,150
0230	120-grit diamond/metal matrix		2.40	13.333		31	515	150	696	975
0240	200-grit diamond/metal matrix	↓	2.80	11.429		31	445	129	605	840
0300	Spray on dye or stain (1 coat)	1 Cefi	16	.500		271	21		292	330
0400	Spray on densifier/hardener (2 coats)	"	8	1	↓	296	42.50		338.50	390

03 35 Concrete Finishing

03 35 43 – Polished Concrete Finishing

03 35 43.10 Polished Concrete Floors	Crew	Daily Output	Labor-Hours	Unit	Material	2012 Bare Costs Labor	Equipment	Total	Total Incl O&P
0410	Auto scrubbing after 2nd coat, when dry	J-4B	16	.500	M.S.F.		17.55	11.15	28.70	39.50
0500	Honing and edging, wet, including wet vac pick-up and auto									
0505	scrubbing between grit changes									
0510	100-grit diamond/resin matrix	J-4A	2.80	11.429	M.S.F.	31	445	129	605	840
0520	200-grit diamond/resin matrix	"	2.80	11.429	"	31	445	129	605	840
0530	Dry, including dry vacuum pick-up system, final hand sweeping									
0540	400-grit diamond/resin matrix	J-4A	2.80	11.429	M.S.F.	31	445	129	605	840
0600	Polishing and edging, dry, including dry vac pick-up and hand									
0605	sweeping between grit changes									
0610	800-grit diamond/resin matrix	J-4A	2.80	11.429	M.S.F.	31	445	129	605	840
0620	1500-grit diamond/resin matrix		2.80	11.429		31	445	129	605	840
0630	3000-grit diamond/resin matrix		2.80	11.429		31	445	129	605	840
0700	Auto scrubbing after final polishing step	J-4B	16	.500			17.55	11.15	28.70	39.50

03 37 Specialty Placed Concrete

03 37 13 – Shotcrete

03 37 13.30 Gunite (Dry-Mix)

		Crew	Daily Output	Labor-Hours	Unit	Material	Labor	Equipment	Total	Total Incl O&P
0010	**GUNITE (DRY-MIX)**									
0020	Typical in place, 1" layers, no mesh included	C-16	2000	.028	S.F.	.33	1.09	.15	1.57	2.19
0100	Mesh for gunite 2 x 2, #12	2 Rodm	800	.020		.65	.98		1.63	2.29
0150	#4 reinforcing bars @ 6" each way	"	500	.032		1.64	1.57		3.21	4.31
0300	Typical in place, including mesh, 2" thick, minimum	C-16	1000	.056		1.31	2.19	.30	3.80	5.10
0350	Maximum		500	.112		1.31	4.38	.60	6.29	8.70
0500	4" thick, minimum		750	.075		1.96	2.92	.40	5.28	7
0550	Maximum		350	.160		1.96	6.25	.86	9.07	12.55
0900	Prepare old walls, no scaffolding, minimum	C-10	1000	.024			.96		.96	1.43
0950	Maximum	"	275	.087			3.48		3.48	5.20
1100	For high finish requirement or close tolerance, add, minimum						50%			
1150	Maximum						110%			

03 37 13.60 Shotcrete (Wet-Mix)

		Crew	Daily Output	Labor-Hours	Unit	Material	Labor	Equipment	Total	Total Incl O&P
0010	**SHOTCRETE (WET-MIX)**									
0020	Wet mix, placed @ up to 12 C.Y. per hour, 3000 psi	C-8C	80	.600	C.Y.	106	23	5.35	134.35	158
0100	Up to 35 C.Y. per hour	C-8E	240	.200	"	95.50	7.65	2.06	105.21	119
1010	Fiber reinforced, 1" thick	C-8C	1740	.028	S.F.	.79	1.06	.25	2.10	2.76
1020	2" thick		900	.053		1.58	2.06	.47	4.11	5.40
1030	3" thick		825	.058		2.38	2.24	.52	5.14	6.60
1040	4" thick		750	.064		3.17	2.47	.57	6.21	7.85

03 39 Concrete Curing

03 39 13 – Water Concrete Curing

03 39 13.50 Water Curing

		Crew	Daily Output	Labor-Hours	Unit	Material	Labor	Equipment	Total	Total Incl O&P
0010	**WATER CURING**									
0015	With burlap, 4 uses assumed, 7.5 oz.	2 Clab	55	.291	C.S.F.	11.80	10.20		22	28.50
0100	10 oz.	"	55	.291	"	21	10.20		31.20	39
0400	Curing blankets, 1" to 2" thick, buy, minimum				S.F.	.34			.34	.37
0450	Maximum				"	.51			.51	.56

03 39 Concrete Curing

03 39 23 – Membrane Concrete Curing

03 39 23.13 Chemical Compound Membrane Concrete Curing	Crew	Daily Output	Labor-Hours	Unit	Material	2012 Bare Costs		Total	Total Incl O&P
						Labor	Equipment		
0010 **CHEMICAL COMPOUND MEMBRANE CONCRETE CURING**									
0300 Sprayed membrane curing compound	2 Clab	95	.168	C.S.F.	7.45	5.90		13.35	17.30
0700 Curing compound, solvent based, 400 S.F./gal., 55 gallon lots				Gal.	20			20	22
0720 5 gallon lots					27			27	29.50
0800 Curing compound, water based, 250 S.F./gal., 55 gallon lots					19.80			19.80	22
0820 5 gallon lots					23			23	25.50

03 39 23.23 Sheet Membrane Concrete Curing

	Crew	Daily Output	Labor-Hours	Unit	Material	Labor	Equipment	Total	Total Incl O&P
0010 **SHEET MEMBRANE CONCRETE CURING**									
0200 Curing blanket, burlap/poly, 2-ply	2 Clab	70	.229	C.S.F.	17.10	8		25.10	31

03 41 Precast Structural Concrete

03 41 05 – Precast Concrete Members

03 41 05.10 Precast Beams

		Crew	Daily Output	Labor-Hours	Unit	Material	Labor	Equipment	Total	Total Incl O&P
0010 **PRECAST BEAMS**	R034105-30									
0011 L-shaped, 20' span, 12" x 20"		C-11	32	2.250	Ea.	1,900	109	57	2,066	2,350
1000 Inverted tee beams, add to above, small beams						15%				
1050 Large beams						20%				
1200 Rectangular, 20' span, 12" x 20"		C-11	32	2.250		1,425	109	57	1,591	1,825
1250 18" x 36"			24	3		2,625	146	76	2,847	3,200
1300 24" x 44"			22	3.273		3,625	159	83	3,867	4,375
1400 30' span, 12" x 36"			24	3		3,300	146	76	3,522	3,950
1450 18" x 44"			20	3.600		4,725	175	91.50	4,991.50	5,600
1500 24" x 52"			16	4.500		6,300	218	114	6,632	7,450
1600 40' span, 12" x 52"			20	3.600		6,100	175	91.50	6,366.50	7,100
1650 18" x 52"			16	4.500		7,275	218	114	7,607	8,500
1700 24" x 52"			12	6		8,425	291	152	8,868	9,925
2000 "T" shaped, 20' span, 12" x 20"			32	2.250		2,475	109	57	2,641	2,975
2050 18" x 36"			24	3		4,475	146	76	4,697	5,250
2100 24" x 44"			22	3.273		6,200	159	83	6,442	7,200
2200 30' span, 12" x 36"			24	3		5,575	146	76	5,797	6,450
2250 18" x 44"			20	3.600		8,000	175	91.50	8,266.50	9,200
2300 24" x 52"			16	4.500		10,700	218	114	11,032	12,300
2500 40' span, 12" x 52"			20	3.600		10,300	175	91.50	10,566.50	11,700
2550 18" x 52"			16	4.500		12,400	218	114	12,732	14,100
2600 24" x 52"			12	6		14,300	291	152	14,743	16,400

03 41 05.15 Precast Columns

		Crew	Daily Output	Labor-Hours	Unit	Material	Labor	Equipment	Total	Total Incl O&P
0010 **PRECAST COLUMNS**	R034105-30									
0020 Rectangular to 12' high, small columns		C-11	120	.600	L.F.	100	29	15.25	144.25	177
0050 Large columns			96	.750		173	36.50	19.05	228.55	275
0300 24' high, small columns			192	.375		100	18.20	9.55	127.75	152
0350 Large columns			144	.500		173	24.50	12.70	210.20	247

03 41 05.25 Precast Joists

		Crew	Daily Output	Labor-Hours	Unit	Material	Labor	Equipment	Total	Total Incl O&P
0010 **PRECAST JOISTS**	R034105-30									
0015 40 psf L.L., 6" deep for 12' spans		C-12	600	.080	L.F.	16.65	3.48	1.02	21.15	25
0050 8" deep for 16' spans			575	.083		27.50	3.64	1.06	32.20	37
0100 10" deep for 20' spans			550	.087		48.50	3.80	1.11	53.41	60.50
0150 12" deep for 24' spans			525	.091		66.50	3.98	1.17	71.65	80.50

03 41 Precast Structural Concrete

03 41 13 – Precast Concrete Hollow Core Planks

03 41 13.50 Precast Slab Planks

		Crew	Daily Output	Labor-Hours	Unit	Material	2012 Bare Costs Labor	Equipment	Total	Total Incl O&P
0010	**PRECAST SLAB PLANKS** R034105-30									
0020	Prestressed roof/floor members, grouted, solid, 4" thick	C-11	2400	.030	S.F.	6.05	1.46	.76	8.27	10
0050	6" thick		2800	.026		6.50	1.25	.65	8.40	10.05
0100	Hollow, 8" thick		3200	.023		7.15	1.09	.57	8.81	10.35
0150	10" thick		3600	.020		7.40	.97	.51	8.88	10.40
0200	12" thick		4000	.018		7.90	.87	.46	9.23	10.70

03 41 16 – Precast Concrete Slabs

03 41 16.20 Precast Concrete Channel Slabs

		Crew	Daily Output	Labor-Hours	Unit	Material	2012 Bare Costs Labor	Equipment	Total	Total Incl O&P
0010	**PRECAST CONCRETE CHANNEL SLABS**									
0335	Lightweight concrete channel slab, long runs, 2-3/4" thick	C-12	1575	.030	S.F.	8.25	1.33	.39	9.97	11.55
0375	3-3/4" thick		1550	.031		8.55	1.35	.39	10.29	11.90
0475	4-3/4" thick		1525	.031		9.25	1.37	.40	11.02	12.75
1275	Short pieces, 2-3/4" thick		785	.061		12.40	2.66	.78	15.84	18.55
1375	3-3/4" thick		770	.062		12.85	2.71	.79	16.35	19.15
1475	4-3/4" thick		762	.063		13.90	2.74	.80	17.44	20.50

03 41 16.50 Precast Lightweight Concrete Plank

		Crew	Daily Output	Labor-Hours	Unit	Material	2012 Bare Costs Labor	Equipment	Total	Total Incl O&P
0010	**PRECAST LIGHTWEIGHT CONCRETE PLANK**									
0015	Lightweight plank, nailable, T&G, 2" thick	C-12	1800	.027	S.F.	7.25	1.16	.34	8.75	10.15
0150	For premium ceiling finish, add				"	10%				
0200	For sloping roofs, slope over 4 in 12, add						25%			
0250	Slope over 6 in 12, add						150%			

03 41 23 – Precast Concrete Stairs

03 41 23.50 Precast Stairs

		Crew	Daily Output	Labor-Hours	Unit	Material	2012 Bare Costs Labor	Equipment	Total	Total Incl O&P
0010	**PRECAST STAIRS**									
0020	Precast concrete treads on steel stringers, 3' wide	C-12	75	.640	Riser	129	28	8.15	165.15	194
0300	Front entrance, 5' wide with 48" platform, 2 risers		16	3	Flight	485	131	38.50	654.50	770
0350	5 risers		12	4		765	174	51	990	1,175
0500	6' wide, 2 risers		15	3.200		535	139	41	715	850
0550	5 risers		11	4.364		845	190	55.50	1,090.50	1,275
0700	7' wide, 2 risers		14	3.429		685	149	43.50	877.50	1,025
0750	5 risers		10	4.800		1,125	209	61	1,395	1,650
1200	Basement entrance stairs, steel bulkhead doors, minimum	B-51	22	2.182		1,400	77	9.60	1,486.60	1,675
1250	Maximum	"	11	4.364		2,350	154	19.20	2,523.20	2,825

03 41 33 – Precast Structural Pretensioned Concrete

03 41 33.60 Tees

		Crew	Daily Output	Labor-Hours	Unit	Material	2012 Bare Costs Labor	Equipment	Total	Total Incl O&P
0010	**TEES** R034105-30									
0020	Quad tee, short spans, roof	C-11	7200	.010	S.F.	7.90	.49	.25	8.64	9.75
0050	Floor		7200	.010		7.90	.49	.25	8.64	9.75
0200	Double tee, floor members, 60' span		8400	.009		9.40	.42	.22	10.04	11.30
0250	80' span		8000	.009		12.15	.44	.23	12.82	14.35
0300	Roof members, 30' span		4800	.015		7.90	.73	.38	9.01	10.40
0350	50' span		6400	.011		8.65	.55	.29	9.49	10.75
0400	Wall members, up to 55' high		3600	.020		11.70	.97	.51	13.18	15.10
0500	Single tee roof members, 40' span		3200	.023		10.65	1.09	.57	12.31	14.20
0550	80' span		5120	.014		12.10	.68	.36	13.14	14.85
0600	100' span		6000	.012		17.70	.58	.30	18.58	21
0650	120' span		6000	.012		19.35	.58	.30	20.23	23
1000	Double tees, floor members									
1100	Lightweight, 20" x 8' wide, 45' span	C-11	20	3.600	Ea.	3,125	175	91.50	3,391.50	3,825
1150	24" x 8' wide, 50' span		18	4		3,475	194	102	3,771	4,275

03 41 Precast Structural Concrete

03 41 33 – Precast Structural Pretensioned Concrete

03 41 33.60 Tees

		Crew	Daily Output	Labor-Hours	Unit	Material	2012 Bare Costs Labor	2012 Bare Costs Equipment	Total	Total Incl O&P
1200	32" x 10' wide, 60' span	C-11	16	4.500	Ea.	5,200	218	114	5,532	6,225
1250	Standard weight, 12" x 8' wide, 20' span		22	3.273		1,250	159	83	1,492	1,750
1300	16" x 8' wide, 25' span		20	3.600		1,575	175	91.50	1,841.50	2,125
1350	18" x 8' wide, 30' span		20	3.600		1,900	175	91.50	2,166.50	2,475
1400	20" x 8' wide, 45' span		18	4		2,825	194	102	3,121	3,575
1450	24" x 8' wide, 50' span		16	4.500		3,150	218	114	3,482	3,975
1500	32" x 10' wide, 60' span	↓	14	5.143	↓	4,725	249	131	5,105	5,775
2000	Roof members									
2050	Lightweight, 20" x 8' wide, 40' span	C-11	20	3.600	Ea.	2,775	175	91.50	3,041.50	3,450
2100	24" x 8' wide, 50' span		18	4		3,475	194	102	3,771	4,275
2150	32" x 10' wide, 60' span		16	4.500		5,200	218	114	5,532	6,225
2200	Standard weight, 12" x 8' wide, 30' span		22	3.273		1,900	159	83	2,142	2,450
2250	16" x 8' wide, 30' span		20	3.600		1,975	175	91.50	2,241.50	2,575
2300	18" x 8' wide, 30' span		20	3.600		2,075	175	91.50	2,341.50	2,700
2350	20" x 8' wide, 40' span		18	4		2,525	194	102	2,821	3,225
2400	24" x 8' wide, 50' span		16	4.500		3,150	218	114	3,482	3,975
2450	32" x 10' wide, 60' span	↓	14	5.143	↓	4,725	249	131	5,105	5,775

03 41 36 – Precast Structural Post-Tensioned Concrete

03 41 36.50 Post-Tensioned Jobs

			Crew	Daily Output	Labor-Hours	Unit	Material	2012 Bare Costs Labor	2012 Bare Costs Equipment	Total	Total Incl O&P
0010	**POST-TENSIONED JOBS**	R034105-30									
0100	Post-tensioned in place, small job	R034136-90	C-17B	8.50	9.647	C.Y.	1,150	440	47	1,637	2,000
0200	Large job		"	10	8.200	"	870	375	40	1,285	1,575

03 45 Precast Architectural Concrete

03 45 13 – Faced Architectural Precast Concrete

03 45 13.50 Precast Wall Panels

			Crew	Daily Output	Labor-Hours	Unit	Material	2012 Bare Costs Labor	2012 Bare Costs Equipment	Total	Total Incl O&P
0010	**PRECAST WALL PANELS**	R034513-10									
0050	Uninsulated 4" thick, smooth gray										
0150	Low rise, 4' x 8' x 4" thick		C-11	320	.225	S.F.	21.50	10.90	5.70	38.10	48.50
0210	8' x 8', 4" thick			576	.125		21	6.05	3.18	30.23	37.50
0250	8' x 16' x 4" thick			1024	.070		21	3.41	1.79	26.20	31
0600	High rise, 4' x 8' x 4" thick			288	.250		21.50	12.15	6.35	40	51.50
0650	8' x 8' x 4" thick			512	.141		21	6.80	3.57	31.37	39
0700	8' x 16' x 4" thick			768	.094		21	4.55	2.38	27.93	33.50
0750	20' x 10', 6" thick, smooth gray		↓	1400	.051		35.50	2.49	1.31	39.30	45
0800	Insulated panel, 2" polystyrene, add						1			1	1.10
0850	2" urethane, add						.78			.78	.86
1200	Finishes, white, add						2.87			2.87	3.15
1250	Exposed aggregate, add						1.88			1.88	2.07
1300	Granite faced, domestic, add						29.50			29.50	32.50
1350	Brick faced, modular, red, add					↓	8.75			8.75	9.65
2200	Fiberglass reinforced cement with urethane core										
2210	R20, 8' x 8', minimum		E-2	750	.075	S.F.	22	3.60	2	27.60	32.50
2220	Maximum		"	600	.093	"	33	4.50	2.49	39.99	46.50

03 47 Site-Cast Concrete

03 47 13 – Tilt-Up Concrete

03 47 13.50 Tilt-Up Wall Panels

		Crew	Daily Output	Labor-Hours	Unit	Material	2012 Bare Costs Labor	Equipment	Total	Total Incl O&P
0010	**TILT-UP WALL PANELS** R034713-20									
0015	Wall panel construction, walls only, 5-1/2" thick	C-14	1600	.090	S.F.	5.25	3.89	.97	10.11	12.85
0100	7-1/2" thick		1550	.093		6.55	4.01	1	11.56	14.50
0500	Walls and columns, 5-1/2" thick walls, 12" x 12" columns		1565	.092		7.90	3.97	.99	12.86	15.95
0550	7-1/2" thick wall, 12" x 12" columns		1370	.105		9.70	4.54	1.13	15.37	18.90
0800	Columns only, site precast, minimum		200	.720	L.F.	13.15	31	7.75	51.90	71
0850	Maximum		105	1.371	"	16.45	59	14.80	90.25	126

03 48 Precast Concrete Specialties

03 48 43 – Precast Concrete Trim

03 48 43.40 Precast Lintels

		Crew	Daily Output	Labor-Hours	Unit	Material	2012 Bare Costs Labor	Equipment	Total	Total Incl O&P
0010	**PRECAST LINTELS**, smooth gray, prestressed, stock units only									
0800	4" wide, 8" high, x 4' long	D-10	28	1.143	Ea.	22.50	49.50	16	88	118
0850	8' long		24	1.333		57.50	58	18.70	134.20	172
1000	6" wide, 8" high, x 4' long		26	1.231		33	53.50	17.25	103.75	135
1050	10' long		22	1.455		88	63	20.50	171.50	215
1200	8" wide, 8" high, x 4' long		24	1.333		41	58	18.70	117.70	154
1250	12' long		20	1.600		134	69.50	22.50	226	278
1275	For custom sizes, types, colors, or finishes of precast lintels, add					150%				

03 48 43.90 Precast Window Sills

		Crew	Daily Output	Labor-Hours	Unit	Material	2012 Bare Costs Labor	Equipment	Total	Total Incl O&P
0010	**PRECAST WINDOW SILLS**									
0600	Precast concrete, 4" tapers to 3", 9" wide	D-1	70	.229	L.F.	12.25	9.10		21.35	27.50
0650	11" wide		60	.267		16	10.65		26.65	33.50
0700	13" wide, 3 1/2" tapers to 2 1/2", 12" wall		50	.320		16	12.75		28.75	37

03 51 Cast Roof Decks

03 51 13 – Cementitious Wood Fiber Decks

03 51 13.50 Cementitious/Wood Fiber Planks

		Crew	Daily Output	Labor-Hours	Unit	Material	2012 Bare Costs Labor	Equipment	Total	Total Incl O&P
0010	**CEMENTITIOUS/WOOD FIBER PLANKS** R051223-50									
0050	Plank, beveled edge, 1" thick	2 Carp	1000	.016	S.F.	2.39	.71		3.10	3.72
0100	1-1/2" thick		975	.016		3.15	.72		3.87	4.58
0150	T & G, 2" thick		950	.017		3.60	.74		4.34	5.10
0200	2-1/2" thick		925	.017		4.22	.76		4.98	5.80
0250	3" thick		900	.018		4.85	.78		5.63	6.55
1000	Bulb tee, sub-purlin and grout, 6' span, add	E-1	5000	.005		2.22	.23	.02	2.47	2.87
1100	8' span	"	4200	.006		2.50	.28	.03	2.81	3.25

03 51 16 – Gypsum Concrete Roof Decks

03 51 16.50 Gypsum Roof Deck

		Crew	Daily Output	Labor-Hours	Unit	Material	2012 Bare Costs Labor	Equipment	Total	Total Incl O&P
0010	**GYPSUM ROOF DECK**									
1000	Poured gypsum, 2" thick	C-8	6000	.009	S.F.	1.34	.36	.12	1.82	2.15
1100	3" thick	"	4800	.012	"	2.01	.46	.15	2.62	3.06

03 52 Lightweight Concrete Roof Insulation

03 52 16 – Lightweight Insulating Concrete

03 52 16.13 Lightweight Cellular Insulating Concrete		Crew	Daily Output	Labor-Hours	Unit	Material	2012 Bare Costs Labor	Equipment	Total	Total Incl O&P
0010	**LIGHTWEIGHT CELLULAR INSULATING CONCRETE** R035216-10									
0020	Portland cement and foaming agent G	C-8	50	1.120	C.Y.	144	44	14.30	202.30	240

03 52 16.16 Lightweight Aggregate Insulating Concrete

		Crew	Daily Output	Labor-Hours	Unit	Material	Labor	Equipment	Total	Total Incl O&P
0010	**LIGHTWEIGHT AGGREGATE INSULATING CONCRETE** R035216-10									
0100	Poured vermiculite or perlite, field mix,									
0110	1:6 field mix G	C-8	50	1.120	C.Y.	246	44	14.30	304.30	350
0200	Ready mix, 1:6 mix, roof fill, 2" thick G		10000	.006	S.F.	1.36	.22	.07	1.65	1.91
0250	3" thick G	↓	7700	.007		2.05	.28	.09	2.42	2.78
0400	Expanded volcanic glass rock, with binder, minimum G	2 Carp	1500	.011		.40	.47		.87	1.16
0450	Maximum G	"	1200	.013	↓	1.19	.59		1.78	2.21

03 54 Cast Underlayment

03 54 13 – Gypsum Cement Underlayment

03 54 13.50 Poured Gypsum Underlayment

		Crew	Daily Output	Labor-Hours	Unit	Material	Labor	Equipment	Total	Total Incl O&P
0010	**POURED GYPSUM UNDERLAYMENT**									
0400	Underlayment, gypsum based, self-leveling 2500 psi, pumped, 1/2"	C-8	24000	.002	S.F.	.33	.09	.03	.45	.54
0500	3/4"		20000	.003		.50	.11	.04	.65	.76
0600	1"	↓	16000	.004		.67	.14	.04	.85	1
1400	Hand placed, 1/2"	C-18	450	.020		.33	.71	.13	1.17	1.60
1500	3/4"	"	300	.030	↓	.50	1.06	.19	1.75	2.39

03 54 16 – Hydraulic Cement Underlayment

03 54 16.50 Cement Underlayment

		Crew	Daily Output	Labor-Hours	Unit	Material	Labor	Equipment	Total	Total Incl O&P
0010	**CEMENT UNDERLAYMENT**									
2510	Underlayment, P.C based self-leveling, 4100 psi, pumped, 1/4"	C-8	20000	.003	S.F.	1.43	.11	.04	1.58	1.78
2520	1/2"		19000	.003		2.85	.12	.04	3.01	3.35
2530	3/4"		18000	.003		4.28	.12	.04	4.44	4.92
2540	1"		17000	.003		5.70	.13	.04	5.87	6.50
2550	1-1/2"	↓	15000	.004		8.55	.15	.05	8.75	9.65
2560	Hand placed, 1/2"	C-18	450	.020		2.85	.71	.13	3.69	4.37
2610	Topping, P.C. based self-level/dry 6100 psi, pumped, 1/4"	C-8	20000	.003		2.16	.11	.04	2.31	2.58
2620	1/2"		19000	.003		4.32	.12	.04	4.48	4.96
2630	3/4"		18000	.003		6.45	.12	.04	6.61	7.30
2660	1"		17000	.003		8.65	.13	.04	8.82	9.75
2670	1-1/2"	↓	15000	.004		12.95	.15	.05	13.15	14.50
2680	Hand placed, 1/2"	C-18	450	.020	↓	4.32	.71	.13	5.16	6

03 62 Non-Shrink Grouting

03 62 13 – Non-Metallic Non-Shrink Grouting

03 62 13.50 Grout, Non-Metallic Non-shrink

		Crew	Daily Output	Labor-Hours	Unit	Material	Labor	Equipment	Total	Total Incl O&P
0010	**GROUT, NON-METALLIC NON-SHRINK**									
0300	Non-shrink, non-metallic, 1" deep	1 Cefi	35	.229	S.F.	6.25	9.70		15.95	21
0350	2" deep	"	25	.320	"	12.50	13.55		26.05	33.50

03 62 16 – Metallic Non-Shrink Grouting

03 62 16.50 Grout, Metallic Non-Shrink

		Crew	Daily Output	Labor-Hours	Unit	Material	Labor	Equipment	Total	Total Incl O&P
0010	**GROUT, METALLIC NON-SHRINK**									
0020	Column & machine bases, non-shrink, metallic, 1" deep	1 Cefi	35	.229	S.F.	10	9.70		19.70	25.50
0050	2" deep	"	25	.320	"	20	13.55		33.55	42

03 63 Epoxy Grouting

03 63 05 - Grouting of Dowels and Fasteners

03 63 05.10 Epoxy Only

	Crew	Daily Output	Labor-Hours	Unit	Material	2012 Bare Costs Labor	Equipment	Total	Total Incl O&P
0010 **EPOXY ONLY**									
1500 Chemical anchoring, epoxy cartridge, excludes layout, drilling, fastener									
1530 For fastener 3/4" diam. x 6" embedment	2 Skwk	72	.222	Ea.	5.10	10.10		15.20	21
1535 1" diam. x 8" embedment		66	.242		7.65	11		18.65	25.50
1540 1-1/4" diam. x 10" embedment		60	.267		15.30	12.10		27.40	35.50
1545 1-3/4" diam. x 12" embedment		54	.296		25.50	13.45		38.95	49
1550 14" embedment		48	.333		30.50	15.15		45.65	57
1555 2" diam. x 12" embedment		42	.381		41	17.30		58.30	71.50
1560 18" embedment		32	.500		51	22.50		73.50	91

03 81 Concrete Cutting

03 81 13 - Flat Concrete Sawing

03 81 13.50 Concrete Floor/Slab Cutting

	Crew	Daily Output	Labor-Hours	Unit	Material	2012 Bare Costs Labor	Equipment	Total	Total Incl O&P
0010 **CONCRETE FLOOR/SLAB CUTTING**									
0300 Saw cut concrete slabs, plain, up to 3" deep	B-89	1060	.015	L.F.	.12	.60	.41	1.13	1.48
0320 Each additional inch of depth		3180	.005		.04	.20	.14	.38	.49
0400 Mesh reinforced, up to 3" deep		980	.016		.14	.65	.44	1.23	1.61
0420 Each additional inch of depth		2940	.005		.05	.22	.15	.42	.54
0500 Rod reinforced, up to 3" deep		800	.020		.17	.79	.54	1.50	1.98
0520 Each additional inch of depth		2400	.007		.06	.26	.18	.50	.66

03 81 16 - Track Mounted Concrete Wall Sawing

03 81 16.50 Concrete Wall Cutting

	Crew	Daily Output	Labor-Hours	Unit	Material	2012 Bare Costs Labor	Equipment	Total	Total Incl O&P
0010 **CONCRETE WALL CUTTING**									
0800 Concrete walls, hydraulic saw, plain, per inch of depth	B-89B	250	.064	L.F.	.44	2.53	2.95	5.92	7.55
0820 Rod reinforcing, per inch of depth	"	150	.107	"	.63	4.22	4.92	9.77	12.50

03 82 Concrete Boring

03 82 13 - Concrete Core Drilling

03 82 13.10 Core Drilling

	Crew	Daily Output	Labor-Hours	Unit	Material	2012 Bare Costs Labor	Equipment	Total	Total Incl O&P
0010 **CORE DRILLING**									
0020 Reinf. conc slab, up to 6" thick, incl. bit, layout & set up									
0100 1" diameter core	B-89A	17	.941	Ea.	.18	38	6.50	44.68	66
0150 Each added inch thick in same hole, add		1440	.011		.03	.45	.08	.56	.80
0200 2" diameter core		16.50	.970		.29	39	6.70	45.99	67.50
0250 Each added inch thick in same hole, add		1080	.015		.05	.60	.10	.75	1.08
0300 3" diameter core		16	1		.41	40.50	6.90	47.81	70
0350 Each added inch thick in same hole, add		720	.022		.07	.89	.15	1.11	1.62
0500 4" diameter core		15	1.067		.49	43	7.35	50.84	74.50
0550 Each added inch thick in same hole, add		480	.033		.49	1.34	.23	2.06	2.86
0700 6" diameter core		14	1.143		.81	46	7.85	54.66	80.50
0750 Each added inch thick in same hole, add		360	.044		.13	1.79	.31	2.23	3.25
0900 8" diameter core		13	1.231		1.03	49.50	8.50	59.03	87
0950 Each added inch thick in same hole, add		288	.056		.17	2.24	.38	2.79	4.06
1100 10" diameter core		12	1.333		1.50	53.50	9.20	64.20	95
1150 Each added inch thick in same hole, add		240	.067		.25	2.68	.46	3.39	4.92
1300 12" diameter core		11	1.455		1.65	58.50	10	70.15	103
1350 Each added inch thick in same hole, add		206	.078		.27	3.13	.54	3.94	5.70
1500 14" diameter core		10	1.600		2.07	64.50	11	77.57	114
1550 Each added inch thick in same hole, add		180	.089		.34	3.58	.61	4.53	6.55

03 82 Concrete Boring

03 82 13 – Concrete Core Drilling

03 82 13.10 Core Drilling

		Crew	Daily Output	Labor-Hours	Unit	Material	2012 Bare Costs Labor	Equipment	Total	Total Incl O&P
1700	18" diameter core	B-89A	9	1.778	Ea.	2.75	71.50	12.25	86.50	127
1750	Each added inch thick in same hole, add		144	.111		.46	4.47	.77	5.70	8.25
1754	24" diameter core		8	2		3.89	80.50	13.80	98.19	143
1756	Each added inch thick in same hole, add		120	.133		.65	5.35	.92	6.92	10
1760	For horizontal holes, add to above						20%	20%		
1770	Prestressed hollow core plank, 8" thick									
1780	1" diameter core	B-89A	17.50	.914	Ea.	.25	37	6.30	43.55	64
1790	Each added inch thick in same hole, add		3840	.004		.03	.17	.03	.23	.32
1794	2" diameter core		17.25	.928		.39	37.50	6.40	44.29	65
1796	Each added inch thick in same hole, add		2880	.006		.05	.22	.04	.31	.44
1800	3" diameter core		17	.941		.54	38	6.50	45.04	66.50
1810	Each added inch thick in same hole, add		1920	.008		.07	.34	.06	.47	.65
1820	4" diameter core		16.50	.970		.65	39	6.70	46.35	68
1830	Each added inch thick in same hole, add		1280	.013		.08	.50	.09	.67	.96
1840	6" diameter core		15.50	1.032		1.08	41.50	7.10	49.68	73
1850	Each added inch thick in same hole, add		960	.017		.13	.67	.11	.91	1.32
1860	8" diameter core		15	1.067		1.37	43	7.35	51.72	75.50
1870	Each added inch thick in same hole, add		768	.021		.17	.84	.14	1.15	1.64
1880	10" diameter core		14	1.143		2	46	7.85	55.85	82
1890	Each added inch thick in same hole, add		640	.025		.25	1.01	.17	1.43	2.01
1900	12" diameter core		13.50	1.185		2.20	47.50	8.15	57.85	85
1910	Each added inch thick in same hole, add		548	.029		.27	1.18	.20	1.65	2.33
3010	Bits for core drill, diamond, premium, 1" diameter					73.50			73.50	81
3020	2" diameter					116			116	128
3030	3" diameter					163			163	179
3040	4" diameter					195			195	215
3060	6" diameter					325			325	355
3080	8" diameter					410			410	455
3110	10" diameter					600			600	660
3120	12" diameter					660			660	725
3140	14" diameter					825			825	910
3180	18" diameter					1,100			1,100	1,200
3240	24" diameter					1,550			1,550	1,700

03 82 16 – Concrete Drilling

03 82 16.10 Concrete Impact Drilling

		Crew	Daily Output	Labor-Hours	Unit	Material	2012 Bare Costs Labor	Equipment	Total	Total Incl O&P
0010	**CONCRETE IMPACT DRILLING**									
0050	Up to 4" deep in conc/brick floor/wall, incl. bit & layout, no anchor									
0100	Holes, 1/4" diameter	1 Carp	75	.107	Ea.	.05	4.70		4.75	7.30
0150	For each additional inch of depth, add		430	.019		.01	.82		.83	1.27
0200	3/8" diameter		63	.127		.05	5.60		5.65	8.65
0250	For each additional inch of depth, add		340	.024		.01	1.04		1.05	1.61
0300	1/2" diameter		50	.160		.05	7.05		7.10	10.90
0350	For each additional inch of depth, add		250	.032		.01	1.41		1.42	2.18
0400	5/8" diameter		48	.167		.08	7.35		7.43	11.40
0450	For each additional inch of depth, add		240	.033		.02	1.47		1.49	2.28
0500	3/4" diameter		45	.178		.10	7.85		7.95	12.15
0550	For each additional inch of depth, add		220	.036		.03	1.60		1.63	2.50
0600	7/8" diameter		43	.186		.13	8.20		8.33	12.75
0650	For each additional inch of depth, add		210	.038		.03	1.68		1.71	2.63
0700	1" diameter		40	.200		.15	8.80		8.95	13.70
0750	For each additional inch of depth, add		190	.042		.04	1.86		1.90	2.90
0800	1-1/4" diameter		38	.211		.23	9.30		9.53	14.55

03 82 Concrete Boring

03 82 16 – Concrete Drilling

03 82 16.10 Concrete Impact Drilling	Crew	Daily Output	Labor-Hours	Unit	Material	2012 Bare Costs Labor	Equipment	Total	Total Incl O&P
0850	For each additional inch of depth, add	1 Carp	180	.044	Ea.	.06	1.96		2.02	3.08
0900	1-1/2" diameter		35	.229		.35	10.10		10.45	15.90
0950	For each additional inch of depth, add		165	.048		.09	2.14		2.23	3.39
1000	For ceiling installations, add						40%			

Division Notes

		CREW	DAILY OUTPUT	LABOR-HOURS	UNIT	2012 BARE COSTS				TOTAL INCL O&P
						MAT.	LABOR	EQUIP.	TOTAL	

Estimating Tips

04 05 00 Common Work Results for Masonry

- The terms *mortar* and *grout* are often used interchangeably, and incorrectly. Mortar is used to bed masonry units, seal the entry of air and moisture, provide architectural appearance, and allow for size variations in the units. Grout is used primarily in reinforced masonry construction and is used to bond the masonry to the reinforcing steel. Common mortar types are M(2500 psi), S(1800 psi), N(750 psi), and O(350 psi), and conform to ASTM C270. Grout is either fine or coarse and conforms to ASTM C476, and in-place strengths generally exceed 2500 psi. Mortar and grout are different components of masonry construction and are placed by entirely different methods. An estimator should be aware of their unique uses and costs.

- Mortar is included in all assembled masonry line items. The mortar cost, part of the assembled masonry material cost, includes all ingredients, all labor, and all equipment required. Please see reference number R040513-10.

- Waste, specifically the loss/droppings of mortar and the breakage of brick and block, is included in all masonry assemblies in this division. A factor of 25% is added for mortar and 3% for brick and concrete masonry units.

- Scaffolding or staging is not included in any of the Division 4 costs. Refer to Subdivision 01 54 23 for scaffolding and staging costs.

04 20 00 Unit Masonry

- The most common types of unit masonry are brick and concrete masonry. The major classifications of brick are building brick (ASTM C62), facing brick (ASTM C216), glazed brick, fire brick, and pavers. Many varieties of texture and appearance can exist within these classifications, and the estimator would be wise to check local custom and availability within the project area. For repair and remodeling jobs, matching the existing brick may be the most important criteria.

- Brick and concrete block are priced by the piece and then converted into a price per square foot of wall. Openings less than two square feet are generally ignored by the estimator because any savings in units used is offset by the cutting and trimming required.

- It is often difficult and expensive to find and purchase small lots of historic brick. Costs can vary widely. Many design issues affect costs, selection of mortar mix, and repairs or replacement of masonry materials. Cleaning techniques must be reflected in the estimate.

- All masonry walls, whether interior or exterior, require bracing. The cost of bracing walls during construction should be included by the estimator, and this bracing must remain in place until permanent bracing is complete. Permanent bracing of masonry walls is accomplished by masonry itself, in the form of pilasters or abutting wall corners, or by anchoring the walls to the structural frame. Accessories in the form of anchors, anchor slots, and ties are used, but their supply and installation can be by different trades. For instance, anchor slots on spandrel beams and columns are supplied and welded in place by the steel fabricator, but the ties from the slots into the masonry are installed by the bricklayer. Regardless of the installation method, the estimator must be certain that these accessories are accounted for in pricing.

Reference Numbers

Reference numbers are shown in shaded boxes at the beginning of some major classifications. These numbers refer to related items in the Reference Section. The reference information may be an estimating procedure, an alternate pricing method, or technical information.

Note: Not all subdivisions listed here necessarily appear in this publication.

04 01 Maintenance of Masonry

04 01 20 – Maintenance of Unit Masonry

04 01 20.20 Pointing Masonry

04 01 20.20 Pointing Masonry	Crew	Daily Output	Labor-Hours	Unit	Material	2012 Bare Costs Labor	2012 Bare Costs Equipment	Total	Total Incl O&P
0010 **POINTING MASONRY**									
0300 Cut and repoint brick, hard mortar, running bond	1 Bric	80	.100	S.F.	.53	4.41		4.94	7.25
0320 Common bond		77	.104		.53	4.58		5.11	7.50
0360 Flemish bond		70	.114		.56	5.05		5.61	8.20
0400 English bond		65	.123		.56	5.45		6.01	8.80
0600 Soft old mortar, running bond		100	.080		.53	3.53		4.06	5.95
0620 Common bond		96	.083		.53	3.67		4.20	6.15
0640 Flemish bond		90	.089		.56	3.92		4.48	6.50
0680 English bond		82	.098		.56	4.30		4.86	7.10
0700 Stonework, hard mortar		140	.057	L.F.	.70	2.52		3.22	4.58
0720 Soft old mortar		160	.050	"	.70	2.21		2.91	4.10
1000 Repoint, mask and grout method, running bond		95	.084	S.F.	.70	3.71		4.41	6.35
1020 Common bond		90	.089		.70	3.92		4.62	6.65
1040 Flemish bond		86	.093		.74	4.10		4.84	7
1060 English bond		77	.104		.74	4.58		5.32	7.70
2000 Scrub coat, sand grout on walls, minimum		120	.067		2.85	2.94		5.79	7.60
2020 Maximum		98	.082		3.96	3.60		7.56	9.80

04 01 20.30 Pointing CMU

04 01 20.30 Pointing CMU	Crew	Daily Output	Labor-Hours	Unit	Material	2012 Bare Costs Labor	2012 Bare Costs Equipment	Total	Total Incl O&P
0010 **POINTING CMU**									
0300 Cut and repoint block, hard mortar, running bond	1 Bric	190	.042	S.F.	.22	1.86		2.08	3.04
0310 Stacked bond		200	.040		.22	1.76		1.98	2.90
0600 Soft old mortar, running bond		230	.035		.22	1.53		1.75	2.56
0610 Stacked bond		245	.033		.22	1.44		1.66	2.41

04 01 20.40 Sawing Masonry

04 01 20.40 Sawing Masonry	Crew	Daily Output	Labor-Hours	Unit	Material	2012 Bare Costs Labor	2012 Bare Costs Equipment	Total	Total Incl O&P
0010 **SAWING MASONRY**									
0050 Brick or block by hand, per inch depth	A-1	125	.064	L.F.		2.25	.53	2.78	4.04

04 01 30 – Unit Masonry Cleaning

04 01 30.20 Cleaning Masonry

04 01 30.20 Cleaning Masonry	Crew	Daily Output	Labor-Hours	Unit	Material	2012 Bare Costs Labor	2012 Bare Costs Equipment	Total	Total Incl O&P
0010 **CLEANING MASONRY**									
0200 Chemical cleaning, new construction, brush and wash, minimum	D-1	1000	.016	S.F.	.05	.64		.69	1.02
0220 Average		800	.020		.08	.80		.88	1.29
0240 Maximum		600	.027		.10	1.06		1.16	1.73
0260 Light restoration, minimum		800	.020		.11	.80		.91	1.32
0270 Average		400	.040		.17	1.59		1.76	2.59
0280 Maximum		330	.048		.22	1.93		2.15	3.16
0300 Heavy restoration, minimum		600	.027		.10	1.06		1.16	1.72
0310 Average		400	.040		.15	1.59		1.74	2.58
0320 Maximum		250	.064		.20	2.55		2.75	4.07
0400 High pressure water only, minimum	C-29	500	.016			.56	.13	.69	1
0420 Average		375	.021			.75	.17	.92	1.33
0440 Maximum		250	.032			1.12	.25	1.37	2.01
2000 Steam cleaning, minimum	A-1H	750	.011			.37	.09	.46	.68
2020 Average		625	.013			.45	.11	.56	.81
2040 Maximum		375	.021			.75	.18	.93	1.35
4000 Add for masking doors and windows	1 Clab	800	.010		.05	.35		.40	.60
4200 Add for pedestrian protection				Job					10%

04 01 30.40 Masonry Building Cleaning

04 01 30.40 Masonry Building Cleaning	Crew	Daily Output	Labor-Hours	Unit	Material	2012 Bare Costs Labor	2012 Bare Costs Equipment	Total	Total Incl O&P
0010 **MASONRY BUILDING CLEANING**									
0020 Steam clean building, not incl scaffolding, minimum	A-1H	750	.011	S.F.		.37	.09	.46	.68
0100 Maximum		375	.021			.75	.18	.93	1.35
0300 Common face brick		438	.018			.64	.15	.79	1.16

04 01 Maintenance of Masonry

04 01 30 – Unit Masonry Cleaning

04 01 30.40 Masonry Building Cleaning

		Crew	Daily Output	Labor-Hours	Unit	Material	2012 Bare Costs Labor	2012 Bare Costs Equipment	Total	Total Incl O&P
0400	Wire cut face brick	A-1H	312	.026	S.F.		.90	.21	1.11	1.62

04 01 30.60 Brick Washing

			Crew	Daily Output	Labor-Hours	Unit	Material	Labor	Equipment	Total	Total Incl O&P
0010	**BRICK WASHING**	R040130-10									
0012	Acid cleanser, smooth brick surface		1 Bric	560	.014	S.F.	.04	.63		.67	1
0050	Rough brick			400	.020		.06	.88		.94	1.40
0060	Stone, acid wash			600	.013		.07	.59		.66	.97
1000	Muriatic acid, price per gallon in 5 gallon lots					Gal.	9			9	9.90

04 05 Common Work Results for Masonry

04 05 05 – Selective Masonry Demolition

04 05 05.10 Selective Demolition

			Crew	Daily Output	Labor-Hours	Unit	Material	Labor	Equipment	Total	Total Incl O&P
0010	**SELECTIVE DEMOLITION**	R024119-10									
0200	Bond beams, 8" block with #4 bar		2 Clab	32	.500	L.F.		17.55		17.55	27
0300	Concrete block walls, unreinforced, 2" thick			1200	.013	S.F.		.47		.47	.72
0310	4" thick			1150	.014			.49		.49	.75
0320	6" thick			1100	.015			.51		.51	.79
0330	8" thick			1050	.015			.53		.53	.82
0340	10" thick			1000	.016			.56		.56	.86
0360	12" thick			950	.017			.59		.59	.91
0380	Reinforced alternate courses, 2" thick			1130	.014			.50		.50	.76
0390	4" thick			1080	.015			.52		.52	.80
0400	6" thick			1035	.015			.54		.54	.83
0410	8" thick			990	.016			.57		.57	.87
0420	10" thick			940	.017			.60		.60	.92
0430	12" thick			890	.018			.63		.63	.97
0440	Reinforced alternate courses & vertically 48" OC, 4" thick			900	.018			.62		.62	.96
0450	6" thick			850	.019			.66		.66	1.02
0460	8" thick			800	.020			.70		.70	1.08
0480	10" thick			750	.021			.75		.75	1.15
0490	12" thick			700	.023			.80		.80	1.23
1000	Chimney, 16" x 16", soft old mortar		1 Clab	55	.145	C.F.		5.10		5.10	7.85
1020	Hard mortar			40	.200			7		7	10.80
1030	16" x 20", soft old mortar			55	.145			5.10		5.10	7.85
1040	Hard mortar			40	.200			7		7	10.80
1050	16" x 24", soft old mortar			55	.145			5.10		5.10	7.85
1060	Hard mortar			40	.200			7		7	10.80
1080	20" x 20", soft old mortar			55	.145			5.10		5.10	7.85
1100	Hard mortar			40	.200			7		7	10.80
1110	20" x 24", soft old mortar			55	.145			5.10		5.10	7.85
1120	Hard mortar			40	.200			7		7	10.80
1140	20" x 32", soft old mortar			55	.145			5.10		5.10	7.85
1160	Hard mortar			40	.200			7		7	10.80
1200	48" x 48", soft old mortar			55	.145			5.10		5.10	7.85
1220	Hard mortar			40	.200			7		7	10.80
1250	Metal, high temp steel jacket, 24" diameter		E-2	130	.431	V.L.F.		21	11.50	32.50	48
1260	60" diameter		"	60	.933			45	25	70	105
1280	Flue lining, up to 12" x 12"		1 Clab	200	.040			1.40		1.40	2.16
1282	Up to 24" x 24"			150	.053			1.87		1.87	2.88
2000	Columns, 8" x 8", soft old mortar			48	.167			5.85		5.85	9
2020	Hard mortar			40	.200			7		7	10.80
2060	16" x 16", soft old mortar			16	.500			17.55		17.55	27

04 05 05 – Selective Masonry Demolition

04 05 05.10 Selective Demolition	Crew	Daily Output	Labor-Hours	Unit	Material	2012 Bare Costs Labor	Equipment	Total	Total Incl O&P
2100	Hard mortar	1 Clab	14	.571	V.L.F.		20		20	31
2140	24" x 24", soft old mortar		8	1			35		35	54
2160	Hard mortar		6	1.333			47		47	72
2200	36" x 36", soft old mortar		4	2			70		70	108
2220	Hard mortar		3	2.667			93.50		93.50	144
2230	Alternate pricing method, soft old mortar		30	.267	C.F.		9.35		9.35	14.40
2240	Hard mortar		23	.348	"		12.20		12.20	18.80
3000	Copings, precast or masonry, to 8" wide									
3020	Soft old mortar	1 Clab	180	.044	L.F.		1.56		1.56	2.40
3040	Hard mortar	"	160	.050	"		1.76		1.76	2.70
3100	To 12" wide									
3120	Soft old mortar	1 Clab	160	.050	L.F.		1.76		1.76	2.70
3140	Hard mortar	"	140	.057	"		2.01		2.01	3.09
4000	Fireplace, brick, 30" x 24" opening									
4020	Soft old mortar	1 Clab	2	4	Ea.		140		140	216
4040	Hard mortar		1.25	6.400			225		225	345
4100	Stone, soft old mortar		1.50	5.333			187		187	288
4120	Hard mortar		1	8			281		281	430
5000	Veneers, brick, soft old mortar		140	.057	S.F.		2.01		2.01	3.09
5020	Hard mortar		125	.064			2.25		2.25	3.46
5100	Granite and marble, 2" thick		180	.044			1.56		1.56	2.40
5120	4" thick		170	.047			1.65		1.65	2.54
5140	Stone, 4" thick		180	.044			1.56		1.56	2.40
5160	8" thick		175	.046			1.60		1.60	2.47
5400	Alternate pricing method, stone, 4" thick		60	.133	C.F.		4.68		4.68	7.20
5420	8" thick		85	.094	"		3.30		3.30	5.10

04 05 13 – Masonry Mortaring

04 05 13.10 Cement

		Crew	Daily Output	Labor-Hours	Unit	Material	Labor	Equipment	Total	Total Incl O&P
0010	**CEMENT**									
0100	Masonry, 70 lb. bag, T.L. lots				Bag	9.85			9.85	10.85
0150	L.T.L. lots					10.45			10.45	11.50
0200	White, 70 lb. bag, T.L. lots					15.55			15.55	17.15
0250	L.T.L. lots					17.75			17.75	19.55

04 05 13.20 Lime

		Crew	Daily Output	Labor-Hours	Unit	Material	Labor	Equipment	Total	Total Incl O&P
0010	**LIME**									
0020	Masons, hydrated, 50 lb. bag, T.L. lots				Bag	9.10			9.10	10
0050	L.T.L. lots					10			10	11
0200	Finish, double hydrated, 50 lb. bag, T.L. lots					9.90			9.90	10.90
0250	L.T.L. lots					10.90			10.90	12

04 05 13.23 Surface Bonding Masonry Mortaring

		Crew	Daily Output	Labor-Hours	Unit	Material	Labor	Equipment	Total	Total Incl O&P
0010	**SURFACE BONDING MASONRY MORTARING**									
0020	Gray or white colors, not incl. block work	1 Bric	540	.015	S.F.	.12	.65		.77	1.13

04 05 13.30 Mortar

		Crew	Daily Output	Labor-Hours	Unit	Material	Labor	Equipment	Total	Total Incl O&P
0010	**MORTAR** R040513-10									
0020	With masonry cement									
0100	Type M, 1:1:6 mix	1 Brhe	143	.056	C.F.	5.15	1.99		7.14	8.65
0200	Type N, 1:3 mix	"	143	.056	"	4.50	1.99		6.49	7.95
2000	With portland cement and lime									
2100	Type M, 1:1/4:3 mix	1 Brhe	143	.056	C.F.	8.80	1.99		10.79	12.70
2200	Type N, 1:1:6 mix, 750 psi		143	.056		6.80	1.99		8.79	10.45
2300	Type O, 1:2:9 mix (Pointing Mortar)		143	.056		7.50	1.99		9.49	11.25

04 05 13 – Masonry Mortaring

04 05 13.30 Mortar

		Crew	Daily Output	Labor-Hours	Unit	Material	2012 Bare Costs Labor	2012 Bare Costs Equipment	Total	Total Incl O&P
2650	Pre-mixed, type S or N				C.F.	4.94			4.94	5.45
2700	Mortar for glass block	1 Brhe	143	.056	↓	10.35	1.99		12.34	14.40
2900	Mortar for Fire Brick, 80 lb. bag, T.L. Lots				Bag	25.50			25.50	28

04 05 13.91 Masonry Restoration Mortaring

		Crew	Daily Output	Labor-Hours	Unit	Material	2012 Bare Costs Labor	2012 Bare Costs Equipment	Total	Total Incl O&P
0010	**MASONRY RESTORATION MORTARING**									
0020	Masonry restoration mix				Lb.	1.97			1.97	2.17
0050	White				"	2.27			2.27	2.50

04 05 13.93 Mortar Pigments

		Crew	Daily Output	Labor-Hours	Unit	Material	2012 Bare Costs Labor	2012 Bare Costs Equipment	Total	Total Incl O&P
0010	**MORTAR PIGMENTS**, 50 lb. bags (2 bags per M bricks) R040513-10									
0020	Color admixture, range 2 to 10 lb. per bag of cement, minimum				Lb.	3.70			3.70	4.07
0050	Average					5.70			5.70	6.25
0100	Maximum				↓	12.20			12.20	13.40

04 05 13.95 Sand

		Crew	Daily Output	Labor-Hours	Unit	Material	2012 Bare Costs Labor	2012 Bare Costs Equipment	Total	Total Incl O&P
0010	**SAND**, screened and washed at pit									
0020	For mortar, per ton				Ton	16.60			16.60	18.25
0050	With 10 mile haul					27			27	29.50
0100	With 30 mile haul				↓	36			36	39.50
0200	Screened and washed, at the pit				C.Y.	23			23	25.50
0250	With 10 mile haul					37.50			37.50	41
0300	With 30 mile haul				↓	50			50	55

04 05 13.98 Mortar Admixtures

		Crew	Daily Output	Labor-Hours	Unit	Material	2012 Bare Costs Labor	2012 Bare Costs Equipment	Total	Total Incl O&P
0010	**MORTAR ADMIXTURES**									
0020	Waterproofing admixture, per quart (1 qt. to 2 bags of masonry cement)				Qt.	9.50			9.50	10.45

04 05 16 – Masonry Grouting

04 05 16.30 Grouting

		Crew	Daily Output	Labor-Hours	Unit	Material	2012 Bare Costs Labor	2012 Bare Costs Equipment	Total	Total Incl O&P
0010	**GROUTING** R040513-10									
0011	Bond beams & lintels, 8" deep, 6" thick, 0.15 C.F. per L.F.	D-4	1480	.022	L.F.	.64	.87	.09	1.60	2.10
0020	8" thick, 0.2 C.F. per L.F.		1400	.023		.99	.91	.09	1.99	2.56
0050	10" thick, 0.25 C.F. per L.F.		1200	.027		1.06	1.07	.11	2.24	2.89
0060	12" thick, 0.3 C.F. per L.F.	↓	1040	.031	↓	1.27	1.23	.12	2.62	3.39
0200	Concrete block cores, solid, 4" thk., by hand, 0.067 C.F./S.F. of wall	D-8	1100	.036	S.F.	.28	1.48		1.76	2.54
0210	6" thick, pumped, 0.175 C.F. per S.F.	D-4	720	.044		.74	1.78	.18	2.70	3.68
0250	8" thick, pumped, 0.258 C.F. per S.F.		680	.047		1.09	1.88	.19	3.16	4.25
0300	10" thick, pumped, 0.340 C.F. per S.F.		660	.048		1.44	1.94	.19	3.57	4.72
0350	12" thick, pumped, 0.422 C.F. per S.F.		640	.050		1.79	2.	.20	3.99	5.20
0500	Cavity walls, 2" space, pumped, 0.167 C.F./S.F. of wall		1700	.019		.71	.75	.07	1.53	2
0550	3" space, 0.250 C.F./S.F.		1200	.027		1.06	1.07	.11	2.24	2.89
0600	4" space, 0.333 C.F. per S.F.		1150	.028		1.41	1.11	.11	2.63	3.35
0700	6" space, 0.500 C.F. per S.F.		800	.040	↓	2.12	1.60	.16	3.88	4.93
0800	Door frames, 3' x 7' opening, 2.5 C.F. per opening		60	.533	Opng.	10.60	21.50	2.12	34.22	46
0850	6' x 7' opening, 3.5 C.F. per opening		45	.711	"	14.80	28.50	2.83	46.13	62.50
2000	Grout, C476, for bond beams, lintels and CMU cores	↓	350	.091	C.F.	4.23	3.66	.36	8.25	10.55

04 05 19 – Masonry Anchorage and Reinforcing

04 05 19.05 Anchor Bolts

		Crew	Daily Output	Labor-Hours	Unit	Material	2012 Bare Costs Labor	2012 Bare Costs Equipment	Total	Total Incl O&P
0010	**ANCHOR BOLTS**									
0015	Installed in fresh grout in CMU bond beams or filled cores, no templates									
0020	Hooked, with nut and washer, 1/2" diam., 8" long	1 Bric	132	.061	Ea.	1.39	2.67		4.06	5.55
0030	12" long		131	.061		1.54	2.69		4.23	5.75
0040	5/8" diameter, 8" long		129	.062		3	2.74		5.74	7.45
0050	12" long	↓	127	.063	↓	3.70	2.78		6.48	8.25

04 05 19 – Masonry Anchorage and Reinforcing

04 05 19.05 Anchor Bolts		Crew	Daily Output	Labor-Hours	Unit	Material	2012 Bare Costs Labor	Equipment	Total	Total Incl O&P
0060	3/4" diameter, 8" long	1 Bric	127	.063	Ea.	3.70	2.78		6.48	8.25
0070	12" long	↓	125	.064	↓	4.62	2.82		7.44	9.35

04 05 19.16 Masonry Anchors

0010	**MASONRY ANCHORS**									
0020	For brick veneer, galv., corrugated, 7/8" x 7", 22 Ga.	1 Bric	10.50	.762	C	13	33.50		46.50	65
0100	24 Ga.		10.50	.762		9.10	33.50		42.60	60.50
0150	16 Ga.		10.50	.762		25.50	33.50		59	78.50
0200	Buck anchors, galv., corrugated, 16 gauge, 2" bend, 8" x 2"		10.50	.762		49	33.50		82.50	105
0250	8" x 3"		10.50	.762		51	33.50		84.50	107
0660	Cavity wall, Z-type, galvanized, 6" long, 1/8" diam.		10.50	.762		23	33.50		56.50	75.50
0670	3/16" diameter		10.50	.762		21	33.50		54.50	74
0680	1/4" diameter		10.50	.762		38	33.50		71.50	92
0850	8" long, 3/16" diameter		10.50	.762		23.50	33.50		57	76
0855	1/4" diameter		10.50	.762		46	33.50		79.50	101
1000	Rectangular type, galvanized, 1/4" diameter, 2" x 6"		10.50	.762		67.50	33.50		101	125
1050	4" x 6"		10.50	.762		82.50	33.50		116	141
1100	3/16" diameter, 2" x 6"		10.50	.762		38.50	33.50		72	93
1150	4" x 6"		10.50	.762		44	33.50		77.50	99
1500	Rigid partition anchors, plain, 8" long, 1" x 1/8"		10.50	.762		200	33.50		233.50	271
1550	1" x 1/4"		10.50	.762		235	33.50		268.50	310
1580	1-1/2" x 1/8"		10.50	.762		220	33.50		253.50	293
1600	1-1/2" x 1/4"		10.50	.762		275	33.50		308.50	355
1650	2" x 1/8"		10.50	.762		260	33.50		293.50	335
1700	2" x 1/4"	↓	10.50	.762	↓	241	33.50		274.50	315

04 05 19.26 Masonry Reinforcing Bars

0010	**MASONRY REINFORCING BARS**	R040519-50								
0015	Steel bars A615, placed horiz., #3 & #4 bars	1 Bric	450	.018	Lb.	.49	.78		1.27	1.72
0020	#5 & #6 bars		800	.010		.49	.44		.93	1.21
0050	Placed vertical, #3 & #4 bars		350	.023		.49	1.01		1.50	2.06
0060	#5 & #6 bars		650	.012	↓	.49	.54		1.03	1.36
0200	Joint reinforcing, regular truss, to 6" wide, mill std galvanized		30	.267	C.L.F.	19.20	11.75		30.95	39
0250	12" wide		20	.400		22.50	17.65		40.15	51
0400	Cavity truss with drip section, to 6" wide		30	.267		19.10	11.75		30.85	39
0450	12" wide	↓	20	.400	↓	22	17.65		39.65	50.50

04 05 23 – Masonry Accessories

04 05 23.13 Masonry Control and Expansion Joints

0010	**MASONRY CONTROL AND EXPANSION JOINTS**									
0020	Rubber, for double wythe 8" minimum wall (Brick/CMU)	1 Bric	400	.020	L.F.	1.88	.88		2.76	3.40
0025	"T" shaped		320	.025		1.15	1.10		2.25	2.94
0030	Cross-shaped for CMU units		280	.029		1.48	1.26		2.74	3.53
0050	PVC, for double wythe 8" minimum wall (Brick/CMU)		400	.020		1.38	.88		2.26	2.85
0120	"T" shaped		320	.025		.71	1.10		1.81	2.45
0160	Cross-shaped for CMU units	↓	280	.029	↓	.95	1.26		2.21	2.95

04 05 23.19 Masonry Cavity Drainage, Weepholes, and Vents

0010	**MASONRY CAVITY DRAINAGE, WEEPHOLES, AND VENTS**									
0020	Extruded aluminum, 4" deep, 2-3/8" x 8-1/8"	1 Bric	30	.267	Ea.	31.50	11.75		43.25	52.50
0050	5" x 8-1/8"		25	.320		41	14.10		55.10	67
0100	2-1/4" x 25"		25	.320		71.50	14.10		85.60	101
0150	5" x 16-1/2"		22	.364		62.50	16.05		78.55	92.50
0200	6" x 16-1/2"		22	.364		83.50	16.05		99.55	116
0250	7-3/4" x 16-1/2"	↓	20	.400	↓	71	17.65		88.65	105

04 05 Common Work Results for Masonry

04 05 23 – Masonry Accessories

04 05 23.19 Masonry Cavity Drainage, Weepholes, and Vents

		Crew	Daily Output	Labor-Hours	Unit	Material	2012 Bare Costs Labor	Equipment	Total	Total Incl O&P
0400	For baked enamel finish, add				Ea.	35%				
0500	For cast aluminum, painted, add					60%				
1000	Stainless steel ventilators, 6" x 6"	1 Bric	25	.320		126	14.10		140.10	160
1050	8" x 8"		24	.333		132	14.70		146.70	168
1100	12" x 12"		23	.348		153	15.35		168.35	191
1150	12" x 6"		24	.333		134	14.70		148.70	169
1200	Foundation block vent, galv., 1-1/4" thk, 8" high, 16" long, no damper		30	.267		24.50	11.75		36.25	45
1250	For damper, add					8.20			8.20	9

04 05 23.95 Wall Plugs

		Crew	Daily Output	Labor-Hours	Unit	Material	2012 Bare Costs Labor	Equipment	Total	Total Incl O&P
0010	**WALL PLUGS** (for nailing to brickwork)									
0020	26 ga., galvanized, plain	1 Bric	10.50	.762	C	31	33.50		64.50	84.50
0050	Wood filled	"	10.50	.762	"	104	33.50		137.50	165

04 21 Clay Unit Masonry

04 21 13 – Brick Masonry

04 21 13.13 Brick Veneer Masonry

		Crew	Daily Output	Labor-Hours	Unit	Material	2012 Bare Costs Labor	Equipment	Total	Total Incl O&P
0010	**BRICK VENEER MASONRY**, T.L. lots, excl. scaff., grout & reinforcing R042110-20									
0015	Material costs incl. 3% brick and 25% mortar waste									
0020	Standard, select common, 4" x 2-2/3" x 8" (6.75/S.F.)	D-8	1.50	26.667	M	535	1,075		1,610	2,250
0050	Red, 4" x 2-2/3" x 8", running bond		1.50	26.667		620	1,075		1,695	2,325
0100	Full header every 6th course (7.88/S.F.) R042110-50		1.45	27.586		620	1,125		1,745	2,375
0150	English, full header every 2nd course (10.13/S.F.)		1.40	28.571		615	1,175		1,790	2,425
0200	Flemish, alternate header every course (9.00/S.F.)		1.40	28.571		620	1,175		1,795	2,425
0250	Flemish, alt. header every 6th course (7.13/S.F.)		1.45	27.586		620	1,125		1,745	2,375
0300	Full headers throughout (13.50/S.F.)		1.40	28.571		615	1,175		1,790	2,425
0350	Rowlock course (13.50/S.F.)		1.35	29.630		615	1,200		1,815	2,500
0400	Rowlock stretcher (4.50/S.F.)		1.40	28.571		625	1,175		1,800	2,450
0450	Soldier course (6.75/S.F.)		1.40	28.571		620	1,175		1,795	2,425
0500	Sailor course (4.50/S.F.)		1.30	30.769		625	1,250		1,875	2,600
0601	Buff or gray face, running bond, (6.75/S.F.)		1.50	26.667		620	1,075		1,695	2,325
0700	Glazed face, 4" x 2-2/3" x 8", running bond		1.40	28.571		1,775	1,175		2,950	3,725
0750	Full header every 6th course (7.88/S.F.)		1.35	29.630		1,700	1,200		2,900	3,700
1000	Jumbo, 6" x 4" x 12",(3.00/S.F.)		1.30	30.769		1,775	1,250		3,025	3,875
1051	Norman, 4" x 2-2/3" x 12" (4.50/S.F.)		1.45	27.586		1,150	1,125		2,275	2,975
1100	Norwegian, 4" x 3-1/5" x 12" (3.75/S.F.)		1.40	28.571		1,150	1,175		2,325	3,000
1150	Economy, 4" x 4" x 8" (4.50 per S.F.)		1.40	28.571		900	1,175		2,075	2,750
1201	Engineer, 4" x 3-1/5" x 8", (5.63/S.F.)		1.45	27.586		630	1,125		1,755	2,400
1251	Roman, 4" x 2" x 12", (6.00/S.F.)		1.50	26.667		985	1,075		2,060	2,725
1300	S.C.R. 6" x 2-2/3" x 12", (4.50/S.F.)		1.40	28.571		1,175	1,175		2,350	3,050
1350	Utility, 4" x 4" x 12" (3.00/S.F.)		1.35	29.630		1,550	1,200		2,750	3,525
1360	For less than truck load lots, add					15%			15%	
1400	For battered walls, add						30%			
1450	For corbels, add						75%			
1500	For curved walls, add						30%			
1550	For pits and trenches, deduct						20%			
1999	Alternate method of figuring by square foot									
2000	Standard, sel. common, 4" x 2-2/3" x 8", (6.75/S.F.)	D-8	230	.174	S.F.	3.61	7.10		10.71	14.65
2020	Standard, red, 4" x 2-2/3" x 8", running bond (6.75/S.F.)		220	.182		4.19	7.40		11.59	15.75
2050	Full header every 6th course (7.88/S.F.)		185	.216		4.88	8.80		13.68	18.65
2100	English, full header every 2nd course (10.13/S.F.)		140	.286		6.25	11.65		17.90	24.50
2150	Flemish, alternate header every course (9.00/S.F.)		150	.267		5.55	10.85		16.40	22.50

04 21 Clay Unit Masonry

04 21 13 – Brick Masonry

04 21 13.13 Brick Veneer Masonry

		Crew	Daily Output	Labor-Hours	Unit	Material	2012 Bare Costs Labor	Equipment	Total	Total Incl O&P
2200	Flemish, alt. header every 6th course (7.13/S.F.)	D-8	205	.195	S.F.	4.42	7.95		12.37	16.85
2250	Full headers throughout (13.50/S.F.)		105	.381		8.30	15.50		23.80	32.50
2300	Rowlock course (13.50/S.F.)		100	.400		8.30	16.30		24.60	33.50
2350	Rowlock stretcher (4.50/S.F.)		310	.129		2.81	5.25		8.06	11.05
2400	Soldier course (6.75/S.F.)		200	.200		4.19	8.15		12.34	16.90
2450	Sailor course (4.50/S.F.)		290	.138		2.81	5.60		8.41	11.60
2600	Buff or gray face, running bond, (6.75/S.F.)		220	.182		4.43	7.40		11.83	16
2700	Glazed face brick, running bond		210	.190		11.45	7.75		19.20	24.50
2750	Full header every 6th course (7.88/S.F.)		170	.235		13.35	9.60		22.95	29
3000	Jumbo, 6" x 4" x 12" running bond (3.00/S.F.)		435	.092		4.90	3.74		8.64	11.05
3050	Norman, 4" x 2-2/3" x 12" running bond, (4.5/S.F.)		320	.125		5.75	5.10		10.85	14
3100	Norwegian, 4" x 3-1/5" x 12" (3.75/S.F.)		375	.107		4.15	4.34		8.49	11.10
3150	Economy, 4" x 4" x 8" (4.50/S.F.)		310	.129		4.02	5.25		9.27	12.35
3200	Engineer, 4" x 3-1/5" x 8" (5.63/S.F.)		260	.154		3.54	6.25		9.79	13.35
3250	Roman, 4" x 2" x 12" (6.00/S.F.)		250	.160		5.75	6.50		12.25	16.20
3300	SCR, 6" x 2-2/3" x 12" (4.50/S.F.)		310	.129		5.25	5.25		10.50	13.70
3350	Utility, 4" x 4" x 12" (3.00/S.F.)		450	.089		4.51	3.62		8.13	10.40
3360	For less than truck load lots, add				M	15%			15%	
3370	For battered walls, add						30%			
3380	For corbels, add						75%			
3400	For cavity wall construction, add						15%			
3450	For stacked bond, add						10%			
3500	For interior veneer construction, add						15%			
3510	For pits and trenches, deduct						20%			
3550	For curved walls, add						30%			

04 21 13.14 Thin Brick Veneer

		Crew	Daily Output	Labor-Hours	Unit	Material	2012 Bare Costs Labor	Equipment	Total	Total Incl O&P
0010	**THIN BRICK VENEER**									
0015	Material costs incl. 3% brick and 25% mortar waste									
0020	On & incl. metal panel support sys, modular, 2-2/3" x 5/8" x 8", red	D-7	92	.174	S.F.	8.75	6.35		15.10	18.95
0100	Closure, 4" x 5/8" x 8"		110	.145		8.75	5.30		14.05	17.45
0110	Norman, 2-2/3" x 5/8" x 12"		110	.145		8.95	5.30		14.25	17.65
0120	Utility, 4" x 5/8" x 12"		125	.128		8.95	4.66		13.61	16.70
0130	Emperor, 4" x 3/4" x 16"		175	.091		10.15	3.33		13.48	16.10
0140	Super emperor, 8" x 3/4" x 16"		195	.082		10.15	2.99		13.14	15.60
0150	For L shaped corners with 4" return, add				L.F.	9.25			9.25	10.20
0200	On masonry/plaster back-up, modular, 2-2/3" x 5/8" x 8", red	D-7	137	.117	S.F.	3.95	4.25		8.20	10.60
0210	Closure, 4" x 5/8" x 8"		165	.097		3.95	3.53		7.48	9.55
0220	Norman, 2-2/3" x 5/8" x 12"		165	.097		4.15	3.53		7.68	9.75
0230	Utility, 4" x 5/8" x 12"		185	.086		4.15	3.15		7.30	9.20
0240	Emperor, 4" x 3/4" x 16"		260	.062		5.35	2.24		7.59	9.20
0250	Super emperor, 8" x 3/4" x 16"		285	.056		5.35	2.05		7.40	8.90
0260	For L shaped corners with 4" return, add				L.F.	9.25			9.25	10.20
0270	For embedment into pre-cast concrete panels, add				S.F.	14.40			14.40	15.85

04 21 13.15 Chimney

		Crew	Daily Output	Labor-Hours	Unit	Material	2012 Bare Costs Labor	Equipment	Total	Total Incl O&P
0010	**CHIMNEY**, excludes foundation, scaffolding, grout and reinforcing									
0100	Brick, 16" x 16", 8" flue	D-1	18.20	.879	V.L.F.	21.50	35		56.50	76.50
0150	16" x 20" with one 8" x 12" flue		16	1		34	40		74	97.50
0200	16" x 24" with two 8" x 8" flues		14	1.143		50.50	45.50		96	125
0250	20" x 20" with one 12" x 12" flue		13.70	1.168		38.50	46.50		85	113
0300	20" x 24" with two 8" x 12" flues		12	1.333		57	53		110	143
0350	20" x 32" with two 12" x 12" flues		10	1.600		68	64		132	172
1800	Metal, high temp. steel jacket, factory lining, 24" diam.	E-2	65	.862		210	41.50	23	274.50	330

04 21 Clay Unit Masonry

04 21 13 – Brick Masonry

04 21 13.15 Chimney

		Crew	Daily Output	Labor-Hours	Unit	Material	2012 Bare Costs Labor	2012 Bare Costs Equipment	Total	Total Incl O&P
1900	60" diameter	E-2	30	1.867	V.L.F.	765	90	50	905	1,050
2100	Poured concrete, brick lining, 200' high x 10' diam.					7,275			7,275	8,000
2800	500' x 20' diameter					12,700			12,700	14,000

04 21 13.18 Columns

		Crew	Daily Output	Labor-Hours	Unit	Material	Labor	Equipment	Total	Total Incl O&P
0010	**COLUMNS**, solid, excludes scaffolding, grout and reinforcing R042110-10									
0050	Brick, 8" x 8", 9 brick per V.L.F.	D-1	56	.286	V.L.F.	5.45	11.40		16.85	23
0100	12" x 8", 13.5 brick per V.L.F.		37	.432		8.15	17.25		25.40	35
0200	12" x 12", 20 brick per V.L.F.		25	.640		12.10	25.50		37.60	52
0300	16" x 12", 27 brick per V.L.F.		19	.842		16.35	33.50		49.85	68.50
0400	16" x 16", 36 brick per V.L.F.		14	1.143		22	45.50		67.50	93
0500	20" x 16", 45 brick per V.L.F.		11	1.455		27	58		85	118
0600	20" x 20", 56 brick per V.L.F.		9	1.778		34	71		105	145
0700	24" x 20", 68 brick per V.L.F.		7	2.286		41	91		132	184
0800	24" x 24", 81 brick per V.L.F.		6	2.667		49	106		155	214
1000	36" x 36", 182 brick per V.L.F.		3	5.333		110	213		323	440

04 21 13.30 Oversized Brick

		Crew	Daily Output	Labor-Hours	Unit	Material	Labor	Equipment	Total	Total Incl O&P
0010	**OVERSIZED BRICK**, excludes scaffolding, grout and reinforcing									
0100	Veneer, 4" x 2.25" x 16"	D-8	387	.103	S.F.	6.45	4.21		10.66	13.40
0105	4" x 2.75" x 16"		412	.097		5.80	3.95		9.75	12.30
0110	4" x 4" x 16"		460	.087		5.30	3.54		8.84	11.20
0120	4" x 8" x 16"		533	.075		6.10	3.05		9.15	11.35
0125	Loadbearing, 6" x 4" x 16", grouted and reinforced		387	.103		9.05	4.21		13.26	16.30
0130	8" x 4" x 16", grouted and reinforced		327	.122		9.45	4.98		14.43	17.85
0135	6" x 8" x 16", grouted and reinforced		440	.091		8.70	3.70		12.40	15.15
0140	8" x 8" x 16", grouted and reinforced		400	.100		9.55	4.07		13.62	16.65
0145	Curtainwall/reinforced veneer, 6" x 4" x 16"		387	.103		13.20	4.21		17.41	21
0150	8" x 4" x 16"		327	.122		15.70	4.98		20.68	25
0155	6" x 8" x 16"		440	.091		13.30	3.70		17	20.50
0160	8" x 8" x 16"		400	.100		15.80	4.07		19.87	23.50
0200	For 1 to 3 slots in face, add					15%				
0210	For 4 to 7 slots in face, add					25%				
0220	For bond beams, add					20%				
0230	For bullnose shapes, add					20%				
0240	For open end knockout, add					10%				
0250	For white or gray color group, add					10%				
0260	For 135 degree corner, add					250%				

04 21 13.35 Common Building Brick

		Crew	Daily Output	Labor-Hours	Unit	Material	Labor	Equipment	Total	Total Incl O&P
0010	**COMMON BUILDING BRICK**, C62, TL lots, material only R042110-20									
0020	Standard, minimum				M	350			350	385
0050	Average (select)				"	430			430	475

04 21 13.40 Structural Brick

		Crew	Daily Output	Labor-Hours	Unit	Material	Labor	Equipment	Total	Total Incl O&P
0010	**STRUCTURAL BRICK** C652, Grade SW, incl. mortar, scaffolding not incl.									
0100	Standard unit, 4-5/8" x 2-3/4" x 9-5/8"	D-8	245	.163	S.F.	5.15	6.65		11.80	15.75
0120	Bond beam		225	.178		9.65	7.25		16.90	21.50
0140	V cut bond beam		225	.178		10.45	7.25		17.70	22.50
0160	Stretcher quoin, 5-5/8" x 2-3/4" x 9-5/8"		245	.163		10.10	6.65		16.75	21
0180	Corner quoin		245	.163		11.30	6.65		17.95	22.50
0200	Corner, 45 deg, 4-5/8" x 2-3/4" x 10-7/16"		235	.170		11.70	6.95		18.65	23.50

04 21 13.45 Face Brick

		Crew	Daily Output	Labor-Hours	Unit	Material	Labor	Equipment	Total	Total Incl O&P
0010	**FACE BRICK** Material Only, C216, TL lots R042110-20									
0300	Standard modular, 4" x 2-2/3" x 8", minimum				M	515			515	565

04 21 Clay Unit Masonry

04 21 13 – Brick Masonry

04 21 13.45 Face Brick

		Crew	Daily Output	Labor-Hours	Unit	Material	2012 Bare Costs Labor	Equipment	Total	Total Incl O&P
0350	Maximum				M	675			675	745
0450	Economy, 4" x 4" x 8", minimum					775			775	855
0500	Maximum					1,000			1,000	1,100
0510	Economy, 4" x 4" x 12", minimum					1,225			1,225	1,350
0520	Maximum					1,600			1,600	1,750
0550	Jumbo, 6" x 4" x 12", minimum					1,475			1,475	1,625
0600	Maximum					1,925			1,925	2,125
0610	Jumbo, 8" x 4" x 12", minimum					1,475			1,475	1,625
0620	Maximum					1,925			1,925	2,125
0650	Norwegian, 4" x 3-1/5" x 12", minimum					980			980	1,075
0700	Maximum					1,275			1,275	1,400
0710	Norwegian, 6" x 3-1/5" x 12", minimum					1,425			1,425	1,575
0720	Maximum					1,875			1,875	2,050
0850	Standard glazed, plain colors, 4" x 2-2/3" x 8", minimum					1,550			1,550	1,725
0900	Maximum					2,025			2,025	2,225
1000	Deep trim shades, 4" x 2-2/3" x 8", minimum					1,925			1,925	2,125
1050	Maximum					2,150			2,150	2,375
1080	Jumbo utility, 4" x 4" x 12"					1,350			1,350	1,500
1120	4" x 8" x 8"					1,675			1,675	1,850
1140	4" x 8" x 16"					4,950			4,950	5,450
1260	Engineer, 4" x 3-1/5" x 8", minimum					520			520	570
1270	Maximum					700			700	770
1350	King, 4" x 2-3/4" x 10", minimum					460			460	505
1360	Maximum					560			560	620
1400	Norman, 4" x 2-3/4" x 12"					460			460	505
1450	Roman, 4" x 2" x 12"					460			460	505
1500	SCR, 6" x 2-2/3" x 12"					460			460	505
1550	Double, 4" x 5-1/3" x 8"					460			460	505
1600	Triple, 4" x 5-1/3" x 12"					460			460	505
1770	Standard modular, double glazed, 4" x 2-2/3" x 8"					2,275			2,275	2,500
1850	Jumbo, colored glazed ceramic, 6" x 4" x 12"					2,400			2,400	2,650
2050	Jumbo utility, glazed, 4" x 4" x 12"					4,500			4,500	4,950
2100	4" x 8" x 8"					5,300			5,300	5,825
2150	4" x 16" x 8"					6,200			6,200	6,825
2170	For less than truck load lots, add					15			15	16.50
2180	For buff or gray brick, add					16			16	17.60

04 21 26 – Glazed Structural Clay Tile Masonry

04 21 26.10 Structural Facing Tile

		Crew	Daily Output	Labor-Hours	Unit	Material	2012 Bare Costs Labor	Equipment	Total	Total Incl O&P
0010	**STRUCTURAL FACING TILE**, std. colors, excl. scaffolding, grout, reinforcing									
0020	6T series, 5-1/3" x 12", 2.3 pieces per S.F., glazed 1 side, 2" thick	D-8	225	.178	S.F.	8.40	7.25		15.65	20
0100	4" thick		220	.182		11.50	7.40		18.90	24
0150	Glazed 2 sides		195	.205		14.95	8.35		23.30	29
0250	6" thick		210	.190		17.10	7.75		24.85	30.50
0300	Glazed 2 sides		185	.216		20.50	8.80		29.30	36
0400	8" thick		180	.222		22.50	9.05		31.55	38.50
0500	Special shapes, group 1		400	.100	Ea.	7.20	4.07		11.27	14.05
0550	Group 2		375	.107		11.65	4.34		15.99	19.40
0600	Group 3		350	.114		15.05	4.65		19.70	23.50
0650	Group 4		325	.123		31	5		36	42
0700	Group 5		300	.133		37.50	5.45		42.95	49
0750	Group 6		275	.145		51	5.90		56.90	65
1000	Fire rated, 4" thick, 1 hr. rating		210	.190	S.F.	16.40	7.75		24.15	30

04 21 Clay Unit Masonry

04 21 26 – Glazed Structural Clay Tile Masonry

04 21 26.10 Structural Facing Tile		Crew	Daily Output	Labor-Hours	Unit	Material	2012 Bare Costs Labor	Equipment	Total	Total Incl O&P
1300	Acoustic, 4" thick	D-8	210	.190	S.F.	34	7.75		41.75	49
2000	8W series, 8" x 16", 1.125 pieces per S.F.									
2050	2" thick, glazed 1 side	D-8	360	.111	S.F.	9.75	4.52		14.27	17.60
2100	4" thick, glazed 1 side		345	.116		12.95	4.72		17.67	21.50
2150	Glazed 2 sides		325	.123		15	5		20	24
2200	6" thick, glazed 1 side		330	.121		22	4.93		26.93	31.50
2250	8" thick, glazed 1 side		310	.129	▼	22.50	5.25		27.75	33
2500	Special shapes, group 1		300	.133	Ea.	13.50	5.45		18.95	23
2550	Group 2		280	.143		18.05	5.80		23.85	28.50
2600	Group 3		260	.154		19.20	6.25		25.45	30.50
2650	Group 4		250	.160		39	6.50		45.50	52.50
2700	Group 5		240	.167		38	6.80		44.80	52
2750	Group 6		230	.174	▼	81.50	7.10		88.60	101
3000	4" thick, glazed 1 side		345	.116	S.F.	12.35	4.72		17.07	21
3100	Acoustic, 4" thick	▼	345	.116	"	16.05	4.72		20.77	25
3120	4W series, 8" x 8", 2.25 pieces per S.F.									
3125	2" thick, glazed 1 side	D-8	360	.111	S.F.	8.75	4.52		13.27	16.45
3130	4" thick, glazed 1 side		345	.116		10.35	4.72		15.07	18.55
3135	Glazed 2 sides		325	.123		14.25	5		19.25	23
3140	6" thick, glazed 1 side		330	.121		14.80	4.93		19.73	23.50
3150	8" thick, glazed 1 side		310	.129	▼	21.50	5.25		26.75	31.50
3155	Special shapes, group I		300	.133	Ea.	6.90	5.45		12.35	15.75
3160	Group II	▼	280	.143	"	7.90	5.80		13.70	17.50
3200	For designer colors, add					25%				
3300	For epoxy mortar joints, add				S.F.	1.58			1.58	1.74

04 21 29 – Terra Cotta Masonry

04 21 29.10 Terra Cotta Masonry Components		Crew	Daily Output	Labor-Hours	Unit	Material	2012 Bare Costs Labor	Equipment	Total	Total Incl O&P
0010	**TERRA COTTA MASONRY COMPONENTS**									
0020	Coping, split type, not glazed, 9" wide	D-1	90	.178	L.F.	10.40	7.10		17.50	22
0100	13" wide		80	.200		15.75	7.95		23.70	29.50
0200	Coping, split type, glazed, 9" wide		90	.178		17.40	7.10		24.50	30
0250	13" wide	▼	80	.200	▼	22.50	7.95		30.45	37
0500	Partition or back-up blocks, scored, in C.L. lots									
0700	Non-load bearing 12" x 12", 3" thick, special order	D-8	550	.073	S.F.	26	2.96		28.96	33
0750	4" thick, standard		500	.080		7.10	3.26		10.36	12.70
0800	6" thick		450	.089		8.70	3.62		12.32	15
0850	8" thick		400	.100		11.05	4.07		15.12	18.35
1000	Load bearing, 12" x 12", 4" thick, in walls		500	.080		7.60	3.26		10.86	13.25
1050	In floors		750	.053		7.60	2.17		9.77	11.65
1200	6" thick, in walls		450	.089		8.70	3.62		12.32	15
1250	In floors		675	.059		8.70	2.41		11.11	13.20
1400	8" thick, in walls		400	.100		11.05	4.07		15.12	18.35
1450	In floors		575	.070		11.05	2.83		13.88	16.50
1600	10" thick, in walls, special order		350	.114		23.50	4.65		28.15	32.50
1650	In floors, special order		500	.080		23.50	3.26		26.76	30.50
1800	12" thick, in walls, special order		300	.133		28	5.45		33.45	38.50
1850	In floors, special order	▼	450	.089		28	3.62		31.62	36
2000	For reinforcing with steel rods, add to above					15%	5%			
2100	For smooth tile instead of scored, add					3.74			3.74	4.11
2200	For L.C.L. quantities, add				▼	10%	10%			

04 21 Clay Unit Masonry

04 21 29 – Terra Cotta Masonry

04 21 29.20 Terra Cotta Tile		Crew	Daily Output	Labor-Hours	Unit	Material	2012 Bare Costs Labor	Equipment	Total	Total Incl O&P
0010	**TERRA COTTA TILE**, on walls, dry set, 1/2" thick									
0100	Square, hexagonal or lattice shapes, unglazed	1 Tilf	135	.059	S.F.	6.35	2.41		8.76	10.50
0300	Glazed, plain colors		130	.062		8.75	2.51		11.26	13.30
0400	Intense colors	↓	125	.064	↓	10.20	2.61		12.81	15.05

04 22 Concrete Unit Masonry

04 22 10 – Concrete Masonry Units

04 22 10.11 Autoclave Aerated Concrete Block

			Crew	Daily Output	Labor-Hours	Unit	Material	Labor	Equipment	Total	Total Incl O&P
0010	**AUTOCLAVE AERATED CONCRETE BLOCK**, excl. scaffolding, grout & reinforcing										
0050	Solid, 4" x 12" x 24", incl mortar	G	D-8	600	.067	S.F.	1.39	2.71		4.10	5.65
0060	6" x 12" x 24"	G		600	.067		2.08	2.71		4.79	6.40
0070	8" x 8" x 24"	G		575	.070		2.78	2.83		5.61	7.35
0080	10" x 12" x 24"	G		575	.070		3.40	2.83		6.23	8
0090	12" x 12" x 24"	G	↓	550	.073	↓	4.16	2.96		7.12	9.05

04 22 10.12 Chimney Block

		Crew	Daily Output	Labor-Hours	Unit	Material	Labor	Equipment	Total	Total Incl O&P
0010	**CHIMNEY BLOCK**, excludes scaffolding, grout and reinforcing									
0220	1 piece, with 8" x 8" flue, 16" x 16"	D-1	28	.571	V.L.F.	17.55	23		40.55	54
0230	2 piece, 16" x 16"		26	.615		15.05	24.50		39.55	53.50
0240	2 piece, with 8" x 12" flue, 16" x 20"	↓	24	.667	↓	30	26.50		56.50	73

04 22 10.14 Concrete Block, Back-Up

			Crew	Daily Output	Labor-Hours	Unit	Material	Labor	Equipment	Total	Total Incl O&P
0010	**CONCRETE BLOCK, BACK-UP**, C90, 2000 psi	R042210-20									
0020	Normal weight, 8" x 16" units, tooled joint 1 side										
0050	Not-reinforced, 2000 psi, 2" thick		D-8	475	.084	S.F.	1.39	3.43		4.82	6.75
0200	4" thick			460	.087		1.58	3.54		5.12	7.10
0300	6" thick			440	.091		2.19	3.70		5.89	8
0350	8" thick			400	.100		2.49	4.07		6.56	8.90
0400	10" thick		↓	330	.121		3.16	4.93		8.09	10.95
0450	12" thick		D-9	310	.155		3.56	6.15		9.71	13.20
1000	Reinforced, alternate courses, 4" thick		D-8	450	.089		1.72	3.62		5.34	7.35
1100	6" thick			430	.093		2.34	3.79		6.13	8.25
1150	8" thick			395	.101		2.65	4.12		6.77	9.10
1200	10" thick		↓	320	.125		3.31	5.10		8.41	11.35
1250	12" thick		D-9	300	.160	↓	3.71	6.40		10.11	13.75

04 22 10.16 Concrete Block, Bond Beam

		Crew	Daily Output	Labor-Hours	Unit	Material	Labor	Equipment	Total	Total Incl O&P
0010	**CONCRETE BLOCK, BOND BEAM**, C90, 2000 psi									
0020	Not including grout or reinforcing									
0125	Regular block, 6" thick	D-8	584	.068	L.F.	2.34	2.79		5.13	6.80
0130	8" high, 8" thick	"	565	.071		2.52	2.88		5.40	7.10
0150	12" thick	D-9	510	.094		3.53	3.75		7.28	9.55
0525	Lightweight, 6" thick	D-8	592	.068		2.55	2.75		5.30	6.95
0530	8" high, 8" thick	"	575	.070		3.09	2.83		5.92	7.65
0550	12" thick	D-9	520	.092	↓	4.15	3.68		7.83	10.10
2000	Including grout and 2 #5 bars									
2100	Regular block, 8" high, 8" thick	D-8	300	.133	L.F.	4.55	5.45		10	13.20
2150	12" thick	D-9	250	.192		6.10	7.65		13.75	18.30
2500	Lightweight, 8" high, 8" thick	D-8	305	.131		5.15	5.35		10.50	13.70
2550	12" thick	D-9	255	.188	↓	6.75	7.50		14.25	18.75

04 22 10 – Concrete Masonry Units

04 22 10.18 Concrete Block, Column	Crew	Daily Output	Labor-Hours	Unit	Material	2012 Bare Costs Labor	Equipment	Total	Total Incl O&P
0010 **CONCRETE BLOCK, COLUMN** or pilaster									
0050 Including vertical reinforcing (4-#4 bars) and grout									
0160 1 piece unit, 16" x 16"	D-1	26	.615	V.L.F.	18.10	24.50		42.60	57
0170 2 piece units, 16" x 20"		24	.667		24	26.50		50.50	66
0180 20" x 20"		22	.727		35	29		64	82.50
0190 22" x 24"		18	.889		49	35.50		84.50	107
0200 20" x 32"		14	1.143		53	45.50		98.50	128

04 22 10.19 Concrete Block, Insulation Inserts	Crew	Daily Output	Labor-Hours	Unit	Material	2012 Bare Costs Labor	Equipment	Total	Total Incl O&P
0010 **CONCRETE BLOCK, INSULATION INSERTS**									
0100 Styrofoam, plant installed, add to block prices									
0200 8" x 16" units, 6" thick				S.F.	1.14			1.14	1.25
0250 8" thick					1.14			1.14	1.25
0300 10" thick					1.93			1.93	2.12
0350 12" thick					2			2	2.20
0500 8" x 8" units, 8" thick					.94			.94	1.03
0550 12" thick					1.14			1.14	1.25

04 22 10.23 Concrete Block, Decorative	Crew	Daily Output	Labor-Hours	Unit	Material	2012 Bare Costs Labor	Equipment	Total	Total Incl O&P
0010 **CONCRETE BLOCK, DECORATIVE**, C90, 2000 psi									
0020 Embossed, simulated brick face									
0100 8" x 16" units, 4" thick	D-8	400	.100	S.F.	3.54	4.07		7.61	10.05
0200 8" thick		340	.118		4.85	4.79		9.64	12.60
0250 12" thick		300	.133		6.35	5.45		11.80	15.20
0400 Embossed both sides									
0500 8" thick	D-8	300	.133	S.F.	5.45	5.45		10.90	14.20
0550 12" thick	"	275	.145	"	6.85	5.90		12.75	16.50
1000 Fluted high strength									
1100 8" x 16" x 4" thick, flutes 1 side,	D-8	345	.116	S.F.	4.20	4.72		8.92	11.75
1150 Flutes 2 sides		335	.119		5.10	4.86		9.96	12.95
1200 8" thick		300	.133		6.60	5.45		12.05	15.45
1250 For special colors, add					.41			.41	.45
1400 Deep grooved, smooth face									
1450 8" x 16" x 4" thick	D-8	345	.116	S.F.	2.75	4.72		7.47	10.15
1500 8" thick	"	300	.133	"	4.70	5.45		10.15	13.35
2000 Formblock, incl. inserts & reinforcing									
2100 8" x 16" x 8" thick	D-8	345	.116	S.F.	4.77	4.72		9.49	12.40
2150 12" thick	"	310	.129	"	5.95	5.25		11.20	14.50
2500 Ground face									
2600 8" x 16" x 4" thick	D-8	345	.116	S.F.	3.35	4.72		8.07	10.85
2650 6" thick		325	.123		4.83	5		9.83	12.85
2700 8" thick		300	.133		6.40	5.45		11.85	15.20
2750 12" thick	D-9	265	.181		5.80	7.20		13	17.25
2900 For special colors, add, minimum					15%				
2950 For special colors, add, maximum					45%				
4000 Slump block									
4100 4" face height x 16" x 4" thick	D-1	165	.097	S.F.	3.53	3.86		7.39	9.75
4150 6" thick		160	.100		4.85	3.99		8.84	11.35
4200 8" thick		155	.103		5.05	4.11		9.16	11.75
4250 10" thick		140	.114		10	4.55		14.55	17.90
4300 12" thick		130	.123		10.25	4.90		15.15	18.65
4400 6" face height x 16" x 6" thick		155	.103		4.51	4.11		8.62	11.15
4450 8" thick		150	.107		6.65	4.25		10.90	13.70
4500 10" thick		130	.123		10.40	4.90		15.30	18.85

04 22 10.23 Concrete Block, Decorative		Crew	Daily Output	Labor-Hours	Unit	Material	2012 Bare Costs Labor	Equipment	Total	Total Incl O&P
4550	12" thick	D-1	120	.133	S.F.	10.70	5.30		16	19.80
5000	Split rib profile units, 1" deep ribs, 8 ribs									
5100	8" x 16" x 4" thick	D-8	345	.116	S.F.	3.49	4.72		8.21	11
5150	6" thick		325	.123		4.01	5		9.01	11.95
5200	8" thick		300	.133		4.66	5.45		10.11	13.35
5250	12" thick	D-9	275	.175		5.50	6.95		12.45	16.50
5400	For special deeper colors, 4" thick, add					1.15			1.15	1.27
5450	12" thick, add					1.19			1.19	1.31
5600	For white, 4" thick, add					1.15			1.15	1.27
5650	6" thick, add					1.14			1.14	1.26
5700	8" thick, add					1.20			1.20	1.32
5750	12" thick, add					1.23			1.23	1.35
6000	Split face									
6100	8" x 16" x 4" thick	D-8	350	.114	S.F.	3.16	4.65		7.81	10.45
6150	6" thick		325	.123		3.65	5		8.65	11.55
6200	8" thick		300	.133		4.22	5.45		9.67	12.85
6250	12" thick	D-9	270	.178		5.10	7.10		12.20	16.35
6300	For scored, add					.35			.35	.39
6400	For special deeper colors, 4" thick, add					.67			.67	.73
6450	6" thick, add					.66			.66	.72
6500	8" thick, add					.68			.68	.75
6550	12" thick, add					.70			.70	.77
6650	For white, 4" thick, add					1.14			1.14	1.26
6700	6" thick, add					1.13			1.13	1.24
6750	8" thick, add					1.16			1.16	1.28
6800	12" thick, add					1.19			1.19	1.31
7000	Scored ground face, 2 to 5 scores									
7100	8" x 16" x 4" thick	D-8	340	.118	S.F.	4.94	4.79		9.73	12.70
7150	6" thick		310	.129		5.65	5.25		10.90	14.15
7200	8" thick		290	.138		6.60	5.60		12.20	15.80
7250	12" thick	D-9	265	.181		8.20	7.20		15.40	19.95
8000	Hexagonal face profile units, 8" x 16" units									
8100	4" thick, hollow	D-8	340	.118	S.F.	3.27	4.79		8.06	10.85
8200	Solid		340	.118		4.20	4.79		8.99	11.85
8300	6" thick, hollow		310	.129		3.50	5.25		8.75	11.80
8350	8" thick, hollow		290	.138		4.75	5.60		10.35	13.70
8500	For stacked bond, add						26%			
8550	For high rise construction, add per story	D-8	67.80	.590	M.S.F.		24		24	36.50
8600	For scored block, add					10%				
8650	For honed or ground face, per face, add				Ea.	.38			.38	.42
8700	For honed or ground end, per end, add				"	2.98			2.98	3.28
8750	For bullnose block, add					10%				
8800	For special color, add					13%				

04 22 10.24 Concrete Block, Exterior		Crew	Daily Output	Labor-Hours	Unit	Material	2012 Bare Costs Labor	Equipment	Total	Total Incl O&P
0010	**CONCRETE BLOCK, EXTERIOR**, C90, 2000 psi									
0020	Reinforced alt courses, tooled joints 2 sides									
0100	Normal weight, 8" x 16" x 6" thick	D-8	395	.101	S.F.	2.57	4.12		6.69	9
0200	8" thick		360	.111		3.34	4.52		7.86	10.50
0250	10" thick		290	.138		4.30	5.60		9.90	13.25
0300	12" thick	D-9	250	.192		4.27	7.65		11.92	16.25
0500	Lightweight, 8" x 16" x 6" thick	D-8	450	.089		2.74	3.62		6.36	8.45
0600	8" thick		430	.093		3.48	3.79		7.27	9.55

04 22 Concrete Unit Masonry

04 22 10 – Concrete Masonry Units

04 22 10.24 Concrete Block, Exterior

		Crew	Daily Output	Labor-Hours	Unit	Material	2012 Bare Costs Labor	2012 Bare Costs Equipment	Total	Total Incl O&P
0650	10" thick	D-8	395	.101	S.F.	4.33	4.12		8.45	10.95
0700	12" thick	D-9	350	.137	↓	6.35	5.45		11.80	15.25

04 22 10.26 Concrete Block Foundation Wall

		Crew	Daily Output	Labor-Hours	Unit	Material	Labor	Equipment	Total	Total Incl O&P
0010	**CONCRETE BLOCK FOUNDATION WALL**, C90/C145									
0050	Normal-weight, cut joints, horiz joint reinf, no vert reinf									
0200	Hollow, 8" x 16" x 6" thick	D-8	455	.088	S.F.	2.65	3.58		6.23	8.30
0250	8" thick		425	.094		2.96	3.83		6.79	9.05
0300	10" thick	↓	350	.114		3.63	4.65		8.28	11
0350	12" thick	D-9	300	.160		4.04	6.40		10.44	14.10
0500	Solid, 8" x 16" block, 6" thick	D-8	440	.091		2.52	3.70		6.22	8.35
0550	8" thick	"	415	.096		3.90	3.92		7.82	10.20
0600	12" thick	D-9	350	.137	↓	5.45	5.45		10.90	14.25

04 22 10.28 Concrete Block, High Strength

		Crew	Daily Output	Labor-Hours	Unit	Material	Labor	Equipment	Total	Total Incl O&P
0010	**CONCRETE BLOCK, HIGH STRENGTH**									
0050	Hollow, reinforced alternate courses, 8" x 16" units									
0200	3500 psi, 4" thick	D-8	440	.091	S.F.	2.25	3.70		5.95	8.10
0250	6" thick		395	.101		2.48	4.12		6.60	8.95
0300	8" thick	↓	360	.111		3.25	4.52		7.77	10.45
0350	12" thick	D-9	250	.192		4.15	7.65		11.80	16.10
0500	5000 psi, 4" thick	D-8	440	.091		2.47	3.70		6.17	8.30
0550	6" thick		395	.101		3.06	4.12		7.18	9.55
0600	8" thick	↓	360	.111		4.07	4.52		8.59	11.35
0650	12" thick	D-9	300	.160	↓	6.05	6.40		12.45	16.30
1000	For 75% solid block, add					30%				
1050	For 100% solid block, add					50%				

04 22 10.30 Concrete Block, Interlocking

		Crew	Daily Output	Labor-Hours	Unit	Material	Labor	Equipment	Total	Total Incl O&P
0010	**CONCRETE BLOCK, INTERLOCKING**									
0100	Not including grout or reinforcing									
0200	8" x 16" units, 2,000 psi, 8" thick	D-1	245	.065	S.F.	2.72	2.60		5.32	6.90
0300	12" thick		220	.073		4.03	2.90		6.93	8.80
0350	16" thick	↓	185	.086		6.05	3.45		9.50	11.85
0400	Including grout & reinforcing, 8" thick	D-4	245	.131		7.35	5.25	.52	13.12	16.55
0450	12" thick		220	.145		8.90	5.80	.58	15.28	19.20
0500	16" thick	↓	185	.173	↓	11.10	6.90	.69	18.69	23.50

04 22 10.32 Concrete Block, Lintels

		Crew	Daily Output	Labor-Hours	Unit	Material	Labor	Equipment	Total	Total Incl O&P
0010	**CONCRETE BLOCK, LINTELS**, C90, normal weight									
0100	Including grout and horizontal reinforcing									
0200	8" x 8" x 8", 1 #4 bar	D-4	300	.107	L.F.	4.87	4.27	.42	9.56	12.25
0250	2 #4 bars		295	.108		5.10	4.34	.43	9.87	12.65
0400	8" x 16" x 8", 1 #4 bar		275	.116		3.38	4.66	.46	8.50	11.30
0450	2 #4 bars		270	.119		3.60	4.74	.47	8.81	11.65
1000	12" x 8" x 8", 1 #4 bar		275	.116		6.80	4.66	.46	11.92	15.05
1100	2 #4 bars		270	.119		7.05	4.74	.47	12.26	15.40
1150	2 #5 bars		270	.119		7.30	4.74	.47	12.51	15.65
1200	2 #6 bars		265	.121		7.60	4.83	.48	12.91	16.20
1500	12" x 16" x 8", 1 #4 bar		250	.128		10.80	5.10	.51	16.41	20
1600	2 #3 bars		245	.131		10.80	5.25	.52	16.57	20.50
1650	2 #4 bars		245	.131		11	5.25	.52	16.77	20.50
1700	2 #5 bars	↓	240	.133	↓	11.25	5.35	.53	17.13	21

04 22 10.34 Concrete Block, Partitions	Crew	Daily Output	Labor-Hours	Unit	Material	2012 Bare Costs Labor	Equipment	Total	Total Incl O&P
0010 **CONCRETE BLOCK, PARTITIONS**, excludes scaffolding R042210-20									
0100 Acoustical slotted block									
0200 4" thick, type A-1	D-8	315	.127	S.F.	3.79	5.15		8.94	11.95
0210 8" thick		275	.145		4.81	5.90		10.71	14.25
0250 8" thick, type Q		275	.145		8.45	5.90		14.35	18.25
0260 4" thick, type RSC		315	.127		5.90	5.15		11.05	14.25
0270 6" thick		295	.136		5.90	5.50		11.40	14.80
0280 8" thick		275	.145		5.90	5.90		11.80	15.40
0290 12" thick		250	.160		5.90	6.50		12.40	16.30
0300 8" thick, type RSR		275	.145		5.90	5.90		11.80	15.40
0400 8" thick, type RSC/RF		275	.145		6.60	5.90		12.50	16.20
0410 10" thick		260	.154		7.10	6.25		13.35	17.25
0420 12" thick		250	.160		7.60	6.50		14.10	18.20
0430 12" thick, type RSC/RF-4		250	.160		9.15	6.50		15.65	19.90
0500 NRC .60 type R, 8" thick		265	.151		8	6.15		14.15	18.10
0600 NRC .65 type RR, 8" thick		265	.151		7.35	6.15		13.50	17.35
0700 NRC .65 type 4R-RF, 8" thick		265	.151		8.25	6.15		14.40	18.35
0710 NRC .70 type R, 12" thick		245	.163		8.65	6.65		15.30	19.55
1000 Lightweight block, tooled joints, 2 sides, hollow									
1100 Not reinforced, 8" x 16" x 4" thick	D-8	440	.091	S.F.	1.64	3.70		5.34	7.40
1150 6" thick		410	.098		2.37	3.97		6.34	8.60
1200 8" thick		385	.104		2.88	4.23		7.11	9.55
1250 10" thick		370	.108		3.48	4.40		7.88	10.50
1300 12" thick	D-9	350	.137		4.33	5.45		9.78	13
2000 Not reinforced, 8" x 24" x 4" thick, hollow		460	.104		1.21	4.16		5.37	7.65
2100 6" thick		440	.109		1.71	4.35		6.06	8.45
2150 8" thick		415	.116		2.12	4.61		6.73	9.30
2200 10" thick		385	.125		2.56	4.97		7.53	10.30
2250 12" thick		365	.132		3.17	5.25		8.42	11.40
2800 Solid, not reinforced, 8" x 16" x 2" thick	D-8	440	.091		1.46	3.70		5.16	7.20
2900 4" thick		420	.095		2.36	3.88		6.24	8.45
2950 6" thick		390	.103		2.63	4.17		6.80	9.20
3000 8" thick		365	.110		3.38	4.46		7.84	10.45
3050 10" thick		350	.114		4.22	4.65		8.87	11.65
3100 12" thick	D-9	330	.145		6.25	5.80		12.05	15.60
4000 Regular block, tooled joints, 2 sides, hollow									
4100 Not reinforced, 8" x 16" x 4" thick	D-8	430	.093	S.F.	1.49	3.79		5.28	7.35
4150 6" thick		400	.100		2.10	4.07		6.17	8.45
4200 8" thick		375	.107		2.40	4.34		6.74	9.20
4250 10" thick		360	.111		3.07	4.52		7.59	10.25
4300 12" thick	D-9	340	.141		3.47	5.65		9.12	12.30
4500 Reinforced alternate courses, 8" x 16" x 4" thick	D-8	425	.094		1.63	3.83		5.46	7.60
4550 6" thick		395	.101		2.25	4.12		6.37	8.65
4600 8" thick		370	.108		2.56	4.40		6.96	9.45
4650 10" thick		355	.113		3.49	4.59		8.08	10.80
4700 12" thick	D-9	335	.143		3.62	5.70		9.32	12.60
4900 Solid, not reinforced, 2" thick	D-8	435	.092		1.34	3.74		5.08	7.10
5000 3" thick		430	.093		1.60	3.79		5.39	7.45
5050 4" thick		415	.096		1.81	3.92		5.73	7.90
5100 6" thick		385	.104		1.98	4.23		6.21	8.55
5150 8" thick		360	.111		3.34	4.52		7.86	10.55
5200 12" thick	D-9	325	.148		4.87	5.90		10.77	14.25

04 22 Concrete Unit Masonry

04 22 10 – Concrete Masonry Units

04 22 10.34 Concrete Block, Partitions

		Crew	Daily Output	Labor-Hours	Unit	Material	2012 Bare Costs Labor	Equipment	Total	Total Incl O&P
5500	Solid, reinforced alternate courses, 4" thick	D-8	420	.095	S.F.	1.92	3.88		5.80	7.95
5550	6" thick		380	.105		2.08	4.28		6.36	8.75
5600	8" thick		355	.113		3.45	4.59		8.04	10.75
5650	12" thick	D-9	320	.150		4.43	6		10.43	13.90

04 22 10.38 Concrete Brick

		Crew	Daily Output	Labor-Hours	Unit	Material	2012 Bare Costs Labor	Equipment	Total	Total Incl O&P
0010	**CONCRETE BRICK**, C55, grade N, type 1									
0100	Regular, 4 x 2-1/4 x 8	D-8	660	.061	Ea.	.50	2.47		2.97	4.28
0125	Rusticated, 4 x 2-1/4 x 8		660	.061		.56	2.47		3.03	4.35
0150	Frog, 4 x 2-1/4 x 8		660	.061		.54	2.47		3.01	4.32
0200	Double, 4 x 4-7/8 x 8		535	.075		.88	3.04		3.92	5.55

04 22 10.42 Concrete Block, Screen Block

		Crew	Daily Output	Labor-Hours	Unit	Material	2012 Bare Costs Labor	Equipment	Total	Total Incl O&P
0010	**CONCRETE BLOCK, SCREEN BLOCK**									
0200	8" x 16", 4" thick	D-8	330	.121	S.F.	1.97	4.93		6.90	9.60
0300	8" thick		270	.148		2.99	6.05		9.04	12.40
0350	12" x 12", 4" thick		290	.138		2.21	5.60		7.81	10.95
0500	8" thick		250	.160		3.06	6.50		9.56	13.20

04 22 10.44 Glazed Concrete Block

		Crew	Daily Output	Labor-Hours	Unit	Material	2012 Bare Costs Labor	Equipment	Total	Total Incl O&P
0010	**GLAZED CONCRETE BLOCK** C744									
0100	Single face, 8" x 16" units, 2" thick	D-8	360	.111	S.F.	9.20	4.52		13.72	17
0200	4" thick		345	.116		9.45	4.72		14.17	17.55
0250	6" thick		330	.121		11.55	4.93		16.48	20
0300	8" thick		310	.129		9.95	5.25		15.20	18.90
0350	10" thick		295	.136		12.50	5.50		18	22
0400	12" thick	D-9	280	.171		13.30	6.85		20.15	25
0700	Double face, 8" x 16" units, 4" thick	D-8	340	.118		13.85	4.79		18.64	22.50
0750	6" thick		320	.125		16.25	5.10		21.35	25.50
0800	8" thick		300	.133		17.05	5.45		22.50	27
1000	Jambs, bullnose or square, single face, 8" x 16", 2" thick		315	.127	Ea.	17.50	5.15		22.65	27
1050	4" thick		285	.140	"	20.50	5.70		26.20	31
1200	Caps, bullnose or square, 8" x 16", 2" thick		420	.095	L.F.	17.80	3.88		21.68	25.50
1250	4" thick		380	.105		18.50	4.28		22.78	27
1500	Cove base, 8" x 16", 2" thick		315	.127		13.60	5.15		18.75	23
1550	4" thick		285	.140		8.65	5.70		14.35	18.20
1600	6" thick		265	.151		9.30	6.15		15.45	19.50
1650	8" thick		245	.163		9.75	6.65		16.40	21

04 23 Glass Unit Masonry

04 23 13 – Vertical Glass Unit Masonry

04 23 13.10 Glass Block

		Crew	Daily Output	Labor-Hours	Unit	Material	2012 Bare Costs Labor	Equipment	Total	Total Incl O&P
0010	**GLASS BLOCK**									
0100	Plain, 4" thick, under 1,000 S.F., 6" x 6"	D-8	115	.348	S.F.	28.50	14.15		42.65	53
0150	8" x 8"		160	.250		15.80	10.20		26	32.50
0160	end block		160	.250		52.50	10.20		62.70	73
0170	90 deg corner		160	.250		56.50	10.20		66.70	77.50
0180	45 deg corner		160	.250		49	10.20		59.20	69
0200	12" x 12"		175	.229		24	9.30		33.30	40.50
0210	4" x 8"		160	.250		16.20	10.20		26.40	33
0220	6" x 8"		160	.250		15.65	10.20		25.85	32.50
0300	1,000 to 5,000 S.F., 6" x 6"		135	.296		28	12.05		40.05	48.50
0350	8" x 8"		190	.211		15.50	8.55		24.05	30

04 23 Glass Unit Masonry

04 23 13 – Vertical Glass Unit Masonry

04 23 13.10 Glass Block		Crew	Daily Output	Labor-Hours	Unit	Material	2012 Bare Costs Labor	Equipment	Total	Total Incl O&P
0400	12" x 12"	D-8	215	.186	S.F.	23.50	7.55		31.05	37.50
0410	4" x 8"		215	.186		6.40	7.55		13.95	18.50
0420	6" x 8"		215	.186		6.10	7.55		13.65	18.15
0500	Over 5,000 S.F., 6" x 6"		145	.276		27	11.25		38.25	47
0550	8" x 8"		215	.186		15	7.55		22.55	28
0600	12" x 12"		240	.167		23	6.80		29.80	35.50
0610	4" x 8"		240	.167		6	6.80		12.80	16.85
0620	6" x 8"	↓	240	.167	↓	6.10	6.80		12.90	16.95
0700	For solar reflective blocks, add					100%				
1000	Thinline, plain, 3-1/8" thick, under 1,000 S.F., 6" x 6"	D-8	115	.348	S.F.	19.90	14.15		34.05	43.50
1050	8" x 8"		160	.250		11	10.20		21.20	27.50
1200	Over 5,000 S.F., 6" x 6"		145	.276		18.70	11.25		29.95	37.50
1250	8" x 8"	↓	215	.186	↓	10.50	7.55		18.05	23
1400	For cleaning block after installation (both sides), add	↓	1000	.040	↓	.11	1.63		1.74	2.58
4000	Accessories									
4100	Anchors, 20 ga. galv., 1-3/4" wide x 24" long				Ea.	3.88			3.88	4.27
4200	Emulsion asphalt				Gal.	8.10			8.10	8.90
4300	Expansion joint, fiberglass				L.F.	.57			.57	.63
4400	Steel mesh, double galvanized				"	.95			.95	1.05

04 24 Adobe Unit Masonry

04 24 16 – Manufactured Adobe Unit Masonry

04 24 16.06 Adobe Brick

			Crew	Daily Output	Labor-Hours	Unit	Material	2012 Bare Costs Labor	Equipment	Total	Total Incl O&P
0010	**ADOBE BRICK**, Semi-stabilized, with cement mortar										
0060	Brick, 10" x 4" x 14", 2.6/S.F.	G	D-8	560	.071	S.F.	3.84	2.91		6.75	8.60
0080	12" x 4" x 16", 2.3/S.F.	G		580	.069		6.10	2.81		8.91	10.95
0100	10" x 4" x 16", 2.3/S.F.	G		590	.068		5.45	2.76		8.21	10.15
0120	8" x 4" x 16", 2.3/S.F.	G		560	.071		4.41	2.91		7.32	9.25
0140	4" x 4" x 16", 2.3/S.F.	G		540	.074		3.98	3.01		6.99	8.90
0160	6" x 4" x 16", 2.3/S.F.	G		540	.074		3.85	3.01		6.86	8.80
0180	4" x 4" x 12", 3.0/S.F.	G		520	.077		3.82	3.13		6.95	8.95
0200	8" x 4" x 12", 3.0/S.F.	G	↓	520	.077	↓	3.80	3.13		6.93	8.90

04 25 Unit Masonry Panels

04 25 20 – Pre-Fabricated Masonry Panels

04 25 20.10 Brick and Epoxy Mortar Panels

		Crew	Daily Output	Labor-Hours	Unit	Material	2012 Bare Costs Labor	Equipment	Total	Total Incl O&P
0010	**BRICK AND EPOXY MORTAR PANELS**									
0020	Prefabricated brick & epoxy mortar, 4" thick, minimum	C-11	775	.093	S.F.	4.95	4.51	2.36	11.82	15.85
0100	Maximum	"	500	.144		6.05	7	3.66	16.71	23
0200	For 2" concrete back-up, add					50%				
0300	For 1" urethane & 3" concrete back-up, add				↓	70%				

04 27 Multiple-Wythe Unit Masonry

04 27 10 – Multiple-Wythe Masonry

04 27 10.20 Cavity Walls

		Crew	Daily Output	Labor-Hours	Unit	Material	2012 Bare Costs Labor	2012 Bare Costs Equipment	Total	Total Incl O&P
0010	**CAVITY WALLS**, brick and CMU, includes joint reinforcing and ties									
0200	4" face brick, 4" block	D-8	165	.242	S.F.	5.80	9.85		15.65	21.50
0400	6" block	↓	145	.276		6.30	11.25		17.55	24
0600	8" block	↓	125	.320	↓	6.50	13		19.50	27

04 27 10.30 Brick Walls

			Crew	Daily Output	Labor-Hours	Unit	Material	2012 Bare Costs Labor	2012 Bare Costs Equipment	Total	Total Incl O&P
0010	**BRICK WALLS**, including mortar, excludes scaffolding	R042110-20									
0020	Estimating by number of brick										
0140	Face brick, 4" thick wall, 6.75 brick/S.F.		D-8	1.45	27.586	M	610	1,125		1,735	2,375
0150	Common brick, 4" thick wall, 6.75 brick/S.F.			1.60	25		440	1,025		1,465	2,000
0204	8" thick, 13.50 bricks per S.F.			1.80	22.222		460	905		1,365	1,875
0250	12" thick, 20.25 bricks per S.F.			1.90	21.053		465	855		1,320	1,800
0304	16" thick, 27.00 bricks per S.F.			2	20		470	815		1,285	1,750
0500	Reinforced, face brick, 4" thick wall, 6.75 brick/S.F.			1.40	28.571		635	1,175		1,810	2,450
0520	Common brick, 4" thick wall, 6.75 brick/S.F.			1.55	25.806		465	1,050		1,515	2,100
0550	8" thick, 13.50 bricks per S.F.			1.75	22.857		485	930		1,415	1,925
0600	12" thick, 20.25 bricks per S.F.			1.85	21.622		485	880		1,365	1,850
0650	16" thick, 27.00 bricks per S.F.		↓	1.95	20.513	↓	495	835		1,330	1,800
0790	Alternate method of figuring by square foot										
0800	Face brick, 4" thick wall, 6.75 brick/S.F.		D-8	215	.186	S.F.	4.12	7.55		11.67	16
0850	Common brick, 4" thick wall, 6.75 brick/S.F.			240	.167		2.98	6.80		9.78	13.55
0900	8" thick, 13.50 bricks per S.F.			135	.296		6.20	12.05		18.25	25
1000	12" thick, 20.25 bricks per S.F.			95	.421		9.35	17.15		26.50	36.50
1050	16" thick, 27.00 bricks per S.F.			75	.533		12.70	21.50		34.20	47
1200	Reinforced, face brick, 4" thick wall, 6.75 brick/S.F.			210	.190		4.29	7.75		12.04	16.40
1220	Common brick, 4" thick wall, 6.75 brick/S.F.			235	.170		3.15	6.95		10.10	13.90
1250	8" thick, 13.50 bricks per S.F.			130	.308		6.55	12.50		19.05	26
1300	12" thick, 20.25 bricks per S.F.			90	.444		9.85	18.10		27.95	38.50
1350	16" thick, 27.00 bricks per S.F.		↓	70	.571	↓	13.35	23.50		36.85	49.50

04 27 10.40 Steps

		Crew	Daily Output	Labor-Hours	Unit	Material	2012 Bare Costs Labor	2012 Bare Costs Equipment	Total	Total Incl O&P
0010	**STEPS**									
0012	Entry steps, select common brick	D-1	.30	53.333	M	430	2,125		2,555	3,675

04 41 Dry-Placed Stone

04 41 10 – Dry Placed Stone

04 41 10.10 Rough Stone Wall

			Crew	Daily Output	Labor-Hours	Unit	Material	2012 Bare Costs Labor	2012 Bare Costs Equipment	Total	Total Incl O&P
0011	**ROUGH STONE WALL**, Dry										
0012	Dry laid (no mortar), under 18" thick	G	D-1	60	.267	C.F.	11.65	10.65		22.30	29
0100	Random fieldstone, under 18" thick	G	D-12	60	.533		11.65	21.50		33.15	45.50
0150	Over 18" thick	G	"	63	.508	↓	14	20.50		34.50	46.50
0500	Field stone veneer	G	D-8	120	.333	S.F.	10.25	13.55		23.80	32
0510	Valley stone veneer	G		120	.333		10.25	13.55		23.80	32
0520	River stone veneer	G	↓	120	.333	↓	10.25	13.55		23.80	32
0600	Rubble stone walls, in mortar bed, up to 18" thick	G	D-11	75	.320	C.F.	14.10	13.40		27.50	36

04 43 Stone Masonry

04 43 10 – Masonry with Natural and Processed Stone

04 43 10.05 Ashlar Veneer	Crew	Daily Output	Labor-Hours	Unit	Material	2012 Bare Costs Labor	2012 Bare Costs Equipment	Total	Total Incl O&P
0011	**ASHLAR VENEER** 4" + or - thk, random or random rectangular									
0150	Sawn face, split joints, low priced stone	D-8	140	.286	S.F.	8.55	11.65		20.20	27
0200	Medium priced stone		130	.308		10.35	12.50		22.85	30.50
0300	High priced stone		120	.333		14.80	13.55		28.35	37
0600	Seam face, split joints, medium price stone		125	.320		7.55	13		20.55	28
0700	High price stone		120	.333		12.20	13.55		25.75	34
1000	Split or rock face, split joints, medium price stone		125	.320		8.40	13		21.40	29
1100	High price stone		120	.333		14	13.55		27.55	36

04 43 10.10 Bluestone	Crew	Daily Output	Labor-Hours	Unit	Material	Labor	Equipment	Total	Total Incl O&P
0010	**BLUESTONE**, cut to size									
0500	Sills, natural cleft, 10" wide to 6' long, 1-1/2" thick	D-11	70	.343	L.F.	13.35	14.40		27.75	36
0550	2" thick	"	63	.381		14.60	15.95		30.55	40
1000	Stair treads, natural cleft, 12" wide, 6' long, 1-1/2" thick	D-10	115	.278		30	12.10	3.90	46	55.50
1050	2" thick		105	.305		30	13.25	4.27	47.52	57.50
1100	Smooth finish, 1-1/2" thick		115	.278		15	12.10	3.90	31	39
1300	Thermal finish, 1-1/2" thick		115	.278		15	12.10	3.90	31	39
1350	2" thick		105	.305		15	13.25	4.27	32.52	41

04 43 10.45 Granite	Crew	Daily Output	Labor-Hours	Unit	Material	Labor	Equipment	Total	Total Incl O&P
0010	**GRANITE**, cut to size									
0050	Veneer, polished face, 3/4" to 1-1/2" thick									
0150	Low price, gray, light gray, etc.	D-10	130	.246	S.F.	26.50	10.70	3.45	40.65	49.50
0220	High price, red, black, etc.	"	130	.246	"	42	10.70	3.45	56.15	66.50
0300	1-1/2" to 2-1/2" thick, veneer									
0350	Low price, gray, light gray, etc.	D-10	130	.246	S.F.	28.50	10.70	3.45	42.65	51.50
0550	High price, red, black, etc.	"	130	.246	"	52.50	10.70	3.45	66.65	77.50
0700	2-1/2" to 4" thick, veneer									
0750	Low price, gray, light gray, etc.	D-10	110	.291	S.F.	38.50	12.65	4.08	55.23	66
0950	High price, red, black, etc.	"	110	.291		63	12.65	4.08	79.73	93
1000	For bush hammered finish, deduct					5%				
1050	Coarse rubbed finish, deduct					10%				
1100	Honed finish, deduct					5%				
1150	Thermal finish, deduct					18%				
2450	For radius under 5', add				L.F.	100%				
2500	Steps, copings, etc., finished on more than one surface									
2550	Minimum	D-10	50	.640	C.F.	91	28	8.95	127.95	152
2600	Maximum	"	50	.640	"	146	28	8.95	182.95	212
2800	Pavers, 4" x 4" x 4" blocks, split face and joints									
2850	Minimum	D-11	80	.300	S.F.	13.10	12.60		25.70	33.50
2900	Maximum	"	80	.300		29	12.60		41.60	51
4000	Soffits, 2" thick, minimum	D-13	35	1.371		37.50	58	12.80	108.30	143
4100	Maximum		35	1.371		91.50	58	12.80	162.30	202
4200	4" thick, minimum		35	1.371		63	58	12.80	133.80	171
4300	Maximum		35	1.371		119	58	12.80	189.80	233

04 43 10.50 Lightweight Natural Stone	Crew	Daily Output	Labor-Hours	Unit	Material	Labor	Equipment	Total	Total Incl O&P
0011	**LIGHTWEIGHT NATURAL STONE** Lava type									
0100	Veneer, rubble face, sawed back, irregular shapes G	D-10	130	.246	S.F.	6.30	10.70	3.45	20.45	27
0200	Sawed face and back, irregular shapes G	"	130	.246	"	6.30	10.70	3.45	20.45	27

04 43 10.55 Limestone	Crew	Daily Output	Labor-Hours	Unit	Material	Labor	Equipment	Total	Total Incl O&P
0010	**LIMESTONE**, cut to size									
0020	Veneer facing panels									
0500	Texture finish, light stick, 4-1/2" thick, 5' x 12'	D-4	300	.107	S.F.	48	4.27	.42	52.69	60
0750	5" thick, 5' x 14' panels	D-10	275	.116		52.50	5.05	1.63	59.18	67

04 43 Stone Masonry

04 43 10 – Masonry with Natural and Processed Stone

04 43 10.55 Limestone		Crew	Daily Output	Labor-Hours	Unit	Material	2012 Bare Costs Labor	Equipment	Total	Total Incl O&P
1000	Sugarcube finish, 2" Thick, 3' x 5' panels	D-10	275	.116	S.F.	35	5.05	1.63	41.68	48
1050	3" Thick, 4' x 9' panels		275	.116		38	5.05	1.63	44.68	51.50
1200	4" Thick, 5' x 11' panels		275	.116		41	5.05	1.63	47.68	54.50
1400	Sugarcube, textured finish, 4-1/2" thick, 5' x 12'		275	.116		45	5.05	1.63	51.68	59
1450	5" thick, 5' x 14' panels		275	.116	↓	53.50	5.05	1.63	60.18	68
2000	Coping, sugarcube finish, top & 2 sides		30	1.067	C.F.	68	46.50	14.95	129.45	161
2100	Sills, lintels, jambs, trim, stops, sugarcube finish, average		20	1.600		68	69.50	22.50	160	205
2150	Detailed		20	1.600	↓	68	69.50	22.50	160	205
2300	Steps, extra hard, 14" wide, 6" rise	↓	50	.640	L.F.	41.50	28	8.95	78.45	98
3000	Quoins, plain finish, 6" x 12" x 12"	D-12	25	1.280	Ea.	38	51.50		89.50	120
3050	6" x 16" x 24"	"	25	1.280	"	50.50	51.50		102	134

04 43 10.60 Marble										
0011	**MARBLE**, ashlar, split face, 4" + or - thick, random									
0040	Lengths 1' to 4' & heights 2" to 7-1/2", average	D-8	175	.229	S.F.	17.75	9.30		27.05	33.50
0100	Base, polished, 3/4" or 7/8" thick, polished, 6" high	D-10	65	.492	L.F.	11.80	21.50	6.90	40.20	53
0300	Carvings or bas relief, from templates, average		80	.400	S.F.	144	17.35	5.60	166.95	190
0350	Maximum	↓	80	.400	"	335	17.35	5.60	357.95	400
0600	Columns, cornices, mouldings, etc.									
0650	Hand or special machine cut, average	D-10	35	.914	C.F.	144	39.50	12.80	196.30	232
0700	Maximum	"	35	.914	"	310	39.50	12.80	362.30	415
1000	Facing, polished finish, cut to size, 3/4" to 7/8" thick									
1050	Average	D-10	130	.246	S.F.	21	10.70	3.45	35.15	43
1100	Maximum		130	.246		36.50	10.70	3.45	50.65	60
1300	1-1/4" thick, average		125	.256		27.50	11.10	3.59	42.19	51.50
1350	Maximum		125	.256		46.50	11.10	3.59	61.19	72
1500	2" thick, average		120	.267		40	11.55	3.74	55.29	65.50
1550	Maximum	↓	120	.267	↓	69	11.55	3.74	84.29	97.50
2200	Window sills, 6" x 3/4" thick	D-1	85	.188	L.F.	10.45	7.50		17.95	23
2500	Flooring, polished tiles, 12" x 12" x 3/8" thick									
2510	Thin set, average	D-11	90	.267	S.F.	11.20	11.20		22.40	29
2600	Maximum		90	.267		99.50	11.20		110.70	126
2700	Mortar bed, average		65	.369		10.60	15.50		26.10	35
2740	Maximum	↓	65	.369		99.50	15.50		115	133
2780	Travertine, 3/8" thick, average	D-10	130	.246		9.25	10.70	3.45	23.40	30
2790	Maximum	"	130	.246		25.50	10.70	3.45	39.65	48
2800	Patio tile, non-slip, 1/2" thick, flame finish	D-11	75	.320	↓	15.25	13.40		28.65	37.50
2900	Shower or toilet partitions, 7/8" thick partitions									
3050	3/4" or 1-1/4" thick stiles, polished 2 sides, average	D-11	75	.320	S.F.	45.50	13.40		58.90	70.50
3201	Soffits, add to above prices				"	20%	100%			
3210	Stairs, risers, 7/8" thick x 6" high	D-10	115	.278	L.F.	15.25	12.10	3.90	31.25	39.50
3360	Treads, 12" wide x 1-1/4" thick	"	115	.278	"	41.50	12.10	3.90	57.50	68
3500	Thresholds, 3' long, 7/8" thick, 4" to 5" wide, plain	D-12	24	1.333	Ea.	19.85	54		73.85	103
3550	Beveled		24	1.333	"	22.50	54		76.50	106
3700	Window stools, polished, 7/8" thick, 5" wide	↓	85	.376	L.F.	20.50	15.20		35.70	45.50

04 43 10.75 Sandstone or Brownstone										
0011	**SANDSTONE OR BROWNSTONE**									
0100	Sawed face veneer, 2-1/2" thick, to 2' x 4' panels	D-10	130	.246	S.F.	21	10.70	3.45	35.15	43
0150	4" thick, to 3'-6" x 8' panels		100	.320		21	13.90	4.48	39.38	49
0300	Split face, random sizes	↓	100	.320	↓	14	13.90	4.48	32.38	41.50
0350	Cut stone trim (limestone)									
0360	Ribbon stone, 4" thick, 5' pieces	D-8	120	.333	Ea.	172	13.55		185.55	210
0370	Cove stone, 4" thick, 5' pieces	↓	105	.381	↓	173	15.50		188.50	214

109

04 43 10 – Masonry with Natural and Processed Stone

04 43 10.75 Sandstone or Brownstone	Crew	Daily Output	Labor-Hours	Unit	Material	2012 Bare Costs Labor	Equipment	Total	Total Incl O&P
0380 Cornice stone, 10" to 12" wide	D-8	90	.444	Ea.	214	18.10		232.10	263
0390 Band stone, 4" thick, 5' pieces		145	.276		110	11.25		121.25	138
0410 Window and door trim, 3" to 4" wide		160	.250		93.50	10.20		103.70	118
0420 Key stone, 18" long	↓	60	.667	↓	98.50	27		125.50	149

04 43 10.80 Slate

	Crew	Daily Output	Labor-Hours	Unit	Material	2012 Bare Costs Labor	Equipment	Total	Total Incl O&P
0010 **SLATE**									
0040 Pennsylvania - blue gray to black									
0050 Vermont - unfading green, mottled green & purple, gray & purple									
0100 Virginia - blue black									
0200 Exterior paving, natural cleft, 1" thick									
0500 24" x 24", Pennsylvania	D-12	120	.267	S.F.	13	10.75		23.75	30.50
0550 Vermont		120	.267		31	10.75		41.75	50.50
0600 Virginia	↓	120	.267	↓	20.50	10.75		31.25	39
1000 Interior flooring, natural cleft, 1/2" thick									
1300 24" x 24" Pennsylvania	D-12	120	.267	S.F.	7.75	10.75		18.50	25
1350 Vermont		120	.267		25	10.75		35.75	44
1400 Virginia	↓	120	.267	↓	14.75	10.75		25.50	32.50
2000 Facing panels, 1-1/4" thick, to 4' x 4' panels									
2100 Natural cleft finish, Pennsylvania	D-10	180	.178	S.F.	34	7.70	2.49	44.19	51.50
2110 Vermont		180	.178		27	7.70	2.49	37.19	44.50
2120 Virginia	↓	180	.178		33	7.70	2.49	43.19	51
2150 Sand rubbed finish, surface, add					10.15			10.15	11.20
2200 Honed finish, add					7.35			7.35	8.10
2500 Ribbon, natural cleft finish, 1" thick, to 9 S.F.	D-10	80	.400		13.10	17.35	5.60	36.05	46.50
2700 1-1/2" thick		78	.410		17.05	17.80	5.75	40.60	52
2850 2" thick	↓	76	.421	↓	20.50	18.25	5.90	44.65	56.50
3500 Stair treads, sand finish, 1" thick x 12" wide									
3600 3 L.F. to 6 L.F.	D-10	120	.267	L.F.	24	11.55	3.74	39.29	47.50
3700 Ribbon, sand finish, 1" thick x 12" wide									
3750 To 6 L.F.	D-10	120	.267	L.F.	20	11.55	3.74	35.29	43.50
4000 Stools or sills, sand finish, 1" thick, 6" wide	D-12	160	.200		11.50	8.05		19.55	25
4200 10" wide		90	.356		17.75	14.35		32.10	41
4400 2" thick, 6" wide		140	.229		18.50	9.20		27.70	34.50
4600 10" wide	↓	90	.356		29	14.35		43.35	53.50
4800 For lengths over 3', add				↓	25%				

04 43 10.85 Window Sill

	Crew	Daily Output	Labor-Hours	Unit	Material	2012 Bare Costs Labor	Equipment	Total	Total Incl O&P
0010 **WINDOW SILL**									
0020 Bluestone, thermal top, 10" wide, 1-1/2" thick	D-1	85	.188	S.F.	16.05	7.50		23.55	29
0050 2" thick		75	.213	"	18.75	8.50		27.25	33.50
0100 Cut stone, 5" x 8" plain		48	.333	L.F.	11.90	13.30		25.20	33
0200 Face brick on edge, brick, 8" wide		80	.200		5.45	7.95		13.40	18
0400 Marble, 9" wide, 1" thick		85	.188		8.50	7.50		16	20.50
0900 Slate, colored, unfading, honed, 12" wide, 1" thick		85	.188		17.10	7.50		24.60	30
0950 2" thick	↓	70	.229	↓	24	9.10		33.10	40

04 51 Flue Liner Masonry

04 51 10 – Clay Flue Lining

04 51 10.10 Flue Lining	Crew	Daily Output	Labor-Hours	Unit	Material	2012 Bare Costs Labor	Equipment	Total	Total Incl O&P
0010 **FLUE LINING**, including mortar									
0020 Clay, 8" x 8"	D-1	125	.128	V.L.F.	5	5.10		10.10	13.20
0100 8" x 12"		103	.155		7.55	6.20		13.75	17.65
0200 12" x 12"		93	.172		9.75	6.85		16.60	21
0300 12" x 18"		84	.190		19	7.60		26.60	32.50
0400 18" x 18"		75	.213		22.50	8.50		31	38
0500 20" x 20"		66	.242		36	9.65		45.65	54.50
0600 24" x 24"		56	.286		47	11.40		58.40	69
1000 Round, 18" diameter		66	.242		30.50	9.65		40.15	48
1100 24" diameter	↓	47	.340	↓	68	13.55		81.55	95

04 54 Refractory Brick Masonry

04 54 10 – Refractory Brick Work

04 54 10.10 Fire Brick

04 54 10.10 Fire Brick	Crew	Daily Output	Labor-Hours	Unit	Material	2012 Bare Costs Labor	Equipment	Total	Total Incl O&P
0010 **FIRE BRICK**									
0012 Low duty, 2000°F, 9" x 2-1/2" x 4-1/2"	D-1	.60	26.667	M	1,425	1,075		2,500	3,150
0050 High duty, 3000°F	"	.60	26.667	"	2,100	1,075		3,175	3,925

04 54 10.20 Fire Clay

04 54 10.20 Fire Clay	Crew	Daily Output	Labor-Hours	Unit	Material	2012 Bare Costs Labor	Equipment	Total	Total Incl O&P
0010 **FIRE CLAY**									
0020 Gray, high duty, 100 lb. bag				Bag	59			59	65
0050 100 lb. drum, premixed (400 brick per drum)				Drum	73			73	80.50

04 57 Masonry Fireplaces

04 57 10 – Brick or Stone Fireplaces

04 57 10.10 Fireplace

04 57 10.10 Fireplace	Crew	Daily Output	Labor-Hours	Unit	Material	2012 Bare Costs Labor	Equipment	Total	Total Incl O&P
0010 **FIREPLACE**									
0100 Brick fireplace, not incl. foundations or chimneys									
0110 30" x 29" opening, incl. chamber, plain brickwork	D-1	.40	40	Ea.	535	1,600		2,135	3,000
0200 Fireplace box only (110 brick)	"	2	8	"	176	320		496	675
0300 For elaborate brickwork and details, add					35%	35%			
0400 For hearth, brick & stone, add	D-1	2	8	Ea.	197	320		517	695
0410 For steel angle, damper, cleanouts, add		4	4		138	159		297	395
0600 Plain brickwork, incl. metal circulator		.50	32	↓	1,025	1,275		2,300	3,050
0800 Face brick only, standard size, 8" x 2-2/3" x 4"	↓	.30	53.333	M	535	2,125		2,660	3,800

04 71 Manufactured Brick Masonry

04 71 10 – Simulated or Manufactured Brick

04 71 10.10 Simulated Brick

04 71 10.10 Simulated Brick	Crew	Daily Output	Labor-Hours	Unit	Material	2012 Bare Costs Labor	Equipment	Total	Total Incl O&P
0010 **SIMULATED BRICK**									
0020 Aluminum, baked on colors	1 Carp	200	.040	S.F.	3.53	1.76		5.29	6.60
0050 Fiberglass panels		200	.040		3.87	1.76		5.63	6.95
0100 Urethane pieces cemented in mastic		150	.053		6.70	2.35		9.05	10.95
0150 Vinyl siding panels	↓	200	.040		2.67	1.76		4.43	5.65
0160 Cement base, brick, incl. mastic	D-1	100	.160	↓	4.22	6.40		10.62	14.30
0170 Corner		50	.320	V.L.F.	10.55	12.75		23.30	31
0180 Stone face, incl. mastic		100	.160	S.F.	10.20	6.40		16.60	21
0190 Corner	↓	50	.320	V.L.F.	11.25	12.75		24	31.50

04 72 Cast Stone Masonry

04 72 10 – Cast Stone Masonry Features

04 72 10.10 Coping	Crew	Daily Output	Labor-Hours	Unit	Material	2012 Bare Costs Labor	Equipment	Total	Total Incl O&P
0010 **COPING**, stock units									
0050 Precast concrete, 10" wide, 4" tapers to 3-1/2", 8" wall	D-1	75	.213	L.F.	18.15	8.50		26.65	33
0100 12" wide, 3-1/2" tapers to 3", 10" wall		70	.229		19.60	9.10		28.70	35.50
0110 14" wide, 4" tapers to 3-1/2", 12" wall		65	.246		21	9.80		30.80	38.50
0150 16" wide, 4" tapers to 3-1/2", 14" wall		60	.267		23	10.65		33.65	41
0250 Precast concrete corners		40	.400	Ea.	29.50	15.95		45.45	56.50
0300 Limestone for 12" wall, 4" thick		90	.178	L.F.	15.35	7.10		22.45	27.50
0350 6" thick		80	.200		23	7.95		30.95	37.50
0500 Marble, to 4" thick, no wash, 9" wide		90	.178		23	7.10		30.10	36
0550 12" wide		80	.200		35	7.95		42.95	50.50
0700 Terra cotta, 9" wide		90	.178		5.60	7.10		12.70	16.85
0750 12" wide		80	.200		9.20	7.95		17.15	22
0800 Aluminum, for 12" wall		80	.200		12.50	7.95		20.45	26

04 72 20 – Cultured Stone Veneer

04 72 20.10 Cultured Stone Veneer Components

	Crew	Daily Output	Labor-Hours	Unit	Material	2012 Bare Costs Labor	Equipment	Total	Total Incl O&P
0010 **CULTURED STONE VENEER COMPONENTS**									
0110 On wood frame and sheathing substrate, random sized cobbles, corner stones	D-8	70	.571	V.L.F.	12.20	23.50		35.70	48.50
0120 Field stones		140	.286	S.F.	8.85	11.65		20.50	27.50
0130 Random sized flats, corner stones		70	.571	V.L.F.	11.95	23.50		35.45	48
0140 Field stones		140	.286	S.F.	10.10	11.65		21.75	28.50
0150 Horizontal lined ledgestones, corner stones		75	.533	V.L.F.	12.20	21.50		33.70	46.50
0160 Field stones		150	.267	S.F.	8.85	10.85		19.70	26
0170 Random shaped flats, corner stones		65	.615	V.L.F.	12.20	25		37.20	51.50
0180 Field stones		150	.267	S.F.	8.85	10.85		19.70	26
0190 Random shaped/textured face, corner stones		65	.615	V.L.F.	12.20	25		37.20	51.50
0200 Field stones		130	.308	S.F.	8.85	12.50		21.35	28.50
0210 Random shaped river rock, corner stones		65	.615	V.L.F.	12.20	25		37.20	51.50
0220 Field stones		130	.308	S.F.	8.85	12.50		21.35	28.50
0240 On concrete or CMU substrate, random sized cobbles, corner stones		70	.571	V.L.F.	11.50	23.50		35	47.50
0250 Field stones		140	.286	S.F.	8.50	11.65		20.15	27
0260 Random sized flats, corner stones		70	.571	V.L.F.	11.30	23.50		34.80	47.50
0270 Field stones		140	.286	S.F.	9.75	11.65		21.40	28.50
0280 Horizontal lined ledgestones, corner stones		75	.533	V.L.F.	11.50	21.50		33	45.50
0290 Field stones		150	.267	S.F.	8.50	10.85		19.35	26
0300 Random shaped flats, corner stones		70	.571	V.L.F.	11.50	23.50		35	47.50
0310 Field stones		140	.286	S.F.	8.50	11.65		20.15	27
0320 Random shaped/textured face, corner stones		65	.615	V.L.F.	11.50	25		36.50	50.50
0330 Field stones		130	.308	S.F.	8.50	12.50		21	28.50
0340 Random shaped river rock, corner stones		65	.615	V.L.F.	11.50	25		36.50	50.50
0350 Field stones		130	.308	S.F.	8.50	12.50		21	28.50
0360 Cultured stone veneer, #15 felt weather resistant barrier	1 Clab	3700	.002	Sq.	5.65	.08		5.73	6.35
0370 Expanded metal lath, diamond, 2.5 lb./S.Y., galvanized	1 Lath	85	.094	S.Y.	3.02	3.69		6.71	8.70
0390 Water table or window sill, 18" long	1 Bric	80	.100	Ea.	10.60	4.41		15.01	18.30

04 73 Manufactured Stone Masonry

04 73 20 – Simulated or Manufactured Stone

04 73 20.10 Simulated Stone	Crew	Daily Output	Labor-Hours	Unit	Material	2012 Bare Costs Labor	Equipment	Total	Total Incl O&P
0010 **SIMULATED STONE**									
0100 Insulated fiberglass panels, 5/8" ply backer	L-4	200	.120	S.F.	11.40	4.96		16.36	20

Division Notes

	CREW	DAILY OUTPUT	LABOR-HOURS	UNIT	2012 BARE COSTS				TOTAL INCL O&P
					MAT.	LABOR	EQUIP.	TOTAL	

Estimating Tips

05 05 00 Common Work Results for Metals

- Nuts, bolts, washers, connection angles, and plates can add a significant amount to both the tonnage of a structural steel job and the estimated cost. As a rule of thumb, add 10% to the total weight to account for these accessories.

- Type 2 steel construction, commonly referred to as "simple construction," consists generally of field-bolted connections with lateral bracing supplied by other elements of the building, such as masonry walls or x-bracing. The estimator should be aware, however, that shop connections may be accomplished by welding or bolting. The method may be particular to the fabrication shop and may have an impact on the estimated cost.

05 12 23 Structural Steel

- Steel items can be obtained from two sources: a fabrication shop or a metals service center. Fabrication shops can fabricate items under more controlled conditions than can crews in the field. They are also more efficient and can produce items more economically. Metal service centers serve as a source of long mill shapes to both fabrication shops and contractors.

- Most line items in this structural steel subdivision, and most items in 05 50 00 Metal Fabrications, are indicated as being shop fabricated. The bare material cost for these shop fabricated items is the "Invoice Cost" from the shop and includes the mill base price of steel plus mill extras, transportation to the shop, shop drawings and detailing where warranted, shop fabrication and handling, sandblasting and a shop coat of primer paint, all necessary structural bolts, and delivery to the job site. The bare labor cost and bare equipment cost for these shop fabricated items is for field installation or erection.

- Line items in Subdivision 05 12 23.40 Lightweight Framing, and other items scattered in Division 5, are indicated as being field fabricated. The bare material cost for these field fabricated items is the "Invoice Cost" from the metals service center and includes the mill base price of steel plus mill extras, transportation to the metals service center, material handling, and delivery of long lengths of mill shapes to the job site. Material costs for structural bolts and welding rods should be added to the estimate. The bare labor cost and bare equipment cost for these items is for both field fabrication and field installation or erection, and include time for cutting, welding and drilling in the fabricated metal items. Drilling into concrete and fasteners to fasten field fabricated items to other work are not included and should be added to the estimate.

05 20 00 Steel Joist Framing

- In any given project the total weight of open web steel joists is determined by the loads to be supported and the design. However, economies can be realized in minimizing the amount of labor used to place the joists. This is done by maximizing the joist spacing, and therefore minimizing the number of joists required to be installed on the job. Certain spacings and locations may be required by the design, but in other cases maximizing the spacing and keeping it as uniform as possible will keep the costs down.

05 30 00 Steel Decking

- The takeoff and estimating of metal deck involves more than simply the area of the floor or roof and the type of deck specified or shown on the drawings. Many different sizes and types of openings may exist. Small openings for individual pipes or conduits may be drilled after the floor/roof is installed, but larger openings may require special deck lengths as well as reinforcing or structural support. The estimator should determine who will be supplying this reinforcing. Additionally, some deck terminations are part of the deck package, such as screed angles and pour stops, and others will be part of the steel contract, such as angles attached to structural members and cast-in-place angles and plates. The estimator must ensure that all pieces are accounted for in the complete estimate.

05 50 00 Metal Fabrications

- The most economical steel stairs are those that use common materials, standard details, and most importantly, a uniform and relatively simple method of field assembly. Commonly available A36 channels and plates are very good choices for the main stringers of the stairs, as are angles and tees for the carrier members. Risers and treads are usually made by specialty shops, and it is most economical to use a typical detail in as many places as possible. The stairs should be pre-assembled and shipped directly to the site. The field connections should be simple and straightforward to be accomplished efficiently, and with minimum equipment and labor.

Reference Numbers

Reference numbers are shown in shaded boxes at the beginning of some major classifications. These numbers refer to related items in the Reference Section. The reference information may be an estimating procedure, an alternate pricing method, or technical information. *Note:* Not all subdivisions listed here necessarily appear in this publication.

05 01 Maintenance of Metals

05 01 10 – Maintenance of Structural Metal Framing

05 01 10.51 Cleaning of Structural Metal Framing	Crew	Daily Output	Labor-Hours	Unit	Material	2012 Bare Costs Labor	Equipment	Total	Total Incl O&P
0010 **CLEANING OF STRUCTURAL METAL FRAMING**									
6125 Steel surface treatments, PDCA guidelines									
6170 Wire brush, hand (SSPC-SP2)	1 Psst	400	.020	S.F.	.02	.79		.81	1.44
6180 Power tool (SSPC-SP3)	"	700	.011		.06	.45		.51	.88
6215 Pressure washing, up to 5000 psi, 5000-15,000 S.F./day	1 Pord	10000	.001			.03		.03	.05
6220 Steam cleaning, 600 psi @ 300 F, 1250 – 2500 S.F./day		2000	.004			.15		.15	.23
6225 Water blasting, up to 25,000 psi, 1750 – 3500 S.F./day	↓	2500	.003			.12		.12	.18
6230 Brush-off blast (SSPC-SP7)	E-11	1750	.018		.25	.72	.12	1.09	1.61
6235 Com'l blast (SSPC-SP6), loose scale, fine powder rust, 2.0#/S.F. sand		1200	.027		.50	1.06	.18	1.74	2.50
6240 Tight mill scale, little/no rust, 3.0#/S.F. sand		1000	.032		.74	1.27	.21	2.22	3.16
6245 Exist coat blistered/pitted, 4.0#/S.F. sand		875	.037		.99	1.45	.24	2.68	3.76
6250 Exist coat badly pitted/nodules, 6.7#/S.F. sand		825	.039		1.66	1.54	.26	3.46	4.67
6255 Near white blast (SSPC-SP10), loose scale, fine rust, 5.6#/S.F. sand		450	.071		1.39	2.82	.47	4.68	6.75
6260 Tight mill scale, little/no rust, 6.9#/S.F. sand		325	.098		1.71	3.90	.65	6.26	9.10
6265 Exist coat blistered/pitted, 9.0#/S.F. sand		225	.142		2.23	5.65	.94	8.82	12.90
6270 Exist coat badly pitted/nodules, 11.3#/S.F. sand	↓	150	.213	↓	2.80	8.45	1.41	12.66	18.65

05 05 Common Work Results for Metals

05 05 05 – Selective Metals Demolition

05 05 05.10 Selective Demolition, Metals

	Crew	Daily Output	Labor-Hours	Unit	Material	2012 Bare Costs Labor	Equipment	Total	Total Incl O&P
0010 **SELECTIVE DEMOLITION, METALS** R024119-10									
0015 Excludes shores, bracing, cutting, loading, hauling, dumping									
0020 Remove nuts only up to 3/4" diameter	1 Sswk	480	.017	Ea.		.82		.82	1.47
0030 7/8" to 1-1/4" diameter		240	.033			1.64		1.64	2.94
0040 1-3/8" to 2" diameter		160	.050			2.47		2.47	4.41
0060 Unbolt and remove structural bolts up to 3/4" diameter		240	.033			1.64		1.64	2.94
0070 7/8" to 2" diameter		160	.050			2.47		2.47	4.41
0140 Light weight framing members, remove whole or cut up, up to 20 lb.	↓	240	.033			1.64		1.64	2.94
0150 21 – 40 lb.	2 Sswk	210	.076			3.76		3.76	6.70
0160 41 – 80 lb.	3 Sswk	180	.133			6.60		6.60	11.75
0170 81 – 120 lb.	4 Sswk	150	.213			10.55		10.55	18.85
0230 Structural members, remove whole or cut up, up to 500 lb.	E-19	48	.500			24	19.80	43.80	63.50
0240 1/4 – 2 tons	E-18	36	1.111			54.50	26.50	81	124
0250 2 – 5 tons	E-24	30	1.067			52	23	75	115
0260 5 – 10 tons	E-20	24	2.667			129	48.50	177.50	276
0270 10 – 15 tons	E-2	18	3.111			150	83	233	350
0340 Fabricated item, remove whole or cut up, up to 20 lb.	1 Sswk	96	.083			4.11		4.11	7.35
0350 21 – 40 lb.	2 Sswk	84	.190			9.40		9.40	16.80
0360 41 – 80 lb.	3 Sswk	72	.333			16.45		16.45	29.50
0370 81 – 120 lb.	4 Sswk	60	.533			26.50		26.50	47
0380 121 – 500 lb.	E-19	48	.500			24	19.80	43.80	63.50
0390 501 – 1000 lb.	"	36	.667	↓		32.50	26.50	59	84
0500 Steel roof decking, uncovered, bare	B-2	5000	.008	S.F.		.28		.28	.44

05 05 13 – Shop-Applied Coatings for Metal

05 05 13.50 Paints and Protective Coatings

	Crew	Daily Output	Labor-Hours	Unit	Material	2012 Bare Costs Labor	Equipment	Total	Total Incl O&P
0010 **PAINTS AND PROTECTIVE COATINGS**									
5900 Galvanizing structural steel in shop, under 1 ton				Ton	540			540	595
5950 1 ton to 20 tons					495			495	545
6000 Over 20 tons				↓	450			450	495

05 05 21 – Fastening Methods for Metal

05 05 21.10 Cutting Steel	Crew	Daily Output	Labor-Hours	Unit	Material	2012 Bare Costs Labor	Equipment	Total	Total Incl O&P
0010 **CUTTING STEEL**									
0020 Hand burning, incl. preparation, torch cutting & grinding, no staging									
0100 Steel to 1/2" thick	E-25	320	.025	L.F.		1.28	.30	1.58	2.63
0150 3/4" thick		260	.031			1.58	.37	1.95	3.23
0200 1" thick	↓	200	.040	↓		2.05	.49	2.54	4.20

05 05 21.15 Drilling Steel	Crew	Daily Output	Labor-Hours	Unit	Material	2012 Bare Costs Labor	Equipment	Total	Total Incl O&P
0010 **DRILLING STEEL**									
1910 Drilling & layout for steel, up to 1/4" deep, no anchor									
1920 Holes, 1/4" diameter	1 Sswk	112	.071	Ea.	.07	3.53		3.60	6.40
1925 For each additional 1/4" depth, add		336	.024		.07	1.18		1.25	2.18
1930 3/8" diameter		104	.077		.08	3.80		3.88	6.90
1935 For each additional 1/4" depth, add		312	.026		.08	1.27		1.35	2.35
1940 1/2" diameter		96	.083		.09	4.11		4.20	7.45
1945 For each additional 1/4" depth, add		288	.028		.09	1.37		1.46	2.54
1950 5/8" diameter		88	.091		.12	4.49		4.61	8.15
1955 For each additional 1/4" depth, add		264	.030		.12	1.50		1.62	2.80
1960 3/4" diameter		80	.100		.15	4.94		5.09	9
1965 For each additional 1/4" depth, add		240	.033		.15	1.64		1.79	3.11
1970 7/8" diameter		72	.111		.20	5.50		5.70	10
1975 For each additional 1/4" depth, add		216	.037		.20	1.83		2.03	3.49
1980 1" diameter		64	.125		.21	6.15		6.36	11.30
1985 For each additional 1/4" depth, add	↓	192	.042	↓	.21	2.06		2.27	3.91
1990 For drilling up, add						40%			

05 05 21.90 Welding Steel	Crew	Daily Output	Labor-Hours	Unit	Material	2012 Bare Costs Labor	Equipment	Total	Total Incl O&P
0010 **WELDING STEEL**, Structural R050521-20									
0020 Field welding, 1/8" E6011, cost per welder, no oper. engineer	E-14	8	1	Hr.	4.22	51.50	15.20	70.92	113
0200 With 1/2 operating engineer	E-13	8	1.500		4.22	73.50	15.20	92.92	147
0300 With 1 operating engineer	E-12	8	2	↓	4.22	96	15.20	115.42	180
0500 With no operating engineer, 2# weld rod per ton	E-14	8	1	Ton	4.22	51.50	15.20	70.92	113
0600 8# E6011 per ton	"	2	4		16.85	205	61	282.85	450
0800 With one operating engineer per welder, 2# E6011 per ton	E-12	8	2		4.22	96	15.20	115.42	180
0900 8# E6011 per ton	"	2	8	↓	16.85	385	61	462.85	720
1200 Continuous fillet, stick welding, incl. equipment									
1300 Single pass, 1/8" thick, 0.1#/L.F.	E-14	150	.053	L.F.	.21	2.74	.81	3.76	6
1400 3/16" thick, 0.2#/L.F.		75	.107		.42	5.50	1.62	7.54	12.05
1500 1/4" thick, 0.3#/L.F.		50	.160		.63	8.20	2.43	11.26	18.10
1610 5/16" thick, 0.4#/L.F.		38	.211		.84	10.80	3.20	14.84	24
1800 3 passes, 3/8" thick, 0.5#/L.F.		30	.267		1.05	13.70	4.05	18.80	30
2010 4 passes, 1/2" thick, 0.7#/L.F.		22	.364		1.48	18.65	5.55	25.68	41
2200 5 to 6 passes, 3/4" thick, 1.3#/L.F.		12	.667		2.74	34	10.15	46.89	75
2400 8 to 11 passes, 1" thick, 2.4#/L.F.	▼	6	1.333		5.05	68.50	20.50	94.05	150
2600 For all position welding, add, minimum						20%			
2700 Maximum						300%			
2900 For semi-automatic welding, deduct, minimum						5%			
3000 Maximum				▼		15%			
4000 Cleaning and welding plates, bars, or rods									
4010 to existing beams, columns, or trusses	E-14	12	.667	L.F.	1.05	34	10.15	45.20	73.50

05 05 23 – Metal Fastenings

05 05 23.05 Anchor Bolts

	Crew	Daily Output	Labor-Hours	Unit	Material	2012 Bare Costs Labor	Equipment	Total	Total Incl O&P
0010 **ANCHOR BOLTS**									
0015 Made from recycled materials G									

05 05 23.05 Anchor Bolts

		Crew	Daily Output	Labor-Hours	Unit	Material	2012 Bare Costs Labor	Equipment	Total	Total Incl O&P
0025	Single bolts installed in fresh concrete, no templates									
0030	Hooked w/nut and washer, 1/2" diameter, 8" long	G	1 Carp	132	.061	Ea.	1.39	2.67		4.06	5.65
0040	12" long	G		131	.061		1.54	2.69		4.23	5.85
0050	5/8" diameter, 8" long	G		129	.062		3	2.74		5.74	7.50
0060	12" long	G		127	.063		3.70	2.78		6.48	8.35
0070	3/4" diameter, 8" long	G		127	.063		3.70	2.78		6.48	8.35
0080	12" long	G		125	.064		4.62	2.82		7.44	9.45
0090	2-bolt pattern, including job-built 2-hole template, per set									
0100	J-type, incl. hex nut & washer, 1/2" diameter x 6" long	G	1 Carp	21	.381	Set	4.55	16.80		21.35	31
0110	12" long	G		21	.381		5.15	16.80		21.95	31.50
0120	18" long	G		21	.381		6.10	16.80		22.90	32.50
0130	3/4" diameter x 8" long	G		20	.400		9.50	17.65		27.15	37.50
0140	12" long	G		20	.400		11.35	17.65		29	39.50
0150	18" long	G		20	.400		14.10	17.65		31.75	42.50
0160	1" diameter x 12" long	G		19	.421		19.90	18.55		38.45	50.50
0170	18" long	G		19	.421		23.50	18.55		42.05	54.50
0180	24" long	G		19	.421		28.50	18.55		47.05	59.50
0190	36" long	G		18	.444		38	19.60		57.60	72
0200	1-1/2" diameter x 18" long	G		17	.471		60	21		81	98
0210	24" long	G		16	.500		71.50	22		93.50	113
0300	L-type, incl. hex nut & washer, 3/4" diameter x 12" long	G		20	.400		10.75	17.65		28.40	39
0310	18" long	G		20	.400		13.15	17.65		30.80	41.50
0320	24" long	G		20	.400		15.60	17.65		33.25	44
0330	30" long	G		20	.400		19.25	17.65		36.90	48
0340	36" long	G		20	.400		21.50	17.65		39.15	51
0350	1" diameter x 12" long	G		19	.421		16.95	18.55		35.50	47
0360	18" long	G		19	.421		20.50	18.55		39.05	51
0370	24" long	G		19	.421		25	18.55		43.55	56
0380	30" long	G		19	.421		29	18.55		47.55	60.50
0390	36" long	G		18	.444		33	19.60		52.60	66.50
0400	42" long	G		18	.444		39.50	19.60		59.10	73.50
0410	48" long	G		18	.444		44	19.60		63.60	78.50
0420	1-1/4" diameter x 18" long	G		18	.444		30	19.60		49.60	63
0430	24" long	G		18	.444		35	19.60		54.60	69
0440	30" long	G		17	.471		40.50	21		61.50	76.50
0450	36" long	G		17	.471		45.50	21		66.50	82.50
0460	42" long	G	2 Carp	32	.500		51.50	22		73.50	90.50
0470	48" long	G		32	.500		58.50	22		80.50	98
0480	54" long	G		31	.516		68.50	23		91.50	110
0490	60" long	G		31	.516		75	23		98	118
0500	1-1/2" diameter x 18" long	G		33	.485		43.50	21.50		65	81
0510	24" long	G		32	.500		50.50	22		72.50	89.50
0520	30" long	G		31	.516		57	23		80	97.50
0530	36" long	G		30	.533		65	23.50		88.50	108
0540	42" long	G		30	.533		74	23.50		97.50	117
0550	48" long	G		29	.552		82.50	24.50		107	129
0560	54" long	G		28	.571		100	25		125	149
0570	60" long	G		28	.571		110	25		135	160
0580	1-3/4" diameter x 18" long	G		31	.516		61.50	23		84.50	103
0590	24" long	G		30	.533		72	23.50		95.50	115
0600	30" long	G		29	.552		83.50	24.50		108	129
0610	36" long	G		28	.571		95	25		120	143
0620	42" long	G		27	.593		106	26		132	157

05 05 23 – Metal Fastenings

05 05 23.05 Anchor Bolts			Crew	Daily Output	Labor-Hours	Unit	Material	2012 Bare Costs Labor	Equipment	Total	Total Incl O&P
0630	48" long	G	2 Carp	26	.615	Set	116	27		143	170
0640	54" long	G		26	.615		144	27		171	200
0650	60" long	G		25	.640		155	28		183	215
0660	2" diameter x 24" long	G		27	.593		96.50	26		122.50	146
0670	30" long	G		27	.593		108	26		134	159
0680	36" long	G		26	.615		119	27		146	173
0690	42" long	G		25	.640		132	28		160	190
0700	48" long	G		24	.667		152	29.50		181.50	212
0710	54" long	G		23	.696		180	30.50		210.50	245
0720	60" long	G		23	.696		194	30.50		224.50	260
0730	66" long	G		22	.727		207	32		239	278
0740	72" long	G		21	.762		227	33.50		260.50	300
1000	4-bolt pattern, including job-built 4-hole template, per set										
1100	J-type, incl. hex nut & washer, 1/2" diameter x 6" long	G	1 Carp	19	.421	Set	7	18.55		25.55	36
1110	12" long	G		19	.421		8.25	18.55		26.80	37.50
1120	18" long	G		18	.444		10.10	19.60		29.70	41
1130	3/4" diameter x 8" long	G		17	.471		16.85	21		37.85	50.50
1140	12" long	G		17	.471		20.50	21		41.50	54.50
1150	18" long	G		17	.471		26	21		47	60.50
1160	1" diameter x 12" long	G		16	.500		38	22		60	75.50
1170	18" long	G		15	.533		45	23.50		68.50	85.50
1180	24" long	G		15	.533		54.50	23.50		78	96
1190	36" long	G		15	.533		74	23.50		97.50	118
1200	1-1/2" diameter x 18" long	G		13	.615		118	27		145	172
1210	24" long	G		12	.667		140	29.50		169.50	200
1300	L-type, incl. hex nut & washer, 3/4" diameter x 12" long	G		17	.471		19.35	21		40.35	53.50
1310	18" long	G		17	.471		24	21		45	58.50
1320	24" long	G		17	.471		29	21		50	64
1330	30" long	G		16	.500		36.50	22		58.50	74
1340	36" long	G		16	.500		41	22		63	79.50
1350	1" diameter x 12" long	G		16	.500		32	22		54	69
1360	18" long	G		15	.533		39	23.50		62.50	79
1370	24" long	G		15	.533		48	23.50		71.50	88.50
1380	30" long	G		15	.533		56	23.50		79.50	97.50
1390	36" long	G		15	.533		64	23.50		87.50	107
1400	42" long	G		14	.571		77	25		102	124
1410	48" long	G		14	.571		86.50	25		111.50	134
1420	1-1/4" diameter x 18" long	G		14	.571		58	25		83	103
1430	24" long	G		14	.571		68.50	25		93.50	114
1440	30" long	G		13	.615		79	27		106	129
1450	36" long	G		13	.615		89.50	27		116.50	140
1460	42" long	G	2 Carp	25	.640		101	28		129	155
1470	48" long	G		24	.667		115	29.50		144.50	171
1480	54" long	G		23	.696		135	30.50		165.50	195
1490	60" long	G		23	.696		148	30.50		178.50	209
1500	1-1/2" diameter x 18" long	G		25	.640		85	28		113	137
1510	24" long	G		24	.667		99	29.50		128.50	154
1520	30" long	G		23	.696		112	30.50		142.50	170
1530	36" long	G		22	.727		128	32		160	191
1540	42" long	G		22	.727		146	32		178	210
1550	48" long	G		21	.762		163	33.50		196.50	231
1560	54" long	G		20	.800		198	35.50		233.50	273
1570	60" long	G		20	.800		217	35.50		252.50	294

05 05 23.05 Anchor Bolts

		Crew	Daily Output	Labor-Hours	Unit	Material	2012 Bare Costs Labor	2012 Bare Costs Equipment	Total	Total Incl O&P
1580	1-3/4" diameter x 18" long	G	2 Carp	22	.727	Set	121	32		153	183
1590	24" long	G		21	.762		142	33.50		175.50	208
1600	30" long	G		21	.762		165	33.50		198.50	233
1610	36" long	G		20	.800		187	35.50		222.50	261
1620	42" long	G		19	.842		210	37		247	288
1630	48" long	G		18	.889		231	39		270	315
1640	54" long	G		18	.889		286	39		325	375
1650	60" long	G		17	.941		310	41.50		351.50	405
1660	2" diameter x 24" long	G		19	.842		191	37		228	267
1670	30" long	G		18	.889		215	39		254	297
1680	36" long	G		18	.889		236	39		275	320
1690	42" long	G		17	.941		263	41.50		304.50	355
1700	48" long	G		16	1		300	44		344	400
1710	54" long	G		15	1.067		360	47		407	470
1720	60" long	G		15	1.067		385	47		432	500
1730	66" long	G		14	1.143		410	50.50		460.50	535
1740	72" long	G		14	1.143		450	50.50		500.50	575
1990	For galvanized, add					Ea.	75%				

05 05 23.10 Bolts and Hex Nuts

		Crew	Daily Output	Labor-Hours	Unit	Material	2012 Bare Costs Labor	2012 Bare Costs Equipment	Total	Total Incl O&P
0010	**BOLTS & HEX NUTS**, Steel, A307										
0100	1/4" diameter, 1/2" long	G	1 Sswk	140	.057	Ea.	.07	2.82		2.89	5.15
0200	1" long	G		140	.057		.08	2.82		2.90	5.15
0300	2" long	G		130	.062		.11	3.04		3.15	5.55
0400	3" long	G		130	.062		.16	3.04		3.20	5.65
0500	4" long	G		120	.067		.18	3.29		3.47	6.10
0600	3/8" diameter, 1" long	G		130	.062		.15	3.04		3.19	5.60
0700	2" long	G		130	.062		.19	3.04		3.23	5.65
0800	3" long	G		120	.067		.25	3.29		3.54	6.20
0900	4" long	G		120	.067		.32	3.29		3.61	6.25
1000	5" long	G		115	.070		.39	3.43		3.82	6.60
1100	1/2" diameter, 1-1/2" long	G		120	.067		.40	3.29		3.69	6.35
1200	2" long	G		120	.067		.46	3.29		3.75	6.40
1300	4" long	G		115	.070		.74	3.43		4.17	6.95
1400	6" long	G		110	.073		1.03	3.59		4.62	7.55
1500	8" long	G		105	.076		1.35	3.76		5.11	8.20
1600	5/8" diameter, 1-1/2" long	G		120	.067		.96	3.29		4.25	6.95
1700	2" long	G		120	.067		1.07	3.29		4.36	7.05
1800	4" long	G		115	.070		1.55	3.43		4.98	7.85
1900	6" long	G		110	.073		2	3.59		5.59	8.60
2000	8" long	G		105	.076		2.99	3.76		6.75	10
2100	10" long	G		100	.080		3.76	3.95		7.71	11.20
2200	3/4" diameter, 2" long	G		120	.067		1.13	3.29		4.42	7.15
2300	4" long	G		110	.073		1.61	3.59		5.20	8.15
2400	6" long	G		105	.076		2.07	3.76		5.83	9
2500	8" long	G		95	.084		3.12	4.16		7.28	10.90
2600	10" long	G		85	.094		4.09	4.64		8.73	12.80
2700	12" long	G		80	.100		4.79	4.94		9.73	14.10
2800	1" diameter, 3" long	G		105	.076		2.81	3.76		6.57	9.80
2900	6" long	G		90	.089		4.06	4.39		8.45	12.30
3000	12" long	G		75	.107		7.20	5.25		12.45	17.35
3100	For galvanized, add						75%				
3200	For stainless, add						350%				

05 05 Common Work Results for Metals

05 05 23 – Metal Fastenings

05 05 23.15 Chemical Anchors

		Crew	Daily Output	Labor-Hours	Unit	Material	2012 Bare Costs Labor	Equipment	Total	Total Incl O&P
0010	**CHEMICAL ANCHORS**									
0020	Includes layout & drilling									
1430	Chemical anchor, w/rod & epoxy cartridge, 3/4" diam. x 9-1/2" long	B-89A	27	.593	Ea.	8.55	24	4.08	36.63	51
1435	1" diameter x 11-3/4" long		24	.667		15.50	27	4.59	47.09	63.50
1440	1-1/4" diameter x 14" long		21	.762		32	30.50	5.25	67.75	88.50
1445	1-3/4" diameter x 15" long		20	.800		60	32	5.50	97.50	122
1450	18" long		17	.941		72	38	6.50	116.50	145
1455	2" diameter x 18" long		16	1		87	40.50	6.90	134.40	166
1460	24" long		15	1.067		113	43	7.35	163.35	198

05 05 23.20 Expansion Anchors

			Crew	Daily Output	Labor-Hours	Unit	Material	2012 Bare Costs Labor	Equipment	Total	Total Incl O&P
0010	**EXPANSION ANCHORS**										
0100	Anchors for concrete, brick or stone, no layout and drilling										
0200	Expansion shields, zinc, 1/4" diameter, 1-5/16" long, single	G	1 Carp	90	.089	Ea.	.42	3.92		4.34	6.50
0300	1-3/8" long, double	G		85	.094		.54	4.15		4.69	7
0400	3/8" diameter, 1-1/2" long, single	G		85	.094		.73	4.15		4.88	7.20
0500	2" long, double	G		80	.100		1.17	4.41		5.58	8.10
0600	1/2" diameter, 2-1/16" long, single	G		80	.100		1.40	4.41		5.81	8.35
0700	2-1/2" long, double	G		75	.107		1.82	4.70		6.52	9.25
0800	5/8" diameter, 2-5/8" long, single	G		75	.107		2.54	4.70		7.24	10.05
0900	2-3/4" long, double	G		70	.114		2.59	5.05		7.64	10.60
1000	3/4" diameter, 2-3/4" long, single	G		70	.114		2.79	5.05		7.84	10.80
1100	3-15/16" long, double	G		65	.123		5	5.45		10.45	13.85
2100	Hollow wall anchors for gypsum wall board, plaster or tile										
2300	1/8" diameter, short	G	1 Carp	160	.050	Ea.	.20	2.21		2.41	3.61
2400	Long	G		150	.053		.23	2.35		2.58	3.87
2500	3/16" diameter, short	G		150	.053		.36	2.35		2.71	4.02
2600	Long	G		140	.057		.49	2.52		3.01	4.42
2700	1/4" diameter, short	G		140	.057		.53	2.52		3.05	4.46
2800	Long	G		130	.062		.53	2.71		3.24	4.76
3000	Toggle bolts, bright steel, 1/8" diameter, 2" long	G		85	.094		.18	4.15		4.33	6.60
3100	4" long	G		80	.100		.22	4.41		4.63	7.05
3200	3/16" diameter, 3" long	G		80	.100		.22	4.41		4.63	7.05
3300	6" long	G		75	.107		.32	4.70		5.02	7.60
3400	1/4" diameter, 3" long	G		75	.107		.28	4.70		4.98	7.55
3500	6" long	G		70	.114		.45	5.05		5.50	8.25
3600	3/8" diameter, 3" long	G		70	.114		.67	5.05		5.72	8.50
3700	6" long	G		60	.133		1.18	5.90		7.08	10.35
3800	1/2" diameter, 4" long	G		60	.133		1.69	5.90		7.59	10.90
3900	6" long	G		50	.160		2.08	7.05		9.13	13.15
4000	Nailing anchors										
4100	Nylon nailing anchor, 1/4" diameter, 1" long		1 Carp	3.20	2.500	C	13.45	110		123.45	185
4200	1-1/2" long			2.80	2.857		16.60	126		142.60	212
4300	2" long			2.40	3.333		19.50	147		166.50	248
4400	Metal nailing anchor, 1/4" diameter, 1" long	G		3.20	2.500		15.75	110		125.75	187
4500	1-1/2" long	G		2.80	2.857		22	126		148	218
4600	2" long	G		2.40	3.333		26.50	147		173.50	255
5000	Screw anchors for concrete, masonry,										
5100	stone & tile, no layout or drilling included										
5700	Lag screw shields, 1/4" diameter, short	G	1 Carp	90	.089	Ea.	.32	3.92		4.24	6.40
5800	Long	G		85	.094		.39	4.15		4.54	6.85
5900	3/8" diameter, short	G		85	.094		.62	4.15		4.77	7.10
6000	Long	G		80	.100		.76	4.41		5.17	7.65

05 05 23.20 Expansion Anchors

		Crew	Daily Output	Labor-Hours	Unit	Material	2012 Bare Costs Labor	2012 Bare Costs Equipment	Total	Total Incl O&P
6100	1/2" diameter, short	G 1 Carp	80	.100	Ea.	1.02	4.41		5.43	7.90
6200	Long	G	75	.107		1.24	4.70		5.94	8.60
6300	5/8" diameter, short	G	70	.114		1.63	5.05		6.68	9.55
6400	Long	G	65	.123		2.13	5.45		7.58	10.70
6600	Lead, #6 & #8, 3/4" long	G	260	.031		.16	1.36		1.52	2.27
6700	#10 - #14, 1-1/2" long	G	200	.040		.27	1.76		2.03	3.01
6800	#16 & #18, 1-1/2" long	G	160	.050		.32	2.21		2.53	3.74
6900	Plastic, #6 & #8, 3/4" long		260	.031		.04	1.36		1.40	2.13
7000	#8 & #10, 7/8" long		240	.033		.04	1.47		1.51	2.30
7100	#10 & #12, 1" long		220	.036		.05	1.60		1.65	2.53
7200	#14 & #16, 1-1/2" long	▼	160	.050	▼	.07	2.21		2.28	3.47
8000	Wedge anchors, not including layout or drilling									
8050	Carbon steel, 1/4" diameter, 1-3/4" long	G 1 Carp	150	.053	Ea.	.36	2.35		2.71	4.02
8100	3-1/4" long	G	140	.057		.48	2.52		3	4.41
8150	3/8" diameter, 2-1/4" long	G	145	.055		.44	2.43		2.87	4.22
8200	5" long	G	140	.057		.77	2.52		3.29	4.73
8250	1/2" diameter, 2-3/4" long	G	140	.057		.89	2.52		3.41	4.86
8300	7" long	G	125	.064		1.53	2.82		4.35	6
8350	5/8" diameter, 3-1/2" long	G	130	.062		1.60	2.71		4.31	5.95
8400	8-1/2" long	G	115	.070		3.40	3.07		6.47	8.45
8450	3/4" diameter, 4-1/4" long	G	115	.070		2.38	3.07		5.45	7.35
8500	10" long	G	95	.084		5.40	3.71		9.11	11.65
8550	1" diameter, 6" long	G	100	.080		8.60	3.53		12.13	14.90
8575	9" long	G	85	.094		11.20	4.15		15.35	18.70
8600	12" long	G	75	.107		12.05	4.70		16.75	20.50
8650	1-1/4" diameter, 9" long	G	70	.114		24	5.05		29.05	34
8700	12" long	G ▼	60	.133	▼	30.50	5.90		36.40	42.50
8750	For type 303 stainless steel, add					350%				
8800	For type 316 stainless steel, add					450%				
8950	Self-drilling concrete screw, hex washer head, 3/16" diam. x 1-3/4" long	G 1 Carp	300	.027	Ea.	.19	1.18		1.37	2.02
8960	2-1/4" long	G	250	.032		.20	1.41		1.61	2.39
8970	Phillips flat head, 3/16" diam. x 1-3/4" long	G	300	.027		.18	1.18		1.36	2.01
8980	2-1/4" long	G ▼	250	.032	▼	.20	1.41		1.61	2.39

05 05 23.25 High Strength Bolts

		Crew	Daily Output	Labor-Hours	Unit	Material	2012 Bare Costs Labor	2012 Bare Costs Equipment	Total	Total Incl O&P
0010	**HIGH STRENGTH BOLTS** R050523-10									
0020	A325 Type 1, structural steel, bolt-nut-washer set									
0100	1/2" diameter x 1-1/2" long	G 1 Sswk	130	.062	Ea.	.93	3.04		3.97	6.50
0120	2" long	G	125	.064		1.01	3.16		4.17	6.75
0150	3" long	G	120	.067		1.41	3.29		4.70	7.45
0170	5/8" diameter x 1-1/2" long	G	125	.064		1.65	3.16		4.81	7.45
0180	2" long	G	120	.067		1.77	3.29		5.06	7.85
0190	3" long	G	115	.070		2.18	3.43		5.61	8.55
0200	3/4" diameter x 2" long	G	120	.067		2.67	3.29		5.96	8.85
0220	3" long	G	115	.070		3.20	3.43		6.63	9.65
0250	4" long	G	110	.073		3.91	3.59		7.50	10.70
0300	6" long	G	105	.076		5.05	3.76		8.81	12.30
0350	8" long	G	95	.084		10	4.16		14.16	18.45
0360	7/8" diameter x 2" long	G	115	.070		3.73	3.43		7.16	10.25
0365	3" long	G	110	.073		4.41	3.59		8	11.25
0370	4" long	G	105	.076		5.30	3.76		9.06	12.55
0380	6" long	G	100	.080		6.75	3.95		10.70	14.50
0390	8" long	G ▼	90	.089	▼	10.70	4.39		15.09	19.65

05 05 Common Work Results for Metals

05 05 23 – Metal Fastenings

05 05 23.25 High Strength Bolts

			Crew	Daily Output	Labor-Hours	Unit	Material	2012 Bare Costs Labor	Equipment	Total	Total Incl O&P
0400	1" diameter x 2" long	G	1 Sswk	105	.076	Ea.	4.67	3.76		8.43	11.85
0420	3" long	G		100	.080		5.30	3.95		9.25	12.85
0450	4" long	G		95	.084		6	4.16		10.16	14.05
0500	6" long	G		90	.089		8	4.39		12.39	16.65
0550	8" long	G		85	.094		13.90	4.64		18.54	23.50
0600	1-1/4" diameter x 3" long	G		85	.094		10.75	4.64		15.39	20
0650	4" long	G		80	.100		2.64	4.94		7.58	11.75
0700	6" long	G		75	.107		15.25	5.25		20.50	26
0750	8" long	G		70	.114		19.50	5.65		25.15	31.50
1020	A490, bolt-nut-washer set										
1170	5/8" diameter x 1-1/2" long	G	1 Sswk	125	.064	Ea.	2.17	3.16		5.33	8.05
1180	2" long	G		120	.067		2.56	3.29		5.85	8.70
1190	3" long	G		115	.070		3.10	3.43		6.53	9.55
1200	3/4" diameter x 2" long	G		120	.067		3.62	3.29		6.91	9.90
1220	3" long	G		115	.070		4.28	3.43		7.71	10.85
1250	4" long	G		110	.073		4.98	3.59		8.57	11.90
1300	6" long	G		105	.076		7.30	3.76		11.06	14.75
1350	8" long	G		95	.084		12.45	4.16		16.61	21
1360	7/8" diameter x 2" long	G		115	.070		5.45	3.43		8.88	12.15
1365	3" long	G		110	.073		6.40	3.59		9.99	13.45
1370	4" long	G		105	.076		8	3.76		11.76	15.50
1380	6" long	G		100	.080		11.25	3.95		15.20	19.40
1390	8" long	G		90	.089		16.35	4.39		20.74	26
1400	1" diameter x 2" long	G		105	.076		6.45	3.76		10.21	13.80
1420	3" long	G		100	.080		7.85	3.95		11.80	15.65
1450	4" long	G		95	.084		9.05	4.16		13.21	17.40
1500	6" long	G		90	.089		12.15	4.39		16.54	21.50
1550	8" long	G		85	.094		19.35	4.64		23.99	30
1600	1-1/4" diameter x 3" long	G		85	.094		14.75	4.64		19.39	24.50
1650	4" long	G		80	.100		16.90	4.94		21.84	27.50
1700	6" long	G		75	.107		23	5.25		28.25	35
1750	8" long	G		70	.114		30	5.65		35.65	43

05 05 23.30 Lag Screws

			Crew	Daily Output	Labor-Hours	Unit	Material	Labor	Equipment	Total	Total Incl O&P
0010	**LAG SCREWS**										
0020	Steel, 1/4" diameter, 2" long	G	1 Carp	200	.040	Ea.	.07	1.76		1.83	2.79
0100	3/8" diameter, 3" long	G		150	.053		.25	2.35		2.60	3.90
0200	1/2" diameter, 3" long	G		130	.062		.49	2.71		3.20	4.72
0300	5/8" diameter, 3" long	G		120	.067		1.01	2.94		3.95	5.65

05 05 23.35 Machine Screws

			Crew	Daily Output	Labor-Hours	Unit	Material	Labor	Equipment	Total	Total Incl O&P
0010	**MACHINE SCREWS**										
0020	Steel, round head, #8 x 1" long	G	1 Carp	4.80	1.667	C	3.76	73.50		77.26	117
0110	#8 x 2" long	G		2.40	3.333		5.85	147		152.85	232
0200	#10 x 1" long	G		4	2		5.85	88		93.85	142
0300	#10 x 2" long	G		2	4		7.95	176		183.95	280

05 05 23.40 Machinery Anchors

			Crew	Daily Output	Labor-Hours	Unit	Material	Labor	Equipment	Total	Total Incl O&P
0010	**MACHINERY ANCHORS**, heavy duty, incl. sleeve, floating base nut,										
0020	lower stud & coupling nut, fiber plug, connecting stud, washer & nut.										
0030	For flush mounted embedment in poured concrete heavy equip. pads.										
0200	Stud & bolt, 1/2" diameter	G	E-16	40	.400	Ea.	68.50	20	3.04	91.54	115
0300	5/8" diameter	G		35	.457		76	23	3.47	102.47	128
0500	3/4" diameter	G		30	.533		87.50	27	4.05	118.55	149
0600	7/8" diameter	G		25	.640		95.50	32	4.86	132.36	168

05 05 23 – Metal Fastenings

05 05 23.40 Machinery Anchors		Crew	Daily Output	Labor-Hours	Unit	Material	2012 Bare Costs Labor	Equipment	Total	Total Incl O&P
0800	1" diameter	G	E-16	20	.800	Ea.	100	40.50	6.10	146.60	189
0900	1-1/4" diameter	G	↓	15	1.067	↓	133	53.50	8.10	194.60	252

05 05 23.50 Powder Actuated Tools and Fasteners		Crew	Daily Output	Labor-Hours	Unit	Material	2012 Bare Costs Labor	Equipment	Total	Total Incl O&P
0010	**POWDER ACTUATED TOOLS & FASTENERS**									
0020	Stud driver, .22 caliber, buy, minimum				Ea.	95			95	105
0100	Maximum				"	370			370	405
0300	Powder charges for above, low velocity				C	5.10			5.10	5.65
0400	Standard velocity					7			7	7.70
0600	Drive pins & studs, 1/4" & 3/8" diam., to 3" long, minimum	G	1 Carp	4.80	1.667		4.75	73.50		78.25	118
0700	Maximum	G	"	4	2	↓	11.40	88		99.40	149

05 05 23.55 Rivets		Crew	Daily Output	Labor-Hours	Unit	Material	2012 Bare Costs Labor	Equipment	Total	Total Incl O&P
0010	**RIVETS**									
0100	Aluminum rivet & mandrel, 1/2" grip length x 1/8" diameter	G	1 Carp	4.80	1.667	C	6.65	73.50		80.15	120
0200	3/16" diameter	G		4	2		11.75	88		99.75	149
0300	Aluminum rivet, steel mandrel, 1/8" diameter	G		4.80	1.667		9.10	73.50		82.60	123
0400	3/16" diameter	G		4	2		14.60	88		102.60	152
0500	Copper rivet, steel mandrel, 1/8" diameter	G		4.80	1.667		8.35	73.50		81.85	122
0800	Stainless rivet & mandrel, 1/8" diameter	G		4.80	1.667		25	73.50		98.50	141
0900	3/16" diameter	G		4	2		43	88		131	184
1000	Stainless rivet, steel mandrel, 1/8" diameter	G		4.80	1.667		22.50	73.50		96	138
1100	3/16" diameter	G		4	2		31.50	88		119.50	171
1200	Steel rivet and mandrel, 1/8" diameter	G		4.80	1.667		8.05	73.50		81.55	122
1300	3/16" diameter	G	↓	4	2	↓	11.20	88		99.20	148
1400	Hand riveting tool, minimum					Ea.	65.50			65.50	72
1500	Maximum						271			271	298
1600	Power riveting tool, minimum						475			475	525
1700	Maximum					↓	2,650			2,650	2,925

05 05 23.70 Structural Blind Bolts		Crew	Daily Output	Labor-Hours	Unit	Material	2012 Bare Costs Labor	Equipment	Total	Total Incl O&P
0010	**STRUCTURAL BLIND BOLTS**									
0100	1/4" diameter x 1/4" grip	G	1 Sswk	240	.033	Ea.	1.24	1.64		2.88	4.30
0150	1/2" grip	G		216	.037		1.33	1.83		3.16	4.73
0200	3/8" diameter x 1/2" grip	G		232	.034		1.75	1.70		3.45	4.97
0250	3/4" grip	G		208	.038		1.83	1.90		3.73	5.40
0300	1/2" diameter x 1/2" grip	G		224	.036		5.15	1.76		6.91	8.80
0350	3/4" grip	G		200	.040		5.60	1.97		7.57	9.70
0400	5/8" diameter x 3/4" grip	G		216	.037		8.25	1.83		10.08	12.35
0450	1" grip	G	↓	192	.042	↓	9.50	2.06		11.56	14.15

05 05 23.80 Vibration and Bearing Pads		Crew	Daily Output	Labor-Hours	Unit	Material	2012 Bare Costs Labor	Equipment	Total	Total Incl O&P
0010	**VIBRATION & BEARING PADS**									
0300	Laminated synthetic rubber impregnated cotton duck, 1/2" thick		2 Sswk	24	.667	S.F.	66	33		99	132
0400	1" thick			20	.800		130	39.50		169.50	214
0600	Neoprene bearing pads, 1/2" thick			24	.667		25.50	33		58.50	87
0700	1" thick			20	.800		50.50	39.50		90	126
0900	Fabric reinforced neoprene, 5000 psi, 1/2" thick			24	.667		11	33		44	71
1000	1" thick			20	.800		22	39.50		61.50	94.50
1200	Felt surfaced vinyl pads, cork and sisal, 5/8" thick			24	.667		28	33		61	89.50
1300	1" thick			20	.800		50.50	39.50		90	126
1500	Teflon bonded to 10 ga. carbon steel, 1/32" layer			24	.667		49.50	33		82.50	114
1600	3/32" layer			24	.667		74	33		107	141
1800	Bonded to 10 ga. stainless steel, 1/32" layer			24	.667		87.50	33		120.50	156
1900	3/32" layer		↓	24	.667	↓	115	33		148	186
2100	Circular machine leveling pad & stud					Kip	6.20			6.20	6.80

05 05 23 – Metal Fastenings

05 05 23.85 Weld Shear Connectors

		Crew	Daily Output	Labor-Hours	Unit	Material	2012 Bare Costs Labor	Equipment	Total	Total Incl O&P
0010	**WELD SHEAR CONNECTORS**									
0020	3/4" diameter, 3-3/16" long [G]	E-10	960	.017	Ea.	.51	.84	.40	1.75	2.50
0030	3-3/8" long [G]		950	.017		.53	.85	.41	1.79	2.56
0200	3-7/8" long [G]		945	.017		.57	.85	.41	1.83	2.60
0300	4-3/16" long [G]		935	.017		.60	.86	.41	1.87	2.65
0500	4-7/8" long [G]		930	.017		.67	.87	.42	1.96	2.75
0600	5-3/16" long [G]		920	.017		.70	.88	.42	2	2.80
0800	5-3/8" long [G]		910	.018		.70	.89	.42	2.01	2.82
0900	6-3/16" long [G]		905	.018		.77	.89	.43	2.09	2.91
1000	7-3/16" long [G]		895	.018		.96	.90	.43	2.29	3.14
1100	8-3/16" long [G]		890	.018		1.05	.91	.43	2.39	3.25
1500	7/8" diameter, 3-11/16" long [G]		920	.017		.82	.88	.42	2.12	2.94
1600	4-3/16" long [G]		910	.018		.88	.89	.42	2.19	3.02
1700	5-3/16" long [G]		905	.018		1	.89	.43	2.32	3.16
1800	6-3/16" long [G]		895	.018		1.12	.90	.43	2.45	3.31
1900	7-3/16" long [G]		890	.018		1.24	.91	.43	2.58	3.47
2000	8-3/16" long [G]		880	.018		1.36	.92	.44	2.72	3.61

05 05 23.87 Weld Studs

		Crew	Daily Output	Labor-Hours	Unit	Material	2012 Bare Costs Labor	Equipment	Total	Total Incl O&P
0010	**WELD STUDS**									
0020	1/4" diameter, 2-11/16" long [G]	E-10	1120	.014	Ea.	.33	.72	.34	1.39	2.03
0100	4-1/8" long [G]		1080	.015		.31	.75	.36	1.42	2.06
0200	3/8" diameter, 4-1/8" long [G]		1080	.015		.36	.75	.36	1.47	2.12
0300	6-1/8" long [G]		1040	.015		.47	.77	.37	1.61	2.30
0400	1/2" diameter, 2-1/8" long [G]		1040	.015		.34	.77	.37	1.48	2.16
0500	3-1/8" long [G]		1025	.016		.41	.79	.38	1.58	2.27
0600	4-1/8" long [G]		1010	.016		.48	.80	.38	1.66	2.38
0700	5-5/16" long [G]		990	.016		.59	.81	.39	1.79	2.53
0800	6-1/8" long [G]		975	.016		.64	.83	.40	1.87	2.62
0900	8-1/8" long [G]		960	.017		.90	.84	.40	2.14	2.93
1000	5/8" diameter, 2-11/16" long [G]		1000	.016		.58	.81	.39	1.78	2.50
1010	4-3/16" long [G]		990	.016		.72	.81	.39	1.92	2.67
1100	6-9/16" long [G]		975	.016		.94	.83	.40	2.17	2.95
1200	8-3/16" long [G]		960	.017		1.26	.84	.40	2.50	3.33

05 05 23.90 Welding Rod

		Crew	Daily Output	Labor-Hours	Unit	Material	2012 Bare Costs Labor	Equipment	Total	Total Incl O&P
0010	**WELDING ROD**									
0020	Steel, type 6011, 1/8" diam., less than 500#				Lb.	2.11			2.11	2.32
0100	500# to 2,000#					1.90			1.90	2.09
0200	2,000# to 5,000#					1.79			1.79	1.96
0300	5/32" diameter, less than 500#					2.10			2.10	2.31
0310	500# to 2,000#					1.89			1.89	2.08
0320	2,000# to 5,000#					1.78			1.78	1.95
0400	3/16" diam., less than 500#					2.26			2.26	2.49
0500	500# to 2,000#					2.04			2.04	2.24
0600	2,000# to 5,000#					1.92			1.92	2.11
0620	Steel, type 6010, 1/8" diam., less than 500#					2.21			2.21	2.43
0630	500# to 2,000#					1.99			1.99	2.19
0640	2,000# to 5,000#					1.87			1.87	2.06
0650	Steel, type 7018 Low Hydrogen, 1/8" diam., less than 500#					2.66			2.66	2.93
0660	500# to 2,000#					2.40			2.40	2.64
0670	2,000# to 5,000#					2.26			2.26	2.48
0700	Steel, type 7024 Jet Weld, 1/8" diam., less than 500#					2.50			2.50	2.75
0710	500# to 2,000#					2.25			2.25	2.48

05 05 Common Work Results for Metals

05 05 23 – Metal Fastenings

05 05 23.90 Welding Rod

		Crew	Daily Output	Labor-Hours	Unit	Material	2012 Bare Costs Labor	Equipment	Total	Total Incl O&P
0720	2,000# to 5,000#				Lb.	2.12			2.12	2.33
1550	Aluminum, type 4043 TIG, 1/8" diam., less than 10#					4.74			4.74	5.20
1560	10# to 60#					4.27			4.27	4.70
1570	Over 60#					4.01			4.01	4.42
1600	Aluminum, type 5356 TIG, 1/8" diam., less than 10#					5.45			5.45	6
1610	10# to 60#					4.90			4.90	5.40
1620	Over 60#					4.61			4.61	5.05
1900	Cast iron, type 8 Nickel, 1/8" diam., less than 500#					27.50			27.50	30
1910	500# to 1,000#					24.50			24.50	27
1920	Over 1,000#					23			23	25.50
2000	Stainless steel, type 316/316L, 1/8" diam., less than 500#					7.55			7.55	8.30
2100	500# to 1000#					6.80			6.80	7.50
2220	Over 1000#					6.40			6.40	7.05

05 12 Structural Steel Framing

05 12 23 – Structural Steel for Buildings

05 12 23.05 Canopy Framing

			Crew	Daily Output	Labor-Hours	Unit	Material	Labor	Equipment	Total	Total Incl O&P
0010	**CANOPY FRAMING**										
0020	6" and 8" members, shop fabricated	G	E-4	3000	.011	Lb.	1.50	.53	.04	2.07	2.64

05 12 23.10 Ceiling Supports

			Crew	Daily Output	Labor-Hours	Unit	Material	Labor	Equipment	Total	Total Incl O&P
0010	**CEILING SUPPORTS**										
1000	Entrance door/folding partition supports, shop fabricated	G	E-4	60	.533	L.F.	25	26.50	2.03	53.53	77
1100	Linear accelerator door supports	G		14	2.286		114	114	8.70	236.70	340
1200	Lintels or shelf angles, hung, exterior hot dipped galv.	G		267	.120		17.05	5.95	.46	23.46	30
1250	Two coats primer paint instead of galv.	G		267	.120		14.80	5.95	.46	21.21	27.50
1400	Monitor support, ceiling hung, expansion bolted	G		4	8	Ea.	395	400	30.50	825.50	1,175
1450	Hung from pre-set inserts	G		6	5.333		425	266	20.50	711.50	970
1600	Motor supports for overhead doors	G		4	8		202	400	30.50	632.50	970
1700	Partition support for heavy folding partitions, without pocket	G		24	1.333	L.F.	57	66.50	5.05	128.55	187
1750	Supports at pocket only	G		12	2.667		114	133	10.15	257.15	375
2000	Rolling grilles & fire door supports	G		34	.941		49	47	3.58	99.58	141
2100	Spider-leg light supports, expansion bolted to ceiling slab	G		8	4	Ea.	163	199	15.20	377.20	550
2150	Hung from pre-set inserts	G		12	2.667	"	175	133	10.15	318.15	440
2400	Toilet partition support	G		36	.889	L.F.	57	44.50	3.38	104.88	145
2500	X-ray travel gantry support	G		12	2.667	"	195	133	10.15	338.15	465

05 12 23.15 Columns, Lightweight

			Crew	Daily Output	Labor-Hours	Unit	Material	Labor	Equipment	Total	Total Incl O&P
0010	**COLUMNS, LIGHTWEIGHT**										
1000	Lightweight units (lally), 3-1/2" diameter		E-2	780	.072	L.F.	7.50	3.46	1.92	12.88	16.30
1050	4" diameter		"	900	.062	"	9	3	1.66	13.66	16.90
5800	Adjustable jack post, 8' maximum height, 2-3/4" diameter	G				Ea.	44.50			44.50	49
5850	4" diameter	G				"	71.50			71.50	78.50

05 12 23.17 Columns, Structural

			Crew	Daily Output	Labor-Hours	Unit	Material	Labor	Equipment	Total	Total Incl O&P
0010	**COLUMNS, STRUCTURAL**	R051223-10									
0015	Made from recycled materials	G									
0020	Shop fab'd for 100-ton, 1-2 story project, bolted connections										
0800	Steel, concrete filled, extra strong pipe, 3-1/2" diameter		E-2	660	.085	L.F.	41.50	4.09	2.27	47.86	55
0830	4" diameter			780	.072		46	3.46	1.92	51.38	58.50
0890	5" diameter			1020	.055		55	2.65	1.47	59.12	66
0930	6" diameter			1200	.047		72.50	2.25	1.25	76	85
0940	8" diameter			1100	.051		72.50	2.46	1.36	76.32	85.50

05 12 Structural Steel Framing

05 12 23 – Structural Steel for Buildings

05 12 23.17 Columns, Structural

		Crew	Daily Output	Labor-Hours	Unit	Material	2012 Bare Costs Labor	Equipment	Total	Total Incl O&P
1100	For galvanizing, add				Lb.	.25			.25	.27
1300	For web ties, angles, etc., add per added lb.	1 Sswk	945	.008		1.25	.42		1.67	2.13
1500	Steel pipe, extra strong, no concrete, 3" to 5" diameter [G]	E-2	16000	.004		1.25	.17	.09	1.51	1.77
1600	6" to 12" diameter [G]		14000	.004	↓	1.25	.19	.11	1.55	1.83
1700	Steel pipe, extra strong, no concrete, 3" diameter x 12'-0" [G]		60	.933	Ea.	154	45	25	224	274
1750	4" diameter x 12'-0" [G]		58	.966		225	46.50	26	297.50	355
1800	6" diameter x 12'-0" [G]		54	1.037		430	50	27.50	507.50	585
1850	8" diameter x 14'-0" [G]		50	1.120		760	54	30	844	960
1900	10" diameter x 16'-0" [G]		48	1.167		1,100	56.50	31	1,187.50	1,325
1950	12" diameter x 18'-0" [G]		45	1.244	↓	1,475	60	33.50	1,568.50	1,775
3300	Structural tubing, square, A500GrB, 4" to 6" square, light section [G]		11270	.005	Lb.	1.25	.24	.13	1.62	1.94
3600	Heavy section [G]	↓	32000	.002	"	1.25	.08	.05	1.38	1.57
4000	Concrete filled, add				L.F.	4.03			4.03	4.43
4500	Structural tubing, sq, 4" x 4" x 1/4" x 12'-0" [G]	E-2	58	.966	Ea.	206	46.50	26	278.50	335
4550	6" x 6" x 1/4" x 12'-0" [G]		54	1.037		340	50	27.50	417.50	485
4600	8" x 8" x 3/8" x 14'-0" [G]		50	1.120		730	54	30	814	930
4650	10" x 10" x 1/2" x 16'-0" [G]		48	1.167	↓	1,350	56.50	31	1,437.50	1,625
5100	Structural tubing, rect, 5" to 6" wide, light section [G]		8000	.007	Lb.	1.25	.34	.19	1.78	2.17
5200	Heavy section [G]		12000	.005		1.25	.23	.12	1.60	1.91
5300	7" to 10" wide, light section [G]		15000	.004		1.25	.18	.10	1.53	1.80
5400	Heavy section [G]		18000	.003	↓	1.25	.15	.08	1.48	1.73
5500	Structural tubing, rect, 5" x 3" x 1/4" x 12'-0" [G]		58	.966	Ea.	200	46.50	26	272.50	330
5550	6" x 4" x 5/16" x 12'-0" [G]		54	1.037		315	50	27.50	392.50	460
5600	8" x 4" x 3/8" x 12'-0" [G]		54	1.037		455	50	27.50	532.50	615
5650	10" x 6" x 3/8" x 14'-0" [G]		50	1.120		730	54	30	814	930
5700	12" x 8" x 1/2" x 16'-0" [G]		48	1.167	↓	1,350	56.50	31	1,437.50	1,600
6800	W Shape, A992 steel, 2 tier, W8 x 24 [G]		1080	.052	L.F.	33	2.50	1.39	36.89	42.50
6850	W8 x 31 [G]		1080	.052		42.50	2.50	1.39	46.39	53
6900	W8 x 48 [G]		1032	.054		66	2.62	1.45	70.07	78.50
6950	W8 x 67 [G]		984	.057		92	2.75	1.52	96.27	107
7000	W10 x 45 [G]		1032	.054		62	2.62	1.45	66.07	74
7050	W10 x 68 [G]		984	.057		93.50	2.75	1.52	97.77	109
7100	W10 x 112 [G]		960	.058		154	2.82	1.56	158.38	176
7150	W12 x 50 [G]		1032	.054		69	2.62	1.45	73.07	81.50
7200	W12 x 87 [G]		984	.057		120	2.75	1.52	124.27	138
7250	W12 x 120 [G]		960	.058		165	2.82	1.56	169.38	189
7300	W12 x 190 [G]		912	.061		261	2.96	1.64	265.60	294
7350	W14 x 74 [G]		984	.057		102	2.75	1.52	106.27	118
7400	W14 x 120 [G]		960	.058		165	2.82	1.56	169.38	189
7450	W14 x 176 [G]	↓	912	.061	↓	242	2.96	1.64	246.60	273
8090	For projects 75 to 99 tons, add				All	10%				
8092	50 to 74 tons, add					20%				
8094	25 to 49 tons, add					30%	10%			
8096	10 to 24 tons, add					50%	25%			
8098	2 to 9 tons, add					75%	50%			
8099	Less than 2 tons, add				↓	100%	100%			

05 12 23.20 Curb Edging

		Crew	Daily Output	Labor-Hours	Unit	Material	2012 Bare Costs Labor	Equipment	Total	Total Incl O&P
0010	**CURB EDGING**									
0020	Steel angle w/anchors, shop fabricated, on forms, 1" x 1", 0.8#/L.F. [G]	E-4	350	.091	L.F.	1.58	4.56	.35	6.49	10.25
0100	2" x 2" angles, 3.92#/L.F. [G]		330	.097		6.25	4.83	.37	11.45	15.95
0200	3" x 3" angles, 6.1#/L.F. [G]		300	.107		9.85	5.30	.41	15.56	21
0300	4" x 4" angles, 8.2#/L.F. [G]	↓	275	.116	↓	13	5.80	.44	19.24	25

05 12 Structural Steel Framing

05 12 23 - Structural Steel for Buildings

05 12 23.20 Curb Edging		Crew	Daily Output	Labor-Hours	Unit	Material	2012 Bare Costs Labor	Equipment	Total	Total Incl O&P
1000	6" x 4" angles, 12.3#/L.F.	G	E-4	250	.128	L.F.	19.15	6.40	.49	26.04	33
1050	Steel channels with anchors, on forms, 3" channel, 5#/L.F.	G		290	.110		7.90	5.50	.42	13.82	18.95
1100	4" channel, 5.4#/L.F.	G		270	.119		8.50	5.90	.45	14.85	20.50
1200	6" channel, 8.2#/L.F.	G		255	.125		13	6.25	.48	19.73	26
1300	8" channel, 11.5#/L.F.	G		225	.142		17.95	7.10	.54	25.59	33
1400	10" channel, 15.3#/L.F.	G		180	.178		23.50	8.85	.68	33.03	42.50
1500	12" channel, 20.7#/L.F.	G		140	.229		32	11.40	.87	44.27	56.50
2000	For curved edging, add						35%	10%			

05 12 23.40 Lightweight Framing		Crew	Daily Output	Labor-Hours	Unit	Material	2012 Bare Costs Labor	Equipment	Total	Total Incl O&P
0010	**LIGHTWEIGHT FRAMING** R051223-35										
0015	Made from recycled materials	G									
0200	For load-bearing steel studs see Section 05 41 13.30										
0400	Angle framing, field fabricated, 4" and larger R051223-45	G	E-3	440	.055	Lb.	.73	2.73	.28	3.74	6
0450	Less than 4" angles	G		265	.091	"	.75	4.53	.46	5.74	9.45
0460	1/2" x 1/2" x 1/8"	G		200	.120	L.F.	.15	6	.61	6.76	11.60
0462	3/4" x 3/4" x 1/8"	G		160	.150		.42	7.50	.76	8.68	14.70
0464	1" x 1" x 1/8"	G		135	.178		.60	8.90	.90	10.40	17.55
0466	1-1/4" x 1-1/4" x 3/16"	G		115	.209		1.11	10.45	1.06	12.62	21
0468	1-1/2" x 1-1/2" x 3/16"	G		100	.240		1.35	12	1.22	14.57	24.50
0470	2" x 2" x 1/4"	G		90	.267		2.39	13.35	1.35	17.09	28
0472	2-1/2" x 2-1/2" x 1/4"	G		72	.333		3.08	16.65	1.69	21.42	35
0474	3" x 2" x 3/8"	G		65	.369		4.43	18.45	1.87	24.75	40
0476	3" x 3" x 3/8"	G		57	.421		5.40	21	2.13	28.53	46
0600	Channel framing, field fabricated, 8" and larger	G		500	.048	Lb.	.75	2.40	.24	3.39	5.40
0650	Less than 8" channels	G		335	.072	"	.75	3.58	.36	4.69	7.65
0660	C2 x 1.78	G		115	.209	L.F.	1.34	10.45	1.06	12.85	21.50
0662	C3 x 4.1	G		80	.300		3.08	15	1.52	19.60	32
0664	C4 x 5.4	G		66	.364		4.05	18.20	1.84	24.09	39
0666	C5 x 6.7	G		57	.421		5.05	21	2.13	28.18	45.50
0668	C6 x 8.2	G		55	.436		5.95	22	2.21	30.16	48
0670	C7 x 9.8	G		40	.600		7.35	30	3.04	40.39	65
0672	C8 x 11.5	G		36	.667		8.65	33.50	3.38	45.53	72.50
0710	Structural bar tee, field fabricated, 3/4" x 3/4" x 1/8"	G		160	.150		.42	7.50	.76	8.68	14.70
0712	1" x 1" x 1/8"	G		135	.178		.60	8.90	.90	10.40	17.55
0714	1-1/2" x 1-1/2" x 1/4"	G		114	.211		1.76	10.55	1.07	13.38	22
0716	2" x 2" x 1/4"	G		89	.270		2.39	13.50	1.37	17.26	28
0718	2-1/2" x 2-1/2" x 3/8"	G		72	.333		4.43	16.65	1.69	22.77	36.50
0720	3" x 3" x 3/8"	G		57	.421		5.40	21	2.13	28.53	46
0730	Structural zee, field fabricated, 1-1/4" x 1-3/4" x 1-3/4"	G		114	.211		.57	10.55	1.07	12.19	20.50
0732	2-11/16" x 3" x 2-11/16"	G		114	.211		1.34	10.55	1.07	12.96	21.50
0734	3-1/16" x 4" x 3-1/16"	G		133	.180		2.02	9.05	.91	11.98	19.40
0736	3-1/4" x 5" x 3-1/4"	G		133	.180		2.75	9.05	.91	12.71	20
0738	3-1/2" x 6" x 3-1/2"	G		160	.150		4.15	7.50	.76	12.41	18.80
0740	Junior beam, field fabricated, 3"	G		80	.300		4.28	15	1.52	20.80	33.50
0742	4"	G		72	.333		5.80	16.65	1.69	24.14	38
0744	5"	G		67	.358		7.50	17.90	1.82	27.22	42.50
0746	6"	G		62	.387		9.40	19.35	1.96	30.71	47
0748	7"	G		57	.421		11.50	21	2.13	34.63	52.50
0750	8"	G		53	.453		13.80	22.50	2.30	38.60	58
1000	Continuous slotted channel framing system, shop fab, minimum	G	2 Sswk	2400	.007	Lb.	3.88	.33		4.21	4.85
1200	Maximum	G	"	1600	.010		4.38	.49		4.87	5.70
1300	Cross bracing, rods, shop fabricated, 3/4" diameter	G	E-3	700	.034		1.50	1.72	.17	3.39	4.91

05 12 23.40 Lightweight Framing		Crew	Daily Output	Labor-Hours	Unit	Material	2012 Bare Costs		Total	Total Incl O&P
							Labor	Equipment		
1310	7/8" diameter	G	E-3	850	.028	Lb.	1.50	1.41	.14	3.05	4.34
1320	1" diameter	G		1000	.024		1.50	1.20	.12	2.82	3.93
1330	Angle, 5" x 5" x 3/8"	G		2800	.009		1.50	.43	.04	1.97	2.47
1350	Hanging lintels, shop fabricated, average	G		850	.028		1.50	1.41	.14	3.05	4.34
1380	Roof frames, shop fabricated, 3'-0" square, 5' span	G	E-2	4200	.013		1.50	.64	.36	2.50	3.14
1400	Tie rod, not upset, 1-1/2" to 4" diameter, with turnbuckle	G	2 Sswk	800	.020		1.63	.99		2.62	3.56
1420	No turnbuckle	G		700	.023		1.56	1.13		2.69	3.74
1500	Upset, 1-3/4" to 4" diameter, with turnbuckle	G		800	.020		1.63	.99		2.62	3.56
1520	No turnbuckle	G		700	.023		1.56	1.13		2.69	3.74

05 12 23.45 Lintels		Crew	Daily Output	Labor-Hours	Unit	Material	Labor	Equipment	Total	Total Incl O&P
0010	**LINTELS**										
0015	Made from recycled materials	G									
0020	Plain steel angles, shop fabricated, under 500 lb.	G	1 Bric	550	.015	Lb.	.96	.64		1.60	2.03
0100	500 to 1000 lb.	G		640	.013		.94	.55		1.49	1.86
0200	1,000 to 2,000 lb.	G		640	.013		.91	.55		1.46	1.83
0300	2,000 to 4,000 lb.	G		640	.013		.89	.55		1.44	1.81
0500	For built-up angles and plates, add to above	G					.31			.31	.34
0700	For engineering, add to above						.13			.13	.14
0900	For galvanizing, add to above, under 500 lb.						.30			.30	.33
0950	500 to 2,000 lb.						.27			.27	.30
1000	Over 2,000 lb.						.25			.25	.27
2000	Steel angles, 3-1/2" x 3", 1/4" thick, 2'-6" long	G	1 Bric	47	.170	Ea.	13.50	7.50		21	26
2100	4'-6" long	G		26	.308		24.50	13.55		38.05	47
2600	4" x 3-1/2", 1/4" thick, 5'-0" long	G		21	.381		31	16.80		47.80	59.50
2700	9'-0" long	G		12	.667		56	29.50		85.50	106

05 12 23.60 Pipe Support Framing		Crew	Daily Output	Labor-Hours	Unit	Material	Labor	Equipment	Total	Total Incl O&P
0010	**PIPE SUPPORT FRAMING**										
0020	Under 10#/L.F., shop fabricated	G	E-4	3900	.008	Lb.	1.68	.41	.03	2.12	2.60
0200	10.1 to 15#/L.F.	G		4300	.007		1.65	.37	.03	2.05	2.51
0400	15.1 to 20#/L.F.	G		4800	.007		1.63	.33	.03	1.99	2.41
0600	Over 20#/L.F.	G		5400	.006		1.60	.30	.02	1.92	2.31

05 12 23.65 Plates		Crew	Daily Output	Labor-Hours	Unit	Material	Labor	Equipment	Total	Total Incl O&P
0010	**PLATES** R051223-80										
0015	Made from recycled materials	G									
0020	For connections & stiffener plates, shop fabricated										
0050	1/8" thick (5.1 lb./S.F.)	G				S.F.	6.40			6.40	7
0100	1/4" thick (10.2 lb./S.F.)	G					12.75			12.75	14.05
0300	3/8" thick (15.3 lb./S.F.)	G					19.15			19.15	21
0400	1/2" thick (20.4 lb./S.F.)	G					25.50			25.50	28
0450	3/4" thick (30.6 lb./S.F.)	G					38.50			38.50	42
0500	1" thick (40.8 lb./S.F.)	G					51			51	56
2000	Steel plate, warehouse prices, no shop fabrication										
2100	1/4" thick (10.2 lb./S.F.)	G				S.F.	8.65			8.65	9.50

05 12 23.70 Stressed Skin Steel Roof and Ceiling System		Crew	Daily Output	Labor-Hours	Unit	Material	Labor	Equipment	Total	Total Incl O&P
0010	**STRESSED SKIN STEEL ROOF & CEILING SYSTEM**										
0020	Double panel flat roof, spans to 100'	G	E-2	1150	.049	S.F.	10	2.35	1.30	13.65	16.45
0100	Double panel convex roof, spans to 200'	G		960	.058		16.25	2.82	1.56	20.63	24.50
0200	Double panel arched roof, spans to 300'	G		760	.074		25	3.56	1.97	30.53	36

05 12 23.75 Structural Steel Members										
0010	**STRUCTURAL STEEL MEMBERS** R051223-10										
0015	Made from recycled materials	G									

05 12 23.75 Structural Steel Members		Crew	Daily Output	Labor-Hours	Unit	Material	2012 Bare Costs Labor	Equipment	Total	Total Incl O&P
0020	Shop fab'd for 100-ton, 1-2 story project, bolted connections									
0100	W 6 x 9	G	E-2	600	.093	L.F.	12.40	4.50	2.49	19.39	24
0120	x 15	G		600	.093		20.50	4.50	2.49	27.49	33
0140	x 20	G		600	.093		27.50	4.50	2.49	34.49	41
0300	W 8 x 10	G		600	.093		13.75	4.50	2.49	20.74	25.50
0320	x 15	G		600	.093		20.50	4.50	2.49	27.49	33
0350	x 21	G		600	.093		29	4.50	2.49	35.99	42.50
0360	x 24	G		550	.102		33	4.91	2.72	40.63	48
0370	x 28	G		550	.102		38.50	4.91	2.72	46.13	54
0500	x 31	G		550	.102		42.50	4.91	2.72	50.13	58.50
0520	x 35	G		550	.102		48	4.91	2.72	55.63	64.50
0540	x 48	G		550	.102		66	4.91	2.72	73.63	84
0600	W 10 x 12	G		600	.093		16.50	4.50	2.49	23.49	28.50
0620	x 15	G		600	.093		20.50	4.50	2.49	27.49	33
0700	x 22	G		600	.093		30.50	4.50	2.49	37.49	44
0720	x 26	G		600	.093		36	4.50	2.49	42.99	50
0740	x 33	G		550	.102		45.50	4.91	2.72	53.13	61.50
0900	x 49	G		550	.102		67.50	4.91	2.72	75.13	85.50
1100	W 12 x 16	G		880	.064		22	3.07	1.70	26.77	31
1300	x 22	G		880	.064		30.50	3.07	1.70	35.27	40.50
1500	x 26	G		880	.064		36	3.07	1.70	40.77	46.50
1520	x 35	G		810	.069		48	3.34	1.85	53.19	60.50
1560	x 50	G		750	.075		69	3.60	2	74.60	84
1580	x 58	G		750	.075		80	3.60	2	85.60	96
1700	x 72	G		640	.088		99	4.22	2.34	105.56	119
1740	x 87	G		640	.088		120	4.22	2.34	126.56	142
1900	W 14 x 26	G		990	.057		36	2.73	1.51	40.24	46
2100	x 30	G		900	.062		41.50	3	1.66	46.16	52.50
2300	x 34	G		810	.069		47	3.34	1.85	52.19	59
2320	x 43	G		810	.069		59	3.34	1.85	64.19	72.50
2340	x 53	G		800	.070		73	3.38	1.87	78.25	88
2360	x 74	G		760	.074		102	3.56	1.97	107.53	120
2380	x 90	G		740	.076		124	3.65	2.02	129.67	144
2500	x 120	G		720	.078		165	3.75	2.08	170.83	191
2700	W 16 x 26	G		1000	.056		36	2.70	1.50	40.20	46
2900	x 31	G		900	.062		42.50	3	1.66	47.16	54
3100	x 40	G		800	.070		55	3.38	1.87	60.25	68.50
3120	x 50	G		800	.070		69	3.38	1.87	74.25	83.50
3140	x 67	G		760	.074		92	3.56	1.97	97.53	109
3300	W 18 x 35	G	E-5	960	.083		48	4.07	1.69	53.76	62
3500	x 40	G		960	.083		55	4.07	1.69	60.76	69.50
3520	x 46	G		960	.083		63.50	4.07	1.69	69.26	78.50
3700	x 50	G		912	.088		69	4.28	1.77	75.05	85
3900	x 55	G		912	.088		75.50	4.28	1.77	81.55	92.50
3920	x 65	G		900	.089		89.50	4.34	1.80	95.64	108
3940	x 76	G		900	.089		105	4.34	1.80	111.14	125
3960	x 86	G		900	.089		118	4.34	1.80	124.14	140
3980	x 106	G		900	.089		146	4.34	1.80	152.14	170
4100	W 21 x 44	G		1064	.075		60.50	3.67	1.52	65.69	74.50
4300	x 50	G		1064	.075		69	3.67	1.52	74.19	83.50
4500	x 62	G		1036	.077		85.50	3.77	1.56	90.83	102
4700	x 68	G		1036	.077		93.50	3.77	1.56	98.83	111
4720	x 83	G		1000	.080		114	3.90	1.62	119.52	135

05 12 23 – Structural Steel for Buildings

05 12 23.75 Structural Steel Members			Crew	Daily Output	Labor-Hours	Unit	Material	2012 Bare Costs Labor	Equipment	Total	Total Incl O&P
4740	x 93	G	E-5	1000	.080	L.F.	128	3.90	1.62	133.52	150
4760	x 101	G		1000	.080		139	3.90	1.62	144.52	162
4780	x 122	G		1000	.080		168	3.90	1.62	173.52	194
4900	W 24 x 55	G		1110	.072		75.50	3.52	1.46	80.48	90.50
5100	x 62	G		1110	.072		85.50	3.52	1.46	90.48	102
5300	x 68	G		1110	.072		93.50	3.52	1.46	98.48	111
5500	x 76	G		1110	.072		105	3.52	1.46	109.98	123
5700	x 84	G		1080	.074		116	3.61	1.50	121.11	135
5720	x 94	G		1080	.074		129	3.61	1.50	134.11	150
5740	x 104	G		1050	.076		143	3.72	1.54	148.26	165
5760	x 117	G		1050	.076		161	3.72	1.54	166.26	185
5780	x 146	G		1050	.076		201	3.72	1.54	206.26	229
5800	W 27 x 84	G		1190	.067		116	3.28	1.36	120.64	134
5900	x 94	G		1190	.067		129	3.28	1.36	133.64	149
5920	x 114	G		1150	.070		157	3.39	1.41	161.80	179
5940	x 146	G		1150	.070		201	3.39	1.41	205.80	228
5960	x 161	G		1150	.070		221	3.39	1.41	225.80	251
6100	W 30 x 99	G		1200	.067		136	3.25	1.35	140.60	157
6300	x 108	G		1200	.067		149	3.25	1.35	153.60	170
6500	x 116	G		1160	.069		160	3.37	1.40	164.77	182
6520	x 132	G		1160	.069		182	3.37	1.40	186.77	207
6540	x 148	G		1160	.069		204	3.37	1.40	208.77	231
6560	x 173	G		1120	.071		238	3.49	1.45	242.94	270
6580	x 191	G		1120	.071		263	3.49	1.45	267.94	297
6700	W 33 x 118	G		1176	.068		162	3.32	1.38	166.70	185
6900	x 130	G		1134	.071		179	3.44	1.43	183.87	205
7100	x 141	G		1134	.071		194	3.44	1.43	198.87	221
7120	x 169	G		1100	.073		232	3.55	1.47	237.02	264
7140	x 201	G		1100	.073		276	3.55	1.47	281.02	315
7300	W 36 x 135	G		1170	.068		186	3.34	1.38	190.72	211
7500	x 150	G		1170	.068		206	3.34	1.38	210.72	234
7600	x 170	G		1150	.070		234	3.39	1.41	238.80	264
7700	x 194	G		1125	.071		267	3.47	1.44	271.91	300
7900	x 231	G		1125	.071		320	3.47	1.44	324.91	360
7920	x 262	G		1035	.077		360	3.77	1.56	365.33	405
8100	x 302	G		1035	.077		415	3.77	1.56	420.33	465
8490	For projects 75 to 99 tons, add						10%				
8492	50 to 74 tons, add						20%				
8494	25 to 49 tons, add						30%	10%			
8496	10 to 24 tons, add						50%	25%			
8498	2 to 9 tons, add						75%	50%			
8499	Less than 2 tons, add						100%	100%			

05 12 23.77 Structural Steel Projects

			Crew	Daily Output	Labor-Hours	Unit	Material	Labor	Equipment	Total	Total Incl O&P
0010	**STRUCTURAL STEEL PROJECTS** R050516-30										
0015	Made from recycled materials	G									
0020	Shop fab'd for 100-ton, 1-2 story project, bolted connections										
0200	Apartments, nursing homes, etc., 1 to 2 stories R050523-10	G	E-5	10.30	7.767	Ton	2,500	380	157	3,037	3,575
0300	3 to 6 stories	G	"	10.10	7.921		2,550	385	160	3,095	3,650
0400	7 to 15 stories R051223-10	G	E-6	14.20	9.014		2,600	440	126	3,166	3,750
0500	Over 15 stories	G	"	13.90	9.209		2,700	450	129	3,279	3,900
0700	Offices, hospitals, etc., steel bearing, 1 to 2 stories R051223-20	G	E-5	10.30	7.767		2,500	380	157	3,037	3,575
0800	3 to 6 stories	G	E-6	14.40	8.889		2,550	435	124	3,109	3,700

05 12 23.77 Structural Steel Projects		Crew	Daily Output	Labor-Hours	Unit	Material	2012 Bare Costs Labor	Equipment	Total	Total Incl O&P
0900	7 to 15 stories R051223-25	G	E-6	14.20	9.014	Ton	2,600	440	126	3,166	3,750
1000	Over 15 stories	G	↓	13.90	9.209		2,700	450	129	3,279	3,900
1100	For multi-story masonry wall bearing construction, add R051223-30							30%			
1300	Industrial bldgs., 1 story, beams & girders, steel bearing	G	E-5	12.90	6.202		2,500	305	125	2,930	3,425
1400	Masonry bearing	G	"	10	8	↓	2,500	390	162	3,052	3,600
1500	Industrial bldgs., 1 story, under 10 tons,										
1510	steel from warehouse, trucked	G	E-2	7.50	7.467	Ton	3,000	360	200	3,560	4,150
1600	1 story with roof trusses, steel bearing	G	E-5	10.60	7.547		2,950	370	153	3,473	4,050
1700	Masonry bearing	G	"	8.30	9.639		2,950	470	195	3,615	4,275
1900	Monumental structures, banks, stores, etc., minimum	G	E-6	13	9.846		2,500	480	138	3,118	3,725
2000	Maximum	G	"	9	14.222		4,150	695	199	5,044	6,000
2200	Churches, minimum	G	E-5	11.60	6.897		2,325	335	140	2,800	3,300
2300	Maximum	G	"	5.20	15.385		3,100	750	310	4,160	5,050
2800	Power stations, fossil fuels, minimum	G	E-6	11	11.636		2,500	570	163	3,233	3,925
2900	Maximum	G		5.70	22.456		3,750	1,100	315	5,165	6,375
2950	Nuclear fuels, non-safety steel, minimum	G		7	18.286		2,500	895	256	3,651	4,575
3000	Maximum	G		5.50	23.273		3,750	1,125	325	5,200	6,450
3040	Safety steel, minimum	G		2.50	51.200		3,650	2,500	715	6,865	9,175
3070	Maximum	G	↓	1.50	85.333		4,800	4,175	1,200	10,175	13,900
3100	Roof trusses, minimum	G	E-5	13	6.154		3,500	300	124	3,924	4,500
3200	Maximum	G		8.30	9.639		4,250	470	195	4,915	5,700
3210	Schools, minimum	G		14.50	5.517		2,500	269	112	2,881	3,350
3220	Maximum	G	↓	8.30	9.639		3,650	470	195	4,315	5,050
3400	Welded construction, simple commercial bldgs., 1 to 2 stories	G	E-7	7.60	10.526		2,550	515	229	3,294	3,950
3500	7 to 15 stories	G	E-9	8.30	15.422		2,950	755	254	3,959	4,825
3700	Welded rigid frame, 1 story, minimum	G	E-7	15.80	5.063		2,600	247	110	2,957	3,400
3800	Maximum	G	"	5.50	14.545	↓	3,375	710	315	4,400	5,300
3810	Fabrication shop costs (included in project material cost, above)										
3820	Mini mill base price, A992	G				Ton	860			860	945
3830	Mill extra for delivery to shop						265			265	292
3840	Shop extra for shop drawings and detailing						285			285	315
3850	Shop fabricating and handling						835			835	920
3860	Shop sandblasting and primer coat of paint						145			145	160
3870	Shop delivery to the job site						110			110	121
3880	Total material cost, shop fabricated, primed, delivered					↓	2,500			2,500	2,750
3900	High strength steel mill spec extras:										
3950	A529, A572 (50 ksi) and A36: same as A992 steel (no extra)										
4000	Add to A992 price for A572 (60, 65 ksi)	G				Ton	100			100	110
4100	A242 and A588 Weathering	G				"	85			85	93.50
4200	Mill size extras for W-Shapes: 0 to 30 plf: no extra charge										
4210	Member sizes 31 to 65 plf, deduct	G				Ton	.01			.01	.01
4220	Member sizes 66 to 100 plf, deduct	G					5.60			5.60	6.15
4230	Member sizes 101 to 387 plf, add	G				↓	56			56	61.50
4300	Column base plates, light, up to 150 lb.	G	2 Sswk	2000	.008	Lb.	1.38	.39		1.77	2.22
4400	Heavy, over 150 lb.	G	E-2	7500	.007	"	1.44	.36	.20	2	2.42
4600	Castellated beams, light sections, to 50#/L.F., minimum	G		10.70	5.234	Ton	2,625	253	140	3,018	3,500
4700	Maximum	G		7	8		2,875	385	214	3,474	4,075
4900	Heavy sections, over 50# per L.F., minimum	G		11.70	4.786		2,750	231	128	3,109	3,550
5000	Maximum	G	↓	7.80	7.179		3,000	345	192	3,537	4,100
5390	For projects 75 to 99 tons, add						10%				
5392	50 to 74 tons, add						20%				
5394	25 to 49 tons, add						30%	10%			
5396	10 to 24 tons, add						50%	25%			

05 12 Structural Steel Framing

05 12 23 – Structural Steel for Buildings

05 12 23.77 Structural Steel Projects	Crew	Daily Output	Labor-Hours	Unit	Material	2012 Bare Costs Labor	Equipment	Total	Total Incl O&P
5398	2 to 9 tons, add				Ton	75%	50%			
5399	Less than 2 tons, add					100%	100%			

05 12 23.80 Subpurlins

	05 12 23.80 Subpurlins	Crew	Daily Output	Labor-Hours	Unit	Material	Labor	Equipment	Total	Total Incl O&P
0010	**SUBPURLINS** R051223-50									
0015	Made from recycled materials [G]									
0020	Bulb tees, shop fabricated, painted, 32-5/8" O.C., 40 psf L.L.									
0100	Type 178, max 8'-9" span, 2.15 plf, 2" high x 1-5/8" wide [G]	E-1	4200	.006	S.F.	1.72	.28	.03	2.03	2.39
0200	Type 218, max 10'-2" span, 3.19 plf, 2-1/8" high x 2-1/8" wide [G]	"	3100	.008		2	.38	.04	2.42	2.88
1420	For 24-5/8" spacing, add					33%	33%			
1430	For 48-5/8" spacing, deduct					50%	50%			

05 14 Structural Aluminum Framing

05 14 23 – Non-Exposed Structural Aluminum Framing

05 14 23.05 Aluminum Shapes

	05 14 23.05 Aluminum Shapes	Crew	Daily Output	Labor-Hours	Unit	Material	Labor	Equipment	Total	Total Incl O&P
0010	**ALUMINUM SHAPES**									
0015	Made from recycled materials [G]									
0020	Structural shapes, 1" to 10" members, under 1 ton [G]	E-2	4000	.014	Lb.	4.18	.68	.37	5.23	6.15
0050	1 to 5 tons [G]		4300	.013		3.82	.63	.35	4.80	5.65
0100	Over 5 tons [G]		4600	.012		3.65	.59	.33	4.57	5.40
0300	Extrusions, over 5 tons, stock shapes [G]		1330	.042		3.39	2.03	1.13	6.55	8.45
0400	Custom shapes [G]		1330	.042		3.39	2.03	1.13	6.55	8.45

05 15 Wire Rope Assemblies

05 15 16 – Steel Wire Rope Assemblies

05 15 16.05 Accessories for Steel Wire Rope

	05 15 16.05 Accessories for Steel Wire Rope	Crew	Daily Output	Labor-Hours	Unit	Material	Labor	Equipment	Total	Total Incl O&P
0010	**ACCESSORIES FOR STEEL WIRE ROPE**									
0015	Made from recycled materials [G]									
1500	Thimbles, heavy duty, 1/4" [G]	E-17	160	.100	Ea.	.51	5.05		5.56	9.55
1510	1/2" [G]		160	.100		2.23	5.05		7.28	11.45
1520	3/4" [G]		105	.152		5.05	7.65		12.70	19.25
1530	1" [G]		52	.308		10.15	15.50		25.65	38.50
1540	1-1/4" [G]		38	.421		15.60	21		36.60	55
1550	1-1/2" [G]		13	1.231		44	62		106	159
1560	1-3/4" [G]		8	2		90.50	101		191.50	280
1570	2" [G]		6	2.667		132	134		266	385
1580	2-1/4" [G]		4	4		178	201		379	555
1600	Clips, 1/4" diameter [G]		160	.100		2.58	5.05		7.63	11.85
1610	3/8" diameter [G]		160	.100		2.83	5.05		7.88	12.10
1620	1/2" diameter [G]		160	.100		4.55	5.05		9.60	14
1630	3/4" diameter [G]		102	.157		7.40	7.90		15.30	22
1640	1" diameter [G]		64	.250		12.30	12.60		24.90	36
1650	1-1/4" diameter [G]		35	.457		20	23		43	63
1670	1-1/2" diameter [G]		26	.615		27	31		58	85.50
1680	1-3/4" diameter [G]		16	1		63.50	50.50		114	160
1690	2" diameter [G]		12	1.333		70.50	67		137.50	198
1700	2-1/4" diameter [G]		10	1.600		104	80.50		184.50	258
1800	Sockets, open swage, 1/4" diameter [G]		160	.100		43.50	5.05		48.55	56.50
1810	1/2" diameter [G]		77	.208		62.50	10.45		72.95	87
1820	3/4" diameter [G]		19	.842		97	42.50		139.50	183

05 15 Wire Rope Assemblies

05 15 16 – Steel Wire Rope Assemblies

05 15 16.05 Accessories for Steel Wire Rope		Crew	Daily Output	Labor-Hours	Unit	Material	2012 Bare Costs Labor	Equipment	Total	Total Incl O&P
1830	1" diameter	G	E-17	9	1.778	Ea.	174	89.50		263.50	350
1840	1-1/4" diameter	G		5	3.200		241	161		402	555
1850	1-1/2" diameter	G		3	5.333		530	269		799	1,050
1860	1-3/4" diameter	G		3	5.333		940	269		1,209	1,500
1870	2" diameter	G		1.50	10.667		1,425	535		1,960	2,500
1900	Closed swage, 1/4" diameter	G		160	.100		25.50	5.05		30.55	37
1910	1/2" diameter	G		104	.154		44	7.75		51.75	62.50
1920	3/4" diameter	G		32	.500		66	25		91	118
1930	1" diameter	G		15	1.067		116	53.50		169.50	224
1940	1-1/4" diameter	G		7	2.286		174	115		289	400
1950	1-1/2" diameter	G		4	4		315	201		516	705
1960	1-3/4" diameter	G		3	5.333		465	269		734	990
1970	2" diameter	G		2	8		905	405		1,310	1,725
2000	Open spelter, galv., 1/4" diameter	G		160	.100		55	5.05		60.05	70
2010	1/2" diameter	G		70	.229		57.50	11.50		69	84
2020	3/4" diameter	G		26	.615		86.50	31		117.50	151
2030	1" diameter	G		10	1.600		240	80.50		320.50	410
2040	1-1/4" diameter	G		5	3.200		345	161		506	670
2050	1-1/2" diameter	G		4	4		730	201		931	1,150
2060	1-3/4" diameter	G		2	8		1,275	405		1,680	2,125
2070	2" diameter	G		1.20	13.333		1,475	670		2,145	2,800
2080	2-1/2" diameter	G		1	16		2,700	805		3,505	4,400
2100	Closed spelter, galv., 1/4" diameter	G		160	.100		46.50	5.05		51.55	60
2110	1/2" diameter	G		88	.182		49.50	9.15		58.65	71
2120	3/4" diameter	G		30	.533		75.50	27		102.50	131
2130	1" diameter	G		13	1.231		160	62		222	287
2140	1-1/4" diameter	G		7	2.286		256	115		371	490
2150	1-1/2" diameter	G		6	2.667		550	134		684	845
2160	1-3/4" diameter	G		2.80	5.714		735	288		1,023	1,325
2170	2" diameter	G		2	8		905	405		1,310	1,725
2200	Jaw & jaw turnbuckles, 1/4" x 4"	G		160	.100		16.35	5.05		21.40	27
2250	1/2" x 6"	G		96	.167		20.50	8.40		28.90	37.50
2260	1/2" x 9"	G		77	.208		27.50	10.45		37.95	49
2270	1/2" x 12"	G		66	.242		31	12.20		43.20	56
2300	3/4" x 6"	G		38	.421		40.50	21		61.50	82.50
2310	3/4" x 9"	G		30	.533		45	27		72	97.50
2320	3/4" x 12"	G		28	.571		57.50	29		86.50	115
2330	3/4" x 18"	G		23	.696		69	35		104	139
2350	1" x 6"	G		17	.941		78.50	47.50		126	171
2360	1" x 12"	G		13	1.231		86	62		148	206
2370	1" x 18"	G		10	1.600		129	80.50		209.50	286
2380	1" x 24"	G		9	1.778		142	89.50		231.50	315
2400	1-1/4" x 12"	G		7	2.286		145	115		260	365
2410	1-1/4" x 18"	G		6.50	2.462		179	124		303	420
2420	1-1/4" x 24"	G		5.60	2.857		241	144		385	520
2450	1-1/2" x 12"	G		5.20	3.077		425	155		580	740
2460	1-1/2" x 18"	G		4	4		450	201		651	860
2470	1-1/2" x 24"	G		3.20	5		610	252		862	1,125
2500	1-3/4" x 18"	G		3.20	5		920	252		1,172	1,450
2510	1-3/4" x 24"	G		2.80	5.714		1,050	288		1,338	1,675
2550	2" x 24"	G		1.60	10		1,425	505		1,930	2,450

05 15 Wire Rope Assemblies

05 15 16 – Steel Wire Rope Assemblies

05 15 16.50 Steel Wire Rope		Crew	Daily Output	Labor-Hours	Unit	Material	2012 Bare Costs Labor	Equipment	Total	Total Incl O&P
0010	**STEEL WIRE ROPE**									
0015	Made from recycled materials	G								
0020	6 x 19, bright, fiber core, 5000' rolls, 1/2" diameter	G				L.F.	.80			.80	.88
0050	Steel core	G					1.06			1.06	1.16
0100	Fiber core, 1" diameter	G					2.71			2.71	2.98
0150	Steel core	G					3.09			3.09	3.40
0300	6 x 19, galvanized, fiber core, 1/2" diameter	G					1.18			1.18	1.30
0350	Steel core	G					1.35			1.35	1.49
0400	Fiber core, 1" diameter	G					3.47			3.47	3.82
0450	Steel core	G					3.64			3.64	4
0500	6 x 7, bright, IPS, fiber core, <500 L.F. w/acc., 1/4" diameter	G	E-17	6400	.003		1.39	.13		1.52	1.75
0510	1/2" diameter	G		2100	.008		3.38	.38		3.76	4.41
0520	3/4" diameter	G		960	.017		6.10	.84		6.94	8.25
0550	6 x 19, bright, IPS, IWRC, <500 L.F. w/acc., 1/4" diameter	G		5760	.003		.89	.14		1.03	1.23
0560	1/2" diameter	G		1730	.009		1.44	.47		1.91	2.41
0570	3/4" diameter	G		770	.021		2.50	1.05		3.55	4.62
0580	1" diameter	G		420	.038		4.23	1.92		6.15	8.10
0590	1-1/4" diameter	G		290	.055		7	2.78		9.78	12.65
0600	1-1/2" diameter	G		192	.083		8.65	4.20		12.85	17
0610	1-3/4" diameter	G	E-18	240	.167		13.75	8.20	3.96	25.91	33.50
0620	2" diameter	G		160	.250		17.65	12.30	5.95	35.90	47.50
0630	2-1/4" diameter	G		160	.250		23.50	12.30	5.95	41.75	54
0650	6 x 37, bright, IPS, IWRC, <500 L.F. w/acc., 1/4" diameter	G	E-17	6400	.003		1.06	.13		1.19	1.40
0660	1/2" diameter	G		1730	.009		1.80	.47		2.27	2.81
0670	3/4" diameter	G		770	.021		2.91	1.05		3.96	5.05
0680	1" diameter	G		430	.037		4.62	1.87		6.49	8.45
0690	1-1/4" diameter	G		290	.055		7	2.78		9.78	12.60
0700	1-1/2" diameter	G		190	.084		10	4.24		14.24	18.60
0710	1-3/4" diameter	G	E-18	260	.154		15.85	7.55	3.66	27.06	34.50
0720	2" diameter	G		200	.200		20.50	9.85	4.75	35.10	45
0730	2-1/4" diameter	G		160	.250		27	12.30	5.95	45.25	58
0800	6 x 19 & 6 x 37, swaged, 1/2" diameter	G	E-17	1220	.013		2.25	.66		2.91	3.66
0810	9/16" diameter	G		1120	.014		2.62	.72		3.34	4.17
0820	5/8" diameter	G		930	.017		3.11	.87		3.98	4.97
0830	3/4" diameter	G		640	.025		3.96	1.26		5.22	6.60
0840	7/8" diameter	G		480	.033		4.99	1.68		6.67	8.50
0850	1" diameter	G		350	.046		6.10	2.30		8.40	10.80
0860	1-1/8" diameter	G		288	.056		7.50	2.80		10.30	13.25
0870	1-1/4" diameter	G		230	.070		9.05	3.50		12.55	16.25
0880	1-3/8" diameter	G		192	.083		10.45	4.20		14.65	19
0890	1-1/2" diameter	G	E-18	300	.133		12.75	6.55	3.17	22.47	29

05 15 16.60 Galvanized Steel Wire Rope and Accessories

05 15 16.60		Crew	Daily Output	Labor-Hours	Unit	Material	Labor	Equipment	Total	Total Incl O&P
0010	**GALVANIZED STEEL WIRE ROPE & ACCESSORIES**									
0015	Made from recycled materials	G								
3000	Aircraft cable, galvanized, 7 x 7 x 1/8"	G	E-17	5000	.003	L.F.	.19	.16		.35	.50
3100	Clamps, 1/8"	"		125	.128	Ea.	1.85	6.45		8.30	13.55

05 15 16.70 Temporary Cable Safety Railing

05 15 16.70		Crew	Daily Output	Labor-Hours	Unit	Material	Labor	Equipment	Total	Total Incl O&P
0010	**TEMPORARY CABLE SAFETY RAILING**, Each 100' strand incl.									
0020	2 eyebolts, 1 turnbuckle, 100' cable, 2 thimbles, 6 clips									
0025	Made from recycled materials	G								
0100	One strand using 1/4" cable & accessories	G	2 Sswk	4	4	C.L.F.	196	197		393	570
0200	1/2" cable & accessories	G	"	2	8	"	410	395		805	1,150

05 21 13 – Deep Longspan Steel Joist Framing

05 21 13.50 Deep Longspan Joists		Crew	Daily Output	Labor-Hours	Unit	Material	2012 Bare Costs Labor	Equipment	Total	Total Incl O&P
0010	**DEEP LONGSPAN JOISTS**									
3010	DLH series, 40-ton job lots, bolted cross bridging, shop primer									
3015	Made from recycled materials	G								
3020	Spans to 144' (shipped in 2 pieces), minimum	G	E-7	16	5	Ton	1,650	244	109	2,003	2,375
3040	Average	G		13	6.154		1,775	300	134	2,209	2,625
3100	Maximum	G		11	7.273		2,150	355	158	2,663	3,175
3200	52DLH11, 26 lb./L.F.	G		2000	.040	L.F.	22.50	1.95	.87	25.32	29
3220	52DLH16, 45 lb./L.F.	G		2000	.040		40	1.95	.87	42.82	48.50
3240	56DLH11, 26 lb./L.F.	G		2000	.040		23	1.95	.87	25.82	30
3260	56DLH16, 46 lb./L.F.	G		2000	.040		41	1.95	.87	43.82	49.50
3280	60DLH12, 29 lb./L.F.	G		2000	.040		26	1.95	.87	28.82	33
3300	60DLH17, 52 lb./L.F.	G		2000	.040		46.50	1.95	.87	49.32	55.50
3320	64DLH12, 31 lb./L.F.	G		2200	.036		27.50	1.77	.79	30.06	34.50
3340	64DLH17, 52 lb./L.F.	G		2200	.036		46.50	1.77	.79	49.06	55
3360	68DLH13, 37 lb./L.F.	G		2200	.036		33	1.77	.79	35.56	40.50
3380	68DLH18, 61 lb./L.F.	G		2200	.036		54.50	1.77	.79	57.06	64
3400	72DLH14, 41 lb./L.F.	G		2200	.036		36.50	1.77	.79	39.06	44
3420	72DLH19, 70 lb./L.F.	G		2200	.036		62.50	1.77	.79	65.06	72.50
3500	For less than 40-ton job lots									
3502	For 30 to 39 tons, add						10%			
3504	20 to 29 tons, add						20%			
3506	10 to 19 tons, add						30%			
3507	5 to 9 tons, add						50%	25%		
3508	1 to 4 tons, add						75%	50%		
3509	Less than 1 ton, add						100%	100%		
4010	SLH series, 40-ton job lots, bolted cross bridging, shop primer									
4020	Spans to 200' (shipped in 3 pieces), minimum	G	E-7	16	5	Ton	1,625	244	109	1,978	2,350
4040	Average	G		13	6.154		1,825	300	134	2,259	2,700
4060	Maximum	G		11	7.273		2,175	355	158	2,688	3,200
4200	80SLH15, 40 lb./L.F.	G		1500	.053	L.F.	36.50	2.60	1.16	40.26	46.50
4220	80SLH20, 75 lb./L.F.	G		1500	.053		68.50	2.60	1.16	72.26	81.50
4240	88SLH16, 46 lb./L.F.	G		1500	.053		42	2.60	1.16	45.76	52.50
4260	88SLH21, 89 lb./L.F.	G		1500	.053		81.50	2.60	1.16	85.26	95.50
4280	96SLH17, 52 lb./L.F.	G		1500	.053		47.50	2.60	1.16	51.26	58.50
4300	96SLH22, 102 lb./L.F.	G		1500	.053		93.50	2.60	1.16	97.26	109
4320	104SLH18, 59 lb./L.F.	G		1800	.044		54	2.17	.97	57.14	64.50
4340	104SLH23, 109 lb./L.F.	G		1800	.044		100	2.17	.97	103.14	115
4360	112SLH19, 67 lb./L.F.	G		1800	.044		61.50	2.17	.97	64.64	72.50
4380	112SLH24, 131 lb./L.F.	G		1800	.044		120	2.17	.97	123.14	137
4400	120SLH20, 77 lb./L.F.	G		1800	.044		70.50	2.17	.97	73.64	82.50
4420	120SLH25, 152 lb./L.F.	G		1800	.044		139	2.17	.97	142.14	158
6100	For less than 40-ton job lots									
6102	For 30 to 39 tons, add						10%			
6104	20 to 29 tons, add						20%			
6106	10 to 19 tons, add						30%			
6107	5 to 9 tons, add						50%	25%		
6108	1 to 4 tons, add						75%	50%		
6109	Less than 1 ton, add						100%	100%		

05 21 16 – Longspan Steel Joist Framing

05 21 16.50 Longspan Joists

0010	**LONGSPAN JOISTS**									
2000	LH series, 40-ton job lots, bolted cross bridging, shop primer									

05 21 Steel Joist Framing

05 21 16 – Longspan Steel Joist Framing

05 21 16.50 Longspan Joists

		Crew	Daily Output	Labor-Hours	Unit	Material	2012 Bare Costs Labor	Equipment	Total	Total Incl O&P
2015	Made from recycled materials G									
2020	Spans to 96', minimum G	E-7	16	5	Ton	1,550	244	109	1,903	2,250
2040	Average G		13	6.154		1,700	300	134	2,134	2,550
2080	Maximum G		11	7.273	↓	2,000	355	158	2,513	3,000
2200	18LH04, 12 lb./L.F. G		1400	.057	L.F.	10.15	2.79	1.24	14.18	17.40
2220	18LH08, 19 lb./L.F. G		1400	.057		16.10	2.79	1.24	20.13	24
2240	20LH04, 12 lb./L.F. G		1400	.057		10.15	2.79	1.24	14.18	17.40
2260	20LH08, 19 lb./L.F. G		1400	.057		16.10	2.79	1.24	20.13	24
2280	24LH05, 13 lb./L.F. G		1400	.057		11	2.79	1.24	15.03	18.30
2300	24LH10, 23 lb./L.F. G		1400	.057		19.50	2.79	1.24	23.53	27.50
2320	28LH06, 16 lb./L.F. G		1800	.044		13.55	2.17	.97	16.69	19.75
2340	28LH11, 25 lb./L.F. G		1800	.044		21	2.17	.97	24.14	28.50
2360	32LH08, 17 lb./L.F. G		1800	.044		14.40	2.17	.97	17.54	20.50
2380	32LH13, 30 lb./L.F. G		1800	.044		25.50	2.17	.97	28.64	33
2400	36LH09, 21 lb./L.F. G		1800	.044		17.80	2.17	.97	20.94	24.50
2420	36LH14, 36 lb./L.F. G		1800	.044		30.50	2.17	.97	33.64	38.50
2440	40LH10, 21 lb./L.F. G		2200	.036		17.80	1.77	.79	20.36	23.50
2460	40LH15, 36 lb./L.F. G		2200	.036		30.50	1.77	.79	33.06	37.50
2480	44LH11, 22 lb./L.F. G		2200	.036		18.65	1.77	.79	21.21	24.50
2500	44LH16, 42 lb./L.F. G		2200	.036		35.50	1.77	.79	38.06	43
2520	48LH11, 22 lb./L.F. G		2200	.036		18.65	1.77	.79	21.21	24.50
2540	48LH16, 42 lb./L.F. G	↓	2200	.036	↓	35.50	1.77	.79	38.06	43
2600	For less than 40-ton job lots									
2602	For 30 to 39 tons, add					10%				
2604	20 to 29 tons, add					20%				
2606	10 to 19 tons, add					30%				
2607	5 to 9 tons, add					50%	25%			
2608	1 to 4 tons, add					75%	50%			
2609	Less than 1 ton, add					100%	100%			
6000	For welded cross bridging, add						30%			

05 21 19 – Open Web Steel Joist Framing

05 21 19.10 Open Web Joists

		Crew	Daily Output	Labor-Hours	Unit	Material	2012 Bare Costs Labor	Equipment	Total	Total Incl O&P
0010	**OPEN WEB JOISTS**									
0015	Made from recycled materials G									
0020	K series, 40-ton lots, horiz. bridging, spans to 30', shop primer, minimum G	E-7	15	5.333	Ton	1,350	260	116	1,726	2,075
0050	Average G		12	6.667		1,525	325	145	1,995	2,400
0080	Maximum G		9	8.889	↓	1,825	435	193	2,453	3,000
0130	8K1, 5.1 lb./L.F. G		1200	.067	L.F.	3.90	3.25	1.45	8.60	11.55
0140	10K1, 5.0 lb./L.F. G		1200	.067		3.83	3.25	1.45	8.53	11.45
0160	12K3, 5.7 lb./L.F. G		1500	.053		4.36	2.60	1.16	8.12	10.60
0180	14K3, 6.0 lb./L.F. G		1500	.053		4.59	2.60	1.16	8.35	10.85
0200	16K3, 6.3 lb./L.F. G		1800	.044		4.82	2.17	.97	7.96	10.15
0220	16K6, 8.1 lb./L.F. G		1800	.044		6.20	2.17	.97	9.34	11.65
0240	18K5, 7.7 lb./L.F. G		2000	.040		5.90	1.95	.87	8.72	10.85
0260	18K9, 10.2 lb./L.F. G		2000	.040	↓	7.80	1.95	.87	10.62	12.95
0410	Span 30' to 50', minimum G		17	4.706	Ton	1,325	230	102	1,657	2,000
0440	Average G		17	4.706		1,500	230	102	1,832	2,175
0460	Maximum G		10	8	↓	1,600	390	174	2,164	2,625
0500	20K5, 8.2 lb./L.F. G		2000	.040	L.F.	6.15	1.95	.87	8.97	11.10
0520	20K9, 10.8 lb./L.F. G		2000	.040		8.10	1.95	.87	10.92	13.25
0540	22K5, 8.8 lb./L.F. G		2000	.040		6.60	1.95	.87	9.42	11.60
0560	22K9, 11.3 lb./L.F. G	↓	2000	.040	↓	8.50	1.95	.87	11.32	13.65

05 21 Steel Joist Framing

05 21 19 – Open Web Steel Joist Framing

05 21 19.10 Open Web Joists

			Crew	Daily Output	Labor-Hours	Unit	Material	2012 Bare Costs Labor	Equipment	Total	Total Incl O&P
0580	24K6, 9.7 lb./L.F.	G	E-7	2200	.036	L.F.	7.30	1.77	.79	9.86	11.95
0600	24K10, 13.1 lb./L.F.	G		2200	.036		9.85	1.77	.79	12.41	14.75
0620	26K6, 10.6 lb./L.F.	G		2200	.036		7.95	1.77	.79	10.51	12.70
0640	26K10, 13.8 lb./L.F.	G		2200	.036		10.35	1.77	.79	12.91	15.35
0660	28K8, 12.7 lb./L.F.	G		2400	.033		9.55	1.63	.72	11.90	14.10
0680	28K12, 17.1 lb./L.F.	G		2400	.033		12.85	1.63	.72	15.20	17.70
0700	30K8, 13.2 lb./L.F.	G		2400	.033		9.90	1.63	.72	12.25	14.50
0720	30K12, 17.6 lb./L.F.	G		2400	.033		13.20	1.63	.72	15.55	18.10
0800	For less than 40-ton job lots										
0802	For 30 to 39 tons, add						10%				
0804	20 to 29 tons, add						20%				
0806	10 to 19 tons, add						30%				
0807	5 to 9 tons, add						50%	25%			
0808	1 to 4 tons, add						75%	50%			
0809	Less than 1 ton, add						100%	100%			
1010	CS series, 40-ton job lots, horizontal bridging, shop primer										
1020	Spans to 30', minimum	G	E-7	15	5.333	Ton	1,425	260	116	1,801	2,125
1040	Average	G		12	6.667		1,575	325	145	2,045	2,450
1060	Maximum	G		9	8.889		1,850	435	193	2,478	3,025
1100	10CS2, 7.5 lb./L.F.	G		1200	.067	L.F.	5.90	3.25	1.45	10.60	13.75
1120	12CS2, 8.0 lb./L.F.	G		1500	.053		6.30	2.60	1.16	10.06	12.70
1140	14CS2, 8.0 lb./L.F.	G		1500	.053		6.30	2.60	1.16	10.06	12.70
1160	16CS2, 8.5 lb./L.F.	G		1800	.044		6.70	2.17	.97	9.84	12.20
1180	16CS4, 14.5 lb./L.F.	G		1800	.044		11.40	2.17	.97	14.54	17.40
1200	18CS2, 9.0 lb./L.F.	G		2000	.040		7.05	1.95	.87	9.87	12.15
1220	18CS4, 15.0 lb./L.F.	G		2000	.040		11.80	1.95	.87	14.62	17.30
1240	20CS2, 9.5 lb./L.F.	G		2000	.040		7.45	1.95	.87	10.27	12.55
1260	20CS4, 16.5 lb./L.F.	G		2000	.040		12.95	1.95	.87	15.77	18.60
1280	22CS2, 10.0 lb./L.F.	G		2000	.040		7.85	1.95	.87	10.67	13
1300	22CS4, 16.5 lb./L.F.	G		2000	.040		12.95	1.95	.87	15.77	18.60
1320	24CS2, 10.0 lb./L.F.	G		2200	.036		7.85	1.77	.79	10.41	12.60
1340	24CS4, 16.5 lb./L.F.	G		2200	.036		12.95	1.77	.79	15.51	18.20
1360	26CS2, 10.0 lb./L.F.	G		2200	.036		7.85	1.77	.79	10.41	12.60
1380	26CS4, 16.5 lb./L.F.	G		2200	.036		12.95	1.77	.79	15.51	18.20
1400	28CS2, 10.5 lb./L.F.	G		2400	.033		8.25	1.63	.72	10.60	12.65
1420	28CS4, 16.5 lb./L.F.	G		2400	.033		12.95	1.63	.72	15.30	17.85
1440	30CS2, 11.0 lb./L.F.	G		2400	.033		8.65	1.63	.72	11	13.10
1460	30CS4, 16.5 lb./L.F.	G		2400	.033		12.95	1.63	.72	15.30	17.85
1500	For less than 40-ton job lots										
1502	For 30 to 39 tons, add						10%				
1504	20 to 29 tons, add						20%				
1506	10 to 19 tons, add						30%				
1507	5 to 9 tons, add						50%	25%			
1508	1 to 4 tons, add						75%	50%			
1509	Less than 1 ton, add						100%	100%			
6200	For shop prime paint other than mfrs. standard, add						20%				
6300	For bottom chord extensions, add per chord	G				Ea.	33			33	36.50
6400	Individual steel bearing plate, 6" x 6" x 1/4" with J-hook	G	1 Bric	160	.050	"	7.50	2.21		9.71	11.60

05 21 23 – Steel Joist Girder Framing

05 21 23.50 Joist Girders

0010	**JOIST GIRDERS**										
0015	Made from recycled materials	G									

05 21 Steel Joist Framing

05 21 23 – Steel Joist Girder Framing

05 21 23.50 Joist Girders

	05 21 23.50 Joist Girders	Crew	Daily Output	Labor-Hours	Unit	Material	Labor	Equipment	Total	Total Incl O&P
							2012 Bare Costs			
7000	Joist girders, 40-ton job lots, shop primer, minimum [G]	E-5	15	5.333	Ton	1,400	260	108	1,768	2,100
7020	Average [G]		13	6.154		1,525	300	124	1,949	2,325
7040	Maximum [G]	↓	11	7.273	↓	1,600	355	147	2,102	2,550
7100	For less than 40-ton job lots									
7102	For 30 to 39 tons, add					10%				
7104	20 to 29 tons, add					20%				
7106	10 to 19 tons, add					30%				
7107	5 to 9 tons, add					50%	25%			
7108	1 to 4 tons, add					75%	50%			
7109	Less than 1 ton, add					100%	100%			
8000	Trusses, 40-ton job lots, shop fabricated WT chords, shop primer, average [G]	E-5	11	7.273	Ton	5,000	355	147	5,502	6,275
8100	For less than 40-ton job lots									
8102	For 30 to 39 tons, add					10%				
8104	20 to 29 tons, add					20%				
8106	10 to 19 tons, add					30%				
8107	5 to 9 tons, add					50%	25%			
8108	1 to 4 tons, add					75%	50%			
8109	Less than 1 ton, add					100%	100%			

05 31 Steel Decking

05 31 13 – Steel Floor Decking

05 31 13.50 Floor Decking

	05 31 13.50 Floor Decking	Crew	Daily Output	Labor-Hours	Unit	Material	Labor	Equipment	Total	Total Incl O&P
							2012 Bare Costs			
0010	**FLOOR DECKING** R053100-10									
0015	Made from recycled materials [G]									
5100	Non-cellular composite decking, galvanized, 1-1/2" deep, 16 gauge [G]	E-4	3500	.009	S.F.	2.95	.46	.03	3.44	4.09
5120	18 gauge [G]		3650	.009		2.39	.44	.03	2.86	3.45
5140	20 gauge [G]		3800	.008		1.90	.42	.03	2.35	2.88
5200	2" deep, 22 gauge [G]		3860	.008		1.65	.41	.03	2.09	2.59
5300	20 gauge [G]		3600	.009		1.83	.44	.03	2.30	2.84
5400	18 gauge [G]		3380	.009		2.34	.47	.04	2.85	3.45
5500	16 gauge [G]		3200	.010		2.91	.50	.04	3.45	4.13
5700	3" deep, 22 gauge [G]		3200	.010		1.80	.50	.04	2.34	2.91
5800	20 gauge [G]		3000	.011		2.01	.53	.04	2.58	3.20
5900	18 gauge [G]		2850	.011		2.49	.56	.04	3.09	3.79
6000	16 gauge [G]	↓	2700	.012	↓	3.31	.59	.05	3.95	4.75

05 31 23 – Steel Roof Decking

05 31 23.50 Roof Decking

	05 31 23.50 Roof Decking	Crew	Daily Output	Labor-Hours	Unit	Material	Labor	Equipment	Total	Total Incl O&P
							2012 Bare Costs			
0010	**ROOF DECKING**									
0015	Made from recycled materials [G]									
2100	Open type, 1-1/2" deep, Type B, wide rib, galv., 22 gauge, under 50 sq. [G]	E-4	4500	.007	S.F.	1.74	.35	.03	2.12	2.57
2200	50-500 squares [G]		4900	.007		1.35	.33	.02	1.70	2.10
2400	Over 500 squares		5100	.006		1.25	.31	.02	1.58	1.97
2600	20 gauge, under 50 squares [G]		3865	.008		2.05	.41	.03	2.49	3.02
2650	50-500 squares [G]		4170	.008		1.64	.38	.03	2.05	2.51
2700	Over 500 squares [G]		4300	.007		1.47	.37	.03	1.87	2.31
2900	18 gauge, under 50 squares [G]		3800	.008		2.65	.42	.03	3.10	3.71
2950	50-500 squares [G]		4100	.008		2.12	.39	.03	2.54	3.06
3000	Over 500 squares [G]		4300	.007		1.91	.37	.03	2.31	2.79
3050	16 gauge, under 50 squares [G]		3700	.009		3.57	.43	.03	4.03	4.73
3060	50-500 squares [G]	↓	4000	.008	↓	2.85	.40	.03	3.28	3.88

05 31 Steel Decking

05 31 23 – Steel Roof Decking

05 31 23.50 Roof Decking		Crew	Daily Output	Labor-Hours	Unit	Material	2012 Bare Costs Labor	Equipment	Total	Total Incl O&P
3100	Over 500 squares	**G** E-4	4200	.008	S.F.	2.57	.38	.03	2.98	3.53
3200	3" deep, Type N, 22 gauge, under 50 squares	**G**	3600	.009		2.59	.44	.03	3.06	3.68
3250	50-500 squares	**G**	3800	.008		2.07	.42	.03	2.52	3.07
3260	over 500 squares	**G**	4000	.008		1.86	.40	.03	2.29	2.79
3300	20 gauge, under 50 squares	**G**	3400	.009		2.78	.47	.04	3.29	3.94
3350	50-500 squares	**G**	3600	.009		2.22	.44	.03	2.69	3.28
3360	over 500 squares	**G**	3800	.008		2	.42	.03	2.45	2.99
3400	18 gauge, under 50 squares	**G**	3200	.010		3.61	.50	.04	4.15	4.90
3450	50-500 squares	**G**	3400	.009		2.88	.47	.04	3.39	4.05
3460	over 500 squares	**G**	3600	.009		2.60	.44	.03	3.07	3.69
3500	16 gauge, under 50 squares	**G**	3000	.011		4.76	.53	.04	5.33	6.25
3550	50-500 squares	**G**	3200	.010		3.81	.50	.04	4.35	5.10
3560	over 500 squares	**G**	3400	.009		3.43	.47	.04	3.94	4.65
3700	4-1/2" deep, Type J, 20 gauge, over 50 squares	**G**	2700	.012		4.48	.59	.05	5.12	6.05
3800	18 gauge	**G**	2460	.013		5.75	.65	.05	6.45	7.55
3900	16 gauge	**G**	2350	.014		4.31	.68	.05	5.04	6
4100	6" deep, Type H, 18 gauge, over 50 squares	**G**	2000	.016		8.25	.80	.06	9.11	10.60
4200	16 gauge	**G**	1930	.017		6.15	.83	.06	7.04	8.35
4300	14 gauge	**G**	1860	.017		7.95	.86	.07	8.88	10.30
4500	7-1/2" deep, Type H, 18 gauge, over 50 squares	**G**	1690	.019		9.05	.94	.07	10.06	11.70
4600	16 gauge	**G**	1590	.020		6.80	1	.08	7.88	9.30
4700	14 gauge	**G**	1490	.021		8.70	1.07	.08	9.85	11.60
4800	For painted instead of galvanized, deduct					5%				
5000	For acoustical perforated with fiberglass insulation, add				S.F.	25%				
5100	For type F intermediate rib instead of type B wide rib, add	**G**				25%				
5150	For type A narrow rib instead of type B wide rib, add	**G**				25%				

05 31 33 – Steel Form Decking

05 31 33.50 Form Decking		Crew	Daily Output	Labor-Hours	Unit	Material	2012 Bare Costs Labor	Equipment	Total	Total Incl O&P
0010	**FORM DECKING**									
0015	Made from recycled materials	**G**								
6100	Slab form, steel, 28 gauge, 9/16" deep, Type UFS, uncoated	**G** E-4	4000	.008	S.F.	1.32	.40	.03	1.75	2.19
6200	Galvanized	**G**	4000	.008		1.17	.40	.03	1.60	2.03
6220	24 gauge, 1" deep, Type UF1X, uncoated	**G**	3900	.008		1.28	.41	.03	1.72	2.16
6240	Galvanized	**G**	3900	.008		1.50	.41	.03	1.94	2.41
6300	24 gauge, 1-5/16" deep, Type UFX, uncoated	**G**	3800	.008		1.35	.42	.03	1.80	2.28
6400	Galvanized	**G**	3800	.008		1.59	.42	.03	2.04	2.54
6500	22 gauge, 1-5/16" deep, uncoated	**G**	3700	.009		1.70	.43	.03	2.16	2.67
6600	Galvanized	**G**	3700	.009		1.73	.43	.03	2.19	2.71
6700	22 gauge, 2" deep, uncoated	**G**	3600	.009		2.23	.44	.03	2.70	3.29
6800	Galvanized	**G**	3600	.009		2.19	.44	.03	2.66	3.24
7000	Sheet metal edge closure form, 12" wide with 2 bends, galvanized									
7100	18 gauge	**G** E-14	360	.022	L.F.	3.59	1.14	.34	5.07	6.35
7200	16 gauge	**G** "	360	.022	"	4.88	1.14	.34	6.36	7.75

05 35 Raceway Decking Assemblies

05 35 13 – Steel Cellular Decking

05 35 13.50 Cellular Decking		Crew	Daily Output	Labor-Hours	Unit	Material	2012 Bare Costs Labor	Equipment	Total	Total Incl O&P
0010	**CELLULAR DECKING**									
0015	Made from recycled materials	G								
0200	Cellular units, galv, 1-1/2" deep, Type BC, 20-20 gauge, over 15 squares	G	E-4	1460	.022	S.F.	6.95	1.09	.08	8.12	9.65
0250	18-20 gauge	G		1420	.023		7.85	1.12	.09	9.06	10.75
0300	18-18 gauge	G		1390	.023		8.05	1.15	.09	9.29	11.05
0320	16-18 gauge	G		1360	.024		9.60	1.17	.09	10.86	12.80
0340	16-16 gauge	G		1330	.024		10.70	1.20	.09	11.99	14.05
0400	3" deep, Type NC, galvanized, 20-20 gauge	G		1375	.023		7.60	1.16	.09	8.85	10.55
0500	18-20 gauge	G		1350	.024		9.20	1.18	.09	10.47	12.35
0600	18-18 gauge	G		1290	.025		9.15	1.24	.09	10.48	12.40
0700	16-18 gauge	G		1230	.026		10.35	1.30	.10	11.75	13.80
0800	16-16 gauge	G		1150	.028		11.25	1.39	.11	12.75	15
1000	4-1/2" deep, Type JC, galvanized, 18-20 gauge	G		1100	.029		10.60	1.45	.11	12.16	14.40
1100	18-18 gauge	G		1040	.031		10.55	1.53	.12	12.20	14.45
1200	16-18 gauge	G		980	.033		11.90	1.63	.12	13.65	16.15
1300	16-16 gauge	G		935	.034		12.95	1.71	.13	14.79	17.45
1500	For acoustical deck, add						15%				
1700	For cells used for ventilation, add						15%				
1900	For multi-story or congested site, add							50%			
8000	Metal deck and trench, 2" thick, 20 gauge, combination										
8010	60% cellular, 40% non-cellular, inserts and trench	G	R-4	1100	.036	S.F.	13.85	1.82	.11	15.78	18.50

05 41 Structural Metal Stud Framing

05 41 13 – Load-Bearing Metal Stud Framing

05 41 13.05 Bracing

			Crew	Daily Output	Labor-Hours	Unit	Material	Labor	Equipment	Total	Total Incl O&P
0010	**BRACING**, shear wall X-bracing, per 10' x 10' bay, one face										
0015	Made of recycled materials	G									
0120	Metal strap, 20 ga x 4" wide	G	2 Carp	18	.889	Ea.	19.80	39		58.80	82.50
0130	6" wide	G		18	.889		33	39		72	96.50
0160	18 ga x 4" wide	G		16	1		34	44		78	106
0170	6" wide	G		16	1		50	44		94	123
0410	Continuous strap bracing, per horizontal row on both faces										
0420	Metal strap, 20 ga x 2" wide, studs 12" O.C.	G	1 Carp	7	1.143	C.L.F.	59.50	50.50		110	143
0430	16" O.C.	G		8	1		59.50	44		103.50	134
0440	24" O.C.	G		10	.800		59.50	35.50		95	120
0450	18 ga x 2" wide, studs 12" O.C.	G		6	1.333		81.50	59		140.50	180
0460	16" O.C.	G		7	1.143		81.50	50.50		132	167
0470	24" O.C.	G		8	1		81.50	44		125.50	158

05 41 13.10 Bridging

			Crew	Daily Output	Labor-Hours	Unit	Material	Labor	Equipment	Total	Total Incl O&P
0010	**BRIDGING**, solid between studs w/1-1/4" leg track, per stud bay										
0015	Made from recycled materials	G									
0200	Studs 12" O.C., 18 ga x 2-1/2" wide	G	1 Carp	125	.064	Ea.	.99	2.82		3.81	5.45
0210	3-5/8" wide	G		120	.067		1.20	2.94		4.14	5.85
0220	4" wide	G		120	.067		1.26	2.94		4.20	5.90
0230	6" wide	G		115	.070		1.65	3.07		4.72	6.55
0240	8" wide	G		110	.073		2.05	3.21		5.26	7.20
0300	16 ga x 2-1/2" wide	G		115	.070		1.26	3.07		4.33	6.10
0310	3-5/8" wide	G		110	.073		1.53	3.21		4.74	6.60
0320	4" wide	G		110	.073		1.62	3.21		4.83	6.70
0330	6" wide	G		105	.076		2.08	3.36		5.44	7.45
0340	8" wide	G		100	.080		2.61	3.53		6.14	8.30

05 41 13 – Load-Bearing Metal Stud Framing

05 41 13.10 Bridging

		Crew	Daily Output	Labor-Hours	Unit	Material	2012 Bare Costs Labor	2012 Bare Costs Equipment	Total	Total Incl O&P
1200	Studs 16" O.C., 18 ga x 2-1/2" wide	G 1 Carp	125	.064	Ea.	1.27	2.82		4.09	5.75
1210	3-5/8" wide	G	120	.067		1.53	2.94		4.47	6.20
1220	4" wide	G	120	.067		1.62	2.94		4.56	6.30
1230	6" wide	G	115	.070		2.11	3.07		5.18	7.05
1240	8" wide	G	110	.073		2.62	3.21		5.83	7.80
1300	16 ga x 2-1/2" wide	G	115	.070		1.62	3.07		4.69	6.50
1310	3-5/8" wide	G	110	.073		1.96	3.21		5.17	7.10
1320	4" wide	G	110	.073		2.08	3.21		5.29	7.20
1330	6" wide	G	105	.076		2.67	3.36		6.03	8.10
1340	8" wide	G	100	.080		3.35	3.53		6.88	9.15
2200	Studs 24" O.C., 18 ga x 2-1/2" wide	G	125	.064		1.84	2.82		4.66	6.35
2210	3-5/8" wide	G	120	.067		2.22	2.94		5.16	6.95
2220	4" wide	G	120	.067		2.34	2.94		5.28	7.10
2230	6" wide	G	115	.070		3.06	3.07		6.13	8.10
2240	8" wide	G	110	.073		3.80	3.21		7.01	9.10
2300	16 ga x 2-1/2" wide	G	115	.070		2.34	3.07		5.41	7.30
2310	3-5/8" wide	G	110	.073		2.84	3.21		6.05	8.05
2320	4" wide	G	110	.073		3.01	3.21		6.22	8.25
2330	6" wide	G	105	.076		3.87	3.36		7.23	9.40
2340	8" wide	G	100	.080		4.85	3.53		8.38	10.80
3000	Continuous bridging, per row									
3100	16 ga x 1-1/2" channel thru studs 12" O.C.	G 1 Carp	6	1.333	C.L.F.	54	59		113	150
3110	16" O.C.	G	7	1.143		54	50.50		104.50	137
3120	24" O.C.	G	8.80	.909		54	40		94	121
4100	2" x 2" angle x 18 ga, studs 12" O.C.	G	7	1.143		82.50	50.50		133	169
4110	16" O.C.	G	9	.889		82.50	39		121.50	152
4120	24" O.C.	G	12	.667		82.50	29.50		112	136
4200	16 ga, studs 12" O.C.	G	5	1.600		106	70.50		176.50	225
4210	16" O.C.	G	7	1.143		106	50.50		156.50	194
4220	24" O.C.	G	10	.800		106	35.50		141.50	171

05 41 13.25 Framing, Boxed Headers/Beams

		Crew	Daily Output	Labor-Hours	Unit	Material	2012 Bare Costs Labor	2012 Bare Costs Equipment	Total	Total Incl O&P
0010	**FRAMING, BOXED HEADERS/BEAMS**	G								
0015	Made from recycled materials	G								
0200	Double, 18 ga x 6" deep	G 2 Carp	220	.073	L.F.	5.70	3.21		8.91	11.20
0210	8" deep	G	210	.076		6.25	3.36		9.61	12.05
0220	10" deep	G	200	.080		7.65	3.53		11.18	13.85
0230	12" deep	G	190	.084		8.35	3.71		12.06	14.90
0300	16 ga x 8" deep	G	180	.089		7.25	3.92		11.17	14
0310	10" deep	G	170	.094		8.75	4.15		12.90	16
0320	12" deep	G	160	.100		9.50	4.41		13.91	17.25
0400	14 ga x 10" deep	G	140	.114		10.10	5.05		15.15	18.85
0410	12" deep	G	130	.123		11.05	5.45		16.50	20.50
1210	Triple, 18 ga x 8" deep	G	170	.094		9.10	4.15		13.25	16.40
1220	10" deep	G	165	.097		11	4.28		15.28	18.70
1230	12" deep	G	160	.100		12.05	4.41		16.46	20
1300	16 ga x 8" deep	G	145	.110		10.50	4.87		15.37	19.05
1310	10" deep	G	140	.114		12.65	5.05		17.70	21.50
1320	12" deep	G	135	.119		13.75	5.25		19	23
1400	14 ga x 10" deep	G	115	.139		13.75	6.15		19.90	24.50
1410	12" deep	G	110	.145		15.20	6.40		21.60	26.50

05 41 13 — Load-Bearing Metal Stud Framing

05 41 13.30 Framing, Stud Walls		Crew	Daily Output	Labor-Hours	Unit	Material	2012 Bare Costs Labor	Equipment	Total	Total Incl O&P
0010	**FRAMING, STUD WALLS** w/top & bottom track, no openings,									
0020	Headers, beams, bridging or bracing									
0025	Made from recycled materials	G								
4100	8' high walls, 18 ga x 2-1/2" wide, studs 12" O.C.	G	2 Carp	54	.296	L.F.	9.40	13.05		22.45	30.50
4110	16" O.C.	G		77	.208		7.55	9.15		16.70	22.50
4120	24" O.C.	G		107	.150		5.65	6.60		12.25	16.35
4130	3-5/8" wide, studs 12" O.C.	G		53	.302		11.25	13.30		24.55	33
4140	16" O.C.	G		76	.211		9	9.30		18.30	24
4150	24" O.C.	G		105	.152		6.75	6.70		13.45	17.80
4160	4" wide, studs 12" O.C.	G		52	.308		11.75	13.55		25.30	34
4170	16" O.C.	G		74	.216		9.40	9.55		18.95	25
4180	24" O.C.	G		103	.155		7.10	6.85		13.95	18.35
4190	6" wide, studs 12" O.C.	G		51	.314		14.90	13.85		28.75	38
4200	16" O.C.	G		73	.219		11.95	9.65		21.60	28
4210	24" O.C.	G		101	.158		9	7		16	20.50
4220	8" wide, studs 12" O.C.	G		50	.320		18.15	14.10		32.25	41.50
4230	16" O.C.	G		72	.222		14.60	9.80		24.40	31
4240	24" O.C.	G		100	.160		11	7.05		18.05	23
4300	16 ga x 2-1/2" wide, studs 12" O.C.	G		47	.340		11.15	15		26.15	35.50
4310	16" O.C.	G		68	.235		8.85	10.40		19.25	25.50
4320	24" O.C.	G		94	.170		6.55	7.50		14.05	18.75
4330	3-5/8" wide, studs 12" O.C.	G		46	.348		13.30	15.35		28.65	38
4340	16" O.C.	G		66	.242		10.55	10.70		21.25	28
4350	24" O.C.	G		92	.174		7.80	7.65		15.45	20.50
4360	4" wide, studs 12" O.C.	G		45	.356		14.10	15.70		29.80	39.50
4370	16" O.C.	G		65	.246		11.15	10.85		22	29
4380	24" O.C.	G		90	.178		8.25	7.85		16.10	21
4390	6" wide, studs 12" O.C.	G		44	.364		17.55	16.05		33.60	44
4400	16" O.C.	G		64	.250		13.95	11.05		25	32.50
4410	24" O.C.	G		88	.182		10.35	8		18.35	23.50
4420	8" wide, studs 12" O.C.	G		43	.372		21.50	16.40		37.90	49.50
4430	16" O.C.	G		63	.254		17.20	11.20		28.40	36
4440	24" O.C.	G		86	.186		12.80	8.20		21	26.50
5100	10' high walls, 18 ga x 2-1/2" wide, studs 12" O.C.	G		54	.296		11.30	13.05		24.35	32.50
5110	16" O.C.	G		77	.208		8.95	9.15		18.10	24
5120	24" O.C.	G		107	.150		6.60	6.60		13.20	17.40
5130	3-5/8" wide, studs 12" O.C.	G		53	.302		13.50	13.30		26.80	35.50
5140	16" O.C.	G		76	.211		10.70	9.30		20	26
5150	24" O.C.	G		105	.152		7.90	6.70		14.60	19
5160	4" wide, studs 12" O.C.	G		52	.308		14.10	13.55		27.65	36.50
5170	16" O.C.	G		74	.216		11.20	9.55		20.75	27
5180	24" O.C.	G		103	.155		8.25	6.85		15.10	19.65
5190	6" wide, studs 12" O.C.	G		51	.314		17.85	13.85		31.70	41
5200	16" O.C.	G		73	.219		14.15	9.65		23.80	30.50
5210	24" O.C.	G		101	.158		10.50	7		17.50	22.50
5220	8" wide, studs 12" O.C.	G		50	.320		21.50	14.10		35.60	45.50
5230	16" O.C.	G		72	.222		17.25	9.80		27.05	34
5240	24" O.C.	G		100	.160		12.80	7.05		19.85	25
5300	16 ga x 2-1/2" wide, studs 12" O.C.	G		47	.340		13.50	15		28.50	38
5310	16" O.C.	G		68	.235		10.60	10.40		21	27.50
5320	24" O.C.	G		94	.170		7.70	7.50		15.20	20
5330	3-5/8" wide, studs 12" O.C.	G		46	.348		16.10	15.35		31.45	41

05 41 13.30 Framing, Stud Walls		Crew	Daily Output	Labor-Hours	Unit	Material	2012 Bare Costs Labor	Equipment	Total	Total Incl O&P
5340	16" O.C.	G	2 Carp	66	.242	L.F.	12.65	10.70		23.35	30.50
5350	24" O.C.	G		92	.174		9.20	7.65		16.85	22
5360	4" wide, studs 12" O.C.	G		45	.356		17	15.70		32.70	42.50
5370	16" O.C.	G		65	.246		13.35	10.85		24.20	31.50
5380	24" O.C.	G		90	.178		9.70	7.85		17.55	22.50
5390	6" wide, studs 12" O.C.	G		44	.364		21	16.05		37.05	48
5400	16" O.C.	G		64	.250		16.65	11.05		27.70	35.50
5410	24" O.C.	G		88	.182		12.15	8		20.15	25.50
5420	8" wide, studs 12" O.C.	G		43	.372		26	16.40		42.40	54
5430	16" O.C.	G		63	.254		20.50	11.20		31.70	40
5440	24" O.C.	G		86	.186		15	8.20		23.20	29
6190	12' high walls, 18 ga x 6" wide, studs 12" O.C.	G		41	.390		21	17.20		38.20	49.50
6200	16" O.C.	G		58	.276		16.35	12.15		28.50	36.50
6210	24" O.C.	G		81	.198		11.95	8.70		20.65	26.50
6220	8" wide, studs 12" O.C.	G		40	.400		25.50	17.65		43.15	55
6230	16" O.C.	G		57	.281		19.90	12.40		32.30	41
6240	24" O.C.	G		80	.200		14.60	8.80		23.40	29.50
6390	16 ga x 6" wide, studs 12" O.C.	G		35	.457		24.50	20		44.50	58
6400	16" O.C.	G		51	.314		19.35	13.85		33.20	43
6410	24" O.C.	G		70	.229		13.95	10.10		24.05	31
6420	8" wide, studs 12" O.C.	G		34	.471		30.50	21		51.50	65.50
6430	16" O.C.	G		50	.320		24	14.10		38.10	48
6440	24" O.C.	G		69	.232		17.20	10.25		27.45	34.50
6530	14 ga x 3-5/8" wide, studs 12" O.C.	G		34	.471		23.50	21		44.50	58
6540	16" O.C.	G		48	.333		18.50	14.70		33.20	43
6550	24" O.C.	G		65	.246		13.30	10.85		24.15	31.50
6560	4" wide, studs 12" O.C.	G		33	.485		25	21.50		46.50	60.50
6570	16" O.C.	G		47	.340		19.45	15		34.45	44.50
6580	24" O.C.	G		64	.250		14	11.05		25.05	32.50
6730	12 ga x 3-5/8" wide, studs 12" O.C.	G		31	.516		33	23		56	71
6740	16" O.C.	G		43	.372		25.50	16.40		41.90	53.50
6750	24" O.C.	G		59	.271		17.85	11.95		29.80	38
6760	4" wide, studs 12" O.C.	G		30	.533		35	23.50		58.50	74.50
6770	16" O.C.	G		42	.381		27	16.80		43.80	55.50
6780	24" O.C.	G		58	.276		19	12.15		31.15	39.50
7390	16' high walls, 16 ga x 6" wide, studs 12" O.C.	G		33	.485		32	21.50		53.50	68
7400	16" O.C.	G		48	.333		24.50	14.70		39.20	49.50
7410	24" O.C.	G		67	.239		17.55	10.55		28.10	35.50
7420	8" wide, studs 12" O.C.	G		32	.500		39.50	22		61.50	77.50
7430	16" O.C.	G		47	.340		30.50	15		45.50	56.50
7440	24" O.C.	G		66	.242		21.50	10.70		32.20	40.50
7560	14 ga x 4" wide, studs 12" O.C.	G		31	.516		32	23		55	70.50
7570	16" O.C.	G		45	.356		25	15.70		40.70	51.50
7580	24" O.C.	G		61	.262		17.65	11.55		29.20	37
7590	6" wide, studs 12" O.C.	G		30	.533		40.50	23.50		64	80.50
7600	16" O.C.	G		44	.364		31.50	16.05		47.55	59
7610	24" O.C.	G		60	.267		22.50	11.75		34.25	42.50
7760	12 ga x 4" wide, studs 12" O.C.	G		29	.552		45.50	24.50		70	87.50
7770	16" O.C.	G		40	.400		35	17.65		52.65	65.50
7780	24" O.C.	G		55	.291		24.50	12.85		37.35	46.50
7790	6" wide, studs 12" O.C.	G		28	.571		57.50	25		82.50	103
7800	16" O.C.	G		39	.410		44	18.10		62.10	76.50
7810	24" O.C.	G		54	.296		31	13.05		44.05	54

05 41 Structural Metal Stud Framing

05 41 13 – Load-Bearing Metal Stud Framing

05 41 13.30 Framing, Stud Walls

		Crew	Daily Output	Labor-Hours	Unit	Material	2012 Bare Costs Labor	Equipment	Total	Total Incl O&P
8590	20' high walls, 14 ga x 6" wide, studs 12" O.C.	G	2 Carp	29	.552	L.F.	50	24.50		74.50	92.50
8600	16" O.C.	G		42	.381		38.50	16.80		55.30	68
8610	24" O.C.	G		57	.281		27	12.40		39.40	48.50
8620	8" wide, studs 12" O.C.	G		28	.571		54	25		79	98.50
8630	16" O.C.	G		41	.390		41.50	17.20		58.70	72.50
8640	24" O.C.	G		56	.286		29.50	12.60		42.10	52
8790	12 ga x 6" wide, studs 12" O.C.	G		27	.593		71	26		97	118
8800	16" O.C.	G		37	.432		54	19.05		73.05	89
8810	24" O.C.	G		51	.314		37.50	13.85		51.35	62.50
8820	8" wide, studs 12" O.C.	G		26	.615		86	27		113	137
8830	16" O.C.	G		36	.444		65.50	19.60		85.10	102
8840	24" O.C.	G		50	.320		45.50	14.10		59.60	71.50

05 42 Cold-Formed Metal Joist Framing

05 42 13 – Cold-Formed Metal Floor Joist Framing

05 42 13.05 Bracing

		Crew	Daily Output	Labor-Hours	Unit	Material	2012 Bare Costs Labor	Equipment	Total	Total Incl O&P
0010	**BRACING**, continuous, per row, top & bottom										
0015	Made from recycled materials	G									
0120	Flat strap, 20 ga x 2" wide, joists at 12" O.C.	G	1 Carp	4.67	1.713	C.L.F.	62	75.50		137.50	185
0130	16" O.C.	G		5.33	1.501		60	66		126	168
0140	24" O.C.	G		6.66	1.201		58	53		111	145
0150	18 ga x 2" wide, joists at 12" O.C.	G		4	2		80.50	88		168.50	225
0160	16" O.C.	G		4.67	1.713		79	75.50		154.50	203
0170	24" O.C.	G		5.33	1.501		77.50	66		143.50	188

05 42 13.10 Bridging

		Crew	Daily Output	Labor-Hours	Unit	Material	2012 Bare Costs Labor	Equipment	Total	Total Incl O&P
0010	**BRIDGING**, solid between joists w/1-1/4" leg track, per joist bay										
0015	Made from recycled materials	G									
0230	Joists 12" O.C., 18 ga track x 6" wide	G	1 Carp	80	.100	Ea.	1.65	4.41		6.06	8.60
0240	8" wide	G		75	.107		2.05	4.70		6.75	9.50
0250	10" wide	G		70	.114		2.55	5.05		7.60	10.55
0260	12" wide	G		65	.123		2.90	5.45		8.35	11.55
0330	16 ga track x 6" wide	G		70	.114		2.08	5.05		7.13	10.05
0340	8" wide	G		65	.123		2.61	5.45		8.06	11.20
0350	10" wide	G		60	.133		3.23	5.90		9.13	12.60
0360	12" wide	G		55	.145		3.71	6.40		10.11	13.95
0440	14 ga track x 8" wide	G		60	.133		3.26	5.90		9.16	12.65
0450	10" wide	G		55	.145		4.04	6.40		10.44	14.30
0460	12" wide	G		50	.160		4.67	7.05		11.72	16
0550	12 ga track x 10" wide	G		45	.178		5.90	7.85		13.75	18.55
0560	12" wide	G		40	.200		6.05	8.80		14.85	20
1230	16" O.C., 18 ga track x 6" wide	G		80	.100		2.11	4.41		6.52	9.10
1240	8" wide	G		75	.107		2.62	4.70		7.32	10.15
1250	10" wide	G		70	.114		3.27	5.05		8.32	11.35
1260	12" wide	G		65	.123		3.71	5.45		9.16	12.45
1330	16 ga track x 6" wide	G		70	.114		2.67	5.05		7.72	10.70
1340	8" wide	G		65	.123		3.35	5.45		8.80	12.05
1350	10" wide	G		60	.133		4.14	5.90		10.04	13.60
1360	12" wide	G		55	.145		4.75	6.40		11.15	15.10
1440	14 ga track x 8" wide	G		60	.133		4.17	5.90		10.07	13.65
1450	10" wide	G		55	.145		5.20	6.40		11.60	15.55
1460	12" wide	G		50	.160		6	7.05		13.05	17.45

05 42 Cold-Formed Metal Joist Framing

05 42 13 – Cold-Formed Metal Floor Joist Framing

05 42 13.10 Bridging

		Crew	Daily Output	Labor-Hours	Unit	Material	2012 Bare Costs Labor	Equipment	Total	Total Incl O&P
1550	12 ga track x 10" wide	G 1 Carp	45	.178	Ea.	7.55	7.85		15.40	20.50
1560	12" wide	G	40	.200		7.75	8.80		16.55	22
2230	24" O.C., 18 ga track x 6" wide	G	80	.100		3.06	4.41		7.47	10.15
2240	8" wide	G	75	.107		3.80	4.70		8.50	11.40
2250	10" wide	G	70	.114		4.73	5.05		9.78	12.95
2260	12" wide	G	65	.123		5.35	5.45		10.80	14.25
2330	16 ga track x 6" wide	G	70	.114		3.87	5.05		8.92	12
2340	8" wide	G	65	.123		4.85	5.45		10.30	13.70
2350	10" wide	G	60	.133		6	5.90		11.90	15.65
2360	12" wide	G	55	.145		6.85	6.40		13.25	17.40
2440	14 ga track x 8" wide	G	60	.133		6.05	5.90		11.95	15.70
2450	10" wide	G	55	.145		7.50	6.40		13.90	18.10
2460	12" wide	G	50	.160		8.65	7.05		15.70	20.50
2550	12 ga track x 10" wide	G	45	.178		10.95	7.85		18.80	24
2560	12" wide	G	40	.200		11.25	8.80		20.05	26

05 42 13.25 Framing, Band Joist

		Crew	Daily Output	Labor-Hours	Unit	Material	Labor	Equipment	Total	Total Incl O&P
0010	**FRAMING, BAND JOIST** (track) fastened to bearing wall	G								
0015	Made from recycled materials	G								
0220	18 ga track x 6" deep	G 2 Carp	1000	.016	L.F.	1.34	.71		2.05	2.57
0230	8" deep	G	920	.017		1.67	.77		2.44	3.02
0240	10" deep	G	860	.019		2.08	.82		2.90	3.55
0320	16 ga track x 6" deep	G	900	.018		1.70	.78		2.48	3.08
0330	8" deep	G	840	.019		2.13	.84		2.97	3.63
0340	10" deep	G	780	.021		2.64	.90		3.54	4.29
0350	12" deep	G	740	.022		3.02	.95		3.97	4.80
0430	14 ga track x 8" deep	G	750	.021		2.66	.94		3.60	4.37
0440	10" deep	G	720	.022		3.30	.98		4.28	5.15
0450	12" deep	G	700	.023		3.81	1.01		4.82	5.75
0540	12 ga track x 10" deep	G	670	.024		4.82	1.05		5.87	6.90
0550	12" deep	G	650	.025		4.95	1.09		6.04	7.10

05 42 13.30 Framing, Boxed Headers/Beams

		Crew	Daily Output	Labor-Hours	Unit	Material	Labor	Equipment	Total	Total Incl O&P
0010	**FRAMING, BOXED HEADERS/BEAMS**	G								
0015	Made from recycled materials	G								
0200	Double, 18 ga x 6" deep	G 2 Carp	220	.073	L.F.	5.70	3.21		8.91	11.20
0210	8" deep	G	210	.076		6.25	3.36		9.61	12.05
0220	10" deep	G	200	.080		7.65	3.53		11.18	13.85
0230	12" deep	G	190	.084		8.35	3.71		12.06	14.90
0300	16 ga x 8" deep	G	180	.089		7.25	3.92		11.17	14
0310	10" deep	G	170	.094		8.75	4.15		12.90	16
0320	12" deep	G	160	.100		9.50	4.41		13.91	17.25
0400	14 ga x 10" deep	G	140	.114		10.10	5.05		15.15	18.85
0410	12" deep	G	130	.123		11.05	5.45		16.50	20.50
0500	12 ga x 10" deep	G	110	.145		13.30	6.40		19.70	24.50
0510	12" deep	G	100	.160		14.70	7.05		21.75	27
1210	Triple, 18 ga x 8" deep	G	170	.094		9.10	4.15		13.25	16.40
1220	10" deep	G	165	.097		11	4.28		15.28	18.70
1230	12" deep	G	160	.100		12.05	4.41		16.46	20
1300	16 ga x 8" deep	G	145	.110		10.50	4.87		15.37	19.05
1310	10" deep	G	140	.114		12.65	5.05		17.70	21.50
1320	12" deep	G	135	.119		13.75	5.25		19	23
1400	14 ga x 10" deep	G	115	.139		14.65	6.15		20.80	25.50
1410	12" deep	G	110	.145		16.10	6.40		22.50	27.50

05 42 Cold-Formed Metal Joist Framing

05 42 13 – Cold-Formed Metal Floor Joist Framing

05 42 13.30 Framing, Boxed Headers/Beams

05 42 13.30 Framing, Boxed Headers/Beams		Crew	Daily Output	Labor-Hours	Unit	Material	2012 Bare Costs Labor	Equipment	Total	Total Incl O&P
1500	12 ga x 10" deep	G	2 Carp	90	.178	L.F.	19.45	7.85		27.30	33.50
1510	12" deep	G	↓	85	.188	↓	21.50	8.30		29.80	37

05 42 13.40 Framing, Joists

05 42 13.40 Framing, Joists		Crew	Daily Output	Labor-Hours	Unit	Material	2012 Bare Costs Labor	Equipment	Total	Total Incl O&P
0010	**FRAMING, JOISTS**, no band joists (track), web stiffeners, headers,										
0020	Beams, bridging or bracing										
0025	Made from recycled materials	G									
0030	Joists (2" flange) and fasteners, materials only										
0220	18 ga x 6" deep	G				L.F.	1.75			1.75	1.93
0230	8" deep	G					2.06			2.06	2.26
0240	10" deep	G					2.43			2.43	2.67
0320	16 ga x 6" deep	G					2.14			2.14	2.36
0330	8" deep	G					2.56			2.56	2.82
0340	10" deep	G					3			3	3.30
0350	12" deep	G					3.40			3.40	3.74
0430	14 ga x 8" deep	G					3.21			3.21	3.53
0440	10" deep	G					3.72			3.72	4.09
0450	12" deep	G					4.22			4.22	4.64
0540	12 ga x 10" deep	G					5.40			5.40	5.95
0550	12" deep	G				↓	6.15			6.15	6.75
1010	Installation of joists to band joists, beams & headers, labor only										
1220	18 ga x 6" deep		2 Carp	110	.145	Ea.		6.40		6.40	9.85
1230	8" deep			90	.178			7.85		7.85	12.05
1240	10" deep			80	.200			8.80		8.80	13.55
1320	16 ga x 6" deep			95	.168			7.45		7.45	11.45
1330	8" deep			70	.229			10.10		10.10	15.50
1340	10" deep			60	.267			11.75		11.75	18.10
1350	12" deep			55	.291			12.85		12.85	19.75
1430	14 ga x 8" deep			65	.246			10.85		10.85	16.70
1440	10" deep			45	.356			15.70		15.70	24
1450	12" deep			35	.457			20		20	31
1540	12 ga x 10" deep			40	.400			17.65		17.65	27
1550	12" deep		↓	30	.533	↓		23.50		23.50	36

05 42 13.45 Framing, Web Stiffeners

05 42 13.45 Framing, Web Stiffeners		Crew	Daily Output	Labor-Hours	Unit	Material	2012 Bare Costs Labor	Equipment	Total	Total Incl O&P
0010	**FRAMING, WEB STIFFENERS** at joist bearing, fabricated from										
0020	Stud piece (1-5/8" flange) to stiffen joist (2" flange)										
0025	Made from recycled materials	G									
2120	For 6" deep joist, with 18 ga x 2-1/2" stud	G	1 Carp	120	.067	Ea.	2.07	2.94		5.01	6.80
2130	3-5/8" stud	G		110	.073		2.30	3.21		5.51	7.45
2140	4" stud	G		105	.076		2.22	3.36		5.58	7.60
2150	6" stud	G		100	.080		2.43	3.53		5.96	8.10
2160	8" stud	G		95	.084		2.49	3.71		6.20	8.45
2220	8" deep joist, with 2-1/2" stud	G		120	.067		2.27	2.94		5.21	7
2230	3-5/8" stud	G		110	.073		2.48	3.21		5.69	7.65
2240	4" stud	G		105	.076		2.43	3.36		5.79	7.80
2250	6" stud	G		100	.080		2.66	3.53		6.19	8.40
2260	8" stud	G		95	.084		2.86	3.71		6.57	8.85
2320	10" deep joist, with 2-1/2" stud	G		110	.073		3.20	3.21		6.41	8.45
2330	3-5/8" stud	G		100	.080		3.53	3.53		7.06	9.35
2340	4" stud	G		95	.084		3.50	3.71		7.21	9.55
2350	6" stud	G		90	.089		3.78	3.92		7.70	10.20
2360	8" stud	G		85	.094		3.84	4.15		7.99	10.65
2420	12" deep joist, with 2-1/2" stud	G	↓	110	.073	↓	3.38	3.21		6.59	8.65

05 42 Cold-Formed Metal Joist Framing

05 42 13 – Cold-Formed Metal Floor Joist Framing

05 42 13.45 Framing, Web Stiffeners

		Crew	Daily Output	Labor-Hours	Unit	Material	2012 Bare Costs Labor	Equipment	Total	Total Incl O&P
2430	3-5/8" stud	G	1 Carp	100	.080	Ea.	3.70	3.53		7.23	9.50
2440	4" stud	G		95	.084		3.63	3.71		7.34	9.70
2450	6" stud	G		90	.089		3.97	3.92		7.89	10.40
2460	8" stud	G		85	.094		4.27	4.15		8.42	11.10
3130	For 6" deep joist, with 16 ga x 3-5/8" stud	G		100	.080		2.42	3.53		5.95	8.10
3140	4" stud	G		95	.084		2.41	3.71		6.12	8.35
3150	6" stud	G		90	.089		2.61	3.92		6.53	8.90
3160	8" stud	G		85	.094		2.78	4.15		6.93	9.45
3230	8" deep joist, with 3-5/8" stud	G		100	.080		2.68	3.53		6.21	8.40
3240	4" stud	G		95	.084		2.64	3.71		6.35	8.60
3250	6" stud	G		90	.089		2.89	3.92		6.81	9.25
3260	8" stud	G		85	.094		3.12	4.15		7.27	9.85
3330	10" deep joist, with 3-5/8" stud	G		85	.094		3.67	4.15		7.82	10.45
3340	4" stud	G		80	.100		3.76	4.41		8.17	10.95
3350	6" stud	G		75	.107		4.03	4.70		8.73	11.70
3360	8" stud	G		70	.114		4.24	5.05		9.29	12.40
3430	12" deep joist, with 3-5/8" stud	G		85	.094		4	4.15		8.15	10.80
3440	4" stud	G		80	.100		3.94	4.41		8.35	11.15
3450	6" stud	G		75	.107		4.32	4.70		9.02	12
3460	8" stud	G		70	.114		4.66	5.05		9.71	12.90
4230	For 8" deep joist, with 14 ga x 3-5/8" stud	G		90	.089		3.48	3.92		7.40	9.90
4240	4" stud	G		85	.094		3.54	4.15		7.69	10.30
4250	6" stud	G		80	.100		3.84	4.41		8.25	11
4260	8" stud	G		75	.107		3.61	4.70		8.31	11.20
4330	10" deep joist, with 3-5/8" stud	G		75	.107		4.88	4.70		9.58	12.60
4340	4" stud	G		70	.114		4.83	5.05		9.88	13.05
4350	6" stud	G		65	.123		5.30	5.45		10.75	14.20
4360	8" stud	G		60	.133		4.88	5.90		10.78	14.40
4430	12" deep joist, with 3-5/8" stud	G		75	.107		5.20	4.70		9.90	12.95
4440	4" stud	G		70	.114		5.30	5.05		10.35	13.55
4450	6" stud	G		65	.123		5.75	5.45		11.20	14.65
4460	8" stud	G		60	.133		5.40	5.90		11.30	15
5330	For 10" deep joist, with 12 ga x 3-5/8" stud	G		65	.123		5.15	5.45		10.60	14.05
5340	4" stud	G		60	.133		5.30	5.90		11.20	14.85
5350	6" stud	G		55	.145		5.85	6.40		12.25	16.25
5360	8" stud	G		50	.160		6.35	7.05		13.40	17.85
5430	12" deep joist, with 3-5/8" stud	G		65	.123		5.75	5.45		11.20	14.65
5440	4" stud	G		60	.133		5.55	5.90		11.45	15.15
5450	6" stud	G		55	.145		6.35	6.40		12.75	16.85
5460	8" stud	G		50	.160		7.25	7.05		14.30	18.85

05 42 23 – Cold-Formed Metal Roof Joist Framing

05 42 23.05 Framing, Bracing

		Crew	Daily Output	Labor-Hours	Unit	Material	2012 Bare Costs Labor	Equipment	Total	Total Incl O&P
0010	**FRAMING, BRACING**										
0015	Made from recycled materials	G									
0020	Continuous bracing, per row										
0100	16 ga x 1-1/2" channel thru rafters/trusses @ 16" O.C.	G	1 Carp	4.50	1.778	C.L.F.	54	78.50		132.50	181
0120	24" O.C.	G		6	1.333		54	59		113	150
0300	2" x 2" angle x 18 ga, rafters/trusses @ 16" O.C.	G		6	1.333		82.50	59		141.50	182
0320	24" O.C.	G		8	1		82.50	44		126.50	159
0400	16 ga, rafters/trusses @ 16" O.C.	G		4.50	1.778		106	78.50		184.50	237
0420	24" O.C.	G		6.50	1.231		106	54.50		160.50	200

05 42 Cold-Formed Metal Joist Framing

05 42 23 – Cold-Formed Metal Roof Joist Framing

05 42 23.10 Framing, Bridging

		Crew	Daily Output	Labor-Hours	Unit	Material	2012 Bare Costs Labor	Equipment	Total	Total Incl O&P
0010	**FRAMING, BRIDGING**									
0015	Made from recycled materials [G]									
0020	Solid, between rafters w/1-1/4" leg track, per rafter bay									
1200	Rafters 16" O.C., 18 ga x 4" deep [G]	1 Carp	60	.133	Ea.	1.62	5.90		7.52	10.85
1210	6" deep [G]		57	.140		2.11	6.20		8.31	11.80
1220	8" deep [G]		55	.145		2.62	6.40		9.02	12.75
1230	10" deep [G]		52	.154		3.27	6.80		10.07	14.05
1240	12" deep [G]		50	.160		3.71	7.05		10.76	14.95
2200	24" O.C., 18 ga x 4" deep [G]		60	.133		2.34	5.90		8.24	11.60
2210	6" deep [G]		57	.140		3.06	6.20		9.26	12.85
2220	8" deep [G]		55	.145		3.80	6.40		10.20	14
2230	10" deep [G]		52	.154		4.73	6.80		11.53	15.65
2240	12" deep [G]		50	.160		5.35	7.05		12.40	16.75

05 42 23.50 Framing, Parapets

		Crew	Daily Output	Labor-Hours	Unit	Material	2012 Bare Costs Labor	Equipment	Total	Total Incl O&P
0010	**FRAMING, PARAPETS**									
0015	Made from recycled materials [G]									
0100	3' high installed on 1st story, 18 ga x 4" wide studs, 12" O.C. [G]	2 Carp	100	.160	L.F.	5.90	7.05		12.95	17.35
0110	16" O.C. [G]		150	.107		5.05	4.70		9.75	12.80
0120	24" O.C. [G]		200	.080		4.16	3.53		7.69	10
0200	6" wide studs, 12" O.C. [G]		100	.160		7.55	7.05		14.60	19.15
0210	16" O.C. [G]		150	.107		6.45	4.70		11.15	14.35
0220	24" O.C. [G]		200	.080		5.35	3.53		8.88	11.35
1100	Installed on 2nd story, 18 ga x 4" wide studs, 12" O.C. [G]		95	.168		5.90	7.45		13.35	17.95
1110	16" O.C. [G]		145	.110		5.05	4.87		9.92	13.05
1120	24" O.C. [G]		190	.084		4.16	3.71		7.87	10.25
1200	6" wide studs, 12" O.C. [G]		95	.168		7.55	7.45		15	19.75
1210	16" O.C. [G]		145	.110		6.45	4.87		11.32	14.60
1220	24" O.C. [G]		190	.084		5.35	3.71		9.06	11.60
2100	Installed on gable, 18 ga x 4" wide studs, 12" O.C. [G]		85	.188		5.90	8.30		14.20	19.25
2110	16" O.C. [G]		130	.123		5.05	5.45		10.50	13.90
2120	24" O.C. [G]		170	.094		4.16	4.15		8.31	10.95
2200	6" wide studs, 12" O.C. [G]		85	.188		7.55	8.30		15.85	21
2210	16" O.C. [G]		130	.123		6.45	5.45		11.90	15.45
2220	24" O.C. [G]		170	.094		5.35	4.15		9.50	12.30

05 42 23.60 Framing, Roof Rafters

		Crew	Daily Output	Labor-Hours	Unit	Material	2012 Bare Costs Labor	Equipment	Total	Total Incl O&P
0010	**FRAMING, ROOF RAFTERS**									
0015	Made from recycled materials [G]									
0100	Boxed ridge beam, double, 18 ga x 6" deep [G]	2 Carp	160	.100	L.F.	5.70	4.41		10.11	13.05
0110	8" deep [G]		150	.107		6.25	4.70		10.95	14.15
0120	10" deep [G]		140	.114		7.65	5.05		12.70	16.15
0130	12" deep [G]		130	.123		8.35	5.45		13.80	17.55
0200	16 ga x 6" deep [G]		150	.107		6.45	4.70		11.15	14.35
0210	8" deep [G]		140	.114		7.25	5.05		12.30	15.70
0220	10" deep [G]		130	.123		8.75	5.45		14.20	17.95
0230	12" deep [G]		120	.133		9.50	5.90		15.40	19.50
1100	Rafters, 2" flange, material only, 18 ga x 6" deep [G]					1.75			1.75	1.93
1110	8" deep [G]					2.06			2.06	2.26
1120	10" deep [G]					2.43			2.43	2.67
1130	12" deep [G]					2.80			2.80	3.08
1200	16 ga x 6" deep [G]					2.14			2.14	2.36
1210	8" deep [G]					2.56			2.56	2.82
1220	10" deep [G]					3			3	3.30

05 42 Cold-Formed Metal Joist Framing

05 42 23 – Cold-Formed Metal Roof Joist Framing

05 42 23.60 Framing, Roof Rafters

		Crew	Daily Output	Labor-Hours	Unit	Material	2012 Bare Costs Labor	Equipment	Total	Total Incl O&P
1230	12" deep	G			L.F.	3.40			3.40	3.74
2100	Installation only, ordinary rafter to 4:12 pitch, 18 ga x 6" deep	2 Carp	35	.457	Ea.		20		20	31
2110	8" deep		30	.533			23.50		23.50	36
2120	10" deep		25	.640			28		28	43.50
2130	12" deep		20	.800			35.50		35.50	54.50
2200	16 ga x 6" deep		30	.533			23.50		23.50	36
2210	8" deep		25	.640			28		28	43.50
2220	10" deep		20	.800			35.50		35.50	54.50
2230	12" deep		15	1.067			47		47	72.50
8100	Add to labor, ordinary rafters on steep roofs						25%			
8110	Dormers & complex roofs						50%			
8200	Hip & valley rafters to 4:12 pitch						25%			
8210	Steep roofs						50%			
8220	Dormers & complex roofs						75%			
8300	Hip & valley jack rafters to 4:12 pitch						50%			
8310	Steep roofs						75%			
8320	Dormers & complex roofs						100%			

05 42 23.70 Framing, Soffits and Canopies

		Crew	Daily Output	Labor-Hours	Unit	Material	2012 Bare Costs Labor	Equipment	Total	Total Incl O&P
0010	**FRAMING, SOFFITS & CANOPIES**									
0015	Made from recycled materials	G								
0130	Continuous ledger track @ wall, studs @ 16" O.C., 18 ga x 4" wide	G 2 Carp	535	.030	L.F.	1.08	1.32		2.40	3.22
0140	6" wide	G	500	.032		1.41	1.41		2.82	3.72
0150	8" wide	G	465	.034		1.75	1.52		3.27	4.25
0160	10" wide	G	430	.037		2.18	1.64		3.82	4.92
0230	Studs @ 24" O.C., 18 ga x 4" wide	G	800	.020		1.03	.88		1.91	2.49
0240	6" wide	G	750	.021		1.34	.94		2.28	2.93
0250	8" wide	G	700	.023		1.67	1.01		2.68	3.39
0260	10" wide	G	650	.025		2.08	1.09		3.17	3.96
1000	Horizontal soffit and canopy members, material only									
1030	1-5/8" flange studs, 18 ga x 4" deep	G			L.F.	1.40			1.40	1.54
1040	6" deep	G				1.76			1.76	1.94
1050	8" deep	G				2.14			2.14	2.35
1140	2" flange joists, 18 ga x 6" deep	G				2			2	2.20
1150	8" deep	G				2.35			2.35	2.59
1160	10" deep	G				2.77			2.77	3.05
4030	Installation only, 18 ga, 1-5/8" flange x 4" deep	2 Carp	130	.123	Ea.		5.45		5.45	8.35
4040	6" deep		110	.145			6.40		6.40	9.85
4050	8" deep		90	.178			7.85		7.85	12.05
4140	2" flange, 18 ga x 6" deep		110	.145			6.40		6.40	9.85
4150	8" deep		90	.178			7.85		7.85	12.05
4160	10" deep		80	.200			8.80		8.80	13.55
6010	Clips to attach facia to rafter tails, 2" x 2" x 18 ga angle	G 1 Carp	120	.067		.98	2.94		3.92	5.60
6020	16 ga angle	G "	100	.080		1.25	3.53		4.78	6.80

05 44 Cold-Formed Metal Trusses

05 44 13 – Cold-Formed Metal Roof Trusses

05 44 13.60 Framing, Roof Trusses		Crew	Daily Output	Labor-Hours	Unit	Material	2012 Bare Costs Labor	Equipment	Total	Total Incl O&P
0010	**FRAMING, ROOF TRUSSES**									
0015	Made from recycled materials [G]									
0020	Fabrication of trusses on ground, Fink (W) or King Post, to 4:12 pitch									
0120	18 ga x 4" chords, 16' span [G]	2 Carp	12	1.333	Ea.	65.50	59		124.50	163
0130	20' span [G]		11	1.455		82	64		146	189
0140	24' span [G]		11	1.455		98.50	64		162.50	207
0150	28' span [G]		10	1.600		115	70.50		185.50	235
0160	32' span [G]		10	1.600		131	70.50		201.50	253
0250	6" chords, 28' span [G]		9	1.778		144	78.50		222.50	279
0260	32' span [G]		9	1.778		165	78.50		243.50	300
0270	36' span [G]		8	2		185	88		273	340
0280	40' span [G]		8	2		206	88		294	360
1120	5:12 to 8:12 pitch, 18 ga x 4" chords, 16' span [G]		10	1.600		75	70.50		145.50	192
1130	20' span [G]		9	1.778		93.50	78.50		172	224
1140	24' span [G]		9	1.778		112	78.50		190.50	245
1150	28' span [G]		8	2		131	88		219	280
1160	32' span [G]		8	2		150	88		238	300
1250	6" chords, 28' span [G]		7	2.286		165	101		266	335
1260	32' span [G]		7	2.286		188	101		289	360
1270	36' span [G]		6	2.667		212	118		330	415
1280	40' span [G]		6	2.667		235	118		353	440
2120	9:12 to 12:12 pitch, 18 ga x 4" chords, 16' span [G]		8	2		93.50	88		181.50	239
2130	20' span [G]		7	2.286		117	101		218	284
2140	24' span [G]		7	2.286		140	101		241	310
2150	28' span [G]		6	2.667		164	118		282	360
2160	32' span [G]		6	2.667		187	118		305	385
2250	6" chords, 28' span [G]		5	3.200		206	141		347	445
2260	32' span [G]		5	3.200		235	141		376	475
2270	36' span [G]		4	4		265	176		441	560
2280	40' span [G]		4	4		294	176		470	595
5120	Erection only of roof trusses, to 4:12 pitch, 16' span	F-6	48	.833			34.50	12.75	47.25	66.50
5130	20' span		46	.870			36	13.30	49.30	69.50
5140	24' span		44	.909			37.50	13.90	51.40	73
5150	28' span		42	.952			39.50	14.55	54.05	76
5160	32' span		40	1			41	15.30	56.30	80
5170	36' span		38	1.053			43.50	16.10	59.60	84
5180	40' span		36	1.111			46	17	63	88.50
5220	5:12 to 8:12 pitch, 16' span		42	.952			39.50	14.55	54.05	76
5230	20' span		40	1			41	15.30	56.30	80
5240	24' span		38	1.053			43.50	16.10	59.60	84
5250	28' span		36	1.111			46	17	63	88.50
5260	32' span		34	1.176			48.50	18	66.50	94.50
5270	36' span		32	1.250			51.50	19.15	70.65	100
5280	40' span		30	1.333			55	20.50	75.50	107
5320	9:12 to 12:12 pitch, 16' span		36	1.111			46	17	63	88.50
5330	20' span		34	1.176			48.50	18	66.50	94.50
5340	24' span		32	1.250			51.50	19.15	70.65	100
5350	28' span		30	1.333			55	20.50	75.50	107
5360	32' span		28	1.429			59	22	81	114
5370	36' span		26	1.538			63.50	23.50	87	123
5380	40' span		24	1.667			68.50	25.50	94	133

05 51 Metal Stairs

05 51 13 – Metal Pan Stairs

05 51 13.50 Pan Stairs

		Crew	Daily Output	Labor-Hours	Unit	Material	2012 Bare Costs Labor	Equipment	Total	Total Incl O&P
0010	**PAN STAIRS**, shop fabricated, steel stringers									
0015	Made from recycled materials	G								
0200	Cement fill metal pan, picket rail, 3'-6" wide	G	E-4	35	.914	Riser	465	45.50	3.47	513.97	595
0300	4'-0" wide	G		30	1.067		525	53	4.05	582.05	680
0350	Wall rail, both sides, 3'-6" wide			53	.604		355	30	2.29	387.29	445
1500	Landing, steel pan, conventional	G		160	.200	S.F.	62	9.95	.76	72.71	86.50
1600	Pre-erected	G		255	.125	"	110	6.25	.48	116.73	133
1700	Pre-erected, steel pan tread, 3'-6" wide, 2 line pipe rail	G	E-2	87	.644	Riser	515	31	17.20	563.20	635

05 51 16 – Metal Floor Plate Stairs

05 51 16.50 Floor Plate Stairs

		Crew	Daily Output	Labor-Hours	Unit	Material	2012 Bare Costs Labor	Equipment	Total	Total Incl O&P
0010	**FLOOR PLATE STAIRS**, shop fabricated, steel stringers									
0015	Made from recycled materials	G								
0400	Cast iron tread and pipe rail, 3'-6" wide	G	E-4	35	.914	Riser	500	45.50	3.47	548.97	635
0500	Checkered plate tread, industrial, 3'-6" wide	G		28	1.143		310	57	4.34	371.34	445
0550	Circular, for tanks, 3'-0" wide	G		33	.970		345	48.50	3.68	397.18	470
0600	For isolated stairs, add							100%			
0800	Custom steel stairs, 3'-6" wide, minimum	G	E-4	35	.914		465	45.50	3.47	513.97	595
0810	Average	G		30	1.067		620	53	4.05	677.05	780
0900	Maximum	G		20	1.600		775	80	6.10	861.10	1,000
1100	For 4' wide stairs, add						5%	5%			
1300	For 5' wide stairs, add						10%	10%			

05 51 19 – Metal Grating Stairs

05 51 19.50 Grating Stairs

		Crew	Daily Output	Labor-Hours	Unit	Material	2012 Bare Costs Labor	Equipment	Total	Total Incl O&P
0010	**GRATING STAIRS**, shop fabricated, steel stringers, safety nosing on treads									
0015	Made from recycled materials	G								
0020	Grating tread and pipe railing, 3'-6" wide	G	E-4	35	.914	Riser	310	45.50	3.47	358.97	425
0100	4'-0" wide	G	"	30	1.067	"	405	53	4.05	462.05	545

05 51 23 – Metal Fire Escapes

05 51 23.25 Fire Escapes

		Crew	Daily Output	Labor-Hours	Unit	Material	2012 Bare Costs Labor	Equipment	Total	Total Incl O&P
0010	**FIRE ESCAPES**, shop fabricated									
0200	2' wide balcony, 1" x 1/4" bars 1-1/2" O.C., with railing	G	2 Sswk	10	1.600	L.F.	55	79		134	202
0400	1st story cantilevered stair, standard, with railing	G		.50	32	Ea.	2,275	1,575		3,850	5,325
0500	Cable counterweighted, with railing			.40	40	"	2,100	1,975		4,075	5,850
0700	36" x 40" platform & fixed stair, with railing	G		.40	40	Flight	1,000	1,975		2,975	4,625
0900	For 3'-6" wide escapes, add to above						100%	150%			

05 51 23.50 Fire Escape Stairs

		Crew	Daily Output	Labor-Hours	Unit	Material	2012 Bare Costs Labor	Equipment	Total	Total Incl O&P
0010	**FIRE ESCAPE STAIRS**, portable									
0100	Portable ladder					Ea.	127			127	139

05 51 33 – Metal Ladders

05 51 33.13 Vertical Metal Ladders

		Crew	Daily Output	Labor-Hours	Unit	Material	2012 Bare Costs Labor	Equipment	Total	Total Incl O&P
0010	**VERTICAL METAL LADDERS**, shop fabricated									
0015	Made from recycled materials	G								
0020	Steel, 20" wide, bolted to concrete, with cage	G	E-4	50	.640	V.L.F.	65	32	2.43	99.43	131
0100	Without cage	G		85	.376		36	18.75	1.43	56.18	74.50
0300	Aluminum, bolted to concrete, with cage	G		50	.640		111	32	2.43	145.43	183
0400	Without cage	G		85	.376		48	18.75	1.43	68.18	88

05 51 33.16 Inclined Metal Ladders

		Crew	Daily Output	Labor-Hours	Unit	Material	2012 Bare Costs Labor	Equipment	Total	Total Incl O&P
0010	**INCLINED METAL LADDERS**, shop fabricated									
0015	Made from recycled materials	G								
3900	Industrial ships ladder, steel, 24" W, grating treads, 2 line pipe rail	G	E-4	30	1.067	Riser	187	53	4.05	244.05	305

05 51 Metal Stairs

05 51 33 – Metal Ladders

05 51 33.16 Inclined Metal Ladders

			Crew	Daily Output	Labor-Hours	Unit	Material	2012 Bare Costs Labor	2012 Bare Costs Equipment	Total	Total Incl O&P
4000	Aluminum	G	E-4	30	1.067	Riser	206	53	4.05	263.05	325

05 51 33.23 Alternating Tread Ladders

			Crew	Daily Output	Labor-Hours	Unit	Material	2012 Bare Costs Labor	2012 Bare Costs Equipment	Total	Total Incl O&P
0010	**ALTERNATING TREAD LADDERS**, shop fabricated										
0015	Made from recycled materials	G									
1350	Alternating tread stair, 56/68°, steel, standard paint color	G	2 Sswk	50	.320	V.L.F.	258	15.80		273.80	310
1360	Non-standard paint color	G		50	.320		292	15.80		307.80	350
1370	Galvanized steel	G		50	.320		300	15.80		315.80	360
1380	Stainless steel	G		50	.320		435	15.80		450.80	505
1390	68°, aluminum	G		50	.320		315	15.80		330.80	380

05 52 Metal Railings

05 52 13 – Pipe and Tube Railings

05 52 13.50 Railings, Pipe

			Crew	Daily Output	Labor-Hours	Unit	Material	2012 Bare Costs Labor	2012 Bare Costs Equipment	Total	Total Incl O&P
0010	**RAILINGS, PIPE**, shop fab'd, 3'-6" high, posts @ 5' O.C.										
0015	Made from recycled materials	G									
0020	Aluminum, 2 rail, satin finish, 1-1/4" diameter	G	E-4	160	.200	L.F.	35	9.95	.76	45.71	57
0030	Clear anodized	G		160	.200		43	9.95	.76	53.71	66
0040	Dark anodized	G		160	.200		48	9.95	.76	58.71	71.50
0080	1-1/2" diameter, satin finish	G		160	.200		41.50	9.95	.76	52.21	64
0090	Clear anodized	G		160	.200		46.50	9.95	.76	57.21	69.50
0100	Dark anodized	G		160	.200		51	9.95	.76	61.71	74.50
0140	Aluminum, 3 rail, 1-1/4" diam., satin finish	G		137	.234		53	11.65	.89	65.54	80.50
0150	Clear anodized	G		137	.234		66	11.65	.89	78.54	94.50
0160	Dark anodized	G		137	.234		73	11.65	.89	85.54	102
0200	1-1/2" diameter, satin finish	G		137	.234		63	11.65	.89	75.54	91.50
0210	Clear anodized	G		137	.234		72	11.65	.89	84.54	101
0220	Dark anodized	G		137	.234		78.50	11.65	.89	91.04	108
0500	Steel, 2 rail, on stairs, primed, 1-1/4" diameter	G		160	.200		23.50	9.95	.76	34.21	44.50
0520	1-1/2" diameter	G		160	.200		25.50	9.95	.76	36.21	46.50
0540	Galvanized, 1-1/4" diameter	G		160	.200		32	9.95	.76	42.71	53.50
0560	1-1/2" diameter	G		160	.200		36	9.95	.76	46.71	58
0580	Steel, 3 rail, primed, 1-1/4" diameter	G		137	.234		35	11.65	.89	47.54	60.50
0600	1-1/2" diameter	G		137	.234		36.50	11.65	.89	49.04	62
0620	Galvanized, 1-1/4" diameter	G		137	.234		49	11.65	.89	61.54	76
0640	1-1/2" diameter	G		137	.234		57	11.65	.89	69.54	84.50
0700	Stainless steel, 2 rail, 1-1/4" diam. #4 finish	G		137	.234		103	11.65	.89	115.54	135
0720	High polish	G		137	.234		167	11.65	.89	179.54	206
0740	Mirror polish	G		137	.234		210	11.65	.89	222.54	253
0760	Stainless steel, 3 rail, 1-1/2" diam., #4 finish	G		120	.267		156	13.30	1.01	170.31	197
0770	High polish	G		120	.267		258	13.30	1.01	272.31	310
0780	Mirror finish	G		120	.267		315	13.30	1.01	329.31	370
0900	Wall rail, alum. pipe, 1-1/4" diam., satin finish	G		213	.150		19.75	7.50	.57	27.82	35.50
0905	Clear anodized	G		213	.150		24.50	7.50	.57	32.57	40.50
0910	Dark anodized	G		213	.150		29	7.50	.57	37.07	46
0915	1-1/2" diameter, satin finish	G		213	.150		22	7.50	.57	30.07	38
0920	Clear anodized	G		213	.150		27.50	7.50	.57	35.57	44.50
0925	Dark anodized	G		213	.150		34	7.50	.57	42.07	51.50
0930	Steel pipe, 1-1/4" diameter, primed	G		213	.150		14	7.50	.57	22.07	29.50
0935	Galvanized	G		213	.150		20.50	7.50	.57	28.57	36.50
0940	1-1/2" diameter	G		176	.182		14.50	9.05	.69	24.24	33
0945	Galvanized	G		213	.150		20.50	7.50	.57	28.57	36.50

05 52 Metal Railings

05 52 13 – Pipe and Tube Railings

05 52 13.50 Railings, Pipe

		Crew	Daily Output	Labor-Hours	Unit	Material	2012 Bare Costs Labor	Equipment	Total	Total Incl O&P
0955	Stainless steel pipe, 1-1/2" diam., #4 finish	G	E-4	107	.299	L.F.	82.50	14.90	1.14	98.54	119
0960	High polish	G		107	.299		168	14.90	1.14	184.04	213
0965	Mirror polish	G	↓	107	.299	↓	199	14.90	1.14	215.04	247
2000	2-line pipe rail (1-1/2" T&B) with 1/2" pickets @ 4-1/2" O.C.,										
2005	attached handrail on brackets										
2010	42" high aluminum, satin finish, straight & level	G	E-4	120	.267	L.F.	193	13.30	1.01	207.31	237
2050	42" high steel, primed, straight & level	G	"	120	.267		119	13.30	1.01	133.31	156
4000	For curved and level rails, add						10%	10%			
4100	For sloped rails for stairs, add					↓	30%	30%			

05 52 16 – Industrial Railings

05 52 16.50 Railings, Industrial

		Crew	Daily Output	Labor-Hours	Unit	Material	2012 Bare Costs Labor	Equipment	Total	Total Incl O&P
0010	**RAILINGS, INDUSTRIAL**, shop fab'd, 3'-6" high, posts @ 5' O.C.										
0020	2 rail, 3'-6" high, 1-1/2" pipe	G	E-4	255	.125	L.F.	25.50	6.25	.48	32.23	39.50
0100	2" angle rail	G	"	255	.125		23.50	6.25	.48	30.23	37.50
0200	For 4" high kick plate, 10 gauge, add	G					5.30			5.30	5.85
0300	1/4" thick, add	G					6.90			6.90	7.60
0500	For curved level rails, add						10%	10%			
0550	For sloped rails for stairs, add					↓	30%	30%			

05 53 Metal Gratings

05 53 13 – Bar Gratings

05 53 13.10 Floor Grating, Aluminum

		Crew	Daily Output	Labor-Hours	Unit	Material	2012 Bare Costs Labor	Equipment	Total	Total Incl O&P
0010	**FLOOR GRATING, ALUMINUM**, field fabricated from panels										
0015	Made from recycled materials	G									
0110	Bearing bars @ 1-3/16" O.C., cross bars @ 4" O.C.,										
0111	Up to 300 S.F., 1" x 1/8" bar	G	E-4	900	.036	S.F.	23.50	1.77	.14	25.41	29.50
0112	Over 300 S.F.	G		850	.038		21.50	1.88	.14	23.52	27
0113	1-1/4" x 1/8" bar, up to 300 S.F.	G		800	.040		27.50	1.99	.15	29.64	34
0114	Over 300 S.F.	G		1000	.032		25	1.60	.12	26.72	30.50
0122	1-1/4" x 3/16" bar, up to 300 S.F.	G		750	.043		36	2.13	.16	38.29	43.50
0124	Over 300 S.F.	G		1000	.032		32.50	1.60	.12	34.22	39
0132	1-1/2" x 3/16" bar, up to 300 S.F.	G		700	.046		41.50	2.28	.17	43.95	50
0134	Over 300 S.F.	G		1000	.032		37.50	1.60	.12	39.22	44.50
0136	1-3/4" x 3/16" bar, up to 300 S.F.	G		500	.064		50	3.19	.24	53.43	61
0138	Over 300 S.F.	G		1000	.032		45.50	1.60	.12	47.22	53
0146	2-1/4" x 3/16" bar, up to 300 S.F.	G		600	.053		69	2.66	.20	71.86	80.50
0148	Over 300 S.F.	G		1000	.032		62.50	1.60	.12	64.22	72
0162	Cross bars @ 2" O.C., 1" x 1/8", up to 300 S.F.	G		600	.053		30.50	2.66	.20	33.36	38.50
0164	Over 300 S.F.	G		1000	.032		27.50	1.60	.12	29.22	33.50
0172	1-1/4" x 3/16" bar, up to 300 S.F.	G		600	.053		40.50	2.66	.20	43.36	50
0174	Over 300 S.F.	G		1000	.032		37	1.60	.12	38.72	43.50
0182	1-1/2" x 3/16" bar, up to 300 S.F.	G		600	.053		45.50	2.66	.20	48.36	55
0184	Over 300 S.F.	G		1000	.032		41.50	1.60	.12	43.22	48.50
0186	1-3/4" x 3/16" bar, up to 300 S.F.	G		600	.053		55.50	2.66	.20	58.36	66
0188	Over 300 S.F.	G	↓	1000	.032	↓	50.50	1.60	.12	52.22	58.50
0200	For straight cuts, add					L.F.	3.90			3.90	4.29
0300	For curved cuts, add						4.80			4.80	5.30
0400	For straight banding, add	G					5			5	5.50
0500	For curved banding, add	G					6			6	6.60
0600	For aluminum checkered plate nosings, add	G				↓	6.50			6.50	7.15

05 53 Metal Gratings

05 53 13 – Bar Gratings

05 53 13.10 Floor Grating, Aluminum

		Crew	Daily Output	Labor-Hours	Unit	Material	2012 Bare Costs Labor	Equipment	Total	Total Incl O&P
0700	For straight toe plate, add G				L.F.	9.80			9.80	10.80
0800	For curved toe plate, add G					11.45			11.45	12.60
1000	For cast aluminum abrasive nosings, add G				↓	9.55			9.55	10.50
1200	Expanded aluminum, .65# per S.F. G	E-4	1050	.030	S.F.	11.65	1.52	.12	13.29	15.65
1400	Extruded I bars are 10% less than 3/16" bars									
1600	Heavy duty, all extruded plank, 3/4" deep, 1.8# per S.F. G	E-4	1100	.029	S.F.	30	1.45	.11	31.56	35.50
1700	1-1/4" deep, 2.9# per S.F. G		1000	.032		35	1.60	.12	36.72	41.50
1800	1-3/4" deep, 4.2# per S.F. G		925	.035		42	1.72	.13	43.85	49
1900	2-1/4" deep, 5.0# per S.F. G	↓	875	.037	↓	50	1.82	.14	51.96	58.50
2100	For safety serrated surface, add					15%				

05 53 13.70 Floor Grating, Steel

		Crew	Daily Output	Labor-Hours	Unit	Material	2012 Bare Costs Labor	Equipment	Total	Total Incl O&P
0010	**FLOOR GRATING, STEEL**, field fabricated from panels									
0015	Made from recycled materials G									
0300	Platforms, to 12' high, rectangular G	E-4	3150	.010	Lb.	3.20	.51	.04	3.75	4.47
0400	Circular G	"	2300	.014	"	4	.69	.05	4.74	5.70
0410	Painted bearing bars @ 1-3/16"									
0412	Cross bars @ 4" O.C., 3/4" x 1/8" bar, up to 300 S.F. G	E-2	500	.112	S.F.	8.55	5.40	2.99	16.94	22
0414	Over 300 S.F. G		750	.075		7.80	3.60	2	13.40	16.95
0422	1-1/4" x 3/16", up to 300 S.F. G		400	.140		13.20	6.75	3.74	23.69	30
0424	Over 300 S.F. G		600	.093		12	4.50	2.49	18.99	23.50
0432	1-1/2" x 3/16", up to 300 S.F. G		400	.140		15.55	6.75	3.74	26.04	33
0434	Over 300 S.F. G		600	.093		14.15	4.50	2.49	21.14	26
0436	1-3/4" x 3/16", up to 300 S.F. G		400	.140		20.50	6.75	3.74	30.99	38
0438	Over 300 S.F. G		600	.093		18.65	4.50	2.49	25.64	31
0452	2-1/4" x 3/16", up to 300 S.F. G		300	.187		25.50	9	4.99	39.49	49
0454	Over 300 S.F. G		450	.124		23	6	3.33	32.33	39.50
0462	Cross bars @ 2" O.C., 3/4" x 1/8", up to 300 S.F. G		500	.112		13.65	5.40	2.99	22.04	27.50
0464	Over 300 S.F. G		750	.075		11.35	3.60	2	16.95	21
0472	1-1/4" x 3/16", up to 300 S.F. G		400	.140		17.85	6.75	3.74	28.34	35.50
0474	Over 300 S.F. G		600	.093		14.85	4.50	2.49	21.84	27
0482	1-1/2" x 3/16", up to 300 S.F. G		400	.140		20	6.75	3.74	30.49	37.50
0484	Over 300 S.F. G		600	.093		16.75	4.50	2.49	23.74	29
0486	1-3/4" x 3/16", up to 300 S.F. G		400	.140		30.50	6.75	3.74	40.99	49
0488	Over 300 S.F. G		600	.093		25	4.50	2.49	31.99	38
0502	2-1/4" x 3/16", up to 300 S.F. G		300	.187		29.50	9	4.99	43.49	53.50
0504	Over 300 S.F. G	↓	450	.124		24.50	6	3.33	33.83	41
0690	For galvanized grating, add				↓	25%				
0800	For straight cuts, add				L.F.	5.60			5.60	6.15
0900	For curved cuts, add					7.10			7.10	7.80
1000	For straight banding, add G					6			6	6.60
1100	For curved banding, add G					7.90			7.90	8.70
1200	For checkered plate nosings, add G					6.95			6.95	7.65
1300	For straight toe or kick plate, add G					12.45			12.45	13.70
1400	For curved toe or kick plate, add G					14.10			14.10	15.50
1500	For abrasive nosings, add G				↓	9.25			9.25	10.20
1510	For stair treads, see Section 05 55 13.50									
1600	For safety serrated surface, minimum, add					15%				
1700	Maximum, add					25%				
2000	Stainless steel gratings, close spaced, 1" x 1/8" bars, up to 300 S.F. G	E-4	450	.071	S.F.	106	3.54	.27	109.81	124
2100	Standard spacing, 3/4" x 1/8" bars G		500	.064		75.50	3.19	.24	78.93	89
2200	1-1/4" x 3/16" bars G		400	.080		116	3.99	.30	120.29	135
2400	Expanded steel grating, at ground, 3.0# per S.F. G	↓	900	.036	↓	7.30	1.77	.14	9.21	11.35

05 53 Metal Gratings

05 53 13 – Bar Gratings

05 53 13.70 Floor Grating, Steel

		Crew	Daily Output	Labor-Hours	Unit	Material	2012 Bare Costs Labor	Equipment	Total	Total Incl O&P
2500	3.14# per S.F.	G	E-4	900	.036	S.F.	7.85	1.77	.14	9.76	11.95
2600	4.0# per S.F.	G		850	.038		9.70	1.88	.14	11.72	14.15
2650	4.27# per S.F.	G		850	.038		10.40	1.88	.14	12.42	14.95
2700	5.0# per S.F.	G		800	.040		12.35	1.99	.15	14.49	17.35
2800	6.25# per S.F.	G		750	.043		16	2.13	.16	18.29	21.50
2900	7.0# per S.F.	G		700	.046		17.70	2.28	.17	20.15	23.50
3100	For flattened expanded steel grating, add						8%				
3300	For elevated installation above 15', add							15%			

05 53 16 – Plank Gratings

05 53 16.50 Grating Planks

		Crew	Daily Output	Labor-Hours	Unit	Material	2012 Bare Costs Labor	Equipment	Total	Total Incl O&P
0010	**GRATING PLANKS**, field fabricated from planks										
0020	Aluminum, 9-1/2" wide, 14 ga., 2" rib	G	E-4	950	.034	L.F.	28	1.68	.13	29.81	34
0200	Galvanized steel, 9-1/2" wide, 14 ga., 2-1/2" rib	G		950	.034		15.15	1.68	.13	16.96	19.80
0300	4" rib	G		950	.034		17.80	1.68	.13	19.61	22.50
0500	12 gauge, 2-1/2" rib	G		950	.034		26	1.68	.13	27.81	31.50
0600	3" rib	G		950	.034		21	1.68	.13	22.81	26
0800	Stainless steel, type 304, 16 ga., 2" rib	G		950	.034		41	1.68	.13	42.81	48
0900	Type 316	G		950	.034		50.50	1.68	.13	52.31	58.50

05 53 19 – Floor Grating Frame

05 53 19.30 Grating Frame

		Crew	Daily Output	Labor-Hours	Unit	Material	2012 Bare Costs Labor	Equipment	Total	Total Incl O&P
0010	**GRATING FRAME**, field fabricated										
0020	Aluminum, for gratings 1" to 1-1/2" deep	G	1 Sswk	70	.114	L.F.	3.36	5.65		9.01	13.80
0100	For each corner, add	G				Ea.	5.05			5.05	5.55

05 54 Metal Floor Plates

05 54 13 – Floor Plates

05 54 13.20 Checkered Plates

		Crew	Daily Output	Labor-Hours	Unit	Material	2012 Bare Costs Labor	Equipment	Total	Total Incl O&P
0010	**CHECKERED PLATES**, steel, field fabricated										
0015	Made from recycled materials	G									
0020	1/4" & 3/8", 2000 to 5000 S.F., bolted	G	E-4	2900	.011	Lb.	.98	.55	.04	1.57	2.11
0100	Welded	G		4400	.007	"	.93	.36	.03	1.32	1.70
0300	Pit or trench cover and frame, 1/4" plate, 2' to 3' wide	G		100	.320	S.F.	13.75	15.95	1.22	30.92	45
0400	For galvanizing, add	G				Lb.	.28			.28	.31
0500	Platforms, 1/4" plate, no handrails included, rectangular	G	E-4	4200	.008		3.20	.38	.03	3.61	4.23
0600	Circular	G	"	2500	.013		4	.64	.05	4.69	5.60

05 54 13.70 Trench Covers

		Crew	Daily Output	Labor-Hours	Unit	Material	2012 Bare Costs Labor	Equipment	Total	Total Incl O&P
0010	**TRENCH COVERS**, field fabricated										
0020	Cast iron grating with bar stops and angle frame, to 18" wide	G	1 Sswk	20	.400	L.F.	268	19.75		287.75	330
0100	Frame only (both sides of trench), 1" grating	G		45	.178		1.54	8.75		10.29	17.40
0150	2" grating	G		35	.229		3.70	11.30		15	24
0200	Aluminum, stock units, including frames and										
0210	3/8" plain cover plate, 4" opening	G	E-4	205	.156	L.F.	22	7.80	.59	30.39	39
0300	6" opening	G		185	.173		26.50	8.60	.66	35.76	45.50
0400	10" opening	G		170	.188		36	9.40	.72	46.12	57
0500	16" opening	G		155	.206		49.50	10.30	.78	60.58	74
0700	Add per inch for additional widths to 24"	G					2.06			2.06	2.26
0900	For custom fabrication, add						50%				
1100	For 1/4" plain cover plate, deduct						12%				
1500	For cover recessed for tile, 1/4" thick, deduct						12%				
1600	3/8" thick, add						5%				

156

05 54 Metal Floor Plates

05 54 13 – Floor Plates

05 54 13.70 Trench Covers		Crew	Daily Output	Labor-Hours	Unit	Material	2012 Bare Costs Labor	Equipment	Total	Total Incl O&P
1800	For checkered plate cover, 1/4" thick, deduct				L.F.	12%				
1900	3/8" thick, add					2%				
2100	For slotted or round holes in cover, 1/4" thick, add					3%				
2200	3/8" thick, add					4%				
2300	For abrasive cover, add					12%				

05 55 Metal Stair Treads and Nosings

05 55 13 – Metal Stair Treads

05 55 13.50 Stair Treads			Crew	Daily Output	Labor-Hours	Unit	Material	2012 Bare Costs Labor	Equipment	Total	Total Incl O&P
0010	**STAIR TREADS**										
0020	Aluminum grating, 3' long, 1-1/2" x 3/16" rect. bars, 6" wide	G	1 Sswk	24	.333	Ea.	62	16.45		78.45	97.50
0100	12" wide	G		22	.364		91.50	17.95		109.45	133
0200	1-1/2" x 3/16" I-bars, 6" wide	G		24	.333		52	16.45		68.45	87
0300	12" wide	G		22	.364		84	17.95		101.95	124
0400	For abrasive nosings, add	G					16.30			16.30	17.90
0500	For narrow mesh, add						60%				
0600	Stair treads, not incl. stringers, see also Section 03 15 05.85										
0700	Cast aluminum, abrasive, 3' long x 12" wide, 5/16" thick	G	1 Sswk	15	.533	Ea.	138	26.50		164.50	198
0800	3/8" thick	G		15	.533		148	26.50		174.50	209
0900	1/2" thick	G		15	.533		160	26.50		186.50	223
1000	Cast bronze, abrasive, 3/8" thick	G		8	1		625	49.50		674.50	780
1100	1/2" thick	G		8	1		820	49.50		869.50	990
1200	Cast iron, abrasive, 3' long x 12" wide, 3/8" thick	G		15	.533		120	26.50		146.50	179
1300	1/2" thick	G		15	.533		140	26.50		166.50	201
2000	Steel grating, painted, 3' long, 1-1/4" x 3/16" bars, 6" wide	G		20	.400		50.50	19.75		70.25	91
2010	12" wide	G		18	.444		58.50	22		80.50	104
2100	Add for abrasive nosing, 3' long, painted	G					15.80			15.80	17.35
2200	Galvanized	G					21			21	23
2300	Painting 3' treads, in shop, nonstandard paint						2.78			2.78	3.06
2400	Added coats, standard paint						1.19			1.19	1.31
2500	Expanded steel, 2-1/2" deep, 9" x 3' long, 18 gauge	G	1 Sswk	20	.400		28	19.75		47.75	66
2600	14 gauge	G	"	20	.400		34	19.75		53.75	73

05 56 Metal Castings

05 56 13 – Metal Construction Castings

05 56 13.50 Construction Castings			Crew	Daily Output	Labor-Hours	Unit	Material	2012 Bare Costs Labor	Equipment	Total	Total Incl O&P
0010	**CONSTRUCTION CASTINGS**										
0020	Manhole covers and frames, see Section 33 44 13.13										
0100	Column bases, cast iron, 16" x 16", approx. 65 lb.	G	E-4	46	.696	Ea.	145	34.50	2.64	182.14	225
0200	32" x 32", approx. 256 lb.	G	"	23	1.391		540	69.50	5.30	614.80	725
0400	Cast aluminum for wood columns, 8" x 8"	G	1 Carp	32	.250		43	11.05		54.05	64.50
0500	12" x 12"	G	"	32	.250		93.50	11.05		104.55	120
0600	Miscellaneous C.I. castings, light sections, less than 150 lb.	G	E-4	3200	.010	Lb.	2.24	.50	.04	2.78	3.39
1100	Heavy sections, more than 150 lb.	G		4200	.008		1.09	.38	.03	1.50	1.91
1300	Special low volume items	G		3200	.010		3.89	.50	.04	4.43	5.20
1500	For ductile iron, add						100%				

05 58 Formed Metal Fabrications

05 58 21 – Formed Chain

05 58 21.05 Alloy Steel Chain

05 58 21.05 Alloy Steel Chain		Crew	Daily Output	Labor-Hours	Unit	Material	2012 Bare Costs Labor	Equipment	Total	Total Incl O&P
0010	**ALLOY STEEL CHAIN**, Grade 80, for lifting									
0015	Self-colored, cut lengths, 1/4"	G	E-17	4	4	C.L.F.	700	201		901	1,125
0020	3/8"	G		2	8		885	405		1,290	1,700
0030	1/2"	G		1.20	13.333		1,425	670		2,095	2,750
0040	5/8"	G		.72	22.222		2,150	1,125		3,275	4,375
0050	3/4"	G	E-18	.48	83.333		2,925	4,100	1,975	9,000	12,500
0060	7/8"	G		.40	100		4,925	4,925	2,375	12,225	16,600
0070	1"	G		.35	114		6,675	5,625	2,725	15,025	20,100
0080	1-1/4"	G		.24	166		13,600	8,200	3,950	25,750	33,600
0110	Hook, Grade 80, Clevis slip, 1/4"	G				Ea.	34.50			34.50	38
0120	3/8"	G					45			45	49.50
0130	1/2"	G					71.50			71.50	79
0140	5/8"	G					89.50			89.50	98.50
0150	3/4"	G					149			149	164
0160	Hook, Grade 80, eye/sling w/hammerlock coupling, 15 Ton	G					320			320	355
0170	22 Ton	G					870			870	955
0180	37 Ton	G					2,400			2,400	2,650

05 58 23 – Formed Metal Guards

05 58 23.90 Window Guards

05 58 23.90 Window Guards		Crew	Daily Output	Labor-Hours	Unit	Material	2012 Bare Costs Labor	Equipment	Total	Total Incl O&P
0010	**WINDOW GUARDS**, shop fabricated									
0015	Expanded metal, steel angle frame, permanent	G	E-4	350	.091	S.F.	21.50	4.56	.35	26.41	32
0025	Steel bars, 1/2" x 1/2", spaced 5" O.C.	G	"	290	.110	"	15	5.50	.42	20.92	27
0030	Hinge mounted, add	G				Opng.	43.50			43.50	48
0040	Removable type, add	G				"	27.50			27.50	30.50
0050	For galvanized guards, add					S.F.	35%				
0070	For pivoted or projected type, add						105%	40%			
0100	Mild steel, stock units, economy	G	E-4	405	.079		5.90	3.94	.30	10.14	13.90
0200	Deluxe	G		405	.079		12	3.94	.30	16.24	20.50
0400	Woven wire, stock units, 3/8" channel frame, 3' x 5' opening	G		40	.800	Opng.	160	40	3.04	203.04	251
0500	4' x 6' opening	G		38	.842		255	42	3.20	300.20	360
0800	Basket guards for above, add	G					220			220	242
1000	Swinging guards for above, add	G					75			75	82.50

05 58 25 – Formed Lamp Posts

05 58 25.40 Lamp Posts

05 58 25.40 Lamp Posts		Crew	Daily Output	Labor-Hours	Unit	Material	2012 Bare Costs Labor	Equipment	Total	Total Incl O&P
0010	**LAMP POSTS**									
0020	Aluminum, 7' high, stock units, post only	G	1 Carp	16	.500	Ea.	67.50	22		89.50	109
0100	Mild steel, plain	G	"	16	.500	"	65	22		87	106

05 71 Decorative Metal Stairs

05 71 13 – Fabricated Metal Spiral Stairs

05 71 13.50 Spiral Stairs

05 71 13.50 Spiral Stairs		Crew	Daily Output	Labor-Hours	Unit	Material	2012 Bare Costs Labor	Equipment	Total	Total Incl O&P
0010	**SPIRAL STAIRS**									
1805	Shop fabricated, custom ordered									
1810	Aluminum, 5'-0" diameter, plain units	G	E-4	45	.711	Riser	560	35.50	2.70	598.20	680
1820	Fancy units	G		45	.711		1,050	35.50	2.70	1,088.20	1,250
1900	Cast iron, 4'-0" diameter, plain units	G		45	.711		505	35.50	2.70	543.20	620
1920	Fancy Units	G		25	1.280		690	64	4.86	758.86	880
2000	Steel, industrial checkered plate, 4' diameter	G		45	.711		505	35.50	2.70	543.20	620
2200	Stock units, 6'-0" diameter	G		40	.800		610	40	3.04	653.04	745
3100	Spiral stair kits, 12 stacking risers to fit exact floor height									

05 71 Decorative Metal Stairs

05 71 13 – Fabricated Metal Spiral Stairs

05 71 13.50 Spiral Stairs

		Crew	Daily Output	Labor-Hours	Unit	Material	2012 Bare Costs Labor	Equipment	Total	Total Incl O&P
3110	Steel, flat metal treads, primed, 3'-6" diameter	**G** 2 Carp	1.60	10	Flight	1,250	440		1,690	2,025
3120	4'-0" diameter	**G**	1.45	11.034		1,425	485		1,910	2,325
3130	4'-6" diameter	**G**	1.35	11.852		1,575	525		2,100	2,525
3140	5'-0" diameter	**G**	1.25	12.800		1,700	565		2,265	2,750
3210	Galvanized, 3'-6" diameter	**G**	1.60	10		1,475	440		1,915	2,300
3220	4'-0" diameter	**G**	1.45	11.034		1,700	485		2,185	2,625
3230	4'-6" diameter	**G**	1.35	11.852		1,900	525		2,425	2,875
3240	5'-0" diameter	**G**	1.25	12.800		2,050	565		2,615	3,125
3310	Checkered plate tread, primed, 3'-6" diameter	**G**	1.45	11.034		1,475	485		1,960	2,375
3320	4'-0" diameter	**G**	1.35	11.852		1,675	525		2,200	2,625
3330	4'-6" diameter	**G**	1.25	12.800		1,825	565		2,390	2,875
3340	5'-0" diameter	**G**	1.15	13.913		2,000	615		2,615	3,125
3410	Galvanized, 3'-6" diameter	**G**	1.45	11.034		1,775	485		2,260	2,700
3420	4'-0" diameter	**G**	1.35	11.852		2,000	525		2,525	3,000
3430	4'-6" diameter	**G**	1.25	12.800		2,175	565		2,740	3,275
3440	5'-0" diameter	**G**	1.15	13.913		2,375	615		2,990	3,575
3510	Red oak covers on flat metal treads, 3'-6" diameter		1.35	11.852		2,550	525		3,075	3,625
3520	4'-0" diameter		1.25	12.800		2,825	565		3,390	3,975
3530	4'-6" diameter		1.15	13.913		3,025	615		3,640	4,300
3540	5'-0" diameter		1.05	15.238		3,325	670		3,995	4,675

05 73 Decorative Metal Railings

05 73 16 – Wire Rope Decorative Metal Railings

05 73 16.10 Cable Railings

		Crew	Daily Output	Labor-Hours	Unit	Material	2012 Bare Costs Labor	Equipment	Total	Total Incl O&P
0010	**CABLE RAILINGS**, with 316 stainless steel 1 x 19 cable, 3/16" diameter									
0015	Made from recycled materials	**G**								
0100	1-3/4" diameter stainless steel posts x 42" high, cables 4" OC	**G** 2 Sswk	25	.640	L.F.	37	31.50		68.50	97

05 73 23 – Ornamental Railings

05 73 23.50 Railings, Ornamental

		Crew	Daily Output	Labor-Hours	Unit	Material	2012 Bare Costs Labor	Equipment	Total	Total Incl O&P
0010	**RAILINGS, ORNAMENTAL**, 3'-6" high, posts @ 6' O.C.									
0020	Bronze or stainless, hand forged, plain	**G** 2 Sswk	24	.667	L.F.	101	33		134	171
0100	Fancy	**G**	18	.889		174	44		218	271
0200	Aluminum, panelized, plain	**G**	24	.667		16.50	33		49.50	77
0300	Fancy	**G**	18	.889		33	44		77	115
0400	Wrought iron, hand forged, plain	**G**	24	.667		67.50	33		100.50	134
0500	Fancy	**G**	18	.889		125	44		169	216
0550	Steel, panelized, plain	**G**	24	.667		16.90	33		49.90	77.50
0560	Fancy	**G**	18	.889		25	44		69	106
0600	Composite metal/wood/glass, plain		18	.889		112	44		156	202
0700	Fancy		12	1.333		224	66		290	365

05 75 Decorative Formed Metal

05 75 13 - Columns

05 75 13.10 Aluminum Columns

		Crew	Daily Output	Labor-Hours	Unit	Material	2012 Bare Costs Labor	Equipment	Total	Total Incl O&P
0010	**ALUMINUM COLUMNS**									
0015	Made from recycled materials	G								
0020	Aluminum, extruded, stock units, no cap or base, 6" diameter	G	E-4	240	.133	L.F.	10.65	6.65	.51	17.81	24
0100	8" diameter	G		170	.188		14.10	9.40	.72	24.22	33
0200	10" diameter	G		150	.213		19.20	10.65	.81	30.66	41
0300	12" diameter	G		140	.229		33.50	11.40	.87	45.77	58.50
0400	15" diameter	G		120	.267		48	13.30	1.01	62.31	77.50
0410	Caps and bases, plain, 6" diameter	G				Set	22.50			22.50	24.50
0420	8" diameter	G					28			28	31
0430	10" diameter	G					39			39	43
0440	12" diameter	G					55.50			55.50	61
0450	15" diameter	G					110			110	121
0460	Caps, ornamental, plain	G					340			340	370
0470	Fancy	G					1,675			1,675	1,850
0500	For square columns, add to column prices above					L.F.	50%				
0700	Residential, flat, 8' high, plain	G	E-4	20	1.600	Ea.	97.50	80	6.10	183.60	257
0720	Fancy	G		20	1.600		190	80	6.10	276.10	360
0740	Corner type, plain	G		20	1.600		168	80	6.10	254.10	335
0760	Fancy	G		20	1.600		330	80	6.10	416.10	515

05 75 13.20 Columns, Ornamental

			Crew	Daily Output	Labor-Hours	Unit	Material	2012 Bare Costs Labor	Equipment	Total	Total Incl O&P
0010	**COLUMNS, ORNAMENTAL**, shop fabricated	R051223-10									
6400	Mild steel, flat, 9" wide, stock units, painted, plain	G	E-4	160	.200	V.L.F.	9.15	9.95	.76	19.86	28.50
6450	Fancy	G		160	.200		17.75	9.95	.76	28.46	38
6500	Corner columns, painted, plain	G		160	.200		15.75	9.95	.76	26.46	36
6550	Fancy	G		160	.200		31	9.95	.76	41.71	53

Estimating Tips

06 05 00 Common Work Results for Wood, Plastics, and Composites

- Common to any wood-framed structure are the accessory connector items such as screws, nails, adhesives, hangers, connector plates, straps, angles, and hold-downs. For typical wood-framed buildings, such as residential projects, the aggregate total for these items can be significant, especially in areas where seismic loading is a concern. For floor and wall framing, the material cost is based on 10 to 25 lbs. per MBF. Hold-downs, hangers, and other connectors should be taken off by the piece.

  Included with material costs are fasteners for a normal installation. RSMeans engineers use manufacturer's recommendations, written specifications, and/or standard construction practice for size and spacing of fasteners. Prices for various fasteners are shown for informational purposes only. Adjustments should be made if unusual fastening conditions exist.

06 10 00 Carpentry

- Lumber is a traded commodity and therefore sensitive to supply and demand in the marketplace. Even in "budgetary" estimating of wood-framed projects, it is advisable to call local suppliers for the latest market pricing.

- Common quantity units for wood-framed projects are "thousand board feet" (MBF). A board foot is a volume of wood, 1" x 1' x 1', or 144 cubic inches. Board-foot quantities are generally calculated using nominal material dimensions—dressed sizes are ignored. Board foot per lineal foot of any stick of lumber can be calculated by dividing the nominal cross-sectional area by 12. As an example, 2,000 lineal feet of 2 x 12 equates to 4 MBF by dividing the nominal area, 2 x 12, by 12, which equals 2, and multiplying by 2,000 to give 4,000 board feet. This simple rule applies to all nominal dimensioned lumber.

- Waste is an issue of concern at the quantity takeoff for any area of construction. Framing lumber is sold in even foot lengths, i.e., 10', 12', 14', 16', and depending on spans, wall heights, and the grade of lumber, waste is inevitable. A rule of thumb for lumber waste is 5%–10% depending on material quality and the complexity of the framing.

- Wood in various forms and shapes is used in many projects, even where the main structural framing is steel, concrete, or masonry. Plywood as a back-up partition material and 2x boards used as blocking and cant strips around roof edges are two common examples. The estimator should ensure that the costs of all wood materials are included in the final estimate.

06 20 00 Finish Carpentry

- It is necessary to consider the grade of workmanship when estimating labor costs for erecting millwork and interior finish. In practice, there are three grades: premium, custom, and economy. The RSMeans daily output for base and case moldings is in the range of 200 to 250 L.F. per carpenter per day. This is appropriate for most average custom-grade projects. For premium projects, an adjustment to productivity of 25%–50% should be made, depending on the complexity of the job.

Reference Numbers

Reference numbers are shown in shaded boxes at the beginning of some major classifications. These numbers refer to related items in the Reference Section. The reference information may be an estimating procedure, an alternate pricing method, or technical information.

Note: Not all subdivisions listed here necessarily appear in this publication.

Division 6 - Wood, Plastics, and Composites

06 05 05 – Selective Wood and Plastics Demolition

06 05 05.10 Selective Demolition Wood Framing		Crew	Daily Output	Labor-Hours	Unit	Material	2012 Bare Costs Labor	Equipment	Total	Total Incl O&P
0010	**SELECTIVE DEMOLITION WOOD FRAMING** R024119-10									
0100	Timber connector, nailed, small	1 Clab	96	.083	Ea.		2.92		2.92	4.50
0110	Medium		60	.133			4.68		4.68	7.20
0120	Large		48	.167			5.85		5.85	9
0130	Bolted, small		48	.167			5.85		5.85	9
0140	Medium		32	.250			8.80		8.80	13.50
0150	Large		24	.333			11.70		11.70	18
2958	Beams, 2" x 6"	2 Clab	1100	.015	L.F.		.51		.51	.79
2960	2" x 8"		825	.019			.68		.68	1.05
2965	2" x 10"		665	.024			.84		.84	1.30
2970	2" x 12"		550	.029			1.02		1.02	1.57
2972	2" x 14"		470	.034			1.19		1.19	1.84
2975	4" x 8"	B-1	413	.058			2.08		2.08	3.20
2980	4" x 10"		330	.073			2.60		2.60	4
2985	4" x 12"		275	.087			3.12		3.12	4.80
3000	6" x 8"		275	.087			3.12		3.12	4.80
3040	6" x 10"		220	.109			3.90		3.90	6
3080	6" x 12"		185	.130			4.64		4.64	7.15
3120	8" x 12"		140	.171			6.15		6.15	9.45
3160	10" x 12"		110	.218			7.80		7.80	12
3162	Alternate pricing method		1.10	21.818	M.B.F.		780		780	1,200
3170	Blocking, in 16" OC wall framing, 2" x 4"	1 Clab	600	.013	L.F.		.47		.47	.72
3172	2" x 6"		400	.020			.70		.70	1.08
3174	In 24" OC wall framing, 2" x 4"		600	.013			.47		.47	.72
3176	2" x 6"		400	.020			.70		.70	1.08
3178	Alt method, wood blocking removal from wood framing		.40	20	M.B.F.		700		700	1,075
3179	Wood blocking removal from steel framing		.36	22.222	"		780		780	1,200
3180	Bracing, let in, 1" x 3", studs 16" OC		1050	.008	L.F.		.27		.27	.41
3181	Studs 24" OC		1080	.007			.26		.26	.40
3182	1" x 4", studs 16" OC		1050	.008			.27		.27	.41
3183	Studs 24" OC		1080	.007			.26		.26	.40
3184	1" x 6", studs 16" OC		1050	.008			.27		.27	.41
3185	Studs 24" OC		1080	.007			.26		.26	.40
3186	2" x 3", studs 16" OC		800	.010			.35		.35	.54
3187	Studs 24" OC		830	.010			.34		.34	.52
3188	2" x 4", studs 16" OC		800	.010			.35		.35	.54
3189	Studs 24" OC		830	.010			.34		.34	.52
3190	2" x 6", studs 16" OC		800	.010			.35		.35	.54
3191	Studs 24" OC		830	.010			.34		.34	.52
3192	2" x 8", studs 16" OC		800	.010			.35		.35	.54
3193	Studs 24" OC		830	.010			.34		.34	.52
3194	"T" shaped metal bracing, studs at 16" OC		1060	.008			.27		.27	.41
3195	Studs at 24" OC		1200	.007			.23		.23	.36
3196	Metal straps, studs at 16" OC		1200	.007			.23		.23	.36
3197	Studs at 24" OC		1240	.006			.23		.23	.35
3200	Columns, round, 8' to 14' tall		40	.200	Ea.		7		7	10.80
3202	Dimensional lumber sizes	2 Clab	1.10	14.545	M.B.F.		510		510	785
3250	Blocking, between joists	1 Clab	320	.025	Ea.		.88		.88	1.35
3252	Bridging, metal strap, between joists		320	.025	Pr.		.88		.88	1.35
3254	Wood, between joists		320	.025	"		.88		.88	1.35
3260	Door buck, studs, header & access., 8' high 2" x 4" wall, 3' wide		32	.250	Ea.		8.80		8.80	13.50
3261	4' wide		32	.250			8.80		8.80	13.50
3262	5' wide		32	.250			8.80		8.80	13.50

06 05 05 – Selective Wood and Plastics Demolition

06 05 05.10 Selective Demolition Wood Framing	Crew	Daily Output	Labor-Hours	Unit	Material	2012 Bare Costs Labor	Equipment	Total	Total Incl O&P
3263	6' wide	1 Clab	32	.250	Ea.		8.80		8.80	13.50
3264	8' wide		30	.267			9.35		9.35	14.40
3265	10' wide		30	.267			9.35		9.35	14.40
3266	12' wide		30	.267			9.35		9.35	14.40
3267	2" x 6" wall, 3' wide		32	.250			8.80		8.80	13.50
3268	4' wide		32	.250			8.80		8.80	13.50
3269	5' wide		32	.250			8.80		8.80	13.50
3270	6' wide		32	.250			8.80		8.80	13.50
3271	8' wide		30	.267			9.35		9.35	14.40
3272	10' wide		30	.267			9.35		9.35	14.40
3273	12' wide		30	.267			9.35		9.35	14.40
3274	Window buck, studs, header & access, 8' high 2" x 4" wall, 2' wide		24	.333			11.70		11.70	18
3275	3' wide		24	.333			11.70		11.70	18
3276	4' wide		24	.333			11.70		11.70	18
3277	5' wide		24	.333			11.70		11.70	18
3278	6' wide		24	.333			11.70		11.70	18
3279	7' wide		24	.333			11.70		11.70	18
3280	8' wide		22	.364			12.75		12.75	19.65
3281	10' wide		22	.364			12.75		12.75	19.65
3282	12' wide		22	.364			12.75		12.75	19.65
3283	2" x 6" wall, 2' wide		24	.333			11.70		11.70	18
3284	3' wide		24	.333			11.70		11.70	18
3285	4' wide		24	.333			11.70		11.70	18
3286	5' wide		24	.333			11.70		11.70	18
3287	6' wide		24	.333			11.70		11.70	18
3288	7' wide		24	.333			11.70		11.70	18
3289	8' wide		22	.364			12.75		12.75	19.65
3290	10' wide		22	.364			12.75		12.75	19.65
3291	12' wide		22	.364	▼		12.75		12.75	19.65
3400	Fascia boards, 1" x 6"		500	.016	L.F.		.56		.56	.86
3440	1" x 8"		450	.018			.62		.62	.96
3480	1" x 10"		400	.020			.70		.70	1.08
3490	2" x 6"		450	.018			.62		.62	.96
3500	2" x 8"		400	.020			.70		.70	1.08
3510	2" x 10"		350	.023	▼		.80		.80	1.23
3610	Furring, on wood walls or ceiling		4000	.002	S.F.		.07		.07	.11
3620	On masonry or concrete walls or ceiling		1200	.007	"		.23		.23	.36
3800	Headers over openings, 2 @ 2" x 6"		110	.073	L.F.		2.55		2.55	3.93
3840	2 @ 2" x 8"		100	.080			2.81		2.81	4.32
3880	2 @ 2" x 10"		90	.089	▼		3.12		3.12	4.80
3885	Alternate pricing method		.26	30.651	M.B.F.		1,075		1,075	1,650
3920	Joists, 1" x 4"		1250	.006	L.F.		.22		.22	.35
3930	1" x 6"		1135	.007			.25		.25	.38
3940	1" x 8"		1000	.008			.28		.28	.43
3950	1" x 10"		895	.009			.31		.31	.48
3960	1" x 12"		765	.010			.37		.37	.56
4200	2" x 4"	2 Clab	1000	.016			.56		.56	.86
4230	2" x 6"		970	.016			.58		.58	.89
4240	2" x 8"		940	.017			.60		.60	.92
4250	2" x 10"		910	.018			.62		.62	.95
4280	2" x 12"		880	.018			.64		.64	.98
4281	2" x 14"		850	.019			.66		.66	1.02
4282	Composite joists, 9-1/2"	▼	960	.017	▼		.59		.59	.90

06 05 05.10 Selective Demolition Wood Framing	Crew	Daily Output	Labor-Hours	Unit	Material	2012 Bare Costs Labor	Equipment	Total	Total Incl O&P
4283	11-7/8"	2 Clab	930	.017	L.F.		.60		.60	.93
4284	14"		897	.018			.63		.63	.96
4285	16"		865	.019			.65		.65	1
4290	Wood joists, alternate pricing method		1.50	10.667	M.B.F.		375		375	575
4500	Open web joist, 12" deep		500	.032	L.F.		1.12		1.12	1.73
4505	14" deep		475	.034			1.18		1.18	1.82
4510	16" deep		450	.036			1.25		1.25	1.92
4520	18" deep		425	.038			1.32		1.32	2.03
4530	24" deep		400	.040			1.40		1.40	2.16
4550	Ledger strips, 1" x 2"	1 Clab	1200	.007			.23		.23	.36
4560	1" x 3"		1200	.007			.23		.23	.36
4570	1" x 4"		1200	.007			.23		.23	.36
4580	2" x 2"		1100	.007			.26		.26	.39
4590	2" x 4"		1000	.008			.28		.28	.43
4600	2" x 6"		1000	.008			.28		.28	.43
4601	2" x 8" or 2" x 10"		800	.010			.35		.35	.54
4602	4" x 6"		600	.013			.47		.47	.72
4604	4" x 8"		450	.018			.62		.62	.96
5400	Posts, 4" x 4"	2 Clab	800	.020			.70		.70	1.08
5405	4" x 6"		550	.029			1.02		1.02	1.57
5410	4" x 8"		440	.036			1.28		1.28	1.96
5425	4" x 10"		390	.041			1.44		1.44	2.22
5430	4" x 12"		350	.046			1.60		1.60	2.47
5440	6" x 6"		400	.040			1.40		1.40	2.16
5445	6" x 8"		350	.046			1.60		1.60	2.47
5450	6" x 10"		320	.050			1.76		1.76	2.70
5455	6" x 12"		290	.055			1.94		1.94	2.98
5480	8" x 8"		300	.053			1.87		1.87	2.88
5500	10" x 10"		240	.067			2.34		2.34	3.60
5660	Tongue and groove floor planks		2	8	M.B.F.		281		281	430
5682	Rafters, ordinary, 16" OC, 2" x 4"		880	.018	S.F.		.64		.64	.98
5683	2" x 6"		840	.019			.67		.67	1.03
5684	2" x 8"		820	.020			.68		.68	1.05
5685	2" x 10"		820	.020			.68		.68	1.05
5686	2" x 12"		810	.020			.69		.69	1.07
5687	24" OC, 2" x 4"		1170	.014			.48		.48	.74
5688	2" x 6"		1117	.014			.50		.50	.77
5689	2" x 8"		1091	.015			.51		.51	.79
5690	2" x 10"		1091	.015			.51		.51	.79
5691	2" x 12"		1077	.015			.52		.52	.80
5795	Rafters, ordinary, 2" x 4" (alternate method)		862	.019	L.F.		.65		.65	1
5800	2" x 6" (alternate method)		850	.019			.66		.66	1.02
5840	2" x 8" (alternate method)		837	.019			.67		.67	1.03
5855	2" x 10" (alternate method)		825	.019			.68		.68	1.05
5865	2" x 12" (alternate method)		812	.020			.69		.69	1.06
5870	Sill plate, 2" x 4"	1 Clab	1170	.007			.24		.24	.37
5871	2" x 6"		780	.010			.36		.36	.55
5872	2" x 8"		586	.014			.48		.48	.74
5873	Alternate pricing method		.78	10.256	M.B.F.		360		360	555
5885	Ridge board, 1" x 4"	2 Clab	900	.018	L.F.		.62		.62	.96
5886	1" x 6"		875	.018			.64		.64	.99
5887	1" x 8"		850	.019			.66		.66	1.02
5888	1" x 10"		825	.019			.68		.68	1.05

06 05 05 – Selective Wood and Plastics Demolition

06 05 05.10 Selective Demolition Wood Framing	Crew	Daily Output	Labor-Hours	Unit	Material	2012 Bare Costs Labor	2012 Bare Costs Equipment	Total	Total Incl O&P
5889	1" x 12"	2 Clab	800	.020	L.F.		.70		.70	1.08
5890	2" x 4"		900	.018			.62		.62	.96
5892	2" x 6"		875	.018			.64		.64	.99
5894	2" x 8"		850	.019			.66		.66	1.02
5896	2" x 10"		825	.019			.68		.68	1.05
5898	2" x 12"		800	.020			.70		.70	1.08
6050	Rafter tie, 1" x 4"		1250	.013			.45		.45	.69
6052	1" x 6"		1135	.014			.49		.49	.76
6054	2" x 4"		1000	.016			.56		.56	.86
6056	2" x 6"		970	.016			.58		.58	.89
6070	Sleepers, on concrete, 1" x 2"	1 Clab	4700	.002			.06		.06	.09
6075	1" x 3"		4000	.002			.07		.07	.11
6080	2" x 4"		3000	.003			.09		.09	.14
6085	2" x 6"		2600	.003			.11		.11	.17
6086	Sheathing from roof, 5/16"	2 Clab	1600	.010	S.F.		.35		.35	.54
6088	3/8"		1525	.010			.37		.37	.57
6090	1/2"		1400	.011			.40		.40	.62
6092	5/8"		1300	.012			.43		.43	.66
6094	3/4"		1200	.013			.47		.47	.72
6096	Board sheathing from roof		1400	.011			.40		.40	.62
6100	Sheathing, from walls, 1/4"		1200	.013			.47		.47	.72
6110	5/16"		1175	.014			.48		.48	.74
6120	3/8"		1150	.014			.49		.49	.75
6130	1/2"		1125	.014			.50		.50	.77
6140	5/8"		1100	.015			.51		.51	.79
6150	3/4"		1075	.015			.52		.52	.80
6152	Board sheathing from walls		1500	.011			.37		.37	.58
6158	Subfloor, with boards		1050	.015			.53		.53	.82
6160	Plywood, 1/2" thick		768	.021			.73		.73	1.12
6162	5/8" thick		760	.021			.74		.74	1.14
6164	3/4" thick		750	.021			.75		.75	1.15
6165	1-1/8" thick		720	.022			.78		.78	1.20
6166	Underlayment, particle board, 3/8" thick	1 Clab	780	.010			.36		.36	.55
6168	1/2" thick		768	.010			.37		.37	.56
6170	5/8" thick		760	.011			.37		.37	.57
6172	3/4" thick		750	.011			.37		.37	.58
6200	Stairs and stringers, minimum	2 Clab	40	.400	Riser		14.05		14.05	21.50
6240	Maximum	"	26	.615	"		21.50		21.50	33
6300	Components, tread	1 Clab	110	.073	Ea.		2.55		2.55	3.93
6320	Riser		80	.100	"		3.51		3.51	5.40
6390	Stringer, 2" x 10"		260	.031	L.F.		1.08		1.08	1.66
6400	2" x 12"		260	.031			1.08		1.08	1.66
6410	3" x 10"		250	.032			1.12		1.12	1.73
6420	3" x 12"		250	.032			1.12		1.12	1.73
6590	Wood studs, 2" x 3"	2 Clab	3076	.005			.18		.18	.28
6600	2" x 4"		2000	.008			.28		.28	.43
6640	2" x 6"		1600	.010			.35		.35	.54
6720	Wall framing, including studs, plates and blocking, 2" x 4"	1 Clab	600	.013	S.F.		.47		.47	.72
6740	2" x 6"		480	.017	"		.59		.59	.90
6750	Headers, 2" x 4"		1125	.007	L.F.		.25		.25	.38
6755	2" x 6"		1125	.007			.25		.25	.38
6760	2" x 8"		1050	.008			.27		.27	.41
6765	2" x 10"		1050	.008			.27		.27	.41

06 05 05 – Selective Wood and Plastics Demolition

06 05 05.10 Selective Demolition Wood Framing

		Crew	Daily Output	Labor-Hours	Unit	Material	2012 Bare Costs Labor	2012 Bare Costs Equipment	Total	Total Incl O&P
6770	2" x 12"	1 Clab	1000	.008	L.F.		.28		.28	.43
6780	4" x 10"		525	.015			.53		.53	.82
6785	4" x 12"		500	.016			.56		.56	.86
6790	6" x 8"		560	.014			.50		.50	.77
6795	6" x 10"		525	.015			.53		.53	.82
6797	6" x 12"		500	.016			.56		.56	.86
8000	Soffit, T & G wood		520	.015	S.F.		.54		.54	.83
8010	Hardboard, vinyl or aluminum		640	.013			.44		.44	.68
8030	Plywood	2 Carp	315	.051			2.24		2.24	3.45
9500	See Section 02 41 19.23 for rubbish handling									

06 05 05.20 Selective Demolition Millwork and Trim

		Crew	Daily Output	Labor-Hours	Unit	Material	2012 Bare Costs Labor	2012 Bare Costs Equipment	Total	Total Incl O&P
0010	**SELECTIVE DEMOLITION MILLWORK AND TRIM** R024119-10									
1000	Cabinets, wood, base cabinets, per L.F.	2 Clab	80	.200	L.F.		7		7	10.80
1020	Wall cabinets, per L.F.	"	80	.200	"		7		7	10.80
1060	Remove and reset, base cabinets	2 Carp	18	.889	Ea.		39		39	60.50
1070	Wall cabinets	"	20	.800	"		35.50		35.50	54.50
1100	Steel, painted, base cabinets	2 Clab	60	.267	L.F.		9.35		9.35	14.40
1120	Wall cabinets	"	60	.267	"		9.35		9.35	14.40
1200	Casework, large area		320	.050	S.F.		1.76		1.76	2.70
1220	Selective		200	.080	"		2.81		2.81	4.32
1500	Counter top, minimum		200	.080	L.F.		2.81		2.81	4.32
1510	Maximum		120	.133			4.68		4.68	7.20
1550	Remove and reset, minimum	2 Carp	50	.320			14.10		14.10	21.50
1560	Maximum	"	40	.400			17.65		17.65	27
2000	Paneling, 4' x 8' sheets	2 Clab	2000	.008	S.F.		.28		.28	.43
2100	Boards, 1" x 4"		700	.023			.80		.80	1.23
2120	1" x 6"		750	.021			.75		.75	1.15
2140	1" x 8"		800	.020			.70		.70	1.08
3000	Trim, baseboard, to 6" wide		1200	.013	L.F.		.47		.47	.72
3040	Greater than 6" and up to 12" wide		1000	.016			.56		.56	.86
3080	Remove and reset, minimum	2 Carp	400	.040			1.76		1.76	2.71
3090	Maximum	"	300	.053			2.35		2.35	3.62
3100	Ceiling trim	2 Clab	1000	.016			.56		.56	.86
3120	Chair rail		1200	.013			.47		.47	.72
3140	Railings with balusters		240	.067			2.34		2.34	3.60
3160	Wainscoting		700	.023	S.F.		.80		.80	1.23

06 05 23 – Wood, Plastic, and Composite Fastenings

06 05 23.10 Nails

		Crew	Daily Output	Labor-Hours	Unit	Material	2012 Bare Costs Labor	2012 Bare Costs Equipment	Total	Total Incl O&P
0010	**NAILS**, material only, based upon 50# box purchase									
0020	Copper nails, plain				Lb.	10			10	11
0400	Stainless steel, plain					7.80			7.80	8.60
0500	Box, 3d to 20d, bright					1.25			1.25	1.38
0520	Galvanized					1.77			1.77	1.95
0600	Common, 3d to 60d, plain					1.30			1.30	1.43
0700	Galvanized					1.56			1.56	1.72
0800	Aluminum					5.80			5.80	6.40
1000	Annular or spiral thread, 4d to 60d, plain					1.80			1.80	1.98
1200	Galvanized					3			3	3.30
1400	Drywall nails, plain					3.59			3.59	3.95
1600	Galvanized					2.19			2.19	2.41
1800	Finish nails, 4d to 10d, plain					1.36			1.36	1.50
2000	Galvanized					1.79			1.79	1.97

06 05 23 – Wood, Plastic, and Composite Fastenings

06 05 23.10 Nails

		Crew	Daily Output	Labor-Hours	Unit	Material	2012 Bare Costs Labor	Equipment	Total	Total Incl O&P
2100	Aluminum				Lb.	5.80			5.80	6.40
2300	Flooring nails, hardened steel, 2d to 10d, plain					3.10			3.10	3.41
2400	Galvanized					4			4	4.40
2500	Gypsum lath nails, 1-1/8", 13 ga. flathead, blued					3.59			3.59	3.95
2600	Masonry nails, hardened steel, 3/4" to 3" long, plain					2.20			2.20	2.42
2700	Galvanized					3.75			3.75	4.13
2900	Roofing nails, threaded, galvanized					1.63			1.63	1.79
3100	Aluminum					4.80			4.80	5.30
3300	Compressed lead head, threaded, galvanized					2.47			2.47	2.72
3600	Siding nails, plain shank, galvanized					2			2	2.20
3800	Aluminum					5.80			5.80	6.40
5000	Add to prices above for cement coating					.11			.11	.12
5200	Zinc or tin plating					.14			.14	.15
5500	Vinyl coated sinkers, 8d to 16d					.65			.65	.72

06 05 23.40 Sheet Metal Screws

		Crew	Daily Output	Labor-Hours	Unit	Material	2012 Bare Costs Labor	Equipment	Total	Total Incl O&P
0010	**SHEET METAL SCREWS**									
0020	Steel, standard, #8 x 3/4", plain				C	2.92			2.92	3.21
0100	Galvanized					2.92			2.92	3.21
0300	#10 x 1", plain					4.03			4.03	4.43
0400	Galvanized					4.03			4.03	4.43
0600	With washers, #14 x 1", plain					12.35			12.35	13.60
0700	Galvanized					12.35			12.35	13.60
0900	#14 x 2", plain					16.90			16.90	18.60
1000	Galvanized					16.90			16.90	18.60
1500	Self-drilling, with washers, (pinch point) #8 x 3/4", plain					7.80			7.80	8.55
1600	Galvanized					7.80			7.80	8.55
1800	#10 x 3/4", plain					8.90			8.90	9.80
1900	Galvanized					8.90			8.90	9.80
3000	Stainless steel w/aluminum or neoprene washers, #14 x 1", plain					30			30	33
3100	#14 x 2", plain					40			40	44

06 05 23.50 Wood Screws

		Crew	Daily Output	Labor-Hours	Unit	Material	2012 Bare Costs Labor	Equipment	Total	Total Incl O&P
0010	**WOOD SCREWS**									
0020	#8 x 1" long, steel				C	3.41			3.41	3.75
0100	Brass					10.50			10.50	11.55
0200	#8, 2" long, steel					5.25			5.25	5.80
0300	Brass					19.95			19.95	22
0400	#10, 1" long, steel					3.65			3.65	4.02
0500	Brass					14.15			14.15	15.60
0600	#10, 2" long, steel					5.85			5.85	6.45
0700	Brass					24.50			24.50	27
0800	#10, 3" long, steel					9.80			9.80	10.75
1000	#12, 2" long, steel					8.05			8.05	8.85
1100	Brass					29.50			29.50	32
1500	#12, 3" long, steel					11.65			11.65	12.80
2000	#12, 4" long, steel					21.50			21.50	23.50

06 05 23.60 Timber Connectors

		Crew	Daily Output	Labor-Hours	Unit	Material	2012 Bare Costs Labor	Equipment	Total	Total Incl O&P
0010	**TIMBER CONNECTORS**									
0020	Add up cost of each part for total cost of connection									
0100	Connector plates, steel, with bolts, straight	2 Carp	75	.213	Ea.	28	9.40		37.40	45
0110	Tee, 7 gauge		50	.320		32	14.10		46.10	56.50
0120	T- Strap, 14 gauge, 12" x 8" x 2"		50	.320		32	14.10		46.10	56.50
0150	Anchor plates, 7 ga, 9" x 7"		75	.213		28	9.40		37.40	45

06 05 23.60 Timber Connectors		Crew	Daily Output	Labor-Hours	Unit	Material	2012 Bare Costs Labor	Equipment	Total	Total Incl O&P
0200	Bolts, machine, sq. hd. with nut & washer, 1/2" diameter, 4" long	1 Carp	140	.057	Ea.	.74	2.52		3.26	4.69
0300	7-1/2" long		130	.062		1.34	2.71		4.05	5.65
0500	3/4" diameter, 7-1/2" long		130	.062		3.12	2.71		5.83	7.60
0610	Machine bolts, w/nut, washer, 3/4" diam., 15" L, HD's & beam hangers		95	.084		5.80	3.71		9.51	12.10
0800	Drilling bolt holes in timber, 1/2" diameter		450	.018	Inch		.78		.78	1.21
0900	1" diameter		350	.023	"		1.01		1.01	1.55
1100	Framing anchor, angle, 3" x 3" x 1-1/2", 12 ga		175	.046	Ea.	2.40	2.02		4.42	5.75
1150	Framing anchors, 18 gauge, 4-1/2" x 2-3/4"		175	.046		2.40	2.02		4.42	5.75
1160	Framing anchors, 18 gauge, 4-1/2" x 3"		175	.046		2.40	2.02		4.42	5.75
1170	Clip anchors plates, 18 gauge, 12" x 1-1/8"		175	.046		2.40	2.02		4.42	5.75
1250	Holdowns, 3 gauge base, 10 gauge body		8	1		23	44		67	93
1260	Holdowns, 7 gauge 11-1/16" x 3-1/4"		8	1		23	44		67	93
1270	Holdowns, 7 gauge 14-3/8" x 3-1/8"		8	1		23	44		67	93
1275	Holdowns, 12 gauge 8" x 2-1/2"		8	1		23	44		67	93
1300	Joist and beam hangers, 18 ga. galv., for 2" x 4" joist		175	.046		.67	2.02		2.69	3.84
1400	2" x 6" to 2" x 10" joist		165	.048		1.28	2.14		3.42	4.70
1600	16 ga. galv., 3" x 6" to 3" x 10" joist		160	.050		2.72	2.21		4.93	6.40
1700	3" x 10" to 3" x 14" joist		160	.050		4.52	2.21		6.73	8.35
1800	4" x 6" to 4" x 10" joist		155	.052		2.82	2.28		5.10	6.60
1900	4" x 10" to 4" x 14" joist		155	.052		4.62	2.28		6.90	8.60
2000	Two-2" x 6" to two-2" x 10" joists		150	.053		3.85	2.35		6.20	7.85
2100	Two-2" x 10" to two-2" x 14" joists		150	.053		4.30	2.35		6.65	8.35
2300	3/16" thick, 6" x 8" joist		145	.055		59.50	2.43		61.93	69
2400	6" x 10" joist		140	.057		62	2.52		64.52	72.50
2500	6" x 12" joist		135	.059		64.50	2.61		67.11	75
2700	1/4" thick, 6" x 14" joist		130	.062		67.50	2.71		70.21	78
2900	Plywood clips, extruded aluminum H clip, for 3/4" panels					.22			.22	.24
3000	Galvanized 18 ga. back-up clip					.17			.17	.19
3200	Post framing, 16 ga. galv. for 4" x 4" base, 2 piece	1 Carp	130	.062		15.05	2.71		17.76	21
3300	Cap		130	.062		21	2.71		23.71	27
3500	Rafter anchors, 18 ga. galv., 1-1/2" wide, 5-1/4" long		145	.055		.45	2.43		2.88	4.24
3600	10-3/4" long		145	.055		1.34	2.43		3.77	5.20
3800	Shear plates, 2-5/8" diameter		120	.067		2.19	2.94		5.13	6.95
3900	4" diameter		115	.070		5.20	3.07		8.27	10.40
4000	Sill anchors, embedded in concrete or block, 25-1/2" long		115	.070		11.65	3.07		14.72	17.55
4100	Spike grids, 3" x 6"		120	.067		.88	2.94		3.82	5.50
4400	Split rings, 2-1/2" diameter		120	.067		1.80	2.94		4.74	6.50
4500	4" diameter		110	.073		2.73	3.21		5.94	7.95
4550	Tie plate, 20 gauge, 7" x 3 1/8"		110	.073		2.73	3.21		5.94	7.95
4560	Tie plate, 20 gauge, 5" x 4 1/8"		110	.073		2.73	3.21		5.94	7.95
4575	Twist straps, 18 gauge, 12" x 1 1/4"		110	.073		2.73	3.21		5.94	7.95
4580	Twist straps, 18 gauge, 16" x 1 1/4"		110	.073		2.73	3.21		5.94	7.95
4600	Strap ties, 20 ga., 2 -1/16" wide, 12 13/16" long		180	.044		.87	1.96		2.83	3.98
4700	Strap ties, 16 ga., 1-3/8" wide, 12" long		180	.044		.87	1.96		2.83	3.98
4800	21-5/8" x 1-1/4"		160	.050		2.72	2.21		4.93	6.40
5000	Toothed rings, 2-5/8" or 4" diameter		90	.089		1.63	3.92		5.55	7.85
5200	Truss plates, nailed, 20 gauge, up to 32' span		17	.471	Truss	11.85	21		32.85	45
5400	Washers, 2" x 2" x 1/8"				Ea.	.37			.37	.41
5500	3" x 3" x 3/16"				"	.98			.98	1.08
6000	Angles and gussets, painted									
6012	7 ga., 3-1/4" x 3-1/4" x 2-1/2" long	1 Carp	1.90	4.211	C	1,025	186		1,211	1,400
6014	3-1/4" x 3-1/4" x 5" long		1.90	4.211		1,975	186		2,161	2,450
6016	3-1/4" x 3-1/4" x 7-1/2" long		1.85	4.324		3,750	191		3,941	4,425

06 05 23 – Wood, Plastic, and Composite Fastenings

06 05 23.60 Timber Connectors	Crew	Daily Output	Labor-Hours	Unit	Material	2012 Bare Costs Labor	Equipment	Total	Total Incl O&P
6018 5-3/4" x 5-3/4" x 2-1/2" long	1 Carp	1.85	4.324	C	2,425	191		2,616	2,950
6020 5-3/4" x 5-3/4" x 5" long		1.85	4.324		3,850	191		4,041	4,525
6022 5-3/4" x 5-3/4" x 7-1/2" long		1.80	4.444		5,675	196		5,871	6,550
6024 3 ga., 4-1/4" x 4-1/4" x 3" long		1.85	4.324		2,575	191		2,766	3,125
6026 4-1/4" x 4-1/4" x 6" long		1.85	4.324		5,525	191		5,716	6,375
6028 4-1/4" x 4-1/4" x 9" long		1.80	4.444		6,200	196		6,396	7,125
6030 7-1/4" x 7-1/4" x 3" long		1.80	4.444		4,450	196		4,646	5,175
6032 7-1/4" x 7-1/4" x 6" long		1.80	4.444		6,000	196		6,196	6,900
6034 7-1/4" x 7-1/4" x 9" long	↓	1.75	4.571	↓	13,400	202		13,602	15,100
6036 Gussets									
6038 7 ga., 8-1/8" x 8-1/8" x 2-3/4" long	1 Carp	1.80	4.444	C	4,250	196		4,446	4,975
6040 3 ga., 9-3/4" x 9-3/4" x 3-1/4" long	"	1.80	4.444	"	5,875	196		6,071	6,775
6101 Beam hangers, polymer painted									
6102 Bolted, 3 ga., (W x H x L)									
6104 3-1/4" x 9" x 12" top flange	1 Carp	1	8	C	18,000	355		18,355	20,300
6106 5-1/4" x 9" x 12" top flange		1	8		18,800	355		19,155	21,200
6108 5-1/4" x 11" x 11-3/4" top flange		1	8		21,400	355		21,755	24,000
6110 6-7/8" x 9" x 12" top flange		1	8		19,500	355		19,855	21,900
6112 6-7/8" x 11" x 13-1/2" top flange		1	8		22,500	355		22,855	25,200
6114 8-7/8" x 11" x 15-1/2" top flange	↓	1	8	↓	24,000	355		24,355	26,900
6116 Nailed, 3 ga., (W x H x L)									
6118 3-1/4" x 10-1/2" x 10" top flange	1 Carp	1.80	4.444	C	18,800	196		18,996	21,000
6120 3-1/4" x 10-1/2" x 12" top flange		1.80	4.444		18,800	196		18,996	21,000
6122 5-1/4" x 9-1/2" x 10" top flange		1.80	4.444		19,200	196		19,396	21,400
6124 5-1/4" x 9-1/2" x 12" top flange		1.80	4.444		19,200	196		19,396	21,400
6128 6-7/8" x 8-1/2" x 12" top flange	↓	1.80	4.444	↓	19,700	196		19,896	21,900
6134 Saddle hangers, glu-lam (W x H x L)									
6136 3-1/4" x 10-1/2" x 5-1/4" x 6" saddle	1 Carp	.50	16	C	14,100	705		14,805	16,600
6138 3-1/4" x 10-1/2" x 6-7/8" x 6" saddle		.50	16		14,800	705		15,505	17,400
6140 3-1/4" x 10-1/2" x 8-7/8" x 6" saddle		.50	16		15,600	705		16,305	18,200
6142 3-1/4" x 19-1/2" x 5-1/4" x 10-1/8" saddle		.40	20		14,100	880		14,980	16,900
6144 3-1/4" x 19-1/2" x 6-7/8" x 10-1/8" saddle		.40	20		14,800	880		15,680	17,700
6146 3-1/4" x 19-1/2" x 8-7/8" x 10-1/8" saddle		.40	20		15,600	880		16,480	18,500
6148 5-1/4" x 9-1/2" x 5-1/4" x 12" saddle		.50	16		16,800	705		17,505	19,600
6150 5-1/4" x 9-1/2" x 6-7/8" x 9" saddle		.50	16		18,400	705		19,105	21,300
6152 5-1/4" x 10-1/2" x spec x 12" saddle		.50	16		20,000	705		20,705	23,100
6154 5-1/4" x 18" x 5-1/4" x 12-1/8" saddle		.40	20		16,800	880		17,680	19,900
6156 5-1/4" x 18" x 6-7/8" x 12-1/8" saddle		.40	20		18,400	880		19,280	21,600
6158 5-1/4" x 18" x spec x 12-1/8" saddle		.40	20		20,000	880		20,880	23,400
6160 6-7/8" x 8-1/2" x 6-7/8" x 12" saddle		.50	16		20,100	705		20,805	23,200
6162 6-7/8" x 8-1/2" x 8-7/8" x 12" saddle		.50	16		20,800	705		21,505	23,900
6164 6-7/8" x 10-1/2" x spec x 12" saddle		.50	16		20,100	705		20,805	23,200
6166 6-7/8" x 18" x 6-7/8" x 13-3/4" saddle		.40	20		20,100	880		20,980	23,500
6168 6-7/8" x 18" x 8-7/8" x 13-3/4" saddle		.40	20		20,800	880		21,680	24,200
6170 6-7/8" x 18" x spec x 13-3/4" saddle		.40	20		22,400	880		23,280	26,000
6172 8-7/8" x 18" x spec x 15-3/4" saddle	↓	.40	20	↓	35,000	880		35,880	39,900
6201 Beam and purlin hangers, galvanized, 12 ga.									
6202 Purlin or joist size, 3" x 8"	1 Carp	1.70	4.706	C	1,875	208		2,083	2,375
6204 3" x 10"		1.70	4.706		2,000	208		2,208	2,525
6206 3" x 12"		1.65	4.848		2,300	214		2,514	2,850
6208 3" x 14"		1.65	4.848		2,450	214		2,664	3,025
6210 3" x 16"		1.65	4.848		2,600	214		2,814	3,175
6212 4" x 8"	↓	1.65	4.848		1,875	214		2,089	2,375

06 05 23.60 Timber Connectors		Crew	Daily Output	Labor-Hours	Unit	Material	2012 Bare Costs Labor	Equipment	Total	Total Incl O&P
6214	4" x 10"	1 Carp	1.65	4.848	C	2,025	214		2,239	2,550
6216	4" x 12"		1.60	5		2,400	221		2,621	2,975
6218	4" x 14"		1.60	5		2,550	221		2,771	3,150
6220	4" x 16"		1.60	5		2,675	221		2,896	3,300
6222	6" x 8"		1.60	5		2,400	221		2,621	3,000
6224	6" x 10"		1.55	5.161		2,450	228		2,678	3,050
6226	6" x 12"		1.55	5.161		4,200	228		4,428	4,975
6228	6" x 14"		1.50	5.333		4,450	235		4,685	5,250
6230	6" x 16"	↓	1.50	5.333	↓	4,725	235		4,960	5,525
6250	Beam seats									
6252	Beam size, 5-1/4" wide									
6254	5" x 7" x 1/4"	1 Carp	1.80	4.444	C	7,100	196		7,296	8,100
6256	6" x 7" x 3/8"		1.80	4.444		7,950	196		8,146	9,050
6258	7" x 7" x 3/8"		1.80	4.444		8,500	196		8,696	9,650
6260	8" x 7" x 3/8"	↓	1.80	4.444	↓	10,000	196		10,196	11,300
6262	Beam size, 6-7/8" wide									
6264	5" x 9" x 1/4"	1 Carp	1.80	4.444	C	8,475	196		8,671	9,625
6266	6" x 9" x 3/8"		1.80	4.444		11,100	196		11,296	12,500
6268	7" x 9" x 3/8"		1.80	4.444		11,200	196		11,396	12,600
6270	8" x 9" x 3/8"	↓	1.80	4.444	↓	13,200	196		13,396	14,900
6272	Special beams, over 6-7/8" wide									
6274	5" x 10" x 3/8"	1 Carp	1.80	4.444	C	11,500	196		11,696	12,900
6276	6" x 10" x 3/8"		1.80	4.444		13,400	196		13,596	15,000
6278	7" x 10" x 3/8"		1.80	4.444		14,000	196		14,196	15,700
6280	8" x 10" x 3/8"		1.75	4.571		15,000	202		15,202	16,800
6282	5-1/4" x 12" x 5/16"		1.75	4.571		11,600	202		11,802	13,100
6284	6-1/2" x 12" x 3/8"		1.75	4.571		19,200	202		19,402	21,500
6286	5-1/4" x 16" x 5/16"		1.70	4.706		17,100	208		17,308	19,100
6288	6-1/2" x 16" x 3/8"		1.70	4.706		22,300	208		22,508	24,800
6290	5-1/4" x 20" x 5/16"		1.70	4.706		20,100	208		20,308	22,400
6292	6-1/2" x 20" x 3/8"	↓	1.65	4.848	↓	26,200	214		26,414	29,100
6300	Column bases									
6302	4 x 4, 16 ga.	1 Carp	1.80	4.444	C	715	196		911	1,075
6306	7 ga.		1.80	4.444		2,625	196		2,821	3,175
6308	4 x 6, 16 ga.		1.80	4.444		1,675	196		1,871	2,125
6312	7 ga.		1.80	4.444		2,725	196		2,921	3,300
6314	6 x 6, 16 ga.		1.75	4.571		1,900	202		2,102	2,375
6318	7 ga.		1.75	4.571		3,725	202		3,927	4,400
6320	6 x 8, 7 ga.		1.70	4.706		2,900	208		3,108	3,525
6322	6 x 10, 7 ga.		1.70	4.706		3,125	208		3,333	3,750
6324	6 x 12, 7 ga.		1.70	4.706		3,375	208		3,583	4,050
6326	8 x 8, 7 ga.		1.65	4.848		5,700	214		5,914	6,600
6330	8 x 10, 7 ga.		1.65	4.848		6,825	214		7,039	7,850
6332	8 x 12, 7 ga.		1.60	5		7,425	221		7,646	8,500
6334	10 x 10, 3 ga.		1.60	5		7,575	221		7,796	8,675
6336	10 x 12, 3 ga.		1.60	5		8,700	221		8,921	9,925
6338	12 x 12, 3 ga.	↓	1.55	5.161	↓	9,450	228		9,678	10,800
6350	Column caps, painted, 3 ga.									
6352	3-1/4" x 3-5/8"	1 Carp	1.80	4.444	C	9,600	196		9,796	10,900
6354	3-1/4" x 5-1/2"		1.80	4.444		9,600	196		9,796	10,900
6356	3-5/8" x 3-5/8"		1.80	4.444		7,850	196		8,046	8,950
6358	3-5/8" x 5-1/2"		1.80	4.444		7,850	196		8,046	8,950
6360	5-1/4" x 5-1/2"	↓	1.75	4.571	↓	10,300	202		10,502	11,600

06 05 23 – Wood, Plastic, and Composite Fastenings

06 05 23.60 Timber Connectors		Crew	Daily Output	Labor-Hours	Unit	Material	2012 Bare Costs Labor	Equipment	Total	Total Incl O&P
6362	5-1/4" x 7-1/2"	1 Carp	1.75	4.571	C	10,300	202		10,502	11,600
6364	5-1/2" x 3-5/8"		1.75	4.571		11,100	202		11,302	12,500
6366	5-1/2" x 5-1/2"		1.75	4.571		11,100	202		11,302	12,500
6368	5-1/2" x 7-1/2"		1.70	4.706		11,100	208		11,308	12,500
6370	6-7/8" x 5-1/2"		1.70	4.706		11,600	208		11,808	13,000
6372	6-7/8" x 6-7/8"		1.70	4.706		11,600	208		11,808	13,000
6374	6-7/8" x 7-1/2"		1.70	4.706		11,600	208		11,808	13,000
6376	7-1/2" x 5-1/2"		1.65	4.848		12,100	214		12,314	13,600
6378	7-1/2" x 7-1/2"		1.65	4.848		12,100	214		12,314	13,600
6380	8-7/8" x 5-1/2"		1.60	5		12,800	221		13,021	14,300
6382	8-7/8" x 7-1/2"		1.60	5		12,800	221		13,021	14,300
6384	9-1/2" x 5-1/2"	↓	1.60	5	↓	17,300	221		17,521	19,300
6400	Floor tie anchors, polymer paint									
6402	10 ga., 3" x 37-1/2"	1 Carp	1.80	4.444	C	4,600	196		4,796	5,350
6404	3-1/2" x 45-1/2"		1.75	4.571		4,825	202		5,027	5,600
6406	3 ga., 3-1/2" x 56"	↓	1.70	4.706	↓	8,500	208		8,708	9,675
6410	Girder hangers									
6412	6" wall thickness, 4" x 6"	1 Carp	1.80	4.444	C	2,575	196		2,771	3,150
6414	4" x 8"		1.80	4.444		2,875	196		3,071	3,475
6416	8" wall thickness, 4" x 6"		1.80	4.444		3,000	196		3,196	3,575
6418	4" x 8"	↓	1.80	4.444	↓	2,975	196		3,171	3,575
6420	Hinge connections, polymer painted									
6422	3/4" thick top plate									
6424	5-1/4" x 12" w/5" x 5" top	1 Carp	1	8	C	33,000	355		33,355	36,800
6426	5-1/4" x 15" w/6" x 6" top		.80	10		35,100	440		35,540	39,300
6428	5-1/4" x 18" w/7" x 7" top		.70	11.429		36,900	505		37,405	41,300
6430	5-1/4" x 26" w/9" x 9" top	↓	.60	13.333	↓	39,200	590		39,790	44,100
6432	1" thick top plate									
6434	6-7/8" x 14" w/5" x 5" top	1 Carp	.80	10	C	40,200	440		40,640	44,900
6436	6-7/8" x 17" w/6" x 6" top		.80	10		44,600	440		45,040	49,800
6438	6-7/8" x 21" w/7" x 7" top		.70	11.429		48,700	505		49,205	54,500
6440	6-7/8" x 31" w/9" x 9" top	↓	.60	13.333	↓	53,500	590		54,090	59,500
6442	1-1/4" thick top plate									
6444	8-7/8" x 16" w/5" x 5" top	1 Carp	.60	13.333	C	50,000	590		50,590	56,000
6446	8-7/8" x 21" w/6" x 6" top		.50	16		55,000	705		55,705	61,500
6448	8-7/8" x 26" w/7" x 7" top		.40	20		62,500	880		63,380	70,000
6450	8-7/8" x 39" w/9" x 9" top	↓	.30	26.667	↓	78,000	1,175		79,175	87,500
6460	Holddowns									
6462	Embedded along edge									
6464	26" long, 12 ga.	1 Carp	.90	8.889	C	1,250	390		1,640	1,975
6466	35" long, 12 ga.		.85	9.412		1,675	415		2,090	2,475
6468	35" long, 10 ga.	↓	.85	9.412	↓	1,325	415		1,740	2,100
6470	Embedded away from edge									
6472	Medium duty, 12 ga.									
6474	18-1/2" long	1 Carp	.95	8.421	C	760	370		1,130	1,400
6476	23-3/4" long		.90	8.889		885	390		1,275	1,575
6478	28" long		.85	9.412		905	415		1,320	1,625
6480	35" long	↓	.85	9.412	↓	1,250	415		1,665	2,025
6482	Heavy duty, 10 ga.									
6484	28" long	1 Carp	.85	9.412	C	1,625	415		2,040	2,425
6486	35" long	"	.85	9.412	"	1,775	415		2,190	2,600
6490	Surface mounted (W x H)									
6492	2-1/2" x 5-3/4", 7 ga.	1 Carp	1	8	C	1,850	355		2,205	2,575

06 05 23.60 Timber Connectors	Crew	Daily Output	Labor-Hours	Unit	Material	2012 Bare Costs Labor	Equipment	Total	Total Incl O&P
6494	2-1/2" x 8", 12 ga.	1 Carp	1	8	C	1,125	355		1,480	1,800
6496	2-7/8" x 6-3/8", 7 ga.		1	8		4,075	355		4,430	5,025
6498	2-7/8" x 12-1/2", 3 ga.		1	8		4,175	355		4,530	5,150
6500	3-3/16" x 9-3/8", 10 ga.		1	8		2,800	355		3,155	3,625
6502	3-1/2" x 11-5/8", 3 ga.		1	8		5,100	355		5,455	6,150
6504	3-1/2" x 14-3/4", 3 ga.		1	8		3,400	355		3,755	4,275
6506	3-1/2" x 16-1/2", 3 ga.		1	8		7,775	355		8,130	9,100
6508	3-1/2" x 20-1/2", 3 ga.		.90	8.889		7,950	390		8,340	9,350
6510	3-1/2" x 24-1/2", 3 ga.		.90	8.889		10,100	390		10,490	11,700
6512	4-1/4" x 20-3/4", 3 ga.		.90	8.889		6,850	390		7,240	8,125
6520	Joist hangers									
6522	Sloped, field adjustable, 18 ga.									
6524	2" x 6"	1 Carp	1.65	4.848	C	500	214		714	880
6526	2" x 8"		1.65	4.848		900	214		1,114	1,325
6528	2" x 10" and up		1.65	4.848		1,500	214		1,714	1,975
6530	3" x 10" and up		1.60	5		1,125	221		1,346	1,575
6532	4" x 10" and up		1.55	5.161		1,350	228		1,578	1,850
6536	Skewed 45°, 16 ga.									
6538	2" x 4"	1 Carp	1.75	4.571	C	800	202		1,002	1,200
6540	2" x 6" or 2" x 8"		1.65	4.848		815	214		1,029	1,225
6542	2" x 10" or 2" x 12"		1.65	4.848		935	214		1,149	1,350
6544	2" x 14" or 2" x 16"		1.60	5		1,675	221		1,896	2,175
6546	(2) 2" x 6" or (2) 2" x 8"		1.60	5		1,500	221		1,721	2,000
6548	(2) 2" x 10" or (2) 2" x 12"		1.55	5.161		1,625	228		1,853	2,125
6550	(2) 2" x 14" or (2) 2" x 16"		1.50	5.333		2,550	235		2,785	3,150
6552	4" x 6" or 4" x 8"		1.60	5		1,250	221		1,471	1,750
6554	4" x 10" or 4" x 12"		1.55	5.161		1,475	228		1,703	1,975
6556	4" x 14" or 4" x 16"		1.55	5.161		2,275	228		2,503	2,850
6560	Skewed 45°, 14 ga.									
6562	(2) 2" x 6" or (2) 2" x 8"	1 Carp	1.60	5	C	1,725	221		1,946	2,250
6564	(2) 2" x 10" or (2) 2" x 12"		1.55	5.161		2,400	228		2,628	3,000
6566	(2) 2" x 14" or (2) 2" x 16"		1.50	5.333		3,525	235		3,760	4,225
6568	4" x 6" or 4" x 8"		1.60	5		2,075	221		2,296	2,625
6570	4" x 10" or 4" x 12"		1.55	5.161		2,175	228		2,403	2,750
6572	4" x 14" or 4" x 16"		1.55	5.161		2,925	228		3,153	3,575
6590	Joist hangers, heavy duty 12 ga., galvanized									
6592	2" x 4"	1 Carp	1.75	4.571	C	1,175	202		1,377	1,575
6594	2" x 6"		1.65	4.848		1,275	214		1,489	1,725
6595	2" x 6", 16 gauge		1.65	4.848		1,275	214		1,489	1,725
6596	2" x 8"		1.65	4.848		1,950	214		2,164	2,450
6597	2" x 8", 16 gauge		1.65	4.848		1,950	214		2,164	2,450
6598	2" x 10"		1.65	4.848		2,000	214		2,214	2,525
6600	2" x 12"		1.65	4.848		2,450	214		2,664	3,025
6602	2" x 14"		1.65	4.848		2,550	214		2,764	3,125
6604	2" x 16"		1.65	4.848		2,700	214		2,914	3,300
6606	3" x 4"		1.65	4.848		1,625	214		1,839	2,100
6608	3" x 6"		1.65	4.848		2,175	214		2,389	2,725
6610	3" x 8"		1.65	4.848		2,200	214		2,414	2,750
6612	3" x 10"		1.60	5		2,550	221		2,771	3,175
6614	3" x 12"		1.60	5		3,075	221		3,296	3,725
6616	3" x 14"		1.60	5		3,600	221		3,821	4,300
6618	3" x 16"		1.60	5		3,950	221		4,171	4,700
6620	(2) 2" x 4"		1.75	4.571		1,900	202		2,102	2,400

06 05 Common Work Results for Wood, Plastics, and Composites

06 05 23 – Wood, Plastic, and Composite Fastenings

06 05 23.60 Timber Connectors	Crew	Daily Output	Labor-Hours	Unit	Material	2012 Bare Costs Labor	Equipment	Total	Total Incl O&P
6622 (2) 2" x 6"	1 Carp	1.60	5	C	2,300	221		2,521	2,875
6624 (2) 2" x 8"		1.60	5		2,350	221		2,571	2,925
6626 (2) 2" x 10"		1.55	5.161		2,550	228		2,778	3,150
6628 (2) 2" x 12"		1.55	5.161		3,250	228		3,478	3,925
6630 (2) 2" x 14"		1.50	5.333		3,275	235		3,510	3,950
6632 (2) 2" x 16"		1.50	5.333		3,300	235		3,535	4,000
6634 4" x 4"		1.65	4.848		1,375	214		1,589	1,850
6636 4" x 6"		1.60	5		1,525	221		1,746	2,025
6638 4" x 8"		1.60	5		1,750	221		1,971	2,275
6640 4" x 10"		1.55	5.161		2,150	228		2,378	2,725
6642 4" x 12"		1.55	5.161		2,300	228		2,528	2,875
6644 4" x 14"		1.55	5.161		2,750	228		2,978	3,375
6646 4" x 16"		1.55	5.161		3,000	228		3,228	3,650
6648 (3) 2" x 10"		1.50	5.333		3,325	235		3,560	4,000
6650 (3) 2" x 12"		1.50	5.333		3,700	235		3,935	4,425
6652 (3) 2" x 14"		1.45	5.517		3,975	243		4,218	4,750
6654 (3) 2" x 16"		1.45	5.517		4,050	243		4,293	4,825
6656 6" x 6"		1.60	5		1,775	221		1,996	2,325
6658 6" x 8"		1.60	5		1,850	221		2,071	2,375
6660 6" x 10"		1.55	5.161		2,200	228		2,428	2,775
6662 6" x 12"		1.55	5.161		2,500	228		2,728	3,100
6664 6" x 14"		1.50	5.333		3,150	235		3,385	3,800
6666 6" x 16"		1.50	5.333		3,700	235		3,935	4,425
6690 Knee braces, galvanized, 12 ga.									
6692 Beam depth, 10" x 15" x 5' long	1 Carp	1.80	4.444	C	4,800	196		4,996	5,600
6694 15" x 22-1/2" x 7' long		1.70	4.706		5,525	208		5,733	6,400
6696 22-1/2" x 28-1/2" x 8' long		1.60	5		5,925	221		6,146	6,875
6698 28-1/2" x 36" x 10' long		1.55	5.161		6,175	228		6,403	7,150
6700 36" x 42" x 12' long		1.50	5.333		6,825	235		7,060	7,850
6710 Mudsill anchors									
6714 2" x 4" or 3" x 4"	1 Carp	115	.070	C	1,225	3.07		1,228.07	1,350
6716 2" x 6" or 3" x 6"		115	.070		1,225	3.07		1,228.07	1,350
6718 Block wall, 13-1/4" long		115	.070		850	3.07		853.07	940
6720 21-1/4" long		115	.070		127	3.07		130.07	144
6730 Post bases, 12 ga. galvanized									
6732 Adjustable, 3-9/16" x 3-9/16"	1 Carp	1.30	6.154	C	980	271		1,251	1,500
6734 3-9/16" x 5-1/2"		1.30	6.154		1,275	271		1,546	1,825
6736 4" x 4"		1.30	6.154		715	271		986	1,200
6738 4" x 6"		1.30	6.154		1,675	271		1,946	2,250
6740 5-1/2" x 5-1/2"		1.30	6.154		2,475	271		2,746	3,150
6742 6" x 6"		1.30	6.154		1,900	271		2,171	2,500
6744 Elevated, 3-9/16" x 3-1/4"		1.30	6.154		1,075	271		1,346	1,600
6746 5-1/2" x 3-5/16"		1.30	6.154		1,525	271		1,796	2,100
6748 5-1/2" x 5"		1.30	6.154		2,275	271		2,546	2,925
6750 Regular, 3-9/16" x 3-3/8"		1.30	6.154		815	271		1,086	1,325
6752 4" x 3-3/8"		1.30	6.154		1,150	271		1,421	1,675
6754 18 gauge, 5-1/4" x 3-1/8"		1.30	6.154		1,200	271		1,471	1,750
6755 5-1/2" x 3-3/8"		1.30	6.154		1,200	271		1,471	1,750
6756 5-1/2" x 5-3/8"		1.30	6.154		1,725	271		1,996	2,325
6758 6" x 3-3/8"		1.30	6.154		2,075	271		2,346	2,700
6760 6" x 5-3/8"		1.30	6.154		2,300	271		2,571	2,975
6762 Post combination cap/bases									
6764 3-9/16" x 3-9/16"	1 Carp	1.20	6.667	C	465	294		759	960

173

06 05 23.60 Timber Connectors		Crew	Daily Output	Labor-Hours	Unit	Material	2012 Bare Costs Labor	2012 Bare Costs Equipment	Total	Total Incl O&P
6766	3-9/16" x 5-1/2"	1 Carp	1.20	6.667	C	1,050	294		1,344	1,600
6768	4" x 4"		1.20	6.667		2,100	294		2,394	2,750
6770	5-1/2" x 5-1/2"		1.20	6.667		1,150	294		1,444	1,725
6772	6" x 6"		1.20	6.667		3,825	294		4,119	4,675
6774	7-1/2" x 7-1/2"		1.20	6.667		4,200	294		4,494	5,075
6776	8" x 8"		1.20	6.667		4,350	294		4,644	5,225
6790	Post-beam connection caps									
6792	Beam size 3-9/16"									
6794	12 ga. post, 4" x 4"	1 Carp	1	8	C	2,650	355		3,005	3,475
6796	4" x 6"		1	8		3,550	355		3,905	4,450
6798	4" x 8"		1	8		5,225	355		5,580	6,300
6800	16 ga. post, 4" x 4"		1	8		1,200	355		1,555	1,875
6802	4" x 6"		1	8		2,000	355		2,355	2,750
6804	4" x 8"		1	8		3,375	355		3,730	4,250
6805	18 ga. post, 2-7/8" x 3"		1	8		3,375	355		3,730	4,250
6806	Beam size 5-1/2"									
6808	12 ga. post, 6" x 4"	1 Carp	1	8	C	3,475	355		3,830	4,375
6810	6" x 6"		1	8		5,475	355		5,830	6,575
6812	6" x 8"		1	8		3,775	355		4,130	4,700
6816	16 ga. post, 6" x 4"		1	8		1,875	355		2,230	2,625
6818	6" x 6"		1	8		2,000	355		2,355	2,750
6820	Beam size 7-1/2"									
6822	12 ga. post, 8" x 4"	1 Carp	1	8	C	4,800	355		5,155	5,850
6824	8" x 6"		1	8		5,050	355		5,405	6,100
6826	8" x 8"		1	8		7,575	355		7,930	8,900
6840	Purlin anchors, embedded									
6842	Heavy duty, 10 ga.									
6844	Straight, 28" long	1 Carp	1.60	5	C	1,325	221		1,546	1,800
6846	35" long		1.50	5.333		1,625	235		1,860	2,150
6848	Twisted, 28" long		1.60	5		1,325	221		1,546	1,800
6850	35" long		1.50	5.333		1,625	235		1,860	2,150
6852	Regular duty, 12 ga.									
6854	Straight, 18-1/2" long	1 Carp	1.80	4.444	C	755	196		951	1,125
6856	23-3/4" long		1.70	4.706		950	208		1,158	1,375
6858	29" long		1.60	5		970	221		1,191	1,425
6860	35" long		1.50	5.333		1,325	235		1,560	1,825
6862	Twisted, 18" long		1.80	4.444		755	196		951	1,125
6866	28" long		1.60	5		900	221		1,121	1,325
6868	35" long		1.50	5.333		1,325	235		1,560	1,825
6870	Straight, plastic coated									
6872	23-1/2" long	1 Carp	1.60	5	C	1,850	221		2,071	2,400
6874	26-7/8" long		1.60	5		2,175	221		2,396	2,750
6876	32-1/2" long		1.50	5.333		2,325	235		2,560	2,900
6878	35-7/8" long		1.50	5.333		2,425	235		2,660	3,000
6890	Purlin hangers, painted									
6892	12 ga., 2" x 6"	1 Carp	1.80	4.444	C	1,700	196		1,896	2,175
6894	2" x 8"		1.80	4.444		1,850	196		2,046	2,350
6896	2" x 10"		1.80	4.444		2,000	196		2,196	2,500
6898	2" x 12"		1.75	4.571		2,150	202		2,352	2,675
6900	2" x 14"		1.75	4.571		2,300	202		2,502	2,825
6902	2" x 16"		1.75	4.571		2,450	202		2,652	2,975
6904	3" x 6"		1.70	4.706		1,725	208		1,933	2,225
6906	3" x 8"		1.70	4.706		1,875	208		2,083	2,375

06 05 23 – Wood, Plastic, and Composite Fastenings

06 05 23.60 Timber Connectors		Crew	Daily Output	Labor-Hours	Unit	Material	2012 Bare Costs Labor	Equipment	Total	Total Incl O&P
6908	3" x 10"	1 Carp	1.70	4.706	C	2,000	208		2,208	2,525
6910	3" x 12"		1.65	4.848		2,300	214		2,514	2,850
6912	3" x 14"		1.65	4.848		2,450	214		2,664	3,025
6914	3" x 16"		1.65	4.848		2,600	214		2,814	3,175
6916	4" x 6"		1.65	4.848		1,725	214		1,939	2,225
6918	4" x 8"		1.65	4.848		1,875	214		2,089	2,375
6920	4" x 10"		1.65	4.848		2,025	214		2,239	2,550
6922	4" x 12"		1.60	5		2,400	221		2,621	2,975
6924	4" x 14"		1.60	5		2,550	221		2,771	3,150
6926	4" x 16"		1.60	5		2,675	221		2,896	3,300
6928	6" x 6"		1.60	5		2,275	221		2,496	2,850
6930	6" x 8"		1.60	5		2,400	221		2,621	3,000
6932	6" x 10"		1.55	5.161		2,450	228		2,678	3,050
6934	double 2" x 6"		1.70	4.706		1,875	208		2,083	2,375
6936	double 2" x 8"		1.70	4.706		2,000	208		2,208	2,525
6938	double 2" x 10"		1.70	4.706		2,150	208		2,358	2,700
6940	double 2" x 12"		1.65	4.848		2,300	214		2,514	2,850
6942	double 2" x 14"		1.65	4.848		2,450	214		2,664	3,025
6944	double 2" x 16"		1.65	4.848		2,600	214		2,814	3,175
6960	11 ga., 4" x 6"		1.65	4.848		3,400	214		3,614	4,075
6962	4" x 8"		1.65	4.848		3,675	214		3,889	4,350
6964	4" x 10"		1.65	4.848		3,925	214		4,139	4,625
6966	6" x 6"		1.60	5		3,450	221		3,671	4,150
6968	6" x 8"		1.60	5		3,700	221		3,921	4,425
6970	6" x 10"		1.55	5.161		3,950	228		4,178	4,700
6972	6" x 12"		1.55	5.161		4,200	228		4,428	4,975
6974	6" x 14"		1.55	5.161		4,450	228		4,678	5,250
6976	6" x 16"		1.50	5.333		4,725	235		4,960	5,525
6978	7 ga., 8" x 6"		1.60	5		3,750	221		3,971	4,475
6980	8" x 8"		1.60	5		4,000	221		4,221	4,750
6982	8" x 10"		1.55	5.161		4,250	228		4,478	5,025
6984	8" x 12"		1.55	5.161		4,500	228		4,728	5,300
6986	8" x 14"		1.50	5.333		4,750	235		4,985	5,575
6988	8" x 16"		1.50	5.333		5,000	235		5,235	5,850
7000	Strap connectors, galvanized									
7002	12 ga., 2-1/16" x 36"	1 Carp	1.55	5.161	C	1,075	228		1,303	1,550
7004	2-1/16" x 47"		1.50	5.333		1,500	235		1,735	2,000
7005	10 ga., 2-1/16" x 72"		1.50	5.333		1,500	235		1,735	2,000
7006	7 ga., 2-1/16" x 34"		1.55	5.161		2,700	228		2,928	3,325
7008	2-1/16" x 45"		1.50	5.333		3,500	235		3,735	4,200
7010	3 ga., 3" x 32"		1.55	5.161		4,525	228		4,753	5,350
7012	3" x 41"		1.55	5.161		4,700	228		4,928	5,525
7014	3" x 50"		1.50	5.333		7,175	235		7,410	8,250
7016	3" x 59"		1.50	5.333		8,750	235		8,985	9,975
7018	3-1/2" x 68"		1.45	5.517		8,875	243		9,118	10,200
7030	Tension ties									
7032	19-1/8" long, 16 ga., 3/4" anchor bolt	1 Carp	1.80	4.444	C	1,250	196		1,446	1,675
7034	20" long, 12 ga., 1/2" anchor bolt		1.80	4.444		1,625	196		1,821	2,075
7036	20" long, 12 ga., 3/4" anchor bolt		1.80	4.444		1,625	196		1,821	2,075
7038	27-3/4" long, 12 ga., 3/4" anchor bolt		1.75	4.571		2,850	202		3,052	3,450
7050	Truss connectors, galvanized									
7052	Adjustable hanger									
7054	18 ga., 2" x 6"	1 Carp	1.65	4.848	C	490	214		704	865

06 05 23 – Wood, Plastic, and Composite Fastenings

06 05 23.60 Timber Connectors	Crew	Daily Output	Labor-Hours	Unit	Material	2012 Bare Costs Labor	Equipment	Total	Total Incl O&P
7056	4" x 6"	1 Carp	1.65	4.848	C	640	214		854	1,025
7058	16 ga., 4" x 10"		1.60	5		940	221		1,161	1,375
7060	(2) 2" x 10"	↓	1.60	5	↓	940	221		1,161	1,375
7062	Connectors to plate									
7064	16 ga., 2" x 4" plate	1 Carp	1.80	4.444	C	480	196		676	830
7066	2" x 6" plate	"	1.80	4.444	"	620	196		816	985
7068	Hip jack connector									
7070	14 ga.	1 Carp	1.50	5.333	C	2,525	235		2,760	3,125

06 05 23.70 Rough Hardware

0010	ROUGH HARDWARE, average percent of carpentry material									
0020	Minimum				Job	.50%				
0200	Maximum					1.50%				
0210	In seismic or hurricane areas, up to				↓	10%				

06 05 23.80 Metal Bracing

0010	METAL BRACING									
0302	Let-in, "T" shaped, 22 ga. galv. steel, studs at 16" O.C.	1 Carp	580	.014	L.F.	.75	.61		1.36	1.77
0402	Studs at 24" O.C.		600	.013		.75	.59		1.34	1.73
0502	16 ga. galv. steel straps, studs at 16" O.C.		600	.013		.97	.59		1.56	1.97
0602	Studs at 24" O.C.	↓	620	.013	↓	.97	.57		1.54	1.95

06 11 Wood Framing

06 11 10 – Framing with Dimensional, Engineered or Composite Lumber

06 11 10.01 Forest Stewardship Council Certification

0010	FOREST STEWARDSHIP COUNCIL CERTIFICATION									
0020	For Forest Stewardship Council (FSC) cert dimension lumber, add G					65%				

06 11 10.02 Blocking

0010	BLOCKING									
2600	Miscellaneous, to wood construction									
2620	2" x 4"	1 Carp	.17	47.059	M.B.F.	490	2,075		2,565	3,750
2625	Pneumatic nailed		.21	38.095		490	1,675		2,165	3,125
2660	2" x 8"		.27	29.630		565	1,300		1,865	2,625
2665	Pneumatic nailed	↓	.33	24.242	↓	565	1,075		1,640	2,275
2720	To steel construction									
2740	2" x 4"	1 Carp	.14	57.143	M.B.F.	490	2,525		3,015	4,425
2780	2" x 8"	"	.21	38.095	"	565	1,675		2,240	3,200

06 11 10.04 Wood Bracing

0010	WOOD BRACING									
0012	Let-in, with 1" x 6" boards, studs @ 16" O.C.	1 Carp	150	.053	L.F.	.72	2.35		3.07	4.41
0202	Studs @ 24" O.C.	"	230	.035	"	.72	1.53		2.25	3.15

06 11 10.06 Bridging

0010	BRIDGING									
0012	Wood, for joists 16" O.C., 1" x 3"	1 Carp	130	.062	Pr.	.51	2.71		3.22	4.74
0017	Pneumatic nailed		170	.047		.51	2.08		2.59	3.75
0102	2" x 3" bridging		130	.062		.50	2.71		3.21	4.73
0107	Pneumatic nailed		170	.047		.50	2.08		2.58	3.74
0302	Steel, galvanized, 18 ga., for 2" x 10" joists at 12" O.C.		130	.062		1.72	2.71		4.43	6.05
0352	16" O.C.		135	.059		1.85	2.61		4.46	6.05
0402	24" O.C.		140	.057		1.85	2.52		4.37	5.90
0602	For 2" x 14" joists at 16" O.C.		130	.062		1.85	2.71		4.56	6.20
0902	Compression type, 16" O.C., 2" x 8" joists	↓	200	.040	↓	1.78	1.76		3.54	4.67

06 11 Wood Framing

06 11 10 - Framing with Dimensional, Engineered or Composite Lumber

06 11 10.06 **Bridging**		Crew	Daily Output	Labor-Hours	Unit	Material	2012 Bare Costs Labor	2012 Bare Costs Equipment	Total	Total Incl O&P
1002	2" x 12" joists	1 Carp	200	.040	Pr.	1.78	1.76		3.54	4.67

06 11 10.10 Beam and Girder Framing

		Crew	Daily Output	Labor-Hours	Unit	Material	Labor	Equipment	Total	Total Incl O&P
0010	**BEAM AND GIRDER FRAMING** R061110-30									
3500	Single, 2" x 6"	2 Carp	.70	22.857	M.B.F.	535	1,000		1,535	2,125
3505	Pneumatic nailed		.81	19.704		535	870		1,405	1,900
3520	2" x 8"		.86	18.605		565	820		1,385	1,875
3525	Pneumatic nailed		1	16.048		565	710		1,275	1,725
3540	2" x 10"		1	16		605	705		1,310	1,750
3545	Pneumatic nailed		1.16	13.793		605	610		1,215	1,600
3560	2" x 12"		1.10	14.545		640	640		1,280	1,675
3565	Pneumatic nailed		1.28	12.539		640	555		1,195	1,550
3580	2" x 14"		1.17	13.675		725	605		1,330	1,725
3585	Pneumatic nailed		1.36	11.791		725	520		1,245	1,600
3600	3" x 8"		1.10	14.545		1,050	640		1,690	2,150
3620	3" x 10"		1.25	12.800		1,075	565		1,640	2,050
3640	3" x 12"		1.35	11.852		1,100	525		1,625	2,000
3660	3" x 14"		1.40	11.429		1,100	505		1,605	1,975
3680	4" x 8"	F-3	2.66	15.038		1,275	675	230	2,180	2,675
3700	4" x 10"		3.16	12.658		1,275	570	194	2,039	2,475
3720	4" x 12"		3.60	11.111		1,375	500	170	2,045	2,475
3740	4" x 14"		3.96	10.101		1,525	455	155	2,135	2,550
4000	Double, 2" x 6"	2 Carp	1.25	12.800		535	565		1,100	1,450
4005	Pneumatic nailed		1.45	11.034		535	485		1,020	1,325
4020	2" x 8"		1.60	10		565	440		1,005	1,300
4025	Pneumatic nailed		1.86	8.621		565	380		945	1,200
4040	2" x 10"		1.92	8.333		605	370		975	1,225
4045	Pneumatic nailed		2.23	7.185		605	315		920	1,150
4060	2" x 12"		2.20	7.273		640	320		960	1,200
4065	Pneumatic nailed		2.55	6.275		640	277		917	1,125
4080	2" x 14"		2.45	6.531		725	288		1,013	1,250
4085	Pneumatic nailed		2.84	5.634		725	248		973	1,175
5000	Triple, 2" x 6"		1.65	9.697		535	430		965	1,250
5005	Pneumatic nailed		1.91	8.377		535	370		905	1,150
5020	2" x 8"		2.10	7.619		565	335		900	1,125
5025	Pneumatic nailed		2.44	6.568		565	290		855	1,075
5040	2" x 10"		2.50	6.400		605	282		887	1,100
5045	Pneumatic nailed		2.90	5.517		605	243		848	1,050
5060	2" x 12"		2.85	5.614		640	248		888	1,075
5065	Pneumatic nailed		3.31	4.840		640	213		853	1,025
5080	2" x 14"		3.15	5.079		725	224		949	1,150
5085	Pneumatic nailed		3.35	4.770		725	210		935	1,125

06 11 10.12 Ceiling Framing

		Crew	Daily Output	Labor-Hours	Unit	Material	Labor	Equipment	Total	Total Incl O&P
0010	**CEILING FRAMING**									
6400	Suspended, 2" x 3"	2 Carp	.50	32	M.B.F.	665	1,400		2,065	2,900
6450	2" x 4"		.59	27.119		490	1,200		1,690	2,400
6500	2" x 6"		.80	20		535	880		1,415	1,925
6550	2" x 8"		.86	18.605		565	820		1,385	1,875

06 11 10.14 Posts and Columns

		Crew	Daily Output	Labor-Hours	Unit	Material	Labor	Equipment	Total	Total Incl O&P
0010	**POSTS AND COLUMNS**									
0400	4" x 4"	2 Carp	.52	30.769	M.B.F.	1,225	1,350		2,575	3,450
0420	4" x 6"		.55	29.091		1,250	1,275		2,525	3,350
0440	4" x 8"		.59	27.119		1,275	1,200		2,475	3,250

06 11 Wood Framing

06 11 10 - Framing with Dimensional, Engineered or Composite Lumber

06 11 10.14 Posts and Columns

		Crew	Daily Output	Labor-Hours	Unit	Material	2012 Bare Costs Labor	Equipment	Total	Total Incl O&P
0460	6" x 6"	2 Carp	.65	24.615	M.B.F.	1,200	1,075		2,275	3,000
0480	6" x 8"		.70	22.857		1,200	1,000		2,200	2,875
0500	6" x 10"	↓	.75	21.333	↓	1,300	940		2,240	2,875

06 11 10.18 Joist Framing

		Crew	Daily Output	Labor-Hours	Unit	Material	2012 Bare Costs Labor	Equipment	Total	Total Incl O&P
0010	**JOIST FRAMING** R061110-30									
2650	Joists, 2" x 4"	2 Carp	.83	19.277	M.B.F.	490	850		1,340	1,850
2655	Pneumatic nailed		.96	16.667		490	735		1,225	1,675
2680	2" x 6"		1.25	12.800		535	565		1,100	1,450
2685	Pneumatic nailed		1.44	11.111		535	490		1,025	1,350
2700	2" x 8"		1.46	10.959		565	485		1,050	1,375
2705	Pneumatic nailed		1.68	9.524		565	420		985	1,275
2720	2" x 10"		1.49	10.738		605	475		1,080	1,400
2725	Pneumatic nailed		1.71	9.357		605	415		1,020	1,300
2740	2" x 12"		1.75	9.143		640	405		1,045	1,325
2745	Pneumatic nailed		2.01	7.960		640	350		990	1,250
2760	2" x 14"		1.79	8.939		725	395		1,120	1,400
2765	Pneumatic nailed		2.06	7.767		725	345		1,070	1,325
2780	3" x 6"		1.39	11.511		900	510		1,410	1,775
2790	3" x 8"		1.90	8.421		1,050	370		1,420	1,750
2800	3" x 10"		1.95	8.205		1,075	360		1,435	1,725
2820	3" x 12"		1.80	8.889		1,100	390		1,490	1,800
2840	4" x 6"		1.60	10		1,250	440		1,690	2,050
2850	4" x 8"		4.15	3.855		1,275	170		1,445	1,650
2860	4" x 10"		2	8		1,275	355		1,630	1,950
2880	4" x 12"		1.80	8.889	↓	1,375	390		1,765	2,125
3000	Composite wood joist 9-1/2" deep		.90	17.778	M.L.F.	1,625	785		2,410	2,975
3010	11-1/2" deep		.88	18.182		1,825	800		2,625	3,225
3020	14" deep		.82	19.512		2,525	860		3,385	4,100
3030	16" deep		.78	20.513		3,125	905		4,030	4,825
4000	Open web joist 12" deep		.88	18.182		2,900	800		3,700	4,425
4010	14" deep		.82	19.512		3,150	860		4,010	4,800
4020	16" deep		.78	20.513		3,200	905		4,105	4,925
4030	18" deep		.74	21.622		3,350	955		4,305	5,150
6000	Composite rim joist, 1-1/4" x 9-1/2"		.90	17.778		1,750	785		2,535	3,125
6010	1-1/4" x 11-1/2"		.88	18.182		2,250	800		3,050	3,700
6020	1-1/4" x 14-1/2"		.82	19.512		2,525	860		3,385	4,100
6030	1-1/4" x 16-1/2"	↓	.78	20.513	↓	3,100	905		4,005	4,800

06 11 10.24 Miscellaneous Framing

		Crew	Daily Output	Labor-Hours	Unit	Material	2012 Bare Costs Labor	Equipment	Total	Total Incl O&P
0010	**MISCELLANEOUS FRAMING**									
8500	Firestops, 2" x 4"	2 Carp	.51	31.373	M.B.F.	490	1,375		1,865	2,675
8505	Pneumatic nailed		.62	25.806		490	1,150		1,640	2,300
8520	2" x 6"		.60	26.667		535	1,175		1,710	2,375
8525	Pneumatic nailed		.73	21.858		535	965		1,500	2,050
8540	2" x 8"		.60	26.667		565	1,175		1,740	2,425
8560	2" x 12"		.70	22.857		640	1,000		1,640	2,250
8600	Nailers, treated, wood construction, 2" x 4"		.53	30.189		500	1,325		1,825	2,600
8605	Pneumatic nailed		.64	25.157		500	1,100		1,600	2,250
8620	2" x 6"		.75	21.333		630	940		1,570	2,150
8625	Pneumatic nailed		.90	17.778		630	785		1,415	1,900
8640	2" x 8"		.93	17.204		660	760		1,420	1,900
8645	Pneumatic nailed		1.12	14.337		660	630		1,290	1,700
8660	Steel construction, 2" x 4"	↓	.50	32	↓	500	1,400		1,900	2,725

178

06 11 Wood Framing

06 11 10 – Framing with Dimensional, Engineered or Composite Lumber

06 11 10.24 Miscellaneous Framing

		Crew	Daily Output	Labor-Hours	Unit	Material	Labor	2012 Bare Costs Equipment	Total	Total Incl O&P
8680	2" x 6"	2 Carp	.70	22.857	M.B.F.	630	1,000		1,630	2,250
8700	2" x 8"		.87	18.391		660	810		1,470	1,975
8760	Rough bucks, treated, for doors or windows, 2" x 6"		.40	40		630	1,775		2,405	3,425
8765	Pneumatic nailed		.48	33.333		630	1,475		2,105	2,950
8780	2" x 8"		.51	31.373		660	1,375		2,035	2,850
8785	Pneumatic nailed		.61	26.144		660	1,150		1,810	2,500
8800	Stair stringers, 2" x 10"		.22	72.727		605	3,200		3,805	5,600
8820	2" x 12"		.26	61.538		640	2,725		3,365	4,875
8840	3" x 10"		.31	51.613		1,075	2,275		3,350	4,675
8860	3" x 12"		.38	42.105		1,100	1,850		2,950	4,050
8870	Laminated structural lumber, 1-1/4" x 11-1/2"		130	.123	L.F.	2.25	5.45		7.70	10.80
8880	1-1/4" x 14-1/2"		130	.123	"	2.53	5.45		7.98	11.15

06 11 10.26 Partitions

		Crew	Daily Output	Labor-Hours	Unit	Material	Labor	2012 Bare Costs Equipment	Total	Total Incl O&P
0010	**PARTITIONS**									
0020	Single bottom and double top plate, no waste, std. & better lumber									
0180	2" x 4" studs, 8' high, studs 12" O.C.	2 Carp	80	.200	L.F.	3.60	8.80		12.40	17.50
0185	12" O.C., pneumatic nailed		96	.167		3.60	7.35		10.95	15.25
0200	16" O.C.		100	.160		2.95	7.05		10	14.10
0205	16" O.C., pneumatic nailed		120	.133		2.95	5.90		8.85	12.30
0300	24" O.C.		125	.128		2.29	5.65		7.94	11.20
0305	24" O.C., pneumatic nailed		150	.107		2.29	4.70		6.99	9.75
0380	10' high, studs 12" O.C.		80	.200		4.26	8.80		13.06	18.25
0385	12" O.C., pneumatic nailed		96	.167		4.26	7.35		11.61	16
0400	16" O.C.		100	.160		3.44	7.05		10.49	14.65
0405	16" O.C., pneumatic nailed		120	.133		3.44	5.90		9.34	12.85
0500	24" O.C.		125	.128		2.62	5.65		8.27	11.60
0505	24" O.C., pneumatic nailed		150	.107		2.62	4.70		7.32	10.15
0580	12' high, studs 12" O.C.		65	.246		4.91	10.85		15.76	22
0585	12" O.C., pneumatic nailed		78	.205		4.91	9.05		13.96	19.30
0600	16" O.C.		80	.200		3.93	8.80		12.73	17.85
0605	16" O.C., pneumatic nailed		96	.167		3.93	7.35		11.28	15.60
0700	24" O.C.		100	.160		2.95	7.05		10	14.10
0705	24" O.C., pneumatic nailed		120	.133		2.95	5.90		8.85	12.30
0780	2" x 6" studs, 8' high, studs 12" O.C.		70	.229		5.85	10.10		15.95	22
0785	12" O.C., pneumatic nailed		84	.190		5.85	8.40		14.25	19.35
0800	16" O.C.		90	.178		4.80	7.85		12.65	17.35
0805	16" O.C., pneumatic nailed		108	.148		4.80	6.55		11.35	15.35
0900	24" O.C.		115	.139		3.73	6.15		9.88	13.55
0905	24" O.C., pneumatic nailed		138	.116		3.73	5.10		8.83	11.95
0980	10' high, studs 12" O.C.		70	.229		6.95	10.10		17.05	23
0985	12" O.C., pneumatic nailed		84	.190		6.95	8.40		15.35	20.50
1000	16" O.C.		90	.178		5.60	7.85		13.45	18.20
1005	16" O.C., pneumatic nailed		108	.148		5.60	6.55		12.15	16.20
1100	24" O.C.		115	.139		4.26	6.15		10.41	14.15
1105	24" O.C., pneumatic nailed		138	.116		4.26	5.10		9.36	12.55
1180	12' high, studs 12" O.C.		55	.291		8	12.85		20.85	28.50
1185	12" O.C., pneumatic nailed		66	.242		8	10.70		18.70	25.50
1200	16" O.C.		70	.229		6.40	10.10		16.50	22.50
1205	16" O.C., pneumatic nailed		84	.190		6.40	8.40		14.80	19.95
1300	24" O.C.		90	.178		4.80	7.85		12.65	17.35
1305	24" O.C., pneumatic nailed		108	.148		4.80	6.55		11.35	15.35
1400	For horizontal blocking, 2" x 4", add		600	.027		.33	1.18		1.51	2.17

06 11 Wood Framing

06 11 10 – Framing with Dimensional, Engineered or Composite Lumber

	06 11 10.26 **Partitions**	Crew	Daily Output	Labor-Hours	Unit	Material	2012 Bare Costs Labor	2012 Bare Costs Equipment	Total	Total Incl O&P
1500	2" x 6", add	2 Carp	600	.027	L.F.	.53	1.18		1.71	2.40
1600	For openings, add	▼	250	.064	▼		2.82		2.82	4.34
1700	Headers for above openings, material only, add				M.B.F.	565			565	620

	06 11 10.28 **Porch or Deck Framing**									
0010	**PORCH OR DECK FRAMING**									
0100	Treated lumber, posts or columns, 4" x 4"	2 Carp	390	.041	L.F.	.82	1.81		2.63	3.68
0110	4" x 6"		275	.058		2.11	2.57		4.68	6.25
0120	4" x 8"		220	.073		4.32	3.21		7.53	9.70
0130	Girder, single, 4" x 4"		675	.024		.82	1.05		1.87	2.51
0140	4" x 6"		600	.027		2.11	1.18		3.29	4.13
0150	4" x 8"		525	.030		4.32	1.34		5.66	6.85
0160	Double, 2" x 4"		625	.026		.70	1.13		1.83	2.51
0170	2" x 6"		600	.027		1.31	1.18		2.49	3.25
0180	2" x 8"		575	.028		1.82	1.23		3.05	3.89
0190	2" x 10"		550	.029		2.41	1.28		3.69	4.62
0200	2" x 12"		525	.030		3.22	1.34		4.56	5.60
0210	Triple, 2" x 4"		575	.028		1.05	1.23		2.28	3.05
0220	2" x 6"		550	.029		1.96	1.28		3.24	4.13
0230	2" x 8"		525	.030		2.73	1.34		4.07	5.05
0240	2" x 10"		500	.032		3.61	1.41		5.02	6.15
0250	2" x 12"		475	.034		4.83	1.49		6.32	7.60
0260	Ledger, bolted 4' O.C., 2" x 4"		400	.040		.46	1.76		2.22	3.21
0270	2" x 6"		395	.041		.75	1.79		2.54	3.58
0280	2" x 8"		390	.041		1	1.81		2.81	3.88
0300	2" x 12"		380	.042		1.69	1.86		3.55	4.72
0310	Joists, 2" x 4"		1250	.013		.35	.56		.91	1.26
0320	2" x 6"		1250	.013		.66	.56		1.22	1.59
0330	2" x 8"		1100	.015		.91	.64		1.55	2
0340	2" x 10"		900	.018		1.21	.78		1.99	2.54
0350	2" x 12"	▼	875	.018		1.45	.81		2.26	2.84
0360	Railings and trim , 1" x 4"	1 Carp	300	.027		.41	1.18		1.59	2.26
0370	2" x 2"		300	.027		.35	1.18		1.53	2.20
0380	2" x 4"		300	.027		.33	1.18		1.51	2.18
0390	2" x 6"		300	.027	▼	.63	1.18		1.81	2.50
0400	Decking, 1" x 4"		275	.029	S.F.	1.64	1.28		2.92	3.77
0410	2" x 4"		300	.027		1.14	1.18		2.32	3.07
0420	2" x 6"		320	.025		1.37	1.10		2.47	3.21
0430	5/4" x 6"	▼	320	.025	▼	1.92	1.10		3.02	3.81
0440	Balusters, square, 2" x 2"	2 Carp	660	.024	L.F.	.36	1.07		1.43	2.03
0450	Turned, 2" x 2"		420	.038		.48	1.68		2.16	3.11
0460	Stair stringer, 2" x 10"		130	.123		1.21	5.45		6.66	9.70
0470	2" x 12"		130	.123		1.45	5.45		6.90	9.95
0480	Stair treads, 1" x 4"		140	.114		1.65	5.05		6.70	9.55
0490	2" x 4"		140	.114		.35	5.05		5.40	8.15
0500	2" x 6"		160	.100		.53	4.41		4.94	7.40
0510	5/4" x 6"		160	.100	▼	.90	4.41		5.31	7.80
0520	Turned handrail post, 4" x 4"		64	.250	Ea.	32	11.05		43.05	52
0530	Lattice panel, 4' x 8', 1/2"		1600	.010	S.F.	.80	.44		1.24	1.56
0535	3/4"		1600	.010	"	1.20	.44		1.64	2
0540	Cedar, posts or columns, 4" x 4"		390	.041	L.F.	2.83	1.81		4.64	5.90
0550	4" x 6"		275	.058		5.25	2.57		7.82	9.70
0560	4" x 8"	▼	220	.073	▼	8.85	3.21		12.06	14.70

06 11 Wood Framing

06 11 10 – Framing with Dimensional, Engineered or Composite Lumber

06 11 10.28 Porch or Deck Framing

		Crew	Daily Output	Labor-Hours	Unit	Material	2012 Bare Costs Labor	2012 Bare Costs Equipment	Total	Total Incl O&P
0800	Decking, 1" x 4"	2 Carp	550	.029	L.F.	1.81	1.28		3.09	3.96
0810	2" x 4"		600	.027		3.62	1.18		4.80	5.80
0820	2" x 6"		640	.025		6.30	1.10		7.40	8.65
0830	5/4" x 6"		640	.025		4.08	1.10		5.18	6.20
0840	Railings and trim, 1" x 4"		600	.027		1.81	1.18		2.99	3.80
0860	2" x 4"		600	.027		3.62	1.18		4.80	5.80
0870	2" x 6"		600	.027		6.30	1.18		7.48	8.75
0920	Stair treads, 1" x 4"		140	.114		1.81	5.05		6.86	9.75
0930	2" x 4"		140	.114		3.62	5.05		8.67	11.75
0940	2" x 6"		160	.100		6.30	4.41		10.71	13.75
0950	5/4" x 6"		160	.100		4.08	4.41		8.49	11.30
0980	Redwood, posts or columns, 4" x 4"		390	.041		6.40	1.81		8.21	9.85
0990	4" x 6"		275	.058		11.20	2.57		13.77	16.25
1000	4" x 8"	▼	220	.073	▼	20.50	3.21		23.71	27.50
1240	Redwood decking, 1" x 4"	1 Carp	275	.029	S.F.	3.97	1.28		5.25	6.35
1260	2" x 6"		340	.024		7.50	1.04		8.54	9.85
1270	5/4" x 6"	▼	320	.025	▼	4.74	1.10		5.84	6.90
1280	Railings and trim, 1" x 4"	2 Carp	600	.027	L.F.	1.17	1.18		2.35	3.09
1310	2" x 6"		600	.027		7.50	1.18		8.68	10.05
1420	Alternative decking, wood/plastic composite, 5/4" x 6"　G		640	.025		3	1.10		4.10	5
1440	1" x 4" square edge fir		550	.029		1.66	1.28		2.94	3.79
1450	1" x 4" tongue and groove fir		450	.036		1.41	1.57		2.98	3.96
1460	1" x 4" mahogany		550	.029		2.51	1.28		3.79	4.73
1462	5/4" x 6" PVC	▼	550	.029	▼	3.01	1.28		4.29	5.30
1465	Framing, porch or deck, alt deck fastening, screws, add	1 Carp	240	.033	S.F.		1.47		1.47	2.26
1470	Accessories, joist hangers, 2" x 4"		160	.050	Ea.	.67	2.21		2.88	4.13
1480	2" x 6" through 2" x 12"	▼	150	.053		1.28	2.35		3.63	5.05
1530	Post footing, incl excav, backfill, tube form & concrete, 4' deep, 8" dia	F-7	12	2.667		11.15	106		117.15	174
1540	10" diameter		11	2.909		16.55	115		131.55	195
1550	12" diameter	▼	10	3.200	▼	22	127		149	219

06 11 10.30 Roof Framing

		Crew	Daily Output	Labor-Hours	Unit	Material	2012 Bare Costs Labor	2012 Bare Costs Equipment	Total	Total Incl O&P
0010	**ROOF FRAMING**									
5250	Composite rafter, 9-1/2" deep	2 Carp	575	.028	L.F.	1.62	1.23		2.85	3.67
5260	11-1/2" deep		575	.028	"	1.82	1.23		3.05	3.90
6070	Fascia boards, 2" x 8"		.30	53.333	M.B.F.	565	2,350		2,915	4,250
6080	2" x 10"		.30	53.333		605	2,350		2,955	4,300
7000	Rafters, to 4 in 12 pitch, 2" x 6"		1	16		535	705		1,240	1,650
7060	2" x 8"		1.26	12.698		565	560		1,125	1,475
7300	Hip and valley rafters, 2" x 6"		.76	21.053		535	930		1,465	2,000
7360	2" x 8"		.96	16.667		565	735		1,300	1,750
7540	Hip and valley jacks, 2" x 6"		.60	26.667		535	1,175		1,710	2,375
7600	2" x 8"	▼	.65	24.615	▼	565	1,075		1,640	2,300
7780	For slopes steeper than 4 in 12, add						30%			
7790	For dormers or complex roofs, add						50%			
7800	Rafter tie, 1" x 4", #3	2 Carp	.27	59.259	M.B.F.	1,225	2,625		3,850	5,375
7820	Ridge board, #2 or better, 1" x 6"		.30	53.333		1,450	2,350		3,800	5,200
7840	1" x 8"		.37	43.243		1,500	1,900		3,400	4,575
7860	1" x 10"		.42	38.095		1,550	1,675		3,225	4,300
7880	2" x 6"		.50	32		535	1,400		1,935	2,750
7900	2" x 8"		.60	26.667		565	1,175		1,740	2,425
7920	2" x 10"		.66	24.242		605	1,075		1,680	2,325
7940	Roof cants, split, 4" x 4"	▼	.86	18.605	▼	1,225	820		2,045	2,600

06 11 Wood Framing

06 11 10 – Framing with Dimensional, Engineered or Composite Lumber

	06 11 10.30 Roof Framing	Crew	Daily Output	Labor-Hours	Unit	Material	2012 Bare Costs Labor	2012 Bare Costs Equipment	Total	Total Incl O&P
7960	6" x 6"	2 Carp	1.80	8.889	M.B.F.	1,200	390		1,590	1,925
7980	Roof curbs, untreated, 2" x 6"		.52	30.769		535	1,350		1,885	2,675
8000	2" x 12"	↓	.80	20	↓	640	880		1,520	2,050

06 11 10.32 Sill and Ledger Framing

		Crew	Daily Output	Labor-Hours	Unit	Material	Labor	Equipment	Total	Total Incl O&P
0010	**SILL AND LEDGER FRAMING**									
4482	Ledgers, nailed, 2" x 4"	2 Carp	.50	32	M.B.F.	490	1,400		1,890	2,725
4484	2" x 6"		.60	26.667		535	1,175		1,710	2,375
4486	Bolted, not including bolts, 3" x 8"		.65	24.615		1,050	1,075		2,125	2,850
4488	3" x 12"		.70	22.857		1,100	1,000		2,100	2,750
4490	Mud sills, redwood, construction grade, 2" x 4"		.59	27.119		3,400	1,200		4,600	5,575
4492	2" x 6"		.78	20.513		3,400	905		4,305	5,150
4500	Sills, 2" x 4"		.40	40		490	1,775		2,265	3,275
4520	2" x 6"		.55	29.091		535	1,275		1,810	2,550
4540	2" x 8"		.67	23.881		565	1,050		1,615	2,250
4600	Treated, 2" x 4"		.36	44.444		500	1,950		2,450	3,575
4620	2" x 6"		.50	32		630	1,400		2,030	2,875
4640	2" x 8"		.60	26.667		660	1,175		1,835	2,525
4700	4" x 4"		.60	26.667		590	1,175		1,765	2,450
4720	4" x 6"		.70	22.857		1,025	1,000		2,025	2,675
4740	4" x 8"		.80	20		1,600	880		2,480	3,100
4760	4" x 10"	↓	.87	18.391	↓	1,750	810		2,560	3,175

06 11 10.34 Sleepers

		Crew	Daily Output	Labor-Hours	Unit	Material	Labor	Equipment	Total	Total Incl O&P
0010	**SLEEPERS**									
0300	On concrete, treated, 1" x 2"	2 Carp	.39	41.026	M.B.F.	1,750	1,800		3,550	4,700
0320	1" x 3"		.50	32		1,700	1,400		3,100	4,050
0340	2" x 4"		.99	16.162		500	715		1,215	1,650
0360	2" x 6"	↓	1.30	12.308	↓	630	545		1,175	1,525

06 11 10.36 Soffit and Canopy Framing

		Crew	Daily Output	Labor-Hours	Unit	Material	Labor	Equipment	Total	Total Incl O&P
0010	**SOFFIT AND CANOPY FRAMING**									
1300	Canopy or soffit framing, 1" x 4"	2 Carp	.30	53.333	M.B.F.	1,225	2,350		3,575	4,975
1340	1" x 8"		.50	32		1,500	1,400		2,900	3,825
1360	2" x 4"		.41	39.024		490	1,725		2,215	3,200
1400	2" x 8"		.67	23.881		565	1,050		1,615	2,250
1420	3" x 4"		.50	32		850	1,400		2,250	3,100
1460	3" x 8"	↓	.60	26.667	↓	1,050	1,175		2,225	2,975

06 11 10.38 Treated Lumber Framing Material

		Crew	Daily Output	Labor-Hours	Unit	Material	Labor	Equipment	Total	Total Incl O&P
0010	**TREATED LUMBER FRAMING MATERIAL**									
0100	2" x 4"				M.B.F.	500			500	550
0110	2" x 6"					630			630	695
0120	2" x 8"					660			660	725
0130	2" x 10"					700			700	770
0140	2" x 12"					780			780	860
0200	4" x 4"					590			590	645
0210	4" x 6"					1,025			1,025	1,125
0220	4" x 8"				↓	1,600			1,600	1,750

06 11 10.40 Wall Framing

		Crew	Daily Output	Labor-Hours	Unit	Material	Labor	Equipment	Total	Total Incl O&P
0010	**WALL FRAMING** R061110-30									
5860	Headers over openings, 2" x 6"	2 Carp	.36	44.444	M.B.F.	535	1,950		2,485	3,600
5865	2" x 6", pneumatic nailed		.43	37.209		535	1,650		2,185	3,100
5880	2" x 8"		.45	35.556		565	1,575		2,140	3,025
5885	2" x 8", pneumatic nailed	↓	.54	29.630	↓	565	1,300		1,865	2,625

06 11 Wood Framing

06 11 10 – Framing with Dimensional, Engineered or Composite Lumber

06 11 10.40 Wall Framing

		Crew	Daily Output	Labor-Hours	Unit	Material	2012 Bare Costs Labor	2012 Bare Costs Equipment	Total	Total Incl O&P
5900	2" x 10"	2 Carp	.53	30.189	M.B.F.	605	1,325		1,930	2,725
5905	2" x 10", pneumatic nailed		.67	23.881		605	1,050		1,655	2,300
5920	2" x 12"		.60	26.667		640	1,175		1,815	2,500
5925	2" x 12", pneumatic nailed		.72	22.222		640	980		1,620	2,200
5940	4" x 12"		.76	21.053		1,375	930		2,305	2,950
5945	4" x 12", pneumatic nailed		.92	17.391		1,375	765		2,140	2,700
5960	6" x 12"		.84	19.048		1,700	840		2,540	3,175
5965	6" x 12", pneumatic nailed		1.01	15.873		1,700	700		2,400	2,950
6000	Plates, untreated, 2" x 3"		.43	37.209		665	1,650		2,315	3,250
6005	2" x 3", pneumatic nailed		.52	30.769		665	1,350		2,015	2,825
6020	2" x 4"		.53	30.189		490	1,325		1,815	2,600
6025	2" x 4", pneumatic nailed		.67	23.881		490	1,050		1,540	2,175
6040	2" x 6"		.75	21.333		535	940		1,475	2,025
6045	2" x 6", pneumatic nailed		.90	17.778		535	785		1,320	1,775
6120	Studs, 8' high wall, 2" x 3"		.60	26.667		665	1,175		1,840	2,525
6125	2" x 3", pneumatic nailed		.72	22.222		665	980		1,645	2,225
6140	2" x 4"		.92	17.391		490	765		1,255	1,725
6145	2" x 4", pneumatic nailed		1.10	14.493		490	640		1,130	1,525
6160	2" x 6"		1	16		535	705		1,240	1,650
6165	2" x 6", pneumatic nailed		1.20	13.333		535	590		1,125	1,500
6180	3" x 4"		.80	20		850	880		1,730	2,275
6185	3" x 4", pneumatic nailed		.96	16.667		850	735		1,585	2,050
8200	For 12' high walls, deduct						5%			
8220	For stub wall, 6' high, add						20%			
8240	3' high, add						40%			
8250	For second story & above, add						5%			
8300	For dormer & gable, add						15%			

06 11 10.42 Furring

		Crew	Daily Output	Labor-Hours	Unit	Material	2012 Bare Costs Labor	2012 Bare Costs Equipment	Total	Total Incl O&P
0010	**FURRING**									
0012	Wood strips, 1" x 2", on walls, on wood	1 Carp	550	.015	L.F.	.25	.64		.89	1.27
0015	On wood, pneumatic nailed		710	.011		.25	.50		.75	1.04
0300	On masonry		495	.016		.25	.71		.96	1.38
0400	On concrete		260	.031		.25	1.36		1.61	2.37
0600	1" x 3", on walls, on wood		550	.015		.34	.64		.98	1.37
0605	On wood, pneumatic nailed		710	.011		.34	.50		.84	1.14
0700	On masonry		495	.016		.34	.71		1.05	1.48
0800	On concrete		260	.031		.34	1.36		1.70	2.47
0850	On ceilings, on wood		350	.023		.34	1.01		1.35	1.93
0855	On wood, pneumatic nailed		450	.018		.34	.78		1.12	1.59
0900	On masonry		320	.025		.34	1.10		1.44	2.08
0950	On concrete		210	.038		.34	1.68		2.02	2.97

06 11 10.44 Grounds

		Crew	Daily Output	Labor-Hours	Unit	Material	2012 Bare Costs Labor	2012 Bare Costs Equipment	Total	Total Incl O&P
0010	**GROUNDS**									
0020	For casework, 1" x 2" wood strips, on wood	1 Carp	330	.024	L.F.	.25	1.07		1.32	1.92
0100	On masonry		285	.028		.25	1.24		1.49	2.18
0200	On concrete		250	.032		.25	1.41		1.66	2.45
0400	For plaster, 3/4" deep, on wood		450	.018		.25	.78		1.03	1.49
0500	On masonry		225	.036		.25	1.57		1.82	2.69
0600	On concrete		175	.046		.25	2.02		2.27	3.38
0700	On metal lath		200	.040		.25	1.76		2.01	2.99

06 12 Structural Panels

06 12 10 – Structural Insulated Panels

06 12 10.10 OSB Faced Panels

		Crew	Daily Output	Labor-Hours	Unit	Material	2012 Bare Costs Labor	Equipment	Total	Total Incl O&P
0010	**OSB FACED PANELS**									
0100	Structural insul. panels, 7/16" OSB both faces, EPS insul, 3-5/8" T [G]	F-3	2075	.019	S.F.	2.79	.86	.30	3.95	4.71
0110	5-5/8" thick [G]		1725	.023		3.13	1.04	.35	4.52	5.40
0120	7-3/8" thick [G]		1425	.028		3.41	1.26	.43	5.10	6.15
0130	9-3/8" thick [G]		1125	.036		3.73	1.59	.54	5.86	7.15
0140	7/16" OSB one face, EPS insul, 3-5/8" thick [G]		2175	.018		2.97	.82	.28	4.07	4.84
0150	5-5/8" thick [G]		1825	.022		3.40	.98	.34	4.72	5.60
0160	7-3/8" thick [G]		1525	.026		3.79	1.18	.40	5.37	6.40
0170	9-3/8" thick [G]		1225	.033		4.22	1.46	.50	6.18	7.45
0190	7/16" OSB - 1/2" GWB faces , EPS insul, 3-5/8" T [G]		2075	.019		2.94	.86	.30	4.10	4.87
0200	5-5/8" thick [G]		1725	.023		3.37	1.04	.35	4.76	5.70
0210	7-3/8" thick [G]		1425	.028		3.75	1.26	.43	5.44	6.55
0220	9-3/8" thick [G]		1125	.036		4.40	1.59	.54	6.53	7.90
0240	7/16" OSB - 1/2" MRGWB faces , EPS insul, 3-5/8" T [G]		2075	.019		2.94	.86	.30	4.10	4.87
0250	5-5/8" thick [G]		1725	.023		3.37	1.04	.35	4.76	5.70
0260	7-3/8" thick [G]		1425	.028		3.75	1.26	.43	5.44	6.55
0270	9-3/8" thick [G]		1125	.036		4.40	1.59	.54	6.53	7.90
0300	For 1/2" GWB added to OSB skin, add [G]					.62			.62	.68
0310	For 1/2" MRGWB added to OSB skin, add [G]					.62			.62	.68
0320	For one T1-11 skin, add to OSB-OSB [G]					1.42			1.42	1.56
0330	For one 19/32" CDX skin, add to OSB-OSB [G]					.72			.72	.79
0500	Structural insulated panel, 7/16" OSB both sides, straw core									
0510	4-3/8" T, walls (w/sill, splines, plates) [G]	F-6	2400	.017	S.F.	6.75	.69	.26	7.70	8.75
0520	Floors (w/splines) [G]		2400	.017		6.75	.69	.26	7.70	8.75
0530	Roof (w/splines) [G]		2400	.017		6.75	.69	.26	7.70	8.75
0550	7-7/8" T, walls (w/sill, splines, plates) [G]		2400	.017		10.20	.69	.26	11.15	12.60
0560	Floors (w/splines) [G]		2400	.017		10.20	.69	.26	11.15	12.60
0570	Roof (w/splines) [G]		2400	.017		10.20	.69	.26	11.15	12.60

06 12 19 – Composite Shearwall Panels

06 12 19.10 Steel and Wood Composite Shearwall Panels

		Crew	Daily Output	Labor-Hours	Unit	Material	2012 Bare Costs Labor	Equipment	Total	Total Incl O&P
0010	**STEEL & WOOD COMPOSITE SHEARWALL PANELS**									
0020	Anchor bolts, 36" long (must be placed in wet concrete)	1 Carp	150	.053	Ea.	33	2.35		35.35	39.50
0030	On concrete, 2 x 4 & 2 x 6 walls, 7' - 10' high, 360 lb. shear, 12" wide	2 Carp	8	2		380	88		468	555
0040	715 lb. shear, 15" wide		8	2		440	88		528	615
0050	1860 lb. shear, 18" wide		8	2		470	88		558	655
0060	2780 lb. shear, 21" wide		8	2		560	88		648	755
0070	3790 lb. shear, 24" wide		8	2		640	88		728	835
0080	2 x 6 walls, 11' to 13' high, 1180 lb. shear, 18" wide		6	2.667		580	118		698	820
0090	1555 lb. shear, 21" wide		6	2.667		725	118		843	980
0100	2280 lb. shear, 24" wide		6	2.667		805	118		923	1,075
0110	For installing above on wood floor frame, add									
0120	Coupler nuts, threaded rods, bolts, shear transfer plate kit	1 Carp	16	.500	Ea.	50	22		72	89
0130	Framing anchors, angle (2 required)	"	96	.083	"	2.40	3.67		6.07	8.30
0140	For blocking see Section 06 11 10.02									
0150	For installing above, first floor to second floor, wood floor frame, add									
0160	Add stack option to first floor wall panel				Ea.	55			55	60.50
0170	Threaded rods, bolts, shear transfer plate kit	1 Carp	16	.500		60	22		82	100
0180	Framing anchors, angle (2 required)	"	96	.083		2.40	3.67		6.07	8.30
0200	For installing stacked panels, balloon framing									
0210	Add stack option to first floor wall panel				Ea.	55			55	60.50
0220	Threaded rods, bolts kit	1 Carp	16	.500	"	35	22		57	72.50

06 13 Heavy Timber Construction

06 13 23 – Heavy Timber Framing

06 13 23.10 Heavy Framing

06 13 23.10 Heavy Framing		Crew	Daily Output	Labor-Hours	Unit	Material	2012 Bare Costs Labor	2012 Bare Costs Equipment	Total	Total Incl O&P
0010	**HEAVY FRAMING**									
0020	Beams, single 6" x 10"	2 Carp	1.10	14.545	M.B.F.	1,400	640		2,040	2,525
0100	Single 8" x 16"		1.20	13.333		1,975	590		2,565	3,075
0200	Built from 2" lumber, multiple 2" x 14"		.90	17.778		725	785		1,510	2,000
0210	Built from 3" lumber, multiple 3" x 6"		.70	22.857		900	1,000		1,900	2,550
0220	Multiple 3" x 8"		.80	20		1,050	880		1,930	2,525
0230	Multiple 3" x 10"		.90	17.778		1,075	785		1,860	2,375
0240	Multiple 3" x 12"		1	16		1,100	705		1,805	2,275
0250	Built from 4" lumber, multiple 4" x 6"		.80	20		1,250	880		2,130	2,725
0260	Multiple 4" x 8"		.90	17.778		1,275	785		2,060	2,600
0270	Multiple 4" x 10"		1	16		1,275	705		1,980	2,475
0280	Multiple 4" x 12"		1.10	14.545		1,375	640		2,015	2,500
0290	Columns, structural grade, 1500f, 4" x 4"		.60	26.667		1,275	1,175		2,450	3,200
0300	6" x 6"		.65	24.615		1,300	1,075		2,375	3,100
0400	8" x 8"		.70	22.857		1,350	1,000		2,350	3,025
0500	10" x 10"		.75	21.333		1,450	940		2,390	3,050
0600	12" x 12"		.80	20		1,525	880		2,405	3,025
0800	Floor planks, 2" thick, T & G, 2" x 6"		1.05	15.238		1,050	670		1,720	2,175
0900	2" x 10"		1.10	14.545		1,050	640		1,690	2,125
1100	3" thick, 3" x 6"		1.05	15.238		1,250	670		1,920	2,400
1200	3" x 10"		1.10	14.545		1,250	640		1,890	2,350
1400	Girders, structural grade, 12" x 12"		.80	20		1,525	880		2,405	3,025
1500	10" x 16"		1	16		2,400	705		3,105	3,725
2300	Roof purlins, 4" thick, structural grade		1.05	15.238		1,275	670		1,945	2,425

06 15 Wood Decking

06 15 16 – Wood Roof Decking

06 15 16.10 Solid Wood Roof Decking

06 15 16.10 Solid Wood Roof Decking		Crew	Daily Output	Labor-Hours	Unit	Material	2012 Bare Costs Labor	2012 Bare Costs Equipment	Total	Total Incl O&P
0010	**SOLID WOOD ROOF DECKING**									
0400	Cedar planks, 3" thick	2 Carp	320	.050	S.F.	7	2.21		9.21	11.05
0500	4" thick		250	.064		9.30	2.82		12.12	14.60
0700	Douglas fir, 3" thick		320	.050		2.94	2.21		5.15	6.60
0800	4" thick		250	.064		3.92	2.82		6.74	8.65
1000	Hemlock, 3" thick		320	.050		3.28	2.21		5.49	7
1100	4" thick		250	.064		4.36	2.82		7.18	9.15
1300	Western white spruce, 3" thick		320	.050		1.92	2.21		4.13	5.50
1400	4" thick		250	.064		2.55	2.82		5.37	7.15

06 15 23 – Laminated Wood Decking

06 15 23.10 Laminated Roof Deck

06 15 23.10 Laminated Roof Deck		Crew	Daily Output	Labor-Hours	Unit	Material	2012 Bare Costs Labor	2012 Bare Costs Equipment	Total	Total Incl O&P
0010	**LAMINATED ROOF DECK**									
0020	Pine or hemlock, 3" thick	2 Carp	425	.038	S.F.	3.83	1.66		5.49	6.75
0100	4" thick		325	.049		5.10	2.17		7.27	8.95
0300	Cedar, 3" thick		425	.038		4.20	1.66		5.86	7.15
0400	4" thick		325	.049		5.35	2.17		7.52	9.20
0600	Fir, 3" thick		425	.038		3.88	1.66		5.54	6.80
0700	4" thick		325	.049		5.15	2.17		7.32	9.05

06 16 Sheathing

06 16 23 – Subflooring

06 16 23.10 Subfloor

			Crew	Daily Output	Labor-Hours	Unit	Material	2012 Bare Costs Labor	Equipment	Total	Total Incl O&P
0010	**SUBFLOOR**	R061636-20									
0011	Plywood, CDX, 1/2" thick		2 Carp	1500	.011	SF Flr.	.47	.47		.94	1.24
0015	Pneumatic nailed			1860	.009		.47	.38		.85	1.10
0100	5/8" thick			1350	.012		.62	.52		1.14	1.48
0105	Pneumatic nailed			1674	.010		.62	.42		1.04	1.33
0200	3/4" thick			1250	.013		.70	.56		1.26	1.64
0205	Pneumatic nailed			1550	.010		.70	.46		1.16	1.47
0300	1-1/8" thick, 2-4-1 including underlayment			1050	.015		1.53	.67		2.20	2.71
0440	With boards, 1" x 6", S4S, laid regular			900	.018		1.57	.78		2.35	2.94
0450	1" x 8", laid regular			1000	.016		1.60	.71		2.31	2.85
0460	Laid diagonal			850	.019		1.60	.83		2.43	3.04
0500	1" x 10", laid regular			1100	.015		1.64	.64		2.28	2.79
0600	Laid diagonal			900	.018		1.64	.78		2.42	3.01
8990	Subfloor adhesive, 3/8" bead		1 Carp	2300	.003	L.F.	.12	.15		.27	.37

06 16 26 – Underlayment

06 16 26.10 Wood Product Underlayment

			Crew	Daily Output	Labor-Hours	Unit	Material	2012 Bare Costs Labor	Equipment	Total	Total Incl O&P
0010	**WOOD PRODUCT UNDERLAYMENT**	R061636-20									
0030	Plywood, underlayment grade, 3/8" thick		2 Carp	1500	.011	SF Flr.	.73	.47		1.20	1.52
0070	Pneumatic nailed			1860	.009		.73	.38		1.11	1.38
0100	1/2" thick			1450	.011		.77	.49		1.26	1.60
0105	Pneumatic nailed			1798	.009		.77	.39		1.16	1.45
0200	5/8" thick			1400	.011		.96	.50		1.46	1.84
0205	Pneumatic nailed			1736	.009		.96	.41		1.37	1.69
0300	3/4" thick			1300	.012		1.17	.54		1.71	2.13
0305	Pneumatic nailed			1612	.010		1.17	.44		1.61	1.96
0500	Particle board, 3/8" thick	G		1500	.011		.35	.47		.82	1.11
0505	Pneumatic nailed	G		1860	.009		.35	.38		.73	.97
0600	1/2" thick	G		1450	.011		.36	.49		.85	1.15
0605	Pneumatic nailed	G		1798	.009		.36	.39		.75	1
0800	5/8" thick	G		1400	.011		.46	.50		.96	1.29
0805	Pneumatic nailed	G		1736	.009		.46	.41		.87	1.14
0900	3/4" thick	G		1300	.012		.65	.54		1.19	1.56
0905	Pneumatic nailed	G		1612	.010		.65	.44		1.09	1.39
0955	Particleboard, 100% recycled straw/wheat, 4' x 8' x 1/4"	G		1450	.011	S.F.	.26	.49		.75	1.04
0960	4' x 8' x 3/8"	G		1450	.011		.39	.49		.88	1.18
0965	4' x 8' x 1/2"	G		1350	.012		.54	.52		1.06	1.39
0970	4' x 8' x 5/8"	G		1300	.012		.64	.54		1.18	1.54
0975	4' x 8' x 3/4"	G		1250	.013		.74	.56		1.30	1.68
0980	4' x 8' x 1"	G		1150	.014		.99	.61		1.60	2.03
0985	4' x 8' x 1-1/4"	G		1100	.015		1.16	.64		1.80	2.27
1100	Hardboard, underlayment grade, 4' x 4', .215" thick	G		1500	.011	SF Flr.	.57	.47		1.04	1.35

06 16 36 – Wood Panel Product Sheathing

06 16 36.10 Sheathing

			Crew	Daily Output	Labor-Hours	Unit	Material	2012 Bare Costs Labor	Equipment	Total	Total Incl O&P
0010	**SHEATHING**	R061636-20									
0012	Plywood on roofs, CDX										
0030	5/16" thick		2 Carp	1600	.010	S.F.	.55	.44		.99	1.28
0035	Pneumatic nailed	R061110-30		1952	.008		.55	.36		.91	1.16
0050	3/8" thick			1525	.010		.47	.46		.93	1.23
0055	Pneumatic nailed			1860	.009		.47	.38		.85	1.10
0100	1/2" thick			1400	.011		.47	.50		.97	1.30
0105	Pneumatic nailed			1708	.009		.47	.41		.88	1.16
0200	5/8" thick			1300	.012		.62	.54		1.16	1.52

06 16 Sheathing

06 16 36 – Wood Panel Product Sheathing

06 16 36.10 Sheathing		Crew	Daily Output	Labor-Hours	Unit	Material	2012 Bare Costs Labor	2012 Bare Costs Equipment	Total	Total Incl O&P
0205	Pneumatic nailed	2 Carp	1586	.010	S.F.	.62	.45		1.07	1.36
0300	3/4" thick		1200	.013		.70	.59		1.29	1.67
0305	Pneumatic nailed		1464	.011		.70	.48		1.18	1.51
0500	Plywood on walls with exterior CDX, 3/8" thick		1200	.013		.47	.59		1.06	1.42
0505	Pneumatic nailed		1488	.011		.47	.47		.94	1.25
0600	1/2" thick		1125	.014		.47	.63		1.10	1.48
0605	Pneumatic nailed		1395	.011		.47	.51		.98	1.30
0700	5/8" thick		1050	.015		.62	.67		1.29	1.71
0705	Pneumatic nailed		1302	.012		.62	.54		1.16	1.51
0800	3/4" thick		975	.016		.70	.72		1.42	1.88
0805	Pneumatic nailed		1209	.013		.70	.58		1.28	1.67
0840	Oriented strand board, 7/16" thick [G]		1400	.011		.39	.50		.89	1.21
0845	Pneumatic nailed [G]		1736	.009		.39	.41		.80	1.06
0846	1/2" thick [G]		1325	.012		.39	.53		.92	1.25
0847	Pneumatic nailed [G]		1643	.010		.39	.43		.82	1.09
0852	5/8" thick [G]		1250	.013		.79	.56		1.35	1.74
0857	Pneumatic nailed [G]		1550	.010		.79	.46		1.25	1.57
1000	For shear wall construction, add						20%			
1200	For structural 1 exterior plywood, add				S.F.	10%				
1400	With boards, on roof 1" x 6" boards, laid horizontal	2 Carp	725	.022		1.57	.97		2.54	3.23
1500	Laid diagonal		650	.025		1.57	1.09		2.66	3.40
1700	1" x 8" boards, laid horizontal		875	.018		1.60	.81		2.41	3
1800	Laid diagonal		725	.022		1.60	.97		2.57	3.26
2000	For steep roofs, add						40%			
2200	For dormers, hips and valleys, add					5%	50%			
2400	Boards on walls, 1" x 6" boards, laid regular	2 Carp	650	.025		1.57	1.09		2.66	3.40
2500	Laid diagonal		585	.027		1.57	1.21		2.78	3.59
2700	1" x 8" boards, laid regular		765	.021		1.60	.92		2.52	3.18
2800	Laid diagonal		650	.025		1.60	1.09		2.69	3.43
2850	Gypsum, weatherproof, 1/2" thick		1125	.014		.38	.63		1.01	1.38
2900	With embedded glass mats		1100	.015		.75	.64		1.39	1.82
3000	Wood fiber, regular, no vapor barrier, 1/2" thick		1200	.013		.56	.59		1.15	1.52
3100	5/8" thick		1200	.013		.71	.59		1.30	1.68
3300	No vapor barrier, in colors, 1/2" thick		1200	.013		.68	.59		1.27	1.65
3400	5/8" thick		1200	.013		.72	.59		1.31	1.69
3600	With vapor barrier one side, white, 1/2" thick		1200	.013		.55	.59		1.14	1.51
3700	Vapor barrier 2 sides, 1/2" thick		1200	.013		.77	.59		1.36	1.75
3800	Asphalt impregnated, 25/32" thick		1200	.013		.28	.59		.87	1.21
3850	Intermediate, 1/2" thick		1200	.013		.20	.59		.79	1.12

06 17 Shop-Fabricated Structural Wood

06 17 33 – Wood I-Joists

06 17 33.10 Wood and Composite I-Joists

0010	WOOD AND COMPOSITE I-JOISTS									
0100	Plywood webs, incl. bridging & blocking, panels 24" O.C.									
1200	15' to 24' span, 50 psf live load	F-5	2400	.013	SF Flr.	1.84	.59		2.43	2.94
1300	55 psf live load		2250	.014		2.08	.63		2.71	3.27
1400	24' to 30' span, 45 psf live load		2600	.012		2.87	.55		3.42	3.99
1500	55 psf live load		2400	.013		3.55	.59		4.14	4.81

187

06 17 Shop-Fabricated Structural Wood

06 17 53 – Shop-Fabricated Wood Trusses

06 17 53.10 Roof Trusses		Crew	Daily Output	Labor-Hours	Unit	Material	2012 Bare Costs Labor	Equipment	Total	Total Incl O&P
0010	**ROOF TRUSSES**									
0100	Fink (W) or King post type, 2'-0" O.C.									
0200	Metal plate connected, 4 in 12 slope									
0210	24' to 29' span	F-3	3000	.013	SF Flr.	1.50	.60	.20	2.30	2.79
0300	30' to 43' span		3000	.013		2.18	.60	.20	2.98	3.54
0400	44' to 60' span		3000	.013		2.36	.60	.20	3.16	3.73
0700	Glued and nailed, add					50%				

06 18 Glued-Laminated Construction

06 18 13 – Glued-Laminated Beams

06 18 13.20 Laminated Framing		Crew	Daily Output	Labor-Hours	Unit	Material	2012 Bare Costs Labor	Equipment	Total	Total Incl O&P
0010	**LAMINATED FRAMING**									
0020	30 lb., short term live load, 15 lb. dead load									
0200	Straight roof beams, 20' clear span, beams 8' O.C.	F-3	2560	.016	SF Flr.	1.99	.70	.24	2.93	3.52
0300	Beams 16' O.C.		3200	.013		1.44	.56	.19	2.19	2.65
0500	40' clear span, beams 8' O.C.		3200	.013		3.80	.56	.19	4.55	5.25
0600	Beams 16' O.C.		3840	.010		3.11	.47	.16	3.74	4.32
0800	60' clear span, beams 8' O.C.	F-4	2880	.017		6.55	.74	.38	7.67	8.75
0900	Beams 16' O.C.	"	3840	.013		4.87	.55	.29	5.71	6.50
1100	Tudor arches, 30' to 40' clear span, frames 8' O.C.	F-3	1680	.024		8.50	1.07	.36	9.93	11.40
1200	Frames 16' O.C.	"	2240	.018		6.70	.80	.27	7.77	8.90
1400	50' to 60' clear span, frames 8' O.C.	F-4	2200	.022		9.20	.97	.50	10.67	12.15
1500	Frames 16' O.C.		2640	.018		7.80	.80	.42	9.02	10.30
1700	Radial arches, 60' clear span, frames 8' O.C.		1920	.025		8.60	1.11	.57	10.28	11.75
1800	Frames 16' O.C.		2880	.017		6.60	.74	.38	7.72	8.80
2000	100' clear span, frames 8' O.C.		1600	.030		8.90	1.33	.69	10.92	12.60
2100	Frames 16' O.C.		2400	.020		7.80	.88	.46	9.14	10.45
2300	120' clear span, frames 8' O.C.		1440	.033		11.85	1.47	.76	14.08	16.10
2400	Frames 16' O.C.		1920	.025		10.80	1.11	.57	12.48	14.20
2600	Bowstring trusses, 20' O.C., 40' clear span	F-3	2400	.017		5.35	.75	.26	6.36	7.30
2700	60' clear span	F-4	3600	.013		4.79	.59	.31	5.69	6.50
2800	100' clear span		4000	.012		6.80	.53	.28	7.61	8.55
2900	120' clear span		3600	.013		7.25	.59	.31	8.15	9.25
3000	For less than 1000 B.F., add					20%				
3050	For over 5000 B.F., deduct					10%				
3100	For premium appearance, add to S.F. prices					5%				
3300	For industrial type, deduct					15%				
3500	For stain and varnish, add					5%				
3900	For 3/4" laminations, add to straight					25%				
4100	Add to curved					15%				
4300	Alternate pricing method: (use nominal footage of									
4310	components). Straight beams, camber less than 6"	F-3	3.50	11.429	M.B.F.	2,950	510	175	3,635	4,225
4400	Columns, including hardware		2	20		3,175	895	305	4,375	5,200
4600	Curved members, radius over 32'		2.50	16		3,250	715	245	4,210	4,950
4700	Radius 10' to 32'		3	13.333		3,225	600	204	4,029	4,675
4900	For complicated shapes, add maximum					100%				
5100	For pressure treating, add to straight					35%				
5200	Add to curved					45%				
6000	Laminated veneer members, southern pine or western species									
6050	1-3/4" wide x 5-1/2" deep	2 Carp	480	.033	L.F.	3.19	1.47		4.66	5.75

06 18 Glued-Laminated Construction

06 18 13 – Glued-Laminated Beams

06 18 13.20 Laminated Framing	Crew	Daily Output	Labor-Hours	Unit	Material	2012 Bare Costs Labor	Equipment	Total	Total Incl O&P
6100	9-1/2" deep	2 Carp	480	.033	L.F.	4.56	1.47		6.03	7.25
6150	14" deep		450	.036		7.30	1.57		8.87	10.45
6200	18" deep		450	.036		9.90	1.57		11.47	13.25
6300	Parallel strand members, southern pine or western species									
6350	1-3/4" wide x 9-1/4" deep	2 Carp	480	.033	L.F.	4.56	1.47		6.03	7.25
6400	11-1/4" deep		450	.036		5.70	1.57		7.27	8.65
6450	14" deep		400	.040		7.30	1.76		9.06	10.75
6500	3-1/2" wide x 9-1/4" deep		480	.033		15.85	1.47		17.32	19.70
6550	11-1/4" deep		450	.036		19.15	1.57		20.72	23.50
6600	14" deep		400	.040		23.50	1.76		25.26	28
6650	7" wide x 9-1/4" deep		450	.036		31.50	1.57		33.07	37.50
6700	11-1/4" deep		420	.038		39.50	1.68		41.18	46
6750	14" deep		400	.040		46.50	1.76		48.26	54
8000	Straight beams									
8102	20' span									
8104	3-1/8" x 9"	F-3	30	1.333	Ea.	138	60	20.50	218.50	266
8106	X 10-1/2"		30	1.333		161	60	20.50	241.50	291
8108	X 12"		30	1.333		184	60	20.50	264.50	315
8110	X 13-1/2"		30	1.333		207	60	20.50	287.50	340
8112	X 15"		29	1.379		230	62	21	313	370
8114	5-1/8" x 10-1/2"		30	1.333		265	60	20.50	345.50	405
8116	X 12"		30	1.333		300	60	20.50	380.50	450
8118	X 13-1/2"		30	1.333		340	60	20.50	420.50	490
8120	X 15"		29	1.379		380	62	21	463	535
8122	X 16-1/2"		29	1.379		415	62	21	498	575
8124	X 18"		29	1.379		455	62	21	538	620
8126	X 19-1/2"		29	1.379		490	62	21	573	660
8128	X 21"		28	1.429		530	64	22	616	700
8130	X 22-1/2"		28	1.429		565	64	22	651	745
8132	X 24"		28	1.429		605	64	22	691	785
8134	6-3/4" x 12"		29	1.379		400	62	21	483	560
8136	X 13-1/2"		29	1.379		450	62	21	533	615
8138	X 15"		29	1.379		500	62	21	583	670
8140	X 16-1/2"		28	1.429		550	64	22	636	720
8142	X 18"		28	1.429		595	64	22	681	775
8144	X 19-1/2"		28	1.429		645	64	22	731	830
8146	X 21"		27	1.481		695	66.50	22.50	784	890
8148	X 22-1/2"		27	1.481		745	66.50	22.50	834	945
8150	X 24"		27	1.481		795	66.50	22.50	884	1,000
8152	X 25-1/2"		27	1.481		845	66.50	22.50	934	1,050
8154	X 27"		26	1.538		895	69	23.50	987.50	1,125
8156	X 28-1/2"		26	1.538		945	69	23.50	1,037.50	1,175
8158	X 30"		26	1.538		995	69	23.50	1,087.50	1,225
8200	30' span									
8250	3-1/8" x 9"	F-3	30	1.333	Ea.	207	60	20.50	287.50	340
8252	X 10-1/2"		30	1.333		242	60	20.50	322.50	380
8254	X 12"		30	1.333		277	60	20.50	357.50	420
8256	X 13-1/2"		30	1.333		310	60	20.50	390.50	455
8258	X 15"		29	1.379		345	62	21	428	500
8260	5-1/8" x 10-1/2"		30	1.333		395	60	20.50	475.50	550
8262	X 12"		30	1.333		455	60	20.50	535.50	615
8264	X 13-1/2"		30	1.333		510	60	20.50	590.50	675
8266	X 15"		29	1.379		565	62	21	648	745

06 18 13 – Glued-Laminated Beams

06 18 13.20 Laminated Framing		Crew	Daily Output	Labor-Hours	Unit	Material	2012 Bare Costs Labor	Equipment	Total	Total Incl O&P
8268	X 16-1/2"	F-3	29	1.379	Ea.	625	62	21	708	805
8270	X 18"		29	1.379		680	62	21	763	870
8272	X 19-1/2"		29	1.379		735	62	21	818	930
8274	X 21"		28	1.429		795	64	22	881	995
8276	X 22-1/2"		28	1.429		850	64	22	936	1,050
8278	X 24"		28	1.429		905	64	22	991	1,125
8280	6-3/4" x 12"		29	1.379		595	62	21	678	775
8282	X 13-1/2"		29	1.379		670	62	21	753	860
8284	X 15"		29	1.379		745	62	21	828	940
8286	X 16-1/2"		28	1.429		820	64	22	906	1,025
8288	X 18"		28	1.429		895	64	22	981	1,100
8290	X 19-1/2"		28	1.429		970	64	22	1,056	1,200
8292	X 21"		27	1.481		1,050	66.50	22.50	1,139	1,275
8294	X 22-1/2"		27	1.481		1,125	66.50	22.50	1,214	1,350
8296	X 24"		27	1.481		1,200	66.50	22.50	1,289	1,450
8298	X 25-1/2"		27	1.481		1,275	66.50	22.50	1,364	1,525
8300	X 27"		26	1.538		1,350	69	23.50	1,442.50	1,600
8302	X 28-1/2"		26	1.538		1,425	69	23.50	1,517.50	1,675
8304	X 30"	▼	26	1.538	▼	1,500	69	23.50	1,592.50	1,775
8400	40' span									
8402	3-1/8" x 9"	F-3	30	1.333	Ea.	277	60	20.50	357.50	420
8404	X 10-1/2"		30	1.333		325	60	20.50	405.50	470
8406	X 12"		30	1.333		370	60	20.50	450.50	520
8408	X 13-1/2"		30	1.333		415	60	20.50	495.50	570
8410	X 15"		29	1.379		460	62	21	543	625
8412	5-1/8" x 10-1/2"		30	1.333		530	60	20.50	610.50	695
8414	X 12"		30	1.333		605	60	20.50	685.50	780
8416	X 13-1/2"		30	1.333		680	60	20.50	760.50	865
8418	X 15"		29	1.379		755	62	21	838	950
8420	X 16-1/2"		29	1.379		830	62	21	913	1,025
8422	X 18"		29	1.379		905	62	21	988	1,125
8424	X 19-1/2"		29	1.379		985	62	21	1,068	1,200
8426	X 21"		28	1.429		1,050	64	22	1,136	1,300
8428	X 22-1/2"		28	1.429		1,125	64	22	1,211	1,375
8430	X 24"		28	1.429		1,200	64	22	1,286	1,450
8432	6-3/4" x 12"		29	1.379		795	62	21	878	995
8434	X 13-1/2"		29	1.379		895	62	21	978	1,100
8436	X 15"		29	1.379		995	62	21	1,078	1,225
8438	X 16-1/2"		28	1.429		1,100	64	22	1,186	1,325
8440	X 18"		28	1.429		1,200	64	22	1,286	1,450
8442	X 19-1/2"		28	1.429		1,300	64	22	1,386	1,550
8444	X 21"		27	1.481		1,400	66.50	22.50	1,489	1,650
8446	X 22-1/2"		27	1.481		1,500	66.50	22.50	1,589	1,775
8448	X 24"		27	1.481		1,600	66.50	22.50	1,689	1,875
8450	X 25-1/2"		27	1.481		1,700	66.50	22.50	1,789	1,975
8452	X 27"		26	1.538		1,800	69	23.50	1,892.50	2,100
8454	X 28-1/2"		26	1.538		1,900	69	23.50	1,992.50	2,200
8456	X 30"	▼	26	1.538	▼	2,000	69	23.50	2,092.50	2,325

06 22 13 – Standard Pattern Wood Trim

06 22 13.15 Moldings, Base	Crew	Daily Output	Labor-Hours	Unit	Material	2012 Bare Costs Labor	Equipment	Total	Total Incl O&P
0010 **MOLDINGS, BASE**									
5100 Classic profile, 5/8" x 5-1/2", finger jointed and primed	1 Carp	250	.032	L.F.	1.26	1.41		2.67	3.55
5105 Poplar		240	.033		1.39	1.47		2.86	3.79
5110 Red oak		220	.036		1.99	1.60		3.59	4.66
5115 Maple		220	.036		3.14	1.60		4.74	5.90
5120 Cherry		220	.036		3.73	1.60		5.33	6.55
5125 3/4 x 7-1/2", finger jointed and primed		250	.032		1.63	1.41		3.04	3.96
5130 Poplar		240	.033		2.16	1.47		3.63	4.63
5135 Red oak		220	.036		2.89	1.60		4.49	5.65
5140 Maple		220	.036		4.18	1.60		5.78	7.05
5145 Cherry		220	.036		4.98	1.60		6.58	7.95
5150 Modern profile, 5/8" x 3-1/2", finger jointed and primed		250	.032		.86	1.41		2.27	3.11
5155 Poplar		240	.033		.86	1.47		2.33	3.20
5160 Red oak		220	.036		1.36	1.60		2.96	3.96
5165 Maple		220	.036		2.20	1.60		3.80	4.89
5170 Cherry		220	.036		2.38	1.60		3.98	5.10
5175 Ogee profile, 7/16" x 3", finger jointed and primed		250	.032		.51	1.41		1.92	2.73
5180 Poplar		240	.033		.56	1.47		2.03	2.87
5185 Red oak		220	.036		.76	1.60		2.36	3.30
5200 9/16" x 3-1/2", finger jointed and primed		250	.032		.51	1.41		1.92	2.73
5205 Pine		240	.033		1.23	1.47		2.70	3.61
5210 Red oak		220	.036		2.32	1.60		3.92	5
5215 9/16" x 4-1/2", red oak		220	.036		3.47	1.60		5.07	6.30
5220 5/8" x 3-1/2", finger jointed and primed		250	.032		.79	1.41		2.20	3.04
5225 Poplar		240	.033		.86	1.47		2.33	3.20
5230 Red oak		220	.036		1.36	1.60		2.96	3.96
5235 Maple		220	.036		2.20	1.60		3.80	4.89
5240 Cherry		220	.036		2.38	1.60		3.98	5.10
5245 5/8" x 4", finger jointed and primed		250	.032		1.02	1.41		2.43	3.29
5250 Poplar		240	.033		1.11	1.47		2.58	3.48
5255 Red oak		220	.036		1.52	1.60		3.12	4.14
5260 Maple		220	.036		2.46	1.60		4.06	5.15
5265 Cherry		220	.036		2.58	1.60		4.18	5.30
5270 Rectangular profile, oak, 3/8" x 1-1/4"		260	.031		1.36	1.36		2.72	3.59
5275 1/2" x 2-1/2"		255	.031		1.91	1.38		3.29	4.23
5280 1/2" x 3-1/2"		250	.032		2.31	1.41		3.72	4.71
5285 1" x 6"		240	.033		2.77	1.47		4.24	5.30
5290 1" x 8"		240	.033		3.68	1.47		5.15	6.30
5295 Pine, 3/8" x 1-3/4"		260	.031		.52	1.36		1.88	2.66
5300 7/16" x 2-1/2"		255	.031		.78	1.38		2.16	2.99
5305 1" x 6"		240	.033		1.06	1.47		2.53	3.42
5310 1" x 8"		240	.033		1.44	1.47		2.91	3.84
5315 Shoe, 1/2" x 3/4", primed		260	.031		.43	1.36		1.79	2.56
5320 Pine		240	.033		.30	1.47		1.77	2.59
5325 Poplar		240	.033		.33	1.47		1.80	2.62
5330 Red oak		220	.036		.44	1.60		2.04	2.95
5335 Maple		220	.036		.61	1.60		2.21	3.14
5340 Cherry		220	.036		.69	1.60		2.29	3.23
5345 11/16" x 1-1/2", pine		240	.033		.68	1.47		2.15	3.01
5350 Caps, 11/16" x 1-3/8", pine		240	.033		.74	1.47		2.21	3.07
5355 3/4" x 1-3/4", finger jointed and primed		260	.031		.62	1.36		1.98	2.77
5360 Poplar		240	.033		.87	1.47		2.34	3.22
5365 Red oak		220	.036		1	1.60		2.60	3.57

06 22 13 – Standard Pattern Wood Trim

		Crew	Daily Output	Labor-Hours	Unit	Material	2012 Bare Costs Labor	Equipment	Total	Total Incl O&P
06 22 13.15 Moldings, Base										
5370	Maple	1 Carp	220	.036	L.F.	1.37	1.60		2.97	3.98
5375	Cherry		220	.036		3.09	1.60		4.69	5.85
5380	Combination base & shoe, 9/16" x 3-1/2" & 1/2" x 3/4", pine		125	.064		1.53	2.82		4.35	6
5385	Three piece oak, 6" high		80	.100		4.21	4.41		8.62	11.45
5390	Including 3/4" x 1" base shoe		70	.114		4.57	5.05		9.62	12.75
5395	Flooring cant strip, 3/4" x 3/4", pre-finished pine		260	.031		.47	1.36		1.83	2.61
5400	For pre-finished, stain and clear coat, add					.40			.40	.44
5405	Clear coat only, add					.35			.35	.39
06 22 13.30 Moldings, Casings										
0010	**MOLDINGS, CASINGS**									
0085	Apron, 9/16" x 2-1/2", pine	1 Carp	250	.032	L.F.	1.63	1.41		3.04	3.96
0090	5/8" x 2-1/2", pine		250	.032		2.17	1.41		3.58	4.56
0110	5/8" x 3-1/2", pine		220	.036		2.17	1.60		3.77	4.86
0300	Band, 11/16" x 1-1/8", pine		270	.030		1	1.31		2.31	3.11
0310	11/16" x 1-1/2", finger jointed and primed		270	.030		.83	1.31		2.14	2.92
0320	Pine		270	.030		1.06	1.31		2.37	3.17
0330	11/16" x 1-3/4", finger jointed and primed		270	.030		.56	1.31		1.87	2.62
0350	Pine		270	.030		1.25	1.31		2.56	3.38
0355	Beaded, 3/4" x 3-1/2", finger jointed and primed		220	.036		1.02	1.60		2.62	3.59
0360	Poplar		220	.036		.96	1.60		2.56	3.52
0365	Red oak		220	.036		1.36	1.60		2.96	3.96
0370	Maple		220	.036		2.20	1.60		3.80	4.89
0375	Cherry		220	.036		2.67	1.60		4.27	5.40
0380	3/4" x 4", finger jointed and primed		220	.036		1.02	1.60		2.62	3.59
0385	Poplar		220	.036		1.24	1.60		2.84	3.83
0390	Red oak		220	.036		1.67	1.60		3.27	4.31
0395	Maple		220	.036		2.30	1.60		3.90	5
0400	Cherry		220	.036		2.90	1.60		4.50	5.65
0405	3/4" x 5-1/2", finger jointed and primed		200	.040		1.02	1.76		2.78	3.83
0410	Poplar		200	.040		1.67	1.76		3.43	4.55
0415	Red oak		200	.040		2.34	1.76		4.10	5.30
0420	Maple		200	.040		3.23	1.76		4.99	6.25
0425	Cherry		200	.040		3.71	1.76		5.47	6.80
0430	Classic profile, 3/4" x 2-3/4", finger jointed and primed		250	.032		.68	1.41		2.09	2.92
0435	Poplar		250	.032		.86	1.41		2.27	3.11
0440	Red oak		250	.032		1.20	1.41		2.61	3.49
0445	Maple		250	.032		1.80	1.41		3.21	4.15
0450	Cherry		250	.032		2.05	1.41		3.46	4.42
0455	Fluted, 3/4" x 3-1/2", poplar		220	.036		.96	1.60		2.56	3.52
0460	Red oak		220	.036		1.36	1.60		2.96	3.96
0465	Maple		220	.036		2.20	1.60		3.80	4.89
0470	Cherry		220	.036		2.67	1.60		4.27	5.40
0475	3/4" x 4", poplar		220	.036		1.24	1.60		2.84	3.83
0480	Red oak		220	.036		3.02	1.60		4.62	5.80
0485	Maple		220	.036		2.30	1.60		3.90	5
0490	Cherry		220	.036		2.90	1.60		4.50	5.65
0495	3/4" x 5-1/2", poplar		200	.040		1.67	1.76		3.43	4.55
0500	Red oak		200	.040		2.34	1.76		4.10	5.30
0505	Maple		200	.040		4.58	1.76		6.34	7.75
0510	Cherry		200	.040		3.71	1.76		5.47	6.80
0515	3/4" x 7-1/2", poplar		190	.042		1.38	1.86		3.24	4.37
0520	Red oak		190	.042		3.24	1.86		5.10	6.40

06 22 13 – Standard Pattern Wood Trim

06 22 13.30 Moldings, Casings	Crew	Daily Output	Labor-Hours	Unit	Material	2012 Bare Costs Labor	Equipment	Total	Total Incl O&P
0525	Maple	1 Carp	190	.042	L.F.	4.89	1.86		6.75	8.25
0530	Cherry		190	.042		5.85	1.86		7.71	9.25
0535	3/4" x 9-1/2", poplar		180	.044		3.56	1.96		5.52	6.95
0540	Red oak		180	.044		5.40	1.96		7.36	8.95
0545	Maple		180	.044		7.75	1.96		9.71	11.55
0550	Cherry		180	.044		8.45	1.96		10.41	12.30
0555	Modern profile, 9/16" x 2-1/4", poplar		250	.032		.49	1.41		1.90	2.71
0560	Red oak		250	.032		.64	1.41		2.05	2.87
0565	11/16" x 2-1/2", finger jointed & primed		250	.032		.58	1.41		1.99	2.81
0570	Pine		250	.032		1.33	1.41		2.74	3.63
0575	3/4" x 2-1/2", poplar		250	.032		.76	1.41		2.17	3
0580	Red oak		250	.032		.98	1.41		2.39	3.25
0585	Maple		250	.032		1.55	1.41		2.96	3.87
0590	Cherry		250	.032		1.96	1.41		3.37	4.32
0595	Mullion, 5/16" x 2", pine		270	.030		.99	1.31		2.30	3.10
0600	9/16" x 2-1/2", fingerjointed and primed		250	.032		.60	1.41		2.01	2.83
0605	Pine		250	.032		1.41	1.41		2.82	3.72
0610	Red oak		250	.032		2.80	1.41		4.21	5.25
0615	1-1/16" x 3-3/4", red oak		220	.036		6.85	1.60		8.45	9.95
0620	Ogee, 7/16" x 2-1/2", poplar		250	.032		.44	1.41		1.85	2.65
0625	Red oak		250	.032		.62	1.41		2.03	2.85
0630	9/16" x 2-1/4", finger jointed and primed		250	.032		.44	1.41		1.85	2.65
0635	Poplar		250	.032		.49	1.41		1.90	2.71
0640	Red oak		250	.032		.64	1.41		2.05	2.87
0645	11/16" x 2-1/2", finger jointed and primed		250	.032		.54	1.41		1.95	2.76
0700	Pine		250	.032		1.32	1.41		2.73	3.62
0730	11/16" x 3-1/2", finger jointed and primed		220	.036		1.82	1.60		3.42	4.47
0750	Pine		220	.036		2.18	1.60		3.78	4.87
0755	3/4" x 2-1/2", finger jointed and primed		250	.032		.62	1.41		2.03	2.85
0760	Poplar		250	.032		.76	1.41		2.17	3
0765	Red oak		250	.032		.98	1.41		2.39	3.25
0770	Maple		250	.032		1.55	1.41		2.96	3.87
0775	Cherry		250	.032		1.96	1.41		3.37	4.32
0780	3/4" x 3-1/2", finger jointed and primed		220	.036		.87	1.60		2.47	3.43
0785	Poplar		220	.036		.96	1.60		2.56	3.52
0790	Red oak		220	.036		1.36	1.60		2.96	3.96
0795	Maple		220	.036		2.20	1.60		3.80	4.89
0800	Cherry		220	.036		2.67	1.60		4.27	5.40
4700	Square profile, 1" x 1", teak		215	.037		2.01	1.64		3.65	4.73
4800	Rectangular profile, 1" x 3", teak		200	.040		5.70	1.76		7.46	9

06 22 13.35 Moldings, Ceilings

		Crew	Daily Output	Labor-Hours	Unit	Material	Labor	Equipment	Total	Total Incl O&P
0010	**MOLDINGS, CEILINGS**									
0600	Bed, stock pine, 9/16" x 1-3/4"	1 Carp	270	.030	L.F.	1.19	1.31		2.50	3.32
0650	9/16" x 2"		240	.033		1.42	1.47		2.89	3.82
1200	Cornice molding, stock pine, 9/16" x 1-3/4"		330	.024		1.19	1.07		2.26	2.95
1300	9/16" x 2-1/4"		300	.027		1.45	1.18		2.63	3.41
2400	Cove scotia, stock pine, 9/16" x 1-3/4"		270	.030		1.11	1.31		2.42	3.23
2500	11/16" x 2-3/4"		255	.031		2.18	1.38		3.56	4.53
2600	Crown, stock pine, 9/16" x 3-5/8"		250	.032		2.68	1.41		4.09	5.10
2700	11/16" x 4-5/8"		220	.036		4.52	1.60		6.12	7.45

06 22 13 – Standard Pattern Wood Trim

06 22 13.40 Moldings, Exterior

06 22 13.40 Moldings, Exterior		Crew	Daily Output	Labor-Hours	Unit	Material	2012 Bare Costs Labor	Equipment	Total	Total Incl O&P
0010	**MOLDINGS, EXTERIOR**									
1500	Cornice, boards, pine, 1" x 2"	1 Carp	330	.024	L.F.	.25	1.07		1.32	1.92
1700	1" x 6"		250	.032		.72	1.41		2.13	2.96
2000	1" x 12"		180	.044		1.63	1.96		3.59	4.81
2200	Three piece, built-up, pine, minimum		80	.100		1.38	4.41		5.79	8.30
2300	Maximum		65	.123		3.93	5.45		9.38	12.65
3000	Corner board, sterling pine, 1" x 4"		200	.040		.80	1.76		2.56	3.59
3100	1" x 6"		200	.040		1.06	1.76		2.82	3.88
3350	Fascia, sterling pine, 1" x 6"		250	.032		1.06	1.41		2.47	3.34
3370	1" x 8"		225	.036		1.44	1.57		3.01	3.99
3400	Trim, back band, 11/16" x 1-1/16"		250	.032		.98	1.41		2.39	3.25
3500	Casing, 11/16" x 4-1/4"		250	.032		3.20	1.41		4.61	5.70
3600	Crown, 11/16" x 4-1/4"		250	.032		4.50	1.41		5.91	7.10
3700	1" x 4" and 1" x 6" railing w/balusters 4" O.C.		22	.364		25.50	16.05		41.55	52.50
3800	Insect screen framing stock, 1-1/16" x 1-3/4"		395	.020		2.38	.89		3.27	3.99
4100	Verge board, sterling pine, 1" x 4"		200	.040		.80	1.76		2.56	3.59
4200	1" x 6"		200	.040		1.06	1.76		2.82	3.88
4300	2" x 6"		165	.048		2.12	2.14		4.26	5.60
4400	2" x 8"		165	.048		2.88	2.14		5.02	6.45
4700	For redwood trim, add					200%				

06 22 13.45 Moldings, Trim

06 22 13.45 Moldings, Trim		Crew	Daily Output	Labor-Hours	Unit	Material	Labor	Equipment	Total	Total Incl O&P
0010	**MOLDINGS, TRIM**									
0200	Astragal, stock pine, 11/16" x 1-3/4"	1 Carp	255	.031	L.F.	1.41	1.38		2.79	3.68
0250	1-5/16" x 2-3/16"		240	.033		2.57	1.47		4.04	5.10
0800	Chair rail, stock pine, 5/8" x 2-1/2"		270	.030		1.68	1.31		2.99	3.86
0900	5/8" x 3-1/2"		240	.033		2.65	1.47		4.12	5.20
1000	Closet pole, stock pine, 1-1/8" diameter		200	.040		1.22	1.76		2.98	4.05
1100	Fir, 1-5/8" diameter		200	.040		2.36	1.76		4.12	5.30
3300	Half round, stock pine, 1/4" x 1/2"		270	.030		.31	1.31		1.62	2.35
3350	1/2" x 1"		255	.031		.73	1.38		2.11	2.93
3400	Handrail, fir, single piece, stock, hardware not included									
3450	1-1/2" x 1-3/4"	1 Carp	80	.100	L.F.	2.50	4.41		6.91	9.55
3470	Pine, 1-1/2" x 1-3/4"		80	.100		2.38	4.41		6.79	9.40
3500	1-1/2" x 2-1/2"		76	.105		2.67	4.64		7.31	10.10
3600	Lattice, stock pine, 1/4" x 1-1/8"		270	.030		.51	1.31		1.82	2.57
3700	1/4" x 1-3/4"		250	.032		.92	1.41		2.33	3.18
3800	Miscellaneous, custom, pine, 1" x 1"		270	.030		.38	1.31		1.69	2.42
3900	1" x 3"		240	.033		1.13	1.47		2.60	3.50
4100	Birch or oak, nominal 1" x 1"		240	.033		.47	1.47		1.94	2.77
4200	Nominal 1" x 3"		215	.037		1.40	1.64		3.04	4.06
4400	Walnut, nominal 1" x 1"		215	.037		.56	1.64		2.20	3.14
4500	Nominal 1" x 3"		200	.040		1.69	1.76		3.45	4.57
4700	Teak, nominal 1" x 1"		215	.037		2.47	1.64		4.11	5.25
4800	Nominal 1" x 3"		200	.040		7.40	1.76		9.16	10.85
4900	Quarter round, stock pine, 1/4" x 1/4"		275	.029		.31	1.28		1.59	2.31
4950	3/4" x 3/4"		255	.031		.70	1.38		2.08	2.90
5600	Wainscot moldings, 1-1/8" x 9/16", 2' high, minimum		76	.105	S.F.	11.55	4.64		16.19	19.85
5700	Maximum		65	.123	"	20.50	5.45		25.95	31

06 22 13.50 Moldings, Window and Door

06 22 13.50 Moldings, Window and Door		Crew	Daily Output	Labor-Hours	Unit	Material	Labor	Equipment	Total	Total Incl O&P
0010	**MOLDINGS, WINDOW AND DOOR**									
2800	Door moldings, stock, decorative, 1-1/8" wide, plain	1 Carp	17	.471	Set	44.50	21		65.50	81
2900	Detailed		17	.471	"	87.50	21		108.50	129

06 22 Millwork

06 22 13 – Standard Pattern Wood Trim

06 22 13.50 Moldings, Window and Door

		Crew	Daily Output	Labor-Hours	Unit	Material	2012 Bare Costs Labor	Equipment	Total	Total Incl O&P
2960	Clear pine door jamb, no stops, 11/16" x 4-9/16"	1 Carp	240	.033	L.F.	4.73	1.47		6.20	7.45
3150	Door trim set, 1 head and 2 sides, pine, 2-1/2 wide		12	.667	Opng.	22.50	29.50		52	69.50
3170	3-1/2" wide		11	.727	"	37	32		69	90
3250	Glass beads, stock pine, 3/8" x 1/2"		275	.029	L.F.	.38	1.28		1.66	2.39
3270	3/8" x 7/8"		270	.030		.50	1.31		1.81	2.56
4850	Parting bead, stock pine, 3/8" x 3/4"		275	.029		.45	1.28		1.73	2.47
4870	1/2" x 3/4"		255	.031		.56	1.38		1.94	2.75
5000	Stool caps, stock pine, 11/16" x 3-1/2"		200	.040		2.65	1.76		4.41	5.65
5100	1-1/16" x 3-1/4"		150	.053		4.15	2.35		6.50	8.20
5300	Threshold, oak, 3' long, inside, 5/8" x 3-5/8"		32	.250	Ea.	9	11.05		20.05	27
5400	Outside, 1-1/2" x 7-5/8"		16	.500	"	38	22		60	76
5900	Window trim sets, including casings, header, stops,									
5910	stool and apron, 2-1/2" wide, minimum	1 Carp	13	.615	Opng.	34.50	27		61.50	79.50
5950	Average		10	.800		40.50	35.50		76	99
6000	Maximum		6	1.333		64.50	59		123.50	162

06 22 13.60 Moldings, Soffits

		Crew	Daily Output	Labor-Hours	Unit	Material	2012 Bare Costs Labor	Equipment	Total	Total Incl O&P
0010	**MOLDINGS, SOFFITS**									
0200	Soffits, pine, 1" x 4"	2 Carp	420	.038	L.F.	.41	1.68		2.09	3.04
0210	1" x 6"		420	.038		.72	1.68		2.40	3.38
0220	1" x 8"		420	.038		1	1.68		2.68	3.69
0230	1" x 10"		400	.040		1.30	1.76		3.06	4.14
0240	1" x 12"		400	.040		1.63	1.76		3.39	4.50
0250	STK cedar, 1" x 4"		420	.038		.48	1.68		2.16	3.12
0260	1" x 6"		420	.038		.83	1.68		2.51	3.50
0270	1" x 8"		420	.038		1.29	1.68		2.97	4.01
0280	1" x 10"		400	.040		1.99	1.76		3.75	4.90
0290	1" x 12"		400	.040		2.69	1.76		4.45	5.65
1000	Exterior AC plywood, 1/4" thick		420	.038	S.F.	.62	1.68		2.30	3.27
1050	3/8" thick		420	.038		.73	1.68		2.41	3.39
1100	1/2" thick		420	.038		.77	1.68		2.45	3.44
1150	Polyvinyl chloride, white, solid	1 Carp	230	.035		2.04	1.53		3.57	4.60
1160	Perforated	"	230	.035		2.04	1.53		3.57	4.60

06 25 Prefinished Paneling

06 25 13 – Prefinished Hardboard Paneling

06 25 13.10 Paneling, Hardboard

			Crew	Daily Output	Labor-Hours	Unit	Material	2012 Bare Costs Labor	Equipment	Total	Total Incl O&P
0010	**PANELING, HARDBOARD**										
0050	Not incl. furring or trim, hardboard, tempered, 1/8" thick	G	2 Carp	500	.032	S.F.	.38	1.41		1.79	2.59
0100	1/4" thick	G		500	.032		.55	1.41		1.96	2.78
0300	Tempered pegboard, 1/8" thick	G		500	.032		.45	1.41		1.86	2.67
0400	1/4" thick	G		500	.032		.64	1.41		2.05	2.87
0600	Untempered hardboard, natural finish, 1/8" thick	G		500	.032		.38	1.41		1.79	2.59
0700	1/4" thick	G		500	.032		.46	1.41		1.87	2.68
0900	Untempered pegboard, 1/8" thick	G		500	.032		.39	1.41		1.80	2.60
1000	1/4" thick	G		500	.032		.42	1.41		1.83	2.63
1200	Plastic faced hardboard, 1/8" thick	G		500	.032		.68	1.41		2.09	2.92
1300	1/4" thick	G		500	.032		.84	1.41		2.25	3.09
1500	Plastic faced pegboard, 1/8" thick	G		500	.032		.62	1.41		2.03	2.85
1600	1/4" thick	G		500	.032		.74	1.41		2.15	2.98
1800	Wood grained, plain or grooved, 1/4" thick, minimum	G		500	.032		.58	1.41		1.99	2.81
1900	Maximum	G		425	.038		1.12	1.66		2.78	3.78

06 25 Prefinished Paneling

06 25 13 – Prefinished Hardboard Paneling

06 25 13.10 Paneling, Hardboard	Crew	Daily Output	Labor-Hours	Unit	Material	2012 Bare Costs Labor	Equipment	Total	Total Incl O&P
2100 Moldings for hardboard, wood or aluminum, minimum	2 Carp	500	.032	L.F.	.39	1.41		1.80	2.60
2200 Maximum		425	.038	"	1.09	1.66		2.75	3.75

06 25 16 – Prefinished Plywood Paneling

06 25 16.10 Paneling, Plywood

		Crew	Daily Output	Labor-Hours	Unit	Material	2012 Bare Costs Labor	Equipment	Total	Total Incl O&P
0010	PANELING, PLYWOOD R061636-20									
2400	Plywood, prefinished, 1/4" thick, 4' x 8' sheets									
2410	with vertical grooves. Birch faced, minimum	2 Carp	500	.032	S.F.	.90	1.41		2.31	3.16
2420	Average		420	.038		1.36	1.68		3.04	4.09
2430	Maximum		350	.046		1.99	2.02		4.01	5.30
2600	Mahogany, African		400	.040		2.54	1.76		4.30	5.50
2700	Philippine (Lauan)		500	.032		1.09	1.41		2.50	3.37
2900	Oak or Cherry, minimum		500	.032		2.12	1.41		3.53	4.50
3000	Maximum		400	.040		3.25	1.76		5.01	6.30
3200	Rosewood		320	.050		4.63	2.21		6.84	8.50
3400	Teak		400	.040		3.25	1.76		5.01	6.30
3600	Chestnut		375	.043		4.82	1.88		6.70	8.20
3800	Pecan		400	.040		2.08	1.76		3.84	5
3900	Walnut, minimum		500	.032		2.78	1.41		4.19	5.25
3950	Maximum		400	.040		5.25	1.76		7.01	8.50
4000	Plywood, prefinished, 3/4" thick, stock grades, minimum		320	.050		1.25	2.21		3.46	4.77
4100	Maximum		224	.071		5.45	3.15		8.60	10.80
4300	Architectural grade, minimum		224	.071		4.02	3.15		7.17	9.25
4400	Maximum		160	.100		6.15	4.41		10.56	13.55
4600	Plywood, "A" face, birch, V.C., 1/2" thick, natural		450	.036		1.91	1.57		3.48	4.51
4700	Select		450	.036		2.08	1.57		3.65	4.70
4900	Veneer core, 3/4" thick, natural		320	.050		2.01	2.21		4.22	5.60
5000	Select		320	.050		2.25	2.21		4.46	5.85
5200	Lumber core, 3/4" thick, natural		320	.050		3.02	2.21		5.23	6.70
5500	Plywood, knotty pine, 1/4" thick, A2 grade		450	.036		1.65	1.57		3.22	4.23
5600	A3 grade		450	.036		2.08	1.57		3.65	4.70
5800	3/4" thick, veneer core, A2 grade		320	.050		2.13	2.21		4.34	5.75
5900	A3 grade		320	.050		2.40	2.21		4.61	6.05
6100	Aromatic cedar, 1/4" thick, plywood		400	.040		2.10	1.76		3.86	5
6200	1/4" thick, particle board		400	.040		1.02	1.76		2.78	3.83

06 25 26 – Panel System

06 25 26.10 Panel Systems

		Crew	Daily Output	Labor-Hours	Unit	Material	2012 Bare Costs Labor	Equipment	Total	Total Incl O&P
0010	PANEL SYSTEMS									
0100	Raised panel, eng. wood core w/wood veneer, std., paint grade	2 Carp	300	.053	S.F.	12.20	2.35		14.55	17
0110	Oak veneer		300	.053		20	2.35		22.35	25.50
0120	Maple veneer		300	.053		24.50	2.35		26.85	30
0130	Cherry veneer		300	.053		29	2.35		31.35	35.50
0300	Class I fire rated, paint grade		300	.053		14.25	2.35		16.60	19.25
0310	Oak veneer		300	.053		23.50	2.35		25.85	29.50
0320	Maple veneer		300	.053		30	2.35		32.35	36.50
0330	Cherry veneer		300	.053		38.50	2.35		40.85	45.50
0510	Beadboard, 5/8" MDF, standard, primed		300	.053		6.60	2.35		8.95	10.85
0520	Oak veneer, unfinished		300	.053		10.05	2.35		12.40	14.65
0530	Maple veneer, unfinished		300	.053		11.20	2.35		13.55	15.90
0610	Rustic paneling, 5/8" MDF, standard, maple veneer, unfinished		300	.053		12.70	2.35		15.05	17.55

06 26 Board Paneling

06 26 13 – Profile Board Paneling

06 26 13.10 Paneling, Boards	Crew	Daily Output	Labor-Hours	Unit	Material	2012 Bare Costs Labor	Equipment	Total	Total Incl O&P
0010 **PANELING, BOARDS**									
6400 Wood board paneling, 3/4" thick, knotty pine	2 Carp	300	.053	S.F.	1.89	2.35		4.24	5.70
6500 Rough sawn cedar		300	.053		2.86	2.35		5.21	6.75
6700 Redwood, clear, 1" x 4" boards		300	.053		4.96	2.35		7.31	9.05
6900 Aromatic cedar, closet lining, boards	↓	275	.058	↓	2	2.57		4.57	6.15

06 43 Wood Stairs and Railings

06 43 13 – Wood Stairs

06 43 13.20 Prefabricated Wood Stairs

	Crew	Daily Output	Labor-Hours	Unit	Material	2012 Bare Costs Labor	Equipment	Total	Total Incl O&P
0010 **PREFABRICATED WOOD STAIRS**									
0100 Box stairs, prefabricated, 3'-0" wide									
0110 Oak treads, up to 14 risers	2 Carp	39	.410	Riser	92.50	18.10		110.60	130
0600 With pine treads for carpet, up to 14 risers	"	39	.410	"	59.50	18.10		77.60	93.50
1100 For 4' wide stairs, add				Flight	25%				
1550 Stairs, prefabricated stair handrail with balusters	1 Carp	30	.267	L.F.	75	11.75		86.75	100
1700 Basement stairs, prefabricated, pine treads									
1710 Pine risers, 3' wide, up to 14 risers	2 Carp	52	.308	Riser	59.50	13.55		73.05	86.50
4000 Residential, wood, oak treads, prefabricated	↓	1.50	10.667	Flight	1,200	470		1,670	2,050
4200 Built in place	↓	.44	36.364	"	1,800	1,600		3,400	4,450
4400 Spiral, oak, 4'-6" diameter, unfinished, prefabricated,									
4500 incl. railing, 9' high	2 Carp	1.50	10.667	Flight	3,625	470		4,095	4,700

06 43 13.40 Wood Stair Parts

	Crew	Daily Output	Labor-Hours	Unit	Material	2012 Bare Costs Labor	Equipment	Total	Total Incl O&P
0010 **WOOD STAIR PARTS**									
0020 Pin top balusters, 1-1/4", oak, 34"	1 Carp	96	.083	Ea.	7.80	3.67		11.47	14.25
0030 38"		96	.083		9.85	3.67		13.52	16.50
0040 42"		96	.083		12.10	3.67		15.77	18.95
0050 Poplar, 34"		96	.083		3.05	3.67		6.72	9
0060 38"		96	.083		3.62	3.67		7.29	9.65
0070 42"		96	.083		8.55	3.67		12.22	15.10
0080 Maple, 34"		96	.083		4.90	3.67		8.57	11.05
0090 38"		96	.083		6.20	3.67		9.87	12.45
0100 42"		96	.083		7.60	3.67		11.27	14
0130 Primed, 34"		96	.083		2.85	3.67		6.52	8.80
0140 38"		96	.083		3.62	3.67		7.29	9.65
0150 42"		96	.083		4.42	3.67		8.09	10.50
0180 Box top balusters, 1-1/4", oak, 34"		60	.133		10.85	5.90		16.75	21
0190 38"		60	.133		12.50	5.90		18.40	23
0200 42"		60	.133		12.95	5.90		18.85	23.50
0210 Poplar, 34"		60	.133		8.65	5.90		14.55	18.55
0220 38"		60	.133		9.50	5.90		15.40	19.50
0230 42"		60	.133		10	5.90		15.90	20
0240 Maple, 34"		60	.133		11.90	5.90		17.80	22
0250 38"		60	.133		13.80	5.90		19.70	24.50
0260 42"		60	.133		14.20	5.90		20.10	24.50
0290 Primed, 34"		60	.133		6.95	5.90		12.85	16.70
0300 38"		60	.133		8	5.90		13.90	17.85
0310 42"		60	.133	↓	8.35	5.90		14.25	18.25
0340 Square balusters, cut from lineal stock, pine, 1-1/16" x 1-1/16"		180	.044	L.F.	1.32	1.96		3.28	4.47
0350 1-5/16" x 1-5/16"		180	.044		1.70	1.96		3.66	4.89
0360 1-5/8" x 1-5/8"		180	.044	↓	2.82	1.96		4.78	6.10
0370 Turned newel, oak, 3-1/2" square, 48" high	↓	8	1	Ea.	94.50	44		138.50	172

06 43 13 – Wood Stairs

06 43 13.40 Wood Stair Parts	Crew	Daily Output	Labor-Hours	Unit	Material	2012 Bare Costs Labor	2012 Bare Costs Equipment	Total	Total Incl O&P
0380	62" high	1 Carp	8	1	Ea.	138	44		182	220
0390	Poplar, 3-1/2" square, 48" high		8	1		52	44		96	125
0400	62" high		8	1		75	44		119	151
0410	Maple, 3-1/2" square, 48" high		8	1		66	44		110	141
0420	62" high		8	1		93	44		137	170
0430	Square newel, oak, 3-1/2" square, 48" high		8	1		50	44		94	123
0440	58" high		8	1		70	44		114	145
0450	Poplar, 3-1/2" square, 48" high		8	1		49	44		93	122
0460	58" high		8	1		54	44		98	128
0470	Maple, 3" square, 48" high		8	1		50	44		94	123
0480	58" high		8	1		62	44		106	136
0490	Railings, oak, minimum		96	.083	L.F.	10.40	3.67		14.07	17.10
0500	Average		96	.083		12.50	3.67		16.17	19.40
0510	Maximum		96	.083		15.50	3.67		19.17	22.50
0520	Maple, minimum		96	.083		15	3.67		18.67	22
0530	Average		96	.083		15.25	3.67		18.92	22.50
0540	Maximum		96	.083		15.50	3.67		19.17	22.50
0550	Oak, for bending rail, minimum		48	.167		21	7.35		28.35	34.50
0560	Average		48	.167		22	7.35		29.35	35.50
0570	Maximum		48	.167		23	7.35		30.35	37
0580	Maple, for bending rail, minimum		48	.167		23	7.35		30.35	36.50
0590	Average		48	.167		25	7.35		32.35	38.50
0600	Maximum		48	.167		27	7.35		34.35	41
0610	Risers, oak, 3/4" x 8", 36" long		80	.100	Ea.	18	4.41		22.41	26.50
0620	42" long		80	.100		21	4.41		25.41	30
0630	48" long		80	.100		24	4.41		28.41	33.50
0640	54" long		80	.100		27	4.41		31.41	36.50
0650	60" long		80	.100		30	4.41		34.41	40
0660	72" long		80	.100		36	4.41		40.41	46.50
0670	Poplar, 3/4" x 8", 36" long		80	.100		11.10	4.41		15.51	19
0680	42" long		80	.100		12.95	4.41		17.36	21
0690	48" long		80	.100		14.80	4.41		19.21	23
0700	54" long		80	.100		16.65	4.41		21.06	25
0710	60" long		80	.100		18.50	4.41		22.91	27.50
0720	72" long		80	.100		22	4.41		26.41	31.50
0730	Pine, 1" x 8", 36" long		80	.100		3	4.41		7.41	10.10
0740	42" long		80	.100		3.50	4.41		7.91	10.65
0750	48" long		80	.100		4	4.41		8.41	11.20
0760	54" long		80	.100		4.50	4.41		8.91	11.75
0770	60" long		80	.100		5	4.41		9.41	12.30
0780	72" long		80	.100		6	4.41		10.41	13.40
0790	Treads, oak, no returns, 1-1/32" x 11-1/2" x 36" long		32	.250		32	11.05		43.05	52
0800	42" long		32	.250		37.50	11.05		48.55	58
0810	48" long		32	.250		42.50	11.05		53.55	64
0820	54" long		32	.250		48	11.05		59.05	70
0830	60" long		32	.250		53.50	11.05		64.55	75.50
0840	72" long		32	.250		64	11.05		75.05	87.50
0850	Mitred return one end, 1-1/32" x 11-1/2" x 36" long		24	.333		46	14.70		60.70	73
0860	42" long		24	.333		53.50	14.70		68.20	81.50
0870	48" long		24	.333		61.50	14.70		76.20	90
0880	54" long		24	.333		69	14.70		83.70	98.50
0890	60" long		24	.333		76.50	14.70		91.20	107
0900	72" long		24	.333		92	14.70		106.70	124

06 43 13 – Wood Stairs

06 43 13.40 Wood Stair Parts

		Crew	Daily Output	Labor-Hours	Unit	Material	2012 Bare Costs Labor	Equipment	Total	Total Incl O&P
0910	Mitred return two ends, 1-1/32" x 11-1/2" x 36" long	1 Carp	12	.667	Ea.	49	29.50		78.50	99
0920	42" long		12	.667		57	29.50		86.50	108
0930	48" long		12	.667		65.50	29.50		95	117
0940	54" long		12	.667		73.50	29.50		103	126
0950	60" long		12	.667		81.50	29.50		111	135
0960	72" long		12	.667		98	29.50		127.50	153
0970	Starting step, oak, 48", bullnose		8	1		192	44		236	279
0980	Double end bullnose		8	1		345	44		389	450
1030	Skirt board, pine, 1" x 10"		55	.145	L.F.	1.30	6.40		7.70	11.30
1040	1" x 12"		52	.154	"	1.63	6.80		8.43	12.25
1050	Oak landing tread, 1-1/16" thick		54	.148	S.F.	8.25	6.55		14.80	19.15
1060	Oak cove molding		96	.083	L.F.	1.45	3.67		5.12	7.25
1070	Oak stringer molding		96	.083	"	2.75	3.67		6.42	8.70
1090	Rail bolt, 5/16" x 3-1/2"		48	.167	Ea.	2.75	7.35		10.10	14.35
1100	5/16" x 4-1/2"		48	.167		2.75	7.35		10.10	14.35
1120	Newel post anchor		16	.500		13.50	22		35.50	49
1130	Tapered plug, 1/2"		240	.033		.30	1.47		1.77	2.59
1140	1"		240	.033		.40	1.47		1.87	2.70

06 43 16 – Wood Railings

06 43 16.10 Wood Handrails and Railings

		Crew	Daily Output	Labor-Hours	Unit	Material	2012 Bare Costs Labor	Equipment	Total	Total Incl O&P
0010	**WOOD HANDRAILS AND RAILINGS**									
0020	Custom design, architectural grade, hardwood, minimum	1 Carp	38	.211	L.F.	9.50	9.30		18.80	25
0100	Maximum		30	.267		50	11.75		61.75	73
0300	Stock interior railing with spindles 4" O.C., 4' long		40	.200		32.50	8.80		41.30	49
0400	8' long		48	.167		32.50	7.35		39.85	47

06 44 19 – Wood Grilles

06 44 19.10 Grilles

		Crew	Daily Output	Labor-Hours	Unit	Material	2012 Bare Costs Labor	Equipment	Total	Total Incl O&P
0010	**GRILLES** and panels, hardwood, sanded									
0020	2' x 4' to 4' x 8', custom designs, unfinished, minimum	1 Carp	38	.211	S.F.	56	9.30		65.30	76.50
0050	Average		30	.267		65	11.75		76.75	89.50
0100	Maximum		19	.421		74	18.55		92.55	110

06 44 33 – Wood Mantels

06 44 33.10 Fireplace Mantels

		Crew	Daily Output	Labor-Hours	Unit	Material	2012 Bare Costs Labor	Equipment	Total	Total Incl O&P
0010	**FIREPLACE MANTELS**									
0015	6" molding, 6' x 3'-6" opening, minimum	1 Carp	5	1.600	Opng.	228	70.50		298.50	360
0100	Maximum		5	1.600		410	70.50		480.50	565
0300	Prefabricated pine, colonial type, stock, deluxe		2	4		1,250	176		1,426	1,675
0400	Economy		3	2.667		530	118		648	765

06 44 33.20 Fireplace Mantel Beam

		Crew	Daily Output	Labor-Hours	Unit	Material	2012 Bare Costs Labor	Equipment	Total	Total Incl O&P
0010	**FIREPLACE MANTEL BEAM**									
0020	Rough texture wood, 4" x 8"	1 Carp	36	.222	L.F.	8.30	9.80		18.10	24
0100	4" x 10"		35	.229	"	10.90	10.10		21	27.50
0300	Laminated hardwood, 2-1/4" x 10-1/2" wide, 6' long		5	1.600	Ea.	110	70.50		180.50	230
0400	8' long		5	1.600	"	150	70.50		220.50	274
0600	Brackets for above, rough sawn		12	.667	Pr.	10	29.50		39.50	56
0700	Laminated		12	.667	"	15	29.50		44.50	61.50

06 44 Ornamental Woodwork

06 44 39 – Wood Posts and Columns

06 44 39.10 Decorative Beams

		Crew	Daily Output	Labor-Hours	Unit	Material	2012 Bare Costs Labor	Equipment	Total	Total Incl O&P
0010	**DECORATIVE BEAMS**									
0020	Rough sawn cedar, non-load bearing, 4" x 4"	2 Carp	180	.089	L.F.	1.45	3.92		5.37	7.65
0100	4" x 6"		170	.094		2.18	4.15		6.33	8.80
0200	4" x 8"		160	.100		2.90	4.41		7.31	10
0300	4" x 10"		150	.107		4.27	4.70		8.97	11.95
0400	4" x 12"		140	.114		5.65	5.05		10.70	13.95
0500	8" x 8"		130	.123		5.80	5.45		11.25	14.75

06 44 39.20 Columns

		Crew	Daily Output	Labor-Hours	Unit	Material	2012 Bare Costs Labor	Equipment	Total	Total Incl O&P
0010	**COLUMNS**									
0050	Aluminum, round colonial, 6" diameter	2 Carp	80	.200	V.L.F.	18.60	8.80		27.40	34
0100	8" diameter		62.25	.257		21.50	11.35		32.85	41
0200	10" diameter		55	.291		25.50	12.85		38.35	48
0250	Fir, stock units, hollow round, 6" diameter		80	.200		20.50	8.80		29.30	36
0300	8" diameter		80	.200		24	8.80		32.80	40
0350	10" diameter		70	.229		33.50	10.10		43.60	52
0400	Solid turned, to 8' high, 3-1/2" diameter		80	.200		9	8.80		17.80	23.50
0500	4-1/2" diameter		75	.213		14	9.40		23.40	30
0600	5-1/2" diameter		70	.229		18.40	10.10		28.50	35.50
0800	Square columns, built-up, 5" x 5"		65	.246		15.50	10.85		26.35	34
0900	Solid, 3-1/2" x 3-1/2"		130	.123		9	5.45		14.45	18.25
1600	Hemlock, tapered, T & G, 12" diam., 10' high		100	.160		34.50	7.05		41.55	49
1700	16' high		65	.246		65.50	10.85		76.35	88.50
1900	10' high, 14" diameter		100	.160		97	7.05		104.05	117
2000	18' high		65	.246		88	10.85		98.85	114
2200	18" diameter, 12' high		65	.246		141	10.85		151.85	172
2300	20' high		50	.320		109	14.10		123.10	142
2500	20" diameter, 14' high		40	.400		159	17.65		176.65	202
2600	20' high		35	.457		173	20		193	221
2800	For flat pilasters, deduct					33%				
3000	For splitting into halves, add				Ea.	105			105	116
4000	Rough sawn cedar posts, 4" x 4"	2 Carp	250	.064	V.L.F.	3.32	2.82		6.14	8
4100	4" x 6"		235	.068		6	3		9	11.20
4200	6" x 6"		220	.073		10.30	3.21		13.51	16.30
4300	8" x 8"		200	.080		20	3.53		23.53	27.50

06 48 Wood Frames

06 48 13 – Exterior Wood Door Frames

06 48 13.10 Exterior Wood Door Frames and Accessories

		Crew	Daily Output	Labor-Hours	Unit	Material	2012 Bare Costs Labor	Equipment	Total	Total Incl O&P
0010	**EXTERIOR WOOD DOOR FRAMES AND ACCESSORIES**									
0400	Exterior frame, incl. ext. trim, pine, 5/4 x 4-9/16" deep	2 Carp	375	.043	L.F.	5.55	1.88		7.43	9
0420	5-3/16" deep		375	.043		8.85	1.88		10.73	12.65
0440	6-9/16" deep		375	.043		7.70	1.88		9.58	11.35
0600	Oak, 5/4 x 4-9/16" deep		350	.046		11.70	2.02		13.72	16
0620	5-3/16" deep		350	.046		12.90	2.02		14.92	17.25
0640	6-9/16" deep		350	.046		15.30	2.02		17.32	19.95
1000	Sills, 8/4 x 8" deep, oak, no horns		100	.160		4.75	7.05		11.80	16.10
1020	2" horns		100	.160		17.10	7.05		24.15	29.50
1040	3" horns		100	.160		17.10	7.05		24.15	29.50
1100	8/4 x 10" deep, oak, no horns		90	.178		5.30	7.85		13.15	17.85
1120	2" horns		90	.178		21	7.85		28.85	35.50

06 48 Wood Frames

06 48 13 – Exterior Wood Door Frames

06 48 13.10 Exterior Wood Door Frames and Accessories	Crew	Daily Output	Labor-Hours	Unit	Material	2012 Bare Costs Labor	Equipment	Total	Total Incl O&P
1140	3" horns	2 Carp	90	.178	L.F.	21	7.85		28.85	35.50
2000	Exterior, colonial, frame & trim, 3' opng., in-swing, minimum		22	.727	Ea.	330	32		362	415
2010	Average		21	.762		525	33.50		558.50	625
2020	Maximum		20	.800		1,125	35.50		1,160.50	1,300
2100	5'-4" opening, in-swing, minimum		17	.941		395	41.50		436.50	500
2120	Maximum		15	1.067		1,125	47		1,172	1,325
2140	Out-swing, minimum		17	.941		395	41.50		436.50	500
2160	Maximum		15	1.067		1,125	47		1,172	1,325
2400	6'-0" opening, in-swing, minimum		16	1		395	44		439	505
2420	Maximum		10	1.600		1,125	70.50		1,195.50	1,350
2460	Out-swing, minimum		16	1		395	44		439	505
2480	Maximum		10	1.600		1,125	70.50		1,195.50	1,350
2600	For two sidelights, add, minimum		30	.533	Opng.	64	23.50		87.50	107
2620	Maximum		20	.800	"	795	35.50		830.50	930
2700	Custom birch frame, 3'-0" opening		16	1	Ea.	220	44		264	310
2750	6'-0" opening		16	1		330	44		374	435
2900	Exterior, modern, plain trim, 3' opng., in-swing, minimum		26	.615		38.50	27		65.50	84.50
2920	Average		24	.667		46	29.50		75.50	95.50
2940	Maximum		22	.727		55	32		87	110

06 48 16 – Interior Wood Door Frames

06 48 16.10 Interior Wood Door Jamb and Frames

		Crew	Daily Output	Labor-Hours	Unit	Material	Labor	Equipment	Total	Total Incl O&P
0010	**INTERIOR WOOD DOOR JAMB AND FRAMES**									
3000	Interior frame, pine, 11/16" x 3-5/8" deep	2 Carp	375	.043	L.F.	5.10	1.88		6.98	8.50
3020	4-9/16" deep		375	.043		5.95	1.88		7.83	9.45
3200	Oak, 11/16" x 3-5/8" deep		350	.046		7.90	2.02		9.92	11.80
3220	4-9/16" deep		350	.046		7.90	2.02		9.92	11.80
3240	5-3/16" deep		350	.046		11.85	2.02		13.87	16.15
3400	Walnut, 11/16" x 3-5/8" deep		350	.046		8.25	2.02		10.27	12.20
3420	4-9/16" deep		350	.046		9	2.02		11.02	13
3440	5-3/16" deep		350	.046		9	2.02		11.02	13
3600	Pocket door frame		16	1	Ea.	74.50	44		118.50	150
3800	Threshold, oak, 5/8" x 3-5/8" deep		200	.080	L.F.	3.56	3.53		7.09	9.35
3820	4-5/8" deep		190	.084		4.13	3.71		7.84	10.25
3840	5-5/8" deep		180	.089		6.50	3.92		10.42	13.20

06 49 Wood Screens and Exterior Wood Shutters

06 49 19 – Exterior Wood Shutters

06 49 19.10 Shutters, Exterior

		Crew	Daily Output	Labor-Hours	Unit	Material	Labor	Equipment	Total	Total Incl O&P
0010	**SHUTTERS, EXTERIOR**									
0012	Aluminum, louvered, 1'-4" wide, 3'-0" long	1 Carp	10	.800	Pr.	62	35.50		97.50	123
0400	6'-8" long		9	.889		124	39		163	198
1000	Pine, louvered, primed, each 1'-2" wide, 3'-3" long		10	.800		107	35.50		142.50	173
1100	4'-7" long		10	.800		145	35.50		180.50	214
1500	Each 1'-6" wide, 3'-3" long		10	.800		114	35.50		149.50	180
1600	4'-7" long		10	.800		160	35.50		195.50	231
1620	Hemlock, louvered, 1'-2" wide, 5'-7" long		10	.800		175	35.50		210.50	248
1630	Each 1'-4" wide, 2'-2" long		10	.800		110	35.50		145.50	175
1670	4'-3" long		10	.800		131	35.50		166.50	199
1690	5'-11" long		10	.800		185	35.50		220.50	258
1700	Door blinds, 6'-9" long, each 1'-3" wide		9	.889		186	39		225	265

06 49 Wood Screens and Exterior Wood Shutters

06 49 19 – Exterior Wood Shutters

06 49 19.10 Shutters, Exterior		Crew	Daily Output	Labor-Hours	Unit	Material	2012 Bare Costs Labor	Equipment	Total	Total Incl O&P
1710	1'-6" wide	1 Carp	9	.889	Pr.	200	39		239	281
1720	Hemlock, solid raised panel, each 1'-4" wide, 3'-3" long		10	.800		177	35.50		212.50	250
1740	4'-3" long		10	.800		224	35.50		259.50	300
1770	5'-11" long		10	.800		300	35.50		335.50	385
1800	Door blinds, 6'-9" long, each 1'-3" wide		9	.889		335	39		374	430
1900	1'-6" wide		9	.889		370	39		409	465
2500	Polystyrene, solid raised panel, each 1'-4" wide, 3'-3" long		10	.800		41.50	35.50		77	100
2700	4'-7" long		10	.800		52	35.50		87.50	112
4500	Polystyrene, louvered, each 1'-2" wide, 3'-3" long		10	.800		28	35.50		63.50	85.50
4750	5'-3" long		10	.800		40	35.50		75.50	98.50
6000	Vinyl, louvered, each 1'-2" x 4'-7" long		10	.800		34.50	35.50		70	92.50
6200	Each 1'-4" x 6'-8" long		9	.889		48.50	39		87.50	114

06 51 Structural Plastic Shapes and Plates

06 51 13 – Plastic Lumber

06 51 13.10 Recycled Plastic Lumber

			Crew	Daily Output	Labor-Hours	Unit	Material	2012 Bare Costs Labor	Equipment	Total	Total Incl O&P
0010	**RECYCLED PLASTIC LUMBER**										
4000	Sheeting, recycled plastic, black or white, 4' x 8' x 1/8"	G	2 Carp	1100	.015	S.F.	1.56	.64		2.20	2.71
4010	4' x 8' x 3/16"	G		1100	.015		2.25	.64		2.89	3.47
4020	4' x 8' x 1/4"	G		950	.017		2.79	.74		3.53	4.21
4030	4' x 8' x 3/8"	G		950	.017		3.89	.74		4.63	5.40
4040	4' x 8' x 1/2"	G		900	.018		4.99	.78		5.77	6.70
4050	4' x 8' x 5/8"	G		900	.018		7	.78		7.78	8.90
4060	4' x 8' x 3/4"	G		850	.019		8.50	.83		9.33	10.65
4070	Add for colors	G				Ea.	5%		5%		6%
8500	100% recycled plastic, var colors, NLB, 2" x 2"	G				L.F.	1.62			1.62	1.78
8510	2" x 4"	G					3.38			3.38	3.72
8520	2" x 6"	G					5.30			5.30	5.80
8530	2" x 8"	G					7.25			7.25	8
8540	2" x 10"	G					10.60			10.60	11.65
8550	5/4" x 4"	G					3.38			3.38	3.72
8560	5/4" x 6"	G					2.99			2.99	3.29
8570	1" x 6"	G					2.58			2.58	2.84
8580	1/2" x 8"	G					2.82			2.82	3.10
8590	2" x 10" T & G	G					7.60			7.60	8.35
8600	3" x 10" T & G	G					12.25			12.25	13.45
8610	Add for premium colors	G					20%		20%		22%

06 52 Plastic Structural Assemblies

06 52 10 – Fiberglass Structural Assemblies

06 52 10.10 Castings, Fiberglass

			Crew	Daily Output	Labor-Hours	Unit	Material	2012 Bare Costs Labor	Equipment	Total	Total Incl O&P
0010	**CASTINGS, FIBERGLASS**										
0100	Angle, 1" x 1" x 1/8" thick		2 Sswk	240	.067	L.F.	1.44	3.29		4.73	7.50
0120	3" x 3" x 1/4" thick			200	.080		6	3.95		9.95	13.65
0140	4" x 4" x 1/4" thick			200	.080		7.70	3.95		11.65	15.55
0160	4" x 4" x 3/8" thick			200	.080		11.85	3.95		15.80	20
0180	6" x 6" x 1/2" thick			160	.100		23.50	4.94		28.44	35
1000	Flat sheet, 1/8" thick			140	.114	S.F.	5.05	5.65		10.70	15.65
1020	1/4" thick			120	.133		9.90	6.60		16.50	22.50

06 52 Plastic Structural Assemblies

06 52 10 – Fiberglass Structural Assemblies

	06 52 10.10 Castings, Fiberglass	Crew	Daily Output	Labor-Hours	Unit	Material	2012 Bare Costs Labor	Equipment	Total	Total Incl O&P
1040	3/8" thick	2 Sswk	100	.160	S.F.	16.85	7.90		24.75	32.50
1060	1/2" thick		80	.200	↓	22	9.85		31.85	42
2000	Handrail, 42" high, 2" diam. rails pickets 5' O.C.		32	.500	L.F.	51	24.50		75.50	100
3000	Round bar, 1/4" diam.		240	.067		.58	3.29		3.87	6.55
3020	1/2" diam.		200	.080		1.01	3.95		4.96	8.15
3040	3/4" diam.		200	.080		2.05	3.95		6	9.30
3060	1" diam.		160	.100		3.62	4.94		8.56	12.85
3080	1-1/4" diam.		160	.100		4.13	4.94		9.07	13.40
3100	1-1/2" diam.		140	.114		5.05	5.65		10.70	15.65
3500	Round tube, 1" diam. x 1/8" thick		240	.067		1.97	3.29		5.26	8.05
3520	2" diam. x 1/4" thick		200	.080		6.35	3.95		10.30	14.05
3540	3" diam. x 1/4" thick		160	.100		10.10	4.94		15.04	19.95
4000	Square bar, 1/2" square		240	.067		3.98	3.29		7.27	10.30
4020	1" square		200	.080		5.10	3.95		9.05	12.65
4040	1-1/2" square		160	.100		10.40	4.94		15.34	20.50
4500	Square tube, 1" x 1" x 1/8" thick		240	.067		2.24	3.29		5.53	8.35
4520	2" x 2" x 1/8" thick		200	.080		4.34	3.95		8.29	11.80
4540	3" x 3" x 1/4" thick		160	.100		11.85	4.94		16.79	22
5000	Threaded rod, 3/8" diam.		320	.050		3.45	2.47		5.92	8.20
5020	1/2" diam.		320	.050		4.06	2.47		6.53	8.90
5040	5/8" diam.		280	.057		4.49	2.82		7.31	10
5060	3/4" diam.		280	.057		5.10	2.82		7.92	10.65
6000	Wide flange beam, 4" x 4" x 1/4" thick		120	.133		10.55	6.60		17.15	23.50
6020	6" x 6" x 1/4" thick		100	.160		19.95	7.90		27.85	36
6040	8" x 8" x 3/8" thick	↓	80	.200	↓	35	9.85		44.85	56

06 52 10.20 Fiberglass Stair Treads

		Crew	Daily Output	Labor-Hours	Unit	Material	2012 Bare Costs Labor	Equipment	Total	Total Incl O&P
0010	**FIBERGLASS STAIR TREADS**									
0100	24" wide	2 Sswk	52	.308	Ea.	39.50	15.20		54.70	70.50
0140	30" wide		52	.308		47	15.20		62.20	79
0180	36" wide		52	.308		55.50	15.20		70.70	88
0220	42" wide	↓	52	.308	↓	65.50	15.20		80.70	99

06 52 10.30 Fiberglass Grating

		Crew	Daily Output	Labor-Hours	Unit	Material	2012 Bare Costs Labor	Equipment	Total	Total Incl O&P
0010	**FIBERGLASS GRATING**									
0100	Molded, green (for mod. corrosive environment)									
0140	1" x 4" mesh, 1" thick	2 Sswk	400	.040	S.F.	11.15	1.97		13.12	15.85
0180	1-1/2" square mesh, 1" thick		400	.040		16.60	1.97		18.57	22
0220	1-1/4" thick		400	.040		13.25	1.97		15.22	18.15
0260	1-1/2" thick		400	.040		24	1.97		25.97	30
0300	2" square mesh, 2" thick	↓	320	.050	↓	22.50	2.47		24.97	29.50
1000	Orange (for highly corrosive environment)									
1040	1" x 4" mesh, 1" thick	2 Sswk	400	.040	S.F.	14.05	1.97		16.02	19
1080	1-1/2" square mesh, 1" thick		400	.040		18.60	1.97		20.57	24
1120	1-1/4" thick		400	.040		19.30	1.97		21.27	25
1160	1-1/2" thick		400	.040		21	1.97		22.97	26.50
1200	2" square mesh, 2" thick	↓	320	.050	↓	25	2.47		27.47	32
3000	Pultruded, green (for mod. corrosive environment)									
3040	1" O.C. bar spacing, 1" thick	2 Sswk	400	.040	S.F.	16.65	1.97		18.62	22
3080	1-1/2" thick		320	.050		18.50	2.47		20.97	25
3120	1-1/2" O.C. bar spacing, 1" thick		400	.040		13.50	1.97		15.47	18.40
3160	1-1/2" thick	↓	400	.040	↓	15.15	1.97		17.12	20
4000	Grating support legs, fixed height, no base				Ea.	43.50			43.50	48
4040	With base				↓	39			39	43

06 52 Plastic Structural Assemblies

06 52 10 – Fiberglass Structural Assemblies

06 52 10.30 Fiberglass Grating	Crew	Daily Output	Labor-Hours	Unit	Material	2012 Bare Costs Labor	Equipment	Total	Total Incl O&P
4080 Adjustable to 60"				Ea.	59			59	65

06 52 10.40 Fiberglass Floor Grating

	06 52 10.40 Fiberglass Floor Grating	Crew	Daily Output	Labor-Hours	Unit	Material	Labor	Equipment	Total	Total Incl O&P
0010	**FIBERGLASS FLOOR GRATING**									
0100	Reinforced polyester, fire retardant, 1" x 4" grid, 1" thick	E-4	510	.063	S.F.	14.25	3.13	.24	17.62	21.50
0200	1-1/2" x 6" mesh, 1-1/2" thick		500	.064		16.65	3.19	.24	20.08	24.50
0300	With grit surface, 1-1/2" x 6" grid, 1-1/2" thick		500	.064		17	3.19	.24	20.43	24.50

06 63 Plastic Railings

06 63 10 – Plastic (PVC) Railings

	06 63 10.10 Plastic Railings	Crew	Daily Output	Labor-Hours	Unit	Material	Labor	Equipment	Total	Total Incl O&P
0010	**PLASTIC RAILINGS**									
0100	Horizontal PVC handrail with balusters, 3-1/2" wide, 36" high	1 Carp	96	.083	L.F.	23.50	3.67		27.17	31
0150	42" high		96	.083		26	3.67		29.67	34.50
0200	Angled PVC handrail with balusters, 3-1/2" wide, 36" high		72	.111		26.50	4.90		31.40	36.50
0250	42" high		72	.111		29.50	4.90		34.40	40
0300	Post sleeve for 4 x 4 post		96	.083		12.05	3.67		15.72	18.90
0400	Post cap for 4 x 4 post, flat profile		48	.167	Ea.	10.40	7.35		17.75	23
0450	Newel post style profile		48	.167		21	7.35		28.35	34.50
0500	Raised corbeled profile		48	.167		23	7.35		30.35	36.50
0550	Post base trim for 4 x 4 post		96	.083		11.50	3.67		15.17	18.30

06 65 Plastic Simulated Wood Trim

06 65 10 – PVC Trim

	06 65 10.10 PVC Trim, Exterior	Crew	Daily Output	Labor-Hours	Unit	Material	Labor	Equipment	Total	Total Incl O&P
0010	**PVC TRIM, EXTERIOR**									
0100	Cornerboards, 5/4" x 6" x 6"	1 Carp	240	.033	L.F.	11.10	1.47		12.57	14.45
0110	Door/window casing, 1" x 4"		200	.040		1.31	1.76		3.07	4.15
0120	1" x 6"		200	.040		2.06	1.76		3.82	4.98
0130	1" x 8"		195	.041		2.72	1.81		4.53	5.75
0140	1" x 10"		195	.041		3.45	1.81		5.26	6.60
0150	1" x 12"		190	.042		4.21	1.86		6.07	7.50
0160	5/4" x 4"		195	.041		1.68	1.81		3.49	4.63
0170	5/4" x 6"		195	.041		2.65	1.81		4.46	5.70
0180	5/4" x 8"		190	.042		3.47	1.86		5.33	6.70
0190	5/4" x 10"		190	.042		4.43	1.86		6.29	7.75
0200	5/4" x 12"		185	.043		4.85	1.91		6.76	8.30
0210	Fascia, 1" x 4"		250	.032		1.31	1.41		2.72	3.61
0220	1" x 6"		250	.032		2.06	1.41		3.47	4.44
0230	1" x 8"		225	.036		2.72	1.57		4.29	5.40
0240	1" x 10"		225	.036		3.45	1.57		5.02	6.20
0250	1" x 12"		200	.040		4.21	1.76		5.97	7.35
0260	5/4" x 4"		240	.033		1.68	1.47		3.15	4.11
0270	5/4" x 6"		240	.033		2.65	1.47		4.12	5.20
0280	5/4" x 8"		215	.037		3.47	1.64		5.11	6.35
0290	5/4" x 10"		215	.037		4.43	1.64		6.07	7.40
0300	5/4" x 12"		190	.042		4.85	1.86		6.71	8.20
0310	Frieze, 1" x 4"		250	.032		1.31	1.41		2.72	3.61
0320	1" x 6"		250	.032		2.06	1.41		3.47	4.44
0330	1" x 8"		225	.036		2.72	1.57		4.29	5.40

06 65 Plastic Simulated Wood Trim

06 65 10 – PVC Trim

06 65 10.10 PVC Trim, Exterior		Crew	Daily Output	Labor-Hours	Unit	Material	2012 Bare Costs Labor	Equipment	Total	Total Incl O&P
0340	1" x 10"	1 Carp	225	.036	L.F.	3.45	1.57		5.02	6.20
0350	1" x 12"		200	.040		4.21	1.76		5.97	7.35
0360	5/4" x 4"		240	.033		1.68	1.47		3.15	4.11
0370	5/4" x 6"		240	.033		2.65	1.47		4.12	5.20
0380	5/4" x 8"		215	.037		3.47	1.64		5.11	6.35
0390	5/4" x 10"		215	.037		4.43	1.64		6.07	7.40
0400	5/4" x 12"		190	.042		4.85	1.86		6.71	8.20
0410	Rake, 1" x 4"		200	.040		1.31	1.76		3.07	4.15
0420	1" x 6"		200	.040		2.06	1.76		3.82	4.98
0430	1" x 8"		190	.042		2.72	1.86		4.58	5.85
0440	1" x 10"		190	.042		3.45	1.86		5.31	6.65
0450	1" x 12"		180	.044		4.21	1.96		6.17	7.65
0460	5/4" x 4"		195	.041		1.68	1.81		3.49	4.63
0470	5/4" x 6"		195	.041		2.65	1.81		4.46	5.70
0480	5/4" x 8"		185	.043		3.47	1.91		5.38	6.75
0490	5/4" x 10"		185	.043		4.43	1.91		6.34	7.80
0500	5/4" x 12"		175	.046		4.85	2.02		6.87	8.45
0510	Rake trim, 1" x 4"		225	.036		1.31	1.57		2.88	3.85
0520	1" x 6"		225	.036		2.06	1.57		3.63	4.68
0560	5/4" x 4"		220	.036		1.68	1.60		3.28	4.32
0570	5/4" x 6"		220	.036		2.65	1.60		4.25	5.40
0610	Soffit, 1" x 4"	2 Carp	420	.038		1.31	1.68		2.99	4.03
0620	1" x 6"		420	.038		2.06	1.68		3.74	4.86
0630	1" x 8"		420	.038		2.72	1.68		4.40	5.60
0640	1" x 10"		400	.040		3.45	1.76		5.21	6.50
0650	1" x 12"		400	.040		4.21	1.76		5.97	7.35
0660	5/4" x 4"		410	.039		1.68	1.72		3.40	4.50
0670	5/4" x 6"		410	.039		2.65	1.72		4.37	5.55
0680	5/4" x 8"		410	.039		3.47	1.72		5.19	6.45
0690	5/4" x 10"		390	.041		4.43	1.81		6.24	7.65
0700	5/4" x 12"		390	.041		4.85	1.81		6.66	8.15

Division Notes

	CREW	DAILY OUTPUT	LABOR-HOURS	UNIT	2012 BARE COSTS				TOTAL INCL O&P
					MAT.	LABOR	EQUIP.	TOTAL	

Estimating Tips

07 10 00 Dampproofing and Waterproofing

- Be sure of the job specifications before pricing this subdivision. The difference in cost between waterproofing and dampproofing can be great. Waterproofing will hold back standing water. Dampproofing prevents the transmission of water vapor. Also included in this section are vapor retarding membranes.

07 20 00 Thermal Protection

- Insulation and fireproofing products are measured by area, thickness, volume or R-value. Specifications may give only what the specific R-value should be in a certain situation. The estimator may need to choose the type of insulation to meet that R-value.

07 30 00 Steep Slope Roofing
07 40 00 Roofing and Siding Panels

- Many roofing and siding products are bought and sold by the square. One square is equal to an area that measures 100 square feet.

This simple change in unit of measure could create a large error if the estimator is not observant. Accessories necessary for a complete installation must be figured into any calculations for both material and labor.

07 50 00 Membrane Roofing
07 60 00 Flashing and Sheet Metal
07 70 00 Roofing and Wall Specialties and Accessories

- The items in these subdivisions compose a roofing system. No one component completes the installation, and all must be estimated. Built-up or single-ply membrane roofing systems are made up of many products and installation trades. Wood blocking at roof perimeters or penetrations, parapet coverings, reglets, roof drains, gutters, downspouts, sheet metal flashing, skylights, smoke vents, and roof hatches all need to be considered along with the roofing material. Several different installation trades will need to work together on the roofing system. Inherent difficulties in the scheduling and coordination of various trades must be accounted for when estimating labor costs.

07 90 00 Joint Protection

- To complete the weather-tight shell, the sealants and caulkings must be estimated. Where different materials meet—at expansion joints, at flashing penetrations, and at hundreds of other locations throughout a construction project—they provide another line of defense against water penetration. Often, an entire system is based on the proper location and placement of caulking or sealants. The detailed drawings that are included as part of a set of architectural plans show typical locations for these materials. When caulking or sealants are shown at typical locations, this means the estimator must include them for all the locations where this detail is applicable. Be careful to keep different types of sealants separate, and remember to consider backer rods and primers if necessary.

Reference Numbers

Reference numbers are shown in shaded boxes at the beginning of some major classifications. These numbers refer to related items in the Reference Section. The reference information may be an estimating procedure, an alternate pricing method, or technical information.

Note: Not all subdivisions listed here necessarily appear in this publication.

Division 7 - Thermal and Moisture Protection

07 01 50 – Maintenance of Membrane Roofing

07 01 50.10 Roof Coatings

		Crew	Daily Output	Labor-Hours	Unit	Material	2012 Bare Costs Labor	Equipment	Total	Total Incl O&P
0010	**ROOF COATINGS**									
0012	Asphalt, brush grade, material only				Gal.	7.60			7.60	8.35
0200	Asphalt base, fibered aluminum coating [G]					10.35			10.35	11.40
0300	Asphalt primer, 5 gallon					6.60			6.60	7.30
0600	Coal tar pitch, 200 lb. barrels				Ton	1,200			1,200	1,325
0700	Tar roof cement, 5 gal. lots				Gal.	15			15	16.50
0800	Glass fibered roof & patching cement, 5 gallon				"	10.35			10.35	11.40
0900	Reinforcing glass membrane, 450 S.F./roll				Ea.	55.50			55.50	61
1000	Neoprene roof coating, 5 gal., 2 gal./sq.				Gal.	30			30	33
1100	Roof patch & flashing cement, 5 gallon					7.90			7.90	8.70
1200	Roof resturant, glass fibered, 3 gal./sq.					7.40			7.40	8.15

07 01 90 – Maintenance of Joint Protection

07 01 90.81 Joint Sealant Replacement

		Crew	Daily Output	Labor-Hours	Unit	Material	2012 Bare Costs Labor	Equipment	Total	Total Incl O&P
0010	**JOINT SEALANT REPLACEMENT**									
0050	Control joints in concrete floors/slabs									
0100	Option 1 for joints with hard dry sealant									
0110	Step 1: Sawcut to remove 95% of old sealant									
0112	1/4" wide x 1/2" deep, with single saw blade	C-27	4800	.003	L.F.	.02	.14	.03	.19	.26
0114	3/8" wide x 3/4" deep, with single saw blade		4000	.004		.03	.17	.04	.24	.32
0116	1/2" wide x 1" deep, with double saw blades		3600	.004		.06	.19	.04	.29	.39
0118	3/4" wide x 1-1/2" deep, with double saw blades		3200	.005		.11	.21	.05	.37	.49
0120	Step 2: Water blast joint faces and edges	C-29	2500	.003			.11	.03	.14	.20
0130	Step 3: Air blast joint faces and edges	C-28	2000	.004			.17	.01	.18	.26
0140	Step 4: Sand blast joint faces and edges	E-11	2000	.016			.63	.11	.74	1.18
0150	Step 5: Air blast joint faces and edges	C-28	2000	.004			.17	.01	.18	.26
0200	Option 2 for joints with soft pliable sealant									
0210	Step 1: Plow joint with rectangular blade	B-62	2600	.009	L.F.		.35	.06	.41	.61
0220	Step 2: Sawcut to re-face joint faces									
0222	1/4" wide x 1/2" deep, with single saw blade	C-27	2400	.007	L.F.	.02	.28	.06	.36	.51
0224	3/8" wide x 3/4" deep, with single saw blade		2000	.008		.04	.34	.07	.45	.62
0226	1/2" wide x 1" deep, with double saw blades		1800	.009		.07	.38	.08	.53	.72
0228	3/4" wide x 1-1/2" deep, with double saw blades		1600	.010		.15	.42	.09	.66	.89
0230	Step 3: Water blast joint faces and edges	C-29	2500	.003			.11	.03	.14	.20
0240	Step 4: Air blast joint faces and edges	C-28	2000	.004			.17	.01	.18	.26
0250	Step 5: Sand blast joint faces and edges	E-11	2000	.016			.63	.11	.74	1.18
0260	Step 6: Air blast joint faces and edges	C-28	2000	.004			.17	.01	.18	.26
0290	For saw cutting new control joints, see Section 03 35 29.35									
8910	For backer rod, see Section 07 91 23.10									
8920	For joint sealant, see Sections 03 15 05.25 or 07 92 13.20									

07 05 Common Work Results for Thermal and Moisture Protection

07 05 05 – Selective Demolition

07 05 05.10 Selective Demo., Thermal and Moist. Protection

			Crew	Daily Output	Labor-Hours	Unit	Material	2012 Bare Costs Labor	Equipment	Total	Total Incl O&P
0010	**SELECTIVE DEMO., THERMAL AND MOISTURE PROTECTION**										
0020	Caulking/sealant, to 1" x 1" joint	R024119-10	1 Clab	600	.013	L.F.		.47		.47	.72
0120	Downspouts, including hangers			350	.023	"		.80		.80	1.23
0220	Flashing, sheet metal			290	.028	S.F.		.97		.97	1.49
0420	Gutters, aluminum or wood, edge hung			240	.033	L.F.		1.17		1.17	1.80
0520	Built-in			100	.080	"		2.81		2.81	4.32
0620	Insulation, air/vapor barrier			3500	.002	S.F.		.08		.08	.12

07 05 05.10 Selective Demo., Thermal and Moist. Protection	Crew	Daily Output	Labor-Hours	Unit	Material	2012 Bare Costs Labor	Equipment	Total	Total Incl O&P
0670	Batts or blankets	1 Clab	1400	.006	C.F.		.20		.20	.31
0720	Foamed or sprayed in place	2 Clab	1000	.016	B.F.		.56		.56	.86
0770	Loose fitting	1 Clab	3000	.003	C.F.		.09		.09	.14
0870	Rigid board		3450	.002	B.F.		.08		.08	.13
1120	Roll roofing, cold adhesive		12	.667	Sq.		23.50		23.50	36
1170	Roof accessories, adjustable metal chimney flashing		9	.889	Ea.		31		31	48
1325	Plumbing vent flashing		32	.250	"		8.80		8.80	13.50
1375	Ridge vent strip, aluminum		310	.026	L.F.		.91		.91	1.39
1620	Skylight to 10 S.F.		8	1	Ea.		35		35	54
2120	Roof edge, aluminum soffit and fascia	▼	570	.014	L.F.		.49		.49	.76
2170	Concrete coping, up to 12" wide	2 Clab	160	.100			3.51		3.51	5.40
2220	Drip edge	1 Clab	1650	.005			.17		.17	.26
2270	Gravel stop		1650	.005			.17		.17	.26
2370	Sheet metal coping, up to 12" wide	▼	240	.033	▼		1.17		1.17	1.80
2470	Roof insulation board, over 2" thick	B-2	7800	.005	B.F.		.18		.18	.28
2520	Up to 2" thick	"	3900	.010	S.F.		.36		.36	.56
2620	Roof ventilation, louvered gable vent	1 Clab	16	.500	Ea.		17.55		17.55	27
2670	Remove, roof hatch	G-3	15	2.133			93		93	142
2675	Rafter vents	1 Clab	960	.008	▼		.29		.29	.45
2720	Soffit vent and/or fascia vent		575	.014	L.F.		.49		.49	.75
2775	Soffit vent strip, aluminum, 3" to 4" wide		160	.050			1.76		1.76	2.70
2820	Roofing accessories, shingle moulding, to 1" x 4"	▼	1600	.005			.18		.18	.27
2870	Cant strip	B-2	2000	.020			.71		.71	1.09
2920	Concrete block walkway	1 Clab	230	.035	▼		1.22		1.22	1.88
3070	Roofing, felt paper, 15#		70	.114	Sq.		4.01		4.01	6.15
3125	#30 felt	▼	30	.267	"		9.35		9.35	14.40
3170	Asphalt shingles, 1 layer	B-2	3500	.011	S.F.		.41		.41	.62
3180	2 layers		1750	.023	"		.81		.81	1.25
3370	Modified bitumen		26	1.538	Sq.		54.50		54.50	84
3420	Built-up, no gravel, 3 ply		25	1.600			57		57	87.50
3470	4 ply		21	1.905	▼		67.50		67.50	104
3620	5 ply		1600	.025	S.F.		.89		.89	1.37
3720	5 ply, with gravel		890	.045			1.60		1.60	2.45
3725	Gravel removal, minimum		5000	.008			.28		.28	.44
3730	Maximum		2000	.020			.71		.71	1.09
3870	Fiberglass sheet		1200	.033			1.18		1.18	1.82
4120	Slate shingles		1900	.021	▼		.75		.75	1.15
4170	Ridge shingles, clay or slate		2000	.020	L.F.		.71		.71	1.09
4320	Single ply membrane, attached at seams		52	.769	Sq.		27.50		27.50	42
4370	Ballasted		75	.533			18.95		18.95	29
4420	Fully adhered	▼	39	1.026	▼		36.50		36.50	56
4550	Roof hatch, 2'-6" x 3'-0"	1 Clab	10	.800	Ea.		28		28	43
4670	Wood shingles	B-2	2200	.018	S.F.		.65		.65	.99
4820	Sheet metal roofing	"	2150	.019			.66		.66	1.02
4970	Siding, horizontal wood clapboards	1 Clab	380	.021			.74		.74	1.14
5025	Exterior insulation finish system	"	120	.067			2.34		2.34	3.60
5070	Tempered hardboard, remove and reset	1 Carp	380	.021			.93		.93	1.43
5120	Tempered hardboard sheet siding	"	375	.021	▼		.94		.94	1.45
5170	Metal, corner strips	1 Clab	850	.009	L.F.		.33		.33	.51
5225	Horizontal strips		444	.018	S.F.		.63		.63	.97
5320	Vertical strips		400	.020			.70		.70	1.08
5520	Wood shingles		350	.023			.80		.80	1.23
5620	Stucco siding	▼	360	.022	▼		.78		.78	1.20

07 05 Common Work Results for Thermal and Moisture Protection

07 05 05 – Selective Demolition

07 05 05.10 Selective Demo., Thermal and Moist. Protection

		Crew	Daily Output	Labor-Hours	Unit	Material	2012 Bare Costs Labor	Equipment	Total	Total Incl O&P
5670	Textured plywood	1 Clab	725	.011	S.F.		.39		.39	.60
5720	Vinyl siding		510	.016	↓		.55		.55	.85
5770	Corner strips		900	.009	L.F.		.31		.31	.48
5870	Wood, boards, vertical	↓	400	.020	S.F.		.70		.70	1.08
5920	Waterproofing, protection/drain board	2 Clab	3900	.004	B.F.		.14		.14	.22
5970	Over 1/2" thick		1750	.009	S.F.		.32		.32	.49
6020	To 1/2" thick	↓	2000	.008	"		.28		.28	.43

07 11 Dampproofing

07 11 13 – Bituminous Dampproofing

07 11 13.10 Bituminous Asphalt Coating

		Crew	Daily Output	Labor-Hours	Unit	Material	2012 Bare Costs Labor	Equipment	Total	Total Incl O&P
0010	**BITUMINOUS ASPHALT COATING**									
0030	Brushed on, below grade, 1 coat	1 Rofc	665	.012	S.F.	.19	.45		.64	.96
0100	2 coat		500	.016		.38	.60		.98	1.42
0300	Sprayed on, below grade, 1 coat		830	.010		.19	.36		.55	.81
0400	2 coat	↓	500	.016	↓	.37	.60		.97	1.41
0500	Asphalt coating, with fibers				Gal.	10.35			10.35	11.40
0600	Troweled on, asphalt with fibers, 1/16" thick	1 Rofc	500	.016	S.F.	.45	.60		1.05	1.50
0700	1/8" thick		400	.020		.80	.75		1.55	2.13
1000	1/2" thick	↓	350	.023	↓	2.59	.86		3.45	4.28

07 11 16 – Cementitious Dampproofing

07 11 16.20 Cementitious Parging

		Crew	Daily Output	Labor-Hours	Unit	Material	2012 Bare Costs Labor	Equipment	Total	Total Incl O&P
0010	**CEMENTITIOUS PARGING**									
0020	Portland cement, 2 coats, 1/2" thick	D-1	250	.064	S.F.	.31	2.55		2.86	4.19
0100	Waterproofed Portland cement, 1/2" thick, 2 coats	"	250	.064	"	3.30	2.55		5.85	7.50

07 12 Built-up Bituminous Waterproofing

07 12 13 – Built-Up Asphalt Waterproofing

07 12 13.20 Membrane Waterproofing

		Crew	Daily Output	Labor-Hours	Unit	Material	2012 Bare Costs Labor	Equipment	Total	Total Incl O&P
0010	**MEMBRANE WATERPROOFING**									
0012	On slabs, 1 ply, felt, mopped	G-1	3000	.019	S.F.	.36	.65	.15	1.16	1.66
0100	On slabs, 1 ply, glass fiber fabric, mopped		2100	.027		.37	.93	.22	1.52	2.21
0300	On slabs, 2 ply, felt, mopped		2500	.022		.72	.78	.18	1.68	2.31
0400	On slabs, 2 ply, glass fiber fabric, mopped		1650	.034		.83	1.19	.27	2.29	3.21
0600	On slabs, 3 ply, felt, mopped		2100	.027		1.09	.93	.22	2.24	2.99
0700	On slabs, 3 ply, glass fiber fabric, mopped	↓	1550	.036		1.12	1.26	.29	2.67	3.68
0710	Asphaltic hardboard protection board, 1/8" thick	2 Rofc	500	.032		.44	1.20		1.64	2.49
1000	1/4" EPS membrane protection board		3500	.005		.34	.17		.51	.66
1050	3/8" thick		3500	.005		.38	.17		.55	.70
1060	1/2" thick		3500	.005	↓	.41	.17		.58	.74
1070	Fiberglass fabric, black, 20/10 mesh	↓	116	.138	Sq.	18.95	5.15		24.10	29.50

07 13 Sheet Waterproofing

07 13 53 – Elastomeric Sheet Waterproofing

07 13 53.10 Elastomeric Sheet Waterproofing and Access.	Crew	Daily Output	Labor-Hours	Unit	Material	2012 Bare Costs Labor	Equipment	Total	Total Incl O&P
0010 **ELASTOMERIC SHEET WATERPROOFING AND ACCESS.**									
0090 EPDM, plain, 45 mils thick	2 Rofc	580	.028	S.F.	1.36	1.03		2.39	3.22
0100 60 mils thick		570	.028		1.45	1.05		2.50	3.35
0300 Nylon reinforced sheets, 45 mils thick		580	.028		1.37	1.03		2.40	3.23
0400 60 mils thick	↓	570	.028	↓	1.47	1.05		2.52	3.37
0600 Vulcanizing splicing tape for above, 2" wide				C.L.F.	50			50	55
0700 4" wide				"	113			113	124
0900 Adhesive, bonding, 60 S.F. per gal.				Gal.	22.50			22.50	24.50
1000 Splicing, 75 S.F. per gal.				"	32			32	35
1200 Neoprene sheets, plain, 45 mils thick	2 Rofc	580	.028	S.F.	1.67	1.03		2.70	3.57
1300 60 mils thick		570	.028		2.35	1.05		3.40	4.35
1500 Nylon reinforced, 45 mils thick		580	.028		1.93	1.03		2.96	3.85
1600 60 mils thick		570	.028		2.81	1.05		3.86	4.85
1800 120 mils thick	↓	500	.032	↓	5.60	1.20		6.80	8.15
1900 Adhesive, splicing, 150 S.F. per gal. per coat				Gal.	32			32	35
2100 Fiberglass reinforced, fluid applied, 1/8" thick	2 Rofc	500	.032	S.F.	1.59	1.20		2.79	3.76
2200 Polyethylene and rubberized asphalt sheets, 60 mils thick		550	.029		.63	1.09		1.72	2.51
2400 Polyvinyl chloride sheets, plain, 10 mils thick		580	.028		.17	1.03		1.20	1.92
2500 20 mils thick		570	.028		.26	1.05		1.31	2.05
2700 30 mils thick	↓	560	.029	↓	.36	1.07		1.43	2.19
3000 Adhesives, trowel grade, 40-100 S.F. per gal.				Gal.	23			23	25.50
3100 Brush grade, 100-250 S.F. per gal.				"	18.50			18.50	20.50
3300 Bitumen modified polyurethane, fluid applied, 55 mils thick	2 Rofc	665	.024	S.F.	.82	.90		1.72	2.41

07 16 Cementitious and Reactive Waterproofing

07 16 16 – Crystalline Waterproofing

07 16 16.20 Cementitious Waterproofing

	Crew	Daily Output	Labor-Hours	Unit	Material	2012 Bare Costs Labor	Equipment	Total	Total Incl O&P
0010 **CEMENTITIOUS WATERPROOFING**									
0020 1/8" application, sprayed on	G-2A	1000	.024	S.F.	2	.80	.63	3.43	4.20
0050 4 coat cementitious metallic slurry	1 Cefi	1.20	6.667	C.S.F.	26.50	282		308.50	445

07 17 Bentonite Waterproofing

07 17 13 – Bentonite Panel Waterproofing

07 17 13.10 Bentonite

	Crew	Daily Output	Labor-Hours	Unit	Material	2012 Bare Costs Labor	Equipment	Total	Total Incl O&P
0010 **BENTONITE**									
0020 Panels, 4' x 4', 3/16" thick	1 Rofc	625	.013	S.F.	1.22	.48		1.70	2.14
0100 Rolls, 3/8" thick, with geotextile fabric both sides	"	550	.015	"	1.35	.54		1.89	2.40
0300 Granular bentonite, 50 lb. bags (.625 C.F.)				Bag	9.90			9.90	10.90
0400 3/8" thick, troweled on	1 Rofc	475	.017	S.F.	.50	.63		1.13	1.60
0500 Drain board, expanded polystyrene, 1-1/2" thick	1 Rohe	1600	.005		.42	.14		.56	.69
0510 2" thick		1600	.005		.56	.14		.70	.85
0520 3" thick		1600	.005		.84	.14		.98	1.15
0530 4" thick		1600	.005		1.12	.14		1.26	1.46
0600 With filter fabric, 1-1/2" thick		1600	.005		.50	.14		.64	.78
0625 2" thick		1600	.005		.64	.14		.78	.94
0650 3" thick		1600	.005		.92	.14		1.06	1.25
0675 4" thick	↓	1600	.005	↓	1.20	.14		1.34	1.55

07 19 Water Repellents

07 19 19 – Silicone Water Repellents

07 19 19.10 Silicone Based Water Repellents		Crew	Daily Output	Labor-Hours	Unit	Material	2012 Bare Costs Labor	Equipment	Total	Total Incl O&P
0010	**SILICONE BASED WATER REPELLENTS**									
0020	Water base liquid, roller applied	2 Rofc	7000	.002	S.F.	.89	.09		.98	1.12
0200	Silicone or stearate, sprayed on CMU, 1 coat	1 Rofc	4000	.002		.33	.07		.40	.49
0300	2 coats	"	3000	.003		.66	.10		.76	.90

07 21 Thermal Insulation

07 21 13 – Board Insulation

07 21 13.10 Rigid Insulation

			Crew	Daily Output	Labor-Hours	Unit	Material	2012 Bare Costs Labor	Equipment	Total	Total Incl O&P
0010	**RIGID INSULATION**, for walls										
0040	Fiberglass, 1.5#/C.F., unfaced, 1" thick, R4.1	G	1 Carp	1000	.008	S.F.	.25	.35		.60	.82
0060	1-1/2" thick, R6.2	G		1000	.008		.37	.35		.72	.95
0080	2" thick, R8.3	G		1000	.008		.41	.35		.76	.99
0120	3" thick, R12.4	G		800	.010		.51	.44		.95	1.24
0370	3#/C.F., unfaced, 1" thick, R4.3	G		1000	.008		.48	.35		.83	1.07
0390	1-1/2" thick, R6.5	G		1000	.008		.72	.35		1.07	1.33
0400	2" thick, R8.7	G		890	.009		.96	.40		1.36	1.67
0420	2-1/2" thick, R10.9	G		800	.010		1.01	.44		1.45	1.79
0440	3" thick, R13	G		800	.010		1.46	.44		1.90	2.29
0520	Foil faced, 1" thick, R4.3	G		1000	.008		.83	.35		1.18	1.45
0540	1-1/2" thick, R6.5	G		1000	.008		1.24	.35		1.59	1.90
0560	2" thick, R8.7	G		890	.009		1.55	.40		1.95	2.32
0580	2-1/2" thick, R10.9	G		800	.010		1.82	.44		2.26	2.68
0600	3" thick, R13	G		800	.010		2	.44		2.44	2.88
1600	Isocyanurate, 4' x 8' sheet, foil faced, both sides										
1610	1/2" thick	G	1 Carp	800	.010	S.F.	.29	.44		.73	1
1620	5/8" thick	G		800	.010		.32	.44		.76	1.03
1630	3/4" thick	G		800	.010		.33	.44		.77	1.04
1640	1" thick	G		800	.010		.51	.44		.95	1.24
1650	1-1/2" thick	G		730	.011		.64	.48		1.12	1.44
1660	2" thick	G		730	.011		.78	.48		1.26	1.60
1670	3" thick	G		730	.011		1.78	.48		2.26	2.70
1680	4" thick	G		730	.011		2	.48		2.48	2.94
1700	Perlite, 1" thick, R2.77	G		800	.010		.30	.44		.74	1.01
1750	2" thick, R5.55	G		730	.011		.60	.48		1.08	1.40
1900	Extruded polystyrene, 25 PSI compressive strength, 1" thick, R5	G		800	.010		.50	.44		.94	1.23
1940	2" thick R10	G		730	.011		1	.48		1.48	1.84
1960	3" thick, R15	G		730	.011		1.44	.48		1.92	2.32
2100	Expanded polystyrene, 1" thick, R3.85	G		800	.010		.28	.44		.72	.99
2120	2" thick, R7.69	G		730	.011		.56	.48		1.04	1.36
2140	3" thick, R11.49	G		730	.011		.84	.48		1.32	1.66

07 21 13.13 Foam Board Insulation

			Crew	Daily Output	Labor-Hours	Unit	Material	2012 Bare Costs Labor	Equipment	Total	Total Incl O&P
0010	**FOAM BOARD INSULATION**										
0600	Polystyrene, expanded, 1" thick, R4	G	1 Carp	680	.012	S.F.	.28	.52		.80	1.11
0700	2" thick, R8	G	"	675	.012	"	.56	.52		1.08	1.42

07 21 16 – Blanket Insulation

07 21 16.10 Blanket Insulation for Floors/Ceilings

			Crew	Daily Output	Labor-Hours	Unit	Material	2012 Bare Costs Labor	Equipment	Total	Total Incl O&P
0010	**BLANKET INSULATION FOR FLOORS/CEILINGS**										
0020	Including spring type wire fasteners										
2000	Fiberglass, blankets or batts, paper or foil backing										
2100	3-1/2" thick, R13	G	1 Carp	700	.011	S.F.	.38	.50		.88	1.20

07 21 Thermal Insulation

07 21 16 – Blanket Insulation

07 21 16.10 Blanket Insulation for Floors/Ceilings

		Crew	Daily Output	Labor-Hours	Unit	Material	2012 Bare Costs Labor	Equipment	Total	Total Incl O&P
2150	6-1/4" thick, R19	G 1 Carp	600	.013	S.F.	.43	.59		1.02	1.37
2210	9-1/2" thick, R30	G	500	.016		.66	.71		1.37	1.82
2220	12" thick, R38	G	475	.017		.72	.74		1.46	1.93
3000	Unfaced, 3-1/2" thick, R13	G	600	.013		.31	.59		.90	1.24
3010	6-1/4" thick, R19	G	500	.016		.35	.71		1.06	1.48
3020	9-1/2" thick, R30	G	450	.018		.62	.78		1.40	1.89
3030	12" thick, R38	G	425	.019		.71	.83		1.54	2.06

07 21 16.20 Blanket Insulation for Walls

		Crew	Daily Output	Labor-Hours	Unit	Material	2012 Bare Costs Labor	Equipment	Total	Total Incl O&P
0010	**BLANKET INSULATION FOR WALLS**									
0020	Kraft faced fiberglass, 3-1/2" thick, R11, 15" wide	G 1 Carp	1350	.006	S.F.	.26	.26		.52	.69
0030	23" wide	G	1600	.005		.26	.22		.48	.63
0060	R13, 11" wide	G	1150	.007		.33	.31		.64	.83
0080	15" wide	G	1350	.006		.33	.26		.59	.76
0100	23" wide	G	1600	.005		.33	.22		.55	.70
0110	R15, 11" wide	G	1150	.007		.43	.31		.74	.94
0120	15" wide	G	1350	.006		.43	.26		.69	.87
0130	23" wide	G	1600	.005		.43	.22		.65	.81
0140	6" thick, R19, 11" wide	G	1150	.007		.38	.31		.69	.89
0160	15" wide	G	1350	.006		.38	.26		.64	.82
0180	23" wide	G	1600	.005		.38	.22		.60	.76
0182	R21, 11" wide	G	1150	.007		.52	.31		.83	1.04
0184	15" wide	G	1350	.006		.52	.26		.78	.97
0186	23" wide	G	1600	.005		.52	.22		.74	.91
0188	9" thick, R30, 11" wide	G	985	.008		.62	.36		.98	1.23
0200	15" wide	G	1150	.007		.62	.31		.93	1.15
0220	23" wide	G	1350	.006		.62	.26		.88	1.08
0230	12" thick, R38, 11" wide	G	985	.008		.72	.36		1.08	1.34
0240	15" wide	G	1150	.007		.72	.31		1.03	1.26
0260	23" wide	G	1350	.006		.72	.26		.98	1.19
0410	Foil faced fiberglass, 3-1/2" thick, R13, 11" wide	G	1150	.007		.34	.31		.65	.84
0420	15" wide	G	1350	.006		.34	.26		.60	.77
0440	23" wide	G	1600	.005		.34	.22		.56	.71
0442	R15, 11" wide	G	1150	.007		.44	.31		.75	.95
0444	15" wide	G	1350	.006		.44	.26		.70	.88
0446	23" wide	G	1600	.005		.44	.22		.66	.82
0448	6" thick, R19, 11" wide	G	1150	.007		.40	.31		.71	.91
0460	15" wide	G	1350	.006		.40	.26		.66	.84
0480	23" wide	G	1600	.005		.40	.22		.62	.78
0482	R21, 11" wide	G	1150	.007		.55	.31		.86	1.08
0484	15" wide	G	1350	.006		.55	.26		.81	1.01
0486	23" wide	G	1600	.005		.55	.22		.77	.95
0488	9" thick, R30, 11" wide	G	985	.008		.65	.36		1.01	1.27
0500	15" wide	G	1150	.007		.65	.31		.96	1.19
0550	23" wide	G	1350	.006		.65	.26		.91	1.12
0560	12" thick, R38, 11" wide	G	985	.008		.75	.36		1.11	1.38
0570	15" wide	G	1150	.007		.75	.31		1.06	1.30
0580	23" wide	G	1350	.006		.75	.26		1.01	1.23
0620	Unfaced fiberglass, 3-1/2" thick, R13, 11" wide	G	1150	.007		.31	.31		.62	.81
0820	15" wide	G	1350	.006		.31	.26		.57	.74
0830	23" wide	G	1600	.005		.31	.22		.53	.68
0832	R15, 11" wide	G	1150	.007		.41	.31		.72	.92
0834	15" wide	G	1350	.006		.41	.26		.67	.85

07 21 16 – Blanket Insulation

07 21 16.20 Blanket Insulation for Walls

			Crew	Daily Output	Labor-Hours	Unit	Material	Labor	Equipment	Total	Total Incl O&P
								2012 Bare Costs			
0836	23" wide	G	1 Carp	1600	.005	S.F.	.41	.22		.63	.79
0838	6" thick, R19, 11" wide	G		1150	.007		.35	.31		.66	.86
0860	15" wide	G		1150	.007		.35	.31		.66	.86
0880	23" wide	G		1350	.006		.35	.26		.61	.79
0882	R21, 11" wide	G		1150	.007		.50	.31		.81	1.02
0886	15" wide	G		1350	.006		.50	.26		.76	.95
0888	23" wide	G		1600	.005		.50	.22		.72	.89
0890	9" thick, R30, 11" wide	G		985	.008		.60	.36		.96	1.21
0900	15" wide	G		1150	.007		.60	.31		.91	1.13
0920	23" wide	G		1350	.006		.60	.26		.86	1.06
0930	12" thick, R38, 11" wide	G		985	.008		.71	.36		1.07	1.33
0940	15" wide	G		1000	.008		.71	.35		1.06	1.32
0960	23" wide	G		1150	.007		.71	.31		1.02	1.25
1300	Mineral fiber batts, kraft faced										
1320	3-1/2" thick, R12	G	1 Carp	1600	.005	S.F.	.38	.22		.60	.76
1340	6" thick, R19	G		1600	.005		.38	.22		.60	.76
1380	10" thick, R30	G		1350	.006		.62	.26		.88	1.08
1850	Friction fit wire insulation supports, 16" O.C.			960	.008	Ea.	.05	.37		.42	.63

07 21 23 – Loose-Fill Insulation

07 21 23.10 Poured Loose-Fill Insulation

			Crew	Daily Output	Labor-Hours	Unit	Material	Labor	Equipment	Total	Total Incl O&P
0010	**POURED LOOSE-FILL INSULATION**										
0020	Cellulose fiber, R3.8 per inch	G	1 Carp	200	.040	C.F.	.53	1.76		2.29	3.30
0040	Ceramic type (perlite), R3.2 per inch	G		200	.040		5	1.76		6.76	8.20
0080	Fiberglass wool, R4 per inch	G		200	.040		.10	1.76		1.86	2.83
0100	Mineral wool, R3 per inch	G		200	.040		.39	1.76		2.15	3.14
0300	Polystyrene, R4 per inch	G		200	.040		3.09	1.76		4.85	6.10
0400	Perlite, R2.7 per inch	G		200	.040		5	1.76		6.76	8.20
0700	Wood fiber, R3.85 per inch	G		200	.040		.70	1.76		2.46	3.48

07 21 23.20 Masonry Loose-Fill Insulation

			Crew	Daily Output	Labor-Hours	Unit	Material	Labor	Equipment	Total	Total Incl O&P
0010	**MASONRY LOOSE-FILL INSULATION**, vermiculite or perlite										
0100	In cores of concrete block, 4" thick wall, .115 C.F./S.F.	G	D-1	4800	.003	S.F.	.58	.13		.71	.83
0200	6" thick wall, .175 C.F./S.F.	G		3000	.005		.88	.21		1.09	1.28
0300	8" thick wall, .258 C.F./S.F.	G		2400	.007		1.29	.27		1.56	1.82
0400	10" thick wall, .340 C.F./S.F.	G		1850	.009		1.70	.34		2.04	2.39
0500	12" thick wall, .422 C.F./S.F.	G		1200	.013		2.11	.53		2.64	3.12
0550	For sand fill, deduct from above	G					70%				
0600	Poured cavity wall, vermiculite or perlite, water repellant	G	D-1	250	.064	C.F.	5	2.55		7.55	9.35
0700	Foamed in place, urethane in 2-5/8" cavity	G	G-2A	1035	.023	S.F.	1.29	.78	.61	2.68	3.35
0800	For each 1" added thickness, add	G	"	2372	.010	"	.42	.34	.27	1.03	1.30

07 21 26 – Blown Insulation

07 21 26.10 Blown Insulation

			Crew	Daily Output	Labor-Hours	Unit	Material	Labor	Equipment	Total	Total Incl O&P
0010	**BLOWN INSULATION** Ceilings, with open access										
0020	Cellulose, 3-1/2" thick, R13	G	G-4	5000	.005	S.F.	.18	.17	.07	.42	.54
0030	5-3/16" thick, R19	G		3800	.006		.27	.23	.09	.59	.75
0050	6-1/2" thick, R22	G		3000	.008		.35	.29	.12	.76	.95
0100	8-11/16" thick, R30	G		2600	.009		.47	.33	.13	.93	1.17
0120	10-7/8" thick, R38	G		1800	.013		.60	.48	.19	1.27	1.60
1000	Fiberglass, 5.5" thick, R11	G		3800	.006		.04	.23	.09	.36	.49
1050	6" thick, R12	G		3000	.008		.05	.29	.12	.46	.63
1100	8.8" thick, R19	G		2200	.011		.06	.39	.16	.61	.84
1200	10" thick, R22	G		1800	.013		.07	.48	.19	.74	1.02

07 21 Thermal Insulation

07 21 26 – Blown Insulation

07 21 26.10 Blown Insulation

		Crew	Daily Output	Labor-Hours	Unit	Material	2012 Bare Costs Labor	2012 Bare Costs Equipment	Total	Total Incl O&P
1300	11.5" thick, R26	G	G-4	1500	.016	S.F.	.09	.57	.23	.89	1.24
1350	13" thick, R30	G		1400	.017		.10	.61	.25	.96	1.32
1450	16" thick, R38	G		1145	.021		.13	.75	.30	1.18	1.62
1500	20" thick, R49	G		920	.026		.17	.93	.38	1.48	2.05

07 21 27 – Reflective Insulation

07 21 27.10 Reflective Insulation Options

		Crew	Daily Output	Labor-Hours	Unit	Material	2012 Bare Costs Labor	2012 Bare Costs Equipment	Total	Total Incl O&P
0010	**REFLECTIVE INSULATION OPTIONS**										
0020	Aluminum foil on reinforced scrim	G	1 Carp	19	.421	C.S.F.	14.20	18.55		32.75	44
0100	Reinforced with woven polyolefin	G		19	.421		18	18.55		36.55	48.50
0500	With single bubble air space, R8.8	G		15	.533		28	23.50		51.50	67
0600	With double bubble air space, R9.8	G		15	.533		29	23.50		52.50	67.50

07 21 29 – Sprayed Insulation

07 21 29.10 Sprayed-On Insulation

		Crew	Daily Output	Labor-Hours	Unit	Material	2012 Bare Costs Labor	2012 Bare Costs Equipment	Total	Total Incl O&P
0010	**SPRAYED-ON INSULATION**										
0020	Fibrous/cementitious, finished wall, 1" thick, R3.7	G	G-2	2050	.012	S.F.	.25	.43	.06	.74	1.01
0100	Attic, 5.2" thick, R19	G		1550	.015	"	.44	.57	.08	1.09	1.44
0200	Fiberglass, R4 per inch, vertical	G		2050	.012	B.F.	.15	.43	.06	.64	.90
0210	Horizontal	G		1550	.015	"	.15	.57	.08	.80	1.13
0300	Closed cell, spray polyurethane foam, 2 pounds per cubic foot density										
0310	1" thick	G	G-2A	6000	.004	S.F.	.42	.13	.11	.66	.80
0320	2" thick	G		3000	.008		.83	.27	.21	1.31	1.59
0330	3" thick	G		2000	.012		1.25	.40	.32	1.97	2.38
0335	3-1/2" thick	G		1715	.014		1.46	.47	.37	2.30	2.76
0340	4" thick	G		1500	.016		1.67	.54	.42	2.63	3.16
0350	5" thick	G		1200	.020		2.08	.67	.53	3.28	3.96
0355	5-1/2" thick	G		1090	.022		2.29	.74	.58	3.61	4.36
0360	6" thick	G		1000	.024		2.50	.80	.63	3.93	4.75

07 22 Roof and Deck Insulation

07 22 16 – Roof Board Insulation

07 22 16.10 Roof Deck Insulation

		Crew	Daily Output	Labor-Hours	Unit	Material	2012 Bare Costs Labor	2012 Bare Costs Equipment	Total	Total Incl O&P
0010	**ROOF DECK INSULATION**										
0020	Fiberboard low density, 1/2" thick R1.39	G	1 Rofc	1000	.008	S.F.	.26	.30		.56	.79
0030	1" thick R2.78	G		800	.010		.46	.37		.83	1.14
0080	1 1/2" thick R4.17	G		800	.010		.70	.37		1.07	1.40
0100	2" thick R5.56	G		800	.010		.93	.37		1.30	1.65
0110	Fiberboard high density, 1/2" thick R1.3	G		1000	.008		.24	.30		.54	.76
0120	1" thick R2.5	G		800	.010		.48	.37		.85	1.16
0130	1-1/2" thick R3.8	G		800	.010		.72	.37		1.09	1.42
0200	Fiberglass, 3/4" thick R2.78	G		1000	.008		.56	.30		.86	1.12
0400	15/16" thick R3.70	G		1000	.008		.74	.30		1.04	1.31
0460	1-1/16" thick R4.17	G		1000	.008		.93	.30		1.23	1.52
0600	1-5/16" thick R5.26	G		1000	.008		1.26	.30		1.56	1.89
0650	2-1/16" thick R8.33	G		800	.010		1.33	.37		1.70	2.09
0700	2-7/16" thick R10	G		800	.010		1.54	.37		1.91	2.32
1650	Perlite, 1/2" thick R1.32	G		1050	.008		.34	.29		.63	.85
1655	3/4" thick R2.08	G		800	.010		.37	.37		.74	1.04
1660	1" thick R2.78	G		800	.010		.46	.37		.83	1.14
1670	1-1/2" thick R4.17	G		800	.010		.48	.37		.85	1.16
1680	2" thick R5.56	G		700	.011		.80	.43		1.23	1.60

07 22 16 – Roof Board Insulation

07 22 16.10 Roof Deck Insulation	Crew	Daily Output	Labor-Hours	Unit	Material	2012 Bare Costs Labor	Equipment	Total	Total Incl O&P
1685	2-1/2" thick R6.67	G	1 Rofc	700	.011	S.F.	.95	.43		1.38	1.77
1690	Tapered for drainage	G		800	.010	B.F.	.73	.37		1.10	1.43
1700	Polyisocyanurate, 2#/C.F. density, 3/4" thick	G		1500	.005	S.F.	.44	.20		.64	.81
1705	1" thick	G		1400	.006		.51	.21		.72	.92
1715	1-1/2" thick	G		1250	.006		.66	.24		.90	1.13
1725	2" thick	G		1100	.007		.78	.27		1.05	1.32
1735	2-1/2" thick	G		1050	.008		1.01	.29		1.30	1.59
1745	3" thick	G		1000	.008		1.23	.30		1.53	1.85
1755	3-1/2" thick	G		1000	.008		1.95	.30		2.25	2.65
1765	Tapered for drainage	G		1400	.006	B.F.	.52	.21		.73	.93
1900	Extruded Polystyrene										
1910	15 PSI compressive strength, 1" thick, R5	G	1 Rofc	1500	.005	S.F.	.44	.20		.64	.81
1920	2" thick, R10	G		1250	.006		.56	.24		.80	1.02
1930	3" thick R15	G		1000	.008		1.15	.30		1.45	1.77
1932	4" thick R20	G		1000	.008		1.52	.30		1.82	2.17
1934	Tapered for drainage	G		1500	.005	B.F.	.51	.20		.71	.89
1940	25 PSI compressive strength, 1" thick R5	G		1500	.005	S.F.	.61	.20		.81	1
1942	2" thick R10	G		1250	.006		1.15	.24		1.39	1.67
1944	3" thick R15	G		1000	.008		1.75	.30		2.05	2.43
1946	4" thick R20	G		1000	.008		2.50	.30		2.80	3.25
1948	Tapered for drainage	G		1500	.005	B.F.	.53	.20		.73	.91
1950	40 psi compressive strength, 1" thick R5	G		1500	.005	S.F.	.47	.20		.67	.85
1952	2" thick R10	G		1250	.006		.88	.24		1.12	1.37
1954	3" thick R15	G		1000	.008		1.28	.30		1.58	1.91
1956	4" thick R20	G		1000	.008		1.70	.30		2	2.37
1958	Tapered for drainage	G		1400	.006	B.F.	.67	.21		.88	1.10
1960	60 PSI compressive strength, 1" thick R5	G		1450	.006	S.F.	.65	.21		.86	1.07
1962	2" thick R10	G		1200	.007		1.25	.25		1.50	1.80
1964	3" thick R15	G		975	.008		2	.31		2.31	2.71
1966	4" thick R20	G		950	.008		2.50	.31		2.81	3.28
1968	Tapered for drainage	G		1400	.006	B.F.	.85	.21		1.06	1.30
2010	Expanded polystyrene, 1#/C.F. density, 3/4" thick R2.89	G		1500	.005	S.F.	.21	.20		.41	.56
2020	1" thick R3.85	G		1500	.005		.28	.20		.48	.64
2100	2" thick R7.69	G		1250	.006		.56	.24		.80	1.02
2110	3" thick R11.49	G		1250	.006		.84	.24		1.08	1.32
2120	4" thick R15.38	G		1200	.007		1.12	.25		1.37	1.65
2130	5" thick R19.23	G		1150	.007		1.40	.26		1.66	1.98
2140	6" thick R23.26	G		1150	.007		1.68	.26		1.94	2.29
2150	Tapered for drainage	G		1500	.005	B.F.	.51	.20		.71	.89
2400	Composites with 2" EPS										
2410	1" fiberboard	G	1 Rofc	950	.008	S.F.	1.32	.31		1.63	1.98
2420	7/16" oriented strand board	G		800	.010		1.06	.37		1.43	1.80
2430	1/2" plywood	G		800	.010		1.31	.37		1.68	2.07
2440	1" perlite	G		800	.010		1.08	.37		1.45	1.82
2450	Composites with 1-1/2" polyisocyanurate										
2460	1" fiberboard	G	1 Rofc	800	.010	S.F.	1.12	.37		1.49	1.86
2470	1" perlite	G		850	.009		1	.35		1.35	1.69
2480	7/16" oriented strand board	G		800	.010		.88	.37		1.25	1.60

07 24 Exterior Insulation and Finish Systems

07 24 13 - Polymer-Based Exterior Insulation and Finish System

07 24 13.10 Exterior Insulation and Finish Systems		Crew	Daily Output	Labor-Hours	Unit	Material	2012 Bare Costs Labor	Equipment	Total	Total Incl O&P
0010	**EXTERIOR INSULATION AND FINISH SYSTEMS**									
0095	Field applied, 1" EPS insulation	G	J-1	390	.103	S.F.	1.97	3.95	.34	6.26	8.45
0100	With 1/2" cement board sheathing	G		268	.149		2.65	5.75	.49	8.89	12.05
0105	2" EPS insulation	G		390	.103		2.25	3.95	.34	6.54	8.75
0110	With 1/2" cement board sheathing	G		268	.149		2.93	5.75	.49	9.17	12.35
0115	3" EPS insulation	G		390	.103		2.53	3.95	.34	6.82	9.05
0120	With 1/2" cement board sheathing	G		268	.149		3.21	5.75	.49	9.45	12.65
0125	4" EPS insulation	G		390	.103		2.81	3.95	.34	7.10	9.35
0130	With 1/2" cement board sheathing	G		268	.149		4.17	5.75	.49	10.41	13.75
0140	Premium finish add			1265	.032		.31	1.22	.10	1.63	2.27
0150	Heavy duty reinforcement add			914	.044		.90	1.69	.14	2.73	3.67
0160	2.5#/S.Y. metal lath substrate add		1 Lath	75	.107	S.Y.	2.37	4.19		6.56	8.75
0170	3.4#/S.Y. metal lath substrate add		"	75	.107	"	2.58	4.19		6.77	9
0180	Color or texture change,		J-1	1265	.032	S.F.	.83	1.22	.10	2.15	2.84
0190	With substrate leveling base coat		1 Plas	530	.015		.83	.61		1.44	1.82
0210	With substrate sealing base coat		1 Pord	1224	.007		.08	.25		.33	.46
0370	V groove shape in panel face					L.F.	.60			.60	.66
0380	U groove shape in panel face					"	.79			.79	.87
0440	For higher than one story, add							25%			

07 25 Weather Barriers

07 25 10 - Weather Barriers or Wraps

07 25 10.10 Weather Barriers

		Crew	Daily Output	Labor-Hours	Unit	Material	2012 Bare Costs Labor	Equipment	Total	Total Incl O&P
0010	**WEATHER BARRIERS**									
0400	Asphalt felt paper, 15#	1 Carp	37	.216	Sq.	5.65	9.55		15.20	21
0401	Per square foot	"	3700	.002	S.F.	.06	.10		.16	.21
0450	Housewrap, exterior, spun bonded polypropylene									
0470	Small roll	1 Carp	3800	.002	S.F.	.13	.09		.22	.29
0480	Large roll	"	4000	.002	"	.13	.09		.22	.28
2100	Asphalt felt roof deck vapor barrier, class 1 metal decks	1 Rofc	37	.216	Sq.	20.50	8.10		28.60	36.50
2200	For all other decks	"	37	.216		15	8.10		23.10	30
2800	Asphalt felt, 50% recycled content, 15 lb., 4 sq per roll	1 Carp	36	.222		5.65	9.80		15.45	21.50
2810	30 lb., 2 sq per roll	"	36	.222		11.30	9.80		21.10	27.50
3000	Building wrap, spunbonded polyethylene	2 Carp	8000	.002	S.F.	.17	.09		.26	.33

07 26 Vapor Retarders

07 26 10 - Above-Grade Vapor Retarders

07 26 10.10 Vapor Retarders

			Crew	Daily Output	Labor-Hours	Unit	Material	2012 Bare Costs Labor	Equipment	Total	Total Incl O&P
0010	**VAPOR RETARDERS**										
0020	Aluminum and kraft laminated, foil 1 side	G	1 Carp	37	.216	Sq.	11.25	9.55		20.80	27
0100	Foil 2 sides	G		37	.216		12.50	9.55		22.05	28.50
0600	Polyethylene vapor barrier, standard, .002" thick	G		37	.216		1.30	9.55		10.85	16.10
0700	.004" thick	G		37	.216		3.23	9.55		12.78	18.20
0900	.006" thick	G		37	.216		4.54	9.55		14.09	19.65
1200	.010" thick	G		37	.216		7	9.55		16.55	22.50
1300	Clear reinforced, fire retardant, .008" thick	G		37	.216		10.65	9.55		20.20	26.50
1350	Cross laminated type, .003" thick	G		37	.216		7.45	9.55		17	23
1400	.004" thick	G		37	.216		8.20	9.55		17.75	23.50
1800	Reinf. waterproof, .002" polyethylene backing, 1 side	G		37	.216		5.95	9.55		15.50	21

07 26 Vapor Retarders

07 26 10 – Above-Grade Vapor Retarders

07 26 10.10 Vapor Retarders	Crew	Daily Output	Labor-Hours	Unit	Material	2012 Bare Costs Labor	Equipment	Total	Total Incl O&P
1900	2 sides	1 Carp	37	.216	Sq.	7.80	9.55		17.35	23.50
2400	Waterproofed kraft with sisal or fiberglass fibers, minimum		37	.216		6.40	9.55		15.95	21.50
2500	Maximum		37	.216		15.85	9.55		25.40	32

07 31 Shingles and Shakes

07 31 13 – Asphalt Shingles

07 31 13.10 Asphalt Roof Shingles

		Crew	Daily Output	Labor-Hours	Unit	Material	2012 Bare Costs Labor	Equipment	Total	Total Incl O&P
0010	**ASPHALT ROOF SHINGLES**									
0100	Standard strip shingles									
0150	Inorganic, class A, 210-235 lb./sq.	1 Rofc	5.50	1.455	Sq.	75.50	54.50		130	174
0155	Pneumatic nailed		7	1.143		75.50	42.50		118	155
0200	Organic, class C, 235-240 lb./sq.		5	1.600		82.50	60		142.50	191
0205	Pneumatic nailed		6.25	1.280		82.50	48		130.50	172
0250	Standard, laminated multi-layered shingles									
0300	Class A, 240-260 lb./sq.	1 Rofc	4.50	1.778	Sq.	92.50	66.50		159	213
0305	Pneumatic nailed		5.63	1.422		92.50	53		145.50	191
0350	Class C, 260-300 lb./square, 4 bundles/square		4	2		135	75		210	274
0355	Pneumatic nailed		5	1.600		135	60		195	249
0400	Premium, laminated multi-layered shingles									
0450	Class A, 260-300 lb., 4 bundles/sq	1 Rofc	3.50	2.286	Sq.	159	85.50		244.50	320
0455	Pneumatic nailed		4.37	1.831		159	68.50		227.50	290
0500	Class C, 300-385 lb./square, 5 bundles/square		3	2.667		165	99.50		264.50	350
0505	Pneumatic nailed		3.75	2.133		165	80		245	315
0800	#15 felt underlayment		64	.125		5.65	4.68		10.33	14.10
0825	#30 felt underlayment		58	.138		11.30	5.15		16.45	21
0850	Self adhering polyethylene and rubberized asphalt underlayment		22	.364		74	13.60		87.60	105
0900	Ridge shingles		330	.024	L.F.	1.90	.91		2.81	3.61
0905	Pneumatic nailed		412.50	.019	"	1.90	.73		2.63	3.31
1000	For steep roofs (7 to 12 pitch or greater), add						50%			

07 31 16 – Metal Shingles

07 31 16.10 Aluminum Shingles

		Crew	Daily Output	Labor-Hours	Unit	Material	2012 Bare Costs Labor	Equipment	Total	Total Incl O&P
0010	**ALUMINUM SHINGLES**									
0020	Mill finish, .019 thick	1 Carp	5	1.600	Sq.	191	70.50		261.50	320
0100	.020" thick	"	5	1.600		214	70.50		284.50	345
0300	For colors, add					20			20	22
0600	Ridge cap, .024" thick	1 Carp	170	.047	L.F.	2.18	2.08		4.26	5.60
0700	End wall flashing, .024" thick		170	.047		1.78	2.08		3.86	5.15
0900	Valley section, .024" thick		170	.047		2.90	2.08		4.98	6.40
1000	Starter strip, .024" thick		400	.020		1.41	.88		2.29	2.91
1200	Side wall flashing, .024" thick		170	.047		1.78	2.08		3.86	5.15
1500	Gable flashing, .024" thick		400	.020		1.41	.88		2.29	2.91

07 31 16.20 Steel Shingles

		Crew	Daily Output	Labor-Hours	Unit	Material	2012 Bare Costs Labor	Equipment	Total	Total Incl O&P
0010	**STEEL SHINGLES**									
0012	Galvanized, 26 gauge	1 Rots	2.20	3.636	Sq.	299	137		436	560
0200	24 gauge	"	2.20	3.636		299	137		436	560
0300	For colored galvanized shingles, add					55			55	60.50
0500	For 1" factory applied polystyrene insulation, add					40			40	44

07 31 Shingles and Shakes

07 31 26 – Slate Shingles

07 31 26.10 Slate Roof Shingles		Crew	Daily Output	Labor-Hours	Unit	Material	2012 Bare Costs Labor	Equipment	Total	Total Incl O&P
0010	**SLATE ROOF SHINGLES**	R073126-20								
0100	Buckingham Virginia black, 3/16" - 1/4" thick	G	1 Rots	1.75	4.571	Sq.	475	172		647	815
0200	1/4" thick	G		1.75	4.571		475	172		647	815
0900	Pennsylvania black, Bangor, #1 clear	G		1.75	4.571		490	172		662	830
1200	Vermont, unfading, green, mottled green	G		1.75	4.571		480	172		652	815
1300	Semi-weathering green & gray	G		1.75	4.571		350	172		522	670
1400	Purple	G		1.75	4.571		425	172		597	755
1500	Black or gray	G		1.75	4.571		460	172		632	795
1600	Red	G		1.75	4.571		1,175	172		1,347	1,575
1700	Variegated purple	G		1.75	4.571		415	172		587	745

07 31 29 – Wood Shingles and Shakes

07 31 29.13 Wood Shingles

07 31 29.13 Wood Shingles		Crew	Daily Output	Labor-Hours	Unit	Material	2012 Bare Costs Labor	Equipment	Total	Total Incl O&P
0010	**WOOD SHINGLES**									
0012	16" No. 1 red cedar shingles, 5" exposure, on roof	1 Carp	2.50	3.200	Sq.	208	141		349	445
0015	Pneumatic nailed		3.25	2.462		208	109		317	395
0200	7-1/2" exposure, on walls		2.05	3.902		139	172		311	420
0205	Pneumatic nailed		2.67	2.996		139	132		271	355
0300	18" No. 1 red cedar perfections, 5-1/2" exposure, on roof		2.75	2.909		228	128		356	450
0305	Pneumatic nailed		3.57	2.241		228	99		327	405
0500	7-1/2" exposure, on walls		2.25	3.556		168	157		325	425
0505	Pneumatic nailed		2.92	2.740		168	121		289	370
0600	Resquared, and rebutted, 5-1/2" exposure, on roof		3	2.667		260	118		378	465
0605	Pneumatic nailed		3.90	2.051		260	90.50		350.50	425
0900	7-1/2" exposure, on walls		2.45	3.265		191	144		335	430
0905	Pneumatic nailed		3.18	2.516		191	111		302	380
1000	Add to above for fire retardant shingles					55			55	60.50
1060	Preformed ridge shingles	1 Carp	400	.020	L.F.	3.60	.88		4.48	5.30
2000	White cedar shingles, 16" long, extras, 5" exposure, on roof		2.40	3.333	Sq.	154	147		301	395
2005	Pneumatic nailed		3.12	2.564		154	113		267	345
2050	5" exposure on walls		2	4		154	176		330	440
2055	Pneumatic nailed		2.60	3.077		154	136		290	380
2100	7-1/2" exposure, on walls		2	4		110	176		286	390
2105	Pneumatic nailed		2.60	3.077		110	136		246	330
2150	"B" grade, 5" exposure on walls		2	4		136	176		312	420
2155	Pneumatic nailed		2.60	3.077		136	136		272	360
2300	For 15# organic felt underlayment on roof, 1 layer, add		64	.125		5.65	5.50		11.15	14.75
2400	2 layers, add		32	.250		11.30	11.05		22.35	29.50
2600	For steep roofs (7/12 pitch or greater), add to above						50%			
2700	Panelized systems, No.1 cedar shingles on 5/16" CDX plywood									
2800	On walls, 8' strips, 7" or 14" exposure	2 Carp	700	.023	S.F.	5.55	1.01		6.56	7.65
3500	On roofs, 8' strips, 7" or 14" exposure	1 Carp	3	2.667	Sq.	555	118		673	790
3505	Pneumatic nailed	"	4	2	"	555	88		643	745

07 31 29.16 Wood Shakes

07 31 29.16 Wood Shakes		Crew	Daily Output	Labor-Hours	Unit	Material	2012 Bare Costs Labor	Equipment	Total	Total Incl O&P
0010	**WOOD SHAKES**									
1100	Hand-split red cedar shakes, 1/2" thick x 24" long, 10" exp. on roof	1 Carp	2.50	3.200	Sq.	233	141		374	475
1105	Pneumatic nailed		3.25	2.462		233	109		342	425
1110	3/4" thick x 24" long, 10" exp. on roof		2.25	3.556		233	157		390	495
1115	Pneumatic nailed		2.92	2.740		233	121		354	440
1200	1/2" thick, 18" long, 8-1/2" exp. on roof		2	4		159	176		335	445
1205	Pneumatic nailed		2.60	3.077		159	136		295	385
1210	3/4" thick x 18" long, 8 1/2" exp. on roof		1.80	4.444		159	196		355	475

07 31 Shingles and Shakes

07 31 29 – Wood Shingles and Shakes

07 31 29.16 Wood Shakes		Crew	Daily Output	Labor-Hours	Unit	Material	2012 Bare Costs Labor	Equipment	Total	Total Incl O&P
1215	Pneumatic nailed	1 Carp	2.34	3.419	Sq.	159	151		310	405
1255	10" exp. on walls		2	4		154	176		330	440
1260	10" exposure on walls, pneumatic nailed		2.60	3.077		154	136		290	380
1700	Add to above for fire retardant shakes, 24" long					55			55	60.50
1800	18" long					55			55	60.50
1810	Ridge shakes	1 Carp	350	.023	L.F.	3.60	1.01		4.61	5.50

07 32 Roof Tiles

07 32 13 – Clay Roof Tiles

07 32 13.10 Clay Tiles

		Crew	Daily Output	Labor-Hours	Unit	Material	2012 Bare Costs Labor	Equipment	Total	Total Incl O&P
0010	**CLAY TILES**, including accessories									
2500	Imported, English shingle, 555 pcs/sq., 6-1/2" x 10-1/2"	3 Rots	5	4.800	Sq.	1,175	180		1,355	1,600
2510	Flat interlocking, 210 pcs/sq., 8" x 14-1/4"		5.50	4.364		585	164		749	920
2520	192 pcs/sq., 9" x 13"		5.50	4.364		610	164		774	945
2530	100 pcs/sq., 13" x 16"		6	4		460	150		610	755
2540	94 pcs/sq., 12-1/2" x 18-1/4"		6	4		485	150		635	785
2550	93 pcs/sq., 12" x 18-1/2"		6	4		485	150		635	785
2560	French, 124 pcs/sq., 10-1/4" x 17-1/2"		6	4		535	150		685	840
2570	120 pcs/sq., 9-3/4" x 17-1/2"		6	4		385	150		535	675
2580	115 pcs/sq., 10-1/4" x 17-1/4"		6	4		485	150		635	785
2590	91 pcs/sq., 12-1/2" x 18-3/4"		6	4		510	150		660	810
2600	French shingle, 495 pcs/sq., 6-3/4" x 12"		5	4.800		1,400	180		1,580	1,850
2610	Gaelic, 190 pcs/sq., 8-3/4" x 12"		5.50	4.364		460	164		624	780
2620	120 pcs/sq., 10-3/4" x 17-1/2"		6	4		560	150		710	865
2630	84 pcs/sq., 13-1/4" x 19"		6	4		460	150		610	755
2640	Handcrafted shingle, 320 pcs/sq., 7-1/4" x 14-3/4"		5	4.800		810	180		990	1,200
2650	Mediterranean S - Roman, 115 pcs/sq., 10" x 17-3/4"		6	4		410	150		560	700
2660	112 pcs/sq., 10" x 17-1/2"		6	4		355	150		505	640
2670	100 pcs/sq., 10-1/2" x 18-1/2"		6	4		254	150		404	530
2680	Mission pan & cover, 260 pcs/sq., 6-1/4" x 17-3/4"		5.50	4.364		435	164		599	755
2690	183 pcs/sq., 8" x 18"		5.50	4.364		535	164		699	865
2700	Pantile, 180 pcs/sq., 9-3/4" x 14"		5.50	4.364		685	164		849	1,025
2710	150 pcs/sq., 10-1/4" x 14-1/2"		6	4		585	150		735	895
2720	115 pcs/sq., 11-1/4" x 18"		6	4		535	150		685	840
3010	#15 felt underlayment	1 Rofc	64	.125		5.65	4.68		10.33	14.10
3020	#30 felt underlayment		58	.138		11.30	5.15		16.45	21
3040	Polyethylene and rubberized asph. underlayment		22	.364		74	13.60		87.60	105

07 32 16 – Concrete Roof Tiles

07 32 16.10 Concrete Tiles

		Crew	Daily Output	Labor-Hours	Unit	Material	2012 Bare Costs Labor	Equipment	Total	Total Incl O&P
0010	**CONCRETE TILES**									
0020	Corrugated, 13" x 16-1/2", 90 per sq., 950 lb. per sq.									
0050	Earthtone colors, nailed to wood deck	1 Rots	1.35	5.926	Sq.	97.50	223		320.50	480
0150	Blues		1.35	5.926		97.50	223		320.50	480
0200	Greens		1.35	5.926		101	223		324	485
0250	Premium colors		1.35	5.926		101	223		324	485
0500	Shakes, 13" x 16-1/2", 90 per sq., 950 lb. per sq.									
0600	All colors, nailed to wood deck	1 Rots	1.50	5.333	Sq.	83	200		283	425
1500	Accessory pieces, ridge & hip, 10" x 16-1/2", 8 lb. each	"	120	.067	Ea.	3.64	2.50		6.14	8.20
1700	Rake, 6-1/2" x 16-3/4", 9 lb. each					3.64			3.64	4
1800	Mansard hip, 10" x 16-1/2", 9.2 lb. each					3.64			3.64	4

07 32 Roof Tiles

07 32 16 – Concrete Roof Tiles

07 32 16.10 Concrete Tiles		Crew	Daily Output	Labor-Hours	Unit	Material	2012 Bare Costs Labor	Equipment	Total	Total Incl O&P
1900	Hip starter, 10" x 16-1/2", 10.5 lb. each				Ea.	11.45			11.45	12.60
2000	3 or 4 way apex, 10" each side, 11.5 lb. each				↓	13.25			13.25	14.55

07 32 19 – Metal Roof Tiles

07 32 19.10 Metal Roof Tiles

		Crew	Daily Output	Labor-Hours	Unit	Material	Labor	Equipment	Total	Total Incl O&P
0010	**METAL ROOF TILES**									
0020	Accessories included, .032" thick aluminum, mission tile	1 Carp	2.50	3.200	Sq.	815	141		956	1,100
0200	Spanish tiles	"	3	2.667	"	560	118		678	795

07 33 Natural Roof Coverings

07 33 63 – Vegetated Roofing

07 33 63.10 Green Roof Systems

			Crew	Daily Output	Labor-Hours	Unit	Material	Labor	Equipment	Total	Total Incl O&P
0010	**GREEN ROOF SYSTEMS**										
0020	Soil mixture for green roof 30% sand, 55% gravel, 15% soil	G									
0100	Hoist and spread soil mixture 4 inch depth up to five stories tall roof	G	B-13B	4000	.014	S.F.	.25	.53	.28	1.06	1.38
0150	6 inch depth	G		2667	.021		.37	.80	.41	1.58	2.08
0200	8 inch depth	G		2000	.028		.50	1.07	.55	2.12	2.79
0250	10 inch depth	G		1600	.035		.62	1.33	.69	2.64	3.49
0300	12 inch depth	G		1335	.042		.75	1.60	.82	3.17	4.17
0310	Alt. man-made soil mix, hoist & spread, 4" deep up to 5 stories tall roof	G		4000	.014		1.50	.53	.28	2.31	2.76
0350	Mobilization 55 ton crane to site	G	1 Eqhv	3.60	2.222	Ea.		106		106	160
0355	Hoisting cost to five stories per day (Avg. 28 picks per day)	G	B-13B	1	56	Day		2,125	1,100	3,225	4,450
0360	Mobilization or demobilization, 100 ton crane to site driver & escort	G	A-3E	2.50	6.400	Ea.		266	55.50	321.50	460
0365	Hoisting cost six to ten stories per day (Avg. 21 picks per day)	G	B-13C	1	56	Day		2,125	1,625	3,750	5,050
0370	Hoist and spread soil mixture 4 inch depth six to ten stories tall roof	G		4000	.014	S.F.	.25	.53	.41	1.19	1.53
0375	6 inch depth	G		2667	.021		.37	.80	.61	1.78	2.30
0380	8 inch depth	G		2000	.028		.50	1.07	.81	2.38	3.08
0385	10 inch depth	G		1600	.035		.62	1.33	1.02	2.97	3.85
0390	12 inch depth	G		1335	.042		.75	1.60	1.22	3.57	4.60
0400	Green roof edging treated lumber 4" x 4" no hoisting included	G	2 Carp	400	.040	L.F.	.82	1.76		2.58	3.61
0410	4" x 6"	G		400	.040		2.11	1.76		3.87	5.05
0420	4" x 8"	G		360	.044		4.32	1.96		6.28	7.80
0430	4" x 6" double stacked	G		300	.053		4.21	2.35		6.56	8.25
0500	Green roof edging redwood lumber 4" x 4" no hoisting included	G		400	.040		6.40	1.76		8.16	9.75
0510	4" x 6"	G		400	.040		11.20	1.76		12.96	15
0520	4" x 8"	G		360	.044		20.50	1.96		22.46	25.50
0530	4" x 6" double stacked	G		300	.053		22.50	2.35		24.85	28
0550	Roof membrane and root barrier, not including insulation:	G									
0560	Fluid applied rubber membrane, reinforced, 215 mil thick	G	G-5	350	.114	S.F.	.26	3.88	.51	4.65	7.35
0570	Root barrier	G	2 Rofc	775	.021		.70	.77		1.47	2.06
0580	Moisture retention barrier and reservoir	G	"	900	.018		2.70	.67		3.37	4.08
0600	Planting sedum, light soil, potted, 2-1/4" diameter, two per S.F.	G	1 Clab	420	.019		5	.67		5.67	6.55
0610	one per S.F.	G	"	840	.010		2.50	.33		2.83	3.26
0630	Planting sedum mat per S.F. including shipping (4000 S.F. min)	G	4 Clab	4000	.008		6	.28		6.28	7.05
0640	Installation sedum mat system (no soil required) per S.F. (4000 S.F. min)	G	"	4000	.008		8.35	.28		8.63	9.65
0645	Note: pricing of sedum mats shipped in full truck loads (4000-5000 S.F.)	G									

07 41 13.10 Aluminum Roof Panels

07 41 13.10 Aluminum Roof Panels	Crew	Daily Output	Labor-Hours	Unit	Material	2012 Bare Costs Labor	Equipment	Total	Total Incl O&P
0010	**ALUMINUM ROOF PANELS**								
0020	Corrugated or ribbed, .0155" thick, natural	G-3	1200	.027	S.F.	.98	1.16		2.14	2.86
0300	Painted		1200	.027		1.43	1.16		2.59	3.35
0400	Corrugated, .018" thick, on steel frame, natural finish		1200	.027		1.29	1.16		2.45	3.20
0600	Painted		1200	.027		1.58	1.16		2.74	3.52
0700	Corrugated, on steel frame, natural, .024" thick		1200	.027		1.83	1.16		2.99	3.79
0800	Painted		1200	.027		2.23	1.16		3.39	4.23
0900	.032" thick, natural		1200	.027		2.36	1.16		3.52	4.38
1200	Painted		1200	.027		3.03	1.16		4.19	5.10
1300	V-Beam, on steel frame construction, .032" thick, natural		1200	.027		2.50	1.16		3.66	4.53
1500	Painted		1200	.027		3.07	1.16		4.23	5.15
1600	.040" thick, natural		1200	.027		3.04	1.16		4.20	5.10
1800	Painted		1200	.027		3.66	1.16		4.82	5.80
1900	.050" thick, natural		1200	.027		3.65	1.16		4.81	5.80
2100	Painted		1200	.027		4.41	1.16		5.57	6.65
2200	For roofing on wood frame, deduct		4600	.007		.08	.30		.38	.55
2400	Ridge cap, .032" thick, natural		800	.040	L.F.	2.88	1.75		4.63	5.85

07 41 13.20 Steel Roofing Panels

07 41 13.20 Steel Roofing Panels	Crew	Daily Output	Labor-Hours	Unit	Material	2012 Bare Costs Labor	Equipment	Total	Total Incl O&P
0010	**STEEL ROOFING PANELS**								
0012	Corrugated or ribbed, on steel framing, 30 ga galv	G-3	1100	.029	S.F.	1.75	1.27		3.02	3.87
0100	28 ga		1050	.030		1.86	1.33		3.19	4.08
0300	26 ga		1000	.032		1.92	1.40		3.32	4.25
0400	24 ga		950	.034		3.03	1.47		4.50	5.60
0600	Colored, 28 ga		1050	.030		1.75	1.33		3.08	3.96
0700	26 ga		1000	.032		2.07	1.40		3.47	4.42
0710	Flat profile, 1-3/4" standing seams, 10" wide, standard finish, 26 ga		1000	.032		4.05	1.40		5.45	6.60
0715	24 ga		950	.034		4.69	1.47		6.16	7.40
0720	22 ga		900	.036		5.80	1.55		7.35	8.70
0725	Zinc aluminum alloy finish, 26 ga		1000	.032		3.17	1.40		4.57	5.65
0730	24 ga		950	.034		3.79	1.47		5.26	6.40
0735	22 ga		900	.036		4.34	1.55		5.89	7.15
0740	12" wide, standard finish, 26 ga		1000	.032		4.04	1.40		5.44	6.60
0745	24 ga		950	.034		5.30	1.47		6.77	8.10
0750	Zinc aluminum alloy finish, 26 ga		1000	.032		4.59	1.40		5.99	7.20
0755	24 ga		950	.034		3.78	1.47		5.25	6.40
0840	Flat profile, 1" x 3/8" batten, 12" wide, standard finish, 26 ga		1000	.032		3.56	1.40		4.96	6.05
0845	24 ga		950	.034		4.17	1.47		5.64	6.85
0850	22 ga		900	.036		4.99	1.55		6.54	7.85
0855	Zinc aluminum alloy finish, 26 ga		1000	.032		3.41	1.40		4.81	5.90
0860	24 ga		950	.034		3.81	1.47		5.28	6.45
0865	22 ga		900	.036		4.40	1.55		5.95	7.20
0870	16-1/2" wide, standard finish, 24 ga		950	.034		4.11	1.47		5.58	6.75
0875	22 ga		900	.036		4.60	1.55		6.15	7.40
0880	Zinc aluminum alloy finish, 24 ga		950	.034		3.59	1.47		5.06	6.20
0885	22 ga		900	.036		4	1.55		5.55	6.75
0890	Flat profile, 2" x 2" batten, 12" wide, standard finish, 26 ga		1000	.032		4.09	1.40		5.49	6.65
0895	24 ga		950	.034		4.88	1.47		6.35	7.60
0900	22 ga		900	.036		6	1.55		7.55	8.95
0905	Zinc aluminum alloy finish, 26 ga		1000	.032		3.81	1.40		5.21	6.35
0910	24 ga		950	.034		4.35	1.47		5.82	7.05
0915	22 ga		900	.036		5.05	1.55		6.60	7.90
0920	16-1/2" wide, standard finish, 24 ga		950	.034		4.49	1.47		5.96	7.20

07 41 Roof Panels

07 41 13 – Metal Roof Panels

07 41 13.20 Steel Roofing Panels	Crew	Daily Output	Labor-Hours	Unit	Material	2012 Bare Costs Labor	Equipment	Total	Total Incl O&P
0925	22 ga	G-3	900	.036	S.F.	5.25	1.55		6.80	8.15
0930	Zinc aluminum alloy finish, 24 ga		950	.034		4.07	1.47		5.54	6.75
0935	22 ga		900	.036	↓	4.63	1.55		6.18	7.45
1200	Ridge, galvanized, 10" wide		800	.040	L.F.	3.18	1.75		4.93	6.15
1210	20" wide	↓	750	.043	"	4.50	1.86		6.36	7.80

07 41 33 – Plastic Roof Panels

07 41 33.10 Fiberglass Panels

		Crew	Daily Output	Labor-Hours	Unit	Material	2012 Bare Costs Labor	Equipment	Total	Total Incl O&P
0010	**FIBERGLASS PANELS**									
0012	Corrugated panels, roofing, 8 oz. per S.F.	G-3	1000	.032	S.F.	1.90	1.40		3.30	4.23
0100	12 oz. per S.F.		1000	.032		3.44	1.40		4.84	5.90
0300	Corrugated siding, 6 oz. per S.F.		880	.036		1.61	1.59		3.20	4.20
0400	8 oz. per S.F.		880	.036		1.90	1.59		3.49	4.52
0500	Fire retardant		880	.036		3.24	1.59		4.83	6
0600	12 oz. siding, textured		880	.036		3.41	1.59		5	6.20
0700	Fire retardant		880	.036		4.53	1.59		6.12	7.40
0900	Flat panels, 6 oz. per S.F., clear or colors		880	.036		2.01	1.59		3.60	4.64
1100	Fire retardant, class A		880	.036		3.28	1.59		4.87	6.05
1300	8 oz. per S.F., clear or colors		880	.036		2.21	1.59		3.80	4.86
1700	Sandwich panels, fiberglass, 1-9/16" thick, panels to 20 S.F.		180	.178		23.50	7.75		31.25	38
1900	As above, but 2-3/4" thick, panels to 100 S.F.	↓	265	.121	↓	17.50	5.25		22.75	27.50

07 42 Wall Panels

07 42 13 – Metal Wall Panels

07 42 13.10 Mansard Panels

		Crew	Daily Output	Labor-Hours	Unit	Material	2012 Bare Costs Labor	Equipment	Total	Total Incl O&P
0010	**MANSARD PANELS**									
0600	Aluminum, stock units, straight surfaces	1 Shee	115	.070	S.F.	3.41	3.63		7.04	9.30
0700	Concave or convex surfaces		75	.107	"	3.74	5.55		9.29	12.60
0800	For framing, to 5' high, add		115	.070	L.F.	3.74	3.63		7.37	9.65
0900	Soffits, to 1' wide	↓	125	.064	S.F.	1.85	3.34		5.19	7.15

07 42 13.20 Aluminum Siding Panels

		Crew	Daily Output	Labor-Hours	Unit	Material	2012 Bare Costs Labor	Equipment	Total	Total Incl O&P
0010	**ALUMINUM SIDING PANELS**									
0012	Corrugated, on steel framing, .019 thick, natural finish	G-3	775	.041	S.F.	1.50	1.80		3.30	4.41
0100	Painted		775	.041		1.61	1.80		3.41	4.53
0400	Farm type, .021" thick on steel frame, natural		775	.041		1.51	1.80		3.31	4.42
0600	Painted		775	.041		1.61	1.80		3.41	4.53
0700	Industrial type, corrugated, on steel, .024" thick, mill		775	.041		2.10	1.80		3.90	5.05
0900	Painted		775	.041		2.26	1.80		4.06	5.25
1000	.032" thick, mill		775	.041		2.42	1.80		4.22	5.40
1200	Painted		775	.041		2.94	1.80		4.74	6
1300	V-Beam, on steel frame, .032" thick, mill		775	.041		2.72	1.80		4.52	5.75
1500	Painted		775	.041		3.12	1.80		4.92	6.20
1600	.040" thick, mill		775	.041		3.31	1.80		5.11	6.40
1800	Painted		775	.041		3.87	1.80		5.67	7
1900	.050" thick, mill		775	.041		3.91	1.80		5.71	7.05
2100	Painted		775	.041		4.66	1.80		6.46	7.90
2200	Ribbed, 3" profile, on steel frame, .032" thick, natural		775	.041		2.35	1.80		4.15	5.35
2400	Painted		775	.041		2.98	1.80		4.78	6.05
2500	.040" thick, natural		775	.041		2.69	1.80		4.49	5.70
2700	Painted		775	.041		3.14	1.80		4.94	6.20
2750	.050" thick, natural	↓	775	.041	↓	3.10	1.80		4.90	6.15

07 42 Wall Panels

07 42 13 – Metal Wall Panels

07 42 13.20 Aluminum Siding Panels

		Crew	Daily Output	Labor-Hours	Unit	Material	2012 Bare Costs Labor	2012 Bare Costs Equipment	Total	Total Incl O&P
2760	Painted	G-3	775	.041	S.F.	3.56	1.80		5.36	6.70
3300	For siding on wood frame, deduct from above	↓	2800	.011	↓	.09	.50		.59	.86
3400	Screw fasteners, aluminum, self tapping, neoprene washer, 1"				M	204			204	224
3600	Stitch screws, self tapping, with neoprene washer, 5/8"				"	153			153	168
3630	Flashing, sidewall, .032" thick	G-3	800	.040	L.F.	2.82	1.75		4.57	5.75
3650	End wall, .040" thick		800	.040		3.30	1.75		5.05	6.30
3670	Closure strips, corrugated, .032" thick		800	.040		.84	1.75		2.59	3.59
3680	Ribbed, 4" or 8", .032" thick		800	.040		.84	1.75		2.59	3.59
3690	V-beam, .040" thick	↓	800	.040	↓	1.13	1.75		2.88	3.91
3800	Horizontal, colored clapboard, 8" wide, plain	2 Carp	515	.031	S.F.	2.21	1.37		3.58	4.54
3810	Insulated		515	.031		2.36	1.37		3.73	4.71
4000	Vertical board & batten, colored, non-insulated	↓	515	.031		1.68	1.37		3.05	3.96
4200	For simulated wood design, add				↓	.10			.10	.11
4300	Corners for above, outside	2 Carp	515	.031	V.L.F.	3.25	1.37		4.62	5.70
4500	Inside corners	"	515	.031	"	1.42	1.37		2.79	3.67
4600	Sandwich panels, 1" insulation, single story	G-3	395	.081	S.F.	7.45	3.53		10.98	13.60
4900	Multi-story	"	345	.093		10.50	4.05		14.55	17.75
5100	For baked enamel finish 1 side, add				↓	.41			.41	.45

07 42 13.30 Steel Siding

		Crew	Daily Output	Labor-Hours	Unit	Material	2012 Bare Costs Labor	2012 Bare Costs Equipment	Total	Total Incl O&P
0010	**STEEL SIDING**									
0020	Beveled, vinyl coated, 8" wide	1 Carp	265	.030	S.F.	2.05	1.33		3.38	4.31
0050	10" wide	"	275	.029		2.19	1.28		3.47	4.38
0080	Galv, corrugated or ribbed, on steel frame, 30 gauge	G-3	800	.040		1.37	1.75		3.12	4.18
0100	28 gauge		795	.040		1.44	1.76		3.20	4.27
0300	26 gauge		790	.041		2.02	1.77		3.79	4.92
0400	24 gauge		785	.041		2.03	1.78		3.81	4.95
0600	22 gauge		770	.042		2.34	1.81		4.15	5.35
0700	Colored, corrugated/ribbed, on steel frame, 10 yr. finish, 28 ga.		800	.040		2.13	1.75		3.88	5
0900	26 gauge		795	.040		2.23	1.76		3.99	5.15
1000	24 gauge		790	.041		2.59	1.77		4.36	5.55
1020	20 gauge		785	.041		3.29	1.78		5.07	6.35
1200	Factory sandwich panel, 26 ga., 1" insulation, galvanized		380	.084		5.45	3.67		9.12	11.60
1300	Colored 1 side		380	.084		6.70	3.67		10.37	12.95
1500	Galvanized 2 sides		380	.084		8.30	3.67		11.97	14.75
1600	Colored 2 sides		380	.084		8.60	3.67		12.27	15.05
1800	Acrylic paint face, regular paint liner	↓	380	.084		6.50	3.67		10.17	12.75
1900	For 2" thick polystyrene, add					1			1	1.10
2000	22 ga, galv, 2" insulation, baked enamel exterior	G-3	360	.089		12.70	3.88		16.58	19.90
2100	Polyvinylidene exterior finish	"	360	.089	↓	13.35	3.88		17.23	20.50

07 44 Faced Panels

07 44 73 – Metal Faced Panels

07 44 73.10 Metal Faced Panels and Accessories

		Crew	Daily Output	Labor-Hours	Unit	Material	2012 Bare Costs Labor	2012 Bare Costs Equipment	Total	Total Incl O&P
0010	**METAL FACED PANELS AND ACCESSORIES**									
0400	Textured aluminum, 4' x 8' x 5/16" plywood backing, single face	2 Shee	375	.043	S.F.	3.67	2.23		5.90	7.45
0600	Double face		375	.043		4.99	2.23		7.22	8.90
0700	4' x 10' x 5/16" plywood backing, single face		375	.043		3.93	2.23		6.16	7.70
0900	Double face		375	.043		5.25	2.23		7.48	9.20
1000	4' x 12' x 5/16" plywood backing, single face		375	.043		3.33	2.23		5.56	7.05
1300	Smooth aluminum, 1/4" plywood panel, fluoropolymer finish, double face		375	.043		5.45	2.23		7.68	9.35
1350	Clear anodized finish, double face	↓	375	.043		9.55	2.23		11.78	13.90

07 44 Faced Panels

07 44 73 – Metal Faced Panels

07 44 73.10 Metal Faced Panels and Accessories	Crew	Daily Output	Labor-Hours	Unit	Material	2012 Bare Costs Labor	Equipment	Total	Total Incl O&P
1400	Double face textured aluminum, structural panel, 1" EPS insulation	2 Shee	375	.043	S.F.	4.92	2.23		7.15	8.80
1500	Accessories, outside corner	1 Shee	175	.046	L.F.	1.90	2.38		4.28	5.70
1600	Inside corner		175	.046		1.34	2.38		3.72	5.10
1800	Batten mounting clip		200	.040		.49	2.09		2.58	3.72
1900	Low profile batten		480	.017		.61	.87		1.48	2
2100	High profile batten		480	.017		1.41	.87		2.28	2.88
2200	Water table		200	.040		2.12	2.09		4.21	5.50
2400	Horizontal joint connector		200	.040		1.67	2.09		3.76	5
2500	Corner cap		200	.040		1.84	2.09		3.93	5.20
2700	H - moulding		480	.017		1.25	.87		2.12	2.71

07 46 Siding

07 46 23 – Wood Siding

07 46 23.10 Wood Board Siding	Crew	Daily Output	Labor-Hours	Unit	Material	2012 Bare Costs Labor	Equipment	Total	Total Incl O&P
0010	**WOOD BOARD SIDING**									
3200	Wood, cedar bevel, A grade, 1/2" x 6"	1 Carp	295	.027	S.F.	4.35	1.20		5.55	6.65
3300	1/2" x 8"		330	.024		6.30	1.07		7.37	8.60
3500	3/4" x 10", clear grade		375	.021		5.90	.94		6.84	7.95
3600	"B" grade		375	.021		3.79	.94		4.73	5.60
3800	Cedar, rough sawn, 1" x 4", A grade, natural		220	.036		3.88	1.60		5.48	6.75
3900	Stained		220	.036		4.25	1.60		5.85	7.15
4100	1" x 12", board & batten, #3 & Btr., natural		420	.019		4.02	.84		4.86	5.70
4200	Stained		420	.019		4.25	.84		5.09	5.95
4400	1" x 8" channel siding, #3 & Btr., natural		330	.024		2.97	1.07		4.04	4.90
4500	Stained		330	.024		2.71	1.07		3.78	4.62
4700	Redwood, clear, beveled, vertical grain, 1/2" x 4"		220	.036		3.85	1.60		5.45	6.70
4750	1/2" x 6"		295	.027		3.85	1.20		5.05	6.10
4800	1/2" x 8"		330	.024		4.10	1.07		5.17	6.15
5000	3/4" x 10"		375	.021		4.12	.94		5.06	6
5200	Channel siding, 1" x 10", B grade		375	.021		4.12	.94		5.06	6
5250	Redwood, T&G boards, B grade, 1" x 4"		220	.036		2.54	1.60		4.14	5.25
5270	1" x 8"		330	.024		4.10	1.07		5.17	6.15
5400	White pine, rough sawn, 1" x 8", natural		330	.024		2	1.07		3.07	3.84
5500	Stained		330	.024		2.20	1.07		3.27	4.06

07 46 29 – Plywood Siding

07 46 29.10 Plywood Siding Options	Crew	Daily Output	Labor-Hours	Unit	Material	2012 Bare Costs Labor	Equipment	Total	Total Incl O&P
0010	**PLYWOOD SIDING OPTIONS**									
0900	Plywood, medium density overlaid, 3/8" thick	2 Carp	750	.021	S.F.	1.15	.94		2.09	2.72
1000	1/2" thick		700	.023		1.26	1.01		2.27	2.94
1100	3/4" thick		650	.025		1.65	1.09		2.74	3.49
1600	Texture 1-11, cedar, 5/8" thick, natural		675	.024		2.43	1.05		3.48	4.28
1700	Factory stained		675	.024		2.69	1.05		3.74	4.57
1900	Texture 1-11, fir, 5/8" thick, natural		675	.024		1.06	1.05		2.11	2.78
2000	Factory stained		675	.024		1.32	1.05		2.37	3.06
2050	Texture 1-11, S.Y.P., 5/8" thick, natural		675	.024		1.19	1.05		2.24	2.92
2100	Factory stained		675	.024		1.24	1.05		2.29	2.97
2200	Rough sawn cedar, 3/8" thick, natural		675	.024		1.20	1.05		2.25	2.93
2300	Factory stained		675	.024		1.46	1.05		2.51	3.22
2500	Rough sawn fir, 3/8" thick, natural		675	.024		.76	1.05		1.81	2.45
2600	Factory stained		675	.024		1.02	1.05		2.07	2.73

07 46 Siding

07 46 29 – Plywood Siding

07 46 29.10 Plywood Siding Options	Crew	Daily Output	Labor-Hours	Unit	Material	2012 Bare Costs Labor	Equipment	Total	Total Incl O&P
2800	Redwood, textured siding, 5/8" thick	2 Carp	675	.024	S.F.	1.92	1.05		2.97	3.72
3000	Polyvinyl chloride coated, 3/8" thick	↓	750	.021	↓	1.11	.94		2.05	2.67

07 46 33 – Plastic Siding

07 46 33.10 Vinyl Siding

		Crew	Daily Output	Labor-Hours	Unit	Material	2012 Bare Costs Labor	Equipment	Total	Total Incl O&P
0010	**VINYL SIDING**									
3995	Clapboard profile, woodgrain texture, .048 thick, double 4	2 Carp	495	.032	S.F.	.85	1.43		2.28	3.13
4000	Double 5		550	.029		.85	1.28		2.13	2.91
4005	Single 8		495	.032		.98	1.43		2.41	3.27
4010	Single 10		550	.029		1.18	1.28		2.46	3.27
4015	.044 thick, double 4		495	.032		.82	1.43		2.25	3.10
4020	Double 5		550	.029		.82	1.28		2.10	2.88
4025	.042 thick, double 4		495	.032		.82	1.43		2.25	3.10
4030	Double 5		550	.029		.82	1.28		2.10	2.88
4035	Cross sawn texture, .040 thick, double 4		495	.032		.58	1.43		2.01	2.83
4040	Double 5		550	.029		.58	1.28		1.86	2.61
4045	Smooth texture, .042 thick, double 4		495	.032		.68	1.43		2.11	2.94
4050	Double 5		550	.029		.73	1.28		2.01	2.78
4055	Single 8		495	.032		.94	1.43		2.37	3.23
4060	Cedar texture, .044 thick, double 4		495	.032		.87	1.43		2.30	3.15
4065	Double 6		600	.027		.87	1.18		2.05	2.77
4070	Dutch lap profile, woodgrain texture, .048 thick, double 5		550	.029		.85	1.28		2.13	2.91
4075	.044 thick, double 4.5		525	.030		.87	1.34		2.21	3.03
4080	.042 thick, double 4.5		525	.030		.74	1.34		2.08	2.89
4085	.040 thick, double 4.5		525	.030		.58	1.34		1.92	2.71
4090	Shingle profile, random grooves, double 7		400	.040		2.73	1.76		4.49	5.70
4095	Triple 5		400	.040		2.72	1.76		4.48	5.70
4100	Shake profile, 10" wide		400	.040		2.91	1.76		4.67	5.90
4105	Vertical pattern, .046 thick, double 5		550	.029		1.22	1.28		2.50	3.32
4110	.044 thick, triple 3		550	.029		1.31	1.28		2.59	3.42
4115	.040 thick, triple 4		550	.029		1.25	1.28		2.53	3.35
4120	.040 thick, triple 2.66		550	.029		1.41	1.28		2.69	3.53
4125	Insulation, fan folded extruded polystyrene, 1/4"		2000	.008		.27	.35		.62	.84
4130	3/8"		2000	.008	↓	.30	.35		.65	.87
4135	Accessories, J channel, 5/8" pocket		700	.023	L.F.	.38	1.01		1.39	1.96
4140	3/4" pocket		695	.023		.40	1.02		1.42	2
4145	1-1/4" pocket		680	.024		.59	1.04		1.63	2.25
4150	Flexible, 3/4" pocket		600	.027		1.92	1.18		3.10	3.92
4155	Under sill finish trim		500	.032		.41	1.41		1.82	2.62
4160	Vinyl starter strip		700	.023		.42	1.01		1.43	2.01
4165	Aluminum starter strip		700	.023		.34	1.01		1.35	1.92
4170	Window casing, 2-1/2" wide, 3/4" pocket		510	.031		1.22	1.38		2.60	3.47
4175	Outside corner, woodgrain finish, 4" face, 3/4" pocket		700	.023		1.58	1.01		2.59	3.29
4180	5/8" pocket		700	.023		1.66	1.01		2.67	3.38
4185	Smooth finish, 4" face, 3/4" pocket		700	.023		1.58	1.01		2.59	3.29
4190	7/8" pocket		690	.023		1.55	1.02		2.57	3.28
4195	1-1/4" pocket		700	.023		1.11	1.01		2.12	2.78
4200	Soffit and fascia, 1' overhang, solid		120	.133		3.35	5.90		9.25	12.75
4205	Vented		120	.133		3.35	5.90		9.25	12.75
4207	18" overhang, solid		110	.145		3.96	6.40		10.36	14.20
4208	Vented		110	.145		3.96	6.40		10.36	14.20
4210	2' overhang, solid		100	.160		4.57	7.05		11.62	15.90
4215	Vented	↓	100	.160	↓	4.57	7.05		11.62	15.90

07 46 Siding

07 46 33 – Plastic Siding

07 46 33.10 Vinyl Siding

	07 46 33.10 Vinyl Siding	Crew	Daily Output	Labor-Hours	Unit	Material	2012 Bare Costs Labor	2012 Bare Costs Equipment	Total	Total Incl O&P
4217	3' overhang, solid	2 Carp	100	.160	L.F.	5.80	7.05		12.85	17.20
4218	Vented	↓	100	.160	↓	5.80	7.05		12.85	17.20
4220	Colors for siding and soffits, add				S.F.	.16			.16	.18
4225	Colors for accessories and trim, add				L.F.	.32			.32	.35

07 46 46 – Fiber Cement Siding

07 46 46.10 Fiber Cement Siding

		Crew	Daily Output	Labor-Hours	Unit	Material	Labor	Equipment	Total	Total Incl O&P
0010	**FIBER CEMENT SIDING**									
0020	Lap siding, 5/16" thick, 6" wide, 4-3/4" exposure, smooth texture	2 Carp	415	.039	S.F.	1.08	1.70		2.78	3.81
0025	Woodgrain texture		415	.039		1.08	1.70		2.78	3.81
0030	7-1/2" wide, 6-1/4" exposure, smooth texture		425	.038		1.04	1.66		2.70	3.69
0035	Woodgrain texture		425	.038		1.04	1.66		2.70	3.69
0040	8" wide, 6-3/4" exposure, smooth texture		425	.038		1.24	1.66		2.90	3.92
0045	Roughsawn texture		425	.038		1.24	1.66		2.90	3.92
0050	9-1/2" wide, 8-1/4" exposure, smooth texture		440	.036		.93	1.60		2.53	3.49
0055	Woodgrain texture		440	.036		.93	1.60		2.53	3.49
0060	12" wide, 10-3/8" exposure, smooth texture		455	.035		1.21	1.55		2.76	3.72
0065	Woodgrain texture		455	.035		1.21	1.55		2.76	3.72
0070	Panel siding, 5/16" thick, smooth texture		750	.021		1.17	.94		2.11	2.74
0075	Stucco texture		750	.021		1.17	.94		2.11	2.74
0080	Grooved woodgrain texture		750	.021		1.17	.94		2.11	2.74
0085	V - grooved woodgrain texture		750	.021	↓	1.17	.94		2.11	2.74
0090	Wood starter strip	↓	400	.040	L.F.	.43	1.76		2.19	3.18

07 46 73 – Soffit

07 46 73.10 Soffit Options

		Crew	Daily Output	Labor-Hours	Unit	Material	Labor	Equipment	Total	Total Incl O&P
0010	**SOFFIT OPTIONS**									
0012	Aluminum, residential, .020" thick	1 Carp	210	.038	S.F.	1.96	1.68		3.64	4.75
0100	Baked enamel on steel, 16 or 18 gauge		105	.076		5.80	3.36		9.16	11.55
0300	Polyvinyl chloride, white, solid		230	.035		2.04	1.53		3.57	4.60
0400	Perforated	↓	230	.035		2.04	1.53		3.57	4.60
0500	For colors, add				↓	.12			.12	.13

07 51 Built-Up Bituminous Roofing

07 51 13 – Built-Up Asphalt Roofing

07 51 13.10 Built-Up Roofing Components

		Crew	Daily Output	Labor-Hours	Unit	Material	Labor	Equipment	Total	Total Incl O&P
0010	**BUILT-UP ROOFING COMPONENTS**									
0012	Asphalt saturated felt, #30, 2 square per roll	1 Rofc	58	.138	Sq.	11.30	5.15		16.45	21
0200	#15, 4 sq per roll, plain or perforated, not mopped		58	.138		5.65	5.15		10.80	14.90
0300	Roll roofing, smooth, #65		15	.533		9.65	19.95		29.60	44
0500	#90		12	.667		34.50	25		59.50	80
0520	Mineralized		12	.667		34.50	25		59.50	80
0540	D.C. (Double coverage), 19" selvage edge	↓	10	.800	↓	44	30		74	98
0580	Adhesive (lap cement)				Gal.	5.50			5.50	6.05
0600	Steep, flat or dead level asphalt, 10 ton lots, bulk				Ton	455			455	500
0800	Packaged				"	780			780	860

07 51 13.13 Cold-Applied Built-Up Asphalt Roofing

		Crew	Daily Output	Labor-Hours	Unit	Material	Labor	Equipment	Total	Total Incl O&P
0010	**COLD-APPLIED BUILT-UP ASPHALT ROOFING**									
0020	3 ply system, installation only (components listed below)	G-5	50	.800	Sq.		27	3.58	30.58	49.50
0100	Spunbond poly. fabric, 1.35 oz./S.Y., 36"W, 10.8 sq./roll				Ea.	175			175	193
0200	49" wide, 14.6 sq./roll					244			244	268
0300	2.10 oz./S.Y., 36" wide, 10.8 sq./roll				↓	266			266	293

07 51 Built-Up Bituminous Roofing

07 51 13 – Built-Up Asphalt Roofing

07 51 13.13 Cold-Applied Built-Up Asphalt Roofing

		Daily Output	Labor-Hours	Unit	Material	2012 Bare Costs Labor	Equipment	Total	Total Incl O&P
0400	49" wide, 14.6 sq./roll			Ea.	360			360	395
0500	Base & finish coat, 3 gal./sq., 5 gal./can			Gal.	7.60			7.60	8.35
0600	Coating, ceramic granules, 1/2 sq./bag			Ea.	15.15			15.15	16.65
0700	Aluminum, 2 gal./sq.			Gal.	12.30			12.30	13.55
0800	Emulsion, fibered or non-fibered, 4 gal./sq.			"	5.20			5.20	5.70

07 51 13.20 Built-Up Roofing Systems

		Crew	Daily Output	Labor-Hours	Unit	Material	Labor	Equipment	Total	Total Incl O&P
0010	**BUILT-UP ROOFING SYSTEMS** R075113-20									
0120	Asphalt flood coat with gravel/slag surfacing, not including									
0140	Insulation, flashing or wood nailers									
0200	Asphalt base sheet, 3 plies #15 asphalt felt, mopped	G-1	22	2.545	Sq.	88	89	20.50	197.50	269
0350	On nailable decks		21	2.667		92.50	93	21.50	207	282
0500	4 plies #15 asphalt felt, mopped		20	2.800		122	98	22.50	242.50	325
0550	On nailable decks		19	2.947		109	103	24	236	320
0700	Coated glass base sheet, 2 plies glass (type IV), mopped		22	2.545		90.50	89	20.50	200	271
0850	3 plies glass, mopped		20	2.800		108	98	22.50	228.50	310
0950	On nailable decks		19	2.947		102	103	24	229	310
1100	4 plies glass fiber felt (type IV), mopped		20	2.800		133	98	22.50	253.50	335
1150	On nailable decks		19	2.947		119	103	24	246	330
1200	Coated & saturated base sheet, 3 plies #15 asph. felt, mopped		20	2.800		98	98	22.50	218.50	297
1250	On nailable decks		19	2.947		91.50	103	24	218.50	299
1300	4 plies #15 asphalt felt, mopped		22	2.545		114	89	20.50	223.50	298
2000	Asphalt flood coat, smooth surface									
2200	Asphalt base sheet & 3 plies #15 asphalt felt, mopped	G-1	24	2.333	Sq.	94.50	81.50	18.85	194.85	262
2400	On nailable decks		23	2.435		88	85	19.65	192.65	261
2600	4 plies #15 asphalt felt, mopped		24	2.333		111	81.50	18.85	211.35	280
2700	On nailable decks		23	2.435		104	85	19.65	208.65	280
2900	Coated glass fiber base sheet, mopped, and 2 plies of									
2910	glass fiber felt (type IV)	G-1	25	2.240	Sq.	86	78.50	18.10	182.60	245
3100	On nailable decks		24	2.333		81	81.50	18.85	181.35	247
3200	3 plies, mopped		23	2.435		104	85	19.65	208.65	279
3300	On nailable decks		22	2.545		97	89	20.50	206.50	279
3800	4 plies glass fiber felt (type IV), mopped		23	2.435		122	85	19.65	226.65	299
3900	On nailable decks		22	2.545		115	89	20.50	224.50	298
4000	Coated & saturated base sheet, 3 plies #15 asph. felt, mopped		24	2.333		93.50	81.50	18.85	193.85	261
4200	On nailable decks		23	2.435		87	85	19.65	191.65	260
4300	4 plies #15 organic felt, mopped		22	2.545		110	89	20.50	219.50	293
4500	Coal tar pitch with gravel/slag surfacing									
4600	4 plies #15 tarred felt, mopped	G-1	21	2.667	Sq.	186	93	21.50	300.50	385
4800	3 plies glass fiber felt (type IV), mopped	"	19	2.947	"	154	103	24	281	370
5000	Coated glass fiber base sheet, and 2 plies of									
5010	glass fiber felt, (type IV), mopped	G-1	19	2.947	Sq.	159	103	24	286	375
5300	On nailable decks		18	3.111		138	109	25	272	360
5600	4 plies glass fiber felt (type IV), mopped		21	2.667		213	93	21.50	327.50	415
5800	On nailable decks		20	2.800		193	98	22.50	313.50	400

07 51 13.30 Cants

		Crew	Daily Output	Labor-Hours	Unit	Material	Labor	Equipment	Total	Total Incl O&P
0010	**CANTS**									
0012	Lumber, treated, 4" x 4" cut diagonally	1 Rofc	325	.025	L.F.	2.02	.92		2.94	3.76
0100	Foamglass		325	.025		2.25	.92		3.17	4.02
0300	Mineral or fiber, trapezoidal, 1" x 4" x 48"		325	.025		.19	.92		1.11	1.75
0400	1-1/2" x 5-5/8" x 48"		325	.025		.31	.92		1.23	1.88

07 51 Built-Up Bituminous Roofing

07 51 13 – Built-Up Asphalt Roofing

07 51 13.40 Felts

		Crew	Daily Output	Labor-Hours	Unit	Material	2012 Bare Costs Labor	Equipment	Total	Total Incl O&P
0010	**FELTS**									
0012	Glass fibered roofing felt, #15, not mopped	1 Rofc	58	.138	Sq.	6.90	5.15		12.05	16.25
0300	Base sheet, #45, channel vented		58	.138		32.50	5.15		37.65	44
0400	#50, coated		58	.138		16.55	5.15		21.70	27
0500	Cap, mineral surfaced		58	.138		45	5.15		50.15	58
0600	Flashing membrane, #65		16	.500		9.65	18.70		28.35	42
0800	Coal tar fibered, #15, no mopping		58	.138		14.05	5.15		19.20	24
0900	Asphalt felt, #15, 4 sq per roll, no mopping		58	.138		5.65	5.15		10.80	14.90
1100	#30, 2 sq per roll		58	.138		11.30	5.15		16.45	21
1200	Double coated, #33		58	.138		9.95	5.15		15.10	19.60
1400	#40, base sheet		58	.138		10.95	5.15		16.10	20.50
1450	Coated and saturated		58	.138		10	5.15		15.15	19.65
1500	Tarred felt, organic, #15, 4 sq rolls		58	.138		11.55	5.15		16.70	21.50
1550	#30, 2 sq roll		58	.138		23	5.15		28.15	34
1700	Add for mopping above felts, per ply, asphalt, 24 lb. per sq.	G-1	192	.292		9.35	10.20	2.36	21.91	30
1800	Coal tar mopping, 30 lb. per sq.		186	.301		18	10.55	2.43	30.98	40
1900	Flood coat, with asphalt, 60 lb. per sq.		60	.933		23.50	32.50	7.55	63.55	88.50
2000	With coal tar, 75 lb. per sq.		56	1		45	35	8.10	88.10	117

07 51 13.50 Walkways for Built-Up Roofs

		Crew	Daily Output	Labor-Hours	Unit	Material	2012 Bare Costs Labor	Equipment	Total	Total Incl O&P
0010	**WALKWAYS FOR BUILT-UP ROOFS**									
0020	Asphalt impregnated, 3' x 6' x 1/2" thick	1 Rofc	400	.020	S.F.	1.73	.75		2.48	3.15
0100	3' x 3' x 3/4" thick	"	400	.020		4.99	.75		5.74	6.75
0300	Concrete patio blocks, 2" thick, natural	1 Clab	115	.070		3.73	2.44		6.17	7.85
0400	Colors	"	115	.070		3.77	2.44		6.21	7.90

07 52 Modified Bituminous Membrane Roofing

07 52 13 – Atactic-Polypropylene-Modified Bituminous Membrane Roofing

07 52 13.10 APP Modified Bituminous Membrane

		Crew	Daily Output	Labor-Hours	Unit	Material	2012 Bare Costs Labor	Equipment	Total	Total Incl O&P
0010	**APP MODIFIED BITUMINOUS MEMBRANE** R075213-30									
0020	Base sheet, #15 glass fiber felt, nailed to deck	1 Rofc	58	.138	Sq.	7.85	5.15		13	17.25
0030	Spot mopped to deck	G-1	295	.190		11.60	6.65	1.53	19.78	25.50
0040	Fully mopped to deck	"	192	.292		16.25	10.20	2.36	28.81	37.50
0050	#15 organic felt, nailed to deck	1 Rofc	58	.138		6.60	5.15		11.75	15.90
0060	Spot mopped to deck	G-1	295	.190		10.35	6.65	1.53	18.53	24
0070	Fully mopped to deck	"	192	.292		15	10.20	2.36	27.56	36
2100	APP mod., smooth surf. cap sheet, poly. reinf., torched, 160 mils	G-5	2100	.019	S.F.	.52	.65	.09	1.26	1.75
2150	170 mils		2100	.019		.55	.65	.09	1.29	1.79
2200	Granule surface cap sheet, poly. reinf., torched, 180 mils		2000	.020		.62	.68	.09	1.39	1.92
2250	Smooth surface flashing, torched, 160 mils		1260	.032		.52	1.08	.14	1.74	2.54
2300	170 mils		1260	.032		.55	1.08	.14	1.77	2.58
2350	Granule surface flashing, torched, 180 mils		1260	.032		.62	1.08	.14	1.84	2.65
2400	Fibrated aluminum coating	1 Rofc	3800	.002		.10	.08		.18	.24

07 52 16 – Styrene-Butadiene-Styrene Modified Bituminous Membrane Roofing

07 52 16.10 SBS Modified Bituminous Membrane

		Crew	Daily Output	Labor-Hours	Unit	Material	2012 Bare Costs Labor	Equipment	Total	Total Incl O&P
0010	**SBS MODIFIED BITUMINOUS MEMBRANE**									
0080	SBS modified, granule surf cap sheet, polyester rein., mopped									
0600	150 mils	G-1	2000	.028	S.F.	.58	.98	.23	1.79	2.53
1100	160 mils		2000	.028		.77	.98	.23	1.98	2.74
1500	Glass fiber reinforced, mopped, 160 mils		2000	.028		.49	.98	.23	1.70	2.43
1600	Smooth surface cap sheet, mopped, 145 mils		2100	.027		.49	.93	.22	1.64	2.34

07 52 Modified Bituminous Membrane Roofing

07 52 16 – Styrene-Butadiene-Styrene Modified Bituminous Membrane Roofing

07 52 16.10 SBS Modified Bituminous Membrane	Crew	Daily Output	Labor-Hours	Unit	Material	2012 Bare Costs Labor	Equipment	Total	Total Incl O&P
1700	Smooth surface flashing, 145 mils	G-1	1260	.044	S.F.	.49	1.55	.36	2.40	3.54
1800	150 mils		1260	.044		.48	1.55	.36	2.39	3.53
1900	Granular surface flashing, 150 mils		1260	.044		.58	1.55	.36	2.49	3.64
2000	160 mils		1260	.044		.77	1.55	.36	2.68	3.85

07 53 Elastomeric Membrane Roofing

07 53 16 – Chlorosulfonate-Polyethylene Roofing

07 53 16.10 Chlorosulfonated Polyethylene Roofing

		Crew	Daily Output	Labor-Hours	Unit	Material	2012 Bare Costs Labor	Equipment	Total	Total Incl O&P
0010	**CHLOROSULFONATED POLYETHYLENE ROOFING**									
0800	Chlorosulfonated polyethylene-hypalon (CSPE), 45 mils,									
0900	0.29 P.S.F., fully adhered	G-5	26	1.538	Sq.	189	52.50	6.90	248.40	305
1100	Loose-laid & ballasted with stone (10 P.S.F.)		51	.784		197	26.50	3.51	227.01	265
1200	Mechanically attached		35	1.143		190	39	5.10	234.10	280
1300	Plates with adhesive attachment		35	1.143		187	39	5.10	231.10	276

07 53 23 – Ethylene-Propylene-Diene-Monomer Roofing

07 53 23.20 Ethylene-Propylene-Diene-Monomer Roofing

		Crew	Daily Output	Labor-Hours	Unit	Material	2012 Bare Costs Labor	Equipment	Total	Total Incl O&P
0010	**ETHYLENE-PROPYLENE-DIENE-MONOMER ROOFING (E.P.D.M.)**									
3500	Ethylene-propylene-diene-monomer (EPDM), 45 mils, 0.28 P.S.F.									
3600	Loose-laid & ballasted with stone (10 P.S.F.)	G-5	51	.784	Sq.	67.50	26.50	3.51	97.51	123
3700	Mechanically attached		35	1.143		63.50	39	5.10	107.60	140
3800	Fully adhered with adhesive		26	1.538		99.50	52.50	6.90	158.90	204
4500	60 mils, 0.40 P.S.F.									
4600	Loose-laid & ballasted with stone (10 P.S.F.)	G-5	51	.784	Sq.	78	26.50	3.51	108.01	134
4700	Mechanically attached		35	1.143		73	39	5.10	117.10	151
4800	Fully adhered with adhesive		26	1.538		109	52.50	6.90	168.40	215
4810	45 mil, .28 PSF, membrane only					39.50			39.50	43.50
4820	60 mil, .40 PSF, membrane only					48			48	53
4850	Seam tape for membrane, 3" x 100' roll				Ea.	50			50	55
4900	Batten strips, 10' sections					4.25			4.25	4.68
4910	Cover tape for batten strips, 6" x 100' roll					113			113	124
4930	Plate anchors				M	110			110	121
4970	Adhesive for fully adhered systems, 60 S.F./gal.				Gal.	22.50			22.50	24.50

07 53 29 – Polyisobutylene Roofing

07 53 29.10 Polyisobutylene Roofing

		Crew	Daily Output	Labor-Hours	Unit	Material	2012 Bare Costs Labor	Equipment	Total	Total Incl O&P
0010	**POLYISOBUTYLENE ROOFING**									
7500	Polyisobutylene (PIB), 100 mils, 0.57 P.S.F.									
7600	Loose-laid & ballasted with stone/gravel (10 P.S.F.)	G-5	51	.784	Sq.	181	26.50	3.51	211.01	247
7700	Partially adhered with adhesive		35	1.143		227	39	5.10	271.10	320
7800	Hot asphalt attachment		35	1.143		217	39	5.10	261.10	310
7900	Fully adhered with contact cement		26	1.538		234	52.50	6.90	293.40	355

07 54 Thermoplastic Membrane Roofing

07 54 19 – Polyvinyl-Chloride Roofing

07 54 19.10 Polyvinyl-Chloride Roofing (P.V.C.)	Crew	Daily Output	Labor-Hours	Unit	Material	2012 Bare Costs Labor	Equipment	Total	Total Incl O&P
0010 **POLYVINYL-CHLORIDE ROOFING (P.V.C.)**									
8200 Heat welded seams									
8700 Reinforced, 48 mils, 0.33 P.S.F.									
8750 Loose-laid & ballasted with stone/gravel (12 P.S.F.)	G-5	51	.784	Sq.	100	26.50	3.51	130.01	158
8800 Mechanically attached		35	1.143		95.50	39	5.10	139.60	176
8850 Fully adhered with adhesive		26	1.538		125	52.50	6.90	184.40	233
8860 Reinforced, 60 mils, .40 P.S.F.									
8870 Loose-laid & ballasted with stone/gravel (12 P.S.F.)	G-5	51	.784	Sq.	102	26.50	3.51	132.01	161
8880 Mechanically attached		35	1.143		97.50	39	5.10	141.60	178
8890 Fully adhered with adhesive		26	1.538		127	52.50	6.90	186.40	235

07 54 23 – Thermoplastic-Polyolefin Roofing

07 54 23.10 Thermoplastic Polyolefin Roofing (T.P.O)

	Crew	Daily Output	Labor-Hours	Unit	Material	Labor	Equipment	Total	Total Incl O&P
0010 **THERMOPLASTIC POLYOLEFIN ROOFING (T.P.O.)**									
0100 45 mils, loose laid & ballasted with stone(1/2 ton/sq.)	G-5	51	.784	Sq.	82	26.50	3.51	112.01	138
0120 Fully adhered		25	1.600		73.50	54.50	7.15	135.15	180
0140 Mechanically attached		34	1.176		73.50	40	5.25	118.75	153
0160 Self adhered		35	1.143		88	39	5.10	132.10	168
0180 60 mil membrane, heat welded seams, ballasted		50	.800		88.50	27	3.58	119.08	146
0200 Fully adhered		25	1.600		79	54.50	7.15	140.65	186
0220 Mechanically attached		34	1.176		83.50	40	5.25	128.75	165
0240 Self adhered		35	1.143		107	39	5.10	151.10	188

07 56 Fluid-Applied Roofing

07 56 10 – Fluid-Applied Roofing Elastomers

07 56 10.10 Elastomeric Roofing

	Crew	Daily Output	Labor-Hours	Unit	Material	Labor	Equipment	Total	Total Incl O&P
0010 **ELASTOMERIC ROOFING**									
0110 Acrylic rubber, fluid applied, 20 mils thick	G-5	2000	.020	S.F.	2.04	.68	.09	2.81	3.48
0120 50 mils, reinforced		1200	.033		3.16	1.13	.15	4.44	5.55
0130 For walking surface, add		900	.044		.96	1.51	.20	2.67	3.81
0300 Hypalon neoprene, fluid applied, 20 mil thick, not-reinforced	G-1	1135	.049		2.57	1.72	.40	4.69	6.15
0600 Non-woven polyester, reinforced		960	.058		2.61	2.04	.47	5.12	6.80
0700 5 coat neoprene deck, 60 mil thick, under 10,000 S.F.		325	.172		5.75	6	1.39	13.14	17.95
0900 Over 10,000 S.F.		625	.090		5.35	3.13	.72	9.20	11.95
1300 Vinyl plastic traffic deck, sprayed, 2 to 4 mils thick		625	.090		1.68	3.13	.72	5.53	7.90
1500 Vinyl and neoprene membrane traffic deck		1550	.036		1.77	1.26	.29	3.32	4.39

07 57 Coated Foamed Roofing

07 57 13 – Sprayed Polyurethane Foam Roofing

07 57 13.10 Sprayed Polyurethane Foam Roofing (S.P.F.)

	Crew	Daily Output	Labor-Hours	Unit	Material	Labor	Equipment	Total	Total Incl O&P
0010 **SPRAYED POLYURETHANE FOAM ROOFING (S.P.F.)**									
0100 Primer for metal substrate (when required)	G-2A	3000	.008	S.F.	.42	.27	.21	.90	1.13
0200 Primer for non-metal substrate (when required)		3000	.008		.16	.27	.21	.64	.85
0300 Closed cell spray, polyurethane foam, 3 lb. per C.F. density, 1", R6.7		15000	.002		.57	.05	.04	.66	.76
0400 2", R13.4		13125	.002		1.14	.06	.05	1.25	1.40
0500 3", R18.6		11485	.002		1.70	.07	.05	1.82	2.04
0550 4", R24.8		10080	.002		2.27	.08	.06	2.41	2.70
0700 Spray-on silicone coating		2500	.010		.92	.32	.25	1.49	1.81
0800 Warranty 5-20 year manufacturer's									.15
0900 Warranty 20 year, no dollar limit									.20

07 58 Roll Roofing

07 58 10 - Asphalt Roll Roofing

07 58 10.10 Roll Roofing

		Crew	Daily Output	Labor-Hours	Unit	Material	2012 Bare Costs Labor	Equipment	Total	Total Incl O&P
0010	**ROLL ROOFING**									
0100	Asphalt, mineral surface									
0200	1 ply #15 organic felt, 1 ply mineral surfaced									
0300	Selvage roofing, lap 19", nailed & mopped	G-1	27	2.074	Sq.	62.50	72.50	16.75	151.75	209
0400	3 plies glass fiber felt (type IV), 1 ply mineral surfaced									
0500	Selvage roofing, lapped 19", mopped	G-1	25	2.240	Sq.	102	78.50	18.10	198.60	263
0600	Coated glass fiber base sheet, 2 plies of glass fiber									
0700	Felt (type IV), 1 ply mineral surfaced selvage									
0800	Roofing, lapped 19", mopped	G-1	25	2.240	Sq.	112	78.50	18.10	208.60	274
0900	On nailable decks	"	24	2.333	"	102	81.50	18.85	202.35	270
1000	3 plies glass fiber felt (type III), 1 ply mineral surfaced									
1100	Selvage roofing, lapped 19", mopped	G-1	25	2.240	Sq.	102	78.50	18.10	198.60	263

07 61 Sheet Metal Roofing

07 61 13 - Standing Seam Sheet Metal Roofing

07 61 13.10 Standing Seam Sheet Metal Roofing, Field Fab.

		Crew	Daily Output	Labor-Hours	Unit	Material	2012 Bare Costs Labor	Equipment	Total	Total Incl O&P
0010	**STANDING SEAM SHEET METAL ROOFING, FIELD FABRICATED**									
0400	Copper standing seam roofing, over 10 squares, 16 oz., 125 lb. per sq.	1 Shee	1.30	6.154	Sq.	985	320		1,305	1,575
0600	18 oz., 140 lb. per sq.		1.20	6.667		1,100	350		1,450	1,725
0700	20 oz., 150 lb. per sq.		1.10	7.273		1,225	380		1,605	1,925
1200	For abnormal conditions or small areas, add					25%	100%			
1300	For lead-coated copper, add					25%				

07 61 16 - Batten Seam Sheet Metal Roofing

07 61 16.10 Batten Seam Sheet Metal Roofing, Field Fabricated

		Crew	Daily Output	Labor-Hours	Unit	Material	2012 Bare Costs Labor	Equipment	Total	Total Incl O&P
0010	**BATTEN SEAM SHEET METAL ROOFING, FIELD FABRICATED**									
0012	Copper batten seam roofing, over 10 sq., 16 oz., 130 lb. per sq.	1 Shee	1.10	7.273	Sq.	1,250	380		1,630	1,950
0020	Lead batten seam roofing, 5 lb. per S.F.		1.20	6.667		620	350		970	1,200
0100	Zinc/copper alloy batten seam roofing, .020 thick		1.20	6.667		1,100	350		1,450	1,750
0200	Copper roofing, batten seam, over 10 sq., 18 oz., 145 lb. per sq.		1	8		1,400	415		1,815	2,150
0300	20 oz., 160 lb. per sq.		1	8		1,550	415		1,965	2,350
0500	Stainless steel batten seam roofing, type 304, 28 gauge		1.20	6.667		625	350		975	1,225
0600	26 gauge		1.15	6.957		630	365		995	1,250
0800	Zinc, copper alloy roofing, batten seam, .027" thick		1.15	6.957		1,550	365		1,915	2,275
0900	.032" thick		1.10	7.273		1,875	380		2,255	2,650
1000	.040" thick		1.05	7.619		2,100	395		2,495	2,900

07 61 19 - Flat Seam Sheet Metal Roofing

07 61 19.10 Flat Seam Sheet Metal Roofing, Field Fabricated

		Crew	Daily Output	Labor-Hours	Unit	Material	2012 Bare Costs Labor	Equipment	Total	Total Incl O&P
0010	**FLAT SEAM SHEET METAL ROOFING, FIELD FABRICATED**									
0900	Copper flat seam roofing, over 10 squares, 16 oz., 115 lb./sq.	1 Shee	1.20	6.667	Sq.	915	350		1,265	1,525
1000	20 oz., 145 lb./sq.		1.10	7.273		1,150	380		1,530	1,825
1100	Lead flat seam roofing, 5 lb. per S.F.		1.30	6.154		615	320		935	1,175

07 65 Flexible Flashing

07 65 10 – Sheet Metal Flashing

07 65 10.10 Sheet Metal Flashing and Counter Flashing	Crew	Daily Output	Labor-Hours	Unit	Material	2012 Bare Costs Labor	Equipment	Total	Total Incl O&P
0010 **SHEET METAL FLASHING AND COUNTER FLASHING**									
0011 Including up to 4 bends									
0020 Aluminum, mill finish, .013" thick	1 Rofc	145	.055	S.F.	.73	2.06		2.79	4.26
0030 .016" thick		145	.055		.84	2.06		2.90	4.38
0060 .019" thick		145	.055		1.07	2.06		3.13	4.64
0100 .032" thick		145	.055		1.25	2.06		3.31	4.84
0200 .040" thick		145	.055		2.14	2.06		4.20	5.80
0300 .050" thick		145	.055		2.25	2.06		4.31	5.95
0325 Mill finish 5" x 7" step flashing, .016" thick		1920	.004	Ea.	.18	.16		.34	.46
0350 Mill finish 12" x 12" step flashing, .016" thick		1600	.005	"	.62	.19		.81	.99
0400 Painted finish, add				S.F.	.34			.34	.37
1000 Mastic-coated 2 sides, .005" thick	1 Rofc	330	.024		1.67	.91		2.58	3.36
1100 .016" thick		330	.024		1.84	.91		2.75	3.54
1600 Copper, 16 oz., sheets, under 1000 lb.		115	.070		7.85	2.60		10.45	13
1700 Over 4000 lb.		155	.052		7.85	1.93		9.78	11.90
1900 20 oz. sheets, under 1000 lb.		110	.073		9.85	2.72		12.57	15.40
2000 Over 4000 lb.		145	.055		9.35	2.06		11.41	13.75
2200 24 oz. sheets, under 1000 lb.		105	.076		12.10	2.85		14.95	18.15
2300 Over 4000 lb.		135	.059		11.50	2.22		13.72	16.35
2500 32 oz. sheets, under 1000 lb.		100	.080		16.15	2.99		19.14	23
2600 Over 4000 lb.		130	.062		15.35	2.30		17.65	21
2700 W shape for valleys, 16 oz., 24" wide		100	.080	L.F.	15.85	2.99		18.84	22.50
5800 Lead, 2.5 lb. per S.F., up to 12" wide		135	.059	S.F.	5.40	2.22		7.62	9.65
5900 Over 12" wide		135	.059		4.43	2.22		6.65	8.60
8900 Stainless steel sheets, 32 ga, .010" thick		155	.052		4	1.93		5.93	7.65
9000 28 ga, .015" thick		155	.052		5.65	1.93		7.58	9.45
9100 26 ga, .018" thick		155	.052		7.25	1.93		9.18	11.25
9200 24 ga, .025" thick		155	.052		8.90	1.93		10.83	13.05
9290 For mechanically keyed flashing, add					40%				
9320 Steel sheets, galvanized, 20 gauge	1 Rofc	130	.062	S.F.	1.43	2.30		3.73	5.45
9340 30 gauge		160	.050		1.33	1.87		3.20	4.60
9400 Terne coated stainless steel, .015" thick, 28 ga		155	.052		7.35	1.93		9.28	11.35
9500 .018" thick, 26 ga		155	.052		8.30	1.93		10.23	12.35
9600 Zinc and copper alloy (brass), .020" thick		155	.052		11.10	1.93		13.03	15.45
9700 .027" thick		155	.052		15.60	1.93		17.53	20.50
9800 .032" thick		155	.052		18.80	1.93		20.73	23.50
9900 .040" thick		155	.052		21	1.93		22.93	26

07 65 12 – Fabric and Mastic Flashings

07 65 12.10 Fabric and Mastic Flashing and Counter Flashing

	Crew	Daily Output	Labor-Hours	Unit	Material	Labor	Equipment	Total	Total Incl O&P
0010 **FABRIC AND MASTIC FLASHING AND COUNTER FLASHING**									
1300 Asphalt flashing cement, 5 gallon				Gal.	10.50			10.50	11.55
4900 Fabric, asphalt-saturated cotton, specification grade	1 Rofc	35	.229	S.Y.	2.96	8.55		11.51	17.60
5000 Utility grade		35	.229		1.42	8.55		9.97	15.90
5300 Close-mesh fabric, saturated, 17 oz. per S.Y.		35	.229		2.15	8.55		10.70	16.70
5500 Fiberglass, resin-coated		35	.229		1.05	8.55		9.60	15.50
8500 Shower pan, bituminous membrane, 7 oz.		155	.052	S.F.	2	1.93		3.93	5.45

07 65 13 – Laminated Sheet Flashing

07 65 13.10 Laminated Sheet Flashing

	Crew	Daily Output	Labor-Hours	Unit	Material	Labor	Equipment	Total	Total Incl O&P
0010 **LAMINATED SHEET FLASHING**, Including up to 4 bends									
0500 Fabric-backed 2 sides, .004" thick	1 Rofc	330	.024	S.F.	1.41	.91		2.32	3.07
0700 .005" thick		330	.024		1.67	.91		2.58	3.36
0750 Mastic-backed, self adhesive		460	.017		3.46	.65		4.11	4.90

07 65 Flexible Flashing

07 65 13 – Laminated Sheet Flashing

07 65 13.10 Laminated Sheet Flashing	Crew	Daily Output	Labor-Hours	Unit	Material	2012 Bare Costs Labor	Equipment	Total	Total Incl O&P
0800	Mastic-coated 2 sides, .004" thick	1 Rofc	330	.024	S.F.	1.41	.91		2.32	3.07
2800	Copper, paperbacked 1 side, 2 oz.		330	.024		1.67	.91		2.58	3.36
2900	3 oz.		330	.024		1.97	.91		2.88	3.69
3100	Paperbacked 2 sides, 2 oz.		330	.024		1.72	.91		2.63	3.41
3150	3 oz.		330	.024		2.07	.91		2.98	3.80
3200	5 oz.		330	.024		2.69	.91		3.60	4.48
3250	7 oz.		330	.024		4.81	.91		5.72	6.80
3400	Mastic-backed 2 sides, copper, 2 oz.		330	.024		1.65	.91		2.56	3.34
3500	3 oz.		330	.024		2	.91		2.91	3.72
3700	5 oz.		330	.024		3	.91		3.91	4.82
3800	Fabric-backed 2 sides, copper, 2 oz.		330	.024		1.93	.91		2.84	3.64
4000	3 oz.		330	.024		2.19	.91		3.10	3.93
4100	5 oz.		330	.024		3.13	.91		4.04	4.96
4300	Copper-clad stainless steel, .015" thick, under 500 lb.		115	.070		5.80	2.60		8.40	10.75
4400	Over 2000 lb.		155	.052		5.55	1.93		7.48	9.35
4600	.018" thick, under 500 lb.		100	.080		6.55	2.99		9.54	12.20
4700	Over 2000 lb.		145	.055		6.20	2.06		8.26	10.30
6100	Lead-coated copper, fabric-backed, 2 oz.		330	.024		2.64	.91		3.55	4.42
6200	5 oz.		330	.024		3.04	.91		3.95	4.86
6400	Mastic-backed 2 sides, 2 oz.		330	.024		2.06	.91		2.97	3.79
6500	5 oz.		330	.024		2.55	.91		3.46	4.33
6700	Paperbacked 1 side, 2 oz.		330	.024		1.78	.91		2.69	3.48
6800	3 oz.		330	.024		2.09	.91		3	3.82
7000	Paperbacked 2 sides, 2 oz.		330	.024		1.84	.91		2.75	3.54
7100	5 oz.		330	.024		3	.91		3.91	4.82
8550	3 ply copper and fabric, 3 oz.		155	.052		2.19	1.93		4.12	5.65
8600	7 oz.		155	.052		3.25	1.93		5.18	6.80
8700	Lead on copper and fabric, 5 oz.		155	.052		3.04	1.93		4.97	6.60
8800	7 oz.		155	.052		5.25	1.93		7.18	9.05
9300	Stainless steel, paperbacked 2 sides, .005" thick		330	.024		3.62	.91		4.53	5.50

07 65 19 – Plastic Sheet Flashing

07 65 19.10 Plastic Sheet Flashing and Counter Flashing

		Crew	Daily Output	Labor-Hours	Unit	Material	Labor	Equipment	Total	Total Incl O&P
0010	**PLASTIC SHEET FLASHING AND COUNTER FLASHING**									
7300	Polyvinyl chloride, black, .010" thick	1 Rofc	285	.028	S.F.	.21	1.05		1.26	1.99
7400	.020" thick		285	.028		.30	1.05		1.35	2.09
7600	.030" thick		285	.028		.39	1.05		1.44	2.19
7700	.056" thick		285	.028		.94	1.05		1.99	2.79
7900	Black or white for exposed roofs, .060" thick		285	.028		2	1.05		3.05	3.96
8060	PVC tape, 5" x 45 mils, for joint covers, 100 L.F./roll				Ea.	136			136	150
8850	Polyvinyl chloride, .030" thick	1 Rofc	160	.050	S.F.	.44	1.87		2.31	3.62

07 65 23 – Rubber Sheet Flashing

07 65 23.10 Rubber Sheet Flashing and Counterflashing

		Crew	Daily Output	Labor-Hours	Unit	Material	Labor	Equipment	Total	Total Incl O&P
0010	**RUBBER SHEET FLASHING AND COUNTERFLASHING**									
8100	Rubber, butyl, 1/32" thick	1 Rofc	285	.028	S.F.	1.75	1.05		2.80	3.69
8200	1/16" thick		285	.028		2.38	1.05		3.43	4.38
8300	Neoprene, cured, 1/16" thick		285	.028		1.92	1.05		2.97	3.87
8400	1/8" thick		285	.028		4	1.05		5.05	6.15

07 71 Roof Specialties

07 71 19 – Manufactured Gravel Stops and Fascias

07 71 19.10 Gravel Stop

		Crew	Daily Output	Labor-Hours	Unit	Material	2012 Bare Costs Labor	Equipment	Total	Total Incl O&P
0010	**GRAVEL STOP**									
0020	Aluminum, .050" thick, 4" face height, mill finish	1 Shee	145	.055	L.F.	6	2.88		8.88	11
0080	Duranodic finish		145	.055		6.50	2.88		9.38	11.55
0100	Painted		145	.055		6.85	2.88		9.73	11.90
0300	6" face height		135	.059		6.45	3.09		9.54	11.80
0350	Duranodic finish		135	.059		6.95	3.09		10.04	12.35
0400	Painted		135	.059		8.10	3.09		11.19	13.60
0600	8" face height		125	.064		7.30	3.34		10.64	13.10
0650	Duranodic finish		125	.064		8	3.34		11.34	13.90
0700	Painted		125	.064		8.15	3.34		11.49	14.05
0900	12" face height, .080 thick, 2 piece		100	.080		10.15	4.17		14.32	17.55
0950	Duranodic finish		100	.080		10.05	4.17		14.22	17.40
1000	Painted		100	.080		11.90	4.17		16.07	19.45
1350	Galv steel, 24 ga., 4" leg, plain, with continuous cleat, 4" face		145	.055		6	2.88		8.88	11
1360	6" face height		145	.055		6	2.88		8.88	11
1500	Polyvinyl chloride, 6" face height		135	.059		4.90	3.09		7.99	10.10
1600	9" face height		125	.064		5.80	3.34		9.14	11.45
1800	Stainless steel, 24 ga., 6" face height		135	.059		9.85	3.09		12.94	15.50
1900	12" face height		100	.080		19.70	4.17		23.87	28
2100	20 ga., 6" face height		135	.059		11.90	3.09		14.99	17.75
2200	12" face height		100	.080		24	4.17		28.17	32.50

07 71 19.30 Fascia

		Crew	Daily Output	Labor-Hours	Unit	Material	2012 Bare Costs Labor	Equipment	Total	Total Incl O&P
0010	**FASCIA**									
0100	Aluminum, reverse board and batten, .032" thick, colored, no furring incl	1 Shee	145	.055	S.F.	6.25	2.88		9.13	11.30
0300	Steel, galv and enameled, stock, no furring, long panels		145	.055		4.20	2.88		7.08	9
0600	Short panels		115	.070		5.70	3.63		9.33	11.85

07 71 23 – Manufactured Gutters and Downspouts

07 71 23.10 Downspouts

		Crew	Daily Output	Labor-Hours	Unit	Material	2012 Bare Costs Labor	Equipment	Total	Total Incl O&P
0010	**DOWNSPOUTS**									
0020	Aluminum 2" x 3", .020" thick, embossed	1 Shee	190	.042	L.F.	.97	2.20		3.17	4.42
0100	Enameled		190	.042		1.26	2.20		3.46	4.74
0300	Enameled, .024" thick, 2" x 3"		180	.044		2.05	2.32		4.37	5.80
0400	3" x 4"		140	.057		2.75	2.98		5.73	7.55
0600	Round, corrugated aluminum, 3" diameter, .020" thick		190	.042		1.72	2.20		3.92	5.25
0700	4" diameter, .025" thick		140	.057		2.65	2.98		5.63	7.45
0900	Wire strainer, round, 2" diameter		155	.052	Ea.	3.92	2.69		6.61	8.40
1000	4" diameter		155	.052		7.20	2.69		9.89	12.05
1200	Rectangular, perforated, 2" x 3"		145	.055		2.30	2.88		5.18	6.90
1300	3" x 4"		145	.055		3.32	2.88		6.20	8.05
1500	Copper, round, 16 oz., stock, 2" diameter		190	.042	L.F.	8.10	2.20		10.30	12.25
1600	3" diameter		190	.042		8.10	2.20		10.30	12.25
1800	4" diameter		145	.055		10.85	2.88		13.73	16.30
1900	5" diameter		130	.062		16.50	3.21		19.71	23
2100	Rectangular, corrugated copper, stock, 2" x 3"		190	.042		8.45	2.20		10.65	12.60
2200	3" x 4"		145	.055		11.10	2.88		13.98	16.60
2400	Rectangular, plain copper, stock, 2" x 3"		190	.042		7.40	2.20		9.60	11.50
2500	3" x 4"		145	.055		9.90	2.88		12.78	15.30
2700	Wire strainers, rectangular, 2" x 3"		145	.055	Ea.	7.90	2.88		10.78	13.05
2800	3" x 4"		145	.055		7.90	2.88		10.78	13.05
3000	Round, 2" diameter		145	.055		7.90	2.88		10.78	13.05
3100	3" diameter		145	.055		7.90	2.88		10.78	13.05
3300	4" diameter		145	.055		14.25	2.88		17.13	20

07 71 Roof Specialties

07 71 23 – Manufactured Gutters and Downspouts

07 71 23.10 Downspouts

		Crew	Daily Output	Labor-Hours	Unit	Material	2012 Bare Costs Labor	2012 Bare Costs Equipment	Total	Total Incl O&P
3400	5" diameter	1 Shee	115	.070	Ea.	19.75	3.63		23.38	27
3600	Lead-coated copper, round, stock, 2" diameter		190	.042	L.F.	34	2.20		36.20	41
3700	3" diameter		190	.042		30	2.20		32.20	36.50
3900	4" diameter		145	.055		31	2.88		33.88	39
4200	6" diameter, corrugated		105	.076		30	3.97		33.97	39
4300	Rectangular, corrugated, stock, 2" x 3"		190	.042		12.70	2.20		14.90	17.30
4500	Plain, stock, 2" x 3"		190	.042		16.35	2.20		18.55	21.50
4600	3" x 4"		145	.055		18.20	2.88		21.08	24.50
4800	Steel, galvanized, round, corrugated, 2" or 3" diameter, 28 gauge		190	.042		1.84	2.20		4.04	5.35
4900	4" diameter, 28 gauge		145	.055		2.54	2.88		5.42	7.20
5100	5" diameter, 28 gauge		130	.062		3.61	3.21		6.82	8.85
5200	26 gauge		130	.062		4	3.21		7.21	9.30
5400	6" diameter, 28 gauge		105	.076		4	3.97		7.97	10.45
5500	26 gauge		105	.076		5.80	3.97		9.77	12.45
5700	Rectangular, corrugated, 28 gauge, 2" x 3"		190	.042		1.28	2.20		3.48	4.76
5800	3" x 4"		145	.055		1.82	2.88		4.70	6.40
6000	Rectangular, plain, 28 gauge, galvanized, 2" x 3"		190	.042		1.28	2.20		3.48	4.76
6100	3" x 4"		145	.055		2.44	2.88		5.32	7.05
6300	Epoxy painted, 24 gauge, corrugated, 2" x 3"		190	.042		2	2.20		4.20	5.55
6400	3" x 4"		145	.055		2.50	2.88		5.38	7.15
6600	Wire strainers, rectangular, 2" x 3"		145	.055	Ea.	12	2.88		14.88	17.60
6700	3" x 4"		145	.055		14	2.88		16.88	19.80
6900	Round strainers, 2" or 3" diameter		145	.055		3.12	2.88		6	7.80
7000	4" diameter		145	.055		5.80	2.88		8.68	10.75
7800	Stainless steel tubing, schedule 5, 2" x 3" or 3" diameter		190	.042	L.F.	37.50	2.20		39.70	45
7900	3" x 4" or 4" diameter		145	.055		48	2.88		50.88	57.50
8100	4" x 5" or 5" diameter		135	.059		99	3.09		102.09	114
8200	Vinyl, rectangular, 2" x 3"		210	.038		2.08	1.99		4.07	5.30
8300	Round, 2-1/2"		220	.036		1.41	1.90		3.31	4.44

07 71 23.20 Downspout Elbows

		Crew	Daily Output	Labor-Hours	Unit	Material	2012 Bare Costs Labor	2012 Bare Costs Equipment	Total	Total Incl O&P
0010	**DOWNSPOUT ELBOWS**									
0020	Aluminum, 2" x 3", embossed	1 Shee	100	.080	Ea.	1.02	4.17		5.19	7.45
0100	Enameled		100	.080		1.02	4.17		5.19	7.45
0200	3" x 4", .025" thick, embossed		100	.080		3.50	4.17		7.67	10.20
0300	Enameled		100	.080		2.02	4.17		6.19	8.55
0400	Round corrugated, 3", embossed, .020" thick		100	.080		2.74	4.17		6.91	9.35
0500	4", .025" thick		100	.080		5.80	4.17		9.97	12.70
0600	Copper, 16 oz. round, 2" diameter		100	.080		8.50	4.17		12.67	15.70
0700	3" diameter		100	.080		11.40	4.17		15.57	18.90
0800	4" diameter		100	.080		12.90	4.17		17.07	20.50
1000	2" x 3" corrugated		100	.080		8.50	4.17		12.67	15.70
1100	3" x 4" corrugated		100	.080		14	4.17		18.17	22
1300	Vinyl, 2-1/2" diameter, 45° or 75°		100	.080		3.98	4.17		8.15	10.75
1400	Tee Y junction		75	.107		12.90	5.55		18.45	22.50

07 71 23.30 Gutters

		Crew	Daily Output	Labor-Hours	Unit	Material	2012 Bare Costs Labor	2012 Bare Costs Equipment	Total	Total Incl O&P
0010	**GUTTERS**									
0012	Aluminum, stock units, 5" K type, .027" thick, plain	1 Shee	120	.067	L.F.	2.67	3.48		6.15	8.25
0100	Enameled		120	.067		2.58	3.48		6.06	8.15
0300	5" K type type, .032" thick, plain		120	.067		2.89	3.48		6.37	8.45
0400	Enameled		120	.067		3.26	3.48		6.74	8.90
0700	Copper, half round, 16 oz., stock units, 4" wide		120	.067		9.90	3.48		13.38	16.20
0900	5" wide		120	.067		9	3.48		12.48	15.20

07 71 Roof Specialties

07 71 23 – Manufactured Gutters and Downspouts

07 71 23.30 Gutters

		Crew	Daily Output	Labor-Hours	Unit	Material	2012 Bare Costs Labor	Equipment	Total	Total Incl O&P
1000	6" wide	1 Shee	115	.070	L.F.	14.60	3.63		18.23	21.50
1200	K type, 16 oz., stock, 4" wide		120	.067		13.20	3.48		16.68	19.80
1300	5" wide		120	.067		11.70	3.48		15.18	18.15
1500	Lead coated copper, half round, stock, 4" wide		120	.067		10.40	3.48		13.88	16.75
1600	6" wide		115	.070		22	3.63		25.63	30
1800	K type, stock, 4" wide		120	.067		11.50	3.48		14.98	17.95
1900	5" wide		120	.067		14.45	3.48		17.93	21
2100	Stainless steel, half round or box, stock, 4" wide		120	.067		5.90	3.48		9.38	11.75
2200	5" wide		120	.067		6.35	3.48		9.83	12.25
2400	Steel, galv, half round or box, 28 ga, 5" wide, plain		120	.067		3.89	3.48		7.37	9.60
2500	Enameled		120	.067		2.02	3.48		5.50	7.50
2700	26 ga, stock, 5" wide		120	.067		2.18	3.48		5.66	7.70
2800	6" wide	↓	120	.067		2.88	3.48		6.36	8.45
3000	Vinyl, O.G., 4" wide	1 Carp	110	.073		1.08	3.21		4.29	6.10
3100	5" wide		110	.073		1.38	3.21		4.59	6.45
3200	4" half round, stock units	↓	110	.073	↓	1.08	3.21		4.29	6.10
3250	Joint connectors				Ea.	2.88			2.88	3.17
3300	Wood, clear treated cedar, fir or hemlock, 3" x 4"	1 Carp	100	.080	L.F.	10	3.53		13.53	16.45
3400	4" x 5"	"	100	.080	"	13.50	3.53		17.03	20.50

07 71 23.35 Gutter Guard

		Crew	Daily Output	Labor-Hours	Unit	Material	2012 Bare Costs Labor	Equipment	Total	Total Incl O&P
0010	**GUTTER GUARD**									
0020	6" wide strip, aluminum mesh	1 Carp	500	.016	L.F.	2.85	.71		3.56	4.23
0100	Vinyl mesh	"	500	.016	"	2.58	.71		3.29	3.93

07 71 26 – Reglets

07 71 26.10 Reglets and Accessories

		Crew	Daily Output	Labor-Hours	Unit	Material	2012 Bare Costs Labor	Equipment	Total	Total Incl O&P
0010	**REGLETS AND ACCESSORIES**									
0020	Aluminum, .025" thick, in concrete parapet	1 Carp	225	.036	L.F.	1.77	1.57		3.34	4.36
0100	Copper, 10 oz.		225	.036		2.87	1.57		4.44	5.55
0300	16 oz.		225	.036		5.60	1.57		7.17	8.55
0400	Galvanized steel, 24 gauge		225	.036		1.03	1.57		2.60	3.54
0600	Stainless steel, .020" thick		225	.036		3.45	1.57		5.02	6.20
0700	Zinc and copper alloy, 20 oz.	↓	225	.036		2.87	1.57		4.44	5.55
0900	Counter flashing for above, 12" wide, .032" aluminum	1 Shee	150	.053		1.78	2.78		4.56	6.20
1000	Copper, 10 oz.		150	.053		5.40	2.78		8.18	10.20
1200	16 oz.		150	.053		5.25	2.78		8.03	10.05
1300	Galvanized steel, .020" thick		150	.053		1.28	2.78		4.06	5.65
1500	Stainless steel, .020" thick		150	.053		5.15	2.78		7.93	9.90
1600	Zinc and copper alloy, 20 oz.	↓	150	.053	↓	5.25	2.78		8.03	10

07 71 29 – Manufactured Roof Expansion Joints

07 71 29.10 Expansion Joints

		Crew	Daily Output	Labor-Hours	Unit	Material	2012 Bare Costs Labor	Equipment	Total	Total Incl O&P
0010	**EXPANSION JOINTS**									
0300	Butyl or neoprene center with foam insulation, metal flanges									
0400	Aluminum, .032" thick for openings to 2-1/2"	1 Rofc	165	.048	L.F.	16	1.81		17.81	20.50
0600	For joint openings to 3-1/2"		165	.048		16	1.81		17.81	20.50
0610	For joint openings to 5"		165	.048		16.40	1.81		18.21	21
0620	For joint openings to 8"		165	.048		18.20	1.81		20.01	23
0700	Copper, 16 oz. for openings to 2-1/2"		165	.048		24.50	1.81		26.31	30
0900	For joint openings to 3-1/2"		165	.048		24.50	1.81		26.31	30
0910	For joint openings to 5"		165	.048		25.50	1.81		27.31	31
0920	For joint openings to 8"		165	.048		27	1.81		28.81	33
1000	Galvanized steel, 26 ga. for openings to 2-1/2"	↓	165	.048	↓	13	1.81		14.81	17.35

07 71 Roof Specialties

07 71 29 – Manufactured Roof Expansion Joints

07 71 29.10 Expansion Joints		Crew	Daily Output	Labor-Hours	Unit	Material	2012 Bare Costs Labor	Equipment	Total	Total Incl O&P
1200	For joint openings to 3-1/2"	1 Rofc	165	.048	L.F.	13	1.81		14.81	17.35
1210	For joint openings to 5"		165	.048		13.90	1.81		15.71	18.35
1220	For joint openings to 8"		165	.048		17.30	1.81		19.11	22
1300	Lead-coated copper, 16 oz. for openings to 2-1/2"		165	.048		33.50	1.81		35.31	40
1500	For joint openings to 3-1/2"		165	.048		40	1.81		41.81	47
1600	Stainless steel, .018", for openings to 2-1/2"		165	.048		17.90	1.81		19.71	22.50
1800	For joint openings to 3-1/2"		165	.048		17.90	1.81		19.71	22.50
1810	For joint openings to 5"		165	.048		18.60	1.81		20.41	23.50
1820	For joint openings to 8"		165	.048		20.50	1.81		22.31	25.50
1900	Neoprene, double-seal type with thick center, 4-1/2" wide		125	.064		14	2.39		16.39	19.40
1950	Polyethylene bellows, with galv steel flat flanges		100	.080		5.40	2.99		8.39	10.95
1960	With galvanized angle flanges		100	.080		6	2.99		8.99	11.60
2000	Roof joint with extruded aluminum cover, 2"	1 Shee	115	.070		37.50	3.63		41.13	47
2100	Roof joint, plastic curbs, foam center, standard	1 Rofc	100	.080		13.70	2.99		16.69	20
2200	Large		100	.080		18.25	2.99		21.24	25
2300	Transitions, regular, minimum		10	.800	Ea.	120	30		150	182
2350	Maximum		4	2		150	75		225	291
2400	Large, minimum		9	.889		173	33		206	247
2450	Maximum		3	2.667		182	99.50		281.50	365
2500	Roof to wall joint with extruded aluminum cover	1 Shee	115	.070	L.F.	31.50	3.63		35.13	40.50
2700	Wall joint, closed cell foam on PVC cover, 9" wide	1 Rofc	125	.064		4.90	2.39		7.29	9.40
2800	12" wide	"	115	.070		5.50	2.60		8.10	10.40

07 71 43 – Drip Edge

07 71 43.10 Drip Edge, Rake Edge, Ice Belts		Crew	Daily Output	Labor-Hours	Unit	Material	Labor	Equipment	Total	Total Incl O&P
0010	**DRIP EDGE, RAKE EDGE, ICE BELTS**									
0020	Aluminum, .016" thick, 5" wide, mill finish	1 Carp	400	.020	L.F.	.38	.88		1.26	1.78
0100	White finish		400	.020		.38	.88		1.26	1.78
0200	8" wide, mill finish		400	.020		.46	.88		1.34	1.87
0300	Ice belt, 28" wide, mill finish		100	.080		3.85	3.53		7.38	9.70
0310	Vented, mill finish		400	.020		2.76	.88		3.64	4.40
0320	Painted finish		400	.020		2.86	.88		3.74	4.51
0400	Galvanized, 5" wide		400	.020		.48	.88		1.36	1.89
0500	8" wide, mill finish		400	.020		.69	.88		1.57	2.12
0510	Rake edge, aluminum, 1-1/2" x 1-1/2"		400	.020		.28	.88		1.16	1.67
0520	3-1/2" x 1-1/2"		400	.020		.38	.88		1.26	1.78

07 72 Roof Accessories

07 72 23 – Relief Vents

07 72 23.10 Roof Vents		Crew	Daily Output	Labor-Hours	Unit	Material	Labor	Equipment	Total	Total Incl O&P
0010	**ROOF VENTS**									
0020	Mushroom shape, for built-up roofs, aluminum	1 Rofc	30	.267	Ea.	88.50	9.95		98.45	114
0100	PVC, 6" high	"	30	.267	"	28.50	9.95		38.45	48

07 72 26 – Ridge Vents

07 72 26.10 Ridge Vents and Accessories		Crew	Daily Output	Labor-Hours	Unit	Material	Labor	Equipment	Total	Total Incl O&P
0010	**RIDGE VENTS AND ACCESSORIES**									
0100	Aluminum strips, mill finish	1 Rofc	160	.050	L.F.	2.62	1.87		4.49	6
0150	Painted finish		160	.050	"	3.74	1.87		5.61	7.25
0200	Connectors		48	.167	Ea.	3.68	6.25		9.93	14.50
0300	End caps		48	.167	"	1.37	6.25		7.62	11.95
0400	Galvanized strips		160	.050	L.F.	2.19	1.87		4.06	5.55

07 72 Roof Accessories

07 72 26 – Ridge Vents

07 72 26.10 Ridge Vents and Accessories	Crew	Daily Output	Labor-Hours	Unit	Material	2012 Bare Costs Labor	Equipment	Total	Total Incl O&P
0430	Molded polyethylene, shingles not included	1 Rofc	160	.050	L.F.	2.99	1.87		4.86	6.45
0440	End plugs		48	.167	Ea.	1.37	6.25		7.62	11.95
0450	Flexible roll, shingles not included		160	.050	L.F.	2.15	1.87		4.02	5.50
2300	Ridge vent strip, mill finish	1 Shee	155	.052	"	2.95	2.69		5.64	7.35

07 72 33 – Roof Hatches

07 72 33.10 Roof Hatch Options

		Crew	Daily Output	Labor-Hours	Unit	Material	Labor	Equipment	Total	Total Incl O&P
0010	ROOF HATCH OPTIONS									
0500	2'-6" x 3', aluminum curb and cover	G-3	10	3.200	Ea.	1,050	140		1,190	1,400
0520	Galvanized steel curb and aluminum cover		10	3.200		640	140		780	920
0540	Galvanized steel curb and cover		10	3.200		590	140		730	865
0600	2'-6" x 4'-6", aluminum curb and cover		9	3.556		920	155		1,075	1,225
0800	Galvanized steel curb and aluminum cover		9	3.556		650	155		805	950
0900	Galvanized steel curb and cover		9	3.556		890	155		1,045	1,225
1100	4' x 4' aluminum curb and cover		8	4		1,800	175		1,975	2,250
1120	Galvanized steel curb and aluminum cover		8	4		1,800	175		1,975	2,250
1140	Galvanized steel curb and cover		8	4		1,625	175		1,800	2,050
1200	2'-6" x 8'-0", aluminum curb and cover		6.60	4.848		1,800	212		2,012	2,300
1400	Galvanized steel curb and aluminum cover		6.60	4.848		1,800	212		2,012	2,300
1500	Galvanized steel curb and cover		6.60	4.848		1,300	212		1,512	1,750
1800	For plexiglass panels, 2'-6" x 3'-0", add to above					440			440	485

07 72 36 – Smoke Vents

07 72 36.10 Smoke Hatches

		Crew	Daily Output	Labor-Hours	Unit	Material	Labor	Equipment	Total	Total Incl O&P
0010	SMOKE HATCHES									
0200	For 3'-0" long, add to roof hatches from Section 07 72 33.10				Ea.	25%	5%			
0250	For 4'-0" long, add to roof hatches from Section 07 72 33.10					20%	5%			
0300	For 8'-0" long, add to roof hatches from Section 07 72 33.10					10%	5%			

07 72 36.20 Smoke Vent Options

		Crew	Daily Output	Labor-Hours	Unit	Material	Labor	Equipment	Total	Total Incl O&P
0010	SMOKE VENT OPTIONS									
0100	4' x 4' aluminum cover and frame	G-3	13	2.462	Ea.	2,050	107		2,157	2,425
0200	Galvanized steel cover and frame		13	2.462		1,800	107		1,907	2,150
0300	4' x 8' aluminum cover and frame		8	4		2,800	175		2,975	3,350
0400	Galvanized steel cover and frame		8	4		2,375	175		2,550	2,900

07 72 53 – Snow Guards

07 72 53.10 Snow Guard Options

		Crew	Daily Output	Labor-Hours	Unit	Material	Labor	Equipment	Total	Total Incl O&P
0010	SNOW GUARD OPTIONS									
0100	Slate & asphalt shingle roofs, fastened with nails	1 Rofc	160	.050	Ea.	9.40	1.87		11.27	13.50
0200	Standing seam metal roofs, fastened with set screws		48	.167		15.60	6.25		21.85	27.50
0300	Surface mount for metal roofs, fastened with solder		48	.167		6.05	6.25		12.30	17.10
0400	Double rail pipe type, including pipe		130	.062	L.F.	22.50	2.30		24.80	29

07 72 73 – Pitch Pockets

07 72 73.10 Pitch Pockets, Variable Sizes

		Crew	Daily Output	Labor-Hours	Unit	Material	Labor	Equipment	Total	Total Incl O&P
0010	PITCH POCKETS, VARIABLE SIZES									
0100	Adjustable, 4" to 7", welded corners, 4" deep	1 Rofc	48	.167	Ea.	15	6.25		21.25	27
0200	Side extenders, 6"	"	240	.033	"	2	1.25		3.25	4.29

07 72 80 – Vents

07 72 80.30 Vent Options

		Crew	Daily Output	Labor-Hours	Unit	Material	Labor	Equipment	Total	Total Incl O&P
0010	VENT OPTIONS									
0020	Plastic, for insulated decks, 1 per M.S.F., minimum	1 Rofc	40	.200	Ea.	19	7.50		26.50	33.50
0100	Maximum		20	.400		40	14.95		54.95	69
0300	Aluminum		30	.267		19	9.95		28.95	37.50

07 72 Roof Accessories

07 72 80 – Vents

07 72 80.30 Vent Options		Crew	Daily Output	Labor-Hours	Unit	Material	2012 Bare Costs Labor	Equipment	Total	Total Incl O&P
0800	Polystyrene baffles, 12" wide for 16" O.C. rafter spacing	1 Carp	90	.089	Ea.	.40	3.92		4.32	6.50
0900	For 24" O.C. rafter spacing	"	110	.073	↓	.75	3.21		3.96	5.75

07 81 Applied Fireproofing

07 81 16 – Cementitious Fireproofing

07 81 16.10 Sprayed Cementitious Fireproofing

		Crew	Daily Output	Labor-Hours	Unit	Material	2012 Bare Costs Labor	Equipment	Total	Total Incl O&P
0010	**SPRAYED CEMENTITIOUS FIREPROOFING**									
0050	Not including canvas protection, normal density									
0100	Per 1" thick, on flat plate steel	G-2	3000	.008	S.F.	.53	.30	.04	.87	1.08
0200	Flat decking		2400	.010		.53	.37	.05	.95	1.20
0400	Beams		1500	.016		.53	.59	.09	1.21	1.57
0500	Corrugated or fluted decks		1250	.019		.79	.71	.10	1.60	2.05
0700	Columns, 1-1/8" thick		1100	.022		.59	.81	.12	1.52	2
0800	2-3/16" thick		700	.034		1.20	1.27	.18	2.65	3.44
0900	For canvas protection, add	↓	5000	.005	↓	.07	.18	.03	.28	.38
1500	Intumescent epoxy fireproofing on wire mesh, 3/16" thick									
1550	1 hour rating, exterior use	G-2	136	.176	S.F.	7.30	6.55	.94	14.79	19
1600	Magnesium oxychloride, 35# to 40# density, 1/4" thick		3000	.008		1.53	.30	.04	1.87	2.18
1650	1/2" thick		2000	.012		3.09	.44	.06	3.59	4.14
1700	60# to 70# density, 1/4" thick		3000	.008		2.04	.30	.04	2.38	2.74
1750	1/2" thick		2000	.012		4.10	.44	.06	4.60	5.25
2000	Vermiculite cement, troweled or sprayed, 1/4" thick		3000	.008		1.39	.30	.04	1.73	2.03
2050	1/2" thick	↓	2000	.012	↓	2.77	.44	.06	3.27	3.79

07 84 Firestopping

07 84 13 – Penetration Firestopping

07 84 13.10 Firestopping

		Crew	Daily Output	Labor-Hours	Unit	Material	2012 Bare Costs Labor	Equipment	Total	Total Incl O&P
0010	**FIRESTOPPING** R078413-30									
0100	Metallic piping, non insulated									
0110	Through walls, 2" diameter	1 Carp	16	.500	Ea.	14.65	22		36.65	50
0120	4" diameter		14	.571		22.50	25		47.50	63.50
0130	6" diameter		12	.667		30	29.50		59.50	78
0140	12" diameter		10	.800		53	35.50		88.50	113
0150	Through floors, 2" diameter		32	.250		8.85	11.05		19.90	26.50
0160	4" diameter		28	.286		12.75	12.60		25.35	33.50
0170	6" diameter		24	.333		16.75	14.70		31.45	41
0180	12" diameter	↓	20	.400	↓	28.50	17.65		46.15	58
0190	Metallic piping, insulated									
0200	Through walls, 2" diameter	1 Carp	16	.500	Ea.	20.50	22		42.50	57
0210	4" diameter		14	.571		28.50	25		53.50	70
0220	6" diameter		12	.667		36	29.50		65.50	84.50
0230	12" diameter		10	.800		59	35.50		94.50	120
0240	Through floors, 2" diameter		32	.250		15	11.05		26.05	33.50
0250	4" diameter		28	.286		18.85	12.60		31.45	40
0260	6" diameter		24	.333		23	14.70		37.70	47.50
0270	12" diameter	↓	20	.400	↓	28.50	17.65		46.15	58
0280	Non metallic piping, non insulated									
0290	Through walls, 2" diameter	1 Carp	12	.667	Ea.	60	29.50		89.50	112
0300	4" diameter	↓	10	.800	↓	75.50	35.50		111	138

07 84 13 – Penetration Firestopping

07 84 13.10 Firestopping	Crew	Daily Output	Labor-Hours	Unit	Material	2012 Bare Costs Labor	Equipment	Total	Total Incl O&P
0310	6" diameter	1 Carp	8	1	Ea.	105	44		149	184
0330	Through floors, 2" diameter		16	.500		47	22		69	86
0340	4" diameter		6	1.333		58.50	59		117.50	155
0350	6" diameter		6	1.333		70.50	59		129.50	168
0370	Ductwork, insulated & non insulated, round									
0380	Through walls, 6" diameter	1 Carp	12	.667	Ea.	30.50	29.50		60	79
0390	12" diameter		10	.800		61	35.50		96.50	122
0400	18" diameter		8	1		99	44		143	177
0410	Through floors, 6" diameter		16	.500		16.75	22		38.75	52.50
0420	12" diameter		14	.571		30.50	25		55.50	73
0430	18" diameter		12	.667		53	29.50		82.50	104
0440	Ductwork, insulated & non insulated, rectangular									
0450	With stiffener/closure angle, through walls, 6" x 12"	1 Carp	8	1	Ea.	25.50	44		69.50	96
0460	12" x 24"		6	1.333		33.50	59		92.50	128
0470	24" x 48"		4	2		96.50	88		184.50	242
0480	With stiffener/closure angle, through floors, 6" x 12"		10	.800		13.75	35.50		49.25	69.50
0490	12" x 24"		8	1		25	44		69	95.50
0500	24" x 48"		6	1.333		48.50	59		107.50	144
0510	Multi trade openings									
0520	Through walls, 6" x 12"	1 Carp	2	4	Ea.	53	176		229	330
0530	12" x 24"	"	1	8		215	355		570	780
0540	24" x 48"	2 Carp	1	16		860	705		1,565	2,025
0550	48" x 96"	"	.75	21.333		3,450	940		4,390	5,250
0560	Through floors, 6" x 12"	1 Carp	2	4		53	176		229	330
0570	12" x 24"	"	1	8		215	355		570	780
0580	24" x 48"	2 Carp	.75	21.333		860	940		1,800	2,400
0590	48" x 96"	"	.50	32		3,450	1,400		4,850	5,975
0600	Structural penetrations, through walls									
0610	Steel beams, W8 x 10	1 Carp	8	1	Ea.	33.50	44		77.50	105
0620	W12 x 14		6	1.333		53	59		112	149
0630	W21 x 44		5	1.600		107	70.50		177.50	227
0640	W36 x 135		3	2.667		260	118		378	465
0650	Bar joists, 18" deep		6	1.333		49	59		108	145
0660	24" deep		6	1.333		61	59		120	158
0670	36" deep		5	1.600		91.50	70.50		162	210
0680	48" deep		4	2		107	88		195	254
0690	Construction joints, floor slab at exterior wall									
0700	Precast, brick, block or drywall exterior									
0710	2" wide joint	1 Carp	125	.064	L.F.	7.60	2.82		10.42	12.75
0720	4" wide joint	"	75	.107	"	15.25	4.70		19.95	24
0730	Metal panel, glass or curtain wall exterior									
0740	2" wide joint	1 Carp	40	.200	L.F.	18.05	8.80		26.85	33.50
0750	4" wide joint	"	25	.320	"	25	14.10		39.10	49
0760	Floor slab to drywall partition									
0770	Flat joint	1 Carp	100	.080	L.F.	7.50	3.53		11.03	13.70
0780	Fluted joint		50	.160		15.25	7.05		22.30	27.50
0790	Etched fluted joint		75	.107		9.90	4.70		14.60	18.15
0800	Floor slab to concrete/masonry partition									
0810	Flat joint	1 Carp	75	.107	L.F.	16.75	4.70		21.45	25.50
0820	Fluted joint	"	50	.160	"	19.85	7.05		26.90	33
0830	Concrete/CMU wall joints									
0840	1" wide	1 Carp	100	.080	L.F.	9.15	3.53		12.68	15.50
0850	2" wide		75	.107		16.75	4.70		21.45	25.50

07 84 Firestopping

07 84 13 – Penetration Firestopping

07 84 13.10 Firestopping		Crew	Daily Output	Labor-Hours	Unit	Material	2012 Bare Costs Labor	Equipment	Total	Total Incl O&P
0860	4" wide	1 Carp	50	.160	L.F.	32	7.05		39.05	46
0870	Concrete/CMU floor joints									
0880	1" wide	1 Carp	200	.040	L.F.	4.59	1.76		6.35	7.75
0890	2" wide		150	.053		8.40	2.35		10.75	12.80
0900	4" wide	↓	100	.080	↓	16	3.53		19.53	23

07 91 Preformed Joint Seals

07 91 13 – Compression Seals

07 91 13.10 Compression Seals

		Crew	Daily Output	Labor-Hours	Unit	Material	2012 Bare Costs Labor	Equipment	Total	Total Incl O&P
0010	**COMPRESSION SEALS**									
4900	O-ring type cord, 1/4"	1 Bric	472	.017	L.F.	.48	.75		1.23	1.66
4910	1/2"		440	.018		1.79	.80		2.59	3.18
4920	3/4"		424	.019		3.62	.83		4.45	5.25
4930	1"		408	.020		6.95	.86		7.81	8.95
4940	1-1/4"		384	.021		13.75	.92		14.67	16.50
4950	1-1/2"		368	.022		17.40	.96		18.36	20.50
4960	1-3/4"		352	.023		36.50	1		37.50	41.50
4970	2"	↓	344	.023	↓	64.50	1.03		65.53	72

07 91 16 – Joint Gaskets

07 91 16.10 Joint Gaskets

		Crew	Daily Output	Labor-Hours	Unit	Material	2012 Bare Costs Labor	Equipment	Total	Total Incl O&P
0010	**JOINT GASKETS**									
4400	Joint gaskets, neoprene, closed cell w/adh, 1/8" x 3/8"	1 Bric	240	.033	L.F.	.27	1.47		1.74	2.52
4500	1/4" x 3/4"		215	.037		.59	1.64		2.23	3.13
4700	1/2" x 1"		200	.040		1.31	1.76		3.07	4.10
4800	3/4" x 1-1/2"	↓	165	.048	↓	1.45	2.14		3.59	4.83

07 91 23 – Backer Rods

07 91 23.10 Backer Rods

		Crew	Daily Output	Labor-Hours	Unit	Material	2012 Bare Costs Labor	Equipment	Total	Total Incl O&P
0010	**BACKER RODS**									
0030	Backer rod, polyethylene, 1/4" diameter	1 Bric	4.60	1.739	C.L.F.	2.75	76.50		79.25	119
0050	1/2" diameter		4.60	1.739		5.65	76.50		82.15	122
0070	3/4" diameter		4.60	1.739		7.30	76.50		83.80	124
0090	1" diameter	↓	4.60	1.739	↓	13.65	76.50		90.15	131

07 91 26 – Joint Fillers

07 91 26.10 Joint Fillers

		Crew	Daily Output	Labor-Hours	Unit	Material	2012 Bare Costs Labor	Equipment	Total	Total Incl O&P
0010	**JOINT FILLERS**									
4360	Butyl rubber filler, 1/4" x 1/4"	1 Bric	290	.028	L.F.	.14	1.22		1.36	2
4365	1/2" x 1/2"		250	.032		.57	1.41		1.98	2.76
4370	1/2" x 3/4"		210	.038		.86	1.68		2.54	3.48
4375	3/4" x 3/4"		230	.035		1.28	1.53		2.81	3.73
4380	1" x 1"	↓	180	.044	↓	1.71	1.96		3.67	4.84
4390	For coloring, add					12%				
4980	Polyethylene joint backing, 1/4" x 2"	1 Bric	2.08	3.846	C.L.F.	12	170		182	269
4990	1/4" x 6"		1.28	6.250	"	28	276		304	445
5600	Silicone, room temp vulcanizing foam seal, 1/4" x 1/2"		1312	.006	L.F.	.31	.27		.58	.76
5610	1/2" x 1/2"		656	.012		.63	.54		1.17	1.50
5620	1/2" x 3/4"		442	.018		.94	.80		1.74	2.25
5630	3/4" x 3/4"		328	.024		1.41	1.08		2.49	3.17
5640	1/8" x 1"		1312	.006		.31	.27		.58	.76
5650	1/8" x 3"	↓	442	.018	↓	.94	.80		1.74	2.25

07 91 Preformed Joint Seals

07 91 26 – Joint Fillers

07 91 26.10 Joint Fillers		Crew	Daily Output	Labor-Hours	Unit	Material	2012 Bare Costs Labor	Equipment	Total	Total Incl O&P
5670	1/4" x 3"	1 Bric	295	.027	L.F.	1.88	1.20		3.08	3.88
5680	1/4" x 6"		148	.054		3.77	2.38		6.15	7.75
5690	1/2" x 6"		82	.098		7.55	4.30		11.85	14.80
5700	1/2" x 9"		52.50	.152		11.30	6.70		18	22.50
5710	1/2" x 12"		33	.242		15.05	10.70		25.75	33

07 92 Joint Sealants

07 92 13 – Elastomeric Joint Sealants

07 92 13.20 Caulking and Sealant Options

		Crew	Daily Output	Labor-Hours	Unit	Material	2012 Bare Costs Labor	Equipment	Total	Total Incl O&P
0010	**CAULKING AND SEALANT OPTIONS**									
0050	Latex acrylic based, bulk				Gal.	29.50			29.50	32.50
0055	Bulk in place 1/4" x 1/4" bead	1 Bric	300	.027	L.F.	.09	1.18		1.27	1.88
0060	1/4" x 3/8"		294	.027		.16	1.20		1.36	1.98
0065	1/4" x 1/2"		288	.028		.21	1.23		1.44	2.08
0075	3/8" x 3/8"		284	.028		.23	1.24		1.47	2.13
0080	3/8" x 1/2"		280	.029		.31	1.26		1.57	2.24
0085	3/8" x 5/8"		276	.029		.39	1.28		1.67	2.36
0095	3/8" x 3/4"		272	.029		.46	1.30		1.76	2.47
0100	1/2" x 1/2"		275	.029		.41	1.28		1.69	2.40
0105	1/2" x 5/8"		269	.030		.52	1.31		1.83	2.55
0110	1/2" x 3/4"		263	.030		.62	1.34		1.96	2.71
0115	1/2" x 7/8"		256	.031		.72	1.38		2.10	2.88
0120	1/2" x 1"		250	.032		.83	1.41		2.24	3.04
0125	3/4" x 3/4"		244	.033		.93	1.45		2.38	3.21
0130	3/4" x 1"		225	.036		1.24	1.57		2.81	3.74
0135	1" x 1"		200	.040		1.66	1.76		3.42	4.48
0190	Cartridges				Gal.	25.50			25.50	28
0200	11 fl. oz. cartridge				Ea.	2.20			2.20	2.42
0500	1/4" x 1/2"	1 Bric	288	.028	L.F.	.18	1.23		1.41	2.05
0600	1/2" x 1/2"		275	.029		.36	1.28		1.64	2.34
0800	3/4" x 3/4"		244	.033		.81	1.45		2.26	3.07
0900	3/4" x 1"		225	.036		1.08	1.57		2.65	3.56
1000	1" x 1"		200	.040		1.35	1.76		3.11	4.14
1400	Butyl based, bulk				Gal.	28.50			28.50	31.50
1500	Cartridges				"	32			32	35
1700	1/4" x 1/2", 154 L.F./gal.	1 Bric	288	.028	L.F.	.19	1.23		1.42	2.05
1800	1/2" x 1/2", 77 L.F./gal.	"	275	.029	"	.37	1.28		1.65	2.35
2300	Polysulfide compounds, 1 component, bulk				Gal.	56			56	61.50
2600	1 or 2 component, in place, 1/4" x 1/4", 308 L.F./gal.	1 Bric	300	.027	L.F.	.18	1.18		1.36	1.98
2700	1/2" x 1/4", 154 L.F./gal.		288	.028		.36	1.23		1.59	2.25
2900	3/4" x 3/8", 68 L.F./gal.		272	.029		.82	1.30		2.12	2.87
3000	1" x 1/2", 38 L.F./gal.		250	.032		1.47	1.41		2.88	3.75
3200	Polyurethane, 1 or 2 component				Gal.	51			51	56
3500	Bulk, in place, 1/4" x 1/4"	1 Bric	300	.027	L.F.	.17	1.18		1.35	1.96
3655	1/2" x 1/4"		288	.028		.33	1.23		1.56	2.21
3800	3/4" x 3/8"		272	.029		.75	1.30		2.05	2.79
3900	1" x 1/2"		250	.032		1.33	1.41		2.74	3.59
4100	Silicone rubber, bulk				Gal.	42			42	46
4200	Cartridges				"	41			41	45

07 92 Joint Sealants

07 92 16 – Rigid Joint Sealants

07 92 16.10 Rigid Joint Sealants

		Crew	Daily Output	Labor-Hours	Unit	Material	2012 Bare Costs Labor	Equipment	Total	Total Incl O&P
0010	**RIGID JOINT SEALANTS**									
5800	Tapes, sealant, P.V.C. foam adhesive, 1/16" x 1/4"				C.L.F.	6			6	6.60
5900	1/16" x 1/2"					8			8	8.80
5950	1/16" x 1"					15			15	16.50
6000	1/8" x 1/2"					9			9	9.90

07 92 19 – Acoustical Joint Sealants

07 92 19.10 Acoustical Sealant

		Crew	Daily Output	Labor-Hours	Unit	Material	2012 Bare Costs Labor	Equipment	Total	Total Incl O&P
0010	**ACOUSTICAL SEALANT**									
0020	Acoustical sealant, elastomeric, cartridges				Ea.	6.25			6.25	6.90
0025	In place, 1/4" x 1/4"	1 Bric	300	.027	L.F.	.26	1.18		1.44	2.06
0030	1/4" x 1/2"		288	.028		.51	1.23		1.74	2.41
0035	1/2" x 1/2"		275	.029		1.02	1.28		2.30	3.06
0040	1/2" x 3/4"		263	.030		1.53	1.34		2.87	3.72
0045	3/4" x 3/4"		244	.033		2.30	1.45		3.75	4.71
0050	1" x 1"		200	.040		4.09	1.76		5.85	7.15

07 95 Expansion Control

07 95 13 – Expansion Joint Cover Assemblies

07 95 13.50 Expansion Joint Assemblies

		Crew	Daily Output	Labor-Hours	Unit	Material	2012 Bare Costs Labor	Equipment	Total	Total Incl O&P
0010	**EXPANSION JOINT ASSEMBLIES**									
0200	Floor cover assemblies, 1" space, aluminum	1 Sswk	38	.211	L.F.	19.60	10.40		30	40
0300	Bronze		38	.211		46.50	10.40		56.90	69.50
0500	2" space, aluminum		38	.211		24	10.40		34.40	45
0600	Bronze		38	.211		50	10.40		60.40	73.50
0800	Wall and ceiling assemblies, 1" space, aluminum		38	.211		11.70	10.40		22.10	31.50
0900	Bronze		38	.211		43.50	10.40		53.90	66
1100	2" space, aluminum		38	.211		19.90	10.40		30.30	40.50
1200	Bronze		38	.211		53.50	10.40		63.90	77
1400	Floor to wall assemblies, 1" space, aluminum		38	.211		17.85	10.40		28.25	38.50
1500	Bronze or stainless		38	.211		52.50	10.40		62.90	76
1700	Gym floor angle covers, aluminum, 3" x 3" angle		46	.174		15.45	8.60		24.05	32.50
1800	3" x 4" angle		46	.174		21.50	8.60		30.10	39
2000	Roof closures, aluminum, flat roof, low profile, 1" space		57	.140		35.50	6.95		42.45	51.50
2100	High profile		57	.140		51.50	6.95		58.45	69
2300	Roof to wall, low profile, 1" space		57	.140		19.75	6.95		26.70	34
2400	High profile		57	.140		25.50	6.95		32.45	41

Estimating Tips

08 10 00 Doors and Frames

All exterior doors should be addressed for their energy conservation (insulation and seals).

- Most metal doors and frames look alike, but there may be significant differences among them. When estimating these items, be sure to choose the line item that most closely compares to the specification or door schedule requirements regarding:
  — type of metal
  — metal gauge
  — door core material
  — fire rating
  — finish
- Wood and plastic doors vary considerably in price. The primary determinant is the veneer material. Lauan, birch, and oak are the most common veneers. Other variables include the following:
  — hollow or solid core
  — fire rating
  — flush or raised panel
  — finish

08 30 00 Specialty Doors and Frames

- There are many varieties of special doors, and they are usually priced per each. Add frames, hardware, or operators required for a complete installation.

08 40 00 Entrances, Storefronts, and Curtain Walls

- Glazed curtain walls consist of the metal tube framing and the glazing material. The cost data in this subdivision is presented for the metal tube framing alone or the composite wall. If your estimate requires a detailed takeoff of the framing, be sure to add the glazing cost.

08 50 00 Windows

- Most metal windows are delivered preglazed. However, some metal windows are priced without glass. Refer to 08 80 00 Glazing for glass pricing. The grade C indicates commercial grade windows, usually ASTM C-35.
- All wood windows and vinyl are priced preglazed. The glazing is insulating glass. Some wood windows may have single pane float glass. Add the cost of screens and grills if required, and not already included.

08 70 00 Hardware

- Hardware costs add considerably to the cost of a door. The most efficient method to determine the hardware requirements for a project is to review the door schedule.

- Door hinges are priced by the pair, with most doors requiring 1-1/2 pairs per door. The hinge prices do not include installation labor because it is included in door installation. Hinges are classified according to the frequency of use.

08 80 00 Glazing

- Different openings require different types of glass. The most common types are:
  — float
  — tempered
  — insulating
  — impact-resistant
  — ballistic-resistant
- Most exterior windows are glazed with insulating glass. Entrance doors and window walls, where the glass is less than 18" from the floor, are generally glazed with tempered glass. Interior windows and some residential windows are glazed with float glass.
- Coastal communities require the use of impact-resistant glass.
- The insulation or 'u' value is a strong consideration, along with solar heat gain, to determine total energy efficiency.

Reference Numbers

Reference numbers are shown in shaded boxes at the beginning of some major classifications. These numbers refer to related items in the Reference Section. The reference information may be an estimating procedure, an alternate pricing method, or technical information.

Note: Not all subdivisions listed here necessarily appear in this publication.

08 05 05 – Selective Windows and Doors Demolition

08 05 05.10 Selective Demolition Doors

		Crew	Daily Output	Labor-Hours	Unit	Material	2012 Bare Costs Labor	Equipment	Total	Total Incl O&P
0010	**SELECTIVE DEMOLITION DOORS** R024119-10									
0200	Doors, exterior, 1-3/4" thick, single, 3' x 7' high	1 Clab	16	.500	Ea.		17.55		17.55	27
0220	Double, 6' x 7' high		12	.667			23.50		23.50	36
0500	Interior, 1-3/8" thick, single, 3' x 7' high		20	.400			14.05		14.05	21.50
0520	Double, 6' x 7' high		16	.500			17.55		17.55	27
0700	Bi-folding, 3' x 6'-8" high		20	.400			14.05		14.05	21.50
0720	6' x 6'-8" high		18	.444			15.60		15.60	24
0900	Bi-passing, 3' x 6'-8" high		16	.500			17.55		17.55	27
0940	6' x 6'-8" high		14	.571			20		20	31
1500	Remove and reset, minimum	1 Carp	8	1			44		44	68
1520	Maximum		6	1.333			59		59	90.50
2000	Frames, including trim, metal		8	1			44		44	68
2200	Wood	2 Carp	32	.500			22		22	34
3000	Special doors, counter doors		6	2.667			118		118	181
3100	Double acting		10	1.600			70.50		70.50	109
3200	Floor door (trap type), or access type		8	2			88		88	136
3300	Glass, sliding, including frames		12	1.333			59		59	90.50
3400	Overhead, commercial, 12' x 12' high		4	4			176		176	271
3440	up to 20' x 16' high		3	5.333			235		235	360
3445	up to 35' x 30' high		1	16			705		705	1,075
3500	Residential, 9' x 7' high		8	2			88		88	136
3540	16' x 7' high		7	2.286			101		101	155
3600	Remove and reset, minimum		4	4			176		176	271
3620	Maximum		2.50	6.400			282		282	435
3700	Roll-up grille		5	3.200			141		141	217
3800	Revolving door		2	8			355		355	545
3900	Storefront swing door		3	5.333			235		235	360
6600	Demo flexible transparent strip entrance	3 Shee	115	.209	SF Surf		10.90		10.90	16.60
7100	Remove double swing pneumatic doors, openers and sensors	2 Skwk	.50	32	Opng.		1,450		1,450	2,250
7110	Remove automatic operators, industrial, sliding doors, to 12' wide	"	.40	40	"		1,825		1,825	2,800
7550	Hangar door demo	2 Sswk	330	.048	S.F.		2.39		2.39	4.28
7570	Remove shock absorbing door	"	1.90	8.421	Opng.		415		415	745

08 05 05.20 Selective Demolition of Windows

		Crew	Daily Output	Labor-Hours	Unit	Material	2012 Bare Costs Labor	Equipment	Total	Total Incl O&P
0010	**SELECTIVE DEMOLITION OF WINDOWS** R024119-10									
0200	Aluminum, including trim, to 12 S.F.	1 Clab	16	.500	Ea.		17.55		17.55	27
0240	To 25 S.F.		11	.727			25.50		25.50	39.50
0280	To 50 S.F.		5	1.600			56		56	86.50
0320	Storm windows/screens, to 12 S.F.		27	.296			10.40		10.40	16
0360	To 25 S.F.		21	.381			13.35		13.35	20.50
0400	To 50 S.F.		16	.500			17.55		17.55	27
0600	Glass, minimum		200	.040	S.F.		1.40		1.40	2.16
0620	Maximum		150	.053	"		1.87		1.87	2.88
1000	Steel, including trim, to 12 S.F.		13	.615	Ea.		21.50		21.50	33
1020	To 25 S.F.		9	.889			31		31	48
1040	To 50 S.F.		4	2			70		70	108
2000	Wood, including trim, to 12 S.F.		22	.364			12.75		12.75	19.65
2020	To 25 S.F.		18	.444			15.60		15.60	24
2060	To 50 S.F.		13	.615			21.50		21.50	33
2065	To 180 S.F.		8	1			35		35	54
4300	Remove bay/bow window	2 Carp	6	2.667			118		118	181
4410	Remove skylight, plstc domes, flush/curb mtd	G-3	395	.081	S.F.		3.53		3.53	5.40
5020	Remove and reset window, minimum	1 Carp	6	1.333	Ea.		59		59	90.50

08 05 Common Work Results for Openings

08 05 05 – Selective Windows and Doors Demolition

08 05 05.20 Selective Demolition of Windows

08 05 05.20 Selective Demolition of Windows	Crew	Daily Output	Labor-Hours	Unit	Material	2012 Bare Costs Labor	Equipment	Total	Total Incl O&P
5040	Average	1 Carp	4	2	Ea.		88		88	136
5080	Maximum	↓	2	4	↓		176		176	271

08 11 Metal Doors and Frames

08 11 63 – Metal Screen and Storm Doors and Frames

08 11 63.23 Aluminum Screen and Storm Doors and Frames

		Crew	Daily Output	Labor-Hours	Unit	Material	2012 Bare Costs Labor	Equipment	Total	Total Incl O&P
0010	**ALUMINUM SCREEN AND STORM DOORS AND FRAMES**									
0020	Combination storm and screen									
0420	Clear anodic coating, 2'-8" wide	2 Carp	14	1.143	Ea.	161	50.50		211.50	255
0440	3'-0" wide	"	14	1.143	"	161	50.50		211.50	255
0500	For 7' door height, add					8%				
1020	Mill finish, 2'-8" wide	2 Carp	14	1.143	Ea.	231	50.50		281.50	330
1040	3'-0" wide	"	14	1.143		253	50.50		303.50	355
1100	For 7'-0" door, add					8%				
1520	White painted, 2'-8" wide	2 Carp	14	1.143		283	50.50		333.50	390
1540	3'-0" wide	"	14	1.143		299	50.50		349.50	410
1600	For 7'-0" door, add					8%				
2000	Wood door & screen, see Section 08 14 33.20									

08 11 74 – Sliding Metal Grilles

08 11 74.10 Rolling Grille Supports

		Crew	Daily Output	Labor-Hours	Unit	Material	2012 Bare Costs Labor	Equipment	Total	Total Incl O&P
0010	**ROLLING GRILLE SUPPORTS**									
0020	Rolling grille supports, overhead framed	E-4	36	.889	L.F.	24.50	44.50	3.38	72.38	110

08 12 Metal Frames

08 12 13 – Hollow Metal Frames

08 12 13.13 Standard Hollow Metal Frames

			Crew	Daily Output	Labor-Hours	Unit	Material	2012 Bare Costs Labor	Equipment	Total	Total Incl O&P
0010	**STANDARD HOLLOW METAL FRAMES**										
0020	16 ga., up to 5-3/4" jamb depth										
0025	6'-8" high, 3'-0" wide, single	G	2 Carp	16	1	Ea.	143	44		187	225
0028	3'-6" wide, single	G		16	1		150	44		194	232
0030	4'-0" wide, single	G		16	1		150	44		194	232
0040	6'-0" wide, double	G		14	1.143		204	50.50		254.50	300
0045	8'-0" wide, double	G		14	1.143		213	50.50		263.50	310
0100	7'-0" high, 3'-0" wide, single	G		16	1		148	44		192	231
0110	3'-6" wide, single	G		16	1		155	44		199	238
0112	4'-0" wide, single	G		16	1		155	44		199	238
0140	6'-0" wide, double	G		14	1.143		194	50.50		244.50	292
0145	8'-0" wide, double	G		14	1.143		229	50.50		279.50	330
1000	16 ga., up to 4-7/8" deep, 7'-0" H, 3'-0" W, single	G		16	1		163	44		207	247
1140	6'-0" wide, double	G		14	1.143		184	50.50		234.50	280
1200	16 ga., 8-3/4" deep, 7'-0" H, 3'-0" W, single	G		16	1		199	44		243	287
1240	6'-0" wide, double	G		14	1.143		234	50.50		284.50	335
2800	14 ga., up to 3-7/8" deep, 7'-0" high, 3'-0" wide, single	G		16	1		177	44		221	263
2840	6'-0" wide, double	G		14	1.143		213	50.50		263.50	310
3000	14 ga., up to 5-3/4" deep, 6'-8" high, 3'-0" wide, single	G		16	1		149	44		193	232
3002	3'-6" wide, single	G		16	1		193	44		237	280
3005	4'-0" wide, single	G		16	1		193	44		237	280
3600	up to 5-3/4" jamb depth, 7'-0" high, 4'-0" wide, single	G		15	1.067		185	47		232	276
3620	6'-0" wide, double	G	↓	12	1.333	↓	235	59		294	350

08 12 Metal Frames

08 12 13 – Hollow Metal Frames

08 12 13.13 Standard Hollow Metal Frames

		Crew	Daily Output	Labor-Hours	Unit	Material	2012 Bare Costs Labor	Equipment	Total	Total Incl O&P
3640	8'-0" wide, double	G 2 Carp	12	1.333	Ea.	246	59		305	360
3700	8'-0" high, 4'-0" wide, single	G	15	1.067		235	47		282	330
3740	8'-0" wide, double	G	12	1.333		290	59		349	410
4000	6-3/4" deep, 7'-0" high, 4'-0" wide, single	G	15	1.067		216	47		263	310
4020	6'-0" wide, double	G	12	1.333		268	59		327	385
4040	8'-0", wide double	G	12	1.333		276	59		335	395
4100	8'-0" high, 4'-0" wide, single	G	15	1.067		276	47		323	380
4140	8'-0" wide, double	G	12	1.333		320	59		379	445
4400	8-3/4" deep, 7'-0" high, 4'-0" wide, single	G	15	1.067		252	47		299	350
4440	8'-0" wide, double	G	12	1.333		315	59		374	435
4500	8'-0" high, 4'-0" wide, single	G	15	1.067		283	47		330	385
4540	8'-0" wide, double	G	12	1.333		350	59		409	475
4900	For welded frames, add					63			63	69.50
5400	14 ga., "B" label, up to 5-3/4" deep, 7'-0" high, 4'-0" wide, single	G 2 Carp	15	1.067		204	47		251	298
5440	8'-0" wide, double	G	12	1.333		267	59		326	385
5800	6-3/4" deep, 7'-0" high, 4'-0" wide, single	G	15	1.067		218	47		265	315
5840	8'-0" wide, double	G	12	1.333		380	59		439	505
6200	8-3/4" deep, 7'-0" high, 4'-0" wide, single	G	15	1.067		305	47		352	410
6240	8'-0" wide, double	G	12	1.333		355	59		414	480
6300	For "A" label use same price as "B" label									
6400	For baked enamel finish, add					30%	15%			
6500	For galvanizing, add					20%				
6600	For hospital stop, add				Ea.	295			295	325
6620	For hospital stop, stainless steel add				"	380			380	420
7900	Transom lite frames, fixed, add	2 Carp	155	.103	S.F.	54	4.55		58.55	66.50
8000	Movable, add	"	130	.123	"	68	5.45		73.45	83.50

08 12 13.25 Channel Metal Frames

		Crew	Daily Output	Labor-Hours	Unit	Material	2012 Bare Costs Labor	Equipment	Total	Total Incl O&P
0010	**CHANNEL METAL FRAMES**									
0020	Steel channels with anchors and bar stops									
0100	6" channel @ 8.2#/L.F., 3' x 7' door, weighs 150#	G E-4	13	2.462	Ea.	225	123	9.35	357.35	475
0200	8" channel @ 11.5#/L.F., 6' x 8' door, weighs 275#	G	9	3.556		415	177	13.50	605.50	785
0300	8' x 12' door, weighs 400#	G	6.50	4.923		600	245	18.70	863.70	1,125
0400	10" channel @ 15.3#/L.F., 10' x 10' door, weighs 500#	G	6	5.333		750	266	20.50	1,036.50	1,325
0500	12' x 12' door, weighs 600#	G	5.50	5.818		900	290	22	1,212	1,525
0600	12" channel @ 20.7#/L.F., 12' x 12' door, weighs 825#	G	4.50	7.111		1,250	355	27	1,632	2,025
0700	12' x 16' door, weighs 1000#	G	4	8		1,500	400	30.50	1,930.50	2,400
0800	For frames without bar stops, light sections, deduct					15%				
0900	Heavy sections, deduct					10%				

08 13 Metal Doors

08 13 13 – Hollow Metal Doors

08 13 13.13 Standard Hollow Metal Doors

		Crew	Daily Output	Labor-Hours	Unit	Material	2012 Bare Costs Labor	Equipment	Total	Total Incl O&P
0010	**STANDARD HOLLOW METAL DOORS** R081313-20									
0015	Flush, full panel, hollow core									
0017	When noted doors are prepared but do not include glass or louvers									
0020	1-3/8" thick, 20 ga., 2'-0" x 6'-8"	G 2 Carp	20	.800	Ea.	335	35.50		370.50	425
0040	2'-8" x 6'-8"	G	18	.889		350	39		389	445
0060	3'-0" x 6'-8"	G	17	.941		350	41.50		391.50	450
0100	3'-0" x 7'-0"	G	17	.941		360	41.50		401.50	460
0120	For vision lite, add					93			93	102
0140	For narrow lite, add					101			101	111

08 13 Metal Doors

08 13 13 – Hollow Metal Doors

08 13 13.13 Standard Hollow Metal Doors		Crew	Daily Output	Labor-Hours	Unit	Material	2012 Bare Costs Labor	Equipment	Total	Total Incl O&P
0320	Half glass, 20 ga., 2'-0" x 6'-8"	G	2 Carp	20	.800	Ea.	490	35.50		525.50	595
0340	2'-8" x 6'-8"	G		18	.889		515	39		554	625
0360	3'-0" x 6'-8"	G		17	.941		510	41.50		551.50	625
0400	3'-0" x 7'-0"	G		17	.941		615	41.50		656.50	745
0410	1-3/8" thick, 18 ga., 2'-0" x 6'-8"	G		20	.800		400	35.50		435.50	495
0420	3'-0" x 6'-8"	G		17	.941		405	41.50		446.50	510
0425	3'-0" x 7'-0"	G		17	.941		410	41.50		451.50	520
0450	For vision lite, add						93			93	102
0452	For narrow lite, add						101			101	111
0460	Half glass, 18 ga., 2'-0" x 6'-8"	G	2 Carp	20	.800		555	35.50		590.50	670
0465	2'-8" x 6'-8"	G		18	.889		575	39		614	695
0470	3'-0" x 6'-8"	G		17	.941		565	41.50		606.50	685
0475	3'-0" x 7'-0"	G		17	.941		565	41.50		606.50	685
0500	Hollow core, 1-3/4" thick, full panel, 20 ga., 2'-8" x 6'-8"	G		18	.889		415	39		454	515
0520	3'-0" x 6'-8"	G		17	.941		420	41.50		461.50	525
0640	3'-0" x 7'-0"	G		17	.941		435	41.50		476.50	540
0680	4'-0" x 7'-0"	G		15	1.067		630	47		677	770
0700	4'-0" x 8'-0"	G		13	1.231		735	54.50		789.50	895
1000	18 ga., 2'-8" x 6'-8"	G		17	.941		490	41.50		531.50	605
1020	3'-0" x 6'-8"	G		16	1		475	44		519	595
1120	3'-0" x 7'-0"	G		17	.941		595	41.50		636.50	715
1180	4'-0" x 7'-0"	G		14	1.143		630	50.50		680.50	775
1200	4'-0" x 8'-0"	G		17	.941		735	41.50		776.50	875
1212	For vision lite, add						93			93	102
1214	For narrow lite, add						101			101	111
1230	Half glass, 20 ga., 2'-8" x 6'-8"	G	2 Carp	20	.800		575	35.50		610.50	690
1240	3'-0" x 6'-8"	G		18	.889		575	39		614	695
1260	3'-0" x 7'-0"	G		18	.889		595	39		634	710
1320	18 ga., 2'-8" x 6'-8"	G		18	.889		640	39		679	765
1340	3'-0" x 6'-8"	G		17	.941		635	41.50		676.50	760
1360	3'-0" x 7'-0"	G		17	.941		645	41.50		686.50	775
1380	4'-0" x 7'-0"	G		15	1.067		785	47		832	940
1400	4'-0" x 8'-0"	G		14	1.143		890	50.50		940.50	1,050
1500	Flush full panel, 16 gauge, steel hollow core										
1505	1-3/8" same price as 1-3/4"										
1520	2'-0" x 6'-8"		2 Carp	20	.800	Ea.	555	35.50		590.50	665
1530	2'-8" x 6'-8"			20	.800		555	35.50		590.50	670
1540	3'-0" x 6'-8"			20	.800		550	35.50		585.50	660
1560	2'-8" x 7'-0"			18	.889		575	39		614	690
1570	3'-0" x 7'-0"			18	.889		560	39		599	675
1580	3'-6" x 7'-0"			18	.889		655	39		694	780
1590	4'-0" x 7'-0"			18	.889		725	39		764	855
1600	2'-8" x 8'-0"			18	.889		700	39		739	830
1620	3'-0" x 8'-0"			18	.889		720	39		759	850
1630	3'-6" x 8'-0"			18	.889		800	39		839	940
1640	4'-0" x 8'-0"			18	.889		850	39		889	990
1720	Insulated, 1-3/4" thick, full panel, 18 ga., 3'-0" x 6'-8"	G		15	1.067		475	47		522	600
1740	2'-8" x 7'-0"	G		16	1		505	44		549	625
1760	3'-0" x 7'-0"	G		15	1.067		490	47		537	615
1800	4'-0" x 8'-0"	G		13	1.231		740	54.50		794.50	895
1805	For vision lite, add						93			93	102
1810	For narrow lite, add						101			101	111
1820	Half glass, 18 ga., 3'-0" x 6'-8"	G	2 Carp	16	1		635	44		679	765

08 13 13.13 Standard Hollow Metal Doors

		Crew	Daily Output	Labor-Hours	Unit	Material	2012 Bare Costs Labor	Equipment	Total	Total Incl O&P
1840	2'-8" x 7'-0" G	2 Carp	17	.941	Ea.	660	41.50		701.50	790
1860	3'-0" x 7'-0" G		16	1		685	44		729	820
1900	4'-0" x 8'-0" G		14	1.143		670	50.50		720.50	815
8100	For bottom louver, add					274			274	300
8110	For baked enamel finish, add					30%	15%			
8120	For galvanizing, add					20%				

08 13 13.15 Metal Fire Doors

		Crew	Daily Output	Labor-Hours	Unit	Material	2012 Bare Costs Labor	Equipment	Total	Total Incl O&P
0010	**METAL FIRE DOORS** R081313-20									
0015	Steel, flush, "B" label, 90 minute									
0020	Full panel, 20 ga., 2'-0" x 6'-8"	2 Carp	20	.800	Ea.	420	35.50		455.50	515
0040	2'-8" x 6'-8"		18	.889		435	39		474	540
0060	3'-0" x 6'-8"		17	.941		435	41.50		476.50	545
0080	3'-0" x 7'-0"		17	.941		455	41.50		496.50	565
0140	18 ga., 3'-0" x 6'-8"		16	1		495	44		539	615
0160	2'-8" x 7'-0"		17	.941		520	41.50		561.50	635
0180	3'-0" x 7'-0"		16	1		505	44		549	625
0200	4'-0" x 7'-0"		15	1.067		650	47		697	790
0220	For "A" label, 3 hour, 18 ga., use same price as "B" label									
0240	For vision lite, add				Ea.	156			156	172
0520	Flush, "B" label 90 min., egress core, 20 ga., 2'-0" x 6'-8"	2 Carp	18	.889		655	39		694	780
0540	2'-8" x 6'-8"		17	.941		665	41.50		706.50	795
0560	3'-0" x 6'-8"		16	1		665	44		709	800
0580	3'-0" x 7'-0"		16	1		685	44		729	820
0640	Flush, "A" label 3 hour, egress core, 18 ga., 3'-0" x 6'-8"		15	1.067		720	47		767	870
0660	2'-8" x 7'-0"		16	1		750	44		794	890
0680	3'-0" x 7'-0"		15	1.067		740	47		787	890
0700	4'-0" x 7'-0"		14	1.143		885	50.50		935.50	1,050

08 13 13.20 Residential Steel Doors

		Crew	Daily Output	Labor-Hours	Unit	Material	2012 Bare Costs Labor	Equipment	Total	Total Incl O&P
0010	**RESIDENTIAL STEEL DOORS**									
0020	Prehung, insulated, exterior									
0030	Embossed, full panel, 2'-8" x 6'-8" G	2 Carp	17	.941	Ea.	298	41.50		339.50	395
0040	3'-0" x 6'-8" G		15	1.067		243	47		290	340
0060	3'-0" x 7'-0" G		15	1.067		330	47		377	440
0070	5'-4" x 6'-8", double G		8	2		625	88		713	820
0220	Half glass, 2'-8" x 6'-8"		17	.941		320	41.50		361.50	415
0240	3'-0" x 6'-8"		16	1		320	44		364	420
0260	3'-0" x 7'-0"		16	1		370	44		414	475
0270	5'-4" x 6'-8", double		8	2		650	88		738	850
1320	Flush face, full panel, 2'-8" x 6'-8"		16	1		248	44		292	340
1340	3'-0" x 6'-8"		15	1.067		248	47		295	345
1360	3'-0" x 7'-0"		15	1.067		296	47		343	400
1380	5'-4" x 6'-8", double		8	2		495	88		583	680
1420	Half glass, 2'-8" x 6'-8"		17	.941		325	41.50		366.50	420
1440	3'-0" x 6'-8"		16	1		325	44		369	425
1460	3'-0" x 7'-0"		16	1		350	44		394	455
1480	5'-4" x 6'-8", double		8	2		635	88		723	830
1500	Sidelight, full lite, 1'-0" x 6'-8" with grille					244			244	268
1510	1'-0" x 6'-8", low e					264			264	290
1520	1'-0" x 6'-8", half lite					254			254	280
1530	1'-0" x 6'-8", half lite, low e					262			262	289
2300	Interior, residential, closet, bi-fold, 6'-8" x 2'-0" wide	2 Carp	16	1		161	44		205	245
2330	3'-0" wide		16	1		200	44		244	288

08 13 Metal Doors

08 13 13 – Hollow Metal Doors

08 13 13.20 Residential Steel Doors		Crew	Daily Output	Labor-Hours	Unit	Material	2012 Bare Costs Labor	Equipment	Total	Total Incl O&P
2360	4'-0" wide	2 Carp	15	1.067	Ea.	260	47		307	360
2400	5'-0" wide		14	1.143		315	50.50		365.50	430
2420	6'-0" wide		13	1.231		330	54.50		384.50	450

08 13 13.25 Doors Hollow Metal

		Crew	Daily Output	Labor-Hours	Unit	Material	Labor	Equipment	Total	Total Incl O&P
0010	**DOORS HOLLOW METAL**									
0500	Exterior, commercial, flush, 20 ga., 1-3/4" x 7'-0" x 2'-6" wide	2 Carp	15	1.067	Ea.	390	47		437	505
0530	2'-8" wide		15	1.067		435	47		482	550
0560	3'-0" wide		14	1.143		435	50.50		485.50	555
1000	18 ga., 1-3/4" x 7'-0" x 2'-6" wide		15	1.067		495	47		542	615
1030	2'-8" wide		15	1.067		500	47		547	625
1060	3'-0" wide		14	1.143		485	50.50		535.50	615
1500	16 ga., 1-3/4 x 7'-0" x 2'-6" wide		15	1.067		560	47		607	690
1530	2'-8" wide		15	1.067		565	47		612	700
1560	3'-0" wide		14	1.143		550	50.50		600.50	685
1590	3'-6" wide		14	1.143		645	50.50		695.50	785
2900	Fire door, "A" label, 1-3/4" x 7'-0" x 2'-6" wide		15	1.067		645	47		692	785
2930	2'-8" wide		15	1.067		655	47		702	795
2960	3'-0" wide		14	1.143		640	50.50		690.50	785
2990	3'-6" wide		14	1.143		730	50.50		780.50	885
3100	"B" label, 2'-6" wide		15	1.067		585	47		632	715
3130	2'-8" wide		15	1.067		590	47		637	720
3160	3'-0" wide		14	1.143		580	50.50		630.50	720

08 13 16 – Aluminum Doors

08 13 16.10 Commercial Aluminum Doors

		Crew	Daily Output	Labor-Hours	Unit	Material	Labor	Equipment	Total	Total Incl O&P
0010	**COMMERCIAL ALUMINUM DOORS**, flush, no glazing									
5000	Flush panel doors, pair of 2'-6" x 7'-0"	2 Sswk	2	8	Pr.	1,375	395		1,770	2,225
5050	3'-0" x 7'-0", single		2.50	6.400	Ea.	780	315		1,095	1,425
5100	Pair of 3'-0" x 7'-0"		2	8	Pr.	1,475	395		1,870	2,325
5150	3'-6" x 7'-0", single		2.50	6.400	Ea.	935	315		1,250	1,600

08 13 73 – Sliding Metal Doors

08 13 73.10 Steel Sliding Doors

		Crew	Daily Output	Labor-Hours	Unit	Material	Labor	Equipment	Total	Total Incl O&P
0010	**STEEL SLIDING DOORS**									
0020	Up to 50' x 18', electric, standard duty, minimum	L-5	360	.156	S.F.	29	7.70	1.94	38.64	47
0100	Maximum		340	.165		45.50	8.15	2.05	55.70	66.50
0500	Heavy duty, minimum		297	.189		37	9.30	2.35	48.65	59.50
0600	Maximum		277	.202		92.50	10	2.52	105.02	122

08 14 Wood Doors

08 14 13 – Carved Wood Doors

08 14 13.10 Types of Wood Doors, Carved

		Crew	Daily Output	Labor-Hours	Unit	Material	Labor	Equipment	Total	Total Incl O&P
0010	**TYPES OF WOOD DOORS, CARVED**									
3000	Solid wood, 1-3/4" thick stile and rail									
3020	Mahogany, 3'-0" x 7'-0", minimum	2 Carp	14	1.143	Ea.	965	50.50		1,015.50	1,125
3030	Maximum		10	1.600		1,750	70.50		1,820.50	2,025
3040	3'-6" x 8'-0", minimum		10	1.600		1,250	70.50		1,320.50	1,475
3050	Maximum		8	2		2,400	88		2,488	2,775
3100	Pine, 3'-0" x 7'-0", minimum		14	1.143		490	50.50		540.50	615
3110	Maximum		10	1.600		825	70.50		895.50	1,025
3120	3'-6" x 8'-0", minimum		10	1.600		900	70.50		970.50	1,100
3130	Maximum		8	2		1,825	88		1,913	2,150

08 14 Wood Doors

08 14 13 – Carved Wood Doors

08 14 13.10 Types of Wood Doors, Carved

		Crew	Daily Output	Labor-Hours	Unit	Material	2012 Bare Costs Labor	Equipment	Total	Total Incl O&P
3200	Red oak, 3'-0" x 7'-0", minimum	2 Carp	14	1.143	Ea.	1,650	50.50		1,700.50	1,900
3210	Maximum		10	1.600		2,200	70.50		2,270.50	2,525
3220	3'-6" x 8'-0", minimum		10	1.600		2,500	70.50		2,570.50	2,850
3230	Maximum		8	2		3,200	88		3,288	3,650
4000	Hand carved door, mahogany									
4020	3'-0" x 7'-0", minimum	2 Carp	14	1.143	Ea.	1,675	50.50		1,725.50	1,925
4030	Maximum		11	1.455		3,650	64		3,714	4,125
4040	3'-6" x 8'-0", minimum		10	1.600		2,500	70.50		2,570.50	2,850
4050	Maximum		8	2		3,600	88		3,688	4,075
4200	Rose wood, 3'-0" x 7'-0", minimum		14	1.143		5,100	50.50		5,150.50	5,675
4210	Maximum		11	1.455		13,600	64		13,664	15,100
4220	3'-6" x 8'-0", minimum		10	1.600		5,800	70.50		5,870.50	6,475
4225	Maximum		10	1.600		8,800	70.50		8,870.50	9,775
4280	For 6'-8" high door, deduct from 7'-0" door					34			34	37.50
4400	For custom finish, add					450			450	495
4600	Side light, mahogany, 7'-0" x 1'-6" wide, minimum	2 Carp	18	.889		1,000	39		1,039	1,150
4610	Maximum		14	1.143		2,625	50.50		2,675.50	2,975
4620	8'-0" x 1'-6" wide, minimum		14	1.143		1,700	50.50		1,750.50	1,950
4630	Maximum		10	1.600		2,000	70.50		2,070.50	2,300
4640	Side light, oak, 7'-0" x 1'-6" wide, minimum		18	.889		1,100	39		1,139	1,250
4650	Maximum		14	1.143		2,000	50.50		2,050.50	2,275
4660	8'-0" x 1'-6" wide, minimum		14	1.143		1,000	50.50		1,050.50	1,175
4670	Maximum		10	1.600		2,000	70.50		2,070.50	2,300

08 14 16 – Flush Wood Doors

08 14 16.09 Smooth Wood Doors

		Crew	Daily Output	Labor-Hours	Unit	Material	2012 Bare Costs Labor	Equipment	Total	Total Incl O&P
0010	**SMOOTH WOOD DOORS**									
0015	Flush, interior, hollow core									
0025	Lauan face, 1-3/8", 3'-0" x 6'-8"	2 Carp	17	.941	Ea.	53.50	41.50		95	123
0030	4'-0" x 6'-8"		16	1		126	44		170	207
0080	1-3/4", 2'-0" x 6'-8"		17	.941		39	41.50		80.50	107
0108	3'-0" x 7'-0"		16	1		99.50	44		143.50	177
0112	Pair of 3'-0" x 7'-0"		9	1.778	Pr.	192	78.50		270.50	330
0140	Birch face, 1-3/8", 2'-6" x 6'-8"		17	.941	Ea.	84	41.50		125.50	157
0180	3'-0" x 6'-8"		17	.941		94.50	41.50		136	168
0200	4'-0" x 6'-8"		16	1		158	44		202	242
0202	1-3/4", 2'-0" x 6'-8"		17	.941		48	41.50		89.50	117
0210	3'-0" x 7'-0"		16	1		108	44		152	187
0214	Pair of 3'-0" x 7'-0"		9	1.778	Pr.	207	78.50		285.50	350
0220	Oak face, 1-3/8", 2'-0" x 6'-8"		17	.941	Ea.	96	41.50		137.50	170
0280	3'-0" x 6'-8"		17	.941		110	41.50		151.50	185
0300	4'-0" x 6'-8"		16	1		134	44		178	216
0305	1-3/4", 2'-6" x 6'-8"		17	.941		106	41.50		147.50	180
0310	3'-0" x 7'-0"		16	1		196	44		240	283
0320	Walnut face, 1-3/8", 2'-0" x 6'-8"		17	.941		178	41.50		219.50	259
0340	2'-6" x 6'-8"		17	.941		184	41.50		225.50	267
0380	3'-0" x 6'-8"		17	.941		192	41.50		233.50	275
0400	4'-0" x 6'-8"		16	1		213	44		257	300
0430	For 7'-0" high, add					25			25	27.50
0440	For 8'-0" high, add					35			35	38.50
0480	For prefinishing, clear, add					40			40	44
0500	For prefinishing, stain, add					50			50	55
1320	M.D. overlay on hardboard, 1-3/8", 2'-0" x 6'-8"	2 Carp	17	.941		110	41.50		151.50	185

08 14 Wood Doors

08 14 16 – Flush Wood Doors

08 14 16.09 Smooth Wood Doors		Crew	Daily Output	Labor-Hours	Unit	Material	2012 Bare Costs Labor	Equipment	Total	Total Incl O&P
1340	2'-6" x 6'-8"	2 Carp	17	.941	Ea.	110	41.50		151.50	185
1380	3'-0" x 6'-8"		17	.941		120	41.50		161.50	196
1400	4'-0" x 6'-8"	↓	16	1		170	44		214	255
1420	For 7'-0" high, add					15			15	16.50
1440	For 8'-0" high, add					30			30	33
1720	H.P. plastic laminate, 1-3/8", 2'-0" x 6'-8"	2 Carp	16	1		232	44		276	325
1740	2'-6" x 6'-8"		16	1		232	44		276	325
1780	3'-0" x 6'-8"		15	1.067		281	47		328	385
1800	4'-0" x 6'-8"	↓	14	1.143		365	50.50		415.50	485
1820	For 7'-0" high, add					15			15	16.50
1840	For 8'-0" high, add					30			30	33
2020	Particle core, lauan face, 1-3/8", 2'-6" x 6'-8"	2 Carp	15	1.067		90	47		137	172
2040	3'-0" x 6'-8"		14	1.143		92	50.50		142.50	179
2080	3'-0" x 7'-0"		13	1.231		96.50	54.50		151	190
2085	4'-0" x 7'-0"		12	1.333		120	59		179	223
2120	Birch face, 1-3/8", 2'-6" x 6'-8"		15	1.067		100	47		147	183
2140	3'-0" x 6'-8"		14	1.143		110	50.50		160.50	199
2180	3'-0" x 7'-0"		13	1.231		120	54.50		174.50	216
2200	4'-0" x 7'-0"		12	1.333		135	59		194	240
2205	1-3/4", 3'-0" x 7'-0"		13	1.231		120	54.50		174.50	216
2220	Oak face, 1-3/8", 2'-6" x 6'-8"		15	1.067		113	47		160	197
2240	3'-0" x 6'-8"		14	1.143		125	50.50		175.50	216
2280	3'-0" x 7'-0"		13	1.231		130	54.50		184.50	227
2300	4'-0" x 7'-0"		12	1.333		152	59		211	258
2305	1-3/4" 3'-0" x 7'-0"		13	1.231		159	54.50		213.50	259
2320	Walnut face, 1-3/8", 2'-0" x 6'-8"		15	1.067		120	47		167	205
2340	2'-6" x 6'-8"		14	1.143		135	50.50		185.50	227
2380	3'-0" x 6'-8"		13	1.231		152	54.50		206.50	251
2400	4'-0" x 6'-8"	↓	12	1.333		200	59		259	310
2440	For 8'-0" high, add					40			40	44
2460	For 8'-0" high walnut, add					35			35	38.50
2480	For solid wood core, add					39			39	43
2720	For prefinishing, clear, add					35			35	38.50
2740	For prefinishing, stain, add					52			52	57
3320	M.D. overlay on hardboard, 1-3/8", 2'-6" x 6'-8"	2 Carp	14	1.143		106	50.50		156.50	195
3340	3'-0" x 6'-8"		13	1.231		112	54.50		166.50	207
3380	3'-0" x 7'-0"		12	1.333		114	59		173	216
3400	4'-0" x 7'-0"	↓	10	1.600		150	70.50		220.50	274
3440	For 8'-0" height, add					35			35	38.50
3460	For solid wood core, add					40			40	44
3720	H.P. plastic laminate, 1-3/8", 2'-6" x 6'-8"	2 Carp	13	1.231		155	54.50		209.50	255
3740	3'-0" x 6'-8"		12	1.333		180	59		239	289
3780	3'-0" x 7'-0"		11	1.455		185	64		249	305
3800	4'-0" x 7'-0"	↓	8	2		222	88		310	380
3840	For 8'-0" height, add					35			35	38.50
3860	For solid wood core, add					40			40	44
4000	Exterior, flush, solid core, birch, 1-3/4" x 2'-6" x 7'-0"	2 Carp	15	1.067		165	47		212	255
4020	2'-8" wide		15	1.067		153	47		200	241
4040	3'-0" wide		14	1.143		197	50.50		247.50	294
4100	Oak faced 1-3/4" x 2'-6" x 7'-0"		15	1.067		185	47		232	277
4120	2'-8" wide		15	1.067		200	47		247	293
4140	3'-0" wide		14	1.143		210	50.50		260.50	310
4200	Walnut faced, 1-3/4" x 2'-6" x 7'-0"		15	1.067	↓	270	47		317	370

08 14 16 – Flush Wood Doors

08 14 16.09 Smooth Wood Doors		Crew	Daily Output	Labor-Hours	Unit	Material	2012 Bare Costs Labor	Equipment	Total	Total Incl O&P
4220	2'-8" wide	2 Carp	15	1.067	Ea.	285	47		332	390
4240	3'-0" wide	↓	14	1.143		295	50.50		345.50	405
4300	For 6'-8" high door, deduct from 7'-0" door					16			16	17.60
5000	Wood doors, for vision lite, add					93			93	102
5010	Wood doors, for narrow lite, add					101			101	111
5015	Wood doors, for bottom (or top) louver, add				↓	274			274	300

08 14 16.20 Wood Fire Doors										
0010	**WOOD FIRE DOORS**									
0020	Particle core, 7 face plys, "B" label,									
0040	1 hour, birch face, 1-3/4" x 2'-6" x 6'-8"	2 Carp	14	1.143	Ea.	405	50.50		455.50	525
0080	3'-0" x 6'-8"		13	1.231		400	54.50		454.50	525
0090	3'-0" x 7'-0"		12	1.333		435	59		494	570
0100	4'-0" x 7'-0"		12	1.333		540	59		599	685
0140	Oak face, 2'-6" x 6'-8"		14	1.143		340	50.50		390.50	455
0180	3'-0" x 6'-8"		13	1.231		400	54.50		454.50	525
0190	3'-0" x 7'-0"		12	1.333		450	59		509	585
0200	4'-0" x 7'-0"		12	1.333		510	59		569	650
0240	Walnut face, 2'-6" x 6'-8"		14	1.143		460	50.50		510.50	585
0280	3'-0" x 6'-8"		13	1.231		480	54.50		534.50	615
0290	3'-0" x 7'-0"		12	1.333		500	59		559	640
0300	4'-0" x 7'-0"		12	1.333		620	59		679	770
0440	M.D. overlay on hardboard, 2'-6" x 6'-8"		15	1.067		310	47		357	415
0480	3'-0" x 6'-8"		14	1.143		370	50.50		420.50	485
0490	3'-0" x 7'-0"		13	1.231		390	54.50		444.50	515
0500	4'-0" x 7'-0"		12	1.333		410	59		469	540
0740	90 minutes, birch face, 1-3/4" x 2'-6" x 6'-8"		14	1.143		315	50.50		365.50	425
0780	3'-0" x 6'-8"		13	1.231		310	54.50		364.50	425
0790	3'-0" x 7'-0"		12	1.333		380	59		439	510
0800	4'-0" x 7'-0"		12	1.333		490	59		549	630
0840	Oak face, 2'-6" x 6'-8"		14	1.143		430	50.50		480.50	555
0880	3'-0" x 6'-8"		13	1.231		440	54.50		494.50	570
0890	3'-0" x 7'-0"		12	1.333		455	59		514	590
0900	4'-0" x 7'-0"		12	1.333		590	59		649	740
0940	Walnut face, 2'-6" x 6'-8"		14	1.143		400	50.50		450.50	520
0980	3'-0" x 6'-8"		13	1.231		410	54.50		464.50	535
0990	3'-0" x 7'-0"		12	1.333		470	59		529	605
1000	4'-0" x 7'-0"		12	1.333		620	59		679	770
1140	M.D. overlay on hardboard, 2'-6" x 6'-8"		15	1.067		355	47		402	465
1180	3'-0" x 6'-8"		14	1.143		375	50.50		425.50	495
1190	3'-0" x 7'-0"	↓	13	1.231		405	54.50		459.50	530
1200	4'-0" x 7'-0"	↓	12	1.333		455	59		514	590
1240	For 8'-0" height, add					70			70	77
1260	For 8'-0" height walnut, add				↓	85			85	93.50
2200	Custom architectural "B" label, flush, 1-3/4" thick, birch,									
2210	Solid core									
2220	2'-6" x 7'-0"	2 Carp	15	1.067	Ea.	300	47		347	405
2260	3'-0" x 7'-0"		14	1.143		310	50.50		360.50	420
2300	4'-0" x 7'-0"		13	1.231		400	54.50		454.50	525
2420	4'-0" x 8'-0"	↓	11	1.455		430	64		494	575
2480	For oak veneer, add					50%				
2500	For walnut veneer, add				↓	75%				

08 14 Wood Doors

08 14 23 – Clad Wood Doors

08 14 23.13 Metal-Faced Wood Doors

		Crew	Daily Output	Labor-Hours	Unit	Material	2012 Bare Costs Labor	2012 Bare Costs Equipment	Total	Total Incl O&P
0010	**METAL-FACED WOOD DOORS**									
0020	Interior, flush type, 3' x 7'	2 Carp	4.30	3.721	Opng.	350	164		514	635

08 14 23.20 Tin Clad Wood Doors

		Crew	Daily Output	Labor-Hours	Unit	Material	2012 Bare Costs Labor	2012 Bare Costs Equipment	Total	Total Incl O&P
0010	**TIN CLAD WOOD DOORS**									
0020	3 ply, 6' x 7', double sliding, doors only	2 Carp	1	16	Opng.	2,000	705		2,705	3,275
1000	For electric operator, add	1 Elec	2	4	"	3,250	206		3,456	3,875

08 14 33 – Stile and Rail Wood Doors

08 14 33.10 Wood Doors Paneled

		Crew	Daily Output	Labor-Hours	Unit	Material	2012 Bare Costs Labor	2012 Bare Costs Equipment	Total	Total Incl O&P
0010	**WOOD DOORS PANELED**									
0020	Interior, six panel, hollow core, 1-3/8" thick									
0040	Molded hardboard, 2'-0" x 6'-8"	2 Carp	17	.941	Ea.	60	41.50		101.50	130
0060	2'-6" x 6'-8"		17	.941		62	41.50		103.50	132
0070	2'-8" x 6'-8"		17	.941		65	41.50		106.50	136
0080	3'-0" x 6'-8"		17	.941		68	41.50		109.50	139
0140	Embossed print, molded hardboard, 2'-0" x 6'-8"		17	.941		62	41.50		103.50	132
0160	2'-6" x 6'-8"		17	.941		62	41.50		103.50	132
0180	3'-0" x 6'-8"		17	.941		68	41.50		109.50	139
0540	Six panel, solid, 1-3/8" thick, pine, 2'-0" x 6'-8"		15	1.067		150	47		197	238
0560	2'-6" x 6'-8"		14	1.143		165	50.50		215.50	260
0580	3'-0" x 6'-8"		13	1.231		140	54.50		194.50	238
1020	Two panel, bored rail, solid, 1-3/8" thick, pine, 1'-6" x 6'-8"		16	1		265	44		309	360
1040	2'-0" x 6'-8"		15	1.067		350	47		397	460
1060	2'-6" x 6'-8"		14	1.143		395	50.50		445.50	515
1340	Two panel, solid, 1-3/8" thick, fir, 2'-0" x 6'-8"		15	1.067		155	47		202	244
1360	2'-6" x 6'-8"		14	1.143		205	50.50		255.50	305
1380	3'-0" x 6'-8"		13	1.231		410	54.50		464.50	535
1740	Five panel, solid, 1-3/8" thick, fir, 2'-0" x 6'-8"		15	1.067		275	47		322	380
1760	2'-6" x 6'-8"		14	1.143		415	50.50		465.50	535
1780	3'-0" x 6'-8"		13	1.231		415	54.50		469.50	540

08 14 33.20 Wood Doors Residential

		Crew	Daily Output	Labor-Hours	Unit	Material	2012 Bare Costs Labor	2012 Bare Costs Equipment	Total	Total Incl O&P
0010	**WOOD DOORS RESIDENTIAL**									
0200	Exterior, combination storm & screen, pine									
0260	2'-8" wide	2 Carp	10	1.600	Ea.	300	70.50		370.50	440
0280	3'-0" wide		9	1.778		310	78.50		388.50	460
0300	7'-1" x 3'-0" wide		9	1.778		330	78.50		408.50	485
0400	Full lite, 6'-9" x 2'-6" wide		11	1.455		320	64		384	450
0420	2'-8" wide		10	1.600		320	70.50		390.50	460
0440	3'-0" wide		9	1.778		325	78.50		403.50	475
0500	7'-1" x 3'-0" wide		9	1.778		355	78.50		433.50	510
0700	Dutch door, pine, 1-3/4" x 2'-8" x 6'-8", minimum		12	1.333		715	59		774	875
0720	Maximum		10	1.600		900	70.50		970.50	1,100
0800	3'-0" wide, minimum		12	1.333		715	59		774	875
0820	Maximum		10	1.600		900	70.50		970.50	1,100
1000	Entrance door, colonial, 1-3/4" x 6'-8" x 2'-8" wide		16	1		485	44		529	605
1020	6 panel pine, 3'-0" wide		15	1.067		445	47		492	565
1100	8 panel pine, 2'-8" wide		16	1		575	44		619	705
1120	3'-0" wide		15	1.067		560	47		607	690
1200	For tempered safety glass lites, (min of 2) add					76.50			76.50	84.50
1300	Flush, birch, solid core, 1-3/4" x 6'-8" x 2'-8" wide	2 Carp	16	1		110	44		154	189
1320	3'-0" wide		15	1.067		115	47		162	199
1350	7'-0" x 2'-8" wide		16	1		119	44		163	198

08 14 33.20 Wood Doors Residential		Crew	Daily Output	Labor-Hours	Unit	Material	2012 Bare Costs Labor	Equipment	Total	Total Incl O&P
1360	3'-0" wide	2 Carp	15	1.067	Ea.	130	47		177	216
1380	For tempered safety glass lites, add					103			103	113
2700	Interior, closet, bi-fold, w/hardware, no frame or trim incl.									
2720	Flush, birch, 6'-6" or 6'-8" x 2'-6" wide	2 Carp	13	1.231	Ea.	67.50	54.50		122	158
2740	3'-0" wide		13	1.231		71.50	54.50		126	162
2760	4'-0" wide		12	1.333		109	59		168	211
2780	5'-0" wide		11	1.455		105	64		169	215
2800	6'-0" wide		10	1.600		128	70.50		198.50	250
3000	Raised panel pine, 6'-6" or 6'-8" x 2'-6" wide		13	1.231		190	54.50		244.50	293
3020	3'-0" wide		13	1.231		278	54.50		332.50	390
3040	4'-0" wide		12	1.333		305	59		364	425
3060	5'-0" wide		11	1.455		365	64		429	500
3080	6'-0" wide		10	1.600		400	70.50		470.50	550
3200	Louvered, pine 6'-6" or 6'-8" x 2'-6" wide		13	1.231		142	54.50		196.50	240
3220	3'-0" wide		13	1.231		206	54.50		260.50	310
3240	4'-0" wide		12	1.333		233	59		292	345
3260	5'-0" wide		11	1.455		260	64		324	385
3280	6'-0" wide		10	1.600		287	70.50		357.50	425
4400	Bi-passing closet, incl. hardware and frame, no trim incl.									
4420	Flush, lauan, 6'-8" x 4'-0" wide	2 Carp	12	1.333	Opng.	176	59		235	285
4440	5'-0" wide		11	1.455		187	64		251	305
4460	6'-0" wide		10	1.600		209	70.50		279.50	340
4600	Flush, birch, 6'-8" x 4'-0" wide		12	1.333		220	59		279	335
4620	5'-0" wide		11	1.455		212	64		276	330
4640	6'-0" wide		10	1.600		256	70.50		326.50	390
4800	Louvered, pine, 6'-8" x 4'-0" wide		12	1.333		405	59		464	535
4820	5'-0" wide		11	1.455		395	64		459	535
4840	6'-0" wide		10	1.600		525	70.50		595.50	685
5000	Paneled, pine, 6'-8" x 4'-0" wide		12	1.333		505	59		564	650
5020	5'-0" wide		11	1.455		410	64		474	550
5040	6'-0" wide		10	1.600		600	70.50		670.50	770
5042	8'-0" wide		12	1.333		1,025	59		1,084	1,225
6100	Folding accordion, closet, including track and frame									
6120	Vinyl, 2 layer, stock	2 Carp	10	1.600	Ea.	65.50	70.50		136	181
6140	Woven mahogany and vinyl, stock		10	1.600		56.50	70.50		127	172
6160	Wood slats with vinyl overlay, stock		10	1.600		150	70.50		220.50	274
6180	Economy vinyl, stock		10	1.600		39	70.50		109.50	152
6200	Rigid PVC		10	1.600		52	70.50		122.50	166
7310	Passage doors, flush, no frame included									
7320	Hardboard, hollow core, 1-3/8" x 6'-8" x 1'-6" wide	2 Carp	18	.889	Ea.	43	39		82	108
7330	2'-0" wide		18	.889		43	39		82	108
7340	2'-6" wide		18	.889		47.50	39		86.50	113
7350	2'-8" wide		18	.889		49	39		88	114
7360	3'-0" wide		17	.941		52	41.50		93.50	121
7420	Lauan, hollow core, 1-3/8" x 6'-8" x 1'-6" wide		18	.889		32	39		71	95.50
7440	2'-0" wide		18	.889		34.50	39		73.50	98.50
7450	2'-4" wide		18	.889		38.50	39		77.50	103
7460	2'-6" wide		18	.889		38.50	39		77.50	103
7480	2'-8" wide		18	.889		40	39		79	104
7500	3'-0" wide		17	.941		42	41.50		83.50	110
7700	Birch, hollow core, 1-3/8" x 6'-8" x 1'-6" wide		18	.889		39.50	39		78.50	104
7720	2'-0" wide		18	.889		42	39		81	107
7740	2'-6" wide		18	.889		49.50	39		88.50	115

08 14 Wood Doors

08 14 33 – Stile and Rail Wood Doors

08 14 33.20 Wood Doors Residential		Crew	Daily Output	Labor-Hours	Unit	Material	2012 Bare Costs Labor	Equipment	Total	Total Incl O&P
7760	2'-8" wide	2 Carp	18	.889	Ea.	50.50	39		89.50	116
7780	3'-0" wide		17	.941		52	41.50		93.50	121
8000	Pine louvered, 1-3/8" x 6'-8" x 1'-6" wide		19	.842		102	37		139	169
8020	2'-0" wide		18	.889		128	39		167	202
8040	2'-6" wide		18	.889		140	39		179	214
8060	2'-8" wide		18	.889		147	39		186	223
8080	3'-0" wide		17	.941		173	41.50		214.50	255
8300	Pine paneled, 1-3/8" x 6'-8" x 1'-6" wide		19	.842		128	37		165	197
8320	2'-0" wide		18	.889		152	39		191	228
8330	2'-4" wide		18	.889		180	39		219	259
8340	2'-6" wide		18	.889		180	39		219	259
8360	2'-8" wide		18	.889		184	39		223	263
8380	3'-0" wide		17	.941		203	41.50		244.50	287

08 14 40 – Interior Cafe Doors

08 14 40.10 Cafe Style Doors

		Crew	Daily Output	Labor-Hours	Unit	Material	Labor	Equipment	Total	Total Incl O&P
0010	**CAFE STYLE DOORS**									
6520	Interior cafe doors, 2'-6" opening, stock, panel pine	2 Carp	16	1	Ea.	192	44		236	279
6540	3'-0" opening	"	16	1	"	199	44		243	287
6550	Louvered pine									
6560	2'-6" opening	2 Carp	16	1	Ea.	166	44		210	250
8000	3'-0" opening		16	1		180	44		224	266
8010	2'-6" opening, hardwood		16	1		247	44		291	340
8020	3'-0" opening		16	1		272	44		316	365

08 16 Composite Doors

08 16 13 – Fiberglass Doors

08 16 13.10 Entrance Doors, Fiberous Glass

			Crew	Daily Output	Labor-Hours	Unit	Material	Labor	Equipment	Total	Total Incl O&P
0010	**ENTRANCE DOORS, FIBEROUS GLASS**										
0020	Exterior, fiberglass, door, 2'-8" wide x 6'-8" high	G	2 Carp	15	1.067	Ea.	259	47		306	360
0040	3'-0" wide x 6'-8" high	G		15	1.067		259	47		306	360
0060	3'-0" wide x 7'-0" high	G		15	1.067		460	47		507	580
0080	3'-0" wide x 6'-8" high, with two lites	G		15	1.067		294	47		341	400
0100	3'-0" wide x 8'-0" high, with two lites	G		15	1.067		495	47		542	620
0110	Half glass, 3'-0" wide x 6'-8" high	G		15	1.067		305	47		352	410
0120	3'-0" wide x 6'-8" high, low e	G		15	1.067		330	47		377	440
0130	3'-0" wide x 8'-0" high	G		15	1.067		580	47		627	715
0140	3'-0" wide x 8'-0" high, low e	G		15	1.067		640	47		687	775
0150	Side lights, 1'-0" wide x 6'-8" high,	G					251			251	276
0160	1'-0" wide x 6'-8" high, low e	G					256			256	281
0180	1'-0" wide x 6'-8" high, full glass	G					275			275	305
0190	1'-0" wide x 6'-8" high, low e	G					284			284	315

08 16 14 – French Doors

08 16 14.10 Exterior Doors With Glass Lites

		Crew	Daily Output	Labor-Hours	Unit	Material	Labor	Equipment	Total	Total Incl O&P
0010	**EXTERIOR DOORS WITH GLASS LITES**									
0020	French, Fir, 1-3/4", 3'-0"wide x 6'-8" high	2 Carp	12	1.333	Ea.	600	59		659	750
0025	Double		12	1.333		1,200	59		1,259	1,425
0030	Maple, 1-3/4", 3'-0"wide x 6'-8" high		12	1.333		675	59		734	835
0035	Double		12	1.333		1,350	59		1,409	1,575
0040	Cherry, 1-3/4", 3'-0"wide x 6'-8" high		12	1.333		790	59		849	955
0045	Double		12	1.333		1,575	59		1,634	1,825

08 16 Composite Doors

08 16 14 – French Doors

08 16 14.10 Exterior Doors With Glass Lites	Crew	Daily Output	Labor-Hours	Unit	Material	2012 Bare Costs Labor	Equipment	Total	Total Incl O&P
0100	Mahogany, 1-3/4", 3'-0"wide x 8'-0" high	2 Carp	10	1.600	Ea.	700	70.50		770.50	880
0105	Double		10	1.600		1,400	70.50		1,470.50	1,650
0110	Fir, 1-3/4", 3'-0"wide x 8'-0" high		10	1.600		1,100	70.50		1,170.50	1,300
0115	Double		10	1.600		2,200	70.50		2,270.50	2,525
0120	Oak, 1-3/4", 3'-0"wide x 8'-0" high		10	1.600		1,825	70.50		1,895.50	2,100
0125	Double	↓	10	1.600	↓	3,650	70.50		3,720.50	4,100

08 17 Integrated Door Opening Assemblies

08 17 23 – Integrated Wood Door Opening Assemblies

08 17 23.10 Pre-Hung Doors

08 17 23.10 Pre-Hung Doors	Crew	Daily Output	Labor-Hours	Unit	Material	2012 Bare Costs Labor	Equipment	Total	Total Incl O&P
0010	**PRE-HUNG DOORS**									
0300	Exterior, wood, comb. storm & screen, 6'-9" x 2'-6" wide	2 Carp	15	1.067	Ea.	296	47		343	400
0320	2'-8" wide		15	1.067		296	47		343	400
0340	3'-0" wide	↓	15	1.067	↓	305	47		352	410
0360	For 7'-0" high door, add				↓	30			30	33
1600	Entrance door, flush, birch, solid core									
1620	4-5/8" solid jamb, 1-3/4" x 6'-8" x 2'-8" wide	2 Carp	16	1	Ea.	289	44		333	390
1640	3'-0" wide		16	1		330	44		374	435
1642	5-5/8" jamb	↓	16	1	↓	325	44		369	430
1680	For 7'-0" high door, add				↓	25			25	27.50
2000	Entrance door, colonial, 6 panel pine									
2020	4-5/8" solid jamb, 1-3/4" x 6'-8" x 2'-8" wide	2 Carp	16	1	Ea.	635	44		679	770
2040	3'-0" wide	"	16	1		665	44		709	805
2060	For 7'-0" high door, add					53.50			53.50	58.50
2200	For 5-5/8" solid jamb, add				↓	41.50			41.50	45.50
4000	Interior, passage door, 4-5/8" solid jamb									
4400	Lauan, flush, solid core, 1-3/8" x 6'-8" x 2'-6" wide	2 Carp	17	.941	Ea.	185	41.50		226.50	267
4420	2'-8" wide		17	.941		185	41.50		226.50	267
4440	3'-0" wide		16	1		200	44		244	287
4600	Hollow core, 1-3/8" x 6'-8" x 2'-6" wide		17	.941		125	41.50		166.50	201
4620	2'-8" wide		17	.941		124	41.50		165.50	200
4640	3'-0" wide	↓	16	1		140	44		184	221
4700	For 7'-0" high door, add					35			35	38.50
5000	Birch, flush, solid core, 1-3/8" x 6'-8" x 2'-6" wide	2 Carp	17	.941		247	41.50		288.50	335
5020	2'-8" wide		17	.941		200	41.50		241.50	283
5040	3'-0" wide		16	1		277	44		321	375
5200	Hollow core, 1-3/8" x 6'-8" x 2'-6" wide		17	.941		187	41.50		228.50	269
5220	2'-8" wide		17	.941		161	41.50		202.50	241
5240	3'-0" wide	↓	16	1		204	44		248	292
5280	For 7'-0" high door, add					30			30	33
5500	Hardboard paneled, 1-3/8" x 6'-8" x 2'-6" wide	2 Carp	17	.941		144	41.50		185.50	222
5520	2'-8" wide		17	.941		152	41.50		193.50	231
5540	3'-0" wide		16	1		149	44		193	232
6000	Pine paneled, 1-3/8" x 6'-8" x 2'-6" wide		17	.941		254	41.50		295.50	345
6020	2'-8" wide		17	.941		273	41.50		314.50	365
6040	3'-0" wide	↓	16	1		281	44		325	380
6500	For 5-5/8" solid jamb, add					22.50			22.50	25
6520	For split jamb, deduct				↓	24			24	26.50

08 31 Access Doors and Panels

08 31 13 – Access Doors and Frames

08 31 13.10 Types of Framed Access Doors	Crew	Daily Output	Labor-Hours	Unit	Material	2012 Bare Costs Labor	Equipment	Total	Total Incl O&P
0010 **TYPES OF FRAMED ACCESS DOORS**									
1000 Fire rated door with lock									
1100 Metal, 12" x 12"	1 Carp	10	.800	Ea.	165	35.50		200.50	237
1150 18" x 18"		9	.889		220	39		259	305
1200 24" x 24"		9	.889		325	39		364	420
1250 24" x 36"		8	1		335	44		379	435
1300 24" x 48"		8	1		430	44		474	545
1350 36" x 36"		7.50	1.067		495	47		542	620
1400 48" x 48"		7.50	1.067		635	47		682	770
1600 Stainless steel, 12" x 12"		10	.800		282	35.50		317.50	365
1650 18" x 18"		9	.889		410	39		449	510
1700 24" x 24"		9	.889		500	39		539	610
1750 24" x 36"	▼	8	1	▼	640	44		684	775
2000 Flush door for finishing									
2100 Metal 8" x 8"	1 Carp	10	.800	Ea.	35	35.50		70.50	93
2150 12" x 12"	"	10	.800	"	40	35.50		75.50	98.50
3000 Recessed door for acoustic tile									
3100 Metal, 12" x 12"	1 Carp	4.50	1.778	Ea.	80	78.50		158.50	209
3150 12" x 24"		4.50	1.778		90	78.50		168.50	220
3200 24" x 24"		4	2		121	88		209	269
3250 24" x 36"	▼	4	2	▼	170	88		258	325
4000 Recessed door for drywall									
4100 Metal 12" x 12"	1 Carp	6	1.333	Ea.	80	59		139	179
4150 12" x 24"		5.50	1.455		114	64		178	224
4200 24" x 36"	▼	5	1.600	▼	182	70.50		252.50	310
6000 Standard door									
6100 Metal, 8" x 8"	1 Carp	10	.800	Ea.	35	35.50		70.50	93
6150 12" x 12"		10	.800		40	35.50		75.50	98.50
6200 18" x 18"		9	.889		62	39		101	129
6250 24" x 24"		9	.889		75	39		114	143
6300 24" x 36"		8	1		115	44		159	195
6350 36" x 36"		8	1		140	44		184	222
6500 Stainless steel, 8" x 8"		10	.800		80	35.50		115.50	143
6550 12" x 12"		10	.800		107	35.50		142.50	173
6600 18" x 18"		9	.889		198	39		237	279
6650 24" x 24"	▼	9	.889		260	39		299	345
7000 Ceiling hatches, 2'-6" x 2'-6", swing up, single leaf, st fr & cover	G-3	11	2.909		580	127		707	835
7010 Aluminum cover	"	11	2.909	▼	560	127		687	810

08 31 13.20 Bulkhead/Cellar Doors

	Crew	Daily Output	Labor-Hours	Unit	Material	Labor	Equipment	Total	Total Incl O&P
0010 **BULKHEAD/CELLAR DOORS**									
0020 Steel, not incl. sides, 44" x 62"	1 Carp	5.50	1.455	Ea.	560	64		624	715
0100 52" x 73"		5.10	1.569		770	69		839	950
0500 With sides and foundation plates, 57" x 45" x 24"		4.70	1.702		810	75		885	1,000
0600 42" x 49" x 51"	▼	4.30	1.860	▼	915	82		997	1,125

08 31 13.30 Commercial Floor Doors

	Crew	Daily Output	Labor-Hours	Unit	Material	Labor	Equipment	Total	Total Incl O&P
0010 **COMMERCIAL FLOOR DOORS**									
0020 Aluminum tile, steel frame, one leaf, 2' x 2' opng.	2 Sswk	3.50	4.571	Opng.	800	226		1,026	1,275
0050 3'-6" x 3'-6" opening		3.50	4.571		1,600	226		1,826	2,150
0500 Double leaf, 4' x 4' opening		3	5.333		1,675	263		1,938	2,325
0550 5' x 5' opening	▼	3	5.333	▼	3,000	263		3,263	3,775

08 31 Access Doors and Panels

08 31 13 – Access Doors and Frames

08 31 13.35 Industrial Floor Doors

		Crew	Daily Output	Labor-Hours	Unit	Material	2012 Bare Costs Labor	Equipment	Total	Total Incl O&P
0010	**INDUSTRIAL FLOOR DOORS**									
0020	Steel 300 psf L.L., single leaf, 2' x 2', 175#	2 Sswk	6	2.667	Opng.	760	132		892	1,075
0050	3' x 3' opening, 300#		5.50	2.909		1,075	144		1,219	1,425
0300	Double leaf, 4' x 4' opening, 455#		5	3.200		1,900	158		2,058	2,375
0350	5' x 5' opening, 645#		4.50	3.556		2,750	175		2,925	3,350
1000	Aluminum, 300 psf L.L., single leaf, 2' x 2', 60#		6	2.667		750	132		882	1,050
1050	3' x 3' opening, 100#		5.50	2.909		1,300	144		1,444	1,675
1500	Double leaf, 4' x 4' opening, 160#		5	3.200		2,100	158		2,258	2,575
1550	5' x 5' opening, 235#		4.50	3.556		2,800	175		2,975	3,400
2000	Aluminum, 150 psf L.L., single leaf, 2' x 2', 60#		6	2.667		700	132		832	1,000
2050	3' x 3' opening, 95#		5.50	2.909		1,000	144		1,144	1,350
2500	Double leaf, 4' x 4' opening, 150#		5	3.200		1,450	158		1,608	1,875
2550	5' x 5' opening, 230#		4.50	3.556		1,950	175		2,125	2,450

08 31 13.40 Kennel Doors

		Crew	Daily Output	Labor-Hours	Unit	Material	2012 Bare Costs Labor	Equipment	Total	Total Incl O&P
0010	**KENNEL DOORS**									
0020	2 way, swinging type, 13" x 19" opening	2 Carp	11	1.455	Opng.	90	64		154	198
0100	17" x 29" opening		11	1.455		110	64		174	220
0200	9" x 9" opening, electronic with accessories		11	1.455		144	64		208	257

08 32 Sliding Glass Doors

08 32 13 – Sliding Aluminum-Framed Glass Doors

08 32 13.10 Sliding Aluminum Doors

		Crew	Daily Output	Labor-Hours	Unit	Material	2012 Bare Costs Labor	Equipment	Total	Total Incl O&P
0010	**SLIDING ALUMINUM DOORS**									
0350	Aluminum, 5/8" tempered insulated glass, 6' wide									
0400	Premium	2 Carp	4	4	Ea.	1,575	176		1,751	2,000
0450	Economy		4	4		885	176		1,061	1,250
0500	8' wide, premium		3	5.333		1,650	235		1,885	2,175
0550	Economy		3	5.333		1,425	235		1,660	1,900
0600	12' wide, premium		2.50	6.400		2,975	282		3,257	3,700
0650	Economy		2.50	6.400		1,375	282		1,657	1,925
4000	Aluminum, baked on enamel, temp glass, 6'-8" x 10'-0" wide		4	4		1,050	176		1,226	1,425
4020	Insulating glass, 6'-8" x 6'-0" wide		4	4		915	176		1,091	1,275
4040	8'-0" wide		3	5.333		1,075	235		1,310	1,525
4060	10'-0" wide		2	8		1,325	355		1,680	2,000
4080	Anodized, temp glass, 6'-8" x 6'-0" wide		4	4		455	176		631	770
4100	8'-0" wide		3	5.333		575	235		810	995
4120	10'-0" wide		2	8		645	355		1,000	1,250
4200	Vinyl clad, anodized, temp glass, 6'-8" x 6'-0" wide		4	4		420	176		596	730
4240	8'-0" wide		3	5.333		585	235		820	1,000
4260	10'-0" wide		2	8		635	355		990	1,250
4280	Insulating glass, 6'-8" x 6'-0" wide		4	4		1,125	176		1,301	1,500
4300	8'-0" wide		3	5.333		935	235		1,170	1,375
4320	10'-0" wide		4	4		1,675	176		1,851	2,100

08 32 19 – Sliding Wood-Framed Glass Doors

08 32 19.15 Sliding Glass Vinyl-Clad Wood Doors

			Crew	Daily Output	Labor-Hours	Unit	Material	2012 Bare Costs Labor	Equipment	Total	Total Incl O&P
0010	**SLIDING GLASS VINYL-CLAD WOOD DOORS**										
0020	Glass, sliding vinyl clad, insul. glass, 6'-0" x 6'-8"	G	2 Carp	4	4	Ea.	1,500	176		1,676	1,950
0025	6'-0" x 6'-10" high	G		4	4	Opng.	1,650	176		1,826	2,100
0030	6'-0" x 8'-0" high	G		4	4	Ea.	1,975	176		2,151	2,450
0050	5'-0" x 6'-8" high	G		4	4	"	1,500	176		1,676	1,925

08 32 Sliding Glass Doors

08 32 19 – Sliding Wood-Framed Glass Doors

08 32 19.15 Sliding Glass Vinyl-Clad Wood Doors		Crew	Daily Output	Labor-Hours	Unit	Material	2012 Bare Costs Labor	Equipment	Total	Total Incl O&P
0100	8'-0" x 6'-10" high	[G] 2 Carp	4	4	Opng.	2,025	176		2,201	2,500
0500	4 leaf, 9'-0" x 6'-10" high	[G]	3	5.333		3,250	235		3,485	3,925
0600	12'-0" x 6'-10" high	[G]	3	5.333		3,875	235		4,110	4,625

08 33 Coiling Doors and Grilles

08 33 13 – Coiling Counter Doors

08 33 13.10 Counter Doors, Coiling Type

		Crew	Daily Output	Labor-Hours	Unit	Material	Labor	Equipment	Total	Total Incl O&P
0010	**COUNTER DOORS, COILING TYPE**									
0020	Manual, incl. frm and hdwe, galv. stl., 4' roll-up, 6' long	2 Carp	2	8	Opng.	1,175	355		1,530	1,850
0300	Galvanized steel, UL label		1.80	8.889		1,225	390		1,615	1,950
0600	Stainless steel, 4' high roll-up, 6' long		2	8		2,100	355		2,455	2,850
0700	10' long		1.80	8.889		2,500	390		2,890	3,350
2000	Aluminum, 4' high, 4' long		2.20	7.273		1,375	320		1,695	2,000
2020	6' long		2	8		1,500	355		1,855	2,200
2040	8' long		1.90	8.421		1,725	370		2,095	2,475
2060	10' long		1.80	8.889		1,825	390		2,215	2,600
2080	14' long		1.40	11.429		2,575	505		3,080	3,600
2100	6' high, 4' long		2	8		1,550	355		1,905	2,250
2120	6' long		1.60	10		1,600	440		2,040	2,425
2140	10' long		1.40	11.429		2,100	505		2,605	3,100

08 33 16 – Coiling Counter Grilles

08 33 16.10 Coiling Grilles

		Crew	Daily Output	Labor-Hours	Unit	Material	Labor	Equipment	Total	Total Incl O&P
0010	**COILING GRILLES**									
0015	Aluminum, manual, incl. frame, mill finish									
0020	Top coiling, 4' high, 4' long	2 Sswk	3.20	5	Opng.	1,500	247		1,747	2,100
0030	6' long		3.20	5		1,750	247		1,997	2,375
0040	8' long		2.40	6.667		1,950	330		2,280	2,750
0050	12' long		2.40	6.667		2,375	330		2,705	3,200
0060	16' long		1.60	10		2,525	495		3,020	3,650
0070	6' high, 4' long		3.20	5		1,750	247		1,997	2,375
0080	6' long		3.20	5		1,750	247		1,997	2,375
0090	8' long		2.40	6.667		1,750	330		2,080	2,525
0100	12' long		1.60	10		2,300	495		2,795	3,425
0110	16' long		1.20	13.333		2,750	660		3,410	4,200
0200	Side coiling, 8' high, 12' long		.60	26.667		2,300	1,325		3,625	4,875
0220	18' long		.50	32		4,025	1,575		5,600	7,250
0240	24' long		.40	40		3,675	1,975		5,650	7,575
0260	12' high, 12' long		.50	32		3,350	1,575		4,925	6,500
0280	18' long		.40	40		4,950	1,975		6,925	8,950
0300	24' long		.28	57.143		6,625	2,825		9,450	12,300

08 33 23 – Overhead Coiling Doors

08 33 23.10 Rolling Service Doors

		Crew	Daily Output	Labor-Hours	Unit	Material	Labor	Equipment	Total	Total Incl O&P
0010	**ROLLING SERVICE DOORS** Steel, manual, 20 ga., incl. hardware									
0050	8' x 8' high	2 Sswk	1.60	10	Ea.	1,100	495		1,595	2,075
0100	10' x 10' high		1.40	11.429		1,825	565		2,390	3,000
0200	20' x 10' high		1	16		3,050	790		3,840	4,750
0300	12' x 12' high		1.20	13.333		1,850	660		2,510	3,200
0400	20' x 12' high		.90	17.778		2,100	875		2,975	3,875
0500	14' x 14' high		.80	20		3,000	985		3,985	5,075
0600	20' x 16' high		.60	26.667		3,450	1,325		4,775	6,150

08 33 Coiling Doors and Grilles

08 33 23 – Overhead Coiling Doors

08 33 23.10 Rolling Service Doors		Crew	Daily Output	Labor-Hours	Unit	Material	2012 Bare Costs Labor	Equipment	Total	Total Incl O&P
0700	10' x 20' high	2 Sswk	.50	32	Ea.	2,450	1,575		4,025	5,500
1000	12' x 12', crank operated, crank on door side		.80	20		1,650	985		2,635	3,600
1100	Crank thru wall		.70	22.857		1,900	1,125		3,025	4,125
1300	For vision panel, add					330			330	365
1400	For 22 ga., deduct				S.F.	1.30			1.30	1.43
1600	3' x 7' pass door within rolling steel door, new construction				Ea.	1,775			1,775	1,975
1700	Existing construction	2 Sswk	2	8		1,900	395		2,295	2,775
2000	Class A fire doors, manual, 20 ga., 8' x 8' high		1.40	11.429		1,475	565		2,040	2,625
2100	10' x 10' high		1.10	14.545		2,025	720		2,745	3,500
2200	20' x 10' high		.80	20		4,125	985		5,110	6,325
2300	12' x 12' high		1	16		3,175	790		3,965	4,900
2400	20' x 12' high		.80	20		4,475	985		5,460	6,700
2500	14' x 14' high		.60	26.667		3,475	1,325		4,800	6,175
2600	20' x 16' high		.50	32		5,625	1,575		7,200	9,000
2700	10' x 20' high		.40	40		4,350	1,975		6,325	8,300
3000	For 18 ga. doors, add				S.F.	1.45			1.45	1.60
3300	For enamel finish, add				"	1.70			1.70	1.87
3600	For safety edge bottom bar, pneumatic, add				L.F.	21			21	23
3700	Electric, add					39.50			39.50	43.50
4000	For weatherstripping, extruded rubber, jambs, add					13.40			13.40	14.75
4100	Hood, add					8.25			8.25	9.05
4200	Sill, add					4.73			4.73	5.20
4500	Motor operators, to 14' x 14' opening	2 Sswk	5	3.200	Ea.	1,050	158		1,208	1,450
4600	Over 14' x 14', jack shaft type	"	5	3.200		1,100	158		1,258	1,500
4700	For fire door, additional fusible link, add					26			26	28.50

08 34 Special Function Doors

08 34 13 – Cold Storage Doors

08 34 13.10 Doors for Cold Area Storage		Crew	Daily Output	Labor-Hours	Unit	Material	2012 Bare Costs Labor	Equipment	Total	Total Incl O&P
0010	**DOORS FOR COLD AREA STORAGE**									
0020	Single, 20 ga. galvanized steel									
0300	Horizontal sliding, 5' x 7', manual operation, 3.5" thick	2 Carp	2	8	Ea.	3,200	355		3,555	4,075
0400	4" thick		2	8		3,225	355		3,580	4,100
0500	6" thick		2	8		3,400	355		3,755	4,300
0800	5' x 7', power operation, 2" thick		1.90	8.421		5,525	370		5,895	6,650
0900	4" thick		1.90	8.421		5,650	370		6,020	6,800
1000	6" thick		1.90	8.421		6,450	370		6,820	7,675
1300	9' x 10', manual operation, 2" insulation		1.70	9.412		4,500	415		4,915	5,600
1400	4" insulation		1.70	9.412		4,600	415		5,015	5,725
1500	6" insulation		1.70	9.412		5,550	415		5,965	6,750
1800	Power operation, 2" insulation		1.60	10		7,625	440		8,065	9,050
1900	4" insulation		1.60	10		7,850	440		8,290	9,325
2000	6" insulation		1.70	9.412		8,850	415		9,265	10,400
2300	For stainless steel face, add					25%				
3000	Hinged, lightweight, 3' x 7'-0", 2" thick	2 Carp	2	8	Ea.	1,400	355		1,755	2,100
3050	4" thick		1.90	8.421		1,750	370		2,120	2,500
3300	Aluminum doors, 3' x 7'-0", 4" thick		1.90	8.421		1,325	370		1,695	2,050
3350	6" thick		1.40	11.429		2,325	505		2,830	3,350
3600	Stainless steel, 3' x 7'-0", 4" thick		1.90	8.421		1,725	370		2,095	2,475
3650	6" thick		1.40	11.429		2,925	505		3,430	4,000
3900	Painted, 3' x 7'-0", 4" thick		1.90	8.421		1,250	370		1,620	1,950

08 34 Special Function Doors

08 34 13 – Cold Storage Doors

08 34 13.10 Doors for Cold Area Storage

08 34 13.10 Doors for Cold Area Storage	Crew	Daily Output	Labor-Hours	Unit	Material	2012 Bare Costs Labor	Equipment	Total	Total Incl O&P
3950	6" thick	2 Carp	1.40	11.429	Ea.	2,325	505		2,830	3,350
5000	Bi-parting, electric operated									
5010	6' x 8' opening, galv. faces, 4" thick for cooler	2 Carp	.80	20	Opng.	7,200	880		8,080	9,275
5050	For freezer, 4" thick		.80	20		7,950	880		8,830	10,100
5300	For door buck framing and door protection, add		2.50	6.400		635	282		917	1,125
6000	Galvanized batten door, galvanized hinges, 4' x 7'		2	8		1,800	355		2,155	2,525
6050	6' x 8'		1.80	8.889		2,450	390		2,840	3,300
6500	Fire door, 3 hr., 6' x 8', single slide		.80	20		8,275	880		9,155	10,500
6550	Double, bi-parting		.70	22.857		12,200	1,000		13,200	15,000

08 34 16 – Hangar Doors

08 34 16.10 Aircraft Hangar Doors

		Crew	Daily Output	Labor-Hours	Unit	Material	Labor	Equipment	Total	Total Incl O&P
0010	**AIRCRAFT HANGAR DOORS**									
0020	Bi-fold, ovhd., 20 PSF wind load, incl. elec. oper.									
0100	12' high x 40'	2 Sswk	240	.067	S.F.	16	3.29		19.29	23.50
0200	16' high x 60'		230	.070		19.20	3.43		22.63	27
0300	20' high x 80'		220	.073		21	3.59		24.59	29.50

08 34 36 – Darkroom Doors

08 34 36.10 Various Types of Darkroom Doors

		Crew	Daily Output	Labor-Hours	Unit	Material	Labor	Equipment	Total	Total Incl O&P
0010	**VARIOUS TYPES OF DARKROOM DOORS**									
0015	Revolving, standard, 2 way, 36" diameter	2 Carp	3.10	5.161	Opng.	2,250	228		2,478	2,825
0020	41" diameter		3.10	5.161		2,650	228		2,878	3,275
0050	3 way, 51" diameter		1.40	11.429		3,075	505		3,580	4,150
1000	4 way, 49" diameter		1.40	11.429		3,625	505		4,130	4,775
2000	Hinged safety, 2 way, 41" diameter		2.30	6.957		2,825	305		3,130	3,600
2500	3 way, 51" diameter		1.40	11.429		3,625	505		4,130	4,775
3000	Pop out safety, 2 way, 41" diameter		3.10	5.161		3,600	228		3,828	4,300
4000	3 way, 51" diameter		1.40	11.429		3,550	505		4,055	4,675
5000	Wheelchair-type, pop out, 51" diameter		1.40	11.429		3,525	505		4,030	4,650
5020	72" diameter		.90	17.778		7,225	785		8,010	9,150
9300	For complete darkrooms, see Section 13 21 53.50									

08 34 53 – Security Doors and Frames

08 34 53.20 Steel Door

		Crew	Daily Output	Labor-Hours	Unit	Material	Labor	Equipment	Total	Total Incl O&P
0010	**STEEL DOOR** with ballistic core and welded frame both 14 gauge									
0050	Flush, UL 752 Level 3, 1-3/4", 3'-0" x 6'-8"	2 Carp	1.50	10.667	Opng.	2,200	470		2,670	3,150
0055	1-3/4", 3'-6" x 6'-8"		1.50	10.667		2,300	470		2,770	3,250
0060	1-3/4", 4'-0" x 6'-8"		1.20	13.333		2,500	590		3,090	3,650
0100	UL 752 Level 8, 1-3/4", 3'-0" x 6'-8"		1.50	10.667		9,600	470		10,070	11,300
0105	1-3/4", 3'-6" x 6'-8"		1.50	10.667		9,800	470		10,270	11,500
0110	1-3/4", 4'-0" x 6'-8"		1.20	13.333		11,100	590		11,690	13,100
0120	UL 752 Level 3, 1-3/4", 3'-0" x 7'-0"		1.50	10.667		2,225	470		2,695	3,175
0125	1-3/4", 3'-6" x 7'-0"		1.50	10.667		2,325	470		2,795	3,300
0130	1-3/4", 4'-0" x 7'-0"		1.20	13.333		2,525	590		3,115	3,675
0150	UL 752 Level 8, 1-3/4", 3'-0" x 7'-0"		1.50	10.667		9,700	470		10,170	11,400
0155	1-3/4", 3'-6" x 7'-0"		1.50	10.667		9,800	470		10,270	11,500
0160	1-3/4", 4'-0" x 7'-0"		1.20	13.333		11,200	590		11,790	13,300
1000	Safe Room sliding door and hardware, 1-3/4", 3'-0" x 7'-0" UL 752 Level 3		.50	32		22,900	1,400		24,300	27,400
1050	Safe Room swinging door and hardware, 1-3/4", 3'-0" x 7'-0" UL 752 Level 3		.50	32		27,500	1,400		28,900	32,500

08 34 53.30 Wood Ballistic Doors

		Crew	Daily Output	Labor-Hours	Unit	Material	Labor	Equipment	Total	Total Incl O&P
0010	**WOOD BALLISTIC DOORS** with frames and hardware									
0050	Wood, 1-3/4", 3'-0" x 7'-0" UL 752 Level 3	2 Carp	1.50	10.667	Opng.	2,125	470		2,595	3,050

08 34 56 – Security Gates

08 34 56.10 Gates	Crew	Daily Output	Labor-Hours	Unit	Material	2012 Bare Costs Labor	Equipment	Total	Total Incl O&P
0010 **GATES**									
0015 Gates include mounting hardware									
0500 Wood, security gate, driveway, dual, 10' wide	H-4	.80	25	Opng.	3,900	1,025		4,925	5,875
0505 12' wide		.80	25		4,300	1,025		5,325	6,300
0510 15' wide		.80	25		5,100	1,025		6,125	7,175
0600 Steel, security gate, driveway, single, 10' wide		.80	25		2,150	1,025		3,175	3,925
0605 12' wide		.80	25		2,225	1,025		3,250	4,000
0620 Steel, security gate, driveway, dual, 12' wide		.80	25		2,050	1,025		3,075	3,850
0625 14' wide		.80	25		2,200	1,025		3,225	4,000
0630 16' wide		.80	25		2,525	1,025		3,550	4,350
0700 Aluminum, security gate, driveway, dual, 10' wide		.80	25		2,200	1,025		3,225	4,000
0705 12' wide		.80	25		2,600	1,025		3,625	4,425
0710 16' wide	↓	.80	25	↓	3,450	1,025		4,475	5,375
1000 Security gate, driveway, opener 12 VDC				Ea.	700			700	770
1010 Wireless					1,700			1,700	1,875
1020 Security gate, driveway, opener 24 VDC					1,100			1,100	1,200
1030 Wireless					1,300			1,300	1,425
1040 Security gate, driveway, opener 12 VDC, solar panel 10 watt	1 Elec	2	4	↓	200	206		406	530
1050 20 watt	"	2	4		320	206		526	660

08 34 59 – Vault Doors and Day Gates

08 34 59.10 Secure Storage Doors

	Crew	Daily Output	Labor-Hours	Unit	Material	Labor	Equipment	Total	Total Incl O&P
0010 **SECURE STORAGE DOORS**									
0020 Door and frame, 32" x 78", clear opening									
0100 1 hour test, 32" door, weighs 750 lb.	2 Sswk	1.50	10.667	Opng.	3,075	525		3,600	4,325
0200 2 hour test, 32" door, weighs 950 lb.		1.30	12.308		3,300	605		3,905	4,700
0250 40" door, weighs 1130 lb.		1	16		3,675	790		4,465	5,425
0300 4 hour test, 32" door, weighs 1025 lb.		1.20	13.333		3,550	660		4,210	5,075
0350 40" door, weighs 1140 lb.	↓	.90	17.778		3,900	875		4,775	5,850
0500 For stainless steel front, including frame, add to above					1,900			1,900	2,100
0550 Back, add				↓	1,900			1,900	2,100
0600 For time lock, two movement, add	1 Elec	2	4	Ea.	2,250	206		2,456	2,775
0650 Three movement, add	"	2	4		2,200	206		2,406	2,725
0800 Day gate, painted, steel, 32" wide	2 Sswk	1.50	10.667		1,850	525		2,375	3,000
0850 40" wide		1.40	11.429		2,050	565		2,615	3,250
0900 Aluminum, 32" wide		1.50	10.667		3,100	525		3,625	4,350
0950 40" wide	↓	1.40	11.429	↓	3,300	565		3,865	4,625
2050 Security vault door, class I, 3' wide, 3 1/2" thick	E-24	.19	166	Opng.	15,300	8,100	3,625	27,025	34,900
2100 Class II, 3' wide, 7" thick		.19	166		18,200	8,100	3,625	29,925	38,000
2150 Class III, 9R, 3' wide, 10" thick, minimum		.13	250	↓	22,800	12,200	5,450	40,450	52,000
2160 Class V, type 1, 40" door		2.48	12.903	Ea.	6,800	630	281	7,711	8,850
2170 Class V, type 2, 40" door	↓	2.48	12.903		6,775	630	281	7,686	8,825
2180 Day gate for class V vault	2 Sswk	2	8	↓	1,825	395		2,220	2,725

08 34 63 – Detention Doors and Frames

08 34 63.13 Steel Detention Doors and Frames

	Crew	Daily Output	Labor-Hours	Unit	Material	Labor	Equipment	Total	Total Incl O&P
0010 **STEEL DETENTION DOORS AND FRAMES**									
0500 Rolling cell door, bar front, 7/8" bars, 4" O.C., 7' H, 5' W, with hardware [G]	E-4	2	16	Ea.	5,175	800	61	6,036	7,175
0550 Actuator for rolling cell door, bar front	2 Skwk	2	8		4,200	365		4,565	5,175
1000 Doors & frames, 3' x 7', complete, with hardware, single plate	E-4	4	8		4,525	400	30.50	4,955.50	5,725
1650 Double plate	"	4	8	↓	5,550	400	30.50	5,980.50	6,875

08 34 Special Function Doors

08 34 73 – Sound Control Door Assemblies

08 34 73.10 Acoustical Doors	Crew	Daily Output	Labor-Hours	Unit	Material	2012 Bare Costs Labor	Equipment	Total	Total Incl O&P
0010 **ACOUSTICAL DOORS**									
0020 Including framed seals, 3' x 7', wood, 27 STC rating	2 Carp	1.50	10.667	Ea.	950	470		1,420	1,775
0100 Steel, 40 STC rating		1.50	10.667		3,000	470		3,470	4,025
0200 45 STC rating		1.50	10.667		3,500	470		3,970	4,575
0300 48 STC rating		1.50	10.667		4,100	470		4,570	5,225
0400 52 STC rating	↓	1.50	10.667	↓	4,600	470		5,070	5,775

08 36 Panel Doors

08 36 13 – Sectional Doors

08 36 13.10 Overhead Commercial Doors

	Crew	Daily Output	Labor-Hours	Unit	Material	Labor	Equipment	Total	Total Incl O&P
0010 **OVERHEAD COMMERCIAL DOORS**									
1000 Stock, sectional, heavy duty, wood, 1-3/4" thick, 8' x 8' high	2 Carp	2	8	Ea.	840	355		1,195	1,475
1100 10' x 10' high		1.80	8.889		1,200	390		1,590	1,925
1200 12' x 12' high		1.50	10.667		1,750	470		2,220	2,650
1300 Chain hoist, 14' x 14' high		1.30	12.308		2,700	545		3,245	3,800
1400 12' x 16' high		1	16		2,650	705		3,355	4,000
1500 20' x 8' high		1.30	12.270		2,425	540		2,965	3,475
1600 20' x 16' high		.65	24.615		4,875	1,075		5,950	7,050
1800 Center mullion openings, 8' high		4	4		1,200	176		1,376	1,600
1900 20' high	↓	2	8	↓	2,000	355		2,355	2,750
2100 For medium duty custom door, deduct					5%	5%			
2150 For medium duty stock doors, deduct					10%	5%			
2300 Fiberglass and aluminum, heavy duty, sectional, 12' x 12' high	2 Carp	1.50	10.667	Ea.	2,675	470		3,145	3,650
2450 Chain hoist, 20' x 20' high		.50	32		6,575	1,400		7,975	9,425
2600 Steel, 24 ga. sectional, manual, 8' x 8' high		2	8		830	355		1,185	1,450
2650 10' x 10' high		1.80	8.889		1,050	390		1,440	1,750
2700 12' x 12' high		1.50	10.667		1,275	470		1,745	2,125
2800 Chain hoist, 20' x 14' high	↓	.70	22.857	↓	3,550	1,000		4,550	5,450
2850 For 1-1/4" rigid insulation and 26 ga. galv.									
2860 back panel, add				S.F.	4.50			4.50	4.95
2900 For electric trolley operator, 1/3 H.P., to 12' x 12', add	1 Carp	2	4	Ea.	925	176		1,101	1,300
2950 Over 12' x 12', 1/2 H.P., add	"	1	8	"	1,075	355		1,430	1,750

08 36 13.20 Residential Garage Doors

	Crew	Daily Output	Labor-Hours	Unit	Material	Labor	Equipment	Total	Total Incl O&P
0010 **RESIDENTIAL GARAGE DOORS**									
0050 Hinged, wood, custom, double door, 9' x 7'	2 Carp	4	4	Ea.	700	176		876	1,050
0070 16' x 7'		3	5.333		1,050	235		1,285	1,500
0200 Overhead, sectional, incl. hardware, fiberglass, 9' x 7', standard		5.28	3.030		865	134		999	1,150
0220 Deluxe		5.28	3.030		1,100	134		1,234	1,400
0300 16' x 7', standard		6	2.667		1,550	118		1,668	1,875
0320 Deluxe		6	2.667		2,100	118		2,218	2,475
0500 Hardboard, 9' x 7', standard		8	2		580	88		668	775
0520 Deluxe		8	2		800	88		888	1,025
0600 16' x 7', standard		6	2.667		1,200	118		1,318	1,500
0620 Deluxe		6	2.667		1,400	118		1,518	1,725
0700 Metal, 9' x 7', standard		5.28	3.030		640	134		774	910
0720 Deluxe		8	2		875	88		963	1,100
0800 16' x 7', standard		3	5.333		775	235		1,010	1,225
0820 Deluxe		6	2.667		1,375	118		1,493	1,675
0900 Wood, 9' x 7', standard		8	2		900	88		988	1,125
0920 Deluxe	↓	8	2	↓	2,100	88		2,188	2,425

08 36 Panel Doors

08 36 13 – Sectional Doors

08 36 13.20 Residential Garage Doors		Crew	Daily Output	Labor-Hours	Unit	Material	2012 Bare Costs Labor	Equipment	Total	Total Incl O&P
1000	16' x 7', standard	2 Carp	6	2.667	Ea.	1,600	118		1,718	1,925
1020	Deluxe	↓	6	2.667		2,975	118		3,093	3,450
1800	Door hardware, sectional	1 Carp	4	2		320	88		408	485
1810	Door tracks only		4	2		159	88		247	310
1820	One side only	↓	7	1.143		110	50.50		160.50	199
3000	Swing-up, including hardware, fiberglass, 9' x 7', standard	2 Carp	8	2		980	88		1,068	1,200
3020	Deluxe		8	2		1,000	88		1,088	1,225
3100	16' x 7', standard		6	2.667		1,225	118		1,343	1,525
3120	Deluxe		6	2.667		1,500	118		1,618	1,825
3200	Hardboard, 9' x 7', standard		8	2		500	88		588	685
3220	Deluxe		8	2		600	88		688	795
3300	16' x 7', standard		6	2.667		620	118		738	860
3320	Deluxe		6	2.667		800	118		918	1,050
3400	Metal, 9' x 7', standard		8	2		420	88		508	600
3420	Deluxe		8	2		950	88		1,038	1,175
3500	16' x 7', standard		6	2.667		685	118		803	935
3520	Deluxe		6	2.667		1,000	118		1,118	1,275
3600	Wood, 9' x 7', standard		8	2		600	88		688	795
3620	Deluxe		8	2		1,050	88		1,138	1,275
3700	16' x 7', standard		6	2.667		800	118		918	1,050
3720	Deluxe	↓	6	2.667		2,000	118		2,118	2,375
3900	Door hardware only, swing up	1 Carp	4	2		163	88		251	315
3920	One side only		7	1.143		80	50.50		130.50	166
4000	For electric operator, economy, add		8	1		410	44		454	520
4100	Deluxe, including remote control	↓	8	1	↓	600	44		644	730
4500	For transmitter/receiver control , add to operator				Total	110			110	121
4600	Transmitters, additional				"	55			55	60.50

08 36 19 – Multi-Leaf Vertical Lift Doors

08 36 19.10 Sectional Vertical Lift Doors

		Crew	Daily Output	Labor-Hours	Unit	Material	2012 Bare Costs Labor	Equipment	Total	Total Incl O&P
0010	**SECTIONAL VERTICAL LIFT DOORS**									
0020	Motorized, 14 ga. stl, incl., frm and ctrl pnl									
0050	16' x 16' high	L-10	.50	48	Ea.	21,500	2,375	1,225	25,100	29,100
0100	10' x 20' high		1.30	18.462		31,800	915	470	33,185	37,100
0120	15' x 20' high		1.30	18.462		39,400	915	470	40,785	45,400
0140	20' x 20' high		1	24		45,900	1,200	610	47,710	53,000
0160	25' x 20' high		1	24		52,500	1,200	610	54,310	60,000
0170	32' x 24' high		.75	32		46,400	1,575	815	48,790	54,500
0180	20' x 25' high		1	24		54,000	1,200	610	55,810	61,500
0200	25' x 25' high		.70	34.286		62,500	1,700	875	65,075	73,000
0220	25' x 30' high		.70	34.286		68,000	1,700	875	70,575	78,500
0240	30' x 30' high		.70	34.286		79,000	1,700	875	81,575	90,500
0260	35' x 30' high	↓	.70	34.286	↓	89,000	1,700	875	91,575	102,000

08 36 23 – Telescoping Vertical Lift Doors

08 36 23.10 Telescoping Steel Doors

		Crew	Daily Output	Labor-Hours	Unit	Material	2012 Bare Costs Labor	Equipment	Total	Total Incl O&P
0010	**TELESCOPING STEEL DOORS**									
1000	Overhead, .03" thick, electric operated, 10' x 10'	E-3	.80	30	Ea.	14,600	1,500	152	16,252	18,900
2000	20' x 10'		.60	40		18,700	2,000	203	20,903	24,400
3000	20' x 16'	↓	.40	60	↓	19,000	3,000	305	22,305	26,600

08 38 Traffic Doors

08 38 13 – Flexible Strip Doors

08 38 13.10 Flexible Transparent Strip Doors		Crew	Daily Output	Labor-Hours	Unit	Material	2012 Bare Costs Labor	Equipment	Total	Total Incl O&P
0010	**FLEXIBLE TRANSPARENT STRIP DOORS**									
0100	12" strip width, 2/3 overlap	3 Shee	135	.178	SF Surf	7.55	9.25		16.80	22.50
0200	Full overlap		115	.209		9.50	10.90		20.40	27
0220	8" strip width, 1/2 overlap		140	.171		6.10	8.95		15.05	20.50
0240	Full overlap	↓	120	.200	↓	7.70	10.45		18.15	24.50
0300	Add for suspension system, header mount				L.F.	9			9	9.90
0400	Wall mount				"	9.30			9.30	10.25

08 38 19 – Rigid Traffic Doors

08 38 19.20 Double Acting Swing Doors

		Crew	Daily Output	Labor-Hours	Unit	Material	Labor	Equipment	Total	Total Incl O&P
0010	**DOUBLE ACTING SWING DOORS**									
1000	.063" aluminum, 7'-0" high, 4'-0" wide	2 Carp	4.20	3.810	Pr.	2,100	168		2,268	2,550
1025	6'-0" wide		4	4		2,200	176		2,376	2,700
1050	6'-8" wide	↓	4	4	↓	2,400	176		2,576	2,925
2000	Solid core wood, 3/4" thick, metal frame, stainless steel									
2010	base plate, 7' high opening, 4' wide	2 Carp	4	4	Pr.	2,600	176		2,776	3,125
2050	7' wide	"	3.80	4.211	"	2,800	186		2,986	3,350

08 38 19.30 Shock Absorbing Doors

		Crew	Daily Output	Labor-Hours	Unit	Material	Labor	Equipment	Total	Total Incl O&P
0010	**SHOCK ABSORBING DOORS**									
0020	Rigid, no frame, 1-1/2" thick, 5' x 7'	2 Sswk	1.90	8.421	Opng.	1,400	415		1,815	2,300
0100	8' x 8'		1.80	8.889		2,000	440		2,440	2,975
0500	Flexible, no frame, insulated, .16" thick, economy, 5' x 7'		2	8		1,750	395		2,145	2,625
0600	Deluxe		1.90	8.421		2,625	415		3,040	3,625
1000	8' x 8' opening, economy		2	8		2,750	395		3,145	3,725
1100	Deluxe	↓	1.90	8.421	↓	3,500	415		3,915	4,600

08 41 Entrances and Storefronts

08 41 13 – Aluminum-Framed Entrances and Storefronts

08 41 13.20 Tube Framing

		Crew	Daily Output	Labor-Hours	Unit	Material	Labor	Equipment	Total	Total Incl O&P
0010	**TUBE FRAMING**, For window walls and store fronts, aluminum stock									
0050	Plain tube frame, mill finish, 1-3/4" x 1-3/4"	2 Glaz	103	.155	L.F.	9.45	6.65		16.10	20.50
0150	1-3/4" x 4"		98	.163		12.95	7		19.95	25
0200	1-3/4" x 4-1/2"		95	.168		14.60	7.20		21.80	27
0250	2" x 6"		89	.180		22	7.70		29.70	35.50
0350	4" x 4"		87	.184		25	7.85		32.85	39.50
0400	4-1/2" x 4-1/2"		85	.188		26.50	8.05		34.55	41
0450	Glass bead		240	.067		2.80	2.85		5.65	7.40
1000	Flush tube frame, mill finish, 1/4" glass, 1-3/4" x 4", open header		80	.200		12.40	8.55		20.95	26.50
1050	Open sill		82	.195		10.10	8.35		18.45	24
1100	Closed back header		83	.193		17.75	8.25		26	32
1150	Closed back sill	↓	85	.188	↓	16.95	8.05		25	31
1160	Tube fmg, spandrel cover both sides, alum 1" wide	1 Sswk	85	.094	S.F.	96	4.64		100.64	113
1170	Tube fmg, spandrel cover both sides, alum 2" wide	"	85	.094	"	37	4.64		41.64	49
1200	Vertical mullion, one piece	2 Glaz	75	.213	L.F.	18.60	9.10		27.70	34.50
1250	Two piece		73	.219		19.75	9.35		29.10	35.50
1300	90° or 180° vertical corner post		75	.213		30	9.10		39.10	47
1400	1-3/4" x 4-1/2", open header		80	.200		15.10	8.55		23.65	29.50
1450	Open sill		82	.195		12.60	8.35		20.95	26.50
1500	Closed back header		83	.193		18	8.25		26.25	32.50
1550	Closed back sill		85	.188		17.75	8.05		25.80	31.50
1600	Vertical mullion, one piece	↓	75	.213	↓	19.80	9.10		28.90	36

08 41 Entrances and Storefronts

08 41 13 – Aluminum-Framed Entrances and Storefronts

08 41 13.20 Tube Framing		Crew	Daily Output	Labor-Hours	Unit	Material	2012 Bare Costs Labor	Equipment	Total	Total Incl O&P
1650	Two piece	2 Glaz	73	.219	L.F.	21	9.35		30.35	37
1700	90° or 180° vertical corner post		75	.213		21.50	9.10		30.60	37.50
2000	Flush tube frame, mil fin.,ins. glass w/thml brk, 2" x 4-1/2", open header		75	.213		15.55	9.10		24.65	31
2050	Open sill		77	.208		13.10	8.90		22	28
2100	Closed back header		78	.205		14.70	8.75		23.45	29.50
2150	Closed back sill		80	.200		15.75	8.55		24.30	30.50
2200	Vertical mullion, one piece		70	.229		16.30	9.75		26.05	32.50
2250	Two piece		68	.235		17.60	10.05		27.65	34.50
2300	90° or 180° vertical corner post		70	.229		16.75	9.75		26.50	33.50
5000	Flush tube frame, mill fin., thermal brk., 2-1/4" x 4-1/2", open header		74	.216		16.20	9.25		25.45	32
5050	Open sill		75	.213		14.40	9.10		23.50	29.50
5100	Vertical mullion, one piece		69	.232		16.75	9.90		26.65	33.50
5150	Two piece		67	.239		20.50	10.20		30.70	38
5200	90° or 180° vertical corner post		69	.232		18.05	9.90		27.95	35
6980	Door stop (snap in)	↓	380	.042	↓	3.09	1.80		4.89	6.15
7000	For joints, 90°, clip type, add				Ea.	24.50			24.50	27
7050	Screw spline joint, add					23			23	25
7100	For joint other than 90°, add				↓	48			48	52.50
8000	For bronze anodized aluminum, add					15%				
8020	For black finish, add					30%				
8050	For stainless steel materials, add					350%				
8100	For monumental grade, add					53%				
8150	For steel stiffener, add	2 Glaz	200	.080	L.F.	11	3.42		14.42	17.30
8200	For 2 to 5 stories, add per story				Story		8%			

08 41 19 – Stainless-Steel-Framed Entrances and Storefronts

08 41 19.10 Stainless-Steel and Glass Entrance Unit

		Crew	Daily Output	Labor-Hours	Unit	Material	2012 Bare Costs Labor	Equipment	Total	Total Incl O&P
0010	**STAINLESS-STEEL AND GLASS ENTRANCE UNIT**, narrow stiles									
0020	3' x 7' opening, including hardware, minimum	2 Sswk	1.60	10	Opng.	6,475	495		6,970	7,975
0050	Average		1.40	11.429		6,875	565		7,440	8,550
0100	Maximum	↓	1.20	13.333		7,425	660		8,085	9,325
1000	For solid bronze entrance units, statuary finish, add					64%				
1100	Without statuary finish, add				↓	45%				
2000	Balanced doors, 3' x 7', economy	2 Sswk	.90	17.778	Ea.	8,675	875		9,550	11,100
2100	Premium	"	.70	22.857	"	14,900	1,125		16,025	18,400

08 41 26 – All-Glass Entrances and Storefronts

08 41 26.10 Window Walls Aluminum, Stock

		Crew	Daily Output	Labor-Hours	Unit	Material	2012 Bare Costs Labor	Equipment	Total	Total Incl O&P
0010	**WINDOW WALLS ALUMINUM, STOCK**, including glazing									
0020	Minimum	H-2	160	.150	S.F.	43.50	6.05		49.55	57
0050	Average		140	.171		60	6.90		66.90	76.50
0100	Maximum	↓	110	.218	↓	165	8.75		173.75	194
0500	For translucent sandwich wall systems, see Section 07 41 33.10									
0850	Cost of the above walls depends on material,									
0860	finish, repetition, and size of units.									
0870	The larger the opening, the lower the S.F. cost									
1200	Double glazed acoustical window wall for airports,									
1220	including 1" thick glass with 2" x 4-1/2" tube frame	H-2	40	.600	S.F.	113	24		137	161

08 42 Entrances

08 42 26 – All-Glass Entrances

08 42 26.10 Swinging Glass Doors

		Daily Output	Labor-Hours	Unit	Material	2012 Bare Costs Labor	Equipment	Total	Total Incl O&P
0010	**SWINGING GLASS DOORS**	Crew							
0020	Including hardware, 1/2" thick, tempered, 3' x 7' opening	2 Glaz	2	8	Opng.	2,175	340		2,515	2,900
0100	6' x 7' opening	"	1.40	11.429	"	4,200	490		4,690	5,375

08 42 29 – Automatic Entrances

08 42 29.23 Sliding Automatic Entrances

		Crew	Daily Output	Labor-Hours	Unit	Material	Labor	Equipment	Total	Total Incl O&P
0010	**SLIDING AUTOMATIC ENTRANCES** 12' x 7'-6" opng., 5' x 7' door, 2 way traffic									
0020	Mat or electronic activated, panic pushout, incl. operator & hardware,									
0030	not including glass or glazing	2 Glaz	.70	22.857	Opng.	8,200	975		9,175	10,500

08 42 33 – Revolving Door Entrances

08 42 33.10 Circular Rotating Entrance Doors

		Crew	Daily Output	Labor-Hours	Unit	Material	Labor	Equipment	Total	Total Incl O&P
0010	**CIRCULAR ROTATING ENTRANCE DOORS**, Aluminum									
0020	6'-10" to 7' high, stock units, minimum	4 Sswk	.75	42.667	Opng.	20,100	2,100		22,200	25,900
0050	Average		.60	53.333		24,200	2,625		26,825	31,300
0100	Maximum		.45	71.111		41,500	3,500		45,000	52,000
1000	Stainless steel		.30	105		39,400	5,225		44,625	52,500
1100	Solid bronze		.15	213		46,400	10,500		56,900	70,000
1500	For automatic controls, add	2 Elec	2	8		14,600	410		15,010	16,700

08 42 36 – Balanced Door Entrances

08 42 36.10 Balanced Entrance Doors

		Crew	Daily Output	Labor-Hours	Unit	Material	Labor	Equipment	Total	Total Incl O&P
0010	**BALANCED ENTRANCE DOORS**									
0020	Hardware & frame, alum. & glass, 3' x 7', econ.	2 Sswk	.90	17.778	Ea.	6,300	875		7,175	8,500
0150	Premium	"	.70	22.857	"	7,700	1,125		8,825	10,500

08 43 Storefronts

08 43 13 – Aluminum-Framed Storefronts

08 43 13.10 Aluminum-Framed Entrance Doors

		Crew	Daily Output	Labor-Hours	Unit	Material	Labor	Equipment	Total	Total Incl O&P
0010	**ALUMINUM-FRAMED ENTRANCE DOORS**									
0015	Standard hardware and glass stops but no glazing									
0020	Entrance door, 3' x 7' opening, clear anodized finish	2 Sswk	7	2.286	Opng.	365	113		478	600
0040	Bronze finish		7	2.286		380	113		493	620
0060	Black finish		7	2.286		425	113		538	670
0500	6' x 7' opening, clear finish		6	2.667		750	132		882	1,050
0520	Bronze finish		6	2.667		785	132		917	1,100
0540	Black finish		6	2.667		875	132		1,007	1,200
1000	With 3' high transom above, 3' x 7' opening, clear finish		5.50	2.909		445	144		589	745
1050	Bronze finish		5.50	2.909		460	144		604	765
1100	Black finish		5.50	2.909		495	144		639	800
1300	3'-6" x 7'-0" opening, clear finish		5.50	2.909		330	144		474	615
1320	Bronze finish		5.50	2.909		335	144		479	625
1340	Black finish		5.50	2.909		355	144		499	645
1500	6' x 7' opening, clear finish		5.50	2.909		560	144		704	870
1550	Bronze finish		5.50	2.909		575	144		719	885
1600	Black finish		5.50	2.909		635	144		779	950

08 43 13.20 Storefront Systems

		Crew	Daily Output	Labor-Hours	Unit	Material	Labor	Equipment	Total	Total Incl O&P
0010	**STOREFRONT SYSTEMS**, aluminum frame clear 3/8" plate glass									
0020	incl. 3' x 7' door with hardware (400 sq. ft. max. wall)									
0500	Wall height to 12' high, commercial grade	2 Glaz	150	.107	S.F.	20.50	4.56		25.06	29.50
0600	Institutional grade		130	.123		25.50	5.25		30.75	36
0700	Monumental grade		115	.139		38.50	5.95		44.45	51

08 43 Storefronts

08 43 13 – Aluminum-Framed Storefronts

08 43 13.20 Storefront Systems	Crew	Daily Output	Labor-Hours	Unit	Material	2012 Bare Costs Labor	Equipment	Total	Total Incl O&P
1000	6' x 7' door with hardware, commercial grade	2 Glaz	135	.119	S.F.	20.50	5.05		25.55	30
1100	Institutional grade		115	.139		28	5.95		33.95	40
1200	Monumental grade		100	.160		52	6.85		58.85	68
1500	For bronze anodized finish, add					15%				
1600	For black anodized finish, add					36%				
1700	For stainless steel framing, add to monumental					78%				

08 43 29 – Sliding Storefronts

08 43 29.10 Sliding Panels

		Crew	Daily Output	Labor-Hours	Unit	Material	2012 Bare Costs Labor	Equipment	Total	Total Incl O&P
0010	**SLIDING PANELS**									
0020	Mall fronts, aluminum & glass, 15' x 9' high	2 Glaz	1.30	12.308	Opng.	3,275	525		3,800	4,400
0100	24' x 9' high		.70	22.857		4,650	975		5,625	6,600
0200	48' x 9' high, with fixed panels		.90	17.778		8,700	760		9,460	10,700
0500	For bronze finish, add					17%				

08 44 Curtain Wall and Glazed Assemblies

08 44 13 – Glazed Aluminum Curtain Walls

08 44 13.10 Glazed Curtain Walls

		Crew	Daily Output	Labor-Hours	Unit	Material	2012 Bare Costs Labor	Equipment	Total	Total Incl O&P
0010	**GLAZED CURTAIN WALLS**, aluminum, stock, including glazing									
0020	Minimum	H-1	205	.156	S.F.	34	7.20		41.20	49.50
0050	Average, single glazed		195	.164		50	7.55		57.55	67.50
0150	Average, double glazed		180	.178		64	8.20		72.20	84
0200	Maximum		160	.200		172	9.20		181.20	204

08 45 Translucent Wall and Roof Assemblies

08 45 10 – Translucent Roof Assemblies

08 45 10.10 Skyroofs

		Crew	Daily Output	Labor-Hours	Unit	Material	2012 Bare Costs Labor	Equipment	Total	Total Incl O&P
0010	**SKYROOFS**, Translucent panels, 2-3/4" thick									
0020	Under 5000 S.F.	G-3	395	.081	SF Hor.	34	3.53		37.53	43
0100	Over 5000 S.F.		465	.069		31.50	3		34.50	39.50
0300	Continuous vaulted, semi-circular, to 8' wide, double glazed [G]		145	.221		52.50	9.65		62.15	72.50
0400	Single glazed		160	.200		35.50	8.75		44.25	52.50
0600	To 20' wide, single glazed		175	.183		46	8		54	62.50
0700	Over 20' wide, single glazed		200	.160		45	7		52	60
0900	Motorized opening type, single glazed, 1/3 opening		145	.221		46	9.65		55.65	65.50
1000	Full opening		130	.246		52.50	10.75		63.25	74
1200	Pyramid type units, self-supporting, to 30' clear opening,									
1300	square or circular, single glazed, minimum	G-3	200	.160	SF Hor.	27	7		34	40
1310	Average		165	.194		37	8.45		45.45	53.50
1400	Maximum		130	.246		53	10.75		63.75	75
1500	Grid type, 4' to 10' modules, single glass glazed, minimum		200	.160		36	7		43	50
1550	Maximum		128	.250		56	10.90		66.90	78
1600	Preformed acrylic, minimum		300	.107		43.50	4.65		48.15	54.50
1650	Maximum		175	.183		59.50	8		67.50	77.50
1800	Skyroofs, dome type, self-supporting, to 100' clear opening, circular									
1900	Rise to span ratio = 0.20									
1920	Minimum	G-3	197	.162	SF Hor.	15.70	7.10		22.80	28
1950	Maximum		113	.283		52.50	12.35		64.85	77
2100	Rise to span ratio = 0.33, minimum		169	.189		31.50	8.25		39.75	47
2150	Maximum		101	.317		62	13.80		75.80	89.50

08 45 Translucent Wall and Roof Assemblies

08 45 10 – Translucent Roof Assemblies

08 45 10.10 Skyroofs		Crew	Daily Output	Labor-Hours	Unit	Material	2012 Bare Costs Labor	Equipment	Total	Total Incl O&P
2200	Rise to span ratio = 0.50, minimum	G-3	148	.216	SF Hor.	47.50	9.45		56.95	67
2250	Maximum		87	.368		69	16.05		85.05	101
2400	Ridge units, continuous, to 8' wide, double	**G**	130	.246		120	10.75		130.75	148
2500	Single	**G**	200	.160		83	7		90	102
2700	Ridge and furrow units, over 4' O.C., double, minimum	**G**	200	.160		26	7		33	39
2750	Maximum	**G**	120	.267		51	11.65		62.65	74
2800	Single, minimum		214	.150		22	6.50		28.50	34.50
2850	Maximum		153	.209	▼	47	9.15		56.15	65.50
3000	Rolling roof, translucent panels, flat roof, minimum		253	.126	S.F.	20	5.50		25.50	30.50
3030	Maximum	**G**	160	.200		40	8.75		48.75	57.50
3100	Lean-to skyroof, long span, double, minimum	**G**	197	.162		26.50	7.10		33.60	40.50
3150	Maximum	**G**	101	.317		53.50	13.80		67.30	80
3300	Single, minimum		321	.100		20.50	4.35		24.85	29
3350	Maximum	▼	160	.200	▼	33.50	8.75		42.25	50

08 51 Metal Windows

08 51 13 – Aluminum Windows

08 51 13.10 Aluminum Sash

		Crew	Daily Output	Labor-Hours	Unit	Material	2012 Bare Costs Labor	Equipment	Total	Total Incl O&P
0010	**ALUMINUM SASH**									
0020	Stock, grade C, glaze & trim not incl., casement	2 Sswk	200	.080	S.F.	38	3.95		41.95	49
0050	Double hung		200	.080		38.50	3.95		42.45	49.50
0100	Fixed casement		200	.080		17	3.95		20.95	26
0150	Picture window		200	.080		18.05	3.95		22	27
0200	Projected window		200	.080		35	3.95		38.95	45.50
0250	Single hung		200	.080		16.20	3.95		20.15	25
0300	Sliding		200	.080	▼	21	3.95		24.95	30
1000	Mullions for above, tubular		240	.067	L.F.	6.05	3.29		9.34	12.55
2000	Custom aluminum sash, grade HC, glazing not included, minimum		200	.080	S.F.	39	3.95		42.95	50
2100	Maximum	▼	85	.188	"	51	9.30		60.30	72.50

08 51 13.20 Aluminum Windows

		Crew	Daily Output	Labor-Hours	Unit	Material	2012 Bare Costs Labor	Equipment	Total	Total Incl O&P
0010	**ALUMINUM WINDOWS**, incl. frame and glazing, commercial grade									
1000	Stock units, casement, 3'-1" x 3'-2" opening	2 Sswk	10	1.600	Ea.	365	79		444	540
1050	Add for storms					118			118	130
1600	Projected, with screen, 3'-1" x 3'-2" opening	2 Sswk	10	1.600		345	79		424	520
1700	Add for storms					115			115	127
2000	4'-5" x 5'-3" opening	2 Sswk	8	2		390	98.50		488.50	605
2100	Add for storms					123			123	135
2500	Enamel finish windows, 3'-1" x 3'-2"	2 Sswk	10	1.600		350	79		429	525
2600	4'-5" x 5'-3"		8	2		395	98.50		493.50	610
3000	Single hung, 2' x 3' opening, enameled, standard glazed		10	1.600		203	79		282	365
3100	Insulating glass		10	1.600		247	79		326	415
3300	2'-8" x 6'-8" opening, standard glazed		8	2		360	98.50		458.50	570
3400	Insulating glass		8	2		460	98.50		558.50	685
3700	3'-4" x 5'-0" opening, standard glazed		9	1.778		294	87.50		381.50	480
3800	Insulating glass		9	1.778		325	87.50		412.50	515
3890	Awning type, 3' x 3' opening standard glass		14	1.143		415	56.50		471.50	555
3900	Insulating glass		14	1.143		440	56.50		496.50	585
3910	3' x 4' opening, standard glass		10	1.600		485	79		564	675
3920	Insulating glass		10	1.600		560	79		639	755
3930	3' x 5'-4" opening, standard glass		10	1.600		585	79		664	785
3940	Insulating glass	▼	10	1.600	▼	690	79		769	900

08 51 Metal Windows

08 51 13 – Aluminum Windows

08 51 13.20 Aluminum Windows	Crew	Daily Output	Labor-Hours	Unit	Material	2012 Bare Costs Labor	Equipment	Total	Total Incl O&P
3950	4' x 5'-4" opening, standard glass	2 Sswk	9	1.778	Ea.	640	87.50		727.50	860
3960	Insulating glass		9	1.778		770	87.50		857.50	1,000
4000	Sliding aluminum, 3' x 2' opening, standard glazed		10	1.600		215	79		294	380
4100	Insulating glass		10	1.600		230	79		309	395
4300	5' x 3' opening, standard glazed		9	1.778		330	87.50		417.50	515
4400	Insulating glass		9	1.778		385	87.50		472.50	575
4600	8' x 4' opening, standard glazed		6	2.667		345	132		477	615
4700	Insulating glass		6	2.667		560	132		692	850
5000	9' x 5' opening, standard glazed		4	4		525	197		722	935
5100	Insulating glass		4	4		840	197		1,037	1,275
5500	Sliding, with thermal barrier and screen, 6' x 4', 2 track		8	2		715	98.50		813.50	965
5700	4 track		8	2		900	98.50		998.50	1,175
6000	For above units with bronze finish, add					15%				
6200	For installation in concrete openings, add					8%				

08 51 23 – Steel Windows

08 51 23.10 Steel Sash

08 51 23.10 Steel Sash	Crew	Daily Output	Labor-Hours	Unit	Material	Labor	Equipment	Total	Total Incl O&P
0010	STEEL SASH Custom units, glazing and trim not included	R085123-10								
0100	Casement, 100% vented	2 Sswk	200	.080	S.F.	64.50	3.95		68.45	77.50
0200	50% vented		200	.080		52.50	3.95		56.45	65
0300	Fixed		200	.080		28	3.95		31.95	38
1000	Projected, commercial, 40% vented		200	.080		50	3.95		53.95	62
1100	Intermediate, 50% vented		200	.080		57	3.95		60.95	69.50
1500	Industrial, horizontally pivoted		200	.080		52	3.95		55.95	64
1600	Fixed		200	.080		30.50	3.95		34.45	40.50
2000	Industrial security sash, 50% vented		200	.080		56.50	3.95		60.45	69
2100	Fixed		200	.080		46	3.95		49.95	57.50
2500	Picture window		200	.080		29.50	3.95		33.45	39.50
3000	Double hung		200	.080		57.50	3.95		61.45	70.50
5000	Mullions for above, open interior face		240	.067	L.F.	10.10	3.29		13.39	17.05
5100	With interior cover		240	.067	"	16.85	3.29		20.14	24.50

08 51 23.20 Steel Windows

08 51 23.20 Steel Windows	Crew	Daily Output	Labor-Hours	Unit	Material	Labor	Equipment	Total	Total Incl O&P
0010	STEEL WINDOWS Stock, including frame, trim and insul. glass									
0020	See Section 13 34 19.50									
1000	Custom units, double hung, 2'-8" x 4'-6" opening	R085123-10	2 Sswk	12	1.333	Ea.	700	66		766	890
1100	2'-4" x 3'-9" opening		12	1.333		580	66		646	755
1500	Commercial projected, 3'-9" x 5'-5" opening		10	1.600		1,200	79		1,279	1,475
1600	6'-9" x 4'-1" opening		7	2.286		1,600	113		1,713	1,950
2000	Intermediate projected, 2'-9" x 4'-1" opening		12	1.333		680	66		746	870
2100	4'-1" x 5'-5" opening		10	1.600		1,400	79		1,479	1,675

08 51 66 – Metal Window Screens

08 51 66.10 Screens

08 51 66.10 Screens	Crew	Daily Output	Labor-Hours	Unit	Material	Labor	Equipment	Total	Total Incl O&P
0010	SCREENS									
0020	For metal sash, aluminum or bronze mesh, flat screen	2 Sswk	1200	.013	S.F.	4.20	.66		4.86	5.80
0500	Wicket screen, inside window		1000	.016		6.50	.79		7.29	8.55
0800	Security screen, aluminum frame with stainless steel cloth		1200	.013		23.50	.66		24.16	26.50
0900	Steel grate, painted, on steel frame		1600	.010		12.80	.49		13.29	15
1000	For solar louvers, add		160	.100		24	4.94		28.94	35.50
4000	See Section 05 58 23.90									

08 52 Wood Windows

08 52 10 – Plain Wood Windows

08 52 10.20 Awning Window

		Crew	Daily Output	Labor-Hours	Unit	Material	2012 Bare Costs Labor	2012 Bare Costs Equipment	Total	Total Incl O&P
0010	**AWNING WINDOW**, Including frame, screens and grilles									
0100	Average quality, 34" x 22", double insulated glass	1 Carp	10	.800	Ea.	269	35.50		304.50	350
0200	Low E glass		10	.800		267	35.50		302.50	350
0300	40" x 28", double insulated glass		9	.889		310	39		349	405
0400	Low E Glass		9	.889		340	39		379	430
0500	48" x 36", double insulated glass		8	1		465	44		509	580
0600	Low E glass		8	1		490	44		534	605
4000	Impact windows, minimum, add					60%				
4010	Impact windows, maximum, add					160%				

08 52 10.40 Casement Window

			Crew	Daily Output	Labor-Hours	Unit	Material	2012 Bare Costs Labor	2012 Bare Costs Equipment	Total	Total Incl O&P
0010	**CASEMENT WINDOW**, including frame, screen and grilles										
0100	Avg. quality, bldrs. model, 2'-0" x 3'-0" H, dbl. insulated glass	G	1 Carp	10	.800	Ea.	284	35.50		319.50	370
0150	Low E glass	G		10	.800		259	35.50		294.50	340
0200	2'-0" x 4'-6" high, double insulated glass	G		9	.889		355	39		394	450
0250	Low E glass	G		9	.889		370	39		409	465
0260	Casement 4'-2" x 4'-2" double insulated glass	G		11	.727		875	32		907	1,000
0270	4'-0" x 4'-0" Low E glass	G		11	.727		535	32		567	640
0290	6'-4" x 5'-7" Low E glass	G		9	.889		1,125	39		1,164	1,275
0300	2'-4" x 6'-0" high, double insulated glass	G		8	1		430	44		474	545
0350	Low E glass	G		8	1		475	44		519	595
0522	Vinyl clad, premium, double insulated glass, 2'-0" x 3'-0"	G		10	.800		266	35.50		301.50	350
0524	2'-0" x 4'-0"	G		9	.889		310	39		349	400
0525	2'-0" x 5'-0"	G		8	1		355	44		399	460
0528	2'-0" x 6'-0"	G		8	1		375	44		419	480
0600	3'-0" x 5'-0"			8	1		650	44		694	785
0700	4'-0" x 3'-0"			8	1		715	44		759	860
0710	4'-0" x 4'-0"	G		8	1		615	44		659	745
0720	4'-8" x 4'-0"			8	1		675	44		719	815
0730	4'-8" x 5'-0"			6	1.333		775	59		834	940
0740	4'-8" x 6'-0"			6	1.333		870	59		929	1,050
0750	6'-0" x 4'-0"			6	1.333		790	59		849	960
0800	6'-0" x 5'-0"			6	1.333		885	59		944	1,075
0900	5'-6" x 5'-6"	G	2 Carp	15	1.067		1,400	47		1,447	1,600
2000	Bay, casement units, 8' x 5', w/screens, dbl. insul. glass			2.50	6.400	Opng.	1,600	282		1,882	2,175
2100	Low E glass			2.50	6.400	"	1,675	282		1,957	2,275
8190	For installation, add per leaf					Ea.		15%			
8200	For multiple leaf units, deduct for stationary sash										
8220	2' high					Ea.	23			23	25.50
8240	4'-6" high						26			26	28.50
8260	6' high						34.50			34.50	38
8300	Impact windows, minimum, add						60%				
8310	Impact windows, maximum, add						160%				

08 52 10.50 Double Hung

			Crew	Daily Output	Labor-Hours	Unit	Material	2012 Bare Costs Labor	2012 Bare Costs Equipment	Total	Total Incl O&P
0010	**DOUBLE HUNG**, Including frame, screens and grilles R085216-10										
0100	Avg. quality, bldrs. model, 2'-0" x 3'-0" high, low E insul. glass	G	1 Carp	10	.800	Ea.	209	35.50		244.50	284
0200	3'-0" x 4'-0" high, double insulated glass	G		9	.889		270	39		309	360
0300	4'-0" x 4'-6" high, low E insulated glass	G		8	1		315	44		359	415
8000	Impact windows, minimum, add						60%				
8010	Impact windows, maximum, add						160%				

08 52 10.55 Picture Window

		Crew	Daily Output	Labor-Hours	Unit	Material	2012 Bare Costs Labor	2012 Bare Costs Equipment	Total	Total Incl O&P
0010	**PICTURE WINDOW**, Including frame and grilles									
0100	Average quality, bldrs. model, 3'-6" x 4'-0" high, dbl. insulated glass	2 Carp	12	1.333	Ea.	400	59		459	525

08 52 Wood Windows

08 52 10 – Plain Wood Windows

08 52 10.55 Picture Window		Crew	Daily Output	Labor-Hours	Unit	Material	2012 Bare Costs Labor	Equipment	Total	Total Incl O&P
0150	Low E glass	2 Carp	12	1.333	Ea.	410	59		469	540
0200	4'-0" x 4'-6" high, double insulated glass		11	1.455		500	64		564	650
0250	Low E glass		11	1.455		530	64		594	685
0300	5'-0" x 4'-0" high, double insulated glass		11	1.455		580	64		644	740
0350	Low E glass		11	1.455		605	64		669	765
0400	6'-0" x 4'-6" high, double insulated glass		10	1.600		625	70.50		695.50	800
0450	Low E glass		10	1.600		635	70.50		705.50	810

08 52 10.65 Wood Sash

		Crew	Daily Output	Labor-Hours	Unit	Material	Labor	Equipment	Total	Total Incl O&P
0010	**WOOD SASH**, Including glazing but not trim									
0050	Custom, 5'-0" x 4'-0", 1" dbl. glazed, 3/16" thick lites	2 Carp	3.20	5	Ea.	226	221		447	590
0100	1/4" thick lites		5	3.200		241	141		382	480
0200	1" thick, triple glazed		5	3.200		405	141		546	660
0300	7'-0" x 4'-6" high, 1" double glazed, 3/16" thick lites		4.30	3.721		415	164		579	705
0400	1/4" thick lites		4.30	3.721		465	164		629	760
0500	1" thick, triple glazed		4.30	3.721		530	164		694	835
0600	8'-6" x 5'-0" high, 1" double glazed, 3/16" thick lites		3.50	4.571		555	202		757	920
0700	1/4" thick lites		3.50	4.571		605	202		807	975
0800	1" thick, triple glazed		3.50	4.571		615	202		817	985
0900	Window frames only, based on perimeter length				L.F.	3.95			3.95	4.35
1200	Window sill, stock, per lineal foot					8.45			8.45	9.30
1250	Casing, stock					3.24			3.24	3.56

08 52 10.70 Sliding Windows

		Crew	Daily Output	Labor-Hours	Unit	Material	Labor	Equipment	Total	Total Incl O&P
0010	**SLIDING WINDOWS**									
0100	Average quality, bldrs. model, 3'-0" x 3'-0" high, double insulated [G]	1 Carp	10	.800	Ea.	280	35.50		315.50	365
0120	Low E glass [G]		10	.800		305	35.50		340.50	390
0200	4'-0" x 3'-6" high, double insulated [G]		9	.889		355	39		394	450
0220	Low E glass [G]		9	.889		355	39		394	450
0300	6'-0" x 5'-0" high, double insulated [G]		8	1		485	44		529	605
0320	Low E glass [G]		8	1		525	44		569	645

08 52 13 – Metal-Clad Wood Windows

08 52 13.10 Awning Windows, Metal-Clad

		Crew	Daily Output	Labor-Hours	Unit	Material	Labor	Equipment	Total	Total Incl O&P
0010	**AWNING WINDOWS, METAL-CLAD**									
2000	Metal clad, awning deluxe, double insulated glass, 34" x 22"	1 Carp	9	.889	Ea.	245	39		284	330
2050	36" x 25"		9	.889		272	39		311	360
2100	40" x 22"		9	.889		289	39		328	380
2150	40" x 30"		9	.889		340	39		379	435
2200	48" x 28"		8	1		345	44		389	450
2250	60" x 36"		8	1		370	44		414	480

08 52 13.20 Casement Windows, Metal-Clad

		Crew	Daily Output	Labor-Hours	Unit	Material	Labor	Equipment	Total	Total Incl O&P
0010	**CASEMENT WINDOWS, METAL-CLAD**									
0100	Metal clad, deluxe, dbl. insul. glass, 2'-0" x 3'-0" high [G]	1 Carp	10	.800	Ea.	279	35.50		314.50	360
0120	2'-0" x 4'-0" high [G]		9	.889		315	39		354	405
0130	2'-0" x 5'-0" high [G]		8	1		325	44		369	430
0140	2'-0" x 6'-0" high [G]		8	1		365	44		409	470
0300	Metal clad, casement, bldrs mdl, 6'-0" x 4'-0", dbl. insltd gls, 3 panels	2 Carp	10	1.600		1,200	70.50		1,270.50	1,425
0310	9'-0" x 4'-0", 4 panels		8	2		1,550	88		1,638	1,825
0320	10'-0" x 5'-0", 5 panels		7	2.286		2,100	101		2,201	2,475
0330	12'-0" x 6'-0", 6 panels		6	2.667		2,700	118		2,818	3,125

08 52 13.30 Double-Hung Windows, Metal-clad

		Crew	Daily Output	Labor-Hours	Unit	Material	Labor	Equipment	Total	Total Incl O&P
0010	**DOUBLE-HUNG WINDOWS, METAL-CLAD**									
0100	Metal clad, deluxe, dbl. insul. glass, 2'-6" x 3'-0" high [G]	1 Carp	10	.800	Ea.	272	35.50		307.50	355

08 52 Wood Windows

08 52 13 – Metal-Clad Wood Windows

08 52 13.30 Double-Hung Windows, Metal-clad

		Crew	Daily Output	Labor-Hours	Unit	Material	2012 Bare Costs Labor	Equipment	Total	Total Incl O&P
0120	3'-0" x 3'-6" high	**G** 1 Carp	10	.800	Ea.	315	35.50		350.50	400
0140	3'-0" x 4'-0" high	**G**	9	.889		325	39		364	420
0160	3'-0" x 4'-6" high	**G**	9	.889		345	39		384	440
0180	3'-0" x 5'-0" high	**G**	8	1		370	44		414	480
0200	3'-6" x 6'-0" high	**G**	8	1		450	44		494	565

08 52 13.35 Picture and Sliding Windows Metal-Clad

		Crew	Daily Output	Labor-Hours	Unit	Material	2012 Bare Costs Labor	Equipment	Total	Total Incl O&P
0010	**PICTURE AND SLIDING WINDOWS METAL-CLAD**									
2000	Metal clad, dlx picture, dbl. insul. glass, 4'-0" x 4'-0" high	2 Carp	12	1.333	Ea.	375	59		434	500
2100	4'-0" x 6'-0" high		11	1.455		545	64		609	700
2200	5'-0" x 6'-0" high		10	1.600		610	70.50		680.50	780
2300	6'-0" x 6'-0" high		10	1.600		695	70.50		765.50	875
2400	Metal clad, dlx sliding, double insulated glass, 3'-0" x 3'-0" high	**G** 1 Carp	10	.800		330	35.50		365.50	415
2420	4'-0" x 3'-6" high	**G**	9	.889		400	39		439	500
2440	5'-0" x 4'-0" high	**G**	9	.889		480	39		519	585
2460	6'-0" x 5'-0" high	**G**	8	1		725	44		769	865

08 52 13.40 Bow and Bay Windows, Metal-Clad

		Crew	Daily Output	Labor-Hours	Unit	Material	2012 Bare Costs Labor	Equipment	Total	Total Incl O&P
0010	**BOW AND BAY WINDOWS, METAL-CLAD**									
0100	Metal clad, deluxe, dbl. insul. glass, 8'-0" x 5'-0" high, 4 panels	2 Carp	10	1.600	Ea.	1,675	70.50		1,745.50	1,925
0120	10'-0" x 5'-0" high, 5 panels		8	2		1,800	88		1,888	2,100
0140	10'-0" x 6'-0" high, 5 panels		7	2.286		2,100	101		2,201	2,475
0160	12'-0" x 6'-0" high, 6 panels		6	2.667		2,925	118		3,043	3,400
0400	Double hung, bldrs. model, bay, 8' x 4' high, dbl. insulated glass		10	1.600		1,325	70.50		1,395.50	1,550
0440	Low E glass		10	1.600		1,425	70.50		1,495.50	1,675
0460	9'-0" x 5'-0" high, double insulated glass		6	2.667		1,425	118		1,543	1,750
0480	Low E glass		6	2.667		1,500	118		1,618	1,825
0500	Metal clad, deluxe, dbl. insul. glass, 7'-0" x 4'-0" high		10	1.600		1,275	70.50		1,345.50	1,500
0520	8'-0" x 4'-0" high		8	2		1,300	88		1,388	1,575
0540	8'-0" x 5'-0" high		7	2.286		1,350	101		1,451	1,650
0560	9'-0" x 5'-0" high		6	2.667		1,450	118		1,568	1,750

08 52 16 – Plastic-Clad Wood Windows

08 52 16.10 Bow Window

		Crew	Daily Output	Labor-Hours	Unit	Material	2012 Bare Costs Labor	Equipment	Total	Total Incl O&P
0010	**BOW WINDOW** Including frames, screens, and grilles									
0020	End panels operable									
1000	Bow type, casement, wood, bldrs mdl, 8' x 5' dbl. insltd glass, 4 panel	2 Carp	10	1.600	Ea.	1,500	70.50		1,570.50	1,750
1050	Low E glass		10	1.600		1,325	70.50		1,395.50	1,550
1100	10'-0" x 5'-0", double insulated glass, 6 panels		6	2.667		1,350	118		1,468	1,675
1200	Low E glass, 6 panels		6	2.667		1,450	118		1,568	1,775
1300	Vinyl clad, bldrs model, double insulated glass, 6'-0" x 4'-0", 3 panel		10	1.600		1,025	70.50		1,095.50	1,225
1340	9'-0" x 4'-0", 4 panel		8	2		1,350	88		1,438	1,625
1380	10'-0" x 6'-0", 5 panels		7	2.286		2,250	101		2,351	2,625
1420	12'-0" x 6'-0", 6 panels		6	2.667		2,925	118		3,043	3,400
2000	Bay window, 8' x 5', dbl. insul glass		10	1.600		1,875	70.50		1,945.50	2,175
2050	Low E glass		10	1.600		2,275	70.50		2,345.50	2,600
2100	12'-0" x 6'-0", double insulated glass, 6 panels		6	2.667		2,350	118		2,468	2,750
2200	Low E glass		6	2.667		2,350	118		2,468	2,775
2280	6'-0" x 4'-0"		11	1.455		1,225	64		1,289	1,450
2300	Vinyl clad, premium, double insulated glass, 8'-0" x 5'-0"		10	1.600		1,750	70.50		1,820.50	2,025
2340	10'-0" x 5'-0"		8	2		2,275	88		2,363	2,650
2380	10'-0" x 6'-0"		7	2.286		2,600	101		2,701	3,000
2420	12'-0" x 6'-0"		6	2.667		3,200	118		3,318	3,700
3300	Vinyl clad, premium, double insulated glass, 7'-0" x 4'-6"		10	1.600		1,375	70.50		1,445.50	1,600
3340	8'-0" x 4'-6"		8	2		1,400	88		1,488	1,650

08 52 Wood Windows

08 52 16 – Plastic-Clad Wood Windows

08 52 16.10 Bow Window	Crew	Daily Output	Labor-Hours	Unit	Material	Labor	2012 Bare Costs Equipment	Total	Total Incl O&P
3380	8'-0" x 5'-0"	2 Carp	7	2.286	Ea.	1,450	101		1,551	1,750
3420	9'-0" x 5'-0"	↓	6	2.667	↓	1,500	118		1,618	1,825

08 52 16.15 Awning Window Vinyl-Clad

		Crew	Daily Output	Labor-Hours	Unit	Material	Labor	Equipment	Total	Total Incl O&P
0010	**AWNING WINDOW VINYL-CLAD** Including frames, screens, and grilles									
0240	Vinyl clad, 34" x 22"	1 Carp	10	.800	Ea.	262	35.50		297.50	345
0280	36" x 28"		9	.889		305	39		344	395
0300	36" x 36"		9	.889		340	39		379	435
0340	40" x 22"		10	.800		287	35.50		322.50	370
0360	48" x 28"		8	1		365	44		409	470
0380	60" x 36"	↓	8	1	↓	515	44		559	635

08 52 16.30 Palladian Windows

		Crew	Daily Output	Labor-Hours	Unit	Material	Labor	Equipment	Total	Total Incl O&P
0010	**PALLADIAN WINDOWS**									
0020	Vinyl clad, double insulated glass, including frame and grilles									
0040	3'-2" x 2'-6" high	2 Carp	11	1.455	Ea.	1,250	64		1,314	1,475
0060	3'-2" x 4'-10"		11	1.455		1,750	64		1,814	2,000
0080	3'-2" x 6'-4"		10	1.600		1,700	70.50		1,770.50	1,975
0100	4'-0" x 4'-0"	↓	10	1.600		1,525	70.50		1,595.50	1,775
0120	4'-0" x 5'-4"	3 Carp	10	2.400		1,875	106		1,981	2,225
0140	4'-0" x 6'-0"		9	2.667		1,950	118		2,068	2,325
0160	4'-0" x 7'-4"		9	2.667		2,125	118		2,243	2,500
0180	5'-5" x 4'-10"		9	2.667		2,275	118		2,393	2,675
0200	5'-5" x 6'-10"		9	2.667		2,575	118		2,693	3,025
0220	5'-5" x 7'-9"		9	2.667		2,800	118		2,918	3,250
0240	6'-0" x 7'-11"		8	3		3,475	132		3,607	4,025
0260	8'-0" x 6'-0"	↓	8	3	↓	3,075	132		3,207	3,575

08 52 16.35 Double-Hung Window

			Crew	Daily Output	Labor-Hours	Unit	Material	Labor	Equipment	Total	Total Incl O&P
0010	**DOUBLE-HUNG WINDOW** Including frames, screens, and grilles										
0300	Vinyl clad, premium, double insulated glass, 2'-6" x 3'-0"	G	1 Carp	10	.800	Ea.	300	35.50		335.50	385
0305	2'-6" x 4'-0"	G		10	.800		355	35.50		390.50	445
0400	3'-0" x 3'-6"	G		10	.800		330	35.50		365.50	420
0500	3'-0" x 4'-0"	G		9	.889		390	39		429	490
0600	3'-0" x 4'-6"	G		9	.889		405	39		444	505
0700	3'-0" x 5'-0"	G		8	1		440	44		484	555
0790	3'-4" x 5'-0"	G		8	1		445	44		489	560
0800	3'-6" x 6'-0"	G		8	1		485	44		529	600
0820	4'-0" x 5'-0"	G		7	1.143		550	50.50		600.50	685
0830	4'-0" x 6'-0"	G	↓	7	1.143	↓	685	50.50		735.50	835

08 52 16.40 Transom Windows

		Crew	Daily Output	Labor-Hours	Unit	Material	Labor	Equipment	Total	Total Incl O&P
0010	**TRANSOM WINDOWS**									
1000	Vinyl clad, premium, dbl. insul. glass, 4'-0" x 4'-0"	2 Carp	12	1.333	Ea.	495	59		554	635
1100	4'-0" x 6'-0"		11	1.455		935	64		999	1,125
1200	5'-0" x 6'-0"		10	1.600		1,050	70.50		1,120.50	1,250
1300	6'-0" x 6'-0"	↓	10	1.600	↓	1,050	70.50		1,120.50	1,250

08 52 16.70 Vinyl Clad, Premium, Dbl. Insulated Glass

			Crew	Daily Output	Labor-Hours	Unit	Material	Labor	Equipment	Total	Total Incl O&P
0010	**VINYL CLAD, PREMIUM, DBL. INSULATED GLASS**										
1000	Sliding, 3'-0" x 3'-0"	G	1 Carp	10	.800	Ea.	605	35.50		640.50	720
1050	4'-0" x 3'-6"	G		9	.889		685	39		724	815
1100	5'-0" x 4'-0"	G		9	.889		900	39		939	1,050
1150	6'-0" x 5'-0"	G	↓	8	1	↓	1,125	44		1,169	1,325

08 52 Wood Windows

08 52 50 – Window Accessories

08 52 50.10 Window Grille or Muntin		Crew	Daily Output	Labor-Hours	Unit	Material	2012 Bare Costs Labor	Equipment	Total	Total Incl O&P
0010	**WINDOW GRILLE OR MUNTIN**, snap in type									
0020	Standard pattern interior grilles									
2000	Wood, awning window, glass size 28" x 16" high	1 Carp	30	.267	Ea.	26.50	11.75		38.25	47
2060	44" x 24" high		32	.250		37.50	11.05		48.55	58.50
2100	Casement, glass size, 20" x 36" high		30	.267		30.50	11.75		42.25	51.50
2180	20" x 56" high		32	.250		42	11.05		53.05	63
2200	Double hung, glass size, 16" x 24" high		24	.333	Set	50	14.70		64.70	78
2280	32" x 32" high		34	.235	"	131	10.40		141.40	160
2500	Picture, glass size, 48" x 48" high		30	.267	Ea.	117	11.75		128.75	146
2580	60" x 68" high		28	.286	"	180	12.60		192.60	217
2600	Sliding, glass size, 14" x 36" high		24	.333	Set	34.50	14.70		49.20	60.50
2680	36" x 36" high		22	.364	"	42.50	16.05		58.55	71

08 52 66 – Wood Window Screens

08 52 66.10 Wood Screens

0010	**WOOD SCREENS**									
0020	Over 3 S.F., 3/4" frames	2 Carp	375	.043	S.F.	4.80	1.88		6.68	8.20
0100	1-1/8" frames	"	375	.043	"	8.10	1.88		9.98	11.80

08 52 69 – Wood Storm Windows

08 52 69.10 Storm Windows

0010	**STORM WINDOWS**, aluminum residential									
0300	Basement, mill finish, incl. fiberglass screen									
0320	1'-10" x 1'-0" high	G	2 Carp	30	.533	Ea.	35	23.50		58.50	74.50
0340	2'-9" x 1'-6" high	G	"	30	.533	"	38	23.50		61.50	78
1600	Double-hung, combination, storm & screen									
2000	Average quality, clear anodic coating, 2'-0" x 3'-5" high	G	2 Carp	30	.533	Ea.	85	23.50		108.50	130
2020	2'-6" x 5'-0" high	G		28	.571		113	25		138	163
2040	4'-0" x 6'-0" high	G		25	.640		126	28		154	183
2400	White painted, 2'-0" x 3'-5" high	G		30	.533		85	23.50		108.50	130
2420	2'-6" x 5'-0" high	G		28	.571		90	25		115	138
2440	4'-0" x 6'-0" high	G		25	.640		98	28		126	152
2600	Mill finish, 2'-0" x 3'-5" high	G		30	.533		75	23.50		98.50	119
2620	2'-6" x 5'-0" high	G		28	.571		85	25		110	133
2640	4'-0" x 6'-8" high	G		25	.640		99	28		127	153

08 53 Plastic Windows

08 53 13 – Vinyl Windows

08 53 13.20 Vinyl Single Hung Windows

0010	**VINYL SINGLE HUNG WINDOWS**, insulated glass									
0100	Grids, low E, J fin, ext. jambs, 21" x 53"	G	2 Carp	18	.889	Ea.	198	39		237	279
0110	21" x 57"	G		17	.941		195	41.50		236.50	279
0120	21" x 65"	G		16	1		200	44		244	288
0130	25" x 41"	G		20	.800		188	35.50		223.50	262
0140	25" x 49"	G		18	.889		198	39		237	279
0150	25" x 57"	G		17	.941		201	41.50		242.50	285
0160	25" x 65"	G		16	1		207	44		251	296
0170	29" x 41"	G		18	.889		192	39		231	272
0180	29" x 53"	G		18	.889		202	39		241	283
0190	29" x 57"	G		17	.941		206	41.50		247.50	291
0200	29" x 65"	G		16	1		212	44		256	300

08 53 13 – Vinyl Windows

08 53 13.20 Vinyl Single Hung Windows		Crew	Daily Output	Labor-Hours	Unit	Material	2012 Bare Costs Labor	Equipment	Total	Total Incl O&P
0210	33" x 41"	G 2 Carp	20	.800	Ea.	198	35.50		233.50	273
0220	33" x 53"	G	18	.889		209	39		248	291
0230	33" x 57"	G	17	.941		212	41.50		253.50	297
0240	33" x 65"	G	16	1		218	44		262	310
0250	37" x 41"	G	20	.800		206	35.50		241.50	282
0260	37" x 53"	G	18	.889		218	39		257	300
0270	37" x 57"	G	17	.941		221	41.50		262.50	305
0280	37" x 65"	G	16	1		232	44		276	325

08 53 13.30 Vinyl Double Hung Windows		Crew	Daily Output	Labor-Hours	Unit	Material	2012 Bare Costs Labor	Equipment	Total	Total Incl O&P
0010	**VINYL DOUBLE HUNG WINDOWS**, insulated glass									
0100	Grids, low E, J fin, ext. jambs, 21" x 53"	G 2 Carp	18	.889	Ea.	215	39		254	298
0102	21" x 37"	G	18	.889		197	39		236	278
0104	21" x 41"	G	18	.889		201	39		240	282
0106	21" x 49"	G	18	.889		221	39		260	305
0110	21" x 57"	G	17	.941		218	41.50		259.50	305
0120	21" x 65"	G	16	1		228	44		272	320
0128	25" x 37"	G	20	.800		206	35.50		241.50	282
0130	25" x 41"	G	20	.800		209	35.50		244.50	285
0140	25" x 49"	G	18	.889		213	39		252	295
0145	25" x 53"	G	18	.889		221	39		260	305
0150	25" x 57"	G	17	.941		221	41.50		262.50	305
0160	25" x 65"	G	16	1		231	44		275	320
0162	25" x 69"	G	16	1		238	44		282	330
0164	25" x 77"	G	16	1		248	44		292	340
0168	29" x 37"	G	18	.889		211	39		250	293
0170	29" x 41"	G	18	.889		214	39		253	296
0172	29" x 49"	G	18	.889		221	39		260	305
0180	29" x 53"	G	18	.889		225	39		264	310
0190	29" x 57"	G	17	.941		229	41.50		270.50	315
0200	29" x 65"	G	16	1		236	44		280	330
0202	29" x 69"	G	16	1		243	44		287	335
0205	29" x 77"	G	16	1		254	44		298	345
0208	33" x 37"	G	20	.800		215	35.50		250.50	292
0210	33" x 41"	G	20	.800		219	35.50		254.50	296
0215	33" x 49"	G	20	.800		227	35.50		262.50	305
0220	33" x 53"	G	18	.889		231	39		270	315
0230	33" x 57"	G	17	.941		235	41.50		276.50	325
0240	33" x 65"	G	16	1		239	44		283	330
0242	33" x 69"	G	16	1		245	44		289	340
0246	33" x 77"	G	16	1		251	44		295	345
0250	37" x 41"	G	20	.800		223	35.50		258.50	300
0255	37" x 49"	G	20	.800		231	35.50		266.50	310
0260	37" x 53"	G	18	.889		239	39		278	325
0270	37" x 57"	G	17	.941		243	41.50		284.50	330
0280	37" x 65"	G	16	1		255	44		299	350
0282	37" x 69"	G	16	1		262	44		306	355
0286	37" x 77"	G	16	1		270	44		314	365
0300	Solid vinyl, average quality, double insulated glass, 2'-0" x 3'-0"	G 1 Carp	10	.800		287	35.50		322.50	370
0310	3'-0" x 4'-0"	G	9	.889		203	39		242	284
0320	4'-0" x 4'-6"	G	8	1		305	44		349	405
0330	Premium, double insulated glass, 2'-6" x 3'-0"	G	10	.800		239	35.50		274.50	320
0340	3'-0" x 3'-6"	G	9	.889		263	39		302	350

08 53 Plastic Windows

08 53 13 – Vinyl Windows

08 53 13.30 Vinyl Double Hung Windows		Crew	Daily Output	Labor-Hours	Unit	Material	2012 Bare Costs Labor	Equipment	Total	Total Incl O&P
0350	3'-0" x 4"-0"	G 1 Carp	9	.889	Ea.	288	39		327	375
0360	3'-0" x 4'-6"	G	9	.889		295	39		334	385
0370	3'-0" x 5'-0"	G	8	1		315	44		359	420
0380	3'-6" x 6"-0"	G	8	1		355	44		399	465

08 53 13.40 Vinyl Casement Windows										
0010	**VINYL CASEMENT WINDOWS**, insulated glass									
0015	Grids, low E, J fin, extension jambs, screens									
0100	One lite, 21" x 41"	G 2 Carp	20	.800	Ea.	271	35.50		306.50	355
0110	21" x 47"	G	20	.800		294	35.50		329.50	380
0120	21" x 53"	G	20	.800		315	35.50		350.50	405
0128	24" x 35"	G	19	.842		263	37		300	345
0130	24" x 41"	G	19	.842		283	37		320	365
0140	24" x 47"	G	19	.842		305	37		342	390
0150	24" x 53"	G	19	.842		325	37		362	415
0158	28" x 35"	G	19	.842		278	37		315	360
0160	28" x 41"	G	19	.842		298	37		335	385
0170	28" x 47"	G	19	.842		320	37		357	405
0180	28" x 53"	G	19	.842		350	37		387	440
0184	28" x 59"	G	19	.842		350	37		387	440
0188	Two lites, 33" x 35"	G	18	.889		430	39		469	530
0190	33" x 41"	G	18	.889		455	39		494	565
0200	33" x 47"	G	18	.889		490	39		529	595
0210	33" x 53"	G	18	.889		520	39		559	630
0212	33" x 59"	G	18	.889		550	39		589	665
0215	33" x 72"	G	18	.889		570	39		609	685
0220	41" x 41"	G	18	.889		495	39		534	605
0230	41" x 47"	G	18	.889		530	39		569	640
0240	41" x 53"	G	17	.941		560	41.50		601.50	680
0242	41" x 59"	G	17	.941		590	41.50		631.50	710
0246	41" x 72"	G	17	.941		615	41.50		656.50	740
0250	47" x 41"	G	17	.941		500	41.50		541.50	615
0260	47" x 47"	G	17	.941		530	41.50		571.50	650
0270	47" x 53"	G	17	.941		560	41.50		601.50	685
0272	47" x 59"	G	17	.941		615	41.50		656.50	740
0280	56" x 41"	G	15	1.067		535	47		582	665
0290	56" x 47"	G	15	1.067		560	47		607	695
0300	56" x 53"	G	15	1.067		610	47		657	745
0302	56" x 59"	G	15	1.067		635	47		682	775
0310	56" x 72"	G	15	1.067		690	47		737	835
0340	Solid vinyl, premium, double insulated glass, 2'-0" x 3'-0" high	G 1 Carp	10	.800		259	35.50		294.50	340
0360	2'-0" x 4'-0" high	G	9	.889		289	39		328	380
0380	2'-0" x 5'-0" high	G	8	1		310	44		354	410

08 53 13.50 Vinyl Picture Windows										
0010	**VINYL PICTURE WINDOWS**, insulated glass									
0100	Grids, low E, J fin, ext. jambs, 33" x 47"	2 Carp	12	1.333	Ea.	250	59		309	365
0110	35" x 71"		12	1.333		335	59		394	460
0120	41" x 47"		12	1.333		270	59		329	390
0130	41" x 71"		12	1.333		350	59		409	475
0140	47" x 47"		12	1.333		305	59		364	425
0150	47" x 71"		11	1.455		380	64		444	520
0160	53" x 47"		11	1.455		330	64		394	465
0170	53" x 71"		11	1.455		345	64		409	480

08 53 Plastic Windows

08 53 13 – Vinyl Windows

08 53 13.50 Vinyl Picture Windows

		Crew	Daily Output	Labor-Hours	Unit	Material	2012 Bare Costs Labor	Equipment	Total	Total Incl O&P
0180	59" x 47"	2 Carp	11	1.455	Ea.	370	64		434	505
0190	59" x 71"		11	1.455		380	64		444	520
0200	71" x 47"		10	1.600		390	70.50		460.50	540
0210	71" x 71"	↓	10	1.600	↓	405	70.50		475.50	555

08 54 Composite Windows

08 54 13 – Fiberglass Windows

08 54 13.10 Fiberglass Single Hung Windows

			Crew	Daily Output	Labor-Hours	Unit	Material	2012 Bare Costs Labor	Equipment	Total	Total Incl O&P
0010	**FIBERGLASS SINGLE HUNG WINDOWS**										
0100	Grids, low E, 18" x 24"	G	2 Carp	18	.889	Ea.	335	39		374	430
0110	18" x 40"	G		17	.941		340	41.50		381.50	440
0130	24" x 40"	G		20	.800		360	35.50		395.50	450
0230	36" x 36"	G		17	.941		370	41.50		411.50	470
0250	36" x 48"	G		20	.800		405	35.50		440.50	500
0260	36" x 60"	G		18	.889		445	39		484	550
0280	36" x 72"	G		16	1		470	44		514	585
0290	48" x 40"	G	↓	16	1	↓	470	44		514	585

08 56 Special Function Windows

08 56 63 – Detention Windows

08 56 63.13 Visitor Cubicle Windows

		Crew	Daily Output	Labor-Hours	Unit	Material	2012 Bare Costs Labor	Equipment	Total	Total Incl O&P
0010	**VISITOR CUBICLE WINDOWS**									
4000	Visitor cubicle, vision panel, no intercom	E-4	2	16	Ea.	3,150	800	61	4,011	4,975

08 62 Unit Skylights

08 62 13 – Domed Unit Skylights

08 62 13.20 Skylights

			Crew	Daily Output	Labor-Hours	Unit	Material	2012 Bare Costs Labor	Equipment	Total	Total Incl O&P
0010	**SKYLIGHTS**, Plastic domes, flush or curb mounted										
0300	Nominal size under 10 S.F., double	G	G-3	130	.246	S.F.	29.50	10.75		40.25	49
0400	Single			160	.200		26	8.75		34.75	42
0600	10 S.F. to 20 S.F., double	G		315	.102		25	4.43		29.43	34.50
0700	Single			395	.081		25	3.53		28.53	33
0900	20 S.F. to 30 S.F., double	G		395	.081		23.50	3.53		27.03	31.50
1000	Single			465	.069		19.25	3		22.25	25.50
1200	30 S.F. to 65 S.F., double	G		465	.069		24	3		27	31
1300	Single		↓	610	.052		17.60	2.29		19.89	23
1500	For insulated 4" curbs, double, add						27%				
1600	Single, add					↓	30%				
1800	For integral insulated 9" curbs, double, add						30%				
1900	Single, add						40%				
2120	Ventilating insulated plexiglass dome with										
2130	curb mounting, 36" x 36"	G	G-3	12	2.667	Ea.	460	116		576	685
2150	52" x 52"	G		12	2.667		640	116		756	885
2160	28" x 52"	G		10	3.200		470	140		610	730
2170	36" x 52"	G	↓	10	3.200		520	140		660	785
2180	For electric opening system, add	G				↓	300			300	330
2300	Insulated safety glass with aluminum frame	G	G-3	160	.200	S.F.	90	8.75		98.75	112

08 62 Unit Skylights

08 62 13 – Domed Unit Skylights

08 62 13.20 Skylights

	08 62 13.20 Skylights		Crew	Daily Output	Labor-Hours	Unit	Material	2012 Bare Costs Labor	Equipment	Total	Total Incl O&P
2400	Sandwich panels, fiberglass, for walls, 1-9/16" thick, to 250 S.F.	G	G-3	200	.160	S.F.	17	7		24	29.50
2700	As above, but for roofs, 2-3/4" thick, to 250 S.F.	G		295	.108		24.50	4.73		29.23	34.50
2800	250 S.F. and up	G		330	.097		20	4.23		24.23	28.50

08 63 Metal-Framed Skylights

08 63 23 – Ridge Metal-Framed Skylights

08 63 23.10 Prefabricated

			Crew	Daily Output	Labor-Hours	Unit	Material	2012 Bare Costs Labor	Equipment	Total	Total Incl O&P
0010	**PREFABRICATED** glass block with metal frame										
0020	Minimum	G	G-3	265	.121	S.F.	56	5.25		61.25	69.50
0100	Maximum	G	"	160	.200	"	108	8.75		116.75	132

08 71 Door Hardware

08 71 13 – Automatic Door Operators

08 71 13.10 Automatic Openers Commercial

			Crew	Daily Output	Labor-Hours	Unit	Material	2012 Bare Costs Labor	Equipment	Total	Total Incl O&P
0010	**AUTOMATIC OPENERS COMMERCIAL**										
0020	Pneumatic, incl opener, motion sens, control box, tubing, compressor										
0050	For single swing door, per opening		2 Skwk	.80	20	Ea.	4,375	910		5,285	6,200
0100	Pair, per opening			.50	32	Opng.	7,100	1,450		8,550	10,100
1000	For single sliding door, per opening			.60	26.667		4,800	1,200		6,000	7,150
1300	Bi-parting pair			.50	32		7,200	1,450		8,650	10,200
1420	Electronic door opener incl motion sens, 12 V control box, motor										
1450	For single swing door, per opening		2 Skwk	.80	20	Opng.	3,550	910		4,460	5,300
1500	Pair, per opening			.50	32		5,800	1,450		7,250	8,625
1600	For single sliding door, per opening			.60	26.667		3,850	1,200		5,050	6,125
1700	Bi-parting pair			.50	32		5,900	1,450		7,350	8,750
1750	Handicap actuator buttons, 2, including 12 V DC wiring, add		1 Carp	1.50	5.333	Pr.	450	235		685	855

08 71 13.20 Automatic Openers Industrial

			Crew	Daily Output	Labor-Hours	Unit	Material	2012 Bare Costs Labor	Equipment	Total	Total Incl O&P
0010	**AUTOMATIC OPENERS INDUSTRIAL**										
0015	Sliding doors up to 6' wide		2 Skwk	.60	26.667	Opng.	5,675	1,200		6,875	8,100
0200	To 12' wide		"	.40	40	"	6,800	1,825		8,625	10,300
0400	Over 12' wide, add per L.F. of excess					L.F.	780			780	860
1000	Swing doors, to 5' wide		2 Skwk	.80	20	Ea.	3,350	910		4,260	5,100
1860	Add for controls, wall pushbutton, 3 button			4	4		225	182		407	530
1870	Ceiling pull cord			4.30	3.721		185	169		354	465

08 71 20 – Hardware

08 71 20.10 Bolts, Flush

			Crew	Daily Output	Labor-Hours	Unit	Material	2012 Bare Costs Labor	Equipment	Total	Total Incl O&P
0010	**BOLTS, FLUSH**										
0020	Standard, concealed		1 Carp	7	1.143	Ea.	23	50.50		73.50	103
0800	Automatic fire exit		"	5	1.600		299	70.50		369.50	440
1600	For electric release, add		1 Elec	3	2.667		125	137		262	345
3000	Barrel, brass, 2" long		1 Carp	40	.200		7.70	8.80		16.50	22
3020	4" long			40	.200		8.50	8.80		17.30	23
3060	6" long			40	.200		15	8.80		23.80	30

08 71 20.15 Hardware

			Crew	Daily Output	Labor-Hours	Unit	Material	2012 Bare Costs Labor	Equipment	Total	Total Incl O&P
0009	**HARDWARE**										
0010	Average percentage for hardware, total job cost										
0025	Minimum					Job					2%
0050	Maximum										500%

08 71 20 – Hardware

08 71 20.15 Hardware		Crew	Daily Output	Labor-Hours	Unit	Material	2012 Bare Costs Labor	Equipment	Total	Total Incl O&P
0500	Total hardware for building, average distribution				Job	85%	15%			
1000	Door hardware, apartment, interior	1 Carp	4	2	Door	515	88		603	705
1300	Average, door hardware, motel/hotel interior, with access card		4	2		635	88		723	830
1500	Hospital bedroom, minimum		4	2		605	88		693	800
2000	Maximum		3	2.667		720	118		838	970
2100	Pocket door		6	1.333	Ea.	134	59		193	238
2250	School, single exterior, incl. lever, incl. panic device		3	2.667	Door	1,125	118		1,243	1,400
2500	Single interior, regular use, lever included		3	2.667		610	118		728	850
2550	Average, door hdwe, school, classroom, ANSI F84, lever handle		3	2.667		775	118		893	1,025
2600	Average, door hdwe, school, classroom, ANSI F88, incl. lever		3	2.667		820	118		938	1,075
2850	Stairway, single interior		3	2.667		535	118		653	770
3100	Double exterior, with panic device		2	4	Pr.	2,075	176		2,251	2,575
6020	Add for fire alarm door holder, electro-magnetic	1 Elec	4	2	Ea.	99.50	103		202.50	264

08 71 20.20 Door Protectors

0010	**DOOR PROTECTORS**									
0020	1-3/4" x 3/4" U channel	2 Carp	80	.200	L.F.	25	8.80		33.80	41
0021	1-3/4" x 1-1/4" U channel		80	.200	"	15	8.80		23.80	30
1000	Tear drop, spring-stl, 8" high x 19" long		15	1.067	Ea.	95	47		142	178
1010	8" high x 32" long		15	1.067		116	47		163	201
1100	Tear drop, stainless stl, 8" high x 19" long		15	1.067		245	47		292	345
1200	8" high x 32" long		15	1.067		305	47		352	410

08 71 20.30 Door Closers

0010	**DOOR CLOSERS**									
0020	Adjustable backcheck, 3 way mount, all sizes, regular arm	1 Carp	6	1.333	Ea.	184	59		243	293
0040	Hold open arm		6	1.333		234	59		293	350
0100	Fusible link		6.50	1.231		158	54.50		212.50	258
0200	Non sized, regular arm		6	1.333		168	59		227	276
0240	Hold open arm		6	1.333		204	59		263	315
0400	4 way mount, non sized, regular arm		6	1.333		224	59		283	335
0440	Hold open arm		6	1.333		234	59		293	350
2000	Backcheck and adjustable power, hinge face mount									
2010	All sizes, regular arm	1 Carp	6.50	1.231	Ea.	214	54.50		268.50	320
2040	Hold open arm		6.50	1.231		249	54.50		303.50	360
2400	Top jamb mount, all sizes, regular arm		6	1.333		204	59		263	315
2440	Hold open arm		6	1.333		214	59		273	325
2800	Top face mount, all sizes, regular arm		6.50	1.231		198	54.50		252.50	300
2840	Hold open arm		6.50	1.231		219	54.50		273.50	325
4000	Backcheck, overhead concealed, all sizes, regular arm		5.50	1.455		220	64		284	340
4040	Concealed arm		5	1.600		231	70.50		301.50	365
4400	Compact overhead, concealed, all sizes, regular arm		5.50	1.455		350	64		414	485
4440	Concealed arm		5	1.600		390	70.50		460.50	540
4800	Concealed in door, all sizes, regular arm		5.50	1.455		360	64		424	495
4840	Concealed arm		5	1.600		380	70.50		450.50	530
4900	Floor concealed, all sizes, single acting		2.20	3.636		385	160		545	670
4940	Double acting		2.20	3.636		325	160		485	605
5000	For cast aluminum cylinder, deduct					35			35	38.50
5040	For delayed action, add					46			46	50.50
5080	For fusible link arm, add					35			35	38.50
5120	For shock absorbing arm, add					50			50	55
5160	For spring power adjustment, add					40			40	44
6000	Closer-holder, hinge face mount, all sizes, exposed arm	1 Carp	6.50	1.231		168	54.50		222.50	269
7000	Electronic closer-holder, hinge facemount, concealed arm		5	1.600		295	70.50		365.50	435

08 71 Door Hardware

08 71 20 – Hardware

08 71 20.30 Door Closers

08 71 20.30 Door Closers	Crew	Daily Output	Labor-Hours	Unit	Material	2012 Bare Costs Labor	Equipment	Total	Total Incl O&P
7400 With built-in detector	1 Carp	5	1.600	Ea.	730	70.50		800.50	915
8000 Surface mounted, stand. duty, parallel arm, primed, traditional		6	1.333		197	59		256	310
8030 Light duty		6	1.333		127	59		186	231
8050 Heavy duty		6	1.333		227	59		286	340
8100 Standard duty, parallel arm, modern		6	1.333		230	59		289	345
8150 Heavy duty		6	1.333		247	59		306	360

08 71 20.31 Door Closers

08 71 20.31 Door Closers	Crew	Daily Output	Labor-Hours	Unit	Material	2012 Bare Costs Labor	Equipment	Total	Total Incl O&P
0010 **DOOR CLOSERS**									
0015 Door closer, rack and pinion	1 Carp	6.50	1.231	Ea.	172	54.50		226.50	273
1520 Door, single acting, standard arm		1	8		203	355		558	770
1522 Hold open arm		1	8		221	355		576	790
1526 Frame, single acting, standard arm		1	8		340	355		695	920
1530 Hold open arm		1	8		360	355		715	940
1534 Double acting, standard arm		1	8		460	355		815	1,050
1536 Hold open arm		1	8		470	355		825	1,050
1554 Floor, center hung, single acting, bottom arm		1	8		325	355		680	905
1558 Double acting		1	8		340	355		695	915
1560 Offset hung, single acting, bottom arm		1	8		380	355		735	960
6500 Electro magnetic closer/holder									
6510 Single point, no detector	1 Carp	1	8	Ea.	360	355		715	940
6515 Including detector		1	8		685	355		1,040	1,300
6520 Multi-point, no detector		1	8		785	355		1,140	1,400
6524 Including detector		1	8		795	355		1,150	1,425
6550 Electric automatic operators									
6555 Operator	1 Carp	1	8	Ea.	1,550	355		1,905	2,250
6570 Wall plate actuator		1	8		199	355		554	765
8010 Light duty, regular arm		1	8		113	355		468	670
8032 Parallel arm		1	8		117	355		472	675
8034 Hold open arm		1	8		123	355		478	680
8036 Fusible link arm		1	8		179	355		534	740
8040 Medium duty, regular arm		1	8		144	355		499	705
8042 Extra duty parallel arm		1	8		153	355		508	715
8044 Hold open arm		1	8		160	355		515	720
8046 Positive stop arm		1	8		173	355		528	735
8052 Heavy duty, regular arm		1	8		175	355		530	735
8054 Top jamb mount		1	8		175	355		530	735
8056 Extra duty parallel arm		1	8		175	355		530	735
8058 Hold open arm		1	8		191	355		546	755
8060 Positive stop arm		1	8		193	355		548	755
8062 Fusible link arm		1	8		222	355		577	790
8080 Universal heavy duty, regular arm		1	8		190	355		545	755
8084 Parallel arm		1	8		190	355		545	755
8088 Extra duty, parallel arm		1	8		195	355		550	760
8090 Hold open arm		1	8		278	355		633	850
8094 Positive stop arm		1	8		282	355		637	855
8098 For delayed action add		1	8		31	355		386	580

08 71 20.35 Panic Devices

08 71 20.35 Panic Devices	Crew	Daily Output	Labor-Hours	Unit	Material	2012 Bare Costs Labor	Equipment	Total	Total Incl O&P
0010 **PANIC DEVICES**									
0015 For rim locks, single door exit only	1 Carp	6	1.333	Ea.	530	59		589	675
0020 Outside key and pull		5	1.600		575	70.50		645.50	740
0200 Bar and vertical rod, exit only		5	1.600		735	70.50		805.50	920
0210 Outside key and pull		4	2		810	88		898	1,025

08 71 Door Hardware

08 71 20 - Hardware

08 71 20.35 Panic Devices

		Crew	Daily Output	Labor-Hours	Unit	Material	2012 Bare Costs Labor	Equipment	Total	Total Incl O&P
0400	Bar and concealed rod	1 Carp	4	2	Ea.	600	88		688	795
0600	Touch bar, exit only		6	1.333		510	59		569	650
0610	Outside key and pull		5	1.600		605	70.50		675.50	775
0700	Touch bar and vertical rod, exit only		5	1.600		670	70.50		740.50	845
0710	Outside key and pull		4	2		780	88		868	995
1000	Mortise, bar, exit only		4	2		550	88		638	740
1600	Touch bar, exit only		4	2		625	88		713	820
2000	Narrow stile, rim mounted, bar, exit only		6	1.333		655	59		714	810
2010	Outside key and pull		5	1.600		700	70.50		770.50	880
2200	Bar and vertical rod, exit only		5	1.600		680	70.50		750.50	860
2210	Outside key and pull		4	2		680	88		768	885
2400	Bar and concealed rod, exit only		3	2.667		765	118		883	1,025
3000	Mortise, bar, exit only		4	2		610	88		698	805
3600	Touch bar, exit only		4	2		880	88		968	1,100

08 71 20.40 Lockset

		Crew	Daily Output	Labor-Hours	Unit	Material	2012 Bare Costs Labor	Equipment	Total	Total Incl O&P
0010	**LOCKSET**, Standard duty									
0020	Non-keyed, passage, w/sect.trim	1 Carp	12	.667	Ea.	57.50	29.50		87	109
0100	Privacy		12	.667		62	29.50		91.50	113
0400	Keyed, single cylinder function		10	.800		92.50	35.50		128	157
0420	Hotel (see also Section 08 71 20.15)		8	1		135	44		179	217
0500	Lever handled, keyed, single cylinder function		10	.800		159	35.50		194.50	230
1000	Heavy duty with sectional trim, non-keyed, passages		12	.667		145	29.50		174.50	205
1100	Privacy		12	.667		185	29.50		214.50	249
1400	Keyed, single cylinder function		10	.800		220	35.50		255.50	297
1420	Hotel		8	1		325	44		369	430
1600	Communicating		10	.800		250	35.50		285.50	330
1690	For re-core cylinder, add					45			45	49.50
1700	Residential, interior door, minimum	1 Carp	16	.500		30	22		52	67
1720	Maximum		8	1		80	44		124	156
1800	Exterior, minimum		14	.571		45	25		70	88.50
1810	Average		8	1		72	44		116	148
1820	Maximum		8	1		180	44		224	266
3800	Cipher lockset w/key pad (security item)		13	.615		845	27		872	970
3900	Cipher lockset, with dial for swinging doors (security item)		13	.615		1,800	27		1,827	2,050
3920	with dial for swinging doors & drill resistant plate (security item)		12	.667		2,200	29.50		2,229.50	2,475
3950	Cipher lockset with dial for safe/vault door (security item)		12	.667		1,500	29.50		1,529.50	1,700
3980	Keyless, pushbutton type									
4000	Residential/light commercial, deadbolt, standard	1 Carp	9	.889	Ea.	150	39		189	226
4010	Heavy duty		9	.889		180	39		219	259
4020	Industrial, heavy duty, with deadbolt		9	.889		325	39		364	420
4030	Key override		9	.889		365	39		404	460
4040	Lever activated handle		9	.889		385	39		424	485
4050	Key override		9	.889		400	39		439	500
4060	Double sided pushbutton type		8	1		755	44		799	900
4070	Key override		8	1		790	44		834	940

08 71 20.41 Dead Locks

		Crew	Daily Output	Labor-Hours	Unit	Material	2012 Bare Costs Labor	Equipment	Total	Total Incl O&P
0010	**DEAD LOCKS**									
0011	Mortise heavy duty outside key (security item)	1 Carp	9	.889	Ea.	275	39		314	365
0020	Double cylinder		9	.889		270	39		309	360
0100	Medium duty, outside key		10	.800		150	35.50		185.50	220
0110	Double cylinder		10	.800		165	35.50		200.50	237
1000	Tubular, standard duty, outside key		10	.800		66	35.50		101.50	127

08 71 Door Hardware

08 71 20 – Hardware

08 71 20.41 Dead Locks	Crew	Daily Output	Labor-Hours	Unit	Material	2012 Bare Costs Labor	2012 Bare Costs Equipment	Total	Total Incl O&P
1010	Double cylinder	1 Carp	10	.800	Ea.	84	35.50		119.50	147
1200	Night latch, outside key	↓	10	.800	↓	83	35.50		118.50	146

08 71 20.42 Mortise Locksets

		Crew	Daily Output	Labor-Hours	Unit	Material	Labor	Equipment	Total	Total Incl O&P
0010	**MORTISE LOCKSETS**, Comm., wrought knobs & full escutcheon trim									
0015	Assumes mortise is cut									
0020	Non-keyed, passage, minimum	1 Carp	9	.889	Ea.	194	39		233	274
0030	Maximum		8	1		300	44		344	400
0040	Privacy, minimum		9	.889		200	39		239	281
0050	Maximum		8	1		325	44		369	430
0100	Keyed, office/entrance/apartment, minimum		8	1		230	44		274	320
0110	Maximum		7	1.143		385	50.50		435.50	505
0120	Single cylinder, typical, minimum		8	1		198	44		242	286
0130	Maximum		7	1.143		360	50.50		410.50	475
0200	Hotel, minimum (see also Section 08 71 20.15)		7	1.143		250	50.50		300.50	355
0210	Maximum (see also Section 08 71 20.15)		6	1.333		400	59		459	530
0300	Communication, double cylinder, minimum		8	1		225	44		269	315
0310	Maximum		7	1.143		300	50.50		350.50	410
1000	Wrought knobs and sectional trim, non-keyed, passage, minimum		10	.800		130	35.50		165.50	198
1010	Maximum		9	.889		245	39		284	330
1040	Privacy, minimum		10	.800		145	35.50		180.50	215
1050	Maximum		9	.889		265	39		304	355
1100	Keyed, entrance, office/apartment, minimum		9	.889		220	39		259	305
1110	Maximum		8	1		315	44		359	420
1120	Single cylinder, typical, minimum		9	.889		225	39		264	310
1130	Maximum	↓	8	1	↓	305	44		349	405
2000	Cast knobs and full escutcheon trim									
2010	Non-keyed, passage, minimum	1 Carp	9	.889	Ea.	275	39		314	365
2020	Maximum		8	1		425	44		469	540
2040	Privacy, minimum		9	.889		320	39		359	410
2050	Maximum		8	1		450	44		494	565
2120	Keyed, single cylinder, typical, minimum		8	1		330	44		374	435
2130	Maximum		7	1.143		500	50.50		550.50	630
3000	Cast knob and sectional trim, non-keyed, passage, minimum		10	.800		210	35.50		245.50	286
3010	Maximum		10	.800		400	35.50		435.50	495
3040	Privacy, minimum		10	.800		225	35.50		260.50	305
3050	Maximum		10	.800		400	35.50		435.50	495
3100	Keyed, office/entrance/apartment, minimum		9	.889		255	39		294	340
3110	Maximum		9	.889		420	39		459	520
3120	Single cylinder, typical, minimum		9	.889		260	39		299	345
3130	Maximum	↓	9	.889		510	39		549	620
3190	For re-core cylinder, add				↓	45			45	49.50

08 71 20.45 Peepholes

		Crew	Daily Output	Labor-Hours	Unit	Material	Labor	Equipment	Total	Total Incl O&P
0010	**PEEPHOLES**									
2010	Peephole	1 Carp	32	.250	Ea.	15.80	11.05		26.85	34.50
2020	Peephole, wide view	"	32	.250	"	17	11.05		28.05	35.50

08 71 20.50 Door Stops

		Crew	Daily Output	Labor-Hours	Unit	Material	Labor	Equipment	Total	Total Incl O&P
0010	**DOOR STOPS**									
0020	Holder & bumper, floor or wall	1 Carp	32	.250	Ea.	35.50	11.05		46.55	56
1300	Wall bumper, 4" diameter, with rubber pad, aluminum		32	.250		10.95	11.05		22	29
1600	Door bumper, floor type, aluminum		32	.250		7.10	11.05		18.15	25
1900	Plunger type, door mounted	↓	32	.250	↓	27	11.05		38.05	46.50

08 71 Door Hardware

08 71 20 – Hardware

08 71 20.55 Push-Pull Plates

		Crew	Daily Output	Labor-Hours	Unit	Material	2012 Bare Costs Labor	2012 Bare Costs Equipment	Total	Total Incl O&P
0010	**PUSH-PULL PLATES**									
0100	Push plate, .050 thick, 4" x 16", aluminum	1 Carp	12	.667	Ea.	10.20	29.50		39.70	56.50
0500	Bronze		12	.667		21.50	29.50		51	68.50
1500	Pull handle and push bar, aluminum		11	.727		131	32		163	194
2000	Bronze		10	.800		169	35.50		204.50	240
3000	Push plate both sides, aluminum		14	.571		19.95	25		44.95	61
3500	Bronze		13	.615		44	27		71	90.50
4000	Door pull, designer style, cast aluminum, minimum		12	.667		71.50	29.50		101	124
5000	Maximum		8	1		370	44		414	475
6000	Cast bronze, minimum		12	.667		87	29.50		116.50	141
7000	Maximum		8	1		405	44		449	515
8000	Walnut, minimum		12	.667		67	29.50		96.50	119
9000	Maximum		8	1		370	44		414	475

08 71 20.60 Entrance Locks

		Crew	Daily Output	Labor-Hours	Unit	Material	Labor	Equipment	Total	Total Incl O&P
0010	**ENTRANCE LOCKS**									
0015	Cylinder, grip handle deadlocking latch	1 Carp	9	.889	Ea.	132	39		171	206
0020	Deadbolt		8	1		160	44		204	244
0100	Push and pull plate, dead bolt		8	1		150	44		194	233
0900	For handicapped lever, add					170			170	187

08 71 20.65 Thresholds

		Crew	Daily Output	Labor-Hours	Unit	Material	Labor	Equipment	Total	Total Incl O&P
0010	**THRESHOLDS**									
0011	Threshold 3' long saddles aluminum	1 Carp	48	.167	L.F.	5.60	7.35		12.95	17.50
0100	Aluminum, 8" wide, 1/2" thick		12	.667	Ea.	41.50	29.50		71	90.50
0500	Bronze		60	.133	L.F.	42	5.90		47.90	55
0600	Bronze, panic threshold, 5" wide, 1/2" thick		12	.667	Ea.	71.50	29.50		101	124
0700	Rubber, 1/2" thick, 5-1/2" wide		20	.400		42	17.65		59.65	73
0800	2-3/4" wide		20	.400		30.50	17.65		48.15	60.50
1950	ADA Compliant Thresholds									
2300	Threshold, aluminum 4" wide x 36" long	1 Carp	12	.667		33	29.50		62.50	81.50
2310	4" wide x 48" long		12	.667		41	29.50		70.50	90
2360	6" wide x 36" long		12	.667		53	29.50		82.50	104
2370	6" wide x 48" long		12	.667		68	29.50		97.50	120

08 71 20.70 Floor Checks

		Crew	Daily Output	Labor-Hours	Unit	Material	Labor	Equipment	Total	Total Incl O&P
0010	**FLOOR CHECKS**									
0020	For over 3' wide doors single acting	1 Carp	2.50	3.200	Ea.	520	141		661	785
0500	Double acting	"	2.50	3.200	"	670	141		811	950

08 71 20.75 Door Hardware Accessories

		Crew	Daily Output	Labor-Hours	Unit	Material	Labor	Equipment	Total	Total Incl O&P
0010	**DOOR HARDWARE ACCESSORIES**									
0050	Door closing coordinator, 36" (for paired openings up to 56")	1 Carp	8	1	Ea.	94	44		138	171
0060	48" (for paired openings up to 84")		8	1		101	44		145	179
0070	56" (for paired openings up to 96")		8	1		111	44		155	190

08 71 20.80 Hasps

		Crew	Daily Output	Labor-Hours	Unit	Material	Labor	Equipment	Total	Total Incl O&P
0010	**HASPS**, steel assembly									
0015	3"	1 Carp	26	.308	Ea.	4.77	13.55		18.32	26.50
0020	4-1/2"		13	.615		6.55	27		33.55	49
0040	6"		12.50	.640		9.30	28		37.30	54

08 71 20.90 Hinges

		Crew	Daily Output	Labor-Hours	Unit	Material	Labor	Equipment	Total	Total Incl O&P
0010	**HINGES** R087120-10									
0012	Full mortise, avg. freq., steel base, USP, 4-1/2" x 4-1/2"				Pr.	27.50			27.50	30
0100	5" x 5", USP					43.50			43.50	48
0200	6" x 6", USP					91.50			91.50	101

08 71 20.90 Hinges

		Crew	Daily Output	Labor-Hours	Unit	Material	2012 Bare Costs Labor	Equipment	Total	Total Incl O&P
0400	Brass base, 4-1/2" x 4-1/2", US10				Pr.	54.50			54.50	60
0500	5" x 5", US10					78			78	85.50
0600	6" x 6", US10					132			132	145
0800	Stainless steel base, 4-1/2" x 4-1/2", US32					82			82	90
0900	For non removable pin, add (security item)				Ea.	4.89			4.89	5.40
0910	For floating pin, driven tips, add					3.30			3.30	3.63
0930	For hospital type tip on pin, add					14.35			14.35	15.80
0940	For steeple type tip on pin, add					13.25			13.25	14.60
0950	Full mortise, high frequency, steel base, 3-1/2" x 3-1/2", US26D				Pr.	29.50			29.50	32.50
1000	4-1/2" x 4-1/2", USP					63			63	69.50
1100	5" x 5", USP					59.50			59.50	65.50
1200	6" x 6", USP					144			144	159
1400	Brass base, 3-1/2" x 3-1/2", US4					52			52	57.50
1430	4-1/2" x 4-1/2", US10					85.50			85.50	94
1500	5" x 5", US10					125			125	138
1600	6" x 6", US10					180			180	198
1800	Stainless steel base, 4-1/2" x 4-1/2", US32					127			127	140
1810	5" x 4-1/2", US32					182			182	200
1930	For hospital type tip on pin, add				Ea.	13.15			13.15	14.45
1950	Full mortise, low frequency, steel base, 3-1/2" x 3-1/2", US26D				Pr.	12.10			12.10	13.30
2000	4-1/2" x 4-1/2", USP					13.05			13.05	14.35
2100	5" x 5", USP					32			32	35.50
2200	6" x 6", USP					64.50			64.50	71
2300	4-1/2" x 4-1/2", US3					18.85			18.85	20.50
2310	5" x 5", US3					45			45	49.50
2400	Brass bass, 4-1/2" x 4-1/2", US10					45.50			45.50	50
2500	5" x 5", US10					68.50			68.50	75
2800	Stainless steel base, 4-1/2" x 4-1/2", US32					77			77	84.50

08 71 20.91 Special Hinges

		Crew	Daily Output	Labor-Hours	Unit	Material	2012 Bare Costs Labor	Equipment	Total	Total Incl O&P
0010	**SPECIAL HINGES**									
0015	Paumelle, high frequency									
0020	Steel base, 6" x 4-1/2", US10				Pr.	145			145	160
0100	Brass base, 5" x 4-1/2", US10				Ea.	208			208	229
0200	Paumelle, average frequency, steel base, 4-1/2" x 3-1/2", US10				Pr.	98.50			98.50	108
0400	Olive knuckle, low frequency, brass base, 6" x 4-1/2", US10				Ea.	152			152	167
1000	Electric hinge with concealed conductor, average frequency									
1010	Steel base, 4-1/2" x 4-1/2", US26D				Pr.	315			315	345
1100	Bronze base, 4-1/2" x 4-1/2", US26D				"	320			320	350
1200	Electric hinge with concealed conductor, high frequency									
1210	Steel base, 4-1/2" x 4-1/2", US26D				Pr.	241			241	265
1600	Double weight, 800 lb., steel base, removable pin, 5" x 6", USP					132			132	145
1700	Steel base-welded pin, 5" x 6", USP					166			166	183
1800	Triple weight, 2000 lb., steel base, welded pin, 5" x 6", USP					151			151	166
2000	Pivot reinf., high frequency, steel base, 7-3/4" door plate, USP					189			189	208
2200	Bronze base, 7-3/4" door plate, US10					232			232	255
3000	Swing clear, full mortise, full or half surface, high frequency,									
3010	Steel base, 5" high, USP				Pr.	159			159	175
3200	Swing clear, full mortise, average frequency									
3210	Steel base, 4-1/2" high, USP				Pr.	127			127	140
4000	Wide throw, average frequency, steel base, 4-1/2" x 6", USP					97			97	107
4200	High frequency, steel base, 4-1/2" x 6", USP					148			148	163
4600	Spring hinge, single acting, 6" flange, steel				Ea.	54.50			54.50	60

08 71 Door Hardware

08 71 20 – Hardware

08 71 20.91 Special Hinges	Crew	Daily Output	Labor-Hours	Unit	Material	2012 Bare Costs Labor	Equipment	Total	Total Incl O&P
4700	Brass				Ea.	96			96	105
4900	Double acting, 6" flange, steel				↓	97			97	107
4950	Brass				↓	159			159	175
8000	Continuous hinges									
8010	Steel, piano, 2" x 72"	1 Carp	20	.400	Ea.	22	17.65		39.65	51
8020	Brass, piano, 1-1/16" x 30"		30	.267		8	11.75		19.75	27
8030	Acrylic, piano, 1-3/4" x 12"		40	.200		15	8.80		23.80	30
9000	Continuous hinge, steel, full mortise, heavy duty, 96 inch	↓	2	4	↓	565	176		741	895

08 71 20.95 Kick Plates	Crew	Daily Output	Labor-Hours	Unit	Material	2012 Bare Costs Labor	Equipment	Total	Total Incl O&P
0010	**KICK PLATES**									
0020	Stainless steel, .050, 16 ga, 8" x 28", US32	1 Carp	15	.533	Ea.	35	23.50		58.50	74.50
0030	8" x 30"		15	.533		36	23.50		59.50	75.50
0040	8" x 34"		15	.533		40	23.50		63.50	80
0050	10" x 28"		15	.533		67	23.50		90.50	110
0060	10" x 30"		15	.533		72	23.50		95.50	115
0070	10" x 34"		15	.533		81	23.50		104.50	125
0080	Mop/Kick, 4" x 28"		15	.533		30	23.50		53.50	69
0090	4" x 30"		15	.533		32	23.50		55.50	71
0100	4" x 34"		15	.533		36	23.50		59.50	75.50
0110	6" x 28"		15	.533		38	23.50		61.50	78
0120	6" x 30"		15	.533		41	23.50		64.50	81
0130	6" x 34"		15	.533		47	23.50		70.50	87.50
0500	Bronze, .050", 8" x 28"		15	.533		48.50	23.50		72	89
0510	8" x 30"		15	.533		63	23.50		86.50	106
0520	8" x 34"		15	.533		71	23.50		94.50	114
0530	10" x 28"		15	.533		73	23.50		96.50	117
0540	10" x 30"		15	.533		78	23.50		101.50	122
0550	10" x 34"		15	.533		89	23.50		112.50	134
0560	Mop/Kick, 4" x 28"		15	.533		31	23.50		54.50	70
0570	4" x 30"		15	.533		34	23.50		57.50	73.50
0580	4" x 34"		15	.533		35	23.50		58.50	74.50
0590	6" x 28"		15	.533		44	23.50		67.50	84.50
0600	6" x 30"		15	.533		50	23.50		73.50	91
0610	6" x 34"		15	.533		54	23.50		77.50	95.50
1000	Acrylic, .125", 8" x 26"		15	.533		28	23.50		51.50	67
1010	8" x 36"		15	.533		38	23.50		61.50	78
1020	8" x 42"		15	.533		45	23.50		68.50	85.50
1030	10" x 26"		15	.533		35	23.50		58.50	74.50
1040	10" x 36"		15	.533		48	23.50		71.50	89
1050	10" x 42"		15	.533		68.50	23.50		92	112
1060	Mop/Kick, 4" x 26"		15	.533		17	23.50		40.50	54.50
1070	4" x 36"		15	.533		24	23.50		47.50	62.50
1080	4" x 42"		15	.533		27	23.50		50.50	65.50
1090	6" x 26"		15	.533		23	23.50		46.50	61.50
1100	6" x 36"		15	.533		34	23.50		57.50	73.50
1110	6" x 42"		15	.533		39	23.50		62.50	79
1220	Brass, .050", 8" x 26"		15	.533		56	23.50		79.50	97.50
1230	8" x 36"		15	.533		75	23.50		98.50	119
1240	8" x 42"		15	.533		86	23.50		109.50	131
1250	10" x 26"		15	.533		71	23.50		94.50	114
1260	10" x 36"		15	.533		91	23.50		114.50	136
1270	10" x 42"	↓	15	.533	↓	105	23.50		128.50	152

08 71 Door Hardware

08 71 20 – Hardware

08 71 20.95 Kick Plates

		Crew	Daily Output	Labor-Hours	Unit	Material	2012 Bare Costs Labor	Equipment	Total	Total Incl O&P
1320	Mop/Kick, 4" x 26"	1 Carp	15	.533	Ea.	28	23.50		51.50	67
1330	4" x 36"		15	.533		39	23.50		62.50	79
1340	4" x 42"		15	.533		44	23.50		67.50	84.50
1350	6" x 26"		15	.533		38	23.50		61.50	78
1360	6" x 36"		15	.533		48	23.50		71.50	89
1370	6" x 42"		15	.533		55	23.50		78.50	96.50
1800	Aluminum, .050", 8" x 26"		15	.533		25	23.50		48.50	63.50
1810	8" x 36"		15	.533		30.50	23.50		54	69.50
1820	8" x 42"		15	.533		35.50	23.50		59	75
1830	10" x 26"		15	.533		27.50	23.50		51	66.50
1840	10" x 36"		15	.533		38	23.50		61.50	78
1850	10" x 42"		15	.533		44.50	23.50		68	85
1860	Mop/Kick, 4" x 26"		15	.533		11	23.50		34.50	48
1870	4" x 36"		15	.533		15.50	23.50		39	53
1880	4" x 42"		15	.533		17	23.50		40.50	54.50
1890	6" x 26"		15	.533		16.50	23.50		40	54
1900	6" x 36"		15	.533		23	23.50		46.50	61.50
1910	6" x 42"		15	.533		27	23.50		50.50	65.50

08 71 21 – Astragals

08 71 21.10 Exterior Mouldings, Astragals

		Crew	Daily Output	Labor-Hours	Unit	Material	2012 Bare Costs Labor	Equipment	Total	Total Incl O&P
0010	**EXTERIOR MOULDINGS, ASTRAGALS**									
0400	One piece, overlapping cadmium plated steel, flat, 3/16" x 2"	1 Carp	90	.089	L.F.	3.78	3.92		7.70	10.20
0600	Prime coated steel, flat, 1/8" x 3"		90	.089		4.95	3.92		8.87	11.50
0800	Stainless steel, flat, 3/32" x 1-5/8"		90	.089		18.80	3.92		22.72	26.50
1000	Aluminum, flat, 1/8" x 2"		90	.089		3.78	3.92		7.70	10.20
1200	Nail on, "T" extrusion		120	.067		1.05	2.94		3.99	5.70
1300	Vinyl bulb insert		105	.076		1.47	3.36		4.83	6.75
1600	Screw on, "T" extrusion		90	.089		5.45	3.92		9.37	12.05
1700	Vinyl insert		75	.107		3.65	4.70		8.35	11.25
2000	"L" extrusion, neoprene bulbs		75	.107		2.20	4.70		6.90	9.65
2100	Neoprene sponge insert		75	.107		6.35	4.70		11.05	14.25
2200	Magnetic		75	.107		10.50	4.70		15.20	18.80
2400	Spring hinged security seal, with cam		75	.107		6.75	4.70		11.45	14.70
2600	Spring loaded locking bolt, vinyl insert		45	.178		9.10	7.85		16.95	22
2800	Neoprene sponge strip, "Z" shaped, aluminum		60	.133		4.45	5.90		10.35	13.95
2900	Solid neoprene strip, nail on aluminum strip		90	.089		3.65	3.92		7.57	10.05
3000	One piece stile protection									
3020	Neoprene fabric loop, nail on aluminum strips	1 Carp	60	.133	L.F.	.95	5.90		6.85	10.10
3110	Flush mounted aluminum extrusion, 1/2" x 1-1/4"		60	.133		3.51	5.90		9.41	12.90
3140	3/4" x 1-3/8"		60	.133		4.46	5.90		10.36	13.95
3160	1-1/8" x 1-3/4"		60	.133		7.05	5.90		12.95	16.80
3300	Mortise, 9/16" x 3/4"		60	.133		3.80	5.90		9.70	13.25
3320	13/16" x 1-3/8"		60	.133		4.10	5.90		10	13.55
3600	Spring bronze strip, nail on type		105	.076		3.30	3.36		6.66	8.80
3620	Screw on, with retainer		75	.107		2.65	4.70		7.35	10.15
3800	Flexible stainless steel housing, pile insert, 1/2" door		105	.076		7.20	3.36		10.56	13.05
3820	3/4" door		105	.076		8.05	3.36		11.41	14
4000	Extruded aluminum retainer, flush mount, pile insert		105	.076		2.55	3.36		5.91	7.95
4080	Mortise, felt insert		90	.089		4.50	3.92		8.42	11
4160	Mortise with spring, pile insert		90	.089		3.57	3.92		7.49	10
4400	Rigid vinyl retainer, mortise, pile insert		105	.076		2.55	3.36		5.91	7.95
4600	Wool pile filler strip, aluminum backing		105	.076		2.60	3.36		5.96	8

08 71 Door Hardware

08 71 21 – Astragals

08 71 21.10 Exterior Mouldings, Astragals

		Crew	Daily Output	Labor-Hours	Unit	Material	2012 Bare Costs Labor	Equipment	Total	Total Incl O&P
5000	Two piece overlapping astragal, extruded aluminum retainer									
5010	Pile insert	1 Carp	60	.133	L.F.	3.30	5.90		9.20	12.70
5020	Vinyl bulb insert		60	.133		2.20	5.90		8.10	11.45
5040	Vinyl flap insert		60	.133		6.55	5.90		12.45	16.25
5060	Solid neoprene flap insert		60	.133		6.50	5.90		12.40	16.20
5080	Hypalon rubber flap insert		60	.133		6.60	5.90		12.50	16.30
5090	Snap on cover, pile insert		60	.133		7.55	5.90		13.45	17.35
5400	Magnetic aluminum, surface mounted		60	.133		25	5.90		30.90	36.50
5500	Interlocking aluminum, 5/8" x 1" neoprene bulb insert		45	.178		4.15	7.85		12	16.60
5600	Adjustable aluminum, 9/16" x 21/32", pile insert		45	.178		19.15	7.85		27	33
5790	For vinyl bulb, deduct					.70			.70	.77
5800	Magnetic, adjustable, 9/16" x 21/32"	1 Carp	45	.178		24.50	7.85		32.35	39
6000	Two piece stile protection									
6010	Cloth backed rubber loop, 1" gap, nail on aluminum strips	1 Carp	45	.178	L.F.	4.30	7.85		12.15	16.80
6040	Screw on aluminum strips		45	.178		6.50	7.85		14.35	19.20
6100	1-1/2" gap, screw on aluminum extrusion		45	.178		5.80	7.85		13.65	18.45
6240	Vinyl fabric loop, slotted aluminum extrusion, 1" gap		45	.178		2.15	7.85		10	14.40
6300	1-1/4" gap		45	.178		6.15	7.85		14	18.80

08 71 25 – Weatherstripping

08 71 25.10 Mechanical Seals, Weatherstripping

		Crew	Daily Output	Labor-Hours	Unit	Material	2012 Bare Costs Labor	Equipment	Total	Total Incl O&P
0010	**MECHANICAL SEALS, WEATHERSTRIPPING**									
1000	Doors, wood frame, interlocking, for 3' x 7' door, zinc	1 Carp	3	2.667	Opng.	20	118		138	203
1100	Bronze		3	2.667		29	118		147	213
1300	6' x 7' opening, zinc		2	4		20	176		196	293
1400	Bronze		2	4		36	176		212	310
1700	Wood frame, spring type, bronze									
1800	3' x 7' door	1 Carp	7.60	1.053	Opng.	19.70	46.50		66.20	93
1900	6' x 7' door	"	7	1.143	"	29	50.50		79.50	109
2200	Metal frame, spring type, bronze									
2300	3' x 7' door	1 Carp	3	2.667	Opng.	38.50	118		156.50	223
2400	6' x 7' door	"	2.50	3.200	"	49	141		190	271
2500	For stainless steel, spring type, add					133%				
2700	Metal frame, extruded sections, 3' x 7' door, aluminum	1 Carp	3	2.667	Opng.	46	118		164	232
2800	Bronze		3	2.667		120	118		238	315
3100	6' x 7' door, aluminum		1.50	5.333		58	235		293	425
3200	Bronze		1.50	5.333		137	235		372	510
3500	Threshold weatherstripping									
3650	Door sweep, flush mounted, aluminum	1 Carp	25	.320	Ea.	15	14.10		29.10	38
3700	Vinyl		25	.320		18	14.10		32.10	41.50
5000	Garage door bottom weatherstrip, 12' aluminum, clear		14	.571		22	25		47	63.50
5010	Bronze		14	.571		86	25		111	134
5050	Bottom protection, Rubber		14	.571		31	25		56	73
5100	Threshold		14	.571		92.50	25		117.50	141

08 74 Access Control Hardware

08 74 13 – Card Key Access Control Hardware

08 74 13.50 Card Key Access	Crew	Daily Output	Labor-Hours	Unit	Material	2012 Bare Costs Labor	2012 Bare Costs Equipment	Total	Total Incl O&P
0010 **CARD KEY ACCESS**, does not include door hardware									
0020 Card type, 1 time zone, minimum				Ea.	665			665	730
0040 Maximum					1,775			1,775	1,950
0060 3 time zones, minimum					2,200			2,200	2,425
0080 Maximum				↓	2,450			2,450	2,700
0100 System with printer, and control console, 3 zones				Total	9,150			9,150	10,100
0120 6 zones				"	12,000			12,000	13,200
0140 For each door, minimum, add				Ea.	1,500			1,500	1,650
0160 Maximum, add				"	2,100			2,100	2,300

08 74 19 – Biometric Identity Access Control Hardware

08 74 19.50 Biometric Identity Access	Crew	Daily Output	Labor-Hours	Unit	Material	2012 Bare Costs Labor	2012 Bare Costs Equipment	Total	Total Incl O&P
0010 **BIOMETRIC IDENTITY ACCESS**									
0220 Hand geometry scanner, mem of 512 users, excl striker/power	1 Elec	3	2.667	Ea.	2,100	137		2,237	2,500
0230 Memory upgrade for, adds 9,700 user profiles		8	1		300	51.50		351.50	405
0240 Adds 32,500 user profiles		8	1		600	51.50		651.50	735
0250 Prison type, memory of 256 users, excl striker, power		3	2.667		2,600	137		2,737	3,050
0260 Memory upgrade for, adds 3,300 user profiles		8	1		250	51.50		301.50	350
0270 Adds 9,700 user profiles		8	1		460	51.50		511.50	580
0280 Adds 27,900 user profiles		8	1		610	51.50		661.50	745
0290 All weather, mem of 512 users, excl striker/power		3	2.667		3,900	137		4,037	4,500
0300 Facial & fingerprint scanner, combination unit, excl striker/power		3	2.667		4,200	137		4,337	4,825
0310 Access for, for initial setup, excl striker/power		3	2.667	↓	1,100	137		1,237	1,400

08 75 Window Hardware

08 75 30 – Weatherstripping

08 75 30.10 Mechanical Weather Seals	Crew	Daily Output	Labor-Hours	Unit	Material	2012 Bare Costs Labor	2012 Bare Costs Equipment	Total	Total Incl O&P
0010 **MECHANICAL WEATHER SEALS**, Window, double hung, 3' X 5'									
0020 Zinc	1 Carp	7.20	1.111	Opng.	13	49		62	90
0100 Bronze		7.20	1.111		28	49		77	107
0500 As above but heavy duty, zinc		4.60	1.739		20	76.50		96.50	140
0600 Bronze	↓	4.60	1.739	↓	35	76.50		111.50	157

08 79 Hardware Accessories

08 79 13 – Key Storage Equipment

08 79 13.10 Key Cabinets	Crew	Daily Output	Labor-Hours	Unit	Material	2012 Bare Costs Labor	2012 Bare Costs Equipment	Total	Total Incl O&P
0010 **KEY CABINETS**									
0020 Wall mounted, 60 key capacity	1 Carp	20	.400	Ea.	103	17.65		120.65	140
0200 Drawer type, 600 key capacity	1 Clab	15	.533		725	18.70		743.70	830
0300 2,400 key capacity		20	.400		4,200	14.05		4,214.05	4,650
0400 Tray type, 20 key capacity		50	.160		46	5.60		51.60	59
0500 50 key capacity	↓	40	.200	↓	102	7		109	123

08 79 20 – Door Accessories

08 79 20.10 Door Hardware Accessories	Crew	Daily Output	Labor-Hours	Unit	Material	2012 Bare Costs Labor	2012 Bare Costs Equipment	Total	Total Incl O&P
0010 **DOOR HARDWARE ACCESSORIES**									
0140 Door bolt, surface, 4"	1 Carp	32	.250	Ea.	11.70	11.05		22.75	30
0160 Door latch	"	12	.667	"	8.45	29.50		37.95	54.50
0200 Sliding closet door									
0220 Track and hanger, single	1 Carp	10	.800	Ea.	56	35.50		91.50	116

08 79 Hardware Accessories

08 79 20 – Door Accessories

08 79 20.10 Door Hardware Accessories	Crew	Daily Output	Labor-Hours	Unit	Material	2012 Bare Costs Labor	Equipment	Total	Total Incl O&P
0240	Double	1 Carp	8	1	Ea.	75	44		119	151
0260	Door guide, single		48	.167		25	7.35		32.35	39
0280	Double		48	.167		34	7.35		41.35	49
0600	Deadbolt and lock cover plate, brass or stainless steel		30	.267		28	11.75		39.75	49
0620	Hole cover plate, brass or chrome		35	.229		7.50	10.10		17.60	24
2240	Mortise lockset, passage, lever handle		9	.889		198	39		237	279
4000	Security chain, standard		18	.444		7.90	19.60		27.50	38.50

08 81 Glass Glazing

08 81 10 – Float Glass

08 81 10.10 Various Types and Thickness of Float Glass

08 81 10.10 Various Types and Thickness of Float Glass	Crew	Daily Output	Labor-Hours	Unit	Material	2012 Bare Costs Labor	Equipment	Total	Total Incl O&P
0010	VARIOUS TYPES AND THICKNESS OF FLOAT GLASS R088110-10									
0020	3/16" Plain	2 Glaz	130	.123	S.F.	4.71	5.25		9.96	13.15
0200	Tempered, clear		130	.123		6.55	5.25		11.80	15.20
0300	Tinted		130	.123		7.45	5.25		12.70	16.15
0600	1/4" thick, clear, plain		120	.133		5.50	5.70		11.20	14.70
0700	Tinted		120	.133		8.80	5.70		14.50	18.35
0800	Tempered, clear		120	.133		8.25	5.70		13.95	17.75
0900	Tinted		120	.133		10.20	5.70		15.90	19.85
1600	3/8" thick, clear, plain		75	.213		9.60	9.10		18.70	24.50
1700	Tinted		75	.213		15.10	9.10		24.20	30.50
1800	Tempered, clear		75	.213		15.85	9.10		24.95	31
1900	Tinted		75	.213		17.60	9.10		26.70	33
2200	1/2" thick, clear, plain		55	.291		16.25	12.45		28.70	37
2300	Tinted		55	.291		26.50	12.45		38.95	48
2400	Tempered, clear		55	.291		23.50	12.45		35.95	44.50
2500	Tinted		55	.291		24.50	12.45		36.95	45.50
2800	5/8" thick, clear, plain		45	.356		26.50	15.20		41.70	52
2900	Tempered, clear		45	.356		30	15.20		45.20	56
3200	3/4" thick, clear, plain		35	.457		34	19.55		53.55	67
3300	Tempered, clear		35	.457		39.50	19.55		59.05	73
3600	1" thick, clear, plain		30	.533		56.50	23		79.50	97
8900	For low emissivity coating for 3/16" & 1/4" only, add to above					18%				

08 81 13 – Decorative Glass Glazing

08 81 13.10 Beveled Glass

08 81 13.10 Beveled Glass	Crew	Daily Output	Labor-Hours	Unit	Material	2012 Bare Costs Labor	Equipment	Total	Total Incl O&P
0010	BEVELED GLASS, with design patterns									
0020	Minimum	2 Glaz	150	.107	S.F.	58	4.56		62.56	71
0050	Average		125	.128		129	5.45		134.45	150
0100	Maximum		100	.160		229	6.85		235.85	261
0102	Maximum, per L.F. for 30" cabinet	1 Glaz	40	.200	L.F.	570	8.55		578.55	645

08 81 13.20 Faceted Glass

08 81 13.20 Faceted Glass	Crew	Daily Output	Labor-Hours	Unit	Material	2012 Bare Costs Labor	Equipment	Total	Total Incl O&P
0010	FACETED GLASS, Color tinted 3/4" thick									
0020	Minimum	2 Glaz	95	.168	S.F.	46.50	7.20		53.70	62
0100	Maximum	"	75	.213	"	81.50	9.10		90.60	103

08 81 13.30 Sandblasted Glass

08 81 13.30 Sandblasted Glass	Crew	Daily Output	Labor-Hours	Unit	Material	2012 Bare Costs Labor	Equipment	Total	Total Incl O&P
0010	SANDBLASTED GLASS, float glass									
0020	1/8" thick	2 Glaz	160	.100	S.F.	10.60	4.28		14.88	18.15
0100	3/16" thick		130	.123		11.70	5.25		16.95	21
0500	1/4" thick		120	.133		12.20	5.70		17.90	22
0600	3/8" thick		75	.213		13.10	9.10		22.20	28

08 81 Glass Glazing

08 81 20 – Vision Panels

08 81 20.10 Full Vision

08 81 20.10 Full Vision		Crew	Daily Output	Labor-Hours	Unit	Material	2012 Bare Costs Labor	Equipment	Total	Total Incl O&P
0010	**FULL VISION**, window system with 3/4" glass mullions									
0020	Up to 10' high	H-2	130	.185	S.F.	60.50	7.40		67.90	78
0100	10' to 20' high, minimum		110	.218		64	8.75		72.75	84
0150	Average		100	.240		69	9.65		78.65	90.50
0200	Maximum		80	.300		78	12.05		90.05	104

08 81 25 – Glazing Variables

08 81 25.10 Applications of Glazing

08 81 25.10 Applications of Glazing		Crew	Daily Output	Labor-Hours	Unit	Material	2012 Bare Costs Labor	Equipment	Total	Total Incl O&P
0010	**APPLICATIONS OF GLAZING** R088110-10									
0500	For high rise glazing, exterior, add per S.F. per story				S.F.					.46
0600	For glass replacement, add				"		100%			
0700	For gasket settings, add				L.F.	5.60			5.60	6.15
0900	For sloped glazing, add				S.F.		26%			
2000	Fabrication, polished edges, 1/4" thick				Inch	.52			.52	.57
2100	1/2" thick					1.25			1.25	1.38
2500	Mitered edges, 1/4" thick					1.25			1.25	1.38
2600	1/2" thick					2.05			2.05	2.26

08 81 30 – Insulating Glass

08 81 30.10 Reduce Heat Transfer Glass

08 81 30.10 Reduce Heat Transfer Glass		Crew	Daily Output	Labor-Hours	Unit	Material	2012 Bare Costs Labor	Equipment	Total	Total Incl O&P
0010	**REDUCE HEAT TRANSFER GLASS** R088110-10									
0015	2 lites 1/8" float, 1/2" thk under 15 S.F.									
0020	Clear G	2 Glaz	95	.168	S.F.	9.45	7.20		16.65	21.50
0100	Tinted G		95	.168		13.20	7.20		20.40	25.50
0200	2 lites 3/16" float, for 5/8" thk unit, 15 to 30 S.F., clear G		90	.178		13.20	7.60		20.80	26
0300	Tinted G		90	.178		13.50	7.60		21.10	26.50
0400	1" thk, dbl. glazed, 1/4" float, 30-70 S.F., clear G		75	.213		15.95	9.10		25.05	31.50
0500	Tinted G		75	.213		22.50	9.10		31.60	39
0600	1" thick double glazed, 1/4" float, 1/4" wire		75	.213		22	9.10		31.10	38.50
0700	1/4" float, 1/4" tempered		75	.213		18.90	9.10		28	35
0800	1/4" wire, 1/4" tempered		75	.213		28	9.10		37.10	45
0900	Both lites, 1/4" wire		75	.213		36	9.10		45.10	53.50
2000	Both lites, light & heat reflective G		85	.188		30	8.05		38.05	45
2500	Heat reflective, film inside, 1" thick unit, clear G		85	.188		26.50	8.05		34.55	41
2600	Tinted G		85	.188		27.50	8.05		35.55	42.50
3000	Film on weatherside, clear, 1/2" thick unit G		95	.168		18.90	7.20		26.10	32
3100	5/8" thick unit G		90	.178		18.70	7.60		26.30	32
3200	1" thick unit G		85	.188		26	8.05		34.05	40.50

08 81 35 – Translucent Glass

08 81 35.10 Obscure Glass

08 81 35.10 Obscure Glass		Crew	Daily Output	Labor-Hours	Unit	Material	2012 Bare Costs Labor	Equipment	Total	Total Incl O&P
0010	**OBSCURE GLASS**									
0020	1/8" thick, Minimum	2 Glaz	140	.114	S.F.	11.20	4.89		16.09	19.70
0100	Maximum		125	.128		13.20	5.45		18.65	23
0300	7/32" thick, minimum		120	.133		12.20	5.70		17.90	22
0400	Maximum		105	.152		15.40	6.50		21.90	27

08 81 35.20 Patterned Glass

08 81 35.20 Patterned Glass		Crew	Daily Output	Labor-Hours	Unit	Material	2012 Bare Costs Labor	Equipment	Total	Total Incl O&P
0010	**PATTERNED GLASS**, colored									
0020	1/8" thick, Minimum	2 Glaz	140	.114	S.F.	9.10	4.89		13.99	17.40
0100	Maximum		125	.128		15.30	5.45		20.75	25
0300	7/32" thick, minimum		120	.133		11.25	5.70		16.95	21
0400	Maximum		105	.152		16.30	6.50		22.80	28

08 81 Glass Glazing

08 81 45 – Sheet Glass

08 81 45.10 Window Glass, Sheet

		Crew	Daily Output	Labor-Hours	Unit	Material	2012 Bare Costs Labor	Equipment	Total	Total Incl O&P
0010	**WINDOW GLASS, SHEET** gray									
0020	1/8" thick	2 Glaz	160	.100	S.F.	5.40	4.28		9.68	12.45
0200	1/4" thick	"	130	.123	"	6.60	5.25		11.85	15.20

08 81 50 – Spandrel Glass

08 81 50.10 Glass for Non Vision Areas

		Crew	Daily Output	Labor-Hours	Unit	Material	2012 Bare Costs Labor	Equipment	Total	Total Incl O&P
0010	**GLASS FOR NON VISION AREAS**, 1/4" thick standard colors									
0020	Up to 1000 S.F.	2 Glaz	110	.145	S.F.	15.85	6.20		22.05	27
0200	1,000 to 2,000 S.F.	"	120	.133	"	14.75	5.70		20.45	25
0300	For custom colors, add				Total	10%				
0500	For 3/8" thick, add				S.F.	10.90			10.90	12
1000	For double coated, 1/4" thick, add					4.08			4.08	4.49
1200	For insulation on panels, add					6.70			6.70	7.35
2000	Panels, insulated, with aluminum backed fiberglass, 1" thick	2 Glaz	120	.133		16.20	5.70		21.90	26.50
2100	2" thick	"	120	.133		19.20	5.70		24.90	29.50
2500	With galvanized steel backing, add					5.60			5.60	6.15

08 81 55 – Window Glass

08 81 55.10 Sheet Glass

		Crew	Daily Output	Labor-Hours	Unit	Material	2012 Bare Costs Labor	Equipment	Total	Total Incl O&P
0010	**SHEET GLASS** (window), clear float, stops, putty bed									
0015	1/8" thick, clear float	2 Glaz	480	.033	S.F.	3.47	1.42		4.89	6
0500	3/16" thick, clear		480	.033		5.50	1.42		6.92	8.20
0600	Tinted		480	.033		7.20	1.42		8.62	10.05
0700	Tempered		480	.033		8.80	1.42		10.22	11.85

08 81 65 – Wire Glass

08 81 65.10 Glass Reinforced With Wire

		Crew	Daily Output	Labor-Hours	Unit	Material	2012 Bare Costs Labor	Equipment	Total	Total Incl O&P
0010	**GLASS REINFORCED WITH WIRE**									
0012	1/4" thick rough obscure	2 Glaz	135	.119	S.F.	21	5.05		26.05	30.50
1000	Polished wire, 1/4" thick, diamond, clear		135	.119		24.50	5.05		29.55	34.50
1500	Pinstripe, obscure		135	.119		30	5.05		35.05	40.50

08 83 Mirrors

08 83 13 – Mirrored Glass Glazing

08 83 13.10 Mirrors

		Crew	Daily Output	Labor-Hours	Unit	Material	2012 Bare Costs Labor	Equipment	Total	Total Incl O&P
0010	**MIRRORS**, No frames, wall type, 1/4" plate glass, polished edge									
0100	Up to 5 S.F.	2 Glaz	125	.128	S.F.	9.50	5.45		14.95	18.75
0200	Over 5 S.F.		160	.100		9.25	4.28		13.53	16.70
0500	Door type, 1/4" plate glass, up to 12 S.F.		160	.100		8.35	4.28		12.63	15.70
1000	Float glass, up to 10 S.F., 1/8" thick		160	.100		5.50	4.28		9.78	12.55
1100	3/16" thick		150	.107		6.75	4.56		11.31	14.35
1500	12" x 12" wall tiles, square edge, clear		195	.082		2	3.51		5.51	7.50
1600	Veined		195	.082		4.95	3.51		8.46	10.75
2000	1/4" thick, stock sizes, one way transparent		125	.128		18.45	5.45		23.90	29
2010	Bathroom, unframed, laminated		160	.100		13	4.28		17.28	21

08 83 13.15 Reflective Glass

			Crew	Daily Output	Labor-Hours	Unit	Material	2012 Bare Costs Labor	Equipment	Total	Total Incl O&P
0010	**REFLECTIVE GLASS**										
0100	1/4" float with fused metallic oxide fixed	G	2 Glaz	115	.139	S.F.	16.30	5.95		22.25	27
0500	1/4" float glass with reflective applied coating	G	"	115	.139	"	13.20	5.95		19.15	23.50

08 84 Plastic Glazing

08 84 10 – Plexiglass Glazing

08 84 10.10 Plexiglass Acrylic	Crew	Daily Output	Labor-Hours	Unit	Material	2012 Bare Costs Labor	Equipment	Total	Total Incl O&P
0010 **PLEXIGLASS ACRYLIC**, clear, masked,									
0020 1/8" thick, cut sheets	2 Glaz	170	.094	S.F.	5.30	4.02		9.32	11.95
0200 Full sheets		195	.082		3.50	3.51		7.01	9.15
0500 1/4" thick, cut sheets		165	.097		9.20	4.15		13.35	16.40
0600 Full sheets		185	.086		6	3.70		9.70	12.20
0900 3/8" thick, cut sheets		155	.103		16.85	4.41		21.26	25.50
1000 Full sheets		180	.089		9.10	3.80		12.90	15.75
1300 1/2" thick, cut sheets		135	.119		19.40	5.05		24.45	29
1400 Full sheets		150	.107		18.80	4.56		23.36	27.50
1700 3/4" thick, cut sheets		115	.139		69	5.95		74.95	85
1800 Full sheets		130	.123		40	5.25		45.25	52
2100 1" thick, cut sheets		105	.152		77.50	6.50		84	95.50
2200 Full sheets		125	.128		48	5.45		53.45	61.50
3000 Colored, 1/8" thick, cut sheets		170	.094		16	4.02		20.02	23.50
3200 Full sheets		195	.082		10.30	3.51		13.81	16.65
3500 1/4" thick, cut sheets		165	.097		18	4.15		22.15	26
3600 Full sheets		185	.086		12.20	3.70		15.90	19
4000 Mirrors, untinted, cut sheets, 1/8" thick		185	.086		8.25	3.70		11.95	14.70
4200 1/4" thick		180	.089		13.20	3.80		17	20.50

08 84 20 – Polycarbonate

08 84 20.10 Thermoplastic

	Crew	Daily Output	Labor-Hours	Unit	Material	Labor	Equipment	Total	Total Incl O&P
0010 **THERMOPLASTIC**, clear, masked, cut sheets									
0020 1/8" thick	2 Glaz	170	.094	S.F.	9.70	4.02		13.72	16.75
0500 3/16" thick		165	.097		11.10	4.15		15.25	18.50
1000 1/4" thick		155	.103		12.50	4.41		16.91	20.50
1500 3/8" thick		150	.107		22	4.56		26.56	31.50

08 84 30 – Vinyl Glass

08 84 30.10 Vinyl Sheets

	Crew	Daily Output	Labor-Hours	Unit	Material	Labor	Equipment	Total	Total Incl O&P
0010 **VINYL SHEETS**, Steel mesh reinforced, stock sizes									
0020 .090" thick	2 Glaz	170	.094	S.F.	13.30	4.02		17.32	21
0500 .120" thick		170	.094		16.30	4.02		20.32	24
1000 .250" thick		155	.103		22	4.41		26.41	30.50
1500 For non-standard sizes, add					20%				

08 87 Glazing Surface Films

08 87 13 – Solar Control Films

08 87 13.10 Solar Films On Glass

		Crew	Daily Output	Labor-Hours	Unit	Material	Labor	Equipment	Total	Total Incl O&P
0010 **SOLAR FILMS ON GLASS** (glass not included)										
2000 Minimum	G	2 Glaz	180	.089	S.F.	6.60	3.80		10.40	13
2050 Maximum	G	"	225	.071	"	14.70	3.04		17.74	21

08 87 23 – Safety and Security Films

08 87 23.13 Safety Films

		Crew	Daily Output	Labor-Hours	Unit	Material	Labor	Equipment	Total	Total Incl O&P
0010 **SAFETY FILMS**										
2010 Window protection film, controls blast damage	G	2 Glaz	80	.200	S.F.	6.60	8.55		15.15	20

08 87 23.16 Security Films

		Crew	Daily Output	Labor-Hours	Unit	Material	Labor	Equipment	Total	Total Incl O&P
0010 **SECURITY FILMS**, clear, 32000 psi tensile strength, adhered to glass										
0100 .002" thick, daylight installation		H-2	950	.025	S.F.	2.25	1.02		3.27	4.03
0150 .004" thick, daylight installation	R088110-10		800	.030		2.80	1.21		4.01	4.92
0200 .006" thick, daylight installation			700	.034		3	1.38		4.38	5.40

08 87 Glazing Surface Films

08 87 23 – Safety and Security Films

08 87 23.16 Security Films		Crew	Daily Output	Labor-Hours	Unit	Material	2012 Bare Costs Labor	2012 Bare Costs Equipment	Total	Total Incl O&P
0210	Install for anchorage	H-2	600	.040	S.F.	3.33	1.61		4.94	6.10
0400	.007" thick, daylight installation		600	.040		4	1.61		5.61	6.85
0410	Install for anchorage		500	.048		4.44	1.93		6.37	7.85
0500	.008" thick, daylight installation		500	.048		4.50	1.93		6.43	7.90
0510	Install for anchorage		500	.048		5	1.93		6.93	8.45
0600	.015" thick, daylight installation		400	.060		8	2.41		10.41	12.45
0610	Install for anchorage		400	.060		5	2.41		7.41	9.15
0900	Security film anchorage, mechanical attachment and cover plate	H-3	370	.043	L.F.	9.45	1.64		11.09	12.90
0950	Security film anchorage, wet glaze structural caulking	1 Glaz	225	.036	"	.84	1.52		2.36	3.22
1000	Adhered security film removal	1 Clab	275	.029	S.F.		1.02		1.02	1.57

08 88 Special Function Glazing

08 88 40 – Acoustical Glass Units

08 88 40.10 Sound Reduction Units

		Crew	Daily Output	Labor-Hours	Unit	Material	2012 Bare Costs Labor	2012 Bare Costs Equipment	Total	Total Incl O&P
0010	**SOUND REDUCTION UNITS**, 1 lite at 3/8", 1 lite at 3/16"									
0020	For 1" thick	2 Glaz	100	.160	S.F.	33	6.85		39.85	46.50
0100	For 4" thick	"	80	.200	"	56	8.55		64.55	74.50

08 88 56 – Ballistics-Resistant Glazing

08 88 56.10 Laminated Glass

		Crew	Daily Output	Labor-Hours	Unit	Material	2012 Bare Costs Labor	2012 Bare Costs Equipment	Total	Total Incl O&P
0010	**LAMINATED GLASS**									
0020	Clear float .03" vinyl 1/4"	2 Glaz	90	.178	S.F.	11.75	7.60		19.35	24.50
0100	3/8" thick		78	.205		21.50	8.75		30.25	37
0200	.06" vinyl, 1/2" thick		65	.246		24.50	10.50		35	43
1000	5/8" thick		90	.178		28.50	7.60		36.10	43
2000	Bullet-resisting, 1-3/16" thick, to 15 S.F.		16	1		90	43		133	164
2100	Over 15 S.F.		16	1		104	43		147	179
2500	2-1/4" thick, to 15 S.F.		12	1.333		140	57		197	241
2600	Over 15 S.F.		12	1.333		130	57		187	230
2700	Level 2 (.357 magnum)		12	1.333		68	57		125	162
2750	Level 3A (.44 magnum)		12	1.333		73.50	57		130.50	168
2800	Level 4 (AK-47)		12	1.333		96	57		153	193
2850	Level 5 (M-16)		12	1.333		107	57		164	205
2900	Level 3 (7.62 Armor Piercing)		12	1.333		127	57		184	227

08 91 Louvers

08 91 19 – Fixed Louvers

08 91 19.10 Aluminum Louvers

		Crew	Daily Output	Labor-Hours	Unit	Material	2012 Bare Costs Labor	2012 Bare Costs Equipment	Total	Total Incl O&P
0010	**ALUMINUM LOUVERS**									
0020	Aluminum with screen, residential, 8" x 8"	1 Carp	38	.211	Ea.	13.50	9.30		22.80	29
0100	12" x 12"		38	.211		15	9.30		24.30	31
0200	12" x 18"		35	.229		19	10.10		29.10	36.50
0250	14" x 24"		30	.267		26	11.75		37.75	46.50
0300	18" x 24"		27	.296		29	13.05		42.05	52
0500	24" x 30"		24	.333		40	14.70		54.70	66.50
0700	Triangle, adjustable, small		20	.400		37.50	17.65		55.15	68.50
0800	Large		15	.533		46.50	23.50		70	87
1200	Extruded aluminum, see Section 23 37 15.40									
2100	Midget, aluminum, 3/4" deep, 1" diameter	1 Carp	85	.094	Ea.	.72	4.15		4.87	7.20
2150	3" diameter		60	.133		2	5.90		7.90	11.25

08 91 Louvers

08 91 19 – Fixed Louvers

08 91 19.10 Aluminum Louvers		Crew	Daily Output	Labor-Hours	Unit	Material	2012 Bare Costs		Total	Total Incl O&P
							Labor	Equipment		
2200	4" diameter	1 Carp	50	.160	Ea.	3.64	7.05		10.69	14.85
2250	6" diameter	↓	30	.267	↓	3.85	11.75		15.60	22.50

08 91 26 – Door Louvers

08 91 26.10 Steel Louvers, 18 Gauge, Fixed Blade

		Crew	Daily Output	Labor-Hours	Unit	Material	Labor	Equipment	Total	Total Incl O&P
0010	**STEEL LOUVERS, 18 GAUGE, FIXED BLADE**									
0050	12" x 12", with enamel or powder coat	1 Carp	20	.400	Ea.	69	17.65		86.65	103
0055	18" x 12"		20	.400		83	17.65		100.65	119
0060	18" x 18"		20	.400		84	17.65		101.65	120
0065	24" x 12"		20	.400		92	17.65		109.65	128
0070	24" x 18"		20	.400		126	17.65		143.65	166
0075	24" x 24"		20	.400		131	17.65		148.65	171
0100	12" x 12", galvanized		20	.400		60	17.65		77.65	93
0105	18" x 12"		20	.400		67	17.65		84.65	101
0115	24" x 12"		20	.400		83	17.65		100.65	119
0125	24" x 24"	↓	20	.400	↓	121	17.65		138.65	160

08 95 Vents

08 95 13 – Soffit Vents

08 95 13.10 Wall Louvers

		Crew	Daily Output	Labor-Hours	Unit	Material	Labor	Equipment	Total	Total Incl O&P
0010	**WALL LOUVERS**									
2330	Soffit vent, continuous, 3" wide, aluminum, mill finish	1 Carp	200	.040	L.F.	.52	1.76		2.28	3.28
2340	Baked enamel finish		200	.040	"	.57	1.76		2.33	3.34
2400	Under eaves vent, aluminum, mill finish, 16" x 4"		48	.167	Ea.	1.84	7.35		9.19	13.30
2500	16" x 8"	↓	48	.167	"	2.14	7.35		9.49	13.65

08 95 16 – Wall Vents

08 95 16.10 Louvers

		Crew	Daily Output	Labor-Hours	Unit	Material	Labor	Equipment	Total	Total Incl O&P
0010	**LOUVERS**									
0020	Redwood, 2'-0" diameter, full circle	1 Carp	16	.500	Ea.	174	22		196	225
0100	Half circle		16	.500		192	22		214	245
0200	Octagonal		16	.500		132	22		154	179
0300	Triangular, 5/12 pitch, 5'-0" at base		16	.500		194	22		216	247
7000	Vinyl gable vent, 8" x 8"		38	.211		12.60	9.30		21.90	28
7020	12" x 12"		38	.211		24	9.30		33.30	41
7080	12" x 18"		35	.229		31	10.10		41.10	49.50
7200	18" x 24"	↓	30	.267	↓	38	11.75		49.75	60

Division Notes

	CREW	DAILY OUTPUT	LABOR-HOURS	UNIT	2012 BARE COSTS				TOTAL INCL O&P
					MAT.	LABOR	EQUIP.	TOTAL	

Estimating Tips

General

- Room Finish Schedule: A complete set of plans should contain a room finish schedule. If one is not available, it would be well worth the time and effort to obtain one.

09 20 00 Plaster and Gypsum Board

- Lath is estimated by the square yard plus a 5% allowance for waste. Furring, channels, and accessories are measured by the linear foot. An extra foot should be allowed for each accessory miter or stop.

- Plaster is also estimated by the square yard. Deductions for openings vary by preference, from zero deduction to 50% of all openings over 2 feet in width. The estimator should allow one extra square foot for each linear foot of horizontal interior or exterior angle located below the ceiling level. Also, double the areas of small radius work.

- Drywall accessories, studs, track, and acoustical caulking are all measured by the linear foot. Drywall taping is figured by the square foot. Gypsum wallboard is estimated by the square foot. No material deductions should be made for door or window openings under 32 S.F.

09 60 00 Flooring

- Tile and terrazzo areas are taken off on a square foot basis. Trim and base materials are measured by the linear foot. Accent tiles are listed per each. Two basic methods of installation are used. Mud set is approximately 30% more expensive than thin set. In terrazzo work, be sure to include the linear footage of embedded decorative strips, grounds, machine rubbing, and power cleanup.

- Wood flooring is available in strip, parquet, or block configuration. The latter two types are set in adhesives with quantities estimated by the square foot. The laying pattern will influence labor costs and material waste. In addition to the material and labor for laying wood floors, the estimator must make allowances for sanding and finishing these areas unless the flooring is prefinished.

- Sheet flooring is measured by the square yard. Roll widths vary, so consideration should be given to use the most economical width, as waste must be figured into the total quantity. Consider also the installation methods available, direct glue down or stretched.

09 70 00 Wall Finishes

- Wall coverings are estimated by the square foot. The area to be covered is measured, length by height of wall above baseboards, to calculate the square footage of each wall. This figure is divided by the number of square feet in the single roll which is being used. Deduct, in full, the areas of openings such as doors and windows. Where a pattern match is required allow 25%–30% waste.

09 80 00 Acoustic Treatment

- Acoustical systems fall into several categories. The takeoff of these materials should be by the square foot of area with a 5% allowance for waste. Do not forget about scaffolding, if applicable, when estimating these systems.

09 90 00 Painting and Coating

- A major portion of the work in painting involves surface preparation. Be sure to include cleaning, sanding, filling, and masking costs in the estimate.

- Protection of adjacent surfaces is not included in painting costs. When considering the method of paint application, an important factor is the amount of protection and masking required. These must be estimated separately and may be the determining factor in choosing the method of application.

Reference Numbers

Reference numbers are shown in shaded boxes at the beginning of some major classifications. These numbers refer to related items in the Reference Section. The reference information may be an estimating procedure, an alternate pricing method, or technical information.

Note: Not all subdivisions listed here necessarily appear in this publication.

09 01 Maintenance of Finishes

09 01 60 – Maintenance of Flooring

09 01 60.10 Carpet Maintenance		Crew	Daily Output	Labor-Hours	Unit	Material	2012 Bare Costs Labor	Equipment	Total	Total Incl O&P
0010	**CARPET MAINTENANCE**									
0020	Steam clean, per cleaning, minimum	1 Clab	3000	.003	S.F.	.07	.09		.16	.22
0500	Maximum	"	2000	.004	"	.12	.14		.26	.35

09 01 70 – Maintenance of Wall Finishes

09 01 70.10 Gypsum Wallboard Repairs		Crew	Daily Output	Labor-Hours	Unit	Material	2012 Bare Costs Labor	Equipment	Total	Total Incl O&P
0010	**GYPSUM WALLBOARD REPAIRS**									
0100	Fill and sand, pin/nail holes	1 Carp	960	.008	Ea.		.37		.37	.57
0110	Screw head pops		480	.017			.74		.74	1.13
0120	Dents, up to 2" square		48	.167		.01	7.35		7.36	11.30
0130	2" to 4" square		24	.333		.03	14.70		14.73	22.50
0140	Cut square, patch, sand and finish, holes, up to 2" square		12	.667		.04	29.50		29.54	45
0150	2" to 4" square		11	.727		.09	32		32.09	49.50
0160	4" to 8" square		10	.800		.21	35.50		35.71	54.50
0170	8" to 12" square		8	1		.40	44		44.40	68.50
0180	12" to 32" square		6	1.333		1.31	59		60.31	92
0210	16" by 48"		5	1.600		2.17	70.50		72.67	111
0220	32" by 48"		4	2		3.20	88		91.20	140
0230	48" square		3.50	2.286		4.48	101		105.48	160
0240	60" square		3.20	2.500		7.60	110		117.60	178
0500	Skim coat surface with joint compound		1600	.005	S.F.	.03	.22		.25	.37
0510	Prepare, retape and refinish joints		60	.133	L.F.	.59	5.90		6.49	9.70

09 05 Common Work Results for Finishes

09 05 05 – Selective Finishes Demolition

09 05 05.10 Selective Demolition, Ceilings		Crew	Daily Output	Labor-Hours	Unit	Material	2012 Bare Costs Labor	Equipment	Total	Total Incl O&P
0010	**SELECTIVE DEMOLITION, CEILINGS** R024119-10									
0200	Ceiling, drywall, furred and nailed or screwed	2 Clab	800	.020	S.F.		.70		.70	1.08
0220	On metal frame		760	.021			.74		.74	1.14
0240	On suspension system, including system		720	.022			.78		.78	1.20
1000	Plaster, lime and horse hair, on wood lath, incl. lath		700	.023			.80		.80	1.23
1020	On metal lath		570	.028			.99		.99	1.52
1100	Gypsum, on gypsum lath		720	.022			.78		.78	1.20
1120	On metal lath		500	.032			1.12		1.12	1.73
1200	Suspended ceiling, mineral fiber, 2' x 2' or 2' x 4'		1500	.011			.37		.37	.58
1250	On suspension system, incl. system		1200	.013			.47		.47	.72
1500	Tile, wood fiber, 12" x 12", glued		900	.018			.62		.62	.96
1540	Stapled		1500	.011			.37		.37	.58
1580	On suspension system, incl. system		760	.021			.74		.74	1.14
2000	Wood, tongue and groove, 1" x 4"		1000	.016			.56		.56	.86
2040	1" x 8"		1100	.015			.51		.51	.79
2400	Plywood or wood fiberboard, 4' x 8' sheets		1200	.013			.47		.47	.72

09 05 05.20 Selective Demolition, Flooring		Crew	Daily Output	Labor-Hours	Unit	Material	2012 Bare Costs Labor	Equipment	Total	Total Incl O&P
0010	**SELECTIVE DEMOLITION, FLOORING** R024119-10									
0200	Brick with mortar	2 Clab	475	.034	S.F.		1.18		1.18	1.82
0400	Carpet, bonded, including surface scraping		2000	.008			.28		.28	.43
0440	Scrim applied		8000	.002			.07		.07	.11
0480	Tackless		9000	.002			.06		.06	.10
0550	Carpet tile, releasable adhesive		5000	.003			.11		.11	.17
0560	Permanent adhesive		1850	.009			.30		.30	.47
0600	Composition, acrylic or epoxy		400	.040			1.40		1.40	2.16

09 05 05 – Selective Finishes Demolition

09 05 05.20 Selective Demolition, Flooring

		Crew	Daily Output	Labor-Hours	Unit	Material	2012 Bare Costs Labor	Equipment	Total	Total Incl O&P
0700	Concrete, scarify skin	A-1A	225	.036	S.F.		1.61	.94	2.55	3.53
0800	Resilient, sheet goods	2 Clab	1400	.011			.40		.40	.62
0820	For gym floors	"	900	.018			.62		.62	.96
0850	Vinyl or rubber cove base	1 Clab	1000	.008	L.F.		.28		.28	.43
0860	Vinyl or rubber cove base, molded corner	"	1000	.008	Ea.		.28		.28	.43
0870	For glued and caulked installation, add to labor						50%			
0900	Vinyl composition tile, 12" x 12"	2 Clab	1000	.016	S.F.		.56		.56	.86
2000	Tile, ceramic, thin set		675	.024			.83		.83	1.28
2020	Mud set		625	.026			.90		.90	1.38
2200	Marble, slate, thin set		675	.024			.83		.83	1.28
2220	Mud set		625	.026			.90		.90	1.38
2600	Terrazzo, thin set		450	.036			1.25		1.25	1.92
2620	Mud set		425	.038			1.32		1.32	2.03
2640	Terrazzo, cast in place		300	.053			1.87		1.87	2.88
3000	Wood, block, on end	1 Carp	400	.020			.88		.88	1.36
3200	Parquet		450	.018			.78		.78	1.21
3400	Strip flooring, interior, 2-1/4" x 25/32" thick		325	.025			1.09		1.09	1.67
3500	Exterior, porch flooring, 1" x 4"		220	.036			1.60		1.60	2.47
3800	Subfloor, tongue and groove, 1" x 6"		325	.025			1.09		1.09	1.67
3820	1" x 8"		430	.019			.82		.82	1.26
3840	1" x 10"		520	.015			.68		.68	1.04
4000	Plywood, nailed		600	.013			.59		.59	.90
4100	Glued and nailed		400	.020			.88		.88	1.36
4200	Hardboard, 1/4" thick		760	.011			.46		.46	.71
8000	Remove flooring, bead blast, minimum	A-1A	1000	.008			.36	.21	.57	.79
8100	Maximum		400	.020			.91	.53	1.44	1.98
8150	Mastic only		1500	.005			.24	.14	.38	.52

09 05 05.30 Selective Demolition, Walls and Partitions

		Crew	Daily Output	Labor-Hours	Unit	Material	2012 Bare Costs Labor	Equipment	Total	Total Incl O&P
0010	**SELECTIVE DEMOLITION, WALLS AND PARTITIONS** R024119-10									
0020	Walls, concrete, reinforced	B-39	120	.400	C.F.		14.80	1.74	16.54	24.50
0025	Plain	"	160	.300			11.10	1.31	12.41	18.50
0100	Brick, 4" to 12" thick	B-9	220	.182			6.45	.95	7.40	11
0200	Concrete block, 4" thick		1000	.040	S.F.		1.42	.21	1.63	2.41
0280	8" thick		810	.049			1.75	.26	2.01	2.98
0300	Exterior stucco 1" thick over mesh		3200	.013			.44	.07	.51	.75
1000	Drywall, nailed or screwed	1 Clab	1000	.008			.28		.28	.43
1010	2 layers		400	.020			.70		.70	1.08
1020	Glued and nailed		900	.009			.31		.31	.48
1500	Fiberboard, nailed		900	.009			.31		.31	.48
1520	Glued and nailed		800	.010			.35		.35	.54
1568	Plenum barrier, sheet lead		300	.027			.94		.94	1.44
2000	Movable walls, metal, 5' high		300	.027			.94		.94	1.44
2020	8' high		400	.020			.70		.70	1.08
2200	Metal or wood studs, finish 2 sides, fiberboard	B-1	520	.046			1.65		1.65	2.54
2250	Lath and plaster		260	.092			3.30		3.30	5.10
2300	Plasterboard (drywall)		520	.046			1.65		1.65	2.54
2350	Plywood		450	.053			1.91		1.91	2.93
2800	Paneling, 4' x 8' sheets	1 Clab	475	.017			.59		.59	.91
3000	Plaster, lime and horsehair, on wood lath		400	.020			.70		.70	1.08
3020	On metal lath		335	.024			.84		.84	1.29
3400	Gypsum or perlite, on gypsum lath		410	.020			.68		.68	1.05
3420	On metal lath		300	.027			.94		.94	1.44

09 05 Common Work Results for Finishes

09 05 05 – Selective Finishes Demolition

09 05 05.30 Selective Demolition, Walls and Partitions

		Crew	Daily Output	Labor-Hours	Unit	Material	2012 Bare Costs Labor	Equipment	Total	Total Incl O&P
3450	Plaster, interior gypsum, acoustic, or cement	1 Clab	60	.133	S.Y.		4.68		4.68	7.20
3500	Stucco, on masonry		145	.055			1.94		1.94	2.98
3510	Commercial 3-coat		80	.100			3.51		3.51	5.40
3520	Interior stucco		25	.320			11.25		11.25	17.30
3600	Plywood, one side	B-1	1500	.016	S.F.		.57		.57	.88
3750	Terra cotta block and plaster, to 6" thick	"	175	.137			4.91		4.91	7.55
3760	Tile, ceramic, on walls, thin set	1 Clab	300	.027			.94		.94	1.44
3765	Mud set		250	.032			1.12		1.12	1.73
3800	Toilet partitions, slate or marble		5	1.600	Ea.		56		56	86.50
3820	Metal or plastic		8	1	"		35		35	54
5000	Wallcovering, vinyl	1 Pape	700	.011	S.F.		.44		.44	.66
5010	With release agent		1500	.005			.21		.21	.31
5025	Wallpaper, 2 layers or less, by hand		250	.032			1.23		1.23	1.84
5035	3 layers or more		165	.048			1.87		1.87	2.79
5040	Designer		480	.017			.64		.64	.96

09 21 Plaster and Gypsum Board Assemblies

09 21 13 – Plaster Assemblies

09 21 13.10 Plaster Partition Wall

		Crew	Daily Output	Labor-Hours	Unit	Material	2012 Bare Costs Labor	Equipment	Total	Total Incl O&P
0010	**PLASTER PARTITION WALL**									
0400	Stud walls, 3.4 lb. metal lath, 3 coat gypsum plaster, 2 sides									
0600	2" x 4" wood studs, 16" O.C.	J-2	315	.152	S.F.	3.41	5.90	.42	9.73	13
0700	2-1/2" metal studs, 25 ga., 12" O.C.		325	.148		3.15	5.70	.40	9.25	12.40
0800	3-5/8" metal studs, 25 ga., 16" O.C.		320	.150		3.17	5.80	.41	9.38	12.60
0900	Gypsum lath, 2 coat vermiculite plaster, 2 sides									
1000	2" x 4" wood studs, 16" O.C.	J-2	355	.135	S.F.	3.80	5.20	.37	9.37	12.40
1200	2-1/2" metal studs, 25 ga., 12" O.C.		365	.132		3.37	5.10	.36	8.83	11.70
1300	3-5/8" metal studs, 25 ga., 16" O.C.		360	.133		3.46	5.15	.36	8.97	11.90

09 21 16 – Gypsum Board Assemblies

09 21 16.23 Gypsum Board Shaft Wall Assemblies

		Crew	Daily Output	Labor-Hours	Unit	Material	2012 Bare Costs Labor	Equipment	Total	Total Incl O&P
0010	**GYPSUM BOARD SHAFT WALL ASSEMBLIES**									
0020	Cavity type on 25 ga. J-track & C-H studs, 24" O.C.									
0030	1" thick coreboard wall liner on shaft side									
0040	2-hour assembly with double layer									
0060	5/8" fire rated gypsum board on room side	2 Carp	220	.073	S.F.	1.54	3.21		4.75	6.60
0100	3-hour assembly with triple layer									
0300	5/8" fire rated gypsum board on room side	2 Carp	180	.089	S.F.	1.29	3.92		5.21	7.45
0400	4-hour assembly, 1" coreboard, 5/8" fire rated gypsum board									
0600	and 3/4" galv. metal furring channels, 24" O.C., with									
0700	Double layer 5/8" fire rated gypsum board on room side	2 Carp	110	.145	S.F.	1.25	6.40		7.65	11.25
0900	For taping & finishing, add per side	1 Carp	1050	.008	"	.05	.34		.39	.57
1000	For insulation, see Section 07 21									
5200	For work over 8' high, add	2 Carp	3060	.005	S.F.		.23		.23	.35
5300	For distribution cost over 3 stories high, add per story	"	6100	.003	"		.12		.12	.18

09 21 16.33 Partition Wall

		Crew	Daily Output	Labor-Hours	Unit	Material	2012 Bare Costs Labor	Equipment	Total	Total Incl O&P
0010	**PARTITION WALL** Stud wall, 8' to 12' high									
0050	1/2", interior, gypsum board, std, tape & finish 2 sides									
0500	Installed on and incl., 2" x 4" wood studs, 16" O.C.	2 Carp	310	.052	S.F.	.90	2.28		3.18	4.49
1000	Metal studs, NLB, 25 ga., 16" O.C., 3-5/8" wide		350	.046		.90	2.02		2.92	4.09
1200	6" wide		330	.048		1.08	2.14		3.22	4.48

09 21 Plaster and Gypsum Board Assemblies

09 21 16 – Gypsum Board Assemblies

09 21 16.33 Partition Wall	Crew	Daily Output	Labor-Hours	Unit	Material	2012 Bare Costs Labor	Equipment	Total	Total Incl O&P
1400 Water resistant, on 2" x 4" wood studs, 16" O.C.	2 Carp	310	.052	S.F.	1.16	2.28		3.44	4.78
1600 Metal studs, NLB, 25 ga., 16" O.C., 3-5/8" wide		350	.046		1.16	2.02		3.18	4.38
1800 6" wide		330	.048		1.34	2.14		3.48	4.77
2000 Fire res., 2 layers, 1-1/2 hr., on 2" x 4" wood studs, 16" O.C.		210	.076		1.50	3.36		4.86	6.80
2200 Metal studs, NLB, 25 ga., 16" O.C., 3-5/8" wide		250	.064		1.50	2.82		4.32	6
2400 6" wide		230	.070		1.68	3.07		4.75	6.55
2600 Fire & water res., 2 layers, 1-1/2 hr., 2" x 4" studs, 16" O.C.		210	.076		1.50	3.36		4.86	6.80
2800 Metal studs, NLB, 25 ga., 16" O.C., 3-5/8" wide		250	.064		1.50	2.82		4.32	6
3000 6" wide	▼	230	.070	▼	1.68	3.07		4.75	6.55
3200 5/8", interior, gypsum board, standard, tape & finish 2 sides									
3400 Installed on and including 2" x 4" wood studs, 16" O.C.	2 Carp	300	.053	S.F.	.96	2.35		3.31	4.68
3600 24" O.C.		330	.048		.88	2.14		3.02	4.26
3800 Metal studs, NLB, 25 ga., 16" O.C., 3-5/8" wide		340	.047		.96	2.08		3.04	4.25
4000 6" wide		320	.050		1.14	2.21		3.35	4.65
4200 24" O.C., 3-5/8" wide		360	.044		.86	1.96		2.82	3.97
4400 6" wide		340	.047		1	2.08		3.08	4.29
4800 Water resistant, on 2" x 4" wood studs, 16" O.C.		300	.053		1.32	2.35		3.67	5.05
5000 24" O.C.		330	.048		1.24	2.14		3.38	4.65
5200 Metal studs, NLB, 25 ga. 16" O.C., 3-5/8" wide		340	.047		1.32	2.08		3.40	4.64
5400 6" wide		320	.050		1.50	2.21		3.71	5.05
5600 24" O.C., 3-5/8" wide		360	.044		1.22	1.96		3.18	4.37
5800 6" wide		340	.047		1.36	2.08		3.44	4.68
6000 Fire resistant, 2 layers, 2 hr., on 2" x 4" wood studs, 16" O.C.		205	.078		1.46	3.44		4.90	6.90
6200 24" O.C.		235	.068		1.46	3		4.46	6.20
6400 Metal studs, NLB, 25 ga., 16" O.C., 3-5/8" wide		245	.065		1.59	2.88		4.47	6.20
6600 6" wide		225	.071		1.72	3.14		4.86	6.70
6800 24" O.C., 3-5/8" wide		265	.060		1.44	2.66		4.10	5.70
7000 6" wide		245	.065		1.58	2.88		4.46	6.15
7200 Fire & water resistant, 2 layers, 2 hr., 2" x 4" studs, 16" O.C.		205	.078		1.54	3.44		4.98	7
7400 24" O.C.		235	.068		1.46	3		4.46	6.20
7600 Metal studs, NLB, 25 ga., 16" O.C., 3-5/8" wide		245	.065		1.54	2.88		4.42	6.10
7800 6" wide		225	.071		1.72	3.14		4.86	6.70
8000 24" O.C., 3-5/8" wide		265	.060		1.44	2.66		4.10	5.70
8200 6" wide	▼	245	.065	▼	1.58	2.88		4.46	6.15
8600 1/2" blueboard, mesh tape both sides									
8620 Installed on and including 2" x 4" wood studs, 16" O.C.	2 Carp	300	.053	S.F.	.96	2.35		3.31	4.68
8640 Metal studs, NLB, 25 ga., 16" O.C., 3-5/8" wide		340	.047		.96	2.08		3.04	4.25
8660 6" wide	▼	320	.050	▼	1.14	2.21		3.35	4.65
9000 Exterior, 1/2" gypsum sheathing, 1/2" gypsum finished, interior,									
9100 including foil faced insulation, metal studs, 20 ga.									
9200 16" O.C., 3-5/8" wide	2 Carp	290	.055	S.F.	1.54	2.43		3.97	5.45
9400 6" wide	▼	270	.059	▼	1.77	2.61		4.38	5.95
9600 Partitions, for work over 8' high, add	▼	1530	.010	▼		.46		.46	.71

09 22 Supports for Plaster and Gypsum Board

09 22 03 – Fastening Methods for Finishes

09 22 03.20 Drilling Plaster/Drywall

		Crew	Daily Output	Labor-Hours	Unit	Material	2012 Bare Costs Labor	Equipment	Total	Total Incl O&P
0010	**DRILLING PLASTER/DRYWALL**									
1100	Drilling & layout for drywall/plaster walls, up to 1" deep, no anchor									
1200	Holes, 1/4" diameter	1 Carp	150	.053	Ea.	.01	2.35		2.36	3.63
1300	3/8" diameter		140	.057		.01	2.52		2.53	3.89
1400	1/2" diameter		130	.062		.01	2.71		2.72	4.19
1500	3/4" diameter		120	.067		.01	2.94		2.95	4.53
1600	1" diameter		110	.073		.02	3.21		3.23	4.95
1700	1-1/4" diameter		100	.080		.03	3.53		3.56	5.50
1800	1-1/2" diameter		90	.089		.04	3.92		3.96	6.10
1900	For ceiling installations, add						40%			

09 22 13 – Metal Furring

09 22 13.13 Metal Channel Furring

		Crew	Daily Output	Labor-Hours	Unit	Material	2012 Bare Costs Labor	Equipment	Total	Total Incl O&P
0010	**METAL CHANNEL FURRING**									
0030	Beams and columns, 7/8" channels, galvanized, 12" O.C.	1 Lath	155	.052	S.F.	.37	2.03		2.40	3.38
0050	16" O.C.		170	.047		.31	1.85		2.16	3.05
0070	24" O.C.		185	.043		.20	1.70		1.90	2.71
0100	Ceilings, on steel, 7/8" channels, galvanized, 12" O.C.		210	.038		.34	1.50		1.84	2.56
0300	16" O.C.		290	.028		.31	1.08		1.39	1.93
0400	24" O.C.		420	.019		.20	.75		.95	1.32
0600	1-5/8" channels, galvanized, 12" O.C.		190	.042		.45	1.65		2.10	2.93
0700	16" O.C.		260	.031		.41	1.21		1.62	2.22
0900	24" O.C.		390	.021		.27	.81		1.08	1.48
0930	7/8" channels with sound isolation clips, 12" O.C.		120	.067		1.40	2.62		4.02	5.40
0940	16" O.C.		100	.080		1.93	3.14		5.07	6.75
0950	24" O.C.		165	.048		1.27	1.90		3.17	4.18
0960	1-5/8" channels, galvanized, 12" O.C.		110	.073		1.52	2.85		4.37	5.85
0970	16" O.C.		100	.080		2.04	3.14		5.18	6.85
0980	24" O.C.		155	.052		1.34	2.03		3.37	4.44
1000	Walls, 7/8" channels, galvanized, 12" O.C.		235	.034		.34	1.34		1.68	2.33
1200	16" O.C.		265	.030		.31	1.19		1.50	2.08
1300	24" O.C.		350	.023		.20	.90		1.10	1.54
1500	1-5/8" channels, galvanized, 12" O.C.		210	.038		.45	1.50		1.95	2.69
1600	16" O.C.		240	.033		.41	1.31		1.72	2.37
1800	24" O.C.		305	.026		.27	1.03		1.30	1.81
1920	7/8" channels with sound isolation clips, 12" O.C.		125	.064		1.40	2.51		3.91	5.25
1940	16" O.C.		100	.080		1.93	3.14		5.07	6.75
1950	24" O.C.		150	.053		1.27	2.09		3.36	4.46
1960	1-5/8" channels, galvanized, 12" O.C.		115	.070		1.52	2.73		4.25	5.70
1970	16" O.C.		95	.084		2.04	3.31		5.35	7.10
1980	24" O.C.		140	.057		1.34	2.24		3.58	4.76

09 22 16 – Non-Structural Metal Framing

09 22 16.13 Non-Structural Metal Stud Framing

		Crew	Daily Output	Labor-Hours	Unit	Material	2012 Bare Costs Labor	Equipment	Total	Total Incl O&P
0010	**NON-STRUCTURAL METAL STUD FRAMING**									
1600	Non-load bearing, galv., 8' high, 25 ga. 1-5/8" wide, 16" O.C.	1 Carp	619	.013	S.F.	.29	.57		.86	1.19
1610	24" O.C.		950	.008		.21	.37		.58	.81
1620	2-1/2" wide, 16" O.C.		613	.013		.33	.58		.91	1.25
1630	24" O.C.		938	.009		.25	.38		.63	.85
1640	3-5/8" wide, 16" O.C.		600	.013		.39	.59		.98	1.33
1650	24" O.C.		925	.009		.29	.38		.67	.91
1660	4" wide, 16" O.C.		594	.013		.45	.59		1.04	1.40
1670	24" O.C.		925	.009		.34	.38		.72	.96
1680	6" wide, 16" O.C.		588	.014		.58	.60		1.18	1.56

09 22 16 – Non-Structural Metal Framing

09 22 16.13 Non-Structural Metal Stud Framing	Crew	Daily Output	Labor-Hours	Unit	Material	2012 Bare Costs Labor	Equipment	Total	Total Incl O&P
1690	24" O.C.	1 Carp	906	.009	S.F.	.43	.39		.82	1.08
1700	20 ga. studs, 1-5/8" wide, 16" O.C.		494	.016		.44	.71		1.15	1.58
1710	24" O.C.		763	.010		.33	.46		.79	1.07
1720	2-1/2" wide, 16" O.C.		488	.016		.50	.72		1.22	1.66
1730	24" O.C.		750	.011		.38	.47		.85	1.14
1740	3-5/8" wide, 16" O.C.		481	.017		.58	.73		1.31	1.77
1750	24" O.C.		738	.011		.43	.48		.91	1.22
1760	4" wide, 16" O.C.		475	.017		.69	.74		1.43	1.89
1770	24" O.C.		738	.011		.51	.48		.99	1.31
1780	6" wide, 16" O.C.		469	.017		.83	.75		1.58	2.07
1790	24" O.C.		725	.011		.62	.49		1.11	1.43
2000	Non-load bearing, galv., 10' high, 25 ga. 1-5/8" wide, 16" O.C.		495	.016		.27	.71		.98	1.40
2100	24" O.C.		760	.011		.20	.46		.66	.93
2200	2-1/2" wide, 16" O.C.		490	.016		.31	.72		1.03	1.45
2250	24" O.C.		750	.011		.23	.47		.70	.97
2300	3-5/8" wide, 16" O.C.		480	.017		.37	.74		1.11	1.54
2350	24" O.C.		740	.011		.27	.48		.75	1.03
2400	4" wide, 16" O.C.		475	.017		.43	.74		1.17	1.61
2450	24" O.C.		740	.011		.31	.48		.79	1.07
2500	6" wide, 16" O.C.		470	.017		.55	.75		1.30	1.76
2550	24" O.C.		725	.011		.40	.49		.89	1.19
2600	20 ga. studs, 1-5/8" wide, 16" O.C.		395	.020		.42	.89		1.31	1.83
2650	24" O.C.		610	.013		.31	.58		.89	1.23
2700	2-1/2" wide, 16" O.C.		390	.021		.48	.90		1.38	1.92
2750	24" O.C.		600	.013		.35	.59		.94	1.29
2800	3-5/8" wide, 16" OC		385	.021		.55	.92		1.47	2.01
2850	24" O.C.		590	.014		.41	.60		1.01	1.37
2900	4" wide, 16" O.C.		380	.021		.65	.93		1.58	2.15
2950	24" O.C.		590	.014		.48	.60		1.08	1.45
3000	6" wide, 16" O.C.		375	.021		.78	.94		1.72	2.31
3050	24" O.C.		580	.014		.58	.61		1.19	1.57
3060	Non-load bearing, galv., 12' high, 25 ga. 1-5/8" wide, 16" O.C.		413	.019		.26	.85		1.11	1.59
3070	24" O.C.		633	.013		.19	.56		.75	1.07
3080	2-1/2" wide, 16" O.C.		408	.020		.30	.86		1.16	1.66
3090	24" O.C.		625	.013		.22	.56		.78	1.11
3100	3-5/8" wide, 16" O.C.		400	.020		.35	.88		1.23	1.75
3110	24" O.C.		617	.013		.26	.57		.83	1.16
3120	4" wide, 16" O.C.		396	.020		.41	.89		1.30	1.82
3130	24" O.C.		617	.013		.30	.57		.87	1.21
3140	6" wide, 16" O.C.		392	.020		.53	.90		1.43	1.96
3150	24" O.C.		604	.013		.38	.58		.96	1.32
3160	20 ga. studs, 1-5/8" wide, 16" O.C.		329	.024		.40	1.07		1.47	2.09
3170	24" O.C.		508	.016		.29	.69		.98	1.39
3180	2-1/2" wide, 16" O.C.		325	.025		.46	1.09		1.55	2.17
3190	24" O.C.		500	.016		.33	.71		1.04	1.46
3200	3-5/8" wide, 16" O.C.		321	.025		.53	1.10		1.63	2.27
3210	24" O.C.		492	.016		.38	.72		1.10	1.52
3220	4" wide, 16" O.C.		317	.025		.62	1.11		1.73	2.39
3230	24" O.C.		492	.016		.45	.72		1.17	1.60
3240	6" wide, 16" O.C.		313	.026		.75	1.13		1.88	2.55
3250	24" O.C.	▼	483	.017	▼	.54	.73		1.27	1.72
5000	Load bearing studs, see Section 05 41 13.30									

09 22 26 – Suspension Systems

09 22 26.13 Ceiling Suspension Systems	Crew	Daily Output	Labor-Hours	Unit	Material	2012 Bare Costs Labor	Equipment	Total	Total Incl O&P
0010 **CEILING SUSPENSION SYSTEMS** for gypsum board or plaster									
8000 Suspended ceilings, including carriers									
8200 1-1/2" carriers, 24" O.C. with:									
8300 7/8" channels, 16" O.C.	1 Lath	275	.029	S.F.	.53	1.14		1.67	2.27
8320 24" O.C.		310	.026		.43	1.01		1.44	1.97
8400 1-5/8" channels, 16" O.C.		205	.039		.64	1.53		2.17	2.95
8420 24" O.C.		250	.032		.50	1.26		1.76	2.39
8600 2" carriers, 24" O.C. with:									
8700 7/8" channels, 16" O.C.	1 Lath	250	.032	S.F.	.56	1.26		1.82	2.45
8720 24" O.C.		285	.028		.45	1.10		1.55	2.12
8800 1-5/8" channels, 16" O.C.		190	.042		.66	1.65		2.31	3.15
8820 24" O.C.		225	.036		.52	1.40		1.92	2.62

09 22 36 – Lath

09 22 36.13 Gypsum Lath

	Crew	Daily Output	Labor-Hours	Unit	Material	2012 Bare Costs Labor	Equipment	Total	Total Incl O&P
0010 **GYPSUM LATH** R092000-50									
0020 Plain or perforated, nailed, 3/8" thick	1 Lath	85	.094	S.Y.	3.87	3.69		7.56	9.65
0100 1/2" thick		80	.100		4.50	3.93		8.43	10.70
0300 Clipped to steel studs, 3/8" thick		75	.107		3.87	4.19		8.06	10.40
0400 1/2" thick		70	.114		4.50	4.49		8.99	11.55
1500 For ceiling installations, add		216	.037			1.45		1.45	2.13
1600 For columns and beams, add		170	.047			1.85		1.85	2.71

09 22 36.23 Metal Lath

	Crew	Daily Output	Labor-Hours	Unit	Material	2012 Bare Costs Labor	Equipment	Total	Total Incl O&P
0010 **METAL LATH** R092000-50									
0020 Diamond, expanded, 2.5 lb. per S.Y., painted				S.Y.	3.44			3.44	3.78
0100 Galvanized					3.02			3.02	3.32
0300 3.4 lb. per S.Y., painted					3.60			3.60	3.96
0400 Galvanized					4.46			4.46	4.91
0600 For 15# asphalt sheathing paper, add					.51			.51	.56
0900 Flat rib, 1/8" high, 2.75 lb., painted					3.14			3.14	3.45
1000 Foil backed					2.83			2.83	3.11
1200 3.4 lb. per S.Y., painted					4			4	4.40
1300 Galvanized					3.50			3.50	3.85
1500 For 15# asphalt sheathing paper, add					.51			.51	.56
1800 High rib, 3/8" high, 3.4 lb. per S.Y., painted					4.84			4.84	5.30
1900 Galvanized					4.84			4.84	5.30
2400 3/4" high, painted, .60 lb. per S.F.				S.F.	.50			.50	.55
2500 .75 lb. per S.F.				"	1.08			1.08	1.19
2800 Stucco mesh, painted, 3.6 lb.				S.Y.	3.51			3.51	3.86
3000 K-lath, perforated, absorbent paper, regular					4.42			4.42	4.86
3100 Heavy duty					5.20			5.20	5.75
3300 Waterproof, heavy duty, grade B backing					5.10			5.10	5.60
3400 Fire resistant backing					5.65			5.65	6.20
3600 2.5 lb. diamond painted, on wood framing, on walls	1 Lath	85	.094		3.44	3.69		7.13	9.20
3700 On ceilings		75	.107		3.44	4.19		7.63	9.95
3900 3.4 lb. diamond painted, on wood framing, on walls		80	.100		4	3.93		7.93	10.15
4000 On ceilings		70	.114		4	4.49		8.49	11
4200 3.4 lb. diamond painted, wired to steel framing		75	.107		4	4.19		8.19	10.55
4300 On ceilings		60	.133		4	5.25		9.25	12.10
4500 Columns and beams, wired to steel		40	.200		4	7.85		11.85	15.90
4600 Cornices, wired to steel		35	.229		4	8.95		12.95	17.55
4800 Screwed to steel studs, 2.5 lb.		80	.100		3.44	3.93		7.37	9.55

09 22 36.23 Metal Lath

		Crew	Daily Output	Labor-Hours	Unit	Material	2012 Bare Costs Labor	2012 Bare Costs Equipment	Total	Total Incl O&P
4900	3.4 lb.	1 Lath	75	.107	S.Y.	3.60	4.19		7.79	10.10
5100	Rib lath, painted, wired to steel, on walls, 2.5 lb.		75	.107		3.14	4.19		7.33	9.60
5200	3.4 lb.		70	.114		4.84	4.49		9.33	11.90
5400	4.0 lb.		65	.123		5.70	4.83		10.53	13.35
5500	For self-furring lath, add					.11			.11	.12
5700	Suspended ceiling system, incl. 3.4 lb. diamond lath, painted	1 Lath	15	.533		3.39	21		24.39	34
5800	Galvanized	"	15	.533		6.50	21		27.50	37.50
6000	Hollow metal stud partitions, 3.4 lb. painted lath both sides									
6010	Non-load bearing, 25 ga., w/rib lath 2-1/2" studs, 12" O.C.	1 Lath	20.30	.394	S.Y.	13.25	15.45		28.70	37
6300	16" O.C.		21.10	.379		12.45	14.90		27.35	35.50
6350	24" O.C.		22.70	.352		11.75	13.85		25.60	33.50
6400	3-5/8" studs, 16" O.C.		19.50	.410		13	16.10		29.10	38
6600	24" O.C.		20.40	.392		12.10	15.40		27.50	36
6700	4" studs, 16" O.C.		20.40	.392		13.50	15.40		28.90	37.50
6900	24" O.C.		21.60	.370		12.50	14.55		27.05	35.50
7000	6" studs, 16" O.C.		19.50	.410		14.65	16.10		30.75	39.50
7100	24" O.C.		21.10	.379		13.30	14.90		28.20	36.50
7200	L.B. partitions, 16 ga., w/rib lath, 2-1/2" studs, 16" O.C.		20	.400		14.50	15.70		30.20	39
7300	3-5/8" studs, 16 ga.		19.70	.406		16.35	15.95		32.30	41.50
7500	4" studs, 16 ga.		19.50	.410		17	16.10		33.10	42
7600	6" studs, 16 ga.		18.70	.428		19.95	16.80		36.75	46.50

09 22 36.43 Security Mesh

		Crew	Daily Output	Labor-Hours	Unit	Material	2012 Bare Costs Labor	2012 Bare Costs Equipment	Total	Total Incl O&P
0010	**SECURITY MESH**, expanded metal, flat, screwed to framing									
0100	On walls, 3/4", 1.76 lb./S.F.	2 Carp	1500	.011	S.F.	1.82	.47		2.29	2.72
0110	1-1/2", 1.14 lb./S.F.		1600	.010		1.54	.44		1.98	2.37
0200	On ceilings, 3/4", 1.76 lb./S.F.		1350	.012		1.82	.52		2.34	2.80
0210	1-1/2", 1.14 lb./S.F.		1450	.011		1.54	.49		2.03	2.44

09 22 36.83 Accessories, Plaster

		Crew	Daily Output	Labor-Hours	Unit	Material	2012 Bare Costs Labor	2012 Bare Costs Equipment	Total	Total Incl O&P
0010	**ACCESSORIES, PLASTER**									
0020	Casing bead, expanded flange, galvanized	1 Lath	2.70	2.963	C.L.F.	53	116		169	230
0200	Foundation weep screed, galvanized	"	2.70	2.963		52	116		168	228
0900	Channels, cold rolled, 16 ga., 3/4" deep, galvanized					34			34	37.50
1200	1-1/2" deep, 16 ga., galvanized					45.50			45.50	50
1620	Corner bead, expanded bullnose, 3/4" radius, #10, galvanized	1 Lath	2.60	3.077		27.50	121		148.50	207
1650	#1, galvanized		2.55	3.137		46.50	123		169.50	232
1670	Expanded wing, 2-3/4" wide, #1, galvanized		2.65	3.019		34	118		152	212
1700	Inside corner (corner rite), 3" x 3", painted		2.60	3.077		27	121		148	207
1750	Strip-ex, 4" wide, painted		2.55	3.137		34.50	123		157.50	219
1800	Expansion joint, 3/4" grounds, limited expansion, galv., 1 piece		2.70	2.963		92	116		208	272
2100	Extreme expansion, galvanized, 2 piece		2.60	3.077		149	121		270	340

09 23 Gypsum Plastering

09 23 13 – Acoustical Gypsum Plastering

09 23 13.10 Perlite or Vermiculite Plaster

	Crew	Daily Output	Labor-Hours	Unit	Material	2012 Bare Costs Labor	2012 Bare Costs Equipment	Total	Total Incl O&P
0010 **PERLITE OR VERMICULITE PLASTER** R092000-50									
0020 In 100 lb. bags, under 200 bags				Bag	14.55			14.55	16.05
0100 Over 200 bags				"	13.30			13.30	14.65
0300 2 coats, no lath included, on walls	J-1	92	.435	S.Y.	3.78	16.75	1.42	21.95	30.50
0400 On ceilings	"	79	.506		3.78	19.50	1.66	24.94	35
0600 On and incl. 3/8" gypsum lath, on metal studs	J-2	84	.571		8.60	22	1.56	32.16	44
0700 On ceilings	"	70	.686		8.60	26.50	1.87	36.97	51
0900 3 coats, no lath included, on walls	J-1	74	.541		6.30	21	1.77	29.07	40
1000 On ceilings	"	63	.635		6.30	24.50	2.08	32.88	45.50
1200 On and incl. painted metal lath, on metal studs	J-2	72	.667		11.15	26	1.82	38.97	53
1300 On ceilings		61	.787		11.15	30.50	2.15	43.80	60
1500 On and incl. suspended metal lath ceiling		37	1.297		9.70	50	3.54	63.24	89.50
1700 For irregular or curved surfaces, add to above						30%			
1800 For columns and beams, add to above						50%			
1900 For soffits, add to ceiling prices						40%			

09 23 20 – Gypsum Plaster

09 23 20.10 Gypsum Plaster On Walls and Ceilings

	Crew	Daily Output	Labor-Hours	Unit	Material	2012 Bare Costs Labor	2012 Bare Costs Equipment	Total	Total Incl O&P
0010 **GYPSUM PLASTER ON WALLS AND CEILINGS** R092000-50									
0020 80# bag, less than 1 ton				Bag	14.80			14.80	16.30
0100 Over 1 ton				"	13.05			13.05	14.35
0300 2 coats, no lath included, on walls	J-1	105	.381	S.Y.	3.57	14.65	1.25	19.47	27.50
0400 On ceilings	"	92	.435		3.57	16.75	1.42	21.74	30.50
0600 On and incl. 3/8" gypsum lath on steel, on walls	J-2	97	.495		7.45	19.10	1.35	27.90	38
0700 On ceilings	"	83	.578		7.45	22.50	1.58	31.53	43.50
0900 3 coats, no lath included, on walls	J-1	87	.460		5.15	17.70	1.50	24.35	34
1000 On ceilings	"	78	.513		5.15	19.75	1.68	26.58	37
1200 On and including painted metal lath, on wood studs	J-2	86	.558		10.05	21.50	1.52	33.07	44.50
1300 On ceilings	"	76.50	.627		10.05	24	1.71	35.76	49
1600 For irregular or curved surfaces, add						30%			
1800 For columns & beams, add						50%			

09 23 20.20 Gauging Plaster

	Crew	Daily Output	Labor-Hours	Unit	Material	2012 Bare Costs Labor	2012 Bare Costs Equipment	Total	Total Incl O&P
0010 **GAUGING PLASTER** R092000-50									
0020 100 lb. bags, less than 1 ton				Bag	18.20			18.20	20
0100 Over 1 ton				"	15.85			15.85	17.45

09 23 20.30 Keenes Cement

	Crew	Daily Output	Labor-Hours	Unit	Material	2012 Bare Costs Labor	2012 Bare Costs Equipment	Total	Total Incl O&P
0010 **KEENES CEMENT** R092000-50									
0020 In 100 lb. bags, less than 1 ton				Bag	22			22	24
0100 Over 1 ton				"	21			21	23
0300 Finish only, add to plaster prices, standard	J-1	215	.186	S.Y.	2	7.15	.61	9.76	13.55
0400 High quality	"	144	.278	"	2.03	10.70	.91	13.64	19.25

09 24 Cement Plastering

09 24 23 – Cement Stucco

09 24 23.40 Stucco

		Crew	Daily Output	Labor-Hours	Unit	Material	2012 Bare Costs Labor	Equipment	Total	Total Incl O&P
0010	**STUCCO** R092000-50									
0015	3 coats 1" thick, float finish, with mesh, on wood frame	J-2	63	.762	S.Y.	5.80	29.50	2.08	37.38	52.50
0100	On masonry construction, no mesh incl.	J-1	67	.597		2.28	23	1.95	27.23	39
0300	For trowel finish, add	1 Plas	170	.047			1.90		1.90	2.84
0400	For 3/4" thick, on masonry, deduct	J-1	880	.045		.61	1.75	.15	2.51	3.45
0600	For coloring and special finish, add, minimum		685	.058		.39	2.25	.19	2.83	4.01
0700	Maximum		200	.200		1.36	7.70	.65	9.71	13.75
0900	For soffits, add	J-2	155	.310		2.11	11.95	.85	14.91	21
1000	Exterior stucco, with bonding agent, 3 coats, on walls, no mesh incl.	J-1	200	.200		3.54	7.70	.65	11.89	16.15
1200	Ceilings		180	.222		3.54	8.55	.73	12.82	17.50
1300	Beams		80	.500		3.54	19.25	1.64	24.43	34.50
1500	Columns		100	.400		3.54	15.40	1.31	20.25	28.50
1600	Mesh, painted, nailed to wood, 1.8 lb.	1 Lath	60	.133		5	5.25		10.25	13.20
1800	3.6 lb.		55	.145		3.51	5.70		9.21	12.25
1900	Wired to steel, painted, 1.8 lb.		53	.151		5	5.90		10.90	14.20
2100	3.6 lb.		50	.160		3.51	6.30		9.81	13.05

09 26 Veneer Plastering

09 26 13 – Gypsum Veneer Plastering

09 26 13.20 Blueboard

		Crew	Daily Output	Labor-Hours	Unit	Material	2012 Bare Costs Labor	Equipment	Total	Total Incl O&P
0010	**BLUEBOARD** For use with thin coat									
0100	plaster application (see Section 09 26 13.80)									
1000	3/8" thick, on walls or ceilings, standard, no finish included	2 Carp	1900	.008	S.F.	.23	.37		.60	.82
1100	With thin coat plaster finish		875	.018		.33	.81		1.14	1.60
1400	On beams, columns, or soffits, standard, no finish included		675	.024		.26	1.05		1.31	1.90
1450	With thin coat plaster finish		475	.034		.36	1.49		1.85	2.69
3000	1/2" thick, on walls or ceilings, standard, no finish included		1900	.008		.25	.37		.62	.85
3100	With thin coat plaster finish		875	.018		.35	.81		1.16	1.62
3300	Fire resistant, no finish included		1900	.008		.25	.37		.62	.85
3400	With thin coat plaster finish		875	.018		.35	.81		1.16	1.62
3450	On beams, columns, or soffits, standard, no finish included		675	.024		.29	1.05		1.34	1.93
3500	With thin coat plaster finish		475	.034		.38	1.49		1.87	2.71
3700	Fire resistant, no finish included		675	.024		.29	1.05		1.34	1.93
3800	With thin coat plaster finish		475	.034		.38	1.49		1.87	2.71
5000	5/8" thick, on walls or ceilings, fire resistant, no finish included		1900	.008		.25	.37		.62	.85
5100	With thin coat plaster finish		875	.018		.35	.81		1.16	1.62
5500	On beams, columns, or soffits, no finish included		675	.024		.29	1.05		1.34	1.93
5600	With thin coat plaster finish		475	.034		.38	1.49		1.87	2.71
6000	For high ceilings, over 8' high, add		3060	.005			.23		.23	.35
6500	For over 3 stories high, add per story		6100	.003			.12		.12	.18

09 26 13.80 Thin Coat Plaster

		Crew	Daily Output	Labor-Hours	Unit	Material	2012 Bare Costs Labor	Equipment	Total	Total Incl O&P
0010	**THIN COAT PLASTER**									
0012	1 coat veneer, not incl. lath	J-1	3600	.011	S.F.	.10	.43	.04	.57	.79
1000	In 50 lb. bags				Bag	12.95			12.95	14.25

09 28 Backing Boards and Underlayments

09 28 13 – Cementitious Backing Boards

09 28 13.10 Cementitious Backerboard	Crew	Daily Output	Labor-Hours	Unit	Material	2012 Bare Costs Labor	Equipment	Total	Total Incl O&P
0010 **CEMENTITIOUS BACKERBOARD**									
0070 Cementitious backerboard, on floor, 3' x 4' x 1/2" sheets	2 Carp	525	.030	S.F.	.73	1.34		2.07	2.88
0080 3' x 5' x 1/2" sheets		525	.030		.73	1.34	'	2.07	2.87
0090 3' x 6' x 1/2" sheets		525	.030		.68	1.34		2.02	2.82
0100 3' x 4' x 5/8" sheets		525	.030		.98	1.34		2.32	3.15
0110 3' x 5' x 5/8" sheets		525	.030		1	1.34		2.34	3.17
0120 3' x 6' x 5/8" sheets		525	.030		.90	1.34		2.24	3.06
0150 On wall, 3' x 4' x 1/2" sheets		350	.046		.73	2.02		2.75	3.91
0160 3' x 5' x 1/2" sheets		350	.046		.73	2.02		2.75	3.90
0170 3' x 6' x 1/2" sheets		350	.046		.68	2.02		2.70	3.85
0180 3' x 4' x 5/8" sheets		350	.046		.98	2.02		3	4.18
0190 3' x 5' x 5/8" sheets		350	.046		1	2.02		3.02	4.20
0200 3' x 6' x 5/8" sheets		350	.046		.90	2.02		2.92	4.09
0250 On counter, 3' x 4' x 1/2" sheets		180	.089		.73	3.92		4.65	6.85
0260 3' x 5' x 1/2" sheets		180	.089		.73	3.92		4.65	6.85
0270 3' x 6' x 1/2" sheets		180	.089		.68	3.92		4.60	6.80
0300 3' x 4' x 5/8" sheets		180	.089		.98	3.92		4.90	7.15
0310 3' x 5' x 5/8" sheets		180	.089		1	3.92		4.92	7.15
0320 3' x 6' x 5/8" sheets		180	.089		.90	3.92		4.82	7.05

09 29 Gypsum Board

09 29 10 – Gypsum Board Panels

09 29 10.30 Gypsum Board

	Crew	Daily Output	Labor-Hours	Unit	Material	2012 Bare Costs Labor	Equipment	Total	Total Incl O&P
0010 **GYPSUM BOARD** on walls & ceilings R092910-10									
0100 Nailed or screwed to studs unless otherwise noted									
0150 3/8" thick, on walls, standard, no finish included	2 Carp	2000	.008	S.F.	.22	.35		.57	.78
0200 On ceilings, standard, no finish included		1800	.009		.22	.39		.61	.84
0250 On beams, columns, or soffits, no finish included		675	.024		.22	1.05		1.27	1.85
0300 1/2" thick, on walls, standard, no finish included		2000	.008		.22	.35		.57	.78
0350 Taped and finished (level 4 finish)		965	.017		.27	.73		1	1.42
0390 With compound skim coat (level 5 finish)		775	.021		.31	.91		1.22	1.75
0400 Fire resistant, no finish included		2000	.008		.26	.35		.61	.83
0450 Taped and finished (level 4 finish)		965	.017		.31	.73		1.04	1.47
0490 With compound skim coat (level 5 finish)		775	.021		.35	.91		1.26	1.79
0500 Water resistant, no finish included		2000	.008		.35	.35		.70	.93
0550 Taped and finished (level 4 finish)		965	.017		.40	.73		1.13	1.57
0590 With compound skim coat (level 5 finish)		775	.021		.44	.91		1.35	1.89
0600 Prefinished, vinyl, clipped to studs		900	.018		.47	.78		1.25	1.73
0700 Mold resistant, no finish included		2000	.008		.38	.35		.73	.96
0710 Taped and finished (level 4 finish)		965	.017		.43	.73		1.16	1.60
0720 With compound skim coat (level 5 finish)		775	.021		.47	.91		1.38	1.92
1000 On ceilings, standard, no finish included		1800	.009		.22	.39		.61	.84
1050 Taped and finished (level 4 finish)		765	.021		.27	.92		1.19	1.71
1090 With compound skim coat (level 5 finish)		610	.026		.31	1.16		1.47	2.13
1100 Fire resistant, no finish included		1800	.009		.26	.39		.65	.89
1150 Taped and finished (level 4 finish)		765	.021		.31	.92		1.23	1.76
1195 With compound skim coat (level 5 finish)		610	.026		.35	1.16		1.51	2.17
1200 Water resistant, no finish included		1800	.009		.35	.39		.74	.99
1250 Taped and finished (level 4 finish)		765	.021		.40	.92		1.32	1.86
1290 With compound skim coat (level 5 finish)		610	.026		.44	1.16		1.60	2.27
1310 Mold resistant, no finish included		1800	.009		.38	.39		.77	1.02

09 29 Gypsum Board

09 29 10 – Gypsum Board Panels

09 29 10.30 Gypsum Board	Crew	Daily Output	Labor-Hours	Unit	Material	2012 Bare Costs Labor	Equipment	Total	Total Incl O&P
1320	Taped and finished (level 4 finish)	2 Carp	765	.021	S.F.	.43	.92		1.35	1.89
1330	With compound skim coat (level 5 finish)		610	.026		.47	1.16		1.63	2.30
1500	On beams, columns, or soffits, standard, no finish included		675	.024		.25	1.05		1.30	1.89
1550	Taped and finished (level 4 finish)		475	.034		.27	1.49		1.76	2.58
1590	With compound skim coat (level 5 finish)		540	.030		.31	1.31		1.62	2.36
1600	Fire resistant, no finish included		675	.024		.26	1.05		1.31	1.90
1650	Taped and finished (level 4 finish)		475	.034		.31	1.49		1.80	2.63
1690	With compound skim coat (level 5 finish)		540	.030		.35	1.31		1.66	2.40
1700	Water resistant, no finish included		675	.024		.40	1.05		1.45	2.05
1750	Taped and finished (level 4 finish)		475	.034		.40	1.49		1.89	2.73
1790	With compound skim coat (level 5 finish)		540	.030		.44	1.31		1.75	2.50
1800	Mold resistant, no finish included		675	.024		.44	1.05		1.49	2.09
1810	Taped and finished (level 4 finish)		475	.034		.43	1.49		1.92	2.76
1820	With compound skim coat (level 5 finish)		540	.030		.47	1.31		1.78	2.53
2000	5/8" thick, on walls, standard, no finish included		2000	.008		.25	.35		.60	.82
2050	Taped and finished (level 4 finish)		965	.017		.30	.73		1.03	1.46
2090	With compound skim coat (level 5 finish)		775	.021		.34	.91		1.25	1.78
2100	Fire resistant, no finish included		2000	.008		.27	.35		.62	.84
2150	Taped and finished (level 4 finish)		965	.017		.32	.73		1.05	1.48
2195	With compound skim coat (level 5 finish)		775	.021		.36	.91		1.27	1.80
2200	Water resistant, no finish included		2000	.008		.43	.35		.78	1.01
2250	Taped and finished (level 4 finish)		965	.017		.48	.73		1.21	1.65
2290	With compound skim coat (level 5 finish)		775	.021		.52	.91		1.43	1.98
2300	Prefinished, vinyl, clipped to studs		900	.018		.87	.78		1.65	2.17
2510	Mold resistant, no finish included		2000	.008		.39	.35		.74	.97
2520	Taped and finished (level 4 finish)		965	.017		.44	.73		1.17	1.61
2530	With compound skim coat (level 5 finish)		775	.021		.48	.91		1.39	1.93
3000	On ceilings, standard, no finish included		1800	.009		.25	.39		.64	.88
3050	Taped and finished (level 4 finish)		765	.021		.30	.92		1.22	1.75
3090	With compound skim coat (level 5 finish)		615	.026		.34	1.15		1.49	2.15
3100	Fire resistant, no finish included		1800	.009		.27	.39		.66	.90
3150	Taped and finished (level 4 finish)		765	.021		.32	.92		1.24	1.77
3190	With compound skim coat (level 5 finish)		615	.026		.36	1.15		1.51	2.17
3200	Water resistant, no finish included		1800	.009		.43	.39		.82	1.07
3250	Taped and finished (level 4 finish)		765	.021		.48	.92		1.40	1.94
3290	With compound skim coat (level 5 finish)		615	.026		.52	1.15		1.67	2.35
3300	Mold resistant, no finish included		1800	.009		.39	.39		.78	1.03
3310	Taped and finished (level 4 finish)		765	.021		.44	.92		1.36	1.90
3320	With compound skim coat (level 5 finish)		615	.026		.48	1.15		1.63	2.30
3500	On beams, columns, or soffits, no finish included		675	.024		.29	1.05		1.34	1.93
3550	Taped and finished (level 4 finish)		475	.034		.34	1.49		1.83	2.66
3590	With compound skim coat (level 5 finish)		380	.042		.40	1.86		2.26	3.30
3600	Fire resistant, no finish included		675	.024		.31	1.05		1.36	1.95
3650	Taped and finished (level 4 finish)		475	.034		.36	1.49		1.85	2.69
3690	With compound skim coat (level 5 finish)		380	.042		.36	1.86		2.22	3.26
3700	Water resistant, no finish included		675	.024		.49	1.05		1.54	2.15
3750	Taped and finished (level 4 finish)		475	.034		.55	1.49		2.04	2.89
3790	With compound skim coat (level 5 finish)		380	.042		.52	1.86		2.38	3.44
3800	Mold resistant, no finish included		675	.024		.45	1.05		1.50	2.10
3810	Taped and finished (level 4 finish)		475	.034		.50	1.49		1.99	2.84
3820	With compound skim coat (level 5 finish)		380	.042		.48	1.86		2.34	3.39
4000	Fireproofing, beams or columns, 2 layers, 1/2" thick, incl finish		330	.048		.57	2.14		2.71	3.91
4010	Mold resistant		330	.048		.81	2.14		2.95	4.18

09 29 10.30 Gypsum Board	Crew	Daily Output	Labor-Hours	Unit	Material	2012 Bare Costs Labor	Equipment	Total	Total Incl O&P
4050	5/8" thick	2 Carp	300	.053	S.F.	.63	2.35		2.98	4.31
4060	Mold resistant		300	.053		.87	2.35		3.22	4.58
4100	3 layers, 1/2" thick		225	.071		.83	3.14		3.97	5.75
4110	Mold resistant		225	.071		1.19	3.14		4.33	6.10
4150	5/8" thick		210	.076		.95	3.36		4.31	6.20
4160	Mold resistant		210	.076		1.31	3.36		4.67	6.60
5050	For 1" thick coreboard on columns		480	.033		.54	1.47		2.01	2.85
5100	For foil-backed board, add					.14			.14	.15
5200	For work over 8' high, add	2 Carp	3060	.005			.23		.23	.35
5270	For textured spray, add	2 Lath	1600	.010		.04	.39		.43	.62
5300	For distribution cost over 3 stories high, add per story	2 Carp	6100	.003			.12		.12	.18
5350	For finishing inner corners, add		950	.017	L.F.	.10	.74		.84	1.25
5355	For finishing outer corners, add		1250	.013		.22	.56		.78	1.11
5500	For acoustical sealant, add per bead	1 Carp	500	.016		.04	.71		.75	1.14
5550	Sealant, 1 quart tube				Ea.	7.05			7.05	7.80
6000	Gypsum sound dampening panels									
6010	1/2" thick on walls, multi-layer, light weight, no finish included	2 Carp	1500	.011	S.F.	1.36	.47		1.83	2.22
6015	Taped and finished (level 4 finish)		725	.022		1.41	.97		2.38	3.05
6020	With compound skim coat (level 5 finish)		580	.028		1.45	1.22		2.67	3.47
6025	5/8" thick on walls, for wood studs, no finish included		1500	.011		2.37	.47		2.84	3.33
6030	Taped and finished (level 4 finish)		725	.022		2.42	.97		3.39	4.16
6035	With compound skim coat (level 5 finish)		580	.028		2.46	1.22		3.68	4.58
6040	For metal stud, no finish included		1500	.011		2.08	.47		2.55	3.01
6045	Taped and finished (level 4 finish)		725	.022		2.13	.97		3.10	3.84
6050	With compound skim coat (level 5 finish)		580	.028		2.17	1.22		3.39	4.26
6055	Abuse resist, no finish included		1500	.011		2.54	.47		3.01	3.51
6060	Taped and finished (level 4 finish)		725	.022		2.59	.97		3.56	4.34
6065	With compound skim coat (level 5 finish)		580	.028		2.63	1.22		3.85	4.77
6070	Shear rated, no finish included		1500	.011		2.76	.47		3.23	3.76
6075	Taped and finished (level 4 finish)		725	.022		2.81	.97		3.78	4.59
6080	With compound skim coat (level 5 finish)		580	.028		2.85	1.22		4.07	5
6085	For SCIF applications, no finish included		1500	.011		3.05	.47		3.52	4.08
6090	Taped and finished (level 4 finish)		725	.022		3.10	.97		4.07	4.91
6095	With compound skim coat (level 5 finish)		580	.028		3.14	1.22		4.36	5.35
6100	1-3/8" thick on walls, THX Certified, no finish included		1500	.011		6.35	.47		6.82	7.70
6105	Taped and finished (level 4 finish)		725	.022		6.40	.97		7.37	8.55
6110	With compound skim coat (level 5 finish)		580	.028		6.45	1.22		7.67	8.95
6115	5/8" thick on walls, score & snap installation, no finish included		2000	.008		1.62	.35		1.97	2.32
6120	Taped and finished (level 4 finish)		965	.017		1.67	.73		2.40	2.96
6125	With compound skim coat (level 5 finish)		775	.021		1.71	.91		2.62	3.29
7020	5/8" thick on ceilings, for wood joists, no finish included		1200	.013		2.37	.59		2.96	3.51
7025	Taped and finished (level 4 finish)		510	.031		2.42	1.38		3.80	4.79
7030	With compound skim coat (level 5 finish)		410	.039		2.46	1.72		4.18	5.35
7035	For metal joists, no finish included		1200	.013		2.08	.59		2.67	3.19
7040	Taped and finished (level 4 finish)		510	.031		2.13	1.38		3.51	4.47
7045	With compound skim coat (level 5 finish)		410	.039		2.17	1.72		3.89	5.05
7050	Abuse resist, no finish included		1200	.013		2.54	.59		3.13	3.69
7055	Taped and finished (level 4 finish)		510	.031		2.59	1.38		3.97	4.97
7060	With compound skim coat (level 5 finish)		410	.039		2.63	1.72		4.35	5.55
7065	Shear rated, no finish included		1200	.013		2.76	.59		3.35	3.94
7070	Taped and finished (level 4 finish)		510	.031		2.81	1.38		4.19	5.20
7075	With compound skim coat (level 5 finish)		410	.039		2.85	1.72		4.57	5.80
7080	For SCIF applications, no finish included		1200	.013		3.05	.59		3.64	4.26

09 29 Gypsum Board

09 29 10 – Gypsum Board Panels

09 29 10.30 Gypsum Board

		Crew	Daily Output	Labor-Hours	Unit	Material	2012 Bare Costs Labor	2012 Bare Costs Equipment	Total	Total Incl O&P
7085	Taped and finished (level 4 finish)	2 Carp	510	.031	S.F.	3.10	1.38		4.48	5.55
7090	With compound skim coat (level 5 finish)		410	.039		3.14	1.72		4.86	6.10
8010	5/8" thick on ceilings, score & snap installation, no finish included		1600	.010		1.62	.44		2.06	2.46
8015	Taped and finished (level 4 finish)		680	.024		1.67	1.04		2.71	3.43
8020	With compound skim coat (level 5 finish)		545	.029		1.71	1.29		3	3.88

09 29 10.50 High Abuse Gypsum Board

		Crew	Daily Output	Labor-Hours	Unit	Material	2012 Bare Costs Labor	2012 Bare Costs Equipment	Total	Total Incl O&P
0010	**HIGH ABUSE GYPSUM BOARD**, fiber reinforced, nailed or									
0100	screwed to studs unless otherwise noted									
0110	1/2" thick, on walls, no finish included	2 Carp	1800	.009	S.F.	.82	.39		1.21	1.50
0120	Taped and finished (level 4 finish)		870	.018		.87	.81		1.68	2.20
0130	With compound skim coat (level 5 finish)		700	.023		.91	1.01		1.92	2.56
0150	On ceilings, no finish included		1620	.010		.82	.44		1.26	1.57
0160	Taped and finished (level 4 finish)		690	.023		.87	1.02		1.89	2.52
0170	With compound skim coat (level 5 finish)		550	.029		.91	1.28		2.19	2.98
0210	5/8" thick, on walls, no finish included		1800	.009		.72	.39		1.11	1.39
0220	Taped and finished (level 4 finish)		870	.018		.77	.81		1.58	2.09
0230	With compound skim coat (level 5 finish)		700	.023		.81	1.01		1.82	2.45
0250	On ceilings, no finish included		1620	.010		.72	.44		1.16	1.46
0260	Taped and finished (level 4 finish)		690	.023		.77	1.02		1.79	2.41
0270	With compound skim coat (level 5 finish)		550	.029		.81	1.28		2.09	2.87
0310	5/8" thick, on walls, very high impact, no finish included		1800	.009		1.15	.39		1.54	1.87
0320	Taped and finished (level 4 finish)		870	.018		1.20	.81		2.01	2.57
0330	With compound skim coat (level 5 finish)		700	.023		1.24	1.01		2.25	2.92
0350	On ceilings, no finish included		1620	.010		1.15	.44		1.59	1.94
0360	Taped and finished (level 4 finish)		690	.023		1.20	1.02		2.22	2.89
0370	With compound skim coat (level 5 finish)		550	.029		1.24	1.28		2.52	3.34
0400	High abuse, gypsum core, paper face									
0410	1/2" thick, on walls, no finish included	2 Carp	1800	.009	S.F.	.76	.39		1.15	1.44
0420	Taped and finished (level 4 finish)		870	.018		.81	.81		1.62	2.14
0430	With compound skim coat (level 5 finish)		700	.023		.85	1.01		1.86	2.49
0450	On ceilings, no finish included		1620	.010		.76	.44		1.20	1.51
0460	Taped and finished (level 4 finish)		690	.023		.81	1.02		1.83	2.46
0470	With compound skim coat (level 5 finish)		550	.029		.85	1.28		2.13	2.91
0510	5/8" thick, on walls, no finish included		1800	.009		.76	.39		1.15	1.44
0520	Taped and finished (level 4 finish)		870	.018		.81	.81		1.62	2.14
0530	With compound skim coat (level 5 finish)		700	.023		.85	1.01		1.86	2.49
0550	On ceilings, no finish included		1620	.010		.76	.44		1.20	1.51
0560	Taped and finished (level 4 finish)		690	.023		.81	1.02		1.83	2.46
0570	With compound skim coat (level 5 finish)		550	.029		.85	1.28		2.13	2.91
1000	For high ceilings, over 8' high, add		2750	.006			.26		.26	.39
1010	For distribution cost over 3 stories high, add per story		5500	.003			.13		.13	.20

09 29 15 – Gypsum Board Accessories

09 29 15.10 Accessories, Gypsum Board

		Crew	Daily Output	Labor-Hours	Unit	Material	2012 Bare Costs Labor	2012 Bare Costs Equipment	Total	Total Incl O&P
0010	**ACCESSORIES, GYPSUM BOARD**									
0020	Casing bead, galvanized steel	1 Carp	2.90	2.759	C.L.F.	24	122		146	214
0100	Vinyl		3	2.667		21.50	118		139.50	205
0300	Corner bead, galvanized steel, 1" x 1"		4	2		14.50	88		102.50	152
0400	1-1/4" x 1-1/4"		3.50	2.286		15.85	101		116.85	172
0600	Vinyl		4	2		16.75	88		104.75	154
0900	Furring channel, galv. steel, 7/8" deep, standard		2.60	3.077		32.50	136		168.50	245
1000	Resilient		2.55	3.137		37.50	138		175.50	255
1100	J trim, galvanized steel, 1/2" wide		3	2.667		26.50	118		144.50	210

09 29 Gypsum Board

09 29 15 – Gypsum Board Accessories

09 29 15.10 Accessories, Gypsum Board	Crew	Daily Output	Labor-Hours	Unit	Material	2012 Bare Costs Labor	Equipment	Total	Total Incl O&P
1120	5/8" wide	1 Carp	2.95	2.712	C.L.F.	28	120		148	215
1140	L trim, galvanized		3	2.667		23.50	118		141.50	207
1150	U trim, galvanized	▼	2.95	2.712	▼	20.50	120		140.50	207
1160	Screws #6 x 1" A				M	8.45			8.45	9.30
1170	#6 x 1-5/8" A				"	12.40			12.40	13.65
1200	For stud partitions, see Section 05 41 13.30 and 09 22 16.13									
1500	Z stud, galvanized steel, 1-1/2" wide	1 Carp	2.60	3.077	C.L.F.	37.50	136		173.50	251
1600	2" wide	"	2.55	3.137	"	58	138		196	277

09 30 Tiling

09 30 13 – Ceramic Tiling

09 30 13.10 Ceramic Tile

		Crew	Daily Output	Labor-Hours	Unit	Material	2012 Bare Costs Labor	Equipment	Total	Total Incl O&P
0010	**CERAMIC TILE**									
0020	Backsplash, thinset, average grade tiles	1 Tilf	50	.160	S.F.	2.42	6.50		8.92	12.25
0022	Custom grade tiles		50	.160		4.84	6.50		11.34	14.90
0024	Luxury grade tiles		50	.160		9.70	6.50		16.20	20.50
0026	Economy grade tiles	▼	50	.160	▼	2.21	6.50		8.71	12.05
0050	Base, using 1' x 4" high pc. with 1" x 1" tiles, mud set	D-7	82	.195	L.F.	4.95	7.10		12.05	15.90
0100	Thin set	"	128	.125		4.59	4.55		9.14	11.75
0300	For 6" high base, 1" x 1" tile face, add					.74			.74	.81
0400	For 2" x 2" tile face, add to above					.39			.39	.43
0600	Cove base, 4-1/4" x 4-1/4" high, mud set	D-7	91	.176		3.68	6.40		10.08	13.45
0700	Thin set		128	.125		3.71	4.55		8.26	10.80
0900	6" x 4-1/4" high, mud set		100	.160		3.40	5.85		9.25	12.30
1000	Thin set		137	.117		3.27	4.25		7.52	9.85
1200	Sanitary cove base, 6" x 4-1/4" high, mud set		93	.172		3.95	6.25		10.20	13.55
1300	Thin set		124	.129		3.82	4.70		8.52	11.10
1500	6" x 6" high, mud set		84	.190		3.72	6.95		10.67	14.30
1600	Thin set	▼	117	.137	▼	3.59	4.98		8.57	11.25
1800	Bathroom accessories, average (soap dish, tooth brush holder)		82	.195	Ea.	11.85	7.10		18.95	23.50
1900	Bathtub, 5', rec. 4-1/4" x 4-1/4" tile wainscot, adhesive set 6' high		2.90	5.517		156	201		357	465
2100	7' high wainscot		2.50	6.400		179	233		412	535
2200	8' high wainscot		2.20	7.273	▼	190	265		455	600
2400	Bullnose trim, 4-1/4" x 4-1/4", mud set		82	.195	L.F.	3.52	7.10		10.62	14.30
2500	Thin set		128	.125		3.43	4.55		7.98	10.45
2700	6" x 4-1/4" bullnose trim, mud set		84	.190		4.84	6.95		11.79	15.55
2800	Thin set		124	.129	▼	4.75	4.70		9.45	12.10
3000	Floors, natural clay, random or uniform, thin set, color group 1		183	.087	S.F.	3.83	3.19		7.02	8.90
3100	Color group 2		183	.087		5.55	3.19		8.74	10.80
3255	Floors, glazed, thin set, 6" x 6", color group 1		300	.053		4.43	1.94		6.37	7.70
3260	8" x 8" tile		300	.053		4.43	1.94		6.37	7.70
3270	12" x 12" tile		290	.055		5.90	2.01		7.91	9.45
3280	16" x 16" tile		280	.057		6.20	2.08		8.28	9.85
3281	18" x 18" tile		270	.059		6.20	2.16		8.36	9.95
3283	24" x 24" tile		250	.064		6.20	2.33		8.53	10.20
3285	Border, 6" x 12" tile		200	.080		22	2.91		24.91	28.50
3290	3" x 12" tile		200	.080		23.50	2.91		26.41	30
3300	Porcelain type, 1 color, color group 2, 1" x 1"		183	.087		4.89	3.19		8.08	10.10
3310	2" x 2" or 2" x 1", thin set	▼	190	.084		4.33	3.07		7.40	9.25
3350	For random blend, 2 colors, add					.95			.95	1.05
3360	4 colors, add				▼	1.33			1.33	1.46

09 30 Tiling

09 30 13 - Ceramic Tiling

09 30 13.10 Ceramic Tile

		Crew	Daily Output	Labor-Hours	Unit	Material	2012 Bare Costs Labor	2012 Bare Costs Equipment	Total	Total Incl O&P
3370	For color group 3, add				S.F.	.54			.54	.59
3380	For abrasive non-slip tile, add					.52			.52	.57
4300	Specialty tile, 4-1/4" x 4-1/4" x 1/2", decorator finish	D-7	183	.087		10.20	3.19		13.39	15.90
4500	Add for epoxy grout, 1/16" joint, 1" x 1" tile		800	.020		.64	.73		1.37	1.77
4600	2" x 2" tile		820	.020		.60	.71		1.31	1.70
4800	Pregrouted sheets, walls, 4-1/4" x 4-1/4", 6" x 4-1/4"									
4810	and 8-1/2" x 4-1/4", 4 S.F. sheets, silicone grout	D-7	240	.067	S.F.	4.87	2.43		7.30	8.90
5100	Floors, unglazed, 2 S.F. sheets,									
5110	Urethane adhesive	D-7	180	.089	S.F.	4.85	3.24		8.09	10.10
5400	Walls, interior, thin set, 4-1/4" x 4-1/4" tile		190	.084		2.20	3.07		5.27	6.95
5500	6" x 4-1/4" tile		190	.084		2.93	3.07		6	7.75
5700	8-1/2" x 4-1/4" tile		190	.084		4.09	3.07		7.16	9
5800	6" x 6" tile		175	.091		2.55	3.33		5.88	7.70
5810	8" x 8" tile		170	.094		3.64	3.43		7.07	9.05
5820	12" x 12" tile		160	.100		3.53	3.64		7.17	9.25
5830	16" x 16" tile		150	.107		4.53	3.89		8.42	10.70
6000	Decorated wall tile, 4-1/4" x 4-1/4", minimum		270	.059		2.80	2.16		4.96	6.25
6100	Maximum		180	.089		45.50	3.24		48.74	55
6300	Exterior walls, frostproof, mud set, 4-1/4" x 4-1/4"		102	.157		7	5.70		12.70	16.10
6400	1-3/8" x 1-3/8"		93	.172		5	6.25		11.25	14.70
6600	Crystalline glazed, 4-1/4" x 4-1/4", mud set, plain		100	.160		3.58	5.85		9.43	12.50
6700	4-1/4" x 4-1/4", scored tile		100	.160		5.30	5.85		11.15	14.40
6900	6" x 6" plain		93	.172		5.65	6.25		11.90	15.40
7000	For epoxy grout, 1/16" joints, 4-1/4" tile, add		800	.020		.40	.73		1.13	1.51
7200	For tile set in dry mortar, add		1735	.009			.34		.34	.49
7300	For tile set in Portland cement mortar, add		290	.055		.16	2.01		2.17	3.13
9300	Ceramic tiles, recycled glass, standard colors, 2" x 2" thru 6" x 6" [G]		190	.084		21	3.07		24.07	27.50
9310	6" x 6" [G]		175	.091		21	3.33		24.33	28
9320	8" x 8" [G]		170	.094		23	3.43		26.43	30.50
9330	12" x 12" [G]		160	.100		23	3.64		26.64	31
9340	Earthtones, 2" x 2" to 4" x 8" [G]		190	.084		26	3.07		29.07	33
9350	6" x 6" [G]		175	.091		26	3.33		29.33	33.50
9360	8" x 8" [G]		170	.094		27	3.43		30.43	34.50
9370	12" x 12" [G]		160	.100		27	3.64		30.64	35
9380	Deep colors, 2" x 2" to 4" x 8" [G]		190	.084		31	3.07		34.07	38.50
9390	6" x 6" [G]		175	.091		31	3.33		34.33	39
9400	8" x 8" [G]		170	.094		32	3.43		35.43	40.50
9410	12" x 12" [G]		160	.100		32	3.64		35.64	41

09 30 13.20 Ceramic Tile Repairs

		Crew	Daily Output	Labor-Hours	Unit	Material	Labor	Equipment	Total	Total Incl O&P
0010	**CERAMIC TILE REPAIRS**									
1000	Grout removal, carbide tipped, rotary grinder	1 Clab	180	.044	L.F.		1.56		1.56	2.40
1100	Regrout tile 4-1/2 x 4-1/2, or larger, wall	1 Tilf	100	.080	S.F.	.14	3.26		3.40	4.94
1150	Floor		125	.064		.15	2.61		2.76	4
1200	Seal tile and grout		360	.022		.17	.91		1.08	1.52

09 30 13.45 Ceramic Tile Accessories

		Crew	Daily Output	Labor-Hours	Unit	Material	Labor	Equipment	Total	Total Incl O&P
0010	**CERAMIC TILE ACCESSORIES**									
0100	Spacers, 1/8"				Ea.	1.44			1.44	1.58
1310	Sealer for natural stone tile, installed	1 Tilf	650	.012	S.F.	.04	.50		.54	.78

09 30 16 - Quarry Tiling

09 30 16.10 Quarry Tile

		Crew	Daily Output	Labor-Hours	Unit	Material	Labor	Equipment	Total	Total Incl O&P
0010	**QUARRY TILE**									
0100	Base, cove or sanitary, mud set, to 5" high, 1/2" thick	D-7	110	.145	L.F.	5.60	5.30		10.90	13.95

09 30 Tiling

09 30 16 – Quarry Tiling

09 30 16.10 Quarry Tile		Crew	Daily Output	Labor-Hours	Unit	Material	2012 Bare Costs Labor	Equipment	Total	Total Incl O&P
0300	Bullnose trim, red, mud set, 6" x 6" x 1/2" thick	D-7	120	.133	L.F.	4.46	4.86		9.32	12.05
0400	4" x 4" x 1/2" thick		110	.145		4.96	5.30		10.26	13.25
0600	4" x 8" x 1/2" thick, using 8" as edge		130	.123		4.38	4.48		8.86	11.40
0700	Floors, mud set, 1,000 S.F. lots, red, 4" x 4" x 1/2" thick		120	.133	S.F.	6.45	4.86		11.31	14.25
0900	6" x 6" x 1/2" thick		140	.114		6	4.16		10.16	12.70
1000	4" x 8" x 1/2" thick		130	.123		6.45	4.48		10.93	13.70
1300	For waxed coating, add					.73			.73	.80
1500	For non-standard colors, add					.43			.43	.47
1600	For abrasive surface, add					.50			.50	.55
1800	Brown tile, imported, 6" x 6" x 3/4"	D-7	120	.133		7.45	4.86		12.31	15.35
1900	8" x 8" x 1"		110	.145		8.15	5.30		13.45	16.75
2100	For thin set mortar application, deduct		700	.023			.83		.83	1.22
2200	For epoxy grout & mortar, 6" x 6" x 1/2", add		350	.046		1.96	1.67		3.63	4.61
2700	Stair tread, 6" x 6" x 3/4", plain		50	.320		5.15	11.65		16.80	23
2800	Abrasive		47	.340		5.75	12.40		18.15	24.50
3000	Wainscot, 6" x 6" x 1/2", thin set, red		105	.152		4.44	5.55		9.99	13.05
3100	Non-standard colors		105	.152		4.92	5.55		10.47	13.55
3300	Window sill, 6" wide, 3/4" thick		90	.178	L.F.	5.30	6.50		11.80	15.35
3400	Corners		80	.200	Ea.	6.20	7.30		13.50	17.50

09 30 23 – Glass Mosaic Tiling

09 30 23.10 Glass Mosaics		Crew	Daily Output	Labor-Hours	Unit	Material	2012 Bare Costs Labor	Equipment	Total	Total Incl O&P
0010	**GLASS MOSAICS** 3/4" tile on 12" sheets, standard grout									
0300	Color group 1 & 2	D-7	73	.219	S.F.	17.20	8		25.20	30.50
0350	Color group 3		73	.219		19.30	8		27.30	33.50
0400	Color group 4		73	.219		25	8		33	39.50
0450	Color group 5		73	.219		28.50	8		36.50	43.50
0500	Color group 6		73	.219		40.50	8		48.50	56.50
0600	Color group 7		73	.219		43.50	8		51.50	60
0700	Color group 8, golds, silvers & specialties		64	.250		57	9.10		66.10	76

09 30 29 – Metal Tiling

09 30 29.10 Metal Tile		Crew	Daily Output	Labor-Hours	Unit	Material	2012 Bare Costs Labor	Equipment	Total	Total Incl O&P
0010	**METAL TILE** 4' x 4' sheet, 24 ga., tile pattern, nailed									
0200	Stainless steel	2 Carp	512	.031	S.F.	27	1.38		28.38	31.50
0400	Aluminized steel	"	512	.031	"	14.50	1.38		15.88	18.05

09 34 Waterproofing-Membrane Tiling

09 34 13 – Waterproofing-Membrane Ceramic Tiling

09 34 13.10 Ceramic Tile Waterproofing Membrane		Crew	Daily Output	Labor-Hours	Unit	Material	2012 Bare Costs Labor	Equipment	Total	Total Incl O&P
0010	**CERAMIC TILE WATERPROOFING MEMBRANE**									
0020	On floors, including thinset									
0030	Fleece laminated polyethylene grid, 1/8" thick	D-7	250	.064	S.F.	2.24	2.33		4.57	5.90
0040	5/16" thick	"	250	.064	"	2.60	2.33		4.93	6.30
0050	On walls, including thinset									
0060	Fleece laminated polyethylene sheet, 8 mil thick	D-7	480	.033	S.F.	2.24	1.21		3.45	4.24
0070	Accessories, including thinset									
0080	Joint and corner sheet, 4 mils thick, 5" wide	1 Tilf	240	.033	L.F.	1.45	1.36		2.81	3.58
0090	7-1/4" wide		180	.044		1.84	1.81		3.65	4.68
0100	10" wide		120	.067		2.24	2.72		4.96	6.45
0110	Pre-formed corners, inside		32	.250	Ea.	4.73	10.20		14.93	20
0120	Outside		32	.250		7.40	10.20		17.60	23

09 34 Waterproofing-Membrane Tiling

09 34 13 – Waterproofing-Membrane Ceramic Tiling

09 34 13.10 Ceramic Tile Waterproofing Membrane		Crew	Daily Output	Labor-Hours	Unit	Material	2012 Bare Costs Labor	Equipment	Total	Total Incl O&P
0130	2" flanged floor drain with 6" stainless steel grate	1 Tilf	16	.500	Ea.	355	20.50		375.50	420
0140	EPS, sloped shower floor		480	.017	S.F.	4.82	.68		5.50	6.30
0150	Curb	↓	32	.250	L.F.	13.55	10.20		23.75	30

09 51 Acoustical Ceilings

09 51 23 – Acoustical Tile Ceilings

09 51 23.10 Suspended Acoustic Ceiling Tiles

		Crew	Daily Output	Labor-Hours	Unit	Material	2012 Bare Costs Labor	Equipment	Total	Total Incl O&P
0010	**SUSPENDED ACOUSTIC CEILING TILES**, not including									
0100	suspension system									
0300	Fiberglass boards, film faced, 2' x 2' or 2' x 4', 5/8" thick	1 Carp	625	.013	S.F.	.96	.56		1.52	1.93
0400	3/4" thick		600	.013		1.55	.59		2.14	2.61
0500	3" thick, thermal, R11		450	.018		1.57	.78		2.35	2.94
0600	Glass cloth faced fiberglass, 3/4" thick		500	.016		1.49	.71		2.20	2.73
0700	1" thick		485	.016		2.18	.73		2.91	3.52
0820	1-1/2" thick, nubby face		475	.017		2.66	.74		3.40	4.07
1110	Mineral fiber tile, lay-in, 2' x 2' or 2' x 4', 5/8" thick, fine texture		625	.013		.75	.56		1.31	1.70
1115	Rough textured		625	.013		1.36	.56		1.92	2.37
1125	3/4" thick, fine textured		600	.013		1.50	.59		2.09	2.55
1130	Rough textured		600	.013		1.79	.59		2.38	2.87
1135	Fissured		600	.013		2.45	.59		3.04	3.60
1150	Tegular, 5/8" thick, fine textured		470	.017		1.45	.75		2.20	2.75
1155	Rough textured		470	.017		1.90	.75		2.65	3.24
1165	3/4" thick, fine textured		450	.018		2.08	.78		2.86	3.50
1170	Rough textured		450	.018		2.35	.78		3.13	3.80
1175	Fissured	↓	450	.018		3.63	.78		4.41	5.20
1180	For aluminum face, add					6.15			6.15	6.80
1185	For plastic film face, add					.94			.94	1.03
1190	For fire rating, add					.45			.45	.50
1300	Mirror tile, acrylic, 2' x 2'	1 Carp	500	.016		10.95	.71		11.66	13.15
1900	Eggcrate, acrylic, 1/2" x 1/2" x 1/2" cubes		500	.016		1.82	.71		2.53	3.09
2100	Polystyrene eggcrate, 3/8" x 3/8" x 1/2" cubes		510	.016		1.53	.69		2.22	2.74
2200	1/2" x 1/2" x 1/2" cubes		500	.016		1.85	.71		2.56	3.13
2400	Luminous panels, prismatic, acrylic		400	.020		2.22	.88		3.10	3.80
2500	Polystyrene		400	.020		1.14	.88		2.02	2.61
2700	Flat white acrylic		400	.020		3.86	.88		4.74	5.60
2800	Polystyrene		400	.020		2.65	.88		3.53	4.28
3000	Drop pan, white, acrylic		400	.020		5.65	.88		6.53	7.60
3100	Polystyrene		400	.020		4.73	.88		5.61	6.55
3600	Perforated aluminum sheets, .024" thick, corrugated, painted		490	.016		2.29	.72		3.01	3.63
3700	Plain		500	.016		3.78	.71		4.49	5.25
3720	Mineral fiber, 24" x 24" or 48", reveal edge, painted, 5/8" thick		600	.013		1.38	.59		1.97	2.42
3740	3/4" thick		575	.014		2.24	.61		2.85	3.40
5020	66 – 78% recycled content, 3/4" thick **G**		600	.013		1.82	.59		2.41	2.90
5040	Mylar, 42% recycled content, 3/4" thick **G**	↓	600	.013	↓	4.28	.59		4.87	5.60

09 51 23.30 Suspended Ceilings, Complete

		Crew	Daily Output	Labor-Hours	Unit	Material	2012 Bare Costs Labor	Equipment	Total	Total Incl O&P
0010	**SUSPENDED CEILINGS, COMPLETE,** including standard									
0100	suspension system but not incl. 1-1/2" carrier channels									
0600	Fiberglass ceiling board, 2' x 4' x 5/8", plain faced	1 Carp	500	.016	S.F.	1.66	.71		2.37	2.91
0700	Offices, 2' x 4' x 3/4"		380	.021		2.25	.93		3.18	3.90
0800	Mineral fiber, on 15/16" T bar susp. 2' x 2' x 3/4" lay-in board		345	.023		2.37	1.02		3.39	4.17
0810	2' x 4' x 5/8" tile	↓	380	.021	↓	1.45	.93		2.38	3.02

09 51 Acoustical Ceilings

09 51 23 – Acoustical Tile Ceilings

09 51 23.30 Suspended Ceilings, Complete		Crew	Daily Output	Labor-Hours	Unit	Material	2012 Bare Costs Labor	Equipment	Total	Total Incl O&P
0820	Tegular, 2' x 2' x 5/8" tile on 9/16" grid	1 Carp	250	.032	S.F.	2.48	1.41		3.89	4.89
0830	2' x 4' x 3/4" tile		275	.029		2.94	1.28		4.22	5.20
0900	Luminous panels, prismatic, acrylic		255	.031		2.92	1.38		4.30	5.35
1200	Metal pan with acoustic pad, steel		75	.107		4.41	4.70		9.11	12.10
1300	Painted aluminum		75	.107		2.99	4.70		7.69	10.55
1500	Aluminum, degreased finish		75	.107		5.10	4.70		9.80	12.85
1600	Stainless steel		75	.107		9.50	4.70		14.20	17.70
1800	Tile, Z bar suspension, 5/8" mineral fiber tile		150	.053		2.27	2.35		4.62	6.10
1900	3/4" mineral fiber tile	↓	150	.053	↓	2.43	2.35		4.78	6.30
2400	For strip lighting, see Section 26 51 13.50									
2500	For rooms under 500 S.F., add				S.F.		25%			

09 51 53 – Direct-Applied Acoustical Ceilings

09 51 53.10 Ceiling Tile

		Crew	Daily Output	Labor-Hours	Unit	Material	2012 Bare Costs Labor	Equipment	Total	Total Incl O&P
0010	**CEILING TILE**, stapled or cemented									
0100	12" x 12" or 12" x 24", not including furring									
0600	Mineral fiber, vinyl coated, 5/8" thick	1 Carp	300	.027	S.F.	1.72	1.18		2.90	3.70
0700	3/4" thick		300	.027		2.12	1.18		3.30	4.14
0900	Fire rated, 3/4" thick, plain faced		300	.027		1.46	1.18		2.64	3.42
1000	Plastic coated face		300	.027		1.24	1.18		2.42	3.17
1200	Aluminum faced, 5/8" thick, plain		300	.027		1.59	1.18		2.77	3.56
3700	Wall application of above, add	↓	1000	.008			.35		.35	.54
3900	For ceiling primer, add					.13			.13	.14
4000	For ceiling cement, add				↓	.37			.37	.41

09 53 Acoustical Ceiling Suspension Assemblies

09 53 23 – Metal Acoustical Ceiling Suspension Assemblies

09 53 23.30 Ceiling Suspension Systems

		Crew	Daily Output	Labor-Hours	Unit	Material	2012 Bare Costs Labor	Equipment	Total	Total Incl O&P
0010	**CEILING SUSPENSION SYSTEMS** for boards and tile									
0050	Class A suspension system, 15/16" T bar, 2' x 4' grid	1 Carp	800	.010	S.F.	.70	.44		1.14	1.45
0300	2' x 2' grid		650	.012		.87	.54		1.41	1.79
0310	25% recycled steel, 2' x 4' grid	G	800	.010		.76	.44		1.20	1.52
0320	2' x 2' grid	G ↓	650	.012		.96	.54		1.50	1.89
0350	For 9/16" grid, add					.16			.16	.18
0360	For fire rated grid, add					.09			.09	.10
0370	For colored grid, add					.21			.21	.23
0400	Concealed Z bar suspension system, 12" module	1 Carp	520	.015		.73	.68		1.41	1.84
0600	1-1/2" carrier channels, 4' O.C., add	"	470	.017	↓	.13	.75		.88	1.29
0700	Carrier channels for ceilings with									
0900	recessed lighting fixtures, add	1 Carp	460	.017	S.F.	.23	.77		1	1.43
1040	Hanging wire, 12 ga., 4' long		65	.123	C.S.F.	7.45	5.45		12.90	16.55
1080	8' long		65	.123	"	14.90	5.45		20.35	24.50
1200	Seismic bracing, 2-1/2" compression post with four bracing wires	↓	50	.160	Ea.	2.88	7.05		9.93	14

09 54 Specialty Ceilings

09 54 33 – Decorative Panel Ceilings

09 54 33.20 Metal Panel Ceilings		Crew	Daily Output	Labor-Hours	Unit	Material	2012 Bare Costs Labor	Equipment	Total	Total Incl O&P
0010	**METAL PANEL CEILINGS**									
0020	Lay-in or screwed to furring, not including grid									
0100	Tin ceilings, 2' x 2' or 2' x 4', bare steel finish	2 Carp	300	.053	S.F.	2.36	2.35		4.71	6.20
0120	Painted white finish		300	.053	"	3.64	2.35		5.99	7.60
0140	Copper, chrome or brass finish		300	.053	L.F.	6.65	2.35		9	10.90
0200	Cornice molding, 2-1/2" to 3-1/2" wide, 4' long, bare steel finish		200	.080	S.F.	2.25	3.53		5.78	7.90
0220	Painted white finish		200	.080		2.94	3.53		6.47	8.70
0240	Copper, chrome or brass finish		200	.080		3.96	3.53		7.49	9.80
0320	5" to 6-1/2" wide, 4' long, bare steel finish		150	.107		3.19	4.70		7.89	10.75
0340	Painted white finish		150	.107		4.20	4.70		8.90	11.85
0360	Copper, chrome or brass finish		150	.107		6.50	4.70		11.20	14.40
0420	Flat molding, 3-1/2" to 5" wide, 4' long, bare steel finish		250	.064		3.69	2.82		6.51	8.40
0440	Painted white finish		250	.064		4.09	2.82		6.91	8.85
0460	Copper, chrome or brass finish		250	.064		7.35	2.82		10.17	12.45

09 62 Specialty Flooring

09 62 19 – Laminate Flooring

09 62 19.10 Floating Floor

		Crew	Daily Output	Labor-Hours	Unit	Material	2012 Bare Costs Labor	Equipment	Total	Total Incl O&P
0010	**FLOATING FLOOR**									
8300	Floating floor, laminate, wood pattern strip, complete	1 Clab	133	.060	S.F.	4.18	2.11		6.29	7.85
8310	Components, T & G wood composite strips					3.27			3.27	3.59
8320	Film					.13			.13	.14
8330	Foam					.25			.25	.28
8340	Adhesive					.50			.50	.55
8350	Installation kit					.17			.17	.19
8360	Trim, 2" wide x 3' long				L.F.	3.85			3.85	4.24
8370	Reducer moulding				"	5.20			5.20	5.70

09 62 23 – Bamboo Flooring

09 62 23.10 Flooring, Bamboo

			Crew	Daily Output	Labor-Hours	Unit	Material	2012 Bare Costs Labor	Equipment	Total	Total Incl O&P
0010	**FLOORING, BAMBOO**										
8600	Flooring, wood, bamboo strips, unfinished, 5/8" x 4" x 3'	G	1 Carp	255	.031	S.F.	4.44	1.38		5.82	7
8610	5/8" x 4" x 4'	G		275	.029		4.61	1.28		5.89	7
8620	5/8" x 4" x 6'	G		295	.027		5.05	1.20		6.25	7.40
8630	Finished, 5/8" x 4" x 3'	G		255	.031		4.89	1.38		6.27	7.55
8640	5/8" x 4" x 4'	G		275	.029		5.10	1.28		6.38	7.60
8650	5/8" x 4" x 6'	G		295	.027		4.28	1.20		5.48	6.55
8660	Stair treads, unfinished, 1-1/16" x 11-1/2" x 4'	G		18	.444	Ea.	42.50	19.60		62.10	77
8670	Finished, 1-1/16" x 11-1/2" x 4'	G		18	.444		71.50	19.60		91.10	109
8680	Stair risers, unfinished, 5/8" x 7-1/2" x 4'	G		18	.444		15.75	19.60		35.35	47.50
8690	Finished, 5/8" x 7-1/2" x 4'	G		18	.444		30	19.60		49.60	63
8700	Stair nosing, unfinished, 6' long	G		16	.500		35	22		57	72.50
8710	Finished, 6' long	G		16	.500		39.50	22		61.50	77.50

09 63 Masonry Flooring

09 63 13 – Brick Flooring

09 63 13.10 Miscellaneous Brick Flooring

		Crew	Daily Output	Labor-Hours	Unit	Material	2012 Bare Costs Labor	Equipment	Total	Total Incl O&P
0010	**MISCELLANEOUS BRICK FLOORING**									
0020	Acid-proof shales, red, 8" x 3-3/4" x 1-1/4" thick	D-7	.43	37.209	M	760	1,350		2,110	2,825
0050	2-1/4" thick	D-1	.40	40		950	1,600		2,550	3,450
0200	Acid-proof clay brick, 8" x 3-3/4" x 2-1/4" thick [G]		.40	40		890	1,600		2,490	3,375
0250	9" x 4-1/2" x 3" [G]		95	.168	S.F.	4.14	6.70		10.84	14.70
0260	Cast ceramic, pressed, 4" x 8" x 1/2", unglazed	D-7	100	.160		6.25	5.85		12.10	15.45
0270	Glazed		100	.160		8.35	5.85		14.20	17.70
0280	Hand molded flooring, 4" x 8" x 3/4", unglazed		95	.168		8.25	6.15		14.40	18.10
0290	Glazed		95	.168		10.35	6.15		16.50	20.50
0300	8" hexagonal, 3/4" thick, unglazed		85	.188		9.05	6.85		15.90	20
0310	Glazed		85	.188		16.35	6.85		23.20	28
0400	Heavy duty industrial, cement mortar bed, 2" thick, not incl. brick	D-1	80	.200		.83	7.95		8.78	12.95
0450	Acid-proof joints, 1/4" wide	"	65	.246		1.44	9.80		11.24	16.40
0500	Pavers, 8" x 4", 1" to 1-1/4" thick, red	D-7	95	.168		3.64	6.15		9.79	13
0510	Ironspot	"	95	.168		5.15	6.15		11.30	14.65
0540	1-3/8" to 1-3/4" thick, red	D-1	95	.168		3.51	6.70		10.21	14
0560	Ironspot		95	.168		5.10	6.70		11.80	15.75
0580	2-1/4" thick, red		90	.178		3.57	7.10		10.67	14.65
0590	Ironspot		90	.178		5.55	7.10		12.65	16.80
0700	Paver, adobe brick, 6" x 12", 1/2" joint [G]		42	.381		1.38	15.20		16.58	24.50
0710	Mexican red, 12" x 12" [G]	1 Tilf	48	.167		1.64	6.80		8.44	11.80
0720	Saltillo, 12" x 12" [G]	"	48	.167		1.64	6.80		8.44	11.80
0800	For sidewalks and patios with pavers, see Section 32 14 16.10									
0870	For epoxy joints, add	D-1	600	.027	S.F.	2.72	1.06		3.78	4.60
0880	For Furan underlayment, add	"	600	.027		2.25	1.06		3.31	4.09
0890	For waxed surface, steam cleaned, add	A-1H	1000	.008		.20	.28	.07	.55	.72

09 63 40 – Stone Flooring

09 63 40.10 Marble

		Crew	Daily Output	Labor-Hours	Unit	Material	2012 Bare Costs Labor	Equipment	Total	Total Incl O&P
0010	**MARBLE**									
0020	Thin gauge tile, 12" x 6", 3/8", white Carara	D-7	60	.267	S.F.	11.45	9.70		21.15	27
0100	Travertine		60	.267		13.15	9.70		22.85	29
0200	12" x 12" x 3/8", thin set, floors		60	.267		6.25	9.70		15.95	21
0300	On walls		52	.308		9.85	11.20		21.05	27.50
1000	Marble threshold, 4" wide x 36" long x 5/8" thick, white		60	.267	Ea.	6.25	9.70		15.95	21

09 63 40.20 Slate Tile

		Crew	Daily Output	Labor-Hours	Unit	Material	2012 Bare Costs Labor	Equipment	Total	Total Incl O&P
0010	**SLATE TILE**									
0020	Vermont, 6" x 6" x 1/4" thick, thin set	D-7	180	.089	S.F.	7.50	3.24		10.74	13
0200	See also Section 32 14 40.10									

09 64 Wood Flooring

09 64 16 – Wood Block Flooring

09 64 16.10 End Grain Block Flooring

		Crew	Daily Output	Labor-Hours	Unit	Material	2012 Bare Costs Labor	Equipment	Total	Total Incl O&P
0010	**END GRAIN BLOCK FLOORING**									
0020	End grain flooring, coated, 2" thick	1 Carp	295	.027	S.F.	3.52	1.20		4.72	5.70
0400	Natural finish, 1" thick, fir		125	.064		3.64	2.82		6.46	8.35
0600	1-1/2" thick, pine		125	.064		3.57	2.82		6.39	8.25
0700	2" thick, pine		125	.064		5	2.82		7.82	9.85

09 64 Wood Flooring

09 64 19 – Wood Composition Flooring

09 64 19.10 Wood Composition

		Crew	Daily Output	Labor-Hours	Unit	Material	2012 Bare Costs Labor	Equipment	Total	Total Incl O&P
0010	**WOOD COMPOSITION** Gym floors									
0100	2-1/4" x 6-7/8" x 3/8", on 2" grout setting bed	D-7	150	.107	S.F.	6.05	3.89		9.94	12.35
0200	Thin set, on concrete	"	250	.064		5.50	2.33		7.83	9.45
0300	Sanding and finishing, add	1 Carp	200	.040	↓	.81	1.76		2.57	3.60

09 64 23 – Wood Parquet Flooring

09 64 23.10 Wood Parquet

		Crew	Daily Output	Labor-Hours	Unit	Material	2012 Bare Costs Labor	Equipment	Total	Total Incl O&P
0010	**WOOD PARQUET** flooring									
5200	Parquetry, standard, 5/16" thick, not incl. finish, oak, minimum	1 Carp	160	.050	S.F.	4.56	2.21		6.77	8.40
5300	Maximum		100	.080		5.80	3.53		9.33	11.80
5500	Teak, minimum		160	.050		4.86	2.21		7.07	8.75
5600	Maximum		100	.080		8.50	3.53		12.03	14.80
5650	13/16" thick, select grade oak, minimum		160	.050		9.50	2.21		11.71	13.85
5700	Maximum		100	.080		14.40	3.53		17.93	21.50
5800	Custom parquetry, including finish, minimum		100	.080		15.65	3.53		19.18	22.50
5900	Maximum		50	.160		21	7.05		28.05	34
6700	Parquetry, prefinished white oak, 5/16" thick, minimum		160	.050		3.30	2.21		5.51	7
6800	Maximum		100	.080		4.47	3.53		8	10.35
7000	Walnut or teak, parquetry, minimum		160	.050		5.35	2.21		7.56	9.30
7100	Maximum	↓	100	.080	↓	9.35	3.53		12.88	15.75
7200	Acrylic wood parquet blocks, 12" x 12" x 5/16",									
7210	Irradiated, set in epoxy	1 Carp	160	.050	S.F.	7.80	2.21		10.01	12

09 64 29 – Wood Strip and Plank Flooring

09 64 29.10 Wood

		Crew	Daily Output	Labor-Hours	Unit	Material	2012 Bare Costs Labor	Equipment	Total	Total Incl O&P
0010	**WOOD**									
0020	Fir, vertical grain, 1" x 4", not incl. finish, grade B & better	1 Carp	255	.031	S.F.	2.74	1.38		4.12	5.15
0100	C grade & better		255	.031		2.58	1.38		3.96	4.97
4000	Maple, strip, 25/32" x 2-1/4", not incl. finish, select		170	.047		5.20	2.08		7.28	8.90
4100	#2 & better		170	.047		3.99	2.08		6.07	7.60
4300	33/32" x 3-1/4", not incl. finish, #1 grade		170	.047		4.40	2.08		6.48	8.05
4400	#2 & better	↓	170	.047	↓	3.92	2.08		6	7.50
4600	Oak, white or red, 25/32" x 2-1/4", not incl. finish									
4700	#1 common	1 Carp	170	.047	S.F.	1.99	2.08		4.07	5.40
4900	Select quartered, 2-1/4" wide		170	.047		2.99	2.08		5.07	6.50
5000	Clear		170	.047		3.75	2.08		5.83	7.30
6100	Prefinished, white oak, prime grade, 2-1/4" wide		170	.047		4.79	2.08		6.87	8.45
6200	3-1/4" wide		185	.043		5.05	1.91		6.96	8.50
6400	Ranch plank		145	.055		6.50	2.43		8.93	10.90
6500	Hardwood blocks, 9" x 9", 25/32" thick		160	.050		5.75	2.21		7.96	9.75
7400	Yellow pine, 3/4" x 3-1/8", T & G, C & better, not incl. finish	↓	200	.040		1.49	1.76		3.25	4.35
7500	Refinish wood floor, sand, 2 coats poly, wax, soft wood, min.	1 Clab	400	.020		.86	.70		1.56	2.03
7600	Hard wood, max.		130	.062		1.28	2.16		3.44	4.73
7800	Sanding and finishing, 2 coats polyurethane	↓	295	.027	↓	.86	.95		1.81	2.41
7900	Subfloor and underlayment, see Section 06 16									
8015	Transition molding, 2 1/4" wide, 5' long	1 Carp	19.20	.417	Ea.	15	18.40		33.40	45

09 64 66 – Wood Athletic Flooring

09 64 66.10 Gymnasium Flooring

		Crew	Daily Output	Labor-Hours	Unit	Material	2012 Bare Costs Labor	Equipment	Total	Total Incl O&P
0010	**GYMNASIUM FLOORING**									
0600	Gym floor, in mastic, over 2 ply felt, #2 & better									
0700	25/32" thick maple	1 Carp	100	.080	S.F.	4.14	3.53		7.67	10
0900	33/32" thick maple	↓	98	.082	↓	4.99	3.60		8.59	11.05

09 64 Wood Flooring

09 64 66 – Wood Athletic Flooring

09 64 66.10 Gymnasium Flooring		Crew	Daily Output	Labor-Hours	Unit	Material	2012 Bare Costs Labor	Equipment	Total	Total Incl O&P
1000	For 1/2" corkboard underlayment, add	1 Carp	750	.011	S.F.	.95	.47		1.42	1.77
1300	For #1 grade maple, add				↓	.49			.49	.54
1600	Maple flooring, over sleepers, #2 & better									
1700	25/32" thick	1 Carp	85	.094	S.F.	4.54	4.15		8.69	11.40
1900	33/32" thick	"	83	.096		5.30	4.25		9.55	12.35
2000	For #1 grade, add					.53			.53	.58
2200	For 3/4" subfloor, add	1 Carp	350	.023		1.18	1.01		2.19	2.85
2300	With two 1/2" subfloors, 25/32" thick	"	69	.116	↓	5.70	5.10		10.80	14.10
2500	Maple, incl. finish, #2 & btr., 25/32" thick, on rubber									
2600	Sleepers, with two 1/2" subfloors	1 Carp	76	.105	S.F.	6.10	4.64		10.74	13.85
2800	With steel spline, double connection to channels	"	73	.110		6.50	4.83		11.33	14.60
2900	For 33/32" maple, add					.70			.70	.77
3100	For #1 grade maple, add					.53			.53	.58
3500	For termite proofing all of the above, add					.28			.28	.31
3700	Portable hardwood, prefinished panels	1 Carp	83	.096		8.15	4.25		12.40	15.50
3720	Insulated with polystyrene, 1" thick, add		165	.048		.70	2.14		2.84	4.06
3750	Running tracks, Sitka spruce surface, 25/32" x 2-1/4"		62	.129		14.90	5.70		20.60	25
3770	3/4" plywood surface, finished	↓	100	.080	↓	3.53	3.53		7.06	9.35

09 65 Resilient Flooring

09 65 10 – Resilient Tile Underlayment

09 65 10.10 Latex Underlayment

0010	**LATEX UNDERLAYMENT**									
3600	Latex underlayment, 1/8" thk., cementitious for resilient flooring	1 Tilf	160	.050	S.F.	.97	2.04		3.01	4.06
4000	Liquid, fortified				Gal.	29.50			29.50	32.50

09 65 13 – Resilient Base and Accessories

09 65 13.13 Resilient Base

0010	**RESILIENT BASE**									
0800	Base, cove, rubber or vinyl									
1100	Standard colors, 0.080" thick, 2-1/2" high	1 Tilf	315	.025	L.F.	1.05	1.04		2.09	2.68
1150	4" high		315	.025		1.01	1.04		2.05	2.63
1200	6" high		315	.025		1.55	1.04		2.59	3.23
1450	1/8" thick, 2-1/2" high		315	.025		1.05	1.04		2.09	2.68
1500	4" high		315	.025	↓	.72	1.04		1.76	2.31
1600	Corners, 2-1/2" high		315	.025	Ea.	1.70	1.04		2.74	3.39
1630	4" high		315	.025		3.15	1.04		4.19	4.99
1660	6" high	↓	315	.025	↓	3.15	1.04		4.19	4.99

09 65 13.23 Resilient Stair Treads and Risers

0010	**RESILIENT STAIR TREADS AND RISERS**									
0300	Rubber, molded tread, 12" wide, 5/16" thick, black	1 Tilf	115	.070	L.F.	13.50	2.84		16.34	19
0400	Colors		115	.070		14	2.84		16.84	19.55
0600	1/4" thick, black		115	.070		13	2.84		15.84	18.45
0700	Colors		115	.070		13.50	2.84		16.34	19
0900	Grip strip safety tread, colors, 5/16" thick		115	.070		17.60	2.84		20.44	23.50
1000	3/16" thick		120	.067	↓	12.75	2.72		15.47	18.05
1200	Landings, smooth sheet rubber, 1/8" thick		120	.067	S.F.	6.65	2.72		9.37	11.30
1300	3/16" thick		120	.067	"	8.35	2.72		11.07	13.20
1500	Nosings, 3" wide, 3/16" thick, black		140	.057	L.F.	3.69	2.33		6.02	7.50
1600	Colors		140	.057		4.09	2.33		6.42	7.90
1800	Risers, 7" high, 1/8" thick, flat	↓	250	.032	↓	6	1.30		7.30	8.50

09 65 Resilient Flooring

09 65 13 – Resilient Base and Accessories

09 65 13.23 Resilient Stair Treads and Risers

		Crew	Daily Output	Labor-Hours	Unit	Material	2012 Bare Costs Labor	Equipment	Total	Total Incl O&P
1900	Coved	1 Tilf	250	.032	L.F.	8.15	1.30		9.45	10.85
2100	Vinyl, molded tread, 12" wide, colors, 1/8" thick		115	.070		6.65	2.84		9.49	11.45
2200	1/4" thick		115	.070	↓	8.35	2.84		11.19	13.35
2300	Landing material, 1/8" thick		200	.040	S.F.	6.65	1.63		8.28	9.70
2400	Riser, 7" high, 1/8" thick, coved		175	.046	L.F.	3.75	1.86		5.61	6.85
2500	Tread and riser combined, 1/8" thick	↓	80	.100	"	10	4.08		14.08	17

09 65 16 – Resilient Sheet Flooring

09 65 16.10 Rubber and Vinyl Sheet Flooring

			Crew	Daily Output	Labor-Hours	Unit	Material	2012 Bare Costs Labor	Equipment	Total	Total Incl O&P
0010	**RUBBER AND VINYL SHEET FLOORING**										
5500	Linoleum, sheet goods	G	1 Tilf	360	.022	S.F.	4.28	.91		5.19	6.05
5900	Rubber, sheet goods, 36" wide, 1/8" thick			120	.067		7.30	2.72		10.02	12.05
5950	3/16" thick			100	.080		9.90	3.26		13.16	15.70
6000	1/4" thick			90	.089		11.70	3.62		15.32	18.15
8000	Vinyl sheet goods, backed, .065" thick, minimum			250	.032		3.95	1.30		5.25	6.25
8050	Maximum			200	.040		3.83	1.63		5.46	6.60
8100	.080" thick, minimum			230	.035		3.95	1.42		5.37	6.45
8150	Maximum			200	.040		5.65	1.63		7.28	8.60
8200	.125" thick, minimum			230	.035		4.57	1.42		5.99	7.15
8250	Maximum		↓	200	.040	↓	6.85	1.63		8.48	9.95
8700	Adhesive cement, 1 gallon per 200 to 300 S.F.					Gal.	26			26	28.50
8800	Asphalt primer, 1 gallon per 300 S.F.						13.40			13.40	14.75
8900	Emulsion, 1 gallon per 140 S.F.					↓	17.25			17.25	19

09 65 19 – Resilient Tile Flooring

09 65 19.10 Miscellaneous Resilient Tile Flooring

			Crew	Daily Output	Labor-Hours	Unit	Material	2012 Bare Costs Labor	Equipment	Total	Total Incl O&P
0010	**MISCELLANEOUS RESILIENT TILE FLOORING**										
2200	Cork tile, standard finish, 1/8" thick	G	1 Tilf	315	.025	S.F.	7.35	1.04		8.39	9.60
2250	3/16" thick	G		315	.025		6.65	1.04		7.69	8.80
2300	5/16" thick	G		315	.025		8.60	1.04		9.64	11
2350	1/2" thick	G		315	.025		10.50	1.04		11.54	13.05
2500	Urethane finish, 1/8" thick	G		315	.025		9.50	1.04		10.54	11.95
2550	3/16" thick	G		315	.025		10.20	1.04		11.24	12.70
2600	5/16" thick	G		315	.025		11.50	1.04		12.54	14.15
2650	1/2" thick	G		315	.025		15.60	1.04		16.64	18.65
6050	Rubber tile, marbleized colors, 12" x 12", 1/8" thick			400	.020		7	.82		7.82	8.90
6100	3/16" thick			400	.020		10.70	.82		11.52	12.95
6300	Special tile, plain colors, 1/8" thick			400	.020		10.70	.82		11.52	12.95
6350	3/16" thick			400	.020		10.70	.82		11.52	12.95
6410	Raised, radial or square, minimum			400	.020		9.90	.82		10.72	12.10
6430	Maximum			400	.020		13.35	.82		14.17	15.90
6450	For golf course, skating rink, etc., 1/4" thick			275	.029		13.35	1.19		14.54	16.45
6700	Synthetic turf, 3/8" thick		↓	90	.089	↓	6.65	3.62		10.27	12.60
6750	Interlocking 2' x 2' squares, 1/2" thick, not										
6810	cemented, for playgrounds, minimum		1 Tilf	210	.038	S.F.	5	1.55		6.55	7.80
6850	Maximum			190	.042		12.85	1.72		14.57	16.65
7000	Vinyl composition tile, 12" x 12", 1/16" thick			500	.016		.94	.65		1.59	1.99
7050	Embossed			500	.016		2.17	.65		2.82	3.35
7100	Marbleized			500	.016		2.17	.65		2.82	3.35
7150	Solid			500	.016		2.79	.65		3.44	4.03
7200	3/32" thick, embossed			500	.016		1.34	.65		1.99	2.43
7250	Marbleized			500	.016		2.49	.65		3.14	3.70
7300	Solid			500	.016		2.31	.65		2.96	3.50
7350	1/8" thick, marbleized			500	.016	↓	1.85	.65		2.50	3

09 65 Resilient Flooring

09 65 19 – Resilient Tile Flooring

09 65 19.10 Miscellaneous Resilient Tile Flooring	Crew	Daily Output	Labor-Hours	Unit	Material	2012 Bare Costs Labor	Equipment	Total	Total Incl O&P
7400	Solid	1 Tilf	500	.016	S.F.	1.58	.65		2.23	2.70
7450	Conductive		500	.016		6.40	.65		7.05	8
7500	Vinyl tile, 12" x 12", .050" thick, minimum		500	.016		3.25	.65		3.90	4.54
7550	Maximum		500	.016		6.50	.65		7.15	8.10
7600	1/8" thick, minimum		500	.016		4.80	.65		5.45	6.25
7650	Solid colors		500	.016		5.85	.65		6.50	7.40
7700	Marbleized or Travertine pattern		500	.016		5.90	.65		6.55	7.45
7750	Florentine pattern		500	.016		6.30	.65		6.95	7.90
7800	Maximum		500	.016		14	.65		14.65	16.35

09 65 33 – Conductive Resilient Flooring

09 65 33.10 Conductive Rubber and Vinyl Flooring

		Crew	Daily Output	Labor-Hours	Unit	Material	Labor	Equipment	Total	Total Incl O&P
0010	**CONDUCTIVE RUBBER AND VINYL FLOORING**									
1700	Conductive flooring, rubber tile, 1/8" thick	1 Tilf	315	.025	S.F.	6.30	1.04		7.34	8.45
1800	Homogeneous vinyl tile, 1/8" thick	"	315	.025	"	8.40	1.04		9.44	10.75

09 66 Terrazzo Flooring

09 66 13 – Portland Cement Terrazzo Flooring

09 66 13.10 Portland Cement Terrazzo

		Crew	Daily Output	Labor-Hours	Unit	Material	Labor	Equipment	Total	Total Incl O&P
0010	**PORTLAND CEMENT TERRAZZO**, cast-in-place R096613-10									
0020	Cove base, 6" high, 16 ga. zinc	1 Mstz	20	.400	L.F.	3.04	16.15		19.19	27
0100	Curb, 6" high and 6" wide		6	1.333		5.40	54		59.40	85
0300	Divider strip for floors, 14 ga., 1-1/4" deep, zinc		375	.021		1.43	.86		2.29	2.84
0400	Brass		375	.021		2.48	.86		3.34	4
0600	Heavy top strip 1/4" thick, 1-1/4" deep, zinc		300	.027		1.87	1.08		2.95	3.64
0900	Galv. bottoms, brass		300	.027		3.21	1.08		4.29	5.10
1200	For thin set floors, 16 ga., 1/2" x 1/2", zinc		350	.023		.85	.92		1.77	2.30
1300	Brass		350	.023		2.30	.92		3.22	3.89
1500	Floor, bonded to concrete, 1-3/4" thick, gray cement	J-3	75	.213	S.F.	3.15	7.90	3.49	14.54	18.90
1600	White cement, mud set		75	.213		3.51	7.90	3.49	14.90	19.30
1800	Not bonded, 3" total thickness, gray cement		70	.229		3.93	8.50	3.74	16.17	21
1900	White cement, mud set		70	.229		4.58	8.50	3.74	16.82	21.50
2100	For Venetian terrazzo, 1" topping, add					50%	50%			
2200	For heavy duty abrasive terrazzo, add					50%	50%			
2700	Monolithic terrazzo, 1/2" thick									
2710	10' panels	J-3	125	.128	S.F.	3	4.75	2.09	9.84	12.55
3000	Stairs, cast in place, pan filled treads		30	.533	L.F.	3.18	19.80	8.75	31.73	42
3100	Treads and risers		14	1.143	"	5.65	42.50	18.70	66.85	89
3300	For stair landings, add to floor prices						50%			
3400	Stair stringers and fascia	J-3	30	.533	S.F.	5.40	19.80	8.75	33.95	44.50
3600	For abrasive metal nosings on stairs, add		150	.107	L.F.	8.85	3.96	1.75	14.56	17.40
3700	For abrasive surface finish, add		600	.027	S.F.	1.47	.99	.44	2.90	3.55
3900	For raised abrasive strips, add		150	.107	L.F.	1.26	3.96	1.75	6.97	9.10
4000	Wainscot, bonded, 1-1/2" thick		30	.533	S.F.	3.81	19.80	8.75	32.36	43
4200	Epoxy terrazzo, 1/4" thick		40	.400	"	5.35	14.85	6.55	26.75	35
4300	Stone chips, onyx gemstone, per 50 lb. bag				Bag	16.30			16.30	17.95

09 66 13.30 Terrazzo, Precast

		Crew	Daily Output	Labor-Hours	Unit	Material	Labor	Equipment	Total	Total Incl O&P
0010	**TERRAZZO, PRECAST**									
0020	Base, 6" high, straight	1 Mstz	70	.114	L.F.	11.50	4.62		16.12	19.45
0100	Cove		60	.133		12.50	5.40		17.90	21.50
0300	8" high, straight		60	.133		11.15	5.40		16.55	20

09 66 Terrazzo Flooring

09 66 13 – Portland Cement Terrazzo Flooring

09 66 13.30 Terrazzo, Precast

		Crew	Daily Output	Labor-Hours	Unit	Material	2012 Bare Costs Labor	2012 Bare Costs Equipment	Total	Total Incl O&P
0400	Cove	1 Mstz	50	.160	L.F.	16.45	6.45		22.90	27.50
0600	For white cement, add					.44			.44	.48
0700	For 16 ga. zinc toe strip, add					1.68			1.68	1.85
0900	Curbs, 4" x 4" high	1 Mstz	40	.200		31.50	8.10		39.60	46.50
1000	8" x 8" high	"	30	.267		36	10.75		46.75	55.50
1200	Floor tiles, non-slip, 1" thick, 12" x 12"	D-1	60	.267	S.F.	18.90	10.65		29.55	37
1300	1-1/4" thick, 12" x 12"		60	.267		20	10.65		30.65	38
1500	16" x 16"		50	.320		21.50	12.75		34.25	43.50
1600	1-1/2" thick, 16" x 16"		45	.356		19.85	14.15		34	43.50
1800	For Venetian terrazzo, add					5.95			5.95	6.55
1900	For white cement, add					.56			.56	.62
2400	Stair treads, 1-1/2" thick, non-slip, three line pattern	2 Mstz	70	.229	L.F.	40	9.25		49.25	57.50
2500	Nosing and two lines		70	.229		40	9.25		49.25	57.50
2700	2" thick treads, straight		60	.267		44	10.75		54.75	64.50
2800	Curved		50	.320		57.50	12.95		70.45	82.50
3000	Stair risers, 1" thick, to 6" high, straight sections		60	.267		10.05	10.75		20.80	27
3100	Cove		50	.320		14.55	12.95		27.50	35
3300	Curved, 1" thick, to 6" high, vertical		48	.333		19.90	13.45		33.35	42
3400	Cove		38	.421		37.50	17		54.50	66.50
3600	Stair tread and riser, single piece, straight, minimum		60	.267		51	10.75		61.75	72
3700	Maximum		40	.400		66	16.15		82.15	96
3900	Curved tread and riser, minimum		40	.400		71.50	16.15		87.65	102
4000	Maximum		32	.500		90	20		110	129
4200	Stair stringers, notched, 1" thick		25	.640		29	26		55	70
4300	2" thick		22	.727		34.50	29.50		64	81
4500	Stair landings, structural, non-slip, 1-1/2" thick		85	.188	S.F.	32	7.60		39.60	46
4600	3" thick		75	.213		45	8.60		53.60	62
4800	Wainscot, 12" x 12" x 1" tiles	1 Mstz	12	.667		6.70	27		33.70	47
4900	16" x 16" x 1-1/2" tiles	"	8	1		13.95	40.50		54.45	75

09 66 16 – Terrazzo Floor Tile

09 66 16.10 Tile or Terrazzo Base

		Crew	Daily Output	Labor-Hours	Unit	Material	2012 Bare Costs Labor	2012 Bare Costs Equipment	Total	Total Incl O&P
0010	**TILE OR TERRAZZO BASE**									
0020	Scratch coat only	1 Mstz	150	.053	S.F.	.42	2.15		2.57	3.63
0500	Scratch and brown coat only	"	75	.107	"	.80	4.31		5.11	7.25

09 66 33 – Conductive Terrazzo Flooring

09 66 33.10 Conductive Terrazzo

		Crew	Daily Output	Labor-Hours	Unit	Material	2012 Bare Costs Labor	2012 Bare Costs Equipment	Total	Total Incl O&P
0010	**CONDUCTIVE TERRAZZO**									
2400	Bonded conductive floor for hospitals	J-3	90	.178	S.F.	4.54	6.60	2.91	14.05	17.90
2500	Epoxy terrazzo, 1/4" thick, minimum		100	.160		4.99	5.95	2.62	13.56	17.10
2550	Average		75	.213		5.80	7.90	3.49	17.19	22
2600	Maximum		60	.267		6.15	9.90	4.36	20.41	26

09 67 Fluid-Applied Flooring

09 67 20 – Epoxy-Marble Chip Flooring

09 67 20.13 Elastomeric Liquid Flooring

		Crew	Daily Output	Labor-Hours	Unit	Material	2012 Bare Costs Labor	Equipment	Total	Total Incl O&P
0010	**ELASTOMERIC LIQUID FLOORING**									
0020	Cementitious acrylic, 1/4" thick	C-6	520	.092	S.F.	1.58	3.38	.11	5.07	7
0100	3/8" thick	"	450	.107		2	3.91	.12	6.03	8.30
0200	Methyl methacrylate, 1/4" thick	C-8A	3000	.016		5.70	.61		6.31	7.15
0210	1/8" thick	"	3000	.016		3.12	.61		3.73	4.35
0300	Cupric oxychloride, on bond coat, minimum	C-6	480	.100		3.41	3.66	.11	7.18	9.50
0400	Maximum		420	.114		5.70	4.19	.13	10.02	12.80
2400	Mastic, hot laid, 2 coat, 1-1/2" thick, standard, minimum		690	.070		3.95	2.55	.08	6.58	8.35
2500	Maximum		520	.092		5.10	3.38	.11	8.59	10.85
2700	Acid-proof, minimum		605	.079		5.10	2.91	.09	8.10	10.15
2800	Maximum		350	.137		7.05	5	.16	12.21	15.55
3000	Neoprene, trowelled on, 1/4" thick, minimum		545	.088		3.89	3.23	.10	7.22	9.30
3100	Maximum		430	.112		5.35	4.09	.13	9.57	12.30
4300	Polyurethane, with suspended vinyl chips, minimum		1065	.045		7.30	1.65	.05	9	10.65
4500	Maximum	↓	860	.056	↓	10.50	2.04	.06	12.60	14.75

09 67 20.16 Epoxy Terrazzo

		Crew	Daily Output	Labor-Hours	Unit	Material	2012 Bare Costs Labor	Equipment	Total	Total Incl O&P
0010	**EPOXY TERRAZZO**									
1800	Epoxy terrazzo, 1/4" thick, chemical resistant, minimum	J-3	200	.080	S.F.	5.65	2.97	1.31	9.93	12
1900	Maximum		150	.107		8.55	3.96	1.75	14.26	17.15
2100	Conductive, minimum		210	.076		7.50	2.83	1.25	11.58	13.75
2200	Maximum	↓	160	.100	↓	9.85	3.71	1.64	15.20	18.05

09 67 20.19 Polyacrylate Terrazzo

		Crew	Daily Output	Labor-Hours	Unit	Material	2012 Bare Costs Labor	Equipment	Total	Total Incl O&P
0010	**POLYACRYLATE TERRAZZO**									
3150	Polyacrylate, 1/4" thick, minimum	C-6	735	.065	S.F.	3.35	2.39	.07	5.81	7.40
3170	Maximum		480	.100		4.26	3.66	.11	8.03	10.40
3200	3/8" thick, minimum		620	.077		4.32	2.84	.09	7.25	9.20
3220	Maximum		480	.100		6.05	3.66	.11	9.82	12.40
3300	Conductive, 1/4" thick, minimum		450	.107		6.85	3.91	.12	10.88	13.65
3330	Maximum		305	.157		9.45	5.75	.18	15.38	19.40
3350	3/8" thick, minimum		365	.132		9	4.82	.15	13.97	17.40
3370	Maximum		255	.188		12.05	6.90	.21	19.16	24
3450	Granite, conductive, 1/4" thick, minimum		695	.069		8.80	2.53	.08	11.41	13.60
3470	Maximum		420	.114		11.35	4.19	.13	15.67	19
3500	3/8" thick, minimum		695	.069		12.85	2.53	.08	15.46	18.05
3520	Maximum	↓	380	.126	↓	15.50	4.63	.14	20.27	24.50

09 67 20.26 Quartz Flooring

		Crew	Daily Output	Labor-Hours	Unit	Material	2012 Bare Costs Labor	Equipment	Total	Total Incl O&P
0010	**QUARTZ FLOORING**									
0600	Epoxy, with colored quartz chips, broadcast, minimum	C-6	675	.071	S.F.	2.67	2.61	.08	5.36	7
0700	Maximum		490	.098		3.85	3.59	.11	7.55	9.85
0900	Trowelled, minimum		560	.086		3.40	3.14	.10	6.64	8.65
1000	Maximum	↓	480	.100	↓	5.10	3.66	.11	8.87	11.40
1200	Heavy duty epoxy topping, 1/4" thick,									
1300	500 to 1,000 S.F.	C-6	420	.114	S.F.	5.25	4.19	.13	9.57	12.35
1500	1,000 to 2,000 S.F.		450	.107		4.76	3.91	.12	8.79	11.35
1600	Over 10,000 S.F.		480	.100		4.39	3.66	.11	8.16	10.55
3600	Polyester, with colored quartz chips, 1/16" thick, minimum		1065	.045		2.99	1.65	.05	4.69	5.85
3700	Maximum		560	.086		3.95	3.14	.10	7.19	9.25
3900	1/8" thick, minimum		810	.059		3.46	2.17	.07	5.70	7.20
4000	Maximum		675	.071		4.48	2.61	.08	7.17	9
4200	Polyester, heavy duty, compared to epoxy, add	↓	2590	.019	↓	1.44	.68	.02	2.14	2.64

09 67 Fluid-Applied Flooring

09 67 66 – Fluid-Applied Athletic Flooring

09 67 66.10 Polyurethane	Crew	Daily Output	Labor-Hours	Unit	Material	2012 Bare Costs Labor	Equipment	Total	Total Incl O&P
0010 **POLYURETHANE**									
4400 Thermoset, prefabricated in place, indoor									
4500 3/8" thick for basketball, gyms, etc.	1 Tilf	100	.080	S.F.	5.20	3.26		8.46	10.50
4600 1/2" thick for professional sports		95	.084		6.95	3.43		10.38	12.70
4700 Outdoor, 1/4" thick, smooth, for tennis		100	.080		5.40	3.26		8.66	10.75
4800 Rough, for track, 3/8" thick		95	.084		6.25	3.43		9.68	11.95
5000 Poured in place, indoor, with finish, 1/4" thick		80	.100		3.84	4.08		7.92	10.20
5050 3/8" thick		65	.123		4.67	5		9.67	12.50
5100 1/2" thick		50	.160		6.45	6.50		12.95	16.70

09 68 Carpeting

09 68 05 – Carpet Accessories

09 68 05.11 Flooring Transition Strip

	Crew	Daily Output	Labor-Hours	Unit	Material	2012 Bare Costs Labor	Equipment	Total	Total Incl O&P
0010 **FLOORING TRANSITION STRIP**									
0107 Clamp down brass divider, 12' strip, vinyl to carpet	1 Tilf	31.25	.256	Ea.	11.45	10.45		21.90	28
0117 Vinyl to hard surface	"	31.25	.256	"	11.45	10.45		21.90	28

09 68 10 – Carpet Pad

09 68 10.10 Commercial Grade Carpet Pad

	Crew	Daily Output	Labor-Hours	Unit	Material	2012 Bare Costs Labor	Equipment	Total	Total Incl O&P
0010 **COMMERCIAL GRADE CARPET PAD**									
9000 Sponge rubber pad, minimum	1 Tilf	150	.053	S.Y.	4.18	2.17		6.35	7.80
9100 Maximum		150	.053		10.40	2.17		12.57	14.65
9200 Felt pad, minimum		150	.053		3.84	2.17		6.01	7.40
9300 Maximum		150	.053		7.45	2.17		9.62	11.40
9400 Bonded urethane pad, minimum		150	.053		4.57	2.17		6.74	8.25
9500 Maximum		150	.053		8.20	2.17		10.37	12.20
9600 Prime urethane pad, minimum		150	.053		3	2.17		5.17	6.50
9700 Maximum		150	.053		5.25	2.17		7.42	9

09 68 13 – Tile Carpeting

09 68 13.10 Carpet Tile

	Crew	Daily Output	Labor-Hours	Unit	Material	2012 Bare Costs Labor	Equipment	Total	Total Incl O&P
0010 **CARPET TILE**									
0100 Tufted nylon, 18" x 18", hard back, 20 oz.	1 Tilf	80	.100	S.Y.	22	4.08		26.08	30.50
0110 26 oz.		80	.100		38	4.08		42.08	47.50
0200 Cushion back, 20 oz.		80	.100		28	4.08		32.08	36.50
0210 26 oz.		80	.100		43.50	4.08		47.58	54
1100 Tufted, 24" x 24", hard back, 24 oz. nylon		80	.100		28	4.08		32.08	37
1180 35 oz.		80	.100		34.50	4.08		38.58	44
5060 42 oz.		80	.100		45.50	4.08		49.58	56

09 68 16 – Sheet Carpeting

09 68 16.10 Sheet Carpet

	Crew	Daily Output	Labor-Hours	Unit	Material	2012 Bare Costs Labor	Equipment	Total	Total Incl O&P
0010 **SHEET CARPET**									
0700 Nylon, level loop, 26 oz., light to medium traffic	1 Tilf	75	.107	S.Y.	29	4.35		33.35	38.50
0720 28 oz., light to medium traffic		75	.107		30	4.35		34.35	39.50
0900 32 oz., medium traffic		75	.107		36.50	4.35		40.85	46.50
1100 40 oz., medium to heavy traffic		75	.107		53.50	4.35		57.85	65.50
2920 Nylon plush, 30 oz., medium traffic		57	.140		28.50	5.70		34.20	39.50
3000 36 oz., medium traffic		75	.107		37.50	4.35		41.85	48
3100 42 oz., medium to heavy traffic		70	.114		42.50	4.66		47.16	53.50
3200 46 oz., medium to heavy traffic		70	.114		49.50	4.66		54.16	61.50
3300 54 oz., heavy traffic		70	.114		54	4.66		58.66	66.50

09 68 Carpeting

09 68 16 – Sheet Carpeting

09 68 16.10 Sheet Carpet

		Crew	Daily Output	Labor-Hours	Unit	Material	2012 Bare Costs Labor	Equipment	Total	Total Incl O&P
3340	60 oz., heavy traffic	1 Tilf	70	.114	S.Y.	74	4.66		78.66	88.50
3665	Olefin, 24 oz., light to medium traffic		75	.107		17	4.35		21.35	25
3670	26 oz., medium traffic		75	.107		18.40	4.35		22.75	26.50
3680	28 oz., medium to heavy traffic		75	.107		20	4.35		24.35	28.50
3700	32 oz., medium to heavy traffic		75	.107		25	4.35		29.35	34
3730	42 oz., heavy traffic		70	.114		27	4.66		31.66	36.50
4110	Wool, level loop, 40 oz., medium traffic		75	.107		126	4.35		130.35	145
4500	50 oz., medium to heavy traffic		75	.107		127	4.35		131.35	146
4700	Patterned, 32 oz., medium to heavy traffic		70	.114		126	4.66		130.66	146
4900	48 oz., heavy traffic		70	.114		129	4.66		133.66	149
5000	For less than full roll (approx. 1500 S.F.), add					25%				
5100	For small rooms, less than 12' wide, add						25%			
5200	For large open areas (no cuts), deduct						25%			
5600	For bound carpet baseboard, add	1 Tilf	300	.027	L.F.	2.75	1.09		3.84	4.63
5610	For stairs, not incl. price of carpet, add	"	30	.267	Riser		10.85		10.85	15.95
5620	For borders and patterns, add to labor						18%			
8950	For tackless, stretched installation, add padding from 09 68 10.10 to above									
9850	For brand-named specific fiber, add				S.Y.	25%				

09 68 20 – Athletic Carpet

09 68 20.10 Indoor Athletic Carpet

		Crew	Daily Output	Labor-Hours	Unit	Material	2012 Bare Costs Labor	Equipment	Total	Total Incl O&P
0010	**INDOOR ATHLETIC CARPET**									
3700	Polyethylene, in rolls, no base incl., landscape surfaces	1 Tilf	275	.029	S.F.	3.47	1.19		4.66	5.55
3800	Nylon action surface, 1/8" thick		275	.029		3.63	1.19		4.82	5.75
3900	1/4" thick		275	.029		5.25	1.19		6.44	7.50
4000	3/8" thick		275	.029		6.55	1.19		7.74	9
4100	Golf tee surface with foam back		235	.034		6.50	1.39		7.89	9.20
4200	Practice putting, knitted nylon surface		235	.034		5.50	1.39		6.89	8.10
5500	Polyvinyl chloride, sheet goods for gyms, 1/4" thick		80	.100		9.20	4.08		13.28	16.10
5600	3/8" thick		60	.133		10.60	5.45		16.05	19.65

09 69 Access Flooring

09 69 13 – Rigid-Grid Access Flooring

09 69 13.10 Access Floors

		Crew	Daily Output	Labor-Hours	Unit	Material	2012 Bare Costs Labor	Equipment	Total	Total Incl O&P
0010	**ACCESS FLOORS**									
0015	Access floor package including panel, pedestal, stringers & laminate cover									
0100	Computer room, greater than 6,000 S.F.	4 Carp	750	.043	S.F.	7.50	1.88		9.38	11.15
0110	Less than 6,000 S.F.	2 Carp	375	.043		8.15	1.88		10.03	11.90
0120	Office, greater than 6,000 S.F.	4 Carp	1050	.030		5.35	1.34		6.69	7.95
0250	Panels, particle board or steel, 1250# load, no covering, under 6,000 S.F.	2 Carp	600	.027		3.75	1.18		4.93	5.95
0300	Over 6,000 S.F.		640	.025		3.21	1.10		4.31	5.25
0400	Aluminum, 24" panels		500	.032		33	1.41		34.41	38.50
0600	For carpet covering, add					8.75			8.75	9.65
0700	For vinyl floor covering, add					7			7	7.70
0900	For high pressure laminate covering, add					5.45			5.45	6
0910	For snap on stringer system, add	2 Carp	1000	.016		1.44	.71		2.15	2.67
0950	Office applications, steel or concrete panels,									
0960	no covering, over 6,000 S.F.	2 Carp	960	.017	S.F.	10.65	.74		11.39	12.85
1000	Machine cutouts after initial installation	1 Carp	50	.160	Ea.	5.05	7.05		12.10	16.40
1050	Pedestals, 6" to 12"	2 Carp	85	.188		8.35	8.30		16.65	22
1100	Air conditioning grilles, 4" x 12"	1 Carp	17	.471		66	21		87	105

09 69 Access Flooring

09 69 13 – Rigid-Grid Access Flooring

09 69 13.10 Access Floors		Crew	Daily Output	Labor-Hours	Unit	Material	2012 Bare Costs Labor	Equipment	Total	Total Incl O&P
1150	4" x 18"	1 Carp	14	.571	Ea.	90.50	25		115.50	139
1200	Approach ramps, minimum	2 Carp	60	.267	S.F.	25.50	11.75		37.25	46
1300	Maximum	"	40	.400	"	33.50	17.65		51.15	63.50
1500	Handrail, 2 rail, aluminum	1 Carp	15	.533	L.F.	97.50	23.50		121	143

09 72 Wall Coverings

09 72 23 – Wallpapering

09 72 23.10 Wallpaper

	09 72 23.10 Wallpaper		Crew	Daily Output	Labor-Hours	Unit	Material	2012 Bare Costs Labor	Equipment	Total	Total Incl O&P
0010	**WALLPAPER** including sizing; add 10-30 percent waste @ takeoff	R097223-10									
0050	Aluminum foil		1 Pape	275	.029	S.F.	1.02	1.12		2.14	2.80
0100	Copper sheets, .025" thick, vinyl backing			240	.033		5.45	1.28		6.73	7.90
0300	Phenolic backing			240	.033		7.05	1.28		8.33	9.70
0600	Cork tiles, light or dark, 12" x 12" x 3/16"			240	.033		4.47	1.28		5.75	6.85
0700	5/16" thick			235	.034		3.80	1.31		5.11	6.15
0900	1/4" basketweave			240	.033		5.85	1.28		7.13	8.35
1000	1/2" natural, non-directional pattern			240	.033		7.50	1.28		8.78	10.15
1100	3/4" natural, non-directional pattern			240	.033		10.45	1.28		11.73	13.40
1200	Granular surface, 12" x 36", 1/2" thick			385	.021		1.26	.80		2.06	2.59
1300	1" thick			370	.022		1.63	.83		2.46	3.04
1500	Polyurethane coated, 12" x 12" x 3/16" thick			240	.033		3.95	1.28		5.23	6.25
1600	5/16" thick			235	.034		5.65	1.31		6.96	8.15
1800	Cork wallpaper, paperbacked, natural			480	.017		1.99	.64		2.63	3.15
1900	Colors			480	.017		2.78	.64		3.42	4.02
2100	Flexible wood veneer, 1/32" thick, plain woods			100	.080		2.36	3.08		5.44	7.20
2200	Exotic woods		▼	95	.084	▼	3.58	3.24		6.82	8.80
2400	Gypsum-based, fabric-backed, fire resistant										
2500	for masonry walls, minimum, 21 oz./S.Y.		1 Pape	800	.010	S.F.	.87	.39		1.26	1.54
2700	Maximum (small quantities)		"	640	.013		1.37	.48		1.85	2.23
2750	Acrylic, modified, semi-rigid PVC, .028" thick		2 Carp	330	.048		1.13	2.14		3.27	4.53
2800	.040" thick		"	320	.050		1.49	2.21		3.70	5.05
3000	Vinyl wall covering, fabric-backed, lightweight, type 1 (12-15 oz./S.Y.)		1 Pape	640	.013		.82	.48		1.30	1.62
3300	Medium weight, type 2 (20-24 oz./S.Y.)			480	.017		.90	.64		1.54	1.95
3400	Heavy weight, type 3 (28 oz./S.Y.)		▼	435	.018	▼	1.52	.71		2.23	2.73
3600	Adhesive, 5 gal. lots (18 S.Y./gal.)					Gal.	11.10			11.10	12.20
3700	Wallpaper, average workmanship, solid pattern, low cost paper		1 Pape	640	.013	S.F.	.37	.48		.85	1.13
3900	basic patterns (matching required), avg. cost paper			535	.015		.97	.58		1.55	1.93
4000	Paper at $85 per double roll, quality workmanship		▼	435	.018	▼	2.47	.71		3.18	3.78
4100	Linen wall covering, paper backed										
4150	Flame treatment, minimum					S.F.	.97			.97	1.07
4180	Maximum						1.65			1.65	1.82
4190	Grass cloth, natural fabric	G	1 Pape	400	.020		2.99	.77		3.76	4.44
4200	Grass cloths with lining paper, minimum	G		400	.020		.78	.77		1.55	2.01
4300	Maximum	G	▼	350	.023	▼	2.49	.88		3.37	4.06

09 77 33 – Fiberglass Reinforced Panels

09 77 33.10 Fiberglass Reinforced Plastic Panels	Crew	Daily Output	Labor-Hours	Unit	Material	2012 Bare Costs Labor	Equipment	Total	Total Incl O&P
0010 **FIBERGLASS REINFORCED PLASTIC PANELS**, .090" thick									
0020 On walls, adhesive mounted, embossed surface	2 Carp	640	.025	S.F.	1.15	1.10		2.25	2.97
0030 Smooth surface		640	.025		1.41	1.10		2.51	3.25
0040 Fire rated, embossed surface		640	.025		2.05	1.10		3.15	3.96
0050 Nylon rivet mounted, on drywall, embossed surface		480	.033		1.15	1.47		2.62	3.53
0060 Smooth surface		480	.033		1.41	1.47		2.88	3.81
0070 Fire rated, embossed surface		480	.033		2.05	1.47		3.52	4.52
0080 On masonry, embossed surface		320	.050		1.15	2.21		3.36	4.66
0090 Smooth surface		320	.050		1.41	2.21		3.62	4.94
0100 Fire rated, embossed surface		320	.050		2.05	2.21		4.26	5.65
0110 Nylon rivet and adhesive mounted, on drywall, embossed surface		240	.067		1.30	2.94		4.24	5.95
0120 Smooth surface		240	.067		1.30	2.94		4.24	5.95
0130 Fire rated, embossed surface		240	.067		2.26	2.94		5.20	7
0140 On masonry, embossed surface		190	.084		1.30	3.71		5.01	7.15
0150 Smooth surface		190	.084		1.30	3.71		5.01	7.15
0160 Fire rated, embossed surface	↓	190	.084	↓	2.26	3.71		5.97	8.20
0170 For moldings add	1 Carp	250	.032	L.F.	.32	1.41		1.73	2.52
0180 On ceilings, for lay in grid system, embossed surface		400	.020	S.F.	1.15	.88		2.03	2.63
0190 Smooth surface		400	.020		1.41	.88		2.29	2.91
0200 Fire rated, embossed surface	↓	400	.020	↓	2.05	.88		2.93	3.62

09 77 43 – Panel Systems

09 77 43.20 Slatwall Panels and Accessories	Crew	Daily Output	Labor-Hours	Unit	Material	2012 Bare Costs Labor	Equipment	Total	Total Incl O&P
0010 **SLATWALL PANELS AND ACCESSORIES**									
0100 Slatwall panel, 4' x 8' x 3/4" T, MDF, paint grade	1 Carp	500	.016	S.F.	1.32	.71		2.03	2.54
0110 Melamine finish		500	.016		1.87	.71		2.58	3.15
0120 High pressure plastic laminate finish	↓	500	.016		3.25	.71		3.96	4.67
0130 Aluminum channel inserts, add				↓	2.94			2.94	3.23
0200 Accessories, corner forms, 8' L				L.F.	4.28			4.28	4.71
0210 T-connector, 8' L					5.15			5.15	5.70
0220 J-mold, 8' L					1.33			1.33	1.46
0230 Edge cap, 8' L					2.05			2.05	2.26
0240 Finish end cap, 8' L				↓	3.18			3.18	3.50
0300 Display hook, metal, 4" L				Ea.	.44			.44	.48
0310 6" L					.49			.49	.54
0320 8" L					.54			.54	.59
0330 10" L					.58			.58	.64
0340 12" L					.63			.63	.69
0350 Acrylic, 4" L					.91			.91	1
0360 6" L					1.05			1.05	1.16
0370 8" L					1.10			1.10	1.21
0380 10" L					1.12			1.12	1.23
0400 Waterfall hanger, metal, 12" - 16"					4.75			4.75	5.25
0410 Acrylic					9.85			9.85	10.85
0500 Shelf bracket, metal, 8"					4.50			4.50	4.95
0510 10"					4.62			4.62	5.10
0520 12"					4.52			4.52	4.97
0530 14"					4.86			4.86	5.35
0540 16"					6.10			6.10	6.75
0550 Acrylic, 8"					3.04			3.04	3.34
0560 10"					3.37			3.37	3.71
0570 12"					3.72			3.72	4.09
0580 14"				↓	3.99			3.99	4.39

09 77 Special Wall Surfacing

09 77 43 – Panel Systems

09 77 43.20 Slatwall Panels and Accessories	Crew	Daily Output	Labor-Hours	Unit	Material	2012 Bare Costs Labor	Equipment	Total	Total Incl O&P
0600	Shelf, acrylic, 12" x 16" x 1/4"				Ea.	19.90			19.90	22
0610	12" x 24" x 1/4"				↓	30			30	33

09 81 Acoustic Insulation

09 81 16 – Acoustic Blanket Insulation

09 81 16.10 Sound Attenuation Blanket

		Crew	Daily Output	Labor-Hours	Unit	Material	2012 Bare Costs Labor	Equipment	Total	Total Incl O&P
0010	**SOUND ATTENUATION BLANKET**									
0020	Blanket, 1" thick	1 Carp	925	.009	S.F.	.25	.38		.63	.87
0500	1-1/2" thick		920	.009		.25	.38		.63	.87
1000	2" thick		915	.009		.34	.39		.73	.96
1500	3" thick		910	.009		.50	.39		.89	1.15
3000	Thermal or acoustical batt above ceiling, 2" thick		900	.009		.49	.39		.88	1.14
3100	3" thick		900	.009		.74	.39		1.13	1.41
3200	4" thick	↓	900	.009		.92	.39		1.31	1.61
3400	Urethane plastic foam, open cell, on wall, 2" thick	2 Carp	2050	.008		3.05	.34		3.39	3.89
3500	3" thick		1550	.010		4.05	.46		4.51	5.15
3600	4" thick		1050	.015		5.70	.67		6.37	7.30
3700	On ceiling, 2" thick		1700	.009		3.04	.42		3.46	3.98
3800	3" thick		1300	.012		4.05	.54		4.59	5.30
3900	4" thick	↓	900	.018	↓	5.70	.78		6.48	7.45
4000	Nylon matting 0.4" thick, with carbon black spinerette									
4010	plus polyester fabric, on floor	D-7	1600	.010	S.F.	1.30	.36		1.66	1.97
4200	Fiberglass reinf. backer board underlayment, 7/16" thick, on floor	"	1500	.011	"	1.90	.39		2.29	2.66

09 84 Acoustic Room Components

09 84 13 – Fixed Sound-Absorptive Panels

09 84 13.10 Fixed Panels

		Crew	Daily Output	Labor-Hours	Unit	Material	2012 Bare Costs Labor	Equipment	Total	Total Incl O&P
0010	**FIXED PANELS** Perforated steel facing, painted with									
0100	Fiberglass or mineral filler, no backs, 2-1/4" thick, modular									
0200	space units, ceiling or wall hung, white or colored	1 Carp	100	.080	S.F.	8.55	3.53		12.08	14.85
0300	Fiberboard sound deadening panels, 1/2" thick	"	600	.013	"	.32	.59		.91	1.25
0500	Fiberglass panels, 4' x 8' x 1" thick, with									
0600	glass cloth face for walls, cemented	1 Carp	155	.052	S.F.	7.90	2.28		10.18	12.20
0700	1-1/2" thick, dacron covered, inner aluminum frame,									
0710	wall mounted	1 Carp	300	.027	S.F.	8.75	1.18		9.93	11.40
0900	Mineral fiberboard panels, fabric covered, 30" x 108",									
1000	3/4" thick, concealed spline, wall mounted	1 Carp	150	.053	S.F.	6.25	2.35		8.60	10.50

09 84 36 – Sound-Absorbing Ceiling Units

09 84 36.10 Barriers

		Crew	Daily Output	Labor-Hours	Unit	Material	2012 Bare Costs Labor	Equipment	Total	Total Incl O&P
0010	**BARRIERS** Plenum									
0600	Aluminum foil, fiberglass reinf., parallel with joists	1 Carp	275	.029	S.F.	.92	1.28		2.20	2.98
0700	Perpendicular to joists		180	.044		.92	1.96		2.88	4.03
0900	Aluminum mesh, kraft paperbacked		275	.029		.77	1.28		2.05	2.82
0970	Fiberglass batts, kraft faced, 3-1/2" thick		1400	.006		.36	.25		.61	.79
0980	6" thick		1300	.006		.65	.27		.92	1.14
1000	Sheet lead, 1 lb., 1/64" thick, perpendicular to joists		150	.053		4.92	2.35		7.27	9
1100	Vinyl foam reinforced, 1/8" thick, 1.0 lb. per S.F.	↓	150	.053	↓	4.40	2.35		6.75	8.45

09 91 03 – Paint Restoration

09 91 03.20 Sanding

		Crew	Daily Output	Labor-Hours	Unit	Material	2012 Bare Costs Labor	Equipment	Total	Total Incl O&P
0010	**SANDING** and puttying interior trim, compared to R099100-10									
0100	Painting 1 coat, on quality work				L.F.		100%			
0300	Medium work						50%			
0400	Industrial grade						25%			
0500	Surface protection, placement and removal									
0510	Basic drop cloths	1 Pord	6400	.001	S.F.		.05		.05	.07
0520	Masking with paper		800	.010		.05	.38		.43	.63
0530	Volume cover up (using plastic sheathing, or building paper)		16000	.001			.02		.02	.03

09 91 03.30 Exterior Surface Preparation

		Crew	Daily Output	Labor-Hours	Unit	Material	2012 Bare Costs Labor	Equipment	Total	Total Incl O&P
0010	**EXTERIOR SURFACE PREPARATION** R099100-10									
0015	Doors, per side, not incl. frames or trim									
0020	Scrape & sand									
0030	Wood, flush	1 Pord	616	.013	S.F.		.50		.50	.74
0040	Wood, detail		496	.016			.62		.62	.92
0050	Wood, louvered		280	.029			1.09		1.09	1.63
0060	Wood, overhead		616	.013			.50		.50	.74
0070	Wire brush									
0080	Metal, flush	1 Pord	640	.013	S.F.		.48		.48	.72
0090	Metal, detail		520	.015			.59		.59	.88
0100	Metal, louvered		360	.022			.85		.85	1.27
0110	Metal or fibr., overhead		640	.013			.48		.48	.72
0120	Metal, roll up		560	.014			.55		.55	.82
0130	Metal, bulkhead		640	.013			.48		.48	.72
0140	Power wash, based on 2500 lb. operating pressure									
0150	Metal, flush	A-1H	2240	.004	S.F.		.13	.03	.16	.22
0160	Metal, detail		2120	.004			.13	.03	.16	.23
0170	Metal, louvered		2000	.004			.14	.03	.17	.26
0180	Metal or fibr., overhead		2400	.003			.12	.03	.15	.21
0190	Metal, roll up		2400	.003			.12	.03	.15	.21
0200	Metal, bulkhead		2200	.004			.13	.03	.16	.23
0400	Windows, per side, not incl. trim									
0410	Scrape & sand									
0420	Wood, 1-2 lite	1 Pord	320	.025	S.F.		.96		.96	1.43
0430	Wood, 3-6 lite		280	.029			1.09		1.09	1.63
0440	Wood, 7-10 lite		240	.033			1.27		1.27	1.91
0450	Wood, 12 lite		200	.040			1.53		1.53	2.29
0460	Wood, Bay/Bow		320	.025			.96		.96	1.43
0470	Wire brush									
0480	Metal, 1-2 lite	1 Pord	480	.017	S.F.		.64		.64	.95
0490	Metal, 3-6 lite		400	.020			.77		.77	1.14
0500	Metal, Bay/Bow		480	.017			.64		.64	.95
0510	Power wash, based on 2500 lb. operating pressure									
0520	1-2 lite	A-1H	4400	.002	S.F.		.06	.02	.08	.12
0530	3-6 lite		4320	.002			.06	.02	.08	.12
0540	7-10 lite		4240	.002			.07	.02	.09	.12
0550	12 lite		4160	.002			.07	.02	.09	.12
0560	Bay/Bow		4400	.002			.06	.02	.08	.12
0600	Siding, scrape and sand, light=10-30%, med.=30-70%									
0610	Heavy=70-100% of surface to sand									
0650	Texture 1-11, light	1 Pord	480	.017	S.F.		.64		.64	.95
0660	Med.		440	.018			.70		.70	1.04
0670	Heavy		360	.022			.85		.85	1.27

09 91 Painting

09 91 03 – Paint Restoration

09 91 03.30 Exterior Surface Preparation

		Crew	Daily Output	Labor-Hours	Unit	Material	2012 Bare Costs Labor	2012 Bare Costs Equipment	Total	Total Incl O&P
0680	Wood shingles, shakes, light	1 Pord	440	.018	S.F.		.70		.70	1.04
0690	Med.		360	.022			.85		.85	1.27
0700	Heavy		280	.029			1.09		1.09	1.63
0710	Clapboard, light		520	.015			.59		.59	.88
0720	Med.		480	.017			.64		.64	.95
0730	Heavy	▼	400	.020	▼		.77		.77	1.14
0740	Wire brush									
0750	Aluminum, light	1 Pord	600	.013	S.F.		.51		.51	.76
0760	Med.		520	.015			.59		.59	.88
0770	Heavy	▼	440	.018	▼		.70		.70	1.04
0780	Pressure wash, based on 2500 lb. operating pressure									
0790	Stucco	A-1H	3080	.003	S.F.		.09	.02	.11	.16
0800	Aluminum or vinyl		3200	.003			.09	.02	.11	.16
0810	Siding, masonry, brick & block	▼	2400	.003	▼		.12	.03	.15	.21
1300	Miscellaneous, wire brush									
1310	Metal, pedestrian gate	1 Pord	100	.080	S.F.		3.06		3.06	4.58
8000	For chemical washing, see Section 04 01 30									
8010	For steam cleaning, see Section 04 01 30.20									
8020	For sand blasting, see Section 03 35 29.60 and 05 01 10.51									

09 91 03.40 Interior Surface Preparation

		Crew	Daily Output	Labor-Hours	Unit	Material	2012 Bare Costs Labor	2012 Bare Costs Equipment	Total	Total Incl O&P
0010	**INTERIOR SURFACE PREPARATION** R099100-10									
0020	Doors, per side, not incl. frames or trim									
0030	Scrape & sand									
0040	Wood, flush	1 Pord	616	.013	S.F.		.50		.50	.74
0050	Wood, detail		496	.016			.62		.62	.92
0060	Wood, louvered	▼	280	.029	▼		1.09		1.09	1.63
0070	Wire brush									
0080	Metal, flush	1 Pord	640	.013	S.F.		.48		.48	.72
0090	Metal, detail		520	.015			.59		.59	.88
0100	Metal, louvered	▼	360	.022	▼		.85		.85	1.27
0110	Hand wash									
0120	Wood, flush	1 Pord	2160	.004	S.F.		.14		.14	.21
0130	Wood, detailed		2000	.004			.15		.15	.23
0140	Wood, louvered		1360	.006			.22		.22	.34
0150	Metal, flush		2160	.004			.14		.14	.21
0160	Metal, detail		2000	.004			.15		.15	.23
0170	Metal, louvered	▼	1360	.006	▼		.22		.22	.34
0400	Windows, per side, not incl. trim									
0410	Scrape & sand									
0420	Wood, 1-2 lite	1 Pord	360	.022	S.F.		.85		.85	1.27
0430	Wood, 3-6 lite		320	.025			.96		.96	1.43
0440	Wood, 7-10 lite		280	.029			1.09		1.09	1.63
0450	Wood, 12 lite		240	.033			1.27		1.27	1.91
0460	Wood, Bay/Bow	▼	360	.022	▼		.85		.85	1.27
0470	Wire brush									
0480	Metal, 1-2 lite	1 Pord	520	.015	S.F.		.59		.59	.88
0490	Metal, 3-6 lite		440	.018			.70		.70	1.04
0500	Metal, Bay/Bow	▼	520	.015	▼		.59		.59	.88
0600	Walls, sanding, light=10-30%, medium - 30-70%,									
0610	heavy=70-100% of surface to sand									
0650	Walls, sand									
0660	Gypsum board or plaster, light	1 Pord	3077	.003	S.F.		.10		.10	.15

09 91 Painting

09 91 03 – Paint Restoration

09 91 03.40 Interior Surface Preparation

		Crew	Daily Output	Labor-Hours	Unit	Material	2012 Bare Costs Labor	Equipment	Total	Total Incl O&P
0670	Gypsum board or plaster, medium	1 Pord	2160	.004	S.F.		.14		.14	.21
0680	Gypsum board or plaster, heavy		923	.009			.33		.33	.50
0690	Wood, T&G, light		2400	.003			.13		.13	.19
0700	Wood, T&G, med.		1600	.005			.19		.19	.29
0710	Wood, T&G, heavy	↓	800	.010	↓		.38		.38	.57
0720	Walls, wash									
0730	Gypsum board or plaster	1 Pord	3200	.003	S.F.		.10		.10	.14
0740	Wood, T&G		3200	.003			.10		.10	.14
0750	Masonry, brick & block, smooth		2800	.003			.11		.11	.16
0760	Masonry, brick & block, coarse	↓	2000	.004	↓		.15		.15	.23
8000	For chemical washing, see Section 04 01 30									
8010	For steam cleaning, see Section 04 01 30.20									
8020	For sand blasting, see Section 03 35 29.60 and 05 01 10.51									

09 91 13 – Exterior Painting

09 91 13.30 Fences

			Crew	Daily Output	Labor-Hours	Unit	Material	2012 Bare Costs Labor	Equipment	Total	Total Incl O&P
0010	**FENCES**	R099100-20									
0100	Chain link or wire metal, one side, water base										
0110	Roll & brush, first coat		1 Pord	960	.008	S.F.	.07	.32		.39	.55
0120	Second coat			1280	.006		.06	.24		.30	.43
0130	Spray, first coat		↓	2275	.004		.07	.13		.20	.27
0140	Second coat		↓	2600	.003	↓	.07	.12		.19	.25
0150	Picket, water base										
0160	Roll & brush, first coat		1 Pord	865	.009	S.F.	.07	.35		.42	.61
0170	Second coat			1050	.008		.07	.29		.36	.52
0180	Spray, first coat			2275	.004		.07	.13		.20	.28
0190	Second coat		↓	2600	.003	↓	.07	.12		.19	.26
0200	Stockade, water base										
0210	Roll & brush, first coat		1 Pord	1040	.008	S.F.	.07	.29		.36	.52
0220	Second coat			1200	.007		.07	.26		.33	.46
0230	Spray, first coat			2275	.004		.07	.13		.20	.28
0240	Second coat		↓	2600	.003	↓	.07	.12		.19	.26

09 91 13.42 Miscellaneous, Exterior

			Crew	Daily Output	Labor-Hours	Unit	Material	2012 Bare Costs Labor	Equipment	Total	Total Incl O&P
0010	**MISCELLANEOUS, EXTERIOR**	R099100-20									
0015	For painting metals, see Section 09 97 13.23										
0100	Railing, ext., decorative wood, incl. cap & baluster										
0110	Newels & spindles @ 12" O.C.										
0120	Brushwork, stain, sand, seal & varnish										
0130	First coat		1 Pord	90	.089	L.F.	.73	3.40		4.13	5.90
0140	Second coat		"	120	.067	"	.73	2.55		3.28	4.62
0150	Rough sawn wood, 42" high, 2" x 2" verticals, 6" O.C.										
0160	Brushwork, stain, each coat		1 Pord	90	.089	L.F.	.24	3.40		3.64	5.35
0170	Wrought iron, 1" rail, 1/2" sq. verticals										
0180	Brushwork, zinc chromate, 60" high, bars 6" O.C.										
0190	Primer		1 Pord	130	.062	L.F.	.83	2.35		3.18	4.43
0200	Finish coat			130	.062		1.09	2.35		3.44	4.72
0210	Additional coat		↓	190	.042	↓	1.27	1.61		2.88	3.81
0220	Shutters or blinds, single panel, 2' x 4', paint all sides										
0230	Brushwork, primer		1 Pord	20	.400	Ea.	.62	15.30		15.92	23.50
0240	Finish coat, exterior latex			20	.400		.54	15.30		15.84	23.50
0250	Primer & 1 coat, exterior latex			13	.615		1.02	23.50		24.52	36
0260	Spray, primer			35	.229		.90	8.75		9.65	14.05
0270	Finish coat, exterior latex		↓	35	.229	↓	1.14	8.75		9.89	14.30

09 91 13.42 Miscellaneous, Exterior

		Crew	Daily Output	Labor-Hours	Unit	Material	2012 Bare Costs Labor	Equipment	Total	Total Incl O&P
0280	Primer & 1 coat, exterior latex	1 Pord	20	.400	Ea.	.97	15.30		16.27	24
0290	For louvered shutters, add				S.F.	10%				
0300	Stair stringers, exterior, metal									
0310	Roll & brush, zinc chromate, to 14", each coat	1 Pord	320	.025	L.F.	.36	.96		1.32	1.83
0320	Rough sawn wood, 4" x 12"									
0330	Roll & brush, exterior latex, each coat	1 Pord	215	.037	L.F.	.08	1.42		1.50	2.22
0340	Trellis/lattice, 2" x 2" @ 3" O.C. with 2" x 8" supports									
0350	Spray, latex, per side, each coat	1 Pord	475	.017	S.F.	.08	.64		.72	1.05
0450	Decking, ext., sealer, alkyd, brushwork, sealer coat		1140	.007		.09	.27		.36	.50
0460	1st coat		1140	.007		.09	.27		.36	.50
0470	2nd coat		1300	.006		.07	.24		.31	.42
0500	Paint, alkyd, brushwork, primer coat		1140	.007		.09	.27		.36	.50
0510	1st coat		1140	.007		.16	.27		.43	.57
0520	2nd coat		1300	.006		.12	.24		.36	.48
0600	Sand paint, alkyd, brushwork, 1 coat		150	.053		.16	2.04		2.20	3.22

09 91 13.60 Siding Exterior

		Crew	Daily Output	Labor-Hours	Unit	Material	2012 Bare Costs Labor	Equipment	Total	Total Incl O&P
0010	SIDING EXTERIOR, Alkyd (oil base) R099100-10									
0450	Steel siding, oil base, paint 1 coat, brushwork	2 Pord	2015	.008	S.F.	.13	.30		.43	.59
0500	Spray R099100-20		4550	.004		.19	.13		.32	.41
0800	Paint 2 coats, brushwork		1300	.012		.25	.47		.72	.98
1000	Spray		2750	.006		.17	.22		.39	.52
1200	Stucco, rough, oil base, paint 2 coats, brushwork		1300	.012		.25	.47		.72	.98
1400	Roller		1625	.010		.27	.38		.65	.85
1600	Spray		2925	.005		.28	.21		.49	.62
1800	Texture 1-11 or clapboard, oil base, primer coat, brushwork		1300	.012		.12	.47		.59	.83
2000	Spray		4550	.004		.12	.13		.25	.33
2400	Paint 2 coats, brushwork		810	.020		.37	.76		1.13	1.53
2600	Spray		2600	.006		.41	.24		.65	.80
3400	Stain 2 coats, brushwork		950	.017		.16	.64		.80	1.14
4000	Spray		3050	.005		.18	.20		.38	.50
4200	Wood shingles, oil base primer coat, brushwork		1300	.012		.11	.47		.58	.82
4400	Spray		3900	.004		.10	.16		.26	.34
5000	Paint 2 coats, brushwork		810	.020		.31	.76		1.07	1.47
5200	Spray		2275	.007		.29	.27		.56	.72
6500	Stain 2 coats, brushwork		950	.017		.16	.64		.80	1.14
7000	Spray		2660	.006		.22	.23		.45	.58
8000	For latex paint, deduct					10%				
8100	For work over 12' H, from pipe scaffolding, add						15%			
8200	For work over 12' H, from extension ladder, add						25%			
8300	For work over 12' H, from swing staging, add						35%			

09 91 13.62 Siding, Misc.

		Crew	Daily Output	Labor-Hours	Unit	Material	2012 Bare Costs Labor	Equipment	Total	Total Incl O&P
0010	SIDING, MISC., latex paint R099100-10									
0100	Aluminum siding									
0110	Brushwork, primer R099100-20	2 Pord	2275	.007	S.F.	.06	.27		.33	.46
0120	Finish coat, exterior latex		2275	.007		.05	.27		.32	.46
0130	Primer & 1 coat exterior latex		1300	.012		.12	.47		.59	.83
0140	Primer & 2 coats exterior latex		975	.016		.17	.63		.80	1.12
0150	Mineral fiber shingles									
0160	Brushwork, primer	2 Pord	1495	.011	S.F.	.12	.41		.53	.74
0170	Finish coat, industrial enamel		1495	.011		.15	.41		.56	.78
0180	Primer & 1 coat enamel		810	.020		.27	.76		1.03	1.42
0190	Primer & 2 coats enamel		540	.030		.42	1.13		1.55	2.15

09 91 Painting

09 91 13 – Exterior Painting

09 91 13.62 Siding, Misc.

		Crew	Daily Output	Labor-Hours	Unit	Material	2012 Bare Costs Labor	2012 Bare Costs Equipment	Total	Total Incl O&P
0200	Roll, primer	2 Pord	1625	.010	S.F.	.13	.38		.51	.70
0210	Finish coat, industrial enamel		1625	.010		.16	.38		.54	.74
0220	Primer & 1 coat enamel		975	.016		.29	.63		.92	1.26
0230	Primer & 2 coats enamel		650	.025		.45	.94		1.39	1.91
0240	Spray, primer		3900	.004		.10	.16		.26	.34
0250	Finish coat, industrial enamel		3900	.004		.14	.16		.30	.38
0260	Primer & 1 coat enamel		2275	.007		.24	.27		.51	.66
0270	Primer & 2 coats enamel		1625	.010		.37	.38		.75	.97
0280	Waterproof sealer, first coat		4485	.004		.08	.14		.22	.29
0290	Second coat	↓	5235	.003	↓	.07	.12		.19	.26
0300	Rough wood incl. shingles, shakes or rough sawn siding									
0310	Brushwork, primer	2 Pord	1280	.013	S.F.	.12	.48		.60	.86
0320	Finish coat, exterior latex		1280	.013		.09	.48		.57	.82
0330	Primer & 1 coat exterior latex		960	.017		.21	.64		.85	1.19
0340	Primer & 2 coats exterior latex		700	.023		.30	.87		1.17	1.64
0350	Roll, primer		2925	.005		.17	.21		.38	.49
0360	Finish coat, exterior latex		2925	.005		.11	.21		.32	.43
0370	Primer & 1 coat exterior latex		1790	.009		.27	.34		.61	.81
0380	Primer & 2 coats exterior latex		1300	.012		.38	.47		.85	1.12
0390	Spray, primer		3900	.004		.14	.16		.30	.38
0400	Finish coat, exterior latex		3900	.004		.08	.16		.24	.32
0410	Primer & 1 coat exterior latex		2600	.006		.22	.24		.46	.60
0420	Primer & 2 coats exterior latex		2080	.008		.31	.29		.60	.78
0430	Waterproof sealer, first coat		4485	.004		.14	.14		.28	.35
0440	Second coat	↓	4485	.004	↓	.08	.14		.22	.29
0450	Smooth wood incl. butt, T&G, beveled, drop or B&B siding									
0460	Brushwork, primer	2 Pord	2325	.007	S.F.	.09	.26		.35	.49
0470	Finish coat, exterior latex		1280	.013		.09	.48		.57	.82
0480	Primer & 1 coat exterior latex		800	.020		.18	.77		.95	1.34
0490	Primer & 2 coats exterior latex		630	.025		.27	.97		1.24	1.75
0500	Roll, primer		2275	.007		.10	.27		.37	.51
0510	Finish coat, exterior latex		2275	.007		.10	.27		.37	.51
0520	Primer & 1 coat exterior latex		1300	.012		.20	.47		.67	.92
0530	Primer & 2 coats exterior latex		975	.016		.29	.63		.92	1.26
0540	Spray, primer		4550	.004		.08	.13		.21	.29
0550	Finish coat, exterior latex		4550	.004		.08	.13		.21	.29
0560	Primer & 1 coat exterior latex		2600	.006		.16	.24		.40	.53
0570	Primer & 2 coats exterior latex		1950	.008		.24	.31		.55	.74
0580	Waterproof sealer, first coat		5230	.003		.08	.12		.20	.27
0590	Second coat	↓	5980	.003	↓	.08	.10		.18	.24
0600	For oil base paint, add					10%				

09 91 13.70 Doors and Windows, Exterior

		Crew	Daily Output	Labor-Hours	Unit	Material	2012 Bare Costs Labor	2012 Bare Costs Equipment	Total	Total Incl O&P
0010	**DOORS AND WINDOWS, EXTERIOR** R099100-10									
0100	Door frames & trim, only									
0110	Brushwork, primer R099100-20	1 Pord	512	.016	L.F.	.06	.60		.66	.95
0120	Finish coat, exterior latex		512	.016		.07	.60		.67	.96
0130	Primer & 1 coat, exterior latex		300	.027		.12	1.02		1.14	1.67
0140	Primer & 2 coats, exterior latex	↓	265	.030	↓	.19	1.15		1.34	1.94
0150	Doors, flush, both sides, incl. frame & trim									
0160	Roll & brush, primer	1 Pord	10	.800	Ea.	4.26	30.50		34.76	50.50
0170	Finish coat, exterior latex		10	.800		5.10	30.50		35.60	51.50
0180	Primer & 1 coat, exterior latex	↓	7	1.143	↓	9.35	43.50		52.85	76

09 91 13.70 Doors and Windows, Exterior

		Crew	Daily Output	Labor-Hours	Unit	Material	2012 Bare Costs Labor	Equipment	Total	Total Incl O&P
0190	Primer & 2 coats, exterior latex	1 Pord	5	1.600	Ea.	14.45	61		75.45	107
0200	Brushwork, stain, sealer & 2 coats polyurethane	↓	4	2	↓	25.50	76.50		102	142
0210	Doors, French, both sides, 10-15 lite, incl. frame & trim									
0220	Brushwork, primer	1 Pord	6	1.333	Ea.	2.13	51		53.13	79
0230	Finish coat, exterior latex		6	1.333		2.55	51		53.55	79.50
0240	Primer & 1 coat, exterior latex		3	2.667		4.67	102		106.67	158
0250	Primer & 2 coats, exterior latex		2	4		7.05	153		160.05	237
0260	Brushwork, stain, sealer & 2 coats polyurethane	↓	2.50	3.200	↓	9.30	122		131.30	193
0270	Doors, louvered, both sides, incl. frame & trim									
0280	Brushwork, primer	1 Pord	7	1.143	Ea.	4.26	43.50		47.76	70
0290	Finish coat, exterior latex		7	1.143		5.10	43.50		48.60	71
0300	Primer & 1 coat, exterior latex		4	2		9.35	76.50		85.85	124
0310	Primer & 2 coats, exterior latex		3	2.667		14.15	102		116.15	169
0320	Brushwork, stain, sealer & 2 coats polyurethane	↓	4.50	1.778	↓	25.50	68		93.50	130
0330	Doors, panel, both sides, incl. frame & trim									
0340	Roll & brush, primer	1 Pord	6	1.333	Ea.	4.26	51		55.26	81
0350	Finish coat, exterior latex		6	1.333		5.10	51		56.10	82
0360	Primer & 1 coat, exterior latex		3	2.667		9.35	102		111.35	163
0370	Primer & 2 coats, exterior latex		2.50	3.200		14.15	122		136.15	199
0380	Brushwork, stain, sealer & 2 coats polyurethane	↓	3	2.667	↓	25.50	102		127.50	181
0400	Windows, per ext. side, based on 15 S.F.									
0410	1 to 6 lite									
0420	Brushwork, primer	1 Pord	13	.615	Ea.	.84	23.50		24.34	36
0430	Finish coat, exterior latex		13	.615		1	23.50		24.50	36
0440	Primer & 1 coat, exterior latex		8	1		1.85	38.50		40.35	59
0450	Primer & 2 coats, exterior latex		6	1.333		2.79	51		53.79	79.50
0460	Stain, sealer & 1 coat varnish	↓	7	1.143	↓	3.67	43.50		47.17	69.50
0470	7 to 10 lite									
0480	Brushwork, primer	1 Pord	11	.727	Ea.	.84	28		28.84	42.50
0490	Finish coat, exterior latex		11	.727		1	28		29	42.50
0500	Primer & 1 coat, exterior latex		7	1.143		1.85	43.50		45.35	67.50
0510	Primer & 2 coats, exterior latex		5	1.600		2.79	61		63.79	94.50
0520	Stain, sealer & 1 coat varnish	↓	6	1.333	↓	3.67	51		54.67	80.50
0530	12 lite									
0540	Brushwork, primer	1 Pord	10	.800	Ea.	.84	30.50		31.34	47
0550	Finish coat, exterior latex		10	.800		1	30.50		31.50	47
0560	Primer & 1 coat, exterior latex		6	1.333		1.85	51		52.85	78.50
0570	Primer & 2 coats, exterior latex		5	1.600		2.79	61		63.79	94.50
0580	Stain, sealer & 1 coat varnish	↓	6	1.333		3.68	51		54.68	80.50
0590	For oil base paint, add				↓	10%				

09 91 13.80 Trim, Exterior

			Crew	Daily Output	Labor-Hours	Unit	Material	2012 Bare Costs Labor	Equipment	Total	Total Incl O&P
0010	**TRIM, EXTERIOR**	R099100-10									
0100	Door frames & trim (see Doors, interior or exterior)										
0110	Fascia, latex paint, one coat coverage	R099100-20									
0120	1" x 4", brushwork		1 Pord	640	.013	L.F.	.02	.48		.50	.74
0130	Roll			1280	.006		.02	.24		.26	.38
0140	Spray			2080	.004		.02	.15		.17	.24
0150	1" x 6" to 1" x 10", brushwork			640	.013		.07	.48		.55	.80
0160	Roll			1230	.007		.07	.25		.32	.45
0170	Spray			2100	.004		.06	.15		.21	.28
0180	1" x 12", brushwork			640	.013		.07	.48		.55	.80
0190	Roll			1050	.008		.07	.29		.36	.52

09 91 Painting

09 91 13 – Exterior Painting

09 91 13.80 Trim, Exterior

		Crew	Daily Output	Labor-Hours	Unit	Material	2012 Bare Costs Labor	Equipment	Total	Total Incl O&P
0200	Spray	1 Pord	2200	.004	L.F.	.06	.14		.20	.27
0210	Gutters & downspouts, metal, zinc chromate paint									
0220	Brushwork, gutters, 5", first coat	1 Pord	640	.013	L.F.	.38	.48		.86	1.14
0230	Second coat		960	.008		.36	.32		.68	.88
0240	Third coat		1280	.006		.29	.24		.53	.68
0250	Downspouts, 4", first coat		640	.013		.38	.48		.86	1.14
0260	Second coat		960	.008		.36	.32		.68	.88
0270	Third coat		1280	.006		.29	.24		.53	.68
0280	Gutters & downspouts, wood									
0290	Brushwork, gutters, 5", primer	1 Pord	640	.013	L.F.	.06	.48		.54	.78
0300	Finish coat, exterior latex		640	.013		.06	.48		.54	.79
0310	Primer & 1 coat exterior latex		400	.020		.12	.77		.89	1.28
0320	Primer & 2 coats exterior latex		325	.025		.19	.94		1.13	1.62
0330	Downspouts, 4", primer		640	.013		.06	.48		.54	.78
0340	Finish coat, exterior latex		640	.013		.06	.48		.54	.79
0350	Primer & 1 coat exterior latex		400	.020		.12	.77		.89	1.28
0360	Primer & 2 coats exterior latex		325	.025		.10	.94		1.04	1.51
0370	Molding, exterior, up to 14" wide									
0380	Brushwork, primer	1 Pord	640	.013	L.F.	.07	.48		.55	.79
0390	Finish coat, exterior latex		640	.013		.07	.48		.55	.80
0400	Primer & 1 coat exterior latex		400	.020		.15	.77		.92	1.30
0410	Primer & 2 coats exterior latex		315	.025		.15	.97		1.12	1.61
0420	Stain & fill		1050	.008		.10	.29		.39	.55
0430	Shellac		1850	.004		.10	.17		.27	.36
0440	Varnish		1275	.006		.11	.24		.35	.49

09 91 13.90 Walls, Masonry (CMU), Exterior

		Crew	Daily Output	Labor-Hours	Unit	Material	2012 Bare Costs Labor	Equipment	Total	Total Incl O&P
0350	**WALLS, MASONRY (CMU), EXTERIOR** R099100-10									
0360	Concrete masonry units (CMU), smooth surface									
0370	Brushwork, latex, first coat	1 Pord	640	.013	S.F.	.06	.48		.54	.79
0380	Second coat		960	.008		.05	.32		.37	.53
0390	Waterproof sealer, first coat		736	.011		.24	.42		.66	.89
0400	Second coat		1104	.007		.24	.28		.52	.68
0410	Roll, latex, paint, first coat		1465	.005		.07	.21		.28	.39
0420	Second coat		1790	.004		.05	.17		.22	.32
0430	Waterproof sealer, first coat		1680	.005		.24	.18		.42	.54
0440	Second coat		2060	.004		.24	.15		.39	.49
0450	Spray, latex, paint, first coat		1950	.004		.06	.16		.22	.29
0460	Second coat		2600	.003		.05	.12		.17	.23
0470	Waterproof sealer, first coat		2245	.004		.24	.14		.38	.47
0480	Second coat		2990	.003		.24	.10		.34	.42
0490	Concrete masonry unit (CMU), porous									
0500	Brushwork, latex, first coat	1 Pord	640	.013	S.F.	.12	.48		.60	.85
0510	Second coat		960	.008		.06	.32		.38	.55
0520	Waterproof sealer, first coat		736	.011		.24	.42		.66	.89
0530	Second coat		1104	.007		.24	.28		.52	.68
0540	Roll latex, first coat		1465	.005		.09	.21		.30	.41
0550	Second coat		1790	.004		.06	.17		.23	.32
0560	Waterproof sealer, first coat		1680	.005		.24	.18		.42	.54
0570	Second coat		2060	.004		.24	.15		.39	.49
0580	Spray latex, first coat		1950	.004		.07	.16		.23	.30
0590	Second coat		2600	.003		.05	.12		.17	.23
0600	Waterproof sealer, first coat		2245	.004		.24	.14		.38	.47

09 91 Painting

09 91 13 – Exterior Painting

09 91 13.90 Walls, Masonry (CMU), Exterior

		Crew	Daily Output	Labor-Hours	Unit	Material	2012 Bare Costs Labor	Equipment	Total	Total Incl O&P
0610	Second coat	1 Pord	2990	.003	S.F.	.24	.10		.34	.42

09 91 23 – Interior Painting

09 91 23.20 Cabinets and Casework

			Crew	Daily Output	Labor-Hours	Unit	Material	2012 Bare Costs Labor	Equipment	Total	Total Incl O&P
0010	**CABINETS AND CASEWORK**	R099100-10									
1000	Primer coat, oil base, brushwork		1 Pord	650	.012	S.F.	.06	.47		.53	.77
2000	Paint, oil base, brushwork, 1 coat	R099100-20		650	.012		.11	.47		.58	.82
3000	Stain, brushwork, wipe off			650	.012		.08	.47		.55	.79
4000	Shellac, 1 coat, brushwork			650	.012		.08	.47		.55	.79
4500	Varnish, 3 coats, brushwork, sand after 1st coat			325	.025		.29	.94		1.23	1.73
5000	For latex paint, deduct						10%				

09 91 23.33 Doors and Windows, Interior Alkyd (Oil Base)

			Crew	Daily Output	Labor-Hours	Unit	Material	2012 Bare Costs Labor	Equipment	Total	Total Incl O&P
0010	**DOORS AND WINDOWS, INTERIOR ALKYD (OIL BASE)**	R099100-10									
0500	Flush door & frame, 3' x 7', oil, primer, brushwork		1 Pord	10	.800	Ea.	3.08	30.50		33.58	49.50
1000	Paint, 1 coat	R099100-20		10	.800		4.11	30.50		34.61	50.50
1400	Stain, brushwork, wipe off			18	.444		1.68	17		18.68	27.50
1600	Shellac, 1 coat, brushwork			25	.320		1.70	12.25		13.95	20
1800	Varnish, 3 coats, brushwork, sand after 1st coat			9	.889		6.05	34		40.05	57.50
2000	Panel door & frame, 3' x 7', oil, primer, brushwork			6	1.333		2.28	51		53.28	79
2200	Paint, 1 coat			6	1.333		4.11	51		55.11	81
2600	Stain, brushwork, panel door, 3' x 7', not incl. frame			16	.500		1.68	19.15		20.83	30.50
2800	Shellac, 1 coat, brushwork			22	.364		1.70	13.90		15.60	23
3000	Varnish, 3 coats, brushwork, sand after 1st coat			7.50	1.067		6.05	41		47.05	67.50
4400	Windows, including frame and trim, per side										
4600	Colonial type, 6/6 lites, 2' x 3', oil, primer, brushwork		1 Pord	14	.571	Ea.	.36	22		22.36	33
5800	Paint, 1 coat			14	.571		.65	22		22.65	33
6200	3' x 5' opening, 6/6 lites, primer coat, brushwork			12	.667		.90	25.50		26.40	39
6400	Paint, 1 coat			12	.667		1.62	25.50		27.12	40
6800	4' x 8' opening, 6/6 lites, primer coat, brushwork			8	1		1.92	38.50		40.42	59
7000	Paint, 1 coat			8	1		3.46	38.50		41.96	61
8000	Single lite type, 2' x 3', oil base, primer coat, brushwork			33	.242		.36	9.25		9.61	14.25
8200	Paint, 1 coat			33	.242		.65	9.25		9.90	14.55
8600	3' x 5' opening, primer coat, brushwork			20	.400		.90	15.30		16.20	24
8800	Paint, 1 coat			20	.400		1.62	15.30		16.92	25
9200	4' x 8' opening, primer coat, brushwork			14	.571		1.92	22		23.92	34.50
9400	Paint, 1 coat			14	.571		3.46	22		25.46	36.50

09 91 23.35 Doors and Windows, Interior Latex

			Crew	Daily Output	Labor-Hours	Unit	Material	2012 Bare Costs Labor	Equipment	Total	Total Incl O&P
0010	**DOORS & WINDOWS, INTERIOR LATEX**	R099100-10									
0100	Doors, flush, both sides, incl. frame & trim										
0110	Roll & brush, primer	R099100-20	1 Pord	10	.800	Ea.	3.94	30.50		34.44	50.50
0120	Finish coat, latex			10	.800		4.43	30.50		34.93	51
0130	Primer & 1 coat latex			7	1.143		8.35	43.50		51.85	74.50
0140	Primer & 2 coats latex			5	1.600		12.55	61		73.55	105
0160	Spray, both sides, primer			20	.400		4.15	15.30		19.45	27.50
0170	Finish coat, latex			20	.400		4.64	15.30		19.94	28
0180	Primer & 1 coat latex			11	.727		8.85	28		36.85	51.50
0190	Primer & 2 coats latex			8	1		13.25	38.50		51.75	71.50
0200	Doors, French, both sides, 10-15 lite, incl. frame & trim										
0210	Roll & brush, primer		1 Pord	6	1.333	Ea.	1.97	51		52.97	78.50
0220	Finish coat, latex			6	1.333		2.21	51		53.21	79
0230	Primer & 1 coat latex			3	2.667		4.18	102		106.18	158
0240	Primer & 2 coats latex			2	4		6.25	153		159.25	236
0260	Doors, louvered, both sides, incl. frame & trim										

09 91 Painting

09 91 23 – Interior Painting

09 91 23.35 Doors and Windows, Interior Latex	Crew	Daily Output	Labor-Hours	Unit	Material	2012 Bare Costs Labor	Equipment	Total	Total Incl O&P
0270	Roll & brush, primer	1 Pord	7	1.143	Ea.	3.94	43.50		47.44	70
0280	Finish coat, latex		7	1.143		4.43	43.50		47.93	70.50
0290	Primer & 1 coat, latex		4	2		8.15	76.50		84.65	123
0300	Primer & 2 coats, latex		3	2.667		12.80	102		114.80	167
0320	Spray, both sides, primer		20	.400		4.15	15.30		19.45	27.50
0330	Finish coat, latex		20	.400		4.64	15.30		19.94	28
0340	Primer & 1 coat, latex		11	.727		8.85	28		36.85	51.50
0350	Primer & 2 coats, latex		8	1		13.55	38.50		52.05	72
0360	Doors, panel, both sides, incl. frame & trim									
0370	Roll & brush, primer	1 Pord	6	1.333	Ea.	4.15	51		55.15	81
0380	Finish coat, latex		6	1.333		4.43	51		55.43	81.50
0390	Primer & 1 coat, latex		3	2.667		8.35	102		110.35	162
0400	Primer & 2 coats, latex		2.50	3.200		12.80	122		134.80	197
0420	Spray, both sides, primer		10	.800		4.15	30.50		34.65	50.50
0430	Finish coat, latex		10	.800		4.64	30.50		35.14	51
0440	Primer & 1 coat, latex		5	1.600		8.85	61		69.85	101
0450	Primer & 2 coats, latex		4	2		13.55	76.50		90.05	129
0460	Windows, per interior side, based on 15 S.F.									
0470	1 to 6 lite									
0480	Brushwork, primer	1 Pord	13	.615	Ea.	.78	23.50		24.28	36
0490	Finish coat, enamel		13	.615		.87	23.50		24.37	36
0500	Primer & 1 coat enamel		8	1		1.65	38.50		40.15	59
0510	Primer & 2 coats enamel		6	1.333		2.52	51		53.52	79.50
0530	7 to 10 lite									
0540	Brushwork, primer	1 Pord	11	.727	Ea.	.78	28		28.78	42.50
0550	Finish coat, enamel		11	.727		.87	28		28.87	42.50
0560	Primer & 1 coat enamel		7	1.143		1.65	43.50		45.15	67.50
0570	Primer & 2 coats enamel		5	1.600		2.52	61		63.52	94.50
0590	12 lite									
0600	Brushwork, primer	1 Pord	10	.800	Ea.	.78	30.50		31.28	47
0610	Finish coat, enamel		10	.800		.87	30.50		31.37	47
0620	Primer & 1 coat enamel		6	1.333		1.65	51		52.65	78.50
0630	Primer & 2 coats enamel		5	1.600		2.52	61		63.52	94.50
0650	For oil base paint, add					10%				

09 91 23.39 Doors and Windows, Interior Latex, Zero Voc

			Crew	Daily Output	Labor-Hours	Unit	Material	Labor	Equipment	Total	Total Incl O&P
0010	**DOORS & WINDOWS, INTERIOR LATEX, ZERO VOC**										
0100	Doors flush, both sides, incl. frame & trim										
0110	Roll & brush, primer	G	1 Pord	10	.800	Ea.	5.10	30.50		35.60	51.50
0120	Finish coat, latex	G		10	.800		5.80	30.50		36.30	52.50
0130	Primer & 1 coat latex	G		7	1.143		10.90	43.50		54.40	77.50
0140	Primer & 2 coats latex	G		5	1.600		16.35	61		77.35	110
0160	Spray, both sides, primer	G		20	.400		5.35	15.30		20.65	29
0170	Finish coat, latex	G		20	.400		6.10	15.30		21.40	29.50
0180	Primer & 1 coat latex	G		11	.727		11.50	28		39.50	54
0190	Primer & 2 coats latex	G		8	1		17.30	38.50		55.80	76
0200	Doors, French, both sides, 10-15 lite, incl. frame & trim										
0210	Roll & brush, primer	G	1 Pord	6	1.333	Ea.	2.55	51		53.55	79.50
0220	Finish coat, latex	G		6	1.333		2.90	51		53.90	79.50
0230	Primer & 1 coat latex	G		3	2.667		5.45	102		107.45	159
0240	Primer & 2 coats latex	G		2	4		8.20	153		161.20	238
0360	Doors, panel, both sides, incl. frame & trim										
0370	Roll & brush, primer	G	1 Pord	6	1.333	Ea.	5.35	51		56.35	82.50

09 91 23 – Interior Painting

09 91 23.39 Doors and Windows, Interior Latex, Zero Voc	Crew	Daily Output	Labor-Hours	Unit	Material	2012 Bare Costs Labor	Equipment	Total	Total Incl O&P
0380	Finish coat, latex	G	1 Pord	6	1.333	Ea.	5.80	51		56.80	83
0390	Primer & 1 coat, latex	G		3	2.667		10.90	102		112.90	165
0400	Primer & 2 coats, latex	G		2.50	3.200		16.70	122		138.70	201
0420	Spray, both sides, primer	G		10	.800		5.35	30.50		35.85	52
0430	Finish coat, latex	G		10	.800		6.10	30.50		36.60	52.50
0440	Primer & 1 coat, latex	G		5	1.600		11.50	61		72.50	104
0450	Primer & 2 coats, latex	G		4	2		17.70	76.50		94.20	133
0460	Windows, per interior side, based on 15 S.F.										
0470	1 to 6 lite										
0480	Brushwork, primer	G	1 Pord	13	.615	Ea.	1	23.50		24.50	36
0490	Finish coat, enamel	G		13	.615		1.15	23.50		24.65	36.50
0500	Primer & 1 coat enamel	G		8	1		2.15	38.50		40.65	59.50
0510	Primer & 2 coats enamel	G		6	1.333		3.30	51		54.30	80

09 91 23.40 Floors, Interior

			Crew	Daily Output	Labor-Hours	Unit	Material	Labor	Equipment	Total	Total Incl O&P
0010	**FLOORS, INTERIOR**	R099100-10									
0100	Concrete paint, latex										
0110	Brushwork										
0120	1st coat		1 Pord	975	.008	S.F.	.16	.31		.47	.64
0130	2nd coat			1150	.007		.10	.27		.37	.51
0140	3rd coat			1300	.006		.08	.24		.32	.44
0150	Roll										
0160	1st coat		1 Pord	2600	.003	S.F.	.21	.12		.33	.41
0170	2nd coat			3250	.002		.13	.09		.22	.28
0180	3rd coat			3900	.002		.09	.08		.17	.22
0190	Spray										
0200	1st coat		1 Pord	2600	.003	S.F.	.18	.12		.30	.38
0210	2nd coat			3250	.002		.10	.09		.19	.25
0220	3rd coat			3900	.002		.08	.08		.16	.21
0300	Acid stain and sealer										
0310	Stain, one coat		1 Pord	650	.012	S.F.	.11	.47		.58	.82
0320	Two coats			570	.014		.22	.54		.76	1.04
0330	Acrylic sealer, one coat			2600	.003		.18	.12		.30	.38
0340	Two coats			1400	.006		.37	.22		.59	.74

09 91 23.52 Miscellaneous, Interior

			Crew	Daily Output	Labor-Hours	Unit	Material	Labor	Equipment	Total	Total Incl O&P
0010	**MISCELLANEOUS, INTERIOR**	R099100-10									
2400	Floors, conc./wood, oil base, primer/sealer coat, brushwork		2 Pord	1950	.008	S.F.	.07	.31		.38	.55
2450	Roller			5200	.003		.08	.12		.20	.26
2600	Spray			6000	.003		.08	.10		.18	.23
2650	Paint 1 coat, brushwork			1950	.008		.09	.31		.40	.57
2800	Roller			5200	.003		.09	.12		.21	.28
2850	Spray			6000	.003		.10	.10		.20	.26
3000	Stain, wood floor, brushwork, 1 coat			4550	.004		.08	.13		.21	.29
3200	Roller			5200	.003		.08	.12		.20	.27
3250	Spray			6000	.003		.08	.10		.18	.24
3400	Varnish, wood floor, brushwork			4550	.004		.10	.13		.23	.31
3450	Roller			5200	.003		.10	.12		.22	.29
3600	Spray			6000	.003		.11	.10		.21	.27
3650	For dust proofing or anti skid, see Section 03 35 29.30										
3800	Grilles, per side, oil base, primer coat, brushwork		1 Pord	520	.015	S.F.	.12	.59		.71	1.01
3850	Spray			1140	.007		.13	.27		.40	.54
3920	Paint 2 coats, brushwork			325	.025		.42	.94		1.36	1.87
3940	Spray			650	.012		.48	.47		.95	1.23

09 91 23.52 Miscellaneous, Interior

		Crew	Daily Output	Labor-Hours	Unit	Material	2012 Bare Costs Labor	Equipment	Total	Total Incl O&P
5000	Pipe, 1" - 4" diameter, primer or sealer coat, oil base, brushwork	2 Pord	1250	.013	L.F.	.08	.49		.57	.81
5100	Spray		2165	.007		.07	.28		.35	.50
5350	Paint 2 coats, brushwork		775	.021		.20	.79		.99	1.40
5400	Spray		1240	.013		.23	.49		.72	.99
6300	13" - 16" diameter, primer or sealer coat, brushwork		310	.052		.31	1.97		2.28	3.29
6350	Spray		540	.030		.34	1.13		1.47	2.06
6500	Paint 2 coats, brushwork		195	.082		.82	3.14		3.96	5.60
6550	Spray		310	.052		.91	1.97		2.88	3.95
7000	Trim, wood, incl. puttying, under 6" wide									
7200	Primer coat, oil base, brushwork	1 Pord	650	.012	L.F.	.03	.47		.50	.73
7250	Paint, 1 coat, brushwork		650	.012		.05	.47		.52	.76
7450	3 coats		325	.025		.16	.94		1.10	1.58
7500	Over 6" wide, primer coat, brushwork		650	.012		.06	.47		.53	.77
7550	Paint, 1 coat, brushwork		650	.012		.11	.47		.58	.82
7650	3 coats		325	.025		.31	.94		1.25	1.75
8000	Cornice, simple design, primer coat, oil base, brushwork		650	.012	S.F.	.06	.47		.53	.77
8250	Paint, 1 coat		650	.012		.11	.47		.58	.82
8350	Ornate design, primer coat		350	.023		.06	.87		.93	1.38
8400	Paint, 1 coat		350	.023		.11	.87		.98	1.43
8600	Balustrades, primer coat, oil base, brushwork		520	.015		.06	.59		.65	.95
8650	Paint, 1 coat		520	.015		.11	.59		.70	1
8900	Trusses and wood frames, primer coat, oil base, brushwork		800	.010		.06	.38		.44	.64
8950	Spray		1200	.007		.06	.26		.32	.45
9220	Paint 2 coats, brushwork		500	.016		.21	.61		.82	1.15
9240	Spray		600	.013		.23	.51		.74	1.02
9260	Stain, brushwork, wipe off		600	.013		.08	.51		.59	.85
9280	Varnish, 3 coats, brushwork		275	.029		.29	1.11		1.40	1.98
9350	For latex paint, deduct					10%				

09 91 23.62 Electrostatic Painting

		Crew	Daily Output	Labor-Hours	Unit	Material	2012 Bare Costs Labor	Equipment	Total	Total Incl O&P
0010	**ELECTROSTATIC PAINTING**									
0100	In shop									
0200	Flat surfaces (lockers, casework, elevator doors. etc.)									
0300	One coat	1 Pord	200	.040	S.F.	.46	1.53		1.99	2.80
0400	Two coats	"	120	.067	"	.67	2.55		3.22	4.55
0500	Irregular surfaces (furniture, door frames, etc.)									
0600	One coat	1 Pord	150	.053	S.F.	.46	2.04		2.50	3.56
0700	Two coats	"	100	.080	"	.67	3.06		3.73	5.30
0800	On site									
0900	Flat surfaces (lockers, casework, elev doors. etc.)									
1000	One coat	1 Pord	150	.053	S.F.	.46	2.04		2.50	3.56
1100	Two coats	"	100	.080	"	.67	3.06		3.73	5.30
1200	Irregular surfaces (furniture, door frames, etc.)									
1300	One coat	1 Pord	115	.070	S.F.	.46	2.66		3.12	4.49
1400	Two coats	"	70	.114	"	.67	4.37		5.04	7.30

09 91 23.72 Walls and Ceilings, Interior

			Crew	Daily Output	Labor-Hours	Unit	Material	2012 Bare Costs Labor	Equipment	Total	Total Incl O&P
0010	**WALLS AND CEILINGS, INTERIOR**	R099100-10									
0100	Concrete, drywall or plaster, latex , primer or sealer coat	R099100-20									
0200	Smooth finish, brushwork		1 Pord	1150	.007	S.F.	.06	.27		.33	.47
0240	Roller			1350	.006		.06	.23		.29	.41
0280	Spray			2750	.003		.05	.11		.16	.23
0300	Sand finish, brushwork			975	.008		.06	.31		.37	.54
0340	Roller			1150	.007		.06	.27		.33	.47

09 91 23.72 Walls and Ceilings, Interior	Crew	Daily Output	Labor-Hours	Unit	Material	2012 Bare Costs Labor	Equipment	Total	Total Incl O&P
0380	Spray	1 Pord	2275	.004	S.F.	.05	.13		.18	.26
0800	Paint 2 coats, smooth finish, brushwork		680	.012		.12	.45		.57	.81
0840	Roller		800	.010		.12	.38		.50	.71
0880	Spray		1625	.005		.12	.19		.31	.41
0900	Sand finish, brushwork		605	.013		.12	.51		.63	.90
0940	Roller		1020	.008		.12	.30		.42	.59
0980	Spray		1700	.005		.12	.18		.30	.40
1200	Paint 3 coats, smooth finish, brushwork		510	.016		.19	.60		.79	1.10
1240	Roller		650	.012		.19	.47		.66	.90
1280	Spray		1625	.005		.17	.19		.36	.47
1600	Glaze coating, 2 coats, spray, clear		1200	.007		.49	.26		.75	.92
1640	Multicolor	▼	1200	.007	▼	.98	.26		1.24	1.46
1660	Painting walls, complete, including surface prep, primer &									
1670	2 coats finish, on drywall or plaster, with roller	1 Pord	325	.025	S.F.	.19	.94		1.13	1.62
1700	For oil base paint, add					10%				
1800	For ceiling installations, add			▼			25%			
2000	Masonry or concrete block, primer/sealer, latex paint									
2100	Primer, smooth finish, brushwork	1 Pord	1000	.008	S.F.	.10	.31		.41	.57
2110	Roller		1150	.007		.10	.27		.37	.51
2180	Spray		2400	.003		.09	.13		.22	.29
2200	Sand finish, brushwork		850	.009		.10	.36		.46	.65
2210	Roller		975	.008		.10	.31		.41	.58
2280	Spray		2050	.004		.09	.15		.24	.32
2400	Finish coat, smooth finish, brush		1100	.007		.08	.28		.36	.51
2410	Roller		1300	.006		.08	.24		.32	.44
2480	Spray		2400	.003		.07	.13		.20	.27
2500	Sand finish, brushwork		950	.008		.08	.32		.40	.57
2510	Roller		1090	.007		.08	.28		.36	.51
2580	Spray		2040	.004		.07	.15		.22	.30
2800	Primer plus one finish coat, smooth brush		525	.015		.29	.58		.87	1.19
2810	Roller		615	.013		.18	.50		.68	.94
2880	Spray		1200	.007		.16	.26		.42	.56
2900	Sand finish, brushwork		450	.018		.18	.68		.86	1.22
2910	Roller		515	.016		.18	.59		.77	1.09
2980	Spray		1025	.008		.16	.30		.46	.63
3200	Primer plus 2 finish coats, smooth, brush		355	.023		.27	.86		1.13	1.58
3210	Roller		415	.019		.27	.74		1.01	1.39
3280	Spray		800	.010		.23	.38		.61	.83
3300	Sand finish, brushwork		305	.026		.27	1		1.27	1.79
3310	Roller		350	.023		.27	.87		1.14	1.60
3380	Spray		675	.012		.23	.45		.68	.94
3600	Glaze coating, 3 coats, spray, clear		900	.009		.70	.34		1.04	1.28
3620	Multicolor		900	.009		1.13	.34		1.47	1.76
4000	Block filler, 1 coat, brushwork		425	.019		.13	.72		.85	1.22
4100	Silicone, water repellent, 2 coats, spray	▼	2000	.004		.30	.15		.45	.56
4120	For oil base paint, add					10%				
8200	For work 8' - 15' H, add						10%			
8300	For work over 15' H, add						20%			
8400	For light textured surfaces, add						10%			
8410	Heavy textured, add			▼			25%			

09 91 Painting

09 91 23 – Interior Painting

09 91 23.74 Walls and Ceilings, Interior, Zero VOC Latex

		Crew	Daily Output	Labor-Hours	Unit	Material	2012 Bare Costs Labor	Equipment	Total	Total Incl O&P
0010	**WALLS AND CEILINGS, INTERIOR, ZERO VOC LATEX**									
0100	Concrete, dry wall or plaster, latex, primer or sealer coat									
0200	Smooth finish, brushwork [G]	1 Pord	1150	.007	S.F.	.07	.27		.34	.47
0240	Roller [G]		1350	.006		.06	.23		.29	.41
0280	Spray [G]		2750	.003		.05	.11		.16	.22
0300	Sand finish, brushwork [G]		975	.008		.06	.31		.37	.54
0340	Roller [G]		1150	.007		.07	.27		.34	.48
0380	Spray [G]		2275	.004		.05	.13		.18	.26
0800	Paint 2 coats, smooth finish, brushwork [G]		680	.012		.17	.45		.62	.85
0840	Roller [G]		800	.010		.17	.38		.55	.76
0880	Spray [G]		1625	.005		.15	.19		.34	.44
0900	Sand finish, brushwork [G]		605	.013		.16	.51		.67	.94
0940	Roller [G]		1020	.008		.17	.30		.47	.64
0980	Spray [G]		1700	.005		.15	.18		.33	.43
1200	Paint 3 coats, smooth finish, brushwork [G]		510	.016		.24	.60		.84	1.17
1240	Roller [G]		650	.012		.26	.47		.73	.99
1280	Spray [G]		1625	.005		.22	.19		.41	.52
1800	For ceiling installations, add [G]						25%			
8200	For work 8' - 15' H, add						10%			
8300	For work over 15' H, add						20%			

09 91 23.75 Dry Fall Painting

		R099100-10	Crew	Daily Output	Labor-Hours	Unit	Material	Labor	Equipment	Total	Total Incl O&P
0010	**DRY FALL PAINTING**										
0100	Sprayed on walls, gypsum board or plaster										
0220	One coat	R099100-20	1 Pord	2600	.003	S.F.	.05	.12		.17	.24
0250	Two coats			1560	.005		.11	.20		.31	.41
0280	Concrete or textured plaster, one coat			1560	.005		.05	.20		.25	.35
0310	Two coats			1300	.006		.11	.24		.35	.47
0340	Concrete block, one coat			1560	.005		.05	.20		.25	.35
0370	Two coats			1300	.006		.11	.24		.35	.47
0400	Wood, one coat			877	.009		.05	.35		.40	.58
0430	Two coats			650	.012		.11	.47		.58	.82
0440	On ceilings, gypsum board or plaster										
0470	One coat		1 Pord	1560	.005	S.F.	.05	.20		.25	.35
0500	Two coats			1300	.006		.11	.24		.35	.47
0530	Concrete or textured plaster, one coat			1560	.005		.05	.20		.25	.35
0560	Two coats			1300	.006		.11	.24		.35	.47
0570	Structural steel, bar joists or metal deck, one coat			1560	.005		.05	.20		.25	.35
0580	Two coats			1040	.008		.11	.29		.40	.56

09 93 Staining and Transparent Finishing

09 93 23 – Interior Staining and Finishing

09 93 23.10 Varnish

		Crew	Daily Output	Labor-Hours	Unit	Material	Labor	Equipment	Total	Total Incl O&P
0010	**VARNISH**									
0012	1 coat + sealer, on wood trim, brush, no sanding included	1 Pord	400	.020	S.F.	.07	.77		.84	1.22
0100	Hardwood floors, 2 coats, no sanding included, roller	"	1890	.004	"	.15	.16		.31	.41

09 96 23 – Graffiti-Resistant Coatings

09 96 23.10 Graffiti Resistant Treatments

		Daily Output	Labor-Hours	Unit	Material	2012 Bare Costs Labor	2012 Bare Costs Equipment	Total	Total Incl O&P
						Crew			
0010	**GRAFFITI RESISTANT TREATMENTS**, sprayed on walls								
0100	Non-sacrificial, permanent non-stick coating, clear, on metals	1 Pord 2000	.004	S.F.	2.06	.15		2.21	2.50
0200	Concrete	2000	.004		2.35	.15		2.50	2.81
0300	Concrete block	2000	.004		3.03	.15		3.18	3.57
0400	Brick	2000	.004		3.44	.15		3.59	4.01
0500	Stone	2000	.004		3.44	.15		3.59	4.01
0600	Unpainted wood	2000	.004		3.97	.15		4.12	4.60
2000	Semi-permanent cross linking polymer primer, on metals	2000	.004		.40	.15		.55	.67
2100	Concrete	2000	.004		.48	.15		.63	.76
2200	Concrete block	2000	.004		.60	.15		.75	.89
2300	Brick	2000	.004		.48	.15		.63	.76
2400	Stone	2000	.004		.48	.15		.63	.76
2500	Unpainted wood	2000	.004		.67	.15		.82	.96
3000	Top coat, on metals	2000	.004		.43	.15		.58	.70
3100	Concrete	2000	.004		.49	.15		.64	.76
3200	Concrete block	2000	.004		.68	.15		.83	.98
3300	Brick	2000	.004		.57	.15		.72	.85
3400	Stone	2000	.004		.57	.15		.72	.85
3500	Unpainted wood	2000	.004		.68	.15		.83	.98
5000	Sacrificial, water based, on metal	2000	.004		.24	.15		.39	.50
5100	Concrete	2000	.004		.24	.15		.39	.50
5200	Concrete block	2000	.004		.24	.15		.39	.50
5300	Brick	2000	.004		.24	.15		.39	.50
5400	Stone	2000	.004		.24	.15		.39	.50
5500	Unpainted wood	2000	.004		.24	.15		.39	.50
8000	Cleaner for use after treatment								
8100	Towels or wipes, per package of 30			Ea.	.63			.63	.70
8200	Aerosol spray, 24 oz. can			"	18			18	19.80

09 96 53 – Elastomeric Coatings

09 96 53.10 Coatings, Elastomeric

		Crew	Daily Output	Labor-Hours	Unit	Material	Labor	Equipment	Total	Total Incl O&P
0010	**COATINGS, ELASTOMERIC**									
0020	High build, water proof, one coat system									
0100	Concrete, brush	1 Pord	650	.012	S.F.	.27	.47		.74	1
0110	Roll		1650	.005		.27	.19		.46	.58
0120	Spray		2600	.003		.27	.12		.39	.48
0200	Concrete block, brush		600	.013		.32	.51		.83	1.12
0210	Roll		1400	.006		.32	.22		.54	.69
0220	Spray		1900	.004		.32	.16		.48	.60
0300	Stucco, brush		400	.020		.45	.77		1.22	1.63
0310	Roll		1000	.008		.45	.31		.76	.95
0320	Spray		1500	.005		.45	.20		.65	.79

09 96 56 – Epoxy Coatings

09 96 56.20 Wall Coatings

		Crew	Daily Output	Labor-Hours	Unit	Material	Labor	Equipment	Total	Total Incl O&P
0010	**WALL COATINGS**									
0100	Acrylic glazed coatings, minimum	1 Pord	525	.015	S.F.	.30	.58		.88	1.20
0200	Maximum		305	.026		.64	1		1.64	2.20
0300	Epoxy coatings, minimum		525	.015		.39	.58		.97	1.30
0400	Maximum		170	.047		1.18	1.80		2.98	3.99
0600	Exposed aggregate, troweled on, 1/16" to 1/4", minimum		235	.034		.60	1.30		1.90	2.61
0700	Maximum (epoxy or polyacrylate)		130	.062		1.28	2.35		3.63	4.93
0900	1/2" to 5/8" aggregate, minimum		130	.062		1.17	2.35		3.52	4.81
1000	Maximum		80	.100		2.02	3.83		5.85	7.90

09 96 High-Performance Coatings

09 96 56 – Epoxy Coatings

09 96 56.20 Wall Coatings		Crew	Daily Output	Labor-Hours	Unit	Material	2012 Bare Costs Labor	Equipment	Total	Total Incl O&P
1500	Exposed aggregate, sprayed on, 1/8" aggregate, minimum	1 Pord	295	.027	S.F.	.55	1.04		1.59	2.16
1600	Maximum		145	.055		1.02	2.11		3.13	4.28
1800	High build epoxy, 50 mil, minimum		390	.021		.66	.78		1.44	1.90
1900	Maximum		95	.084		1.12	3.22		4.34	6.05
2100	Laminated epoxy with fiberglass, minimum		295	.027		.71	1.04		1.75	2.33
2200	Maximum		145	.055		1.27	2.11		3.38	4.56
2400	Sprayed perlite or vermiculite, 1/16" thick, minimum		2935	.003		.26	.10		.36	.45
2500	Maximum		640	.013		.72	.48		1.20	1.51
2700	Vinyl plastic wall coating, minimum		735	.011		.32	.42		.74	.97
2800	Maximum		240	.033		.79	1.27		2.06	2.78
3000	Urethane on smooth surface, 2 coats, minimum		1135	.007		.26	.27		.53	.69
3100	Maximum		665	.012		.57	.46		1.03	1.32
3600	Ceramic-like glazed coating, cementitious, minimum		440	.018		.46	.70		1.16	1.55
3700	Maximum		345	.023		.78	.89		1.67	2.19
3900	Resin base, minimum		640	.013		.32	.48		.80	1.07
4000	Maximum		330	.024		.52	.93		1.45	1.96

09 97 Special Coatings

09 97 13 – Steel Coatings

09 97 13.23 Exterior Steel Coatings

		Crew	Daily Output	Labor-Hours	Unit	Material	2012 Bare Costs Labor	Equipment	Total	Total Incl O&P
0010	**EXTERIOR STEEL COATINGS** R050516-30									
6100	Cold galvanizing, brush in field	1 Psst	1100	.007	S.F.	.12	.29		.41	.66
6510	Paints & protective coatings, sprayed in field									
6520	Alkyds, primer	2 Psst	3600	.004	S.F.	.07	.17		.24	.40
6540	Gloss topcoats		3200	.005		.08	.20		.28	.45
6560	Silicone alkyd		3200	.005		.12	.20		.32	.50
6610	Epoxy, primer		3000	.005		.25	.21		.46	.65
6630	Intermediate or topcoat		2800	.006		.22	.22		.44	.65
6650	Enamel coat		2800	.006		.28	.22		.50	.71
6700	Epoxy ester, primer		2800	.006		.49	.22		.71	.95
6720	Topcoats		2800	.006		.18	.22		.40	.61
6810	Latex primer		3600	.004		.06	.17		.23	.38
6830	Topcoats		3200	.005		.06	.20		.26	.43
6910	Universal primers, one part, phenolic, modified alkyd		2000	.008		.11	.31		.42	.69
6940	Two part, epoxy spray		2000	.008		.26	.31		.57	.86
7000	Zinc rich primers, self cure, spray, inorganic		1800	.009		.83	.35		1.18	1.54
7010	Epoxy, spray, organic		1800	.009		.22	.35		.57	.87
7020	Above one story, spray painting simple structures, add						25%			
7030	Intricate structures, add						50%			

Estimating Tips

General

- The items in this division are usually priced per square foot or each.
- Many items in Division 10 require some type of support system or special anchors that are not usually furnished with the item. The required anchors must be added to the estimate in the appropriate division.
- Some items in Division 10, such as lockers, may require assembly before installation. Verify the amount of assembly required. Assembly can often exceed installation time.

10 20 00 Interior Specialties

- Support angles and blocking are not included in the installation of toilet compartments, shower/dressing compartments, or cubicles. Appropriate line items from Divisions 5 or 6 may need to be added to support the installations.
- Toilet partitions are priced by the stall. A stall consists of a side wall, pilaster, and door with hardware. Toilet tissue holders and grab bars are extra.
- The required acoustical rating of a folding partition can have a significant impact on costs. Verify the sound transmission coefficient rating of the panel priced to the specification requirements.

- Grab bar installation does not include supplemental blocking or backing to support the required load. When grab bars are installed at an existing facility, provisions must be made to attach the grab bars to solid structure.

Reference Numbers

Reference numbers are shown in shaded boxes at the beginning of some major classifications. These numbers refer to related items in the Reference Section. The reference information may be an estimating procedure, an alternate pricing method, or technical information.

Note: Not all subdivisions listed here necessarily appear in this publication.

10 05 05 – Selective Specialties Demolition

10 05 05.10 Selective Demolition, Specialties	Crew	Daily Output	Labor-Hours	Unit	Material	2012 Bare Costs Labor	Equipment	Total	Total Incl O&P
0010	**SELECTIVE DEMOLITION, SPECIALTIES**								
1100	Boards and panels, wall mounted	2 Clab	15	1.067	Ea.		37.50		37.50	57.50
1200	Cases, for directory and/or bulletin boards, including doors		24	.667			23.50		23.50	36
1850	Shower partitions, cabinet or stall, including base and door	↓	8	2			70		70	108
1855	Shower receptor, terrazzo or concrete	1 Clab	14	.571	↓		20		20	31
1900	Curtain track or rod, hospital type, ceiling mounted or suspended	"	220	.036	L.F.		1.28		1.28	1.96
1910	Toilet cubicles, remove	2 Clab	8	2	Ea.		70		70	108
1930	Urinal screen, remove	1 Clab	12	.667	"		23.50		23.50	36
2650	Wall guard, misc. wall or corner protection	"	320	.025	L.F.		.88		.88	1.35
2750	Access floor, metal panel system, including pedestals, covering	2 Clab	850	.019	S.F.		.66		.66	1.02
3050	Fireplace, prefab, freestanding or wall hung, including hood and screen	1 Clab	2	4	Ea.		140		140	216
3054	Chimney top, simulated brick, 4' high	"	15	.533			18.70		18.70	29
3200	Stove, woodburning, cast iron	2 Clab	2	8			281		281	430
3440	Weathervane, residential	1 Clab	12	.667			23.50		23.50	36
3500	Flagpole, groundset, to 70' high, excluding base/foundation	K-1	1	16			630	264	894	1,250
3555	To 30' high	"	2.50	6.400			251	106	357	500
4300	Letter, signs or plaques, exterior on wall	1 Clab	20	.400			14.05		14.05	21.50
4310	Signs, street, reflective aluminum, including post and bracket		60	.133			4.68		4.68	7.20
4320	Door signs interior on door 6" x 6", selective demolition	↓	20	.400			14.05		14.05	21.50
4550	Turnstiles, manual or electric	2 Clab	2	8	↓		281		281	430
5050	Lockers	1 Clab	15	.533	Opng.		18.70		18.70	29
5250	Cabinets, recessed	Q-12	12	1.333	Ea.		65		65	98
5260	Mail boxes, Horiz., Key Lock, front loading, Remove	1 Carp	34	.235	"		10.40		10.40	15.95
5350	Awning, fabric, including frame	2 Clab	100	.160	S.F.		5.60		5.60	8.65
6050	Partition, woven wire		1400	.011			.40		.40	.62
6100	Folding gate, security, door or window		500	.032			1.12		1.12	1.73
6580	Acoustic air wall		650	.025	↓		.86		.86	1.33
7550	Telephone enclosure, exterior, post mounted		3	5.333	Ea.		187		187	288
8850	Scale, platform, excludes foundation or pit	↓	.25	64	"		2,250		2,250	3,450

10 11 Visual Display Units

10 11 13 – Chalkboards

10 11 13.13 Fixed Chalkboards

		Crew	Daily Output	Labor-Hours	Unit	Material	2012 Bare Costs Labor	Equipment	Total	Total Incl O&P
0010	**FIXED CHALKBOARDS** Porcelain enamel steel									
3900	Wall hung									
4000	Aluminum frame and chalktrough									
4200	3' x 4'	2 Carp	16	1	Ea.	240	44		284	330
4300	3' x 5'		15	1.067		299	47		346	405
4500	4' x 8'		14	1.143		410	50.50		460.50	530
4600	4' x 12'	↓	13	1.231	↓	575	54.50		629.50	715
4700	Wood frame and chalktrough									
4800	3' x 4'	2 Carp	16	1	Ea.	217	44		261	305
5000	3' x 5'		15	1.067		267	47		314	365
5100	4' x 5'		14	1.143		275	50.50		325.50	380
5300	4' x 8'	↓	13	1.231	↓	355	54.50		409.50	475
5400	Liquid chalk, white porcelain enamel, wall hung									
5420	Deluxe units, aluminum trim and chalktrough									
5450	4' x 4'	2 Carp	16	1	Ea.	250	44		294	345
5500	4' x 8'		14	1.143		370	50.50		420.50	485
5550	4' x 12'	↓	12	1.333	↓	530	59		589	675
5700	Wood trim and chalktrough									

10 11 Visual Display Units

10 11 13 – Chalkboards

10 11 13.13 Fixed Chalkboards		Crew	Daily Output	Labor-Hours	Unit	Material	2012 Bare Costs Labor	Equipment	Total	Total Incl O&P
5900	4' x 4'	2 Carp	16	1	Ea.	715	44		759	855
6000	4' x 6'		15	1.067		470	47		517	595
6200	4' x 8'	↓	14	1.143	↓	970	50.50		1,020.50	1,150
6300	Liquid chalk, felt tip markers					1.95			1.95	2.15
6500	Erasers					1.77			1.77	1.95
6600	Board cleaner, 8 oz. bottle				↓	6			6	6.60

10 11 13.23 Modular-Support-Mounted Chalkboards

		Crew	Daily Output	Labor-Hours	Unit	Material	2012 Bare Costs Labor	Equipment	Total	Total Incl O&P
0010	**MODULAR-SUPPORT-MOUNTED CHALKBOARDS**									
0400	Sliding chalkboards									
0450	Vertical, one sliding board with back panel, wall mounted									
0500	8' x 4'	2 Carp	8	2	Ea.	2,225	88		2,313	2,575
0520	8' x 8'		7.50	2.133		3,050	94		3,144	3,500
0540	8' x 12'	↓	7	2.286	↓	4,175	101		4,276	4,750
0600	Two sliding boards, with back panel									
0620	8' x 4'	2 Carp	8	2	Ea.	3,500	88		3,588	3,975
0640	8' x 8'		7.50	2.133		5,025	94		5,119	5,700
0660	8' x 12'	↓	7	2.286	↓	8,400	101		8,501	9,400
0700	Horizontal, two track									
0800	4' x 8', 2 sliding panels	2 Carp	8	2	Ea.	1,950	88		2,038	2,275
0820	4' x 12', 2 sliding panels		7.50	2.133		2,550	94		2,644	2,950
0840	4' x 16', 4 sliding panels	↓	7	2.286	↓	3,425	101		3,526	3,925
0900	Four track, four sliding panels									
0920	4' x 8'	2 Carp	8	2	Ea.	3,150	88		3,238	3,575
0940	4' x 12'		7.50	2.133		4,100	94		4,194	4,675
0960	4' x 16'	↓	7	2.286	↓	5,350	101		5,451	6,025
1200	Vertical, motor operated									
1400	One sliding panel with back panel									
1450	10' x 4'	2 Carp	4	4	Ea.	5,650	176		5,826	6,500
1500	10' x 10'		3.75	4.267		6,825	188		7,013	7,800
1550	10' x 16'	↓	3.50	4.571	↓	8,050	202		8,252	9,150
1700	Two sliding panels with back panel									
1750	10' x 4'	2 Carp	4	4	Ea.	10,000	176		10,176	11,300
1800	10' x 10'		3.75	4.267		11,700	188		11,888	13,100
1850	10' x 16'	↓	3.50	4.571	↓	13,300	202		13,502	15,000
2000	Three sliding panels with back panel									
2100	10' x 4'	2 Carp	4	4	Ea.	14,000	176		14,176	15,700
2150	10' x 10'		3.75	4.267		15,300	188		15,488	17,200
2200	10' x 16'	↓	3.50	4.571	↓	18,500	202		18,702	20,700
2400	For projection screen, glass beaded, add				S.F.	8.10			8.10	8.90
2500	For remote control, 1 panel control, add				Ea.	380			380	420
2600	2 panel control, add				"	655			655	725
2800	For units without back panels, deduct				S.F.	5.10			5.10	5.65
2850	For liquid chalk porcelain panels, add				"	5.50			5.50	6.05
3000	Swing leaf, any comb. of chalkboard & cork, aluminum frame									
3100	Floor style, 6 panels									
3150	30" x 40" panels				Ea.	1,575			1,575	1,750
3200	48" x 40" panels				"	2,650			2,650	2,925
3300	Wall mounted, 6 panels									
3400	30" x 40" panels	2 Carp	16	1	Ea.	1,500	44		1,544	1,725
3450	48" x 40" panels	"	16	1	"	1,850	44		1,894	2,100
3600	Extra panels for swing leaf units									
3700	30" x 40" panels				Ea.	299			299	330

10 11 Visual Display Units

10 11 13 – Chalkboards

10 11 13.23 Modular-Support-Mounted Chalkboards	Crew	Daily Output	Labor-Hours	Unit	Material	2012 Bare Costs Labor	Equipment	Total	Total Incl O&P
3750	48" x 40" panels				Ea.	370			370	405

10 11 13.43 Portable Chalkboards

		Crew	Daily Output	Labor-Hours	Unit	Material	Labor	Equipment	Total	Total Incl O&P
0010	**PORTABLE CHALKBOARDS**									
0100	Freestanding, reversible									
0120	Economy, wood frame, 4' x 6'									
0140	Chalkboard both sides				Ea.	595			595	655
0160	Chalkboard one side, cork other side				"	565			565	620
0200	Standard, lightweight satin finished aluminum, 4' x 6'									
0220	Chalkboard both sides				Ea.	675			675	745
0240	Chalkboard one side, cork other side				"	695			695	765
0300	Deluxe, heavy duty extruded aluminum, 4' x 6'									
0320	Chalkboard both sides				Ea.	1,025			1,025	1,125
0340	Chalkboard one side, cork other side				"	930			930	1,025

10 11 16 – Markerboards

10 11 16.53 Electronic Markerboards

		Crew	Daily Output	Labor-Hours	Unit	Material	Labor	Equipment	Total	Total Incl O&P
0010	**ELECTRONIC MARKERBOARDS**									
0100	Wall hung or free standing, 3' x 4' to 4' x 6'	2 Carp	8	2	S.F.	83.50	88		171.50	228
0150	5' x 6' to 4' x 8'		8	2	"	64.50	88		152.50	207
0500	Interactive projection module for existing whiteboards		8	2	Ea.	2,300	88		2,388	2,675

10 11 23 – Tackboards

10 11 23.10 Fixed Tackboards

		Crew	Daily Output	Labor-Hours	Unit	Material	Labor	Equipment	Total	Total Incl O&P
0010	**FIXED TACKBOARDS**									
0020	Cork sheets, unbacked, no frame, 1/4" thick	2 Carp	290	.055	S.F.	2.07	2.43		4.50	6
0100	1/2" thick		290	.055		4.16	2.43		6.59	8.30
0300	Fabric-face, no frame, on 7/32" cork underlay		290	.055		6.30	2.43		8.73	10.70
0400	On 1/4" cork on 1/4" hardboard		290	.055		8.35	2.43		10.78	12.90
0600	With edges wrapped		290	.055		9.45	2.43		11.88	14.15
0700	On 7/16" fire retardant core		290	.055		6.15	2.43		8.58	10.50
0900	With edges wrapped		290	.055		7.80	2.43		10.23	12.30
1000	Designer fabric only, cut to size					2.70			2.70	2.97
1200	1/4" vinyl cork, on 1/4" hardboard, no frame	2 Carp	290	.055		7.70	2.43		10.13	12.25
1300	On 1/4" coreboard		290	.055		4.18	2.43		6.61	8.35
2000	For map and display rail, economy, add		385	.042	L.F.	3.06	1.83		4.89	6.20
2100	Deluxe, add		350	.046	"	4.70	2.02		6.72	8.25
2120	Prefabricated, 1/4" cork, 3' x 5' with aluminum frame		16	1	Ea.	128	44		172	209
2140	Wood frame		16	1		153	44		197	237
2160	4' x 4' with aluminum frame		16	1		119	44		163	199
2180	Wood frame		16	1		175	44		219	260
2200	4' x 8' with aluminum frame		14	1.143		261	50.50		311.50	365
2210	With wood frame		14	1.143		195	50.50		245.50	293
2220	4' x 12' with aluminum frame		12	1.333		385	59		444	515
2230	Bulletin board case, single glass door, with lock									
2240	36" x 24", economy	2 Carp	12	1.333	Ea.	315	59		374	440
2250	Deluxe		12	1.333		565	59		624	710
2260	42" x 30", economy		12	1.333		360	59		419	485
2270	Deluxe		12	1.333		510	59		569	650
2300	Glass enclosed cabinets, alum., cork panel, hinged doors									
2400	3' x 3', 1 door	2 Carp	12	1.333	Ea.	605	59		664	755
2500	4' x 4', 2 door		11	1.455		1,000	64		1,064	1,200
2600	4' x 7', 3 door		10	1.600		1,775	70.50		1,845.50	2,050
2800	4' x 10', 4 door		8	2		2,350	88		2,438	2,725

10 11 Visual Display Units

10 11 23 – Tackboards

10 11 23.10 Fixed Tackboards

		Crew	Daily Output	Labor-Hours	Unit	Material	2012 Bare Costs Labor	Equipment	Total	Total Incl O&P
2900	For lights, add per door opening	1 Elec	13	.615	Ea.	161	31.50		192.50	226
3100	Horizontal sliding units, 4 doors, 4' x 8', 8' x 4'	2 Carp	9	1.778		1,950	78.50		2,028.50	2,275
3200	4' x 12'		7	2.286		2,550	101		2,651	2,950
3400	8 doors, 4' x 16'		5	3.200		3,450	141		3,591	4,000
3500	4' x 24'		4	4		4,650	176		4,826	5,375

10 11 23.20 Control Boards

		Crew	Daily Output	Labor-Hours	Unit	Material	2012 Bare Costs Labor	Equipment	Total	Total Incl O&P
0010	**CONTROL BOARDS**									
0020	Magnetic, porcelain finish, 18" x 24", framed	2 Carp	8	2	Ea.	199	88		287	355
0100	24" x 36"		7.50	2.133		295	94		389	470
0200	36" x 48"		7	2.286		390	101		491	585
0300	48" x 72"		6	2.667		860	118		978	1,125
0400	48" x 96"		5	3.200		1,075	141		1,216	1,400

10 13 Directories

10 13 10 – Building Directories

10 13 10.10 Directory Boards

		Crew	Daily Output	Labor-Hours	Unit	Material	2012 Bare Costs Labor	Equipment	Total	Total Incl O&P
0010	**DIRECTORY BOARDS**									
0050	Plastic, glass covered, 30" x 20"	2 Carp	3	5.333	Ea.	260	235		495	645
0100	36" x 48"		2	8		845	355		1,200	1,475
0300	Grooved cork, 30" x 20"		3	5.333		405	235		640	805
0400	36" x 48"		2	8		555	355		910	1,150
0600	Black felt, 30" x 20"		3	5.333		229	235		464	610
0700	36" x 48"		2	8		420	355		775	1,000
0900	Outdoor, weatherproof, black plastic, 36" x 24"		2	8		760	355		1,115	1,375
1000	36" x 36"		1.50	10.667		880	470		1,350	1,700
1800	Indoor, economy, open face, 18" x 24"		7	2.286		149	101		250	320
1900	24" x 36"		7	2.286		154	101		255	325
2000	36" x 24"		6	2.667		154	118		272	350
2100	36" x 48"		6	2.667		251	118		369	455
2400	Building directory, alum., black felt panels, 1 door, 24" x 18"		4	4		315	176		491	615
2500	36" x 24"		3.50	4.571		375	202		577	725
2600	48" x 32"		3	5.333		610	235		845	1,025
2700	2 door, 36" x 48"		2.50	6.400		655	282		937	1,150
2800	36" x 60"		2	8		860	355		1,215	1,500
2900	48" x 60"		1	16		970	705		1,675	2,150
3100	For bronze enamel finish, add					15%				
3200	For bronze anodized finish, add					25%				
3400	For illuminated directory, single door unit, add					135			135	148
3500	For 6" header panel, 6 letters per foot, add				L.F.	21.50			21.50	23.50
6050	Building directory, electronic display, alum. frame, wall mounted	2 Carp	32	.500	S.F.	785	22		807	895
6100	Free standing	"	60	.267	"	1,025	11.75		1,036.75	1,175

351

10 14 Signage

10 14 19 – Dimensional Letter Signage

10 14 19.10 Exterior Signs

		Daily Output	Labor-Hours	Unit	Material	2012 Bare Costs Labor	Equipment	Total	Total Incl O&P
0010	**EXTERIOR SIGNS**	Crew							
0020	Letters, 2" high, 3/8" deep, cast bronze	1 Carp	24	.333	Ea.	26	14.70		40.70	51
0140	1/2" deep, cast aluminum		18	.444		26	19.60		45.60	58.50
0160	Cast bronze		32	.250		33.50	11.05		44.55	53.50
0300	6" high, 5/8" deep, cast aluminum		24	.333		28.50	14.70		43.20	54
0400	Cast bronze		24	.333		62	14.70		76.70	91
0600	8" high, 3/4" deep, cast aluminum		14	.571		37.50	25		62.50	80.50
0700	Cast bronze		20	.400		90	17.65		107.65	126
0900	10" high, 1" deep, cast aluminum		18	.444		61	19.60		80.60	97
1000	Bronze		18	.444		116	19.60		135.60	157
1200	12" high, 1-1/4" deep, cast aluminum		12	.667		61.50	29.50		91	113
1500	Cast bronze		18	.444		133	19.60		152.60	176
1600	14" high, 2-5/16" deep, cast aluminum		12	.667		77	29.50		106.50	130
1800	Fabricated stainless steel, 6" high, 2" deep		20	.400		41.50	17.65		59.15	72.50
1900	12" high, 3" deep		18	.444		67	19.60		86.60	104
2100	18" high, 3" deep		12	.667		109	29.50		138.50	164
2200	24" high, 4" deep		10	.800		212	35.50		247.50	288
2700	Acrylic, on high density foam, 12" high, 2" deep		20	.400		17.60	17.65		35.25	46.50
2800	18" high, 2" deep		18	.444		32.50	19.60		52.10	65.50
3900	Plaques, custom, 20" x 30", for up to 450 letters, cast aluminum	2 Carp	4	4		1,125	176		1,301	1,500
4000	Cast bronze		4	4		1,675	176		1,851	2,100
4200	30" x 36", up to 900 letters cast aluminum		3	5.333		2,000	235		2,235	2,550
4300	Cast bronze		3	5.333		3,250	235		3,485	3,925
4500	36" x 48", for up to 1300 letters, cast bronze		2	8		4,425	355		4,780	5,425
4800	Signs, reflective alum. directional signs, dbl. face, 2-way, w/bracket		30	.533		144	23.50		167.50	194
4900	4-way		30	.533		231	23.50		254.50	290
5100	Exit signs, 24 ga. alum., 14" x 12" surface mounted	1 Carp	30	.267		43	11.75		54.75	65
5200	10" x 7"		20	.400		29.50	17.65		47.15	59.50
5400	Bracket mounted, double face, 12" x 10"		30	.267		78	11.75		89.75	104
5500	Sticky back, stock decals, 14" x 10"	1 Clab	50	.160		25.50	5.60		31.10	36.50
6000	Interior elec., wall mount, fiberglass panels, 2 lamps, 6"	1 Elec	8	1		132	51.50		183.50	222
6100	8"	"	8	1		103	51.50		154.50	190
6400	Replacement sign faces, 6" or 8"	1 Clab	50	.160		62.50	5.60		68.10	77

10 14 23 – Panel Signage

10 14 23.13 Engraved Panel Signage

		Crew	Daily Output	Labor-Hours	Unit	Material	2012 Bare Costs Labor	Equipment	Total	Total Incl O&P
0010	**ENGRAVED PANEL SIGNAGE**, interior									
1010	Flexible door sign, adhesive back, w/Braille, 5/8" letters, 4" x 4"	1 Clab	32	.250	Ea.	35	8.80		43.80	52
1050	6" x 6"		32	.250		52.50	8.80		61.30	71
1100	8" x 2"		32	.250		36	8.80		44.80	53
1150	8" x 4"		32	.250		45.50	8.80		54.30	63.50
1200	8" x 8"		32	.250		71.50	8.80		80.30	92
1250	12" x 2"		32	.250		35	8.80		43.80	52
1300	12" x 6"		32	.250		50	8.80		58.80	68.50
1350	12" x 12"		32	.250		143	8.80		151.80	171
1500	Graphic symbols, 2" x 2"		32	.250		12	8.80		20.80	26.50
1550	6" x 6"		32	.250		31	8.80		39.80	47.50
1600	8" x 8"		32	.250		31	8.80		39.80	47.50
2010	Corridor, stock acrylic, 2-sided, with mounting bracket, 2" x 8"	1 Carp	24	.333		26	14.70		40.70	51
2020	2" x 10"		24	.333		35.50	14.70		50.20	61.50
2050	3" x 8"		24	.333		31.50	14.70		46.20	57
2060	3" x 10"		24	.333		44.50	14.70		59.20	71.50
2070	3" x 12"		24	.333		65	14.70		79.70	94

10 14 Signage

10 14 23 – Panel Signage

10 14 23.13 Engraved Panel Signage

		Crew	Daily Output	Labor-Hours	Unit	Material	2012 Bare Costs Labor	Equipment	Total	Total Incl O&P
2100	4" x 8"	1 Carp	24	.333	Ea.	35.50	14.70		50.20	61.50
2110	4" x 10"		24	.333		42	14.70		56.70	68.50
2120	4" x 12"		24	.333		56.50	14.70		71.20	84.50

10 14 53 – Traffic Signage

10 14 53.20 Traffic Signs

		Crew	Daily Output	Labor-Hours	Unit	Material	2012 Bare Costs Labor	Equipment	Total	Total Incl O&P
0010	**TRAFFIC SIGNS**									
0012	Stock, 24" x 24", no posts, .080" alum. reflectorized	B-80	70	.457	Ea.	81.50	17.30	9.55	108.35	127
0100	High intensity		70	.457		81.50	17.30	9.55	108.35	127
0300	30" x 30", reflectorized		70	.457		248	17.30	9.55	274.85	310
0400	High intensity		70	.457		275	17.30	9.55	301.85	340
0600	Guide and directional signs, 12" x 18", reflectorized		70	.457		34.50	17.30	9.55	61.35	75
0700	High intensity		70	.457		57	17.30	9.55	83.85	100
0900	18" x 24", stock signs, reflectorized		70	.457		45.50	17.30	9.55	72.35	87
1000	High intensity		70	.457		50.50	17.30	9.55	77.35	92.50
1200	24" x 24", stock signs, reflectorized		70	.457		55	17.30	9.55	81.85	97.50
1300	High intensity		70	.457		60	17.30	9.55	86.85	103
1500	Add to above for steel posts, galvanized, 10'-0" upright, bolted		200	.160		42	6.05	3.34	51.39	59.50
1600	12'-0" upright, bolted		140	.229		39	8.65	4.77	52.42	61.50
1800	Highway road signs, aluminum, over 20 S. F., reflectorized		350	.091	S.F.	29	3.46	1.91	34.37	39
2000	High intensity		350	.091		29	3.46	1.91	34.37	39
2200	Highway, suspended over road, 80 S.F. min., reflectorized		165	.194		29	7.35	4.05	40.40	47
2300	High intensity		165	.194		29	7.35	4.05	40.40	47

10 17 Telephone Specialties

10 17 16 – Telephone Enclosures

10 17 16.10 Commercial Telephone Enclosures

		Crew	Daily Output	Labor-Hours	Unit	Material	2012 Bare Costs Labor	Equipment	Total	Total Incl O&P
0010	**COMMERCIAL TELEPHONE ENCLOSURES**									
0300	Shelf type, wall hung, minimum	2 Carp	5	3.200	Ea.	995	141		1,136	1,325
0400	Maximum	"	5	3.200	"	3,225	141		3,366	3,775

10 21 Compartments and Cubicles

10 21 13 – Toilet Compartments

10 21 13.13 Metal Toilet Compartments

			Crew	Daily Output	Labor-Hours	Unit	Material	2012 Bare Costs Labor	Equipment	Total	Total Incl O&P
0010	**METAL TOILET COMPARTMENTS**										
0110	Cubicles, ceiling hung										
0200	Powder coated steel		2 Carp	4	4	Ea.	500	176		676	820
0500	Stainless steel		"	4	4		945	176		1,121	1,325
0600	For handicap units, incl. 52" grab bars, add	♿					420			420	460
0900	Floor and ceiling anchored										
1000	Powder coated steel		2 Carp	5	3.200	Ea.	615	141		756	895
1300	Stainless steel		"	5	3.200		1,075	141		1,216	1,425
1400	For handicap units, incl. 52" grab bars, add	♿					300			300	330
1610	Floor anchored										
1700	Powder coated steel		2 Carp	7	2.286	Ea.	590	101		691	805
2000	Stainless steel		"	7	2.286		1,375	101		1,476	1,675
2100	For handicap units, incl. 52" grab bars, add	♿					299			299	330
2200	For juvenile units, deduct						41.50			41.50	45.50
2450	Floor anchored, headrail braced										
2500	Powder coated steel		2 Carp	6	2.667	Ea.	385	118		503	600

10 21 13 – Toilet Compartments

10 21 13.13 Metal Toilet Compartments

		Crew	Daily Output	Labor-Hours	Unit	Material	2012 Bare Costs Labor	Equipment	Total	Total Incl O&P
2804	Stainless steel	2 Carp	4.60	3.478	Ea.	1,025	153		1,178	1,350
2900	For handicap units, incl. 52" grab bars, add					350			350	385
3000	Wall hung partitions, powder coated steel	2 Carp	7	2.286		625	101		726	840
3300	Stainless steel	"	7	2.286		1,650	101		1,751	1,975
3400	For handicap units, incl. 52" grab bars, add					350			350	385
4000	Screens, entrance, floor mounted, 58" high, 48" wide									
4200	Powder coated steel	2 Carp	15	1.067	Ea.	238	47		285	335
4500	Stainless steel	"	15	1.067	"	895	47		942	1,050
4650	Urinal screen, 18" wide									
4704	Powder coated steel	2 Carp	6.15	2.602	Ea.	220	115		335	420
5004	Stainless steel	"	6.15	2.602	"	580	115		695	815
5100	Floor mounted, head rail braced									
5300	Powder coated steel	2 Carp	8	2	Ea.	228	88		316	385
5600	Stainless steel	"	8	2	"	595	88		683	790
5750	Pilaster, flush									
5800	Powder coated steel	2 Carp	10	1.600	Ea.	276	70.50		346.50	415
6100	Stainless steel		10	1.600		670	70.50		740.50	845
6300	Post braced, powder coated steel		10	1.600		153	70.50		223.50	277
6600	Stainless steel		10	1.600		450	70.50		520.50	605
6700	Wall hung, bracket supported									
6800	Powder coated steel	2 Carp	10	1.600	Ea.	153	70.50		223.50	277
7100	Stainless steel		10	1.600		219	70.50		289.50	350
7400	Flange supported, powder coated steel		10	1.600		106	70.50		176.50	225
7700	Stainless steel		10	1.600		310	70.50		380.50	450
7800	Wedge type, powder coated steel		10	1.600		134	70.50		204.50	257
8100	Stainless steel		10	1.600		565	70.50		635.50	730

10 21 13.14 Metal Toilet Compartment Components

		Crew	Daily Output	Labor-Hours	Unit	Material	2012 Bare Costs Labor	Equipment	Total	Total Incl O&P
0010	**METAL TOILET COMPARTMENT COMPONENTS**									
0100	Pilasters									
0110	Overhead braced, powder coated steel, 7" wide x 82" high	2 Carp	22.20	.721	Ea.	72.50	32		104.50	129
0120	Stainless steel		22.20	.721		121	32		153	182
0130	Floor braced, powder coated steel, 7" wide x 70" high		23.30	.687		122	30.50		152.50	181
0140	Stainless steel		23.30	.687		236	30.50		266.50	305
0150	Ceiling hung, powder coated steel, 7" wide x 83" high		13.30	1.203		129	53		182	224
0160	Stainless steel		13.30	1.203		264	53		317	370
0170	Wall hung, powder coated steel, 3" wide x 58" high		18.90	.847		129	37.50		166.50	200
0180	Stainless steel		18.90	.847		183	37.50		220.50	259
0200	Panels									
0210	Powder coated steel, 31" wide x 58" high	2 Carp	18.90	.847	Ea.	128	37.50		165.50	199
0220	Stainless steel		18.90	.847		350	37.50		387.50	445
0230	Powder coated steel, 53" wide x 58" high		18.90	.847		158	37.50		195.50	232
0240	Stainless steel		18.90	.847		450	37.50		487.50	555
0250	Powder coated steel, 63" wide x 58" high		18.90	.847		186	37.50		223.50	263
0260	Stainless steel		18.90	.847		495	37.50		532.50	605
0300	Doors									
0310	Powder coated steel, 24" wide x 58" high	2 Carp	14.10	1.135	Ea.	133	50		183	224
0320	Stainless steel		14.10	1.135		275	50		325	380
0330	Powder coated steel, 26" wide x 58" high		14.10	1.135		136	50		186	226
0340	Stainless steel		14.10	1.135		284	50		334	385
0350	Powder coated steel, 28" wide x 58" high		14.10	1.135		156	50		206	248
0360	Stainless steel		14.10	1.135		315	50		365	425
0370	Powder coated steel, 36" wide x 58" high		14.10	1.135		167	50		217	261

10 21 13 – Toilet Compartments

10 21 13.14 Metal Toilet Compartment Components

		Crew	Daily Output	Labor-Hours	Unit	Material	2012 Bare Costs Labor	Equipment	Total	Total Incl O&P
0380	Stainless steel	2 Carp	14.10	1.135	Ea.	355	50		405	465
0400	Headrails									
0410	For powder coated steel, 62" long	2 Carp	65	.246	Ea.	21	10.85		31.85	40
0420	Stainless steel		65	.246		21	10.85		31.85	40
0430	For powder coated steel, 84" long		50	.320		32	14.10		46.10	56.50
0440	Stainless steel		50	.320		32	14.10		46.10	56.50
0450	For powder coated steel, 120" long		30	.533		42	23.50		65.50	82
0460	Stainless steel		30	.533		42	23.50		65.50	82

10 21 13.16 Plastic-Laminate-Clad Toilet Compartments

		Crew	Daily Output	Labor-Hours	Unit	Material	2012 Bare Costs Labor	Equipment	Total	Total Incl O&P
0010	**PLASTIC-LAMINATE-CLAD TOILET COMPARTMENTS**									
0110	Cubicles, ceiling hung									
0300	Plastic laminate on particle board	2 Carp	4	4	Ea.	540	176		716	865
0600	For handicap units, incl. 52" grab bars, add				"	420			420	460
0900	Floor and ceiling anchored									
1100	Plastic laminate on particle board	2 Carp	5	3.200	Ea.	790	141		931	1,075
1400	For handicap units, incl. 52" grab bars, add				"	300			300	330
1610	Floor mounted									
1800	Plastic laminate on particle board	2 Carp	7	2.286	Ea.	495	101		596	700
2450	Floor mounted, headrail braced									
2600	Plastic laminate on particle board	2 Carp	6	2.667	Ea.	750	118		868	1,000
3400	For handicap units, incl. 52" grab bars, add					350			350	385
4300	Entrance screen, floor mtd., plas. lam., 58" high, 48" wide	2 Carp	15	1.067		610	47		657	745
4800	Urinal screen, 18" wide, ceiling braced, plastic laminate		8	2		174	88		262	325
5400	Floor mounted, headrail braced		8	2		180	88		268	335
5900	Pilaster, flush, plastic laminate		10	1.600		505	70.50		575.50	670
6400	Post braced, plastic laminate		10	1.600		295	70.50		365.50	435
6700	Wall hung, bracket supported									
6900	Plastic laminate on particle board	2 Carp	10	1.600	Ea.	90.50	70.50		161	209
7450	Flange supported									
7500	Plastic laminate on particle board	2 Carp	10	1.600	Ea.	229	70.50		299.50	360

10 21 13.17 Plastic-Laminate Clad Toilet Compartment Components

		Crew	Daily Output	Labor-Hours	Unit	Material	2012 Bare Costs Labor	Equipment	Total	Total Incl O&P
0010	**PLASTIC-LAMINATE CLAD TOILET COMPARTMENT COMPONENTS**									
0100	Pilasters									
0110	Overhead braced, 7" wide x 82" high	2 Carp	22.20	.721	Ea.	92	32		124	150
0130	Floor anchored, 7" wide x 70" high		23.30	.687		92.50	30.50		123	149
0150	Ceiling hung, 7" wide x 83" high		13.30	1.203		98	53		151	190
0180	Wall hung, 3" wide x 58" high		18.90	.847		90.50	37.50		128	157
0200	Panels									
0210	31" wide x 58" high	2 Carp	18.90	.847	Ea.	138	37.50		175.50	210
0230	51" wide x 58" high		18.90	.847		190	37.50		227.50	267
0250	63" wide x 58" high		18.90	.847		221	37.50		258.50	300
0300	Doors									
0310	24" wide x 58" high	2 Carp	14.10	1.135	Ea.	135	50		185	226
0330	26" wide x 58" high		14.10	1.135		140	50		190	231
0350	28" wide x 58" high		14.10	1.135		145	50		195	237
0370	36" wide x 58" high		14.10	1.135		174	50		224	268
0400	Headrails									
0410	62" long	2 Carp	65	.246	Ea.	22.50	10.85		33.35	41
0430	84" long		60	.267		32	11.75		43.75	53
0450	120" long		30	.533		42	23.50		65.50	82

10 21 13.19 Plastic Toilet Compartments

		Crew	Daily Output	Labor-Hours	Unit	Material	2012 Bare Costs Labor	Equipment	Total	Total Incl O&P
0010	**PLASTIC TOILET COMPARTMENTS**									
0110	Cubicles, ceiling hung									
0250	Phenolic	2 Carp	4	4	Ea.	850	176		1,026	1,200
0600	For handicap units, incl. 52" grab bars, add				"	420			420	460
0900	Floor and ceiling anchored									
1050	Phenolic	2 Carp	5	3.200	Ea.	860	141		1,001	1,150
1400	For handicap units, incl. 52" grab bars, add				"	300			300	330
1610	Floor mounted									
1750	Phenolic	2 Carp	7	2.286	Ea.	750	101		851	980
2100	For handicap units, incl. 52" grab bars, add					299			299	330
2200	For juvenile units, deduct					41.50			41.50	45.50
2450	Floor mounted, headrail braced									
2550	Phenolic	2 Carp	6	2.667	Ea.	750	118		868	1,000

10 21 13.20 Plastic Toilet Compartment Components

		Crew	Daily Output	Labor-Hours	Unit	Material	2012 Bare Costs Labor	Equipment	Total	Total Incl O&P
0010	**PLASTIC TOILET COMPARTMENT COMPONENTS**									
0100	Pilasters									
0110	Overhead braced, polymer plastic, 7" wide x 82" high	2 Carp	22.20	.721	Ea.	103	32		135	163
0120	Phenolic		22.20	.721		146	32		178	210
0130	Floor braced, polymer plastic, 7" wide x 70" high		23.30	.687		169	30.50		199.50	233
0140	Phenolic		23.30	.687		138	30.50		168.50	199
0150	Ceiling hung, polymer plastic, 7" wide x 83" high		13.30	1.203		166	53		219	265
0160	Phenolic		13.30	1.203		156	53		209	254
0180	Wall hung, phenolic, 3" wide x 58" high		18.90	.847		93	37.50		130.50	160
0200	Panels									
0210	Polymer plastic, 31" high x 55" high	2 Carp	18.90	.847	Ea.	295	37.50		332.50	385
0220	Phenolic, 31" wide x 58" high		18.90	.847		249	37.50		286.50	330
0230	Polymer plastic, 51" wide x 55" high		18.90	.847		395	37.50		432.50	495
0240	Phenolic, 51" wide x 58" high		18.90	.847		380	37.50		417.50	480
0250	Polymer plastic, 63" wide x 55" high		18.90	.847		540	37.50		577.50	655
0260	Phenolic, 63" wide x 58" high		18.90	.847		400	37.50		437.50	500
0300	Doors									
0310	Polymer plastic, 24" wide x 55" high	2 Carp	14.10	1.135	Ea.	202	50		252	299
0320	Phenolic, 24" wide x 58" high		14.10	1.135		283	50		333	385
0330	Polymer plastic, 26" high x 55" high		14.10	1.135		215	50		265	315
0340	Phenolic, 26" wide x 58" high		14.10	1.135		300	50		350	405
0350	Polymer plastic, 28" wide x 55" high		14.10	1.135		240	50		290	340
0360	Phenolic, 28" wide x 58" high		14.10	1.135		315	50		365	425
0370	Polymer plastic, 36" wide x 55" high		14.10	1.135		280	50		330	385
0380	Phenolic, 36" wide x 58" high		14.10	1.135		395	50		445	510
0400	Headrails									
0410	For polymer plastic, 62" long	2 Carp	65	.246	Ea.	22.50	10.85		33.35	41
0420	Phenolic		65	.246		22.50	10.85		33.35	41
0430	For polymer plastic, 84" long		50	.320		32	14.10		46.10	56.50
0440	Phenolic		50	.320		32	14.10		46.10	56.50
0450	For polymer plastic, 120" long		30	.533		42	23.50		65.50	82
0460	Phenolic		30	.533		42	23.50		65.50	82

10 21 13.40 Stone Toilet Compartments

		Crew	Daily Output	Labor-Hours	Unit	Material	2012 Bare Costs Labor	Equipment	Total	Total Incl O&P
0010	**STONE TOILET COMPARTMENTS**									
0100	Cubicles, ceiling hung, marble	2 Marb	2	8	Ea.	1,525	330		1,855	2,175
0600	For handicap units, incl. 52" grab bars, add					420			420	460
0800	Floor & ceiling anchored, marble	2 Marb	2.50	6.400		1,925	263		2,188	2,500
1400	For handicap units, incl. 52" grab bars, add					300			300	330

10 21 Compartments and Cubicles

10 21 13 – Toilet Compartments

10 21 13.40 Stone Toilet Compartments

		Crew	Daily Output	Labor-Hours	Unit	Material	2012 Bare Costs Labor	Equipment	Total	Total Incl O&P
1600	Floor mounted, marble	2 Marb	3	5.333	Ea.	1,125	219		1,344	1,575
2400	Floor mounted, headrail braced, marble	"	3	5.333		1,100	219		1,319	1,525
2900	For handicap units, incl. 52" grab bars, add					350			350	385
4100	Entrance screen, floor mounted marble, 58" high, 48" wide	2 Marb	9	1.778		710	73		783	890
4600	Urinal screen, 18" wide, ceiling braced, marble	D-1	6	2.667		725	106		831	960
5100	Floor mounted, head rail braced									
5200	Marble	D-1	6	2.667	Ea.	615	106		721	840
5700	Pilaster, flush, marble		9	1.778		810	71		881	995
6200	Post braced, marble		9	1.778		800	71		871	985

10 21 16 – Shower and Dressing Compartments

10 21 16.10 Partitions, Shower

		Crew	Daily Output	Labor-Hours	Unit	Material	2012 Bare Costs Labor	Equipment	Total	Total Incl O&P
0010	**PARTITIONS, SHOWER** floor mounted, no plumbing									
0400	Cabinet, one piece, fiberglass, 32" x 32"	2 Carp	5	3.200	Ea.	505	141		646	770
0420	36" x 36"		5	3.200		570	141		711	840
0440	36" x 48"		5	3.200		1,450	141		1,591	1,825
0460	Acrylic, 32" x 32"		5	3.200		325	141		466	570
0480	36" x 36"		5	3.200		1,025	141		1,166	1,350
0500	36" x 48"		5	3.200		1,475	141		1,616	1,850
0520	Shower door for above, clear plastic, 24" wide	1 Carp	8	1		177	44		221	263
0540	28" wide		8	1		201	44		245	290
0560	Tempered glass, 24" wide		8	1		191	44		235	278
0580	28" wide		8	1		216	44		260	305
2400	Glass stalls, with doors, no receptors, chrome on brass	2 Shee	3	5.333		1,550	278		1,828	2,125
2700	Anodized aluminum	"	4	4		1,075	209		1,284	1,500
2900	Marble shower stall, stock design, with shower door	2 Marb	1.20	13.333		2,325	550		2,875	3,375
3000	With curtain		1.30	12.308		2,050	505		2,555	3,025
3200	Receptors, precast terrazzo, 32" x 32"		14	1.143		297	47		344	395
3300	48" x 34"		9.50	1.684		405	69		474	550
3500	Plastic, simulated terrazzo receptor, 32" x 32"		14	1.143		130	47		177	214
3600	32" x 48"		12	1.333		208	55		263	310
3800	Precast concrete, colors, 32" x 32"		14	1.143		166	47		213	254
3900	48" x 48"		8	2		230	82		312	375
4100	Shower doors, economy plastic, 24" wide	1 Shee	9	.889		134	46.50		180.50	219
4200	Tempered glass door, economy		8	1		252	52		304	355
4400	Folding, tempered glass, aluminum frame		6	1.333		390	69.50		459.50	535
4500	Sliding, tempered glass, 48" opening		6	1.333		540	69.50		609.50	695
4700	Deluxe, tempered glass, chrome on brass frame, minimum		8	1		385	52		437	500
4800	Maximum		1	8		625	415		1,040	1,325
4850	On anodized aluminum frame, minimum		2	4		540	209		749	910
4900	Maximum		1	8		625	415		1,040	1,325
5100	Shower enclosure, tempered glass, anodized alum. frame									
5120	2 panel & door, corner unit, 32" x 32"	1 Shee	2	4	Ea.	1,075	209		1,284	1,500
5140	Neo-angle corner unit, 16" x 24" x 16"	"	2	4		1,150	209		1,359	1,575
5200	Shower surround, 3 wall, polypropylene, 32" x 32"	1 Carp	4	2		415	88		503	590
5220	PVC, 32" x 32"		4	2		355	88		443	530
5240	Fiberglass		4	2		390	88		478	565
5250	2 wall, polypropylene, 32" x 32"		4	2		296	88		384	460
5270	PVC		4	2		370	88		458	540
5290	Fiberglass		4	2		375	88		463	545
5300	Tub doors, tempered glass & frame, minimum	1 Shee	8	1		214	52		266	315
5400	Maximum		6	1.333		500	69.50		569.50	655
5600	Chrome plated, brass frame, minimum		8	1		284	52		336	390

10 21 Compartments and Cubicles

10 21 16 – Shower and Dressing Compartments

10 21 16.10 Partitions, Shower	Crew	Daily Output	Labor-Hours	Unit	Material	2012 Bare Costs Labor	Equipment	Total	Total Incl O&P
5700	Maximum	1 Shee	6	1.333	Ea.	695	69.50		764.50	870
5900	Tub/shower enclosure, temp. glass, alum. frame, minimum		2	4		385	209		594	740
6200	Maximum		1.50	5.333		800	278		1,078	1,300
6500	On chrome-plated brass frame, minimum		2	4		530	209		739	905
6600	Maximum		1.50	5.333		1,125	278		1,403	1,675
6800	Tub surround, 3 wall, polypropylene	1 Carp	4	2		240	88		328	400
6900	PVC		4	2		365	88		453	535
7000	Fiberglass, minimum		4	2		375	88		463	550
7100	Maximum		3	2.667		635	118		753	880

10 21 23 – Cubicle Curtains and Track

10 21 23.16 Cubicle Track and Hardware

		Crew	Daily Output	Labor-Hours	Unit	Material	Labor	Equipment	Total	Total Incl O&P
0010	**CUBICLE TRACK AND HARDWARE**									
0020	Curtain track, box channel, ceiling mounted	1 Carp	135	.059	L.F.	5.70	2.61		8.31	10.30
0100	Suspended	"	100	.080	"	8.20	3.53		11.73	14.50
0300	Curtains, nylon mesh tops, fire resistant, 11 oz. per lineal yard									
0310	Polyester oxford cloth, 9' ceiling height	1 Carp	425	.019	L.F.	12	.83		12.83	14.50
0500	8' ceiling height		425	.019		9.85	.83		10.68	12.10
0700	Designer oxford cloth		425	.019		9.45	.83		10.28	11.70
0800	I.V. track systems									
0820	I.V. track, oval	1 Carp	135	.059	L.F.	5.25	2.61		7.86	9.80
0830	I.V. trolley		32	.250	Ea.	36	11.05		47.05	57
0840	I.V. pendent, (tree, 5 hook)		32	.250	"	129	11.05		140.05	159

10 22 Partitions

10 22 13 – Wire Mesh Partitions

10 22 13.10 Partitions, Woven Wire

		Crew	Daily Output	Labor-Hours	Unit	Material	Labor	Equipment	Total	Total Incl O&P
0010	**PARTITIONS, WOVEN WIRE** for tool or stockroom enclosures									
0100	Channel frame, 1-1/2" diamond mesh, 10 ga. wire, painted									
0300	Wall panels, 4'-0" wide, 7' high	2 Carp	25	.640	Ea.	133	28		161	190
0400	8' high		23	.696		150	30.50		180.50	212
0600	10' high		18	.889		179	39		218	258
0700	For 5' wide panels, add					5%				
0900	Ceiling panels, 10' long, 2' wide	2 Carp	25	.640		136	28		164	194
1000	4' wide		15	1.067		172	47		219	262
1200	Panel with service window & shelf, 5' wide, 7' high		20	.800		415	35.50		450.50	515
1300	8' high		15	1.067		415	47		462	530
1500	Sliding doors, full height, 3' wide, 7' high		6	2.667		430	118		548	650
1600	10' high		5	3.200		465	141		606	725
1800	6' wide sliding door, 7' full height		5	3.200		590	141		731	860
1900	10' high		4	4		730	176		906	1,075
2100	Swinging doors, 3' wide, 7' high, no transom		6	2.667		273	118		391	480
2200	7' high, 3' transom		5	3.200		325	141		466	575

10 22 16 – Folding Gates

10 22 16.10 Security Gates

		Crew	Daily Output	Labor-Hours	Unit	Material	Labor	Equipment	Total	Total Incl O&P
0010	**SECURITY GATES** for roll up type, see Section 08 33 13.10									
0300	Scissors type folding gate, ptd. steel, single, 6-1/2' high, 5-1/2'wide	2 Sswk	4	4	Opng.	229	197		426	605
0350	6-1/2' wide		4	4		242	197		439	620
0400	7-1/2' wide		4	4		215	197		412	590
0600	Double gate, 8' high, 8' wide		2.50	6.400		355	315		670	955
0650	10' wide		2.50	6.400		390	315		705	995

10 22 Partitions

10 22 16 – Folding Gates

10 22 16.10 Security Gates

		Crew	Daily Output	Labor-Hours	Unit	Material	2012 Bare Costs Labor	2012 Bare Costs Equipment	Total	Total Incl O&P
0700	12' wide	2 Sswk	2	8	Opng.	590	395		985	1,350
0750	14' wide		2	8		675	395		1,070	1,450
0900	Door gate, folding steel, 4' wide, 61" high		4	4		129	197		326	495
1000	71" high		4	4		161	197		358	530
1200	81" high		4	4		175	197		372	545
1300	Window gates, 2' to 4' wide, 31" high		4	4		62.50	197		259.50	425
1500	55" high		3.75	4.267		97.50	211		308.50	480
1600	79" high	↓	3.50	4.571	↓	113	226		339	530

10 22 19 – Demountable Partitions

10 22 19.43 Demountable Composite Partitions

		Crew	Daily Output	Labor-Hours	Unit	Material	2012 Bare Costs Labor	2012 Bare Costs Equipment	Total	Total Incl O&P
0010	**DEMOUNTABLE COMPOSITE PARTITIONS**, add for doors									
0100	Do not deduct door openings from total L.F.									
0900	Demountable gypsum system on 2" to 2-1/2"									
1000	steel studs, 9' high, 3" to 3-3/4" thick									
1200	Vinyl clad gypsum	2 Carp	48	.333	L.F.	58	14.70		72.70	86
1300	Fabric clad gypsum		44	.364		144	16.05		160.05	183
1500	Steel clad gypsum	↓	40	.400	↓	156	17.65		173.65	199
1600	1.75 system, aluminum framing, vinyl clad hardboard,									
1800	paper honeycomb core panel, 1-3/4" to 2-1/2" thick									
1900	9' high	2 Carp	48	.333	L.F.	97	14.70		111.70	130
2100	7' high		60	.267		87	11.75		98.75	114
2200	5' high	↓	80	.200	↓	73.50	8.80		82.30	94.50
2250	Unitized gypsum system									
2300	Unitized panel, 9' high, 2" to 2-1/2" thick									
2350	Vinyl clad gypsum	2 Carp	48	.333	L.F.	125	14.70		139.70	160
2400	Fabric clad gypsum	"	44	.364	"	206	16.05		222.05	251
2500	Unitized mineral fiber system									
2510	Unitized panel, 9' high, 2-1/4" thick, aluminum frame									
2550	Vinyl clad mineral fiber	2 Carp	48	.333	L.F.	124	14.70		138.70	160
2600	Fabric clad mineral fiber	"	44	.364	"	185	16.05		201.05	229
2800	Movable steel walls, modular system									
2900	Unitized panels, 9' high, 48" wide									
3100	Baked enamel, pre-finished	2 Carp	60	.267	L.F.	141	11.75		152.75	173
3200	Fabric clad steel		56	.286	"	203	12.60		215.60	243
5310	Trackless wall, cork finish, semi-acoustic, 1-5/8" thick, minimum		325	.049	S.F.	37	2.17		39.17	44
5320	Maximum		190	.084		41	3.71		44.71	50.50
5330	Acoustic, 2" thick, minimum		305	.052		35	2.31		37.31	42
5340	Maximum	↓	225	.071		53.50	3.14		56.64	64
5500	For acoustical partitions, add, minimum					2.22			2.22	2.44
5550	Maximum				↓	10.35			10.35	11.40
5700	For doors, see Sections 08 11 & 08 16									
5800	For door hardware, see Section 08 71									
6100	In-plant modular office system, w/prehung hollow core door									
6200	3" thick polystyrene core panels									
6250	12' x 12', 2 wall	2 Clab	3.80	4.211	Ea.	3,900	148		4,048	4,500
6300	4 wall		1.90	8.421		6,800	296		7,096	7,925
6350	16' x 16', 2 wall		3.60	4.444		5,850	156		6,006	6,675
6400	4 wall	↓	1.80	8.889	↓	8,225	310		8,535	9,500

10 22 Partitions

10 22 23 – Portable Partitions, Screens, and Panels

10 22 23.13 Wall Screens

		Crew	Daily Output	Labor-Hours	Unit	Material	2012 Bare Costs Labor	Equipment	Total	Total Incl O&P
0010	**WALL SCREENS**, divider panels, free standing, fiber core									
0020	Fabric face straight									
0100	3'-0" long, 4'-0" high	2 Carp	100	.160	L.F.	146	7.05		153.05	172
0200	5'-0" high		90	.178		103	7.85		110.85	125
0500	6'-0" high		75	.213		120	9.40		129.40	146
0900	5'-0" long, 4'-0" high		175	.091		90.50	4.03		94.53	106
1000	5'-0" high		150	.107		98.50	4.70		103.20	115
1500	6"-0" high		125	.128		114	5.65		119.65	135
1600	6'-0" long, 5'-0" high		162	.099		95.50	4.36		99.86	112
3200	Economical panels, fabric face, 4'-0" long, 5'-0" high		132	.121		50.50	5.35		55.85	63.50
3250	6'-0" high		112	.143		54	6.30		60.30	68.50
3300	5'-0" long, 5'-0" high		150	.107		43	4.70		47.70	54.50
3350	6'-0" high		125	.128		44	5.65		49.65	56.50
3450	Acoustical panels, 60 to 90 NRC, 3'-0" long, 5'-0" high		90	.178		89.50	7.85		97.35	111
3550	6'-0" high		75	.213		99.50	9.40		108.90	123
3600	5'-0" long, 5'-0" high		150	.107		56.50	4.70		61.20	69.50
3650	6'-0" high		125	.128		65	5.65		70.65	80.50
3700	6'-0" long, 5'-0" high		162	.099		48	4.36		52.36	59
3750	6'-0" high		138	.116		56	5.10		61.10	69.50
3800	Economy acoustical panels, 40 N.R.C., 4'-0" long, 5'-0" high		132	.121		50.50	5.35		55.85	63.50
3850	6'-0" high		112	.143		54	6.30		60.30	68.50
3900	5'-0" long, 6'-0" high		125	.128		44	5.65		49.65	56.50
3950	6'-0" long, 5'-0" high		162	.099		41	4.36		45.36	51.50
4000	Metal chalkboard, 6'-6" high, chalkboard, 1 side		125	.128		116	5.65		121.65	137
4100	Metal chalkboard, 2 sides		120	.133		133	5.90		138.90	155
4300	Tackboard, both sides		123	.130		105	5.75		110.75	125

10 22 33 – Accordion Folding Partitions

10 22 33.10 Partitions, Accordion Folding

		Crew	Daily Output	Labor-Hours	Unit	Material	2012 Bare Costs Labor	Equipment	Total	Total Incl O&P
0010	**PARTITIONS, ACCORDION FOLDING**									
0100	Vinyl covered, over 150 S.F., frame not included									
0300	Residential, 1.25 lb. per S.F., 8' maximum height	2 Carp	300	.053	S.F.	20.50	2.35		22.85	26.50
0400	Commercial, 1.75 lb. per S.F., 8' maximum height		225	.071		23.50	3.14		26.64	31
0600	2 lb. per S.F., 17' maximum height		150	.107		24.50	4.70		29.20	34.50
0700	Industrial, 4 lb. per S.F., 20' maximum height		75	.213		35	9.40		44.40	53
0900	Acoustical, 3 lb. per S.F., 17' maximum height		100	.160		27.50	7.05		34.55	41
1200	5 lb. per S.F., 20' maximum height		95	.168		38	7.45		45.45	53.50
1300	5.5 lb. per S.F., 17' maximum height		90	.178		44.50	7.85		52.35	61
1400	Fire rated, 4.5 psf, 20' maximum height		160	.100		44.50	4.41		48.91	56
1500	Vinyl clad wood or steel, electric operation, 5.0 psf		160	.100		52	4.41		56.41	64
1900	Wood, non-acoustic, birch or mahogany, to 10' high		300	.053		28	2.35		30.35	34

10 22 39 – Folding Panel Partitions

10 22 39.10 Partitions, Folding Panel

		Crew	Daily Output	Labor-Hours	Unit	Material	2012 Bare Costs Labor	Equipment	Total	Total Incl O&P
0010	**PARTITIONS, FOLDING PANEL**, acoustic, wood									
0100	Vinyl faced, to 18' high, 6 psf, minimum	2 Carp	60	.267	S.F.	54	11.75		65.75	77.50
0150	Average		45	.356		64.50	15.70		80.20	95
0200	Maximum		30	.533		83.50	23.50		107	128
0400	Formica or hardwood finish, minimum		60	.267		55.50	11.75		67.25	79.50
0500	Maximum		30	.533		59.50	23.50		83	102
0600	Wood, low acoustical type, 4.5 psf, to 14' high		50	.320		40.50	14.10		54.60	66
1100	Steel, acoustical, 9 to 12 lb. per S.F., vinyl faced, minimum		60	.267		57.50	11.75		69.25	81.50
1200	Maximum		30	.533		70.50	23.50		94	114

10 22 Partitions

10 22 39 – Folding Panel Partitions

	10 22 39.10 Partitions, Folding Panel	Crew	Daily Output	Labor-Hours	Unit	Material	2012 Bare Costs Labor	Equipment	Total	Total Incl O&P
1700	Aluminum framed, acoustical, to 12' high, 5.5 psf, minimum	2 Carp	60	.267	S.F.	39	11.75		50.75	60.50
1800	Maximum		30	.533		47	23.50		70.50	87.50
2000	6.5 lb. per S.F., minimum		60	.267		41	11.75		52.75	63
2100	Maximum	↓	30	.533	↓	50.50	23.50		74	91.50

10 22 43 – Sliding Partitions

	10 22 43.10 Partitions, Sliding	Crew	Daily Output	Labor-Hours	Unit	Material	2012 Bare Costs Labor	Equipment	Total	Total Incl O&P
0010	**PARTITIONS, SLIDING**									
0020	Acoustic air wall, 1-5/8" thick, minimum	2 Carp	375	.043	S.F.	31.50	1.88		33.38	37.50
0100	Maximum		365	.044		54	1.93		55.93	62.50
0300	2-1/4" thick, minimum		360	.044		35.50	1.96		37.46	42
0400	Maximum	↓	330	.048	↓	62	2.14		64.14	71.50
0600	For track type, add to above				L.F.	115			115	127
0700	Overhead track type, acoustical, 3" thick, 11 psf, minimum	2 Carp	350	.046	S.F.	79.50	2.02		81.52	90.50
0800	Maximum	"	300	.053	"	95.50	2.35		97.85	109

10 26 Wall and Door Protection

10 26 13 – Corner Guards

10 26 13.10 Metal Corner Guards

	10 26 13.10 Metal Corner Guards	Crew	Daily Output	Labor-Hours	Unit	Material	2012 Bare Costs Labor	Equipment	Total	Total Incl O&P
0010	**METAL CORNER GUARDS**									
0020	Steel angle w/anchors, 1" x 1" x 1/4", 1.5#/L.F.	2 Carp	160	.100	L.F.	6.85	4.41		11.26	14.35
0100	2" x 2" x 1/4" angles, 3.2#/L.F.		150	.107		9.60	4.70		14.30	17.80
0200	3" x 3" x 5/16" angles, 6.1#/L.F.		140	.114		13.70	5.05		18.75	23
0310	4" x 4" x 5/16" angles on masonry, 8.2#/L.F.	↓	55	.291		16.35	12.85		29.20	37.50
0350	For angles drilled and anchored to masonry, add					15%	120%			
0370	Drilled and anchored to concrete, add					20%	170%			
0400	For galvanized angles, add					35%				
0450	For stainless steel angles, add				↓	100%				
0500	Steel door track/wheel guards, 4' - 0" high	E-4	22	1.455	Ea.	153	72.50	5.55	231.05	305
0800	Pipe bumper for truck doors, 8' long, 6" diameter, filled		20	1.600		725	80	6.10	811.10	950
0900	8" diameter	↓	20	1.600	↓	1,075	80	6.10	1,161.10	1,325

10 26 13.20 Corner Protection

	10 26 13.20 Corner Protection	Crew	Daily Output	Labor-Hours	Unit	Material	2012 Bare Costs Labor	Equipment	Total	Total Incl O&P
0010	**CORNER PROTECTION**									
0100	Stainless steel, 16 ga., adhesive mount, 3-1/2" leg	1 Sswk	80	.100	L.F.	19.60	4.94		24.54	30.50
0200	12 ga. stainless, adhesive mount	"	80	.100		21.50	4.94		26.44	32.50
0300	For screw mount, add						10%			
0500	Vinyl acrylic, adhesive mount, 3" leg	1 Carp	128	.063		8.20	2.76		10.96	13.25
0550	1-1/2" leg		160	.050		4.85	2.21		7.06	8.75
0600	Screw mounted, 3" leg		80	.100		8.40	4.41		12.81	16.05
0650	1-1/2" leg		100	.080		4.59	3.53		8.12	10.50
0700	Clear plastic, screw mounted, 2-1/2"		60	.133		4.69	5.90		10.59	14.20
1000	Vinyl cover, alum. retainer, surface mount, 3" x 3"		48	.167		9.25	7.35		16.60	21.50
1050	2" x 2"		48	.167		9.45	7.35		16.80	21.50
1100	Flush mounted, 3" x 3"		32	.250		20.50	11.05		31.55	39.50
1150	2" x 2"	↓	32	.250	↓	16.70	11.05		27.75	35.50

10 26 16 – Bumper Guards

10 26 16.10 Wallguard

	10 26 16.10 Wallguard	Crew	Daily Output	Labor-Hours	Unit	Material	2012 Bare Costs Labor	Equipment	Total	Total Incl O&P
0010	**WALLGUARD**									
0400	Rub rail, vinyl, adhesive mounted	1 Carp	185	.043	L.F.	7.85	1.91		9.76	11.60
0500	Neoprene, aluminum backing, 1-1/2" x 2"		110	.073		8.80	3.21		12.01	14.65
1000	Trolley rail, PVC, clipped to wall, 5" high	↓	185	.043	↓	8.70	1.91		10.61	12.55

10 26 Wall and Door Protection

10 26 16 – Bumper Guards

10 26 16.10 Wallguard		Crew	Daily Output	Labor-Hours	Unit	Material	2012 Bare Costs Labor	Equipment	Total	Total Incl O&P
1050	8" high	1 Carp	180	.044	L.F.	11.10	1.96		13.06	15.20
1200	Bed bumper, vinyl acrylic, alum. retainer, 21" long		10	.800	Ea.	40.50	35.50		76	99
1300	53" long with aligner		9	.889	"	102	39		141	173
1400	Bumper, vinyl cover, alum. retain., cush. mnt., 1-1/2" x 2-3/4"		80	.100	L.F.	14.20	4.41		18.61	22.50
1500	2" x 4-1/4"		80	.100		20.50	4.41		24.91	30
1600	Surface mounted, 1-3/4" x 3-5/8"		80	.100		11.95	4.41		16.36	19.95
2000	Crash rail, vinyl cover, alum. retainer, 1" x 4"		110	.073		11.05	3.21		14.26	17.10
2100	1" x 8"		90	.089		17.25	3.92		21.17	25
2150	Vinyl inserts, aluminum plate, 1" x 2-1/2"		110	.073		14.90	3.21		18.11	21.50
2200	1" x 5"		90	.089		23	3.92		26.92	31.50
3000	Handrail/bumper, vinyl cover, alum. retainer									
3010	Bracket mounted, flat rail, 5-1/2"	1 Carp	80	.100	L.F.	18.30	4.41		22.71	27
3100	6-1/2"		80	.100		23	4.41		27.41	32
3200	Bronze bracket, 1-3/4" diam. rail		80	.100		16.50	4.41		20.91	25

10 28 Toilet, Bath, and Laundry Accessories

10 28 13 – Toilet Accessories

10 28 13.13 Commercial Toilet Accessories

10 28 13.13 Commercial Toilet Accessories		Crew	Daily Output	Labor-Hours	Unit	Material	2012 Bare Costs Labor	Equipment	Total	Total Incl O&P
0010	**COMMERCIAL TOILET ACCESSORIES**									
0200	Curtain rod, stainless steel, 5' long, 1" diameter	1 Carp	13	.615	Ea.	36.50	27		63.50	82
0300	1-1/4" diameter		13	.615		27.50	27		54.50	72
0350	Chrome, 1" diameter		13	.615		28	27		55	72.50
0360	For vinyl curtain, add		1950	.004	S.F.	.84	.18		1.02	1.20
0400	Diaper changing station, horizontal, wall mounted, plastic		10	.800	Ea.	218	35.50		253.50	294
0500	Dispenser units, combined soap & towel dispensers,									
0510	mirror and shelf, flush mounted	1 Carp	10	.800	Ea.	310	35.50		345.50	395
0600	Towel dispenser and waste receptacle,									
0610	18 gallon capacity	1 Carp	10	.800	Ea.	325	35.50		360.50	415
0800	Grab bar, straight, 1-1/4" diameter, stainless steel, 18" long		24	.333		29	14.70		43.70	54.50
0900	24" long		23	.348		29.50	15.35		44.85	56
1000	30" long		22	.364		31.50	16.05		47.55	59
1100	36" long		20	.400		37	17.65		54.65	67.50
1105	42" long		20	.400		37.50	17.65		55.15	68
1200	1-1/2" diameter, 24" long		23	.348		30	15.35		45.35	56.50
1300	36" long		20	.400		34.50	17.65		52.15	64.50
1310	42" long		18	.444		43	19.60		62.60	77.50
1500	Tub bar, 1-1/4" diameter, 24" x 36"		14	.571		87	25		112	135
1600	Plus vertical arm		12	.667		85.50	29.50		115	139
1900	End tub bar, 1" diameter, 90° angle, 16" x 32"		12	.667		106	29.50		135.50	161
2010	Tub/shower/toilet, 2-wall, 36" x 24"		12	.667		78	29.50		107.50	131
2300	Hand dryer, surface mounted, electric, 115 volt, 20 amp		4	2		475	88		563	655
2400	230 volt, 10 amp		4	2		720	88		808	930
2600	Hat and coat strip, stainless steel, 4 hook, 36" long		24	.333		71	14.70		85.70	101
2700	6 hook, 60" long		20	.400		112	17.65		129.65	150
3000	Mirror, with stainless steel 3/4" square frame, 18" x 24"		20	.400		42.50	17.65		60.15	74
3100	36" x 24"		15	.533		96.50	23.50		120	142
3200	48" x 24"		10	.800		144	35.50		179.50	214
3300	72" x 24"		6	1.333		206	59		265	315
3500	With 5" stainless steel shelf, 18" x 24"		20	.400		181	17.65		198.65	226
3600	36" x 24"		15	.533		200	23.50		223.50	256
3700	48" x 24"		10	.800		236	35.50		271.50	315

10 28 Toilet, Bath, and Laundry Accessories

10 28 13 – Toilet Accessories

10 28 13.13 Commercial Toilet Accessories

	10 28 13.13 Commercial Toilet Accessories	Crew	Daily Output	Labor-Hours	Unit	Material	2012 Bare Costs Labor	Equipment	Total	Total Incl O&P
3800	72" x 24"	1 Carp	6	1.333	Ea.	291	59		350	410
4100	Mop holder strip, stainless steel, 5 holders, 48" long		20	.400		86	17.65		103.65	122
4200	Napkin/tampon dispenser, recessed		15	.533		520	23.50		543.50	610
4250	Napkin receptacle, recessed		6.50	1.231		157	54.50		211.50	257
4300	Robe hook, single, regular		36	.222		21	9.80		30.80	38
4400	Heavy duty, concealed mounting		36	.222		15.20	9.80		25	32
4600	Soap dispenser, chrome, surface mounted, liquid		20	.400		42	17.65		59.65	73.50
4700	Powder		20	.400		54.50	17.65		72.15	87
5000	Recessed stainless steel, liquid		10	.800		123	35.50		158.50	190
5100	Powder		10	.800		180	35.50		215.50	253
5300	Soap tank, stainless steel, 1 gallon		10	.800		200	35.50		235.50	275
5400	5 gallon		5	1.600		275	70.50		345.50	410
5600	Shelf, stainless steel, 5" wide, 18 ga., 24" long		24	.333		64.50	14.70		79.20	93.50
5700	48" long		16	.500		106	22		128	151
5800	8" wide shelf, 18 ga., 24" long		22	.364		68.50	16.05		84.55	100
5900	48" long		14	.571		121	25		146	172
6000	Toilet seat cover dispenser, stainless steel, recessed		20	.400		158	17.65		175.65	201
6050	Surface mounted		15	.533		32	23.50		55.50	71.50
6100	Toilet tissue dispenser, surface mounted, SS, single roll		30	.267		17.05	11.75		28.80	37
6200	Double roll		24	.333		22.50	14.70		37.20	47.50
6240	Plastic, twin/jumbo dbl. roll		24	.333		25.50	14.70		40.20	50.50
6400	Towel bar, stainless steel, 18" long		23	.348		39.50	15.35		54.85	67
6500	30" long		21	.381		111	16.80		127.80	149
6700	Towel dispenser, stainless steel, surface mounted		16	.500		42.50	22		64.50	80.50
6800	Flush mounted, recessed		10	.800		231	35.50		266.50	310
6900	Plastic, touchless, battery operated		16	.500		78.50	22		100.50	121
7000	Towel holder, hotel type, 2 guest size		20	.400		44.50	17.65		62.15	76
7200	Towel shelf, stainless steel, 24" long, 8" wide		20	.400		56.50	17.65		74.15	89
7400	Tumbler holder, tumbler only		30	.267		42	11.75		53.75	64.50
7500	Soap, tumbler & toothbrush		30	.267		19.25	11.75		31	39
7700	Wall urn ash receiver, surface mount, 11" long		12	.667		94.50	29.50		124	149
7800	7-1/2", long		18	.444		116	19.60		135.60	158
8000	Waste receptacles, stainless steel, with top, 13 gallon		10	.800		276	35.50		311.50	360
8100	36 gallon		8	1		365	44		409	470

10 28 16 – Bath Accessories

10 28 16.20 Medicine Cabinets

	10 28 16.20 Medicine Cabinets	Crew	Daily Output	Labor-Hours	Unit	Material	Labor	Equipment	Total	Total Incl O&P
0010	**MEDICINE CABINETS**									
0020	With mirror, st. st. frame, 16" x 22", unlighted	1 Carp	14	.571	Ea.	78	25		103	125
0100	Wood frame		14	.571		121	25		146	173
0300	Sliding mirror doors, 20" x 16" x 4-3/4", unlighted		7	1.143		114	50.50		164.50	203
0400	24" x 19" x 8-1/2", lighted		5	1.600		177	70.50		247.50	305
0600	Triple door, 30" x 32", unlighted, plywood body		7	1.143		320	50.50		370.50	430
0700	Steel body		7	1.143		370	50.50		420.50	485
0900	Oak door, wood body, beveled mirror, single door		7	1.143		175	50.50		225.50	270
1000	Double door		6	1.333		390	59		449	520
1200	Hotel cabinets, stainless, with lower shelf, unlighted		10	.800		200	35.50		235.50	274
1300	Lighted		5	1.600		300	70.50		370.50	440

10 28 23 – Laundry Accessories

10 28 23.13 Built-In Ironing Boards

	10 28 23.13 Built-In Ironing Boards	Crew	Daily Output	Labor-Hours	Unit	Material	Labor	Equipment	Total	Total Incl O&P
0010	**BUILT-IN IRONING BOARDS**									
0020	Including cabinet, board & light, minimum	1 Carp	2	4	Ea.	345	176		521	650
0100	Maximum, see also Section 11 23 23.13	"	1.50	5.333	"	580	235		815	1,000

10 31 Manufactured Fireplaces

10 31 13 – Manufactured Fireplace Chimneys

10 31 13.10 Fireplace Chimneys	Crew	Daily Output	Labor-Hours	Unit	Material	2012 Bare Costs Labor	Equipment	Total	Total Incl O&P
0010 **FIREPLACE CHIMNEYS**									
0500 Chimney dbl. wall, all stainless, over 8'-6", 7" diam., add to fireplace	1 Carp	33	.242	V.L.F.	81	10.70		91.70	105
0600 10" diameter, add to fireplace		32	.250		119	11.05		130.05	147
0700 12" diameter, add to fireplace		31	.258		210	11.40		221.40	249
0800 14" diameter, add to fireplace		30	.267		233	11.75		244.75	274
1000 Simulated brick chimney top, 4' high, 16" x 16"		10	.800	Ea.	425	35.50		460.50	525
1100 24" x 24"		7	1.143	"	520	50.50		570.50	650

10 31 13.20 Chimney Accessories

	Crew	Daily Output	Labor-Hours	Unit	Material	Labor	Equipment	Total	Total Incl O&P
0010 **CHIMNEY ACCESSORIES**									
0020 Chimney screens, galv., 13" x 13" flue	1 Bric	8	1	Ea.	58.50	44		102.50	131
0050 24" x 24" flue		5	1.600		118	70.50		188.50	237
0200 Stainless steel, 13" x 13" flue		8	1		97.50	44		141.50	174
0250 20" x 20" flue		5	1.600		175	70.50		245.50	300
2400 Squirrel and bird screens, galvanized, 8" x 8" flue		16	.500		46.50	22		68.50	84.50
2450 13" x 13" flue		12	.667		50.50	29.50		80	100

10 31 16 – Manufactured Fireplace Forms

10 31 16.10 Fireplace Forms

	Crew	Daily Output	Labor-Hours	Unit	Material	Labor	Equipment	Total	Total Incl O&P
0010 **FIREPLACE FORMS**									
1800 Fireplace forms, no accessories, 32" opening	1 Bric	3	2.667	Ea.	670	118		788	920
1900 36" opening		2.50	3.200		855	141		996	1,150
2000 40" opening		2	4		1,125	176		1,301	1,525
2100 78" opening		1.50	5.333		1,650	235		1,885	2,175

10 31 23 – Prefabricated Fireplaces

10 31 23.10 Fireplace, Prefabricated

	Crew	Daily Output	Labor-Hours	Unit	Material	Labor	Equipment	Total	Total Incl O&P
0010 **FIREPLACE, PREFABRICATED**, free standing or wall hung									
0100 With hood & screen, minimum	1 Carp	1.30	6.154	Ea.	1,200	271		1,471	1,750
0150 Average		1	8		1,425	355		1,780	2,100
0200 Maximum		.90	8.889		2,725	390		3,115	3,600
1500 Simulated logs, gas fired, 40,000 BTU, 2' long, minimum		7	1.143	Set	500	50.50		550.50	630
1600 Maximum		6	1.333		1,100	59		1,159	1,325
1700 Electric, 1,500 BTU, 1'-6" long, minimum		7	1.143		200	50.50		250.50	298
1800 11,500 BTU, maximum		6	1.333		450	59		509	585
2000 Fireplace, built-in, 36" hearth, radiant		1.30	6.154	Ea.	650	271		921	1,125
2100 Recirculating, small fan		1	8		1,025	355		1,380	1,700
2150 Large fan		.90	8.889		2,050	390		2,440	2,850
2200 42" hearth, radiant		1.20	6.667		920	294		1,214	1,475
2300 Recirculating, small fan		.90	8.889		1,250	390		1,640	1,975
2350 Large fan		.80	10		2,050	440		2,490	2,925
2400 48" hearth, radiant		1.10	7.273		1,925	320		2,245	2,600
2500 Recirculating, small fan		.80	10		2,400	440		2,840	3,300
2550 Large fan		.70	11.429		3,700	505		4,205	4,825
3000 See through, including doors		.80	10		3,125	440		3,565	4,125
3200 Corner (2 wall)		1	8		3,175	355		3,530	4,025

10 32 Fireplace Specialties

10 32 13 – Fireplace Dampers

10 32 13.10 Dampers

		Crew	Daily Output	Labor-Hours	Unit	Material	2012 Bare Costs Labor	Equipment	Total	Total Incl O&P
0010	**DAMPERS**									
0800	Damper, rotary control, steel, 30" opening	1 Bric	6	1.333	Ea.	89	59		148	187
0850	Cast iron, 30" opening		6	1.333		97	59		156	196
0880	36" opening		6	1.333		105	59		164	205
0900	48" opening		6	1.333		150	59		209	254
0920	60" opening		6	1.333		335	59		394	460
0950	72" opening		5	1.600		405	70.50		475.50	550
1000	84" opening, special order		5	1.600		865	70.50		935.50	1,050
1050	96" opening, special order		4	2		880	88		968	1,100
1200	Steel plate, poker control, 60" opening		8	1		305	44		349	400
1250	84" opening, special order		5	1.600		560	70.50		630.50	720
1400	"Universal" type, chain operated, 32" x 20" opening		8	1		239	44		283	330
1450	48" x 24" opening	↓	5	1.600	↓	355	70.50		425.50	495

10 32 23 – Fireplace Doors

10 32 23.10 Doors

		Crew	Daily Output	Labor-Hours	Unit	Material	2012 Bare Costs Labor	Equipment	Total	Total Incl O&P
0010	**DOORS**									
0400	Cleanout doors and frames, cast iron, 8" x 8"	1 Bric	12	.667	Ea.	39	29.50		68.50	87.50
0450	12" x 12"		10	.800		108	35.50		143.50	173
0500	18" x 24"		8	1		143	44		187	224
0550	Cast iron frame, steel door, 24" x 30"		5	1.600		310	70.50		380.50	445
1600	Dutch Oven door and frame, cast iron, 12" x 15" opening		13	.615		125	27		152	179
1650	Copper plated, 12" x 15" opening	↓	13	.615	↓	245	27		272	310

10 35 Stoves

10 35 13 – Heating Stoves

10 35 13.10 Woodburning Stoves

		Crew	Daily Output	Labor-Hours	Unit	Material	2012 Bare Costs Labor	Equipment	Total	Total Incl O&P
0010	**WOODBURNING STOVES**									
0015	Cast iron, minimum	2 Carp	1.30	12.308	Ea.	905	545		1,450	1,825
0020	Average		1	16		1,675	705		2,380	2,925
0030	Maximum	↓	.80	20		2,450	880		3,330	4,050
0050	For gas log lighter, add				↓	42.50			42.50	46.50

10 44 Fire Protection Specialties

10 44 13 – Fire Protection Cabinets

10 44 13.53 Fire Equipment Cabinets

		Crew	Daily Output	Labor-Hours	Unit	Material	2012 Bare Costs Labor	Equipment	Total	Total Incl O&P
0010	**FIRE EQUIPMENT CABINETS**, not equipped, 20 ga. steel box									
0040	recessed, D.S. glass in door, box size given									
1000	Portable extinguisher, single, 8" x 12" x 27", alum. door & frame	Q-12	8	2	Ea.	131	97.50		228.50	291
1100	Steel door and frame	"	8	2	"	90.50	97.50		188	247
3000	Hose rack assy., 1-1/2" valve & 100' hose, 24" x 40" x 5-1/2"									
3100	Aluminum door and frame	Q-12	6	2.667	Ea.	305	130		435	530
3200	Steel door and frame		6	2.667		360	130		490	590
3300	Stainless steel door and frame	↓	6	2.667	↓	405	130		535	640
4000	Hose rack assy., 2-1/2" x 1-1/2" valve, 100' hose, 24" x 40" x 8"									
4100	Aluminum door and frame	Q-12	6	2.667	Ea.	305	130		435	530
4200	Steel door and frame		6	2.667		208	130		338	425
4300	Stainless steel door and frame	↓	6	2.667	↓	410	130		540	645
5000	Hose rack assy., 2-1/2" x 1-1/2" valve, 100' hose									
5010	and extinguisher, 30" x 40" x 8"									

10 44 Fire Protection Specialties

10 44 13 – Fire Protection Cabinets

10 44 13.53 Fire Equipment Cabinets

		Crew	Daily Output	Labor-Hours	Unit	Material	2012 Bare Costs Labor	Equipment	Total	Total Incl O&P
5100	Aluminum door and frame	Q-12	5	3.200	Ea.	390	156		546	665
5200	Steel door and frame		5	3.200		260	156		416	520
5300	Stainless steel door and frame		5	3.200		475	156		631	755
8000	Valve cabinet for 2-1/2" FD angle valve, 18" x 18" x 8"									
8100	Aluminum door and frame	Q-12	12	1.333	Ea.	135	65		200	247
8200	Steel door and frame		12	1.333		111	65		176	221
8300	Stainless steel door and frame		12	1.333		182	65		247	298

10 44 16 – Fire Extinguishers

10 44 16.13 Portable Fire Extinguishers

		Crew	Daily Output	Labor-Hours	Unit	Material	2012 Bare Costs Labor	Equipment	Total	Total Incl O&P
0010	**PORTABLE FIRE EXTINGUISHERS**									
0140	CO_2, with hose and "H" horn, 10 lb.				Ea.	248			248	273
0160	15 lb.					315			315	350
0180	20 lb.					355			355	395
1000	Dry chemical, pressurized									
1040	Standard type, portable, painted, 2-1/2 lb.				Ea.	34.50			34.50	38
1060	5 lb.					47			47	51.50
1080	10 lb.					74.50			74.50	81.50
1100	20 lb.					126			126	138
1120	30 lb.					385			385	420
1300	Standard type, wheeled, 150 lb.					2,150			2,150	2,350
2000	ABC all purpose type, portable, 2-1/2 lb.					21			21	23.50
2060	5 lb.					26.50			26.50	29
2080	9-1/2 lb.					44			44	48
2100	20 lb.					79.50			79.50	87.50
5000	Pressurized water, 2-1/2 gallon, stainless steel					121			121	133
5060	With anti-freeze					106			106	116
9400	Installation of extinguishers, 12 or more, on nailable surface	1 Carp	30	.267			11.75		11.75	18.10
9420	On masonry or concrete	"	15	.533			23.50		23.50	36

10 44 16.16 Wheeled Fire Extinguisher Units

		Crew	Daily Output	Labor-Hours	Unit	Material	2012 Bare Costs Labor	Equipment	Total	Total Incl O&P
0010	**WHEELED FIRE EXTINGUISHER UNITS**									
0350	CO_2, portable, with swivel horn									
0360	Wheeled type, cart mounted, 50 lb.				Ea.	1,025			1,025	1,125
0400	100 lb.				"	2,625			2,625	2,875
2200	ABC all purpose type									
2300	Wheeled, 45 lb.				Ea.	735			735	810
2360	150 lb.				"	1,850			1,850	2,025

10 51 Lockers

10 51 13 – Metal Lockers

10 51 13.10 Lockers

		Crew	Daily Output	Labor-Hours	Unit	Material	2012 Bare Costs Labor	Equipment	Total	Total Incl O&P
0011	**LOCKERS** steel, baked enamel, pre-assembled									
0110	Single tier box locker, 12" x 15" x 72"	1 Shee	20	.400	Ea.	220	21		241	273
0120	18" x 15" x 72"		20	.400		241	21		262	297
0130	12" x 18" x 72"		20	.400		255	21		276	310
0140	18" x 18" x 72"		20	.400		265	21		286	325
0410	Double tier, 12" x 15" x 36"		30	.267		225	13.90		238.90	268
0420	18" x 15" x 36"		30	.267		235	13.90		248.90	280
0430	12" x 18" x 36"		30	.267		236	13.90		249.90	281
0440	18" x 18" x 36"		30	.267		236	13.90		249.90	281
0500	Two person, 18" x 15" x 72"		20	.400		268	21		289	325

10 51 Lockers

10 51 13 – Metal Lockers

10 51 13.10 Lockers		Crew	Daily Output	Labor-Hours	Unit	Material	2012 Bare Costs Labor	Equipment	Total	Total Incl O&P
0510	18" x 18" x 72"	1 Shee	20	.400	Ea.	279	21		300	335
0520	Duplex, 15" x 15" x 72"		20	.400		268	21		289	325
0530	15" x 21" x 72"		20	.400		315	21		336	380
0600	5 tier box lockers, minimum		30	.267	Opng.	42.50	13.90		56.40	68
0700	Maximum		24	.333		49.50	17.40		66.90	81
0900	6 tier box lockers, minimum		36	.222		35	11.60		46.60	56
1000	Maximum		30	.267		43	13.90		56.90	68
1100	Wire meshed wardrobe, floor. mtd., open front varsity type		7.50	1.067	Ea.	247	55.50		302.50	355
2400	16-person locker unit with clothing rack									
2500	72 wide x 15" deep x 72" high	1 Shee	15	.533	Ea.	445	28		473	535
2550	18" deep	"	15	.533	"	600	28		628	705
3000	Wall mounted lockers, 4 person, with coat bar									
3100	48" wide x 18" deep x 12" high	1 Shee	20	.400	Ea.	305	21		326	365
3250	Rack w/24 wire mesh baskets		1.50	5.333	Set	390	278		668	850
3260	30 baskets		1.25	6.400		365	335		700	910
3270	36 baskets		.95	8.421		460	440		900	1,175
3280	42 baskets		.80	10		435	520		955	1,275
3300	For built-in lock with 2 keys, add				Ea.	13.15			13.15	14.45
3600	For hanger rods, add					1.86			1.86	2.05
3650	For number plate kit, 100 plates #1 - #100, add	1 Shee	4	2		66	104		170	232
3700	For locker base, closed front panel		90	.089		6.90	4.64		11.54	14.60
3710	End panel, bolted		36	.222		8.25	11.60		19.85	26.50
3800	For sloping top, 12" wide		24	.333		28.50	17.40		45.90	58
3810	15" wide		24	.333		31	17.40		48.40	60.50
3820	18" wide		24	.333		28	17.40		45.40	57.50
3850	Sloping top end panel, 12" deep		72	.111		10.95	5.80		16.75	21
3860	15" deep		72	.111		11.10	5.80		16.90	21
3870	18" deep		72	.111		11.60	5.80		17.40	21.50
3900	For finish end panels, steel, 60" high, 15" deep		12	.667		50	35		85	108
3910	72" high, 12" deep		12	.667		50	35		85	108
3920	18" deep		12	.667		60	35		95	119
5000	For 'ready to assemble' lockers,									
5010	Add to labor						75%			
5020	Deduct from material					20%				

10 51 26 – Plastic Lockers

10 51 26.13 Recycled Plastic Lockers

0011	RECYCLED PLASTIC LOCKERS, 30% recycled										
0110	Single tier box locker, 12" x 12" x 72"	G	1 Shee	8	1	Ea.	420	52		472	545
0120	12" x 15" x 72"	G		8	1		420	52		472	540
0130	12" x 18" x 72"	G		8	1		405	52		457	525
0410	Double tier, 12" x 12" x 72"	G		21	.381		435	19.85		454.85	510
0420	12" x 15" x 72"	G		21	.381		430	19.85		449.85	505
0430	12" x 18" x 72"	G		21	.381		420	19.85		439.85	490

10 51 53 – Locker Room Benches

10 51 53.10 Benches

0010	BENCHES									
2100	Locker bench, laminated maple, top only	1 Shee	100	.080	L.F.	15.40	4.17		19.57	23.50
2200	Pedestals, steel pipe		25	.320	Ea.	41	16.70		57.70	70.50
2250	Plastic, 9.5" top with PVC pedestals		80	.100	L.F.	58	5.20		63.20	72

10 55 Postal Specialties

10 55 23 – Mail Boxes

10 55 23.10 Commercial Mail Boxes	Crew	Daily Output	Labor-Hours	Unit	Material	2012 Bare Costs Labor	2012 Bare Costs Equipment	Total	Total Incl O&P
0010 **COMMERCIAL MAIL BOXES**									
0020 Horiz., key lock, 5"H x 6"W x 15"D, alum., rear load	1 Carp	34	.235	Ea.	35	10.40		45.40	54.50
0100 Front loading		34	.235		35	10.40		45.40	54.50
0200 Double, 5"H x 12"W x 15"D, rear loading		26	.308		67.50	13.55		81.05	95.50
0300 Front loading		26	.308		67	13.55		80.55	94.50
0500 Quadruple, 10"H x 12"W x 15"D, rear loading		20	.400		81.50	17.65		99.15	117
0600 Front loading		20	.400		95.50	17.65		113.15	132
0800 Vertical, front load, 15"H x 5"W x 6"D, alum., per compartment		34	.235		30	10.40		40.40	49
0900 Bronze, duranodic finish		34	.235		40	10.40		50.40	60
1000 Steel, enameled		34	.235		30	10.40		40.40	49
1700 Alphabetical directories, 120 names		10	.800		130	35.50		165.50	198
1800 Letter collection box		6	1.333		660	59		719	820
1830 Lobby collection boxes, aluminum	2 Shee	5	3.200		2,025	167		2,192	2,475
1840 Bronze or stainless	"	4.50	3.556		2,500	185		2,685	3,025
1900 Letter slot, residential	1 Carp	20	.400		89.50	17.65		107.15	126
2000 Post office type		8	1		105	44		149	184
2200 Post office counter window, with grille		2	4		550	176		726	875
2250 Key keeper, single key, aluminum		26	.308		37	13.55		50.55	62
2300 Steel, enameled		26	.308		107	13.55		120.55	139

10 56 Storage Assemblies

10 56 13 – Metal Storage Shelving

10 56 13.10 Shelving

10 56 13.10 Shelving	Crew	Daily Output	Labor-Hours	Unit	Material	2012 Bare Costs Labor	2012 Bare Costs Equipment	Total	Total Incl O&P
0010 **SHELVING**									
0020 Metal, industrial, cross-braced, 3' wide, 12" deep	1 Sswk	175	.046	SF Shlf	7.15	2.26		9.41	11.95
0100 24" deep		330	.024		4.61	1.20		5.81	7.20
0300 4' wide, 12" deep		185	.043		6.20	2.13		8.33	10.60
0400 24" deep		380	.021		4.46	1.04		5.50	6.75
1200 Enclosed sides, cross-braced back, 3' wide, 12" deep		175	.046		11.25	2.26		13.51	16.40
1300 24" deep		290	.028		8.25	1.36		9.61	11.55
1500 Fully enclosed, sides and back, 3' wide, 12" deep		150	.053		13.70	2.63		16.33	19.75
1600 24" deep		255	.031		9.20	1.55		10.75	12.85
1800 4' wide, 12" deep		150	.053		9.25	2.63		11.88	14.90
1900 24" deep		290	.028		7	1.36		8.36	10.15
2200 Wide span, 1600 lb. capacity per shelf, 6' wide, 24" deep		380	.021		6.30	1.04		7.34	8.80
2400 36" deep		440	.018		6.40	.90		7.30	8.65
2600 8' wide, 24" deep		440	.018		6.85	.90		7.75	9.10
2800 36" deep		520	.015		5.55	.76		6.31	7.45
4000 Pallet racks, steel frame 5,000 lb. capacity, 8' long, 36" deep	2 Sswk	450	.036		8.70	1.75		10.45	12.70
4200 42" deep		500	.032		7.85	1.58		9.43	11.45
4400 48" deep		520	.031		7.05	1.52		8.57	10.45

10 56 13.20 Parts Bins

10 56 13.20 Parts Bins	Crew	Daily Output	Labor-Hours	Unit	Material	2012 Bare Costs Labor	2012 Bare Costs Equipment	Total	Total Incl O&P
0010 **PARTS BINS** metal, gray baked enamel finish									
0100 6'-3" high, 3' wide									
0300 12 bins, 18" wide x 12" high, 12" deep	2 Clab	10	1.600	Ea.	315	56		371	435
0400 24" deep		10	1.600		470	56		526	600
0600 72 bins, 6" wide x 6" high, 12" deep		8	2		560	70		630	725
0700 18" deep		8	2		790	70		860	980
1000 7'-3" high, 3' wide									
1200 14 bins, 18" wide x 12" high, 12" deep	2 Clab	10	1.600	Ea.	340	56		396	460
1300 24" deep		10	1.600		450	56		506	580

10 56 Storage Assemblies

10 56 13 – Metal Storage Shelving

10 56 13.20 Parts Bins		Crew	Daily Output	Labor-Hours	Unit	Material	2012 Bare Costs Labor	Equipment	Total	Total Incl O&P
1500	84 bins, 6" wide x 6" high, 12" deep	2 Clab	8	2	Ea.	775	70		845	960
1600	24" deep	↓	8	2	↓	980	70		1,050	1,175

10 57 Wardrobe and Closet Specialties

10 57 13 – Hat and Coat Racks

10 57 13.10 Coat Racks and Wardrobes

		Crew	Daily Output	Labor-Hours	Unit	Material	2012 Bare Costs Labor	Equipment	Total	Total Incl O&P
0010	**COAT RACKS AND WARDROBES**									
0020	Hat & coat rack, floor model, 6 hangers									
0050	Standing, beech wood, 21" x 21" x 72", chrome				Ea.	237			237	260
0100	18 gauge tubular steel, 21" x 21" x 69", wood walnut				"	289			289	320
0500	16 gauge steel frame, 22 gauge steel shelves									
0650	Single pedestal, 30" x 18" x 63"				Ea.	257			257	283
0800	Single face rack, 29" x 18-1/2" x 62"					281			281	310
0900	51" x 18-1/2" x 70"					395			395	435
0910	Double face rack, 39" x 26" x 70"					350			350	385
0920	63" x 26" x 70"				↓	425			425	470
0940	For 2" ball casters, add				Set	88			88	97
1400	Utility hook strips, 3/8" x 2-1/2" x 18", 6 hooks	1 Carp	48	.167	Ea.	58	7.35		65.35	75
1500	34" long, 12 hooks	"	48	.167	"	55.50	7.35		62.85	72.50
1650	Wall mounted racks, 16 gauge steel frame, 22 gauge steel shelves									
1850	12" x 15" x 26", 6 hangers	1 Carp	32	.250	Ea.	155	11.05		166.05	187
2000	12" x 15" x 50", 12 hangers	"	32	.250	"	181	11.05		192.05	216
2150	Wardrobe cabinet, steel, baked enamel finish									
2300	36" x 21" x 78", incl. top shelf & hanger rod				Ea.	282			282	310

10 57 23 – Closet and Utility Shelving

10 57 23.19 Wood Closet and Utility Shelving

		Crew	Daily Output	Labor-Hours	Unit	Material	2012 Bare Costs Labor	Equipment	Total	Total Incl O&P
0010	**WOOD CLOSET AND UTILITY SHELVING**									
0020	Pine, clear grade, no edge band, 1" x 8"	1 Carp	115	.070	L.F.	3.01	3.07		6.08	8.05
0100	1" x 10"		110	.073		3.75	3.21		6.96	9.05
0200	1" x 12"		105	.076		4.52	3.36		7.88	10.10
0600	Plywood, 3/4" thick with lumber edge, 12" wide		75	.107		1.55	4.70		6.25	8.95
0700	24" wide		70	.114	↓	2.72	5.05		7.77	10.75
0900	Bookcase, clear grade pine, shelves 12" O.C., 8" deep, per S.F. shelf		70	.114	S.F.	9.80	5.05		14.85	18.50
1000	12" deep shelves		65	.123	"	14.70	5.45		20.15	24.50
1200	Adjustable closet rod and shelf, 12" wide, 3' long		20	.400	Ea.	10.45	17.65		28.10	38.50
1300	8' long		15	.533	"	19.35	23.50		42.85	57.50
1500	Prefinished shelves with supports, stock, 8" wide		75	.107	L.F.	3.17	4.70		7.87	10.75
1600	10" wide	↓	70	.114	"	4.32	5.05		9.37	12.50

10 71 Exterior Protection

10 71 13 – Exterior Sun Control Devices

10 71 13.19 Rolling Exterior Shutters	Crew	Daily Output	Labor-Hours	Unit	Material	2012 Bare Costs Labor	2012 Bare Costs Equipment	Total	Total Incl O&P
0010 **ROLLING EXTERIOR SHUTTERS**									
0020 Roll-up, manual operation, aluminum, 3' x 4', incl. frame	2 Carp	8	2	Ea.	765	88		853	980
0030 6' x 7'	"	8	2	"	1,600	88		1,688	1,875

10 73 Protective Covers

10 73 13 – Awnings

10 73 13.10 Awnings, Fabric

10 73 13.10 Awnings, Fabric	Crew	Daily Output	Labor-Hours	Unit	Material	2012 Bare Costs Labor	2012 Bare Costs Equipment	Total	Total Incl O&P
0010 **AWNINGS, FABRIC**									
0020 Including acrylic canvas and frame, standard design									
0100 Door and window, slope, 3' high, 4' wide	1 Carp	4.50	1.778	Ea.	705	78.50		783.50	895
0110 6' wide		3.50	2.286		910	101		1,011	1,150
0120 8' wide		3	2.667		1,100	118		1,218	1,400
0200 Quarter round convex, 4' wide		3	2.667		1,100	118		1,218	1,375
0210 6' wide		2.25	3.556		1,425	157		1,582	1,825
0220 8' wide		1.80	4.444		1,750	196		1,946	2,225
0300 Dome, 4' wide		7.50	1.067		425	47		472	540
0310 6' wide		3.50	2.286		950	101		1,051	1,200
0320 8' wide		2	4		1,700	176		1,876	2,125
0350 Elongated dome, 4' wide		1.33	6.015		1,575	265		1,840	2,150
0360 6' wide		1.11	7.207		1,900	320		2,220	2,600
0370 8' wide		1	8		2,225	355		2,580	3,000
1000 Entry or walkway, peak, 12' long, 4' wide	2 Carp	.90	17.778		5,025	785		5,810	6,725
1010 6' wide		.60	26.667		7,725	1,175		8,900	10,300
1020 8' wide		.40	40		10,700	1,775		12,475	14,400
1100 Radius with dome end, 4' wide		1.10	14.545		3,800	640		4,440	5,150
1110 6' wide		.70	22.857		6,125	1,000		7,125	8,275
1120 8' wide		.50	32		8,700	1,400		10,100	11,800
2000 Retractable lateral arm awning, manual									
2010 To 12' wide, 8' - 6" projection	2 Carp	1.70	9.412	Ea.	1,125	415		1,540	1,900
2020 To 14' wide, 8' - 6" projection		1.10	14.545		1,325	640		1,965	2,425
2030 To 19' wide, 8' - 6" projection		.85	18.824		1,800	830		2,630	3,250
2040 To 24' wide, 8' - 6" projection		.67	23.881		2,275	1,050		3,325	4,125
2050 Motor for above, add	1 Carp	2.67	3		970	132		1,102	1,275
3000 Patio/deck canopy with frame									
3010 12' wide, 12' projection	2 Carp	2	8	Ea.	1,600	355		1,955	2,300
3020 16' wide, 14' projection	"	1.20	13.333		2,500	590		3,090	3,650
9000 For fire retardant canvas, add					7%				
9010 For lettering or graphics, add					35%				
9020 For painted or coated acrylic canvas, deduct					8%				
9030 For translucent or opaque vinyl canvas, add					10%				
9040 For 6 or more units, deduct					20%	15%			

10 73 16 – Canopies

10 73 16.20 Metal Canopies

10 73 16.20 Metal Canopies	Crew	Daily Output	Labor-Hours	Unit	Material	2012 Bare Costs Labor	2012 Bare Costs Equipment	Total	Total Incl O&P
0010 **METAL CANOPIES**									
0020 Wall hung, .032", aluminum, prefinished, 8' x 10'	K-2	1.30	18.462	Ea.	2,150	830	203	3,183	4,000
0300 8' x 20'		1.10	21.818		4,275	980	240	5,495	6,675
0500 10' x 10'		1.30	18.462		2,400	830	203	3,433	4,275
0700 10' x 20'		1.10	21.818		4,450	980	240	5,670	6,875
1000 12' x 20'		1	24		5,475	1,075	264	6,814	8,175
1360 12' x 30'		.80	30		8,050	1,350	330	9,730	11,500

10 73 Protective Covers

10 73 16 – Canopies

10 73 16.20 Metal Canopies

10 73 16.20 Metal Canopies		Crew	Daily Output	Labor-Hours	Unit	Material	2012 Bare Costs Labor	Equipment	Total	Total Incl O&P
1700	12' x 40'	K-2	.60	40	Ea.	9,800	1,800	440	12,040	14,400
1900	For free standing units, add				↓	20%	10%			
2300	Aluminum entrance canopies, flat soffit, .032"									
2500	3'-6" x 4'-0", clear anodized	2 Carp	4	4	Ea.	900	176		1,076	1,250
2700	Bronze anodized		4	4		1,575	176		1,751	2,025
3000	Polyurethane painted		4	4		1,275	176		1,451	1,675
3300	4'-6" x 10'-0", clear anodized		2	8		2,475	355		2,830	3,275
3500	Bronze anodized		2	8		3,175	355		3,530	4,025
3700	Polyurethane painted	↓	2	8		2,650	355		3,005	3,475
4000	Wall downspout, 10 L.F., clear anodized	1 Carp	7	1.143		154	50.50		204.50	248
4300	Bronze anodized		7	1.143		269	50.50		319.50	375
4500	Polyurethane painted	↓	7	1.143	↓	231	50.50		281.50	330
7000	Carport, baked vinyl finish, .032", 20' x 10', no foundations, min.	K-2	4	6	Car	3,850	270	66	4,186	4,775
7250	Maximum		2	12	"	7,000	540	132	7,672	8,775
7500	Walkway cover, to 12' wide, stl., vinyl finish, .032",no fndtns., min.		250	.096	S.F.	22.50	4.32	1.06	27.88	33.50
7750	Maximum	↓	200	.120	"	24.50	5.40	1.32	31.22	38

10 74 Manufactured Exterior Specialties

10 74 23 – Cupolas

10 74 23.10 Wood Cupolas

		Crew	Daily Output	Labor-Hours	Unit	Material	Labor	Equipment	Total	Total Incl O&P
0010	**WOOD CUPOLAS**									
0020	Stock units, pine, painted, 18" sq., 28" high, alum. roof	1 Carp	4.10	1.951	Ea.	167	86		253	315
0100	Copper roof		3.80	2.105		233	93		326	400
0300	23" square, 33" high, aluminum roof		3.70	2.162		385	95.50		480.50	570
0400	Copper roof		3.30	2.424		480	107		587	695
0600	30" square, 37" high, aluminum roof		3.70	2.162		510	95.50		605.50	710
0700	Copper roof		3.30	2.424		610	107		717	835
0900	Hexagonal, 31" wide, 46" high, copper roof		4	2		755	88		843	965
1000	36" wide, 50" high, copper roof	↓	3.50	2.286		1,375	101		1,476	1,675
1200	For deluxe stock units, add to above					25%				
1400	For custom built units, add to above				↓	50%	50%			

10 74 29 – Steeples

10 74 29.10 Prefabricated Steeples

		Crew	Daily Output	Labor-Hours	Unit	Material	Labor	Equipment	Total	Total Incl O&P
0010	**PREFABRICATED STEEPLES**									
4000	Steeples, translucent fiberglass, 30" square, 15' high	F-3	2	20	Ea.	6,700	895	305	7,900	9,075
4150	25' high		1.80	22.222		7,825	995	340	9,160	10,500
4350	Opaque fiberglass, 24" square, 14' high		2	20		5,575	895	305	6,775	7,850
4500	28' high	↓	1.80	22.222		5,450	995	340	6,785	7,900
4600	Aluminum, baked finish, 16" square, 14' high					4,850			4,850	5,325
4620	20' high, 3'-6" base					8,125			8,125	8,950
4640	35' high, 8' base					31,400			31,400	34,500
4660	60' high, 14' base					67,000			67,000	73,500
4680	152' high, custom					553,000			553,000	608,500
4700	Porcelain enamel steeples, custom, 40' high	F-3	.50	80		14,100	3,575	1,225	18,900	22,500
4800	60' high	"	.30	133	↓	24,500	5,975	2,050	32,525	38,400

10 74 46 – Window Wells

10 74 46.10 Area Window Wells

		Crew	Daily Output	Labor-Hours	Unit	Material	Labor	Equipment	Total	Total Incl O&P
0010	**AREA WINDOW WELLS**, Galvanized steel									
0020	20 ga., 3'-2" wide, 1' deep	1 Sswk	29	.276	Ea.	14.45	13.60		28.05	40.50
0100	2' deep	↓	23	.348	↓	24	17.15		41.15	56.50

10 74 Manufactured Exterior Specialties

10 74 46 – Window Wells

10 74 46.10 Area Window Wells	Crew	Daily Output	Labor-Hours	Unit	Material	2012 Bare Costs Labor	Equipment	Total	Total Incl O&P
0300	16 ga., 3'-2" wide, 1' deep	1 Sswk	29	.276	Ea.	19.85	13.60		33.45	46.50
0400	3' deep		23	.348		37	17.15		54.15	71.50
0600	Welded grating for above, 15 lb., painted		45	.178		85	8.75		93.75	109
0700	Galvanized		45	.178		116	8.75		124.75	143
0900	Translucent plastic cap for above		60	.133		19	6.60		25.60	33

10 75 Flagpoles

10 75 16 – Ground-Set Flagpoles

10 75 16.10 Flagpoles	Crew	Daily Output	Labor-Hours	Unit	Material	2012 Bare Costs Labor	Equipment	Total	Total Incl O&P
0010	**FLAGPOLES**, ground set									
0050	Not including base or foundation									
0100	Aluminum, tapered, ground set 20' high	K-1	2	8	Ea.	880	315	132	1,327	1,600
0200	25' high		1.70	9.412		1,275	370	156	1,801	2,150
0300	30' high		1.50	10.667		1,250	420	176	1,846	2,200
0400	35' high		1.40	11.429		1,650	450	189	2,289	2,725
0500	40' high		1.20	13.333		2,350	525	220	3,095	3,625
0600	50' high		1	16		2,850	630	264	3,744	4,375
0700	60' high		.90	17.778		5,050	695	294	6,039	6,950
0800	70' high		.80	20		7,650	785	330	8,765	9,975
1100	Counterbalanced, internal halyard, 20' high		1.80	8.889		2,350	350	147	2,847	3,275
1200	30' high		1.50	10.667		2,600	420	176	3,196	3,675
1300	40' high		1.30	12.308		5,675	485	203	6,363	7,225
1400	50' high		1	16		7,900	630	264	8,794	9,925
2820	Aluminum, electronically operated, 30' high		1.40	11.429		4,375	450	189	5,014	5,725
2840	35' high		1.30	12.308		4,725	485	203	5,413	6,175
2860	39' high		1.10	14.545		5,150	570	240	5,960	6,800
2880	45' high		1	16		6,575	630	264	7,469	8,500
2900	50' high		.90	17.778		7,000	695	294	7,989	9,100
3000	Fiberglass, tapered, ground set, 23' high		2	8		695	315	132	1,142	1,400
3100	29'-7" high		1.50	10.667		1,425	420	176	2,021	2,400
3200	36'-1" high		1.40	11.429		1,975	450	189	2,614	3,075
3300	39'-5" high		1.20	13.333		2,225	525	220	2,970	3,500
3400	49'-2" high		1	16		3,975	630	264	4,869	5,625
3500	59' high		.90	17.778		4,800	695	294	5,789	6,675
4300	Steel, direct imbedded installation									
4400	Internal halyard, 20' high	K-1	2.50	6.400	Ea.	1,175	251	106	1,532	1,800
4500	25' high		2.50	6.400		1,775	251	106	2,132	2,450
4600	30' high		2.30	6.957		2,125	273	115	2,513	2,875
4700	40' high		2.10	7.619		2,675	299	126	3,100	3,525
4800	50' high		1.90	8.421		3,500	330	139	3,969	4,500
5000	60' high		1.80	8.889		6,250	350	147	6,747	7,575
5100	70' high		1.60	10		6,800	390	165	7,355	8,250
5200	80' high		1.40	11.429		8,750	450	189	9,389	10,500
5300	90' high		1.20	13.333		15,000	525	220	15,745	17,500
5500	100' high		1	16		15,500	630	264	16,394	18,400
6400	Wood poles, tapered, clear vertical grain fir with tilting									
6410	base, not incl. foundation, 4" butt, 25' high	K-1	1.90	8.421	Ea.	1,375	330	139	1,844	2,175
6800	6" butt, 30' high	"	1.30	12.308	"	2,250	485	203	2,938	3,450
7300	Foundations for flagpoles, including									
7400	excavation and concrete, to 35' high poles	C-1	10	3.200	Ea.	650	134		784	920
7600	40' to 50' high		3.50	9.143		1,200	385		1,585	1,925

372

10 75 Flagpoles

10 75 16 – Ground-Set Flagpoles

10 75 16.10 Flagpoles		Crew	Daily Output	Labor-Hours	Unit	Material	2012 Bare Costs Labor	Equipment	Total	Total Incl O&P
7700	Over 60' high	C-1	2	16	Ea.	1,500	670		2,170	2,675

10 75 23 – Wall-Mounted Flagpoles

10 75 23.10 Flagpoles										
0010	**FLAGPOLES**, structure mounted									
0100	Fiberglass, vertical wall set, 19'-8" long	K-1	1.50	10.667	Ea.	1,300	420	176	1,896	2,250
0200	23' long		1.40	11.429		1,375	450	189	2,014	2,400
0300	26'-3" long		1.30	12.308		1,775	485	203	2,463	2,925
0800	19'-8" long outrigger		1.30	12.308		1,225	485	203	1,913	2,325
1300	Aluminum, vertical wall set, tapered, with base, 20' high		1.20	13.333		975	525	220	1,720	2,125
1400	29'-6" high		1	16		2,250	630	264	3,144	3,725
2400	Outrigger poles with base, 12' long		1.30	12.308		930	485	203	1,618	2,000
2500	14' long	▼	1	16	▼	1,425	630	264	2,319	2,825

10 81 Pest Control Devices

10 81 13 – Bird Control Devices

10 81 13.10 Bird Control Netting										
0010	**BIRD CONTROL NETTING**									
0020	1/8" square mesh	4 Clab	4000	.008	S.F.	.54	.28		.82	1.02
0100	1/4" square mesh		4000	.008		.75	.28		1.03	1.26
0120	1/2" square mesh		4000	.008		.11	.28		.39	.55
0140	5/8" x 3/4" mesh		4000	.008		.07	.28		.35	.51
0160	1-1/4" x 1-1/2" mesh		4000	.008		.13	.28		.41	.57
0200	4" square mesh	▼	4000	.008	▼	.11	.28		.39	.55
1000	Poly clips				Ea.	.02			.02	.02

10 88 Scales

10 88 05 – Commercial Scales

10 88 05.10 Scales										
0010	**SCALES**									
0700	Truck scales, incl. steel weigh bridge,									
0800	not including foundation, pits									
1550	Digital, electronic, 100 ton capacity, steel deck 12' x 10' platform	3 Carp	.20	120	Ea.	12,800	5,300		18,100	22,300
1600	40' x 10' platform		.14	171		26,100	7,550		33,650	40,300
1640	60' x 10' platform		.13	184		32,300	8,150		40,450	48,000
1680	70' x 10' platform	▼	.12	200		36,100	8,825		44,925	53,500
2000	For standard automatic printing device, add					2,600			2,600	2,850
2100	For remote reading electronic system, add					2,350			2,350	2,575
2300	Concrete foundation pits for above, 8' x 6', 5 C.Y. required	C-1	.50	64		980	2,675		3,655	5,200
2400	14' x 6' platform, 10 C.Y. required		.35	91.429		1,450	3,825		5,275	7,450
2600	50' x 10' platform, 30 C.Y. required		.25	128		1,925	5,350		7,275	10,400
2700	70' x 10' platform, 40 C.Y. required	▼	.15	213		4,250	8,925		13,175	18,400
2750	Crane scales, dial, 1 ton capacity					1,050			1,050	1,150
2780	5 ton capacity					1,450			1,450	1,600
2800	Digital, 1 ton capacity					1,800			1,800	1,975
2850	10 ton capacity				▼	4,625			4,625	5,100
2900	Low profile electronic warehouse scale,									
3000	not incl. printer, 4' x 4' platform, 10,000 lb. capacity	2 Carp	.30	53.333	Ea.	1,725	2,350		4,075	5,500
3300	5' x 7' platform, 10,000 lb. capacity		.25	64		4,250	2,825		7,075	9,025
3400	20,000 lb. capacity	▼	.20	80	▼	5,475	3,525		9,000	11,500

10 88 Scales

10 88 05 – Commercial Scales

10 88 05.10 Scales	Crew	Daily Output	Labor-Hours	Unit	Material	2012 Bare Costs Labor	Equipment	Total	Total Incl O&P
3500	For printers, incl. time, date & numbering, add				Ea.	1,350			1,350	1,500
3800	Portable, beam type, capacity 1000#, platform 18" x 24"					770			770	845
3900	Dial type, capacity 2000#, platform 24" x 24"					1,375			1,375	1,500
4000	Digital type, capacity 1000#, platform 24" x 30"					2,175			2,175	2,400
4100	Portable contractor truck scales, 50 ton cap., 40' x 10' platform					33,000			33,000	36,300
4200	60' x 10' platform					30,400			30,400	33,400

Estimating Tips

General

- The items in this division are usually priced per square foot or each. Many of these items are purchased by the owner for installation by the contractor. Check the specifications for responsibilities and include time for receiving, storage, installation, and mechanical and electrical hookups in the appropriate divisions.

- Many items in Division 11 require some type of support system that is not usually furnished with the item. Examples of these systems include blocking for the attachment of casework and support angles for ceiling-hung projection screens. The required blocking or supports must be added to the estimate in the appropriate division.

- Some items in Division 11 may require assembly or electrical hookups. Verify the amount of assembly required or the need for a hard electrical connection and add the appropriate costs.

Reference Numbers

Reference numbers are shown in shaded boxes at the beginning of some major classifications. These numbers refer to related items in the Reference Section. The reference information may be an estimating procedure, an alternate pricing method, or technical information.

Note: Not all subdivisions listed here necessarily appear in this publication.

11 05 05.10 Selective Demolition		Crew	Daily Output	Labor-Hours	Unit	Material	2012 Bare Costs Labor	2012 Bare Costs Equipment	Total	Total Incl O&P
0010	**SELECTIVE DEMOLITION**									
0130	Central vacuum, motor unit, residential or commercial	1 Clab	2	4	Ea.		140		140	216
0210	Vault door and frame	2 Skwk	2	8			365		365	560
0215	Day gate, for vault	"	3	5.333			242		242	375
0380	Bank equipment, teller window, bullet resistant	1 Clab	1.20	6.667	↓		234		234	360
0381	Counter	2 Clab	1.50	10.667	Station		375		375	575
0382	Drive-up window, including drawer and glass		1.50	10.667	"		375		375	575
0383	Thru-wall boxes and chests, selective demolition		2.50	6.400	Ea.		225		225	345
0384	Bullet resistant partitions	↓	20	.800	L.F.		28		28	43
0385	Pneumatic tube system, 2 lane drive-up	L-3	.45	35.556	Ea.		1,700		1,700	2,600
0386	Safety deposit box	1 Clab	50	.160	Opng.		5.60		5.60	8.65
0387	Surveillance system, video, complete	2 Elec	2	8	Ea.		410		410	615
0410	Church equipment, misc moveable fixtures	2 Clab	1	16			560		560	865
0412	Steeple, to 28' high	F-3	3	13.333			600	204	804	1,150
0414	40' to 60' high	"	.80	50	↓		2,250	765	3,015	4,275
0510	Library equipment, bookshelves, wood, to 90" high	1 Clab	20	.400	L.F.		14.05		14.05	21.50
0515	Carrels, hardwood, 36" x 24"	"	9	.889	Ea.		31		31	48
0630	Stage equipment, light control panel	1 Elec	1	8	"		410		410	615
0632	Border lights		40	.200	L.F.		10.30		10.30	15.40
0634	Spotlights	↓	8	1	Ea.		51.50		51.50	77
0636	Telescoping platforms and risers	2 Clab	175	.091	SF Stg.		3.21		3.21	4.94
1020	Barber equipment, hydraulic chair	1 Clab	40	.200	Ea.		7		7	10.80
1030	Checkout counter, supermarket or warehouse conveyor	2 Clab	18	.889			31		31	48
1040	Food cases, refrigerated or frozen	Q-5	6	2.667			134		134	202
1190	Laundry equipment, commercial	L-6	3	4			214		214	320
1360	Movie equipment, lamphouse, to 4000 watt, incl rectifier	1 Elec	4	2			103		103	154
1365	Sound system, incl amplifier	"	1.25	6.400			330		330	495
1410	Air compressor, to 5 H.P.	2 Clab	2.50	6.400	↓		225		225	345
1412	Lubrication equipment, automotive, 3 reel type, incl pump, excl piping	L-4	1	24	Set		995		995	1,525
1414	Booth, spray paint, complete, to 26' long	"	.80	30	"		1,250		1,250	1,925
1560	Parking equipment, cashier booth	B-22	2	15	Ea.		615	99.50	714.50	1,050
1600	Loading dock equipment, dock bumpers, rubber	1 Clab	50	.160	"		5.60		5.60	8.65
1610	Door seal for door perimeter	"	50	.160	L.F.		5.60		5.60	8.65
1620	Platform lifter, fixed, 6' x 8', 5000 lb. capacity	E-16	1.50	10.667	Ea.		535	81	616	1,050
1630	Dock leveller	"	2	8			405	61	466	785
1640	Lights, single or double arm	1 Elec	8	1			51.50		51.50	77
1650	Shelter, fabric, truck or train	1 Clab	1.50	5.333			187		187	288
1790	Waste handling equipment, commercial compactor	L-4	2	12			495		495	765
1792	Commercial or municipal incinerator, gas	"	2	12			495		495	765
1795	Crematory, excluding building	Q-3	.25	128			6,650		6,650	10,000
1910	Detection equipment, cell bar front	E-4	4	8			400	30.50	430.50	750
1912	Cell door and frame		8	4			199	15.20	214.20	370
1914	Prefab cell, 4' to 5' wide, 7' to 8' high, 7' deep		8	4			199	15.20	214.20	370
1916	Cot, bolted, single		40	.800			40	3.04	43.04	75
1918	Visitor cubicle	↓	4	8			400	30.50	430.50	750
2850	Hydraulic gates, canal, flap, knife, slide or sluice, to 18" diameter	L-5A	8	4			198	70.50	268.50	420
2852	19" to 36" diameter		6	5.333			264	94	358	560
2854	37" to 48" diameter		2	16			790	283	1,073	1,650
2856	49" to 60" diameter		1	32			1,575	565	2,140	3,350
2858	Over 60" diameter	↓	.30	106			5,275	1,875	7,150	11,200
3100	Sewage pumping system, prefabricated, to 1000 GPM	C-17D	.20	420			19,300	3,875	23,175	34,100
3110	Sewage treatment, holding tank for recirc chemical water closet	1 Plum	8	1			54.50		54.50	82.50
3900	Wastewater treatment system, to 1500 gallons	B-21	2	14	↓		565	66	631	945

11 05 05 – Selective Equipment Demolition

11 05 05.10 Selective Demolition	Crew	Daily Output	Labor-Hours	Unit	Material	2012 Bare Costs Labor	2012 Bare Costs Equipment	Total	Total Incl O&P
4050	Food storage equipment, walk-in refrigerator/freezer	2 Clab	64	.250	S.F.		8.80		8.80	13.50
4052	Shelving, stainless steel, 4 tier or dunnage rack	1 Clab	12	.667	Ea.		23.50		23.50	36
4100	Food preparation equipment, small countertop		18	.444			15.60		15.60	24
4150	Food delivery carts, heated cabinets		18	.444			15.60		15.60	24
4200	Cooking equipment, commercial range	Q-1	12	1.333			65.50		65.50	99
4250	Hood and ventilation equipment, kitchen exhaust hood, excl fire prot	1 Clab	3	2.667			93.50		93.50	144
4255	Fire protection system	Q-1	3	5.333			262		262	395
4300	Food dispensing equipment, countertop items	1 Clab	15	.533			18.70		18.70	29
4310	Serving counter	"	65	.123	L.F.		4.32		4.32	6.65
4350	Ice machine, ice cube maker, flakers and storage bins, to 2000 lb./day	Q-1	1.60	10	Ea.		490		490	740
4400	Cleaning and disposal, commercial dishwasher, to 50 racks per hour	L-6	1	12			645		645	965
4405	To 275 racks per hour	L-4	1	24			995		995	1,525
4410	Dishwasher hood	2 Clab	5	3.200			112		112	173
4420	Garbage disposal, commercial, to 5 H.P.	L-1	8	2			106		106	159
4540	Water heater, residential, to 80 gal/day	"	5	3.200			170		170	255
4542	Water softener, automatic	2 Plum	10	1.600			87.50		87.50	132
4544	Disappearing stairway, to 15' floor height	2 Clab	6	2.667			93.50		93.50	144
4710	Darkroom equipment, light	L-7	10	2.800			119		119	183
4712	Heavy	"	1.50	18.667			795		795	1,225
4720	Doors	2 Clab	3.50	4.571	Opng.		160		160	247
4830	Bowling alley, complete, incl pinsetter, scorer, counters, misc supplies	4 Clab	.40	80	Lane		2,800		2,800	4,325
4840	Health club equipment, circuit training apparatus	2 Clab	2	8	Set		281		281	430
4842	Squat racks	"	10	1.600	Ea.		56		56	86.50
4860	School equipment, basketball backstop	L-2	2	8			310		310	475
4862	Table and benches, folding, in wall, 14' long	L-4	4	6			248		248	385
4864	Bleachers, telescoping, to 30 tier	F-5	120	.267	Seat		11.90		11.90	18.30
4866	Boxing ring, elevated	L-4	.20	120	Ea.		4,975		4,975	7,675
4867	Boxing ring, floor level	"	2	12			495		495	765
4868	Exercise equipment	1 Clab	6	1.333			47		47	72
4870	Gym divider	L-4	1000	.024	S.F.		.99		.99	1.53
4875	Scoreboard	R-3	2	10	Ea.		510	66	576	840
4880	Shooting range, incl bullet traps, targets, excl structure	L-9	1	36	Point		1,450		1,450	2,325
5200	Vocational shop equipment	2 Clab	8	2	Ea.		70		70	108
6200	Fume hood, incl countertop, excl HVAC	"	6	2.667	L.F.		93.50		93.50	144
7100	Medical sterilizing, distiller, water, steam heated, 50 gal. capacity	1 Plum	2.80	2.857	Ea.		156		156	235
7200	Medical equipment, surgery table, minor	1 Clab	1	8			281		281	430
7210	Surgical lights, doctors office, single or double arm	2 Elec	3	5.333			275		275	410
7300	Physical therapy, table	2 Clab	4	4			140		140	216
7310	Whirlpool bath, fixed, incl mixing valves	1 Plum	4	2			109		109	165
7400	Dental equipment, chair, electric or hydraulic	1 Clab	.75	10.667			375		375	575
7410	Central suction system	1 Plum	2	4			218		218	330
7420	Drill console with accessories	1 Clab	3.20	2.500			88		88	135
7430	X-ray unit	"	4	2			70		70	108
7440	X-ray developer	1 Plum	10	.800			43.50		43.50	66

11 05 10 – Equipment Installation

11 05 10.10 Industrial Equipment Installation

		Crew	Daily Output	Labor-Hours	Unit	Material	Labor	Equipment	Total	Total Incl O&P
0010	**INDUSTRIAL EQUIPMENT INSTALLATION**									
0020	Industrial equipment, minimum	E-2	12	4.667	Ton		225	125	350	520
0200	Maximum	"	2	28	"		1,350	750	2,100	3,150

11 11 Vehicle Service Equipment

11 11 13 – Compressed-Air Vehicle Service Equipment

11 11 13.10 Compressed Air Equipment

11 11 13.10 Compressed Air Equipment	Crew	Daily Output	Labor-Hours	Unit	Material	2012 Bare Costs Labor	Equipment	Total	Total Incl O&P
0010 **COMPRESSED AIR EQUIPMENT**									
0030 Compressors, electric, 1-1/2 H.P., standard controls	L-4	1.50	16	Ea.	420	660		1,080	1,475
0550 Dual controls		1.50	16		765	660		1,425	1,875
0600 5 H.P., 115/230 volt, standard controls		1	24		2,025	995		3,020	3,750
0650 Dual controls	↓	1	24	↓	2,575	995		3,570	4,350

11 11 19 – Vehicle Lubrication Equipment

11 11 19.10 Lubrication Equipment

	Crew	Daily Output	Labor-Hours	Unit	Material	Labor	Equipment	Total	Total Incl O&P
0010 **LUBRICATION EQUIPMENT**									
3000 Lube equipment, 3 reel type, with pumps, not including piping	L-4	.50	48	Set	8,700	1,975		10,675	12,600

11 11 33 – Vehicle Spray Painting Equipment

11 11 33.10 Spray Painting Equipment

	Crew	Daily Output	Labor-Hours	Unit	Material	Labor	Equipment	Total	Total Incl O&P
0010 **SPRAY PAINTING EQUIPMENT**									
4000 Spray painting booth, 26' long, complete	L-4	.40	60	Ea.	17,600	2,475		20,075	23,200

11 12 Parking Control Equipment

11 12 13 – Parking Key and Card Control Units

11 12 13.10 Parking Control Units

	Crew	Daily Output	Labor-Hours	Unit	Material	Labor	Equipment	Total	Total Incl O&P
0010 **PARKING CONTROL UNITS**									
5100 Card reader	1 Elec	2	4	Ea.	1,900	206		2,106	2,400
5120 Proximity with customer display	2 Elec	1	16		5,500	825		6,325	7,275
6000 Parking control software, minimum	1 Elec	.50	16		23,600	825		24,425	27,100
6020 Maximum	"	.20	40	↓	103,000	2,050		105,050	116,500

11 12 16 – Parking Ticket Dispensers

11 12 16.10 Ticket Dispensers

	Crew	Daily Output	Labor-Hours	Unit	Material	Labor	Equipment	Total	Total Incl O&P
0010 **TICKET DISPENSERS**									
5900 Ticket spitter with time/date stamp, standard	2 Elec	2	8	Ea.	6,400	410		6,810	7,650
5920 Mag stripe encoding	"	2	8	"	18,600	410		19,010	21,000

11 12 26 – Parking Fee Collection Equipment

11 12 26.13 Parking Fee Coin Collection Equipment

	Crew	Daily Output	Labor-Hours	Unit	Material	Labor	Equipment	Total	Total Incl O&P
0010 **PARKING FEE COIN COLLECTION EQUIPMENT**									
5200 Cashier booth, average	B-22	1	30	Ea.	10,100	1,225	199	11,524	13,200
5300 Collector station, pay on foot	2 Elec	.20	80		109,500	4,125		113,625	126,500
5320 Credit card only	"	.50	32	↓	20,200	1,650		21,850	24,800

11 12 26.23 Fee Equipment

	Crew	Daily Output	Labor-Hours	Unit	Material	Labor	Equipment	Total	Total Incl O&P
0010 **FEE EQUIPMENT**									
5600 Fee computer	1 Elec	1.50	5.333	Ea.	13,900	275		14,175	15,700

11 12 33 – Parking Gates

11 12 33.13 Parking Gate Equipment

	Crew	Daily Output	Labor-Hours	Unit	Material	Labor	Equipment	Total	Total Incl O&P
0010 **PARKING GATE EQUIPMENT**									
5000 Barrier gate with programmable controller	2 Elec	3	5.333	Ea.	3,525	275		3,800	4,300
5020 Industrial		3	5.333		4,875	275		5,150	5,775
5500 Exit verifier	↓	1	16		17,400	825		18,225	20,400
5700 Full sign, 4" letters	1 Elec	2	4		1,225	206		1,431	1,625
5800 Inductive loop	2 Elec	4	4		169	206		375	495
5950 Vehicle detector, microprocessor based	1 Elec	3	2.667	↓	405	137		542	650

11 13 Loading Dock Equipment

11 13 13 – Loading Dock Bumpers

11 13 13.10 Dock Bumpers

		Crew	Daily Output	Labor-Hours	Unit	Material	2012 Bare Costs Labor	Equipment	Total	Total Incl O&P
0010	**DOCK BUMPERS** Bolts not included									
0020	2" x 6" to 4" x 8", average	1 Carp	.30	26.667	M.B.F.	1,325	1,175		2,500	3,275
0050	Bumpers, rubber blocks 4-1/2" thick, 10" high, 14" long		26	.308	Ea.	51	13.55		64.55	77
0200	24" long		22	.364		95	16.05		111.05	129
0300	36" long		17	.471		104	21		125	147
0500	12" high, 14" long		25	.320		100	14.10		114.10	132
0550	24" long		20	.400		111	17.65		128.65	149
0600	36" long		15	.533		123	23.50		146.50	172
0800	Rubber blocks 6" thick, 10" high, 14" long		22	.364		102	16.05		118.05	138
0850	24" long		18	.444		121	19.60		140.60	163
0900	36" long		13	.615		173	27		200	232
0910	20" high, 11" long		13	.615		147	27		174	203
0920	Extruded rubber bumpers, T section, 22" x 22" x 3" thick		41	.195		61	8.60		69.60	80.50
0940	Molded rubber bumpers, 24" x 12" x 3" thick		20	.400		52.50	17.65		70.15	84.50
1000	Welded installation of above bumpers	E-14	8	1		3.74	51.50	15.20	70.44	113
1100	For drilled anchors, add per anchor	1 Carp	36	.222		6.70	9.80		16.50	22.50
1300	Steel bumpers, see Section 10 26 13.10									

11 13 16 – Loading Dock Seals and Shelters

11 13 16.10 Dock Seals and Shelters

		Crew	Daily Output	Labor-Hours	Unit	Material	2012 Bare Costs Labor	Equipment	Total	Total Incl O&P
0010	**DOCK SEALS AND SHELTERS**									
3600	Door seal for door perimeter, 12" x 12", vinyl covered	1 Carp	26	.308	L.F.	31.50	13.55		45.05	55.50
3900	Folding gates, see Section 10 22 16.10									
6200	Shelters, fabric, for truck or train, scissor arms, minimum	1 Carp	1	8	Ea.	1,650	355		2,005	2,375
6300	Maximum	"	.50	16	"	2,200	705		2,905	3,475

11 13 19 – Stationary Loading Dock Equipment

11 13 19.10 Dock Equipment

		Crew	Daily Output	Labor-Hours	Unit	Material	2012 Bare Costs Labor	Equipment	Total	Total Incl O&P
0010	**DOCK EQUIPMENT**									
2200	Dock boards, heavy duty, 60" x 60", aluminum, 5,000 lb. capacity				Ea.	1,250			1,250	1,375
2700	9,000 lb. capacity					1,325			1,325	1,450
3200	15,000 lb. capacity					1,575			1,575	1,750
4200	Platform lifter, 6' x 6', portable, 3,000 lb. capacity					9,150			9,150	10,100
4250	4,000 lb. capacity					11,300			11,300	12,400
4400	Fixed, 6' x 8', 5,000 lb. capacity	E-16	.70	22.857		9,850	1,150	174	11,174	13,000
4500	Levelers, hinged for trucks, 10 ton capacity, 6' x 8'		1.08	14.815		4,975	745	113	5,833	6,925
4650	7' x 8'		1.08	14.815		5,750	745	113	6,608	7,800
4670	Air bag power operated, 10 ton cap., 6' x 8'		1.08	14.815		5,775	745	113	6,633	7,800
4680	7' x 8'		1.08	14.815		5,800	745	113	6,658	7,825
4700	Hydraulic, 10 ton capacity, 6' x 8'		1.08	14.815		8,650	745	113	9,508	11,000
4800	7' x 8'		1.08	14.815		9,325	745	113	10,183	11,600
5800	Loading dock safety restraints, manual style		1.08	14.815		3,175	745	113	4,033	4,925
5900	Automatic style		1.08	14.815		5,325	745	113	6,183	7,300

11 13 26 – Loading Dock Lights

11 13 26.10 Dock Lights

		Crew	Daily Output	Labor-Hours	Unit	Material	2012 Bare Costs Labor	Equipment	Total	Total Incl O&P
0010	**DOCK LIGHTS**									
5000	Lights for loading docks, single arm, 24" long	1 Elec	3.80	2.105	Ea.	137	109		246	315
5700	Double arm, 60" long	"	3.80	2.105	"	194	109		303	375

11 14 13 – Pedestrian Gates

11 14 13.13 Portable Posts and Railings	Crew	Daily Output	Labor-Hours	Unit	Material	2012 Bare Costs Labor	Equipment	Total	Total Incl O&P
0010 **PORTABLE POSTS AND RAILINGS**									
0020 Portable for pedestrian traffic control, standard, minimum				Ea.	131			131	144
0100 Maximum					186			186	204
0300 Deluxe posts, minimum					207			207	228
0400 Maximum				↓	420			420	460
0600 Ropes for above posts, plastic covered, 1-1/2" diameter				L.F.	17.95			17.95	19.75
0700 Chain core				"	11.75			11.75	12.95
1500 Portable security or safety barrier, black with 7' yellow strap				Ea.	215			215	236
1510 12' yellow strap					250			250	275
1550 Sign holder, standard design				↓	70			70	77

11 14 13.19 Turnstiles

11 14 13.19 Turnstiles	Crew	Daily Output	Labor-Hours	Unit	Material	2012 Bare Costs Labor	Equipment	Total	Total Incl O&P
0010 **TURNSTILES**									
0020 One way, 4 arm, 46" diameter, economy, manual	2 Carp	5	3.200	Ea.	1,525	141		1,666	1,900
0100 Electric		1.20	13.333		1,750	590		2,340	2,825
0300 High security, galv., 5'-5" diameter, 7' high, manual		1	16		6,100	705		6,805	7,800
0350 Electric		.60	26.667		6,900	1,175		8,075	9,375
0420 Three arm, 24" opening, light duty, manual		2	8		2,600	355		2,955	3,400
0450 Heavy duty		1.50	10.667		4,700	470		5,170	5,900
0460 Manual, with registering & controls, light duty		2	8		3,275	355		3,630	4,150
0470 Heavy duty		1.50	10.667		3,950	470		4,420	5,075
0480 Electric, heavy duty	↓	1.10	14.545		4,575	640		5,215	6,025
0500 For coin or token operating, add				↓	730			730	805
1200 One way gate with horizontal bars, 5'-5" diameter									
1300 7' high, recreation or transit type	2 Carp	.80	20	Ea.	5,300	880		6,180	7,175
1500 For electronic counter, add				"	180			180	198

11 16 16 – Safes

11 16 16.10 Commercial Safes	Crew	Daily Output	Labor-Hours	Unit	Material	2012 Bare Costs Labor	Equipment	Total	Total Incl O&P
0010 **COMMERCIAL SAFES**									
0200 Office, 1 hr. rating, 30" x 18" x 18"				Ea.	2,125			2,125	2,325
0250 40" x 18" x 18"					4,025			4,025	4,450
0300 60" x 36" x 18", double door					7,750			7,750	8,525
0600 Data, 1 hr. rating, 27" x 19" x 16"					6,325			6,325	6,975
0700 63" x 34" x 16"					13,400			13,400	14,700
0750 Diskette, 1 hr., 14" x 12" x 11", inside					4,025			4,025	4,425
0800 Money, "B" label, 9" x 14" x 14"					525			525	580
0900 Tool resistive, 24" x 24" x 20"					3,875			3,875	4,250
1050 Tool and torch resistive, 24" x 24" x 20"					7,050			7,050	7,750
1150 Jewelers, 23" x 20" x 18"					8,350			8,350	9,200
1200 63" x 25" x 18"					14,200			14,200	15,700
1300 For handling into building, add, minimum	A-2	8.50	2.824			98.50	25	123.50	179
1400 Maximum	"	.78	30.769	↓		1,075	270	1,345	1,950

11 17 Teller and Service Equipment

11 17 13 – Teller Equipment Systems

11 17 13.10 Bank Equipment

11 17 13.10 Bank Equipment	Crew	Daily Output	Labor-Hours	Unit	Material	2012 Bare Costs Labor	Equipment	Total	Total Incl O&P
0010 **BANK EQUIPMENT**									
0020 Alarm system, police	2 Elec	1.60	10	Ea.	4,675	515		5,190	5,925
0100 With vault alarm	"	.40	40		18,600	2,050		20,650	23,500
0400 Bullet resistant teller window, 44" x 60"	1 Glaz	.60	13.333		3,600	570		4,170	4,825
0500 48" x 60"	"	.60	13.333	↓	4,775	570		5,345	6,150
3000 Counters for banks, frontal only	2 Carp	1	16	Station	1,750	705		2,455	3,000
3100 Complete with steel undercounter	"	.50	32	"	3,400	1,400		4,800	5,925
4600 Door and frame, bullet-resistant, with vision panel, minimum	2 Sswk	1.10	14.545	Ea.	5,500	720		6,220	7,325
4700 Maximum		1.10	14.545		7,475	720		8,195	9,475
4800 Drive-up window, drawer & mike, not incl. glass, minimum		1	16		5,900	790		6,690	7,900
4900 Maximum		.50	32		8,875	1,575		10,450	12,600
5000 Night depository, with chest, minimum		1	16		7,250	790		8,040	9,375
5100 Maximum		.50	32		10,300	1,575		11,875	14,100
5200 Package receiver, painted		3.20	5		1,300	247		1,547	1,875
5300 Stainless steel	↓	3.20	5	↓	2,225	247		2,472	2,875
5400 Partitions, bullet-resistant, 1-3/16" glass, 8' high	2 Carp	10	1.600	L.F.	189	70.50		259.50	315
5450 Acrylic	"	10	1.600	"	360	70.50		430.50	505
5500 Pneumatic tube systems, 2 lane drive-up, complete	L-3	.25	64	Total	24,600	3,075		27,675	31,800
5550 With T.V. viewer	"	.20	80	"	47,600	3,850		51,450	58,500
5570 Safety deposit boxes, minimum	1 Sswk	44	.182	Opng.	54.50	8.95		63.45	76
5580 Maximum, 10" x 15" opening		19	.421		116	21		137	164
5590 Teller locker, average	↓	15	.533	↓	1,500	26.50		1,526.50	1,700
5600 Pass thru, bullet-res. window, painted steel, 24" x 36"	2 Sswk	1.60	10	Ea.	2,450	495		2,945	3,575
5700 48" x 48"		1.20	13.333		2,550	660		3,210	3,975
5800 72" x 40"	↓	.80	20	↓	3,825	985		4,810	6,000
5900 For stainless steel frames, add					20%				
6100 Surveillance system, video camera, complete	2 Elec	1	16	Ea.	9,925	825		10,750	12,100
6110 For each additional camera, add				"	945			945	1,050
6120 CCTV system, see Section 27 41 19.10									
6200 Twenty-four hour teller, single unit,									
6300 automated deposit, cash and memo	L-3	.25	64	Ea.	42,800	3,075		45,875	51,500
7000 Vault front, see Section 08 34 59.10									

11 19 Detention Equipment

11 19 30 – Detention Cell Equipment

11 19 30.10 Cell Equipment

	Crew	Daily Output	Labor-Hours	Unit	Material	Labor	Equipment	Total	Total Incl O&P
0010 **CELL EQUIPMENT**									
3000 Toilet apparatus including wash basin, average	L-8	1.50	13.333	Ea.	3,350	615		3,965	4,650

11 21 Mercantile and Service Equipment

11 21 13 – Cash Registers and Checking Equipment

11 21 13.10 Checkout Counter

	Crew	Daily Output	Labor-Hours	Unit	Material	Labor	Equipment	Total	Total Incl O&P
0010 **CHECKOUT COUNTER**									
0020 Supermarket conveyor, single belt	2 Clab	10	1.600	Ea.	2,975	56		3,031	3,350
0100 Double belt, power take-away		9	1.778		4,300	62.50		4,362.50	4,825
0400 Double belt, power take-away, incl. side scanning		7	2.286		5,050	80		5,130	5,675
0800 Warehouse or bulk type	↓	6	2.667	↓	5,975	93.50		6,068.50	6,725
1000 Scanning system, 2 lanes, w/registers, scan gun & memory				System	15,600			15,600	17,200
1100 10 lanes, single processor, full scan, with scales				"	148,500			148,500	163,500

11 21 Mercantile and Service Equipment

11 21 13 – Cash Registers and Checking Equipment

11 21 13.10 Checkout Counter	Crew	Daily Output	Labor-Hours	Unit	Material	2012 Bare Costs Labor	2012 Bare Costs Equipment	Total	Total Incl O&P
2000	Register, restaurant, minimum				Ea.	695			695	765
2100	Maximum					2,925			2,925	3,225
2150	Store, minimum					695			695	765
2200	Maximum					2,925			2,925	3,225

11 21 53 – Barber and Beauty Shop Equipment

11 21 53.10 Barber Equipment

		Crew	Daily Output	Labor-Hours	Unit	Material	Labor	Equipment	Total	Total Incl O&P
0010	**BARBER EQUIPMENT**									
0020	Chair, hydraulic, movable, minimum	1 Carp	24	.333	Ea.	465	14.70		479.70	535
0050	Maximum	"	16	.500		2,975	22		2,997	3,300
0200	Wall hung styling station with mirrors, minimum	L-2	8	2		395	77.50		472.50	555
0300	Maximum	"	4	4		1,975	155		2,130	2,425
0500	Sink, hair washing basin, rough plumbing not incl.	1 Plum	8	1		425	54.50		479.50	550
1000	Sterilizer, liquid solution for tools					138			138	151
1100	Total equipment, rule of thumb, per chair, minimum	L-8	1	20		1,725	925		2,650	3,325
1150	Maximum	"	1	20		4,400	925		5,325	6,275

11 23 Commercial Laundry and Dry Cleaning Equipment

11 23 13 – Dry Cleaning Equipment

11 23 13.13 Dry Cleaners

		Crew	Daily Output	Labor-Hours	Unit	Material	Labor	Equipment	Total	Total Incl O&P
0010	**DRY CLEANERS** Not incl. rough-in									
2000	Dry cleaners, electric, 20 lb. capacity	L-1	.20	80	Ea.	33,300	4,250		37,550	43,000
2050	25 lb. capacity		.17	94.118		46,500	5,000		51,500	58,500
2100	30 lb. capacity		.15	106		49,000	5,650		54,650	62,500
2150	60 lb. capacity		.09	177		78,000	9,425		87,425	100,000

11 23 16 – Drying and Conditioning Equipment

11 23 16.13 Dryers

		Crew	Daily Output	Labor-Hours	Unit	Material	Labor	Equipment	Total	Total Incl O&P
0010	**DRYERS**, Not including rough-in									
1500	Industrial, 30 lb. capacity	1 Plum	2	4	Ea.	3,025	218		3,243	3,650
1600	50 lb. capacity	"	1.70	4.706		3,250	257		3,507	3,975
4700	Lint collector, ductwork not included, 8,000 to 10,000 CFM	Q-10	.30	80		8,800	3,900		12,700	15,600

11 23 19 – Finishing Equipment

11 23 19.13 Folders and Spreaders

		Crew	Daily Output	Labor-Hours	Unit	Material	Labor	Equipment	Total	Total Incl O&P
0010	**FOLDERS AND SPREADERS**									
3500	Folders, blankets & sheets, minimum	1 Elec	.17	47.059	Ea.	31,600	2,425		34,025	38,400
3700	King size with automatic stacker		.10	80		57,000	4,125		61,125	69,000
3800	For conveyor delivery, add		.45	17.778		13,500	915		14,415	16,300
4900	Spreader feeders, 240V, 2 station	L-6	.70	17.143		54,500	920		55,420	61,500
4920	4 station	"	.35	34.286		66,000	1,825		67,825	75,500

11 23 23 – Commercial Ironing Equipment

11 23 23.13 Irons and Pressers

		Crew	Daily Output	Labor-Hours	Unit	Material	Labor	Equipment	Total	Total Incl O&P
0010	**IRONS AND PRESSERS**									
4500	Ironers, institutional, 110", single roll	1 Elec	.20	40	Ea.	30,700	2,050		32,750	36,900
4800	Pressers, low capacity air operated	L-6	1.75	6.857		9,325	365		9,690	10,900
4820	Hand operated		1.75	6.857		8,625	365		8,990	10,000
4840	Ironer 48", 240V		3.50	3.429		113,500	184		113,684	125,500
6600	Hand operated presser		.70	17.143		6,100	920		7,020	8,100
6620	Mushroom press 115 V		.70	17.143		7,600	920		8,520	9,750

11 23 Commercial Laundry and Dry Cleaning Equipment

11 23 26 – Commercial Washers and Extractors

11 23 26.13 Washers and Extractors	Crew	Daily Output	Labor-Hours	Unit	Material	2012 Bare Costs Labor	Equipment	Total	Total Incl O&P
0010 **WASHERS AND EXTRACTORS**, not including rough-in									
6000 Combination washer/extractor, 20 lb. capacity	L-6	1.50	8	Ea.	5,700	430		6,130	6,925
6100 30 lb. capacity		.80	15		8,975	805		9,780	11,100
6200 50 lb. capacity		.68	17.647		10,500	945		11,445	13,000
6300 75 lb. capacity		.30	40		19,900	2,150		22,050	25,100
6350 125 lb. capacity		.16	75		27,300	4,025		31,325	36,200
6380 Washer extractor/dryer, 110 lb., 240V		1	12		8,775	645		9,420	10,600
6400 Washer extractor, 135 lb., 240V		1	12		25,000	645		25,645	28,500
6450 Pass through		1	12		69,000	645		69,645	77,000
6500 200 lb. washer extractor		1	12		71,000	645		71,645	79,000
6550 Pass through		1	12		74,500	645		75,145	83,000
6600 Extractor, low capacity		1.75	6.857		7,275	365		7,640	8,550

11 23 33 – Coin-Operated Laundry Equipment

11 23 33.13 Coin Operated Washers and Dryers	Crew	Daily Output	Labor-Hours	Unit	Material	2012 Bare Costs Labor	Equipment	Total	Total Incl O&P
0010 **COIN OPERATED WASHERS AND DRYERS**									
0990 Dryer, gas fired									
1000 Commercial, 30 lb. capacity, coin operated, single	1 Plum	3	2.667	Ea.	3,150	146		3,296	3,700
1100 Double stacked	"	2	4		6,825	218		7,043	7,825
4860 Coin dry cleaner 20 lb.	L-6	1.75	6.857		28,500	365		28,865	31,900
5290 Clothes washer									
5300 Commercial, coin operated, average	1 Plum	3	2.667	Ea.	1,250	146		1,396	1,600

11 24 Maintenance Equipment

11 24 19 – Vacuum Cleaning Systems

11 24 19.10 Vacuum Cleaning	Crew	Daily Output	Labor-Hours	Unit	Material	2012 Bare Costs Labor	Equipment	Total	Total Incl O&P
0010 **VACUUM CLEANING**									
0020 Central, 3 inlet, residential	1 Skwk	.90	8.889	Total	1,025	405		1,430	1,775
0200 Commercial		.70	11.429		1,225	520		1,745	2,150
0400 5 inlet system, residential		.50	16		1,450	725		2,175	2,725
0600 7 inlet system, commercial		.40	20		1,675	910		2,585	3,250
0800 9 inlet system, residential		.30	26.667		3,600	1,200		4,800	5,825
4010 Rule of thumb: First 1200 S.F., installed								1,425	1,575
4020 For each additional S.F., add				S.F.					.26

11 26 Unit Kitchens

11 26 13 – Metal Unit Kitchens

11 26 13.10 Commercial Unit Kitchens	Crew	Daily Output	Labor-Hours	Unit	Material	2012 Bare Costs Labor	Equipment	Total	Total Incl O&P
0010 **COMMERCIAL UNIT KITCHENS**									
1500 Combination range, refrigerator and sink, 30" wide, minimum	L-1	2	8	Ea.	1,125	425		1,550	1,875
1550 Maximum		1	16		3,875	850		4,725	5,525
1570 60" wide, average		1.40	11.429		3,375	605		3,980	4,625
1590 72" wide, average		1.20	13.333		4,400	710		5,110	5,925
1600 Office model, 48" wide		2	8		2,725	425		3,150	3,650
1620 Refrigerator and sink only		2.40	6.667		2,850	355		3,205	3,650
1640 Combination range, refrigerator, sink, microwave									
1660 Oven and ice maker	L-1	.80	20	Ea.	4,675	1,050		5,725	6,725

11 27 Photographic Processing Equipment

11 27 13 – Darkroom Processing Equipment

11 27 13.10 Darkroom Equipment

		Crew	Daily Output	Labor-Hours	Unit	Material	2012 Bare Costs Labor	Equipment	Total	Total Incl O&P
0010	**DARKROOM EQUIPMENT**									
0020	Developing sink, 5" deep, 24" x 48"	Q-1	2	8	Ea.	5,625	395		6,020	6,775
0050	48" x 52"		1.70	9.412		5,675	465		6,140	6,925
0200	10" deep, 24" x 48"		1.70	9.412		7,775	465		8,240	9,250
0250	24" x 108"		1.50	10.667		10,200	525		10,725	12,100
0500	Dryers, dehumidified filtered air, 36" x 25" x 68" high	L-7	6	4.667		5,075	199		5,274	5,875
0550	48" x 25" x 68" high		5	5.600		10,000	239		10,239	11,400
2000	Processors, automatic, color print, minimum		4	7		15,700	298		15,998	17,800
2050	Maximum		.60	46.667		24,700	2,000		26,700	30,300
2300	Black and white print, minimum		2	14		11,900	595		12,495	14,000
2350	Maximum		.80	35		61,500	1,500		63,000	70,000
2600	Manual processor, 16" x 20" maximum print size		2	14		9,450	595		10,045	11,300
2650	20" x 24" maximum print size		1	28		8,975	1,200		10,175	11,700
3000	Viewing lights, 20" x 24"		6	4.667		380	199		579	720
3100	20" x 24" with color correction		6	4.667		535	199		734	895
3500	Washers, round, minimum sheet 11" x 14"	Q-1	2	8		3,400	395		3,795	4,350
3550	Maximum sheet 20" x 24"		1	16		3,475	785		4,260	5,000
3800	Square, minimum sheet 20" x 24"		1	16		3,125	785		3,910	4,600
3900	Maximum sheet 50" x 56"		.80	20		4,850	985		5,835	6,800
4500	Combination tank sink, tray sink, washers, with									
4510	Dry side tables, average	Q-1	.45	35.556	Ea.	10,600	1,750		12,350	14,300

11 31 Residential Appliances

11 31 13 – Residential Kitchen Appliances

11 31 13.13 Cooking Equipment

		Crew	Daily Output	Labor-Hours	Unit	Material	2012 Bare Costs Labor	Equipment	Total	Total Incl O&P
0010	**COOKING EQUIPMENT**									
0020	Cooking range, 30" free standing, 1 oven, minimum	2 Clab	10	1.600	Ea.	405	56		461	530
0050	Maximum		4	4		1,775	140		1,915	2,175
0150	2 oven, minimum		10	1.600		1,100	56		1,156	1,275
0200	Maximum		10	1.600		1,175	56		1,231	1,350
0350	Built-in, 30" wide, 1 oven, minimum	1 Elec	6	1.333		650	68.50		718.50	820
0400	Maximum	2 Carp	2	8		1,250	355		1,605	1,925
0500	2 oven, conventional, minimum		4	4		1,350	176		1,526	1,750
0550	1 conventional, 1 microwave, maximum		2	8		1,850	355		2,205	2,575
0700	Free-standing, 1 oven, 21" wide range, minimum	2 Clab	10	1.600		365	56		421	490
0750	21" wide, maximum	"	4	4		360	140		500	610
0900	Countertop cooktops, 4 burner, standard, minimum	1 Elec	6	1.333		275	68.50		343.50	410
0950	Maximum		3	2.667		1,100	137		1,237	1,400
1050	As above, but with grill and griddle attachment, minimum		6	1.333		1,050	68.50		1,118.50	1,250
1100	Maximum		3	2.667		3,325	137		3,462	3,850
1200	Induction cooktop, 30" wide		3	2.667		1,100	137		1,237	1,425
1250	Microwave oven, minimum		4	2		110	103		213	275
1300	Maximum		2	4		420	206		626	770

11 31 13.23 Refrigeration Equipment

		Crew	Daily Output	Labor-Hours	Unit	Material	2012 Bare Costs Labor	Equipment	Total	Total Incl O&P
0010	**REFRIGERATION EQUIPMENT**									
2000	Deep freeze, 15 to 23 C.F., minimum	2 Clab	10	1.600	Ea.	575	56		631	720
2050	Maximum		5	3.200		745	112		857	995
2200	30 C.F., minimum		8	2		840	70		910	1,025
2250	Maximum		3	5.333		890	187		1,077	1,275
5200	Icemaker, automatic, 20 lb. per day	1 Plum	7	1.143		920	62.50		982.50	1,125
5350	51 lb. per day	"	2	4		1,350	218		1,568	1,800

11 31 Residential Appliances

11 31 13 – Residential Kitchen Appliances

11 31 13.23 Refrigeration Equipment		Crew	Daily Output	Labor-Hours	Unit	Material	2012 Bare Costs Labor	2012 Bare Costs Equipment	Total	Total Incl O&P
5500	Refrigerator, no frost, 10 C.F. to 12 C.F., minimum	2 Clab	10	1.600	Ea.	425	56		481	555
5600	Maximum		6	2.667		510	93.50		603.50	705
5750	14 C.F. to 16 C.F., minimum		9	1.778		515	62.50		577.50	660
5800	Maximum		5	3.200		735	112		847	985
5950	18 C.F. to 20 C.F., minimum		8	2		650	70		720	825
6000	Maximum		4	4		1,400	140		1,540	1,775
6150	21 C.F. to 29 C.F., minimum		7	2.286		955	80		1,035	1,175
6200	Maximum		3	5.333		3,175	187		3,362	3,775
6790	Energy-star qualified, 18 C.F., minimum [G]	2 Carp	4	4		450	176		626	765
6795	Maximum [G]		2	8		1,150	355		1,505	1,800
6797	21.7 C.F., minimum [G]		4	4		955	176		1,131	1,325
6799	Maximum [G]		4	4		1,875	176		2,051	2,350

11 31 13.33 Kitchen Cleaning Equipment

		Crew	Daily Output	Labor-Hours	Unit	Material	Labor	Equipment	Total	Total Incl O&P
0010	**KITCHEN CLEANING EQUIPMENT**									
2750	Dishwasher, built-in, 2 cycles, minimum	L-1	4	4	Ea.	212	212		424	555
2800	Maximum		2	8		400	425		825	1,075
2950	4 or more cycles, minimum		4	4		325	212		537	680
2960	Average		4	4		435	212		647	795
3000	Maximum		2	8		1,100	425		1,525	1,875
3100	Energy-star qualified, minimum [G]		4	4		355	212		567	710
3110	Maximum [G]		2	8		1,300	425		1,725	2,075

11 31 13.43 Waste Disposal Equipment

		Crew	Daily Output	Labor-Hours	Unit	Material	Labor	Equipment	Total	Total Incl O&P
0010	**WASTE DISPOSAL EQUIPMENT**									
1750	Compactor, residential size, 4 to 1 compaction, minimum	1 Carp	5	1.600	Ea.	580	70.50		650.50	745
1800	Maximum	"	3	2.667		860	118		978	1,125
3300	Garbage disposal, sink type, minimum	L-1	10	1.600		83	85		168	219
3350	Maximum	"	10	1.600		197	85		282	345

11 31 13.53 Kitchen Ventilation Equipment

		Crew	Daily Output	Labor-Hours	Unit	Material	Labor	Equipment	Total	Total Incl O&P
0010	**KITCHEN VENTILATION EQUIPMENT**									
4150	Hood for range, 2 speed, vented, 30" wide, minimum	L-3	5	3.200	Ea.	57.50	154		211.50	297
4200	Maximum		3	5.333		750	256		1,006	1,225
4300	42" wide, minimum		5	3.200		181	154		335	435
4330	Custom		5	3.200		1,650	154		1,804	2,025
4350	Maximum		3	5.333		2,000	256		2,256	2,600
4500	For ventless hood, 2 speed, add					17.90			17.90	19.70
4650	For vented 1 speed, deduct from maximum					47.50			47.50	52.50

11 31 23 – Residential Laundry Appliances

11 31 23.13 Washers

		Crew	Daily Output	Labor-Hours	Unit	Material	Labor	Equipment	Total	Total Incl O&P
0010	**WASHERS**									
5000	Residential, 4 cycle, average	1 Plum	3	2.667	Ea.	850	146		996	1,150
6650	Washing machine, automatic, minimum		3	2.667		415	146		561	675
6700	Maximum		1	8		1,500	435		1,935	2,300
6750	Energy star qualified, front loading, minimum [G]		3	2.667		550	146		696	830
6760	Maximum [G]		1	8		1,425	435		1,860	2,225
6764	Top loading, minimum [G]		3	2.667		370	146		516	630
6766	Maximum [G]		3	2.667		1,825	146		1,971	2,225

11 31 23.23 Dryers

		Crew	Daily Output	Labor-Hours	Unit	Material	Labor	Equipment	Total	Total Incl O&P
0010	**DRYERS**									
0500	Gas fired residential, 16 lb. capacity, average	1 Plum	3	2.667	Ea.	645	146		791	930
6770	Electric, front loading, energy-star qualified, minimum [G]	L-2	3	5.333		395	206		601	750
6780	Maximum [G]	"	2	8		1,700	310		2,010	2,350

11 31 Residential Appliances

11 31 23 – Residential Laundry Appliances

11 31 23.23 Dryers		Crew	Daily Output	Labor-Hours	Unit	Material	2012 Bare Costs			Total	Total Incl O&P
							Labor	Equipment			
7450	Vent kits for dryers	1 Carp	10	.800	Ea.	41	35.50			76.50	99.50

11 31 33 – Miscellaneous Residential Appliances

11 31 33.13 Sump Pumps

		Crew	Daily Output	Labor-Hours	Unit	Material	Labor	Equipment		Total	Total Incl O&P
0010	**SUMP PUMPS**										
6400	Cellar drainer, pedestal, 1/3 H.P., molded PVC base	1 Plum	3	2.667	Ea.	128	146			274	360
6450	Solid brass	"	2	4	"	272	218			490	630
6460	Sump pump, see also Section 22 14 29.16										

11 31 33.23 Water Heaters

		Crew	Daily Output	Labor-Hours	Unit	Material	Labor	Equipment		Total	Total Incl O&P
0010	**WATER HEATERS**										
6900	Electric, glass lined, 30 gallon, minimum	L-1	5	3.200	Ea.	520	170			690	825
6950	Maximum		3	5.333		720	283			1,003	1,225
7100	80 gallon, minimum		2	8		935	425			1,360	1,675
7150	Maximum		1	16		1,300	850			2,150	2,700
7180	Gas, glass lined, 30 gallon, minimum	2 Plum	5	3.200		805	175			980	1,150
7220	Maximum		3	5.333		1,125	291			1,416	1,675
7260	50 gallon, minimum		2.50	6.400		845	350			1,195	1,450
7300	Maximum		1.50	10.667		1,175	580			1,755	2,150
7310	Water heater, see also Section 22 33 30.13										

11 31 33.43 Air Quality

		Crew	Daily Output	Labor-Hours	Unit	Material	Labor	Equipment		Total	Total Incl O&P
0010	**AIR QUALITY**										
2450	Dehumidifier, portable, automatic, 15 pint				Ea.	162				162	178
2550	40 pint					212				212	233
3550	Heater, electric, built-in, 1250 watt, ceiling type, minimum	1 Elec	4	2		102	103			205	266
3600	Maximum		3	2.667		166	137			303	385
3700	Wall type, minimum		4	2		164	103			267	335
3750	Maximum		3	2.667		176	137			313	400
3900	1500 watt wall type, with blower		4	2		164	103			267	335
3950	3000 watt		3	2.667		335	137			472	575
4850	Humidifier, portable, 8 gallons per day					173				173	191
5000	15 gallons per day					208				208	229

11 33 Retractable Stairs

11 33 10 – Disappearing Stairs

11 33 10.10 Disappearing Stairway

		Crew	Daily Output	Labor-Hours	Unit	Material	Labor	Equipment		Total	Total Incl O&P
0010	**DISAPPEARING STAIRWAY** No trim included										
0100	Custom grade, pine, 8'-6" ceiling, minimum	1 Carp	4	2	Ea.	155	88			243	305
0150	Average		3.50	2.286		239	101			340	420
0200	Maximum		3	2.667		310	118			428	520
0500	Heavy duty, pivoted, from 7'-7" to 12'-10" floor to floor		3	2.667		735	118			853	985
0600	16'-0" ceiling		2	4		1,500	176			1,676	1,925
0800	Economy folding, pine, 8'-6" ceiling		4	2		132	88			220	281
0900	9'-6" ceiling		4	2		155	88			243	305
1100	Automatic electric, aluminum, floor to floor height, 8' to 9'	2 Carp	1	16		8,575	705			9,280	10,500
1400	11' to 12'		.90	17.778		9,275	785			10,060	11,400
1700	14' to 15'		.70	22.857		9,875	1,000			10,875	12,500

11 41 Foodservice Storage Equipment

11 41 13 – Refrigerated Food Storage Cases

11 41 13.10 Refrigerated Food Cases	Crew	Daily Output	Labor-Hours	Unit	Material	2012 Bare Costs Labor	Equipment	Total	Total Incl O&P
0010	**REFRIGERATED FOOD CASES**									
0030	Dairy, multi-deck, 12' long	Q-5	3	5.333	Ea.	11,200	267		11,467	12,700
0100	For rear sliding doors, add					1,725			1,725	1,900
0200	Delicatessen case, service deli, 12' long, single deck	Q-5	3.90	4.103		6,575	206		6,781	7,550
0300	Multi-deck, 18 S.F. shelf display		3	5.333		6,225	267		6,492	7,250
0400	Freezer, self-contained, chest-type, 30 C.F.		3.90	4.103		7,425	206		7,631	8,475
0500	Glass door, upright, 78 C.F.		3.30	4.848		11,900	243		12,143	13,500
0600	Frozen food, chest type, 12' long		3.30	4.848		7,975	243		8,218	9,150
0700	Glass door, reach-in, 5 door		3	5.333		13,900	267		14,167	15,700
0800	Island case, 12' long, single deck		3.30	4.848		7,200	243		7,443	8,300
0900	Multi-deck		3	5.333		8,400	267		8,667	9,650
1000	Meat case, 12' long, single deck		3.30	4.848		6,000	243		6,243	6,975
1050	Multi-deck		3.10	5.161		10,300	259		10,559	11,700
1100	Produce, 12' long, single deck		3.30	4.848		6,775	243		7,018	7,825
1200	Multi-deck		3.10	5.161		8,625	259		8,884	9,875

11 41 13.20 Refrigerated Food Storage Equipment

		Crew	Daily Output	Labor-Hours	Unit	Material	2012 Bare Costs Labor	Equipment	Total	Total Incl O&P
0010	**REFRIGERATED FOOD STORAGE EQUIPMENT**									
2350	Cooler, reach-in, beverage, 6' long	Q-1	6	2.667	Ea.	3,625	131		3,756	4,175
4300	Freezers, reach-in, 44 C.F.		4	4		2,625	197		2,822	3,200
4500	68 C.F.		3	5.333		4,550	262		4,812	5,425
4600	Freezer, pre-fab, 8' x 8' w/refrigeration	2 Carp	.45	35.556		11,000	1,575		12,575	14,500
4620	8' x 12'		.35	45.714		11,800	2,025		13,825	16,000
4640	8' x 16'		.25	64		16,600	2,825		19,425	22,600
4660	8' x 20'		.17	94.118		19,900	4,150		24,050	28,300
4680	Reach-in, 1 compartment	Q-1	4	4		2,275	197		2,472	2,800
4685	Energy star rated [G]	R-18	7.80	3.333		2,325	135		2,460	2,750
4700	2 compartment	Q-1	3	5.333		3,650	262		3,912	4,400
4705	Energy star rated [G]	R-18	6.20	4.194		3,250	169		3,419	3,825
4710	3 compartment	Q-1	3	5.333		4,825	262		5,087	5,725
4715	Energy star rated [G]	R-18	5.60	4.643		4,025	187		4,212	4,700
8320	Refrigerator, reach-in, 1 compartment		7.80	3.333		2,175	135		2,310	2,575
8325	Energy star rated [G]		7.80	3.333		2,325	135		2,460	2,750
8330	2 compartment		6.20	4.194		3,450	169		3,619	4,025
8335	Energy star rated [G]		6.20	4.194		3,250	169		3,419	3,825
8340	3 compartment		5.60	4.643		4,275	187		4,462	4,975
8345	Energy star rated [G]		5.60	4.643		4,025	187		4,212	4,700
8350	Pre-fab, with refrigeration, 8' x 8'	2 Carp	.45	35.556		6,575	1,575		8,150	9,625
8360	8' x 12'		.35	45.714		7,025	2,025		9,050	10,900
8370	8' x 16'		.25	64		11,800	2,825		14,625	17,400
8380	8' x 20'		.17	94.118		14,900	4,150		19,050	22,800
8390	Pass-thru/roll-in, 1 compartment	R-18	7.80	3.333		4,300	135		4,435	4,950
8400	2 compartment		6.24	4.167		6,075	168		6,243	6,950
8410	3 compartment		5.60	4.643		8,375	187		8,562	9,475
8420	Walk-in, alum, door & floor only, no refrig, 6' x 6' x 7'-6"	2 Carp	1.40	11.429		9,175	505		9,680	10,900
8430	10' x 6' x 7'-6"		.55	29.091		13,100	1,275		14,375	16,400
8440	12' x 14' x 7'-6"		.25	64		18,400	2,825		21,225	24,600
8450	12' x 20' x 7'-6"		.17	94.118		22,400	4,150		26,550	31,100
8460	Refrigerated cabinets, mobile					3,725			3,725	4,075
8470	Refrigerator/freezer, reach-in, 1 compartment	R-18	5.60	4.643		5,050	187		5,237	5,850
8480	2 compartment	"	4.80	5.417		7,125	219		7,344	8,150

11 41 Foodservice Storage Equipment

11 41 13 – Refrigerated Food Storage Cases

11 41 13.30 Wine Cellar

		Crew	Daily Output	Labor-Hours	Unit	Material	2012 Bare Costs Labor	2012 Bare Costs Equipment	Total	Total Incl O&P
0010	**WINE CELLAR**, refrigerated, Redwood interior, carpeted, walk-in type									
0020	6'-8" high, including racks									
0200	80" W x 48" D for 900 bottles	2 Carp	1.50	10.667	Ea.	3,275	470		3,745	4,325
0250	80" W x 72" D for 1300 bottles		1.33	12.030		4,125	530		4,655	5,375
0300	80" W x 94" D for 1900 bottles		1.17	13.675		4,900	605		5,505	6,300
0400	80" W x 124" D for 2500 bottles	↓	1	16	↓	6,000	705		6,705	7,675
0600	Portable cabinets, red oak, reach-in temp.& humidity controlled									
0650	26-5/8"W x 26-1/2"D x 68"H for 235 bottles				Ea.	3,450			3,450	3,800
0660	32"W x 21-1/2"D x 73-1/2"H for 144 bottles					2,450			2,450	2,700
0670	32"W x 29-1/2"D x 73-1/2"H for 288 bottles					2,925			2,925	3,225
0680	39-1/2"W x 29-1/2"D x 86-1/2"H for 440 bottles					4,100			4,100	4,500
0690	52-1/2"W x 29-1/2"D x 73-1/2"H for 468 bottles					4,025			4,025	4,425
0700	52-1/2"W x 29-1/2"D x 86-1/2"H for 572 bottles				↓	4,625			4,625	5,075
0730	Portable, red oak, can be built-in with glass door									
0750	23-7/8"W x 24"D x 34-1/2"H for 50 bottles				Ea.	990			990	1,075

11 41 33 – Foodservice Shelving

11 41 33.20 Metal Food Storage Shelving

		Crew	Daily Output	Labor-Hours	Unit	Material	2012 Bare Costs Labor	2012 Bare Costs Equipment	Total	Total Incl O&P
0010	**METAL FOOD STORAGE SHELVING**									
8600	Stainless steel shelving, louvered 4-tier, 20" x 3'	1 Clab	6	1.333	Ea.	1,300	47		1,347	1,525
8605	20" x 4'		6	1.333		1,875	47		1,922	2,150
8610	20" x 6'		6	1.333		2,725	47		2,772	3,075
8615	24" x 3'		6	1.333		1,725	47		1,772	1,975
8620	24" x 4'		6	1.333		2,000	47		2,047	2,275
8625	24" x 6'		6	1.333		2,900	47		2,947	3,275
8630	Flat 4-tier, 20" x 3'		6	1.333		1,175	47		1,222	1,375
8635	20" x 4'		6	1.333		1,425	47		1,472	1,625
8640	20" x 5'		6	1.333		1,650	47		1,697	1,875
8645	24" x 3'		6	1.333		1,325	47		1,372	1,550
8650	24" x 4'		6	1.333		2,200	47		2,247	2,475
8655	24" x 6'		6	1.333		2,625	47		2,672	2,950
8700	Galvanized shelving, louvered 4-tier, 20" x 3'		6	1.333		670	47		717	805
8705	20" x 4'		6	1.333		745	47		792	890
8710	20" x 6'		6	1.333		915	47		962	1,075
8715	24" x 3'		6	1.333		580	47		627	710
8720	24" x 4'		6	1.333		805	47		852	955
8725	24" x 6'		6	1.333		1,150	47		1,197	1,350
8730	Flat 4-tier, 20" x 3'		6	1.333		595	47		642	725
8735	20" x 4'		6	1.333		655	47		702	795
8740	20" x 6'		6	1.333		940	47		987	1,100
8745	24" x 3'		6	1.333		660	47		707	800
8750	24" x 4'		6	1.333		670	47		717	810
8755	24" x 6'		6	1.333		980	47		1,027	1,150
8760	Stainless steel dunnage rack, 24" x 3'		8	1		330	35		365	415
8765	24" x 4'		8	1		420	35		455	515
8770	Galvanized dunnage rack, 24" x 3'		8	1		178	35		213	250
8775	24" x 4'	↓	8	1	↓	198	35		233	272

11 42 Food Preparation Equipment

11 42 10 – Commercial Food Preparation Equipment

11 42 10.10 Choppers, Mixers and Misc. Equipment	Crew	Daily Output	Labor-Hours	Unit	Material	2012 Bare Costs Labor	Equipment	Total	Total Incl O&P
0010 **CHOPPERS, MIXERS AND MISC. EQUIPMENT**									
1700 Choppers, 5 pounds	R-18	7	3.714	Ea.	1,950	150		2,100	2,375
1720 16 pounds		5	5.200		2,300	210		2,510	2,875
1740 35 to 40 pounds		4	6.500		3,425	262		3,687	4,175
1840 Coffee brewer, 5 burners	1 Plum	3	2.667		1,275	146		1,421	1,625
1850 Coffee urn, twin 6 gallon urns		2	4		2,375	218		2,593	2,950
1860 Single, 3 gallon		3	2.667		1,725	146		1,871	2,125
3000 Fast food equipment, total package, minimum	6 Skwk	.08	600		201,000	27,200		228,200	263,500
3100 Maximum	"	.07	685		274,000	31,100		305,100	349,500
3800 Food mixers, bench type, 20 quarts	L-7	7	4		2,475	170		2,645	2,975
3850 40 quarts		5.40	5.185		6,425	221		6,646	7,425
3900 60 quarts		5	5.600		10,600	239		10,839	12,100
4040 80 quarts		3.90	7.179		11,600	305		11,905	13,300
4080 130 quarts		2.20	12.727		16,700	540		17,240	19,200
4100 Floor type, 20 quarts		15	1.867		3,075	79.50		3,154.50	3,500
4120 60 quarts		14	2		9,475	85		9,560	10,500
4140 80 quarts		12	2.333		11,200	99.50		11,299.50	12,600
4160 140 quarts		8.60	3.256		18,600	139		18,739	20,600
6700 Peelers, small	R-18	8	3.250		1,650	131		1,781	2,025
6720 Large	"	6	4.333		4,975	175		5,150	5,750
6800 Pulper/extractor, close coupled, 5 HP	1 Plum	1.90	4.211		3,650	230		3,880	4,375
8580 Slicer with table	R-18	9	2.889		5,075	117		5,192	5,750

11 43 Food Delivery Carts and Conveyors

11 43 13 – Food Delivery Carts

11 43 13.10 Mobile Carts, Racks and Trays

	Crew	Daily Output	Labor-Hours	Unit	Material	2012 Bare Costs Labor	Equipment	Total	Total Incl O&P
0010 **MOBILE CARTS, RACKS AND TRAYS**									
1650 Cabinet, heated, 1 compartment, reach-in	R-18	5.60	4.643	Ea.	3,850	187		4,037	4,500
1655 Pass-thru roll-in		5.60	4.643		5,200	187		5,387	6,000
1660 2 compartment, reach-in		4.80	5.417		8,275	219		8,494	9,425
1670 Mobile					2,875			2,875	3,175
6850 Mobile rack w/pan slide					1,350			1,350	1,475
9180 Tray and silver dispenser, mobile	1 Clab	16	.500		890	17.55		907.55	1,000

11 44 Food Cooking Equipment

11 44 13 – Commercial Ranges

11 44 13.10 Cooking Equipment

	Crew	Daily Output	Labor-Hours	Unit	Material	2012 Bare Costs Labor	Equipment	Total	Total Incl O&P
0010 **COOKING EQUIPMENT**									
0020 Bake oven, gas, one section	Q-1	8	2	Ea.	4,800	98.50		4,898.50	5,425
0300 Two sections		7	2.286		9,175	112		9,287	10,300
0600 Three sections		6	2.667		14,900	131		15,031	16,600
0900 Electric convection, single deck	L-7	4	7		6,425	298		6,723	7,525
1300 Broiler, without oven, standard	Q-1	8	2		3,500	98.50		3,598.50	4,000
1550 Infrared	L-7	4	7		9,350	298		9,648	10,800
4750 Fryer, with twin baskets, modular model	Q-1	7	2.286		1,125	112		1,237	1,400
5000 Floor model, on 6" legs	"	5	3.200		2,125	157		2,282	2,575
5100 Extra single basket, large					97			97	107
5170 Energy star rated, 50 lb. capacity G	R-18	4	6.500		4,100	262		4,362	4,900
5175 85 lb. capacity G	"	4	6.500		6,775	262		7,037	7,850

11 44 Food Cooking Equipment

11 44 13 – Commercial Ranges

11 44 13.10 Cooking Equipment	Crew	Daily Output	Labor-Hours	Unit	Material	2012 Bare Costs Labor	Equipment	Total	Total Incl O&P
5300	Griddle, SS, 24" plate, w/4" legs, elec, 208 V, 3 phase, 3' long	Q-1	7	2.286	Ea.	1,250	112		1,362	1,550
5550	4' long	"	6	2.667		2,200	131		2,331	2,625
6200	Iced tea brewer	1 Plum	3.44	2.326		710	127		837	970
6350	Kettle, w/steam jacket, tilting, w/positive lock, SS, 20 gallons	L-7	7	4		7,375	170		7,545	8,350
6600	60 gallons	"	6	4.667		10,400	199		10,599	11,700
6900	Range, restaurant type, 6 burners and 1 standard oven, 36" wide	Q-1	7	2.286		2,875	112		2,987	3,325
6950	Convection		7	2.286		4,400	112		4,512	5,025
7150	2 standard ovens, 24" griddle, 60" wide		6	2.667		3,775	131		3,906	4,375
7200	1 standard, 1 convection oven		6	2.667		8,800	131		8,931	9,875
7450	Heavy duty, single 34" standard oven, open top		5	3.200		4,650	157		4,807	5,325
7500	Convection oven		5	3.200		4,750	157		4,907	5,450
7700	Griddle top		6	2.667		2,600	131		2,731	3,075
7750	Convection oven	▼	6	2.667		7,375	131		7,506	8,300
8850	Steamer, electric 27 KW	L-7	7	4		8,500	170		8,670	9,600
9100	Electric, 10 KW or gas 100,000 BTU	"	5	5.600		6,250	239		6,489	7,250
9150	Toaster, conveyor type, 16-22 slices per minute					990			990	1,100
9160	Pop-up, 2 slot				▼	610			610	670
9200	For deluxe models of above equipment, add					75%				
9400	Rule of thumb: Equipment cost based									
9410	on kitchen work area									
9420	Office buildings, minimum	L-7	77	.364	S.F.	89.50	15.50		105	122
9450	Maximum		58	.483		151	20.50		171.50	198
9550	Public eating facilities, minimum		77	.364		117	15.50		132.50	153
9600	Maximum		46	.609		191	26		217	250
9750	Hospitals, minimum		58	.483		121	20.50		141.50	165
9800	Maximum	▼	39	.718	▼	222	30.50		252.50	291

11 46 Food Dispensing Equipment

11 46 16 – Service Line Equipment

11 46 16.10 Commercial Food Dispensing Equipment

		Crew	Daily Output	Labor-Hours	Unit	Material	Labor	Equipment	Total	Total Incl O&P
0010	**COMMERCIAL FOOD DISPENSING EQUIPMENT**									
1050	Butter pat dispenser	1 Clab	13	.615	Ea.	945	21.50		966.50	1,075
1100	Bread dispenser, counter top		13	.615		920	21.50		941.50	1,050
1900	Cup and glass dispenser, drop in		4	2		1,050	70		1,120	1,250
1920	Disposable cup, drop in		16	.500		470	17.55		487.55	545
2650	Dish dispenser, drop in, 12"		11	.727		2,275	25.50		2,300.50	2,550
2660	Mobile	▼	10	.800		2,700	28		2,728	3,025
3300	Food warmer, counter, 1.2 KW					510			510	560
3550	1.6 KW					2,050			2,050	2,275
3600	Well, hot food, built-in, rectangular, 12" x 20"	R-30	10	2.600		710	108		818	950
3610	Circular, 7 qt.		10	2.600		375	108		483	575
3620	Refrigerated, 2 compartments		10	2.600		2,875	108		2,983	3,350
3630	3 compartments		9	2.889		3,450	120		3,570	3,975
3640	4 compartments		8	3.250		4,225	135		4,360	4,850
4720	Frost cold plate	▼	9	2.889		18,200	120		18,320	20,200
5700	Hot chocolate dispenser	1 Plum	4	2		1,175	109		1,284	1,475
5750	Ice dispenser 567 pound	Q-1	6	2.667		5,000	131		5,131	5,700
6250	Jet spray dispenser	R-18	4.50	5.778		3,225	233		3,458	3,900
6300	Juice dispenser, concentrate	"	4.50	5.778		2,125	233		2,358	2,700
6690	Milk dispenser, bulk, 2 flavor	R-30	8	3.250		1,700	135		1,835	2,075
6695	3 flavor	"	8	3.250	▼	2,300	135		2,435	2,725

11 46 Food Dispensing Equipment

11 46 16 – Service Line Equipment

11 46 16.10 Commercial Food Dispensing Equipment	Crew	Daily Output	Labor-Hours	Unit	Material	2012 Bare Costs Labor	Equipment	Total	Total Incl O&P
8800	Serving counter, straight	1 Carp	40	.200	L.F.	840	8.80		848.80	940
8820	Curved section	"	30	.267	"	1,025	11.75		1,036.75	1,150
8825	Solid surface, see Section 12 36 61.16									
8860	Sneeze guard with lights, 60" L	1 Clab	16	.500	Ea.	375	17.55		392.55	435
9100	Soft serve ice cream machine, medium	R-18	11	2.364		11,900	95.50		11,995.50	13,200
9110	Large	"	9	2.889	↓	19,200	117		19,317	21,300

11 47 Ice Machines

11 47 10 – Commercial Ice Machines

11 47 10.10 Commercial Ice Equipment	Crew	Daily Output	Labor-Hours	Unit	Material	2012 Bare Costs Labor	Equipment	Total	Total Incl O&P
0010	**COMMERCIAL ICE EQUIPMENT**									
5800	Ice cube maker, 50 pounds per day	Q-1	6	2.667	Ea.	1,475	131		1,606	1,825
5900	250 pounds per day		1.20	13.333		2,625	655		3,280	3,875
6050	500 pounds per day		4	4		2,750	197		2,947	3,325
6060	With bin		1.20	13.333		3,100	655		3,755	4,400
6090	1000 pounds per day, with bin		1	16		4,700	785		5,485	6,350
6100	Ice flakers, 300 pounds per day		1.60	10		2,575	490		3,065	3,600
6120	600 pounds per day		.95	16.842		3,750	830		4,580	5,375
6130	1000 pounds per day		.75	21.333		4,325	1,050		5,375	6,325
6140	2000 pounds per day	↓	.65	24.615		21,300	1,200		22,500	25,200
6160	Ice storage bin, 500 pound capacity	Q-5	1	16		975	800		1,775	2,275
6180	1000 pound	"	.56	28.571	↓	2,500	1,425		3,925	4,875

11 48 Cleaning and Disposal Equipment

11 48 13 – Commercial Dishwashers

11 48 13.10 Dishwashers	Crew	Daily Output	Labor-Hours	Unit	Material	2012 Bare Costs Labor	Equipment	Total	Total Incl O&P
0010	**DISHWASHERS**									
2700	Dishwasher, commercial, rack type									
2720	10 to 12 racks per hour	Q-1	3.20	5	Ea.	4,300	246		4,546	5,100
2730	Energy star rated, 35 to 40 racks/hour G		1.30	12.308		4,725	605		5,330	6,100
2740	50 to 60 racks/hour G	↓	1.30	12.308		10,200	605		10,805	12,100
2800	Automatic, 190 to 230 racks per hour	L-6	.35	34.286		13,500	1,825		15,325	17,600
2820	235 to 275 racks per hour		.25	48		31,900	2,575		34,475	39,000
2840	8,750 to 12,500 dishes per hour	↓	.10	120	↓	49,500	6,425		55,925	64,000
2950	Dishwasher hood, canopy type	L-3A	10	1.200	L.F.	800	57.50		857.50	970
2960	Pant leg type	"	2.50	4.800	Ea.	8,600	231		8,831	9,825
5200	Garbage disposal 1.5 HP, 100 GPH	L-1	4.80	3.333		2,000	177		2,177	2,475
5210	3 HP, 120 GPH		4.60	3.478		2,100	185		2,285	2,575
5220	5 HP, 250 GPH	↓	4.50	3.556	↓	3,375	189		3,564	4,000
6750	Pot sink, 3 compartment	1 Plum	7.25	1.103	L.F.	1,000	60.50		1,060.50	1,200
6760	Pot washer, low temp wash/rinse		1.60	5	Ea.	3,800	273		4,073	4,600
6770	High pressure wash, high temperature rinse	↓	1.20	6.667		36,500	365		36,865	40,700
9170	Trash compactor, small, up to 125 lb. compacted weight	L-4	4	6		23,900	248		24,148	26,700
9175	Large, up to 175 lb. compacted weight	"	3	8	↓	29,200	330		29,530	32,600

11 52 Audio-Visual Equipment

11 52 13 – Projection Screens

		Daily Output	Labor-Hours	Unit	Material	2012 Bare Costs Labor	Equipment	Total	Total Incl O&P
11 52 13.10 Projection Screens, Wall or Ceiling Hung	Crew								
0010	**PROJECTION SCREENS, WALL OR CEILING HUNG**, matte white								
0100	Manually operated, economy	2 Carp	500	.032	S.F.	5.85	1.41		7.26	8.60
0300	Intermediate		450	.036		6.80	1.57		8.37	9.90
0400	Deluxe		400	.040		9.45	1.76		11.21	13.05
0600	Electric operated, matte white, 25 S.F., economy		5	3.200	Ea.	855	141		996	1,150
0700	Deluxe		4	4		1,775	176		1,951	2,225
0900	50 S.F., economy		3	5.333		750	235		985	1,175
1000	Deluxe		2	8		1,975	355		2,330	2,725
1200	Heavy duty, electric operated, 200 S.F.		1.50	10.667		3,900	470		4,370	5,025
1300	400 S.F.		1	16		4,800	705		5,505	6,375
1500	Rigid acrylic in wall, for rear projection, 1/4" thick	2 Glaz	30	.533	S.F.	47	23		70	86.50
1600	1/2" thick (maximum size 10' x 20')	"	25	.640	"	83.50	27.50		111	133

11 52 16 – Projectors

		Crew	Daily Output	Labor-Hours	Unit	Material	2012 Bare Costs Labor	Equipment	Total	Total Incl O&P
11 52 16.10 Movie Equipment										
0010	**MOVIE EQUIPMENT**									
0020	Changeover, minimum				Ea.	465			465	510
0100	Maximum					905			905	995
0400	Film transport, incl. platters and autowind, minimum					5,025			5,025	5,525
0500	Maximum					14,300			14,300	15,700
0800	Lamphouses, incl. rectifiers, xenon, 1,000 watt	1 Elec	2	4		6,650	206		6,856	7,600
0900	1,600 watt		2	4		7,100	206		7,306	8,100
1000	2,000 watt		1.50	5.333		7,625	275		7,900	8,775
1100	4,000 watt		1.50	5.333		9,425	275		9,700	10,800
1400	Lenses, anamorphic, minimum					1,275			1,275	1,400
1500	Maximum					2,875			2,875	3,150
1800	Flat 35 mm, minimum					1,100			1,100	1,225
1900	Maximum					1,725			1,725	1,900
2200	Pedestals, for projectors					1,475			1,475	1,625
2300	Console type					10,700			10,700	11,800
2600	Projector mechanisms, incl. soundhead, 35 mm, minimum					11,100			11,100	12,200
2700	Maximum					15,200			15,200	16,700
3000	Projection screens, rigid, in wall, acrylic, 1/4" thick	2 Glaz	195	.082	S.F.	42	3.51		45.51	51.50
3100	1/2" thick	"	130	.123	"	48.50	5.25		53.75	61.50
3300	Electric operated, heavy duty, 400 S.F.	2 Carp	1	16	Ea.	2,950	705		3,655	4,325
3320	Theater projection screens, matte white, including frames	"	200	.080	S.F.	6.65	3.53		10.18	12.75
3400	Also see Section 11 52 13.10									
3700	Sound systems, incl. amplifier, mono, minimum	1 Elec	.90	8.889	Ea.	3,300	460		3,760	4,300
3800	Dolby/Super Sound, maximum		.40	20		18,100	1,025		19,125	21,500
4100	Dual system, 2 channel, front surround, minimum		.70	11.429		4,625	590		5,215	5,975
4200	Dolby/Super Sound, 4 channel, maximum		.40	20		16,500	1,025		17,525	19,800
4500	Sound heads, 35 mm					5,300			5,300	5,825
4900	Splicer, tape system, minimum					740			740	815
5000	Tape type, maximum					1,325			1,325	1,450
5300	Speakers, recessed behind screen, minimum	1 Elec	2	4		1,050	206		1,256	1,475
5400	Maximum	"	1	8		3,100	410		3,510	4,025
5700	Seating, painted steel, upholstered, minimum	2 Carp	35	.457		132	20		152	176
5800	Maximum	"	28	.571		420	25		445	505
6100	Rewind tables, minimum					2,650			2,650	2,900
6200	Maximum					4,725			4,725	5,175
7000	For automation, varying sophistication, minimum	1 Elec	1	8	System	2,375	410		2,785	3,250
7100	Maximum	2 Elec	.30	53.333	"	5,575	2,750		8,325	10,200

11 52 Audio-Visual Equipment

11 52 16 – Projectors

11 52 16.20 Movie Equipment- Digital	Crew	Daily Output	Labor-Hours	Unit	Material	2012 Bare Costs Labor	2012 Bare Costs Equipment	Total	Total Incl O&P
0010 **MOVIE EQUIPMENT- DIGITAL**									
1000 Digital 2K projection system, 98" DMD	1 Elec	2	4	Ea.	44,000	206		44,206	48,700
1100 OEM lens		2	4		5,375	206		5,581	6,200
2000 Pedestal with power distribution		2	4		2,050	206		2,256	2,550
3000 Software		2	4		1,675	206		1,881	2,150

11 53 Laboratory Equipment

11 53 03 – Laboratory Test Equipment

11 53 03.13 Test Equipment

	Crew	Daily Output	Labor-Hours	Unit	Material	Labor	Equipment	Total	Total Incl O&P
0010 **TEST EQUIPMENT**									
1700 Thermometer, electric, portable				Ea.	450			450	495
1800 Titration unit, four 2000 ml reservoirs				"	5,500			5,500	6,050

11 53 13 – Laboratory Fume Hoods

11 53 13.13 Recirculating Laboratory Fume Hoods

	Crew	Daily Output	Labor-Hours	Unit	Material	Labor	Equipment	Total	Total Incl O&P
0010 **RECIRCULATING LABORATORY FUME HOODS**									
0600 Fume hood, with countertop & base, not including HVAC									
0610 Simple, minimum	2 Carp	5.40	2.963	L.F.	620	131		751	880
0620 Complex, including fixtures		2.40	6.667		1,850	294		2,144	2,475
0630 Special, maximum		1.70	9.412		825	415		1,240	1,550
0670 Service fixtures, average				Ea.	295			295	325
0680 For sink assembly with hot and cold water, add	1 Plum	1.40	5.714		725	310		1,035	1,275
0750 Glove box, fiberglass, bacteriological					16,900			16,900	18,600
0760 Controlled atmosphere					19,400			19,400	21,300
0770 Radioisotope					16,900			16,900	18,600
0780 Carcinogenic					16,900			16,900	18,600

11 53 13.23 Exhaust Hoods

	Crew	Daily Output	Labor-Hours	Unit	Material	Labor	Equipment	Total	Total Incl O&P
0010 **EXHAUST HOODS**									
0650 Ductwork, minimum	2 Shee	1	16	Hood	3,725	835		4,560	5,375
0660 Maximum	"	.50	32	"	5,750	1,675		7,425	8,875

11 53 16 – Laboratory Incubators

11 53 16.13 Incubators

	Crew	Daily Output	Labor-Hours	Unit	Material	Labor	Equipment	Total	Total Incl O&P
0010 **INCUBATORS**									
1000 Incubators, minimum				Ea.	2,725			2,725	2,975
1010 Maximum				"	14,300			14,300	15,700

11 53 19 – Laboratory Sterilizers

11 53 19.13 Sterilizers

	Crew	Daily Output	Labor-Hours	Unit	Material	Labor	Equipment	Total	Total Incl O&P
0010 **STERILIZERS**									
0700 Glassware washer, undercounter, minimum	L-1	1.80	8.889	Ea.	6,000	470		6,470	7,300
0710 Maximum	"	1	16		12,200	850		13,050	14,700
1850 Utensil washer-sanitizer	1 Plum	2	4		11,100	218		11,318	12,500

11 53 23 – Laboratory Refrigerators

11 53 23.13 Refrigerators

	Crew	Daily Output	Labor-Hours	Unit	Material	Labor	Equipment	Total	Total Incl O&P
0010 **REFRIGERATORS**									
1200 Blood bank, 28.6 C.F. emergency signal				Ea.	9,150			9,150	10,100
1210 Reach-in, 16.9 C.F.				"	8,200			8,200	9,000

11 53 Laboratory Equipment

11 53 33 – Emergency Safety Appliances

11 53 33.13 Emergency Equipment	Crew	Daily Output	Labor-Hours	Unit	Material	2012 Bare Costs Labor	Equipment	Total	Total Incl O&P
0010 **EMERGENCY EQUIPMENT**									
1400 Safety equipment, eye wash, hand held				Ea.	405			405	445
1450 Deluge shower				"	740			740	815

11 53 43 – Service Fittings and Accessories

11 53 43.13 Fittings

	Crew	Daily Output	Labor-Hours	Unit	Material	2012 Bare Costs Labor	Equipment	Total	Total Incl O&P
0010 **FITTINGS**									
1600 Sink, one piece plastic, flask wash, hose, free standing	1 Plum	1.60	5	Ea.	1,850	273		2,123	2,425
1610 Epoxy resin sink, 25" x 16" x 10"	"	2	4	"	204	218		422	555
1950 Utility table, acid resistant top with drawers	2 Carp	30	.533	L.F.	151	23.50		174.50	202
8000 Alternate pricing method: as percent of lab furniture									
8050 Installation, not incl. plumbing & duct work				% Furn.				20%	22%
8100 Plumbing, final connections, simple system								9.09%	10%
8110 Moderately complex system								13.64%	15%
8120 Complex system								18.18%	20%
8150 Electrical, simple system								9.09%	10%
8160 Moderately complex system								18.18%	20%
8170 Complex system								31.80%	35%

11 57 Vocational Shop Equipment

11 57 10 – Shop Equipment

11 57 10.10 Vocational School Shop Equipment

	Crew	Daily Output	Labor-Hours	Unit	Material	2012 Bare Costs Labor	Equipment	Total	Total Incl O&P
0010 **VOCATIONAL SCHOOL SHOP EQUIPMENT**									
0020 Benches, work, wood, average	2 Carp	5	3.200	Ea.	530	141		671	795
0100 Metal, average		5	3.200		490	141		631	755
0400 Combination belt & disc sander, 6"		4	4		1,425	176		1,601	1,825
0700 Drill press, floor mounted, 12", 1/2 H.P.		4	4		385	176		561	695
0800 Dust collector, not incl. ductwork, 6" diameter	1 Shee	1.10	7.273		4,175	380		4,555	5,150
0810 Dust collector bag, 20" diameter	"	5	1.600		405	83.50		488.50	570
1000 Grinders, double wheel, 1/2 H.P.	2 Carp	5	3.200		203	141		344	440
1300 Jointer, 4", 3/4 H.P.		4	4		1,275	176		1,451	1,675
1600 Kilns, 16 C.F., to 2000°		4	4		1,500	176		1,676	1,900
1900 Lathe, woodworking, 10", 1/2 H.P.		4	4		830	176		1,006	1,175
2200 Planer, 13" x 6"		4	4		950	176		1,126	1,325
2500 Potter's wheel, motorized		4	4		1,050	176		1,226	1,425
2800 Saws, band, 14", 3/4 H.P.		4	4		860	176		1,036	1,225
3100 Metal cutting band saw, 14"		4	4		2,275	176		2,451	2,775
3400 Radial arm saw, 10", 2 H.P.		4	4		1,300	176		1,476	1,700
3700 Scroll saw, 24"		4	4		590	176		766	920
4000 Table saw, 10", 3 H.P.		4	4		2,275	176		2,451	2,775
4300 Welder AC arc, 30 amp capacity		4	4		3,100	176		3,276	3,700

11 61 Theater and Stage Equipment

11 61 23 – Folding and Portable Stages

11 61 23.10 Portable Stages	Crew	Daily Output	Labor-Hours	Unit	Material	2012 Bare Costs Labor	Equipment	Total	Total Incl O&P
0010 **PORTABLE STAGES**									
1500 Flooring, portable oak parquet, 3' x 3' sections				S.F.	13.40			13.40	14.70
1600 Cart to carry 225 S.F. of flooring				Ea.	400			400	440
5000 Stages, portable with steps, folding legs, stock, 8" high				SF Stg.	31			31	34.50
5100 16" high					24.50			24.50	27
5200 32" high					43			43	47
5300 40" high					59.50			59.50	65
6000 Telescoping platforms, extruded alum., straight, minimum	4 Carp	157	.204		33	9		42	50
6100 Maximum		77	.416		46	18.35		64.35	78.50
6500 Pie-shaped, minimum		150	.213		71	9.40		80.40	93
6600 Maximum		70	.457		79.50	20		99.50	119
6800 For 3/4" plywood covered deck, deduct					4.05			4.05	4.46
7000 Band risers, steel frame, plywood deck, minimum	4 Carp	275	.116		30.50	5.15		35.65	41.50
7100 Maximum	"	138	.232		67	10.25		77.25	89.50
7500 Chairs for above, self-storing, minimum	2 Carp	43	.372	Ea.	107	16.40		123.40	144
7600 Maximum	"	40	.400	"	189	17.65		206.65	235

11 61 33 – Rigging Systems and Controls

11 61 33.10 Controls

	Crew	Daily Output	Labor-Hours	Unit	Material	2012 Bare Costs Labor	Equipment	Total	Total Incl O&P
0010 **CONTROLS**									
0050 Control boards with dimmers and breakers, minimum	1 Elec	1	8	Ea.	12,300	410		12,710	14,100
0100 Average		.50	16		38,600	825		39,425	43,600
0150 Maximum		.20	40		126,000	2,050		128,050	141,500
8000 Rule of thumb: total stage equipment, minimum	4 Carp	100	.320	SF Stg.	97	14.10		111.10	129
8100 Maximum	"	25	1.280	"	545	56.50		601.50	685

11 61 43 – Stage Curtains

11 61 43.10 Curtains

	Crew	Daily Output	Labor-Hours	Unit	Material	2012 Bare Costs Labor	Equipment	Total	Total Incl O&P
0010 **CURTAINS**									
0500 Curtain track, straight, light duty	2 Carp	20	.800	L.F.	27	35.50		62.50	84
0600 Heavy duty		18	.889		61	39		100	128
0700 Curved sections		12	1.333		173	59		232	281
1000 Curtains, velour, medium weight		600	.027	S.F.	7.90	1.18		9.08	10.50
1150 Silica based yarn, inherently fire retardant		50	.320	"	15.20	14.10		29.30	38

11 62 Musical Equipment

11 62 16 – Carillons

11 62 16.10 Bell Tower Equipment

	Crew	Daily Output	Labor-Hours	Unit	Material	2012 Bare Costs Labor	Equipment	Total	Total Incl O&P
0010 **BELL TOWER EQUIPMENT**									
0300 Carillon, 4 octave (48 bells), with keyboard				System	937,000			937,000	1,031,000
0320 2 octave (24 bells)					441,000			441,000	485,000
0340 3 to 4 bell peal, minimum					110,500			110,500	121,500
0360 Maximum					661,500			661,500	727,500
0380 Cast bronze bell, average				Ea.	99,000			99,000	109,000
0400 Electronic, digital, minimum					16,500			16,500	18,200
0410 With keyboard, maximum					82,500			82,500	91,000

11 66 Athletic Equipment

11 66 13 – Exercise Equipment

11 66 13.10 Physical Training Equipment	Crew	Daily Output	Labor-Hours	Unit	Material	2012 Bare Costs Labor	Equipment	Total	Total Incl O&P
0010 **PHYSICAL TRAINING EQUIPMENT**									
0020 Abdominal rack, 2 board capacity				Ea.	510			510	565
0050 Abdominal board, upholstered					685			685	750
0200 Bicycle trainer, minimum					430			430	470
0300 Deluxe, electric					4,300			4,300	4,725
0400 Barbell set, chrome plated steel, 25 lb.					249			249	274
0420 100 lb.					400			400	440
0450 200 lb.					665			665	730
0500 Weight plates, cast iron, per lb.				Lb.	5.20			5.20	5.70
0520 Storage rack, 10 station				Ea.	900			900	990
0600 Circuit training apparatus, 12 machines minimum	2 Clab	1.25	12.800	Set	28,700	450		29,150	32,200
0700 Average		1	16		35,200	560		35,760	39,600
0800 Maximum		.75	21.333		41,700	750		42,450	47,100
0820 Dumbbell set, cast iron, with rack and 5 pair					615			615	675
0900 Squat racks	2 Clab	5	3.200	Ea.	890	112		1,002	1,150
1200 Multi-station gym machine, 5 station					5,000			5,000	5,500
1250 9 station					10,100			10,100	11,100
1280 Rowing machine, hydraulic					1,750			1,750	1,925
1300 Treadmill, manual					1,100			1,100	1,200
1320 Motorized					3,625			3,625	4,000
1340 Electronic					3,700			3,700	4,075
1360 Cardio-testing					4,525			4,525	5,000
1400 Treatment/massage tables, minimum					605			605	670
1420 Deluxe, with accessories					735			735	810
4150 Exercise equipment, bicycle trainer					750			750	825
4180 Chinning bar, adjustable, wall mounted	1 Carp	5	1.600		300	70.50		370.50	440
4200 Exercise ladder, 16' x 1'-7", suspended	L-2	3	5.333		1,700	206		1,906	2,175
4210 High bar, floor plate attached	1 Carp	4	2		1,250	88		1,338	1,500
4240 Parallel bars, adjustable		4	2		2,250	88		2,338	2,625
4270 Uneven parallel bars, adjustable		4	2		2,950	88		3,038	3,375
4280 Wall mounted, adjustable	L-2	1.50	10.667	Set	820	415		1,235	1,550
4300 Rope, ceiling mounted, 18' long	1 Carp	3.66	2.186	Ea.	286	96.50		382.50	465
4330 Side horse, vaulting		5	1.600		1,775	70.50		1,845.50	2,050
4360 Treadmill, motorized, deluxe, training type		5	1.600		3,000	70.50		3,070.50	3,400
4390 Weight lifting multi-station, minimum	2 Clab	1	16		450	560		1,010	1,350
4450 Maximum	"	.50	32		14,700	1,125		15,825	17,900

11 66 23 – Gymnasium Equipment

11 66 23.13 Basketball Equipment

	Crew	Daily Output	Labor-Hours	Unit	Material	2012 Bare Costs Labor	Equipment	Total	Total Incl O&P
0010 **BASKETBALL EQUIPMENT**									
1000 Backstops, wall mtd., 6' extended, fixed, minimum	L-2	1	16	Ea.	1,300	620		1,920	2,375
1100 Maximum		1	16		1,925	620		2,545	3,075
1200 Swing up, minimum		1	16		1,500	620		2,120	2,600
1250 Maximum		1	16		6,125	620		6,745	7,700
1300 Portable, manual, heavy duty, spring operated		1.90	8.421		15,200	325		15,525	17,200
1400 Ceiling suspended, stationary, minimum		.78	20.513		3,650	795		4,445	5,250
1450 Fold up, with accessories, maximum		.40	40		6,975	1,550		8,525	10,100
1600 For electrically operated, add	1 Elec	1	8		2,300	410		2,710	3,150
5800 Wall pads, 1-1/2" thick, standard (not fire rated)	2 Carp	640	.025	S.F.	7.30	1.10		8.40	9.75

11 66 23.19 Boxing Ring

	Crew	Daily Output	Labor-Hours	Unit	Material	2012 Bare Costs Labor	Equipment	Total	Total Incl O&P
0010 **BOXING RING**									
4100 Elevated, 22' x 22'	L-4	.10	240	Ea.	7,725	9,925		17,650	23,800
4110 For cellular plastic foam padding, add		.10	240		975	9,925		10,900	16,400

11 66 Athletic Equipment

11 66 23 – Gymnasium Equipment

11 66 23.19 Boxing Ring

		Crew	Daily Output	Labor-Hours	Unit	Material	2012 Bare Costs Labor	2012 Bare Costs Equipment	Total	Total Incl O&P
4120	Floor level, including posts and ropes only, 20' x 20'	L-4	.80	30	Ea.	3,750	1,250		5,000	6,050
4130	Canvas, 30' x 30'	↓	5	4.800	↓	1,250	199		1,449	1,675

11 66 23.47 Gym Mats

0010	**GYM MATS**									
5500	2" thick, naugahyde covered				S.F.	3.81			3.81	4.19
5600	Vinyl/nylon covered					6.60			6.60	7.30
6000	Wrestling mats, 1" thick, heavy duty				↓	5.60			5.60	6.15

11 66 43 – Interior Scoreboards

11 66 43.10 Scoreboards

		Crew	Daily Output	Labor-Hours	Unit	Material	2012 Bare Costs Labor	2012 Bare Costs Equipment	Total	Total Incl O&P
0010	**SCOREBOARDS**									
7000	Baseball, minimum	R-3	1.30	15.385	Ea.	4,025	785	102	4,912	5,725
7200	Maximum		.05	400		17,700	20,400	2,650	40,750	53,000
7300	Football, minimum		.86	23.256		5,225	1,175	154	6,554	7,700
7400	Maximum		.20	100		15,600	5,100	660	21,360	25,500
7500	Basketball (one side), minimum		2.07	9.662		2,350	495	64	2,909	3,400
7600	Maximum		.30	66.667		3,450	3,400	440	7,290	9,350
7700	Hockey-basketball (four sides), minimum		.25	80		5,550	4,075	530	10,155	12,800
7800	Maximum	↓	.15	133	↓	5,625	6,800	885	13,310	17,300

11 66 53 – Gymnasium Dividers

11 66 53.10 Divider Curtains

		Crew	Daily Output	Labor-Hours	Unit	Material	2012 Bare Costs Labor	2012 Bare Costs Equipment	Total	Total Incl O&P
0010	**DIVIDER CURTAINS**									
4500	Gym divider curtain, mesh top, vinyl bottom, manual	L-4	500	.048	S.F.	9.15	1.99		11.14	13.15
4700	Electric roll up	L-7	400	.070	"	11.15	2.98		14.13	16.80

11 67 Recreational Equipment

11 67 13 – Bowling Alley Equipment

11 67 13.10 Bowling Alleys

		Crew	Daily Output	Labor-Hours	Unit	Material	2012 Bare Costs Labor	2012 Bare Costs Equipment	Total	Total Incl O&P
0010	**BOWLING ALLEYS** Including alley, pinsetter, scorer,									
0020	Counters and misc. supplies, minimum	4 Carp	.20	160	Lane	41,300	7,050		48,350	56,500
0150	Average		.19	168		45,400	7,425		52,825	61,500
0300	Maximum	↓	.18	177		52,000	7,850		59,850	69,500
0400	Combo table ball rack, add					1,150			1,150	1,250
0600	For automatic scorer, add, minimum					8,225			8,225	9,050
0700	Maximum				↓	9,875			9,875	10,900

11 67 23 – Shooting Range Equipment

11 67 23.10 Shooting Range

		Crew	Daily Output	Labor-Hours	Unit	Material	2012 Bare Costs Labor	2012 Bare Costs Equipment	Total	Total Incl O&P
0010	**SHOOTING RANGE** Incl. bullet traps, target provisions, controls,									
0100	Separators, ceiling system, etc. Not incl. structural shell									
0200	Commercial	L-9	.64	56.250	Point	19,500	2,250		21,750	25,100
0300	Law enforcement		.28	128		36,000	5,175		41,175	47,900
0400	National Guard armories		.71	50.704		20,500	2,050		22,550	25,900
0500	Reserve training centers		.71	50.704		15,800	2,050		17,850	20,600
0600	Schools and colleges		.32	112		38,600	4,525		43,125	49,800
0700	Major academies	↓	.19	189		45,500	7,625		53,125	62,000
0800	For acoustical treatment, add					10%	10%			
0900	For lighting, add					28%	25%			
1000	For plumbing, add					5%	5%			
1100	For ventilating system, add, minimum					40%	40%			
1200	Add, average				↓	25%	25%			

11 67 Recreational Equipment

11 67 23 – Shooting Range Equipment

11 67 23.10 Shooting Range		Crew	Daily Output	Labor-Hours	Unit	Material	2012 Bare Costs Labor	Equipment	Total	Total Incl O&P
1300	Add, maximum				Point	35%	35%			

11 68 Play Field Equipment and Structures

11 68 13 – Playground Equipment

11 68 13.10 Free-Standing Playground Equipment

			Crew	Daily Output	Labor-Hours	Unit	Material	Labor	Equipment	Total	Total Incl O&P
0010	**FREE-STANDING PLAYGROUND EQUIPMENT** See also individual items										
0200	Bike rack, 10' long, permanent	G	B-1	12	2	Ea.	405	71.50		476.50	555
0400	Horizontal monkey ladder, 14' long, 6' high			4	6		980	215		1,195	1,400
0590	Parallel bars, 10' long			4	6		279	215		494	635
0600	Posts, tether ball set, 2-3/8" O.D.			12	2		355	71.50		426.50	500
0800	Poles, multiple purpose, 10'-6" long			12	2	Pr.	168	71.50		239.50	295
1000	Ground socket for movable posts, 2-3/8" post			10	2.400		107	86		193	250
1100	3-1/2" post			10	2.400		144	86		230	291
1300	See-saw, spring, steel, 2 units			6	4	Ea.	775	143		918	1,075
1400	4 units			4	6		1,375	215		1,590	1,825
1500	6 units			3	8		1,750	286		2,036	2,375
1700	Shelter, fiberglass golf tee, 3 person			4.60	5.217		3,350	187		3,537	3,975
1900	Slides, stainless steel bed, 12' long, 6' high			3	8		3,350	286		3,636	4,150
2000	20' long, 10' high			2	12		4,475	430		4,905	5,550
2200	Swings, plain seats, 8' high, 4 seats			2	12		1,050	430		1,480	1,800
2300	8 seats			1.30	18.462		1,900	660		2,560	3,100
2500	12' high, 4 seats			2	12		2,175	430		2,605	3,025
2600	8 seats			1.30	18.462		3,750	660		4,410	5,150
2800	Whirlers, 8' diameter			3	8		2,900	286		3,186	3,625
2900	10' diameter			3	8		6,200	286		6,486	7,275

11 68 13.20 Modular Playground

		Crew	Daily Output	Labor-Hours	Unit	Material	Labor	Equipment	Total	Total Incl O&P
0010	**MODULAR PLAYGROUND** Basic components									
0100	Deck, square, steel, 48" x 48"	B-1	1	24	Ea.	500	860		1,360	1,875
0110	Recycled polyurethane		1	24		500	860		1,360	1,875
0120	Triangular, steel, 48" side		1	24		660	860		1,520	2,050
0130	Post, steel, 5" square		18	1.333	L.F.	37	47.50		84.50	115
0140	Aluminum, 2-3/8" square		20	1.200		36.50	43		79.50	106
0150	5" square		18	1.333		39	47.50		86.50	117
0160	Roof, square poly, 54" side		18	1.333	Ea.	1,275	47.50		1,322.50	1,475
0170	Wheelchair transfer module, for 3' high deck		3	8	"	2,850	286		3,136	3,600
0180	Guardrail, pipe, 36" high		60	.400	L.F.	174	14.30		188.30	214
0190	Steps, deck-to-deck, 3 – 8" steps		8	3	Ea.	1,050	107		1,157	1,350
0200	Activity panel, crawl through panel		2	12		570	430		1,000	1,300
0210	Alphabet/spelling panel		2	12		625	430		1,055	1,350
0360	With guardrails		3	8		1,650	286		1,936	2,275
0370	Crawl tunnel, straight, 56" long		4	6		1,200	215		1,415	1,650
0380	90°, 4' long		4	6		1,300	215		1,515	1,750
1200	Slide, tunnel, for 56" high deck		8	3		1,800	107		1,907	2,150
1210	Straight, poly		8	3		2,425	107		2,532	2,850
1220	Stainless steel, 54" high deck		6	4		685	143		828	975
1230	Curved, poly, 40" high deck		6	4		805	143		948	1,100
1240	Spiral slide, 56" - 72" high		5	4.800		3,600	172		3,772	4,225
1300	Ladder, vertical, for 24" - 72" high deck		5	4.800		565	172		737	885
1310	Horizontal, 8' long		5	4.800		675	172		847	1,000
1320	Corkscrew climber, 6' high		3	8		1,050	286		1,336	1,600
1330	Fire pole for 72" high deck		6	4		600	143		743	880

11 68 Play Field Equipment and Structures

11 68 13 – Playground Equipment

11 68 13.20 Modular Playground

		Crew	Daily Output	Labor-Hours	Unit	Material	2012 Bare Costs Labor	2012 Bare Costs Equipment	Total	Total Incl O&P
1340	Bridge, ring climber, 8' long	B-1	4	6	Ea.	1,750	215		1,965	2,250
1350	Suspension	↓	4	6	L.F.	875	215		1,090	1,300

11 68 16 – Play Structures

11 68 16.10 Handball/Squash Court

		Crew	Daily Output	Labor-Hours	Unit	Material	Labor	Equipment	Total	Total Incl O&P
0010	**HANDBALL/SQUASH COURT**, outdoor									
0900	Handball or squash court, outdoor, wood	2 Carp	.50	32	Ea.	4,950	1,400		6,350	7,625
1000	Masonry handball/squash court	D-1	.30	53.333	"	24,000	2,125		26,125	29,600

11 68 16.30 Platform/Paddle Tennis Court

		Crew	Daily Output	Labor-Hours	Unit	Material	Labor	Equipment	Total	Total Incl O&P
0010	**PLATFORM/PADDLE TENNIS COURT** Complete with lighting, etc.									
0100	Aluminum slat deck with aluminum frame	B-1	.08	300	Court	60,000	10,700		70,700	82,500
0500	Aluminum slat deck with wood frame	C-1	.12	266		54,500	11,200		65,700	77,000
0800	Aluminum deck heater, add	B-1	1.18	20.339		2,500	730		3,230	3,875
0900	Douglas fir planking with wood frame 2" x 6" x 30'	C-1	.12	266		51,000	11,200		62,200	73,500
1000	Plywood deck with steel frame		.12	266		51,000	11,200		62,200	73,500
1100	Steel slat deck with wood frame	↓	.12	266	↓	36,300	11,200		47,500	57,000

11 68 33 – Athletic Field Equipment

11 68 33.13 Football Field Equipment

		Crew	Daily Output	Labor-Hours	Unit	Material	Labor	Equipment	Total	Total Incl O&P
0010	**FOOTBALL FIELD EQUIPMENT**									
0020	Goal posts, steel, football, double post	B-1	1.50	16	Pr.	3,225	570		3,795	4,400
0100	Deluxe, single post		1.50	16		2,600	570		3,170	3,750
0300	Football, convertible to soccer		1.50	16		2,350	570		2,920	3,475
0500	Soccer, regulation	↓	2	12	↓	1,200	430		1,630	1,975

11 71 Medical Sterilizing Equipment

11 71 10 – Medical Sterilizers & Distillers

11 71 10.10 Sterilizers and Distillers

		Crew	Daily Output	Labor-Hours	Unit	Material	Labor	Equipment	Total	Total Incl O&P
0010	**STERILIZERS AND DISTILLERS**									
0700	Distiller, water, steam heated, 50 gal. capacity	1 Plum	1.40	5.714	Ea.	19,000	310		19,310	21,400
5600	Sterilizers, floor loading, 26" x 62" x 42", single door, steam					115,000			115,000	126,500
5650	Double door, steam					204,000			204,000	224,000
5800	General purpose, 20" x 20" x 38", single door					13,000			13,000	14,300
6000	Portable, counter top, steam, minimum					3,300			3,300	3,625
6020	Maximum					5,475			5,475	6,025
6050	Portable, counter top, gas, 17" x 15" x 32-1/2"					39,300			39,300	43,200
6150	Manual washer/sterilizer, 16" x 16" x 26"	1 Plum	2	4	↓	54,000	218		54,218	59,500
6200	Steam generators, electric 10 kW to 180 kW, freestanding									
6250	Minimum	1 Elec	3	2.667	Ea.	7,750	137		7,887	8,725
6300	Maximum	"	.70	11.429	↓	28,800	590		29,390	32,600
8200	Bed pan washer-sanitizer	1 Plum	2	4	↓	7,275	218		7,493	8,325

11 72 Examination and Treatment Equipment

11 72 13 – Examination Equipment

11 72 13.13 Examination Equipment	Crew	Daily Output	Labor-Hours	Unit	Material	2012 Bare Costs Labor	Equipment	Total	Total Incl O&P
0010 **EXAMINATION EQUIPMENT**									
0300 Blood pressure unit, mercurial, wall				Ea.	162			162	178
0400 Diagnostic set, wall					735			735	810
4400 Scale, physician's, with height rod					290			290	320

11 72 53 – Treatment Equipment

11 72 53.13 Medical Treatment Equipment

	Crew	Daily Output	Labor-Hours	Unit	Material	Labor	Equipment	Total	Total Incl O&P
0010 **MEDICAL TREATMENT EQUIPMENT**									
6500 Surgery table, minor minimum	1 Sswk	.70	11.429	Ea.	14,100	565		14,665	16,500
6520 Maximum	"	.50	16		26,600	790		27,390	30,700
6700 Surgical lights, doctor's office, single arm	2 Elec	2	8		1,250	410		1,660	2,000
6750 Dual arm	"	1	16		5,925	825		6,750	7,750

11 73 Patient Care Equipment

11 73 10 – Patient Treatment Equipment

11 73 10.10 Treatment Equipment

	Crew	Daily Output	Labor-Hours	Unit	Material	Labor	Equipment	Total	Total Incl O&P
0010 **TREATMENT EQUIPMENT**									
0750 Exam room furnishings, average per room				Ea.	6,650			6,650	7,325
1800 Heat therapy unit, humidified, 26" x 78" x 28"				"	3,500			3,500	3,850
2100 Hubbard tank with accessories, stainless steel,									
2110 125 GPM at 45 psi water pressure				Ea.	26,800			26,800	29,500
2150 For electric overhead hoist, add					2,925			2,925	3,225
2900 K-Module for heat therapy, 20 oz. capacity, 75°F to 110°F					420			420	460
3600 Paraffin bath, 126°F, auto controlled					1,100			1,100	1,200
3900 Parallel bars for walking training, 12'-0"					1,300			1,300	1,425
4600 Station, dietary, medium, with ice					16,500			16,500	18,200
4700 Medicine					7,450			7,450	8,200
7000 Tables, physical therapy, walk off, electric	2 Carp	3	5.333		3,275	235		3,510	3,950
7150 Standard, vinyl top with base cabinets, minimum		3	5.333		1,150	235		1,385	1,600
7200 Maximum		2	8		5,550	355		5,905	6,650
8400 Whirlpool bath, mobile, sst, 18" x 24" x 60"					4,600			4,600	5,050
8450 Fixed, incl. mixing valves	1 Plum	2	4		9,400	218		9,618	10,600

11 73 10.20 Bariatric Equipment

	Crew	Daily Output	Labor-Hours	Unit	Material	Labor	Equipment	Total	Total Incl O&P
0010 **BARIATRIC EQUIPMENT**									
5000 Patient lift, electric operated, arm style									
5110 400 lb. capacity				Ea.	2,150			2,150	2,375
5120 450 lb. capacity					2,975			2,975	3,250
5130 600 lb. capacity					3,825			3,825	4,200
5140 700 lb. capacity					4,925			4,925	5,400
5150 1,000 lb. capacity					9,900			9,900	10,900
5200 Overhead, 4-post, 1,000 lb. capacity					15,900			15,900	17,500
5300 Overhead, track type, 450 lb. capacity, not including track					3,175			3,175	3,500
5500 For fabric sling, add					300			300	330
5550 For digital scale, add					900			900	990

11 74 Dental Equipment

11 74 10 – Dental Office Equipment

11 74 10.10 Diagnostic and Treatment Equipment

		Daily Output	Labor-Hours	Unit	2012 Bare Costs				Total Incl O&P
11 74 10.10 Diagnostic and Treatment Equipment	Crew				Material	Labor	Equipment	Total	
0010 **DIAGNOSTIC AND TREATMENT EQUIPMENT**									
0020 Central suction system, minimum	1 Plum	1.20	6.667	Ea.	1,550	365		1,915	2,250
0100 Maximum	"	.90	8.889		4,250	485		4,735	5,400
0300 Air compressor, minimum	1 Skwk	.80	10		2,775	455		3,230	3,750
0400 Maximum		.50	16		7,350	725		8,075	9,200
0600 Chair, electric or hydraulic, minimum		.50	16		2,300	725		3,025	3,650
0700 Maximum		.25	32		8,125	1,450		9,575	11,200
0800 Doctor's/assistant's stool, minimum					220			220	242
0850 Maximum					665			665	730
1000 Drill console with accessories, minimum	1 Skwk	1.60	5		2,150	227		2,377	2,725
1100 Maximum		1.60	5		5,125	227		5,352	6,000
2000 Light, ceiling mounted, minimum		8	1		1,125	45.50		1,170.50	1,325
2100 Maximum		8	1		1,700	45.50		1,745.50	1,950
2200 Unit light, minimum	2 Skwk	5.33	3.002		620	136		756	890
2210 Maximum		5.33	3.002		1,375	136		1,511	1,700
2220 Track light, minimum		3.20	5		1,500	227		1,727	2,000
2230 Maximum		3.20	5		2,700	227		2,927	3,325
2300 Sterilizers, steam portable, minimum					1,350			1,350	1,500
2350 Maximum					10,400			10,400	11,400
2600 Steam, institutional					3,250			3,250	3,550
2650 Dry heat, electric, portable, 3 trays					1,225			1,225	1,325
2700 Ultra-sonic cleaner, portable, minimum					410			410	450
2750 Maximum (institutional)					1,225			1,225	1,350
3000 X-ray unit, wall, minimum	1 Skwk	4	2		1,825	91		1,916	2,150
3010 Maximum		4	2		3,800	91		3,891	4,325
3100 Panoramic unit		.60	13.333		15,500	605		16,105	18,000
3105 Deluxe, minimum	2 Skwk	1.60	10		11,000	455		11,455	12,900
3110 Maximum	"	1.60	10		42,000	455		42,455	46,900
3500 Developers, X-ray, average	1 Plum	5.33	1.501		4,950	82		5,032	5,575
3600 Maximum	"	5.33	1.501		8,100	82		8,182	9,025

11 76 Operating Room Equipment

11 76 10 – Operating Room Equipment

11 76 10.10 Surgical Equipment

		Daily Output	Labor-Hours	Unit	Material	Labor	Equipment	Total	Total Incl O&P
0010 **SURGICAL EQUIPMENT**									
5000 Scrub, surgical, stainless steel, single station, minimum	1 Plum	3	2.667	Ea.	4,050	146		4,196	4,675
5100 Maximum					6,825			6,825	7,500
6550 Major surgery table, minimum	1 Sswk	.50	16		32,100	790		32,890	36,700
6570 Maximum	"	.50	16		106,000	790		106,790	118,000
6800 Surgical lights, major operating room, dual head, minimum	2 Elec	1	16		15,100	825		15,925	17,800
6850 Maximum	"	1	16		29,700	825		30,525	33,900

11 77 Radiology Equipment

11 77 10 - Radiology Equipment

11 77 10.10 X-Ray Equipment	Crew	Daily Output	Labor-Hours	Unit	Material	2012 Bare Costs Labor	Equipment	Total	Total Incl O&P
0010 **X-RAY EQUIPMENT**									
8700 X-ray, mobile, minimum				Ea.	12,400			12,400	13,600
8750 Maximum					77,000			77,000	84,500
8900 Stationary, minimum					42,700			42,700	47,000
8950 Maximum					222,000			222,000	244,000
9150 Developing processors, minimum					11,900			11,900	13,100
9200 Maximum					31,200			31,200	34,300

11 78 Mortuary Equipment

11 78 13 - Mortuary Refrigerators

11 78 13.10 Mortuary and Autopsy Equipment

	Crew	Daily Output	Labor-Hours	Unit	Material	Labor	Equipment	Total	Total Incl O&P
0010 **MORTUARY AND AUTOPSY EQUIPMENT**									
0015 Autopsy table, standard	1 Plum	1	8	Ea.	9,175	435		9,610	10,800
0020 Deluxe	"	.60	13.333		15,600	730		16,330	18,200
3200 Mortuary refrigerator, end operated, 2 capacity					13,100			13,100	14,400
3300 6 capacity					21,000			21,000	23,100

11 78 16 - Crematorium Equipment

11 78 16.10 Crematory

	Crew	Daily Output	Labor-Hours	Unit	Material	Labor	Equipment	Total	Total Incl O&P
0010 **CREMATORY**									
1500 Crematory, not including building, 1 place	Q-3	.20	160	Ea.	65,000	8,325		73,325	84,000
1750 2 place	"	.10	320	"	93,000	16,600		109,600	127,000

11 82 Solid Waste Handling Equipment

11 82 19 - Packaged Incinerators

11 82 19.10 Packaged Gas Fired Incinerators

	Crew	Daily Output	Labor-Hours	Unit	Material	Labor	Equipment	Total	Total Incl O&P
0010 **PACKAGED GAS FIRED INCINERATORS**									
4400 Incinerator, gas, not incl. chimney, elec. or pipe, 50#/hr., minimum	Q-3	.80	40	Ea.	11,900	2,075		13,975	16,100
4420 Maximum		.70	45.714		38,200	2,375		40,575	45,700
4440 200 lb. per hr., minimum (batch type)		.60	53.333		65,500	2,775		68,275	76,000
4460 Maximum (with feeder)		.50	64		93,000	3,325		96,325	107,000
4480 400 lb. per hr., minimum (batch type)		.30	106		72,500	5,550		78,050	88,500
4500 Maximum (with feeder)		.25	128		98,500	6,650		105,150	118,000
4520 800 lb. per hr., with feeder, minimum		.20	160		118,500	8,325		126,825	143,000
4540 Maximum		.17	188		164,000	9,800		173,800	195,500
4560 1,200 lb. per hr., with feeder, minimum		.15	213		159,500	11,100		170,600	192,000
4580 Maximum		.11	290		190,500	15,100		205,600	232,500
4600 2,000 lb. per hr., with feeder, minimum		.10	320		390,500	16,600		407,100	454,500
4620 Maximum		.05	640		534,500	33,300		567,800	638,000
4700 For heat recovery system, add, minimum		.25	128		75,500	6,650		82,150	93,000
4710 Add, maximum		.11	290		241,000	15,100		256,100	288,000
4720 For automatic ash conveyer, add		.50	64		31,600	3,325		34,925	39,800
4750 Large municipal incinerators, incl. stack, minimum		.25	128	Ton/day	19,200	6,650		25,850	31,200
4850 Maximum		.10	320	"	51,000	16,600		67,600	81,500

11 82 26 - Facility Waste Compactors

11 82 26.10 Compactors

	Crew	Daily Output	Labor-Hours	Unit	Material	Labor	Equipment	Total	Total Incl O&P
0010 **COMPACTORS**									
0020 Compactors, 115 volt, 250#/hr., chute fed	L-4	1	24	Ea.	11,900	995		12,895	14,600
0100 Hand fed		2.40	10		8,725	415		9,140	10,200

11 82 Solid Waste Handling Equipment

11 82 26 – Facility Waste Compactors

11 82 26.10 Compactors

		Daily Output	Labor-Hours	Unit	Material	2012 Bare Costs Labor	2012 Bare Costs Equipment	Total	Total Incl O&P
0300	Multi-bag, 230 volt, 600#/hr., chute fed	L-4	1	24	Ea.	10,800	995		11,795	13,400
0400	Hand fed		1	24		9,275	995		10,270	11,700
0500	Containerized, hand fed, 2 to 6 C.Y. containers, 250#/hr.		1	24		11,500	995		12,495	14,100
0550	For chute fed, add per floor		1	24		1,325	995		2,320	2,975
1000	Heavy duty industrial compactor, 0.5 C.Y. capacity		1	24		7,750	995		8,745	10,100
1050	1.0 C.Y. capacity		1	24		11,800	995		12,795	14,500
1100	3.0 C.Y. capacity		.50	48		16,100	1,975		18,075	20,800
1150	5.0 C.Y. capacity		.50	48		19,900	1,975		21,875	25,000
1200	Combination shredder/compactor (5,000 lb./hr.)		.50	48		39,500	1,975		41,475	46,500
1400	For handling hazardous waste materials, 55 gallon drum packer, std.					18,300			18,300	20,200
1410	55 gallon drum packer w/HEPA filter					22,900			22,900	25,200
1420	55 gallon drum packer w/charcoal & HEPA filter					30,600			30,600	33,600
1430	All of the above made explosion proof, add					13,900			13,900	15,300
5500	Shredder, municipal use, 35 ton per hour					286,500			286,500	315,000
5600	60 ton per hour					610,000			610,000	671,000
5750	Shredder & baler, 50 ton per day					572,000			572,000	629,500
5800	Shredder, industrial, minimum					22,500			22,500	24,800
5850	Maximum					121,000			121,000	133,000
5900	Baler, industrial, minimum					9,025			9,025	9,925
5950	Maximum					527,500			527,500	580,000
6000	Transfer station compactor, with power unit									
6050	and pedestal, not including pit, 50 ton per hour				Ea.	180,500			180,500	198,500

11 91 Religious Equipment

11 91 13 – Baptisteries

11 91 13.10 Baptistry

		Crew	Daily Output	Labor-Hours	Unit	Material	2012 Bare Costs Labor	2012 Bare Costs Equipment	Total	Total Incl O&P
0010	**BAPTISTRY**									
0150	Fiberglass, 3'-6" deep, x 13'-7" long,									
0160	steps at both ends, incl. plumbing, minimum	L-8	1	20	Ea.	4,600	925		5,525	6,475
0200	Maximum	"	.70	28.571		7,500	1,325		8,825	10,300
0250	Add for filter, heater and lights					1,650			1,650	1,800

11 91 23 – Sanctuary Equipment

11 91 23.10 Sanctuary Furnishings

		Crew	Daily Output	Labor-Hours	Unit	Material	2012 Bare Costs Labor	2012 Bare Costs Equipment	Total	Total Incl O&P
0010	**SANCTUARY FURNISHINGS**									
0020	Altar, wood, custom design, plain	1 Carp	1.40	5.714	Ea.	2,150	252		2,402	2,775
0050	Deluxe	"	.20	40		10,400	1,775		12,175	14,100
0070	Granite or marble, average	2 Marb	.50	32		12,300	1,325		13,625	15,500
0090	Deluxe	"	.20	80		35,100	3,300		38,400	43,600
0100	Arks, prefabricated, plain	2 Carp	.80	20		9,000	880		9,880	11,300
0130	Deluxe, maximum	"	.20	80		128,000	3,525		131,525	146,500
0500	Reconciliation room, wood, prefabricated, single, plain	1 Carp	.60	13.333		2,800	590		3,390	3,975
0550	Deluxe		.40	20		7,725	880		8,605	9,850
0650	Double, plain		.40	20		5,625	880		6,505	7,525
0700	Deluxe		.20	40		17,500	1,775		19,275	22,000
1000	Lecterns, wood, plain		5	1.600		835	70.50		905.50	1,025
1100	Deluxe		2	4		5,675	176		5,851	6,500
2000	Pulpits, hardwood, prefabricated, plain		2	4		1,375	176		1,551	1,775
2100	Deluxe		1.60	5		9,350	221		9,571	10,600
2500	Railing, hardwood, average		25	.320	L.F.	192	14.10		206.10	233
3000	Seating, individual, oak, contour, laminated		21	.381	Person	165	16.80		181.80	207

403

11 91 23.10 Sanctuary Furnishings	Crew	Daily Output	Labor-Hours	Unit	Material	2012 Bare Costs Labor	2012 Bare Costs Equipment	Total	Total Incl O&P
3100 Cushion seat	1 Carp	21	.381	Person	149	16.80		165.80	190
3200 Fully upholstered		21	.381		146	16.80		162.80	186
3300 Combination, self-rising		21	.381		375	16.80		391.80	440
3500 For cherry, add					30%				
5000 Wall cross, aluminum, extruded, 2" x 2" section	1 Carp	34	.235	L.F.	160	10.40		170.40	192
5150 4" x 4" section		29	.276		230	12.15		242.15	272
5300 Bronze, extruded, 1" x 2" section		31	.258		315	11.40		326.40	365
5350 2-1/2" x 2-1/2" section		34	.235		480	10.40		490.40	540
5450 Solid bar stock, 1/2" x 3" section		29	.276		625	12.15		637.15	710
5600 Fiberglass, stock		34	.235		130	10.40		140.40	158
5700 Stainless steel, 4" deep, channel section		29	.276		505	12.15		517.15	575
5800 4" deep box section		29	.276		690	12.15		702.15	780

Estimating Tips

General

- The items in this division are usually priced per square foot or each. Most of these items are purchased by the owner and installed by the contractor. Do not assume the items in Division 12 will be purchased and installed by the contractor. Check the specifications for responsibilities and include receiving, storage, installation, and mechanical and electrical hookups in the appropriate divisions.

- Some items in this division require some type of support system that is not usually furnished with the item. Examples of these systems include blocking for the attachment of casework and heavy drapery rods. The required blocking must be added to the estimate in the appropriate division.

Reference Numbers

Reference numbers are shown in shaded boxes at the beginning of some major classifications. These numbers refer to related items in the Reference Section. The reference information may be an estimating procedure, an alternate pricing method, or technical information.

Note: Not all subdivisions listed here necessarily appear in this publication.

Division 12 - Furnishings

12 21 Window Blinds

12 21 13 – Horizontal Louver Blinds

12 21 13.13 Metal Horizontal Louver Blinds

		Crew	Daily Output	Labor-Hours	Unit	Material	2012 Bare Costs Labor	Equipment	Total	Total Incl O&P
0010	**METAL HORIZONTAL LOUVER BLINDS**									
0020	Horizontal, 1" aluminum slats, solid color, stock	1 Carp	590	.014	S.F.	4.99	.60		5.59	6.40
0090	Custom, minimum		590	.014		5.50	.60		6.10	6.95
0100	Maximum		440	.018		7.40	.80		8.20	9.40
0250	2" aluminum slats, custom, minimum		590	.014		8.55	.60		9.15	10.30
0350	Maximum		440	.018		13.80	.80		14.60	16.45
0450	Stock, minimum		590	.014		5	.60		5.60	6.40
0500	Maximum		440	.018		8.15	.80		8.95	10.20
0600	2" steel slats, stock, minimum		590	.014		1.90	.60		2.50	3.01
0630	Maximum		440	.018		5.30	.80		6.10	7.10
0750	Custom, minimum		590	.014		1.72	.60		2.32	2.81
0850	Maximum		400	.020		8.50	.88		9.38	10.70

12 21 13.33 Vinyl Horizontal Louver Blinds

		Crew	Daily Output	Labor-Hours	Unit	Material	2012 Bare Costs Labor	Equipment	Total	Total Incl O&P
0010	**VINYL HORIZONTAL LOUVER BLINDS**									
0100	2" custom composite, 48" wide, 48" high	1 Carp	30	.267	Ea.	89	11.75		100.75	116
0120	72" high		29	.276		133	12.15		145.15	165
0140	96" high		28	.286		167	12.60		179.60	203
0200	60" wide, 60" high		27	.296		109	13.05		122.05	140
0220	72" high		25	.320		127	14.10		141.10	161
0240	96" high		24	.333		176	14.70		190.70	217
0300	72" wide, 72" high		25	.320		183	14.10		197.10	223
0320	96" high		23	.348		255	15.35		270.35	305
0400	96" wide, 96" high		20	.400		310	17.65		327.65	370
1000	2" custom faux wood, 48" wide, 48" high		30	.267		57.50	11.75		69.25	81
1020	72" high		29	.276		75.50	12.15		87.65	102
1040	96" high		28	.286		94	12.60		106.60	122
1300	72" wide, 72" high		25	.320		117	14.10		131.10	151
1320	96" high		23	.348		186	15.35		201.35	228
1400	96" wide, 96" high		20	.400		207	17.65		224.65	254

12 21 16 – Vertical Louver Blinds

12 21 16.13 Metal Vertical Louver Blinds

		Crew	Daily Output	Labor-Hours	Unit	Material	2012 Bare Costs Labor	Equipment	Total	Total Incl O&P
0010	**METAL VERTICAL LOUVER BLINDS**									
1500	Vertical, 3" to 5" PVC or cloth strips, minimum	1 Carp	460	.017	S.F.	9.90	.77		10.67	12.05
1600	Maximum		400	.020		26.50	.88		27.38	30.50
1800	4" aluminum slats, minimum		460	.017		7.95	.77		8.72	9.95
1900	Maximum		400	.020		13	.88		13.88	15.65
1950	Mylar mirror-finish strips, to 8" wide, minimum		460	.017		4.85	.77		5.62	6.55
1970	Maximum		400	.020		11.05	.88		11.93	13.50

12 22 Curtains and Drapes

12 22 16 – Drapery Track and Accessories

12 22 16.10 Drapery Hardware

		Crew	Daily Output	Labor-Hours	Unit	Material	2012 Bare Costs Labor	Equipment	Total	Total Incl O&P
0010	**DRAPERY HARDWARE**									
0030	Standard traverse, per foot, minimum	1 Carp	59	.136	L.F.	5.75	6		11.75	15.55
0100	Maximum		51	.157	"	8.05	6.90		14.95	19.50
4000	Traverse rods, adjustable, 28" to 48"		22	.364	Ea.	22	16.05		38.05	49
4020	48" to 84"		20	.400		28	17.65		45.65	58
4040	66" to 120"		18	.444		34	19.60		53.60	67.50
4060	84" to 156"		16	.500		38.50	22		60.50	76
4080	100" to 180"		14	.571		44	25		69	87.50

12 22 Curtains and Drapes

12 22 16 – Drapery Track and Accessories

	12 22 16.10 Drapery Hardware	Crew	Daily Output	Labor-Hours	Unit	Material	2012 Bare Costs Labor	Equipment	Total	Total Incl O&P
4090	156" to 228"	1 Carp	13	.615	Ea.	49.50	27		76.50	96.50
4100	228" to 312"		13	.615		59.50	27		86.50	108
4200	Double rods, adjustable, 30" to 48"		9	.889		37.50	39		76.50	102
4220	48" to 86"		9	.889		55.50	39		94.50	122
4240	86" to 150"		8	1		60	44		104	134
4260	100" to 180"		7	1.143		63	50.50		113.50	147
4300	Tray and curtain rod, adjustable, 30" to 48"		9	.889		29.50	39		68.50	92.50
4320	48" to 86"		9	.889		41.50	39		80.50	106
4340	86" to 150"		8	1		52.50	44		96.50	126
4360	100" to 180"		7	1.143		61	50.50		111.50	145
4600	Valance, pinch pleated fabric, 12" deep, up to 54" long, minimum					36.50			36.50	40.50
4610	Maximum					91.50			91.50	101
4620	Up to 77" long, minimum					56.50			56.50	62
4630	Maximum					148			148	163
5000	Stationary rods, first 2'					9.35			9.35	10.25
5020	Each additional foot, add				L.F.	3.64			3.64	4

12 22 16.20 Blast Curtains

		Crew	Daily Output	Labor-Hours	Unit	Material	2012 Bare Costs Labor	Equipment	Total	Total Incl O&P
0010	**BLAST CURTAINS** per L.F. horizontal opening width, off-white or gray fabric									
0100	Blast curtains, drapery system, complete, including hardware, minimum	1 Carp	10.25	.780	L.F.	185	34.50		219.50	257
0120	Average		10.25	.780		200	34.50		234.50	273
0140	Maximum		10.25	.780		230	34.50		264.50	305

12 23 Interior Shutters

12 23 10 – Wood Interior Shutters

12 23 10.10 Wood Interior Shutters

		Crew	Daily Output	Labor-Hours	Unit	Material	2012 Bare Costs Labor	Equipment	Total	Total Incl O&P
0010	**WOOD INTERIOR SHUTTERS**, louvered									
0200	Two panel, 27" wide, 36" high	1 Carp	5	1.600	Set	146	70.50		216.50	270
0300	33" wide, 36" high		5	1.600		192	70.50		262.50	320
0500	47" wide, 36" high		5	1.600		253	70.50		323.50	385
1000	Four panel, 27" wide, 36" high		5	1.600		179	70.50		249.50	305
1100	33" wide, 36" high		5	1.600		230	70.50		300.50	360
1300	47" wide, 36" high		5	1.600		305	70.50		375.50	445

12 23 10.13 Wood Panels

		Crew	Daily Output	Labor-Hours	Unit	Material	2012 Bare Costs Labor	Equipment	Total	Total Incl O&P
0010	**WOOD PANELS**									
3000	Wood folding panels with movable louvers, 7" x 20" each	1 Carp	17	.471	Pr.	78	21		99	118
3300	8" x 28" each		17	.471		78	21		99	118
3450	9" x 36" each		17	.471		90	21		111	131
3600	10" x 40" each		17	.471		98	21		119	140
4000	Fixed louver type, stock units, 8" x 20" each		17	.471		92	21		113	133
4150	10" x 28" each		17	.471		78	21		99	118
4300	12" x 36" each		17	.471		92	21		113	133
4450	18" x 40" each		17	.471		132	21		153	177
5000	Insert panel type, stock, 7" x 20" each		17	.471		20	21		41	54
5150	8" x 28" each		17	.471		37	21		58	72.50
5300	9" x 36" each		17	.471		47	21		68	83.50
5450	10" x 40" each		17	.471		50	21		71	87
5600	Raised panel type, stock, 10" x 24" each		17	.471		233	21		254	288
5650	12" x 26" each		17	.471		233	21		254	288
5700	14" x 30" each		17	.471		258	21		279	315
5750	16" x 36" each		17	.471		285	21		306	345

12 23 Interior Shutters

12 23 10 – Wood Interior Shutters

12 23 10.13 Wood Panels	Crew	Daily Output	Labor- Hours	Unit	Material	2012 Bare Costs Labor	Equipment	Total	Total Incl O&P
6000	For custom built pine, add				Pr.	22%				
6500	For custom built hardwood blinds, add				↓	42%				

12 24 Window Shades

12 24 13 – Roller Window Shades

12 24 13.10 Shades

		Crew	Daily Output	Labor- Hours	Unit	Material	Labor	Equipment	Total	Total Incl O&P
0010	**SHADES**									
0020	Basswood, roll-up, stain finish, 3/8" slats	1 Carp	300	.027	S.F.	13.95	1.18		15.13	17.15
0200	7/8" slats		300	.027		13.15	1.18		14.33	16.30
0300	Vertical side slide, stain finish, 3/8" slats		300	.027		18.05	1.18		19.23	21.50
0400	7/8" slats	↓	300	.027		18.05	1.18		19.23	21.50
0500	For fire retardant finishes, add					16%				
0600	For "B" rated finishes, add					20%				
0900	Mylar, single layer, non-heat reflective	1 Carp	685	.012		6.90	.52		7.42	8.35
1000	Double layered, heat reflective		685	.012		9.55	.52		10.07	11.30
1100	Triple layered, heat reflective	↓	685	.012		11.10	.52		11.62	13
1200	For metal roller instead of wood, add per				Shade	4.33			4.33	4.76
1300	Vinyl coated cotton, standard	1 Carp	685	.012	S.F.	3.53	.52		4.05	4.67
1400	Lightproof decorator shades		685	.012		2.21	.52		2.73	3.22
1500	Vinyl, lightweight, 4 gauge		685	.012		.58	.52		1.10	1.43
1600	Heavyweight, 6 gauge		685	.012		1.75	.52		2.27	2.72
1700	Vinyl laminated fiberglass, 6 ga., translucent		685	.012		2.48	.52		3	3.52
1800	Lightproof		685	.012		4.03	.52		4.55	5.20
3000	Woven aluminum, 3/8" thick, lightproof and fireproof	↓	350	.023	↓	5.65	1.01		6.66	7.75

12 32 Manufactured Wood Casework

12 32 16 – Manufactured Plastic-Laminate-Clad Casework

12 32 16.20 Plastic Laminate Casework Doors

		Crew	Daily Output	Labor- Hours	Unit	Material	Labor	Equipment	Total	Total Incl O&P
0010	**PLASTIC LAMINATE CASEWORK DOORS**									
1000	For casework frames, see Section 12 32 23.15									
1100	For casework hardware, see Section 12 32 23.35									
6000	Plastic laminate on particle board									
6100	12" wide, 18" high	1 Carp	25	.320	Ea.	12.20	14.10		26.30	35
6140	30" high		23	.348		20.50	15.35		35.85	46
6500	18" wide, 18" high		24	.333		18.35	14.70		33.05	42.50
6600	30" high	↓	22	.364	↓	30.50	16.05		46.55	58

12 32 16.25 Plastic Laminate Drawer Fronts

		Crew	Daily Output	Labor- Hours	Unit	Material	Labor	Equipment	Total	Total Incl O&P
0010	**PLASTIC LAMINATE DRAWER FRONTS**									
2800	Plastic laminate on particle board front									
3000	4" high, 12" wide	1 Carp	17	.471	Ea.	5.20	21		26.20	37.50
3200	18" wide	"	16	.500	"	7.75	22		29.75	42.50

12 32 23 – Hardwood Casework

12 32 23.10 Manufactured Wood Casework, Stock Units

		Crew	Daily Output	Labor- Hours	Unit	Material	Labor	Equipment	Total	Total Incl O&P
0010	**MANUFACTURED WOOD CASEWORK, STOCK UNITS**									
0300	Built-in drawer units, pine, 18" deep, 32" high, unfinished									
0400	Minimum	2 Carp	53	.302	L.F.	149	13.30		162.30	184
0500	Maximum	"	40	.400	"	185	17.65		202.65	230
0700	Kitchen base cabinets, hardwood, not incl. counter tops,									

12 32 Manufactured Wood Casework

12 32 23 – Hardwood Casework

12 32 23.10 Manufactured Wood Casework, Stock Units

		Crew	Daily Output	Labor-Hours	Unit	Material	2012 Bare Costs Labor	Equipment	Total	Total Incl O&P
0710	24" deep, 35" high, prefinished									
0800	One top drawer, one door below, 12" wide	2 Carp	24.80	.645	Ea.	234	28.50		262.50	300
0840	18" wide		23.30	.687		263	30.50		293.50	335
0880	24" wide		22.30	.717		310	31.50		341.50	390
1000	Four drawers, 12" wide		24.80	.645		246	28.50		274.50	315
1040	18" wide		23.30	.687		275	30.50		305.50	350
1060	24" wide		22.30	.717		310	31.50		341.50	390
1200	Two top drawers, two doors below, 27" wide		22	.727		340	32		372	425
1260	36" wide		20.30	.788		410	35		445	505
1300	48" wide		18.90	.847		470	37.50		507.50	575
1500	Range or sink base, two doors below, 30" wide		21.40	.748		315	33		348	395
1540	36" wide		20.30	.788		355	35		390	445
1580	48" wide		18.90	.847		395	37.50		432.50	495
1800	For sink front units, deduct					140			140	154
2000	Corner base cabinets, 36" wide, standard	2 Carp	18	.889		550	39		589	670
2100	Lazy Susan with revolving door	"	16.50	.970		740	43		783	880
4000	Kitchen wall cabinets, hardwood, 12" deep with two doors									
4050	12" high, 30" wide	2 Carp	24.80	.645	Ea.	210	28.50		238.50	275
4100	36" wide		24	.667		246	29.50		275.50	315
4400	15" high, 30" wide		24	.667		210	29.50		239.50	276
4440	36" wide		22.70	.705		246	31		277	320
4700	24" high, 30" wide		23.30	.687		282	30.50		312.50	355
4720	36" wide		22.70	.705		310	31		341	390
5000	30" high, one door, 12" wide		22	.727		200	32		232	270
5040	18" wide		20.90	.766		234	34		268	310
5060	24" wide		20.30	.788		273	35		308	355
5300	Two doors, 27" wide		19.80	.808		300	35.50		335.50	385
5340	36" wide		18.80	.851		350	37.50		387.50	445
5380	48" wide		18.40	.870		440	38.50		478.50	545
6000	Corner wall, 30" high, 24" wide		18	.889		271	39		310	360
6050	30" wide		17.20	.930		280	41		321	375
6100	36" wide		16.50	.970		296	43		339	390
6500	Revolving Lazy Susan		15.20	1.053		425	46.50		471.50	535
7000	Broom cabinet, 84" high, 24" deep, 18" wide		10	1.600		590	70.50		660.50	760
7500	Oven cabinets, 84" high, 24" deep, 27" wide		8	2		860	88		948	1,075
7750	Valance board trim		396	.040	L.F.	12.40	1.78		14.18	16.40
9000	For deluxe models of all cabinets, add					40%				
9500	For custom built in place, add					25%	10%			
9558	Rule of thumb, kitchen cabinets not including									
9560	appliances & counter top, minimum	2 Carp	30	.533	L.F.	150	23.50		173.50	201
9600	Maximum	"	25	.640	"	320	28		348	395
9610	For metal cabinets, see Section 12 35 70.13									

12 32 23.15 Manufactured Wood Casework Frames

		Crew	Daily Output	Labor-Hours	Unit	Material	2012 Bare Costs Labor	Equipment	Total	Total Incl O&P
0010	**MANUFACTURED WOOD CASEWORK FRAMES**									
0050	Base cabinets, counter storage, 36" high									
0100	One bay, 18" wide	1 Carp	2.70	2.963	Ea.	137	131		268	350
0400	Two bay, 36" wide		2.20	3.636		209	160		369	475
1100	Three bay, 54" wide		1.50	5.333		249	235		484	635
2800	Bookcases, one bay, 7' high, 18" wide		2.40	3.333		161	147		308	405
3500	Two bay, 36" wide		1.60	5		234	221		455	595
4100	Three bay, 54" wide		1.20	6.667		385	294		679	875
5100	Coat racks, one bay, 7' high, 24" wide		4.50	1.778		161	78.50		239.50	298

12 32 Manufactured Wood Casework

12 32 23 – Hardwood Casework

12 32 23.15 Manufactured Wood Casework Frames

		Crew	Daily Output	Labor-Hours	Unit	Material	2012 Bare Costs Labor	Equipment	Total	Total Incl O&P
5300	Two bay, 48" wide	1 Carp	2.75	2.909	Ea.	224	128		352	445
5800	Three bay, 72" wide		2.10	3.810		330	168		498	620
6100	Wall mounted cabinet, one bay, 24" high, 18" wide		3.60	2.222		88.50	98		186.50	248
6800	Two bay, 36" wide		2.20	3.636		129	160		289	390
7400	Three bay, 54" wide		1.70	4.706		161	208		369	495
8400	30" high, one bay, 18" wide		3.60	2.222		96	98		194	257
9000	Two bay, 36" wide		2.15	3.721		128	164		292	395
9400	Three bay, 54" wide		1.60	5		160	221		381	515
9800	Wardrobe, 7' high, single, 24" wide		2.70	2.963		177	131		308	395
9880	Partition & adjustable shelves, 48" wide		1.70	4.706		226	208		434	570
9950	Partition, adjustable shelves & drawers, 48" wide		1.40	5.714		340	252		592	760

12 32 23.20 Manufactured Hardwood Casework Doors

		Crew	Daily Output	Labor-Hours	Unit	Material	2012 Bare Costs Labor	Equipment	Total	Total Incl O&P
0010	**MANUFACTURED HARDWOOD CASEWORK DOORS**									
2000	Glass panel, hardwood frame									
2200	12" wide, 18" high	1 Carp	34	.235	Ea.	25	10.40		35.40	43.50
2600	30" high		32	.250		41.50	11.05		52.55	63
4450	18" wide, 18" high		32	.250		37.50	11.05		48.55	58.50
4550	30" high		29	.276		62.50	12.15		74.65	87.50
5000	Hardwood, raised panel									
5100	12" wide, 18" high	1 Carp	16	.500	Ea.	28	22		50	65
5200	30" high		15	.533		47	23.50		70.50	87.50
5500	18" wide, 18" high		15	.533		42	23.50		65.50	82.50
5600	30" high		14	.571		70.50	25		95.50	117

12 32 23.25 Manufactured Wood Casework Drawer Fronts

		Crew	Daily Output	Labor-Hours	Unit	Material	2012 Bare Costs Labor	Equipment	Total	Total Incl O&P
0010	**MANUFACTURED WOOD CASEWORK DRAWER FRONTS**									
0100	Solid hardwood front									
1000	4" high, 12" wide	1 Carp	17	.471	Ea.	3.41	21		24.41	36
1200	18" wide	"	16	.500	"	5.05	22		27.05	39.50

12 32 23.30 Manufactured Wood Casework Vanities

		Crew	Daily Output	Labor-Hours	Unit	Material	2012 Bare Costs Labor	Equipment	Total	Total Incl O&P
0010	**MANUFACTURED WOOD CASEWORK VANITIES**									
8000	Vanity bases, 2 doors, 30" high, 21" deep, 24" wide	2 Carp	20	.800	Ea.	233	35.50		268.50	310
8050	30" wide		16	1		268	44		312	365
8100	36" wide		13.33	1.200		355	53		408	470
8150	48" wide		11.43	1.400		425	61.50		486.50	565
9000	For deluxe models of all vanities, add to above					40%				
9500	For custom built in place, add to above					25%	10%			

12 32 23.35 Manufactured Wood Casework Hardware

		Crew	Daily Output	Labor-Hours	Unit	Material	2012 Bare Costs Labor	Equipment	Total	Total Incl O&P
0010	**MANUFACTURED WOOD CASEWORK HARDWARE**									
1000	Catches, minimum	1 Carp	235	.034	Ea.	1.15	1.50		2.65	3.58
1040	Maximum	"	80	.100	"	7.20	4.41		11.61	14.70
2000	Door/drawer pulls, handles									
2200	Handles and pulls, projecting, metal, minimum	1 Carp	48	.167	Ea.	4.66	7.35		12.01	16.45
2240	Maximum		24	.333		9.85	14.70		24.55	33.50
2300	Wood, minimum		48	.167		4.90	7.35		12.25	16.70
2340	Maximum		24	.333		9	14.70		23.70	32.50
2600	Flush, metal, minimum		48	.167		4.90	7.35		12.25	16.70
2640	Maximum		24	.333		9	14.70		23.70	32.50
3000	Drawer tracks/glides, minimum		48	.167	Pr.	8.30	7.35		15.65	20.50
3040	Maximum		24	.333		24	14.70		38.70	49
4000	Cabinet hinges, minimum		160	.050		2.82	2.21		5.03	6.50
4040	Maximum		68	.118		10.70	5.20		15.90	19.75

12 35 Specialty Casework

12 35 50 – Educational/Library Casework

12 35 50.13 Educational Casework

		Crew	Daily Output	Labor-Hours	Unit	Material	2012 Bare Costs Labor	Equipment	Total	Total Incl O&P
0010	**EDUCATIONAL CASEWORK**									
5000	School, 24" deep, metal, 84" high units	2 Carp	15	1.067	L.F.	395	47		442	510
5150	Counter height units		20	.800		266	35.50		301.50	350
5450	Wood, custom fabricated, 32" high counter		20	.800		221	35.50		256.50	299
5600	Add for counter top		56	.286		23.50	12.60		36.10	45.50
5800	84" high wall units		15	1.067		430	47		477	550
6000	Laminated plastic finish is same price as wood									

12 35 53 – Laboratory Casework

12 35 53.13 Metal Laboratory Casework

		Crew	Daily Output	Labor-Hours	Unit	Material	2012 Bare Costs Labor	Equipment	Total	Total Incl O&P
0010	**METAL LABORATORY CASEWORK**									
0020	Cabinets, base, door units, metal	2 Carp	18	.889	L.F.	212	39		251	294
0300	Drawer units		18	.889		475	39		514	580
0700	Tall storage cabinets, open, 7' high		20	.800		455	35.50		490.50	555
0900	With glazed doors		20	.800		695	35.50		730.50	820
1300	Wall cabinets, metal, 12-1/2" deep, open		20	.800		149	35.50		184.50	219
1500	With doors		20	.800		310	35.50		345.50	395
6300	Rule of thumb: lab furniture including installation & connection									
6320	High school				S.F.				35	39
6340	College								52	57
6360	Clinical, health care								45	49.50
6380	Industrial								72.50	79.50

12 35 59 – Display Casework

12 35 59.10 Display Cases

		Crew	Daily Output	Labor-Hours	Unit	Material	2012 Bare Costs Labor	Equipment	Total	Total Incl O&P
0010	**DISPLAY CASES** Free standing, all glass									
0020	Aluminum frame, 42" high x 36" x 12" deep	2 Carp	8	2	Ea.	1,225	88		1,313	1,450
0100	70" high x 48" x 18" deep	"	6	2.667		3,575	118		3,693	4,125
0500	For wood bases, add					9%				
0600	For hardwood frames, deduct					8%				
0700	For bronze, baked enamel finish, add					10%				
2000	Wall mounted, glass front, aluminum frame									
2010	Non-illuminated, one section 3' x 4' x 1'-4"	2 Carp	5	3.200	Ea.	2,150	141		2,291	2,575
2100	5' x 4' x 1'-4"		5	3.200		2,475	141		2,616	2,950
2200	6' x 4' x 1'-4"		4	4		2,975	176		3,151	3,550
2500	Two sections, 8' x 4' x 1'-4"		2	8		2,050	355		2,405	2,800
2600	10' x 4' x 1'-4"		2	8		2,075	355		2,430	2,825
3000	Three sections, 16' x 4' x 1'-4"		1.50	10.667		3,900	470		4,370	5,025
3500	For fluorescent lights, add				Section	330			330	365
4000	Table exhibit cases, 2' wide, 3' high, 4' long, flat top	2 Carp	5	3.200	Ea.	1,250	141		1,391	1,600
4100	3' wide, 3' high, 4' long, sloping top	"	3	5.333	"	925	235		1,160	1,375

12 35 70 – Healthcare Casework

12 35 70.13 Hospital Casework

		Crew	Daily Output	Labor-Hours	Unit	Material	2012 Bare Costs Labor	Equipment	Total	Total Incl O&P
0010	**HOSPITAL CASEWORK**									
0500	Base cabinets, laminated plastic	2 Carp	10	1.600	L.F.	254	70.50		324.50	390
1000	Stainless steel	"	10	1.600		465	70.50		535.50	625
1200	For all drawers, add					26.50			26.50	29
1300	Cabinet base trim, 4" high, enameled steel	2 Carp	200	.080		42.50	3.53		46.03	52.50
1400	Stainless steel		200	.080		85	3.53		88.53	99
1450	Countertop, laminated plastic, no backsplash		40	.400		44	17.65		61.65	75
1650	With backsplash		40	.400		54.50	17.65		72.15	87
1800	For sink cutout, add		12.20	1.311	Ea.		58		58	89
1900	Stainless steel counter top		40	.400	L.F.	142	17.65		159.65	183

12 35 Specialty Casework

12 35 70 – Healthcare Casework

12 35 70.13 Hospital Casework

	Crew	Daily Output	Labor-Hours	Unit	Material	2012 Bare Costs Labor	Equipment	Total	Total Incl O&P
2000	For drop-in stainless 43" x 21" sink, add				Ea.	930			930	1,025
2500	Wall cabinets, laminated plastic	2 Carp	15	1.067	L.F.	190	47		237	282
2600	Enameled steel		15	1.067		233	47		280	330
2700	Stainless steel	↓	15	1.067		465	47		512	585
2800	For glass doors, add				↓	32			32	35

12 35 70.16 Nurse Station Casework

	Crew	Daily Output	Labor-Hours	Unit	Material	2012 Bare Costs Labor	Equipment	Total	Total Incl O&P
0010	**NURSE STATION CASEWORK**									
2100	Door type, laminated plastic	2 Carp	10	1.600	L.F.	294	70.50		364.50	435
2200	Enameled steel		10	1.600		281	70.50		351.50	420
2300	Stainless steel	↓	10	1.600		560	70.50		630.50	730
2400	For drawer type, add				↓	239			239	263

12 35 80 – Commercial Kitchen Casework

12 35 80.13 Metal Kitchen Casework

	Crew	Daily Output	Labor-Hours	Unit	Material	2012 Bare Costs Labor	Equipment	Total	Total Incl O&P
0010	**METAL KITCHEN CASEWORK**									
3500	Base cabinets, metal, minimum	2 Carp	30	.533	L.F.	68.50	23.50		92	112
3600	Maximum		25	.640		174	28		202	235
3700	Wall cabinets, metal, minimum		30	.533		68.50	23.50		92	112
3800	Maximum	↓	25	.640	↓	157	28		185	217

12 36 Countertops

12 36 16 – Metal Countertops

12 36 16.10 Stainless Steel Countertops

	Crew	Daily Output	Labor-Hours	Unit	Material	2012 Bare Costs Labor	Equipment	Total	Total Incl O&P
0010	**STAINLESS STEEL COUNTERTOPS**									
3200	Stainless steel, custom	1 Carp	24	.333	S.F.	145	14.70		159.70	183

12 36 19 – Wood Countertops

12 36 19.10 Maple Countertops

	Crew	Daily Output	Labor-Hours	Unit	Material	2012 Bare Costs Labor	Equipment	Total	Total Incl O&P
0010	**MAPLE COUNTERTOPS**									
2900	Solid, laminated, 1-1/2" thick, no splash	1 Carp	28	.286	L.F.	74	12.60		86.60	101
3000	With square splash		28	.286	"	88	12.60		100.60	116
3400	Recessed cutting block with trim, 16" x 20" x 1"	↓	8	1	Ea.	74	44		118	150

12 36 23 – Plastic Countertops

12 36 23.13 Plastic-Laminate-Clad Countertops

	Crew	Daily Output	Labor-Hours	Unit	Material	2012 Bare Costs Labor	Equipment	Total	Total Incl O&P
0010	**PLASTIC-LAMINATE-CLAD COUNTERTOPS**									
0020	Stock, 24" wide w/backsplash, minimum	1 Carp	30	.267	L.F.	9.85	11.75		21.60	29
0100	Maximum		25	.320		21	14.10		35.10	44.50
0300	Custom plastic, 7/8" thick, aluminum molding, no splash		30	.267		22.50	11.75		34.25	43
0400	Cove splash		30	.267		29	11.75		40.75	50
0600	1-1/4" thick, no splash		28	.286		31.50	12.60		44.10	54
0700	Square splash		28	.286		32	12.60		44.60	54.50
0900	Square edge, plastic face, 7/8" thick, no splash		30	.267		28	11.75		39.75	49
1000	With splash	↓	30	.267		36	11.75		47.75	57.50
1200	For stainless channel edge, 7/8" thick, add					3			3	3.30
1300	1-1/4" thick, add					3.50			3.50	3.85
1500	For solid color suede finish, add				↓	3			3	3.30
1700	For end splash, add				Ea.	18			18	19.80
1900	For cut outs, standard, add, minimum	1 Carp	32	.250		4	11.05		15.05	21.50
2000	Maximum		8	1	↓	6	44		50	74.50
2100	Postformed, including backsplash and front edge		30	.267	L.F.	11	11.75		22.75	30
2110	Mitred, add	↓	12	.667	Ea.		29.50		29.50	45

12 36 Countertops

12 36 23 – Plastic Countertops

12 36 23.13 Plastic-Laminate-Clad Countertops

		Crew	Daily Output	Labor-Hours	Unit	Material	2012 Bare Costs Labor	Equipment	Total	Total Incl O&P
2200	Built-in place, 25" wide, plastic laminate	1 Carp	25	.320	L.F.	30	14.10		44.10	54.50

12 36 33 – Tile Countertops

12 36 33.10 Ceramic Tile Countertops

		Crew	Daily Output	Labor-Hours	Unit	Material	2012 Bare Costs Labor	Equipment	Total	Total Incl O&P
0010	**CERAMIC TILE COUNTERTOPS**									
2300	Ceramic tile mosaic	1 Carp	25	.320	L.F.	33	14.10		47.10	58

12 36 40 – Stone Countertops

12 36 40.10 Natural Stone Countertops

		Crew	Daily Output	Labor-Hours	Unit	Material	2012 Bare Costs Labor	Equipment	Total	Total Incl O&P
0010	**NATURAL STONE COUNTERTOPS**									
2500	Marble, stock, with splash, 1/2" thick, minimum	1 Bric	17	.471	L.F.	41	21		62	76.50
2700	3/4" thick, maximum		13	.615		103	27		130	154
2800	Granite, average, 1-1/4" thick, 24" wide, no splash		13.01	.615		135	27		162	190

12 36 53 – Laboratory Countertops

12 36 53.10 Laboratory Countertops and Sinks

		Crew	Daily Output	Labor-Hours	Unit	Material	2012 Bare Costs Labor	Equipment	Total	Total Incl O&P
0010	**LABORATORY COUNTERTOPS AND SINKS**									
0020	Countertops, not incl. base cabinets, acid-proof, minimum	2 Carp	82	.195	S.F.	36.50	8.60		45.10	53.50
0030	Maximum		70	.229		46.50	10.10		56.60	67
0040	Stainless steel		82	.195		135	8.60		143.60	161

12 36 61 – Simulated Stone Countertops

12 36 61.16 Solid Surface Countertops

		Crew	Daily Output	Labor-Hours	Unit	Material	2012 Bare Costs Labor	Equipment	Total	Total Incl O&P
0010	**SOLID SURFACE COUNTERTOPS**, Acrylic polymer									
0020	Pricing for orders of 100 L.F. or greater									
0100	25" wide, solid colors	2 Carp	28	.571	L.F.	44.50	25		69.50	88
0200	Patterned colors		28	.571		56.50	25		81.50	101
0300	Premium patterned colors		28	.571		70.50	25		95.50	117
0400	With silicone attached 4" backsplash, solid colors		27	.593		49	26		75	94
0500	Patterned colors		27	.593		62	26		88	108
0600	Premium patterned colors		27	.593		77	26		103	125
0700	With hard seam attached 4" backsplash, solid colors		23	.696		49	30.50		79.50	101
0800	Patterned colors		23	.696		62	30.50		92.50	115
0900	Premium patterned colors		23	.696		77	30.50		107.50	132
1000	Pricing for order of 51 – 99 L.F.									
1100	25" wide, solid colors	2 Carp	24	.667	L.F.	51.50	29.50		81	102
1200	Patterned colors		24	.667		65	29.50		94.50	117
1300	Premium patterned colors		24	.667		81.50	29.50		111	135
1400	With silicone attached 4" backsplash, solid colors		23	.696		56.50	30.50		87	109
1500	Patterned colors		23	.696		71.50	30.50		102	126
1600	Premium patterned colors		23	.696		89	30.50		119.50	145
1700	With hard seam attached 4" backsplash, solid colors		20	.800		56.50	35.50		92	117
1800	Patterned colors		20	.800		71.50	35.50		107	133
1900	Premium patterned colors		20	.800		89	35.50		124.50	152
2000	Pricing for order of 1 – 50 L.F.									
2100	25" wide, solid colors	2 Carp	20	.800	L.F.	60	35.50		95.50	121
2200	Patterned colors		20	.800		76.50	35.50		112	139
2300	Premium patterned colors		20	.800		95.50	35.50		131	160
2400	With silicone attached 4" backsplash, solid colors		19	.842		66	37		103	130
2500	Patterned colors		19	.842		83.50	37		120.50	149
2600	Premium patterned colors		19	.842		104	37		141	172
2700	With hard seam attached 4" backsplash, solid colors		15	1.067		66	47		113	145
2800	Patterned colors		15	1.067		83.50	47		130.50	165
2900	Premium patterned colors		15	1.067		104	47		151	188

12 36 Countertops

12 36 61.16 Solid Surface Countertops

		Crew	Daily Output	Labor-Hours	Unit	Material	2012 Bare Costs Labor	2012 Bare Costs Equipment	Total	Total Incl O&P
3000	Sinks, pricing for order of 100 or greater units									
3100	Single bowl, hard seamed, solid colors, 13" x 17"	1 Carp	3	2.667	Ea.	300	118		418	510
3200	10" x 15"		7	1.143		139	50.50		189.50	231
3300	Cutouts for sinks		8	1			44		44	68
3400	Sinks, pricing for order of 51 – 99 units									
3500	Single bowl, hard seamed, solid colors, 13" x 17"	1 Carp	2.55	3.137	Ea.	345	138		483	595
3600	10" x 15"		6	1.333		160	59		219	267
3700	Cutouts for sinks		7	1.143			50.50		50.50	77.50
3800	Sinks, pricing for order of 1 – 50 units									
3900	Single bowl, hard seamed, solid colors, 13" x 17"	1 Carp	2	4	Ea.	405	176		581	715
4000	10" x 15"		4.55	1.758		188	77.50		265.50	325
4100	Cutouts for sinks		5.25	1.524			67		67	103
4200	Cooktop cutouts, pricing for 100 or greater units		4	2		22	88		110	161
4300	51 – 99 units		3.40	2.353		25.50	104		129.50	188
4400	1 – 50 units		3	2.667		30	118		148	214

12 36 61.17 Solid Surface Vanity Tops

		Crew	Daily Output	Labor-Hours	Unit	Material	2012 Bare Costs Labor	2012 Bare Costs Equipment	Total	Total Incl O&P
0010	**SOLID SURFACE VANITY TOPS**									
0015	Solid surface, center bowl, 17" x 19"	1 Carp	12	.667	Ea.	175	29.50		204.50	238
0020	19" x 25"		12	.667		180	29.50		209.50	243
0030	19" x 31"		12	.667		210	29.50		239.50	276
0040	19" x 37"		12	.667		244	29.50		273.50	315
0050	22" x 25"		10	.800		320	35.50		355.50	410
0060	22" x 31"		10	.800		375	35.50		410.50	465
0070	22" x 37"		10	.800		435	35.50		470.50	535
0080	22" x 43"		10	.800		495	35.50		530.50	600
0090	22" x 49"		10	.800		550	35.50		585.50	660
0110	22" x 55"		8	1		625	44		669	755
0120	22" x 61"		8	1		715	44		759	855
0220	Double bowl, 22" x 61"		8	1		805	44		849	955
0230	Double bowl, 22" x 73"		8	1		1,400	44		1,444	1,600
0240	For aggregate colors, add					35%				
0250	For faucets and fittings, see Section 22 41 39.10									

12 36 61.19 Quartz Agglomerate Countertops

		Crew	Daily Output	Labor-Hours	Unit	Material	2012 Bare Costs Labor	2012 Bare Costs Equipment	Total	Total Incl O&P
0010	**QUARTZ AGGLOMERATE COUNTERTOPS**									
0100	25" wide, 4" backsplash, color group A, minimum	2 Carp	15	1.067	L.F.	58	47		105	137
0110	Maximum		15	1.067		81	47		128	162
0120	Color group B, minimum		15	1.067		60	47		107	139
0130	Maximum		15	1.067		85	47		132	166
0140	Color group C, minimum		15	1.067		70	47		117	150
0150	Maximum		15	1.067		96	47		143	179
0160	Color group D, minimum		15	1.067		76	47		123	156
0170	Maximum		15	1.067		103	47		150	186

12 46 Furnishing Accessories

12 46 13 – Ash Receptacles

12 46 13.10 Ash/Trash Receivers	Crew	Daily Output	Labor-Hours	Unit	Material	2012 Bare Costs Labor	Equipment	Total	Total Incl O&P
0010 **ASH/TRASH RECEIVERS**									
1000 Ash urn, cylindrical metal									
1020 8" diameter, 20" high	1 Clab	60	.133	Ea.	160	4.68		164.68	183
1060 10" diameter, 26" high	"	60	.133	"	120	4.68		124.68	139
2000 Combination ash/trash urn, metal									
2020 8" diameter, 20" high	1 Clab	60	.133	Ea.	160	4.68		164.68	183
2050 10" diameter, 26" high	"	60	.133	"	120	4.68		124.68	139

12 46 19 – Clocks

12 46 19.50 Wall Clocks

	Crew	Daily Output	Labor-Hours	Unit	Material	2012 Bare Costs Labor	Equipment	Total	Total Incl O&P
0010 **WALL CLOCKS**									
0080 12" diameter, single face	1 Elec	8	1	Ea.	139	51.50		190.50	229
0100 Double face	"	6.20	1.290	"	350	66.50		416.50	485

12 46 33 – Waste Receptacles

12 46 33.13 Trash Receptacles

	Crew	Daily Output	Labor-Hours	Unit	Material	2012 Bare Costs Labor	Equipment	Total	Total Incl O&P
0010 **TRASH RECEPTACLES**									
4000 Trash receptacle, metal									
4020 8" diameter, 15" high	1 Clab	60	.133	Ea.	75	4.68		79.68	89.50
4040 10" diameter, 18" high		60	.133		150	4.68		154.68	172
5040 16" x 8" x 14" high	↓	60	.133	↓	24	4.68		28.68	33.50
5500 Plastic, with lid									
5520 35 gallon	1 Clab	60	.133	Ea.	130	4.68		134.68	150
5540 45 gallon		60	.133		220	4.68		224.68	249
5550 Plastic recycling barrel, w/lid & wheels, 32 gal. **G**		60	.133		80	4.68		84.68	95
5560 65 gal. **G**		60	.133		515	4.68		519.68	570
5570 95 gal. **G**	↓	60	.133	↓	985	4.68		989.68	1,075

12 48 Rugs and Mats

12 48 13 – Entrance Floor Mats and Frames

12 48 13.13 Entrance Floor Mats

	Crew	Daily Output	Labor-Hours	Unit	Material	2012 Bare Costs Labor	Equipment	Total	Total Incl O&P
0010 **ENTRANCE FLOOR MATS**									
0020 Recessed, in-laid black rubber, 3/8" thick, solid	1 Clab	155	.052	S.F.	21	1.81		22.81	26
0050 Perforated		155	.052		14.50	1.81		16.31	18.75
0100 1/2" thick, solid		155	.052		17.40	1.81		19.21	22
0150 Perforated		155	.052		22.50	1.81		24.31	27.50
0200 In colors, 3/8" thick, solid		155	.052		18.75	1.81		20.56	23.50
0250 Perforated		155	.052		19.35	1.81		21.16	24.50
0300 1/2" thick, solid		155	.052		24.50	1.81		26.31	29.50
0350 Perforated		155	.052		25	1.81		26.81	30.50
2000 Recycled rubber tire tile, 12" x 12" x 3/8" thick **G**		125	.064		8.65	2.25		10.90	13
2510 Natural cocoa fiber, 1/2" thick **G**		125	.064		7.65	2.25		9.90	11.85
2520 3/4" thick **G**		125	.064		6.40	2.25		8.65	10.45
2530 1" thick **G**	↓	125	.064	↓	7.05	2.25		9.30	11.20

12 51 Office Furniture

12 51 16 - Case Goods

12 51 16.13 Metal Case Goods

		Crew	Daily Output	Labor-Hours	Unit	Material	2012 Bare Costs Labor	Equipment	Total	Total Incl O&P
0010	**METAL CASE GOODS**									
0020	Desks, 29" high, double pedestal, 30" x 60", metal, minimum				Ea.	540			540	595
0030	Maximum					1,325			1,325	1,475
0600	Desks, single pedestal, 30" x 60", metal, minimum					475			475	520
0620	Maximum					1,075			1,075	1,175
0720	Desks, secretarial, 30" x 60", metal, minimum					480			480	530
0730	Maximum					755			755	835
0740	Return, 20" x 42", minimum					335			335	370
0750	Maximum					500			500	550
0940	59" x 12" x 23" high, steel, minimum					290			290	320
0960	Maximum					355			355	390

12 51 16.16 Wood Case Goods

		Crew	Daily Output	Labor-Hours	Unit	Material	2012 Bare Costs Labor	Equipment	Total	Total Incl O&P
0010	**WOOD CASE GOODS**									
0150	Desk, 29" high, double pedestal, 30" x 60"									
0160	Wood, minimum				Ea.	700			700	770
0180	Maximum				"	2,575			2,575	2,850
0630	Single pedestal, 30" x 60"									
0640	Wood, minimum				Ea.	565			565	620
0650	Maximum					810			810	890
0670	Executive return, 24" x 42", with box, file, wood, minimum					345			345	380
0680	Maximum					810			810	890
0790	Desk, 29" high, secretarial, 30" x 60"									
0800	Wood, minimum				Ea.	470			470	515
0810	Maximum					2,625			2,625	2,875
0820	Return, 20" x 42", minimum					299			299	330
0830	Maximum					1,025			1,025	1,125
0900	Desktop organizer, 72" x 14" x 36" high, wood, minimum					177			177	195
0920	Maximum					425			425	465
1110	Furniture, credenza, 29" high, 18" to 22" x 60" to 72"									
1120	Wood, minimum				Ea.	665			665	735
1140	Maximum				"	2,525			2,525	2,775

12 51 23 - Office Tables

12 51 23.33 Conference Tables

		Crew	Daily Output	Labor-Hours	Unit	Material	2012 Bare Costs Labor	Equipment	Total	Total Incl O&P
0010	**CONFERENCE TABLES**									
6050	Boat, 96" x 42", minimum				Ea.	795			795	875
6150	Maximum					3,225			3,225	3,550
6720	Rectangle, 96" x 42", minimum					1,200			1,200	1,300
6740	Maximum					3,225			3,225	3,550

12 52 Seating

12 52 23 - Office Seating

12 52 23.13 Office Chairs

		Crew	Daily Output	Labor-Hours	Unit	Material	2012 Bare Costs Labor	Equipment	Total	Total Incl O&P
0010	**OFFICE CHAIRS**									
2000	Standard office chair, executive, minimum				Ea.	330			330	365
2150	Maximum					1,825			1,825	2,000
2200	Management, minimum					207			207	227
2250	Maximum					2,275			2,275	2,525
2280	Task, minimum					145			145	160
2290	Maximum					500			500	550
2300	Arm kit, minimum					62.50			62.50	68.50

12 52 Seating

12 52 23 – Office Seating

12 52 23.13 Office Chairs	Crew	Daily Output	Labor-Hours	Unit	Material	2012 Bare Costs Labor	Equipment	Total	Total Incl O&P
2320	Maximum				Ea.	111			111	122

12 54 Hospitality Furniture

12 54 13 – Hotel and Motel Furniture

12 54 13.10 Hotel Furniture

		Crew	Daily Output	Labor-Hours	Unit	Material	Labor	Equipment	Total	Total Incl O&P
0010	**HOTEL FURNITURE**									
0020	Standard quality set, minimum				Room	2,500			2,500	2,750
0200	Maximum				"	8,875			8,875	9,750

12 54 16 – Restaurant Furniture

12 54 16.10 Tables, Folding

		Crew	Daily Output	Labor-Hours	Unit	Material	Labor	Equipment	Total	Total Incl O&P
0010	**TABLES, FOLDING** Laminated plastic tops									
1000	Tubular steel legs with glides									
1020	18" x 60", minimum				Ea.	253			253	279
1040	Maximum					1,350			1,350	1,500
1840	36" x 96", minimum					292			292	320
1860	Maximum					2,725			2,725	3,000
2000	Round, wood stained, plywood top, 60" diameter, minimum					420			420	465
2020	Maximum					1,475			1,475	1,625

12 54 16.20 Furniture, Restaurant

		Crew	Daily Output	Labor-Hours	Unit	Material	Labor	Equipment	Total	Total Incl O&P
0010	**FURNITURE, RESTAURANT**									
0020	Bars, built-in, front bar	1 Carp	5	1.600	L.F.	262	70.50		332.50	400
0200	Back bar	"	5	1.600	"	191	70.50		261.50	320
0300	Booth seating, see Section 12 54 16.70									
2000	Chair, bentwood side chair, metal, minimum				Ea.	87.50			87.50	96.50
2020	Maximum					103			103	113
2600	Upholstered seat & back, arms, minimum					143			143	157
2620	Maximum					405			405	445

12 54 16.70 Booths

		Crew	Daily Output	Labor-Hours	Unit	Material	Labor	Equipment	Total	Total Incl O&P
0010	**BOOTHS**									
1000	Banquet, upholstered seat and back, custom									
1500	Straight, minimum	2 Carp	40	.400	L.F.	186	17.65		203.65	232
1520	Maximum		36	.444		360	19.60		379.60	430
1600	"L" or "U" shape, minimum		35	.457		190	20		210	240
1620	Maximum		30	.533		335	23.50		358.50	405
1800	Upholstered outside finished backs for									
1810	single booths and custom banquets									
1820	Minimum	2 Carp	44	.364	L.F.	21.50	16.05		37.55	48
1840	Maximum	"	40	.400	"	64.50	17.65		82.15	97.50
3000	Fixed seating, one piece plastic chair and									
3010	plastic laminate table top									
3100	Two seat, 24" x 24" table, minimum	F-7	30	1.067	Ea.	750	42		792	890
3120	Maximum		26	1.231		1,075	48.50		1,123.50	1,250
3200	Four seat, 24" x 48" table, minimum		28	1.143		745	45.50		790.50	885
3220	Maximum		24	1.333		1,275	53		1,328	1,475
5000	Mount in floor, wood fiber core with									
5010	plastic laminate face, single booth									
5050	24" wide	F-7	30	1.067	Ea.	287	42		329	380
5100	48" wide	"	28	1.143	"	365	45.50		410.50	470

12 55 Detention Furniture

12 55 13 – Detention Bunks

12 55 13.13 Cots

		Crew	Daily Output	Labor-Hours	Unit	Material	2012 Bare Costs Labor	Equipment	Total	Total Incl O&P
0010	**COTS**									
2500	Bolted, single, painted steel	E-4	20	1.600	Ea.	330	80	6.10	416.10	515
2700	Stainless steel	"	20	1.600	"	960	80	6.10	1,046.10	1,200

12 56 Institutional Furniture

12 56 33 – Classroom Furniture

12 56 33.10 Furniture, School

		Crew	Daily Output	Labor-Hours	Unit	Material	2012 Bare Costs Labor	Equipment	Total	Total Incl O&P
0010	**FURNITURE, SCHOOL**									
0500	Classroom, movable chair & desk type, minimum				Set				73.50	81
0600	Maximum				"				155	171
1000	Chair, molded plastic									
1100	Integral tablet arm, minimum				Ea.	74			74	81.50
1150	Maximum					166			166	183
2000	Desk, single pedestal, top book compartment, minimum					132			132	145
2020	Maximum					154			154	170
2200	Flip top, minimum					193			193	212
2220	Maximum					235			235	259

12 56 43 – Dormitory Furniture

12 56 43.10 Dormitory Furnishings

		Crew	Daily Output	Labor-Hours	Unit	Material	2012 Bare Costs Labor	Equipment	Total	Total Incl O&P
0010	**DORMITORY FURNISHINGS**									
0300	Bunkable bed, twin, minimum				Ea.	390			390	430
0320	Maximum					765			765	845
1000	Chest, four drawer, minimum					340			340	375
1020	Maximum					660			660	725
1050	Built-in, minimum	2 Carp	13	1.231	L.F.	149	54.50		203.50	247
1150	Maximum		10	1.600		275	70.50		345.50	410
1200	Desk top, built-in, laminated plastic, 24" deep, minimum		50	.320		39.50	14.10		53.60	65
1300	Maximum		40	.400		118	17.65		135.65	157
1450	30" deep, minimum		50	.320		50.50	14.10		64.60	77
1550	Maximum		40	.400		221	17.65		238.65	270
1750	Dressing unit, built-in, minimum		12	1.333		223	59		282	335
1850	Maximum		8	2		670	88		758	870
8000	Rule of thumb: total cost for furniture, minimum				Student				2,525	2,800
8050	Maximum				"				4,850	5,350

12 56 51 – Library Furniture

12 56 51.10 Library Furnishings

		Crew	Daily Output	Labor-Hours	Unit	Material	2012 Bare Costs Labor	Equipment	Total	Total Incl O&P
0010	**LIBRARY FURNISHINGS**									
0100	Attendant desk, 36" x 62" x 29" high	1 Carp	16	.500	Ea.	2,375	22		2,397	2,625
0200	Book display, "A" frame display, both sides, 42" x 42" x 60" high		16	.500		965	22		987	1,100
0220	Table with bulletin board, 42" x 24" x 49" high		16	.500		860	22		882	980
0800	Card catalogue, 30 tray unit		16	.500		2,900	22		2,922	3,225
0840	60 tray unit		16	.500		5,275	22		5,297	5,850
0880	72 tray unit	2 Carp	16	1		6,450	44		6,494	7,175
1000	Carrels, single face, initial unit	1 Carp	16	.500		880	22		902	1,000
1500	Double face, initial unit	2 Carp	16	1		1,150	44		1,194	1,325
1710	Carrels, hardwood, 36" x 24", minimum	1 Carp	5	1.600		580	70.50		650.50	750
1720	Maximum	"	4	2		1,325	88		1,413	1,575
2700	Card catalog file, 60 trays, complete					5,950			5,950	6,550
2720	Alternate method: each tray					99			99	109
3800	Charging desk, built-in, with counter, plastic laminated top	1 Carp	7	1.143	L.F.	410	50.50		460.50	535

12 56 Institutional Furniture

12 56 51 – Library Furniture

12 56 51.10 Library Furnishings

	12 56 51.10 Library Furnishings	Crew	Daily Output	Labor-Hours	Unit	Material	2012 Bare Costs Labor	2012 Bare Costs Equipment	Total	Total Incl O&P
4000	Dictionary stand, stationary	1 Carp	16	.500	Ea.	840	22		862	955
4020	Revolving		16	.500		345	22		367	415
4200	Exhibit case, table style, 60" x 28" x 36"		11	.727	↓	2,800	32		2,832	3,150
6010	Bookshelf, metal, 90" high, 10" shelf, double face		11.50	.696	L.F.	122	30.50		152.50	181
6020	Single face	↓	12	.667	"	105	29.50		134.50	160
6050	For 8" shelving, subtract from above					10%				
6060	For 12" shelving, add to above					10%				
6070	For 42" high with countertop, subtract from above					20%				
6100	Mobile compacted shelving, hand crank, 9'-0" high									
6110	Double face, including track, 3' section				Ea.	1,125			1,125	1,250
6150	For electrical operation, add					25%				
6200	Magazine shelving, 82" high, 12" deep, single face	1 Carp	11.50	.696	L.F.	92.50	30.50		123	149
6210	Double face	↓	11.50	.696	"	122	30.50		152.50	181
7000	Tables, card catalog reference, 24" x 60" x 42"	↓	16	.500	Ea.	975	22		997	1,100
7200	Reading table, laminated top, 60" x 36"				"	660			660	725

12 56 70 – Healthcare Furniture

12 56 70.10 Furniture, Hospital

	12 56 70.10 Furniture, Hospital	Crew	Daily Output	Labor-Hours	Unit	Material	Labor	Equipment	Total	Total Incl O&P
0010	**FURNITURE, HOSPITAL**									
0020	Beds, manual, minimum				Ea.	770			770	850
0100	Maximum					2,600			2,600	2,875
0300	Manual and electric beds, minimum					1,075			1,075	1,175
0400	Maximum					3,075			3,075	3,375
0600	All electric hospital beds, minimum					1,375			1,375	1,525
0700	Maximum					4,450			4,450	4,875
0900	Manual, nursing home beds, minimum					875			875	965
1000	Maximum					1,825			1,825	2,000
1020	Overbed table, laminated top, minimum					375			375	410
1040	Maximum				↓	930			930	1,025
1100	Patient wall systems, not incl. plumbing, minimum				Room	1,100			1,100	1,225
1200	Maximum				"	2,050			2,050	2,250
2000	Geriatric chairs, minimum				Ea.	390			390	430
2020	Maximum				"	730			730	800

12 61 Fixed Audience Seating

12 61 13 – Upholstered Audience Seating

12 61 13.13 Auditorium Chairs

	12 61 13.13 Auditorium Chairs	Crew	Daily Output	Labor-Hours	Unit	Material	Labor	Equipment	Total	Total Incl O&P
0010	**AUDITORIUM CHAIRS**									
2000	All veneer construction	2 Carp	22	.727	Ea.	184	32		216	252
2200	Veneer back, padded seat		22	.727		243	32		275	315
2350	Fully upholstered, spring seat	↓	22	.727		232	32		264	305
2450	For tablet arms, add					63.50			63.50	70
2500	For fire retardancy, CATB-133, add				↓	27.50			27.50	30

12 61 13.23 Lecture Hall Seating

	12 61 13.23 Lecture Hall Seating	Crew	Daily Output	Labor-Hours	Unit	Material	Labor	Equipment	Total	Total Incl O&P
0010	**LECTURE HALL SEATING**									
1000	Pedestal type, minimum	2 Carp	22	.727	Ea.	174	32		206	241
1200	Maximum	"	14.50	1.103	"	580	48.50		628.50	715

12 63 Stadium and Arena Seating

12 63 13 - Stadium and Arena Bench Seating

12 63 13.13 Bleachers		Crew	Daily Output	Labor-Hours	Unit	Material	2012 Bare Costs Labor	Equipment	Total	Total Incl O&P
0010	**BLEACHERS**									
3000	Telescoping, manual to 15 tier, minimum	F-5	65	.492	Seat	86	22		108	129
3100	Maximum		60	.533		129	24		153	179
3300	16 to 20 tier, minimum		60	.533		207	24		231	265
3400	Maximum		55	.582		259	26		285	325
3600	21 to 30 tier, minimum		50	.640		215	28.50		243.50	281
3700	Maximum		40	.800		259	35.50		294.50	340
3900	For integral power operation, add, minimum	2 Elec	300	.053		43	2.75		45.75	51.50
4000	Maximum	"	250	.064		69	3.30		72.30	81
5000	Benches, folding, in wall, 14' table, 2 benches	L-4	2	12	Set	730	495		1,225	1,575

12 67 Pews and Benches

12 67 13 - Pews

12 67 13.13 Sanctuary Pews		Crew	Daily Output	Labor-Hours	Unit	Material	2012 Bare Costs Labor	Equipment	Total	Total Incl O&P
0010	**SANCTUARY PEWS**									
1500	Bench type, hardwood, minimum	1 Carp	20	.400	L.F.	83	17.65		100.65	119
1550	Maximum	"	15	.533		165	23.50		188.50	217
1570	For kneeler, add					20.50			20.50	22.50

12 92 Interior Planters and Artificial Plants

12 92 33 - Interior Planters

12 92 33.10 Planters		Crew	Daily Output	Labor-Hours	Unit	Material	2012 Bare Costs Labor	Equipment	Total	Total Incl O&P
0010	**PLANTERS**									
1000	Fiberglass, hanging, 12" diameter, 7" high				Ea.	111			111	122
1500	Rectangular, 48" long, 16" high x 15" wide					650			650	715
1650	60" long, 30" high, 28" wide					1,050			1,050	1,150
2000	Round, 12" diameter, 13" high					134			134	148
2050	25" high					193			193	212
5000	Square, 10" side, 20" high					189			189	208
5100	14" side, 15" high					206			206	227
6000	Metal bowl, 32" diameter, 8" high, minimum					555			555	610
6050	Maximum					745			745	820
8750	Wood, fiberglass liner, square									
8780	14" square, 15" high, minimum				Ea.	360			360	400
8800	Maximum					445			445	490
9400	Plastic cylinder, molded, 10" diameter, 10" high					14.70			14.70	16.15
9500	11" diameter, 11" high					46.50			46.50	51

12 93 Site Furnishings

12 93 23 – Trash and Litter Receptors

12 93 23.10 Trash Receptacles

	12 93 23.10 Trash Receptacles		Crew	Daily Output	Labor-Hours	Unit	Material	2012 Bare Costs Labor	Equipment	Total	Total Incl O&P
0010	**TRASH RECEPTACLES**										
0020	Fiberglass, 2' square, 18" high		2 Clab	30	.533	Ea.	780	18.70		798.70	885
0100	2' square, 2'-6" high			30	.533		550	18.70		568.70	635
0300	Circular, 2' diameter, 18" high			30	.533		465	18.70		483.70	540
0400	2' diameter, 2'-6" high			30	.533		520	18.70		538.70	600
0500	Recycled plastic, var colors, round, 32 gal., 28" x 38" H	G		5	3.200		445	112		557	665
0510	32 gal., 31" x 32" H	G		5	3.200		490	112		602	715

12 93 23.20 Trash Closure

	12 93 23.20 Trash Closure	Crew	Daily Output	Labor-Hours	Unit	Material	Labor	Equipment	Total	Total Incl O&P
0010	**TRASH CLOSURE**									
0020	Steel with pullover cover, 2'-3" wide, 4'-7" high, 6'-2" long	2 Clab	5	3.200	Ea.	1,500	112		1,612	1,825
0100	10'-1" long		4	4		1,925	140		2,065	2,350
0300	Wood, 10' wide, 6' high, 10' long		1.20	13.333		1,675	470		2,145	2,575

12 93 33 – Manufactured Planters

12 93 33.10 Planters

	12 93 33.10 Planters	Crew	Daily Output	Labor-Hours	Unit	Material	Labor	Equipment	Total	Total Incl O&P
0010	**PLANTERS**									
0012	Concrete, sandblasted, precast, 48" diameter, 24" high	2 Clab	15	1.067	Ea.	695	37.50		732.50	820
0100	Fluted, precast, 7' diameter, 36" high		10	1.600		1,425	56		1,481	1,650
0300	Fiberglass, circular, 36" diameter, 24" high		15	1.067		660	37.50		697.50	785
0320	36" diameter, 27" high		12	1.333		780	47		827	930
0330	33" high		15	1.067		835	37.50		872.50	980
0335	24" diameter, 36" high		15	1.067		535	37.50		572.50	645
0340	60" diameter, 39" high		8	2		1,350	70		1,420	1,575
0400	60" diameter, 24" high		10	1.600		775	56		831	935
0600	Square, 24" side, 36" high		15	1.067		680	37.50		717.50	805
0610	24" side, 27" high		12	1.333		680	47		727	815
0620	24" side, 16" high		20	.800		360	28		388	440
0700	48" side, 36" high		15	1.067		990	37.50		1,027.50	1,150
0900	Planter/bench, 72" square, 36" high		5	3.200		1,425	112		1,537	1,750
1000	96" square, 27" high		5	3.200		2,175	112		2,287	2,575
1200	Wood, square, 48" side, 24" high		15	1.067		1,325	37.50		1,362.50	1,525
1300	Circular, 48" diameter, 30" high		10	1.600		910	56		966	1,075
1500	72" diameter, 30" high		10	1.600		1,425	56		1,481	1,625
1600	Planter/bench, 72"		5	3.200		3,150	112		3,262	3,650

12 93 43 – Site Seating and Tables

12 93 43.13 Site Seating

	12 93 43.13 Site Seating	Crew	Daily Output	Labor-Hours	Unit	Material	Labor	Equipment	Total	Total Incl O&P
0010	**SITE SEATING**									
0012	Seating, benches, park, precast conc, w/backs, wood rails, 4' long	2 Clab	5	3.200	Ea.	535	112		647	765
0100	8' long		4	4		1,050	140		1,190	1,375
0300	Fiberglass, without back, one piece, 4' long		10	1.600		585	56		641	730
0400	8' long		7	2.286		800	80		880	1,000
0500	Steel barstock pedestals w/backs, 2" x 3" wood rails, 4' long		10	1.600		1,175	56		1,231	1,375
0510	8' long		7	2.286		1,425	80		1,505	1,700
0520	3" x 8" wood plank, 4' long		10	1.600		1,175	56		1,231	1,375
0530	8' long		7	2.286		1,425	80		1,505	1,700
0540	Backless, 4" x 4" wood plank, 4' square		10	1.600		1,200	56		1,256	1,375
0550	8' long		7	2.286		920	80		1,000	1,125
0600	Aluminum pedestals, with backs, aluminum slats, 8' long		8	2		440	70		510	595
0610	15' long		5	3.200		440	112		552	660
0620	Portable, aluminum slats, 8' long		8	2		440	70		510	595
0630	15' long		5	3.200		480	112		592	705
0800	Cast iron pedestals, back & arms, wood slats, 4' long		8	2		370	70		440	515

12 93 Site Furnishings

12 93 43 – Site Seating and Tables

12 93 43.13 Site Seating		Crew	Daily Output	Labor-Hours	Unit	Material	2012 Bare Costs Labor	Equipment	Total	Total Incl O&P
0820	8' long	2 Clab	5	3.200	Ea.	1,025	112		1,137	1,300
0840	Backless, wood slats, 4' long		8	2		455	70		525	610
0860	8' long		5	3.200		1,075	112		1,187	1,350
1700	Steel frame, fir seat, 10' long		10	1.600		325	56		381	445

Estimating Tips

General

- The items and systems in this division are usually estimated, purchased, supplied, and installed as a unit by one or more subcontractors. The estimator must ensure that all parties are operating from the same set of specifications and assumptions, and that all necessary items are estimated and will be provided. Many times the complex items and systems are covered, but the more common ones, such as excavation or a crane, are overlooked for the very reason that everyone assumes nobody could miss them. The estimator should be the central focus and be able to ensure that all systems are complete.

- Another area where problems can develop in this division is at the interface between systems. The estimator must ensure, for instance, that anchor bolts, nuts, and washers are estimated and included for the air-supported structures and pre-engineered buildings to be bolted to their foundations. Utility supply is a common area where essential items or pieces of equipment can be missed or overlooked due to the fact that each subcontractor may feel it is another's responsibility. The estimator should also be aware of certain items which may be supplied as part of a package but installed by others, and ensure that the installing contractor's estimate includes the cost of installation. Conversely, the estimator must also ensure that items are not costed by two different subcontractors, resulting in an inflated overall estimate.

13 30 00 Special Structures

- The foundations and floor slab, as well as rough mechanical and electrical, should be estimated, as this work is required for the assembly and erection of the structure. Generally, as noted in the book, the pre-engineered building comes as a shell. Pricing is based on the size and structural design parameters stated in the reference section. Additional features, such as windows and doors with their related structural framing, must also be included by the estimator. Here again, the estimator must have a clear understanding of the scope of each portion of the work and all the necessary interfaces.

Reference Numbers

Reference numbers are shown in shaded boxes at the beginning of some major classifications. These numbers refer to related items in the Reference Section. The reference information may be an estimating procedure, an alternate pricing method, or technical information.

Note: Not all subdivisions listed here necessarily appear in this publication.

13 05 05 – Selective Special Construction Demolition

13 05 05.10 Selective Demolition, Air Supported Structures

		Crew	Daily Output	Labor-Hours	Unit	Material	2012 Bare Costs Labor	2012 Bare Costs Equipment	Total	Total Incl O&P
0010	**SELECTIVE DEMOLITION, AIR SUPPORTED STRUCTURES**									
0020	Tank covers, scrim, dbl. layer, vinyl poly w/hdw., blower & controls									
0050	Round and rectangular R024119-10	B-2	9000	.004	S.F.		.16		.16	.24
0100	Warehouse structures									
0120	Poly/vinyl fabric, 28 oz., incl. tension cables & inflation system	4 Clab	9000	.004	SF Flr.		.13		.13	.19
0150	Reinforced vinyl, 12 oz., 3000 S.F.	"	5000	.006			.22		.22	.35
0200	12,000 to 24,000 S.F.	8 Clab	20000	.003			.11		.11	.17
0250	Tedlar vinyl fabric, 28 oz. w/liner, to 3000 S.F.	4 Clab	5000	.006			.22		.22	.35
0300	12,000 to 24,000 S.F.	8 Clab	20000	.003			.11		.11	.17
0350	Greenhouse/shelter, woven polyethylene with liner									
0400	3000 S.F.	4 Clab	5000	.006	SF Flr.		.22		.22	.35
0450	12,000 to 24,000 S.F.	8 Clab	20000	.003			.11		.11	.17
0500	Tennis/gymnasium, poly/vinyl fabric, 28 oz., incl. thermal liner	4 Clab	9000	.004			.13		.13	.19
0600	Stadium/convention center, teflon coated fiberglass, incl. thermal liner	9 Clab	40000	.002			.06		.06	.10
0700	Doors, air lock, 15' long, 10' x 10'	2 Carp	1.50	10.667	Ea.		470		470	725
0720	15' x 15'		.80	20			880		880	1,350
0750	Revolving personnel door, 6' dia. x 6'-6" high		1.50	10.667			470		470	725

13 05 05.20 Selective Demolition, Garden Houses

		Crew	Daily Output	Labor-Hours	Unit	Material	Labor	Equipment	Total	Total Incl O&P
0010	**SELECTIVE DEMOLITION, GARDEN HOUSES** R024119-10									
0020	Prefab, wood, excl foundation, average	2 Clab	400	.040	SF Flr.		1.40		1.40	2.16

13 05 05.25 Selective Demolition, Geodesic Domes

		Crew	Daily Output	Labor-Hours	Unit	Material	Labor	Equipment	Total	Total Incl O&P
0010	**SELECTIVE DEMOLITION, GEODESIC DOMES**									
0050	Shell only, interlocking plywood panels, 30' diameter	F-5	3.20	10	Ea.		445		445	685
0060	34' diameter		2.30	13.913			620		620	955
0070	39' diameter		2	16			715		715	1,100
0080	45' diameter	F-3	2.20	18.182			815	278	1,093	1,550
0090	55' diameter		2	20			895	305	1,200	1,700
0100	60' diameter		2	20			895	305	1,200	1,700
0110	65' diameter		1.60	25			1,125	385	1,510	2,150

13 05 05.30 Selective Demolition, Greenhouses

		Crew	Daily Output	Labor-Hours	Unit	Material	Labor	Equipment	Total	Total Incl O&P
0010	**SELECTIVE DEMOLITION, GREENHOUSES** R024119-10									
0020	Resi-type, free standing, excl. foundations, 9' long x 8' wide	2 Clab	160	.100	SF Flr.		3.51		3.51	5.40
0030	9' long x 11' wide		170	.094			3.30		3.30	5.10
0040	9' long x 14' wide		220	.073			2.55		2.55	3.93
0050	9' long x 17' wide		320	.050			1.76		1.76	2.70
0060	Lean-to type, 4' wide		64	.250			8.80		8.80	13.50
0070	7' wide		120	.133			4.68		4.68	7.20
0080	Geodesic hemisphere, 1/8" plexiglass glazing, 8' diam.		4	4	Ea.		140		140	216
0090	24' diam.		.80	20			700		700	1,075
0100	48' diam.		.40	40			1,400		1,400	2,150

13 05 05.35 Selective Demolition, Hangars

		Crew	Daily Output	Labor-Hours	Unit	Material	Labor	Equipment	Total	Total Incl O&P
0010	**SELECTIVE DEMOLITION, HANGARS**									
0020	T type hangars, prefab, steel , galv roof & walls, incl doors, excl fndtn	E-2	2550	.022	SF Flr.		1.06	.59	1.65	2.47
0030	Circular type, prefab, steel frame, plastic skin, incl foundation, 80' diam	"	.50	112	Total		5,400	3,000	8,400	12,600

13 05 05.45 Selective Demolition, Lightning Protection

		Crew	Daily Output	Labor-Hours	Unit	Material	Labor	Equipment	Total	Total Incl O&P
0010	**SELECTIVE DEMOLITION, LIGHTNING PROTECTION**									
0020	Air terminal & base, copper, 3/8" diam. x 10", to 75' h	1 Clab	16	.500	Ea.		17.55		17.55	27
0030	1/2" diam. x 12", over 75' h		16	.500			17.55		17.55	27
0050	Aluminum, 1/2" diam. x 12", to 75' h		16	.500			17.55		17.55	27
0060	5/8" diam. x 12", over 75' h		16	.500			17.55		17.55	27
0070	Cable, copper, 220 lb. per thousand feet, to 75' high		640	.013	L.F.		.44		.44	.68

13 05 05 – Selective Special Construction Demolition

13 05 05.45 Selective Demolition, Lightning Protection

		Crew	Daily Output	Labor-Hours	Unit	Material	2012 Bare Costs Labor	2012 Bare Costs Equipment	Total	Total Incl O&P
0080	375 lb. per thousand feet, over 75' high	1 Clab	460	.017	L.F.		.61		.61	.94
0090	Aluminum, 101 lb. per thousand feet, to 75' high		560	.014			.50		.50	.77
0100	199 lb. per thousand feet, over 75' high		480	.017			.59		.59	.90
0110	Arrester, 175 V AC, to ground		16	.500	Ea.		17.55		17.55	27
0120	650 V AC, to ground		13	.615	"		21.50		21.50	33

13 05 05.50 Selective Demolition, Pre-Engineered Steel Buildings

		Crew	Daily Output	Labor-Hours	Unit	Material	2012 Bare Costs Labor	2012 Bare Costs Equipment	Total	Total Incl O&P
0010	**SELECTIVE DEMOLITION, PRE-ENGINEERED STEEL BUILDINGS**									
0500	Pre-engd. steel bldgs., rigid frame, clear span & multi post, excl. salvage									
0550	3,500 to 7,500 S.F.	L-10	1000	.024	SF Flr.		1.19	.61	1.80	2.69
0600	7,501 to 12,500 S.F.		1500	.016			.79	.41	1.20	1.80
0650	12,500 S.F. or greater		1650	.015			.72	.37	1.09	1.63
0700	Pre-engd. steel building components									
0710	Entrance canopy, including frame 4' x 4'	E-24	8	4	Ea.		195	87	282	430
0720	4' x 8'	"	7	4.571			222	99.50	321.50	495
0730	H.M. doors, self framing, single leaf	2 Skwk	8	2			91		91	140
0740	Double leaf		5	3.200			145		145	225
0760	Gutter, eave type		600	.027	L.F.		1.21		1.21	1.87
0770	Sash, single slide, double slide or fixed		24	.667	Ea.		30.50		30.50	47
0780	Skylight, fiberglass, to 30 S.F.		16	1			45.50		45.50	70
0785	Roof vents, circular, 12" to 24" diameter		12	1.333			60.50		60.50	93.50
0790	Continuous, 10' long		8	2			91		91	140
0900	Shelters, aluminum frame									
0910	Acrylic glazing, 3' x 9' x 8' high	2 Skwk	2	8	Ea.		365		365	560
0920	9' x 12' x 8' high	"	1.50	10.667	"		485		485	750

13 05 05.60 Selective Demolition, Silos

		Crew	Daily Output	Labor-Hours	Unit	Material	2012 Bare Costs Labor	2012 Bare Costs Equipment	Total	Total Incl O&P
0010	**SELECTIVE DEMOLITION, SILOS**									
0020	Conc stave, indstrl, conical/sloping bott, excl fndtn, 12' diam., 35' h	E-24	.18	177	Ea.		8,650	3,875	12,525	19,200
0030	16' diam., 45' h		.12	266			13,000	5,800	18,800	28,700
0040	25' diam., 75' h		.08	400			19,500	8,725	28,225	43,100
0050	Steel, factory fabricated, 30,000 gal. cap, painted or epoxy lined	L-5	2	28			1,375	350	1,725	2,800

13 05 05.65 Selective Demolition, Sound Control

		Crew	Daily Output	Labor-Hours	Unit	Material	2012 Bare Costs Labor	2012 Bare Costs Equipment	Total	Total Incl O&P
0010	**SELECTIVE DEMOLITION, SOUND CONTROL** R024119-10									
0120	Acoustical enclosure, 4" thick walls & ceiling panels, 8 lb./S.F.	3 Carp	144	.167	SF Surf		7.35		7.35	11.30
0130	10.5 lb./S.F.		128	.188			8.25		8.25	12.70
0140	Reverb chamber, parallel walls, 4" thick		120	.200			8.80		8.80	13.55
0150	Skewed walls, parallel roof, 4" thick		110	.218			9.60		9.60	14.80
0160	Skewed walls/roof, 4" layer/air space		96	.250			11.05		11.05	16.95
0170	Sound-absorbing panels, painted metal, 2'-6" x 8', under 1,000 S.F.		430	.056			2.46		2.46	3.79
0180	Over 1,000 S.F.		480	.050			2.21		2.21	3.39
0190	Flexible transparent curtain, clear	3 Shee	430	.056			2.91		2.91	4.44
0192	50% clear, 50% foam		430	.056			2.91		2.91	4.44
0194	25% clear, 75% foam		430	.056			2.91		2.91	4.44
0196	100% foam		430	.056			2.91		2.91	4.44
0200	Audio-masking sys., incl. speakers, amplfr., signal gnrtr.									
0205	Ceiling mounted, 5,000 S.F.	2 Elec	4800	.003	S.F.		.17		.17	.26
0210	10,000 S.F.		5600	.003			.15		.15	.22
0220	Plenum mounted, 5,000 S.F.		7600	.002			.11		.11	.16
0230	10,000 S.F.		8800	.002			.09		.09	.14

13 05 05.70 Selective Demolition, Special Purpose Rooms

		Crew	Daily Output	Labor-Hours	Unit	Material	2012 Bare Costs Labor	2012 Bare Costs Equipment	Total	Total Incl O&P
0010	**SELECTIVE DEMOLITION, SPECIAL PURPOSE ROOMS** R024119-10									
0100	Audiometric rooms, under 500 S.F. surface	4 Carp	200	.160	SF Surf		7.05		7.05	10.85
0110	Over 500 S.F. surface	"	240	.133	"		5.90		5.90	9.05

13 05 05 – Selective Special Construction Demolition

13 05 05.70 Selective Demolition, Special Purpose Rooms	Crew	Daily Output	Labor-Hours	Unit	Material	2012 Bare Costs Labor	Equipment	Total	Total Incl O&P
0200 Clean rooms, 12' x 12' soft wall, class 100	1 Carp	.30	26.667	Ea.		1,175		1,175	1,800
0210 Class 1000		.30	26.667			1,175		1,175	1,800
0220 Class 10,000		.35	22.857			1,000		1,000	1,550
0230 Class 100,000		.35	22.857			1,000		1,000	1,550
0300 Darkrooms, shell complete, 8' high	2 Carp	220	.073	SF Flr.		3.21		3.21	4.93
0310 12' high		110	.145	"		6.40		6.40	9.85
0350 Darkrooms doors, mini-cylindrical, revolving		4	4	Ea.		176		176	271
0400 Music room, practice modular		140	.114	SF Surf		5.05		5.05	7.75
0500 Refrigeration structures and finishes									
0510 Wall finish, 2 coat portland cement plaster, 1/2" thick	1 Clab	200	.040	S.F.		1.40		1.40	2.16
0520 Fiberglass panels, 1/8" thick		400	.020			.70		.70	1.08
0530 Ceiling finish, polystyrene plastic, 1" to 2" thick		500	.016			.56		.56	.86
0540 4" thick		450	.018			.62		.62	.96
0550 Refrigerator, prefab aluminum walk-in, 7'-6" high, 6' x 6' OD	2 Carp	100	.160	SF Flr.		7.05		7.05	10.85
0560 10' x 10' OD		160	.100			4.41		4.41	6.80
0570 Over 150 S.F.		200	.080			3.53		3.53	5.45
0600 Sauna, prefabricated, including heater & controls, 7' high, to 30 S.F.		120	.133			5.90		5.90	9.05
0610 To 40 S.F.		140	.114			5.05		5.05	7.75
0620 To 60 S.F.		175	.091			4.03		4.03	6.20
0630 To 100 S.F.		220	.073			3.21		3.21	4.93
0640 To 130 S.F.		250	.064			2.82		2.82	4.34
0650 Steam bath, heater, timer, head, single, to 140 C.F.	1 Plum	2.20	3.636	Ea.		199		199	299
0660 To 300 C.F.		2.20	3.636			199		199	299
0670 Steam bath, comm. size, w/blow-down assembly, to 800 C.F.		1.80	4.444			243		243	365
0680 To 2500 C.F.		1.60	5			273		273	410
0690 Steam bath, comm. size, multiple, for motels, apts, 500 C.F., 2 baths		2	4			218		218	330
0700 1,000 C.F., 4 baths		1.40	5.714			310		310	470

13 05 05.75 Selective Demolition, Storage Tanks

		Crew	Daily Output	Labor-Hours	Unit	Material	Labor	Equipment	Total	Total Incl O&P
0010	**SELECTIVE DEMOLITION, STORAGE TANKS**									
0500	Steel tank, single wall, above ground, not incl. fdn., pumps or piping									
0510	Single wall, 275 gallon R024119-10	Q-1	3	5.333	Ea.		262		262	395
0520	550 thru 2,000 gallon	B-34P	2	12			545	299	844	1,150
0530	5,000 thru 10,000 gallon	B-34Q	2	12			550	580	1,130	1,475
0540	15,000 thru 30,000 gallon	B-34S	2	16			780	1,750	2,530	3,075
0600	Steel tank, double wall, above ground not incl. fdn., pumps & piping									
0620	500 thru 2,000 gallon	B-34P	2	12	Ea.		545	299	844	1,150

13 05 05.85 Selective Demolition, Swimming Pool Equip

		Crew	Daily Output	Labor-Hours	Unit	Material	Labor	Equipment	Total	Total Incl O&P
0010	**SELECTIVE DEMOLITION, SWIMMING POOL EQUIP**									
0020	Diving stand, stainless steel, 3 meter	2 Clab	3	5.333	Ea.		187		187	288
0030	1 meter		5	3.200			112		112	173
0040	Diving board, 16' long, aluminum		5.40	2.963			104		104	160
0050	Fiberglass		5.40	2.963			104		104	160
0070	Ladders, heavy duty, stainless steel, 2 tread		14	1.143			40		40	61.50
0080	4 tread		12	1.333			47		47	72
0090	Lifeguard chair, stainless steel, fixed		5	3.200			112		112	173
0100	Slide, tubular, fiberglass, aluminum handrails & ladder, 5', straight		4	4			140		140	216
0110	8', curved		6	2.667			93.50		93.50	144
0120	10', curved		3	5.333			187		187	288
0130	12' straight, with platform		2.50	6.400			225		225	345
0140	Removable access ramp, stainless steel		4	4			140		140	216
0150	Removable stairs, stainless steel, collapsible		4	4			140		140	216

13 05 05 – Selective Special Construction Demolition

13 05 05.90 Selective Demolition, Tension Structures

		Crew	Daily Output	Labor-Hours	Unit	Material	2012 Bare Costs Labor	2012 Bare Costs Equipment	Total	Total Incl O&P
0010	**SELECTIVE DEMOLITION, TENSION STRUCTURES**									
0020	Steel/alum. frame, fabric shell, 60' clear span, 6,000 S.F.	B-41	2000	.022	SF Flr.		.80	.14	.94	1.38
0030	12,000 S.F.		2200	.020			.73	.13	.86	1.26
0040	80' clear span, 20,800 S.F.		2440	.018			.65	.11	.76	1.14
0050	100' clear span, 10,000 S.F.	L-5	4350	.013			.64	.16	.80	1.29
0060	26,000 S.F.	"	4600	.012			.60	.15	.75	1.22

13 05 05.95 Selective Demo, X-Ray/Radio Freq Protection

		Crew	Daily Output	Labor-Hours	Unit	Material	2012 Bare Costs Labor	2012 Bare Costs Equipment	Total	Total Incl O&P
0010	**SELECTIVE DEMO, X-RAY/RADIO FREQ PROTECTION**									
0020	Shielding lead, lined door frame, excl. hdwe., 1/16" thick	1 Clab	4.80	1.667	Ea.		58.50		58.50	90
0030	Lead sheets, 1/16" thick R024119-10	2 Clab	270	.059	S.F.		2.08		2.08	3.20
0040	1/8" thick		240	.067			2.34		2.34	3.60
0050	Lead shielding, 1/4" thick		270	.059			2.08		2.08	3.20
0060	1/2" thick		240	.067			2.34		2.34	3.60
0070	Lead glass, 1/4" thick, 2.0 mm LE, 12" x 16"	2 Glaz	16	1	Ea.		43		43	65
0080	24" x 36"		8	2			85.50		85.50	130
0090	36" x 60"		4	4			171		171	259
0100	Lead glass window frame, with 1/16" lead & voice passage, 36" x 60"		4	4			171		171	259
0110	Lead glass window frame, 24" x 36"		8	2			85.50		85.50	130
0120	Lead gypsum board, 5/8" thick with 1/16" lead	2 Clab	320	.050	S.F.		1.76		1.76	2.70
0130	1/8" lead		280	.057			2.01		2.01	3.09
0140	1/32" lead		400	.040			1.40		1.40	2.16
0150	Butt joints, 1/8" lead or thicker, 2" x 7' long batten strip		480	.033	Ea.		1.17		1.17	1.80
0160	X-ray protection, average radiography room, up to 300 S.F., 1/16" lead, min		.50	32	Total		1,125		1,125	1,725
0170	Maximum		.30	53.333			1,875		1,875	2,875
0180	Deep therapy X-ray room, 250 kV cap, up to 300 S.F., 1/4" lead, min		.20	80			2,800		2,800	4,325
0190	Maximum		.12	133			4,675		4,675	7,200
0880	Radio frequency shielding, prefab or screen-type copper or steel, minimum		360	.044	SF Surf		1.56		1.56	2.40
0890	Average		310	.052			1.81		1.81	2.79
0895	Maximum		290	.055			1.94		1.94	2.98

13 11 Swimming Pools

13 11 13 – Below-Grade Swimming Pools

13 11 13.50 Swimming Pools

		Crew	Daily Output	Labor-Hours	Unit	Material	2012 Bare Costs Labor	2012 Bare Costs Equipment	Total	Total Incl O&P
0010	**SWIMMING POOLS** Residential in-ground, vinyl lined, concrete									
0020	Sides including equipment, sand bottom	B-52	300	.187	SF Surf	17.55	7.60	1.84	26.99	33
0100	Metal or polystyrene sides R131113-20	B-14	410	.117		14.70	4.34	.81	19.85	23.50
0200	Add for vermiculite bottom					1.12			1.12	1.23
0500	Gunite bottom and sides, white plaster finish									
0600	12' x 30' pool	B-52	145	.386	SF Surf	32.50	15.75	3.82	52.07	64
0720	16' x 32' pool		155	.361		29.50	14.75	3.57	47.82	59
0750	20' x 40' pool		250	.224		26.50	9.15	2.21	37.86	45.50
0810	Concrete bottom and sides, tile finish									
0820	12' x 30' pool	B-52	80	.700	SF Surf	33	28.50	6.90	68.40	87.50
0830	16' x 32' pool		95	.589		27.50	24	5.80	57.30	73.50
0840	20' x 40' pool		130	.431		21.50	17.60	4.26	43.36	55.50
1100	Motel, gunite with plaster finish, incl. medium									
1150	capacity filtration & chlorination	B-52	115	.487	SF Surf	40	19.85	4.81	64.66	80.50
1200	Municipal, gunite with plaster finish, incl. high									
1250	capacity filtration & chlorination	B-52	100	.560	SF Surf	52	23	5.55	80.55	98
1350	Add for formed gutters				L.F.	76.50			76.50	84

13 11 Swimming Pools

13 11 13 – Below-Grade Swimming Pools

13 11 13.50 Swimming Pools

		Crew	Daily Output	Labor-Hours	Unit	Material	2012 Bare Costs Labor	2012 Bare Costs Equipment	Total	Total Incl O&P
1360	Add for stainless steel gutters				L.F.	226			226	249
1600	For water heating system, see Section 23 52 28.10									
1700	Filtration and deck equipment only, as % of total				Total				20%	20%
1800	Deck equipment, rule of thumb, 20' x 40' pool				SF Pool				1.18	1.30
1900	5000 S.F. pool				"				1.73	1.90
3000	Painting pools, preparation + 3 coats, 20' x 40' pool, epoxy	2 Pord	.33	48.485	Total	1,700	1,850		3,550	4,625
3100	Rubber base paint, 18 gallons	"	.33	48.485		1,100	1,850		2,950	3,975
3500	42' x 82' pool, 75 gallons, epoxy paint	3 Pord	.14	171		7,150	6,550		13,700	17,700
3600	Rubber base paint	"	.14	171		4,500	6,550		11,050	14,800

13 11 46 – Swimming Pool Accessories

13 11 46.50 Swimming Pool Equipment

		Crew	Daily Output	Labor-Hours	Unit	Material	2012 Bare Costs Labor	2012 Bare Costs Equipment	Total	Total Incl O&P
0010	**SWIMMING POOL EQUIPMENT**									
0020	Diving stand, stainless steel, 3 meter	2 Carp	.40	40	Ea.	11,100	1,775		12,875	14,900
0300	1 meter		2.70	5.926		6,450	261		6,711	7,475
0600	Diving boards, 16' long, aluminum		2.70	5.926		3,725	261		3,986	4,500
0700	Fiberglass		2.70	5.926		2,700	261		2,961	3,375
1200	Ladders, heavy duty, stainless steel, 2 tread		7	2.286		675	101		776	895
1500	4 tread		6	2.667		870	118		988	1,125
1800	Lifeguard chair, stainless steel, fixed		2.70	5.926		3,375	261		3,636	4,125
1900	Portable					2,750			2,750	3,025
2100	Lights, underwater, 12 volt, with transformer, 300 watt	1 Elec	1	8		274	410		684	915
2200	110 volt, 500 watt, standard		1	8		208	410		618	845
2400	Low water cutoff type		1	8		257	410		667	900
2800	Heaters, see Section 23 52 28.10									
3000	Pool covers, reinforced vinyl	3 Clab	1800	.013	S.F.	.35	.47		.82	1.11
3050	Automatic, electric								8.75	9.65
3100	Vinyl water tube	3 Clab	3200	.008		.50	.26		.76	.96
3200	Maximum	"	3000	.008		.63	.28		.91	1.12
3250	Sealed air bubble polyethylene solar blanket, 16 mils					.36			.36	.40
3300	Slides, tubular, fiberglass, aluminum handrails & ladder, 5'-0", straight	2 Carp	1.60	10	Ea.	3,300	440		3,740	4,300
3320	8'-0", curved		3	5.333		6,750	235		6,985	7,775
3400	10'-0", curved		1	16		30,100	705		30,805	34,200
3420	12'-0", straight with platform		1.20	13.333		13,700	590		14,290	15,900
4500	Hydraulic lift, movable pool bottom, single ram									
4520	Under 1,000 S.F. area	L-9	72	.500	S.F.	130	20		150	175
4600	Four ram lift, over 1,000 S.F.	"	109	.330	"	105	13.30		118.30	138
5000	Removable access ramp, stainless steel	2 Clab	2	8	Ea.	10,500	281		10,781	11,900

13 17 Tubs and Pools

13 17 33 – Whirlpool Tubs

13 17 33.10 Whirlpool Bath

		Crew	Daily Output	Labor-Hours	Unit	Material	2012 Bare Costs Labor	2012 Bare Costs Equipment	Total	Total Incl O&P
0010	**WHIRLPOOL BATH**									
6000	Whirlpool, bath with vented overflow, molded fiberglass									
6100	66" x 36" x 24"	Q-1	1	16	Ea.	3,100	785		3,885	4,600

13 18 Ice Rinks

13 18 13 – Ice Rink Floor Systems

13 18 13.50 Ice Skating		Crew	Daily Output	Labor-Hours	Unit	Material	2012 Bare Costs Labor	Equipment	Total	Total Incl O&P
0010	**ICE SKATING** Equipment incl. refrigeration, plumbing & cooling									
0020	coils & concrete slab, 85' x 200' rink									
0300	55° system, 5 mos., 100 ton				Total	575,000			575,000	632,500
0700	90° system, 12 mos., 135 ton				"	650,000			650,000	715,000
1200	Subsoil heating system (recycled from compressor), 85' x 200'	Q-7	.27	118	Ea.	40,000	6,275		46,275	53,500
1300	Subsoil insulation, 2 lb. polystyrene with vapor barrier, 85' x 200'	2 Carp	.14	114	"	30,000	5,050		35,050	40,800

13 18 16 – Ice Rink Dasher Boards

13 18 16.50 Ice Rink Dasher Boards		Crew	Daily Output	Labor-Hours	Unit	Material	2012 Bare Costs Labor	Equipment	Total	Total Incl O&P
0010	**ICE RINK DASHER BOARDS**									
1000	Dasher boards, 1/2" H.D. polyethylene faced steel frame, 3' acrylic									
1020	screen at sides, 5' acrylic ends, 85' x 200'	F-5	.06	533	Ea.	135,000	23,800		158,800	185,000
1100	Fiberglass & aluminum construction, same sides and ends	"	.06	533	"	155,000	23,800		178,800	207,000

13 21 Controlled Environment Rooms

13 21 13 – Clean Rooms

13 21 13.50 Clean Room Components		Crew	Daily Output	Labor-Hours	Unit	Material	2012 Bare Costs Labor	Equipment	Total	Total Incl O&P
0010	**CLEAN ROOM COMPONENTS**									
1100	Clean room, soft wall, 12' x 12', Class 100	1 Carp	.18	44.444	Ea.	16,400	1,950		18,350	21,000
1110	Class 1,000		.18	44.444		13,200	1,950		15,150	17,600
1120	Class 10,000		.21	38.095		11,600	1,675		13,275	15,400
1130	Class 100,000	↓	.21	38.095	↓	10,900	1,675		12,575	14,600
2800	Ceiling grid support, slotted channel struts 4'-0" O.C., ea. way				S.F.				5.90	6.50
3000	Ceiling panel, vinyl coated foil on mineral substrate									
3020	Sealed, non-perforated				S.F.				1.27	1.40
4000	Ceiling panel seal, silicone sealant, 150 L.F./gal.	1 Carp	150	.053	L.F.	.30	2.35		2.65	3.95
4100	Two sided adhesive tape	"	240	.033	"	.13	1.47		1.60	2.40
4200	Clips, one per panel				Ea.	1.04			1.04	1.14
6000	HEPA filter, 2' x 4', 99.97% eff., 3" dp beveled frame (silicone seal)					455			455	500
6040	6" deep skirted frame (channel seal)					390			390	430
6100	99.99% efficient, 3" deep beveled frame (silicone seal)					530			530	585
6140	6" deep skirted frame (channel seal)					405			405	445
6200	99.999% efficient, 3" deep beveled frame (silicone seal)					600			600	660
6240	6" deep skirted frame (channel seal)				↓	420			420	465
7000	Wall panel systems, including channel strut framing									
7020	Polyester coated aluminum, particle board				S.F.				18.20	20
7100	Porcelain coated aluminum, particle board								32	35
7400	Wall panel support, slotted channel struts, to 12' high				↓				16.35	18

13 21 26 – Cold Storage Rooms

13 21 26.50 Refrigeration		Crew	Daily Output	Labor-Hours	Unit	Material	2012 Bare Costs Labor	Equipment	Total	Total Incl O&P
0010	**REFRIGERATION**									
0020	Curbs, 12" high, 4" thick, concrete	2 Carp	58	.276	L.F.	4.61	12.15		16.76	24
1000	Doors, see Section 08 34 13.10									
2400	Finishes, 2 coat portland cement plaster, 1/2" thick	1 Plas	48	.167	S.F.	1.27	6.75		8.02	11.45
2500	For galvanized reinforcing mesh, add	1 Lath	335	.024		.84	.94		1.78	2.30
2700	3/16" thick latex cement	1 Plas	88	.091		2.16	3.67		5.83	7.90
2900	For glass cloth reinforced ceilings, add	"	450	.018		.52	.72		1.24	1.64
3100	Fiberglass panels, 1/8" thick	1 Carp	149.45	.054		2.86	2.36		5.22	6.80
3200	Polystyrene, plastic finish ceiling, 1" thick		274	.029		2.64	1.29		3.93	4.88
3400	2" thick		274	.029		3.01	1.29		4.30	5.30
3500	4" thick	↓	219	.037	↓	3.34	1.61		4.95	6.15

13 21 26 – Cold Storage Rooms

13 21 26.50 Refrigeration	Crew	Daily Output	Labor-Hours	Unit	Material	2012 Bare Costs Labor	Equipment	Total	Total Incl O&P
3800	Floors, concrete, 4" thick	1 Cefi	93	.086	S.F.	1.32	3.64		4.96	6.80
3900	6" thick	"	85	.094	↓	2.07	3.99		6.06	8.15
4000	Insulation, 1" to 6" thick, cork				B.F.	1.22			1.22	1.34
4100	Urethane					.51			.51	.56
4300	Polystyrene, regular					.50			.50	.55
4400	Bead board				↓	.28			.28	.31
4600	Installation of above, add per layer	2 Carp	657.60	.024	S.F.	.40	1.07		1.47	2.09
4700	Wall and ceiling juncture		298.90	.054	L.F.	1.97	2.36		4.33	5.80
4900	Partitions, galvanized sandwich panels, 4" thick, stock		219.20	.073	S.F.	8.25	3.22		11.47	14.05
5000	Aluminum or fiberglass	↓	219.20	.073	"	9.05	3.22		12.27	14.90
5200	Prefab walk-in, 7'-6" high, aluminum, incl. door & floors,									
5210	not incl. partitions or refrigeration, 6' x 6' O.D. nominal	2 Carp	54.80	.292	SF Flr.	148	12.90		160.90	183
5500	10' x 10' O.D. nominal		82.20	.195		119	8.60		127.60	144
5700	12' x 14' O.D. nominal		109.60	.146		107	6.45		113.45	127
5800	12' x 20' O.D. nominal	↓	109.60	.146		93	6.45		99.45	112
6100	For 8'-6" high, add					5%				
6300	Rule of thumb for complete units, w/o doors & refrigeration, cooler	2 Carp	146	.110		134	4.83		138.83	154
6400	Freezer		109.60	.146	↓	158	6.45		164.45	184
6600	Shelving, plated or galvanized, steel wire type		360	.044	SF Hor.	11.75	1.96		13.71	15.90
6700	Slat shelf type	↓	375	.043		14.45	1.88		16.33	18.80
6900	For stainless steel shelving, add				↓	300%				
7000	Vapor barrier, on wood walls	2 Carp	1644	.010	S.F.	.17	.43		.60	.85
7200	On masonry walls	"	1315	.012	"	.44	.54		.98	1.31
7500	For air curtain doors, see Section 23 34 33.10									

13 21 48 – Sound-Conditioned Rooms

13 21 48.10 Anechoic Chambers

		Crew	Daily Output	Labor-Hours	Unit	Material	Labor	Equipment	Total	Total Incl O&P
0010	**ANECHOIC CHAMBERS** Standard units, 7' ceiling heights									
0100	Area for pricing is net inside dimensions									
0300	200 cycles per second cutoff, 25 S.F. floor area				SF Flr.	1,600			1,600	1,750
0400	50 S.F.								1,050	1,150
0600	75 S.F.								1,000	1,100
0700	100 S.F.					1,200			1,200	1,325
0900	For 150 cycles per second cutoff, add to 100 S.F. room									30%
1000	For 100 cycles per second cutoff, add to 100 S.F. room				↓					45%

13 21 48.15 Audiometric Rooms

		Crew	Daily Output	Labor-Hours	Unit	Material	Labor	Equipment	Total	Total Incl O&P
0010	**AUDIOMETRIC ROOMS**									
0020	Under 500 S.F. surface	4 Carp	98	.327	SF Surf	51.50	14.40		65.90	78.50
0100	Over 500 S.F. surface	"	120	.267	"	49	11.75		60.75	71.50

13 21 53 – Darkrooms

13 21 53.50 Darkrooms

		Crew	Daily Output	Labor-Hours	Unit	Material	Labor	Equipment	Total	Total Incl O&P
0010	**DARKROOMS**									
0020	Shell, complete except for door, 64 S.F., 8' high	2 Carp	128	.125	SF Flr.	56	5.50		61.50	70
0100	12' high		64	.250		71.50	11.05		82.55	96
0500	120 S.F. floor, 8' high		120	.133		39	5.90		44.90	52
0600	12' high		60	.267		53.50	11.75		65.25	76.50
0800	240 S.F. floor, 8' high		120	.133		28	5.90		33.90	40
0900	12' high		60	.267	↓	40	11.75		51.75	62
1200	Mini-cylindrical, revolving, unlined, 4' diameter		3.50	4.571	Ea.	2,350	202		2,552	2,875
1400	5'-6" diameter	↓	2.50	6.400		5,600	282		5,882	6,575
1600	Add for lead lining, inner cylinder, 1/32" thick					1,650			1,650	1,825
1700	1/16" thick				↓	4,450			4,450	4,875

13 21 Controlled Environment Rooms

13 21 53 – Darkrooms

13 21 53.50 Darkrooms	Crew	Daily Output	Labor-Hours	Unit	Material	2012 Bare Costs Labor	Equipment	Total	Total Incl O&P
1800	Add for lead lining, inner and outer cylinder, 1/32" thick				Ea.	3,050			3,050	3,375
1900	1/16" thick				↓	6,825			6,825	7,500
2000	For darkroom door, see Section 08 34 36.10									

13 21 56 – Music Rooms

13 21 56.50 Music Rooms

		Crew	Daily Output	Labor-Hours	Unit	Material	Labor	Equipment	Total	Total Incl O&P
0010	**MUSIC ROOMS**									
0020	Practice room, modular, perforated steel, under 500 S.F.	2 Carp	70	.229	SF Surf	31.50	10.10		41.60	50
0100	Over 500 S.F.	"	80	.200	"	26.50	8.80		35.30	43

13 24 Special Activity Rooms

13 24 16 – Saunas

13 24 16.50 Saunas and Heaters

		Crew	Daily Output	Labor-Hours	Unit	Material	Labor	Equipment	Total	Total Incl O&P
0010	**SAUNAS AND HEATERS**									
0020	Prefabricated, incl. heater & controls, 7' high, 6' x 4', C/C	L-7	2.20	12.727	Ea.	5,125	540		5,665	6,475
0050	6' x 4', C/P		2	14		4,300	595		4,895	5,650
0400	6' x 5', C/C		2	14		5,225	595		5,820	6,675
0450	6' x 5', C/P		2	14		4,450	595		5,045	5,825
0600	6' x 6', C/C		1.80	15.556		5,550	665		6,215	7,125
0650	6' x 6', C/P		1.80	15.556		4,750	665		5,415	6,250
0800	6' x 9', C/C		1.60	17.500		6,875	745		7,620	8,700
0850	6' x 9', C/P		1.60	17.500		6,425	745		7,170	8,225
1000	8' x 12', C/C		1.10	25.455		11,200	1,075		12,275	14,000
1050	8' x 12', C/P		1.10	25.455		9,900	1,075		10,975	12,600
1200	8' x 8', C/C		1.40	20		8,525	850		9,375	10,700
1250	8' x 8', C/P		1.40	20		7,750	850		8,600	9,825
1400	8' x 10', C/C		1.20	23.333		9,450	995		10,445	11,900
1450	8' x 10', C/P		1.20	23.333		8,500	995		9,495	10,900
1600	10' x 12', C/C		1	28		11,800	1,200		13,000	14,800
1650	10' x 12', C/P	↓	1	28		10,300	1,200		11,500	13,200
1700	Door only, cedar, 2' x 6', with tempered insulated glass window	2 Carp	3.40	4.706		620	208		828	1,000
1800	Prehung, incl. jambs, pulls & hardware	"	12	1.333		620	59		679	775
2500	Heaters only (incl. above), wall mounted, to 200 C.F.					615			615	675
2750	To 300 C.F.					795			795	875
3000	Floor standing, to 720 C.F., 10,000 watts, w/controls	1 Elec	3	2.667		2,100	137		2,237	2,500
3250	To 1,000 C.F., 16,000 watts	"	3	2.667	↓	2,700	137		2,837	3,150

13 24 26 – Steam Baths

13 24 26.50 Steam Baths and Components

		Crew	Daily Output	Labor-Hours	Unit	Material	Labor	Equipment	Total	Total Incl O&P
0010	**STEAM BATHS AND COMPONENTS**									
0020	Heater, timer & head, single, to 140 C.F.	1 Plum	1.20	6.667	Ea.	2,100	365		2,465	2,850
0500	To 300 C.F.		1.10	7.273		2,425	395		2,820	3,275
1000	Commercial size, with blow-down assembly, to 800 C.F.		.90	8.889		7,800	485		8,285	9,300
1500	To 2500 C.F.	↓	.80	10		8,775	545		9,320	10,500
2000	Multiple, motels, apts., 2 baths, w/blow-down assm., 500 C.F.	Q-1	1.30	12.308		5,900	605		6,505	7,400
2500	4 baths	"	.70	22.857		7,450	1,125		8,575	9,900
2700	Conversion unit for residential tub, including door				↓	4,100			4,100	4,500

13 28 Athletic and Recreational Special Construction

13 28 33 – Athletic and Recreational Court Walls

13 28 33.50 Sport Court	Crew	Daily Output	Labor-Hours	Unit	Material	2012 Bare Costs Labor	Equipment	Total	Total Incl O&P
0010 **SPORT COURT**									
0020 Floors, No. 2 & better maple, 25/32" thick				SF Flr.				6.05	6.65
0100 Walls, laminated plastic bonded to galv. steel studs				SF Wall				7.70	8.45
0150 Laminated fiberglass surfacing, minimum				↓				1.91	2.10
0180 Maximum								2.15	2.37
0300 Squash, regulation court in existing building, minimum				Court	36,800			36,800	40,400
0400 Maximum				"	41,000			41,000	45,000
0450 Rule of thumb for components:									
0470 Walls	3 Carp	.15	160	Court	11,000	7,050		18,050	23,000
0500 Floor	"	.25	96		8,725	4,225		12,950	16,100
0550 Lighting	2 Elec	.60	26.667		2,100	1,375		3,475	4,350
0600 Handball, racquetball court in existing building, minimum	C-1	.20	160		39,800	6,700		46,500	54,000
0800 Maximum	"	.10	320		43,100	13,400		56,500	68,000
0900 Rule of thumb for components: walls	3 Carp	.12	200		12,600	8,825		21,425	27,500
1000 Floor	↓	.25	96		8,725	4,225		12,950	16,100
1100 Ceiling		.33	72.727		4,200	3,200		7,400	9,550
1200 Lighting	2 Elec	.60	26.667	↓	2,200	1,375		3,575	4,475

13 31 Fabric Structures

13 31 13 – Air-Supported Fabric Structures

13 31 13.09 Air Supported Tank Covers

13 31 13.09 Air Supported Tank Covers	Crew	Daily Output	Labor-Hours	Unit	Material	2012 Bare Costs Labor	Equipment	Total	Total Incl O&P
0010 **AIR SUPPORTED TANK COVERS**, vinyl polyester									
0100 Scrim, double layer, with hardware, blower, standby & controls									
0200 Round, 75' diameter	B-2	4500	.009	S.F.	10.10	.32		10.42	11.65
0300 100' diameter		5000	.008		9.20	.28		9.48	10.55
0400 150' diameter		5000	.008		7.30	.28		7.58	8.45
0500 Rectangular, 20' x 20'		4500	.009		23	.32		23.32	26
0600 30' x 40'		4500	.009		23	.32		23.32	26
0700 50' x 60'	↓	4500	.009		23	.32		23.32	26
0800 For single wall construction, deduct, minimum					.78			.78	.86
0900 Maximum					2.31			2.31	2.54
1000 For maximum resistance to atmosphere or cold, add				↓	1.13			1.13	1.24
1100 For average shipping charges, add				Total	1,950			1,950	2,150

13 31 13.13 Single-Walled Air-Supported Structures

13 31 13.13 Single-Walled Air-Supported Structures	Crew	Daily Output	Labor-Hours	Unit	Material	2012 Bare Costs Labor	Equipment	Total	Total Incl O&P
0010 **SINGLE-WALLED AIR-SUPPORTED STRUCTURES** R133113-10									
0020 Site preparation, incl. anchor placement and utilities	B-11B	1000	.016	SF Flr.	1.15	.64	.27	2.06	2.53
0030 For concrete, see Section 03 30 53.40									
0050 Warehouse, polyester/vinyl fabric, 28 oz., over 10 yr. life, welded									
0060 Seams, tension cables, primary & auxiliary inflation system,									
0070 airlock, personnel doors and liner									
0100 5,000 S.F.	4 Clab	5000	.006	SF Flr.	25	.22		25.22	28
0250 12,000 S.F.	"	6000	.005		17.80	.19		17.99	19.90
0400 24,000 S.F.	8 Clab	12000	.005		12.55	.19		12.74	14.10
0500 50,000 S.F.	"	12500	.005	↓	11.60	.18		11.78	13.05
0700 12 oz. reinforced vinyl fabric, 5 yr. life, sewn seams,									
0710 accordion door, including liner									
0750 3000 S.F.	4 Clab	3000	.011	SF Flr.	12	.37		12.37	13.80
0800 12,000 S.F.	"	6000	.005		10.20	.19		10.39	11.55
0850 24,000 S.F.	8 Clab	12000	.005		8.65	.19		8.84	9.80
0950 Deduct for single layer					1.02			1.02	1.12
1000 Add for welded seams				↓	1.01			1.01	1.11

13 31 Fabric Structures

13 31 13 – Air-Supported Fabric Structures

13 31 13.13 Single-Walled Air-Supported Structures		Crew	Daily Output	Labor-Hours	Unit	Material	2012 Bare Costs Labor	Equipment	Total	Total Incl O&P
1050	Add for double layer, welded seams included				SF Flr.	2.08			2.08	2.29
1250	Tedlar/vinyl fabric, 28 oz., with liner, over 10 yr. life,									
1260	incl. overhead and personnel doors									
1300	3000 S.F.	4 Clab	3000	.011	SF Flr.	24	.37		24.37	27
1450	12,000 S.F.	"	6000	.005		17.05	.19		17.24	19.05
1550	24,000 S.F.	8 Clab	12000	.005		13.15	.19		13.34	14.80
1700	Deduct for single layer					1.44			1.44	1.58
2250	Greenhouse/shelter, woven polyethylene with liner, 2 yr. life,									
2260	sewn seams, including doors									
2300	3000 S.F.	4 Clab	3000	.011	SF Flr.	6.60	.37		6.97	7.85
2350	12,000 S.F.	"	6000	.005		6.70	.19		6.89	7.70
2450	24,000 S.F.	8 Clab	12000	.005		5.60	.19		5.79	6.45
2550	Deduct for single layer					.63			.63	.69
2600	Tennis/gymnasium, polyester/vinyl fabric, 28 oz., over 10 yr. life,									
2610	including thermal liner, heat and lights									
2650	7,200 S.F.	4 Clab	6000	.005	SF Flr.	23	.19		23.19	26
2750	13,000 S.F.	"	6500	.005		17.80	.17		17.97	19.85
2850	Over 24,000 S.F.	8 Clab	12000	.005		16.25	.19		16.44	18.20
2860	For low temperature conditions, add					1.13			1.13	1.24
2870	For average shipping charges, add				Total	5,600			5,600	6,150
2900	Thermal liner, translucent reinforced vinyl				SF Flr.	1.13			1.13	1.24
2950	Metalized mylar fabric and mesh, double liner				"	2.31			2.31	2.54
3050	Stadium/convention center, teflon coated fiberglass, heavy weight,									
3060	over 20 yr. life, incl. thermal liner and heating system									
3100	Minimum	9 Clab	26000	.003	SF Flr.	56.50	.10		56.60	62.50
3110	Maximum	"	19000	.004	"	67.50	.13		67.63	74
3400	Doors, air lock, 15' long, 10' x 10'	2 Carp	.80	20	Ea.	20,200	880		21,080	23,700
3600	15' x 15'	"	.50	32		30,000	1,400		31,400	35,200
3700	For each added 5' length, add					5,375			5,375	5,900
3900	Revolving personnel door, 6' diameter, 6'-6" high	2 Carp	.80	20		15,000	880		15,880	17,900
4200	Double wall, self supporting, shell only, minimum				SF Flr.				19.10	21
4300	Maximum				"				35.50	39

13 31 23 – Tensioned Fabric Structures

13 31 23.50 Tension Structures

		Crew	Daily Output	Labor-Hours	Unit	Material	2012 Bare Costs Labor	Equipment	Total	Total Incl O&P
0010	**TENSION STRUCTURES** Rigid steel/alum. frame, vinyl coated poly									
0100	Fabric shell, 60' clear span, not incl. foundations or floors									
0200	6,000 S.F.	B-41	1000	.044	SF Flr.	15.15	1.60	.28	17.03	19.45
0300	12,000 S.F.		1100	.040		14.05	1.45	.25	15.75	17.95
0400	80' clear span, 20,800 S.F.		1220	.036		13.90	1.31	.23	15.44	17.55
0410	100' clear span, 10,000 S.F.	L-5	2175	.026		15.35	1.27	.32	16.94	19.50
0430	26,000 S.F.		2300	.024		14.45	1.20	.30	15.95	18.35
0450	36,000 S.F.		2500	.022		14.10	1.11	.28	15.49	17.75
0460	120' clear span, 24,000 S.F.		3000	.019		16.05	.92	.23	17.20	19.55
0470	150' clear span, 30,000 S.F.		6000	.009		17.30	.46	.12	17.88	19.95
0480	200' clear span, 40,000 S.F.	E-6	8000	.016		19.75	.78	.22	20.75	23
0500	For roll-up door, 12' x 14', add	L-2	1	16	Ea.	5,325	620		5,945	6,800
0600	For personnel doors, add, minimum				SF Flr.	5%				
0700	Add, maximum					15%				
0800	For site work, simple foundation, etc., add, minimum								1.25	1.95
0900	Add, maximum								2.75	3.05

433

13 34 Fabricated Engineered Structures

13 34 13 – Glazed Structures

13 34 13.13 Greenhouses

		Crew	Daily Output	Labor-Hours	Unit	Material	2012 Bare Costs Labor	Equipment	Total	Total Incl O&P
0010	**GREENHOUSES**, Shell only, stock units, not incl. 2' stub walls,									
0020	foundation, floors, heat or compartments									
0300	Residential type, free standing, 8'-6" long x 7'-6" wide	2 Carp	59	.271	SF Flr.	33	11.95		44.95	54.50
0400	10'-6" wide		85	.188		28	8.30		36.30	44
0600	13'-6" wide		108	.148		24.50	6.55		31.05	37
0700	17'-0" wide		160	.100		23.50	4.41		27.91	33
0900	Lean-to type, 3'-10" wide		34	.471		32.50	21		53.50	68
1000	6'-10" wide		58	.276		43.50	12.15		55.65	66.50
1500	Commercial, custom, truss frame, incl. equip., plumbing, elec.,									
1510	benches and controls, under 2,000 S.F., minimum				SF Flr.				27.50	30
1550	Maximum					14.45			14.45	15.90
1700	Over 5,000 S.F., minimum					14.60			14.60	16.05
1750	Maximum								27.50	30
2000	Institutional, custom, rigid frame, including compartments and									
2010	multi-controls, under 500 S.F., minimum				SF Flr.	45			45	49.50
2050	Maximum					35.50			35.50	39
2150	Over 2,000 S.F., minimum					11.70			11.70	12.90
2200	Maximum					59			59	65
2400	Concealed rigid frame, under 500 S.F., minimum								82	90
2450	Maximum								100	110
2550	Over 2,000 S.F., minimum								62	68
2600	Maximum								72	79
2800	Lean-to type, under 500 S.F., minimum								72.50	80
2850	Maximum								109	120
3000	Over 2,000 S.F., minimum								40	44
3050	Maximum								66.50	73
3600	For 1/4" clear plate glass, deduct				SF Surf	3.80			3.80	4.18
3700	For 1/4" tempered glass, add				"	7.85			7.85	8.65
3900	Cooling, 1200 cfm exhaust fan, add				Ea.	239			239	263
4000	7850 cfm					900			900	990
4200	For heaters, 10 MBH, add					195			195	215
4300	60 MBH, add					645			645	710
4500	For benches, 2' x 8', add					150			150	165
4600	4' x 10', add					220			220	242
4800	For controls, add, minimum				Total	210			210	231
4900	Maximum				"	700			700	770
5100	For humidification equipment, add				Ea.	560			560	615
5200	For vinyl shading, add				S.F.	.22			.22	.24
6000	Geodesic hemisphere, 1/8" plexiglass glazing									
6050	8' diameter	2 Carp	2	8	Ea.	4,950	355		5,305	6,000
6150	24' diameter		.35	45.714		11,000	2,025		13,025	15,100
6250	48' diameter		.20	80		45,000	3,525		48,525	55,000

13 34 13.19 Swimming Pool Enclosures

		Crew	Daily Output	Labor-Hours	Unit	Material	2012 Bare Costs Labor	Equipment	Total	Total Incl O&P
0010	**SWIMMING POOL ENCLOSURES** Translucent, free standing									
0020	not including foundations, heat or light									
0200	Economy, minimum	2 Carp	200	.080	SF Hor.	14.80	3.53		18.33	21.50
0300	Maximum		100	.160		45.50	7.05		52.55	61
0400	Deluxe, minimum		100	.160		50.50	7.05		57.55	66.50
0600	Maximum		70	.229		227	10.10		237.10	266
0700	For motorized roof, 40% opening, solid roof, add					11.70			11.70	12.85
0800	Skylight type roof, add					9.35			9.35	10.30
0900	Air-inflated, including blowers and heaters, minimum								3.18	3.50

13 34 Fabricated Engineered Structures

13 34 13 – Glazed Structures

13 34 13.19 Swimming Pool Enclosures	Crew	Daily Output	Labor-Hours	Unit	Material	Labor	Equipment	Total	Total Incl O&P
1000	Maximum				SF Hor.				6.10	6.70

13 34 16 – Grandstands and Bleachers

13 34 16.13 Grandstands

		Crew	Daily Output	Labor-Hours	Unit	Material	Labor	Equipment	Total	Total Incl O&P
0010	**GRANDSTANDS** Permanent, municipal, including foundation									
0050	Steel understructure w/aluminum closed deck, minimum				Seat				118	130
0100	Maximum								205	225
0300	Steel, minimum					55			55	60.50
0400	Maximum					110			110	121
0600	Aluminum, extruded, stock design, minimum								63.50	70
0700	Maximum								105	115
0900	Composite, steel, wood and plastic, stock design, minimum					130			130	143
1000	Maximum					260			260	286

13 34 16.53 Bleachers

		Crew	Daily Output	Labor-Hours	Unit	Material	Labor	Equipment	Total	Total Incl O&P
0010	**BLEACHERS**									
0020	Bleachers, outdoor, portable, 3 to 5 tiers, to 300' long, min	2 Sswk	120	.133	Seat	37.50	6.60		44.10	53
0100	Maximum, less than 15' long, prefabricated		80	.200		41.50	9.85		51.35	63
0200	6 to 20 tiers, minimum, up to 300' long		120	.133		70.50	6.60		77.10	89.50
0300	Max., under 15', (highly prefabricated, on wheels)		80	.200		85.50	9.85		95.35	112
0500	Permanent grandstands, wood seat, steel frame, 24" row									
0600	3 to 15 tiers, minimum	2 Sswk	60	.267	Seat	63.50	13.15		76.65	93
0700	Maximum		48	.333		76	16.45		92.45	113
0900	16 to 30 tiers, minimum		60	.267		191	13.15		204.15	234
0950	Average		55	.291		211	14.35		225.35	258
1000	Maximum		48	.333		244	16.45		260.45	299
1200	Seat backs only, 30" row, fiberglass		160	.100		40	4.94		44.94	53
1300	Steel and wood		160	.100		62	4.94		66.94	77
1400	NOTE: average seating is 1.5' in width									

13 34 19 – Metal Building Systems

13 34 19.50 Pre-Engineered Steel Buildings

		Crew	Daily Output	Labor-Hours	Unit	Material	Labor	Equipment	Total	Total Incl O&P
0010	**PRE-ENGINEERED STEEL BUILDINGS** R133419-10									
0100	Clear span rigid frame, 26 ga. colored roofing and siding									
0150	20' to 29' wide, 10' eave height	E-2	425	.132	SF Flr.	8.50	6.35	3.52	18.37	24
0160	14' eave height		350	.160		9.20	7.70	4.28	21.18	28
0170	16' eave height		320	.175		9.85	8.45	4.68	22.98	30.50
0180	20' eave height		275	.204		10.75	9.85	5.45	26.05	34.50
0190	24' eave height		240	.233		11.85	11.25	6.25	29.35	39
0200	30' to 40' wide, 10' eave height		535	.105		6.55	5.05	2.80	14.40	18.95
0300	14' eave height		450	.124		7.05	6	3.33	16.38	21.50
0400	16' eave height		415	.135		7.50	6.50	3.61	17.61	23.50
0500	20' eave height		360	.156		8.15	7.50	4.16	19.81	26.50
0600	24' eave height		320	.175		8.95	8.45	4.68	22.08	29.50
0700	50' to 100' wide, 10' eave height		770	.073		5.55	3.51	1.94	11	14.25
0900	16' eave height		600	.093		6.40	4.50	2.49	13.39	17.45
1000	20' eave height		490	.114		6.90	5.50	3.06	15.46	20.50
1100	24' eave height		435	.129		7.60	6.20	3.44	17.24	23
1200	Clear span tapered beam frame, 26 ga. colored roofing/siding									
1300	30' to 39' wide, 10' eave height	E-2	535	.105	SF Flr.	7.40	5.05	2.80	15.25	19.90
1400	14' eave height		450	.124		8.20	6	3.33	17.53	23
1500	16' eave height		415	.135		8.60	6.50	3.61	18.71	24.50
1600	20' eave height		360	.156		9.50	7.50	4.16	21.16	28
1700	40' wide, 10' eave height		600	.093		6.60	4.50	2.49	13.59	17.70

13 34 19 – Metal Building Systems

13 34 19.50 Pre-Engineered Steel Buildings		Crew	Daily Output	Labor-Hours	Unit	Material	2012 Bare Costs Labor	Equipment	Total	Total Incl O&P
1800	14' eave height	E-2	510	.110	SF Flr.	7.30	5.30	2.94	15.54	20.50
1900	16' eave height		475	.118		7.60	5.70	3.15	16.45	21.50
2000	20' eave height		415	.135		8.40	6.50	3.61	18.51	24.50
2100	50' to 79' wide, 10' eave height		770	.073		6.20	3.51	1.94	11.65	14.95
2200	14' eave height		675	.083		6.70	4	2.22	12.92	16.70
2300	16' eave height		635	.088		7	4.26	2.36	13.62	17.60
2400	20' eave height		490	.114		7.60	5.50	3.06	16.16	21
2410	80' to 100' wide, 10' eave height		935	.060		5.50	2.89	1.60	9.99	12.75
2420	14' eave height		750	.075		6.05	3.60	2	11.65	15.05
2430	16' eave height		685	.082		6.30	3.95	2.19	12.44	16.10
2440	20' eave height		560	.100		6.75	4.83	2.67	14.25	18.60
2460	101' to 120' wide, 10' eave height		950	.059		5.05	2.84	1.58	9.47	12.15
2470	14' eave height		770	.073		5.60	3.51	1.94	11.05	14.30
2480	16' eave height		675	.083		5.95	4	2.22	12.17	15.85
2490	20' eave height	▼	560	.100	▼	6.35	4.83	2.67	13.85	18.20
2500	Single post 2-span frame, 26 ga. colored roofing and siding									
2600	80' wide, 14' eave height	E-2	740	.076	SF Flr.	5.55	3.65	2.02	11.22	14.60
2700	16' eave height		695	.081		5.90	3.89	2.15	11.94	15.50
2800	20' eave height		625	.090		6.40	4.32	2.40	13.12	17.05
2900	24' eave height		570	.098		7	4.74	2.63	14.37	18.75
3000	100' wide, 14' eave height		835	.067		5.40	3.24	1.79	10.43	13.45
3100	16' eave height		795	.070		5.05	3.40	1.88	10.33	13.45
3200	20' eave height		730	.077		6.10	3.70	2.05	11.85	15.30
3300	24' eave height		670	.084		6.75	4.03	2.23	13.01	16.80
3400	120' wide, 14' eave height		870	.064		6.25	3.11	1.72	11.08	14.10
3500	16' eave height		830	.067		5.60	3.26	1.80	10.66	13.75
3600	20' eave height		765	.073		6.10	3.53	1.96	11.59	14.90
3700	24' eave height	▼	705	.079	▼	6.65	3.83	2.12	12.60	16.25
3800	Double post 3-span frame, 26 ga. colored roofing and siding									
3900	150' wide, 14' eave height	E-2	925	.061	SF Flr.	4.43	2.92	1.62	8.97	11.65
4000	16' eave height		890	.063		4.62	3.04	1.68	9.34	12.15
4100	20' eave height		820	.068		5.05	3.30	1.83	10.18	13.20
4200	24' eave height	▼	765	.073	▼	5.55	3.53	1.96	11.04	14.35
4300	Triple post 4-span frame, 26 ga. colored roofing and siding									
4400	160' wide, 14' eave height	E-2	970	.058	SF Flr.	4.35	2.79	1.54	8.68	11.25
4500	16' eave height		930	.060		4.54	2.91	1.61	9.06	11.75
4600	20' eave height		870	.064		4.48	3.11	1.72	9.31	12.15
4700	24' eave height		815	.069		5.10	3.32	1.84	10.26	13.30
4800	200' wide, 14' eave height		1030	.054		4.01	2.62	1.45	8.08	10.50
4900	16' eave height		995	.056		4.15	2.72	1.50	8.37	10.90
5000	20' eave height		935	.060		4.58	2.89	1.60	9.07	11.75
5100	24' eave height	▼	885	.063	▼	5.15	3.05	1.69	9.89	12.75
5200	Accessory items: add to the basic building cost above									
5250	Eave overhang, 2' wide, 26 ga., with soffit	E-2	360	.156	L.F.	30.50	7.50	4.16	42.16	51
5300	4' wide, without soffit		300	.187		26.50	9	4.99	40.49	50.50
5350	With soffit		250	.224		38.50	10.80	6	55.30	67.50
5400	6' wide, without soffit		250	.224		34.50	10.80	6	51.30	63
5450	With soffit		200	.280	▼	46	13.50	7.50	67	82.50
5500	Entrance canopy, incl. frame, 4' x 4'		25	2.240	Ea.	450	108	60	618	745
5550	4' x 8'		19	2.947	"	525	142	79	746	905
5600	End wall roof overhang, 4' wide, without soffit		850	.066	L.F.	17.10	3.18	1.76	22.04	26
5650	With soffit	▼	500	.112	"	27.50	5.40	2.99	35.89	42.50
5700	Doors, H.M. self-framing, incl. butts, lockset and trim									

13 34 19.50 Pre-Engineered Steel Buildings	Crew	Daily Output	Labor-Hours	Unit	Material	Labor	Equipment	Total	Total Incl O&P
5750	Single leaf, 3070 (3' x 7'), economy	2 Sswk	5	3.200	Opng.	580	158		738	920
5800	Deluxe		4	4		635	197		832	1,050
5825	Glazed		4	4		740	197		937	1,175
5850	3670 (3'-6" x 7')		4	4		815	197		1,012	1,250
5900	4070 (4' x 7')		3	5.333		885	263		1,148	1,450
5950	Double leaf, 6070 (6' x 7')		2	8		1,075	395		1,470	1,875
6000	Glazed		2	8		1,400	395		1,795	2,250
6050	Framing only, for openings, 3' x 7'		4	4		174	197		371	545
6100	10' x 10'		3	5.333		575	263		838	1,100
6150	For windows below, 2020 (2' x 2')		6	2.667		185	132		317	440
6200	4030 (4' x 3')		5	3.200		226	158		384	530
6250	Flashings, 26 ga., corner or eave, painted		240	.067	L.F.	4.31	3.29		7.60	10.65
6300	Galvanized		240	.067		3.91	3.29		7.20	10.20
6350	Rake flashing, painted		240	.067		4.65	3.29		7.94	11
6400	Galvanized		240	.067		4.19	3.29		7.48	10.50
6450	Ridge flashing, 18" wide, painted		240	.067		6.20	3.29		9.49	12.75
6500	Galvanized		240	.067		6.75	3.29		10.04	13.35
6550	Gutter, eave type, 26 ga., painted		320	.050		6.70	2.47		9.17	11.75
6650	Valley type, between buildings, painted		120	.133		12.25	6.60		18.85	25
6710	Insulation, rated .6 lb. density, unfaced 4" thick, R13	2 Carp	2300	.007	S.F.	.37	.31		.68	.88
6720	6" thick, R19		2300	.007		.51	.31		.82	1.03
6730	10" thick, R30		2300	.007		.90	.31		1.21	1.46
6750	Insulation, rated .6 lb. density, poly/scrim/foil (PSF) faced									
6760	4" thick R13	2 Carp	2300	.007	S.F.	.47	.31		.78	.99
6770	6" thick, R19		2300	.007		.61	.31		.92	1.14
6780	9 1/2" thick, R30		2300	.007		.78	.31		1.09	1.33
6800	Insulation, rated .6 lb. density, vinyl faced 1-1/2" thick, R5		2300	.007		.27	.31		.58	.77
6850	3" thick, R10		2300	.007		.31	.31		.62	.81
6900	4" thick, R13		2300	.007		.41	.31		.72	.92
6920	6" thick, R19		2300	.007		.54	.31		.85	1.06
6930	10" thick, R30		2300	.007		1.16	.31		1.47	1.75
6950	Foil/scrim/kraft (FSK) faced, 1-1/2" thick, R5		2300	.007		.29	.31		.60	.79
7000	2" thick, R6		2300	.007		.37	.31		.68	.88
7050	3" thick, R10		2300	.007		.42	.31		.73	.93
7100	4" thick, R13		2300	.007		.43	.31		.74	.94
7110	6" thick, R19		2300	.007		.55	.31		.86	1.08
7120	10" thick, R30		2300	.007		.74	.31		1.05	1.28
7150	Met polyester/scrim/kraft (PSK) facing,1-1/2" thk, R5		2300	.007		.47	.31		.78	.99
7200	2" thick, R6		2300	.007		.55	.31		.86	1.08
7250	3" thick, R11		2300	.007		.63	.31		.94	1.16
7300	4" thick, R13		2300	.007		.72	.31		1.03	1.26
7310	6" thick, R19		2300	.007		.86	.31		1.17	1.42
7320	10" thick, R30		2300	.007		.95	.31		1.26	1.52
7350	Vinyl/scrim/foil (VSF), 1-1/2" thick, R5		2300	.007		.41	.31		.72	.92
7400	2" thick, R6		2300	.007		.52	.31		.83	1.04
7450	3" thick, R10		2300	.007		.56	.31		.87	1.09
7500	4" thick, R13		2300	.007		.69	.31		1	1.23
7510	Vinyl/scrim/vinyl (VSV) 4" thick, R13		2300	.007		.61	.31		.92	1.14
7520	6" thick, R19		2300	.007		.76	.31		1.07	1.31
7530	9 1/2" thick, R30		2300	.007		.77	.31		1.08	1.32
7540	Polyprop/scrim/polyester (PSP), 4", R13		2300	.007		.45	.31		.76	.97
7550	6" thick, R19		2300	.007		.61	.31		.92	1.14
7555	10" thick, R30		2300	.007		1.03	.31		1.34	1.60

13 34 Fabricated Engineered Structures

13 34 19 – Metal Building Systems

13 34 19.50 Pre-Engineered Steel Buildings

		Crew	Daily Output	Labor-Hours	Unit	Material	2012 Bare Costs Labor	2012 Bare Costs Equipment	Total	Total Incl O&P
7560	Polyprop/scrim/kraft/polyester (PSKP), 4" thick, R13	2 Carp	2300	.007	S.F.	.45	.31		.76	.97
7570	6" thick, R19		2300	.007		.62	.31		.93	1.15
7580	10" thick, R19		2300	.007		1.18	.31		1.49	1.77
7585	Vinyl/scrim/polyester (VSP), 4" thick, R13		2300	.007		.44	.31		.75	.95
7590	6" thick, R19		2300	.007		.59	.31		.90	1.12
7600	10" thick, R30		2300	.007		1.06	.31		1.37	1.64
7635	Insulation installation, over the purlin, second layer, up to 4" thick, add						90%			
7640	Insulation installation, between the purlins, up to 4" thick, add						100%			
7650	Sash, single slide, glazed, with screens, 2020 (2' x 2')	E-1	22	1.091	Opng.	121	53	5.55	179.55	229
7700	3030 (3' x 3')		14	1.714		272	83	8.70	363.70	450
7750	4030 (4' x 3')		13	1.846		365	89.50	9.35	463.85	560
7800	6040 (6' x 4')		12	2		725	97	10.15	832.15	975
7850	Double slide sash, 3030 (3' x 3')		14	1.714		211	83	8.70	302.70	385
7900	6040 (6' x 4')		12	2		560	97	10.15	667.15	795
7950	Fixed glass, no screens, 3030 (3' x 3')		14	1.714		210	83	8.70	301.70	380
8000	6040 (6' x 4')		12	2		560	97	10.15	667.15	790
8050	Prefinished storm sash, 3030 (3' x 3')		70	.343		72	16.60	1.74	90.34	109
8100	Siding and roofing, see Sections 07 41 13.00 & 07 42 13.00									
8200	Skylight, fiberglass panels, to 30 S.F.	E-1	10	2.400	Ea.	103	116	12.15	231.15	325
8250	Larger sizes, add for excess over 30 S.F.	"	300	.080	S.F.	3.44	3.88	.41	7.73	10.85
8300	Roof vents, circular with damper, birdscreen									
8350	and operator hardware, painted									
8400	26 ga., 12" diameter	1 Sswk	4	2	Ea.	69	98.50		167.50	253
8450	20" diameter		3	2.667		142	132		274	390
8500	24 ga., 24" diameter		2	4		282	197		479	665
8550	Galvanized		2	4		236	197		433	615
8600	Continuous, 26 ga., 10' long, 9" wide	2 Sswk	4	4		22	197		219	380
8650	12" wide	"	4	4		26.50	197		223.50	385

13 34 23 – Fabricated Structures

13 34 23.10 Comfort Stations

		Crew	Daily Output	Labor-Hours	Unit	Material	2012 Bare Costs Labor	2012 Bare Costs Equipment	Total	Total Incl O&P
0010	**COMFORT STATIONS** Prefab., stock, w/doors, windows & fixt.									
0100	Not incl. interior finish or electrical									
0300	Mobile, on steel frame, minimum				S.F.	48			48	52.50
0350	Maximum					81			81	89
0400	Permanent, including concrete slab, minimum	B-12J	50	.320		199	13.25	16.90	229.15	258
0500	Maximum	"	43	.372		288	15.40	19.65	323.05	360
0600	Alternate pricing method, mobile, minimum				Fixture	2,150			2,150	2,375
0650	Maximum					3,175			3,175	3,500
0700	Permanent, minimum	B-12J	.70	22.857		12,200	945	1,200	14,345	16,300
0750	Maximum	"	.50	32		20,400	1,325	1,700	23,425	26,300

13 34 23.15 Domes

		Crew	Daily Output	Labor-Hours	Unit	Material	2012 Bare Costs Labor	2012 Bare Costs Equipment	Total	Total Incl O&P
0010	**DOMES**									
0020	Domes, rev. alum., elec. drive, for astronomy obsv. shell only, stock units									
0600	10'-6" diameter	2 Carp	.25	64	Ea.	16,000	2,825		18,825	22,000
0900	18'-6" diameter		.17	94.118		43,000	4,150		47,150	53,500
1200	24'-6" diameter		.08	200		77,000	8,825		85,825	98,000
1500	Domes, bulk storage, shell only, dual radius hemisphere, arch, steel									
1600	framing, corrugated steel covering, 150' diameter	E-2	550	.102	SF Flr.	28	4.91	2.72	35.63	42
1700	400' diameter	"	720	.078		22.50	3.75	2.08	28.33	33.50
1800	Wood framing, wood decking, to 400' diameter	F-4	400	.120		20.50	5.30	2.75	28.55	33.50
1900	Radial framed wood (2" x 6"), 1/2" thick									
2000	plywood, asphalt shingles, 50' diameter	F-3	2000	.020	SF Flr.	40.50	.90	.31	41.71	46

13 34 Fabricated Engineered Structures

13 34 23 – Fabricated Structures

13 34 23.15 **Domes**		Crew	Daily Output	Labor-Hours	Unit	Material	2012 Bare Costs Labor	Equipment	Total	Total Incl O&P
2100	60' diameter	F-3	1900	.021	SF Flr.	31	.94	.32	32.26	36
2200	72' diameter		1800	.022		27.50	1	.34	28.84	32.50
2300	116' diameter		1730	.023		24.50	1.04	.35	25.89	29
2400	150' diameter	↓	1500	.027	↓	21	1.20	.41	22.61	25.50

13 34 23.16 Fabricated Control Booths

0010	**FABRICATED CONTROL BOOTHS**									
0100	Guard House, prefab conc w/bullet resistant doors & windows, roof & wiring									
0110	8' x 8', Level III	L-10	1	24	Ea.	24,200	1,200	610	26,010	29,300
0120	8' x 8', Level IV	"	1	24	"	30,300	1,200	610	32,110	36,000

13 34 23.25 Garage Costs

0010	**GARAGE COSTS**									
0020	Public parking, average				Car				17,300	19,000
0100	See also Square Foot Costs in Reference Section									
0300	Residential, wood, 12' x 20', one car prefab shell, stock, economy	2 Carp	1	16	Total	3,775	705		4,480	5,225
0350	Custom		.67	23.881		4,075	1,050		5,125	6,100
0400	Two car, 24' x 20', economy		.67	23.881		8,050	1,050		9,100	10,500
0450	Custom	↓	.50	32	↓	8,625	1,400		10,025	11,700

13 34 23.30 Garden House

0010	**GARDEN HOUSE** Prefab wood, no floors or foundations									
0100	32 to 200 S.F., minimum	2 Carp	200	.080	SF Flr.	32	3.53		35.53	41
0300	Maximum	"	48	.333	"	47	14.70		61.70	74

13 34 23.35 Geodesic Domes

0010	**GEODESIC DOMES** Shell only, interlocking plywood panels R133423-30									
0400	30' diameter	F-5	1.60	20	Ea.	22,100	890		22,990	25,700
0500	33' diameter		1.14	28.070		23,300	1,250		24,550	27,600
0600	40' diameter	↓	1	32		27,100	1,425		28,525	32,000
0700	45' diameter	F-3	1.13	35.556		28,400	1,600	545	30,545	34,300
0750	56' diameter		1	40		51,000	1,800	610	53,410	59,500
0800	60' diameter		1	40		57,500	1,800	610	59,910	67,000
0850	67' diameter	↓	.80	50	↓	81,000	2,250	765	84,015	93,500
1100	Aluminum panel, with 6" insulation									
1200	100' diameter				SF Flr.	41			41	45.50
1300	500' diameter				"	35.50			35.50	39
1600	Aluminum framed, plexiglass closure panels									
1700	40' diameter				SF Flr.	104			104	115
1800	200' diameter				"	83.50			83.50	91.50
2100	Aluminum framed, aluminum closure panels									
2200	40' diameter				SF Flr.	33			33	36
2300	100' diameter					24.50			24.50	27
2400	200' diameter					23.50			23.50	26
2500	For VRP faced bonded fiberglass insulation, add				↓					10
2700	Aluminum framed, fiberglass sandwich panel closure									
2800	6' diameter	2 Carp	150	.107	SF Flr.	36.50	4.70		41.20	48
2900	28' diameter	"	350	.046	"	33.50	2.02		35.52	39.50

13 34 23.45 Kiosks

0010	**KIOSKS**									
0020	Round, 5' diameter, 8' high, 1/4" fiberglass wall				Ea.	5,800			5,800	6,375
0100	1" insulated double wall, fiberglass					6,600			6,600	7,250
0500	Rectangular, 5' x 9', 7'-6" high, 1/4" fiberglass wall					8,450			8,450	9,300
0600	1" insulated double wall, fiberglass				↓	10,000			10,000	11,000

13 34 Fabricated Engineered Structures

13 34 23 – Fabricated Structures

13 34 23.60 Portable Booths

	13 34 23.60 Portable Booths	Crew	Daily Output	Labor-Hours	Unit	Material	2012 Bare Costs Labor	Equipment	Total	Total Incl O&P
0010	**PORTABLE BOOTHS** Prefab. aluminum with doors, windows, ext. roof									
0100	lights wiring & insulation, 15 S.F. building, O.D., painted, minimum				S.F.	296			296	325
0300	30 S.F. building					216			216	238
0400	50 S.F. building					161			161	177
0600	80 S.F. building					131			131	144
0700	100 S.F. building				↓	123			123	135
0900	Acoustical booth, 27 Db @ 1,000 Hz, 15 S.F. floor				Ea.	3,550			3,550	3,925
1000	7' x 7'-6", including light & ventilation					7,325			7,325	8,050
1200	Ticket booth, galv. steel, not incl. foundations., 4' x 4'					5,000			5,000	5,500
1300	4' x 6'				↓	6,500			6,500	7,150

13 34 23.70 Shelters

		Crew	Daily Output	Labor-Hours	Unit	Material	Labor	Equipment	Total	Total Incl O&P
0010	**SHELTERS**									
0020	Aluminum frame, acrylic glazing, 3' x 9' x 8' high	2 Sswk	1.14	14.035	Ea.	3,350	695		4,045	4,950
0100	9' x 12' x 8' high	"	.73	21.918	"	5,000	1,075		6,075	7,425

13 34 43 – Aircraft Hangars

13 34 43.50 Hangars

		Crew	Daily Output	Labor-Hours	Unit	Material	Labor	Equipment	Total	Total Incl O&P
0010	**HANGARS** Prefabricated steel T hangars, Galv. steel roof &									
0100	walls, incl. electric bi-folding doors, 4 or more units,									
0110	not including floors or foundations, minimum	E-2	1275	.044	SF Flr.	16.25	2.12	1.17	19.54	23
0130	Maximum		1063	.053		18.85	2.54	1.41	22.80	26.50
0900	With bottom rolling doors, minimum		1386	.040		15.80	1.95	1.08	18.83	22
1000	Maximum	↓	966	.058	↓	18.35	2.80	1.55	22.70	26.50
1200	Alternate pricing method:									
1300	Galv. roof and walls, electric bi-folding doors, minimum	E-2	1.06	52.830	Plane	20,600	2,550	1,400	24,550	28,600
1500	Maximum		.91	61.538		24,900	2,975	1,650	29,525	34,200
1600	With bottom rolling doors, minimum		1.25	44.800		15,400	2,150	1,200	18,750	21,900
1800	Maximum		.97	57.732	↓	17,900	2,775	1,550	22,225	26,200
2000	Circular type, prefab., steel frame, plastic skin, electric									
2010	door, including foundations, 80' diameter,									
2020	for up to 5 light planes, minimum	E-2	.50	112	Total	88,500	5,400	3,000	96,900	110,000
2200	Maximum	"	.25	224	"	100,500	10,800	6,000	117,300	135,500

13 34 53 – Agricultural Structures

13 34 53.50 Silos

		Crew	Daily Output	Labor-Hours	Unit	Material	Labor	Equipment	Total	Total Incl O&P
0010	**SILOS**									
0500	Steel, factory fab., 30,000 gallon cap., painted, minimum	L-5	1	56	Ea.	21,000	2,775	695	24,470	28,700
0700	Maximum		.50	112		33,400	5,525	1,400	40,325	48,000
0800	Epoxy lined, minimum		1	56		34,300	2,775	695	37,770	43,400
1000	Maximum	↓	.50	112	↓	43,600	5,525	1,400	50,525	59,000

13 34 63 – Natural Fiber Construction

13 34 63.50 Straw Bale Construction

			Crew	Daily Output	Labor-Hours	Unit	Material	Labor	Equipment	Total	Total Incl O&P
0010	**STRAW BALE CONSTRUCTION**										
2000	Straw bales wall, incl. labor & material complete, minimum	G				SF Flr.					130
2010	Maximum	G				"					160
2020	Straw bales in walls w/modified post and beam frame	G	2 Carp	320	.050	S.F.	6.60	2.21		8.81	10.65

13 36 Towers

13 36 13 – Metal Towers

13 36 13.50 Control Towers

		Crew	Daily Output	Labor-Hours	Unit	Material	2012 Bare Costs Labor	Equipment	Total	Total Incl O&P
0010	**CONTROL TOWERS**									
0020	Modular 12' x 10', incl. instruments, min.				Ea.	670,000			670,000	737,000
0100	Maximum								927,500	1,020,000
0500	With standard 40' tower, average					935,000			935,000	1,028,500
1000	Temporary portable control towers, 8' x 12',									
1010	complete with one position communications, minimum				Ea.				266,000	293,000
2000	For fixed facilities, depending on height, minimum								57,000	63,000
2010	Maximum								115,000	127,000

13 42 Building Modules

13 42 63 – Detention Cell Modules

13 42 63.16 Steel Detention Cell Modules

		Crew	Daily Output	Labor-Hours	Unit	Material	2012 Bare Costs Labor	Equipment	Total	Total Incl O&P
0010	**STEEL DETENTION CELL MODULES**									
2000	Cells, prefab., 5' to 6' wide, 7' to 8' high, 7' to 8' deep,									
2010	bar front, cot, not incl. plumbing	E-4	1.50	21.333	Ea.	9,650	1,075	81	10,806	12,600

13 48 Sound, Vibration, and Seismic Control

13 48 13 – Manufactured Sound and Vibration Control Components

13 48 13.50 Audio Masking

		Crew	Daily Output	Labor-Hours	Unit	Material	2012 Bare Costs Labor	Equipment	Total	Total Incl O&P
0010	**AUDIO MASKING**, acoustical enclosure, 4" thick wall and ceiling									
0020	8# per S.F., up to 12' span	3 Carp	72	.333	SF Surf	32	14.70		46.70	58
0300	Better quality panels, 10.5# per S.F.		64	.375		36.50	16.55		53.05	65.50
0400	Reverb-chamber, 4" thick, parallel walls		60	.400		45.50	17.65		63.15	77
0600	Skewed wall, parallel roof, 4" thick panels		55	.436		52	19.25		71.25	86.50
0700	Skewed walls, skewed roof, 4" layers, 4" air space		48	.500		58.50	22		80.50	98.50
0900	Sound-absorbing panels, pntd. mtl., 2'-6" x 8', under 1,000 S.F.		215	.112		12.10	4.92		17.02	21
1100	Over 1000 S.F.		240	.100		11.65	4.41		16.06	19.60
1200	Fabric faced		240	.100		9.45	4.41		13.86	17.15
1500	Flexible transparent curtain, clear	3 Shee	215	.112		7.30	5.80		13.10	16.85
1600	50% foam		215	.112		10.20	5.80		16	20
1700	75% foam		215	.112		10.20	5.80		16	20
1800	100% foam		215	.112		10.20	5.80		16	20
3100	Audio masking system, including speakers, amplification									
3110	and signal generator									
3200	Ceiling mounted, 5,000 S.F.	2 Elec	2400	.007	S.F.	1.25	.34		1.59	1.89
3300	10,000 S.F.		2800	.006		1.01	.29		1.30	1.55
3400	Plenum mounted, 5,000 S.F.		3800	.004		1.06	.22		1.28	1.49
3500	10,000 S.F.		4400	.004		.72	.19		.91	1.07

13 49 Radiation Protection

13 49 13 – Integrated X-Ray Shielding Assemblies

13 49 13.50 Lead Sheets

		Crew	Daily Output	Labor-Hours	Unit	Material	2012 Bare Costs Labor	2012 Bare Costs Equipment	Total	Total Incl O&P
0010	**LEAD SHEETS**									
0300	Lead sheets, 1/16" thick	2 Lath	135	.119	S.F.	10.05	4.65		14.70	17.90
0400	1/8" thick		120	.133		20	5.25		25.25	29.50
0500	Lead shielding, 1/4" thick		135	.119		49	4.65		53.65	61
0550	1/2" thick	↓	120	.133	↓	83.50	5.25		88.75	99
0950	Lead headed nails (average 1 lb. per sheet)				Lb.	7.60			7.60	8.35
1000	Butt joints in 1/8" lead or thicker, 2" batten strip x 7' long	2 Lath	240	.067	Ea.	23.50	2.62		26.12	30
1200	X-ray protection, average radiography or fluoroscopy									
1210	room, up to 300 S.F. floor, 1/16" lead, minimum	2 Lath	.25	64	Total	9,200	2,500		11,700	13,800
1500	Maximum, 7'-0" walls	"	.15	106	"	11,100	4,175		15,275	18,400
1600	Deep therapy X-ray room, 250 kV capacity,									
1800	up to 300 S.F. floor, 1/4" lead, minimum	2 Lath	.08	200	Total	25,700	7,850		33,550	39,800
1900	Maximum, 7'-0" walls	"	.06	266	"	31,700	10,500		42,200	50,500

13 49 19 – Lead-Lined Materials

13 49 19.50 Shielding Lead

		Crew	Daily Output	Labor-Hours	Unit	Material	2012 Bare Costs Labor	2012 Bare Costs Equipment	Total	Total Incl O&P
0010	**SHIELDING LEAD**									
0100	Laminated lead in wood doors, 1/16" thick, no hardware				S.F.	65.50			65.50	72
0200	Lead lined door frame, not incl. hardware,									
0210	1/16" thick lead, butt prepared for hardware	1 Lath	2.40	3.333	Ea.	770	131		901	1,025
0850	Window frame with 1/16" lead and voice passage, 36" x 60"	2 Glaz	2	8		1,550	340		1,890	2,225
0870	24" x 36" frame		4	4	↓	1,275	171		1,446	1,650
0900	Lead gypsum board, 5/8" thick with 1/16" lead		160	.100	S.F.	9.15	4.28		13.43	16.55
0910	1/8" lead	↓	140	.114		22	4.89		26.89	31.50
0930	1/32" lead	2 Lath	200	.080	↓	6.20	3.14		9.34	11.45

13 49 21 – Lead Glazing

13 49 21.50 Lead Glazing

		Crew	Daily Output	Labor-Hours	Unit	Material	2012 Bare Costs Labor	2012 Bare Costs Equipment	Total	Total Incl O&P
0010	**LEAD GLAZING**									
0600	Lead glass, 1/4" thick, 2.0 mm LE, 12" x 16"	2 Glaz	13	1.231	Ea.	305	52.50		357.50	415
0700	24" x 36"		8	2		1,100	85.50		1,185.50	1,350
0800	36" x 60"	↓	2	8	↓	3,375	340		3,715	4,225
2000	X-ray viewing panels, clear lead plastic									
2010	7 mm thick, 0.3 mm LE, 2.3 lb./S.F.	H-3	139	.115	S.F.	224	4.38		228.38	253
2020	12 mm thick, 0.5 mm LE, 3.9 lb./S.F.		82	.195		305	7.40		312.40	345
2030	18 mm thick, 0.8 mm LE, 5.9 lb./S.F.		54	.296		330	11.25		341.25	375
2040	22 mm thick, 1.0 mm LE, 7.2 lb./S.F.		44	.364		335	13.85		348.85	390
2050	35 mm thick, 1.5 mm LE, 11.5 lb./S.F.		28	.571		375	21.50		396.50	450
2060	46 mm thick, 2.0 mm LE, 15.0 lb./S.F.	↓	21	.762	↓	495	29		524	590
2090	For panels 12 S.F. to 48 S.F., add crating charge				Ea.					50

13 49 23 – Integrated RFI/EMI Shielding Assemblies

13 49 23.50 Modular Shielding Partitions

		Crew	Daily Output	Labor-Hours	Unit	Material	2012 Bare Costs Labor	2012 Bare Costs Equipment	Total	Total Incl O&P
0010	**MODULAR SHIELDING PARTITIONS**									
4000	X-ray barriers, modular, panels mounted within framework for									
4002	attaching to floor, wall or ceiling, upper portion is clear lead									
4005	plastic window panels 48"H, lower portion is opaque leaded									
4008	steel panels 36"H, structural supports not incl.									
4010	1-section barrier, 36"W x 84"H overall									
4020	0.5 mm LE panels	H-3	6.40	2.500	Ea.	7,075	95		7,170	7,925
4030	0.8 mm LE panels		6.40	2.500		7,525	95		7,620	8,450
4040	1.0 mm LE panels		5.33	3.002		7,675	114		7,789	8,625
4050	1.5 mm LE panels	↓	5.33	3.002	↓	8,175	114		8,289	9,175
4060	2-section barrier, 72"W x 84"H overall									

13 49 Radiation Protection

13 49 23 – Integrated RFI/EMI Shielding Assemblies

13 49 23.50 Modular Shielding Partitions	Crew	Daily Output	Labor-Hours	Unit	Material	2012 Bare Costs Labor	Equipment	Total	Total Incl O&P
4070	0.5 mm LE panels	H-3	4	4	Ea.	14,400	152		14,552	16,100
4080	0.8 mm LE panels		4	4		15,300	152		15,452	17,000
4090	1.0 mm LE panels		3.56	4.494		15,600	171		15,771	17,500
5000	1.5 mm LE panels		3.20	5		17,700	190		17,890	19,800
5010	3-section barrier, 108"W x 84"H overall									
5020	0.5 mm LE panels	H-3	3.20	5	Ea.	21,700	190		21,890	24,200
5030	0.8 mm LE panels		3.20	5		22,900	190		23,090	25,500
5040	1.0 mm LE panels		2.67	5.993		23,300	228		23,528	25,900
5050	1.5 mm LE panels		2.46	6.504		25,300	247		25,547	28,200
7000	X-ray barriers, mobile, mounted within framework w/casters on									
7005	bottom, clear lead plastic window panels on upper portion,									
7010	opaque on lower, 30"W x 75"H overall, incl. framework									
7020	24"H upper w/0.5 mm LE, 48"H lower w/0.8 mm LE	1 Carp	16	.500	Ea.	3,775	22		3,797	4,175
7030	48"W x 75"H overall, incl. framework									
7040	36"H upper w/0.5 mm LE, 36"H lower w/0.8 mm LE	1 Carp	16	.500	Ea.	6,900	22		6,922	7,625
7050	36"H upper w/1.0 mm LE, 36"H lower w/1.5 mm LE	"	16	.500	"	8,200	22		8,222	9,050
7060	72"W x 75"H overall, incl. framework									
7070	36"H upper w/0.5 mm LE, 36"H lower w/0.8 mm LE	1 Carp	16	.500	Ea.	8,175	22		8,197	9,025
7080	36"H upper w/1.0 mm LE, 36"H lower w/1.5 mm LE	"	16	.500	"	10,200	22		10,222	11,300

13 49 33 – Radio Frequency Shielding

13 49 33.50 Shielding, Radio Frequency

		Crew	Daily Output	Labor-Hours	Unit	Material	2012 Bare Costs Labor	Equipment	Total	Total Incl O&P
0010	**SHIELDING, RADIO FREQUENCY**									
0020	Prefabricated or screen-type copper or steel, minimum	2 Carp	180	.089	SF Surf	30	3.92		33.92	39.50
0100	Average		155	.103		32	4.55		36.55	42.50
0150	Maximum		145	.110		34	4.87		38.87	45

13 53 Meteorological Instrumentation

13 53 09 – Weather Instrumentation

13 53 09.50 Weather Station

		Crew	Daily Output	Labor-Hours	Unit	Material	2012 Bare Costs Labor	Equipment	Total	Total Incl O&P
0010	**WEATHER STATION**									
0020	Remote recording, solar powered, with rain gauge & display, 400 ft range				Ea.	1,675			1,675	1,825
0100	1 mile range				"	1,775			1,775	1,950

Division Notes

	CREW	DAILY OUTPUT	LABOR-HOURS	UNIT	2012 BARE COSTS				TOTAL INCL O&P
					MAT.	LABOR	EQUIP.	TOTAL	

Estimating Tips

General

• Many products in Division 14 will require some type of support or blocking for installation not included with the item itself. Examples are supports for conveyors or tube systems, attachment points for lifts, and footings for hoists or cranes. Add these supports in the appropriate division.

14 10 00 Dumbwaiters
14 20 00 Elevators

• Dumbwaiters and elevators are estimated and purchased in a method similar to buying a car. The manufacturer has a base unit with standard features. Added to this base unit price will be whatever options the owner or specifications require. Increased load capacity, additional vertical travel, additional stops, higher speed, and cab finish options are items to be considered. When developing an estimate for dumbwaiters and elevators, remember that some items needed by the installers may have to be included as part of the general contract.

Examples are:
— shaftway
— rail support brackets
— machine room
— electrical supply
— sill angles
— electrical connections
— pits
— roof penthouses
— pit ladders

Check the job specifications and drawings before pricing.

• Installation of elevators and handicapped lifts in historic structures can require significant additional costs. The associated structural requirements may involve cutting into and repairing finishes, mouldings, flooring, etc. The estimator must account for these special conditions.

14 30 00 Escalators and Moving Walks

• Escalators and moving walks are specialty items installed by specialty contractors. There are numerous options associated with these items. For specific options, contact a manufacturer or contractor. In a method similar to estimating dumbwaiters and elevators, you should verify the extent of general contract work and add items as necessary.

14 40 00 Lifts
14 90 00 Other Conveying Equipment

• Products such as correspondence lifts, chutes, and pneumatic tube systems, as well as other items specified in this subdivision, may require trained installers. The general contractor might not have any choice as to who will perform the installation or when it will be performed. Long lead times are often required for these products, making early decisions in scheduling necessary.

Reference Numbers

Reference numbers are shown in shaded boxes at the beginning of some major classifications. These numbers refer to related items in the Reference Section. The reference information may be an estimating procedure, an alternate pricing method, or technical information.

Note: Not all subdivisions listed here necessarily appear in this publication.

14 11 Manual Dumbwaiters

14 11 10 – Hand Operated Dumbwaiters

14 11 10.20 Manual Dumbwaiters	Crew	Daily Output	Labor-Hours	Unit	Material	2012 Bare Costs Labor	Equipment	Total	Total Incl O&P
0010 **MANUAL DUMBWAITERS**									
0020 2 stop, minimum	2 Elev	.75	21.333	Ea.	2,925	1,500		4,425	5,500
0100 Maximum	↓	.50	32	"	6,650	2,275		8,925	10,700
0300 For each additional stop, add		.75	21.333	Stop	1,050	1,500		2,550	3,450

14 12 Electric Dumbwaiters

14 12 10 – Dumbwaiters

14 12 10.10 Electric Dumbwaiters

	Crew	Daily Output	Labor-Hours	Unit	Material	2012 Bare Costs Labor	Equipment	Total	Total Incl O&P
0010 **ELECTRIC DUMBWAITERS**									
0020 2 stop, minimum	2 Elev	.13	123	Ea.	7,375	8,725		16,100	21,200
0100 Maximum	↓	.11	145	"	22,200	10,300		32,500	39,800
0600 For each additional stop, add		.54	29.630	Stop	3,250	2,100		5,350	6,725
0750 Correspondence lift, 1 floor, 2 stop, 45 lb. capacity	2 Elec	.20	80	Ea.	8,350	4,125		12,475	15,400

14 21 Electric Traction Elevators

14 21 13 – Electric Traction Freight Elevators

14 21 13.10 Electric Traction Freight Elevators and Options

	Crew	Daily Output	Labor-Hours	Unit	Material	2012 Bare Costs Labor	Equipment	Total	Total Incl O&P
0010 **ELECTRIC TRACTION FREIGHT ELEVATORS AND OPTIONS** R142000-10									
0425 Electric freight, base unit, 4000 lb., 200 fpm, 4 stop, std. fin.	2 Elev	.05	320	Ea.	99,000	22,700		121,700	143,000
0450 For 5000 lb. capacity, add					5,500			5,500	6,050
0500 For 6000 lb. capacity, add					16,000			16,000	17,600
0525 For 7000 lb. capacity, add					19,800			19,800	21,700
0550 For 8000 lb. capacity, add					43,400			43,400	47,700
0575 For 10000 lb. capacity, add					30,000			30,000	33,000
0600 For 12000 lb. capacity, add					36,500			36,500	40,200
0625 For 16000 lb. capacity, add					46,500			46,500	51,000
0650 For 20000 lb. capacity, add					55,000			55,000	60,500
0675 For increased speed, 250 fpm, add					20,400			20,400	22,400
0700 300 fpm, geared electric, add					25,800			25,800	28,300
0725 350 fpm, geared electric, add					31,200			31,200	34,300
0750 400 fpm, geared electric, add					36,300			36,300	39,900
0775 500 fpm, gearless electric, add					43,800			43,800	48,100
0800 600 fpm, gearless electric, add					53,000			53,000	58,000
0825 700 fpm, gearless electric, add					63,000			63,000	69,500
0850 800 fpm, gearless electric, add					72,500			72,500	80,000
0875 For class "B" loading, add					5,875			5,875	6,450
0900 For class "C-1" loading, add					8,300			8,300	9,125
0925 For class "C-2" loading, add					9,475			9,475	10,400
0950 For class "C-3" loading, add				↓	12,000			12,000	13,200
0975 For travel over 40 V.L.F., add	2 Elev	7.25	2.207	V.L.F.	775	156		931	1,100
1000 For number of stops over 4, add	"	.27	59.259	Stop	3,925	4,200		8,125	10,600

14 21 23 – Electric Traction Passenger Elevators

14 21 23.10 Electric Traction Passenger Elevators and Options

	Crew	Daily Output	Labor-Hours	Unit	Material	2012 Bare Costs Labor	Equipment	Total	Total Incl O&P
0010 **ELECTRIC TRACTION PASSENGER ELEVATORS AND OPTIONS**									
1625 Electric pass., base unit, 2000 lb., 200 fpm, 4 stop, std. fin.	2 Elev	.05	320	Ea.	84,000	22,700		106,700	126,500
1650 For 2500 lb. capacity, add					3,525			3,525	3,875
1675 For 3000 lb. capacity, add					4,200			4,200	4,600
1700 For 3500 lb. capacity, add					7,025			7,025	7,725
1725 For 4000 lb. capacity, add				↓	6,600			6,600	7,250

14 21 Electric Traction Elevators

14 21 23 – Electric Traction Passenger Elevators

14 21 23.10 Electric Traction Passenger Elevators and Options	Crew	Daily Output	Labor-Hours	Unit	Material	2012 Bare Costs Labor	Equipment	Total	Total Incl O&P
1750	For 4500 lb. capacity, add				Ea.	8,325			8,325	9,175
1775	For 5000 lb. capacity, add					10,100			10,100	11,100
1800	For increased speed, 250 fpm, geared electric, add					6,100			6,100	6,700
1825	300 fpm, geared electric, add					9,325			9,325	10,200
1850	350 fpm, geared electric, add					10,600			10,600	11,700
1875	400 fpm, geared electric, add					13,600			13,600	14,900
1900	500 fpm, gearless electric, add					32,000			32,000	35,200
1925	600 fpm, gearless electric, add					39,600			39,600	43,600
1950	700 fpm, gearless electric, add					41,800			41,800	46,000
1975	800 fpm, gearless electric, add					48,000			48,000	53,000
2000	For travel over 40 V.L.F., add	2 Elev	7.25	2.207	V.L.F.	910	156		1,066	1,225
2025	For number of stops over 4, add		.27	59.259	Stop	6,825	4,200		11,025	13,800
2400	Electric hospital, base unit, 4000 lb., 200 fpm, 4 stop, std fin.		.05	320	Ea.	72,500	22,700		95,200	113,500
2425	For 4500 lb. capacity, add					5,175			5,175	5,675
2450	For 5000 lb. capacity, add					6,225			6,225	6,825
2475	For increased speed, 250 fpm, geared electric, add					6,150			6,150	6,775
2500	300 fpm, geared electric, add					9,200			9,200	10,100
2525	350 fpm, geared electric, add					10,500			10,500	11,600
2550	400 fpm, geared electric, add					13,400			13,400	14,700
2575	500 fpm, gearless electric, add					33,200			33,200	36,500
2600	600 fpm, gearless electric, add					38,300			38,300	42,100
2625	700 fpm, gearless electric, add					43,700			43,700	48,100
2650	800 fpm, gearless electric, add					50,500			50,500	55,500
2675	For travel over 40 V.L.F., add	2 Elev	7.25	2.207	V.L.F.	580	156		736	875
2700	For number of stops over 4, add	"	.27	59.259	Stop	3,875	4,200		8,075	10,600

14 21 33 – Electric Traction Residential Elevators

14 21 33.20 Residential Elevators

		Crew	Daily Output	Labor-Hours	Unit	Material	2012 Bare Costs Labor	Equipment	Total	Total Incl O&P
0010	**RESIDENTIAL ELEVATORS**									
7000	Residential, cab type, 1 floor, 2 stop, minimum	2 Elev	.20	80	Ea.	11,000	5,675		16,675	20,600
7100	Maximum		.10	160		18,600	11,300		29,900	37,500
7200	2 floor, 3 stop, minimum		.12	133		16,400	9,450		25,850	32,200
7300	Maximum		.06	266		26,700	18,900		45,600	57,500

14 24 Hydraulic Elevators

14 24 13 – Hydraulic Freight Elevators

14 24 13.10 Hydraulic Freight Elevators and Options

		Crew	Daily Output	Labor-Hours	Unit	Material	2012 Bare Costs Labor	Equipment	Total	Total Incl O&P
0010	**HYDRAULIC FREIGHT ELEVATORS AND OPTIONS**									
1025	Hydraulic freight, base unit, 2000 lb., 50 fpm, 2 stop, std. fin.	2 Elev	.10	160	Ea.	76,000	11,300		87,300	101,000
1050	For 2500 lb. capacity, add					4,450			4,450	4,900
1075	For 3000 lb. capacity, add					7,825			7,825	8,600
1100	For 3500 lb. capacity, add					14,100			14,100	15,500
1125	For 4000 lb. capacity, add					15,100			15,100	16,600
1150	For 4500 lb. capacity, add					17,800			17,800	19,600
1175	For 5000 lb. capacity, add					22,400			22,400	24,700
1200	For 6000 lb. capacity, add					27,800			27,800	30,600
1225	For 7000 lb. capacity, add					33,800			33,800	37,100
1250	For 8000 lb. capacity, add					37,200			37,200	40,900
1275	For 10000 lb. capacity, add					46,600			46,600	51,500
1300	For 12000 lb. capacity, add					61,000			61,000	67,000
1325	For 16000 lb. capacity, add					83,500			83,500	92,000

14 24 Hydraulic Elevators

14 24 13 – Hydraulic Freight Elevators

14 24 13.10 Hydraulic Freight Elevators and Options	Crew	Daily Output	Labor-Hours	Unit	Material	2012 Bare Costs Labor	Equipment	Total	Total Incl O&P
1350	For 20000 lb. capacity, add				Ea.	101,500			101,500	112,000
1375	For increased speed, 100 fpm, add					1,950			1,950	2,150
1400	125 fpm, add					4,250			4,250	4,675
1425	150 fpm, add					5,100			5,100	5,625
1450	175 fpm, add					8,650			8,650	9,525
1475	For class "B" loading, add					5,850			5,850	6,425
1500	For class "C-1" loading, add					8,275			8,275	9,100
1525	For class "C-2" loading, add					9,450			9,450	10,400
1550	For class "C-3" loading, add					12,000			12,000	13,200
1575	For travel over 20 V.L.F., add	2 Elev	7.25	2.207	V.L.F.	1,525	156		1,681	1,900
1600	For number of stops over 2, add	"	.27	59.259	Stop	4,125	4,200		8,325	10,900

14 24 23 – Hydraulic Passenger Elevators

14 24 23.10 Hydraulic Passenger Elevators and Options

		Crew	Daily Output	Labor-Hours	Unit	Material	2012 Bare Costs Labor	Equipment	Total	Total Incl O&P
0010	**HYDRAULIC PASSENGER ELEVATORS AND OPTIONS**									
2050	Hyd. pass., base unit, 1500 lb., 100 fpm, 2 stop, std. fin.	2 Elev	.10	160	Ea.	44,400	11,300		55,700	66,000
2075	For 2000 lb. capacity, add					1,750			1,750	1,925
2100	For 2500 lb. capacity, add					2,800			2,800	3,075
2125	For 3000 lb. capacity, add					3,725			3,725	4,075
2150	For 3500 lb. capacity, add					5,675			5,675	6,250
2175	For 4000 lb. capacity, add					7,000			7,000	7,700
2200	For 4500 lb. capacity, add					7,900			7,900	8,700
2225	For 5000 lb. capacity, add					10,500			10,500	11,600
2250	For increased speed, 125 fpm, add					2,575			2,575	2,825
2275	150 fpm, add					3,300			3,300	3,625
2300	175 fpm, add					5,000			5,000	5,500
2325	200 fpm, add					7,025			7,025	7,725
2350	For travel over 12 V.L.F., add	2 Elev	7.25	2.207	V.L.F.	835	156		991	1,150
2375	For number of stops over 2, add		.27	59.259	Stop	2,250	4,200		6,450	8,775
2725	Hydraulic hospital, base unit, 4000 lb., 100 fpm, 2 stop, std. fin.		.10	160	Ea.	56,500	11,300		67,800	79,500
2775	For 4500 lb. capacity, add					6,425			6,425	7,050
2800	For 5000 lb. capacity, add					8,500			8,500	9,350
2825	For increased speed, 125 fpm, add					3,000			3,000	3,325
2850	150 fpm, add					3,900			3,900	4,275
2875	175 fpm, add					5,675			5,675	6,250
2900	200 fpm, add					6,850			6,850	7,550
2925	For travel over 12 V.L.F., add	2 Elev	7.25	2.207	V.L.F.	665	156		821	965
2950	For number of stops over 2, add	"	.27	59.259	Stop	3,875	4,200		8,075	10,600

14 27 Custom Elevator Cabs and Doors

14 27 13 – Custom Elevator Cab Finishes

14 27 13.10 Cab Finishes

		Crew	Daily Output	Labor-Hours	Unit	Material	2012 Bare Costs Labor	Equipment	Total	Total Incl O&P
0010	**CAB FINISHES**									
3325	Passenger elevator cab finishes (based on 3500 lb. cab size)									
3350	Acrylic panel ceiling				Ea.	715			715	785
3375	Aluminum eggcrate ceiling					1,075			1,075	1,175
3400	Stainless steel doors					3,075			3,075	3,375
3425	Carpet flooring					580			580	640
3450	Epoxy flooring					440			440	485
3475	Quarry tile flooring					825			825	905
3500	Slate flooring					1,450			1,450	1,600

14 27 Custom Elevator Cabs and Doors

14 27 13 – Custom Elevator Cab Finishes

14 27 13.10 Cab Finishes

		Crew	Daily Output	Labor-Hours	Unit	Material	2012 Bare Costs Labor	2012 Bare Costs Equipment	Total	Total Incl O&P
3525	Textured rubber flooring				Ea.	190			190	208
3550	Stainless steel walls					4,725			4,725	5,200
3575	Stainless steel returns at door					1,075			1,075	1,200
4450	Hospital elevator cab finishes (based on 3500 lb. cab size)									
4475	Aluminum eggcrate ceiling				Ea.	1,050			1,050	1,175
4500	Stainless steel doors					2,575			2,575	2,825
4525	Epoxy flooring					355			355	390
4550	Quarry tile flooring					830			830	910
4575	Textured rubber flooring					630			630	690
4600	Stainless steel walls					4,700			4,700	5,175
4625	Stainless steel returns at door					1,075			1,075	1,200

14 28 Elevator Equipment and Controls

14 28 10 – Elevator Equipment and Control Options

14 28 10.10 Elevator Controls and Doors

		Crew	Daily Output	Labor-Hours	Unit	Material	2012 Bare Costs Labor	2012 Bare Costs Equipment	Total	Total Incl O&P
0010	**ELEVATOR CONTROLS AND DOORS**									
2975	Passenger elevator options									
3000	2 car group automatic controls	2 Elev	.66	24.242	Ea.	5,825	1,725		7,550	9,000
3025	3 car group automatic controls		.44	36.364		11,400	2,575		13,975	16,500
3050	4 car group automatic controls		.33	48.485		23,200	3,425		26,625	30,700
3075	5 car group automatic controls		.26	61.538		42,300	4,350		46,650	53,000
3100	6 car group automatic controls		.22	72.727		86,500	5,150		91,650	102,500
3125	Intercom service		3	5.333		1,225	380		1,605	1,925
3150	Duplex car selective collective		.66	24.242		10,700	1,725		12,425	14,400
3175	Center opening 1 speed doors		2	8		1,750	565		2,315	2,775
3200	Center opening 2 speed doors		2	8		2,450	565		3,015	3,550
3225	Rear opening doors (opposite front)		2	8		2,825	565		3,390	3,950
3250	Side opening 2 speed doors		2	8		4,475	565		5,040	5,775
3275	Automatic emergency power switching		.66	24.242		1,100	1,725		2,825	3,800
3300	Manual emergency power switching		8	2		485	142		627	745
3625	Hall finishes, stainless steel doors					1,825			1,825	2,000
3650	Stainless steel frames					1,425			1,425	1,575
3675	12 month maintenance contract								3,600	3,950
3700	Signal devices, hall lanterns	2 Elev	8	2		530	142		672	795
3725	Position indicators, up to 3		9.40	1.702		570	121		691	810
3750	Position indicators, per each over 3		32	.500		540	35.50		575.50	650
3775	High speed heavy duty door opener					3,775			3,775	4,150
3800	Variable voltage, O.H. gearless machine, min.	2 Elev	.16	100		61,500	7,075		68,575	78,000
3815	Maximum		.07	228		92,000	16,200		108,200	126,000
3825	Basement installed geared machine		.33	48.485		63,000	3,425		66,425	74,000
3850	Freight elevator options									
3875	Doors, bi-parting	2 Elev	.66	24.242	Ea.	8,900	1,725		10,625	12,400
3900	Power operated door and gate	"	.66	24.242		22,300	1,725		24,025	27,200
3925	Finishes, steel plate floor					2,200			2,200	2,425
3950	14 ga. 1/4" x 4' steel plate walls					1,825			1,825	2,000
3975	12 month maintenance contract								3,600	3,950
4000	Signal devices, hall lanterns	2 Elev	8	2		525	142		667	785
4025	Position indicators, up to 3		9.40	1.702		580	121		701	820
4050	Position indicators, per each over 3		32	.500		545	35.50		580.50	655
4075	Variable voltage basement installed geared machine		.66	24.242		18,600	1,725		20,325	23,100
4100	Hospital elevator options									

14 28 10 – Elevator Equipment and Control Options

14 28 10.10 Elevator Controls and Doors	Crew	Daily Output	Labor-Hours	Unit	Material	2012 Bare Costs Labor	Equipment	Total	Total Incl O&P
4125	2 car group automatic controls	2 Elev	.66	24.242	Ea.	5,825	1,725		7,550	9,000
4150	3 car group automatic controls		.44	36.364		11,400	2,575		13,975	16,400
4175	4 car group automatic controls		.33	48.485		23,100	3,425		26,525	30,600
4200	5 car group automatic controls		.26	61.538		42,100	4,350		46,450	53,000
4225	6 car group automatic controls		.22	72.727		86,000	5,150		91,150	102,000
4250	Intercom service		3	5.333		1,225	380		1,605	1,925
4275	Duplex car selective collective		.66	24.242		10,700	1,725		12,425	14,300
4300	Center opening 1 speed doors		2	8		1,750	565		2,315	2,775
4325	Center opening 2 speed doors		2	8		2,375	565		2,940	3,475
4350	Rear opening doors (opposite front)		2	8		3,400	565		3,965	4,575
4375	Side opening 2 speed doors		2	8		4,950	565		5,515	6,300
4400	Automatic emergency power switching		.66	24.242		1,075	1,725		2,800	3,750
4425	Manual emergency power switching	↓	8	2		475	142		617	730
4675	Hall finishes, stainless steel doors					1,825			1,825	2,000
4700	Stainless steel frames					1,425			1,425	1,575
4725	12 month maintenance contract								3,600	3,950
4750	Signal devices, hall lanterns	2 Elev	8	2		520	142		662	780
4775	Position indicators, up to 3		9.40	1.702		570	121		691	805
4800	Position indicators, per each over 3	↓	32	.500		540	35.50		575.50	650
4825	High speed heavy duty door opener					3,775			3,775	4,150
4850	Variable voltage, O.H. gearless machine, min.	2 Elev	.16	100		61,000	7,075		68,075	78,000
4865	Maximum		.07	228		90,500	16,200		106,700	124,000
4875	Basement installed geared machine	↓	.33	48.485	↓	23,400	3,425		26,825	31,000
5000	Drilling for piston, casing included, 18" diameter	B-48	80	.700	V.L.F.	49.50	27.50	34.50	111.50	134

14 31 10 – Glass and Steel Escalators

14 31 10.10 Escalators

		Crew	Daily Output	Labor-Hours	Unit	Material	2012 Bare Costs Labor	Equipment	Total	Total Incl O&P
0010	**ESCALATORS** R143110-10									
1000	Glass, 32" wide x 10' floor to floor height	M-1	.07	457	Ea.	82,000	30,800	660	113,460	137,500
1010	48" wide x 10' floor to floor height		.07	457		88,000	30,800	660	119,460	143,500
1020	32" wide x 15' floor to floor height		.06	533		86,500	35,900	770	123,170	150,000
1030	48" wide x 15' floor to floor height		.06	533		91,000	35,900	770	127,670	155,000
1040	32" wide x 20' floor to floor height		.05	653		91,000	44,000	940	135,940	167,000
1050	48" wide x 20' floor to floor height		.05	653		99,000	44,000	940	143,940	175,500
1060	32" wide x 25' floor to floor height		.04	800		98,500	54,000	1,150	153,650	190,500
1070	48" wide x 25' floor to floor height		.04	800		113,500	54,000	1,150	168,650	207,000
1080	Enameled steel, 32" wide x 10' floor to floor height		.07	457		88,000	30,800	660	119,460	143,500
1090	48" wide x 10' floor to floor height		.07	457		94,500	30,800	660	125,960	151,000
1110	32" wide x 15' floor to floor height		.06	533		94,500	35,900	770	131,170	159,000
1120	48" wide x 15' floor to floor height		.06	533		94,500	35,900	770	131,170	159,000
1130	32" wide x 20' floor to floor height		.05	653		97,500	44,000	940	142,440	174,500
1140	48" wide x 20' floor to floor height		.05	653		104,500	44,000	940	149,440	182,000
1150	32" wide x 25' floor to floor height		.04	800		104,500	54,000	1,150	159,650	197,000
1160	48" wide x 25' floor to floor height		.04	800		104,500	54,000	1,150	159,650	197,000
1170	Stainless steel, 32" wide x 10' floor to floor height		.07	457		104,500	30,800	660	135,960	162,000
1180	48" wide x 10' floor to floor height		.07	457		104,500	30,800	660	135,960	162,000
1500	32" wide x 15' floor to floor height		.06	533		95,500	35,900	770	132,170	160,000
1700	48" wide x 15' floor to floor height		.06	533		97,000	35,900	770	133,670	162,000
2300	32" wide x 25' floor to floor height		.04	800		105,000	54,000	1,150	160,150	197,500
2500	48" wide x 25' floor to floor height	↓	.04	800	↓	126,500	54,000	1,150	181,650	221,500

14 32 Moving Walks

14 32 10 – Moving Walkways

14 32 10.10 Moving Walks		Crew	Daily Output	Labor-Hours	Unit	Material	2012 Bare Costs Labor	Equipment	Total	Total Incl O&P
0010	**MOVING WALKS** R143210-20									
0020	Walk, 27" tread width, minimum	M-1	6.50	4.923	L.F.	850	330	7.10	1,187.10	1,450
0100	300' to 500', maximum		4.43	7.223		1,175	485	10.40	1,670.40	2,050
0300	48" tread width walk, minimum		4.43	7.223		1,925	485	10.40	2,420.40	2,875
0400	100' to 350', maximum		3.82	8.377		2,250	565	12.05	2,827.05	3,325
0600	Ramp, 12° incline, 36" tread width, minimum		5.27	6.072		1,575	410	8.75	1,993.75	2,350
0700	70' to 90' maximum		3.82	8.377		2,250	565	12.05	2,827.05	3,325
0900	48" tread width, minimum		3.57	8.964		2,300	605	12.90	2,917.90	3,450
1000	40' to 70', maximum		2.91	10.997		2,900	740	15.85	3,655.85	4,325

14 42 Wheelchair Lifts

14 42 13 – Inclined Wheelchair Lifts

14 42 13.10 Inclined Wheelchair Lifts and Stairclimbers

		Crew	Daily Output	Labor-Hours	Unit	Material	Labor	Equipment	Total	Total Incl O&P
0010	**INCLINED WHEELCHAIR LIFTS AND STAIRCLIMBERS**									
7700	Stair climber (chair lift), single seat, minimum	2 Elev	1	16	Ea.	5,325	1,125		6,450	7,575
7800	Maximum		.20	80		7,350	5,675		13,025	16,600
8700	Stair lift, minimum		1	16		14,500	1,125		15,625	17,600
8900	Maximum		.20	80		22,900	5,675		28,575	33,700

14 42 16 – Vertical Wheelchair Lifts

14 42 16.10 Wheelchair Lifts

		Crew	Daily Output	Labor-Hours	Unit	Material	Labor	Equipment	Total	Total Incl O&P
0010	**WHEELCHAIR LIFTS**									
8000	Wheelchair lift, minimum	2 Elev	1	16	Ea.	7,325	1,125		8,450	9,750
8500	Maximum	"	.50	32	"	17,300	2,275		19,575	22,400

14 45 Vehicle Lifts

14 45 10 – Hydraulic Vehicle Lifts

14 45 10.10 Hydraulic Lifts

		Crew	Daily Output	Labor-Hours	Unit	Material	Labor	Equipment	Total	Total Incl O&P
0010	**HYDRAULIC LIFTS**									
0200	Double post, 9000 lb. frame	L-4	1.15	20.870	Ea.	3,350	865		4,215	5,000
0400	26,000 lb. frame		.22	109		17,300	4,525		21,825	26,100
2200	Hoists, single post, 8000 lb. capacity, swivel arms		.40	60		5,325	2,475		7,800	9,700
2400	Two posts, adjustable frames, 11,000 lb. capacity		.25	96		6,850	3,975		10,825	13,700
2500	24,000 lb. capacity		.15	160		9,150	6,625		15,775	20,300
2700	7500 lb. capacity, frame supports		.50	48		7,625	1,975		9,600	11,500
2800	Four post, roll on ramp		.50	48		6,850	1,975		8,825	10,600
2810	Hydraulic lifts, above ground, 2 post, clear floor, 6000 lb. cap		2.67	8.989		7,525	370		7,895	8,850
2815	9000 lb. capacity		2.29	10.480		17,800	435		18,235	20,300
2820	15,000 lb. capacity		2	12		20,200	495		20,695	23,000
2825	30,000 lb. capacity		1.60	15		44,600	620		45,220	50,000
2830	4 post, ramp style, 25,000 lb. capacity		2	12		17,400	495		17,895	20,000
2835	35,000 lb. capacity		1	24		81,500	995		82,495	91,000
2840	50,000 lb. capacity		1	24		91,000	995		91,995	101,500
2845	75,000 lb. capacity		1	24		105,500	995		106,495	117,500
2850	For drive thru tracks, add, minimum					1,125			1,125	1,225
2855	Maximum					1,925			1,925	2,100
2860	Ramp extensions, 3' (set of 2)					920			920	1,000
2865	Rolling jack platform					3,200			3,200	3,525
2870	Elec/hyd jacking beam					8,550			8,550	9,425
2880	Scissor lift, portable, 6000 lb. capacity					8,400			8,400	9,250

14 51 Correspondence and Parcel Lifts

14 51 10 – Electric Correspondence and Parcel Lifts

14 51 10.10 Correspondence Lifts

		Crew	Daily Output	Labor-Hours	Unit	Material	2012 Bare Costs Labor	Equipment	Total	Total Incl O&P
0010	**CORRESPONDENCE LIFTS**									
0020	1 floor, 2 stop, 25 lb. capacity, electric	2 Elev	.20	80	Ea.	5,975	5,675		11,650	15,100
0100	Hand, 5 lb. capacity	"	.20	80	"	2,250	5,675		7,925	11,000

14 51 10.20 Parcel Lifts

		Crew	Daily Output	Labor-Hours	Unit	Material	2012 Bare Costs Labor	Equipment	Total	Total Incl O&P
0010	**PARCEL LIFTS**									
0020	20" x 20", 100 lb. capacity, electric, per floor	2 Mill	.25	64	Ea.	9,350	2,950		12,300	14,700

14 91 Facility Chutes

14 91 33 – Laundry and Linen Chutes

14 91 33.10 Chutes

		Crew	Daily Output	Labor-Hours	Unit	Material	2012 Bare Costs Labor	Equipment	Total	Total Incl O&P
0011	**CHUTES**, linen, trash or refuse									
0050	Aluminized steel, 16 ga., 18" diameter	2 Shee	3.50	4.571	Floor	1,700	238		1,938	2,250
0100	24" diameter		3.20	5		1,775	261		2,036	2,350
0200	30" diameter		3	5.333		2,125	278		2,403	2,750
0300	36" diameter		2.80	5.714		2,625	298		2,923	3,350
0400	Galvanized steel, 16 ga., 18" diameter		3.50	4.571		1,000	238		1,238	1,475
0500	24" diameter		3.20	5		1,125	261		1,386	1,650
0600	30" diameter		3	5.333		1,275	278		1,553	1,825
0700	36" diameter		2.80	5.714		1,500	298		1,798	2,100
0800	Stainless steel, 18" diameter		3.50	4.571		3,000	238		3,238	3,675
0900	24" diameter		3.20	5		3,150	261		3,411	3,850
1000	30" diameter		3	5.333		3,750	278		4,028	4,550
1005	36" diameter		2.80	5.714		3,950	298		4,248	4,775
1200	Linen chute bottom collector, aluminized steel		4	4	Ea.	1,400	209		1,609	1,875
1300	Stainless steel		4	4		1,800	209		2,009	2,300
1500	Refuse, bottom hopper, aluminized steel, 18" diameter		3	5.333		1,025	278		1,303	1,550
1600	24" diameter		3	5.333		1,250	278		1,528	1,800
1800	36" diameter		3	5.333		2,500	278		2,778	3,175

14 91 82 – Trash Chutes

14 91 82.10 Trash Chutes and Accessories

		Crew	Daily Output	Labor-Hours	Unit	Material	2012 Bare Costs Labor	Equipment	Total	Total Incl O&P
0010	**TRASH CHUTES AND ACCESSORIES**									
2900	Package chutes, spiral type, minimum	2 Shee	4.50	3.556	Floor	2,425	185		2,610	2,950
3000	Maximum	"	1.50	10.667	"	6,325	555		6,880	7,825

14 92 Pneumatic Tube Systems

14 92 10 – Conventional, Automatic and Computer Controlled Pneumatic Tube Systems

14 92 10.10 Pneumatic Tube Systems

		Crew	Daily Output	Labor-Hours	Unit	Material	2012 Bare Costs Labor	Equipment	Total	Total Incl O&P
0010	**PNEUMATIC TUBE SYSTEMS**									
0020	100' long, single tube, 2 stations, stock									
0100	3" diameter	2 Stpi	.12	133	Total	6,600	7,425		14,025	18,500
0300	4" diameter	"	.09	177	"	7,450	9,900		17,350	23,100
0400	Twin tube, two stations or more, conventional system									
0600	2-1/2" round	2 Stpi	62.50	.256	L.F.	13.30	14.25		27.55	36
0700	3" round		46	.348		15.25	19.35		34.60	46
0900	4" round		49.60	.323		16.90	17.95		34.85	45.50
1000	4" x 7" oval		37.60	.426		24.50	23.50		48	62.50
1050	Add for blower		2	8	System	5,200	445		5,645	6,400
1110	Plus for each round station, add		7.50	2.133	Ea.	585	119		704	825

14 92 10.10 Pneumatic Tube Systems		Crew	Daily Output	Labor-Hours	Unit	Material	2012 Bare Costs Labor	Equipment	Total	Total Incl O&P
1150	Plus for each oval station, add	2 Stpi	7.50	2.133	Ea.	585	119		704	825
1200	Alternate pricing method: base cost, minimum		.75	21.333	Total	6,375	1,200		7,575	8,825
1300	Maximum		.25	64	"	12,700	3,575		16,275	19,300
1500	Plus total system length, add, minimum		93.40	.171	L.F.	8.20	9.55		17.75	23.50
1600	Maximum		37.60	.426	"	24.50	23.50		48	62
1800	Completely automatic system, 4" round, 15 to 50 stations		.29	55.172	Station	20,100	3,075		23,175	26,700
2200	51 to 144 stations		.32	50		15,600	2,775		18,375	21,400
2400	6" round or 4" x 7" oval, 15 to 50 stations		.24	66.667		25,200	3,725		28,925	33,300
2800	51 to 144 stations		.23	69.565		21,100	3,875		24,975	29,100

Division Notes

		CREW	DAILY OUTPUT	LABOR-HOURS	UNIT	2012 BARE COSTS				TOTAL INCL O&P
						MAT.	LABOR	EQUIP.	TOTAL	

Estimating Tips

Pipe for fire protection and all uses is located in Subdivisions 21 11 13 and 22 11 13.

The labor adjustment factors listed in Subdivision 22 01 02.20 also apply to Division 21.

Many, but not all, areas in the U.S. require backflow protection in the fire system. It is advisable to check local building codes for specific requirements.

For your reference, the following is a list of the most applicable Fire Codes and Standards which may be purchased from the NFPA, 1 Batterymarch Park, Quincy, MA 02169-7471.

NFPA 1: Uniform Fire Code

NFPA 10: Portable Fire Extinguishers

NFPA 11: Low-, Medium-, and High-Expansion Foam

NFPA 12: Carbon Dioxide Extinguishing Systems (Also companion 12A)

NFPA 13: Installation of Sprinkler Systems (Also companion 13D, 13E, and 13R)

NFPA 14: Installation of Standpipe and Hose Systems

NFPA 15: Water Spray Fixed Systems for Fire Protection

NFPA 16: Installation of Foam-Water Sprinkler and Foam-Water Spray Systems

NFPA 17: Dry Chemical Extinguishing Systems (Also companion 17A)

NFPA 18: Wetting Agents

NFPA 20: Installation of Stationary Pumps for Fire Protection

NFPA 22: Water Tanks for Private Fire Protection

NFPA 24: Installation of Private Fire Service Mains and their Appurtenances

NFPA 25: Inspection, Testing and Maintenance of Water-Based Fire Protection

Reference Numbers

Reference numbers are shown in shaded boxes at the beginning of some major classifications. These numbers refer to related items in the Reference Section. The reference information may be an estimating procedure, an alternate pricing method, or technical information.

Note: Not all subdivisions listed here necessarily appear in this publication.

Note: **Trade Service,** *in part, has been used as a reference source for some of the material prices used in Division 21.*

21 05 Common Work Results for Fire Suppression

21 05 23 – General-Duty Valves for Water-Based Fire-Suppression Piping

21 05 23.50 General-Duty Valves	Crew	Daily Output	Labor-Hours	Unit	Material	2012 Bare Costs Labor	2012 Bare Costs Equipment	Total	Total Incl O&P
0010 **GENERAL-DUTY VALVES**, for water-based fire suppression									
6200 Valves and components									
6500 Check, swing, C.I. body, brass fittings, auto. ball drip									
6520 4" size	Q-12	3	5.333	Ea.	291	259		550	710
6800 Check, wafer, butterfly type, C.I. body, bronze fittings									
6820 4" size	Q-12	4	4	Ea.	410	195		605	745

21 11 Facility Fire-Suppression Water-Service Piping

21 11 16 – Facility Fire Hydrants

21 11 16.50 Fire Hydrants for Buildings

	Crew	Daily Output	Labor-Hours	Unit	Material	2012 Bare Costs Labor	2012 Bare Costs Equipment	Total	Total Incl O&P
0010 **FIRE HYDRANTS FOR BUILDINGS**									
3750 Hydrants, wall, w/caps, single, flush, polished brass									
3800 2-1/2" x 2-1/2"	Q-12	5	3.200	Ea.	175	156		331	430
3840 2-1/2" x 3"	"	5	3.200	"	355	156		511	625
3950 Double, flush, polished brass									
4000 2-1/2" x 2-1/2" x 4"	Q-12	5	3.200	Ea.	470	156		626	755
4040 2-1/2" x 2-1/2" x 6"	"	4.60	3.478	↓	675	169		844	1,000
4200 For polished chrome, add					10%				
4350 Double, projecting, polished brass									
4400 2-1/2" x 2-1/2" x 4"	Q-12	5	3.200	Ea.	209	156		365	465
4450 2-1/2" x 2-1/2" x 6"	"	4.60	3.478	"	425	169		594	725
4460 Valve control, dbl. flush/projecting hydrant, cap &									
4470 chain, extension rod & cplg., escutcheon, polished brass	Q-12	8	2	Ea.	249	97.50		346.50	420

21 11 19 – Fire-Department Connections

21 11 19.50 Connections for the Fire-Department

	Crew	Daily Output	Labor-Hours	Unit	Material	2012 Bare Costs Labor	2012 Bare Costs Equipment	Total	Total Incl O&P
0010 **CONNECTIONS FOR THE FIRE-DEPARTMENT**									
7140 Standpipe connections, wall, w/plugs & chains									
7160 Single, flush, brass, 2-1/2" x 2-1/2"	Q-12	5	3.200	Ea.	123	156		279	370
7180 2-1/2" x 3"	"	5	3.200	"	135	156		291	385
7240 For polished chrome, add					15%				
7280 Double, flush, polished brass									
7300 2-1/2" x 2-1/2" x 4"	Q-12	5	3.200	Ea.	425	156		581	705
7330 2-1/2" x 2-1/2" x 6"	"	4.60	3.478	"	595	169		764	910
7400 For polished chrome, add					15%				
7440 For sill cock combination, add				Ea.	50.50			50.50	56
7900 Three way, flush, polished brass									
7920 2-1/2" (3) x 4"	Q-12	4.80	3.333	Ea.	1,225	162		1,387	1,600
7930 2-1/2" (3) x 6"	"	4.80	3.333	↓	1,375	162		1,537	1,750
8000 For polished chrome, add					9%				
8020 Three way, projecting, polished brass									
8040 2-1/2"(3) x 4"	Q-12	4.80	3.333	Ea.	985	162		1,147	1,325

21 12 Fire-Suppression Standpipes

21 12 13 – Fire-Suppression Hoses and Nozzles

21 12 13.50 Fire Hoses and Nozzles	Crew	Daily Output	Labor-Hours	Unit	Material	2012 Bare Costs Labor	Equipment	Total	Total Incl O&P
0010 **FIRE HOSES AND NOZZLES**									
0200 Adapters, rough brass, straight hose threads									
0220 One piece, female to male, rocker lugs									
0240 1" x 1"				Ea.	39.50			39.50	43.50
2200 Hose, less couplings									
2260 Synthetic jacket, lined, 300 lb. test, 1-1/2" diameter	Q-12	2600	.006	L.F.	2.39	.30		2.69	3.08
2280 2-1/2" diameter		2200	.007		3.98	.35		4.33	4.91
2360 High strength, 500 lb. test, 1-1/2" diameter		2600	.006		2.47	.30		2.77	3.17
2380 2-1/2" diameter		2200	.007		4.29	.35		4.64	5.25
5600 Nozzles, brass									
5620 Adjustable fog, 3/4" booster line				Ea.	87.50			87.50	96
5630 1" booster line					128			128	141
5640 1-1/2" leader line					94			94	104
5660 2-1/2" direct connection					123			123	135
5780 For chrome plated, add					8%				
5850 Electrical fire, adjustable fog, no shock									
5900 1-1/2"				Ea.	400			400	440
5920 2-1/2"					540			540	595
5980 For polished chrome, add					6%				
6200 Heavy duty, comb. adj. fog and str. stream, with handle									
6210 1" booster line				Ea.	360			360	395

21 12 19 – Fire-Suppression Hose Racks

21 12 19.50 Fire Hose Racks

	Crew	Daily Output	Labor-Hours	Unit	Material	Labor	Equipment	Total	Total Incl O&P
0010 **FIRE HOSE RACKS**									
2600 Hose rack, swinging, for 1-1/2" diameter hose,									
2620 Enameled steel, 50' & 75' lengths of hose	Q-12	20	.800	Ea.	50	39		89	114
2640 100' and 125' lengths of hose	"	20	.800	"	50.50	39		89.50	115

21 12 23 – Fire-Suppression Hose Valves

21 12 23.70 Fire Hose Valves

	Crew	Daily Output	Labor-Hours	Unit	Material	Labor	Equipment	Total	Total Incl O&P
0010 **FIRE HOSE VALVES**									
0080 Wheel handle, 300 lb., 1-1/2"	1 Spri	12	.667	Ea.	89	36		125	152
0090 2-1/2"	"	7	1.143	"	60	62		122	159
0100 For polished brass, add					35%				
0110 For polished chrome, add					50%				

21 13 Fire-Suppression Sprinkler Systems

21 13 13 – Wet-Pipe Sprinkler Systems

21 13 13.50 Wet-Pipe Sprinkler System Components

	Crew	Daily Output	Labor-Hours	Unit	Material	Labor	Equipment	Total	Total Incl O&P
0010 **WET-PIPE SPRINKLER SYSTEM COMPONENTS**									
2600 Sprinkler heads, not including supply piping									
3700 Standard spray, pendent or upright, brass, 135°F to 286°F									
3730 1/2" NPT, 7/16" orifice	1 Spri	16	.500	Ea.	13.15	27		40.15	55.50
3740 1/2" NPT, 1/2" orifice	"	16	.500		8.55	27		35.55	50.50
3860 For wax and lead coating, add					29.50			29.50	32
3880 For wax coating, add					17.50			17.50	19.25
3900 For lead coating, add					19			19	21
3920 For 360°F, same cost									
3930 For 400°F, add				Ea.	55.50			55.50	61
3940 For 500°F, add				"	55.50			55.50	61
4500 Sidewall, horizontal, brass, 135°F to 286°F									

21 13 Fire-Suppression Sprinkler Systems

21 13 13 – Wet-Pipe Sprinkler Systems

21 13 13.50 Wet-Pipe Sprinkler System Components	Crew	Daily Output	Labor-Hours	Unit	Material	2012 Bare Costs Labor	Equipment	Total	Total Incl O&P
4520 1/2" NPT, 1/2" orifice	1 Spri	16	.500	Ea.	21	27		48	64
4540 For 360°F, same cost									
4800 Recessed pendent, brass, 135°F to 286°F									
4820 1/2" NPT, 3/8" orifice	1 Spri	10	.800	Ea.	40	43		83	109
4830 1/2" NPT, 7/16" orifice		10	.800		21	43		64	88
4840 1/2" NPT, 1/2" orifice	↓	10	.800	↓	16.25	43		59.25	83

21 13 16 – Dry-Pipe Sprinkler Systems

21 13 16.50 Dry-Pipe Sprinkler System Components

	Crew	Daily Output	Labor-Hours	Unit	Material	2012 Bare Costs Labor	Equipment	Total	Total Incl O&P
0010 **DRY-PIPE SPRINKLER SYSTEM COMPONENTS**									
0600 Accelerator	1 Spri	8	1	Ea.	25.50	54		79.50	110
2600 Sprinkler heads, not including supply piping									
2640 Dry, pendent, 1/2" orifice, 3/4" or 1" NPT									
2700 15-1/4" to 18" length	1 Spri	14	.571	Ea.	123	31		154	182
2710 18-1/4" to 21" length		13	.615		128	33.50		161.50	190
2720 21-1/4" to 24" length		13	.615		132	33.50		165.50	195
2730 24-1/4" to 27" length	↓	13	.615	↓	137	33.50		170.50	201

21 13 26 – Deluge Fire-Suppression Sprinkler Systems

21 13 26.50 Deluge Fire-Suppression Sprinkler Sys. Comp.

	Crew	Daily Output	Labor-Hours	Unit	Material	2012 Bare Costs Labor	Equipment	Total	Total Incl O&P
0010 **DELUGE FIRE-SUPPRESSION SPRINKLER SYSTEM COMPONENTS**									
6200 Valves and components									
7000 Deluge, assembly, incl. trim, pressure									
7020 operated relief, emergency release, gauges									
7040 2" size	Q-12	2	8	Ea.	3,175	390		3,565	4,050
7060 3" size	"	1.50	10.667	"	3,550	520		4,070	4,675

21 13 39 – Foam-Water Systems

21 13 39.50 Foam-Water System Components

	Crew	Daily Output	Labor-Hours	Unit	Material	2012 Bare Costs Labor	Equipment	Total	Total Incl O&P
0010 **FOAM-WATER SYSTEM COMPONENTS**									
2600 Sprinkler heads, not including supply piping									
3600 Foam-water, pendent or upright, 1/2" NPT	1 Spri	12	.667	Ea.	157	36		193	227

21 21 Carbon-Dioxide Fire-Extinguishing Systems

21 21 16 – Carbon-Dioxide Fire-Extinguishing Equipment

21 21 16.50 CO2 Fire Extinguishing System

	Crew	Daily Output	Labor-Hours	Unit	Material	2012 Bare Costs Labor	Equipment	Total	Total Incl O&P
0010 **CO$_2$ FIRE EXTINGUISHING SYSTEM**									
0042 For detectors and control stations, see Section 28 31 23.50									
1000 Dispersion nozzle, CO$_2$, 3" x 5"	1 Plum	18	.444	Ea.	60	24.50		84.50	103
2000 Extinguisher, CO$_2$ system, high pressure, 75 lb. cylinder	Q-1	6	2.667		1,150	131		1,281	1,450
3000 Electro/mechanical release	L-1	4	4		150	212		362	485
3400 Manual pull station	1 Plum	6	1.333	↓	54	73		127	170

21 22 Clean-Agent Fire-Extinguishing Systems

21 22 16 – Clean-Agent Fire-Extinguishing Equipment

21 22 16.50 FM200 Fire Extinguishing System	Crew	Daily Output	Labor-Hours	Unit	Material	2012 Bare Costs Labor	Equipment	Total	Total Incl O&P
0010 **FM200 FIRE EXTINGUISHING SYSTEM**									
1100 Dispersion nozzle FM200, 1-1/2"	1 Plum	14	.571	Ea.	60	31		91	113
2400 Extinguisher, FM200 system, filled, with mounting bracket									
2540 196 lb. container	Q-1	4	4	Ea.	6,975	197		7,172	7,975
6000 Average FM200 system, minimum				C.F.	1.58			1.58	1.74
6020 Maximum				"	3.15			3.15	3.47

21 31 Centrifugal Fire Pumps

21 31 13 – Electric-Drive, Centrifugal Fire Pumps

21 31 13.50 Electric-Drive Fire Pumps

	Crew	Daily Output	Labor-Hours	Unit	Material	2012 Bare Costs Labor	Equipment	Total	Total Incl O&P
0010 **ELECTRIC-DRIVE FIRE PUMPS** Including controller, fittings and relief valve									
3100 250 GPM, 55 psi, 15 HP, 3550 RPM, 2" pump	Q-13	.70	45.714	Ea.	14,200	2,350		16,550	19,200
3200 500 GPM, 50 psi, 27 HP, 1770 RPM, 4" pump		.68	47.059		14,500	2,425		16,925	19,700
3350 750 GPM, 50 psi, 44 HP, 1770 RPM, 5" pump		.64	50		15,100	2,575		17,675	20,500
3400 750 GPM, 100 psi, 66 HP, 3550 RPM, 4" pump		.58	55.172		17,400	2,850		20,250	23,400
5000 For jockey pump 1", 3 HP, with control, add	Q-12	2	8		2,350	390		2,740	3,150

21 31 16 – Diesel-Drive, Centrifugal Fire Pumps

21 31 16.50 Diesel-Drive Fire Pumps

	Crew	Daily Output	Labor-Hours	Unit	Material	2012 Bare Costs Labor	Equipment	Total	Total Incl O&P
0010 **DIESEL-DRIVE FIRE PUMPS** Including controller, fittings and relief valve									
0050 500 GPM, 50 psi, 27 HP, 4" pump	Q-13	.64	50	Ea.	32,900	2,575		35,475	40,100
0200 750 GPM, 50 psi, 44 HP, 5" pump		.60	53.333		33,900	2,750		36,650	41,400
0400 1,000 GPM, 100 psi, 89 HP, 4" pump		.56	57.143		37,100	2,950		40,050	45,300
0700 2,000 GPM, 100 psi, 167 HP, 6" pump		.34	94.118		47,200	4,850		52,050	59,500
0950 3,500 GPM, 100 psi, 300 HP, 10" pump		.24	133		65,500	6,875		72,375	82,500

Division Notes

	CREW	DAILY OUTPUT	LABOR-HOURS	UNIT	2012 BARE COSTS				TOTAL INCL O&P
					MAT.	LABOR	EQUIP.	TOTAL	

Estimating Tips

22 10 00 Plumbing Piping and Pumps

This subdivision is primarily basic pipe and related materials. The pipe may be used by any of the mechanical disciplines, i.e., plumbing, fire protection, heating, and air conditioning.

Note: CPVC plastic piping approved for tire protection is located in 21 11 13.

- The labor adjustment factors listed in Subdivision 22 01 02.20 apply throughout Divisions 21, 22, and 23. CAUTION: the correct percentage may vary for the same items. For example, the percentage add for the basic pipe installation should be based on the maximum height that the craftsman must install for that particular section. If the pipe is to be located 14′ above the floor but it is suspended on threaded rod from beams, the bottom flange of which is 18′ high (4′ rods), then the height is actually 18′ and the add is 20%. The pipe coverer, however, does not have to go above the 14′, and so his or her add should be 10%.

- Most pipe is priced first as straight pipe with a joint (coupling, weld, etc.) every 10′ and a hanger usually every 10′.

There are exceptions with hanger spacing such as for cast iron pipe (5′) and plastic pipe (3 per 10′). Following each type of pipe there are several lines listing sizes and the amount to be subtracted to delete couplings and hangers. This is for pipe that is to be buried or supported together on trapeze hangers. The reason that the couplings are deleted is that these runs are usually long, and frequently longer lengths of pipe are used. By deleting the couplings, the estimator is expected to look up and add back the correct reduced number of couplings.

- When preparing an estimate, it may be necessary to approximate the fittings. Fittings usually run between 25% and 50% of the cost of the pipe. The lower percentage is for simpler runs, and the higher number is for complex areas, such as mechanical rooms.

- For historic restoration projects, the systems must be as invisible as possible, and pathways must be sought for pipes, conduit, and ductwork. While installations in accessible spaces (such as basements and attics) are relatively straightforward to estimate, labor costs may be more difficult to determine when delivery systems must be concealed.

22 40 00 Plumbing Fixtures

- Plumbing fixture costs usually require two lines: the fixture itself and its "rough-in, supply, and waste."

- In the Assemblies Section (Plumbing D2010) for the desired fixture, the System Components Group at the center of the page shows the fixture on the first line. The rest of the list (fittings, pipe, tubing, etc.) will total up to what we refer to in the Unit Price section as "Rough-in, supply, waste, and vent." Note that for most fixtures we allow a nominal 5′ of tubing to reach from the fixture to a main or riser.

- Remember that gas- and oil-fired units need venting.

Reference Numbers

Reference numbers are shown in shaded boxes at the beginning of some major classifications. These numbers refer to related items in the Reference Section. The reference information may be an estimating procedure, an alternate pricing method, or technical information.

Note: Not all subdivisions listed here necessarily appear in this publication.

Note: **Trade Service,** *in part, has been used as a reference source for some of the material prices used in Division 22.*

22 01 02 – Labor Adjustments

		Crew	Daily Output	Labor-Hours	Unit	Material	2012 Bare Costs Labor	Equipment	Total	Total Incl O&P
22 01 02.10 Boilers, General										
0010	**BOILERS, GENERAL**, Prices do not include flue piping, elec. wiring,									
0020	gas or oil piping, boiler base, pad, or tankless unless noted									
0100	Boiler H.P.: 10 KW = 34 lb./steam/hr. = 33,475 BTU/hr. R235000-50									
0150	To convert SFR to BTU rating: Hot water, 150 x SFR;									
0160	Forced hot water, 180 x SFR; steam, 240 x SFR									
22 01 02.20 Labor Adjustment Factors										
0010	**LABOR ADJUSTMENT FACTORS**, (For Div. 21, 22 and 23) R220102-20									
1000	Add to labor for elevated installation (Above floor level)									
1080	10' to 14.5' high						10%			
1100	15' to 19.5' high						20%			
1120	20' to 24.5' high						25%			
1140	25' to 29.5' high						35%			
1160	30' to 34.5' high						40%			
1180	35' to 39.5' high						50%			
1200	40' and higher						55%			
2000	Add to labor for crawl space									
2100	3' high						40%			
2140	4' high						30%			
3000	Add to labor for multi-story building									
3100	Add per floor for floors 3 thru 19						2%			
3140	Add per floor for floors 20 and up						4%			
4000	Add to labor for working in existing occupied buildings									
4100	Hospital						35%			
4140	Office building						25%			
4180	School						20%			
4220	Factory or warehouse						15%			
4260	Multi dwelling						15%			
5000	Add to labor, miscellaneous									
5100	Cramped shaft						35%			
5140	Congested area						15%			
5180	Excessive heat or cold						30%			
9000	Labor factors, The above are reasonable suggestions, however									
9010	each project should be evaluated for its own peculiarities.									
9100	Other factors to be considered are:									
9140	Movement of material and equipment through finished areas									
9180	Equipment room									
9220	Attic space									
9260	No service road									
9300	Poor unloading/storage area									
9340	Congested site area/heavy traffic									

22 05 Common Work Results for Plumbing

22 05 05 – Selective Plumbing Demolition

22 05 05.10 Plumbing Demolition

		Crew	Daily Output	Labor-Hours	Unit	Material	2012 Bare Costs Labor	Equipment	Total	Total Incl O&P
0010	**PLUMBING DEMOLITION**									
1020	Fixtures, including 10' piping									
1100	Bathtubs, cast iron	1 Plum	4	2	Ea.		109		109	165
1120	Fiberglass		6	1.333			73		73	110
1140	Steel		5	1.600			87.50		87.50	132
1200	Lavatory, wall hung		10	.800			43.50		43.50	66
1220	Counter top		8	1			54.50		54.50	82.50
1300	Sink, single compartment		8	1			54.50		54.50	82.50
1320	Double compartment		7	1.143			62.50		62.50	94
1400	Water closet, floor mounted		8	1			54.50		54.50	82.50
1420	Wall mounted		7	1.143			62.50		62.50	94
1500	Urinal, floor mounted		4	2			109		109	165
1520	Wall mounted		7	1.143			62.50		62.50	94
1600	Water fountains, free standing		8	1			54.50		54.50	82.50
1620	Wall or deck mounted		6	1.333	▼		73		73	110
2000	Piping, metal, up thru 1-1/2" diameter		200	.040	L.F.		2.18		2.18	3.29
2050	2" thru 3-1/2" diameter	▼	150	.053			2.91		2.91	4.39
2100	4" thru 6" diameter	2 Plum	100	.160			8.75		8.75	13.20
2150	8" thru 14" diameter	"	60	.267			14.55		14.55	22
2153	16" thru 20" diameter	Q-18	70	.343			17.80	.53	18.33	27.50
2155	24" thru 26" diameter		55	.436			22.50	.67	23.17	34.50
2156	30" thru 36" diameter	▼	40	.600	▼		31	.92	31.92	48
2212	Deduct for salvage, aluminum scrap				Ton				700	770
2214	Brass scrap								2,450	2,675
2216	Copper scrap								3,200	3,525
2218	Lead scrap								520	570
2220	Steel scrap				▼				180	200
2250	Water heater, 40 gal.	1 Plum	6	1.333	Ea.		73		73	110
9470	Water softener	Q-1	2	8	"		395		395	595

22 05 23 – General-Duty Valves for Plumbing Piping

22 05 23.10 Valves, Brass

		Crew	Daily Output	Labor-Hours	Unit	Material	2012 Bare Costs Labor	Equipment	Total	Total Incl O&P
0010	**VALVES, BRASS**									
0500	Gas cocks, threaded									
0530	1/2"	1 Plum	24	.333	Ea.	13.15	18.20		31.35	42
0540	3/4"		22	.364		15.50	19.85		35.35	47
0550	1"		19	.421		31	23		54	68.50
0560	1-1/4"	▼	15	.533	▼	44.50	29		73.50	93

22 05 23.20 Valves, Bronze

		Crew	Daily Output	Labor-Hours	Unit	Material	2012 Bare Costs Labor	Equipment	Total	Total Incl O&P
0010	**VALVES, BRONZE**									
1020	Angle, 150 lb., rising stem, threaded									
1030	1/8"	1 Plum	24	.333	Ea.	119	18.20		137.20	159
1040	1/4"		24	.333		119	18.20		137.20	159
1050	3/8"		24	.333		119	18.20		137.20	159
1060	1/2"		22	.364		119	19.85		138.85	161
1070	3/4"		20	.400		162	22		184	211
1080	1"		19	.421		234	23		257	292
1100	1-1/2"		13	.615		360	33.50		393.50	445
1110	2"	▼	11	.727	▼	630	39.50		669.50	755
1300	Ball									
1382	2"	1 Plum	9	.889	Ea.	145	48.50		193.50	233
1398	150 psi, threaded									
1400	1/4"	1 Plum	24	.333	Ea.	12.70	18.20		30.90	41.50

463

22 05 23.20 Valves, Bronze		Crew	Daily Output	Labor-Hours	Unit	Material	2012 Bare Costs Labor	2012 Bare Costs Equipment	Total	Total Incl O&P
1430	3/8"	1 Plum	24	.333	Ea.	12.70	18.20		30.90	41.50
1450	1/2"		22	.364		12.70	19.85		32.55	44
1460	3/4"		20	.400		21	22		43	56
1470	1"		19	.421		28.50	23		51.50	66
1480	1-1/4"		15	.533		50	29		79	99
1490	1-1/2"		13	.615		65	33.50		98.50	122
1500	2"	▼	11	.727	▼	78.50	39.50		118	147
1750	Check, swing, class 150, regrinding disc, threaded									
1800	1/8"	1 Plum	24	.333	Ea.	91	18.20		109.20	128
1830	1/4"		24	.333		91	18.20		109.20	128
1840	3/8"		24	.333		88.50	18.20		106.70	125
1850	1/2"		24	.333		107	18.20		125.20	146
1860	3/4"		20	.400		142	22		164	189
1870	1"		19	.421		145	23		168	194
1880	1-1/4"		15	.533		198	29		227	262
1890	1-1/2"		13	.615		292	33.50		325.50	370
1900	2"	▼	11	.727		400	39.50		439.50	495
1910	2-1/2"	Q-1	15	1.067	▼	615	52.50		667.50	755
2000	For 200 lb., add					5%	10%			
2040	For 300 lb., add					15%	15%			
2850	Gate, N.R.S., soldered, 125 psi									
2900	3/8"	1 Plum	24	.333	Ea.	48	18.20		66.20	80.50
2920	1/2"		24	.333		41	18.20		59.20	72.50
2940	3/4"		20	.400		47	22		69	84.50
2950	1"		19	.421		66.50	23		89.50	108
2960	1-1/4"		15	.533		102	29		131	156
2970	1-1/2"		13	.615		114	33.50		147.50	176
2980	2"	▼	11	.727		161	39.50		200.50	237
2990	2-1/2"	Q-1	15	1.067		405	52.50		457.50	525
3000	3"	"	13	1.231	▼	525	60.50		585.50	670
3850	Rising stem, soldered, 300 psi									
3950	1"	1 Plum	19	.421	Ea.	136	23		159	185
3980	2"	"	11	.727		360	39.50		399.50	455
4000	3"	Q-1	13	1.231	▼	1,150	60.50		1,210.50	1,350
4250	Threaded, class 150									
4310	1/4"	1 Plum	24	.333	Ea.	62.50	18.20		80.70	96.50
4320	3/8"		24	.333		62.50	18.20		80.70	96.50
4330	1/2"		24	.333		57.50	18.20		75.70	91
4340	3/4"		20	.400		67	22		89	107
4350	1"		19	.421		90	23		113	134
4360	1-1/4"		15	.533		122	29		151	179
4370	1-1/2"		13	.615		154	33.50		187.50	220
4380	2"	▼	11	.727		208	39.50		247.50	289
4390	2-1/2"	Q-1	15	1.067		485	52.50		537.50	615
4400	3"	"	13	1.231		675	60.50		735.50	835
4500	For 300 psi, threaded, add					100%	15%			
4540	For chain operated type, add				▼	15%				
4850	Globe, class 150, rising stem, threaded									
4920	1/4"	1 Plum	24	.333	Ea.	88.50	18.20		106.70	125
4940	3/8" size		24	.333		88.50	18.20		106.70	125
4950	1/2"		24	.333		88.50	18.20		106.70	125
4960	3/4"		20	.400		119	22		141	164
4970	1"	▼	19	.421	▼	187	23		210	240

22 05 23 – General-Duty Valves for Plumbing Piping

22 05 23.20 Valves, Bronze		Crew	Daily Output	Labor-Hours	Unit	Material	2012 Bare Costs Labor	2012 Bare Costs Equipment	Total	Total Incl O&P
4980	1-1/4"	1 Plum	15	.533	Ea.	298	29		327	375
4990	1-1/2"		13	.615		360	33.50		393.50	450
5000	2"		11	.727		540	39.50		579.50	655
5010	2-1/2"	Q-1	15	1.067		1,100	52.50		1,152.50	1,275
5020	3"	"	13	1.231		1,425	60.50		1,485.50	1,675
5120	For 300 lb. threaded, add					50%	15%			
5600	Relief, pressure & temperature, self-closing, ASME, threaded									
5640	3/4"	1 Plum	28	.286	Ea.	142	15.60		157.60	180
5650	1"		24	.333		217	18.20		235.20	266
5660	1-1/4"		20	.400		420	22		442	495
5670	1-1/2"		18	.444		800	24.50		824.50	915
5680	2"		16	.500		870	27.50		897.50	1,000
5950	Pressure, poppet type, threaded									
6000	1/2"	1 Plum	30	.267	Ea.	37.50	14.55		52.05	63.50
6040	3/4"	"	28	.286	"	40	15.60		55.60	67.50
6400	Pressure, water, ASME, threaded									
6440	3/4"	1 Plum	28	.286	Ea.	80	15.60		95.60	112
6450	1"		24	.333		170	18.20		188.20	215
6460	1-1/4"		20	.400		270	22		292	330
6470	1-1/2"		18	.444		375	24.50		399.50	445
6480	2"		16	.500		540	27.50		567.50	635
6490	2-1/2"		15	.533		2,650	29		2,679	2,975
6900	Reducing, water pressure									
6920	300 psi to 25-75 psi, threaded or sweat									
6940	1/2"	1 Plum	24	.333	Ea.	257	18.20		275.20	310
6950	3/4"		20	.400		257	22		279	315
6960	1"		19	.421		400	23		423	475
6970	1-1/4"		15	.533		715	29		744	830
6980	1-1/2"		13	.615		1,075	33.50		1,108.50	1,225
8350	Tempering, water, sweat connections									
8400	1/2"	1 Plum	24	.333	Ea.	85	18.20		103.20	121
8440	3/4"	"	20	.400	"	104	22		126	148
8650	Threaded connections									
8700	1/2"	1 Plum	24	.333	Ea.	128	18.20		146.20	169
8740	3/4"		20	.400		475	22		497	560
8750	1"		19	.421		535	23		558	625
8760	1-1/4"		15	.533		830	29		859	955
8770	1-1/2"		13	.615		905	33.50		938.50	1,050
8780	2"		11	.727		1,350	39.50		1,389.50	1,550

22 05 23.60 Valves, Plastic

22 05 23.60 Valves, Plastic		Crew	Daily Output	Labor-Hours	Unit	Material	2012 Bare Costs Labor	2012 Bare Costs Equipment	Total	Total Incl O&P
0010	**VALVES, PLASTIC**									
1100	Angle, PVC, threaded									
1110	1/4"	1 Plum	26	.308	Ea.	74	16.80		90.80	107
1120	1/2"		26	.308		74	16.80		90.80	107
1130	3/4"		25	.320		87.50	17.45		104.95	123
1140	1"		23	.348		106	19		125	146
1150	Ball, PVC, socket or threaded, single union									
1230	1/2"	1 Plum	26	.308	Ea.	26.50	16.80		43.30	55
1240	3/4"		25	.320		30.50	17.45		47.95	60
1250	1"		23	.348		38.50	19		57.50	70.50
1260	1-1/4"		21	.381		51.50	21		72.50	88.50
1270	1-1/2"		20	.400		61.50	22		83.50	101

22 05 23 – General-Duty Valves for Plumbing Piping

22 05 23.60 Valves, Plastic

		Crew	Daily Output	Labor-Hours	Unit	Material	2012 Bare Costs Labor	2012 Bare Costs Equipment	Total	Total Incl O&P
1280	2"	1 Plum	17	.471	Ea.	89	25.50		114.50	137
1360	For PVC, flanged, add					100%	15%			
1650	CPVC, socket or threaded, single union									
1700	1/2"	1 Plum	26	.308	Ea.	53	16.80		69.80	84
1720	3/4"		25	.320		66.50	17.45		83.95	100
1730	1"		23	.348		80	19		99	117
1750	1-1/4"		21	.381		127	21		148	172
1760	1-1/2"		20	.400		127	22		149	173
1840	For CPVC, flanged, add					65%	15%			
1880	For true union, socket or threaded, add					50%	5%			
2050	Polypropylene, threaded									
2100	1/4"	1 Plum	26	.308	Ea.	40	16.80		56.80	69.50
2120	3/8"		26	.308		40	16.80		56.80	69.50
2130	1/2"		26	.308		40	16.80		56.80	69.50
2140	3/4"		25	.320		49	17.45		66.45	80
2150	1"		23	.348		56	19		75	90
2160	1-1/4"		21	.381		75	21		96	114
2170	1-1/2"		20	.400		93	22		115	135
2180	2"		17	.471		125	25.50		150.50	176
4850	Foot valve, PVC, socket or threaded									
4900	1/2"	1 Plum	34	.235	Ea.	66.50	12.85		79.35	92.50
4930	3/4"		32	.250		75	13.65		88.65	103
4940	1"		28	.286		97.50	15.60		113.10	131
4950	1-1/4"		27	.296		187	16.20		203.20	231
4960	1-1/2"		26	.308		187	16.80		203.80	232
6350	Y sediment strainer, PVC, socket or threaded									
6400	1/2"	1 Plum	26	.308	Ea.	49	16.80		65.80	79
6440	3/4"		24	.333		52	18.20		70.20	84.50
6450	1"		23	.348		62.50	19		81.50	97
6460	1-1/4"		21	.381		104	21		125	146
6470	1-1/2"		20	.400		104	22		126	147

22 05 48 – Vibration and Seismic Controls for Plumbing Piping and Equipment

22 05 48.10 Seismic Bracing Supports

		Crew	Daily Output	Labor-Hours	Unit	Material	2012 Bare Costs Labor	2012 Bare Costs Equipment	Total	Total Incl O&P
0010	**SEISMIC BRACING SUPPORTS**									
0020	Clamps									
0030	C-clamp, for mounting on steel beam									
0040	3/8" threaded rod	1 Skwk	160	.050	Ea.	3.31	2.27		5.58	7.15
0050	1/2" threaded rod		160	.050		4.12	2.27		6.39	8.05
0060	5/8" threaded rod		160	.050		6.30	2.27		8.57	10.45
0070	3/4" threaded rod		160	.050		8.50	2.27		10.77	12.85
0100	Brackets									
0110	Beam side or wall mallable iron									
0120	3/8" threaded rod	1 Skwk	48	.167	Ea.	3.42	7.55		10.97	15.45
0130	1/2" threaded rod		48	.167		6.05	7.55		13.60	18.35
0140	5/8" threaded rod		48	.167		9.25	7.55		16.80	22
0150	3/4" threaded rod		48	.167		11.20	7.55		18.75	24
0160	7/8" threaded rod		48	.167		13.15	7.55		20.70	26
0170	For concrete installation, add						30%			
0180	Wall, welded steel									
0190	0 size 12" wide 18" deep	1 Skwk	34	.235	Ea.	227	10.70		237.70	267
0200	1 size 18" wide 24" deep		34	.235		270	10.70		280.70	315
0210	2 size 24" wide 30" deep		34	.235		355	10.70		365.70	410

22 05 48 – Vibration and Seismic Controls for Plumbing Piping and Equipment

22 05 48.10 Seismic Bracing Supports		Crew	Daily Output	Labor-Hours	Unit	Material	2012 Bare Costs Labor	2012 Bare Costs Equipment	Total	Total Incl O&P
0300	Rod, carbon steel									
0310	Continuous thread									
0320	1/4" thread	1 Skwk	144	.056	L.F.	1.45	2.52		3.97	5.50
0330	3/8" thread		144	.056		1.55	2.52		4.07	5.60
0340	1/2" thread		144	.056		2.44	2.52		4.96	6.60
0350	5/8" thread		144	.056		3.47	2.52		5.99	7.70
0360	3/4" thread		144	.056		6.10	2.52		8.62	10.65
0370	7/8" thread		144	.056		7.65	2.52		10.17	12.35
0380	For galvanized, add					30%				
0400	Channel, steel									
0410	3/4" x 1-1/2"	1 Skwk	80	.100	L.F.	6.70	4.54		11.24	14.35
0420	1-1/2" x 1-1/2"		70	.114		8.80	5.20		14	17.70
0430	1-7/8" x 1-1/2"		60	.133		30	6.05		36.05	42.50
0440	3" x 1-1/2"		50	.160		31.50	7.25		38.75	46
0450	Spring nuts									
0460	3/8"	1 Skwk	100	.080	Ea.	1.94	3.63		5.57	7.75
0470	1/2"	"	80	.100	"	2	4.54		6.54	9.20
0500	Welding, field									
0510	Cleaning and welding plates, bars, or rods									
0520	To existing beams, columns, or trusses									
0530	1" weld	1 Skwk	144	.056	Ea.	.21	2.52		2.73	4.13
0540	2" weld		72	.111		.36	5.05		5.41	8.20
0550	3" weld		54	.148		.56	6.75		7.31	11
0560	4" weld		36	.222		.77	10.10		10.87	16.45
0570	5" weld		30	.267		.97	12.10		13.07	19.75
0580	6" weld		24	.333		1.10	15.15		16.25	24.50
0600	Vibration absorbers									
0610	Hangers, neoprene flex									
0620	10-120 lb. capacity	1 Skwk	8	1	Ea.	23.50	45.50		69	96
0630	75-550 lb. capacity		8	1		32	45.50		77.50	105
0640	250-1100 lb. capacity		6	1.333		57.50	60.50		118	157
0650	1000-4000 lb. capacity		6	1.333		104	60.50		164.50	209
0660	Spring flex									
0670	60-450 lb. capacity	1 Skwk	8	1	Ea.	66	45.50		111.50	143
0680	85-450 lb. capacity		8	1		88	45.50		133.50	167
0690	600-900 lb. capacity		8	1		91	45.50		136.50	170
0700	1100-1300 lb. capacity		6	1.333		97	60.50		157.50	201
0710	Mounts, neoprene									
0720	135-380 lb. capacity	1 Skwk	7	1.143	Ea.	16.15	52		68.15	98
0730	250-1100 lb. capacity		7	1.143		42.50	52		94.50	127
0740	1000-4000 lb. capacity		5	1.600		95.50	72.50		168	217
0900	Wire support									
0910	Light fixture, recessed (pair)	1 Skwk	96	.083	Ea.	3.55	3.78		7.33	9.75

22 05 76 – Facility Drainage Piping Cleanouts

22 05 76.10 Cleanouts

		Crew	Daily Output	Labor-Hours	Unit	Material	2012 Bare Costs Labor	2012 Bare Costs Equipment	Total	Total Incl O&P
0010	**CLEANOUTS**									
0060	Floor type									
0080	Round or square, scoriated nickel bronze top									
0100	2" pipe size	1 Plum	10	.800	Ea.	167	43.50		210.50	250
0120	3" pipe size		8	1		181	54.50		235.50	282
0140	4" pipe size		6	1.333		251	73		324	385
0980	Round top, recessed for terrazzo									

22 05 76 – Facility Drainage Piping Cleanouts

22 05 76.10 Cleanouts

		Crew	Daily Output	Labor-Hours	Unit	Material	2012 Bare Costs Labor	Equipment	Total	Total Incl O&P
1000	2" pipe size	1 Plum	9	.889	Ea.	167	48.50		215.50	257
1080	3" pipe size		6	1.333		251	73		324	385
1100	4" pipe size		4	2		251	109		360	440
1120	5" pipe size	Q-1	6	2.667		320	131		451	550

22 05 76.20 Cleanout Tees

		Crew	Daily Output	Labor-Hours	Unit	Material	2012 Bare Costs Labor	Equipment	Total	Total Incl O&P
0010	**CLEANOUT TEES**									
0100	Cast iron, B&S, with countersunk plug									
0200	2" pipe size	1 Plum	4	2	Ea.	211	109		320	395
0220	3" pipe size		3.60	2.222		248	121		369	455
0240	4" pipe size		3.30	2.424		310	132		442	540
0280	6" pipe size	Q-1	5	3.200		835	157		992	1,150
0500	For round smooth access cover, same price									
4000	Plastic, tees and adapters. Add plugs									
4010	ABS, DWV									
4020	Cleanout tee, 1-1/2" pipe size	1 Plum	15	.533	Ea.	10.40	29		39.40	55.50
4030	2" pipe size	Q-1	27	.593		11.40	29		40.40	56.50
4040	3" pipe size		21	.762		21.50	37.50		59	80
4050	4" pipe size		16	1		46	49		95	125
4100	Cleanout plug, 1-1/2" pipe size	1 Plum	32	.250		1.80	13.65		15.45	22.50
4110	2" pipe size	Q-1	56	.286		2.37	14.05		16.42	23.50
4120	3" pipe size		36	.444		3.82	22		25.82	37
4130	4" pipe size		30	.533		6.70	26		32.70	47
4180	Cleanout adapter fitting, 1-1/2" pipe size	1 Plum	32	.250		2.85	13.65		16.50	23.50
4190	2" pipe size	Q-1	56	.286		4.38	14.05		18.43	26
4200	3" pipe size		36	.444		11.20	22		33.20	45.50
4210	4" pipe size		30	.533		20.50	26		46.50	62
5000	PVC, DWV									
5010	Cleanout tee, 1-1/2" pipe size	1 Plum	15	.533	Ea.	12.50	29		41.50	58
5020	2" pipe size	Q-1	27	.593		10.05	29		39.05	55
5030	3" pipe size		21	.762		30	37.50		67.50	89.50
5040	4" pipe size		16	1		38.50	49		87.50	117
5090	Cleanout plug, 1-1/2" pipe size	1 Plum	32	.250		1.80	13.65		15.45	22.50
5100	2" pipe size	Q-1	56	.286		2	14.05		16.05	23
5110	3" pipe size		36	.444		3.58	22		25.58	37
5120	4" pipe size		30	.533		5.30	26		31.30	45.50
5130	6" pipe size		24	.667		17.60	33		50.60	69
5170	Cleanout adapter fitting, 1-1/2" pipe size	1 Plum	32	.250		2.40	13.65		16.05	23
5180	2" pipe size	Q-1	56	.286		3.11	14.05		17.16	24.50
5190	3" pipe size		36	.444		8.75	22		30.75	42.50
5200	4" pipe size		30	.533		14.35	26		40.35	55.50
5210	6" pipe size		24	.667		42	33		75	95.50

RSMeans Professional Development Seminars

RSMeans Seminars are approved for Continuing Education Units (CEUs) by both the American Institute of Architects Continuing Education Systems (AIA/CES) and AACE International Certification Board.

Seminar Topics Include:

- Practical Project Management for Construction Professionals
- Scheduling with MSProject for Construction Professionals
- Assessing Scope of Work for Facilities Construction Estimating
- Site Work and Heavy Construction Estimating
- Facilities Construction Estimating
- Job Order Contracting
- Mechanical and Electrical Estimating
- Facilities Maintenance and Repair Estimating
- Unit Price Estimating
- RSMeans Product Training (MeansCostWorks.com and RSMeans CostWorks CD)

Ask us about on-site product training.
Call 1-800-334-3509, option 2.

For more information and a seminar schedule visit
www.reedconstructiondata.com/rsmeans/seminars

RSMeans

22 07 19 – Plumbing Piping Insulation

22 07 19.10 Piping Insulation		Crew	Daily Output	Labor-Hours	Unit	Material	2012 Bare Costs Labor	Equipment	Total	Total Incl O&P
0010	**PIPING INSULATION**									
0100	Rule of thumb, as a percentage of total mechanical costs				Job				10%	
0110	Insulation req'd is based on the surface size/area to be covered									
0600	Pipe covering (price copper tube one size less than IPS)									
6600	Fiberglass, with all service jacket									
6840	1" wall, 1/2" iron pipe size	G	Q-14	240	.067	L.F.	.89	2.93		3.82	5.55
6870	1" iron pipe size	G		220	.073		1.04	3.20		4.24	6.10
6900	2" iron pipe size	G		200	.080		1.32	3.52		4.84	6.95
6920	3" iron pipe size	G		180	.089		1.60	3.91		5.51	7.85
6940	4" iron pipe size	G		150	.107		2.12	4.69		6.81	9.65
7320	2" wall, 1/2" iron pipe size	G		220	.073		2.62	3.20		5.82	7.85
7440	6" iron pipe size	G		100	.160		5.25	7.05		12.30	16.75
7460	8" iron pipe size	G		80	.200		6.40	8.80		15.20	20.50
7480	10" iron pipe size	G		70	.229		7.65	10.05		17.70	24
7490	12" iron pipe size	G		65	.246		8.55	10.85		19.40	26.50
7800	For fiberglass with standard canvas jacket, deduct						5%				
7802	For fittings, add 3 L.F. for each fitting										
7804	plus 4 L.F. for each flange of the fitting										
7810	Finishes										
7812	For .016" aluminum jacket, add	G	Q-14	200	.080	S.F.	.94	3.52		4.46	6.55
7813	For .010" stainless steel, add	G	"	160	.100	"	2.79	4.40		7.19	9.90
7814	For single layer of felt, add						10%	10%			
7816	For roofing paper, 45 lb. to 55 lb., add						25%	10%			
7879	Rubber tubing, flexible closed cell foam										
7880	3/8" wall, 1/4" iron pipe size	G	1 Asbe	120	.067	L.F.	.38	3.26		3.64	5.45
7910	1/2" iron pipe size	G		115	.070		.48	3.40		3.88	5.85
7920	3/4" iron pipe size	G		115	.070		.55	3.40		3.95	5.90
7930	1" iron pipe size	G		110	.073		.63	3.55		4.18	6.25
7950	1-1/2" iron pipe size	G		110	.073		.85	3.55		4.40	6.50
8100	1/2" wall, 1/4" iron pipe size	G		90	.089		.57	4.34		4.91	7.40
8130	1/2" iron pipe size	G		89	.090		.71	4.39		5.10	7.65
8140	3/4" iron pipe size	G		89	.090		.79	4.39		5.18	7.70
8150	1" iron pipe size	G		88	.091		.87	4.44		5.31	7.85
8170	1-1/2" iron pipe size	G		87	.092		1.21	4.49		5.70	8.35
8180	2" iron pipe size	G		86	.093		1.55	4.54		6.09	8.80
8200	3" iron pipe size	G		85	.094		2.18	4.60		6.78	9.55
8300	3/4" wall, 1/4" iron pipe size	G		90	.089		.90	4.34		5.24	7.75
8330	1/2" iron pipe size	G		89	.090		1.17	4.39		5.56	8.15
8340	3/4" iron pipe size	G		89	.090		1.43	4.39		5.82	8.40
8350	1" iron pipe size	G		88	.091		1.63	4.44		6.07	8.70
8370	1-1/2" iron pipe size	G		87	.092		2.46	4.49		6.95	9.70
8380	2" iron pipe size	G		86	.093		2.84	4.54		7.38	10.20
8400	3" iron pipe size	G		85	.094		4.32	4.60		8.92	11.90
8444	1" wall, 1/2" iron pipe size	G		86	.093		2.17	4.54		6.71	9.50
8445	3/4" iron pipe size	G		84	.095		2.63	4.65		7.28	10.15
8446	1" iron pipe size	G		84	.095		3.06	4.65		7.71	10.60
8447	1-1/4" iron pipe size	G		82	.098		3.42	4.77		8.19	11.15
8448	1-1/2" iron pipe size	G		82	.098		3.98	4.77		8.75	11.80
8449	2" iron pipe size	G		80	.100		5.25	4.89		10.14	13.35
8450	2-1/2" iron pipe size	G		80	.100		6.85	4.89		11.74	15.10
8456	Rubber insulation tape, 1/8" x 2" x 30'	G				Ea.	12.05			12.05	13.25

22 11 13 – Facility Water Distribution Piping

		Crew	Daily Output	Labor-Hours	Unit	Material	2012 Bare Costs Labor	Equipment	Total	Total Incl O&P
22 11 13.14 Pipe, Brass										
0010	**PIPE, BRASS**, Plain end									
0900	Field threaded, coupling & clevis hanger assembly 10' O.C.									
0920	Regular weight									
1120	1/2" diameter	1 Plum	48	.167	L.F.	6.35	9.10		15.45	20.50
1140	3/4" diameter		46	.174		8.45	9.50		17.95	23.50
1160	1" diameter		43	.186		12.25	10.15		22.40	29
1180	1-1/4" diameter	Q-1	72	.222		18.65	10.90		29.55	37
1200	1-1/2" diameter		65	.246		22.50	12.10		34.60	43
1220	2" diameter		53	.302		31	14.85		45.85	57
22 11 13.23 Pipe/Tube, Copper										
0010	**PIPE/TUBE, COPPER**, Solder joints									
1000	Type K tubing, couplings & clevis hanger assemblies 10' O.C.									
1100	1/4" diameter	1 Plum	84	.095	L.F.	4.94	5.20		10.14	13.30
1200	1" diameter		66	.121		12.85	6.60		19.45	24
1260	2" diameter		40	.200		32	10.90		42.90	51.50
2000	Type L tubing, couplings & clevis hanger assemblies 10' O.C.									
2100	1/4" diameter	1 Plum	88	.091	L.F.	2.69	4.96		7.65	10.45
2120	3/8" diameter		84	.095		3.73	5.20		8.93	11.95
2140	1/2" diameter		81	.099		4.30	5.40		9.70	12.90
2160	5/8" diameter		79	.101		6.25	5.55		11.80	15.25
2180	3/4" diameter		76	.105		6.60	5.75		12.35	15.90
2200	1" diameter		68	.118		9.85	6.40		16.25	20.50
2220	1-1/4" diameter		58	.138		13.45	7.55		21	26
2240	1-1/2" diameter		52	.154		17.25	8.40		25.65	31.50
2260	2" diameter		42	.190		27	10.40		37.40	45.50
2280	2-1/2" diameter	Q-1	62	.258		42	12.70		54.70	65
2300	3" diameter		56	.286		57	14.05		71.05	84
2320	3-1/2" diameter		43	.372		81	18.30		99.30	117
2340	4" diameter		39	.410		99	20		119	140
2360	5" diameter		34	.471		211	23		234	267
2380	6" diameter	Q-2	40	.600		300	30.50		330.50	375
2400	8" diameter	"	36	.667		485	34		519	585
2410	For other than full hard temper, add					21%				
2590	For silver solder, add						15%			
4000	Type DWV tubing, couplings & clevis hanger assemblies 10' O.C.									
4100	1-1/4" diameter	1 Plum	60	.133	L.F.	11.95	7.30		19.25	24
4120	1-1/2" diameter		54	.148		14.95	8.10		23.05	28.50
4140	2" diameter		44	.182		21	9.95		30.95	38
4160	3" diameter	Q-1	58	.276		39	13.55		52.55	63
4180	4" diameter		40	.400		70	19.65		89.65	107
4200	5" diameter		36	.444		191	22		213	243
4220	6" diameter	Q-2	42	.571		280	29		309	355
22 11 13.44 Pipe, Steel										
0010	**PIPE, STEEL**									
0012	The steel pipe in this section does not include fittings such as ells, tees									
0014	For fittings either add a % (usually 25 to 35%) or see									
0015	the Mechanical or Plumbing Cost Data									
0020	All pipe sizes are to Spec. A-53 unless noted otherwise R221113-50									
0050	Schedule 40, threaded, with couplings, and clevis hanger									
0060	assemblies sized for covering, 10' O.C.									
0540	Black, 1/4" diameter	1 Plum	66	.121	L.F.	6	6.60		12.60	16.60
0550	3/8" diameter		65	.123		6.75	6.70		13.45	17.60

22 11 13 – Facility Water Distribution Piping

22 11 13.44 Pipe, Steel

		Crew	Daily Output	Labor-Hours	Unit	Material	2012 Bare Costs Labor	2012 Bare Costs Equipment	Total	Total Incl O&P
0560	1/2" diameter	1 Plum	63	.127	L.F.	4.38	6.95		11.33	15.25
0570	3/4" diameter		61	.131		5.15	7.15		12.30	16.45
0580	1" diameter		53	.151		7.55	8.25		15.80	21
0590	1-1/4" diameter	Q-1	89	.180		9.50	8.85		18.35	24
0600	1-1/2" diameter		80	.200		11.15	9.85		21	27
0610	2" diameter		64	.250		14.90	12.30		27.20	35
0620	2-1/2" diameter		50	.320		23	15.75		38.75	48.50
0630	3" diameter		43	.372		29.50	18.30		47.80	60
0640	3-1/2" diameter		40	.400		40	19.65		59.65	73.50
0650	4" diameter		36	.444		43.50	22		65.50	80.50
1280	All pipe sizes are to Spec. A-53 unless noted otherwise									
1281	Schedule 40, threaded, with couplings and clevis hanger									
1282	assemblies sized for covering, 10' O. C.									
1290	Galvanized, 1/4" diameter	1 Plum	66	.121	L.F.	8.10	6.60		14.70	18.90
1300	3/8" diameter		65	.123		8.95	6.70		15.65	20
1310	1/2" diameter		63	.127		6.35	6.95		13.30	17.45
1320	3/4" diameter		61	.131		7.60	7.15		14.75	19.15
1330	1" diameter		53	.151		10.40	8.25		18.65	24
1340	1-1/4" diameter	Q-1	89	.180		13.30	8.85		22.15	28
1350	1-1/2" diameter		80	.200		15.65	9.85		25.50	32
1360	2" diameter		64	.250		21	12.30		33.30	41.50
1370	2-1/2" diameter		50	.320		34.50	15.75		50.25	61.50
1380	3" diameter		43	.372		44	18.30		62.30	76
1390	3-1/2" diameter		40	.400		55.50	19.65		75.15	90.50
1400	4" diameter		36	.444		63.50	22		85.50	103
2000	Welded, sch. 40, on yoke & roll hanger assy's, sized for covering, 10' O.C.									
2040	Black, 1" diameter	Q-15	93	.172	L.F.	6.80	8.45	.40	15.65	20.50
2070	2" diameter		61	.262		12.35	12.90	.61	25.86	33.50
2090	3" diameter		43	.372		25.50	18.30	.86	44.66	57
2110	4" diameter		37	.432		35	21.50	1	57.50	71.50
2120	5" diameter		32	.500		38	24.50	1.16	63.66	80
2130	6" diameter	Q-16	36	.667		50.50	34	1.03	85.53	108
2140	8" diameter		29	.828		55	42	1.27	98.27	125
2150	10" diameter		24	1		64	51	1.54	116.54	149
2160	12" diameter		19	1.263		87.50	64.50	1.95	153.95	195

22 11 13.48 Pipe, Fittings and Valves, Steel, Grooved-Joint

		Crew	Daily Output	Labor-Hours	Unit	Material	2012 Bare Costs Labor	2012 Bare Costs Equipment	Total	Total Incl O&P
0010	**PIPE, FITTINGS AND VALVES, STEEL, GROOVED-JOINT**									
0012	Fittings are ductile iron. Steel fittings noted.									
0020	Pipe includes coupling & clevis type hanger assemblies, 10' O.C.									
1000	Schedule 40, black									
1040	3/4" diameter	1 Plum	71	.113	L.F.	6.30	6.15		12.45	16.20
1050	1" diameter		63	.127		6.30	6.95		13.25	17.40
1060	1-1/4" diameter		58	.138		8	7.55		15.55	20
1070	1-1/2" diameter		51	.157		9.25	8.55		17.80	23
1080	2" diameter		40	.200		11.85	10.90		22.75	29.50
1090	2-1/2" diameter	Q-1	57	.281		20.50	13.80		34.30	43.50
1100	3" diameter		50	.320		26	15.75		41.75	52
1110	4" diameter		45	.356		35	17.50		52.50	65
1120	5" diameter		37	.432		38.50	21.50		60	74.50
1130	6" diameter	Q-2	42	.571		53	29		82	103
1800	Galvanized									
1840	3/4" diameter	1 Plum	71	.113	L.F.	8.70	6.15		14.85	18.85

22 11 13 – Facility Water Distribution Piping

22 11 13.48 Pipe, Fittings and Valves, Steel, Grooved-Joint		Crew	Daily Output	Labor-Hours	Unit	Material	2012 Bare Costs Labor	Equipment	Total	Total Incl O&P
1850	1" diameter	1 Plum	63	.127	L.F.	8.25	6.95		15.20	19.55
1860	1-1/4" diameter		58	.138		10.55	7.55		18.10	23
1870	1-1/2" diameter		51	.157		12.25	8.55		20.80	26.50
1880	2" diameter		40	.200		15.80	10.90		26.70	34
1890	2-1/2" diameter	Q-1	57	.281		30	13.80		43.80	54
1900	3" diameter		50	.320		38.50	15.75		54.25	65.50
1910	4" diameter		45	.356		53.50	17.50		71	85.50
1920	5" diameter		37	.432		58	21.50		79.50	96
1930	6" diameter	Q-2	42	.571		65.50	29		94.50	116
3990	Fittings: coupling material required at joints not incl. in fitting price.									
3994	Add 1 selected coupling, material only, per joint for installed price.									
4000	Elbow, 90° or 45°, painted									
4030	3/4" diameter	1 Plum	50	.160	Ea.	46	8.75		54.75	63.50
4040	1" diameter		50	.160		24.50	8.75		33.25	40
4050	1-1/4" diameter		40	.200		24.50	10.90		35.40	43.50
4060	1-1/2" diameter		33	.242		24.50	13.25		37.75	47
4070	2" diameter		25	.320		24.50	17.45		41.95	53.50
4080	2-1/2" diameter	Q-1	40	.400		24.50	19.65		44.15	56.50
4090	3" diameter		33	.485		43	24		67	83.50
4100	4" diameter		25	.640		47	31.50		78.50	99
4110	5" diameter		20	.800		112	39.50		151.50	184
4120	6" diameter	Q-2	25	.960		132	49		181	219
4250	For galvanized elbows, add					26%				
4690	Tee, painted									
4700	3/4" diameter	1 Plum	38	.211	Ea.	49	11.50		60.50	71.50
4740	1" diameter		33	.242		38	13.25		51.25	61.50
4750	1-1/4" diameter		27	.296		38	16.20		54.20	66
4760	1-1/2" diameter		22	.364		38	19.85		57.85	71.50
4770	2" diameter		17	.471		38	25.50		63.50	80.50
4780	2-1/2" diameter	Q-1	27	.593		38	29		67	85.50
4790	3" diameter		22	.727		52	36		88	111
4800	4" diameter		17	.941		79.50	46.50		126	158
4810	5" diameter		13	1.231		185	60.50		245.50	295
4820	6" diameter	Q-2	17	1.412		214	72		286	345
4900	For galvanized tees, add					24%				
4906	Couplings, rigid style, painted									
4908	1" diameter	1 Plum	100	.080	Ea.	18.40	4.37		22.77	26.50
4909	1-1/4" diameter		100	.080		18.40	4.37		22.77	26.50
4910	1-1/2" diameter		67	.119		18.40	6.50		24.90	30
4912	2" diameter		50	.160		20.50	8.75		29.25	35.50
4914	2-1/2" diameter	Q-1	80	.200		23.50	9.85		33.35	40.50
4916	3" diameter		67	.239		27	11.75		38.75	47.50
4918	4" diameter		50	.320		38	15.75		53.75	65.50
4920	5" diameter		40	.400		49	19.65		68.65	83.50
4922	6" diameter	Q-2	50	.480		65	24.50		89.50	109
4940	Flexible, standard, painted									
4950	3/4" diameter	1 Plum	100	.080	Ea.	13.40	4.37		17.77	21.50
4960	1" diameter		100	.080		13.40	4.37		17.77	21.50
4970	1-1/4" diameter		80	.100		17.60	5.45		23.05	27.50
4980	1-1/2" diameter		67	.119		19.20	6.50		25.70	31
4990	2" diameter		50	.160		20.50	8.75		29.25	35.50
5000	2-1/2" diameter	Q-1	80	.200		24	9.85		33.85	41.50
5010	3" diameter		67	.239		27	11.75		38.75	47

22 11 Facility Water Distribution

22 11 13 – Facility Water Distribution Piping

22 11 13.48 Pipe, Fittings and Valves, Steel, Grooved-Joint

		Crew	Daily Output	Labor-Hours	Unit	Material	2012 Bare Costs Labor	Equipment	Total	Total Incl O&P
5020	3-1/2" diameter	Q-1	57	.281	Ea.	38	13.80		51.80	63
5030	4" diameter		50	.320		38.50	15.75		54.25	66
5040	5" diameter	↓	40	.400		58.50	19.65		78.15	94
5050	6" diameter	Q-2	50	.480	↓	69.50	24.50		94	114

22 11 13.64 Pipe, Stainless Steel

		Crew	Daily Output	Labor-Hours	Unit	Material	2012 Bare Costs Labor	Equipment	Total	Total Incl O&P
0010	**PIPE, STAINLESS STEEL**									
3500	Threaded, couplings and clevis hanger assemblies, 10' O.C.									
3520	Schedule 40, type 304									
3540	1/4" diameter	1 Plum	54	.148	L.F.	6.65	8.10		14.75	19.50
3550	3/8" diameter		53	.151		8.45	8.25		16.70	22
3560	1/2" diameter		52	.154		7.55	8.40		15.95	21
3580	1" diameter	↓	45	.178		16.15	9.70		25.85	32.50
3610	2" diameter	Q-1	57	.281		35.50	13.80		49.30	60
3640	4" diameter	Q-2	51	.471		101	24		125	147
3740	For small quantities, add				↓	10%				
4250	Schedule 40, type 316									
4290	1/4" diameter	1 Plum	54	.148	L.F.	8	8.10		16.10	21
4300	3/8" diameter		53	.151		9.10	8.25		17.35	22.50
4310	1/2" diameter		52	.154		11.95	8.40		20.35	26
4320	3/4" diameter		51	.157		14.65	8.55		23.20	29
4330	1" diameter	↓	45	.178		21	9.70		30.70	37.50
4360	2" diameter	Q-1	57	.281		46.50	13.80		60.30	72.50
4390	4" diameter	Q-2	51	.471		150	24		174	201
4490	For small quantities, add				↓	10%				

22 11 13.74 Pipe, Plastic

		Crew	Daily Output	Labor-Hours	Unit	Material	2012 Bare Costs Labor	Equipment	Total	Total Incl O&P
0010	**PIPE, PLASTIC**									
1800	PVC, couplings 10' O.C., clevis hanger assemblies, 3 per 10'									
1820	Schedule 40									
1860	1/2" diameter	1 Plum	54	.148	L.F.	2.83	8.10		10.93	15.30
1870	3/4" diameter		51	.157		3.17	8.55		11.72	16.40
1880	1" diameter		46	.174		3.59	9.50		13.09	18.25
1890	1-1/4" diameter		42	.190		4.33	10.40		14.73	20.50
1900	1-1/2" diameter	↓	36	.222		4.59	12.15		16.74	23.50
1910	2" diameter	Q-1	59	.271		5.75	13.35		19.10	26.50
1920	2-1/2" diameter		56	.286		8.40	14.05		22.45	30.50
1930	3" diameter		53	.302		10.70	14.85		25.55	34.50
1940	4" diameter		48	.333		14.35	16.40		30.75	40.50
1950	5" diameter		43	.372		20	18.30		38.30	49.50
1960	6" diameter	↓	39	.410	↓	26	20		46	59.50
4100	DWV type, schedule 40, couplings 10' O.C., clevis hanger assy's, 3 per 10'									
4210	ABS, schedule 40, foam core type									
4212	Plain end black									
4214	1-1/2" diameter	1 Plum	39	.205	L.F.	2.97	11.20		14.17	20
4216	2" diameter	Q-1	62	.258		3.44	12.70		16.14	23
4218	3" diameter		56	.286		6.15	14.05		20.20	28
4220	4" diameter		51	.314		8.55	15.40		23.95	33
4222	6" diameter	↓	42	.381	↓	17.90	18.70		36.60	47.50
4240	To delete coupling & hangers, subtract									
4244	1-1/2" diam. to 6" diam.					43%	48%			
4400	PVC									
4410	1-1/4" diameter	1 Plum	42	.190	L.F.	3.21	10.40		13.61	19.25
4420	1-1/2" diameter	"	36	.222	↓	3.01	12.15		15.16	21.50

22 11 13.74 Pipe, Plastic	Crew	Daily Output	Labor-Hours	Unit	Material	2012 Bare Costs Labor	Equipment	Total	Total Incl O&P
4460	2" diameter	Q-1	59	.271	L.F.	3.46	13.35		16.81	24
4470	3" diameter		53	.302		6.15	14.85		21	29.50
4480	4" diameter		48	.333		8.30	16.40		24.70	33.50
4490	6" diameter		39	.410		15.65	20		35.65	47.50
5300	CPVC, socket joint, couplings 10' O.C., clevis hanger assemblies, 3 per 10'									
5302	Schedule 40									
5304	1/2" diameter	1 Plum	54	.148	L.F.	3.58	8.10		11.68	16.15
5305	3/4" diameter		51	.157		4.17	8.55		12.72	17.50
5306	1" diameter		46	.174		5.25	9.50		14.75	20
5307	1-1/4" diameter		42	.190		6.60	10.40		17	23
5308	1-1/2" diameter		36	.222		7.60	12.15		19.75	26.50
5309	2" diameter	Q-1	59	.271		9.35	13.35		22.70	30.50
5310	2-1/2" diameter		56	.286		15.65	14.05		29.70	38
5311	3" diameter		53	.302		18.85	14.85		33.70	43
5360	CPVC, threaded, couplings 10' O.C., clevis hanger assemblies, 3 per 10'									
5380	Schedule 40									
5460	1/2" diameter	1 Plum	54	.148	L.F.	4.34	8.10		12.44	16.95
5470	3/4" diameter		51	.157		5.50	8.55		14.05	18.95
5480	1" diameter		46	.174		6.60	9.50		16.10	21.50
5490	1-1/4" diameter		42	.190		7.65	10.40		18.05	24
5500	1-1/2" diameter		36	.222		8.50	12.15		20.65	27.50
5510	2" diameter	Q-1	59	.271		10.45	13.35		23.80	31.50
5520	2-1/2" diameter		56	.286		16.80	14.05		30.85	39.50
5530	3" diameter		53	.302		20.50	14.85		35.35	45
7280	PEX, flexible, no couplings or hangers									
7282	Note: For labor costs add 25% to the couplings and fittings labor total.									
7285	For fittings see section 23 83 16.10 7000									
7300	Non-barrier type, hot/cold tubing rolls									
7310	1/4" diameter x 100'				L.F.	.46			.46	.51
7350	3/8" diameter x 100'					.51			.51	.56
7360	1/2" diameter x 100'					.57			.57	.63
7370	1/2" diameter x 500'					.57			.57	.63
7380	1/2" diameter x 1000'					.57			.57	.63
7400	3/4" diameter x 100'					1.05			1.05	1.16
7410	3/4" diameter x 500'					1.05			1.05	1.16
7420	3/4" diameter x 1000'					1.05			1.05	1.16
7460	1" diameter x 100'					1.80			1.80	1.98
7470	1" diameter x 300'					1.80			1.80	1.98
7480	1" diameter x 500'					1.78			1.78	1.96
7500	1-1/4" diameter x 100'					3.04			3.04	3.34
7510	1-1/4" diameter x 300'					3.04			3.04	3.34
7540	1-1/2" diameter x 100'					4.15			4.15	4.57
7550	1-1/2" diameter x 300'					4.15			4.15	4.57
7596	Most sizes available in red or blue									
7700	Non-barrier type, hot/cold tubing straight lengths									
7710	1/2" diameter x 20'				L.F.	.57			.57	.63
7750	3/4" diameter x 20'					1.05			1.05	1.16
7760	1" diameter x 20'					1.80			1.80	1.98
7770	1-1/4" diameter x 20'					3.04			3.04	3.34
7780	1-1/2" diameter x 20'					4.15			4.15	4.57
7790	2" diameter					8.10			8.10	8.90
7796	Most sizes available in red or blue									

22 11 19 – Domestic Water Piping Specialties

22 11 19.10 Flexible Connectors	Crew	Daily Output	Labor-Hours	Unit	Material	2012 Bare Costs Labor	Equipment	Total	Total Incl O&P
0010 **FLEXIBLE CONNECTORS**, Corrugated, 7/8" O.D., 1/2" I.D.									
0050 Gas, seamless brass, steel fittings									
0200 12" long	1 Plum	36	.222	Ea.	16.30	12.15		28.45	36
0220 18" long		36	.222		20	12.15		32.15	40.50
0240 24" long		34	.235		24	12.85		36.85	46
0280 36" long		32	.250		28.50	13.65		42.15	52
0340 60" long		30	.267		43	14.55		57.55	69.50
2000 Water, copper tubing, dielectric separators									
2100 12" long	1 Plum	36	.222	Ea.	15.95	12.15		28.10	36
2260 24" long	"	34	.235	"	23.50	12.85		36.35	45.50

22 11 19.14 Flexible Metal Hose	Crew	Daily Output	Labor-Hours	Unit	Material	Labor	Equipment	Total	Total Incl O&P
0010 **FLEXIBLE METAL HOSE**, Connectors, standard lengths									
0100 Bronze braided, bronze ends									
0120 3/8" diameter x 12"	1 Stpi	26	.308	Ea.	18.85	17.15		36	47
0160 3/4" diameter x 12"		20	.400		26.50	22.50		49	62.50
0180 1" diameter x 18"		19	.421		34	23.50		57.50	73
0200 1-1/2" diameter x 18"		13	.615		53.50	34.50		88	111
0220 2" diameter x 18"		11	.727		64.50	40.50		105	132

22 11 19.26 Pressure Regulators	Crew	Daily Output	Labor-Hours	Unit	Material	Labor	Equipment	Total	Total Incl O&P
0010 **PRESSURE REGULATORS**									
3000 Steam, high capacity, bronze body, stainless steel trim									
3020 Threaded, 1/2" diameter	1 Stpi	24	.333	Ea.	1,550	18.55		1,568.55	1,750
3030 3/4" diameter		24	.333		1,550	18.55		1,568.55	1,750
3040 1" diameter		19	.421		1,750	23.50		1,773.50	1,950
3060 1-1/4" diameter		15	.533		1,925	29.50		1,954.50	2,175
3080 1-1/2" diameter		13	.615		2,200	34.50		2,234.50	2,475
3100 2" diameter		11	.727		2,700	40.50		2,740.50	3,025
3120 2-1/2" diameter	Q-5	12	1.333		3,375	67		3,442	3,825
3140 3" diameter	"	11	1.455		3,850	73		3,923	4,350
3500 Flanged connection, iron body, 125 lb. W.S.P.									
3520 3" diameter	Q-5	11	1.455	Ea.	4,225	73		4,298	4,750
3540 4" diameter	"	5	3.200	"	5,325	160		5,485	6,100

22 11 19.38 Water Supply Meters	Crew	Daily Output	Labor-Hours	Unit	Material	Labor	Equipment	Total	Total Incl O&P
0010 **WATER SUPPLY METERS**									
2000 Domestic/commercial, bronze									
2020 Threaded									
2060 5/8" diameter, to 20 GPM	1 Plum	16	.500	Ea.	46	27.50		73.50	91.50
2080 3/4" diameter, to 30 GPM		14	.571		83.50	31		114.50	139
2100 1" diameter, to 50 GPM		12	.667		127	36.50		163.50	195
2300 Threaded/flanged									
2340 1-1/2" diameter, to 100 GPM	1 Plum	8	1	Ea.	310	54.50		364.50	425
2360 2" diameter, to 160 GPM	"	6	1.333	"	420	73		493	570
2600 Flanged, compound									
2640 3" diameter, 320 GPM	Q-1	3	5.333	Ea.	2,850	262		3,112	3,550
2660 4" diameter, to 500 GPM		1.50	10.667		4,575	525		5,100	5,825
2680 6" diameter, to 1,000 GPM		1	16		7,300	785		8,085	9,200
2700 8" diameter, to 1,800 GPM		.80	20		11,500	985		12,485	14,100

22 11 19.42 Backflow Preventers

	Crew	Daily Output	Labor-Hours	Unit	Material	Labor	Equipment	Total	Total Incl O&P
0010 **BACKFLOW PREVENTERS**, Includes valves									
0020 and four test cocks, corrosion resistant, automatic operation									
4000 Reduced pressure principle									

22 11 19 – Domestic Water Piping Specialties

22 11 19.42 Backflow Preventers

		Crew	Daily Output	Labor-Hours	Unit	Material	2012 Bare Costs Labor	Equipment	Total	Total Incl O&P
4100	Threaded, bronze, valves are ball									
4120	3/4" pipe size	1 Plum	16	.500	Ea.	420	27.50		447.50	505
4140	1" pipe size		14	.571		455	31		486	545
4150	1-1/4" pipe size		12	.667		835	36.50		871.50	975
4160	1-1/2" pipe size		10	.800		915	43.50		958.50	1,075
4180	2" pipe size	↓	7	1.143	↓	1,025	62.50		1,087.50	1,225
5000	Flanged, bronze, valves are OS&Y									
5060	2-1/2" pipe size	Q-1	5	3.200	Ea.	6,100	157		6,257	6,950
5080	3" pipe size		4.50	3.556		7,100	175		7,275	8,100
5100	4" pipe size	↓	3	5.333		8,325	262		8,587	9,575
5120	6" pipe size	Q-2	3	8	↓	13,300	410		13,710	15,200
5600	Flanged, iron, valves are OS&Y									
5660	2-1/2" pipe size	Q-1	5	3.200	Ea.	3,150	157		3,307	3,700
5680	3" pipe size		4.50	3.556		3,425	175		3,600	4,050
5700	4" pipe size	↓	3	5.333		3,800	262		4,062	4,575
5720	6" pipe size	Q-2	3	8		5,500	410		5,910	6,675
5740	8" pipe size		2	12		9,675	610		10,285	11,500
5760	10" pipe size	↓	1	24	↓	12,900	1,225		14,125	16,100

22 11 19.50 Vacuum Breakers

		Crew	Daily Output	Labor-Hours	Unit	Material	2012 Bare Costs Labor	Equipment	Total	Total Incl O&P
0010	**VACUUM BREAKERS**									
0013	See also backflow preventers Section 22 11 19.42									
1000	Anti-siphon continuous pressure type									
1010	Max. 150 PSI - 210°F									
1020	Bronze body									
1030	1/2" size	1 Stpi	24	.333	Ea.	171	18.55		189.55	216
1040	3/4" size		20	.400		171	22.50		193.50	222
1050	1" size		19	.421		177	23.50		200.50	230
1060	1-1/4" size		15	.533		350	29.50		379.50	430
1070	1-1/2" size		13	.615		430	34.50		464.50	520
1080	2" size	↓	11	.727	↓	445	40.50		485.50	545
1200	Max. 125 PSI with atmospheric vent									
1210	Brass, in-line construction									
1220	1/4" size	1 Stpi	24	.333	Ea.	68	18.55		86.55	103
1230	3/8" size	"	24	.333		68	18.55		86.55	103
1260	For polished chrome finish, add				↓	13%				
2000	Anti-siphon, non-continuous pressure type									
2010	Hot or cold water 125 PSI - 210°F									
2020	Bronze body									
2030	1/4" size	1 Stpi	24	.333	Ea.	43.50	18.55		62.05	75.50
2040	3/8" size		24	.333		43.50	18.55		62.05	75.50
2050	1/2" size		24	.333		49	18.55		67.55	82
2060	3/4" size		20	.400		58	22.50		80.50	97.50
2070	1" size		19	.421		91	23.50		114.50	136
2080	1-1/4" size		15	.533		159	29.50		188.50	220
2090	1-1/2" size		13	.615		187	34.50		221.50	258
2100	2" size		11	.727		291	40.50		331.50	380
2110	2-1/2" size		8	1		840	55.50		895.50	1,000
2120	3" size	↓	6	1.333	↓	1,125	74.50		1,199.50	1,325
2150	For polished chrome finish, add					50%				

22 11 19.54 Water Hammer Arresters/Shock Absorbers

		Crew	Daily Output	Labor-Hours	Unit	Material	2012 Bare Costs Labor	Equipment	Total	Total Incl O&P
0010	**WATER HAMMER ARRESTERS/SHOCK ABSORBERS**									
0490	Copper									

22 11 Facility Water Distribution

22 11 19 – Domestic Water Piping Specialties

22 11 19.54 Water Hammer Arresters/Shock Absorbers	Crew	Daily Output	Labor-Hours	Unit	Material	2012 Bare Costs Labor	Equipment	Total	Total Incl O&P
0500	3/4" male I.P.S. For 1 to 11 fixtures	1 Plum	12	.667	Ea.	21.50	36.50		58	79
0600	1" male I.P.S. For 12 to 32 fixtures		8	1		34	54.50		88.50	120
0700	1-1/4" male I.P.S. For 33 to 60 fixtures		8	1		44	54.50		98.50	131
0800	1-1/2" male I.P.S. For 61 to 113 fixtures		8	1		63	54.50		117.50	152
0900	2" male I.P.S. For 114 to 154 fixtures		8	1		92	54.50		146.50	184
1000	2-1/2" male I.P.S. For 155 to 330 fixtures		4	2		285	109		394	480

22 11 19.64 Hydrants

		Crew	Daily Output	Labor-Hours	Unit	Material	Labor	Equipment	Total	Total Incl O&P
0010	**HYDRANTS**									
0050	Wall type, moderate climate, bronze, encased									
0200	3/4" IPS connection	1 Plum	16	.500	Ea.	650	27.50		677.50	755
0300	1" IPS connection	"	14	.571		745	31		776	865
0500	Anti-siphon type, 3/4" connection					560			560	615
1000	Non-freeze, bronze, exposed									
1100	3/4" IPS connection, 4" to 9" thick wall	1 Plum	14	.571	Ea.	440	31		471	530
1120	10" to 14" thick wall		12	.667		475	36.50		511.50	575
1140	15" to 19" thick wall		12	.667		530	36.50		566.50	640
1160	20" to 24" thick wall		10	.800		575	43.50		618.50	695
1200	For 1" IPS connection, add					15%	10%			
1240	For 3/4" adapter type vacuum breaker, add				Ea.	55.50			55.50	61
1280	For anti-siphon type, add				"	115			115	126
2000	Non-freeze bronze, encased, anti-siphon type									
2100	3/4" IPS connection, 5" to 9" thick wall	1 Plum	14	.571	Ea.	1,125	31		1,156	1,275
2140	15" to 19" thick wall	"	12	.667	"	1,200	36.50		1,236.50	1,375
3000	Ground box type, bronze frame, 3/4" IPS connection									
3080	Non-freeze, all bronze, polished face, set flush									
3100	2 feet depth of bury	1 Plum	8	1	Ea.	830	54.50		884.50	995
3180	6 feet depth of bury		7	1.143		1,075	62.50		1,137.50	1,300
3220	8 feet depth of bury		5	1.600		1,200	87.50		1,287.50	1,450
3400	For 1" IPS connection, add					15%	10%			
3550	For 2" connection, add					445%	24%			
3600	For tapped drain port in box, add					75			75	82.50
5000	Moderate climate, all bronze, polished face									
5020	and scoriated cover, set flush									
5100	3/4" IPS connection	1 Plum	16	.500	Ea.	575	27.50		602.50	670
5120	1" IPS connection	"	14	.571		705	31		736	820
5200	For tapped drain port in box, add					75			75	82.50

22 11 23 – Domestic Water Pumps

22 11 23.10 General Utility Pumps

		Crew	Daily Output	Labor-Hours	Unit	Material	Labor	Equipment	Total	Total Incl O&P
0010	**GENERAL UTILITY PUMPS**									
2000	Single stage									
3000	Double suction,									
3190	75 HP, to 2500 GPM	Q-3	.28	114	Ea.	19,800	5,950		25,750	30,800
3220	100 HP, to 3000 GPM		.26	123		25,200	6,400		31,600	37,400
3240	150 HP, to 4000 GPM		.24	133		35,300	6,925		42,225	49,400

22 13 16.20 Pipe, Cast Iron

	Crew	Daily Output	Labor-Hours	Unit	Material	2012 Bare Costs Labor	Equipment	Total	Total Incl O&P
0010 **PIPE, CAST IRON**, Soil, on clevis hanger assemblies, 5' O.C.									
0020 Single hub, service wt., lead & oakum joints 10' O.C.									
2120 2" diameter	Q-1	63	.254	L.F.	8.50	12.50		21	28
2140 3" diameter		60	.267		11.95	13.10		25.05	33
2160 4" diameter	↓	55	.291		15.45	14.30		29.75	38.50
2180 5" diameter	Q-2	76	.316		21	16.10		37.10	48
2200 6" diameter	"	73	.329		26.50	16.75		43.25	54.50
2220 8" diameter	Q-3	59	.542		41	28		69	87.50
2240 10" diameter		54	.593		66.50	31		97.50	120
2260 12" diameter	↓	48	.667		94	34.50		128.50	156
2320 For service weight, double hub, add					10%				
2340 For extra heavy, single hub, add					48%	4%			
2360 For extra heavy, double hub, add				↓	71%	4%			
2400 Lead for caulking, (1#/dia in)	Q-1	160	.100	Lb.	1.04	4.92		5.96	8.55
2420 Oakum for caulking, (1/8#/dia in)	"	40	.400	"	4.20	19.65		23.85	34
4000 No hub, couplings 10' O.C.									
4100 1-1/2" diameter	Q-1	71	.225	L.F.	8.45	11.10		19.55	26
4120 2" diameter		67	.239		8.65	11.75		20.40	27.50
4160 4" diameter	↓	58	.276	↓	15.70	13.55		29.25	38

22 13 16.50 Shower Drains

	Crew	Daily Output	Labor-Hours	Unit	Material	2012 Bare Costs Labor	Equipment	Total	Total Incl O&P
0010 **SHOWER DRAINS**									
2780 Shower, with strainer, uniform diam. trap, bronze top									
2800 2" and 3" pipe size	Q-1	8	2	Ea.	360	98.50		458.50	545
2820 4" pipe size	"	7	2.286		400	112		512	605
2840 For galvanized body, add				↓	160			160	176

22 13 16.60 Traps

	Crew	Daily Output	Labor-Hours	Unit	Material	2012 Bare Costs Labor	Equipment	Total	Total Incl O&P
0010 **TRAPS**									
0030 Cast iron, service weight									
0050 Running P trap, without vent									
1100 2"	Q-1	16	1	Ea.	122	49		171	208
1140 3"		14	1.143		122	56		178	219
1150 4"	↓	13	1.231		122	60.50		182.50	225
1160 6"	Q-2	17	1.412	↓	530	72		602	695
1180 Running trap, single hub, with vent									
2080 3" pipe size, 3" vent	Q-1	14	1.143	Ea.	96.50	56		152.50	191
2120 4" pipe size, 4" vent	"	13	1.231		131	60.50		191.50	235
2300 For double hub, vent, add					10%	20%			
3000 P trap, B&S, 2" pipe size	Q-1	16	1		29	49		78	106
3040 3" pipe size	"	14	1.143	↓	43	56		99	132
3350 Deep seal trap, B&S									
3400 1-1/4" pipe size	Q-1	14	1.143	Ea.	46	56		102	135
3410 1-1/2" pipe size		14	1.143		46	56		102	135
3420 2" pipe size		14	1.143		41	56		97	130
3440 3" pipe size	↓	12	1.333	↓	53	65.50		118.50	158
4700 Copper, drainage, drum trap									
4800 3" x 5" solid, 1-1/2" pipe size	1 Plum	16	.500	Ea.	256	27.50		283.50	325
4840 3" x 6" swivel, 1-1/2" pipe size	"	16	.500	"	405	27.50		432.50	485
5100 P trap, standard pattern									
5200 1-1/4" pipe size	1 Plum	18	.444	Ea.	189	24.50		213.50	245
5240 1-1/2" pipe size		17	.471		183	25.50		208.50	240
5260 2" pipe size		15	.533		282	29		311	355
5280 3" pipe size	↓	11	.727	↓	675	39.50		714.50	805

22 13 Facility Sanitary Sewerage

22 13 16 – Sanitary Waste and Vent Piping

22 13 16.60 Traps

		Crew	Daily Output	Labor-Hours	Unit	Material	2012 Bare Costs Labor	2012 Bare Costs Equipment	Total	Total Incl O&P
5340	With cleanout, swivel joint and slip joint									
5360	1-1/4" pipe size	1 Plum	18	.444	Ea.	134	24.50		158.50	184
5400	1-1/2" pipe size	"	17	.471	"	278	25.50		303.50	345

22 13 16.80 Vent Flashing and Caps

		Crew	Daily Output	Labor-Hours	Unit	Material	2012 Bare Costs Labor	2012 Bare Costs Equipment	Total	Total Incl O&P
0010	**VENT FLASHING AND CAPS**									
0120	Vent caps									
0140	Cast iron									
0180	2-1/2" - 3-5/8" pipe	1 Plum	21	.381	Ea.	40	21		61	75.50
0190	4" - 4-1/8" pipe	"	19	.421	"	48	23		71	87.50
0900	Vent flashing									
1000	Aluminum with lead ring									
1020	1-1/4" pipe	1 Plum	20	.400	Ea.	11.20	22		33.20	45.50
1030	1-1/2" pipe		20	.400		13.40	22		35.40	47.50
1040	2" pipe		18	.444		12.05	24.50		36.55	50
1050	3" pipe		17	.471		13.30	25.50		38.80	53.50
1060	4" pipe	↓	16	.500	↓	16.05	27.50		43.55	58.50
1350	Copper with neoprene ring									
1400	1-1/4" pipe	1 Plum	20	.400	Ea.	19.95	22		41.95	55
1430	1-1/2" pipe		20	.400		19.95	22		41.95	55
1440	2" pipe		18	.444		21	24.50		45.50	59.50
1450	3" pipe		17	.471		24.50	25.50		50	66
1460	4" pipe	↓	16	.500	↓	27	27.50		54.50	71

22 13 19 – Sanitary Waste Piping Specialties

22 13 19.13 Sanitary Drains

		Crew	Daily Output	Labor-Hours	Unit	Material	2012 Bare Costs Labor	2012 Bare Costs Equipment	Total	Total Incl O&P
0010	**SANITARY DRAINS**									
0400	Deck, auto park, C.I., 13" top									
0440	3", 4", 5", and 6" pipe size	Q-1	8	2	Ea.	1,225	98.50		1,323.50	1,500
0480	For galvanized body, add				"	660			660	725
2000	Floor, medium duty, C.I., deep flange, 7" dia top									
2040	2" and 3" pipe size	Q-1	12	1.333	Ea.	171	65.50		236.50	287
2080	For galvanized body, add					82			82	90
2120	With polished bronze top				↓	267			267	293
2400	Heavy duty, with sediment bucket, C.I., 12" dia. loose grate									
2420	2", 3", 4", 5", and 6" pipe size	Q-1	9	1.778	Ea.	585	87.50		672.50	775
2460	With polished bronze top				"	830			830	915
2500	Heavy duty, cleanout & trap w/bucket, C.I., 15" top									
2540	2", 3", and 4" pipe size	Q-1	6	2.667	Ea.	5,575	131		5,706	6,350
2560	For galvanized body, add					1,425			1,425	1,575
2580	With polished bronze top				↓	6,200			6,200	6,825

22 13 23 – Sanitary Waste Interceptors

22 13 23.10 Interceptors

		Crew	Daily Output	Labor-Hours	Unit	Material	2012 Bare Costs Labor	2012 Bare Costs Equipment	Total	Total Incl O&P
0010	**INTERCEPTORS**									
0150	Grease, fabricated steel, 4 GPM, 8 lb. fat capacity	1 Plum	4	2	Ea.	1,000	109		1,109	1,275
0200	7 GPM, 14 lb. fat capacity		4	2		1,400	109		1,509	1,725
1000	10 GPM, 20 lb. fat capacity		4	2		1,650	109		1,759	2,000
1040	15 GPM, 30 lb. fat capacity		4	2		2,450	109		2,559	2,850
1060	20 GPM, 40 lb. fat capacity	↓	3	2.667		2,975	146		3,121	3,500
1120	50 GPM, 100 lb. fat capacity	Q-1	2	8		8,800	395		9,195	10,300
1160	100 GPM, 200 lb. fat capacity	"	2	8	↓	12,500	395		12,895	14,300
1580	For seepage pan, add					7%				
3000	Hair, cast iron, 1-1/4" and 1-1/2" pipe connection	1 Plum	8	1	Ea.	380	54.50		434.50	500

22 13 Facility Sanitary Sewerage

22 13 23 – Sanitary Waste Interceptors

22 13 23.10 Interceptors

		Crew	Daily Output	Labor-Hours	Unit	Material	2012 Bare Costs Labor	2012 Bare Costs Equipment	Total	Total Incl O&P
3100	For chrome-plated cast iron, add				Ea.	232			232	256
4000	Oil, fabricated steel, 10 GPM, 2" pipe size	1 Plum	4	2		2,250	109		2,359	2,650
4100	15 GPM, 2" or 3" pipe size		4	2		3,100	109		3,209	3,575
4120	20 GPM, 2" or 3" pipe size		3	2.667		3,725	146		3,871	4,325
4220	100 GPM, 3" pipe size	Q-1	2	8		12,500	395		12,895	14,300
6000	Solids, precious metals recovery, C.I., 1-1/4" to 2" pipe	1 Plum	4	2		570	109		679	795
6100	Dental Lab., large, C.I., 1-1/2" to 2" pipe	"	3	2.667		2,000	146		2,146	2,425

22 13 29 – Sanitary Sewerage Pumps

22 13 29.13 Wet-Pit-Mounted, Vertical Sewerage Pumps

		Crew	Daily Output	Labor-Hours	Unit	Material	2012 Bare Costs Labor	2012 Bare Costs Equipment	Total	Total Incl O&P
0010	**WET-PIT-MOUNTED, VERTICAL SEWERAGE PUMPS**									
0020	Controls incl. alarm/disconnect panel w/wire. Excavation not included									
0260	Simplex, 9 GPM at 60 PSIG, 91 gal. tank				Ea.	3,150			3,150	3,450
0300	Unit with manway, 26" I.D., 18" high					3,575			3,575	3,950
0340	26" I.D., 36" high					3,625			3,625	4,000
0380	43" I.D., 4' high					3,750			3,750	4,125
3000	Indoor residential type installation									
3020	Simplex, 9 GPM at 60 PSIG, 91 gal. HDPE tank				Ea.	3,125			3,125	3,450

22 13 29.14 Sewage Ejector Pumps

		Crew	Daily Output	Labor-Hours	Unit	Material	2012 Bare Costs Labor	2012 Bare Costs Equipment	Total	Total Incl O&P
0010	**SEWAGE EJECTOR PUMPS**, With operating and level controls									
0100	Simplex system incl. tank, cover, pump 15' head									
0500	37 gal. PE tank, 12 GPM, 1/2 HP, 2" discharge	Q-1	3.20	5	Ea.	455	246		701	870
0510	3" discharge		3.10	5.161		495	254		749	930
0530	87 GPM, .7 HP, 2" discharge		3.20	5		690	246		936	1,125
0540	3" discharge		3.10	5.161		750	254		1,004	1,200
0600	45 gal. coated stl tank, 12 GPM, 1/2 HP, 2" discharge		3	5.333		805	262		1,067	1,275
0610	3" discharge		2.90	5.517		840	271		1,111	1,325
0630	87 GPM, .7 HP, 2" discharge		3	5.333		1,025	262		1,287	1,525
0640	3" discharge		2.90	5.517		1,100	271		1,371	1,600
0660	134 GPM, 1 HP, 2" discharge		2.80	5.714		1,125	281		1,406	1,650
0680	3" discharge		2.70	5.926		1,175	291		1,466	1,750
0700	70 gal. PE tank, 12 GPM, 1/2 HP, 2" discharge		2.60	6.154		880	300		1,180	1,425
0710	3" discharge		2.40	6.667		935	330		1,265	1,525
0730	87 GPM, 0.7 HP, 2" discharge		2.50	6.400		1,125	315		1,440	1,700
0740	3" discharge		2.30	6.957		1,200	340		1,540	1,850
0760	134 GPM, 1 HP, 2" discharge		2.20	7.273		1,225	355		1,580	1,900
0770	3" discharge		2	8		1,300	395		1,695	2,025

22 14 Facility Storm Drainage

22 14 23 – Storm Drainage Piping Specialties

22 14 23.33 Backwater Valves

		Crew	Daily Output	Labor-Hours	Unit	Material	2012 Bare Costs Labor	2012 Bare Costs Equipment	Total	Total Incl O&P
0010	**BACKWATER VALVES**, C.I. Body									
6980	Bronze gate and automatic flapper valves									
7000	3" and 4" pipe size	Q-1	13	1.231	Ea.	1,725	60.50		1,785.50	2,000
7100	5" and 6" pipe size	"	13	1.231	"	2,675	60.50		2,735.50	3,025
7240	Bronze flapper valve, bolted cover									
7260	2" pipe size	Q-1	16	1	Ea.	510	49		559	635
7300	4" pipe size	"	13	1.231		975	60.50		1,035.50	1,175
7340	6" pipe size	Q-2	17	1.412		1,400	72		1,472	1,650

22 14 Facility Storm Drainage

22 14 26 – Facility Storm Drains

22 14 26.13 Roof Drains

22 14 26.13 Roof Drains	Crew	Daily Output	Labor-Hours	Unit	Material	2012 Bare Costs Labor	Equipment	Total	Total Incl O&P
0010 **ROOF DRAINS**									
0140 Cornice, C.I., 45° or 90° outlet									
0200 3" and 4" pipe size	Q-1	12	1.333	Ea.	360	65.50		425.50	495
0260 For galvanized body, add					64.50			64.50	71
0280 For polished bronze dome, add					50			50	55
3860 Roof, flat metal deck, C.I. body, 12" C.I. dome									
3890 3" pipe size	Q-1	14	1.143	Ea.	330	56		386	450
3920 6" pipe size	"	10	1.600	"	540	78.50		618.50	710
4620 Main, all aluminum, 12" low profile dome									
4640 2", 3" and 4" pipe size	Q-1	14	1.143	Ea.	370	56		426	495

22 14 26.16 Facility Area Drains

22 14 26.16 Facility Area Drains	Crew	Daily Output	Labor-Hours	Unit	Material	2012 Bare Costs Labor	Equipment	Total	Total Incl O&P
0010 **FACILITY AREA DRAINS**									
4980 Scupper floor, oblique strainer, C.I.									
5000 6" x 7" top, 2", 3" and 4" pipe size	Q-1	16	1	Ea.	240	49		289	340
5100 8" x 12" top, 5" and 6" pipe size	"	14	1.143		490	56		546	625
5160 For galvanized body, add					40%				
5200 For polished bronze strainer, add					85%				

22 14 26.19 Facility Trench Drains

22 14 26.19 Facility Trench Drains	Crew	Daily Output	Labor-Hours	Unit	Material	2012 Bare Costs Labor	Equipment	Total	Total Incl O&P
0010 **FACILITY TRENCH DRAINS**									
5980 Trench, floor, hvy duty, modular, C.I., 12" x 12" top									
6000 2", 3", 4", 5", & 6" pipe size	Q-1	8	2	Ea.	760	98.50		858.50	990
6100 For unit with polished bronze top	"	8	2	"	1,125	98.50		1,223.50	1,400
6600 Trench, floor, for cement concrete encasement									
6610 Not including trenching or concrete									
6640 Polyester polymer concrete									
6650 4" internal width, with grate									
6660 Light duty steel grate	Q-1	120	.133	L.F.	32	6.55		38.55	45.50
6670 Medium duty steel grate		115	.139		42.50	6.85		49.35	57
6680 Heavy duty iron grate		110	.145		56.50	7.15		63.65	73.50
6700 12" internal width, with grate									
6770 Heavy duty galvanized grate	Q-1	80	.200	L.F.	154	9.85		163.85	185
6800 Fiberglass									
6810 8" internal width, with grate									
6820 Medium duty galvanized grate	Q-1	115	.139	L.F.	96.50	6.85		103.35	116
6830 Heavy duty iron grate	"	110	.145	"	120	7.15		127.15	143

22 14 29 – Sump Pumps

22 14 29.13 Wet-Pit-Mounted, Vertical Sump Pumps

22 14 29.13 Wet-Pit-Mounted, Vertical Sump Pumps	Crew	Daily Output	Labor-Hours	Unit	Material	2012 Bare Costs Labor	Equipment	Total	Total Incl O&P
0010 **WET-PIT-MOUNTED, VERTICAL SUMP PUMPS**									
0400 Molded PVC base, 21 GPM at 15' head, 1/3 HP	1 Plum	5	1.600	Ea.	128	87.50		215.50	272
0800 Iron base, 21 GPM at 15' head, 1/3 HP		5	1.600		156	87.50		243.50	305
1200 Solid brass, 21 GPM at 15' head, 1/3 HP		5	1.600		272	87.50		359.50	430

22 14 29.16 Submersible Sump Pumps

22 14 29.16 Submersible Sump Pumps	Crew	Daily Output	Labor-Hours	Unit	Material	2012 Bare Costs Labor	Equipment	Total	Total Incl O&P
0010 **SUBMERSIBLE SUMP PUMPS**									
7000 Sump pump, automatic									
7100 Plastic, 1-1/4" discharge, 1/4 HP	1 Plum	6	1.333	Ea.	126	73		199	249
7140 1/3 HP		5	1.600		189	87.50		276.50	340
7160 1/2 HP		5	1.600		207	87.50		294.50	360
7180 1-1/2" discharge, 1/2 HP		4	2		256	109		365	445
7500 Cast iron, 1-1/4" discharge, 1/4 HP		6	1.333		177	73		250	305
7540 1/3 HP		6	1.333		209	73		282	340
7560 1/2 HP		5	1.600		252	87.50		339.50	410

22 31 Domestic Water Softeners

22 31 13 – Residential Domestic Water Softeners

22 31 13.10 Residential Water Softeners	Crew	Daily Output	Labor-Hours	Unit	Material	2012 Bare Costs Labor	Equipment	Total	Total Incl O&P
0010 **RESIDENTIAL WATER SOFTENERS**									
7350 Water softener, automatic, to 30 grains per gallon	2 Plum	5	3.200	Ea.	340	175		515	640
7400 To 100 grains per gallon	"	4	4	"	680	218		898	1,075

22 31 16 – Commercial Domestic Water Softeners

22 31 16.10 Water Softeners

	Crew	Daily Output	Labor-Hours	Unit	Material	2012 Bare Costs Labor	Equipment	Total	Total Incl O&P
0010 **WATER SOFTENERS**									
5800 Softener systems, automatic, intermediate sizes									
5820 available, may be used in multiples.									
6000 Hardness capacity between regenerations and flow									
6100 150,000 grains, 37 GPM cont., 51 GPM peak	Q-1	1.20	13.333	Ea.	4,350	655		5,005	5,775
6200 300,000 grains, 81 GPM cont., 113 GPM peak		1	16		8,925	785		9,710	11,000
6300 750,000 grains, 160 GPM cont., 230 GPM peak		.80	20		11,600	985		12,585	14,300
6400 900,000 grains, 185 GPM cont., 270 GPM peak		.70	22.857		18,700	1,125		19,825	22,300

22 33 Electric Domestic Water Heaters

22 33 13 – Instantaneous Electric Domestic Water Heaters

22 33 13.10 Hot Water Dispensers

	Crew	Daily Output	Labor-Hours	Unit	Material	2012 Bare Costs Labor	Equipment	Total	Total Incl O&P
0010 **HOT WATER DISPENSERS**									
0160 Commercial, 100 cup, 11.3 amp	1 Plum	14	.571	Ea.	249	31		280	320
3180 Household, 60 cup	"	14	.571	"	245	31		276	315

22 33 30 – Residential, Electric Domestic Water Heaters

22 33 30.13 Residential, Small-Capacity Elec. Water Heaters

	Crew	Daily Output	Labor-Hours	Unit	Material	2012 Bare Costs Labor	Equipment	Total	Total Incl O&P
0010 **RESIDENTIAL, SMALL-CAPACITY ELECTRIC DOMESTIC WATER HEATERS**									
1000 Residential, electric, glass lined tank, 5 yr., 10 gal., single element	1 Plum	2.30	3.478	Ea.	385	190		575	710
1040 20 gallon, single element		2.20	3.636		440	199		639	785
1060 30 gallon, double element		2.20	3.636		575	199		774	935
1080 40 gallon, double element		2	4		615	218		833	1,000
1100 52 gallon, double element		2	4		690	218		908	1,075
1180 120 gallon, double element		1.40	5.714		1,450	310		1,760	2,075

22 33 33 – Light-Commercial Electric Domestic Water Heaters

22 33 33.10 Commercial Electric Water Heaters

	Crew	Daily Output	Labor-Hours	Unit	Material	2012 Bare Costs Labor	Equipment	Total	Total Incl O&P
0010 **COMMERCIAL ELECTRIC WATER HEATERS**									
4000 Commercial, 100° rise. NOTE: for each size tank, a range of									
4010 heaters between the ones shown are available									
4020 Electric									
4100 5 gal., 3 kW, 12 GPH, 208 volt	1 Plum	2	4	Ea.	2,625	218		2,843	3,200
4120 10 gal., 6 kW, 25 GPH, 208 volt		2	4		2,900	218		3,118	3,525
4140 50 gal., 9 kW, 37 GPH, 208 volt		1.80	4.444		3,975	243		4,218	4,750
4160 50 gal., 36 kW, 148 GPH, 208 volt		1.80	4.444		6,075	243		6,318	7,075
4300 200 gal., 15 kW, 61 GPH, 480 volt	Q-1	1.70	9.412		18,800	465		19,265	21,300
4320 200 gal., 120 kW , 490 GPH, 480 volt		1.70	9.412		25,600	465		26,065	28,900
4460 400 gal., 30 kW, 123 GPH, 480 volt		1	16		25,500	785		26,285	29,300
5400 Modulating step control, 2-5 steps	1 Elec	5.30	1.509		900	78		978	1,100
5440 6-10 steps		3.20	2.500		1,150	129		1,279	1,475
5460 11-15 steps		2.70	2.963		1,625	153		1,778	2,025
5480 16-20 steps		1.60	5		2,400	258		2,658	3,025

22 34 Fuel-Fired Domestic Water Heaters

22 34 13 – Instantaneous, Tankless, Gas Domestic Water Heaters

22 34 13.10 Instantaneous, Tankless, Gas Water Heaters	Crew	Daily Output	Labor-Hours	Unit	Material	2012 Bare Costs Labor	Equipment	Total	Total Incl O&P
0010 **INSTANTANEOUS, TANKLESS, GAS WATER HEATERS**									
9410 Natural gas/propane, 3.2 GPM **G**	1 Plum	2	4	Ea.	535	218		753	920
9420 6.4 GPM		1.90	4.211		630	230		860	1,050
9430 8.4 GPM		1.80	4.444		755	243		998	1,200
9440 9.5 GPM		1.60	5		930	273		1,203	1,425

22 34 30 – Residential Gas Domestic Water Heaters

22 34 30.13 Residential, Atmos, Gas Domestic Wtr Heaters

	Crew	Daily Output	Labor-Hours	Unit	Material	Labor	Equipment	Total	Total Incl O&P
0010 **RESIDENTIAL, ATMOSPHERIC, GAS DOMESTIC WATER HEATERS**									
2000 Gas fired, foam lined tank, 10 yr., vent not incl.									
2040 30 gallon	1 Plum	2	4	Ea.	895	218		1,113	1,325
2100 75 gallon		1.50	5.333		1,225	291		1,516	1,800
2120 100 gallon		1.30	6.154		1,525	335		1,860	2,175

22 34 36 – Commercial Gas Domestic Water Heaters

22 34 36.13 Commercial, Atmos., Gas Domestic Water Htrs.

	Crew	Daily Output	Labor-Hours	Unit	Material	Labor	Equipment	Total	Total Incl O&P
0010 **COMMERCIAL, ATMOSPHERIC, GAS DOMESTIC WATER HEATERS**									
6000 Gas fired, flush jacket, std. controls, vent not incl.									
6040 75 MBH input, 73 GPH	1 Plum	1.40	5.714	Ea.	1,575	310		1,885	2,200
6060 98 MBH input, 95 GPH		1.40	5.714		4,575	310		4,885	5,525
6080 120 MBH input, 110 GPH		1.20	6.667		4,975	365		5,340	6,000
6180 200 MBH input, 192 GPH		.60	13.333		8,025	730		8,755	9,925
6200 250 MBH input, 245 GPH		.50	16		8,375	875		9,250	10,500
6900 For low water cutoff, add		8	1		315	54.50		369.50	430
6960 For bronze body hot water circulator, add		4	2		1,875	109		1,984	2,250

22 34 46 – Oil-Fired Domestic Water Heaters

22 34 46.10 Residential Oil-Fired Water Heaters

	Crew	Daily Output	Labor-Hours	Unit	Material	Labor	Equipment	Total	Total Incl O&P
0010 **RESIDENTIAL OIL-FIRED WATER HEATERS**									
3000 Oil fired, glass lined tank, 5 yr., vent not included, 30 gallon	1 Plum	2	4	Ea.	665	218		883	1,050
3040 50 gallon		1.80	4.444		1,050	243		1,293	1,525
3060 70 gallon		1.50	5.333		1,275	291		1,566	1,850

22 34 46.20 Commercial Oil-Fired Water Heaters

	Crew	Daily Output	Labor-Hours	Unit	Material	Labor	Equipment	Total	Total Incl O&P
0010 **COMMERCIAL OIL-FIRED WATER HEATERS**									
8000 Oil fired, glass lined, UL listed, std. controls, vent not incl.									
8060 140 gal., 140 MBH input, 134 GPH	Q-1	2.13	7.512	Ea.	14,900	370		15,270	17,000
8080 140 gal., 199 MBH input, 191 GPH		2	8		15,400	395		15,795	17,500
8100 140 gal., 255 MBH input, 247 GPH		1.60	10		15,900	490		16,390	18,100
8160 140 gal., 540 MBH input, 519 GPH		.96	16.667		21,000	820		21,820	24,300
8180 140 gal., 720 MBH input, 691 GPH		.92	17.391		21,400	855		22,255	24,800
8280 201 gal., 1250 MBH input, 1200 GPH	Q-2	1.22	19.672		32,900	1,000		33,900	37,600
8300 201 gal., 1500 MBH input, 1441 GPH	"	1.16	20.690		35,700	1,050		36,750	40,900
8900 For low water cutoff, add	1 Plum	8	1		300	54.50		354.50	415
8960 For bronze body hot water circulator, add	"	4	2		710	109		819	945

22 35 Domestic Water Heat Exchangers

22 35 30 – Water Heating by Steam

22 35 30.10 Water Heating Transfer Package	Crew	Daily Output	Labor-Hours	Unit	Material	2012 Bare Costs Labor	Equipment	Total	Total Incl O&P
0010 **WATER HEATING TRANSFER PACKAGE**, Complete controls,									
0020 expansion tank, converter, air separator									
1000 Hot water, 180°F enter, 200°F leaving, 15# steam									
1010 One pump system, 28 GPM	Q-6	.75	32	Ea.	18,700	1,675		20,375	23,100
1020 35 GPM		.70	34.286		20,300	1,775		22,075	25,100
1040 55 GPM		.65	36.923		23,900	1,925		25,825	29,200
1060 130 GPM		.55	43.636		30,100	2,275		32,375	36,600
1080 255 GPM		.40	60		39,400	3,125		42,525	48,100
1100 550 GPM	↓	.30	80	↓	50,000	4,150		54,150	61,500

22 41 Residential Plumbing Fixtures

22 41 06 – Plumbing Fixtures General

22 41 06.10 Plumbing Fixture Notes

0010 **PLUMBING FIXTURE NOTES**, Incl. trim fittings unless otherwise noted R224000-40									
0080 For rough-in, supply, waste, and vent, see add for each type									
0122 For electric water coolers, see Section 22 47 16.10									
0160 For color, unless otherwise noted, add				Ea.	20%				

22 41 13 – Residential Water Closets, Urinals, and Bidets

22 41 13.40 Water Closets

	Crew	Daily Output	Labor-Hours	Unit	Material	Labor	Equipment	Total	Total Incl O&P
0010 **WATER CLOSETS** R224000-40									
0032 For automatic flush, see Line 22 42 39.10 0972									
0150 Tank type, vitreous china, incl. seat, supply pipe w/stop, 1.6 gpf or noted									
0200 Wall hung									
0400 Two piece, close coupled	Q-1	5.30	3.019	Ea.	605	148		753	890
0960 For rough-in, supply, waste, vent and carrier	"	2.73	5.861	"	1,000	288		1,288	1,525
0999 Floor mounted									
1100 Two piece, close coupled	Q-1	5.30	3.019	Ea.	247	148		395	495
1102 Economy		5.30	3.019		137	148		285	375
1110 Two piece, close coupled, dual flush		5.30	3.019		273	148		421	525
1140 Two piece, close coupled, 1.28 gpf, ADA G	↓	5.30	3.019	↓	295	148		443	550
1960 For color, add					30%				
1980 For rough-in, supply, waste and vent	Q-1	3.05	5.246	Ea.	340	258		598	765

22 41 16 – Residential Lavatories and Sinks

22 41 16.10 Lavatories

	Crew	Daily Output	Labor-Hours	Unit	Material	Labor	Equipment	Total	Total Incl O&P
0010 **LAVATORIES**, With trim, white unless noted otherwise R224000-40									
0500 Vanity top, porcelain enamel on cast iron									
0600 20" x 18"	Q-1	6.40	2.500	Ea.	305	123		428	520
0640 33" x 19" oval		6.40	2.500		625	123		748	870
0720 19" round	↓	6.40	2.500	↓	228	123		351	435
0860 For color, add					25%				
1000 Cultured marble, 19" x 17", single bowl	Q-1	6.40	2.500	Ea.	179	123		302	380
1040 25" x 19", single bowl	"	6.40	2.500	"	204	123		327	410
1580 For color, same price									
1900 Stainless steel, self-rimming, 25" x 22", single bowl, ledge	Q-1	6.40	2.500	Ea.	360	123		483	580
1960 17" x 22", single bowl		6.40	2.500		350	123		473	570
2600 Steel, enameled, 20" x 17", single bowl		5.80	2.759		188	136		324	410
2660 19" round		5.80	2.759		164	136		300	385
2900 Vitreous china, 20" x 16", single bowl		5.40	2.963		260	146		406	505
2960 20" x 17", single bowl		5.40	2.963		176	146		322	415
3020 19" round, single bowl	↓	5.40	2.963	↓	174	146		320	410

22 41 Residential Plumbing Fixtures

22 41 16 – Residential Lavatories and Sinks

22 41 16.10 Lavatories

		Crew	Daily Output	Labor-Hours	Unit	Material	2012 Bare Costs Labor	Equipment	Total	Total Incl O&P
3200	22" x 13", single bowl	Q-1	5.40	2.963	Ea.	267	146		413	515
3560	For color, add					50%				
3580	Rough-in, supply, waste and vent for all above lavatories	Q-1	2.30	6.957	Ea.	420	340		760	975
4000	Wall hung									
4040	Porcelain enamel on cast iron, 16" x 14", single bowl	Q-1	8	2	Ea.	460	98.50		558.50	655
4180	20" x 18", single bowl		8	2		355	98.50		453.50	545
4240	22" x 19", single bowl	↓	8	2	↓	615	98.50		713.50	825
4580	For color, add					30%				
6000	Vitreous china, 18" x 15", single bowl with backsplash	Q-1	7	2.286	Ea.	231	112		343	425
6500	For color, add					30%				
6960	Rough-in, supply, waste and vent for above lavatories	Q-1	1.66	9.639	Ea.	480	475		955	1,250
7000	Pedestal type									
7600	Vitreous china, 27" x 21", white	Q-1	6.60	2.424	Ea.	555	119		674	795
7610	27" x 21", colored		6.60	2.424		670	119		789	915
7620	27" x 21", premium color		6.60	2.424		785	119		904	1,050
7660	26" x 20", white		6.60	2.424		505	119		624	735
7670	26" x 20", colored		6.60	2.424		600	119		719	840
7680	26" x 20", premium color		6.60	2.424		635	119		754	880
7700	24" x 20", white		6.60	2.424		490	119		609	715
7710	24" x 20", colored		6.60	2.424		585	119		704	820
7720	24" x 20", premium color		6.60	2.424		680	119		799	925
7760	21" x 18", white		6.60	2.424		235	119		354	440
7770	21" x 18", colored		6.60	2.424		267	119		386	475
7990	Rough-in, supply, waste and vent for pedestal lavatories	↓	1.66	9.639	↓	480	475		955	1,250

22 41 16.30 Sinks

		Crew	Daily Output	Labor-Hours	Unit	Material	2012 Bare Costs Labor	Equipment	Total	Total Incl O&P
0010	**SINKS**, With faucets and drain R224000-40									
2000	Kitchen, counter top style, P.E. on C.I., 24" x 21" single bowl	Q-1	5.60	2.857	Ea.	257	140		397	495
2100	31" x 22" single bowl		5.60	2.857		525	140		665	785
2200	32" x 21" double bowl		4.80	3.333		286	164		450	560
2300	42" x 21" double bowl		4.80	3.333		1,050	164		1,214	1,400
3000	Stainless steel, self rimming, 19" x 18" single bowl		5.60	2.857		580	140		720	845
3100	25" x 22" single bowl		5.60	2.857		645	140		785	920
4000	Steel, enameled, with ledge, 24" x 21" single bowl		5.60	2.857		440	140		580	695
4100	32" x 21" double bowl	↓	4.80	3.333		400	164		564	685
4960	For color sinks except stainless steel, add					10%				
4980	For rough-in, supply, waste and vent, counter top sinks	Q-1	2.14	7.477	↓	470	365		835	1,075
5000	Kitchen, raised deck, P.E. on C.I.									
5100	32" x 21", dual level, double bowl	Q-1	2.60	6.154	Ea.	345	300		645	835
5700	For color, add					20%				
5790	For rough-in, supply, waste & vent, sinks	Q-1	1.85	8.649	↓	470	425		895	1,150

22 41 19 – Residential Bathtubs

22 41 19.10 Baths

		Crew	Daily Output	Labor-Hours	Unit	Material	2012 Bare Costs Labor	Equipment	Total	Total Incl O&P
0010	**BATHS**									
0100	Tubs, recessed porcelain enamel on cast iron, with trim									
0180	48" x 42"	Q-1	4	4	Ea.	2,225	197		2,422	2,750
0220	72" x 36"		3	5.333		2,500	262		2,762	3,150
2000	Enameled formed steel, 4'-6" long		5.80	2.759		490	136		626	745
4000	Soaking, acrylic with pop-up drain, 66" x 35" x 20" deep		5.50	2.909		1,800	143		1,943	2,200
4100	60" x 48" x 18-1/2" deep		5	3.200		1,025	157		1,182	1,350
9600	Rough-in, supply, waste and vent, for all above tubs, add	↓	2.07	7.729	↓	485	380		865	1,100

22 41 Residential Plumbing Fixtures

22 41 23 – Residential Showers

22 41 23.20 Showers

		Crew	Daily Output	Labor-Hours	Unit	Material	2012 Bare Costs Labor	2012 Bare Costs Equipment	Total	Total Incl O&P
0010	**SHOWERS**									
1500	Stall, with drain only. Add for valve and door/curtain									
3000	Fiberglass, one piece, with 3 walls, 32" x 32" square	Q-1	5.50	2.909	Ea.	545	143		688	810
3100	36" x 36" square		5.50	2.909		555	143		698	830
3250	64" x 65-3/4" x 81-1/2" fold. seat, whlchr.		3.80	4.211		2,050	207		2,257	2,575
4000	Polypropylene, stall only, w/molded-stone floor, 30" x 30"		2	8		500	395		895	1,150
4200	Rough-in, supply, waste and vent for above showers	↓	2.05	7.805	↓	590	385		975	1,225

22 41 23.40 Shower System Components

		Crew	Daily Output	Labor-Hours	Unit	Material	2012 Bare Costs Labor	2012 Bare Costs Equipment	Total	Total Incl O&P
0010	**SHOWER SYSTEM COMPONENTS**									
4500	Receptor only									
4510	For tile, 36" x 36"	1 Plum	4	2	Ea.	325	109		434	525
4520	Fiberglass receptor only, 32" x 32"		8	1		107	54.50		161.50	201
4530	34" x 34"		7.80	1.026		126	56		182	224
4540	36" x 36"	↓	7.60	1.053	↓	132	57.50		189.50	232
4600	Rectangular									
4620	32" x 48"	1 Plum	7.40	1.081	Ea.	149	59		208	252
4630	32" x 54"		7.20	1.111		196	60.50		256.50	305
4640	32" x 60"		7	1.143		208	62.50		270.50	320
5000	Built-in, head, arm, 2.5 GPM valve		4	2		80	109		189	253
5200	Head, arm, by-pass, integral stops, handles	↓	3.60	2.222	↓	216	121		337	420

22 41 36 – Residential Laundry Trays

22 41 36.10 Laundry Sinks

		Crew	Daily Output	Labor-Hours	Unit	Material	2012 Bare Costs Labor	2012 Bare Costs Equipment	Total	Total Incl O&P
0010	**LAUNDRY SINKS**, With trim									
0020	Porcelain enamel on cast iron, black iron frame									
0050	24" x 21", single compartment	Q-1	6	2.667	Ea.	465	131		596	710
0100	26" x 21", single compartment	"	6	2.667	"	450	131		581	695
2000	Molded stone, on wall hanger or legs									
2020	22" x 23", single compartment	Q-1	6	2.667	Ea.	138	131		269	350
2100	45" x 21", double compartment	"	5	3.200	"	259	157		416	520
3000	Plastic, on wall hanger or legs									
3020	18" x 23", single compartment	Q-1	6.50	2.462	Ea.	149	121		270	345
3300	40" x 24", double compartment		5.50	2.909		267	143		410	510
5000	Stainless steel, counter top, 22" x 17" single compartment		6	2.667		61.50	131		192.50	266
5200	33" x 22", double compartment		5	3.200		73.50	157		230.50	320
9600	Rough-in, supply, waste and vent, for all laundry sinks	↓	2.14	7.477	↓	470	365		835	1,075

22 41 39 – Residential Faucets, Supplies and Trim

22 41 39.10 Faucets and Fittings

		Crew	Daily Output	Labor-Hours	Unit	Material	2012 Bare Costs Labor	2012 Bare Costs Equipment	Total	Total Incl O&P
0010	**FAUCETS AND FITTINGS**									
0150	Bath, faucets, diverter spout combination, sweat	1 Plum	8	1	Ea.	86.50	54.50		141	178
0200	For integral stops, IPS unions, add					109			109	120
0420	Bath, press-bal mix valve w/diverter, spout, shower hd, arm/flange	1 Plum	8	1	↓	163	54.50		217.50	262
0810	Bidet									
0812	Fitting, over the rim, swivel spray/pop-up drain	1 Plum	8	1	Ea.	172	54.50		226.50	272
1000	Kitchen sink faucets, top mount, cast spout		10	.800		61.50	43.50		105	134
1100	For spray, add	↓	24	.333	↓	16.15	18.20		34.35	45.50
1300	Single control lever handle									
1310	With pull out spray									
1320	Polished chrome	1 Plum	10	.800	Ea.	171	43.50		214.50	255
2000	Laundry faucets, shelf type, IPS or copper unions		12	.667		49.50	36.50		86	110
2100	Lavatory faucet, centerset, without drain	↓	10	.800	↓	44.50	43.50		88	115
2210	Porcelain cross handles and pop-up drain									

22 41 Residential Plumbing Fixtures

22 41 39 – Residential Faucets, Supplies and Trim

22 41 39.10 Faucets and Fittings

	22 41 39.10 Faucets and Fittings	Crew	Daily Output	Labor-Hours	Unit	Material	2012 Bare Costs Labor	2012 Bare Costs Equipment	Total	Total Incl O&P
2220	Polished chrome	1 Plum	6.66	1.201	Ea.	128	65.50		193.50	240
2230	Polished brass	"	6.66	1.201	"	193	65.50		258.50	310
2260	Single lever handle and pop-up drain									
2280	Satin nickel	1 Plum	6.66	1.201	Ea.	274	65.50		339.50	400
2290	Polished chrome		6.66	1.201		182	65.50		247.50	299
2810	Automatic sensor and operator, with faucet head [G]		6.15	1.301		370	71		441	515
4000	Shower by-pass valve with union		18	.444		68.50	24.50		93	112
4200	Shower thermostatic mixing valve, concealed, with shower head trim kit	↓	8	1	↓	320	54.50		374.50	435
4220	Shower pressure balancing mixing valve,									
4230	With shower head, arm, flange and diverter tub spout									
4240	Chrome	1 Plum	6.14	1.303	Ea.	222	71		293	350
4250	Polished brass		6.14	1.303		279	71		350	410
4260	Satin		6.14	1.303		279	71		350	410
4270	Polished chrome/brass		6.14	1.303		229	71		300	360
5000	Sillcock, compact, brass, IPS or copper to hose	↓	24	.333	↓	13.30	18.20		31.50	42

22 42 Commercial Plumbing Fixtures

22 42 13 – Commercial Water Closets, Urinals, and Bidets

22 42 13.30 Urinals

	22 42 13.30 Urinals		Crew	Daily Output	Labor-Hours	Unit	Material	2012 Bare Costs Labor	2012 Bare Costs Equipment	Total	Total Incl O&P
0010	**URINALS**										
3000	Wall hung, vitreous china, with hanger & self-closing valve										
3100	Siphon jet type		Q-1	3	5.333	Ea.	293	262		555	720
3120	Blowout type			3	5.333		285	262		547	710
3140	Water saving .5 gpf	[G]		3	5.333	↓	655	262		917	1,125
3300	Rough-in, supply, waste & vent			2.83	5.654		585	278		863	1,050
5000	Stall type, vitreous china, includes valve			2.50	6.400		655	315		970	1,200
6980	Rough-in, supply, waste and vent		↓	1.99	8.040	↓	590	395		985	1,250
8000	Waterless (no flush) urinal										
8010	Wall hung										
8014	Fiberglass reinforced polyester										
8020	Standard unit	[G]	Q-1	21.30	.751	Ea.	385	37		422	475
8030	ADA compliant unit	[G]	"	21.30	.751		400	37		437	495
8070	For solid color, add	[G]					60			60	66
8080	For 2" brass flange, (new const.), add	[G]	Q-1	96	.167	↓	19.20	8.20		27.40	33.50
8200	Vitreous china										
8220	ADA compliant unit, 14"	[G]	Q-1	21.30	.751	Ea.	198	37		235	274
8240	ADA compliant unit, 18"	[G]		21.30	.751		320	37		357	405
8250	ADA compliant unit, 15.5"			21.30	.751		272	37		309	355
8270	For solid color, add	[G]					60			60	66
8290	Rough-in, supply, waste & vent	[G]	Q-1	2.92	5.479	↓	530	269		799	985
8400	Trap liquid										
8410	1 quart	[G]				Ea.	15.15			15.15	16.70
8420	1 gallon	[G]				"	56			56	61.50

22 42 13.40 Water Closets

	22 42 13.40 Water Closets		Crew	Daily Output	Labor-Hours	Unit	Material	2012 Bare Costs Labor	2012 Bare Costs Equipment	Total	Total Incl O&P
0010	**WATER CLOSETS**										
3000	Bowl only, with flush valve, seat, 1.6 gpf unless noted										
3100	Wall hung		Q-1	5.80	2.759	Ea.	455	136		591	705
3200	For rough-in, supply, waste and vent, single WC			2.56	6.250		1,075	305		1,380	1,650
3300	Floor mounted			5.80	2.759		315	136		451	555
3350	With wall outlet			5.80	2.759		455	136		591	705
3360	With floor outlet, 1.28 gpf	[G]	↓	5.80	2.759		505	136		641	760

22 42 Commercial Plumbing Fixtures

22 42 13 – Commercial Water Closets, Urinals, and Bidets

22 42 13.40 Water Closets

		Crew	Daily Output	Labor-Hours	Unit	Material	2012 Bare Costs Labor	2012 Bare Costs Equipment	Total	Total Incl O&P
3362	With floor outlet, 1.28 gpf, ADA　G	Q-1	5.80	2.759	Ea.	595	136		731	860
3400	For rough-in, supply, waste and vent, single WC	↓	2.84	5.634	↓	420	277		697	885

22 42 16 – Commercial Lavatories and Sinks

22 42 16.10 Handwasher-Dryer Module

		Crew	Daily Output	Labor-Hours	Unit	Material	Labor	Equipment	Total	Total Incl O&P
0010	**HANDWASHER-DRYER MODULE**									
0030	Wall mounted									
0040	With electric dryer									
0050	Sensor operated	Q-1	8	2	Ea.	3,650	98.50		3,748.50	4,150
0110	Sensor operated (ADA)	↓	8	2		3,650	98.50		3,748.50	4,175
0140	Sensor operated (ADA), surface mounted	↓	8	2	↓	4,400	98.50		4,498.50	5,000
0150	With paper towels									
0180	Sensor operated	Q-1	8	2	Ea.	3,550	98.50		3,648.50	4,050

22 42 16.14 Lavatories

0010	**LAVATORIES**, With trim, white unless noted otherwise									
0020	Commercial lavatories same as residential. See Section 22 41 16.10									

22 42 16.40 Service Sinks

		Crew	Daily Output	Labor-Hours	Unit	Material	Labor	Equipment	Total	Total Incl O&P
0010	**SERVICE SINKS**									
6650	Service, floor, corner, P.E. on C.I., 28" x 28"	Q-1	4.40	3.636	Ea.	880	179		1,059	1,250
6790	For rough-in, supply, waste & vent, floor service sinks		1.64	9.756		1,400	480		1,880	2,250
7000	Service, wall, P.E. on C.I., roll rim, 22" x 18"		4	4		675	197		872	1,025
7100	24" x 20"		4	4		740	197		937	1,100
8600	Vitreous china, 22" x 20"	↓	4	4		540	197		737	890
8960	For stainless steel rim guard, front or one side, add					55			55	60.50
8980	For rough-in, supply, waste & vent, wall service sinks	Q-1	1.30	12.308	↓	1,675	605		2,280	2,725

22 42 23 – Commercial Showers

22 42 23.30 Group Showers

		Crew	Daily Output	Labor-Hours	Unit	Material	Labor	Equipment	Total	Total Incl O&P
0010	**GROUP SHOWERS**									
6000	Group, w/pressure balancing valve, rough-in and rigging not included									
6800	Column, 6 heads, no receptors, less partitions	Q-1	3	5.333	Ea.	3,000	262		3,262	3,700
6900	With stainless steel partitions		1	16		9,250	785		10,035	11,400
7600	5 heads, no receptors, less partitions		3	5.333		2,475	262		2,737	3,125
7620	4 heads (1 handicap) no receptors, less partitions		3	5.333		3,975	262		4,237	4,775
7700	With stainless steel partitions		1	16		7,600	785		8,385	9,525
8000	Wall, 2 heads, no receptors, less partitions		4	4		1,375	197		1,572	1,825
8100	With stainless steel partitions	↓	2	8	↓	3,225	395		3,620	4,150

22 42 33 – Wash Fountains

22 42 33.20 Commercial Wash Fountains

		Crew	Daily Output	Labor-Hours	Unit	Material	Labor	Equipment	Total	Total Incl O&P
0010	**COMMERCIAL WASH FOUNTAINS**									
1900	Group, foot control									
2000	Precast terrazzo, circular, 36" diam., 5 or 6 persons	Q-2	3	8	Ea.	5,450	410		5,860	6,625
2100	54" diameter for 8 or 10 persons		2.50	9.600		6,800	490		7,290	8,225
2400	Semi-circular, 36" diam. for 3 persons		3	8		4,775	410		5,185	5,875
2500	54" diam. for 4 or 5 persons		2.50	9.600		6,050	490		6,540	7,400
2700	Quarter circle (corner), 54" for 3 persons		3.50	6.857		6,000	350		6,350	7,125
3000	Stainless steel, circular, 36" diameter		3.50	6.857		5,450	350		5,800	6,525
3100	54" diameter		2.80	8.571		7,050	435		7,485	8,400
3400	Semi-circular, 36" diameter		3.50	6.857		4,750	350		5,100	5,750
3500	54" diameter	↓	2.80	8.571	↓	6,000	435		6,435	7,250
5610	Group, infrared control, barrier free									
5614	Precast terrazzo									

22 42 Commercial Plumbing Fixtures

22 42 33 – Wash Fountains

22 42 33.20 Commercial Wash Fountains

		Crew	Daily Output	Labor-Hours	Unit	Material	2012 Bare Costs Labor	Equipment	Total	Total Incl O&P
5620	Semi-circular 36" diam. for 3 persons	Q-2	3	8	Ea.	7,150	410		7,560	8,500
5630	46" diam. for 4 persons		2.80	8.571		7,725	435		8,160	9,150
5640	Circular, 54" diam. for 8 persons, button control	↓	2.50	9.600		9,125	490		9,615	10,700
5700	Rough-in, supply, waste and vent for above wash fountains	Q-1	1.82	8.791		675	430		1,105	1,400
6200	Duo for small washrooms, stainless steel		2	8		2,725	395		3,120	3,575
6500	Rough-in, supply, waste & vent for duo fountains	↓	2.02	7.921	↓	400	390		790	1,025

22 42 39 – Commercial Faucets, Supplies, and Trim

22 42 39.10 Faucets and Fittings

		Crew	Daily Output	Labor-Hours	Unit	Material	2012 Bare Costs Labor	Equipment	Total	Total Incl O&P
0010	**FAUCETS AND FITTINGS**									
0840	Flush valves, with vacuum breaker									
0850	Water closet									
0860	Exposed, rear spud	1 Plum	8	1	Ea.	137	54.50		191.50	234
0870	Top spud		8	1		126	54.50		180.50	222
0880	Concealed, rear spud		8	1		180	54.50		234.50	281
0890	Top spud		8	1		149	54.50		203.50	247
0900	Wall hung	↓	8	1	↓	159	54.50		213.50	258
0920	Urinal									
0930	Exposed, stall	1 Plum	8	1	Ea.	126	54.50		180.50	222
0940	Wall, (washout)		8	1		126	54.50		180.50	222
0950	Pedestal, top spud		8	1		129	54.50		183.50	225
0960	Concealed, stall		8	1		145	54.50		199.50	242
0970	Wall (washout)	↓	8	1	↓	156	54.50		210.50	255
0971	Automatic flush sensor and operator for ♿ G									
0972	urinals or water closets, standard G	1 Plum	8	1	Ea.	415	54.50		469.50	545
0980	High efficiency water saving									
0984	Water closets, 1.28 gpf G	1 Plum	8	1	Ea.	395	54.50		449.50	520
0988	Urinals, .5 gpf G	"	8	1	"	395	54.50		449.50	520
2790	Faucets for lavatories									
2800	Self-closing, center set	1 Plum	10	.800	Ea.	131	43.50		174.50	210
2810	Automatic sensor and operator, with faucet head		6.15	1.301		370	71		441	515
3000	Service sink faucet, cast spout, pail hook, hose end	↓	14	.571	↓	80	31		111	135

22 42 39.30 Carriers and Supports

		Crew	Daily Output	Labor-Hours	Unit	Material	2012 Bare Costs Labor	Equipment	Total	Total Incl O&P
0010	**CARRIERS AND SUPPORTS**, For plumbing fixtures									
0500	Drinking fountain, wall mounted									
0600	Plate type with studs, top back plate	1 Plum	7	1.143	Ea.	76.50	62.50		139	178
0700	Top front and back plate		7	1.143		94.50	62.50		157	198
0800	Top & bottom, front & back plates, w/bearing jacks	↓	7	1.143	↓	136	62.50		198.50	243
3000	Lavatory, concealed arm									
3050	Floor mounted, single									
3100	High back fixture	1 Plum	6	1.333	Ea.	425	73		498	575
3200	Flat slab fixture		6	1.333		370	73		443	515
3220	Paraplegic ♿	↓	6	1.333	↓	475	73		548	630
3250	Floor mounted, back to back									
3300	High back fixtures	1 Plum	5	1.600	Ea.	605	87.50		692.50	795
3400	Flat slab fixtures		5	1.600		745	87.50		832.50	950
3430	Paraplegic ♿	↓	5	1.600	↓	690	87.50		777.50	890
3500	Wall mounted, in stud or masonry									
3600	High back fixture	1 Plum	6	1.333	Ea.	251	73		324	385
3700	Flat slab fixture	"	6	1.333	"	217	73		290	350
4600	Sink, floor mounted									
4650	Exposed arm system									
4700	Single heavy fixture	1 Plum	5	1.600	Ea.	217	87.50		304.50	370

22 42 39.30 Carriers and Supports	Crew	Daily Output	Labor-Hours	Unit	Material	2012 Bare Costs Labor	2012 Bare Costs Equipment	Total	Total Incl O&P
4750	Single heavy sink with slab	1 Plum	5	1.600	Ea.	910	87.50		997.50	1,125
4800	Back to back, standard fixtures		5	1.600		520	87.50		607.50	705
4850	Back to back, heavy fixtures		5	1.600		805	87.50		892.50	1,025
4900	Back to back, heavy sink with slab		5	1.600		805	87.50		892.50	1,025
4950	Exposed offset arm system									
5000	Single heavy deep fixture	1 Plum	5	1.600	Ea.	685	87.50		772.50	885
5100	Plate type system									
5200	With bearing jacks, single fixture	1 Plum	5	1.600	Ea.	775	87.50		862.50	985
5300	With exposed arms, single heavy fixture		5	1.600		995	87.50		1,082.50	1,225
5400	Wall mounted, exposed arms, single heavy fixture		5	1.600		360	87.50		447.50	530
6000	Urinal, floor mounted, 2" or 3" coupling, blowout type		6	1.333		455	73		528	610
6100	With fixture or hanger bolts, blowout or washout		6	1.333		320	73		393	465
6200	With bearing plate		6	1.333		360	73		433	505
6300	Wall mounted, plate type system		6	1.333		277	73		350	415
6980	Water closet, siphon jet									
7000	Horizontal, adjustable, caulk									
7040	Single, 4" pipe size	1 Plum	5.33	1.501	Ea.	815	82		897	1,025
7050	4" pipe size, paraplegic		5.33	1.501		815	82		897	1,025
7060	5" pipe size		5.33	1.501		1,025	82		1,107	1,250
7100	Double, 4" pipe size		5	1.600		1,425	87.50		1,512.50	1,675
7110	4" pipe size, paraplegic		5	1.600		1,425	87.50		1,512.50	1,675
7120	5" pipe size		5	1.600		1,725	87.50		1,812.50	2,000
7160	Horizontal, adjustable, extended, caulk									
7180	Single, 4" pipe size	1 Plum	5.33	1.501	Ea.	815	82		897	1,025
7200	5" pipe size		5.33	1.501		1,025	82		1,107	1,250
7240	Double, 4" pipe size		5	1.600		1,425	87.50		1,512.50	1,675
7260	5" pipe size		5	1.600		1,725	87.50		1,812.50	2,000
7400	Vertical, adjustable, caulk or thread									
7440	Single, 4" pipe size	1 Plum	5.33	1.501	Ea.	830	82		912	1,025
7460	5" pipe size		5.33	1.501		1,025	82		1,107	1,250
7480	6" pipe size		5	1.600		1,175	87.50		1,262.50	1,400
7520	Double, 4" pipe size		5	1.600		1,225	87.50		1,312.50	1,475
7540	5" pipe size		5	1.600		1,425	87.50		1,512.50	1,700
7560	6" pipe size		4	2		1,575	109		1,684	1,900
7600	Vertical, adjustable, extended, caulk									
7620	Single, 4" pipe size	1 Plum	5.33	1.501	Ea.	830	82		912	1,025
7640	5" pipe size		5.33	1.501		1,025	82		1,107	1,250
7680	6" pipe size		5	1.600		1,175	87.50		1,262.50	1,400
7720	Double, 4" pipe size		5	1.600		1,225	87.50		1,312.50	1,475
7740	5" pipe size		5	1.600		1,425	87.50		1,512.50	1,700
7760	6" pipe size		4	2		1,575	109		1,684	1,900
7780	Water closet, blow out									
7800	Vertical offset, caulk or thread									
7820	Single, 4" pipe size	1 Plum	5.33	1.501	Ea.	630	82		712	815
7840	Double, 4" pipe size	"	5	1.600	"	1,050	87.50		1,137.50	1,275
7880	Vertical offset, extended, caulk									
7900	Single, 4" pipe size	1 Plum	5.33	1.501	Ea.	770	82		852	970
7920	Double, 4" pipe size	"	5	1.600	"	1,200	87.50		1,287.50	1,450
7960	Vertical, for floor mounted back-outlet									
7980	Single, 4" thread, 2" vent	1 Plum	5.33	1.501	Ea.	540	82		622	720
8000	Double, 4" thread, 2" vent	"	6	1.333	"	1,575	73		1,648	1,850
8040	Vertical, for floor mounted back-outlet, extended									
8060	Single, 4" caulk, 2" vent	1 Plum	6	1.333	Ea.	540	73		613	705

22 42 Commercial Plumbing Fixtures

22 42 39 – Commercial Faucets, Supplies, and Trim

22 42 39.30 Carriers and Supports		Crew	Daily Output	Labor-Hours	Unit	Material	2012 Bare Costs Labor	Equipment	Total	Total Incl O&P
8080	Double, 4" caulk, 2" vent	1 Plum	6	1.333	Ea.	1,575	73		1,648	1,850
8200	Water closet, residential									
8220	Vertical centerline, floor mount									
8240	Single, 3" caulk, 2" or 3" vent	1 Plum	6	1.333	Ea.	490	73		563	650
8260	4" caulk, 2" or 4" vent		6	1.333		630	73		703	805
8280	3" copper sweat, 3" vent		6	1.333		440	73		513	595
8300	4" copper sweat, 4" vent		6	1.333		530	73		603	695
8400	Vertical offset, floor mount									
8420	Single, 3" or 4" caulk, vent	1 Plum	4	2	Ea.	610	109		719	840
8440	3" or 4" copper sweat, vent		5	1.600		630	87.50		717.50	820
8460	Double, 3" or 4" caulk, vent		4	2		1,050	109		1,159	1,325
8480	3" or 4" copper sweat, vent		5	1.600		1,050	87.50		1,137.50	1,275
9000	Water cooler (electric), floor mounted									
9100	Plate type with bearing plate, single	1 Plum	6	1.333	Ea.	310	73		383	450

22 45 Emergency Plumbing Fixtures

22 45 13 – Emergency Showers

22 45 13.10 Emergency Showers

		Crew	Daily Output	Labor-Hours	Unit	Material	2012 Bare Costs Labor	Equipment	Total	Total Incl O&P
0010	**EMERGENCY SHOWERS**, Rough-in not included									
5000	Shower, single head, drench, ball valve, pull, freestanding	Q-1	4	4	Ea.	350	197		547	680
5200	Horizontal or vertical supply		4	4		520	197		717	865
6000	Multi-nozzle, eye/face wash combination		4	4		615	197		812	970
6400	Multi-nozzle, 12 spray, shower only		4	4		1,950	197		2,147	2,450
6600	For freeze-proof, add		6	2.667		440	131		571	685

22 45 16 – Eyewash Equipment

22 45 16.10 Eyewash Safety Equipment

		Crew	Daily Output	Labor-Hours	Unit	Material	2012 Bare Costs Labor	Equipment	Total	Total Incl O&P
0010	**EYEWASH SAFETY EQUIPMENT**, Rough-in not included									
1000	Eye wash fountain									
1400	Plastic bowl, pedestal mounted	Q-1	4	4	Ea.	262	197		459	585
1600	Unmounted		4	4		227	197		424	545
1800	Wall mounted		4	4		420	197		617	760
2000	Stainless steel, pedestal mounted		4	4		405	197		602	740
2200	Unmounted		4	4		262	197		459	585
2400	Wall mounted		4	4		299	197		496	625

22 45 19 – Self-Contained Eyewash Equipment

22 45 19.10 Self-Contained Eyewash Safety Equipment

		Crew	Daily Output	Labor-Hours	Unit	Material	2012 Bare Costs Labor	Equipment	Total	Total Incl O&P
0010	**SELF-CONTAINED EYEWASH SAFETY EQUIPMENT**									
3000	Eye wash, portable, self-contained				Ea.	1,050			1,050	1,150

22 45 26 – Eye/Face Wash Equipment

22 45 26.10 Eye/Face Wash Safety Equipment

		Crew	Daily Output	Labor-Hours	Unit	Material	2012 Bare Costs Labor	Equipment	Total	Total Incl O&P
0010	**EYE/FACE WASH SAFETY EQUIPMENT**, Rough-in not included									
4000	Eye and face wash, combination fountain									
4200	Stainless steel, pedestal mounted	Q-1	4	4	Ea.	895	197		1,092	1,275
4400	Unmounted		4	4		262	197		459	585
4600	Wall mounted		4	4		269	197		466	595

22 47 Drinking Fountains and Water Coolers

22 47 13 – Drinking Fountains

22 47 13.10 Drinking Water Fountains	Crew	Daily Output	Labor-Hours	Unit	Material	2012 Bare Costs Labor	Equipment	Total	Total Incl O&P
0010 **DRINKING WATER FOUNTAINS**, For connection to cold water supply R224000-40									
1000 Wall mounted, non-recessed									
1400 Bronze, with no back	1 Plum	4	2	Ea.	975	109		1,084	1,250
1800 Cast aluminum, enameled, for correctional institutions		4	2		1,375	109		1,484	1,675
2000 Fiberglass, 12" back, single bubbler unit		4	2		1,450	109		1,559	1,775
2040 Dual bubbler		3.20	2.500		1,950	137		2,087	2,350
2400 Precast stone, no back		4	2		760	109		869	1,000
2700 Stainless steel, single bubbler, no back		4	2		920	109		1,029	1,175
2740 With back		4	2		635	109		744	865
2780 Dual handle & wheelchair projection type		4	2		685	109		794	920
2820 Dual level for handicapped type		3.20	2.500		1,400	137		1,537	1,750
3300 Vitreous china									
3340 7" back	1 Plum	4	2	Ea.	610	109		719	835
3940 For vandal-resistant bottom plate, add					115			115	127
3960 For freeze-proof valve system, add	1 Plum	2	4		610	218		828	1,000
3980 For rough-in, supply and waste, add	"	2.21	3.620		345	198		543	680
4000 Wall mounted, semi-recessed									
4200 Poly-marble, single bubbler	1 Plum	4	2	Ea.	925	109		1,034	1,200
4600 Stainless steel, satin finish, single bubbler		4	2		995	109		1,104	1,275
4900 Vitreous china, single bubbler		4	2		840	109		949	1,100
5980 For rough-in, supply and waste, add		1.83	4.372		345	239		584	740
6000 Wall mounted, fully recessed									
6400 Poly-marble, single bubbler	1 Plum	4	2	Ea.	1,525	109		1,634	1,850
6800 Stainless steel, single bubbler		4	2		1,525	109		1,634	1,850
7560 For freeze-proof valve system, add		2	4		730	218		948	1,125
7580 For rough-in, supply and waste, add		1.83	4.372		345	239		584	740
7600 Floor mounted, pedestal type									
7700 Aluminum, architectural style, C.I. base	1 Plum	2	4	Ea.	2,200	218		2,418	2,725
7780 Wheelchair handicap unit		2	4		1,475	218		1,693	1,950
8400 Stainless steel, architectural style		2	4		2,650	218		2,868	3,250
8600 Enameled iron, heavy duty service, 2 bubblers		2	4		2,500	218		2,718	3,100
8660 4 bubblers		2	4		3,825	218		4,043	4,525
8880 For freeze-proof valve system, add		2	4		610	218		828	1,000
8900 For rough-in, supply and waste, add		1.83	4.372		345	239		584	740
9100 Deck mounted									
9500 Stainless steel, circular receptor	1 Plum	4	2	Ea.	415	109		524	620
9760 White enameled steel, 14" x 9" receptor		4	2		360	109		469	560
9860 White enameled cast iron, 24" x 16" receptor		3	2.667		405	146		551	665
9980 For rough-in, supply and waste, add		1.83	4.372		345	239		584	740

22 47 16 – Pressure Water Coolers

22 47 16.10 Electric Water Coolers

	Crew	Daily Output	Labor-Hours	Unit	Material	2012 Bare Costs Labor	Equipment	Total	Total Incl O&P
0010 **ELECTRIC WATER COOLERS** R224000-50									
0100 Wall mounted, non-recessed									
0140 4 GPH	Q-1	4	4	Ea.	680	197		877	1,050
0160 8 GPH, barrier free, sensor operated		4	4		890	197		1,087	1,275
0180 8.2 G.P.H.		4	4		680	197		877	1,050
0600 8 GPH hot and cold water		4	4		1,175	197		1,372	1,600
0640 For stainless steel cabinet, add					84			84	92
1000 Dual height, 8.2 GPH	Q-1	3.80	4.211		1,975	207		2,182	2,475
1040 14.3 GPH	"	3.80	4.211		1,125	207		1,332	1,550
1240 For stainless steel cabinet, add					150			150	165
2600 Wheelchair type, 8 GPH	Q-1	4	4		975	197		1,172	1,375

22 47 Drinking Fountains and Water Coolers

22 47 16 – Pressure Water Coolers

22 47 16.10 Electric Water Coolers

		Crew	Daily Output	Labor-Hours	Unit	Material	2012 Bare Costs Labor	2012 Bare Costs Equipment	Total	Total Incl O&P
3300	Semi-recessed, 8.1 GPH	Q-1	4	4	Ea.	630	197		827	985
3320	12 GPH	↓	4	4	↓	725	197		922	1,100
4600	Floor mounted, flush-to-wall									
4640	4 GPH	1 Plum	3	2.667	Ea.	620	146		766	905
4680	8.2 GPH		3	2.667		655	146		801	940
4720	14.3 GPH		3	2.667		750	146		896	1,050
4960	14 GPH hot and cold water	↓	3	2.667		1,050	146		1,196	1,375
4980	For stainless steel cabinet, add					128			128	141
5000	Dual height, 8.2 GPH	1 Plum	2	4		1,100	218		1,318	1,550
5040	14.3 GPH	"	2	4		1,150	218		1,368	1,600
5120	For stainless steel cabinet, add					186			186	205
9800	For supply, waste & vent, all coolers	1 Plum	2.21	3.620	↓	345	198		543	680

22 51 Swimming Pool Plumbing Systems

22 51 19 – Swimming Pool Water Treatment Equipment

22 51 19.50 Swimming Pool Filtration Equipment

		Crew	Daily Output	Labor-Hours	Unit	Material	2012 Bare Costs Labor	2012 Bare Costs Equipment	Total	Total Incl O&P
0010	**SWIMMING POOL FILTRATION EQUIPMENT**									
0900	Filter system, sand or diatomite type, incl. pump, 6,000 gal./hr.	2 Plum	1.80	8.889	Total	1,775	485		2,260	2,675
1020	Add for chlorination system, 800 S.F. pool	↓	3	5.333	Ea.	105	291		396	555
1040	5,000 S.F. pool	↓	3	5.333	"	1,875	291		2,166	2,525

22 52 Fountain Plumbing Systems

22 52 16 – Fountain Pumps

22 52 16.10 Fountain Water Pumps

		Crew	Daily Output	Labor-Hours	Unit	Material	2012 Bare Costs Labor	2012 Bare Costs Equipment	Total	Total Incl O&P
0010	**FOUNTAIN WATER PUMPS**									
0100	Pump w/controls									
0200	Single phase, 100' cord, 1/2 H.P. pump	2 Skwk	4.40	3.636	Ea.	1,225	165		1,390	1,600
0300	3/4 H.P. pump		4.30	3.721		2,125	169		2,294	2,575
0400	1 H.P. pump		4.20	3.810		2,300	173		2,473	2,800
0500	1-1/2 H.P. pump		4.10	3.902		2,700	177		2,877	3,250
0600	2 H.P. pump		4	4		3,675	182		3,857	4,325
0700	Three phase, 200' cord, 5 H.P. pump		3.90	4.103		5,025	186		5,211	5,825
0800	7-1/2 H.P. pump		3.80	4.211		8,725	191		8,916	9,900
0900	10 H.P. pump		3.70	4.324		9,550	196		9,746	10,800
1000	15 H.P. pump	↓	3.60	4.444	↓	11,500	202		11,702	13,000
2000	DESIGN NOTE: Use two horsepower per surface acre.									

22 52 33 – Fountain Ancillary

22 52 33.10 Fountain Miscellaneous

		Crew	Daily Output	Labor-Hours	Unit	Material	2012 Bare Costs Labor	2012 Bare Costs Equipment	Total	Total Incl O&P
0010	**FOUNTAIN MISCELLANEOUS**									
1100	Nozzles, minimum	2 Skwk	8	2	Ea.	149	91		240	305
1200	Maximum		8	2		320	91		411	490
1300	Lights w/mounting kits, 200 watt		18	.889		1,050	40.50		1,090.50	1,225
1400	300 watt		18	.889		1,275	40.50		1,315.50	1,475
1500	500 watt		18	.889		1,425	40.50		1,465.50	1,650
1600	Color blender	↓	12	1.333	↓	555	60.50		615.50	705

22 66 53 – Laboratory Chemical-Waste and Vent Piping

22 66 53.30 Glass Pipe	Crew	Daily Output	Labor-Hours	Unit	Material	2012 Bare Costs Labor	2012 Bare Costs Equipment	Total	Total Incl O&P
0010 **GLASS PIPE**, Borosilicate, couplings & clevis hanger assemblies, 10' O.C.									
0020 Drainage									
1100 1-1/2" diameter	Q-1	52	.308	L.F.	10	15.10		25.10	34
1120 2" diameter		44	.364		13.05	17.85		30.90	41.50
1140 3" diameter		39	.410		17.65	20		37.65	50
1160 4" diameter		30	.533		31.50	26		57.50	74
1180 6" diameter		26	.615		58.50	30.50		89	110

22 66 53.60 Corrosion Resistant Pipe	Crew	Daily Output	Labor-Hours	Unit	Material	2012 Bare Costs Labor	2012 Bare Costs Equipment	Total	Total Incl O&P
0010 **CORROSION RESISTANT PIPE**, No couplings or hangers									
0020 Iron alloy, drain, mechanical joint									
1000 1-1/2" diameter	Q-1	70	.229	L.F.	46	11.25		57.25	67.50
1100 2" diameter		66	.242		47	11.90		58.90	69.50
1120 3" diameter		60	.267		60.50	13.10		73.60	86.50
1140 4" diameter		52	.308		77.50	15.10		92.60	109
2980 Plastic, epoxy, fiberglass filament wound, B&S joint									
3000 2" diameter	Q-1	62	.258	L.F.	12	12.70		24.70	32.50
3100 3" diameter		51	.314		14	15.40		29.40	39
3120 4" diameter		45	.356		20	17.50		37.50	48.50
3140 6" diameter		32	.500		28	24.50		52.50	68
3980 Polyester, fiberglass filament wound, B&S joint									
4000 2" diameter	Q-1	62	.258	L.F.	13.05	12.70		25.75	33.50
4100 3" diameter		51	.314		17	15.40		32.40	42
4120 4" diameter		45	.356		25	17.50		42.50	54
4140 6" diameter		32	.500		36	24.50		60.50	76.50
4980 Polypropylene, acid resistant, fire retardant, schedule 40									
5000 1-1/2" diameter	Q-1	68	.235	L.F.	6.90	11.55		18.45	25
5100 2" diameter		62	.258		9.45	12.70		22.15	29.50
5120 3" diameter		51	.314		19.30	15.40		34.70	44.50
5140 4" diameter		45	.356		24.50	17.50		42	53.50
5980 Proxylene, fire retardant, Schedule 40									
6000 1-1/2" diameter	Q-1	68	.235	L.F.	10.80	11.55		22.35	29.50
6100 2" diameter		62	.258		14.80	12.70		27.50	35.50
6120 3" diameter		51	.314		27	15.40		42.40	53
6140 4" diameter		45	.356		38	17.50		55.50	68

Estimating Tips

The labor adjustment factors listed in Subdivision 22 01 02.20 also apply to Division 23.

23 10 00 Facility Fuel Systems

- The prices in this subdivision for above- and below-ground storage tanks do not include foundations or hold-down slabs, unless noted. The estimator should refer to Divisions 3 and 31 for foundation system pricing. In addition to the foundations, required tank accessories, such as tank gauges, leak detection devices, and additional manholes and piping, must be added to the tank prices.

23 50 00 Central Heating Equipment

- When estimating the cost of an HVAC system, check to see who is responsible for providing and installing the temperature control system. It is possible to overlook controls, assuming that they would be included in the electrical estimate.

- When looking up a boiler, be careful on specified capacity. Some manufacturers rate their products on output while others use input.

- Include HVAC insulation for pipe, boiler, and duct (wrap and liner).

- Be careful when looking up mechanical items to get the correct pressure rating and connection type (thread, weld, flange).

23 70 00 Central HVAC Equipment

- Combination heating and cooling units are sized by the air conditioning requirements. (See Reference No. R236000-20 for preliminary sizing guide.)

- A ton of air conditioning is nominally 400 CFM.

- Rectangular duct is taken off by the linear foot for each size, but its cost is usually estimated by the pound. Remember that SMACNA standards now base duct on internal pressure.

- Prefabricated duct is estimated and purchased like pipe: straight sections and fittings.

- Note that cranes or other lifting equipment are not included on any lines in Division 23. For example, if a crane is required to lift a heavy piece of pipe into place high above a gym floor, or to put a rooftop unit on the roof of a four-story building, etc., it must be added. Due to the potential for extreme variation—from nothing additional required to a major crane or helicopter—we feel that including a nominal amount for "lifting contingency" would be useless and detract from the accuracy of the estimate. When using equipment rental cost data from RSMeans, do not forget to include the cost of the operator(s).

Reference Numbers

Reference numbers are shown in shaded boxes at the beginning of some major classifications. These numbers refer to related items in the Reference Section. The reference information may be an estimating procedure, an alternate pricing method, or technical information.

Note: Not all subdivisions listed here necessarily appear in this publication.

Note: **Trade Service,** *in part, has been used as a reference source for some of the material prices used in Division 23.*

23 05 02 – HVAC General

23 05 02.10 Air Conditioning, General	Crew	Daily Output	Labor-Hours	Unit	Material	2012 Bare Costs Labor	Equipment	Total	Total Incl O&P
0010 **AIR CONDITIONING, GENERAL** Prices are for standard efficiencies (SEER 13)									
0020 for upgrade to SEER 14 add					20%				

23 05 05 – Selective HVAC Demolition

23 05 05.10 HVAC Demolition

	Crew	Daily Output	Labor-Hours	Unit	Material	Labor	Equipment	Total	Total Incl O&P
0010 **HVAC DEMOLITION**									
0100 Air conditioner, split unit, 3 ton	Q-5	2	8	Ea.		400		400	605
0150 Package unit, 3 ton	Q-6	3	8	"		415		415	625
0298 Boilers									
0300 Electric, up thru 148 kW	Q-19	2	12	Ea.		605		605	915
0310 150 thru 518 kW	"	1	24			1,225		1,225	1,825
0320 550 thru 2000 kW	Q-21	.40	80			4,150		4,150	6,250
0330 2070 kW and up	"	.30	106			5,525		5,525	8,325
0340 Gas and/or oil, up thru 150 MBH	Q-7	2.20	14.545			770		770	1,175
0350 160 thru 2000 MBH		.80	40			2,125		2,125	3,200
0360 2100 thru 4500 MBH		.50	64			3,400		3,400	5,125
0370 4600 thru 7000 MBH		.30	106			5,650		5,650	8,525
0380 7100 thru 12,000 MBH		.16	200			10,600		10,600	16,000
0390 12,200 thru 25,000 MBH	▼	.12	266	▼		14,100		14,100	21,300
1000 Ductwork, 4" high, 8" wide	1 Clab	200	.040	L.F.		1.40		1.40	2.16
1100 6" high, 8" wide		165	.048			1.70		1.70	2.62
1200 10" high, 12" wide		125	.064			2.25		2.25	3.46
1300 12"-14" high, 16"-18" wide		85	.094			3.30		3.30	5.10
1400 18" high, 24" wide		67	.119			4.19		4.19	6.45
1500 30" high, 36" wide		56	.143			5		5	7.70
1540 72" wide	▼	50	.160	▼		5.60		5.60	8.65
3000 Mechanical equipment, light items. Unit is weight, not cooling.	Q-5	.90	17.778	Ton		890		890	1,350
3600 Heavy items	"	1.10	14.545	"		730		730	1,100
3700 Deduct for salvage (when applicable), minimum				Job				73	80
3710 Maximum				"				455	500

23 05 23 – General-Duty Valves for HVAC Piping

23 05 23.30 Valves, Iron Body

	Crew	Daily Output	Labor-Hours	Unit	Material	Labor	Equipment	Total	Total Incl O&P
0010 **VALVES, IRON BODY**									
1020 Butterfly, wafer type, gear actuator, 200 lb.									
1030 2"	1 Plum	14	.571	Ea.	172	31		203	237
1040 2-1/2"	Q-1	9	1.778		178	87.50		265.50	325
1050 3"		8	2		184	98.50		282.50	350
1060 4"	▼	5	3.200		209	157		366	465
1070 5"	Q-2	5	4.800		278	245		523	675
1080 6"	"	5	4.800	▼	315	245		560	715
1650 Gate, 125 lb., N.R.S.									
2150 Flanged									
2200 2"	1 Plum	5	1.600	Ea.	610	87.50		697.50	800
2240 2-1/2"	Q-1	5	3.200		625	157		782	920
2260 3"		4.50	3.556		700	175		875	1,025
2280 4"	▼	3	5.333		1,000	262		1,262	1,500
2300 6"	Q-2	3	8	▼	1,700	410		2,110	2,500
3550 OS&Y, 125 lb., flanged									
3600 2"	1 Plum	5	1.600	Ea.	315	87.50		402.50	475
3660 3"	Q-1	4.50	3.556		445	175		620	750
3680 4"	"	3	5.333		635	262		897	1,100
3700 6"	Q-2	3	8	▼	1,050	410		1,460	1,775
3900 For 175 lb., flanged, add					200%	10%			

23 05 23 – General-Duty Valves for HVAC Piping

23 05 23.30 Valves, Iron Body		Crew	Daily Output	Labor-Hours	Unit	Material	2012 Bare Costs Labor	Equipment	Total	Total Incl O&P
5450	Swing check, 125 lb., threaded									
5470	1"	1 Plum	13	.615	Ea.	575	33.50		608.50	685
5500	2"	"	11	.727		780	39.50		819.50	915
5540	2-1/2"	Q-1	15	1.067		675	52.50		727.50	825
5550	3"		13	1.231		750	60.50		810.50	915
5560	4"		10	1.600		1,300	78.50		1,378.50	1,550
5950	Flanged									
6000	2"	1 Plum	5	1.600	Ea.	335	87.50		422.50	500
6040	2-1/2"	Q-1	5	3.200		340	157		497	605
6050	3"		4.50	3.556		345	175		520	640
6060	4"		3	5.333		540	262		802	985
6070	6"	Q-2	3	8		920	410		1,330	1,625

23 05 23.80 Valves, Steel

23 05 23.80 Valves, Steel		Crew	Daily Output	Labor-Hours	Unit	Material	2012 Bare Costs Labor	Equipment	Total	Total Incl O&P
0010	**VALVES, STEEL**									
0800	Cast									
1350	Check valve, swing type, 150 lb., flanged									
1370	1"	1 Plum	10	.800	Ea.	390	43.50		433.50	495
1400	2"	"	8	1		660	54.50		714.50	810
1440	2-1/2"	Q-1	5	3.200		765	157		922	1,075
1450	3"		4.50	3.556		780	175		955	1,125
1460	4"		3	5.333		1,125	262		1,387	1,625
1540	For 300 lb., flanged, add					50%	15%			
1548	For 600 lb., flanged, add					110%	20%			
1950	Gate valve, 150 lb., flanged									
2000	2"	1 Plum	8	1	Ea.	710	54.50		764.50	865
2040	2-1/2"	Q-1	5	3.200		1,000	157		1,157	1,325
2050	3"		4.50	3.556		1,000	175		1,175	1,375
2060	4"		3	5.333		1,250	262		1,512	1,775
2070	6"	Q-2	3	8		1,850	410		2,260	2,650
3650	Globe valve, 150 lb., flanged									
3700	2"	1 Plum	8	1	Ea.	890	54.50		944.50	1,075
3740	2-1/2"	Q-1	5	3.200		1,125	157		1,282	1,475
3750	3"		4.50	3.556		1,125	175		1,300	1,525
3760	4"		3	5.333		1,650	262		1,912	2,225
3770	6"	Q-2	3	8		2,600	410		3,010	3,475
5150	Forged									
5650	Check valve, class 800, horizontal, socket									
5698	Threaded									
5700	1/4"	1 Plum	24	.333	Ea.	94.50	18.20		112.70	132
5720	3/8"		24	.333		94.50	18.20		112.70	132
5730	1/2"		24	.333		94.50	18.20		112.70	132
5740	3/4"		20	.400		101	22		123	144
5750	1"		19	.421		119	23		142	166
5760	1-1/4"		15	.533		233	29		262	300

23 05 93 – Testing, Adjusting, and Balancing for HVAC

23 05 93.10 Balancing, Air

23 05 93.10 Balancing, Air		Crew	Daily Output	Labor-Hours	Unit	Material	2012 Bare Costs Labor	Equipment	Total	Total Incl O&P
0010	**BALANCING, AIR** (Subcontractor's quote incl. material and labor)									
0900	Heating and ventilating equipment									
1000	Centrifugal fans, utility sets				Ea.				340	375
1100	Heating and ventilating unit								515	565
1200	In-line fan								515	565
1300	Propeller and wall fan								97	107

23 05 Common Work Results for HVAC

23 05 93 – Testing, Adjusting, and Balancing for HVAC

23 05 93.10 Balancing, Air		Crew	Daily Output	Labor-Hours	Unit	Material	2012 Bare Costs Labor	Equipment	Total	Total Incl O&P
1400	Roof exhaust fan				Ea.				230	251
2000	Air conditioning equipment, central station								745	815
2100	Built-up low pressure unit								685	755
2200	Built-up high pressure unit								800	880
2500	Multi-zone A.C. and heating unit								515	565
2600	For each zone over one, add								114	126
2700	Package A.C. unit								286	315
2800	Rooftop heating and cooling unit								400	440
3000	Supply, return, exhaust, registers & diffusers, avg. height ceiling								68	75
3100	High ceiling								103	113
3200	Floor height								57	63

23 05 93.20 Balancing, Water

		Crew	Daily Output	Labor-Hours	Unit	Material	2012 Bare Costs Labor	Equipment	Total	Total Incl O&P
0010	**BALANCING, WATER** (Subcontractor's quote incl. material and labor)									
0050	Air cooled condenser				Ea.				215	235
0080	Boiler								430	475
0100	Cabinet unit heater								73.50	80.50
0200	Chiller								520	570
0300	Convector								61	67
0500	Cooling tower								400	435
0600	Fan coil unit, unit ventilator								110	121
0700	Fin tube and radiant panels								122	134
0800	Main and duct re-heat coils								113	124
0810	Heat exchanger								113	124
1000	Pumps								269	296
1100	Unit heater								85.50	94

23 07 HVAC Insulation

23 07 13 – Duct Insulation

23 07 13.10 Duct Thermal Insulation

			Crew	Daily Output	Labor-Hours	Unit	Material	2012 Bare Costs Labor	Equipment	Total	Total Incl O&P
0010	**DUCT THERMAL INSULATION**										
0100	Rule of thumb, as a percentage of total mechanical costs					Job				10%	
0110	Insulation req'd is based on the surface size/area to be covered										
3000	Ductwork										
3020	Blanket type, fiberglass, flexible										
3140	FSK vapor barrier wrap, .75 lb. density										
3160	1" thick	G	Q-14	350	.046	S.F.	.18	2.01		2.19	3.33
3170	1-1/2" thick	G		320	.050		.21	2.20		2.41	3.66
3180	2" thick	G		300	.053		.26	2.35		2.61	3.94
3190	3" thick	G		260	.062		.37	2.71		3.08	4.63
3200	4" thick	G		242	.066		.60	2.91		3.51	5.20
3210	Vinyl jacket, same as FSK										
3280	Unfaced, 1 lb. density										
3310	1" thick	G	Q-14	360	.044	S.F.	.19	1.95		2.14	3.25
3320	1-1/2" thick	G		330	.048		.23	2.13		2.36	3.57
3330	2" thick	G		310	.052		.31	2.27		2.58	3.88
3400	FSK facing, 1 lb. density										
3420	1-1/2" thick	G	Q-14	310	.052	S.F.	.28	2.27		2.55	3.85
3430	2" thick	G	"	300	.053	"	.33	2.35		2.68	4.01
3450	FSK facing, 1.5 lb. density										
3470	1-1/2" thick	G	Q-14	300	.053	S.F.	.37	2.35		2.72	4.06
3480	2" thick	G	"	290	.055	"	.48	2.43		2.91	4.31

23 07 HVAC Insulation

23 07 13 – Duct Insulation

23 07 13.10 Duct Thermal Insulation

23 07 13.10 Duct Thermal Insulation		Crew	Daily Output	Labor-Hours	Unit	Material	2012 Bare Costs Labor	Equipment	Total	Total Incl O&P
3795	Finishes									
3800	Stainless steel woven mesh	Q-14	100	.160	S.F.	.70	7.05		7.75	11.70
3810	For .010" stainless steel, add		160	.100		2.79	4.40		7.19	9.90
3820	18 oz. fiberglass cloth, pasted on		170	.094		.67	4.14		4.81	7.20
3900	8 oz. canvas, pasted on		180	.089		.19	3.91		4.10	6.30
3940	For .016" aluminum jacket, add		200	.080		.94	3.52		4.46	6.55
7878	Contact cement, quart can				Ea.	10.50			10.50	11.55

23 07 16 – HVAC Equipment Insulation

23 07 16.10 HVAC Equipment Thermal Insulation

23 07 16.10 HVAC Equipment Thermal Insulation			Crew	Daily Output	Labor-Hours	Unit	Material	2012 Bare Costs Labor	Equipment	Total	Total Incl O&P
0010	**HVAC EQUIPMENT THERMAL INSULATION**										
0100	Rule of thumb, as a percentage of total mechanical costs					Job				10%	
0110	Insulation req'd is based on the surface size/area to be covered										
1000	Boiler, 1-1/2" calcium silicate only	G	Q-14	110	.145	S.F.	3.50	6.40		9.90	13.80
1020	Plus 2" fiberglass	G	"	80	.200	"	4.47	8.80		13.27	18.60
2000	Breeching, 2" calcium silicate										
2020	Rectangular	G	Q-14	42	.381	S.F.	6.85	16.75		23.60	33.50
2040	Round	G	"	38.70	.413	"	7.15	18.20		25.35	36.50

23 09 Instrumentation and Control for HVAC

23 09 33 – Electric and Electronic Control System for HVAC

23 09 33.10 Electronic Control Systems

23 09 33.10 Electronic Control Systems		Crew	Daily Output	Labor-Hours	Unit	Material	2012 Bare Costs Labor	Equipment	Total	Total Incl O&P
0010	**ELECTRONIC CONTROL SYSTEMS**									
0020	For electronic costs, add to Section 23 09 43.10				Ea.					15%

23 09 43 – Pneumatic Control System for HVAC

23 09 43.10 Pneumatic Control Systems

23 09 43.10 Pneumatic Control Systems			Crew	Daily Output	Labor-Hours	Unit	Material	2012 Bare Costs Labor	Equipment	Total	Total Incl O&P
0010	**PNEUMATIC CONTROL SYSTEMS**										
0011	Including a nominal 50 ft. of tubing. Add control panelboard if req'd.										
0100	Heating and ventilating, split system										
0200	Mixed air control, economizer cycle, panel readout, tubing										
0220	Up to 10 tons	G	Q-19	.68	35.294	Ea.	3,800	1,775		5,575	6,875
0240	For 10 to 20 tons	G		.63	37.915		4,075	1,925		6,000	7,350
0260	For over 20 tons	G		.58	41.096		4,425	2,075		6,500	7,975
0300	Heating coil, hot water, 3 way valve,										
0320	Freezestat, limit control on discharge, readout		Q-5	.69	23.088	Ea.	2,825	1,150		3,975	4,850
0500	Cooling coil, chilled water, room										
0520	Thermostat, 3 way valve		Q-5	2	8	Ea.	1,250	400		1,650	1,975
0600	Cooling tower, fan cycle, damper control,										
0620	Control system including water readout in/out at panel		Q-19	.67	35.821	Ea.	5,000	1,825		6,825	8,225
1000	Unit ventilator, day/night operation,										
1100	freezestat, ASHRAE, cycle 2		Q-19	.91	26.374	Ea.	2,775	1,325		4,100	5,050
2000	Compensated hot water from boiler, valve control,										
2100	readout and reset at panel, up to 60 GPM		Q-19	.55	43.956	Ea.	5,175	2,225		7,400	9,050
2120	For 120 GPM			.51	47.059		5,550	2,375		7,925	9,675
2140	For 240 GPM			.49	49.180		5,800	2,500		8,300	10,200
3000	Boiler room combustion air, damper to 5 S.F., controls			1.37	17.582		2,500	890		3,390	4,100
3500	Fan coil, heating and cooling valves, 4 pipe control system			3	8		1,125	405		1,530	1,850
3600	Heat exchanger system controls			.86	27.907		2,425	1,400		3,825	4,800
4000	Pneumatic thermostat, including controlling room radiator valve		Q-5	2.43	6.593		755	330		1,085	1,325
4060	Pump control system		Q-19	3	8		1,150	405		1,555	1,875
4500	Air supply for pneumatic control system										

23 09 Instrumentation and Control for HVAC

23 09 43 – Pneumatic Control System for HVAC

23 09 43.10 Pneumatic Control Systems	Crew	Daily Output	Labor-Hours	Unit	Material	2012 Bare Costs Labor	Equipment	Total	Total Incl O&P
4600	Tank mounted duplex compressor, starter, alternator,									
4620	piping, dryer, PRV station and filter									
4630	1/2 HP	Q-19	.68	35.139	Ea.	9,300	1,775		11,075	12,900
4660	1-1/2 HP		.58	41.739		11,400	2,100		13,500	15,700
4690	5 HP		.42	57.143		27,000	2,900		29,900	34,100

23 13 Facility Fuel-Storage Tanks

23 13 13 – Facility Underground Fuel-Oil, Storage Tanks

23 13 13.09 Single-Wall Steel Fuel-Oil Tanks

		Crew	Daily Output	Labor-Hours	Unit	Material	2012 Bare Costs Labor	Equipment	Total	Total Incl O&P
0010	**SINGLE-WALL STEEL FUEL-OIL TANKS**									
5000	Tanks, steel ugnd, sti-p3, not incl hold-down bars									
5500	Excavation, pad, pumps and piping not included									
5510	Single wall, 500 gallon capacity, 7 gauge shell	Q-5	2.70	5.926	Ea.	2,050	297		2,347	2,700
5520	1,000 gallon capacity, 7 gauge shell	"	2.50	6.400		3,475	320		3,795	4,300
5530	2,000 gallon capacity, 1/4" thick shell	Q-7	4.60	6.957		5,625	370		5,995	6,750
5535	2,500 gallon capacity, 7 gauge shell	Q-5	3	5.333		6,225	267		6,492	7,250
5540	5,000 gallon capacity, 1/4" thick shell	Q-7	3.20	10		10,500	530		11,030	12,300
5580	15,000 gallon capacity, 5/16" thick shell		1.70	18.824		18,100	1,000		19,100	21,400
5600	20,000 gallon capacity, 5/16" thick shell		1.50	21.333		33,000	1,125		34,125	38,000
5610	25,000 gallon capacity, 3/8" thick shell		1.30	24.615		32,300	1,300		33,600	37,500
5620	30,000 gallon capacity, 3/8" thick shell		1.10	29.091		34,900	1,550		36,450	40,700
5630	40,000 gallon capacity, 3/8" thick shell		.90	35.556		35,600	1,875		37,475	42,100
5640	50,000 gallon capacity, 3/8" thick shell		.80	40		39,500	2,125		41,625	46,700

23 13 13.23 Glass-Fiber-Reinfcd-Plastic, Fuel-Oil, Storage

		Crew	Daily Output	Labor-Hours	Unit	Material	2012 Bare Costs Labor	Equipment	Total	Total Incl O&P
0010	**GLASS-FIBER-REINFCD-PLASTIC, UNDERGRND FUEL-OIL, STORAGE**									
0210	Fiberglass, underground, single wall, U.L. listed, not including									
0220	manway or hold-down strap									
0240	2,000 gallon capacity	Q-7	4.57	7.002	Ea.	6,700	370		7,070	7,925
0250	4,000 gallon capacity		3.55	9.014		9,525	480		10,005	11,200
0260	6,000 gallon capacity		2.67	11.985		10,600	635		11,235	12,600
0280	10,000 gallon capacity		2	16		14,200	850		15,050	16,900
0284	15,000 gallon capacity		1.68	19.048		26,300	1,000		27,300	30,500
0290	20,000 gallon capacity		1.45	22.069		28,900	1,175		30,075	33,600
0500	For manway, fittings and hold-downs, add					20%	15%			
1020	Fiberglass, underground, double wall, U.L. listed									
1030	includes manways, not incl. hold-down straps									
1040	600 gallon capacity	Q-5	2.42	6.612	Ea.	7,875	330		8,205	9,150
1050	1,000 gallon capacity	"	2.25	7.111		10,600	355		10,955	12,100
1060	2,500 gallon capacity	Q-7	4.16	7.692		15,100	410		15,510	17,200
1070	3,000 gallon capacity		3.90	8.205		16,400	435		16,835	18,700
1080	4,000 gallon capacity		3.64	8.791		16,500	465		16,965	18,900
1090	6,000 gallon capacity		2.42	13.223		21,800	700		22,500	25,100
1100	8,000 gallon capacity		2.08	15.385		24,700	815		25,515	28,400
1110	10,000 gallon capacity		1.82	17.582		27,000	935		27,935	31,100
1120	12,000 gallon capacity		1.70	18.824		33,100	1,000		34,100	37,900
2210	Fiberglass, underground, single wall, U.L. listed, including									
2220	hold-down straps, no manways									
2240	2,000 gallon capacity	Q-7	3.55	9.014	Ea.	7,100	480		7,580	8,525
2250	4,000 gallon capacity		2.90	11.034		9,925	585		10,510	11,800
2260	6,000 gallon capacity		2	16		11,400	850		12,250	13,800
2280	10,000 gallon capacity		1.60	20		15,000	1,050		16,050	18,100

23 13 Facility Fuel-Storage Tanks

23 13 13 – Facility Underground Fuel-Oil, Storage Tanks

23 13 13.23 Glass-Fiber-Reinfcd-Plastic, Fuel-Oil, Storage

		Crew	Daily Output	Labor-Hours	Unit	Material	2012 Bare Costs Labor	2012 Bare Costs Equipment	Total	Total Incl O&P
2284	15,000 gallon capacity	Q-7	1.39	23.022	Ea.	27,100	1,225		28,325	31,800
2290	20,000 gallon capacity	↓	1.14	28.070	↓	30,100	1,500		31,600	35,400
3020	Fiberglass, underground, double wall, U.L. listed									
3030	includes manways and hold-down straps									
3040	600 gallon capacity	Q-5	1.86	8.602	Ea.	8,275	430		8,705	9,750
3050	1,000 gallon capacity	"	1.70	9.412		11,000	470		11,470	12,800
3060	2,500 gallon capacity	Q-7	3.29	9.726		15,500	515		16,015	17,900
3070	3,000 gallon capacity		3.13	10.224		16,800	540		17,340	19,300
3080	4,000 gallon capacity		2.93	10.922		16,900	580		17,480	19,500
3090	6,000 gallon capacity		1.86	17.204		22,600	915		23,515	26,200
3100	8,000 gallon capacity		1.65	19.394		25,500	1,025		26,525	29,700
3110	10,000 gallon capacity		1.48	21.622		27,800	1,150		28,950	32,300
3120	12,000 gallon capacity	↓	1.40	22.857	↓	33,900	1,200		35,100	39,100

23 13 23 – Facility Aboveground Fuel-Oil, Storage Tanks

23 13 23.13 Vertical, Steel, Abvground Fuel-Oil, Stor. Tanks

		Crew	Daily Output	Labor-Hours	Unit	Material	2012 Bare Costs Labor	2012 Bare Costs Equipment	Total	Total Incl O&P
0010	**VERTICAL, STEEL, ABOVEGROUND FUEL-OIL, STORAGE TANKS**									
4000	Fixed roof oil storage tanks, steel, (1 BBL=42 gal. w/foundation 3'D x 1'W)									
4200	5,000 barrels				Ea.				194,000	213,500
4300	24,000 barrels								333,500	367,000
4500	56,000 barrels								729,000	802,000
4600	110,000 barrels								1,060,000	1,166,000
4800	143,000 barrels								1,250,000	1,375,000
4900	225,000 barrels								1,360,000	1,496,000
5100	Floating roof gasoline tanks, steel, 5,000 barrels (w/foundation 3'D x 1'W)								204,000	225,000
5200	25,000 barrels								381,000	419,000
5400	55,000 barrels								839,000	923,000
5500	100,000 barrels								1,253,000	1,379,000
5700	150,000 barrels								1,532,000	1,685,000
5800	225,000 barrels								2,300,000	2,783,000

23 13 23.16 Horizontal, Stl, Abvgrd Fuel-Oil, Storage Tanks

		Crew	Daily Output	Labor-Hours	Unit	Material	2012 Bare Costs Labor	2012 Bare Costs Equipment	Total	Total Incl O&P
0010	**HORIZONTAL, STEEL, ABOVEGROUND FUEL-OIL, STORAGE TANKS**									
3000	Steel, storage, above ground, including cradles, coating,									
3020	fittings, not including foundation, pumps or piping									
3040	Single wall, 275 gallon	Q-5	5	3.200	Ea.	445	160		605	730
3060	550 gallon	"	2.70	5.926		3,100	297		3,397	3,850
3080	1,000 gallon	Q-7	5	6.400		5,075	340		5,415	6,100
3100	1,500 gallon		4.75	6.737		7,250	355		7,605	8,525
3120	2,000 gallon		4.60	6.957		8,800	370		9,170	10,200
3140	5,000 gallon		3.20	10		19,100	530		19,630	21,800
3150	10,000 gallon		2	16		29,000	850		29,850	33,200
3160	15,000 gallon		1.70	18.824		37,300	1,000		38,300	42,500
3170	20,000 gallon		1.45	22.069		48,200	1,175		49,375	55,000
3180	25,000 gallon		1.30	24.615		56,500	1,300		57,800	64,000
3190	30,000 gallon	↓	1.10	29.091		67,500	1,550		69,050	76,500
3320	Double wall, 500 gallon capacity	Q-5	2.40	6.667		2,375	335		2,710	3,125
3330	2000 gallon capacity	Q-7	4.15	7.711		9,075	410		9,485	10,600
3340	4000 gallon capacity		3.60	8.889		16,200	470		16,670	18,500
3350	6000 gallon capacity		2.40	13.333		19,100	705		19,805	22,100
3360	8000 gallon capacity		2	16		24,500	850		25,350	28,300
3370	10000 gallon capacity		1.80	17.778		27,500	945		28,445	31,600
3380	15000 gallon capacity		1.50	21.333		41,700	1,125		42,825	47,600
3390	20000 gallon capacity	↓	1.30	24.615	↓	47,500	1,300		48,800	54,500

23 13 Facility Fuel-Storage Tanks

23 13 23 – Facility Aboveground Fuel-Oil, Storage Tanks

23 13 23.16 Horizontal, Stl, Abvgrd Fuel-Oil, Storage Tanks	Crew	Daily Output	Labor-Hours	Unit	Material	2012 Bare Costs Labor	Equipment	Total	Total Incl O&P
3400	25000 gallon capacity	Q-7	1.15	27.826	Ea.	57,500	1,475		58,975	65,500
3410	30000 gallon capacity	↓	1	32	↓	63,500	1,700		65,200	72,000

23 21 Hydronic Piping and Pumps

23 21 20 – Hydronic HVAC Piping Specialties

23 21 20.10 Air Control

		Crew	Daily Output	Labor-Hours	Unit	Material	Labor	Equipment	Total	Total Incl O&P
0010	**AIR CONTROL**									
0030	Air separator, with strainer									
0040	2" diameter	Q-5	6	2.667	Ea.	905	134		1,039	1,200
0080	2-1/2" diameter		5	3.200		1,025	160		1,185	1,375
0100	3" diameter		4	4		1,550	201		1,751	2,025
0120	4" diameter	↓	3	5.333		2,250	267		2,517	2,875
0130	5" diameter	Q-6	3.60	6.667		2,875	345		3,220	3,700
0140	6" diameter	"	3.40	7.059	↓	3,450	365		3,815	4,350

23 21 20.18 Automatic Air Vent

		Crew	Daily Output	Labor-Hours	Unit	Material	Labor	Equipment	Total	Total Incl O&P
0010	**AUTOMATIC AIR VENT**									
0020	Cast iron body, stainless steel internals, float type									
0060	1/2" NPT inlet, 300 psi	1 Stpi	12	.667	Ea.	93.50	37		130.50	159
0220	3/4" NPT inlet, 250 psi	"	10	.800		299	44.50		343.50	395
0340	1-1/2" NPT inlet, 250 psi	Q-5	12	1.333	↓	925	67		992	1,125

23 21 20.42 Expansion Joints

		Crew	Daily Output	Labor-Hours	Unit	Material	Labor	Equipment	Total	Total Incl O&P
0010	**EXPANSION JOINTS**									
0100	Bellows type, neoprene cover, flanged spool									
0140	6" face to face, 1-1/4" diameter	1 Stpi	11	.727	Ea.	255	40.50		295.50	340
0160	1-1/2" diameter	"	10.60	.755		255	42		297	345
0180	2" diameter	Q-5	13.30	1.203		258	60.50		318.50	375
0190	2-1/2" diameter		12.40	1.290		267	64.50		331.50	390
0200	3" diameter		11.40	1.404		299	70.50		369.50	435
0480	10" face to face, 2" diameter		13	1.231		370	61.50		431.50	505
0500	2-1/2" diameter		12	1.333		390	67		457	530
0520	3" diameter		11	1.455		400	73		473	550
0540	4" diameter		8	2		455	100		555	650
0560	5" diameter		7	2.286		540	115		655	770
0580	6" diameter	↓	6	2.667	↓	560	134		694	815

23 21 20.46 Expansion Tanks

		Crew	Daily Output	Labor-Hours	Unit	Material	Labor	Equipment	Total	Total Incl O&P
0010	**EXPANSION TANKS**									
1507	Underground fuel-oil storage tanks, see Section 23 13 13									
1512	Tank leak detection systems, see Section 28 33 33.50									
2000	Steel, liquid expansion, ASME, painted, 15 gallon capacity	Q-5	17	.941	Ea.	595	47		642	725
2020	24 gallon capacity		14	1.143		635	57.50		692.50	785
2040	30 gallon capacity		12	1.333		665	67		732	835
2060	40 gallon capacity		10	1.600		780	80		860	980
2080	60 gallon capacity		8	2		935	100		1,035	1,175
2100	80 gallon capacity		7	2.286		1,000	115		1,115	1,275
2120	100 gallon capacity		6	2.667		1,350	134		1,484	1,700
3000	Steel ASME expansion, rubber diaphragm, 19 gal. cap. accept.		12	1.333		2,450	67		2,517	2,800
3020	31 gallon capacity		8	2		2,600	100		2,700	3,000
3040	61 gallon capacity		6	2.667		3,650	134		3,784	4,225
3080	119 gallon capacity		4	4		3,925	201		4,126	4,625
3100	158 gallon capacity	↓	3.80	4.211		5,475	211		5,686	6,350

23 21 20 – Hydronic HVAC Piping Specialties

23 21 20.46 Expansion Tanks	Crew	Daily Output	Labor-Hours	Unit	Material	2012 Bare Costs Labor	Equipment	Total	Total Incl O&P
3140	317 gallon capacity	Q-5	2.80	5.714	Ea.	8,250	286		8,536	9,500
3180	528 gallon capacity	▼	2.40	6.667	▼	13,400	335		13,735	15,200

23 21 20.58 Hydronic Heating Control Valves

0010	**HYDRONIC HEATING CONTROL VALVES**								
0100	Radiator supply, 1/2" diameter	1 Stpi	24	.333	Ea.	59.50	18.55		78.05	93
0120	3/4" diameter		20	.400		62.50	22.50		85	102
0140	1" diameter		19	.421		77.50	23.50		101	121
0160	1-1/4" diameter	▼	15	.533	▼	111	29.50		140.50	167
0500	For low pressure steam, add					25%				

23 21 20.70 Steam Traps

0010	**STEAM TRAPS**								
0030	Cast iron body, threaded								
0040	Inverted bucket								
0050	1/2" pipe size	1 Stpi	12	.667	Ea.	136	37		173	205
0070	3/4" pipe size		10	.800		236	44.50		280.50	325
0100	1" pipe size		9	.889		365	49.50		414.50	475
0120	1-1/4" pipe size	▼	8	1	▼	550	55.50		605.50	690
1000	Float & thermostatic, 15 psi								
1010	3/4" pipe size	1 Stpi	16	.500	Ea.	109	28		137	162
1020	1" pipe size		15	.533		131	29.50		160.50	189
1040	1-1/2" pipe size		9	.889		231	49.50		280.50	330
1060	2" pipe size	▼	6	1.333	▼	425	74.50		499.50	580

23 21 20.76 Strainers, Y Type, Bronze Body

0010	**STRAINERS, Y TYPE, BRONZE BODY**								
0050	Screwed, 125 lb., 1/4" pipe size	1 Stpi	24	.333	Ea.	22	18.55		40.55	52
0070	3/8" pipe size		24	.333		29	18.55		47.55	59.50
0100	1/2" pipe size		20	.400		29	22.50		51.50	65
0140	1" pipe size		17	.471		38.50	26		64.50	81.50
0160	1-1/2" pipe size		14	.571		83	32		115	140
0180	2" pipe size		13	.615		110	34.50		144.50	173
0182	3" pipe size	▼	12	.667		835	37		872	975
0200	300 lb., 2-1/2" pipe size	Q-5	17	.941		485	47		532	605
0220	3" pipe size		16	1		960	50		1,010	1,125
0240	4" pipe size	▼	15	1.067	▼	2,200	53.50		2,253.50	2,475
0500	For 300 lb. rating 1/4" thru 2", add					15%				
1000	Flanged, 150 lb., 1-1/2" pipe size	1 Stpi	11	.727	Ea.	450	40.50		490.50	555
1020	2" pipe size	"	8	1		610	55.50		665.50	755
1030	2-1/2" pipe size	Q-5	5	3.200		905	160		1,065	1,225
1040	3" pipe size		4.50	3.556		1,125	178		1,303	1,500
1060	4" pipe size	▼	3	5.333		1,700	267		1,967	2,250
1100	6" pipe size	Q-6	3	8		3,225	415		3,640	4,175
1106	8" pipe size	"	2.60	9.231	▼	3,550	480		4,030	4,625
1500	For 300 lb. rating, add					40%				

23 21 20.78 Strainers, Y Type, Iron Body

0010	**STRAINERS, Y TYPE, IRON BODY**								
0050	Screwed, 250 lb., 1/4" pipe size	1 Stpi	20	.400	Ea.	10.55	22.50		33.05	45
0070	3/8" pipe size		20	.400		10.55	22.50		33.05	45
0100	1/2" pipe size		20	.400		10.55	22.50		33.05	45
0140	1" pipe size		16	.500		17.30	28		45.30	61
0160	1-1/2" pipe size		12	.667		28.50	37		65.50	87
0180	2" pipe size	▼	8	1		42.50	55.50		98	131
0220	3" pipe size	Q-5	11	1.455	▼	275	73		348	415

23 21 20 – Hydronic HVAC Piping Specialties

23 21 20.78 Strainers, Y Type, Iron Body		Crew	Daily Output	Labor-Hours	Unit	Material	2012 Bare Costs		Total	Total Incl O&P
							Labor	Equipment		
0240	4" pipe size	Q-5	5	3.200	Ea.	465	160		625	755
0500	For galvanized body, add					50%				
1000	Flanged, 125 lb., 1-1/2" pipe size	1 Stpi	11	.727	Ea.	147	40.50		187.50	223
1020	2" pipe size	"	8	1		108	55.50		163.50	203
1040	3" pipe size	Q-5	4.50	3.556		143	178		321	425
1060	4" pipe size	"	3	5.333		260	267		527	690
1080	5" pipe size	Q-6	3.40	7.059		405	365		770	1,000
1100	6" pipe size	"	3	8		495	415		910	1,175
1500	For 250 lb. rating, add					20%				
2000	For galvanized body, add					50%				
2500	For steel body, add					40%				

23 21 20.88 Venturi Flow

		Crew	Daily Output	Labor-Hours	Unit	Material	Labor	Equipment	Total	Total Incl O&P
0010	**VENTURI FLOW**, Measuring device									
0050	1/2" diameter	1 Stpi	24	.333	Ea.	281	18.55		299.55	340
0120	1" diameter		19	.421		276	23.50		299.50	340
0140	1-1/4" diameter		15	.533		340	29.50		369.50	420
0160	1-1/2" diameter		13	.615		355	34.50		389.50	440
0180	2" diameter		11	.727		365	40.50		405.50	465
0220	3" diameter	Q-5	14	1.143		515	57.50		572.50	655
0240	4" diameter	"	11	1.455		775	73		848	960
0280	6" diameter	Q-6	3.50	6.857		1,125	355		1,480	1,800
0500	For meter, add					2,125			2,125	2,350

23 21 23 – Hydronic Pumps

23 21 23.13 In-Line Centrifugal Hydronic Pumps

		Crew	Daily Output	Labor-Hours	Unit	Material	Labor	Equipment	Total	Total Incl O&P
0010	**IN-LINE CENTRIFUGAL HYDRONIC PUMPS**									
0600	Bronze, sweat connections, 1/40 HP, in line									
0640	3/4" size	Q-1	16	1	Ea.	206	49		255	300
1000	Flange connection, 3/4" to 1-1/2" size									
1040	1/12 HP	Q-1	6	2.667	Ea.	530	131		661	785
1060	1/8 HP		6	2.667		915	131		1,046	1,200
1100	1/3 HP		6	2.667		1,025	131		1,156	1,325
1140	2" size, 1/6 HP		5	3.200		1,325	157		1,482	1,675
1180	2-1/2" size, 1/4 HP		5	3.200		1,675	157		1,832	2,075
2000	Cast iron, flange connection									
2040	3/4" to 1-1/2" size, in line, 1/12 HP	Q-1	6	2.667	Ea.	345	131		476	580
2100	1/3 HP		6	2.667		640	131		771	905
2140	2" size, 1/6 HP		5	3.200		700	157		857	1,000
2180	2-1/2" size, 1/4 HP		5	3.200		905	157		1,062	1,225
2220	3" size, 1/4 HP		4	4		920	197		1,117	1,300
2600	For non-ferrous impeller, add					3%				

23 21 29 – Automatic Condensate Pump Units

23 21 29.10 Condensate Removal Pump System

			Crew	Daily Output	Labor-Hours	Unit	Material	Labor	Equipment	Total	Total Incl O&P
0010	**CONDENSATE REMOVAL PUMP SYSTEM**										
0020	Pump with 1 gal. ABS tank	G									
0100	115 V	G									
0120	1/50 HP, 200 GPH	G	1 Stpi	12	.667	Ea.	185	37		222	260
0140	1/18 HP, 270 GPH	G		10	.800		205	44.50		249.50	293
0160	1/5 HP, 450 GPH	G		8	1		445	55.50		500.50	570
0200	230 V	G									
0240	1/18 HP, 270 GPH	G	1 Stpi	10	.800	Ea.	192	44.50		236.50	278
0260	1/5 HP, 450 GPH	G	"	8	1	"	485	55.50		540.50	620

23 22 Steam and Condensate Piping and Pumps

23 22 13 – Steam and Condensate Heating Piping

23 22 13.23 Aboveground Steam and Condensate Piping	Crew	Daily Output	Labor-Hours	Unit	Material	2012 Bare Costs Labor	Equipment	Total	Total Incl O&P
0010 **ABOVEGROUND STEAM AND CONDENSATE HEATING PIPING**									
0020 Condensate meter									
0100 500 lb. per hour	1 Stpi	14	.571	Ea.	2,350	32		2,382	2,650
0140 1500 lb. per hour	"	7	1.143	"	2,550	63.50		2,613.50	2,900

23 22 23 – Steam Condensate Pumps

23 22 23.10 Condensate Return System

	Crew	Daily Output	Labor-Hours	Unit	Material	2012 Bare Costs Labor	Equipment	Total	Total Incl O&P
0010 **CONDENSATE RETURN SYSTEM**									
2000 Simplex									
2010 With pump, motor, CI receiver, float switch									
2020 3/4 HP, 15 GPM	Q-1	1.80	8.889	Ea.	5,825	435		6,260	7,050
2100 Duplex									
2110 With 2 pumps and motors, CI receiver, float switch, alternator									
2120 3/4 HP, 15 GPM, 15 gal. CI rcvr	Q-1	1.40	11.429	Ea.	6,350	560		6,910	7,825
2130 1 HP, 25 GPM		1.20	13.333		7,625	655		8,280	9,375
2140 1-1/2 HP, 45 GPM		1	16		8,850	785		9,635	10,900
2150 1-1/2 HP, 60 GPM		1	16		9,925	785		10,710	12,100

23 31 HVAC Ducts and Casings

23 31 13 – Metal Ducts

23 31 13.13 Rectangular Metal Ducts

	Crew	Daily Output	Labor-Hours	Unit	Material	2012 Bare Costs Labor	Equipment	Total	Total Incl O&P
0010 **RECTANGULAR METAL DUCTS** R233100-20									
0020 Fabricated rectangular, includes fittings, joints, supports,									
0030 allowance for flexible connections, no insulation									
0031 NOTE: Fabrication and installation are combined									
0040 as LABOR cost. Approx. 25% fittings assumed.									
0050 Add to labor for elevated installation									
0051 of fabricated ductwork									
0052 10' to 15' high						6%			
0053 15' to 20' high						12%			
0054 20' to 25' high						15%			
0055 25' to 30' high						21%			
0056 30' to 35' high						24%			
0057 35' to 40' high						30%			
0058 Over 40' high						33%			
0072 For duct insulation see Line 23 07 13.10 3000									
0100 Aluminum, alloy 3003-H14, under 100 lb.	Q-10	75	.320	Lb.	3.58	15.55		19.13	27.50
0110 100 to 500 lb.		80	.300		2.34	14.60		16.94	25
0120 500 to 1,000 lb.		95	.253		2.21	12.30		14.51	21
0140 1,000 to 2,000 lb.		120	.200		2.13	9.75		11.88	17.20
0150 2,000 to 5,000 lb.		130	.185		2.13	9		11.13	16.05
0160 Over 5,000 lb.		145	.166		2.13	8.05		10.18	14.65
0500 Galvanized steel, under 200 lb.		235	.102		.70	4.97		5.67	8.35
0520 200 to 500 lb.		245	.098		.69	4.77		5.46	8
0540 500 to 1,000 lb.		255	.094		.68	4.58		5.26	7.75
0560 1,000 to 2,000 lb.		265	.091		.67	4.41		5.08	7.45
0570 2,000 to 5,000 lb.		275	.087		.66	4.25		4.91	7.20
0580 Over 5,000 lb.		285	.084		.66	4.10		4.76	7
1000 Stainless steel, type 304, under 100 lb.		165	.145		6.35	7.10		13.45	17.75
1020 100 to 500 lb.		175	.137		4.05	6.65		10.70	14.60
1030 500 to 1,000 lb.		190	.126		2.95	6.15		9.10	12.60

23 31 HVAC Ducts and Casings

23 31 13 – Metal Ducts

23 31 13.13 Rectangular Metal Ducts

		Crew	Daily Output	Labor-Hours	Unit	Material	2012 Bare Costs Labor	Equipment	Total	Total Incl O&P
1040	1,000 to 2,000 lb.	Q-10	200	.120	Lb.	2.89	5.85		8.74	12.10
1050	2,000 to 5,000 lb.		225	.107		2.39	5.20		7.59	10.55
1060	Over 5,000 lb.		235	.102		2.35	4.97		7.32	10.20
1100	For medium pressure ductwork, add						15%			
1200	For high pressure ductwork, add						40%			
1210	For welded ductwork, add						85%			
1220	For 30% fittings, add						11%			
1224	For 40% fittings, add						34%			
1228	For 50% fittings, add						56%			
1232	For 60% fittings, add						79%			
1236	For 70% fittings, add						101%			
1240	For 80% fittings, add						124%			
1244	For 90% fittings, add						147%			
1248	For 100% fittings, add						169%			
1252	Note: Fittings add includes time for detailing and installation.									

23 31 13.19 Metal Duct Fittings

		Crew	Daily Output	Labor-Hours	Unit	Material	2012 Bare Costs Labor	Equipment	Total	Total Incl O&P
0010	**METAL DUCT FITTINGS**									
2000	Fabrics for flexible connections, with metal edge	1 Shee	100	.080	L.F.	2.17	4.17		6.34	8.75
2100	Without metal edge	"	160	.050	"	1.64	2.61		4.25	5.80

23 31 16 – Nonmetal Ducts

23 31 16.13 Fibrous-Glass Ducts

		Crew	Daily Output	Labor-Hours	Unit	Material	2012 Bare Costs Labor	Equipment	Total	Total Incl O&P
0010	**FIBROUS-GLASS DUCTS** R233100-20									
3490	Rigid fiberglass duct board, foil reinf. kraft facing									
3500	Rectangular, 1" thick, alum. faced, (FRK), std. weight	Q-10	350	.069	SF Surf	.77	3.34		4.11	5.95

23 33 Air Duct Accessories

23 33 13 – Dampers

23 33 13.13 Volume-Control Dampers

		Crew	Daily Output	Labor-Hours	Unit	Material	2012 Bare Costs Labor	Equipment	Total	Total Incl O&P
0010	**VOLUME-CONTROL DAMPERS**									
5990	Multi-blade dampers, opposed blade, 8" x 6"	1 Shee	24	.333	Ea.	21	17.40		38.40	50
5994	8" x 8"		22	.364		22	18.95		40.95	53
5996	10" x 10"		21	.381		25.50	19.85		45.35	59
6000	12" x 12"		21	.381		28.50	19.85		48.35	62
6020	12" x 18"		18	.444		38.50	23		61.50	78
6030	14" x 10"		20	.400		28	21		49	63
6031	14" x 14"		17	.471		34	24.50		58.50	75
6033	16" x 12"		17	.471		34	24.50		58.50	75
6035	16" x 16"		16	.500		42.50	26		68.50	86.50
6037	18" x 16"		15	.533		47	28		75	94
6038	18" x 18"		15	.533		50.50	28		78.50	98
6070	20" x 16"		14	.571		50.50	30		80.50	101
6072	20" x 20"		13	.615		60.50	32		92.50	116
6074	22" x 18"		14	.571		60.50	30		90.50	112
6076	24" x 16"		11	.727		59.50	38		97.50	124
6078	24" x 20"		8	1		70.50	52		122.50	157
6080	24" x 24"		8	1		82.50	52		134.50	170
6110	26" x 26"		6	1.333		90.50	69.50		160	206
6133	30" x 30"	Q-9	6.60	2.424		130	114		244	315
6135	32" x 32"		6.40	2.500		151	117		268	345
6180	48" x 36"		5.60	2.857		249	134		383	480

23 33 Air Duct Accessories

23 33 13 – Dampers

23 33 13.13 Volume-Control Dampers

		Crew	Daily Output	Labor-Hours	Unit	Material	2012 Bare Costs Labor	2012 Bare Costs Equipment	Total	Total Incl O&P
8000	Multi-blade dampers, parallel blade									
8100	8" x 8"	1 Shee	24	.333	Ea.	82	17.40		99.40	117
8140	16" x 10"		20	.400		103	21		124	146
8200	24" x 16"		11	.727		134	38		172	206
8260	30" x 18"		7	1.143		186	59.50		245.50	295

23 33 13.16 Fire Dampers

		Crew	Daily Output	Labor-Hours	Unit	Material	Labor	Equipment	Total	Total Incl O&P
0010	**FIRE DAMPERS**									
3000	Fire damper, curtain type, 1-1/2 hr. rated, vertical, 6" x 6"	1 Shee	24	.333	Ea.	24	17.40		41.40	53
3020	8" x 6"		22	.364		24	18.95		42.95	55.50
3240	16" x 14"		18	.444		44	23		67	84
3400	24" x 20"		8	1		54	52		106	139

23 33 13.28 Splitter Damper Assembly

		Crew	Daily Output	Labor-Hours	Unit	Material	Labor	Equipment	Total	Total Incl O&P
0009	**SPLITTER DAMPER ASSEMBLY**									
0010	Self locking, 1' rod	1 Shee	24	.333	Ea.	23.50	17.40		40.90	52
7020	3' rod		22	.364		30	18.95		48.95	62
7040	4' rod		20	.400		33.50	21		54.50	69
7060	6' rod		18	.444		41	23		64	80.50

23 33 19 – Duct Silencers

23 33 19.10 Duct Silencers

		Crew	Daily Output	Labor-Hours	Unit	Material	Labor	Equipment	Total	Total Incl O&P
0009	**DUCT SILENCERS**									
0010	Silencers, noise control for air flow, duct				MCFM	57			57	63

23 33 33 – Duct-Mounting Access Doors

23 33 33.13 Duct Access Doors

		Crew	Daily Output	Labor-Hours	Unit	Material	Labor	Equipment	Total	Total Incl O&P
0010	**DUCT ACCESS DOORS**									
1000	Duct access door, insulated, 6" x 6"	1 Shee	14	.571	Ea.	16.05	30		46.05	63
1020	10" x 10"		11	.727		18.45	38		56.45	78.50
1040	12" x 12"		10	.800		20	41.50		61.50	85.50
1050	12" x 18"		9	.889		37	46.50		83.50	111
1070	18" x 18"		8	1		32.50	52		84.50	115
1074	24" x 18"		8	1		44	52		96	128

23 33 46 – Flexible Ducts

23 33 46.10 Flexible Air Ducts

		Crew	Daily Output	Labor-Hours	Unit	Material	Labor	Equipment	Total	Total Incl O&P
0010	**FLEXIBLE AIR DUCTS** R233100-20									
1280	Add to labor for elevated installation									
1282	of prefabricated (purchased) ductwork									
1283	10' to 15' high						10%			
1284	15' to 20' high						20%			
1285	20' to 25' high						25%			
1286	25' to 30' high						35%			
1287	30' to 35' high						40%			
1288	35' to 40' high						50%			
1289	Over 40' high						55%			
1300	Flexible, coated fiberglass fabric on corr. resist. metal helix									
1400	pressure to 12" (WG) UL-181									
1500	Non-insulated, 3" diameter	Q-9	400	.040	L.F.	1.12	1.88		3	4.09
1540	5" diameter		320	.050		1.12	2.35		3.47	4.81
1560	6" diameter		280	.057		1.40	2.68		4.08	5.65
1580	7" diameter		240	.067		1.46	3.13		4.59	6.40
1600	8" diameter		200	.080		1.76	3.75		5.51	7.65
1640	10" diameter		160	.100		2.28	4.69		6.97	9.65

23 33 46 – Flexible Ducts

23 33 46.10 Flexible Air Ducts		Crew	Daily Output	Labor-Hours	Unit	Material	2012 Bare Costs Labor	Equipment	Total	Total Incl O&P
1660	12" diameter	Q-9	120	.133	L.F.	2.73	6.25		8.98	12.55
1900	Insulated, 1" thick, PE jacket, 3" diameter [G]		380	.042		2.22	1.98		4.20	5.45
1910	4" diameter [G]		340	.047		2.22	2.21		4.43	5.80
1920	5" diameter [G]		300	.053		2.23	2.50		4.73	6.25
1940	6" diameter [G]		260	.062		2.50	2.89		5.39	7.15
1960	7" diameter [G]		220	.073		2.73	3.41		6.14	8.20
1980	8" diameter [G]		180	.089		3.01	4.17		7.18	9.65
2020	10" diameter [G]		140	.114		3.66	5.35		9.01	12.25
2040	12" diameter [G]		100	.160		4.20	7.50		11.70	16.05

23 33 53 – Duct Liners

23 33 53.10 Duct Liner Board		Crew	Daily Output	Labor-Hours	Unit	Material	2012 Bare Costs Labor	Equipment	Total	Total Incl O&P
0010	**DUCT LINER BOARD**									
3340	Board type fiberglass liner, FSK, 1-1/2 lb. density									
3344	1" thick	Q-14	150	.107	S.F.	.60	4.69		5.29	7.95
3345	1-1/2" thick		130	.123		.66	5.40		6.06	9.20
3346	2" thick		120	.133		.77	5.85		6.62	10
3348	3" thick		110	.145		.99	6.40		7.39	11.05
3350	4" thick		100	.160		1.22	7.05		8.27	12.30
3356	3 lb. density, 1" thick		150	.107		.77	4.69		5.46	8.15
3358	1-1/2" thick		130	.123		.97	5.40		6.37	9.50
3360	2" thick		120	.133		1.18	5.85		7.03	10.45
3362	2-1/2" thick		110	.145		1.38	6.40		7.78	11.45
3364	3" thick		100	.160		1.59	7.05		8.64	12.70
3366	4" thick		90	.178		2	7.80		9.80	14.40
3370	6 lb. density, 1" thick		140	.114		1.09	5.05		6.14	9.05
3374	1-1/2" thick		120	.133		1.46	5.85		7.31	10.75
3378	2" thick		100	.160		1.82	7.05		8.87	12.95
3490	Board type, fiberglass liner, 3 lb. density									
3680	No finish									
3700	1" thick [G]	Q-14	170	.094	S.F.	.43	4.14		4.57	6.90
3710	1-1/2" thick [G]		140	.114		.64	5.05		5.69	8.55
3720	2" thick [G]		130	.123		.86	5.40		6.26	9.40
3940	Board type, non-fibrous foam									
3950	Temperature, bacteria and fungi resistant									
3960	1" thick [G]	Q-14	150	.107	S.F.	2.25	4.69		6.94	9.80
3970	1-1/2" thick [G]		130	.123		2.95	5.40		8.35	11.70
3980	2" thick [G]		120	.133		3.60	5.85		9.45	13.10

23 34 HVAC Fans

23 34 13 – Axial HVAC Fans

23 34 13.10 Axial Flow HVAC Fans		Crew	Daily Output	Labor-Hours	Unit	Material	2012 Bare Costs Labor	Equipment	Total	Total Incl O&P
0010	**AXIAL FLOW HVAC FANS** R233400-10									
0020	Air conditioning and process air handling									
0030	Axial flow, compact, low sound, 2.5" S.P.									
0050	3,800 CFM, 5 HP	Q-20	3.40	5.882	Ea.	4,800	281		5,081	5,700
0080	6,400 CFM, 5 HP		2.80	7.143		5,350	340		5,690	6,425
0100	10,500 CFM, 7-1/2 HP		2.40	8.333		6,675	400		7,075	7,925
0120	15,600 CFM, 10 HP		1.60	12.500		8,400	600		9,000	10,200
1500	Vaneaxial, low pressure, 2000 CFM, 1/2 HP		3.60	5.556		2,025	266		2,291	2,625
1520	4,000 CFM, 1 HP		3.20	6.250		2,375	299		2,674	3,050

23 34 HVAC Fans

23 34 13 – Axial HVAC Fans

23 34 13.10 Axial Flow HVAC Fans	Crew	Daily Output	Labor-Hours	Unit	Material	2012 Bare Costs Labor	Equipment	Total	Total Incl O&P
1540	8,000 CFM, 2 HP	Q-20	2.80	7.143	Ea.	3,000	340		3,340	3,825

23 34 14 – Blower HVAC Fans

23 34 14.10 Blower Type HVAC Fans

	23 34 14.10 Blower Type HVAC Fans	Crew	Daily Output	Labor-Hours	Unit	Material	2012 Bare Costs Labor	Equipment	Total	Total Incl O&P
0010	**BLOWER TYPE HVAC FANS**									
2500	Ceiling fan, right angle, extra quiet, 0.10" S.P.									
2520	95 CFM	Q-20	20	1	Ea.	264	48		312	365
2540	210 CFM		19	1.053		310	50.50		360.50	420
2560	385 CFM		18	1.111		395	53		448	515
2580	885 CFM		16	1.250		780	60		840	950
2600	1,650 CFM		13	1.538		1,075	73.50		1,148.50	1,300
2620	2,960 CFM		11	1.818		1,450	87		1,537	1,700
2640	For wall or roof cap, add	1 Shee	16	.500		264	26		290	330
2660	For straight thru fan, add					10%				
2680	For speed control switch, add	1 Elec	16	.500		144	26		170	198
7500	Utility set, steel construction, pedestal, 1/4" S.P.									
7520	Direct drive, 150 CFM, 1/8 HP	Q-20	6.40	3.125	Ea.	805	150		955	1,100
7540	485 CFM, 1/6 HP		5.80	3.448		1,025	165		1,190	1,375
7560	1950 CFM, 1/2 HP		4.80	4.167		1,200	199		1,399	1,600
7580	2410 CFM, 3/4 HP		4.40	4.545		2,200	218		2,418	2,725
7600	3328 CFM, 1-1/2 HP		3	6.667		2,450	320		2,770	3,150
7680	V-belt drive, drive cover, 3 phase									
7700	800 CFM, 1/4 HP	Q-20	6	3.333	Ea.	900	160		1,060	1,225
7720	1,300 CFM, 1/3 HP		5	4		940	191		1,131	1,325
7740	2,000 CFM, 1 HP		4.60	4.348		1,125	208		1,333	1,550
7760	2,900 CFM, 3/4 HP		4.20	4.762		1,500	228		1,728	2,000

23 34 16 – Centrifugal HVAC Fans

23 34 16.10 Centrifugal Type HVAC Fans

	23 34 16.10 Centrifugal Type HVAC Fans	Crew	Daily Output	Labor-Hours	Unit	Material	2012 Bare Costs Labor	Equipment	Total	Total Incl O&P
0010	**CENTRIFUGAL TYPE HVAC FANS**									
0200	In-line centrifugal, supply/exhaust booster									
0220	aluminum wheel/hub, disconnect switch, 1/4" S.P.									
0240	500 CFM, 10" diameter connection	Q-20	3	6.667	Ea.	1,150	320		1,470	1,750
0260	1,380 CFM, 12" diameter connection		2	10		1,225	480		1,705	2,075
0280	1,520 CFM, 16" diameter connection		2	10		1,325	480		1,805	2,175
0300	2,560 CFM, 18" diameter connection		1	20		1,450	955		2,405	3,050
0320	3,480 CFM, 20" diameter connection		.80	25		1,725	1,200		2,925	3,700
0326	5,080 CFM, 20" diameter connection		.75	26.667		1,875	1,275		3,150	3,975
3500	Centrifugal, airfoil, motor and drive, complete									
3520	1000 CFM, 1/2 HP	Q-20	2.50	8	Ea.	1,725	385		2,110	2,475
3540	2,000 CFM, 1 HP		2	10		1,950	480		2,430	2,875
3560	4,000 CFM, 3 HP		1.80	11.111		2,500	530		3,030	3,550
3580	8,000 CFM, 7-1/2 HP		1.40	14.286		3,750	685		4,435	5,175
3600	12,000 CFM, 10 HP		1	20		5,000	955		5,955	6,950
4500	Corrosive fume resistant, plastic									
4600	roof ventilators, centrifugal, V belt drive, motor									
4620	1/4" S.P., 250 CFM, 1/4 HP	Q-20	6	3.333	Ea.	4,150	160		4,310	4,825
4640	895 CFM, 1/3 HP		5	4		4,525	191		4,716	5,275
4660	1630 CFM, 1/2 HP		4	5		5,350	239		5,589	6,250
4680	2240 CFM, 1 HP		3	6.667		5,575	320		5,895	6,600
5000	Utility set, centrifugal, V belt drive, motor									
5020	1/4" S.P., 1200 CFM, 1/4 HP	Q-20	6	3.333	Ea.	1,450	160		1,610	1,825
5040	1520 CFM, 1/3 HP		5	4		1,775	191		1,966	2,275
5060	1850 CFM, 1/2 HP		4	5		1,775	239		2,014	2,350

23 34 HVAC Fans

23 34 16 – Centrifugal HVAC Fans

23 34 16.10 Centrifugal Type HVAC Fans	Crew	Daily Output	Labor-Hours	Unit	Material	2012 Bare Costs Labor	Equipment	Total	Total Incl O&P
5080	2180 CFM, 3/4 HP	Q-20	3	6.667	Ea.	2,125	320		2,445	2,800
5100	1/2" S.P., 3600 CFM, 1 HP		2	10		2,450	480		2,930	3,400
5120	4250 CFM, 1-1/2 HP		1.60	12.500		2,675	600		3,275	3,850
5140	4800 CFM, 2 HP	↓	1.40	14.286	↓	3,000	685		3,685	4,350
7000	Roof exhauster, centrifugal, aluminum housing, 12" galvanized									
7020	curb, bird screen, back draft damper, 1/4" S.P.									
7100	Direct drive, 320 CFM, 11" sq. damper	Q-20	7	2.857	Ea.	625	137		762	900
7120	600 CFM, 11" sq. damper		6	3.333		785	160		945	1,100
7140	815 CFM, 13" sq. damper		5	4		785	191		976	1,150
7160	1450 CFM, 13" sq. damper		4.20	4.762		1,275	228		1,503	1,750
7180	2050 CFM, 16" sq. damper		4	5		1,525	239		1,764	2,050
7200	V-belt drive, 1650 CFM, 12" sq. damper		6	3.333		1,125	160		1,285	1,500
7220	2750 CFM, 21" sq. damper		5	4		1,350	191		1,541	1,775
7230	3500 CFM, 21" sq. damper		4.50	4.444		1,525	213		1,738	2,000
7240	4910 CFM, 23" sq. damper		4	5		1,850	239		2,089	2,400
7260	8525 CFM, 28" sq. damper		3	6.667		2,475	320		2,795	3,200
7280	13,760 CFM, 35" sq. damper		2	10		3,325	480		3,805	4,375
7300	20,558 CFM, 43" sq. damper	↓	1	20		6,925	955		7,880	9,075
7320	For 2 speed winding, add					15%				
7340	For explosionproof motor, add					495			495	545
7360	For belt driven, top discharge, add				↓	15%				
8500	Wall exhausters, centrifugal, auto damper, 1/8" S.P.									
8520	Direct drive, 610 CFM, 1/20 HP	Q-20	14	1.429	Ea.	385	68.50		453.50	525
8540	796 CFM, 1/12 HP		13	1.538		755	73.50		828.50	940
8560	822 CFM, 1/6 HP		12	1.667		975	80		1,055	1,200
8580	1,320 CFM, 1/4 HP	↓	12	1.667	↓	1,100	80		1,180	1,325
9500	V-belt drive, 3 phase									
9520	2,800 CFM, 1/4 HP	Q-20	9	2.222	Ea.	1,750	106		1,856	2,075
9540	3,740 CFM, 1/2 HP	"	8	2.500	"	1,800	120		1,920	2,175

23 34 23 – HVAC Power Ventilators

23 34 23.10 HVAC Power Circulators and Ventilators

0010	**HVAC POWER CIRCULATORS AND VENTILATORS**									
3000	Paddle blade air circulator, 3 speed switch									
3020	42", 5,000 CFM high, 3000 CFM low	G	1 Elec	2.40	3.333	Ea.	154	172		326	425
3040	52", 6,500 CFM high, 4000 CFM low	G	"	2.20	3.636	"	161	187		348	455
3100	For antique white motor, same cost									
3200	For brass plated motor, same cost									
3300	For light adaptor kit, add	G				Ea.	38.50			38.50	42.50
6000	Propeller exhaust, wall shutter, 1/4" S.P.									
6020	Direct drive, two speed									
6100	375 CFM, 1/10 HP		Q-20	10	2	Ea.	450	95.50		545.50	640
6120	730 CFM, 1/7 HP			9	2.222		505	106		611	715
6140	1000 CFM, 1/8 HP			8	2.500		700	120		820	950
6200	4720 CFM, 1 HP		↓	5	4	↓	1,125	191		1,316	1,525
6300	V-belt drive, 3 phase									
6320	6175 CFM, 3/4 HP		Q-20	5	4	Ea.	945	191		1,136	1,350
6340	7500 CFM, 3/4 HP			5	4		1,000	191		1,191	1,400
6360	10,100 CFM, 1 HP			4.50	4.444		1,100	213		1,313	1,525
6380	14,300 CFM, 1-1/2 HP		↓	4	5	↓	1,375	239		1,614	1,875
6650	Residential, bath exhaust, grille, back draft damper									
6660	50 CFM		Q-20	24	.833	Ea.	49	40		89	115
6670	110 CFM		↓	22	.909	↓	82	43.50		125.50	156

23 34 23 – HVAC Power Ventilators

23 34 23.10 HVAC Power Circulators and Ventilators		Crew	Daily Output	Labor-Hours	Unit	Material	2012 Bare Costs Labor	Equipment	Total	Total Incl O&P
6680	Light combination, squirrel cage, 100 watt, 70 CFM	Q-20	24	.833	Ea.	92.50	40		132.50	163
6700	Light/heater combination, ceiling mounted									
6710	70 CFM, 1450 watt	Q-20	24	.833	Ea.	174	40		214	252
6800	Heater combination, recessed, 70 CFM		24	.833		55	40		95	121
6820	With 2 infrared bulbs		23	.870		86	41.50		127.50	158
6900	Kitchen exhaust, grille, complete, 160 CFM		22	.909		92	43.50		135.50	167
6910	180 CFM		20	1		78	48		126	159
6920	270 CFM		18	1.111		141	53		194	236
6930	350 CFM		16	1.250		110	60		170	212
6940	Residential roof jacks and wall caps									
6944	Wall cap with back draft damper									
6946	3" & 4" dia. round duct	1 Shee	11	.727	Ea.	22.50	38		60.50	82.50
6948	6" dia. round duct	"	11	.727	"	55.50	38		93.50	119
6958	Roof jack with bird screen and back draft damper									
6960	3" & 4" dia. round duct	1 Shee	11	.727	Ea.	22.50	38		60.50	82.50
6962	3-1/4" x 10" rectangular duct	"	10	.800	"	42	41.50		83.50	110
6980	Transition									
6982	3-1/4" x 10" to 6" dia. round	1 Shee	20	.400	Ea.	25.50	21		46.50	60

23 34 33 – Air Curtains

23 34 33.10 Air Barrier Curtains

		Crew	Daily Output	Labor-Hours	Unit	Material	2012 Bare Costs Labor	Equipment	Total	Total Incl O&P
0010	**AIR BARRIER CURTAINS**, Incl. motor starters, transformers,									
0050	and door switches									
2450	Conveyor openings or service windows									
3000	Service window, 5' high x 25" wide	2 Shee	5	3.200	Ea.	350	167		517	640
3100	Environmental separation									
3110	Door heights up to 8', low profile, super quiet									
3120	Unheated, variable speed									
3130	36" wide	2 Shee	4	4	Ea.	620	209		829	1,000
3134	42" wide		3.80	4.211		650	220		870	1,050
3138	48" wide		3.60	4.444		675	232		907	1,100
3142	60" wide		3.40	4.706		705	245		950	1,150
3146	72" wide	Q-3	4.60	6.957		895	360		1,255	1,525
3150	96" wide		4.40	7.273		1,350	380		1,730	2,075
3154	120" wide		4.20	7.619		1,450	395		1,845	2,200
3158	144" wide		4	8		1,800	415		2,215	2,600
3200	Door heights up to 10'									
3210	Unheated									
3230	36" wide	2 Shee	3.80	4.211	Ea.	620	220		840	1,025
3234	42" wide		3.60	4.444		640	232		872	1,050
3238	48" wide		3.40	4.706		660	245		905	1,100
3242	60" wide		3.20	5		1,000	261		1,261	1,500
3246	72" wide	Q-3	4.40	7.273		1,075	380		1,455	1,750
3250	96" wide		4.20	7.619		1,275	395		1,670	2,000
3254	120" wide		4	8		1,775	415		2,190	2,575
3258	144" wide		3.80	8.421		1,925	440		2,365	2,775
3300	Door heights up to 12'									
3310	Unheated									
3334	42" wide	2 Shee	3.40	4.706	Ea.	875	245		1,120	1,350
3338	48" wide		3.20	5		885	261		1,146	1,375
3342	60" wide		3	5.333		905	278		1,183	1,425
3346	72" wide	Q-3	4.20	7.619		1,650	395		2,045	2,400
3350	96" wide		4	8		1,800	415		2,215	2,600

23 34 HVAC Fans

23 34 33 – Air Curtains

	23 34 33.10 Air Barrier Curtains	Crew	Daily Output	Labor-Hours	Unit	Material	2012 Bare Costs Labor	2012 Bare Costs Equipment	Total	Total Incl O&P
3354	120" wide	Q-3	3.80	8.421	Ea.	2,225	440		2,665	3,100
3358	144" wide	↓	3.60	8.889	↓	2,375	460		2,835	3,300
3400	Door heights up to 16'									
3410	Unheated									
3438	48" wide	2 Shee	3	5.333	Ea.	985	278		1,263	1,500
3442	60" wide	"	2.80	5.714		1,025	298		1,323	1,575
3446	72" wide	Q-3	3.80	8.421		1,925	440		2,365	2,775
3450	96" wide		3.60	8.889		2,025	460		2,485	2,925
3454	120" wide		3.40	9.412		2,575	490		3,065	3,575
3458	144" wide	↓	3.20	10	↓	2,650	520		3,170	3,700
3470	Heated, electric									
3474	48" wide	2 Shee	2.90	5.517	Ea.	1,925	288		2,213	2,575
3478	60" wide	"	2.70	5.926		1,950	310		2,260	2,625
3482	72" wide	Q-3	3.70	8.649		3,325	450		3,775	4,350
3486	96" wide		3.50	9.143		3,450	475		3,925	4,525
3490	120" wide		3.30	9.697		4,625	505		5,130	5,850
3494	144" wide	↓	3.10	10.323	↓	4,725	535		5,260	6,000

23 37 Air Outlets and Inlets

23 37 13 – Diffusers, Registers, and Grilles

23 37 13.10 Diffusers

		Crew	Daily Output	Labor-Hours	Unit	Material	2012 Bare Costs Labor	2012 Bare Costs Equipment	Total	Total Incl O&P
0010	**DIFFUSERS**, Aluminum, opposed blade damper unless noted									
0100	Ceiling, linear, also for sidewall									
0500	Perforated, 24" x 24" lay-in panel size, 6" x 6"	1 Shee	16	.500	Ea.	151	26		177	206
0520	8" x 8"		15	.533		159	28		187	218
0530	9" x 9"		14	.571		161	30		191	223
0540	10" x 10"		14	.571		162	30		192	224
0560	12" x 12"		12	.667		168	35		203	238
0590	16" x 16"		11	.727		189	38		227	265
0600	18" x 18"		10	.800		202	41.50		243.50	286
0610	20" x 20"		10	.800		218	41.50		259.50	305
0620	24" x 24"		9	.889		239	46.50		285.50	335
1000	Rectangular, 1 to 4 way blow, 6" x 6"		16	.500		50	26		76	95
1010	8" x 8"		15	.533		58	28		86	106
1014	9" x 9"		15	.533		64.50	28		92.50	114
1016	10" x 10"		15	.533		80	28		108	130
1020	12" x 6"		15	.533		71.50	28		99.50	121
1040	12" x 9"		14	.571		75.50	30		105.50	129
1060	12" x 12"		12	.667		84	35		119	146
1070	14" x 6"		13	.615		77.50	32		109.50	135
1074	14" x 14"		12	.667		128	35		163	193
1150	18" x 18"		9	.889		136	46.50		182.50	220
1160	21" x 21"		8	1		215	52		267	315
1170	24" x 12"		10	.800		163	41.50		204.50	244
1500	Round, butterfly damper, steel, diffuser size, 6" diameter		18	.444		24.50	23		47.50	62.50
1520	8" diameter		16	.500		26	26		52	68.50
1540	10" diameter		14	.571		32	30		62	81
1560	12" diameter		12	.667		42	35		77	99.50
1580	14" diameter		10	.800		52	41.50		93.50	121
2000	T bar mounting, 24" x 24" lay-in frame, 6" x 6"		16	.500		45.50	26		71.50	90
2020	8" x 8"	↓	14	.571	↓	47	30		77	97.50

23 37 Air Outlets and Inlets

23 37 13 – Diffusers, Registers, and Grilles

23 37 13.10 Diffusers

		Crew	Daily Output	Labor-Hours	Unit	Material	2012 Bare Costs Labor	2012 Bare Costs Equipment	Total	Total Incl O&P
2040	12" x 12"	1 Shee	12	.667	Ea.	55.50	35		90.50	114
2060	16" x 16"		11	.727		75	38		113	141
2080	18" x 18"	↓	10	.800		83	41.50		124.50	155
6000	For steel diffusers instead of aluminum, deduct				↓	10%				

23 37 13.30 Grilles

		Crew	Daily Output	Labor-Hours	Unit	Material	2012 Bare Costs Labor	2012 Bare Costs Equipment	Total	Total Incl O&P
0010	**GRILLES**									
0020	Aluminum									
1000	Air return, steel, 6" x 6"	1 Shee	26	.308	Ea.	18.15	16.05		34.20	44.50
1020	10" x 6"		24	.333		18.15	17.40		35.55	46.50
1080	16" x 8"		22	.364		26.50	18.95		45.45	58
1100	12" x 12"		22	.364		26.50	18.95		45.45	58
1120	24" x 12"		18	.444		34.50	23		57.50	73.50
1220	24" x 18"		16	.500		42.50	26		68.50	86.50
1280	36" x 24"		14	.571		73	30		103	126
3000	Filter grille with filter, 12" x 12"		24	.333		51	17.40		68.40	83
3020	18" x 12"		20	.400		68	21		89	107
3040	24" x 18"		18	.444		80.50	23		103.50	124
3060	24" x 24"	↓	16	.500		94	26		120	143
6000	For steel grilles instead of aluminum in above, deduct				↓	10%				

23 37 13.60 Registers

		Crew	Daily Output	Labor-Hours	Unit	Material	2012 Bare Costs Labor	2012 Bare Costs Equipment	Total	Total Incl O&P
0010	**REGISTERS**									
0980	Air supply									
1000	Ceiling/wall, O.B. damper, anodized aluminum									
1010	One or two way deflection, adj. curved face bars									
1020	8" x 4"	1 Shee	26	.308	Ea.	34.50	16.05		50.55	62.50
1120	12" x 12"		18	.444		63.50	23		86.50	106
1240	20" x 6"		18	.444		56.50	23		79.50	98
1340	24" x 8"		13	.615		76.50	32		108.50	133
1350	24" x 18"	↓	12	.667		142	35		177	209
2700	Above registers in steel instead of aluminum, deduct				↓	10%				
4000	Floor, toe operated damper, enameled steel									
4020	4" x 8"	1 Shee	32	.250	Ea.	27.50	13.05		40.55	50
4100	8" x 10"		22	.364		33.50	18.95		52.45	66
4140	10" x 10"		20	.400		40	21		61	76
4220	14" x 14"		16	.500		141	26		167	195
4240	14" x 20"	↓	15	.533	↓	188	28		216	250
4980	Air return									
5000	Ceiling or wall, fixed 45° face blades									
5010	Adjustable O.B. damper, anodized aluminum									
5020	4" x 8"	1 Shee	26	.308	Ea.	28.50	16.05		44.55	56
5060	6" x 10"		19	.421		28.50	22		50.50	65
5280	24" x 24"		11	.727		94.50	38		132.50	162
5300	24" x 36"	↓	8	1	↓	151	52		203	246
6000	For steel construction instead of aluminum, deduct					10%				

23 37 15 – Louvers

23 37 15.40 HVAC Louvers

		Crew	Daily Output	Labor-Hours	Unit	Material	2012 Bare Costs Labor	2012 Bare Costs Equipment	Total	Total Incl O&P
0010	**HVAC LOUVERS**									
0100	Aluminum, extruded, with screen, mill finish									
1002	Brick vent, see also Section 04 05 23.19									
1100	Standard, 4" deep, 8" wide, 5" high	1 Shee	24	.333	Ea.	34.50	17.40		51.90	64
1200	Modular, 4" deep, 7-3/4" wide, 5" high		24	.333		36	17.40		53.40	66
1300	Speed brick, 4" deep, 11-5/8" wide, 3-7/8" high	↓	24	.333	↓	36	17.40		53.40	66

23 37 15 – Louvers

23 37 15.40 HVAC Louvers	Crew	Daily Output	Labor-Hours	Unit	Material	2012 Bare Costs Labor	Equipment	Total	Total Incl O&P
1400	Fuel oil brick, 4" deep, 8" wide, 5" high	1 Shee	24	.333	Ea.	61.50	17.40		78.90	94.50
2000	Cooling tower and mechanical equip., screens, light weight		40	.200	S.F.	15.40	10.45		25.85	33
2020	Standard weight		35	.229		41	11.90		52.90	63
2500	Dual combination, automatic, intake or exhaust		20	.400		56	21		77	93.50
2520	Manual operation		20	.400		41.50	21		62.50	78
2540	Electric or pneumatic operation		20	.400		41.50	21		62.50	78
2560	Motor, for electric or pneumatic		14	.571	Ea.	480	30		510	575
3000	Fixed blade, continuous line									
3100	Mullion type, stormproof	1 Shee	28	.286	S.F.	41.50	14.90		56.40	68.50
3200	Stormproof		28	.286		41.50	14.90		56.40	68.50
3300	Vertical line		28	.286		49.50	14.90		64.40	76.50
3500	For damper to use with above, add					50%	30%			
3520	Motor, for damper, electric or pneumatic	1 Shee	14	.571	Ea.	480	30		510	575
4000	Operating, 45°, manual, electric or pneumatic		24	.333	S.F.	50	17.40		67.40	81.50
4100	Motor, for electric or pneumatic		14	.571	Ea.	480	30		510	575
4200	Penthouse, roof		56	.143	S.F.	24.50	7.45		31.95	38.50
4300	Walls		40	.200		58	10.45		68.45	80
5000	Thinline, under 4" thick, fixed blade		40	.200		24	10.45		34.45	42.50
5010	Finishes, applied by mfr. at additional cost, available in colors									
5020	Prime coat only, add				S.F.	3.30			3.30	3.63
5040	Baked enamel finish coating, add					6.10			6.10	6.70
5060	Anodized finish, add					6.60			6.60	7.25
5080	Duranodic finish, add					12			12	13.20
5100	Fluoropolymer finish coating, add					18.90			18.90	21
9980	For small orders (under 10 pieces), add					25%				

23 37 23 – HVAC Gravity Ventilators

23 37 23.10 HVAC Gravity Air Ventilators

		Crew	Daily Output	Labor-Hours	Unit	Material	2012 Bare Costs Labor	Equipment	Total	Total Incl O&P
0010	**HVAC GRAVITY AIR VENTILATORS**, Incl. base and damper									
1280	Rotary ventilators, wind driven, galvanized									
1300	4" neck diameter	Q-9	20	.800	Ea.	50	37.50		87.50	112
1340	6" neck diameter		16	1		56	47		103	133
1400	12" neck diameter		10	1.600		81	75		156	204
1500	24" neck diameter, 3,100 CFM		8	2		236	94		330	405
1540	36" neck diameter, 5,500 CFM		6	2.667		695	125		820	955
2000	Stationary, gravity, syphon, galvanized									
2160	6" neck diameter, 66 CFM	Q-9	16	1	Ea.	53	47		100	130
2240	12" neck diameter, 160 CFM		10	1.600		91	75		166	214
2340	24" neck diameter, 900 CFM		8	2		248	94		342	415
2380	36" neck diameter, 2,000 CFM		6	2.667		775	125		900	1,050
4200	Stationary mushroom, aluminum, 16" orifice diameter		10	1.600		565	75		640	735
4220	26" orifice diameter		6.15	2.602		830	122		952	1,100
4230	30" orifice diameter		5.71	2.802		1,225	132		1,357	1,550
4240	38" orifice diameter		5	3.200		1,750	150		1,900	2,150
4250	42" orifice diameter		4.70	3.404		2,300	160		2,460	2,800
4260	50" orifice diameter		4.44	3.604		2,750	169		2,919	3,275
5000	Relief vent									
5500	Rectangular, aluminum, galvanized curb									
5510	intake/exhaust, 0.033" SP									
5580	500 CFM, 12" x 12"	Q-9	8.60	1.860	Ea.	645	87.50		732.50	845
5600	600 CFM, 12" x 16"		8	2		730	94		824	945
5640	1000 CFM, 12" x 24"		6.60	2.424		820	114		934	1,075
5680	3000 CFM, 20" x 42"		4	4		1,450	188		1,638	1,850

23 37 Air Outlets and Inlets

23 37 23 – HVAC Gravity Ventilators

23 37 23.10 HVAC Gravity Air Ventilators	Crew	Daily Output	Labor-Hours	Unit	Material	2012 Bare Costs Labor	Equipment	Total	Total Incl O&P
5880 Size is throat area, volume is at 500 fpm									

23 38 Ventilation Hoods

23 38 13 – Commercial-Kitchen Hoods

23 38 13.10 Hood and Ventilation Equipment

		Crew	Daily Output	Labor-Hours	Unit	Material	Labor	Equipment	Total	Total Incl O&P
0010	**HOOD AND VENTILATION EQUIPMENT**									
2970	Exhaust hood, sst, gutter on all sides, 4' x 4' x 2'	1 Carp	1.80	4.444	Ea.	4,675	196		4,871	5,425
2980	4' x 4' x 7'	"	1.60	5		7,425	221		7,646	8,525
7950	Hood fire protection system, minimum	Q-1	3	5.333		4,400	262		4,662	5,250
8050	Maximum	"	1	16		35,300	785		36,085	40,100

23 41 Particulate Air Filtration

23 41 13 – Panel Air Filters

23 41 13.10 Panel Type Air Filters

			Crew	Daily Output	Labor-Hours	Unit	Material	Labor	Equipment	Total	Total Incl O&P
0010	**PANEL TYPE AIR FILTERS**										
2950	Mechanical media filtration units										
3000	High efficiency type, with frame, non-supported	G				MCFM	45			45	49.50
3100	Supported type	G				"	60			60	66
5500	Throwaway glass or paper media type					Ea.	3.35			3.35	3.69

23 41 16 – Renewable-Media Air Filters

23 41 16.10 Disposable Media Air Filters

		Crew	Daily Output	Labor-Hours	Unit	Material	Labor	Equipment	Total	Total Incl O&P
0010	**DISPOSABLE MEDIA AIR FILTERS**									
5000	Renewable disposable roll				MCFM	11.50			11.50	12.65

23 41 19 – Washable Air Filters

23 41 19.10 Permanent Air Filters

			Crew	Daily Output	Labor-Hours	Unit	Material	Labor	Equipment	Total	Total Incl O&P
0010	**PERMANENT AIR FILTERS**										
4500	Permanent washable	G				MCFM	20			20	22

23 41 23 – Extended Surface Filters

23 41 23.10 Expanded Surface Filters

			Crew	Daily Output	Labor-Hours	Unit	Material	Labor	Equipment	Total	Total Incl O&P
0010	**EXPANDED SURFACE FILTERS**										
4000	Medium efficiency, extended surface	G				MCFM	5.50			5.50	6.05

23 42 Gas-Phase Air Filtration

23 42 13 – Activated-Carbon Air Filtration

23 42 13.10 Charcoal Type Air Filtration

		Crew	Daily Output	Labor-Hours	Unit	Material	Labor	Equipment	Total	Total Incl O&P
0010	**CHARCOAL TYPE AIR FILTRATION**									
0050	Activated charcoal type, full flow				MCFM	600			600	660
0060	Full flow, impregnated media 12" deep					225			225	248
0070	HEPA filter & frame for field erection					300			300	330
0080	HEPA filter-diffuser, ceiling install.					300			300	330

23 43 Electronic Air Cleaners

23 43 13 – Washable Electronic Air Cleaners

23 43 13.10 Electronic Air Cleaners	Crew	Daily Output	Labor-Hours	Unit	Material	2012 Bare Costs Labor	Equipment	Total	Total Incl O&P
0010 **ELECTRONIC AIR CLEANERS**									
2000 Electronic air cleaner, duct mounted									
2150 1000 CFM	1 Shee	4	2	Ea.	380	104		484	580
2200 1200 CFM		3.80	2.105		430	110		540	640
2250 1400 CFM		3.60	2.222		435	116		551	655

23 51 Breechings, Chimneys, and Stacks

23 51 13 – Draft Control Devices

23 51 13.13 Draft-Induction Fans

	Crew	Daily Output	Labor-Hours	Unit	Material	2012 Bare Costs Labor	Equipment	Total	Total Incl O&P
0010 **DRAFT-INDUCTION FANS**									
1000 Breeching installation									
1800 Hot gas, 600°F, variable pitch pulley and motor									
1860 8" diam. inlet, 1/4 H.P., 1phase, 1120 CFM	Q-9	4	4	Ea.	2,675	188		2,863	3,225
1900 12" diam. inlet, 3/4 H.P., 3 phase, 2960 CFM		3	5.333		3,650	250		3,900	4,400
1980 24" diam. inlet, 7-1/2 H.P., 3 phase, 17,760 CFM		.80	20		10,500	940		11,440	12,900
2300 For multi-blade damper at fan inlet, add					20%				

23 51 23 – Gas Vents

23 51 23.10 Gas Chimney Vents

	Crew	Daily Output	Labor-Hours	Unit	Material	2012 Bare Costs Labor	Equipment	Total	Total Incl O&P
0010 **GAS CHIMNEY VENTS**, Prefab metal, U.L. listed									
0020 Gas, double wall, galvanized steel									
0080 3" diameter	Q-9	72	.222	V.L.F.	4.65	10.45		15.10	21
0100 4" diameter		68	.235		6.35	11.05		17.40	24
0120 5" diameter		64	.250		7.20	11.75		18.95	26
0140 6" diameter		60	.267		9.05	12.50		21.55	29
0160 7" diameter		56	.286		10.40	13.40		23.80	32
0180 8" diameter		52	.308		15.80	14.45		30.25	39.50
0200 10" diameter		48	.333		30	15.65		45.65	57
0220 12" diameter		44	.364		36	17.05		53.05	65.50
0260 16" diameter		40	.400		86	18.75		104.75	124
0300 20" diameter	Q-10	36	.667		125	32.50		157.50	188

23 51 26 – All-Fuel Vent Chimneys

23 51 26.30 All-Fuel Vent Chimneys, Double Wall, St. Stl.

	Crew	Daily Output	Labor-Hours	Unit	Material	2012 Bare Costs Labor	Equipment	Total	Total Incl O&P
0010 **ALL-FUEL VENT CHIMNEYS, DOUBLE WALL, STAINLESS STEEL**									
7800 6" diameter	Q-9	60	.267	V.L.F.	62	12.50		74.50	87
7804 8" diameter		52	.308		71	14.45		85.45	100
7806 10" diameter		48	.333		79	15.65		94.65	111
7808 12" diameter		44	.364		90.50	17.05		107.55	126
7810 14" diameter		42	.381		102	17.90		119.90	140

23 51 33 – Insulated Sectional Chimneys

23 51 33.10 Prefabricated Insulated Sectional Chimneys

	Crew	Daily Output	Labor-Hours	Unit	Material	2012 Bare Costs Labor	Equipment	Total	Total Incl O&P
0010 **PREFABRICATED INSULATED SECTIONAL CHIMNEYS**									
9000 High temp. (2000°F), steel jacket, acid resistant refractory lining									
9010 11 ga. galvanized jacket, U.L. listed									
9020 Straight section, 48" long, 10" diameter	Q-10	13.30	1.805	Ea.	400	88		488	575
9040 18" diameter		7.40	3.243		615	158		773	915
9070 36" diameter		2.70	8.889		1,325	435		1,760	2,100
9090 48" diameter	Q-11	2.70	11.852		1,925	590		2,515	3,000
9120 Tee section, 10" diameter	Q-10	4.40	5.455		855	265		1,120	1,350
9140 18" diameter		2.40	10		1,225	485		1,710	2,100

23 51 Breechings, Chimneys, and Stacks

23 51 33 – Insulated Sectional Chimneys

23 51 33.10 Prefabricated Insulated Sectional Chimneys	Crew	Daily Output	Labor-Hours	Unit	Material	2012 Bare Costs Labor	Equipment	Total	Total Incl O&P
9170	36" diameter	Q-10	.80	30	Ea.	3,975	1,450		5,425	6,600
9190	48" diameter	Q-11	.90	35.556		6,775	1,775		8,550	10,200
9220	Cleanout pier section, 10" diameter	Q-10	3.50	6.857		655	335		990	1,225
9240	18" diameter		1.90	12.632		885	615		1,500	1,900
9270	36" diameter		.75	32		1,625	1,550		3,175	4,150
9290	48" diameter	Q-11	.75	42.667		2,475	2,125		4,600	5,950
9320	For drain, add					53%				
9330	Elbow, 30° and 45°, 10" diameter	Q-10	6.60	3.636		555	177		732	880
9350	18" diameter		3.70	6.486		900	315		1,215	1,475
9380	36" diameter		1.30	18.462		2,225	900		3,125	3,825
9400	48" diameter	Q-11	1.35	23.704		3,900	1,175		5,075	6,075
9430	For 60° and 90° elbow, add					112%				
9440	End cap, 10" diameter	Q-10	26	.923		955	45		1,000	1,125
9460	18" diameter		15	1.600		1,350	78		1,428	1,600
9490	36" diameter		5	4.800		2,350	234		2,584	2,950
9510	48" diameter	Q-11	5.30	6.038		3,475	300		3,775	4,250
9540	Increaser (1 diameter), 10" diameter	Q-10	6.60	3.636		475	177		652	790
9560	18" diameter		3.70	6.486		740	315		1,055	1,300
9590	36" diameter		1.30	18.462		1,450	900		2,350	2,975
9592	42" diameter	Q-11	1.60	20		1,600	995		2,595	3,275
9594	48" diameter		1.35	23.704		1,925	1,175		3,100	3,900
9596	54" diameter		1.10	29.091		2,325	1,450		3,775	4,750
9600	For expansion joints, add to straight section					7%				
9610	For 1/4" hot rolled steel jacket, add					157%				
9620	For 2950°F very high temperature, add					89%				
9630	26 ga. aluminized jacket, straight section, 48" long									
9640	10" diameter	Q-10	15.30	1.569	V.L.F.	237	76.50		313.50	375
9660	18" diameter		8.50	2.824		365	137		502	615
9690	36" diameter		3.10	7.742		845	375		1,220	1,500
9810	Draw band, galv. stl., 11 gauge, 10" diameter		32	.750	Ea.	65	36.50		101.50	127
9820	12" diameter		30	.800		69	39		108	135
9830	18" diameter		26	.923		86	45		131	163
9860	36" diameter		18	1.333		146	65		211	259
9880	48" diameter	Q-11	20	1.600		251	79.50		330.50	400

23 52 Heating Boilers

23 52 13 – Electric Boilers

23 52 13.10 Electric Boilers, ASME

		Crew	Daily Output	Labor-Hours	Unit	Material	2012 Bare Costs Labor	Equipment	Total	Total Incl O&P
0010	**ELECTRIC BOILERS, ASME**, Standard controls and trim									
1000	Steam, 6 KW, 20.5 MBH	Q-19	1.20	20	Ea.	3,625	1,000		4,625	5,525
1160	60 KW, 205 MBH		1	24		6,250	1,225		7,475	8,700
1220	112 KW, 382 MBH		.75	32		8,550	1,625		10,175	11,800
1280	222 KW, 758 MBH		.55	43.636		13,500	2,200		15,700	18,200
1380	518 KW, 1768 MBH	Q-21	.36	88.889		22,800	4,600		27,400	32,100
1480	814 KW, 2778 MBH		.25	128		31,900	6,650		38,550	45,100
1600	2,340 KW, 7984 MBH		.16	200		65,500	10,400		75,900	87,500
2000	Hot water, 7.5 KW, 25.6 MBH	Q-19	1.30	18.462		3,825	935		4,760	5,600
2100	90 KW, 307 MBH		1.10	21.818		5,575	1,100		6,675	7,775
2220	296 KW, 1010 MBH		.55	43.636		11,900	2,200		14,100	16,400
2500	1036 KW, 3536 MBH	Q-21	.34	94.118		27,700	4,875		32,575	37,900
2680	2400 KW, 8191 MBH		.25	128		58,000	6,650		64,650	73,500

23 52 Heating Boilers

23 52 13 – Electric Boilers

23 52 13.10 Electric Boilers, ASME	Crew	Daily Output	Labor- Hours	Unit	Material	2012 Bare Costs Labor	Equipment	Total	Total Incl O&P
2820 3600 KW, 12,283 MBH	Q-21	.16	200	Ea.	81,500	10,400		91,900	105,000

23 52 23 – Cast-Iron Boilers

23 52 23.20 Gas-Fired Boilers

		Daily	Labor-			2012 Bare Costs			Total
	Crew	Output	Hours	Unit	Material	Labor	Equipment	Total	Incl O&P
0010 **GAS-FIRED BOILERS**, Natural or propane, standard controls, packaged									
1000 Cast iron, with insulated jacket									
2000 Steam, gross output, 81 MBH	Q-7	1.40	22.857	Ea.	1,950	1,200		3,150	3,975
2080 203 MBH		.90	35.556		3,100	1,875		4,975	6,275
2180 400 MBH		.56	56.838		5,200	3,025		8,225	10,300
2240 765 MBH		.43	74.419		10,300	3,950		14,250	17,400
2320 1,875 MBH		.30	106		16,400	5,650		22,050	26,500
2440 4,720 MBH		.15	207		73,500	11,000		84,500	97,000
2480 6,100 MBH		.13	246		81,000	13,100		94,100	109,000
2540 6,970 MBH		.10	320		88,500	17,000		105,500	123,000
3000 Hot water, gross output, 80 MBH		1.46	21.918		1,775	1,175		2,950	3,725
3140 320 MBH		.80	40		4,200	2,125		6,325	7,825
3260 1,088 MBH		.40	80		12,500	4,250		16,750	20,100
3360 2,856 MBH		.20	160		22,100	8,475		30,575	37,100
3380 3,264 MBH		.18	179		23,500	9,525		33,025	40,300
3480 6,100 MBH		.13	250		81,000	13,300		94,300	109,500
3540 6,970 MBH		.09	359		88,500	19,100		107,600	126,500
7000 For tankless water heater, add					10%				

23 52 23.30 Gas/Oil Fired Boilers

0010 **GAS/OIL FIRED BOILERS**, Combination with burners and controls, packaged									
1000 Cast iron with insulated jacket									
2000 Steam, gross output, 720 MBH	Q-7	.43	74.074	Ea.	13,300	3,925		17,225	20,600
2080 1,600 MBH		.30	107		18,900	5,675		24,575	29,400
2140 2,700 MBH		.19	165		25,400	8,800		34,200	41,300
2280 5,520 MBH		.14	235		82,000	12,500		94,500	109,000
2340 6,390 MBH		.11	296		88,000	15,700		103,700	120,000
2380 6,970 MBH		.09	372		93,000	19,700		112,700	132,500
2900 Hot water, gross output									
2910 200 MBH	Q-6	.62	39.024	Ea.	8,925	2,025		10,950	12,900
2920 300 MBH		.49	49.080		8,925	2,550		11,475	13,700
2930 400 MBH		.41	57.971		10,400	3,025		13,425	16,100
2940 500 MBH		.36	67.039		11,300	3,475		14,775	17,700
3000 584 MBH	Q-7	.44	72.072		12,400	3,825		16,225	19,500
3060 1,460 MBH		.28	113		24,800	6,000		30,800	36,300
3160 4,088 MBH		.16	195		52,000	10,300		62,300	73,000
3300 13,500 MBH, 403.3 BHP		.04	727		161,500	38,600		200,100	235,500

23 52 23.40 Oil-Fired Boilers

0010 **OIL-FIRED BOILERS**, Standard controls, flame retention burner, packaged									
1000 Cast iron, with insulated flush jacket									
2000 Steam, gross output, 109 MBH	Q-7	1.20	26.667	Ea.	2,025	1,425		3,450	4,350
2060 207 MBH		.90	35.556		2,775	1,875		4,650	5,900
2180 1,084 MBH		.38	85.106		9,400	4,525		13,925	17,100
2280 3,000 MBH		.19	170		21,600	9,025		30,625	37,300
2380 5,520 MBH		.14	235		68,500	12,500		81,000	94,500
2460 6,970 MBH		.09	363		88,000	19,300		107,300	126,000
3000 Hot water, same price as steam									
4000 For tankless coil in smaller sizes, add				Ea.	15%				

23 52 Heating Boilers

23 52 26 - Steel Boilers

23 52 26.40 Oil-Fired Boilers

23 52 26.40 Oil-Fired Boilers	Crew	Daily Output	Labor-Hours	Unit	Material	2012 Bare Costs Labor	Equipment	Total	Total Incl O&P
0010 **OIL-FIRED BOILERS**, Standard controls, flame retention burner									
5000 Steel, with insulated flush jacket									
7000 Hot water, gross output, 103 MBH	Q-6	1.60	15	Ea.	1,800	780		2,580	3,150
7120 420 MBH		.70	34.483		5,500	1,800		7,300	8,750
7320 3,150 MBH	↓	.13	184		37,100	9,600		46,700	55,500
7340 For tankless coil in steam or hot water, add				↓	7%				

23 52 28 - Swimming Pool Boilers

23 52 28.10 Swimming Pool Heaters

23 52 28.10 Swimming Pool Heaters	Crew	Daily Output	Labor-Hours	Unit	Material	Labor	Equipment	Total	Total Incl O&P
0010 **SWIMMING POOL HEATERS**, Not including wiring, external									
0020 piping, base or pad,									
0160 Gas fired, input, 155 MBH	Q-6	1.50	16	Ea.	1,700	830		2,530	3,100
0200 199 MBH		1	24		1,800	1,250		3,050	3,850
0280 500 MBH		.40	60		7,650	3,125		10,775	13,100
0400 1,800 MBH	↓	.14	171		16,600	8,900		25,500	31,700
2000 Electric, 12 KW, 4,800 gallon pool	Q-19	3	8		2,050	405		2,455	2,850
2020 15 KW, 7,200 gallon pool		2.80	8.571		2,075	435		2,510	2,950
2040 24 KW, 9,600 gallon pool		2.40	10		2,400	505		2,905	3,400
2100 57 KW, 24,000 gallon pool	↓	1.20	20	↓	3,525	1,000		4,525	5,400

23 54 Furnaces

23 54 13 - Electric-Resistance Furnaces

23 54 13.10 Electric Furnaces

23 54 13.10 Electric Furnaces	Crew	Daily Output	Labor-Hours	Unit	Material	Labor	Equipment	Total	Total Incl O&P
0010 **ELECTRIC FURNACES**, Hot air, blowers, std. controls									
0011 not including gas, oil or flue piping									
1000 Electric, UL listed									
1020 10.2 MBH	Q-20	5	4	Ea.	475	191		666	810
1040 17.1 MBH		4.80	4.167		495	199		694	850
1060 27.3 MBH		4.60	4.348		600	208		808	975
1100 34.1 MBH	↓	4.40	4.545	↓	635	218		853	1,025

23 54 16 - Fuel-Fired Furnaces

23 54 16.13 Gas-Fired Furnaces

23 54 16.13 Gas-Fired Furnaces	Crew	Daily Output	Labor-Hours	Unit	Material	Labor	Equipment	Total	Total Incl O&P
0010 **GAS-FIRED FURNACES**									
3000 Gas, AGA certified, upflow, direct drive models									
3020 45 MBH input	Q-9	4	4	Ea.	620	188		808	965
3040 60 MBH input		3.80	4.211		645	198		843	1,000
3060 75 MBH input		3.60	4.444		665	209		874	1,050
3100 100 MBH input	↓	3.20	5	↓	720	235		955	1,150

23 54 16.16 Oil-Fired Furnaces

23 54 16.16 Oil-Fired Furnaces	Crew	Daily Output	Labor-Hours	Unit	Material	Labor	Equipment	Total	Total Incl O&P
0010 **OIL-FIRED FURNACES**									
6000 Oil, UL listed, atomizing gun type burner									
6020 56 MBH output	Q-9	3.60	4.444	Ea.	1,475	209		1,684	1,950
6030 84 MBH output		3.50	4.571		1,850	215		2,065	2,375
6040 95 MBH output		3.40	4.706		1,875	221		2,096	2,400
6060 134 MBH output		3.20	5		2,175	235		2,410	2,750
6080 151 MBH output	↓	3	5.333	↓	2,250	250		2,500	2,850

23 55 Fuel-Fired Heaters

23 55 13 - Fuel-Fired Duct Heaters

23 55 13.16 Gas-Fired Duct Heaters

23 55 13.16 Gas-Fired Duct Heaters		Crew	Daily Output	Labor-Hours	Unit	Material	2012 Bare Costs Labor	Equipment	Total	Total Incl O&P
0010	**GAS-FIRED DUCT HEATERS**, Includes burner, controls, stainless steel									
0020	heat exchanger. Gas fired, electric ignition									
0030	Indoor installation									
0100	120 MBH output	Q-5	4	4	Ea.	2,125	201		2,326	2,650
0130	200 MBH output		2.70	5.926		2,850	297		3,147	3,575
0140	240 MBH output		2.30	6.957		2,975	350		3,325	3,775
0180	320 MBH output		1.60	10		3,550	500		4,050	4,675
0300	For powered venter and adapter, add					460			460	510
0502	For required flue pipe, see Section 23 51 23.10									
1000	Outdoor installation, with power venter									
1020	75 MBH output	Q-5	4	4	Ea.	2,775	201		2,976	3,350
1060	120 MBH output		4	4		3,525	201		3,726	4,200
1100	187 MBH output		3	5.333		5,050	267		5,317	5,950
1140	300 MBH output		1.80	8.889		6,100	445		6,545	7,375
1180	450 MBH output		1.40	11.429		7,450	575		8,025	9,075

23 55 33 - Fuel-Fired Unit Heaters

23 55 33.13 Oil-Fired Unit Heaters

23 55 33.13 Oil-Fired Unit Heaters		Crew	Daily Output	Labor-Hours	Unit	Material	Labor	Equipment	Total	Total Incl O&P
0010	**OIL-FIRED UNIT HEATERS**, Cabinet, grilles, fan, ctrl, burner, no piping									
6000	Oil fired, suspension mounted, 94 MBH output	Q-5	4	4	Ea.	3,400	201		3,601	4,025
6040	140 MBH output		3	5.333		3,625	267		3,892	4,375
6060	184 MBH output		3	5.333		3,900	267		4,167	4,700

23 55 33.16 Gas-Fired Unit Heaters

23 55 33.16 Gas-Fired Unit Heaters		Crew	Daily Output	Labor-Hours	Unit	Material	Labor	Equipment	Total	Total Incl O&P
0010	**GAS-FIRED UNIT HEATERS**, Cabinet, grilles, fan, ctrls., burner, no piping									
0022	thermostat, no piping. For flue see Section 23 51 23.10									
1000	Gas fired, floor mounted									
1100	60 MBH output	Q-5	10	1.600	Ea.	830	80		910	1,025
1140	100 MBH output		8	2		915	100		1,015	1,150
1180	180 MBH output		6	2.667		1,300	134		1,434	1,650
2000	Suspension mounted, propeller fan, 20 MBH output		8.50	1.882		985	94.50		1,079.50	1,225
2040	60 MBH output		7	2.286		1,200	115		1,315	1,500
2060	80 MBH output		6	2.667		1,225	134		1,359	1,550
2100	130 MBH output		5	3.200		1,500	160		1,660	1,900
2240	320 MBH output		2	8		2,650	400		3,050	3,500
2500	For powered venter and adapter, add					355			355	390
5000	Wall furnace, 17.5 MBH output	Q-5	6	2.667		715	134		849	985
5020	24 MBH output		5	3.200		835	160		995	1,150
5040	35 MBH output		4	4		960	201		1,161	1,350

23 56 Solar Energy Heating Equipment

23 56 16 - Packaged Solar Heating Equipment

23 56 16.40 Solar Heating Systems

23 56 16.40 Solar Heating Systems			Crew	Daily Output	Labor-Hours	Unit	Material	Labor	Equipment	Total	Total Incl O&P
0010	**SOLAR HEATING SYSTEMS**	R235616-60									
0020	System/Package prices, not including connecting										
0030	pipe, insulation, or special heating/plumbing fixtures										
0500	Hot water, standard package, low temperature										
0540	1 collector, circulator, fittings, 65 gal. tank	G	Q-1	.50	32	Ea.	1,625	1,575		3,200	4,150
0580	2 collectors, circulator, fittings, 120 gal. tank	G		.40	40		2,375	1,975		4,350	5,575
0620	3 collectors, circulator, fittings, 120 gal. tank	G		.34	47.059		3,225	2,325		5,550	7,050
0700	Medium temperature package										
0720	1 collector, circulator, fittings, 80 gal. tank	G	Q-1	.50	32	Ea.	2,375	1,575		3,950	4,975

23 56 Solar Energy Heating Equipment

23 56 16 – Packaged Solar Heating Equipment

23 56 16.40 Solar Heating Systems		Crew	Daily Output	Labor-Hours	Unit	Material	2012 Bare Costs Labor	Equipment	Total	Total Incl O&P
0740	2 collectors, circulator, fittings, 120 gal. tank	G Q-1	.40	40	Ea.	3,425	1,975		5,400	6,750
0780	3 collectors, circulator, fittings, 120 gal. tank	G ↓	.30	53.333		4,450	2,625		7,075	8,850
0980	For each additional 120 gal. tank, add	G			↓	1,550			1,550	1,700

23 56 19 – Solar Heating Components

23 56 19.50 Solar Heating Ancillary

		Crew	Daily Output	Labor-Hours	Unit	Material	2012 Bare Costs Labor	Equipment	Total	Total Incl O&P
0010	**SOLAR HEATING ANCILLARY**									
2300	Circulators, air									
2310	Blowers									
2400	Reversible fan, 20" diameter, 2 speed	G Q-9	18	.889	Ea.	113	41.50		154.50	188
2870	1/12 HP, 30 GPM	G Q-1	10	1.600	"	345	78.50		423.50	500
3000	Collector panels, air with aluminum absorber plate									
3010	Wall or roof mount									
3040	Flat black, plastic glazing									
3080	4' x 8'	G Q-9	6	2.667	Ea.	660	125		785	920
3200	Flush roof mount, 10' to 16' x 22" wide	G "	96	.167	L.F.	470	7.80		477.80	525
3300	Collector panels, liquid with copper absorber plate									
3330	Alum. frame, 4' x 8', 5/32" single glazing	G Q-1	9.50	1.684	Ea.	995	83		1,078	1,225
3390	Alum. frame, 4' x 10', 5/32" single glazing	G	6	2.667		1,150	131		1,281	1,450
3450	Flat black, alum. frame, 3.5' x 7.5'	G	9	1.778		690	87.50		777.50	890
3500	4' x 8'	G	5.50	2.909		865	143		1,008	1,175
3520	4' x 10'	G	10	1.600		970	78.50		1,048.50	1,200
3540	4' x 12.5'	G	5	3.200		1,175	157		1,332	1,500
3600	Liquid, full wetted, plastic, alum. frame, 4' x 10'	G	5	3.200		234	157		391	495
3650	Collector panel mounting, flat roof or ground rack	G	7	2.286	↓	228	112		340	420
3670	Roof clamps	G ↓	70	.229	Set	2.46	11.25		13.71	19.65
3700	Roof strap, teflon	G 1 Plum	205	.039	L.F.	19.95	2.13		22.08	25
3900	Differential controller with two sensors									
3930	Thermostat, hard wired	G 1 Plum	8	1	Ea.	83	54.50		137.50	174
4100	Five station with digital read-out	G "	3	2.667	"	229	146		375	470
4300	Heat exchanger									
4580	Fluid to fluid package includes two circulating pumps									
4590	expansion tank, check valve, relief valve									
4600	controller, high temperature cutoff and sensors	G Q-1	2.50	6.400	Ea.	725	315		1,040	1,275
4650	Heat transfer fluid									
4700	Propylene glycol, inhibited anti-freeze	G 1 Plum	28	.286	Gal.	14.50	15.60		30.10	39.50
8250	Water storage tank with heat exchanger and electric element									
8300	80 gal. with 2" x 2 lb. density insulation	G 1 Plum	1.60	5	Ea.	1,100	273		1,373	1,600
8380	120 gal. with 2" x 2 lb. density insulation	G	1.40	5.714		1,250	310		1,560	1,850
8400	120 gal. with 2" x 2 lb. density insul., 40 S.F. heat coil	G ↓	1.40	5.714	↓	1,575	310		1,885	2,200

23 57 Heat Exchangers for HVAC

23 57 16 – Steam-to-Water Heat Exchangers

23 57 16.10 Shell/Tube Type Steam-to-Water Heat Exch.

		Crew	Daily Output	Labor-Hours	Unit	Material	2012 Bare Costs Labor	Equipment	Total	Total Incl O&P
0010	**SHELL AND TUBE TYPE STEAM-TO-WATER HEAT EXCHANGERS**									
0016	Shell & tube type, 2 or 4 pass, 3/4" O.D. copper tubes,									
0020	C.I. heads, C.I. tube sheet, steel shell									
0100	Hot water 40°F to 180°F, by steam at 10 PSI									
0120	8 GPM	Q-5	6	2.667	Ea.	1,925	134		2,059	2,325
0140	10 GPM		5	3.200		2,900	160		3,060	3,450
0160	40 GPM	↓	4	4	↓	4,500	201		4,701	5,250

521

23 57 Heat Exchangers for HVAC

23 57 16 – Steam-to-Water Heat Exchangers

23 57 16.10 Shell/Tube Type Steam-to-Water Heat Exch.

		Crew	Daily Output	Labor-Hours	Unit	Material	2012 Bare Costs Labor	Equipment	Total	Total Incl O&P
0180	64 GPM	Q-5	2	8	Ea.	6,900	400		7,300	8,200
0200	96 GPM	↓	1	16		9,250	800		10,050	11,400
0220	120 GPM	Q-6	1.50	16	↓	12,100	830		12,930	14,600

23 57 19 – Liquid-to-Liquid Heat Exchangers

23 57 19.13 Plate-Type, Liquid-to-Liquid Heat Exchangers

		Crew	Daily Output	Labor-Hours	Unit	Material	2012 Bare Costs Labor	Equipment	Total	Total Incl O&P
0010	**PLATE-TYPE, LIQUID-TO-LIQUID HEAT EXCHANGERS**									
3000	Plate type,									
3100	400 GPM	Q-6	.80	30	Ea.	33,600	1,550		35,150	39,300
3120	800 GPM	"	.50	48		58,000	2,500		60,500	67,500
3140	1200 GPM	Q-7	.34	94.118		86,000	5,000		91,000	102,000
3160	1800 GPM	"	.24	133	↓	114,000	7,075		121,075	136,000

23 57 19.16 Shell-Type, Liquid-to-Liquid Heat Exchangers

		Crew	Daily Output	Labor-Hours	Unit	Material	2012 Bare Costs Labor	Equipment	Total	Total Incl O&P
0010	**SHELL-TYPE, LIQUID-TO-LIQUID HEAT EXCHANGERS**									
1000	Hot water 40°F to 140°F, by water at 200°F									
1020	7 GPM	Q-5	6	2.667	Ea.	2,375	134		2,509	2,800
1040	16 GPM		5	3.200		3,350	160		3,510	3,950
1060	34 GPM		4	4		5,100	201		5,301	5,900
1100	74 GPM	↓	1.50	10.667	↓	9,225	535		9,760	10,900

23 62 Packaged Compressor and Condenser Units

23 62 13 – Packaged Air-Cooled Refrigerant Compressor and Condenser Units

23 62 13.10 Packaged Air-Cooled Refrig. Condensing Units

		Crew	Daily Output	Labor-Hours	Unit	Material	2012 Bare Costs Labor	Equipment	Total	Total Incl O&P
0010	**PACKAGED AIR-COOLED REFRIGERANT CONDENSING UNITS**									
0020	Condensing unit									
0030	Air cooled, compressor, standard controls									
0050	1.5 ton	Q-5	2.50	6.400	Ea.	1,125	320		1,445	1,700
0500	5 ton	↓	.60	26.667		2,250	1,325		3,575	4,525
0600	10 ton	↓	.50	32		4,950	1,600		6,550	7,875
0700	20 ton	Q-6	.40	60	↓	11,200	3,125		14,325	17,000

23 63 Refrigerant Condensers

23 63 13 – Air-Cooled Refrigerant Condensers

23 63 13.10 Air-Cooled Refrig. Condensers

		Crew	Daily Output	Labor-Hours	Unit	Material	2012 Bare Costs Labor	Equipment	Total	Total Incl O&P
0010	**AIR-COOLED REFRIG. CONDENSERS**, Ratings for 30°F TD, R-22.									
0080	Air cooled, belt drive, propeller fan									
0240	50 ton	Q-6	.69	34.985	Ea.	10,400	1,825		12,225	14,200
0280	59 ton		.58	41.308		12,400	2,150		14,550	17,000
0320	73 ton		.47	51.173		16,300	2,650		18,950	21,900
0360	86 ton		.40	60.302		18,800	3,125		21,925	25,400
0380	88 ton	↓	.39	61.697	↓	20,200	3,200		23,400	27,000
1550	Air cooled, direct drive, propeller fan									
1590	1 ton	Q-5	3.80	4.211	Ea.	685	211		896	1,075
1600	1-1/2 ton		3.60	4.444		840	223		1,063	1,250
1620	2 ton		3.20	5		900	251		1,151	1,375
1640	5 ton		2	8		1,825	400		2,225	2,625
1660	10 ton		1.40	11.429		2,825	575		3,400	3,975
1690	16 ton		1.10	14.545		3,975	730		4,705	5,475
1720	26 ton	↓	.84	19.002		5,150	955		6,105	7,100

23 63 Refrigerant Condensers

23 63 13 – Air-Cooled Refrigerant Condensers

23 63 13.10 Air-Cooled Refrig. Condensers		Crew	Daily Output	Labor-Hours	Unit	Material	2012 Bare Costs Labor	Equipment	Total	Total Incl O&P
1760	41 ton	Q-6	.77	31.008	Ea.	9,700	1,600		11,300	13,100
1800	63 ton	"	.55	44.037	↓	15,300	2,300		17,600	20,300

23 64 Packaged Water Chillers

23 64 13 – Absorption Water Chillers

23 64 13.16 Indirect-Fired Absorption Water Chillers

		Crew	Daily Output	Labor-Hours	Unit	Material	Labor	Equipment	Total	Total Incl O&P
0010	**INDIRECT-FIRED ABSORPTION WATER CHILLERS**									
0020	Steam or hot water, water cooled									
0050	100 ton	Q-7	.13	240	Ea.	125,000	12,700		137,700	156,500
0400	420 ton	"	.10	323	"	347,000	17,100		364,100	407,500

23 64 16 – Centrifugal Water Chillers

23 64 16.10 Centrifugal Type Water Chillers

		Crew	Daily Output	Labor-Hours	Unit	Material	Labor	Equipment	Total	Total Incl O&P
0010	**CENTRIFUGAL TYPE WATER CHILLERS**, With standard controls									
0020	Centrifugal liquid chiller, water cooled									
0030	not including water tower									
0100	2000 ton (twin 1000 ton units)	Q-7	.07	477	Ea.	865,500	25,300		890,800	990,000

23 64 19 – Reciprocating Water Chillers

23 64 19.10 Reciprocating Type Water Chillers

		Crew	Daily Output	Labor-Hours	Unit	Material	Labor	Equipment	Total	Total Incl O&P
0010	**RECIPROCATING TYPE WATER CHILLERS**, With standard controls									
0494	Water chillers, integral air cooled condenser									
0600	100 ton cooling	Q-7	.25	129	Ea.	71,500	6,850		78,350	89,000
0980	Water cooled, multiple compressor, semi-hermetic, tower not incl.									
1000	15 ton cooling	Q-6	.36	65.934	Ea.	17,500	3,425		20,925	24,500
1020	25 ton cooling	Q-7	.41	78.049		18,900	4,150		23,050	27,100
1060	35 ton cooling		.31	101		22,600	5,400		28,000	33,000
1100	50 ton cooling		.28	113		32,300	6,050		38,350	44,600
1160	100 ton cooling		.18	179		58,500	9,525		68,025	79,000
1180	125 ton cooling		.16	196		66,500	10,400		76,900	88,500
1200	145 ton cooling	↓	.16	202	↓	77,500	10,700		88,200	101,500
1451	Water cooled, dual compressors, semi-hermetic, tower not incl.									
1500	80 ton cooling	Q-7	.14	222	Ea.	28,800	11,800		40,600	49,400
1520	100 ton cooling		.14	228		38,800	12,100		50,900	61,000
1540	120 ton cooling	↓	.14	231	↓	47,100	12,300		59,400	70,500

23 64 23 – Scroll Water Chillers

23 64 23.10 Scroll Water Chillers

		Crew	Daily Output	Labor-Hours	Unit	Material	Labor	Equipment	Total	Total Incl O&P
0010	**SCROLL WATER CHILLERS**, With standard controls									
0490	Packaged w/integral air cooled condenser, 15 ton cool	Q-7	.37	86.486	Ea.	20,400	4,575		24,975	29,400
0500	20 ton cooling		.34	94.118		26,400	5,000		31,400	36,500
0520	40 ton cooling	↓	.30	108	↓	34,100	5,725		39,825	46,200
0680	Scroll water cooled, single compressor, hermetic, tower not incl.									
0700	2 ton cooling	Q-5	.57	28.070	Ea.	3,400	1,400		4,800	5,850
0710	5 ton cooling		.57	28.070		4,050	1,400		5,450	6,575
0740	8 ton cooling	↓	.31	52.117		5,575	2,625		8,200	10,100
0760	10 ton cooling	Q-6	.36	67.039		6,425	3,475		9,900	12,300
0800	20 ton cooling	Q-7	.38	83.990		11,500	4,450		15,950	19,300
0820	30 ton cooling	"	.33	96.096	↓	12,900	5,100		18,000	21,800

23 64 Packaged Water Chillers

23 64 26 – Rotary-Screw Water Chillers

23 64 26.10 Rotary-Screw Type Water Chillers	Crew	Daily Output	Labor-Hours	Unit	Material	2012 Bare Costs Labor	Equipment	Total	Total Incl O&P
0010 **ROTARY-SCREW TYPE WATER CHILLERS**, With standard controls									
0110 Screw, liquid chiller, air cooled, insulated evaporator									
0120 130 ton	Q-7	.14	228	Ea.	86,500	12,100		98,600	113,500
0124 160 ton		.13	246		106,500	13,100		119,600	136,500
0128 180 ton		.13	250		119,500	13,300		132,800	151,500
0132 210 ton		.12	258		132,000	13,700		145,700	165,500
0136 270 ton		.12	266		150,500	14,100		164,600	187,500
0140 320 ton		.12	275		189,000	14,600		203,600	230,000
0200 Packaged unit, water cooled, not incl. tower									
0210 80 ton	Q-7	.14	223	Ea.	43,400	11,900		55,300	65,500
0240 200 ton		.13	251		82,000	13,400		95,400	110,000
0270 350 ton		.12	275		162,000	14,600		176,600	200,000
1450 Water cooled, tower not included									
1580 150 ton cooling, screw compressors	Q-7	.13	240	Ea.	67,000	12,800		79,800	92,500
1620 200 ton cooling, screw compressors		.13	250		92,000	13,300		105,300	121,500
1660 291 ton cooling, screw compressors		.12	260		96,000	13,800		109,800	127,000

23 65 Cooling Towers

23 65 13 – Forced-Draft Cooling Towers

23 65 13.10 Forced-Draft Type Cooling Towers	Crew	Daily Output	Labor-Hours	Unit	Material	2012 Bare Costs Labor	Equipment	Total	Total Incl O&P
0010 **FORCED-DRAFT TYPE COOLING TOWERS**, Packaged units									
0070 Galvanized steel									
0080 Induced draft, crossflow									
0100 Vertical, belt drive, 61 tons	Q-6	90	.267	TonAC	197	13.85		210.85	237
0150 100 ton		100	.240		190	12.50		202.50	228
0200 115 ton		109	.220		165	11.45		176.45	199
0250 131 ton		120	.200		198	10.40		208.40	234
0260 162 ton		132	.182		160	9.45		169.45	191
1500 Induced air, double flow									
1900 Vertical, gear drive, 167 ton	Q-6	126	.190	TonAC	156	9.90		165.90	186
2000 297 ton		129	.186		96	9.65		105.65	121
2100 582 ton		132	.182		53.50	9.45		62.95	73
2150 849 ton		142	.169		72	8.80		80.80	92.50
2200 1016 ton		150	.160		72.50	8.30		80.80	92.50
3000 For higher capacities, use multiples									
3500 For pumps and piping, add	Q-6	38	.632	TonAC	98.50	33		131.50	158
4000 For absorption systems, add				"	75%	75%			
5000 Fiberglass tower on galvanized steel support structure									
5010 Draw thru									
5100 100 ton	Q-6	1.40	17.143	Ea.	10,100	890		10,990	12,500
5120 120 ton		1.20	20		11,700	1,050		12,750	14,500
5140 140 ton		1	24		12,700	1,250		13,950	15,900
5160 160 ton		.80	30		14,100	1,550		15,650	17,900
5180 180 ton		.65	36.923		16,000	1,925		17,925	20,500
5200 200 ton		.48	50		18,100	2,600		20,700	23,800
5300 For stainless steel support structure, add					30%				
5360 For higher capacities, use multiples of each size									
6000 Stainless steel									
6010 Induced draft, crossflow, horizontal, belt drive									
6100 57 ton	Q-6	1.50	16	Ea.	25,300	830		26,130	29,100

23 65 Cooling Towers

23 65 13 – Forced-Draft Cooling Towers

23 65 13.10 Forced-Draft Type Cooling Towers	Crew	Daily Output	Labor-Hours	Unit	Material	2012 Bare Costs Labor	2012 Bare Costs Equipment	Total	Total Incl O&P
6120	91 ton	Q-6	.99	24.242	Ea.	30,600	1,250		31,850	35,500
6140	111 ton		.43	55.814		38,900	2,900		41,800	47,100
6160	126 ton		.22	109		38,900	5,675		44,575	51,500

23 73 Indoor Central-Station Air-Handling Units

23 73 13 – Modular Indoor Central-Station Air-Handling Units

23 73 13.10 Air Handling Units

		Crew	Daily Output	Labor-Hours	Unit	Material	2012 Bare Costs Labor	2012 Bare Costs Equipment	Total	Total Incl O&P
0010	**AIR HANDLING UNITS**, Built-Up									
0100	With cooling/heating coil section, filters, mixing box									
0880	Single zone, horizontal/vertical									
0890	Constant volume									
0900	1600 CFM	Q-5	1.20	13.333	Ea.	3,425	670		4,095	4,775
0920	5000 CFM	Q-6	1.40	17.143		8,925	890		9,815	11,200
0940	11,500 CFM		1	24		18,100	1,250		19,350	21,800
0970	22,000 CFM		.60	40		34,600	2,075		36,675	41,200
1000	40,000 CFM		.30	80		53,000	4,150		57,150	65,000
1140	60,000 CFM	Q-7	.18	177		75,000	9,425		84,425	96,000

23 73 39 – Indoor, Direct Gas-Fired Heating and Ventilating Units

23 73 39.10 Make-Up Air Unit

		Crew	Daily Output	Labor-Hours	Unit	Material	2012 Bare Costs Labor	2012 Bare Costs Equipment	Total	Total Incl O&P
0010	**MAKE-UP AIR UNIT**									
0020	Indoor suspension, natural/LP gas, direct fired,									
0032	standard control. For flue see Section 23 51 23.10									
0040	70°F temperature rise, MBH is input									
0100	75 MBH input	Q-6	3.60	6.667	Ea.	2,400	345		2,745	3,175
0160	150 MBH input		3	8		2,850	415		3,265	3,750
0220	225 MBH input		2.40	10		3,300	520		3,820	4,400
0300	400 MBH input		1.60	15		7,450	780		8,230	9,350
0600	For discharge louver assembly, add					5%				
0700	For filters, add					10%				
0800	For air shut-off damper section, add					30%				

23 74 Packaged Outdoor HVAC Equipment

23 74 33 – Dedicated Outdoor-Air Units

23 74 33.10 Roof Top Air Conditioners

			Crew	Daily Output	Labor-Hours	Unit	Material	2012 Bare Costs Labor	2012 Bare Costs Equipment	Total	Total Incl O&P
0010	**ROOF TOP AIR CONDITIONERS**, Standard controls, curb, economizer										
1000	Single zone, electric cool, gas heat										
1100	3 ton cooling, 60 MBH heating	R236000-20	Q-5	.70	22.857	Ea.	3,600	1,150		4,750	5,675
1120	4 ton cooling, 95 MBH heating			.61	26.403		4,250	1,325		5,575	6,675
1140	5 ton cooling, 112 MBH heating			.56	28.521		4,975	1,425		6,400	7,625
1145	6 ton cooling, 140 MBH heating			.52	30.769		5,825	1,550		7,375	8,750
1150	7.5 ton cooling, 170 MBH heating			.50	32.258		7,650	1,625		9,275	10,900
1160	10 ton cooling, 200 MBH heating		Q-6	.67	35.982		10,600	1,875		12,475	14,400
1170	12.5 ton cooling, 230 MBH heating			.63	37.975		13,400	1,975		15,375	17,700
1190	17.5 ton cooling, 330 MBH heating			.52	45.889		18,800	2,375		21,175	24,300
1200	20 ton cooling, 360 MBH heating		Q-7	.67	47.976		21,900	2,550		24,450	28,000
1210	25 ton cooling, 450 MBH heating			.56	57.554		25,000	3,050		28,050	32,100
1220	30 ton cooling, 540 MBH heating			.47	68.376		25,500	3,625		29,125	33,600
1240	40 ton cooling, 675 MBH heating			.35	91.168		33,400	4,825		38,225	44,000
2000	Multizone, electric cool, gas heat, economizer										

23 74 Packaged Outdoor HVAC Equipment

23 74 33 – Dedicated Outdoor-Air Units

23 74 33.10 Roof Top Air Conditioners	Crew	Daily Output	Labor-Hours	Unit	Material	2012 Bare Costs Labor	Equipment	Total	Total Incl O&P
2100	15 ton cooling, 360 MBH heating	Q-7	.61	52.545	Ea.	59,500	2,775		62,275	69,500
2120	20 ton cooling, 360 MBH heating		.53	60.038		64,000	3,175		67,175	75,500
2200	40 ton cooling, 540 MBH heating		.28	113		112,500	6,050		118,550	133,000
2210	50 ton cooling, 540 MBH heating		.23	142		140,000	7,550		147,550	165,500
2220	70 ton cooling, 1500 MBH heating		.16	198		151,500	10,500		162,000	182,500
2240	80 ton cooling, 1500 MBH heating		.14	228		173,000	12,100		185,100	209,000
2260	90 ton cooling, 1500 MBH heating		.13	256		181,500	13,600		195,100	220,000
2280	105 ton cooling, 1500 MBH heating		.11	290		200,000	15,400		215,400	243,500
2400	For hot water heat coil, deduct					5%				
2500	For steam heat coil, deduct					2%				
2600	For electric heat, deduct					3%	5%			

23 81 Decentralized Unitary HVAC Equipment

23 81 13 – Packaged Terminal Air-Conditioners

23 81 13.10 Packaged Cabinet Type Air-Conditioners

23 81 13.10 Packaged Cabinet Type Air-Conditioners	Crew	Daily Output	Labor-Hours	Unit	Material	Labor	Equipment	Total	Total Incl O&P
0010	**PACKAGED CABINET TYPE AIR-CONDITIONERS**, Cabinet, wall sleeve,									
0100	louver, electric heat, thermostat, manual changeover, 208 V									
0200	6,000 BTUH cooling, 8800 BTU heat	Q-5	6	2.667	Ea.	720	134		854	990
0220	9,000 BTUH cooling, 13,900 BTU heat		5	3.200		785	160		945	1,100
0240	12,000 BTUH cooling, 13,900 BTU heat		4	4		850	201		1,051	1,225
0260	15,000 BTUH cooling, 13,900 BTU heat		3	5.333		1,050	267		1,317	1,550
0500	For hot water coil, increase heat by 10%, add					5%	10%			
1000	For steam, increase heat output by 30%, add					8%	10%			

23 81 19 – Self-Contained Air-Conditioners

23 81 19.20 Self-Contained Single Package

23 81 19.20 Self-Contained Single Package	Crew	Daily Output	Labor-Hours	Unit	Material	Labor	Equipment	Total	Total Incl O&P
0010	**SELF-CONTAINED SINGLE PACKAGE**									
0100	Air cooled, for free blow or duct, not incl. remote condenser									
0200	3 ton cooling	Q-5	1	16	Ea.	3,475	800		4,275	5,025
0220	5 ton cooling	Q-6	1.20	20		4,150	1,050		5,200	6,150
0240	10 ton cooling	Q-7	1	32		7,000	1,700		8,700	10,300
0260	20 ton cooling		.90	35.556		15,600	1,875		17,475	20,100
0280	30 ton cooling		.80	40		23,600	2,125		25,725	29,200
0340	60 ton cooling	Q-8	.40	80		53,500	4,250	92	57,842	65,500
0490	For duct mounting no price change									
0500	For steam heating coils, add				Ea.	10%	10%			
1000	Water cooled for free blow or duct, not including tower									
1010	Constant volume									
1100	3 ton cooling	Q-6	1	24	Ea.	3,475	1,250		4,725	5,700
1120	5 ton cooling	"	1	24		4,550	1,250		5,800	6,900
1140	10 ton cooling	Q-7	.90	35.556		8,550	1,875		10,425	12,300
1160	20 ton cooling		.80	40		26,800	2,125		28,925	32,700
1180	30 ton cooling		.70	45.714		35,500	2,425		37,925	42,800

23 81 23 – Computer-Room Air-Conditioners

23 81 23.10 Computer Room Units

23 81 23.10 Computer Room Units	Crew	Daily Output	Labor-Hours	Unit	Material	Labor	Equipment	Total	Total Incl O&P
0010	**COMPUTER ROOM UNITS**									
1000	Air cooled, includes remote condenser but not									
1020	interconnecting tubing or refrigerant									
1080	3 ton	Q-5	.50	32	Ea.	17,800	1,600		19,400	22,000
1120	5 ton		.45	35.556		19,100	1,775		20,875	23,700
1160	6 ton		.30	53.333		33,600	2,675		36,275	40,900

23 81 Decentralized Unitary HVAC Equipment

23 81 23 – Computer-Room Air-Conditioners

23 81 23.10 Computer Room Units	Crew	Daily Output	Labor-Hours	Unit	Material	2012 Bare Costs Labor	2012 Bare Costs Equipment	Total	Total Incl O&P
1200	8 ton	Q-5	.27	59.259	Ea.	35,600	2,975		38,575	43,700
1240	10 ton		.25	64		37,200	3,200		40,400	45,800
1260	12 ton		.24	66.667		38,600	3,350		41,950	47,500
1280	15 ton		.22	72.727		40,900	3,650		44,550	50,500
1290	18 ton		.20	80		46,600	4,000		50,600	57,000
1300	20 ton	Q-6	.26	92.308		49,000	4,800		53,800	61,000
1320	22 ton		.24	100		49,700	5,200		54,900	62,500
1360	30 ton		.21	114		61,000	5,950		66,950	76,000
2200	Chilled water, for connection to									
2220	existing chiller system of adequate capacity									
2260	5 ton	Q-5	.74	21.622	Ea.	13,600	1,075		14,675	16,600

23 81 43 – Air-Source Unitary Heat Pumps

23 81 43.10 Air-Source Heat Pumps

		Crew	Daily Output	Labor-Hours	Unit	Material	Labor	Equipment	Total	Total Incl O&P
0010	**AIR-SOURCE HEAT PUMPS**, Not including interconnecting tubing									
1000	Air to air, split system, not including curbs, pads, fan coil and ductwork									
1012	Outside condensing unit only, for fan coil see Section 23 82 19.10									
1020	2 ton cooling, 8.5 MBH heat @ 0°F	Q-5	2	8	Ea.	2,125	400		2,525	2,925
1060	5 ton cooling, 27 MBH heat @ 0°F		.50	32		3,425	1,600		5,025	6,200
1080	7.5 ton cooling, 33 MBH heat @ 0°F		.45	35.556		5,925	1,775		7,700	9,200
1100	10 ton cooling, 50 MBH heat @ 0°F	Q-6	.64	37.500		8,250	1,950		10,200	12,000
1120	15 ton cooling, 64 MBH heat @ 0°F		.50	48		12,200	2,500		14,700	17,200
1130	20 ton cooling, 85 MBH heat @ 0°F		.35	68.571		16,000	3,575		19,575	23,000
1140	25 ton cooling, 119 MBH heat @ 0°F		.25	96		19,300	5,000		24,300	28,700
1500	Single package, not including curbs, pads, or plenums									
1520	2 ton cooling, 6.5 MBH heat @ 0°F	Q-5	1.50	10.667	Ea.	2,750	535		3,285	3,825
1580	4 ton cooling, 13 MBH heat @ 0°F		.96	16.667		3,850	835		4,685	5,500
1640	7.5 ton cooling, 35 MBH heat @ 0°F		.40	40		6,850	2,000		8,850	10,600

23 81 46 – Water-Source Unitary Heat Pumps

23 81 46.10 Water Source Heat Pumps

		Crew	Daily Output	Labor-Hours	Unit	Material	Labor	Equipment	Total	Total Incl O&P
0010	**WATER SOURCE HEAT PUMPS**, Not incl. connecting tubing or water source									
2000	Water source to air, single package									
2100	1 ton cooling, 13 MBH heat @ 75°F	Q-5	2	8	Ea.	1,700	400		2,100	2,450
2140	2 ton cooling, 19 MBH heat @ 75°F		1.70	9.412		1,875	470		2,345	2,775
2220	5 ton cooling, 29 MBH heat @ 75°F		.90	17.778		2,825	890		3,715	4,450
3960	For supplementary heat coil, add					10%				
4000	For increase in capacity thru use									
4020	of solar collector, size boiler at 60%									

23 82 Convection Heating and Cooling Units

23 82 16 – Air Coils

23 82 16.10 Flanged Coils

		Crew	Daily Output	Labor-Hours	Unit	Material	Labor	Equipment	Total	Total Incl O&P
0010	**FLANGED COILS**									
0500	Chilled water cooling, 6 rows, 24" x 48"	Q-5	3.20	5	Ea.	3,850	251		4,101	4,600
1000	Direct expansion cooling, 6 rows, 24" x 48"		2.80	5.714		4,175	286		4,461	5,025
1500	Hot water heating, 1 row, 24" x 48"		4	4		1,525	201		1,726	2,000
2000	Steam heating, 1 row, 24" x 48"		3.06	5.229		2,175	262		2,437	2,800

23 82 16.20 Duct Heaters

		Crew	Daily Output	Labor-Hours	Unit	Material	Labor	Equipment	Total	Total Incl O&P
0010	**DUCT HEATERS**, Electric, 480 V, 3 Ph.									
0020	Finned tubular insert, 500°F									
0100	8" wide x 6" high, 4.0 kW	Q-20	16	1.250	Ea.	740	60		800	905

23 82 Convection Heating and Cooling Units

23 82 16 – Air Coils

23 82 16.20 Duct Heaters	Crew	Daily Output	Labor-Hours	Unit	Material	2012 Bare Costs Labor	Equipment	Total	Total Incl O&P
0120	12" high, 8.0 kW	Q-20	15	1.333	Ea.	1,225	64		1,289	1,450
0140	18" high, 12.0 kW		14	1.429		1,725	68.50		1,793.50	2,000
0160	24" high, 16.0 kW		13	1.538		2,225	73.50		2,298.50	2,525
0180	30" high, 20.0 kW		12	1.667		2,700	80		2,780	3,100
0300	12" wide x 6" high, 6.7 kW		15	1.333		785	64		849	960
0360	24" high, 26.7 kW		12	1.667		2,275	80		2,355	2,650
0700	24" wide x 6" high, 17.8 kW		13	1.538		935	73.50		1,008.50	1,125
0760	24" high, 71.1 kW		10	2		2,850	95.50		2,945.50	3,300
8000	To obtain BTU multiply kW by 3413									

23 82 19 – Fan Coil Units

23 82 19.10 Fan Coil Air Conditioning

		Crew	Daily Output	Labor-Hours	Unit	Material	2012 Bare Costs Labor	Equipment	Total	Total Incl O&P
0010	**FAN COIL AIR CONDITIONING** Cabinet mounted, filters, controls									
0100	Chilled water, 1/2 ton cooling	Q-5	8	2	Ea.	710	100		810	930
0120	1 ton cooling		6	2.667		815	134		949	1,100
0140	1.5 ton cooling		5.50	2.909		920	146		1,066	1,225
0150	2 ton cooling		5.25	3.048		1,150	153		1,303	1,500
0180	3 ton cooling		4	4		1,725	201		1,926	2,200
0262	For hot water coil, add					40%	10%			
0940	Direct expansion, for use w/air cooled condensing unit, 1.5 ton cooling	Q-5	5	3.200	Ea.	690	160		850	1,000
1000	5 ton cooling	"	3	5.333		1,325	267		1,592	1,850
1040	10 ton cooling	Q-6	2.60	9.231		2,975	480		3,455	4,000
1060	20 ton cooling	"	.70	34.286		5,575	1,775		7,350	8,850

23 82 19.20 Heating and Ventilating Units

		Crew	Daily Output	Labor-Hours	Unit	Material	2012 Bare Costs Labor	Equipment	Total	Total Incl O&P
0010	**HEATING AND VENTILATING UNITS**, Classroom units									
0020	Includes filter, heating/cooling coils, standard controls									
0080	750 CFM, 2 tons cooling	Q-6	2	12	Ea.	3,900	625		4,525	5,250
0120	1250 CFM, 3 tons cooling		1.40	17.143		4,750	890		5,640	6,575
0140	1500 CFM, 4 tons cooling		.80	30		5,100	1,550		6,650	7,950
0500	For electric heat, add					35%				
1000	For no cooling, deduct					25%	10%			

23 82 27 – Infrared Units

23 82 27.10 Infrared Type Heating Units

		Crew	Daily Output	Labor-Hours	Unit	Material	2012 Bare Costs Labor	Equipment	Total	Total Incl O&P
0010	**INFRARED TYPE HEATING UNITS**									
0020	Gas fired, unvented, electric ignition, 100% shutoff.									
0030	Piping and wiring not included									
0060	Input, 15 MBH	Q-5	7	2.286	Ea.	520	115		635	745
0120	45 MBH		5	3.200		1,500	160		1,660	1,900
0160	60 MBH		4	4		1,600	201		1,801	2,075
0240	120 MBH		2	8		1,975	400		2,375	2,750

23 82 29 – Radiators

23 82 29.10 Hydronic Heating

		Crew	Daily Output	Labor-Hours	Unit	Material	2012 Bare Costs Labor	Equipment	Total	Total Incl O&P
0010	**HYDRONIC HEATING**, Terminal units, not incl. main supply pipe									
1000	Radiation									
1100	Panel, baseboard, C.I., including supports, no covers	Q-5	46	.348	L.F.	35.50	17.45		52.95	65.50
3000	Radiators, cast iron									
3100	Free standing or wall hung, 6 tube, 25" high	Q-5	96	.167	Section	37.50	8.35		45.85	54
3200	4 tube, 19" high	"	96	.167	"	26	8.35		34.35	41.50
3250	Adj. brackets, 2 per wall radiator up to 30 sections	1 Stpi	32	.250	Ea.	36.50	13.95		50.45	61
9500	To convert SFR to BTU rating: Hot water, 150 x SFR									
9510	Forced hot water, 180 x SFR; steam, 240 x SFR									

23 82 Convection Heating and Cooling Units

23 82 33 – Convectors

23 82 33.10 Convector Units

	23 82 33.10 Convector Units	Crew	Daily Output	Labor-Hours	Unit	Material	2012 Bare Costs Labor	Equipment	Total	Total Incl O&P
0010	**CONVECTOR UNITS**, Terminal units, not incl. main supply pipe									
2204	Convector, multifin, 2 pipe w/cabinet									
2210	17" H x 24" L	Q-5	10	1.600	Ea.	99.50	80		179.50	231
2214	17" H x 36" L		8.60	1.860		149	93.50		242.50	305
2218	17" H x 48" L		7.40	2.162		199	108		307	380
2222	21" H x 24" L		9	1.778		99.50	89		188.50	244
2226	21" H x 36" L		8.20	1.951		149	98		247	310
2228	21" H x 48" L		6.80	2.353		199	118		317	395
2240	For knob operated damper, add					140%				
2241	For metal trim strips, add	Q-5	64	.250	Ea.	19.25	12.55		31.80	40
2243	For snap-on inlet grille, add					10%	10%			
2245	For hinged access door, add	Q-5	64	.250	Ea.	35.50	12.55		48.05	58
2246	For air chamber, auto-venting, add	"	58	.276	"	6.85	13.85		20.70	28.50

23 82 36 – Finned-Tube Radiation Heaters

23 82 36.10 Finned Tube Radiation

	23 82 36.10 Finned Tube Radiation	Crew	Daily Output	Labor-Hours	Unit	Material	2012 Bare Costs Labor	Equipment	Total	Total Incl O&P
0010	**FINNED TUBE RADIATION**, Terminal units, not incl. main supply pipe									
1150	Fin tube, wall hung, 14" slope top cover, with damper									
1200	1-1/4" copper tube, 4-1/4" alum. fin	Q-5	38	.421	L.F.	42.50	21		63.50	78.50
1250	1-1/4" steel tube, 4-1/4" steel fin	"	36	.444	"	39	22.50		61.50	76.50
1500	Note: fin tube may also require corners, caps, etc.									

23 82 39 – Unit Heaters

23 82 39.16 Propeller Unit Heaters

	23 82 39.16 Propeller Unit Heaters	Crew	Daily Output	Labor-Hours	Unit	Material	2012 Bare Costs Labor	Equipment	Total	Total Incl O&P
0010	**PROPELLER UNIT HEATERS**									
3950	Unit heaters, propeller, 115 V 2 psi steam, 60°F entering air									
4000	Horizontal, 12 MBH	Q-5	12	1.333	Ea.	345	67		412	480
4060	43.9 MBH		8	2		520	100		620	725
4140	96.8 MBH		6	2.667		760	134		894	1,025
4180	157.6 MBH		4	4		1,000	201		1,201	1,400
4240	286.9 MBH		2	8		1,575	400		1,975	2,325
4250	326.0 MBH		1.90	8.421		1,825	420		2,245	2,625
4260	364 MBH		1.80	8.889		1,975	445		2,420	2,850
4270	404 MBH		1.60	10		2,500	500		3,000	3,500
4300	For vertical diffuser, add					174			174	191
4310	Vertical flow, 40 MBH	Q-5	11	1.455		735	73		808	920
4314	58.5 MBH		8	2		765	100		865	990
4326	131.0 MBH		4	4		1,150	201		1,351	1,575
4346	297.0 MBH		1.80	8.889		2,200	445		2,645	3,100
4354	420 MBH, (460 V)	Q-6	1.80	13.333		3,375	695		4,070	4,775
4358	500 MBH, (460 V)		1.71	14.035		4,700	730		5,430	6,275
4362	570 MBH, (460 V)		1.40	17.143		5,975	890		6,865	7,925
4366	620 MBH, (460 V)		1.30	18.462		6,525	960		7,485	8,625
4370	960 MBH, (460 V)		1.10	21.818		7,200	1,125		8,325	9,600

23 83 Radiant Heating Units

23 83 16 – Radiant-Heating Hydronic Piping

23 83 16.10 Radiant Floor Heating		Crew	Daily Output	Labor-Hours	Unit	Material	2012 Bare Costs Labor	Equipment	Total	Total Incl O&P
0010	**RADIANT FLOOR HEATING**									
0100	Tubing, PEX (cross-linked polyethylene)									
0110	Oxygen barrier type for systems with ferrous materials									
0120	1/2"	Q-5	800	.020	L.F.	.99	1		1.99	2.60
0130	3/4"		535	.030		1.40	1.50		2.90	3.80
0140	1"	↓	400	.040	↓	2.19	2.01		4.20	5.45
0200	Non barrier type for ferrous free systems									
0210	1/2"	Q-5	800	.020	L.F.	.57	1		1.57	2.14
0220	3/4"		535	.030		1.05	1.50		2.55	3.42
0230	1"	↓	400	.040	↓	1.80	2.01		3.81	5
1000	Manifolds									
1110	Brass									
1120	With supply and return valves, flow meter, thermometer,									
1122	auto air vent and drain/fill valve.									
1130	1", 2 circuit	Q-5	14	1.143	Ea.	245	57.50		302.50	355
1140	1", 3 circuit		13.50	1.185		280	59.50		339.50	400
1150	1", 4 circuit		13	1.231		305	61.50		366.50	430
1154	1", 5 circuit		12.50	1.280		365	64		429	495
1158	1", 6 circuit		12	1.333		395	67		462	535
1162	1", 7 circuit		11.50	1.391		430	70		500	580
1166	1", 8 circuit		11	1.455		480	73		553	635
1172	1", 9 circuit		10.50	1.524		515	76.50		591.50	680
1174	1", 10 circuit		10	1.600		555	80		635	730
1178	1", 11 circuit		9.50	1.684		575	84.50		659.50	760
1182	1", 12 circuit	↓	9	1.778	↓	635	89		724	835
1610	Copper manifold header, (cut to size)									
1620	1" header, 12 – 1/2" sweat outlets	Q-5	3.33	4.805	Ea.	91	241		332	465
1630	1-1/4" header, 12 – 1/2" sweat outlets		3.20	5		106	251		357	495
1640	1-1/4" header, 12 – 3/4" sweat outlets		3	5.333		114	267		381	530
1650	1-1/2" header, 12 – 3/4" sweat outlets		3.10	5.161		139	259		398	545
1660	2" header, 12 – 3/4" sweat outlets	↓	2.90	5.517	↓	201	277		478	635
3000	Valves									
3110	Thermostatic zone valve actuator with end switch	Q-5	40	.400	Ea.	36.50	20		56.50	70
3114	Thermostatic zone valve actuator	"	36	.444	"	81	22.50		103.50	123
3120	Motorized straight zone valve with operator complete									
3130	3/4"	Q-5	35	.457	Ea.	125	23		148	173
3140	1"		32	.500		137	25		162	189
3150	1-1/4"	↓	29.60	.541	↓	165	27		192	223
3500	4 Way mixing valve, manual, brass									
3530	1"	Q-5	13.30	1.203	Ea.	126	60.50		186.50	230
3540	1-1/4"		11.40	1.404		137	70.50		207.50	257
3550	1-1/2"		11	1.455		203	73		276	335
3560	2"		10.60	1.509		288	75.50		363.50	430
3800	Mixing valve motor, 4 way for valves, 1" and 1-1/4"		34	.471		310	23.50		333.50	375
3810	Mixing valve motor, 4 way for valves, 1-1/2" and 2"	↓	30	.533	↓	355	26.50		381.50	430
5000	Radiant floor heating, zone control panel									
5120	4 Zone actuator valve control, expandable	Q-5	20	.800	Ea.	163	40		203	240
5130	6 Zone actuator valve control, expandable		18	.889		199	44.50		243.50	286
6070	Thermal track, straight panel for long continuous runs, 5.333 S.F.		40	.400		20	20		40	52
6080	Thermal track, utility panel, for direction reverse at run end, 5.333 S.F.		40	.400		20	20		40	52
6090	Combination panel, for direction reverse plus straight run, 5.333 S.F.	↓	40	.400	↓	20	20		40	52
7000	PEX tubing fittings									
7100	Compression type									

23 83 Radiant Heating Units

23 83 16 – Radiant-Heating Hydronic Piping

23 83 16.10 Radiant Floor Heating		Crew	Daily Output	Labor-Hours	Unit	Material	2012 Bare Costs Labor	Equipment	Total	Total Incl O&P
7116	Coupling									
7120	1/2" x 1/2"	1 Stpi	27	.296	Ea.	6.40	16.50		22.90	32
7124	3/4" x 3/4"	"	23	.348	"	10.60	19.35		29.95	40.50
7130	Adapter									
7132	1/2" x female sweat 1/2"	1 Stpi	27	.296	Ea.	4.33	16.50		20.83	30
7134	1/2" x female sweat 3/4"		26	.308		4.88	17.15		22.03	31.50
7136	5/8" x female sweat 3/4"		24	.333		7	18.55		25.55	35.50
7140	Elbow									
7142	1/2" x female sweat 1/2"	1 Stpi	27	.296	Ea.	6.65	16.50		23.15	32.50
7144	1/2" x female sweat 3/4"		26	.308		7.80	17.15		24.95	34.50
7146	5/8" x female sweat 3/4"		24	.333		8.75	18.55		27.30	37.50
7200	Insert type									
7206	PEX x male NPT									
7210	1/2" x 1/2"	1 Stpi	29	.276	Ea.	2.48	15.35		17.83	25.50
7220	3/4" x 3/4"		27	.296		3.67	16.50		20.17	29
7230	1" x 1"		26	.308		6.05	17.15		23.20	32.50
7300	PEX coupling									
7310	1/2" x 1/2"	1 Stpi	30	.267	Ea.	.94	14.85		15.79	23.50
7320	3/4" x 3/4"		29	.276		1.17	15.35		16.52	24.50
7330	1" x 1"		28	.286		2.20	15.90		18.10	26.50
7400	PEX stainless crimp ring									
7410	1/2" x 1/2"	1 Stpi	86	.093	Ea.	.34	5.20		5.54	8.15
7420	3/4" x 3/4"		84	.095		.47	5.30		5.77	8.50
7430	1" x 1"		82	.098		.68	5.45		6.13	8.95

23 83 33 – Electric Radiant Heaters

23 83 33.10 Electric Heating

		Crew	Daily Output	Labor-Hours	Unit	Material	2012 Bare Costs Labor	Equipment	Total	Total Incl O&P
0010	**ELECTRIC HEATING**, not incl. conduit or feed wiring									
1100	Rule of thumb: Baseboard units, including control	1 Elec	4.40	1.818	kW	98	93.50		191.50	248
1300	Baseboard heaters, 2' long, 375 watt		8	1	Ea.	40	51.50		91.50	121
1400	3' long, 500 watt		8	1		46	51.50		97.50	128
1600	4' long, 750 watt		6.70	1.194		64	61.50		125.50	163
1800	5' long, 935 watt		5.70	1.404		68	72.50		140.50	183
2000	6' long, 1125 watt		5	1.600		74	82.50		156.50	205
2400	8' long, 1500 watt		4	2		92	103		195	255
2950	Wall heaters with fan, 120 to 277 volt									
3600	Thermostats, integral	1 Elec	16	.500	Ea.	30	26		56	71.50
3800	Line voltage, 1 pole	"	8	1	"	31	51.50		82.50	111

23 84 Humidity Control Equipment

23 84 13 – Humidifiers

23 84 13.10 Humidifier Units

		Crew	Daily Output	Labor-Hours	Unit	Material	2012 Bare Costs Labor	Equipment	Total	Total Incl O&P
0010	**HUMIDIFIER UNITS**									
0520	Steam, room or duct, filter, regulators, auto. controls, 220 V									
0540	11 lb. per hour	Q-5	6	2.667	Ea.	2,450	134		2,584	2,900
0560	22 lb. per hour		5	3.200		2,700	160		2,860	3,225
0580	33 lb. per hour		4	4		2,775	201		2,976	3,350
0600	50 lb. per hour		4	4		3,400	201		3,601	4,050
0620	100 lb. per hour		3	5.333		4,075	267		4,342	4,875

Division Notes

	CREW	DAILY OUTPUT	LABOR-HOURS	UNIT	2012 BARE COSTS				TOTAL INCL O&P
					MAT.	LABOR	EQUIP.	TOTAL	

Estimating Tips

26 05 00 Common Work Results for Electrical

- Conduit should be taken off in three main categories—power distribution, branch power, and branch lighting—so the estimator can concentrate on systems and components, therefore making it easier to ensure all items have been accounted for.

- For cost modifications for elevated conduit installation, add the percentages to labor according to the height of installation, and only to the quantities exceeding the different height levels, not to the total conduit quantities.

- Remember that aluminum wiring of equal ampacity is larger in diameter than copper and may require larger conduit.

- If more than three wires at a time are being pulled, deduct percentages from the labor hours of that grouping of wires.

- When taking off grounding systems, identify separately the type and size of wire, and list each unique type of ground connection.

- The estimator should take the weights of materials into consideration when completing a takeoff. Topics to consider include: How will the materials be supported? What methods of support are available? How high will the support structure have to reach? Will the final support structure be able to withstand the total burden? Is the support material included or separate from the fixture, equipment, and material specified?

- Do not overlook the costs for equipment used in the installation. If scaffolding or highlifts are available in the field, contractors may use them in lieu of the proposed ladders and rolling staging.

26 20 00 Low-Voltage Electrical Transmission

- Supports and concrete pads may be shown on drawings for the larger equipment, or the support system may be only a piece of plywood for the back of a panelboard. In either case, it must be included in the costs.

26 40 00 Electrical and Cathodic Protection

- When taking off cathodic protections systems, identify the type and size of cable, and list each unique type of anode connection.

26 50 00 Lighting

- Fixtures should be taken off room by room, using the fixture schedule, specifications, and the ceiling plan. For large concentrations of lighting fixtures in the same area, deduct the percentages from labor hours.

Reference Numbers

Reference numbers are shown in shaded boxes at the beginning of some major classifications. These numbers refer to related items in the Reference Section. The reference information may be an estimating procedure, an alternate pricing method, or technical information.

Note: Not all subdivisions listed here necessarily appear in this publication.

Note: **Trade Service,** *in part, has been used as a reference source for some of the material prices used in Division 26.*

26 05 05 – Selective Electrical Demolition

26 05 05.10 Electrical Demolition	Crew	Daily Output	Labor-Hours	Unit	Material	2012 Bare Costs Labor	2012 Bare Costs Equipment	Total	Total Incl O&P
0010	**ELECTRICAL DEMOLITION**									
0020	Conduit to 15' high, including fittings & hangers								
0100	Rigid galvanized steel, 1/2" to 1" diameter	1 Elec	242	.033	L.F.		1.70		1.70	2.55
0120	1-1/4" to 2"	"	200	.040			2.06		2.06	3.08
0140	2-1/2" to 3-1/2"	2 Elec	302	.053			2.73		2.73	4.08
0160	4" to 6"	"	160	.100			5.15		5.15	7.70
0200	Electric metallic tubing (EMT), 1/2" to 1"	1 Elec	394	.020			1.05		1.05	1.56
0220	1-1/4" to 1-1/2"	"	326	.025			1.27		1.27	1.89
0260	3-1/2" to 4"	2 Elec	310	.052			2.66		2.66	3.98
0270	Armored cable, (BX) avg. 50' runs								
0280	#14, 2 wire	1 Elec	690	.012	L.F.		.60		.60	.89
0290	#14, 3 wire		571	.014			.72		.72	1.08
0300	#12, 2 wire		605	.013			.68		.68	1.02
0310	#12, 3 wire		514	.016			.80		.80	1.20
0320	#10, 2 wire		514	.016			.80		.80	1.20
0330	#10, 3 wire		425	.019			.97		.97	1.45
0340	#8, 3 wire		342	.023			1.21		1.21	1.80
0350	Non metallic sheathed cable (Romex)								
0360	#14, 2 wire	1 Elec	720	.011	L.F.		.57		.57	.86
0370	#14, 3 wire		657	.012			.63		.63	.94
0380	#12, 2 wire		629	.013			.66		.66	.98
0390	#10, 3 wire		450	.018			.92		.92	1.37
0400	Wiremold raceway, including fittings & hangers								
0420	No. 3000	1 Elec	250	.032	L.F.		1.65		1.65	2.47
0440	No. 4000		217	.037			1.90		1.90	2.84
0460	No. 6000		166	.048			2.48		2.48	3.71
0465	Telephone/power pole		12	.667	Ea.		34.50		34.50	51.50
0470	Non-metallic, straight section		480	.017	L.F.		.86		.86	1.28
0500	Channels, steel, including fittings & hangers								
0520	3/4" x 1-1/2"	1 Elec	308	.026	L.F.		1.34		1.34	2
0540	1-1/2" x 1-1/2"		269	.030			1.53		1.53	2.29
0560	1-1/2" x 1-7/8"		229	.035			1.80		1.80	2.69
0600	Copper bus duct, indoor, 3 phase								
0610	Including hangers & supports								
0620	225 amp	2 Elec	135	.119	L.F.		6.10		6.10	9.15
0640	400 amp		106	.151			7.80		7.80	11.65
0660	600 amp		86	.186			9.60		9.60	14.35
0680	1000 amp		60	.267			13.75		13.75	20.50
0700	1600 amp		40	.400			20.50		20.50	31
0720	3000 amp		10	1.600			82.50		82.50	123
1300	Transformer, dry type, 1 phase, incl. removal of								
1320	supports, wire & conduit terminations								
1340	1 kVA	1 Elec	7.70	1.039	Ea.		53.50		53.50	80
1420	75 kVA	2 Elec	2.50	6.400	"		330		330	495
1440	3 phase to 600V, primary								
1460	3 kVA	1 Elec	3.85	2.078	Ea.		107		107	160
1520	75 kVA	2 Elec	2.70	5.926			305		305	455
1550	300 kVA	R-3	1.80	11.111			565	73.50	638.50	930
1570	750 kVA	"	1.10	18.182			925	120	1,045	1,525
1800	Wire, THW-THWN-THHN, removed from								
1810	in place conduit, to 15' high								
1830	#14	1 Elec	65	.123	C.L.F.		6.35		6.35	9.50
1840	#12		55	.145			7.50		7.50	11.20

26 05 05 – Selective Electrical Demolition

26 05 05.10 Electrical Demolition

		Crew	Daily Output	Labor-Hours	Unit	Material	2012 Bare Costs Labor	2012 Bare Costs Equipment	Total	Total Incl O&P
1850	#10	1 Elec	45.50	.176	C.L.F.		9.05		9.05	13.55
1860	#8		40.40	.198			10.20		10.20	15.25
1870	#6		32.60	.245			12.65		12.65	18.90
1880	#4	2 Elec	53	.302			15.55		15.55	23.50
1890	#3		50	.320			16.50		16.50	24.50
1900	#2		44.60	.359			18.50		18.50	27.50
1910	1/0		33.20	.482			25		25	37
1920	2/0		29.20	.548			28.50		28.50	42
1930	3/0		25	.640			33		33	49.50
1940	4/0		22	.727			37.50		37.50	56
1950	250 kcmil		20	.800			41		41	61.50
1960	300 kcmil		19	.842			43.50		43.50	65
1970	350 kcmil		18	.889			46		46	68.50
1980	400 kcmil		17	.941			48.50		48.50	72.50
1990	500 kcmil		16.20	.988			51		51	76
2000	Interior fluorescent fixtures, incl. supports									
2010	& whips, to 15' high									
2100	Recessed drop-in 2' x 2', 2 lamp	2 Elec	35	.457	Ea.		23.50		23.50	35
2120	2' x 4', 2 lamp		33	.485			25		25	37.50
2140	2' x 4', 4 lamp		30	.533			27.50		27.50	41
2160	4' x 4', 4 lamp		20	.800			41		41	61.50
2180	Surface mount, acrylic lens & hinged frame									
2200	1' x 4', 2 lamp	2 Elec	44	.364	Ea.		18.75		18.75	28
2220	2' x 2', 2 lamp		44	.364			18.75		18.75	28
2260	2' x 4', 4 lamp		33	.485			25		25	37.50
2280	4' x 4', 4 lamp		23	.696			36		36	53.50
2300	Strip fixtures, surface mount									
2320	4' long, 1 lamp	2 Elec	53	.302	Ea.		15.55		15.55	23.50
2340	4' long, 2 lamp		50	.320			16.50		16.50	24.50
2360	8' long, 1 lamp		42	.381			19.65		19.65	29.50
2380	8' long, 2 lamp		40	.400			20.50		20.50	31
2400	Pendant mount, industrial, incl. removal									
2410	of chain or rod hangers, to 15' high									
2420	4' long, 2 lamp	2 Elec	35	.457	Ea.		23.50		23.50	35
2440	8' long, 2 lamp	"	27	.593	"		30.50		30.50	45.50

26 05 13 – Medium-Voltage Cables

26 05 13.16 Medium-Voltage, Single Cable

		Crew	Daily Output	Labor-Hours	Unit	Material	2012 Bare Costs Labor	2012 Bare Costs Equipment	Total	Total Incl O&P
0010	**MEDIUM-VOLTAGE, SINGLE CABLE** Splicing & terminations not included									
0040	Copper, XLP shielding, 5 kV, #6	2 Elec	4.40	3.636	C.L.F.	162	187		349	460
0050	#4		4.40	3.636		210	187		397	510
0100	#2		4	4		248	206		454	585
0200	#1		4	4		285	206		491	625
0400	1/0		3.80	4.211		315	217		532	670
0600	2/0		3.60	4.444		395	229		624	775
0800	4/0		3.20	5		515	258		773	950
1000	250 kcmil	3 Elec	4.50	5.333		635	275		910	1,100
1200	350 kcmil		3.90	6.154		820	315		1,135	1,375
1400	500 kcmil		3.60	6.667		1,000	345		1,345	1,625
1600	15 kV, ungrounded neutral, #1	2 Elec	4	4		345	206		551	690
1800	1/0		3.80	4.211		415	217		632	780
2000	2/0		3.60	4.444		470	229		699	860
2200	4/0		3.20	5		625	258		883	1,075

26 05 Common Work Results for Electrical

26 05 13 – Medium-Voltage Cables

	26 05 13.16 Medium-Voltage, Single Cable	Crew	Daily Output	Labor-Hours	Unit	Material	2012 Bare Costs Labor	Equipment	Total	Total Incl O&P
2400	250 kcmil	3 Elec	4.50	5.333	C.L.F.	695	275		970	1,175
2600	350 kcmil		3.90	6.154		875	315		1,190	1,450
2800	500 kcmil	▼	3.60	6.667	▼	1,075	345		1,420	1,700

26 05 19 – Low-Voltage Electrical Power Conductors and Cables

26 05 19.20 Armored Cable

	26 05 19.20 Armored Cable	Crew	Daily Output	Labor-Hours	Unit	Material	2012 Bare Costs Labor	Equipment	Total	Total Incl O&P
0010	**ARMORED CABLE**									
0050	600 volt, copper (BX), #14, 2 conductor, solid	1 Elec	2.40	3.333	C.L.F.	71.50	172		243.50	335
0100	3 conductor, solid		2.20	3.636		111	187		298	400
0150	#12, 2 conductor, solid		2.30	3.478		72.50	179		251.50	350
0200	3 conductor, solid		2	4		117	206		323	440
0250	#10, 2 conductor, solid		2	4		133	206		339	455
0300	3 conductor, solid		1.60	5		184	258		442	585
0350	#8, 3 conductor, solid		1.30	6.154		345	315		660	855
0400	3 conductor with PVC jacket, in cable tray, #6	▼	3.10	2.581		390	133		523	630
0450	#4	2 Elec	5.40	2.963		500	153		653	780
0500	#2		4.60	3.478		665	179		844	1,000
0550	#1		4	4		905	206		1,111	1,300
0600	1/0		3.60	4.444		1,075	229		1,304	1,525
0650	2/0		3.40	4.706		1,300	243		1,543	1,800
0700	3/0		3.20	5		1,500	258		1,758	2,025
0750	4/0	▼	3	5.333		1,725	275		2,000	2,300
0800	250 kcmil	3 Elec	3.60	6.667		1,950	345		2,295	2,650
0850	350 kcmil		3.30	7.273		2,600	375		2,975	3,400
0900	500 kcmil	▼	3	8	▼	3,375	410		3,785	4,350
1050	5 kV, copper, 3 conductor with PVC jacket,									
1060	non-shielded, in cable tray, #4	2 Elec	380	.042	L.F.	7.65	2.17		9.82	11.70
1100	#2		360	.044		10	2.29		12.29	14.40
1200	#1		300	.053		12.70	2.75		15.45	18.05
1400	1/0		290	.055		14.65	2.84		17.49	20.50
1600	2/0		260	.062		16.95	3.17		20.12	23.50
2000	4/0	▼	240	.067		22.50	3.44		25.94	30
2100	250 kcmil	3 Elec	330	.073		31	3.75		34.75	39.50
2150	350 kcmil		315	.076		38	3.93		41.93	48
2200	500 kcmil	▼	270	.089	▼	50.50	4.58		55.08	62.50
2400	15 kV, copper, 3 conductor with PVC jacket galv steel armored									
2500	grounded neutral, in cable tray, #2	2 Elec	300	.053	L.F.	16.10	2.75		18.85	22
2600	#1		280	.057		17.15	2.95		20.10	23.50
2800	1/0		260	.062		19.75	3.17		22.92	26
2900	2/0		220	.073		26	3.75		29.75	34
3000	4/0	▼	190	.084		29.50	4.34		33.84	39
3100	250 kcmil	3 Elec	270	.089		33	4.58		37.58	43.50
3150	350 kcmil		240	.100		39	5.15		44.15	50
3200	500 kcmil	▼	210	.114	▼	52	5.90		57.90	66.50
3400	15 kV, copper, 3 conductor with PVC jacket,									
3450	ungrounded neutral, in cable tray, #2	2 Elec	260	.062	L.F.	17.35	3.17		20.52	24
3500	#1		230	.070		19.25	3.59		22.84	26.50
3600	1/0		200	.080		22	4.12		26.12	30
3700	2/0		190	.084		27	4.34		31.34	36.50
3800	4/0	▼	160	.100		32.50	5.15		37.65	43.50
4000	250 kcmil	3 Elec	210	.114		38	5.90		43.90	51
4050	350 kcmil		195	.123		50	6.35		56.35	64.50
4100	500 kcmil	▼	180	.133	▼	61	6.85		67.85	77.50

26 05 19.20 Armored Cable

		Crew	Daily Output	Labor-Hours	Unit	Material	2012 Bare Costs Labor	Equipment	Total	Total Incl O&P
9010	600 volt, copper (MC) steel clad, #14, 2 wire	1 Elec	2.40	3.333	C.L.F.	72.50	172		244.50	335
9020	3 wire		2.20	3.636		116	187		303	405
9040	#12, 2 wire		2.30	3.478		73	179		252	350
9050	3 wire		2	4		128	206		334	450
9070	#10, 2 wire		2	4		158	206		364	485
9080	3 wire		1.60	5		221	258		479	630
9100	#8, 2 wire, stranded		1.80	4.444		296	229		525	665
9110	3 wire, stranded		1.30	6.154		415	315		730	930
9200	600 volt, copper (MC) aluminum clad, #14, 2 wire		2.65	3.019		72.50	156		228.50	315
9210	3 wire		2.45	3.265		116	168		284	380
9220	4 wire		2.20	3.636		161	187		348	455
9230	#12, 2 wire		2.55	3.137		73	162		235	325
9240	3 wire		2.20	3.636		128	187		315	420
9250	4 wire		2	4		173	206		379	500
9260	#10, 2 wire		2.20	3.636		158	187		345	455
9270	3 wire		1.80	4.444		221	229		450	585
9280	4 wire		1.55	5.161		345	266		611	780

26 05 19.35 Cable Terminations

		Crew	Daily Output	Labor-Hours	Unit	Material	2012 Bare Costs Labor	Equipment	Total	Total Incl O&P
0010	**CABLE TERMINATIONS**									
0015	Wire connectors, screw type, #22 to #14	1 Elec	260	.031	Ea.	.09	1.59		1.68	2.47
0020	#18 to #12		240	.033		.11	1.72		1.83	2.69
0025	#18 to #10		240	.033		.17	1.72		1.89	2.76
0030	Screw-on connectors, insulated, #18 to #12		240	.033		.13	1.72		1.85	2.71
0035	#16 to #10		230	.035		.18	1.79		1.97	2.88
0040	#14 to #8		210	.038		.39	1.96		2.35	3.37
0045	#12 to #6		180	.044		.64	2.29		2.93	4.12
0050	Terminal lugs, solderless, #16 to #10		50	.160		.61	8.25		8.86	13
0100	#8 to #4		30	.267		.94	13.75		14.69	21.50
0150	#2 to #1		22	.364		1.26	18.75		20.01	29.50
0200	1/0 to 2/0		16	.500		2.23	26		28.23	41
0250	3/0		12	.667		4.31	34.50		38.81	56
0300	4/0		11	.727		4.31	37.50		41.81	60.50
0350	250 kcmil		9	.889		4.31	46		50.31	73
0400	350 kcmil		7	1.143		5.50	59		64.50	94
0450	500 kcmil		6	1.333		10.70	68.50		79.20	115
1600	Crimp 1 hole lugs, copper or aluminum, 600 volt									
1620	#14	1 Elec	60	.133	Ea.	.46	6.85		7.31	10.75
1630	#12		50	.160		.63	8.25		8.88	13.05
1640	#10		45	.178		.63	9.15		9.78	14.40
1780	#8		36	.222		2.63	11.45		14.08	20
1800	#6		30	.267		3.46	13.75		17.21	24.50
2000	#4		27	.296		4.70	15.25		19.95	28
2200	#2		24	.333		7.40	17.20		24.60	33.50
2400	#1		20	.400		7.60	20.50		28.10	39.50
2600	2/0		15	.533		9.45	27.50		36.95	51.50
2800	3/0		12	.667		11.10	34.50		45.60	63.50
3000	4/0		11	.727		12.35	37.50		49.85	69.50
3200	250 kcmil		9	.889		14.55	46		60.55	84.50
3400	300 kcmil		8	1		16.80	51.50		68.30	95.50
3500	350 kcmil		7	1.143		17.25	59		76.25	107
3600	400 kcmil		6.50	1.231		21	63.50		84.50	118
3800	500 kcmil		6	1.333		24.50	68.50		93	130

26 05 19.50 Mineral Insulated Cable

		Crew	Daily Output	Labor-Hours	Unit	Material	2012 Bare Costs Labor	Equipment	Total	Total Incl O&P
0010	**MINERAL INSULATED CABLE** 600 volt									
0100	1 conductor, #12	1 Elec	1.60	5	C.L.F.	385	258		643	805
0200	#10		1.60	5		495	258		753	930
0400	#8		1.50	5.333		550	275		825	1,025
0500	#6		1.40	5.714		655	295		950	1,150
0600	#4	2 Elec	2.40	6.667		880	345		1,225	1,475
0800	#2		2.20	7.273		1,250	375		1,625	1,925
0900	#1		2.10	7.619		1,425	395		1,820	2,150
1000	1/0		2	8		1,675	410		2,085	2,475
1100	2/0		1.90	8.421		2,000	435		2,435	2,850
1200	3/0		1.80	8.889		2,400	460		2,860	3,325
1400	4/0		1.60	10		2,775	515		3,290	3,825
1410	250 kcmil	3 Elec	2.40	10		3,125	515		3,640	4,225
1420	350 kcmil		1.95	12.308		3,575	635		4,210	4,875
1430	500 kcmil		1.95	12.308		4,600	635		5,235	6,000

26 05 19.55 Non-Metallic Sheathed Cable

		Crew	Daily Output	Labor-Hours	Unit	Material	2012 Bare Costs Labor	Equipment	Total	Total Incl O&P
0010	**NON-METALLIC SHEATHED CABLE** 600 volt									
0100	Copper with ground wire, (Romex)									
0150	#14, 2 conductor	1 Elec	2.70	2.963	C.L.F.	29.50	153		182.50	261
0200	3 conductor		2.40	3.333		41	172		213	300
0250	#12, 2 conductor		2.50	3.200		45	165		210	297
0300	3 conductor		2.20	3.636		63	187		250	350
0350	#10, 2 conductor		2.20	3.636		71	187		258	360
0400	3 conductor		1.80	4.444		100	229		329	450
0450	#8, 3 conductor		1.50	5.333		160	275		435	585
0500	#6, 3 conductor		1.40	5.714		259	295		554	725
0550	SE type SER aluminum cable, 3 RHW and									
0600	1 bare neutral, 3 #8 & 1 #8	1 Elec	1.60	5	C.L.F.	140	258		398	540
0650	3 #6 & 1 #6	"	1.40	5.714		159	295		454	615
0700	3 #4 & 1 #6	2 Elec	2.40	6.667		148	345		493	680
0750	3 #2 & 1 #4		2.20	7.273		263	375		638	850
0800	3 #1/0 & 1 #2		2	8		395	410		805	1,050
0850	3 #2/0 & 1 #1		1.80	8.889		470	460		930	1,200
0900	3 #4/0 & 1 #2/0		1.60	10		665	515		1,180	1,500

26 05 19.90 Wire

			Crew	Daily Output	Labor-Hours	Unit	Material	2012 Bare Costs Labor	Equipment	Total	Total Incl O&P
0010	**WIRE**	R260519-92									
0020	600 volt, copper type THW, solid, #14		1 Elec	13	.615	C.L.F.	8.20	31.50		39.70	56.50
0030	#12			11	.727		12.40	37.50		49.90	69.50
0040	#10			10	.800		19.55	41		60.55	83
0050	Stranded, #14	R260533-22		13	.615		9.25	31.50		40.75	57.50
0100	#12			11	.727		14.20	37.50		51.70	71.50
0120	#10			10	.800		22.50	41		63.50	86
0140	#8			8	1		37	51.50		88.50	118
0160	#6			6.50	1.231		63	63.50		126.50	164
0180	#4		2 Elec	10.60	1.509		98.50	78		176.50	225
0200	#3			10	1.600		124	82.50		206.50	259
0220	#2			9	1.778		157	91.50		248.50	310
0240	#1			8	2		198	103		301	370
0260	1/0			6.60	2.424		247	125		372	460
0280	2/0			5.80	2.759		310	142		452	555
0300	3/0			5	3.200		390	165		555	675
0350	4/0			4.40	3.636		495	187		682	825

26 05 19 – Low-Voltage Electrical Power Conductors and Cables

26 05 19.90 Wire		Crew	Daily Output	Labor-Hours	Unit	Material	2012 Bare Costs Labor	2012 Bare Costs Equipment	Total	Total Incl O&P
0400	250 kcmil	3 Elec	6	4	C.L.F.	580	206		786	945
0420	300 kcmil		5.70	4.211		690	217		907	1,075
0450	350 kcmil		5.40	4.444		845	229		1,074	1,275
0480	400 kcmil		5.10	4.706		970	243		1,213	1,450
0490	500 kcmil		4.80	5		1,150	258		1,408	1,625
0540	600 volt, aluminum type THHN, stranded, #6	1 Elec	8	1		28	51.50		79.50	108
0560	#4	2 Elec	13	1.231		34.50	63.50		98	133
0580	#2		10.60	1.509		47	78		125	168
0600	#1		9	1.778		68.50	91.50		160	213
0620	1/0		8	2		82.50	103		185.50	245
0640	2/0		7.20	2.222		97.50	115		212.50	278
0680	3/0		6.60	2.424		121	125		246	320
0700	4/0		6.20	2.581		134	133		267	345
0720	250 kcmil	3 Elec	8.70	2.759		164	142		306	395
0740	300 kcmil		8.10	2.963		227	153		380	475
0760	350 kcmil		7.50	3.200		231	165		396	500
0780	400 kcmil		6.90	3.478		270	179		449	565
0800	500 kcmil		6	4		297	206		503	635
0850	600 kcmil		5.70	4.211		375	217		592	740
0880	700 kcmil		5.10	4.706		435	243		678	845
0900	750 kcmil		4.80	5		500	258		758	935
0920	600 volt, copper type THWN-THHN, solid, #14	1 Elec	13	.615		8.20	31.50		39.70	56.50
0940	#12		11	.727		12.40	37.50		49.90	69.50
0960	#10		10	.800		19.55	41		60.55	83
1000	Stranded, #14		13	.615		9.35	31.50		40.85	58
1200	#12		11	.727		14.35	37.50		51.85	72
1250	#10		10	.800		22	41		63	86
1300	#8		8	1		33.50	51.50		85	114
1350	#6		6.50	1.231		57.50	63.50		121	158
1400	#4	2 Elec	10.60	1.509		90	78		168	215

26 05 23 – Control-Voltage Electrical Power Cables

26 05 23.10 Control Cable

		Crew	Daily Output	Labor-Hours	Unit	Material	Labor	Equipment	Total	Total Incl O&P
0010	**CONTROL CABLE**									
0020	600 volt, copper, #14 THWN wire with PVC jacket, 2 wires	1 Elec	9	.889	C.L.F.	23	46		69	94
0100	4 wires		7	1.143		39.50	59		98.50	132
0200	6 wires		6	1.333		74.50	68.50		143	185
0300	8 wires		5.30	1.509		91	78		169	216
0400	10 wires		4.80	1.667		107	86		193	246
0500	12 wires		4.30	1.860		125	96		221	281
0600	14 wires		3.80	2.105		149	109		258	325
0700	16 wires		3.50	2.286		153	118		271	345
0800	18 wires		3.30	2.424		169	125		294	375
0900	20 wires		3	2.667		206	137		343	430
1000	22 wires		2.80	2.857		212	147		359	455

26 05 26 – Grounding and Bonding for Electrical Systems

26 05 26.80 Grounding

		Crew	Daily Output	Labor-Hours	Unit	Material	Labor	Equipment	Total	Total Incl O&P
0010	**GROUNDING**									
0030	Rod, copper clad, 8' long, 1/2" diameter	1 Elec	5.50	1.455	Ea.	18	75		93	132
0050	3/4" diameter		5.30	1.509		32.50	78		110.50	152
0080	10' long, 1/2" diameter		4.80	1.667		21.50	86		107.50	152
0100	3/4" diameter		4.40	1.818		38.50	93.50		132	183
0130	15' long, 3/4" diameter		4	2		83.50	103		186.50	246

26 05 26 – Grounding and Bonding for Electrical Systems

26 05 26.80 Grounding

		Crew	Daily Output	Labor-Hours	Unit	Material	2012 Bare Costs Labor	Equipment	Total	Total Incl O&P
0390	Bare copper wire, stranded, #8	1 Elec	11	.727	C.L.F.	32	37.50		69.50	91.50
0400	#6	↓	10	.800		56.50	41		97.50	124
0600	#2	2 Elec	10	1.600		151	82.50		233.50	289
0800	3/0		6.60	2.424		335	125		460	550
1000	4/0	↓	5.70	2.807		420	145		565	680
1200	250 kcmil	3 Elec	7.20	3.333	↓	500	172		672	805
1800	Water pipe ground clamps, heavy duty									
2000	Bronze, 1/2" to 1" diameter	1 Elec	8	1	Ea.	29.50	51.50		81	110
2100	1-1/4" to 2" diameter		8	1		38.50	51.50		90	120
2200	2-1/2" to 3" diameter		6	1.333		67.50	68.50		136	178
2800	Brazed connections, #6 wire		12	.667		15.85	34.50		50.35	69
3000	#2 wire		10	.800		21	41		62	85
3100	3/0 wire		8	1		32	51.50		83.50	112
3200	4/0 wire		7	1.143		36.50	59		95.50	128
3400	250 kcmil wire		5	1.600		42.50	82.50		125	170
3600	500 kcmil wire	↓	4	2	↓	52.50	103		155.50	212

26 05 33 – Raceway and Boxes for Electrical Systems

26 05 33.13 Conduit

		Crew	Daily Output	Labor-Hours	Unit	Material	2012 Bare Costs Labor	Equipment	Total	Total Incl O&P
0010	**CONDUIT** To 15' high, includes 2 terminations, 2 elbows, R260533-22									
0020	11 beam clamps, and 11 couplings per 100 L.F.									
0300	Aluminum, 1/2" diameter	1 Elec	100	.080	L.F.	2.30	4.12		6.42	8.70
0500	3/4" diameter		90	.089		3.09	4.58		7.67	10.25
0700	1" diameter		80	.100		4.10	5.15		9.25	12.20
1000	1-1/4" diameter		70	.114		5.50	5.90		11.40	14.85
1030	1-1/2" diameter		65	.123		6.60	6.35		12.95	16.80
1050	2" diameter		60	.133		9	6.85		15.85	20
1070	2-1/2" diameter		50	.160		14.65	8.25		22.90	28.50
1100	3" diameter	2 Elec	90	.178		20.50	9.15		29.65	36
1130	3-1/2" diameter		80	.200		26.50	10.30		36.80	45
1140	4" diameter	↓	70	.229		32.50	11.80		44.30	53
1750	Rigid galvanized steel, 1/2" diameter	1 Elec	90	.089		2.92	4.58		7.50	10.05
1770	3/4" diameter		80	.100		3.10	5.15		8.25	11.10
1800	1" diameter		65	.123		4.18	6.35		10.53	14.10
1830	1-1/4" diameter		60	.133		5.90	6.85		12.75	16.75
1850	1-1/2" diameter		55	.145		6.80	7.50		14.30	18.70
1870	2" diameter		45	.178		8.75	9.15		17.90	23.50
1900	2-1/2" diameter	↓	35	.229		17	11.80		28.80	36.50
1930	3" diameter	2 Elec	50	.320		21.50	16.50		38	48
1950	3-1/2" diameter		44	.364		26.50	18.75		45.25	57
1970	4" diameter	↓	40	.400		31.50	20.50		52	65.50
2500	Steel, intermediate conduit (IMC), 1/2" diameter	1 Elec	100	.080		1.99	4.12		6.11	8.35
2530	3/4" diameter		90	.089		2.38	4.58		6.96	9.45
2550	1" diameter		70	.114		3.32	5.90		9.22	12.45
2570	1-1/4" diameter		65	.123		4.42	6.35		10.77	14.35
2600	1-1/2" diameter		60	.133		5.35	6.85		12.20	16.15
2630	2" diameter		50	.160		6.80	8.25		15.05	19.85
2650	2-1/2" diameter		40	.200		13.60	10.30		23.90	30.50
2670	3" diameter	2 Elec	60	.267		18.25	13.75		32	40.50
2700	3-1/2" diameter		54	.296		23	15.25		38.25	48.50
2730	4" diameter	↓	50	.320		25.50	16.50		42	53
5000	Electric metallic tubing (EMT), 1/2" diameter	1 Elec	170	.047		.70	2.43		3.13	4.40
5020	3/4" diameter	↓	130	.062	↓	1.12	3.17		4.29	5.95

26 05 33.13 Conduit		Crew	Daily Output	Labor-Hours	Unit	Material	2012 Bare Costs Labor	Equipment	Total	Total Incl O&P
5040	1" diameter	1 Elec	115	.070	L.F.	1.94	3.59		5.53	7.50
5060	1-1/4" diameter		100	.080		3.22	4.12		7.34	9.70
5080	1-1/2" diameter		90	.089		4.21	4.58		8.79	11.50
5100	2" diameter		80	.100		5.45	5.15		10.60	13.70
5120	2-1/2" diameter		60	.133		13.05	6.85		19.90	24.50
5140	3" diameter	2 Elec	100	.160		15.45	8.25		23.70	29.50
5160	3-1/2" diameter		90	.178		19.70	9.15		28.85	35
5180	4" diameter		80	.200		21.50	10.30		31.80	39.50
9900	Add to labor for higher elevated installation									
9910	15' to 20' high, add						10%			
9920	20' to 25' high, add						20%			
9930	25' to 30' high, add						25%			
9940	30' to 35' high, add						30%			
9950	35' to 40' high, add						35%			
9960	Over 40' high, add						40%			
9980	Allow. for cond. ftngs., 5% min.-20% max.									

26 05 33.16 Outlet Boxes

26 05 33.16 Outlet Boxes		Crew	Daily Output	Labor-Hours	Unit	Material	2012 Bare Costs Labor	Equipment	Total	Total Incl O&P
0010	**OUTLET BOXES**									
0020	Pressed steel, octagon, 4"	1 Elec	20	.400	Ea.	4.24	20.50		24.74	35.50
0060	Covers, blank		64	.125		1.77	6.45		8.22	11.60
0100	Extension rings		40	.200		7.05	10.30		17.35	23
0150	Square, 4"		20	.400		2.90	20.50		23.40	34
0200	Extension rings		40	.200		7.10	10.30		17.40	23
0250	Covers, blank		64	.125		1.93	6.45		8.38	11.75
0300	Plaster rings		64	.125		3.90	6.45		10.35	13.95
0650	Switchbox		27	.296		6.80	15.25		22.05	30.50
1100	Concrete, floor, 1 gang		5.30	1.509		124	78		202	252
2000	Poke-thru fitting, fire rated, for 3-3/4" floor		6.80	1.176		120	60.50		180.50	223
2040	For 7" floor		6.80	1.176		124	60.50		184.50	228
2100	Pedestal, 15 amp, duplex receptacle & blank plate		5.25	1.524		135	78.50		213.50	265
2120	Duplex receptacle and telephone plate		5.25	1.524		135	78.50		213.50	265
2140	Pedestal, 20 amp, duplex recept. & phone plate		5	1.600		136	82.50		218.50	272
2200	Abandonment plate		32	.250		38.50	12.90		51.40	61.50

26 05 33.18 Pull Boxes

26 05 33.18 Pull Boxes		Crew	Daily Output	Labor-Hours	Unit	Material	2012 Bare Costs Labor	Equipment	Total	Total Incl O&P
0010	**PULL BOXES**									
0100	Sheet metal, pull box, NEMA 1, type SC, 6" W x 6" H x 4" D	1 Elec	8	1	Ea.	15.50	51.50		67	94
0200	8" W x 8" H x 4" D		8	1		21	51.50		72.50	101
0300	10" W x 12" H x 6" D		5.30	1.509		25.50	78		103.50	144
0400	16" W x 20" H x 8" D		4	2		86.50	103		189.50	250
0500	20" W x 24" H x 8" D		3.20	2.500		94	129		223	296
0600	24" W x 36" H x 8" D		2.70	2.963		145	153		298	385
0650	Pull box, hinged , NEMA 1, 6" W x 6" H x 4" D		8	1		9.95	51.50		61.45	88
0800	12" W x 16" H x 6" D		4.70	1.702		32.50	87.50		120	167
1000	20" W x 20" H x 6" D		3.60	2.222		65	115		180	243
1200	20" W x 20" H x 8" D		3.20	2.500		130	129		259	335
1400	24" W x 36" H x 8" D		2.70	2.963		208	153		361	455
1600	24" W x 42" H x 8" D		2	4		315	206		521	655
2100	Pull box, NEMA 3R, type SC, raintight & weatherproof									
2150	6" L x 6" W x 6" D	1 Elec	10	.800	Ea.	26	41		67	90
2200	8" L x 6" W x 6" D		8	1		32.50	51.50		84	113
2250	10" L x 6" W x 6" D		7	1.143		43.50	59		102.50	136
2300	12" L x 12" W x 6" D		5	1.600		61	82.50		143.50	191

26 05 33 – Raceway and Boxes for Electrical Systems

26 05 33.18 Pull Boxes

		Crew	Daily Output	Labor-Hours	Unit	Material	2012 Bare Costs Labor	Equipment	Total	Total Incl O&P
2350	16" L x 16" W x 6" D	1 Elec	4.50	1.778	Ea.	122	91.50		213.50	271
2400	20" L x 20" W x 6" D		4	2		167	103		270	340
2450	24" L x 18" W x 8" D		3	2.667		173	137		310	395
2500	24" L x 24" W x 10" D		2.50	3.200		222	165		387	490
2550	30" L x 24" W x 12" D		2	4		445	206		651	800
2600	36" L x 36" W x 12" D		1.50	5.333		590	275		865	1,050
2800	Cast iron, pull boxes for surface mounting									
3000	NEMA 4, watertight & dust tight									
3050	6" L x 6" W x 6" D	1 Elec	4	2	Ea.	252	103		355	430
3100	8" L x 6" W x 6" D		3.20	2.500		273	129		402	495
3150	10" L x 6" W x 6" D		2.50	3.200		340	165		505	620
3200	12" L x 12" W x 6" D		2.30	3.478		600	179		779	930
3250	16" L x 16" W x 6" D		1.30	6.154		1,575	315		1,890	2,200
3300	20" L x 20" W x 6" D		.80	10		2,925	515		3,440	3,975
3350	24" L x 18" W x 8" D		.70	11.429		3,050	590		3,640	4,250
3400	24" L x 24" W x 10" D		.50	16		4,700	825		5,525	6,375
3450	30" L x 24" W x 12" D		.40	20		7,850	1,025		8,875	10,200
3500	36" L x 36" W x 12" D		.20	40		9,725	2,050		11,775	13,800
6000	J.I.C. wiring boxes, NEMA 12, dust tight & drip tight									
6050	6" L x 8" W x 4" D	1 Elec	10	.800	Ea.	60	41		101	128
6100	8" L x 10" W x 4" D		8	1		75	51.50		126.50	160
6150	12" L x 14" W x 6" D		5.30	1.509		126	78		204	254
6200	14" L x 16" W x 6" D		4.70	1.702		157	87.50		244.50	305
6250	16" L x 20" W x 6" D		4.40	1.818		310	93.50		403.50	480
6300	24" L x 30" W x 6" D		3.20	2.500		455	129		584	695
6350	24" L x 30" W x 8" D		2.90	2.759		490	142		632	750
6400	24" L x 36" W x 8" D		2.70	2.963		540	153		693	825
6450	24" L x 42" W x 8" D		2.30	3.478		595	179		774	920
6500	24" L x 48" W x 8" D		2	4		645	206		851	1,025

26 05 33.23 Wireway

		Crew	Daily Output	Labor-Hours	Unit	Material	2012 Bare Costs Labor	Equipment	Total	Total Incl O&P
0010	**WIREWAY** to 15' high									
0100	Screw cover, NEMA 1 w/fittings and supports, 2-1/2" x 2-1/2"	1 Elec	45	.178	L.F.	11.45	9.15		20.60	26.50
0200	4" x 4"	"	40	.200		12.65	10.30		22.95	29.50
0400	6" x 6"	2 Elec	60	.267		19.75	13.75		33.50	42
0600	8" x 8"	"	40	.400		33	20.50		53.50	67.50
4475	Screw cover, NEMA 3R w/fittings and supports, 4" x 4"	1 Elec	36	.222		27	11.45		38.45	46.50
4480	6" x 6"	2 Elec	55	.291		34	15		49	60
4485	8" x 8"		36	.444		55	23		78	94.50
4490	12" x 12"		18	.889		77	46		123	153
9980	Allow. for wireway ftngs., 5% min.-20% max.									

26 05 36 – Cable Trays for Electrical Systems

26 05 36.10 Cable Tray Ladder Type

		Crew	Daily Output	Labor-Hours	Unit	Material	2012 Bare Costs Labor	Equipment	Total	Total Incl O&P
0010	**CABLE TRAY LADDER TYPE** w/ftngs. & supports, 4" dp., to 15' elev.									
0160	Galvanized steel tray									
0170	4" rung spacing, 6" wide	2 Elec	98	.163	L.F.	17.80	8.40		26.20	32
0200	12" wide		86	.186		21.50	9.60		31.10	38
0400	18" wide		82	.195		25	10.05		35.05	42.50
0600	24" wide		78	.205		28.50	10.55		39.05	47.50
3200	Aluminum tray, 4" deep, 6" rung spacing, 6" wide		134	.119		17.65	6.15		23.80	28.50
3220	12" wide		124	.129		19.75	6.65		26.40	32
3230	18" wide		114	.140		22	7.25		29.25	35
3240	24" wide		106	.151		25.50	7.80		33.30	39.50

26 05 36 – Cable Trays for Electrical Systems

26 05 36.10 Cable Tray Ladder Type	Crew	Daily Output	Labor-Hours	Unit	Material	2012 Bare Costs Labor	Equipment	Total	Total Incl O&P
9980	Allow. for tray ftngs., 5% min.-20% max.								

26 05 39 – Underfloor Raceways for Electrical Systems

26 05 39.30 Conduit In Concrete Slab

		Crew	Daily Output	Labor-Hours	Unit	Material	Labor	Equipment	Total	Total Incl O&P
0010	**CONDUIT IN CONCRETE SLAB** Including terminations,									
0020	fittings and supports									
3230	PVC, schedule 40, 1/2" diameter	1 Elec	270	.030	L.F.	.80	1.53		2.33	3.16
3250	3/4" diameter		230	.035		.97	1.79		2.76	3.74
3270	1" diameter		200	.040		1.30	2.06		3.36	4.51
3300	1-1/4" diameter		170	.047		1.82	2.43		4.25	5.65
3330	1-1/2" diameter		140	.057		2.19	2.95		5.14	6.80
3350	2" diameter		120	.067		2.80	3.44		6.24	8.25
4350	Rigid galvanized steel, 1/2" diameter		200	.040		2.65	2.06		4.71	6
4400	3/4" diameter		170	.047		2.83	2.43		5.26	6.75
4450	1" diameter		130	.062		3.91	3.17		7.08	9.05
4500	1-1/4" diameter		110	.073		5.35	3.75		9.10	11.50
4600	1-1/2" diameter		100	.080		6.25	4.12		10.37	13.05
4800	2" diameter		90	.089		7.95	4.58		12.53	15.60

26 05 39.40 Conduit In Trench

		Crew	Daily Output	Labor-Hours	Unit	Material	Labor	Equipment	Total	Total Incl O&P
0010	**CONDUIT IN TRENCH** Includes terminations and fittings									
0020	Does not include excavation or backfill, see Section 31 23 16.00									
0200	Rigid galvanized steel, 2" diameter	1 Elec	150	.053	L.F.	7.65	2.75		10.40	12.50
0400	2-1/2" diameter	"	100	.080		14.90	4.12		19.02	22.50
0600	3" diameter	2 Elec	160	.100		19.25	5.15		24.40	28.50
0800	3-1/2" diameter		140	.114		25	5.90		30.90	36.50
1000	4" diameter		100	.160		28	8.25		36.25	43.50
1200	5" diameter		80	.200		57	10.30		67.30	78.50
1400	6" diameter		60	.267		84	13.75		97.75	113

26 05 43 – Underground Ducts and Raceways for Electrical Systems

26 05 43.10 Trench Duct

		Crew	Daily Output	Labor-Hours	Unit	Material	Labor	Equipment	Total	Total Incl O&P
0010	**TRENCH DUCT** Steel with cover									
0020	Standard adjustable, depths to 4"									
0100	Straight, single compartment, 9" wide	2 Elec	40	.400	L.F.	114	20.50		134.50	157
0200	12" wide		32	.500		130	26		156	182
0400	18" wide		26	.615		173	31.50		204.50	238
0600	24" wide		22	.727		222	37.50		259.50	300
0800	30" wide		20	.800		268	41		309	355
1000	36" wide		16	1		320	51.50		371.50	430
1200	Horizontal elbow, 9" wide		5.40	2.963	Ea.	425	153		578	695
1400	12" wide		4.60	3.478		490	179		669	805
1600	18" wide		4	4		625	206		831	1,000
1800	24" wide		3.20	5		880	258		1,138	1,350
2000	30" wide		2.60	6.154		1,175	315		1,490	1,775
2200	36" wide		2.40	6.667		1,525	345		1,870	2,225
2400	Vertical elbow, 9" wide		5.40	2.963		148	153		301	390
2600	12" wide		4.60	3.478		160	179		339	445
2800	18" wide		4	4		184	206		390	515
3000	24" wide		3.20	5		229	258		487	635
3200	30" wide		2.60	6.154		252	315		567	755
3400	36" wide		2.40	6.667		278	345		623	820
3600	Cross, 9" wide		4	4		695	206		901	1,075
3800	12" wide		3.20	5		735	258		993	1,200

26 05 43.10 Trench Duct		Crew	Daily Output	Labor-Hours	Unit	Material	2012 Bare Costs Labor	2012 Bare Costs Equipment	Total	Total Incl O&P
4000	18" wide	2 Elec	2.60	6.154	Ea.	880	315		1,195	1,450
4200	24" wide		2.20	7.273		1,125	375		1,500	1,775
4400	30" wide		2	8		1,450	410		1,860	2,225
4600	36" wide		1.80	8.889		1,800	460		2,260	2,650
4800	End closure, 9" wide		14.40	1.111		43.50	57.50		101	133
5000	12" wide		12	1.333		49.50	68.50		118	158
5200	18" wide		10	1.600		76	82.50		158.50	207
5400	24" wide		8	2		100	103		203	264
5600	30" wide		6.60	2.424		126	125		251	325
5800	36" wide		5.80	2.759		149	142		291	375
6000	Tees, 9" wide		4	4		420	206		626	775
6200	12" wide		3.60	4.444		490	229		719	880
6400	18" wide		3.20	5		625	258		883	1,075
6600	24" wide		3	5.333		900	275		1,175	1,400
6800	30" wide		2.60	6.154		1,175	315		1,490	1,775
7000	36" wide		2	8		1,525	410		1,935	2,325
7200	Riser, and cabinet connector, 9" wide		5.40	2.963		184	153		337	430
7400	12" wide		4.60	3.478		215	179		394	505
7600	18" wide		4	4		264	206		470	600
7800	24" wide		3.20	5		320	258		578	735
8000	30" wide		2.60	6.154		365	315		680	880
8200	36" wide		2	8		425	410		835	1,075
8400	Insert assembly, cell to conduit adapter, 1-1/4"	1 Elec	16	.500		73	26		99	119

26 05 43.20 Underfloor Duct

	26 05 43.20 Underfloor Duct	Crew	Daily Output	Labor-Hours	Unit	Material	2012 Bare Costs Labor	2012 Bare Costs Equipment	Total	Total Incl O&P
0010	**UNDERFLOOR DUCT**									
0100	Duct, 1-3/8" x 3-1/8" blank, standard	2 Elec	160	.100	L.F.	13.95	5.15		19.10	23
0200	1-3/8" x 7-1/4" blank, super duct		120	.133		28	6.85		34.85	41
0400	7/8" or 1-3/8" insert type, 24" O.C., 1-3/8" x 3-1/8", std.		140	.114		18.70	5.90		24.60	29.50
0600	1-3/8" x 7-1/4", super duct		100	.160		32.50	8.25		40.75	48.50
0800	Junction box, single duct, 1 level, 3-1/8"	1 Elec	4	2	Ea.	420	103		523	615
1000	Junction box, single duct, 1 level, 7-1/4"		2.70	2.963		490	153		643	765
1200	1 level, 2 duct, 3-1/8"		3.20	2.500		560	129		689	810
1400	Junction box, 1 level, 2 duct, 7-1/4"		2.30	3.478		1,450	179		1,629	1,850
1600	Triple duct, 3-1/8"		2.30	3.478		950	179		1,129	1,325
1800	Insert to conduit adapter, 3/4" & 1"		32	.250		34.50	12.90		47.40	57
2000	Support, single cell		27	.296		51.50	15.25		66.75	80
2200	Super duct		16	.500		51.50	26		77.50	95.50
2400	Double cell		16	.500		51.50	26		77.50	95.50
2600	Triple cell		11	.727		51.50	37.50		89	113
2800	Vertical elbow, standard duct		10	.800		92.50	41		133.50	164
3000	Super duct		8	1		92.50	51.50		144	179
3200	Cabinet connector, standard duct		32	.250		69.50	12.90		82.40	96
3400	Super duct		27	.296		69.50	15.25		84.75	99.50
3600	Conduit adapter, 1" to 1-1/4"		32	.250		69.50	12.90		82.40	96
3800	2" to 1-1/4"		27	.296		83.50	15.25		98.75	115
4000	Outlet, low tension (tele, computer, etc.)		8	1		97.50	51.50		149	184
4200	High tension, receptacle (120 volt)		8	1		97.50	51.50		149	184

26 05 80 – Wiring Connections

26 05 80.10 Motor Connections

	26 05 80.10 Motor Connections	Crew	Daily Output	Labor-Hours	Unit	Material	2012 Bare Costs Labor	2012 Bare Costs Equipment	Total	Total Incl O&P
0010	**MOTOR CONNECTIONS**									
0020	Flexible conduit and fittings, 115 volt, 1 phase, up to 1 HP motor	1 Elec	8	1	Ea.	10.30	51.50		61.80	88.50
0120	230 volt, 10 HP motor, 3 phase		4.20	1.905		17.55	98		115.55	166

26 05 Common Work Results for Electrical

26 05 80 – Wiring Connections

26 05 80.10 Motor Connections

		Crew	Daily Output	Labor-Hours	Unit	Material	2012 Bare Costs Labor	2012 Bare Costs Equipment	Total	Total Incl O&P
0200	25 HP motor	1 Elec	2.70	2.963	Ea.	36	153		189	268
0400	50 HP motor		2.20	3.636		89	187		276	380
0600	100 HP motor		1.50	5.333		217	275		492	650

26 05 90 – Residential Applications

26 05 90.10 Residential Wiring

		Crew	Daily Output	Labor-Hours	Unit	Material	2012 Bare Costs Labor	2012 Bare Costs Equipment	Total	Total Incl O&P
0010	**RESIDENTIAL WIRING**									
0020	20' avg. runs and #14/2 wiring incl. unless otherwise noted									
1000	Service & panel, includes 24' SE-AL cable, service eye, meter,									
1010	Socket, panel board, main bkr., ground rod, 15 or 20 amp									
1020	1-pole circuit breakers, and misc. hardware									
1100	100 amp, with 10 branch breakers	1 Elec	1.19	6.723	Ea.	590	345		935	1,175
1110	With PVC conduit and wire		.92	8.696		670	450		1,120	1,400
1120	With RGS conduit and wire		.73	10.959		845	565		1,410	1,775
1150	150 amp, with 14 branch breakers		1.03	7.767		915	400		1,315	1,600
1170	With PVC conduit and wire		.82	9.756		1,075	505		1,580	1,925
1180	With RGS conduit and wire		.67	11.940		1,400	615		2,015	2,475
1200	200 amp, with 18 branch breakers	2 Elec	1.80	8.889		1,200	460		1,660	2,000
1220	With PVC conduit and wire		1.46	10.959		1,375	565		1,940	2,350
1230	With RGS conduit and wire		1.24	12.903		1,825	665		2,490	3,025
1800	Lightning surge suppressor for above services, add	1 Elec	32	.250		59.50	12.90		72.40	84.50
2000	Switch devices									
2100	Single pole, 15 amp, Ivory, with a 1-gang box, cover plate,									
2110	Type NM (Romex) cable	1 Elec	17.10	.468	Ea.	13.95	24		37.95	51.50
2120	Type MC (BX) cable		14.30	.559		30	29		59	76
2130	EMT & wire		5.71	1.401		38	72		110	150
2150	3-way, #14/3, type NM cable		14.55	.550		18.60	28.50		47.10	63
2170	Type MC cable		12.31	.650		40.50	33.50		74	94.50
2180	EMT & wire		5	1.600		42	82.50		124.50	170
2200	4-way, #14/3, type NM cable		14.55	.550		28	28.50		56.50	73.50
2220	Type MC cable		12.31	.650		50	33.50		83.50	105
2230	EMT & wire		5	1.600		51.50	82.50		134	180
2250	S.P., 20 amp, #12/2, type NM cable		13.33	.600		21.50	31		52.50	70
2270	Type MC cable		11.43	.700		35	36		71	92.50
2280	EMT & wire		4.85	1.649		47	85		132	179
2290	S.P. rotary dimmer, 600W, no wiring		17	.471		24	24.50		48.50	63
2300	S.P. rotary dimmer, 600W, type NM cable		14.55	.550		30	28.50		58.50	75.50
2320	Type MC cable		12.31	.650		46	33.50		79.50	101
2330	EMT & wire		5	1.600		55.50	82.50		138	185
2350	3-way rotary dimmer, type NM cable		13.33	.600		37.50	31		68.50	87
2370	Type MC cable		11.43	.700		53.50	36		89.50	113
2380	EMT & wire		4.85	1.649		63	85		148	197
2400	Interval timer wall switch, 20 amp, 1-30 min., #12/2									
2410	Type NM cable	1 Elec	14.55	.550	Ea.	59	28.50		87.50	107
2420	Type MC cable		12.31	.650		65.50	33.50		99	122
2430	EMT & wire		5	1.600		84	82.50		166.50	216
2500	Decorator style									
2510	S.P., 15 amp, type NM cable	1 Elec	17.10	.468	Ea.	18.85	24		42.85	56.50
2520	Type MC cable		14.30	.559		35	29		64	81.50
2530	EMT & wire		5.71	1.401		43	72		115	156
2550	3-way, #14/3, type NM cable		14.55	.550		23.50	28.50		52	68.50
2570	Type MC cable		12.31	.650		45	33.50		78.50	99.50
2580	EMT & wire		5	1.600		47	82.50		129.50	175

26 05 90.10 Residential Wiring	Crew	Daily Output	Labor-Hours	Unit	Material	2012 Bare Costs Labor	Equipment	Total	Total Incl O&P
2600 4-way, #14/3, type NM cable	1 Elec	14.55	.550	Ea.	33	28.50		61.50	79
2620 Type MC cable		12.31	.650		54.50	33.50		88	110
2630 EMT & wire		5	1.600		56.50	82.50		139	185
2650 S.P., 20 amp, #12/2, type NM cable		13.33	.600		26.50	31		57.50	75
2670 Type MC cable		11.43	.700		40	36		76	98
2680 EMT & wire		4.85	1.649		52	85		137	184
2700 S.P., slide dimmer, type NM cable		17.10	.468		37.50	24		61.50	77.50
2720 Type MC cable		14.30	.559		54	29		83	103
2730 EMT & wire		5.71	1.401		63.50	72		135.50	178
2750 S.P., touch dimmer, type NM cable		17.10	.468		32.50	24		56.50	71.50
2770 Type MC cable		14.30	.559		48.50	29		77.50	96.50
2780 EMT & wire		5.71	1.401		58.50	72		130.50	172
2800 3-way touch dimmer, type NM cable		13.33	.600		52.50	31		83.50	104
2820 Type MC cable		11.43	.700		68.50	36		104.50	130
2830 EMT & wire	▼	4.85	1.649	▼	78	85		163	213
3000 Combination devices									
3100 S.P. switch/15 amp recpt., Ivory, 1-gang box, plate									
3110 Type NM cable	1 Elec	11.43	.700	Ea.	26.50	36		62.50	83.50
3120 Type MC cable		10	.800		43	41		84	109
3130 EMT & wire		4.40	1.818		52.50	93.50		146	198
3150 S.P. switch/pilot light, type NM cable		11.43	.700		27.50	36		63.50	84.50
3170 Type MC cable		10	.800		44	41		85	110
3180 EMT & wire		4.43	1.806		53.50	93		146.50	198
3190 2-S.P. switches, 2-#14/2, no wiring		14	.571		8.90	29.50		38.40	54
3200 2-S.P. switches, 2-#14/2, type NM cables		10	.800		31.50	41		72.50	96
3220 Type MC cable		8.89	.900		56	46.50		102.50	132
3230 EMT & wire		4.10	1.951		56.50	101		157.50	212
3250 3-way switch/15 amp recpt., #14/3, type NM cable		10	.800		35.50	41		76.50	101
3270 Type MC cable		8.89	.900		57	46.50		103.50	133
3280 EMT & wire		4.10	1.951		59	101		160	215
3300 2-3 way switches, 2-#14/3, type NM cables		8.89	.900		46.50	46.50		93	121
3320 Type MC cable		8	1		82.50	51.50		134	168
3330 EMT & wire		4	2		67	103		170	228
3350 S.P. switch/20 amp recpt., #12/2, type NM cable		10	.800		39.50	41		80.50	105
3370 Type MC cable		8.89	.900		46.50	46.50		93	121
3380 EMT & wire	▼	4.10	1.951	▼	65	101		166	222
3400 Decorator style									
3410 S.P. switch/15 amp recpt., type NM cable	1 Elec	11.43	.700	Ea.	31.50	36		67.50	89
3420 Type MC cable		10	.800		48	41		89	114
3430 EMT & wire		4.40	1.818		57.50	93.50		151	203
3450 S.P. switch/pilot light, type NM cable		11.43	.700		32.50	36		68.50	90
3470 Type MC cable		10	.800		48.50	41		89.50	115
3480 EMT & wire		4.40	1.818		58.50	93.50		152	204
3500 2-S.P. switches, 2-#14/2, type NM cables		10	.800		36.50	41		77.50	102
3520 Type MC cable		8.89	.900		61	46.50		107.50	137
3530 EMT & wire		4.10	1.951		61.50	101		162.50	218
3550 3-way/15 amp recpt., #14/3, type NM cable		10	.800		40.50	41		81.50	106
3570 Type MC cable		8.89	.900		62	46.50		108.50	138
3580 EMT & wire		4.10	1.951		64	101		165	221
3650 2-3 way switches, 2-#14/3, type NM cables		8.89	.900		51.50	46.50		98	127
3670 Type MC cable		8	1		87	51.50		138.50	173
3680 EMT & wire		4	2		72	103		175	233
3700 S.P. switch/20 amp recpt., #12/2, type NM cable	▼	10	.800	▼	44.50	41		85.50	111

26 05 90.10 Residential Wiring		Crew	Daily Output	Labor-Hours	Unit	Material	2012 Bare Costs Labor	Equipment	Total	Total Incl O&P
3720	Type MC cable	1 Elec	8.89	.900	Ea.	51.50	46.50		98	126
3730	EMT & wire	↓	4.10	1.951	↓	70	101		171	227
4000	Receptacle devices									
4010	Duplex outlet, 15 amp recpt., Ivory, 1-gang box, plate									
4015	Type NM cable	1 Elec	14.55	.550	Ea.	12.45	28.50		40.95	56
4020	Type MC cable		12.31	.650		28.50	33.50		62	81.50
4030	EMT & wire		5.33	1.501		36.50	77.50		114	157
4050	With #12/2, type NM cable		12.31	.650		15.55	33.50		49.05	67
4070	Type MC cable		10.67	.750		29	38.50		67.50	89.50
4080	EMT & wire		4.71	1.699		41	87.50		128.50	176
4100	20 amp recpt., #12/2, type NM cable		12.31	.650		25	33.50		58.50	77.50
4120	Type MC cable		10.67	.750		38.50	38.50		77	100
4130	EMT & wire	↓	4.71	1.699	↓	50.50	87.50		138	187
4140	For GFI see Section 26 05 90.10 line 4300 below									
4150	Decorator style, 15 amp recpt., type NM cable	1 Elec	14.55	.550	Ea.	17.30	28.50		45.80	61.50
4170	Type MC cable		12.31	.650		33.50	33.50		67	87
4180	EMT & wire		5.33	1.501		41.50	77.50		119	162
4200	With #12/2, type NM cable		12.31	.650		20.50	33.50		54	72.50
4220	Type MC cable		10.67	.750		33.50	38.50		72	95
4230	EMT & wire		4.71	1.699		45.50	87.50		133	181
4250	20 amp recpt. #12/2, type NM cable		12.31	.650		30	33.50		63.50	83
4270	Type MC cable		10.67	.750		43	38.50		81.50	106
4280	EMT & wire		4.71	1.699		55	87.50		142.50	192
4300	GFI, 15 amp recpt., type NM cable		12.31	.650		48	33.50		81.50	103
4320	Type MC cable		10.67	.750		64.50	38.50		103	129
4330	EMT & wire		4.71	1.699		72.50	87.50		160	211
4350	GFI with #12/2, type NM cable		10.67	.750		51.50	38.50		90	115
4370	Type MC cable		9.20	.870		64.50	45		109.50	138
4380	EMT & wire		4.21	1.900		76.50	98		174.50	230
4400	20 amp recpt., #12/2 type NM cable		10.67	.750		53.50	38.50		92	117
4420	Type MC cable		9.20	.870		66.50	45		111.50	141
4430	EMT & wire		4.21	1.900		78.50	98		176.50	233
4500	Weather-proof cover for above receptacles, add	↓	32	.250	↓	5.65	12.90		18.55	25.50
4550	Air conditioner outlet, 20 amp-240 volt recpt.									
4560	30' of #12/2, 2 pole circuit breaker									
4570	Type NM cable	1 Elec	10	.800	Ea.	65.50	41		106.50	134
4580	Type MC cable		9	.889		82	46		128	159
4590	EMT & wire		4	2		90	103		193	253
4600	Decorator style, type NM cable		10	.800		70.50	41		111.50	139
4620	Type MC cable		9	.889		86.50	46		132.50	164
4630	EMT & wire	↓	4	2	↓	95	103		198	258
4650	Dryer outlet, 30 amp-240 volt recpt., 20' of #10/3									
4660	2 pole circuit breaker									
4670	Type NM cable	1 Elec	6.41	1.248	Ea.	73	64.50		137.50	177
4680	Type MC cable		5.71	1.401		81.50	72		153.50	198
4690	EMT & wire	↓	3.48	2.299	↓	86.50	119		205.50	272
4700	Range outlet, 50 amp-240 volt recpt., 30' of #8/3									
4710	Type NM cable	1 Elec	4.21	1.900	Ea.	105	98		203	262
4720	Type MC cable		4	2		167	103		270	340
4730	EMT & wire		2.96	2.703		119	139		258	340
4750	Central vacuum outlet, Type NM cable		6.40	1.250		71	64.50		135.50	175
4770	Type MC cable		5.71	1.401		94	72		166	211
4780	EMT & wire	↓	3.48	2.299	↓	98	119		217	285

26 05 90.10 Residential Wiring	Crew	Daily Output	Labor-Hours	Unit	Material	2012 Bare Costs Labor	Equipment	Total	Total Incl O&P
4800	30 amp-110 volt locking recpt., #10/2 circ. bkr.								
4810	Type NM cable	1 Elec	6.20	1.290	Ea.	82.50	66.50		149	190
4820	Type MC cable		5.40	1.481		108	76.50		184.50	233
4830	EMT & wire	↓	3.20	2.500	↓	109	129		238	310
4900	Low voltage outlets									
4910	Telephone recpt., 20' of 4/C phone wire	1 Elec	26	.308	Ea.	12.15	15.85		28	37
4920	TV recpt., 20' of RG59U coax wire, F type connector	"	16	.500	"	20.50	26		46.50	61
4950	Door bell chime, transformer, 2 buttons, 60' of bellwire									
4970	Economy model	1 Elec	11.50	.696	Ea.	76.50	36		112.50	138
4980	Custom model		11.50	.696		119	36		155	185
4990	Luxury model, 3 buttons	↓	9.50	.842	↓	320	43.50		363.50	420
6000	Lighting outlets									
6050	Wire only (for fixture), type NM cable	1 Elec	32	.250	Ea.	8.05	12.90		20.95	28
6070	Type MC cable		24	.333		17.30	17.20		34.50	44.50
6080	EMT & wire		10	.800		24	41		65	88
6100	Box (4"), and wire (for fixture), type NM cable		25	.320		20.50	16.50		37	47
6120	Type MC cable		20	.400		29.50	20.50		50	63.50
6130	EMT & wire	↓	11	.727	↓	36	37.50		73.50	96
6200	Fixtures (use with lines 6050 or 6100 above)									
6210	Canopy style, economy grade	1 Elec	40	.200	Ea.	37	10.30		47.30	56.50
6220	Custom grade		40	.200		57	10.30		67.30	78
6250	Dining room chandelier, economy grade		19	.421		92.50	21.50		114	135
6260	Custom grade		19	.421		272	21.50		293.50	330
6270	Luxury grade		15	.533		610	27.50		637.50	715
6310	Kitchen fixture (fluorescent), economy grade		30	.267		68.50	13.75		82.25	96
6320	Custom grade		25	.320		188	16.50		204.50	232
6350	Outdoor, wall mounted, economy grade		30	.267		35	13.75		48.75	59
6360	Custom grade		30	.267		120	13.75		133.75	153
6370	Luxury grade		25	.320		272	16.50		288.50	325
6410	Outdoor PAR floodlights, 1 lamp, 150 watt		20	.400		37.50	20.50		58	72.50
6420	2 lamp, 150 watt each		20	.400		62.50	20.50		83	99.50
6425	Motion sensing, 2 lamp, 150 watt each		20	.400		75	20.50		95.50	114
6430	For infrared security sensor, add		32	.250		126	12.90		138.90	157
6450	Outdoor, quartz-halogen, 300 watt flood		20	.400		45.50	20.50		66	81
6600	Recessed downlight, round, pre-wired, 50 or 75 watt trim		30	.267		46.50	13.75		60.25	71.50
6610	With shower light trim		30	.267		57.50	13.75		71.25	83.50
6620	With wall washer trim		28	.286		69.50	14.75		84.25	98
6630	With eye-ball trim	↓	28	.286		69.50	14.75		84.25	98
6640	For direct contact with insulation, add					2.47			2.47	2.72
6700	Porcelain lamp holder	1 Elec	40	.200		4.83	10.30		15.13	20.50
6710	With pull switch		40	.200		5.65	10.30		15.95	21.50
6750	Fluorescent strip, 1-20 watt tube, wrap around diffuser, 24"		24	.333		58	17.20		75.20	89.50
6760	1-34 watt tube, 48"		24	.333		74.50	17.20		91.70	108
6770	2-34 watt tubes, 48"		20	.400		88.50	20.50		109	129
6780	With residential ballast		20	.400		99.50	20.50		120	140
6800	Bathroom heat lamp, 1-250 watt		28	.286		51	14.75		65.75	78
6810	2-250 watt lamps	↓	28	.286	↓	81.50	14.75		96.25	112
6820	For timer switch, see Section 26 05 90.10 line 2400									
6900	Outdoor post lamp, incl. post, fixture, 35' of #14/2									
6910	Type NMC cable	1 Elec	3.50	2.286	Ea.	207	118		325	405
6920	Photo-eye, add		27	.296		37.50	15.25		52.75	64.50
6950	Clock dial time switch, 24 hr., w/enclosure, type NM cable		11.43	.700		72	36		108	133
6970	Type MC cable	↓	11	.727	↓	88	37.50		125.50	153

26 05 90.10 Residential Wiring	Crew	Daily Output	Labor-Hours	Unit	Material	2012 Bare Costs Labor	Equipment	Total	Total Incl O&P
6980　　　　EMT & wire	1 Elec	4.85	1.649	Ea.	96	85		181	233
7000　Alarm systems									
7050　　　Smoke detectors, box, #14/3, type NM cable	1 Elec	14.55	.550	Ea.	37	28.50		65.50	83
7070　　　　Type MC cable	↓	12.31	.650		52.50	33.50		86	108
7080　　　　EMT & wire	↓	5	1.600		54	82.50		136.50	183
7090　　　For relay output to security system, add				↓	14			14	15.40
8000　Residential equipment									
8050　　　Disposal hook-up, incl. switch, outlet box, 3' of flex									
8060　　　　20 amp-1 pole circ. bkr., and 25' of #12/2									
8070　　　　　Type NM cable	1 Elec	10	.800	Ea.	36.50	41		77.50	102
8080　　　　　Type MC cable	↓	8	1		51	51.50		102.50	133
8090　　　　　EMT & wire	↓	5	1.600	↓	64	82.50		146.50	193
8100　　　Trash compactor or dishwasher hook-up, incl. outlet box,									
8110　　　　3' of flex, 15 amp-1 pole circ. bkr., and 25' of #14/2									
8120　　　　　Type NM cable	1 Elec	10	.800	Ea.	27	41		68	91
8130　　　　　Type MC cable	↓	8	1		45.50	51.50		97	127
8140　　　　　EMT & wire	↓	5	1.600	↓	55	82.50		137.50	184
8150　　　Hot water sink dispensor hook-up, use line 8100									
8200　　　Vent/exhaust fan hook-up, type NM cable	1 Elec	32	.250	Ea.	8.05	12.90		20.95	28
8220　　　　Type MC cable	↓	24	.333		17.30	17.20		34.50	44.50
8230　　　　EMT & wire	↓	10	.800	↓	24	41		65	88
8250　　　Bathroom vent fan, 50 CFM (use with above hook-up)									
8260　　　　Economy model	1 Elec	15	.533	Ea.	27	27.50		54.50	70.50
8270　　　　Low noise model	↓	15	.533		36.50	27.50		64	81
8280　　　　Custom model	↓	12	.667	↓	126	34.50		160.50	190
8300　　　Bathroom or kitchen vent fan, 110 CFM									
8310　　　　Economy model	1 Elec	15	.533	Ea.	74.50	27.50		102	123
8320　　　　Low noise model	"	15	.533	"	92.50	27.50		120	143
8350　　　Paddle fan, variable speed (w/o lights)									
8360　　　　Economy model (AC motor)	1 Elec	10	.800	Ea.	124	41		165	198
8362　　　　　With light kit		10	.800		162	41		203	240
8370　　　　Custom model (AC motor)		10	.800		195	41		236	277
8372　　　　　With light kit		10	.800		234	41		275	320
8380　　　　Luxury model (DC motor)		8	1		385	51.50		436.50	500
8382　　　　　With light kit		8	1		425	51.50		476.50	545
8390　　　　Remote speed switch for above, add	↓	12	.667	↓	32	34.50		66.50	87
8500　　　Whole house exhaust fan, ceiling mount, 36", variable speed									
8510　　　　Remote switch, incl. shutters, 20 amp-1 pole circ. bkr.									
8520　　　　30' of #12/2, type NM cable	1 Elec	4	2	Ea.	1,200	103		1,303	1,450
8530　　　　Type MC cable		3.50	2.286		1,200	118		1,318	1,500
8540　　　　EMT & wire	↓	3	2.667	↓	1,225	137		1,362	1,550
8600　　　Whirlpool tub hook-up, incl. timer switch, outlet box									
8610　　　　3' of flex, 20 amp-1 pole GFI circ. bkr.									
8620　　　　30' of #12/2, type NM cable	1 Elec	5	1.600	Ea.	133	82.50		215.50	269
8630　　　　Type MC cable		4.20	1.905		142	98		240	305
8640　　　　EMT & wire	↓	3.40	2.353	↓	154	121		275	350
8650　　　Hot water heater hook-up, incl. 1-2 pole circ. bkr., box;									
8660　　　　3' of flex, 20' of #10/2, type NM cable	1 Elec	5	1.600	Ea.	37.50	82.50		120	165
8670　　　　Type MC cable		4.20	1.905		58	98		156	211
8680　　　　EMT & wire	↓	3.40	2.353	↓	58	121		179	245
9000　Heating/air conditioning									
9050　　　Furnace/boiler hook-up, incl. firestat, local on-off switch									
9060　　　　Emergency switch, and 40' of type NM cable	1 Elec	4	2	Ea.	62	103		165	223

26 05 Common Work Results for Electrical

26 05 90 – Residential Applications

26 05 90.10 Residential Wiring	Crew	Daily Output	Labor-Hours	Unit	Material	2012 Bare Costs Labor	Equipment	Total	Total Incl O&P
9070	Type MC cable	1 Elec	3.50	2.286	Ea.	87	118		205	272
9080	EMT & wire	↓	1.50	5.333	↓	101	275		376	520
9100	Air conditioner hook-up, incl. local 60 amp disc. switch									
9110	3' sealtite, 40 amp, 2 pole circuit breaker									
9130	40' of #8/2, type NM cable	1 Elec	3.50	2.286	Ea.	234	118		352	435
9140	Type MC cable	↓	3	2.667		330	137		467	565
9150	EMT & wire	↓	1.30	6.154	↓	273	315		588	775
9200	Heat pump hook-up, 1-40 & 1-100 amp 2 pole circ. bkr.									
9210	Local disconnect switch, 3' sealtite									
9220	40' of #8/2 & 30' of #3/2									
9230	Type NM cable	1 Elec	1.30	6.154	Ea.	615	315		930	1,150
9240	Type MC cable		1.08	7.407		760	380		1,140	1,400
9250	EMT & wire	↓	.94	8.511	↓	670	440		1,110	1,400
9500	Thermostat hook-up, using low voltage wire									
9520	Heating only, 25' of #18-3	1 Elec	24	.333	Ea.	10.95	17.20		28.15	37.50
9530	Heating/cooling, 25' of #18-4	"	20	.400	"	13.20	20.50		33.70	45.50

26 09 Instrumentation and Control for Electrical Systems

26 09 13 – Electrical Power Monitoring

26 09 13.10 Switchboard Instruments

		Crew	Daily Output	Labor-Hours	Unit	Material	2012 Bare Costs Labor	Equipment	Total	Total Incl O&P
0010	**SWITCHBOARD INSTRUMENTS** 3 phase, 4 wire									
0100	AC indicating, ammeter & switch	1 Elec	8	1	Ea.	1,925	51.50		1,976.50	2,200
0200	Voltmeter & switch		8	1		1,925	51.50		1,976.50	2,200
0300	Wattmeter		8	1		3,825	51.50		3,876.50	4,275
0400	AC recording, ammeter		4	2		6,800	103		6,903	7,625
0500	Voltmeter		4	2		6,800	103		6,903	7,625
0600	Ground fault protection, zero sequence		2.70	2.963		6,000	153		6,153	6,825
0700	Ground return path		2.70	2.963		6,000	153		6,153	6,825
0800	3 current transformers, 5 to 800 amp		2	4		2,800	206		3,006	3,375
0900	1000 to 1500 amp		1.30	6.154		4,025	315		4,340	4,900
1200	2000 to 4000 amp		1	8		4,750	410		5,160	5,850
1300	Fused potential transformer, maximum 600 volt	↓	8	1	↓	1,050	51.50		1,101.50	1,225

26 12 Medium-Voltage Transformers

26 12 19 – Pad-Mounted, Liquid-Filled, Medium-Voltage Transformers

26 12 19.10 Transformer, Oil-Filled

		Crew	Daily Output	Labor-Hours	Unit	Material	2012 Bare Costs Labor	Equipment	Total	Total Incl O&P
0010	**TRANSFORMER, OIL-FILLED** primary delta or Y,									
0050	Pad mounted 5 kV or 15 kV, with taps, 277/480 V secondary, 3 phase									
0100	150 kVA	R-3	.65	30.769	Ea.	9,725	1,575	204	11,504	13,300
0200	300 kVA		.45	44.444		12,000	2,275	294	14,569	16,900
0300	500 kVA		.40	50		18,800	2,550	330	21,680	24,900
0400	750 kVA		.38	52.632		21,700	2,675	350	24,725	28,200
0500	1000 kVA		.26	76.923		25,700	3,925	510	30,135	34,700
0600	1500 kVA		.23	86.957		30,500	4,425	575	35,500	40,900
0700	2000 kVA		.20	100		38,500	5,100	660	44,260	51,000
0800	3750 kVA	↓	.16	125	↓	72,000	6,375	830	79,205	90,000

26 22 Low-Voltage Transformers

26 22 13 – Low-Voltage Distribution Transformers

26 22 13.10 Transformer, Dry-Type	Crew	Daily Output	Labor-Hours	Unit	Material	2012 Bare Costs Labor	Equipment	Total	Total Incl O&P
0010 **TRANSFORMER, DRY-TYPE**									
0050 Single phase, 240/480 volt primary, 120/240 volt secondary									
0100 1 kVA	1 Elec	2	4	Ea.	257	206		463	590
0300 2 kVA		1.60	5		385	258		643	810
0500 3 kVA		1.40	5.714		475	295		770	965
0700 5 kVA		1.20	6.667		660	345		1,005	1,250
0900 7.5 kVA	2 Elec	2.20	7.273		915	375		1,290	1,550
1100 10 kVA		1.60	10		1,175	515		1,690	2,050
1300 15 kVA		1.20	13.333		1,275	685		1,960	2,425
1500 25 kVA		1	16		1,725	825		2,550	3,125
1700 37.5 kVA		.80	20		2,225	1,025		3,250	4,000
1900 50 kVA		.70	22.857		2,650	1,175		3,825	4,675
2100 75 kVA		.65	24.615		3,525	1,275		4,800	5,775
2190 480 V primary 120/240 V secondary, nonvent., 15 kVA		1.20	13.333		1,275	685		1,960	2,425
2200 25 kVA		.90	17.778		1,875	915		2,790	3,450
2210 37 kVA		.75	21.333		2,225	1,100		3,325	4,100
2220 50 kVA		.65	24.615		2,650	1,275		3,925	4,825
2300 3 phase, 480 volt primary 120/208 volt secondary									
2310 Ventilated, 3 kVA	1 Elec	1	8	Ea.	805	410		1,215	1,500
2700 6 kVA		.80	10		1,100	515		1,615	1,975
2900 9 kVA		.70	11.429		1,250	590		1,840	2,250
3100 15 kVA	2 Elec	1.10	14.545		1,275	750		2,025	2,525
3300 30 kVA		.90	17.778		1,300	915		2,215	2,800
3500 45 kVA		.80	20		1,550	1,025		2,575	3,250
3700 75 kVA		.70	22.857		2,350	1,175		3,525	4,325
3900 112.5 kVA	R-3	.90	22.222		3,100	1,125	147	4,372	5,275
4100 150 kVA		.85	23.529		4,050	1,200	156	5,406	6,425
4300 225 kVA		.65	30.769		5,500	1,575	204	7,279	8,625
4500 300 kVA		.55	36.364		6,950	1,850	241	9,041	10,700
4700 500 kVA		.45	44.444		11,500	2,275	294	14,069	16,300
4800 750 kVA		.35	57.143		20,200	2,925	380	23,505	27,000

26 24 Switchboards and Panelboards

26 24 13 – Switchboards

26 24 13.10 Incoming Switchboards

	Crew	Daily Output	Labor-Hours	Unit	Material	2012 Bare Costs Labor	Equipment	Total	Total Incl O&P
0010 **INCOMING SWITCHBOARDS** main service section									
0100 Aluminum bus bars, not including CT's or PT's									
0200 No main disconnect, includes CT compartment									
0300 120/208 volt, 4 wire, 600 amp	2 Elec	1	16	Ea.	4,025	825		4,850	5,650
0400 800 amp		.88	18.182		4,025	935		4,960	5,825
0500 1000 amp		.80	20		4,850	1,025		5,875	6,875
0600 1200 amp		.72	22.222		4,850	1,150		6,000	7,025
0700 1600 amp		.66	24.242		4,850	1,250		6,100	7,200
0800 2000 amp		.62	25.806		5,225	1,325		6,550	7,725
1000 3000 amp		.56	28.571		6,875	1,475		8,350	9,775
2000 Fused switch & CT compartment									
2100 120/208 volt, 4 wire, 400 amp	2 Elec	1.12	14.286	Ea.	2,700	735		3,435	4,075
2200 600 amp		.94	17.021		3,200	875		4,075	4,825
2300 800 amp		.84	19.048		10,900	980		11,880	13,500
2400 1200 amp		.68	23.529		14,200	1,225		15,425	17,400
2900 Pressure switch & CT compartment									

26 24 Switchboards and Panelboards

26 24 13 – Switchboards

26 24 13.10 Incoming Switchboards

		Crew	Daily Output	Labor-Hours	Unit	Material	2012 Bare Costs Labor	2012 Bare Costs Equipment	Total	Total Incl O&P
3000	120/208 volt, 4 wire, 800 amp	2 Elec	.80	20	Ea.	9,800	1,025		10,825	12,400
3100	1200 amp		.66	24.242		19,000	1,250		20,250	22,700
3200	1600 amp		.62	25.806		20,100	1,325		21,425	24,200
3300	2000 amp		.56	28.571		21,400	1,475		22,875	25,800
4400	Circuit breaker, molded case & CT compartment									
4600	3 pole, 4 wire, 600 amp	2 Elec	.94	17.021	Ea.	8,425	875		9,300	10,600
4800	800 amp		.84	19.048		10,100	980		11,080	12,600
5000	1200 amp		.68	23.529		13,800	1,225		15,025	16,900
5100	Copper bus bars, not incl. CT's or PT's, add, minimum					15%				

26 24 13.30 Distribution Switchboards Section

		Crew	Daily Output	Labor-Hours	Unit	Material	2012 Bare Costs Labor	2012 Bare Costs Equipment	Total	Total Incl O&P
0010	**DISTRIBUTION SWITCHBOARDS SECTION**									
0100	Aluminum bus bars, not including breakers									
0200	120/208 or 277/480 volt, 4 wire, 600 amp	2 Elec	1	16	Ea.	2,225	825		3,050	3,675
0300	800 amp		.88	18.182		2,925	935		3,860	4,625
0400	1000 amp		.80	20		2,925	1,025		3,950	4,775
0500	1200 amp		.72	22.222		3,625	1,150		4,775	5,700
0600	1600 amp		.66	24.242		4,175	1,250		5,425	6,450
0700	2000 amp		.62	25.806		4,875	1,325		6,200	7,375
0800	2500 amp		.60	26.667		5,500	1,375		6,875	8,100
0900	3000 amp		.56	28.571		6,675	1,475		8,150	9,550
0950	4000 amp		.52	30.769		9,750	1,575		11,325	13,100

26 24 13.40 Switchboards Feeder Section

		Crew	Daily Output	Labor-Hours	Unit	Material	2012 Bare Costs Labor	2012 Bare Costs Equipment	Total	Total Incl O&P
0010	**SWITCHBOARDS FEEDER SECTION** group mounted devices									
0030	Circuit breakers									
0160	FA frame, 15 to 60 amp, 240 volt, 1 pole	1 Elec	8	1	Ea.	119	51.50		170.50	208
0280	FA frame, 70 to 100 amp, 240 volt, 1 pole		7	1.143		185	59		244	292
0420	KA frame, 70 to 225 amp		3.20	2.500		1,175	129		1,304	1,500
0430	LA frame, 125 to 400 amp		2.30	3.478		2,650	179		2,829	3,200
0460	MA frame, 450 to 600 amp		1.60	5		4,725	258		4,983	5,575
0470	700 to 800 amp		1.30	6.154		6,150	315		6,465	7,250
0480	MAL frame, 1000 amp		1	8		6,375	410		6,785	7,650
0490	PA frame, 1200 amp		.80	10		13,000	515		13,515	15,100
0500	Branch circuit, fusible switch, 600 volt, double 30/30 amp		4	2		890	103		993	1,125
0550	60/60 amp		3.20	2.500		915	129		1,044	1,200
0600	100/100 amp		2.70	2.963		1,150	153		1,303	1,500
0650	Single, 30 amp		5.30	1.509		740	78		818	930
0700	60 amp		4.70	1.702		740	87.50		827.50	945
0750	100 amp		4	2		1,075	103		1,178	1,350
0800	200 amp		2.70	2.963		1,425	153		1,578	1,800
0850	400 amp		2.30	3.478		2,600	179		2,779	3,150
0900	600 amp		1.80	4.444		3,200	229		3,429	3,850
0950	800 amp		1.30	6.154		5,350	315		5,665	6,375
1000	1200 amp		.80	10		6,125	515		6,640	7,525

26 24 16 – Panelboards

26 24 16.20 Panelboard and Load Center Circuit Breakers

		Crew	Daily Output	Labor-Hours	Unit	Material	2012 Bare Costs Labor	2012 Bare Costs Equipment	Total	Total Incl O&P
0010	**PANELBOARD AND LOAD CENTER CIRCUIT BREAKERS**									
0050	Bolt-on, 10,000 amp IC, 120 volt, 1 pole									
0100	15 to 50 amp	1 Elec	10	.800	Ea.	14.80	41		55.80	78
0200	60 amp		8	1		21.50	51.50		73	101
0300	70 amp		8	1		28	51.50		79.50	108
0350	240 volt, 2 pole									
0400	15 to 50 amp	1 Elec	8	1	Ea.	32.50	51.50		84	113

26 24 16 — Panelboards

26 24 16.20 Panelboard and Load Center Circuit Breakers	Crew	Daily Output	Labor- Hours	Unit	Material	2012 Bare Costs Labor	2012 Bare Costs Equipment	Total	Total Incl O&P
0500	60 amp	1 Elec	7.50	1.067	Ea.	58	55		113	146
0600	80 to 100 amp		5	1.600		83.50	82.50		166	215
0700	3 pole, 15 to 60 amp		6.20	1.290		103	66.50		169.50	213
0800	70 amp		5	1.600		130	82.50		212.50	265
0900	80 to 100 amp		3.60	2.222		147	115		262	335
1000	22,000 amp I.C., 240 volt, 2 pole, 70 – 225 amp		2.70	2.963		635	153		788	925
1100	3 pole, 70 – 225 amp		2.30	3.478		705	179		884	1,050
1200	14,000 amp I.C., 277 volts, 1 pole, 15 – 30 amp		8	1		39	51.50		90.50	120
1300	22,000 amp I.C., 480 volts, 2 pole, 70 – 225 amp		2.70	2.963		635	153		788	925
1400	3 pole, 70 – 225 amp		2.30	3.478		780	179		959	1,125

26 24 16.30 Panelboards Commercial Applications

		Crew	Daily Output	Labor- Hours	Unit	Material	Labor	Equipment	Total	Total Incl O&P
0010	**PANELBOARDS COMMERCIAL APPLICATIONS**									
0050	NQOD, w/20 amp 1 pole bolt-on circuit breakers									
0100	3 wire, 120/240 volts, 100 amp main lugs									
0150	10 circuits	1 Elec	1	8	Ea.	545	410		955	1,225
0200	14 circuits		.88	9.091		665	470		1,135	1,425
0250	18 circuits		.75	10.667		720	550		1,270	1,625
0300	20 circuits		.65	12.308		810	635		1,445	1,850
0350	225 amp main lugs, 24 circuits	2 Elec	1.20	13.333		910	685		1,595	2,025
0400	30 circuits		.90	17.778		1,050	915		1,965	2,525
0450	36 circuits		.80	20		1,200	1,025		2,225	2,875
0500	38 circuits		.72	22.222		1,275	1,150		2,425	3,125
0550	42 circuits		.66	24.242		1,350	1,250		2,600	3,350
0600	4 wire, 120/208 volts, 100 amp main lugs, 12 circuits	1 Elec	1	8		640	410		1,050	1,325
0650	16 circuits		.75	10.667		735	550		1,285	1,625
0700	20 circuits		.65	12.308		850	635		1,485	1,875
0750	24 circuits		.60	13.333		915	685		1,600	2,025
0800	30 circuits		.53	15.094		1,050	780		1,830	2,325
0850	225 amp main lugs, 32 circuits	2 Elec	.90	17.778		1,175	915		2,090	2,675
0900	34 circuits		.84	19.048		1,200	980		2,180	2,800
0950	36 circuits		.80	20		1,225	1,025		2,250	2,900
1000	42 circuits		.68	23.529		1,375	1,225		2,600	3,350
1200	NEHB, w/20 amp, 1 pole bolt-on circuit breakers									
1250	4 wire, 277/480 volts, 100 amp main lugs, 12 circuits	1 Elec	.88	9.091	Ea.	1,200	470		1,670	2,000
1300	20 circuits	"	.60	13.333		1,775	685		2,460	2,975
1350	225 amp main lugs, 24 circuits	2 Elec	.90	17.778		2,000	915		2,915	3,575
1400	30 circuits		.80	20		2,400	1,025		3,425	4,200
1450	36 circuits		.72	22.222		2,800	1,150		3,950	4,775
1600	NQOD panel, w/20 amp, 1 pole, circuit breakers									
1650	3 wire, 120/240 volt with main circuit breaker									
1700	100 amp main, 12 circuits	1 Elec	.80	10	Ea.	775	515		1,290	1,625
1750	20 circuits	"	.60	13.333		985	685		1,670	2,100
1800	225 amp main, 30 circuits	2 Elec	.68	23.529		1,850	1,225		3,075	3,875
1850	42 circuits		.52	30.769		2,150	1,575		3,725	4,725
1900	400 amp main, 30 circuits		.54	29.630		2,550	1,525		4,075	5,075
1950	42 circuits		.50	32		2,850	1,650		4,500	5,600
2000	4 wire, 120/208 volts with main circuit breaker									
2050	100 amp main, 24 circuits	1 Elec	.47	17.021	Ea.	1,150	875		2,025	2,550
2100	30 circuits	"	.40	20		1,300	1,025		2,325	2,975
2200	225 amp main, 32 circuits	2 Elec	.72	22.222		2,150	1,150		3,300	4,075
2250	42 circuits		.56	28.571		2,350	1,475		3,825	4,775
2300	400 amp main, 42 circuits		.48	33.333		3,150	1,725		4,875	6,050

26 24 Switchboards and Panelboards

26 24 16 – Panelboards

26 24 16.30 Panelboards Commercial Applications

		Crew	Daily Output	Labor-Hours	Unit	Material	2012 Bare Costs Labor	Equipment	Total	Total Incl O&P
2350	600 amp main, 42 circuits	2 Elec	.40	40	Ea.	4,675	2,050		6,725	8,225
2400	NEHB, with 20 amp, 1 pole circuit breaker									
2450	4 wire, 277/480 volts with main circuit breaker									
2500	100 amp main, 24 circuits	1 Elec	.42	19.048	Ea.	2,300	980		3,280	4,025
2550	30 circuits	"	.38	21.053		2,700	1,075		3,775	4,600
2600	225 amp main, 30 circuits	2 Elec	.72	22.222		3,400	1,150		4,550	5,425
2650	42 circuits	"	.56	28.571		4,175	1,475		5,650	6,800

26 24 19 – Motor-Control Centers

26 24 19.40 Motor Starters and Controls

		Crew	Daily Output	Labor-Hours	Unit	Material	2012 Bare Costs Labor	Equipment	Total	Total Incl O&P
0010	**MOTOR STARTERS AND CONTROLS**									
0050	Magnetic, FVNR, with enclosure and heaters, 480 volt									
0100	5 HP, size 0	1 Elec	2.30	3.478	Ea.	259	179		438	555
0200	10 HP, size 1	"	1.60	5		291	258		549	705
0300	25 HP, size 2	2 Elec	2.20	7.273		545	375		920	1,150
0400	50 HP, size 3		1.80	8.889		890	460		1,350	1,675
0500	100 HP, size 4		1.20	13.333		1,975	685		2,660	3,200
0600	200 HP, size 5		.90	17.778		4,625	915		5,540	6,475
0700	Combination, with motor circuit protectors, 5 HP, size 0	1 Elec	1.80	4.444		815	229		1,044	1,225
0800	10 HP, size 1	"	1.30	6.154		845	315		1,160	1,400
0900	25 HP, size 2	2 Elec	2	8		1,175	410		1,585	1,925
1000	50 HP, size 3		1.32	12.121		1,700	625		2,325	2,800
1200	100 HP, size 4		.80	20		3,725	1,025		4,750	5,650
1400	Combination, with fused switch, 5 HP, size 0	1 Elec	1.80	4.444		620	229		849	1,025
1600	10 HP, size 1	"	1.30	6.154		665	315		980	1,200
1800	25 HP, size 2	2 Elec	2	8		1,075	410		1,485	1,800
2000	50 HP, size 3		1.32	12.121		1,825	625		2,450	2,925
2200	100 HP, size 4		.80	20		3,175	1,025		4,200	5,050

26 25 Enclosed Bus Assemblies

26 25 13 – Bus Duct/Busway and Fittings

26 25 13.40 Copper Bus Duct

		Crew	Daily Output	Labor-Hours	Unit	Material	2012 Bare Costs Labor	Equipment	Total	Total Incl O&P
0010	**COPPER BUS DUCT** 10 ft. long									
0050	Indoor 3 pole 4 wire, plug-in, straight section, 225 amp	2 Elec	40	.400	L.F.	225	20.50		245.50	279
1000	400 amp		32	.500		225	26		251	287
1500	600 amp		26	.615		225	31.50		256.50	296
2400	800 amp		20	.800		267	41		308	355
2450	1000 amp		18	.889		295	46		341	395
2500	1350 amp		16	1		400	51.50		451.50	515
2510	1600 amp		12	1.333		455	68.50		523.50	605
2520	2000 amp		10	1.600		575	82.50		657.50	755
2550	Feeder, 600 amp		28	.571		197	29.50		226.50	261
2600	800 amp		22	.727		239	37.50		276.50	320
2700	1000 amp		20	.800		267	41		308	355
2800	1350 amp		18	.889		375	46		421	480
2900	1600 amp		14	1.143		430	59		489	560
3000	2000 amp		12	1.333		550	68.50		618.50	710
3100	Elbows, 225 amp		4	4	Ea.	1,325	206		1,531	1,775
3200	400 amp		3.60	4.444		1,325	229		1,554	1,825
3300	600 amp		3.20	5		1,325	258		1,583	1,850
3400	800 amp		2.80	5.714		1,450	295		1,745	2,050

26 25 Enclosed Bus Assemblies

26 25 13 – Bus Duct/Busway and Fittings

26 25 13.40 Copper Bus Duct	Crew	Daily Output	Labor-Hours	Unit	Material	2012 Bare Costs Labor	Equipment	Total	Total Incl O&P
3500	1000 amp	2 Elec	2.60	6.154	Ea.	1,625	315		1,940	2,250
3600	1350 amp		2.40	6.667		1,800	345		2,145	2,500
3700	1600 amp		2.20	7.273		1,950	375		2,325	2,700
3800	2000 amp		1.80	8.889		2,425	460		2,885	3,325
4000	End box, 225 amp		34	.471		180	24.50		204.50	235
4100	400 amp		32	.500		180	26		206	237
4200	600 amp		28	.571		180	29.50		209.50	242
4300	800 amp		26	.615		180	31.50		211.50	246
4400	1000 amp		24	.667		180	34.50		214.50	250
4500	1350 amp		22	.727		171	37.50		208.50	244
4600	1600 amp		20	.800		171	41		212	250
4700	2000 amp		18	.889		210	46		256	300
4800	Cable tap box end, 225 amp		3.20	5		1,475	258		1,733	1,975
5000	400 amp		2.60	6.154		1,475	315		1,790	2,075
5100	600 amp		2.20	7.273		1,475	375		1,850	2,150
5200	800 amp		2	8		1,675	410		2,085	2,475
5300	1000 amp		1.60	10		1,800	515		2,315	2,775
5400	1350 amp		1.40	11.429		2,150	590		2,740	3,225
5500	1600 amp		1.20	13.333		2,400	685		3,085	3,650
5600	2000 amp		1	16		2,675	825		3,500	4,175
5700	Switchboard stub, 225 amp		5.40	2.963		1,350	153		1,503	1,700
5800	400 amp		4.60	3.478		1,350	179		1,529	1,750
5900	600 amp		4	4		1,350	206		1,556	1,775
6000	800 amp		3.20	5		1,625	258		1,883	2,175
6100	1000 amp		3	5.333		1,900	275		2,175	2,475
6200	1350 amp		2.60	6.154		2,325	315		2,640	3,025
6300	1600 amp		2.40	6.667		2,625	345		2,970	3,400
6400	2000 amp		2	8		3,175	410		3,585	4,100
6490	Tee fittings, 225 amp		2.40	6.667		1,825	345		2,170	2,550
6500	400 amp		2	8		1,825	410		2,235	2,650
6600	600 amp		1.80	8.889		1,825	460		2,285	2,700
6700	800 amp		1.60	10		2,125	515		2,640	3,100
6800	1350 amp		1.20	13.333		2,900	685		3,585	4,200
7000	1600 amp		1	16		3,300	825		4,125	4,850
7100	2000 amp		.80	20		3,900	1,025		4,925	5,850
7200	Plug-in fusible switches w/3 fuses, 600 volt, 3 pole, 30 amp	1 Elec	4	2		820	103		923	1,050
7300	60 amp		3.60	2.222		925	115		1,040	1,200
7400	100 amp		2.70	2.963		1,400	153		1,553	1,775
7500	200 amp	2 Elec	3.20	5		2,525	258		2,783	3,150
7600	400 amp		1.40	11.429		7,375	590		7,965	8,975
7700	600 amp		.90	17.778		8,375	915		9,290	10,600
7800	800 amp		.66	24.242		11,700	1,250		12,950	14,800
7900	1200 amp		.50	32		22,000	1,650		23,650	26,700
8000	Plug-in circuit breakers, molded case, 15 to 50 amp	1 Elec	4.40	1.818		780	93.50		873.50	995
8100	70 to 100 amp	"	3.10	2.581		865	133		998	1,150
8200	150 to 225 amp	2 Elec	3.40	4.706		2,350	243		2,593	2,950
8300	250 to 400 amp		1.40	11.429		4,100	590		4,690	5,400
8400	500 to 600 amp		1	16		5,550	825		6,375	7,325
8500	700 to 800 amp		.64	25		6,850	1,300		8,150	9,450
8600	900 to 1000 amp		.56	28.571		9,800	1,475		11,275	13,000
8700	1200 amp		.44	36.364		11,800	1,875		13,675	15,800

26 27 Low-Voltage Distribution Equipment

26 27 16 – Electrical Cabinets and Enclosures

26 27 16.10 Cabinets	Crew	Daily Output	Labor-Hours	Unit	Material	2012 Bare Costs Labor	Equipment	Total	Total Incl O&P
0010 **CABINETS**									
7000 Cabinets, current transformer									
7050 Single door, 24" H x 24" W x 10" D	1 Elec	1.60	5	Ea.	134	258		392	530
7100 30" H x 24" W x 10" D		1.30	6.154		137	315		452	625
7150 36" H x 24" W x 10" D		1.10	7.273		140	375		515	715
7200 30" H x 30" W x 10" D		1	8		146	410		556	775
7250 36" H x 30" W x 10" D		.90	8.889		198	460		658	905
7300 36" H x 36" W x 10" D		.80	10		211	515		726	1,000
7500 Double door, 48" H x 36" W x 10" D		.60	13.333		435	685		1,120	1,500
7550 24" H x 24" W x 12" D		1	8		178	410		588	810

26 27 23 – Indoor Service Poles

26 27 23.40 Surface Raceway

	Crew	Daily Output	Labor-Hours	Unit	Material	2012 Bare Costs Labor	Equipment	Total	Total Incl O&P
0010 **SURFACE RACEWAY**									
0090 Metal, straight section									
0100 No. 500	1 Elec	100	.080	L.F.	1.03	4.12		5.15	7.30
0110 No. 700		100	.080		1.16	4.12		5.28	7.45
0400 No. 1500, small pancake		90	.089		2.08	4.58		6.66	9.15
0600 No. 2000, base & cover, blank		90	.089		2.13	4.58		6.71	9.20
0800 No. 3000, base & cover, blank		75	.107		4.10	5.50		9.60	12.70
1000 No. 4000, base & cover, blank		65	.123		6.70	6.35		13.05	16.85
1200 No. 6000, base & cover, blank		50	.160		11.10	8.25		19.35	24.50
2400 Fittings, elbows, No. 500		40	.200	Ea.	1.85	10.30		12.15	17.45
2800 Elbow cover, No. 2000		40	.200		3.50	10.30		13.80	19.25
2880 Tee, No. 500		42	.190		3.56	9.80		13.36	18.60
2900 No. 2000		27	.296		11.70	15.25		26.95	36
3000 Switch box, No. 500		16	.500		12.40	26		38.40	52
3400 Telephone outlet, No. 1500		16	.500		14.55	26		40.55	54.50
3600 Junction box, No. 1500		16	.500		9.95	26		35.95	49.50
3800 Plugmold wired sections, No. 2000									
4000 1 circuit, 6 outlets, 3 ft. long	1 Elec	8	1	Ea.	36	51.50		87.50	117
4100 2 circuits, 8 outlets, 6 ft. long	"	5.30	1.509	"	59.50	78		137.50	182

26 27 26 – Wiring Devices

26 27 26.10 Low Voltage Switching

	Crew	Daily Output	Labor-Hours	Unit	Material	2012 Bare Costs Labor	Equipment	Total	Total Incl O&P
0010 **LOW VOLTAGE SWITCHING**									
3600 Relays, 120 V or 277 V standard	1 Elec	12	.667	Ea.	41	34.50		75.50	96.50
3800 Flush switch, standard		40	.200		11.30	10.30		21.60	28
4000 Interchangeable		40	.200		14.75	10.30		25.05	31.50
4100 Surface switch, standard		40	.200		7.95	10.30		18.25	24
4200 Transformer 115 V to 25 V		12	.667		128	34.50		162.50	192
4400 Master control, 12 circuit, manual		4	2		124	103		227	290
4500 25 circuit, motorized		4	2		137	103		240	305
4600 Rectifier, silicon		12	.667		44.50	34.50		79	101
4800 Switchplates, 1 gang, 1, 2 or 3 switch, plastic		80	.100		4.90	5.15		10.05	13.10
5000 Stainless steel		80	.100		11.15	5.15		16.30	19.95
5400 2 gang, 3 switch, stainless steel		53	.151		22.50	7.80		30.30	36
5500 4 switch, plastic		53	.151		10.10	7.80		17.90	23
5800 3 gang, 9 switch, stainless steel		32	.250		63.50	12.90		76.40	89.50

26 27 26.20 Wiring Devices Elements

	Crew	Daily Output	Labor-Hours	Unit	Material	2012 Bare Costs Labor	Equipment	Total	Total Incl O&P
0010 **WIRING DEVICES ELEMENTS**									
0200 Toggle switch, quiet type, single pole, 15 amp	1 Elec	40	.200	Ea.	5.40	10.30		15.70	21.50
0600 3 way, 15 amp		23	.348		8.20	17.95		26.15	36

26 27 Low-Voltage Distribution Equipment

26 27 26 – Wiring Devices

26 27 26.20 Wiring Devices Elements

26 27 26.20 Wiring Devices Elements		Crew	Daily Output	Labor-Hours	Unit	Material	2012 Bare Costs Labor	2012 Bare Costs Equipment	Total	Total Incl O&P
0900	4 way, 15 amp	1 Elec	15	.533	Ea.	14.80	27.50		42.30	57.50
1650	Dimmer switch, 120 volt, incandescent, 600 watt, 1 pole [G]		16	.500		14	26		40	54
2460	Receptacle, duplex, 120 volt, grounded, 15 amp		40	.200		1.42	10.30		11.72	16.95
2470	20 amp		27	.296		10.95	15.25		26.20	35
2490	Dryer, 30 amp		15	.533		6.75	27.50		34.25	48.50
2500	Range, 50 amp		11	.727		15	37.50		52.50	72.50
2600	Wall plates, stainless steel, 1 gang		80	.100		2.50	5.15		7.65	10.45
2800	2 gang		53	.151		4.25	7.80		12.05	16.35
3200	Lampholder, keyless		26	.308		13.60	15.85		29.45	38.50
3400	Pullchain with receptacle	↓	22	.364	↓	20	18.75		38.75	50

26 27 73 – Door Chimes

26 27 73.10 Doorbell System

26 27 73.10 Doorbell System		Crew	Daily Output	Labor-Hours	Unit	Material	2012 Bare Costs Labor	2012 Bare Costs Equipment	Total	Total Incl O&P
0010	**DOORBELL SYSTEM**, incl. transformer, button & signal									
0100	6" bell	1 Elec	4	2	Ea.	112	103		215	277
0200	Buzzer	"	4	2	"	89.50	103		192.50	252

26 28 Low-Voltage Circuit Protective Devices

26 28 16 – Enclosed Switches and Circuit Breakers

26 28 16.10 Circuit Breakers

26 28 16.10 Circuit Breakers		Crew	Daily Output	Labor-Hours	Unit	Material	2012 Bare Costs Labor	2012 Bare Costs Equipment	Total	Total Incl O&P
0010	**CIRCUIT BREAKERS** (in enclosure)									
0100	Enclosed (NEMA 1), 600 volt, 3 pole, 30 amp	1 Elec	3.20	2.500	Ea.	485	129		614	730
0200	60 amp		2.80	2.857		600	147		747	875
0400	100 amp		2.30	3.478		685	179		864	1,025
0600	225 amp	↓	1.50	5.333		1,575	275		1,850	2,150
0700	400 amp	2 Elec	1.60	10		2,700	515		3,215	3,750
0800	600 amp		1.20	13.333		3,900	685		4,585	5,325
1000	800 amp	↓	.94	17.021	↓	5,100	875		5,975	6,900

26 28 16.20 Safety Switches

26 28 16.20 Safety Switches		Crew	Daily Output	Labor-Hours	Unit	Material	2012 Bare Costs Labor	2012 Bare Costs Equipment	Total	Total Incl O&P
0010	**SAFETY SWITCHES**									
0100	General duty 240 volt, 3 pole NEMA 1, fusible, 30 amp	1 Elec	3.20	2.500	Ea.	119	129		248	325
0200	60 amp		2.30	3.478		202	179		381	490
0300	100 amp		1.90	4.211		345	217		562	705
0400	200 amp	↓	1.30	6.154		745	315		1,060	1,300
0500	400 amp	2 Elec	1.80	8.889		1,875	460		2,335	2,750
0600	600 amp	"	1.20	13.333	↓	3,500	685		4,185	4,875
2900	Heavy duty, 240 volt, 3 pole NEMA 1 fusible									
2910	30 amp	1 Elec	3.20	2.500	Ea.	192	129		321	405
3000	60 amp		2.30	3.478		325	179		504	630
3300	100 amp		1.90	4.211		515	217		732	890
3500	200 amp	↓	1.30	6.154		880	315		1,195	1,450
3700	400 amp	2 Elec	1.80	8.889		2,250	460		2,710	3,175
3900	600 amp	"	1.20	13.333	↓	3,900	685		4,585	5,300

26 29 Low-Voltage Controllers

26 29 13 – Enclosed Controllers

26 29 13.20 Control Stations		Crew	Daily Output	Labor-Hours	Unit	Material	2012 Bare Costs Labor	Equipment	Total	Total Incl O&P
0010	**CONTROL STATIONS**									
0050	NEMA 1, heavy duty, stop/start	1 Elec	8	1	Ea.	135	51.50		186.50	226
0100	Stop/start, pilot light		6.20	1.290		184	66.50		250.50	305
0200	Hand/off/automatic		6.20	1.290		100	66.50		166.50	210
0400	Stop/start/reverse	↓	5.30	1.509	↓	182	78		260	315

26 32 Packaged Generator Assemblies

26 32 13 – Engine Generators

26 32 13.13 Diesel-Engine-Driven Generator Sets

		Crew	Daily Output	Labor-Hours	Unit	Material	2012 Bare Costs Labor	Equipment	Total	Total Incl O&P
0010	**DIESEL-ENGINE-DRIVEN GENERATOR SETS**									
2000	Diesel engine, including battery, charger,									
2010	muffler, automatic transfer switch & day tank, 30 kW	R-3	.55	36.364	Ea.	22,700	1,850	241	24,791	28,000
2100	50 kW		.42	47.619		26,900	2,425	315	29,640	33,600
2200	75 kW		.35	57.143		34,900	2,925	380	38,205	43,200
2300	100 kW		.31	64.516		38,800	3,300	425	42,525	48,100
2400	125 kW		.29	68.966		40,900	3,525	455	44,880	51,000
2500	150 kW		.26	76.923		46,900	3,925	510	51,335	58,000
2600	175 kW		.25	80		51,000	4,075	530	55,605	62,500
2700	200 kW		.24	83.333		52,500	4,250	550	57,300	65,000
2800	250 kW		.23	86.957		62,000	4,425	575	67,000	76,000
2900	300 kW		.22	90.909		67,000	4,625	600	72,225	81,500
3000	350 kW		.20	100		76,000	5,100	660	81,760	92,000
3100	400 kW		.19	105		93,500	5,375	695	99,570	112,000
3200	500 kW	↓	.18	111	↓	118,000	5,675	735	124,410	139,000

26 32 13.16 Gas-Engine-Driven Generator Sets

		Crew	Daily Output	Labor-Hours	Unit	Material	2012 Bare Costs Labor	Equipment	Total	Total Incl O&P
0010	**GAS-ENGINE-DRIVEN GENERATOR SETS**									
0020	Gas or gasoline operated, includes battery,									
0050	charger, muffler & transfer switch									
0200	3 phase 4 wire, 277/480 volt, 7.5 kW	R-3	.83	24.096	Ea.	8,550	1,225	160	9,935	11,500
0300	11.5 kW		.71	28.169		12,100	1,425	186	13,711	15,700
0400	20 kW		.63	31.746		14,300	1,625	210	16,135	18,400
0500	35 kW		.55	36.364		17,000	1,850	241	19,091	21,700
0600	80 kW		.40	50		27,900	2,550	330	30,780	34,900
0700	100 kW		.33	60.606		30,600	3,100	400	34,100	38,800
0800	125 kW		.28	71.429		62,500	3,650	475	66,625	75,000
0900	185 kW	↓	.25	80	↓	83,000	4,075	530	87,605	97,500

26 35 Power Filters and Conditioners

26 35 13 – Capacitors

26 35 13.10 Capacitors Indoor

		Crew	Daily Output	Labor-Hours	Unit	Material	2012 Bare Costs Labor	Equipment	Total	Total Incl O&P
0010	**CAPACITORS INDOOR**									
0020	240 volts, single & 3 phase, 0.5 kVAR	1 Elec	2.70	2.963	Ea.	395	153		548	665
0100	1.0 kVAR		2.70	2.963		475	153		628	755
0150	2.5 kVAR		2	4		535	206		741	900
0200	5.0 kVAR		1.80	4.444		640	229		869	1,050
0250	7.5 kVAR		1.60	5		745	258		1,003	1,200
0300	10 kVAR		1.50	5.333		900	275		1,175	1,400
0350	15 kVAR		1.30	6.154		1,225	315		1,540	1,825
0400	20 kVAR		1.10	7.273		1,550	375		1,925	2,250

26 35 Power Filters and Conditioners

26 35 13 – Capacitors

26 35 13.10 Capacitors Indoor	Crew	Daily Output	Labor-Hours	Unit	Material	2012 Bare Costs Labor	Equipment	Total	Total Incl O&P
0450	25 kVAR	1 Elec	1	8	Ea.	1,800	410		2,210	2,625
1000	480 volts, single & 3 phase, 1 kVAR		2.70	2.963		360	153		513	625
1050	2 kVAR		2.70	2.963		415	153		568	685
1100	5 kVAR		2	4		520	206		726	885
1150	7.5 kVAR		2	4		560	206		766	930
1200	10 kVAR		2	4		630	206		836	1,000
1250	15 kVAR		2	4		775	206		981	1,175
1300	20 kVAR		1.60	5		865	258		1,123	1,325
1350	30 kVAR		1.50	5.333		1,075	275		1,350	1,600
1400	40 kVAR		1.20	6.667		1,375	345		1,720	2,050
1450	50 kVAR		1.10	7.273		1,575	375		1,950	2,275

26 42 Cathodic Protection

26 42 16 – Passive Cathodic Protection for Underground Storage Tank

26 42 16.50 Cathodic Protection Wiring Methods

		Crew	Daily Output	Labor-Hours	Unit	Material	2012 Bare Costs Labor	Equipment	Total	Total Incl O&P
0010	**CATHODIC PROTECTION WIRING METHODS**									
1000	Anodes, magnesium type, 9 #	R-15	18.50	2.595	Ea.	38.50	131	17.50	187	258
1010	17 #		13	3.692		70	186	25	281	385
1020	32 #		10	4.800		122	242	32.50	396.50	535
1030	48 #		7.20	6.667		160	335	45	540	730
1100	Graphite type w/epoxy cap, 3" x 60" (32 #)	R-22	8.40	4.438		125	194		319	430
1110	4" x 80" (68 #)		6	6.213		237	272		509	670
1120	6" x 72" (80 #)		5.20	7.169		1,475	315		1,790	2,100
1130	6" x 36" (45 #)		9.60	3.883		745	170		915	1,075
2000	Rectifiers, silicon type, air cooled, 28 V/10 A	R-19	3.50	5.714		2,275	295		2,570	2,950
2010	20 V/20 A		3.50	5.714		2,325	295		2,620	3,000
2100	Oil immersed, 28 V/10 A		3	6.667		3,100	345		3,445	3,925
2110	20 V/20 A		3	6.667		3,125	345		3,470	3,950
3000	Anode backfill, coke breeze	R-22	3850	.010	Lb.	.22	.42		.64	.88
4000	Cable, HMWPE, No. 8		2.40	15.533	M.L.F.	490	680		1,170	1,575
4010	No. 6		2.40	15.533		730	680		1,410	1,825
4020	No. 4		2.40	15.533		1,100	680		1,780	2,250
4030	No. 2		2.40	15.533		1,725	680		2,405	2,925
4040	No. 1		2.20	16.945		2,350	740		3,090	3,725
4050	No. 1/0		2.20	16.945		2,925	740		3,665	4,325
4060	No. 2/0		2.20	16.945		4,525	740		5,265	6,125
4070	No. 4/0		2	18.640		5,875	815		6,690	7,700
5000	Test station, 7 terminal box, flush curb type w/lockable cover	R-19	12	1.667	Ea.	70.50	86		156.50	207
5010	Reference cell, 2" dia PVC conduit, cplg., plug, set flush	"	4.80	4.167	"	148	215		363	485

26 51 13 – Interior Lighting Fixtures, Lamps, and Ballasts

26 51 13.50 Interior Lighting Fixtures	Crew	Daily Output	Labor-Hours	Unit	Material	2012 Bare Costs Labor	Equipment	Total	Total Incl O&P
0010 **INTERIOR LIGHTING FIXTURES** Including lamps, mounting									
0030 hardware and connections									
0100 Fluorescent, C.W. lamps, troffer, recess mounted in grid, RS									
0130 Grid ceiling mount									
0200 Acrylic lens, 1'W x 4'L, two 40 watt	1 Elec	5.70	1.404	Ea.	54	72.50		126.50	168
0210 1'W x 4'L, three 40 watt		5.40	1.481		62.50	76.50		139	183
0300 2'W x 2'L, two U40 watt		5.70	1.404		59.50	72.50		132	174
0400 2'W x 4'L, two 40 watt		5.30	1.509		58	78		136	180
0500 2'W x 4'L, three 40 watt		5	1.600		64	82.50		146.50	194
0600 2'W x 4'L, four 40 watt		4.70	1.702		67	87.50		154.50	205
0700 4'W x 4'L, four 40 watt	2 Elec	6.40	2.500		340	129		469	570
0800 4'W x 4'L, six 40 watt		6.20	2.581		350	133		483	585
0900 4'W x 4'L, eight 40 watt		5.80	2.759		365	142		507	615
0910 Acrylic lens, 1'W x 4'L, two 32 watt T8 G	1 Elec	5.70	1.404		67	72.50		139.50	182
0930 2'W x 2'L, two U32 watt T8 G		5.70	1.404		80	72.50		152.50	196
0940 2'W x 4'L, two 32 watt T8 G		5.30	1.509		72.50	78		150.50	196
0950 2'W x 4'L, three 32 watt T8 G		5	1.600		76.50	82.50		159	207
0960 2'W x 4'L, four 32 watt T8 G		4.70	1.702		78.50	87.50		166	217
1000 Surface mounted, RS									
1030 Acrylic lens with hinged & latched door frame									
1100 1'W x 4'L, two 40 watt	1 Elec	7	1.143	Ea.	74	59		133	170
1110 1'W x 4'L, three 40 watt		6.70	1.194		76.50	61.50		138	176
1200 2'W x 2'L, two U40 watt		7	1.143		79.50	59		138.50	175
1300 2'W x 4'L, two 40 watt		6.20	1.290		91	66.50		157.50	200
1400 2'W x 4'L, three 40 watt		5.70	1.404		92	72.50		164.50	209
1500 2'W x 4'L, four 40 watt		5.30	1.509		94.50	78		172.50	220
1600 4'W x 4'L, four 40 watt	2 Elec	7.20	2.222		465	115		580	680
1700 4'W x 4'L, six 40 watt		6.60	2.424		505	125		630	740
1800 4'W x 4'L, eight 40 watt		6.20	2.581		525	133		658	780
1900 2'W x 8'L, four 40 watt		6.40	2.500		185	129		314	395
2000 2'W x 8'L, eight 40 watt		6.20	2.581		199	133		332	420
2100 Strip fixture									
2130 Surface mounted									
2200 4' long, one 40 watt, RS	1 Elec	8.50	.941	Ea.	27	48.50		75.50	102
2300 4' long, two 40 watt, RS		8	1		38	51.50		89.50	119
2400 4' long, one 40 watt, SL		8	1		46	51.50		97.50	128
2500 4' long, two 40 watt, SL		7	1.143		62.50	59		121.50	157
2600 8' long, one 75 watt, SL	2 Elec	13.40	1.194		47	61.50		108.50	144
2700 8' long, two 75 watt, SL	"	12.40	1.290		57	66.50		123.50	162
2800 4' long, two 60 watt, HO	1 Elec	6.70	1.194		91.50	61.50		153	193
2900 8' long, two 110 watt, HO	2 Elec	10.60	1.509		96.50	78		174.50	222
2950 High bay pendent mounted, 16" W x 4' L, four 54 watt, T5HO G		8.90	1.798		224	92.50		316.50	385
2952 2' W x 4' L, six 54 watt, T5HO G		8.50	1.882		300	97		397	475
2954 2' W x 4' L, six 32 watt, T8 G		8.50	1.882		175	97		272	340
3000 Strip, pendent mounted, industrial, white porcelain enamel									
3100 4' long, two 40 watt, RS	1 Elec	5.70	1.404	Ea.	52	72.50		124.50	165
3200 4' long, two 60 watt, HO	"	5	1.600		81	82.50		163.50	212
3300 8' long, two 75 watt, SL	2 Elec	8.80	1.818		96	93.50		189.50	246
3400 8' long, two 110 watt, HO	"	8	2		123	103		226	289
3470 Troffer, air handling, 2'W x 4'L with four 32 watt T8 G	1 Elec	4	2		110	103		213	275
3480 2'W x 2'L with two U32 watt T8 G		5.50	1.455		106	75		181	229
3490 Air connector insulated, 5" diameter		20	.400		66	20.50		86.50	104
3500 6" diameter		20	.400		67.50	20.50		88	105

26 51 Interior Lighting

26 51 13 – Interior Lighting Fixtures, Lamps, and Ballasts

	26 51 13.50 Interior Lighting Fixtures	Crew	Daily Output	Labor-Hours	Unit	Material	2012 Bare Costs Labor	Equipment	Total	Total Incl O&P
3510	Troffer parabolic lay-in, 1'W x 4'L with one 32 W T8 **G**	1 Elec	5.70	1.404	Ea.	112	72.50		184.50	231
3520	1'W x 4'L with two 32 W T8 **G**		5.30	1.509		134	78		212	263
3525	2'W x 2'L with two U32 W T8 **G**		5.70	1.404		113	72.50		185.50	232
3530	2'W x 4'L with three 32 W T8 **G**		5	1.600		125	82.50		207.50	261
4450	Incandescent, high hat can, round alzak reflector, prewired									
4470	100 watt	1 Elec	8	1	Ea.	70	51.50		121.50	154
4480	150 watt		8	1		102	51.50		153.50	189
4500	300 watt		6.70	1.194		236	61.50		297.50	350
4600	Square glass lens with metal trim, prewired									
4630	100 watt	1 Elec	6.70	1.194	Ea.	54	61.50		115.50	152
4700	200 watt		6.70	1.194		95	61.50		156.50	197
4800	300 watt		5.70	1.404		142	72.50		214.50	264
4900	Ceiling/wall, surface mounted, metal cylinder, 75 watt		10	.800		55	41		96	122
4920	150 watt		10	.800		79	41		120	149
5200	Ceiling, surface mounted, opal glass drum									
5300	8", one 60 watt lamp	1 Elec	10	.800	Ea.	43.50	41		84.50	110
5400	10", two 60 watt lamps		8	1		49	51.50		100.50	131
5500	12", four 60 watt lamps		6.70	1.194		69	61.50		130.50	168
6010	Vapor tight, incandescent, ceiling mounted, 200 watt		6.20	1.290		81	66.50		147.50	189
6100	Fluorescent, surface mounted, 2 lamps, 4'L, RS, 40 watt		3.20	2.500		113	129		242	315
6850	Vandalproof, surface mounted, fluorescent, two 32 watt T8 **G**		3.20	2.500		248	129		377	465
6860	Incandescent, one 150 watt		8	1		84	51.50		135.50	170
7500	Ballast replacement, by weight of ballast, to 15' high									
7520	Indoor fluorescent, less than 2 lb.	1 Elec	10	.800	Ea.	25.50	41		66.50	89.50
7540	Two 40W, watt reducer, 2 to 5 lb.		9.40	.851		40.50	44		84.50	110
7560	Two F96 slimline, over 5 lb.		8	1		75.50	51.50		127	160
7580	Vaportite ballast, less than 2 lb.		9.40	.851		25.50	44		69.50	93.50
7600	2 lb. to 5 lb.		8.90	.899		40.50	46.50		87	114
7620	Over 5 lb.		7.60	1.053		75.50	54.50		130	164
7630	Electronic ballast for two tubes		8	1		36.50	51.50		88	117
7640	Dimmable ballast one lamp **G**		8	1		104	51.50		155.50	191
7650	Dimmable ballast two-lamp **G**		7.60	1.053		102	54.50		156.50	193

26 52 Emergency Lighting

26 52 13 – Emergency Lighting Equipments

26 52 13.10 Emergency Lighting and Battery Units

		Crew	Daily Output	Labor-Hours	Unit	Material	Labor	Equipment	Total	Total Incl O&P
0010	**EMERGENCY LIGHTING AND BATTERY UNITS**									
0300	Emergency light units, battery operated									
0350	Twin sealed beam light, 25 watt, 6 volt each									
0500	Lead battery operated	1 Elec	4	2	Ea.	124	103		227	290
0700	Nickel cadmium battery operated		4	2		560	103		663	770
0900	Self-contained fluorescent lamp pack		10	.800		154	41		195	231

26 53 Exit Signs

26 53 13 – Exit Lighting

26 53 13.10 Exit Lighting Fixtures	Crew	Daily Output	Labor-Hours	Unit	Material	2012 Bare Costs Labor	Equipment	Total	Total Incl O&P
0010 **EXIT LIGHTING FIXTURES**									
0080 Exit light ceiling or wall mount, incandescent, single face	1 Elec	8	1	Ea.	38	51.50		89.50	119
0100 Double face		6.70	1.194		38	61.50		99.50	134
0200 L.E.D. standard, single face Ⓖ		8	1		74.50	51.50		126	159
0220 Double face Ⓖ		6.70	1.194		74.50	61.50		136	174
0230 L.E.D. vandal-resistant, single face Ⓖ		7.27	1.100		223	56.50		279.50	330
0240 L.E.D. w/battery unit, single face Ⓖ		4.40	1.818		128	93.50		221.50	281
0260 Double face Ⓖ		4	2		130	103		233	297
0262 L.E.D. w/battery unit, vandal-resistant, single face Ⓖ		4.40	1.818		370	93.50		463.50	550
0270 Combination emergency light units and exit sign		4	2		195	103		298	370

26 54 Classified Location Lighting

26 54 13 – Classified Lighting

26 54 13.20 Explosionproof

	Crew	Daily Output	Labor-Hours	Unit	Material	2012 Bare Costs Labor	Equipment	Total	Total Incl O&P
0010 **EXPLOSIONPROOF**, incl lamps, mounting hardware and connections									
6510 Incandescent, ceiling mounted, 200 watt	1 Elec	4	2	Ea.	1,050	103		1,153	1,300
6600 Fluorescent, RS, 4' long, ceiling mounted, two 40 watt	"	2.70	2.963	"	2,800	153		2,953	3,300

26 55 Special Purpose Lighting

26 55 61 – Theatrical Lighting

26 55 61.10 Lights

	Crew	Daily Output	Labor-Hours	Unit	Material	2012 Bare Costs Labor	Equipment	Total	Total Incl O&P
0010 **LIGHTS**									
2000 Lights, border, quartz, reflector, vented,									
2100 colored or white	1 Elec	20	.400	L.F.	173	20.50		193.50	221
2500 Spotlight, follow spot, with transformer, 2,100 watt	"	4	2	Ea.	3,075	103		3,178	3,525
2600 For no transformer, deduct					910			910	1,000
3000 Stationary spot, fresnel quartz, 6" lens	1 Elec	4	2		131	103		234	298
3100 8" lens		4	2		231	103		334	410
3500 Ellipsoidal quartz, 1,000W, 6" lens		4	2		335	103		438	525
3600 12" lens		4	2		600	103		703	815
4000 Strobe light, 1 to 15 flashes per second, quartz		3	2.667		750	137		887	1,025
4500 Color wheel, portable, five hole, motorized		4	2		194	103		297	370

26 56 Exterior Lighting

26 56 13 – Lighting Poles and Standards

26 56 13.10 Lighting Poles

	Crew	Daily Output	Labor-Hours	Unit	Material	2012 Bare Costs Labor	Equipment	Total	Total Incl O&P
0010 **LIGHTING POLES**									
2800 Light poles, anchor base									
2820 not including concrete bases									
2840 Aluminum pole, 8' high	1 Elec	4	2	Ea.	680	103		783	905
3000 20' high	R-3	2.90	6.897		900	350	45.50	1,295.50	1,575
3200 30' high		2.60	7.692		1,650	390	51	2,091	2,450
3400 35' high		2.30	8.696		1,875	445	57.50	2,377.50	2,775
3600 40' high		2	10		2,125	510	66	2,701	3,200
3800 Bracket arms, 1 arm	1 Elec	8	1		122	51.50		173.50	211
4000 2 arms		8	1		245	51.50		296.50	345
4200 3 arms		5.30	1.509		380	78		458	530

26 56 Exterior Lighting

26 56 13 – Lighting Poles and Standards

26 56 13.10 Lighting Poles

26 56 13.10 Lighting Poles		Crew	Daily Output	Labor-Hours	Unit	Material	2012 Bare Costs Labor	Equipment	Total	Total Incl O&P
4400	4 arms	1 Elec	5.30	1.509	Ea.	505	78		583	670
4500	Steel pole, galvanized, 8' high	↓	3.80	2.105		610	109		719	830
4600	20' high	R-3	2.60	7.692		1,075	390	51	1,516	1,850
4800	30' high		2.30	8.696		1,175	445	57.50	1,677.50	2,000
5000	35' high		2.20	9.091		1,375	465	60	1,900	2,275
5200	40' high	↓	1.70	11.765		1,700	600	78	2,378	2,850
5400	Bracket arms, 1 arm	1 Elec	8	1		172	51.50		223.50	266
5600	2 arms		8	1		267	51.50		318.50	370
5800	3 arms		5.30	1.509		289	78		367	435
6000	4 arms	↓	5.30	1.509	↓	400	78		478	560

26 56 19 – Roadway Lighting

26 56 19.20 Roadway Area Luminaire

		Crew	Daily Output	Labor-Hours	Unit	Material	2012 Bare Costs Labor	Equipment	Total	Total Incl O&P
0010	**ROADWAY AREA LUMINAIRE**									
2650	Roadway area luminaire, low pressure sodium, 135 watt	1 Elec	2	4	Ea.	625	206		831	995
2700	180 watt	"	2	4		660	206		866	1,025
2750	Metal halide, 400 watt	2 Elec	4.40	3.636		535	187		722	870
2760	1000 watt		4	4		605	206		811	975
2780	High pressure sodium, 400 watt		4.40	3.636		555	187		742	890
2790	1000 watt	↓	4	4	↓	635	206		841	1,000

26 56 23 – Area Lighting

26 56 23.10 Exterior Fixtures

		Crew	Daily Output	Labor-Hours	Unit	Material	2012 Bare Costs Labor	Equipment	Total	Total Incl O&P
0010	**EXTERIOR FIXTURES** With lamps									
0200	Wall mounted, incandescent, 100 watt	1 Elec	8	1	Ea.	33	51.50		84.50	114
0400	Quartz, 500 watt		5.30	1.509		57.50	78		135.50	180
1100	Wall pack, low pressure sodium, 35 watt		4	2		221	103		324	395
1150	55 watt	↓	4	2	↓	262	103		365	440

26 56 36 – Flood Lighting

26 56 36.20 Floodlights

		Crew	Daily Output	Labor-Hours	Unit	Material	2012 Bare Costs Labor	Equipment	Total	Total Incl O&P
0010	**FLOODLIGHTS** with ballast and lamp,									
1400	Pole mounted, pole not included									
1950	Metal halide, 175 watt	1 Elec	2.70	2.963	Ea.	325	153		478	590
2000	400 watt	2 Elec	4.40	3.636		405	187		592	725
2200	1000 watt	"	4	4		555	206		761	920
2340	High pressure sodium, 70 watt	1 Elec	2.70	2.963		245	153		398	500
2400	400 watt	2 Elec	4.40	3.636		365	187		552	680
2600	1000 watt	"	4	4	↓	625	206		831	1,000

26 61 Lighting Systems and Accessories

26 61 23 – Lamps Applications

26 61 23.10 Lamps

		Crew	Daily Output	Labor-Hours	Unit	Material	2012 Bare Costs Labor	Equipment	Total	Total Incl O&P
0010	**LAMPS**									
0080	Fluorescent, rapid start, cool white, 2' long, 20 watt	1 Elec	1	8	C	370	410		780	1,025
0100	4' long, 40 watt		.90	8.889		315	460		775	1,025
0200	Slimline, 4' long, 40 watt		.90	8.889		895	460		1,355	1,675
0210	4' long, 30 watt energy saver [G]		.90	8.889		895	460		1,355	1,675
0400	High output, 4' long, 60 watt		.90	8.889		1,025	460		1,485	1,800
0410	8' long, 95 watt energy saver [G]		.80	10		1,025	515		1,540	1,925
0500	8' long, 110 watt		.80	10		1,025	515		1,540	1,925
0512	2' long, T5, 14 watt energy saver [G]	↓	1	8	↓	1,400	410		1,810	2,175

26 61 Lighting Systems and Accessories

26 61 23 – Lamps Applications

26 61 23.10 Lamps		Crew	Daily Output	Labor-Hours	Unit	Material	2012 Bare Costs Labor	Equipment	Total	Total Incl O&P
0514	3' long, T5, 21 watt energy saver	G	1 Elec	.90	8.889	C	1,400	460		1,860	2,225
0516	4' long, T5, 28 watt energy saver	G		.90	8.889		1,225	460		1,685	2,000
0517	4' long, T5, 54 watt energy saver	G		.90	8.889		1,775	460		2,235	2,625
0560	Twin tube compact lamp	G		.90	8.889		585	460		1,045	1,325
0570	Double twin tube compact lamp	G		.80	10		1,300	515		1,815	2,225
0600	Mercury vapor, mogul base, deluxe white, 100 watt			.30	26.667		3,175	1,375		4,550	5,525
0700	250 watt			.30	26.667		4,900	1,375		6,275	7,450
0800	400 watt			.30	26.667		7,050	1,375		8,425	9,800
0900	1000 watt			.20	40		9,200	2,050		11,250	13,200
1000	Metal halide, mogul base, 175 watt			.30	26.667		4,175	1,375		5,550	6,650
1200	400 watt			.30	26.667		7,775	1,375		9,150	10,600
1300	1000 watt			.20	40		10,800	2,050		12,850	15,000
1350	High pressure sodium, 70 watt			.30	26.667		5,050	1,375		6,425	7,600
1380	250 watt			.30	26.667		5,750	1,375		7,125	8,375
1400	400 watt			.30	26.667		5,900	1,375		7,275	8,550
1450	1000 watt			.20	40		16,300	2,050		18,350	21,000
3000	Guards, fluorescent lamp, 4' long			1	8		1,175	410		1,585	1,900
3200	8' long			.90	8.889		2,325	460		2,785	3,250

26 71 Electrical Machines

26 71 13 – Motors Applications

26 71 13.20 Motors

		Crew	Daily Output	Labor-Hours	Unit	Material	2012 Bare Costs Labor	Equipment	Total	Total Incl O&P
0010	**MOTORS** 230/460 volts, 60 HZ									
0050	Dripproof, premium efficiency, 1.15 service factor									
0060	1800 RPM, 1/4 HP	1 Elec	5.33	1.501	Ea.	188	77.50		265.50	325
0070	1/3 HP		5.33	1.501		193	77.50		270.50	330
0080	1/2 HP		5.33	1.501		288	77.50		365.50	430
0090	3/4 HP		5.33	1.501		320	77.50		397.50	465
0100	1 HP		4.50	1.778		330	91.50		421.50	500
0250	5 HP		4.50	1.778		455	91.50		546.50	635
0350	10 HP		4	2		720	103		823	945
0450	20 HP	2 Elec	5.20	3.077		1,150	159		1,309	1,500

Estimating Tips

27 20 00 Data Communications

27 30 00 Voice Communications

27 40 00 Audio-Video Communications

When estimating material costs for special systems, it is always prudent to obtain manufacturers' quotations for equipment prices and special installation requirements which will affect the total costs.

Reference Numbers

Reference numbers are shown in shaded boxes at the beginning of some major classifications. These numbers refer to related items in the Reference Section. The reference information may be an estimating procedure, an alternate pricing method, or technical information.

Note: Not all subdivisions listed here necessarily appear in this publication.

Note: **Trade Service,** *in part, has been used as a reference source for some of the material prices used in Division 27.*

27 13 Communications Backbone Cabling

27 13 23 – Communications Optical Fiber Backbone Cabling

27 13 23.13 Communications Optical Fiber		Crew	Daily Output	Labor-Hours	Unit	Material	2012 Bare Costs Labor	Equipment	Total	Total Incl O&P
0010	**COMMUNICATIONS OPTICAL FIBER**									
0040	Specialized tools & techniques cause installation costs to vary.									
0070	Fiber optic, cable, bulk simplex, minimum	1 Elec	8	1	C.L.F.	23	51.50		74.50	103
0080	Bulk plenum quad, maximum	"	2.29	3.493	"	68	180		248	345
0150	Jumper				Ea.	34.50			34.50	38
0200	Pigtail					31			31	34
0300	Connector	1 Elec	24	.333		20	17.20		37.20	47.50
0350	Finger splice		32	.250		32	12.90		44.90	54.50
0400	Transceiver (low cost bi-directional)		8	1		410	51.50		461.50	530
0450	Rack housing, 4 rack spaces, 12 panels (144 fibers)		2	4		490	206		696	850
0500	Patch panel, 12 ports		6	1.333		242	68.50		310.50	370

27 41 Audio-Video Systems

27 41 19 – Portable Audio-Video Equipment

27 41 19.10 T.V. Systems

		Crew	Daily Output	Labor-Hours	Unit	Material	2012 Bare Costs Labor	Equipment	Total	Total Incl O&P
0010	**T.V. SYSTEMS**, not including rough-in wires, cables & conduits									
0100	Master TV antenna system									
0200	VHF reception & distribution, 12 outlets	1 Elec	6	1.333	Outlet	136	68.50		204.50	253
0400	30 outlets		10	.800		174	41		215	253
0600	100 outlets		13	.615		146	31.50		177.50	209
0800	VHF & UHF reception & distribution, 12 outlets		6	1.333		205	68.50		273.50	330
1000	30 outlets		10	.800		138	41		179	214
1200	100 outlets		13	.615		136	31.50		167.50	198
1400	School and deluxe systems, 12 outlets		2.40	3.333		270	172		442	555
1600	30 outlets		4	2		237	103		340	415
1800	80 outlets		5.30	1.509		227	78		305	365

27 51 Distributed Audio-Video Communications Systems

27 51 16 – Public Address and Mass Notification Systems

27 51 16.10 Public Address System

		Crew	Daily Output	Labor-Hours	Unit	Material	2012 Bare Costs Labor	Equipment	Total	Total Incl O&P
0010	**PUBLIC ADDRESS SYSTEM**									
0100	Conventional, office	1 Elec	5.33	1.501	Speaker	131	77.50		208.50	261
0200	Industrial	"	2.70	2.963	"	254	153		407	505

27 51 19 – Sound Masking Systems

27 51 19.10 Sound System

		Crew	Daily Output	Labor-Hours	Unit	Material	2012 Bare Costs Labor	Equipment	Total	Total Incl O&P
0010	**SOUND SYSTEM**, not including rough-in wires, cables & conduits									
0100	Components, projector outlet	1 Elec	8	1	Ea.	57	51.50		108.50	140
0200	Microphone		4	2		78	103		181	240
0400	Speakers, ceiling or wall		8	1		115	51.50		166.50	203
0600	Trumpets		4	2		213	103		316	390
0800	Privacy switch		8	1		85	51.50		136.50	171
1000	Monitor panel		4	2		380	103		483	570
1200	Antenna, AM/FM		4	2		132	103		235	299
1400	Volume control		8	1		86	51.50		137.50	172
1600	Amplifier, 250 watts		1	8		1,225	410		1,635	1,975
1800	Cabinets		1	8		825	410		1,235	1,525
2000	Intercom, 30 station capacity, master station	2 Elec	2	8		2,250	410		2,660	3,100
2200	Remote station	1 Elec	8	1		159	51.50		210.50	252
2400	Intercom outlets		8	1		93.50	51.50		145	180

27 51 Distributed Audio-Video Communications Systems

27 51 19 – Sound Masking Systems

27 51 19.10 Sound System

		Crew	Daily Output	Labor-Hours	Unit	Material	2012 Bare Costs Labor	Equipment	Total	Total Incl O&P
2600	Handset	1 Elec	4	2	Ea.	310	103		413	495
2800	Emergency call system, 12 zones, annunciator		1.30	6.154		935	315		1,250	1,500
3000	Bell		5.30	1.509		96.50	78		174.50	222
3200	Light or relay		8	1		48	51.50		99.50	130
3400	Transformer		4	2		212	103		315	385
3600	House telephone, talking station		1.60	5		455	258		713	885
3800	Press to talk, release to listen		5.30	1.509		106	78		184	233
4000	System-on button					63.50			63.50	70
4200	Door release	1 Elec	4	2		113	103		216	279
4400	Combination speaker and microphone		8	1		193	51.50		244.50	289
4600	Termination box		3.20	2.500		61	129		190	260
4800	Amplifier or power supply		5.30	1.509		700	78		778	885
5000	Vestibule door unit		16	.500	Name	128	26		154	180
5200	Strip cabinet		27	.296	Ea.	242	15.25		257.25	290
5400	Directory		16	.500	"	114	26		140	164

27 52 Healthcare Communications and Monitoring Systems

27 52 23 – Nurse Call/Code Blue Systems

27 52 23.10 Nurse Call Systems

		Crew	Daily Output	Labor-Hours	Unit	Material	2012 Bare Costs Labor	Equipment	Total	Total Incl O&P
0010	**NURSE CALL SYSTEMS**									
0100	Single bedside call station	1 Elec	8	1	Ea.	226	51.50		277.50	325
0200	Ceiling speaker station		8	1		63.50	51.50		115	147
0400	Emergency call station		8	1		68	51.50		119.50	152
0600	Pillow speaker		8	1		167	51.50		218.50	261
0800	Double bedside call station		4	2		135	103		238	305
1000	Duty station		4	2		145	103		248	315
1200	Standard call button		8	1		82	51.50		133.50	168
1400	Lights, corridor, dome or zone indicator		8	1		46.50	51.50		98	128
1600	Master control station for 20 stations	2 Elec	.65	24.615	Total	3,900	1,275		5,175	6,200

27 53 Distributed Systems

27 53 13 – Clock Systems

27 53 13.50 Clock Equipments

		Crew	Daily Output	Labor-Hours	Unit	Material	2012 Bare Costs Labor	Equipment	Total	Total Incl O&P
0010	**CLOCK EQUIPMENTS**, not including wires & conduits									
0100	Time system components, master controller	1 Elec	.33	24.242	Ea.	1,825	1,250		3,075	3,875
0200	Program bell		8	1		85	51.50		136.50	171
0400	Combination clock & speaker		3.20	2.500		210	129		339	425
0600	Frequency generator		2	4		2,350	206		2,556	2,900
0800	Job time automatic stamp recorder		4	2		460	103		563	660
1600	Master time clock system, clocks & bells, 20 room	4 Elec	.20	160		6,300	8,250		14,550	19,200
1800	50 room	"	.08	400		13,000	20,600		33,600	45,100
2000	Time clock, 100 cards in & out, 1 color	1 Elec	3.20	2.500		840	129		969	1,125
2200	2 colors		3.20	2.500		840	129		969	1,125
2800	Metal rack for 25 cards		7	1.143		39	59		98	131

Division Notes

	CREW	DAILY OUTPUT	LABOR-HOURS	UNIT	2012 BARE COSTS				TOTAL INCL O&P
					MAT.	LABOR	EQUIP.	TOTAL	

Estimating Tips

- When estimating material costs for electronic safety and security systems, it is always prudent to obtain manufacturers' quotations for equipment prices and special installation requirements that affect the total cost.

- Fire alarm systems consist of control panels, annunciator panels, battery with rack, charger, and fire alarm actuating and indicating devices. Some fire alarm systems include speakers, telephone lines, door closer controls, and other components. Be careful not to overlook the costs related to installation for these items.

Also be aware of costs for integrated automation instrumentation and terminal devices, control equipment, control wiring, and programming.

- Security equipment includes items such as CCTV, access control, and other detection and identification systems to perform alert and alarm functions. Be sure to consider the costs related to installation for this security equipment, such as for integrated automation instrumentation and terminal devices, control equipment, control wiring, and programming.

Reference Numbers

Reference numbers are shown in shaded boxes at the beginning of some major classifications. These numbers refer to related items in the Reference Section. The reference information may be an estimating procedure, an alternate pricing method, or technical information.

Note: Not all subdivisions listed here necessarily appear in this publication.

Division 28 – Electronic Safety and Security

28 13 Access Control

28 13 53 – Security Access Detection

28 13 53.13 Security Access Metal Detectors

		Crew	Daily Output	Labor-Hours	Unit	Material	2012 Bare Costs Labor	Equipment	Total	Total Incl O&P
0010	**SECURITY ACCESS METAL DETECTORS**									
0240	Metal detector, hand-held, wand type, unit only				Ea.	89			89	98
0250	Metal detector, walk through portal type, single zone	1 Elec	2	4		3,150	206		3,356	3,750
0260	Multi-zone	"	2	4		4,000	206		4,206	4,700

28 13 53.16 Security Access X-Ray Equipment

		Crew	Daily Output	Labor-Hours	Unit	Material	2012 Bare Costs Labor	Equipment	Total	Total Incl O&P
0010	**SECURITY ACCESS X-RAY EQUIPMENT**									
0290	X-ray machine, desk top, for mail/small packages/letters	1 Elec	4	2	Ea.	3,350	103		3,453	3,850
0300	Conveyor type, incl monitor, minimum		2	4		15,700	206		15,906	17,500
0310	Maximum		2	4		28,000	206		28,206	31,100
0320	X-ray machine, large unit, for airports, incl monitor, min	2 Elec	1	16		39,200	825		40,025	44,300
0330	Maximum	"	.50	32		67,000	1,650		68,650	76,500

28 13 53.23 Security Access Explosive Detection Equipment

		Crew	Daily Output	Labor-Hours	Unit	Material	2012 Bare Costs Labor	Equipment	Total	Total Incl O&P
0010	**SECURITY ACCESS EXPLOSIVE DETECTION EQUIPMENT**									
0270	Explosives detector, walk through portal type	1 Elec	2	4	Ea.	43,300	206		43,506	47,900
0280	Hand-held, battery operated				"				25,500	28,100

28 16 Intrusion Detection

28 16 16 – Intrusion Detection Systems Infrastructure

28 16 16.50 Intrusion Detection

		Crew	Daily Output	Labor-Hours	Unit	Material	2012 Bare Costs Labor	Equipment	Total	Total Incl O&P
0010	**INTRUSION DETECTION**, not including wires & conduits									
0100	Burglar alarm, battery operated, mechanical trigger	1 Elec	4	2	Ea.	272	103		375	455
0200	Electrical trigger		4	2		325	103		428	510
0400	For outside key control, add		8	1		82	51.50		133.50	168
0600	For remote signaling circuitry, add		8	1		122	51.50		173.50	212
0800	Card reader, flush type, standard		2.70	2.963		910	153		1,063	1,225
1000	Multi-code		2.70	2.963		1,175	153		1,328	1,525
1200	Door switches, hinge switch		5.30	1.509		57.50	78		135.50	179
1400	Magnetic switch		5.30	1.509		67.50	78		145.50	191
1600	Exit control locks, horn alarm		4	2		270	103		373	450
1800	Flashing light alarm		4	2		300	103		403	485
2000	Indicating panels, 1 channel		2.70	2.963		360	153		513	625
2200	10 channel	2 Elec	3.20	5		1,225	258		1,483	1,725
2400	20 channel		2	8		2,400	410		2,810	3,275
2600	40 channel		1.14	14.035		4,375	725		5,100	5,875
2800	Ultrasonic motion detector, 12 volt	1 Elec	2.30	3.478		225	179		404	515
3000	Infrared photoelectric detector	"	4	2		186	103		289	360

28 23 Video Surveillance

28 23 13 – Video Surveillance Control and Management Systems

28 23 13.10 Closed Circuit Television System

		Crew	Daily Output	Labor-Hours	Unit	Material	2012 Bare Costs Labor	Equipment	Total	Total Incl O&P
0010	**CLOSED CIRCUIT TELEVISION SYSTEM**									
2000	Surveillance, one station (camera & monitor)	2 Elec	2.60	6.154	Total	1,300	315		1,615	1,925
2200	For additional camera stations, add	1 Elec	2.70	2.963	Ea.	735	153		888	1,025
2400	Industrial quality, one station (camera & monitor)	2 Elec	2.60	6.154	Total	2,725	315		3,040	3,450
2600	For additional camera stations, add	1 Elec	2.70	2.963	Ea.	1,650	153		1,803	2,050
2610	For low light, add		2.70	2.963		1,325	153		1,478	1,700
2620	For very low light, add		2.70	2.963		9,825	153		9,978	11,000
2800	For weatherproof camera station, add		1.30	6.154		1,025	315		1,340	1,600
3000	For pan and tilt, add		1.30	6.154		2,650	315		2,965	3,375

28 23 Video Surveillance

28 23 13 – Video Surveillance Control and Management Systems

28 23 13.10 Closed Circuit Television System

		Crew	Daily Output	Labor-Hours	Unit	Material	2012 Bare Costs Labor	2012 Bare Costs Equipment	Total	Total Incl O&P
3200	For zoom lens - remote control, add, minimum	1 Elec	2	4	Ea.	2,450	206		2,656	2,975
3400	Maximum		2	4		8,925	206		9,131	10,100
3410	For automatic iris for low light, add	↓	2	4	↓	2,125	206		2,331	2,650
3600	Educational T.V. studio, basic 3 camera system, black & white,									
3800	electrical & electronic equip. only, minimum	4 Elec	.80	40	Total	12,700	2,050		14,750	17,100
4000	Maximum (full console)		.28	114		54,000	5,900		59,900	68,500
4100	As above, but color system, minimum		.28	114		71,500	5,900		77,400	87,500
4120	Maximum	↓	.12	266	↓	310,500	13,700		324,200	362,000
4200	For film chain, black & white, add	1 Elec	1	8	Ea.	14,500	410		14,910	16,600
4250	Color, add		.25	32		17,600	1,650		19,250	21,900
4400	For video recorders, add, minimum	↓	1	8		3,050	410		3,460	3,975
4600	Maximum	4 Elec	.40	80	↓	25,400	4,125		29,525	34,100

28 23 23 – Video Surveillance Systems Infrastructure

28 23 23.50 Video Surveillance Equipments

		Crew	Daily Output	Labor-Hours	Unit	Material	2012 Bare Costs Labor	2012 Bare Costs Equipment	Total	Total Incl O&P
0010	**VIDEO SURVEILLANCE EQUIPMENTS**									
0200	Video cameras, wireless, hidden in exit signs, clocks, etc, incl receiver	1 Elec	3	2.667	Ea.	106	137		243	320
0210	Accessories for video recorder, single camera		3	2.667		179	137		316	400
0220	For multiple cameras		3	2.667		1,725	137		1,862	2,100
0230	Video cameras, wireless, for under vehicle searching, complete	↓	2	4	↓	9,975	206		10,181	11,300

28 31 Fire Detection and Alarm

28 31 23 – Fire Detection and Alarm Annunciation Panels and Fire Stations

28 31 23.50 Alarm Panels and Devices

		Crew	Daily Output	Labor-Hours	Unit	Material	2012 Bare Costs Labor	2012 Bare Costs Equipment	Total	Total Incl O&P
0010	**ALARM PANELS AND DEVICES**, not including wires & conduits									
3594	Fire, alarm control panel									
3600	4 zone	2 Elec	2	8	Ea.	390	410		800	1,050
3800	8 zone		1	16		760	825		1,585	2,050
4000	12 zone	↓	.67	23.988		2,325	1,225		3,550	4,400
4020	Alarm device	1 Elec	8	1		231	51.50		282.50	330
4050	Actuating device		8	1		325	51.50		376.50	435
4200	Battery and rack		4	2		400	103		503	595
4400	Automatic charger		8	1		565	51.50		616.50	700
4600	Signal bell		8	1		71	51.50		122.50	155
4800	Trouble buzzer or manual station		8	1		81	51.50		132.50	166
5600	Strobe and horn		5.30	1.509		148	78		226	278
5800	Fire alarm horn		6.70	1.194		59	61.50		120.50	157
6000	Door holder, electro-magnetic		4	2		99.50	103		202.50	264
6200	Combination holder and closer		3.20	2.500		120	129		249	325
6600	Drill switch		8	1		360	51.50		411.50	470
6800	Master box		2.70	2.963		6,225	153		6,378	7,075
7000	Break glass station		8	1		65.50	51.50		117	149
7800	Remote annunciator, 8 zone lamp	↓	1.80	4.444		203	229		432	565
8000	12 zone lamp	2 Elec	2.60	6.154		325	315		640	835
8200	16 zone lamp	"	2.20	7.273	↓	405	375		780	1,000

28 31 43 – Fire Detection Sensors

28 31 43.50 Fire and Heat Detectors

		Crew	Daily Output	Labor-Hours	Unit	Material	2012 Bare Costs Labor	2012 Bare Costs Equipment	Total	Total Incl O&P
0010	**FIRE & HEAT DETECTORS**									
5000	Detector, rate of rise	1 Elec	8	1	Ea.	49.50	51.50		101	132

28 31 Fire Detection and Alarm

28 31 46 – Smoke Detection Sensors

28 31 46.50 Smoke Detectors	Crew	Daily Output	Labor-Hours	Unit	Material	2012 Bare Costs Labor	Equipment	Total	Total Incl O&P
0010 **SMOKE DETECTORS**									
5200 Smoke detector, ceiling type	1 Elec	6.20	1.290	Ea.	127	66.50		193.50	240
5400 Duct type	"	3.20	2.500	"	320	129		449	545

28 33 Gas Detection and Alarm

28 33 33 – Gas Detection Sensors

28 33 33.50 Tank Leak Detection Systems

28 33 33.50 Tank Leak Detection Systems	Crew	Daily Output	Labor-Hours	Unit	Material	2012 Bare Costs Labor	Equipment	Total	Total Incl O&P
0010 **TANK LEAK DETECTION SYSTEMS** Liquid and vapor									
0100 For hydrocarbons and hazardous liquids/vapors									
0120 Controller, data acquisition, incl. printer, modem, RS232 port									
0140 24 channel, for use with all probes				Ea.	4,050			4,050	4,450
0160 9 channel, for external monitoring				"	890			890	980
0200 Probes									
0210 Well monitoring									
0220 Liquid phase detection				Ea.	460			460	505
0230 Hydrocarbon vapor, fixed position					460			460	505
0240 Hydrocarbon vapor, float mounted					460			460	505
0250 Both liquid and vapor hydrocarbon					460			460	505
0300 Secondary containment, liquid phase									
0310 Pipe trench/manway sump				Ea.	800			800	880
0320 Double wall pipe and manual sump					800			800	880
0330 Double wall fiberglass annular space					315			315	345
0340 Double wall steel tank annular space					325			325	360
0500 Accessories									
0510 Modem, non-dedicated phone line				Ea.	294			294	325
0600 Monitoring, internal									
0610 Automatic tank gauge, incl. overfill				Ea.	1,125			1,125	1,225
0620 Product line				"	1,125			1,125	1,225
0700 Monitoring, special									
0710 Cathodic protection				Ea.	700			700	770
0720 Annular space chemical monitor				"	955			955	1,050

28 39 Mass Notification Systems

28 39 10 – Notification Systems

28 39 10.10 Mass Notification System

28 39 10.10 Mass Notification System	Crew	Daily Output	Labor-Hours	Unit	Material	2012 Bare Costs Labor	Equipment	Total	Total Incl O&P
0010 **MASS NOTIFICATION SYSTEM**									
0100 Wireless command center, 10,000 devices	2 Elec	1.33	12.030	Ea.	3,500	620		4,120	4,775
0200 Option, email notification					1,700			1,700	1,875
0210 Remote device supervision & monitor					2,375			2,375	2,600
0300 Antenna VHF or UHF, for medium range	1 Elec	4	2		125	103		228	292
0310 For high-power transmitter		2	4		650	206		856	1,025
0400 Transmitter, 25 watt		4	2		1,975	103		2,078	2,325
0410 40 watt		2.66	3.008		2,625	155		2,780	3,125
0420 100 watt		1.33	6.015		6,575	310		6,885	7,700
0500 Wireless receiver/control module for speaker		8	1		257	51.50		308.50	360
0600 Desktop paging controller, stand alone		4	2		360	103		463	550

Estimating Tips

31 05 00 Common Work Results for Earthwork

- Estimating the actual cost of performing earthwork requires careful consideration of the variables involved. This includes items such as type of soil, whether water will be encountered, dewatering, whether banks need bracing, disposal of excavated earth, and length of haul to fill or spoil sites, etc. If the project has large quantities of cut or fill, consider raising or lowering the site to reduce costs, while paying close attention to the effect on site drainage and utilities.

- If the project has large quantities of fill, creating a borrow pit on the site can significantly lower the costs.

- It is very important to consider what time of year the project is scheduled for completion. Bad weather can create large cost overruns from dewatering, site repair, and lost productivity from cold weather.

- New lines have been added for marine piling.

Reference Numbers

Reference numbers are shown in shaded boxes at the beginning of some major classifications. These numbers refer to related items in the Reference Section. The reference information may be an estimating procedure, an alternate pricing method, or technical information.

Note: Not all subdivisions listed here necessarily appear in this publication.

31 05 Common Work Results for Earthwork

31 05 13 – Soils for Earthwork

31 05 13.10 Borrow

	31 05 13.10 Borrow		Crew	Daily Output	Labor-Hours	Unit	Material	2012 Bare Costs Labor	Equipment	Total	Total Incl O&P
0010	**BORROW**	R312316-40									
0020	Spread, 200 H.P. dozer, no compaction, 2 mi. RT haul										
0200	Common borrow		B-15	600	.047	C.Y.	14.50	1.80	4.02	20.32	23
0700	Screened loam			600	.047		33.50	1.80	4.02	39.32	44
0800	Topsoil, weed free		↓	600	.047		23	1.80	4.02	28.82	32.50
0900	For 5 mile haul, add		B-34B	200	.040	↓		1.41	2.92	4.33	5.35

31 05 16 – Aggregates for Earthwork

31 05 16.10 Borrow

	31 05 16.10 Borrow		Crew	Daily Output	Labor-Hours	Unit	Material	2012 Bare Costs Labor	Equipment	Total	Total Incl O&P
0010	**BORROW**	R312316-40									
0020	Spread, with 200 H.P. dozer, no compaction, 2 mi. RT haul										
0100	Bank run gravel		B-15	600	.047	C.Y.	23	1.80	4.02	28.82	32
0300	Crushed stone (1.40 tons per CY) , 1-1/2"			600	.047		31.50	1.80	4.02	37.32	41.50
0320	3/4"			600	.047		31.50	1.80	4.02	37.32	41.50
0340	1/2"			600	.047		44	1.80	4.02	49.82	55.50
0360	3/8"			600	.047		32.50	1.80	4.02	38.32	43
0400	Sand, washed, concrete			600	.047		36.50	1.80	4.02	42.32	47
0500	Dead or bank sand			600	.047		17	1.80	4.02	22.82	26
0600	Select structural fill		↓	600	.047		21.50	1.80	4.02	27.32	30.50
0900	For 5 mile haul, add		B-34B	200	.040	↓		1.41	2.92	4.33	5.35

31 05 23 – Cement and Concrete for Earthwork

31 05 23.30 Plant Mixed Bituminous Concrete

	31 05 23.30 Plant Mixed Bituminous Concrete	Crew	Daily Output	Labor-Hours	Unit	Material	2012 Bare Costs Labor	Equipment	Total	Total Incl O&P
0010	**PLANT MIXED BITUMINOUS CONCRETE**									
0020	Asphaltic concrete plant mix (145 lb. per C.F.)				Ton	62			62	68
0040	Asphaltic concrete less than 300 tons add trucking costs									
0050	See Section 31 23 23.20 for hauling costs									
0200	All weather patching mix, hot				Ton	68			68	75
0250	Cold patch					68			68	75
0300	Berm mix				↓	68			68	75

31 06 Schedules for Earthwork

31 06 60 – Schedules for Special Foundations and Load Bearing Elements

31 06 60.14 Piling Special Costs

	31 06 60.14 Piling Special Costs	Crew	Daily Output	Labor-Hours	Unit	Material	2012 Bare Costs Labor	Equipment	Total	Total Incl O&P
0010	**PILING SPECIAL COSTS**									
0011	Piling special costs, pile caps, see Section 03 30 53.40									
0500	Cutoffs, concrete piles, plain	1 Pile	5.50	1.455	Ea.		62.50		62.50	100
0600	With steel thin shell, add		38	.211			9		9	14.50
0700	Steel pile or "H" piles		19	.421			18		18	29
0800	Wood piles	↓	38	.211	↓		9		9	14.50
0900	Pre-augering up to 30' deep, average soil, 24" diameter	B-43	180	.267	L.F.		10.30	13.55	23.85	30.50
0920	36" diameter		115	.417			16.10	21	37.10	48
0960	48" diameter		70	.686			26.50	35	61.50	79
0980	60" diameter	↓	50	.960	↓		37	49	86	110
1000	Testing, any type piles, test load is twice the design load									
1050	50 ton design load, 100 ton test				Ea.				14,000	15,500
1100	100 ton design load, 200 ton test								20,000	22,000
1150	150 ton design load, 300 ton test								26,000	28,500
1200	200 ton design load, 400 ton test								28,000	31,000
1250	400 ton design load, 800 ton test				↓				32,000	35,000
1500	Wet conditions, soft damp ground									
1600	Requiring mats for crane, add								40%	40%

31 06 Schedules for Earthwork

31 06 60 – Schedules for Special Foundations and Load Bearing Elements

31 06 60.14 Piling Special Costs

		Crew	Daily Output	Labor-Hours	Unit	Material	2012 Bare Costs Labor	2012 Bare Costs Equipment	Total	Total Incl O&P
1700	Barge mounted driving rig, add								30%	30%

31 06 60.15 Mobilization

		Crew	Daily Output	Labor-Hours	Unit	Material	2012 Bare Costs Labor	2012 Bare Costs Equipment	Total	Total Incl O&P
0010	**MOBILIZATION**									
0020	Set up & remove, air compressor, 600 CFM	A-5	3.30	5.455	Ea.		191	16	207	310
0100	1200 CFM	"	2.20	8.182			287	24	311	465
0200	Crane, with pile leads and pile hammer, 75 ton	B-19	.60	106			4,700	3,050	7,750	10,700
0300	150 ton	"	.36	177			7,850	5,100	12,950	17,900
0500	Drill rig, for caissons, to 36", minimum	B-43	2	24			925	1,225	2,150	2,775
0600	Up to 84"	"	1	48			1,850	2,450	4,300	5,500
0800	Auxiliary boiler, for steam small	A-5	1.66	10.843			380	32	412	620
0900	Large	"	.83	21.687			760	63.50	823.50	1,250
1100	Rule of thumb: complete pile driving set up, small	B-19	.45	142			6,275	4,075	10,350	14,300
1200	Large	"	.27	237			10,500	6,775	17,275	23,900
1500	Mobilization, barge, by tug boat	B-83	25	.640	Mile		26	30	56	73

31 11 Clearing and Grubbing

31 11 10 – Clearing and Grubbing Land

31 11 10.10 Clear and Grub Site

		Crew	Daily Output	Labor-Hours	Unit	Material	2012 Bare Costs Labor	2012 Bare Costs Equipment	Total	Total Incl O&P
0010	**CLEAR AND GRUB SITE**									
0020	Cut & chip light trees to 6" diam.	B-7	1	48	Acre		1,800	1,475	3,275	4,375
0150	Grub stumps and remove	B-30	2	12			470	1,075	1,545	1,875
0200	Cut & chip medium, trees to 12" diam.	B-7	.70	68.571			2,550	2,100	4,650	6,250
0250	Grub stumps and remove	B-30	1	24			935	2,125	3,060	3,775
0300	Cut & chip heavy, trees to 24" diam.	B-7	.30	160			5,975	4,925	10,900	14,600
0350	Grub stumps and remove	B-30	.50	48			1,875	4,250	6,125	7,525
0400	If burning is allowed, reduce cut & chip									40%
3000	Chipping stumps, to 18" deep, 12" diam.	B-86	20	.400	Ea.		18.60	8.55	27.15	37.50
3040	18" diameter		16	.500			23.50	10.65	34.15	47
3080	24" diameter		14	.571			26.50	12.20	38.70	53.50
3100	30" diameter		12	.667			31	14.25	45.25	62.50
3120	36" diameter		10	.800			37	17.05	54.05	75
3160	48" diameter		8	1			46.50	21.50	68	94
5000	Tree thinning, feller buncher, conifer									
5080	Up to 8" diameter	B-93	240	.033	Ea.		1.55	2.40	3.95	4.99
5120	12" diameter		160	.050			2.33	3.61	5.94	7.50
5240	Hardwood, up to 4" diameter		240	.033			1.55	2.40	3.95	4.99
5280	8" diameter		180	.044			2.07	3.21	5.28	6.65
5320	12" diameter		120	.067			3.10	4.81	7.91	10
7000	Tree removal, congested area, aerial lift truck									
7040	8" diameter	B-85	7	5.714	Ea.		214	133	347	470
7080	12" diameter		6	6.667			250	155	405	550
7120	18" diameter		5	8			299	186	485	660
7160	24" diameter		4	10			375	232	607	825
7240	36" diameter		3	13.333			500	310	810	1,100
7280	48" diameter		2	20			750	465	1,215	1,650

31 13 13 – Selective Tree and Shrub Removal

31 13 13.10 Selective Clearing	Crew	Daily Output	Labor-Hours	Unit	Material	2012 Bare Costs Labor	2012 Bare Costs Equipment	Total	Total Incl O&P
0010 **SELECTIVE CLEARING**									
0020 Clearing brush with brush saw	A-1C	.25	32	Acre		1,125	110	1,235	1,850
0100 By hand	1 Clab	.12	66.667			2,350		2,350	3,600
0300 With dozer, ball and chain, light clearing	B-11A	2	8			325	620	945	1,175
0400 Medium clearing		1.50	10.667			435	830	1,265	1,575
0500 With dozer and brush rake, light		10	1.600			65.50	124	189.50	237
0550 Medium brush to 4" diameter		8	2			81.50	155	236.50	295
0600 Heavy brush to 4" diameter		6.40	2.500			102	194	296	370
1000 Brush mowing, tractor w/rotary mower, no removal									
1020 Light density	B-84	2	4	Acre		186	172	358	470
1040 Medium density		1.50	5.333			248	229	477	625
1080 Heavy density		1	8			370	345	715	940

31 13 13.20 Selective Tree Removal

	Crew	Daily Output	Labor-Hours	Unit	Material	2012 Bare Costs Labor	2012 Bare Costs Equipment	Total	Total Incl O&P
0010 **SELECTIVE TREE REMOVAL**									
0011 With tractor, large tract, firm									
0020 level terrain, no boulders, less than 12" diam. trees									
0300 300 HP dozer, up to 400 trees/acre, 0 to 25% hardwoods	B-10M	.75	16	Acre		685	2,300	2,985	3,550
0340 25% to 50% hardwoods		.60	20			855	2,875	3,730	4,450
0370 75% to 100% hardwoods		.45	26.667			1,150	3,825	4,975	5,925
0400 500 trees/acre, 0% to 25% hardwoods		.60	20			855	2,875	3,730	4,450
0440 25% to 50% hardwoods		.48	25			1,075	3,575	4,650	5,550
0470 75% to 100% hardwoods		.36	33.333			1,425	4,775	6,200	7,400
0500 More than 600 trees/acre, 0 to 25% hardwoods		.52	23.077			985	3,300	4,285	5,125
0540 25% to 50% hardwoods		.42	28.571			1,225	4,100	5,325	6,350
0570 75% to 100% hardwoods		.31	38.710			1,650	5,550	7,200	8,600
0900 Large tract clearing per tree									
1500 300 HP dozer, to 12" diameter, softwood	B-10M	320	.038	Ea.		1.60	5.35	6.95	8.35
1550 Hardwood		100	.120			5.15	17.20	22.35	26.50
1600 12" to 24" diameter, softwood		200	.060			2.56	8.60	11.16	13.35
1650 Hardwood		80	.150			6.40	21.50	27.90	33
1700 24" to 36" diameter, softwood		100	.120			5.15	17.20	22.35	26.50
1750 Hardwood		50	.240			10.25	34.50	44.75	53.50
1800 36" to 48" diameter, softwood		70	.171			7.35	24.50	31.85	38
1850 Hardwood		35	.343			14.65	49	63.65	76
2000 Stump removal on site by hydraulic backhoe, 1-1/2 C.Y.									
2040 4" to 6" diameter	B-17	60	.533	Ea.		20	11.55	31.55	43
2050 8" to 12" diameter	B-30	33	.727			28.50	64.50	93	114
2100 14" to 24" diameter		25	.960			37.50	85	122.50	151
2150 26" to 36" diameter		16	1.500			58.50	133	191.50	235
3000 Remove selective trees, on site using chain saws and chipper,									
3050 not incl. stumps, up to 6" diameter	B-7	18	2.667	Ea.		99.50	82	181.50	244
3100 8" to 12" diameter		12	4			149	123	272	365
3150 14" to 24" diameter		10	4.800			179	148	327	435
3200 26" to 36" diameter		8	6			224	185	409	550
3300 Machine load, 2 mile haul to dump, 12" diam. tree	A-3B	8	2			82	132	214	269

31 14 Earth Stripping and Stockpiling

31 14 13 – Soil Stripping and Stockpiling

31 14 13.23 Topsoil Stripping and Stockpiling

	31 14 13.23 Topsoil Stripping and Stockpiling	Crew	Daily Output	Labor-Hours	Unit	Material	2012 Bare Costs Labor	Equipment	Total	Total Incl O&P
0010	**TOPSOIL STRIPPING AND STOCKPILING**									
0020	200 H.P. dozer, ideal conditions	B-10B	2300	.005	C.Y.		.22	.54	.76	.93
0100	Adverse conditions	"	1150	.010			.45	1.08	1.53	1.87
0200	300 H.P. dozer, ideal conditions	B-10M	3000	.004			.17	.57	.74	.89
0300	Adverse conditions	"	1650	.007			.31	1.04	1.35	1.61
0400	400 H.P. dozer, ideal conditions	B-10X	3900	.003			.13	.58	.71	.84
0500	Adverse conditions	"	2000	.006			.26	1.14	1.40	1.64
0600	Clay, dry and soft, 200 H.P. dozer, ideal conditions	B-10B	1600	.008			.32	.78	1.10	1.34
0700	Adverse conditions	"	800	.015			.64	1.55	2.19	2.68
1000	Medium hard, 300 H.P. dozer, ideal conditions	B-10M	2000	.006			.26	.86	1.12	1.33
1100	Adverse conditions	"	1100	.011			.47	1.56	2.03	2.43
1200	Very hard, 400 H.P. dozer, ideal conditions	B-10X	2600	.005			.20	.87	1.07	1.26
1300	Adverse conditions	"	1340	.009	▼		.38	1.70	2.08	2.45
1400	Loam or topsoil, remove and stockpile on site									
1420	6" deep, 200' haul	B-10B	865	.014	C.Y.		.59	1.44	2.03	2.48
1430	300' haul		520	.023			.99	2.39	3.38	4.13
1440	500' haul		225	.053	▼		2.28	5.50	7.78	9.55
1450	Alternate method: 6" deep, 200' haul		5090	.002	S.Y.		.10	.24	.34	.42
1460	500' haul		1325	.009	"		.39	.94	1.33	1.62

31 22 Grading

31 22 16 – Fine Grading

31 22 16.10 Finish Grading

	31 22 16.10 Finish Grading	Crew	Daily Output	Labor-Hours	Unit	Material	Labor	Equipment	Total	Total Incl O&P
0010	**FINISH GRADING**									
0012	Finish grading area to be paved with grader, small area	B-11L	400	.040	S.Y.		1.63	1.53	3.16	4.17
0100	Large area		2000	.008			.33	.31	.64	.84
1100	Fine grade for slab on grade, machine		1040	.015			.63	.59	1.22	1.61
1150	Hand grading	B-18	700	.034	▼		1.23	.06	1.29	1.95
3500	Finish grading lagoon bottoms	B-11L	4	4	M.S.F.		163	153	316	415

31 23 Excavation and Fill

31 23 16 – Excavation

31 23 16.13 Excavating, Trench

	31 23 16.13 Excavating, Trench	Crew	Daily Output	Labor-Hours	Unit	Material	Labor	Equipment	Total	Total Incl O&P
0010	**EXCAVATING, TRENCH**									
0011	Or continuous footing									
0020	Common earth with no sheeting or dewatering included									
0050	1' to 4' deep, 3/8 C.Y. excavator	B-11C	150	.107	B.C.Y.		4.36	2.23	6.59	9.10
0060	1/2 C.Y. excavator	B-11M	200	.080			3.27	1.82	5.09	6.95
0090	4' to 6' deep, 1/2 C.Y. excavator	"	200	.080			3.27	1.82	5.09	6.95
0100	5/8 C.Y. excavator	B-12Q	250	.064			2.65	2.21	4.86	6.45
0110	3/4 C.Y. excavator	B-12F	300	.053			2.21	2.16	4.37	5.75
0300	1/2 C.Y. excavator, truck mounted	B-12J	200	.080			3.32	4.22	7.54	9.70
0500	6' to 10' deep, 3/4 C.Y. excavator	B-12F	225	.071			2.95	2.88	5.83	7.65
0510	1 C.Y. excavator	B-12A	400	.040			1.66	1.87	3.53	4.58
0600	1 C.Y. excavator, truck mounted	B-12K	400	.040			1.66	2.40	4.06	5.15
0610	1-1/2 C.Y. excavator	B-12B	600	.027			1.11	1.60	2.71	3.44
0900	10' to 14' deep, 3/4 C.Y. excavator	B-12F	200	.080			3.32	3.24	6.56	8.60
0910	1 C.Y. excavator	B-12A	360	.044			1.84	2.08	3.92	5.10
1000	1-1/2 C.Y. excavator	B-12B	540	.030	▼		1.23	1.78	3.01	3.83

31 23 16 – Excavation

31 23 16.13 Excavating, Trench	Crew	Daily Output	Labor-Hours	Unit	Material	2012 Bare Costs Labor	Equipment	Total	Total Incl O&P
1300 14' to 20' deep, 1 C.Y. excavator	B-12A	320	.050	B.C.Y.		2.07	2.34	4.41	5.70
1310 1-1/2 C.Y. excavator	B-12B	480	.033			1.38	2	3.38	4.30
1320 2-1/2 C.Y. excavator	B-12S	765	.021			.87	2.16	3.03	3.70
1340 20' to 24' deep, 1 C.Y. excavator	B-12A	288	.056			2.30	2.60	4.90	6.35
1342 1-1/2 C.Y. excavator	B-12B	432	.037			1.54	2.22	3.76	4.79
1344 2-1/2 C.Y. excavator	B-12S	685	.023			.97	2.41	3.38	4.12
1352 4' to 6' deep, 1/2 C.Y. excavator w/trench box	B-13H	188	.085			3.53	5.10	8.63	10.95
1354 5/8 C.Y. excavator	"	235	.068			2.82	4.07	6.89	8.80
1356 3/4 C.Y. excavator	B-13G	282	.057			2.35	2.70	5.05	6.55
1362 6' to 10' deep, 3/4 C.Y. excavator w/trench box	"	212	.075			3.13	3.59	6.72	8.70
1370 1 C.Y. excavator	B-13D	376	.043			1.76	2.29	4.05	5.20
1371 1-1/2 C.Y. excavator	B-13E	564	.028			1.18	1.90	3.08	3.88
1374 10' to 14' deep, 3/4 C.Y. excavator w/trench box	B-13G	188	.085			3.53	4.05	7.58	9.80
1375 1 C.Y. excavator	B-13D	338	.047			1.96	2.55	4.51	5.80
1376 1-1/2 C.Y. excavator	B-13E	508	.032			1.31	2.11	3.42	4.32
1381 14' to 20' deep, 1 C.Y. excavator w/trench box	B-13D	301	.053			2.20	2.86	5.06	6.50
1382 1-1/2 C.Y. excavator	B-13E	451	.035			1.47	2.38	3.85	4.86
1383 2-1/2 C.Y. excavator	B-13J	720	.022			.92	2.45	3.37	4.10
1386 20' to 24' deep, 1 C.Y. excavator w/trench box	B-13D	271	.059			2.45	3.18	5.63	7.20
1387 1-1/2 C.Y. excavator	B-13E	406	.039			1.63	2.64	4.27	5.40
1388 2-1/2 C.Y. excavator	B-13J	645	.025			1.03	2.74	3.77	4.58
1391 Shoring by S.F./day trench wall protected loose mat., 4' W	B-6	3200	.008	SF Wall	.39	.29	.10	.78	.98
1392 Rent shoring per week per S.F. wall protected, loose mat., 4' W					1.07			1.07	1.18
1395 Hydraulic shoring, S.F. trench wall protected stable mat., 4' W	2 Clab	2700	.006		.17	.21		.38	.50
1397 semi-stable material, 4' W	"	2400	.007		.23	.23		.46	.61
1398 Rent hydraulic shoring per day/S.F. wall, stable mat., 4' W					.28			.28	.30
1399 semi-stable material					.34			.34	.37
1400 By hand with pick and shovel 2' to 6' deep, light soil	1 Clab	8	1	B.C.Y.		35		35	54
1500 Heavy soil	"	4	2	"		70		70	108
1700 For tamping backfilled trenches, air tamp, add	A-1G	100	.080	E.C.Y.		2.81	.50	3.31	4.87
1900 Vibrating plate, add	B-18	180	.133	"		4.77	.23	5	7.60
2100 Trim sides and bottom for concrete pours, common earth		1500	.016	S.F.		.57	.03	.60	.91
2300 Hardpan		600	.040	"		1.43	.07	1.50	2.27
2400 Pier and spread footing excavation, add to above				B.C.Y.				30%	30%
3000 Backfill trench, F.E. loader, wheel mtd., 1 C.Y. bucket									
3020 Minimal haul	B-10R	400	.030	L.C.Y.		1.28	.67	1.95	2.68
3040 100' haul	"	200	.060			2.56	1.34	3.90	5.35
3080 2-1/4 C.Y. bucket, minimum haul	B-10T	600	.020			.85	.79	1.64	2.17
3090 100' haul	"	300	.040			1.71	1.57	3.28	4.32
5020 Loam & Sandy clay with no sheeting or dewatering included									
5050 1' to 4' deep, 3/8 C.Y. tractor loader/backhoe	B-11C	162	.099	B.C.Y.		4.03	2.06	6.09	8.40
5060 1/2 C.Y. excavator	B-11M	216	.074			3.02	1.68	4.70	6.45
5080 4' to 6' deep, 1/2 C.Y. excavator	"	216	.074			3.02	1.68	4.70	6.45
5090 5/8 C.Y. excavator	B-12Q	276	.058			2.40	2	4.40	5.85
5100 3/4 C.Y. excavator	B-12F	324	.049			2.05	2	4.05	5.30
5130 1/2 C.Y. excavator, truck mounted	B-12J	216	.074			3.07	3.91	6.98	8.95
5140 6' to 10' deep, 3/4 C.Y. excavator	B-12F	243	.066			2.73	2.66	5.39	7.10
5150 1 C.Y. excavator	B-12A	432	.037			1.54	1.73	3.27	4.25
5160 1 C.Y. excavator, truck mounted	B-12K	432	.037			1.54	2.23	3.77	4.79
5170 1-1/2 C.Y. excavator	B-12B	648	.025			1.02	1.48	2.50	3.19
5190 10' to 14' deep, 3/4 C.Y. excavator	B-12F	216	.074			3.07	3	6.07	7.95
5200 1 C.Y. excavator	B-12A	389	.041			1.70	1.92	3.62	4.71
5210 1-1/2 C.Y. excavator	B-12B	583	.027			1.14	1.65	2.79	3.54

31 23 16 – Excavation

31 23 16.13 Excavating, Trench	Crew	Daily Output	Labor-Hours	Unit	Material	2012 Bare Costs Labor	2012 Bare Costs Equipment	2012 Bare Costs Total	Total Incl O&P
5250	14' to 20' deep, 1 C.Y. excavator	B-12A	346	.046	B.C.Y.		1.92	2.16	4.08	5.30
5260	1-1/2 C.Y. excavator	B-12B	518	.031			1.28	1.85	3.13	3.99
5270	2-1/2 C.Y. excavator	B-12S	826	.019			.80	2	2.80	3.42
5300	20' to 24' deep, 1 C.Y. excavator	B-12A	311	.051			2.13	2.41	4.54	5.90
5310	1-1/2 C.Y. excavator	B-12B	467	.034			1.42	2.06	3.48	4.42
5320	2-1/2 C.Y. excavator	B-12S	740	.022			.90	2.23	3.13	3.82
5352	4' to 6' deep, 1/2 C.Y. excavator w/trench box	B-13H	205	.078			3.24	4.67	7.91	10.05
5354	5/8 C.Y. excavator	"	257	.062			2.58	3.73	6.31	8.05
5356	3/4 C.Y. excavator	B-13G	308	.052			2.15	2.47	4.62	6
5362	6' to 10' deep, 3/4 C.Y. excavator w/trench box	"	231	.069			2.87	3.29	6.16	8
5364	1 C.Y. excavator	B-13D	410	.039			1.62	2.10	3.72	4.77
5366	1-1/2 C.Y. excavator	B-13E	616	.026			1.08	1.74	2.82	3.56
5370	10' to 14' deep, 3/4 C.Y. excavator w/trench box	B-13G	205	.078			3.24	3.71	6.95	9
5372	1 C.Y. excavator	B-13D	370	.043			1.79	2.33	4.12	5.30
5374	1-1/2 C.Y. excavator	B-13E	554	.029			1.20	1.94	3.14	3.95
5382	14' to 20' deep, 1 C.Y. excavator w/trench box	B-13D	329	.049			2.02	2.62	4.64	5.95
5384	1-1/2 C.Y. excavator	B-13E	492	.033			1.35	2.18	3.53	4.45
5386	2-1/2 C.Y. excavator	B-13J	780	.021			.85	2.26	3.11	3.78
5392	20' to 24' deep, 1 C.Y. excavator w/trench box	B-13D	295	.054			2.25	2.92	5.17	6.65
5394	1-1/2 C.Y. excavator	B-13E	444	.036			1.49	2.42	3.91	4.93
5396	2-1/2 C.Y. excavator	B-13J	695	.023			.95	2.54	3.49	4.24
6020	Sand & gravel with no sheeting or dewatering included									
6050	1' to 4' deep, 3/8 C.Y. excavator	B-11C	165	.097	B.C.Y.		3.96	2.02	5.98	8.25
6060	1/2 C.Y. excavator	B-11M	220	.073			2.97	1.65	4.62	6.35
6080	4' to 6' deep, 1/2 C.Y. excavator	"	220	.073			2.97	1.65	4.62	6.35
6090	5/8 C.Y. excavator	B-12Q	275	.058			2.41	2.01	4.42	5.90
6100	3/4 C.Y. excavator	B-12F	330	.048			2.01	1.96	3.97	5.20
6130	1/2 C.Y. excavator, truck mounted	B-12J	220	.073			3.01	3.84	6.85	8.80
6140	6' to 10' deep, 3/4 C.Y. excavator	B-12F	248	.065			2.67	2.61	5.28	6.95
6150	1 C.Y. excavator	B-12A	440	.036			1.51	1.70	3.21	4.16
6160	1 C.Y. excavator, truck mounted	B-12K	440	.036			1.51	2.18	3.69	4.69
6170	1-1/2 C.Y. excavator	B-12B	660	.024			1	1.46	2.46	3.13
6190	10' to 14' deep, 3/4 C.Y. excavator	B-12F	220	.073			3.01	2.94	5.95	7.85
6200	1 C.Y. excavator	B-12A	396	.040			1.67	1.89	3.56	4.63
6210	1-1/2 C.Y. excavator	B-12B	594	.027			1.12	1.62	2.74	3.48
6250	14' to 20' deep, 1 C.Y. excavator	B-12A	352	.045			1.88	2.13	4.01	5.20
6260	1-1/2 C.Y. excavator	B-12B	528	.030			1.26	1.82	3.08	3.91
6270	2-1/2 C.Y. excavator	B-12S	840	.019			.79	1.97	2.76	3.36
6300	20' to 24' deep, 1 C.Y. excavator	B-12A	317	.050			2.09	2.36	4.45	5.80
6310	1-1/2 C.Y. excavator	B-12B	475	.034			1.40	2.02	3.42	4.34
6320	2-1/2 C.Y. excavator	B-12S	755	.021			.88	2.19	3.07	3.75
6352	4' to 6' deep, 1/2 C.Y. excavator w/trench box	B-13H	209	.077			3.17	4.58	7.75	9.90
6354	5/8 C.Y. excavator	"	261	.061			2.54	3.67	6.21	7.90
6356	3/4 C.Y. excavator	B-13G	314	.051			2.11	2.42	4.53	5.85
6362	6' to 10' deep, 3/4 C.Y. excavator w/trench box	"	236	.068			2.81	3.22	6.03	7.85
6364	1 C.Y. excavator	B-13D	418	.038			1.59	2.06	3.65	4.68
6366	1-1/2 C.Y. excavator	B-13E	627	.026			1.06	1.71	2.77	3.49
6370	10' to 14' deep, 3/4 C.Y. excavator w/trench box	B-13G	209	.077			3.17	3.64	6.81	8.85
6372	1 C.Y. excavator	B-13D	376	.043			1.76	2.29	4.05	5.20
6374	1-1/2 C.Y. excavator	B-13E	564	.028			1.18	1.90	3.08	3.88
6382	14' to 20' deep, 1 C.Y. excavator w/trench box	B-13D	334	.048			1.99	2.58	4.57	5.85
6384	1-1/2 C.Y. excavator	B-13E	502	.032			1.32	2.14	3.46	4.36
6386	2-1/2 C.Y. excavator	B-13J	790	.020			.84	2.23	3.07	3.74

31 23 Excavation and Fill

31 23 16 – Excavation

31 23 16.13 Excavating, Trench

		Crew	Daily Output	Labor-Hours	Unit	Material	Labor	Equipment	Total	Total Incl O&P
							2012 Bare Costs			
6392	20' to 24' deep, 1 C.Y. excavator w/trench box	B-13D	301	.053	B.C.Y.		2.20	2.86	5.06	6.50
6394	1-1/2 C.Y. excavator	B-13E	452	.035			1.47	2.38	3.85	4.84
6396	2-1/2 C.Y. excavator	B-13J	710	.023	↓		.93	2.49	3.42	4.16
7020	Dense hard clay with no sheeting or dewatering included									
7050	1' to 4' deep, 3/8 C.Y. excavator	B-11C	132	.121	B.C.Y.		4.95	2.53	7.48	10.35
7060	1/2 C.Y. excavator	B-11M	176	.091			3.71	2.07	5.78	7.90
7080	4' to 6' deep, 1/2 C.Y. excavator	"	176	.091			3.71	2.07	5.78	7.90
7090	5/8 C.Y. excavator	B-12Q	220	.073			3.01	2.51	5.52	7.35
7100	3/4 C.Y. excavator	B-12F	264	.061			2.51	2.45	4.96	6.50
7130	1/2 C.Y. excavator, truck mounted	B-12J	176	.091			3.77	4.80	8.57	11.05
7140	6' to 10' deep, 3/4 C.Y. excavator	B-12F	198	.081			3.35	3.27	6.62	8.70
7150	1 C.Y. excavator	B-12A	352	.045			1.88	2.13	4.01	5.20
7160	1 C.Y. excavator, truck mounted	B-12K	352	.045			1.88	2.73	4.61	5.85
7170	1-1/2 C.Y. excavator	B-12B	528	.030			1.26	1.82	3.08	3.91
7190	10' to 14' deep, 3/4 C.Y. excavator	B-12F	176	.091			3.77	3.68	7.45	9.80
7200	1 C.Y. excavator	B-12A	317	.050			2.09	2.36	4.45	5.80
7210	1-1/2 C.Y. excavator	B-12B	475	.034			1.40	2.02	3.42	4.34
7250	14' to 20' deep, 1 C.Y. excavator	B-12A	282	.057			2.35	2.65	5	6.50
7260	1-1/2 C.Y. excavator	B-12B	422	.038			1.57	2.28	3.85	4.89
7270	2-1/2 C.Y. excavator	B-12S	675	.024			.98	2.45	3.43	4.19
7300	20' to 24' deep, 1 C.Y. excavator	B-12A	254	.063			2.61	2.95	5.56	7.20
7310	1-1/2 C.Y. excavator	B-12B	380	.042			1.75	2.53	4.28	5.45
7320	2-1/2 C.Y. excavator	B-12S	605	.026	↓		1.10	2.73	3.83	4.67

31 23 16.14 Excavating, Utility Trench

		Crew	Daily Output	Labor-Hours	Unit	Material	Labor	Equipment	Total	Total Incl O&P
0010	**EXCAVATING, UTILITY TRENCH**									
0011	Common earth									
0050	Trenching with chain trencher, 12 H.P., operator walking									
0100	4" wide trench, 12" deep	B-53	800	.010	L.F.		.45	.08	.53	.77
0150	18" deep		750	.011			.48	.08	.56	.81
0200	24" deep		700	.011			.51	.09	.60	.87
0300	6" wide trench, 12" deep		650	.012			.55	.10	.65	.94
0350	18" deep		600	.013			.60	.10	.70	1.02
0400	24" deep		550	.015			.65	.11	.76	1.11
0450	36" deep		450	.018			.80	.14	.94	1.35
0600	8" wide trench, 12" deep		475	.017			.75	.13	.88	1.29
0650	18" deep		400	.020			.90	.16	1.06	1.52
0700	24" deep		350	.023			1.02	.18	1.20	1.74
0750	36" deep	↓	300	.027	↓		1.19	.21	1.40	2.03
1000	Backfill by hand including compaction, add									
1050	4" wide trench, 12" deep	A-1G	800	.010	L.F.		.35	.06	.41	.61
1100	18" deep		530	.015			.53	.09	.62	.91
1150	24" deep		400	.020			.70	.12	.82	1.22
1300	6" wide trench, 12" deep		540	.015			.52	.09	.61	.90
1350	18" deep		405	.020			.69	.12	.81	1.21
1400	24" deep		270	.030			1.04	.18	1.22	1.80
1450	36" deep		180	.044			1.56	.28	1.84	2.70
1600	8" wide trench, 12" deep		400	.020			.70	.12	.82	1.22
1650	18" deep		265	.030			1.06	.19	1.25	1.84
1700	24" deep		200	.040			1.40	.25	1.65	2.43
1750	36" deep	↓	135	.059	↓		2.08	.37	2.45	3.61
2000	Chain trencher, 40 H.P. operator riding									
2050	6" wide trench and backfill, 12" deep	B-54	1200	.007	L.F.		.30	.26	.56	.74

31 23 Excavation and Fill

31 23 16 – Excavation

31 23 16.14 Excavating, Utility Trench

		Crew	Daily Output	Labor-Hours	Unit	Material	Labor	Equipment	Total	Total Incl O&P
							2012 Bare Costs			
2100	18" deep	B-54	1000	.008	L.F.		.36	.31	.67	.88
2150	24" deep		975	.008			.37	.32	.69	.90
2200	36" deep		900	.009			.40	.35	.75	.98
2250	48" deep		750	.011			.48	.42	.90	1.18
2300	60" deep		650	.012			.55	.48	1.03	1.36
2400	8" wide trench and backfill, 12" deep		1000	.008			.36	.31	.67	.88
2450	18" deep		950	.008			.38	.33	.71	.93
2500	24" deep		900	.009			.40	.35	.75	.98
2550	36" deep		800	.010			.45	.39	.84	1.11
2600	48" deep		650	.012			.55	.48	1.03	1.36
2700	12" wide trench and backfill, 12" deep		975	.008			.37	.32	.69	.90
2750	18" deep		860	.009			.42	.36	.78	1.03
2800	24" deep		800	.010			.45	.39	.84	1.11
2850	36" deep		725	.011			.49	.43	.92	1.22
3000	16" wide trench and backfill, 12" deep		835	.010			.43	.37	.80	1.06
3050	18" deep		750	.011			.48	.42	.90	1.18
3100	24" deep	▼	700	.011	▼		.51	.45	.96	1.26
3200	Compaction with vibratory plate, add								35%	35%
5100	Hand excavate and trim for pipe bells after trench excavation									
5200	8" pipe	1 Clab	155	.052	L.F.		1.81		1.81	2.79
5300	18" pipe	"	130	.062	"		2.16		2.16	3.32

31 23 16.16 Structural Excavation for Minor Structures

		Crew	Daily Output	Labor-Hours	Unit	Material	Labor	Equipment	Total	Total Incl O&P
0010	**STRUCTURAL EXCAVATION FOR MINOR STRUCTURES** R312316-40									
0015	Hand, pits to 6' deep, sandy soil	1 Clab	8	1	B.C.Y.		35		35	54
0100	Heavy soil or clay		4	2			70		70	108
0300	Pits 6' to 12' deep, sandy soil		5	1.600			56		56	86.50
0500	Heavy soil or clay		3	2.667			93.50		93.50	144
0700	Pits 12' to 18' deep, sandy soil		4	2			70		70	108
0900	Heavy soil or clay		2	4			140		140	216
1100	Hand loading trucks from stock pile, sandy soil		12	.667			23.50		23.50	36
1300	Heavy soil or clay	▼	8	1	▼		35		35	54
1500	For wet or muck hand excavation, add to above								50%	50%
6000	Machine excavation, for spread and mat footings, elevator pits,									
6001	and small building foundations									
6030	Common earth, hydraulic backhoe, 1/2 C.Y. bucket	B-12E	55	.291	B.C.Y.		12.05	6.75	18.80	26
6035	3/4 C.Y. bucket	B-12F	90	.178			7.35	7.20	14.55	19.10
6040	1 C.Y. bucket	B-12A	108	.148			6.15	6.95	13.10	17
6050	1-1/2 C.Y. bucket	B-12B	144	.111			4.61	6.65	11.26	14.35
6060	2 C.Y. bucket	B-12C	200	.080			3.32	6.25	9.57	11.95
6070	Sand and gravel, 3/4 C.Y. bucket	B-12F	100	.160			6.65	6.45	13.10	17.20
6080	1 C.Y. bucket	B-12A	120	.133			5.55	6.25	11.80	15.25
6090	1-1/2 C.Y. bucket	B-12B	160	.100			4.15	6	10.15	12.90
6100	2 C.Y. bucket	B-12C	220	.073			3.01	5.70	8.71	10.85
6110	Clay, till, or blasted rock, 3/4 C.Y. bucket	B-12F	80	.200			8.30	8.10	16.40	21.50
6120	1 C.Y. bucket	B-12A	95	.168			7	7.90	14.90	19.25
6130	1-1/2 C.Y. bucket	B-12B	130	.123			5.10	7.40	12.50	15.90
6140	2 C.Y. bucket	B-12C	175	.091			3.79	7.15	10.94	13.65
6230	Sandy clay & loam, hydraulic backhoe, 1/2 C.Y. bucket	B-12E	60	.267			11.05	6.20	17.25	23.50
6235	3/4 C.Y. bucket	B-12F	98	.163			6.75	6.60	13.35	17.55
6240	1 C.Y. bucket	B-12A	116	.138			5.70	6.45	12.15	15.80
6250	1-1/2 C.Y. bucket	B-12B	156	.103	▼		4.25	6.15	10.40	13.20
9010	For mobilization or demobilization, see Section 01 54 36.50									

31 23 Excavation and Fill

31 23 16 – Excavation

31 23 16.16 Structural Excavation for Minor Structures	Crew	Daily Output	Labor-Hours	Unit	Material	2012 Bare Costs Labor	2012 Bare Costs Equipment	Total	Total Incl O&P
9020	For dewatering, see Section 31 23 19.20									
9022	For larger structures, see Bulk Excavation, Section 31 23 16.42									
9024	For loading onto trucks, add								15%	15%
9026	For hauling, see Section 31 23 23.20									
9030	For sheeting or soldier bms/lagging, see Section 31 52 16.10									
9040	For trench excavation of strip ftgs, see Section 31 23 16.13									

31 23 16.26 Rock Removal

		Crew	Daily Output	Labor-Hours	Unit	Material	Labor	Equipment	Total	Total Incl O&P
0010	**ROCK REMOVAL** R312316-40									
0015	Drilling only rock, 2" hole for rock bolts	B-47	316	.076	L.F.		2.96	4.54	7.50	9.50
0800	2-1/2" hole for pre-splitting		600	.040			1.56	2.39	3.95	5
4600	Quarry operations, 2-1/2" to 3-1/2" diameter		715	.034			1.31	2.01	3.32	4.21

31 23 16.30 Drilling and Blasting Rock

		Crew	Daily Output	Labor-Hours	Unit	Material	Labor	Equipment	Total	Total Incl O&P
0010	**DRILLING AND BLASTING ROCK**									
0020	Rock, open face, under 1500 C.Y.	B-47	225	.107	B.C.Y.	3.05	4.16	6.40	13.61	16.70
0100	Over 1500 C.Y.		300	.080		3.05	3.12	4.78	10.95	13.35
0200	Areas where blasting mats are required, under 1500 C.Y.		175	.137		3.05	5.35	8.20	16.60	20.50
0250	Over 1500 C.Y.		250	.096		3.05	3.74	5.75	12.54	15.35
0300	Bulk drilling and blasting, can vary greatly, average								9.50	12
0500	Pits, average								25	31
1300	Deep hole method, up to 1500 C.Y.	B-47	50	.480		3.05	18.70	28.50	50.25	63.50
1400	Over 1500 C.Y.		66	.364		3.05	14.15	21.50	38.70	49
1900	Restricted areas, up to 1500 C.Y.		13	1.846		3.05	72	110	185.05	234
2000	Over 1500 C.Y.		20	1.200		3.05	47	71.50	121.55	154
2200	Trenches, up to 1500 C.Y.		22	1.091		8.85	42.50	65	116.35	146
2300	Over 1500 C.Y.		26	.923		8.85	36	55	99.85	125
2500	Pier holes, up to 1500 C.Y.		22	1.091		3.05	42.50	65	110.55	140
2600	Over 1500 C.Y.		31	.774		3.05	30	46.50	79.55	100
2800	Boulders under 1/2 C.Y., loaded on truck, no hauling	B-100	80	.150			6.40	11.20	17.60	22
2900	Boulders, drilled, blasted	B-47	100	.240		3.05	9.35	14.35	26.75	33.50
3100	Jackhammer operators with foreman compressor, air tools	B-9	1	40	Day		1,425	209	1,634	2,400
3300	Track drill, compressor, operator and foreman	B-47	1	24	"		935	1,425	2,360	3,000
3500	Blasting caps				Ea.	5.65			5.65	6.20
3700	Explosives					.40			.40	.44
3900	Blasting mats, rent, for first day					135			135	149
4000	Per added day					45			45	49.50
4200	Preblast survey for 6 room house, individual lot, minimum	A-6	2.40	6.667			294	33	327	485
4300	Maximum	"	1.35	11.852			520	58.50	578.50	865
4500	City block within zone of influence, minimum	A-8	25200	.001	S.F.		.06		.06	.09
4600	Maximum	"	15100	.002	"		.10	.01	.11	.16

31 23 16.42 Excavating, Bulk Bank Measure

		Crew	Daily Output	Labor-Hours	Unit	Material	Labor	Equipment	Total	Total Incl O&P
0010	**EXCAVATING, BULK BANK MEASURE** R312316-40									
0011	Common earth piled									
0020	For loading onto trucks, add								15%	15%
0050	For mobilization and demobilization, see Section 01 54 36.50 R312316-45									
0100	For hauling, see Section 31 23 23.20									
0200	Excavator, hydraulic, crawler mtd., 1 C.Y. cap. = 100 C.Y./hr.	B-12A	800	.020	B.C.Y.		.83	.94	1.77	2.29
0250	1-1/2 C.Y. cap. = 125 C.Y./hr.	B-12B	1000	.016			.66	.96	1.62	2.07
0260	2 C.Y. cap. = 165 C.Y./hr.	B-12C	1320	.012			.50	.95	1.45	1.80
0300	3 C.Y. cap. = 260 C.Y./hr.	B-12D	2080	.008			.32	1.06	1.38	1.66
0305	3-1/2 C.Y. cap. = 300 C.Y./hr.	"	2400	.007			.28	.92	1.20	1.43
0310	Wheel mounted, 1/2 C.Y. cap. = 40 C.Y./hr.	B-12E	320	.050			2.07	1.16	3.23	4.43
0360	3/4 C.Y. cap. = 60 C.Y./hr.	B-12F	480	.033			1.38	1.35	2.73	3.58

31 23 16 – Excavation

31 23 16.42 Excavating, Bulk Bank Measure	Crew	Daily Output	Labor-Hours	Unit	Material	2012 Bare Costs Labor	Equipment	Total	Total Incl O&P
0500	Clamshell, 1/2 C.Y. cap. = 20 C.Y./hr.	B-12G	160	.100	B.C.Y.		4.15	4.28	8.43	11
0550	1 C.Y. cap. = 35 C.Y./hr.	B-12H	280	.057			2.37	4.13	6.50	8.15
0950	Dragline, 1/2 C.Y. cap. = 30 C.Y./hr.	B-12I	240	.067			2.76	3.55	6.31	8.10
1000	3/4 C.Y. cap. = 35 C.Y./hr.	"	280	.057			2.37	3.04	5.41	6.95
1050	1-1/2 C.Y. cap. = 65 C.Y./hr.	B-12P	520	.031			1.28	2.21	3.49	4.37
1200	Front end loader, track mtd., 1-1/2 C.Y. cap. = 70 C.Y./hr.	B-10N	560	.021			.92	.87	1.79	2.35
1250	2-1/2 C.Y. cap. = 95 C.Y./hr.	B-100	760	.016			.67	1.18	1.85	2.32
1300	3 C.Y. cap. = 130 C.Y./hr.	B-10P	1040	.012			.49	1.06	1.55	1.92
1350	5 C.Y. cap. = 160 C.Y./hr.	B-10Q	1280	.009			.40	1.13	1.53	1.85
1500	Wheel mounted, 3/4 C.Y. cap. = 45 C.Y./hr.	B-10R	360	.033			1.42	.74	2.16	2.98
1550	1-1/2 C.Y. cap. = 80 C.Y./hr.	B-10S	640	.019			.80	.52	1.32	1.79
1600	2-1/4 C.Y. cap. = 100 C.Y./hr.	B-10T	800	.015			.64	.59	1.23	1.62
1650	5 C.Y. cap. = 185 C.Y./hr.	B-10U	1480	.008			.35	.65	1	1.24
1800	Hydraulic excavator, truck mtd. 1/2 C.Y. = 30 C.Y./hr.	B-12J	240	.067			2.76	3.52	6.28	8.10
1850	48 inch bucket, 1 C.Y. = 45 C.Y./hr.	B-12K	360	.044			1.84	2.67	4.51	5.75
3700	Shovel, 1/2 C.Y. capacity = 55 C.Y./hr.	B-12L	440	.036			1.51	1.60	3.11	4.05
3750	3/4 C.Y. capacity = 85 C.Y./hr.	B-12M	680	.024			.98	1.32	2.30	2.93
3800	1 C.Y. capacity = 120 C.Y./hr.	B-12N	960	.017			.69	1.23	1.92	2.40
3850	1-1/2 C.Y. capacity = 160 C.Y./hr.	B-120	1280	.013			.52	.93	1.45	1.81
3900	3 C.Y. cap. = 250 C.Y./hr.	B-12T	2000	.008			.33	.76	1.09	1.34
4000	For soft soil or sand, deduct								15%	15%
4100	For heavy soil or stiff clay, add								60%	60%
4200	For wet excavation with clamshell or dragline, add								100%	100%
4250	All other equipment, add								50%	50%
4400	Clamshell in sheeting or cofferdam, minimum	B-12H	160	.100			4.15	7.25	11.40	14.25
4450	Maximum	"	60	.267			11.05	19.30	30.35	38
5000	Excavating, bulk bank measure, sandy clay & loam piled									
5020	For loading onto trucks, add								15%	15%
5100	Excavator, hydraulic, crawler mtd., 1 C.Y. cap. = 120 C.Y./hr.	B-12A	960	.017	B.C.Y.		.69	.78	1.47	1.91
5150	1-1/2 C.Y. cap. = 150 C.Y./hr.	B-12B	1200	.013			.55	.80	1.35	1.72
5300	2 C.Y. cap. = 195 C.Y./hr.	B-12C	1560	.010			.43	.80	1.23	1.53
5400	3 C.Y. cap. = 300 C.Y./hr.	B-12D	2400	.007			.28	.92	1.20	1.43
5500	3.5 C.Y. cap. = 350 C.Y./hr.	"	2800	.006			.24	.79	1.03	1.23
5610	Wheel mounted, 1/2 C.Y. cap. = 44 C.Y./hr.	B-12E	352	.045			1.88	1.05	2.93	4.03
5660	3/4 C.Y. cap. = 66 C.Y./hr.	B-12F	528	.030			1.26	1.23	2.49	3.26
8000	For hauling excavated material, see Section 31 23 23.20									

31 23 16.46 Excavating, Bulk, Dozer

		Crew	Daily Output	Labor-Hours	Unit	Material	Labor	Equipment	Total	Total Incl O&P
0010	**EXCAVATING, BULK, DOZER**									
0011	Open site									
2000	80 H.P., 50' haul, sand & gravel	B-10L	460	.026	B.C.Y.		1.11	.96	2.07	2.75
2010	Sandy clay & loam		440	.027			1.17	1.01	2.18	2.88
2020	Common earth		400	.030			1.28	1.11	2.39	3.16
2040	Clay		250	.048			2.05	1.77	3.82	5.05
2200	150' haul, sand & gravel		230	.052			2.23	1.92	4.15	5.50
2210	Sandy clay & loam		220	.055			2.33	2.01	4.34	5.75
2220	Common earth		200	.060			2.56	2.21	4.77	6.30
2240	Clay		125	.096			4.10	3.54	7.64	10.10
2400	300' haul, sand & gravel		120	.100			4.27	3.69	7.96	10.55
2410	Sandy clay & loam		115	.104			4.46	3.85	8.31	11
2420	Common earth		100	.120			5.15	4.42	9.57	12.65
2440	Clay		65	.185			7.90	6.80	14.70	19.45
3000	105 H.P., 50' haul, sand & gravel	B-10W	700	.017			.73	.79	1.52	1.98

31 23 16 – Excavation

31 23 16.46 Excavating, Bulk, Dozer	Crew	Daily Output	Labor-Hours	Unit	Material	2012 Bare Costs Labor	2012 Bare Costs Equipment	Total	Total Incl O&P
3010 Sandy clay & loam	B-10W	680	.018	B.C.Y.		.75	.81	1.56	2.03
3020 Common earth		610	.020			.84	.90	1.74	2.27
3040 Clay		385	.031			1.33	1.43	2.76	3.59
3200 150' haul, sand & gravel		310	.039			1.65	1.78	3.43	4.46
3210 Sandy clay & loam		300	.040			1.71	1.84	3.55	4.61
3220 Common earth		270	.044			1.90	2.04	3.94	5.10
3240 Clay		170	.071			3.02	3.24	6.26	8.15
3300 300' haul, sand & gravel		140	.086			3.66	3.93	7.59	9.90
3310 Sandy clay & loam		135	.089			3.80	4.08	7.88	10.25
3320 Common earth		120	.100			4.27	4.59	8.86	11.55
3340 Clay	▼	100	.120			5.15	5.50	10.65	13.85
4000 200 H.P., 50' haul, sand & gravel	B-10B	1400	.009			.37	.89	1.26	1.54
4010 Sandy clay & loam		1360	.009			.38	.91	1.29	1.58
4020 Common earth		1230	.010			.42	1.01	1.43	1.74
4040 Clay		770	.016			.67	1.61	2.28	2.79
4200 150' haul, sand & gravel		595	.020			.86	2.09	2.95	3.61
4210 Sandy clay & loam		580	.021			.88	2.14	3.02	3.70
4220 Common earth		516	.023			.99	2.41	3.40	4.16
4240 Clay		325	.037			1.58	3.82	5.40	6.60
4400 300' haul, sand & gravel		310	.039			1.65	4.01	5.66	6.90
4410 Sandy clay & loam		300	.040			1.71	4.14	5.85	7.15
4420 Common earth		270	.044			1.90	4.60	6.50	7.95
4440 Clay	▼	170	.071			3.02	7.30	10.32	12.65
5000 300 H.P., 50' haul, sand & gravel	B-10M	1900	.006			.27	.90	1.17	1.41
5010 Sandy clay & loam		1850	.006			.28	.93	1.21	1.44
5020 Common earth		1650	.007			.31	1.04	1.35	1.61
5040 Clay		1025	.012			.50	1.68	2.18	2.60
5200 150' haul, sand & gravel		920	.013			.56	1.87	2.43	2.90
5210 Sandy clay & loam		895	.013			.57	1.92	2.49	2.98
5220 Common earth		800	.015			.64	2.15	2.79	3.33
5240 Clay		500	.024			1.03	3.44	4.47	5.35
5400 300' haul, sand & gravel		470	.026			1.09	3.66	4.75	5.70
5410 Sandy clay & loam		455	.026			1.13	3.78	4.91	5.85
5420 Common earth		410	.029			1.25	4.19	5.44	6.50
5440 Clay	▼	250	.048			2.05	6.85	8.90	10.65
5500 460 H.P., 50' haul, sand & gravel	B-10X	1930	.006			.27	1.18	1.45	1.70
5506 Sandy clay & loam		1880	.006			.27	1.21	1.48	1.74
5510 Common earth		1680	.007			.31	1.35	1.66	1.95
5520 Clay		1050	.011			.49	2.16	2.65	3.12
5530 150' haul, sand & gravel		1290	.009			.40	1.76	2.16	2.54
5535 Sandy clay & loam		1250	.010			.41	1.82	2.23	2.62
5540 Common earth		1120	.011			.46	2.03	2.49	2.92
5550 Clay		700	.017			.73	3.25	3.98	4.68
5560 300' haul, sand & gravel		660	.018			.78	3.44	4.22	4.97
5565 Sandy clay & loam		640	.019			.80	3.55	4.35	5.15
5570 Common earth		575	.021			.89	3.95	4.84	5.70
5580 Clay	▼	350	.034			1.47	6.50	7.97	9.35
6000 700 H.P., 50' haul, sand & gravel	B-10V	3500	.003			.15	1.21	1.36	1.55
6006 Sandy clay & loam		3400	.004			.15	1.25	1.40	1.60
6010 Common earth		3035	.004			.17	1.39	1.56	1.79
6020 Clay		1925	.006			.27	2.20	2.47	2.82
6030 150' haul, sand & gravel		2025	.006			.25	2.09	2.34	2.68
6035 Sandy clay & loam	▼	1960	.006			.26	2.16	2.42	2.78

31 23 16.46 Excavating, Bulk, Dozer

		Crew	Daily Output	Labor-Hours	Unit	Material	Labor	Equipment	Total	Total Incl O&P
6040	Common earth	B-10V	1750	.007	B.C.Y.		.29	2.42	2.71	3.10
6050	Clay		1100	.011			.47	3.85	4.32	4.95
6060	300' haul, sand & gravel		1030	.012			.50	4.11	4.61	5.30
6065	Sandy clay & loam		1005	.012			.51	4.21	4.72	5.40
6070	Common earth		900	.013			.57	4.70	5.27	6
6080	Clay		550	.022			.93	7.70	8.63	9.85

31 23 16.50 Excavation, Bulk, Scrapers

		Crew	Daily Output	Labor-Hours	Unit	Material	Labor	Equipment	Total	Total Incl O&P
0010	**EXCAVATION, BULK, SCRAPERS** R312316-40									
0100	Elev. scraper 11 C.Y., sand & gravel 1500' haul, 1/4 dozer	B-33F	690	.020	B.C.Y.		.88	2.22	3.10	3.78
0150	3000' haul		610	.023			.99	2.52	3.51	4.28
0200	5000' haul		505	.028			1.20	3.04	4.24	5.15
0300	Common earth, 1500' haul		600	.023			1.01	2.56	3.57	4.34
0350	3000' haul		530	.026			1.14	2.90	4.04	4.92
0400	5000' haul		440	.032			1.38	3.49	4.87	5.95
0410	Sandy clay & loam, 1500' haul		648	.022			.93	2.37	3.30	4.02
0420	3000' haul		572	.024			1.06	2.68	3.74	4.56
0430	5000' haul		475	.029			1.28	3.23	4.51	5.50
0500	Clay, 1500' haul		375	.037			1.62	4.09	5.71	6.95
0550	3000' haul		330	.042			1.84	4.65	6.49	7.90
0600	5000' haul		275	.051			2.20	5.60	7.80	9.50
1000	Self propelled scraper, 14 C.Y. 1/4 push dozer, sand									
1050	Sand and gravel, 1500' haul	B-33D	920	.015	B.C.Y.		.66	2.14	2.80	3.35
1100	3000' haul		805	.017			.75	2.44	3.19	3.83
1200	5000' haul		645	.022			.94	3.05	3.99	4.77
1300	Common earth, 1500' haul		800	.018			.76	2.46	3.22	3.85
1350	3000' haul		700	.020			.87	2.81	3.68	4.40
1400	5000' haul		560	.025			1.08	3.51	4.59	5.50
1420	Sandy clay & loam, 1500' haul		864	.016			.70	2.28	2.98	3.56
1430	3000' haul		786	.018			.77	2.50	3.27	3.92
1440	5000' haul		605	.023			1	3.25	4.25	5.10
1500	Clay, 1500' haul		500	.028			1.21	3.93	5.14	6.15
1550	3000' haul		440	.032			1.38	4.47	5.85	7
1600	5000' haul		350	.040			1.73	5.60	7.33	8.80
2000	21 C.Y., 1/4 push dozer, sand & gravel, 1500' haul	B-33E	1180	.012			.51	2.04	2.55	3.02
2100	3000' haul		910	.015			.67	2.64	3.31	3.91
2200	5000' haul		750	.019			.81	3.21	4.02	4.75
2300	Common earth, 1500' haul		1030	.014			.59	2.33	2.92	3.46
2350	3000' haul		790	.018			.77	3.04	3.81	4.51
2400	5000' haul		650	.022			.93	3.70	4.63	5.50
2420	Sandy clay & loam, 1500' haul		1112	.013			.54	2.16	2.70	3.21
2430	3000' haul		854	.016			.71	2.81	3.52	4.18
2440	5000' haul		702	.020			.86	3.42	4.28	5.10
2500	Clay, 1500' haul		645	.022			.94	3.73	4.67	5.50
2550	3000' haul		495	.028			1.22	4.86	6.08	7.20
2600	5000' haul		405	.035			1.50	5.95	7.45	8.80
2700	Towed, 10 C.Y., 1/4 push dozer, sand & gravel, 1500' haul	B-33B	560	.025			1.08	4.08	5.16	6.15
2720	3000' haul		450	.031			1.35	5.10	6.45	7.65
2730	5000' haul		365	.038			1.66	6.25	7.91	9.40
2750	Common earth, 1500' haul		420	.033			1.44	5.45	6.89	8.20
2770	3000' haul		400	.035			1.51	5.70	7.21	8.60
2780	5000' haul		310	.045			1.95	7.40	9.35	11.05
2785	Sandy clay & Loam, 1500' haul		454	.031			1.33	5.05	6.38	7.55

31 23 16 – Excavation

	31 23 16.50 Excavation, Bulk, Scrapers	Crew	Daily Output	Labor-Hours	Unit	Material	2012 Bare Costs Labor	Equipment	Total	Total Incl O&P
2790	3000' haul	B-33B	432	.032	B.C.Y.		1.40	5.30	6.70	7.95
2795	5000' haul		340	.041			1.78	6.75	8.53	10.10
2800	Clay, 1500' haul		315	.044			1.92	7.25	9.17	10.90
2820	3000' haul		300	.047			2.02	7.60	9.62	11.45
2840	5000' haul	↓	225	.062			2.69	10.15	12.84	15.30
2900	15 C.Y., 1/4 push dozer, sand & gravel, 1500' haul	B-33C	800	.018			.76	2.88	3.64	4.32
2920	3000' haul		640	.022			.95	3.60	4.55	5.40
2940	5000' haul		520	.027			1.17	4.43	5.60	6.65
2960	Common earth, 1500' haul		600	.023			1.01	3.84	4.85	5.75
2980	3000' haul		560	.025			1.08	4.11	5.19	6.15
3000	5000' haul		440	.032			1.38	5.25	6.63	7.85
3005	Sandy clay & Loam, 1500' haul		648	.022			.93	3.56	4.49	5.35
3010	3000' haul		605	.023			1	3.81	4.81	5.70
3015	5000' haul		475	.029			1.28	4.85	6.13	7.30
3020	Clay, 1500' haul		450	.031			1.35	5.10	6.45	7.70
3040	3000' haul		420	.033			1.44	5.50	6.94	8.25
3060	5000' haul	↓	320	.044	↓		1.89	7.20	9.09	10.75

31 23 19 – Dewatering

31 23 19.20 Dewatering Systems

		Crew	Daily Output	Labor-Hours	Unit	Material	Labor	Equipment	Total	Total Incl O&P
0010	**DEWATERING SYSTEMS**									
0020	Excavate drainage trench, 2' wide, 2' deep	B-11C	90	.178	C.Y.		7.25	3.71	10.96	15.15
0100	2' wide, 3' deep, with backhoe loader	"	135	.119			4.84	2.47	7.31	10.05
0200	Excavate sump pits by hand, light soil	1 Clab	7.10	1.127			39.50		39.50	61
0300	Heavy soil	"	3.50	2.286	↓		80		80	123
0500	Pumping 8 hr., attended 2 hrs. per day, including 20 L.F.									
0550	of suction hose & 100 L.F. discharge hose									
0600	2" diaphragm pump used for 8 hours	B-10H	4	3	Day		128	18.50	146.50	215
0650	4" diaphragm pump used for 8 hours	B-10I	4	3			128	31	159	228
0800	8 hrs. attended, 2" diaphragm pump	B-10H	1	12			515	74	589	860
0900	3" centrifugal pump	B-10J	1	12			515	83	598	870
1000	4" diaphragm pump	B-10I	1	12			515	124	639	915
1100	6" centrifugal pump	B-10K	1	12	↓		515	345	860	1,150
1300	CMP, incl. excavation 3' deep, 12" diameter	B-6	115	.209	L.F.	10.50	8	2.91	21.41	27
1400	18" diameter		100	.240	"	15.75	9.20	3.34	28.29	35
1600	Sump hole construction, incl. excavation and gravel, pit		1250	.019	C.F.	1	.74	.27	2.01	2.51
1700	With 12" gravel collar, 12" pipe, corrugated, 16 ga.		70	.343	L.F.	24	13.15	4.77	41.92	52
1800	15" pipe, corrugated, 16 ga.		55	.436		31	16.70	6.05	53.75	66
1900	18" pipe, corrugated, 16 ga.		50	.480		36	18.40	6.70	61.10	75
2000	24" pipe, corrugated, 14 ga.		40	.600	↓	43.50	23	8.35	74.85	91.50
2200	Wood lining, up to 4' x 4', add	↓	300	.080	SFCA	16.65	3.07	1.11	20.83	24.50
9950	See Section 31 23 19.40 for wellpoints									
9960	See Section 31 23 19.30 for deep well systems									

31 23 19.30 Wells

		Crew	Daily Output	Labor-Hours	Unit	Material	Labor	Equipment	Total	Total Incl O&P
0010	**WELLS**									
0011	For dewatering 10' to 20' deep, 2' diameter									
0020	with steel casing, minimum	B-6	165	.145	V.L.F.	37	5.55	2.02	44.57	51
0050	Average		98	.245		42	9.40	3.41	54.81	64
0100	Maximum	↓	49	.490	↓	47.50	18.75	6.80	73.05	88
0300	For dewatering pumps see 01 54 33 in Reference Section									
0500	For domestic water wells, see Section 33 21 13.10									

31 23 Excavation and Fill

31 23 19 – Dewatering

31 23 19.40 Wellpoints

	Crew	Daily Output	Labor-Hours	Unit	Material	2012 Bare Costs Labor	2012 Bare Costs Equipment	Total	Total Incl O&P
0010 **WELLPOINTS** R312319-90									
0011 For equipment rental, see 01 54 33 in Reference Section									
0100 Installation and removal of single stage system									
0110 Labor only, .75 labor-hours per L.F.	1 Clab	10.70	.748	LF Hdr		26		26	40.50
0200 2.0 labor-hours per L.F.	"	4	2	"		70		70	108
0400 Pump operation, 4 @ 6 hr. shifts									
0410 Per 24 hour day	4 Eqlt	1.27	25.197	Day		1,125		1,125	1,700
0500 Per 168 hour week, 160 hr. straight, 8 hr. double time		.18	177	Week		7,950		7,950	12,000
0550 Per 4.3 week month		.04	800	Month		35,800		35,800	54,000
0600 Complete installation, operation, equipment rental, fuel &									
0610 removal of system with 2" wellpoints 5' O.C.									
0700 100' long header, 6" diameter, first month	4 Eqlt	3.23	9.907	LF Hdr	155	445		600	840
0800 Thereafter, per month		4.13	7.748		124	345		469	660
1000 200' long header, 8" diameter, first month		6	5.333		140	239		379	515
1100 Thereafter, per month		8.39	3.814		69.50	171		240.50	335
1300 500' long header, 8" diameter, first month		10.63	3.010		54	135		189	263
1400 Thereafter, per month		20.91	1.530		38.50	68.50		107	146
1600 1,000' long header, 10" diameter, first month		11.62	2.754		46.50	123		169.50	237
1700 Thereafter, per month		41.81	.765		23	34.50		57.50	77
1900 Note: above figures include pumping 168 hrs. per week									
1910 and include the pump operator and one stand-by pump.									

31 23 23 – Fill

31 23 23.13 Backfill

	Crew	Daily Output	Labor-Hours	Unit	Material	2012 Bare Costs Labor	2012 Bare Costs Equipment	Total	Total Incl O&P
0010 **BACKFILL** R312323-30									
0015 By hand, no compaction, light soil	1 Clab	14	.571	L.C.Y.		20		20	31
0100 Heavy soil		11	.727	"		25.50		25.50	39.50
0300 Compaction in 6" layers, hand tamp, add to above		20.60	.388	E.C.Y.		13.65		13.65	21
0400 Roller compaction operator walking, add	B-10A	100	.120			5.15	1.72	6.87	9.70
0500 Air tamp, add	B-9D	190	.211			7.45	1.27	8.72	12.90
0600 Vibrating plate, add	A-1D	60	.133			4.68	.54	5.22	7.80
0800 Compaction in 12" layers, hand tamp, add to above	1 Clab	34	.235			8.25		8.25	12.70
0900 Roller compaction operator walking, add	B-10A	150	.080			3.42	1.14	4.56	6.45
1000 Air tamp, add	B-9	285	.140			4.98	.73	5.71	8.45
1100 Vibrating plate, add	A-1E	90	.089			3.12	.45	3.57	5.30
1300 Dozer backfilling, bulk, up to 300' haul, no compaction	B-10B	1200	.010	L.C.Y.		.43	1.04	1.47	1.79
1400 Air tamped, add	B-11B	80	.200	E.C.Y.		8	3.35	11.35	15.85
1600 Compacting backfill, 6" to 12" lifts, vibrating roller	B-10C	800	.015			.64	2.04	2.68	3.21
1700 Sheepsfoot roller	B-10D	750	.016			.68	2.22	2.90	3.48
1900 Dozer backfilling, trench, up to 300' haul, no compaction	B-10B	900	.013	L.C.Y.		.57	1.38	1.95	2.38
2000 Air tamped, add	B-11B	80	.200	E.C.Y.		8	3.35	11.35	15.85
2200 Compacting backfill, 6" to 12" lifts, vibrating roller	B-10C	700	.017			.73	2.33	3.06	3.67
2300 Sheepsfoot roller	B-10D	650	.018			.79	2.56	3.35	4.01
2350 Spreading in 8" layers, small dozer	B-10B	1060	.011	L.C.Y.		.48	1.17	1.65	2.02

31 23 23.14 Backfill, Structural

	Crew	Daily Output	Labor-Hours	Unit	Material	2012 Bare Costs Labor	2012 Bare Costs Equipment	Total	Total Incl O&P
0010 **BACKFILL, STRUCTURAL**									
0011 Dozer or F.E. loader									
0020 From existing stockpile, no compaction									
2000 80 H.P., 50' haul, sand & gravel	B-10L	1100	.011	L.C.Y.		.47	.40	.87	1.15
2010 Sandy clay & loam		1070	.011			.48	.41	.89	1.18
2020 Common earth		975	.012			.53	.45	.98	1.30
2040 Clay		850	.014			.60	.52	1.12	1.49

31 23 23.14 Backfill, Structural

	Crew	Daily Output	Labor-Hours	Unit	Material	2012 Bare Costs Labor	Equipment	Total	Total Incl O&P
2400	300' haul, sand & gravel	B-10L	370	.032	L.C.Y.		1.39	1.20	2.59	3.42
2410	Sandy clay & loam		360	.033			1.42	1.23	2.65	3.51
2420	Common earth		330	.036			1.55	1.34	2.89	3.83
2440	Clay		290	.041			1.77	1.53	3.30	4.36
3000	105 H.P., 50' haul, sand & gravel	B-10W	1350	.009			.38	.41	.79	1.03
3010	Sandy clay & loam		1325	.009			.39	.42	.81	1.05
3020	Common earth		1225	.010			.42	.45	.87	1.13
3040	Clay		1100	.011			.47	.50	.97	1.26
3300	300' haul, sand & gravel		465	.026			1.10	1.18	2.28	2.97
3310	Sandy clay & loam		455	.026			1.13	1.21	2.34	3.04
3320	Common earth		415	.029			1.24	1.33	2.57	3.33
3340	Clay		370	.032			1.39	1.49	2.88	3.74
4000	200 H.P., 50' haul, sand & gravel	B-10B	2500	.005			.21	.50	.71	.86
4010	Sandy clay & loam		2435	.005			.21	.51	.72	.88
4020	Common earth		2200	.005			.23	.56	.79	.97
4040	Clay		1950	.006			.26	.64	.90	1.10
4400	300' haul, sand & gravel		805	.015			.64	1.54	2.18	2.67
4410	Sandy clay & loam		790	.015			.65	1.57	2.22	2.71
4420	Common earth		735	.016			.70	1.69	2.39	2.92
4440	Clay		660	.018			.78	1.88	2.66	3.25
5000	300 H.P., 50' haul, sand & gravel	B-10M	3170	.004			.16	.54	.70	.85
5010	Sandy clay & loam		3110	.004			.16	.55	.71	.86
5020	Common earth		2900	.004			.18	.59	.77	.92
5040	Clay		2700	.004			.19	.64	.83	.99
5400	300' haul, sand & gravel		1500	.008			.34	1.15	1.49	1.78
5410	Sandy clay & loam		1470	.008			.35	1.17	1.52	1.82
5420	Common earth		1350	.009			.38	1.27	1.65	1.98
5440	Clay		1225	.010			.42	1.40	1.82	2.18
6010	For trench backfill, see Section 31 23 16.13 and 31 23 16.14									
6100	For compaction, see Section 31 23 23.24									

31 23 23.16 Fill By Borrow and Utility Bedding

	Crew	Daily Output	Labor-Hours	Unit	Material	2012 Bare Costs Labor	Equipment	Total	Total Incl O&P
0010	**FILL BY BORROW AND UTILITY BEDDING**									
0015	Fill by borrow, load, 1 mile haul, spread with dozer									
0020	for embankments	B-15	1200	.023	L.C.Y.	14.50	.90	2.01	17.41	19.50
0035	Select fill for shoulders & embankments	"	1200	.023	"	21.50	.90	2.01	24.41	27
0040	Fill, for hauling over 1 mile, add to above per C.Y., see Section 31 23 23.20				Mile				1.26	1.62
0049	Utility bedding, for pipe & conduit, not incl. compaction									
0050	Crushed or screened bank run gravel	B-6	150	.160	L.C.Y.	26	6.15	2.23	34.38	40.50
0100	Crushed stone 3/4" to 1/2"		150	.160		31.50	6.15	2.23	39.88	46.50
0200	Sand, dead or bank		150	.160		17	6.15	2.23	25.38	30.50
0500	Compacting bedding in trench	A-1D	90	.089	E.C.Y.		3.12	.36	3.48	5.20
0600	If material source exceeds 2 miles, add for extra mileage.									
0610	See Section 31 23 23.20 for hauling mileage add.									

31 23 23.17 General Fill

	Crew	Daily Output	Labor-Hours	Unit	Material	2012 Bare Costs Labor	Equipment	Total	Total Incl O&P
0010	**GENERAL FILL**									
0011	Spread dumped material, no compaction									
0020	By dozer, no compaction	B-10B	1000	.012	L.C.Y.		.51	1.24	1.75	2.15
0100	By hand	1 Clab	12	.667	"		23.50		23.50	36
0500	Gravel fill, compacted, under floor slabs, 4" deep	B-37	10000	.005	S.F.	.43	.18	.01	.62	.76
0600	6" deep		8600	.006		.64	.21	.02	.87	1.04
0700	9" deep		7200	.007		1.06	.25	.02	1.33	1.57
0800	12" deep		6000	.008		1.49	.30	.02	1.81	2.12

31 23 Excavation and Fill

31 23 23 – Fill

31 23 23.17 General Fill

		Crew	Daily Output	Labor-Hours	Unit	Material	2012 Bare Costs Labor	2012 Bare Costs Equipment	Total	Total Incl O&P
1000	Alternate pricing method, 4" deep	B-37	120	.400	E.C.Y.	32	14.80	1.24	48.04	59
1100	6" deep		160	.300		32	11.10	.93	44.03	53
1200	9" deep		200	.240		32	8.90	.75	41.65	49.50
1300	12" deep		220	.218		32	8.10	.68	40.78	48
1500	For fill under exterior paving, see Section 32 11 23.23									
1600	For flowable fill, see Section 03 31 05.35									

31 23 23.20 Hauling

		Crew	Daily Output	Labor-Hours	Unit	Material	2012 Bare Costs Labor	2012 Bare Costs Equipment	Total	Total Incl O&P
0010	**HAULING**									
0011	Excavated or borrow, loose cubic yards									
0012	no loading equipment, including hauling, waiting, loading/dumping									
0013	time per cycle (wait, load, travel, unload or dump & return)									
0014	8 C.Y. truck, 15 MPH ave, cycle 0.5 miles, 10 min. wait/Ld./Uld.	B-34A	320	.025	L.C.Y.		.88	1.12	2	2.57
0016	cycle 1 mile		272	.029			1.04	1.31	2.35	3.02
0018	cycle 2 miles		208	.038			1.36	1.72	3.08	3.95
0020	cycle 4 miles		144	.056			1.96	2.48	4.44	5.70
0022	cycle 6 miles		112	.071			2.52	3.19	5.71	7.35
0024	cycle 8 miles		88	.091			3.21	4.06	7.27	9.35
0026	20 MPH ave, cycle 0.5 mile		336	.024			.84	1.06	1.90	2.45
0028	cycle 1 mile		296	.027			.95	1.21	2.16	2.78
0030	cycle 2 miles		240	.033			1.18	1.49	2.67	3.42
0032	cycle 4 miles		176	.045			1.60	2.03	3.63	4.66
0034	cycle 6 miles		136	.059			2.08	2.63	4.71	6.05
0036	cycle 8 miles		112	.071			2.52	3.19	5.71	7.35
0044	25 MPH ave, cycle 4 miles		192	.042			1.47	1.86	3.33	4.28
0046	cycle 6 miles		160	.050			1.77	2.23	4	5.15
0048	cycle 8 miles		128	.063			2.21	2.79	5	6.40
0050	30 MPH ave, cycle 4 miles		216	.037			1.31	1.65	2.96	3.80
0052	cycle 6 miles		176	.045			1.60	2.03	3.63	4.66
0054	cycle 8 miles		144	.056			1.96	2.48	4.44	5.70
0114	15 MPH ave, cycle 0.5 mile, 15 min. wait/Ld./Uld.		224	.036			1.26	1.60	2.86	3.66
0116	cycle 1 mile		200	.040			1.41	1.79	3.20	4.11
0118	cycle 2 miles		168	.048			1.68	2.13	3.81	4.89
0120	cycle 4 miles		120	.067			2.35	2.98	5.33	6.85
0122	cycle 6 miles		96	.083			2.94	3.72	6.66	8.55
0124	cycle 8 miles		80	.100			3.53	4.47	8	10.25
0126	20 MPH ave, cycle 0.5 mile		232	.034			1.22	1.54	2.76	3.54
0128	cycle 1 mile		208	.038			1.36	1.72	3.08	3.95
0130	cycle 2 miles		184	.043			1.53	1.94	3.47	4.47
0132	cycle 4 miles		144	.056			1.96	2.48	4.44	5.70
0134	cycle 6 miles		112	.071			2.52	3.19	5.71	7.35
0136	cycle 8 miles		96	.083			2.94	3.72	6.66	8.55
0144	25 MPH ave, cycle 4 miles		152	.053			1.86	2.35	4.21	5.40
0146	cycle 6 miles		128	.063			2.21	2.79	5	6.40
0148	cycle 8 miles		112	.071			2.52	3.19	5.71	7.35
0150	30 MPH ave, cycle 4 miles		168	.048			1.68	2.13	3.81	4.89
0152	cycle 6 miles		144	.056			1.96	2.48	4.44	5.70
0154	cycle 8 miles		120	.067			2.35	2.98	5.33	6.85
0214	15 MPH ave, cycle 0.5 mile, 20 min wait/Ld./Uld.		176	.045			1.60	2.03	3.63	4.66
0216	cycle 1 mile		160	.050			1.77	2.23	4	5.15
0218	cycle 2 miles		136	.059			2.08	2.63	4.71	6.05
0220	cycle 4 miles		104	.077			2.72	3.44	6.16	7.90
0222	cycle 6 miles		88	.091			3.21	4.06	7.27	9.35

31 23 23.20 Hauling		Crew	Daily Output	Labor-Hours	Unit	Material	2012 Bare Costs Labor	Equipment	Total	Total Incl O&P
0224	cycle 8 miles	B-34A	72	.111	L.C.Y.		3.92	4.96	8.88	11.40
0226	20 MPH ave, cycle 0.5 mile		176	.045			1.60	2.03	3.63	4.66
0228	cycle 1 mile		168	.048			1.68	2.13	3.81	4.89
0230	cycle 2 miles		144	.056			1.96	2.48	4.44	5.70
0232	cycle 4 miles		120	.067			2.35	2.98	5.33	6.85
0234	cycle 6 miles		96	.083			2.94	3.72	6.66	8.55
0236	cycle 8 miles		88	.091			3.21	4.06	7.27	9.35
0244	25 MPH ave, cycle 4 miles		128	.063			2.21	2.79	5	6.40
0246	cycle 6 miles		112	.071			2.52	3.19	5.71	7.35
0248	cycle 8 miles		96	.083			2.94	3.72	6.66	8.55
0250	30 MPH ave, cycle 4 miles		136	.059			2.08	2.63	4.71	6.05
0252	cycle 6 miles		120	.067			2.35	2.98	5.33	6.85
0254	cycle 8 miles		104	.077			2.72	3.44	6.16	7.90
0314	15 MPH ave, cycle 0.5 mile, 25 min wait/Ld./Uld.		144	.056			1.96	2.48	4.44	5.70
0316	cycle 1 mile		128	.063			2.21	2.79	5	6.40
0318	cycle 2 miles		112	.071			2.52	3.19	5.71	7.35
0320	cycle 4 miles		96	.083			2.94	3.72	6.66	8.55
0322	cycle 6 miles		80	.100			3.53	4.47	8	10.25
0324	cycle 8 miles		64	.125			4.41	5.60	10.01	12.85
0326	20 MPH ave, cycle 0.5 mile		144	.056			1.96	2.48	4.44	5.70
0328	cycle 1 mile		136	.059			2.08	2.63	4.71	6.05
0330	cycle 2 miles		120	.067			2.35	2.98	5.33	6.85
0332	cycle 4 miles		104	.077			2.72	3.44	6.16	7.90
0334	cycle 6 miles		88	.091			3.21	4.06	7.27	9.35
0336	cycle 8 miles		80	.100			3.53	4.47	8	10.25
0344	25 MPH ave, cycle 4 miles		112	.071			2.52	3.19	5.71	7.35
0346	cycle 6 miles		96	.083			2.94	3.72	6.66	8.55
0348	cycle 8 miles		88	.091			3.21	4.06	7.27	9.35
0350	30 MPH ave, cycle 4 miles		112	.071			2.52	3.19	5.71	7.35
0352	cycle 6 miles		104	.077			2.72	3.44	6.16	7.90
0354	cycle 8 miles		96	.083			2.94	3.72	6.66	8.55
0414	15 MPH ave, cycle 0.5 mile, 30 min wait/Ld./Uld.		120	.067			2.35	2.98	5.33	6.85
0416	cycle 1 mile		112	.071			2.52	3.19	5.71	7.35
0418	cycle 2 miles		96	.083			2.94	3.72	6.66	8.55
0420	cycle 4 miles		80	.100			3.53	4.47	8	10.25
0422	cycle 6 miles		72	.111			3.92	4.96	8.88	11.40
0424	cycle 8 miles		64	.125			4.41	5.60	10.01	12.85
0426	20 MPH ave, cycle 0.5 mile		120	.067			2.35	2.98	5.33	6.85
0428	cycle 1 mile		112	.071			2.52	3.19	5.71	7.35
0430	cycle 2 miles		104	.077			2.72	3.44	6.16	7.90
0432	cycle 4 miles		88	.091			3.21	4.06	7.27	9.35
0434	cycle 6 miles		80	.100			3.53	4.47	8	10.25
0436	cycle 8 miles		72	.111			3.92	4.96	8.88	11.40
0444	25 MPH ave, cycle 4 miles		96	.083			2.94	3.72	6.66	8.55
0446	cycle 6 miles		88	.091			3.21	4.06	7.27	9.35
0448	cycle 8 miles		80	.100			3.53	4.47	8	10.25
0450	30 MPH ave, cycle 4 miles		96	.083			2.94	3.72	6.66	8.55
0452	cycle 6 miles		88	.091			3.21	4.06	7.27	9.35
0454	cycle 8 miles		80	.100			3.53	4.47	8	10.25
0514	15 MPH ave, cycle 0.5 mile, 35 min wait/Ld./Uld.		104	.077			2.72	3.44	6.16	7.90
0516	cycle 1 mile		96	.083			2.94	3.72	6.66	8.55
0518	cycle 2 miles		88	.091			3.21	4.06	7.27	9.35
0520	cycle 4 miles		72	.111			3.92	4.96	8.88	11.40

31 23 23 - Fill

31 23 23.20 Hauling		Crew	Daily Output	Labor-Hours	Unit	Material	2012 Bare Costs Labor	2012 Bare Costs Equipment	Total	Total Incl O&P
0522	cycle 6 miles	B-34A	64	.125	L.C.Y.		4.41	5.60	10.01	12.85
0524	cycle 8 miles		56	.143			5.05	6.40	11.45	14.65
0526	20 MPH ave, cycle 0.5 mile		104	.077			2.72	3.44	6.16	7.90
0528	cycle 1 mile		96	.083			2.94	3.72	6.66	8.55
0530	cycle 2 miles		96	.083			2.94	3.72	6.66	8.55
0532	cycle 4 miles		80	.100			3.53	4.47	8	10.25
0534	cycle 6 miles		72	.111			3.92	4.96	8.88	11.40
0536	cycle 8 miles		64	.125			4.41	5.60	10.01	12.85
0544	25 MPH ave, cycle 4 miles		88	.091			3.21	4.06	7.27	9.35
0546	cycle 6 miles		80	.100			3.53	4.47	8	10.25
0548	cycle 8 miles		72	.111			3.92	4.96	8.88	11.40
0550	30 MPH ave, cycle 4 miles		88	.091			3.21	4.06	7.27	9.35
0552	cycle 6 miles		80	.100			3.53	4.47	8	10.25
0554	cycle 8 miles		72	.111			3.92	4.96	8.88	11.40
1014	12 C.Y. truck, cycle 0.5 mile, 15 MPH ave, 15 min. wait/Ld./Uld.	B-34B	336	.024			.84	1.74	2.58	3.19
1016	cycle 1 mile		300	.027			.94	1.95	2.89	3.57
1018	cycle 2 miles		252	.032			1.12	2.32	3.44	4.25
1020	cycle 4 miles		180	.044			1.57	3.25	4.82	5.95
1022	cycle 6 miles		144	.056			1.96	4.06	6.02	7.45
1024	cycle 8 miles		120	.067			2.35	4.87	7.22	8.90
1025	cycle 10 miles		96	.083			2.94	6.10	9.04	11.15
1026	20 MPH ave, cycle 0.5 mile		348	.023			.81	1.68	2.49	3.08
1028	cycle 1 mile		312	.026			.91	1.87	2.78	3.43
1030	cycle 2 miles		276	.029			1.02	2.12	3.14	3.88
1032	cycle 4 miles		216	.037			1.31	2.71	4.02	4.96
1034	cycle 6 miles		168	.048			1.68	3.48	5.16	6.40
1036	cycle 8 miles		144	.056			1.96	4.06	6.02	7.45
1038	cycle 10 miles		120	.067			2.35	4.87	7.22	8.90
1040	25 MPH ave, cycle 4 miles		228	.035			1.24	2.56	3.80	4.70
1042	cycle 6 miles		192	.042			1.47	3.05	4.52	5.60
1044	cycle 8 miles		168	.048			1.68	3.48	5.16	6.40
1046	cycle 10 miles		144	.056			1.96	4.06	6.02	7.45
1050	30 MPH ave, cycle 4 miles		252	.032			1.12	2.32	3.44	4.25
1052	cycle 6 miles		216	.037			1.31	2.71	4.02	4.96
1054	cycle 8 miles		180	.044			1.57	3.25	4.82	5.95
1056	cycle 10 miles		156	.051			1.81	3.75	5.56	6.85
1060	35 MPH ave, cycle 4 miles		264	.030			1.07	2.21	3.28	4.06
1062	cycle 6 miles		228	.035			1.24	2.56	3.80	4.70
1064	cycle 8 miles		204	.039			1.38	2.87	4.25	5.25
1066	cycle 10 miles		180	.044			1.57	3.25	4.82	5.95
1068	cycle 20 miles		120	.067			2.35	4.87	7.22	8.90
1069	cycle 30 miles		84	.095			3.36	6.95	10.31	12.75
1070	cycle 40 miles		72	.111			3.92	8.10	12.02	14.90
1072	40 MPH ave, cycle 6 miles		240	.033			1.18	2.44	3.62	4.46
1074	cycle 8 miles		216	.037			1.31	2.71	4.02	4.96
1076	cycle 10 miles		192	.042			1.47	3.05	4.52	5.60
1078	cycle 20 miles		120	.067			2.35	4.87	7.22	8.90
1080	cycle 30 miles		96	.083			2.94	6.10	9.04	11.15
1082	cycle 40 miles		72	.111			3.92	8.10	12.02	14.90
1084	cycle 50 miles		60	.133			4.71	9.75	14.46	17.85
1094	45 MPH ave, cycle 8 miles		216	.037			1.31	2.71	4.02	4.96
1096	cycle 10 miles		204	.039			1.38	2.87	4.25	5.25
1098	cycle 20 miles		132	.061			2.14	4.43	6.57	8.10

31 23 23 – Fill

31 23 23.20 **Hauling**	Crew	Daily Output	Labor-Hours	Unit	Material	2012 Bare Costs Labor	Equipment	Total	Total Incl O&P
1100	cycle 30 miles	B-34B	108	.074	L.C.Y.		2.61	5.40	8.01	9.90
1102	cycle 40 miles		84	.095			3.36	6.95	10.31	12.75
1104	cycle 50 miles		72	.111			3.92	8.10	12.02	14.90
1106	50 MPH ave, cycle 10 miles		216	.037			1.31	2.71	4.02	4.96
1108	cycle 20 miles		144	.056			1.96	4.06	6.02	7.45
1110	cycle 30 miles		108	.074			2.61	5.40	8.01	9.90
1112	cycle 40 miles		84	.095			3.36	6.95	10.31	12.75
1114	cycle 50 miles		72	.111			3.92	8.10	12.02	14.90
1214	15 MPH ave, cycle 0.5 mile, 20 min. wait/Ld./Uld.		264	.030			1.07	2.21	3.28	4.06
1216	cycle 1 mile		240	.033			1.18	2.44	3.62	4.46
1218	cycle 2 miles		204	.039			1.38	2.87	4.25	5.25
1220	cycle 4 miles		156	.051			1.81	3.75	5.56	6.85
1222	cycle 6 miles		132	.061			2.14	4.43	6.57	8.10
1224	cycle 8 miles		108	.074			2.61	5.40	8.01	9.90
1225	cycle 10 miles		96	.083			2.94	6.10	9.04	11.15
1226	20 MPH ave, cycle 0.5 mile		264	.030			1.07	2.21	3.28	4.06
1228	cycle 1 mile		252	.032			1.12	2.32	3.44	4.25
1230	cycle 2 miles		216	.037			1.31	2.71	4.02	4.96
1232	cycle 4 miles		180	.044			1.57	3.25	4.82	5.95
1234	cycle 6 miles		144	.056			1.96	4.06	6.02	7.45
1236	cycle 8 miles		132	.061			2.14	4.43	6.57	8.10
1238	cycle 10 miles		108	.074			2.61	5.40	8.01	9.90
1240	25 MPH ave, cycle 4 miles		192	.042			1.47	3.05	4.52	5.60
1242	cycle 6 miles		168	.048			1.68	3.48	5.16	6.40
1244	cycle 8 miles		144	.056			1.96	4.06	6.02	7.45
1246	cycle 10 miles		132	.061			2.14	4.43	6.57	8.10
1250	30 MPH ave, cycle 4 miles		204	.039			1.38	2.87	4.25	5.25
1252	cycle 6 miles		180	.044			1.57	3.25	4.82	5.95
1254	cycle 8 miles		156	.051			1.81	3.75	5.56	6.85
1256	cycle 10 miles		144	.056			1.96	4.06	6.02	7.45
1260	35 MPH ave, cycle 4 miles		216	.037			1.31	2.71	4.02	4.96
1262	cycle 6 miles		192	.042			1.47	3.05	4.52	5.60
1264	cycle 8 miles		168	.048			1.68	3.48	5.16	6.40
1266	cycle 10 miles		156	.051			1.81	3.75	5.56	6.85
1268	cycle 20 miles		108	.074			2.61	5.40	8.01	9.90
1269	cycle 30 miles		72	.111			3.92	8.10	12.02	14.90
1270	cycle 40 miles		60	.133			4.71	9.75	14.46	17.85
1272	40 MPH ave, cycle 6 miles		192	.042			1.47	3.05	4.52	5.60
1274	cycle 8 miles		180	.044			1.57	3.25	4.82	5.95
1276	cycle 10 miles		156	.051			1.81	3.75	5.56	6.85
1278	cycle 20 miles		108	.074			2.61	5.40	8.01	9.90
1280	cycle 30 miles		84	.095			3.36	6.95	10.31	12.75
1282	cycle 40 miles		72	.111			3.92	8.10	12.02	14.90
1284	cycle 50 miles		60	.133			4.71	9.75	14.46	17.85
1294	45 MPH ave, cycle 8 miles		180	.044			1.57	3.25	4.82	5.95
1296	cycle 10 miles		168	.048			1.68	3.48	5.16	6.40
1298	cycle 20 miles		120	.067			2.35	4.87	7.22	8.90
1300	cycle 30 miles		96	.083			2.94	6.10	9.04	11.15
1302	cycle 40 miles		72	.111			3.92	8.10	12.02	14.90
1304	cycle 50 miles		60	.133			4.71	9.75	14.46	17.85
1306	50 MPH ave, cycle 10 miles		180	.044			1.57	3.25	4.82	5.95
1308	cycle 20 miles		132	.061			2.14	4.43	6.57	8.10
1310	cycle 30 miles		96	.083			2.94	6.10	9.04	11.15

31 23 23.20 Hauling		Crew	Daily Output	Labor-Hours	Unit	Material	2012 Bare Costs Labor	2012 Bare Costs Equipment	Total	Total Incl O&P
1312	cycle 40 miles	B-34B	84	.095	L.C.Y.		3.36	6.95	10.31	12.75
1314	cycle 50 miles		72	.111			3.92	8.10	12.02	14.90
1414	15 MPH ave, cycle 0.5 mile, 25 min. wait/Ld./Uld.		204	.039			1.38	2.87	4.25	5.25
1416	cycle 1 mile		192	.042			1.47	3.05	4.52	5.60
1418	cycle 2 miles		168	.048			1.68	3.48	5.16	6.40
1420	cycle 4 miles		132	.061			2.14	4.43	6.57	8.10
1422	cycle 6 miles		120	.067			2.35	4.87	7.22	8.90
1424	cycle 8 miles		96	.083			2.94	6.10	9.04	11.15
1425	cycle 10 miles		84	.095			3.36	6.95	10.31	12.75
1426	20 MPH ave, cycle 0.5 mile		216	.037			1.31	2.71	4.02	4.96
1428	cycle 1 mile		204	.039			1.38	2.87	4.25	5.25
1430	cycle 2 miles		180	.044			1.57	3.25	4.82	5.95
1432	cycle 4 miles		156	.051			1.81	3.75	5.56	6.85
1434	cycle 6 miles		132	.061			2.14	4.43	6.57	8.10
1436	cycle 8 miles		120	.067			2.35	4.87	7.22	8.90
1438	cycle 10 miles		96	.083			2.94	6.10	9.04	11.15
1440	25 MPH ave, cycle 4 miles		168	.048			1.68	3.48	5.16	6.40
1442	cycle 6 miles		144	.056			1.96	4.06	6.02	7.45
1444	cycle 8 miles		132	.061			2.14	4.43	6.57	8.10
1446	cycle 10 miles		108	.074			2.61	5.40	8.01	9.90
1450	30 MPH ave, cycle 4 miles		168	.048			1.68	3.48	5.16	6.40
1452	cycle 6 miles		156	.051			1.81	3.75	5.56	6.85
1454	cycle 8 miles		132	.061			2.14	4.43	6.57	8.10
1456	cycle 10 miles		120	.067			2.35	4.87	7.22	8.90
1460	35 MPH ave, cycle 4 miles		180	.044			1.57	3.25	4.82	5.95
1462	cycle 6 miles		156	.051			1.81	3.75	5.56	6.85
1464	cycle 8 miles		144	.056			1.96	4.06	6.02	7.45
1466	cycle 10 miles		132	.061			2.14	4.43	6.57	8.10
1468	cycle 20 miles		96	.083			2.94	6.10	9.04	11.15
1469	cycle 30 miles		72	.111			3.92	8.10	12.02	14.90
1470	cycle 40 miles		60	.133			4.71	9.75	14.46	17.85
1472	40 MPH ave, cycle 6 miles		168	.048			1.68	3.48	5.16	6.40
1474	cycle 8 miles		156	.051			1.81	3.75	5.56	6.85
1476	cycle 10 miles		144	.056			1.96	4.06	6.02	7.45
1478	cycle 20 miles		96	.083			2.94	6.10	9.04	11.15
1480	cycle 30 miles		84	.095			3.36	6.95	10.31	12.75
1482	cycle 40 miles		60	.133			4.71	9.75	14.46	17.85
1484	cycle 50 miles		60	.133			4.71	9.75	14.46	17.85
1494	45 MPH ave, cycle 8 miles		156	.051			1.81	3.75	5.56	6.85
1496	cycle 10 miles		144	.056			1.96	4.06	6.02	7.45
1498	cycle 20 miles		108	.074			2.61	5.40	8.01	9.90
1500	cycle 30 miles		84	.095			3.36	6.95	10.31	12.75
1502	cycle 40 miles		72	.111			3.92	8.10	12.02	14.90
1504	cycle 50 miles		60	.133			4.71	9.75	14.46	17.85
1506	50 MPH ave, cycle 10 miles		156	.051			1.81	3.75	5.56	6.85
1508	cycle 20 miles		120	.067			2.35	4.87	7.22	8.90
1510	cycle 30 miles		96	.083			2.94	6.10	9.04	11.15
1512	cycle 40 miles		72	.111			3.92	8.10	12.02	14.90
1514	cycle 50 miles		60	.133			4.71	9.75	14.46	17.85
1614	15 MPH, cycle 0.5 mile, 30 min. wait/Ld./Uld.		180	.044			1.57	3.25	4.82	5.95
1616	cycle 1 mile		168	.048			1.68	3.48	5.16	6.40
1618	cycle 2 miles		144	.056			1.96	4.06	6.02	7.45
1620	cycle 4 miles		120	.067			2.35	4.87	7.22	8.90

31 23 23.20 Hauling		Crew	Daily Output	Labor-Hours	Unit	Material	2012 Bare Costs Labor	2012 Bare Costs Equipment	Total	Total Incl O&P
1622	cycle 6 miles	B-34B	108	.074	L.C.Y.		2.61	5.40	8.01	9.90
1624	cycle 8 miles		84	.095			3.36	6.95	10.31	12.75
1625	cycle 10 miles		84	.095			3.36	6.95	10.31	12.75
1626	20 MPH ave, cycle 0.5 mile		180	.044			1.57	3.25	4.82	5.95
1628	cycle 1 mile		168	.048			1.68	3.48	5.16	6.40
1630	cycle 2 miles		156	.051			1.81	3.75	5.56	6.85
1632	cycle 4 miles		132	.061			2.14	4.43	6.57	8.10
1634	cycle 6 miles		120	.067			2.35	4.87	7.22	8.90
1636	cycle 8 miles		108	.074			2.61	5.40	8.01	9.90
1638	cycle 10 miles		96	.083			2.94	6.10	9.04	11.15
1640	25 MPH ave, cycle 4 miles		144	.056			1.96	4.06	6.02	7.45
1642	cycle 6 miles		132	.061			2.14	4.43	6.57	8.10
1644	cycle 8 miles		108	.074			2.61	5.40	8.01	9.90
1646	cycle 10 miles		108	.074			2.61	5.40	8.01	9.90
1650	30 MPH ave, cycle 4 miles		144	.056			1.96	4.06	6.02	7.45
1652	cycle 6 miles		132	.061			2.14	4.43	6.57	8.10
1654	cycle 8 miles		120	.067			2.35	4.87	7.22	8.90
1656	cycle 10 miles		108	.074			2.61	5.40	8.01	9.90
1660	35 MPH ave, cycle 4 miles		156	.051			1.81	3.75	5.56	6.85
1662	cycle 6 miles		144	.056			1.96	4.06	6.02	7.45
1664	cycle 8 miles		132	.061			2.14	4.43	6.57	8.10
1666	cycle 10 miles		120	.067			2.35	4.87	7.22	8.90
1668	cycle 20 miles		84	.095			3.36	6.95	10.31	12.75
1669	cycle 30 miles		72	.111			3.92	8.10	12.02	14.90
1670	cycle 40 miles		60	.133			4.71	9.75	14.46	17.85
1672	40 MPH, cycle 6 miles		144	.056			1.96	4.06	6.02	7.45
1674	cycle 8 miles		132	.061			2.14	4.43	6.57	8.10
1676	cycle 10 miles		120	.067			2.35	4.87	7.22	8.90
1678	cycle 20 miles		96	.083			2.94	6.10	9.04	11.15
1680	cycle 30 miles		72	.111			3.92	8.10	12.02	14.90
1682	cycle 40 miles		60	.133			4.71	9.75	14.46	17.85
1684	cycle 50 miles		48	.167			5.90	12.20	18.10	22.50
1694	45 MPH ave, cycle 8 miles		144	.056			1.96	4.06	6.02	7.45
1696	cycle 10 miles		132	.061			2.14	4.43	6.57	8.10
1698	cycle 20 miles		96	.083			2.94	6.10	9.04	11.15
1700	cycle 30 miles		84	.095			3.36	6.95	10.31	12.75
1702	cycle 40 miles		60	.133			4.71	9.75	14.46	17.85
1704	cycle 50 miles		60	.133			4.71	9.75	14.46	17.85
1706	50 MPH ave, cycle 10 miles		132	.061			2.14	4.43	6.57	8.10
1708	cycle 20 miles		108	.074			2.61	5.40	8.01	9.90
1710	cycle 30 miles		84	.095			3.36	6.95	10.31	12.75
1712	cycle 40 miles		72	.111			3.92	8.10	12.02	14.90
1714	cycle 50 miles		60	.133			4.71	9.75	14.46	17.85
2000	Hauling, 8 C.Y. truck, small project cost per hour	B-34A	8	1	Hr.		35.50	44.50	80	103
2100	12 C.Y. Truck	B-34B	8	1			35.50	73	108.50	134
2150	16.5 C.Y. Truck	B-34C	8	1			35.50	80.50	116	142
2175	18 C.Y. 8 wheel Truck	B-34I	8	1			35.50	93	128.50	156
2200	20 C.Y. Truck	B-34D	8	1			35.50	82.50	118	145
2300	Grading at dump, or embankment if required, by dozer	B-10B	1000	.012	L.C.Y.		.51	1.24	1.75	2.15
2310	Spotter at fill or cut, if required	1 Clab	8	1	Hr.		35		35	54
9014	18 C.Y. truck, 8 wheels,15 min. wait/Ld./Uld.,15 MPH, cycle 0.5 mi.	B-34I	504	.016	L.C.Y.		.56	1.48	2.04	2.48
9016	cycle 1 mile		450	.018			.63	1.66	2.29	2.77
9018	cycle 2 miles		378	.021			.75	1.97	2.72	3.30

31 23 Excavation and Fill

31 23 23 – Fill

31 23 23.20 Hauling		Crew	Daily Output	Labor-Hours	Unit	Material	2012 Bare Costs		Total	Total Incl O&P
							Labor	Equipment		
9020	cycle 4 miles	B-34I	270	.030	L.C.Y.		1.05	2.76	3.81	4.63
9022	cycle 6 miles		216	.037			1.31	3.45	4.76	5.75
9024	cycle 8 miles		180	.044			1.57	4.14	5.71	6.95
9025	cycle 10 miles		144	.056			1.96	5.15	7.11	8.70
9026	20 MPH ave, cycle 0.5 mile		522	.015			.54	1.43	1.97	2.39
9028	cycle 1 mile		468	.017			.60	1.59	2.19	2.67
9030	cycle 2 miles		414	.019			.68	1.80	2.48	3.01
9032	cycle 4 miles		324	.025			.87	2.30	3.17	3.85
9034	cycle 6 miles		252	.032			1.12	2.96	4.08	4.95
9036	cycle 8 miles		216	.037			1.31	3.45	4.76	5.75
9038	cycle 10 miles		180	.044			1.57	4.14	5.71	6.95
9040	25 MPH ave, cycle 4 miles		342	.023			.83	2.18	3.01	3.65
9042	cycle 6 miles		288	.028			.98	2.59	3.57	4.34
9044	cycle 8 miles		252	.032			1.12	2.96	4.08	4.95
9046	cycle 10 miles		216	.037			1.31	3.45	4.76	5.75
9050	30 MPH ave, cycle 4 miles		378	.021			.75	1.97	2.72	3.30
9052	cycle 6 miles		324	.025			.87	2.30	3.17	3.85
9054	cycle 8 miles		270	.030			1.05	2.76	3.81	4.63
9056	cycle 10 miles		234	.034			1.21	3.18	4.39	5.35
9060	35 MPH ave, cycle 4 miles		396	.020			.71	1.88	2.59	3.15
9062	cycle 6 miles		342	.023			.83	2.18	3.01	3.65
9064	cycle 8 miles		288	.028			.98	2.59	3.57	4.34
9066	cycle 10 miles		270	.030			1.05	2.76	3.81	4.63
9068	cycle 20 miles		162	.049			1.74	4.60	6.34	7.70
9070	cycle 30 miles		126	.063			2.24	5.90	8.14	9.90
9072	cycle 40 miles		90	.089			3.14	8.30	11.44	13.85
9074	40 MPH ave, cycle 6 miles		360	.022			.78	2.07	2.85	3.47
9076	cycle 8 miles		324	.025			.87	2.30	3.17	3.85
9078	cycle 10 miles		288	.028			.98	2.59	3.57	4.34
9080	cycle 20 miles		180	.044			1.57	4.14	5.71	6.95
9082	cycle 30 miles		144	.056			1.96	5.15	7.11	8.70
9084	cycle 40 miles		108	.074			2.61	6.90	9.51	11.55
9086	cycle 50 miles		90	.089			3.14	8.30	11.44	13.85
9094	45 MPH ave, cycle 8 miles		324	.025			.87	2.30	3.17	3.85
9096	cycle 10 miles		306	.026			.92	2.43	3.35	4.08
9098	cycle 20 miles		198	.040			1.43	3.76	5.19	6.30
9100	cycle 30 miles		144	.056			1.96	5.15	7.11	8.70
9102	cycle 40 miles		126	.063			2.24	5.90	8.14	9.90
9104	cycle 50 miles		108	.074			2.61	6.90	9.51	11.55
9106	50 MPH ave, cycle 10 miles		324	.025			.87	2.30	3.17	3.85
9108	cycle 20 miles		216	.037			1.31	3.45	4.76	5.75
9110	cycle 30 miles		162	.049			1.74	4.60	6.34	7.70
9112	cycle 40 miles		126	.063			2.24	5.90	8.14	9.90
9114	cycle 50 miles		108	.074			2.61	6.90	9.51	11.55
9214	20 min. wait/Ld./Uld.,15 MPH, cycle 0.5 mi.		396	.020			.71	1.88	2.59	3.15
9216	cycle 1 mile		360	.022			.78	2.07	2.85	3.47
9218	cycle 2 miles		306	.026			.92	2.43	3.35	4.08
9220	cycle 4 miles		234	.034			1.21	3.18	4.39	5.35
9222	cycle 6 miles		198	.040			1.43	3.76	5.19	6.30
9224	cycle 8 miles		162	.049			1.74	4.60	6.34	7.70
9225	cycle 10 miles		144	.056			1.96	5.15	7.11	8.70
9226	20 MPH ave, cycle 0.5 mile		396	.020			.71	1.88	2.59	3.15
9228	cycle 1 mile		378	.021			.75	1.97	2.72	3.30

31 23 23.20 Hauling		Crew	Daily Output	Labor-Hours	Unit	Material	2012 Bare Costs Labor	2012 Bare Costs Equipment	Total	Total Incl O&P
9230	cycle 2 miles	B-34I	324	.025	L.C.Y.		.87	2.30	3.17	3.85
9232	cycle 4 miles		270	.030			1.05	2.76	3.81	4.63
9234	cycle 6 miles		216	.037			1.31	3.45	4.76	5.75
9236	cycle 8 miles		198	.040			1.43	3.76	5.19	6.30
9238	cycle 10 miles		162	.049			1.74	4.60	6.34	7.70
9240	25 MPH ave, cycle 4 miles		288	.028			.98	2.59	3.57	4.34
9242	cycle 6 miles		252	.032			1.12	2.96	4.08	4.95
9244	cycle 8 miles		216	.037			1.31	3.45	4.76	5.75
9246	cycle 10 miles		198	.040			1.43	3.76	5.19	6.30
9250	30 MPH ave, cycle 4 miles		306	.026			.92	2.43	3.35	4.08
9252	cycle 6 miles		270	.030			1.05	2.76	3.81	4.63
9254	cycle 8 miles		234	.034			1.21	3.18	4.39	5.35
9256	cycle 10 miles		216	.037			1.31	3.45	4.76	5.75
9260	35 MPH ave, cycle 4 miles		324	.025			.87	2.30	3.17	3.85
9262	cycle 6 miles		288	.028			.98	2.59	3.57	4.34
9264	cycle 8 miles		252	.032			1.12	2.96	4.08	4.95
9266	cycle 10 miles		234	.034			1.21	3.18	4.39	5.35
9268	cycle 20 miles		162	.049			1.74	4.60	6.34	7.70
9270	cycle 30 miles		108	.074			2.61	6.90	9.51	11.55
9272	cycle 40 miles		90	.089			3.14	8.30	11.44	13.85
9274	40 MPH ave, cycle 6 miles		288	.028			.98	2.59	3.57	4.34
9276	cycle 8 miles		270	.030			1.05	2.76	3.81	4.63
9278	cycle 10 miles		234	.034			1.21	3.18	4.39	5.35
9280	cycle 20 miles		162	.049			1.74	4.60	6.34	7.70
9282	cycle 30 miles		126	.063			2.24	5.90	8.14	9.90
9284	cycle 40 miles		108	.074			2.61	6.90	9.51	11.55
9286	cycle 50 miles		90	.089			3.14	8.30	11.44	13.85
9294	45 MPH ave, cycle 8 miles		270	.030			1.05	2.76	3.81	4.63
9296	cycle 10 miles		252	.032			1.12	2.96	4.08	4.95
9298	cycle 20 miles		180	.044			1.57	4.14	5.71	6.95
9300	cycle 30 miles		144	.056			1.96	5.15	7.11	8.70
9302	cycle 40 miles		108	.074			2.61	6.90	9.51	11.55
9304	cycle 50 miles		90	.089			3.14	8.30	11.44	13.85
9306	50 MPH ave, cycle 10 miles		270	.030			1.05	2.76	3.81	4.63
9308	cycle 20 miles		198	.040			1.43	3.76	5.19	6.30
9310	cycle 30 miles		144	.056			1.96	5.15	7.11	8.70
9312	cycle 40 miles		126	.063			2.24	5.90	8.14	9.90
9314	cycle 50 miles		108	.074			2.61	6.90	9.51	11.55
9414	25 min. wait/Ld./Uld.,15 MPH, cycle 0.5 mi.		306	.026			.92	2.43	3.35	4.08
9416	cycle 1 mile		288	.028			.98	2.59	3.57	4.34
9418	cycle 2 miles		252	.032			1.12	2.96	4.08	4.95
9420	cycle 4 miles		198	.040			1.43	3.76	5.19	6.30
9422	cycle 6 miles		180	.044			1.57	4.14	5.71	6.95
9424	cycle 8 miles		144	.056			1.96	5.15	7.11	8.70
9425	cycle 10 miles		126	.063			2.24	5.90	8.14	9.90
9426	20 MPH ave, cycle 0.5 mile		324	.025			.87	2.30	3.17	3.85
9428	cycle 1 mile		306	.026			.92	2.43	3.35	4.08
9430	cycle 2 miles		270	.030			1.05	2.76	3.81	4.63
9432	cycle 4 miles		234	.034			1.21	3.18	4.39	5.35
9434	cycle 6 miles		198	.040			1.43	3.76	5.19	6.30
9436	cycle 8 miles		180	.044			1.57	4.14	5.71	6.95
9438	cycle 10 miles		144	.056			1.96	5.15	7.11	8.70
9440	25 MPH ave, cycle 4 miles		252	.032			1.12	2.96	4.08	4.95

31 23 23 — Fill

31 23 23.20 Hauling	Crew	Daily Output	Labor-Hours	Unit	Material	2012 Bare Costs Labor	2012 Bare Costs Equipment	Total	Total Incl O&P
9442	cycle 6 miles	B-34I	216	.037	L.C.Y.		1.31	3.45	4.76	5.75
9444	cycle 8 miles		198	.040			1.43	3.76	5.19	6.30
9446	cycle 10 miles		180	.044			1.57	4.14	5.71	6.95
9450	30 MPH ave, cycle 4 miles		252	.032			1.12	2.96	4.08	4.95
9452	cycle 6 miles		234	.034			1.21	3.18	4.39	5.35
9454	cycle 8 miles		198	.040			1.43	3.76	5.19	6.30
9456	cycle 10 miles		180	.044			1.57	4.14	5.71	6.95
9460	35 MPH ave, cycle 4 miles		270	.030			1.05	2.76	3.81	4.63
9462	cycle 6 miles		234	.034			1.21	3.18	4.39	5.35
9464	cycle 8 miles		216	.037			1.31	3.45	4.76	5.75
9466	cycle 10 miles		198	.040			1.43	3.76	5.19	6.30
9468	cycle 20 miles		144	.056			1.96	5.15	7.11	8.70
9470	cycle 30 miles		108	.074			2.61	6.90	9.51	11.55
9472	cycle 40 miles		90	.089			3.14	8.30	11.44	13.85
9474	40 MPH ave, cycle 6 miles		252	.032			1.12	2.96	4.08	4.95
9476	cycle 8 miles		234	.034			1.21	3.18	4.39	5.35
9478	cycle 10 miles		216	.037			1.31	3.45	4.76	5.75
9480	cycle 20 miles		144	.056			1.96	5.15	7.11	8.70
9482	cycle 30 miles		126	.063			2.24	5.90	8.14	9.90
9484	cycle 40 miles		90	.089			3.14	8.30	11.44	13.85
9486	cycle 50 miles		90	.089			3.14	8.30	11.44	13.85
9494	45 MPH ave, cycle 8 miles		234	.034			1.21	3.18	4.39	5.35
9496	cycle 10 miles		216	.037			1.31	3.45	4.76	5.75
9498	cycle 20 miles		162	.049			1.74	4.60	6.34	7.70
9500	cycle 30 miles		126	.063			2.24	5.90	8.14	9.90
9502	cycle 40 miles		108	.074			2.61	6.90	9.51	11.55
9504	cycle 50 miles		90	.089			3.14	8.30	11.44	13.85
9506	50 MPH ave, cycle 10 miles		234	.034			1.21	3.18	4.39	5.35
9508	cycle 20 miles		180	.044			1.57	4.14	5.71	6.95
9510	cycle 30 miles		144	.056			1.96	5.15	7.11	8.70
9512	cycle 40 miles		108	.074			2.61	6.90	9.51	11.55
9514	cycle 50 miles		90	.089			3.14	8.30	11.44	13.85
9614	30 min. wait/Ld./Uld.,15 MPH, cycle 0.5 mi.		270	.030			1.05	2.76	3.81	4.63
9616	cycle 1 mile		252	.032			1.12	2.96	4.08	4.95
9618	cycle 2 miles		216	.037			1.31	3.45	4.76	5.75
9620	cycle 4 miles		180	.044			1.57	4.14	5.71	6.95
9622	cycle 6 miles		162	.049			1.74	4.60	6.34	7.70
9624	cycle 8 miles		126	.063			2.24	5.90	8.14	9.90
9625	cycle 10 miles		126	.063			2.24	5.90	8.14	9.90
9626	20 MPH ave, cycle 0.5 mile		270	.030			1.05	2.76	3.81	4.63
9628	cycle 1 mile		252	.032			1.12	2.96	4.08	4.95
9630	cycle 2 miles		234	.034			1.21	3.18	4.39	5.35
9632	cycle 4 miles		198	.040			1.43	3.76	5.19	6.30
9634	cycle 6 miles		180	.044			1.57	4.14	5.71	6.95
9636	cycle 8 miles		162	.049			1.74	4.60	6.34	7.70
9638	cycle 10 miles		144	.056			1.96	5.15	7.11	8.70
9640	25 MPH ave, cycle 4 miles		216	.037			1.31	3.45	4.76	5.75
9642	cycle 6 miles		198	.040			1.43	3.76	5.19	6.30
9644	cycle 8 miles		180	.044			1.57	4.14	5.71	6.95
9646	cycle 10 miles		162	.049			1.74	4.60	6.34	7.70
9650	30 MPH ave, cycle 4 miles		216	.037			1.31	3.45	4.76	5.75
9652	cycle 6 miles		198	.040			1.43	3.76	5.19	6.30
9654	cycle 8 miles		180	.044			1.57	4.14	5.71	6.95

31 23 Excavation and Fill

31 23 23 – Fill

31 23 23.20 Hauling		Crew	Daily Output	Labor-Hours	Unit	Material	2012 Bare Costs Labor	2012 Bare Costs Equipment	Total	Total Incl O&P
9656	cycle 10 miles	B-34I	162	.049	L.C.Y.		1.74	4.60	6.34	7.70
9660	35 MPH ave, cycle 4 miles		234	.034			1.21	3.18	4.39	5.35
9662	cycle 6 miles		216	.037			1.31	3.45	4.76	5.75
9664	cycle 8 miles		198	.040			1.43	3.76	5.19	6.30
9666	cycle 10 miles		180	.044			1.57	4.14	5.71	6.95
9668	cycle 20 miles		126	.063			2.24	5.90	8.14	9.90
9670	cycle 30 miles		108	.074			2.61	6.90	9.51	11.55
9672	cycle 40 miles		90	.089			3.14	8.30	11.44	13.85
9674	40 MPH ave, cycle 6 miles		216	.037			1.31	3.45	4.76	5.75
9676	cycle 8 miles		198	.040			1.43	3.76	5.19	6.30
9678	cycle 10 miles		180	.044			1.57	4.14	5.71	6.95
9680	cycle 20 miles		144	.056			1.96	5.15	7.11	8.70
9682	cycle 30 miles		108	.074			2.61	6.90	9.51	11.55
9684	cycle 40 miles		90	.089			3.14	8.30	11.44	13.85
9686	cycle 50 miles		72	.111			3.92	10.35	14.27	17.35
9694	45 MPH ave, cycle 8 miles		216	.037			1.31	3.45	4.76	5.75
9696	cycle 10 miles		198	.040			1.43	3.76	5.19	6.30
9698	cycle 20 miles		144	.056			1.96	5.15	7.11	8.70
9700	cycle 30 miles		126	.063			2.24	5.90	8.14	9.90
9702	cycle 40 miles		108	.074			2.61	6.90	9.51	11.55
9704	cycle 50 miles		90	.089			3.14	8.30	11.44	13.85
9706	50 MPH ave, cycle 10 miles		198	.040			1.43	3.76	5.19	6.30
9708	cycle 20 miles		162	.049			1.74	4.60	6.34	7.70
9710	cycle 30 miles		126	.063			2.24	5.90	8.14	9.90
9712	cycle 40 miles		108	.074			2.61	6.90	9.51	11.55
9714	cycle 50 miles	↓	90	.089	↓		3.14	8.30	11.44	13.85

31 23 23.24 Compaction, Structural

			Crew	Daily Output	Labor-Hours	Unit	Material	Labor	Equipment	Total	Total Incl O&P
0010	**COMPACTION, STRUCTURAL**	R312323-30									
0020	Steel wheel tandem roller, 5 tons		B-10E	8	1.500	Hr.		64	18.65	82.65	118
0100	10 tons		B-10F	8	1.500	"		64	28.50	92.50	129
0300	Sheepsfoot or wobbly wheel roller, 8" lifts, common fill		B-10G	1300	.009	E.C.Y.		.39	.85	1.24	1.53
0400	Select fill		"	1500	.008			.34	.73	1.07	1.33
0600	Vibratory plate, 8" lifts, common fill		A-1D	200	.040			1.40	.16	1.56	2.34
0700	Select fill		"	216	.037	↓		1.30	.15	1.45	2.16

31 25 Erosion and Sedimentation Controls

31 25 14 – Stabilization Measures for Erosion and Sedimentation Control

31 25 14.16 Rolled Erosion Control Mats and Blankets

			Crew	Daily Output	Labor-Hours	Unit	Material	Labor	Equipment	Total	Total Incl O&P
0010	**ROLLED EROSION CONTROL MATS AND BLANKETS**										
0020	Jute mesh, 100 S.Y. per roll, 4' wide, stapled	G	B-80A	2400	.010	S.Y.	.60	.35	.11	1.06	1.32
0070	Paper biodegradable mesh	G	B-1	2500	.010		.12	.34		.46	.66
0080	Paper mulch	G	B-64	20000	.001		.09	.03	.02	.14	.16
0100	Plastic netting, stapled, 2" x 1" mesh, 20 mil	G	B-1	2500	.010		.34	.34		.68	.90
0200	Polypropylene mesh, stapled, 6.5 oz./S.Y.	G		2500	.010		1.32	.34		1.66	1.98
0300	Tobacco netting, or jute mesh #2, stapled	G	↓	2500	.010	↓	.12	.34		.46	.66
1000	Silt fence, polypropylene, 3' high, ideal conditions	G	2 Clab	1600	.010	L.F.	.25	.35		.60	.82
1100	Adverse conditions	G	"	950	.017	"	.25	.59		.84	1.19
1200	Place and remove hay bales	G	A-2	3	8	Ton	265	279	70.50	614.50	795
1250	Hay bales, staked	G	"	2500	.010	L.F.	10.60	.33	.08	11.01	12.25

31 31 Soil Treatment

31 31 16 – Termite Control

31 31 16.13 Chemical Termite Control	Crew	Daily Output	Labor-Hours	Unit	Material	2012 Bare Costs Labor	Equipment	Total	Total Incl O&P
0010 **CHEMICAL TERMITE CONTROL**									
0020 Slab and walls, residential	1 Skwk	1200	.007	SF Flr.	.33	.30		.63	.83
0100 Commercial, minimum		2496	.003		.33	.15		.48	.59
0200 Maximum		1645	.005		.50	.22		.72	.88
0400 Insecticides for termite control, minimum		14.20	.563	Gal.	59	25.50		84.50	105
0500 Maximum		11	.727	"	101	33		134	162

31 32 Soil Stabilization

31 32 13 – Soil Mixing Stabilization

31 32 13.30 Calcium Chloride

	Crew	Daily Output	Labor-Hours	Unit	Material	Labor	Equipment	Total	Total Incl O&P
0010 **CALCIUM CHLORIDE**									
0020 Calcium chloride delivered, 100 lb. bags, truckload lots				Ton	360			360	395
0030 Solution, 4 lb. flake per gallon, tank truck delivery				Gal.	1.44			1.44	1.58

31 33 Rock Stabilization

31 33 13 – Rock Bolting and Grouting

31 33 13.10 Rock Bolting

	Crew	Daily Output	Labor-Hours	Unit	Material	Labor	Equipment	Total	Total Incl O&P
0010 **ROCK BOLTING**									
2020 Hollow core, prestressable anchor, 1" diameter, 5' long	2 Skwk	32	.500	Ea.	104	22.50		126.50	150
2025 10' long		24	.667		209	30.50		239.50	277
2060 2" diameter, 5' long		32	.500		400	22.50		422.50	470
2065 10' long		24	.667		780	30.50		810.50	905
2100 Super high-tensile, 3/4" diameter, 5' long		32	.500		23.50	22.50		46	61
2105 10' long		24	.667		47.50	30.50		78	99
2160 2" diameter, 5' long		32	.500		211	22.50		233.50	267
2165 10' long		24	.667		420	30.50		450.50	510
4400 Drill hole for rock bolt, 1-3/4" diam., 5' long (for 3/4" bolt)	B-56	17	.941			37.50	83.50	121	149
4405 10' long		9	1.778			71	158	229	282
4420 2" diameter, 5' long (for 1" bolt)		13	1.231			49	109	158	195
4425 10' long		7	2.286			91.50	203	294.50	360
4460 3-1/2" diameter, 5' long (for 2" bolt)		10	1.600			64	142	206	254
4465 10' long		5	3.200			128	284	412	505

31 36 Gabions

31 36 13 – Gabion Boxes

31 36 13.10 Gabion Box Systems

	Crew	Daily Output	Labor-Hours	Unit	Material	Labor	Equipment	Total	Total Incl O&P
0010 **GABION BOX SYSTEMS**									
0400 Gabions, galvanized steel mesh mats or boxes, stone filled, 6" deep	B-13	200	.280	S.Y.	29.50	10.65	3.49	43.64	52.50
0500 9" deep		163	.344		37	13.10	4.28	54.38	65
0600 12" deep		153	.366		54	13.95	4.56	72.51	86
0700 18" deep		102	.549		71.50	21	6.85	99.35	118
0800 36" deep		60	.933		116	35.50	11.60	163.10	195

31 37 Riprap

31 37 13 – Machined Riprap

31 37 13.10 Riprap and Rock Lining

		Crew	Daily Output	Labor-Hours	Unit	Material	2012 Bare Costs Labor	2012 Bare Costs Equipment	Total	Total Incl O&P
0010	**RIPRAP AND ROCK LINING**									
0011	Random, broken stone									
0100	Machine placed for slope protection	B-12G	62	.258	L.C.Y.	31.50	10.70	11.05	53.25	63.50
0110	3/8 to 1/4 C.Y. pieces, grouted	B-13	80	.700	S.Y.	67.50	26.50	8.70	102.70	125
0200	18" minimum thickness, not grouted	"	53	1.057	"	19.75	40	13.15	72.90	97.50
0300	Dumped, 50 lb. average	B-11A	800	.020	Ton	23.50	.82	1.55	25.87	28.50
0350	100 lb. average		700	.023		23.50	.93	1.78	26.21	29
0370	300 lb. average	↓	600	.027	↓	23.50	1.09	2.07	26.66	29.50

31 41 Shoring

31 41 13 – Timber Shoring

31 41 13.10 Building Shoring

		Crew	Daily Output	Labor-Hours	Unit	Material	2012 Bare Costs Labor	2012 Bare Costs Equipment	Total	Total Incl O&P
0010	**BUILDING SHORING**									
0020	Shoring, existing building, with timber, no salvage allowance	B-51	2.20	21.818	M.B.F.	640	770	96	1,506	1,975
1000	On cribbing with 35 ton screw jacks, per box and jack	"	3.60	13.333	Jack	66	470	58.50	594.50	860

31 41 16 – Sheet Piling

31 41 16.10 Sheet Piling Systems

			Crew	Daily Output	Labor-Hours	Unit	Material	2012 Bare Costs Labor	2012 Bare Costs Equipment	Total	Total Incl O&P
0010	**SHEET PILING SYSTEMS**										
0020	Sheet piling steel, not incl. wales, 22 psf, 15' excav., left in place		B-40	10.81	5.920	Ton	1,525	261	320	2,106	2,425
0100	Drive, extract & salvage	R314116-40		6	10.667		530	470	575	1,575	1,950
0300	20' deep excavation, 27 psf, left in place			12.95	4.942		1,525	218	266	2,009	2,300
0400	Drive, extract & salvage	R314116-45		6.55	9.771		530	430	525	1,485	1,850
0600	25' deep excavation, 38 psf, left in place			19	3.368		1,525	149	182	1,856	2,100
0700	Drive, extract & salvage			10.50	6.095		530	269	330	1,129	1,375
0900	40' deep excavation, 38 psf, left in place			21.20	3.019		1,525	133	163	1,821	2,075
1000	Drive, extract & salvage			12.25	5.224	↓	530	230	282	1,042	1,250
1200	15' deep excavation, 22 psf, left in place			983	.065	S.F.	17.65	2.87	3.51	24.03	28
1300	Drive, extract & salvage			545	.117		5.95	5.20	6.35	17.50	21.50
1500	20' deep excavation, 27 psf, left in place			960	.067		22	2.94	3.59	28.53	33
1600	Drive, extract & salvage			485	.132		7.75	5.80	7.10	20.65	25.50
1800	25' deep excavation, 38 psf, left in place			1000	.064		32.50	2.82	3.45	38.77	44
1900	Drive, extract & salvage		↓	553	.116	↓	10.65	5.10	6.25	22	26.50
2100	Rent steel sheet piling and wales, first month					Ton	305			305	335
2200	Per added month						30.50			30.50	33.50
2300	Rental piling left in place, add to rental						1,225			1,225	1,325
2500	Wales, connections & struts, 2/3 salvage						485			485	535
2700	High strength piling, 50,000 psi, add						70			70	77
2800	55,000 psi, add						79.50			79.50	87.50
3000	Tie rod, not upset, 1-1/2" to 4" diameter with turnbuckle						2,200			2,200	2,425
3100	No turnbuckle						1,725			1,725	1,875
3300	Upset, 1-3/4" to 4" diameter with turnbuckle						2,500			2,500	2,750
3400	No turnbuckle					↓	2,175			2,175	2,375
3600	Lightweight, 18" to 28" wide, 7 ga., 9.22 psf, and										
3610	9 ga., 8.6 psf, minimum					Lb.	.91			.91	1
3700	Average						.99			.99	1.09
3750	Maximum					↓	1.13			1.13	1.24
3900	Wood, solid sheeting, incl. wales, braces and spacers,										
3910	drive, extract & salvage, 8' deep excavation		B-31	330	.121	S.F.	1.56	4.52	.59	6.67	9.30
4000	10' deep, 50 S.F./hr. in & 150 S.F./hr. out			300	.133		1.61	4.97	.65	7.23	10.15
4100	12' deep, 45 S.F./hr. in & 135 S.F./hr. out		↓	270	.148	↓	1.66	5.55	.72	7.93	11.10

31 41 Shoring

31 41 16 – Sheet Piling

31 41 16.10 Sheet Piling Systems	Crew	Daily Output	Labor-Hours	Unit	Material	2012 Bare Costs Labor	Equipment	Total	Total Incl O&P
4200	14' deep, 42 S.F./hr. in & 126 S.F./hr. out	B-31	250	.160	S.F.	1.71	5.95	.78	8.44	11.95
4300	16' deep, 40 S.F./hr. in & 120 S.F./hr. out		240	.167		1.76	6.20	.81	8.77	12.40
4400	18' deep, 38 S.F./hr. in & 114 S.F./hr. out		230	.174		1.82	6.50	.85	9.17	12.95
4500	20' deep, 35 S.F./hr. in & 105 S.F./hr. out		210	.190		1.88	7.10	.93	9.91	14.05
4520	Left in place, 8' deep, 55 S.F./hr.		440	.091		2.82	3.39	.44	6.65	8.80
4540	10' deep, 50 S.F./hr.		400	.100		2.96	3.73	.49	7.18	9.55
4560	12' deep, 45 S.F./hr.		360	.111		3.13	4.14	.54	7.81	10.45
4565	14' deep, 42 S.F./hr.		335	.119		3.31	4.45	.58	8.34	11.15
4570	16' deep, 40 S.F./hr.		320	.125		3.52	4.66	.61	8.79	11.70
4580	18' deep, 38 S.F./hr.		305	.131		3.75	4.89	.64	9.28	12.40
4590	20' deep, 35 S.F./hr.		280	.143		4.02	5.35	.70	10.07	13.40
4700	Alternate pricing, left in place, 8' deep		1.76	22.727	M.B.F.	635	850	111	1,596	2,125
4800	Drive, extract and salvage, 8' deep		1.32	30.303	"	565	1,125	148	1,838	2,525
5000	For treated lumber add cost of treatment to lumber									

31 43 Concrete Raising

31 43 13 – Pressure Grouting

31 43 13.13 Concrete Pressure Grouting

		Crew	Daily Output	Labor-Hours	Unit	Material	Labor	Equipment	Total	Total Incl O&P
0010	**CONCRETE PRESSURE GROUTING**									
0020	Grouting, pressure, cement & sand, 1:1 mix, minimum	B-61	124	.323	Bag	14.10	12.05	2.64	28.79	37
0100	Maximum		51	.784	"	14.10	29.50	6.40	50	67.50
0200	Cement and sand, 1:1 mix, minimum		250	.160	C.F.	28	6	1.31	35.31	41.50
0300	Maximum		100	.400		42.50	14.95	3.27	60.72	73
0400	Epoxy cement grout, minimum		137	.292		560	10.95	2.39	573.34	635
0500	Maximum		57	.702		560	26.50	5.75	592.25	660
0700	Alternate pricing method: (Add for materials)									
0710	5 person crew and equipment	B-61	1	40	Day		1,500	325	1,825	2,650

31 45 Vibroflotation and Densification

31 45 13 – Vibroflotation

31 45 13.10 Vibroflotation Densification

		Crew	Daily Output	Labor-Hours	Unit	Material	Labor	Equipment	Total	Total Incl O&P
0010	**VIBROFLOTATION DENSIFICATION** R314513-90									
0900	Vibroflotation compacted sand cylinder, minimum	B-60	750	.075	V.L.F.		3.05	2.82	5.87	7.75
0950	Maximum		325	.172			7.05	6.50	13.55	17.85
1100	Vibro replacement compacted stone cylinder, minimum		500	.112			4.57	4.23	8.80	11.60
1150	Maximum		250	.224			9.15	8.45	17.60	23
1300	Mobilization and demobilization, minimum		.47	119	Total		4,875	4,500	9,375	12,400
1400	Maximum		.14	400	"		16,300	15,100	31,400	41,400

31 46 Needle Beams

31 46 13 – Cantilever Needle Beams

31 46 13.10 Needle Beams	Crew	Daily Output	Labor-Hours	Unit	Material	2012 Bare Costs Labor	Equipment	Total	Total Incl O&P
0010 **NEEDLE BEAMS**									
0011 Incl. wood shoring 10' x 10' opening									
0400 Block, concrete, 8" thick	B-9	7.10	5.634	Ea.	48	200	29.50	277.50	395
0420 12" thick		6.70	5.970		56	212	31	299	420
0800 Brick, 4" thick with 8" backup block		5.70	7.018		56	249	36.50	341.50	485
1000 Brick, solid, 8" thick		6.20	6.452		48	229	33.50	310.50	440
1040 12" thick		4.90	8.163		56	290	42.50	388.50	555
1080 16" thick		4.50	8.889		71.50	315	46.50	433	615
2000 Add for additional floors of shoring	B-1	6	4		48	143		191	273

31 48 Underpinning

31 48 13 – Underpinning Piers

31 48 13.10 Underpinning Foundations	Crew	Daily Output	Labor-Hours	Unit	Material	2012 Bare Costs Labor	Equipment	Total	Total Incl O&P
0010 **UNDERPINNING FOUNDATIONS**									
0011 Including excavation,									
0020 forming, reinforcing, concrete and equipment									
0100 5' to 16' below grade, 100 to 500 C.Y.	B-52	2.30	24.348	C.Y.	272	995	241	1,508	2,100
0200 Over 500 C.Y.		2.50	22.400		245	915	221	1,381	1,900
0400 16' to 25' below grade, 100 to 500 C.Y.		2	28		299	1,150	277	1,726	2,375
0500 Over 500 C.Y.		2.10	26.667		283	1,100	263	1,646	2,275
0700 26' to 40' below grade, 100 to 500 C.Y.		1.60	35		325	1,425	345	2,095	2,925
0800 Over 500 C.Y.		1.80	31.111		299	1,275	305	1,879	2,625
0900 For under 50 C.Y., add					10%	40%			
1000 For 50 C.Y. to 100 C.Y., add					5%	20%			

31 52 Cofferdams

31 52 16 – Timber Cofferdams

31 52 16.10 Cofferdams	Crew	Daily Output	Labor-Hours	Unit	Material	2012 Bare Costs Labor	Equipment	Total	Total Incl O&P
0010 **COFFERDAMS**									
0011 Incl. mobilization and temporary sheeting									
0080 Soldier beams & lagging H piles with 3" wood sheeting									
0090 horizontal between piles, including removal of wales & braces									
0100 No hydrostatic head, 15' deep, 1 line of braces, minimum	B-50	545	.206	S.F.	6.40	8.65	4.26	19.31	25.50
0200 Maximum		495	.226		7.10	9.50	4.69	21.29	28
0400 15' to 22' deep with 2 lines of braces, 10" H, minimum		360	.311		7.50	13.10	6.45	27.05	36
0500 Maximum		330	.339		8.50	14.25	7.05	29.80	39.50
0700 23' to 35' deep with 3 lines of braces, 12" H, minimum		325	.345		9.80	14.50	7.15	31.45	41.50
0800 Maximum		295	.380		10.65	15.95	7.85	34.45	45.50
1000 36' to 45' deep with 4 lines of braces, 14" H, minimum		290	.386		11	16.25	8	35.25	46.50
1100 Maximum		265	.423		11.60	17.75	8.75	38.10	50.50
1300 No hydrostatic head, left in place, 15' dp., 1 line of braces, min.		635	.176		8.50	7.40	3.65	19.55	25
1400 Maximum		575	.195		9.10	8.20	4.04	21.34	27.50
1600 15' to 22' deep with 2 lines of braces, minimum		455	.246		12.75	10.35	5.10	28.20	36
1700 Maximum		415	.270		14.20	11.35	5.60	31.15	39.50
1900 23' to 35' deep with 3 lines of braces, minimum		420	.267		15.20	11.20	5.55	31.95	40.50
2000 Maximum		380	.295		16.80	12.40	6.10	35.30	44.50
2200 36' to 45' deep with 4 lines of braces, minimum		385	.291		18.25	12.25	6.05	36.55	46
2300 Maximum		350	.320		21.50	13.45	6.65	41.60	52
2350 Lagging only, 3" thick wood between piles 8' O.C., minimum	B-46	400	.120		1.42	4.71	.11	6.24	9.15

31 52 Cofferdams

31 52 16 – Timber Cofferdams

31 52 16.10 Cofferdams	Crew	Daily Output	Labor-Hours	Unit	Material	2012 Bare Costs Labor	Equipment	Total	Total Incl O&P
2370	Maximum	B-46	250	.192	S.F.	2.13	7.55	.17	9.85	14.45
2400	Open sheeting no bracing, for trenches to 10' deep, min.		1736	.028		.64	1.09	.03	1.76	2.44
2450	Maximum		1510	.032		.71	1.25	.03	1.99	2.78
2500	Tie-back method, add to open sheeting, add, minimum								20%	20%
2550	Maximum								60%	60%
2700	Tie-backs only, based on tie-backs total length, minimum	B-46	86.80	.553	L.F.	14	21.50	.50	36	50.50
2750	Maximum		38.50	1.247	"	24.50	49	1.13	74.63	105
3500	Tie-backs only, typical average, 25' long		2	24	Ea.	615	945	22	1,582	2,175
3600	35' long		1.58	30.380	"	820	1,200	27.50	2,047.50	2,800

31 56 Slurry Walls

31 56 23 – Lean Concrete Slurry Walls

31 56 23.20 Slurry Trench	Crew	Daily Output	Labor-Hours	Unit	Material	2012 Bare Costs Labor	Equipment	Total	Total Incl O&P
0010	**SLURRY TRENCH**									
0011	Excavated slurry trench in wet soils									
0020	backfilled with 3000 PSI concrete, no reinforcing steel									
0050	Minimum	C-7	333	.216	C.F.	7.55	8.25	3.54	19.34	24.50
0100	Maximum		200	.360	"	12.65	13.70	5.90	32.25	41.50
0200	Alternate pricing method, minimum		150	.480	S.F.	15.10	18.30	7.85	41.25	53.50
0300	Maximum		120	.600		22.50	23	9.80	55.30	71
0500	Reinforced slurry trench, minimum	B-48	177	.316		11.35	12.50	15.65	39.50	49
0600	Maximum	"	69	.812		38	32	40	110	135
0800	Haul for disposal, 2 mile haul, excavated material, add	B-34B	99	.081	C.Y.		2.85	5.90	8.75	10.85
0900	Haul bentonite castings for disposal, add	"	40	.200	"		7.05	14.60	21.65	27

31 62 Driven Piles

31 62 13 – Concrete Piles

31 62 13.23 Prestressed Concrete Piles	Crew	Daily Output	Labor-Hours	Unit	Material	2012 Bare Costs Labor	Equipment	Total	Total Incl O&P
0010	**PRESTRESSED CONCRETE PILES**, 200 piles									
0020	Unless specified otherwise, not incl. pile caps or mobilization									
2200	Precast, prestressed, 50' long, 12" diam., 2-3/8" wall	B-19	720	.089	V.L.F.	21.50	3.92	2.54	27.96	33
2300	14" diameter, 2-1/2" wall		680	.094		29	4.15	2.69	35.84	41.50
2500	16" diameter, 3" wall		640	.100		41.50	4.41	2.86	48.77	55.50
2600	18" diameter, 3" wall	B-19A	600	.107		52	4.71	3.72	60.43	69
2800	20" diameter, 3-1/2" wall		560	.114		62.50	5.05	3.99	71.54	81.50
2900	24" diameter, 3-1/2" wall		520	.123		71	5.45	4.30	80.75	91
3100	Precast, prestressed, 40' long, 10" thick, square	B-19	700	.091		15.85	4.03	2.62	22.50	26.50
3200	12" thick, square		680	.094		20	4.15	2.69	26.84	31.50
3400	14" thick, square		600	.107		23.50	4.71	3.05	31.26	36.50
3500	Octagonal		640	.100		32	4.41	2.86	39.27	45
3700	16" thick, square		560	.114		37	5.05	3.27	45.32	52
3800	Octagonal		600	.107		38.50	4.71	3.05	46.26	53.50
4000	18" thick, square	B-19A	520	.123		44.50	5.45	4.30	54.25	62
4100	Octagonal	B-19	560	.114		45.50	5.05	3.27	53.82	61.50
4300	20" thick, square	B-19A	480	.133		55.50	5.90	4.65	66.05	75.50
4400	Octagonal	B-19	520	.123		50.50	5.45	3.52	59.47	68
4600	24" thick, square	B-19A	440	.145		78.50	6.40	5.10	90	102
4700	Octagonal	B-19	480	.133		72.50	5.90	3.82	82.22	93
4730	Precast, prestressed, 60' long, 10" thick, square		700	.091		16.65	4.03	2.62	23.30	27.50

31 62 Driven Piles

31 62 13 – Concrete Piles

31 62 13.23 Prestressed Concrete Piles	Crew	Daily Output	Labor-Hours	Unit	Material	2012 Bare Costs Labor	Equipment	Total	Total Incl O&P
4740	12" thick, square (60' long)	B-19	680	.094	V.L.F.	21	4.15	2.69	27.84	32.50
4750	Mobilization for 10,000 L.F. pile job, add		3300	.019			.86	.55	1.41	1.95
4800	25,000 L.F. pile job, add	▼	8500	.008	▼		.33	.22	.55	.76

31 62 16 – Steel Piles

31 62 16.13 Steel Piles

		Crew	Daily Output	Labor-Hours	Unit	Material	2012 Bare Costs Labor	Equipment	Total	Total Incl O&P
0010	**STEEL PILES**									
0100	Step tapered, round, concrete filled									
0110	8" tip, 60 ton capacity, 30' depth	B-19	760	.084	V.L.F.	10.50	3.71	2.41	16.62	20
0120	60' depth		740	.086		11.90	3.82	2.48	18.20	22
0130	80' depth		700	.091		12.30	4.03	2.62	18.95	23
0150	10" tip, 90 ton capacity, 30' depth		700	.091		12.60	4.03	2.62	19.25	23
0160	60' depth		690	.093		13	4.09	2.65	19.74	23.50
0170	80' depth		670	.096		14	4.21	2.73	20.94	25
0190	12" tip, 120 ton capacity, 30' depth		660	.097		17.05	4.28	2.78	24.11	28.50
0200	60' depth, 12" diameter		630	.102		17	4.48	2.91	24.39	29
0210	80' depth		590	.108		15.70	4.78	3.10	23.58	28
0250	"H" Sections, 50' long, HP8 x 36		640	.100		19.30	4.41	2.86	26.57	31
0400	HP10 X 42		610	.105		22.50	4.63	3	30.13	35.50
0500	HP10 X 57		610	.105		30.50	4.63	3	38.13	44
0700	HP12 X 53	▼	590	.108		29	4.78	3.10	36.88	42.50
0800	HP12 X 74	B-19A	590	.108		40.50	4.78	3.79	49.07	56
1000	HP14 X 73		540	.119		39.50	5.25	4.14	48.89	56.50
1100	HP14 X 89		540	.119		49	5.25	4.14	58.39	67
1300	HP14 X 102		510	.125		56	5.55	4.38	65.93	75
1400	HP14 X 117	▼	510	.125	▼	64	5.55	4.38	73.93	84
1600	Splice on standard points, not in leads, 8" or 10"	1 Sswl	5	1.600	Ea.	127	79		206	280
1700	12" or 14"		4	2		185	98.50		283.50	380
1900	Heavy duty points, not in leads, 10" wide		4	2		196	98.50		294.50	390
2100	14" wide	▼	3.50	2.286	▼	254	113		367	480

31 62 19 – Timber Piles

31 62 19.10 Wood Piles

		Crew	Daily Output	Labor-Hours	Unit	Material	2012 Bare Costs Labor	Equipment	Total	Total Incl O&P
0010	**WOOD PILES**									
0011	Friction or end bearing, not including									
0050	mobilization or demobilization									
0100	Untreated piles, up to 30' long, 12" butts, 8" points	B-19	625	.102	V.L.F.	20	4.52	2.93	27.45	32.50
0200	30' to 39' long, 12" butts, 8" points		700	.091		20	4.03	2.62	26.65	31
0300	40' to 49' long, 12" butts, 7" points		720	.089		20	3.92	2.54	26.46	31
0400	50' to 59' long, 13"butts, 7" points		800	.080		20	3.53	2.29	25.82	30
0500	60' to 69' long, 13" butts, 7" points		840	.076		22.50	3.36	2.18	28.04	32.50
0600	70' to 80' long, 13" butts, 6" points	▼	840	.076	▼	25	3.36	2.18	30.54	35
0800	Treated piles, 12 lb. per C.F.,									
0810	friction or end bearing, ASTM class B									
1000	Up to 30' long, 12" butts, 8" points	B-19	625	.102	V.L.F.	18.60	4.52	2.93	26.05	31
1100	30' to 39' long, 12" butts, 8" points		700	.091		20	4.03	2.62	26.65	31
1200	40' to 49' long, 12" butts, 7" points		720	.089		20	3.92	2.54	26.46	31
1300	50' to 59' long, 13" butts, 7" points	▼	800	.080		21	3.53	2.29	26.82	31
1400	60' to 69' long, 13" butts, 6" points	B-19A	840	.076		30	3.36	2.66	36.02	41
1500	70' to 80' long, 13" butts, 6" points	"	840	.076	▼	32	3.36	2.66	38.02	43
1600	Treated piles, C.C.A., 2.5# per C.F.									
1610	8" butts, 10' long	B-19	400	.160	V.L.F.	20	7.05	4.58	31.63	38
1620	11" to 16' long		500	.128		20	5.65	3.66	29.31	35
1630	17' to 20' long	▼	575	.111	▼	20	4.91	3.19	28.10	33

31 62 Driven Piles

31 62 19 – Timber Piles

31 62 19.10 Wood Piles

		Crew	Daily Output	Labor-Hours	Unit	Material	2012 Bare Costs Labor	Equipment	Total	Total Incl O&P
1640	10" butts, 10' to 16' long	B-19	500	.128	V.L.F.	20	5.65	3.66	29.31	35
1650	17' to 20' long		575	.111		20	4.91	3.19	28.10	33
1660	21' to 40' long		700	.091		20	4.03	2.62	26.65	31
1670	12" butts, 10' to 20' long		575	.111		20	4.91	3.19	28.10	33
1680	21' to 35' long		650	.098		20	4.34	2.82	27.16	32
1690	36' to 40' long		700	.091		20	4.03	2.62	26.65	31
1695	14" butts. to 40' long		700	.091		25	4.03	2.62	31.65	36.50
1700	Boot for pile tip, minimum	1 Pile	27	.296	Ea.	24.50	12.70		37.20	47.50
1800	Maximum		21	.381		73.50	16.30		89.80	107
2000	Point for pile tip, minimum		20	.400		24.50	17.10		41.60	54.50
2100	Maximum		15	.533		88.50	23		111.50	134
2300	Splice for piles over 50' long, minimum	B-46	35	1.371		61.50	54	1.25	116.75	154
2400	Maximum		20	2.400		73.50	94.50	2.18	170.18	232
2600	Concrete encasement with wire mesh and tube		331	.145	V.L.F.	11.55	5.70	.13	17.38	22
2700	Mobilization for 10,000 L.F. pile job, add	B-19	3300	.019			.86	.55	1.41	1.95
2800	25,000 L.F. pile job, add	"	8500	.008			.33	.22	.55	.76

31 62 23 – Composite Piles

31 62 23.13 Concrete-Filled Steel Piles

		Crew	Daily Output	Labor-Hours	Unit	Material	2012 Bare Costs Labor	Equipment	Total	Total Incl O&P
0010	**CONCRETE-FILLED STEEL PILES** no mobilization or demobilization									
2600	Pipe piles, 50' lg. 8" diam., 29 lb. per L.F., no concrete	B-19	500	.128	V.L.F.	20.50	5.65	3.66	29.81	35.50
2700	Concrete filled		460	.139		21.50	6.15	3.98	31.63	37.50
2900	10" diameter, 34 lb. per L.F., no concrete		500	.128		25.50	5.65	3.66	34.81	41
3000	Concrete filled		450	.142		27.50	6.25	4.07	37.82	45
3200	12" diameter, 44 lb. per L.F., no concrete		475	.135		31	5.95	3.86	40.81	48
3300	Concrete filled		415	.154		32	6.80	4.41	43.21	51
3500	14" diameter, 46 lb. per L.F., no concrete		430	.149		33.50	6.55	4.26	44.31	52
3600	Concrete filled		355	.180		34.50	7.95	5.15	47.60	56
3800	16" diameter, 52 lb. per L.F., no concrete		385	.166		37	7.35	4.76	49.11	58
3900	Concrete filled		335	.191		39.50	8.45	5.45	53.40	62
4100	18" diameter, 59 lb. per L.F., no concrete		355	.180		49	7.95	5.15	62.10	71.50
4200	Concrete filled		310	.206		55	9.10	5.90	70	81.50
4400	Splices for pipe piles, not in leads, 8" diameter	1 Sswl	4.67	1.713	Ea.	91	84.50		175.50	251
4500	14" diameter		3.79	2.111		119	104		223	315
4600	16" diameter		3.03	2.640		147	130		277	395
4800	Points, standard, 8" diameter		4.61	1.735		149	85.50		234.50	315
4900	14" diameter		4.05	1.975		207	97.50		304.50	400
5000	16" diameter		3.37	2.374		252	117		369	485
5200	Points, heavy duty, 10" diameter		2.89	2.768		295	137		432	570
5300	14" or 16" diameter		2.02	3.960		470	195		665	865
5500	For reinforcing steel, add		1150	.007	Lb.	.83	.34		1.17	1.52
5700	For thick wall sections, add				"	.93			.93	1.02

31 63 Bored Piles

31 63 26 – Drilled Caissons

31 63 26.13 Fixed End Caisson Piles	Crew	Daily Output	Labor-Hours	Unit	Material	2012 Bare Costs Labor	Equipment	Total	Total Incl O&P
0010 **FIXED END CAISSON PILES** R316326-60									
0015 Including excavation, concrete, 50 lb. reinforcing									
0020 per C.Y., not incl. mobilization, boulder removal, disposal									
0100 Open style, machine drilled, to 50' deep, in stable ground, no									
0110 casings or ground water, 18" diam., 0.065 C.Y./L.F.	B-43	200	.240	V.L.F.	8.20	9.25	12.20	29.65	36.50
0200 24" diameter, 0.116 C.Y./L.F.		190	.253		14.65	9.75	12.85	37.25	45
0300 30" diameter, 0.182 C.Y./L.F.		150	.320		23	12.35	16.25	51.60	62.50
0400 36" diameter, 0.262 C.Y./L.F.		125	.384		33	14.80	19.55	67.35	80.50
0500 48" diameter, 0.465 C.Y./L.F.		100	.480		59	18.50	24.50	102	120
0600 60" diameter, 0.727 C.Y./L.F.		90	.533		92	20.50	27	139.50	163
0700 72" diameter, 1.05 C.Y./L.F.		80	.600		133	23	30.50	186.50	215
0800 84" diameter, 1.43 C.Y./L.F.		75	.640		181	24.50	32.50	238	273
1000 For bell excavation and concrete, add									
1020 4' bell diameter, 24" shaft, 0.444 C.Y.	B-43	20	2.400	Ea.	45.50	92.50	122	260	325
1040 6' bell diameter, 30" shaft, 1.57 C.Y.		5.70	8.421		160	325	430	915	1,150
1060 8' bell diameter, 36" shaft, 3.72 C.Y.		2.40	20		380	770	1,025	2,175	2,725
1080 9' bell diameter, 48" shaft, 4.48 C.Y.		2	24		455	925	1,225	2,605	3,275
1100 10' bell diameter, 60" shaft, 5.24 C.Y.		1.70	28.235		535	1,100	1,425	3,060	3,850
1120 12' bell diameter, 72" shaft, 8.74 C.Y.		1	48		890	1,850	2,450	5,190	6,475
1140 14' bell diameter, 84" shaft, 13.6 C.Y.		.70	68.571		1,375	2,650	3,475	7,500	9,400
1200 Open style, machine drilled, to 50' deep, in wet ground, pulled									
1300 casing and pumping, 18" diameter, 0.065 C.Y./L.F.	B-48	160	.350	V.L.F.	8.20	13.80	17.35	39.35	49
1400 24" diameter, 0.116 C.Y./L.F.		125	.448		14.65	17.70	22	54.35	67.50
1500 30" diameter, 0.182 C.Y./L.F.		85	.659		23	26	32.50	81.50	101
1600 36" diameter, 0.262 C.Y./L.F.		60	.933		33	37	46	116	144
1700 48" diameter, 0.465 C.Y./L.F.	B-49	55	1.600		59	65.50	63	187.50	235
1800 60" diameter, 0.727 C.Y./L.F.		35	2.514		92	103	99	294	370
1900 72" diameter, 1.05 C.Y./L.F.		30	2.933		133	120	116	369	460
2000 84" diameter, 1.43 C.Y./L.F.		25	3.520		181	144	139	464	575
2100 For bell excavation and concrete, add									
2120 4' bell diameter, 24" shaft, 0.444 C.Y.	B-48	19.80	2.828	Ea.	45.50	112	140	297.50	375
2140 6' bell diameter, 30" shaft, 1.57 C.Y.		5.70	9.825		160	390	485	1,035	1,300
2160 8' bell diameter, 36" shaft, 3.72 C.Y.		2.40	23.333		380	920	1,150	2,450	3,100
2180 9' bell diameter, 48" shaft, 4.48 C.Y.	B-49	3.30	26.667		455	1,100	1,050	2,605	3,325
2200 10' bell diameter, 60" shaft, 5.24 C.Y.		2.80	31.429		535	1,300	1,250	3,085	3,950
2220 12' bell diameter, 72" shaft, 8.74 C.Y.		1.60	55		890	2,250	2,175	5,315	6,825
2240 14' bell diameter, 84" shaft, 13.6 C.Y.		1	88		1,375	3,600	3,475	8,450	10,900
2300 Open style, machine drilled, to 50' deep, in soft rocks and									
2400 medium hard shales, 18" diameter, 0.065 C.Y./L.F.	B-49	50	1.760	V.L.F.	8.20	72	69.50	149.70	197
2500 24" diameter, 0.116 C.Y./L.F.		30	2.933		14.65	120	116	250.65	330
2600 30" diameter, 0.182 C.Y./L.F.		20	4.400		23	180	173	376	495
2700 36" diameter, 0.262 C.Y./L.F.		15	5.867		33	240	231	504	660
2800 48" diameter, 0.465 C.Y./L.F.		10	8.800		59	360	345	764	1,000
2900 60" diameter, 0.727 C.Y./L.F.		7	12.571		92	515	495	1,102	1,425
3000 72" diameter, 1.05 C.Y./L.F.		6	14.667		133	600	580	1,313	1,700
3100 84" diameter, 1.43 C.Y./L.F.		5	17.600		181	720	695	1,596	2,075
3200 For bell excavation and concrete, add									
3220 4' bell diameter, 24" shaft, 0.444 C.Y.	B-49	10.90	8.073	Ea.	45.50	330	320	695.50	910
3240 6' bell diameter, 30" shaft, 1.57 C.Y.		3.10	28.387		160	1,175	1,125	2,460	3,200
3260 8' bell diameter, 36" shaft, 3.72 C.Y.		1.30	67.692		380	2,775	2,675	5,830	7,625
3280 9' bell diameter, 48" shaft, 4.48 C.Y.		1.10	80		455	3,275	3,150	6,880	9,025
3300 10' bell diameter, 60" shaft, 5.24 C.Y.		.90	97.778		535	4,000	3,850	8,385	11,000
3320 12' bell diameter, 72" shaft, 8.74 C.Y.		.60	146		890	6,000	5,775	12,665	16,600

31 63 Bored Piles

31 63 26 – Drilled Caissons

31 63 26.13 Fixed End Caisson Piles

	Crew	Daily Output	Labor-Hours	Unit	Material	2012 Bare Costs Labor	Equipment	Total	Total Incl O&P
3340	14' bell diameter, 84" shaft, 13.6 C.Y.	B-49	.40	220	Ea.	1,375	9,025	8,675	19,075	25,000
3600	For rock excavation, sockets, add, minimum		120	.733	C.F.		30	29	59	78
3650	Average		95	.926			38	36.50	74.50	98.50
3700	Maximum		48	1.833			75	72.50	147.50	196
3900	For 50' to 100' deep, add				V.L.F.				7%	7%
4000	For 100' to 150' deep, add								25%	25%
4100	For 150' to 200' deep, add								30%	30%
4200	For casings left in place, add				Lb.	1.04			1.04	1.14
4300	For other than 50 lb. reinf. per C.Y., add or deduct				"	1.03			1.03	1.13
4400	For steel "I" beam cores, add	B-49	8.30	10.602	Ton	2,000	435	420	2,855	3,325
4500	Load and haul excess excavation, 2 miles	B-34B	178	.045	L.C.Y.		1.59	3.28	4.87	6
4600	For mobilization, 50 mile radius, rig to 36"	B-43	2	24	Ea.		925	1,225	2,150	2,775
4650	Rig to 84"	B-48	1.75	32			1,275	1,575	2,850	3,675
4700	For low headroom, add								50%	
5000	Bottom inspection	1 Skwk	1.20	6.667			305		305	470

31 63 26.16 Concrete Caissons for Marine Construction

	Crew	Daily Output	Labor-Hours	Unit	Material	2012 Bare Costs Labor	Equipment	Total	Total Incl O&P
0010	**CONCRETE CAISSONS FOR MARINE CONSTRUCTION**									
0100	Caissons, incl. mobilization and demobilization, up to 50 miles									
0200	Uncased shafts, 30 to 80 tons cap., 17" diam., 10' depth	B-44	88	.727	V.L.F.	20.50	31.50	20	72	94.50
0300	25' depth		165	.388		14.55	16.80	10.80	42.15	54.50
0400	80-150 ton capacity, 22" diameter, 10' depth		80	.800		25.50	34.50	22.50	82.50	107
0500	20' depth		130	.492		20.50	21.50	13.70	55.70	71
0700	Cased shafts, 10 to 30 ton capacity, 10-5/8" diam., 20' depth		175	.366		14.55	15.85	10.15	40.55	52.50
0800	30' depth		240	.267		13.60	11.55	7.40	32.55	41.50
0850	30 to 60 ton capacity, 12" diameter, 20' depth		160	.400		20.50	17.35	11.10	48.95	62.50
0900	40' depth		230	.278		15.70	12.05	7.75	35.50	44.50
1000	80 to 100 ton capacity, 16" diameter, 20' depth		160	.400		29	17.35	11.10	57.45	72
1100	40' depth		230	.278		27	12.05	7.75	46.80	57.50
1200	110 to 140 ton capacity, 17-5/8" diameter, 20' depth		160	.400		31.50	17.35	11.10	59.95	74.50
1300	40' depth		230	.278		29	12.05	7.75	48.80	59.50
1400	140 to 175 ton capacity, 19" diameter, 20' depth		130	.492		34	21.50	13.70	69.20	86
1500	40' depth		210	.305		31.50	13.20	8.50	53.20	65
1700	Over 30' long, L.F. cost tends to be lower									
1900	Maximum depth is about 90'									

31 63 29 – Drilled Concrete Piers and Shafts

31 63 29.13 Uncased Drilled Concrete Piers

	Crew	Daily Output	Labor-Hours	Unit	Material	2012 Bare Costs Labor	Equipment	Total	Total Incl O&P
0010	**UNCASED DRILLED CONCRETE PIERS**									
0020	Unless specified otherwise, not incl. pile caps or mobilization									
0050	Cast in place augered piles, no casing or reinforcing									
0060	8" diameter	B-43	540	.089	V.L.F.	3.92	3.43	4.52	11.87	14.55
0065	10" diameter		480	.100		6.25	3.86	5.10	15.21	18.35
0070	12" diameter		420	.114		8.80	4.41	5.80	19.01	23
0075	14" diameter		360	.133		11.85	5.15	6.80	23.80	28.50
0080	16" diameter		300	.160		15.95	6.15	8.15	30.25	36
0085	18" diameter		240	.200		19.75	7.70	10.15	37.60	44.50
0100	Cast in place, thin wall shell pile, straight sided,									
0110	not incl. reinforcing, 8" diam., 16 ga., 5.8 lb./L.F.	B-19	700	.091	V.L.F.	9.05	4.03	2.62	15.70	19.20
0200	10" diameter, 16 ga. corrugated, 7.3 lb./L.F.		650	.098		11.85	4.34	2.82	19.01	23
0300	12" diameter, 16 ga. corrugated, 8.7 lb./L.F.		600	.107		15.40	4.71	3.05	23.16	27.50
0400	14" diameter, 16 ga. corrugated, 10.0 lb./L.F.		550	.116		18.10	5.15	3.33	26.58	31.50
0500	16" diameter, 16 ga. corrugated, 11.6 lb./L.F.		500	.128		22	5.65	3.66	31.31	37.50
0800	Cast in place friction pile, 50' long, fluted,									

31 63 29 – Drilled Concrete Piers and Shafts

31 63 29.13 Uncased Drilled Concrete Piers	Crew	Daily Output	Labor-Hours	Unit	Material	2012 Bare Costs Labor	Equipment	Total	Total Incl O&P
0810	tapered steel, 4000 psi concrete, no reinforcing								
0900	12" diameter, 7 ga.	B-19	600	.107	V.L.F.	28.50	4.71	3.05	36.26	42.50
1000	14" diameter, 7 ga.		560	.114		31	5.05	3.27	39.32	46
1100	16" diameter, 7 ga.		520	.123		36.50	5.45	3.52	45.47	53
1200	18" diameter, 7 ga.	▼	480	.133	▼	43	5.90	3.82	52.72	61
1300	End bearing, fluted, constant diameter,								
1320	4000 psi concrete, no reinforcing								
1340	12" diameter, 7 ga.	B-19	600	.107	V.L.F.	30	4.71	3.05	37.76	44
1360	14" diameter, 7 ga.		560	.114		37.50	5.05	3.27	45.82	52.50
1380	16" diameter, 7 ga.		520	.123		43.50	5.45	3.52	52.47	60
1400	18" diameter, 7 ga.	▼	480	.133	▼	48	5.90	3.82	57.72	66

31 63 29.20 Cast In Place Piles, Adds

31 63 29.20 Cast In Place Piles, Adds	Crew	Daily Output	Labor-Hours	Unit	Material	2012 Bare Costs Labor	Equipment	Total	Total Incl O&P
0010	**CAST IN PLACE PILES, ADDS**									
1500	For reinforcing steel, add				Lb.	.98			.98	1.08
1700	For ball or pedestal end, add	B-19	11	5.818	C.Y.	144	257	167	568	745
1900	For lengths above 60', concrete, add	"	11	5.818	"	150	257	167	574	755
2000	For steel thin shell, pipe only				Lb.	1.17			1.17	1.29

Estimating Tips

32 01 00 Operations and Maintenance of Exterior Improvements

- Recycling of asphalt pavement is becoming very popular and is an alternative to removal and replacement. It can be a good value engineering proposal if removed pavement can be recycled, either at the project site or at another site that is reasonably close to the project site. Two new sections on repair of flexible and rigid pavement have been added.

32 10 00 Bases, Ballasts, and Paving

- When estimating paving, keep in mind the project schedule. If an asphaltic paving project is in a colder climate and runs through to the spring, consider placing the base course in the autumn and then topping it in the spring, just prior to completion. This could save considerable costs in spring repair. Keep in mind that prices for asphalt and concrete are generally higher in the cold seasons. New lines have been added for pavement markings including tactile warning systems and new fence lines.

32 90 00 Planting

- The timing of planting and guarantee specifications often dictate the costs for establishing tree and shrub growth and a stand of grass or ground cover. Establish the work performance schedule to coincide with the local planting season. Maintenance and growth guarantees can add from 20%–100% to the total landscaping cost. The cost to replace trees and shrubs can be as high as 5% of the total cost, depending on the planting zone, soil conditions, and time of year.

Reference Numbers

Reference numbers are shown in shaded boxes at the beginning of some major classifications. These numbers refer to related items in the Reference Section. The reference information may be an estimating procedure, an alternate pricing method, or technical information.

Note: Not all subdivisions listed here necessarily appear in this publication.

Division 32 - Exterior Improvements

32 01 Operation and Maintenance of Exterior Improvements

32 01 13 – Flexible Paving Surface Treatment

32 01 13.61 Slurry Seal (Latex Modified)

		Crew	Daily Output	Labor-Hours	Unit	Material	2012 Bare Costs Labor	Equipment	Total	Total Incl O&P
0010	**SLURRY SEAL (LATEX MODIFIED)**									
3780	Rubberized asphalt (latex) seal	B-45	5000	.003	S.Y.	2.70	.13	.17	3	3.35

32 01 13.64 Sand Seal

0010	**SAND SEAL**									
2080	Sand sealing, sharp sand, asphalt emulsion, small area	B-91	10000	.006	S.Y.	1.40	.26	.22	1.88	2.18
2120	Roadway or large area	"	18000	.004	"	1.20	.15	.12	1.47	1.67

32 01 13.66 Fog Seal

0010	**FOG SEAL**									
0012	Sealcoating, 2 coat coal tar pitch emulsion over 10,000 S.Y.	B-45	5000	.003	S.Y.	1	.13	.17	1.30	1.48
0030	1000 to 10,000 S.Y.	"	3000	.005		1	.22	.28	1.50	1.73
0100	Under 1000 S.Y.	B-1	1050	.023		1	.82		1.82	2.36
0300	Petroleum resistant, over 10,000 S.Y.	B-45	5000	.003		1.28	.13	.17	1.58	1.79
0320	1000 to 10,000 S.Y.	"	3000	.005		1.28	.22	.28	1.78	2.04
0400	Under 1000 S.Y.	B-1	1050	.023		1.28	.82		2.10	2.67
0600	Non-skid pavement renewal, over 10,000 S.Y.	B-45	5000	.003		1.40	.13	.17	1.70	1.92
0620	1000 to 10,000 S.Y.	"	3000	.005		1.40	.22	.28	1.90	2.17
0700	Under 1000 S.Y.	B-1	1050	.023		1.40	.82		2.22	2.80
0800	Prepare and clean surface for above	A-2	8545	.003			.10	.02	.12	.18
1000	Hand seal asphalt curbing	B-1	4420	.005	L.F.	.84	.19		1.03	1.22
1900	Asphalt surface treatment, single course, small area									
1901	0.30 gal/S.Y. asphalt material, 20#/S.Y. aggregate	B-91	5000	.013	S.Y.	1.92	.53	.43	2.88	3.38
1910	Roadway or large area		10000	.006		1.77	.26	.22	2.25	2.58
1950	Asphalt surface treatment, dbl. course for small area		3000	.021		3.84	.88	.72	5.44	6.35
1960	Roadway or large area		6000	.011		3.46	.44	.36	4.26	4.86
1980	Asphalt surface treatment, single course, for shoulders		7500	.009		1.68	.35	.29	2.32	2.70

32 06 Schedules for Exterior Improvements

32 06 10 – Schedules for Bases, Ballasts, and Paving

32 06 10.10 Sidewalks, Driveways and Patios

		Crew	Daily Output	Labor-Hours	Unit	Material	2012 Bare Costs Labor	Equipment	Total	Total Incl O&P
0010	**SIDEWALKS, DRIVEWAYS AND PATIOS** No base									
0020	Asphaltic concrete, 2" thick	B-37	720	.067	S.Y.	6.55	2.47	.21	9.23	11.20
0100	2-1/2" thick	"	660	.073	"	8.35	2.69	.23	11.27	13.55
0300	Concrete, 3000 psi, CIP, 6 x 6 - W1.4 x W1.4 mesh,									
0310	broomed finish, no base, 4" thick	B-24	600	.040	S.F.	1.70	1.62		3.32	4.33
0350	5" thick		545	.044		2.27	1.78		4.05	5.20
0400	6" thick		510	.047		2.64	1.91		4.55	5.80
0450	For bank run gravel base, 4" thick, add	B-18	2500	.010		.53	.34	.02	.89	1.13
0520	8" thick, add	"	1600	.015		1.07	.54	.03	1.64	2.04
0550	Exposed aggregate finish, add to above, minimum	B-24	1875	.013		.13	.52		.65	.93
0600	Maximum	"	455	.053		.43	2.14		2.57	3.72
1000	Crushed stone, 1" thick, white marble	2 Clab	1700	.009		.24	.33		.57	.78
1050	Bluestone	"	1700	.009		.18	.33		.51	.71
1700	Redwood, prefabricated, 4' x 4' sections	2 Carp	316	.051		4.74	2.23		6.97	8.65
1750	Redwood planks, 1" thick, on sleepers	"	240	.067		4.74	2.94		7.68	9.70
2250	Stone dust, 4" thick	B-62	900	.027	S.Y.	3.66	1.02	.18	4.86	5.80

32 06 10.20 Steps

0010	**STEPS**									
0011	Incl. excav., borrow & concrete base as required									
0100	Brick steps	B-24	35	.686	LF Riser	11.70	28		39.70	55
0200	Railroad ties	2 Clab	25	.640		3.10	22.50		25.60	38

32 06 Schedules for Exterior Improvements

32 06 10 - Schedules for Bases, Ballasts, and Paving

32 06 10.20 Steps		Crew	Daily Output	Labor-Hours	Unit	Material	2012 Bare Costs		Total	Total Incl O&P
							Labor	Equipment		
0300	Bluestone treads, 12" x 2" or 12" x 1-1/2"	B-24	30	.800	LF Riser	28	32.50		60.50	79.50
0500	Concrete, cast in place, see Section 03 30 53.40									
0600	Precast concrete, see Section 03 41 23.50									
4025	Steel edge strips, incl. stakes, 1/4" x 5"	B-1	390	.062	L.F.	2.50	2.20		4.70	6.15
4050	Edging, landscape timber or railroad ties, 6" x 8"	2 Carp	170	.094	"	3.97	4.15		8.12	10.75

32 11 Base Courses

32 11 23 - Aggregate Base Courses

32 11 23.23 Base Course Drainage Layers

		Crew	Daily Output	Labor-Hours	Unit	Material	2012 Bare Costs		Total	Total Incl O&P
							Labor	Equipment		
0010	**BASE COURSE DRAINAGE LAYERS**									
0011	For roadways and large areas									
0050	Crushed 3/4" stone base, compacted, 3" deep	B-36C	5200	.008	S.Y.	3.04	.33	.72	4.09	4.63
0100	6" deep		5000	.008		6.10	.34	.75	7.19	8.05
0200	9" deep		4600	.009		9.15	.37	.81	10.33	11.50
0300	12" deep		4200	.010		12.20	.40	.89	13.49	15
0301	Crushed 1-1/2" stone base, compacted to 4" deep	B-36B	6000	.011		3.98	.44	.72	5.14	5.85
0302	6" deep		5400	.012		5.95	.49	.80	7.24	8.15
0303	8" deep		4500	.014		7.95	.58	.97	9.50	10.70
0304	12" deep		3800	.017		11.95	.69	1.14	13.78	15.45
0350	Bank run gravel, spread and compacted									
0370	6" deep	B-32	6000	.005	S.Y.	4.44	.23	.35	5.02	5.60
0390	9" deep		4900	.007		6.65	.29	.42	7.36	8.25
0400	12" deep		4200	.008		8.90	.33	.50	9.73	10.85
6000	Stabilization fabric, polypropylene, 6 oz./S.Y.	B-6	10000	.002		1.14	.09	.03	1.26	1.43
6900	For small and irregular areas, add						50%	50%		
7000	Prepare and roll sub-base, small areas to 2500 S.Y.	B-32A	1500	.016	S.Y.		.68	.83	1.51	1.96
8000	Large areas over 2500 S.Y.	"	3500	.007			.29	.36	.65	.83
8050	For roadways	B-32	4000	.008			.35	.52	.87	1.10

32 11 26 - Asphaltic Base Courses

32 11 26.19 Bituminous-Stabilized Base Courses

		Crew	Daily Output	Labor-Hours	Unit	Material	2012 Bare Costs		Total	Total Incl O&P
							Labor	Equipment		
0010	**BITUMINOUS-STABILIZED BASE COURSES**									
0020	And large paved areas									
0700	Liquid application to gravel base, asphalt emulsion	B-45	6000	.003	Gal.	6.70	.11	.14	6.95	7.65
0800	Prime and seal, cut back asphalt		6000	.003	"	7.90	.11	.14	8.15	9
1000	Macadam penetration crushed stone, 2 gal. per S.Y., 4" thick		6000	.003	S.Y.	13.40	.11	.14	13.65	15.05
1100	6" thick, 3 gal. per S.Y.		4000	.004		20	.16	.21	20.37	22.50
1200	8" thick, 4 gal. per S.Y.		3000	.005		27	.22	.28	27.50	30
8900	For small and irregular areas, add						50%	50%		

32 12 Flexible Paving

32 12 16 – Asphalt Paving

32 12 16.13 Plant-Mix Asphalt Paving

		Crew	Daily Output	Labor-Hours	Unit	Material	2012 Bare Costs Labor	Equipment	Total	Total Incl O&P
0010	**PLANT-MIX ASPHALT PAVING**									
0020	And large paved areas with no hauling included									
0025	See Section 31 23 23.20 for hauling costs									
0080	Binder course, 1-1/2" thick	B-25	7725	.011	S.Y.	5.05	.44	.33	5.82	6.65
0120	2" thick		6345	.014		6.75	.53	.40	7.68	8.65
0130	2-1/2" thick		5620	.016		8.45	.60	.45	9.50	10.65
0160	3" thick		4905	.018		10.10	.69	.52	11.31	12.75
0170	3-1/2" thick		4520	.019		11.80	.75	.56	13.11	14.75
0200	4" thick		4140	.021		13.50	.82	.61	14.93	16.75
0300	Wearing course, 1" thick	B-25B	10575	.009		3.34	.35	.26	3.95	4.51
0340	1-1/2" thick		7725	.012		5.60	.49	.36	6.45	7.35
0380	2" thick		6345	.015		7.55	.59	.44	8.58	9.70
0420	2-1/2" thick		5480	.018		9.30	.68	.50	10.48	11.85
0460	3" thick		4900	.020		11.10	.77	.56	12.43	14
0470	3-1/2" thick		4520	.021		13.05	.83	.61	14.49	16.30
0480	4" thick		4140	.023		14.90	.91	.67	16.48	18.45
0500	Open graded friction course	B-25C	5000	.010		2.70	.38	.44	3.52	4.04
0800	Alternate method of figuring paving costs									
0810	Binder course, 1-1/2" thick	B-25	630	.140	Ton	62	5.35	4.03	71.38	80.50
0811	2" thick		690	.128		62	4.90	3.68	70.58	79.50
0812	3" thick		800	.110		62	4.22	3.17	69.39	78
0813	4" thick		900	.098		62	3.75	2.82	68.57	77
0850	Wearing course, 1" thick	B-25B	575	.167		68	6.50	4.81	79.31	90.50
0851	1-1/2" thick		630	.152		68	5.95	4.39	78.34	89
0852	2" thick		690	.139		68	5.45	4.01	77.46	87.50
0853	2-1/2" thick		765	.125		68	4.90	3.61	76.51	86.50
0854	3" thick		800	.120		68	4.69	3.46	76.15	86
1000	Pavement replacement over trench, 2" thick	B-37	90	.533	S.Y.	6.95	19.75	1.66	28.36	40
1050	4" thick		70	.686		13.80	25.50	2.13	41.43	56.50
1080	6" thick		55	.873		22	32.50	2.71	57.21	76.50

32 12 16.14 Asphaltic Concrete Paving

		Crew	Daily Output	Labor-Hours	Unit	Material	2012 Bare Costs Labor	Equipment	Total	Total Incl O&P
0011	**ASPHALTIC CONCRETE PAVING**, parking lots & driveways									
0015	No asphalt hauling included									
0018	Use 6.05 C. Y. per inch per M.S.F. for hauling									
0020	6" stone base, 2" binder course, 1" topping	B-25C	9000	.005	S.F.	1.70	.21	.25	2.16	2.46
0025	2" binder course, 2" topping		9000	.005		2.11	.21	.25	2.57	2.91
0030	3" binder course, 2" topping		9000	.005		2.49	.21	.25	2.95	3.32
0035	4" binder course, 2" topping		9000	.005		2.86	.21	.25	3.32	3.73
0040	1.5" binder course, 1" topping		9000	.005		1.51	.21	.25	1.97	2.25
0042	3" binder course, 1" topping		9000	.005		2.07	.21	.25	2.53	2.87
0045	3" binder course, 3" topping		9000	.005		2.90	.21	.25	3.36	3.78
0050	4" binder course, 3" topping		9000	.005		3.27	.21	.25	3.73	4.19
0055	4" binder course, 4" topping		9000	.005		3.68	.21	.25	4.14	4.64
0300	Binder course, 1-1/2" thick		35000	.001		.56	.05	.06	.67	.77
0400	2" thick		25000	.002		.73	.08	.09	.90	1.02
0500	3" thick		15000	.003		1.13	.13	.15	1.41	1.59
0600	4" thick		10800	.004		1.48	.17	.21	1.86	2.12
0800	Sand finish course, 3/4" thick		41000	.001		.32	.05	.05	.42	.48
0900	1" thick		34000	.001		.39	.06	.07	.52	.58
1000	Fill pot holes, hot mix, 2" thick	B-16	4200	.008		.84	.27	.14	1.25	1.49
1100	4" thick		3500	.009		1.23	.33	.17	1.73	2.03
1120	6" thick		3100	.010		1.65	.37	.19	2.21	2.59

32 12 Flexible Paving

32 12 16 – Asphalt Paving

32 12 16.14 Asphaltic Concrete Paving		Crew	Daily Output	Labor-Hours	Unit	Material	2012 Bare Costs Labor	Equipment	Total	Total Incl O&P
1140	Cold patch, 2" thick	B-51	3000	.016	S.F.	.85	.57	.07	1.49	1.89
1160	4" thick		2700	.018		1.62	.63	.08	2.33	2.83
1180	6" thick	↓	1900	.025	↓	2.52	.89	.11	3.52	4.26

32 13 Rigid Paving

32 13 13 – Concrete Paving

32 13 13.23 Concrete Paving Surface Treatment

		Crew	Daily Output	Labor-Hours	Unit	Material	2012 Bare Costs Labor	Equipment	Total	Total Incl O&P
0010	**CONCRETE PAVING SURFACE TREATMENT**									
0015	Including joints, finishing and curing									
0020	Fixed form, 12' pass, unreinforced, 6" thick	B-26	3000	.029	S.Y.	20.50	1.15	1.05	22.70	25.50
0100	8" thick		2750	.032		28	1.26	1.14	30.40	33.50
0110	8" thick, small area		1375	.064		28	2.52	2.28	32.80	37
0200	9" thick		2500	.035		32	1.38	1.26	34.64	38.50
0300	10" thick		2100	.042		35	1.65	1.50	38.15	42.50
0310	10" thick, small area		1050	.084		35	3.29	2.99	41.28	47
0400	12" thick		1800	.049		40.50	1.92	1.75	44.17	49.50
0410	Conc pavement, w/jt,fnsh&curing,fix form,24' pass,unreinforced,6"T		6000	.015		19.50	.58	.52	20.60	23
0430	8" thick		5500	.016		26.50	.63	.57	27.70	31
0440	9" thick		5000	.018		30.50	.69	.63	31.82	35.50
0450	10" thick		4200	.021		33.50	.82	.75	35.07	39
0460	12" thick		3600	.024		39	.96	.87	40.83	45.50
0470	15" thick		3000	.029		51	1.15	1.05	53.20	59
0500	Fixed form 12' pass 15" thick	↓	1500	.059	↓	51	2.31	2.09	55.40	62.50
0510	For small irregular areas, add				%	10%	100%	100%		
0520	Welded wire fabric, sheets for rigid paving 2.33 lb./S.Y.	2 Rodm	389	.041	S.Y.	1.24	2.02		3.26	4.59
0530	Reinforcing steel for rigid paving 12 lb./S.Y.		666	.024		6.20	1.18		7.38	8.70
0540	Reinforcing steel for rigid paving 18 lb./S.Y.	↓	444	.036		9.25	1.77		11.02	13.05
0620	Slip form, 12' pass, unreinforced, 6" thick	B-26A	5600	.016		19.80	.62	.59	21.01	23.50
0624	8" thick		5300	.017		27	.65	.62	28.27	31
0626	9" thick		4820	.018		31	.72	.68	32.40	36
0628	10" thick		4050	.022		34	.85	.81	35.66	39.50
0630	12" thick		3470	.025		39.50	1	.95	41.45	46
0632	15" thick		2890	.030		50	1.20	1.14	52.34	57.50
0640	Slip form, 24' pass, unreinforced, 6" thick		11200	.008		19.45	.31	.29	20.05	22.50
0644	8" thick		10600	.008		26	.33	.31	26.64	30
0646	9" thick		9640	.009		30	.36	.34	30.70	34
0648	10" thick		8100	.011		33	.43	.41	33.84	37.50
0650	12" thick		6940	.013		38.50	.50	.47	39.47	44
0652	15" thick	↓	5780	.015		48.50	.60	.57	49.67	55
0700	Finishing, broom finish small areas	2 Cefi	120	.133			5.65		5.65	8.30
1000	Curing, with sprayed membrane by hand	2 Clab	1500	.011	↓	.67	.37		1.04	1.32
1650	For integral coloring, see Section 03 05 13.20									

32 14 Unit Paving

32 14 13 – Precast Concrete Unit Paving

32 14 13.13 Interlocking Precast Concrete Unit Paving

		Crew	Daily Output	Labor-Hours	Unit	Material	2012 Bare Costs Labor	Equipment	Total	Total Incl O&P
0010	**INTERLOCKING PRECAST CONCRETE UNIT PAVING**									
0020	"V" blocks for retaining soil	D-1	205	.078	S.F.	9.50	3.11		12.61	15.15

32 14 13.16 Precast Concrete Unit Paving Slabs

		Crew	Daily Output	Labor-Hours	Unit	Material	2012 Bare Costs Labor	Equipment	Total	Total Incl O&P
0010	**PRECAST CONCRETE UNIT PAVING SLABS**									
0710	Precast concrete patio blocks, 2-3/8" thick, colors, 8" x 16"	D-1	265	.060	S.F.	2	2.41		4.41	5.85
0750	Exposed local aggregate, natural	2 Bric	250	.064		7.10	2.82		9.92	12.05
0800	Colors		250	.064		7.10	2.82		9.92	12.05
0850	Exposed granite or limestone aggregate		250	.064		7.20	2.82		10.02	12.15
0900	Exposed white tumblestone aggregate		250	.064		6.15	2.82		8.97	11

32 14 13.18 Precast Concrete Plantable Pavers

		Crew	Daily Output	Labor-Hours	Unit	Material	2012 Bare Costs Labor	Equipment	Total	Total Incl O&P
0010	**PRECAST CONCRETE PLANTABLE PAVERS** (50% grass)									
0015	Subgrade preparation and grass planting not included									
0100	Precast concrete plantable pavers with topsoil, 24" x 16"	B-63	800	.050	S.F.	4.39	1.85	.20	6.44	7.90
0200	Less than 600 Square Feet or irregular area	"	500	.080	"	4.39	2.96	.32	7.67	9.70
0300	3/4" crushed stone base for plantable pavers, 6 inch depth	B-62	1000	.024	S.Y.	6.10	.92	.16	7.18	8.30
0400	8 inch depth		900	.027		8.15	1.02	.18	9.35	10.75
0500	10 inch depth		800	.030		10.15	1.15	.20	11.50	13.15
0600	12 inch depth		700	.034		12.20	1.31	.23	13.74	15.65
0700	Hydro seeding plantable pavers	B-81A	20	.800	M.S.F.	19.60	28	19.25	66.85	85
0800	Apply fertilizer and seed to plantable pavers	1 Clab	8	1	"	34.50	35		69.50	92

32 14 16 – Brick Unit Paving

32 14 16.10 Brick Paving

		Crew	Daily Output	Labor-Hours	Unit	Material	2012 Bare Costs Labor	Equipment	Total	Total Incl O&P
0010	**BRICK PAVING**									
0012	4" x 8" x 1-1/2", without joints (4.5 brick/S.F.)	D-1	110	.145	S.F.	2.86	5.80		8.66	11.90
0100	Grouted, 3/8" joint (3.9 brick/S.F.)		90	.178		3.38	7.10		10.48	14.40
0200	4" x 8" x 2-1/4", without joints (4.5 bricks/S.F.)		110	.145		3.74	5.80		9.54	12.85
0300	Grouted, 3/8" joint (3.9 brick/S.F.)		90	.178		3.45	7.10		10.55	14.50
0500	Bedding, asphalt, 3/4" thick	B-25	5130	.017		.62	.66	.49	1.77	2.23
0540	Course washed sand bed, 1" thick	B-18	5000	.005		.36	.17	.01	.54	.67
0580	Mortar, 1" thick	D-1	300	.053		.66	2.13		2.79	3.93
0620	2" thick		200	.080		1.31	3.19		4.50	6.25
1500	Brick on 1" thick sand bed laid flat, 4.5 per S.F.		100	.160		3.03	6.40		9.43	13
2000	Brick pavers, laid on edge, 7.2 per S.F.		70	.229		2.80	9.10		11.90	16.85
2500	For 4" thick concrete bed and joints, add		595	.027		1.12	1.07		2.19	2.86
2800	For steam cleaning, add	A-1H	950	.008		.08	.30	.07	.45	.62

32 14 23 – Asphalt Unit Paving

32 14 23.10 Asphalt Blocks

		Crew	Daily Output	Labor-Hours	Unit	Material	2012 Bare Costs Labor	Equipment	Total	Total Incl O&P
0010	**ASPHALT BLOCKS**									
0020	Rectangular, 6" x 12" x 1-1/4", w/bed & neopr. adhesive	D-1	135	.119	S.F.	9.05	4.72		13.77	17.10
0100	3" thick		130	.123		12.65	4.90		17.55	21.50
0300	Hexagonal tile, 8" wide, 1-1/4" thick		135	.119		9.05	4.72		13.77	17.10
0400	2" thick		130	.123		12.65	4.90		17.55	21.50
0500	Square, 8" x 8", 1-1/4" thick		135	.119		9.05	4.72		13.77	17.10
0600	2" thick		130	.123		12.65	4.90		17.55	21.50
0900	For exposed aggregate (ground finish) add					.63			.63	.69
0910	For colors, add					.48			.48	.53

32 14 40 – Stone Paving

32 14 40.10 Stone Pavers

		Crew	Daily Output	Labor-Hours	Unit	Material	2012 Bare Costs Labor	Equipment	Total	Total Incl O&P
0010	**STONE PAVERS**									
1100	Flagging, bluestone, irregular, 1" thick,	D-1	81	.198	S.F.	5.75	7.85		13.60	18.25

32 14 Unit Paving

32 14 40 – Stone Paving

32 14 40.10 Stone Pavers

		Crew	Daily Output	Labor-Hours	Unit	Material	2012 Bare Costs Labor	2012 Bare Costs Equipment	Total	Total Incl O&P
1150	Snapped random rectangular, 1" thick	D-1	92	.174	S.F.	8.70	6.95		15.65	20
1200	1-1/2" thick		85	.188		10.45	7.50		17.95	23
1250	2" thick		83	.193		12.20	7.70		19.90	25
1300	Slate, natural cleft, irregular, 3/4" thick		92	.174		9.60	6.95		16.55	21
1350	Random rectangular, gauged, 1/2" thick		105	.152		21	6.05		27.05	32
1400	Random rectangular, butt joint, gauged, 1/4" thick	↓	150	.107	↓	22.50	4.25		26.75	31
1500	For interior setting, add								25%	25%
1550	Granite blocks, 3-1/2" x 3-1/2" x 3-1/2"	D-1	92	.174	S.F.	10	6.95		16.95	21.50
1600	4" to 12" long, 3" to 5" wide, 3" to 5" thick		98	.163		8.35	6.50		14.85	19
1650	6" to 15" long, 3" to 6" wide, 3" to 5" thick	↓	105	.152	↓	4.45	6.05		10.50	14.05

32 16 Curbs, Gutters, Sidewalks, and Driveways

32 16 13 – Curbs and Gutters

32 16 13.13 Cast-in-Place Concrete Curbs and Gutters

		Crew	Daily Output	Labor-Hours	Unit	Material	2012 Bare Costs Labor	2012 Bare Costs Equipment	Total	Total Incl O&P
0010	**CAST-IN-PLACE CONCRETE CURBS AND GUTTERS**									
0290	Forms only, no concrete									
0300	Concrete, wood forms, 6" x 18", straight	C-2	500	.096	L.F.	2.15	4.12		6.27	8.70
0400	6" x 18", radius	"	200	.240	"	2.25	10.30		12.55	18.30
0402	Forms and concrete complete									
0404	Concrete, wood forms, 6" x 18", straight & concrete	C-2A	500	.096	L.F.	5	4.09		9.09	11.75
0406	6" x 18", radius		200	.240		5.10	10.25		15.35	21.50
0410	Steel forms, 6" x 18", straight		700	.069		4.42	2.92		7.34	9.35
0411	6" x 18", radius	↓	400	.120		6.60	5.10		11.70	15.05
0415	Machine formed, 6" x 18", straight	B-69A	2000	.024		6.15	.93	.38	7.46	8.60
0416	6" x 18", radius	"	900	.053	↓	6.25	2.06	.85	9.16	10.95
0421	Curb and gutter, straight									
0422	with 6" high curb and 6" thick gutter, wood forms									
0430	24" wide, .055 C.Y. per L.F.	C-2A	375	.128	L.F.	13.35	5.45		18.80	23
0435	30" wide, .066 C.Y. per L.F.		340	.141		14.85	6		20.85	25.50
0440	Steel forms, 24" wide, straight		700	.069		11.05	2.92		13.97	16.60
0441	Radius		300	.160		11.05	6.80		17.85	22.50
0442	30" wide, straight		700	.069		13.05	2.92		15.97	18.80
0443	Radius	↓	300	.160		13.05	6.80		19.85	25
0445	Machine formed, 24" wide, straight	B-69A	2000	.024		5.70	.93	.38	7.01	8.15
0446	Radius		900	.053		5.70	2.06	.85	8.61	10.35
0447	30" wide, straight		2000	.024		6.65	.93	.38	7.96	9.15
0448	Radius	↓	900	.053	↓	6.65	2.06	.85	9.56	11.35

32 16 13.23 Precast Concrete Curbs and Gutters

		Crew	Daily Output	Labor-Hours	Unit	Material	2012 Bare Costs Labor	2012 Bare Costs Equipment	Total	Total Incl O&P
0010	**PRECAST CONCRETE CURBS AND GUTTERS**									
0550	Precast, 6" x 18", straight	B-29	700	.080	L.F.	8.75	3.05	1.21	13.01	15.65
0600	6" x 18", radius	"	325	.172	"	19.50	6.55	2.60	28.65	34.50

32 16 13.33 Asphalt Curbs

		Crew	Daily Output	Labor-Hours	Unit	Material	2012 Bare Costs Labor	2012 Bare Costs Equipment	Total	Total Incl O&P
0010	**ASPHALT CURBS**									
0012	Curbs, asphaltic, machine formed, 8" wide, 6" high, 40 L.F./ton	B-27	1000	.032	L.F.	1.65	1.14	.28	3.07	3.88
0100	8" wide, 8" high, 30 L.F. per ton		900	.036		2.20	1.27	.31	3.78	4.71
0150	Asphaltic berm, 12" W, 3"-6" H, 35 L.F./ton, before pavement	↓	700	.046		.04	1.63	.40	2.07	2.98
0200	12" W, 1-1/2" to 4" H, 60 L.F. per ton, laid with pavement	B-2	1050	.038	↓	.02	1.35		1.37	2.10

32 16 13.43 Stone Curbs

		Crew	Daily Output	Labor-Hours	Unit	Material	2012 Bare Costs Labor	2012 Bare Costs Equipment	Total	Total Incl O&P
0010	**STONE CURBS**									
1000	Granite, split face, straight, 5" x 16"	D-13	500	.096	L.F.	12.75	4.05	.90	17.70	21

32 16 Curbs, Gutters, Sidewalks, and Driveways

32 16 13 – Curbs and Gutters

32 16 13.43 Stone Curbs

		Crew	Daily Output	Labor-Hours	Unit	Material	2012 Bare Costs Labor	Equipment	Total	Total Incl O&P
1100	6" x 18"	D-13	450	.107	L.F.	16.75	4.50	1	22.25	26.50
1300	Radius curbing, 6" x 18", over 10' radius	B-29	260	.215	↓	20.50	8.20	3.25	31.95	38.50
1400	Corners, 2' radius		80	.700	Ea.	69	26.50	10.55	106.05	128
1600	Edging, 4-1/2" x 12", straight		300	.187	L.F.	6.40	7.10	2.81	16.31	21
1800	Curb inlets, (guttermouth) straight	↓	41	1.366	Ea.	153	52	20.50	225.50	270
2000	Indian granite (belgian block)									
2100	Jumbo, 10-1/2" x 7-1/2" x 4", grey	D-1	150	.107	L.F.	6.50	4.25		10.75	13.55
2150	Pink		150	.107		6.70	4.25		10.95	13.80
2200	Regular, 9" x 4-1/2" x 4-1/2", grey		160	.100		6.60	3.99		10.59	13.25
2250	Pink		160	.100		5.95	3.99		9.94	12.50
2300	Cubes, 4" x 4" x 4", grey		175	.091		3.83	3.64		7.47	9.70
2350	Pink		175	.091		3.23	3.64		6.87	9.05
2400	6" x 6" x 6", pink	↓	155	.103	↓	11.20	4.11		15.31	18.50
2500	Alternate pricing method for indian granite									
2550	Jumbo, 10-1/2" x 7-1/2" x 4" (30 lb), grey				Ton	370			370	405
2600	Pink					390			390	430
2650	Regular, 9" x 4-1/2" x 4-1/2" (20 lb), grey					465			465	510
2700	Pink					415			415	455
2750	Cubes, 4" x 4" x 4" (5 lb), grey					465			465	510
2800	Pink					415			415	455
2850	6" x 6" x 6" (25 lb), pink					435			435	480
2900	For pallets, add				↓	20			20	22

32 17 Paving Specialties

32 17 13 – Parking Bumpers

32 17 13.13 Metal Parking Bumpers

		Crew	Daily Output	Labor-Hours	Unit	Material	2012 Bare Costs Labor	Equipment	Total	Total Incl O&P
0010	**METAL PARKING BUMPERS**									
0015	Bumper rails for garages, 12 Ga. rail, 6" wide, with steel									
0020	posts 12'-6" O.C., minimum	E-4	190	.168	L.F.	16.15	8.40	.64	25.19	33.50
0030	Average		165	.194		20	9.65	.74	30.39	40
0100	Maximum		140	.229		24	11.40	.87	36.27	48
0300	12" channel rail, minimum		160	.200		20	9.95	.76	30.71	40.50
0400	Maximum	↓	120	.267	↓	30.50	13.30	1.01	44.81	58.50
1300	Pipe bollards, conc filled/paint, 8' L x 4' D hole, 6" diam.	B-6	20	1.200	Ea.	705	46	16.70	767.70	865
1400	8" diam.		15	1.600		1,025	61.50	22.50	1,109	1,250
1500	12" diam.		12	2		1,150	76.50	28	1,254.50	1,425
1592	Bollards, steel, 3' H, retractable, incl hydraulic controls, min		4	6		44,000	230	83.50	44,313.50	48,800
1594	Max	↓	2	12		46,000	460	167	46,627	51,500
2030	Folding with individual padlocks	B-2	50	.800	↓	560	28.50		588.50	660
8000	Parking lot control, see Section 11 12 13.10									
8900	Security bollards, SS, lighted, hyd, incl. controls, group of 3	L-7	.06	509	Ea.	47,000	21,700		68,700	84,500
8910	Group of 5	"	.04	682	"	63,000	29,100		92,100	114,000

32 17 13.16 Plastic Parking Bumpers

		Crew	Daily Output	Labor-Hours	Unit	Material	2012 Bare Costs Labor	Equipment	Total	Total Incl O&P
0010	**PLASTIC PARKING BUMPERS**									
1200	Thermoplastic, 6" x 10" x 6'-0"	B-2	120	.333	Ea.	150	11.85		161.85	183

32 17 13.19 Precast Concrete Parking Bumpers

		Crew	Daily Output	Labor-Hours	Unit	Material	2012 Bare Costs Labor	Equipment	Total	Total Incl O&P
0010	**PRECAST CONCRETE PARKING BUMPERS**									
1000	Wheel stops, precast concrete incl. dowels, 6" x 10" x 6'-0"	B-2	120	.333	Ea.	59.50	11.85		71.35	83.50
1100	8" x 13" x 6'-0"	"	120	.333	"	87.50	11.85		99.35	115

32 17 Paving Specialties

32 17 13 – Parking Bumpers

32 17 13.26 Wood Parking Bumpers	Crew	Daily Output	Labor-Hours	Unit	Material	2012 Bare Costs Labor	2012 Bare Costs Equipment	Total	Total Incl O&P
0010 **WOOD PARKING BUMPERS**									
0020 Parking barriers, timber w/saddles, treated type									
0100 4" x 4" for cars	B-2	520	.077	L.F.	2.26	2.73		4.99	6.70
0200 6" x 6" for trucks		520	.077	"	4.61	2.73		7.34	9.25
0600 Flexible fixed stanchion, 2' high, 3" diameter		100	.400	Ea.	37	14.20		51.20	62.50

32 17 23 – Pavement Markings

32 17 23.13 Painted Pavement Markings

32 17 23.13 Painted Pavement Markings	Crew	Daily Output	Labor-Hours	Unit	Material	2012 Bare Costs Labor	2012 Bare Costs Equipment	Total	Total Incl O&P
0010 **PAINTED PAVEMENT MARKINGS**									
0020 Acrylic waterborne, white or yellow, 4" wide, less than 3000 L.F.	B-78	20000	.002	L.F.	.12	.08	.03	.23	.29
0200 6" wide, less than 3000 L.F.		11000	.004		.18	.15	.05	.38	.49
0500 8" wide, less than 3000 L.F.		10000	.005		.24	.17	.05	.46	.59
0600 12" wide, less than 3000 L.F.		4000	.012		.36	.42	.14	.92	1.20
0620 Arrows or gore lines		2300	.021	S.F.	.15	.74	.23	1.12	1.56
0640 Temporary paint, white or yellow, less than 3000 L.F.		15000	.003	L.F.	.05	.11	.04	.20	.27
0660 Removal	1 Clab	300	.027			.94		.94	1.44
0680 Temporary tape	2 Clab	1500	.011		.42	.37		.79	1.04
0710 Thermoplastic, white or yellow, 4" wide, less than 6000 L.F.	B-79	15000	.003		.28	.09	.09	.46	.55
0730 6" wide, less than 6000 L.F.		14000	.003		.42	.10	.09	.61	.72
0740 8" wide, less than 6000 L.F.		12000	.003		.56	.12	.11	.79	.92
0750 12" wide, less than 6000 L.F.		6000	.007		.83	.24	.22	1.29	1.51
0760 Arrows		660	.061	S.F.	.56	2.14	1.97	4.67	6.05
0770 Gore lines		2500	.016		.56	.57	.52	1.65	2.06
0780 Letters		660	.061		.56	2.14	1.97	4.67	6.05
1000 Airport painted markings									
1050 Traffic safety flashing truck for airport painting	A-2B	1	8	Day		275	211	486	645
1100 Painting, white or yellow, taxiway markings	B-78	4000	.012	S.F.	.24	.42	.14	.80	1.06
1110 with 12 lb. beads per 100 S.F.		4000	.012		.52	.42	.14	1.08	1.37
1200 Runway markings		3500	.014		.24	.48	.15	.87	1.17
1210 with 12 lb. beads per 100 S.F.		3500	.014		.52	.48	.15	1.15	1.48
1300 Pavement location or direction signs		2500	.019		.24	.68	.22	1.14	1.54
1310 with 12 lb. beads per 100 S.F.		2500	.019		.52	.68	.22	1.42	1.85
1350 Mobilization airport pavement painting		4	12	Ea.		425	135	560	800
1400 Paint markings or pavement signs removal daytime	B-78B	400	.045	S.F.		1.63	.86	2.49	3.44
1500 Removal nighttime		335	.054	"		1.94	1.03	2.97	4.11
1600 Mobilization pavement paint removal		4	4.500	Ea.		163	86	249	345

32 17 23.14 Pavement Parking Markings

32 17 23.14 Pavement Parking Markings	Crew	Daily Output	Labor-Hours	Unit	Material	2012 Bare Costs Labor	2012 Bare Costs Equipment	Total	Total Incl O&P
0010 **PAVEMENT PARKING MARKINGS**									
0790 Layout of pavement marking	A-2	25000	.001	L.F.		.03	.01	.04	.06
0800 Lines on pvmt, parking stall, paint, white, 4" wide	B-78B	400	.045	Stall	4.40	1.63	.86	6.89	8.30
0825 Parking stall, small quantities	2 Pord	80	.200		8.80	7.65		16.45	21
0830 Lines on pvmt, parking stall, thermoplastic, white, 4" wide	B-79	300	.133		12	4.71	4.33	21.04	25
1000 Street letters and numbers	B-78B	1600	.011	S.F.	.64	.41	.21	1.26	1.56

32 18 Athletic and Recreational Surfacing

32 18 13 – Synthetic Grass Surfacing

32 18 13.10 Artificial Grass Surfacing

		Crew	Daily Output	Labor-Hours	Unit	Material	2012 Bare Costs Labor	2012 Bare Costs Equipment	Total	Total Incl O&P
0010	**ARTIFICIAL GRASS SURFACING**									
0015	Not including asphalt base or drainage,									
0020	but including cushion pad, over 50,000 S.F.									
0200	1/2" pile and 5/16" cushion pad, standard	C-17	3200	.025	S.F.	10	1.15		11.15	12.75
0300	Deluxe		2560	.031		14.80	1.43		16.23	18.50
0500	1/2" pile and 5/8" cushion pad, standard		2844	.028		14.45	1.29		15.74	17.90
0600	Deluxe		2327	.034		15.90	1.57		17.47	19.95
0800	For asphaltic concrete base, 2-1/2" thick,									
0900	with 6" crushed stone sub-base, add	B-25	12000	.007	S.F.	1.60	.28	.21	2.09	2.42

32 18 23 – Athletic Surfacing

32 18 23.33 Running Track Surfacing

		Crew	Daily Output	Labor-Hours	Unit	Material	2012 Bare Costs Labor	2012 Bare Costs Equipment	Total	Total Incl O&P
0010	**RUNNING TRACK SURFACING**									
0020	Running track, asphalt, incl base, 3" thick	B-37	300	.160	S.Y.	23	5.95	.50	29.45	34.50
0100	Surface, latex rubber system, 3/8" thick, black	B-20	125	.192		10.30	7.55		17.85	23
0150	Colors		125	.192		18.30	7.55		25.85	31.50
0300	Urethane rubber system, 3/8" thick, black		120	.200		27.50	7.85		35.35	42
0400	Color coating		115	.209		33.50	8.20		41.70	49.50

32 18 23.53 Tennis Court Surfacing

		Crew	Daily Output	Labor-Hours	Unit	Material	2012 Bare Costs Labor	2012 Bare Costs Equipment	Total	Total Incl O&P
0010	**TENNIS COURT SURFACING**									
0020	Tennis court, asphalt, incl. base, 2-1/2" thick, one court	B-37	450	.107	S.Y.	35	3.95	.33	39.28	45
0200	Two courts		675	.071		14.35	2.63	.22	17.20	20
0300	Clay courts		360	.133		35.50	4.94	.41	40.85	47
0400	Pulverized natural greenstone with 4" base, fast dry		250	.192		38	7.10	.60	45.70	53
0800	Rubber-acrylic base resilient pavement		600	.080		53.50	2.96	.25	56.71	64
1000	Colored sealer, acrylic emulsion, 3 coats	2 Clab	800	.020		5.80	.70		6.50	7.45
1100	3 coat, 2 colors	"	900	.018		8.10	.62		8.72	9.90
1200	For preparing old courts, add	1 Clab	825	.010			.34		.34	.52
1400	Posts for nets, 3-1/2" diameter with eye bolts	B-1	3.40	7.059	Pr.	228	252		480	640
1500	With pulley & reel		3.40	7.059	"	325	252		577	750
1700	Net, 42' long, nylon thread with binder		50	.480	Ea.	300	17.15		317.15	355
1800	All metal		6.50	3.692	"	530	132		662	790
2000	Paint markings on asphalt, 2 coats	1 Pord	1.78	4.494	Court	144	172		316	415
2200	Complete court with fence, etc., asphaltic conc., minimum	B-37	.20	240		27,000	8,900	745	36,645	44,100
2300	Maximum		.16	300		54,000	11,100	935	66,035	77,500
2800	Clay courts, minimum		.20	240		26,000	8,900	745	35,645	43,000
2900	Maximum		.16	300		42,000	11,100	935	54,035	64,000

32 31 Fences and Gates

32 31 13 – Chain Link Fences and Gates

32 31 13.20 Fence, Chain Link Industrial

		Crew	Daily Output	Labor-Hours	Unit	Material	2012 Bare Costs Labor	2012 Bare Costs Equipment	Total	Total Incl O&P
0010	**FENCE, CHAIN LINK INDUSTRIAL**									
0011	Schedule 40, including concrete									
0020	3 strands barb wire, 2" post @ 10' O.C., set in concrete, 6' H									
0200	9 ga. wire, galv. steel, in concrete	B-80C	240	.100	L.F.	18.50	3.49	.91	22.90	27
0248	Fence, add for vinyl coated fabric				S.F.	.68			.68	.75
0300	Aluminized steel	B-80C	240	.100	L.F.	18.90	3.49	.91	23.30	27.50
0500	6 ga. wire, galv. steel		240	.100		20	3.49	.91	24.40	28.50
0600	Aluminized steel		240	.100		23	3.49	.91	27.40	32
0800	6 ga. wire, 6' high but omit barbed wire, galv. steel		250	.096		18.55	3.35	.87	22.77	26.50
0900	Aluminized steel, in concrete		250	.096		22.50	3.35	.87	26.72	30.50

32 31 Fences and Gates

32 31 13 - Chain Link Fences and Gates

32 31 13.20 Fence, Chain Link Industrial

	Crew	Daily Output	Labor-Hours	Unit	Material	2012 Bare Costs Labor	Equipment	Total	Total Incl O&P
0920	8' H, 6 ga. wire, 2-1/2" line post, galv. steel, in concrete	B-80C	180	.133	L.F.	29.50	4.65	1.21	35.36	41
0940	Aluminized steel, in concrete		180	.133	↓	36	4.65	1.21	41.86	48
1100	Add for corner posts, 3" diam., galv. steel, in concrete		40	.600	Ea.	86	21	5.45	112.45	133
1200	Aluminized steel, in concrete		40	.600		99	21	5.45	125.45	147
1300	Add for braces, galv. steel		80	.300		30	10.45	2.73	43.18	52
1350	Aluminized steel		80	.300		39	10.45	2.73	52.18	62
1400	Gate for 6' high fence, 1-5/8" frame, 3' wide, galv. steel		10	2.400		188	83.50	22	293.50	360
1500	Aluminized steel, in concrete	↓	10	2.400	↓	188	83.50	22	293.50	360
2000	5'-0" high fence, 9 ga., no barbed wire, 2" line post, in concrete									
2010	10' O.C., 1-5/8" top rail, in concrete									
2100	Galvanized steel, in concrete	B-80C	300	.080	L.F.	18.20	2.79	.73	21.72	25
2200	Aluminized steel, in concrete		300	.080	"	18.90	2.79	.73	22.42	26
2400	Gate, 4' wide, 5' high, 2" frame, galv. steel, in concrete		10	2.400	Ea.	82.50	83.50	22	188	243
2500	Aluminized steel, in concrete		10	2.400	"	96.50	83.50	22	202	258
3100	Overhead slide gate, chain link, 6' high, to 18' wide, in concrete	↓	38	.632	L.F.	92.50	22	5.75	120.25	142
3110	Cantilever type, in concrete	B-80	48	.667		65.50	25	13.90	104.40	126
3120	8' high, in concrete		24	1.333		147	50.50	28	225.50	270
3130	10' high, in concrete	↓	18	1.778	↓	180	67.50	37	284.50	340
5000	Double swing gates, incl. posts & hardware, in concrete									
5010	5' high, 12' opening, in concrete	B-80C	3.40	7.059	Opng.	253	246	64.50	563.50	725
5020	20' opening, in concrete		2.80	8.571		292	299	78	669	860
5060	6' high, 12' opening, in concrete		3.20	7.500		233	261	68.50	562.50	730
5070	20' opening, in concrete		2.60	9.231		270	320	84	674	885
5080	8' high, 12' opening, in concrete	B-80	2.13	15.002		270	570	315	1,155	1,500
5090	20' opening, in concrete		1.45	22.069		320	835	460	1,615	2,125
5100	10' high, 12' opening, in concrete		1.31	24.427		2,000	925	510	3,435	4,150
5110	20' opening, in concrete		1.03	31.068		2,175	1,175	650	4,000	4,900
5120	12' high, 12' opening, in concrete		1.05	30.476		1,425	1,150	635	3,210	4,000
5130	20' opening, in concrete	↓	.85	37.647	↓	2,025	1,425	785	4,235	5,275
5190	For aluminized steel add					20%				
7075	Fence, for small jobs 100 L.F. or less fence w/or wo gate, add				S.F.	20%			20%	

32 31 13.25 Fence, Chain Link Residential

	Crew	Daily Output	Labor-Hours	Unit	Material	2012 Bare Costs Labor	Equipment	Total	Total Incl O&P
0010	**FENCE, CHAIN LINK RESIDENTIAL**									
0011	Schedule 20, 11 gauge wire, 1-5/8" post									
0020	10' O.C., 1-3/8" top rail, 2" corner post, galv. stl. 3' high	B-80C	500	.048	L.F.	2.13	1.67	.44	4.24	5.40
0050	4' high		400	.060		4.86	2.09	.55	7.50	9.15
0100	6' high		200	.120	↓	6.35	4.18	1.09	11.62	14.55
0150	Add for gate 3' wide, 1-3/8" frame, 3' high		12	2	Ea.	69	69.50	18.20	156.70	203
0170	4' high		10	2.400		74	83.50	22	179.50	234
0190	6' high		10	2.400		105	83.50	22	210.50	268
0200	Add for gate 4' wide, 1-3/8" frame, 3' high		9	2.667		74	93	24.50	191.50	250
0220	4' high		9	2.667		78	93	24.50	195.50	255
0240	6' high		8	3	↓	113	105	27.50	245.50	315
0350	Aluminized steel, 11 ga. wire, 3' high		500	.048	L.F.	5.70	1.67	.44	7.81	9.35
0380	4' high		400	.060		7.30	2.09	.55	9.94	11.85
0400	6' high		200	.120	↓	10.55	4.18	1.09	15.82	19.20
0450	Add for gate 3' wide, 1-3/8" frame, 3' high		12	2	Ea.	83	69.50	18.20	170.70	219
0470	4' high		10	2.400		99.50	83.50	22	205	262
0490	6' high		10	2.400		124	83.50	22	229.50	288
0500	Add for gate 4' wide, 1-3/8" frame, 3' high		10	2.400		97	83.50	22	202.50	259
0520	4' high		9	2.667		116	93	24.50	233.50	297
0540	6' high		8	3	↓	137	105	27.50	269.50	340

32 31 Fences and Gates

32 31 13 – Chain Link Fences and Gates

32 31 13.25 Fence, Chain Link Residential

		Crew	Daily Output	Labor-Hours	Unit	Material	2012 Bare Costs Labor	Equipment	Total	Total Incl O&P
0620	Vinyl covered, 9 ga. wire, 3' high	B-80C	500	.048	L.F.	4.95	1.67	.44	7.06	8.50
0640	4' high		400	.060		5.90	2.09	.55	8.54	10.30
0660	6' high		200	.120		9.30	4.18	1.09	14.57	17.80
0720	Add for gate 3' wide, 1-3/8" frame, 3' high		12	2	Ea.	45.50	69.50	18.20	133.20	177
0740	4' high		10	2.400		51.50	83.50	22	157	209
0760	6' high		10	2.400		115	83.50	22	220.50	279
0780	Add for gate 4' wide, 1-3/8" frame, 3' high		10	2.400		74	83.50	22	179.50	234
0800	4' high		9	2.667		84	93	24.50	201.50	261
0820	6' high		8	3		121	105	27.50	253.50	325
7076	Fence, for small jobs 100 L.F. fence or less w/or wo gate, add				S.F.	20%			20%	

32 31 13.26 Tennis Court Fences and Gates

		Crew	Daily Output	Labor-Hours	Unit	Material	Labor	Equipment	Total	Total Incl O&P
0010	**TENNIS COURT FENCES AND GATES**									
0860	Tennis courts, 11 ga. wire, 2-1/2" post set									
0870	in concrete, 10' O.C., 1-5/8" top rail									
0900	10' high	B-80	190	.168	L.F.	16.60	6.35	3.51	26.46	32
0920	12' high		170	.188	"	18.60	7.10	3.93	29.63	35.50
1000	Add for gate 4' wide, 1-5/8" frame 7' high		10	3.200	Ea.	183	121	67	371	460
1040	Aluminized steel, 11 ga. wire 10' high		190	.168	L.F.	21.50	6.35	3.51	31.36	37.50
1100	12' high		170	.188	"	21.50	7.10	3.93	32.53	39
1140	Add for gate 4' wide, 1-5/8" frame, 7' high		10	3.200	Ea.	213	121	67	401	495
1250	Vinyl covered, 9 ga. wire, 10' high		190	.168	L.F.	19.60	6.35	3.51	29.46	35
1300	12' high		170	.188	"	25	7.10	3.93	36.03	42.50
1310	Fence, CL, tennis ct, transom gate, single, galv, 4' x 7' x 3'	B-80A	8.72	2.752	Ea.	180	96.50	30.50	307	380
1400	Add for gate 4' wide, 1-5/8" frame, 7' high	B-80	10	3.200	"	288	121	67	476	575

32 31 13.33 Chain Link Backstops

		Crew	Daily Output	Labor-Hours	Unit	Material	Labor	Equipment	Total	Total Incl O&P
0010	**CHAIN LINK BACKSTOPS**									
0015	Backstops, baseball, prefabricated, 30' wide, 12' high & 1 overhang	B-1	1	24	Ea.	4,650	860		5,510	6,450
0100	40' wide, 12' high & 2 overhangs	"	.75	32		6,450	1,150		7,600	8,850
0300	Basketball, steel, single goal	B-13	3.04	18.421		1,450	700	229	2,379	2,925
0400	Double goal	"	1.92	29.167		2,125	1,100	365	3,590	4,425
0600	Tennis, wire mesh with pair of ends	B-1	2.48	9.677	Set	2,500	345		2,845	3,275
0700	Enclosed court	"	1.30	18.462	Ea.	9,750	660		10,410	11,700

32 31 13.53 High-Security Chain Link Fences, Gates and Systems

		Crew	Daily Output	Labor-Hours	Unit	Material	Labor	Equipment	Total	Total Incl O&P
0010	**HIGH-SECURITY CHAIN LINK FENCES, GATES AND SYSTEMS**									
0100	Fence, chain link, security, 7' H, standard FE-7, incl excavation & posts	B-80C	480	.050	L.F.	41	1.74	.46	43.20	48
0200	Fence, barbed wire, security, 7' high, with 3 wire barbed wire arm	"	400	.060	"	6.20	2.09	.55	8.84	10.60
0300	Complete systems, including material and installation									
0310	Taunt wire fence detection system				M.L.F.				24,600	27,000
0410	Microwave fence detection system								40,500	44,500
0510	Passive magnetic fence detection system								19,100	21,000
0610	Infrared fence detection system								12,700	14,000
0710	Strain relief fence detection system								24,600	27,000
0810	Electro-shock fence detection system								35,200	38,700
0910	Photo-electric fence detection system								16,000	17,600

32 31 19 – Decorative Metal Fences and Gates

32 31 19.10 Decorative Fence

		Crew	Daily Output	Labor-Hours	Unit	Material	Labor	Equipment	Total	Total Incl O&P
0010	**DECORATIVE FENCE**									
5300	Tubular picket, steel, 6' sections, 1-9/16" posts, 4' high	B-80C	300	.080	L.F.	29	2.79	.73	32.52	37
5400	2" posts, 5' high		240	.100		44	3.49	.91	48.40	54.50
5600	2" posts, 6' high		200	.120		50.50	4.18	1.09	55.77	63.50
5700	Staggered picket 1-9/16" posts, 4' high		300	.080		29	2.79	.73	32.52	37

32 31 Fences and Gates

32 31 19 – Decorative Metal Fences and Gates

32 31 19.10 Decorative Fence

	32 31 19.10 Decorative Fence	Crew	Daily Output	Labor-Hours	Unit	Material	2012 Bare Costs Labor	Equipment	Total	Total Incl O&P
5800	2" posts, 5' high	B-80C	240	.100	L.F.	44	3.49	.91	48.40	54.50
5900	2" posts, 6' high	↓	200	.120	↓	50.50	4.18	1.09	55.77	63.50
6200	Gates, 4' high, 3' wide	B-1	10	2.400	Ea.	250	86		336	405
6300	5' high, 3' wide		10	2.400		264	86		350	420
6400	6' high, 3' wide		10	2.400		315	86		401	480
6500	4' wide	↓	10	2.400	↓	345	86		431	510

32 31 26 – Wire Fences and Gates

32 31 26.10 Fences, Misc. Metal

	32 31 26.10 Fences, Misc. Metal	Crew	Daily Output	Labor-Hours	Unit	Material	2012 Bare Costs Labor	Equipment	Total	Total Incl O&P
0010	**FENCES, MISC. METAL**									
0012	Chicken wire, posts @ 4', 1" mesh, 4' high	B-80C	410	.059	L.F.	4.30	2.04	.53	6.87	8.45
0100	2" mesh, 6' high		350	.069		2.32	2.39	.62	5.33	6.90
0200	Galv. steel, 12 ga., 2" x 4" mesh, posts 5' O.C., 3' high		300	.080		2.59	2.79	.73	6.11	7.90
0300	5' high		300	.080		3.96	2.79	.73	7.48	9.45
0400	14 ga., 1" x 2" mesh, 3' high		300	.080		2.93	2.79	.73	6.45	8.30
0500	5' high		300	.080	↓	4.60	2.79	.73	8.12	10.10
1000	Kennel fencing, 1-1/2" mesh, 6' long, 3'-6" wide, 6'-2" high	2 Clab	4	4	Ea.	490	140		630	750
1050	12' long		4	4		635	140		775	915
1200	Top covers, 1-1/2" mesh, 6' long		15	1.067		163	37.50		200.50	237
1250	12' long	↓	12	1.333	↓	237	47		284	335
1300	For kennel doors, see Section 08 31 13.40									
4500	Security fence, prison grade, set in concrete, 12' high	B-80	25	1.280	L.F.	54	48.50	26.50	129	163
4600	16' high	"	20	1.600	"	73.50	60.50	33.50	167.50	210

32 31 26.20 Wire Fencing, General

	32 31 26.20 Wire Fencing, General	Crew	Daily Output	Labor-Hours	Unit	Material	2012 Bare Costs Labor	Equipment	Total	Total Incl O&P
0010	**WIRE FENCING, GENERAL**									
0015	Barbed wire, galvanized, domestic steel, hi-tensile 15-1/2 ga.				M.L.F.	139			139	153
0020	Standard, 12-3/4 ga.					160			160	176
0210	Barbless wire, 2-strand galvanized, 12-1/2 ga.				↓	160			160	176
0500	Helical razor ribbon, stainless steel, 18" dia x 18" spacing				C.L.F.	440			440	485
0600	Hardware cloth galv., 1/4" mesh, 23 ga., 2' wide				C.S.F.	89.50			89.50	98.50
0700	3' wide					89.50			89.50	98.50
0900	1/2" mesh, 19 ga., 2' wide					89.50			89.50	98.50
1000	4' wide					67			67	73.50
1200	Chain link fabric, steel, 2" mesh, 6 ga, galvanized					239			239	263
1300	9 ga, galvanized					91.50			91.50	101
1350	Vinyl coated					102			102	112
1360	Aluminized					105			105	116
1400	2-1/4" mesh, 11.5 ga, galvanized					57.50			57.50	63
1600	1-3/4" mesh (tennis courts), 11.5 ga (core), vinyl coated					57			57	63
1700	9 ga, galvanized					88			88	96.50
2100	Welded wire fabric, galvanized, 1" x 2", 14 ga.					65.50			65.50	72
2200	2" x 4", 12-1/2 ga.				↓	65.50			65.50	72

32 31 29 – Wood Fences and Gates

32 31 29.20 Fence, Wood Rail

	32 31 29.20 Fence, Wood Rail	Crew	Daily Output	Labor-Hours	Unit	Material	2012 Bare Costs Labor	Equipment	Total	Total Incl O&P
0010	**FENCE, WOOD RAIL**									
0012	Picket, No. 2 cedar, Gothic, 2 rail, 3' high	B-1	160	.150	L.F.	6.85	5.35		12.20	15.80
0050	Gate, 3'-6" wide	B-80C	9	2.667	Ea.	59	93	24.50	176.50	233
0400	3 rail, 4' high		150	.160	L.F.	7.90	5.60	1.46	14.96	18.85
0500	Gate, 3'-6" wide		9	2.667	Ea.	71	93	24.50	188.50	247
0600	Open rail, rustic, No. 1 cedar, 2 rail, 3' high		160	.150	L.F.	6.10	5.25	1.37	12.72	16.20
0650	Gate, 3' wide		9	2.667	Ea.	68.50	93	24.50	186	244
0700	3 rail, 4' high	↓	150	.160	L.F.	7	5.60	1.46	14.06	17.85

32 31 Fences and Gates

32 31 29 – Wood Fences and Gates

32 31 29.20 Fence, Wood Rail	Crew	Daily Output	Labor-Hours	Unit	Material	2012 Bare Costs Labor	Equipment	Total	Total Incl O&P
0900	Gate, 3' wide	B-80C	9	2.667	Ea.	88	93	24.50	205.50	266
1200	Stockade, No. 2 cedar, treated wood rails, 6' high		160	.150	L.F.	8.55	5.25	1.37	15.17	18.90
1250	Gate, 3' wide		9	2.667	Ea.	70	93	24.50	187.50	246
1300	No. 1 cedar, 3-1/4" cedar rails, 6' high		160	.150	L.F.	19.70	5.25	1.37	26.32	31
1500	Gate, 3' wide		9	2.667	Ea.	174	93	24.50	291.50	360
1520	Open rail, split, No. 1 cedar, 2 rail, 3' high		160	.150	L.F.	6.15	5.25	1.37	12.77	16.30
1540	3 rail, 4'-0" high		150	.160		7.45	5.60	1.46	14.51	18.35
3300	Board, shadow box, 1" x 6", treated pine, 6' high		160	.150		11.60	5.25	1.37	18.22	22.50
3400	No. 1 cedar, 6' high		150	.160		23	5.60	1.46	30.06	35
3900	Basket weave, No. 1 cedar, 6' high		160	.150		35	5.25	1.37	41.62	48
3950	Gate, 3'-6" wide	B-1	8	3	Ea.	180	107		287	365
4000	Treated pine, 6' high		150	.160	L.F.	17.55	5.70		23.25	28
4200	Gate, 3'-6" wide		9	2.667	Ea.	92.50	95.50		188	249
5000	Fence rail, redwood, 2" x 4", merch grade 8'		2400	.010	L.F.	2.03	.36		2.39	2.78
5050	Select grade, 8'		2400	.010	"	4.52	.36		4.88	5.50
6000	Fence post, select redwood, earthpacked & treated, 4" x 4" x 6'		96	.250	Ea.	12.75	8.95		21.70	28
6010	4" x 4" x 8'		96	.250		15.15	8.95		24.10	30.50
6020	Set in concrete, 4" x 4" x 6'		50	.480		16.40	17.15		33.55	44.50
6030	4" x 4" x 8'		50	.480		19.55	17.15		36.70	48
6040	Wood post, 4' high, set in concrete, incl. concrete		50	.480		9.15	17.15		26.30	36.50
6050	Earth packed		96	.250		6.25	8.95		15.20	20.50
6060	6' high, set in concrete, incl. concrete		50	.480		11.70	17.15		28.85	39.50
6070	Earth packed		96	.250		8.55	8.95		17.50	23

32 32 Retaining Walls

32 32 13 – Cast-in-Place Concrete Retaining Walls

32 32 13.10 Retaining Walls, Cast Concrete

		Crew	Daily Output	Labor-Hours	Unit	Material	2012 Bare Costs Labor	Equipment	Total	Total Incl O&P
0010	**RETAINING WALLS, CAST CONCRETE**									
1800	Concrete gravity wall with vertical face including excavation & backfill									
1850	No reinforcing									
1900	6' high, level embankment	C-17C	36	2.306	L.F.	70.50	106	16.20	192.70	258
2000	33° slope embankment		32	2.594		86.50	119	18.20	223.70	300
2200	8' high, no surcharge		27	3.074		92.50	141	21.50	255	345
2300	33° slope embankment		24	3.458		112	159	24.50	295.50	395
2500	10' high, level embankment		19	4.368		133	200	30.50	363.50	490
2600	33° slope embankment		18	4.611		184	212	32.50	428.50	565
2800	Reinforced concrete cantilever, incl. excavation, backfill & reinf.									
2900	6' high, 33° slope embankment	C-17C	35	2.371	L.F.	68	109	16.65	193.65	261
3000	8' high, 33° slope embankment		29	2.862		78.50	131	20	229.50	310
3100	10' high, 33° slope embankment		20	4.150		102	190	29	321	440
3200	20' high, 500 lb. per L.F. surcharge		7.50	11.067		305	510	77.50	892.50	1,200
3500	Concrete cribbing, incl. excavation and backfill									
3700	12' high, open face	B-13	210	.267	S.F.	34	10.15	3.32	47.47	56.50
3900	Closed face	"	210	.267	"	32	10.15	3.32	45.47	54
4100	Concrete filled slurry trench, see Section 31 56 23.20									

32 32 23 – Segmental Retaining Walls

32 32 23.13 Segmental Conc. Unit Masonry Retaining Walls

0010	**SEGMENTAL CONC. UNIT MASONRY RETAINING WALLS**									
7100	Segmental Retaining Wall system, incl pins, and void fill									
7120	base not included									

32 32 Retaining Walls

32 32 23 – Segmental Retaining Walls

32 32 23.13 Segmental Conc. Unit Masonry Retaining Walls	Crew	Daily Output	Labor-Hours	Unit	Material	2012 Bare Costs Labor	2012 Bare Costs Equipment	Total	Total Incl O&P
7140	Large unit, 8" high x 18" wide x 20" deep, 3 plane split	B-62	300	.080	S.F.	12.20	3.07	.53	15.80	18.65
7150	Straight split		300	.080		12.10	3.07	.53	15.70	18.55
7160	Medium, lt.wt., 8" high x 8" wide x 12" deep, 3 plane split		400	.060		9.05	2.30	.40	11.75	13.95
7170	Straight split		400	.060		8.55	2.30	.40	11.25	13.35
7180	Small unit, 4" x 18" x 10" deep, 3 plane split		400	.060		12.95	2.30	.40	15.65	18.20
7190	Straight split		400	.060		13.20	2.30	.40	15.90	18.45
7200	Cap unit, 3 plane split		300	.080		12.75	3.07	.53	16.35	19.25
7210	Cap unit, straight split	▼	300	.080	▼	12.75	3.07	.53	16.35	19.25
7260	For reinforcing, add								4.70	5.15
8000	For higher walls, add components as necessary									

32 32 26 – Metal Crib Retaining Walls

32 32 26.10 Metal Bin Retaining Walls

		Crew	Daily Output	Labor-Hours	Unit	Material	2012 Bare Costs Labor	2012 Bare Costs Equipment	Total	Total Incl O&P
0010	**METAL BIN RETAINING WALLS**									
0011	Aluminized steel bin, excavation									
0020	and backfill not included, 10' wide									
0100	4' high, 5.5' deep	B-13	650	.086	S.F.	27.50	3.28	1.07	31.85	36
0200	8' high, 5.5' deep		615	.091		31.50	3.47	1.13	36.10	41
0300	10' high, 7.7' deep		580	.097		33.50	3.68	1.20	38.38	43.50
0400	12' high, 7.7' deep		530	.106		36	4.02	1.32	41.34	47
0500	16' high, 7.7' deep		515	.109		38	4.14	1.35	43.49	50
0600	16' high, 9.9' deep		500	.112		40	4.27	1.39	45.66	52
0700	20' high, 9.9' deep		470	.119		45	4.54	1.48	51.02	58
0800	20' high, 12.1' deep		460	.122		39.50	4.64	1.52	45.66	52.50
0900	24' high, 12.1' deep		455	.123		42	4.69	1.53	48.22	55.50
1000	24' high, 14.3' deep		450	.124		48.50	4.74	1.55	54.79	62.50
1100	28' high, 14.3' deep	▼	440	.127	▼	50.50	4.85	1.58	56.93	64.50
1300	For plain galvanized bin type walls, deduct					10%				

32 32 36 – Gabion Retaining Walls

32 32 36.10 Stone Gabion Retaining Walls

		Crew	Daily Output	Labor-Hours	Unit	Material	2012 Bare Costs Labor	2012 Bare Costs Equipment	Total	Total Incl O&P
0010	**STONE GABION RETAINING WALLS**									
4300	Stone filled gabions, not incl. excavation,									
4310	Stone, delivered, 3' wide									
4350	Galvanized, 6' high, 33° slope embankment	B-13	49	1.143	L.F.	41	43.50	14.25	98.75	127
4500	Highway surcharge		27	2.074		81.50	79	26	186.50	240
4600	9' high, up to 33° slope embankment		24	2.333		92	89	29	210	269
4700	Highway surcharge		16	3.500		143	133	43.50	319.50	410
4900	12' high, up to 33° slope embankment		14	4		143	152	50	345	445
5000	Highway surcharge	▼	11	5.091	▼	204	194	63.50	461.50	590
5950	For PVC coating, add					20%				

32 32 60 – Stone Retaining Walls

32 32 60.10 Retaining Walls, Stone

		Crew	Daily Output	Labor-Hours	Unit	Material	2012 Bare Costs Labor	2012 Bare Costs Equipment	Total	Total Incl O&P
0010	**RETAINING WALLS, STONE**									
0015	Including excavation, concrete footing and									
0020	stone 3' below grade. Price is exposed face area.									
0200	Decorative random stone, to 6' high, 1'-6" thick, dry set	D-1	35	.457	S.F.	43	18.20		61.20	75
0300	Mortar set		40	.400		48	15.95		63.95	77
0500	Cut stone, to 6' high, 1'-6" thick, dry set		35	.457		43	18.20		61.20	75
0600	Mortar set		40	.400		46	15.95		61.95	74.50
0800	Random stone, 6' to 10' high, 2' thick, dry set		45	.356		46	14.15		60.15	72
0900	Mortar set		50	.320		51	12.75		63.75	75.50
1100	Cut stone, 6' to 10' high, 2' thick, dry set	▼	45	.356	▼	46	14.15		60.15	72

32 32 Retaining Walls

32 32 60 – Stone Retaining Walls

32 32 60.10 Retaining Walls, Stone	Crew	Daily Output	Labor-Hours	Unit	Material	2012 Bare Costs Labor	Equipment	Total	Total Incl O&P
1200 Mortar set	D-1	50	.320	S.F.	51	12.75		63.75	75.50

32 34 Fabricated Bridges

32 34 20 – Fabricated Pedestrian Bridges

32 34 20.10 Bridges, Pedestrian

	32 34 20.10 Bridges, Pedestrian	Crew	Daily Output	Labor-Hours	Unit	Material	2012 Bare Costs Labor	Equipment	Total	Total Incl O&P
0010	**BRIDGES, PEDESTRIAN**									
0011	Spans over streams, roadways, etc.									
0020	including erection, not including foundations									
0050	Precast concrete, complete in place, 8' wide, 60' span	E-2	215	.260	S.F.	108	12.55	6.95	127.50	148
0100	100' span		185	.303		118	14.60	8.10	140.70	164
0150	120' span		160	.350		129	16.90	9.35	155.25	180
0200	150' span		145	.386		134	18.65	10.30	162.95	190
0300	Steel, trussed or arch spans, compl. in place, 8' wide, 40' span		320	.175		108	8.45	4.68	121.13	139
0400	50' span		395	.142		97	6.85	3.79	107.64	123
0500	60' span		465	.120		97	5.80	3.22	106.02	120
0600	80' span		570	.098		115	4.74	2.63	122.37	138
0700	100' span		465	.120		162	5.80	3.22	171.02	191
0800	120' span		365	.153		205	7.40	4.10	216.50	242
0900	150' span		310	.181		218	8.70	4.83	231.53	260
1000	160' span		255	.220		218	10.60	5.85	234.45	265
1100	10' wide, 80' span		640	.088		120	4.22	2.34	126.56	142
1200	120' span		415	.135		156	6.50	3.61	166.11	186
1300	150' span		445	.126		175	6.05	3.36	184.41	206
1400	200' span		205	.273		186	13.20	7.30	206.50	236
1600	Wood, laminated type, complete in place, 80' span	C-12	203	.236		64	10.30	3.01	77.31	89.50
1700	130' span	"	153	.314		66.50	13.65	4	84.15	99

32 35 Screening Devices

32 35 16 – Sound Barriers

32 35 16.10 Traffic Barriers, Highway Sound Barriers

	32 35 16.10 Traffic Barriers, Highway Sound Barriers	Crew	Daily Output	Labor-Hours	Unit	Material	2012 Bare Costs Labor	Equipment	Total	Total Incl O&P
0010	**TRAFFIC BARRIERS, HIGHWAY SOUND BARRIERS**									
0020	Highway sound barriers, not including footing									
0100	Precast concrete, concrete columns @ 30' OC, 8" T, 8' H	C-12	400	.120	L.F.	122	5.25	1.53	128.78	144
0110	12' H		265	.181		182	7.90	2.31	192.21	216
0120	16' H		200	.240		243	10.45	3.06	256.51	287
0130	20' H		160	.300		305	13.05	3.83	321.88	360
0400	Lt. Wt. composite panel, cementitious face, St. posts @ 12' OC, 8' H	B-80B	190	.168		145	6.30	1.22	152.52	171
0410	12' H		125	.256		218	9.60	1.86	229.46	257
0420	16' H		95	.337		290	12.65	2.45	305.10	340
0430	20' H		75	.427		365	16	3.10	384.10	430

32 84 23.10 Sprinkler Irrigation System	Crew	Daily Output	Labor-Hours	Unit	Material	2012 Bare Costs Labor	Equipment	Total	Total Incl O&P
0010 **SPRINKLER IRRIGATION SYSTEM**									
0011 For lawns									
0100 Golf course with fully automatic system	C-17	.05	1600	9 holes	98,000	73,500		171,500	221,500
0200 24' diam. head at 15' O.C incl. piping, auto oper., minimum	B-20	70	.343	Head	22	13.45		35.45	45
0300 Maximum		40	.600		54.50	23.50		78	96.50
0500 60' diam. head at 40' O.C incl. piping, auto oper., minimum		28	.857		69.50	33.50		103	128
0600 Maximum		23	1.043		192	41		233	274
0800 Residential system, custom, 1" supply		2000	.012	S.F.	.34	.47		.81	1.10
0900 1-1/2" supply		1800	.013	"	.39	.52		.91	1.24
1020 Pop up spray head w/risers, hi-pop, full circle pattern, 4"	2 Skwk	76	.211	Ea.	4.31	9.55		13.86	19.55
1030 1/2 circle pattern, 4"		76	.211		4.30	9.55		13.85	19.55
1040 6", full circle pattern		76	.211		9.55	9.55		19.10	25.50
1050 1/2 circle pattern, 6"		76	.211		9.55	9.55		19.10	25.50
1060 12", full circle pattern		76	.211		11.05	9.55		20.60	27
1070 1/2 circle pattern, 12"		76	.211		11.05	9.55		20.60	27
1080 Pop up bubbler head w/risers, hi-pop bubbler head, 4"		76	.211		4.24	9.55		13.79	19.45
1090 6"		76	.211		9.20	9.55		18.75	25
1100 12"		76	.211		11.10	9.55		20.65	27
1110 Impact full/part circle sprinklers, 28'-54' 25-60 PSI		37	.432		18.55	19.65		38.20	51
1120 Spaced 37'-49' @ 25-50 PSI		37	.432		30.50	19.65		50.15	64
1130 Spaced 43'-61' @ 30-60 PSI		37	.432		85	19.65		104.65	124
1140 Spaced 54'-78' @ 40-80 PSI		37	.432		121	19.65		140.65	164
1145 Impact rotor pop-up full/part commercial circle sprinklers									
1150 Spaced 42'-65' 35-80 PSI	2 Skwk	25	.640	Ea.	10.50	29		39.50	56.50
1160 Spaced 48'-76' 45-85 PSI	"	25	.640	"	8.50	29		37.50	54.50
1165 Impact rotor pop-up part. circle comm., 53'-75', 55-100 PSI, w/accessories									
1170 Plastic case, metal cover	2 Skwk	25	.640	Ea.	82.50	29		111.50	136
1180 Rubber cover		25	.640		62.50	29		91.50	114
1190 Iron case, metal cover		22	.727		133	33		166	197
1200 Rubber cover		22	.727		143	33		176	208
1250 Plastic case, 2 nozzle, metal cover		25	.640		130	29		159	188
1260 Rubber cover		25	.640		130	29		159	188
1270 Iron case, 2 nozzle, metal cover		22	.727		186	33		219	255
1280 Rubber cover		22	.727		186	33		219	255
1282 Impact rotor pop-up full circle commercial, 39'-99', 30-100 PSI									
1284 Plastic case, metal cover	2 Skwk	25	.640	Ea.	140	29		169	199
1286 Rubber cover		25	.640		158	29		187	218
1288 Iron case, metal cover		22	.727		203	33		236	274
1290 Rubber cover		22	.727		210	33		243	282
1292 Plastic case, 2 nozzle, metal cover		22	.727		150	33		183	216
1294 Rubber cover		22	.727		150	33		183	216
1296 Iron case, 2 nozzle, metal cover		20	.800		198	36.50		234.50	273
1298 Rubber cover		20	.800		205	36.50		241.50	282
1305 Electric remote control valve, plastic, 3/4"		18	.889		15.95	40.50		56.45	80
1310 1"		18	.889		18.35	40.50		58.85	82.50
1320 1-1/2"		18	.889		73	40.50		113.50	143
1330 2"		18	.889		113	40.50		153.50	187
1335 Quick coupling valves, brass, locking cover									
1340 Inlet coupling valve, 3/4"	2 Skwk	18.75	.853	Ea.	28.50	38.50		67	91.50
1350 1"		18.75	.853		45.50	38.50		84	111
1360 Controller valve boxes, 6" round boxes		18.75	.853		9.75	38.50		48.25	71
1370 10" round boxes		14.25	1.123		15.50	51		66.50	96
1380 12" square box		9.75	1.641		19.50	74.50		94	137

32 84 Planting Irrigation

32 84 23 – Underground Sprinklers

32 84 23.10 Sprinkler Irrigation System	Crew	Daily Output	Labor-Hours	Unit	Material	2012 Bare Costs Labor	Equipment	Total	Total Incl O&P
1388 Electromech. control, 14 day 3-60 min, auto start to 23/day									
1390 4 station	2 Skwk	1.04	15.385	Ea.	180	700		880	1,275
1400 7 station		.64	25		143	1,125		1,268	1,900
1410 12 station		.40	40		188	1,825		2,013	3,000
1420 Dual programs, 18 station		.24	66.667		340	3,025		3,365	5,050
1430 23 station		.16	100		570	4,550		5,120	7,650
1435 Backflow preventer, bronze, 0-175 PSI, w/valves, test cocks									
1440 3/4"	2 Skwk	2	8	Ea.	100	365		465	670
1450 1"		2	8		121	365		486	695
1460 1-1/2"		2	8		305	365		670	895
1470 2"		2	8		291	365		656	880
1475 Pressure vacuum breaker, brass, 15-150 PSI									
1480 3/4"	2 Skwk	2	8	Ea.	58.50	365		423.50	625
1490 1"		2	8		69	365		434	635
1500 1-1/2"		2	8		260	365		625	845
1510 2"		2	8		282	365		647	870

32 91 Planting Preparation

32 91 13 – Soil Preparation

32 91 13.16 Mulching

	Crew	Daily Output	Labor-Hours	Unit	Material	2012 Bare Costs Labor	Equipment	Total	Total Incl O&P
0010 **MULCHING**									
0100 Aged barks, 3" deep, hand spread	1 Clab	100	.080	S.Y.	2.92	2.81		5.73	7.55
0150 Skid steer loader	B-63	13.50	2.963	M.S.F.	325	110	11.80	446.80	535
0200 Hay, 1" deep, hand spread	1 Clab	475	.017	S.Y.	.33	.59		.92	1.27
0250 Power mulcher, small	B-64	180	.089	M.S.F.	36.50	3.09	1.96	41.55	47.50
0350 Large	B-65	530	.030	"	36.50	1.05	.94	38.49	43
0400 Humus peat, 1" deep, hand spread	1 Clab	700	.011	S.Y.	2.80	.40		3.20	3.70
0450 Push spreader	"	2500	.003	"	2.80	.11		2.91	3.25
0550 Tractor spreader	B-66	700	.011	M.S.F.	310	.51	.35	310.86	340
0600 Oat straw, 1" deep, hand spread	1 Clab	475	.017	S.Y.	.52	.59		1.11	1.48
0650 Power mulcher, small	B-64	180	.089	M.S.F.	58	3.09	1.96	63.05	70.50
0700 Large	B-65	530	.030	"	58	1.05	.94	59.99	66
0750 Add for asphaltic emulsion	B-45	1770	.009	Gal.	5.25	.37	.47	6.09	6.85
0800 Peat moss, 1" deep, hand spread	1 Clab	900	.009	S.Y.	2.80	.31		3.11	3.56
0850 Push spreader	"	2500	.003	"	2.80	.11		2.91	3.25
0950 Tractor spreader	B-66	700	.011	M.S.F.	310	.51	.35	310.86	340
1000 Polyethylene film, 6 mil.	2 Clab	2000	.008	S.Y.	.24	.28		.52	.69
1100 Redwood nuggets, 3" deep, hand spread	1 Clab	150	.053	"	3	1.87		4.87	6.20
1150 Skid steer loader	B-63	13.50	2.963	M.S.F.	335	110	11.80	456.80	545
1200 Stone mulch, hand spread, ceramic chips, economy	1 Clab	125	.064	S.Y.	7	2.25		9.25	11.15
1250 Deluxe	"	95	.084	"	11	2.96		13.96	16.65
1300 Granite chips	B-1	10	2.400	C.Y.	36.50	86		122.50	172
1400 Marble chips		10	2.400		132	86		218	277
1600 Pea gravel		28	.857		72	30.50		102.50	126
1700 Quartz		10	2.400		172	86		258	320
1800 Tar paper, 15 lb. felt	1 Clab	800	.010	S.Y.	.51	.35		.86	1.10
1900 Wood chips, 2" deep, hand spread	"	220	.036	"	1.75	1.28		3.03	3.89
1950 Skid steer loader	B-63	20.30	1.970	M.S.F.	194	73	7.85	274.85	335

32 91 13.26 Planting Beds

	Crew	Daily Output	Labor-Hours	Unit	Material	2012 Bare Costs Labor	Equipment	Total	Total Incl O&P
0010 **PLANTING BEDS**									
0100 Backfill planting pit, by hand, on site topsoil	2 Clab	18	.889	C.Y.		31		31	48

626

32 91 Planting Preparation

32 91 13 – Soil Preparation

32 91 13.26 Planting Beds

		Crew	Daily Output	Labor-Hours	Unit	Material	2012 Bare Costs Labor	Equipment	Total	Total Incl O&P
0200	Prepared planting mix, by hand	2 Clab	24	.667	C.Y.		23.50		23.50	36
0300	Skid steer loader, on site topsoil	B-62	340	.071			2.71	.47	3.18	4.65
0400	Prepared planting mix	"	410	.059			2.24	.39	2.63	3.86
1000	Excavate planting pit, by hand, sandy soil	2 Clab	16	1			35		35	54
1100	Heavy soil or clay	"	8	2			70		70	108
1200	1/2 C.Y. backhoe, sandy soil	B-11C	150	.107			4.36	2.23	6.59	9.10
1300	Heavy soil or clay	"	115	.139			5.70	2.91	8.61	11.85
2000	Mix planting soil, incl. loam, manure, peat, by hand	2 Clab	60	.267		40	9.35		49.35	58.50
2100	Skid steer loader	B-62	150	.160		40	6.15	1.06	47.21	54.50
3000	Pile sod, skid steer loader	"	2800	.009	S.Y.		.33	.06	.39	.56
3100	By hand	2 Clab	400	.040			1.40		1.40	2.16
4000	Remove sod, F.E. loader	B-10S	2000	.006			.26	.17	.43	.57
4100	Sod cutter	B-12K	3200	.005			.21	.30	.51	.65
4200	By hand	2 Clab	240	.067			2.34		2.34	3.60

32 91 19 – Landscape Grading

32 91 19.13 Topsoil Placement and Grading

		Crew	Daily Output	Labor-Hours	Unit	Material	2012 Bare Costs Labor	Equipment	Total	Total Incl O&P
0010	**TOPSOIL PLACEMENT AND GRADING**									
0400	Spread from pile to rough finish grade, F.E. loader, 1.5 C.Y.	B-10S	200	.060	C.Y.		2.56	1.67	4.23	5.75
0500	Up to 200' radius, by hand	1 Clab	14	.571			20		20	31
0600	Top dress by hand, 1 C.Y. for 600 S.F.	"	11.50	.696		33.50	24.50		58	74.50
0700	Furnish and place, truck dumped, screened, 4" deep	B-10S	1300	.009	S.Y.	4.21	.39	.26	4.86	5.50
0800	6" deep	"	820	.015	"	5.40	.63	.41	6.44	7.35

32 92 Turf and Grasses

32 92 19 – Seeding

32 92 19.13 Mechanical Seeding

		Crew	Daily Output	Labor-Hours	Unit	Material	2012 Bare Costs Labor	Equipment	Total	Total Incl O&P
0010	**MECHANICAL SEEDING** R329219-50									
0020	Mechanical seeding, 215 lb./acre	B-66	1.50	5.333	Acre	605	239	165	1,009	1,200
0100	44 lb./M.S.Y.	"	2500	.003	S.Y.	.16	.14	.10	.40	.51
0300	Fine grading and seeding incl. lime, fertilizer & seed,									
0310	with equipment	B-14	1000	.048	S.Y.	.40	1.78	.33	2.51	3.54
0600	Limestone hand push spreader, 50 lb. per M.S.F.	1 Clab	180	.044	M.S.F.	3.30	1.56		4.86	6.05
0800	Grass seed hand push spreader, 4.5 lb. per M.S.F.	"	180	.044	"	17.80	1.56		19.36	22
1000	Hydro or air seeding for large areas, incl. seed and fertilizer	B-81	8900	.003	S.Y.	.40	.11	.07	.58	.68
1100	With wood fiber mulch added	"	8900	.003	"	1.66	.11	.07	1.84	2.07
1300	Seed only, over 100 lb., field seed, minimum				Lb.	2.28			2.28	2.51
1400	Maximum					2.87			2.87	3.16
1500	Lawn seed, minimum					2.17			2.17	2.39
1600	Maximum					2.90			2.90	3.19
1800	Aerial operations, seeding only, field seed	B-58	50	.480	Acre	740	18.40	59	817.40	910
1900	Lawn seed		50	.480		705	18.40	59	782.40	870
2100	Seed and liquid fertilizer, field seed		50	.480		810	18.40	59	887.40	985
2200	Lawn seed		50	.480		775	18.40	59	852.40	950

32 92 23 – Sodding

32 92 23.10 Sodding Systems

		Crew	Daily Output	Labor-Hours	Unit	Material	2012 Bare Costs Labor	Equipment	Total	Total Incl O&P
0010	**SODDING SYSTEMS**									
0020	Sodding, 1" deep, bluegrass sod, on level ground, over 8 M.S.F.	B-63	22	1.818	M.S.F.	175	67.50	7.25	249.75	305
0200	4 M.S.F.		17	2.353		240	87	9.40	336.40	405
0300	1000 S.F.		13.50	2.963		295	110	11.80	416.80	505
0500	Sloped ground, over 8 M.S.F.		6	6.667		175	247	26.50	448.50	605

32 92 Turf and Grasses

32 92 23 – Sodding

32 92 23.10 Sodding Systems	Crew	Daily Output	Labor-Hours	Unit	Material	2012 Bare Costs Labor	Equipment	Total	Total Incl O&P
0600	4 M.S.F.	B-63	5	8	M.S.F.	240	296	32	568	755
0700	1000 S.F.		4	10		295	370	40	705	935
1000	Bent grass sod, on level ground, over 6 M.S.F.		20	2		600	74	8	682	780
1100	3 M.S.F.		18	2.222		665	82.50	8.85	756.35	865
1200	Sodding 1000 S.F. or less		14	2.857		875	106	11.40	992.40	1,150
1500	Sloped ground, over 6 M.S.F.		15	2.667		600	99	10.65	709.65	825
1600	3 M.S.F.		13.50	2.963		665	110	11.80	786.80	910
1700	1000 S.F.	↓	12	3.333	↓	875	123	13.30	1,011.30	1,175

32 93 Plants

32 93 10 – General Planting Costs

32 93 10.12 Travel

		Crew	Daily Output	Labor-Hours	Unit	Material	Labor	Equipment	Total	Total Incl O&P
0010	**TRAVEL** to all nursery items, for 10 to 20 miles, add									
0015	10 to 20 miles one way, add				All					5%
0100	30 to 50 miles one way, add				"					10%

32 93 13 – Ground Covers

32 93 13.10 Ground Cover Plants

		Crew	Daily Output	Labor-Hours	Unit	Material	Labor	Equipment	Total	Total Incl O&P
0010	**GROUND COVER PLANTS**									
0012	Plants, pachysandra, in prepared beds	B-1	15	1.600	C	31	57		88	122
0200	Vinca minor, 1 yr., bare root, in prepared beds		12	2	"	39	71.50		110.50	153
0600	Stone chips, in 50 lb. bags, Georgia marble		520	.046	Bag	5	1.65		6.65	8.05
0700	Onyx gemstone		260	.092		18	3.30		21.30	25
0800	Quartz		260	.092	↓	6.90	3.30		10.20	12.65
0900	Pea gravel, truckload lots	↓	28	.857	Ton	30.50	30.50		61	80.50

32 93 33 – Shrubs

32 93 33.10 Shrubs and Trees

		Crew	Daily Output	Labor-Hours	Unit	Material	Labor	Equipment	Total	Total Incl O&P
0010	**SHRUBS AND TREES**									
0011	Evergreen, in prepared beds, B & B									
0100	Arborvitae pyramidal, 4'-5'	B-17	30	1.067	Ea.	52	40	23	115	144
0150	Globe, 12"-15"	B-1	96	.250		21.50	8.95		30.45	37.50
0300	Cedar, blue, 8'-10'	B-17	18	1.778		220	67	38.50	325.50	385
0500	Hemlock, Canadian, 2-1/2'-3'	B-1	36	.667		36	24		60	76
0550	Holly, Savannah, 8' - 10' H		9.68	2.479		163	88.50		251.50	315
0600	Juniper, andorra, 18"-24"		80	.300		20	10.75		30.75	38.50
0620	Wiltoni, 15"-18"	↓	80	.300		24.50	10.75		35.25	43.50
0640	Skyrocket, 4-1/2'-5'	B-17	55	.582		139	22	12.55	173.55	200
0660	Blue pfitzer, 2'-2-1/2'	B-1	44	.545		39	19.50		58.50	73
0680	Ketleerie, 2-1/2'-3'		50	.480		40	17.15		57.15	70.50
0700	Pine, black, 2-1/2'-3'		50	.480		53	17.15		70.15	85
0720	Mugo, 18"-24"	↓	60	.400		45	14.30		59.30	71.50
0740	White, 4'-5'	B-17	75	.427		40	16.05	9.20	65.25	78.50
0800	Spruce, blue, 18"-24"	B-1	60	.400		49	14.30		63.30	76
0840	Norway, 4'-5'	B-17	75	.427		75	16.05	9.20	100.25	117
0900	Yew, denisforma, 12"-15"	B-1	60	.400		34.50	14.30		48.80	60
1000	Capitata, 18"-24"		30	.800		18	28.50		46.50	64
1100	Hicksi, 2'-2-1/2'	↓	30	.800	↓	32	28.50		60.50	79.50

32 93 33.20 Shrubs

		Crew	Daily Output	Labor-Hours	Unit	Material	Labor	Equipment	Total	Total Incl O&P
0010	**SHRUBS**									
0011	Broadleaf Evergreen, planted in prepared beds									
0100	Andromeda, 15"-18", container	B-1	96	.250	Ea.	18	8.95		26.95	33.50

32 93 Plants

32 93 33 – Shrubs

32 93 33.20 Shrubs

		Crew	Daily Output	Labor-Hours	Unit	Material	2012 Bare Costs Labor	Equipment	Total	Total Incl O&P
0200	Azalea, 15" - 18", container	B-1	96	.250	Ea.	22	8.95		30.95	38
0300	Barberry, 9"-12", container		130	.185		13	6.60		19.60	24.50
0400	Boxwood, 15"-18", B&B		96	.250		30	8.95		38.95	47
0500	Euonymus, emerald gaiety, 12" to 15", container		115	.209		22	7.45		29.45	35.50
0600	Holly, 15"-18", B & B		96	.250		17	8.95		25.95	32.50
0900	Mount laurel, 18" - 24", B & B		80	.300		55	10.75		65.75	77
1000	Paxistema, 9 – 12" high		130	.185		20	6.60		26.60	32
1100	Rhododendron, 18"-24", container		48	.500		28.50	17.90		46.40	59
1200	Rosemary, 1 gal. container		600	.040		10.60	1.43		12.03	13.85
2000	Deciduous, planted in prepared beds, amelanchier, 2'-3', B & B		57	.421		119	15.05		134.05	154
2100	Azalea, 15"-18", B & B		96	.250		17.50	8.95		26.45	33
2300	Bayberry, 2'-3', B & B		57	.421		22	15.05		37.05	47
2600	Cotoneaster, 15"-18", B & B		80	.300		10.60	10.75		21.35	28
2800	Dogwood, 3'-4', B & B	B-17	40	.800		27	30	17.30	74.30	94.50
2900	Euonymus, alatus compacta, 15" to 18", container	B-1	80	.300		22	10.75		32.75	40.50
3200	Forsythia, 2'-3', container	"	60	.400		22	14.30		36.30	46
3300	Hibiscus, 3'-4', B & B	B-17	75	.427		41.50	16.05	9.20	66.75	80
3400	Honeysuckle, 3'-4', B & B	B-1	60	.400		20.50	14.30		34.80	44.50
3500	Hydrangea, 2'-3', B & B	"	57	.421		27	15.05		42.05	52.50
3600	Lilac, 3'-4', B & B	B-17	40	.800		19.65	30	17.30	66.95	86.50
3900	Privet, bare root, 18"-24"	B-1	80	.300		22	10.75		32.75	40.50
4100	Quince, 2'-3', B & B	"	57	.421		23	15.05		38.05	48.50
4200	Russian olive, 3'-4', B & B	B-17	75	.427		25	16.05	9.20	50.25	62
4400	Spirea, 3'-4', B & B	B-1	70	.343		21.50	12.25		33.75	43
4500	Viburnum, 3'-4', B & B	B-17	40	.800		21.50	30	17.30	68.80	88.50

32 93 43 – Trees

32 93 43.20 Trees

			Crew	Daily Output	Labor-Hours	Unit	Material	2012 Bare Costs Labor	Equipment	Total	Total Incl O&P
0010	**TREES**										
0011	Deciduous, in prep. beds, balled & burlapped (B&B)										
0100	Ash, 2" caliper	G	B-17	8	4	Ea.	200	150	86.50	436.50	545
0200	Beech, 5'-6'	G		50	.640		180	24	13.85	217.85	250
0300	Birch, 6'-8', 3 stems	G		20	1.600		210	60	34.50	304.50	360
0500	Crabapple, 6'-8'	G		20	1.600		99	60	34.50	193.50	239
0600	Dogwood, 4'-5'	G		40	.800		138	30	17.30	185.30	216
0700	Eastern redbud 4'-5'	G		40	.800		117	30	17.30	164.30	193
0800	Elm, 8'-10'	G		20	1.600		137	60	34.50	231.50	281
0900	Ginkgo, 6'-7'	G		24	1.333		177	50	29	256	305
1000	Hawthorn, 8'-10', 1" caliper	G		20	1.600		173	60	34.50	267.50	320
1100	Honeylocust, 10'-12', 1-1/2" caliper	G		10	3.200		162	120	69	351	435
1300	Larch, 8'	G		32	1		106	37.50	21.50	165	199
1400	Linden, 8'-10', 1" caliper	G		20	1.600		129	60	34.50	223.50	272
1500	Magnolia, 4'-5'	G		20	1.600		94	60	34.50	188.50	233
1600	Maple, red, 8'-10', 1-1/2" caliper	G		10	3.200		191	120	69	380	470
1700	Mountain ash, 8'-10', 1" caliper	G		16	2		184	75	43	302	365
1800	Oak, 2-1/2"-3" caliper	G		6	5.333		264	200	115	579	720
2100	Planetree, 9'-11', 1-1/4" caliper	G		10	3.200		122	120	69	311	395
2200	Plum, 6'-8', 1" caliper	G		20	1.600		98	60	34.50	192.50	238
2300	Poplar, 9'-11', 1-1/4" caliper	G		10	3.200		101	120	69	290	370
2500	Sumac, 2'-3'	G		75	.427		23	16.05	9.20	48.25	60
2700	Tulip, 5'-6'	G		40	.800		59	30	17.30	106.30	130
2800	Willow, 6'-8', 1" caliper	G		20	1.600		67.50	60	34.50	162	204

32 94 Planting Accessories

32 94 13 – Landscape Edging

32 94 13.20 Edging	Crew	Daily Output	Labor-Hours	Unit	Material	2012 Bare Costs Labor	2012 Bare Costs Equipment	Total	Total Incl O&P
0010 **EDGING**									
0050 Aluminum alloy, including stakes, 1/8" x 4", mill finish	B-1	390	.062	L.F.	3.94	2.20		6.14	7.70
0051 Black paint		390	.062		4.57	2.20		6.77	8.45
0052 Black anodized		390	.062		5.30	2.20		7.50	9.20
0100 Brick, set horizontally, 1-1/2 bricks per L.F.	D-1	370	.043		1.35	1.72		3.07	4.09
0150 Set vertically, 3 bricks per L.F.	"	135	.119		3	4.72		7.72	10.45
0200 Corrugated aluminum, roll, 4" wide	1 Carp	650	.012		1.84	.54		2.38	2.86
0250 6" wide	"	550	.015		2.30	.64		2.94	3.52
0600 Railroad ties, 6" x 8"	2 Carp	170	.094		3.97	4.15		8.12	10.75
0650 7" x 9"		136	.118		4.41	5.20		9.61	12.85
0750 Redwood 2" x 4"		330	.048		2.26	2.14		4.40	5.80
0800 Steel edge strips, incl. stakes, 1/4" x 5"	B-1	390	.062		2.50	2.20		4.70	6.15
0850 3/16" x 4"	"	390	.062		1.98	2.20		4.18	5.55

32 94 50 – Tree Guying

32 94 50.10 Tree Guying Systems	Crew	Daily Output	Labor-Hours	Unit	Material	2012 Bare Costs Labor	2012 Bare Costs Equipment	Total	Total Incl O&P
0010 **TREE GUYING SYSTEMS**									
0015 Tree guying Including stakes, guy wire and wrap									
0100 Less than 3" caliper, 2 stakes	2 Clab	35	.457	Ea.	14	16.05		30.05	40
0200 3" to 4" caliper, 3 stakes	"	21	.762	"	18.95	26.50		45.45	62
1000 Including arrowhead anchor, cable, turnbuckles and wrap									
1100 Less than 3" caliper, 3" anchors	2 Clab	20	.800	Ea.	44	28		72	91.50
1200 3" to 6" caliper, 4" anchors		15	1.067		37.50	37.50		75	99
1300 6" caliper, 6" anchors		12	1.333		44	47		91	121
1400 8" caliper, 8" anchors		9	1.778		126	62.50		188.50	234

32 96 Transplanting

32 96 23 – Plant and Bulb Transplanting

32 96 23.23 Planting

	Crew	Daily Output	Labor-Hours	Unit	Material	2012 Bare Costs Labor	2012 Bare Costs Equipment	Total	Total Incl O&P
0010 **PLANTING**									
0012 Moving shrubs on site, 12" ball	B-62	28	.857	Ea.		33	5.70	38.70	56.50
0100 24" ball	"	22	1.091	"		42	7.25	49.25	72

32 96 23.43 Moving Trees

	Crew	Daily Output	Labor-Hours	Unit	Material	2012 Bare Costs Labor	2012 Bare Costs Equipment	Total	Total Incl O&P
0290 **MOVING TREES**, On site									
0300 Moving trees on site, 36" ball	B-6	3.75	6.400	Ea.		245	89	334	475
0400 60" ball	"	1	24	"		920	335	1,255	1,775

Estimating Tips

33 10 00 Water Utilities
33 30 00 Sanitary Sewerage Utilities
33 40 00 Storm Drainage Utilities

• Never assume that the water, sewer, and drainage lines will go in at the early stages of the project. Consider the site access needs before dividing the site in half with open trenches, loose pipe, and machinery obstructions. Always inspect the site to establish that the site drawings are complete. Check off all existing utilities on your drawings as you locate them. Be especially careful with underground utilities because appurtenances are sometimes buried during regrading or repaving operations. If you find any discrepancies, mark up the site plan for further research. Differing site conditions can be very costly if discovered later in the project.

• See also Section 33 01 00 for restoration of pipe where removal/replacement may be undesirable. Use of new types of piping materials can reduce the overall project cost. Owners/design engineers should consider the installing contractor as a valuable source of current information on utility products and local conditions that could lead to significant cost savings.

Reference Numbers

Reference numbers are shown in shaded boxes at the beginning of some major classifications. These numbers refer to related items in the Reference Section. The reference information may be an estimating procedure, an alternate pricing method, or technical information.

Note: Not all subdivisions listed here necessarily appear in this publication.

Division 33 – Utilities

Note: **Trade Service,** *in part, has been used as a reference source for some of the material prices used in Division 33.*

33 01 Operation and Maintenance of Utilities

33 01 10 – Operation and Maintenance of Water Utilities

33 01 10.10 Corrosion Resistance

		Crew	Daily Output	Labor-Hours	Unit	Material	2012 Bare Costs Labor	2012 Bare Costs Equipment	Total	Total Incl O&P
0010	**CORROSION RESISTANCE**									
0012	Wrap & coat, add to pipe, 4" diameter				L.F.	2.24			2.24	2.46
0040	6" diameter					3.36			3.36	3.70
0060	8" diameter					4.36			4.36	4.80
0100	12" diameter					6.60			6.60	7.25
0200	24" diameter					12.70			12.70	13.95
0500	Coating, bituminous, per diameter inch, 1 coat, add					.55			.55	.61
0540	3 coat					1.65			1.65	1.82
0560	Coal tar epoxy, per diameter inch, 1 coat, add					.28			.28	.31
0600	3 coat					.84			.84	.92

33 01 30 – Operation and Maintenance of Sewer Utilities

33 01 30.72 Relining Sewers

		Crew	Daily Output	Labor-Hours	Unit	Material	2012 Bare Costs Labor	2012 Bare Costs Equipment	Total	Total Incl O&P
0010	**RELINING SEWERS**									
0011	With cement incl. bypass & cleaning									
0020	Less than 10,000 L.F., urban, 6" to 10"	C-17E	130	.615	L.F.	9.20	28	.70	37.90	54.50
0200	24" to 36"		90	.889		14.65	40.50	1	56.15	80
0300	48" to 72"		80	1		23.50	46	1.13	70.63	98

33 05 Common Work Results for Utilities

33 05 16 – Utility Structures

33 05 16.13 Precast Concrete Utility Boxes

		Crew	Daily Output	Labor-Hours	Unit	Material	2012 Bare Costs Labor	2012 Bare Costs Equipment	Total	Total Incl O&P
0010	**PRECAST CONCRETE UTILITY BOXES**, 6" thick									
0050	5' x 10' x 6' high, I.D.	B-13	2	28	Ea.	2,475	1,075	350	3,900	4,725
0100	6' x 10' x 6' high, I.D.		2	28		2,575	1,075	350	4,000	4,825
0150	5' x 12' x 6' high, I.D.		2	28		2,725	1,075	350	4,150	5,000
0200	6' x 12' x 6' high, I.D.		1.80	31.111		3,050	1,175	385	4,610	5,575
0250	6' x 13' x 6' high, I.D.		1.50	37.333		4,000	1,425	465	5,890	7,075
0300	8' x 14' x 7' high, I.D.		1	56		4,325	2,125	695	7,145	8,775
0350	Hand hole, precast concrete, 1-1/2" thick									
0400	1'-0" x 2'-0" x 1'-9", I.D., light duty	B-1	4	6	Ea.	390	215		605	760
0450	4'-6" x 3'-2" x 2'-0", O.D., heavy duty	B-6	3	8	"	1,400	305	111	1,816	2,150

33 05 23 – Trenchless Utility Installation

33 05 23.19 Microtunneling

		Crew	Daily Output	Labor-Hours	Unit	Material	2012 Bare Costs Labor	2012 Bare Costs Equipment	Total	Total Incl O&P
0010	**MICROTUNNELING**									
0011	Not including excavation, backfill, shoring,									
0020	or dewatering, average 50'/day, slurry method									
0100	24" to 48" outside diameter, minimum				L.F.				830	915
0110	Adverse conditions, add				%				50%	50%
1000	Rent microtunneling machine, average monthly lease				Month				92,000	101,000
1010	Operating technician				Day				610	680
1100	Mobilization and demobilization, minimum				Job				40,000	45,000
1110	Maximum				"				420,000	462,000

33 05 23.20 Horizontal Boring

		Crew	Daily Output	Labor-Hours	Unit	Material	2012 Bare Costs Labor	2012 Bare Costs Equipment	Total	Total Incl O&P
0010	**HORIZONTAL BORING**									
0011	Casing only, 100' minimum,									
0020	not incl. jacking pits or dewatering									
0100	Roadwork, 1/2" thick wall, 24" diameter casing	B-42	20	3.200	L.F.	146	126	63.50	335.50	430
0200	36" diameter		16	4		220	158	79	457	575
0300	48" diameter		15	4.267		294	168	84.50	546.50	680
0500	Railroad work, 24" diameter		15	4.267		146	168	84.50	398.50	520

33 05 Common Work Results for Utilities

33 05 23 – Trenchless Utility Installation

33 05 23.20 Horizontal Boring	Crew	Daily Output	Labor-Hours	Unit	Material	2012 Bare Costs Labor	Equipment	Total	Total Incl O&P
0600	36" diameter	B-42	14	4.571	L.F.	220	181	90.50	491.50	625
0700	48" diameter	↓	12	5.333		294	211	106	611	770
0900	For ledge, add				↓					20%

33 05 26 – Utility Identification

33 05 26.10 Utility Accessories

0010	UTILITY ACCESSORIES									
0400	Underground tape, detectable, reinforced, alum. foil core, 2"	1 Clab	150	.053	C.L.F.	2.05	1.87		3.92	5.15
0500	6"	"	140	.057	"	5.60	2.01		7.61	9.25

33 11 Water Utility Distribution Piping

33 11 13 – Public Water Utility Distribution Piping

33 11 13.15 Water Supply, Ductile Iron Pipe

33 11 13.15 Water Supply, Ductile Iron Pipe	Crew	Daily Output	Labor-Hours	Unit	Material	2012 Bare Costs Labor	Equipment	Total	Total Incl O&P
0010	WATER SUPPLY, DUCTILE IRON PIPE R331113-80									
0020	Not including excavation or backfill									
2000	Pipe, class 50 water piping, 18' lengths									
2020	Mechanical joint, 4" diameter	B-21A	200	.200	L.F.	12.60	8.75	2.24	23.59	29.50
2040	6" diameter		160	.250		14.75	10.90	2.80	28.45	36
2060	8" diameter		133.33	.300		21.50	13.10	3.36	37.96	47
2080	10" diameter		114.29	.350		26	15.30	3.92	45.22	56
2100	12" diameter		105.26	.380		31.50	16.60	4.26	52.36	64
2120	14" diameter		100	.400		49	17.45	4.48	70.93	85
2140	16" diameter		72.73	.550		50.50	24	6.15	80.65	99
2160	18" diameter		68.97	.580		53.50	25.50	6.50	85.50	105
2170	20" diameter		57.14	.700		56.50	30.50	7.85	94.85	117
2180	24" diameter		47.06	.850		61.50	37	9.55	108.05	135
3000	Tyton, push-on joint, 4" diameter		400	.100		11.40	4.37	1.12	16.89	20.50
3020	6" diameter		333.33	.120		15.80	5.25	1.35	22.40	27
3040	8" diameter		200	.200		22	8.75	2.24	32.99	40
3060	10" diameter		181.82	.220		31.50	9.60	2.47	43.57	52
3080	12" diameter		160	.250		33	10.90	2.80	46.70	56
3100	14" diameter		133.33	.300		36.50	13.10	3.36	52.96	63.50
3120	16" diameter		114.29	.350		49.50	15.30	3.92	68.72	82
3140	18" diameter		100	.400		55	17.45	4.48	76.93	92
3160	20" diameter		88.89	.450		57.50	19.65	5.05	82.20	98.50
3180	24" diameter	↓	76.92	.520	↓	72	22.50	5.85	100.35	120
8000	Fittings, mechanical joint									
8006	90° bend, 4" diameter	B-20A	16	2	Ea.	251	85.50		336.50	405
8020	6" diameter		12.80	2.500		375	107		482	570
8040	8" diameter	↓	10.67	2.999		665	128		793	925
8060	10" diameter	B-21A	11.43	3.500		890	153	39	1,082	1,250
8080	12" diameter		10.53	3.799		1,250	166	42.50	1,458.50	1,675
8100	14" diameter		10	4		1,675	175	45	1,895	2,175
8120	16" diameter		7.27	5.502		2,175	240	61.50	2,476.50	2,825
8140	18" diameter		6.90	5.797		2,925	253	65	3,243	3,675
8160	20" diameter		5.71	7.005		3,675	305	78.50	4,058.50	4,600
8180	24" diameter	↓	4.70	8.511		5,675	370	95.50	6,140.50	6,925
8200	Wye or tee, 4" diameter	B-20A	10.67	2.999		405	128		533	640
8220	6" diameter		8.53	3.751		615	160		775	920
8240	8" diameter	↓	7.11	4.501		975	192		1,167	1,375
8260	10" diameter	B-21A	7.62	5.249	↓	1,400	229	59	1,688	1,975

33 11 13 - Public Water Utility Distribution Piping

33 11 13.15 Water Supply, Ductile Iron Pipe

		Crew	Daily Output	Labor-Hours	Unit	Material	Labor	Equipment	Total	Total Incl O&P
8280	12" diameter	B-21A	7.02	5.698	Ea.	1,850	249	64	2,163	2,475
8300	14" diameter		6.67	5.997		2,750	262	67	3,079	3,500
8320	16" diameter		4.85	8.247		3,550	360	92.50	4,002.50	4,550
8340	18" diameter		4.60	8.696		4,400	380	97.50	4,877.50	5,500
8360	20" diameter		3.81	10.499		6,200	460	118	6,778	7,650
8380	24" diameter		3.14	12.739		10,500	555	143	11,198	12,500
8450	Decreaser, 6" x 4" diameter	B-20A	14.22	2.250		330	96		426	510
8460	8" x 6" diameter	"	11.64	2.749		485	117		602	710
8470	10" x 6" diameter	B-21A	13.33	3.001		590	131	33.50	754.50	885
8480	12" x 6" diameter		12.70	3.150		730	138	35.50	903.50	1,050
8490	16" x 6" diameter		10	4		1,300	175	45	1,520	1,750
8500	20" x 6" diameter		8.42	4.751		2,750	207	53.50	3,010.50	3,400
8550	Piping, butterfly valves, cast iron									
8560	4" diameter	B-20	6	4	Ea.	800	157		957	1,125
8570	6" diameter	"	5	4.800		850	188		1,038	1,225
8580	8" diameter	B-21	4	7		1,125	283	33	1,441	1,725
8590	10" diameter		3.50	8		1,700	325	38	2,063	2,400
8600	12" diameter		3	9.333		2,425	375	44	2,844	3,300
8610	14" diameter		2	14		4,100	565	66	4,731	5,475
8620	16" diameter		2	14		4,825	565	66	5,456	6,250

33 11 13.25 Water Supply, Polyvinyl Chloride Pipe

		Crew	Daily Output	Labor-Hours	Unit	Material	Labor	Equipment	Total	Total Incl O&P
0010	**WATER SUPPLY, POLYVINYL CHLORIDE PIPE**									
0020	Not including excavation or backfill, unless specified									
2100	PVC pipe, Class 150, 1-1/2" diameter	Q-1A	750	.013	L.F.	.33	.73		1.06	1.47
2120	2" diameter		686	.015		.45	.80		1.25	1.71
2140	2-1/2" diameter		500	.020		.65	1.10		1.75	2.38
2160	3" diameter	B-20	430	.056		.93	2.19		3.12	4.39
3010	AWWA C905, PR 100, DR 25									
3030	14" diameter	B-20A	213	.150	L.F.	14	6.40		20.40	25
3040	16" diameter		200	.160		14.20	6.80		21	26
3050	18" diameter		160	.200		24	8.55		32.55	39.50
3060	20" diameter		133	.241		30	10.25		40.25	48.50
3070	24" diameter		107	.299		43	12.75		55.75	66.50
3080	30" diameter		80	.400		80	17.05		97.05	114
3090	36" diameter		80	.400		124	17.05		141.05	162
3100	42" diameter		60	.533		166	22.50		188.50	218
3200	48" diameter		60	.533		217	22.50		239.50	273
4520	Pressure pipe Class 150, SDR 18, AWWA C900, 4" diameter		380	.084		2.34	3.59		5.93	8
4530	6" diameter		316	.101		5.70	4.32		10.02	12.85
4540	8" diameter		264	.121		9.50	5.15		14.65	18.30
4550	10" diameter		220	.145		12.45	6.20		18.65	23
4560	12" diameter		186	.172		17.60	7.35		24.95	30.50
8000	Fittings with rubber gasket									
8003	Class 150, D.R. 18									
8006	90° Bend , 4" diameter	B-20	100	.240	Ea.	49.50	9.40		58.90	69
8020	6" diameter		90	.267		88	10.45		98.45	113
8040	8" diameter		80	.300		158	11.75		169.75	191
8060	10" diameter		50	.480		330	18.80		348.80	390
8080	12" diameter		30	.800		425	31.50		456.50	515
8100	Tee, 4" diameter		90	.267		58	10.45		68.45	79.50
8120	6" diameter		80	.300		145	11.75		156.75	178
8140	8" diameter		70	.343		190	13.45		203.45	230

33 11 Water Utility Distribution Piping

33 11 13 – Public Water Utility Distribution Piping

33 11 13.25 Water Supply, Polyvinyl Chloride Pipe		Crew	Daily Output	Labor-Hours	Unit	Material	2012 Bare Costs Labor	Equipment	Total	Total Incl O&P
8160	10" diameter	B-20	40	.600	Ea.	565	23.50		588.50	655
8180	12" diameter		20	1.200		710	47		757	855
8200	45° Bend, 4" diameter		100	.240		47	9.40		56.40	66
8220	6" diameter		90	.267		85.50	10.45		95.95	110
8240	8" diameter		50	.480		167	18.80		185.80	212
8260	10" diameter		50	.480		330	18.80		348.80	390
8280	12" diameter		30	.800		420	31.50		451.50	510
8300	Reducing tee 6" x 4"		100	.240		129	9.40		138.40	157
8320	8" x 6"		90	.267		248	10.45		258.45	288
8330	10" x 6"		90	.267		350	10.45		360.45	400
8340	10" x 8"		90	.267		365	10.45		375.45	415
8350	12" x 6"		90	.267		430	10.45		440.45	485
8360	12" x 8"		90	.267		470	10.45		480.45	530
8400	Tapped service tee (threaded type) 6" x 6" x 3/4"		100	.240		95	9.40		104.40	120
8430	6" x 6" x 1"		90	.267		95	10.45		105.45	121
8440	6" x 6" x 1-1/2"		90	.267		95	10.45		105.45	121
8450	6" x 6" x 2"		90	.267		95	10.45		105.45	121
8460	8" x 8" x 3/4"		90	.267		140	10.45		150.45	170
8470	8" x 8" x 1"		90	.267		140	10.45		150.45	170
8480	8" x 8" x 1-1/2"		90	.267		140	10.45		150.45	170
8490	8" x 8" x 2"		90	.267		140	10.45		150.45	170
8500	Repair coupling 4"		100	.240		25.50	9.40		34.90	42.50
8520	6" diameter		90	.267		40	10.45		50.45	60
8540	8" diameter		50	.480		96	18.80		114.80	135
8560	10" diameter		50	.480		202	18.80		220.80	251
8580	12" diameter		50	.480		295	18.80		313.80	355
8600	Plug end 4"		100	.240		22.50	9.40		31.90	39.50
8620	6" diameter		90	.267		40.50	10.45		50.95	60.50
8640	8" diameter		50	.480		69.50	18.80		88.30	105
8660	10" diameter		50	.480		96.50	18.80		115.30	135
8680	12" diameter		50	.480		118	18.80		136.80	159

33 12 Water Utility Distribution Equipment

33 12 19 – Water Utility Distribution Fire Hydrants

33 12 19.10 Fire Hydrants

		Crew	Daily Output	Labor-Hours	Unit	Material	2012 Bare Costs Labor	Equipment	Total	Total Incl O&P
0010	**FIRE HYDRANTS**									
0020	Mechanical joints unless otherwise noted									
1000	Fire hydrants, two way; excavation and backfill not incl.									
1100	4-1/2" valve size, depth 2'-0"	B-21	10	2.800	Ea.	1,450	113	13.25	1,576.25	1,800
1120	2'-6"		10	2.800		1,525	113	13.25	1,651.25	1,875
1140	3'-0"		10	2.800		1,650	113	13.25	1,776.25	2,000
1300	7'-0"		6	4.667		1,900	189	22	2,111	2,400
2400	Lower barrel extensions with stems, 1'-0"	B-20	14	1.714		380	67		447	525
2480	3'-0"	"	12	2		740	78.50		818.50	935

33 16 Water Utility Storage Tanks

33 16 13 – Aboveground Water Utility Storage Tanks

33 16 13.13 Steel Water Storage Tanks

		Crew	Daily Output	Labor-Hours	Unit	Material	2012 Bare Costs Labor	Equipment	Total	Total Incl O&P
0010	**STEEL WATER STORAGE TANKS**									
0910	Steel, ground level, ht./diam. less than 1, not incl. fdn., 100,000 gallons				Ea.				180,000	217,500
1000	250,000 gallons								263,000	288,500
1200	500,000 gallons								372,000	408,500
1250	750,000 gallons								479,500	527,500
1300	1,000,000 gallons								588,000	647,000
1500	2,000,000 gallons								1,022,500	1,023,000
1600	4,000,000 gallons								1,890,500	2,079,000
1800	6,000,000 gallons								2,758,500	3,034,500
1850	8,000,000 gallons								3,625,500	3,988,000
1910	10,000,000 gallons				▼				4,501,000	4,950,500
2100	Steel standpipes, ht./diam. more than 1, 100' to overflow, no fdn.									
2200	500,000 gallons				Ea.				487,500	535,000
2400	750,000 gallons								644,000	708,000
2500	1,000,000 gallons								945,000	1,040,000
2700	1,500,000 gallons								1,558,500	1,714,000
2800	2,000,000 gallons				▼				2,074,000	2,281,000

33 16 13.16 Prestressed Conc. Aboveground Water Utility Storage Tanks

		Crew	Daily Output	Labor-Hours	Unit	Material	2012 Bare Costs Labor	Equipment	Total	Total Incl O&P
0010	**PRESTRESSED CONC. ABOVEGROUND WATER UTILITY STORAGE TANKS**									
0020	Not including fdn., pipe or pumps, 250,000 gallons				Ea.				267,500	295,000
0100	500,000 gallons								436,500	480,000
0300	1,000,000 gallons								633,000	696,000
0400	2,000,000 gallons								960,000	1,056,000
0600	4,000,000 gallons								1,528,000	1,680,000
0700	6,000,000 gallons								2,030,000	2,232,000
0750	8,000,000 gallons								2,619,000	2,880,000
0800	10,000,000 gallons				▼				3,164,000	3,848,000

33 16 13.23 Plastic-Coated Fabric Pillow Water Tanks

		Crew	Daily Output	Labor-Hours	Unit	Material	2012 Bare Costs Labor	Equipment	Total	Total Incl O&P
0010	**PLASTIC-COATED FABRIC PILLOW WATER TANKS**									
7000	Water tanks, vinyl coated fabric pillow tanks, freestanding, 5,000 gallons	4 Clab	4	8	Ea.	4,600	281		4,881	5,475
7100	Supporting embankment not included, 25,000 gallons	6 Clab	2	24		10,500	840		11,340	12,900
7200	50,000 gallons	8 Clab	1.50	42.667		18,500	1,500		20,000	22,700
7300	100,000 gallons	9 Clab	.90	80		40,000	2,800		42,800	48,300
7400	150,000 gallons		.50	144		57,500	5,050		62,550	71,500
7500	200,000 gallons		.40	180		70,000	6,325		76,325	86,500
7600	250,000 gallons	▼	.30	240	▼	100,000	8,425		108,425	123,000

33 16 19 – Elevated Water Utility Storage Tanks

33 16 19.50 Elevated Water Storage Tanks

		Crew	Daily Output	Labor-Hours	Unit	Material	2012 Bare Costs Labor	Equipment	Total	Total Incl O&P
0010	**ELEVATED WATER STORAGE TANKS**									
0011	Not incl. pipe, pumps or foundation									
3000	Elevated water tanks, 100' to bottom capacity line, incl. painting									
3010	50,000 gallons				Ea.				167,000	184,000
3300	100,000 gallons								255,000	280,000
3400	250,000 gallons								706,000	776,000
3600	500,000 gallons								1,255,000	1,380,000
3700	750,000 gallons								1,523,000	1,675,000
3900	1,000,000 gallons				▼				2,180,000	2,400,000

33 21 13 – Public Water Supply Wells

33 21 13.10 Wells and Accessories	Crew	Daily Output	Labor-Hours	Unit	Material	2012 Bare Costs Labor	Equipment	Total	Total Incl O&P
0010 **WELLS & ACCESSORIES**									
0011 Domestic									
0100 Drilled, 4" to 6" diameter	B-23	120	.333	L.F.		11.85	22.50	34.35	43
0200 8" diameter	"	95.20	.420	"		14.90	28.50	43.40	54.50
0400 Gravel pack well, 40' deep, incl. gravel & casing, complete									
0500 24" diameter casing x 18" diameter screen	B-23	.13	307	Total	36,200	10,900	20,800	67,900	79,500
0600 36" diameter casing x 18" diameter screen		.12	333	"	37,400	11,800	22,500	71,700	84,000
0800 Observation wells, 1-1/4" riser pipe	↓	163	.245	V.L.F.	18.20	8.70	16.60	43.50	51.50
0900 For flush Buffalo roadway box, add	1 Skwk	16.60	.482	Ea.	49.50	22		71.50	88.50
1200 Test well, 2-1/2" diameter, up to 50' deep (15 to 50 GPM)	B-23	1.51	26.490	"	745	940	1,800	3,485	4,250
1300 Over 50' deep, add	"	121.80	.328	L.F.	19.85	11.65	22	53.50	64.50
1500 Pumps, installed in wells to 100' deep, 4" submersible									
1510 1/2 H.P.	Q-1	3.22	4.969	Ea.	420	244		664	830
1520 3/4 H.P.		2.66	6.015		625	296		921	1,125
1600 1 H.P.	↓	2.29	6.987		680	345		1,025	1,275
1700 1-1/2 H.P.	Q-22	1.60	10		780	490	385	1,655	2,025
1800 2 H.P.		1.33	12.030		850	590	460	1,900	2,325
1900 3 H.P.		1.14	14.035		1,325	690	535	2,550	3,125
2000 5 H.P.		1.14	14.035		2,225	690	535	3,450	4,100
3000 Pump, 6" submersible, 25' to 150' deep, 25 H.P., 249 to 297 GPM		.89	17.978		7,350	885	690	8,925	10,200
3100 25' to 500' deep, 30 H.P., 100 to 300 GPM	↓	.73	21.918	↓	8,650	1,075	840	10,565	12,000
8000 Steel well casing	B-23A	3020	.008	Lb.	.78	.31	.85	1.94	2.28
9950 See Section 31 23 19.40 for wellpoints									
9960 See Section 31 23 19.30 for drainage wells									

33 21 13.20 Water Supply Wells, Pumps

	Crew	Daily Output	Labor-Hours	Unit	Material	Labor	Equipment	Total	Total Incl O&P
0010 **WATER SUPPLY WELLS, PUMPS**									
0011 With pressure control									
1000 Deep well, jet, 42 gal. galvanized tank									
1040 3/4 HP	1 Plum	.80	10	Ea.	1,350	545		1,895	2,300
3000 Shallow well, jet, 30 gal. galvanized tank									
3040 1/2 HP	1 Plum	2	4	Ea.	1,050	218		1,268	1,500

33 31 13 – Public Sanitary Utility Sewerage Piping

33 31 13.15 Sewage Collection, Concrete Pipe

	Crew	Daily Output	Labor-Hours	Unit	Material	Labor	Equipment	Total	Total Incl O&P
0010 **SEWAGE COLLECTION, CONCRETE PIPE**									
0020 See Section 33 41 13.60 for sewage/drainage collection, concrete pipe									

33 31 13.25 Sewage Collection, Polyvinyl Chloride Pipe

	Crew	Daily Output	Labor-Hours	Unit	Material	Labor	Equipment	Total	Total Incl O&P
0010 **SEWAGE COLLECTION, POLYVINYL CHLORIDE PIPE**									
0020 Not including excavation or backfill									
2000 20' lengths, S.D.R. 35, B&S, 4" diameter	B-20	375	.064	L.F.	1.64	2.51		4.15	5.65
2040 6" diameter		350	.069		3.45	2.69		6.14	7.95
2080 13' lengths , S.D.R. 35, B&S, 8" diameter	↓	335	.072		7.85	2.81		10.66	12.95
2120 10" diameter	B-21	330	.085		12.70	3.43	.40	16.53	19.65
2160 12" diameter		320	.088		14.70	3.54	.41	18.65	22
2200 15" diameter	↓	240	.117		15.15	4.72	.55	20.42	24.50
4000 Piping, DWV PVC, no exc./bkfill., 10' L, Sch 40, 4" diameter	B-20	375	.064		4.09	2.51		6.60	8.35
4010 6" diameter		350	.069		8.45	2.69		11.14	13.45
4020 8" diameter	↓	335	.072	↓	15.30	2.81		18.11	21

33 36 Utility Septic Tanks

33 36 13 – Utility Septic Tank and Effluent Wet Wells

33 36 13.10 Septic Tanks	Crew	Daily Output	Labor-Hours	Unit	Material	2012 Bare Costs Labor	Equipment	Total	Total Incl O&P
0010 **SEPTIC TANKS**									
0011 Not including excavation or piping									
0015 Septic tanks, precast, 1,000 gallon	B-21	8	3.500	Ea.	1,025	142	16.55	1,183.55	1,350
0060 1,500 gallon		7	4		1,575	162	18.90	1,755.90	2,025
0100 2,000 gallon	↓	5	5.600		2,000	226	26.50	2,252.50	2,575
0200 5,000 gallon	B-13	3.50	16		9,100	610	199	9,909	11,100
0300 15,000 gallon, 4 piece	B-13B	1.70	32.941		20,900	1,250	645	22,795	25,600
0400 25,000 gallon, 4 piece		1.10	50.909		40,500	1,950	1,000	43,450	48,700
0500 40,000 gallon, 4 piece	↓	.80	70		52,000	2,675	1,375	56,050	62,500
0520 50,000 gallon, 5 piece	B-13C	.60	93.333		60,000	3,550	2,725	66,275	74,500
0540 75,000 gallon, cast in place	C-14C	.25	448		73,000	18,900	108	92,008	109,500
0560 100,000 gallon	"	.15	746		90,000	31,600	179	121,779	148,000
0600 High density polyethylene, 1,000 gallon	B-21	6	4.667		1,100	189	22	1,311	1,525
0700 1,500 gallon		4	7		1,400	283	33	1,716	2,000
0900 Galley, 4' x 4' x 4'	↓	16	1.750		260	71	8.30	339.30	405
1000 Distribution boxes, concrete, 7 outlets	2 Clab	16	1		75	35		110	137
1100 9 outlets		8	2		460	70		530	615
1120 Distribution boxes, HDPE, 5 outlets		20	.800		45	28		73	92.50
1125 8 outlets		10	1.600		56	56		112	148
1150 Leaching field chambers, 13' x 3'-7" x 1'-4", standard	B-13	16	3.500		470	133	43.50	646.50	770
1190 Distribution boxes, Outlet Flow Leveler	1 Clab	50	.160		2.95	5.60		8.55	11.90
1200 Heavy duty, 8' x 4' x 1'-6"	B-13	14	4		278	152	50	480	595
1300 13' x 3'-9" x 1'-6"		12	4.667		1,050	178	58	1,286	1,475
1350 20' x 4' x 1'-6"	↓	5	11.200		1,100	425	139	1,664	2,025
1400 Leaching pit, precast concrete, 3' diameter, 3' deep	B-21	8	3.500		660	142	16.55	818.55	960
1500 6' diameter, 3' section		4.70	5.957		815	241	28	1,084	1,300
2000 Velocity reducing pit, precast conc., 6' diameter, 3' deep	↓	4.70	5.957	↓	1,550	241	28	1,819	2,100
2200 Excavation for septic tank, 3/4 C.Y. backhoe	B-12F	145	.110	C.Y.		4.57	4.46	9.03	11.85
2400 4' trench for disposal field, 3/4 C.Y. backhoe	"	335	.048	L.F.		1.98	1.93	3.91	5.15
2600 Gravel fill, run of bank	B-6	150	.160	C.Y.	20.50	6.15	2.23	28.88	35
2800 Crushed stone, 3/4"	"	150	.160	"	39.50	6.15	2.23	47.88	55
3000 Effluent filter, 4" diameter	1 Skwk	8	1	Ea.	50	45.50		95.50	125
3020 Effluent filter, 6" diameter	"	7	1.143	"	200	52		252	300

33 41 Storm Utility Drainage Piping

33 41 13 – Public Storm Utility Drainage Piping

33 41 13.40 Piping, Storm Drainage, Corrugated Metal

	Crew	Daily Output	Labor-Hours	Unit	Material	2012 Bare Costs Labor	Equipment	Total	Total Incl O&P
0010 **PIPING, STORM DRAINAGE, CORRUGATED METAL**									
0020 Not including excavation or backfill									
2000 Corrugated metal pipe, galvanized									
2020 Bituminous coated with paved invert, 20' lengths									
2040 8" diameter, 16 ga.	B-14	330	.145	L.F.	11.10	5.40	1.01	17.51	21.50
2060 10" diameter, 16 ga.		260	.185		11.55	6.85	1.29	19.69	24.50
2080 12" diameter, 16 ga.		210	.229		14	8.45	1.59	24.04	30
2100 15" diameter, 16 ga.		200	.240		19.45	8.90	1.67	30.02	37
2120 18" diameter, 16 ga.		190	.253		21	9.35	1.76	32.11	39.50
2140 24" diameter, 14 ga.	↓	160	.300		25	11.10	2.09	38.19	47
2160 30" diameter, 14 ga.	B-13	120	.467		28	17.75	5.80	51.55	64.50
2180 36" diameter, 12 ga.		120	.467		35	17.75	5.80	58.55	72
2200 48" diameter, 12 ga.	↓	100	.560		55	21.50	6.95	83.45	101
2220 60" diameter, 10 ga.	B-13B	75	.747	↓	85	28.50	14.65	128.15	153

33 41 13 – Public Storm Utility Drainage Piping

33 41 13.40 Piping, Storm Drainage, Corrugated Metal

		Crew	Daily Output	Labor-Hours	Unit	Material	Labor	Equipment	Total	Total Incl O&P
							2012 Bare Costs			
2240	72" diameter, 8 ga.	B-13B	45	1.244	L.F.	95	47.50	24.50	167	205
2500	Galvanized, uncoated, 20' lengths									
2520	8" diameter, 16 ga.	B-14	355	.135	L.F.	8.30	5	.94	14.24	17.90
2540	10" diameter, 16 ga.		280	.171		8.80	6.35	1.19	16.34	21
2560	12" diameter, 16 ga.		220	.218		9.80	8.10	1.52	19.42	25
2580	15" diameter, 16 ga.		220	.218		12.90	8.10	1.52	22.52	28.50
2600	18" diameter, 16 ga.		205	.234		15	8.65	1.63	25.28	31.50
2620	24" diameter, 14 ga.	↓	175	.274		19.20	10.15	1.91	31.26	38.50
2640	30" diameter, 14 ga.	B-13	130	.431		24.50	16.40	5.35	46.25	58
2660	36" diameter, 12 ga.		130	.431		31.50	16.40	5.35	53.25	65.50
2680	48" diameter, 12 ga.	↓	110	.509		46.50	19.40	6.35	72.25	87.50
2690	60" diameter, 10 ga.	B-13B	78	.718	↓	69	27.50	14.10	110.60	134
2780	End sections, 8" diameter	B-14	35	1.371	Ea.	65.50	51	9.55	126.05	161
2785	10" diameter		35	1.371		69.50	51	9.55	130.05	165
2790	12" diameter		35	1.371		103	51	9.55	163.55	202
2800	18" diameter	↓	30	1.600		125	59.50	11.15	195.65	241
2810	24" diameter	B-13	25	2.240		188	85.50	28	301.50	370
2820	30" diameter		25	2.240		297	85.50	28	410.50	485
2825	36" diameter		20	2.800		455	107	35	597	700
2830	48" diameter		10	5.600		900	213	69.50	1,182.50	1,400
2835	60" diameter	B-13B	5	11.200		1,550	425	220	2,195	2,600
2840	72" diameter	"	4	14	↓	1,850	535	275	2,660	3,175

33 41 13.60 Sewage/Drainage Collection, Concrete Pipe

		Crew	Daily Output	Labor-Hours	Unit	Material	Labor	Equipment	Total	Total Incl O&P
0010	**SEWAGE/DRAINAGE COLLECTION, CONCRETE PIPE**									
0020	Not including excavation or backfill									
1000	Non-reinforced pipe, extra strength, B&S or T&G joints									
1010	6" diameter	B-14	265.04	.181	L.F.	6.40	6.70	1.26	14.36	18.75
1020	8" diameter		224	.214		7.05	7.95	1.49	16.49	21.50
1030	10" diameter		216	.222		7.80	8.25	1.55	17.60	23
1040	12" diameter		200	.240		9.55	8.90	1.67	20.12	26
1050	15" diameter		180	.267		13.05	9.90	1.86	24.81	31.50
1060	18" diameter		144	.333		15.40	12.35	2.32	30.07	38.50
1070	21" diameter		112	.429		18.95	15.85	2.98	37.78	49
1080	24" diameter	↓	100	.480	↓	25.50	17.80	3.34	46.64	59
2000	Reinforced culvert, class 3, no gaskets									
2010	12" diameter	B-14	150	.320	L.F.	11.75	11.85	2.23	25.83	33.50
2020	15" diameter		150	.320		16	11.85	2.23	30.08	38
2030	18" diameter		132	.364		20	13.45	2.53	35.98	45.50
2035	21" diameter		120	.400		24	14.80	2.78	41.58	52
2040	24" diameter	↓	100	.480		28	17.80	3.34	49.14	62
2045	27" diameter	B-13	92	.609		41	23	7.60	71.60	89
2050	30" diameter		88	.636		49	24	7.90	80.90	99.50
2060	36" diameter	↓	72	.778		59	29.50	9.70	98.20	121
2070	42" diameter	B-13B	72	.778		78	29.50	15.30	122.80	148
2080	48" diameter		64	.875		97.50	33.50	17.20	148.20	177
2090	60" diameter		48	1.167		125	44.50	23	192.50	231
2100	72" diameter		40	1.400		204	53.50	27.50	285	335
2120	84" diameter		32	1.750		293	66.50	34.50	394	465
2140	96" diameter	↓	24	2.333		350	89	46	485	570
2200	With gaskets, class 3, 12" diameter	B-21	168	.167		12.90	6.75	.79	20.44	25.50
2220	15" diameter		160	.175		17.60	7.10	.83	25.53	31
2230	18" diameter	↓	152	.184	↓	22	7.45	.87	30.32	37

33 41 Storm Utility Drainage Piping

33 41 13 – Public Storm Utility Drainage Piping

33 41 13.60 Sewage/Drainage Collection, Concrete Pipe	Crew	Daily Output	Labor-Hours	Unit	Material	2012 Bare Costs Labor	Equipment	Total	Total Incl O&P
2240	24" diameter	B-21	136	.206	L.F.	34.50	8.30	.97	43.77	52
2260	30" diameter	B-13	88	.636		56	24	7.90	87.90	107
2270	36" diameter	"	72	.778		67.50	29.50	9.70	106.70	131
2290	48" diameter	B-13B	64	.875		109	33.50	17.20	159.70	190
2310	72" diameter	"	40	1.400		222	53.50	27.50	303	355
2330	Flared ends, 6'-1" long, 12" diameter	B-21	190	.147		37.50	5.95	.70	44.15	51
2340	15" diameter		155	.181		40.50	7.30	.85	48.65	56.50
2400	6'-2" long, 18" diameter		122	.230		52	9.30	1.09	62.39	73
2420	24" diameter		88	.318		56	12.85	1.51	70.36	83
2440	36" diameter	B-13	60	.933		129	35.50	11.60	176.10	209
3080	Radius pipe, add to pipe prices, 12" to 60" diameter					50%				
3090	Over 60" diameter, add					20%				
3500	Reinforced elliptical, 8' lengths, C507 class 3									
3520	14" x 23" inside, round equivalent 18" diameter	B-21	82	.341	L.F.	55.50	13.80	1.62	70.92	84.50
3530	24" x 38" inside, round equivalent 30" diameter	B-13	58	.966		82.50	37	12	131.50	160
3540	29" x 45" inside, round equivalent 36" diameter		52	1.077		99	41	13.40	153.40	186
3550	38" x 60" inside, round equivalent 48" diameter		38	1.474		164	56	18.35	238.35	287
3560	48" x 76" inside, round equivalent 60" diameter		26	2.154		208	82	27	317	385
3570	58" x 91" inside, round equivalent 72" diameter		22	2.545		315	97	31.50	443.50	530
3780	Concrete slotted pipe, class 4 mortar joint									
3800	12" diameter	B-21	168	.167	L.F.	35.50	6.75	.79	43.04	50.50
3840	18" diameter	"	152	.184	"	41	7.45	.87	49.32	57.50
3900	Concrete slotted pipe, Class 4 O-ring joint									
3940	12" diameter	B-21	168	.167	L.F.	35.50	6.75	.79	43.04	50.50
3960	18" diameter	"	152	.184	"	41	7.45	.87	49.32	57.50

33 42 Culverts

33 42 16 – Concrete Culverts

33 42 16.15 Oval Arch Culverts

		Crew	Daily Output	Labor-Hours	Unit	Material	2012 Bare Costs Labor	Equipment	Total	Total Incl O&P
0010	**OVAL ARCH CULVERTS**									
3000	Corrugated galvanized or aluminum, coated & paved									
3020	17" x 13", 16 ga., 15" equivalent	B-14	200	.240	L.F.	17	8.90	1.67	27.57	34
3040	21" x 15", 16 ga., 18" equivalent		150	.320		20.50	11.85	2.23	34.58	43
3060	28" x 20", 14 ga., 24" equivalent		125	.384		28	14.20	2.67	44.87	56
3080	35" x 24", 14 ga., 30" equivalent		100	.480		34	17.80	3.34	55.14	68.50
3100	42" x 29", 12 ga., 36" equivalent	B-13	100	.560		39	21.50	6.95	67.45	83
3120	49" x 33", 12 ga., 42" equivalent		90	.622		49	23.50	7.75	80.25	98.50
3140	57" x 38", 12 ga., 48" equivalent		75	.747		58	28.50	9.30	95.80	118
3160	Steel, plain oval arch culverts, plain									
3180	17" x 13", 16 ga., 15" equivalent	B-14	225	.213	L.F.	12	7.90	1.48	21.38	27
3200	21" x 15", 16 ga., 18" equivalent		175	.274		15.25	10.15	1.91	27.31	34.50
3220	28" x 20", 14 ga., 24" equivalent		150	.320		22	11.85	2.23	36.08	45
3240	35" x 24", 14 ga., 30" equivalent	B-13	108	.519		27	19.75	6.45	53.20	67
3260	42" x 29", 12 ga., 36" equivalent		108	.519		34.50	19.75	6.45	60.70	75
3280	49" x 33", 12 ga., 42" equivalent		92	.609		44	23	7.60	74.60	92.50
3300	57" x 38", 12 ga., 48" equivalent		75	.747		53	28.50	9.30	90.80	112
3320	End sections, 17" x 13"		22	2.545	Ea.	130	97	31.50	258.50	325
3340	42" x 29"		17	3.294	"	360	125	41	526	630
3360	Multi-plate arch, steel	B-20	1690	.014	Lb.	1.25	.56		1.81	2.24

33 44 13 - Utility Area Drains

33 44 13.13 Catchbasins

		Crew	Daily Output	Labor-Hours	Unit	Material	2012 Bare Costs Labor	2012 Bare Costs Equipment	Total	Total Incl O&P
0010	**CATCHBASINS**									
0011	Not including footing & excavation									
1600	Frames & covers, C.I., 24" square, 500 lb.	B-6	7.80	3.077	Ea.	320	118	43	481	575
1700	26" D shape, 600 lb.		7	3.429		480	131	47.50	658.50	785
1800	Light traffic, 18" diameter, 100 lb.		10	2.400		115	92	33.50	240.50	305
1900	24" diameter, 300 lb.		8.70	2.759		192	106	38.50	336.50	415
2000	36" diameter, 900 lb.		5.80	4.138		560	159	57.50	776.50	920
2100	Heavy traffic, 24" diameter, 400 lb.		7.80	3.077		250	118	43	411	500
2200	36" diameter, 1150 lb.		3	8		800	305	111	1,216	1,475
2300	Mass. State standard, 26" diameter, 475 lb.		7	3.429		270	131	47.50	448.50	550
2400	30" diameter, 620 lb.		7	3.429		350	131	47.50	528.50	640
2500	Watertight, 24" diameter, 350 lb.		7.80	3.077		237	118	43	398	485
2600	26" diameter, 500 lb.		7	3.429		390	131	47.50	568.50	685
2700	32" diameter, 575 lb.	↓	6	4	↓	750	153	55.50	958.50	1,125
2800	3 piece cover & frame, 10" deep,									
2900	1200 lb., for heavy equipment	B-6	3	8	Ea.	1,000	305	111	1,416	1,700
3000	Raised for paving 1-1/4" to 2" high									
3100	4 piece expansion ring									
3200	20" to 26" diameter	1 Clab	3	2.667	Ea.	147	93.50		240.50	305
3300	30" to 36" diameter	"	3	2.667	"	205	93.50		298.50	370
3320	Frames and covers, existing, raised for paving, 2", including									
3340	row of brick, concrete collar, up to 12" wide frame	B-6	18	1.333	Ea.	46.50	51	18.55	116.05	150
3360	20" to 26" wide frame		11	2.182		69.50	83.50	30.50	183.50	238
3380	30" to 36" wide frame	↓	9	2.667		86	102	37	225	292
3400	Inverts, single channel brick	D-1	3	5.333		100	213		313	430
3500	Concrete		5	3.200		78	128		206	279
3600	Triple channel, brick		2	8		153	320		473	650
3700	Concrete	↓	3	5.333	↓	132	213		345	465

33 46 16 - Subdrainage Piping

33 46 16.25 Piping, Subdrainage, Corrugated Metal

		Crew	Daily Output	Labor-Hours	Unit	Material	2012 Bare Costs Labor	2012 Bare Costs Equipment	Total	Total Incl O&P
0010	**PIPING, SUBDRAINAGE, CORRUGATED METAL**									
0021	Not including excavation and backfill									
2010	Aluminum, perforated									
2020	6" diameter, 18 ga.	B-14	380	.126	L.F.	6	4.68	.88	11.56	14.70
2200	8" diameter, 16 ga.		370	.130		11.10	4.81	.90	16.81	20.50
2220	10" diameter, 16 ga.		360	.133		13.90	4.94	.93	19.77	24
2240	12" diameter, 16 ga.		285	.168		15.55	6.25	1.17	22.97	28
2260	18" diameter, 16 ga.	↓	205	.234	↓	23.50	8.65	1.63	33.78	40.50
3000	Uncoated galvanized, perforated									
3020	6" diameter, 18 ga.	B-20	380	.063	L.F.	5.50	2.48		7.98	9.85
3200	8" diameter, 16 ga.	"	370	.065		8.30	2.54		10.84	13.05
3220	10" diameter, 16 ga.	B-21	360	.078		8.80	3.14	.37	12.31	14.95
3240	12" diameter, 16 ga.		285	.098		9.80	3.97	.46	14.23	17.35
3260	18" diameter, 16 ga.	↓	205	.137	↓	15	5.50	.65	21.15	25.50
4000	Steel, perforated, asphalt coated									
4020	6" diameter 18 ga.	B-20	380	.063	L.F.	5.50	2.48		7.98	9.85
4030	8" diameter 18 ga.	"	370	.065		8.30	2.54		10.84	13.05
4040	10" diameter 16 ga.	B-21	360	.078		8.80	3.14	.37	12.31	14.95
4050	12" diameter 16 ga.	↓	285	.098	↓	9.80	3.97	.46	14.23	17.40

33 46 Subdrainage

33 46 16 – Subdrainage Piping

33 46 16.25 Piping, Subdrainage, Corrugated Metal	Crew	Daily Output	Labor-Hours	Unit	Material	2012 Bare Costs Labor	Equipment	Total	Total Incl O&P
4060	18" diameter 16 ga.	B-21	205	.137	L.F.	15	5.50	.65	21.15	25.50

33 46 16.40 Piping, Subdrainage, Polyvinyl Chloride									
0010	**PIPING, SUBDRAINAGE, POLYVINYL CHLORIDE**								
0020	Perforated, price as solid pipe, Section 33 31 13.25								

33 49 Storm Drainage Structures

33 49 13 – Storm Drainage Manholes, Frames, and Covers

33 49 13.10 Storm Drainage Manholes, Frames and Covers

33 49 13.10 Storm Drainage Manholes, Frames and Covers		Crew	Daily Output	Labor-Hours	Unit	Material	2012 Bare Costs Labor	Equipment	Total	Total Incl O&P
0010	**STORM DRAINAGE MANHOLES, FRAMES & COVERS**									
0020	Excludes footing, excavation, backfill (See line items for frame & cover)									
0050	Brick, 4' inside diameter, 4' deep	D-1	1	16	Ea.	410	640		1,050	1,425
0100	6' deep		.70	22.857		570	910		1,480	2,000
0150	8' deep		.50	32	↓	730	1,275		2,005	2,725
0200	For depths over 8', add		4	4	V.L.F.	85	159		244	335
0400	Concrete blocks (radial), 4' I.D., 4' deep		1.50	10.667	Ea.	375	425		800	1,050
0500	6' deep		1	16		500	640		1,140	1,525
0600	8' deep		.70	22.857	↓	630	910		1,540	2,075
0700	For depths over 8', add	↓	5.50	2.909	V.L.F.	66	116		182	248
0800	Concrete, cast in place, 4' x 4', 8" thick, 4' deep	C-14H	2	24	Ea.	445	1,050	13.70	1,508.70	2,100
0900	6' deep		1.50	32		635	1,400	18.25	2,053.25	2,875
1000	8' deep		1	48	↓	925	2,075	27.50	3,027.50	4,250
1100	For depths over 8', add	↓	8	6	V.L.F.	101	261	3.42	365.42	515
1110	Precast, 4' I.D., 4' deep	B-22	4.10	7.317	Ea.	775	299	48.50	1,122.50	1,375
1120	6' deep		3	10		900	410	66	1,376	1,700
1130	8' deep		2	15	↓	1,025	615	99.50	1,739.50	2,175
1140	For depths over 8', add	↓	16	1.875	V.L.F.	121	76.50	12.40	209.90	265
1150	5' I.D., 4' deep	B-6	3	8	Ea.	1,625	305	111	2,041	2,400
1160	6' deep		2	12		2,025	460	167	2,652	3,100
1170	8' deep		1.50	16	↓	2,475	615	223	3,313	3,900
1180	For depths over 8', add		12	2	V.L.F.	305	76.50	28	409.50	485
1190	6' I.D., 4' deep		2	12	Ea.	2,100	460	167	2,727	3,200
1200	6' deep		1.50	16		2,525	615	223	3,363	3,950
1210	8' deep		1	24	↓	3,100	920	335	4,355	5,200
1220	For depths over 8', add	↓	8	3	V.L.F.	360	115	42	517	620
1250	Slab tops, precast, 8" thick									
1300	4' diameter manhole	B-6	8	3	Ea.	272	115	42	429	520
1400	5' diameter manhole		7.50	3.200		425	123	44.50	592.50	705
1500	6' diameter manhole	↓	7	3.429		635	131	47.50	813.50	955
3800	Steps, heavyweight cast iron, 7" x 9"	1 Bric	40	.200		18.05	8.80		26.85	33
3900	8" x 9"		40	.200		21.50	8.80		30.30	37.50
3928	12" x 10-1/2"		40	.200		25.50	8.80		34.30	41.50
4000	Standard sizes, galvanized steel		40	.200		24	8.80		32.80	40
4100	Aluminum	↓	40	.200	↓	24	8.80		32.80	40

33 51 Natural-Gas Distribution

33 51 13 – Natural-Gas Piping

33 51 13.10 Piping, Gas Service and Distribution, P.E.	Crew	Daily Output	Labor-Hours	Unit	Material	2012 Bare Costs Labor	Equipment	Total	Total Incl O&P
0010 **PIPING, GAS SERVICE AND DISTRIBUTION, POLYETHYLENE**									
0020 Not including excavation or backfill									
1000 60 psi coils, compression coupling @ 100', 1/2" diameter, SDR 11	B-20A	608	.053	L.F.	.91	2.24		3.15	4.41
1010 1" diameter, SDR 11		544	.059		2.03	2.51		4.54	6.05
1040 1-1/4" diameter, SDR 11		544	.059		2.04	2.51		4.55	6.05
1100 2" diameter, SDR 11		488	.066		3.36	2.80		6.16	7.95
1160 3" diameter, SDR 11		408	.078		4.37	3.34		7.71	9.90
1500 60 PSI 40' joints with coupling, 3" diameter, SDR 11	B-21A	408	.098		4.37	4.28	1.10	9.75	12.50
1540 4" diameter, SDR 11		352	.114		9.15	4.96	1.27	15.38	19
1600 6" diameter, SDR 11		328	.122		26.50	5.30	1.37	33.17	38.50
1640 8" diameter, SDR 11		272	.147		32.50	6.40	1.65	40.55	47

33 52 Liquid Fuel Distribution

33 52 16 – Gasoline Distribution

33 52 16.13 Gasoline Piping

	Crew	Daily Output	Labor-Hours	Unit	Material	2012 Bare Costs Labor	Equipment	Total	Total Incl O&P
0010 **GASOLINE PIPING**									
0020 Primary containment pipe, fiberglass-reinforced									
0030 Plastic pipe 15' & 30' lengths									
0040 2" diameter	Q-6	425	.056	L.F.	5.60	2.94		8.54	10.65
0050 3" diameter		400	.060		8.60	3.12		11.72	14.15
0060 4" diameter		375	.064		11.25	3.33		14.58	17.40
0100 Fittings									
0110 Elbows, 90° & 45°, bell-ends, 2"	Q-6	24	1	Ea.	41	52		93	124
0120 3" diameter		22	1.091		51.50	56.50		108	143
0130 4" diameter		20	1.200		66	62.50		128.50	167
0200 Tees, bell ends, 2"		21	1.143		57	59.50		116.50	152
0210 3" diameter		18	1.333		60.50	69.50		130	172
0220 4" diameter		15	1.600		79.50	83		162.50	212
0230 Flanges bell ends, 2"		24	1		31.50	52		83.50	113
0240 3" diameter		22	1.091		37	56.50		93.50	126
0250 4" diameter		20	1.200		42.50	62.50		105	141
0260 Sleeve couplings, 2"		21	1.143		13.50	59.50		73	104
0270 3" diameter		18	1.333		17.10	69.50		86.60	124
0280 4" diameter		15	1.600		24.50	83		107.50	152
0290 Threaded adapters 2"		21	1.143		17.10	59.50		76.60	108
0300 3" diameter		18	1.333		32.50	69.50		102	141
0310 4" diameter		15	1.600		36	83		119	165
0320 Reducers, 2"		27	.889		26	46		72	98
0330 3" diameter		22	1.091		26	56.50		82.50	114
0340 4" diameter		20	1.200		35	62.50		97.50	133
1010 Gas station product line for secondary containment (double wall)									
1100 Fiberglass reinforced plastic pipe 25' lengths									
1120 Pipe, plain end, 3" diameter	Q-6	375	.064	L.F.	6.75	3.33		10.08	12.45
1130 4" diameter		350	.069		7.90	3.56		11.46	14.10
1140 5" diameter		325	.074		13.80	3.84		17.64	21
1150 6" diameter		300	.080		14.20	4.16		18.36	22
1200 Fittings									
1230 Elbows, 90° & 45°, 3" diameter	Q-6	18	1.333	Ea.	56.50	69.50		126	167
1240 4" diameter		16	1.500		105	78		183	234
1250 5" diameter		14	1.714		280	89		369	445
1260 6" diameter		12	2		320	104		424	505

33 52 Liquid Fuel Distribution

33 52 16 – Gasoline Distribution

33 52 16.13 Gasoline Piping		Crew	Daily Output	Labor-Hours	Unit	Material	2012 Bare Costs		Total	Total Incl O&P
							Labor	Equipment		
1270	Tees, 3" diameter	Q-6	15	1.600	Ea.	86	83		169	220
1280	4" diameter		12	2		131	104		235	300
1290	5" diameter		9	2.667		300	139		439	540
1300	6" diameter		6	4		365	208		573	720
1310	Couplings, 3" diameter		18	1.333		55	69.50		124.50	166
1320	4" diameter		16	1.500		115	78		193	245
1330	5" diameter		14	1.714		250	89		339	410
1340	6" diameter		12	2		310	104		414	495
1350	Cross-over nipples, 3" diameter		18	1.333		10	69.50		79.50	116
1360	4" diameter		16	1.500		12	78		90	131
1370	5" diameter		14	1.714		15	89		104	151
1380	6" diameter		12	2		18	104		122	177
1400	Telescoping, reducers, concentric 4" x 3"		18	1.333		45	69.50		114.50	155
1410	5" x 4"		17	1.412		90	73.50		163.50	210
1420	6" x 5"		16	1.500		220	78		298	360

33 71 Electrical Utility Transmission and Distribution

33 71 16 – Electrical Utility Poles

33 71 16.33 Wood Electrical Utility Poles

		Crew	Daily Output	Labor-Hours	Unit	Material	Labor	Equipment	Total	Total Incl O&P
0010	**WOOD ELECTRICAL UTILITY POLES**									
0011	Excludes excavation, backfill and cast-in-place concrete									
6200	Electric & tel sitework, 20' high, treated wd., see Section 26 56 13.10	R-3	3.10	6.452	Ea.	267	330	42.50	639.50	835
6400	25' high		2.90	6.897		325	350	45.50	720.50	935
6600	30' high		2.60	7.692		425	390	51	866	1,100
6800	35' high		2.40	8.333		525	425	55	1,005	1,275
7000	40' high		2.30	8.696		745	445	57.50	1,247.50	1,550
7200	45' high		1.70	11.765		890	600	78	1,568	1,975
7400	Cross arms with hardware & insulators									
7600	4' long	1 Elec	2.50	3.200	Ea.	143	165		308	405
7800	5' long		2.40	3.333		166	172		338	440
8000	6' long		2.20	3.636		190	187		377	490

33 71 19 – Electrical Underground Ducts and Manholes

33 71 19.17 Electric and Telephone Underground

		Crew	Daily Output	Labor-Hours	Unit	Material	Labor	Equipment	Total	Total Incl O&P
0010	**ELECTRIC AND TELEPHONE UNDERGROUND**									
0011	Not including excavation									
0200	backfill and cast in place concrete									
0400	Hand holes, precast concrete, with concrete cover									
0600	2' x 2' x 3' deep	R-3	2.40	8.333	Ea.	395	425	55	875	1,125
0800	3' x 3' x 3' deep		1.90	10.526		510	535	69.50	1,114.50	1,450
1000	4' x 4' x 4' deep		1.40	14.286		1,200	730	94.50	2,024.50	2,525
1200	Manholes, precast with iron racks & pulling irons, C.I. frame									
1400	and cover, 4' x 6' x 7' deep	B-13	2	28	Ea.	4,000	1,075	350	5,425	6,400
1600	6' x 8' x 7' deep		1.90	29.474		4,500	1,125	365	5,990	7,075
1800	6' x 10' x 7' deep		1.80	31.111		5,050	1,175	385	6,610	7,775
4200	Underground duct, banks ready for concrete fill, min. of 7.5"									
4400	between conduits, center to center									
4580	PVC, type EB, 1 @ 2" diameter	2 Elec	480	.033	L.F.	.37	1.72		2.09	2.98
4600	2 @ 2" diameter		240	.067		.74	3.44		4.18	5.95
4800	4 @ 2" diameter		120	.133		1.48	6.85		8.33	11.90
5000	2 @ 3" diameter		200	.080		1.03	4.12		5.15	7.30

33 71 Electrical Utility Transmission and Distribution

33 71 19 – Electrical Underground Ducts and Manholes

33 71 19.17 Electric and Telephone Underground	Crew	Daily Output	Labor-Hours	Unit	Material	2012 Bare Costs Labor	Equipment	Total	Total Incl O&P
5200	4 @ 3" diameter	2 Elec	100	.160	L.F.	2.05	8.25		10.30	14.60
5400	2 @ 4" diameter		160	.100		1.51	5.15		6.66	9.35
5600	4 @ 4" diameter		80	.200		3.02	10.30		13.32	18.75
5800	6 @ 4" diameter		54	.296		4.54	15.25		19.79	28
6200	Rigid galvanized steel, 2 @ 2" diameter		180	.089		15.15	4.58		19.73	23.50
6400	4 @ 2" diameter		90	.178		30.50	9.15		39.65	47
6800	2 @ 3" diameter		100	.160		38	8.25		46.25	54
7000	4 @ 3" diameter		50	.320		75.50	16.50		92	108
7200	2 @ 4" diameter		70	.229		55	11.80		66.80	78
7400	4 @ 4" diameter		34	.471		110	24.50		134.50	158
7600	6 @ 4" diameter		22	.727		164	37.50		201.50	237

33 81 Communications Structures

33 81 13 – Communications Transmission Towers

33 81 13.10 Radio Towers

		Crew	Daily Output	Labor-Hours	Unit	Material	2012 Bare Costs Labor	Equipment	Total	Total Incl O&P
0010	**RADIO TOWERS**									
0020	Guyed, 50' H, 40 lb. sec., 70 MPH basic wind spd.	2 Sswk	1	16	Ea.	2,900	790		3,690	4,575
0100	Wind load 90 MPH basic wind speed	"	1	16		3,450	790		4,240	5,175
0300	190' high, 40 lb. section, wind load 70 MPH basic wind speed	K-2	.33	72.727		9,000	3,275	800	13,075	16,400
0400	200' high, 70 lb. section, wind load 90 MPH basic wind speed		.33	72.727		15,000	3,275	800	19,075	23,000
0600	300' high, 70 lb. section, wind load 70 MPH basic wind speed		.20	120		24,000	5,400	1,325	30,725	37,100
0700	270' high, 90 lb. section, wind load 90 MPH basic wind speed		.20	120		26,000	5,400	1,325	32,725	39,300
0800	400' high, 100 lb. section, wind load 70 MPH basic wind speed		.14	171		36,000	7,725	1,900	45,625	55,000
0900	Self-supporting, 60' high, wind load 70 MPH basic wind speed		.80	30		4,500	1,350	330	6,180	7,650
0910	60' high, wind load 90 MPH basic wind speed		.45	53.333		5,500	2,400	590	8,490	10,800
1000	120' high, wind load 70 MPH basic wind speed		.40	60		10,500	2,700	660	13,860	17,000
1200	190' high, wind load 90 MPH basic wind speed		.20	120		29,500	5,400	1,325	36,225	43,200
2000	For states west of Rocky Mountains, add for shipping					10%				

Division Notes

	CREW	DAILY OUTPUT	LABOR-HOURS	UNIT	2012 BARE COSTS				TOTAL INCL O&P
					MAT.	LABOR	EQUIP.	TOTAL	

Estimating Tips

34 11 00 Rail Tracks

This subdivision includes items that may involve either repair of existing, or construction of new, railroad tracks. Additional preparation work, such as the roadbed earthwork, would be found in Division 31. Additional new construction siding and turnouts are found in Subdivision 34 72. Maintenance of railroads is found under 34 01 23 Operation and Maintenance of Railways.

34 41 13 Traffic Signals

This subdivision includes traffic signal systems. Other traffic control devices such as traffic signs are found in Subdivision 10 14 53 Traffic Signage.

34 71 13 Vehicle Barriers

This subdivision includes security vehicle barriers, guide and guard rails, crash barriers, and delineators. The actual maintenance and construction of concrete and asphalt pavement is found in Division 32.

Reference Numbers

Reference numbers are shown in shaded boxes at the beginning of some major classifications. These numbers refer to related items in the Reference Section. The reference information may be an estimating procedure, an alternate pricing method, or technical information.

Note: Not all subdivisions listed here necessarily appear in this publication.

34 01 Operation and Maintenance of Transportation

34 01 23 – Operation and Maintenance of Railways

34 01 23.51 Maintenance of Railroads		Crew	Daily Output	Labor-Hours	Unit	Material	2012 Bare Costs Labor	Equipment	Total	Total Incl O&P
0010	**MAINTENANCE OF RAILROADS**									
0400	Resurface and realign existing track	B-14	200	.240	L.F.		8.90	1.67	10.57	15.50
0600	For crushed stone ballast, add	"	500	.096	"	11.95	3.56	.67	16.18	19.35

34 11 Rail Tracks

34 11 13 – Track Rails

34 11 13.23 Heavy Rail Track

		Crew	Daily Output	Labor-Hours	Unit	Material	2012 Bare Costs Labor	Equipment	Total	Total Incl O&P
0010	**HEAVY RAIL TRACK** R347216-10									
1000	Rail, 100 lb. prime grade				L.F.	31.50			31.50	35
1500	Relay rail				"	15.85			15.85	17.45

34 11 33 – Track Cross Ties

34 11 33.13 Concrete Track Cross Ties

		Crew	Daily Output	Labor-Hours	Unit	Material	2012 Bare Costs Labor	Equipment	Total	Total Incl O&P
0010	**CONCRETE TRACK CROSS TIES**									
1400	Ties, concrete, 8'-6" long, 30" O.C.	B-14	80	.600	Ea.	157	22	4.18	183.18	211

34 11 33.16 Timber Track Cross Ties

		Crew	Daily Output	Labor-Hours	Unit	Material	2012 Bare Costs Labor	Equipment	Total	Total Incl O&P
0010	**TIMBER TRACK CROSS TIES**									
1600	Wood, pressure treated, 6" x 8" x 8'-6", C.L. lots	B-14	90	.533	Ea.	47	19.75	3.71	70.46	86.50
1700	L.C.L. lots		90	.533		49.50	19.75	3.71	72.96	89
1900	Heavy duty, 7" x 9" x 8'-6", C.L. lots		70	.686		52	25.50	4.77	82.27	101
2000	L.C.L. lots		70	.686		52	25.50	4.77	82.27	101

34 11 33.17 Timber Switch Ties

		Crew	Daily Output	Labor-Hours	Unit	Material	2012 Bare Costs Labor	Equipment	Total	Total Incl O&P
0010	**TIMBER SWITCH TIES**									
1200	Switch timber, for a #8 switch, pressure treated	B-14	3.70	12.973	M.B.F.	2,825	480	90.50	3,395.50	3,925
1300	Complete set of timbers, 3.7 M.B.F. for #8 switch	"	1	48	Total	10,900	1,775	335	13,010	15,100

34 11 93 – Track Appurtenances and Accessories

34 11 93.50 Track Accessories

		Crew	Daily Output	Labor-Hours	Unit	Material	2012 Bare Costs Labor	Equipment	Total	Total Incl O&P
0010	**TRACK ACCESSORIES**									
0020	Car bumpers, test	B-14	2	24	Ea.	3,550	890	167	4,607	5,450
0100	Heavy duty R347216-20		2	24		6,700	890	167	7,757	8,925
0200	Derails hand throw (sliding)		10	4.800		1,100	178	33.50	1,311.50	1,525
0300	Hand throw with standard timbers, open stand & target		8	6		1,200	222	42	1,464	1,700
2400	Wheel stops, fixed		18	2.667	Pr.	870	99	18.55	987.55	1,125
2450	Hinged		14	3.429	"	1,175	127	24	1,326	1,525

34 11 93.60 Track Material

		Crew	Daily Output	Labor-Hours	Unit	Material	2012 Bare Costs Labor	Equipment	Total	Total Incl O&P
0010	**TRACK MATERIAL**									
0020	Track bolts				Ea.	3.63			3.63	3.99
0100	Joint bars				Pr.	84			84	92.50
0200	Spikes				Ea.	1.20			1.20	1.32
0300	Tie plates				"	14.05			14.05	15.50

34 41 Roadway Signaling and Control Equipment

34 41 13 – Traffic Signals

34 41 13.10 Traffic Signals Systems

		Crew	Daily Output	Labor-Hours	Unit	Material	2012 Bare Costs Labor	2012 Bare Costs Equipment	Total	Total Incl O&P
0010	**TRAFFIC SIGNALS SYSTEMS**									
0020	Mid block pedestrian crosswalk with pushbutton and mast arms	R-11	.30	186	Total	85,000	9,100	2,650	96,750	110,000
0600	Traffic signals, school flashing system, solar powered, remote controlled	"	1	56	Signal	19,600	2,725	795	23,120	26,600
1000	Intersection traffic signals, LED, mast, programmable, no lane control									
1010	Includes all labor, material and equip. for complete installation	R-11	1	56	Signal	195,000	2,725	795	198,520	219,500
1100	Intersection traffic signals, LED, mast, programmable, R/L lane control									
1110	Includes all labor, material and equip. for complete installation	R-11	1	56	Signal	252,000	2,725	795	255,520	282,000
1200	Add protective/permissive left turns to existing traffic light									
1210	Includes all labor, material and equip. for complete installation	R-11	1	56	Signal	79,000	2,725	795	82,520	92,000
1300	Replace existing light heads with LED heads wire hung									
1310	Includes all labor, material and equip. for complete installation	R-11	1	56	Signal	53,000	2,725	795	56,520	63,500
1400	Replace existing light heads with LED heads mast arm hung									
1410	Includes all labor, material and equip. for complete installation	R-11	1	56	Signal	95,000	2,725	795	98,520	109,500

34 71 Roadway Construction

34 71 13 – Vehicle Barriers

34 71 13.17 Security Vehicle Barriers

		Crew	Daily Output	Labor-Hours	Unit	Material	2012 Bare Costs Labor	2012 Bare Costs Equipment	Total	Total Incl O&P
0010	**SECURITY VEHICLE BARRIERS**									
0020	Security planters excludes filling material									
0100	Concrete security planter, exposed aggregate 36" dia x 30" high	B-11M	8	2	Ea.	455	81.50	45.50	582	675
0200	48" dia x 36" high		8	2		565	81.50	45.50	692	795
0300	53" dia x 18" high		8	2		495	81.50	45.50	622	720
0400	72" dia x 18" high with seats		8	2		1,075	81.50	45.50	1,202	1,350
0450	84" dia x 36" high		8	2		1,275	81.50	45.50	1,402	1,575
0500	36" x 36" x 24" high square		8	2		385	81.50	45.50	512	600
0600	36" x 36" x 30" high square		8	2		450	81.50	45.50	577	670
0700	48" L x 24" W x 30" H rectangle		8	2		445	81.50	45.50	572	665
0800	72" L x 24" W x 30" H rectangle		8	2		550	81.50	45.50	677	780
0900	96" L x 24" W x 30" H rectangle		8	2		680	81.50	45.50	807	925
0950	Decorative geometric concrete barrier, 96" L x 24" W x 36" H		8	2		640	81.50	45.50	767	880
1000	Concrete security planter, filling with washed sand or gravel <1 C.Y.		8	2		34.50	81.50	45.50	161.50	212
1050	2 C.Y. or less		6	2.667		69.50	109	60.50	239	310
1200	Jersey concrete barrier, 10' L x 2' by 0.5' W x 30" H	B-21B	16	2.500		385	95	38.50	518.50	615
1300	10 or more same site		24	1.667		385	63.50	25.50	474	550
1400	10' L x 2' by 0.5' W x 32" H		16	2.500		415	95	38.50	548.50	645
1500	10 or more same site		24	1.667		415	63.50	25.50	504	580
1600	20' L x 2' by 0.5' W x 30" H		12	3.333		580	127	51	758	890
1700	10 or more same site		18	2.222		580	84.50	34	698.50	805
1800	20' L x 2' by 0.5' W x 32" H		12	3.333		710	127	51	888	1,025
1900	10 or more same site		18	2.222		710	84.50	34	828.50	945
2000	GFRC decorative security barrier per 10 feet section including concrete		4	10		2,200	380	153	2,733	3,175
2100	Per 12 feet section including concrete		4	10		2,650	380	153	3,183	3,675
2210	GFRC decorative security barrier will stop 30 MPH, 4000 lb. vehicle									
2300	High security barrier base prep per 12 feet section on bare ground	B-11C	4	4	Ea.	28.50	163	83.50	275	375
2310	GFRC barrier base prep does not include haul away of excavated matl									
2400	GFRC decorative high security barrier per 12 feet section w/concrete	B-21B	3	13.333	Ea.	5,350	505	204	6,059	6,875
2410	GFRC decorative high security barrier will stop 50 MPH, 15000 lb. vehicle									
2500	GFRC decorative impaler security barrier per 10 feet section w/prep	B-6	4	6	Ea.	2,075	230	83.50	2,388.50	2,725
2600	Per 12 feet section w/prep	"	4	6	"	2,475	230	83.50	2,788.50	3,175
2610	Impaler barrier should stop 50 MPH, 15000 lb. vehicle w/some penetr									
2700	Pipe bollards, steel, concrete filled/painted, 8' L x 4' D hole, 8" diam.	B-6	10	2.400	Ea.	875	92	33.50	1,000.50	1,150

34 71 13 – Vehicle Barriers

34 71 13.17 Security Vehicle Barriers		Crew	Daily Output	Labor-Hours	Unit	Material	2012 Bare Costs Labor	Equipment	Total	Total Incl O&P
2710	Schedule 80 concrete bollards will stop 4000 lb. vehicle @ 30 MPH									
2800	GFRC decorative jersey barrier cover per 10 feet section excludes soil	B-6	8	3	Ea.	880	115	42	1,037	1,200
2900	Per 12 feet section excludes soil		8	3		1,050	115	42	1,207	1,375
3000	GFRC decorative 8" diameter bollard cover		12	2		470	76.50	28	574.50	665
3100	GFRC decorative barrier face 10 foot section excludes earth backing		10	2.400		900	92	33.50	1,025.50	1,175
3200	Drop arm crash barrier, 15000 lb. vehicle @ 50 MPH									
3205	Includes all material, labor for complete installation									
3210	12' width				Ea.				62,000	68,000
3310	24' width								83,000	91,000
3410	Wedge crash barrier, 10' width								98,000	108,000
3510	12.5' width								109,000	120,000
3520	15' width								121,000	133,000
3610	Sliding crash barrier, 12' width								175,000	192,000
3710	Sliding roller crash barrier, 20' width								194,000	213,000
3810	Sliding cantilever crash barrier, 20' width								194,000	213,000
3890	Note: Raised bollard crash barriers should be used w/tire shredders									
3910	Raised bollard crash barrier, 10' width				Ea.				38,000	42,000
4010	12' width								44,000	48,000
4110	Raised bollard crash barrier, 10' width, solar powered								46,000	51,000
4210	12' width								51,000	56,000
4310	In ground tire shredder, 16' width								43,000	47,000

34 71 13.26 Vehicle Guide Rails

		Crew	Daily Output	Labor-Hours	Unit	Material	Labor	Equipment	Total	Total Incl O&P
0010	**VEHICLE GUIDE RAILS**									
0012	Corrugated stl, galv. stl posts, 6'-3" O.C.	B-80	850	.038	L.F.	20	1.42	.79	22.21	25
0200	End sections, galvanized, flared		50	.640	Ea.	88	24	13.35	125.35	149
0300	Wrap around end		50	.640	"	128	24	13.35	165.35	193
0400	Timber guide rail, 4" x 8" with 6" x 8" wood posts, treated		960	.033	L.F.	10.35	1.26	.70	12.31	14.05
0600	Cable guide rail, 3 at 3/4" cables, steel posts, single face		900	.036		9.90	1.35	.74	11.99	13.75
0700	Wood posts		950	.034		11.85	1.27	.70	13.82	15.70
0900	Guide rail, steel box beam, 6" x 6"		120	.267		31	10.10	5.55	46.65	55.50
1100	Median barrier, steel box beam, 6" x 8"		215	.149		41	5.65	3.11	49.76	57
1400	Resilient guide fence and light shield, 6' high	B-2	130	.308		34	10.90		44.90	54.50
1500	Concrete posts, individual, 6'-5", triangular	B-80	110	.291	Ea.	64	11	6.05	81.05	94
1550	Square	"	110	.291	"	70.50	11	6.05	87.55	101
2000	Median, precast concrete, 3'-6" high, 2' wide, single face	B-29	380	.147	L.F.	49.50	5.60	2.22	57.32	65.50
2200	Double face	"	340	.165	"	57	6.25	2.48	65.73	75
2400	Speed bumps, thermoplastic, 10-1/2" x 2-1/4" x 48" long	B-2	120	.333	Ea.	138	11.85		149.85	170
3030	Impact barrier, UTMCD, barrel type	B-16	30	1.067	"	470	38	19.50	527.50	595

34 71 19 – Vehicle Delineators

34 71 19.13 Fixed Vehicle Delineators

		Crew	Daily Output	Labor-Hours	Unit	Material	Labor	Equipment	Total	Total Incl O&P
0010	**FIXED VEHICLE DELINEATORS**									
0020	Crash barriers									
0100	Traffic channelizing pavement markers, layout only	A-7	2000	.012	Ea.		.58	.04	.62	.93
0110	13" x 7-1/2" x 2-1/2" high, non-plowable install	2 Clab	96	.167		22.50	5.85		28.35	34
0200	8" x 8" x 3-1/4" high, non-plowable, install		96	.167		21	5.85		26.85	32.50
0230	4" x 4" x 3/4" high, non-plowable, install		120	.133		1.22	4.68		5.90	8.55
0240	9-1/4" x 5-7/8" x 1/4" high, plowable, concrete pvmt.	A-2A	70	.343		19	11.95	5.15	36.10	45
0250	9-1/4" x 5-7/8" x 1/4" high, plowable, asphalt pvmt.	"	120	.200		4	6.95	2.99	13.94	18.35
0300	Barrier and curb delineators, reflectorized, 2" x 4"	2 Clab	150	.107		2.22	3.74		5.96	8.20
0310	3" x 5"	"	150	.107		4.25	3.74		7.99	10.45
0500	Rumble strip, polycarbonate									
0510	24" x 3-1/2" x 1/2" high	2 Clab	50	.320	Ea.	7.85	11.25		19.10	26

34 72 Railway Construction

34 72 16 – Railway Siding

34 72 16.50 Railroad Sidings		Crew	Daily Output	Labor-Hours	Unit	Material	2012 Bare Costs Labor	Equipment	Total	Total Incl O&P
0010	**RAILROAD SIDINGS** R347216-10									
0800	Siding, yard spur, level grade									
0820	100 lb. new rail	B-14	57	.842	L.F.	119	31	5.85	155.85	185
1002	Steel ties in concrete, incl. fasteners & plates									
1020	100 lb. new rail	B-14	22	2.182	L.F.	166	81	15.20	262.20	325

34 72 16.60 Railroad Turnouts		Crew	Daily Output	Labor-Hours	Unit	Material	2012 Bare Costs Labor	Equipment	Total	Total Incl O&P
0010	**RAILROAD TURNOUTS**									
2200	Turnout, #8 complete, w/rails, plates, bars, frog, switch point,									
2250	timbers, and ballast to 6" below bottom of ties									
2280	90 lb. rails	B-13	.25	224	Ea.	36,400	8,525	2,800	47,725	56,000
2290	90 lb. relay rails		.25	224		24,500	8,525	2,800	35,825	43,000
2300	100 lb. rails		.25	224		40,600	8,525	2,800	51,925	60,500
2310	100 lb. relay rails		.25	224		27,300	8,525	2,800	38,625	46,100
2320	110 lb. rails		.25	224		44,700	8,525	2,800	56,025	65,000
2330	110 lb. relay rails		.25	224		30,100	8,525	2,800	41,425	49,200
2340	115 lb. rails		.25	224		49,000	8,525	2,800	60,325	70,000
2350	115 lb. relay rails		.25	224		32,900	8,525	2,800	44,225	52,500
2360	132 lb. rails		.25	224		56,000	8,525	2,800	67,325	77,500
2370	132 lb. relay rails		.25	224		37,100	8,525	2,800	48,425	57,000

Division Notes

	CREW	DAILY OUTPUT	LABOR-HOURS	UNIT	2012 BARE COSTS				TOTAL INCL O&P
					MAT.	LABOR	EQUIP.	TOTAL	

Estimating Tips

35 20 16 Hydraulic Gates

This subdivision includes various types of gates that are commonly used in waterway and canal construction. Various earthwork items and structural support is found in Division 31, and concrete work in Division 3.

35 20 23 Dredging

This subdivision includes barge and shore dredging systems for rivers, canals, and channels.

35 31 00 Shoreline Protection

This subdivision includes breakwaters, bulkheads, and revetments for ocean and river inlets. Additional earthwork may be required from Division 31, and concrete work from Division 3.

35 41 00 Levees

Information on levee construction, including estimated cost of clay cone material.

35 49 00 Waterway Structures

This subdivision includes breakwaters and bulkheads for canals.

35 51 00 Floating Construction

This section includes floating piers, docks, and dock accessories. Fixed Pier Timber Construction is found in 06 13 33. Driven piles are found in Division 31, as well as sheet piling, cofferdams, and riprap.

Reference Numbers

Reference numbers are shown in shaded boxes at the beginning of some major classifications. These numbers refer to related items in the Reference Section. The reference information may be an estimating procedure, an alternate pricing method, or technical information.

Note: Not all subdivisions listed here necessarily appear in this publication.

Division 35 - Waterway and Marine

35 20 Waterway and Marine Construction and Equipment

35 20 23 – Dredging

35 20 23.13 Mechanical Dredging

		Crew	Daily Output	Labor-Hours	Unit	Material	2012 Bare Costs Labor	2012 Bare Costs Equipment	Total	Total Incl O&P
0010	**MECHANICAL DREDGING**									
0020	Dredging mobilization and demobilization, add to below, minimum	B-8	.53	120	Total		4,725	5,600	10,325	13,400
0100	Maximum	"	.10	640	"		25,000	29,700	54,700	70,500
0300	Barge mounted clamshell excavation into scows									
0310	Dumped 20 miles at sea, minimum	B-57	310	.155	B.C.Y.		6.20	5.55	11.75	15.55
0400	Maximum	"	213	.225	"		9.05	8.05	17.10	22.50
0500	Barge mounted dragline or clamshell, hopper dumped,									
0510	pumped 1000' to shore dump, minimum	B-57	340	.141	B.C.Y.		5.65	5.05	10.70	14.20
0525	All pumping uses 2000 gallons of water per cubic yard									
0600	Maximum	B-57	243	.198	B.C.Y.		7.95	7.05	15	19.90

35 20 23.23 Hydraulic Dredging

		Crew	Daily Output	Labor-Hours	Unit	Material	2012 Bare Costs Labor	2012 Bare Costs Equipment	Total	Total Incl O&P
0010	**HYDRAULIC DREDGING**									
1000	Hydraulic method, pumped 1000' to shore dump, minimum	B-57	460	.104	B.C.Y.		4.19	3.74	7.93	10.50
1100	Maximum		310	.155			6.20	5.55	11.75	15.55
1400	Into scows dumped 20 miles, minimum		425	.113			4.54	4.04	8.58	11.35
1500	Maximum		243	.198			7.95	7.05	15	19.90
1600	For inland rivers and canals in South, deduct								30%	30%

35 51 Floating Construction

35 51 13 – Floating Piers

35 51 13.23 Floating Wood Piers

		Crew	Daily Output	Labor-Hours	Unit	Material	2012 Bare Costs Labor	2012 Bare Costs Equipment	Total	Total Incl O&P
0010	**FLOATING WOOD PIERS**									
0020	Polyethylene encased polystyrene, no pilings included	F-3	330	.121	S.F.	40.50	5.45	1.85	47.80	55
0200	Pile supported, shore constructed, bare, 3" decking		130	.308		27.50	13.80	4.71	46.01	56.50
0250	4" decking		120	.333		28	14.95	5.10	48.05	59.50
0400	Floating, small boat, prefab, no shore facilities, minimum		250	.160		14	7.15	2.45	23.60	29
0500	Maximum		150	.267		45.50	11.95	4.08	61.53	73
0700	Per slip, minimum (180 S.F. each)		1.59	25.157	Ea.	2,925	1,125	385	4,435	5,375
0800	Maximum		1.40	28.571	"	9,400	1,275	435	11,110	12,800

Estimating Tips

Products such as conveyors, material handling cranes and hoists, as well as other items specified in this division, require trained installers. The general contractor may not have any choice as to who will perform the installation or when it will be performed. Long lead times are often required for these products, making early decisions in purchasing and scheduling necessary.

The installation of this type of equipment may require the embedment of mounting hardware during construction of floors, structural walls, or interior walls/partitions. Electrical connections will require coordination with the electrical contractor.

Reference Numbers

Reference numbers are shown in shaded boxes at the beginning of some major classifications. These numbers refer to related items in the Reference Section. The reference information may be an estimating procedure, an alternate pricing method, or technical information.

Note: Not all subdivisions listed here necessarily appear in this publication.

41 21 Conveyors

41 21 23 – Piece Material Conveyors

41 21 23.16 Container Piece Material Conveyors	Crew	Daily Output	Labor-Hours	Unit	Material	2012 Bare Costs Labor	Equipment	Total	Total Incl O&P
0010 **CONTAINER PIECE MATERIAL CONVEYORS**									
0020 Gravity fed, 2" rollers, 3" O.C.									
0050 10' sections with 2 supports, 600 lb. capacity, 18" wide				Ea.	450			450	495
0100 24" wide					515			515	570
0150 1400 lb. capacity, 18" wide					415			415	455
0200 24" wide					460			460	505
0350 Horizontal belt, center drive and takeup, 60 fpm									
0400 16" belt, 26.5' length	2 Mill	.50	32	Ea.	3,150	1,475		4,625	5,650
0450 24" belt, 41.5' length		.40	40		4,900	1,850		6,750	8,100
0500 61.5' length		.30	53.333		6,525	2,450		8,975	10,800
0600 Inclined belt, 10' rise with horizontal loader and									
0620 End idler assembly, 27.5' length, 18" belt	2 Mill	.30	53.333	Ea.	9,200	2,450		11,650	13,700
0700 24" belt	"	.15	106	"	10,300	4,900		15,200	18,600
3600 Monorail, overhead, manual, channel type									
3700 125 lb. per L.F.	1 Mill	26	.308	L.F.	10.25	14.15		24.40	32.50
3900 500 lb. per L.F.	"	21	.381	"	9.25	17.55		26.80	36
4000 Trolleys for above, 2 wheel, 125 lb. capacity				Ea.	100			100	110
4200 4 wheel, 250 lb. capacity					248			248	272
4300 8 wheel, 1,000 lb. capacity					340			340	375

41 22 Cranes and Hoists

41 22 13 – Cranes

41 22 13.10 Crane Rail

		Crew	Daily Output	Labor-Hours	Unit	Material	Labor	Equipment	Total	Total Incl O&P
0010	**CRANE RAIL**									
0020	Box beam bridge, no equipment included	E-4	3400	.009	Lb.	1.19	.47	.04	1.70	2.19
0200	Running track only, 104 lb. per yard	"	5600	.006	"	.62	.28	.02	.92	1.21

41 22 13.13 Bridge Cranes

		Crew	Daily Output	Labor-Hours	Unit	Material	Labor	Equipment	Total	Total Incl O&P
0010	**BRIDGE CRANES**									
0100	1 girder, 20' span, 3 ton	M-3	1	34	Ea.	23,800	1,825	130	25,755	29,100
0125	5 ton		1	34		26,000	1,825	130	27,955	31,500
0150	7.5 ton		1	34		31,000	1,825	130	32,955	37,000
0175	10 ton		.80	42.500		41,000	2,275	162	43,437	48,600
0200	15 ton		.80	42.500		52,500	2,275	162	54,937	61,500
0225	30' span, 3 ton		1	34		24,700	1,825	130	26,655	30,100
0250	5 ton		1	34		27,200	1,825	130	29,155	32,800
0275	7.5 ton		1	34		32,600	1,825	130	34,555	38,700
0300	10 ton		.80	42.500		42,400	2,275	162	44,837	50,000
0325	15 ton		.80	42.500		55,000	2,275	162	57,437	64,000
0350	2 girder, 40' span, 3 ton	M-4	.50	72		40,900	3,825	330	45,055	51,000
0375	5 ton		.50	72		42,800	3,825	330	46,955	53,000
0400	7.5 ton		.50	72		46,900	3,825	330	51,055	57,500
0425	10 ton		.40	90		55,000	4,775	415	60,190	68,000
0450	15 ton		.40	90		74,500	4,775	415	79,690	89,500
0475	25 ton		.30	120		88,000	6,350	555	94,905	107,000
0500	50' span, 3 ton		.50	72		46,500	3,825	330	50,655	57,000
0525	5 ton		.50	72		48,400	3,825	330	52,555	59,000
0550	7.5 ton		.50	72		52,000	3,825	330	56,155	63,000
0575	10 ton		.40	90		59,500	4,775	415	64,690	73,000
0600	15 ton		.40	90		78,000	4,775	415	83,190	93,500
0625	25 ton		.30	120		92,500	6,350	555	99,405	111,500

41 22 Cranes and Hoists

41 22 23 – Hoists

41 22 23.10 Material Handling	Crew	Daily Output	Labor-Hours	Unit	Material	2012 Bare Costs Labor	Equipment	Total	Total Incl O&P
0010 **MATERIAL HANDLING**, cranes, hoists and lifts									
1500 Cranes, portable hydraulic, floor type, 2,000 lb. capacity				Ea.	2,475			2,475	2,700
1600 4,000 lb. capacity					4,150			4,150	4,575
1800 Movable gantry type, 12' to 15' range, 2,000 lb. capacity					3,600			3,600	3,975
1900 6,000 lb. capacity					5,225			5,225	5,750
2100 Hoists, electric overhead, chain, hook hung, 15' lift, 1 ton cap.					2,325			2,325	2,575
2200 3 ton capacity					2,725			2,725	3,000
2500 5 ton capacity					6,625			6,625	7,275
2600 For hand-pushed trolley, add					15%				
2700 For geared trolley, add					30%				
2800 For motor trolley, add					75%				
3000 For lifts over 15', 1 ton, add				L.F.	23.50			23.50	25.50
3100 5 ton, add				"	52			52	57
3300 Lifts, scissor type, portable, electric, 36" high, 2,000 lb.				Ea.	4,275			4,275	4,700
3400 48" high, 4,000 lb.				"	4,675			4,675	5,150

Division Notes

	CREW	DAILY OUTPUT	LABOR-HOURS	UNIT	2012 BARE COSTS				TOTAL INCL O&P
					MAT.	LABOR	EQUIP.	TOTAL	

Estimating Tips

This section involves equipment and construction costs for air noise and odor pollution control systems.

These systems may be interrelated and care must be taken that the complete systems are estimated. For example, air pollution equipment may include dust and air-entrained particles that have to be collected. The vacuum systems could be noisy, requiring silencers to reduce noise pollution, and the collected solids have to be disposed of to prevent solid pollution.

Reference Numbers

Reference numbers are shown in shaded boxes at the beginning of some major classifications. These numbers refer to related items in the Reference Section. The reference information may be an estimating procedure, an alternate pricing method, or technical information.

Note: Not all subdivisions listed here necessarily appear in this publication.

Division 44 - Pollution Control Equipment

44 11 16.10 Dust Collection Systems	Crew	Daily Output	Labor-Hours	Unit	Material	2012 Bare Costs Labor	Equipment	Total	Total Incl O&P
0010 **DUST COLLECTION SYSTEMS** Commercial/Industrial									
0120 Central vacuum units									
0130 Includes stand, filters and motorized shaker									
0200 500 CFM, 10" inlet, 2 HP	Q-20	2.40	8.333	Ea.	3,900	400		4,300	4,900
0220 1000 CFM, 10" inlet, 3 HP		2.20	9.091		4,100	435		4,535	5,175
0240 1500 CFM, 10" inlet, 5 HP		2	10		4,325	480		4,805	5,475
0260 3000 CFM, 13" inlet, 10 HP		1.50	13.333		12,200	640		12,840	14,400
0280 5000 CFM, 16" inlet, 2 @ 10 HP	↓	1	20	↓	12,900	955		13,855	15,700
1000 Vacuum tubing, galvanized									
1100 2-1/8" OD, 16 ga.	Q-9	440	.036	L.F.	2.80	1.71		4.51	5.70
1110 2-1/2" OD, 16 ga.		420	.038		3.23	1.79		5.02	6.30
1120 3" OD, 16 ga.		400	.040		3.69	1.88		5.57	6.90
1130 3-1/2" OD, 16 ga.		380	.042		5.35	1.98		7.33	8.85
1140 4" OD, 16 ga.		360	.044		6	2.09		8.09	9.80
1150 5" OD, 14 ga.		320	.050		10.75	2.35		13.10	15.45
1160 6" OD, 14 ga.		280	.057		12.45	2.68		15.13	17.80
1170 8" OD, 14 ga.		200	.080		18.75	3.75		22.50	26
1180 10" OD, 12 ga.		160	.100		29.50	4.69		34.19	39.50
1190 12" OD, 12 ga.		120	.133		32.50	6.25		38.75	45.50
1200 14" OD, 12 ga.	↓	80	.200	↓	39.50	9.40		48.90	58
1940 Hose, flexible wire reinforced rubber									
1956 3" dia.	Q-9	400	.040	L.F.	8.20	1.88		10.08	11.85
1960 4" dia.		360	.044		9.50	2.09		11.59	13.65
1970 5" dia.		320	.050		11.90	2.35		14.25	16.70
1980 6" dia.	↓	280	.057	↓	12.90	2.68		15.58	18.25
2000 90° Elbow, slip fit									
2110 2-1/8" dia.	Q-9	70	.229	Ea.	11.70	10.75		22.45	29
2120 2-1/2" dia.		65	.246		16.35	11.55		27.90	35.50
2130 3" dia.		60	.267		22	12.50		34.50	43.50
2140 3-1/2" dia.		55	.291		27.50	13.65		41.15	51
2150 4" dia.		50	.320		34	15		49	60.50
2160 5" dia.		45	.356		64.50	16.70		81.20	96.50
2170 6" dia.		40	.400		89	18.75		107.75	127
2180 8" dia.	↓	30	.533	↓	169	25		194	224
2400 45° Elbow, slip fit									
2410 2-1/8" dia.	Q-9	70	.229	Ea.	10.25	10.75		21	27.50
2420 2-1/2" dia.		65	.246		15.10	11.55		26.65	34
2430 3" dia.		60	.267		18.20	12.50		30.70	39
2440 3-1/2" dia.		55	.291		23	13.65		36.65	46
2450 4" dia.		50	.320		29.50	15		44.50	55.50
2460 5" dia.		45	.356		50	16.70		66.70	80.50
2470 6" dia.		40	.400		68	18.75		86.75	103
2480 8" dia.	↓	35	.457	↓	133	21.50		154.50	179
2800 90° TY, slip fit thru 6" dia.									
2810 2-1/8" dia.	Q-9	42	.381	Ea.	23	17.90		40.90	52.50
2820 2-1/2" dia.		39	.410		30	19.25		49.25	62.50
2830 3" dia.		36	.444		39.50	21		60.50	75.50
2840 3-1/2" dia.		33	.485		50	23		73	89.50
2850 4" dia.		30	.533		72	25		97	117
2860 5" dia.		27	.593		137	28		165	194
2870 6" dia.	↓	24	.667		186	31.50		217.50	253
2880 8" dia., butt end					395			395	435
2890 10" dia., butt end				↓	715			715	785

44 11 16.10 Dust Collection Systems		Crew	Daily Output	Labor-Hours	Unit	Material	2012 Bare Costs Labor	Equipment	Total	Total Incl O&P
2900	12" dia., butt end				Ea.	715			715	785
2910	14" dia., butt end					1,300			1,300	1,425
2920	6" x 4" dia., butt end					114			114	125
2930	8" x 4" dia., butt end					215			215	237
2940	10" x 4" dia., butt end					282			282	310
2950	12" x 4" dia., butt end					325			325	355
3100	90° Elbow, butt end segmented									
3110	8" dia., butt end, segmented				Ea.	169			169	186
3120	10" dia., butt end, segmented					420			420	460
3130	12" dia., butt end, segmented					530			530	585
3140	14" dia., butt end, segmented					665			665	735
3200	45° Elbow, butt end segmented									
3210	8" dia., butt end, segmented				Ea.	133			133	146
3220	10" dia., butt end, segmented					305			305	335
3230	12" dia., butt end, segmented					310			310	340
3240	14" dia., butt end, segmented					380			380	420
3400	All butt end fittings require one coupling per joint.									
3410	Labor for fitting included with couplings.									
3460	Compression coupling, galvanized, neoprene gasket									
3470	2-1/8" dia.	Q-9	44	.364	Ea.	13.55	17.05		30.60	41
3480	2-1/2" dia.		44	.364		13.55	17.05		30.60	41
3490	3" dia.		38	.421		19.55	19.75		39.30	51.50
3500	3-1/2" dia.		35	.457		22	21.50		43.50	57
3510	4" dia.		33	.485		24	23		47	61
3520	5" dia.		29	.552		27.50	26		53.50	69.50
3530	6" dia.		26	.615		27.50	29		56.50	74.50
3540	8" dia.		22	.727		57.50	34		91.50	115
3550	10" dia.		20	.800		77.50	37.50		115	142
3560	12" dia.		18	.889		96	41.50		137.50	169
3570	14" dia.		16	1		122	47		169	207
3800	Air gate valves, galvanized									
3810	2-1/8" dia.	Q-9	30	.533	Ea.	198	25		223	256
3820	2-1/2" dia.		28	.571		219	27		246	282
3830	3" dia.		26	.615		239	29		268	305
3840	4" dia.		23	.696		300	32.50		332.50	380
3850	6" dia.		18	.889		405	41.50		446.50	510

Division Notes

	CREW	DAILY OUTPUT	LABOR-HOURS	UNIT	2012 BARE COSTS				TOTAL INCL O&P
					MAT.	LABOR	EQUIP.	TOTAL	

Estimating Tips

This new division contains information about water and wastewater equipment and systems, which was formerly located in Division 44.

The main areas of focus are total wastewater treatment plants and components of wastewater treatment plants. In addition, there are assemblies such as sewage treatment lagoons that can be found in some publications under G30 Site Mechanical Utilities.

Also included in this section are oil/water separators for wastewater treatment.

New sections added for this year are for activated sludge cells and air diffuser systems for wastewater treatment.

Reference Numbers

Reference numbers are shown in shaded boxes at the beginning of some major classifications. These numbers refer to related items in the Reference Section. The reference information may be an estimating procedure, an alternate pricing method, or technical information.

Note: Not all subdivisions listed here necessarily appear in this publication.

Division 46 - Water & Wastewater Equipment

46 07 53 – Packaged Wastewater Treatment Equipment

46 07 53.10 Biological Packaged Water Treatment Plants	Crew	Daily Output	Labor-Hours	Unit	Material	2012 Bare Costs Labor	Equipment	Total	Total Incl O&P
0010 **BIOLOGICAL PACKAGED WATER TREATMENT PLANTS**									
0011 Not including fencing or external piping									
0020 Steel packaged, blown air aeration plants									
0100 1,000 GPD				Gal.				95	109
0200 5,000 GPD								57	66
0300 15,000 GPD								48	55
0400 30,000 GPD								43	50
0500 50,000 GPD								38	44
0600 100,000 GPD								33	38
0700 200,000 GPD								28	32
0800 500,000 GPD								24	28
1000 Concrete, extended aeration, primary and secondary treatment									
1010 10,000 GPD				Gal.				48	55
1100 30,000 GPD								43	50
1200 50,000 GPD								38	44
1400 100,000 GPD								33	38
1500 500,000 GPD								24	28
1700 Municipal wastewater treatment facility									
1720 1.0 MGD				Gal.				9.70	11
1740 1.5 MGD								9.20	10.60
1760 2.0 MGD								8.75	10
1780 3.0 MGD								6.80	7.80
1800 5.0 MGD								5.80	6.70
2000 Holding tank system, not incl. excavation or backfill									
2010 Recirculating chemical water closet	2 Plum	4	4	Ea.	450	218		668	825
2100 For voltage converter, add	"	16	1		245	54.50		299.50	355
2200 For high level alarm, add	1 Plum	7.80	1.026		140	56		196	239

46 07 53.20 Wastewater Treatment System

	Crew	Daily Output	Labor-Hours	Unit	Material	2012 Bare Costs Labor	Equipment	Total	Total Incl O&P
0010 **WASTEWATER TREATMENT SYSTEM**									
0020 Fiberglass, 1,000 gallon	B-21	1.29	21.705	Ea.	3,375	880	103	4,358	5,200
0100 1,500 gallon	B-21	1.03	27.184	Ea.	7,400	1,100	129	8,629	10,000

Reference Section

All the reference information is in one section, making it easy to find what you need to know ...and easy to use the book on a daily basis. This section is visually identified by a vertical gray bar on the page edges.

In this Reference Section, we've included Equipment Rental Costs, a listing of rental and operating costs; Crew Listings, a full listing of all crews and equipment, and their costs; Historical Cost Indexes for cost comparisons over time; City Cost Indexes and Location Factors for adjusting costs to the region you are in; Reference Tables, where you will find explanations, estimating information and procedures, or technical data; Change Orders, information on pricing changes to contract documents; Square Foot Costs that allow you to make a rough estimate for the overall cost of a project; and an explanation of all the Abbreviations in the book.

Table of Contents

Estimating Tips

- This section contains the average costs to rent and operate hundreds of pieces of construction equipment. This is useful information when estimating the time and material requirements of any particular operation in order to establish a unit or total cost. Equipment costs include not only rental, but also operating costs for equipment under normal use.

Rental Costs

- Equipment rental rates are obtained from industry sources throughout North America–contractors, suppliers, dealers, manufacturers, and distributors.
- Rental rates vary throughout the country, with larger cities generally having lower rates. Lease plans for new equipment are available for periods in excess of six months, with a percentage of payments applying toward purchase.
- Monthly rental rates vary from 2% to 5% of the purchase price of the equipment depending on the anticipated life of the equipment and its wearing parts.
- Weekly rental rates are about 1/3 the monthly rates, and daily rental rates are about 1/3 the weekly rate.
- Rental rates can also be treated as reimbursement costs for contractor-owned equipment. Owned equipment costs include depreciation, loan payments, interest, taxes, insurance, storage, and major repairs.

Operating Costs

- The operating costs include parts and labor for routine servicing, such as repair and replacement of pumps, filters and worn lines. Normal operating expendables, such as fuel, lubricants, tires and electricity (where applicable), are also included.
- Extraordinary operating expendables with highly variable wear patterns, such as diamond bits and blades, are excluded. These costs can be found as material costs in the Unit Price section.
- The hourly operating costs listed do not include the operator's wages.

Equipment Cost/Day

- Any power equipment required by a crew is shown in the Crew Listings with a daily cost.
- The daily cost of equipment needed by a crew is based on dividing the weekly rental rate by 5 (number of working days in the week), and then adding the hourly operating cost times 8 (the number of hours in a day). This "Equipment Cost/ Day" is shown in the far right column of the Equipment Rental pages.
- If equipment is needed for only one or two days, it is best to develop your own cost by including components for daily rent and hourly operating cost. This is important when the listed Crew for a task does not contain the equipment needed, such as a crane for lifting mechanical heating/cooling equipment up onto a roof.

- If the quantity of work is less than the crew's Daily Output shown for a Unit Price line item that includes a bare unit equipment cost, it is recommended to estimate one day's rental cost and operating cost for equipment shown in the Crew Listing for that line item.

Mobilization/ Demobilization

- The cost to move construction equipment from an equipment yard or rental company to the jobsite and back again is not included in equipment rental costs listed in the Reference section, nor in the bare equipment cost of any Unit Price line item, nor in any equipment costs shown in the Crew listings.
- Mobilization (to the site) and demobilization (from the site) costs can be found in the Unit Price section.
- If a piece of equipment is already at the jobsite, it is not appropriate to utilize mobil./ demob. costs again in an estimate.

01 54 33 | Equipment Rental

			UNIT	HOURLY OPER. COST	RENT PER DAY	RENT PER WEEK	RENT PER MONTH	EQUIPMENT COST/DAY	
10	0010	**CONCRETE EQUIPMENT RENTAL** without operators	R015433 -10						**10**
	0200	Bucket, concrete lightweight, 1/2 C.Y.	Ea.	.75	22	66	198	19.20	
	0300	1 C.Y.		.80	26.50	80	240	22.40	
	0400	1-1/2 C.Y.		1.05	36.50	110	330	30.40	
	0500	2 C.Y.		1.15	43.50	130	390	35.20	
	0580	8 C.Y.		5.80	255	765	2,300	199.40	
	0600	Cart, concrete, self-propelled, operator walking, 10 C.F.		2.85	56.50	170	510	56.80	
	0700	Operator riding, 18 C.F.		4.80	93.50	280	840	94.40	
	0800	Conveyer for concrete, portable, gas, 16" wide, 26' long		11.00	122	365	1,100	161	
	0900	46' long		11.40	147	440	1,325	179.20	
	1000	56' long		11.50	157	470	1,400	186	
	1100	Core drill, electric, 2-1/2 H.P., 1" to 8" bit diameter		1.40	59.50	179	535	47	
	1150	11 H.P., 8" to 18" cores		5.40	112	335	1,000	110.20	
	1200	Finisher, concrete floor, gas, riding trowel, 96" wide		10.90	145	435	1,300	174.20	
	1300	Gas, walk-behind, 3 blade, 36" trowel		1.80	20	60	180	26.40	
	1400	4 blade, 48" trowel		3.55	27	81	243	44.60	
	1500	Float, hand-operated (Bull float) 48" wide		.08	13.65	41	123	8.85	
	1570	Curb builder, 14 H.P., gas, single screw		13.10	245	735	2,200	251.80	
	1590	Double screw		13.75	283	850	2,550	280	
	1600	Floor grinder, concrete and terrazzo, electric, 22" path		1.84	117	350	1,050	84.70	
	1700	Edger, concrete, electric, 7" path		.90	51.50	155	465	38.20	
	1750	Vacuum pick-up system for floor grinders, wet/dry		1.35	81.50	245	735	59.80	
	1800	Mixer, powered, mortar and concrete, gas, 6 C.F., 18 H.P.		7.60	117	350	1,050	130.80	
	1900	10 C.F., 25 H.P.		9.45	142	425	1,275	160.60	
	2000	16 C.F.		9.75	165	495	1,475	177	
	2100	Concrete, stationary, tilt drum, 2 C.Y.		6.55	230	690	2,075	190.40	
	2120	Pump, concrete, truck mounted 4" line 80' boom		23.95	875	2,620	7,850	715.60	
	2140	5" line, 110' boom		31.05	1,150	3,465	10,400	941.40	
	2160	Mud jack, 50 C.F. per hr.		6.55	125	375	1,125	127.40	
	2180	225 C.F. per hr.		8.70	142	425	1,275	154.60	
	2190	Shotcrete pump rig, 12 C.Y./hr.		13.75	220	660	1,975	242	
	2200	35 C.Y./hr.		15.85	237	710	2,125	268.80	
	2600	Saw, concrete, manual, gas, 18 H.P.		5.00	43.50	130	390	66	
	2650	Self-propelled, gas, 30 H.P.		11.00	100	300	900	148	
	2700	Vibrators, concrete, electric, 60 cycle, 2 H.P.		.30	8.65	26	78	7.60	
	2800	3 H.P.		.43	11.65	35	105	10.45	
	2900	Gas engine, 5 H.P.		1.70	16	48	144	23.20	
	3000	8 H.P.		2.30	15	45	135	27.40	
	3050	Vibrating screed, gas engine, 8 H.P.		2.81	71.50	215	645	65.50	
	3120	Concrete transit mixer, 6 x 4, 250 H.P., 8 C.Y., rear discharge		52.20	570	1,705	5,125	758.60	
	3200	Front discharge		61.30	695	2,090	6,275	908.40	
	3300	6 x 6, 285 H.P., 12 C.Y., rear discharge		60.40	655	1,970	5,900	877.20	
	3400	Front discharge	▼	63.10	705	2,115	6,350	927.80	
20	0010	**EARTHWORK EQUIPMENT RENTAL** without operators	R015433 -10						**20**
	0040	Aggregate spreader, push type 8' to 12' wide	Ea.	2.70	25.50	76	228	36.80	
	0045	Tailgate type, 8' wide		2.55	33	99	297	40.20	
	0055	Earth auger, truck-mounted, for fence & sign posts, utility poles		15.30	470	1,405	4,225	403.40	
	0060	For borings and monitoring wells		41.30	645	1,940	5,825	718.40	
	0070	Portable, trailer mounted		2.75	31.50	95	285	41	
	0075	Truck-mounted, for caissons, water wells		89.05	2,875	8,645	25,900	2,441	
	0080	Horizontal boring machine, 12" to 36" diameter, 45 H.P.		22.20	192	575	1,725	292.60	
	0090	12" to 48" diameter, 65 H.P.		30.95	335	1,000	3,000	447.60	
	0095	Auger, for fence posts, gas engine, hand held		.50	6	18	54	7.60	
	0100	Excavator, diesel hydraulic, crawler mounted, 1/2 C.Y. capacity		18.80	370	1,105	3,325	371.40	
	0120	5/8 C.Y. capacity		28.60	540	1,615	4,850	551.80	
	0140	3/4 C.Y. capacity		34.90	615	1,840	5,525	647.20	
	0150	1 C.Y. capacity	▼	42.70	680	2,035	6,100	748.60	

01 54 33 | Equipment Rental

		UNIT	HOURLY OPER. COST	RENT PER DAY	RENT PER WEEK	RENT PER MONTH	EQUIPMENT COST/DAY
20	0200	1-1/2 C.Y. capacity	Ea.	51.80	910	2,730	8,200	960.40
	0300	2 C.Y. capacity		68.70	1,175	3,515	10,500	1,253
	0320	2-1/2 C.Y. capacity		92.40	1,525	4,565	13,700	1,652
	0325	3-1/2 C.Y. capacity		124.00	2,025	6,095	18,300	2,211
	0330	4-1/2 C.Y. capacity		149.80	2,575	7,750	23,300	2,748
	0335	6 C.Y. capacity		188.25	2,875	8,615	25,800	3,229
	0340	7 C.Y. capacity		190.45	2,975	8,900	26,700	3,304
	0342	Excavator attachments, bucket thumbs		3.05	240	720	2,150	168.40
	0345	Grapples		2.75	208	625	1,875	147
	0347	Hydraulic hammer for boom mounting, 5000 ft lb.		13.70	425	1,275	3,825	364.60
	0349	11,000 ft lb.		22.35	755	2,260	6,775	630.80
	0350	Gradall type, truck mounted, 3 ton @ 15' radius, 5/8 C.Y.		39.75	875	2,630	7,900	844
	0370	1 C.Y. capacity		43.65	1,025	3,060	9,175	961.20
	0400	Backhoe-loader, 40 to 45 H.P., 5/8 C.Y. capacity		12.90	240	720	2,150	247.20
	0450	45 H.P. to 60 H.P., 3/4 C.Y. capacity		19.75	293	880	2,650	334
	0460	80 H.P., 1-1/4 C.Y. capacity		21.60	320	955	2,875	363.80
	0470	112 H.P., 1-1/2 C.Y. capacity		35.65	655	1,960	5,875	677.20
	0482	Backhoe-loader attachment, compactor, 20,000 lb.		5.50	142	425	1,275	129
	0485	Hydraulic hammer, 750 ft lb.		3.10	95	285	855	81.80
	0486	Hydraulic hammer, 1200 ft lb.		6.05	217	650	1,950	178.40
	0500	Brush chipper, gas engine, 6" cutter head, 35 H.P.		9.75	107	320	960	142
	0550	Diesel engine, 12" cutter head, 130 H.P.		18.60	225	675	2,025	283.80
	0600	15" cutter head, 165 H.P.		23.35	265	795	2,375	345.80
	0750	Bucket, clamshell, general purpose, 3/8 C.Y.		1.20	38.50	115	345	32.60
	0800	1/2 C.Y.		1.30	45	135	405	37.40
	0850	3/4 C.Y.		1.45	53.50	160	480	43.60
	0900	1 C.Y.		1.50	58.50	175	525	47
	0950	1-1/2 C.Y.		2.45	80	240	720	67.60
	1000	2 C.Y.		2.55	88.50	265	795	73.40
	1010	Bucket, dragline, medium duty, 1/2 C.Y.		.70	23.50	70	210	19.60
	1020	3/4 C.Y.		.70	24.50	74	222	20.40
	1030	1 C.Y.		.75	26.50	80	240	22
	1040	1-1/2 C.Y.		1.15	40	120	360	33.20
	1050	2 C.Y.		1.25	45	135	405	37
	1070	3 C.Y.		1.85	61.50	185	555	51.80
	1200	Compactor, manually guided 2-drum vibratory smooth roller, 7.5 H.P.		6.70	197	590	1,775	171.60
	1250	Rammer/tamper, gas, 8"		2.40	41.50	125	375	44.20
	1260	15"		2.60	48.50	145	435	49.80
	1300	Vibratory plate, gas, 18" plate, 3000 lb. blow		2.30	23	69	207	32.20
	1350	21" plate, 5000 lb. blow		2.85	30	90	270	40.80
	1370	Curb builder/extruder, 14 H.P., gas, single screw		13.10	245	735	2,200	251.80
	1390	Double screw		13.75	283	850	2,550	280
	1500	Disc harrow attachment, for tractor		.42	69.50	209	625	45.15
	1810	Feller buncher, shearing & accumulating trees, 100 H.P.		31.15	545	1,640	4,925	577.20
	1860	Grader, self-propelled, 25,000 lb.		27.85	525	1,580	4,750	538.80
	1910	30,000 lb.		31.10	605	1,810	5,425	610.80
	1920	40,000 lb.		48.75	1,025	3,105	9,325	1,011
	1930	55,000 lb.		64.25	1,500	4,530	13,600	1,420
	1950	Hammer, pavement breaker, self-propelled, diesel, 1000 to 1250 lb.		26.10	345	1,040	3,125	416.80
	2000	1300 to 1500 lb.		39.15	695	2,080	6,250	729.20
	2050	Pile driving hammer, steam or air, 4150 ft lb. @ 225 bpm		10.00	480	1,435	4,300	367
	2100	8750 ft lb. @ 145 bpm		12.00	665	2,000	6,000	496
	2150	15,000 ft lb. @ 60 bpm		13.55	805	2,415	7,250	591.40
	2200	24,450 ft lb. @ 111 bpm		14.60	895	2,680	8,050	652.80
	2250	Leads, 60' high for pile driving hammers up to 20,000 ft lb.		3.15	80.50	242	725	73.60
	2300	90' high for hammers over 20,000 ft lb.		4.75	143	430	1,300	124
	2350	Diesel type hammer, 22,400 ft lb.		27.60	620	1,860	5,575	592.80
	2400	41,300 ft lb.		35.35	675	2,020	6,050	686.80

01 54 33 | Equipment Rental

		UNIT	HOURLY OPER. COST	RENT PER DAY	RENT PER WEEK	RENT PER MONTH	EQUIPMENT COST/DAY	
20	2450	141,000 ft lb.	Ea.	52.60	1,150	3,440	10,300	1,109	20
	2500	Vib. elec. hammer/extractor, 200 kW diesel generator, 34 H.P.		45.65	645	1,940	5,825	753.20	
	2550	80 H.P.		82.40	945	2,840	8,525	1,227	
	2600	150 H.P.		155.30	1,825	5,465	16,400	2,335	
	2800	Log chipper, up to 22" diameter, 600 H.P.		50.70	510	1,525	4,575	710.60	
	2850	Logger, for skidding & stacking logs, 150 H.P.		46.90	815	2,450	7,350	865.20	
	2860	Mulcher, diesel powered, trailer mounted		19.70	215	645	1,925	286.60	
	2900	Rake, spring tooth, with tractor		15.41	335	1,007	3,025	324.70	
	3000	Roller, vibratory, tandem, smooth drum, 20 H.P.		7.90	143	430	1,300	149.20	
	3050	35 H.P.		9.95	247	740	2,225	227.60	
	3100	Towed type vibratory compactor, smooth drum, 50 H.P.		23.60	335	1,000	3,000	388.80	
	3150	Sheepsfoot, 50 H.P.		24.90	370	1,105	3,325	420.20	
	3170	Landfill compactor, 220 H.P.		77.35	1,425	4,305	12,900	1,480	
	3200	Pneumatic tire roller, 80 H.P.		14.15	335	1,000	3,000	313.20	
	3250	120 H.P.		21.25	560	1,680	5,050	506	
	3300	Sheepsfoot vibratory roller, 240 H.P.		58.10	1,050	3,180	9,550	1,101	
	3320	340 H.P.		85.45	1,525	4,575	13,700	1,599	
	3350	Smooth drum vibratory roller, 75 H.P.		22.10	565	1,690	5,075	514.80	
	3400	125 H.P.		28.40	690	2,065	6,200	640.20	
	3410	Rotary mower, brush, 60", with tractor		19.80	310	925	2,775	343.40	
	3420	Rototiller, walk-behind, gas, 5 H.P.		2.23	67.50	203	610	58.45	
	3422	8 H.P.		3.47	103	310	930	89.75	
	3440	Scrapers, towed type, 7 C.Y. capacity		5.40	108	325	975	108.20	
	3450	10 C.Y. capacity		6.15	150	450	1,350	139.20	
	3500	15 C.Y. capacity		6.60	173	520	1,550	156.80	
	3525	Self-propelled, single engine, 14 C.Y. capacity		91.35	1,350	4,030	12,100	1,537	
	3550	Dual engine, 21 C.Y. capacity		131.60	1,525	4,605	13,800	1,974	
	3600	31 C.Y. capacity		181.80	2,350	7,070	21,200	2,868	
	3640	44 C.Y. capacity		230.40	3,200	9,625	28,900	3,768	
	3650	Elevating type, single engine, 11 C.Y. capacity		61.60	1,025	3,060	9,175	1,105	
	3700	22 C.Y. capacity		113.95	1,975	5,935	17,800	2,099	
	3710	Screening plant 110 H.P. w/5' x 10' screen		33.90	435	1,305	3,925	532.20	
	3720	5' x 16' screen		36.15	540	1,625	4,875	614.20	
	3850	Shovel, crawler-mounted, front-loading, 7 C.Y. capacity		200.60	2,750	8,270	24,800	3,259	
	3855	12 C.Y. capacity		268.10	3,500	10,520	31,600	4,249	
	3860	Shovel/backhoe bucket, 1/2 C.Y.		2.30	65	195	585	57.40	
	3870	3/4 C.Y.		2.40	71.50	215	645	62.20	
	3880	1 C.Y.		2.50	80	240	720	68	
	3890	1-1/2 C.Y.		2.60	93.50	280	840	76.80	
	3910	3 C.Y.		2.95	127	380	1,150	99.60	
	3950	Stump chipper, 18" deep, 30 H.P.		7.21	188	565	1,700	170.70	
	4110	Dozer, crawler, torque converter, diesel 80 H.P.		25.30	400	1,200	3,600	442.40	
	4150	105 H.P.		30.85	505	1,520	4,550	550.80	
	4200	140 H.P.		44.30	770	2,315	6,950	817.40	
	4260	200 H.P.		65.75	1,200	3,585	10,800	1,243	
	4310	300 H.P.		86.45	1,700	5,130	15,400	1,718	
	4360	410 H.P.		117.20	2,225	6,670	20,000	2,272	
	4370	500 H.P.		150.10	2,900	8,680	26,000	2,937	
	4380	700 H.P.		222.15	4,100	12,290	36,900	4,235	
	4400	Loader, crawler, torque conv., diesel, 1-1/2 C.Y., 80 H.P.		26.70	460	1,375	4,125	488.60	
	4450	1-1/2 to 1-3/4 C.Y., 95 H.P.		30.40	590	1,775	5,325	598.20	
	4510	1-3/4 to 2-1/4 C.Y., 130 H.P.		47.40	860	2,580	7,750	895.20	
	4530	2-1/2 to 3-1/4 C.Y., 190 H.P.		57.75	1,075	3,220	9,650	1,106	
	4560	3-1/2 to 5 C.Y., 275 H.P.		73.90	1,425	4,260	12,800	1,443	
	4610	Front end loader, 4WD, articulated frame, diesel, 1 to 1-1/4 C.Y., 70 H.P.		16.65	225	675	2,025	268.20	
	4620	1-1/2 to 1-3/4 C.Y., 95 H.P.		20.85	278	835	2,500	333.80	
	4650	1-3/4 to 2 C.Y., 130 H.P.		24.40	365	1,100	3,300	415.20	
	4710	2-1/2 to 3-1/2 C.Y., 145 H.P.		27.55	420	1,260	3,775	472.40	

01 54 33 | Equipment Rental

		UNIT	HOURLY OPER. COST	RENT PER DAY	RENT PER WEEK	RENT PER MONTH	EQUIPMENT COST/DAY	
20	4730	3 to 4-1/2 C.Y., 185 H.P.	Ea.	35.25	550	1,645	4,925	611	20
	4760	5-1/4 to 5-3/4 C.Y., 270 H.P.		56.50	845	2,540	7,625	960	
	4810	7 to 9 C.Y., 475 H.P.		98.35	1,775	5,345	16,000	1,856	
	4870	9 - 11 C.Y., 620 H.P.		132.15	2,500	7,495	22,500	2,556	
	4880	Skid steer loader, wheeled, 10 C.F., 30 H.P. gas		8.95	147	440	1,325	159.60	
	4890	1 C.Y., 78 H.P., diesel		16.90	243	730	2,200	281.20	
	4892	Skid-steer attachment, auger		.55	92	276	830	59.60	
	4893	Backhoe		.70	116	348	1,050	75.20	
	4894	Broom		.74	124	371	1,125	80.10	
	4895	Forks		.25	41	123	370	26.60	
	4896	Grapple		.57	94.50	283	850	61.15	
	4897	Concrete hammer		1.08	179	538	1,625	116.25	
	4898	Tree spade		.94	156	468	1,400	101.10	
	4899	Trencher		.76	126	378	1,125	81.70	
	4900	Trencher, chain, boom type, gas, operator walking, 12 H.P.		4.35	46.50	140	420	62.80	
	4910	Operator riding, 40 H.P.		17.10	293	880	2,650	312.80	
	5000	Wheel type, diesel, 4' deep, 12" wide		73.05	835	2,500	7,500	1,084	
	5100	6' deep, 20" wide		84.15	1,900	5,715	17,100	1,816	
	5150	Chain type, diesel, 5' deep, 8" wide		32.30	560	1,685	5,050	595.40	
	5200	Diesel, 8' deep, 16" wide		134.05	2,950	8,880	26,600	2,848	
	5202	Rock trencher, wheel type, 6" wide x 18" deep		19.05	320	960	2,875	344.40	
	5206	Chain type, 18" wide x 7' deep		103.05	2,825	8,465	25,400	2,517	
	5210	Tree spade, self-propelled		13.73	267	800	2,400	269.85	
	5250	Truck, dump, 2-axle, 12 ton, 8 C.Y. payload, 220 H.P.		27.80	225	675	2,025	357.40	
	5300	Three axle dump, 16 ton, 12 C.Y. payload, 400 H.P.		48.95	320	965	2,900	584.60	
	5310	Four axle dump, 25 ton, 18 C.Y. payload, 450 H.P.		58.00	470	1,405	4,225	745	
	5350	Dump trailer only, rear dump, 16-1/2 C.Y.		5.15	137	410	1,225	123.20	
	5400	20 C.Y.		5.60	155	465	1,400	137.80	
	5450	Flatbed, single axle, 1-1/2 ton rating		21.25	68.50	205	615	211	
	5500	3 ton rating		25.80	96.50	290	870	264.40	
	5550	Off highway rear dump, 25 ton capacity		67.70	1,225	3,645	10,900	1,271	
	5600	35 ton capacity		63.10	1,075	3,260	9,775	1,157	
	5610	50 ton capacity		85.40	1,475	4,425	13,300	1,568	
	5620	65 ton capacity		88.20	1,450	4,330	13,000	1,572	
	5630	100 ton capacity		113.50	1,950	5,815	17,400	2,071	
	6000	Vibratory plow, 25 H.P., walking		7.35	60	180	540	94.80	
40	0010	**GENERAL EQUIPMENT RENTAL** without operators							40
	0150	Aerial lift, scissor type, to 15' high, 1000 lb. capacity, electric	Ea.	3.00	53.50	160	480	56	
	0160	To 25' high, 2000 lb. capacity		3.40	68.50	205	615	68.20	
	0170	Telescoping boom to 40' high, 500 lb. capacity, gas		17.40	310	925	2,775	324.20	
	0180	To 45' high, 500 lb. capacity		18.50	355	1,070	3,200	362	
	0190	To 60' high, 600 lb. capacity		20.80	470	1,405	4,225	447.40	
	0195	Air compressor, portable, 6.5 CFM, electric		.31	12.65	38	114	10.10	
	0196	Gasoline		.77	19	57	171	17.55	
	0200	Towed type, gas engine, 60 CFM		11.55	46.50	140	420	120.40	
	0300	160 CFM		13.35	50	150	450	136.80	
	0400	Diesel engine, rotary screw, 250 CFM		14.00	108	325	975	177	
	0500	365 CFM		18.75	132	395	1,175	229	
	0550	450 CFM		23.70	165	495	1,475	288.60	
	0600	600 CFM		41.45	227	680	2,050	467.60	
	0700	750 CFM		41.60	237	710	2,125	474.80	
	0800	For silenced models, small sizes, add to rent		3%	5%	5%	5%		
	0900	Large sizes, add to rent		5%	7%	7%	7%		
	0930	Air tools, breaker, pavement, 60 lb.	Ea.	.50	9.65	29	87	9.80	
	0940	80 lb.		.50	10.35	31	93	10.20	
	0950	Drills, hand (jackhammer) 65 lb.		.60	16.65	50	150	14.80	
	0960	Track or wagon, swing boom, 4" drifter		55.90	815	2,450	7,350	937.20	
	0970	5" drifter		67.20	915	2,750	8,250	1,088	

R015433 -10

01 54 33 | Equipment Rental

		UNIT	HOURLY OPER. COST	RENT PER DAY	RENT PER WEEK	RENT PER MONTH	EQUIPMENT COST/DAY
0975	Track mounted quarry drill, 6" diameter drill	Ea.	102.85	1,350	4,060	12,200	1,635
0980	Dust control per drill		.63	22	66	198	18.25
0990	Hammer, chipping, 12 lb.		.55	25.50	77	231	19.80
1000	Hose, air with couplings, 50' long, 3/4" diameter		.03	4.67	14	42	3.05
1100	1" diameter		.04	6.35	19	57	4.10
1200	1-1/2" diameter		.06	9.65	29	87	6.30
1300	2" diameter		.07	12	36	108	7.75
1400	2-1/2" diameter		.11	19	57	171	12.30
1410	3" diameter		.14	23	69	207	14.90
1450	Drill, steel, 7/8" x 2'		.05	8.65	26	78	5.60
1460	7/8" x 6'		.05	9	27	81	5.80
1520	Moil points		.02	4	12	36	2.55
1525	Pneumatic nailer w/accessories		.48	31.50	95	285	22.85
1530	Sheeting driver for 60 lb. breaker		.04	6	18	54	3.90
1540	For 90 lb. breaker		.12	8	24	72	5.75
1550	Spade, 25 lb.		.45	6.65	20	60	7.60
1560	Tamper, single, 35 lb.		.55	36.50	109	325	26.20
1570	Triple, 140 lb.		.82	54.50	164	490	39.35
1580	Wrenches, impact, air powered, up to 3/4" bolt		.40	11.35	34	102	10
1590	Up to 1-1/4" bolt		.50	23.50	70	210	18
1600	Barricades, barrels, reflectorized, 1 to 99 barrels		.03	4.60	13.80	41.50	3
1610	100 to 200 barrels		.02	3.53	10.60	32	2.30
1620	Barrels with flashers, 1 to 99 barrels		.03	5.25	15.80	47.50	3.40
1630	100 to 200 barrels		.03	4.20	12.60	38	2.75
1640	Barrels with steady burn type C lights		.04	7	21	63	4.50
1650	Illuminated board, trailer mounted, with generator		3.25	133	400	1,200	106
1670	Portable barricade, stock, with flashers, 1 to 6 units		.03	5.25	15.80	47.50	3.40
1680	25 to 50 units		.03	4.90	14.70	44	3.20
1685	Butt fusion machine, wheeled, 1.5 HP electric, 2" - 8" diameter pipe		2.56	167	500	1,500	120.50
1690	Tracked, 20 HP diesel, 4"-12" diameter pipe		10.43	490	1,465	4,400	376.45
1695	83 HP diesel, 8" - 24" diameter pipe		27.52	975	2,930	8,800	806.15
1700	Carts, brick, hand powered, 1000 lb. capacity		.44	72.50	218	655	47.10
1800	Gas engine, 1500 lb., 7-1/2' lift		4.04	115	345	1,025	101.30
1822	Dehumidifier, medium, 6 lb./hr., 150 CFM		.93	60	180	540	43.45
1824	Large, 18 lb./hr., 600 CFM		1.89	122	366	1,100	88.30
1830	Distributor, asphalt, trailer mounted, 2000 gal., 38 H.P. diesel		9.80	330	990	2,975	276.40
1840	3000 gal., 38 H.P. diesel		11.25	360	1,080	3,250	306
1850	Drill, rotary hammer, electric		.82	26.50	80	240	22.55
1860	Carbide bit, 1-1/2" diameter, add to electric rotary hammer		.02	4.97	14.90	44.50	3.15
1865	Rotary, crawler, 250 H.P.		131.70	2,050	6,175	18,500	2,289
1870	Emulsion sprayer, 65 gal., 5 H.P. gas engine		2.85	97	291	875	81
1880	200 gal., 5 H.P. engine		7.25	162	485	1,450	155
1900	Floor auto-scrubbing machine, walk-behind, 28" path		3.76	247	740	2,225	178.10
1930	Floodlight, mercury vapor, or quartz, on tripod, 1000 watt		.30	16	48	144	12
1940	2000 watt		.55	30.50	91	273	22.60
1950	Floodlights, trailer mounted with generator, 1 - 300 watt light		3.35	71.50	215	645	69.80
1960	2 - 1000 watt lights		4.35	96.50	290	870	92.80
2000	4 - 300 watt lights		4.10	91.50	275	825	87.80
2005	Foam spray rig, incl. box trailer, compressor, generator, proportioner		27.23	475	1,430	4,300	503.85
2020	Forklift, straight mast, 12' lift, 5000 lb., 2 wheel drive, gas		22.00	198	595	1,775	295
2040	21' lift, 5000 lb., 4 wheel drive, diesel		17.90	243	730	2,200	289.20
2050	For rough terrain, 42' lift, 35' reach, 9000 lb., 110 H.P.		25.10	470	1,405	4,225	481.80
2060	For plant, 4 ton capacity, 80 H.P., 2 wheel drive, gas		13.40	98.50	295	885	166.20
2080	10 ton capacity, 120 H.P., 2 wheel drive, diesel		19.95	172	515	1,550	262.60
2100	Generator, electric, gas engine, 1.5 kW to 3 kW		3.10	11.35	34	102	31.60
2200	5 kW		4.00	15	45	135	41
2300	10 kW		7.50	36.50	110	330	82
2400	25 kW		8.80	85	255	765	121.40

40

01 54 33 | Equipment Rental

		UNIT	HOURLY OPER. COST	RENT PER DAY	RENT PER WEEK	RENT PER MONTH	EQUIPMENT COST/DAY	
40	2500	Diesel engine, 20 kW	Ea.	10.10	71.50	215	645	123.80	40
	2600	50 kW		18.15	98.50	295	885	204.20	
	2700	100 kW		35.05	138	415	1,250	363.40	
	2800	250 kW		69.05	243	730	2,200	698.40	
	2850	Hammer, hydraulic, for mounting on boom, to 500 ft lb.		2.40	75	225	675	64.20	
	2860	1000 ft lb.		4.15	125	375	1,125	108.20	
	2900	Heaters, space, oil or electric, 50 MBH		1.84	8	24	72	19.50	
	3000	100 MBH		3.32	11.35	34	102	33.35	
	3100	300 MBH		10.62	36.50	110	330	106.95	
	3150	500 MBH		21.42	48.50	145	435	200.35	
	3200	Hose, water, suction with coupling, 20' long, 2" diameter		.02	3	9	27	1.95	
	3210	3" diameter		.03	4.33	13	39	2.85	
	3220	4" diameter		.03	5	15	45	3.25	
	3230	6" diameter		.11	17.65	53	159	11.50	
	3240	8" diameter		.20	33.50	100	300	21.60	
	3250	Discharge hose with coupling, 50' long, 2" diameter		.02	2.67	8	24	1.75	
	3260	3" diameter		.03	4.67	14	42	3.05	
	3270	4" diameter		.04	7.35	22	66	4.70	
	3280	6" diameter		.11	18.65	56	168	12.10	
	3290	8" diameter		.20	33.50	100	300	21.60	
	3295	Insulation blower		.67	6	18	54	8.95	
	3300	Ladders, extension type, 16' to 36' long		.14	24	72	216	15.50	
	3400	40' to 60' long		.20	32.50	98	294	21.20	
	3405	Lance for cutting concrete		2.17	104	313	940	79.95	
	3407	Lawn mower, rotary, 22", 5 H.P.		2.14	63.50	190	570	55.10	
	3408	48" self propelled		2.88	75	225	675	68.05	
	3410	Level, electronic, automatic, with tripod and leveling rod		1.64	109	328	985	78.70	
	3430	Laser type, for pipe and sewer line and grade		.69	46	138	415	33.10	
	3440	Rotating beam for interior control		1.15	76.50	230	690	55.20	
	3460	Builder's optical transit, with tripod and rod		.09	15.35	46	138	9.90	
	3500	Light towers, towable, with diesel generator, 2000 watt		4.10	91.50	275	825	87.80	
	3600	4000 watt		4.35	96.50	290	870	92.80	
	3700	Mixer, powered, plaster and mortar, 6 C.F., 7 H.P.		2.30	19.65	59	177	30.20	
	3800	10 C.F., 9 H.P.		2.45	30.50	92	276	38	
	3850	Nailer, pneumatic		.49	32.50	98	294	23.50	
	3900	Paint sprayers complete, 8 CFM		.73	48.50	145	435	34.85	
	4000	17 CFM		1.35	89.50	269	805	64.60	
	4020	Pavers, bituminous, rubber tires, 8' wide, 50 H.P., diesel		29.45	490	1,465	4,400	528.60	
	4030	10' wide, 150 H.P.		96.65	1,800	5,395	16,200	1,852	
	4050	Crawler, 8' wide, 100 H.P., diesel		83.45	1,800	5,410	16,200	1,750	
	4060	10' wide, 150 H.P.		103.55	2,225	6,645	19,900	2,157	
	4070	Concrete paver, 12' to 24' wide, 250 H.P.		88.70	1,550	4,660	14,000	1,642	
	4080	Placer-spreader-trimmer, 24' wide, 300 H.P.		127.05	2,525	7,575	22,700	2,531	
	4100	Pump, centrifugal gas pump, 1-1/2" diameter, 65 GPM		3.70	48.50	145	435	58.60	
	4200	2" diameter, 130 GPM		4.85	53.50	160	480	70.80	
	4300	3" diameter, 250 GPM		5.10	55	165	495	73.80	
	4400	6" diameter, 1500 GPM		25.55	173	520	1,550	308.40	
	4500	Submersible electric pump, 1-1/4" diameter, 55 GPM		.31	16.35	49	147	12.30	
	4600	1-1/2" diameter, 83 GPM		.35	18.65	56	168	14	
	4700	2" diameter, 120 GPM		1.35	23.50	70	210	24.80	
	4800	3" diameter, 300 GPM		2.25	41.50	125	375	43	
	4900	4" diameter, 560 GPM		9.65	158	475	1,425	172.20	
	5000	6" diameter, 1590 GPM		14.15	213	640	1,925	241.20	
	5100	Diaphragm pump, gas, single, 1-1/2" diameter		1.15	49.50	148	445	38.80	
	5200	2" diameter		3.95	61.50	185	555	68.60	
	5300	3" diameter		3.95	61.50	185	555	68.60	
	5400	Double, 4" diameter		5.90	107	320	960	111.20	
	5450	Pressure washer 5 GPM, 3000 psi		4.25	48.50	145	435	63	

01 54 33 | Equipment Rental

		UNIT	HOURLY OPER. COST	RENT PER DAY	RENT PER WEEK	RENT PER MONTH	EQUIPMENT COST/DAY
5500	Trash pump, self-priming, gas, 2" diameter	Ea.	4.05	21	63	189	45
5600	Diesel, 4" diameter		7.75	73.50	220	660	106
5650	Diesel, 6" diameter		20.50	140	420	1,250	248
5655	Grout Pump		21.80	177	530	1,600	280.40
5700	Salamanders, L.P. gas fired, 100,000 Btu		3.33	12.65	38	114	34.25
5705	50,000 Btu		2.50	10.35	31	93	26.20
5720	Sandblaster, portable, open top, 3 C.F. capacity		.55	26.50	80	240	20.40
5730	6 C.F. capacity		.90	40	120	360	31.20
5740	Accessories for above		.12	20.50	62	186	13.35
5750	Sander, floor		.59	19.65	59	177	16.50
5760	Edger		.54	21.50	64	192	17.10
5800	Saw, chain, gas engine, 18" long		1.95	19.65	59	177	27.40
5900	Hydraulic powered, 36" long		.70	63.50	190	570	43.60
5950	60" long		.70	65	195	585	44.60
6000	Masonry, table mounted, 14" diameter, 5 H.P.		1.25	56.50	170	510	44
6050	Portable cut-off, 8 H.P.		2.10	32.50	98	294	36.40
6100	Circular, hand held, electric, 7-1/4" diameter		.14	4.67	14	42	3.90
6200	12" diameter		.21	7.65	23	69	6.30
6250	Wall saw, w/hydraulic power, 10 H.P.		7.00	60	180	540	92
6275	Shot blaster, walk-behind, 20" wide		4.65	290	870	2,600	211.20
6280	Sidewalk broom, walk-behind		2.09	60.50	182	545	53.10
6300	Steam cleaner, 100 gallons per hour		3.35	66.50	200	600	66.80
6310	200 gallons per hour		4.65	81.50	245	735	86.20
6340	Tar Kettle/Pot, 400 gallons		5.50	78.50	235	705	91
6350	Torch, cutting, acetylene-oxygen, 150' hose, excludes gases		.30	14	42	126	10.80
6360	Hourly operating cost includes tips and gas		10.80				86.40
6410	Toilet, portable chemical		.12	20	60	180	12.95
6420	Recycle flush type		.14	24	72	216	15.50
6430	Toilet, fresh water flush, garden hose,		.17	28.50	85	255	18.35
6440	Hoisted, non-flush, for high rise		.14	23.50	71	213	15.30
6465	Tractor, farm with attachment		18.25	275	825	2,475	311
6500	Trailers, platform, flush deck, 2 axle, 25 ton capacity		5.20	112	335	1,000	108.60
6600	40 ton capacity		6.75	155	465	1,400	147
6700	3 axle, 50 ton capacity		7.25	172	515	1,550	161
6800	75 ton capacity		9.05	225	675	2,025	207.40
6810	Trailer mounted cable reel for high voltage line work		5.14	245	734	2,200	187.90
6820	Trailer mounted cable tensioning rig		10.22	485	1,460	4,375	373.75
6830	Cable pulling rig		68.64	2,725	8,180	24,500	2,185
6900	Water tank trailer, engine driven discharge, 5000 gallons		6.75	145	435	1,300	141
6925	10,000 gallons		9.20	202	605	1,825	194.60
6950	Water truck, off highway, 6000 gallons		75.60	800	2,400	7,200	1,085
7010	Tram car for high voltage line work, powered, 2 conductor		4.44	133	399	1,200	115.30
7020	Transit (builder's level) with tripod		.09	15.35	46	138	9.90
7030	Trench box, 3000 lb., 6' x 8'		.56	93	279	835	60.30
7040	7200 lb., 6' x 20'		1.05	175	525	1,575	113.40
7050	8000 lb., 8' x 16'		1.05	175	525	1,575	113.40
7060	9500 lb., 8' x 20'		1.19	199	597	1,800	128.90
7065	11,000 lb., 8' x 24'		1.25	209	627	1,875	135.40
7070	12,000 lb., 10' x 20'		1.36	227	680	2,050	146.90
7100	Truck, pickup, 3/4 ton, 2 wheel drive		11.50	58.50	175	525	127
7200	4 wheel drive		11.75	75	225	675	139
7250	Crew carrier, 9 passenger		16.30	86.50	260	780	182.40
7290	Flat bed truck, 20,000 lb. GVW		17.60	125	375	1,125	215.80
7300	Tractor, 4 x 2, 220 H.P.		24.60	197	590	1,775	314.80
7410	330 H.P.		36.35	270	810	2,425	452.80
7500	6 x 4, 380 H.P.		41.65	315	945	2,825	522.20
7600	450 H.P.		50.75	380	1,145	3,425	635
7610	Tractor, with A frame, boom and winch, 225 H.P.		27.40	272	815	2,450	382.20

01 54 33 | Equipment Rental

		UNIT	HOURLY OPER. COST	RENT PER DAY	RENT PER WEEK	RENT PER MONTH	EQUIPMENT COST/DAY	
40	7620	Vacuum truck, hazardous material, 2500 gallons	Ea.	11.55	298	895	2,675	271.40	**40**
	7625	5,000 gallons		13.40	415	1,250	3,750	357.20	
	7650	Vacuum, HEPA, 16 gallon, wet/dry		.74	18	54	162	16.70	
	7655	55 gallon, wet/dry		.59	27	81	243	20.90	
	7660	Water tank, portable		.16	27	80.70	242	17.40	
	7690	Sewer/catch basin vacuum, 14 C.Y., 1500 gallons		17.81	625	1,880	5,650	518.50	
	7700	Welder, electric, 200 amp		2.14	16.35	49	147	26.90	
	7800	300 amp		3.11	20	60	180	36.90	
	7900	Gas engine, 200 amp		11.70	23.50	70	210	107.60	
	8000	300 amp		13.35	24.50	74	222	121.60	
	8100	Wheelbarrow, any size		.08	12.65	38	114	8.25	
	8200	Wrecking ball, 4000 lb.		2.25	70	210	630	60	
50	0010	**HIGHWAY EQUIPMENT RENTAL** without operators							**50**
	0050	Asphalt batch plant, portable drum mixer, 100 ton/hr.	Ea.	70.30	1,425	4,310	12,900	1,424	
	0060	200 ton/hr.		78.75	1,525	4,570	13,700	1,544	
	0070	300 ton/hr.		91.25	1,800	5,375	16,100	1,805	
	0100	Backhoe attachment, long stick, up to 185 H.P., 10.5' long		.34	22.50	67	201	16.10	
	0140	Up to 250 H.P., 12' long		.36	24	72	216	17.30	
	0180	Over 250 H.P., 15' long		.50	33	99	297	23.80	
	0200	Special dipper arm, up to 100 H.P., 32' long		1.02	67.50	203	610	48.75	
	0240	Over 100 H.P., 33' long		1.27	84.50	253	760	60.75	
	0280	Catch basin/sewer cleaning truck, 3 ton, 9 C.Y., 1000 gal.		37.35	395	1,185	3,550	535.80	
	0300	Concrete batch plant, portable, electric, 200 C.Y./hr.		27.69	510	1,530	4,600	527.50	
	0520	Grader/dozer attachment, ripper/scarifier, rear mounted, up to 135 H.P.		3.35	65	195	585	65.80	
	0540	Up to 180 H.P.		3.95	83.50	250	750	81.60	
	0580	Up to 250 H.P.		4.35	95	285	855	91.80	
	0700	Pavement removal bucket, for hydraulic excavator, up to 90 H.P.		1.80	51.50	155	465	45.40	
	0740	Up to 200 H.P.		2.00	71.50	215	645	59	
	0780	Over 200 H.P.		2.15	85	255	765	68.20	
	0900	Aggregate spreader, self-propelled, 187 H.P.		46.10	690	2,065	6,200	781.80	
	1000	Chemical spreader, 3 C.Y.		3.00	43.50	130	390	50	
	1900	Hammermill, traveling, 250 H.P.		68.15	1,950	5,880	17,600	1,721	
	2000	Horizontal borer, 3" diameter, 13 H.P. gas driven		5.65	53.50	160	480	77.20	
	2150	Horizontal directional drill, 20,000 lb. thrust, 78 H.P. diesel		28.00	665	1,990	5,975	622	
	2160	30,000 lb. thrust, 115 H.P.		34.65	1,025	3,045	9,125	886.20	
	2170	50,000 lb. thrust, 170 H.P.		49.60	1,300	3,895	11,700	1,176	
	2190	Mud trailer for HDD, 1500 gallons, 175 H.P., gas		27.15	153	460	1,375	309.20	
	2200	Hydromulcher, diesel, 3000 gallon, for truck mounting		19.10	252	755	2,275	303.80	
	2300	Gas, 600 gallon		7.55	100	300	900	120.40	
	2400	Joint & crack cleaner, walk behind, 25 H.P.		3.40	50	150	450	57.20	
	2500	Filler, trailer mounted, 400 gallons, 20 H.P.		8.55	213	640	1,925	196.40	
	3000	Paint striper, self-propelled, 40 gallon, 22 H.P.		6.70	158	475	1,425	148.60	
	3100	120 gallon, 120 H.P.		20.45	410	1,225	3,675	408.60	
	3200	Post drivers, 6" I-Beam frame, for truck mounting		15.15	405	1,210	3,625	363.20	
	3400	Road sweeper, self-propelled, 8' wide, 90 H.P.		35.10	565	1,700	5,100	620.80	
	3450	Road sweeper, vacuum assisted, 4 C.Y., 220 gallons		63.20	620	1,855	5,575	876.60	
	4000	Road mixer, self-propelled, 130 H.P.		43.95	770	2,305	6,925	812.60	
	4100	310 H.P.		76.25	2,125	6,360	19,100	1,882	
	4220	Cold mix paver, incl. pug mill and bitumen tank, 165 H.P.		90.00	2,150	6,460	19,400	2,012	
	4240	Pavement brush, towed		3.10	95	285	855	81.80	
	4250	Paver, asphalt, wheel or crawler, 130 H.P., diesel		89.70	2,125	6,390	19,200	1,996	
	4300	Paver, road widener, gas 1' to 6', 67 H.P.		44.75	875	2,620	7,850	882	
	4400	Diesel, 2' to 14', 88 H.P.		56.90	1,050	3,155	9,475	1,086	
	4600	Slipform pavers, curb and gutter, 2 track, 75 H.P.		40.75	725	2,175	6,525	761	
	4700	4 track, 165 H.P.		36.65	735	2,200	6,600	733.20	
	4800	Median barrier, 215 H.P.		41.40	755	2,260	6,775	783.20	
	4901	Trailer, low bed, 75 ton capacity		9.75	225	675	2,025	213	
	5000	Road planer, walk behind, 10" cutting width, 10 H.P.		3.20	32.50	98	294	45.20	

R015433 -10

01 54 33 | Equipment Rental

		UNIT	HOURLY OPER. COST	RENT PER DAY	RENT PER WEEK	RENT PER MONTH	EQUIPMENT COST/DAY	
50	5100	Self-propelled, 12" cutting width, 64 H.P.	Ea.	8.90	112	335	1,000	138.20	**50**
	5120	Traffic line remover, metal ball blaster, truck mounted, 115 H.P.		44.60	740	2,220	6,650	800.80	
	5140	Grinder, truck mounted, 115 H.P.		50.10	805	2,410	7,225	882.80	
	5160	Walk-behind, 11 H.P.		3.65	53.50	160	480	61.20	
	5200	Pavement profiler, 4' to 6' wide, 450 H.P.		227.55	3,350	10,035	30,100	3,827	
	5300	8' to 10' wide, 750 H.P.		353.65	4,375	13,150	39,500	5,459	
	5400	Roadway plate, steel, 1" x 8' x 20'		.07	12.35	37	111	7.95	
	5600	Stabilizer, self-propelled, 150 H.P.		43.25	610	1,825	5,475	711	
	5700	310 H.P.		84.85	1,500	4,530	13,600	1,585	
	5800	Striper, thermal, truck mounted, 120 gallon paint, 460 H.P.		54.85	495	1,485	4,450	735.80	
	5900	Thermal paint heating kettle, 115 gallons		3.64	25.50	77	231	44.50	
	6000	Tar kettle, 330 gallon, trailer mounted		5.15	60	180	540	77.20	
	7000	Tunnel locomotive, diesel, 8 to 12 ton		29.20	570	1,710	5,125	575.60	
	7005	Electric, 10 ton		25.40	650	1,955	5,875	594.20	
	7010	Muck cars, 1/2 C.Y. capacity		2.00	23.50	71	213	30.20	
	7020	1 C.Y. capacity		2.20	31.50	95	285	36.60	
	7030	2 C.Y. capacity		2.30	36.50	110	330	40.40	
	7040	Side dump, 2 C.Y. capacity		2.50	43.50	130	390	46	
	7050	3 C.Y. capacity		3.40	50	150	450	57.20	
	7060	5 C.Y. capacity		4.85	63.50	190	570	76.80	
	7100	Ventilating blower for tunnel, 7-1/2 H.P.		1.75	50	150	450	44	
	7110	10 H.P.		1.88	51.50	155	465	46.05	
	7120	20 H.P.		2.68	67.50	202	605	61.85	
	7140	40 H.P.		4.17	95	285	855	90.35	
	7160	60 H.P.		6.38	147	440	1,325	139.05	
	7175	75 H.P.		8.35	196	587	1,750	184.20	
	7180	200 H.P.		15.40	293	880	2,650	299.20	
	7800	Windrow loader, elevating		51.85	1,325	3,960	11,900	1,207	
60	0010	**LIFTING AND HOISTING EQUIPMENT RENTAL** without operators							**60**
	0120	Aerial lift truck, 2 person, to 80'	Ea.	26.70	705	2,120	6,350	637.60	
	0140	Boom work platform, 40' snorkel		15.80	270	810	2,425	288.40	
	0150	Crane, flatbed mounted, 3 ton capacity		14.55	193	580	1,750	232.40	
	0200	Crane, climbing, 106' jib, 6000 lb. capacity, 410 fpm		29.95	1,550	4,670	14,000	1,174	
	0300	101' jib, 10,250 lb. capacity, 270 fpm		36.20	1,975	5,920	17,800	1,474	
	0500	Tower, static, 130' high, 106' jib, 6200 lb. capacity at 400 fpm		33.60	1,800	5,400	16,200	1,349	
	0600	Crawler mounted, lattice boom, 1/2 C.Y., 15 tons at 12' radius		33.19	635	1,910	5,725	647.50	
	0700	3/4 C.Y., 20 tons at 12' radius		44.25	795	2,390	7,175	832	
	0800	1 C.Y., 25 tons at 12' radius		59.00	1,075	3,190	9,575	1,110	
	0900	1-1/2 C.Y., 40 tons at 12' radius		59.00	1,075	3,215	9,650	1,115	
	1000	2 C.Y., 50 tons at 12' radius		62.90	1,250	3,770	11,300	1,257	
	1100	3 C.Y., 75 tons at 12' radius		67.55	1,475	4,415	13,200	1,423	
	1200	100 ton capacity, 60' boom		76.50	1,700	5,080	15,200	1,628	
	1300	165 ton capacity, 60' boom		96.25	1,975	5,940	17,800	1,958	
	1400	200 ton capacity, 70' boom		118.45	2,475	7,440	22,300	2,436	
	1500	350 ton capacity, 80' boom		166.45	3,750	11,245	33,700	3,581	
	1600	Truck mounted, lattice boom, 6 x 4, 20 tons at 10' radius		35.08	1,125	3,350	10,100	950.65	
	1700	25 tons at 10' radius		37.95	1,225	3,650	11,000	1,034	
	1800	8 x 4, 30 tons at 10' radius		41.11	1,300	3,880	11,600	1,105	
	1900	40 tons at 12' radius		43.92	1,350	4,060	12,200	1,163	
	2000	60 tons at 15' radius		49.48	1,425	4,300	12,900	1,256	
	2050	82 tons at 15' radius		55.38	1,525	4,590	13,800	1,361	
	2100	90 tons at 15' radius		62.13	1,675	5,000	15,000	1,497	
	2200	115 tons at 15' radius		70.13	1,875	5,590	16,800	1,679	
	2300	150 tons at 18' radius		81.45	1,950	5,885	17,700	1,829	
	2350	165 tons at 18' radius		82.43	2,075	6,240	18,700	1,907	
	2400	Truck mounted, hydraulic, 12 ton capacity		36.75	530	1,590	4,775	612	
	2500	25 ton capacity		39.15	640	1,920	5,750	697.20	
	2550	33 ton capacity		39.70	655	1,970	5,900	711.60	

Reference boxes: R015433 -10, R015433 -15, R312316 -45

01 54 33 | Equipment Rental

		UNIT	HOURLY OPER. COST	RENT PER DAY	RENT PER WEEK	RENT PER MONTH	EQUIPMENT COST/DAY	
60	2560	40 ton capacity	Ea.	51.70	770	2,305	6,925	874.60	60
	2600	55 ton capacity		67.65	930	2,795	8,375	1,100	
	2700	80 ton capacity		88.65	1,400	4,225	12,700	1,554	
	2720	100 ton capacity		84.25	1,450	4,365	13,100	1,547	
	2740	120 ton capacity		96.15	1,575	4,690	14,100	1,707	
	2760	150 ton capacity		114.30	2,075	6,190	18,600	2,152	
	2800	Self-propelled, 4 x 4, with telescoping boom, 5 ton		15.70	232	695	2,075	264.60	
	2900	12-1/2 ton capacity		28.15	370	1,115	3,350	448.20	
	3000	15 ton capacity		28.85	390	1,175	3,525	465.80	
	3050	20 ton capacity		31.05	450	1,355	4,075	519.40	
	3100	25 ton capacity		32.65	505	1,520	4,550	565.20	
	3150	40 ton capacity		40.25	570	1,710	5,125	664	
	3200	Derricks, guy, 20 ton capacity, 60' boom, 75' mast		23.51	380	1,144	3,425	416.90	
	3300	100' boom, 115' mast		37.04	655	1,970	5,900	690.30	
	3400	Stiffleg, 20 ton capacity, 70' boom, 37' mast		25.93	495	1,490	4,475	505.45	
	3500	100' boom, 47' mast		39.91	795	2,380	7,150	795.30	
	3550	Helicopter, small, lift to 1250 lb. maximum, w/pilot		97.18	3,075	9,240	27,700	2,625	
	3600	Hoists, chain type, overhead, manual, 3/4 ton		.10	.33	1	3	1	
	3900	10 ton		.70	6	18	54	9.20	
	4000	Hoist and tower, 5000 lb. capacity, portable electric, 40' high		3.96	220	659	1,975	163.50	
	4100	For each added 10' section, add		.10	17.35	52	156	11.20	
	4200	Hoist and single tubular tower, 5000 lb. electric, 100' high		5.43	305	920	2,750	227.45	
	4300	For each added 6'-6" section, add		.18	29.50	89	267	19.25	
	4400	Hoist and double tubular tower, 5000 lb., 100' high		5.89	340	1,013	3,050	249.70	
	4500	For each added 6'-6" section, add		.20	33	99	297	21.40	
	4550	Hoist and tower, mast type, 6000 lb., 100' high		6.25	350	1,051	3,150	260.20	
	4570	For each added 10' section, add		.12	20.50	62	186	13.35	
	4600	Hoist and tower, personnel, electric, 2000 lb., 100' @ 125 fpm		14.61	930	2,790	8,375	674.90	
	4700	3000 lb., 100' @ 200 fpm		16.68	1,050	3,170	9,500	767.45	
	4800	3000 lb., 150' @ 300 fpm		18.58	1,175	3,550	10,700	858.65	
	4900	4000 lb., 100' @ 300 fpm		19.09	1,200	3,620	10,900	876.70	
	5000	6000 lb., 100' @ 275 fpm		20.27	1,275	3,790	11,400	920.15	
	5100	For added heights up to 500', add	L.F.	.01	1.67	5	15	1.10	
	5200	Jacks, hydraulic, 20 ton	Ea.	.05	2	6	18	1.60	
	5500	100 ton		.40	11.65	35	105	10.20	
	6100	Jacks, hydraulic, climbing w/50' jackrods, control console, 30 ton capacity		1.89	126	378	1,125	90.70	
	6150	For each added 10' jackrod section, add		.05	3.33	10	30	2.40	
	6300	50 ton capacity		3.05	203	609	1,825	146.20	
	6350	For each added 10' jackrod section, add		.06	4	12	36	2.90	
	6500	125 ton capacity		8.00	535	1,600	4,800	384	
	6550	For each added 10' jackrod section, add		.55	36.50	109	325	26.20	
	6600	Cable jack, 10 ton capacity with 200' cable		1.59	106	317	950	76.10	
	6650	For each added 50' of cable, add		.19	12.35	37	111	8.90	
70	0010	**WELLPOINT EQUIPMENT RENTAL** without operators [R015433 -10]							70
	0020	Based on 2 months rental							
	0100	Combination jetting & wellpoint pump, 60 H.P. diesel	Ea.	15.87	315	938	2,825	314.55	
	0200	High pressure gas jet pump, 200 H.P., 300 psi	"	36.61	267	802	2,400	453.30	
	0300	Discharge pipe, 8" diameter	L.F.	.01	.51	1.53	4.59	.40	
	0350	12" diameter		.01	.75	2.24	6.70	.55	
	0400	Header pipe, flows up to 150 GPM, 4" diameter		.01	.46	1.38	4.14	.35	
	0500	400 GPM, 6" diameter		.01	.54	1.63	4.89	.40	
	0600	800 GPM, 8" diameter		.01	.75	2.24	6.70	.55	
	0700	1500 GPM, 10" diameter		.01	.79	2.36	7.10	.55	
	0800	2500 GPM, 12" diameter		.02	1.49	4.47	13.40	1.05	
	0900	4500 GPM, 16" diameter		.03	1.91	5.72	17.15	1.40	
	0950	For quick coupling aluminum and plastic pipe, add		.03	1.97	5.92	17.75	1.40	
	1100	Wellpoint, 25' long, with fittings & riser pipe, 1-1/2" or 2" diameter	Ea.	.06	3.94	11.81	35.50	2.85	

01 54 33 | Equipment Rental

			UNIT	HOURLY OPER. COST	RENT PER DAY	RENT PER WEEK	RENT PER MONTH	EQUIPMENT COST/DAY	
70	1200	Wellpoint pump, diesel powered, 4" suction, 20 H.P.	Ea.	6.89	180	541	1,625	163.30	**70**
	1300	6" suction, 30 H.P.		9.35	224	671	2,025	209	
	1400	8" suction, 40 H.P.		12.64	305	920	2,750	285.10	
	1500	10" suction, 75 H.P.		19.15	360	1,075	3,225	368.20	
	1600	12" suction, 100 H.P.		27.47	570	1,710	5,125	561.75	
	1700	12" suction, 175 H.P.		.36	630	1,890	5,675	700.90	
80	0010	**MARINE EQUIPMENT RENTAL** without operators	Ea.						**80**
	0200	Barge, 400 Ton, 30' wide x 90' long [R015433 -10]		16.65	1,075	3,230	9,700	779.20	
	0240	800 Ton, 45' wide x 90' long		20.15	1,300	3,915	11,700	944.20	
	2000	Tugboat, diesel, 100 H.P.		31.60	215	645	1,925	381.80	
	2040	250 H.P.		64.00	395	1,180	3,550	748	
	2080	380 H.P.		131.50	1,175	3,505	10,500	1,753	

Crew A-1

	Bare Costs Hr.	Daily	Incl. Subs O&P Hr.	Daily	Cost Per Labor-Hour Bare Costs	Incl. O&P
1 Building Laborer	$35.10	$280.80	$54.00	$432.00	$35.10	$54.00
1 Concrete saw, gas manual		66.00		72.60	8.25	9.07
8 L.H., Daily Totals		$346.80		$504.60	$43.35	$63.08

Crew A-1A

	Bare Costs Hr.	Daily	Incl. Subs O&P Hr.	Daily	Cost Per Labor-Hour Bare Costs	Incl. O&P
1 Skilled Worker	$45.40	$363.20	$70.20	$561.60	$45.40	$70.20
1 Shot Blaster, 20"		211.20		232.32	26.40	29.04
8 L.H., Daily Totals		$574.40		$793.92	$71.80	$99.24

Crew A-1B

	Bare Costs Hr.	Daily	Incl. Subs O&P Hr.	Daily	Cost Per Labor-Hour Bare Costs	Incl. O&P
1 Building Laborer	$35.10	$280.80	$54.00	$432.00	$35.10	$54.00
1 Concrete Saw		148.00		162.80	18.50	20.35
8 L.H., Daily Totals		$428.80		$594.80	$53.60	$74.35

Crew A-1C

	Bare Costs Hr.	Daily	Incl. Subs O&P Hr.	Daily	Cost Per Labor-Hour Bare Costs	Incl. O&P
1 Building Laborer	$35.10	$280.80	$54.00	$432.00	$35.10	$54.00
1 Chain Saw, gas, 18"		27.40		30.14	3.42	3.77
8 L.H., Daily Totals		$308.20		$462.14	$38.52	$57.77

Crew A-1D

	Bare Costs Hr.	Daily	Incl. Subs O&P Hr.	Daily	Cost Per Labor-Hour Bare Costs	Incl. O&P
1 Building Laborer	$35.10	$280.80	$54.00	$432.00	$35.10	$54.00
1 Vibrating plate, gas, 18"		32.20		35.42	4.03	4.43
8 L.H., Daily Totals		$313.00		$467.42	$39.13	$58.43

Crew A-1E

	Bare Costs Hr.	Daily	Incl. Subs O&P Hr.	Daily	Cost Per Labor-Hour Bare Costs	Incl. O&P
1 Building Laborer	$35.10	$280.80	$54.00	$432.00	$35.10	$54.00
1 Vibratory Plate, gas, 21"		40.80		44.88	5.10	5.61
8 L.H., Daily Totals		$321.60		$476.88	$40.20	$59.61

Crew A-1F

	Bare Costs Hr.	Daily	Incl. Subs O&P Hr.	Daily	Cost Per Labor-Hour Bare Costs	Incl. O&P
1 Building Laborer	$35.10	$280.80	$54.00	$432.00	$35.10	$54.00
1 Rammer/tamper, gas, 8"		44.20		48.62	5.53	6.08
8 L.H., Daily Totals		$325.00		$480.62	$40.63	$60.08

Crew A-1G

	Bare Costs Hr.	Daily	Incl. Subs O&P Hr.	Daily	Cost Per Labor-Hour Bare Costs	Incl. O&P
1 Building Laborer	$35.10	$280.80	$54.00	$432.00	$35.10	$54.00
1 Rammer/tamper, gas, 15"		49.80		54.78	6.22	6.85
8 L.H., Daily Totals		$330.60		$486.78	$41.33	$60.85

Crew A-1H

	Bare Costs Hr.	Daily	Incl. Subs O&P Hr.	Daily	Cost Per Labor-Hour Bare Costs	Incl. O&P
1 Building Laborer	$35.10	$280.80	$54.00	$432.00	$35.10	$54.00
1 Exterior Steam Cleaner		66.80		73.48	8.35	9.19
8 L.H., Daily Totals		$347.60		$505.48	$43.45	$63.19

Crew A-1J

	Bare Costs Hr.	Daily	Incl. Subs O&P Hr.	Daily	Cost Per Labor-Hour Bare Costs	Incl. O&P
1 Building Laborer	$35.10	$280.80	$54.00	$432.00	$35.10	$54.00
1 Cultivator, Walk-Behind, 5 H.P.		58.45		64.30	7.31	8.04
8 L.H., Daily Totals		$339.25		$496.30	$42.41	$62.04

Crew A-1K

	Bare Costs Hr.	Daily	Incl. Subs O&P Hr.	Daily	Cost Per Labor-Hour Bare Costs	Incl. O&P
1 Building Laborer	$35.10	$280.80	$54.00	$432.00	$35.10	$54.00
1 Cultivator, Walk-Behind, 8 H.P.		89.75		98.72	11.22	12.34
8 L.H., Daily Totals		$370.55		$530.73	$46.32	$66.34

Crew A-1M

	Bare Costs Hr.	Daily	Incl. Subs O&P Hr.	Daily	Cost Per Labor-Hour Bare Costs	Incl. O&P
1 Building Laborer	$35.10	$280.80	$54.00	$432.00	$35.10	$54.00
1 Snow Blower, Walk-Behind		53.10		58.41	6.64	7.30
8 L.H., Daily Totals		$333.90		$490.41	$41.74	$61.30

Crew A-2

	Bare Costs Hr.	Daily	Incl. Subs O&P Hr.	Daily	Cost Per Labor-Hour Bare Costs	Incl. O&P
2 Laborers	$35.10	$561.60	$54.00	$864.00	$34.85	$53.37
1 Truck Driver (light)	34.35	274.80	52.10	416.80		
1 Flatbed Truck, Gas, 1.5 Ton		211.00		232.10	8.79	9.67
24 L.H., Daily Totals		$1047.40		$1512.90	$43.64	$63.04

Crew A-2A

	Bare Costs Hr.	Daily	Incl. Subs O&P Hr.	Daily	Cost Per Labor-Hour Bare Costs	Incl. O&P
2 Laborers	$35.10	$561.60	$54.00	$864.00	$34.85	$53.37
1 Truck Driver (light)	34.35	274.80	52.10	416.80		
1 Flatbed Truck, Gas, 1.5 Ton		211.00		232.10		
1 Concrete Saw		148.00		162.80	14.96	16.45
24 L.H., Daily Totals		$1195.40		$1675.70	$49.81	$69.82

Crew A-2B

	Bare Costs Hr.	Daily	Incl. Subs O&P Hr.	Daily	Cost Per Labor-Hour Bare Costs	Incl. O&P
1 Truck Driver (light)	$34.35	$274.80	$52.10	$416.80	$34.35	$52.10
1 Flatbed Truck, Gas, 1.5 Ton		211.00		232.10	26.38	29.01
8 L.H., Daily Totals		$485.80		$648.90	$60.73	$81.11

Crew A-3A

	Bare Costs Hr.	Daily	Incl. Subs O&P Hr.	Daily	Cost Per Labor-Hour Bare Costs	Incl. O&P
1 Truck Driver (light)	$34.35	$274.80	$52.10	$416.80	$34.35	$52.10
1 Pickup truck, 4 x 4, 3/4 ton		139.00		152.90	17.38	19.11
8 L.H., Daily Totals		$413.80		$569.70	$51.73	$71.21

Crew A-3B

	Bare Costs Hr.	Daily	Incl. Subs O&P Hr.	Daily	Cost Per Labor-Hour Bare Costs	Incl. O&P
1 Equip. Oper. (med.)	$46.55	$372.40	$70.25	$562.00	$40.92	$61.90
1 Truck Driver (heav.)	35.30	282.40	53.55	428.40		
1 Dump Truck, 12 C.Y., 400 H.P.		584.60		643.06		
1 F.E. Loader, W.M.,2.5 C.Y.		472.40		519.64	66.06	72.67
16 L.H., Daily Totals		$1711.80		$2153.10	$106.99	$134.57

Crew A-3C

	Bare Costs Hr.	Daily	Incl. Subs O&P Hr.	Daily	Cost Per Labor-Hour Bare Costs	Incl. O&P
1 Equip. Oper. (light)	$44.75	$358.00	$67.55	$540.40	$44.75	$67.55
1 Loader, Skid Steer, 78 H.P.		281.20		309.32	35.15	38.66
8 L.H., Daily Totals		$639.20		$849.72	$79.90	$106.22

Crew A-3D

	Bare Costs Hr.	Daily	Incl. Subs O&P Hr.	Daily	Cost Per Labor-Hour Bare Costs	Incl. O&P
1 Truck Driver, Light	$34.35	$274.80	$52.10	$416.80	$34.35	$52.10
1 Pickup truck, 4 x 4, 3/4 ton		139.00		152.90		
1 Flatbed Trailer, 25 Ton		108.60		119.46	30.95	34.05
8 L.H., Daily Totals		$522.40		$689.16	$65.30	$86.14

Crew A-3E

	Bare Costs Hr.	Daily	Incl. Subs O&P Hr.	Daily	Cost Per Labor-Hour Bare Costs	Incl. O&P
1 Equip. Oper. (crane)	$47.80	$382.40	$72.15	$577.20	$41.55	$62.85
1 Truck Driver (heavy)	35.30	282.40	53.55	428.40		
1 Pickup truck, 4 x 4, 3/4 ton		139.00		152.90	8.69	9.56
16 L.H., Daily Totals		$803.80		$1158.50	$50.24	$72.41

Crew A-3F

	Bare Costs Hr.	Daily	Incl. Subs O&P Hr.	Daily	Cost Per Labor-Hour Bare Costs	Incl. O&P
1 Equip. Oper. (crane)	$47.80	$382.40	$72.15	$577.20	$41.55	$62.85
1 Truck Driver (heavy)	35.30	282.40	53.55	428.40		
1 Pickup truck, 4 x 4, 3/4 ton		139.00		152.90		
1 Truck Tractor, 6x4, 380 H.P.		522.20		574.42		
1 Lowbed Trailer, 75 Ton		213.00		234.30	54.64	60.10
16 L.H., Daily Totals		$1539.00		$1967.22	$96.19	$122.95

Crew No.	Bare Costs		Incl. Subs O&P		Cost Per Labor-Hour	
Crew A-3G	Hr.	Daily	Hr.	Daily	Bare Costs	Incl. O&P
1 Equip. Oper. (crane)	$47.80	$382.40	$72.15	$577.20	$41.55	$62.85
1 Truck Driver (heavy)	35.30	282.40	53.55	428.40		
1 Pickup truck, 4 x 4, 3/4 ton		139.00		152.90		
1 Truck Tractor, 6x4, 450 H.P.		635.00		698.50		
1 Lowbed Trailer, 75 Ton		213.00		234.30	61.69	67.86
16 L.H., Daily Totals		$1651.80		$2091.30	$103.24	$130.71
Crew A-3H	Hr.	Daily	Hr.	Daily	Bare Costs	Incl. O&P
1 Equip. Oper. (crane)	$47.80	$382.40	$72.15	$577.20	$47.80	$72.15
1 Hyd. crane, 12 Ton (daily)		824.00		906.40	103.00	113.30
8 L.H., Daily Totals		$1206.40		$1483.60	$150.80	$185.45
Crew A-3I	Hr.	Daily	Hr.	Daily	Bare Costs	Incl. O&P
1 Equip. Oper. (crane)	$47.80	$382.40	$72.15	$577.20	$47.80	$72.15
1 Hyd. crane, 25 Ton (daily)		953.20		1048.52	119.15	131.07
8 L.H., Daily Totals		$1335.60		$1625.72	$166.95	$203.22
Crew A-3J	Hr.	Daily	Hr.	Daily	Bare Costs	Incl. O&P
1 Equip. Oper. (crane)	$47.80	$382.40	$72.15	$577.20	$47.80	$72.15
1 Hyd. crane, 40 Ton (daily)		1184.00		1302.40	148.00	162.80
8 L.H., Daily Totals		$1566.40		$1879.60	$195.80	$234.95
Crew A-3K	Hr.	Daily	Hr.	Daily	Bare Costs	Incl. O&P
1 Equip. Oper. (crane)	$47.80	$382.40	$72.15	$577.20	$44.52	$67.20
1 Equip. Oper. Oiler	41.25	330.00	62.25	498.00		
1 Hyd. crane, 55 Ton (daily)		1471.00		1618.10		
1 P/U Truck, 3/4 Ton (daily)		152.00		167.20	101.44	111.58
16 L.H., Daily Totals		$2335.40		$2860.50	$145.96	$178.78
Crew A-3L	Hr.	Daily	Hr.	Daily	Bare Costs	Incl. O&P
1 Equip. Oper. (crane)	$47.80	$382.40	$72.15	$577.20	$44.52	$67.20
1 Equip. Oper. Oiler	41.25	330.00	62.25	498.00		
1 Hyd. crane, 80 Ton (daily)		2119.00		2330.90		
1 P/U Truck, 3/4 Ton (daily)		152.00		167.20	141.94	156.13
16 L.H., Daily Totals		$2983.40		$3573.30	$186.46	$223.33
Crew A-3M	Hr.	Daily	Hr.	Daily	Bare Costs	Incl. O&P
1 Equip. Oper. (crane)	$47.80	$382.40	$72.15	$577.20	$44.52	$67.20
1 Equip. Oper. Oiler	41.25	330.00	62.25	498.00		
1 Hyd. crane, 100 Ton (daily)		2129.00		2341.90		
1 P/U Truck, 3/4 Ton (daily)		152.00		167.20	142.56	156.82
16 L.H., Daily Totals		$2993.40		$3584.30	$187.09	$224.02
Crew A-3N	Hr.	Daily	Hr.	Daily	Bare Costs	Incl. O&P
1 Equip. Oper. (crane)	$47.80	$382.40	$72.15	$577.20	$47.80	$72.15
1 Tower crane (monthly)		1005.00		1105.50	125.63	138.19
8 L.H., Daily Totals		$1387.40		$1682.70	$173.43	$210.34
Crew A-3P	Hr.	Daily	Hr.	Daily	Bare Costs	Incl. O&P
1 Equip. Oper., Light	$44.75	$358.00	$67.55	$540.40	$44.75	$67.55
1 A.T. Forklift, 42' lift		481.80		529.98	60.23	66.25
8 L.H., Daily Totals		$839.80		$1070.38	$104.97	$133.80
Crew A-4	Hr.	Daily	Hr.	Daily	Bare Costs	Incl. O&P
2 Carpenters	$44.10	$705.60	$67.85	$1085.60	$42.15	$64.30
1 Painter, Ordinary	38.25	306.00	57.20	457.60		
24 L.H., Daily Totals		$1011.60		$1543.20	$42.15	$64.30

Crew No.	Bare Costs		Incl. Subs O&P		Cost Per Labor-Hour	
Crew A-5	Hr.	Daily	Hr.	Daily	Bare Costs	Incl. O&P
2 Laborers	$35.10	$561.60	$54.00	$864.00	$35.02	$53.79
.25 Truck Driver (light)	34.35	68.70	52.10	104.20		
.25 Flatbed Truck, Gas, 1.5 Ton		52.75		58.02	2.93	3.22
18 L.H., Daily Totals		$683.05		$1026.22	$37.95	$57.01
Crew A-6	Hr.	Daily	Hr.	Daily	Bare Costs	Incl. O&P
1 Instrument Man	$45.40	$363.20	$70.20	$561.60	$44.08	$67.47
1 Rodman/Chainman	42.75	342.00	64.75	518.00		
1 Level, electronic		78.70		86.57	4.92	5.41
16 L.H., Daily Totals		$783.90		$1166.17	$48.99	$72.89
Crew A-7	Hr.	Daily	Hr.	Daily	Bare Costs	Incl. O&P
1 Chief Of Party	$56.40	$451.20	$87.15	$697.20	$48.18	$74.03
1 Instrument Man	45.40	363.20	70.20	561.60		
1 Rodman/Chainman	42.75	342.00	64.75	518.00		
1 Level, electronic		78.70		86.57	3.28	3.61
24 L.H., Daily Totals		$1235.10		$1863.37	$51.46	$77.64
Crew A-8	Hr.	Daily	Hr.	Daily	Bare Costs	Incl. O&P
1 Chief of Party	$56.40	$451.20	$87.15	$697.20	$46.83	$71.71
1 Instrument Man	45.40	363.20	70.20	561.60		
2 Rodmen/Chainmen	42.75	684.00	64.75	1036.00		
1 Level, electronic		78.70		86.57	2.46	2.71
32 L.H., Daily Totals		$1577.10		$2381.37	$49.28	$74.42
Crew A-9	Hr.	Daily	Hr.	Daily	Bare Costs	Incl. O&P
1 Asbestos Foreman	$49.35	$394.80	$76.90	$615.20	$48.91	$76.20
7 Asbestos Workers	48.85	2735.60	76.10	4261.60		
64 L.H., Daily Totals		$3130.40		$4876.80	$48.91	$76.20
Crew A-10A	Hr.	Daily	Hr.	Daily	Bare Costs	Incl. O&P
1 Asbestos Foreman	$49.35	$394.80	$76.90	$615.20	$49.02	$76.37
2 Asbestos Workers	48.85	781.60	76.10	1217.60		
24 L.H., Daily Totals		$1176.40		$1832.80	$49.02	$76.37
Crew A-10B	Hr.	Daily	Hr.	Daily	Bare Costs	Incl. O&P
1 Asbestos Foreman	$49.35	$394.80	$76.90	$615.20	$48.98	$76.30
3 Asbestos Workers	48.85	1172.40	76.10	1826.40		
32 L.H., Daily Totals		$1567.20		$2441.60	$48.98	$76.30
Crew A-10C	Hr.	Daily	Hr.	Daily	Bare Costs	Incl. O&P
3 Asbestos Workers	$48.85	$1172.40	$76.10	$1826.40	$48.85	$76.10
1 Flatbed Truck, Gas, 1.5 Ton		211.00		232.10	8.79	9.67
24 L.H., Daily Totals		$1383.40		$2058.50	$57.64	$85.77
Crew A-10D	Hr.	Daily	Hr.	Daily	Bare Costs	Incl. O&P
2 Asbestos Workers	$48.85	$781.60	$76.10	$1217.60	$46.69	$71.65
1 Equip. Oper. (crane)	47.80	382.40	72.15	577.20		
1 Equip. Oper. Oiler	41.25	330.00	62.25	498.00		
1 Hydraulic Crane, 33 Ton		711.60		782.76	22.24	24.46
32 L.H., Daily Totals		$2205.60		$3075.56	$68.92	$96.11
Crew A-11	Hr.	Daily	Hr.	Daily	Bare Costs	Incl. O&P
1 Asbestos Foreman	$49.35	$394.80	$76.90	$615.20	$48.91	$76.20
7 Asbestos Workers	48.85	2735.60	76.10	4261.60		
2 Chip. Hammers, 12 Lb., Elec.		39.60		43.56	.62	.68
64 L.H., Daily Totals		$3170.00		$4920.36	$49.53	$76.88

Crew No.	Bare Costs Hr.	Daily	Incl. Subs O&P Hr.	Daily	Cost Per Labor-Hour Bare Costs	Incl. O&P
Crew A-12	Hr.	Daily	Hr.	Daily	Bare Costs	Incl. O&P
1 Asbestos Foreman	$49.35	$394.80	$76.90	$615.20	$48.91	$76.20
7 Asbestos Workers	48.85	2735.60	76.10	4261.60		
1 Trk-mtd vac, 14 CY, 1500 Gal.		518.50		570.35		
1 Flatbed Truck, 20,000 GVW		215.80		237.38	11.47	12.62
64 L.H., Daily Totals		$3864.70		$5684.53	$60.39	$88.82
Crew A-13	Hr.	Daily	Hr.	Daily	Bare Costs	Incl. O&P
1 Equip. Oper. (light)	$44.75	$358.00	$67.55	$540.40	$44.75	$67.55
1 Trk-mtd vac, 14 CY, 1500 Gal.		518.50		570.35		
1 Flatbed Truck, 20,000 GVW		215.80		237.38	91.79	100.97
8 L.H., Daily Totals		$1092.30		$1348.13	$136.54	$168.52
Crew B-1	Hr.	Daily	Hr.	Daily	Bare Costs	Incl. O&P
1 Labor Foreman (outside)	$37.10	$296.80	$57.10	$456.80	$35.77	$55.03
2 Laborers	35.10	561.60	54.00	864.00		
24 L.H., Daily Totals		$858.40		$1320.80	$35.77	$55.03
Crew B-1A	Hr.	Daily	Hr.	Daily	Bare Costs	Incl. O&P
1 Labor Foreman (outside)	$37.10	$296.80	$57.10	$456.80	$35.77	$55.03
2 Laborers	35.10	561.60	54.00	864.00		
2 Cutting Torches		21.60		23.76		
2 Sets of Gases		172.80		190.08	8.10	8.91
24 L.H., Daily Totals		$1052.80		$1534.64	$43.87	$63.94
Crew B-1B	Hr.	Daily	Hr.	Daily	Bare Costs	Incl. O&P
1 Labor Foreman (outside)	$37.10	$296.80	$57.10	$456.80	$38.77	$59.31
2 Laborers	35.10	561.60	54.00	864.00		
1 Equip. Oper. (crane)	47.80	382.40	72.15	577.20		
2 Cutting Torches		21.60		23.76		
2 Sets of Gases		172.80		190.08		
1 Hyd. Crane, 12 Ton		612.00		673.20	25.20	27.72
32 L.H., Daily Totals		$2047.20		$2785.04	$63.98	$87.03
Crew B-2	Hr.	Daily	Hr.	Daily	Bare Costs	Incl. O&P
1 Labor Foreman (outside)	$37.10	$296.80	$57.10	$456.80	$35.50	$54.62
4 Laborers	35.10	1123.20	54.00	1728.00		
40 L.H., Daily Totals		$1420.00		$2184.80	$35.50	$54.62
Crew B-3	Hr.	Daily	Hr.	Daily	Bare Costs	Incl. O&P
1 Labor Foreman (outside)	$37.10	$296.80	$57.10	$456.80	$37.41	$57.08
2 Laborers	35.10	561.60	54.00	864.00		
1 Equip. Oper. (med.)	46.55	372.40	70.25	562.00		
2 Truck Drivers (heavy)	35.30	564.80	53.55	856.80		
1 Crawler Loader, 3 C.Y.		1106.00		1216.60		
2 Dump Trucks, 12 C.Y., 400 H.P.		1169.20		1286.12	47.40	52.14
48 L.H., Daily Totals		$4070.80		$5242.32	$84.81	$109.22
Crew B-3A	Hr.	Daily	Hr.	Daily	Bare Costs	Incl. O&P
4 Laborers	$35.10	$1123.20	$54.00	$1728.00	$37.39	$57.25
1 Equip. Oper. (med.)	46.55	372.40	70.25	562.00		
1 Hyd. Excavator, 1.5 C.Y.		960.40		1056.44	24.01	26.41
40 L.H., Daily Totals		$2456.00		$3346.44	$61.40	$83.66
Crew B-3B	Hr.	Daily	Hr.	Daily	Bare Costs	Incl. O&P
2 Laborers	$35.10	$561.60	$54.00	$864.00	$38.01	$57.95
1 Equip. Oper. (med.)	46.55	372.40	70.25	562.00		
1 Truck Driver (heavy)	35.30	282.40	53.55	428.40		
1 Backhoe Loader, 80 H.P.		363.80		400.18		
1 Dump Truck, 12 C.Y., 400 H.P.		584.60		643.06	29.64	32.60
32 L.H., Daily Totals		$2164.80		$2897.64	$67.65	$90.55
Crew B-3C	Hr.	Daily	Hr.	Daily	Bare Costs	Incl. O&P
3 Laborers	$35.10	$842.40	$54.00	$1296.00	$37.96	$58.06
1 Equip. Oper. (med.)	46.55	372.40	70.25	562.00		
1 Crawler Loader, 4 C.Y.		1443.00		1587.30	45.09	49.60
32 L.H., Daily Totals		$2657.80		$3445.30	$83.06	$107.67
Crew B-4	Hr.	Daily	Hr.	Daily	Bare Costs	Incl. O&P
1 Labor Foreman (outside)	$37.10	$296.80	$57.10	$456.80	$35.47	$54.44
4 Laborers	35.10	1123.20	54.00	1728.00		
1 Truck Driver (heavy)	35.30	282.40	53.55	428.40		
1 Truck Tractor, 220 H.P.		314.80		346.28		
1 Flatbed Trailer, 40 Ton		147.00		161.70	9.62	10.58
48 L.H., Daily Totals		$2164.20		$3121.18	$45.09	$65.02
Crew B-5	Hr.	Daily	Hr.	Daily	Bare Costs	Incl. O&P
1 Labor Foreman (outside)	$37.10	$296.80	$57.10	$456.80	$38.66	$59.09
4 Laborers	35.10	1123.20	54.00	1728.00		
2 Equip. Oper. (med.)	46.55	744.80	70.25	1124.00		
1 Air Compressor, 250 cfm		177.00		194.70		
2 Breakers, Pavement, 60 lb.		19.60		21.56		
2 -50' Air Hoses, 1.5"		12.60		13.86		
1 Crawler Loader, 3 C.Y.		1106.00		1216.60	23.49	25.83
56 L.H., Daily Totals		$3480.00		$4755.52	$62.14	$84.92
Crew B-5A	Hr.	Daily	Hr.	Daily	Bare Costs	Incl. O&P
1 Labor Foreman (outside)	$37.10	$296.80	$57.10	$456.80	$38.01	$58.02
6 Laborers	35.10	1684.80	54.00	2592.00		
2 Equip. Oper. (med.)	46.55	744.80	70.25	1124.00		
1 Equip. Oper. (light)	44.75	358.00	67.55	540.40		
2 Truck Drivers (heavy)	35.30	564.80	53.55	856.80		
1 Air Compressor, 365 cfm		229.00		251.90		
2 Breakers, Pavement, 60 lb.		19.60		21.56		
8 -50' Air Hoses, 1"		32.80		36.08		
2 Dump Trucks, 8 C.Y., 220 H.P.		714.80		786.28	10.38	11.41
96 L.H., Daily Totals		$4645.40		$6665.82	$48.39	$69.44
Crew B-5B	Hr.	Daily	Hr.	Daily	Bare Costs	Incl. O&P
1 Powderman	$45.40	$363.20	$70.20	$561.60	$40.73	$61.89
2 Equip. Oper. (med.)	46.55	744.80	70.25	1124.00		
3 Truck Drivers (heavy)	35.30	847.20	53.55	1285.20		
1 F.E. Loader, W.M.,2.5 C.Y.		472.40		519.64		
3 Dump Trucks, 12 C.Y., 400 H.P.		1753.80		1929.18		
1 Air Compressor, 365 CFM		229.00		251.90	51.15	56.27
48 L.H., Daily Totals		$4410.40		$5671.52	$91.88	$118.16
Crew B-5C	Hr.	Daily	Hr.	Daily	Bare Costs	Incl. O&P
3 Laborers	$35.10	$842.40	$54.00	$1296.00	$38.94	$59.22
1 Equip. Oper. (med.)	46.55	372.40	70.25	562.00		
2 Truck Drivers (heav.)	35.30	564.80	53.55	856.80		
1 Equip. Oper. (crane)	47.80	382.40	72.15	577.20		
1 Equip. Oper. Oiler	41.25	330.00	62.25	498.00		
2 Dump Trucks, 12 C.Y., 400 H.P.		1169.20		1286.12		
1 Crawler Loader, 4 C.Y.		1443.00		1587.30		
1 S.P. Crane, 4x4, 25 Ton		565.20		621.72	49.65	54.61
64 L.H., Daily Totals		$5669.40		$7285.14	$88.58	$113.83

681

Crew No.	Bare Costs		Incl. Subs O&P		Cost Per Labor-Hour	

Crew B-5D

Crew B-5D	Hr.	Daily	Hr.	Daily	Bare Costs	Incl. O&P
1 Labor Foreman (outside)	$37.10	$296.80	$57.10	$456.80	$38.24	$58.39
4 Laborers	35.10	1123.20	54.00	1728.00		
2 Equip. Oper. (med.)	46.55	744.80	70.25	1124.00		
1 Truck Driver (heavy)	35.30	282.40	53.55	428.40		
1 Air Compressor, 250 cfm		177.00		194.70		
2 Breakers, Pavement, 60 lb.		19.60		21.56		
2 -50' Air Hoses, 1.5"		12.60		13.86		
1 Crawler Loader, 3 C.Y.		1106.00		1216.60		
1 Dump Truck, 12 C.Y., 400 H.P.		584.60		643.06	29.68	32.65
64 L.H., Daily Totals		$4347.00		$5826.98	$67.92	$91.05

Crew B-6	Hr.	Daily	Hr.	Daily	Bare Costs	Incl. O&P
2 Laborers	$35.10	$561.60	$54.00	$864.00	$38.32	$58.52
1 Equip. Oper. (light)	44.75	358.00	67.55	540.40		
1 Backhoe Loader, 48 H.P.		334.00		367.40	13.92	15.31
24 L.H., Daily Totals		$1253.60		$1771.80	$52.23	$73.83

Crew B-6A	Hr.	Daily	Hr.	Daily	Bare Costs	Incl. O&P
.5 Labor Foreman (outside)	$37.10	$148.40	$57.10	$228.40	$40.08	$61.12
1 Laborer	35.10	280.80	54.00	432.00		
1 Equip. Oper. (med.)	46.55	372.40	70.25	562.00		
1 Vacuum Trk.,5000 Gal.		357.20		392.92	17.86	19.65
20 L.H., Daily Totals		$1158.80		$1615.32	$57.94	$80.77

Crew B-6B	Hr.	Daily	Hr.	Daily	Bare Costs	Incl. O&P
2 Labor Foremen (outside)	$37.10	$593.60	$57.10	$913.60	$35.77	$55.03
4 Laborers	35.10	1123.20	54.00	1728.00		
1 S.P. Crane, 4x4, 5 Ton		264.60		291.06		
1 Flatbed Truck, Gas, 1.5 Ton		211.00		232.10		
1 Butt Fusion Mach., 4"-12" diam.		376.45		414.10	17.75	19.53
48 L.H., Daily Totals		$2568.85		$3578.86	$53.52	$74.56

Crew B-6C	Hr.	Daily	Hr.	Daily	Bare Costs	Incl. O&P
2 Labor Foremen (outside)	$37.10	$593.60	$57.10	$913.60	$35.77	$55.03
4 Laborers	35.10	1123.20	54.00	1728.00		
1 S.P. Crane, 4x4, 12 Ton		448.20		493.02		
1 Flatbed Truck, Gas, 3 Ton		264.40		290.84		
1 Butt Fusion Mach., 8"-24" diam.		806.15		886.76	31.64	34.80
48 L.H., Daily Totals		$3235.55		$4312.23	$67.41	$89.84

Crew B-7	Hr.	Daily	Hr.	Daily	Bare Costs	Incl. O&P
1 Labor Foreman (outside)	$37.10	$296.80	$57.10	$456.80	$37.34	$57.23
4 Laborers	35.10	1123.20	54.00	1728.00		
1 Equip. Oper. (med.)	46.55	372.40	70.25	562.00		
1 Brush Chipper, 12", 130 H.P.		283.80		312.18		
1 Crawler Loader, 3 C.Y.		1106.00		1216.60		
2 Chain Saws, gas, 36" Long		87.20		95.92	30.77	33.85
48 L.H., Daily Totals		$3269.40		$4371.50	$68.11	$91.07

Crew B-7A	Hr.	Daily	Hr.	Daily	Bare Costs	Incl. O&P
2 Laborers	$35.10	$561.60	$54.00	$864.00	$38.32	$58.52
1 Equip. Oper. (light)	44.75	358.00	67.55	540.40		
1 Rake w/Tractor		324.70		357.17		
2 Chain Saws, gas, 18"		54.80		60.28	15.81	17.39
24 L.H., Daily Totals		$1299.10		$1821.85	$54.13	$75.91

Crew No.	Bare Costs		Incl. Subs O&P		Cost Per Labor-Hour	

Crew B-7B	Hr.	Daily	Hr.	Daily	Bare Costs	Incl. O&P
1 Labor Foreman (outside)	$37.10	$296.80	$57.10	$456.80	$37.05	$56.70
4 Laborers	35.10	1123.20	54.00	1728.00		
1 Equip. Oper. (med.)	46.55	372.40	70.25	562.00		
1 Truck Driver (heavy)	35.30	282.40	53.55	428.40		
1 Brush Chipper, 12", 130 H.P.		283.80		312.18		
1 Crawler Loader, 3 C.Y.		1106.00		1216.60		
2 Chain Saws, gas, 36" Long		87.20		95.92		
1 Dump Truck, 8 C.Y., 220 H.P.		357.40		393.14	32.76	36.03
56 L.H., Daily Totals		$3909.20		$5193.04	$69.81	$92.73

Crew B-7C	Hr.	Daily	Hr.	Daily	Bare Costs	Incl. O&P
1 Labor Foreman (outside)	$37.10	$296.80	$57.10	$456.80	$37.05	$56.70
4 Laborers	35.10	1123.20	54.00	1728.00		
1 Equip. Oper. (med.)	46.55	372.40	70.25	562.00		
1 Truck Driver (heavy)	35.30	282.40	53.55	428.40		
1 Brush Chipper, 12", 130 H.P.		283.80		312.18		
1 Crawler Loader, 3 C.Y.		1106.00		1216.60		
2 Chain Saws, gas, 36" Long		87.20		95.92		
1 Dump Truck, 12 C.Y., 400 H.P.		584.60		643.06	36.81	40.50
56 L.H., Daily Totals		$4136.40		$5442.96	$73.86	$97.20

Crew B-8	Hr.	Daily	Hr.	Daily	Bare Costs	Incl. O&P
1 Labor Foreman (outside)	$37.10	$296.80	$57.10	$456.80	$39.03	$59.37
2 Laborers	35.10	561.60	54.00	864.00		
2 Equip. Oper. (med.)	46.55	744.80	70.25	1124.00		
1 Equip. Oper. Oiler	41.25	330.00	62.25	498.00		
2 Truck Drivers (heavy)	35.30	564.80	53.55	856.80		
1 Hyd. Crane, 25 Ton		697.20		766.92		
1 Crawler Loader, 3 C.Y.		1106.00		1216.60		
2 Dump Trucks, 12 C.Y., 400 H.P.		1169.20		1286.12	46.44	51.09
64 L.H., Daily Totals		$5470.40		$7069.24	$85.48	$110.46

Crew B-9	Hr.	Daily	Hr.	Daily	Bare Costs	Incl. O&P
1 Labor Foreman (outside)	$37.10	$296.80	$57.10	$456.80	$35.50	$54.62
4 Laborers	35.10	1123.20	54.00	1728.00		
1 Air Compressor, 250 cfm		177.00		194.70		
2 Breakers, Pavement, 60 lb.		19.60		21.56		
2 -50' Air Hoses, 1.5"		12.60		13.86	5.23	5.75
40 L.H., Daily Totals		$1629.20		$2414.92	$40.73	$60.37

Crew B-9A	Hr.	Daily	Hr.	Daily	Bare Costs	Incl. O&P
2 Laborers	$35.10	$561.60	$54.00	$864.00	$35.17	$53.85
1 Truck Driver (heavy)	35.30	282.40	53.55	428.40		
1 Water Tank Trailer, 5000 Gal.		141.00		155.10		
1 Truck Tractor, 220 H.P.		314.80		346.28		
2 -50' Discharge Hoses, 3"		6.10		6.71	19.25	21.17
24 L.H., Daily Totals		$1305.90		$1800.49	$54.41	$75.02

Crew B-9B	Hr.	Daily	Hr.	Daily	Bare Costs	Incl. O&P
2 Laborers	$35.10	$561.60	$54.00	$864.00	$35.17	$53.85
1 Truck Driver (heavy)	35.30	282.40	53.55	428.40		
2 -50' Discharge Hoses, 3"		6.10		6.71		
1 Water Tank Trailer, 5000 Gal.		141.00		155.10		
1 Truck Tractor, 220 H.P.		314.80		346.28		
1 Pressure Washer		63.00		69.30	21.87	24.06
24 L.H., Daily Totals		$1368.90		$1869.79	$57.04	$77.91

Crews

Crew No.	Bare Costs Hr.	Daily	Incl. Subs O&P Hr.	Daily	Cost Per Labor-Hour Bare Costs	Incl. O&P
Crew B-9D	Hr.	Daily	Hr.	Daily	Bare Costs	Incl. O&P
1 Labor Foreman (Outside)	$37.10	$296.80	$57.10	$456.80	$35.50	$54.62
4 Common Laborers	35.10	1123.20	54.00	1728.00		
1 Air Compressor, 250 cfm		177.00		194.70		
2 -50' Air Hoses, 1.5"		12.60		13.86		
2 Air Powered Tampers		52.40		57.64	6.05	6.66
40 L.H., Daily Totals		$1662.00		$2451.00	$41.55	$61.27
Crew B-10	Hr.	Daily	Hr.	Daily	Bare Costs	Incl. O&P
1 Equip. Oper. (med.)	$46.55	$372.40	$70.25	$562.00	$42.73	$64.83
.5 Laborer	35.10	140.40	54.00	216.00		
12 L.H., Daily Totals		$512.80		$778.00	$42.73	$64.83
Crew B-10A	Hr.	Daily	Hr.	Daily	Bare Costs	Incl. O&P
1 Equip. Oper. (med.)	$46.55	$372.40	$70.25	$562.00	$42.73	$64.83
.5 Laborer	35.10	140.40	54.00	216.00		
1 Roller, 2-Drum, W.B., 7.5 H.P.		171.60		188.76	14.30	15.73
12 L.H., Daily Totals		$684.40		$966.76	$57.03	$80.56
Crew B-10B	Hr.	Daily	Hr.	Daily	Bare Costs	Incl. O&P
1 Equip. Oper. (med.)	$46.55	$372.40	$70.25	$562.00	$42.73	$64.83
.5 Laborer	35.10	140.40	54.00	216.00		
1 Dozer, 200 H.P.		1243.00		1367.30	103.58	113.94
12 L.H., Daily Totals		$1755.80		$2145.30	$146.32	$178.78
Crew B-10C	Hr.	Daily	Hr.	Daily	Bare Costs	Incl. O&P
1 Equip. Oper. (med.)	$46.55	$372.40	$70.25	$562.00	$42.73	$64.83
.5 Laborer	35.10	140.40	54.00	216.00		
1 Dozer, 200 H.P.		1243.00		1367.30		
1 Vibratory Roller, Towed, 23 Ton		388.80		427.68	135.98	149.58
12 L.H., Daily Totals		$2144.60		$2572.98	$178.72	$214.41
Crew B-10D	Hr.	Daily	Hr.	Daily	Bare Costs	Incl. O&P
1 Equip. Oper. (med.)	$46.55	$372.40	$70.25	$562.00	$42.73	$64.83
.5 Laborer	35.10	140.40	54.00	216.00		
1 Dozer, 200 H.P.		1243.00		1367.30		
1 Sheepsft. Roller, Towed		420.20		462.22	138.60	152.46
12 L.H., Daily Totals		$2176.00		$2607.52	$181.33	$217.29
Crew B-10E	Hr.	Daily	Hr.	Daily	Bare Costs	Incl. O&P
1 Equip. Oper. (med.)	$46.55	$372.40	$70.25	$562.00	$42.73	$64.83
.5 Laborer	35.10	140.40	54.00	216.00		
1 Tandem Roller, 5 Ton		149.20		164.12	12.43	13.68
12 L.H., Daily Totals		$662.00		$942.12	$55.17	$78.51
Crew B-10F	Hr.	Daily	Hr.	Daily	Bare Costs	Incl. O&P
1 Equip. Oper. (med.)	$46.55	$372.40	$70.25	$562.00	$42.73	$64.83
.5 Laborer	35.10	140.40	54.00	216.00		
1 Tandem Roller, 10 Ton		227.60		250.36	18.97	20.86
12 L.H., Daily Totals		$740.40		$1028.36	$61.70	$85.70
Crew B-10G	Hr.	Daily	Hr.	Daily	Bare Costs	Incl. O&P
1 Equip. Oper. (med.)	$46.55	$372.40	$70.25	$562.00	$42.73	$64.83
.5 Laborer	35.10	140.40	54.00	216.00		
1 Sheepsft. Roll., 240 H.P.		1101.00		1211.10	91.75	100.93
12 L.H., Daily Totals		$1613.80		$1989.10	$134.48	$165.76

Crew No.	Bare Costs Hr.	Daily	Incl. Subs O&P Hr.	Daily	Cost Per Labor-Hour Bare Costs	Incl. O&P
Crew B-10H	Hr.	Daily	Hr.	Daily	Bare Costs	Incl. O&P
1 Equip. Oper. (med.)	$46.55	$372.40	$70.25	$562.00	$42.73	$64.83
.5 Laborer	35.10	140.40	54.00	216.00		
1 Diaphragm Water Pump, 2"		68.60		75.46		
1 -20' Suction Hose, 2"		1.95		2.15		
2 -50' Discharge Hoses, 2"		3.50		3.85	6.17	6.79
12 L.H., Daily Totals		$586.85		$859.46	$48.90	$71.62
Crew B-10I	Hr.	Daily	Hr.	Daily	Bare Costs	Incl. O&P
1 Equip. Oper. (med.)	$46.55	$372.40	$70.25	$562.00	$42.73	$64.83
.5 Laborer	35.10	140.40	54.00	216.00		
1 Diaphragm Water Pump, 4"		111.20		122.32		
1 -20' Suction Hose, 4"		3.25		3.58		
2 -50' Discharge Hoses, 4"		9.40		10.34	10.32	11.35
12 L.H., Daily Totals		$636.65		$914.24	$53.05	$76.19
Crew B-10J	Hr.	Daily	Hr.	Daily	Bare Costs	Incl. O&P
1 Equip. Oper. (med.)	$46.55	$372.40	$70.25	$562.00	$42.73	$64.83
.5 Laborer	35.10	140.40	54.00	216.00		
1 Centrifugal Water Pump, 3"		73.80		81.18		
1 -20' Suction Hose, 3"		2.85		3.13		
2 -50' Discharge Hoses, 3"		6.10		6.71	6.90	7.59
12 L.H., Daily Totals		$595.55		$869.02	$49.63	$72.42
Crew B-10K	Hr.	Daily	Hr.	Daily	Bare Costs	Incl. O&P
1 Equip. Oper. (med.)	$46.55	$372.40	$70.25	$562.00	$42.73	$64.83
.5 Laborer	35.10	140.40	54.00	216.00		
1 Centr. Water Pump, 6"		308.40		339.24		
1 -20' Suction Hose, 6"		11.50		12.65		
2 -50' Discharge Hoses, 6"		24.20		26.62	28.68	31.54
12 L.H., Daily Totals		$856.90		$1156.51	$71.41	$96.38
Crew B-10L	Hr.	Daily	Hr.	Daily	Bare Costs	Incl. O&P
1 Equip. Oper. (med.)	$46.55	$372.40	$70.25	$562.00	$42.73	$64.83
.5 Laborer	35.10	140.40	54.00	216.00		
1 Dozer, 80 H.P.		442.40		486.64	36.87	40.55
12 L.H., Daily Totals		$955.20		$1264.64	$79.60	$105.39
Crew B-10M	Hr.	Daily	Hr.	Daily	Bare Costs	Incl. O&P
1 Equip. Oper. (med.)	$46.55	$372.40	$70.25	$562.00	$42.73	$64.83
.5 Laborer	35.10	140.40	54.00	216.00		
1 Dozer, 300 H.P.		1718.00		1889.80	143.17	157.48
12 L.H., Daily Totals		$2230.80		$2667.80	$185.90	$222.32
Crew B-10N	Hr.	Daily	Hr.	Daily	Bare Costs	Incl. O&P
1 Equip. Oper. (med.)	$46.55	$372.40	$70.25	$562.00	$42.73	$64.83
.5 Laborer	35.10	140.40	54.00	216.00		
1 F.E. Loader, T.M., 1.5 C.Y		488.60		537.46	40.72	44.79
12 L.H., Daily Totals		$1001.40		$1315.46	$83.45	$109.62
Crew B-10O	Hr.	Daily	Hr.	Daily	Bare Costs	Incl. O&P
1 Equip. Oper. (med.)	$46.55	$372.40	$70.25	$562.00	$42.73	$64.83
.5 Laborer	35.10	140.40	54.00	216.00		
1 F.E. Loader, T.M., 2.25 C.Y.		895.20		984.72	74.60	82.06
12 L.H., Daily Totals		$1408.00		$1762.72	$117.33	$146.89
Crew B-10P	Hr.	Daily	Hr.	Daily	Bare Costs	Incl. O&P
1 Equip. Oper. (med.)	$46.55	$372.40	$70.25	$562.00	$42.73	$64.83
.5 Laborer	35.10	140.40	54.00	216.00		
1 Crawler Loader, 3 C.Y.		1106.00		1216.60	92.17	101.38
12 L.H., Daily Totals		$1618.80		$1994.60	$134.90	$166.22

Crew B-10Q

Crew No.	Bare Costs Hr.	Daily	Incl. Subs O&P Hr.	Daily	Cost Per L.H. Bare Costs	Incl. O&P
1 Equip. Oper. (med.)	$46.55	$372.40	$70.25	$562.00	$42.73	$64.83
.5 Laborer	35.10	140.40	54.00	216.00		
1 Crawler Loader, 4 C.Y.		1443.00		1587.30	120.25	132.28
12 L.H., Daily Totals		$1955.80		$2365.30	$162.98	$197.11

Crew B-10R

Crew No.	Bare Costs Hr.	Daily	Incl. Subs O&P Hr.	Daily	Cost Per L.H. Bare Costs	Incl. O&P
1 Equip. Oper. (med.)	$46.55	$372.40	$70.25	$562.00	$42.73	$64.83
.5 Laborer	35.10	140.40	54.00	216.00		
1 F.E. Loader, W.M., 1 C.Y.		268.20		295.02	22.35	24.59
12 L.H., Daily Totals		$781.00		$1073.02	$65.08	$89.42

Crew B-10S

Crew No.	Bare Costs Hr.	Daily	Incl. Subs O&P Hr.	Daily	Cost Per L.H. Bare Costs	Incl. O&P
1 Equip. Oper. (med.)	$46.55	$372.40	$70.25	$562.00	$42.73	$64.83
.5 Laborer	35.10	140.40	54.00	216.00		
1 F.E. Loader, W.M., 1.5 C.Y.		333.80		367.18	27.82	30.60
12 L.H., Daily Totals		$846.60		$1145.18	$70.55	$95.43

Crew B-10T

Crew No.	Bare Costs Hr.	Daily	Incl. Subs O&P Hr.	Daily	Cost Per L.H. Bare Costs	Incl. O&P
1 Equip. Oper. (med.)	$46.55	$372.40	$70.25	$562.00	$42.73	$64.83
.5 Laborer	35.10	140.40	54.00	216.00		
1 F.E. Loader, W.M., 2.5 C.Y.		472.40		519.64	39.37	43.30
12 L.H., Daily Totals		$985.20		$1297.64	$82.10	$108.14

Crew B-10U

Crew No.	Bare Costs Hr.	Daily	Incl. Subs O&P Hr.	Daily	Cost Per L.H. Bare Costs	Incl. O&P
1 Equip. Oper. (med.)	$46.55	$372.40	$70.25	$562.00	$42.73	$64.83
.5 Laborer	35.10	140.40	54.00	216.00		
1 F.E. Loader, W.M., 5.5 C.Y.		960.00		1056.00	80.00	88.00
12 L.H., Daily Totals		$1472.80		$1834.00	$122.73	$152.83

Crew B-10V

Crew No.	Bare Costs Hr.	Daily	Incl. Subs O&P Hr.	Daily	Cost Per L.H. Bare Costs	Incl. O&P
1 Equip. Oper. (med.)	$46.55	$372.40	$70.25	$562.00	$42.73	$64.83
.5 Laborer	35.10	140.40	54.00	216.00		
1 Dozer, 700 H.P.		4235.00		4658.50	352.92	388.21
12 L.H., Daily Totals		$4747.80		$5436.50	$395.65	$453.04

Crew B-10W

Crew No.	Bare Costs Hr.	Daily	Incl. Subs O&P Hr.	Daily	Cost Per L.H. Bare Costs	Incl. O&P
1 Equip. Oper. (med.)	$46.55	$372.40	$70.25	$562.00	$42.73	$64.83
.5 Laborer	35.10	140.40	54.00	216.00		
1 Dozer, 105 H.P.		550.80		605.88	45.90	50.49
12 L.H., Daily Totals		$1063.60		$1383.88	$88.63	$115.32

Crew B-10X

Crew No.	Bare Costs Hr.	Daily	Incl. Subs O&P Hr.	Daily	Cost Per L.H. Bare Costs	Incl. O&P
1 Equip. Oper. (med.)	$46.55	$372.40	$70.25	$562.00	$42.73	$64.83
.5 Laborer	35.10	140.40	54.00	216.00		
1 Dozer, 410 H.P.		2272.00		2499.20	189.33	208.27
12 L.H., Daily Totals		$2784.80		$3277.20	$232.07	$273.10

Crew B-10Y

Crew No.	Bare Costs Hr.	Daily	Incl. Subs O&P Hr.	Daily	Cost Per L.H. Bare Costs	Incl. O&P
1 Equip. Oper. (med.)	$46.55	$372.40	$70.25	$562.00	$42.73	$64.83
.5 Laborer	35.10	140.40	54.00	216.00		
1 Vibr. Roller, Towed, 12 Ton		514.80		566.28	42.90	47.19
12 L.H., Daily Totals		$1027.60		$1344.28	$85.63	$112.02

Crew B-11A

Crew No.	Bare Costs Hr.	Daily	Incl. Subs O&P Hr.	Daily	Cost Per L.H. Bare Costs	Incl. O&P
1 Equipment Oper. (med.)	$46.55	$372.40	$70.25	$562.00	$40.83	$62.13
1 Laborer	35.10	280.80	54.00	432.00		
1 Dozer, 200 H.P.		1243.00		1367.30	77.69	85.46
16 L.H., Daily Totals		$1896.20		$2361.30	$118.51	$147.58

Crew B-11B

Crew No.	Bare Costs Hr.	Daily	Incl. Subs O&P Hr.	Daily	Cost Per L.H. Bare Costs	Incl. O&P
1 Equipment Oper. (light)	$44.75	$358.00	$67.55	$540.40	$39.92	$60.77
1 Laborer	35.10	280.80	54.00	432.00		
1 Air Powered Tamper		26.20		28.82		
1 Air Compressor, 365 cfm		229.00		251.90		
2 -50' Air Hoses, 1.5"		12.60		13.86	16.74	18.41
16 L.H., Daily Totals		$906.60		$1266.98	$56.66	$79.19

Crew B-11C

Crew No.	Bare Costs Hr.	Daily	Incl. Subs O&P Hr.	Daily	Cost Per L.H. Bare Costs	Incl. O&P
1 Equipment Oper. (med.)	$46.55	$372.40	$70.25	$562.00	$40.83	$62.13
1 Laborer	35.10	280.80	54.00	432.00		
1 Backhoe Loader, 48 H.P.		334.00		367.40	20.88	22.96
16 L.H., Daily Totals		$987.20		$1361.40	$61.70	$85.09

Crew B-11J

Crew No.	Bare Costs Hr.	Daily	Incl. Subs O&P Hr.	Daily	Cost Per L.H. Bare Costs	Incl. O&P
1 Equipment Oper. (med.)	$46.55	$372.40	$70.25	$562.00	$40.83	$62.13
1 Laborer	35.10	280.80	54.00	432.00		
1 Grader, 30,000 Lbs.		610.80		671.88		
1 Ripper, beam & 1 shank		81.60		89.76	43.27	47.60
16 L.H., Daily Totals		$1345.60		$1755.64	$84.10	$109.73

Crew B-11K

Crew No.	Bare Costs Hr.	Daily	Incl. Subs O&P Hr.	Daily	Cost Per L.H. Bare Costs	Incl. O&P
1 Equipment Oper. (med.)	$46.55	$372.40	$70.25	$562.00	$40.83	$62.13
1 Laborer	35.10	280.80	54.00	432.00		
1 Trencher, Chain Type, 8' D		2848.00		3132.80	178.00	195.80
16 L.H., Daily Totals		$3501.20		$4126.80	$218.82	$257.93

Crew B-11L

Crew No.	Bare Costs Hr.	Daily	Incl. Subs O&P Hr.	Daily	Cost Per L.H. Bare Costs	Incl. O&P
1 Equipment Oper. (med.)	$46.55	$372.40	$70.25	$562.00	$40.83	$62.13
1 Laborer	35.10	280.80	54.00	432.00		
1 Grader, 30,000 Lbs.		610.80		671.88	38.17	41.99
16 L.H., Daily Totals		$1264.00		$1665.88	$79.00	$104.12

Crew B-11M

Crew No.	Bare Costs Hr.	Daily	Incl. Subs O&P Hr.	Daily	Cost Per L.H. Bare Costs	Incl. O&P
1 Equipment Oper. (med.)	$46.55	$372.40	$70.25	$562.00	$40.83	$62.13
1 Laborer	35.10	280.80	54.00	432.00		
1 Backhoe Loader, 80 H.P.		363.80		400.18	22.74	25.01
16 L.H., Daily Totals		$1017.00		$1394.18	$63.56	$87.14

Crew B-11N

Crew No.	Bare Costs Hr.	Daily	Incl. Subs O&P Hr.	Daily	Cost Per L.H. Bare Costs	Incl. O&P
1 Labor Foreman (outside)	$37.10	$296.80	$57.10	$456.80	$38.00	$57.66
2 Equipment Operators (med.)	46.55	744.80	70.25	1124.00		
6 Truck Drivers (hvy.)	35.30	1694.40	53.55	2570.40		
1 F.E. Loader, W.M., 5.5 C.Y.		960.00		1056.00		
1 Dozer, 410 H.P.		2272.00		2499.20		
6 Dump Trucks, Off Hwy., 50 Ton		9408.00		10348.80	175.56	193.11
72 L.H., Daily Totals		$15376.00		$18055.20	$213.56	$250.77

Crew B-11Q

Crew No.	Bare Costs Hr.	Daily	Incl. Subs O&P Hr.	Daily	Cost Per L.H. Bare Costs	Incl. O&P
1 Equipment Operator (med.)	$46.55	$372.40	$70.25	$562.00	$42.73	$64.83
.5 Laborer	35.10	140.40	54.00	216.00		
1 Dozer, 140 H.P.		817.40		899.14	68.12	74.93
12 L.H., Daily Totals		$1330.20		$1677.14	$110.85	$139.76

Crew B-11R

Crew No.	Bare Costs Hr.	Daily	Incl. Subs O&P Hr.	Daily	Cost Per L.H. Bare Costs	Incl. O&P
1 Equipment Operator (med.)	$46.55	$372.40	$70.25	$562.00	$42.73	$64.83
.5 Laborer	35.10	140.40	54.00	216.00		
1 Dozer, 200 H.P.		1243.00		1367.30	103.58	113.94
12 L.H., Daily Totals		$1755.80		$2145.30	$146.32	$178.78

Crews

Crew No.	Bare Costs Hr.	Daily	Incl. Subs O&P Hr.	Daily	Cost Per Labor-Hour Bare Costs	Incl. O&P
Crew B-11S	Hr.	Daily	Hr.	Daily	Bare Costs	Incl. O&P
1 Equipment Operator (med.)	$46.55	$372.40	$70.25	$562.00	$42.73	$64.83
.5 Laborer	35.10	140.40	54.00	216.00		
1 Dozer, 300 H.P.		1718.00		1889.80		
1 Ripper, beam & 1 shank		81.60		89.76	149.97	164.96
12 L.H., Daily Totals		$2312.40		$2757.56	$192.70	$229.80
Crew B-11T	Hr.	Daily	Hr.	Daily	Bare Costs	Incl. O&P
1 Equipment Operator (med.)	$46.55	$372.40	$70.25	$562.00	$42.73	$64.83
.5 Laborer	35.10	140.40	54.00	216.00		
1 Dozer, 410 H.P.		2272.00		2499.20		
1 Ripper, beam & 2 shanks		91.80		100.98	196.98	216.68
12 L.H., Daily Totals		$2876.60		$3378.18	$239.72	$281.51
Crew B-11U	Hr.	Daily	Hr.	Daily	Bare Costs	Incl. O&P
1 Equipment Operator (med.)	$46.55	$372.40	$70.25	$562.00	$42.73	$64.83
.5 Laborer	35.10	140.40	54.00	216.00		
1 Dozer, 520 H.P.		2937.00		3230.70	244.75	269.23
12 L.H., Daily Totals		$3449.80		$4008.70	$287.48	$334.06
Crew B-11V	Hr.	Daily	Hr.	Daily	Bare Costs	Incl. O&P
3 Laborers	$35.10	$842.40	$54.00	$1296.00	$35.10	$54.00
1 Roller, 2-Drum, W.B., 7.5 H.P.		171.60		188.76	7.15	7.87
24 L.H., Daily Totals		$1014.00		$1484.76	$42.25	$61.87
Crew B-11W	Hr.	Daily	Hr.	Daily	Bare Costs	Incl. O&P
1 Equipment Operator (med.)	$46.55	$372.40	$70.25	$562.00	$36.22	$54.98
1 Common Laborer	35.10	280.80	54.00	432.00		
10 Truck Drivers (hvy.)	35.30	2824.00	53.55	4284.00		
1 Dozer, 200 H.P.		1243.00		1367.30		
1 Vibratory Roller, Towed, 23 Ton		388.60		427.68		
10 Dump Trucks, 8 C.Y., 220 H.P.		3574.00		3931.40	54.23	59.65
96 L.H., Daily Totals		$8683.00		$11004.38	$90.45	$114.63
Crew B-11Y	Hr.	Daily	Hr.	Daily	Bare Costs	Incl. O&P
1 Labor Foreman (Outside)	$37.10	$296.80	$57.10	$456.80	$39.14	$59.76
5 Common Laborers	35.10	1404.00	54.00	2160.00		
3 Equipment Operators (med.)	46.55	1117.20	70.25	1686.00		
1 Dozer, 80 H.P.		442.40		486.64		
2 Roller, 2-Drum, W.B., 7.5 H.P.		343.20		377.52		
4 Vibratory Plates, gas, 21"		163.20		179.52	13.18	14.50
72 L.H., Daily Totals		$3766.80		$5346.48	$52.32	$74.26
Crew B-12A	Hr.	Daily	Hr.	Daily	Bare Costs	Incl. O&P
1 Equip. Oper. (crane)	$47.80	$382.40	$72.15	$577.20	$41.45	$63.08
1 Laborer	35.10	280.80	54.00	432.00		
1 Hyd. Excavator, 1 C.Y.		748.60		823.46	46.79	51.47
16 L.H., Daily Totals		$1411.80		$1832.66	$88.24	$114.54
Crew B-12B	Hr.	Daily	Hr.	Daily	Bare Costs	Incl. O&P
1 Equip. Oper. (crane)	$47.80	$382.40	$72.15	$577.20	$41.45	$63.08
1 Laborer	35.10	280.80	54.00	432.00		
1 Hyd. Excavator, 1.5 C.Y.		960.40		1056.44	60.02	66.03
16 L.H., Daily Totals		$1623.60		$2065.64	$101.47	$129.10
Crew B-12C	Hr.	Daily	Hr.	Daily	Bare Costs	Incl. O&P
1 Equip. Oper. (crane)	$47.80	$382.40	$72.15	$577.20	$41.45	$63.08
1 Laborer	35.10	280.80	54.00	432.00		
1 Hyd. Excavator, 2 C.Y.		1253.00		1378.30	78.31	86.14
16 L.H., Daily Totals		$1916.20		$2387.50	$119.76	$149.22
Crew B-12D	Hr.	Daily	Hr.	Daily	Bare Costs	Incl. O&P
1 Equip. Oper. (crane)	$47.80	$382.40	$72.15	$577.20	$41.45	$63.08
1 Laborer	35.10	280.80	54.00	432.00		
1 Hyd. Excavator, 3.5 C.Y.		2211.00		2432.10	138.19	152.01
16 L.H., Daily Totals		$2874.20		$3441.30	$179.64	$215.08
Crew B-12E	Hr.	Daily	Hr.	Daily	Bare Costs	Incl. O&P
1 Equip. Oper. (crane)	$47.80	$382.40	$72.15	$577.20	$41.45	$63.08
1 Laborer	35.10	280.80	54.00	432.00		
1 Hyd. Excavator, .5 C.Y.		371.40		408.54	23.21	25.53
16 L.H., Daily Totals		$1034.60		$1417.74	$64.66	$88.61
Crew B-12F	Hr.	Daily	Hr.	Daily	Bare Costs	Incl. O&P
1 Equip. Oper. (crane)	$47.80	$382.40	$72.15	$577.20	$41.45	$63.08
1 Laborer	35.10	280.80	54.00	432.00		
1 Hyd. Excavator, .75 C.Y.		647.20		711.92	40.45	44.49
16 L.H., Daily Totals		$1310.40		$1721.12	$81.90	$107.57
Crew B-12G	Hr.	Daily	Hr.	Daily	Bare Costs	Incl. O&P
1 Equip. Oper. (crane)	$47.80	$382.40	$72.15	$577.20	$41.45	$63.08
1 Laborer	35.10	280.80	54.00	432.00		
1 Crawler Crane, 15 Ton		647.50		712.25		
1 Clamshell Bucket, .5 C.Y.		37.40		41.14	42.81	47.09
16 L.H., Daily Totals		$1348.10		$1762.59	$84.26	$110.16
Crew B-12H	Hr.	Daily	Hr.	Daily	Bare Costs	Incl. O&P
1 Equip. Oper. (crane)	$47.80	$382.40	$72.15	$577.20	$41.45	$63.08
1 Laborer	35.10	280.80	54.00	432.00		
1 Crawler Crane, 25 Ton		1110.00		1221.00		
1 Clamshell Bucket, 1 C.Y.		47.00		51.70	72.31	79.54
16 L.H., Daily Totals		$1820.20		$2281.90	$113.76	$142.62
Crew B-12I	Hr.	Daily	Hr.	Daily	Bare Costs	Incl. O&P
1 Equip. Oper. (crane)	$47.80	$382.40	$72.15	$577.20	$41.45	$63.08
1 Laborer	35.10	280.80	54.00	432.00		
1 Crawler Crane, 20 Ton		832.00		915.20		
1 Dragline Bucket, .75 C.Y.		20.40		22.44	53.27	58.60
16 L.H., Daily Totals		$1515.60		$1946.84	$94.72	$121.68
Crew B-12J	Hr.	Daily	Hr.	Daily	Bare Costs	Incl. O&P
1 Equip. Oper. (crane)	$47.80	$382.40	$72.15	$577.20	$41.45	$63.08
1 Laborer	35.10	280.80	54.00	432.00		
1 Gradall, 5/8 C.Y.		844.00		928.40	52.75	58.02
16 L.H., Daily Totals		$1507.20		$1937.60	$94.20	$121.10
Crew B-12K	Hr.	Daily	Hr.	Daily	Bare Costs	Incl. O&P
1 Equip. Oper. (crane)	$47.80	$382.40	$72.15	$577.20	$41.45	$63.08
1 Laborer	35.10	280.80	54.00	432.00		
1 Gradall, 3 Ton, 1 C.Y.		961.20		1057.32	60.08	66.08
16 L.H., Daily Totals		$1624.40		$2066.52	$101.53	$129.16
Crew B-12L	Hr.	Daily	Hr.	Daily	Bare Costs	Incl. O&P
1 Equip. Oper. (crane)	$47.80	$382.40	$72.15	$577.20	$41.45	$63.08
1 Laborer	35.10	280.80	54.00	432.00		
1 Crawler Crane, 15 Ton		647.50		712.25		
1 F.E. Attachment, .5 C.Y.		57.40		63.14	44.06	48.46
16 L.H., Daily Totals		$1368.10		$1784.59	$85.51	$111.54

Crew No.	Bare Costs Hr.	Daily	Incl. Subs O&P Hr.	Daily	Cost Per Labor-Hour Bare Costs	Incl. O&P
Crew B-12M					Bare Costs	Incl. O&P
1 Equip. Oper. (crane)	$47.80	$382.40	$72.15	$577.20	$41.45	$63.08
1 Laborer	35.10	280.80	54.00	432.00		
1 Crawler Crane, 20 Ton		832.00		915.20		
1 F.E. Attachment, .75 C.Y.		62.20		68.42	55.89	61.48
16 L.H., Daily Totals		$1557.40		$1992.82	$97.34	$124.55
Crew B-12N					Bare Costs	Incl. O&P
1 Equip. Oper. (crane)	$47.80	$382.40	$72.15	$577.20	$41.45	$63.08
1 Laborer	35.10	280.80	54.00	432.00		
1 Crawler Crane, 25 Ton		1110.00		1221.00		
1 F.E. Attachment, 1 C.Y.		68.00		74.80	73.63	80.99
16 L.H., Daily Totals		$1841.20		$2305.00	$115.08	$144.06
Crew B-12O					Bare Costs	Incl. O&P
1 Equip. Oper. (crane)	$47.80	$382.40	$72.15	$577.20	$41.45	$63.08
1 Laborer	35.10	280.80	54.00	432.00		
1 Crawler Crane, 40 Ton		1115.00		1226.50		
1 F.E. Attachment, 1.5 C.Y.		76.80		84.48	74.49	81.94
16 L.H., Daily Totals		$1855.00		$2320.18	$115.94	$145.01
Crew B-12P					Bare Costs	Incl. O&P
1 Equip. Oper. (crane)	$47.80	$382.40	$72.15	$577.20	$41.45	$63.08
1 Laborer	35.10	280.80	54.00	432.00		
1 Crawler Crane, 40 Ton		1115.00		1226.50		
1 Dragline Bucket, 1.5 C.Y.		33.20		36.52	71.76	78.94
16 L.H., Daily Totals		$1811.40		$2272.22	$113.21	$142.01
Crew B-12Q					Bare Costs	Incl. O&P
1 Equip. Oper. (crane)	$47.80	$382.40	$72.15	$577.20	$41.45	$63.08
1 Laborer	35.10	280.80	54.00	432.00		
1 Hyd. Excavator, 5/8 C.Y.		551.80		606.98	34.49	37.94
16 L.H., Daily Totals		$1215.00		$1616.18	$75.94	$101.01
Crew B-12S					Bare Costs	Incl. O&P
1 Equip. Oper. (crane)	$47.80	$382.40	$72.15	$577.20	$41.45	$63.08
1 Laborer	35.10	280.80	54.00	432.00		
1 Hyd. Excavator, 2.5 C.Y.		1652.00		1817.20	103.25	113.58
16 L.H., Daily Totals		$2315.20		$2826.40	$144.70	$176.65
Crew B-12T					Bare Costs	Incl. O&P
1 Equip. Oper. (crane)	$47.80	$382.40	$72.15	$577.20	$41.45	$63.08
1 Laborer	35.10	280.80	54.00	432.00		
1 Crawler Crane, 75 Ton		1423.00		1565.30		
1 F.E. Attachment, 3 C.Y.		99.60		109.56	95.16	104.68
16 L.H., Daily Totals		$2185.80		$2684.06	$136.61	$167.75
Crew B-12V					Bare Costs	Incl. O&P
1 Equip. Oper. (crane)	$47.80	$382.40	$72.15	$577.20	$41.45	$63.08
1 Laborer	35.10	280.80	54.00	432.00		
1 Crawler Crane, 75 Ton		1423.00		1565.30		
1 Dragline Bucket, 3 C.Y.		51.80		56.98	92.17	101.39
16 L.H., Daily Totals		$2138.00		$2631.48	$133.63	$164.47
Crew B-12Y					Bare Costs	Incl. O&P
1 Equip. Oper. (crane)	$47.80	$382.40	$72.15	$577.20	$39.33	$60.05
2 Laborers	35.10	561.60	54.00	864.00		
1 Hyd. Excavator, 3.5 C.Y.		2211.00		2432.10	92.13	101.34
24 L.H., Daily Totals		$3155.00		$3873.30	$131.46	$161.39

Crew No.	Bare Costs Hr.	Daily	Incl. Subs O&P Hr.	Daily	Cost Per Labor-Hour Bare Costs	Incl. O&P
Crew B-12Z					Bare Costs	Incl. O&P
1 Equip. Oper. (crane)	$47.80	$382.40	$72.15	$577.20	$39.33	$60.05
2 Laborers	35.10	561.60	54.00	864.00		
1 Hyd. Excavator, 2.5 C.Y.		1652.00		1817.20	68.83	75.72
24 L.H., Daily Totals		$2596.00		$3258.40	$108.17	$135.77
Crew B-13					Bare Costs	Incl. O&P
1 Labor Foreman (outside)	$37.10	$296.80	$57.10	$456.80	$38.08	$58.21
4 Laborers	35.10	1123.20	54.00	1728.00		
1 Equip. Oper. (crane)	47.80	382.40	72.15	577.20		
1 Equip. Oper. Oiler	41.25	330.00	62.25	498.00		
1 Hyd. Crane, 25 Ton		697.20		766.92	12.45	13.70
56 L.H., Daily Totals		$2829.60		$4026.92	$50.53	$71.91
Crew B-13A					Bare Costs	Incl. O&P
1 Labor Foreman (outside)	$37.10	$296.80	$57.10	$456.80	$38.71	$58.96
2 Laborers	35.10	561.60	54.00	864.00		
2 Equipment Operators (med.)	46.55	744.80	70.25	1124.00		
2 Truck Drivers (heavy)	35.30	564.80	53.55	856.80		
1 Crawler Crane, 75 Ton		1423.00		1565.30		
1 Crawler Loader, 4 C.Y.		1443.00		1587.30		
2 Dump Trucks, 8 C.Y., 220 H.P.		714.80		786.28	63.94	70.34
56 L.H., Daily Totals		$5748.80		$7240.48	$102.66	$129.29
Crew B-13B					Bare Costs	Incl. O&P
1 Labor Foreman (outside)	$37.10	$296.80	$57.10	$456.80	$38.08	$58.21
4 Laborers	35.10	1123.20	54.00	1728.00		
1 Equip. Oper. (crane)	47.80	382.40	72.15	577.20		
1 Equip. Oper. Oiler	41.25	330.00	62.25	498.00		
1 Hyd. Crane, 55 Ton		1100.00		1210.00	19.64	21.61
56 L.H., Daily Totals		$3232.40		$4470.00	$57.72	$79.82
Crew B-13C					Bare Costs	Incl. O&P
1 Labor Foreman (outside)	$37.10	$296.80	$57.10	$456.80	$38.08	$58.21
4 Laborers	35.10	1123.20	54.00	1728.00		
1 Equip. Oper. (crane)	47.80	382.40	72.15	577.20		
1 Equip. Oper. Oiler	41.25	330.00	62.25	498.00		
1 Crawler Crane, 100 Ton		1628.00		1790.80	29.07	31.98
56 L.H., Daily Totals		$3760.40		$5050.80	$67.15	$90.19
Crew B-13D					Bare Costs	Incl. O&P
1 Laborer	$35.10	$280.80	$54.00	$432.00	$41.45	$63.08
1 Equip. Oper. (crane)	47.80	382.40	72.15	577.20		
1 Hyd. Excavator, 1 C.Y.		748.60		823.46		
1 Trench Box		113.40		124.74	53.88	59.26
16 L.H., Daily Totals		$1525.20		$1957.40	$95.33	$122.34
Crew B-13E					Bare Costs	Incl. O&P
1 Laborer	$35.10	$280.80	$54.00	$432.00	$41.45	$63.08
1 Equip. Oper. (crane)	47.80	382.40	72.15	577.20		
1 Hyd. Excavator, 1.5 C.Y.		960.40		1056.44		
1 Trench Box		113.40		124.74	67.11	73.82
16 L.H., Daily Totals		$1737.00		$2190.38	$108.56	$136.90
Crew B-13F					Bare Costs	Incl. O&P
1 Laborer	$35.10	$280.80	$54.00	$432.00	$41.45	$63.08
1 Equip. Oper. (crane)	47.80	382.40	72.15	577.20		
1 Hyd. Excavator, 3.5 C.Y.		2211.00		2432.10		
1 Trench Box		113.40		124.74	145.28	159.80
16 L.H., Daily Totals		$2987.60		$3566.04	$186.72	$222.88

Crew B-13G

Crew No.	Bare Costs Hr.	Daily	Incl. Subs O&P Hr.	Daily	Cost Per LH Bare Costs	Incl. O&P
1 Laborer	$35.10	$280.80	$54.00	$432.00	$41.45	$63.08
1 Equip. Oper. (crane)	47.80	382.40	72.15	577.20		
1 Hyd. Excavator, .75 C.Y.		647.20		711.92		
1 Trench Box		113.40		124.74	47.54	52.29
16 L.H., Daily Totals		$1423.80		$1845.86	$88.99	$115.37

Crew B-13H

Crew No.	Bare Costs Hr.	Daily	Incl. Subs O&P Hr.	Daily	Cost Per LH Bare Costs	Incl. O&P
1 Laborer	$35.10	$280.80	$54.00	$432.00	$41.45	$63.08
1 Equip. Oper. (crane)	47.80	382.40	72.15	577.20		
1 Gradall, 5/8 C.Y.		844.00		928.40		
1 Trench Box		113.40		124.74	59.84	65.82
16 L.H., Daily Totals		$1620.60		$2062.34	$101.29	$128.90

Crew B-13I

Crew No.	Bare Costs Hr.	Daily	Incl. Subs O&P Hr.	Daily	Cost Per LH Bare Costs	Incl. O&P
1 Laborer	$35.10	$280.80	$54.00	$432.00	$41.45	$63.08
1 Equip. Oper. (crane)	47.80	382.40	72.15	577.20		
1 Gradall, 3 Ton, 1 C.Y.		961.20		1057.32		
1 Trench Box		113.40		124.74	67.16	73.88
16 L.H., Daily Totals		$1737.80		$2191.26	$108.61	$136.95

Crew B-13J

Crew No.	Bare Costs Hr.	Daily	Incl. Subs O&P Hr.	Daily	Cost Per LH Bare Costs	Incl. O&P
1 Laborer	$35.10	$280.80	$54.00	$432.00	$41.45	$63.08
1 Equip. Oper. (crane)	47.80	382.40	72.15	577.20		
1 Hyd. Excavator, 2.5 C.Y.		1652.00		1817.20		
1 Trench Box		113.40		124.74	110.34	121.37
16 L.H., Daily Totals		$2428.60		$2951.14	$151.79	$184.45

Crew B-14

Crew No.	Bare Costs Hr.	Daily	Incl. Subs O&P Hr.	Daily	Cost Per LH Bare Costs	Incl. O&P
1 Labor Foreman (outside)	$37.10	$296.80	$57.10	$456.80	$37.04	$56.77
4 Laborers	35.10	1123.20	54.00	1728.00		
1 Equip. Oper. (light)	44.75	358.00	67.55	540.40		
1 Backhoe Loader, 48 H.P.		334.00		367.40	6.96	7.65
48 L.H., Daily Totals		$2112.00		$3092.60	$44.00	$64.43

Crew B-14A

Crew No.	Bare Costs Hr.	Daily	Incl. Subs O&P Hr.	Daily	Cost Per LH Bare Costs	Incl. O&P
1 Equip. Oper. (crane)	$47.80	$382.40	$72.15	$577.20	$43.57	$66.10
.5 Laborer	35.10	140.40	54.00	216.00		
1 Hyd. Excavator, 4.5 C.Y.		2748.00		3022.80	229.00	251.90
12 L.H., Daily Totals		$3270.80		$3816.00	$272.57	$318.00

Crew B-14B

Crew No.	Bare Costs Hr.	Daily	Incl. Subs O&P Hr.	Daily	Cost Per LH Bare Costs	Incl. O&P
1 Equip. Oper. (crane)	$47.80	$382.40	$72.15	$577.20	$43.57	$66.10
.5 Laborer	35.10	140.40	54.00	216.00		
1 Hyd. Excavator, 6 C.Y.		3229.00		3551.90	269.08	295.99
12 L.H., Daily Totals		$3751.80		$4345.10	$312.65	$362.09

Crew B-14C

Crew No.	Bare Costs Hr.	Daily	Incl. Subs O&P Hr.	Daily	Cost Per LH Bare Costs	Incl. O&P
1 Equip. Oper. (crane)	$47.80	$382.40	$72.15	$577.20	$43.57	$66.10
.5 Laborer	35.10	140.40	54.00	216.00		
1 Hyd. Excavator, 7 C.Y.		3304.00		3634.40	275.33	302.87
12 L.H., Daily Totals		$3826.80		$4427.60	$318.90	$368.97

Crew B-14F

Crew No.	Bare Costs Hr.	Daily	Incl. Subs O&P Hr.	Daily	Cost Per LH Bare Costs	Incl. O&P
1 Equip. Oper. (crane)	$47.80	$382.40	$72.15	$577.20	$43.57	$66.10
.5 Laborer	35.10	140.40	54.00	216.00		
1 Hyd. Shovel, 7 C.Y.		3259.00		3584.90	271.58	298.74
12 L.H., Daily Totals		$3781.80		$4378.10	$315.15	$364.84

Crew B-14G

Crew No.	Bare Costs Hr.	Daily	Incl. Subs O&P Hr.	Daily	Cost Per LH Bare Costs	Incl. O&P
1 Equip. Oper. (crane)	$47.80	$382.40	$72.15	$577.20	$43.57	$66.10
.5 Laborer	35.10	140.40	54.00	216.00		
1 Hyd. Shovel, 12 C.Y.		4249.00		4673.90	354.08	389.49
12 L.H., Daily Totals		$4771.80		$5467.10	$397.65	$455.59

Crew B-14J

Crew No.	Bare Costs Hr.	Daily	Incl. Subs O&P Hr.	Daily	Cost Per LH Bare Costs	Incl. O&P
1 Equip. Oper. (med.)	$46.55	$372.40	$70.25	$562.00	$42.73	$64.83
.5 Laborer	35.10	140.40	54.00	216.00		
1 F.E. Loader, 8 C.Y.		1856.00		2041.60	154.67	170.13
12 L.H., Daily Totals		$2368.80		$2819.60	$197.40	$234.97

Crew B-14K

Crew No.	Bare Costs Hr.	Daily	Incl. Subs O&P Hr.	Daily	Cost Per LH Bare Costs	Incl. O&P
1 Equip. Oper. (med.)	$46.55	$372.40	$70.25	$562.00	$42.73	$64.83
.5 Laborer	35.10	140.40	54.00	216.00		
1 F.E. Loader, 10 C.Y.		2556.00		2811.60	213.00	234.30
12 L.H., Daily Totals		$3068.80		$3589.60	$255.73	$299.13

Crew B-15

Crew No.	Bare Costs Hr.	Daily	Incl. Subs O&P Hr.	Daily	Cost Per LH Bare Costs	Incl. O&P
1 Equipment Oper. (med.)	$46.55	$372.40	$70.25	$562.00	$38.49	$58.39
.5 Laborer	35.10	140.40	54.00	216.00		
2 Truck Drivers (heavy)	35.30	564.80	53.55	856.80		
2 Dump Trucks, 12 C.Y., 400 H.P.		1169.20		1286.12		
1 Dozer, 200 H.P.		1243.00		1367.30	86.15	94.77
28 L.H., Daily Totals		$3489.80		$4288.22	$124.64	$153.15

Crew B-16

Crew No.	Bare Costs Hr.	Daily	Incl. Subs O&P Hr.	Daily	Cost Per LH Bare Costs	Incl. O&P
1 Labor Foreman (outside)	$37.10	$296.80	$57.10	$456.80	$35.65	$54.66
2 Laborers	35.10	561.60	54.00	864.00		
1 Truck Driver (heavy)	35.30	282.40	53.55	428.40		
1 Dump Truck, 12 C.Y., 400 H.P.		584.60		643.06	18.27	20.10
32 L.H., Daily Totals		$1725.40		$2392.26	$53.92	$74.76

Crew B-17

Crew No.	Bare Costs Hr.	Daily	Incl. Subs O&P Hr.	Daily	Cost Per LH Bare Costs	Incl. O&P
2 Laborers	$35.10	$561.60	$54.00	$864.00	$37.56	$57.27
1 Equip. Oper. (light)	44.75	358.00	67.55	540.40		
1 Truck Driver (heavy)	35.30	282.40	53.55	428.40		
1 Backhoe Loader, 48 H.P.		334.00		367.40		
1 Dump Truck, 8 C.Y., 220 H.P.		357.40		393.14	21.61	23.77
32 L.H., Daily Totals		$1893.40		$2593.34	$59.17	$81.04

Crew B-17A

Crew No.	Bare Costs Hr.	Daily	Incl. Subs O&P Hr.	Daily	Cost Per LH Bare Costs	Incl. O&P
2 Labor Foremen (outside)	$37.10	$593.60	$57.10	$913.60	$37.76	$58.17
6 Laborers	35.10	1684.80	54.00	2592.00		
1 Skilled Worker Foreman (out)	47.40	379.20	73.30	586.40		
1 Skilled Worker	45.40	363.20	70.20	561.60		
80 L.H., Daily Totals		$3020.80		$4653.60	$37.76	$58.17

Crew B-17B

Crew No.	Bare Costs Hr.	Daily	Incl. Subs O&P Hr.	Daily	Cost Per LH Bare Costs	Incl. O&P
2 Laborers	$35.10	$561.60	$54.00	$864.00	$37.56	$57.27
1 Equip. Oper. (light)	44.75	358.00	67.55	540.40		
1 Truck Driver (heavy)	35.30	282.40	53.55	428.40		
1 Backhoe Loader, 48 H.P.		334.00		367.40		
1 Dump Truck, 12 C.Y., 400 H.P.		584.60		643.06	28.71	31.58
32 L.H., Daily Totals		$2120.60		$2843.26	$66.27	$88.85

Crew B-18

Crew No.	Bare Costs Hr.	Daily	Incl. Subs O&P Hr.	Daily	Cost Per LH Bare Costs	Incl. O&P
1 Labor Foreman (outside)	$37.10	$296.80	$57.10	$456.80	$35.77	$55.03
2 Laborers	35.10	561.60	54.00	864.00		
1 Vibratory Plate, gas, 21"		40.80		44.88	1.70	1.87
24 L.H., Daily Totals		$899.20		$1365.68	$37.47	$56.90

Crew No.	Bare Costs		Incl. Subs O&P		Cost Per Labor-Hour	

Crew B-19	Hr.	Daily	Hr.	Daily	Bare Costs	Incl. O&P
1 Pile Driver Foreman (outside)	$44.80	$358.40	$72.00	$576.00	$44.11	$69.22
4 Pile Drivers	42.80	1369.60	68.80	2201.60		
2 Equip. Oper. (crane)	47.80	764.80	72.15	1154.40		
1 Equip. Oper. Oiler	41.25	330.00	62.25	498.00		
1 Crawler Crane, 40 Ton		1115.00		1226.50		
1 Lead, 90' high		124.00		136.40		
1 Hammer, Diesel, 22k ft-lb		592.80		652.08	28.62	31.48
64 L.H., Daily Totals		$4654.60		$6444.98	$72.73	$100.70

Crew B-19A	Hr.	Daily	Hr.	Daily	Bare Costs	Incl. O&P
1 Pile Driver Foreman (outside)	$44.80	$358.40	$72.00	$576.00	$44.11	$69.22
4 Pile Drivers	42.80	1369.60	68.80	2201.60		
2 Equip. Oper. (crane)	47.80	764.80	72.15	1154.40		
1 Equip. Oper. Oiler	41.25	330.00	62.25	498.00		
1 Crawler Crane, 75 Ton		1423.00		1565.30		
1 Lead, 90' high		124.00		136.40		
1 Hammer, Diesel, 41k ft-lb		686.80		755.48	34.90	38.39
64 L.H., Daily Totals		$5056.60		$6887.18	$79.01	$107.61

Crew B-19B	Hr.	Daily	Hr.	Daily	Bare Costs	Incl. O&P
1 Pile Driver Foreman (outside)	$44.80	$358.40	$72.00	$576.00	$44.11	$69.22
4 Pile Drivers	42.80	1369.60	68.80	2201.60		
2 Equip. Oper. (crane)	47.80	764.80	72.15	1154.40		
1 Equip. Oper. Oiler	41.25	330.00	62.25	498.00		
1 Crawler Crane, 40 Ton		1115.00		1226.50		
1 Lead, 90' high		124.00		136.40		
1 Hammer, Diesel, 22k ft-lb		592.80		652.08		
1 Barge, 400 Ton		779.20		857.12	40.80	44.88
64 L.H., Daily Totals		$5433.80		$7302.10	$84.90	$114.10

Crew B-19C	Hr.	Daily	Hr.	Daily	Bare Costs	Incl. O&P
1 Pile Driver Foreman (outside)	$44.80	$358.40	$72.00	$576.00	$44.11	$69.22
4 Pile Drivers	42.80	1369.60	68.80	2201.60		
2 Equip. Oper. (crane)	47.80	764.80	72.15	1154.40		
1 Equip. Oper. Oiler	41.25	330.00	62.25	498.00		
1 Crawler Crane, 75 Ton		1423.00		1565.30		
1 Lead, 90' high		124.00		136.40		
1 Hammer, Diesel, 41k ft-lb		686.80		755.48		
1 Barge, 400 Ton		779.20		857.12	47.08	51.79
64 L.H., Daily Totals		$5835.80		$7744.30	$91.18	$121.00

Crew B-20	Hr.	Daily	Hr.	Daily	Bare Costs	Incl. O&P
1 Labor Foreman (outside)	$37.10	$296.80	$57.10	$456.80	$39.20	$60.43
1 Skilled Worker	45.40	363.20	70.20	561.60		
1 Laborer	35.10	280.80	54.00	432.00		
24 L.H., Daily Totals		$940.80		$1450.40	$39.20	$60.43

Crew B-20A	Hr.	Daily	Hr.	Daily	Bare Costs	Incl. O&P
1 Labor Foreman (outside)	$37.10	$296.80	$57.10	$456.80	$42.63	$64.84
1 Laborer	35.10	280.80	54.00	432.00		
1 Plumber	54.60	436.80	82.35	658.80		
1 Plumber Apprentice	43.70	349.60	65.90	527.20		
32 L.H., Daily Totals		$1364.00		$2074.80	$42.63	$64.84

Crew B-21	Hr.	Daily	Hr.	Daily	Bare Costs	Incl. O&P
1 Labor Foreman (outside)	$37.10	$296.80	$57.10	$456.80	$40.43	$62.11
1 Skilled Worker	45.40	363.20	70.20	561.60		
1 Laborer	35.10	280.80	54.00	432.00		
.5 Equip. Oper. (crane)	47.80	191.20	72.15	288.60		
.5 S.P. Crane, 4x4, 5 Ton		132.30		145.53	4.72	5.20
28 L.H., Daily Totals		$1264.30		$1884.53	$45.15	$67.30

Crew B-21A	Hr.	Daily	Hr.	Daily	Bare Costs	Incl. O&P
1 Labor Foreman (outside)	$37.10	$296.80	$57.10	$456.80	$43.66	$66.30
1 Laborer	35.10	280.80	54.00	432.00		
1 Plumber	54.60	436.80	82.35	658.80		
1 Plumber Apprentice	43.70	349.60	65.90	527.20		
1 Equip. Oper. (crane)	47.80	382.40	72.15	577.20		
1 S.P. Crane, 4x4, 12 Ton		448.20		493.02	11.21	12.33
40 L.H., Daily Totals		$2194.60		$3145.02	$54.87	$78.63

Crew B-21B	Hr.	Daily	Hr.	Daily	Bare Costs	Incl. O&P
1 Labor Foreman (outside)	$37.10	$296.80	$57.10	$456.80	$38.04	$58.25
3 Laborers	35.10	842.40	54.00	1296.00		
1 Equip. Oper. (crane)	47.80	382.40	72.15	577.20		
1 Hyd. Crane, 12 Ton		612.00		673.20	15.30	16.83
40 L.H., Daily Totals		$2133.60		$3003.20	$53.34	$75.08

Crew B-21C	Hr.	Daily	Hr.	Daily	Bare Costs	Incl. O&P
1 Labor Foreman (outside)	$37.10	$296.80	$57.10	$456.80	$38.08	$58.21
4 Laborers	35.10	1123.20	54.00	1728.00		
1 Equip. Oper. (crane)	47.80	382.40	72.15	577.20		
1 Equip. Oper. Oiler	41.25	330.00	62.25	498.00		
2 Cutting Torches		21.60		23.76		
2 Sets of Gases		172.80		190.08		
1 Lattice Boom Crane, 90 Ton		1497.00		1646.70	30.20	33.22
56 L.H., Daily Totals		$3823.80		$5120.54	$68.28	$91.44

Crew B-22	Hr.	Daily	Hr.	Daily	Bare Costs	Incl. O&P
1 Labor Foreman (outside)	$37.10	$296.80	$57.10	$456.80	$40.92	$62.78
1 Skilled Worker	45.40	363.20	70.20	561.60		
1 Laborer	35.10	280.80	54.00	432.00		
.75 Equip. Oper. (crane)	47.80	286.80	72.15	432.90		
.75 S.P. Crane, 4x4, 5 Ton		198.45		218.29	6.62	7.28
30 L.H., Daily Totals		$1426.05		$2101.59	$47.53	$70.05

Crew B-22A	Hr.	Daily	Hr.	Daily	Bare Costs	Incl. O&P
1 Labor Foreman (outside)	$37.10	$296.80	$57.10	$456.80	$40.10	$61.49
1 Skilled Worker	45.40	363.20	70.20	561.60		
2 Laborers	35.10	561.60	54.00	864.00		
1 Equipment Oper. (crane)	47.80	382.40	72.15	577.20		
1 S.P. Crane, 4x4, 5 Ton		264.60		291.06		
1 Butt Fusion Mach., 4"-12" diam.		376.45		414.10	16.03	17.63
40 L.H., Daily Totals		$2245.05		$3164.76	$56.13	$79.12

Crew B-22B	Hr.	Daily	Hr.	Daily	Bare Costs	Incl. O&P
1 Labor Foreman (outside)	$37.10	$296.80	$57.10	$456.80	$40.10	$61.49
1 Skilled Worker	45.40	363.20	70.20	561.60		
2 Laborers	35.10	561.60	54.00	864.00		
1 Equip. Oper. (crane)	47.80	382.40	72.15	577.20		
1 S.P. Crane, 4x4, 5 Ton		264.60		291.06		
1 Butt Fusion Mach., 8"-24" diam.		806.15		886.76	26.77	29.45
40 L.H., Daily Totals		$2674.75		$3637.43	$66.87	$90.94

Crew B-22C	Hr.	Daily	Hr.	Daily	Bare Costs	Incl. O&P
1 Skilled Worker	$45.40	$363.20	$70.20	$561.60	$40.25	$62.10
1 Laborer	35.10	280.80	54.00	432.00		
1 Butt Fusion Mach., 2"-8" diam.		120.50		132.55	7.53	8.28
16 L.H., Daily Totals		$764.50		$1126.15	$47.78	$70.38

Crew No.	Bare Costs		Incl. Subs O&P		Cost Per Labor-Hour	
Crew B-23	**Hr.**	**Daily**	**Hr.**	**Daily**	**Bare Costs**	**Incl. O&P**
1 Labor Foreman (outside)	$37.10	$296.80	$57.10	$456.80	$35.50	$54.62
4 Laborers	35.10	1123.20	54.00	1728.00		
1 Drill Rig, Truck-Mounted		2441.00		2685.10		
1 Flatbed Truck, Gas, 3 Ton		264.40		290.84	67.64	74.40
40 L.H., Daily Totals		$4125.40		$5160.74	$103.14	$129.02
Crew B-23A	**Hr.**	**Daily**	**Hr.**	**Daily**	**Bare Costs**	**Incl. O&P**
1 Labor Foreman (outside)	$37.10	$296.80	$57.10	$456.80	$39.58	$60.45
1 Laborer	35.10	280.80	54.00	432.00		
1 Equip. Operator (med.)	46.55	372.40	70.25	562.00		
1 Drill Rig, Truck-Mounted		2441.00		2685.10		
1 Pickup Truck, 3/4 Ton		127.00		139.70	107.00	117.70
24 L.H., Daily Totals		$3518.00		$4275.60	$146.58	$178.15
Crew B-23B	**Hr.**	**Daily**	**Hr.**	**Daily**	**Bare Costs**	**Incl. O&P**
1 Labor Foreman (outside)	$37.10	$296.80	$57.10	$456.80	$39.58	$60.45
1 Laborer	35.10	280.80	54.00	432.00		
1 Equip. Operator (med.)	46.55	372.40	70.25	562.00		
1 Drill Rig, Truck-Mounted		2441.00		2685.10		
1 Pickup Truck, 3/4 Ton		127.00		139.70		
1 Centr. Water Pump, 6"		308.40		339.24	119.85	131.84
24 L.H., Daily Totals		$3826.40		$4614.84	$159.43	$192.29
Crew B-24	**Hr.**	**Daily**	**Hr.**	**Daily**	**Bare Costs**	**Incl. O&P**
1 Cement Finisher	$42.35	$338.80	$62.30	$498.40	$40.52	$61.38
1 Laborer	35.10	280.80	54.00	432.00		
1 Carpenter	44.10	352.80	67.85	542.80		
24 L.H., Daily Totals		$972.40		$1473.20	$40.52	$61.38
Crew B-25	**Hr.**	**Daily**	**Hr.**	**Daily**	**Bare Costs**	**Incl. O&P**
1 Labor Foreman (outside)	$37.10	$296.80	$57.10	$456.80	$38.40	$58.71
7 Laborers	35.10	1965.60	54.00	3024.00		
3 Equip. Oper. (med.)	46.55	1117.20	70.25	1686.00		
1 Asphalt Paver, 130 H.P.		1996.00		2195.60		
1 Tandem Roller, 10 Ton		227.60		250.36		
1 Roller, Pneum. Whl, 12 Ton		313.20		344.52	28.83	31.71
88 L.H., Daily Totals		$5916.40		$7957.28	$67.23	$90.42
Crew B-25B	**Hr.**	**Daily**	**Hr.**	**Daily**	**Bare Costs**	**Incl. O&P**
1 Labor Foreman (outside)	$37.10	$296.80	$57.10	$456.80	$39.08	$59.67
7 Laborers	35.10	1965.60	54.00	3024.00		
4 Equip. Oper. (med.)	46.55	1489.60	70.25	2248.00		
1 Asphalt Paver, 130 H.P.		1996.00		2195.60		
2 Tandem Rollers, 10 Ton		455.20		500.72		
1 Roller, Pneum. Whl, 12 Ton		313.20		344.52	28.80	31.68
96 L.H., Daily Totals		$6516.40		$8769.64	$67.88	$91.35
Crew B-25C	**Hr.**	**Daily**	**Hr.**	**Daily**	**Bare Costs**	**Incl. O&P**
1 Labor Foreman (outside)	$37.10	$296.80	$57.10	$456.80	$39.25	$59.93
3 Laborers	35.10	842.40	54.00	1296.00		
2 Equip. Oper. (med.)	46.55	744.80	70.25	1124.00		
1 Asphalt Paver, 130 H.P.		1996.00		2195.60		
1 Tandem Roller, 10 Ton		227.60		250.36	46.33	50.96
48 L.H., Daily Totals		$4107.60		$5322.76	$85.58	$110.89

Crew No.	Bare Costs		Incl. Subs O&P		Cost Per Labor-Hour	
Crew B-26	**Hr.**	**Daily**	**Hr.**	**Daily**	**Bare Costs**	**Incl. O&P**
1 Labor Foreman (outside)	$37.10	$296.80	$57.10	$456.80	$39.30	$60.21
6 Laborers	35.10	1684.80	54.00	2592.00		
2 Equip. Oper. (med.)	46.55	744.80	70.25	1124.00		
1 Rodman (reinf.)	49.10	392.80	78.40	627.20		
1 Cement Finisher	42.35	338.80	62.30	498.40		
1 Grader, 30,000 Lbs.		610.80		671.88		
1 Paving Mach. & Equip.		2531.00		2784.10	35.70	39.27
88 L.H., Daily Totals		$6599.80		$8754.38	$75.00	$99.48
Crew B-26A	**Hr.**	**Daily**	**Hr.**	**Daily**	**Bare Costs**	**Incl. O&P**
1 Labor Foreman (outside)	$37.10	$296.80	$57.10	$456.80	$39.30	$60.21
6 Laborers	35.10	1684.80	54.00	2592.00		
2 Equip. Oper. (med.)	46.55	744.80	70.25	1124.00		
1 Rodman (reinf.)	49.10	392.80	78.40	627.20		
1 Cement Finisher	42.35	338.80	62.30	498.40		
1 Grader, 30,000 Lbs.		610.80		671.88		
1 Paving Mach. & Equip.		2531.00		2784.10		
1 Concrete Saw		148.00		162.80	37.38	41.12
88 L.H., Daily Totals		$6747.80		$8917.18	$76.68	$101.33
Crew B-26B	**Hr.**	**Daily**	**Hr.**	**Daily**	**Bare Costs**	**Incl. O&P**
1 Labor Foreman (outside)	$37.10	$296.80	$57.10	$456.80	$39.90	$61.05
6 Laborers	35.10	1684.80	54.00	2592.00		
3 Equip. Oper. (med.)	46.55	1117.20	70.25	1686.00		
1 Rodman (reinf.)	49.10	392.80	78.40	627.20		
1 Cement Finisher	42.35	338.80	62.30	498.40		
1 Grader, 30,000 Lbs.		610.80		671.88		
1 Paving Mach. & Equip.		2531.00		2784.10		
1 Concrete Pump, 110' Boom		941.40		1035.54	42.53	46.79
96 L.H., Daily Totals		$7913.60		$10351.92	$82.43	$107.83
Crew B-27	**Hr.**	**Daily**	**Hr.**	**Daily**	**Bare Costs**	**Incl. O&P**
1 Labor Foreman (outside)	$37.10	$296.80	$57.10	$456.80	$35.60	$54.77
3 Laborers	35.10	842.40	54.00	1296.00		
1 Berm Machine		280.00		308.00	8.75	9.63
32 L.H., Daily Totals		$1419.20		$2060.80	$44.35	$64.40
Crew B-28	**Hr.**	**Daily**	**Hr.**	**Daily**	**Bare Costs**	**Incl. O&P**
2 Carpenters	$44.10	$705.60	$67.85	$1085.60	$41.10	$63.23
1 Laborer	35.10	280.80	54.00	432.00		
24 L.H., Daily Totals		$986.40		$1517.60	$41.10	$63.23
Crew B-29	**Hr.**	**Daily**	**Hr.**	**Daily**	**Bare Costs**	**Incl. O&P**
1 Labor Foreman (outside)	$37.10	$296.80	$57.10	$456.80	$38.08	$58.21
4 Laborers	35.10	1123.20	54.00	1728.00		
1 Equip. Oper. (crane)	47.80	382.40	72.15	577.20		
1 Equip. Oper. Oiler	41.25	330.00	62.25	498.00		
1 Gradall, 5/8 C.Y.		844.00		928.40	15.07	16.58
56 L.H., Daily Totals		$2976.40		$4188.40	$53.15	$74.79
Crew B-30	**Hr.**	**Daily**	**Hr.**	**Daily**	**Bare Costs**	**Incl. O&P**
1 Equip. Oper. (med.)	$46.55	$372.40	$70.25	$562.00	$39.05	$59.12
2 Truck Drivers (heavy)	35.30	564.80	53.55	856.80		
1 Hyd. Excavator, 1.5 C.Y.		960.40		1056.44		
2 Dump Trucks, 12 C.Y., 400 H.P.		1169.20		1286.12	88.73	97.61
24 L.H., Daily Totals		$3066.80		$3761.36	$127.78	$156.72

Crew No.	Bare Costs		Incl. Subs O&P		Cost Per Labor-Hour	
Crew B-31	Hr.	Daily	Hr.	Daily	Bare Costs	Incl. O&P
1 Labor Foreman (outside)	$37.10	$296.80	$57.10	$456.80	$37.30	$57.39
3 Laborers	35.10	842.40	54.00	1296.00		
1 Carpenter	44.10	352.80	67.85	542.80		
1 Air Compressor, 250 cfm		177.00		194.70		
1 Sheeting Driver		5.75		6.33		
2 -50' Air Hoses, 1.5"		12.60		13.86	4.88	5.37
40 L.H., Daily Totals		$1687.35		$2510.49	$42.18	$62.76
Crew B-32	Hr.	Daily	Hr.	Daily	Bare Costs	Incl. O&P
1 Laborer	$35.10	$280.80	$54.00	$432.00	$43.69	$66.19
3 Equip. Oper. (med.)	46.55	1117.20	70.25	1686.00		
1 Grader, 30,000 Lbs.		610.80		671.88		
1 Tandem Roller, 10 Ton		227.60		250.36		
1 Dozer, 200 H.P.		1243.00		1367.30	65.04	71.55
32 L.H., Daily Totals		$3479.40		$4407.54	$108.73	$137.74
Crew B-32A	Hr.	Daily	Hr.	Daily	Bare Costs	Incl. O&P
1 Laborer	$35.10	$280.80	$54.00	$432.00	$42.73	$64.83
2 Equip. Oper. (med.)	46.55	744.80	70.25	1124.00		
1 Grader, 30,000 Lbs.		610.80		671.88		
1 Roller, Vibratory, 25 Ton		640.20		704.22	52.13	57.34
24 L.H., Daily Totals		$2276.60		$2932.10	$94.86	$122.17
Crew B-32B	Hr.	Daily	Hr.	Daily	Bare Costs	Incl. O&P
1 Laborer	$35.10	$280.80	$54.00	$432.00	$42.73	$64.83
2 Equip. Oper. (med.)	46.55	744.80	70.25	1124.00		
1 Dozer, 200 H.P.		1243.00		1367.30		
1 Roller, Vibratory, 25 Ton		640.20		704.22	78.47	86.31
24 L.H., Daily Totals		$2908.80		$3627.52	$121.20	$151.15
Crew B-32C	Hr.	Daily	Hr.	Daily	Bare Costs	Incl. O&P
1 Labor Foreman (outside)	$37.10	$296.80	$57.10	$456.80	$41.16	$62.64
2 Laborers	35.10	561.60	54.00	864.00		
3 Equip. Oper. (med.)	46.55	1117.20	70.25	1686.00		
1 Grader, 30,000 Lbs.		610.80		671.88		
1 Tandem Roller, 10 Ton		227.60		250.36		
1 Dozer, 200 H.P.		1243.00		1367.30	43.36	47.70
48 L.H., Daily Totals		$4057.00		$5296.34	$84.52	$110.34
Crew B-33A	Hr.	Daily	Hr.	Daily	Bare Costs	Incl. O&P
1 Equip. Oper. (med.)	$46.55	$372.40	$70.25	$562.00	$43.28	$65.61
.5 Laborer	35.10	140.40	54.00	216.00		
.25 Equip. Oper. (med.)	46.55	93.10	70.25	140.50		
1 Scraper, Towed, 7 C.Y.		108.20		119.02		
1.25 Dozers, 300 H.P.		2147.50		2362.25	161.12	177.23
14 L.H., Daily Totals		$2861.60		$3399.77	$204.40	$242.84
Crew B-33B	Hr.	Daily	Hr.	Daily	Bare Costs	Incl. O&P
1 Equip. Oper. (med.)	$46.55	$372.40	$70.25	$562.00	$43.28	$65.61
.5 Laborer	35.10	140.40	54.00	216.00		
.25 Equip. Oper. (med.)	46.55	93.10	70.25	140.50		
1 Scraper, Towed, 10 C.Y.		139.20		153.12		
1.25 Dozers, 300 H.P.		2147.50		2362.25	163.34	179.67
14 L.H., Daily Totals		$2892.60		$3433.87	$206.61	$245.28

Crew No.	Bare Costs		Incl. Subs O&P		Cost Per Labor-Hour	
Crew B-33C	Hr.	Daily	Hr.	Daily	Bare Costs	Incl. O&P
1 Equip. Oper. (med.)	$46.55	$372.40	$70.25	$562.00	$43.28	$65.61
.5 Laborer	35.10	140.40	54.00	216.00		
.25 Equip. Oper. (med.)	46.55	93.10	70.25	140.50		
1 Scraper, Towed, 15 C.Y.		156.80		172.48		
1.25 Dozers, 300 H.P.		2147.50		2362.25	164.59	181.05
14 L.H., Daily Totals		$2910.20		$3453.23	$207.87	$246.66
Crew B-33D	Hr.	Daily	Hr.	Daily	Bare Costs	Incl. O&P
1 Equip. Oper. (med.)	$46.55	$372.40	$70.25	$562.00	$43.28	$65.61
.5 Laborer	35.10	140.40	54.00	216.00		
.25 Equip. Oper. (med.)	46.55	93.10	70.25	140.50		
1 S.P. Scraper, 14 C.Y.		1537.00		1690.70		
.25 Dozer, 300 H.P.		429.50		472.45	140.46	154.51
14 L.H., Daily Totals		$2572.40		$3081.65	$183.74	$220.12
Crew B-33E	Hr.	Daily	Hr.	Daily	Bare Costs	Incl. O&P
1 Equip. Oper. (med.)	$46.55	$372.40	$70.25	$562.00	$43.28	$65.61
.5 Laborer	35.10	140.40	54.00	216.00		
.25 Equip. Oper. (med.)	46.55	93.10	70.25	140.50		
1 S.P. Scraper, 21 C.Y.		1974.00		2171.40		
.25 Dozer, 300 H.P.		429.50		472.45	171.68	188.85
14 L.H., Daily Totals		$3009.40		$3562.35	$214.96	$254.45
Crew B-33F	Hr.	Daily	Hr.	Daily	Bare Costs	Incl. O&P
1 Equip. Oper. (med.)	$46.55	$372.40	$70.25	$562.00	$43.28	$65.61
.5 Laborer	35.10	140.40	54.00	216.00		
.25 Equip. Oper. (med.)	46.55	93.10	70.25	140.50		
1 Elev. Scraper, 11 C.Y.		1105.00		1215.50		
.25 Dozer, 300 H.P.		429.50		472.45	109.61	120.57
14 L.H., Daily Totals		$2140.40		$2606.45	$152.89	$186.18
Crew B-33G	Hr.	Daily	Hr.	Daily	Bare Costs	Incl. O&P
1 Equip. Oper. (med.)	$46.55	$372.40	$70.25	$562.00	$43.28	$65.61
.5 Laborer	35.10	140.40	54.00	216.00		
.25 Equip. Oper. (med.)	46.55	93.10	70.25	140.50		
1 Elev. Scraper, 22 C.Y.		2099.00		2308.90		
.25 Dozer, 300 H.P.		429.50		472.45	180.61	198.67
14 L.H., Daily Totals		$3134.40		$3699.85	$223.89	$264.27
Crew B-33H	Hr.	Daily	Hr.	Daily	Bare Costs	Incl. O&P
.5 Laborer	$35.10	$140.40	$54.00	$216.00	$43.28	$65.61
1 Equipment Operator (med.)	46.55	372.40	70.25	562.00		
.25 Equipment Operator (med.)	46.55	93.10	70.25	140.50		
1 S.P. Scraper, 44 C.Y.		3768.00		4144.80		
.25 Dozer, 410 H.P.		568.00		624.80	309.71	340.69
14 L.H., Daily Totals		$4941.90		$5688.10	$352.99	$406.29
Crew B-33J	Hr.	Daily	Hr.	Daily	Bare Costs	Incl. O&P
1 Equipment Operator (med.)	$46.55	$372.40	$70.25	$562.00	$46.55	$70.25
1 S.P. Scraper, 14 C.Y.		1537.00		1690.70	192.13	211.34
8 L.H., Daily Totals		$1909.40		$2252.70	$238.68	$281.59
Crew B-33K	Hr.	Daily	Hr.	Daily	Bare Costs	Incl. O&P
1 Equipment Operator (med.)	$46.55	$372.40	$70.25	$562.00	$43.28	$65.61
.25 Equipment Operator (med.)	46.55	93.10	70.25	140.50		
.5 Laborer	35.10	140.40	54.00	216.00		
1 S.P. Scraper, 31 C.Y.		2868.00		3154.80		
.25 Dozer, 410 H.P.		568.00		624.80	245.43	269.97
14 L.H., Daily Totals		$4041.90		$4698.10	$288.71	$335.58

Crew No.	Bare Costs Hr.	Bare Costs Daily	Incl. Subs O&P Hr.	Incl. Subs O&P Daily	Cost Per Labor-Hour Bare Costs	Cost Per Labor-Hour Incl. O&P
Crew B-34A	Hr.	Daily	Hr.	Daily	Bare Costs	Incl. O&P
1 Truck Driver (heavy)	$35.30	$282.40	$53.55	$428.40	$35.30	$53.55
1 Dump Truck, 8 C.Y., 220 H.P.		357.40		393.14	44.67	49.14
8 L.H., Daily Totals		$639.80		$821.54	$79.97	$102.69
Crew B-34B	Hr.	Daily	Hr.	Daily	Bare Costs	Incl. O&P
1 Truck Driver (heavy)	$35.30	$282.40	$53.55	$428.40	$35.30	$53.55
1 Dump Truck, 12 C.Y., 400 H.P.		584.60		643.06	73.08	80.38
8 L.H., Daily Totals		$867.00		$1071.46	$108.38	$133.93
Crew B-34C	Hr.	Daily	Hr.	Daily	Bare Costs	Incl. O&P
1 Truck Driver (heavy)	$35.30	$282.40	$53.55	$428.40	$35.30	$53.55
1 Truck Tractor, 6x4, 380 H.P.		522.20		574.42		
1 Dump Trailer, 16.5 C.Y.		123.20		135.52	80.67	88.74
8 L.H., Daily		$927.80		$1138.34	$115.97	$142.29
Crew B-34D	Hr.	Daily	Hr.	Daily	Bare Costs	Incl. O&P
1 Truck Driver (heavy)	$35.30	$282.40	$53.55	$428.40	$35.30	$53.55
1 Truck Tractor, 6x4, 380 H.P.		522.20		574.42		
1 Dump Trailer, 20 C.Y.		137.80		151.58	82.50	90.75
8 L.H., Daily		$942.40		$1154.40	$117.80	$144.30
Crew B-34E	Hr.	Daily	Hr.	Daily	Bare Costs	Incl. O&P
1 Truck Driver (heavy)	$35.30	$282.40	$53.55	$428.40	$35.30	$53.55
1 Dump Truck, Off Hwy., 25 Ton		1271.00		1398.10	158.88	174.76
8 L.H., Daily Totals		$1553.40		$1826.50	$194.18	$228.31
Crew B-34F	Hr.	Daily	Hr.	Daily	Bare Costs	Incl. O&P
1 Truck Driver (heavy)	$35.30	$282.40	$53.55	$428.40	$35.30	$53.55
1 Dump Truck, Off Hwy., 35 Ton		1157.00		1272.70	144.63	159.09
8 L.H., Daily Totals		$1439.40		$1701.10	$179.93	$212.64
Crew B-34G	Hr.	Daily	Hr.	Daily	Bare Costs	Incl. O&P
1 Truck Driver (heavy)	$35.30	$282.40	$53.55	$428.40	$35.30	$53.55
1 Dump Truck, Off Hwy., 50 Ton		1568.00		1724.80	196.00	215.60
8 L.H., Daily Totals		$1850.40		$2153.20	$231.30	$269.15
Crew B-34H	Hr.	Daily	Hr.	Daily	Bare Costs	Incl. O&P
1 Truck Driver (heavy)	$35.30	$282.40	$53.55	$428.40	$35.30	$53.55
1 Dump Truck, Off Hwy., 65 Ton		1572.00		1729.20	196.50	216.15
8 L.H., Daily Totals		$1854.40		$2157.60	$231.80	$269.70
Crew B-34I	Hr.	Daily	Hr.	Daily	Bare Costs	Incl. O&P
1 Truck Driver (heavy)	$35.30	$282.40	$53.55	$428.40	$35.30	$53.55
1 Dump Truck, 18 C.Y., 450 H.P.		745.00		819.50	93.13	102.44
8 L.H., Daily Totals		$1027.40		$1247.90	$128.43	$155.99
Crew B-34J	Hr.	Daily	Hr.	Daily	Bare Costs	Incl. O&P
1 Truck Driver (heavy)	$35.30	$282.40	$53.55	$428.40	$35.30	$53.55
1 Dump Truck, Off Hwy., 100 Ton		2071.00		2278.10	258.88	284.76
8 L.H., Daily Totals		$2353.40		$2706.50	$294.18	$338.31
Crew B-34K	Hr.	Daily	Hr.	Daily	Bare Costs	Incl. O&P
1 Truck Driver (heavy)	$35.30	$282.40	$53.55	$428.40	$35.30	$53.55
1 Truck Tractor, 6x4, 450 H.P.		635.00		698.50		
1 Lowbed Trailer, 75 Ton		213.00		234.30	106.00	116.60
8 L.H., Daily Totals		$1130.40		$1361.20	$141.30	$170.15

Crew No.	Bare Costs Hr.	Bare Costs Daily	Incl. Subs O&P Hr.	Incl. Subs O&P Daily	Cost Per Labor-Hour Bare Costs	Cost Per Labor-Hour Incl. O&P
Crew B-34L	Hr.	Daily	Hr.	Daily	Bare Costs	Incl. O&P
1 Equip. Oper. (light)	$44.75	$358.00	$67.55	$540.40	$44.75	$67.55
1 Flatbed Truck, Gas, 1.5 Ton		211.00		232.10	26.38	29.01
8 L.H., Daily Totals		$569.00		$772.50	$71.13	$96.56
Crew B-34M	Hr.	Daily	Hr.	Daily	Bare Costs	Incl. O&P
1 Equip. Oper. (light)	$44.75	$358.00	$67.55	$540.40	$44.75	$67.55
1 Flatbed Truck, Gas, 3 Ton		264.40		290.84	33.05	36.35
8 L.H., Daily Totals		$622.40		$831.24	$77.80	$103.91
Crew B-34N	Hr.	Daily	Hr.	Daily	Bare Costs	Incl. O&P
1 Truck Driver (heavy)	$35.30	$282.40	$53.55	$428.40	$35.30	$53.55
1 Dump Truck, 8 C.Y., 220 H.P.		357.40		393.14		
1 Flatbed Trailer, 40 Ton		147.00		161.70	63.05	69.36
8 L.H., Daily Totals		$786.80		$983.24	$98.35	$122.91
Crew B-34P	Hr.	Daily	Hr.	Daily	Bare Costs	Incl. O&P
1 Pipe Fitter	$55.70	$445.60	$84.00	$672.00	$45.53	$68.78
1 Truck Driver (light)	34.35	274.80	52.10	416.80		
1 Equip. Oper. (med.)	46.55	372.40	70.25	562.00		
1 Flatbed Truck, Gas, 3 Ton		264.40		290.84		
1 Backhoe Loader, 48 H.P.		334.00		367.40	24.93	27.43
24 L.H., Daily Totals		$1691.20		$2309.04	$70.47	$96.21
Crew B-34Q	Hr.	Daily	Hr.	Daily	Bare Costs	Incl. O&P
1 Pipe Fitter	$55.70	$445.60	$84.00	$672.00	$45.95	$69.42
1 Truck Driver (light)	34.35	274.80	52.10	416.80		
1 Eqip. Oper. (crane)	47.80	382.40	72.15	577.20		
1 Flatbed Trailer, 25 Ton		108.60		119.46		
1 Dump Truck, 8 C.Y., 220 H.P.		357.40		393.14		
1 Hyd. Crane, 25 Ton		697.20		766.92	48.47	53.31
24 L.H., Daily Totals		$2266.00		$2945.52	$94.42	$122.73
Crew B-34R	Hr.	Daily	Hr.	Daily	Bare Costs	Incl. O&P
1 Pipe Fitter	$55.70	$445.60	$84.00	$672.00	$45.95	$69.42
1 Truck Driver (light)	34.35	274.80	52.10	416.80		
1 Eqip. Oper. (crane)	47.80	382.40	72.15	577.20		
1 Flatbed Trailer, 25 Ton		108.60		119.46		
1 Dump Truck, 8 C.Y., 220 H.P.		357.40		393.14		
1 Hyd. Crane, 25 Ton		697.20		766.92		
1 Hyd. Excavator, 1 C.Y.		748.60		823.46	79.66	87.62
24 L.H., Daily Totals		$3014.60		$3768.98	$125.61	$157.04
Crew B-34S	Hr.	Daily	Hr.	Daily	Bare Costs	Incl. O&P
2 Pipe Fitters	$55.70	$891.20	$84.00	$1344.00	$48.63	$73.42
1 Truck Driver (heavy)	35.30	282.40	53.55	428.40		
1 Eqip. Oper. (crane)	47.80	382.40	72.15	577.20		
1 Flatbed Trailer, 40 Ton		147.00		161.70		
1 Truck Tractor, 6x4, 380 H.P.		522.20		574.42		
1 Hyd. Crane, 80 Ton		1554.00		1709.40		
1 Hyd. Excavator, 2 C.Y.		1253.00		1378.30	108.63	119.49
32 L.H., Daily Totals		$5032.20		$6173.42	$157.26	$192.92
Crew B-34T	Hr.	Daily	Hr.	Daily	Bare Costs	Incl. O&P
2 Pipe Fitters	$55.70	$891.20	$84.00	$1344.00	$48.63	$73.42
1 Truck Driver (heavy)	35.30	282.40	53.55	428.40		
1 Eqip. Oper. (crane)	47.80	382.40	72.15	577.20		
1 Flatbed Trailer, 40 Ton		147.00		161.70		
1 Truck Tractor, 6x4, 380 H.P.		522.20		574.42		
1 Hyd. Crane, 80 Ton		1554.00		1709.40	69.47	76.42
32 L.H., Daily Totals		$3779.20		$4795.12	$118.10	$149.85

Crew No.	Bare Costs		Incl. Subs O&P		Cost Per Labor-Hour	

Crew B-35

Crew B-35	Hr.	Daily	Hr.	Daily	Bare Costs	Incl. O&P
1 Labor Foreman (outside)	$37.10	$296.80	$57.10	$456.80	$43.54	$66.34
1 Skilled Worker	45.40	363.20	70.20	561.60		
1 Welder (plumber)	54.60	436.80	82.35	658.80		
1 Laborer	35.10	280.80	54.00	432.00		
1 Equip. Oper. (crane)	47.80	382.40	72.15	577.20		
1 Equip. Oper. Oiler	41.25	330.00	62.25	498.00		
1 Welder, electric, 300 amp		36.90		40.59		
1 Hyd. Excavator, .75 C.Y.		647.20		711.92	14.25	15.68
48 L.H., Daily Totals		$2774.10		$3936.91	$57.79	$82.02

Crew B-35A	Hr.	Daily	Hr.	Daily	Bare Costs	Incl. O&P
1 Labor Foreman (outside)	$37.10	$296.80	$57.10	$456.80	$42.34	$64.58
2 Laborers	35.10	561.60	54.00	864.00		
1 Skilled Worker	45.40	363.20	70.20	561.60		
1 Welder (plumber)	54.60	436.80	82.35	658.80		
1 Equip. Oper. (crane)	47.80	382.40	72.15	577.20		
1 Equip. Oper. Oiler	41.25	330.00	62.25	498.00		
1 Welder, gas engine, 300 amp		121.60		133.76		
1 Crawler Crane, 75 Ton		1423.00		1565.30	27.58	30.34
56 L.H., Daily Totals		$3915.40		$5315.46	$69.92	$94.92

Crew B-36	Hr.	Daily	Hr.	Daily	Bare Costs	Incl. O&P
1 Labor Foreman (outside)	$37.10	$296.80	$57.10	$456.80	$40.08	$61.12
2 Laborers	35.10	561.60	54.00	864.00		
2 Equip. Oper. (med.)	46.55	744.80	70.25	1124.00		
1 Dozer, 200 H.P.		1243.00		1367.30		
1 Aggregate Spreader		36.80		40.48		
1 Tandem Roller, 10 Ton		227.60		250.36	37.69	41.45
40 L.H., Daily Totals		$3110.60		$4102.94	$77.77	$102.57

Crew B-36A	Hr.	Daily	Hr.	Daily	Bare Costs	Incl. O&P
1 Labor Foreman (outside)	$37.10	$296.80	$57.10	$456.80	$41.93	$63.73
2 Laborers	35.10	561.60	54.00	864.00		
4 Equip. Oper. (med.)	46.55	1489.60	70.25	2248.00		
1 Dozer, 200 H.P.		1243.00		1367.30		
1 Aggregate Spreader		36.80		40.48		
1 Tandem Roller, 10 Ton		227.60		250.36		
1 Roller, Pneum. Whl, 12 Ton		313.20		344.52	32.51	35.76
56 L.H., Daily Totals		$4168.60		$5571.46	$74.44	$99.49

Crew B-36B	Hr.	Daily	Hr.	Daily	Bare Costs	Incl. O&P
1 Labor Foreman (outside)	$37.10	$296.80	$57.10	$456.80	$41.10	$62.46
2 Laborers	35.10	561.60	54.00	864.00		
4 Equip. Oper. (med.)	46.55	1489.60	70.25	2248.00		
1 Truck Driver, Heavy	35.30	282.40	53.55	428.40		
1 Grader, 30,000 Lbs.		610.80		671.88		
1 F.E. Loader, crl, 1.5 C.Y.		598.20		658.02		
1 Dozer, 300 H.P.		1718.00		1889.80		
1 Roller, Vibratory, 25 Ton		640.20		704.22		
1 Truck Tractor, 6x4, 450 H.P.		635.00		698.50		
1 Water Tank Trailer, 5000 Gal.		141.00		155.10	67.86	74.65
64 L.H., Daily Totals		$6973.60		$8774.72	$108.96	$137.11

Crew B-36C	Hr.	Daily	Hr.	Daily	Bare Costs	Incl. O&P
1 Labor Foreman (outside)	$37.10	$296.80	$57.10	$456.80	$42.41	$64.28
3 Equip. Oper. (med.)	46.55	1117.20	70.25	1686.00		
1 Truck Driver, Heavy	35.30	282.40	53.55	428.40		
1 Grader, 30,000 Lbs.		610.80		671.88		
1 Dozer, 300 H.P.		1718.00		1889.80		
1 Roller, Vibratory, 25 Ton		640.20		704.22		
1 Truck Tractor, 6x4, 450 H.P.		635.00		698.50		
1 Water Tank Trailer, 5000 Gal.		141.00		155.10	93.63	102.99
40 L.H., Daily Totals		$5441.40		$6690.70	$136.04	$167.27

Crew B-36D	Hr.	Daily	Hr.	Daily	Bare Costs	Incl. O&P
1 Labor Foreman (outside)	$37.10	$296.80	$57.10	$456.80	$44.19	$66.96
3 Equip. Oper. (medium)	46.55	1117.20	70.25	1686.00		
1 Grader, 30,000 Lbs.		610.80		671.88		
1 Dozer, 300 H.P.		1718.00		1889.80		
1 Roller, Vibratory, 25 Ton		640.20		704.22	92.78	102.06
32 L.H., Daily Totals		$4383.00		$5408.70	$136.97	$169.02

Crew B-37	Hr.	Daily	Hr.	Daily	Bare Costs	Incl. O&P
1 Labor Foreman (outside)	$37.10	$296.80	$57.10	$456.80	$37.04	$56.77
4 Laborers	35.10	1123.20	54.00	1728.00		
1 Equip. Oper. (light)	44.75	358.00	67.55	540.40		
1 Tandem Roller, 5 Ton		149.20		164.12	3.11	3.42
48 L.H., Daily Totals		$1927.20		$2889.32	$40.15	$60.19

Crew B-37A	Hr.	Daily	Hr.	Daily	Bare Costs	Incl. O&P
2 Laborers	$35.10	$561.60	$54.00	$864.00	$34.85	$53.37
1 Truck Driver (light)	34.35	274.80	52.10	416.80		
1 Flatbed Truck, Gas, 1.5 Ton		211.00		232.10		
1 Tar Kettle, T.M.		77.20		84.92	12.01	13.21
24 L.H., Daily Totals		$1124.60		$1597.82	$46.86	$66.58

Crew B-37B	Hr.	Daily	Hr.	Daily	Bare Costs	Incl. O&P
3 Laborers	$35.10	$842.40	$54.00	$1296.00	$34.91	$53.52
1 Truck Driver (light)	34.35	274.80	52.10	416.80		
1 Flatbed Truck, Gas, 1.5 Ton		211.00		232.10		
1 Tar Kettle, T.M.		77.20		84.92	9.01	9.91
32 L.H., Daily Totals		$1405.40		$2029.82	$43.92	$63.43

Crew B-37C	Hr.	Daily	Hr.	Daily	Bare Costs	Incl. O&P
2 Laborers	$35.10	$561.60	$54.00	$864.00	$34.73	$53.05
2 Truck Drivers (light)	34.35	549.60	52.10	833.60		
2 Flatbed Trucks, Gas, 1.5 Ton		422.00		464.20		
1 Tar Kettle, T.M.		77.20		84.92	15.60	17.16
32 L.H., Daily Totals		$1610.40		$2246.72	$50.33	$70.21

Crew B-37D	Hr.	Daily	Hr.	Daily	Bare Costs	Incl. O&P
1 Laborer	$35.10	$280.80	$54.00	$432.00	$34.73	$53.05
1 Truck Driver (light)	34.35	274.80	52.10	416.80		
1 Pickup Truck, 3/4 Ton		127.00		139.70	7.94	8.73
16 L.H., Daily Totals		$682.60		$988.50	$42.66	$61.78

Crews

Crew B-37E

Crew No.	Bare Costs Hr.	Daily	Incl. Subs O&P Hr.	Daily	Cost Per Labor-Hour Bare Costs	Incl. O&P
3 Laborers	$35.10	$842.40	$54.00	$1296.00	$37.90	$57.71
1 Equip. Oper. (light)	44.75	358.00	67.55	540.40		
1 Equip. Oper. (medium)	46.55	372.40	70.25	562.00		
2 Truck Drivers (light)	34.35	549.60	52.10	833.60		
4 Barrels w/ Flasher		13.60		14.96		
1 Concrete Saw		148.00		162.80		
1 Rotary Hammer Drill		22.55		24.81		
1 Hammer Drill Bit		3.15		3.46		
1 Loader, Skid Steer, 30 H.P., gas		159.60		175.56		
1 Conc. Hammer Attach.		116.25		127.88		
1 Vibrating plate, gas, 18"		32.20		35.42		
2 Flatbed Trucks, Gas, 1.5 Ton		422.00		464.20	16.38	18.02
56 L.H., Daily Totals		$3039.75		$4241.09	$54.28	$75.73

Crew B-37F

Crew No.	Bare Costs Hr.	Daily	Incl. Subs O&P Hr.	Daily	Cost Per Labor-Hour Bare Costs	Incl. O&P
3 Laborers	$35.10	$842.40	$54.00	$1296.00	$34.91	$53.52
1 Truck Driver (light)	34.35	274.80	52.10	416.80		
4 Barrels w/ Flasher		13.60		14.96		
1 Concrete Mixer, 10 C.F.		160.60		176.66		
1 Air Compressor, 60 cfm		120.40		132.44		
1 -50' Air Hose, 3/4"		3.05		3.36		
1 Spade (chipper)		7.60		8.36		
1 Flatbed Truck, Gas, 1.5 Ton		211.00		232.10	16.13	17.75
32 L.H., Daily Totals		$1633.45		$2280.68	$51.05	$71.27

Crew B-37G

Crew No.	Bare Costs Hr.	Daily	Incl. Subs O&P Hr.	Daily	Cost Per Labor-Hour Bare Costs	Incl. O&P
1 Labor Foreman (outside)	$37.10	$296.80	$57.10	$456.80	$37.04	$56.77
4 Laborers	35.10	1123.20	54.00	1728.00		
1 Equip. Oper. (light)	44.75	358.00	67.55	540.40		
1 Berm Machine		280.00		308.00		
1 Tandem Roller, 5 Ton		149.20		164.12	8.94	9.84
48 L.H., Daily Totals		$2207.20		$3197.32	$45.98	$66.61

Crew B-38

Crew No.	Bare Costs Hr.	Daily	Incl. Subs O&P Hr.	Daily	Cost Per Labor-Hour Bare Costs	Incl. O&P
1 Labor Foreman (outside)	$37.10	$296.80	$57.10	$456.80	$39.72	$60.58
2 Laborers	35.10	561.60	54.00	864.00		
1 Equip. Oper. (light)	44.75	358.00	67.55	540.40		
1 Equip. Oper. (med.)	46.55	372.40	70.25	562.00		
1 Backhoe Loader, 48 H.P.		334.00		367.40		
1 Hyd.Hammer, (1200 lb.)		178.40		196.24		
1 F.E. Loader, W.M., 4 C.Y.		611.00		672.10		
1 Pvmt. Rem. Bucket		59.00		64.90	29.56	32.52
40 L.H., Daily Totals		$2771.20		$3723.84	$69.28	$93.10

Crew B-39

Crew No.	Bare Costs Hr.	Daily	Incl. Subs O&P Hr.	Daily	Cost Per Labor-Hour Bare Costs	Incl. O&P
1 Labor Foreman (outside)	$37.10	$296.80	$57.10	$456.80	$37.04	$56.77
4 Laborers	35.10	1123.20	54.00	1728.00		
1 Equip. Oper. (light)	44.75	358.00	67.55	540.40		
1 Air Compressor, 250 cfm		177.00		194.70		
2 Breakers, Pavement, 60 lb.		19.60		21.56		
2 -50' Air Hoses, 1.5"		12.60		13.86	4.36	4.79
48 L.H., Daily Totals		$1987.20		$2955.32	$41.40	$61.57

Crew B-40

Crew No.	Bare Costs Hr.	Daily	Incl. Subs O&P Hr.	Daily	Cost Per Labor-Hour Bare Costs	Incl. O&P
1 Pile Driver Foreman (outside)	$44.80	$358.40	$72.00	$576.00	$44.11	$69.22
4 Pile Drivers	42.80	1369.60	68.80	2201.60		
2 Equip. Oper. (crane)	47.80	764.80	72.15	1154.40		
1 Equip. Oper. Oiler	41.25	330.00	62.25	498.00		
1 Crawler Crane, 40 Ton		1115.00		1226.50		
1 Vibratory Hammer & Gen.		2335.00		2568.50	53.91	59.30
64 L.H., Daily Totals		$6272.80		$8225.00	$98.01	$128.52

Crew B-40B

Crew No.	Bare Costs Hr.	Daily	Incl. Subs O&P Hr.	Daily	Cost Per Labor-Hour Bare Costs	Incl. O&P
1 Labor Foreman (outside)	$37.10	$296.80	$57.10	$456.80	$38.58	$58.92
3 Laborers	35.10	842.40	54.00	1296.00		
1 Equip. Oper. (crane)	47.80	382.40	72.15	577.20		
1 Equip. Oper. Oiler	41.25	330.00	62.25	498.00		
1 Lattice Boom Crane, 40 Ton		1163.00		1279.30	24.23	26.65
48 L.H., Daily Totals		$3014.60		$4107.30	$62.80	$85.57

Crew B-41

Crew No.	Bare Costs Hr.	Daily	Incl. Subs O&P Hr.	Daily	Cost Per Labor-Hour Bare Costs	Incl. O&P
1 Labor Foreman (outside)	$37.10	$296.80	$57.10	$456.80	$36.32	$55.76
4 Laborers	35.10	1123.20	54.00	1728.00		
.25 Equip. Oper. (crane)	47.80	95.60	72.15	144.30		
.25 Equip. Oper. Oiler	41.25	82.50	62.25	124.50		
.25 Crawler Crane, 40 Ton		278.75		306.63	6.34	6.97
44 L.H., Daily Totals		$1876.85		$2760.22	$42.66	$62.73

Crew B-42

Crew No.	Bare Costs Hr.	Daily	Incl. Subs O&P Hr.	Daily	Cost Per Labor-Hour Bare Costs	Incl. O&P
1 Labor Foreman (outside)	$37.10	$296.80	$57.10	$456.80	$39.49	$61.97
4 Laborers	35.10	1123.20	54.00	1728.00		
1 Equip. Oper. (crane)	47.80	382.40	72.15	577.20		
1 Equip. Oper. Oiler	41.25	330.00	62.25	498.00		
1 Welder	49.35	394.80	88.25	706.00		
1 Hyd. Crane, 25 Ton		697.20		766.92		
1 Welder, gas engine, 300 amp		121.60		133.76		
1 Horz. Boring Csg. Mch.		447.60		492.36	19.79	21.77
64 L.H., Daily Totals		$3793.60		$5359.04	$59.27	$83.74

Crew B-43

Crew No.	Bare Costs Hr.	Daily	Incl. Subs O&P Hr.	Daily	Cost Per Labor-Hour Bare Costs	Incl. O&P
1 Labor Foreman (outside)	$37.10	$296.80	$57.10	$456.80	$38.58	$58.92
3 Laborers	35.10	842.40	54.00	1296.00		
1 Equip. Oper. (crane)	47.80	382.40	72.15	577.20		
1 Equip. Oper. Oiler	41.25	330.00	62.25	498.00		
1 Drill Rig, Truck-Mounted		2441.00		2685.10	50.85	55.94
48 L.H., Daily Totals		$4292.60		$5513.10	$89.43	$114.86

Crew B-44

Crew No.	Bare Costs Hr.	Daily	Incl. Subs O&P Hr.	Daily	Cost Per Labor-Hour Bare Costs	Incl. O&P
1 Pile Driver Foreman (outside)	$44.80	$358.40	$72.00	$576.00	$43.34	$68.19
4 Pile Drivers	42.80	1369.60	68.80	2201.60		
2 Equip. Oper. (crane)	47.80	764.80	72.15	1154.40		
1 Laborer	35.10	280.80	54.00	432.00		
1 Crawler Crane, 40 Ton		1115.00		1226.50		
1 Lead, 60' high		73.60		80.96		
1 Hammer, diesel, 15K ft.-lbs.		591.40		650.54	27.81	30.59
64 L.H., Daily Totals		$4553.60		$6322.00	$71.15	$98.78

Crew B-45

Crew No.	Bare Costs Hr.	Daily	Incl. Subs O&P Hr.	Daily	Cost Per Labor-Hour Bare Costs	Incl. O&P
1 Equip. Oper. (med.)	$46.55	$372.40	$70.25	$562.00	$40.92	$61.90
1 Truck Driver (heavy)	35.30	282.40	53.55	428.40		
1 Dist. Tanker, 3000 Gallon		306.00		336.60		
1 Truck Tractor, 6x4, 380 H.P.		522.20		574.42	51.76	56.94
16 L.H., Daily Totals		$1483.00		$1901.42	$92.69	$118.84

Crew B-46

Crew No.	Bare Costs Hr.	Daily	Incl. Subs O&P Hr.	Daily	Cost Per Labor-Hour Bare Costs	Incl. O&P
1 Pile Driver Foreman (outside)	$44.80	$358.40	$72.00	$576.00	$39.28	$61.93
2 Pile Drivers	42.80	684.80	68.80	1100.80		
3 Laborers	35.10	842.40	54.00	1296.00		
1 Chain Saw, gas, 36" Long		43.60		47.96	.91	1.00
48 L.H., Daily Totals		$1929.20		$3020.76	$40.19	$62.93

Crew No.	Bare Costs		Incl. Subs O&P		Cost Per Labor-Hour	

Crew B-47

Crew B-47	Hr.	Daily	Hr.	Daily	Bare Costs	Incl. O&P
1 Blast Foreman (outside)	$37.10	$296.80	$57.10	$456.80	$38.98	$59.55
1 Driller	35.10	280.80	54.00	432.00		
1 Equip. Oper. (light)	44.75	358.00	67.55	540.40		
1 Air Track Drill, 4"		937.20		1030.92		
1 Air Compressor, 600 cfm		467.60		514.36		
2 -50' Air Hoses, 3"		29.80		32.78	59.77	65.75
24 L.H., Daily Totals		$2370.20		$3007.26	$98.76	$125.30

Crew B-47A	Hr.	Daily	Hr.	Daily	Bare Costs	Incl. O&P
1 Drilling Foreman (outside)	$37.10	$296.80	$57.10	$456.80	$42.05	$63.83
1 Equip. Oper. (heavy)	47.80	382.40	72.15	577.20		
1 Oiler	41.25	330.00	62.25	498.00		
1 Air Track Drill, 5"		1088.00		1196.80	45.33	49.87
24 L.H., Daily Totals		$2097.20		$2728.80	$87.38	$113.70

Crew B-47C	Hr.	Daily	Hr.	Daily	Bare Costs	Incl. O&P
1 Laborer	$35.10	$280.80	$54.00	$432.00	$39.92	$60.77
1 Equip. Oper. (light)	44.75	358.00	67.55	540.40		
1 Air Compressor, 750 cfm		474.80		522.28		
2 -50' Air Hoses, 3"		29.80		32.78		
1 Air Track Drill, 4"		937.20		1030.92	90.11	99.12
16 L.H., Daily Totals		$2080.60		$2558.38	$130.04	$159.90

Crew B-47E	Hr.	Daily	Hr.	Daily	Bare Costs	Incl. O&P
1 Labor Foreman (outside)	$37.10	$296.80	$57.10	$456.80	$35.60	$54.77
3 Laborers	35.10	842.40	54.00	1296.00		
1 Flatbed Truck, Gas, 3 Ton		264.40		290.84	8.26	9.09
32 L.H., Daily Totals		$1403.60		$2043.64	$43.86	$63.86

Crew B-47G	Hr.	Daily	Hr.	Daily	Bare Costs	Incl. O&P
1 Labor Foreman (outside)	$37.10	$296.80	$57.10	$456.80	$38.01	$58.16
2 Laborers	35.10	561.60	54.00	864.00		
1 Equip. Oper. (light)	44.75	358.00	67.55	540.40		
1 Air Track Drill, 4"		937.20		1030.92		
1 Air Compressor, 600 cfm		467.60		514.36		
2 -50' Air Hoses, 3"		29.80		32.78		
1 Gunite Pump Rig		280.40		308.44	53.59	58.95
32 L.H., Daily Totals		$2931.40		$3747.70	$91.61	$117.12

Crew B-47H	Hr.	Daily	Hr.	Daily	Bare Costs	Incl. O&P
1 Skilled Worker Foreman (out)	$47.40	$379.20	$73.30	$586.40	$45.90	$70.97
3 Skilled Workers	45.40	1089.60	70.20	1684.80		
1 Flatbed Truck, Gas, 3 Ton		264.40		290.84	8.26	9.09
32 L.H., Daily Totals		$1733.20		$2562.04	$54.16	$80.06

Crew B-48	Hr.	Daily	Hr.	Daily	Bare Costs	Incl. O&P
1 Labor Foreman (outside)	$37.10	$296.80	$57.10	$456.80	$39.46	$60.15
3 Laborers	35.10	842.40	54.00	1296.00		
1 Equip. Oper. (crane)	47.80	382.40	72.15	577.20		
1 Equip. Oper. Oiler	41.25	330.00	62.25	498.00		
1 Equip. Oper. (light)	44.75	358.00	67.55	540.40		
1 Centr. Water Pump, 6"		308.40		339.24		
1 -20' Suction Hose, 6"		11.50		12.65		
1 -50' Discharge Hose, 6"		12.10		13.31		
1 Drill Rig, Truck-Mounted		2441.00		2685.10	49.52	54.47
56 L.H., Daily Totals		$4982.60		$6418.70	$88.97	$114.62

Crew B-49	Hr.	Daily	Hr.	Daily	Bare Costs	Incl. O&P
1 Labor Foreman (outside)	$37.10	$296.80	$57.10	$456.80	$40.99	$63.00
3 Laborers	35.10	842.40	54.00	1296.00		
2 Equip. Oper. (crane)	47.80	764.80	72.15	1154.40		
2 Equip. Oper. Oilers	41.25	660.00	62.25	996.00		
1 Equip. Oper. (light)	44.75	358.00	67.55	540.40		
2 Pile Drivers	42.80	684.80	68.80	1100.80		
1 Hyd. Crane, 25 Ton		697.20		766.92		
1 Centr. Water Pump, 6"		308.40		339.24		
1 -20' Suction Hose, 6"		11.50		12.65		
1 -50' Discharge Hose, 6"		12.10		13.31		
1 Drill Rig, Truck-Mounted		2441.00		2685.10	39.43	43.38
88 L.H., Daily Totals		$7077.00		$9361.62	$80.42	$106.38

Crew B-50	Hr.	Daily	Hr.	Daily	Bare Costs	Incl. O&P
2 Pile Driver Foremen (outside)	$44.80	$716.80	$72.00	$1152.00	$42.04	$66.10
6 Pile Drivers	42.80	2054.40	68.80	3302.40		
2 Equip. Oper. (crane)	47.80	764.80	72.15	1154.40		
1 Equip. Oper. Oiler	41.25	330.00	62.25	498.00		
3 Laborers	35.10	842.40	54.00	1296.00		
1 Crawler Crane, 40 Ton		1115.00		1226.50		
1 Lead, 60' high		73.60		80.96		
1 Hammer, diesel, 15K ft.-lbs.		591.40		650.54		
1 Air Compressor, 600 cfm		467.60		514.36		
2 -50' Air Hoses, 3"		29.80		32.78		
1 Chain Saw, gas, 36" Long		43.60		47.96	20.72	22.80
112 L.H., Daily Totals		$7029.40		$9955.90	$62.76	$88.89

Crew B-51	Hr.	Daily	Hr.	Daily	Bare Costs	Incl. O&P
1 Labor Foreman (outside)	$37.10	$296.80	$57.10	$456.80	$35.31	$54.20
4 Laborers	35.10	1123.20	54.00	1728.00		
1 Truck Driver (light)	34.35	274.80	52.10	416.80		
1 Flatbed Truck, Gas, 1.5 Ton		211.00		232.10	4.40	4.84
48 L.H., Daily Totals		$1905.80		$2833.70	$39.70	$59.04

Crew B-52	Hr.	Daily	Hr.	Daily	Bare Costs	Incl. O&P
1 Carpenter Foreman (outside)	$46.10	$368.80	$70.95	$567.60	$40.81	$62.49
1 Carpenter	44.10	352.80	67.85	542.80		
3 Laborers	35.10	842.40	54.00	1296.00		
1 Cement Finisher	42.35	338.80	62.30	498.40		
.5 Rodman (reinf.)	49.10	196.40	78.40	313.60		
.5 Equip. Oper. (med.)	46.55	186.20	70.25	281.00		
.5 Crawler Loader, 3 C.Y.		553.00		608.30	9.88	10.86
56 L.H., Daily Totals		$2838.40		$4107.70	$50.69	$73.35

Crew B-53	Hr.	Daily	Hr.	Daily	Bare Costs	Incl. O&P
1 Equip. Oper. (light)	$44.75	$358.00	$67.55	$540.40	$44.75	$67.55
1 Trencher, Chain, 12 H.P.		62.80		69.08	7.85	8.63
8 L.H., Daily Totals		$420.80		$609.48	$52.60	$76.19

Crew B-54	Hr.	Daily	Hr.	Daily	Bare Costs	Incl. O&P
1 Equip. Oper. (light)	$44.75	$358.00	$67.55	$540.40	$44.75	$67.55
1 Trencher, Chain, 40 H.P.		312.80		344.08	39.10	43.01
8 L.H., Daily Totals		$670.80		$884.48	$83.85	$110.56

Crew B-54A	Hr.	Daily	Hr.	Daily	Bare Costs	Incl. O&P
.17 Labor Foreman (outside)	$37.10	$50.46	$57.10	$77.66	$45.18	$68.34
1 Equipment Operator (med.)	46.55	372.40	70.25	562.00		
1 Wheel Trencher, 67 H.P.		1084.00		1192.40	115.81	127.39
9.36 L.H., Daily Totals		$1506.86		$1832.06	$160.99	$195.73

Crew B-54B

Crew No.	Bare Costs Hr.	Daily	Incl. Subs O&P Hr.	Daily	Cost Per Labor-Hour Bare Costs	Incl. O&P
.25 Labor Foreman (outside)	$37.10	$74.20	$57.10	$114.20	$44.66	$67.62
1 Equipment Operator (med.)	46.55	372.40	70.25	562.00		
1 Wheel Trencher, 150 H.P.		1816.00		1997.60	181.60	199.76
10 L.H., Daily Totals		$2262.60		$2673.80	$226.26	$267.38

Crew B-54C

Crew No.	Bare Costs Hr.	Daily	Incl. Subs O&P Hr.	Daily	Cost Per Labor-Hour Bare Costs	Incl. O&P
1 Laborer	$35.10	$280.80	$54.00	$432.00	$40.83	$62.13
1 Equipment Operator (med.)	46.55	372.40	70.25	562.00		
1 Wheel Trencher, 67 H.P.		1084.00		1192.40	67.75	74.53
16 L.H., Daily Totals		$1737.20		$2186.40	$108.58	$136.65

Crew B-54D

Crew No.	Bare Costs Hr.	Daily	Incl. Subs O&P Hr.	Daily	Cost Per Labor-Hour Bare Costs	Incl. O&P
1 Laborer	$35.10	$280.80	$54.00	$432.00	$40.83	$62.13
1 Equipment Operator (med.)	46.55	372.40	70.25	562.00		
1 Rock trencher, 6" width		344.40		378.84	21.52	23.68
16 L.H., Daily Totals		$997.60		$1372.84	$62.35	$85.80

Crew B-54E

Crew No.	Bare Costs Hr.	Daily	Incl. Subs O&P Hr.	Daily	Cost Per Labor-Hour Bare Costs	Incl. O&P
1 Laborer	$35.10	$280.80	$54.00	$432.00	$40.83	$62.13
1 Equipment Operator (med.)	46.55	372.40	70.25	562.00		
1 Rock trencher, 18" width		2517.00		2768.70	157.31	173.04
16 L.H., Daily Totals		$3170.20		$3762.70	$198.14	$235.17

Crew B-55

Crew No.	Bare Costs Hr.	Daily	Incl. Subs O&P Hr.	Daily	Cost Per Labor-Hour Bare Costs	Incl. O&P
2 Laborers	$35.10	$561.60	$54.00	$864.00	$34.85	$53.37
1 Truck Driver (light)	34.35	274.80	52.10	416.80		
1 Truck-mounted earth auger		718.40		790.24		
1 Flatbed Truck, Gas, 3 Ton		264.40		290.84	40.95	45.05
24 L.H., Daily Totals		$1819.20		$2361.88	$75.80	$98.41

Crew B-56

Crew No.	Bare Costs Hr.	Daily	Incl. Subs O&P Hr.	Daily	Cost Per Labor-Hour Bare Costs	Incl. O&P
1 Laborer	$35.10	$280.80	$54.00	$432.00	$39.92	$60.77
1 Equip. Oper. (light)	44.75	358.00	67.55	540.40		
1 Air Track Drill, 4"		937.20		1030.92		
1 Air Compressor, 600 cfm		467.60		514.36		
1 -50' Air Hose, 3"		14.90		16.39	88.73	97.60
16 L.H., Daily Totals		$2058.50		$2534.07	$128.66	$158.38

Crew B-57

Crew No.	Bare Costs Hr.	Daily	Incl. Subs O&P Hr.	Daily	Cost Per Labor-Hour Bare Costs	Incl. O&P
1 Labor Foreman (outside)	$37.10	$296.80	$57.10	$456.80	$40.18	$61.17
2 Laborers	35.10	561.60	54.00	864.00		
1 Equip. Oper. (crane)	47.80	382.40	72.15	577.20		
1 Equip. Oper. (light)	44.75	358.00	67.55	540.40		
1 Equip. Oper. Oiler	41.25	330.00	62.25	498.00		
1 Crawler Crane, 25 Ton		1110.00		1221.00		
1 Clamshell Bucket, 1 C.Y.		47.00		51.70		
1 Centr. Water Pump, 6"		308.40		339.24		
1 -20' Suction Hose, 6"		11.50		12.65		
20 -50' Discharge Hoses, 6"		242.00		266.20	35.81	39.39
48 L.H., Daily Totals		$3647.70		$4827.19	$75.99	$100.57

Crew B-58

Crew No.	Bare Costs Hr.	Daily	Incl. Subs O&P Hr.	Daily	Cost Per Labor-Hour Bare Costs	Incl. O&P
2 Laborers	$35.10	$561.60	$54.00	$864.00	$38.32	$58.52
1 Equip. Oper. (light)	44.75	358.00	67.55	540.40		
1 Backhoe Loader, 48 H.P.		334.00		367.40		
1 Small Helicopter, w/pilot		2625.00		2887.50	123.29	135.62
24 L.H., Daily Totals		$3878.60		$4659.30	$161.61	$194.14

Crew B-59

Crew No.	Bare Costs Hr.	Daily	Incl. Subs O&P Hr.	Daily	Cost Per Labor-Hour Bare Costs	Incl. O&P
1 Truck Driver (heavy)	$35.30	$282.40	$53.55	$428.40	$35.30	$53.55
1 Truck Tractor, 220 H.P.		314.80		346.28		
1 Water Tank Trailer, 5000 Gal.		141.00		155.10	56.98	62.67
8 L.H., Daily Totals		$738.20		$929.78	$92.28	$116.22

Crew B-59A

Crew No.	Bare Costs Hr.	Daily	Incl. Subs O&P Hr.	Daily	Cost Per Labor-Hour Bare Costs	Incl. O&P
2 Laborers	$35.10	$561.60	$54.00	$864.00	$35.17	$53.85
1 Truck Driver (heavy)	35.30	282.40	53.55	428.40		
1 Water Tank Trailer, 5000 Gal.		141.00		155.10		
1 Truck Tractor, 220 H.P.		314.80		346.28	18.99	20.89
24 L.H., Daily Totals		$1299.80		$1793.78	$54.16	$74.74

Crew B-60

Crew No.	Bare Costs Hr.	Daily	Incl. Subs O&P Hr.	Daily	Cost Per Labor-Hour Bare Costs	Incl. O&P
1 Labor Foreman (outside)	$37.10	$296.80	$57.10	$456.80	$40.84	$62.09
2 Laborers	35.10	561.60	54.00	864.00		
1 Equip. Oper. (crane)	47.80	382.40	72.15	577.20		
2 Equip. Oper. (light)	44.75	716.00	67.55	1080.80		
1 Equip. Oper. Oiler	41.25	330.00	62.25	498.00		
1 Crawler Crane, 40 Ton		1115.00		1226.50		
1 Lead, 60' high		73.60		80.96		
1 Hammer, diesel, 15K ft.-lbs.		591.40		650.54		
1 Backhoe Loader, 48 H.P.		334.00		367.40	37.75	41.52
56 L.H., Daily Totals		$4400.80		$5802.20	$78.59	$103.61

Crew B-61

Crew No.	Bare Costs Hr.	Daily	Incl. Subs O&P Hr.	Daily	Cost Per Labor-Hour Bare Costs	Incl. O&P
1 Labor Foreman (outside)	$37.10	$296.80	$57.10	$456.80	$37.43	$57.33
3 Laborers	35.10	842.40	54.00	1296.00		
1 Equip. Oper. (light)	44.75	358.00	67.55	540.40		
1 Cement Mixer, 2 C.Y.		190.40		209.44		
1 Air Compressor, 160 cfm		136.80		150.48	8.18	9.00
40 L.H., Daily Totals		$1824.40		$2653.12	$45.61	$66.33

Crew B-62

Crew No.	Bare Costs Hr.	Daily	Incl. Subs O&P Hr.	Daily	Cost Per Labor-Hour Bare Costs	Incl. O&P
2 Laborers	$35.10	$561.60	$54.00	$864.00	$38.32	$58.52
1 Equip. Oper. (light)	44.75	358.00	67.55	540.40		
1 Loader, Skid Steer, 30 H.P.,gas		159.60		175.56	6.65	7.32
24 L.H., Daily Totals		$1079.20		$1579.96	$44.97	$65.83

Crew B-63

Crew No.	Bare Costs Hr.	Daily	Incl. Subs O&P Hr.	Daily	Cost Per Labor-Hour Bare Costs	Incl. O&P
4 Laborers	$35.10	$1123.20	$54.00	$1728.00	$37.03	$56.71
1 Equip. Oper. (light)	44.75	358.00	67.55	540.40		
1 Loader, Skid Steer, 30 H.P.,gas		159.60		175.56	3.99	4.39
40 L.H., Daily Totals		$1640.80		$2443.96	$41.02	$61.10

Crew B-64

Crew No.	Bare Costs Hr.	Daily	Incl. Subs O&P Hr.	Daily	Cost Per Labor-Hour Bare Costs	Incl. O&P
1 Laborer	$35.10	$280.80	$54.00	$432.00	$34.73	$53.05
1 Truck Driver (light)	34.35	274.80	52.10	416.80		
1 Power Mulcher (small)		142.00		156.20		
1 Flatbed Truck, Gas, 1.5 Ton		211.00		232.10	22.06	24.27
16 L.H., Daily Totals		$908.60		$1237.10	$56.79	$77.32

Crew B-65

Crew No.	Bare Costs Hr.	Daily	Incl. Subs O&P Hr.	Daily	Cost Per Labor-Hour Bare Costs	Incl. O&P
1 Laborer	$35.10	$280.80	$54.00	$432.00	$34.73	$53.05
1 Truck Driver (light)	34.35	274.80	52.10	416.80		
1 Power Mulcher (large)		286.60		315.26		
1 Flatbed Truck, Gas, 1.5 Ton		211.00		232.10	31.10	34.21
16 L.H., Daily Totals		$1053.20		$1396.16	$65.83	$87.26

Crews

Crew B-66

Crew No.	Bare Costs Hr.	Daily	Incl. Subs O&P Hr.	Daily	Bare Costs	Incl. O&P
1 Equip. Oper. (light)	$44.75	$358.00	$67.55	$540.40	$44.75	$67.55
1 Loader-Backhoe		247.20		271.92	30.90	33.99
8 L.H., Daily Totals		$605.20		$812.32	$75.65	$101.54

Crew B-67

Crew No.	Bare Costs Hr.	Daily	Incl. Subs O&P Hr.	Daily	Bare Costs	Incl. O&P
1 Millwright	$46.05	$368.40	$67.85	$542.80	$45.40	$67.70
1 Equip. Oper. (light)	44.75	358.00	67.55	540.40		
1 Forklift, R/T, 4,000 Lb.		289.20		318.12	18.07	19.88
16 L.H., Daily Totals		$1015.60		$1401.32	$63.48	$87.58

Crew B-68

Crew No.	Bare Costs Hr.	Daily	Incl. Subs O&P Hr.	Daily	Bare Costs	Incl. O&P
2 Millwrights	$46.05	$736.80	$67.85	$1085.60	$45.62	$67.75
1 Equip. Oper. (light)	44.75	358.00	67.55	540.40		
1 Forklift, R/T, 4,000 Lb.		289.20		318.12	12.05	13.26
24 L.H., Daily Totals		$1384.00		$1944.12	$57.67	$81.00

Crew B-69

Crew No.	Bare Costs Hr.	Daily	Incl. Subs O&P Hr.	Daily	Bare Costs	Incl. O&P
1 Labor Foreman (outside)	$37.10	$296.80	$57.10	$456.80	$38.58	$58.92
3 Laborers	35.10	842.40	54.00	1296.00		
1 Equip Oper. (crane)	47.80	382.40	72.15	577.20		
1 Equip Oper. Oiler	41.25	330.00	62.25	498.00		
1 Hyd. Crane, 80 Ton		1554.00		1709.40	32.38	35.61
48 L.H., Daily Totals		$3405.60		$4537.40	$70.95	$94.53

Crew B-69A

Crew No.	Bare Costs Hr.	Daily	Incl. Subs O&P Hr.	Daily	Bare Costs	Incl. O&P
1 Labor Foreman (outside)	$37.10	$296.80	$57.10	$456.80	$38.55	$58.61
3 Laborers	35.10	842.40	54.00	1296.00		
1 Equip. Oper. (med.)	46.55	372.40	70.25	562.00		
1 Concrete Finisher	42.35	338.80	62.30	498.40		
1 Curb/Gutter Paver, 2-Track		761.00		837.10	15.85	17.44
48 L.H., Daily Totals		$2611.40		$3650.30	$54.40	$76.05

Crew B-69B

Crew No.	Bare Costs Hr.	Daily	Incl. Subs O&P Hr.	Daily	Bare Costs	Incl. O&P
1 Labor Foreman (outside)	$37.10	$296.80	$57.10	$456.80	$38.55	$58.61
3 Laborers	35.10	842.40	54.00	1296.00		
1 Equip. Oper. (med.)	46.55	372.40	70.25	562.00		
1 Cement Finisher	42.35	338.80	62.30	498.40		
1 Curb/Gutter Paver, 4-Track		733.20		806.52	15.28	16.80
48 L.H., Daily Totals		$2583.60		$3619.72	$53.83	$75.41

Crew B-70

Crew No.	Bare Costs Hr.	Daily	Incl. Subs O&P Hr.	Daily	Bare Costs	Incl. O&P
1 Labor Foreman (outside)	$37.10	$296.80	$57.10	$456.80	$40.29	$61.41
3 Laborers	35.10	842.40	54.00	1296.00		
3 Equip. Oper. (med.)	46.55	1117.20	70.25	1686.00		
1 Grader, 30,000 Lbs.		610.80		671.88		
1 Ripper, beam & 1 shank		81.60		89.76		
1 Road Sweeper, S.P., 8' wide		620.80		682.88		
1 F.E. Loader, W.M., 1.5 C.Y.		333.80		367.18	29.41	32.35
56 L.H., Daily Totals		$3903.40		$5250.50	$69.70	$93.76

Crew B-70A

Crew No.	Bare Costs Hr.	Daily	Incl. Subs O&P Hr.	Daily	Bare Costs	Incl. O&P
1 Laborer	$35.10	$280.80	$54.00	$432.00	$44.26	$67.00
4 Equip. Oper. (med.)	46.55	1489.60	70.25	2248.00		
1 Grader, 40,000 Lbs.		1011.00		1112.10		
1 F.E. Loader, W.M., 2.5 C.Y.		472.40		519.64		
1 Dozer, 80 H.P.		442.40		486.64		
1 Roller, Pneum. Whl, 12 Ton		313.20		344.52	55.98	61.57
40 L.H., Daily Totals		$4009.40		$5142.90	$100.24	$128.57

Crew B-71

Crew No.	Bare Costs Hr.	Daily	Incl. Subs O&P Hr.	Daily	Bare Costs	Incl. O&P
1 Labor Foreman (outside)	$37.10	$296.80	$57.10	$456.80	$40.29	$61.41
3 Laborers	35.10	842.40	54.00	1296.00		
3 Equip. Oper. (med.)	46.55	1117.20	70.25	1686.00		
1 Pvmt. Profiler, 750 H.P.		5459.00		6004.90		
1 Road Sweeper, S.P., 8' wide		620.80		682.88		
1 F.E. Loader, W.M., 1.5 C.Y.		333.80		367.18	114.53	125.98
56 L.H., Daily Totals		$8670.00		$10493.76	$154.82	$187.39

Crew B-72

Crew No.	Bare Costs Hr.	Daily	Incl. Subs O&P Hr.	Daily	Bare Costs	Incl. O&P
1 Labor Foreman (outside)	$37.10	$296.80	$57.10	$456.80	$41.08	$62.51
3 Laborers	35.10	842.40	54.00	1296.00		
4 Equip. Oper. (med.)	46.55	1489.60	70.25	2248.00		
1 Pvmt. Profiler, 750 H.P.		5459.00		6004.90		
1 Hammermill, 250 H.P.		1721.00		1893.10		
1 Windrow Loader		1207.00		1327.70		
1 Mix Paver 165 H.P.		2012.00		2213.20		
1 Roller, Pneum. Whl, 12 Ton		313.20		344.52	167.38	184.12
64 L.H., Daily Totals		$13341.00		$15784.22	$208.45	$246.63

Crew B-73

Crew No.	Bare Costs Hr.	Daily	Incl. Subs O&P Hr.	Daily	Bare Costs	Incl. O&P
1 Labor Foreman (outside)	$37.10	$296.80	$57.10	$456.80	$42.51	$64.54
2 Laborers	35.10	561.60	54.00	864.00		
5 Equip. Oper. (med.)	46.55	1862.00	70.25	2810.00		
1 Road Mixer, 310 H.P.		1882.00		2070.20		
1 Tandem Roller, 10 Ton		227.60		250.36		
1 Hammermill, 250 H.P.		1721.00		1893.10		
1 Grader, 30,000 Lbs.		610.80		671.88		
.5 F.E. Loader, W.M., 1.5 C.Y.		166.90		183.59		
.5 Truck Tractor, 220 H.P.		157.40		173.14		
.5 Water Tank Trailer, 5000 Gal.		70.50		77.55	75.57	83.12
64 L.H., Daily Totals		$7556.60		$9450.62	$118.07	$147.67

Crew B-74

Crew No.	Bare Costs Hr.	Daily	Incl. Subs O&P Hr.	Daily	Bare Costs	Incl. O&P
1 Labor Foreman (outside)	$37.10	$296.80	$57.10	$456.80	$41.13	$62.40
1 Laborer	35.10	280.80	54.00	432.00		
4 Equip. Oper. (med.)	46.55	1489.60	70.25	2248.00		
2 Truck Drivers (heavy)	35.30	564.80	53.55	856.80		
1 Grader, 30,000 Lbs.		610.80		671.88		
1 Ripper, beam & 1 shank		81.60		89.76		
2 Stabilizers, 310 H.P.		3170.00		3487.00		
1 Flatbed Truck, Gas, 3 Ton		264.40		290.84		
1 Chem. Spreader, Towed		50.00		55.00		
1 Roller, Vibratory, 25 Ton		640.20		704.22		
1 Water Tank Trailer, 5000 Gal.		141.00		155.10		
1 Truck Tractor, 220 H.P.		314.80		346.28	82.39	90.63
64 L.H., Daily Totals		$7904.80		$9793.68	$123.51	$153.03

Crew B-75

Crew No.	Bare Costs Hr.	Daily	Incl. Subs O&P Hr.	Daily	Bare Costs	Incl. O&P
1 Labor Foreman (outside)	$37.10	$296.80	$57.10	$456.80	$41.96	$63.66
1 Laborer	35.10	280.80	54.00	432.00		
4 Equip. Oper. (med.)	46.55	1489.60	70.25	2248.00		
1 Truck Driver (heavy)	35.30	282.40	53.55	428.40		
1 Grader, 30,000 Lbs.		610.80		671.88		
1 Ripper, beam & 1 shank		81.60		89.76		
2 Stabilizers, 310 H.P.		3170.00		3487.00		
1 Dist. Tanker, 3000 Gallon		306.00		336.60		
1 Truck Tractor, 6x4, 380 H.P.		522.20		574.42		
1 Roller, Vibratory, 25 Ton		640.20		704.22	95.19	104.71
56 L.H., Daily Totals		$7680.40		$9429.08	$137.15	$168.38

Crew No.	Bare Costs		Incl. Subs O&P		Cost Per Labor-Hour	
Crew B-76	**Hr.**	**Daily**	**Hr.**	**Daily**	**Bare Costs**	**Incl. O&P**
1 Dock Builder Foreman (outside)	$44.80	$358.40	$72.00	$576.00	$43.96	$69.17
5 Dock Builders	42.80	1712.00	68.80	2752.00		
2 Equip. Oper. (crane)	47.80	764.80	72.15	1154.40		
1 Equip. Oper. Oiler	41.25	330.00	62.25	498.00		
1 Crawler Crane, 50 Ton		1257.00		1382.70		
1 Barge, 400 Ton		779.20		857.12		
1 Hammer, diesel, 15K ft.-lbs.		591.40		650.54		
1 Lead, 60' high		73.60		80.96		
1 Air Compressor, 600 cfm		467.60		514.36		
2 -50' Air Hoses, 3"		29.80		32.78	44.42	48.87
72 L.H., Daily Totals		$6363.80		$8498.86	$88.39	$118.04
Crew B-76A	**Hr.**	**Daily**	**Hr.**	**Daily**	**Bare Costs**	**Incl. O&P**
1 Labor Foreman (outside)	$37.10	$296.80	$57.10	$456.80	$37.71	$57.69
5 Laborers	35.10	1404.00	54.00	2160.00		
1 Equip. Oper. (crane)	47.80	382.40	72.15	577.20		
1 Equip. Oper. Oiler	41.25	330.00	62.25	498.00		
1 Crawler Crane, 50 Ton		1257.00		1382.70		
1 Barge, 400 Ton		779.20		857.12	31.82	35.00
64 L.H., Daily Totals		$4449.40		$5931.82	$69.52	$92.68
Crew B-77	**Hr.**	**Daily**	**Hr.**	**Daily**	**Bare Costs**	**Incl. O&P**
1 Labor Foreman (outside)	$37.10	$296.80	$57.10	$456.80	$35.35	$54.24
3 Laborers	35.10	842.40	54.00	1296.00		
1 Truck Driver (light)	34.35	274.80	52.10	416.80		
1 Crack Cleaner, 25 H.P.		57.20		62.92		
1 Crack Filler, Trailer Mtd.		196.40		216.04		
1 Flatbed Truck, Gas, 3 Ton		264.40		290.84	12.95	14.24
40 L.H., Daily Totals		$1932.00		$2739.40	$48.30	$68.48
Crew B-78	**Hr.**	**Daily**	**Hr.**	**Daily**	**Bare Costs**	**Incl. O&P**
1 Labor Foreman (outside)	$37.10	$296.80	$57.10	$456.80	$35.31	$54.20
4 Laborers	35.10	1123.20	54.00	1728.00		
1 Truck Driver (light)	34.35	274.80	52.10	416.80		
1 Paint Striper, S.P., 40 Gallon		148.60		163.46		
1 Flatbed Truck, Gas, 3 Ton		264.40		290.84		
1 Pickup Truck, 3/4 Ton		127.00		139.70	11.25	12.38
48 L.H., Daily Totals		$2234.80		$3195.60	$46.56	$66.58
Crew B-78A	**Hr.**	**Daily**	**Hr.**	**Daily**	**Bare Costs**	**Incl. O&P**
1 Equip. Oper. (light)	$44.75	$358.00	$67.55	$540.40	$44.75	$67.55
1 Line Rem. (metal balls) 115 H.P.		800.80		880.88	100.10	110.11
8 L.H., Daily Totals		$1158.80		$1421.28	$144.85	$177.66
Crew B-78B	**Hr.**	**Daily**	**Hr.**	**Daily**	**Bare Costs**	**Incl. O&P**
2 Laborers	$35.10	$561.60	$54.00	$864.00	$36.17	$55.51
.25 Equip. Oper. (light)	44.75	89.50	67.55	135.10		
1 Pickup Truck, 3/4 Ton		127.00		139.70		
1 Line Rem., 11 H.P.,walk behind		61.20		67.32		
.25 Road Sweeper, S.P., 8' wide		155.20		170.72	19.08	20.99
18 L.H., Daily Totals		$994.50		$1376.84	$55.25	$76.49
Crew B-78C	**Hr.**	**Daily**	**Hr.**	**Daily**	**Bare Costs**	**Incl. O&P**
1 Labor Foreman (outside)	$37.10	$296.80	$57.10	$456.80	$35.31	$54.20
4 Laborers	35.10	1123.20	54.00	1728.00		
1 Truck Driver (light)	34.35	274.80	52.10	416.80		
1 Paint Striper, T.M., 120 Gal.		735.80		809.38		
1 Flatbed Truck, Gas, 3 Ton		264.40		290.84		
1 Pickup Truck, 3/4 Ton		127.00		139.70	23.48	25.83
48 L.H., Daily Totals		$2822.00		$3841.52	$58.79	$80.03

Crew No.	Bare Costs		Incl. Subs O&P		Cost Per Labor-Hour	
Crew B-78D	**Hr.**	**Daily**	**Hr.**	**Daily**	**Bare Costs**	**Incl. O&P**
2 Labor Foremen (Outside)	$37.10	$593.60	$57.10	$913.60	$35.42	$54.43
7 Laborers	35.10	1965.60	54.00	3024.00		
1 Truck Driver (light)	34.35	274.80	52.10	416.80		
1 Paint Striper, T.M., 120 Gal.		735.80		809.38		
1 Flatbed Truck, Gas, 3 Ton		264.40		290.84		
3 Pickup Trucks, 3/4 Ton		381.00		419.10		
1 Air Compressor, 60 cfm		120.40		132.44		
1 -50' Air Hose, 3/4"		3.05		3.36		
1 Breakers, Pavement, 60 lb.		9.80		10.78	18.93	20.82
80 L.H., Daily Totals		$4348.45		$6020.30	$54.36	$75.25
Crew B-78E	**Hr.**	**Daily**	**Hr.**	**Daily**	**Bare Costs**	**Incl. O&P**
2 Labor Foremen (Outside)	$37.10	$593.60	$57.10	$913.60	$35.37	$54.36
9 Laborers	35.10	2527.20	54.00	3888.00		
1 Truck Driver (light)	34.35	274.80	52.10	416.80		
1 Paint Striper, T.M., 120 Gal.		735.80		809.38		
1 Flatbed Truck, Gas, 3 Ton		264.40		290.84		
4 Pickup Trucks, 3/4 Ton		508.00		558.80		
2 Air Compressor, 60 cfm		240.80		264.88		
2 -50' Air Hose, 3/4"		6.10		6.71		
2 Breakers, Pavement, 60 lb.		19.60		21.56	18.49	20.34
96 L.H., Daily Totals		$5170.30		$7170.57	$53.86	$74.69
Crew B-78F	**Hr.**	**Daily**	**Hr.**	**Daily**	**Bare Costs**	**Incl. O&P**
2 Labor Foremen (Outside)	$37.10	$593.60	$57.10	$913.60	$35.33	$54.31
11 Laborers	35.10	3088.80	54.00	4752.00		
1 Truck Driver (light)	34.35	274.80	52.10	416.80		
1 Paint Striper, T.M., 120 Gal.		735.80		809.38		
1 Flatbed Truck, Gas, 3 Ton		264.40		290.84		
7 Pickup Trucks, 3/4 Ton		889.00		977.90		
3 Air Compressor, 60 cfm		361.20		397.32		
3 -50' Air Hose, 3/4"		9.15		10.07		
3 Breakers, Pavement, 60 lb.		29.40		32.34	20.44	22.48
112 L.H., Daily Totals		$6246.15		$8600.25	$55.77	$76.79
Crew B-79	**Hr.**	**Daily**	**Hr.**	**Daily**	**Bare Costs**	**Incl. O&P**
1 Labor Foreman (outside)	$37.10	$296.80	$57.10	$456.80	$35.35	$54.24
3 Laborers	35.10	842.40	54.00	1296.00		
1 Truck Driver (light)	34.35	274.80	52.10	416.80		
1 Paint Striper, T.M., 120 Gal.		735.80		809.38		
1 Heating Kettle, 115 Gallon		44.50		48.95		
1 Flatbed Truck, Gas, 3 Ton		264.40		290.84		
2 Pickup Trucks, 3/4 Ton		254.00		279.40	32.47	35.71
40 L.H., Daily Totals		$2712.70		$3598.17	$67.82	$89.95
Crew B-79A	**Hr.**	**Daily**	**Hr.**	**Daily**	**Bare Costs**	**Incl. O&P**
1.5 Equip. Oper. (light)	$44.75	$537.00	$67.55	$810.60	$44.75	$67.55
.5 Line Remov. (grinder) 115 H.P.		441.40		485.54		
1 Line Rem. (metal balls) 115 H.P.		800.80		880.88	103.52	113.87
12 L.H., Daily Totals		$1779.20		$2177.02	$148.27	$181.42
Crew B-79B	**Hr.**	**Daily**	**Hr.**	**Daily**	**Bare Costs**	**Incl. O&P**
1 Laborer	$35.10	$280.80	$54.00	$432.00	$35.10	$54.00
1 Set of Gases		86.40		95.04	10.80	11.88
8 L.H., Daily Totals		$367.20		$527.04	$45.90	$65.88

697

Crew No.	Bare Costs Hr.	Daily	Incl. Subs O&P Hr.	Daily	Bare Costs	Incl. O&P
Crew B-79C	Hr.	Daily	Hr.	Daily	Bare Costs	Incl. O&P
1 Labor Foreman (outside)	$37.10	$296.80	$57.10	$456.80	$35.28	$54.17
5 Laborers	35.10	1404.00	54.00	2160.00		
1 Truck Driver (light)	34.35	274.80	52.10	416.80		
1 Paint Striper, T.M., 120 Gal.		735.80		809.38		
1 Heating Kettle, 115 Gallon		44.50		48.95		
1 Flatbed Truck, Gas, 3 Ton		264.40		290.84		
3 Pickup Trucks, 3/4 Ton		381.00		419.10		
1 Air Compressor, 60 cfm		120.40		132.44		
1 -50' Air Hose, 3/4"		3.05		3.36		
1 Breakers, Pavement, 60 lb.		9.80		10.78	27.84	30.62
56 L.H., Daily Totals		$3534.55		$4748.44	$63.12	$84.79
Crew B-79D	Hr.	Daily	Hr.	Daily	Bare Costs	Incl. O&P
2 Labor Foremen (Outside)	$37.10	$593.60	$57.10	$913.60	$35.51	$54.54
5 Laborers	35.10	1404.00	54.00	2160.00		
1 Truck Driver (light)	34.35	274.80	52.10	416.80		
1 Paint Striper, T.M., 120 Gal.		735.80		809.38		
1 Heating Kettle, 115 Gallon		44.50		48.95		
1 Flatbed Truck, Gas, 3 Ton		264.40		290.84		
4 Pickup Trucks, 3/4 Ton		508.00		558.80		
1 Air Compressor, 60 cfm		120.40		132.44		
1 -50' Air Hose, 3/4"		3.05		3.36		
1 Breakers, Pavement, 60 lb.		9.80		10.78	26.34	28.98
64 L.H., Daily Totals		$3958.35		$5344.94	$61.85	$83.51
Crew B-79E	Hr.	Daily	Hr.	Daily	Bare Costs	Incl. O&P
2 Labor Foremen (Outside)	$37.10	$593.60	$57.10	$913.60	$35.42	$54.43
7 Laborers	35.10	1965.60	54.00	3024.00		
1 Truck Driver (light)	34.35	274.80	52.10	416.80		
1 Paint Striper, T.M., 120 Gal.		735.80		809.38		
1 Heating Kettle, 115 Gallon		44.50		48.95		
1 Flatbed Truck, Gas, 3 Ton		264.40		290.84		
5 Pickup Trucks, 3/4 Ton		635.00		698.50		
2 Air Compressors, 60 cfm		240.80		264.88		
2 -50' Air Hoses, 3/4"		6.10		6.71		
2 Breakers, Pavement, 60 lb.		19.60		21.56	24.33	26.76
80 L.H., Daily Totals		$4780.20		$6495.22	$59.75	$81.19
Crew B-80	Hr.	Daily	Hr.	Daily	Bare Costs	Incl. O&P
1 Labor Foreman (outside)	$37.10	$296.80	$57.10	$456.80	$37.83	$57.69
1 Laborer	35.10	280.80	54.00	432.00		
1 Truck Driver (light)	34.35	274.80	52.10	416.80		
1 Equip. Oper. (light)	44.75	358.00	67.55	540.40		
1 Flatbed Truck, Gas, 3 Ton		264.40		290.84		
1 Earth Auger, Truck-Mtd.		403.40		443.74	20.87	22.96
32 L.H., Daily Totals		$1878.20		$2580.58	$58.69	$80.64
Crew B-80A	Hr.	Daily	Hr.	Daily	Bare Costs	Incl. O&P
3 Laborers	$35.10	$842.40	$54.00	$1296.00	$35.10	$54.00
1 Flatbed Truck, Gas, 3 Ton		264.40		290.84	11.02	12.12
24 L.H., Daily Totals		$1106.80		$1586.84	$46.12	$66.12
Crew B-80B	Hr.	Daily	Hr.	Daily	Bare Costs	Incl. O&P
3 Laborers	$35.10	$842.40	$54.00	$1296.00	$37.51	$57.39
1 Equip. Oper. (light)	44.75	358.00	67.55	540.40		
1 Crane, Flatbed Mounted, 3 Ton		232.40		255.64	7.26	7.99
32 L.H., Daily Totals		$1432.80		$2092.04	$44.77	$65.38

Crew No.	Bare Costs Hr.	Daily	Incl. Subs O&P Hr.	Daily	Bare Costs	Incl. O&P
Crew B-80C	Hr.	Daily	Hr.	Daily	Bare Costs	Incl. O&P
2 Laborers	$35.10	$561.60	$54.00	$864.00	$34.85	$53.37
1 Truck Driver (light)	34.35	274.80	52.10	416.80		
1 Flatbed Truck, Gas, 1.5 Ton		211.00		232.10		
1 Manual fence post auger, gas		7.60		8.36	9.11	10.02
24 L.H., Daily Totals		$1055.00		$1521.26	$43.96	$63.39
Crew B-81	Hr.	Daily	Hr.	Daily	Bare Costs	Incl. O&P
1 Laborer	$35.10	$280.80	$54.00	$432.00	$38.98	$59.27
1 Equip. Oper. (med.)	46.55	372.40	70.25	562.00		
1 Truck Driver (heavy)	35.30	282.40	53.55	428.40		
1 Hydromulcher, T.M., 3000 Gal.		303.80		334.18		
1 Truck Tractor, 220 H.P.		314.80		346.28	25.77	28.35
24 L.H., Daily Totals		$1554.20		$2102.86	$64.76	$87.62
Crew B-81A	Hr.	Daily	Hr.	Daily	Bare Costs	Incl. O&P
1 Laborer	$35.10	$280.80	$54.00	$432.00	$34.73	$53.05
1 Truck Driver (light)	34.35	274.80	52.10	416.80		
1 Hydromulcher, T.M., 600 Gal.		120.40		132.44		
1 Flatbed Truck, Gas, 3 Ton		264.40		290.84	24.05	26.45
16 L.H., Daily Totals		$940.40		$1272.08	$58.77	$79.50
Crew B-82	Hr.	Daily	Hr.	Daily	Bare Costs	Incl. O&P
1 Laborer	$35.10	$280.80	$54.00	$432.00	$39.92	$60.77
1 Equip. Oper. (light)	44.75	358.00	67.55	540.40		
1 Horiz. Borer, 6 H.P.		77.20		84.92	4.83	5.31
16 L.H., Daily Totals		$716.00		$1057.32	$44.75	$66.08
Crew B-82A	Hr.	Daily	Hr.	Daily	Bare Costs	Incl. O&P
1 Laborer	$35.10	$280.80	$54.00	$432.00	$39.92	$60.77
1 Equip. Oper. (light)	44.75	358.00	67.55	540.40		
1 Flatbed Truck, Gas, 3 Ton		264.40		290.84		
1 Flatbed Trailer, 25 Ton		108.60		119.46		
1 Horiz. Dir. Drill, 20k lb. thrust		622.00		684.20	62.19	68.41
16 L.H., Daily Totals		$1633.80		$2066.90	$102.11	$129.18
Crew B-82B	Hr.	Daily	Hr.	Daily	Bare Costs	Incl. O&P
2 Laborers	$35.10	$561.60	$54.00	$864.00	$38.32	$58.52
1 Equip. Oper. (light)	44.75	358.00	67.55	540.40		
1 Flatbed Truck, Gas, 3 Ton		264.40		290.84		
1 Flatbed Trailer, 25 Ton		108.60		119.46		
1 Horiz. Dir. Drill, 30k lb. thrust		886.20		974.82	52.47	57.71
24 L.H., Daily Totals		$2178.80		$2789.52	$90.78	$116.23
Crew B-82C	Hr.	Daily	Hr.	Daily	Bare Costs	Incl. O&P
2 Laborers	$35.10	$561.60	$54.00	$864.00	$38.32	$58.52
1 Equip. Oper. (light)	44.75	358.00	67.55	540.40		
1 Flatbed Truck, Gas, 3 Ton		264.40		290.84		
1 Flatbed Trailer, 25 Ton		108.60		119.46		
1 Horiz. Dir. Drill, 50k lb. thrust		1176.00		1293.60	64.54	71.00
24 L.H., Daily Totals		$2468.60		$3108.30	$102.86	$129.51
Crew B-82D	Hr.	Daily	Hr.	Daily	Bare Costs	Incl. O&P
1 Equip. Oper. (light)	$44.75	$358.00	$67.55	$540.40	$44.75	$67.55
1 Mud Trailer for HDD, 1500 gallon		309.20		340.12	38.65	42.52
8 L.H., Daily Totals		$667.20		$880.52	$83.40	$110.07

Crew No.	Bare Costs Hr.	Daily	Incl. Subs O&P Hr.	Daily	Cost Per Labor-Hour Bare Costs	Incl. O&P
Crew B-83	Hr.	Daily	Hr.	Daily	Bare Costs	Incl. O&P
1 Tugboat Captain	$46.55	$372.40	$70.25	$562.00	$40.83	$62.13
1 Tugboat Hand	35.10	280.80	54.00	432.00		
1 Tugboat, 250 H.P.		748.00		822.80	46.75	51.42
16 L.H., Daily Totals		$1401.20		$1816.80	$87.58	$113.55

Crew B-84	Hr.	Daily	Hr.	Daily	Bare Costs	Incl. O&P
1 Equip. Oper. (med.)	$46.55	$372.40	$70.25	$562.00	$46.55	$70.25
1 Rotary Mower/Tractor		343.40		377.74	42.92	47.22
8 L.H., Daily Totals		$715.80		$939.74	$89.47	$117.47

Crew B-85	Hr.	Daily	Hr.	Daily	Bare Costs	Incl. O&P
3 Laborers	$35.10	$842.40	$54.00	$1296.00	$37.43	$57.16
1 Equip. Oper. (med.)	46.55	372.40	70.25	562.00		
1 Truck Driver (heavy)	35.30	282.40	53.55	428.40		
1 Aerial Lift Truck, 80'		637.60		701.36		
1 Brush Chipper, 12", 130 H.P.		283.80		312.18		
1 Pruning Saw, Rotary		6.30		6.93	23.19	25.51
40 L.H., Daily Totals		$2424.90		$3306.87	$60.62	$82.67

Crew B-86	Hr.	Daily	Hr.	Daily	Bare Costs	Incl. O&P
1 Equip. Oper. (med.)	$46.55	$372.40	$70.25	$562.00	$46.55	$70.25
1 Stump Chipper, S.P.		170.70		187.77	21.34	23.47
8 L.H., Daily Totals		$543.10		$749.77	$67.89	$93.72

Crew B-86A	Hr.	Daily	Hr.	Daily	Bare Costs	Incl. O&P
1 Equip. Oper. (med.)	$46.55	$372.40	$70.25	$562.00	$46.55	$70.25
1 Grader, 30,000 Lbs.		610.80		671.88	76.35	83.98
8 L.H., Daily Totals		$983.20		$1233.88	$122.90	$154.24

Crew B-86B	Hr.	Daily	Hr.	Daily	Bare Costs	Incl. O&P
1 Equip. Oper. (med.)	$46.55	$372.40	$70.25	$562.00	$46.55	$70.25
1 Dozer, 200 H.P.		1243.00		1367.30	155.38	170.91
8 L.H., Daily Totals		$1615.40		$1929.30	$201.93	$241.16

Crew B-87	Hr.	Daily	Hr.	Daily	Bare Costs	Incl. O&P
1 Laborer	$35.10	$280.80	$54.00	$432.00	$44.26	$67.00
4 Equip. Oper. (med.)	46.55	1489.60	70.25	2248.00		
2 Feller Bunchers, 100 H.P.		1154.40		1269.84		
1 Log Chipper, 22" Tree		710.60		781.66		
1 Dozer, 105 H.P.		550.80		605.88		
1 Chain Saw, gas, 36" Long		43.60		47.96	61.48	67.63
40 L.H., Daily Totals		$4229.80		$5385.34	$105.75	$134.63

Crew B-88	Hr.	Daily	Hr.	Daily	Bare Costs	Incl. O&P
1 Laborer	$35.10	$280.80	$54.00	$432.00	$44.91	$67.93
6 Equip. Oper. (med.)	46.55	2234.40	70.25	3372.00		
2 Feller Bunchers, 100 H.P.		1154.40		1269.84		
1 Log Chipper, 22" Tree		710.60		781.66		
2 Log Skidders, 50 H.P.		1730.40		1903.44		
1 Dozer, 105 H.P.		550.80		605.88		
1 Chain Saw, gas, 36" Long		43.60		47.96	74.82	82.30
56 L.H., Daily Totals		$6705.00		$8412.78	$119.73	$150.23

Crew B-89	Hr.	Daily	Hr.	Daily	Bare Costs	Incl. O&P
1 Equip. Oper. (light)	$44.75	$358.00	$67.55	$540.40	$39.55	$59.83
1 Truck Driver (light)	34.35	274.80	52.10	416.80		
1 Flatbed Truck, Gas, 3 Ton		264.40		290.84		
1 Concrete Saw		148.00		162.80		
1 Water Tank, 65 Gal.		17.40		19.14	26.86	29.55
16 L.H., Daily Totals		$1062.60		$1429.98	$66.41	$89.37

Crew No.	Bare Costs Hr.	Daily	Incl. Subs O&P Hr.	Daily	Cost Per Labor-Hour Bare Costs	Incl. O&P
Crew B-89A	Hr.	Daily	Hr.	Daily	Bare Costs	Incl. O&P
1 Skilled Worker	$45.40	$363.20	$70.20	$561.60	$40.25	$62.10
1 Laborer	35.10	280.80	54.00	432.00		
1 Core Drill (large)		110.20		121.22	6.89	7.58
16 L.H., Daily Totals		$754.20		$1114.82	$47.14	$69.68

Crew B-89B	Hr.	Daily	Hr.	Daily	Bare Costs	Incl. O&P
1 Equip. Oper. (light)	$44.75	$358.00	$67.55	$540.40	$39.55	$59.83
1 Truck Driver, Light	34.35	274.80	52.10	416.80		
1 Wall Saw, Hydraulic, 10 H.P.		92.00		101.20		
1 Generator, Diesel, 100 kW		363.40		399.74		
1 Water Tank, 65 Gal.		17.40		19.14		
1 Flatbed Truck, Gas, 3 Ton		264.40		290.84	46.08	50.68
16 L.H., Daily Totals		$1370.00		$1768.12	$85.63	$110.51

Crew B-90	Hr.	Daily	Hr.	Daily	Bare Costs	Incl. O&P
1 Labor Foreman (outside)	$37.10	$296.80	$57.10	$456.80	$37.81	$57.66
3 Laborers	35.10	842.40	54.00	1296.00		
2 Equip. Oper. (light)	44.75	716.00	67.55	1080.80		
2 Truck Drivers (heavy)	35.30	564.80	53.55	856.80		
1 Road Mixer, 310 H.P.		1882.00		2070.20		
1 Dist. Truck, 2000 Gal.		276.40		304.04	33.73	37.10
64 L.H., Daily Totals		$4578.40		$6064.64	$71.54	$94.76

Crew B-90A	Hr.	Daily	Hr.	Daily	Bare Costs	Incl. O&P
1 Labor Foreman (outside)	$37.10	$296.80	$57.10	$456.80	$41.93	$63.73
2 Laborers	35.10	561.60	54.00	864.00		
4 Equip. Oper. (med.)	46.55	1489.60	70.25	2248.00		
2 Graders, 30,000 Lbs.		1221.60		1343.76		
1 Tandem Roller, 10 Ton		227.60		250.36		
1 Roller, Pneum. Whl, 12 Ton		313.20		344.52	31.47	34.62
56 L.H., Daily Totals		$4110.40		$5507.44	$73.40	$98.35

Crew B-90B	Hr.	Daily	Hr.	Daily	Bare Costs	Incl. O&P
1 Labor Foreman (outside)	$37.10	$296.80	$57.10	$456.80	$41.16	$62.64
2 Laborers	35.10	561.60	54.00	864.00		
3 Equip. Oper. (med.)	46.55	1117.20	70.25	1686.00		
1 Roller, Pneum. Whl, 12 Ton		313.20		344.52		
1 Road Mixer, 310 H.P.		1882.00		2070.20	45.73	50.31
48 L.H., Daily Totals		$4170.80		$5421.52	$86.89	$112.95

Crew B-90C	Hr.	Daily	Hr.	Daily	Bare Costs	Incl. O&P
1 Labor Foreman (outside)	$37.10	$296.80	$57.10	$456.80	$38.46	$58.59
4 Laborers	35.10	1123.20	54.00	1728.00		
3 Equip. Oper. (med.)	46.55	1117.20	70.25	1686.00		
3 Truck Drivers (heavy)	35.30	847.20	53.55	1285.20		
3 Road Mixers, 310 H.P.		5646.00		6210.60	64.16	70.58
88 L.H., Daily Totals		$9030.40		$11366.60	$102.62	$129.17

Crew B-90D	Hr.	Daily	Hr.	Daily	Bare Costs	Incl. O&P
1 Labor Foreman (outside)	$37.10	$296.80	$57.10	$456.80	$37.94	$57.88
6 Laborers	35.10	1684.80	54.00	2592.00		
3 Equip. Oper. (med.)	46.55	1117.20	70.25	1686.00		
3 Truck Drivers (heavy)	35.30	847.20	53.55	1285.20		
3 Road Mixers, 310 H.P.		5646.00		6210.60	54.29	59.72
104 L.H., Daily Totals		$9592.00		$12230.60	$92.23	$117.60

Crew No.	Bare Costs Hr.	Daily	Incl. Subs O&P Hr.	Daily	Cost Per Labor-Hour Bare Costs	Incl. O&P
Crew B-91						
1 Labor Foreman (outside)	$37.10	$296.80	$57.10	$456.80	$41.10	$62.46
2 Laborers	35.10	561.60	54.00	864.00		
4 Equip. Oper. (med.)	46.55	1489.60	70.25	2248.00		
1 Truck Driver (heavy)	35.30	282.40	53.55	428.40		
1 Dist. Tanker, 3000 Gallon		306.00		336.60		
1 Truck Tractor, 6x4, 380 H.P.		522.20		574.42		
1 Aggreg. Spreader, S.P.		781.80		859.98		
1 Roller, Pneum. Whl, 12 Ton		313.20		344.52		
1 Tandem Roller, 10 Ton		227.60		250.36	33.61	36.97
64 L.H., Daily Totals		$4781.20		$6363.08	$74.71	$99.42
Crew B-91B						
1 Laborer	$35.10	$280.80	$54.00	$432.00	$40.83	$62.13
1 Equipment Oper. (med.)	46.55	372.40	70.25	562.00		
1 Road Sweeper, Vac. Assist.		876.60		964.26	54.79	60.27
16 L.H., Daily Totals		$1529.80		$1958.26	$95.61	$122.39
Crew B-91C						
1 Laborer	$35.10	$280.80	$54.00	$432.00	$34.73	$53.05
1 Truck Driver (light)	34.35	274.80	52.10	416.80		
1 Catch Basin Cleaning Truck		535.80		589.38	33.49	36.84
16 L.H., Daily Totals		$1091.40		$1438.18	$68.21	$89.89
Crew B-91D						
1 Labor Foreman (outside)	$37.10	$296.80	$57.10	$456.80	$39.69	$60.42
5 Laborers	35.10	1404.00	54.00	2160.00		
5 Equip. Oper. (med.)	46.55	1862.00	70.25	2810.00		
2 Truck Drivers (heavy)	35.30	564.80	53.55	856.80		
1 Aggreg. Spreader, S.P.		781.80		859.98		
2 Truck Tractor, 6x4, 380 H.P.		1044.40		1148.84		
2 Dist. Tanker, 3000 Gallon		612.00		673.20		
2 Pavement brush, towed		163.60		179.96		
2 Roller, Pneum. Whl, 12 Ton		626.40		689.04	31.04	34.14
104 L.H., Daily Totals		$7355.80		$9834.62	$70.73	$94.56
Crew B-92						
1 Labor Foreman (outside)	$37.10	$296.80	$57.10	$456.80	$35.60	$54.77
3 Laborers	35.10	842.40	54.00	1296.00		
1 Crack Cleaner, 25 H.P.		57.20		62.92		
1 Air Compressor, 60 cfm		120.40		132.44		
1 Tar Kettle, T.M.		77.20		84.92		
1 Flatbed Truck, Gas, 3 Ton		264.40		290.84	16.23	17.85
32 L.H., Daily Totals		$1658.40		$2323.92	$51.83	$72.62
Crew B-93						
1 Equip. Oper. (med.)	$46.55	$372.40	$70.25	$562.00	$46.55	$70.25
1 Feller Buncher, 100 H.P.		577.20		634.92	72.15	79.36
8 L.H., Daily Totals		$949.60		$1196.92	$118.70	$149.62
Crew B-94A						
1 Laborer	$35.10	$280.80	$54.00	$432.00	$35.10	$54.00
1 Diaphragm Water Pump, 2"		68.60		75.46		
1 -20' Suction Hose, 2"		1.95		2.15		
2 -50' Discharge Hoses, 2"		3.50		3.85	9.26	10.18
8 L.H., Daily Totals		$354.85		$513.46	$44.36	$64.18

Crew No.	Bare Costs Hr.	Daily	Incl. Subs O&P Hr.	Daily	Cost Per Labor-Hour Bare Costs	Incl. O&P
Crew B-94B						
1 Laborer	$35.10	$280.80	$54.00	$432.00	$35.10	$54.00
1 Diaphragm Water Pump, 4"		111.20		122.32		
1 -20' Suction Hose, 4"		3.25		3.58		
2 -50' Discharge Hoses, 4"		9.40		10.34	15.48	17.03
8 L.H., Daily Totals		$404.65		$568.24	$50.58	$71.03
Crew B-94C						
1 Laborer	$35.10	$280.80	$54.00	$432.00	$35.10	$54.00
1 Centrifugal Water Pump, 3"		73.80		81.18		
1 -20' Suction Hose, 3"		2.85		3.13		
2 -50' Discharge Hoses, 3"		6.10		6.71	10.34	11.38
8 L.H., Daily Totals		$363.55		$523.02	$45.44	$65.38
Crew B-94D						
1 Laborer	$35.10	$280.80	$54.00	$432.00	$35.10	$54.00
1 Centr. Water Pump, 6"		308.40		339.24		
1 -20' Suction Hose, 6"		11.50		12.65		
2 -50' Discharge Hoses, 6"		24.20		26.62	43.01	47.31
8 L.H., Daily Totals		$624.90		$810.51	$78.11	$101.31
Crew C-1						
3 Carpenters	$44.10	$1058.40	$67.85	$1628.40	$41.85	$64.39
1 Laborer	35.10	280.80	54.00	432.00		
32 L.H., Daily Totals		$1339.20		$2060.40	$41.85	$64.39
Crew C-2						
1 Carpenter Foreman (outside)	$46.10	$368.80	$70.95	$567.60	$42.93	$66.06
4 Carpenters	44.10	1411.20	67.85	2171.20		
1 Laborer	35.10	280.80	54.00	432.00		
48 L.H., Daily Totals		$2060.80		$3170.80	$42.93	$66.06
Crew C-2A						
1 Carpenter Foreman (outside)	$46.10	$368.80	$70.95	$567.60	$42.64	$65.13
3 Carpenters	44.10	1058.40	67.85	1628.40		
1 Cement Finisher	42.35	338.80	62.30	498.40		
1 Laborer	35.10	280.80	54.00	432.00		
48 L.H., Daily Totals		$2046.80		$3126.40	$42.64	$65.13
Crew C-3						
1 Rodman Foreman (outside)	$51.10	$408.80	$81.60	$652.80	$45.31	$71.34
4 Rodmen (reinf.)	49.10	1571.20	78.40	2508.80		
1 Equip. Oper. (light)	44.75	358.00	67.55	540.40		
2 Laborers	35.10	561.60	54.00	864.00		
3 Stressing Equipment		30.60		33.66		
.5 Grouting Equipment		77.30		85.03	1.69	1.85
64 L.H., Daily Totals		$3007.50		$4684.69	$46.99	$73.20
Crew C-4						
1 Rodman Foreman (outside)	$51.10	$408.80	$81.60	$652.80	$49.60	$79.20
3 Rodmen (reinf.)	49.10	1178.40	78.40	1881.60		
3 Stressing Equipment		30.60		33.66	.96	1.05
32 L.H., Daily Totals		$1617.80		$2568.06	$50.56	$80.25
Crew C-4A						
2 Rodmen (reinf.)	$49.10	$785.60	$78.40	$1254.40	$49.10	$78.40
4 Stressing Equipment		40.80		44.88	2.55	2.81
16 L.H., Daily Totals		$826.40		$1299.28	$51.65	$81.20

Crew No.	Bare Costs		Incl. Subs O&P		Cost Per Labor-Hour	

Crew C-5

Crew C-5	Hr.	Daily	Hr.	Daily	Bare Costs	Incl. O&P
1 Rodman Foreman (outside)	$51.10	$408.80	$81.60	$652.80	$48.08	$75.66
4 Rodmen (reinf.)	49.10	1571.20	78.40	2508.80		
1 Equip. Oper. (crane)	47.80	382.40	72.15	577.20		
1 Equip. Oper. Oiler	41.25	330.00	62.25	498.00		
1 Hyd. Crane, 25 Ton		697.20		766.92	12.45	13.70
56 L.H., Daily Totals		$3389.60		$5003.72	$60.53	$89.35

Crew C-6	Hr.	Daily	Hr.	Daily	Bare Costs	Incl. O&P
1 Labor Foreman (outside)	$37.10	$296.80	$57.10	$456.80	$36.64	$55.90
4 Laborers	35.10	1123.20	54.00	1728.00		
1 Cement Finisher	42.35	338.80	62.30	498.40		
2 Gas Engine Vibrators		54.80		60.28	1.14	1.26
48 L.H., Daily Totals		$1813.60		$2743.48	$37.78	$57.16

Crew C-7	Hr.	Daily	Hr.	Daily	Bare Costs	Incl. O&P
1 Labor Foreman (outside)	$37.10	$296.80	$57.10	$456.80	$38.08	$57.99
5 Laborers	35.10	1404.00	54.00	2160.00		
1 Cement Finisher	42.35	338.80	62.30	498.40		
1 Equip. Oper. (med.)	46.55	372.40	70.25	562.00		
1 Equip. Oper. (oiler)	41.25	330.00	62.25	498.00		
2 Gas Engine Vibrators		54.80		60.28		
1 Concrete Bucket, 1 C.Y.		22.40		24.64		
1 Hyd. Crane, 55 Ton		1100.00		1210.00	16.35	17.98
72 L.H., Daily Totals		$3919.20		$5470.12	$54.43	$75.97

Crew C-7A	Hr.	Daily	Hr.	Daily	Bare Costs	Incl. O&P
1 Labor Foreman (outside)	$37.10	$296.80	$57.10	$456.80	$35.40	$54.27
5 Laborers	35.10	1404.00	54.00	2160.00		
2 Truck Drivers (Heavy)	35.30	564.80	53.55	856.80		
2 Conc. Transit Mixers		1855.60		2041.16	28.99	31.89
64 L.H., Daily Totals		$4121.20		$5514.76	$64.39	$86.17

Crew C-7B	Hr.	Daily	Hr.	Daily	Bare Costs	Incl. O&P
1 Labor Foreman (outside)	$37.10	$296.80	$57.10	$456.80	$37.71	$57.69
5 Laborers	35.10	1404.00	54.00	2160.00		
1 Equipment Oper. (crane)	47.80	382.40	72.15	577.20		
1 Equipment Oiler	41.25	330.00	62.25	498.00		
1 Conc. Bucket, 2 C.Y.		35.20		38.72		
1 Lattice Boom Crane, 165 Ton		1907.00		2097.70	30.35	33.38
64 L.H., Daily Totals		$4355.40		$5828.42	$68.05	$91.07

Crew C-7C	Hr.	Daily	Hr.	Daily	Bare Costs	Incl. O&P
1 Labor Foreman (outside)	$37.10	$296.80	$57.10	$456.80	$38.21	$58.45
5 Laborers	35.10	1404.00	54.00	2160.00		
2 Equipment Operators (med.)	46.55	744.80	70.25	1124.00		
2 F.E. Loaders, W.M., 4 C.Y.		1222.00		1344.20	19.09	21.00
64 L.H., Daily Totals		$3667.60		$5085.00	$57.31	$79.45

Crew C-7D	Hr.	Daily	Hr.	Daily	Bare Costs	Incl. O&P
1 Labor Foreman (outside)	$37.10	$296.80	$57.10	$456.80	$37.02	$56.76
5 Laborers	35.10	1404.00	54.00	2160.00		
1 Equip. Oper. (med.)	46.55	372.40	70.25	562.00		
1 Concrete Conveyer		186.00		204.60	3.32	3.65
56 L.H., Daily Totals		$2259.20		$3383.40	$40.34	$60.42

Crew C-8	Hr.	Daily	Hr.	Daily	Bare Costs	Incl. O&P
1 Labor Foreman (outside)	$37.10	$296.80	$57.10	$456.80	$39.09	$59.14
3 Laborers	35.10	842.40	54.00	1296.00		
2 Cement Finishers	42.35	677.60	62.30	996.80		
1 Equip. Oper. (med.)	46.55	372.40	70.25	562.00		
1 Concrete Pump (small)		715.60		787.16	12.78	14.06
56 L.H., Daily Totals		$2904.80		$4098.76	$51.87	$73.19

Crew C-8A	Hr.	Daily	Hr.	Daily	Bare Costs	Incl. O&P
1 Labor Foreman (outside)	$37.10	$296.80	$57.10	$456.80	$37.85	$57.28
3 Laborers	35.10	842.40	54.00	1296.00		
2 Cement Finishers	42.35	677.60	62.30	996.80		
48 L.H., Daily Totals		$1816.80		$2749.60	$37.85	$57.28

Crew C-8B	Hr.	Daily	Hr.	Daily	Bare Costs	Incl. O&P
1 Labor Foreman (outside)	$37.10	$296.80	$57.10	$456.80	$37.79	$57.87
3 Laborers	35.10	842.40	54.00	1296.00		
1 Equip. Oper. (med.)	46.55	372.40	70.25	562.00		
1 Vibrating Power Screed		65.50		72.05		
1 Roller, Vibratory, 25 Ton		640.20		704.22		
1 Dozer, 200 H.P.		1243.00		1367.30	48.72	53.59
40 L.H., Daily Totals		$3460.30		$4458.37	$86.51	$111.46

Crew C-8C	Hr.	Daily	Hr.	Daily	Bare Costs	Incl. O&P
1 Labor Foreman (outside)	$37.10	$296.80	$57.10	$456.80	$38.55	$58.61
3 Laborers	35.10	842.40	54.00	1296.00		
1 Cement Finisher	42.35	338.80	62.30	498.40		
1 Equip. Oper. (med.)	46.55	372.40	70.25	562.00		
1 Shotcrete Rig, 12 C.Y./hr		242.00		266.20		
1 Air Compressor, 160 cfm		136.80		150.48		
4 -50' Air Hoses, 1"		16.40		18.04		
4 -50' Air Hoses, 2"		31.00		34.10	8.88	9.77
48 L.H., Daily Totals		$2276.60		$3282.02	$47.43	$68.38

Crew C-8D	Hr.	Daily	Hr.	Daily	Bare Costs	Incl. O&P
1 Labor Foreman (outside)	$37.10	$296.80	$57.10	$456.80	$39.83	$60.24
1 Laborer	35.10	280.80	54.00	432.00		
1 Cement Finisher	42.35	338.80	62.30	498.40		
1 Equipment Oper. (light)	44.75	358.00	67.55	540.40		
1 Air Compressor, 250 cfm		177.00		194.70		
2 -50' Air Hoses, 1"		8.20		9.02	5.79	6.37
32 L.H., Daily Totals		$1459.60		$2131.32	$45.61	$66.60

Crew C-8E	Hr.	Daily	Hr.	Daily	Bare Costs	Incl. O&P
1 Labor Foreman (outside)	$37.10	$296.80	$57.10	$456.80	$38.25	$58.16
3 Laborers	35.10	842.40	54.00	1296.00		
1 Cement Finisher	42.35	338.80	62.30	498.40		
1 Equipment Oper. (light)	44.75	358.00	67.55	540.40		
1 Shotcrete Rig, 35 C.Y./hr		268.80		295.68		
1 Air Compressor, 250 cfm		177.00		194.70		
4 -50' Air Hoses, 1"		16.40		18.04		
4 -50' Air Hoses, 2"		31.00		34.10	10.28	11.30
48 L.H., Daily Totals		$2329.20		$3334.12	$48.52	$69.46

Crew C-10	Hr.	Daily	Hr.	Daily	Bare Costs	Incl. O&P
1 Laborer	$35.10	$280.80	$54.00	$432.00	$39.93	$59.53
2 Cement Finishers	42.35	677.60	62.30	996.80		
24 L.H., Daily Totals		$958.40		$1428.80	$39.93	$59.53

Crew C-10B

Crew No.	Bare Costs Hr.	Daily	Incl. Subs O&P Hr.	Daily	Cost Per Labor-Hour Bare Costs	Incl. O&P
3 Laborers	$35.10	$842.40	$54.00	$1296.00	$38.00	$57.32
2 Cement Finishers	42.35	677.60	62.30	996.80		
1 Concrete Mixer, 10 C.F.		160.60		176.66		
2 Trowels, 48" Walk-Behind		89.20		98.12	6.25	6.87
40 L.H., Daily Totals		$1769.80		$2567.58	$44.24	$64.19

Crew C-10C

Crew No.	Bare Costs Hr.	Daily	Incl. Subs O&P Hr.	Daily	Cost Per Labor-Hour Bare Costs	Incl. O&P
1 Laborer	$35.10	$280.80	$54.00	$432.00	$39.93	$59.53
2 Cement Finishers	42.35	677.60	62.30	996.80		
1 Trowel, 48" Walk-Behind		44.60		49.06	1.86	2.04
24 L.H., Daily Totals		$1003.00		$1477.86	$41.79	$61.58

Crew C-10D

Crew No.	Bare Costs Hr.	Daily	Incl. Subs O&P Hr.	Daily	Cost Per Labor-Hour Bare Costs	Incl. O&P
1 Laborer	$35.10	$280.80	$54.00	$432.00	$39.93	$59.53
2 Cement Finishers	42.35	677.60	62.30	996.80		
1 Vibrating Power Screed		65.50		72.05		
1 Trowel, 48" Walk-Behind		44.60		49.06	4.59	5.05
24 L.H., Daily Totals		$1068.50		$1549.91	$44.52	$64.58

Crew C-10E

Crew No.	Bare Costs Hr.	Daily	Incl. Subs O&P Hr.	Daily	Cost Per Labor-Hour Bare Costs	Incl. O&P
1 Laborer	$35.10	$280.80	$54.00	$432.00	$39.93	$59.53
2 Cement Finishers	42.35	677.60	62.30	996.80		
1 Vibrating Power Screed		65.50		72.05		
1 Cement Trowel, 96" Ride-On		174.20		191.62	9.99	10.99
24 L.H., Daily Totals		$1198.10		$1692.47	$49.92	$70.52

Crew C-11

Crew No.	Bare Costs Hr.	Daily	Incl. Subs O&P Hr.	Daily	Cost Per Labor-Hour Bare Costs	Incl. O&P
1 Struc. Steel Foreman (outside)	$51.35	$410.80	$91.80	$734.40	$48.50	$83.97
6 Struc. Steel Workers	49.35	2368.80	88.25	4236.00		
1 Equip. Oper. (crane)	47.80	382.40	72.15	577.20		
1 Equip. Oper. Oiler	41.25	330.00	62.25	498.00		
1 Lattice Boom Crane, 150 Ton		1829.00		2011.90	25.40	27.94
72 L.H., Daily Totals		$5321.00		$8057.50	$73.90	$111.91

Crew C-12

Crew No.	Bare Costs Hr.	Daily	Incl. Subs O&P Hr.	Daily	Cost Per Labor-Hour Bare Costs	Incl. O&P
1 Carpenter Foreman (outside)	$46.10	$368.80	$70.95	$567.60	$43.55	$66.78
3 Carpenters	44.10	1058.40	67.85	1628.40		
1 Laborer	35.10	280.80	54.00	432.00		
1 Equip. Oper. (crane)	47.80	382.40	72.15	577.20		
1 Hyd. Crane, 12 Ton		612.00		673.20	12.75	14.03
48 L.H., Daily Totals		$2702.40		$3878.40	$56.30	$80.80

Crew C-13

Crew No.	Bare Costs Hr.	Daily	Incl. Subs O&P Hr.	Daily	Cost Per Labor-Hour Bare Costs	Incl. O&P
1 Struc. Steel Worker	$49.35	$394.80	$88.25	$706.00	$47.60	$81.45
1 Welder	49.35	394.80	88.25	706.00		
1 Carpenter	44.10	352.80	67.85	542.80		
1 Welder, gas engine, 300 amp		121.60		133.76	5.07	5.57
24 L.H., Daily Totals		$1264.00		$2088.56	$52.67	$87.02

Crew C-14

Crew No.	Bare Costs Hr.	Daily	Incl. Subs O&P Hr.	Daily	Cost Per Labor-Hour Bare Costs	Incl. O&P
1 Carpenter Foreman (outside)	$46.10	$368.80	$70.95	$567.60	$43.17	$66.60
5 Carpenters	44.10	1764.00	67.85	2714.00		
4 Laborers	35.10	1123.20	54.00	1728.00		
4 Rodmen (reinf.)	49.10	1571.20	78.40	2508.80		
2 Cement Finishers	42.35	677.60	62.30	996.80		
1 Equip. Oper. (crane)	47.80	382.40	72.15	577.20		
1 Equip. Oper. Oiler	41.25	330.00	62.25	498.00		
1 Hyd. Crane, 80 Ton		1554.00		1709.40	10.79	11.87
144 L.H., Daily Totals		$7771.20		$11299.80	$53.97	$78.47

Crew C-14A

Crew No.	Bare Costs Hr.	Daily	Incl. Subs O&P Hr.	Daily	Cost Per Labor-Hour Bare Costs	Incl. O&P
1 Carpenter Foreman (outside)	$46.10	$368.80	$70.95	$567.60	$44.29	$68.43
16 Carpenters	44.10	5644.80	67.85	8684.80		
4 Rodmen (reinf.)	49.10	1571.20	78.40	2508.80		
2 Laborers	35.10	561.60	54.00	864.00		
1 Cement Finisher	42.35	338.80	62.30	498.40		
1 Equip. Oper. (med.)	46.55	372.40	70.25	562.00		
1 Gas Engine Vibrator		27.40		30.14		
1 Concrete Pump (small)		715.60		787.16	3.71	4.09
200 L.H., Daily Totals		$9600.60		$14502.90	$48.00	$72.51

Crew C-14B

Crew No.	Bare Costs Hr.	Daily	Incl. Subs O&P Hr.	Daily	Cost Per Labor-Hour Bare Costs	Incl. O&P
1 Carpenter Foreman (outside)	$46.10	$368.80	$70.95	$567.60	$44.21	$68.19
16 Carpenters	44.10	5644.80	67.85	8684.80		
4 Rodmen (reinf.)	49.10	1571.20	78.40	2508.80		
2 Laborers	35.10	561.60	54.00	864.00		
2 Cement Finishers	42.35	677.60	62.30	996.80		
1 Equip. Oper. (med.)	46.55	372.40	70.25	562.00		
1 Gas Engine Vibrator		27.40		30.14		
1 Concrete Pump (small)		715.60		787.16	3.57	3.93
208 L.H., Daily Totals		$9939.40		$15001.30	$47.79	$72.12

Crew C-14C

Crew No.	Bare Costs Hr.	Daily	Incl. Subs O&P Hr.	Daily	Cost Per Labor-Hour Bare Costs	Incl. O&P
1 Carpenter Foreman (outside)	$46.10	$368.80	$70.95	$567.60	$42.26	$65.22
6 Carpenters	44.10	2116.80	67.85	3256.80		
2 Rodmen (reinf.)	49.10	785.60	78.40	1254.40		
4 Laborers	35.10	1123.20	54.00	1728.00		
1 Cement Finisher	42.35	338.80	62.30	498.40		
1 Gas Engine Vibrator		27.40		30.14	.24	.27
112 L.H., Daily Totals		$4760.60		$7335.34	$42.51	$65.49

Crew C-14D

Crew No.	Bare Costs Hr.	Daily	Incl. Subs O&P Hr.	Daily	Cost Per Labor-Hour Bare Costs	Incl. O&P
1 Carpenter Foreman (outside)	$46.10	$368.80	$70.95	$567.60	$43.89	$67.58
18 Carpenters	44.10	6350.40	67.85	9770.40		
2 Rodmen (reinf.)	49.10	785.60	78.40	1254.40		
2 Laborers	35.10	561.60	54.00	864.00		
1 Cement Finisher	42.35	338.80	62.30	498.40		
1 Equip. Oper. (med.)	46.55	372.40	70.25	562.00		
1 Gas Engine Vibrator		27.40		30.14		
1 Concrete Pump (small)		715.60		787.16	3.71	4.09
200 L.H., Daily Totals		$9520.60		$14334.10	$47.60	$71.67

Crew C-14E

Crew No.	Bare Costs Hr.	Daily	Incl. Subs O&P Hr.	Daily	Cost Per Labor-Hour Bare Costs	Incl. O&P
1 Carpenter Foreman (outside)	$46.10	$368.80	$70.95	$567.60	$43.49	$67.69
2 Carpenters	44.10	705.60	67.85	1085.60		
4 Rodmen (reinf.)	49.10	1571.20	78.40	2508.80		
3 Laborers	35.10	842.40	54.00	1296.00		
1 Cement Finisher	42.35	338.80	62.30	498.40		
1 Gas Engine Vibrator		27.40		30.14	.31	.34
88 L.H., Daily Totals		$3854.20		$5986.54	$43.80	$68.03

Crew C-14F

Crew No.	Bare Costs Hr.	Daily	Incl. Subs O&P Hr.	Daily	Cost Per Labor-Hour Bare Costs	Incl. O&P
1 Labor Foreman (outside)	$37.10	$296.80	$57.10	$456.80	$40.16	$59.88
2 Laborers	35.10	561.60	54.00	864.00		
6 Cement Finishers	42.35	2032.80	62.30	2990.40		
1 Gas Engine Vibrator		27.40		30.14	.38	.42
72 L.H., Daily Totals		$2918.60		$4341.34	$40.54	$60.30

Crew No.	Bare Costs		Incl. Subs O&P		Cost Per Labor-Hour	
Crew C-14G	Hr.	Daily	Hr.	Daily	Bare Costs	Incl. O&P
1 Labor Foreman (outside)	$37.10	$296.80	$57.10	$456.80	$39.53	$59.19
2 Laborers	35.10	561.60	54.00	864.00		
4 Cement Finishers	42.35	1355.20	62.30	1993.60		
1 Gas Engine Vibrator		27.40		30.14	.49	.54
56 L.H., Daily Totals		$2241.00		$3344.54	$40.02	$59.72
Crew C-14H	Hr.	Daily	Hr.	Daily	Bare Costs	Incl. O&P
1 Carpenter Foreman (outside)	$46.10	$368.80	$70.95	$567.60	$43.48	$66.89
2 Carpenters	44.10	705.60	67.85	1085.60		
1 Rodman (reinf.)	49.10	392.80	78.40	627.20		
1 Laborer	35.10	280.80	54.00	432.00		
1 Cement Finisher	42.35	338.80	62.30	498.40		
1 Gas Engine Vibrator		27.40		30.14	.57	.63
48 L.H., Daily Totals		$2114.20		$3240.94	$44.05	$67.52
Crew C-14L	Hr.	Daily	Hr.	Daily	Bare Costs	Incl. O&P
1 Carpenter Foreman (outside)	$46.10	$368.80	$70.95	$567.60	$41.12	$63.03
6 Carpenters	44.10	2116.80	67.85	3256.80		
4 Laborers	35.10	1123.20	54.00	1728.00		
1 Cement Finisher	42.35	338.80	62.30	498.40		
1 Gas Engine Vibrator		27.40		30.14	.29	.31
96 L.H., Daily Totals		$3975.00		$6080.94	$41.41	$63.34
Crew C-15	Hr.	Daily	Hr.	Daily	Bare Costs	Incl. O&P
1 Carpenter Foreman (outside)	$46.10	$368.80	$70.95	$567.60	$41.49	$63.52
2 Carpenters	44.10	705.60	67.85	1085.60		
3 Laborers	35.10	842.40	54.00	1296.00		
2 Cement Finishers	42.35	677.60	62.30	996.80		
1 Rodman (reinf.)	49.10	392.80	78.40	627.20		
72 L.H., Daily Totals		$2987.20		$4573.20	$41.49	$63.52
Crew C-16	Hr.	Daily	Hr.	Daily	Bare Costs	Incl. O&P
1 Labor Foreman (outside)	$37.10	$296.80	$57.10	$456.80	$39.09	$59.14
3 Laborers	35.10	842.40	54.00	1296.00		
2 Cement Finishers	42.35	677.60	62.30	996.80		
1 Equip. Oper. (med.)	46.55	372.40	70.25	562.00		
1 Gunite Pump Rig		280.40		308.44		
2 -50' Air Hoses, 3/4"		6.10		6.71		
2 -50' Air Hoses, 2"		15.50		17.05	5.39	5.93
56 L.H., Daily Totals		$2491.20		$3643.80	$44.49	$65.07
Crew C-17	Hr.	Daily	Hr.	Daily	Bare Costs	Incl. O&P
2 Skilled Worker Foremen (out)	$47.40	$758.40	$73.30	$1172.80	$45.80	$70.82
8 Skilled Workers	45.40	2905.60	70.20	4492.80		
80 L.H., Daily Totals		$3664.00		$5665.60	$45.80	$70.82
Crew C-17A	Hr.	Daily	Hr.	Daily	Bare Costs	Incl. O&P
2 Skilled Worker Foremen (out)	$47.40	$758.40	$73.30	$1172.80	$45.82	$70.84
8 Skilled Workers	45.40	2905.60	70.20	4492.80		
.125 Equip. Oper. (crane)	47.80	47.80	72.15	72.15		
.125 Hyd. Crane, 80 Ton		194.25		213.68	2.40	2.64
81 L.H., Daily Totals		$3906.05		$5951.43	$48.22	$73.47
Crew C-17B	Hr.	Daily	Hr.	Daily	Bare Costs	Incl. O&P
2 Skilled Worker Foremen (out)	$47.40	$758.40	$73.30	$1172.80	$45.85	$70.85
8 Skilled Workers	45.40	2905.60	70.20	4492.80		
.25 Equip. Oper. (crane)	47.80	95.60	72.15	144.30		
.25 Hyd. Crane, 80 Ton		388.50		427.35		
.25 Trowel, 48" Walk-Behind		11.15		12.27	4.87	5.36
82 L.H., Daily Totals		$4159.25		$6249.52	$50.72	$76.21

Crew No.	Bare Costs		Incl. Subs O&P		Cost Per Labor-Hour	
Crew C-17C	Hr.	Daily	Hr.	Daily	Bare Costs	Incl. O&P
2 Skilled Worker Foremen (out)	$47.40	$758.40	$73.30	$1172.80	$45.87	$70.87
8 Skilled Workers	45.40	2905.60	70.20	4492.80		
.375 Equip. Oper. (crane)	47.80	143.40	72.15	216.45		
.375 Hyd. Crane, 80 Ton		582.75		641.02	7.02	7.72
83 L.H., Daily Totals		$4390.15		$6523.07	$52.89	$78.59
Crew C-17D	Hr.	Daily	Hr.	Daily	Bare Costs	Incl. O&P
2 Skilled Worker Foremen (out)	$47.40	$758.40	$73.30	$1172.80	$45.90	$70.88
8 Skilled Workers	45.40	2905.60	70.20	4492.80		
.5 Equip. Oper. (crane)	47.80	191.20	72.15	288.60		
.5 Hyd. Crane, 80 Ton		777.00		854.70	9.25	10.18
84 L.H., Daily Totals		$4632.20		$6808.90	$55.15	$81.06
Crew C-17E	Hr.	Daily	Hr.	Daily	Bare Costs	Incl. O&P
2 Skilled Worker Foremen (out)	$47.40	$758.40	$73.30	$1172.80	$45.80	$70.82
8 Skilled Workers	45.40	2905.60	70.20	4492.80		
1 Hyd. Jack with Rods		90.70		99.77	1.13	1.25
80 L.H., Daily Totals		$3754.70		$5765.37	$46.93	$72.07
Crew C-18	Hr.	Daily	Hr.	Daily	Bare Costs	Incl. O&P
.125 Labor Foreman (outside)	$37.10	$37.10	$57.10	$57.10	$35.32	$54.34
1 Laborer	35.10	280.80	54.00	432.00		
1 Concrete Cart, 10 C.F.		56.80		62.48	6.31	6.94
9 L.H., Daily Totals		$374.70		$551.58	$41.63	$61.29
Crew C-19	Hr.	Daily	Hr.	Daily	Bare Costs	Incl. O&P
.125 Labor Foreman (outside)	$37.10	$37.10	$57.10	$57.10	$35.32	$54.34
1 Laborer	35.10	280.80	54.00	432.00		
1 Concrete Cart, 18 C.F.		94.40		103.84	10.49	11.54
9 L.H., Daily Totals		$412.30		$592.94	$45.81	$65.88
Crew C-20	Hr.	Daily	Hr.	Daily	Bare Costs	Incl. O&P
1 Labor Foreman (outside)	$37.10	$296.80	$57.10	$456.80	$37.69	$57.46
5 Laborers	35.10	1404.00	54.00	2160.00		
1 Cement Finisher	42.35	338.80	62.30	498.40		
1 Equip. Oper. (med.)	46.55	372.40	70.25	562.00		
2 Gas Engine Vibrators		54.80		60.28		
1 Concrete Pump (small)		715.60		787.16	12.04	13.24
64 L.H., Daily Totals		$3182.40		$4524.64	$49.73	$70.70
Crew C-21	Hr.	Daily	Hr.	Daily	Bare Costs	Incl. O&P
1 Labor Foreman (outside)	$37.10	$296.80	$57.10	$456.80	$37.69	$57.46
5 Laborers	35.10	1404.00	54.00	2160.00		
1 Cement Finisher	42.35	338.80	62.30	498.40		
1 Equip. Oper. (med.)	46.55	372.40	70.25	562.00		
2 Gas Engine Vibrators		54.80		60.28		
1 Concrete Conveyer		186.00		204.60	3.76	4.14
64 L.H., Daily Totals		$2652.80		$3942.08	$41.45	$61.60
Crew C-22	Hr.	Daily	Hr.	Daily	Bare Costs	Incl. O&P
1 Rodman Foreman (outside)	$51.10	$408.80	$81.60	$652.80	$49.26	$78.48
4 Rodmen (reinf.)	49.10	1571.20	78.40	2508.80		
.125 Equip. Oper. (crane)	47.80	47.80	72.15	72.15		
.125 Equip. Oper. Oiler	41.25	41.25	62.25	62.25		
.125 Hyd. Crane, 25 Ton		87.15		95.86	2.08	2.28
42 L.H., Daily Totals		$2156.20		$3391.86	$51.34	$80.76

Crew No.	Bare Costs		Incl. Subs O&P		Cost Per Labor-Hour	
Crew C-23	Hr.	Daily	Hr.	Daily	Bare Costs	Incl. O&P
2 Skilled Worker Foremen (out)	$47.40	$758.40	$73.30	$1172.80	$45.63	$70.22
6 Skilled Workers	45.40	2179.20	70.20	3369.60		
1 Equip. Oper. (crane)	47.80	382.40	72.15	577.20		
1 Equip. Oper. Oiler	41.25	330.00	62.25	498.00		
1 Lattice Boom Crane, 90 Ton		1497.00		1646.70	18.71	20.58
80 L.H., Daily Totals		$5147.00		$7264.30	$64.34	$90.80
Crew C-23A	Hr.	Daily	Hr.	Daily	Bare Costs	Incl. O&P
1 Labor Foreman (outside)	$37.10	$296.80	$57.10	$456.80	$39.27	$59.90
2 Laborers	35.10	561.60	54.00	864.00		
1 Equip. Oper. (crane)	47.80	382.40	72.15	577.20		
1 Equip. Oper. Oiler	41.25	330.00	62.25	498.00		
1 Crawler Crane, 100 Ton		1628.00		1790.80		
3 Conc. Buckets, 8 C.Y.		598.20		658.02	55.66	61.22
40 L.H., Daily Totals		$3797.00		$4844.82	$94.92	$121.12
Crew C-24	Hr.	Daily	Hr.	Daily	Bare Costs	Incl. O&P
2 Skilled Worker Foremen (out)	$47.40	$758.40	$73.30	$1172.80	$45.63	$70.22
6 Skilled Workers	45.40	2179.20	70.20	3369.60		
1 Equip. Oper. (crane)	47.80	382.40	72.15	577.20		
1 Equip. Oper. Oiler	41.25	330.00	62.25	498.00		
1 Lattice Boom Crane, 150 Ton		1829.00		2011.90	22.86	25.15
80 L.H., Daily Totals		$5479.00		$7629.50	$68.49	$95.37
Crew C-25	Hr.	Daily	Hr.	Daily	Bare Costs	Incl. O&P
2 Rodmen (reinf.)	$49.10	$785.60	$78.40	$1254.40	$38.48	$62.55
2 Rodmen Helpers	27.85	445.60	46.70	747.20		
32 L.H., Daily Totals		$1231.20		$2001.60	$38.48	$62.55
Crew C-27	Hr.	Daily	Hr.	Daily	Bare Costs	Incl. O&P
2 Cement Finishers	$42.35	$677.60	$62.30	$996.80	$42.35	$62.30
1 Concrete Saw		148.00		162.80	9.25	10.18
16 L.H., Daily Totals		$825.60		$1159.60	$51.60	$72.47
Crew C-28	Hr.	Daily	Hr.	Daily	Bare Costs	Incl. O&P
1 Cement Finisher	$42.35	$338.80	$62.30	$498.40	$42.35	$62.30
1 Portable Air Compressor, Gas		17.55		19.31	2.19	2.41
8 L.H., Daily Totals		$356.35		$517.71	$44.54	$64.71
Crew C-29	Hr.	Daily	Hr.	Daily	Bare Costs	Incl. O&P
1 Laborer	$35.10	$280.80	$54.00	$432.00	$35.10	$54.00
1 Pressure Washer		63.00		69.30	7.88	8.66
8 L.H., Daily Totals		$343.80		$501.30	$42.98	$62.66
Crew C-30	Hr.	Daily	Hr.	Daily	Bare Costs	Incl. O&P
1 Laborer	$35.10	$280.80	$54.00	$432.00	$35.10	$54.00
1 Concrete Mixer, 10 C.F.		160.60		176.66	20.07	22.08
8 L.H., Daily Totals		$441.40		$608.66	$55.17	$76.08
Crew D-1	Hr.	Daily	Hr.	Daily	Bare Costs	Incl. O&P
1 Bricklayer	$44.10	$352.80	$66.60	$532.80	$39.85	$60.17
1 Bricklayer Helper	35.60	284.80	53.75	430.00		
16 L.H., Daily Totals		$637.60		$962.80	$39.85	$60.17
Crew D-2	Hr.	Daily	Hr.	Daily	Bare Costs	Incl. O&P
3 Bricklayers	$44.10	$1058.40	$66.60	$1598.40	$41.01	$62.04
2 Bricklayer Helpers	35.60	569.60	53.75	860.00		
.5 Carpenter	44.10	176.40	67.85	271.40		
44 L.H., Daily Totals		$1804.40		$2729.80	$41.01	$62.04

Crew No.	Bare Costs		Incl. Subs O&P		Cost Per Labor-Hour	
Crew D-3	Hr.	Daily	Hr.	Daily	Bare Costs	Incl. O&P
3 Bricklayers	$44.10	$1058.40	$66.60	$1598.40	$40.86	$61.76
2 Bricklayer Helpers	35.60	569.60	53.75	860.00		
.25 Carpenter	44.10	88.20	67.85	135.70		
42 L.H., Daily Totals		$1716.20		$2594.10	$40.86	$61.76
Crew D-4	Hr.	Daily	Hr.	Daily	Bare Costs	Incl. O&P
1 Bricklayer	$44.10	$352.80	$66.60	$532.80	$40.01	$60.41
2 Bricklayer Helpers	35.60	569.60	53.75	860.00		
1 Equip. Oper. (light)	44.75	358.00	67.55	540.40		
1 Grout Pump, 50 C.F./hr.		127.40		140.14	3.98	4.38
32 L.H., Daily Totals		$1407.80		$2073.34	$43.99	$64.79
Crew D-5	Hr.	Daily	Hr.	Daily	Bare Costs	Incl. O&P
1 Bricklayer	$44.10	$352.80	$66.60	$532.80	$44.10	$66.60
8 L.H., Daily Totals		$352.80		$532.80	$44.10	$66.60
Crew D-6	Hr.	Daily	Hr.	Daily	Bare Costs	Incl. O&P
3 Bricklayers	$44.10	$1058.40	$66.60	$1598.40	$40.02	$60.48
3 Bricklayer Helpers	35.60	854.40	53.75	1290.00		
.25 Carpenter	44.10	88.20	67.85	135.70		
50 L.H., Daily Totals		$2001.00		$3024.10	$40.02	$60.48
Crew D-7	Hr.	Daily	Hr.	Daily	Bare Costs	Incl. O&P
1 Tile Layer	$40.75	$326.00	$59.85	$478.80	$36.42	$53.50
1 Tile Layer Helper	32.10	256.80	47.15	377.20		
16 L.H., Daily Totals		$582.80		$856.00	$36.42	$53.50
Crew D-8	Hr.	Daily	Hr.	Daily	Bare Costs	Incl. O&P
3 Bricklayers	$44.10	$1058.40	$66.60	$1598.40	$40.70	$61.46
2 Bricklayer Helpers	35.60	569.60	53.75	860.00		
40 L.H., Daily Totals		$1628.00		$2458.40	$40.70	$61.46
Crew D-9	Hr.	Daily	Hr.	Daily	Bare Costs	Incl. O&P
3 Bricklayers	$44.10	$1058.40	$66.60	$1598.40	$39.85	$60.17
3 Bricklayer Helpers	35.60	854.40	53.75	1290.00		
48 L.H., Daily Totals		$1912.80		$2888.40	$39.85	$60.17
Crew D-10	Hr.	Daily	Hr.	Daily	Bare Costs	Incl. O&P
1 Bricklayer Foreman (outside)	$46.10	$368.80	$69.60	$556.80	$43.40	$65.53
1 Bricklayer	44.10	352.80	66.60	532.80		
1 Bricklayer Helper	35.60	284.80	53.75	430.00		
1 Equip. Oper. (crane)	47.80	382.40	72.15	577.20		
1 S.P. Crane, 4x4, 12 Ton		448.20		493.02	14.01	15.41
32 L.H., Daily Totals		$1837.00		$2589.82	$57.41	$80.93
Crew D-11	Hr.	Daily	Hr.	Daily	Bare Costs	Incl. O&P
1 Bricklayer Foreman (outside)	$46.10	$368.80	$69.60	$556.80	$41.93	$63.32
1 Bricklayer	44.10	352.80	66.60	532.80		
1 Bricklayer Helper	35.60	284.80	53.75	430.00		
24 L.H., Daily Totals		$1006.40		$1519.60	$41.93	$63.32
Crew D-12	Hr.	Daily	Hr.	Daily	Bare Costs	Incl. O&P
1 Bricklayer Foreman (outside)	$46.10	$368.80	$69.60	$556.80	$40.35	$60.92
1 Bricklayer	44.10	352.80	66.60	532.80		
2 Bricklayer Helpers	35.60	569.60	53.75	860.00		
32 L.H., Daily Totals		$1291.20		$1949.60	$40.35	$60.92

Crew D-13

Crew No.	Bare Costs Hr.	Bare Costs Daily	Incl. Subs O&P Hr.	Incl. Subs O&P Daily	Cost Per Labor-Hour Bare Costs	Cost Per Labor-Hour Incl. O&P
1 Bricklayer Foreman (outside)	$46.10	$368.80	$69.60	$556.80	$42.22	$63.95
1 Bricklayer	44.10	352.80	66.60	532.80		
2 Bricklayer Helpers	35.60	569.60	53.75	860.00		
1 Carpenter	44.10	352.80	67.85	542.80		
1 Equip. Oper. (crane)	47.80	382.40	72.15	577.20		
1 S.P. Crane, 4x4, 12 Ton		448.20		493.02	9.34	10.27
48 L.H., Daily Totals		$2474.60		$3562.62	$51.55	$74.22

Crew E-1

Crew No.	Bare Costs Hr.	Bare Costs Daily	Incl. Subs O&P Hr.	Incl. Subs O&P Daily	Cost Per Labor-Hour Bare Costs	Cost Per Labor-Hour Incl. O&P
1 Welder Foreman (outside)	$51.35	$410.80	$91.80	$734.40	$48.48	$82.53
1 Welder	49.35	394.80	88.25	706.00		
1 Equip. Oper. (light)	44.75	358.00	67.55	540.40		
1 Welder, gas engine, 300 amp		121.60		133.76	5.07	5.57
24 L.H., Daily Totals		$1285.20		$2114.56	$53.55	$88.11

Crew E-2

Crew No.	Bare Costs Hr.	Bare Costs Daily	Incl. Subs O&P Hr.	Incl. Subs O&P Daily	Cost Per Labor-Hour Bare Costs	Cost Per Labor-Hour Incl. O&P
1 Struc. Steel Foreman (outside)	$51.35	$410.80	$91.80	$734.40	$48.26	$82.74
4 Struc. Steel Workers	49.35	1579.20	88.25	2824.00		
1 Equip. Oper. (crane)	47.80	382.40	72.15	577.20		
1 Equip. Oper. Oiler	41.25	330.00	62.25	498.00		
1 Lattice Boom Crane, 90 Ton		1497.00		1646.70	26.73	29.41
56 L.H., Daily Totals		$4199.40		$6280.30	$74.99	$112.15

Crew E-3

Crew No.	Bare Costs Hr.	Bare Costs Daily	Incl. Subs O&P Hr.	Incl. Subs O&P Daily	Cost Per Labor-Hour Bare Costs	Cost Per Labor-Hour Incl. O&P
1 Struc. Steel Foreman (outside)	$51.35	$410.80	$91.80	$734.40	$50.02	$89.43
1 Struc. Steel Worker	49.35	394.80	88.25	706.00		
1 Welder	49.35	394.80	88.25	706.00		
1 Welder, gas engine, 300 amp		121.60		133.76	5.07	5.57
24 L.H., Daily Totals		$1322.00		$2280.16	$55.08	$95.01

Crew E-4

Crew No.	Bare Costs Hr.	Bare Costs Daily	Incl. Subs O&P Hr.	Incl. Subs O&P Daily	Cost Per Labor-Hour Bare Costs	Cost Per Labor-Hour Incl. O&P
1 Struc. Steel Foreman (outside)	$51.35	$410.80	$91.80	$734.40	$49.85	$89.14
3 Struc. Steel Workers	49.35	1184.40	88.25	2118.00		
1 Welder, gas engine, 300 amp		121.60		133.76	3.80	4.18
32 L.H., Daily Totals		$1716.80		$2986.16	$53.65	$93.32

Crew E-5

Crew No.	Bare Costs Hr.	Bare Costs Daily	Incl. Subs O&P Hr.	Incl. Subs O&P Daily	Cost Per Labor-Hour Bare Costs	Cost Per Labor-Hour Incl. O&P
2 Struc. Steel Foremen (outside)	$51.35	$821.60	$91.80	$1468.80	$48.78	$84.75
5 Struc. Steel Workers	49.35	1974.00	88.25	3530.00		
1 Equip. Oper. (crane)	47.80	382.40	72.15	577.20		
1 Welder	49.35	394.80	88.25	706.00		
1 Equip. Oper. Oiler	41.25	330.00	62.25	498.00		
1 Lattice Boom Crane, 90 Ton		1497.00		1646.70		
1 Welder, gas engine, 300 amp		121.60		133.76	20.23	22.26
80 L.H., Daily Totals		$5521.40		$8560.46	$69.02	$107.01

Crew E-6

Crew No.	Bare Costs Hr.	Bare Costs Daily	Incl. Subs O&P Hr.	Incl. Subs O&P Daily	Cost Per Labor-Hour Bare Costs	Cost Per Labor-Hour Incl. O&P
3 Struc. Steel Foremen (outside)	$51.35	$1232.40	$91.80	$2203.20	$48.83	$84.99
9 Struc. Steel Workers	49.35	3553.20	88.25	6354.00		
1 Equip. Oper. (crane)	47.80	382.40	72.15	577.20		
1 Welder	49.35	394.80	88.25	706.00		
1 Equip. Oper. Oiler	41.25	330.00	62.25	498.00		
1 Equip. Oper. (light)	44.75	358.00	67.55	540.40		
1 Lattice Boom Crane, 90 Ton		1497.00		1646.70		
1 Welder, gas engine, 300 amp		121.60		133.76		
1 Air Compressor, 160 cfm		136.80		150.48		
2 Impact Wrenches		36.00		39.60	14.00	15.39
128 L.H., Daily Totals		$8042.20		$12849.34	$62.83	$100.39

Crew E-7

Crew No.	Bare Costs Hr.	Bare Costs Daily	Incl. Subs O&P Hr.	Incl. Subs O&P Daily	Cost Per Labor-Hour Bare Costs	Cost Per Labor-Hour Incl. O&P
1 Struc. Steel Foreman (outside)	$51.35	$410.80	$91.80	$734.40	$48.78	$84.75
4 Struc. Steel Workers	49.35	1579.20	88.25	2824.00		
1 Equip. Oper. (crane)	47.80	382.40	72.15	577.20		
1 Equip. Oper. Oiler	41.25	330.00	62.25	498.00		
1 Welder Foreman (outside)	51.35	410.80	91.80	734.40		
2 Welders	49.35	789.60	88.25	1412.00		
1 Lattice Boom Crane, 90 Ton		1497.00		1646.70		
2 Welders, gas engine, 300 amp		243.20		267.52	21.75	23.93
80 L.H., Daily Totals		$5643.00		$8694.22	$70.54	$108.68

Crew E-8

Crew No.	Bare Costs Hr.	Bare Costs Daily	Incl. Subs O&P Hr.	Incl. Subs O&P Daily	Cost Per Labor-Hour Bare Costs	Cost Per Labor-Hour Incl. O&P
1 Struc. Steel Foreman (outside)	$51.35	$410.80	$91.80	$734.40	$48.56	$83.97
4 Struc. Steel Workers	49.35	1579.20	88.25	2824.00		
1 Welder Foreman (outside)	51.35	410.80	91.80	734.40		
4 Welders	49.35	1579.20	88.25	2824.00		
1 Equip. Oper. (crane)	47.80	382.40	72.15	577.20		
1 Equip. Oper. Oiler	41.25	330.00	62.25	498.00		
1 Equip. Oper. (light)	44.75	358.00	67.55	540.40		
1 Lattice Boom Crane, 90 Ton		1497.00		1646.70		
4 Welders, gas engine, 300 amp		486.40		535.04	19.07	20.98
104 L.H., Daily Totals		$7033.80		$10914.14	$67.63	$104.94

Crew E-9

Crew No.	Bare Costs Hr.	Bare Costs Daily	Incl. Subs O&P Hr.	Incl. Subs O&P Daily	Cost Per Labor-Hour Bare Costs	Cost Per Labor-Hour Incl. O&P
2 Struc. Steel Foremen (outside)	$51.35	$821.60	$91.80	$1468.80	$48.83	$84.99
5 Struc. Steel Workers	49.35	1974.00	88.25	3530.00		
1 Welder Foreman (outside)	51.35	410.80	91.80	734.40		
5 Welders	49.35	1974.00	88.25	3530.00		
1 Equip. Oper. (crane)	47.80	382.40	72.15	577.20		
1 Equip. Oper. Oiler	41.25	330.00	62.25	498.00		
1 Equip. Oper. (light)	44.75	358.00	67.55	540.40		
1 Lattice Boom Crane, 90 Ton		1497.00		1646.70		
5 Welders, gas engine, 300 amp		608.00		668.80	16.45	18.09
128 L.H., Daily Totals		$8355.80		$13194.30	$65.28	$103.08

Crew E-10

Crew No.	Bare Costs Hr.	Bare Costs Daily	Incl. Subs O&P Hr.	Incl. Subs O&P Daily	Cost Per Labor-Hour Bare Costs	Cost Per Labor-Hour Incl. O&P
1 Welder Foreman (outside)	$51.35	$410.80	$91.80	$734.40	$50.35	$90.03
1 Welder	49.35	394.80	88.25	706.00		
1 Welder, gas engine, 300 amp		121.60		133.76		
1 Flatbed Truck, Gas, 3 Ton		264.40		290.84	24.13	26.54
16 L.H., Daily Totals		$1191.60		$1865.00	$74.47	$116.56

Crew E-11

Crew No.	Bare Costs Hr.	Bare Costs Daily	Incl. Subs O&P Hr.	Incl. Subs O&P Daily	Cost Per Labor-Hour Bare Costs	Cost Per Labor-Hour Incl. O&P
2 Painters, Struc. Steel	$39.30	$628.80	$71.15	$1138.40	$39.61	$65.96
1 Building Laborer	35.10	280.80	54.00	432.00		
1 Equip. Oper. (light)	44.75	358.00	67.55	540.40		
1 Air Compressor, 250 cfm		177.00		194.70		
1 Sandblaster, portable, 3 C.F.		20.40		22.44		
1 Set Sand Blasting Accessories		13.35		14.69	6.59	7.24
32 L.H., Daily Totals		$1478.35		$2342.63	$46.20	$73.21

Crew E-12

Crew No.	Bare Costs Hr.	Bare Costs Daily	Incl. Subs O&P Hr.	Incl. Subs O&P Daily	Cost Per Labor-Hour Bare Costs	Cost Per Labor-Hour Incl. O&P
1 Welder Foreman (outside)	$51.35	$410.80	$91.80	$734.40	$48.05	$79.67
1 Equip. Oper. (light)	44.75	358.00	67.55	540.40		
1 Welder, gas engine, 300 amp		121.60		133.76	7.60	8.36
16 L.H., Daily Totals		$890.40		$1408.56	$55.65	$88.03

Crew E-13

Crew No.	Bare Costs Hr.	Bare Costs Daily	Incl. Subs O&P Hr.	Incl. Subs O&P Daily	Cost Per Labor-Hour Bare Costs	Cost Per Labor-Hour Incl. O&P
1 Welder Foreman (outside)	$51.35	$410.80	$91.80	$734.40	$49.15	$83.72
.5 Equip. Oper. (light)	44.75	179.00	67.55	270.20		
1 Welder, gas engine, 300 amp		121.60		133.76	10.13	11.15
12 L.H., Daily Totals		$711.40		$1138.36	$59.28	$94.86

Crew E-14

Crew No.	Bare Costs Hr.	Daily	Incl. Subs O&P Hr.	Daily	Bare Costs	Incl. O&P
1 Welder Foreman (outside)	$51.35	$410.80	$91.80	$734.40	$51.35	$91.80
1 Welder, gas engine, 300 amp		121.60		133.76	15.20	16.72
8 L.H., Daily Totals		$532.40		$868.16	$66.55	$108.52

Crew E-16

Crew No.	Bare Costs Hr.	Daily	Incl. Subs O&P Hr.	Daily	Bare Costs	Incl. O&P
1 Welder Foreman (outside)	$51.35	$410.80	$91.80	$734.40	$50.35	$90.03
1 Welder	49.35	394.80	88.25	706.00		
1 Welder, gas engine, 300 amp		121.60		133.76	7.60	8.36
16 L.H., Daily Totals		$927.20		$1574.16	$57.95	$98.39

Crew E-17

Crew No.	Bare Costs Hr.	Daily	Incl. Subs O&P Hr.	Daily	Bare Costs	Incl. O&P
1 Struc. Steel Foreman (outside)	$51.35	$410.80	$91.80	$734.40	$50.35	$90.03
1 Structural Steel Worker	49.35	394.80	88.25	706.00		
16 L.H., Daily Totals		$805.60		$1440.40	$50.35	$90.03

Crew E-18

Crew No.	Bare Costs Hr.	Daily	Incl. Subs O&P Hr.	Daily	Bare Costs	Incl. O&P
1 Struc. Steel Foreman (outside)	$51.35	$410.80	$91.80	$734.40	$49.19	$85.36
3 Structural Steel Workers	49.35	1184.40	88.25	2118.00		
1 Equipment Operator (med.)	46.55	372.40	70.25	562.00		
1 Lattice Boom Crane, 20 Ton		950.65		1045.71	23.77	26.14
40 L.H., Daily Totals		$2918.25		$4460.11	$72.96	$111.50

Crew E-19

Crew No.	Bare Costs Hr.	Daily	Incl. Subs O&P Hr.	Daily	Bare Costs	Incl. O&P
1 Structural Steel Worker	$49.35	$394.80	$88.25	$706.00	$48.48	$82.53
1 Struc. Steel Foreman (outside)	51.35	410.80	91.80	734.40		
1 Equip. Oper. (light)	44.75	358.00	67.55	540.40		
1 Lattice Boom Crane, 20 Ton		950.65		1045.71	39.61	43.57
24 L.H., Daily Totals		$2114.25		$3026.51	$88.09	$126.10

Crew E-20

Crew No.	Bare Costs Hr.	Daily	Incl. Subs O&P Hr.	Daily	Bare Costs	Incl. O&P
1 Struc. Steel Foreman (outside)	$51.35	$410.80	$91.80	$734.40	$48.39	$83.43
5 Structural Steel Workers	49.35	1974.00	88.25	3530.00		
1 Equip. Oper. (crane)	47.80	382.40	72.15	577.20		
1 Oiler	41.25	330.00	62.25	498.00		
1 Lattice Boom Crane, 40 Ton		1163.00		1279.30	18.17	19.99
64 L.H., Daily Totals		$4260.20		$6618.90	$66.57	$103.42

Crew E-22

Crew No.	Bare Costs Hr.	Daily	Incl. Subs O&P Hr.	Daily	Bare Costs	Incl. O&P
1 Skilled Worker Foreman (out)	$47.40	$379.20	$73.30	$586.40	$46.07	$71.23
2 Skilled Workers	45.40	726.40	70.20	1123.20		
24 L.H., Daily Totals		$1105.60		$1709.60	$46.07	$71.23

Crew E-24

Crew No.	Bare Costs Hr.	Daily	Incl. Subs O&P Hr.	Daily	Bare Costs	Incl. O&P
3 Structural Steel Workers	$49.35	$1184.40	$88.25	$2118.00	$48.65	$83.75
1 Equipment Operator (med.)	46.55	372.40	70.25	562.00		
1 Hyd. Crane, 25 Ton		697.20		766.92	21.79	23.97
32 L.H., Daily Totals		$2254.00		$3446.92	$70.44	$107.72

Crew E-25

Crew No.	Bare Costs Hr.	Daily	Incl. Subs O&P Hr.	Daily	Bare Costs	Incl. O&P
1 Welder Foreman (outside)	$51.35	$410.80	$91.80	$734.40	$51.35	$91.80
1 Cutting Torch		10.80		11.88		
1 Set of Gases		86.40		95.04	12.15	13.37
8 L.H., Daily Totals		$508.00		$841.32	$63.50	$105.17

Crew F-3

Crew No.	Bare Costs Hr.	Daily	Incl. Subs O&P Hr.	Daily	Bare Costs	Incl. O&P
4 Carpenters	$44.10	$1411.20	$67.85	$2171.20	$44.84	$68.71
1 Equip. Oper. (crane)	47.80	382.40	72.15	577.20		
1 Hyd. Crane, 12 Ton		612.00		673.20	15.30	16.83
40 L.H., Daily Totals		$2405.60		$3421.60	$60.14	$85.54

Crew F-4

Crew No.	Bare Costs Hr.	Daily	Incl. Subs O&P Hr.	Daily	Bare Costs	Incl. O&P
4 Carpenters	$44.10	$1411.20	$67.85	$2171.20	$44.24	$67.63
1 Equip. Oper. (crane)	47.80	382.40	72.15	577.20		
1 Equip. Oper. Oiler	41.25	330.00	62.25	498.00		
1 Hyd. Crane, 55 Ton		1100.00		1210.00	22.92	25.21
48 L.H., Daily Totals		$3223.60		$4456.40	$67.16	$92.84

Crew F-5

Crew No.	Bare Costs Hr.	Daily	Incl. Subs O&P Hr.	Daily	Bare Costs	Incl. O&P
1 Carpenter Foreman (outside)	$46.10	$368.80	$70.95	$567.60	$44.60	$68.63
3 Carpenters	44.10	1058.40	67.85	1628.40		
32 L.H., Daily Totals		$1427.20		$2196.00	$44.60	$68.63

Crew F-6

Crew No.	Bare Costs Hr.	Daily	Incl. Subs O&P Hr.	Daily	Bare Costs	Incl. O&P
2 Carpenters	$44.10	$705.60	$67.85	$1085.60	$41.24	$63.17
2 Building Laborers	35.10	561.60	54.00	864.00		
1 Equip. Oper. (crane)	47.80	382.40	72.15	577.20		
1 Hyd. Crane, 12 Ton		612.00		673.20	15.30	16.83
40 L.H., Daily Totals		$2261.60		$3200.00	$56.54	$80.00

Crew F-7

Crew No.	Bare Costs Hr.	Daily	Incl. Subs O&P Hr.	Daily	Bare Costs	Incl. O&P
2 Carpenters	$44.10	$705.60	$67.85	$1085.60	$39.60	$60.92
2 Building Laborers	35.10	561.60	54.00	864.00		
32 L.H., Daily Totals		$1267.20		$1949.60	$39.60	$60.92

Crew G-1

Crew No.	Bare Costs Hr.	Daily	Incl. Subs O&P Hr.	Daily	Bare Costs	Incl. O&P
1 Roofer Foreman (outside)	$39.40	$315.20	$66.05	$528.40	$34.96	$58.61
4 Roofers, Composition	37.40	1196.80	62.70	2006.40		
2 Roofer Helpers	27.85	445.60	46.70	747.20		
1 Application Equipment		179.20		197.12		
1 Tar Kettle/Pot		91.00		100.10		
1 Crew Truck		182.40		200.64	8.08	8.89
56 L.H., Daily Totals		$2410.20		$3779.86	$43.04	$67.50

Crew G-2

Crew No.	Bare Costs Hr.	Daily	Incl. Subs O&P Hr.	Daily	Bare Costs	Incl. O&P
1 Plasterer	$40.35	$322.80	$60.40	$483.20	$37.07	$55.97
1 Plasterer Helper	35.75	286.00	53.50	428.00		
1 Building Laborer	35.10	280.80	54.00	432.00		
1 Grout Pump, 50 C.F./hr.		127.40		140.14	5.31	5.84
24 L.H., Daily Totals		$1017.00		$1483.34	$42.38	$61.81

Crew G-2A

Crew No.	Bare Costs Hr.	Daily	Incl. Subs O&P Hr.	Daily	Bare Costs	Incl. O&P
1 Roofer, composition	$37.40	$299.20	$62.70	$501.60	$33.45	$54.47
1 Roofer Helper	27.85	222.80	46.70	373.60		
1 Building Laborer	35.10	280.80	54.00	432.00		
1 Foam spray rig, trailer-mtd.		503.85		554.24		
1 Pickup Truck, 3/4 Ton		127.00		139.70	26.29	28.91
24 L.H., Daily Totals		$1433.65		$2001.14	$59.74	$83.38

Crew G-3

Crew No.	Bare Costs Hr.	Daily	Incl. Subs O&P Hr.	Daily	Bare Costs	Incl. O&P
2 Sheet Metal Workers	$52.15	$834.40	$79.50	$1272.00	$43.63	$66.75
2 Building Laborers	35.10	561.60	54.00	864.00		
32 L.H., Daily Totals		$1396.00		$2136.00	$43.63	$66.75

Crew G-4

Crew No.	Bare Costs Hr.	Daily	Incl. Subs O&P Hr.	Daily	Bare Costs	Incl. O&P
1 Labor Foreman (outside)	$37.10	$296.80	$57.10	$456.80	$35.77	$55.03
2 Building Laborers	35.10	561.60	54.00	864.00		
1 Flatbed Truck, Gas, 1.5 Ton		211.00		232.10		
1 Air Compressor, 160 cfm		136.80		150.48	14.49	15.94
24 L.H., Daily Totals		$1206.20		$1703.38	$50.26	$70.97

Crew G-5

	Bare Costs Hr.	Daily	Incl. Subs O&P Hr.	Daily	Bare Costs	Incl. O&P
1 Roofer Foreman (outside)	$39.40	$315.20	$66.05	$528.40	$33.98	$56.97
2 Roofers, Composition	37.40	598.40	62.70	1003.20		
2 Roofer Helpers	27.85	445.60	46.70	747.20		
1 Application Equipment		179.20		197.12	4.48	4.93
40 L.H., Daily Totals		$1538.40		$2475.92	$38.46	$61.90

Crew G-6A

	Bare Costs Hr.	Daily	Incl. Subs O&P Hr.	Daily	Bare Costs	Incl. O&P
2 Roofers Composition	$37.40	$598.40	$62.70	$1003.20	$37.40	$62.70
1 Small Compressor, Electric		10.10		11.11		
2 Pneumatic Nailers		45.70		50.27	3.49	3.84
16 L.H., Daily Totals		$654.20		$1064.58	$40.89	$66.54

Crew G-7

	Bare Costs Hr.	Daily	Incl. Subs O&P Hr.	Daily	Bare Costs	Incl. O&P
1 Carpenter	$44.10	$352.80	$67.85	$542.80	$44.10	$67.85
1 Small Compressor, Electric		10.10		11.11		
1 Pneumatic Nailer		22.85		25.14	4.12	4.53
8 L.H., Daily Totals		$385.75		$579.04	$48.22	$72.38

Crew H-1

	Bare Costs Hr.	Daily	Incl. Subs O&P Hr.	Daily	Bare Costs	Incl. O&P
2 Glaziers	$42.75	$684.00	$64.75	$1036.00	$46.05	$76.50
2 Struc. Steel Workers	49.35	789.60	88.25	1412.00		
32 L.H., Daily Totals		$1473.60		$2448.00	$46.05	$76.50

Crew H-2

	Bare Costs Hr.	Daily	Incl. Subs O&P Hr.	Daily	Bare Costs	Incl. O&P
2 Glaziers	$42.75	$684.00	$64.75	$1036.00	$40.20	$61.17
1 Building Laborer	35.10	280.80	54.00	432.00		
24 L.H., Daily Totals		$964.80		$1468.00	$40.20	$61.17

Crew H-3

	Bare Costs Hr.	Daily	Incl. Subs O&P Hr.	Daily	Bare Costs	Incl. O&P
1 Glazier	$42.75	$342.00	$64.75	$518.00	$38.02	$57.98
1 Helper	33.30	266.40	51.20	409.60		
16 L.H., Daily Totals		$608.40		$927.60	$38.02	$57.98

Crew H-4

	Bare Costs Hr.	Daily	Incl. Subs O&P Hr.	Daily	Bare Costs	Incl. O&P
1 Carpenter	$44.10	$352.80	$67.85	$542.80	$41.27	$63.03
1 Carpenter Helper	33.30	266.40	51.20	409.60		
.5 Electrician	51.55	206.20	77.05	308.20		
20 L.H., Daily Totals		$825.40		$1260.60	$41.27	$63.03

Crew J-1

	Bare Costs Hr.	Daily	Incl. Subs O&P Hr.	Daily	Bare Costs	Incl. O&P
3 Plasterers	$40.35	$968.40	$60.40	$1449.60	$38.51	$57.64
2 Plasterer Helpers	35.75	572.00	53.50	856.00		
1 Mixing Machine, 6 C.F.		130.80		143.88	3.27	3.60
40 L.H., Daily Totals		$1671.20		$2449.48	$41.78	$61.24

Crew J-2

	Bare Costs Hr.	Daily	Incl. Subs O&P Hr.	Daily	Bare Costs	Incl. O&P
3 Plasterers	$40.35	$968.40	$60.40	$1449.60	$38.63	$57.63
2 Plasterer Helpers	35.75	572.00	53.50	856.00		
1 Lather	39.25	314.00	57.60	460.80		
1 Mixing Machine, 6 C.F.		130.80		143.88	2.73	3.00
48 L.H., Daily Totals		$1985.20		$2910.28	$41.36	$60.63

Crew J-3

	Bare Costs Hr.	Daily	Incl. Subs O&P Hr.	Daily	Bare Costs	Incl. O&P
1 Terrazzo Worker	$40.40	$323.20	$59.35	$474.80	$37.08	$54.48
1 Terrazzo Helper	33.75	270.00	49.60	396.80		
1 Floor Grinder, 22" path		84.70		93.17		
1 Terrazzo Mixer		177.00		194.70	16.36	17.99
16 L.H., Daily Totals		$854.90		$1159.47	$53.43	$72.47

Crew J-4

	Bare Costs Hr.	Daily	Incl. Subs O&P Hr.	Daily	Bare Costs	Incl. O&P
2 Cement Finishers	$42.35	$677.60	$62.30	$996.80	$39.93	$59.53
1 Laborer	35.10	280.80	54.00	432.00		
1 Floor Grinder, 22" path		84.70		93.17		
1 Floor Edger, 7" path		38.20		42.02		
1 Vacuum Pick-Up System		59.80		65.78	7.61	8.37
24 L.H., Daily Totals		$1141.10		$1629.77	$47.55	$67.91

Crew J-4A

	Bare Costs Hr.	Daily	Incl. Subs O&P Hr.	Daily	Bare Costs	Incl. O&P
2 Cement Finishers	$42.35	$677.60	$62.30	$996.80	$38.73	$58.15
2 Laborers	35.10	561.60	54.00	864.00		
1 Floor Grinder, 22" path		84.70		93.17		
1 Floor Edger, 7" path		38.20		42.02		
1 Vacuum Pick-Up System		59.80		65.78		
1 Floor Auto Scrubber		178.10		195.91	11.28	12.40
32 L.H., Daily Totals		$1600.00		$2257.68	$50.00	$70.55

Crew J-4B

	Bare Costs Hr.	Daily	Incl. Subs O&P Hr.	Daily	Bare Costs	Incl. O&P
1 Laborer	$35.10	$280.80	$54.00	$432.00	$35.10	$54.00
1 Floor Auto Scrubber		178.10		195.91	22.26	24.49
8 L.H., Daily Totals		$458.90		$627.91	$57.36	$78.49

Crew J-6

	Bare Costs Hr.	Daily	Incl. Subs O&P Hr.	Daily	Bare Costs	Incl. O&P
2 Painters	$38.25	$612.00	$57.20	$915.20	$39.09	$58.99
1 Building Laborer	35.10	280.80	54.00	432.00		
1 Equip. Oper. (light)	44.75	358.00	67.55	540.40		
1 Air Compressor, 250 cfm		177.00		194.70		
1 Sandblaster, portable, 3 C.F.		20.40		22.44		
1 Set Sand Blasting Accessories		13.35		14.69	6.59	7.24
32 L.H., Daily Totals		$1461.55		$2119.43	$45.67	$66.23

Crew K-1

	Bare Costs Hr.	Daily	Incl. Subs O&P Hr.	Daily	Bare Costs	Incl. O&P
1 Carpenter	$44.10	$352.80	$67.85	$542.80	$39.23	$59.98
1 Truck Driver (light)	34.35	274.80	52.10	416.80		
1 Flatbed Truck, Gas, 3 Ton		264.40		290.84	16.52	18.18
16 L.H., Daily Totals		$892.00		$1250.44	$55.75	$78.15

Crew K-2

	Bare Costs Hr.	Daily	Incl. Subs O&P Hr.	Daily	Bare Costs	Incl. O&P
1 Struc. Steel Foreman (outside)	$51.35	$410.80	$91.80	$734.40	$45.02	$77.38
1 Struc. Steel Worker	49.35	394.80	88.25	706.00		
1 Truck Driver (light)	34.35	274.80	52.10	416.80		
1 Flatbed Truck, Gas, 3 Ton		264.40		290.84	11.02	12.12
24 L.H., Daily Totals		$1344.80		$2148.04	$56.03	$89.50

Crew L-1

	Bare Costs Hr.	Daily	Incl. Subs O&P Hr.	Daily	Bare Costs	Incl. O&P
1 Electrician	$51.55	$412.40	$77.05	$616.40	$53.08	$79.70
1 Plumber	54.60	436.80	82.35	658.80		
16 L.H., Daily Totals		$849.20		$1275.20	$53.08	$79.70

Crew L-2

	Bare Costs Hr.	Daily	Incl. Subs O&P Hr.	Daily	Bare Costs	Incl. O&P
1 Carpenter	$44.10	$352.80	$67.85	$542.80	$38.70	$59.52
1 Carpenter Helper	33.30	266.40	51.20	409.60		
16 L.H., Daily Totals		$619.20		$952.40	$38.70	$59.52

Crew L-3

	Bare Costs Hr.	Daily	Incl. Subs O&P Hr.	Daily	Bare Costs	Incl. O&P
1 Carpenter	$44.10	$352.80	$67.85	$542.80	$47.98	$73.06
.5 Electrician	51.55	206.20	77.05	308.20		
.5 Sheet Metal Worker	52.15	208.60	79.50	318.00		
16 L.H., Daily Totals		$767.60		$1169.00	$47.98	$73.06

Crew No.	Bare Costs		Incl. Subs O&P		Cost Per Labor-Hour	
Crew L-3A	Hr.	Daily	Hr.	Daily	Bare Costs	Incl. O&P
1 Carpenter Foreman (outside)	$46.10	$368.80	$70.95	$567.60	$48.12	$73.80
.5 Sheet Metal Worker	52.15	208.60	79.50	318.00		
12 L.H., Daily Totals		$577.40		$885.60	$48.12	$73.80
Crew L-4	Hr.	Daily	Hr.	Daily	Bare Costs	Incl. O&P
2 Skilled Workers	$45.40	$726.40	$70.20	$1123.20	$41.37	$63.87
1 Helper	33.30	266.40	51.20	409.60		
24 L.H., Daily Totals		$992.80		$1532.80	$41.37	$63.87
Crew L-5	Hr.	Daily	Hr.	Daily	Bare Costs	Incl. O&P
1 Struc. Steel Foreman (outside)	$51.35	$410.80	$91.80	$734.40	$49.41	$86.46
5 Struc. Steel Workers	49.35	1974.00	88.25	3530.00		
1 Equip. Oper. (crane)	47.80	382.40	72.15	577.20		
1 Hyd. Crane, 25 Ton		697.20		766.92	12.45	13.70
56 L.H., Daily Totals		$3464.40		$5608.52	$61.86	$100.15
Crew L-5A	Hr.	Daily	Hr.	Daily	Bare Costs	Incl. O&P
1 Struc. Steel Foreman (outside)	$51.35	$410.80	$91.80	$734.40	$49.46	$85.11
2 Structural Steel Workers	49.35	789.60	88.25	1412.00		
1 Equip. Oper. (crane)	47.80	382.40	72.15	577.20		
1 S.P. Crane, 4x4, 25 Ton		565.20		621.72	17.66	19.43
32 L.H., Daily Totals		$2148.00		$3345.32	$67.13	$104.54
Crew L-5B	Hr.	Daily	Hr.	Daily	Bare Costs	Incl. O&P
1 Struc. Steel Foreman (outside)	$51.35	$410.80	$91.80	$734.40	$50.40	$80.53
2 Structural Steel Workers	49.35	789.60	88.25	1412.00		
2 Electricians	51.55	824.80	77.05	1232.80		
2 Steamfitters/Pipefitters	55.70	891.20	84.00	1344.00		
1 Equip. Oper. (crane)	47.80	382.40	72.15	577.20		
1 Equip. Oper. Oiler	41.25	330.00	62.25	498.00		
1 Hyd. Crane, 80 Ton		1554.00		1709.40	21.58	23.74
72 L.H., Daily Totals		$5182.80		$7507.80	$71.98	$104.28
Crew L-6	Hr.	Daily	Hr.	Daily	Bare Costs	Incl. O&P
1 Plumber	$54.60	$436.80	$82.35	$658.80	$53.58	$80.58
.5 Electrician	51.55	206.20	77.05	308.20		
12 L.H., Daily Totals		$643.00		$967.00	$53.58	$80.58
Crew L-7	Hr.	Daily	Hr.	Daily	Bare Costs	Incl. O&P
2 Carpenters	$44.10	$705.60	$67.85	$1085.60	$42.59	$65.21
1 Building Laborer	35.10	280.80	54.00	432.00		
.5 Electrician	51.55	206.20	77.05	308.20		
28 L.H., Daily Totals		$1192.60		$1825.80	$42.59	$65.21
Crew L-8	Hr.	Daily	Hr.	Daily	Bare Costs	Incl. O&P
2 Carpenters	$44.10	$705.60	$67.85	$1085.60	$46.20	$70.75
.5 Plumber	54.60	218.40	82.35	329.40		
20 L.H., Daily Totals		$924.00		$1415.00	$46.20	$70.75
Crew L-9	Hr.	Daily	Hr.	Daily	Bare Costs	Incl. O&P
1 Labor Foreman (inside)	$35.60	$284.80	$54.80	$438.40	$40.21	$64.35
2 Building Laborers	35.10	561.60	54.00	864.00		
1 Struc. Steel Worker	49.35	394.80	88.25	706.00		
.5 Electrician	51.55	206.20	77.05	308.20		
36 L.H., Daily Totals		$1447.40		$2316.60	$40.21	$64.35

Crew No.	Bare Costs		Incl. Subs O&P		Cost Per Labor-Hour	
Crew L-10	Hr.	Daily	Hr.	Daily	Bare Costs	Incl. O&P
1 Struc. Steel Foreman (outside)	$51.35	$410.80	$91.80	$734.40	$49.50	$84.07
1 Structural Steel Worker	49.35	394.80	88.25	706.00		
1 Equip. Oper. (crane)	47.80	382.40	72.15	577.20		
1 Hyd. Crane, 12 Ton		612.00		673.20	25.50	28.05
24 L.H., Daily Totals		$1800.00		$2690.80	$75.00	$112.12
Crew L-11	Hr.	Daily	Hr.	Daily	Bare Costs	Incl. O&P
2 Wreckers	$35.10	$561.60	$59.40	$950.40	$40.69	$64.63
1 Equip. Oper. (crane)	47.80	382.40	72.15	577.20		
1 Equip. Oper. (light)	44.75	358.00	67.55	540.40		
1 Hyd. Excavator, 2.5 C.Y.		1652.00		1817.20		
1 Loader, Skid Steer, 78 H.P.		281.20		309.32	60.41	66.45
32 L.H., Daily Totals		$3235.20		$4194.52	$101.10	$131.08
Crew M-1	Hr.	Daily	Hr.	Daily	Bare Costs	Incl. O&P
3 Elevator Constructors	$70.85	$1700.40	$106.15	$2547.60	$67.31	$100.85
1 Elevator Apprentice	56.70	453.60	84.95	679.60		
5 Hand Tools		46.00		50.60	1.44	1.58
32 L.H., Daily Totals		$2200.00		$3277.80	$68.75	$102.43
Crew M-3	Hr.	Daily	Hr.	Daily	Bare Costs	Incl. O&P
1 Electrician Foreman (outside)	$53.55	$428.40	$80.05	$640.40	$53.61	$80.64
1 Common Laborer	35.10	280.80	54.00	432.00		
.25 Equipment Operator, Med.	46.55	93.10	70.25	140.50		
1 Elevator Constructor	70.85	566.80	106.15	849.20		
1 Elevator Apprentice	56.70	453.60	84.95	679.60		
.25 S.P. Crane, 4x4, 20 Ton		129.85		142.84	3.82	4.20
34 L.H., Daily Totals		$1952.55		$2884.53	$57.43	$84.84
Crew M-4	Hr.	Daily	Hr.	Daily	Bare Costs	Incl. O&P
1 Electrician Foreman (outside)	$53.55	$428.40	$80.05	$640.40	$52.99	$79.72
1 Common Laborer	35.10	280.80	54.00	432.00		
.25 Equipment Operator, Crane	47.80	95.60	72.15	144.30		
.25 Equipment Operator, Oiler	41.25	82.50	62.25	124.50		
1 Elevator Constructor	70.85	566.80	106.15	849.20		
1 Elevator Apprentice	56.70	453.60	84.95	679.60		
.25 S.P. Crane, 4x4, 40 Ton		166.00		182.60	4.61	5.07
36 L.H., Daily Totals		$2073.70		$3052.60	$57.60	$84.79
Crew Q-1	Hr.	Daily	Hr.	Daily	Bare Costs	Incl. O&P
1 Plumber	$54.60	$436.80	$82.35	$658.80	$49.15	$74.13
1 Plumber Apprentice	43.70	349.60	65.90	527.20		
16 L.H., Daily Totals		$786.40		$1186.00	$49.15	$74.13
Crew Q-1A	Hr.	Daily	Hr.	Daily	Bare Costs	Incl. O&P
.25 Plumber Foreman (outside)	$56.60	$113.20	$85.35	$170.70	$55.00	$82.95
1 Plumber	54.60	436.80	82.35	658.80		
10 L.H., Daily Totals		$550.00		$829.50	$55.00	$82.95
Crew Q-1C	Hr.	Daily	Hr.	Daily	Bare Costs	Incl. O&P
1 Plumber	$54.60	$436.80	$82.35	$658.80	$48.28	$72.83
1 Plumber Apprentice	43.70	349.60	65.90	527.20		
1 Equip. Oper. (medium)	46.55	372.40	70.25	562.00		
1 Trencher, Chain Type, 8' D		2848.00		3132.80	118.67	130.53
24 L.H., Daily Totals		$4006.80		$4880.80	$166.95	$203.37
Crew Q-2	Hr.	Daily	Hr.	Daily	Bare Costs	Incl. O&P
2 Plumbers	$54.60	$873.60	$82.35	$1317.60	$50.97	$76.87
1 Plumber Apprentice	43.70	349.60	65.90	527.20		
24 L.H., Daily Totals		$1223.20		$1844.80	$50.97	$76.87

Crews

Crew Q-3

Crew No.	Bare Hr.	Bare Daily	Incl. Subs O&P Hr.	Incl. Subs O&P Daily	Cost Per LH Bare Costs	Cost Per LH Incl. O&P
1 Plumber Foreman (inside)	$55.10	$440.80	$83.10	$664.80	$52.00	$78.42
2 Plumbers	54.60	873.60	82.35	1317.60		
1 Plumber Apprentice	43.70	349.60	65.90	527.20		
32 L.H., Daily Totals		$1664.00		$2509.60	$52.00	$78.42

Crew Q-4

Crew No.	Bare Hr.	Bare Daily	Incl. Subs O&P Hr.	Incl. Subs O&P Daily	Cost Per LH Bare Costs	Cost Per LH Incl. O&P
1 Plumber Foreman (inside)	$55.10	$440.80	$83.10	$664.80	$52.00	$78.42
1 Plumber	54.60	436.80	82.35	658.80		
1 Welder (plumber)	54.60	436.80	82.35	658.80		
1 Plumber Apprentice	43.70	349.60	65.90	527.20		
1 Welder, electric, 300 amp		36.90		40.59	1.15	1.27
32 L.H., Daily Totals		$1700.90		$2550.19	$53.15	$79.69

Crew Q-5

Crew No.	Bare Hr.	Bare Daily	Incl. Subs O&P Hr.	Incl. Subs O&P Daily	Cost Per LH Bare Costs	Cost Per LH Incl. O&P
1 Steamfitter	$55.70	$445.60	$84.00	$672.00	$50.13	$75.60
1 Steamfitter Apprentice	44.55	356.40	67.20	537.60		
16 L.H., Daily Totals		$802.00		$1209.60	$50.13	$75.60

Crew Q-6

Crew No.	Bare Hr.	Bare Daily	Incl. Subs O&P Hr.	Incl. Subs O&P Daily	Cost Per LH Bare Costs	Cost Per LH Incl. O&P
2 Steamfitters	$55.70	$891.20	$84.00	$1344.00	$51.98	$78.40
1 Steamfitter Apprentice	44.55	356.40	67.20	537.60		
24 L.H., Daily Totals		$1247.60		$1881.60	$51.98	$78.40

Crew Q-7

Crew No.	Bare Hr.	Bare Daily	Incl. Subs O&P Hr.	Incl. Subs O&P Daily	Cost Per LH Bare Costs	Cost Per LH Incl. O&P
1 Steamfitter Foreman (inside)	$56.20	$449.60	$84.75	$678.00	$53.04	$79.99
2 Steamfitters	55.70	891.20	84.00	1344.00		
1 Steamfitter Apprentice	44.55	356.40	67.20	537.60		
32 L.H., Daily Totals		$1697.20		$2559.60	$53.04	$79.99

Crew Q-8

Crew No.	Bare Hr.	Bare Daily	Incl. Subs O&P Hr.	Incl. Subs O&P Daily	Cost Per LH Bare Costs	Cost Per LH Incl. O&P
1 Steamfitter Foreman (inside)	$56.20	$449.60	$84.75	$678.00	$53.04	$79.99
1 Steamfitter	55.70	445.60	84.00	672.00		
1 Welder (steamfitter)	55.70	445.60	84.00	672.00		
1 Steamfitter Apprentice	44.55	356.40	67.20	537.60		
1 Welder, electric, 300 amp		36.90		40.59	1.15	1.27
32 L.H., Daily Totals		$1734.10		$2600.19	$54.19	$81.26

Crew Q-9

Crew No.	Bare Hr.	Bare Daily	Incl. Subs O&P Hr.	Incl. Subs O&P Daily	Cost Per LH Bare Costs	Cost Per LH Incl. O&P
1 Sheet Metal Worker	$52.15	$417.20	$79.50	$636.00	$46.92	$71.53
1 Sheet Metal Apprentice	41.70	333.60	63.55	508.40		
16 L.H., Daily Totals		$750.80		$1144.40	$46.92	$71.53

Crew Q-10

Crew No.	Bare Hr.	Bare Daily	Incl. Subs O&P Hr.	Incl. Subs O&P Daily	Cost Per LH Bare Costs	Cost Per LH Incl. O&P
2 Sheet Metal Workers	$52.15	$834.40	$79.50	$1272.00	$48.67	$74.18
1 Sheet Metal Apprentice	41.70	333.60	63.55	508.40		
24 L.H., Daily Totals		$1168.00		$1780.40	$48.67	$74.18

Crew Q-11

Crew No.	Bare Hr.	Bare Daily	Incl. Subs O&P Hr.	Incl. Subs O&P Daily	Cost Per LH Bare Costs	Cost Per LH Incl. O&P
1 Sheet Metal Foreman (inside)	$52.65	$421.20	$80.25	$642.00	$49.66	$75.70
2 Sheet Metal Workers	52.15	834.40	79.50	1272.00		
1 Sheet Metal Apprentice	41.70	333.60	63.55	508.40		
32 L.H., Daily Totals		$1589.20		$2422.40	$49.66	$75.70

Crew Q-12

Crew No.	Bare Hr.	Bare Daily	Incl. Subs O&P Hr.	Incl. Subs O&P Daily	Cost Per LH Bare Costs	Cost Per LH Incl. O&P
1 Sprinkler Installer	$54.05	$432.40	$81.55	$652.40	$48.65	$73.40
1 Sprinkler Apprentice	43.25	346.00	65.25	522.00		
16 L.H., Daily Totals		$778.40		$1174.40	$48.65	$73.40

Crew Q-13

Crew No.	Bare Hr.	Bare Daily	Incl. Subs O&P Hr.	Incl. Subs O&P Daily	Cost Per LH Bare Costs	Cost Per LH Incl. O&P
1 Sprinkler Foreman (inside)	$54.55	$436.40	$82.30	$658.40	$51.48	$77.66
2 Sprinkler Installers	54.05	864.80	81.55	1304.80		
1 Sprinkler Apprentice	43.25	346.00	65.25	522.00		
32 L.H., Daily Totals		$1647.20		$2485.20	$51.48	$77.66

Crew Q-14

Crew No.	Bare Hr.	Bare Daily	Incl. Subs O&P Hr.	Incl. Subs O&P Daily	Cost Per LH Bare Costs	Cost Per LH Incl. O&P
1 Asbestos Worker	$48.85	$390.80	$76.10	$608.80	$43.98	$68.50
1 Asbestos Apprentice	39.10	312.80	60.90	487.20		
16 L.H., Daily Totals		$703.60		$1096.00	$43.98	$68.50

Crew Q-15

Crew No.	Bare Hr.	Bare Daily	Incl. Subs O&P Hr.	Incl. Subs O&P Daily	Cost Per LH Bare Costs	Cost Per LH Incl. O&P
1 Plumber	$54.60	$436.80	$82.35	$658.80	$49.15	$74.13
1 Plumber Apprentice	43.70	349.60	65.90	527.20		
1 Welder, electric, 300 amp		36.90		40.59	2.31	2.54
16 L.H., Daily Totals		$823.30		$1226.59	$51.46	$76.66

Crew Q-16

Crew No.	Bare Hr.	Bare Daily	Incl. Subs O&P Hr.	Incl. Subs O&P Daily	Cost Per LH Bare Costs	Cost Per LH Incl. O&P
2 Plumbers	$54.60	$873.60	$82.35	$1317.60	$50.97	$76.87
1 Plumber Apprentice	43.70	349.60	65.90	527.20		
1 Welder, electric, 300 amp		36.90		40.59	1.54	1.69
24 L.H., Daily Totals		$1260.10		$1885.39	$52.50	$78.56

Crew Q-17

Crew No.	Bare Hr.	Bare Daily	Incl. Subs O&P Hr.	Incl. Subs O&P Daily	Cost Per LH Bare Costs	Cost Per LH Incl. O&P
1 Steamfitter	$55.70	$445.60	$84.00	$672.00	$50.13	$75.60
1 Steamfitter Apprentice	44.55	356.40	67.20	537.60		
1 Welder, electric, 300 amp		36.90		40.59	2.31	2.54
16 L.H., Daily Totals		$838.90		$1250.19	$52.43	$78.14

Crew Q-17A

Crew No.	Bare Hr.	Bare Daily	Incl. Subs O&P Hr.	Incl. Subs O&P Daily	Cost Per LH Bare Costs	Cost Per LH Incl. O&P
1 Steamfitter	$55.70	$445.60	$84.00	$672.00	$49.35	$74.45
1 Steamfitter Apprentice	44.55	356.40	67.20	537.60		
1 Equip. Oper. (crane)	47.80	382.40	72.15	577.20		
1 Hyd. Crane, 12 Ton		612.00		673.20		
1 Welder, electric, 300 amp		36.90		40.59	27.04	29.74
24 L.H., Daily Totals		$1833.30		$2500.59	$76.39	$104.19

Crew Q-18

Crew No.	Bare Hr.	Bare Daily	Incl. Subs O&P Hr.	Incl. Subs O&P Daily	Cost Per LH Bare Costs	Cost Per LH Incl. O&P
2 Steamfitters	$55.70	$891.20	$84.00	$1344.00	$51.98	$78.40
1 Steamfitter Apprentice	44.55	356.40	67.20	537.60		
1 Welder, electric, 300 amp		36.90		40.59	1.54	1.69
24 L.H., Daily Totals		$1284.50		$1922.19	$53.52	$80.09

Crew Q-19

Crew No.	Bare Hr.	Bare Daily	Incl. Subs O&P Hr.	Incl. Subs O&P Daily	Cost Per LH Bare Costs	Cost Per LH Incl. O&P
1 Steamfitter	$55.70	$445.60	$84.00	$672.00	$50.60	$76.08
1 Steamfitter Apprentice	44.55	356.40	67.20	537.60		
1 Electrician	51.55	412.40	77.05	616.40		
24 L.H., Daily Totals		$1214.40		$1826.00	$50.60	$76.08

Crew Q-20

Crew No.	Bare Hr.	Bare Daily	Incl. Subs O&P Hr.	Incl. Subs O&P Daily	Cost Per LH Bare Costs	Cost Per LH Incl. O&P
1 Sheet Metal Worker	$52.15	$417.20	$79.50	$636.00	$47.85	$72.63
1 Sheet Metal Apprentice	41.70	333.60	63.55	508.40		
.5 Electrician	51.55	206.20	77.05	308.20		
20 L.H., Daily Totals		$957.00		$1452.60	$47.85	$72.63

Crew Q-21

Crew No.	Bare Hr.	Bare Daily	Incl. Subs O&P Hr.	Incl. Subs O&P Daily	Cost Per LH Bare Costs	Cost Per LH Incl. O&P
2 Steamfitters	$55.70	$891.20	$84.00	$1344.00	$51.88	$78.06
1 Steamfitter Apprentice	44.55	356.40	67.20	537.60		
1 Electrician	51.55	412.40	77.05	616.40		
32 L.H., Daily Totals		$1660.00		$2498.00	$51.88	$78.06

Crew No.	Bare Costs		Incl. Subs O&P		Cost Per Labor-Hour	
Crew Q-22	Hr.	Daily	Hr.	Daily	Bare Costs	Incl. O&P
1 Plumber	$54.60	$436.80	$82.35	$658.80	$49.15	$74.13
1 Plumber Apprentice	43.70	349.60	65.90	527.20		
1 Hyd. Crane, 12 Ton		612.00		673.20	38.25	42.08
16 L.H., Daily Totals		$1398.40		$1859.20	$87.40	$116.20
Crew Q-22A	Hr.	Daily	Hr.	Daily	Bare Costs	Incl. O&P
1 Plumber	$54.60	$436.80	$82.35	$658.80	$45.30	$68.60
1 Plumber Apprentice	43.70	349.60	65.90	527.20		
1 Laborer	35.10	280.80	54.00	432.00		
1 Equip. Oper. (crane)	47.80	382.40	72.15	577.20		
1 Hyd. Crane, 12 Ton		612.00		673.20	19.13	21.04
32 L.H., Daily Totals		$2061.60		$2868.40	$64.42	$89.64
Crew Q-23	Hr.	Daily	Hr.	Daily	Bare Costs	Incl. O&P
1 Plumber Foreman (outside)	$56.60	$452.80	$85.35	$682.80	$52.58	$79.32
1 Plumber	54.60	436.80	82.35	658.80		
1 Equip. Oper. (med.)	46.55	372.40	70.25	562.00		
1 Lattice Boom Crane, 20 Ton		950.65		1045.71	39.61	43.57
24 L.H., Daily Totals		$2212.65		$2949.32	$92.19	$122.89
Crew R-1	Hr.	Daily	Hr.	Daily	Bare Costs	Incl. O&P
1 Electrician Foreman	$52.05	$416.40	$77.80	$622.40	$45.55	$68.56
3 Electricians	51.55	1237.20	77.05	1849.20		
2 Helpers	33.30	532.80	51.20	819.20		
48 L.H., Daily Totals		$2186.40		$3290.80	$45.55	$68.56
Crew R-1A	Hr.	Daily	Hr.	Daily	Bare Costs	Incl. O&P
1 Electrician	$51.55	$412.40	$77.05	$616.40	$42.42	$64.13
1 Helper	33.30	266.40	51.20	409.60		
16 L.H., Daily Totals		$678.80		$1026.00	$42.42	$64.13
Crew R-2	Hr.	Daily	Hr.	Daily	Bare Costs	Incl. O&P
1 Electrician Foreman	$52.05	$416.40	$77.80	$622.40	$45.87	$69.07
3 Electricians	51.55	1237.20	77.05	1849.20		
2 Helpers	33.30	532.80	51.20	819.20		
1 Equip. Oper. (crane)	47.80	382.40	72.15	577.20		
1 S.P. Crane, 4x4, 5 Ton		264.60		291.06	4.72	5.20
56 L.H., Daily Totals		$2833.40		$4159.06	$50.60	$74.27
Crew R-3	Hr.	Daily	Hr.	Daily	Bare Costs	Incl. O&P
1 Electrician Foreman	$52.05	$416.40	$77.80	$622.40	$51.00	$76.37
1 Electrician	51.55	412.40	77.05	616.40		
.5 Equip. Oper. (crane)	47.80	191.20	72.15	288.60		
.5 S.P. Crane, 4x4, 5 Ton		132.30		145.53	6.62	7.28
20 L.H., Daily Totals		$1152.30		$1672.93	$57.62	$83.65
Crew R-4	Hr.	Daily	Hr.	Daily	Bare Costs	Incl. O&P
1 Struc. Steel Foreman (outside)	$51.35	$410.80	$91.80	$734.40	$50.19	$86.72
3 Struc. Steel Workers	49.35	1184.40	88.25	2118.00		
1 Electrician	51.55	412.40	77.05	616.40		
1 Welder, gas engine, 300 amp		121.60		133.76	3.04	3.34
40 L.H., Daily Totals		$2129.20		$3602.56	$53.23	$90.06

Crew No.	Bare Costs		Incl. Subs O&P		Cost Per Labor-Hour	
Crew R-5	Hr.	Daily	Hr.	Daily	Bare Costs	Incl. O&P
1 Electrician Foreman	$52.05	$416.40	$77.80	$622.40	$44.96	$67.72
4 Electrician Linemen	51.55	1649.60	77.05	2465.60		
2 Electrician Operators	51.55	824.80	77.05	1232.80		
4 Electrician Groundmen	33.30	1065.60	51.20	1638.40		
1 Crew Truck		182.40		200.64		
1 Flatbed Truck, 20,000 GVW		215.80		237.38		
1 Pickup Truck, 3/4 Ton		127.00		139.70		
.2 Hyd. Crane, 55 Ton		220.00		242.00		
.2 Hyd. Crane, 12 Ton		122.40		134.64		
.2 Earth Auger, Truck-Mtd.		80.68		88.75		
1 Tractor w/Winch		382.20		420.42	15.12	16.63
88 L.H., Daily Totals		$5286.88		$7422.73	$60.08	$84.35
Crew R-6	Hr.	Daily	Hr.	Daily	Bare Costs	Incl. O&P
1 Electrician Foreman	$52.05	$416.40	$77.80	$622.40	$44.96	$67.72
4 Electrician Linemen	51.55	1649.60	77.05	2465.60		
2 Electrician Operators	51.55	824.80	77.05	1232.80		
4 Electrician Groundmen	33.30	1065.60	51.20	1638.40		
1 Crew Truck		182.40		200.64		
1 Flatbed Truck, 20,000 GVW		215.80		237.38		
1 Pickup Truck, 3/4 Ton		127.00		139.70		
.2 Hyd. Crane, 55 Ton		220.00		242.00		
.2 Hyd. Crane, 12 Ton		122.40		134.64		
.2 Earth Auger, Truck-Mtd.		80.68		88.75		
1 Tractor w/Winch		382.20		420.42		
3 Cable Trailers		563.70		620.07		
.5 Tensioning Rig		186.88		205.56		
.5 Cable Pulling Rig		1092.50		1201.75	36.06	39.67
88 L.H., Daily Totals		$7129.95		$9450.11	$81.02	$107.39
Crew R-7	Hr.	Daily	Hr.	Daily	Bare Costs	Incl. O&P
1 Electrician Foreman	$52.05	$416.40	$77.80	$622.40	$36.42	$55.63
5 Electrician Groundmen	33.30	1332.00	51.20	2048.00		
1 Crew Truck		182.40		200.64	3.80	4.18
48 L.H., Daily Totals		$1930.80		$2871.04	$40.23	$59.81
Crew R-8	Hr.	Daily	Hr.	Daily	Bare Costs	Incl. O&P
1 Electrician Foreman	$52.05	$416.40	$77.80	$622.40	$45.55	$68.56
3 Electrician Linemen	51.55	1237.20	77.05	1849.20		
2 Electrician Groundmen	33.30	532.80	51.20	819.20		
1 Pickup Truck, 3/4 Ton		127.00		139.70		
1 Crew Truck		182.40		200.64	6.45	7.09
48 L.H., Daily Totals		$2495.80		$3631.14	$52.00	$75.65
Crew R-9	Hr.	Daily	Hr.	Daily	Bare Costs	Incl. O&P
1 Electrician Foreman	$52.05	$416.40	$77.80	$622.40	$42.49	$64.22
1 Electrician Lineman	51.55	412.40	77.05	616.40		
2 Electrician Operators	51.55	824.80	77.05	1232.80		
4 Electrician Groundmen	33.30	1065.60	51.20	1638.40		
1 Pickup Truck, 3/4 Ton		127.00		139.70		
1 Crew Truck		182.40		200.64	4.83	5.32
64 L.H., Daily Totals		$3028.60		$4450.34	$47.32	$69.54
Crew R-10	Hr.	Daily	Hr.	Daily	Bare Costs	Incl. O&P
1 Electrician Foreman	$52.05	$416.40	$77.80	$622.40	$48.59	$72.87
4 Electrician Linemen	51.55	1649.60	77.05	2465.60		
1 Electrician Groundman	33.30	266.40	51.20	409.60		
1 Crew Truck		182.40		200.64		
3 Tram Cars		345.90		380.49	11.01	12.11
48 L.H., Daily Totals		$2860.70		$4078.73	$59.60	$84.97

Crew No.	Bare Costs		Incl. Subs O&P		Cost Per Labor-Hour	
Crew R-11	Hr.	Daily	Hr.	Daily	Bare Costs	Incl. O&P
1 Electrician Foreman	$52.05	$416.40	$77.80	$622.40	$48.74	$73.16
4 Electricians	51.55	1649.60	77.05	2465.60		
1 Equip. Oper. (crane)	47.80	382.40	72.15	577.20		
1 Common Laborer	35.10	280.80	54.00	432.00		
1 Crew Truck		182.40		200.64		
1 Hyd. Crane, 12 Ton		612.00		673.20	14.19	15.60
56 L.H., Daily Totals		$3523.60		$4971.04	$62.92	$88.77
Crew R-12	Hr.	Daily	Hr.	Daily	Bare Costs	Incl. O&P
1 Carpenter Foreman (inside)	$44.60	$356.80	$68.65	$549.20	$41.57	$64.96
4 Carpenters	44.10	1411.20	67.85	2171.20		
4 Common Laborers	35.10	1123.20	54.00	1728.00		
1 Equip. Oper. (med.)	46.55	372.40	70.25	562.00		
1 Steel Worker	49.35	394.80	88.25	706.00		
1 Dozer, 200 H.P.		1243.00		1367.30		
1 Pickup Truck, 3/4 Ton		127.00		139.70	15.57	17.13
88 L.H., Daily Totals		$5028.40		$7223.40	$57.14	$82.08
Crew R-13	Hr.	Daily	Hr.	Daily	Bare Costs	Incl. O&P
1 Electrician Foreman	$52.05	$416.40	$77.80	$622.40	$49.50	$74.14
3 Electricians	51.55	1237.20	77.05	1849.20		
.25 Equip. Oper. (crane)	47.80	95.60	72.15	144.30		
1 Equipment Oiler	41.25	330.00	62.25	498.00		
.25 Hydraulic Crane, 33 Ton		177.90		195.69	4.24	4.66
42 L.H., Daily Totals		$2257.10		$3309.59	$53.74	$78.80
Crew R-15	Hr.	Daily	Hr.	Daily	Bare Costs	Incl. O&P
1 Electrician Foreman	$52.05	$416.40	$77.80	$622.40	$50.50	$75.59
4 Electricians	51.55	1649.60	77.05	2465.60		
1 Equipment Oper. (light)	44.75	358.00	67.55	540.40		
1 Aerial Lift Truck		324.20		356.62	6.75	7.43
48 L.H., Daily Totals		$2748.20		$3985.02	$57.25	$83.02
Crew R-18	Hr.	Daily	Hr.	Daily	Bare Costs	Incl. O&P
.25 Electrician Foreman	$52.05	$104.10	$77.80	$155.60	$40.36	$61.20
1 Electrician	51.55	412.40	77.05	616.40		
2 Helpers	33.30	532.80	51.20	819.20		
26 L.H., Daily Totals		$1049.30		$1591.20	$40.36	$61.20
Crew R-19	Hr.	Daily	Hr.	Daily	Bare Costs	Incl. O&P
.5 Electrician Foreman	$52.05	$208.20	$77.80	$311.20	$51.65	$77.20
2 Electricians	51.55	824.80	77.05	1232.80		
20 L.H., Daily Totals		$1033.00		$1544.00	$51.65	$77.20
Crew R-21	Hr.	Daily	Hr.	Daily	Bare Costs	Incl. O&P
1 Electrician Foreman	$52.05	$416.40	$77.80	$622.40	$51.55	$77.07
3 Electricians	51.55	1237.20	77.05	1849.20		
.1 Equip. Oper. (med.)	46.55	37.24	70.25	56.20		
.1 S.P. Crane, 4x4, 25 Ton		56.52		62.17	1.72	1.90
32.8 L.H., Daily Totals		$1747.36		$2589.97	$53.27	$78.96
Crew R-22	Hr.	Daily	Hr.	Daily	Bare Costs	Incl. O&P
.66 Electrician Foreman	$52.05	$274.82	$77.80	$410.78	$43.79	$66.06
2 Helpers	33.30	532.80	51.20	819.20		
2 Electricians	51.55	824.80	77.05	1232.80		
37.28 L.H., Daily Totals		$1632.42		$2462.78	$43.79	$66.06

Crew No.	Bare Costs		Incl. Subs O&P		Cost Per Labor-Hour	
Crew R-30	Hr.	Daily	Hr.	Daily	Bare Costs	Incl. O&P
.25 Electrician Foreman (outside)	$53.55	$107.10	$80.05	$160.10	$41.58	$63.10
1 Electrician	51.55	412.40	77.05	616.40		
2 Laborers, (Semi-Skilled)	35.10	561.60	54.00	864.00		
26 L.H., Daily Totals		$1081.10		$1640.50	$41.58	$63.10
Crew R-31	Hr.	Daily	Hr.	Daily	Bare Costs	Incl. O&P
1 Electrician	$51.55	$412.40	$77.05	$616.40	$51.55	$77.05
1 Core Drill, Electric, 2.5 H.P.		47.00		51.70	5.88	6.46
8 L.H., Daily Totals		$459.40		$668.10	$57.42	$83.51
Crew W-41E	Hr.	Daily	Hr.	Daily	Bare Costs	Incl. O&P
.5 Plumber Foreman (outside)	$56.60	$226.40	$85.35	$341.40	$47.20	$71.61
1 Plumber	54.60	436.80	82.35	658.80		
1 Laborer	35.10	280.80	54.00	432.00		
20 L.H., Daily Totals		$944.00		$1432.20	$47.20	$71.61

Historical Cost Indexes

The table below lists both the RSMeans Historical Cost Index based on Jan. 1, 1993 = 100 as well as the computed value of an index based on Jan. 1, 2012 costs. Since the Jan. 1, 2012 figure is estimated, space is left to write in the actual index figures as they become available through either the quarterly *Means Construction Cost Indexes* or as printed in the *Engineering News-Record*. To compute the actual index based on Jan. 1, 2012 = 100, divide the Historical Cost Index for a particular year by the actual Jan. 1, 2012 Construction Cost Index. Space has been left to advance the index figures as the year progresses.

Year	Historical Cost Index Jan. 1, 1993 = 100		Current Index Based on Jan. 1, 2012 = 100		Year	Historical Cost Index Jan. 1, 1993 = 100	Current Index Based on Jan. 1, 2012 = 100		Year	Historical Cost Index Jan. 1, 1993 = 100	Current Index Based on Jan. 1, 2012 = 100	
	Est.	Actual	Est.	Actual		Actual	Est.	Actual		Actual	Est.	Actual
Oct 2012*					July 1997	112.8	58.5		July 1979	57.8	30.0	
July 2012*					1996	110.2	57.2		1978	53.5	27.7	
April 2012*					1995	107.6	55.8		1977	49.5	25.7	
Jan 2012*	192.8		100.0	100.0	1994	104.4	54.1		1976	46.9	24.3	
July 2011		185.7	96.3		1993	101.7	52.7		1975	44.8	23.2	
2010		183.5	95.2		1992	99.4	51.6		1974	41.4	21.5	
2009		180.1	93.4		1991	96.8	50.2		1973	37.7	19.6	
2008		180.4	93.6		1990	94.3	48.9		1972	34.8	18.0	
2007		169.4	87.9		1989	92.1	47.8		1971	32.1	16.6	
2006		162.0	84.0		1988	89.9	46.6		1970	28.7	14.9	
2005		151.6	78.6		1987	87.7	45.5		1969	26.9	14.0	
2004		143.7	74.5		1986	84.2	43.7		1968	24.9	12.9	
2003		132.0	68.5		1985	82.6	42.8		1967	23.5	12.2	
2002		128.7	66.8		1984	82.0	42.5		1966	22.7	11.8	
2001		125.1	64.9		1983	80.2	41.6		1965	21.7	11.3	
2000		120.9	62.7		1982	76.1	39.5		1964	21.2	11.0	
1999		117.6	61.0		1981	70.0	36.3		1963	20.7	10.7	
1998		115.1	59.7		1980	62.9	32.6		1962	20.2	10.5	

Adjustments to Costs

The Historical Cost Index can be used to convert National Average building costs at a particular time to the approximate building costs for some other time.

Example:

Estimate and compare construction costs for different years in the same city.

To estimate the National Average construction cost of a building in 1970, knowing that it cost $900,000 in 2012:

INDEX in 1970 = 28.7

INDEX in 2012 = 192.8

Note: The City Cost Indexes for Canada can be used to convert U.S. National averages to local costs in Canadian dollars.

Time Adjustment using the Historical Cost Indexes:

$$\frac{\text{Index for Year A}}{\text{Index for Year B}} \times \text{Cost in Year B} = \text{Cost in Year A}$$

$$\frac{\text{INDEX } 1970}{\text{INDEX } 2012} \times \text{Cost } 2012 = \text{Cost } 1970$$

$$\frac{28.7}{192.8} \times \$900,000 = .149 \times \$900,000 = \$134,100$$

The construction cost of the building in 1970 is $134,100.

*Historical Cost Index updates and other resources are provided on the following website.
http://www.reedconstructiondata.com/rsmeans/chgnotice/176028

How to Use the City Cost Indexes

What you should know before you begin

RSMeans City Cost Indexes (CCI) are an extremely useful tool to use when you want to compare costs from city to city and region to region.

This publication contains average construction cost indexes for 731 U.S. and Canadian cities covering over 930 three-digit zip code locations, as listed directly under each city.

Keep in mind that a City Cost Index number is a percentage ratio of a specific city's cost to the national average cost of the same item at a stated time period.

In other words, these index figures represent relative construction factors (or, if you prefer, multipliers) for Material and Installation costs, as well as the weighted average for Total In Place costs for each CSI MasterFormat division. Installation costs include both labor and equipment rental costs. When estimating equipment rental rates only, for a specific location, use 015433 CONTRACTOR EQUIPMENT index.

The 30 City Average Index is the average of 30 major U.S. cities and serves as a National Average.

Index figures for both material and installation are based on the 30 major city average of 100 and represent the cost relationship as of July 1, 2011. The index for each division is computed from representative material and labor quantities for that division. The weighted average for each city is a weighted total of the components listed above it, but does not include relative productivity between trades or cities.

As changes occur in local material prices, labor rates, and equipment rental rates, (including fuel costs) the impact of these changes should be accurately measured by the change in the City Cost Index for each particular city (as compared to the 30 City Average).

Therefore, if you know (or have estimated) building costs in one city today, you can easily convert those costs to expected building costs in another city.

In addition, by using the Historical Cost Index, you can easily convert National Average building costs at a particular time to the approximate building costs for some other time. The City Cost Indexes can then be applied to calculate the costs for a particular city.

Quick Calculations

Location Adjustment Using the City Cost Indexes:

$$\frac{\text{Index for City A}}{\text{Index for City B}} \times \text{Cost in City B} = \text{Cost in City A}$$

Time Adjustment for the National Average Using the Historical Cost Index:

$$\frac{\text{Index for Year A}}{\text{Index for Year B}} \times \text{Cost in Year B} = \text{Cost in Year A}$$

Adjustment from the National Average:

$$\frac{\text{Index for City A}}{100} \times \text{National Average Cost} = \text{Cost in City A}$$

Since each of the other RSMeans publications contains many different items, any *one* item multiplied by the particular city index may give incorrect results. However, the larger the number of items compiled, the closer the results should be to actual costs for that particular city.

The City Cost Indexes for Canadian cities are calculated using Canadian material and equipment prices and labor rates, in Canadian dollars. Therefore, indexes for Canadian cities can be used to convert U.S. National Average prices to local costs in Canadian dollars.

How to use this section

1. Compare costs from city to city.

In using the RSMeans Indexes, remember that an index number is not a fixed number but a ratio: It's a percentage ratio of a building component's cost at any stated time to the National Average cost of that same component at the same time period. Put in the form of an equation:

$$\frac{\text{Specific City Cost}}{\text{National Average Cost}} \times 100 = \text{City Index Number}$$

Therefore, when making cost comparisons between cities, do not subtract one city's index number from the index number of another city and read the result as a percentage difference. Instead, divide one city's index number by that of the other city. The resulting number may then be used as a multiplier to calculate cost differences from city to city.

The formula used to find cost differences between cities for the purpose of comparison is as follows:

$$\frac{\text{City A Index}}{\text{City B Index}} \times \text{City B Cost (Known)} = \text{City A Cost (Unknown)}$$

In addition, you can use RSMeans CCI to calculate and compare costs division by division between cities using the same basic formula. (Just be sure that you're comparing similar divisions.)

2. Compare a specific city's construction costs with the National Average.

When you're studying construction location feasibility, it's advisable to compare a prospective project's cost index with an index of the National Average cost.

For example, divide the weighted average index of construction costs of a specific city by that of the 30 City Average, which = 100.

$$\frac{\text{City Index}}{100} = \% \text{ of National Average}$$

As a result, you get a ratio that indicates the relative cost of construction in that city in comparison with the National Average.

3. Convert U.S. National Average to actual costs in Canadian City.

$$\frac{\text{Index for Canadian City}}{100} \times \text{National Average Cost} = \text{Cost in Canadian City in \$ CAN}$$

4. Adjust construction cost data based on a National Average.

When you use a source of construction cost data which is based on a National Average (such as RSMeans cost data publications), it is necessary to adjust those costs to a specific location.

$$\frac{\text{City Index}}{100} \quad \text{x} \quad \frac{\text{"Book" Cost Based on}}{\text{National Average Costs}} \quad = \quad \frac{\text{City Cost}}{\text{(Unknown)}}$$

5. When applying the City Cost Indexes to demolition projects, use the appropriate division installation index. For example, for removal of existing doors and windows, use Division 8 (Openings) index.

What you might like to know about how we developed the Indexes

The information presented in the CCI is organized according to the Construction Specifications Institute (CSI) MasterFormat 2010.

To create a reliable index, RSMeans researched the building type most often constructed in the United States and Canada. Because it was concluded that no one type of building completely represented the building construction industry, nine different types of buildings were combined to create a composite model.

The exact material, labor, and equipment quantities are based on detailed analyses of these nine building types, then each quantity is weighted in proportion to expected usage. These various material items, labor hours, and equipment rental rates are thus combined to form a composite building representing as closely as possible the actual usage of materials, labor and equipment used in the North American Building Construction Industry.

The following structures were chosen to make up that composite model:

1. Factory, 1 story
2. Office, 2–4 story
3. Store, Retail
4. Town Hall, 2–3 story
5. High School, 2–3 story
6. Hospital, 4–8 story
7. Garage, Parking
8. Apartment, 1–3 story
9. Hotel/Motel, 2–3 story

For the purposes of ensuring the timeliness of the data, the components of the index for the composite model have been streamlined. They currently consist of:

- specific quantities of 66 commonly used construction materials;
- specific labor-hours for 21 building construction trades; and
- specific days of equipment rental for 6 types of construction equipment (normally used to install the 66 material items by the 21 trades.) Fuel costs and routine maintenance costs are included in the equipment cost.

A sophisticated computer program handles the updating of all costs for each city on a quarterly basis. Material and equipment price quotations are gathered quarterly from 731 cities in the United States and Canada. These prices and the latest negotiated labor wage rates for 21 different building trades are used to compile the quarterly update of the City Cost Index.

The 30 major U.S. cities used to calculate the National Average are:

Atlanta, GA	Memphis, TN
Baltimore, MD	Milwaukee, WI
Boston, MA	Minneapolis, MN
Buffalo, NY	Nashville, TN
Chicago, IL	New Orleans, LA
Cincinnati, OH	New York, NY
Cleveland, OH	Philadelphia, PA
Columbus, OH	Phoenix, AZ
Dallas, TX	Pittsburgh, PA
Denver, CO	St. Louis, MO
Detroit, MI	San Antonio, TX
Houston, TX	San Diego, CA
Indianapolis, IN	San Francisco, CA
Kansas City, MO	Seattle, WA
Los Angeles, CA	Washington, DC

What the CCI does not indicate

The weighted average for each city is a total of the divisional components weighted to reflect typical usage, but it does not include the productivity variations between trades or cities.

In addition, the CCI does not take into consideration factors such as the following:

- managerial efficiency
- competitive conditions
- automation
- restrictive union practices
- unique local requirements
- regional variations due to specific building codes

City Cost Indexes

Table 1

DIVISION		UNITED STATES 30 CITY AVERAGE			ALABAMA — ANNISTON 362			BIRMINGHAM 350-352			BUTLER 369			DECATUR 356			DOTHAN 363		
		MAT.	INST.	TOTAL	MAT.	INST.	TOTAL	MAT.	INST.	TOTAL	MAT.	INST.	TOTAL	MAT.	INST.	TOTAL	MAT.	INST.	TOTAL
015433	CONTRACTOR EQUIPMENT		100.0	100.0		100.7	100.7		100.8	100.8		98.0	98.0		100.7	100.7		98.0	98.0
0241, 31 - 34	SITE & INFRASTRUCTURE, DEMOLITION	100.0	100.0	100.0	87.6	93.4	91.6	85.6	94.0	91.4	103.5	87.5	92.4	80.6	92.0	88.5	100.8	87.5	91.6
0310	Concrete Forming & Accessories	100.0	100.0	100.0	91.9	49.9	55.4	92.3	72.4	75.0	88.6	45.2	50.8	96.0	48.1	54.3	96.8	44.6	51.4
0320	Concrete Reinforcing	100.0	100.0	100.0	93.3	83.3	88.3	93.3	83.7	88.5	98.6	48.8	73.5	93.3	78.1	85.7	98.6	48.2	73.2
0330	Cast-in-Place Concrete	100.0	100.0	100.0	92.5	51.8	76.5	98.7	66.6	86.0	90.2	57.3	77.2	91.9	64.9	81.3	90.2	48.2	73.7
03	CONCRETE	100.0	100.0	100.0	95.3	58.4	77.3	94.3	73.5	84.2	96.1	51.9	74.6	91.1	61.2	76.5	95.5	48.4	72.6
04	MASONRY	100.0	100.0	100.0	92.0	70.5	78.9	94.8	79.1	85.2	96.9	51.3	69.1	93.0	54.2	69.3	98.1	42.0	63.8
05	METALS	100.0	100.0	100.0	93.6	91.3	92.9	94.5	93.1	94.0	92.6	78.9	88.1	95.8	87.9	93.2	92.7	77.7	87.7
06	WOOD, PLASTICS & COMPOSITES	100.0	100.0	100.0	94.1	45.4	66.0	96.0	71.0	81.5	89.3	45.0	63.7	99.0	45.6	68.1	100.4	45.0	68.4
07	THERMAL & MOISTURE PROTECTION	100.0	100.0	100.0	93.1	55.8	78.0	95.1	82.0	89.8	93.2	59.5	79.5	92.9	61.3	80.1	93.1	51.8	76.3
08	OPENINGS	100.0	100.0	100.0	93.8	51.1	83.1	97.6	73.3	91.4	93.9	46.8	82.0	97.9	52.6	86.4	93.9	46.8	82.1
0920	Plaster & Gypsum Board	100.0	100.0	100.0	88.2	44.1	56.7	97.1	70.5	78.1	86.0	43.7	55.8	93.2	44.4	58.3	96.0	43.7	58.7
0950, 0980	Ceilings & Acoustic Treatment	100.0	100.0	100.0	89.3	44.1	59.4	94.9	70.5	78.8	89.3	43.7	59.2	93.8	44.4	61.2	89.3	43.7	58.2
0960	Flooring	100.0	100.0	100.0	101.8	42.1	84.3	104.5	63.1	92.4	107.6	58.6	93.2	104.5	53.0	89.4	114.6	28.8	89.4
0970, 0990	Wall Finishes & Painting/Coating	100.0	100.0	100.0	96.6	33.2	57.5	96.6	58.6	73.1	96.6	56.2	71.7	96.6	59.8	73.9	96.6	56.2	71.7
09	FINISHES	100.0	100.0	100.0	95.4	45.5	67.6	98.5	68.8	81.9	99.4	48.4	70.9	97.1	49.4	70.5	102.3	42.4	68.9
COVERS	DIVS. 10 - 14, 25, 28, 41, 43, 44, 46	100.0	100.0	100.0	100.0	73.9	94.8	100.0	88.7	97.8	100.0	44.7	89.0	100.0	46.7	89.4	100.0	44.8	89.0
21, 22, 23	FIRE SUPPRESSION, PLUMBING & HVAC	100.0	100.0	100.0	99.2	59.4	83.0	99.9	65.9	86.1	96.7	35.7	72.0	99.9	43.7	77.2	96.7	35.6	72.0
26, 27, 3370	ELECTRICAL, COMMUNICATIONS & UTIL.	100.0	100.0	100.0	96.6	60.9	78.2	103.0	62.5	82.1	98.5	42.7	69.7	98.2	65.6	81.4	97.3	61.0	78.6
MF2010	WEIGHTED AVERAGE	100.0	100.0	100.0	96.0	64.5	82.1	97.6	74.6	87.4	96.5	52.0	76.9	96.5	60.1	80.5	96.6	52.0	76.9

Table 2 — ALABAMA

DIVISION		EVERGREEN 364			GADSDEN 359			HUNTSVILLE 357-358			JASPER 355			MOBILE 365-366			MONTGOMERY 360-361		
		MAT.	INST.	TOTAL	MAT.	INST.	TOTAL	MAT.	INST.	TOTAL	MAT.	INST.	TOTAL	MAT.	INST.	TOTAL	MAT.	INST.	TOTAL
015433	CONTRACTOR EQUIPMENT		98.0	98.0		100.7	100.7		100.7	100.7		100.7	100.7		98.0	98.0		98.0	98.0
0241, 31 - 34	SITE & INFRASTRUCTURE, DEMOLITION	104.4	87.8	92.8	86.3	92.9	90.9	80.4	93.1	89.1	85.7	92.7	90.6	95.1	88.5	90.5	93.6	88.4	90.0
0310	Concrete Forming & Accessories	85.4	46.9	51.9	88.7	43.9	49.7	96.0	64.3	68.4	93.6	36.4	43.9	96.1	56.9	62.0	94.0	45.9	52.2
0320	Concrete Reinforcing	98.7	48.8	73.5	98.9	82.5	90.6	93.3	78.6	85.9	93.3	82.4	87.8	96.3	81.7	89.0	96.3	82.3	89.3
0330	Cast-in-Place Concrete	90.2	59.9	78.3	92.0	51.0	75.8	89.6	63.3	79.2	101.9	45.9	79.8	94.5	69.3	84.6	96.5	48.6	77.6
03	CONCRETE	96.3	53.6	75.5	95.1	55.3	75.7	89.9	67.9	79.2	98.6	50.3	75.0	92.9	67.3	80.4	93.7	55.4	75.1
04	MASONRY	96.9	56.2	72.0	91.4	45.4	63.3	94.5	60.3	73.6	88.0	43.6	60.9	95.2	55.7	71.1	93.3	41.6	61.7
05	METALS	92.7	78.9	88.1	93.7	90.5	92.6	95.8	89.5	93.7	93.6	89.8	92.4	94.6	91.6	93.6	94.5	90.1	93.0
06	WOOD, PLASTICS & COMPOSITES	85.9	45.0	62.3	89.9	41.1	61.7	99.0	65.2	79.4	96.0	33.5	59.9	99.1	56.1	74.3	95.6	45.0	66.4
07	THERMAL & MOISTURE PROTECTION	93.1	60.1	79.7	93.0	66.1	82.1	92.8	74.7	85.4	93.0	47.8	74.7	92.8	73.3	84.9	92.4	64.8	81.2
08	OPENINGS	93.9	46.8	82.0	93.8	49.9	82.8	97.9	63.2	89.1	93.8	48.8	82.5	97.9	60.7	88.5	97.9	54.3	86.9
0920	Plaster & Gypsum Board	85.1	43.7	55.5	85.5	39.7	52.8	93.2	64.5	72.7	88.7	31.9	48.1	93.2	55.2	66.1	97.1	43.7	59.0
0950, 0980	Ceilings & Acoustic Treatment	89.3	43.7	59.2	89.3	39.7	56.6	95.8	64.5	75.2	89.3	31.9	51.4	95.8	55.2	69.0	94.9	43.7	61.1
0960	Flooring	105.0	58.6	91.4	100.0	47.1	84.5	104.5	56.6	90.4	102.6	42.1	84.9	113.6	58.6	97.5	112.0	31.0	88.2
0970, 0990	Wall Finishes & Painting/Coating	96.6	56.2	71.7	96.6	64.0	76.5	96.6	61.6	75.0	96.6	40.7	62.1	100.1	57.2	73.6	96.6	56.2	71.7
09	FINISHES	98.6	49.6	71.2	94.4	44.9	66.8	97.6	62.2	77.8	95.4	36.5	62.5	102.6	56.1	76.7	101.8	42.8	68.9
COVERS	DIVS. 10 - 14, 25, 28, 41, 43, 44, 46	100.0	46.4	89.3	100.0	80.4	96.1	100.0	85.1	97.0	100.0	42.9	88.6	100.0	81.6	96.3	100.0	80.3	96.1
21, 22, 23	FIRE SUPPRESSION, PLUMBING & HVAC	96.7	38.2	73.0	101.8	38.6	76.2	99.9	55.5	81.9	101.8	56.4	83.4	99.9	64.6	85.6	100.0	36.4	74.2
26, 27, 3370	ELECTRICAL, COMMUNICATIONS & UTIL.	95.9	42.7	68.5	98.3	62.5	79.9	99.4	65.6	82.0	97.7	62.5	79.6	99.4	60.7	79.5	98.5	64.7	81.0
MF2010	WEIGHTED AVERAGE	96.2	53.4	77.3	96.6	57.6	79.4	96.6	68.1	84.1	96.8	57.6	79.5	97.7	67.5	84.4	97.4	56.6	79.4

Table 3 — ALABAMA / ALASKA

DIVISION		ALABAMA — PHENIX CITY 368			SELMA 367			TUSCALOOSA 354			ALASKA — ANCHORAGE 995-996			FAIRBANKS 997			JUNEAU 998		
		MAT.	INST.	TOTAL	MAT.	INST.	TOTAL	MAT.	INST.	TOTAL	MAT.	INST.	TOTAL	MAT.	INST.	TOTAL	MAT.	INST.	TOTAL
015433	CONTRACTOR EQUIPMENT		98.0	98.0		98.0	98.0		100.7	100.7		114.5	114.5		114.5	114.5		114.5	114.5
0241, 31 - 34	SITE & INFRASTRUCTURE, DEMOLITION	108.1	88.4	94.5	100.6	88.3	92.2	80.8	93.1	89.3	125.9	129.7	128.5	118.0	129.7	126.1	125.5	129.7	128.4
0310	Concrete Forming & Accessories	91.9	40.7	47.4	89.8	45.1	50.9	95.9	49.5	55.6	139.3	117.2	120.1	142.3	117.3	120.6	138.5	117.2	119.9
0320	Concrete Reinforcing	98.6	65.1	81.7	98.6	82.1	90.3	93.3	82.4	87.9	148.6	108.8	128.5	143.5	108.9	126.0	111.7	108.8	110.3
0330	Cast-in-Place Concrete	90.2	49.2	74.0	90.2	48.4	73.7	93.3	51.0	76.6	144.5	116.5	133.5	136.6	116.9	128.8	144.4	116.5	133.4
03	CONCRETE	98.7	50.1	75.1	95.0	54.9	75.5	91.7	57.9	75.2	131.5	114.7	123.3	119.6	114.9	117.3	122.1	114.7	118.5
04	MASONRY	96.9	43.4	64.2	100.2	40.8	64.0	93.3	48.5	66.0	193.3	121.7	149.6	195.8	121.7	150.6	178.8	121.7	144.0
05	METALS	92.6	83.8	89.7	92.6	89.0	91.4	94.9	90.5	93.4	120.1	101.7	114.0	117.1	101.8	112.1	117.1	101.6	112.0
06	WOOD, PLASTICS & COMPOSITES	93.8	37.6	61.4	90.9	45.0	64.4	99.0	47.2	69.0	116.7	116.0	116.3	121.5	116.0	118.4	116.7	116.0	116.3
07	THERMAL & MOISTURE PROTECTION	93.5	64.7	81.8	93.0	54.7	77.4	92.9	68.2	82.9	191.0	116.3	160.6	179.7	117.6	154.4	181.2	116.3	154.8
08	OPENINGS	93.8	46.1	81.8	93.8	54.3	83.9	97.9	60.4	88.4	119.9	112.9	118.1	116.4	113.0	115.5	115.7	112.9	115.0
0920	Plaster & Gypsum Board	89.6	36.1	51.4	87.3	43.7	56.2	93.2	46.0	59.5	143.4	116.3	124.1	150.2	116.3	126.0	130.9	116.3	120.5
0950, 0980	Ceilings & Acoustic Treatment	89.3	36.1	54.2	89.3	43.7	59.2	95.8	46.0	62.9	136.0	116.3	123.0	122.5	116.3	118.4	116.2	116.3	116.3
0960	Flooring	110.1	34.0	86.9	108.3	29.8	85.3	104.5	39.3	85.4	160.1	128.4	150.8	150.4	128.4	143.9	150.6	128.4	144.1
0970, 0990	Wall Finishes & Painting/Coating	96.6	56.2	71.7	96.6	56.2	71.7	96.6	48.4	66.9	150.9	116.9	129.9	150.4	125.8	135.2	150.9	116.9	129.9
09	FINISHES	101.1	38.7	66.3	99.4	42.6	67.7	97.5	46.3	69.0	150.9	119.6	133.4	143.6	120.5	130.8	140.5	119.6	128.8
COVERS	DIVS. 10 - 14, 25, 28, 41, 43, 44, 46	100.0	79.9	96.0	100.0	44.7	89.0	100.0	82.3	96.5	100.0	108.7	101.7	100.0	108.7	101.7	100.0	108.7	101.7
21, 22, 23	FIRE SUPPRESSION, PLUMBING & HVAC	96.7	36.8	72.4	96.7	35.9	72.1	99.9	36.2	74.1	100.3	104.5	102.0	100.3	108.2	103.5	100.3	104.0	101.8
26, 27, 3370	ELECTRICAL, COMMUNICATIONS & UTIL.	98.0	73.9	85.5	97.0	42.7	69.0	98.9	62.5	80.1	133.3	116.7	124.7	143.5	116.7	129.6	129.3	116.7	122.8
MF2010	WEIGHTED AVERAGE	97.1	55.9	78.9	96.3	51.9	76.7	96.6	58.6	79.9	124.5	114.0	119.9	122.4	115.0	119.1	120.3	113.9	117.5

City Cost Indexes

		ALASKA			ARIZONA															
		KETCHIKAN			CHAMBERS			FLAGSTAFF			GLOBE			KINGMAN			MESA/TEMPE		
	DIVISION	999			865			860			855			864			852		
		MAT.	INST.	TOTAL	MAT.	INST.	TOTAL	MAT.	INST.	TOTAL	MAT.	INST.	TOTAL	MAT.	INST.	TOTAL	MAT.	INST.	TOTAL
015433	CONTRACTOR EQUIPMENT		114.5	114.5		94.0	94.0		94.0	94.0		95.9	95.9		94.0	94.0		95.9	95.9
0241, 31 - 34	SITE & INFRASTRUCTURE, DEMOLITION	176.4	129.7	144.2	64.0	96.6	86.5	82.4	98.5	93.5	96.6	98.8	98.1	64.0	98.6	87.9	86.8	100.4	96.2
0310	Concrete Forming & Accessories	134.3	117.2	119.4	96.5	45.5	52.2	101.3	68.1	72.4	99.2	45.4	52.4	94.8	67.3	70.9	101.8	67.2	71.7
0320	Concrete Reinforcing	112.2	108.8	110.5	98.0	78.0	87.9	97.8	85.2	91.5	102.6	77.9	90.2	98.1	85.2	91.6	103.3	85.2	94.1
0330	Cast-in-Place Concrete	276.4	116.7	213.4	89.3	54.1	75.4	89.4	75.5	83.9	86.2	54.2	73.6	89.0	71.9	82.3	86.9	71.9	81.0
03	CONCRETE	203.0	114.7	159.9	94.8	55.0	75.4	112.9	73.8	93.9	105.5	55.0	80.9	94.4	72.3	83.6	97.5	72.3	85.2
04	MASONRY	204.8	121.7	154.1	92.9	41.2	61.3	93.0	62.8	74.6	107.4	41.1	66.9	92.9	61.3	73.6	107.5	61.3	79.3
05	METALS	117.3	101.6	112.1	94.0	68.6	85.6	94.5	76.8	88.7	90.6	68.8	83.4	94.7	76.6	88.7	90.9	76.8	86.3
06	WOOD, PLASTICS & COMPOSITES	112.9	116.0	114.7	94.6	43.4	65.0	99.9	68.3	81.6	88.8	43.5	62.6	90.6	68.3	77.7	92.0	68.4	78.3
07	THERMAL & MOISTURE PROTECTION	182.8	116.3	155.8	96.4	50.9	77.9	98.2	66.3	85.2	102.6	48.4	80.6	96.4	62.1	82.5	101.8	62.8	86.0
08	OPENINGS	118.6	112.9	117.2	104.8	50.8	91.2	105.0	73.2	97.0	97.7	50.9	85.9	105.0	72.0	96.7	97.8	72.0	91.3
0920	Plaster & Gypsum Board	137.9	116.3	122.5	91.3	41.7	55.9	94.8	67.3	75.2	97.8	41.7	57.7	85.2	67.3	72.4	100.6	67.3	76.8
0950, 0980	Ceilings & Acoustic Treatment	117.4	116.3	116.7	112.4	41.7	65.7	113.3	67.3	82.9	88.9	41.7	61.1	113.3	67.3	82.9	98.9	67.3	78.0
0960	Flooring	150.4	128.4	143.9	92.7	41.4	77.7	95.1	41.6	79.4	98.1	41.4	81.4	91.6	56.2	81.2	99.5	51.9	85.5
0970, 0990	Wall Finishes & Painting/Coating	150.4	116.9	129.7	93.2	38.9	59.7	93.2	57.7	71.3	97.8	38.9	61.5	93.2	57.7	71.3	97.8	57.7	73.1
09	FINISHES	146.4	119.6	131.4	97.4	42.8	67.0	100.7	61.4	78.8	98.1	42.9	67.3	96.5	63.5	78.1	97.7	62.7	78.2
COVERS	DIVS. 10 - 14, 25, 28, 41, 43, 44, 46	100.0	108.7	101.7	100.0	77.7	95.6	100.0	82.9	96.6	100.0	77.9	95.6	100.0	82.3	96.5	100.0	82.6	96.5
21, 22, 23	FIRE SUPPRESSION, PLUMBING & HVAC	98.6	104.0	100.8	96.4	74.0	87.3	100.2	83.0	93.2	94.3	73.7	86.0	96.4	82.0	90.6	100.2	76.9	90.8
26, 27, 3370	ELECTRICAL, COMMUNICATIONS & UTIL.	143.4	116.7	129.6	101.4	68.5	84.4	100.4	62.3	80.7	96.6	63.7	79.7	101.4	62.3	81.2	93.6	62.3	77.4
MF2010	WEIGHTED AVERAGE	134.1	113.9	125.2	96.5	63.0	81.7	100.3	73.9	88.7	97.4	62.5	82.0	96.4	73.4	86.3	97.4	72.4	86.3

		ARIZONA												ARKANSAS					
		PHOENIX			PRESCOTT			SHOW LOW			TUCSON			BATESVILLE			CAMDEN		
	DIVISION	850,853			863			859			856 - 857			725			717		
		MAT.	INST.	TOTAL	MAT.	INST.	TOTAL	MAT.	INST.	TOTAL	MAT.	INST.	TOTAL	MAT.	INST.	TOTAL	MAT.	INST.	TOTAL
015433	CONTRACTOR EQUIPMENT		96.4	96.4		94.0	94.0		95.9	95.9		95.9	95.9		88.3	88.3		88.3	88.3
0241, 31 - 34	SITE & INFRASTRUCTURE, DEMOLITION	87.3	100.9	96.7	70.3	96.6	88.4	98.8	98.8	98.8	82.3	100.7	95.0	71.0	84.4	80.2	71.7	83.9	80.1
0310	Concrete Forming & Accessories	102.7	70.4	74.6	97.8	45.3	52.1	105.3	45.7	53.4	102.0	70.0	74.2	83.5	47.6	52.3	83.3	32.4	39.0
0320	Concrete Reinforcing	101.5	85.4	93.4	97.8	78.2	87.9	103.3	78.1	90.6	88.5	85.2	85.3	95.4	64.5	79.8	96.7	64.3	80.4
0330	Cast-in-Place Concrete	87.0	75.7	82.5	89.3	54.0	75.4	86.2	54.3	73.6	89.4	75.5	84.0	80.2	47.4	67.3	82.2	40.0	65.6
03	CONCRETE	97.1	75.1	86.3	99.7	54.9	77.9	107.6	55.2	82.0	95.8	74.7	85.5	85.2	51.5	68.8	86.9	42.2	65.1
04	MASONRY	94.3	65.9	77.0	93.0	46.4	64.5	107.4	42.7	67.9	92.4	62.8	74.3	105.5	43.1	67.4	113.6	33.2	64.5
05	METALS	92.5	78.2	87.7	94.6	68.7	86.0	90.4	69.2	83.4	91.7	77.0	86.8	96.9	66.9	87.0	97.0	66.4	86.9
06	WOOD, PLASTICS & COMPOSITES	92.9	71.2	80.3	95.8	43.4	65.5	96.2	43.5	65.8	92.3	71.2	80.1	86.9	48.3	64.6	86.7	31.5	54.8
07	THERMAL & MOISTURE PROTECTION	101.5	67.0	87.4	96.9	50.2	77.9	102.8	50.8	81.7	102.9	62.7	86.6	96.2	46.4	76.0	96.0	37.7	72.3
08	OPENINGS	98.9	74.8	92.9	105.0	50.8	91.3	96.7	50.9	85.2	94.0	74.8	89.2	95.1	46.9	83.0	91.3	41.3	78.7
0920	Plaster & Gypsum Board	103.2	70.2	79.6	91.5	41.7	55.9	103.3	41.7	59.3	107.3	70.2	80.8	76.7	47.2	55.6	76.7	29.9	43.3
0950, 0980	Ceilings & Acoustic Treatment	106.1	70.2	82.4	111.5	41.7	65.4	98.9	41.7	61.1	99.8	70.2	80.2	88.4	47.2	61.2	88.4	29.9	49.7
0960	Flooring	99.8	54.2	86.4	93.6	41.4	78.3	101.3	47.2	85.4	90.5	41.6	76.1	102.7	61.9	90.7	102.6	41.5	84.6
0970, 0990	Wall Finishes & Painting/Coating	97.8	64.0	76.9	93.2	38.9	59.7	97.8	38.9	61.5	98.4	57.7	73.3	96.2	43.4	63.6	96.2	49.9	67.6
09	FINISHES	99.7	66.1	81.0	98.0	42.8	67.2	99.9	43.9	68.7	95.9	63.2	77.6	89.7	49.5	67.3	89.8	35.1	59.3
COVERS	DIVS. 10 - 14, 25, 28, 41, 43, 44, 46	100.0	83.5	96.7	100.0	77.7	95.6	100.0	77.9	95.6	100.0	83.4	96.7	100.0	42.7	88.6	100.0	38.4	87.7
21, 22, 23	FIRE SUPPRESSION, PLUMBING & HVAC	100.2	83.1	93.2	100.2	71.2	88.4	94.3	74.1	86.1	100.2	73.5	89.4	94.3	53.6	77.8	94.3	50.9	76.7
26, 27, 3370	ELECTRICAL, COMMUNICATIONS & UTIL.	100.7	70.9	85.3	100.1	62.2	80.5	93.9	63.6	78.3	95.6	59.9	77.2	98.4	63.4	80.3	94.9	60.4	77.1
MF2010	WEIGHTED AVERAGE	97.9	76.6	88.5	98.2	62.1	82.2	97.5	62.9	82.2	96.1	72.1	85.5	94.2	56.1	77.4	94.1	50.1	74.7

		ARKANSAS																	
		FAYETTEVILLE			FORT SMITH			HARRISON			HOT SPRINGS			JONESBORO			LITTLE ROCK		
	DIVISION	727			729			726			719			724			720 - 722		
		MAT.	INST.	TOTAL	MAT.	INST.	TOTAL	MAT.	INST.	TOTAL	MAT.	INST.	TOTAL	MAT.	INST.	TOTAL	MAT.	INST.	TOTAL
015433	CONTRACTOR EQUIPMENT		88.3	88.3		88.3	88.3		88.3	88.3		88.3	88.3		106.3	106.3		88.3	88.3
0241, 31 - 34	SITE & INFRASTRUCTURE, DEMOLITION	70.4	85.9	81.1	75.1	85.8	82.5	75.5	84.4	81.6	74.9	85.3	82.1	94.8	99.4	97.9	82.1	85.8	84.6
0310	Concrete Forming & Accessories	79.5	40.9	45.9	96.7	58.2	63.2	87.6	47.6	52.8	81.0	37.7	43.4	86.9	51.1	55.7	89.3	62.6	66.1
0320	Concrete Reinforcing	95.4	69.4	82.3	96.6	70.1	83.2	95.0	64.4	79.6	94.9	64.5	79.6	92.3	66.8	79.4	96.7	65.2	80.9
0330	Cast-in-Place Concrete	80.2	51.7	69.0	91.7	66.6	81.8	89.0	45.4	71.8	84.1	40.5	66.9	87.4	57.5	75.6	92.1	66.6	82.1
03	CONCRETE	85.0	50.9	68.4	93.1	63.8	78.8	92.5	50.7	72.1	90.4	44.7	68.2	89.8	57.6	74.1	92.9	64.9	79.2
04	MASONRY	95.0	43.7	63.7	100.3	52.0	70.8	105.9	41.3	66.4	85.2	35.4	54.8	96.4	44.3	64.6	97.1	52.0	69.6
05	METALS	96.9	69.0	87.7	99.3	70.7	89.8	98.1	66.6	87.7	96.9	66.9	87.0	93.8	80.0	89.2	95.4	69.1	86.7
06	WOOD, PLASTICS & COMPOSITES	82.9	37.7	56.8	102.5	60.6	78.3	92.0	48.3	66.8	84.0	37.9	57.4	90.3	51.3	67.8	97.2	66.4	79.4
07	THERMAL & MOISTURE PROTECTION	97.1	47.9	77.1	97.4	56.0	80.6	96.5	45.1	75.6	96.2	40.0	73.4	100.9	51.9	81.0	96.5	56.6	80.3
08	OPENINGS	95.1	46.4	82.8	96.0	57.0	86.2	96.0	47.6	83.8	91.3	43.2	79.1	98.2	55.4	87.4	96.0	60.2	87.0
0920	Plaster & Gypsum Board	74.5	36.3	47.2	82.8	59.9	66.5	81.4	47.2	57.0	75.4	36.5	47.6	90.7	50.0	61.6	88.4	65.9	72.3
0950, 0980	Ceilings & Acoustic Treatment	88.4	36.3	54.0	92.0	59.9	70.8	92.0	47.2	62.4	88.4	36.5	54.1	92.5	50.0	64.4	93.8	65.9	75.4
0960	Flooring	99.6	61.9	88.5	110.3	63.7	96.6	105.4	61.9	92.6	101.3	61.9	89.7	73.6	55.9	68.4	111.7	63.7	97.6
0970, 0990	Wall Finishes & Painting/Coating	96.2	31.0	55.9	96.2	56.7	71.8	96.2	43.4	63.6	96.2	49.9	67.6	86.3	51.0	64.5	96.2	58.2	72.8
09	FINISHES	88.6	42.4	62.8	93.7	59.3	74.5	92.2	49.5	68.4	89.6	42.5	63.3	86.8	51.4	67.0	95.4	62.9	77.3
COVERS	DIVS. 10 - 14, 25, 28, 41, 43, 44, 46	100.0	52.6	90.6	100.0	77.1	95.4	100.0	53.1	90.7	100.0	39.4	87.9	100.0	48.7	89.8	100.0	77.7	95.6
21, 22, 23	FIRE SUPPRESSION, PLUMBING & HVAC	94.3	52.2	77.2	100.3	51.6	80.5	94.3	51.3	76.9	94.3	50.4	76.5	100.5	54.5	81.9	100.2	59.3	83.6
26, 27, 3370	ELECTRICAL, COMMUNICATIONS & UTIL.	92.3	54.4	72.7	95.7	68.6	81.7	97.0	39.4	67.3	97.1	67.2	81.6	103.4	63.4	82.8	100.5	69.7	84.6
MF2010	WEIGHTED AVERAGE	92.9	54.1	75.8	97.1	62.5	81.8	95.5	52.3	76.5	93.4	52.9	75.5	96.8	60.7	80.9	97.0	65.0	82.9

ARKANSAS / CALIFORNIA

DIVISION		PINE BLUFF 716 MAT.	INST.	TOTAL	RUSSELLVILLE 728 MAT.	INST.	TOTAL	TEXARKANA 718 MAT.	INST.	TOTAL	WEST MEMPHIS 723 MAT.	INST.	TOTAL	ALHAMBRA 917-918 MAT.	INST.	TOTAL	ANAHEIM 928 MAT.	INST.	TOTAL
015433	CONTRACTOR EQUIPMENT		88.3	88.3		88.3	88.3		88.9	88.9		106.3	106.3		99.3	99.3		100.8	100.8
0241, 31 - 34	SITE & INFRASTRUCTURE, DEMOLITION	76.8	85.8	83.0	72.2	84.3	80.5	89.7	86.3	87.4	101.5	99.4	100.0	97.9	109.3	105.8	96.8	108.4	104.8
0310	Concrete Forming & Accessories	80.7	62.5	64.9	84.1	56.5	60.1	86.5	42.6	48.3	91.7	51.3	56.5	121.6	119.9	120.1	107.0	121.9	119.9
0320	Concrete Reinforcing	96.6	65.2	80.8	96.1	64.3	80.1	96.2	64.5	80.2	92.3	66.8	79.4	109.7	118.2	114.0	95.9	118.1	107.1
0330	Cast-in-Place Concrete	84.1	66.6	77.2	84.1	47.3	69.6	91.7	55.5	73.4	91.6	57.5	78.2	89.2	118.3	100.7	92.9	120.8	103.9
03	CONCRETE	91.2	64.8	78.3	88.4	55.4	72.3	89.6	48.5	69.6	96.9	57.8	77.8	100.0	118.0	108.8	99.7	118.9	109.5
04	MASONRY	122.3	52.0	79.4	99.5	39.5	62.9	100.3	35.0	60.5	81.7	44.3	58.9	116.0	118.2	117.4	80.3	116.4	102.4
05	METALS	97.8	69.0	88.3	96.9	66.5	86.9	90.9	67.0	83.0	92.8	80.2	88.6	83.5	101.6	89.5	100.7	101.8	101.1
06	WOOD, PLASTICS & COMPOSITES	83.5	66.4	73.6	88.1	60.6	72.2	91.5	44.6	64.4	96.0	51.3	70.2	96.8	117.7	108.9	97.0	120.3	110.5
07	THERMAL & MOISTURE PROTECTION	96.3	56.6	80.2	97.3	46.1	76.5	97.1	46.7	76.6	101.4	51.9	81.2	101.2	117.4	107.8	101.6	121.3	109.6
08	OPENINGS	91.3	60.2	83.4	95.1	56.2	85.3	96.5	47.6	84.2	98.2	55.4	87.4	91.5	119.7	98.6	103.4	121.3	107.9
0920	Plaster & Gypsum Board	74.9	65.9	68.5	76.7	59.9	64.7	79.7	43.4	53.8	93.4	50.0	62.4	98.5	118.3	112.7	103.2	120.9	115.8
0950, 0980	Ceilings & Acoustic Treatment	88.4	65.9	73.5	88.4	59.9	69.6	93.8	43.4	60.5	90.5	50.0	63.7	92.8	118.3	109.7	107.5	120.9	116.3
0960	Flooring	101.0	63.7	90.0	102.3	61.9	90.4	103.4	53.4	88.7	76.2	55.9	70.2	97.2	105.0	99.4	108.9	111.4	109.6
0970, 0990	Wall Finishes & Painting/Coating	96.2	58.2	72.8	96.2	36.9	59.6	96.2	32.0	56.6	86.3	51.0	64.5	108.1	108.5	108.4	102.8	106.7	105.2
09	FINISHES	89.5	62.9	74.7	89.9	56.0	71.0	92.6	42.7	64.8	88.2	51.4	67.7	97.3	115.9	107.7	104.4	118.3	112.2
COVERS	DIVS. 10 - 14, 25, 28, 41, 43, 44, 46	100.0	77.7	95.6	100.0	44.0	88.9	100.0	35.2	87.1	100.0	48.7	89.8	100.0	109.9	102.0	100.0	110.5	102.1
21, 22, 23	FIRE SUPPRESSION, PLUMBING & HVAC	100.3	54.0	81.5	94.3	53.4	77.8	100.3	57.3	82.9	94.5	66.3	83.1	94.1	111.1	101.0	100.1	112.0	104.9
26, 27, 3370	ELECTRICAL, COMMUNICATIONS & UTIL.	95.0	69.7	82.0	95.7	47.5	70.8	97.0	38.2	66.7	105.0	60.3	81.9	112.7	115.8	114.3	92.2	105.4	99.0
MF2010	WEIGHTED AVERAGE	96.7	63.8	82.2	94.1	55.4	77.0	95.9	51.3	76.2	95.8	62.8	81.2	97.1	113.5	104.3	98.9	112.8	105.1

CALIFORNIA

DIVISION		BAKERSFIELD 932-933 MAT.	INST.	TOTAL	BERKELEY 947 MAT.	INST.	TOTAL	EUREKA 955 MAT.	INST.	TOTAL	FRESNO 936-938 MAT.	INST.	TOTAL	INGLEWOOD 903-905 MAT.	INST.	TOTAL	LONG BEACH 906-908 MAT.	INST.	TOTAL
015433	CONTRACTOR EQUIPMENT		98.7	98.7		98.5	98.5		98.4	98.4		98.7	98.7		96.5	96.5		96.5	96.5
0241, 31 - 34	SITE & INFRASTRUCTURE, DEMOLITION	99.1	106.0	103.9	120.8	105.3	110.1	110.3	102.8	105.2	102.0	105.5	104.4	91.4	103.8	100.0	98.8	103.8	102.3
0310	Concrete Forming & Accessories	105.5	121.4	119.3	119.6	148.1	144.4	117.0	132.1	130.1	103.0	133.1	129.2	111.2	119.8	118.7	106.4	119.8	118.1
0320	Concrete Reinforcing	108.0	118.0	113.0	95.0	119.4	107.3	104.7	118.8	111.8	91.5	118.7	105.2	107.6	118.1	112.9	106.6	118.1	112.4
0330	Cast-in-Place Concrete	94.6	119.7	104.5	128.5	121.8	125.9	100.7	116.2	106.8	97.9	117.0	105.4	88.4	118.7	100.3	100.7	118.7	107.8
03	CONCRETE	100.2	119.1	109.4	114.5	131.6	122.8	110.2	122.7	116.3	101.3	123.4	112.1	97.2	118.2	107.4	107.0	118.2	112.5
04	MASONRY	99.4	115.9	109.4	130.4	132.7	131.8	103.8	127.3	118.1	102.6	122.1	114.5	76.5	118.3	102.0	85.4	118.3	105.5
05	METALS	96.1	100.9	97.7	105.9	101.0	104.3	100.6	100.0	100.4	100.7	101.6	101.0	91.1	102.0	94.7	90.9	102.0	94.6
06	WOOD, PLASTICS & COMPOSITES	92.1	120.4	108.4	118.6	153.8	138.9	110.4	137.2	125.9	103.5	137.2	123.0	99.9	117.6	110.1	93.6	117.6	107.5
07	THERMAL & MOISTURE PROTECTION	100.5	111.9	105.1	114.4	135.6	123.0	105.7	115.6	109.7	95.3	115.0	103.3	99.4	118.7	107.2	99.6	119.0	107.5
08	OPENINGS	100.9	116.8	104.9	101.4	141.2	111.4	102.2	114.6	105.3	102.7	126.0	108.6	93.5	119.6	100.1	93.5	119.6	100.1
0920	Plaster & Gypsum Board	94.5	120.9	113.3	110.4	154.9	142.2	108.4	138.2	129.7	97.4	138.2	126.6	103.7	118.3	114.2	99.2	118.3	112.9
0950, 0980	Ceilings & Acoustic Treatment	108.0	120.9	116.5	109.4	154.9	139.4	112.9	138.2	129.6	110.5	138.2	128.8	100.7	118.3	112.3	100.7	118.3	112.3
0960	Flooring	109.1	97.4	105.7	115.8	126.8	119.1	113.5	122.0	116.0	115.5	139.8	122.6	100.0	105.0	101.5	97.3	105.0	99.6
0970, 0990	Wall Finishes & Painting/Coating	104.9	93.8	98.0	103.9	141.5	127.1	104.2	48.7	70.0	120.2	111.0	114.5	99.7	108.5	105.2	99.7	108.5	105.2
09	FINISHES	105.4	114.7	110.6	111.0	145.5	130.2	110.9	123.9	118.1	109.8	133.8	123.2	102.5	115.8	109.9	101.8	115.8	109.6
COVERS	DIVS. 10 - 14, 25, 28, 41, 43, 44, 46	100.0	107.6	101.5	100.0	124.0	104.8	100.0	120.3	104.0	100.0	120.3	104.0	100.0	109.6	101.9	100.0	109.6	101.9
21, 22, 23	FIRE SUPPRESSION, PLUMBING & HVAC	100.1	105.5	102.3	94.3	147.5	115.9	94.1	103.5	97.9	100.2	113.2	105.5	93.7	111.0	100.7	93.7	111.0	100.7
26, 27, 3370	ELECTRICAL, COMMUNICATIONS & UTIL.	101.6	101.2	101.4	106.6	138.5	123.1	99.8	116.8	108.6	91.7	100.4	96.2	103.4	115.9	109.8	103.1	115.9	109.7
MF2010	WEIGHTED AVERAGE	100.0	109.5	104.2	105.5	133.2	117.7	101.6	114.2	107.2	100.5	115.7	107.2	95.3	113.1	103.2	96.9	113.2	104.1

CALIFORNIA

DIVISION		LOS ANGELES 900-902 MAT.	INST.	TOTAL	MARYSVILLE 959 MAT.	INST.	TOTAL	MODESTO 953 MAT.	INST.	TOTAL	MOJAVE 935 MAT.	INST.	TOTAL	OAKLAND 946 MAT.	INST.	TOTAL	OXNARD 930 MAT.	INST.	TOTAL
015433	CONTRACTOR EQUIPMENT		100.1	100.1		98.4	98.4		98.4	98.4		98.7	98.7		98.5	98.5		97.6	97.6
0241, 31 - 34	SITE & INFRASTRUCTURE, DEMOLITION	98.8	107.5	104.8	106.2	105.0	105.4	100.5	105.0	103.6	95.5	106.0	102.7	127.4	105.3	112.2	102.3	104.1	103.5
0310	Concrete Forming & Accessories	108.2	121.9	120.1	107.6	133.0	129.7	104.2	133.2	129.5	116.2	116.2	116.2	109.1	148.1	143.0	108.1	121.9	120.1
0320	Concrete Reinforcing	108.4	118.3	113.4	104.7	118.5	111.7	108.5	118.7	113.7	110.0	117.9	114.0	97.1	119.4	108.4	108.0	118.0	113.0
0330	Cast-in-Place Concrete	95.5	119.4	104.9	112.6	117.0	114.3	100.8	117.1	107.2	84.7	119.6	98.5	121.8	121.8	121.8	98.6	120.0	107.0
03	CONCRETE	102.8	119.4	110.9	112.2	123.4	117.7	103.3	123.6	113.2	93.8	116.8	105.0	113.9	131.7	122.6	102.3	119.4	110.7
04	MASONRY	90.2	120.1	108.5	104.8	119.1	113.5	103.8	123.0	115.5	100.7	115.1	109.5	138.3	132.7	134.9	104.1	114.5	110.5
05	METALS	98.1	103.5	99.9	100.1	101.6	100.6	97.3	102.1	98.7	99.7	100.7	98.7	99.9	101.0	100.3	95.9	101.3	97.7
06	WOOD, PLASTICS & COMPOSITES	99.6	120.2	111.5	98.2	137.2	120.7	94.4	137.2	119.1	102.8	113.6	109.0	106.6	153.8	133.9	98.1	120.4	111.0
07	THERMAL & MOISTURE PROTECTION	98.8	121.0	107.8	105.2	118.0	110.4	104.8	117.7	110.0	101.2	110.4	105.0	112.1	135.6	121.6	104.7	116.9	109.7
08	OPENINGS	100.6	121.3	105.8	101.4	125.9	107.6	100.2	126.6	106.8	96.7	113.1	100.9	101.4	141.2	111.4	99.8	121.4	105.2
0920	Plaster & Gypsum Board	103.7	120.9	115.9	100.4	138.2	127.4	103.6	138.2	128.3	109.9	113.9	112.7	103.8	154.9	140.3	103.2	120.9	115.8
0950, 0980	Ceilings & Acoustic Treatment	110.8	120.9	117.4	112.0	138.2	129.3	107.5	138.2	127.8	112.0	113.9	113.2	112.3	154.9	140.4	112.5	120.9	118.0
0960	Flooring	97.7	111.4	101.7	108.6	110.5	109.2	109.1	116.7	111.3	115.9	97.4	110.5	110.2	126.8	115.1	107.5	111.4	108.7
0970, 0990	Wall Finishes & Painting/Coating	98.8	108.5	104.8	104.2	115.2	111.0	104.2	118.4	112.9	104.2	93.8	97.7	103.9	141.5	127.1	104.2	99.7	101.4
09	FINISHES	104.2	118.4	112.1	107.7	129.4	119.8	106.6	130.8	120.1	110.0	110.6	110.3	109.9	145.5	129.7	107.2	117.6	113.0
COVERS	DIVS. 10 - 14, 25, 28, 41, 43, 44, 46	100.0	110.2	102.0	100.0	120.3	104.0	100.0	120.3	104.0	100.0	106.8	101.3	100.0	124.0	104.8	100.0	110.7	102.1
21, 22, 23	FIRE SUPPRESSION, PLUMBING & HVAC	100.2	111.9	104.9	94.1	108.9	100.1	100.1	113.2	105.4	94.1	100.9	96.9	100.2	147.5	119.4	100.1	112.0	104.9
26, 27, 3370	ELECTRICAL, COMMUNICATIONS & UTIL.	101.7	122.2	112.3	96.4	103.8	100.2	98.9	110.5	104.9	90.3	98.2	94.4	105.8	143.6	125.3	96.2	107.5	102.0
MF2010	WEIGHTED AVERAGE	100.1	115.5	106.9	100.9	114.4	106.8	100.7	116.9	107.9	97.0	106.9	101.3	106.3	133.9	118.5	100.2	112.3	105.5

City Cost Indexes

CALIFORNIA

DIVISION		PALM SPRINGS 922			PALO ALTO 943			PASADENA 910-912			REDDING 960			RICHMOND 948			RIVERSIDE 925		
		MAT.	INST.	TOTAL	MAT.	INST.	TOTAL	MAT.	INST.	TOTAL	MAT.	INST.	TOTAL	MAT.	INST.	TOTAL	MAT.	INST.	TOTAL
015433	CONTRACTOR EQUIPMENT		99.6	99.6		98.5	98.5		99.3	99.3		98.4	98.4		98.5	98.5		99.6	99.6
0241, 31 - 34	SITE & INFRASTRUCTURE, DEMOLITION	88.3	106.5	100.8	116.5	105.3	108.8	94.7	109.3	104.8	106.2	105.0	105.4	126.8	105.3	112.0	95.1	106.5	102.9
0310	Concrete Forming & Accessories	103.5	119.9	117.7	107.3	140.1	135.9	111.8	119.9	118.9	107.4	133.0	129.7	122.0	147.7	144.4	107.1	121.9	119.9
0320	Concrete Reinforcing	110.2	118.0	114.1	95.0	119.2	107.2	110.6	118.2	114.4	104.7	118.5	111.7	95.0	119.2	107.2	107.0	118.0	112.6
0330	Cast-in-Place Concrete	88.5	120.7	101.2	108.7	121.6	113.8	84.7	118.3	97.9	111.6	117.0	113.7	125.0	121.6	123.7	96.3	120.8	105.9
03	CONCRETE	95.0	118.8	106.6	102.8	128.0	115.1	95.7	118.1	106.6	111.2	123.4	117.1	116.0	131.4	123.5	101.0	119.7	110.1
04	MASONRY	78.1	115.4	100.8	109.6	128.8	121.3	100.8	118.2	111.4	104.8	119.1	113.5	130.1	128.8	129.3	79.1	115.4	101.2
05	METALS	101.3	101.5	101.4	97.4	100.3	98.3	83.6	101.6	89.5	100.4	101.6	100.8	97.4	100.4	98.4	100.9	101.7	101.2
06	WOOD, PLASTICS & COMPOSITES	91.9	117.8	106.9	104.0	143.6	126.9	84.4	117.7	103.7	97.9	137.2	120.6	122.2	153.8	140.4	97.0	120.3	110.5
07	THERMAL & MOISTURE PROTECTION	101.3	119.3	108.6	111.6	128.7	118.6	100.9	117.9	107.8	105.2	117.5	110.2	112.2	133.9	121.0	101.8	120.7	109.5
08	OPENINGS	98.8	118.2	103.7	101.4	134.7	109.8	91.5	119.7	98.6	102.4	125.9	108.3	101.4	140.2	111.2	101.9	121.3	106.8
0920	Plaster & Gypsum Board	97.3	118.3	112.3	101.5	144.3	132.1	91.2	118.3	110.6	100.7	138.2	127.5	112.4	154.9	142.8	102.3	120.9	115.6
0950, 0980	Ceilings & Acoustic Treatment	104.8	118.3	113.7	110.3	144.3	132.8	92.8	118.3	109.7	118.0	138.2	131.3	110.3	154.9	139.7	109.8	120.9	117.1
0960	Flooring	111.1	105.0	109.3	109.3	120.8	112.7	91.7	105.0	95.6	108.6	110.5	109.1	117.7	120.8	118.6	112.9	111.4	112.5
0970, 0990	Wall Finishes & Painting/Coating	100.9	117.3	111.0	103.9	139.2	125.6	108.1	108.5	108.4	104.2	115.2	111.0	103.9	139.2	125.6	100.9	106.7	104.5
09	FINISHES	102.9	116.9	110.7	108.0	138.2	124.8	94.6	115.9	106.5	108.9	129.4	120.3	112.6	144.2	130.2	105.6	118.3	112.7
COVERS	DIVS. 10 - 14, 25, 28, 41, 43, 44, 46	100.0	110.2	102.0	100.0	122.9	104.5	100.0	109.9	102.0	100.0	120.3	104.0	100.0	124.0	104.8	100.0	110.5	102.1
21, 22, 23	FIRE SUPPRESSION, PLUMBING & HVAC	94.1	107.3	99.5	94.3	141.2	113.3	94.1	111.1	101.0	100.1	108.9	103.6	94.3	136.5	111.4	100.1	104.9	102.1
26, 27, 3370	ELECTRICAL, COMMUNICATIONS & UTIL.	95.9	104.4	100.3	105.7	142.0	124.4	109.7	115.8	112.9	99.3	103.8	101.6	106.2	135.0	121.1	92.4	104.4	98.6
MF2010	WEIGHTED AVERAGE	96.4	110.9	102.8	101.3	129.9	113.9	95.2	113.6	103.0	102.8	114.4	107.9	104.7	129.7	115.7	98.9	112.4	104.9

CALIFORNIA

DIVISION		SACRAMENTO 942,956 - 958			SALINAS 939			SAN BERNARDINO 923 - 924			SAN DIEGO 919 - 921			SAN FRANCISCO 940 - 941			SAN JOSE 951		
		MAT.	INST.	TOTAL	MAT.	INST.	TOTAL	MAT.	INST.	TOTAL	MAT.	INST.	TOTAL	MAT.	INST.	TOTAL	MAT.	INST.	TOTAL
015433	CONTRACTOR EQUIPMENT		98.0	98.0		98.7	98.7		99.6	99.6		99.3	99.3		109.1	109.1		99.1	99.1
0241, 31 - 34	SITE & INFRASTRUCTURE, DEMOLITION	100.9	110.7	107.6	116.7	105.7	109.1	74.3	106.5	96.5	100.8	103.0	102.3	129.7	111.0	116.8	135.8	100.2	111.2
0310	Concrete Forming & Accessories	107.4	136.1	132.3	110.7	136.4	133.0	110.5	119.8	118.6	108.2	110.6	110.3	108.7	148.9	143.7	110.0	147.9	142.9
0320	Concrete Reinforcing	89.7	118.8	104.4	108.5	119.1	113.9	107.0	118.0	112.5	107.2	118.0	112.6	110.6	119.7	115.2	95.5	119.4	107.6
0330	Cast-in-Place Concrete	103.0	118.1	109.0	97.4	117.4	105.3	66.5	120.7	87.8	92.8	106.6	98.2	124.9	123.1	124.2	117.1	121.0	118.6
03	CONCRETE	103.2	124.7	113.7	111.5	125.2	118.1	75.6	118.8	96.7	100.5	110.0	105.1	117.5	133.1	125.1	109.8	131.7	120.5
04	MASONRY	111.6	123.0	118.5	100.4	128.2	117.4	85.8	115.4	103.8	92.0	113.7	105.2	138.8	135.9	137.0	132.7	132.8	132.7
05	METALS	95.6	96.0	95.7	100.3	103.2	101.2	100.9	101.4	101.0	98.7	102.7	100.0	106.5	110.8	107.9	95.7	107.7	99.6
06	WOOD, PLASTICS & COMPOSITES	99.3	140.7	123.2	102.3	140.4	124.3	100.6	117.8	110.6	97.9	107.2	103.3	106.6	153.9	134.0	106.5	153.5	133.6
07	THERMAL & MOISTURE PROTECTION	118.7	122.1	120.1	102.0	125.8	111.7	100.4	119.3	108.1	106.5	105.1	106.0	114.2	137.1	123.5	101.2	136.7	115.6
08	OPENINGS	114.1	128.5	117.7	101.2	133.9	109.4	98.9	118.2	103.7	99.5	111.9	102.6	105.9	141.3	114.8	93.1	141.0	105.1
0920	Plaster & Gypsum Board	97.8	141.5	129.0	104.5	141.5	130.9	104.1	118.3	114.3	98.8	107.2	104.8	106.8	154.9	141.1	97.5	154.9	138.5
0950, 0980	Ceilings & Acoustic Treatment	110.3	141.5	130.9	112.0	141.5	131.5	107.5	118.3	114.6	100.0	107.2	104.8	120.2	154.9	143.1	102.6	154.9	137.1
0960	Flooring	109.4	116.7	111.5	109.9	126.8	114.9	115.1	105.0	112.1	100.1	111.4	103.4	110.2	126.8	115.1	101.4	126.8	108.9
0970, 0990	Wall Finishes & Painting/Coating	101.1	118.4	111.8	105.3	141.5	127.7	100.9	104.9	103.4	104.5	109.4	107.5	103.9	151.4	133.2	104.7	141.5	127.3
09	FINISHES	106.3	132.9	121.1	110.0	137.0	125.0	104.1	115.6	110.5	100.3	110.2	105.9	112.1	146.7	131.4	104.7	145.3	127.3
COVERS	DIVS. 10 - 14, 25, 28, 41, 43, 44, 46	100.0	121.3	104.2	100.0	120.7	104.1	100.0	107.1	101.4	100.0	108.4	101.7	100.0	124.5	104.9	100.0	123.4	104.7
21, 22, 23	FIRE SUPPRESSION, PLUMBING & HVAC	100.1	116.8	106.9	94.1	120.0	104.6	94.1	107.3	99.5	100.1	110.4	104.3	100.2	169.6	128.3	100.1	146.7	118.9
26, 27, 3370	ELECTRICAL, COMMUNICATIONS & UTIL.	101.1	113.3	107.4	91.7	125.8	109.3	95.9	103.6	99.9	96.0	101.0	98.6	105.9	155.0	131.2	102.1	152.2	128.0
MF2010	WEIGHTED AVERAGE	102.9	118.6	109.9	100.5	122.7	110.3	94.3	110.5	101.4	99.3	107.8	103.0	108.5	142.3	123.4	103.2	135.1	117.3

CALIFORNIA

DIVISION		SAN LUIS OBISPO 934			SAN MATEO 944			SAN RAFAEL 949			SANTA ANA 926 - 927			SANTA BARBARA 931			SANTA CRUZ 950		
		MAT.	INST.	TOTAL	MAT.	INST.	TOTAL	MAT.	INST.	TOTAL	MAT.	INST.	TOTAL	MAT.	INST.	TOTAL	MAT.	INST.	TOTAL
015433	CONTRACTOR EQUIPMENT		98.7	98.7		98.5	98.5		98.5	98.5		99.6	99.6		98.7	98.7		99.1	99.1
0241, 31 - 34	SITE & INFRASTRUCTURE, DEMOLITION	108.1	106.0	106.7	124.1	105.3	111.1	114.7	110.7	112.0	86.7	106.5	100.3	102.3	106.0	104.9	135.1	100.0	110.9
0310	Concrete Forming & Accessories	117.8	119.6	119.4	112.9	147.8	143.2	115.2	147.5	143.3	110.8	119.8	118.6	109.1	121.8	120.1	110.1	136.5	133.1
0320	Concrete Reinforcing	110.0	117.9	114.0	95.0	119.4	107.3	95.5	119.3	107.5	110.7	118.0	114.4	108.0	118.0	113.0	118.3	119.2	118.7
0330	Cast-in-Place Concrete	104.7	119.7	110.6	121.0	121.7	121.3	140.9	120.6	132.9	85.0	120.7	99.0	98.2	119.8	106.7	116.3	119.3	117.5
03	CONCRETE	110.0	118.3	114.1	112.3	131.4	121.6	134.4	130.8	132.6	92.6	118.8	105.4	102.2	119.4	110.6	112.9	126.1	119.4
04	MASONRY	102.3	115.3	110.1	129.8	132.2	131.3	106.7	132.2	122.2	75.1	115.7	99.9	100.9	115.1	109.6	136.8	124.8	131.7
05	METALS	98.3	100.9	99.2	97.2	100.6	98.4	98.8	97.7	98.4	101.0	101.4	101.1	96.3	101.3	97.9	102.6	106.6	103.9
06	WOOD, PLASTICS & COMPOSITES	105.0	117.9	112.5	111.6	153.8	135.9	106.4	153.5	133.6	102.2	117.8	111.2	98.1	120.4	111.0	106.5	140.5	126.1
07	THERMAL & MOISTURE PROTECTION	102.1	115.9	107.7	112.0	134.8	121.3	116.6	131.7	122.7	101.6	119.3	108.8	101.4	116.2	107.4	101.1	128.6	112.3
08	OPENINGS	99.0	115.5	103.2	101.4	140.2	111.2	111.3	140.1	118.5	98.1	118.2	103.1	100.9	121.4	106.1	94.3	134.0	104.3
0920	Plaster & Gypsum Board	111.8	118.3	116.5	107.4	154.9	141.3	109.6	154.9	141.9	105.0	118.3	114.5	103.2	120.9	115.8	104.0	141.5	130.8
0950, 0980	Ceilings & Acoustic Treatment	112.0	118.3	116.2	110.3	154.9	139.7	118.4	154.9	142.5	107.5	118.3	114.6	112.5	120.9	118.0	109.3	141.5	130.6
0960	Flooring	116.9	97.4	111.1	112.3	120.8	114.8	123.2	120.8	122.5	115.7	105.0	112.5	109.1	97.4	105.7	105.9	126.8	112.0
0970, 0990	Wall Finishes & Painting/Coating	104.2	99.7	101.4	103.9	139.2	125.6	99.9	139.2	124.1	100.9	104.9	103.4	104.2	99.7	101.4	104.9	141.5	127.5
09	FINISHES	111.7	112.9	112.4	110.2	143.9	129.0	112.8	144.0	130.2	105.7	115.6	111.2	107.9	115.3	112.0	108.3	137.0	124.3
COVERS	DIVS. 10 - 14, 25, 28, 41, 43, 44, 46	100.0	117.5	103.5	100.0	124.0	104.8	100.0	123.3	104.6	100.0	110.2	102.0	100.0	110.7	102.1	100.0	121.0	104.2
21, 22, 23	FIRE SUPPRESSION, PLUMBING & HVAC	94.1	107.3	99.5	94.3	138.4	112.2	94.2	146.1	115.3	94.1	107.4	99.5	100.1	112.0	104.9	100.1	120.0	108.2
26, 27, 3370	ELECTRICAL, COMMUNICATIONS & UTIL.	90.3	98.1	94.3	105.7	141.1	124.0	102.3	116.2	109.5	95.9	105.0	100.6	89.3	108.2	99.0	101.0	125.8	113.8
MF2010	WEIGHTED AVERAGE	99.7	109.4	104.0	103.8	131.3	115.9	106.0	129.5	116.4	96.1	110.8	102.6	99.5	112.3	105.1	105.0	122.8	112.9

		CALIFORNIA																COLORADO		
	DIVISION	SANTA ROSA			STOCKTON			SUSANVILLE			VALLEJO			VAN NUYS			ALAMOSA		
		954			952			961			945			913 - 916			811		
		MAT.	INST.	TOTAL	MAT.	INST.	TOTAL	MAT.	INST.	TOTAL	MAT.	INST.	TOTAL	MAT.	INST.	TOTAL	MAT.	INST.	TOTAL
015433	CONTRACTOR EQUIPMENT		99.0	99.0		98.4	98.4		98.4	98.4		98.5	98.5		99.3	99.3		95.2	95.2
0241, 31 - 34	SITE & INFRASTRUCTURE, DEMOLITION	100.9	105.1	103.8	100.2	105.0	103.5	113.2	105.0	107.5	100.3	110.6	107.4	112.4	109.3	110.3	131.0	91.6	103.8
0310	Concrete Forming & Accessories	104.5	146.9	141.4	107.9	133.3	130.0	108.7	133.0	129.8	106.1	146.6	141.4	117.6	119.9	119.6	105.9	71.5	76.0
0320	Concrete Reinforcing	105.7	119.6	112.7	108.5	118.8	113.7	104.7	118.5	111.7	96.7	119.5	108.2	110.6	118.2	114.4	105.8	80.7	93.2
0330	Cast-in-Place Concrete	110.6	118.7	113.8	98.2	117.1	105.6	101.6	117.0	107.6	112.2	119.3	115.0	89.3	118.3	100.7	99.6	82.0	92.7
03	CONCRETE	111.7	130.4	120.8	102.2	123.6	112.6	112.1	123.4	117.6	108.8	130.0	119.1	108.8	118.0	113.3	113.2	77.2	95.6
04	MASONRY	102.8	133.5	121.5	103.8	123.0	115.5	103.2	119.1	112.9	80.2	133.5	112.7	116.0	118.2	117.4	126.0	75.6	95.2
05	METALS	101.3	104.9	102.5	97.5	102.2	99.0	99.5	101.6	100.2	98.8	98.0	98.5	82.6	101.6	88.9	96.0	82.7	91.6
06	WOOD, PLASTICS & COMPOSITES	93.2	153.3	127.9	99.7	137.2	121.4	99.8	137.2	121.4	96.0	153.5	129.2	92.1	117.7	106.9	93.2	71.8	80.8
07	THERMAL & MOISTURE PROTECTION	102.1	133.9	115.0	104.9	119.8	111.0	105.8	117.5	110.5	114.2	133.6	122.1	101.8	117.4	108.2	106.1	80.8	95.8
08	OPENINGS	99.8	140.9	110.2	100.2	126.6	106.8	102.3	125.9	108.3	113.7	141.0	120.6	91.3	119.7	98.5	95.6	78.0	91.2
0920	Plaster & Gypsum Board	99.1	154.9	138.9	103.6	138.2	128.3	101.8	138.2	127.8	103.7	154.9	140.3	96.1	118.3	112.0	81.6	70.7	73.8
0950, 0980	Ceilings & Acoustic Treatment	107.5	154.9	138.8	112.5	138.2	129.5	112.0	138.2	129.3	120.4	154.9	143.2	91.0	118.3	109.0	108.8	70.7	83.6
0960	Flooring	111.7	120.4	114.3	109.1	116.7	111.3	109.1	110.5	109.5	118.0	126.8	120.6	94.9	105.0	97.8	111.5	57.4	95.6
0970, 0990	Wall Finishes & Painting/Coating	100.9	141.5	126.0	104.2	118.4	112.9	104.2	115.2	111.0	100.8	141.5	125.9	108.1	108.5	108.4	107.9	24.9	56.7
09	FINISHES	105.0	143.5	126.5	107.7	130.8	120.6	108.9	129.4	120.3	109.3	144.7	129.0	97.2	115.9	107.6	106.9	63.5	82.7
COVERS	DIVS. 10 - 14, 25, 28, 41, 43, 44, 46	100.0	122.0	104.4	100.0	120.3	104.0	100.0	120.3	104.0	100.0	122.6	104.5	100.0	109.9	102.0	100.0	90.2	98.0
21, 22, 23	FIRE SUPPRESSION, PLUMBING & HVAC	94.1	168.2	124.1	100.1	113.2	105.4	94.1	108.9	100.1	100.2	130.4	112.4	94.1	111.1	101.0	94.1	74.6	86.2
26, 27, 3370	ELECTRICAL, COMMUNICATIONS & UTIL.	96.2	120.9	108.9	98.9	112.2	105.8	99.7	103.8	101.8	98.1	129.6	114.4	109.7	115.8	112.9	95.7	75.1	85.1
MF2010	WEIGHTED AVERAGE	100.3	135.0	115.6	100.7	117.2	108.0	101.5	114.4	107.2	102.2	128.2	113.7	98.1	113.5	104.9	101.5	76.7	90.6

		COLORADO																		
	DIVISION	BOULDER			COLORADO SPRINGS			DENVER			DURANGO			FORT COLLINS			FORT MORGAN		
		803			808 - 809			800 - 802			813			805			807		
		MAT.	INST.	TOTAL	MAT.	INST.	TOTAL	MAT.	INST.	TOTAL	MAT.	INST.	TOTAL	MAT.	INST.	TOTAL	MAT.	INST.	TOTAL
015433	CONTRACTOR EQUIPMENT		96.2	96.2		94.7	94.7		99.1	99.1		95.2	95.2		96.2	96.2		96.2	96.2
0241, 31 - 34	SITE & INFRASTRUCTURE, DEMOLITION	91.9	97.2	95.6	93.3	95.4	94.7	92.4	103.6	100.1	123.9	91.6	101.6	103.6	96.8	98.9	94.1	96.8	96.0
0310	Concrete Forming & Accessories	104.0	80.0	83.1	95.1	80.4	82.3	100.4	80.7	83.2	111.3	71.8	77.0	101.4	79.1	82.0	104.3	79.5	82.7
0320	Concrete Reinforcing	98.9	80.8	89.8	98.1	85.2	91.6	98.1	85.2	91.6	105.8	80.7	93.2	99.0	79.6	89.2	99.1	80.7	89.8
0330	Cast-in-Place Concrete	103.2	82.4	95.0	106.0	85.8	98.0	100.3	85.5	94.5	114.5	82.1	101.8	116.5	81.2	102.6	101.3	81.3	93.4
03	CONCRETE	102.3	81.3	92.0	105.5	83.4	94.8	100.0	83.5	91.9	116.2	77.4	97.3	112.8	80.1	96.9	100.9	80.5	90.9
04	MASONRY	92.8	76.1	82.6	93.5	76.1	82.9	95.5	76.1	83.7	113.4	75.6	90.3	110.8	55.2	76.9	107.6	75.8	88.2
05	METALS	99.5	86.0	95.0	102.6	88.5	98.0	105.2	88.7	99.7	96.0	83.0	91.7	100.7	81.1	94.2	99.2	82.9	93.8
06	WOOD, PLASTICS & COMPOSITES	97.6	82.1	88.7	88.7	82.2	85.0	95.6	82.2	87.8	101.0	71.8	84.1	95.4	82.1	87.7	97.8	82.1	88.7
07	THERMAL & MOISTURE PROTECTION	105.0	78.8	94.4	105.9	83.1	96.6	104.2	77.0	93.2	106.0	80.8	95.7	105.4	67.9	90.2	105.0	82.4	95.8
08	OPENINGS	101.5	83.6	96.9	106.1	84.9	100.8	106.3	84.8	100.9	103.4	78.0	97.0	101.4	83.6	96.9	101.4	83.6	96.9
0920	Plaster & Gypsum Board	108.6	81.8	89.4	89.1	81.8	83.9	104.6	81.8	88.3	93.4	70.7	77.2	102.7	81.8	87.8	109.1	81.8	89.6
0950, 0980	Ceilings & Acoustic Treatment	99.5	81.8	87.8	106.8	81.8	90.3	105.9	81.8	90.0	108.8	70.7	83.6	99.5	81.8	87.8	99.5	81.8	87.8
0960	Flooring	109.4	87.8	103.1	100.7	71.5	92.1	105.8	88.8	100.8	116.4	57.4	99.1	105.7	57.4	91.5	109.9	57.4	94.5
0970, 0990	Wall Finishes & Painting/Coating	103.0	70.2	82.8	102.7	42.7	65.7	103.0	79.8	88.7	107.9	24.9	56.7	103.0	42.4	65.6	103.0	56.4	74.3
09	FINISHES	104.2	80.5	91.0	101.0	74.7	86.3	103.9	81.9	91.6	108.8	63.5	83.5	103.0	71.4	85.4	104.3	73.0	86.8
COVERS	DIVS. 10 - 14, 25, 28, 41, 43, 44, 46	100.0	90.4	98.1	100.0	90.8	98.2	100.0	90.4	98.1	100.0	90.1	98.0	100.0	90.4	98.1	100.0	90.4	98.1
21, 22, 23	FIRE SUPPRESSION, PLUMBING & HVAC	94.1	80.2	88.4	100.2	74.8	89.9	99.9	80.5	92.1	94.1	89.1	92.1	100.0	80.1	91.9	94.1	80.1	88.4
26, 27, 3370	ELECTRICAL, COMMUNICATIONS & UTIL.	98.9	84.0	91.2	102.4	84.7	93.3	104.1	87.0	95.3	95.1	73.7	84.1	98.9	86.8	92.6	99.3	86.8	92.8
MF2010	WEIGHTED AVERAGE	98.6	82.9	91.7	101.6	81.7	92.8	101.7	84.7	94.2	101.9	79.6	92.1	102.6	79.1	92.2	99.3	82.0	91.7

		COLORADO																		
	DIVISION	GLENWOOD SPRINGS			GOLDEN			GRAND JUNCTION			GREELEY			MONTROSE			PUEBLO		
		816			804			815			806			814			810		
		MAT.	INST.	TOTAL	MAT.	INST.	TOTAL	MAT.	INST.	TOTAL	MAT.	INST.	TOTAL	MAT.	INST.	TOTAL	MAT.	INST.	TOTAL
015433	CONTRACTOR EQUIPMENT		97.9	97.9		96.2	96.2		97.9	97.9		96.2	96.2		96.5	96.5		95.2	95.2
0241, 31 - 34	SITE & INFRASTRUCTURE, DEMOLITION	140.5	98.5	111.5	105.6	97.1	99.7	123.1	98.2	105.9	90.5	96.2	94.5	133.7	94.7	106.8	114.8	94.1	100.5
0310	Concrete Forming & Accessories	103.3	79.6	82.7	97.4	79.5	81.8	110.1	79.2	83.2	99.6	83.0	85.2	102.9	79.4	82.4	107.8	80.7	84.2
0320	Concrete Reinforcing	104.7	80.7	92.6	99.1	80.7	89.8	105.0	80.7	92.7	98.9	79.3	89.0	104.6	80.7	92.5	101.2	85.2	93.1
0330	Cast-in-Place Concrete	99.5	81.5	92.4	101.3	81.4	93.5	110.2	80.8	98.6	97.4	60.3	82.8	99.6	81.4	92.4	98.8	86.6	94.0
03	CONCRETE	117.9	80.7	99.8	110.1	80.5	95.7	112.6	80.1	96.8	97.6	74.6	86.4	109.6	80.4	95.4	102.7	83.9	93.5
04	MASONRY	100.2	75.8	85.3	110.4	75.8	89.3	133.2	75.6	98.0	104.7	51.6	72.3	106.6	75.6	87.7	96.7	76.1	84.1
05	METALS	95.7	83.5	91.7	99.4	83.0	94.0	97.3	82.2	92.3	100.7	80.0	93.8	94.9	82.5	90.8	98.9	89.3	95.7
06	WOOD, PLASTICS & COMPOSITES	88.7	82.2	85.0	90.7	82.1	85.7	98.8	82.2	89.2	92.9	88.2	90.2	89.9	82.4	85.6	95.5	82.5	88.0
07	THERMAL & MOISTURE PROTECTION	105.9	81.9	96.2	105.9	76.4	93.9	104.9	70.8	91.0	104.8	63.3	87.9	106.1	81.9	96.3	104.4	82.6	95.6
08	OPENINGS	102.5	83.6	97.8	101.4	83.6	96.9	103.4	83.6	98.4	101.4	86.9	97.8	103.6	83.7	98.6	97.5	85.0	94.4
0920	Plaster & Gypsum Board	113.3	81.8	90.8	100.0	81.8	87.0	125.1	81.8	94.2	100.9	88.1	91.8	80.7	81.8	81.5	86.6	81.8	83.1
0950, 0980	Ceilings & Acoustic Treatment	107.9	81.8	90.6	99.5	81.8	87.8	107.9	81.8	90.6	99.5	88.1	92.0	108.8	81.8	90.9	117.8	81.8	94.0
0960	Flooring	110.6	52.8	93.6	103.3	57.4	89.8	115.8	57.4	98.7	104.5	57.4	90.7	113.9	47.5	94.4	112.6	88.8	105.6
0970, 0990	Wall Finishes & Painting/Coating	107.9	56.4	76.2	103.0	70.2	82.8	107.9	70.2	84.6	103.0	26.4	55.8	107.9	24.9	56.7	107.9	39.6	65.8
09	FINISHES	110.7	72.3	89.3	102.9	75.3	87.5	111.4	74.6	90.9	101.5	73.3	85.8	107.5	67.7	85.3	107.7	77.7	91.0
COVERS	DIVS. 10 - 14, 25, 28, 41, 43, 44, 46	100.0	90.6	98.1	100.0	90.4	98.1	100.0	90.5	98.1	100.0	91.1	98.2	100.0	90.9	98.2	100.0	91.4	98.3
21, 22, 23	FIRE SUPPRESSION, PLUMBING & HVAC	94.1	88.9	92.0	94.1	79.7	88.3	100.0	88.6	95.4	100.0	80.0	91.9	94.1	88.9	92.0	100.0	74.8	89.8
26, 27, 3370	ELECTRICAL, COMMUNICATIONS & UTIL.	92.7	73.8	82.9	99.3	83.9	91.4	94.8	57.2	75.4	98.9	86.8	92.6	94.8	59.9	76.8	95.7	75.1	85.1
MF2010	WEIGHTED AVERAGE	101.6	82.2	93.1	100.7	81.7	92.3	104.2	79.6	93.4	100.0	78.2	90.4	100.8	79.3	91.3	100.4	80.8	91.8

719

Section 1

	DIVISION	COLORADO SALIDA 812			CONNECTICUT BRIDGEPORT 066			BRISTOL 060			HARTFORD 061			MERIDEN 064			NEW BRITAIN 060		
		MAT.	INST.	TOTAL	MAT.	INST.	TOTAL	MAT.	INST.	TOTAL	MAT.	INST.	TOTAL	MAT.	INST.	TOTAL	MAT.	INST.	TOTAL
015433	CONTRACTOR EQUIPMENT		96.5	96.5		100.4	100.4		100.4	100.4		100.4	100.4		100.9	100.9		100.4	100.4
0241, 31 - 34	SITE & INFRASTRUCTURE, DEMOLITION	123.6	95.0	103.9	111.8	103.3	105.9	111.0	103.3	105.7	103.6	103.3	103.4	109.2	104.0	105.6	111.2	103.3	105.7
0310	Concrete Forming & Accessories	110.1	79.2	83.3	99.3	121.7	118.8	99.3	121.6	118.7	97.3	121.6	118.4	99.0	121.6	118.6	99.6	121.6	118.7
0320	Concrete Reinforcing	104.3	80.7	92.4	103.6	128.8	116.3	103.6	128.8	116.3	103.6	128.8	116.3	103.6	128.8	116.3	103.6	128.8	116.3
0330	Cast-in-Place Concrete	114.1	81.4	101.2	107.8	125.0	114.6	101.0	125.0	110.4	107.0	125.0	114.1	97.1	125.0	108.1	102.6	125.0	115.7
03	CONCRETE	111.5	80.4	96.3	110.6	124.0	117.1	107.1	123.9	115.3	110.0	123.9	116.8	105.2	123.9	114.3	108.0	123.9	115.7
04	MASONRY	134.8	75.6	98.7	104.1	129.8	119.8	96.7	129.8	116.9	97.2	129.8	117.1	96.3	129.8	116.7	98.0	129.8	117.4
05	METALS	94.6	82.4	90.6	99.1	125.2	107.7	99.1	125.0	107.7	103.9	125.0	110.9	96.5	125.0	105.9	95.7	125.0	105.4
06	WOOD, PLASTICS & COMPOSITES	96.5	82.4	88.3	98.6	120.3	111.1	98.6	120.3	111.1	97.1	120.3	110.5	98.6	120.3	111.1	98.6	120.3	111.1
07	THERMAL & MOISTURE PROTECTION	105.0	81.9	95.6	101.2	126.6	111.6	101.3	123.7	110.4	102.6	123.7	111.2	101.3	123.7	110.4	101.3	123.7	110.4
08	OPENINGS	95.7	83.7	92.7	102.4	130.1	109.4	102.4	130.1	109.4	103.1	130.1	109.9	105.1	130.1	111.4	102.4	130.1	109.4
0920	Plaster & Gypsum Board	81.2	81.8	81.6	97.8	120.3	113.8	97.8	120.3	113.8	95.9	120.3	113.3	99.6	120.3	114.4	97.8	120.3	113.8
0950, 0980	Ceilings & Acoustic Treatment	108.8	81.8	90.9	102.0	120.3	114.1	102.0	120.3	114.1	100.2	120.3	113.5	106.3	120.3	115.5	102.0	120.3	114.1
0960	Flooring	119.1	47.5	98.1	94.5	134.4	106.2	94.5	134.4	106.2	94.5	134.4	106.2	94.5	134.4	106.2	94.5	134.4	106.2
0970, 0990	Wall Finishes & Painting/Coating	107.9	24.9	56.7	90.1	117.5	107.0	90.1	117.5	107.0	90.1	117.5	107.0	90.1	117.5	107.0	90.1	117.5	107.0
09	FINISHES	107.9	67.7	85.5	101.7	123.2	113.7	101.8	123.2	113.7	99.9	123.2	112.9	102.9	123.2	114.2	101.8	123.2	113.7
COVERS	DIVS. 10 - 14, 25, 28, 41, 43, 44, 46	100.0	91.0	98.2	100.0	108.6	101.7	100.0	108.6	101.7	100.0	108.6	101.7	100.0	108.6	101.7	100.0	108.6	101.7
21, 22, 23	FIRE SUPPRESSION, PLUMBING & HVAC	94.1	74.6	86.2	100.0	114.7	106.0	100.0	114.7	106.0	100.1	114.7	106.0	94.1	114.7	102.4	100.0	114.7	106.0
26, 27, 3370	ELECTRICAL, COMMUNICATIONS & UTIL.	95.1	75.1	84.8	102.2	109.9	106.2	102.2	109.6	106.0	99.2	110.5	105.1	102.1	109.6	106.0	102.3	109.6	106.1
MF2010	WEIGHTED AVERAGE	101.3	78.3	91.2	102.2	118.7	109.5	101.5	118.6	109.0	101.2	118.7	109.4	99.7	118.6	108.0	101.1	118.6	108.8

Section 2

	DIVISION	CONNECTICUT NEW HAVEN 065			NEW LONDON 063			NORWALK 068			STAMFORD 069			WATERBURY 067			WILLIMANTIC 062		
		MAT.	INST.	TOTAL	MAT.	INST.	TOTAL	MAT.	INST.	TOTAL	MAT.	INST.	TOTAL	MAT.	INST.	TOTAL	MAT.	INST.	TOTAL
015433	CONTRACTOR EQUIPMENT		100.9	100.9		100.9	100.9		100.4	100.4		100.4	100.4		100.4	100.4		100.4	100.4
0241, 31 - 34	SITE & INFRASTRUCTURE, DEMOLITION	111.1	104.0	106.2	102.8	104.0	103.6	111.6	103.3	105.9	112.3	103.3	106.1	111.4	103.3	105.8	111.9	103.3	105.9
0310	Concrete Forming & Accessories	99.0	121.6	118.6	99.0	121.6	118.6	99.3	122.1	119.1	99.3	122.1	119.1	99.3	121.6	118.7	99.3	121.5	118.6
0320	Concrete Reinforcing	103.6	128.8	116.3	81.2	128.8	105.2	103.6	129.0	116.4	103.6	129.0	116.4	103.6	128.8	116.3	103.6	128.8	116.3
0330	Cast-in-Place Concrete	104.4	125.0	112.5	88.9	125.0	103.1	106.0	126.5	114.1	107.8	126.5	115.2	107.8	125.0	114.6	100.7	124.9	110.2
03	CONCRETE	123.2	123.9	123.5	94.7	123.9	108.9	109.7	124.7	117.0	110.6	124.7	117.4	110.6	123.9	117.0	107.0	123.8	115.2
04	MASONRY	96.9	129.8	117.0	95.3	129.8	116.3	96.4	131.3	117.7	97.2	131.3	118.0	97.2	129.8	117.1	96.5	129.8	116.8
05	METALS	95.9	125.0	105.5	95.6	125.0	105.3	99.1	125.7	107.9	99.1	125.7	107.9	99.1	125.0	107.7	98.9	124.9	107.5
06	WOOD, PLASTICS & COMPOSITES	98.6	120.3	111.1	98.6	120.3	111.1	98.6	120.3	111.1	98.6	120.3	111.1	98.6	120.3	111.1	98.6	120.3	111.1
07	THERMAL & MOISTURE PROTECTION	101.4	123.6	110.4	101.3	123.7	110.4	101.4	127.3	111.9	101.3	127.3	111.9	101.3	123.6	110.4	101.5	123.3	110.4
08	OPENINGS	102.4	130.1	109.4	105.6	130.1	111.8	102.4	130.1	109.4	102.4	130.1	109.4	102.4	130.1	109.4	105.6	130.1	111.8
0920	Plaster & Gypsum Board	97.8	120.3	113.8	97.8	120.3	113.8	97.8	120.3	113.8	97.8	120.3	113.8	97.8	120.3	113.8	97.8	120.3	113.8
0950, 0980	Ceilings & Acoustic Treatment	102.0	120.3	114.1	100.0	120.3	113.4	102.0	120.3	114.1	102.0	120.3	114.1	102.0	120.3	114.1	100.0	120.3	113.4
0960	Flooring	94.5	134.4	106.2	94.5	134.4	106.2	94.5	134.4	106.2	94.5	134.4	106.2	94.5	134.4	106.2	94.5	119.7	101.9
0970, 0990	Wall Finishes & Painting/Coating	90.1	117.5	107.0	90.1	117.5	107.0	90.1	117.5	107.0	90.1	117.5	107.0	90.1	117.5	107.0	90.1	117.5	107.0
09	FINISHES	101.8	123.2	113.7	100.6	123.2	113.2	101.8	123.2	113.7	101.9	123.2	113.7	101.6	123.2	113.6	101.5	120.7	112.2
COVERS	DIVS. 10 - 14, 25, 28, 41, 43, 44, 46	100.0	108.6	101.7	100.0	108.6	101.7	100.0	108.6	101.7	100.0	108.8	101.7	100.0	108.6	101.7	100.0	108.6	101.7
21, 22, 23	FIRE SUPPRESSION, PLUMBING & HVAC	100.0	114.7	106.0	94.1	114.7	102.4	100.0	114.8	106.0	100.0	114.8	106.0	100.0	114.7	106.0	100.0	114.7	106.0
26, 27, 3370	ELECTRICAL, COMMUNICATIONS & UTIL.	102.1	109.6	106.0	98.7	109.6	104.3	102.2	162.3	133.2	102.2	162.3	133.2	101.6	109.9	105.9	102.2	110.5	106.5
MF2010	WEIGHTED AVERAGE	102.8	118.6	109.8	97.6	118.6	106.9	101.8	126.3	112.6	101.9	126.3	112.7	101.8	118.6	109.2	101.7	118.4	109.1

Section 3

	DIVISION	D.C. WASHINGTON 200 - 205			DELAWARE DOVER 199			NEWARK 197			WILMINGTON 198			FLORIDA DAYTONA BEACH 321			FORT LAUDERDALE 333		
		MAT.	INST.	TOTAL	MAT.	INST.	TOTAL	MAT.	INST.	TOTAL	MAT.	INST.	TOTAL	MAT.	INST.	TOTAL	MAT.	INST.	TOTAL
015433	CONTRACTOR EQUIPMENT		103.2	103.2		116.9	116.9		116.9	116.9		117.0	117.0		98.0	98.0		90.9	90.9
0241, 31 - 34	SITE & INFRASTRUCTURE, DEMOLITION	109.6	92.6	97.9	98.2	111.9	107.6	98.6	111.9	107.8	88.8	112.2	104.9	116.5	89.6	97.9	99.7	77.9	84.6
0310	Concrete Forming & Accessories	97.4	81.1	83.2	98.3	102.7	102.1	100.4	102.7	102.4	101.4	102.7	102.5	96.7	72.1	75.3	95.3	72.8	75.7
0320	Concrete Reinforcing	98.5	89.7	94.1	96.0	102.1	99.1	96.8	102.1	99.5	96.8	102.1	99.5	96.3	77.0	86.6	96.3	76.7	86.4
0330	Cast-in-Place Concrete	129.8	90.0	114.1	95.5	101.9	98.0	85.8	101.9	92.2	92.6	101.9	96.3	91.8	73.9	84.7	96.2	81.0	90.2
03	CONCRETE	115.7	87.0	101.7	97.3	103.2	100.2	92.7	103.2	97.8	96.2	103.2	99.6	91.6	74.8	83.4	93.8	77.5	85.8
04	MASONRY	101.4	81.4	89.2	103.8	96.8	99.5	102.1	96.8	98.9	108.2	96.8	101.2	93.6	69.4	78.8	93.9	72.2	80.6
05	METALS	96.5	106.6	99.8	103.9	116.9	108.2	103.9	116.9	108.2	100.8	116.9	106.1	96.1	92.2	94.8	96.0	93.1	95.1
06	WOOD, PLASTICS & COMPOSITES	101.9	79.9	89.2	97.2	102.6	100.3	100.1	102.6	101.6	102.5	102.6	102.5	100.0	73.6	84.7	95.9	71.4	81.8
07	THERMAL & MOISTURE PROTECTION	100.2	84.9	94.0	97.9	112.4	103.8	97.9	112.4	103.8	98.4	112.4	104.1	95.1	77.8	88.1	95.1	85.4	91.2
08	OPENINGS	103.0	88.8	99.4	95.6	110.4	99.3	95.6	110.4	99.3	95.3	110.4	99.1	100.2	70.5	92.7	97.9	69.5	90.7
0920	Plaster & Gypsum Board	108.6	79.3	87.7	105.9	102.5	103.5	107.3	102.5	103.9	107.8	102.5	104.0	93.2	73.2	78.9	92.3	71.0	77.1
0950, 0980	Ceilings & Acoustic Treatment	105.8	79.3	88.3	104.2	102.5	103.1	104.2	102.5	103.1	99.5	102.5	101.5	95.8	73.2	80.9	95.8	71.0	79.4
0960	Flooring	115.0	97.2	109.8	97.1	107.6	100.1	97.3	107.6	100.3	96.9	107.6	100.0	117.7	77.1	105.8	117.7	72.6	104.5
0970, 0990	Wall Finishes & Painting/Coating	121.0	86.6	99.8	98.1	102.8	101.0	98.1	102.8	101.0	98.1	102.8	101.0	111.2	78.1	90.8	105.7	74.0	86.2
09	FINISHES	107.0	83.9	94.1	100.1	103.1	101.8	100.6	103.1	102.0	99.4	103.1	101.5	107.1	73.4	88.3	103.9	72.1	86.2
COVERS	DIVS. 10 - 14, 25, 28, 41, 43, 44, 46	100.0	99.4	99.9	100.0	93.7	98.7	100.0	93.7	98.7	100.0	93.7	98.7	100.0	78.3	95.7	100.0	84.8	97.0
21, 22, 23	FIRE SUPPRESSION, PLUMBING & HVAC	100.2	94.2	97.7	99.9	113.4	105.4	100.0	113.4	105.5	100.1	113.4	105.5	99.9	77.0	90.6	99.9	70.6	88.1
26, 27, 3370	ELECTRICAL, COMMUNICATIONS & UTIL.	98.8	106.0	102.5	96.4	110.1	103.5	100.3	110.1	105.3	98.6	110.1	104.5	98.0	58.4	77.6	98.0	74.3	85.8
MF2010	WEIGHTED AVERAGE	102.4	92.8	98.2	99.5	107.9	103.2	99.4	107.9	103.2	99.1	108.0	103.0	98.8	75.2	88.4	98.1	75.9	88.3

		FLORIDA															
	DIVISION	FORT MYERS			GAINESVILLE			JACKSONVILLE			LAKELAND			MELBOURNE			MIAMI		
		339,341			326,344			320,322			338			329			330 - 332,340		
		MAT.	INST.	TOTAL	MAT.	INST.	TOTAL	MAT.	INST.	TOTAL	MAT.	INST.	TOTAL	MAT.	INST.	TOTAL	MAT.	INST.	TOTAL
015433	CONTRACTOR EQUIPMENT		98.0	98.0		98.0	98.0		98.0	98.0		98.0	98.0		98.0	98.0		90.9	90.9
0241, 31 - 34	SITE & INFRASTRUCTURE, DEMOLITION	113.5	88.9	96.5	127.3	88.6	100.6	116.5	88.9	97.5	115.7	89.1	97.3	124.9	88.8	100.0	101.6	77.6	85.1
0310	Concrete Forming & Accessories	91.1	79.5	81.1	92.0	57.5	62.0	96.5	57.9	62.9	87.6	80.1	81.1	93.1	73.8	76.3	99.7	73.2	76.6
0320	Concrete Reinforcing	97.4	98.6	98.0	102.3	67.6	84.8	96.3	67.7	81.9	99.9	99.5	99.7	97.4	77.0	87.1	96.3	76.7	86.5
0330	Cast-in-Place Concrete	100.4	71.6	89.1	105.3	65.3	89.6	92.7	71.4	84.3	102.7	72.9	90.9	110.6	76.4	97.1	98.1	82.0	91.8
03	CONCRETE	94.7	81.3	88.2	102.6	63.6	83.6	92.0	65.9	79.3	96.6	82.1	89.6	103.3	76.4	90.2	95.0	78.0	86.7
04	MASONRY	87.5	65.4	74.0	109.0	64.6	81.9	93.6	64.7	76.0	105.5	80.1	90.0	92.7	73.7	81.1	94.4	76.0	83.2
05	METALS	98.2	100.5	99.0	95.1	87.1	92.4	94.7	87.5	92.3	98.1	101.2	99.1	104.2	92.3	100.3	100.9	91.9	97.9
06	WOOD, PLASTICS & COMPOSITES	92.8	82.3	86.7	94.0	55.0	71.5	100.0	55.0	74.0	88.6	82.3	85.0	95.6	73.6	82.9	101.6	71.4	84.2
07	THERMAL & MOISTURE PROTECTION	94.9	84.5	90.7	95.4	64.7	82.9	95.3	65.3	83.1	94.9	89.3	92.6	95.5	80.3	89.3	103.6	79.5	93.8
08	OPENINGS	98.8	79.0	93.8	98.2	55.6	87.4	100.2	58.9	89.8	98.7	79.7	93.9	99.3	73.9	92.9	97.9	69.5	90.7
0920	Plaster & Gypsum Board	87.3	82.2	83.7	88.2	54.1	63.8	93.2	54.1	65.3	85.1	82.2	83.0	88.2	73.2	77.5	90.6	71.0	76.6
0950, 0980	Ceilings & Acoustic Treatment	89.3	82.2	84.6	89.3	54.1	66.0	95.8	54.1	68.2	89.3	82.2	84.6	91.3	73.2	79.3	93.1	71.0	78.5
0960	Flooring	114.1	57.7	97.5	114.6	45.2	94.2	117.7	67.6	103.0	111.4	59.2	96.1	114.8	77.1	103.7	125.3	76.4	111.0
0970, 0990	Wall Finishes & Painting/Coating	111.2	64.1	82.1	111.2	64.1	82.1	111.2	65.3	82.9	111.2	64.1	82.1	111.2	96.9	102.4	105.7	74.0	86.2
09	FINISHES	103.7	74.1	87.2	105.5	54.8	77.2	107.1	59.4	80.5	102.8	74.4	86.9	105.5	76.6	89.4	105.5	73.3	87.6
COVERS	DIVS. 10 - 14, 25, 28, 41, 43, 44, 46	100.0	78.6	95.7	100.0	76.6	95.3	100.0	74.6	95.0	100.0	78.6	95.7	100.0	79.8	96.0	100.0	85.4	97.1
21, 22, 23	FIRE SUPPRESSION, PLUMBING & HVAC	96.7	66.2	84.4	98.8	68.6	86.6	99.9	68.6	87.3	96.7	86.7	92.7	99.9	79.2	91.5	99.9	71.3	88.3
26, 27, 3370	ELECTRICAL, COMMUNICATIONS & UTIL.	100.3	65.5	82.3	98.3	76.3	87.0	97.7	66.1	81.4	98.3	65.4	81.3	99.4	68.0	83.2	99.9	76.9	88.0
MF2010	WEIGHTED AVERAGE	98.1	75.8	88.3	100.4	69.7	86.9	98.7	69.4	85.7	99.0	82.0	91.5	101.5	78.2	91.2	99.6	76.7	89.6

		FLORIDA															
	DIVISION	ORLANDO			PANAMA CITY			PENSACOLA			SARASOTA			ST. PETERSBURG			TALLAHASSEE		
		327 - 328,347			324			325			342			337			323		
		MAT.	INST.	TOTAL	MAT.	INST.	TOTAL	MAT.	INST.	TOTAL	MAT.	INST.	TOTAL	MAT.	INST.	TOTAL	MAT.	INST.	TOTAL
015433	CONTRACTOR EQUIPMENT		98.0	98.0		98.0	98.0		98.0	98.0		98.0	98.0		98.0	98.0		98.0	98.0
0241, 31 - 34	SITE & INFRASTRUCTURE, DEMOLITION	116.7	88.7	97.4	131.7	87.8	101.4	130.9	88.3	101.5	117.9	88.8	97.8	117.3	88.4	97.4	112.6	88.0	95.6
0310	Concrete Forming & Accessories	96.3	76.4	79.0	95.8	46.6	53.0	93.8	55.0	60.0	96.7	79.6	81.9	94.8	54.1	59.4	98.6	46.8	53.5
0320	Concrete Reinforcing	96.3	76.3	86.2	100.6	77.1	88.7	103.2	77.6	90.3	96.3	99.4	97.9	99.9	86.2	93.0	96.3	67.5	81.8
0330	Cast-in-Place Concrete	115.0	73.8	98.7	97.4	59.3	82.4	120.3	68.9	100.0	104.9	72.7	92.2	103.8	67.2	89.4	100.2	58.8	83.9
03	CONCRETE	104.5	76.5	90.8	100.2	58.4	79.8	111.1	65.6	88.9	98.2	81.8	90.2	98.1	66.2	82.6	95.9	56.6	76.8
04	MASONRY	99.2	69.4	81.0	98.6	49.9	68.9	116.3	57.5	80.4	94.5	80.1	85.7	148.1	51.7	89.3	96.8	56.8	72.4
05	METALS	101.6	91.5	98.3	95.9	89.3	93.7	97.0	90.8	95.0	98.8	100.7	99.4	99.0	93.7	97.3	89.8	86.8	88.8
06	WOOD, PLASTICS & COMPOSITES	99.5	79.9	88.2	98.8	44.4	67.3	96.9	54.8	72.6	100.0	82.3	89.8	97.3	53.2	71.9	97.3	43.9	66.5
07	THERMAL & MOISTURE PROTECTION	96.3	78.4	89.0	95.6	59.7	81.0	95.5	65.8	83.4	95.1	89.3	92.8	95.1	61.0	81.2	101.0	75.5	90.6
08	OPENINGS	100.2	72.9	93.3	97.7	49.3	85.5	97.7	60.4	88.3	100.2	78.7	94.8	98.7	63.3	89.8	100.7	49.9	87.9
0920	Plaster & Gypsum Board	104.6	79.8	86.9	91.0	43.1	56.8	91.5	53.8	64.6	93.2	82.2	85.4	90.1	52.2	63.0	107.2	42.6	61.0
0950, 0980	Ceilings & Acoustic Treatment	98.5	79.8	86.1	91.3	43.1	59.4	91.3	53.8	66.5	93.8	82.2	86.1	91.3	52.2	65.5	98.5	42.6	61.6
0960	Flooring	117.9	77.1	105.9	117.2	45.0	96.0	112.0	65.9	98.5	117.7	60.6	100.9	116.3	57.8	99.2	117.7	64.9	102.2
0970, 0990	Wall Finishes & Painting/Coating	111.2	74.0	88.3	111.2	66.3	83.5	111.2	66.3	83.5	111.2	64.1	82.1	111.2	64.1	82.1	111.2	66.3	83.5
09	FINISHES	108.8	76.8	90.9	107.5	47.2	73.9	105.6	57.5	78.8	106.7	74.6	88.8	105.4	54.8	77.2	108.7	50.8	76.4
COVERS	DIVS. 10 - 14, 25, 28, 41, 43, 44, 46	100.0	79.0	95.8	100.0	49.9	90.0	100.0	50.0	90.1	100.0	78.6	95.7	100.0	60.3	92.1	100.0	71.2	94.3
21, 22, 23	FIRE SUPPRESSION, PLUMBING & HVAC	99.9	60.6	84.0	99.9	55.8	82.1	99.9	56.2	82.2	99.8	64.8	85.6	99.9	60.1	83.8	99.9	41.8	76.4
26, 27, 3370	ELECTRICAL, COMMUNICATIONS & UTIL.	100.1	62.8	80.8	96.7	60.1	77.8	100.7	57.2	78.3	98.0	65.4	81.2	98.3	65.4	81.3	103.8	60.1	81.3
MF2010	WEIGHTED AVERAGE	101.8	73.0	89.1	100.1	60.5	82.7	102.6	64.1	85.6	100.0	77.3	90.0	102.5	65.8	86.3	99.4	59.3	81.7

		FLORIDA						GEORGIA									
	DIVISION	TAMPA			WEST PALM BEACH			ALBANY			ATHENS			ATLANTA			AUGUSTA		
		335 - 336,346			334,349			317,398			306			300 - 303,399			308 - 309		
		MAT.	INST.	TOTAL	MAT.	INST.	TOTAL	MAT.	INST.	TOTAL	MAT.	INST.	TOTAL	MAT.	INST.	TOTAL	MAT.	INST.	TOTAL
015433	CONTRACTOR EQUIPMENT		98.0	98.0		90.9	90.9		91.8	91.8		93.2	93.2		93.7	93.7		93.2	93.2
0241, 31 - 34	SITE & INFRASTRUCTURE, DEMOLITION	117.6	89.0	97.9	96.4	77.9	83.6	99.9	79.2	85.6	103.6	92.3	95.8	99.4	94.1	95.7	95.8	92.0	93.2
0310	Concrete Forming & Accessories	97.6	80.2	82.4	98.6	72.4	75.8	96.0	45.8	52.3	93.6	43.5	50.0	96.9	75.3	78.1	94.5	67.0	70.6
0320	Concrete Reinforcing	96.3	99.5	97.9	99.1	76.4	87.6	96.0	83.5	89.7	102.7	82.7	92.6	102.0	84.9	93.4	103.0	75.5	89.1
0330	Cast-in-Place Concrete	101.4	72.9	90.2	91.5	78.0	86.2	98.0	47.8	78.2	111.6	53.1	88.5	111.6	71.7	95.9	105.4	50.0	83.6
03	CONCRETE	96.5	82.2	89.5	90.7	76.2	83.7	94.5	55.1	75.3	103.1	55.0	79.7	101.1	75.9	88.8	96.6	63.0	80.2
04	MASONRY	94.9	80.1	85.9	93.4	70.4	79.4	96.0	44.6	64.7	82.4	56.9	66.8	96.8	67.7	79.1	97.1	41.4	63.1
05	METALS	98.1	101.2	99.1	95.2	92.6	94.3	95.4	86.9	92.6	91.2	75.4	86.0	92.1	79.5	87.9	90.9	72.7	84.9
06	WOOD, PLASTICS & COMPOSITES	101.2	82.3	90.3	100.6	71.4	83.7	99.0	40.3	65.1	93.8	37.3	61.2	97.7	78.0	86.3	95.1	72.7	82.1
07	THERMAL & MOISTURE PROTECTION	95.3	89.3	92.9	94.9	74.1	86.4	92.8	61.4	80.0	94.5	54.7	78.3	94.4	74.2	86.2	94.1	58.5	79.6
08	OPENINGS	100.2	82.1	95.6	97.0	69.5	90.0	97.9	46.6	84.9	91.2	45.2	79.6	97.0	73.7	91.1	91.3	64.4	84.5
0920	Plaster & Gypsum Board	93.2	82.2	85.4	96.0	71.0	78.1	93.0	38.9	54.4	96.5	35.7	53.1	98.7	77.7	83.7	97.4	72.2	79.4
0950, 0980	Ceilings & Acoustic Treatment	95.8	82.2	86.8	89.3	71.0	77.2	94.9	38.9	57.9	95.8	35.7	56.1	95.8	77.7	83.8	96.8	72.2	80.6
0960	Flooring	117.7	59.2	100.5	120.1	69.6	105.3	117.7	37.5	94.2	89.2	56.4	79.5	90.6	68.3	84.1	89.6	41.1	75.4
0970, 0990	Wall Finishes & Painting/Coating	111.2	64.1	82.1	105.7	74.0	86.2	105.7	55.0	74.4	94.6	36.9	59.0	96.6	84.6	89.2	96.6	38.5	60.8
09	FINISHES	107.2	74.4	88.9	103.2	71.5	85.6	103.7	43.2	70.0	94.6	43.5	66.1	94.8	75.0	83.8	94.2	59.9	75.1
COVERS	DIVS. 10 - 14, 25, 28, 41, 43, 44, 46	100.0	79.1	95.9	100.0	84.8	97.0	100.0	80.6	96.1	100.0	42.2	88.5	100.0	85.3	97.1	100.0	53.6	90.8
21, 22, 23	FIRE SUPPRESSION, PLUMBING & HVAC	99.9	86.7	94.6	96.7	67.3	84.8	99.9	68.2	87.1	94.0	72.9	85.4	99.9	73.4	89.2	100.0	65.7	86.1
26, 27, 3370	ELECTRICAL, COMMUNICATIONS & UTIL.	98.0	65.4	81.2	99.4	74.3	86.5	93.5	60.1	76.3	102.3	73.9	87.7	101.7	75.5	88.2	103.0	65.6	83.7
MF2010	WEIGHTED AVERAGE	99.8	82.1	92.0	96.7	74.4	86.9	97.6	61.5	81.7	95.5	64.2	81.7	98.0	76.5	88.5	96.7	64.5	82.5

GEORGIA

DIVISION		COLUMBUS 318 - 319			DALTON 307			GAINESVILLE 305			MACON 310 - 312			SAVANNAH 313 - 314			STATESBORO 304		
		MAT.	INST.	TOTAL	MAT.	INST.	TOTAL	MAT.	INST.	TOTAL	MAT.	INST.	TOTAL	MAT.	INST.	TOTAL	MAT.	INST.	TOTAL
015433	CONTRACTOR EQUIPMENT		91.8	91.8		104.7	104.7		93.2	93.2		101.9	101.9		92.7	92.7		92.6	92.6
0241, 31 - 34	SITE & INFRASTRUCTURE, DEMOLITION	99.8	79.3	85.7	104.0	96.3	98.7	103.0	92.3	95.6	101.1	94.0	96.2	100.8	80.0	86.5	104.5	77.0	85.6
0310	Concrete Forming & Accessories	95.9	57.1	62.2	87.3	49.0	54.0	96.6	45.3	52.0	95.2	54.3	59.6	96.0	50.7	56.6	82.3	42.5	47.6
0320	Concrete Reinforcing	96.3	84.2	90.2	102.2	78.3	90.1	102.5	82.7	92.5	97.6	83.5	90.5	97.2	75.7	86.4	101.8	44.0	72.6
0330	Cast-in-Place Concrete	97.6	56.3	81.3	108.4	47.0	84.2	117.3	52.9	91.9	96.3	50.2	78.1	107.4	49.2	84.5	111.4	61.7	91.8
03	CONCRETE	94.4	63.2	79.2	100.0	55.4	79.3	105.3	55.8	81.2	93.9	59.8	77.3	99.4	56.3	78.4	102.6	51.2	77.5
04	MASONRY	96.0	58.4	73.1	84.2	33.9	53.5	91.4	57.3	70.6	108.2	39.4	66.3	95.1	51.5	68.5	85.6	41.6	58.8
05	METALS	94.9	88.6	92.8	91.9	85.0	89.6	90.4	75.1	85.3	91.1	87.7	90.0	91.4	84.2	89.1	95.4	72.4	87.8
06	WOOD, PLASTICS & COMPOSITES	99.0	55.4	73.8	76.7	50.4	61.5	97.5	40.2	64.4	108.9	54.4	77.4	115.1	47.5	76.1	70.7	41.8	54.0
07	THERMAL & MOISTURE PROTECTION	92.7	65.7	81.7	96.3	52.7	78.6	94.4	57.6	79.4	91.3	63.1	79.9	95.0	58.2	80.1	95.0	51.0	77.1
08	OPENINGS	97.9	56.7	87.5	92.9	47.1	81.3	91.2	42.4	78.9	96.2	55.2	85.9	99.1	48.5	86.4	93.6	33.4	78.4
0920	Plaster & Gypsum Board	93.0	54.5	65.4	80.3	49.2	58.1	98.7	38.7	55.9	101.8	53.4	67.2	92.2	46.4	59.5	82.2	40.4	52.3
0950, 0980	Ceilings & Acoustic Treatment	94.9	54.5	68.2	107.8	49.2	69.1	95.8	38.7	58.1	89.7	53.4	65.7	98.5	46.4	64.1	104.0	40.4	62.0
0960	Flooring	117.7	52.8	98.6	90.0	37.5	74.6	90.6	37.5	75.0	91.6	38.0	75.9	118.3	48.6	97.8	106.1	47.0	88.8
0970, 0990	Wall Finishes & Painting/Coating	105.7	68.2	82.5	87.3	56.3	68.2	96.6	36.9	59.8	108.1	55.0	75.3	107.1	57.2	76.3	94.6	34.6	57.6
09	FINISHES	103.6	56.4	77.3	103.6	47.0	72.0	95.0	40.9	64.8	91.3	50.2	68.4	104.7	50.1	74.2	106.4	41.6	70.3
COVERS	DIVS. 10 - 14, 25, 28, 41, 43, 44, 46	100.0	82.3	96.5	100.0	24.2	84.9	100.0	42.5	88.6	100.0	80.3	96.1	100.0	52.6	90.6	100.0	39.2	87.9
21, 22, 23	FIRE SUPPRESSION, PLUMBING & HVAC	99.9	49.9	79.7	94.1	59.2	79.9	94.0	72.8	85.4	99.9	66.3	86.3	99.9	61.6	84.4	94.5	56.7	79.2
26, 27, 3370	ELECTRICAL, COMMUNICATIONS & UTIL.	93.5	74.0	83.4	113.8	72.0	92.2	102.3	73.9	87.7	92.5	62.1	76.8	95.0	57.4	75.6	102.6	57.4	79.3
MF2010	WEIGHTED AVERAGE	97.5	64.6	83.0	97.6	60.1	81.1	96.1	64.0	82.0	96.4	64.2	82.2	98.1	60.5	81.5	97.4	54.0	78.3

DIVISION		GEORGIA VALDOSTA 316			GEORGIA WAYCROSS 315			HAWAII HILO 967			HAWAII HONOLULU 968			HAWAII STATES & POSS., GUAM 969			IDAHO BOISE 836 - 837		
		MAT.	INST.	TOTAL	MAT.	INST.	TOTAL	MAT.	INST.	TOTAL	MAT.	INST.	TOTAL	MAT.	INST.	TOTAL	MAT.	INST.	TOTAL
015433	CONTRACTOR EQUIPMENT		91.8	91.8		91.8	91.8		99.4	99.4		99.4	99.4		162.1	162.1		99.3	99.3
0241, 31 - 34	SITE & INFRASTRUCTURE, DEMOLITION	110.3	79.2	88.9	106.8	78.0	86.9	124.4	107.4	112.6	127.4	107.4	113.6	166.7	104.8	124.0	77.8	98.3	91.9
0310	Concrete Forming & Accessories	86.0	44.9	50.2	87.8	68.6	71.1	110.2	135.0	131.8	117.2	135.0	132.7	112.4	67.0	72.9	100.7	73.9	77.4
0320	Concrete Reinforcing	98.3	71.4	84.8	98.3	77.2	87.7	109.0	122.1	115.6	108.0	122.1	115.1	192.4	32.0	111.5	99.8	74.9	87.2
0330	Cast-in-Place Concrete	95.9	51.7	78.5	108.3	50.4	85.5	187.8	125.4	163.2	158.4	125.4	145.4	162.8	108.9	141.6	90.1	74.5	84.0
03	CONCRETE	98.4	53.9	76.7	102.6	64.7	84.1	147.4	128.1	138.0	132.9	128.1	130.5	148.0	75.4	112.6	98.8	74.4	86.9
04	MASONRY	102.1	51.1	71.0	103.0	41.5	65.5	121.3	127.0	124.8	114.1	127.0	122.0	173.3	46.0	95.6	122.1	62.5	85.8
05	METALS	94.6	83.1	90.8	93.6	79.9	89.1	107.4	108.1	107.6	128.1	108.1	121.5	138.5	77.6	118.4	99.9	74.8	91.6
06	WOOD, PLASTICS & COMPOSITES	87.0	39.3	59.5	88.8	77.0	82.0	101.4	137.6	122.4	111.7	137.6	126.7	112.0	70.9	88.3	87.9	76.2	81.2
07	THERMAL & MOISTURE PROTECTION	92.9	64.4	81.3	92.7	53.1	76.6	104.7	124.7	112.8	114.5	124.7	118.6	120.8	69.9	100.1	96.0	66.6	84.1
08	OPENINGS	93.2	42.7	80.4	93.2	60.7	85.0	99.7	131.3	107.7	105.0	131.3	111.6	104.5	57.0	92.5	96.3	70.8	89.9
0920	Plaster & Gypsum Board	85.1	37.9	51.4	85.5	76.7	79.2	103.7	138.6	128.6	134.7	138.6	137.5	202.8	59.2	100.2	86.8	75.4	78.7
0950, 0980	Ceilings & Acoustic Treatment	91.3	37.9	56.0	89.3	76.7	81.0	109.3	138.6	128.7	114.4	138.6	130.4	220.9	59.2	114.0	111.7	75.4	87.7
0960	Flooring	110.4	48.7	92.3	111.8	31.3	88.2	133.3	133.6	133.4	157.0	133.6	150.1	156.7	48.6	125.0	96.5	66.5	87.7
0970, 0990	Wall Finishes & Painting/Coating	105.7	55.0	74.4	105.7	42.0	66.4	98.7	144.2	126.8	103.1	144.2	128.5	104.2	37.9	63.3	98.8	41.6	63.5
09	FINISHES	100.9	45.0	69.7	100.4	60.1	77.9	115.8	136.8	127.5	128.5	136.8	133.1	190.9	61.9	118.9	99.3	70.1	83.0
COVERS	DIVS. 10 - 14, 25, 28, 41, 43, 44, 46	100.0	51.7	90.4	100.0	43.0	88.7	100.0	115.6	103.1	100.0	115.6	103.1	100.0	79.7	96.0	100.0	62.2	92.5
21, 22, 23	FIRE SUPPRESSION, PLUMBING & HVAC	99.9	68.1	87.0	96.1	58.1	80.7	100.1	109.7	104.0	100.1	109.7	104.0	102.1	40.0	76.9	100.0	67.6	86.9
26, 27, 3370	ELECTRICAL, COMMUNICATIONS & UTIL.	91.6	59.8	75.2	95.8	60.5	77.6	107.0	127.2	117.4	105.9	127.2	116.9	154.1	43.8	97.2	96.4	73.0	84.3
MF2010	WEIGHTED AVERAGE	97.6	60.8	81.4	97.3	61.4	81.5	110.3	121.2	115.1	113.3	121.2	116.8	131.1	60.8	100.1	99.3	72.4	87.4

DIVISION		IDAHO COEUR D'ALENE 838			IDAHO IDAHO FALLS 834			IDAHO LEWISTON 835			IDAHO POCATELLO 832			IDAHO TWIN FALLS 833			ILLINOIS BLOOMINGTON 617		
		MAT.	INST.	TOTAL	MAT.	INST.	TOTAL	MAT.	INST.	TOTAL	MAT.	INST.	TOTAL	MAT.	INST.	TOTAL	MAT.	INST.	TOTAL
015433	CONTRACTOR EQUIPMENT		93.8	93.8		99.3	99.3		93.8	93.8		99.3	99.3		99.3	99.3		100.8	100.8
0241, 31 - 34	SITE & INFRASTRUCTURE, DEMOLITION	76.2	93.9	88.4	76.3	97.4	90.8	82.2	94.4	90.6	79.6	98.2	92.5	86.4	96.6	93.4	97.8	96.8	97.1
0310	Concrete Forming & Accessories	117.9	57.8	65.6	95.4	41.4	48.4	122.5	59.1	67.4	100.9	73.6	77.2	101.6	29.6	39.0	86.0	120.1	115.6
0320	Concrete Reinforcing	107.4	93.5	100.4	101.8	77.7	89.6	107.4	93.6	100.5	100.2	74.9	87.5	102.1	77.9	89.9	98.6	107.6	103.1
0330	Cast-in-Place Concrete	97.4	86.9	93.3	85.8	52.0	72.5	101.3	87.8	96.0	92.7	74.3	85.4	95.1	39.8	73.3	97.9	114.2	104.3
03	CONCRETE	105.9	75.0	90.9	91.4	52.8	72.6	109.5	76.0	93.1	98.3	74.2	86.6	105.5	43.4	75.2	96.3	115.6	105.7
04	MASONRY	123.7	82.4	98.5	117.2	34.6	66.8	124.1	83.3	99.2	119.8	64.4	86.0	122.3	32.6	67.6	113.8	119.2	117.1
05	METALS	94.2	84.0	90.8	107.5	73.7	96.4	93.6	85.0	90.8	107.6	74.7	96.7	107.6	73.1	96.2	96.9	111.3	101.6
06	WOOD, PLASTICS & COMPOSITES	93.4	50.5	68.6	82.6	39.8	57.9	98.0	50.5	70.6	88.0	76.2	81.2	88.7	28.3	53.8	85.6	119.8	105.4
07	THERMAL & MOISTURE PROTECTION	147.8	73.7	117.7	95.6	44.6	74.9	148.0	73.7	117.8	96.1	61.3	81.9	96.9	38.7	73.2	97.9	113.9	104.4
08	OPENINGS	115.3	56.0	100.4	100.5	46.8	87.0	115.3	58.7	101.0	97.1	65.5	89.1	100.6	38.4	84.9	89.4	107.0	93.8
0920	Plaster & Gypsum Board	154.5	49.0	79.1	81.0	37.8	50.1	155.4	49.0	79.4	83.8	75.4	77.8	84.5	25.9	42.6	97.2	120.3	113.7
0950, 0980	Ceilings & Acoustic Treatment	147.1	49.0	82.3	109.7	37.8	62.2	147.1	49.0	82.3	117.8	75.4	89.8	112.4	25.9	55.3	94.9	120.3	111.7
0960	Flooring	135.3	48.0	109.7	96.2	45.1	81.2	138.5	89.5	124.1	99.7	66.5	90.0	100.9	45.1	84.6	97.4	112.2	101.8
0970, 0990	Wall Finishes & Painting/Coating	116.6	51.9	74.4	98.8	34.2	59.0	116.6	61.9	82.8	116.6	61.9	82.8	98.8	25.5	53.6	96.0	124.2	113.4
09	FINISHES	167.7	54.8	104.7	97.5	40.6	65.8	168.9	63.3	110.0	101.2	70.3	84.0	101.2	30.9	62.0	97.5	119.8	109.9
COVERS	DIVS. 10 - 14, 25, 28, 41, 43, 44, 46	100.0	52.4	90.5	100.0	39.5	88.0	100.0	65.0	93.0	100.0	62.2	92.5	100.0	35.5	87.2	100.0	105.0	101.0
21, 22, 23	FIRE SUPPRESSION, PLUMBING & HVAC	99.6	83.2	93.0	100.9	61.5	84.9	100.8	90.2	96.5	100.0	67.5	86.8	100.0	55.6	82.0	94.0	106.7	99.1
26, 27, 3370	ELECTRICAL, COMMUNICATIONS & UTIL.	87.4	80.3	83.7	87.6	73.0	80.1	85.5	80.8	83.1	93.9	73.0	83.1	88.9	62.9	75.5	94.3	96.7	95.6
MF2010	WEIGHTED AVERAGE	107.2	76.3	93.6	98.9	58.7	81.2	107.9	79.7	95.5	100.3	72.2	87.9	101.3	52.5	79.8	96.3	109.3	102.0

City Cost Indexes

ILLINOIS

DIVISION		CARBONDALE 629 MAT.	INST.	TOTAL	CENTRALIA 628 MAT.	INST.	TOTAL	CHAMPAIGN 618-619 MAT.	INST.	TOTAL	CHICAGO 606-608 MAT.	INST.	TOTAL	DECATUR 625 MAT.	INST.	TOTAL	EAST ST. LOUIS 620-622 MAT.	INST.	TOTAL
015433	CONTRACTOR EQUIPMENT		107.0	107.0		107.0	107.0		101.6	101.6		92.7	92.7		101.6	101.6		107.0	107.0
0241, 31 - 34	SITE & INFRASTRUCTURE, DEMOLITION	104.8	97.4	99.7	105.2	98.7	100.7	107.5	97.7	100.7	111.8	93.7	99.3	91.6	97.7	95.8	107.2	98.6	101.3
0310	Concrete Forming & Accessories	94.0	104.1	102.8	95.9	112.4	110.2	91.7	118.5	115.0	98.0	156.1	148.5	95.8	118.0	115.1	91.7	112.7	110.0
0320	Concrete Reinforcing	97.6	94.6	96.1	97.6	104.5	101.1	98.6	102.5	100.6	99.0	161.1	130.3	95.6	102.9	99.3	97.5	108.5	103.1
0330	Cast-in-Place Concrete	91.6	99.1	94.6	92.1	116.1	101.5	113.5	109.9	112.1	107.8	147.0	123.2	100.0	112.9	105.1	93.6	116.2	102.5
03	CONCRETE	87.4	101.4	94.2	87.9	112.8	100.0	108.7	112.4	110.5	107.0	152.5	126.1	97.5	113.3	105.2	88.8	113.7	101.0
04	MASONRY	75.5	102.6	92.0	75.6	116.7	100.6	137.4	116.3	124.5	98.4	153.1	131.8	72.3	115.4	98.6	75.8	116.6	100.7
05	METALS	96.0	111.1	101.0	96.0	117.9	103.3	96.8	107.3	100.3	95.3	133.9	108.1	99.5	107.8	102.3	97.2	119.8	104.6
06	WOOD, PLASTICS & COMPOSITES	99.2	103.4	101.6	101.6	109.8	106.4	92.0	118.1	107.1	101.5	155.5	132.7	98.8	118.1	110.0	96.8	109.8	104.3
07	THERMAL & MOISTURE PROTECTION	93.1	98.9	95.5	93.2	112.9	101.2	98.6	113.3	104.5	99.7	143.6	117.6	98.7	108.0	102.5	93.2	112.4	101.0
08	OPENINGS	85.7	108.5	91.4	85.7	113.9	92.8	90.1	113.0	95.9	97.5	158.0	112.7	96.0	113.1	100.3	85.7	115.7	93.3
0920	Plaster & Gypsum Board	100.8	103.4	102.7	102.2	110.1	107.8	99.9	118.6	113.3	101.0	157.2	141.1	104.1	118.6	114.5	99.9	110.1	107.2
0950, 0980	Ceilings & Acoustic Treatment	94.9	103.4	100.5	94.9	110.1	104.9	94.9	118.6	110.6	109.9	157.2	141.1	101.2	118.6	112.7	94.9	110.1	104.9
0960	Flooring	116.3	125.8	119.1	117.4	118.6	117.8	100.5	125.8	107.9	97.0	148.5	112.1	104.0	117.6	108.0	115.1	118.6	116.2
0970, 0990	Wall Finishes & Painting/Coating	104.0	101.0	102.1	104.0	108.3	106.6	96.0	112.3	106.1	95.1	161.0	135.8	96.0	110.6	105.0	104.0	108.3	106.6
09	FINISHES	100.9	107.0	104.3	101.4	113.0	107.9	99.6	119.8	110.9	101.2	156.2	131.9	101.7	118.2	110.9	100.5	113.2	107.6
COVERS	DIVS. 10 - 14, 25, 28, 41, 43, 44, 46	100.0	100.5	100.1	100.0	103.9	100.8	100.0	104.7	100.9	100.0	125.2	105.0	100.0	104.4	100.9	100.0	103.9	100.8
21, 22, 23	FIRE SUPPRESSION, PLUMBING & HVAC	93.9	96.7	95.1	93.9	92.8	93.5	94.0	100.5	96.6	100.0	133.9	113.7	99.9	101.5	100.6	99.9	97.1	98.8
26, 27, 3370	ELECTRICAL, COMMUNICATIONS & UTIL.	93.0	101.5	97.4	94.4	105.4	100.1	97.6	95.8	96.7	95.4	133.4	115.0	97.1	90.3	93.6	94.0	105.4	99.8
MF2010	WEIGHTED AVERAGE	93.1	102.0	97.1	93.4	107.0	99.4	99.8	108.1	103.5	99.1	138.9	116.6	97.4	106.3	101.3	95.1	108.3	100.9

ILLINOIS

DIVISION		EFFINGHAM 624 MAT.	INST.	TOTAL	GALESBURG 614 MAT.	INST.	TOTAL	JOLIET 604 MAT.	INST.	TOTAL	KANKAKEE 609 MAT.	INST.	TOTAL	LA SALLE 613 MAT.	INST.	TOTAL	NORTH SUBURBAN 600-603 MAT.	INST.	TOTAL
015433	CONTRACTOR EQUIPMENT		101.6	101.6		100.8	100.8		90.9	90.9		90.9	90.9		100.8	100.8		90.9	90.9
0241, 31 - 34	SITE & INFRASTRUCTURE, DEMOLITION	99.0	97.2	97.7	100.3	96.7	97.8	111.7	92.6	98.5	104.8	92.0	96.0	99.7	97.6	98.2	110.7	91.8	97.7
0310	Concrete Forming & Accessories	100.1	111.5	110.0	91.6	115.7	112.6	100.1	154.6	147.5	94.2	135.5	130.1	104.0	121.0	118.8	99.5	148.9	142.4
0320	Concrete Reinforcing	98.6	99.9	99.3	98.1	107.2	102.7	99.0	146.2	122.8	99.8	140.4	120.3	98.3	133.4	116.0	99.0	155.2	127.3
0330	Cast-in-Place Concrete	99.7	105.3	101.9	100.8	105.2	102.5	107.7	142.3	121.3	100.4	127.1	110.9	100.6	114.4	106.1	107.8	139.8	120.4
03	CONCRETE	98.3	107.2	102.7	99.0	110.5	104.6	101.1	147.4	123.7	95.3	132.6	113.5	99.8	121.1	110.2	101.1	145.7	122.8
04	MASONRY	79.3	104.4	94.6	113.9	113.7	113.8	101.0	146.3	128.6	97.8	132.5	119.0	113.9	117.0	115.8	98.3	142.4	125.2
05	METALS	96.7	105.6	99.6	96.8	110.1	101.2	93.2	125.8	104.0	93.2	121.4	102.5	96.9	125.6	106.4	94.4	128.7	105.7
06	WOOD, PLASTICS & COMPOSITES	101.2	113.3	108.2	91.8	115.0	105.2	103.2	156.1	133.8	96.5	134.4	118.4	105.5	120.3	114.1	101.5	150.6	129.9
07	THERMAL & MOISTURE PROTECTION	98.2	106.0	101.4	98.1	106.6	101.5	99.6	140.1	116.0	98.8	131.9	112.2	98.2	114.9	105.0	100.0	138.5	115.7
08	OPENINGS	89.9	109.7	94.9	89.4	108.8	94.3	95.8	154.3	110.5	87.5	139.4	100.5	89.4	122.7	97.8	95.8	152.1	110.0
0920	Plaster & Gypsum Board	104.0	113.6	110.9	99.9	115.4	111.0	97.8	157.8	140.7	95.5	135.5	124.1	106.3	120.9	116.7	101.0	152.2	137.6
0950, 0980	Ceilings & Acoustic Treatment	94.9	113.6	107.3	94.9	115.4	108.4	109.9	157.8	141.6	109.9	135.5	126.8	94.9	120.9	112.1	109.9	152.2	137.8
0960	Flooring	105.3	125.8	111.3	100.4	112.2	103.9	96.6	138.1	108.8	93.7	138.1	106.7	107.4	116.3	110.0	97.0	138.1	109.1
0970, 0990	Wall Finishes & Painting/Coating	96.0	105.9	102.1	96.0	93.9	94.7	93.2	151.2	129.9	93.2	124.2	112.4	96.0	124.2	113.4	95.1	151.2	129.7
09	FINISHES	100.7	114.7	108.5	98.9	113.6	107.1	100.7	152.6	129.6	99.0	133.0	118.0	101.7	119.6	111.7	101.2	148.4	127.5
COVERS	DIVS. 10 - 14, 25, 28, 41, 43, 44, 46	100.0	72.5	94.5	100.0	100.9	100.2	100.0	123.8	104.7	100.0	114.8	102.9	100.0	101.0	100.2	100.0	117.0	103.4
21, 22, 23	FIRE SUPPRESSION, PLUMBING & HVAC	94.0	101.1	96.9	94.0	102.7	97.5	100.1	133.2	113.5	94.1	127.3	107.5	94.0	124.7	106.4	100.0	123.2	109.4
26, 27, 3370	ELECTRICAL, COMMUNICATIONS & UTIL.	95.0	101.5	98.4	95.3	87.7	91.4	94.7	137.9	117.0	89.5	137.1	114.0	92.4	134.8	114.2	94.5	133.3	114.6
MF2010	WEIGHTED AVERAGE	95.3	103.9	99.1	96.9	104.8	100.4	98.6	136.4	115.3	94.6	127.4	109.1	97.0	120.8	107.5	98.7	132.4	113.5

ILLINOIS

DIVISION		PEORIA 615-616 MAT.	INST.	TOTAL	QUINCY 623 MAT.	INST.	TOTAL	ROCK ISLAND 612 MAT.	INST.	TOTAL	ROCKFORD 610-611 MAT.	INST.	TOTAL	SOUTH SUBURBAN 605 MAT.	INST.	TOTAL	SPRINGFIELD 626-627 MAT.	INST.	TOTAL
015433	CONTRACTOR EQUIPMENT		100.8	100.8		101.6	101.6		100.8	100.8		100.8	100.8		90.9	90.9		101.6	101.6
0241, 31 - 34	SITE & INFRASTRUCTURE, DEMOLITION	100.5	96.8	97.9	97.8	97.3	97.4	98.4	96.1	96.8	99.6	98.0	98.5	110.7	91.8	97.7	99.2	97.8	98.2
0310	Concrete Forming & Accessories	94.2	116.9	113.9	98.0	114.6	112.4	93.1	100.0	99.1	98.2	132.3	127.9	99.5	148.9	142.5	96.5	118.5	115.6
0320	Concrete Reinforcing	95.6	107.3	101.5	98.3	99.8	99.0	98.1	99.8	99.0	90.4	140.0	115.4	99.0	155.2	127.3	95.6	102.9	99.3
0330	Cast-in-Place Concrete	97.7	115.2	104.6	99.8	97.6	99.0	98.6	96.0	97.6	100.1	125.9	110.2	107.8	139.8	120.4	89.9	108.1	97.1
03	CONCRETE	96.3	114.5	105.2	98.0	106.0	101.9	97.1	99.0	98.0	96.9	131.2	113.6	101.1	145.7	122.8	92.5	111.9	101.9
04	MASONRY	114.0	114.4	114.3	99.5	101.1	100.5	113.8	95.6	102.7	88.9	137.3	118.4	98.3	142.4	125.2	77.2	116.5	101.2
05	METALS	99.5	111.0	103.3	96.7	105.3	99.6	96.8	106.1	99.9	99.5	128.8	109.2	94.4	128.7	105.7	102.0	107.9	103.9
06	WOOD, PLASTICS & COMPOSITES	98.8	115.0	108.2	98.7	118.1	109.9	93.3	99.3	96.7	98.8	131.0	117.4	101.5	150.6	129.9	100.1	118.1	110.5
07	THERMAL & MOISTURE PROTECTION	98.8	112.1	104.2	98.2	103.5	100.4	98.1	97.2	97.7	101.3	129.6	112.8	100.0	138.7	115.8	98.4	112.1	104.0
08	OPENINGS	96.1	112.8	100.3	90.8	112.4	96.2	89.4	98.4	91.7	96.1	132.0	105.1	95.8	152.1	110.0	97.1	113.1	101.1
0920	Plaster & Gypsum Board	104.1	115.4	112.2	102.2	118.6	113.9	99.9	99.1	99.4	104.1	131.9	124.0	101.0	152.2	137.6	105.9	118.6	115.0
0950, 0980	Ceilings & Acoustic Treatment	101.2	115.4	110.6	94.9	118.6	110.6	94.9	99.1	97.7	101.2	131.9	121.5	109.9	152.2	137.8	100.3	118.6	112.4
0960	Flooring	104.0	112.2	106.4	104.0	104.0	104.0	101.6	94.7	99.6	104.0	116.3	107.6	97.0	138.1	109.1	104.0	109.0	105.5
0970, 0990	Wall Finishes & Painting/Coating	96.0	124.2	113.4	96.0	108.2	103.5	96.0	97.0	96.6	96.0	131.3	117.8	95.1	151.2	129.7	96.0	108.2	103.5
09	FINISHES	101.8	117.1	110.4	100.1	113.5	107.6	99.1	98.6	98.9	101.8	129.7	117.4	101.2	148.4	127.5	101.5	116.5	109.9
COVERS	DIVS. 10 - 14, 25, 28, 41, 43, 44, 46	100.0	104.7	100.9	100.0	75.5	95.1	100.0	98.1	99.6	100.0	94.4	98.9	100.0	117.0	103.4	100.0	104.7	100.9
21, 22, 23	FIRE SUPPRESSION, PLUMBING & HVAC	99.9	105.7	102.3	94.0	98.3	95.7	94.0	99.4	96.2	100.0	115.6	106.3	100.0	123.2	109.4	99.8	104.8	101.8
26, 27, 3370	ELECTRICAL, COMMUNICATIONS & UTIL.	96.2	97.2	96.7	92.5	81.6	86.9	87.4	98.4	93.1	96.6	131.8	114.8	94.5	133.3	114.6	97.5	92.4	94.9
MF2010	WEIGHTED AVERAGE	99.5	108.3	103.4	96.0	100.1	97.8	95.9	98.9	97.2	98.4	124.3	109.8	98.7	132.4	113.5	97.8	107.1	101.9

723

City Cost Indexes

INDIANA

DIVISION		ANDERSON 460 MAT.	INST.	TOTAL	BLOOMINGTON 474 MAT.	INST.	TOTAL	COLUMBUS 472 MAT.	INST.	TOTAL	EVANSVILLE 476-477 MAT.	INST.	TOTAL	FORT WAYNE 467-468 MAT.	INST.	TOTAL	GARY 463-464 MAT.	INST.	TOTAL
015433	CONTRACTOR EQUIPMENT		96.6	96.6		88.1	88.1		88.1	88.1		116.9	116.9		96.6	96.6		96.6	96.6
0241, 31 - 34	SITE & INFRASTRUCTURE, DEMOLITION	94.9	94.9	94.9	79.4	95.4	90.4	76.4	95.1	89.3	84.4	124.7	112.2	95.4	94.6	94.8	95.4	98.2	97.3
0310	Concrete Forming & Accessories	95.0	81.5	83.3	100.5	82.4	84.8	95.0	79.3	81.4	94.6	80.5	82.3	93.4	74.7	77.1	95.1	110.3	108.3
0320	Concrete Reinforcing	97.9	82.7	90.3	89.6	82.5	86.0	90.0	82.5	86.2	98.1	76.6	87.3	97.9	77.8	87.8	97.9	103.3	100.6
0330	Cast-in-Place Concrete	98.2	79.7	90.9	98.2	80.8	91.3	97.8	74.5	88.6	93.8	87.5	91.3	104.4	75.3	93.0	102.7	107.9	104.7
03	CONCRETE	94.7	81.6	88.3	100.8	81.6	91.4	100.2	78.1	89.4	101.0	82.4	91.9	97.7	76.2	87.2	96.9	108.0	102.3
04	MASONRY	87.4	81.7	83.9	88.8	79.1	82.9	88.6	77.9	82.1	84.5	83.0	83.6	91.2	79.2	83.9	88.7	113.2	103.7
05	METALS	94.2	89.5	92.7	94.6	78.8	89.4	94.7	78.1	89.2	88.3	84.7	87.1	94.2	86.9	91.8	94.2	104.8	97.7
06	WOOD, PLASTICS & COMPOSITES	107.1	81.7	92.4	119.4	82.1	97.9	114.6	78.8	93.9	98.2	79.1	87.1	106.8	74.0	87.8	104.7	108.2	106.8
07	THERMAL & MOISTURE PROTECTION	98.2	80.2	90.9	93.9	82.8	89.3	93.5	82.0	88.8	97.6	84.4	92.2	98.0	82.6	91.7	97.1	109.5	102.1
08	OPENINGS	98.3	82.0	94.2	106.5	82.3	100.4	101.5	80.4	96.2	98.2	78.9	93.3	98.3	74.2	92.2	98.3	111.6	101.7
0920	Plaster & Gypsum Board	99.8	81.6	86.8	97.7	82.2	86.6	94.5	78.8	83.3	93.0	77.8	82.1	98.8	73.6	80.8	92.9	108.9	104.4
0950, 0980	Ceilings & Acoustic Treatment	91.1	81.6	84.8	83.4	82.2	82.6	83.4	78.8	80.3	88.4	77.8	81.4	91.1	73.6	79.5	91.1	108.9	102.9
0960	Flooring	98.3	81.3	93.3	105.9	78.5	97.9	100.9	78.5	94.3	99.1	84.5	94.8	98.3	78.5	92.4	98.3	121.4	105.1
0970, 0990	Wall Finishes & Painting/Coating	98.8	70.0	81.0	91.4	83.3	86.4	91.4	83.3	86.4	96.0	85.8	89.7	98.8	75.2	84.2	98.8	122.6	113.5
09	FINISHES	96.5	80.3	87.5	96.7	81.8	88.4	94.9	79.5	86.3	95.2	81.6	87.6	96.2	75.5	84.7	95.6	113.6	105.7
COVERS	DIVS. 10 - 14, 25, 28, 41, 43, 44, 46	100.0	92.5	98.5	100.0	87.8	97.6	100.0	87.0	97.4	100.0	92.0	98.4	100.0	87.5	97.5	100.0	103.4	100.7
21, 22, 23	FIRE SUPPRESSION, PLUMBING & HVAC	100.0	77.4	90.8	99.8	80.8	92.1	93.8	78.7	87.7	99.9	81.1	92.3	100.0	74.5	89.7	100.0	105.7	102.3
26, 27, 3370	ELECTRICAL, COMMUNICATIONS & UTIL.	88.2	86.3	87.2	99.7	85.7	92.5	98.9	86.3	92.4	95.8	85.6	90.6	89.0	74.1	81.3	100.4	114.3	107.6
MF2010	WEIGHTED AVERAGE	96.1	83.4	90.5	98.4	82.9	91.6	96.0	81.5	89.6	96.0	86.5	91.8	96.7	78.7	88.8	97.6	108.5	102.4

INDIANA

DIVISION		INDIANAPOLIS 461-462 MAT.	INST.	TOTAL	KOKOMO 469 MAT.	INST.	TOTAL	LAFAYETTE 479 MAT.	INST.	TOTAL	LAWRENCEBURG 470 MAT.	INST.	TOTAL	MUNCIE 473 MAT.	INST.	TOTAL	NEW ALBANY 471 MAT.	INST.	TOTAL
015433	CONTRACTOR EQUIPMENT		92.4	92.4		96.6	96.6		88.1	88.1		105.8	105.8		96.4	96.4		96.0	96.0
0241, 31 - 34	SITE & INFRASTRUCTURE, DEMOLITION	95.1	97.9	97.0	91.5	94.8	93.8	77.3	95.2	89.6	74.4	111.8	100.2	79.2	95.5	90.5	71.8	98.1	90.0
0310	Concrete Forming & Accessories	95.7	85.2	86.6	98.2	77.1	79.9	92.9	82.0	83.5	76.1	75.9	78.0	93.2	81.3	82.9	88.8	73.7	75.7
0320	Concrete Reinforcing	97.3	82.8	90.0	88.6	82.6	85.6	89.6	82.7	86.1	88.9	78.8	83.8	99.1	82.7	90.8	90.2	76.5	83.3
0330	Cast-in-Place Concrete	94.5	86.4	91.3	97.3	84.0	92.0	98.3	83.1	92.3	91.9	73.6	84.7	103.2	79.9	94.0	94.9	72.5	86.1
03	CONCRETE	96.5	84.9	90.8	91.6	81.1	86.5	100.3	82.3	91.6	93.6	76.3	85.1	100.1	81.6	91.1	98.6	74.3	86.8
04	MASONRY	94.6	82.9	87.4	87.1	79.6	82.5	93.5	81.5	86.2	73.6	77.8	76.2	90.7	81.6	85.2	80.7	71.0	74.8
05	METALS	96.3	80.9	91.2	90.8	89.2	90.3	93.1	78.9	88.4	90.0	86.5	88.8	96.5	89.8	94.3	91.8	81.8	88.5
06	WOOD, PLASTICS & COMPOSITES	103.8	85.5	93.2	110.1	75.5	90.1	111.9	81.9	94.5	97.0	75.2	84.4	114.3	81.6	95.4	98.7	74.2	84.6
07	THERMAL & MOISTURE PROTECTION	96.2	84.6	91.5	97.8	79.9	90.5	93.5	83.3	89.3	96.8	79.3	89.6	96.4	80.5	90.0	85.6	70.4	79.4
08	OPENINGS	106.5	84.1	100.8	92.4	78.7	89.0	99.8	82.1	95.4	99.9	76.5	94.0	101.3	81.9	96.4	97.7	76.1	92.3
0920	Plaster & Gypsum Board	97.7	85.3	88.8	105.2	75.2	83.8	91.4	81.9	84.6	72.1	74.9	74.1	93.0	81.6	84.9	91.2	73.7	78.7
0950, 0980	Ceilings & Acoustic Treatment	95.6	85.3	88.8	91.1	75.2	80.6	78.9	81.9	80.9	93.8	74.9	81.3	84.3	81.6	82.5	88.4	73.7	78.7
0960	Flooring	98.1	87.5	94.9	102.4	88.8	98.4	99.7	89.1	96.6	73.7	87.5	77.7	99.4	81.3	94.1	97.3	65.8	88.1
0970, 0990	Wall Finishes & Painting/Coating	98.8	83.3	89.3	98.8	75.6	84.5	91.4	98.1	95.6	92.5	74.4	81.3	91.4	70.0	78.2	96.0	87.2	90.6
09	FINISHES	97.5	85.6	90.9	98.1	78.8	87.3	93.2	85.1	88.7	86.4	78.2	81.8	94.1	80.2	86.3	94.7	74.0	83.1
COVERS	DIVS. 10 - 14, 25, 28, 41, 43, 44, 46	100.0	93.8	98.8	100.0	87.1	97.4	100.0	92.2	98.4	100.0	46.4	89.3	100.0	92.0	98.4	100.0	45.8	89.2
21, 22, 23	FIRE SUPPRESSION, PLUMBING & HVAC	99.9	83.3	93.2	94.0	80.7	88.6	93.8	79.6	88.1	94.5	75.6	86.8	99.8	77.3	90.7	94.0	77.0	87.1
26, 27, 3370	ELECTRICAL, COMMUNICATIONS & UTIL.	102.4	86.3	94.1	92.7	79.2	85.7	98.4	83.4	90.7	93.7	75.6	84.4	91.5	80.3	85.7	94.3	76.4	85.0
MF2010	WEIGHTED AVERAGE	99.2	85.6	93.2	93.7	82.2	88.6	95.7	83.2	90.2	92.5	79.6	86.8	97.1	82.6	90.7	93.6	76.6	86.1

INDIANA / IOWA

DIVISION		SOUTH BEND 465-466 MAT.	INST.	TOTAL	TERRE HAUTE 478 MAT.	INST.	TOTAL	WASHINGTON 475 MAT.	INST.	TOTAL	BURLINGTON 526 MAT.	INST.	TOTAL	CARROLL 514 MAT.	INST.	TOTAL	CEDAR RAPIDS 522-524 MAT.	INST.	TOTAL
015433	CONTRACTOR EQUIPMENT		104.8	104.8		116.9	116.9		116.9	116.9		99.8	99.8		99.8	99.8		96.5	96.5
0241, 31 - 34	SITE & INFRASTRUCTURE, DEMOLITION	97.8	95.1	95.9	86.1	125.0	113.0	86.3	122.8	111.5	100.9	96.9	98.2	88.6	97.1	94.4	103.1	96.4	98.5
0310	Concrete Forming & Accessories	98.2	78.8	81.3	95.5	80.5	82.4	96.2	79.3	81.5	99.1	77.1	79.9	88.1	52.7	57.3	104.3	81.6	84.6
0320	Concrete Reinforcing	97.9	77.9	87.8	98.1	82.9	90.5	90.7	51.8	71.1	100.6	82.3	91.4	101.5	81.8	91.5	101.4	84.8	93.1
0330	Cast-in-Place Concrete	95.6	79.5	89.3	90.8	88.9	90.1	98.8	88.2	94.7	111.3	56.9	89.8	111.3	61.8	91.8	111.6	80.4	99.3
03	CONCRETE	89.6	80.2	85.0	103.4	84.1	94.0	108.9	77.3	93.5	104.3	71.9	88.5	103.3	62.7	83.5	104.4	82.4	93.7
04	MASONRY	87.2	80.0	82.8	91.9	82.2	86.0	84.6	83.2	83.7	100.0	68.3	80.7	102.2	74.9	85.6	106.1	80.3	90.3
05	METALS	94.2	99.3	95.9	89.0	88.1	88.7	83.7	70.8	79.5	89.2	91.4	90.0	89.3	91.1	89.9	91.6	93.4	92.2
06	WOOD, PLASTICS & COMPOSITES	104.4	78.5	89.5	100.1	79.4	88.1	100.5	78.9	88.0	97.1	77.9	86.0	85.2	47.7	63.5	102.9	80.6	90.1
07	THERMAL & MOISTURE PROTECTION	92.8	83.8	89.2	97.6	82.7	91.6	97.8	84.5	92.4	106.8	76.9	94.6	107.1	71.5	92.7	107.7	80.3	96.6
08	OPENINGS	91.4	78.1	88.1	98.8	80.8	94.3	95.2	69.1	88.6	91.9	71.2	86.7	97.0	56.1	86.7	98.0	81.7	93.9
0920	Plaster & Gypsum Board	90.8	78.3	81.9	93.0	78.1	82.4	92.5	77.7	81.9	104.2	77.2	84.9	100.1	46.1	61.5	108.9	80.3	88.5
0950, 0980	Ceilings & Acoustic Treatment	85.6	78.3	80.8	88.4	78.1	81.6	82.0	77.7	79.1	109.7	77.2	88.2	109.7	46.1	67.7	113.3	80.3	91.5
0960	Flooring	96.5	83.4	92.7	99.1	87.5	95.7	100.2	80.0	94.3	113.3	39.6	91.7	106.4	34.8	85.4	131.2	82.7	116.8
0970, 0990	Wall Finishes & Painting/Coating	98.3	87.2	91.4	96.0	85.2	89.3	96.0	85.8	89.7	99.0	82.8	89.0	99.0	78.3	86.2	101.0	72.4	83.3
09	FINISHES	93.7	80.5	86.4	95.2	82.1	87.9	94.3	80.7	86.8	107.8	70.8	87.2	103.3	50.3	73.7	114.3	80.6	95.5
COVERS	DIVS. 10 - 14, 25, 28, 41, 43, 44, 46	100.0	71.1	94.2	100.0	94.6	98.9	100.0	92.0	98.4	100.0	87.3	97.5	100.0	69.8	94.0	100.0	88.3	97.7
21, 22, 23	FIRE SUPPRESSION, PLUMBING & HVAC	99.9	77.4	90.8	99.9	79.9	91.8	94.0	80.3	88.4	94.3	77.9	87.7	94.3	71.1	84.9	100.3	81.9	92.8
26, 27, 3370	ELECTRICAL, COMMUNICATIONS & UTIL.	99.7	84.2	91.7	94.0	85.7	89.7	94.5	85.6	89.9	100.8	71.9	85.9	101.5	82.4	91.7	98.4	81.0	89.4
MF2010	WEIGHTED AVERAGE	95.7	82.9	90.1	96.7	86.9	92.4	94.4	83.7	89.7	97.6	77.3	88.6	97.4	72.4	86.4	100.7	84.0	93.3

IOWA

DIVISION		COUNCIL BLUFFS 515			CRESTON 508			DAVENPORT 527 - 528			DECORAH 521			DES MOINES 500 - 503,509			DUBUQUE 520		
		MAT.	INST.	TOTAL	MAT.	INST.	TOTAL	MAT.	INST.	TOTAL	MAT.	INST.	TOTAL	MAT.	INST.	TOTAL	MAT.	INST.	TOTAL
015433	CONTRACTOR EQUIPMENT		96.0	96.0		99.8	99.8		99.8	99.8		99.8	99.8		101.4	101.4		95.3	95.3
0241, 31 - 34	SITE & INFRASTRUCTURE, DEMOLITION	111.3	92.7	98.5	88.5	96.6	94.1	100.8	99.4	99.8	99.4	96.5	97.4	89.8	100.2	96.9	100.9	93.6	95.9
0310	Concrete Forming & Accessories	87.6	72.0	74.0	88.6	68.7	71.3	103.7	88.0	90.1	96.8	46.5	53.0	105.2	78.3	81.8	88.8	70.9	73.2
0320	Concrete Reinforcing	103.5	77.8	90.5	100.7	81.8	91.2	101.4	95.8	98.6	100.6	81.3	90.9	101.4	82.0	91.7	100.0	84.7	92.3
0330	Cast-in-Place Concrete	116.0	76.6	100.5	110.9	65.8	93.2	107.4	86.9	99.3	108.3	58.5	88.7	93.5	90.5	92.3	109.2	87.7	100.8
03	CONCRETE	107.0	75.5	91.7	103.0	71.1	87.5	102.3	89.7	96.2	102.2	58.6	81.0	94.4	84.0	89.3	101.3	80.2	91.0
04	MASONRY	107.6	75.8	88.2	102.2	82.0	89.8	102.7	85.8	92.4	121.3	71.1	90.7	96.6	83.1	88.3	107.0	76.6	88.5
05	METALS	96.9	89.9	94.6	89.2	91.4	89.9	91.6	101.5	94.9	89.4	89.3	89.3	91.7	93.6	92.4	90.3	92.9	91.1
06	WOOD, PLASTICS & COMPOSITES	84.2	71.4	76.8	85.7	65.5	74.0	102.9	87.0	93.8	94.3	39.3	62.6	102.7	76.1	87.4	85.7	67.2	75.0
07	THERMAL & MOISTURE PROTECTION	107.2	69.8	92.0	107.9	78.9	96.1	107.2	85.9	98.6	107.0	54.9	85.8	108.6	80.1	97.0	107.5	72.1	93.1
08	OPENINGS	97.0	76.6	91.9	109.5	66.2	98.6	98.0	89.0	95.7	95.3	51.6	84.3	98.0	82.1	94.0	97.0	76.1	91.7
0920	Plaster & Gypsum Board	100.1	70.8	79.2	100.1	64.5	74.7	108.9	86.7	93.0	102.8	37.5	56.2	97.7	75.4	81.8	100.1	66.4	76.1
0950, 0980	Ceilings & Acoustic Treatment	109.7	70.8	84.0	109.7	64.5	79.8	113.3	86.7	95.7	109.7	37.5	62.0	110.6	75.4	87.4	109.7	66.4	81.1
0960	Flooring	105.4	84.5	99.3	106.6	34.8	85.5	116.0	86.9	107.5	112.8	49.4	94.2	115.6	84.5	106.5	120.5	79.7	108.5
0970, 0990	Wall Finishes & Painting/Coating	96.0	67.5	78.4	99.0	78.3	86.2	99.0	88.6	92.6	99.0	34.7	59.3	99.0	82.5	88.8	100.0	67.2	79.8
09	FINISHES	104.6	73.5	87.3	103.3	62.6	80.6	109.8	87.7	97.5	107.4	44.0	72.0	106.3	79.5	91.3	109.1	71.2	88.0
COVERS	DIVS. 10 - 14, 25, 28, 41, 43, 44, 46	100.0	81.7	96.4	100.0	74.2	94.9	100.0	90.2	98.1	100.0	81.9	96.4	100.0	88.2	97.7	100.0	86.3	97.3
21, 22, 23	FIRE SUPPRESSION, PLUMBING & HVAC	100.3	74.6	89.9	94.3	80.7	88.8	100.3	89.6	96.0	94.3	73.3	85.8	100.2	81.3	92.5	100.3	76.4	90.6
26, 27, 3370	ELECTRICAL, COMMUNICATIONS & UTIL.	103.6	82.0	92.5	96.3	82.4	89.1	95.3	86.7	90.8	98.4	47.1	71.9	107.5	82.5	94.6	102.2	79.1	90.3
MF2010	WEIGHTED AVERAGE	101.7	78.8	91.6	98.1	78.8	89.6	99.6	90.4	95.5	98.5	65.7	84.0	99.1	84.7	92.7	99.9	79.7	91.0

IOWA

DIVISION		FORT DODGE 505			MASON CITY 504			OTTUMWA 525			SHENANDOAH 516			SIBLEY 512			SIOUX CITY 510 - 511		
		MAT.	INST.	TOTAL	MAT.	INST.	TOTAL	MAT.	INST.	TOTAL	MAT.	INST.	TOTAL	MAT.	INST.	TOTAL	MAT.	INST.	TOTAL
015433	CONTRACTOR EQUIPMENT		99.8	99.8		99.8	99.8		95.3	95.3		96.0	96.0		99.8	99.8		99.8	99.8
0241, 31 - 34	SITE & INFRASTRUCTURE, DEMOLITION	96.4	95.4	95.7	96.4	96.5	96.5	101.7	91.7	94.8	109.3	91.8	97.2	112.0	95.3	100.4	113.2	96.1	101.4
0310	Concrete Forming & Accessories	89.2	47.5	52.9	93.5	47.2	53.2	95.0	59.3	64.0	89.1	59.4	63.2	89.4	39.7	46.1	104.3	65.8	70.8
0320	Concrete Reinforcing	100.7	64.4	82.4	100.6	81.3	90.9	100.6	82.6	91.5	103.5	68.7	85.9	103.5	64.3	83.7	101.4	64.7	82.9
0330	Cast-in-Place Concrete	104.1	46.1	81.2	104.1	59.1	86.4	112.0	56.3	90.1	112.2	62.1	92.4	109.9	46.5	84.9	110.6	57.3	89.6
03	CONCRETE	98.4	51.7	75.6	98.6	59.2	79.4	104.0	63.9	84.4	104.3	63.2	84.3	103.2	48.3	76.5	103.8	63.7	84.2
04	MASONRY	100.7	40.0	63.7	114.1	69.9	87.1	103.5	59.4	76.6	107.1	75.8	88.0	126.0	40.3	73.7	99.9	56.3	73.3
05	METALS	89.3	81.0	86.6	89.3	89.6	89.4	89.2	91.1	89.8	95.9	84.7	92.2	89.5	79.8	86.3	91.6	82.5	88.6
06	WOOD, PLASTICS & COMPOSITES	86.1	48.0	64.1	90.1	39.3	60.8	92.0	60.8	74.0	85.7	56.1	68.6	86.4	38.0	58.5	102.9	65.9	81.5
07	THERMAL & MOISTURE PROTECTION	107.3	59.7	87.9	106.8	65.1	89.8	107.6	64.5	90.1	106.5	68.1	90.9	106.8	50.6	84.0	107.2	65.2	90.2
08	OPENINGS	102.2	50.4	89.2	92.8	51.5	82.4	97.0	66.2	89.2	86.8	56.9	79.3	93.1	45.1	81.0	98.0	62.0	88.9
0920	Plaster & Gypsum Board	100.1	46.4	61.7	100.1	37.5	55.4	101.0	59.9	71.6	100.1	55.0	67.9	100.1	36.2	54.4	108.9	64.9	77.4
0950, 0980	Ceilings & Acoustic Treatment	109.7	46.4	67.9	109.7	37.5	62.0	109.7	59.9	76.8	109.7	55.0	73.6	109.7	36.2	61.1	113.3	64.9	81.3
0960	Flooring	108.5	49.4	91.1	111.1	49.4	93.0	124.0	52.5	103.1	106.3	35.9	85.7	107.4	35.6	86.3	116.3	56.2	98.7
0970, 0990	Wall Finishes & Painting/Coating	99.0	62.0	76.2	99.0	31.7	57.5	100.0	81.2	88.4	96.0	67.5	78.4	99.0	62.0	76.2	99.5	62.0	76.4
09	FINISHES	105.5	48.4	73.7	106.3	43.9	71.5	110.4	59.5	82.0	104.7	54.8	76.9	107.4	39.9	69.8	111.9	63.5	84.9
COVERS	DIVS. 10 - 14, 25, 28, 41, 43, 44, 46	100.0	77.8	95.6	100.0	82.3	96.5	100.0	80.1	96.0	100.0	71.0	94.2	100.0	76.8	95.4	100.0	84.2	96.9
21, 22, 23	FIRE SUPPRESSION, PLUMBING & HVAC	94.3	66.8	83.2	94.3	74.4	86.2	94.3	72.6	85.5	94.3	72.1	85.3	94.3	62.7	81.5	100.3	69.3	87.7
26, 27, 3370	ELECTRICAL, COMMUNICATIONS & UTIL.	103.2	44.5	72.9	102.2	59.1	80.0	100.6	67.9	83.7	98.4	82.0	89.9	98.4	44.0	70.3	98.4	74.4	86.0
MF2010	WEIGHTED AVERAGE	97.9	59.7	81.0	97.6	67.8	84.5	98.4	70.7	86.2	98.2	72.3	86.8	98.9	56.5	80.2	100.5	70.7	87.3

IOWA / KANSAS

DIVISION		SPENCER 513			WATERLOO 506 - 507			BELLEVILLE 669			COLBY 677			DODGE CITY 678			EMPORIA 668		
		MAT.	INST.	TOTAL	MAT.	INST.	TOTAL	MAT.	INST.	TOTAL	MAT.	INST.	TOTAL	MAT.	INST.	TOTAL	MAT.	INST.	TOTAL
015433	CONTRACTOR EQUIPMENT		99.8	99.8		99.8	99.8		102.5	102.5		102.5	102.5		102.5	102.5		100.8	100.8
0241, 31 - 34	SITE & INFRASTRUCTURE, DEMOLITION	112.0	95.3	100.5	101.2	96.5	98.0	115.8	93.8	100.6	113.1	93.8	99.8	115.5	93.4	100.3	106.5	91.6	96.2
0310	Concrete Forming & Accessories	94.9	39.7	46.9	104.6	49.7	56.8	94.1	56.6	61.5	100.0	61.1	66.2	93.6	61.0	65.3	86.5	68.5	70.8
0320	Concrete Reinforcing	103.5	64.3	83.7	101.4	84.5	92.9	103.5	55.5	79.3	106.1	55.5	80.6	103.5	55.5	79.4	102.0	56.1	78.9
0330	Cast-in-Place Concrete	109.9	46.5	84.9	111.6	56.5	89.9	118.6	57.9	94.7	114.0	57.9	91.9	116.0	57.6	93.0	114.6	48.5	88.6
03	CONCRETE	103.6	48.3	76.6	104.5	60.0	82.8	116.2	58.2	88.0	113.4	60.2	87.5	114.6	60.0	88.0	109.0	60.4	85.3
04	MASONRY	126.0	40.3	73.7	101.4	75.9	85.8	97.3	58.4	73.6	98.4	58.4	74.0	106.5	60.9	78.7	102.8	68.7	82.0
05	METALS	89.4	79.8	86.3	91.6	92.4	91.9	95.0	78.6	89.5	95.3	78.5	89.8	96.8	77.5	90.4	94.7	79.8	89.8
06	WOOD, PLASTICS & COMPOSITES	91.9	38.0	60.8	103.5	39.3	66.4	93.4	55.4	71.4	100.1	61.5	77.8	92.7	61.5	74.7	85.8	70.8	77.2
07	THERMAL & MOISTURE PROTECTION	107.7	50.6	84.5	107.0	72.1	92.8	99.5	59.3	83.1	99.4	59.9	83.4	99.4	60.9	83.8	98.0	74.3	88.4
08	OPENINGS	106.2	45.1	90.8	93.8	54.1	83.8	95.8	50.6	84.5	95.9	53.9	85.4	95.9	53.9	85.3	93.4	59.0	84.8
0920	Plaster & Gypsum Board	101.0	36.2	54.7	108.9	37.5	57.9	101.7	53.9	67.6	106.3	60.2	73.4	100.8	60.2	71.8	98.5	69.8	78.0
0950, 0980	Ceilings & Acoustic Treatment	109.7	36.2	61.1	113.3	37.5	63.2	94.9	53.9	67.8	94.9	60.2	72.0	94.9	60.2	72.0	94.9	69.8	78.4
0960	Flooring	110.6	35.6	88.6	117.7	79.0	106.4	103.6	41.0	85.2	107.3	41.0	87.8	103.4	41.0	85.1	98.6	38.1	80.8
0970, 0990	Wall Finishes & Painting/Coating	99.0	62.0	76.2	99.5	82.0	88.7	96.0	41.0	62.1	96.0	41.0	62.1	96.0	41.0	62.1	96.0	41.0	62.1
09	FINISHES	108.5	39.9	70.2	110.4	56.3	80.2	101.5	51.9	73.8	102.9	55.5	76.5	101.3	55.5	75.7	98.5	60.4	77.3
COVERS	DIVS. 10 - 14, 25, 28, 41, 43, 44, 46	100.0	76.8	95.4	100.0	83.6	96.7	100.0	43.6	88.8	100.0	44.2	88.9	100.0	44.3	88.9	100.0	43.3	88.7
21, 22, 23	FIRE SUPPRESSION, PLUMBING & HVAC	94.3	62.7	81.5	100.3	81.2	92.6	94.0	70.8	84.6	94.0	67.4	83.2	99.9	67.4	86.8	94.0	71.0	84.7
26, 27, 3370	ELECTRICAL, COMMUNICATIONS & UTIL.	100.1	44.0	71.2	98.4	59.1	78.1	99.9	67.3	83.1	99.9	67.3	83.1	96.7	73.7	84.8	97.1	73.6	84.9
MF2010	WEIGHTED AVERAGE	100.6	56.5	81.2	99.7	72.2	87.6	99.6	65.5	84.6	99.5	65.8	84.6	101.2	66.8	86.0	97.9	69.7	85.4

725

City Cost Indexes

KANSAS

DIVISION		FORT SCOTT 667 MAT.	INST.	TOTAL	HAYS 676 MAT.	INST.	TOTAL	HUTCHINSON 675 MAT.	INST.	TOTAL	INDEPENDENCE 673 MAT.	INST.	TOTAL	KANSAS CITY 660 - 662 MAT.	INST.	TOTAL	LIBERAL 679 MAT.	INST.	TOTAL
015433	CONTRACTOR EQUIPMENT		101.7	101.7		102.5	102.5		102.5	102.5		102.5	102.5		99.2	99.2		102.5	102.5
0241, 31 - 34	SITE & INFRASTRUCTURE, DEMOLITION	102.8	92.0	95.4	119.6	93.8	101.8	95.2	93.9	94.3	116.8	94.0	101.0	95.0	91.8	92.8	119.8	93.9	102.0
0310	Concrete Forming & Accessories	100.7	80.7	83.3	97.7	61.1	65.9	88.8	61.0	64.6	107.8	69.5	74.5	97.6	97.1	97.2	94.1	61.0	65.3
0320	Concrete Reinforcing	101.4	89.1	95.2	103.5	55.5	79.3	103.5	55.6	79.3	102.9	60.5	81.5	98.3	94.4	96.3	105.0	55.6	80.1
0330	Cast-in-Place Concrete	106.3	54.6	86.7	89.9	57.9	77.3	83.3	47.9	69.4	116.5	48.2	89.6	91.1	96.1	93.0	89.9	47.9	73.4
03	CONCRETE	103.9	74.6	89.7	104.1	60.2	82.7	87.8	56.7	72.6	115.8	61.4	89.3	95.6	96.5	96.0	106.1	56.7	82.0
04	MASONRY	104.0	59.4	76.8	106.7	58.4	77.3	97.7	60.6	75.1	95.7	63.2	75.9	106.0	97.5	100.8	104.3	60.6	77.6
05	METALS	94.6	91.1	93.5	94.9	78.5	89.5	94.7	77.4	89.0	94.6	79.9	89.8	102.1	98.7	101.0	95.2	77.4	89.3
06	WOOD, PLASTICS & COMPOSITES	102.0	88.8	94.4	97.5	61.5	76.7	88.1	61.5	72.7	109.2	72.3	87.9	98.3	97.7	97.9	93.4	61.5	75.0
07	THERMAL & MOISTURE PROTECTION	98.8	71.0	87.5	99.7	59.9	83.6	98.4	60.8	83.1	99.5	72.9	88.7	98.2	97.3	97.8	99.9	60.8	84.0
08	OPENINGS	93.4	80.6	90.2	95.8	53.9	85.3	95.8	53.9	85.3	93.4	61.0	85.3	94.8	90.8	93.8	95.5	53.9	85.3
0920	Plaster & Gypsum Board	104.5	88.4	93.0	104.5	60.2	72.9	99.9	60.2	71.6	112.2	71.4	83.0	96.2	97.5	97.1	101.7	60.2	72.1
0950, 0980	Ceilings & Acoustic Treatment	94.9	88.4	90.6	94.9	60.2	72.0	94.9	60.2	72.0	94.9	71.4	79.4	94.9	97.5	96.6	94.9	60.2	72.0
0960	Flooring	115.0	41.3	93.4	106.1	41.0	87.0	100.4	41.0	83.0	112.0	41.0	91.2	91.6	98.3	93.6	103.6	41.0	85.2
0970, 0990	Wall Finishes & Painting/Coating	98.0	41.0	62.8	96.0	41.0	62.1	96.0	41.0	62.1	96.0	41.0	62.1	104.0	71.4	83.9	96.0	41.0	62.1
09	FINISHES	104.2	70.7	85.5	103.0	55.5	76.5	98.4	55.5	74.5	105.2	61.9	81.0	98.4	93.7	95.8	102.3	55.5	76.2
COVERS	DIVS. 10 - 14, 25, 28, 41, 43, 44, 46	100.0	49.7	90.0	100.0	44.2	88.9	100.0	44.3	88.9	100.0	45.5	89.2	100.0	66.4	93.3	100.0	44.3	88.9
21, 22, 23	FIRE SUPPRESSION, PLUMBING & HVAC	94.0	69.1	83.9	94.0	67.4	83.2	94.0	67.4	83.2	94.0	69.4	84.0	99.9	99.0	99.5	94.0	67.4	83.2
26, 27, 3370	ELECTRICAL, COMMUNICATIONS & UTIL.	96.3	73.6	84.6	98.8	67.3	82.5	93.7	67.3	80.1	95.9	73.6	84.4	102.3	95.9	99.0	96.7	69.6	82.7
MF2010	WEIGHTED AVERAGE	97.7	73.9	87.2	98.8	65.8	84.3	94.8	65.4	81.8	99.2	69.5	86.1	99.5	95.4	97.7	98.7	65.8	84.2

KANSAS / KENTUCKY

DIVISION		SALINA 674 MAT.	INST.	TOTAL	TOPEKA 664 - 666 MAT.	INST.	TOTAL	WICHITA 670 - 672 MAT.	INST.	TOTAL	ASHLAND 411 - 412 MAT.	INST.	TOTAL	BOWLING GREEN 421 - 422 MAT.	INST.	TOTAL	CAMPTON 413 - 414 MAT.	INST.	TOTAL
015433	CONTRACTOR EQUIPMENT		102.5	102.5		100.8	100.8		102.5	102.5		98.5	98.5		96.0	96.0		102.2	102.2
0241, 31 - 34	SITE & INFRASTRUCTURE, DEMOLITION	105.0	93.8	97.3	97.0	90.7	92.6	96.9	92.6	93.9	106.0	87.7	93.4	71.9	98.8	90.5	80.4	100.0	93.9
0310	Concrete Forming & Accessories	90.7	51.0	61.4	97.4	42.8	49.9	95.6	46.4	53.9	89.5	100.8	99.4	85.3	80.1	80.8	90.0	87.7	88.0
0320	Concrete Reinforcing	102.9	64.3	83.5	95.6	96.6	96.1	95.6	75.8	85.6	91.7	106.6	99.2	88.9	89.0	88.9	89.8	105.6	97.8
0330	Cast-in-Place Concrete	100.7	51.5	81.3	90.6	47.7	73.7	88.6	50.6	73.6	86.4	101.2	92.2	85.7	99.8	91.2	95.6	63.7	83.0
03	CONCRETE	101.7	57.8	80.3	92.9	56.2	75.0	91.8	54.8	73.8	94.5	102.6	98.4	92.7	88.7	90.8	96.8	82.8	90.0
04	MASONRY	121.5	58.4	83.0	100.2	56.8	73.7	93.8	53.1	69.0	90.6	104.8	99.3	92.4	73.6	80.9	90.3	73.0	79.7
05	METALS	96.6	82.2	91.8	102.0	96.1	100.0	102.0	84.1	96.1	90.9	110.1	97.3	92.5	86.6	90.6	91.8	94.2	92.6
06	WOOD, PLASTICS & COMPOSITES	89.5	55.4	69.8	94.5	39.6	62.8	96.0	45.2	66.6	80.0	99.3	91.2	93.9	78.7	85.1	92.6	94.7	93.8
07	THERMAL & MOISTURE PROTECTION	99.0	59.3	82.8	100.0	63.3	85.0	98.3	54.6	80.6	89.9	98.9	93.6	85.5	81.3	83.8	97.9	72.3	87.5
08	OPENINGS	95.8	51.5	84.6	96.1	56.5	86.1	96.1	53.0	85.2	96.9	96.7	96.9	97.8	77.0	92.5	99.5	92.4	97.7
0920	Plaster & Gypsum Board	99.9	53.9	67.1	98.6	37.6	55.0	98.3	43.3	59.0	59.3	99.3	87.9	87.1	78.4	80.8	87.1	93.9	91.9
0950, 0980	Ceilings & Acoustic Treatment	94.9	53.9	67.8	95.8	37.6	57.4	94.9	43.3	60.9	83.3	99.3	93.9	88.4	78.4	81.7	88.4	93.9	92.0
0960	Flooring	101.7	64.9	90.9	105.0	77.2	96.8	103.8	66.0	92.7	78.5	104.9	86.3	95.1	90.1	93.6	96.5	43.2	80.9
0970, 0990	Wall Finishes & Painting/Coating	96.0	41.0	62.1	96.0	55.4	70.9	96.0	50.6	67.9	98.3	96.5	96.6	96.0	69.9	79.9	96.0	69.0	79.3
09	FINISHES	99.8	56.5	75.6	100.0	48.6	71.4	99.6	49.9	71.9	82.5	101.1	92.9	93.4	80.8	86.4	94.1	78.4	85.4
COVERS	DIVS. 10 - 14, 25, 28, 41, 43, 44, 46	100.0	54.1	90.9	100.0	51.0	90.2	100.0	50.3	90.1	100.0	99.4	99.9	100.0	61.4	92.3	100.0	60.3	92.1
21, 22, 23	FIRE SUPPRESSION, PLUMBING & HVAC	99.9	67.5	86.8	99.9	57.4	82.7	99.7	63.6	85.1	93.9	91.3	92.9	99.9	83.9	93.5	94.1	80.9	88.7
26, 27, 3370	ELECTRICAL, COMMUNICATIONS & UTIL.	96.4	73.7	84.7	102.0	73.6	87.4	99.9	73.7	86.4	92.2	97.3	94.8	94.8	83.7	89.1	92.2	70.3	80.9
MF2010	WEIGHTED AVERAGE	100.0	66.9	85.4	99.2	64.5	83.9	98.4	64.1	83.3	93.3	98.5	95.6	95.0	83.5	90.0	94.5	81.2	88.6

KENTUCKY

DIVISION		CORBIN 407 - 409 MAT.	INST.	TOTAL	COVINGTON 410 MAT.	INST.	TOTAL	ELIZABETHTOWN 427 MAT.	INST.	TOTAL	FRANKFORT 406 MAT.	INST.	TOTAL	HAZARD 417 - 418 MAT.	INST.	TOTAL	HENDERSON 424 MAT.	INST.	TOTAL
015433	CONTRACTOR EQUIPMENT		102.2	102.2		105.8	105.8		96.0	96.0		102.2	102.2		102.2	102.2		116.9	116.9
0241, 31 - 34	SITE & INFRASTRUCTURE, DEMOLITION	77.8	99.7	92.9	75.9	113.8	102.1	67.0	98.7	88.8	82.9	100.7	95.2	78.3	101.4	94.2	75.4	125.3	109.8
0310	Concrete Forming & Accessories	87.3	69.2	71.5	86.1	96.4	95.1	81.0	79.1	79.3	97.1	75.2	78.1	87.3	88.4	88.3	92.9	83.1	84.4
0320	Concrete Reinforcing	89.4	68.8	79.0	88.4	95.8	92.2	89.4	92.5	91.0	90.7	92.5	91.6	90.2	105.6	98.0	89.1	90.0	89.5
0330	Cast-in-Place Concrete	90.5	61.2	78.9	91.5	94.0	92.5	77.5	76.0	76.9	87.5	71.7	81.3	91.9	87.2	90.1	75.7	89.4	81.1
03	CONCRETE	92.1	66.9	79.8	95.1	95.5	95.3	85.2	80.8	83.0	91.4	77.6	84.7	93.7	91.2	92.5	89.7	86.7	88.2
04	MASONRY	90.5	65.5	75.2	103.6	107.4	105.9	77.0	79.6	78.6	87.4	81.8	84.0	87.7	75.7	80.4	96.3	92.5	94.0
05	METALS	91.7	78.8	87.4	89.9	93.5	91.1	91.7	88.5	90.6	98.7	89.4	95.6	91.8	94.4	92.7	83.5	89.6	85.5
06	WOOD, PLASTICS & COMPOSITES	89.8	73.1	80.1	90.9	89.5	90.1	89.5	78.7	83.3	101.2	73.1	84.9	89.8	94.7	92.6	96.2	80.6	87.2
07	THERMAL & MOISTURE PROTECTION	97.7	67.5	85.4	97.0	97.7	97.3	85.2	78.2	82.4	97.6	76.9	89.2	97.8	71.0	86.9	97.3	93.0	95.5
08	OPENINGS	98.7	64.1	90.0	101.0	88.3	97.8	97.8	82.7	94.0	97.9	79.5	93.3	99.9	82.6	95.5	95.7	83.0	92.5
0920	Plaster & Gypsum Board	86.2	71.6	75.8	69.3	89.6	83.8	86.2	78.4	80.6	97.9	71.6	79.1	86.2	93.9	91.7	89.3	79.4	82.2
0950, 0980	Ceilings & Acoustic Treatment	88.4	71.6	77.3	92.0	89.6	90.4	88.4	78.4	81.7	91.1	71.6	78.2	88.4	93.9	92.0	82.0	79.4	80.3
0960	Flooring	94.9	43.2	79.7	71.3	88.1	76.2	92.5	90.1	91.8	99.5	57.0	87.0	94.9	44.0	79.9	98.3	90.1	95.9
0970, 0990	Wall Finishes & Painting/Coating	96.0	52.1	68.9	92.5	85.5	88.2	96.0	78.3	85.1	96.0	84.1	88.7	96.0	69.0	79.3	96.0	93.8	94.6
09	FINISHES	93.2	62.8	76.3	85.2	93.5	89.9	92.2	80.3	85.6	96.9	72.3	83.2	93.3	79.3	85.5	92.6	85.4	88.6
COVERS	DIVS. 10 - 14, 25, 28, 41, 43, 44, 46	100.0	51.5	90.3	100.0	104.5	100.9	100.0	90.4	98.1	100.0	70.1	94.1	100.0	61.2	92.3	100.0	69.6	93.9
21, 22, 23	FIRE SUPPRESSION, PLUMBING & HVAC	94.0	69.6	84.1	94.5	93.1	93.9	94.4	82.4	89.5	100.0	83.6	93.4	94.1	81.8	89.1	94.4	78.2	87.8
26, 27, 3370	ELECTRICAL, COMMUNICATIONS & UTIL.	92.4	83.7	87.9	96.0	80.3	87.9	92.0	83.7	87.7	100.0	92.6	96.2	92.1	62.0	76.6	94.2	83.9	88.9
MF2010	WEIGHTED AVERAGE	93.7	72.5	84.4	94.4	95.2	94.8	91.4	83.9	88.1	97.2	83.5	91.2	93.9	81.5	88.4	92.4	87.9	90.4

KENTUCKY

DIVISION		LEXINGTON 403-405			LOUISVILLE 400-402			OWENSBORO 423			PADUCAH 420			PIKEVILLE 415-416			SOMERSET 425-426		
		MAT.	INST.	TOTAL	MAT.	INST.	TOTAL	MAT.	INST.	TOTAL	MAT.	INST.	TOTAL	MAT.	INST.	TOTAL	MAT.	INST.	TOTAL
015433	CONTRACTOR EQUIPMENT		102.2	102.2		96.0	96.0		116.9	116.9		116.9	116.9		98.5	98.5		102.2	102.2
0241, 31-34	SITE & INFRASTRUCTURE, DEMOLITION	79.5	101.7	94.8	69.8	98.7	89.7	84.6	125.2	112.6	78.1	124.8	110.3	116.7	87.0	96.2	71.7	100.3	91.4
0310	Concrete Forming & Accessories	98.1	76.1	79.0	93.5	79.1	80.9	91.5	82.6	83.8	89.6	79.3	80.6	97.3	94.3	94.7	87.9	73.9	75.7
0320	Concrete Reinforcing	98.1	92.0	95.0	98.1	92.5	95.3	89.1	89.7	89.4	89.6	86.1	87.8	92.2	106.5	99.4	89.4	91.8	90.6
0330	Cast-in-Place Concrete	92.5	86.7	90.3	94.5	75.9	87.2	88.3	98.0	92.1	80.7	87.5	83.4	94.9	98.7	96.4	75.7	102.8	86.4
03	CONCRETE	95.2	83.1	89.3	95.9	80.7	88.5	101.3	89.4	95.5	94.0	83.6	89.0	107.4	98.7	103.2	80.6	87.5	84.0
04	MASONRY	88.7	70.1	77.4	89.9	79.4	83.5	88.8	83.7	85.7	91.8	91.6	91.7	88.1	101.0	96.0	83.0	66.6	73.0
05	METALS	94.1	89.2	92.5	100.9	88.6	96.9	85.0	89.3	86.4	82.0	87.6	83.8	90.9	108.9	96.8	91.7	88.0	90.5
06	WOOD, PLASTICS & COMPOSITES	103.7	73.1	86.0	101.2	78.7	88.2	94.4	80.6	86.4	92.1	77.7	83.8	88.4	94.0	91.6	90.3	73.1	80.3
07	THERMAL & MOISTURE PROTECTION	97.9	86.3	93.2	92.0	78.2	86.4	97.6	88.8	94.0	97.3	81.8	91.0	90.5	84.0	87.9	97.3	67.5	85.2
08	OPENINGS	98.8	77.0	93.3	98.8	82.7	94.8	95.7	83.8	92.7	94.8	77.4	90.4	97.8	90.9	96.0	98.7	77.3	93.3
0920	Plaster & Gypsum Board	95.9	71.6	78.5	89.0	78.4	81.4	87.9	79.4	81.8	86.5	76.4	79.3	62.0	93.9	84.8	86.2	71.6	75.8
0950, 0980	Ceilings & Acoustic Treatment	93.8	71.6	79.1	91.1	78.4	82.7	82.0	79.4	80.3	82.0	76.4	78.3	83.3	93.9	90.3	88.4	71.6	77.3
0960	Flooring	99.9	73.6	92.2	98.3	90.1	95.9	97.7	90.1	95.4	96.3	62.2	86.3	82.9	104.9	89.3	95.1	43.2	79.9
0970, 0990	Wall Finishes & Painting/Coating	96.0	84.1	88.7	96.0	78.3	85.1	96.0	97.0	96.6	96.0	81.0	86.7	98.3	85.6	90.5	96.0	78.3	85.1
09	FINISHES	97.1	75.7	85.2	94.7	80.9	87.0	92.9	85.2	88.6	91.9	76.0	83.1	85.0	95.9	91.1	92.6	68.3	79.1
COVERS	DIVS. 10-14, 25, 28, 41, 43, 44, 46	100.0	92.6	98.5	100.0	90.3	98.1	100.0	98.7	99.7	100.0	65.5	93.1	100.0	61.0	92.2	100.0	61.8	92.4
21, 22, 23	FIRE SUPPRESSION, PLUMBING & HVAC	99.9	76.7	90.5	100.0	82.3	92.8	99.9	77.2	90.7	94.4	82.5	89.6	93.9	88.8	91.8	94.4	75.1	86.6
26, 27, 3370	ELECTRICAL, COMMUNICATIONS & UTIL.	95.3	83.7	89.3	102.7	80.8	91.4	94.3	83.9	88.9	96.6	80.5	88.3	95.2	78.0	86.3	92.6	83.7	88.0
MF2010	WEIGHTED AVERAGE	96.5	81.8	90.0	97.8	83.5	91.5	95.2	87.9	92.0	92.6	85.7	89.6	95.6	91.7	93.9	91.9	79.1	86.2

LOUISIANA

DIVISION		ALEXANDRIA 713-714			BATON ROUGE 707-708			HAMMOND 704			LAFAYETTE 705			LAKE CHARLES 706			MONROE 712		
		MAT.	INST.	TOTAL	MAT.	INST.	TOTAL	MAT.	INST.	TOTAL	MAT.	INST.	TOTAL	MAT.	INST.	TOTAL	MAT.	INST.	TOTAL
015433	CONTRACTOR EQUIPMENT		88.9	88.9		88.7	88.7		89.4	89.4		89.4	89.4		88.7	88.7		88.9	88.9
0241, 31-34	SITE & INFRASTRUCTURE, DEMOLITION	96.1	85.7	88.9	104.0	85.2	91.1	105.3	86.5	92.3	106.0	86.4	92.5	106.7	85.0	91.8	96.1	85.0	88.5
0310	Concrete Forming & Accessories	82.5	36.4	42.4	93.8	62.0	66.1	79.3	48.7	52.7	95.1	53.7	59.1	96.6	57.3	62.4	82.1	36.7	42.6
0320	Concrete Reinforcing	98.0	57.5	77.6	99.0	59.3	79.0	97.6	61.1	79.2	99.0	59.3	79.0	99.0	58.2	78.5	96.9	57.3	76.9
0330	Cast-in-Place Concrete	95.8	39.4	73.6	94.8	57.7	80.2	97.6	45.1	76.9	97.1	45.7	76.9	102.2	66.5	88.2	95.8	55.8	80.1
03	CONCRETE	95.1	42.8	69.6	95.1	60.5	78.2	95.4	50.7	73.6	96.3	52.8	75.1	99.0	61.3	80.6	94.9	48.4	72.3
04	MASONRY	118.1	49.3	76.2	106.1	51.1	72.5	105.6	52.8	73.4	105.6	50.3	71.8	104.6	56.7	75.4	113.4	45.8	72.1
05	METALS	89.5	71.8	83.7	101.3	70.2	91.1	97.0	70.9	88.4	96.1	70.4	87.6	96.1	70.1	87.5	89.5	69.5	82.9
06	WOOD, PLASTICS & COMPOSITES	86.7	35.1	56.9	100.8	67.8	81.8	86.7	49.9	65.5	107.0	57.0	78.1	105.2	61.1	79.7	86.0	35.5	56.8
07	THERMAL & MOISTURE PROTECTION	97.5	50.9	78.6	95.5	57.9	80.2	97.2	58.8	81.6	97.7	56.4	80.9	96.9	58.9	81.5	97.5	50.1	78.2
08	OPENINGS	98.2	41.5	83.9	101.1	59.6	90.7	96.8	54.3	86.1	101.1	53.8	89.2	101.1	56.8	90.0	98.2	45.4	84.9
0920	Plaster & Gypsum Board	76.2	33.6	45.8	95.1	67.2	75.2	89.4	48.7	60.3	98.8	56.0	68.3	98.8	60.2	71.3	75.7	34.0	45.9
0950, 0980	Ceilings & Acoustic Treatment	91.1	33.6	53.1	102.2	67.2	79.1	99.3	48.7	65.9	97.5	56.0	70.1	98.5	60.2	73.2	91.1	34.0	53.3
0960	Flooring	101.3	66.9	91.2	102.1	66.9	91.8	93.4	66.9	85.6	102.3	66.9	91.9	102.3	76.4	94.7	100.8	59.8	88.8
0970, 0990	Wall Finishes & Painting/Coating	96.2	33.6	57.6	99.9	38.5	62.0	99.9	50.9	69.7	99.9	38.5	62.0	99.9	35.4	60.1	96.2	40.1	61.6
09	FINISHES	91.6	40.7	63.2	100.4	61.1	78.5	96.5	52.3	71.8	99.8	54.5	74.5	100.0	58.7	77.0	91.4	40.3	62.9
COVERS	DIVS. 10-14, 25, 28, 41, 43, 44, 46	100.0	51.0	90.2	100.0	79.6	95.9	100.0	49.0	89.8	100.0	78.1	95.6	100.0	78.9	95.8	100.0	46.8	89.4
21, 22, 23	FIRE SUPPRESSION, PLUMBING & HVAC	100.3	51.2	80.4	100.1	59.1	83.5	94.2	45.9	74.6	100.2	59.4	83.6	100.2	59.8	83.8	100.3	50.1	80.0
26, 27, 3370	ELECTRICAL, COMMUNICATIONS & UTIL.	93.9	56.5	74.6	110.9	61.3	85.3	102.0	56.6	78.6	103.3	67.0	84.5	102.7	67.0	84.3	95.8	57.7	76.1
MF2010	WEIGHTED AVERAGE	97.1	53.5	77.9	101.2	63.0	84.4	97.5	56.1	79.3	99.9	61.5	83.0	100.1	64.1	84.2	97.0	53.6	77.9

LOUISIANA / MAINE

DIVISION		NEW ORLEANS 700-701			SHREVEPORT 710-711			THIBODAUX 703			AUGUSTA 043			BANGOR 044			BATH 045		
		MAT.	INST.	TOTAL	MAT.	INST.	TOTAL	MAT.	INST.	TOTAL	MAT.	INST.	TOTAL	MAT.	INST.	TOTAL	MAT.	INST.	TOTAL
015433	CONTRACTOR EQUIPMENT		89.6	89.6		88.9	88.9		89.4	89.4		100.4	100.4		100.4	100.4		100.4	100.4
0241, 31-34	SITE & INFRASTRUCTURE, DEMOLITION	109.6	88.0	94.7	98.6	84.9	89.1	107.9	86.4	93.1	87.9	100.2	96.4	90.9	97.9	95.7	89.1	100.2	96.7
0310	Concrete Forming & Accessories	98.5	68.9	72.7	96.1	38.8	46.2	91.7	61.6	65.5	98.1	100.3	100.0	93.7	91.0	91.3	90.1	100.3	98.9
0320	Concrete Reinforcing	99.0	61.8	80.3	96.6	57.7	77.0	97.6	61.0	79.1	84.6	101.0	92.9	84.3	101.1	92.8	83.4	100.9	92.2
0330	Cast-in-Place Concrete	101.2	69.9	88.9	95.3	44.9	75.5	104.9	48.0	82.5	95.3	61.9	82.2	83.5	104.4	91.8	83.5	61.9	75.0
03	CONCRETE	98.6	68.4	83.9	94.9	45.7	70.9	100.8	57.4	79.6	105.7	86.6	96.4	99.1	96.8	98.0	99.1	86.5	93.0
04	MASONRY	105.9	62.0	79.1	103.4	47.4	69.2	133.9	48.5	81.8	95.4	60.0	73.8	109.0	59.7	78.9	115.5	56.9	79.7
05	METALS	109.1	74.3	97.6	86.3	69.6	80.7	97.0	70.6	88.3	86.7	85.0	86.1	86.3	80.8	84.5	84.7	84.7	84.7
06	WOOD, PLASTICS & COMPOSITES	101.7	70.5	83.7	101.8	38.5	65.2	94.6	67.0	78.7	97.2	112.0	105.8	92.2	98.8	96.0	87.9	112.0	101.8
07	THERMAL & MOISTURE PROTECTION	97.4	70.4	86.5	96.6	49.1	77.3	97.5	54.5	80.0	100.9	66.1	86.8	101.5	65.7	86.9	101.5	65.3	86.8
08	OPENINGS	101.9	68.4	93.5	96.0	43.5	82.8	102.1	63.6	92.4	99.3	90.9	97.2	99.2	83.5	95.3	99.2	90.7	97.1
0920	Plaster & Gypsum Board	97.0	70.0	77.7	82.8	37.1	50.1	92.6	66.4	73.9	95.2	111.7	107.0	92.4	98.1	96.5	90.2	111.7	105.6
0950, 0980	Ceilings & Acoustic Treatment	97.5	70.0	79.3	92.0	37.1	55.7	99.3	66.4	77.6	94.8	111.7	106.0	94.8	98.1	97.0	92.7	111.7	105.3
0960	Flooring	102.7	66.9	92.2	109.5	64.2	96.2	99.6	46.2	83.9	94.5	57.9	83.7	92.3	64.4	84.1	90.5	57.9	80.9
0970, 0990	Wall Finishes & Painting/Coating	101.3	67.6	80.5	96.2	33.6	57.6	101.3	50.9	70.2	90.1	36.4	57.0	90.1	37.8	57.8	90.1	36.4	57.0
09	FINISHES	100.3	68.6	82.6	95.1	42.1	65.6	99.0	58.3	76.3	97.4	87.6	91.9	96.9	81.6	88.4	95.6	87.6	91.1
COVERS	DIVS. 10-14, 25, 28, 41, 43, 44, 46	100.0	83.6	96.7	100.0	64.9	93.0	100.0	80.1	96.0	100.0	93.7	98.7	100.0	106.1	101.2	100.0	93.7	98.7
21, 22, 23	FIRE SUPPRESSION, PLUMBING & HVAC	100.1	65.8	86.2	100.2	55.1	81.9	94.2	59.6	80.2	99.9	65.4	85.9	100.0	61.4	84.4	94.1	65.4	82.4
26, 27, 3370	ELECTRICAL, COMMUNICATIONS & UTIL.	103.4	71.7	87.0	100.0	66.6	82.8	100.4	69.3	84.4	98.5	80.1	89.0	100.1	80.1	89.8	98.0	80.1	88.8
MF2010	WEIGHTED AVERAGE	102.3	70.5	88.3	96.7	56.3	78.9	100.2	63.4	84.0	97.6	79.8	89.8	97.7	78.9	89.4	96.0	79.5	88.7

MAINE

DIVISION		HOULTON 047			KITTERY 039			LEWISTON 042			MACHIAS 046			PORTLAND 040 - 041			ROCKLAND 048		
		MAT.	INST.	TOTAL	MAT.	INST.	TOTAL	MAT.	INST.	TOTAL	MAT.	INST.	TOTAL	MAT.	INST.	TOTAL	MAT.	INST.	TOTAL
015433	CONTRACTOR EQUIPMENT		100.4	100.4		100.4	100.4		100.4	100.4		100.4	100.4		100.4	100.4		100.4	100.4
0241, 31 - 34	SITE & INFRASTRUCTURE, DEMOLITION	90.8	100.0	97.2	84.7	100.2	95.4	87.5	97.9	94.7	90.1	100.0	97.0	85.0	97.9	93.9	86.0	100.1	95.7
0310	Concrete Forming & Accessories	96.7	99.4	99.0	90.7	100.7	99.4	98.3	91.0	91.9	94.3	99.5	98.8	97.2	91.0	91.8	95.2	99.7	99.1
0320	Concrete Reinforcing	84.3	101.0	92.7	82.9	101.0	92.0	103.6	101.1	102.3	84.3	101.0	92.7	103.6	101.1	102.3	84.3	101.1	92.8
0330	Cast-in-Place Concrete	83.5	60.1	74.3	83.5	62.8	75.4	85.2	104.4	92.8	83.5	60.2	74.4	105.7	104.4	105.2	85.3	61.3	75.8
03	CONCRETE	100.0	85.5	93.0	94.6	87.0	90.9	99.2	96.8	98.0	99.5	85.6	92.7	109.4	96.8	103.2	96.7	86.1	91.5
04	MASONRY	93.4	58.0	71.8	106.2	59.5	77.7	94.0	59.7	73.1	93.4	58.3	72.0	94.3	59.7	73.2	88.0	59.7	70.8
05	METALS	84.9	84.9	84.9	84.7	84.8	84.7	89.3	80.8	86.5	85.0	84.9	84.9	90.6	80.8	87.4	84.8	84.9	84.9
06	WOOD, PLASTICS & COMPOSITES	96.0	112.0	105.3	88.3	112.0	102.0	97.9	98.8	98.4	93.1	112.0	104.0	97.1	98.8	98.1	93.9	112.0	104.4
07	THERMAL & MOISTURE PROTECTION	101.6	63.6	86.1	101.1	66.3	86.9	101.3	65.7	86.8	101.5	63.6	86.1	104.6	65.7	88.8	101.3	64.5	86.3
08	OPENINGS	99.3	89.0	96.7	99.2	90.2	96.9	102.4	83.5	97.7	99.3	89.0	96.7	102.4	83.5	97.7	99.2	89.0	96.7
0920	Plaster & Gypsum Board	95.2	111.7	107.0	90.2	111.7	105.6	98.7	98.1	98.3	93.3	111.7	106.5	99.5	98.1	98.5	93.3	111.7	106.5
0950, 0980	Ceilings & Acoustic Treatment	92.7	111.7	105.3	92.7	111.7	105.3	102.0	98.1	99.4	92.7	111.7	105.3	99.3	98.1	98.5	92.7	111.7	105.3
0960	Flooring	93.5	54.2	82.0	90.6	60.4	81.7	95.1	64.4	86.1	92.6	54.2	81.3	94.5	64.4	85.6	92.9	54.2	81.5
0970, 0990	Wall Finishes & Painting/Coating	90.1	37.8	57.8	90.1	36.4	57.0	90.1	37.8	57.8	90.1	37.8	57.8	90.1	37.8	57.8	90.1	37.8	57.8
09	FINISHES	97.2	86.6	91.3	95.2	88.3	91.3	99.6	81.6	89.6	96.6	86.6	91.0	98.3	81.6	89.0	96.3	86.8	91.0
COVERS	DIVS. 10 - 14, 25, 28, 41, 43, 44, 46	100.0	92.9	98.6	100.0	95.2	99.1	100.0	106.1	101.2	100.0	93.0	98.6	100.0	106.1	101.2	100.0	93.2	98.7
21, 22, 23	FIRE SUPPRESSION, PLUMBING & HVAC	94.1	64.3	82.0	94.1	66.0	82.7	100.0	61.4	84.4	94.1	64.4	82.0	99.9	61.4	84.3	94.1	64.7	82.2
26, 27, 3370	ELECTRICAL, COMMUNICATIONS & UTIL.	102.3	80.1	90.9	98.0	80.1	88.8	102.3	80.1	90.9	102.3	80.1	90.9	103.9	80.1	91.6	102.2	80.1	90.8
MF2010	WEIGHTED AVERAGE	95.7	78.9	88.3	94.8	80.1	88.3	98.1	78.9	89.7	95.5	79.0	88.2	99.5	78.9	90.4	94.7	79.4	88.0

		MAINE			MARYLAND														
DIVISION		WATERVILLE 049			ANNAPOLIS 214			BALTIMORE 210 - 212			COLLEGE PARK 207 - 208			CUMBERLAND 215			EASTON 216		
		MAT.	INST.	TOTAL	MAT.	INST.	TOTAL	MAT.	INST.	TOTAL	MAT.	INST.	TOTAL	MAT.	INST.	TOTAL	MAT.	INST.	TOTAL
015433	CONTRACTOR EQUIPMENT		100.4	100.4		98.6	98.6		102.0	102.0		103.1	103.1		98.6	98.6		98.6	98.6
0241, 31 - 34	SITE & INFRASTRUCTURE, DEMOLITION	91.0	100.2	97.3	99.7	90.5	93.4	100.2	94.4	96.2	106.9	92.8	97.2	91.2	90.2	90.5	98.2	87.6	90.9
0310	Concrete Forming & Accessories	89.7	100.3	98.9	94.1	71.0	74.0	96.9	75.3	78.1	85.5	71.8	73.6	88.5	77.8	79.2	86.7	72.4	74.3
0320	Concrete Reinforcing	84.3	101.0	92.7	99.8	82.5	91.1	99.8	82.5	91.1	98.5	80.9	89.6	88.2	74.7	81.4	87.4	80.7	84.0
0330	Cast-in-Place Concrete	83.5	61.9	75.0	140.3	79.5	116.4	124.9	80.3	107.3	131.0	80.5	111.1	111.5	84.1	100.7	123.8	50.6	95.0
03	CONCRETE	100.8	86.6	93.8	120.9	77.2	99.6	113.4	79.4	96.8	115.7	77.8	97.2	101.2	80.3	91.0	110.1	67.4	89.3
04	MASONRY	103.1	60.0	76.8	101.1	75.8	85.7	102.2	75.8	86.1	115.1	70.5	87.9	100.2	78.5	86.9	114.3	45.4	72.3
05	METALS	84.9	85.0	84.9	94.9	94.7	94.9	96.1	95.3	95.8	83.9	96.9	88.2	92.5	91.3	92.1	92.8	88.8	91.4
06	WOOD, PLASTICS & COMPOSITES	87.3	112.0	101.6	90.0	70.8	78.9	92.8	76.5	83.4	87.6	71.1	78.1	83.3	76.3	79.2	81.3	79.5	80.2
07	THERMAL & MOISTURE PROTECTION	101.6	66.1	87.1	97.1	79.7	90.0	97.4	80.5	90.5	100.5	77.8	91.3	96.8	77.3	88.9	96.9	61.6	82.5
08	OPENINGS	99.3	90.9	97.2	93.9	78.5	90.0	96.1	81.6	92.4	97.2	75.0	91.6	95.1	77.5	90.7	93.1	73.4	88.2
0920	Plaster & Gypsum Board	90.2	111.7	105.6	98.3	70.3	78.3	99.9	76.0	82.9	99.7	70.3	78.7	94.5	75.9	81.2	94.0	79.2	83.4
0950, 0980	Ceilings & Acoustic Treatment	92.7	111.7	105.3	98.9	70.3	80.0	99.8	76.0	84.1	95.8	70.3	78.9	99.8	75.9	84.0	99.8	79.2	86.2
0960	Flooring	90.2	57.9	80.7	93.8	80.9	90.0	93.8	80.9	90.0	107.1	80.9	99.4	89.8	82.2	87.6	89.1	54.1	78.8
0970, 0990	Wall Finishes & Painting/Coating	90.1	36.4	57.0	96.8	83.5	88.6	96.8	83.5	88.6	121.0	78.7	94.9	96.8	77.3	84.8	96.8	78.7	85.6
09	FINISHES	95.8	87.6	91.2	96.0	73.2	83.3	96.6	76.6	85.4	101.4	73.5	85.8	94.1	78.3	85.3	94.3	71.3	81.4
COVERS	DIVS. 10 - 14, 25, 28, 41, 43, 44, 46	100.0	93.7	98.7	100.0	86.1	97.2	100.0	87.1	97.4	100.0	82.2	96.5	100.0	89.8	98.0	100.0	81.1	96.2
21, 22, 23	FIRE SUPPRESSION, PLUMBING & HVAC	94.1	65.4	82.4	99.9	83.7	93.3	99.9	83.8	93.4	94.3	86.1	91.0	93.9	72.0	85.0	93.9	69.5	84.0
26, 27, 3370	ELECTRICAL, COMMUNICATIONS & UTIL.	102.3	80.1	90.9	96.2	93.2	94.7	99.4	93.8	96.5	98.3	98.9	98.6	97.1	82.0	89.3	96.6	66.1	80.9
MF2010	WEIGHTED AVERAGE	96.1	79.8	88.9	100.2	83.2	92.7	100.2	84.7	93.3	98.5	84.2	92.2	95.8	80.2	89.0	97.5	70.3	85.5

		MARYLAND															MASSACHUSETTS		
DIVISION		ELKTON 219			HAGERSTOWN 217			SALISBURY 218			SILVER SPRING 209			WALDORF 206			BOSTON 020 - 022, 024		
		MAT.	INST.	TOTAL	MAT.	INST.	TOTAL	MAT.	INST.	TOTAL	MAT.	INST.	TOTAL	MAT.	INST.	TOTAL	MAT.	INST.	TOTAL
015433	CONTRACTOR EQUIPMENT		98.6	98.6		98.6	98.6		98.6	98.6		96.1	96.1		96.1	96.1		106.0	106.0
0241, 31 - 34	SITE & INFRASTRUCTURE, DEMOLITION	85.3	88.3	87.4	89.2	91.1	90.5	98.2	87.7	90.9	94.3	86.0	88.6	101.6	86.0	90.9	95.2	107.4	103.6
0310	Concrete Forming & Accessories	92.1	85.8	86.6	87.8	78.1	79.3	99.2	54.9	60.7	93.9	70.1	73.2	100.6	70.2	74.1	101.8	142.2	136.9
0320	Concrete Reinforcing	87.4	104.7	96.1	88.2	74.7	81.4	87.4	67.5	77.4	97.4	76.0	86.6	98.0	76.0	86.9	102.5	149.5	126.2
0330	Cast-in-Place Concrete	100.2	78.9	91.8	106.2	84.2	97.5	123.8	49.2	94.5	134.1	79.4	112.5	150.3	79.4	122.4	107.8	149.9	124.4
03	CONCRETE	92.9	87.6	90.3	97.1	80.5	89.0	110.9	56.8	84.5	114.3	75.6	95.5	126.1	75.7	101.5	113.4	145.1	128.9
04	MASONRY	100.1	62.3	77.0	106.0	78.5	89.2	114.1	50.0	75.0	114.2	68.7	86.4	97.9	68.7	80.1	112.0	162.5	142.8
05	METALS	92.8	101.2	95.6	92.7	91.5	92.3	92.8	83.9	89.9	87.9	93.3	89.7	87.9	93.4	89.7	101.2	129.1	110.4
06	WOOD, PLASTICS & COMPOSITES	87.5	92.3	90.3	82.4	76.4	78.9	96.6	58.5	74.6	93.8	70.4	80.3	101.0	70.4	83.3	101.1	141.6	124.5
07	THERMAL & MOISTURE PROTECTION	96.5	75.1	87.8	96.5	76.1	88.2	97.1	66.8	84.8	103.1	81.4	94.3	103.6	81.4	94.6	101.6	150.1	121.3
08	OPENINGS	93.1	84.4	90.9	93.1	76.3	88.9	93.4	64.7	86.2	90.3	74.6	86.4	91.1	74.6	87.0	98.5	143.3	109.8
0920	Plaster & Gypsum Board	96.7	92.4	93.7	94.0	76.0	81.2	102.7	57.5	70.4	105.4	70.3	80.3	109.0	70.3	81.4	101.7	142.2	130.7
0950, 0980	Ceilings & Acoustic Treatment	99.8	92.4	94.9	100.8	76.0	84.4	99.8	57.5	71.9	105.8	70.3	82.3	105.8	70.3	82.3	101.0	142.2	128.2
0960	Flooring	91.6	61.2	82.7	89.5	82.2	87.4	95.3	67.8	87.2	113.9	80.9	104.2	118.0	80.9	107.1	95.0	178.2	119.4
0970, 0990	Wall Finishes & Painting/Coating	96.8	78.7	85.6	96.8	78.7	85.6	96.8	78.7	85.6	128.5	78.7	97.8	128.5	78.7	97.8	92.8	153.9	130.4
09	FINISHES	94.5	81.1	87.0	94.0	78.5	85.3	97.1	59.2	76.0	101.6	71.7	84.9	103.6	72.3	86.1	99.4	150.6	128.0
COVERS	DIVS. 10 - 14, 25, 28, 41, 43, 44, 46	100.0	64.7	93.0	100.0	89.8	98.0	100.0	41.7	88.4	100.0	79.7	96.0	100.0	79.7	96.0	100.0	120.2	104.0
21, 22, 23	FIRE SUPPRESSION, PLUMBING & HVAC	93.9	83.3	89.6	99.9	84.4	93.6	93.9	64.3	81.9	94.3	84.8	90.4	94.3	84.8	90.4	100.1	128.6	111.6
26, 27, 3370	ELECTRICAL, COMMUNICATIONS & UTIL.	98.4	93.2	95.7	96.9	82.0	89.2	95.4	71.4	83.0	95.4	98.9	97.2	93.0	98.9	96.0	101.5	132.0	117.2
MF2010	WEIGHTED AVERAGE	94.7	84.3	90.1	96.8	82.9	90.7	97.9	65.3	83.5	97.7	82.3	90.9	98.5	82.4	91.4	102.2	136.8	117.5

MASSACHUSETTS

DIVISION		BROCKTON 023			BUZZARDS BAY 025			FALL RIVER 027			FITCHBURG 014			FRAMINGHAM 017			GREENFIELD 013		
		MAT.	INST.	TOTAL	MAT.	INST.	TOTAL	MAT.	INST.	TOTAL	MAT.	INST.	TOTAL	MAT.	INST.	TOTAL	MAT.	INST.	TOTAL
015433	CONTRACTOR EQUIPMENT		102.2	102.2		102.2	102.2		103.1	103.1		100.4	100.4		101.6	101.6		100.4	100.4
0241, 31 - 34	SITE & INFRASTRUCTURE, DEMOLITION	92.5	104.0	100.4	82.7	103.9	97.3	91.6	104.0	100.2	84.5	103.9	97.9	81.0	103.6	96.6	88.4	102.4	98.1
0310	Concrete Forming & Accessories	102.4	127.9	124.6	100.1	127.2	123.7	102.4	127.5	124.2	94.8	127.1	122.9	101.1	127.7	124.3	93.3	108.3	106.3
0320	Concrete Reinforcing	103.6	149.4	126.7	83.0	121.5	102.4	103.6	121.5	112.6	83.0	142.0	112.7	83.0	149.0	116.3	86.5	125.7	106.2
0330	Cast-in-Place Concrete	102.6	152.3	122.2	85.3	152.0	111.6	99.3	152.6	120.3	89.4	145.9	111.6	89.4	139.7	109.2	91.9	124.2	104.6
03	CONCRETE	110.5	139.5	124.7	93.3	133.9	113.1	108.9	134.2	121.2	91.7	135.4	113.0	94.6	135.0	114.3	95.7	116.5	105.8
04	MASONRY	107.9	158.8	139.0	99.3	158.8	135.6	108.4	158.8	139.1	95.5	151.9	129.9	101.7	152.0	132.4	99.8	125.9	115.7
05	METALS	98.0	126.9	107.6	93.0	115.2	100.3	98.0	115.6	103.8	92.9	120.7	102.1	93.0	126.1	103.9	95.1	109.4	99.8
06	WOOD, PLASTICS & COMPOSITES	100.0	124.7	114.3	97.2	124.7	113.1	100.0	125.0	114.4	93.7	124.5	111.5	99.7	124.5	114.0	92.0	106.0	100.1
07	THERMAL & MOISTURE PROTECTION	101.6	145.9	119.6	101.0	141.2	117.3	101.6	140.4	117.4	101.0	137.8	116.0	101.1	142.8	118.1	101.1	118.2	108.0
08	OPENINGS	96.6	133.8	105.9	92.1	120.5	99.3	96.6	120.6	102.6	101.1	131.7	108.8	92.6	133.6	102.9	101.2	111.9	103.9
0920	Plaster & Gypsum Board	95.7	124.8	116.5	91.3	124.8	115.2	95.7	124.8	116.5	92.4	124.6	115.4	95.6	124.8	116.5	93.7	105.5	102.1
0950, 0980	Ceilings & Acoustic Treatment	103.0	124.8	117.4	93.7	124.8	114.3	103.0	124.8	117.4	92.7	124.6	113.8	92.7	124.8	113.9	100.0	105.5	103.6
0960	Flooring	95.5	178.2	119.7	94.2	178.2	118.8	95.3	178.2	119.6	92.3	178.2	117.5	93.9	178.2	118.7	91.6	140.6	106.0
0970, 0990	Wall Finishes & Painting/Coating	92.3	139.1	121.2	92.3	139.1	121.2	92.3	139.1	121.2	90.1	139.1	120.4	91.2	139.1	120.8	90.1	109.4	102.0
09	FINISHES	99.4	138.1	121.0	95.6	138.1	119.3	99.3	138.3	121.1	95.0	138.0	119.0	95.6	137.9	119.2	96.9	114.2	106.5
COVERS	DIVS. 10 - 14, 25, 28, 41, 43, 44, 46	100.0	117.1	103.4	100.0	117.1	103.4	100.0	117.7	103.5	100.0	107.0	101.4	100.0	116.6	103.3	100.0	105.3	101.1
21, 22, 23	FIRE SUPPRESSION, PLUMBING & HVAC	100.0	111.9	104.8	94.1	111.6	101.2	100.0	111.7	104.7	95.0	115.0	103.1	95.0	124.6	107.0	95.0	102.0	97.8
26, 27, 3370	ELECTRICAL, COMMUNICATIONS & UTIL.	100.6	98.8	99.7	97.6	98.8	98.3	100.5	98.8	99.7	102.8	101.8	102.3	99.1	128.4	114.2	102.8	93.0	97.7
MF2010	WEIGHTED AVERAGE	100.8	124.7	111.4	94.9	122.1	106.9	100.6	122.2	110.1	96.1	123.3	108.1	95.5	129.4	110.7	97.3	108.4	102.2

MASSACHUSETTS

DIVISION		HYANNIS 026			LAWRENCE 019			LOWELL 018			NEW BEDFORD 027			PITTSFIELD 012			SPRINGFIELD 010 - 011		
		MAT.	INST.	TOTAL	MAT.	INST.	TOTAL	MAT.	INST.	TOTAL	MAT.	INST.	TOTAL	MAT.	INST.	TOTAL	MAT.	INST.	TOTAL
015433	CONTRACTOR EQUIPMENT		102.2	102.2		102.2	102.2		100.4	100.4		103.1	103.1		100.4	100.4		100.4	100.4
0241, 31 - 34	SITE & INFRASTRUCTURE, DEMOLITION	88.8	103.9	99.2	93.1	104.0	100.6	92.4	103.9	100.3	90.0	104.0	99.7	93.4	102.2	99.4	92.8	102.4	99.5
0310	Concrete Forming & Accessories	95.0	127.2	123.1	102.1	128.0	124.6	99.4	128.0	124.3	102.4	127.5	124.2	99.3	106.5	105.5	99.6	108.4	107.2
0320	Concrete Reinforcing	83.0	121.5	102.4	102.7	135.5	119.2	103.6	135.2	119.5	103.6	121.5	112.6	85.8	108.6	97.3	103.6	125.7	114.7
0330	Cast-in-Place Concrete	93.6	152.0	116.6	103.5	146.8	120.5	94.1	146.8	114.8	87.4	152.6	113.1	102.6	121.8	110.2	97.9	124.3	108.3
03	CONCRETE	100.4	133.9	116.7	110.8	135.0	122.6	101.4	134.8	117.7	102.9	134.2	118.2	102.9	111.7	107.2	103.3	116.6	109.8
04	MASONRY	106.9	158.8	138.6	107.1	158.8	138.7	94.7	151.9	129.6	106.8	158.8	138.5	95.4	121.6	111.4	95.0	126.1	114.0
05	METALS	94.5	115.2	101.3	95.5	121.4	104.0	95.4	118.6	103.1	98.0	115.6	103.8	95.3	102.4	97.6	98.0	109.4	101.7
06	WOOD, PLASTICS & COMPOSITES	91.2	124.7	110.6	100.0	124.7	114.3	99.5	124.7	114.1	100.0	125.0	114.4	99.5	106.0	103.3	99.5	106.0	103.3
07	THERMAL & MOISTURE PROTECTION	101.2	141.2	117.5	101.5	145.1	119.2	101.4	143.2	118.4	101.5	140.4	117.3	101.4	116.3	107.5	101.4	118.3	108.2
08	OPENINGS	92.8	120.5	99.8	96.6	130.0	105.0	102.4	130.0	109.4	96.6	120.6	102.6	102.4	107.3	103.6	102.4	111.9	104.8
0920	Plaster & Gypsum Board	87.6	124.8	114.2	98.7	124.8	117.4	98.7	124.8	117.4	95.7	124.8	116.5	98.7	105.5	103.6	98.7	105.5	103.6
0950, 0980	Ceilings & Acoustic Treatment	95.8	124.8	115.0	102.0	124.8	117.1	102.0	124.8	117.1	103.0	124.8	117.4	102.0	105.5	104.3	102.0	105.5	104.3
0960	Flooring	92.0	178.2	117.3	94.5	178.2	119.0	94.5	178.2	119.0	95.3	178.2	119.6	94.7	140.6	108.1	94.3	140.6	107.9
0970, 0990	Wall Finishes & Painting/Coating	92.3	139.1	121.2	90.2	139.1	120.4	90.1	139.1	120.4	92.3	139.1	121.2	90.1	109.4	102.0	91.7	109.4	102.6
09	FINISHES	95.5	138.1	119.3	99.0	138.1	120.8	99.0	138.1	120.8	99.2	138.3	121.0	99.1	113.1	106.9	99.1	114.2	107.5
COVERS	DIVS. 10 - 14, 25, 28, 41, 43, 44, 46	100.0	117.1	103.4	100.0	117.2	103.4	100.0	117.2	103.4	100.0	117.7	103.5	100.0	103.8	100.8	100.0	105.4	101.1
21, 22, 23	FIRE SUPPRESSION, PLUMBING & HVAC	100.0	111.6	104.7	100.0	119.0	107.7	100.0	124.5	109.9	100.0	111.7	104.7	100.0	97.6	99.1	100.0	102.1	100.9
26, 27, 3370	ELECTRICAL, COMMUNICATIONS & UTIL.	98.1	98.8	98.5	101.8	128.4	115.5	102.3	124.2	113.6	101.4	98.8	100.1	102.3	93.0	97.5	102.3	94.9	98.5
MF2010	WEIGHTED AVERAGE	98.0	122.1	108.6	100.5	129.0	113.1	99.4	128.5	112.3	99.9	122.2	109.7	99.6	105.3	102.1	100.0	108.7	103.9

DIVISION		MASSACHUSETTS WORCESTER 015 - 016			MICHIGAN ANN ARBOR 481			BATTLE CREEK 490			BAY CITY 487			DEARBORN 481			DETROIT 482		
		MAT.	INST.	TOTAL	MAT.	INST.	TOTAL	MAT.	INST.	TOTAL	MAT.	INST.	TOTAL	MAT.	INST.	TOTAL	MAT.	INST.	TOTAL
015433	CONTRACTOR EQUIPMENT		100.4	100.4		110.6	110.6		103.2	103.2		110.6	110.6		110.6	110.6		98.9	98.9
0241, 31 - 34	SITE & INFRASTRUCTURE, DEMOLITION	92.8	103.9	100.4	74.7	96.5	89.8	84.8	88.5	87.4	66.5	95.7	86.6	74.5	96.6	89.7	89.3	98.2	95.4
0310	Concrete Forming & Accessories	99.8	127.1	123.6	100.7	110.6	109.3	98.6	89.8	90.9	100.8	89.2	90.7	100.5	114.6	112.8	101.3	114.6	112.9
0320	Concrete Reinforcing	103.6	144.3	124.1	99.7	123.6	111.8	98.1	91.1	94.6	99.7	122.8	111.4	99.7	123.8	111.9	99.0	123.8	111.5
0330	Cast-in-Place Concrete	97.3	145.9	116.5	81.9	112.4	93.9	92.4	102.5	96.4	78.4	93.3	84.2	80.1	111.5	92.5	87.9	111.6	97.2
03	CONCRETE	103.0	135.8	119.0	90.3	113.9	101.8	95.2	93.8	94.5	88.5	97.6	92.9	89.3	115.4	102.0	93.2	114.3	103.5
04	MASONRY	94.5	151.9	129.5	97.9	108.5	104.4	96.0	86.8	90.3	97.5	88.5	92.0	97.8	111.9	106.4	96.9	111.9	106.1
05	METALS	98.0	121.6	105.8	93.8	119.0	102.1	93.0	87.3	91.1	94.5	115.9	101.5	93.9	119.4	102.3	97.9	102.2	99.3
06	WOOD, PLASTICS & COMPOSITES	99.9	124.5	114.1	97.8	111.1	105.5	100.1	89.6	94.1	97.8	88.4	92.4	97.8	115.7	108.1	98.2	115.7	108.3
07	THERMAL & MOISTURE PROTECTION	101.4	137.8	116.2	101.3	110.0	104.8	93.7	87.4	91.2	99.1	96.0	97.8	99.9	114.7	105.9	98.3	114.8	105.0
08	OPENINGS	102.4	132.4	110.0	95.1	109.8	98.8	92.7	83.8	90.5	95.1	95.6	95.2	95.1	112.3	99.4	96.6	112.3	100.6
0920	Plaster & Gypsum Board	98.7	124.6	117.2	104.7	110.4	108.8	94.6	85.4	88.0	104.7	87.0	92.1	104.7	115.1	112.2	104.7	115.1	112.2
0950, 0980	Ceilings & Acoustic Treatment	102.0	124.6	116.9	93.7	110.4	104.7	93.8	85.4	88.2	94.7	87.0	89.6	93.7	115.1	107.9	94.7	115.1	108.2
0960	Flooring	94.5	170.9	116.9	94.2	117.1	100.9	98.7	96.2	98.0	94.0	80.5	90.0	93.8	117.8	100.8	93.9	117.8	100.9
0970, 0990	Wall Finishes & Painting/Coating	90.1	139.1	120.4	92.8	102.0	98.5	96.0	82.6	87.7	92.8	83.3	87.0	92.8	102.1	98.5	94.3	102.1	99.1
09	FINISHES	99.0	136.5	119.9	94.8	111.1	103.9	96.5	90.1	92.9	94.5	86.2	89.9	94.7	114.3	105.6	96.2	114.3	106.3
COVERS	DIVS. 10 - 14, 25, 28, 41, 43, 44, 46	100.0	110.4	102.1	100.0	106.8	101.4	100.0	103.0	100.6	100.0	97.1	99.4	100.0	107.6	101.5	100.0	107.6	101.5
21, 22, 23	FIRE SUPPRESSION, PLUMBING & HVAC	100.0	115.1	106.1	100.0	100.2	100.1	99.9	90.9	96.3	100.0	87.9	95.1	100.0	110.5	104.3	100.1	111.8	104.8
26, 27, 3370	ELECTRICAL, COMMUNICATIONS & UTIL.	102.3	102.6	102.4	95.6	110.4	103.2	94.8	83.4	88.9	94.4	91.3	92.8	95.6	109.2	102.6	96.7	109.1	103.1
MF2010	WEIGHTED AVERAGE	100.0	123.5	110.4	95.8	108.1	101.2	96.0	89.2	93.0	95.2	93.6	94.5	95.7	111.4	102.6	97.4	110.1	103.0

MICHIGAN

DIVISION		FLINT 484 - 485			GAYLORD 497			GRAND RAPIDS 493,495			IRON MOUNTAIN 498 - 499			JACKSON 492			KALAMAZOO 491		
		MAT.	INST.	TOTAL	MAT.	INST.	TOTAL	MAT.	INST.	TOTAL	MAT.	INST.	TOTAL	MAT.	INST.	TOTAL	MAT.	INST.	TOTAL
015433	CONTRACTOR EQUIPMENT		110.6	110.6		105.6	105.6		103.2	103.2		94.9	94.9		105.6	105.6		103.2	103.2
0241, 31 - 34	SITE & INFRASTRUCTURE, DEMOLITION	64.3	95.9	86.1	79.0	86.7	84.3	86.3	88.4	87.7	87.5	94.9	92.6	101.3	88.4	92.4	85.1	88.5	87.4
0310	Concrete Forming & Accessories	103.5	91.4	93.0	95.9	56.1	61.2	98.9	81.0	83.4	88.0	86.1	86.4	93.4	88.0	88.7	98.6	89.5	90.7
0320	Concrete Reinforcing	99.7	123.1	111.5	91.5	92.7	92.1	98.1	90.8	94.4	91.3	94.4	92.8	88.9	122.9	106.0	98.1	91.0	94.6
0330	Cast-in-Place Concrete	82.5	95.4	87.6	92.2	71.4	84.0	98.4	95.5	97.3	109.2	89.4	101.4	92.1	94.5	93.0	94.3	102.4	97.5
03	CONCRETE	90.7	99.4	95.0	91.6	69.8	81.0	98.2	87.5	93.0	101.2	88.9	95.2	96.8	97.4	97.1	98.4	93.6	96.1
04	MASONRY	98.0	96.2	96.9	106.1	82.9	91.9	91.8	81.9	85.8	92.5	88.8	90.3	86.7	93.2	90.7	94.6	86.8	89.8
05	METALS	93.9	116.5	101.4	94.1	100.6	96.3	94.2	86.3	91.6	93.5	90.7	92.6	94.3	113.8	100.8	93.0	87.0	91.0
06	WOOD, PLASTICS & COMPOSITES	100.9	89.7	94.5	92.5	48.8	67.3	97.2	79.1	86.8	87.3	86.1	86.6	91.3	86.3	88.4	100.1	89.6	94.1
07	THERMAL & MOISTURE PROTECTION	99.1	99.3	99.2	92.2	70.8	83.5	95.5	77.1	88.0	95.4	84.1	90.8	91.6	96.6	93.6	93.7	87.4	91.2
08	OPENINGS	95.1	96.3	95.4	95.0	53.9	84.7	96.6	82.2	92.9	100.6	73.4	93.7	94.0	87.2	92.3	92.7	83.8	90.5
0920	Plaster & Gypsum Board	106.5	88.4	93.5	94.0	46.0	59.7	95.5	74.6	80.6	58.1	86.1	78.1	92.5	84.6	86.8	94.6	85.4	88.0
0950, 0980	Ceilings & Acoustic Treatment	93.7	88.4	90.2	92.0	46.0	61.6	92.9	74.6	80.8	91.8	86.1	88.0	92.9	84.6	87.4	93.8	85.4	88.2
0960	Flooring	94.0	83.2	90.8	91.8	50.4	79.7	98.9	70.5	90.5	112.8	96.8	108.1	90.5	69.4	84.3	98.7	75.8	92.0
0970, 0990	Wall Finishes & Painting/Coating	92.8	90.0	91.1	92.5	45.6	63.5	96.0	81.3	86.9	111.5	58.8	79.0	92.5	75.7	82.1	96.0	82.6	87.7
09	FINISHES	94.0	89.0	91.2	96.7	51.8	71.7	96.6	78.4	86.5	99.3	85.3	91.5	98.0	82.4	89.3	96.5	86.2	90.7
COVERS	DIVS. 10 - 14, 25, 28, 41, 43, 44, 46	100.0	98.2	99.6	100.0	91.8	98.4	100.0	101.6	100.3	100.0	93.7	98.7	100.0	98.0	99.6	100.0	103.0	100.6
21, 22, 23	FIRE SUPPRESSION, PLUMBING & HVAC	100.0	94.1	97.6	94.4	82.8	89.7	99.9	83.2	93.1	94.3	86.8	91.3	94.4	92.0	93.4	99.9	83.0	93.1
26, 27, 3370	ELECTRICAL, COMMUNICATIONS & UTIL.	95.6	95.0	95.3	92.9	49.0	70.3	99.1	75.1	86.7	99.2	87.6	93.2	96.8	110.3	103.8	94.7	86.9	90.7
MF2010	WEIGHTED AVERAGE	95.5	97.1	96.2	94.7	72.8	85.1	97.2	83.0	90.9	96.7	87.8	92.8	94.3	95.9	95.0	96.3	87.4	92.4

MICHIGAN / MINNESOTA

DIVISION		LANSING 488 - 489			MUSKEGON 494			ROYAL OAK 480,483			SAGINAW 486			TRAVERSE CITY 496			BEMIDJI 566		
		MAT.	INST.	TOTAL	MAT.	INST.	TOTAL	MAT.	INST.	TOTAL	MAT.	INST.	TOTAL	MAT.	INST.	TOTAL	MAT.	INST.	TOTAL
015433	CONTRACTOR EQUIPMENT		110.6	110.6		103.2	103.2		96.4	96.4		110.6	110.6		94.9	94.9		98.7	98.7
0241, 31 - 34	SITE & INFRASTRUCTURE, DEMOLITION	83.2	95.8	91.9	82.9	88.4	86.7	79.0	96.0	90.7	67.4	95.7	86.9	74.3	94.6	88.3	98.4	98.6	98.6
0310	Concrete Forming & Accessories	104.2	88.4	90.5	99.1	88.1	89.5	97.1	111.7	109.8	100.7	89.0	90.3	88.0	71.1	73.3	90.8	94.2	93.7
0320	Concrete Reinforcing	99.7	122.9	111.4	98.8	90.5	94.6	90.4	120.5	105.6	99.7	122.8	111.3	92.7	90.2	91.4	103.5	109.4	106.5
0330	Cast-in-Place Concrete	87.0	93.7	89.6	92.1	96.5	93.8	71.7	110.4	87.0	80.9	93.2	85.8	85.2	63.7	76.7	105.6	106.6	106.0
03	CONCRETE	93.1	97.5	95.2	93.6	90.9	92.3	78.0	111.9	94.5	89.8	97.5	93.6	83.4	72.7	78.2	99.5	102.4	100.9
04	MASONRY	91.7	93.8	92.9	93.1	76.2	82.8	93.0	111.9	104.5	99.5	88.5	92.8	90.7	82.8	85.8	100.3	107.1	104.4
05	METALS	92.5	115.9	100.2	90.7	85.9	89.1	97.0	99.9	98.0	93.9	115.7	101.1	93.5	88.8	91.9	92.6	122.5	102.5
06	WOOD, PLASTICS & COMPOSITES	100.3	86.5	92.3	96.1	88.5	91.7	93.2	112.2	104.2	93.6	88.4	90.6	87.3	69.5	77.0	72.8	90.9	83.3
07	THERMAL & MOISTURE PROTECTION	100.2	92.0	96.1	92.9	78.2	86.9	97.7	114.0	104.4	99.8	96.0	98.3	94.5	71.4	85.1	108.5	99.7	104.9
08	OPENINGS	95.1	94.5	94.9	92.0	87.7	90.9	95.3	110.1	99.0	92.9	95.6	93.6	100.6	64.9	91.6	97.7	111.9	101.3
0920	Plaster & Gypsum Board	97.4	85.0	88.6	75.1	84.2	81.6	101.5	111.5	108.6	104.7	87.0	92.1	58.1	69.0	65.9	105.6	91.0	95.1
0950, 0980	Ceilings & Acoustic Treatment	93.7	85.0	88.0	94.7	84.2	87.8	92.9	111.5	105.2	93.7	87.0	89.3	91.8	69.0	76.7	137.6	91.0	106.8
0960	Flooring	104.0	83.2	97.9	97.9	65.3	88.3	91.8	116.4	99.0	94.2	80.5	90.2	112.8	50.4	94.5	109.8	126.5	114.7
0970, 0990	Wall Finishes & Painting/Coating	103.9	84.1	91.7	95.3	81.3	86.7	94.3	102.0	99.0	92.8	83.3	87.0	111.5	45.6	70.8	96.0	101.0	99.1
09	FINISHES	98.5	86.3	91.7	93.4	83.0	87.6	93.8	111.9	103.9	94.4	86.2	89.9	98.2	63.8	79.0	112.4	100.2	105.6
COVERS	DIVS. 10 - 14, 25, 28, 41, 43, 44, 46	100.0	97.6	99.5	100.0	102.8	100.6	100.0	102.7	100.5	100.0	97.1	99.4	100.0	90.3	98.1	100.0	95.7	99.1
21, 22, 23	FIRE SUPPRESSION, PLUMBING & HVAC	99.9	92.2	96.8	99.8	85.2	93.9	94.5	109.6	100.6	100.0	87.5	95.0	94.3	82.4	89.5	94.5	88.1	91.9
26, 27, 3370	ELECTRICAL, COMMUNICATIONS & UTIL.	98.7	93.0	95.8	95.2	80.1	87.4	97.6	94.0	95.8	92.7	91.3	92.0	94.6	76.7	85.3	104.5	104.0	104.2
MF2010	WEIGHTED AVERAGE	96.4	95.1	95.9	94.9	84.9	90.5	93.4	106.2	99.1	95.0	93.5	94.4	93.6	78.7	87.0	98.7	101.3	99.9

MINNESOTA

DIVISION		BRAINERD 564			DETROIT LAKES 565			DULUTH 556 - 558			MANKATO 560			MINNEAPOLIS 553 - 555			ROCHESTER 559		
		MAT.	INST.	TOTAL	MAT.	INST.	TOTAL	MAT.	INST.	TOTAL	MAT.	INST.	TOTAL	MAT.	INST.	TOTAL	MAT.	INST.	TOTAL
015433	CONTRACTOR EQUIPMENT		101.4	101.4		98.7	98.7		101.2	101.2		101.4	101.4		104.7	104.7		101.2	101.2
0241, 31 - 34	SITE & INFRASTRUCTURE, DEMOLITION	96.6	103.5	101.3	96.4	99.1	98.2	98.8	102.3	101.2	92.9	104.1	100.6	101.3	107.6	105.7	100.3	102.1	101.6
0310	Concrete Forming & Accessories	94.0	97.0	96.6	88.1	96.9	95.7	101.1	116.4	114.4	101.0	104.7	104.2	103.2	129.6	126.1	103.8	110.6	109.7
0320	Concrete Reinforcing	102.3	109.6	106.0	103.5	109.4	106.5	98.8	109.9	104.4	102.1	119.8	111.0	99.0	120.2	109.7	98.8	119.9	109.4
0330	Cast-in-Place Concrete	114.8	111.6	113.6	102.6	110.8	105.8	107.4	109.8	108.4	105.8	110.7	107.7	120.3	119.0	119.8	117.9	103.0	112.0
03	CONCRETE	103.9	105.4	104.6	97.0	105.0	100.9	102.3	113.5	107.7	99.0	110.4	104.6	110.5	124.5	117.3	107.7	110.5	109.1
04	MASONRY	123.6	117.6	119.9	123.6	114.8	118.2	108.6	121.3	116.3	111.9	114.4	113.4	108.1	129.1	120.9	107.3	117.0	113.2
05	METALS	93.8	122.5	103.3	92.6	122.3	102.4	99.6	125.1	108.0	93.6	128.2	105.1	102.4	131.9	112.1	99.5	130.1	109.6
06	WOOD, PLASTICS & COMPOSITES	93.7	90.7	92.0	70.2	90.9	82.2	110.4	116.3	113.8	101.9	102.6	102.3	113.6	127.8	121.8	113.6	109.4	111.2
07	THERMAL & MOISTURE PROTECTION	107.6	112.3	109.5	108.3	112.5	110.0	109.0	119.9	113.5	108.0	103.2	106.0	108.8	132.6	118.5	109.1	104.6	107.3
08	OPENINGS	85.7	111.8	92.2	97.6	111.9	101.2	93.9	121.1	100.7	90.4	121.3	98.2	97.2	135.7	106.9	93.9	125.7	101.9
0920	Plaster & Gypsum Board	86.5	91.0	89.7	104.7	91.0	94.9	94.3	117.2	110.7	90.6	103.2	99.6	102.2	129.0	121.3	101.7	110.1	107.7
0950, 0980	Ceilings & Acoustic Treatment	67.9	91.0	83.1	137.6	91.0	106.8	92.0	117.2	108.7	67.9	103.2	91.2	93.8	129.0	117.1	92.0	110.1	103.9
0960	Flooring	108.6	126.5	113.9	108.4	126.5	113.7	105.9	127.8	112.4	110.7	126.5	115.4	103.7	126.5	110.4	106.1	126.5	112.1
0970, 0990	Wall Finishes & Painting/Coating	90.2	101.0	96.9	96.0	101.0	99.1	88.5	111.6	102.7	101.7	105.1	103.8	97.6	132.1	118.9	92.0	105.1	100.1
09	FINISHES	94.1	101.9	98.5	111.7	102.1	106.4	97.0	118.4	108.9	95.7	108.6	102.9	98.4	129.7	115.8	98.2	113.1	106.5
COVERS	DIVS. 10 - 14, 25, 28, 41, 43, 44, 46	100.0	97.9	99.6	100.0	98.2	99.6	100.0	105.4	101.1	100.0	98.1	99.6	100.0	110.2	102.0	100.0	104.1	100.8
21, 22, 23	FIRE SUPPRESSION, PLUMBING & HVAC	93.7	92.0	93.0	94.5	91.7	93.3	100.1	101.0	100.5	93.7	91.7	92.9	100.0	121.6	108.8	100.0	99.4	99.8
26, 27, 3370	ELECTRICAL, COMMUNICATIONS & UTIL.	102.1	108.9	105.6	104.1	70.5	86.8	99.5	108.9	104.3	108.7	89.0	98.5	101.6	113.1	107.6	100.6	89.1	94.7
MF2010	WEIGHTED AVERAGE	97.6	105.3	101.0	99.4	99.3	99.4	100.1	112.0	105.4	97.7	104.6	100.7	102.1	123.0	111.3	100.9	107.6	103.8

MINNESOTA / MISSISSIPPI

DIVISION		SAINT PAUL 550 - 551			ST. CLOUD 563			THIEF RIVER FALLS 567			WILLMAR 562			WINDOM 561			BILOXI 395		
		MAT.	INST.	TOTAL	MAT.	INST.	TOTAL	MAT.	INST.	TOTAL	MAT.	INST.	TOTAL	MAT.	INST.	TOTAL	MAT.	INST.	TOTAL
015433	CONTRACTOR EQUIPMENT		101.2	101.2		101.4	101.4		98.7	98.7		101.4	101.4		101.4	101.4		98.5	98.5
0241, 31 - 34	SITE & INFRASTRUCTURE, DEMOLITION	101.0	102.9	102.3	91.1	105.4	101.0	97.1	98.6	98.1	90.8	103.2	99.4	84.1	101.9	96.4	101.8	88.9	92.9
0310	Concrete Forming & Accessories	97.8	129.3	125.2	91.3	124.9	120.5	91.4	93.4	93.2	91.2	96.2	95.5	95.2	87.9	88.9	94.7	56.1	61.2
0320	Concrete Reinforcing	95.4	120.1	107.8	102.3	120.0	111.2	103.9	109.3	106.7	101.9	119.6	110.8	101.9	118.7	110.4	96.3	60.1	78.1
0330	Cast-in-Place Concrete	120.3	118.4	119.5	101.3	117.9	107.8	104.7	91.3	99.4	102.9	89.3	97.5	89.0	67.0	80.4	106.6	55.2	86.4
03	CONCRETE	112.0	124.1	117.9	94.9	121.9	108.0	98.2	96.8	97.5	95.0	99.3	97.1	85.6	87.8	86.6	98.8	58.2	79.0
04	MASONRY	117.2	129.1	124.4	108.6	126.5	119.5	100.2	107.1	104.4	113.0	117.6	115.8	123.1	90.1	102.9	95.6	45.7	65.1
05	METALS	99.0	131.6	109.8	94.5	129.5	106.0	92.8	121.8	102.3	93.6	127.0	104.6	93.5	124.6	103.8	92.5	83.7	89.6
06	WOOD, PLASTICS & COMPOSITES	106.6	127.6	118.7	91.3	122.2	109.2	73.6	90.9	83.6	91.1	91.0	91.0	94.9	88.1	91.0	99.3	59.6	76.4
07	THERMAL & MOISTURE PROTECTION	110.3	132.4	119.3	107.8	122.8	113.9	109.4	101.5	106.2	107.6	109.6	108.4	107.6	86.6	99.0	92.8	57.0	78.3
08	OPENINGS	91.7	135.6	102.7	90.4	132.7	101.1	97.7	111.9	101.3	88.0	96.2	90.0	92.1	94.6	92.8	97.9	60.7	88.5
0920	Plaster & Gypsum Board	91.2	128.9	118.1	86.5	123.4	112.9	105.2	91.0	95.0	86.5	91.2	89.9	86.5	88.2	87.8	97.3	58.8	69.8
0950, 0980	Ceilings & Acoustic Treatment	89.3	128.9	115.4	67.9	123.4	104.6	137.6	91.0	106.8	67.9	91.2	83.3	67.9	88.2	81.3	95.8	58.8	71.4
0960	Flooring	100.0	126.5	107.8	104.6	126.5	111.1	109.4	126.5	114.4	106.5	126.5	112.4	109.2	126.5	114.3	117.7	55.3	99.4
0970, 0990	Wall Finishes & Painting/Coating	97.6	127.6	116.1	101.7	132.1	120.5	96.0	101.0	99.1	96.0	101.0	99.1	96.0	105.1	101.6	105.7	45.8	68.7
09	FINISHES	95.0	129.0	114.0	93.2	126.2	111.6	112.2	100.2	105.5	93.4	99.6	96.8	93.5	95.0	94.3	104.4	55.1	76.9
COVERS	DIVS. 10 - 14, 25, 28, 41, 43, 44, 46	100.0	109.9	102.0	100.0	104.5	100.9	100.0	95.6	99.1	100.0	97.8	99.6	100.0	92.9	98.6	100.0	59.9	92.0
21, 22, 23	FIRE SUPPRESSION, PLUMBING & HVAC	100.0	117.1	106.9	99.7	121.3	108.4	94.5	87.7	91.7	93.7	110.0	100.3	93.7	86.1	90.6	99.9	50.4	79.9
26, 27, 3370	ELECTRICAL, COMMUNICATIONS & UTIL.	98.7	113.1	106.1	102.1	113.1	107.8	101.7	70.5	85.6	102.1	90.5	96.1	108.7	89.0	98.5	98.8	59.0	78.3
MF2010	WEIGHTED AVERAGE	101.1	121.5	110.1	97.6	120.8	107.9	98.3	95.8	97.2	96.0	105.1	100.0	96.4	93.8	95.2	98.3	60.2	81.5

MISSISSIPPI

DIVISION		CLARKSDALE 386			COLUMBUS 397			GREENVILLE 387			GREENWOOD 389			JACKSON 390 - 392			LAUREL 394		
		MAT.	INST.	TOTAL	MAT.	INST.	TOTAL	MAT.	INST.	TOTAL	MAT.	INST.	TOTAL	MAT.	INST.	TOTAL	MAT.	INST.	TOTAL
015433	CONTRACTOR EQUIPMENT		98.5	98.5		98.5	98.5		98.5	98.5		98.5	98.5		98.5	98.5		98.5	98.5
0241, 31 - 34	SITE & INFRASTRUCTURE, DEMOLITION	99.2	88.7	91.9	100.5	89.0	92.6	105.7	88.9	94.1	102.4	88.4	92.8	97.3	88.9	91.5	107.1	88.8	94.5
0310	Concrete Forming & Accessories	86.1	52.4	56.8	84.3	54.5	58.4	83.0	67.7	69.7	94.6	52.3	57.8	94.0	61.1	65.4	84.4	56.3	59.9
0320	Concrete Reinforcing	103.8	44.2	73.8	103.2	45.2	74.0	104.4	64.0	84.0	103.8	55.8	79.6	96.3	59.1	77.6	103.8	41.5	72.4
0330	Cast-in-Place Concrete	103.9	56.1	85.1	108.7	61.4	90.1	107.0	58.3	87.8	111.8	55.7	89.7	90.7	59.1	78.2	106.3	57.8	87.2
03	CONCRETE	97.0	53.8	76.0	100.6	56.8	79.2	102.0	65.1	84.0	103.3	55.7	80.1	90.8	61.5	76.5	102.6	55.7	79.8
04	MASONRY	95.4	47.6	66.2	122.5	54.6	81.1	141.8	51.1	86.4	96.0	47.4	66.4	96.6	51.1	68.8	118.2	50.9	77.1
05	METALS	89.6	76.2	85.2	89.5	77.1	85.4	90.7	85.6	89.0	89.6	77.7	85.7	92.4	83.5	89.5	89.6	76.1	85.1
06	WOOD, PLASTICS & COMPOSITES	87.3	54.4	68.3	85.4	55.5	68.1	84.1	72.4	77.4	99.3	54.4	73.4	98.0	63.3	78.0	86.4	58.0	70.0
07	THERMAL & MOISTURE PROTECTION	92.7	49.8	75.2	92.8	54.3	77.1	93.0	63.0	80.8	93.0	49.7	75.4	94.6	62.0	81.4	92.9	56.1	77.9
08	OPENINGS	97.1	52.4	85.9	97.1	52.5	85.9	97.1	65.1	89.1	97.1	53.8	86.2	98.3	58.8	88.3	93.7	53.3	83.5
0920	Plaster & Gypsum Board	85.5	53.5	62.6	85.1	54.5	63.3	85.1	72.0	75.7	96.0	53.5	65.6	91.2	62.7	70.8	85.1	57.2	65.2
0950, 0980	Ceilings & Acoustic Treatment	89.3	53.5	65.6	89.3	54.5	66.3	91.3	72.0	78.6	89.3	53.5	65.6	93.1	62.7	73.0	89.3	57.2	68.1
0960	Flooring	111.4	55.3	95.0	110.2	62.1	96.1	109.4	55.3	93.5	117.7	55.3	99.4	118.1	55.3	99.7	108.4	55.3	92.8
0970, 0990	Wall Finishes & Painting/Coating	105.7	48.6	70.5	105.7	55.8	74.9	105.7	69.4	83.3	105.7	48.6	70.5	105.7	69.4	83.3	105.7	64.2	80.1
09	FINISHES	99.5	52.8	73.5	99.4	56.4	75.4	100.0	66.6	81.4	103.0	52.8	75.0	103.4	61.3	79.9	99.5	57.5	76.1
COVERS	DIVS. 10 - 14, 25, 28, 41, 43, 44, 46	100.0	60.3	92.1	100.0	61.5	92.3	100.0	63.2	92.7	100.0	60.2	92.1	100.0	62.2	92.5	100.0	40.4	88.1
21, 22, 23	FIRE SUPPRESSION, PLUMBING & HVAC	97.4	44.3	75.9	96.7	40.3	73.8	99.9	51.1	80.1	97.4	42.2	75.0	99.9	56.5	82.3	96.7	49.2	77.5
26, 27, 3370	ELECTRICAL, COMMUNICATIONS & UTIL.	98.1	49.3	72.9	96.3	52.2	73.5	98.1	66.6	81.9	98.1	46.2	71.3	99.5	66.6	82.5	97.8	59.5	78.1
MF2010	WEIGHTED AVERAGE	96.4	55.5	78.4	97.8	56.9	79.8	100.3	65.1	84.7	97.6	55.1	78.8	97.4	64.4	82.9	97.9	58.8	80.6

MISSISSIPPI / MISSOURI

DIVISION		MCCOMB 396			MERIDIAN 393			TUPELO 388			BOWLING GREEN 633			CAPE GIRARDEAU 637			CHILLICOTHE 646		
		MAT.	INST.	TOTAL	MAT.	INST.	TOTAL	MAT.	INST.	TOTAL	MAT.	INST.	TOTAL	MAT.	INST.	TOTAL	MAT.	INST.	TOTAL
015433	CONTRACTOR EQUIPMENT		98.5	98.5		98.5	98.5		98.5	98.5		106.1	106.1		106.1	106.1		102.2	102.2
0241, 31 - 34	SITE & INFRASTRUCTURE, DEMOLITION	92.8	88.6	89.9	97.0	89.2	91.6	96.8	88.6	91.1	91.4	93.6	92.9	92.7	93.5	93.3	100.3	90.9	93.8
0310	Concrete Forming & Accessories	84.3	54.2	58.1	81.9	67.2	69.1	83.5	54.5	58.3	91.8	87.1	87.7	85.7	75.0	76.4	90.3	90.0	90.0
0320	Concrete Reinforcing	104.4	43.6	73.8	103.2	59.1	81.0	101.6	55.8	78.5	100.9	96.9	98.9	102.1	83.3	92.7	97.5	100.1	98.8
0330	Cast-in-Place Concrete	94.4	55.6	79.1	101.0	61.6	85.4	103.9	57.6	85.7	94.6	75.6	87.1	93.7	82.5	89.2	93.0	65.8	82.3
03	CONCRETE	90.5	54.2	72.8	94.8	65.1	80.3	96.5	57.3	77.4	96.3	86.1	91.4	95.5	80.6	88.3	99.8	83.9	92.1
04	MASONRY	123.9	46.9	76.9	95.3	55.3	70.9	131.6	50.6	82.2	124.4	95.9	107.0	121.5	77.3	94.5	98.6	90.8	93.8
05	METALS	89.7	74.3	84.6	90.6	83.7	88.3	89.5	77.9	85.7	97.9	110.8	102.1	98.9	104.4	100.8	95.5	96.5	95.8
06	WOOD, PLASTICS & COMPOSITES	85.4	57.4	69.2	83.0	69.3	75.1	84.7	55.5	67.8	88.9	87.4	88.0	82.6	72.3	76.6	91.9	96.0	94.3
07	THERMAL & MOISTURE PROTECTION	92.4	48.9	74.7	92.6	64.5	81.1	92.6	55.8	77.7	101.0	98.4	99.9	100.9	81.1	92.8	97.6	84.9	92.4
08	OPENINGS	97.2	54.1	86.3	97.1	67.9	89.8	97.1	53.7	86.2	99.0	83.1	95.0	99.0	74.3	92.7	90.8	84.5	89.2
0920	Plaster & Gypsum Board	85.1	56.5	64.7	85.1	68.8	73.4	85.1	54.5	63.3	97.3	87.0	89.9	95.5	71.3	78.2	103.1	95.7	97.8
0950, 0980	Ceilings & Acoustic Treatment	89.3	56.5	67.6	91.3	68.8	76.4	89.3	54.5	66.3	90.1	87.0	88.0	90.1	71.3	77.7	89.5	95.7	93.6
0960	Flooring	110.2	55.3	94.1	108.4	55.3	92.8	109.7	55.3	93.7	101.2	95.8	99.6	97.5	88.0	94.7	93.5	45.0	79.3
0970, 0990	Wall Finishes & Painting/Coating	105.7	44.6	68.0	105.7	80.3	90.0	105.7	61.9	78.7	102.0	103.4	102.8	102.0	72.7	83.9	94.3	47.3	65.3
09	FINISHES	98.6	54.0	73.7	98.9	67.0	81.1	98.9	55.7	74.8	98.5	89.4	93.4	97.1	75.3	84.9	96.4	79.5	87.0
COVERS	DIVS. 10 - 14, 25, 28, 41, 43, 44, 46	100.0	63.9	92.8	100.0	64.3	92.9	100.0	61.5	92.3	100.0	89.5	97.9	100.0	90.4	98.1	100.0	55.7	91.2
21, 22, 23	FIRE SUPPRESSION, PLUMBING & HVAC	96.7	37.1	72.6	99.9	58.7	83.2	97.7	46.4	76.9	93.9	86.7	91.0	99.9	94.9	97.8	94.2	44.0	73.9
26, 27, 3370	ELECTRICAL, COMMUNICATIONS & UTIL.	94.8	52.5	73.0	97.8	59.2	77.9	97.8	52.3	74.4	99.7	82.9	91.0	99.7	107.9	103.9	95.9	34.9	64.5
MF2010	WEIGHTED AVERAGE	96.3	54.6	77.9	96.7	66.1	83.2	98.0	57.9	80.4	98.4	90.5	94.9	99.7	89.6	95.3	96.1	70.0	84.6

MISSOURI

DIVISION		COLUMBIA 652			FLAT RIVER 636			HANNIBAL 634			HARRISONVILLE 647			JEFFERSON CITY 650 - 651			JOPLIN 648		
		MAT.	INST.	TOTAL	MAT.	INST.	TOTAL	MAT.	INST.	TOTAL	MAT.	INST.	TOTAL	MAT.	INST.	TOTAL	MAT.	INST.	TOTAL
015433	CONTRACTOR EQUIPMENT		107.0	107.0		106.1	106.1		106.1	106.1		102.2	102.2		107.0	107.0		105.7	105.7
0241, 31 - 34	SITE & INFRASTRUCTURE, DEMOLITION	105.3	95.9	98.8	94.1	93.6	93.8	89.0	93.4	92.0	91.7	95.8	94.5	101.7	95.8	97.6	100.4	98.5	99.1
0310	Concrete Forming & Accessories	87.3	79.5	80.5	97.1	87.1	88.4	90.2	77.3	79.0	87.6	104.1	101.9	96.7	79.3	81.6	101.1	73.5	77.1
0320	Concrete Reinforcing	99.3	115.6	107.5	102.1	105.8	104.0	100.3	96.8	98.5	97.1	111.9	104.5	98.1	106.7	102.4	100.6	72.3	86.3
0330	Cast-in-Place Concrete	89.5	82.6	86.8	97.7	94.2	96.3	89.6	75.1	83.9	95.3	106.7	99.8	90.0	82.5	87.1	100.7	66.1	87.1
03	CONCRETE	86.1	88.8	87.4	99.3	94.2	96.8	92.5	81.6	87.2	96.4	106.7	101.4	87.5	87.0	87.2	99.6	71.6	86.0
04	MASONRY	133.3	85.6	104.2	122.0	78.8	95.6	115.2	95.9	103.4	93.1	107.9	102.1	96.0	85.6	89.7	92.4	84.8	87.7
05	METALS	94.5	119.5	102.7	97.8	113.8	103.1	97.9	109.6	101.7	95.9	112.3	101.4	93.9	115.8	101.1	98.6	88.4	95.2
06	WOOD, PLASTICS & COMPOSITES	92.4	77.1	83.6	95.6	87.4	90.9	87.1	75.1	80.2	88.6	103.7	97.3	102.4	77.1	87.8	102.8	74.5	86.4
07	THERMAL & MOISTURE PROTECTION	93.3	86.0	90.3	101.2	93.8	98.2	100.8	92.3	97.4	96.9	107.7	101.3	93.4	86.0	90.4	97.4	72.2	87.1
08	OPENINGS	90.6	94.5	91.6	99.0	85.7	95.6	99.0	76.4	93.3	91.2	106.0	94.9	87.8	92.1	88.9	92.0	72.4	87.1
0920	Plaster & Gypsum Board	99.9	76.3	83.1	102.7	87.0	91.5	95.9	74.3	80.5	97.1	103.6	101.8	104.2	76.3	84.3	110.2	73.5	84.0
0950, 0980	Ceilings & Acoustic Treatment	94.9	76.3	82.6	90.1	87.0	88.0	90.1	74.3	79.7	89.5	103.6	98.8	94.0	76.3	82.3	90.4	73.5	79.2
0960	Flooring	112.4	95.9	107.5	104.5	88.0	99.6	100.3	95.8	99.0	89.0	107.3	94.4	120.0	95.9	112.9	118.8	42.2	96.3
0970, 0990	Wall Finishes & Painting/Coating	104.0	79.4	88.8	102.0	62.6	77.7	102.0	103.4	102.8	99.2	107.0	104.0	104.0	79.4	88.8	93.8	37.0	58.7
09	FINISHES	99.5	81.1	89.2	100.2	83.4	90.9	97.8	82.1	89.1	93.9	104.8	100.0	102.4	81.1	90.5	103.0	64.4	81.5
COVERS	DIVS. 10 - 14, 25, 28, 41, 43, 44, 46	100.0	95.3	99.1	100.0	92.6	98.5	100.0	54.7	91.0	100.0	97.6	99.5	100.0	95.3	99.1	100.0	62.0	92.4
21, 22, 23	FIRE SUPPRESSION, PLUMBING & HVAC	99.9	95.2	98.0	93.9	86.7	91.0	93.9	86.6	91.0	94.1	103.3	97.8	99.9	89.5	95.7	100.2	48.0	79.0
26, 27, 3370	ELECTRICAL, COMMUNICATIONS & UTIL.	95.6	88.0	91.7	104.2	107.9	106.1	98.4	82.9	90.4	102.6	101.7	102.1	99.3	88.0	93.4	93.7	69.0	81.0
MF2010	WEIGHTED AVERAGE	97.7	92.4	95.4	99.4	92.9	96.5	97.3	87.1	92.8	95.7	104.5	99.6	96.2	90.5	93.7	98.2	70.4	85.9

MISSOURI

DIVISION		KANSAS CITY 640 - 641			KIRKSVILLE 635			POPLAR BLUFF 639			ROLLA 654 - 655			SEDALIA 653			SIKESTON 638		
		MAT.	INST.	TOTAL	MAT.	INST.	TOTAL	MAT.	INST.	TOTAL	MAT.	INST.	TOTAL	MAT.	INST.	TOTAL	MAT.	INST.	TOTAL
015433	CONTRACTOR EQUIPMENT		103.4	103.4		98.5	98.5		100.9	100.9		107.0	107.0		99.2	99.2		100.9	100.9
0241, 31 - 34	SITE & INFRASTRUCTURE, DEMOLITION	93.2	98.2	96.7	90.3	89.5	89.8	77.2	92.9	88.1	104.8	94.0	97.3	96.7	91.8	93.3	80.7	93.2	89.3
0310	Concrete Forming & Accessories	100.1	107.9	106.9	83.5	77.9	78.6	83.7	72.6	74.0	93.9	90.4	90.8	91.3	79.8	81.3	84.7	73.6	75.0
0320	Concrete Reinforcing	95.6	116.5	106.1	101.2	86.0	93.5	104.2	83.3	93.7	99.8	67.8	83.7	98.3	104.7	101.5	103.6	83.2	93.3
0330	Cast-in-Place Concrete	93.7	108.0	99.3	97.6	75.1	88.8	75.1	79.7	77.0	91.6	85.5	89.2	95.7	83.2	90.8	80.3	82.0	81.0
03	CONCRETE	95.6	109.7	102.5	110.1	79.3	95.1	85.7	78.1	82.0	87.8	85.8	86.8	99.2	86.5	93.0	89.4	79.3	84.5
04	MASONRY	95.1	109.7	104.0	128.1	87.4	103.2	119.9	71.6	90.4	109.8	78.8	90.8	115.7	86.0	97.5	119.6	76.2	93.1
05	METALS	106.5	114.9	109.3	97.4	96.4	97.1	98.0	96.5	97.5	93.8	102.3	96.6	92.4	106.7	97.1	98.3	95.7	97.5
06	WOOD, PLASTICS & COMPOSITES	102.0	107.6	105.2	75.9	77.1	76.6	75.1	72.3	73.5	99.1	93.2	95.7	92.0	77.1	83.4	76.5	72.3	74.1
07	THERMAL & MOISTURE PROTECTION	97.0	109.6	102.1	103.3	85.8	98.6	105.9	80.9	95.7	93.4	91.0	92.4	98.8	93.0	96.4	106.1	79.4	95.2
08	OPENINGS	98.2	109.8	101.2	103.3	74.9	96.2	104.3	73.6	96.6	90.6	87.3	89.8	93.9	87.1	92.2	104.3	67.1	94.9
0920	Plaster & Gypsum Board	106.1	107.6	107.2	90.4	76.3	80.3	90.9	71.3	76.9	100.8	92.9	95.1	93.9	76.3	81.4	92.7	71.3	77.5
0950, 0980	Ceilings & Acoustic Treatment	96.7	107.6	103.9	88.3	76.3	80.4	90.1	71.3	77.7	94.9	92.9	93.6	94.9	76.3	82.6	90.1	71.3	77.7
0960	Flooring	94.3	109.0	98.6	77.5	95.9	82.9	92.2	81.4	89.1	116.3	95.9	110.3	88.4	97.1	90.9	92.7	81.4	89.4
0970, 0990	Wall Finishes & Painting/Coating	99.2	114.7	108.7	97.2	79.4	86.2	96.7	72.1	81.5	104.0	86.4	93.1	104.0	105.7	105.0	96.7	57.6	72.6
09	FINISHES	98.1	108.6	104.0	96.7	80.2	87.5	97.0	72.7	83.4	100.9	90.4	95.0	97.4	84.5	90.2	97.7	71.6	83.1
COVERS	DIVS. 10 - 14, 25, 28, 41, 43, 44, 46	100.0	101.5	100.3	100.0	53.1	90.7	100.0	88.4	97.7	100.0	57.1	91.5	100.0	89.7	97.9	100.0	90.0	98.0
21, 22, 23	FIRE SUPPRESSION, PLUMBING & HVAC	100.0	110.9	104.4	93.9	86.1	90.7	93.9	91.3	92.9	93.9	86.9	91.1	93.9	96.0	94.7	93.9	84.4	90.1
26, 27, 3370	ELECTRICAL, COMMUNICATIONS & UTIL.	104.3	102.0	103.1	98.6	82.8	90.5	98.9	107.8	103.5	94.1	87.9	90.9	95.7	101.6	98.8	97.9	107.8	103.0
MF2010	WEIGHTED AVERAGE	100.1	108.0	103.6	100.4	83.8	93.1	97.0	86.8	92.5	95.2	87.9	92.0	96.6	92.8	94.9	97.5	85.5	92.2

DIVISION		MISSOURI SPRINGFIELD 656 - 658			ST. JOSEPH 644 - 645			ST. LOUIS 630 - 631			MONTANA BILLINGS 590 - 591			BUTTE 597			GREAT FALLS 594		
		MAT.	INST.	TOTAL	MAT.	INST.	TOTAL	MAT.	INST.	TOTAL	MAT.	INST.	TOTAL	MAT.	INST.	TOTAL	MAT.	INST.	TOTAL
015433	CONTRACTOR EQUIPMENT		101.7	101.7		102.2	102.2		107.1	107.1		98.9	98.9		98.7	98.7		98.7	98.7
0241, 31 - 34	SITE & INFRASTRUCTURE, DEMOLITION	99.2	93.6	95.4	94.7	94.0	94.2	92.6	97.5	96.0	98.6	97.0	97.5	109.4	96.3	100.4	113.3	97.1	102.1
0310	Concrete Forming & Accessories	98.8	74.1	77.4	99.9	87.6	89.2	97.1	103.4	102.6	101.9	65.9	70.6	90.3	64.0	67.5	101.7	67.4	71.8
0320	Concrete Reinforcing	95.6	115.1	105.4	94.4	111.1	102.8	93.4	107.6	100.5	101.4	80.2	90.7	110.0	80.3	95.0	101.4	80.2	90.7
0330	Cast-in-Place Concrete	97.2	72.4	87.5	93.6	104.1	97.7	93.7	106.3	98.7	115.2	66.3	95.9	126.8	67.9	103.6	134.0	64.4	106.5
03	CONCRETE	96.3	82.1	89.4	95.4	98.4	96.9	94.9	106.0	100.3	106.3	69.7	88.5	110.4	69.5	90.5	115.5	69.7	93.2
04	MASONRY	89.5	84.5	86.4	94.9	93.9	94.3	101.5	109.4	106.3	125.4	69.9	91.5	120.7	71.7	90.8	125.3	73.9	93.9
05	METALS	98.4	106.2	101.0	102.5	110.5	105.1	103.8	118.1	108.5	110.2	89.8	103.5	103.0	89.9	98.7	106.8	89.8	101.2
06	WOOD, PLASTICS & COMPOSITES	99.0	73.3	84.2	102.8	85.1	92.5	95.5	101.7	99.1	97.8	66.2	79.6	86.9	63.6	73.4	98.9	66.2	80.0
07	THERMAL & MOISTURE PROTECTION	97.5	80.1	90.5	97.5	95.5	96.7	100.9	106.7	103.3	107.4	69.0	91.8	107.3	69.5	91.9	107.8	68.4	91.8
08	OPENINGS	96.1	81.5	92.4	96.8	96.4	96.7	98.2	108.6	100.9	96.6	66.3	89.0	94.6	66.1	87.4	98.0	65.6	89.8
0920	Plaster & Gypsum Board	102.2	72.3	80.9	111.8	84.4	92.2	103.9	101.7	102.3	96.9	65.5	74.4	100.6	62.7	73.6	108.9	65.5	77.9
0950, 0980	Ceilings & Acoustic Treatment	94.9	72.3	80.0	95.8	84.4	88.3	95.5	101.7	99.6	106.2	65.5	79.3	111.5	62.7	79.3	113.3	65.5	81.7
0960	Flooring	113.8	42.2	92.8	97.5	97.1	97.4	104.2	99.3	102.8	109.2	73.6	98.7	106.4	77.7	98.0	114.8	77.7	103.9
0970, 0990	Wall Finishes & Painting/Coating	98.0	52.9	70.2	94.3	88.8	90.9	102.0	106.7	104.9	97.5	54.5	71.0	96.0	54.0	70.1	96.0	51.3	68.4
09	FINISHES	103.2	65.4	82.1	99.2	88.7	93.4	101.4	102.4	101.9	104.5	65.7	82.8	105.0	65.0	82.7	109.2	67.0	85.7
COVERS	DIVS. 10 - 14, 25, 28, 41, 43, 44, 46	100.0	93.4	98.7	100.0	96.8	99.4	100.0	102.3	100.5	100.0	74.9	95.0	100.0	74.7	95.0	100.0	76.3	95.3
21, 22, 23	FIRE SUPPRESSION, PLUMBING & HVAC	99.9	70.8	88.1	100.2	87.3	94.9	99.9	101.7	100.6	100.2	71.4	88.5	100.3	68.4	87.4	100.3	73.5	89.4
26, 27, 3370	ELECTRICAL, COMMUNICATIONS & UTIL.	99.8	72.4	85.7	102.7	76.5	89.2	102.2	108.3	105.3	97.3	76.7	86.7	104.9	72.6	88.2	97.5	73.5	85.1
MF2010	WEIGHTED AVERAGE	98.5	79.9	90.3	99.3	91.8	96.0	100.0	105.7	102.5	103.5	74.7	90.8	103.5	73.5	90.3	105.0	75.3	91.9

MONTANA

DIVISION		HAVRE 595			HELENA 596			KALISPELL 599			MILES CITY 593			MISSOULA 598			WOLF POINT 592		
		MAT.	INST.	TOTAL	MAT.	INST.	TOTAL	MAT.	INST.	TOTAL	MAT.	INST.	TOTAL	MAT.	INST.	TOTAL	MAT.	INST.	TOTAL
015433	CONTRACTOR EQUIPMENT		98.7	98.7		98.7	98.7		98.7	98.7		98.7	98.7		98.7	98.7		98.7	98.7
0241, 31 - 34	SITE & INFRASTRUCTURE, DEMOLITION	118.8	96.7	103.6	101.9	96.2	98.0	98.2	96.0	96.7	105.9	96.4	99.3	89.2	96.2	94.0	127.4	96.5	106.1
0310	Concrete Forming & Accessories	84.2	57.2	60.7	101.9	66.6	71.2	93.2	55.2	60.1	100.4	55.7	61.5	93.2	62.3	66.4	93.7	50.7	56.3
0320	Concrete Reinforcing	110.8	64.3	87.4	104.9	64.0	84.3	112.9	80.0	96.3	110.4	64.3	87.2	111.8	80.2	95.9	112.0	64.3	88.0
0330	Cast-in-Place Concrete	136.5	62.6	107.5	104.0	63.3	88.0	110.1	60.6	90.6	120.5	62.6	97.7	93.4	62.6	81.3	135.0	63.1	106.6
03	CONCRETE	118.9	61.7	91.0	101.0	65.9	83.9	99.8	62.9	81.8	107.2	60.9	84.6	87.9	66.8	77.7	122.5	58.8	91.5
04	MASONRY	121.6	72.0	91.3	116.6	68.5	87.2	119.6	66.9	87.4	127.2	67.2	90.6	145.2	70.2	99.4	128.3	67.9	91.5
05	METALS	99.4	82.0	93.7	106.0	81.3	97.8	99.3	88.2	95.6	98.6	81.9	93.1	99.8	89.0	96.3	98.7	81.8	93.1
06	WOOD, PLASTICS & COMPOSITES	79.6	54.1	64.9	98.9	66.2	80.0	89.8	52.4	68.2	96.4	54.5	72.2	89.8	62.4	74.0	88.9	47.5	65.0
07	THERMAL & MOISTURE PROTECTION	107.6	65.6	90.5	107.6	70.4	92.5	106.9	66.5	90.5	107.2	63.8	89.5	106.5	68.9	91.2	108.0	62.8	89.6
08	OPENINGS	94.7	54.5	84.6	97.4	60.0	88.0	94.7	58.1	85.5	94.2	54.7	84.2	94.7	63.3	86.8	94.2	50.7	83.2
0920	Plaster & Gypsum Board	96.1	52.9	65.3	97.7	65.5	74.7	100.6	51.2	65.3	107.8	53.4	68.9	100.6	61.5	72.7	102.3	46.2	62.2
0950, 0980	Ceilings & Acoustic Treatment	111.5	52.9	72.8	110.6	65.5	80.8	111.5	51.2	71.7	109.7	53.4	72.5	111.5	61.5	78.5	109.7	46.2	67.7
0960	Flooring	103.1	77.7	95.6	114.8	61.9	99.3	108.5	77.7	99.4	114.6	73.6	102.5	108.5	77.7	99.4	110.2	73.6	99.4
0970, 0990	Wall Finishes & Painting/Coating	96.0	51.3	68.4	96.0	45.6	64.9	96.0	39.4	61.1	96.0	47.3	66.0	96.0	50.2	67.7	96.0	51.3	68.4
09	FINISHES	104.3	59.4	79.2	106.6	63.7	82.7	104.9	56.7	78.0	107.6	57.3	79.6	104.2	63.5	81.5	107.4	53.7	77.4
COVERS	DIVS. 10 - 14, 25, 28, 41, 43, 44, 46	100.0	74.3	94.9	100.0	76.3	95.3	100.0	73.4	94.7	100.0	72.7	94.6	100.0	74.1	94.9	100.0	72.1	94.5
21, 22, 23	FIRE SUPPRESSION, PLUMBING & HVAC	94.3	72.1	85.3	100.2	69.9	87.9	94.3	68.1	83.7	94.3	69.6	84.3	100.3	67.6	87.0	94.3	70.0	84.5
26, 27, 3370	ELECTRICAL, COMMUNICATIONS & UTIL.	97.5	66.0	81.2	101.9	73.5	87.3	101.7	71.9	86.3	97.5	68.8	82.7	102.7	71.9	86.8	97.5	64.0	80.2
MF2010	WEIGHTED AVERAGE	102.0	70.2	88.0	102.6	72.0	89.1	99.6	70.1	86.6	100.7	69.1	86.8	100.8	72.2	88.2	103.1	67.5	87.4

NEBRASKA

DIVISION		ALLIANCE 693			COLUMBUS 686			GRAND ISLAND 688			HASTINGS 689			LINCOLN 683 - 685			MCCOOK 690		
		MAT.	INST.	TOTAL	MAT.	INST.	TOTAL	MAT.	INST.	TOTAL	MAT.	INST.	TOTAL	MAT.	INST.	TOTAL	MAT.	INST.	TOTAL
015433	CONTRACTOR EQUIPMENT		97.0	97.0		100.8	100.8		100.8	100.8		100.8	100.8		100.8	100.8		100.8	100.8
0241, 31 - 34	SITE & INFRASTRUCTURE, DEMOLITION	99.4	97.5	98.1	98.7	90.9	93.3	103.4	92.1	95.6	102.3	91.3	94.7	91.0	92.1	91.7	103.2	91.1	94.9
0310	Concrete Forming & Accessories	88.2	56.3	60.5	96.3	59.1	64.0	95.9	76.0	78.6	98.8	74.3	77.4	100.2	60.3	65.5	93.7	56.6	61.5
0320	Concrete Reinforcing	113.6	47.6	80.3	104.7	46.4	75.3	104.1	75.5	89.7	104.1	47.1	75.4	95.6	74.9	85.2	106.1	75.0	90.4
0330	Cast-in-Place Concrete	108.2	58.5	88.6	110.2	55.4	88.6	116.7	69.8	98.2	116.7	64.1	96.0	87.2	72.3	81.3	116.7	58.2	93.7
03	CONCRETE	117.8	56.1	87.7	104.1	56.6	81.0	108.7	74.4	92.0	108.8	66.4	88.2	91.3	68.3	80.1	108.8	61.8	85.9
04	MASONRY	110.1	66.8	83.7	115.3	71.6	88.7	108.6	79.0	90.6	117.7	88.6	99.9	96.8	69.7	80.2	105.1	66.9	81.8
05	METALS	100.6	65.3	88.9	92.7	74.6	86.8	94.5	86.3	91.8	95.2	76.2	88.9	99.6	85.3	94.9	95.5	84.2	91.7
06	WOOD, PLASTICS & COMPOSITES	85.2	55.5	68.0	96.4	60.3	75.6	95.8	76.6	84.8	99.3	76.8	86.3	98.5	55.8	73.8	93.9	55.5	71.7
07	THERMAL & MOISTURE PROTECTION	104.5	67.8	89.6	98.3	64.8	84.7	98.4	75.3	89.0	98.5	86.7	93.7	97.9	70.6	86.8	98.4	65.0	84.8
08	OPENINGS	89.9	51.7	80.3	89.8	56.2	81.4	89.8	72.7	85.5	89.8	65.2	83.6	95.3	57.3	85.7	89.9	58.7	82.0
0920	Plaster & Gypsum Board	91.3	54.1	64.7	100.4	59.0	70.8	99.9	76.0	82.8	101.7	76.0	83.4	101.6	54.4	67.9	99.9	54.1	67.2
0950, 0980	Ceilings & Acoustic Treatment	99.2	54.1	69.4	94.9	59.0	71.2	94.9	76.0	82.4	94.9	76.0	82.4	97.6	54.4	69.0	94.9	54.1	67.9
0960	Flooring	103.9	86.0	98.6	102.6	105.1	103.3	102.4	112.3	105.3	103.8	105.1	104.2	104.0	87.5	99.2	101.5	86.0	96.9
0970, 0990	Wall Finishes & Painting/Coating	173.8	57.7	102.2	96.0	67.2	78.2	96.0	71.3	80.7	96.0	67.2	78.2	96.0	82.1	87.4	96.0	50.3	67.8
09	FINISHES	103.3	61.1	79.7	100.1	67.4	81.8	100.3	81.4	89.8	100.9	78.6	88.5	100.9	66.7	81.9	100.9	60.2	77.8
COVERS	DIVS. 10 - 14, 25, 28, 41, 43, 44, 46	100.0	68.0	93.6	100.0	77.2	95.5	100.0	82.6	96.5	100.0	81.2	96.3	100.0	80.2	96.1	100.0	67.3	93.5
21, 22, 23	FIRE SUPPRESSION, PLUMBING & HVAC	94.2	59.4	80.1	94.0	72.3	85.2	99.9	78.2	91.1	94.0	75.4	86.5	99.8	78.2	91.1	94.0	72.8	85.4
26, 27, 3370	ELECTRICAL, COMMUNICATIONS & UTIL.	92.7	73.7	82.9	92.3	81.9	86.9	91.0	67.3	78.8	90.4	88.8	89.6	99.8	67.3	83.0	94.8	73.7	83.9
MF2010	WEIGHTED AVERAGE	99.7	65.8	84.8	96.8	71.7	85.7	98.7	78.4	89.8	97.8	79.3	89.6	98.0	73.6	87.2	97.6	70.7	85.7

NEBRASKA / NEVADA

DIVISION		NORFOLK 687			NORTH PLATTE 691			OMAHA 680 - 681			VALENTINE 692			CARSON CITY 897			ELKO 898		
		MAT.	INST.	TOTAL	MAT.	INST.	TOTAL	MAT.	INST.	TOTAL	MAT.	INST.	TOTAL	MAT.	INST.	TOTAL	MAT.	INST.	TOTAL
015433	CONTRACTOR EQUIPMENT		92.2	92.2		100.8	100.8		92.2	92.2		95.1	95.1		99.3	99.3		99.3	99.3
0241, 31 - 34	SITE & INFRASTRUCTURE, DEMOLITION	81.1	90.5	87.6	104.3	91.3	95.3	81.5	91.1	88.1	86.2	95.0	92.3	69.7	101.3	91.5	56.0	101.0	87.0
0310	Concrete Forming & Accessories	83.1	73.4	74.7	95.8	74.1	76.9	91.5	73.1	75.5	84.3	55.9	59.6	101.5	97.6	98.1	108.0	84.0	87.1
0320	Concrete Reinforcing	104.8	49.1	76.7	105.5	76.1	90.7	100.1	75.5	87.7	106.1	46.1	75.9	105.0	117.5	111.3	106.7	91.8	99.2
0330	Cast-in-Place Concrete	111.4	63.3	92.4	116.7	65.5	96.5	93.6	78.7	87.7	103.2	55.9	84.6	94.6	87.8	91.9	95.4	83.5	90.7
03	CONCRETE	103.0	65.7	84.9	108.9	72.2	91.0	93.9	75.7	85.0	105.5	54.9	80.8	100.1	97.9	99.0	97.9	85.5	91.9
04	MASONRY	122.3	69.5	90.1	92.9	88.6	90.3	102.3	79.9	88.6	105.4	67.0	81.9	116.9	76.6	92.3	119.2	69.9	89.1
05	METALS	96.0	68.0	86.8	94.7	86.2	91.9	98.9	78.6	92.2	106.5	66.3	93.3	94.1	103.2	97.1	93.2	90.7	92.4
06	WOOD, PLASTICS & COMPOSITES	80.0	76.3	77.9	95.7	76.8	84.8	89.6	72.5	79.8	79.7	55.0	65.5	87.7	100.4	95.1	96.7	84.4	89.6
07	THERMAL & MOISTURE PROTECTION	97.9	73.7	88.1	98.4	80.6	91.2	94.1	80.5	88.6	98.8	67.6	86.1	99.6	87.0	94.5	100.9	75.9	90.7
08	OPENINGS	91.3	65.7	84.9	89.1	72.0	84.8	98.0	72.4	91.6	91.4	52.9	81.7	97.0	112.6	100.9	99.8	82.2	95.4
0920	Plaster & Gypsum Board	100.7	76.0	83.1	99.9	76.0	82.8	104.9	72.1	81.5	102.6	54.1	67.9	90.4	100.3	97.5	94.5	83.8	86.9
0950, 0980	Ceilings & Acoustic Treatment	109.8	76.0	87.5	94.9	76.0	82.4	111.6	72.1	85.5	113.1	54.1	74.1	109.7	100.3	103.5	112.4	83.8	93.5
0960	Flooring	130.7	105.1	123.2	102.4	105.1	103.2	135.4	85.8	120.8	133.2	83.6	118.6	101.1	82.0	95.5	106.6	53.5	91.0
0970, 0990	Wall Finishes & Painting/Coating	172.0	67.2	107.3	96.0	67.2	78.2	172.0	68.9	108.4	172.0	69.6	108.8	99.8	83.3	89.6	98.8	83.3	89.2
09	FINISHES	121.5	78.6	97.6	100.3	78.6	88.2	123.9	74.7	96.5	123.8	61.5	89.0	100.8	93.4	96.7	101.9	78.3	88.7
COVERS	DIVS. 10 - 14, 25, 28, 41, 43, 44, 46	100.0	80.0	96.0	100.0	70.8	94.2	100.0	81.0	96.2	100.0	65.2	93.1	100.0	97.1	99.4	100.0	67.8	93.6
21, 22, 23	FIRE SUPPRESSION, PLUMBING & HVAC	93.8	75.3	86.3	99.9	73.7	89.3	99.7	78.0	90.9	93.7	57.5	79.0	99.9	84.2	93.5	97.2	83.8	91.8
26, 27, 3370	ELECTRICAL, COMMUNICATIONS & UTIL.	91.2	82.5	86.7	93.0	81.9	87.3	99.2	82.8	90.8	89.9	92.0	91.0	91.1	98.5	94.9	93.9	98.4	96.3
MF2010	WEIGHTED AVERAGE	98.5	75.1	88.2	98.1	79.4	89.9	100.0	79.2	90.8	99.7	67.6	85.6	97.8	93.5	95.9	97.2	85.3	92.0

Table 1

DIVISION		NEVADA									NEW HAMPSHIRE								
		ELY 893			LAS VEGAS 889-891			RENO 894-895			CHARLESTON 036			CLAREMONT 037			CONCORD 032-033		
		MAT.	INST.	TOTAL	MAT.	INST.	TOTAL	MAT.	INST.	TOTAL	MAT.	INST.	TOTAL	MAT.	INST.	TOTAL	MAT.	INST.	TOTAL
015433	CONTRACTOR EQUIPMENT		99.3	99.3		99.3	99.3		99.3	99.3		100.4	100.4		100.4	100.4		100.4	100.4
0241, 31-34	SITE & INFRASTRUCTURE, DEMOLITION	60.8	101.2	88.6	63.1	103.8	91.1	60.7	101.3	88.7	85.6	96.0	92.8	79.1	96.0	90.7	91.9	98.0	96.1
0310	Concrete Forming & Accessories	101.8	104.9	104.5	103.6	114.8	113.3	98.9	97.6	97.7	88.6	43.4	49.3	93.8	43.4	50.0	95.5	87.1	88.2
0320	Concrete Reinforcing	105.5	85.0	95.1	97.2	125.4	111.4	99.7	124.1	112.0	82.9	46.7	64.6	82.9	46.7	64.6	82.9	91.9	87.4
0330	Cast-in-Place Concrete	102.5	103.1	102.7	99.2	110.8	103.8	108.4	87.8	100.3	100.0	50.2	80.4	91.9	50.2	75.5	108.8	86.5	100.0
03	CONCRETE	106.0	100.3	103.3	101.3	114.9	107.9	106.0	99.0	102.6	104.8	47.3	76.8	96.2	47.3	72.4	107.6	87.7	97.9
04	MASONRY	124.0	79.1	96.6	111.6	106.7	108.6	118.3	76.6	92.8	90.1	37.8	58.2	90.1	37.8	58.2	93.0	80.2	85.2
05	METALS	93.1	90.2	92.2	100.3	110.4	103.6	94.6	105.8	98.3	92.3	62.2	82.3	92.3	62.2	82.3	93.0	88.7	91.6
06	WOOD, PLASTICS & COMPOSITES	88.6	106.8	99.1	87.8	114.8	103.4	84.3	100.4	93.6	86.7	45.0	62.6	92.5	45.0	65.0	96.2	91.2	93.3
07	THERMAL & MOISTURE PROTECTION	101.5	87.3	95.7	115.1	101.8	109.7	101.0	87.0	95.3	100.8	41.5	76.7	100.6	41.5	76.6	99.4	84.9	93.5
08	OPENINGS	99.7	92.5	97.9	98.6	122.2	104.6	97.4	108.0	100.1	99.1	40.7	84.4	100.3	40.7	85.3	100.3	81.9	95.7
0920	Plaster & Gypsum Board	90.9	106.9	102.3	90.9	115.2	108.3	83.8	100.3	95.6	90.2	42.5	56.1	91.5	42.5	56.5	94.2	90.3	91.4
0950, 0980	Ceilings & Acoustic Treatment	112.4	106.9	108.8	121.4	115.2	117.3	117.8	100.3	106.3	92.7	42.5	59.6	92.7	42.5	59.6	91.8	90.3	90.8
0960	Flooring	104.0	59.9	91.0	94.9	103.4	97.4	100.7	82.0	95.2	89.2	33.6	72.9	91.6	33.6	74.6	89.2	93.7	90.5
0970, 0990	Wall Finishes & Painting/Coating	98.8	118.3	110.8	101.6	122.1	114.2	98.8	83.3	89.2	90.1	48.2	64.3	90.1	48.2	64.3	90.1	100.6	96.6
09	FINISHES	101.3	98.9	100.0	100.7	113.8	108.0	100.5	93.4	96.5	93.7	42.6	65.2	93.9	42.6	65.3	94.9	90.6	92.5
COVERS	DIVS. 10-14, 25, 28, 41, 43, 44, 46	100.0	66.9	93.4	100.0	93.2	98.7	100.0	97.1	99.4	100.0	38.6	87.8	100.0	38.6	87.8	100.0	62.1	92.5
21, 22, 23	FIRE SUPPRESSION, PLUMBING & HVAC	97.2	104.8	100.3	100.1	108.5	103.5	100.0	84.2	93.6	94.1	39.3	71.9	94.1	39.4	71.9	99.9	83.5	93.2
26, 27, 3370	ELECTRICAL, COMMUNICATIONS & UTIL.	94.2	112.0	103.4	98.5	116.6	107.8	94.5	98.5	96.5	99.5	34.1	65.8	99.5	34.1	65.8	98.3	59.4	78.2
MF2010	WEIGHTED AVERAGE	98.5	98.1	98.3	99.8	110.7	104.6	98.7	93.7	96.5	96.3	47.1	74.6	95.3	47.1	74.1	98.6	82.4	91.5

Table 2

DIVISION		NEW HAMPSHIRE															NEW JERSEY		
		KEENE 034			LITTLETON 035			MANCHESTER 031			NASHUA 030			PORTSMOUTH 038			ATLANTIC CITY 082,084		
		MAT.	INST.	TOTAL	MAT.	INST.	TOTAL	MAT.	INST.	TOTAL	MAT.	INST.	TOTAL	MAT.	INST.	TOTAL	MAT.	INST.	TOTAL
015433	CONTRACTOR EQUIPMENT		100.4	100.4		100.4	100.4		100.4	100.4		100.4	100.4		100.4	100.4		98.7	98.7
0241, 31-34	SITE & INFRASTRUCTURE, DEMOLITION	94.3	96.2	95.6	79.0	95.7	90.5	89.6	98.1	95.4	94.9	98.1	97.1	88.6	99.5	96.1	98.9	103.0	101.7
0310	Concrete Forming & Accessories	92.6	41.5	48.1	102.2	46.3	53.6	97.6	87.7	89.3	89.7	92.6	92.3	89.7	92.6	92.3	107.0	124.2	121.7
0320	Concrete Reinforcing	82.9	55.6	69.1	83.6	33.8	58.5	103.6	92.0	97.7	103.6	92.0	99.7	82.9	92.1	87.5	79.0	114.8	97.0
0330	Cast-in-Place Concrete	100.4	51.7	81.2	90.1	52.4	75.3	112.0	108.4	110.5	95.1	108.4	100.3	90.2	115.8	100.3	85.3	130.0	102.9
03	CONCRETE	104.5	48.6	77.3	95.4	46.9	71.8	112.5	95.5	104.2	104.2	95.5	100.0	95.5	100.4	97.9	97.1	123.4	109.9
04	MASONRY	93.9	40.0	61.0	101.2	43.4	65.9	95.6	80.2	86.2	94.7	80.2	85.9	90.6	92.3	91.6	97.3	127.6	115.8
05	METALS	93.0	66.3	84.1	93.0	57.0	81.1	98.0	89.3	95.1	98.0	89.3	95.1	94.5	92.1	93.7	92.7	104.6	96.6
06	WOOD, PLASTICS & COMPOSITES	90.9	40.1	61.6	100.8	45.9	69.1	96.2	91.2	93.3	96.1	91.2	94.7	87.9	91.2	89.8	109.4	123.9	117.8
07	THERMAL & MOISTURE PROTECTION	101.2	43.5	77.7	100.7	47.3	79.0	100.7	88.2	95.6	101.5	88.2	96.1	101.2	112.4	105.7	101.2	124.8	110.8
08	OPENINGS	97.6	45.4	84.5	101.4	38.5	85.6	102.4	81.9	97.2	102.4	81.9	97.2	103.1	77.2	96.6	100.0	121.6	105.4
0920	Plaster & Gypsum Board	90.6	37.6	52.7	103.3	43.5	60.6	102.2	90.3	93.7	98.7	90.3	92.7	90.2	90.3	90.2	102.9	124.0	118.0
0950, 0980	Ceilings & Acoustic Treatment	92.7	37.6	56.3	92.7	43.5	60.2	98.4	90.3	93.0	102.0	90.3	94.2	94.8	90.3	91.8	92.7	124.0	113.4
0960	Flooring	91.1	55.1	80.6	100.4	33.6	80.8	95.4	93.7	94.9	94.5	93.7	94.2	89.6	93.7	90.8	98.1	133.1	108.4
0970, 0990	Wall Finishes & Painting/Coating	90.1	48.2	64.3	90.1	63.2	73.5	90.1	100.6	96.6	90.1	100.6	96.6	90.1	100.6	96.6	90.1	121.1	109.2
09	FINISHES	95.8	44.4	67.1	97.8	46.0	68.9	98.5	90.6	94.1	99.7	90.6	94.6	95.1	93.5	94.2	100.0	125.9	114.4
COVERS	DIVS. 10-14, 25, 28, 41, 43, 44, 46	100.0	52.8	90.6	100.0	40.4	88.1	100.0	62.2	92.5	100.0	62.2	92.5	100.0	66.2	93.3	100.0	110.4	102.1
21, 22, 23	FIRE SUPPRESSION, PLUMBING & HVAC	94.1	42.6	73.2	94.1	65.6	82.5	99.9	83.5	93.2	100.0	83.5	93.3	100.0	89.6	95.8	99.6	123.3	109.2
26, 27, 3370	ELECTRICAL, COMMUNICATIONS & UTIL.	99.5	34.1	65.8	100.7	59.3	79.3	100.3	81.4	90.6	102.1	81.4	91.5	100.0	81.5	90.5	97.2	136.9	117.7
MF2010	WEIGHTED AVERAGE	96.9	49.4	76.0	96.5	56.6	78.9	100.7	86.7	94.5	100.2	86.7	94.3	97.8	91.2	94.9	98.1	122.1	108.7

Table 3

DIVISION		NEW JERSEY																	
		CAMDEN 081			DOVER 078			ELIZABETH 072			HACKENSACK 076			JERSEY CITY 073			LONG BRANCH 077		
		MAT.	INST.	TOTAL	MAT.	INST.	TOTAL	MAT.	INST.	TOTAL	MAT.	INST.	TOTAL	MAT.	INST.	TOTAL	MAT.	INST.	TOTAL
015433	CONTRACTOR EQUIPMENT		98.7	98.7		100.4	100.4		100.4	100.4		100.4	100.4		98.7	98.7		98.3	98.3
0241, 31-34	SITE & INFRASTRUCTURE, DEMOLITION	99.4	103.5	102.3	111.1	104.1	106.3	115.9	104.1	107.8	111.3	104.6	106.7	99.4	104.5	102.9	105.2	104.4	104.7
0310	Concrete Forming & Accessories	99.6	124.3	121.1	96.2	125.0	121.2	106.6	125.0	122.6	96.2	125.0	121.2	99.6	125.0	121.7	100.1	124.7	121.5
0320	Concrete Reinforcing	103.6	116.4	110.1	79.8	123.6	101.9	79.8	123.6	101.9	79.8	123.6	101.9	103.6	123.6	113.7	79.8	123.6	101.9
0330	Cast-in-Place Concrete	82.7	131.7	102.0	106.0	130.9	115.8	91.0	130.9	106.8	103.6	131.0	114.4	82.7	131.0	101.7	91.9	129.9	106.9
03	CONCRETE	98.0	124.3	110.8	105.8	125.9	115.6	100.6	125.0	112.9	103.8	125.9	114.6	98.0	125.8	111.5	102.8	125.3	113.7
04	MASONRY	89.2	127.6	112.6	93.0	128.1	114.4	108.0	128.1	120.2	96.7	128.1	115.8	86.5	128.1	111.9	100.9	127.6	117.1
05	METALS	97.8	104.7	100.1	92.7	111.9	99.1	94.2	111.9	100.1	92.8	111.9	99.1	97.9	109.6	101.8	92.8	109.4	98.3
06	WOOD, PLASTICS & COMPOSITES	99.5	123.9	113.6	98.9	123.9	113.4	112.4	123.9	119.1	98.9	123.9	113.4	99.5	123.9	113.6	100.5	123.9	114.0
07	THERMAL & MOISTURE PROTECTION	101.1	124.5	110.6	101.6	126.4	111.6	101.7	126.4	111.7	101.3	123.9	110.5	101.1	126.4	111.4	101.2	125.0	110.9
08	OPENINGS	102.4	121.6	107.2	105.4	123.6	110.0	103.4	123.6	108.5	102.6	123.6	107.9	102.4	123.6	107.8	98.4	123.6	104.8
0920	Plaster & Gypsum Board	98.7	124.0	116.8	95.2	124.0	115.8	102.4	124.0	118.0	95.2	124.0	115.8	98.7	124.0	116.8	97.4	124.0	116.4
0950, 0980	Ceilings & Acoustic Treatment	102.0	124.0	116.5	92.7	124.0	113.4	94.8	124.0	114.1	92.7	124.0	113.4	102.0	124.0	116.5	92.7	124.0	113.4
0960	Flooring	94.5	133.1	105.8	93.4	164.3	114.2	98.7	164.3	117.9	93.4	164.3	114.2	94.5	164.3	114.9	94.7	164.3	115.1
0970, 0990	Wall Finishes & Painting/Coating	90.1	121.1	109.2	90.0	123.1	110.4	90.0	123.1	110.4	90.0	123.1	110.4	90.0	123.1	110.4	90.1	121.1	109.2
09	FINISHES	100.2	125.9	114.6	97.8	131.5	116.6	101.4	131.5	118.2	97.6	131.5	116.5	100.2	131.5	117.7	98.8	131.1	116.9
COVERS	DIVS. 10-14, 25, 28, 41, 43, 44, 46	100.0	110.5	102.1	100.0	110.3	102.1	100.0	110.3	102.1	100.0	110.4	102.1	100.0	110.4	102.1	100.0	110.3	102.1
21, 22, 23	FIRE SUPPRESSION, PLUMBING & HVAC	100.0	124.9	110.1	99.6	125.7	110.2	100.0	125.6	110.4	99.6	125.7	110.2	100.0	125.7	110.4	99.6	124.7	109.8
26, 27, 3370	ELECTRICAL, COMMUNICATIONS & UTIL.	102.6	136.9	120.3	98.6	137.0	118.4	99.3	137.0	118.8	98.6	136.9	118.4	103.8	136.9	120.9	98.3	132.7	116.0
MF2010	WEIGHTED AVERAGE	99.4	122.6	109.6	99.7	124.6	110.7	100.6	124.5	111.2	99.4	124.5	110.5	99.4	124.4	110.4	99.0	123.3	109.7

City Cost Indexes

NEW JERSEY

DIVISION		NEW BRUNSWICK 088 - 089			NEWARK 070 - 071			PATERSON 074 - 075			POINT PLEASANT 087			SUMMIT 079			TRENTON 085 - 086		
		MAT.	INST.	TOTAL	MAT.	INST.	TOTAL	MAT.	INST.	TOTAL	MAT.	INST.	TOTAL	MAT.	INST.	TOTAL	MAT.	INST.	TOTAL
015433	CONTRACTOR EQUIPMENT		98.3	98.3		100.4	100.4		100.4	100.4		98.3	98.3		100.4	100.4		98.3	98.3
0241, 31 - 34	SITE & INFRASTRUCTURE, DEMOLITION	114.2	104.4	107.5	115.2	104.1	107.5	113.0	104.6	107.2	115.7	104.4	107.9	113.1	104.1	106.9	96.1	104.4	101.8
0310	Concrete Forming & Accessories	102.1	124.9	121.9	96.0	125.0	121.2	98.1	125.0	121.5	97.5	124.4	120.9	98.8	125.0	121.6	96.7	124.6	121.0
0320	Concrete Reinforcing	79.8	123.6	101.9	103.6	123.6	113.7	103.6	123.6	113.7	79.8	123.5	101.8	79.8	123.6	101.9	103.6	119.0	111.3
0330	Cast-in-Place Concrete	105.3	130.2	115.1	109.9	131.0	118.2	105.4	131.0	115.5	105.3	129.8	115.0	88.0	131.0	104.9	101.5	129.7	112.7
03	CONCRETE	115.3	125.5	120.3	111.4	125.9	118.5	109.3	125.9	117.4	115.0	125.1	119.9	97.4	125.9	111.3	107.3	124.4	115.6
04	MASONRY	95.5	128.0	115.3	98.2	128.1	116.4	93.5	128.1	114.6	86.9	127.6	111.7	95.2	128.1	115.3	85.8	127.6	111.3
05	METALS	92.8	109.5	98.3	97.9	111.9	102.5	93.0	111.9	99.3	92.8	109.2	98.2	92.7	111.9	99.1	93.0	107.2	97.7
06	WOOD, PLASTICS & COMPOSITES	103.6	123.9	115.3	98.2	123.9	113.1	101.3	123.9	114.4	97.2	123.9	112.6	102.4	123.9	114.8	96.6	123.9	112.4
07	THERMAL & MOISTURE PROTECTION	101.4	125.2	111.1	101.3	126.4	111.5	101.6	123.9	110.7	101.4	125.0	111.0	102.0	126.4	111.9	97.2	124.6	108.3
08	OPENINGS	94.5	123.6	101.8	94.5	123.6	101.8	108.3	123.6	112.1	96.5	124.5	103.6	110.6	123.6	113.9	103.0	122.1	107.8
0920	Plaster & Gypsum Board	99.7	124.0	117.1	96.2	124.0	116.1	98.7	124.0	116.8	95.2	124.0	115.8	97.4	124.0	116.4	96.5	124.0	116.1
0950, 0980	Ceilings & Acoustic Treatment	92.7	124.0	113.4	98.4	124.0	115.3	102.0	124.0	116.5	92.7	124.0	113.4	92.7	124.0	113.4	99.3	124.0	115.6
0960	Flooring	95.9	164.3	116.0	94.7	164.3	115.1	94.5	164.3	114.9	93.4	133.1	105.1	94.8	164.3	115.2	94.8	164.3	115.2
0970, 0990	Wall Finishes & Painting/Coating	90.1	123.1	110.5	90.0	123.1	110.4	90.0	123.1	110.4	90.1	121.1	109.2	90.0	123.1	110.4	90.1	121.1	109.2
09	FINISHES	100.5	131.5	117.8	99.0	131.5	117.2	100.4	131.5	117.8	99.3	125.9	114.1	98.9	131.5	117.1	98.8	131.1	116.9
COVERS	DIVS. 10 - 14, 25, 28, 41, 43, 44, 46	100.0	110.2	102.0	100.0	110.3	102.1	100.0	110.4	102.1	100.0	107.1	101.4	100.0	110.3	102.1	100.0	110.3	102.1
21, 22, 23	FIRE SUPPRESSION, PLUMBING & HVAC	99.6	125.0	109.9	100.1	126.1	110.6	100.0	125.7	110.4	99.6	124.5	109.7	99.6	126.1	110.3	100.2	124.7	110.1
26, 27, 3370	ELECTRICAL, COMMUNICATIONS & UTIL.	98.0	138.0	118.7	105.5	136.9	121.7	103.8	137.0	120.9	97.2	132.7	115.5	99.3	137.0	118.8	105.1	137.6	121.9
MF2010	WEIGHTED AVERAGE	100.1	124.3	110.8	102.7	124.6	112.4	101.4	124.5	111.6	99.7	122.5	109.7	99.6	124.7	110.7	99.6	123.6	110.2

DIVISION		NEW JERSEY VINELAND 080,083			NEW MEXICO														
					ALBUQUERQUE 870 - 872			CARRIZOZO 883			CLOVIS 881			FARMINGTON 874			GALLUP 873		
		MAT.	INST.	TOTAL	MAT.	INST.	TOTAL	MAT.	INST.	TOTAL	MAT.	INST.	TOTAL	MAT.	INST.	TOTAL	MAT.	INST.	TOTAL
015433	CONTRACTOR EQUIPMENT		98.7	98.7		111.1	111.1		111.1	111.1		111.1	111.1		111.1	111.1		111.1	111.1
0241, 31 - 34	SITE & INFRASTRUCTURE, DEMOLITION	104.1	103.5	103.7	78.9	106.9	98.2	100.5	106.9	104.9	87.9	106.9	101.0	85.1	106.9	100.1	93.6	106.9	102.8
0310	Concrete Forming & Accessories	95.3	124.3	120.5	99.4	65.8	70.2	99.4	65.8	70.2	99.4	65.6	70.0	99.4	65.8	70.2	99.4	65.8	70.2
0320	Concrete Reinforcing	79.0	114.8	97.0	100.4	71.1	85.6	107.7	71.1	89.2	108.9	71.0	89.8	109.7	71.1	90.2	105.0	71.1	87.9
0330	Cast-in-Place Concrete	91.9	130.0	106.9	101.4	73.4	90.4	94.3	73.4	86.1	94.3	73.4	86.0	102.3	73.4	90.9	96.3	73.4	87.3
03	CONCRETE	102.3	123.5	112.6	102.6	70.5	87.0	116.2	70.5	93.9	105.5	70.4	88.4	106.2	70.5	88.8	111.7	70.5	91.6
04	MASONRY	85.9	127.6	111.3	107.7	62.9	80.4	104.4	62.9	79.1	104.5	62.9	79.1	114.7	62.9	83.1	104.5	62.9	79.2
05	METALS	92.7	104.7	96.6	100.4	88.2	96.4	97.3	88.2	94.3	96.9	88.1	94.0	98.2	88.2	94.9	97.3	88.2	94.3
06	WOOD, PLASTICS & COMPOSITES	94.4	123.9	111.5	87.9	66.1	75.3	88.0	66.1	75.4	88.0	66.1	75.4	88.0	66.1	75.4	88.0	66.1	75.4
07	THERMAL & MOISTURE PROTECTION	101.0	124.8	110.7	100.6	73.4	89.5	102.2	73.4	90.5	101.0	73.4	89.8	100.9	73.4	89.7	101.9	73.4	90.3
08	OPENINGS	96.0	121.6	102.5	97.1	68.8	90.0	95.8	68.8	89.0	96.0	68.8	89.2	100.1	68.8	92.2	100.1	68.8	92.2
0920	Plaster & Gypsum Board	93.3	124.0	115.2	89.3	64.7	71.7	81.2	64.7	69.4	81.2	64.7	69.4	81.2	64.7	69.4	81.2	64.7	69.4
0950, 0980	Ceilings & Acoustic Treatment	92.7	124.0	113.4	112.0	64.7	80.7	108.8	64.7	79.7	108.8	64.7	79.7	108.8	64.7	79.7	108.8	64.7	79.7
0960	Flooring	92.6	133.1	104.4	98.4	69.1	89.8	100.7	69.1	91.4	100.7	69.1	91.4	100.7	69.1	91.4	100.7	69.1	91.4
0970, 0990	Wall Finishes & Painting/Coating	90.1	121.1	109.2	103.7	51.7	71.6	98.8	51.7	69.8	98.8	51.7	69.8	98.8	51.7	69.8	98.8	51.7	69.8
09	FINISHES	97.7	125.9	113.5	100.3	64.6	80.4	101.2	64.6	80.8	99.7	64.6	80.2	99.3	64.6	80.0	100.6	64.6	80.5
COVERS	DIVS. 10 - 14, 25, 28, 41, 43, 44, 46	100.0	110.5	102.1	100.0	80.2	96.1	100.0	80.2	96.1	100.0	80.2	96.1	100.0	80.2	96.1	100.0	80.2	96.1
21, 22, 23	FIRE SUPPRESSION, PLUMBING & HVAC	99.6	124.8	109.8	100.1	68.2	87.2	96.6	68.2	85.1	96.6	67.8	84.9	100.0	68.2	87.1	96.4	68.2	85.0
26, 27, 3370	ELECTRICAL, COMMUNICATIONS & UTIL.	97.2	136.9	117.7	89.5	71.1	80.0	89.5	71.1	80.0	87.2	71.1	78.9	87.8	71.1	79.2	87.1	71.1	78.8
MF2010	WEIGHTED AVERAGE	97.6	122.4	108.6	98.7	73.6	87.7	99.4	73.6	88.1	97.4	73.5	86.9	99.4	73.6	88.0	98.8	73.6	87.7

NEW MEXICO

DIVISION		LAS CRUCES 880			LAS VEGAS 877			ROSWELL 882			SANTA FE 875			SOCORRO 878			TRUTH/CONSEQUENCES 879		
		MAT.	INST.	TOTAL	MAT.	INST.	TOTAL	MAT.	INST.	TOTAL	MAT.	INST.	TOTAL	MAT.	INST.	TOTAL	MAT.	INST.	TOTAL
015433	CONTRACTOR EQUIPMENT		87.3	87.3		111.1	111.1		111.1	111.1		111.1	111.1		111.1	111.1		87.3	87.3
0241, 31 - 34	SITE & INFRASTRUCTURE, DEMOLITION	90.5	86.0	87.4	84.4	106.9	99.9	90.0	106.9	101.7	83.4	106.9	99.6	81.0	106.9	98.9	101.7	86.0	90.9
0310	Concrete Forming & Accessories	96.3	64.6	68.7	99.4	65.8	70.2	99.4	65.8	70.2	98.6	65.8	70.1	99.4	65.8	70.2	96.3	64.6	68.7
0320	Concrete Reinforcing	103.6	70.9	87.1	106.7	71.1	88.8	108.9	71.1	89.8	107.7	71.1	89.2	108.9	71.1	89.8	102.4	70.9	86.5
0330	Cast-in-Place Concrete	89.0	65.8	79.9	99.5	73.4	89.2	94.3	73.4	86.1	103.6	73.4	91.7	97.5	73.4	88.0	106.9	65.8	90.7
03	CONCRETE	85.1	67.0	76.3	103.5	70.5	87.4	106.2	70.5	88.8	104.8	70.5	88.1	102.3	70.5	86.8	96.9	67.0	82.4
04	MASONRY	100.5	62.6	77.4	104.8	62.9	79.3	115.3	62.9	83.3	104.7	62.9	79.2	104.7	62.9	79.2	102.1	62.6	78.0
05	METALS	98.1	81.7	92.7	97.0	88.2	94.1	98.2	88.2	94.9	98.2	88.2	94.9	97.3	88.2	94.3	96.9	81.7	91.9
06	WOOD, PLASTICS & COMPOSITES	79.3	65.0	71.0	88.0	66.1	75.4	88.0	66.1	75.4	87.5	66.1	75.2	88.0	66.1	75.4	79.3	65.0	71.0
07	THERMAL & MOISTURE PROTECTION	87.1	68.7	79.6	100.5	73.4	89.5	101.1	73.4	89.8	100.2	73.4	89.3	100.4	73.4	89.4	87.6	68.7	79.9
08	OPENINGS	90.1	68.2	84.6	96.0	68.8	89.2	95.8	68.8	89.0	96.0	68.8	89.1	95.8	68.8	89.0	90.0	68.2	84.5
0920	Plaster & Gypsum Board	83.3	64.7	70.0	81.2	64.7	69.4	81.2	64.7	69.4	89.2	64.7	71.7	81.2	64.7	69.4	83.3	64.7	70.0
0950, 0980	Ceilings & Acoustic Treatment	97.8	64.7	75.9	108.8	64.7	79.7	108.8	64.7	79.7	109.7	64.7	80.0	108.8	64.7	79.7	97.8	64.7	75.9
0960	Flooring	133.3	69.1	114.4	100.7	69.1	91.4	100.7	69.1	91.4	100.7	69.1	91.4	100.7	69.1	91.4	133.3	69.1	114.4
0970, 0990	Wall Finishes & Painting/Coating	90.0	51.7	66.4	98.8	51.7	69.8	98.8	51.7	69.8	98.8	51.7	69.8	98.8	51.7	69.8	90.0	51.7	66.4
09	FINISHES	112.8	63.8	85.5	99.1	64.6	79.9	99.8	64.6	80.2	100.6	64.6	80.5	99.0	64.6	79.9	113.6	63.8	85.8
COVERS	DIVS. 10 - 14, 25, 28, 41, 43, 44, 46	100.0	77.4	95.5	100.0	80.2	96.1	100.0	80.2	96.1	100.0	80.2	96.1	100.0	80.2	96.1	100.0	77.4	95.5
21, 22, 23	FIRE SUPPRESSION, PLUMBING & HVAC	100.3	67.9	87.2	96.4	68.2	85.0	100.0	68.2	87.1	99.9	68.2	87.1	96.4	68.2	85.0	96.3	67.9	84.8
26, 27, 3370	ELECTRICAL, COMMUNICATIONS & UTIL.	88.4	66.3	77.0	89.5	71.1	80.0	88.6	71.1	79.6	96.6	71.1	83.4	87.6	71.1	79.1	91.2	71.1	80.8
MF2010	WEIGHTED AVERAGE	96.1	69.6	84.4	97.3	73.6	86.8	99.3	73.6	88.0	99.4	73.6	88.0	96.8	73.6	86.6	97.0	70.3	85.3

DIVISION		NEW MEXICO – TUCUMCARI 884			NEW YORK – ALBANY 120-122			NEW YORK – BINGHAMTON 137-139			NEW YORK – BRONX 104			NEW YORK – BROOKLYN 112			NEW YORK – BUFFALO 140-142		
		MAT.	INST.	TOTAL	MAT.	INST.	TOTAL	MAT.	INST.	TOTAL	MAT.	INST.	TOTAL	MAT.	INST.	TOTAL	MAT.	INST.	TOTAL
015433	CONTRACTOR EQUIPMENT		111.1	111.1		113.6	113.6		114.2	114.2		113.9	113.9		114.5	114.5		96.5	96.5
0241, 31 - 34	SITE & INFRASTRUCTURE, DEMOLITION	87.6	106.9	100.9	70.5	106.9	95.6	93.4	92.7	92.9	114.2	124.5	121.3	119.7	128.0	125.4	97.5	97.3	97.3
0310	Concrete Forming & Accessories	99.4	65.6	70.0	97.8	95.5	95.8	102.1	82.7	85.2	101.7	176.5	166.8	108.5	176.2	167.4	102.0	115.7	113.9
0320	Concrete Reinforcing	106.7	71.0	88.7	98.3	91.8	95.0	97.3	95.1	96.2	95.9	179.8	138.2	98.8	187.6	143.5	102.5	101.8	102.1
0330	Cast-in-Place Concrete	94.3	73.4	86.0	76.6	102.0	86.6	98.4	94.5	96.8	89.2	164.2	118.8	100.4	162.5	124.9	103.1	117.8	108.9
03	CONCRETE	104.7	70.4	88.0	93.9	97.8	95.8	96.7	90.8	93.8	96.2	171.0	132.6	108.5	171.5	139.2	106.4	113.0	109.6
04	MASONRY	116.0	62.9	83.6	92.4	100.8	97.6	107.9	89.5	96.7	93.8	175.1	143.4	117.3	175.0	152.5	106.8	116.8	112.9
05	METALS	96.9	88.1	94.0	100.3	104.0	101.5	96.9	114.2	102.6	104.9	142.9	117.5	105.2	141.8	117.3	101.5	94.7	99.2
06	WOOD, PLASTICS & COMPOSITES	88.0	66.1	75.4	96.1	94.3	95.1	106.3	81.2	91.8	103.1	178.3	146.6	107.9	177.9	148.3	99.6	116.5	109.3
07	THERMAL & MOISTURE PROTECTION	101.0	73.4	89.7	88.6	93.7	90.7	103.4	83.3	95.2	104.1	164.3	128.6	105.8	163.1	129.1	100.9	108.4	104.0
08	OPENINGS	95.8	68.8	89.0	94.3	87.7	92.6	89.3	80.3	87.0	85.0	172.8	107.1	88.7	172.1	109.7	90.7	103.7	94.0
0920	Plaster & Gypsum Board	81.2	64.7	69.4	101.2	93.9	96.0	112.3	80.2	89.3	103.6	180.4	158.4	109.8	180.4	160.2	109.2	116.7	114.6
0950, 0980	Ceilings & Acoustic Treatment	108.8	64.7	79.7	94.4	93.9	94.1	97.1	80.2	85.9	83.5	180.4	147.5	93.6	180.4	150.9	105.9	116.7	113.1
0960	Flooring	100.7	69.1	91.4	91.7	103.9	95.3	103.7	89.6	99.6	99.0	177.0	121.9	110.9	177.0	130.3	102.3	116.7	106.5
0970, 0990	Wall Finishes & Painting/Coating	98.8	51.7	69.8	89.3	78.4	82.6	90.8	81.8	85.2	106.1	152.7	134.8	121.2	152.7	140.6	98.1	112.6	107.1
09	FINISHES	99.7	64.6	80.1	95.0	95.3	95.2	97.4	83.5	89.7	101.2	175.0	142.3	113.7	174.7	147.7	102.0	116.7	110.2
COVERS	DIVS. 10 - 14, 25, 28, 41, 43, 44, 46	100.0	80.2	96.1	100.0	93.6	98.7	100.0	94.3	98.9	100.0	134.3	106.8	100.0	133.5	106.7	100.0	106.5	101.3
21, 22, 23	FIRE SUPPRESSION, PLUMBING & HVAC	96.6	67.8	84.9	100.0	92.1	96.8	100.4	80.7	92.4	100.1	166.8	127.1	99.7	166.7	126.9	99.9	92.8	97.0
26, 27, 3370	ELECTRICAL, COMMUNICATIONS & UTIL.	89.5	71.1	80.0	99.2	91.6	95.3	100.7	90.2	95.3	96.9	173.3	136.3	100.9	173.3	138.3	98.6	99.7	99.2
MF2010	WEIGHTED AVERAGE	98.1	73.5	87.3	96.7	96.4	96.5	98.4	89.2	94.4	98.9	163.6	127.4	103.4	163.8	130.0	100.3	104.0	101.9

DIVISION		NEW YORK – ELMIRA 148-149			NEW YORK – FAR ROCKAWAY 116			NEW YORK – FLUSHING 113			NEW YORK – GLENS FALLS 128			NEW YORK – HICKSVILLE 115,117,118			NEW YORK – JAMAICA 114		
		MAT.	INST.	TOTAL	MAT.	INST.	TOTAL	MAT.	INST.	TOTAL	MAT.	INST.	TOTAL	MAT.	INST.	TOTAL	MAT.	INST.	TOTAL
015433	CONTRACTOR EQUIPMENT		115.2	115.2		114.5	114.5		114.5	114.5		113.6	113.6		114.5	114.5		114.5	114.5
0241, 31 - 34	SITE & INFRASTRUCTURE, DEMOLITION	93.1	93.6	93.4	123.6	128.0	126.7	123.6	128.0	126.7	62.4	106.2	92.6	112.3	126.6	122.2	117.5	128.0	124.8
0310	Concrete Forming & Accessories	87.4	85.3	85.6	95.4	176.2	165.7	99.1	176.2	166.2	86.7	88.8	88.5	92.1	146.0	139.0	99.1	176.2	166.2
0320	Concrete Reinforcing	103.7	85.9	94.8	98.8	187.6	143.5	100.5	187.6	144.4	97.9	88.6	93.2	98.8	179.5	139.5	98.8	187.6	143.5
0330	Cast-in-Place Concrete	96.4	94.5	95.6	108.5	162.5	129.8	108.5	162.5	129.8	74.8	99.7	84.6	92.2	153.5	116.4	100.4	162.5	124.9
03	CONCRETE	100.2	90.4	95.4	114.7	171.5	142.4	115.2	171.5	142.6	86.9	93.4	90.0	100.4	153.5	126.3	107.9	171.5	138.9
04	MASONRY	103.2	91.0	95.7	121.4	175.0	154.1	115.4	175.0	151.8	93.8	96.8	95.6	111.9	160.6	141.6	119.3	175.0	153.3
05	METALS	96.8	113.1	102.2	105.2	141.8	117.3	105.2	141.8	117.3	96.4	102.7	98.5	106.7	136.5	116.5	105.2	141.8	117.3
06	WOOD, PLASTICS & COMPOSITES	83.9	84.4	84.2	92.0	177.9	141.7	96.6	177.9	143.6	83.6	87.5	85.9	88.6	145.9	121.7	96.6	177.9	143.6
07	THERMAL & MOISTURE PROTECTION	101.8	84.8	94.9	105.8	163.1	129.1	105.8	163.1	129.1	88.0	90.6	89.0	105.4	149.6	123.3	105.6	163.1	129.0
08	OPENINGS	91.6	79.9	88.7	87.1	172.1	108.5	87.1	172.1	108.5	88.0	83.3	86.8	87.1	152.9	103.7	87.1	172.1	108.5
0920	Plaster & Gypsum Board	103.6	83.7	89.4	98.4	180.4	156.9	101.1	180.4	157.7	95.8	86.9	89.5	97.9	147.3	133.2	101.1	180.4	157.7
0950, 0980	Ceilings & Acoustic Treatment	101.4	83.7	89.7	82.7	180.4	147.2	82.7	180.4	147.2	91.8	86.9	88.6	81.7	147.3	125.0	82.7	180.4	147.2
0960	Flooring	91.5	89.6	90.9	105.5	177.0	126.5	107.0	177.0	127.5	83.3	103.9	89.4	104.4	149.4	117.6	107.0	177.0	127.5
0970, 0990	Wall Finishes & Painting/Coating	96.6	77.9	85.0	121.2	152.7	140.6	121.2	152.7	140.6	89.3	81.9	84.7	121.2	152.7	140.6	121.2	152.7	140.6
09	FINISHES	96.8	85.1	90.3	108.9	174.7	145.6	109.6	174.7	145.9	90.6	90.7	90.7	107.2	146.7	129.2	109.1	174.7	145.7
COVERS	DIVS. 10 - 14, 25, 28, 41, 43, 44, 46	100.0	87.8	97.6	100.0	133.5	106.7	100.0	133.5	106.7	100.0	91.5	98.3	100.0	123.0	104.6	100.0	133.5	106.7
21, 22, 23	FIRE SUPPRESSION, PLUMBING & HVAC	94.2	86.3	91.0	93.8	166.7	123.3	93.8	166.7	123.3	94.3	84.6	90.4	99.7	148.7	119.6	93.8	166.7	123.3
26, 27, 3370	ELECTRICAL, COMMUNICATIONS & UTIL.	97.3	93.5	95.3	108.1	173.3	141.7	108.1	173.3	141.7	96.2	91.6	93.9	100.1	141.0	121.2	99.0	173.3	137.4
MF2010	WEIGHTED AVERAGE	96.7	91.0	94.2	103.1	163.8	129.8	102.9	163.8	129.7	92.3	92.6	92.5	101.3	145.6	120.8	101.1	163.8	128.7

DIVISION		NEW YORK – JAMESTOWN 147			NEW YORK – KINGSTON 124			NEW YORK – LONG ISLAND CITY 111			NEW YORK – MONTICELLO 127			NEW YORK – MOUNT VERNON 105			NEW YORK – NEW ROCHELLE 108		
		MAT.	INST.	TOTAL	MAT.	INST.	TOTAL	MAT.	INST.	TOTAL	MAT.	INST.	TOTAL	MAT.	INST.	TOTAL	MAT.	INST.	TOTAL
015433	CONTRACTOR EQUIPMENT		93.5	93.5		114.5	114.5		114.5	114.5		114.5	114.5		113.9	113.9		113.9	113.9
0241, 31 - 34	SITE & INFRASTRUCTURE, DEMOLITION	94.3	92.9	93.3	119.2	124.7	123.0	121.0	128.0	125.9	114.5	124.7	121.5	122.3	119.2	120.2	121.1	119.2	119.8
0310	Concrete Forming & Accessories	87.5	81.0	81.8	88.4	108.3	105.7	103.2	176.2	166.7	95.0	108.4	106.6	92.6	129.3	124.5	106.7	129.3	126.4
0320	Concrete Reinforcing	103.9	97.6	100.7	98.2	134.7	116.6	98.8	187.6	143.5	97.5	134.7	116.2	94.6	178.4	136.9	94.7	178.4	136.9
0330	Cast-in-Place Concrete	100.0	93.5	97.5	101.1	130.7	112.8	103.5	162.5	126.8	94.7	130.7	108.9	99.7	129.7	111.5	99.7	129.7	111.5
03	CONCRETE	103.4	88.4	96.1	108.8	120.6	114.6	111.0	171.5	140.5	103.6	120.7	111.9	105.7	137.6	121.3	105.1	137.6	121.0
04	MASONRY	111.8	90.8	99.0	109.1	135.1	124.9	114.0	175.0	151.2	102.1	135.1	122.2	99.7	127.3	116.5	99.7	127.3	116.5
05	METALS	94.1	89.2	92.5	105.2	115.1	108.4	105.2	141.8	117.3	105.1	115.2	108.4	104.7	132.6	113.9	105.0	132.6	114.1
06	WOOD, PLASTICS & COMPOSITES	82.5	78.0	79.9	84.9	102.6	95.1	102.0	177.9	145.9	92.5	102.6	98.3	93.2	132.4	115.9	110.3	132.4	123.1
07	THERMAL & MOISTURE PROTECTION	101.2	85.8	95.0	107.5	136.0	119.1	105.7	163.1	129.1	107.1	136.0	118.9	105.0	137.9	118.3	105.1	137.9	118.4
08	OPENINGS	91.5	81.4	88.9	92.3	119.6	99.2	87.1	172.1	108.5	87.2	119.6	95.3	85.0	145.6	100.2	85.0	145.6	100.3
0920	Plaster & Gypsum Board	98.0	77.1	83.0	94.2	102.6	100.2	104.8	180.4	158.8	96.5	102.6	100.9	98.5	133.0	123.2	111.2	133.0	126.8
0950, 0980	Ceilings & Acoustic Treatment	98.6	77.1	84.4	76.4	102.6	93.7	82.7	180.4	147.2	76.4	102.6	93.7	81.7	133.0	115.6	81.7	133.0	115.6
0960	Flooring	94.5	92.3	93.9	97.7	75.9	91.3	108.8	177.0	128.8	100.4	75.9	93.2	91.1	172.3	114.9	97.9	172.3	119.8
0970, 0990	Wall Finishes & Painting/Coating	98.1	78.2	85.9	118.7	106.3	111.0	121.2	152.7	140.6	118.7	108.8	112.6	104.1	152.7	134.1	104.1	152.7	134.1
09	FINISHES	96.7	82.1	88.5	104.2	99.9	101.8	110.3	174.7	146.2	104.8	100.2	102.2	98.7	139.1	121.2	102.0	139.1	122.7
COVERS	DIVS. 10 - 14, 25, 28, 41, 43, 44, 46	100.0	93.3	98.7	100.0	108.4	101.7	100.0	133.5	106.7	100.0	108.4	101.7	100.0	117.2	103.4	100.0	117.2	103.4
21, 22, 23	FIRE SUPPRESSION, PLUMBING & HVAC	94.1	82.2	89.3	94.2	115.7	102.9	99.7	166.7	126.9	94.2	118.1	103.9	94.3	117.9	103.8	94.3	117.9	103.8
26, 27, 3370	ELECTRICAL, COMMUNICATIONS & UTIL.	96.0	86.4	91.1	98.0	115.7	107.1	99.6	173.3	137.6	98.0	115.7	107.1	95.2	157.5	127.4	95.2	157.5	127.4
MF2010	WEIGHTED AVERAGE	96.9	86.4	92.3	100.9	117.4	108.2	103.0	163.8	129.8	99.4	118.0	107.6	98.6	133.0	113.8	99.0	133.0	114.0

NEW YORK

DIVISION		NEW YORK 100 - 102			NIAGARA FALLS 143			PLATTSBURGH 129			POUGHKEEPSIE 125 - 126			QUEENS 110			RIVERHEAD 119		
		MAT.	INST.	TOTAL	MAT.	INST.	TOTAL	MAT.	INST.	TOTAL	MAT.	INST.	TOTAL	MAT.	INST.	TOTAL	MAT.	INST.	TOTAL
015433	CONTRACTOR EQUIPMENT		114.3	114.3		93.5	93.5		99.5	99.5		114.5	114.5		114.5	114.5		114.5	114.5
0241, 31 - 34	SITE & INFRASTRUCTURE, DEMOLITION	123.0	125.1	124.5	96.5	94.7	95.3	85.4	102.9	97.5	115.3	124.3	121.5	115.7	128.0	124.2	113.2	126.8	122.6
0310	Concrete Forming & Accessories	106.0	184.9	174.6	87.4	117.3	113.5	91.0	86.4	87.0	88.4	153.4	144.9	92.3	176.2	165.3	96.5	146.8	140.3
0320	Concrete Reinforcing	101.5	179.8	141.0	102.5	102.3	102.4	102.6	88.2	95.4	98.2	134.9	116.7	100.5	187.6	144.4	100.6	176.6	138.9
0330	Cast-in-Place Concrete	100.4	180.1	131.8	103.4	122.6	111.0	91.6	95.4	93.1	97.9	128.2	109.9	95.4	162.5	121.8	93.7	154.7	117.8
03	CONCRETE	106.4	180.1	142.3	105.7	115.4	110.4	100.5	89.5	95.2	105.9	139.8	122.4	103.6	171.5	136.7	101.1	153.7	126.7
04	MASONRY	103.9	178.2	149.2	119.6	123.4	121.9	89.5	90.7	90.2	102.2	129.4	118.8	108.3	175.0	149.0	117.2	162.8	145.1
05	METALS	119.0	143.1	126.9	96.9	93.5	95.8	101.0	83.5	95.2	105.2	116.8	109.0	105.2	141.8	117.3	107.1	135.4	116.5
06	WOOD, PLASTICS & COMPOSITES	107.5	187.8	153.9	82.4	115.3	101.4	90.5	85.1	87.4	84.9	164.5	130.9	88.7	177.9	140.3	93.7	145.9	123.8
07	THERMAL & MOISTURE PROTECTION	104.1	166.8	129.6	101.3	110.7	105.1	101.1	88.2	95.9	107.4	140.3	120.8	105.4	163.1	128.9	106.0	150.5	124.1
08	OPENINGS	90.2	178.0	112.3	91.5	103.2	94.4	96.7	81.2	92.8	92.3	151.4	107.2	87.1	172.1	108.5	87.1	152.2	103.5
0920	Plaster & Gypsum Board	112.0	190.1	167.8	98.0	115.5	110.5	117.0	83.9	93.4	94.2	166.5	145.9	97.9	180.4	156.8	99.6	147.3	133.6
0950, 0980	Ceilings & Acoustic Treatment	102.5	190.1	160.4	98.6	115.5	109.8	104.9	83.9	91.0	76.4	166.5	135.9	82.7	180.4	147.2	82.6	147.3	125.3
0960	Flooring	100.3	177.0	122.8	94.4	117.0	101.0	107.8	103.9	106.7	97.7	145.8	111.8	104.4	177.0	125.7	105.5	149.4	118.4
0970, 0990	Wall Finishes & Painting/Coating	106.1	155.0	136.2	98.1	113.1	107.4	116.5	78.8	93.2	118.7	106.3	111.0	121.2	152.7	140.6	121.2	152.7	140.6
09	FINISHES	107.5	181.6	148.8	96.8	117.8	108.5	101.2	88.9	94.3	103.9	149.2	129.2	107.6	174.7	145.0	107.8	147.2	129.8
COVERS	DIVS. 10 - 14, 25, 28, 41, 43, 44, 46	100.0	137.9	107.5	100.0	104.2	100.8	100.0	51.4	90.3	100.0	113.5	102.7	100.0	133.5	106.7	100.0	123.8	104.7
21, 22, 23	FIRE SUPPRESSION, PLUMBING & HVAC	100.0	168.4	127.7	94.1	99.4	96.2	94.2	84.6	90.3	94.2	117.2	103.5	99.7	166.7	126.9	99.9	149.8	120.1
26, 27, 3370	ELECTRICAL, COMMUNICATIONS & UTIL.	105.0	173.3	140.3	94.7	98.8	96.8	93.9	85.7	89.7	98.0	115.7	107.1	100.1	173.3	137.9	101.7	141.0	122.0
MF2010	WEIGHTED AVERAGE	104.8	166.9	132.2	97.9	106.1	101.5	96.9	87.0	92.6	100.1	128.4	112.6	101.4	163.8	128.9	102.1	146.1	121.5

NEW YORK

DIVISION		ROCHESTER 144 - 146			SCHENECTADY 123			STATEN ISLAND 103			SUFFERN 109			SYRACUSE 130 - 132			UTICA 133 - 135		
		MAT.	INST.	TOTAL	MAT.	INST.	TOTAL	MAT.	INST.	TOTAL	MAT.	INST.	TOTAL	MAT.	INST.	TOTAL	MAT.	INST.	TOTAL
015433	CONTRACTOR EQUIPMENT		115.6	115.6		113.6	113.6		113.9	113.9		113.9	113.9		113.6	113.6		113.6	113.6
0241, 31 - 34	SITE & INFRASTRUCTURE, DEMOLITION	74.5	107.6	97.3	71.9	106.9	96.0	127.1	124.5	125.3	117.7	120.3	119.5	92.4	106.6	102.2	70.2	104.2	93.7
0310	Concrete Forming & Accessories	103.5	91.7	93.3	102.0	95.1	96.0	92.1	176.5	165.6	100.6	126.4	123.1	101.0	88.0	89.7	102.5	80.6	83.5
0320	Concrete Reinforcing	103.4	89.2	96.2	96.9	91.8	94.3	95.9	179.8	138.2	94.7	135.0	115.0	98.3	95.5	96.9	98.3	87.1	92.6
0330	Cast-in-Place Concrete	96.8	96.6	96.7	90.4	101.4	94.7	99.7	164.2	125.1	96.7	126.8	108.5	91.3	95.8	93.1	83.6	87.5	85.1
03	CONCRETE	112.7	93.9	103.5	100.8	97.4	99.2	107.4	171.0	138.4	102.1	127.4	114.4	99.7	93.0	96.5	97.7	85.3	91.7
04	MASONRY	103.3	96.3	99.0	91.9	99.8	96.7	106.6	175.1	148.4	99.0	123.0	113.6	98.9	101.8	100.6	90.6	88.2	89.1
05	METALS	98.0	104.2	100.1	100.4	104.0	101.6	102.8	142.9	116.0	102.8	118.0	107.8	100.3	105.0	101.8	98.3	101.6	99.4
06	WOOD, PLASTICS & COMPOSITES	96.6	91.7	93.8	102.5	94.3	97.8	91.7	178.3	141.7	102.8	132.6	120.0	102.5	85.6	92.7	102.5	82.4	90.9
07	THERMAL & MOISTURE PROTECTION	96.7	93.9	95.5	89.1	93.2	90.8	104.4	164.3	128.8	104.9	134.4	116.9	99.5	96.2	98.2	88.9	90.0	89.4
08	OPENINGS	94.2	86.1	92.2	94.3	87.7	92.6	85.0	172.8	107.1	85.0	136.9	98.1	92.3	79.9	89.2	94.3	78.6	90.4
0920	Plaster & Gypsum Board	114.8	91.4	98.1	106.3	93.9	97.5	98.6	180.4	157.0	102.6	133.3	124.5	106.3	84.8	91.0	106.3	81.6	88.7
0950, 0980	Ceilings & Acoustic Treatment	106.9	91.4	96.7	97.1	93.9	95.0	83.5	180.4	147.5	81.7	133.3	115.7	97.1	84.8	89.0	97.1	81.6	86.9
0960	Flooring	90.3	103.0	94.1	91.0	103.9	94.8	95.2	177.0	119.2	94.3	149.4	110.4	93.1	95.0	93.7	91.0	90.4	90.8
0970, 0990	Wall Finishes & Painting/Coating	97.9	95.5	96.4	89.3	78.4	82.6	106.1	152.7	134.8	104.1	114.4	110.5	96.6	98.5	97.8	89.3	98.5	95.0
09	FINISHES	102.9	94.4	98.2	96.0	95.1	95.5	100.8	175.0	142.2	99.6	126.9	114.9	97.9	90.0	93.5	96.0	84.2	89.4
COVERS	DIVS. 10 - 14, 25, 28, 41, 43, 44, 46	100.0	94.1	98.8	100.0	93.3	98.7	100.0	134.3	106.8	100.0	115.6	103.1	100.0	95.7	99.2	100.0	87.2	97.4
21, 22, 23	FIRE SUPPRESSION, PLUMBING & HVAC	100.0	87.1	94.8	100.2	91.5	96.7	100.1	166.8	127.1	94.3	112.0	101.4	100.2	88.3	95.3	100.2	78.4	91.3
26, 27, 3370	ELECTRICAL, COMMUNICATIONS & UTIL.	99.2	89.6	94.2	100.9	91.6	96.1	96.9	173.3	136.3	103.2	123.8	113.8	100.9	94.3	97.5	99.0	94.3	96.6
MF2010	WEIGHTED AVERAGE	100.0	94.0	97.3	97.8	96.1	97.0	100.8	163.6	128.5	98.7	122.0	109.0	98.9	94.5	97.0	96.8	88.2	93.0

DIVISION		NEW YORK									NORTH CAROLINA								
		WATERTOWN 136			WHITE PLAINS 106			YONKERS 107			ASHEVILLE 287 - 288			CHARLOTTE 281 - 282			DURHAM 277		
		MAT.	INST.	TOTAL	MAT.	INST.	TOTAL	MAT.	INST.	TOTAL	MAT.	INST.	TOTAL	MAT.	INST.	TOTAL	MAT.	INST.	TOTAL
015433	CONTRACTOR EQUIPMENT		113.6	113.6		113.9	113.9		113.9	113.9		94.7	94.7		94.7	94.7		99.9	99.9
0241, 31 - 34	SITE & INFRASTRUCTURE, DEMOLITION	78.2	106.5	97.7	113.8	119.2	117.5	122.5	119.1	120.1	108.4	75.5	85.7	109.7	75.6	86.2	108.6	84.0	91.6
0310	Concrete Forming & Accessories	88.2	90.3	90.1	105.3	129.3	126.2	105.7	129.3	126.2	92.5	37.4	44.6	96.6	39.2	46.7	94.8	38.0	45.4
0320	Concrete Reinforcing	98.9	95.6	97.2	94.7	178.4	136.9	98.5	178.4	138.8	93.6	41.0	67.1	94.0	38.3	65.9	95.2	47.9	71.4
0330	Cast-in-Place Concrete	97.3	94.7	96.3	88.5	129.7	104.8	99.0	129.8	111.1	118.0	45.0	89.3	124.5	47.0	93.9	119.9	45.3	90.5
03	CONCRETE	110.4	93.6	102.2	95.6	137.6	116.1	105.3	137.6	121.1	109.0	42.7	76.7	111.9	43.6	78.6	109.7	44.3	77.8
04	MASONRY	91.6	101.4	97.6	98.5	127.3	116.1	103.4	127.3	118.0	87.1	38.2	57.3	93.2	41.8	61.8	92.4	34.4	57.0
05	METALS	98.3	105.0	100.5	104.3	132.6	113.7	114.4	132.6	120.4	87.2	72.4	82.3	91.2	71.6	84.7	100.4	75.5	92.2
06	WOOD, PLASTICS & COMPOSITES	85.2	89.4	87.6	108.6	132.4	122.4	108.4	132.4	122.3	94.5	36.9	61.3	99.7	39.0	64.6	96.9	38.3	63.1
07	THERMAL & MOISTURE PROTECTION	89.2	96.3	92.1	104.8	137.9	118.2	105.1	137.9	118.4	99.9	41.2	76.1	97.3	42.7	75.1	100.7	43.2	77.3
08	OPENINGS	94.3	85.7	92.1	85.1	145.6	100.3	88.2	145.6	102.7	93.2	36.7	79.0	97.3	38.1	82.4	97.3	40.5	83.0
0920	Plaster & Gypsum Board	97.7	88.8	91.3	106.2	133.0	125.4	111.4	133.0	126.9	104.3	34.8	54.7	100.7	36.9	55.1	108.3	36.2	56.8
0950, 0980	Ceilings & Acoustic Treatment	97.1	88.8	91.6	81.7	133.0	115.6	100.7	133.0	122.1	93.7	34.8	54.8	96.4	36.9	57.1	97.3	36.2	57.0
0960	Flooring	83.8	91.2	85.9	96.3	172.3	118.6	95.9	172.3	118.3	103.7	43.6	86.1	106.7	43.8	88.2	107.3	43.6	88.6
0970, 0990	Wall Finishes & Painting/Coating	89.3	88.9	89.1	104.1	152.7	134.1	104.1	152.7	134.1	114.8	33.7	64.8	114.8	38.5	67.7	114.8	35.4	65.8
09	FINISHES	93.7	90.1	91.7	100.1	139.1	121.8	105.6	139.1	124.3	100.8	37.9	65.7	101.5	39.8	67.1	103.2	38.4	67.0
COVERS	DIVS. 10 - 14, 25, 28, 41, 43, 44, 46	100.0	95.7	99.2	100.0	117.2	103.4	100.0	120.4	104.1	100.0	64.5	92.9	100.0	65.0	93.0	100.0	72.4	94.5
21, 22, 23	FIRE SUPPRESSION, PLUMBING & HVAC	100.2	83.8	93.5	100.3	117.9	107.4	100.3	117.9	107.4	100.2	36.9	74.6	99.9	37.1	74.5	100.3	35.8	74.2
26, 27, 3370	ELECTRICAL, COMMUNICATIONS & UTIL.	100.9	91.7	96.2	95.2	157.5	127.4	103.3	157.5	131.3	96.6	36.6	65.7	97.2	47.6	71.6	97.5	46.3	71.1
MF2010	WEIGHTED AVERAGE	98.5	93.5	96.3	98.8	133.0	113.9	103.5	133.1	116.5	97.8	45.5	74.7	99.4	47.8	76.7	100.8	47.9	77.5

NORTH CAROLINA

	DIVISION	ELIZABETH CITY 279			FAYETTEVILLE 283			GASTONIA 280			GREENSBORO 270,272 - 274			HICKORY 286			KINSTON 285		
		MAT.	INST.	TOTAL	MAT.	INST.	TOTAL	MAT.	INST.	TOTAL	MAT.	INST.	TOTAL	MAT.	INST.	TOTAL	MAT.	INST.	TOTAL
015433	CONTRACTOR EQUIPMENT		103.9	103.9		99.9	99.9		94.7	94.7		99.9	99.9		99.9	99.9		99.9	99.9
0241, 31 - 34	SITE & INFRASTRUCTURE, DEMOLITION	114.3	85.0	94.1	107.4	84.1	91.3	108.0	75.5	85.6	108.5	84.1	91.7	107.6	84.0	91.4	106.2	84.1	90.9
0310	Concrete Forming & Accessories	82.0	33.0	39.4	92.1	38.8	45.7	98.4	38.3	46.1	94.5	38.6	45.9	88.9	36.6	43.4	85.5	38.8	44.8
0320	Concrete Reinforcing	93.1	39.9	66.3	97.3	48.1	72.5	94.0	40.9	67.3	94.0	56.2	75.0	93.6	18.6	55.8	93.1	18.1	55.3
0330	Cast-in-Place Concrete	120.1	36.9	87.4	123.7	48.2	94.0	115.3	52.1	90.4	119.0	46.5	90.5	117.9	46.1	89.6	113.8	39.3	84.5
03	CONCRETE	109.9	37.7	74.7	111.4	45.7	79.4	107.4	45.5	77.2	109.0	44.6	78.6	108.7	38.6	74.6	105.4	37.2	72.2
04	MASONRY	105.0	25.7	56.6	90.6	39.1	59.2	91.7	36.1	57.8	90.5	35.0	56.6	76.8	39.3	53.9	82.8	31.5	51.5
05	METALS	89.2	71.9	83.5	104.3	75.6	94.9	88.1	72.4	82.9	94.5	78.6	89.2	87.3	66.1	80.3	86.3	65.2	79.3
06	WOOD, PLASTICS & COMPOSITES	81.9	34.2	54.3	93.5	38.3	61.6	102.4	38.3	65.3	96.6	38.3	62.9	89.3	36.7	58.9	86.3	39.6	59.3
07	THERMAL & MOISTURE PROTECTION	100.0	32.5	72.6	99.8	41.7	76.2	100.2	41.5	76.3	100.5	40.8	76.2	100.3	40.2	75.8	100.1	33.8	73.2
08	OPENINGS	94.2	32.1	78.5	93.3	40.5	80.0	97.3	38.2	82.4	97.3	42.8	83.6	93.3	29.8	77.3	93.4	31.4	77.7
0920	Plaster & Gypsum Board	101.8	31.3	51.4	109.0	36.2	57.0	110.4	36.2	57.4	110.4	36.2	57.4	104.3	34.6	54.5	104.5	37.6	56.7
0950, 0980	Ceilings & Acoustic Treatment	97.3	31.3	53.7	94.6	36.2	56.0	97.3	36.2	56.9	97.3	36.2	56.9	93.7	34.6	54.6	97.3	37.6	57.9
0960	Flooring	98.0	23.7	76.2	103.8	43.6	86.2	107.3	43.6	88.6	107.3	40.3	87.6	103.5	33.6	83.0	100.7	19.2	76.8
0970, 0990	Wall Finishes & Painting/Coating	114.8	43.6	70.9	114.8	33.7	64.8	114.8	38.7	67.9	114.8	32.8	64.2	114.8	32.8	64.2	114.8	29.9	62.4
09	FINISHES	100.1	32.2	62.2	101.5	38.7	66.5	103.3	39.1	67.5	103.4	38.1	67.0	101.0	35.3	64.3	100.7	34.6	63.8
COVERS	DIVS. 10 - 14, 25, 28, 41, 43, 44, 46	100.0	72.8	94.6	100.0	73.1	94.7	100.0	64.5	92.9	100.0	64.7	93.0	100.0	64.5	92.9	100.0	73.6	94.7
21, 22, 23	FIRE SUPPRESSION, PLUMBING & HVAC	94.3	29.0	67.8	100.2	36.9	74.5	100.2	36.6	74.4	100.2	37.0	74.6	94.3	30.4	68.4	94.3	31.0	68.6
26, 27, 3370	ELECTRICAL, COMMUNICATIONS & UTIL.	97.1	29.0	62.0	95.6	50.5	72.3	96.1	34.0	64.1	96.4	36.8	65.7	94.2	30.8	61.5	94.0	47.9	70.2
MF2010	WEIGHTED AVERAGE	97.8	40.6	72.6	100.7	49.4	78.1	98.6	45.4	75.1	99.7	47.3	76.6	95.5	42.3	72.1	95.2	43.8	72.6

NORTH CAROLINA / NORTH DAKOTA

	DIVISION	MURPHY 289			RALEIGH 275 - 276			ROCKY MOUNT 278			WILMINGTON 284			WINSTON-SALEM 271			BISMARCK 585		
		MAT.	INST.	TOTAL	MAT.	INST.	TOTAL	MAT.	INST.	TOTAL	MAT.	INST.	TOTAL	MAT.	INST.	TOTAL	MAT.	INST.	TOTAL
015433	CONTRACTOR EQUIPMENT		94.7	94.7		99.9	99.9		99.9	99.9		94.7	94.7		99.9	99.9		98.7	98.7
0241, 31 - 34	SITE & INFRASTRUCTURE, DEMOLITION	110.1	75.3	86.1	109.1	84.1	91.9	112.0	84.2	92.8	109.5	75.7	86.2	108.8	84.0	91.7	100.1	97.4	98.2
0310	Concrete Forming & Accessories	99.0	23.9	33.7	94.9	39.5	46.7	87.7	31.6	38.9	93.7	44.1	50.5	95.8	37.7	45.2	99.8	42.5	49.9
0320	Concrete Reinforcing	93.2	17.9	55.2	94.0	48.1	70.9	93.1	26.3	59.5	94.3	46.4	70.2	94.0	38.0	65.8	110.0	85.6	97.7
0330	Cast-in-Place Concrete	122.0	35.2	87.8	127.6	51.5	97.6	117.5	38.5	86.4	117.5	48.0	90.1	122.0	44.0	91.3	101.1	49.3	80.7
03	CONCRETE	112.2	29.2	71.8	113.3	47.1	81.1	110.4	35.1	73.7	108.8	47.7	79.0	110.6	41.9	77.1	100.3	54.3	77.9
04	MASONRY	79.4	25.4	46.4	91.7	40.0	60.2	83.2	33.1	52.6	77.1	40.0	54.5	90.7	36.0	57.3	104.3	59.6	77.0
05	METALS	85.3	64.5	78.4	92.2	75.5	86.7	88.4	65.4	80.8	87.0	74.9	83.0	92.2	71.1	85.2	96.5	86.1	93.0
06	WOOD, PLASTICS & COMPOSITES	103.1	22.1	56.3	96.9	38.9	63.4	88.6	28.9	54.1	96.2	44.1	66.1	96.6	37.6	62.5	82.3	36.9	56.0
07	THERMAL & MOISTURE PROTECTION	100.1	30.6	71.9	99.5	42.0	76.1	100.5	31.8	72.6	99.9	46.1	78.0	100.5	40.5	76.1	107.4	52.6	85.1
08	OPENINGS	93.2	24.5	75.9	94.1	40.6	80.6	93.4	25.6	76.3	93.4	43.9	80.9	97.3	37.5	82.2	98.0	48.4	85.5
0920	Plaster & Gypsum Board	109.3	19.5	45.1	100.7	36.9	55.1	103.7	26.5	48.6	106.0	42.2	60.4	110.4	35.5	56.9	98.3	35.2	53.2
0950, 0980	Ceilings & Acoustic Treatment	93.7	19.5	44.7	96.4	36.9	57.1	94.6	26.5	49.6	94.6	42.2	60.0	97.3	35.5	56.5	144.8	35.2	72.4
0960	Flooring	107.6	24.6	83.3	107.3	43.6	88.6	102.0	22.3	78.6	104.5	45.5	87.2	107.3	43.6	88.6	115.3	75.5	103.6
0970, 0990	Wall Finishes & Painting/Coating	114.8	18.1	55.1	114.8	37.2	66.9	114.8	40.0	68.6	114.8	40.2	68.8	114.8	34.4	65.2	96.0	31.0	55.9
09	FINISHES	102.8	22.7	58.1	101.8	39.7	67.2	101.1	30.3	61.6	101.4	43.9	69.3	103.4	38.2	67.1	114.9	45.3	76.1
COVERS	DIVS. 10 - 14, 25, 28, 41, 43, 44, 46	100.0	61.8	92.4	100.0	73.5	94.7	100.0	72.9	94.6	100.0	74.7	95.0	100.0	64.4	92.9	100.0	81.2	96.3
21, 22, 23	FIRE SUPPRESSION, PLUMBING & HVAC	94.3	18.5	63.6	100.0	35.3	73.8	94.3	35.7	70.6	100.2	38.3	75.1	100.2	35.7	74.1	100.3	60.7	84.3
26, 27, 3370	ELECTRICAL, COMMUNICATIONS & UTIL.	97.5	25.7	60.5	100.6	36.5	67.5	99.1	38.8	68.0	96.8	36.9	65.9	96.4	40.1	67.4	98.1	71.7	84.5
MF2010	WEIGHTED AVERAGE	96.4	33.2	68.5	99.8	47.6	76.8	96.8	42.5	72.9	97.4	48.5	75.8	99.5	46.0	75.9	100.7	64.4	84.7

NORTH DAKOTA

	DIVISION	DEVILS LAKE 583			DICKINSON 586			FARGO 580 - 581			GRAND FORKS 582			JAMESTOWN 584			MINOT 587		
		MAT.	INST.	TOTAL	MAT.	INST.	TOTAL	MAT.	INST.	TOTAL	MAT.	INST.	TOTAL	MAT.	INST.	TOTAL	MAT.	INST.	TOTAL
015433	CONTRACTOR EQUIPMENT		98.7	98.7		98.7	98.7		98.7	98.7		98.7	98.7		98.7	98.7		98.7	98.7
0241, 31 - 34	SITE & INFRASTRUCTURE, DEMOLITION	109.6	95.4	99.8	119.7	94.1	102.1	100.7	97.4	98.4	114.2	94.1	100.4	108.5	94.1	98.6	111.2	97.4	101.7
0310	Concrete Forming & Accessories	103.3	36.9	45.6	93.3	36.1	43.6	101.0	43.2	50.7	96.9	36.1	44.0	94.6	35.8	43.4	93.0	60.8	65.0
0320	Concrete Reinforcing	110.0	85.9	97.9	111.0	42.0	76.2	101.4	85.7	93.5	108.3	85.8	96.9	110.6	49.8	80.0	112.0	85.9	98.9
0330	Cast-in-Place Concrete	122.8	47.9	93.3	111.1	46.3	85.6	100.2	51.2	80.9	111.1	46.3	85.6	121.3	46.2	91.7	111.1	48.3	86.4
03	CONCRETE	114.0	51.4	83.5	113.0	42.2	78.5	107.8	55.3	82.2	110.1	50.2	80.9	112.6	43.4	78.9	108.8	62.1	86.1
04	MASONRY	112.8	65.2	83.8	114.8	59.0	80.8	106.0	59.6	77.7	107.3	64.8	81.3	125.1	34.7	69.9	106.5	64.2	80.7
05	METALS	96.5	85.1	92.8	96.4	64.8	86.0	98.8	86.4	94.7	96.5	80.5	91.2	96.5	65.9	86.4	96.8	86.7	93.4
06	WOOD, PLASTICS & COMPOSITES	85.5	34.0	55.7	75.2	34.0	51.4	82.1	37.4	56.3	78.9	34.0	53.0	76.9	34.0	52.1	74.9	61.3	67.1
07	THERMAL & MOISTURE PROTECTION	108.2	50.8	84.9	108.8	48.4	84.2	105.7	53.7	84.6	108.5	50.8	85.0	108.1	43.0	81.6	108.2	56.4	87.2
08	OPENINGS	98.0	42.3	84.0	97.9	31.9	81.3	97.9	48.7	85.5	98.0	42.3	83.9	98.0	34.1	81.9	98.1	61.7	88.9
0920	Plaster & Gypsum Board	118.9	32.2	57.0	109.9	32.2	54.4	98.3	35.7	53.6	111.7	32.2	54.9	111.2	32.2	54.8	109.9	60.4	74.5
0950, 0980	Ceilings & Acoustic Treatment	145.7	32.2	70.7	145.7	32.2	70.7	144.8	35.7	72.8	145.7	32.2	70.7	145.7	32.2	70.7	145.7	60.4	89.4
0960	Flooring	119.5	37.6	95.4	111.2	37.6	89.6	114.9	75.5	103.3	113.6	37.6	91.3	112.2	37.6	90.3	111.0	84.2	103.1
0970, 0990	Wall Finishes & Painting/Coating	96.0	23.7	51.4	96.0	23.7	51.4	96.0	73.8	82.3	96.0	23.7	51.4	96.0	23.7	51.4	96.0	26.0	52.8
09	FINISHES	119.5	34.1	71.9	117.1	34.1	70.8	114.9	50.4	78.9	117.4	34.1	70.9	116.4	34.1	70.5	116.1	60.9	85.3
COVERS	DIVS. 10 - 14, 25, 28, 41, 43, 44, 46	100.0	34.9	87.1	100.0	35.1	87.1	100.0	81.3	96.3	100.0	35.0	87.1	100.0	79.1	95.8	100.0	84.0	96.8
21, 22, 23	FIRE SUPPRESSION, PLUMBING & HVAC	94.5	63.0	81.7	94.5	70.5	84.7	100.3	66.0	86.4	100.4	35.6	74.2	94.5	37.5	71.4	100.4	60.7	84.3
26, 27, 3370	ELECTRICAL, COMMUNICATIONS & UTIL.	96.3	38.0	66.3	105.1	71.7	87.8	99.7	66.9	82.8	99.9	56.9	77.7	96.3	38.0	66.2	103.1	80.4	91.4
MF2010	WEIGHTED AVERAGE	101.7	57.0	82.0	102.7	58.8	83.3	102.1	65.7	86.0	102.8	53.1	80.9	101.9	46.4	77.4	102.7	70.2	88.4

		NORTH DAKOTA			OHIO														
DIVISION		WILLISTON			AKRON			ATHENS			CANTON			CHILLICOTHE			CINCINNATI		
		588			442 - 443			457			446 - 447			456			451 - 452		
		MAT.	INST.	TOTAL	MAT.	INST.	TOTAL	MAT.	INST.	TOTAL	MAT.	INST.	TOTAL	MAT.	INST.	TOTAL	MAT.	INST.	TOTAL
015433	CONTRACTOR EQUIPMENT		98.7	98.7		97.0	97.0		91.7	91.7		97.0	97.0		101.6	101.6		101.4	101.4
0241, 31 - 34	SITE & INFRASTRUCTURE, DEMOLITION	112.2	94.1	99.7	91.7	104.6	100.6	103.8	94.8	97.6	91.8	104.1	100.3	89.0	106.6	101.2	86.1	106.1	99.9
0310	Concrete Forming & Accessories	98.4	36.1	44.2	100.3	92.6	93.6	91.7	89.0	89.4	100.3	82.7	85.0	93.3	95.4	95.1	95.0	82.2	83.8
0320	Concrete Reinforcing	113.2	42.0	77.3	93.4	92.0	92.7	93.4	87.6	90.4	93.4	77.3	85.3	90.6	81.0	85.8	96.0	83.2	89.6
0330	Cast-in-Place Concrete	111.1	46.3	85.6	95.5	98.4	96.6	108.5	95.2	103.3	96.4	96.0	96.2	98.5	94.3	96.8	90.7	86.6	89.1
03	CONCRETE	110.2	42.2	77.1	95.1	93.8	94.5	102.6	90.3	96.6	95.6	85.9	90.9	96.9	92.1	94.6	91.6	84.0	87.9
04	MASONRY	101.2	59.0	75.5	90.4	95.6	93.6	76.1	91.1	85.2	91.1	86.5	88.3	82.9	100.4	93.6	82.5	87.3	85.4
05	METALS	96.6	64.8	86.1	92.7	82.5	89.3	100.5	79.3	93.5	92.7	76.2	87.2	92.8	85.1	90.3	95.2	84.7	91.7
06	WOOD, PLASTICS & COMPOSITES	80.2	34.0	53.5	102.2	91.4	96.0	83.8	90.3	87.5	102.7	80.9	90.1	96.2	94.1	95.0	98.5	80.0	87.8
07	THERMAL & MOISTURE PROTECTION	108.4	48.4	84.0	107.2	96.9	103.0	99.8	93.0	97.0	108.2	92.7	101.9	100.9	96.4	99.1	99.3	91.1	95.9
08	OPENINGS	98.0	31.9	81.4	112.2	91.5	107.0	101.0	86.3	97.3	105.9	78.0	98.9	92.9	86.1	91.2	101.7	79.7	96.1
0920	Plaster & Gypsum Board	111.7	32.2	54.9	91.0	90.8	90.8	91.1	89.6	90.1	92.3	80.0	83.5	91.7	94.2	93.5	93.8	79.6	83.7
0950, 0980	Ceilings & Acoustic Treatment	145.7	32.2	70.7	94.7	90.8	92.1	103.6	89.6	94.4	94.7	80.0	85.0	97.7	94.2	95.4	98.6	79.6	86.1
0960	Flooring	114.7	37.6	92.1	100.4	92.1	98.0	127.7	100.1	119.6	100.6	82.1	95.2	103.9	103.6	103.8	104.8	90.6	100.7
0970, 0990	Wall Finishes & Painting/Coating	96.0	23.7	51.4	100.8	102.3	101.7	103.1	96.1	98.8	100.8	83.3	90.0	100.1	92.5	95.4	100.1	86.1	91.5
09	FINISHES	117.6	34.1	71.0	98.1	93.3	95.4	103.0	92.7	97.3	98.3	82.1	89.3	99.6	97.1	98.2	100.1	83.7	90.9
COVERS	DIVS. 10 - 14, 25, 28, 41, 43, 44, 46	100.0	35.1	87.1	100.0	94.0	98.8	100.0	56.1	91.3	100.0	91.5	98.3	100.0	92.0	98.4	100.0	92.4	98.5
21, 22, 23	FIRE SUPPRESSION, PLUMBING & HVAC	94.5	70.5	84.7	100.0	97.0	98.8	94.0	55.0	78.2	100.0	83.9	93.5	94.4	93.4	94.0	99.9	83.0	93.1
26, 27, 3370	ELECTRICAL, COMMUNICATIONS & UTIL.	100.2	71.7	85.5	97.9	94.5	96.1	99.3	102.6	101.0	97.2	90.9	93.9	98.9	85.2	91.8	97.7	80.4	88.8
MF2010	WEIGHTED AVERAGE	101.1	58.8	82.4	98.7	94.5	96.9	98.0	83.1	91.4	98.2	86.3	93.0	95.2	93.4	94.4	96.9	85.8	92.0

		OHIO																	
DIVISION		CLEVELAND			COLUMBUS			DAYTON			HAMILTON			LIMA			LORAIN		
		441			430 - 432			453 - 454			450			458			440		
		MAT.	INST.	TOTAL	MAT.	INST.	TOTAL	MAT.	INST.	TOTAL	MAT.	INST.	TOTAL	MAT.	INST.	TOTAL	MAT.	INST.	TOTAL
015433	CONTRACTOR EQUIPMENT		97.3	97.3		95.9	95.9		96.7	96.7		101.6	101.6		94.2	94.2		97.0	97.0
0241, 31 - 34	SITE & INFRASTRUCTURE, DEMOLITION	91.7	104.9	100.8	93.2	100.9	98.5	84.9	106.0	99.4	84.8	106.2	99.5	97.6	94.5	95.5	91.1	103.4	99.6
0310	Concrete Forming & Accessories	100.4	97.9	98.2	100.6	87.4	89.1	95.0	81.6	83.4	95.0	81.7	83.4	91.6	85.9	86.6	100.4	89.7	89.5
0320	Concrete Reinforcing	93.8	92.3	93.0	103.7	83.0	93.3	96.0	81.2	88.6	96.0	83.2	89.6	93.4	81.4	87.4	93.4	92.1	92.8
0330	Cast-in-Place Concrete	93.6	106.4	98.7	88.9	94.0	90.9	84.4	87.6	85.6	90.4	86.0	88.7	99.7	97.4	98.8	90.9	104.2	96.1
03	CONCRETE	94.3	99.0	96.6	93.5	88.5	91.1	88.4	83.4	85.9	91.5	83.6	87.6	95.8	88.7	92.4	92.8	93.8	93.3
04	MASONRY	95.0	103.4	100.1	91.4	95.0	93.6	82.0	83.7	83.1	82.4	86.0	84.6	104.8	84.9	92.6	87.1	97.6	93.5
05	METALS	94.3	84.9	91.2	94.3	81.2	90.0	94.4	77.9	88.9	94.5	84.7	91.2	100.5	81.1	94.1	93.3	83.0	89.9
06	WOOD, PLASTICS & COMPOSITES	101.2	94.9	97.6	98.6	84.9	90.7	99.9	80.3	88.6	98.5	80.0	87.8	83.7	85.2	84.6	102.2	85.3	92.5
07	THERMAL & MOISTURE PROTECTION	105.9	107.7	106.6	100.7	95.1	98.5	104.2	89.0	96.8	101.0	90.7	96.8	99.3	95.8	97.9	108.1	101.0	105.3
08	OPENINGS	101.7	93.4	99.6	103.4	81.6	97.9	101.8	79.9	96.3	99.1	79.7	94.2	101.0	80.0	95.7	105.9	88.2	101.4
0920	Plaster & Gypsum Board	90.0	94.4	93.1	96.1	84.5	87.8	93.8	80.0	83.9	93.8	79.6	83.7	91.1	84.5	86.4	91.0	84.5	86.3
0950, 0980	Ceilings & Acoustic Treatment	92.9	94.4	93.9	95.5	84.5	88.2	99.7	80.0	86.7	98.6	79.6	86.1	102.6	84.5	90.6	94.7	84.5	88.0
0960	Flooring	100.2	105.0	101.6	95.2	90.0	93.7	107.7	83.0	100.5	104.8	90.6	100.7	126.8	90.8	116.3	100.6	105.0	101.9
0970, 0990	Wall Finishes & Painting/Coating	100.8	106.5	104.3	99.0	92.5	95.0	100.1	85.9	91.4	100.1	86.1	91.5	103.2	84.9	91.9	100.8	106.5	104.3
09	FINISHES	97.6	99.7	98.8	97.3	88.0	92.1	101.1	82.0	90.4	100.0	83.4	90.7	102.1	86.4	93.3	98.1	92.6	95.1
COVERS	DIVS. 10 - 14, 25, 28, 41, 43, 44, 46	100.0	98.3	99.7	100.0	94.7	98.9	100.0	91.9	98.4	100.0	92.0	98.4	100.0	92.9	98.6	100.0	95.2	99.0
21, 22, 23	FIRE SUPPRESSION, PLUMBING & HVAC	99.9	98.8	99.5	99.9	93.9	97.5	100.6	84.8	94.2	100.4	82.4	93.1	94.0	86.2	90.9	100.0	84.7	93.8
26, 27, 3370	ELECTRICAL, COMMUNICATIONS & UTIL.	97.5	107.4	102.6	98.7	88.1	93.2	96.3	83.8	89.9	96.7	81.9	89.1	99.7	80.8	89.9	97.4	89.9	93.5
MF2010	WEIGHTED AVERAGE	97.9	99.8	98.7	97.8	90.6	94.6	96.7	85.2	91.6	96.6	85.6	91.7	98.4	86.1	93.0	97.7	91.3	94.9

		OHIO																	
DIVISION		MANSFIELD			MARION			SPRINGFIELD			STEUBENVILLE			TOLEDO			YOUNGSTOWN		
		448 - 449			433			455			439			434 - 436			444 - 445		
		MAT.	INST.	TOTAL	MAT.	INST.	TOTAL	MAT.	INST.	TOTAL	MAT.	INST.	TOTAL	MAT.	INST.	TOTAL	MAT.	INST.	TOTAL
015433	CONTRACTOR EQUIPMENT		97.0	97.0		95.6	95.6		96.7	96.7		100.3	100.3		98.3	98.3		97.0	97.0
0241, 31 - 34	SITE & INFRASTRUCTURE, DEMOLITION	87.9	104.4	99.3	89.2	99.8	96.5	85.2	104.6	98.6	134.7	109.2	117.1	92.4	101.8	98.9	91.6	105.0	100.8
0310	Concrete Forming & Accessories	91.1	86.6	87.2	97.5	84.4	86.1	95.0	84.7	86.1	100.5	86.7	88.5	100.5	95.9	96.5	100.3	86.0	87.9
0320	Concrete Reinforcing	85.0	77.9	81.4	95.6	83.0	89.3	96.0	81.2	88.6	93.1	85.4	89.2	103.7	89.8	96.7	93.4	86.9	90.1
0330	Cast-in-Place Concrete	88.4	93.0	90.2	81.2	91.1	85.1	86.6	87.7	87.1	88.2	93.1	90.1	88.9	103.4	94.6	94.6	98.5	96.1
03	CONCRETE	87.8	86.7	87.2	85.8	86.1	86.0	89.6	84.7	87.2	89.7	88.1	88.9	93.5	97.0	95.2	94.7	90.0	92.4
04	MASONRY	89.6	97.4	94.3	92.9	96.1	94.8	82.2	84.0	83.3	79.8	91.2	86.7	99.5	100.1	99.8	90.7	92.7	91.9
05	METALS	93.5	77.1	88.1	93.4	79.2	88.7	94.3	77.6	88.8	90.1	79.6	86.6	94.1	87.6	91.9	92.7	80.3	88.6
06	WOOD, PLASTICS & COMPOSITES	91.2	85.3	87.8	94.9	84.9	89.2	101.5	84.7	91.8	90.9	85.1	87.6	98.6	95.1	96.6	102.2	83.8	91.6
07	THERMAL & MOISTURE PROTECTION	107.6	95.8	102.8	100.3	84.8	94.0	104.0	89.6	98.1	112.3	94.3	105.0	102.5	103.6	102.9	108.3	94.9	102.9
08	OPENINGS	105.1	79.6	98.7	97.3	80.2	93.0	99.5	80.4	94.7	98.3	82.9	94.4	101.2	91.4	98.7	105.9	84.7	100.6
0920	Plaster & Gypsum Board	84.9	84.5	84.6	93.8	84.5	87.1	93.8	84.5	87.1	94.4	84.1	87.1	96.1	94.9	95.2	91.0	83.0	85.3
0950, 0980	Ceilings & Acoustic Treatment	95.6	84.5	88.3	95.5	84.5	88.2	99.7	84.5	89.6	92.3	84.1	86.9	95.5	94.9	95.1	94.7	83.0	87.0
0960	Flooring	95.7	104.5	98.3	93.8	73.0	87.7	107.7	83.0	100.5	122.3	95.1	114.3	94.5	101.8	96.6	100.6	93.7	98.6
0970, 0990	Wall Finishes & Painting/Coating	100.8	91.2	94.9	99.0	50.1	68.8	100.1	85.9	91.4	113.1	95.5	102.2	99.0	102.9	101.4	100.8	93.5	96.3
09	FINISHES	95.9	90.3	92.8	96.1	78.9	86.5	101.1	84.6	91.9	114.7	88.6	100.2	97.0	97.5	97.3	98.2	87.6	92.3
COVERS	DIVS. 10 - 14, 25, 28, 41, 43, 44, 46	100.0	92.4	98.5	100.0	57.9	91.6	100.0	92.5	98.5	100.0	96.3	99.3	100.0	98.5	99.7	100.0	92.3	98.5
21, 22, 23	FIRE SUPPRESSION, PLUMBING & HVAC	94.0	83.4	89.7	94.0	89.0	92.0	100.6	85.0	94.3	94.3	91.2	93.0	100.0	99.2	99.7	100.0	92.1	96.8
26, 27, 3370	ELECTRICAL, COMMUNICATIONS & UTIL.	95.0	80.6	87.6	92.6	80.6	86.4	96.3	84.0	89.9	87.8	111.5	100.0	98.7	104.1	101.5	97.4	86.1	91.5
MF2010	WEIGHTED AVERAGE	95.2	87.4	91.7	93.9	85.4	90.2	96.6	85.8	91.8	96.1	93.5	95.0	97.9	98.3	98.1	98.1	90.1	94.6

OHIO / OKLAHOMA

DIVISION		ZANESVILLE 437-438 MAT.	INST.	TOTAL	ARDMORE 734 MAT.	INST.	TOTAL	CLINTON 736 MAT.	INST.	TOTAL	DURANT 747 MAT.	INST.	TOTAL	ENID 737 MAT.	INST.	TOTAL	GUYMON 739 MAT.	INST.	TOTAL
015433	CONTRACTOR EQUIPMENT		95.6	95.6		81.8	81.8		81.0	81.0		81.0	81.0		81.0	81.0		81.0	81.0
0241, 31 - 34	SITE & INFRASTRUCTURE, DEMOLITION	92.0	101.1	98.2	98.9	90.4	93.0	100.5	89.0	92.5	93.8	88.8	90.3	101.9	89.0	93.0	105.1	87.6	93.0
0310	Concrete Forming & Accessories	94.8	84.5	85.9	91.3	39.6	46.3	90.2	34.3	41.5	83.8	39.5	45.3	93.4	33.6	41.4	96.7	21.9	31.6
0320	Concrete Reinforcing	95.0	84.9	89.9	96.4	77.7	87.0	96.9	77.7	87.2	97.3	61.6	79.3	96.3	77.7	86.9	96.9	27.3	61.8
0330	Cast-in-Place Concrete	85.6	90.1	87.3	97.9	43.1	76.3	94.6	43.1	74.3	91.5	43.0	72.4	94.6	45.4	75.2	94.6	31.2	69.6
03	CONCRETE	89.2	86.2	87.7	96.9	48.5	73.3	96.3	46.1	71.8	91.5	45.4	69.1	96.7	46.7	72.3	99.5	27.0	64.1
04	MASONRY	90.3	82.2	85.4	97.2	55.8	72.0	125.5	55.8	83.0	95.8	63.6	76.2	106.1	55.8	75.4	100.3	22.2	52.6
05	METALS	94.8	80.4	90.0	91.0	69.6	84.0	91.1	69.6	84.0	91.1	63.2	81.8	92.5	69.6	85.0	91.6	40.7	74.8
06	WOOD, PLASTICS & COMPOSITES	91.3	84.9	87.6	97.2	38.2	63.1	96.2	31.0	58.6	88.4	38.2	59.4	99.9	30.2	59.6	103.5	20.0	55.3
07	THERMAL & MOISTURE PROTECTION	100.5	89.1	95.8	97.5	59.2	81.9	97.6	58.5	81.7	97.2	57.0	80.9	97.6	58.4	81.7	97.9	28.2	69.6
08	OPENINGS	97.3	82.1	93.5	94.3	48.7	82.8	94.4	44.8	81.9	94.4	44.6	81.8	94.3	44.1	81.7	94.5	21.4	76.1
0920	Plaster & Gypsum Board	91.1	84.5	86.4	79.5	36.9	49.1	79.0	29.5	43.7	74.9	36.9	47.8	80.4	28.6	43.4	81.2	18.2	36.2
0950, 0980	Ceilings & Acoustic Treatment	95.5	84.5	88.2	83.8	36.9	52.9	83.8	29.5	48.0	83.8	36.9	52.9	83.8	28.6	47.4	86.5	18.2	41.4
0960	Flooring	92.0	87.4	90.7	107.2	45.2	89.0	105.8	43.1	87.4	102.0	64.8	91.1	108.0	43.1	88.9	110.0	25.5	85.2
0970, 0990	Wall Finishes & Painting/Coating	99.0	78.4	86.3	96.2	54.1	70.2	96.2	54.1	70.2	96.2	54.1	70.2	96.2	54.1	70.2	96.2	19.3	48.7
09	FINISHES	95.6	84.2	89.2	92.7	40.3	63.5	92.5	35.7	60.8	90.3	44.4	64.7	93.4	35.2	60.9	95.1	21.9	54.3
COVERS	DIVS. 10 - 14, 25, 28, 41, 43, 44, 46	100.0	88.0	97.6	100.0	72.6	94.5	100.0	71.8	94.4	100.0	72.7	94.6	100.0	71.7	94.4	100.0	69.8	94.0
21, 22, 23	FIRE SUPPRESSION, PLUMBING & HVAC	94.0	88.9	92.0	94.3	61.3	80.9	94.3	61.3	80.9	94.3	61.4	81.0	100.3	61.3	84.5	94.3	23.1	65.5
26, 27, 3370	ELECTRICAL, COMMUNICATIONS & UTIL.	93.1	81.1	86.9	94.4	68.6	81.1	95.5	68.6	81.6	97.1	51.7	73.7	95.5	68.6	81.6	97.1	16.4	55.5
MF2010	WEIGHTED AVERAGE	94.5	86.1	90.8	95.0	60.1	79.6	96.5	58.8	79.8	94.1	57.7	78.1	97.3	58.7	80.3	96.2	31.0	67.5

OKLAHOMA

DIVISION		LAWTON 735 MAT.	INST.	TOTAL	McALESTER 745 MAT.	INST.	TOTAL	MIAMI 743 MAT.	INST.	TOTAL	MUSKOGEE 744 MAT.	INST.	TOTAL	OKLAHOMA CITY 730 - 731 MAT.	INST.	TOTAL	PONCA CITY 746 MAT.	INST.	TOTAL
015433	CONTRACTOR EQUIPMENT		81.8	81.8		81.0	81.0		88.9	88.9		88.9	88.9		82.1	82.1		81.0	81.0
0241 - 34	SITE & INFRASTRUCTURE, DEMOLITION	97.6	90.4	92.6	86.9	89.1	88.4	88.4	86.3	87.0	88.6	85.0	86.1	96.1	90.9	92.5	94.4	89.0	90.6
0310	Concrete Forming & Accessories	96.5	45.2	51.9	82.0	40.9	46.2	93.6	69.2	72.4	97.6	30.1	38.9	95.7	39.3	46.6	89.6	39.9	46.4
0320	Concrete Reinforcing	96.6	77.7	87.1	96.9	40.9	68.7	95.4	77.8	86.5	96.3	31.1	63.5	96.6	77.7	87.1	96.3	77.7	86.9
0330	Cast-in-Place Concrete	91.5	45.5	73.3	80.2	45.5	66.5	84.0	44.9	68.6	85.0	35.9	65.7	92.6	48.1	75.0	93.9	38.4	72.0
03	CONCRETE	93.0	51.8	72.9	82.3	43.0	63.2	86.7	63.0	75.1	88.4	33.7	61.7	93.5	50.1	72.3	93.5	47.0	70.9
04	MASONRY	100.2	55.8	73.1	114.9	56.4	79.2	98.6	57.0	73.2	117.7	45.3	73.5	99.8	57.5	74.0	91.1	56.4	69.9
05	METALS	96.1	69.7	87.4	91.0	53.9	78.8	91.0	81.7	87.9	92.5	56.2	80.5	98.0	69.6	88.6	91.0	69.3	83.8
06	WOOD, PLASTICS & COMPOSITES	102.5	45.8	69.8	86.0	41.4	60.3	99.8	77.5	86.9	103.8	29.9	61.1	101.8	36.9	64.3	95.6	38.8	62.8
07	THERMAL & MOISTURE PROTECTION	97.4	60.0	82.2	96.9	55.9	80.2	97.2	60.5	82.3	97.3	41.5	74.6	94.2	59.8	80.2	97.4	58.4	81.5
08	OPENINGS	96.0	52.6	85.1	94.3	40.2	80.7	94.3	70.5	88.3	94.3	29.0	77.9	96.0	47.8	83.9	94.3	49.0	82.9
0920	Plaster & Gypsum Board	82.8	44.8	55.6	73.6	40.2	49.8	80.4	77.3	78.2	82.8	28.2	43.8	84.9	35.6	49.7	79.0	37.5	49.4
0950, 0980	Ceilings & Acoustic Treatment	92.0	44.8	60.8	83.8	40.2	55.0	83.8	77.3	79.5	92.0	28.2	49.8	92.0	35.6	54.7	83.8	37.5	53.2
0960	Flooring	110.3	43.1	90.6	100.8	43.1	83.9	108.8	64.8	95.9	111.4	41.7	90.9	110.3	43.1	90.6	105.4	43.1	87.1
0970, 0990	Wall Finishes & Painting/Coating	96.2	54.1	70.2	96.2	39.8	61.4	96.2	81.7	87.2	96.2	28.9	54.7	96.2	54.1	70.2	96.2	54.1	70.2
09	FINISHES	95.5	44.4	67.0	89.1	40.4	61.9	92.1	70.8	80.2	95.1	32.0	59.9	95.5	39.6	64.3	91.9	40.4	63.2
COVERS	DIVS. 10 - 14, 25, 28, 41, 43, 44, 46	100.0	73.5	94.7	100.0	73.1	94.7	100.0	77.8	95.6	100.0	71.3	94.3	100.0	73.0	94.6	100.0	72.8	94.6
21, 22, 23	FIRE SUPPRESSION, PLUMBING & HVAC	100.3	61.3	84.5	94.3	30.2	68.3	94.3	60.2	80.5	100.3	23.4	69.1	100.2	62.1	84.7	94.3	59.9	80.4
26, 27, 3370	ELECTRICAL, COMMUNICATIONS & UTIL.	97.1	68.6	82.4	95.5	66.4	80.5	97.0	66.5	81.2	95.1	28.0	60.5	98.7	68.6	83.2	95.1	62.6	78.3
MF2010	WEIGHTED AVERAGE	97.5	61.4	81.6	93.6	50.6	74.6	93.8	67.9	82.3	96.7	39.3	71.4	97.8	60.6	81.4	94.1	58.7	78.5

OKLAHOMA / OREGON

DIVISION		POTEAU 749 MAT.	INST.	TOTAL	SHAWNEE 748 MAT.	INST.	TOTAL	TULSA 740 - 741 MAT.	INST.	TOTAL	WOODWARD 738 MAT.	INST.	TOTAL	BEND 977 MAT.	INST.	TOTAL	EUGENE 974 MAT.	INST.	TOTAL
015433	CONTRACTOR EQUIPMENT		88.3	88.3		81.0	81.0		88.9	88.9		81.0	81.0		98.7	98.7		98.7	98.7
0241, 31 - 34	SITE & INFRASTRUCTURE, DEMOLITION	73.2	85.1	81.4	97.5	88.9	91.6	94.7	86.4	89.0	100.9	89.0	92.7	113.0	103.8	106.6	101.4	103.8	103.0
0310	Concrete Forming & Accessories	87.7	42.1	48.1	83.7	38.6	44.5	97.5	39.1	46.7	90.3	34.2	41.5	111.3	94.4	96.6	108.0	94.3	96.1
0320	Concrete Reinforcing	97.4	77.7	87.5	96.3	45.0	70.4	96.6	77.6	87.0	96.3	77.7	86.9	100.8	104.5	102.7	105.2	104.5	104.8
0330	Cast-in-Place Concrete	84.1	44.4	68.4	97.0	42.6	75.5	92.7	44.4	73.7	94.6	43.1	74.3	99.4	104.7	101.5	96.2	104.7	99.6
03	CONCRETE	88.8	50.8	70.3	95.4	41.8	69.2	93.6	49.5	72.1	96.5	46.1	72.0	109.1	99.7	104.5	100.7	99.6	100.2
04	MASONRY	99.5	56.4	73.2	116.2	55.2	79.0	99.4	57.6	73.9	93.9	55.8	70.7	106.6	108.5	107.8	103.9	108.5	106.7
05	METALS	91.0	81.3	87.8	90.9	55.7	79.3	95.7	80.4	90.6	91.2	69.6	84.0	86.8	98.1	90.5	87.5	98.0	90.9
06	WOOD, PLASTICS & COMPOSITES	92.7	41.5	63.1	88.3	38.0	59.2	103.1	38.0	65.5	96.4	30.9	58.6	102.0	91.3	95.8	98.0	91.3	94.1
07	THERMAL & MOISTURE PROTECTION	97.3	56.3	80.6	97.4	59.0	81.8	97.3	54.8	80.0	97.6	58.4	81.7	105.3	95.4	101.2	104.5	92.4	99.6
08	OPENINGS	94.3	51.0	83.4	94.3	37.4	80.0	96.0	48.0	83.9	94.4	44.7	81.9	97.8	99.5	98.2	98.1	99.5	98.5
0920	Plaster & Gypsum Board	77.7	40.2	50.9	74.9	36.7	47.6	82.8	36.6	49.8	79.8	29.4	43.8	102.2	90.9	94.1	99.9	90.9	93.5
0950, 0980	Ceilings & Acoustic Treatment	83.8	40.2	55.0	83.8	36.7	52.7	92.0	36.6	55.4	86.5	29.4	48.8	96.6	90.9	92.8	97.6	90.9	93.2
0960	Flooring	104.7	44.8	93.0	102.0	34.3	82.1	110.1	44.8	91.0	105.9	45.2	88.1	109.2	108.2	108.9	107.4	108.2	107.6
0970, 0990	Wall Finishes & Painting/Coating	96.2	54.1	70.2	96.2	36.0	59.0	96.2	42.7	63.2	96.2	54.1	70.2	108.7	69.0	84.2	108.7	69.0	84.2
09	FINISHES	89.7	46.4	65.5	90.6	36.1	60.2	94.9	39.3	63.9	93.3	36.0	61.4	105.7	93.4	98.8	103.8	93.4	98.0
COVERS	DIVS. 10 - 14, 25, 28, 41, 43, 44, 46	100.0	73.4	94.7	100.0	72.4	94.5	100.0	73.4	94.7	100.0	71.8	94.4	100.0	102.9	100.6	100.0	102.9	100.6
21, 22, 23	FIRE SUPPRESSION, PLUMBING & HVAC	94.3	59.9	80.4	94.3	60.6	80.6	100.3	47.3	78.8	94.3	61.3	80.9	94.1	97.6	95.5	100.1	97.6	99.1
26, 27, 3370	ELECTRICAL, COMMUNICATIONS & UTIL.	95.2	66.5	80.4	97.2	68.6	82.4	97.1	36.5	65.9	97.0	68.6	82.3	100.5	93.9	97.1	99.1	93.9	96.4
MF2010	WEIGHTED AVERAGE	93.2	61.4	79.2	95.7	56.5	78.4	97.4	53.5	78.0	95.1	58.8	79.1	98.8	98.6	98.7	98.6	98.5	98.6

City Cost Indexes

OREGON

DIVISION		KLAMATH FALLS 976			MEDFORD 975			PENDLETON 978			PORTLAND 970 - 972			SALEM 973			VALE 979		
		MAT.	INST.	TOTAL	MAT.	INST.	TOTAL	MAT.	INST.	TOTAL	MAT.	INST.	TOTAL	MAT.	INST.	TOTAL	MAT.	INST.	TOTAL
015433	CONTRACTOR EQUIPMENT		98.7	98.7		98.7	98.7		96.2	96.2		98.7	98.7		98.7	98.7		96.2	96.2
0241, 31 - 34	SITE & INFRASTRUCTURE, DEMOLITION	117.6	103.8	108.1	110.0	103.8	105.7	107.8	97.3	100.6	104.0	103.8	103.9	96.6	103.8	101.5	94.7	97.3	96.5
0310	Concrete Forming & Accessories	104.9	94.3	95.6	104.1	94.3	95.5	104.4	94.5	95.8	109.1	94.6	96.5	106.9	94.4	96.1	110.3	93.7	95.8
0320	Concrete Reinforcing	100.8	104.5	102.7	102.6	104.5	103.5	100.0	104.5	102.3	106.0	104.5	105.2	106.2	104.5	105.3	97.5	104.4	101.0
0330	Cast-in-Place Concrete	99.4	104.7	101.5	99.3	104.7	101.4	100.1	106.1	102.5	98.9	104.8	101.2	88.2	104.7	94.7	78.9	104.7	88.6
03	CONCRETE	111.9	99.6	105.9	106.7	99.6	103.2	93.1	100.2	96.6	102.2	99.8	101.0	96.8	99.7	98.2	78.2	99.7	88.6
04	MASONRY	120.1	108.5	113.0	100.9	108.5	105.5	110.4	108.6	109.3	104.5	108.5	107.0	108.7	108.5	108.6	108.7	108.6	108.6
05	METALS	86.8	97.9	90.5	87.1	97.9	90.6	92.6	98.5	94.5	88.3	98.2	91.6	87.8	98.1	91.2	92.5	97.6	94.2
06	WOOD, PLASTICS & COMPOSITES	93.4	91.3	92.2	92.3	91.3	91.7	94.8	91.4	92.9	98.9	91.3	94.5	95.2	91.3	93.0	102.9	91.4	96.3
07	THERMAL & MOISTURE PROTECTION	105.5	93.9	100.8	105.1	93.9	100.6	97.3	94.2	96.0	104.5	98.0	101.8	102.5	95.4	99.6	96.7	91.9	94.7
08	OPENINGS	97.8	99.5	98.2	100.7	99.5	100.4	93.2	99.6	94.8	95.7	99.5	96.7	97.6	99.5	98.1	93.1	92.3	92.9
0920	Plaster & Gypsum Board	97.3	90.9	92.7	96.4	90.9	92.5	80.4	90.9	87.9	100.2	90.9	93.5	94.4	90.9	91.9	87.2	90.9	89.8
0950, 0980	Ceilings & Acoustic Treatment	104.8	90.9	95.6	109.8	90.9	97.3	67.6	90.9	83.0	99.7	90.9	93.9	99.9	90.9	93.9	67.6	90.9	83.0
0960	Flooring	105.9	108.2	106.6	105.2	108.2	106.1	74.1	108.2	84.1	104.9	108.2	105.9	107.6	108.2	107.8	76.1	108.2	85.5
0970, 0990	Wall Finishes & Painting/Coating	108.7	61.3	79.4	108.7	61.3	79.4	98.3	69.0	80.2	108.3	69.0	84.0	108.7	69.0	84.2	98.3	69.0	80.2
09	FINISHES	106.5	92.6	98.7	106.3	92.6	98.7	72.6	93.5	84.3	103.6	93.4	97.9	103.2	93.4	97.7	72.8	93.5	84.3
COVERS	DIVS. 10 - 14, 25, 28, 41, 43, 44, 46	100.0	102.9	100.6	100.0	102.9	100.6	100.0	95.8	99.2	100.0	103.0	100.6	100.0	102.9	100.6	100.0	103.2	100.6
21, 22, 23	FIRE SUPPRESSION, PLUMBING & HVAC	94.1	97.6	95.5	100.1	97.6	99.1	95.8	106.9	100.3	100.1	100.9	100.4	100.1	97.6	99.1	95.8	94.7	87.3
26, 27, 3370	ELECTRICAL, COMMUNICATIONS & UTIL.	99.2	83.4	91.1	102.6	83.4	92.7	91.4	94.3	92.9	99.4	100.2	99.8	103.5	93.9	98.6	91.4	76.7	83.8
MF2010	WEIGHTED AVERAGE	99.8	97.0	98.6	100.1	97.0	98.7	94.1	100.0	96.7	98.8	100.3	99.5	98.6	98.6	98.6	91.9	90.5	91.3

PENNSYLVANIA

DIVISION		ALLENTOWN 181			ALTOONA 166			BEDFORD 155			BRADFORD 167			BUTLER 160			CHAMBERSBURG 172		
		MAT.	INST.	TOTAL	MAT.	INST.	TOTAL	MAT.	INST.	TOTAL	MAT.	INST.	TOTAL	MAT.	INST.	TOTAL	MAT.	INST.	TOTAL
015433	CONTRACTOR EQUIPMENT		113.6	113.6		113.6	113.6		110.9	110.9		113.6	113.6		113.6	113.6		112.9	112.9
0241, 31 - 34	SITE & INFRASTRUCTURE, DEMOLITION	91.4	105.8	101.4	95.6	105.6	102.5	109.6	102.5	104.7	90.6	105.0	100.5	86.7	107.1	100.8	84.9	103.1	97.4
0310	Concrete Forming & Accessories	100.4	113.5	111.8	86.4	79.5	80.4	85.6	82.0	82.4	88.4	80.2	81.2	87.8	93.7	92.9	86.1	81.1	81.8
0320	Concrete Reinforcing	98.3	111.7	105.0	95.3	90.1	92.7	96.3	81.9	89.1	97.3	90.3	93.8	96.0	110.0	103.0	95.7	102.2	99.0
0330	Cast-in-Place Concrete	82.8	104.8	91.5	92.2	84.8	89.3	107.1	72.8	93.6	88.1	89.2	88.6	81.5	96.8	87.5	89.5	72.5	82.8
03	CONCRETE	94.4	110.9	102.4	90.1	84.8	87.5	101.7	80.1	91.2	96.3	86.7	91.6	82.1	99.0	90.4	102.0	83.6	93.0
04	MASONRY	95.7	104.8	101.3	97.8	68.1	79.7	107.4	85.2	93.9	95.0	83.6	88.1	100.0	96.2	97.7	96.0	90.3	92.5
05	METALS	100.6	123.5	108.2	94.4	110.6	99.7	96.9	105.0	99.6	98.5	110.2	102.4	94.1	121.9	103.3	98.5	114.9	103.9
06	WOOD, PLASTICS & COMPOSITES	101.9	115.4	109.7	81.3	81.7	81.5	82.7	81.7	82.1	87.6	78.4	82.3	82.6	91.5	87.7	91.0	81.7	85.6
07	THERMAL & MOISTURE PROTECTION	99.5	117.2	106.7	98.9	90.5	95.4	104.8	88.5	98.2	99.5	91.2	96.1	98.5	100.1	99.2	98.7	75.5	89.3
08	OPENINGS	92.3	114.6	97.9	86.8	85.4	86.4	94.6	83.2	91.8	92.7	87.3	91.3	86.8	103.0	90.9	89.1	83.4	87.7
0920	Plaster & Gypsum Board	103.9	115.6	112.3	96.6	80.9	85.4	95.4	80.9	85.0	97.2	77.5	83.1	96.6	91.0	92.6	100.8	80.9	86.6
0950, 0980	Ceilings & Acoustic Treatment	88.9	115.6	106.6	93.5	80.9	85.2	89.7	80.9	83.9	91.8	77.5	82.4	94.5	91.0	92.2	91.8	80.9	84.6
0960	Flooring	93.1	98.0	94.6	86.9	92.2	88.4	88.4	92.3	89.5	87.0	100.1	90.8	87.7	84.8	86.9	90.8	48.3	78.3
0970, 0990	Wall Finishes & Painting/Coating	96.6	74.4	82.9	92.6	106.9	101.4	97.3	86.0	90.3	96.6	92.3	93.9	92.6	106.9	101.4	96.6	86.0	90.1
09	FINISHES	95.8	106.8	101.9	94.6	84.0	88.7	94.2	82.9	87.9	93.9	83.3	88.0	94.4	92.2	93.2	94.4	75.3	83.7
COVERS	DIVS. 10 - 14, 25, 28, 41, 43, 44, 46	100.0	106.2	101.2	100.0	95.9	99.2	100.0	99.0	99.8	100.0	98.9	99.8	100.0	102.9	100.6	100.0	96.2	99.2
21, 22, 23	FIRE SUPPRESSION, PLUMBING & HVAC	100.2	112.2	105.0	99.7	83.7	93.2	94.1	88.4	91.7	94.3	90.3	92.7	93.8	95.0	94.3	94.3	89.9	92.5
26, 27, 3370	ELECTRICAL, COMMUNICATIONS & UTIL.	99.9	97.7	98.7	89.8	107.8	99.1	92.6	107.8	100.5	93.5	107.8	100.9	90.4	102.0	96.4	92.1	85.8	88.8
MF2010	WEIGHTED AVERAGE	97.9	109.2	102.9	94.8	90.6	93.0	97.2	91.7	94.8	95.5	93.8	94.7	92.3	100.5	95.9	95.5	89.5	92.9

PENNSYLVANIA

DIVISION		DOYLESTOWN 189			DUBOIS 158			ERIE 164 - 165			GREENSBURG 156			HARRISBURG 170 - 171			HAZLETON 182		
		MAT.	INST.	TOTAL	MAT.	INST.	TOTAL	MAT.	INST.	TOTAL	MAT.	INST.	TOTAL	MAT.	INST.	TOTAL	MAT.	INST.	TOTAL
015433	CONTRACTOR EQUIPMENT		94.9	94.9		110.9	110.9		113.6	113.6		110.9	110.9		112.9	112.9		113.6	113.6
0241, 31 - 34	SITE & INFRASTRUCTURE, DEMOLITION	104.8	91.7	95.8	115.4	102.6	106.6	92.2	106.2	101.8	105.1	104.7	104.8	81.2	104.7	97.4	84.5	106.0	99.3
0310	Concrete Forming & Accessories	85.3	128.2	122.6	85.0	83.0	83.2	100.0	87.4	89.0	91.4	93.7	93.4	93.7	90.7	91.1	83.2	88.3	87.6
0320	Concrete Reinforcing	95.1	138.9	117.2	95.7	90.5	93.1	97.3	92.0	94.6	95.7	110.0	102.9	98.3	102.6	100.5	95.4	106.5	101.0
0330	Cast-in-Place Concrete	78.3	89.1	82.5	103.2	91.4	98.6	90.7	84.2	88.1	99.4	96.4	98.2	87.2	96.3	90.8	78.2	91.6	83.5
03	CONCRETE	90.2	116.4	103.0	102.6	88.6	95.8	89.0	88.4	88.7	96.6	98.8	97.7	94.5	96.1	95.3	87.3	94.2	90.6
04	MASONRY	99.0	128.9	117.3	107.4	86.0	94.4	87.8	90.1	89.2	118.0	96.2	104.7	95.5	93.5	94.3	108.2	96.8	101.2
05	METALS	98.1	127.7	107.9	96.9	109.8	101.1	94.6	110.9	100.0	96.8	120.7	104.7	102.7	118.6	107.9	100.3	120.3	106.9
06	WOOD, PLASTICS & COMPOSITES	83.3	130.2	110.4	81.6	81.7	81.6	97.9	85.9	91.0	89.0	91.4	90.4	95.9	90.0	92.5	82.0	86.8	84.8
07	THERMAL & MOISTURE PROTECTION	97.8	132.0	111.7	105.1	94.8	100.9	99.2	93.5	96.9	104.7	99.8	102.7	105.2	108.9	106.7	99.1	105.7	101.8
08	OPENINGS	94.1	139.9	105.6	94.6	89.0	93.2	87.0	88.0	87.2	94.6	102.9	96.7	92.3	94.4	92.8	93.0	92.9	93.0
0920	Plaster & Gypsum Board	94.8	130.9	120.6	94.0	80.9	84.6	103.9	85.2	90.5	96.5	91.0	92.5	100.9	89.4	92.7	95.3	86.1	88.7
0950, 0980	Ceilings & Acoustic Treatment	88.1	130.9	116.4	89.7	80.9	83.9	88.9	85.2	86.5	88.8	91.0	90.2	93.5	89.4	90.8	90.0	86.1	87.4
0960	Flooring	76.9	133.3	93.4	88.1	100.1	91.7	92.3	96.1	93.4	91.8	71.2	85.7	93.7	93.7	93.7	84.3	96.7	87.9
0970, 0990	Wall Finishes & Painting/Coating	96.1	72.7	81.7	97.3	106.9	103.2	100.7	90.2	94.2	97.3	106.9	103.2	96.6	89.0	91.9	96.6	100.5	99.0
09	FINISHES	86.8	122.4	106.6	94.6	87.9	90.8	96.4	88.9	92.2	94.7	90.5	92.4	95.1	90.8	92.7	92.0	89.1	90.4
COVERS	DIVS. 10 - 14, 25, 28, 41, 43, 44, 46	100.0	74.2	94.9	100.0	99.0	99.8	100.0	101.1	100.2	100.0	102.8	100.6	100.0	98.2	99.6	100.0	96.5	99.3
21, 22, 23	FIRE SUPPRESSION, PLUMBING & HVAC	93.8	126.9	107.2	94.1	88.6	91.8	99.7	93.0	97.0	94.1	92.8	93.6	100.0	92.2	96.9	94.3	98.3	95.9
26, 27, 3370	ELECTRICAL, COMMUNICATIONS & UTIL.	92.7	117.9	105.7	93.2	107.8	100.8	91.7	87.4	89.5	93.2	107.8	100.8	96.7	85.8	91.1	94.3	85.8	89.9
MF2010	WEIGHTED AVERAGE	94.7	120.1	105.9	97.6	94.5	96.2	94.6	93.6	94.1	97.2	100.3	98.5	97.7	96.1	97.0	95.1	97.2	96.0

PENNSYLVANIA

DIVISION		INDIANA 157 MAT.	INST.	TOTAL	JOHNSTOWN 159 MAT.	INST.	TOTAL	KITTANNING 162 MAT.	INST.	TOTAL	LANCASTER 175-176 MAT.	INST.	TOTAL	LEHIGH VALLEY 180 MAT.	INST.	TOTAL	MONTROSE 188 MAT.	INST.	TOTAL
015433	CONTRACTOR EQUIPMENT		110.9	110.9		110.9	110.9		113.6	113.6		112.9	112.9		113.6	113.6		113.6	113.6
0241, 31-34	SITE & INFRASTRUCTURE, DEMOLITION	103.1	103.6	103.4	110.2	103.7	105.8	89.4	107.1	101.6	77.0	104.6	96.1	88.6	105.6	100.3	87.5	103.9	98.8
0310	Concrete Forming & Accessories	86.1	86.2	86.2	85.0	83.3	83.5	87.8	93.8	93.0	87.9	89.8	89.6	94.3	112.4	110.1	84.0	87.6	87.2
0320	Concrete Reinforcing	94.9	110.0	102.6	96.3	109.8	103.1	96.0	110.1	103.1	95.4	102.5	99.0	95.4	111.7	103.6	100.0	106.3	103.1
0330	Cast-in-Place Concrete	97.4	96.3	97.0	108.0	89.9	100.9	84.7	96.8	89.5	76.2	95.5	83.8	84.7	103.5	92.1	83.0	90.6	86.0
03	CONCRETE	94.3	95.5	94.9	102.5	91.9	97.3	84.6	99.1	91.7	89.6	95.5	92.5	93.5	109.9	101.5	92.2	93.3	92.7
04	MASONRY	103.3	97.9	100.0	104.1	85.6	92.8	103.7	97.9	100.2	101.3	92.4	95.9	95.7	102.7	100.0	95.7	97.8	97.0
05	METALS	96.9	120.5	104.7	96.9	118.7	104.1	94.2	122.2	103.4	98.5	118.1	105.0	100.3	122.7	107.7	98.6	115.0	104.0
06	WOOD, PLASTICS & COMPOSITES	83.4	81.7	82.4	81.6	81.7	81.6	82.6	91.5	87.7	93.5	90.0	91.5	94.2	115.4	106.5	82.7	86.9	85.1
07	THERMAL & MOISTURE PROTECTION	104.6	97.8	101.8	104.8	93.3	100.2	98.6	100.6	99.4	98.3	97.3	97.9	99.4	105.2	101.7	99.1	94.3	97.1
08	OPENINGS	94.6	94.9	94.7	94.6	91.2	93.8	86.8	100.2	90.2	89.1	98.1	91.4	93.0	114.6	98.5	93.2	92.9	90.2
0920	Plaster & Gypsum Board	95.8	80.9	85.2	93.7	80.9	84.6	96.6	91.0	92.6	103.1	89.4	93.3	98.5	115.6	110.7	95.8	86.2	88.9
0950, 0980	Ceilings & Acoustic Treatment	89.7	80.9	83.9	88.8	80.9	83.6	94.5	91.0	92.2	91.8	89.4	90.2	90.0	115.6	106.9	91.8	86.2	88.1
0960	Flooring	88.9	100.1	92.2	88.1	92.3	89.4	87.7	102.4	92.0	91.8	93.7	92.3	89.8	98.0	92.2	84.9	60.6	77.8
0970, 0990	Wall Finishes & Painting/Coating	97.3	106.9	103.2	97.3	106.9	103.2	92.6	106.9	101.4	96.6	60.2	74.2	96.6	70.0	80.2	96.6	100.5	99.0
09	FINISHES	93.8	89.1	91.2	93.8	86.4	89.7	94.6	95.3	95.0	94.3	87.3	90.4	94.3	104.8	100.1	92.9	83.7	87.8
COVERS	DIVS. 10-14, 25, 28, 41, 43, 44, 46	100.0	101.7	100.3	100.0	99.2	99.8	100.0	102.9	100.6	100.0	97.6	99.5	100.0	102.1	100.4	100.0	97.1	99.4
21, 22, 23	FIRE SUPPRESSION, PLUMBING & HVAC	94.1	92.8	93.5	94.1	89.0	92.0	93.8	97.7	95.4	94.3	91.6	93.2	94.3	111.1	101.1	94.3	99.6	96.4
26, 27, 3370	ELECTRICAL, COMMUNICATIONS & UTIL.	93.2	107.8	100.8	93.2	107.8	100.8	89.8	107.9	99.1	93.5	46.8	69.4	94.3	138.5	117.1	93.5	94.4	93.9
MF2010	WEIGHTED AVERAGE	96.0	99.1	97.4	97.2	95.8	96.6	92.8	102.4	97.0	94.3	89.7	92.3	95.6	113.5	103.5	94.5	97.0	95.6

PENNSYLVANIA

DIVISION		NEW CASTLE 161 MAT.	INST.	TOTAL	NORRISTOWN 194 MAT.	INST.	TOTAL	OIL CITY 163 MAT.	INST.	TOTAL	PHILADELPHIA 190-191 MAT.	INST.	TOTAL	PITTSBURGH 150-152 MAT.	INST.	TOTAL	POTTSVILLE 179 MAT.	INST.	TOTAL
015433	CONTRACTOR EQUIPMENT		113.6	113.6		98.2	98.2		113.6	113.6		97.6	97.6		112.1	112.1		112.9	112.9
0241, 31-34	SITE & INFRASTRUCTURE, DEMOLITION	87.2	107.0	100.9	91.9	99.2	96.9	85.6	105.2	99.1	98.6	98.3	98.4	108.7	106.8	107.4	79.9	104.8	97.1
0310	Concrete Forming & Accessories	87.8	96.2	95.1	87.0	132.2	126.3	87.8	81.4	82.3	102.8	134.7	130.6	98.5	97.4	97.5	80.2	88.4	87.3
0320	Concrete Reinforcing	94.7	93.7	94.2	95.5	146.3	121.1	96.0	93.5	94.7	98.5	146.3	122.6	96.8	110.5	103.7	94.6	102.0	98.3
0330	Cast-in-Place Concrete	82.3	95.9	87.6	81.1	128.6	99.8	79.9	93.7	85.4	99.4	128.6	110.9	103.2	96.6	100.6	81.1	78.3	80.0
03	CONCRETE	82.5	96.8	89.4	87.6	133.2	109.8	80.9	89.4	85.0	100.0	134.3	116.7	99.9	100.6	100.3	93.4	88.8	91.2
04	MASONRY	99.3	96.2	97.4	109.3	129.0	121.3	99.8	89.9	93.8	96.2	129.1	116.3	97.1	103.6	101.0	95.4	94.4	94.8
05	METALS	94.2	114.0	100.7	99.7	133.0	110.7	94.2	111.8	100.0	103.5	133.2	113.3	98.3	121.7	106.0	98.7	118.1	105.1
06	WOOD, PLASTICS & COMPOSITES	82.6	95.8	90.3	84.3	132.2	112.0	82.6	78.4	80.2	102.7	135.5	121.7	98.2	95.8	96.8	84.3	86.8	85.7
07	THERMAL & MOISTURE PROTECTION	98.6	97.7	98.2	97.5	132.5	111.7	98.5	93.8	96.6	98.4	132.9	112.4	104.9	102.2	103.8	98.3	105.9	101.4
08	OPENINGS	86.8	97.5	89.5	89.6	143.1	103.1	86.8	80.3	85.2	98.1	145.0	109.9	97.6	106.0	99.7	89.2	91.5	89.8
0920	Plaster & Gypsum Board	96.6	95.4	95.8	90.8	133.0	121.0	96.6	77.5	83.0	98.6	136.5	125.7	103.1	95.4	97.6	97.6	86.1	89.4
0950, 0980	Ceilings & Acoustic Treatment	94.5	95.4	95.1	99.7	133.0	121.7	94.5	77.5	83.3	99.7	136.5	124.0	89.7	95.4	93.5	91.8	86.1	88.0
0960	Flooring	87.7	60.2	79.6	91.9	139.3	105.8	87.7	100.1	91.4	99.8	141.2	112.0	95.3	107.4	98.9	87.8	93.7	89.6
0970, 0990	Wall Finishes & Painting/Coating	92.6	106.9	101.4	99.6	146.7	128.7	92.6	106.9	101.4	99.6	146.7	128.7	97.3	114.0	107.6	96.6	100.5	99.0
09	FINISHES	94.5	90.3	92.1	95.6	134.9	117.5	94.3	85.7	89.5	99.2	137.2	120.4	96.9	100.3	98.8	92.8	90.4	91.4
COVERS	DIVS. 10-14, 25, 28, 41, 43, 44, 46	100.0	103.3	100.7	100.0	115.0	103.0	100.0	100.0	100.0	100.0	119.0	103.8	100.0	103.3	100.7	100.0	94.5	98.9
21, 22, 23	FIRE SUPPRESSION, PLUMBING & HVAC	93.8	94.7	94.2	93.9	128.3	107.9	93.8	92.3	93.2	99.9	128.4	111.5	100.0	100.3	100.1	94.3	99.5	96.4
26, 27, 3370	ELECTRICAL, COMMUNICATIONS & UTIL.	90.4	91.4	90.9	94.7	142.8	119.5	92.4	102.0	97.4	99.1	142.8	121.7	95.5	107.9	101.9	91.7	90.2	90.9
MF2010	WEIGHTED AVERAGE	92.4	97.5	94.6	95.2	130.2	110.6	92.4	94.7	93.4	99.9	130.9	113.6	99.1	104.6	101.5	94.2	96.8	95.4

PENNSYLVANIA

DIVISION		READING 195-196 MAT.	INST.	TOTAL	SCRANTON 184-185 MAT.	INST.	TOTAL	STATE COLLEGE 168 MAT.	INST.	TOTAL	STROUDSBURG 183 MAT.	INST.	TOTAL	SUNBURY 178 MAT.	INST.	TOTAL	UNIONTOWN 154 MAT.	INST.	TOTAL
015433	CONTRACTOR EQUIPMENT		116.9	116.9		113.6	113.6		112.9	112.9		113.6	113.6		113.6	113.6		110.9	110.9
0241, 31-34	SITE & INFRASTRUCTURE, DEMOLITION	96.3	110.9	106.4	91.8	106.1	101.7	82.1	104.4	97.4	86.3	104.0	98.5	90.3	105.8	101.0	103.6	104.7	104.4
0310	Concrete Forming & Accessories	102.3	86.0	88.1	100.6	89.2	90.7	85.2	79.4	80.2	89.1	88.8	88.8	92.3	87.5	88.1	79.8	93.7	91.9
0320	Concrete Reinforcing	96.8	102.1	99.5	98.3	106.5	102.4	96.5	90.8	93.6	98.6	118.6	108.7	97.3	102.6	100.0	95.7	110.0	102.9
0330	Cast-in-Place Concrete	72.7	96.7	82.1	86.5	92.6	88.9	83.4	68.3	77.4	81.5	75.1	79.0	88.7	95.3	91.3	97.4	96.4	97.0
03	CONCRETE	86.3	94.2	90.1	96.2	95.0	95.6	96.3	79.2	88.0	91.0	90.8	90.9	96.9	94.4	95.7	94.0	98.8	96.4
04	MASONRY	97.8	95.5	96.4	96.1	98.3	97.4	101.4	81.5	89.3	92.5	102.3	98.5	95.0	92.0	93.1	120.8	96.2	105.8
05	METALS	99.9	118.9	106.2	102.7	120.6	108.6	98.3	111.5	102.7	100.4	120.2	106.9	98.4	118.4	105.0	96.6	120.7	104.6
06	WOOD, PLASTICS & COMPOSITES	102.8	82.8	91.3	101.9	86.9	93.2	89.7	81.7	85.1	88.5	86.9	87.6	91.8	86.8	88.9	75.5	91.4	84.7
07	THERMAL & MOISTURE PROTECTION	97.9	108.9	102.4	99.4	98.0	98.9	98.5	91.5	95.7	99.3	87.1	94.3	99.4	104.9	101.6	104.5	99.8	102.6
08	OPENINGS	95.3	94.0	95.0	92.3	92.9	92.5	89.1	85.4	88.2	93.0	89.7	92.2	89.3	91.7	89.9	94.6	102.9	96.7
0920	Plaster & Gypsum Board	103.0	82.1	88.0	106.3	86.2	92.0	98.7	80.9	86.0	96.6	86.2	89.2	97.3	86.1	89.3	91.0	91.0	91.0
0950, 0980	Ceilings & Acoustic Treatment	88.7	82.1	84.3	97.1	86.2	89.9	89.0	80.9	83.7	88.1	86.2	86.9	89.0	86.1	87.1	88.8	91.0	90.2
0960	Flooring	97.3	93.7	96.2	93.1	110.2	98.1	90.3	102.1	93.8	87.5	53.8	77.6	88.7	91.8	89.6	85.4	100.1	89.7
0970, 0990	Wall Finishes & Painting/Coating	98.1	100.9	99.8	96.6	100.5	99.0	96.6	106.9	103.0	96.6	68.1	79.0	96.6	100.5	99.0	97.3	106.9	103.2
09	FINISHES	96.5	87.6	91.5	97.9	93.6	95.5	93.1	86.0	89.1	92.8	79.9	85.6	93.9	89.5	91.4	92.0	94.6	93.5
COVERS	DIVS. 10-14, 25, 28, 41, 43, 44, 46	100.0	99.6	99.9	100.0	97.1	99.4	100.0	93.1	98.6	100.0	65.9	93.2	100.0	93.6	98.7	100.0	102.8	100.6
21, 22, 23	FIRE SUPPRESSION, PLUMBING & HVAC	100.0	107.9	103.2	100.2	100.5	100.3	94.3	85.7	90.8	94.3	102.2	97.5	94.3	91.3	93.1	94.1	92.8	93.6
26, 27, 3370	ELECTRICAL, COMMUNICATIONS & UTIL.	101.3	90.2	95.6	99.9	94.5	97.1	92.6	107.8	100.5	94.3	142.6	119.2	92.0	86.7	89.3	90.2	107.8	99.3
MF2010	WEIGHTED AVERAGE	97.6	100.0	98.6	98.6	99.5	99.0	95.0	91.8	93.6	94.9	103.0	98.5	95.1	95.1	95.1	96.4	100.8	98.3

PENNSYLVANIA

DIVISION		WASHINGTON 153			WELLSBORO 169			WESTCHESTER 193			WILKES-BARRE 186 - 187			WILLIAMSPORT 177			YORK 173 - 174		
		MAT.	INST.	TOTAL	MAT.	INST.	TOTAL	MAT.	INST.	TOTAL	MAT.	INST.	TOTAL	MAT.	INST.	TOTAL	MAT.	INST.	TOTAL
015433	CONTRACTOR EQUIPMENT		110.9	110.9		113.6	113.6		98.2	98.2		113.6	113.6		113.6	113.6		112.9	112.9
0241, 31 - 34	SITE & INFRASTRUCTURE, DEMOLITION	103.7	104.7	104.4	94.7	103.8	101.0	98.0	96.9	97.2	84.3	106.1	99.3	81.6	104.4	97.3	80.5	104.7	97.2
0310	Concrete Forming & Accessories	86.2	93.8	92.9	87.8	85.9	86.1	93.5	129.2	124.6	91.5	89.2	89.5	89.1	61.9	65.4	83.3	87.9	87.3
0320	Concrete Reinforcing	95.7	110.1	103.0	96.5	106.1	101.4	94.6	112.9	103.8	97.3	106.5	102.0	96.5	68.5	82.4	97.3	102.6	100.0
0330	Cast-in-Place Concrete	97.4	96.5	97.1	87.4	88.5	87.8	89.7	128.0	104.8	78.2	92.6	83.9	75.0	74.0	74.6	81.5	95.7	87.1
03	CONCRETE	94.4	99.0	96.6	98.7	91.8	95.3	95.0	125.4	109.8	88.1	95.0	91.4	84.9	69.3	77.3	94.6	94.7	94.7
04	MASONRY	103.1	97.9	99.9	101.1	92.0	95.6	103.4	129.0	119.0	108.8	98.3	102.4	86.9	92.3	90.2	96.1	92.4	93.9
05	METALS	96.6	121.0	104.6	98.4	115.2	104.0	99.7	117.0	105.4	98.6	120.7	105.9	98.5	101.5	99.5	100.1	118.6	106.2
06	WOOD, PLASTICS & COMPOSITES	83.5	91.4	88.1	87.1	86.8	86.9	91.6	130.1	113.8	90.9	86.7	88.5	88.5	54.2	68.7	87.8	86.8	87.2
07	THERMAL & MOISTURE PROTECTION	104.6	100.2	102.8	99.7	92.0	96.6	97.8	131.9	111.7	99.0	106.3	102.0	98.9	101.4	99.9	98.4	108.1	102.4
08	OPENINGS	94.5	102.9	96.7	92.6	92.9	92.7	89.6	126.7	99.0	89.3	97.5	91.4	89.3	56.8	81.1	89.1	92.6	90.0
0920	Plaster & Gypsum Board	95.6	91.0	92.3	96.4	86.1	89.1	92.7	130.9	120.0	97.6	86.0	89.4	97.6	52.5	65.4	99.0	86.1	89.8
0950, 0980	Ceilings & Acoustic Treatment	88.8	91.0	90.2	89.0	86.1	87.1	99.7	130.9	120.3	91.8	86.0	88.0	91.8	52.5	65.9	90.8	86.1	87.7
0960	Flooring	88.9	100.1	92.2	86.7	53.2	76.9	95.4	141.2	108.9	88.4	96.7	90.9	87.5	52.1	77.1	89.2	93.7	90.5
0970, 0990	Wall Finishes & Painting/Coating	97.3	106.9	103.2	96.6	100.5	99.0	99.6	146.7	128.7	96.6	103.2	100.7	96.6	103.2	100.7	96.6	89.0	91.9
09	FINISHES	93.6	95.0	94.4	93.6	81.2	86.7	97.4	131.9	116.6	93.8	91.1	92.3	93.4	63.5	76.7	93.1	88.7	90.6
COVERS	DIVS. 10 - 14, 25, 28, 41, 43, 44, 46	100.0	102.8	100.6	100.0	93.6	98.7	100.0	114.7	102.9	100.0	97.1	99.4	100.0	93.3	98.7	100.0	97.5	99.5
21, 22, 23	FIRE SUPPRESSION, PLUMBING & HVAC	94.1	97.9	95.6	94.3	90.6	92.8	93.9	125.8	106.8	94.3	99.2	96.3	94.3	90.7	92.8	100.2	91.7	96.7
26, 27, 3370	ELECTRICAL, COMMUNICATIONS & UTIL.	92.6	107.8	100.5	93.5	84.0	88.6	94.6	120.8	108.1	94.3	85.8	89.9	92.5	49.5	70.3	93.5	85.8	89.5
MF2010	WEIGHTED AVERAGE	95.9	102.1	98.6	96.1	92.4	94.5	96.1	122.8	107.9	94.8	98.2	96.3	93.0	79.6	87.1	96.3	95.2	95.8

DIVISION		PUERTO RICO SAN JUAN 009			RHODE ISLAND NEWPORT 028			PROVIDENCE 029			SOUTH CAROLINA AIKEN 298			BEAUFORT 299			CHARLESTON 294		
		MAT.	INST.	TOTAL	MAT.	INST.	TOTAL	MAT.	INST.	TOTAL	MAT.	INST.	TOTAL	MAT.	INST.	TOTAL	MAT.	INST.	TOTAL
015433	CONTRACTOR EQUIPMENT		90.2	90.2		101.9	101.9		101.9	101.9		99.5	99.5		99.5	99.5		99.5	99.5
0241, 31 - 34	SITE & INFRASTRUCTURE, DEMOLITION	131.5	90.1	103.0	87.5	102.9	98.1	88.6	102.9	98.5	128.4	86.4	99.4	122.8	84.5	96.4	103.7	84.1	90.2
0310	Concrete Forming & Accessories	95.4	17.4	27.6	102.3	117.2	115.2	101.0	117.2	115.0	96.3	71.7	74.9	95.2	39.4	46.6	94.3	41.1	48.1
0320	Concrete Reinforcing	186.3	12.2	98.6	103.6	112.7	108.2	103.6	112.7	108.2	95.1	69.3	82.1	94.2	27.3	60.5	94.0	58.9	76.3
0330	Cast-in-Place Concrete	107.8	31.0	77.6	83.5	117.3	96.8	102.6	117.3	108.4	93.1	73.1	85.2	93.1	48.6	75.6	109.2	49.3	85.6
03	CONCRETE	115.0	22.2	69.8	100.9	115.9	108.2	110.4	115.9	113.1	110.3	72.8	92.0	107.6	42.4	75.8	104.1	49.0	77.2
04	MASONRY	84.4	16.3	42.8	101.1	122.3	114.0	104.6	122.3	115.4	81.2	65.0	71.3	97.2	34.8	59.1	98.6	38.8	62.1
05	METALS	111.9	35.1	86.5	98.0	108.9	101.6	98.0	108.9	101.6	86.1	86.6	86.2	86.1	69.5	80.6	87.9	77.0	84.3
06	WOOD, PLASTICS & COMPOSITES	93.4	17.2	49.4	99.9	117.8	110.2	99.1	117.8	110.0	99.8	72.9	84.2	97.8	40.1	64.5	96.6	40.1	64.0
07	THERMAL & MOISTURE PROTECTION	129.6	21.4	85.6	101.5	114.5	106.8	99.1	114.5	105.3	101.3	70.2	88.7	101.0	37.8	75.3	100.1	42.3	76.6
08	OPENINGS	146.8	14.6	113.5	96.6	115.7	101.4	97.2	115.7	101.9	93.2	68.7	87.1	93.3	38.8	79.6	97.3	43.8	83.8
0920	Plaster & Gypsum Board	115.3	14.4	43.3	95.1	117.7	111.2	93.0	117.7	110.6	108.4	71.9	82.3	111.3	38.1	59.0	113.4	38.1	59.6
0950, 0980	Ceilings & Acoustic Treatment	243.9	14.4	92.3	99.2	117.7	111.4	98.5	117.7	111.2	93.7	71.9	79.3	97.3	38.1	58.2	97.3	38.1	58.2
0960	Flooring	197.1	13.7	143.3	95.3	134.8	106.9	94.9	134.8	106.6	106.2	71.6	96.0	107.8	53.6	91.9	107.3	61.0	93.7
0970, 0990	Wall Finishes & Painting/Coating	193.0	12.6	81.7	92.3	120.8	109.9	92.3	120.8	109.9	114.8	74.1	89.7	114.8	36.6	66.5	114.8	41.3	69.4
09	FINISHES	207.2	16.4	100.8	98.5	121.4	111.2	97.7	121.4	110.9	105.4	72.0	86.7	106.5	42.0	70.5	104.2	44.3	70.8
COVERS	DIVS. 10 - 14, 25, 28, 41, 43, 44, 46	100.0	17.0	83.5	100.0	103.1	100.6	100.0	103.1	100.6	100.0	77.6	95.5	100.0	46.7	89.4	100.0	70.3	94.1
21, 22, 23	FIRE SUPPRESSION, PLUMBING & HVAC	102.8	13.3	66.5	100.0	108.3	103.4	99.9	108.3	103.3	94.3	67.7	83.5	94.3	25.4	66.4	100.2	39.9	75.8
26, 27, 3370	ELECTRICAL, COMMUNICATIONS & UTIL.	130.6	12.6	69.7	101.4	99.0	100.2	99.7	99.0	99.3	94.3	70.0	81.8	97.7	32.2	63.9	96.2	94.3	95.2
MF2010	WEIGHTED AVERAGE	121.3	24.1	78.4	99.2	111.1	104.5	100.2	111.1	105.0	96.8	72.8	86.2	97.5	42.5	73.3	98.4	57.4	80.3

SOUTH CAROLINA / SOUTH DAKOTA

DIVISION		COLUMBIA 290 - 292			FLORENCE 295			GREENVILLE 296			ROCK HILL 297			SPARTANBURG 293			SOUTH DAKOTA ABERDEEN 574		
		MAT.	INST.	TOTAL	MAT.	INST.	TOTAL	MAT.	INST.	TOTAL	MAT.	INST.	TOTAL	MAT.	INST.	TOTAL	MAT.	INST.	TOTAL
015433	CONTRACTOR EQUIPMENT		99.5	99.5		99.5	99.5		99.5	99.5		99.5	99.5		99.5	99.5		98.7	98.7
0241, 31 - 34	SITE & INFRASTRUCTURE, DEMOLITION	101.5	84.0	89.5	115.4	84.1	93.8	109.7	83.7	91.8	107.1	83.6	90.9	109.5	83.7	91.7	98.8	94.9	96.1
0310	Concrete Forming & Accessories	96.0	40.2	47.5	83.0	40.3	45.9	93.9	40.2	47.2	92.5	37.9	45.0	96.8	40.2	47.6	100.2	38.9	46.9
0320	Concrete Reinforcing	94.0	58.7	76.2	93.7	58.8	76.1	93.6	46.9	70.1	94.4	46.7	70.4	93.6	46.9	70.1	108.5	40.4	74.1
0330	Cast-in-Place Concrete	112.1	49.9	87.6	93.0	49.1	75.7	93.0	49.0	75.7	93.0	44.1	73.7	93.0	49.0	75.7	105.9	45.7	82.2
03	CONCRETE	105.6	48.7	77.9	102.2	48.5	76.1	100.8	46.3	74.2	98.9	43.5	71.9	101.0	46.3	74.3	103.1	43.0	73.8
04	MASONRY	92.9	36.8	58.6	81.3	38.8	55.4	79.1	38.8	54.5	105.1	34.1	61.8	81.3	38.8	55.4	108.2	56.7	76.8
05	METALS	87.9	76.3	84.1	86.9	76.6	83.5	86.9	72.3	82.1	86.1	71.8	81.4	86.9	72.3	82.1	102.9	66.1	90.8
06	WOOD, PLASTICS & COMPOSITES	99.0	39.7	64.7	83.1	39.7	58.0	96.6	39.7	63.7	94.9	37.7	61.8	100.4	39.7	65.3	96.9	38.2	63.0
07	THERMAL & MOISTURE PROTECTION	98.8	41.0	75.3	100.4	42.2	76.8	100.3	42.2	76.7	100.2	39.9	75.7	100.4	42.2	76.7	107.2	50.3	84.0
08	OPENINGS	97.3	43.5	83.7	93.3	43.5	80.8	93.2	40.7	80.0	93.3	39.8	79.8	93.3	40.7	80.0	93.8	39.0	80.0
0920	Plaster & Gypsum Board	100.7	37.6	55.6	101.0	37.6	55.7	106.6	37.6	57.3	105.7	35.6	55.6	108.9	37.6	58.0	111.3	36.6	57.9
0950, 0980	Ceilings & Acoustic Treatment	96.4	37.6	57.6	94.6	37.6	57.0	93.7	37.6	56.6	93.7	35.6	55.3	93.7	37.6	56.6	112.9	36.6	62.5
0960	Flooring	104.0	45.7	86.8	98.5	45.7	83.0	104.9	59.9	91.7	104.0	46.5	87.1	106.4	59.9	92.7	115.7	52.7	97.2
0970, 0990	Wall Finishes & Painting/Coating	112.2	41.3	68.5	114.8	41.3	69.4	114.8	41.3	69.4	114.8	41.3	69.4	114.8	41.3	69.4	96.0	39.1	60.9
09	FINISHES	100.8	41.0	67.5	100.0	41.5	67.8	102.9	43.9	70.0	102.2	39.5	67.2	103.6	43.9	70.3	109.1	41.1	71.1
COVERS	DIVS. 10 - 14, 25, 28, 41, 43, 44, 46	100.0	70.1	94.0	100.0	70.1	94.0	100.0	70.1	94.0	100.0	69.3	93.9	100.0	70.1	94.0	100.0	43.8	88.8
21, 22, 23	FIRE SUPPRESSION, PLUMBING & HVAC	99.9	35.3	73.7	100.2	35.3	73.9	100.2	33.9	73.3	94.3	32.9	69.4	100.2	33.9	73.3	100.2	42.7	76.9
26, 27, 3370	ELECTRICAL, COMMUNICATIONS & UTIL.	95.2	48.2	71.0	94.2	45.9	69.3	96.2	49.6	72.2	96.2	49.6	72.2	96.2	49.6	72.2	100.3	54.4	76.6
MF2010	WEIGHTED AVERAGE	97.8	49.3	76.4	96.6	49.3	75.8	96.6	49.0	75.6	96.0	47.2	74.5	96.8	49.0	75.7	101.6	52.2	79.8

743

City Cost Indexes

SOUTH DAKOTA

DIVISION		MITCHELL 573			MOBRIDGE 576			PIERRE 575			RAPID CITY 577			SIOUX FALLS 570-571			WATERTOWN 572		
		MAT.	INST.	TOTAL	MAT.	INST.	TOTAL	MAT.	INST.	TOTAL	MAT.	INST.	TOTAL	MAT.	INST.	TOTAL	MAT.	INST.	TOTAL
015433	CONTRACTOR EQUIPMENT		98.7	98.7		98.7	98.7		98.7	98.7		98.7	98.7		99.6	99.6		98.7	98.7
0241, 31 - 34	SITE & INFRASTRUCTURE, DEMOLITION	95.7	94.9	95.1	95.6	94.9	95.1	94.0	94.9	94.7	96.9	95.0	95.6	94.6	96.6	96.0	95.5	94.9	95.1
0310	Concrete Forming & Accessories	99.5	39.3	47.1	91.1	39.1	45.8	98.9	40.0	47.7	106.9	38.1	47.0	99.6	42.6	50.0	88.1	38.9	45.3
0320	Concrete Reinforcing	107.8	49.2	78.2	110.6	40.4	75.2	108.0	59.1	83.3	101.4	59.2	80.1	101.4	59.3	80.2	104.9	40.5	72.4
0330	Cast-in-Place Concrete	102.9	44.4	79.9	102.9	45.7	80.4	88.8	44.0	71.2	102.1	44.6	79.5	94.5	45.0	75.0	102.9	46.7	80.4
03	CONCRETE	100.8	44.2	73.2	100.7	43.1	72.6	93.7	46.5	70.7	99.9	45.8	73.5	94.5	48.0	71.8	99.6	43.0	72.0
04	MASONRY	96.8	54.5	71.0	105.7	56.7	75.8	103.8	54.8	73.9	105.7	55.9	75.3	101.3	56.6	74.0	131.0	58.9	87.0
05	METALS	101.9	66.3	90.2	102.0	66.1	90.1	102.9	74.6	93.6	104.7	75.1	94.9	104.9	75.4	95.1	101.9	66.6	90.3
06	WOOD, PLASTICS & COMPOSITES	96.0	38.8	62.9	86.3	38.5	58.7	94.8	39.0	62.6	100.2	35.4	62.7	95.6	41.7	64.5	83.3	38.2	57.2
07	THERMAL & MOISTURE PROTECTION	107.0	48.0	83.0	107.1	50.3	84.0	106.9	48.3	83.1	107.5	49.8	84.1	109.4	51.2	85.7	106.8	50.9	84.1
08	OPENINGS	92.3	39.5	79.0	95.2	38.5	80.9	96.9	44.9	83.8	98.0	43.0	84.1	98.2	46.4	85.2	92.3	38.7	78.8
0920	Plaster & Gypsum Board	109.3	37.1	57.8	103.1	36.8	55.8	98.3	37.4	54.8	110.5	33.7	55.6	108.3	40.2	59.6	101.1	36.6	55.0
0950, 0980	Ceilings & Acoustic Treatment	109.2	37.1	61.6	112.9	36.8	62.6	108.3	37.4	61.5	114.7	33.7	61.2	112.0	40.2	64.6	109.2	36.6	61.2
0960	Flooring	115.1	52.7	96.8	109.6	52.7	92.9	114.8	37.5	92.1	114.8	75.4	103.2	114.8	78.2	104.1	108.1	52.7	91.8
0970, 0990	Wall Finishes & Painting/Coating	96.0	42.8	63.2	96.0	43.9	63.9	96.0	47.0	65.8	96.0	47.0	65.8	96.0	47.0	65.8	96.0	39.1	60.9
09	FINISHES	107.7	41.8	70.9	106.2	41.8	70.2	105.9	39.5	68.9	109.0	45.0	73.3	108.0	49.1	75.2	104.6	41.1	69.2
COVERS	DIVS. 10 - 14, 25, 28, 41, 43, 44, 46	100.0	40.6	88.2	100.0	43.8	88.8	100.0	74.2	94.9	100.0	74.0	94.8	100.0	74.5	94.9	100.0	43.7	88.8
21, 22, 23	FIRE SUPPRESSION, PLUMBING & HVAC	94.2	42.3	73.2	94.2	42.8	73.4	100.1	42.4	76.7	100.2	42.9	77.0	100.2	41.2	76.3	94.2	42.8	73.4
26, 27, 3370	ELECTRICAL, COMMUNICATIONS & UTIL.	98.7	46.7	71.9	100.3	47.6	73.2	100.1	52.1	75.3	97.0	52.1	73.8	99.2	63.1	80.6	97.9	47.6	72.0
MF2010	WEIGHTED AVERAGE	98.6	51.0	77.6	99.3	51.4	78.2	100.1	53.8	79.7	101.4	54.6	80.7	100.7	57.1	81.5	99.8	51.6	78.5

TENNESSEE

DIVISION		CHATTANOOGA 373 - 374			COLUMBIA 384			COOKEVILLE 385			JACKSON 383			JOHNSON CITY 376			KNOXVILLE 377 - 379		
		MAT.	INST.	TOTAL	MAT.	INST.	TOTAL	MAT.	INST.	TOTAL	MAT.	INST.	TOTAL	MAT.	INST.	TOTAL	MAT.	INST.	TOTAL
015433	CONTRACTOR EQUIPMENT		103.8	103.8		98.0	98.0		98.0	98.0		104.0	104.0		97.8	97.8		97.8	97.8
0241, 31 - 34	SITE & INFRASTRUCTURE, DEMOLITION	102.1	98.6	99.7	89.3	87.3	87.9	95.5	86.7	89.4	98.8	97.1	97.6	110.0	87.5	94.5	87.2	88.0	87.7
0310	Concrete Forming & Accessories	96.6	57.1	62.3	83.4	61.3	64.1	83.5	37.4	43.4	89.7	47.5	53.0	84.5	40.7	46.4	95.3	43.8	50.5
0320	Concrete Reinforcing	88.5	62.6	75.5	90.9	62.6	76.7	90.9	62.6	76.6	90.9	62.9	76.8	89.1	61.0	75.0	88.5	61.1	74.7
0330	Cast-in-Place Concrete	97.3	62.1	83.4	92.7	53.3	77.2	105.0	45.2	81.5	102.6	51.5	82.4	78.3	62.7	72.1	91.4	50.1	75.1
03	CONCRETE	93.3	61.6	77.9	93.9	60.4	77.6	103.7	47.0	76.1	95.7	53.7	75.3	98.9	54.2	77.1	90.6	51.3	71.4
04	MASONRY	103.9	51.6	72.0	121.4	56.6	81.8	115.6	46.2	73.3	121.6	49.9	77.8	118.8	47.3	75.2	81.4	55.0	65.3
05	METALS	96.2	87.4	93.3	90.6	88.0	89.7	90.7	86.9	89.4	92.9	87.6	91.1	93.5	86.1	91.0	96.8	86.3	93.3
06	WOOD, PLASTICS & COMPOSITES	106.2	58.8	78.8	74.0	62.8	67.6	74.2	35.7	52.0	90.5	48.2	66.0	79.2	39.2	56.1	92.4	39.2	61.7
07	THERMAL & MOISTURE PROTECTION	95.6	59.6	81.0	90.4	61.1	78.5	90.8	42.8	73.5	92.6	51.5	75.8	91.1	55.5	76.6	89.2	59.3	77.0
08	OPENINGS	102.1	57.3	90.8	92.0	57.6	83.4	92.0	44.0	79.9	101.5	53.9	89.5	97.5	46.9	84.8	94.3	47.0	82.4
0920	Plaster & Gypsum Board	76.1	58.0	63.2	82.3	62.1	67.9	82.3	34.2	47.9	83.9	47.0	57.6	98.3	37.8	55.1	105.8	37.8	57.3
0950, 0980	Ceilings & Acoustic Treatment	99.7	58.0	72.2	82.9	62.1	69.2	82.9	34.2	50.7	91.8	47.0	62.2	95.8	37.8	57.5	96.7	37.8	57.8
0960	Flooring	98.1	59.3	86.7	92.8	21.5	71.9	92.9	61.4	83.7	91.7	42.7	77.3	94.7	42.6	79.4	99.5	58.4	87.4
0970, 0990	Wall Finishes & Painting/Coating	104.4	55.7	74.3	94.6	35.2	57.9	94.6	38.4	59.9	96.0	52.3	69.0	101.6	44.4	66.3	101.6	51.2	70.5
09	FINISHES	96.1	57.4	74.5	94.7	51.5	70.6	95.3	40.7	64.8	95.0	46.3	67.8	100.1	40.3	66.8	92.6	46.1	66.7
COVERS	DIVS. 10 - 14, 25, 28, 41, 43, 44, 46	100.0	44.4	88.9	100.0	50.7	90.2	100.0	44.2	88.9	100.0	46.5	89.3	100.0	46.6	89.4	100.0	49.3	89.9
21, 22, 23	FIRE SUPPRESSION, PLUMBING & HVAC	100.2	63.3	85.2	96.9	77.4	89.0	96.9	72.1	86.9	100.0	71.6	88.5	99.9	56.3	82.2	99.9	66.7	86.5
26, 27, 3370	ELECTRICAL, COMMUNICATIONS & UTIL.	102.4	69.3	85.3	95.1	61.0	77.5	96.8	61.9	78.8	102.5	62.1	81.7	92.3	49.2	70.1	97.9	58.1	77.3
MF2010	WEIGHTED AVERAGE	99.0	66.2	84.6	95.7	67.1	83.1	97.0	60.2	80.8	99.2	63.8	83.6	98.7	56.7	80.2	95.4	61.4	80.4

TENNESSEE / TEXAS

DIVISION		MCKENZIE 382			MEMPHIS 375,380 - 381			NASHVILLE 370 - 372			ABILENE 795 - 796			AMARILLO 790 - 791			AUSTIN 786 - 787		
		MAT.	INST.	TOTAL	MAT.	INST.	TOTAL	MAT.	INST.	TOTAL	MAT.	INST.	TOTAL	MAT.	INST.	TOTAL	MAT.	INST.	TOTAL
015433	CONTRACTOR EQUIPMENT		98.0	98.0		102.2	102.2		104.1	104.1		88.9	88.9		88.9	88.9		88.6	88.6
0241, 31 - 34	SITE & INFRASTRUCTURE, DEMOLITION	95.0	87.0	89.5	90.5	93.3	92.5	96.6	99.5	98.6	95.1	86.0	88.8	95.3	87.1	89.6	97.0	86.8	90.0
0310	Concrete Forming & Accessories	90.8	40.3	46.9	95.4	60.4	64.9	97.0	64.0	68.3	95.1	40.2	47.4	95.6	54.1	59.5	90.2	53.4	58.2
0320	Concrete Reinforcing	91.1	62.9	76.9	97.3	63.5	80.3	92.6	63.0	77.7	93.8	51.8	72.6	93.8	50.0	71.7	88.8	48.8	68.7
0330	Cast-in-Place Concrete	102.8	60.0	85.9	90.8	62.7	79.8	86.1	66.3	78.3	96.1	48.7	77.4	95.3	55.4	79.6	92.4	49.3	75.4
03	CONCRETE	102.3	53.5	78.5	91.9	63.4	78.0	88.9	66.1	77.8	94.8	46.5	71.3	94.4	54.7	75.0	87.5	51.9	70.2
04	MASONRY	119.9	48.8	76.5	96.6	63.7	76.6	86.2	59.2	69.7	103.1	57.5	75.3	103.1	58.3	75.8	99.7	47.6	67.9
05	METALS	90.7	87.9	89.8	97.8	89.5	95.0	97.8	88.7	94.8	99.3	68.9	89.2	99.3	68.6	89.2	95.5	64.8	85.4
06	WOOD, PLASTICS & COMPOSITES	82.6	38.5	57.1	100.8	61.2	78.0	99.9	65.0	79.8	99.6	38.7	64.4	98.1	56.5	74.1	94.3	56.6	72.5
07	THERMAL & MOISTURE PROTECTION	90.8	49.4	74.0	90.2	64.8	79.9	93.9	61.9	80.9	97.4	51.0	78.5	97.4	54.2	79.8	86.7	51.1	72.2
08	OPENINGS	92.0	45.8	80.4	101.4	61.7	91.4	99.6	64.1	90.6	92.6	42.1	79.9	92.6	53.7	82.8	98.2	54.2	87.1
0920	Plaster & Gypsum Board	85.5	37.0	50.9	91.7	60.4	69.3	92.1	64.3	72.3	82.8	37.3	50.3	84.4	55.7	63.9	81.5	55.8	63.1
0950, 0980	Ceilings & Acoustic Treatment	82.9	37.0	52.6	99.8	60.4	73.7	94.7	64.3	74.6	92.0	37.3	55.9	92.0	55.7	68.0	93.7	55.8	68.6
0960	Flooring	96.1	41.2	80.0	94.8	37.8	78.0	99.8	67.3	90.2	110.3	77.2	100.6	110.1	77.2	100.5	95.9	41.4	79.3
0970, 0990	Wall Finishes & Painting/Coating	94.6	52.3	68.5	99.3	55.3	72.2	102.2	68.9	81.7	94.7	51.8	68.2	94.7	51.1	67.8	92.2	42.1	61.3
09	FINISHES	96.5	40.5	65.2	96.3	55.1	73.3	98.3	64.9	79.7	94.9	47.5	68.5	95.2	58.1	74.5	90.4	49.9	67.8
COVERS	DIVS. 10 - 14, 25, 28, 41, 43, 44, 46	100.0	29.5	86.0	100.0	81.1	96.2	100.0	81.8	96.4	100.0	76.2	95.3	100.0	71.3	94.3	100.0	78.0	95.6
21, 22, 23	FIRE SUPPRESSION, PLUMBING & HVAC	96.9	71.8	86.8	100.0	70.2	87.9	100.0	77.3	90.8	100.3	45.2	78.0	100.2	51.2	80.3	100.1	55.3	82.0
26, 27, 3370	ELECTRICAL, COMMUNICATIONS & UTIL.	96.5	66.8	81.2	103.7	65.9	84.2	95.6	61.9	78.2	97.3	50.5	73.1	99.5	67.1	82.8	96.4	64.5	79.9
MF2010	WEIGHTED AVERAGE	97.1	61.8	81.6	98.2	69.6	85.6	96.8	72.2	86.0	97.9	54.2	78.6	98.1	61.0	81.7	96.1	58.8	79.6

City Cost Indexes

TEXAS

DIVISION		BEAUMONT 776 - 777			BROWNWOOD 768			BRYAN 778			CHILDRESS 792			CORPUS CHRISTI 783 - 784			DALLAS 752 - 753		
		MAT.	INST.	TOTAL	MAT.	INST.	TOTAL	MAT.	INST.	TOTAL	MAT.	INST.	TOTAL	MAT.	INST.	TOTAL	MAT.	INST.	TOTAL
015433	CONTRACTOR EQUIPMENT		89.8	89.8		88.9	88.9		89.8	89.8		88.9	88.9		95.7	95.7		98.0	98.0
0241, 31 - 34	SITE & INFRASTRUCTURE, DEMOLITION	86.6	86.3	86.4	101.1	85.5	90.3	78.2	87.9	84.9	106.1	85.9	92.1	137.0	82.6	99.5	109.8	87.3	94.2
0310	Concrete Forming & Accessories	101.5	50.1	56.8	93.6	34.1	41.8	81.2	41.6	46.8	93.7	61.5	65.7	96.4	38.1	45.7	93.9	64.3	68.1
0320	Concrete Reinforcing	91.2	42.7	66.7	94.1	49.1	71.4	93.7	48.9	71.1	93.9	49.3	71.4	88.2	47.3	67.6	101.0	52.3	76.5
0330	Cast-in-Place Concrete	86.3	53.1	73.2	103.7	43.8	80.1	69.1	66.0	67.9	98.6	50.1	79.5	102.5	46.2	80.3	96.5	56.2	80.6
03	CONCRETE	93.0	50.7	72.4	102.2	41.5	72.6	78.1	52.6	65.7	103.2	55.9	80.2	95.3	44.5	70.5	97.3	60.6	79.4
04	MASONRY	104.5	58.5	76.5	132.2	50.2	82.2	143.7	57.2	90.9	107.8	51.5	73.4	88.8	49.3	64.7	102.6	58.4	75.6
05	METALS	97.0	65.3	86.5	96.9	62.5	85.6	96.5	68.2	87.1	96.7	64.4	86.0	95.1	75.2	88.5	100.4	79.9	93.6
06	WOOD, PLASTICS & COMPOSITES	111.8	50.2	76.2	98.3	30.2	59.0	78.2	34.9	53.2	99.0	66.6	80.3	113.9	37.3	69.7	98.2	67.2	80.3
07	THERMAL & MOISTURE PROTECTION	103.6	56.8	84.6	97.5	45.7	76.4	96.1	56.9	80.1	98.0	50.3	78.6	95.4	46.1	75.4	92.6	64.0	81.0
08	OPENINGS	95.9	46.8	83.5	88.6	37.5	75.7	97.6	43.0	83.9	89.6	55.9	81.1	105.0	38.6	88.3	100.7	58.4	90.1
0920	Plaster & Gypsum Board	95.1	49.3	62.4	81.7	28.6	43.8	83.6	33.4	47.8	81.7	66.1	70.5	86.3	35.8	50.2	88.7	66.6	72.9
0950, 0980	Ceilings & Acoustic Treatment	99.7	49.3	66.4	88.4	28.6	48.9	93.0	33.4	53.7	88.4	66.1	73.6	93.7	35.8	55.4	94.6	66.6	76.1
0960	Flooring	109.1	74.6	99.0	109.2	46.8	90.9	82.1	55.2	74.2	108.2	42.8	89.0	109.0	66.0	96.4	99.3	50.0	84.8
0970, 0990	Wall Finishes & Painting/Coating	91.8	50.2	66.1	94.7	30.2	54.9	89.0	62.4	72.6	94.7	39.9	60.9	104.7	59.6	76.8	103.0	53.2	72.3
09	FINISHES	91.2	54.3	70.6	94.3	35.2	61.3	79.8	44.5	60.1	94.8	56.8	73.6	98.7	44.9	68.7	97.7	60.8	77.1
COVERS	DIVS. 10 - 14, 25, 28, 41, 43, 44, 46	100.0	79.6	95.9	100.0	39.2	87.9	100.0	78.8	95.8	100.0	73.4	94.7	100.0	77.0	95.4	100.0	82.1	96.4
21, 22, 23	FIRE SUPPRESSION, PLUMBING & HVAC	100.1	59.3	83.6	94.3	48.4	75.7	94.1	66.4	82.9	94.3	52.7	77.5	100.1	40.1	75.8	100.0	63.7	85.3
26, 27, 3370	ELECTRICAL, COMMUNICATIONS & UTIL.	94.8	72.6	83.4	93.3	45.4	68.6	92.9	67.4	79.8	97.3	45.1	70.3	92.7	54.7	73.1	95.5	67.6	81.1
MF2010	WEIGHTED AVERAGE	97.2	62.0	81.7	97.7	49.0	76.2	94.3	61.7	79.9	97.3	57.2	79.6	98.9	52.3	78.4	99.4	66.8	85.0

TEXAS

DIVISION		DEL RIO 788			DENTON 762			EASTLAND 764			EL PASO 798 - 799,885			FORT WORTH 760 - 761			GALVESTON 775		
		MAT.	INST.	TOTAL	MAT.	INST.	TOTAL	MAT.	INST.	TOTAL	MAT.	INST.	TOTAL	MAT.	INST.	TOTAL	MAT.	INST.	TOTAL
015433	CONTRACTOR EQUIPMENT		88.6	88.6		94.7	94.7		88.9	88.9		88.9	88.9		88.9	88.9		98.5	98.5
0241, 31 - 34	SITE & INFRASTRUCTURE, DEMOLITION	117.8	85.1	95.2	101.7	78.7	85.9	104.0	85.4	91.2	92.4	85.4	87.6	97.1	86.4	89.7	107.8	85.7	92.6
0310	Concrete Forming & Accessories	91.2	37.7	44.7	99.0	33.5	42.0	94.2	37.1	44.5	94.4	45.0	51.4	91.3	63.7	67.3	89.8	61.9	65.5
0320	Concrete Reinforcing	88.8	44.3	66.4	95.5	50.3	72.7	94.3	49.0	71.5	93.8	47.2	70.3	93.8	52.2	72.8	93.1	64.0	78.4
0330	Cast-in-Place Concrete	110.4	43.7	84.1	80.2	45.7	66.6	109.6	44.5	83.9	89.8	41.8	70.9	99.0	52.1	80.5	91.7	67.1	82.0
03	CONCRETE	112.3	42.1	78.1	80.7	42.9	62.3	106.9	43.0	75.8	91.6	45.3	69.0	96.0	58.2	77.6	94.4	65.6	80.4
04	MASONRY	103.6	51.9	72.0	140.8	48.6	84.6	101.6	50.1	70.2	100.8	47.9	68.5	97.3	58.3	73.5	99.9	57.2	73.9
05	METALS	94.0	59.6	82.6	96.5	75.6	89.6	96.7	62.9	85.6	99.0	63.4	87.3	96.3	68.7	87.2	98.0	87.1	94.4
06	WOOD, PLASTICS & COMPOSITES	93.7	35.9	60.3	107.9	33.1	64.7	105.3	34.0	64.1	99.2	47.5	69.3	100.9	67.1	81.4	95.3	61.8	76.0
07	THERMAL & MOISTURE PROTECTION	92.3	45.2	73.2	96.0	45.6	75.5	97.9	44.9	76.3	93.6	53.8	77.4	98.3	54.2	80.3	95.3	65.2	83.1
08	OPENINGS	99.0	34.2	82.6	102.7	38.2	86.4	76.9	39.5	67.5	92.6	43.6	80.3	86.7	58.3	79.5	102.3	61.6	92.0
0920	Plaster & Gypsum Board	80.9	34.4	47.7	85.3	31.4	46.8	81.7	32.4	46.5	87.4	46.4	58.1	85.4	66.6	72.0	89.9	61.1	69.3
0950, 0980	Ceilings & Acoustic Treatment	89.2	34.4	53.0	92.9	31.4	52.3	88.4	32.4	51.4	92.0	46.4	61.9	93.8	66.6	75.8	95.7	61.1	72.8
0960	Flooring	92.8	35.7	76.0	101.7	44.8	85.0	143.0	46.8	114.7	110.3	65.3	97.1	144.3	42.6	114.5	95.7	55.2	83.8
0970, 0990	Wall Finishes & Painting/Coating	92.2	26.5	51.7	104.7	30.5	59.0	96.2	31.5	56.3	94.7	36.0	58.5	96.2	52.7	69.3	99.2	48.5	67.9
09	FINISHES	90.6	35.8	60.0	90.7	34.6	59.4	104.7	37.5	67.2	95.1	47.9	68.8	106.0	59.2	79.9	89.3	58.8	72.3
COVERS	DIVS. 10 - 14, 25, 28, 41, 43, 44, 46	100.0	39.0	87.9	100.0	37.8	87.6	100.0	40.3	88.1	100.0	77.9	95.6	100.0	81.8	96.4	100.0	83.6	96.7
21, 22, 23	FIRE SUPPRESSION, PLUMBING & HVAC	94.1	46.3	74.7	94.3	39.8	72.2	94.3	48.7	75.8	100.2	35.4	73.9	100.2	59.1	83.5	94.1	65.9	82.7
26, 27, 3370	ELECTRICAL, COMMUNICATIONS & UTIL.	94.1	31.2	61.7	95.8	45.5	69.9	93.2	45.4	68.5	93.9	53.3	72.9	93.4	62.4	77.4	94.5	67.3	80.4
MF2010	WEIGHTED AVERAGE	98.0	46.5	75.3	97.1	47.8	75.4	96.5	49.7	75.9	96.8	51.2	76.7	97.2	63.2	82.2	96.5	68.2	84.0

TEXAS

DIVISION		GIDDINGS 789			GREENVILLE 754			HOUSTON 770 - 772			HUNTSVILLE 773			LAREDO 780			LONGVIEW 756		
		MAT.	INST.	TOTAL	MAT.	INST.	TOTAL	MAT.	INST.	TOTAL	MAT.	INST.	TOTAL	MAT.	INST.	TOTAL	MAT.	INST.	TOTAL
015433	CONTRACTOR EQUIPMENT		88.6	88.6		95.6	95.6		98.4	98.4		89.8	89.8		88.6	88.6		90.5	90.5
0241, 31 - 34	SITE & INFRASTRUCTURE, DEMOLITION	103.3	85.4	90.9	102.5	82.2	88.5	106.0	85.5	91.9	94.1	86.0	88.5	96.9	86.6	89.8	97.8	89.3	91.9
0310	Concrete Forming & Accessories	89.4	57.6	61.7	86.6	28.2	35.8	91.7	62.0	65.9	87.7	57.1	61.1	91.2	38.3	45.2	83.7	29.0	36.2
0320	Concrete Reinforcing	89.4	46.2	67.6	101.6	25.6	63.3	92.9	64.5	78.6	94.0	48.3	70.9	88.8	47.2	67.9	100.4	23.4	61.6
0330	Cast-in-Place Concrete	93.5	44.1	74.1	87.9	38.2	68.3	88.9	67.1	80.3	95.0	48.5	76.7	78.9	59.2	71.2	102.2	35.8	76.0
03	CONCRETE	91.2	51.4	71.8	90.1	33.2	62.4	92.1	65.8	79.3	100.4	53.3	77.4	85.8	48.3	67.5	105.5	31.7	69.5
04	MASONRY	111.6	52.0	75.2	160.5	47.4	91.5	99.7	67.2	79.9	142.0	43.8	82.1	97.3	47.4	66.9	156.1	43.7	87.5
05	METALS	93.4	61.9	83.0	97.7	63.6	86.4	101.0	88.0	96.7	96.4	63.8	85.6	96.5	63.8	85.7	91.2	51.2	78.0
06	WOOD, PLASTICS & COMPOSITES	92.7	61.8	74.8	89.6	27.3	53.6	97.5	61.8	76.9	86.1	60.4	71.2	93.7	37.2	61.1	83.9	29.5	52.5
07	THERMAL & MOISTURE PROTECTION	92.9	49.4	75.2	92.7	40.8	71.6	94.9	67.5	83.8	97.0	47.4	76.9	91.3	49.4	74.3	93.9	36.2	70.5
08	OPENINGS	98.2	50.2	86.1	99.5	26.9	81.2	104.5	61.6	93.7	97.6	49.3	85.5	98.8	39.2	83.8	90.5	26.9	74.5
0920	Plaster & Gypsum Board	79.5	61.1	66.4	82.8	25.5	41.8	92.8	61.1	70.2	87.2	59.7	67.6	82.2	35.8	49.1	80.6	27.9	42.9
0950, 0980	Ceilings & Acoustic Treatment	89.2	61.1	70.6	90.1	25.5	47.4	101.1	61.1	74.7	93.0	59.7	71.0	93.7	35.8	55.4	87.3	27.9	48.1
0960	Flooring	92.8	35.7	76.0	95.0	49.1	81.5	97.0	55.2	84.7	85.6	35.7	70.9	92.7	66.0	84.9	100.6	36.1	81.6
0970, 0990	Wall Finishes & Painting/Coating	92.2	31.5	54.8	103.0	25.9	55.5	99.2	62.4	76.5	89.0	32.3	54.0	92.2	56.1	69.9	94.6	30.3	54.9
09	FINISHES	89.2	52.0	68.4	94.1	31.4	59.2	97.1	60.4	76.6	82.3	51.7	65.2	89.9	44.4	64.5	99.7	30.4	61.0
COVERS	DIVS. 10 - 14, 25, 28, 41, 43, 44, 46	100.0	45.2	89.1	100.0	36.1	87.3	100.0	83.6	96.7	100.0	45.4	89.1	100.0	74.9	95.0	100.0	22.9	84.7
21, 22, 23	FIRE SUPPRESSION, PLUMBING & HVAC	94.1	60.1	80.3	94.1	45.7	74.5	100.0	66.6	86.5	94.1	61.0	80.7	100.0	38.8	75.2	94.1	28.9	67.6
26, 27, 3370	ELECTRICAL, COMMUNICATIONS & UTIL.	91.1	45.1	67.3	92.2	28.1	59.1	96.5	67.5	81.5	92.9	45.1	68.2	94.3	62.2	77.7	92.4	45.6	68.3
MF2010	WEIGHTED AVERAGE	95.1	56.2	77.9	98.6	43.3	74.2	99.1	69.8	86.2	97.5	55.9	79.2	95.8	52.7	76.8	98.6	40.4	72.9

745

		TEXAS																
		LUBBOCK			LUFKIN			MCALLEN			MCKINNEY			MIDLAND			ODESSA		
		793 - 794			759			785			750			797			797		
DIVISION		MAT.	INST.	TOTAL	MAT.	INST.	TOTAL	MAT.	INST.	TOTAL	MAT.	INST.	TOTAL	MAT.	INST.	TOTAL	MAT.	INST.	TOTAL
015433	CONTRACTOR EQUIPMENT		97.2	97.2		90.5	90.5		95.7	95.7		95.6	95.6		97.2	97.2		88.9	88.9
0241, 31 - 34	SITE & INFRASTRUCTURE, DEMOLITION	120.9	84.3	95.7	92.5	90.3	91.0	142.0	82.5	101.0	98.5	82.8	87.6	124.3	84.3	96.7	95.2	86.3	89.1
0310	Concrete Forming & Accessories	93.2	52.6	57.9	86.6	58.7	62.3	95.8	35.7	43.5	85.7	37.6	43.9	96.6	39.6	47.0	95.0	39.5	46.7
0320	Concrete Reinforcing	95.0	51.8	73.2	102.1	60.8	81.3	88.3	47.2	67.6	101.6	50.3	75.7	96.1	51.1	73.4	93.8	51.0	72.2
0330	Cast-in-Place Concrete	96.3	49.0	77.6	91.4	45.2	73.2	111.4	44.5	85.1	82.4	46.4	68.2	102.3	46.3	80.3	96.1	45.3	76.1
03	CONCRETE	93.6	52.9	73.7	97.9	55.0	77.0	101.2	42.9	72.8	85.6	44.9	65.8	98.1	46.1	72.8	94.8	44.9	70.5
04	MASONRY	102.4	55.9	74.0	119.7	43.8	73.4	103.4	49.3	70.4	172.7	49.9	97.8	120.6	43.6	73.7	103.1	43.6	66.8
05	METALS	102.6	80.4	95.3	97.8	67.5	87.8	93.9	74.8	87.6	97.6	74.6	90.0	100.8	79.2	93.7	98.6	67.4	88.3
06	WOOD, PLASTICS & COMPOSITES	98.7	56.7	74.4	92.0	61.7	74.5	112.6	34.8	67.7	88.3	37.7	59.1	103.0	39.3	66.2	99.6	39.1	64.7
07	THERMAL & MOISTURE PROTECTION	88.0	53.3	73.9	93.7	47.1	74.8	95.4	44.9	74.9	92.6	47.5	74.2	88.2	44.9	70.6	97.4	44.0	75.7
08	OPENINGS	102.2	50.7	89.2	73.0	54.6	68.4	103.3	37.7	86.8	99.5	40.6	84.7	101.1	40.9	85.9	92.6	40.8	79.6
0920	Plaster & Gypsum Board	83.3	55.7	63.6	79.8	61.1	66.5	86.3	33.2	48.4	81.4	36.3	49.2	84.8	37.7	51.2	82.8	37.7	50.6
0950, 0980	Ceilings & Acoustic Treatment	93.8	55.7	68.6	80.1	61.1	67.5	93.7	33.2	53.7	90.1	36.3	54.5	91.1	37.7	55.9	92.0	37.7	56.2
0960	Flooring	102.8	40.5	84.5	135.1	39.2	107.0	108.5	66.0	96.0	94.6	36.6	77.6	104.3	50.0	88.3	110.3	50.0	92.6
0970, 0990	Wall Finishes & Painting/Coating	105.2	51.1	71.8	94.6	39.0	60.3	104.7	29.8	58.5	103.0	42.5	65.7	105.2	51.1	71.8	94.7	51.1	67.8
09	FINISHES	97.7	50.3	71.2	107.8	53.8	77.7	99.1	40.1	66.2	93.5	37.9	62.5	98.1	42.0	66.8	94.9	41.9	65.3
COVERS	DIVS. 10 - 14, 25, 28, 41, 43, 44, 46	100.0	77.9	95.6	100.0	74.6	94.9	100.0	74.8	95.0	100.0	41.7	88.4	100.0	77.6	95.5	100.0	77.3	95.5
21, 22, 23	FIRE SUPPRESSION, PLUMBING & HVAC	99.9	44.3	77.3	94.1	61.0	80.7	94.1	30.4	68.3	94.1	43.4	73.6	93.9	44.5	73.9	100.3	44.4	77.6
26, 27, 3370	ELECTRICAL, COMMUNICATIONS & UTIL.	95.9	51.5	73.0	93.7	45.1	68.6	91.9	35.5	62.8	92.3	46.9	68.9	95.9	41.9	68.0	97.4	41.8	68.8
MF2010	WEIGHTED AVERAGE	99.6	56.8	80.7	95.8	58.3	79.3	98.6	46.6	75.7	98.5	50.2	77.2	99.4	51.4	78.2	97.8	50.2	76.8

		TEXAS																	
		PALESTINE			SAN ANGELO			SAN ANTONIO			TEMPLE			TEXARKANA			TYLER		
		758			769			781 - 782			765			755			757		
DIVISION		MAT.	INST.	TOTAL	MAT.	INST.	TOTAL	MAT.	INST.	TOTAL	MAT.	INST.	TOTAL	MAT.	INST.	TOTAL	MAT.	INST.	TOTAL
015433	CONTRACTOR EQUIPMENT		90.5	90.5		88.9	88.9		91.0	91.0		88.9	88.9		90.5	90.5		90.5	90.5
0241, 31 - 34	SITE & INFRASTRUCTURE, DEMOLITION	99.6	90.1	93.1	97.3	86.8	90.1	96.6	91.3	92.9	85.6	85.2	85.3	86.6	90.1	89.0	97.2	90.1	92.3
0310	Concrete Forming & Accessories	79.0	57.8	60.6	93.9	37.4	44.7	91.2	56.8	61.3	96.8	35.2	43.2	92.5	35.2	42.6	88.0	34.1	41.1
0320	Concrete Reinforcing	99.6	49.1	74.2	94.0	51.0	72.3	95.1	48.1	71.4	94.1	48.8	71.3	99.5	52.0	75.6	100.4	52.2	76.1
0330	Cast-in-Place Concrete	83.5	43.8	67.9	97.8	43.5	76.4	77.4	67.5	73.5	80.2	48.9	67.8	84.2	44.6	68.6	100.4	43.5	78.0
03	CONCRETE	99.0	51.9	76.1	97.6	43.2	71.1	86.0	59.7	73.2	83.4	43.7	64.1	90.9	42.9	67.5	105.0	42.1	74.3
04	MASONRY	115.0	50.4	75.6	128.8	43.0	76.4	97.2	59.3	74.1	140.2	48.8	84.4	177.1	46.4	97.3	166.5	49.3	94.9
05	METALS	97.5	62.4	86.0	97.1	64.6	86.4	97.2	68.3	87.7	96.8	63.5	85.8	91.1	67.2	83.2	97.4	67.7	87.6
06	WOOD, PLASTICS & COMPOSITES	83.1	61.7	70.8	98.5	37.4	63.2	93.7	54.9	71.3	107.6	34.0	65.1	94.2	33.5	59.1	93.5	31.6	57.7
07	THERMAL & MOISTURE PROTECTION	94.2	48.2	75.5	97.3	43.7	75.5	91.0	63.9	80.0	96.9	45.3	75.9	93.5	48.8	75.4	94.0	53.2	77.4
08	OPENINGS	72.9	51.6	67.6	88.6	39.8	76.3	100.7	53.0	88.7	74.8	36.2	65.1	90.5	38.7	77.4	72.9	37.7	64.0
0920	Plaster & Gypsum Board	75.3	61.1	65.2	81.7	36.0	49.1	82.2	53.9	62.0	81.7	32.4	46.5	84.7	32.0	47.1	79.8	30.0	44.3
0950, 0980	Ceilings & Acoustic Treatment	80.1	61.1	67.5	88.4	36.0	53.8	93.7	53.9	67.4	88.4	32.4	51.4	87.3	32.0	50.8	80.1	30.0	47.0
0960	Flooring	126.5	60.6	107.1	109.3	35.7	87.7	92.7	66.2	84.9	145.6	42.9	115.5	108.5	61.0	94.6	137.0	38.1	108.0
0970, 0990	Wall Finishes & Painting/Coating	94.6	31.5	55.7	94.7	51.1	67.8	92.2	56.1	69.9	96.2	34.7	58.3	94.6	40.6	61.3	94.6	40.6	61.3
09	FINISHES	105.5	54.6	78.1	94.0	37.7	62.6	89.9	57.4	71.8	104.0	36.0	66.1	101.8	39.5	67.0	108.7	33.8	66.9
COVERS	DIVS. 10 - 14, 25, 28, 41, 43, 44, 46	100.0	42.8	88.6	100.0	74.8	95.0	100.0	81.0	96.2	100.0	37.3	87.5	100.0	74.9	95.0	100.0	74.6	94.9
21, 22, 23	FIRE SUPPRESSION, PLUMBING & HVAC	94.1	57.7	79.3	94.3	43.7	73.8	99.9	66.1	86.2	94.3	34.9	70.3	94.1	34.4	69.9	94.1	59.2	79.9
26, 27, 3370	ELECTRICAL, COMMUNICATIONS & UTIL.	89.9	45.1	66.8	97.3	47.5	71.6	94.7	62.3	78.0	94.4	67.4	80.5	93.6	63.5	78.1	92.4	54.0	72.6
MF2010	WEIGHTED AVERAGE	95.2	56.6	78.2	97.3	49.6	76.3	96.2	65.0	82.4	95.1	49.4	75.0	98.0	51.0	77.2	99.0	54.4	79.4

		TEXAS															UTAH		
		VICTORIA			WACO			WAXAHACHIE			WHARTON			WICHITA FALLS			LOGAN		
		779			766 - 767			751			774			763			843		
DIVISION		MAT.	INST.	TOTAL	MAT.	INST.	TOTAL	MAT.	INST.	TOTAL	MAT.	INST.	TOTAL	MAT.	INST.	TOTAL	MAT.	INST.	TOTAL
015433	CONTRACTOR EQUIPMENT		97.4	97.4		88.9	88.9		95.6	95.6		98.5	98.5		88.9	88.9		98.6	98.6
0241, 31 - 34	SITE & INFRASTRUCTURE, DEMOLITION	112.8	81.1	90.9	94.0	86.4	88.7	100.6	83.3	88.7	118.9	83.9	94.8	94.7	86.0	88.7	90.7	98.4	96.0
0310	Concrete Forming & Accessories	90.1	36.4	43.4	95.5	40.5	47.7	85.7	59.2	62.6	85.1	37.9	44.0	95.4	38.4	45.8	105.4	60.2	66.1
0320	Concrete Reinforcing	89.4	31.2	60.1	93.8	48.8	71.1	101.6	50.4	75.8	93.0	53.1	72.9	93.8	51.0	72.2	100.2	75.9	87.9
0330	Cast-in-Place Concrete	103.0	42.0	79.0	86.7	53.2	73.5	87.0	42.4	69.4	105.8	52.1	84.7	92.7	45.8	74.2	86.0	72.1	80.5
03	CONCRETE	101.7	39.6	71.4	90.1	47.7	69.4	89.2	53.1	71.7	105.7	47.7	77.4	93.1	44.6	69.5	105.3	67.8	87.0
04	MASONRY	116.7	35.4	67.1	99.9	55.8	73.0	161.1	50.0	93.3	101.0	46.2	67.5	100.3	57.5	74.2	106.8	62.0	79.4
05	METALS	96.6	70.9	88.1	99.2	66.7	88.5	97.7	76.6	90.7	98.0	78.0	91.4	99.1	68.9	89.1	99.9	76.8	92.3
06	WOOD, PLASTICS & COMPOSITES	99.2	37.2	63.4	106.1	36.4	65.8	88.3	66.6	75.8	89.5	33.8	57.3	106.1	36.3	65.8	79.9	58.4	67.5
07	THERMAL & MOISTURE PROTECTION	98.9	40.9	75.3	97.9	48.7	77.9	92.7	47.5	74.3	95.6	50.7	77.4	97.9	52.0	79.2	99.6	63.7	85.0
08	OPENINGS	102.1	36.4	85.6	86.7	37.5	74.3	99.5	56.3	88.7	102.3	38.3	86.2	86.7	41.3	75.3	91.7	57.6	83.1
0920	Plaster & Gypsum Board	87.8	35.7	50.6	82.8	34.9	48.6	82.0	66.1	70.6	85.8	32.1	47.5	82.8	34.8	48.5	81.4	57.0	64.0
0950, 0980	Ceilings & Acoustic Treatment	96.6	35.7	56.4	92.0	34.9	54.3	91.9	66.1	74.8	95.7	32.1	53.7	92.0	34.8	54.2	109.7	57.0	74.9
0960	Flooring	95.1	39.2	78.7	144.3	46.4	115.6	94.6	39.2	78.3	93.2	46.8	79.6	145.2	77.2	125.3	100.7	50.0	85.8
0970, 0990	Wall Finishes & Painting/Coating	99.4	39.0	62.1	96.2	34.7	58.3	103.0	39.0	63.5	99.2	32.3	57.9	98.7	51.1	69.3	98.8	62.0	76.1
09	FINISHES	87.5	37.3	59.5	105.1	39.8	68.7	94.2	54.7	72.2	89.0	38.0	60.5	105.5	46.0	72.3	100.6	57.8	76.8
COVERS	DIVS. 10 - 14, 25, 28, 41, 43, 44, 46	100.0	38.4	87.7	100.0	78.3	95.7	100.0	42.2	88.5	100.0	40.6	88.2	100.0	68.7	93.8	100.0	57.0	91.5
21, 22, 23	FIRE SUPPRESSION, PLUMBING & HVAC	94.1	32.8	69.3	100.3	51.3	80.4	94.1	42.5	73.2	94.1	61.0	80.7	100.3	46.6	78.5	100.0	71.3	88.4
26, 27, 3370	ELECTRICAL, COMMUNICATIONS & UTIL.	99.4	41.2	69.4	97.5	67.6	82.0	92.3	67.6	79.5	98.1	45.5	71.0	99.2	57.9	77.9	94.5	73.7	83.8
MF2010	WEIGHTED AVERAGE	98.6	44.0	74.5	97.4	56.4	79.4	98.5	57.3	80.3	98.5	54.0	78.9	98.0	54.8	79.0	99.2	70.0	86.3

UTAH / VERMONT

DIVISION		OGDEN 842,844 MAT.	INST.	TOTAL	PRICE 845 MAT.	INST.	TOTAL	PROVO 846 - 847 MAT.	INST.	TOTAL	SALT LAKE CITY 840 - 841 MAT.	INST.	TOTAL	BELLOWS FALLS 051 MAT.	INST.	TOTAL	BENNINGTON 052 MAT.	INST.	TOTAL
015433	CONTRACTOR EQUIPMENT		98.6	98.6		97.6	97.6		97.6	97.6		98.6	98.6		100.4	100.4		100.4	100.4
0241, 31 - 34	SITE & INFRASTRUCTURE, DEMOLITION	78.7	98.4	92.3	88.4	96.7	94.1	87.1	96.7	93.7	78.3	98.3	92.1	81.9	97.8	92.8	81.3	97.5	92.5
0310	Concrete Forming & Accessories	105.5	60.2	66.1	107.4	51.0	58.4	106.6	60.1	66.1	107.9	60.1	66.3	99.3	48.8	55.4	97.2	53.0	58.8
0320	Concrete Reinforcing	99.9	75.9	87.8	107.5	75.7	91.5	108.4	75.8	92.0	102.1	75.8	88.9	82.9	63.7	73.2	82.9	63.6	73.2
0330	Cast-in-Place Concrete	87.3	72.1	81.3	86.0	61.7	76.4	86.0	72.0	80.5	95.4	72.0	86.2	96.9	71.2	86.8	96.9	67.9	85.5
03	CONCRETE	95.8	67.8	82.3	106.6	60.1	84.0	105.2	67.7	86.9	114.2	67.7	91.5	100.9	60.0	81.0	100.7	60.7	81.2
04	MASONRY	100.3	62.0	76.9	112.2	62.0	81.6	112.3	62.0	81.6	114.0	62.0	82.3	93.6	60.6	73.5	102.3	60.5	76.8
05	METALS	100.4	76.8	92.6	97.1	76.1	90.2	98.1	76.7	91.1	105.0	76.7	95.6	92.5	69.6	85.0	92.5	68.7	84.7
06	WOOD, PLASTICS & COMPOSITES	79.9	58.4	67.5	82.5	47.1	62.1	81.2	58.4	68.0	81.5	58.4	68.2	99.5	38.4	64.2	96.7	44.9	66.8
07	THERMAL & MOISTURE PROTECTION	98.5	63.7	84.4	101.4	59.4	84.4	101.4	63.7	86.1	105.1	63.7	88.3	100.7	59.6	84.0	100.7	54.7	82.0
08	OPENINGS	91.7	57.6	83.1	96.2	50.3	84.6	96.2	57.6	86.5	93.7	57.6	84.6	99.2	41.1	84.5	99.2	43.4	85.1
0920	Plaster & Gypsum Board	81.4	57.0	64.0	83.7	45.3	56.3	81.9	57.0	64.1	84.4	57.0	64.8	96.5	35.8	53.1	94.2	42.5	57.3
0950, 0980	Ceilings & Acoustic Treatment	109.7	57.0	74.9	109.7	45.3	67.2	109.7	57.0	74.9	103.6	57.0	72.8	92.7	35.8	55.1	92.7	42.5	59.5
0960	Flooring	98.5	50.0	84.2	101.8	37.1	82.8	101.4	50.0	86.3	102.1	50.0	86.8	94.5	31.9	76.1	93.4	31.9	75.3
0970, 0990	Wall Finishes & Painting/Coating	98.8	62.0	76.1	98.8	39.6	62.3	98.8	50.6	69.1	101.9	50.6	70.2	90.1	28.1	51.9	90.1	28.1	51.9
09	FINISHES	98.6	57.8	75.9	101.8	46.1	70.7	101.3	56.6	76.4	99.4	56.6	75.5	95.6	42.0	65.7	95.0	45.9	67.6
COVERS	DIVS. 10 - 14, 25, 28, 41, 43, 44, 46	100.0	57.0	91.5	100.0	49.5	90.0	100.0	57.0	91.4	100.0	57.0	91.4	100.0	45.9	89.2	100.0	46.5	89.4
21, 22, 23	FIRE SUPPRESSION, PLUMBING & HVAC	100.0	71.3	88.4	97.0	68.8	85.6	100.0	71.3	88.4	100.1	71.3	88.4	94.1	95.7	94.7	94.1	89.5	92.2
26, 27, 3370	ELECTRICAL, COMMUNICATIONS & UTIL.	94.8	73.7	83.9	100.0	73.7	86.4	95.4	63.5	78.9	97.8	63.5	80.1	102.0	94.8	98.3	102.0	64.9	82.8
MF2010	WEIGHTED AVERAGE	97.3	70.0	85.3	99.5	65.9	84.7	99.7	68.3	85.8	101.6	68.4	86.9	96.5	72.7	86.0	96.8	67.8	84.0

VERMONT

DIVISION		BRATTLEBORO 053 MAT.	INST.	TOTAL	BURLINGTON 054 MAT.	INST.	TOTAL	GUILDHALL 059 MAT.	INST.	TOTAL	MONTPELIER 056 MAT.	INST.	TOTAL	RUTLAND 057 MAT.	INST.	TOTAL	ST. JOHNSBURY 058 MAT.	INST.	TOTAL
015433	CONTRACTOR EQUIPMENT		100.4	100.4		100.4	100.4		100.4	100.4		100.4	100.4		100.4	100.4		100.4	100.4
0241, 31 - 34	SITE & INFRASTRUCTURE, DEMOLITION	82.9	97.8	93.1	83.8	96.5	92.5	80.9	95.8	91.1	84.8	96.5	92.9	84.8	96.5	92.8	80.9	96.0	91.3
0310	Concrete Forming & Accessories	99.6	48.7	55.3	96.8	51.3	57.3	97.6	43.8	50.8	101.5	51.4	57.9	99.8	51.4	57.7	96.5	43.9	50.7
0320	Concrete Reinforcing	82.0	63.7	72.8	103.6	82.7	93.1	83.6	37.7	60.4	82.5	82.7	82.6	103.6	82.7	93.1	82.0	42.9	62.3
0330	Cast-in-Place Concrete	100.0	68.3	87.5	105.7	98.9	103.0	94.0	54.9	78.6	111.9	98.9	106.8	95.0	98.9	96.5	94.0	55.2	78.7
03	CONCRETE	103.3	59.0	81.7	109.4	74.2	92.2	98.3	47.4	73.5	109.5	74.2	92.3	104.1	74.2	89.6	98.0	48.4	73.8
04	MASONRY	101.3	60.6	76.4	104.9	49.1	70.8	101.5	39.0	63.4	89.6	49.1	64.9	85.0	49.1	63.1	126.1	39.1	73.0
05	METALS	92.5	69.5	84.9	99.7	78.9	92.8	92.5	58.5	81.3	92.5	78.9	88.0	98.0	78.9	91.7	92.5	59.0	81.5
06	WOOD, PLASTICS & COMPOSITES	99.8	38.4	64.3	94.7	50.7	69.3	96.2	44.9	66.6	98.8	50.7	71.0	99.9	50.7	71.5	92.3	44.9	64.9
07	THERMAL & MOISTURE PROTECTION	100.8	53.6	81.6	100.7	59.7	84.0	100.6	41.4	76.5	100.1	60.1	83.9	100.9	60.1	84.3	100.5	41.4	76.5
08	OPENINGS	99.2	41.1	84.5	102.4	52.5	89.9	99.2	38.2	83.8	99.2	52.5	87.4	102.4	52.5	89.9	99.2	38.2	83.8
0920	Plaster & Gypsum Board	96.5	35.8	53.1	98.1	48.5	62.7	102.4	42.5	59.6	105.6	48.5	64.8	97.3	48.5	62.5	105.2	42.5	60.4
0950, 0980	Ceilings & Acoustic Treatment	92.7	35.8	55.1	94.8	48.5	64.2	92.7	42.5	59.5	91.8	48.5	63.2	97.5	48.5	65.1	92.7	42.5	59.5
0960	Flooring	94.6	31.9	76.2	94.7	57.7	83.8	97.7	31.9	78.4	101.3	57.7	88.5	94.5	57.7	83.7	101.3	31.9	80.9
0970, 0990	Wall Finishes & Painting/Coating	90.1	28.1	51.9	90.1	32.7	54.7	90.1	21.8	48.0	90.1	32.7	54.7	90.1	32.7	54.7	90.1	22.9	48.7
09	FINISHES	95.7	42.0	65.8	96.4	49.3	70.1	97.1	39.4	64.9	98.5	49.3	71.1	96.8	49.3	70.3	98.5	39.6	65.6
COVERS	DIVS. 10 - 14, 25, 28, 41, 43, 44, 46	100.0	45.9	89.2	100.0	70.6	94.1	100.0	38.6	87.8	100.0	70.6	94.1	100.0	70.6	94.1	100.0	38.6	87.8
21, 22, 23	FIRE SUPPRESSION, PLUMBING & HVAC	94.1	97.8	95.6	99.9	62.1	84.5	94.1	60.4	80.4	93.9	62.1	81.0	100.0	62.1	84.6	94.1	60.4	80.4
26, 27, 3370	ELECTRICAL, COMMUNICATIONS & UTIL.	102.0	94.8	98.3	99.4	64.9	81.6	102.0	64.9	82.8	100.7	64.9	82.2	102.0	64.9	82.9	102.0	64.9	82.8
MF2010	WEIGHTED AVERAGE	97.2	72.9	86.5	100.6	65.4	85.1	96.6	55.0	78.3	97.4	65.4	83.3	99.2	65.4	84.3	97.9	55.2	79.1

VERMONT / VIRGINIA

DIVISION		WHITE RIVER JCT. 050 MAT.	INST.	TOTAL	ALEXANDRIA 223 MAT.	INST.	TOTAL	ARLINGTON 222 MAT.	INST.	TOTAL	BRISTOL 242 MAT.	INST.	TOTAL	CHARLOTTESVILLE 229 MAT.	INST.	TOTAL	CULPEPER 227 MAT.	INST.	TOTAL
015433	CONTRACTOR EQUIPMENT		100.4	100.4		101.1	101.1		99.9	99.9		99.9	99.9		104.0	104.0		99.9	99.9
0241, 31 - 34	SITE & INFRASTRUCTURE, DEMOLITION	85.4	95.9	92.7	118.6	88.3	97.7	130.3	85.1	99.2	113.8	84.7	93.7	119.2	86.7	96.8	116.8	85.5	95.2
0310	Concrete Forming & Accessories	94.8	42.1	49.0	89.8	75.3	77.2	89.3	71.8	74.1	85.7	44.1	49.5	84.2	50.2	54.6	81.7	70.4	71.9
0320	Concrete Reinforcing	82.9	37.6	60.1	83.6	79.9	81.7	94.4	71.6	82.9	94.3	68.5	81.3	93.7	69.4	81.5	94.4	65.7	79.9
0330	Cast-in-Place Concrete	100.0	54.7	82.1	122.1	83.8	107.0	118.8	79.2	103.2	118.3	48.9	91.0	123.1	57.0	97.0	121.6	66.3	99.8
03	CONCRETE	105.2	46.6	76.6	111.6	80.1	96.2	115.8	75.0	95.9	112.2	52.2	83.0	113.3	57.9	86.3	110.3	69.0	90.2
04	MASONRY	113.5	37.8	67.3	93.7	72.5	80.7	107.7	66.1	82.3	95.8	41.5	62.7	121.9	48.0	76.8	109.7	61.8	80.5
05	METALS	92.5	58.7	81.4	95.8	94.2	95.3	94.5	81.8	90.3	93.3	85.4	90.7	93.6	88.0	91.8	93.7	80.6	89.4
06	WOOD, PLASTICS & COMPOSITES	93.7	43.2	64.5	95.7	74.4	83.4	92.9	74.4	82.2	85.9	42.4	60.8	84.5	48.3	63.6	83.2	74.4	78.1
07	THERMAL & MOISTURE PROTECTION	100.9	36.6	74.7	99.0	80.9	91.6	101.0	68.6	87.8	100.5	48.1	79.2	100.1	53.5	81.2	100.3	62.0	84.7
08	OPENINGS	99.2	39.5	84.1	97.3	75.4	91.8	95.2	69.4	88.7	98.6	48.4	86.0	96.6	53.3	85.7	96.9	65.0	88.9
0920	Plaster & Gypsum Board	92.4	40.7	55.5	110.4	73.5	84.0	107.8	73.5	83.3	103.4	40.5	58.4	103.4	45.9	62.3	103.7	73.5	82.1
0950, 0980	Ceilings & Acoustic Treatment	92.7	40.7	58.4	97.3	73.5	81.6	94.6	73.5	80.6	93.7	40.5	58.5	93.7	45.9	62.1	94.6	73.5	80.6
0960	Flooring	92.3	31.9	74.6	107.3	86.1	101.1	105.6	63.4	93.2	101.8	39.2	83.4	100.3	46.2	84.4	100.3	63.4	89.5
0970, 0990	Wall Finishes & Painting/Coating	90.1	19.1	46.3	130.0	81.1	99.8	130.0	81.1	99.8	114.8	43.8	71.0	114.8	62.0	82.2	130.0	55.1	83.8
09	FINISHES	94.8	37.9	63.1	104.4	77.4	89.3	104.5	71.9	86.3	100.8	42.4	68.2	100.4	49.9	72.2	101.1	68.8	83.1
COVERS	DIVS. 10 - 14, 25, 28, 41, 43, 44, 46	100.0	38.2	87.7	100.0	89.3	97.9	100.0	70.2	94.1	100.0	73.2	94.7	100.0	73.1	94.6	100.0	59.7	92.0
21, 22, 23	FIRE SUPPRESSION, PLUMBING & HVAC	94.1	60.0	80.2	100.2	88.8	95.6	100.2	84.6	93.9	94.3	53.8	77.9	94.3	56.9	79.1	94.3	62.8	81.5
26, 27, 3370	ELECTRICAL, COMMUNICATIONS & UTIL.	102.0	64.9	82.8	96.3	102.5	99.5	94.0	98.3	96.2	96.0	37.6	65.9	95.9	71.0	83.1	98.4	55.4	76.2
MF2010	WEIGHTED AVERAGE	98.0	54.4	78.8	100.6	86.0	94.2	101.6	79.9	92.0	98.6	54.3	79.1	100.0	62.8	83.6	99.3	67.0	85.1

		VIRGINIA																
DIVISION		FAIRFAX 220-221			FARMVILLE 239			FREDERICKSBURG 224-225			GRUNDY 246			HARRISONBURG 228			LYNCHBURG 245		
		MAT.	INST.	TOTAL	MAT.	INST.	TOTAL	MAT.	INST.	TOTAL	MAT.	INST.	TOTAL	MAT.	INST.	TOTAL	MAT.	INST.	TOTAL
015433	CONTRACTOR EQUIPMENT		99.9	99.9		104.0	104.0		99.9	99.9		99.9	99.9		99.9	99.9		99.9	99.9
0241, 31-34	SITE & INFRASTRUCTURE, DEMOLITION	129.2	85.9	99.4	114.8	86.0	94.9	116.2	85.8	95.3	111.3	83.6	92.2	125.8	84.8	97.5	112.2	84.9	93.3
0310	Concrete Forming & Accessories	84.3	72.3	73.9	95.1	45.7	52.1	84.3	50.0	54.5	88.6	38.1	44.7	80.7	44.2	49.0	85.7	63.1	66.1
0320	Concrete Reinforcing	94.4	79.8	87.0	94.3	52.9	73.5	95.1	78.8	86.9	93.0	53.5	73.1	94.4	48.1	71.1	93.7	69.1	81.3
0330	Cast-in-Place Concrete	118.8	80.1	103.5	123.0	54.9	96.2	120.6	53.8	94.3	118.3	49.3	91.1	118.7	60.0	95.6	118.3	59.0	94.1
03	CONCRETE	115.5	77.6	97.0	112.4	51.9	82.9	109.9	58.4	84.8	111.0	46.4	79.5	113.3	52.5	83.7	110.9	64.5	88.3
04	MASONRY	107.6	66.4	82.4	104.7	47.3	69.7	108.8	46.0	70.5	99.2	50.0	69.2	105.8	52.1	73.0	114.0	54.2	77.5
05	METALS	93.7	95.1	94.2	93.6	77.5	88.3	93.7	91.6	93.0	93.3	71.5	86.1	93.6	83.7	90.4	93.5	86.7	91.3
06	WOOD, PLASTICS & COMPOSITES	85.9	74.4	79.2	97.7	46.7	68.2	85.9	47.7	63.8	88.6	37.2	58.9	82.3	41.4	58.6	85.9	66.5	74.7
07	THERMAL & MOISTURE PROTECTION	100.9	68.9	87.9	100.1	48.5	79.1	100.3	51.6	80.5	100.4	47.8	79.0	100.7	61.4	84.7	100.3	55.7	82.2
08	OPENINGS	95.2	76.6	90.5	96.9	45.6	84.0	96.6	55.2	86.2	98.6	36.5	83.0	96.9	51.9	85.6	96.9	61.5	88.0
0920	Plaster & Gypsum Board	103.7	73.5	82.1	109.3	44.2	62.8	103.7	45.9	62.4	103.4	35.1	54.6	103.4	39.4	57.7	103.4	65.3	76.2
0950, 0980	Ceilings & Acoustic Treatment	94.6	73.5	80.6	93.7	44.2	61.0	94.6	45.9	62.4	93.7	35.1	55.0	93.7	39.4	57.8	93.7	65.3	74.9
0960	Flooring	102.1	66.2	91.6	107.3	63.1	94.3	102.1	77.5	94.9	103.3	35.3	83.3	100.0	77.5	93.4	101.8	43.7	84.7
0970, 0990	Wall Finishes & Painting/Coating	130.0	81.1	99.8	114.8	40.8	69.2	130.0	67.7	91.6	114.8	37.0	66.8	130.0	53.5	82.8	114.8	43.8	71.0
09	FINISHES	102.9	72.4	85.9	102.8	48.7	72.6	101.6	55.7	76.0	101.0	37.1	65.4	101.7	50.2	73.0	100.6	57.9	76.8
COVERS	DIVS. 10-14, 25, 28, 41, 43, 44, 46	100.0	70.2	94.1	100.0	49.1	89.9	100.0	74.0	94.8	100.0	47.1	89.5	100.0	72.9	94.6	100.0	76.6	95.3
21, 22, 23	FIRE SUPPRESSION, PLUMBING & HVAC	94.3	84.9	90.5	94.3	36.7	70.9	94.3	82.3	89.4	94.3	41.1	72.7	94.3	53.7	77.8	94.3	56.4	78.9
26, 27, 3370	ELECTRICAL, COMMUNICATIONS & UTIL.	97.1	98.3	97.7	93.3	49.9	70.9	94.2	91.3	92.7	96.0	47.4	70.9	96.2	83.0	89.4	97.0	55.9	75.8
MF2010	WEIGHTED AVERAGE	100.1	82.0	92.1	98.9	52.3	78.4	98.8	71.8	86.9	98.6	49.7	77.0	99.6	63.0	83.4	99.3	63.6	83.5

		VIRGINIA																
DIVISION		NEWPORT NEWS 236			NORFOLK 233-235			PETERSBURG 238			PORTSMOUTH 237			PULASKI 243			RICHMOND 230-232		
		MAT.	INST.	TOTAL	MAT.	INST.	TOTAL	MAT.	INST.	TOTAL	MAT.	INST.	TOTAL	MAT.	INST.	TOTAL	MAT.	INST.	TOTAL
015433	CONTRACTOR EQUIPMENT		104.0	104.0		104.6	104.6		104.0	104.0		103.9	103.9		99.9	99.9		104.0	104.0
0241, 31-34	SITE & INFRASTRUCTURE, DEMOLITION	113.0	87.4	95.3	111.3	88.3	95.4	117.7	87.7	97.0	111.5	86.7	94.4	110.7	83.6	92.0	110.4	87.7	94.7
0310	Concrete Forming & Accessories	94.5	66.1	69.8	95.9	66.2	70.1	88.5	59.9	63.3	84.9	54.3	58.3	88.6	38.3	44.8	94.0	59.6	64.1
0320	Concrete Reinforcing	94.0	72.3	83.1	94.0	72.3	83.1	93.7	70.1	81.8	93.7	69.8	81.6	93.0	53.5	73.1	94.0	70.1	82.0
0330	Cast-in-Place Concrete	119.6	60.2	96.2	130.7	60.3	102.9	127.0	55.2	98.7	118.4	59.7	95.3	118.3	49.3	91.1	115.2	55.2	91.6
03	CONCRETE	109.3	66.5	88.4	114.9	66.6	91.4	115.2	61.5	89.1	108.0	60.6	84.9	111.0	46.5	79.6	107.1	61.5	84.9
04	MASONRY	99.5	54.1	71.8	105.6	54.1	74.2	113.2	55.6	78.1	105.0	53.9	73.8	92.5	50.0	66.6	97.3	55.6	71.9
05	METALS	95.8	89.9	93.9	94.8	90.0	93.2	93.7	89.5	92.3	94.8	87.2	92.3	93.4	71.6	86.2	98.0	89.5	95.2
06	WOOD, PLASTICS & COMPOSITES	96.6	70.3	81.4	98.6	70.3	82.2	89.0	62.1	73.5	85.7	54.5	67.7	88.6	37.2	58.9	95.7	62.1	76.3
07	THERMAL & MOISTURE PROTECTION	100.1	56.1	82.2	97.5	56.1	80.6	100.1	56.3	82.3	100.1	53.8	81.3	100.4	47.8	79.0	98.7	56.3	81.5
08	OPENINGS	97.3	64.9	89.1	97.3	64.9	89.1	96.6	60.8	87.6	97.4	55.7	86.9	98.6	36.5	83.0	97.3	60.8	88.1
0920	Plaster & Gypsum Board	110.4	68.5	80.5	101.0	68.5	77.8	103.7	60.1	72.6	104.1	52.3	67.1	103.4	35.1	54.6	114.0	60.1	75.5
0950, 0980	Ceilings & Acoustic Treatment	97.3	68.5	78.3	97.3	68.5	78.3	94.6	60.1	71.8	97.3	52.3	67.6	93.7	35.1	55.0	100.6	60.1	74.0
0960	Flooring	107.3	52.9	91.3	106.7	52.9	90.9	102.8	74.6	94.5	99.5	66.9	90.0	103.3	35.3	83.3	106.7	74.6	97.3
0970, 0990	Wall Finishes & Painting/Coating	114.8	44.3	71.3	114.8	67.8	85.8	114.8	62.0	82.2	114.8	67.8	85.8	114.8	37.0	66.8	114.8	62.0	82.2
09	FINISHES	103.5	61.7	80.2	101.8	64.4	80.9	101.3	62.7	79.8	100.5	57.4	76.5	101.0	37.1	65.4	104.0	62.7	81.0
COVERS	DIVS. 10-14, 25, 28, 41, 43, 44, 46	100.0	78.3	95.7	100.0	78.3	95.7	100.0	73.3	94.7	100.0	76.6	95.3	100.0	47.1	89.5	100.0	73.3	94.7
21, 22, 23	FIRE SUPPRESSION, PLUMBING & HVAC	100.2	63.9	85.5	100.0	63.7	85.3	94.3	55.0	78.4	100.2	62.1	84.8	94.3	41.4	72.9	100.0	55.0	81.8
26, 27, 3370	ELECTRICAL, COMMUNICATIONS & UTIL.	96.0	60.4	77.6	101.2	59.9	79.9	96.2	71.0	83.2	94.3	59.8	76.5	96.0	51.2	72.9	99.3	71.0	84.7
MF2010	WEIGHTED AVERAGE	100.4	67.3	85.8	101.4	67.6	86.5	99.9	66.0	85.0	99.9	64.4	84.3	98.3	50.3	77.1	100.6	66.0	85.3

		VIRGINIA									WASHINGTON								
DIVISION		ROANOKE 240-241			STAUNTON 244			WINCHESTER 226			CLARKSTON 994			EVERETT 982			OLYMPIA 985		
		MAT.	INST.	TOTAL	MAT.	INST.	TOTAL	MAT.	INST.	TOTAL	MAT.	INST.	TOTAL	MAT.	INST.	TOTAL	MAT.	INST.	TOTAL
015433	CONTRACTOR EQUIPMENT		99.9	99.9		104.0	104.0		99.9	99.9		91.0	91.0		104.1	104.1		104.1	104.1
0241, 31-34	SITE & INFRASTRUCTURE, DEMOLITION	110.3	84.8	92.7	115.4	86.5	95.5	124.3	84.9	97.1	96.4	90.4	92.2	92.9	112.9	106.7	94.4	112.9	107.1
0310	Concrete Forming & Accessories	94.3	62.9	67.0	88.1	52.4	57.0	82.9	49.2	53.6	124.9	70.5	77.6	113.7	99.4	101.2	103.5	99.5	100.0
0320	Concrete Reinforcing	94.0	69.2	81.5	93.7	51.9	72.6	93.7	83.8	88.7	106.6	84.5	95.5	116.6	103.0	109.7	117.2	103.0	110.1
0330	Cast-in-Place Concrete	134.4	59.7	105.0	123.0	53.2	95.5	118.7	56.3	94.1	102.5	82.4	94.6	101.0	105.8	102.9	91.9	106.0	97.5
03	CONCRETE	116.7	64.3	91.2	112.6	54.3	84.2	112.7	59.7	86.9	99.5	77.4	88.7	98.4	101.8	100.0	94.4	101.9	98.0
04	MASONRY	100.6	55.8	73.3	110.2	54.3	76.1	103.4	49.9	70.8	104.1	79.6	89.2	122.3	99.2	108.2	112.2	99.3	104.3
05	METALS	95.6	86.6	92.6	93.6	82.3	89.9	93.7	90.2	92.6	87.3	80.3	84.9	100.8	92.9	98.2	95.8	93.0	94.9
06	WOOD, PLASTICS & COMPOSITES	96.6	66.5	79.2	88.6	52.9	68.0	84.5	47.7	63.2	101.0	67.8	81.8	106.4	99.0	102.1	93.4	99.0	96.6
07	THERMAL & MOISTURE PROTECTION	100.0	54.7	81.6	100.1	52.3	80.6	100.8	58.7	83.7	151.2	76.5	120.8	102.4	99.0	101.1	102.1	96.8	99.9
08	OPENINGS	97.3	61.5	88.3	96.9	51.0	85.4	98.7	56.5	88.1	106.4	71.0	97.5	101.2	100.2	101.0	101.0	100.2	100.8
0920	Plaster & Gypsum Board	110.4	65.3	78.2	103.4	50.6	65.7	103.7	45.9	62.4	130.2	66.6	84.8	115.5	98.9	103.6	105.1	98.9	100.7
0950, 0980	Ceilings & Acoustic Treatment	97.3	65.3	76.2	93.7	50.6	65.2	94.6	45.9	62.4	99.3	66.6	77.7	107.8	98.9	101.9	101.2	98.9	99.7
0960	Flooring	107.3	47.1	89.6	102.7	41.3	84.7	101.4	77.5	94.4	103.2	49.4	87.4	116.5	95.9	110.5	109.2	96.1	105.3
0970, 0990	Wall Finishes & Painting/Coating	114.8	43.8	71.0	114.8	30.6	62.8	130.0	47.1	78.8	110.1	61.7	80.2	105.5	74.8	86.6	105.5	84.1	92.3
09	FINISHES	103.3	58.7	78.4	100.8	48.6	71.7	102.2	53.4	75.0	116.9	64.5	87.7	108.7	96.0	101.6	104.0	97.1	100.1
COVERS	DIVS. 10-14, 25, 28, 41, 43, 44, 46	100.0	76.5	95.3	100.0	76.3	95.3	100.0	55.1	91.1	100.0	81.0	96.2	100.0	96.6	99.3	100.0	101.0	100.2
21, 22, 23	FIRE SUPPRESSION, PLUMBING & HVAC	100.2	47.5	78.9	94.3	55.4	78.5	94.3	67.4	83.4	94.4	87.9	91.8	100.0	90.9	96.3	99.9	90.9	96.2
26, 27, 3370	ELECTRICAL, COMMUNICATIONS & UTIL.	96.0	40.4	67.3	94.9	43.9	68.6	94.6	45.6	69.3	99.9	92.2	96.0	103.8	82.3	92.7	98.1	97.0	97.6
MF2010	WEIGHTED AVERAGE	101.2	59.8	82.9	99.2	58.2	81.1	99.4	62.2	83.0	100.4	81.3	92.0	102.2	95.6	99.3	99.4	97.9	98.7

City Cost Indexes

WASHINGTON

Division		RICHLAND 993 MAT.	INST.	TOTAL	SEATTLE 980-981,987 MAT.	INST.	TOTAL	SPOKANE 990-992 MAT.	INST.	TOTAL	TACOMA 983-984 MAT.	INST.	TOTAL	VANCOUVER 986 MAT.	INST.	TOTAL	WENATCHEE 988 MAT.	INST.	TOTAL
015433	CONTRACTOR EQUIPMENT		91.0	91.0		103.9	103.9		91.0	91.0		104.1	104.1		98.8	98.8		104.1	104.1
0241, 31 - 34	SITE & INFRASTRUCTURE, DEMOLITION	98.0	91.3	93.4	97.0	111.1	106.7	97.4	91.3	93.1	96.1	112.9	107.7	107.4	100.2	102.5	107.0	111.5	110.1
0310	Concrete Forming & Accessories	125.0	79.9	85.7	105.2	100.3	100.9	129.2	79.7	86.1	105.2	99.5	100.2	106.3	86.6	89.2	106.6	78.0	81.7
0320	Concrete Reinforcing	102.3	84.6	93.4	115.1	103.1	109.1	103.0	84.6	93.7	115.1	103.0	109.0	116.1	102.7	109.4	116.1	91.2	103.6
0330	Cast-in-Place Concrete	102.7	84.5	95.5	106.0	106.3	106.1	106.7	84.4	97.9	103.9	106.0	104.7	116.0	100.3	109.8	106.0	79.4	95.5
03	CONCRETE	99.0	82.4	90.9	101.2	102.5	101.8	101.4	82.2	92.1	100.1	101.9	101.0	110.1	94.3	102.4	107.3	81.4	94.5
04	MASONRY	105.0	81.9	90.9	116.4	99.3	106.0	105.6	81.9	91.2	116.2	99.3	105.9	117.3	95.8	104.1	119.0	81.3	96.0
05	METALS	87.6	82.9	86.1	102.5	95.1	100.1	89.9	82.7	87.6	102.5	93.0	99.4	100.0	93.5	97.8	100.0	86.7	95.6
06	WOOD, PLASTICS & COMPOSITES	101.2	78.7	88.2	97.9	99.0	98.5	109.7	78.7	91.8	96.5	99.0	98.0	88.4	83.7	85.7	97.6	76.5	85.4
07	THERMAL & MOISTURE PROTECTION	152.4	78.6	122.4	102.3	99.1	101.0	148.9	79.5	120.6	102.2	96.8	100.0	102.7	91.2	98.0	102.8	79.8	93.5
08	OPENINGS	109.0	72.5	99.8	103.4	100.2	102.6	109.6	72.9	100.3	101.8	100.2	101.4	99.5	90.0	97.1	101.3	72.9	94.1
0920	Plaster & Gypsum Board	130.2	77.9	92.8	109.6	98.9	101.9	130.6	77.9	92.9	112.5	98.9	102.8	110.2	83.4	91.0	114.1	75.7	86.7
0950, 0980	Ceilings & Acoustic Treatment	104.4	77.9	86.9	110.3	98.9	102.7	101.4	77.9	85.8	110.5	98.9	102.8	105.2	83.4	90.8	103.0	75.7	85.0
0960	Flooring	103.4	83.9	97.7	109.1	103.9	107.6	103.5	89.3	99.4	109.6	96.1	105.6	115.2	107.0	112.8	112.7	65.6	98.9
0970, 0990	Wall Finishes & Painting/Coating	110.1	61.7	80.2	105.5	84.1	92.3	110.1	61.7	80.3	105.5	84.1	92.3	108.9	65.2	82.0	105.5	61.7	78.5
09	FINISHES	118.3	78.3	96.0	106.6	98.7	102.2	117.9	79.4	96.4	107.0	97.1	101.5	104.9	87.7	95.3	107.6	73.2	88.4
COVERS	DIVS. 10 - 14, 25, 28, 41, 43, 44, 46	100.0	66.9	93.4	100.0	101.0	100.2	100.0	66.8	93.4	100.0	101.0	100.2	100.0	64.1	92.9	100.0	82.3	96.5
21, 22, 23	FIRE SUPPRESSION, PLUMBING & HVAC	100.4	100.1	100.3	100.0	122.1	109.0	100.3	89.0	95.7	100.1	90.9	96.4	100.2	93.0	97.3	94.1	85.4	90.6
26, 27, 3370	ELECTRICAL, COMMUNICATIONS & UTIL.	97.1	93.1	95.1	103.6	107.9	105.8	95.1	79.1	86.8	103.6	97.0	100.2	108.5	98.0	103.1	104.4	80.2	91.9
MF2010	WEIGHTED AVERAGE	102.1	86.8	95.4	102.5	106.3	104.2	102.5	82.7	93.7	102.2	97.9	100.3	103.5	93.0	98.8	101.8	83.6	93.8

WASHINGTON / WEST VIRGINIA

Division		YAKIMA 989 MAT.	INST.	TOTAL	BECKLEY 258-259 MAT.	INST.	TOTAL	BLUEFIELD 247-248 MAT.	INST.	TOTAL	BUCKHANNON 262 MAT.	INST.	TOTAL	CHARLESTON 250-253 MAT.	INST.	TOTAL	CLARKSBURG 263-264 MAT.	INST.	TOTAL
015433	CONTRACTOR EQUIPMENT		104.1	104.1		99.9	99.9		99.9	99.9		99.9	99.9		99.9	99.9		99.9	99.9
0241, 31 - 34	SITE & INFRASTRUCTURE, DEMOLITION	98.9	112.3	108.2	104.9	86.9	92.5	104.7	86.9	92.4	111.5	86.7	94.4	102.7	88.4	92.9	112.1	86.7	94.6
0310	Concrete Forming & Accessories	105.5	95.1	96.5	84.6	86.5	86.3	85.8	86.2	86.1	85.2	84.5	84.6	99.5	88.5	89.9	82.9	84.4	84.2
0320	Concrete Reinforcing	115.7	91.3	103.4	92.7	84.2	88.4	92.7	66.3	79.4	93.3	83.1	88.2	94.0	84.4	89.2	93.3	83.7	88.4
0330	Cast-in-Place Concrete	111.0	84.2	100.4	115.7	102.7	110.6	115.7	102.6	110.5	115.4	100.0	109.3	107.0	100.9	104.6	126.4	100.0	116.0
03	CONCRETE	104.9	90.3	97.8	106.5	92.3	99.6	106.6	88.8	97.9	109.5	90.3	100.2	103.3	92.6	98.1	114.6	90.4	102.8
04	MASONRY	109.5	81.3	92.3	99.4	89.6	93.5	94.2	89.6	91.4	105.6	86.9	94.2	96.2	91.5	93.3	108.1	86.9	95.2
05	METALS	100.7	87.0	96.2	93.6	95.2	94.1	93.6	88.7	91.9	93.7	95.7	94.4	95.8	96.4	96.0	93.7	95.8	94.4
06	WOOD, PLASTICS & COMPOSITES	97.0	99.0	98.2	86.0	87.8	87.1	87.7	87.8	87.8	87.0	83.3	84.9	102.6	87.8	94.1	83.9	83.3	83.6
07	THERMAL & MOISTURE PROTECTION	102.3	81.4	93.8	100.2	87.2	94.9	100.2	87.2	94.9	100.5	86.7	94.9	97.3	87.7	93.4	100.4	86.7	94.8
08	OPENINGS	101.3	81.7	96.4	98.0	81.2	93.7	98.8	77.1	93.4	98.9	78.5	93.8	98.5	81.2	94.2	98.9	80.5	94.2
0920	Plaster & Gypsum Board	112.3	98.9	102.7	101.5	87.3	91.4	102.4	87.3	91.6	103.0	82.7	88.5	100.5	87.3	91.1	100.7	82.7	87.8
0950, 0980	Ceilings & Acoustic Treatment	105.0	98.9	101.0	91.9	87.3	88.9	91.9	87.3	88.9	93.7	82.7	86.4	95.5	87.3	90.1	93.7	82.7	86.4
0960	Flooring	110.8	60.7	96.1	98.9	110.4	102.3	99.5	110.4	102.7	99.2	104.1	100.6	107.3	110.4	108.2	98.0	104.1	99.8
0970, 0990	Wall Finishes & Painting/Coating	105.5	61.7	78.5	114.8	36.9	66.7	114.8	36.9	66.7	114.8	89.3	99.0	114.8	89.5	99.2	114.8	89.3	99.0
09	FINISHES	106.4	85.5	94.8	98.6	86.4	91.8	98.9	86.4	91.9	99.9	88.4	93.5	101.2	92.7	96.5	99.2	88.4	93.2
COVERS	DIVS. 10 - 14, 25, 28, 41, 43, 44, 46	100.0	98.2	99.6	100.0	57.3	91.5	100.0	57.2	91.5	100.0	94.6	98.9	100.0	95.6	99.1	100.0	94.6	98.9
21, 22, 23	FIRE SUPPRESSION, PLUMBING & HVAC	100.1	100.2	100.1	94.3	80.1	88.6	94.3	79.7	88.4	94.3	88.0	91.7	100.0	82.2	92.8	94.3	87.7	91.6
26, 27, 3370	ELECTRICAL, COMMUNICATIONS & UTIL.	106.6	93.1	99.6	92.4	90.9	91.6	95.0	71.5	82.9	96.3	91.5	93.8	97.2	90.9	93.9	96.3	91.5	93.8
MF2010	WEIGHTED AVERAGE	102.5	92.6	98.1	97.3	86.6	92.6	97.4	82.6	90.9	98.8	89.0	94.5	99.2	89.5	94.9	99.4	89.1	94.9

WEST VIRGINIA

Division		GASSAWAY 266 MAT.	INST.	TOTAL	HUNTINGTON 255-257 MAT.	INST.	TOTAL	LEWISBURG 249 MAT.	INST.	TOTAL	MARTINSBURG 254 MAT.	INST.	TOTAL	MORGANTOWN 265 MAT.	INST.	TOTAL	PARKERSBURG 261 MAT.	INST.	TOTAL
015433	CONTRACTOR EQUIPMENT		99.9	99.9		99.9	99.9		99.9	99.9		99.9	99.9		99.9	99.9		99.9	99.9
0241, 31 - 34	SITE & INFRASTRUCTURE, DEMOLITION	108.5	87.2	93.8	109.3	89.4	95.5	122.4	86.9	97.9	108.8	84.9	92.3	105.6	87.0	92.8	114.4	88.4	96.5
0310	Concrete Forming & Accessories	84.7	84.6	84.6	95.4	91.3	91.8	83.2	86.2	85.8	84.6	76.9	77.9	83.2	84.1	84.0	86.9	86.0	86.1
0320	Concrete Reinforcing	93.3	84.4	88.8	94.0	90.6	92.3	93.3	66.3	79.7	92.7	74.1	83.3	93.3	83.7	88.4	92.7	83.0	87.8
0330	Cast-in-Place Concrete	120.8	100.1	112.6	126.2	105.1	117.9	115.8	102.6	110.6	121.2	89.7	108.8	115.3	99.9	109.2	117.9	91.5	107.5
03	CONCRETE	110.4	90.6	100.8	112.7	96.4	104.7	115.9	88.8	102.7	110.5	81.5	96.4	106.3	90.2	98.5	111.7	88.0	100.1
04	MASONRY	109.2	90.1	97.6	99.9	91.8	95.0	98.6	89.6	93.1	101.2	79.5	87.9	128.6	86.9	103.2	84.3	84.7	84.5
05	METALS	93.7	86.4	91.3	95.9	98.6	96.8	93.7	88.7	92.0	93.9	85.3	91.1	93.8	95.5	94.4	94.4	95.5	94.8
06	WOOD, PLASTICS & COMPOSITES	86.2	83.3	84.5	96.6	89.9	92.7	84.2	87.8	86.3	86.0	78.1	81.5	84.2	83.3	83.7	87.1	84.9	85.8
07	THERMAL & MOISTURE PROTECTION	100.2	87.5	95.1	100.3	88.4	95.5	101.1	87.2	95.5	100.4	72.7	89.2	100.3	86.7	94.8	100.1	86.0	94.4
08	OPENINGS	96.8	78.8	92.2	97.3	83.8	93.9	98.9	77.1	93.4	100.2	66.9	91.8	100.3	80.5	95.3	97.9	79.4	93.2
0920	Plaster & Gypsum Board	102.1	82.7	88.2	109.3	89.5	95.2	100.7	87.3	91.1	102.1	77.3	84.4	100.7	82.7	87.8	103.4	84.3	89.7
0950, 0980	Ceilings & Acoustic Treatment	93.7	82.7	86.4	93.7	89.5	90.9	93.7	87.3	89.5	93.7	77.3	82.9	93.7	82.7	86.4	93.7	84.3	87.5
0960	Flooring	99.0	110.4	102.3	107.1	105.1	106.5	98.1	110.4	101.7	98.9	56.2	86.4	98.1	104.1	99.9	102.3	96.6	100.6
0970, 0990	Wall Finishes & Painting/Coating	114.8	89.3	99.2	114.8	83.2	95.3	114.8	36.9	66.7	114.8	43.6	70.9	114.8	89.3	99.0	114.8	88.4	98.5
09	FINISHES	99.4	89.7	94.0	102.3	92.9	97.1	100.5	86.4	92.6	99.4	70.1	83.0	98.7	88.4	93.0	100.9	88.0	93.7
COVERS	DIVS. 10 - 14, 25, 28, 41, 43, 44, 46	100.0	94.6	98.9	100.0	96.9	99.4	100.0	57.2	91.5	100.0	55.4	91.1	100.0	66.6	93.4	100.0	95.1	99.0
21, 22, 23	FIRE SUPPRESSION, PLUMBING & HVAC	94.3	88.8	92.1	100.2	84.4	93.8	94.3	80.0	88.5	94.3	82.6	89.6	94.3	87.6	91.6	100.2	87.8	95.1
26, 27, 3370	ELECTRICAL, COMMUNICATIONS & UTIL.	96.3	90.9	93.5	96.0	93.9	94.9	92.4	71.5	81.6	98.0	81.6	89.6	96.5	91.5	93.9	96.4	90.3	93.3
MF2010	WEIGHTED AVERAGE	98.7	89.8	94.8	100.6	91.4	96.6	99.1	82.7	91.9	98.9	79.1	90.2	99.5	88.2	94.5	99.6	88.4	94.7

WEST VIRGINIA / WISCONSIN

DIVISION		PETERSBURG 268 MAT.	INST.	TOTAL	ROMNEY 267 MAT.	INST.	TOTAL	WHEELING 260 MAT.	INST.	TOTAL	BELOIT 535 MAT.	INST.	TOTAL	EAU CLAIRE 547 MAT.	INST.	TOTAL	GREEN BAY 541 - 543 MAT.	INST.	TOTAL
015433	CONTRACTOR EQUIPMENT		99.9	99.9		99.9	99.9		99.9	99.9		100.9	100.9		100.8	100.8		98.7	98.7
0241, 31 - 34	SITE & INFRASTRUCTURE, DEMOLITION	104.9	87.3	92.8	108.0	87.3	93.7	115.1	87.6	96.1	97.8	106.2	103.6	98.6	103.4	101.9	102.8	99.5	100.5
0310	Concrete Forming & Accessories	86.0	79.7	80.5	82.6	79.8	80.1	88.6	84.5	85.1	103.3	94.6	95.7	101.9	96.3	97.0	107.3	95.6	97.1
0320	Concrete Reinforcing	92.7	74.6	83.6	93.3	71.8	82.5	92.1	83.7	87.9	102.0	118.4	110.3	100.1	101.7	100.9	98.1	88.4	93.2
0330	Cast-in-Place Concrete	115.3	94.0	106.9	120.8	94.0	110.2	117.9	100.1	110.9	103.5	92.8	99.3	103.6	93.5	99.6	107.2	98.0	103.6
03	CONCRETE	106.4	84.5	95.7	110.3	84.1	97.5	111.7	90.5	101.3	99.7	98.6	99.2	101.1	96.6	98.9	104.0	95.4	99.9
04	MASONRY	101.2	86.9	92.5	99.0	83.7	89.7	108.9	86.9	95.5	104.6	93.9	98.1	92.6	97.9	95.8	124.4	97.2	107.8
05	METALS	93.8	91.8	93.2	93.9	90.8	92.9	94.6	95.9	95.0	95.9	108.3	100.0	95.7	102.8	98.1	97.9	96.9	97.6
06	WOOD, PLASTICS & COMPOSITES	88.0	78.1	82.3	83.3	78.1	80.3	88.6	83.3	85.5	101.5	97.0	98.9	108.3	97.1	101.8	109.3	97.1	102.2
07	THERMAL & MOISTURE PROTECTION	100.3	75.1	90.1	100.4	74.2	89.8	100.5	87.3	95.2	105.8	91.4	99.9	106.3	84.9	97.6	107.8	85.2	98.6
08	OPENINGS	100.3	73.8	93.6	100.2	73.1	93.4	98.8	80.5	94.2	102.1	105.3	102.9	99.6	95.9	98.7	97.8	94.6	97.0
0920	Plaster & Gypsum Board	103.0	77.3	84.6	100.3	77.3	83.8	103.4	82.7	88.6	100.3	97.3	98.1	105.2	97.3	99.5	96.6	97.3	97.1
0950, 0980	Ceilings & Acoustic Treatment	93.7	77.3	82.9	93.7	77.3	82.9	93.7	82.7	86.4	87.5	97.3	93.9	101.5	97.3	98.7	93.4	97.3	96.0
0960	Flooring	100.0	104.1	101.2	97.9	97.8	97.9	103.3	104.1	103.5	104.5	117.5	108.3	96.3	113.3	101.3	115.9	113.3	115.1
0970, 0990	Wall Finishes & Painting/Coating	114.8	89.3	99.0	114.8	89.3	99.0	114.8	89.3	99.0	91.9	97.4	95.3	88.4	91.1	90.1	98.0	80.3	87.1
09	FINISHES	99.5	85.4	91.6	98.9	84.1	90.6	101.2	88.4	94.1	100.3	99.3	99.7	99.7	99.3	99.5	104.0	97.9	100.6
COVERS	DIVS. 10 - 14, 25, 28, 41, 43, 44, 46	100.0	66.0	93.2	100.0	94.0	98.8	100.0	96.2	99.2	100.0	89.4	97.9	100.0	91.8	98.4	100.0	91.5	98.3
21, 22, 23	FIRE SUPPRESSION, PLUMBING & HVAC	94.3	88.7	92.0	94.3	88.6	92.0	100.2	87.7	95.2	99.9	89.1	95.5	100.2	84.8	93.9	100.4	82.9	93.3
26, 27, 3370	ELECTRICAL, COMMUNICATIONS & UTIL.	99.6	81.7	90.4	98.8	81.7	90.0	93.7	91.5	92.6	101.1	87.5	94.1	103.7	90.2	96.7	98.4	82.8	90.3
MF2010	WEIGHTED AVERAGE	98.5	84.9	92.5	98.7	85.1	92.7	100.7	89.2	95.7	100.0	96.0	98.3	99.7	94.3	97.3	101.8	91.6	97.3

WISCONSIN

DIVISION		KENOSHA 531 MAT.	INST.	TOTAL	LA CROSSE 546 MAT.	INST.	TOTAL	LANCASTER 538 MAT.	INST.	TOTAL	MADISON 537 MAT.	INST.	TOTAL	MILWAUKEE 530,532 MAT.	INST.	TOTAL	NEW RICHMOND 540 MAT.	INST.	TOTAL
015433	CONTRACTOR EQUIPMENT		98.8	98.8		100.8	100.8		100.9	100.9		100.9	100.9		89.3	89.3		101.4	101.4
0241, 31 - 34	SITE & INFRASTRUCTURE, DEMOLITION	104.7	103.6	103.9	91.8	103.5	99.9	97.2	106.6	103.7	93.3	106.7	102.5	98.6	97.2	97.6	93.7	103.7	100.6
0310	Concrete Forming & Accessories	107.0	103.3	103.7	90.2	96.4	95.6	102.7	95.4	96.4	102.8	97.2	97.9	105.4	120.2	118.3	98.0	103.1	102.4
0320	Concrete Reinforcing	101.8	101.1	101.5	99.7	90.6	95.1	103.4	90.4	96.9	102.0	90.8	96.4	102.0	101.7	101.8	97.1	101.6	99.3
0330	Cast-in-Place Concrete	112.7	97.4	106.7	93.2	91.0	92.3	102.9	96.5	100.4	94.6	97.2	95.6	101.1	113.3	105.9	107.9	81.0	97.3
03	CONCRETE	104.5	100.9	102.7	92.3	93.7	93.0	99.5	95.0	97.3	95.2	96.1	95.6	98.8	113.4	105.9	99.3	95.3	97.4
04	MASONRY	101.9	107.0	105.0	91.7	99.8	96.7	104.7	99.6	101.6	102.0	102.1	102.1	104.5	122.3	115.4	119.1	94.3	104.0
05	METALS	96.6	102.4	98.5	95.6	97.6	96.3	93.4	94.5	93.8	98.5	96.5	97.8	99.2	96.4	98.3	95.8	101.9	97.8
06	WOOD, PLASTICS & COMPOSITES	102.0	102.5	102.3	95.1	97.1	96.2	100.9	97.0	98.6	98.8	97.0	97.7	103.3	120.5	113.2	97.7	109.0	104.2
07	THERMAL & MOISTURE PROTECTION	105.4	96.3	101.7	105.9	84.5	97.2	105.6	79.8	95.1	102.4	90.8	97.7	104.3	115.0	108.3	107.8	92.1	101.4
08	OPENINGS	97.5	105.6	99.5	99.6	83.8	95.6	97.2	84.6	94.0	102.1	97.5	100.9	104.3	115.3	107.1	87.7	102.4	91.4
0920	Plaster & Gypsum Board	88.1	103.0	98.8	100.1	97.3	98.1	99.0	97.3	97.8	102.7	97.3	98.8	101.7	121.4	115.7	87.2	109.8	103.3
0950, 0980	Ceilings & Acoustic Treatment	87.5	103.0	97.7	99.7	97.3	98.1	82.9	97.3	92.4	92.2	97.3	95.5	92.0	121.4	111.4	67.0	109.8	95.3
0960	Flooring	123.0	112.2	119.8	89.2	113.3	96.2	103.7	106.2	104.5	98.6	106.2	100.8	107.5	119.4	111.0	108.5	113.3	109.9
0970, 0990	Wall Finishes & Painting/Coating	100.5	103.6	102.4	88.4	72.5	78.6	91.9	70.0	78.4	91.9	97.4	95.3	93.9	117.5	108.5	101.7	91.1	95.2
09	FINISHES	105.2	104.5	105.3	96.1	97.7	97.0	98.9	94.4	96.4	99.4	99.2	99.3	102.5	120.6	112.6	94.5	103.7	99.6
COVERS	DIVS. 10 - 14, 25, 28, 41, 43, 44, 46	100.0	99.3	99.9	100.0	92.4	98.5	100.0	51.0	90.3	100.0	92.2	98.5	100.0	104.2	100.8	100.0	89.5	97.9
21, 22, 23	FIRE SUPPRESSION, PLUMBING & HVAC	100.1	94.8	97.9	100.2	85.5	94.3	93.9	85.3	90.4	99.8	94.6	97.7	99.9	103.7	101.4	93.7	82.2	89.1
26, 27, 3370	ELECTRICAL, COMMUNICATIONS & UTIL.	101.7	99.4	100.5	104.0	90.4	97.0	100.9	90.4	95.5	99.4	94.1	96.7	102.6	100.1	101.3	102.0	90.4	96.0
MF2010	WEIGHTED AVERAGE	100.8	101.0	100.9	98.1	93.2	95.9	97.6	91.5	94.9	99.3	97.2	98.4	100.9	108.3	104.2	97.3	94.4	96.0

WISCONSIN

DIVISION		OSHKOSH 549 MAT.	INST.	TOTAL	PORTAGE 539 MAT.	INST.	TOTAL	RACINE 534 MAT.	INST.	TOTAL	RHINELANDER 545 MAT.	INST.	TOTAL	SUPERIOR 548 MAT.	INST.	TOTAL	WAUSAU 544 MAT.	INST.	TOTAL
015433	CONTRACTOR EQUIPMENT		98.7	98.7		100.9	100.9		100.9	100.9		98.7	98.7		101.4	101.4		98.7	98.7
0241, 31 - 34	SITE & INFRASTRUCTURE, DEMOLITION	93.9	98.5	97.0	87.1	106.6	100.6	97.6	107.3	104.3	107.9	99.6	102.2	90.2	104.1	99.7	89.3	99.5	96.3
0310	Concrete Forming & Accessories	92.2	84.6	85.6	96.3	96.3	96.3	103.6	103.3	103.3	89.9	95.5	94.8	96.5	92.7	93.2	91.6	95.6	95.1
0320	Concrete Reinforcing	98.2	77.7	87.9	103.5	90.5	97.0	102.0	101.1	101.6	98.4	92.7	95.5	97.1	93.7	95.4	98.4	90.6	94.4
0330	Cast-in-Place Concrete	99.4	88.3	95.0	88.3	98.7	92.4	101.5	97.2	99.8	112.7	98.7	107.2	101.6	96.9	99.8	92.6	98.0	94.7
03	CONCRETE	94.0	85.1	89.7	87.6	96.2	91.8	98.7	100.8	99.7	106.2	96.3	101.4	93.8	94.7	94.2	89.0	95.8	92.3
04	MASONRY	105.6	87.9	94.8	103.6	99.5	101.1	104.6	107.0	106.0	122.5	99.1	108.2	118.4	104.0	109.6	105.1	97.2	100.3
05	METALS	96.0	89.8	94.0	94.1	96.4	94.5	97.5	102.4	99.1	95.9	97.9	96.5	96.7	99.7	97.7	95.8	97.8	96.4
06	WOOD, PLASTICS & COMPOSITES	91.8	86.2	88.5	93.0	97.0	95.3	101.7	102.5	102.2	89.6	97.1	93.9	96.2	91.2	93.3	91.3	97.1	94.6
07	THERMAL & MOISTURE PROTECTION	106.9	78.3	95.3	105.1	89.3	98.7	105.8	95.9	101.8	107.7	83.5	97.9	107.5	95.1	102.5	106.8	84.2	97.6
08	OPENINGS	93.2	83.2	90.7	97.4	91.6	95.9	102.1	105.6	103.0	93.2	86.8	91.6	87.5	92.3	88.7	93.4	95.2	93.9
0920	Plaster & Gypsum Board	85.7	86.0	85.9	93.6	97.3	96.2	100.3	103.0	102.2	85.7	97.3	94.0	87.0	91.5	90.2	85.7	97.3	94.0
0950, 0980	Ceilings & Acoustic Treatment	93.4	86.0	88.5	84.7	97.3	93.0	87.5	103.0	97.7	93.4	97.3	96.0	67.9	91.5	83.5	93.4	97.3	96.0
0960	Flooring	106.5	99.7	104.5	99.6	112.7	103.5	104.5	112.2	106.7	105.6	113.3	107.9	110.0	122.5	113.6	106.3	113.3	108.4
0970, 0990	Wall Finishes & Painting/Coating	96.0	68.3	78.9	91.9	70.0	78.4	91.9	108.7	102.2	96.0	63.6	76.0	90.2	104.0	96.5	96.0	80.3	86.3
09	FINISHES	99.1	85.2	91.4	96.7	96.4	96.5	100.3	105.9	103.4	100.1	95.5	97.5	94.0	98.9	96.7	98.7	97.9	98.3
COVERS	DIVS. 10 - 14, 25, 28, 41, 43, 44, 46	100.0	64.1	92.8	100.0	65.9	93.2	100.0	99.3	99.9	100.0	56.9	91.4	100.0	89.2	97.8	100.0	91.5	98.3
21, 22, 23	FIRE SUPPRESSION, PLUMBING & HVAC	94.5	74.6	86.4	93.9	90.2	92.4	99.9	94.8	97.8	94.5	84.8	90.5	93.7	92.3	93.1	94.5	83.6	90.0
26, 27, 3370	ELECTRICAL, COMMUNICATIONS & UTIL.	102.6	76.2	89.0	104.8	93.7	99.0	100.9	97.3	99.1	101.9	81.4	91.3	107.0	96.9	101.8	103.7	81.4	92.2
MF2010	WEIGHTED AVERAGE	97.2	82.6	90.8	96.1	94.5	95.4	100.1	101.0	100.5	99.8	90.5	95.7	97.1	96.9	97.0	96.5	91.7	94.4

City Cost Indexes

WYOMING

DIVISION		CASPER 826			CHEYENNE 820			NEWCASTLE 827			RAWLINS 823			RIVERTON 825			ROCK SPRINGS 829-831		
		MAT.	INST.	TOTAL	MAT.	INST.	TOTAL	MAT.	INST.	TOTAL	MAT.	INST.	TOTAL	MAT.	INST.	TOTAL	MAT.	INST.	TOTAL
015433	CONTRACTOR EQUIPMENT		99.3	99.3		99.3	99.3		99.3	99.3		99.3	99.3		99.3	99.3		99.3	99.3
0241, 31 - 34	SITE & INFRASTRUCTURE, DEMOLITION	96.2	97.4	97.0	88.5	97.4	94.7	79.6	96.4	91.2	94.4	96.4	95.8	87.6	96.4	93.6	82.9	96.4	92.2
0310	Concrete Forming & Accessories	103.0	41.3	49.3	104.6	56.0	62.3	96.0	43.8	50.6	99.8	43.8	51.1	95.0	43.8	50.5	101.4	43.8	51.3
0320	Concrete Reinforcing	106.8	64.1	85.3	100.8	64.9	82.7	108.3	64.7	86.3	107.9	64.7	86.1	108.9	64.7	86.6	108.9	66.3	87.4
0330	Cast-in-Place Concrete	92.9	74.7	85.7	91.3	74.9	84.8	92.2	51.9	76.3	92.3	51.9	76.4	92.3	51.9	76.4	92.2	51.9	76.3
03	CONCRETE	99.6	58.0	79.3	98.3	64.7	81.9	99.0	51.3	75.8	111.4	51.3	82.2	106.5	51.3	79.6	99.5	51.6	76.2
04	MASONRY	101.3	35.8	61.4	98.1	49.4	68.4	94.8	44.3	64.0	94.8	44.3	64.0	94.8	44.3	64.0	152.6	44.6	86.5
05	METALS	97.6	68.4	87.9	98.8	69.8	89.2	93.9	67.7	85.2	93.9	67.7	85.2	94.0	67.7	85.3	94.7	68.3	86.0
06	WOOD, PLASTICS & COMPOSITES	91.8	38.4	60.9	93.5	57.8	72.9	84.3	44.8	61.5	87.9	44.8	63.0	83.3	44.8	61.0	91.2	44.8	64.4
07	THERMAL & MOISTURE PROTECTION	97.8	47.5	77.3	99.5	53.1	80.6	100.8	46.6	78.8	102.5	46.6	79.8	101.9	46.6	79.4	100.9	46.6	78.8
08	OPENINGS	96.3	43.2	83.0	98.6	53.8	87.3	102.8	46.7	88.7	102.5	46.7	88.4	102.7	46.7	88.6	103.2	47.1	89.1
0920	Plaster & Gypsum Board	100.7	36.4	54.8	89.0	56.4	65.7	83.7	42.9	54.6	84.1	42.9	54.7	83.7	42.9	54.6	92.8	42.9	57.2
0950, 0980	Ceilings & Acoustic Treatment	114.2	36.4	62.8	106.7	56.4	73.5	108.8	42.9	65.3	108.8	42.9	65.3	108.8	42.9	65.3	108.8	42.9	65.3
0960	Flooring	99.6	34.6	80.5	104.1	55.5	89.8	97.6	48.8	83.3	100.6	48.8	85.4	97.0	48.8	82.8	102.9	48.8	87.1
0970, 0990	Wall Finishes & Painting/Coating	98.8	45.4	65.9	102.3	45.4	67.2	98.1	29.1	55.5	98.1	29.1	55.5	98.1	29.1	55.5	98.1	29.1	55.5
09	FINISHES	105.1	39.1	68.3	104.1	54.8	76.6	98.5	42.4	67.2	101.2	42.4	68.4	99.4	42.4	67.6	101.1	42.4	68.4
COVERS	DIVS. 10 - 14, 25, 28, 41, 43, 44, 46	100.0	74.3	94.9	100.0	76.5	95.3	100.0	57.2	91.5	100.0	57.2	91.5	100.0	57.2	91.5	100.0	57.2	91.5
21, 22, 23	FIRE SUPPRESSION, PLUMBING & HVAC	100.0	55.0	81.8	100.0	55.0	81.8	97.0	52.6	79.0	97.0	52.6	79.0	97.0	52.6	79.0	100.0	52.7	80.8
26, 27, 3370	ELECTRICAL, COMMUNICATIONS & UTIL.	94.9	66.9	80.5	97.6	72.9	84.9	95.1	66.8	80.5	95.1	66.8	80.5	95.1	66.9	80.5	93.4	66.9	79.7
MF2010	WEIGHTED AVERAGE	98.9	57.7	80.7	99.1	63.8	83.5	96.9	57.0	79.3	99.0	57.0	80.5	98.1	57.0	80.0	101.0	57.2	81.7

WYOMING / CANADA

DIVISION		SHERIDAN 828			WHEATLAND 822			WORLAND 824			YELLOWSTONE NAT'L PA 821			BARRIE, ONTARIO			BATHURST, NEW BRUNSWICK		
		MAT.	INST.	TOTAL	MAT.	INST.	TOTAL	MAT.	INST.	TOTAL	MAT.	INST.	TOTAL	MAT.	INST.	TOTAL	MAT.	INST.	TOTAL
015433	CONTRACTOR EQUIPMENT		99.3	99.3		99.3	99.3		99.3	99.3		99.3	99.3		101.7	101.7		102.0	102.0
0241, 31 - 34	SITE & INFRASTRUCTURE, DEMOLITION	88.1	97.4	94.5	83.8	96.4	92.5	81.2	96.4	91.7	81.3	96.7	91.9	114.9	101.4	105.6	94.6	97.7	96.7
0310	Concrete Forming & Accessories	102.6	46.2	53.5	97.9	44.2	51.2	97.9	43.8	50.9	97.9	46.1	52.9	123.2	93.6	97.5	103.1	66.6	71.4
0320	Concrete Reinforcing	108.9	64.8	86.7	108.3	64.7	86.3	108.9	64.7	86.6	110.8	69.6	90.0	180.2	87.6	133.5	148.8	59.1	103.6
0330	Cast-in-Place Concrete	95.4	74.8	87.3	96.2	52.4	79.0	92.2	51.9	76.3	92.2	55.2	77.6	167.6	92.3	137.9	138.2	64.1	109.0
03	CONCRETE	106.6	60.3	84.1	103.7	51.7	78.3	99.2	51.3	75.9	99.5	54.4	77.5	153.4	92.1	123.5	135.8	65.0	101.3
04	MASONRY	95.1	48.8	66.9	95.1	45.2	64.7	94.8	44.3	64.0	94.9	50.3	67.7	167.0	102.0	127.3	158.0	67.6	102.8
05	METALS	97.5	69.6	88.3	93.8	67.7	85.2	94.0	67.7	85.3	94.6	69.5	86.3	107.3	92.6	102.5	104.7	73.5	94.4
06	WOOD, PLASTICS & COMPOSITES	94.7	44.8	65.9	85.9	44.8	62.2	86.0	44.8	62.2	86.0	44.8	62.2	117.4	92.2	102.8	94.7	66.9	78.6
07	THERMAL & MOISTURE PROTECTION	101.9	51.5	81.4	101.1	47.5	79.3	100.9	46.6	78.8	100.4	49.7	79.8	111.6	93.1	104.0	105.0	65.1	88.8
08	OPENINGS	103.5	47.6	89.4	101.4	46.7	87.6	103.0	46.7	88.9	95.9	47.9	83.8	92.4	90.3	91.8	87.4	58.3	80.1
0920	Plaster & Gypsum Board	111.2	42.9	62.5	83.7	42.9	54.6	83.7	42.9	54.6	84.0	42.9	54.7	155.0	91.9	109.9	162.1	65.7	93.3
0950, 0980	Ceilings & Acoustic Treatment	111.8	42.9	66.3	108.8	42.9	65.3	108.8	42.9	65.3	109.7	42.9	65.6	95.8	91.9	93.2	96.7	65.7	76.2
0960	Flooring	100.8	48.8	85.5	99.2	48.8	84.4	99.2	48.8	84.4	99.2	48.8	84.4	132.5	101.8	123.5	115.8	46.9	96.1
0970, 0990	Wall Finishes & Painting/Coating	100.4	45.4	66.5	98.1	29.1	55.5	98.1	29.1	55.5	98.1	29.1	55.5	112.8	93.1	100.7	112.0	53.5	75.9
09	FINISHES	106.5	45.4	72.4	99.4	42.6	67.7	99.0	42.4	67.4	99.2	43.9	68.3	118.4	95.4	105.5	112.6	62.1	84.5
COVERS	DIVS. 10 - 14, 25, 28, 41, 43, 44, 46	100.0	75.0	95.0	100.0	57.5	91.5	100.0	57.2	91.5	100.0	59.2	91.9	140.0	79.2	127.9	140.5	69.5	126.4
21, 22, 23	FIRE SUPPRESSION, PLUMBING & HVAC	97.0	55.0	80.0	97.0	53.1	79.2	97.0	52.6	79.0	97.0	55.7	80.3	103.0	104.9	103.8	103.2	73.7	91.3
26, 27, 3370	ELECTRICAL, COMMUNICATIONS & UTIL.	97.8	66.9	81.9	95.1	72.8	83.6	95.1	66.9	80.5	94.1	66.9	80.1	119.3	94.5	106.5	119.9	64.6	91.4
MF2010	WEIGHTED AVERAGE	99.7	60.6	82.4	97.5	58.2	80.2	97.1	57.0	79.4	96.4	59.3	80.1	118.6	96.9	109.1	114.1	70.1	94.7

CANADA

DIVISION		BRANDON, MANITOBA			BRANTFORD, ONTARIO			BRIDGEWATER, NOVA SCOTIA			CALGARY, ALBERTA			CAP-DE-LA-MADELEINE, QUEBEC			CHARLESBOURG, QUEBEC		
		MAT.	INST.	TOTAL	MAT.	INST.	TOTAL	MAT.	INST.	TOTAL	MAT.	INST.	TOTAL	MAT.	INST.	TOTAL	MAT.	INST.	TOTAL
015433	CONTRACTOR EQUIPMENT		103.9	103.9		101.7	101.7		101.7	101.7		106.4	106.4		102.6	102.6		102.6	102.6
0241, 31 - 34	SITE & INFRASTRUCTURE, DEMOLITION	110.7	100.3	103.5	112.7	101.7	105.1	97.0	99.3	98.5	125.9	105.2	111.6	92.7	100.9	98.4	92.7	100.9	98.4
0310	Concrete Forming & Accessories	122.5	75.8	81.9	123.0	101.4	104.2	97.4	75.0	77.9	127.5	96.2	100.3	128.1	92.1	96.8	128.1	92.1	96.8
0320	Concrete Reinforcing	161.8	55.6	108.3	175.1	86.3	130.3	148.8	48.8	98.4	143.8	72.5	107.9	148.8	82.4	115.3	148.8	82.4	115.3
0330	Cast-in-Place Concrete	151.1	79.4	122.8	163.5	114.5	144.2	168.1	74.8	131.3	201.8	106.1	164.1	131.9	101.5	119.9	131.9	101.5	119.9
03	CONCRETE	135.9	73.9	105.7	150.5	103.0	127.4	148.4	70.7	110.5	167.2	95.3	132.2	130.9	93.6	112.7	130.9	93.6	112.7
04	MASONRY	161.7	70.4	106.0	163.3	106.7	128.8	158.8	74.8	107.6	212.7	90.7	138.3	159.0	90.6	117.3	159.0	90.6	117.3
05	METALS	107.5	78.8	98.0	106.7	93.4	102.3	105.8	76.3	96.1	143.7	89.0	125.6	105.0	90.0	100.1	105.0	90.0	100.1
06	WOOD, PLASTICS & COMPOSITES	112.9	76.8	92.0	115.2	100.1	106.5	87.4	74.3	79.8	113.5	96.0	103.4	126.1	92.0	106.4	126.1	92.0	106.4
07	THERMAL & MOISTURE PROTECTION	105.2	75.0	93.0	111.5	98.9	106.3	106.5	73.1	92.9	105.4	94.0	100.5	105.4	93.9	100.7	105.4	93.9	100.7
08	OPENINGS	93.4	67.6	86.9	92.2	96.7	93.4	83.9	67.7	79.9	93.4	85.7	91.4	93.4	84.6	91.2	93.4	84.6	91.2
0920	Plaster & Gypsum Board	130.0	75.6	91.2	155.0	100.1	115.8	155.9	73.4	96.9	171.9	95.4	117.2	185.9	91.5	118.5	185.9	91.5	118.5
0950, 0980	Ceilings & Acoustic Treatment	95.8	75.6	82.5	95.8	100.1	98.6	95.8	73.4	81.0	151.3	95.4	114.4	95.8	91.5	93.0	95.8	91.5	93.0
0960	Flooring	132.5	72.3	114.8	132.5	101.8	123.5	112.1	69.3	99.5	134.6	95.4	123.1	132.5	102.5	123.7	132.5	102.5	123.7
0970, 0990	Wall Finishes & Painting/Coating	112.0	60.8	80.5	112.0	102.2	106.0	112.0	66.5	83.9	111.9	109.7	110.5	112.0	95.0	101.5	112.0	95.0	101.5
09	FINISHES	114.5	74.0	91.9	117.1	102.0	108.7	110.3	73.2	89.6	133.8	98.0	113.8	119.7	94.6	105.7	119.7	94.6	105.7
COVERS	DIVS. 10 - 14, 25, 28, 41, 43, 44, 46	140.5	71.8	126.9	140.5	81.5	128.8	140.5	70.8	126.7	140.5	101.4	132.8	140.5	90.7	130.6	140.5	90.7	130.6
21, 22, 23	FIRE SUPPRESSION, PLUMBING & HVAC	103.2	88.8	97.4	103.2	108.0	105.2	103.2	89.2	97.5	100.5	92.6	97.3	103.8	96.1	100.7	103.8	96.1	100.7
26, 27, 3370	ELECTRICAL, COMMUNICATIONS & UTIL.	122.0	72.8	96.6	117.2	94.0	105.2	123.2	67.3	94.4	123.5	84.2	103.2	116.9	75.9	95.8	116.9	75.9	95.8
MF2010	WEIGHTED AVERAGE	116.2	79.0	99.8	117.7	101.0	110.4	115.6	77.8	98.9	129.8	93.1	113.6	114.8	91.3	104.4	114.8	91.3	104.4

City Cost Indexes

CANADA

DIVISION		CHARLOTTETOWN, PRINCE EDWARD ISLAND			CHICOUTIMI, QUEBEC			CORNER BROOK, NEWFOUNDLAND			CORNWALL, ONTARIO			DALHOUSIE, NEW BRUNSWICK			DARTMOUTH, NOVA SCOTIA		
		MAT.	INST.	TOTAL	MAT.	INST.	TOTAL	MAT.	INST.	TOTAL	MAT.	INST.	TOTAL	MAT.	INST.	TOTAL	MAT.	INST.	TOTAL
015433	CONTRACTOR EQUIPMENT		101.6	101.6		102.6	102.6		103.3	103.3		101.7	101.7		102.0	102.0		101.7	101.7
0241, 31 - 34	SITE & INFRASTRUCTURE, DEMOLITION	108.2	96.8	100.3	92.2	99.7	97.4	112.8	98.3	102.8	110.8	101.2	104.2	94.5	97.7	96.7	104.7	99.3	100.9
0310	Concrete Forming & Accessories	97.8	60.9	65.7	128.1	89.7	94.7	104.5	64.5	69.7	121.0	93.8	97.3	103.1	66.9	71.6	97.5	75.0	77.9
0320	Concrete Reinforcing	155.0	48.9	101.5	112.2	92.1	102.1	148.8	50.8	99.4	175.1	86.0	130.2	148.8	59.1	103.6	155.2	48.8	101.6
0330	Cast-in-Place Concrete	160.5	63.3	122.2	130.1	96.8	117.0	172.3	73.2	133.2	147.0	104.4	130.2	138.2	64.2	109.1	169.8	74.8	132.4
03	CONCRETE	153.8	60.3	108.3	124.3	92.7	108.9	162.3	65.9	115.3	142.1	96.0	119.7	135.8	65.1	101.4	150.3	70.7	111.5
04	MASONRY	164.1	65.2	103.7	159.5	90.6	117.5	159.2	66.7	102.8	162.0	97.8	122.8	155.8	67.6	101.9	170.6	74.8	112.2
05	METALS	111.6	69.4	97.7	104.5	91.1	100.1	107.9	75.0	97.1	106.7	92.0	101.8	104.7	73.8	94.5	108.0	76.3	97.5
06	WOOD, PLASTICS & COMPOSITES	87.7	60.4	71.9	126.1	89.8	105.1	95.7	63.5	77.1	114.1	93.1	102.0	94.7	66.9	78.6	87.4	74.3	79.8
07	THERMAL & MOISTURE PROTECTION	106.6	62.5	88.6	105.4	93.0	100.3	108.6	64.7	90.7	111.3	93.2	103.9	108.9	65.1	91.1	106.5	73.1	92.9
08	OPENINGS	95.7	52.1	84.7	92.9	77.9	89.1	99.7	59.6	89.6	93.4	90.0	92.5	87.4	58.3	80.1	83.9	67.7	79.9
0920	Plaster & Gypsum Board	157.7	59.1	87.2	185.9	89.3	116.9	162.1	62.2	90.7	224.8	92.8	130.5	162.1	65.7	93.3	158.9	73.4	97.8
0950, 0980	Ceilings & Acoustic Treatment	110.3	59.1	76.5	95.8	89.3	91.5	96.7	62.2	73.9	98.5	92.8	94.8	96.7	65.7	76.2	105.8	73.4	84.4
0960	Flooring	112.2	64.8	98.2	132.5	102.5	123.7	116.5	59.0	99.6	132.5	100.3	123.1	115.8	74.3	103.6	112.1	69.3	99.5
0970, 0990	Wall Finishes & Painting/Coating	112.0	44.5	70.4	112.0	101.1	105.3	112.0	64.0	82.4	112.0	95.3	101.7	112.0	53.5	75.9	112.0	66.5	83.9
09	FINISHES	115.2	60.2	84.5	119.7	93.7	105.2	114.0	63.5	85.8	125.8	95.4	108.8	112.6	67.2	87.3	112.9	73.2	90.7
COVERS	DIVS. 10 - 14, 25, 28, 41, 43, 44, 46	140.5	68.9	126.3	140.5	90.1	130.5	140.5	69.8	126.5	140.5	78.8	128.2	140.5	69.4	126.4	140.5	70.8	126.7
21, 22, 23	FIRE SUPPRESSION, PLUMBING & HVAC	103.3	67.4	88.8	103.2	95.6	100.1	103.2	75.9	92.1	103.8	105.8	104.6	103.2	73.7	91.3	103.2	89.2	97.5
26, 27, 3370	ELECTRICAL, COMMUNICATIONS & UTIL.	113.6	54.7	83.2	116.5	83.0	99.2	119.1	60.9	89.1	118.3	95.0	106.3	120.0	61.1	89.6	123.7	67.3	94.6
MF2010	WEIGHTED AVERAGE	118.2	65.4	95.0	113.7	91.6	104.0	119.5	70.5	97.9	117.7	97.2	108.7	114.1	70.3	94.8	117.2	77.8	99.8

CANADA

DIVISION		EDMONTON, ALBERTA			FORT MCMURRAY, ALBERTA			FREDERICTON, NEW BRUNSWICK			GATINEAU, QUEBEC			GRANBY, QUEBEC			HALIFAX, NOVA SCOTIA		
		MAT.	INST.	TOTAL	MAT.	INST.	TOTAL	MAT.	INST.	TOTAL	MAT.	INST.	TOTAL	MAT.	INST.	TOTAL	MAT.	INST.	TOTAL
015433	CONTRACTOR EQUIPMENT		106.4	106.4		103.7	103.7		102.0	102.0		102.6	102.6		102.6	102.6		101.7	101.7
0241, 31 - 34	SITE & INFRASTRUCTURE, DEMOLITION	151.7	105.2	119.6	118.6	102.2	107.3	96.4	97.7	97.3	92.5	100.9	98.3	93.0	100.9	98.5	99.9	99.0	99.3
0310	Concrete Forming & Accessories	129.7	96.2	100.6	121.2	96.4	99.6	120.7	67.2	74.1	128.1	92.0	96.7	128.1	91.9	96.6	97.8	77.3	80.0
0320	Concrete Reinforcing	143.8	72.5	107.9	161.8	72.5	116.8	145.2	59.3	101.9	157.4	82.4	119.6	157.4	82.4	119.6	153.0	61.2	106.7
0330	Cast-in-Place Concrete	210.4	106.1	169.3	217.2	105.3	173.1	132.5	64.3	105.6	130.1	101.4	118.8	134.5	101.4	121.4	172.4	80.6	136.2
03	CONCRETE	171.7	94.3	134.5	175.3	95.1	136.2	133.6	65.3	100.4	131.3	93.5	112.9	133.5	93.5	114.0	151.2	76.1	114.6
04	MASONRY	200.5	90.7	133.5	202.4	90.7	134.2	175.9	69.2	110.8	158.9	90.6	117.2	159.2	90.6	117.3	171.4	82.4	117.1
05	METALS	144.2	89.0	126.0	132.2	89.1	118.0	110.3	74.3	98.4	105.0	89.9	100.0	105.0	89.8	100.0	126.5	81.6	111.7
06	WOOD, PLASTICS & COMPOSITES	110.0	96.0	101.9	109.8	96.0	101.8	115.9	66.9	87.6	126.1	92.0	106.4	126.1	92.0	106.4	87.5	76.2	81.0
07	THERMAL & MOISTURE PROTECTION	125.1	94.0	112.5	118.9	93.5	108.5	109.8	66.1	92.0	105.4	93.9	100.7	105.4	92.3	100.0	107.3	77.1	95.0
08	OPENINGS	93.4	85.7	91.4	93.4	85.7	91.4	87.3	57.1	79.7	93.4	79.7	89.9	93.4	79.7	89.9	83.9	71.4	80.8
0920	Plaster & Gypsum Board	166.3	95.4	115.6	154.6	95.5	112.4	177.8	65.7	97.7	147.1	91.5	107.4	150.1	91.5	108.3	158.0	75.4	99.0
0950, 0980	Ceilings & Acoustic Treatment	158.0	95.4	116.6	105.8	95.5	99.0	107.6	65.7	79.9	95.8	91.5	93.0	95.8	91.5	93.0	111.2	75.4	87.5
0960	Flooring	135.8	95.4	123.9	132.5	95.4	121.6	128.0	77.5	113.2	132.5	102.5	123.7	132.5	102.5	123.7	112.1	91.0	105.9
0970, 0990	Wall Finishes & Painting/Coating	112.0	109.7	110.6	112.0	99.0	104.0	112.0	68.2	84.9	112.0	95.0	101.5	112.0	95.0	101.5	112.0	79.8	92.1
09	FINISHES	137.0	98.0	115.2	120.8	96.8	107.4	120.8	69.4	92.2	115.2	94.6	103.7	115.5	94.6	103.9	114.2	80.4	95.4
COVERS	DIVS. 10 - 14, 25, 28, 41, 43, 44, 46	140.5	101.4	132.8	140.5	101.3	132.7	140.5	69.5	126.4	140.5	90.7	130.6	140.5	90.7	130.6	140.5	71.5	126.8
21, 22, 23	FIRE SUPPRESSION, PLUMBING & HVAC	100.6	92.6	97.4	103.5	104.4	103.9	103.3	83.8	95.4	103.8	96.1	100.7	103.2	96.1	100.4	99.8	78.0	91.0
26, 27, 3370	ELECTRICAL, COMMUNICATIONS & UTIL.	115.6	84.2	99.4	111.0	89.3	99.8	124.5	80.5	101.8	116.9	75.9	95.8	117.6	75.9	96.1	120.9	79.8	99.7
MF2010	WEIGHTED AVERAGE	130.0	93.1	113.7	126.5	92.8	113.0	117.0	75.6	98.8	114.4	91.1	104.2	114.7	91.0	104.3	119.0	80.3	101.9

CANADA

DIVISION		HAMILTON, ONTARIO			HULL, QUEBEC			JOLIETTE, QUEBEC			KAMLOOPS, BRITISH COLUMBIA			KINGSTON, ONTARIO			KITCHENER, ONTARIO		
		MAT.	INST.	TOTAL	MAT.	INST.	TOTAL	MAT.	INST.	TOTAL	MAT.	INST.	TOTAL	MAT.	INST.	TOTAL	MAT.	INST.	TOTAL
015433	CONTRACTOR EQUIPMENT		108.4	108.4		102.6	102.6		102.6	102.6		105.7	105.7		104.0	104.0		103.9	103.9
0241, 31 - 34	SITE & INFRASTRUCTURE, DEMOLITION	116.3	112.8	113.9	92.5	100.9	98.3	93.2	100.9	98.5	117.7	104.3	108.5	110.8	104.9	106.7	100.3	105.1	103.6
0310	Concrete Forming & Accessories	130.8	94.3	99.0	128.1	92.0	96.7	128.1	92.1	96.8	123.5	95.0	98.7	121.2	93.8	97.4	117.6	85.6	89.8
0320	Concrete Reinforcing	162.7	85.9	124.0	157.4	82.4	119.6	148.8	82.4	115.3	116.8	80.3	98.4	175.1	86.0	130.2	110.5	85.8	98.1
0330	Cast-in-Place Concrete	153.0	95.8	130.4	130.1	101.4	118.8	135.6	101.5	122.1	117.3	105.3	112.5	147.0	104.4	130.2	140.8	86.9	119.5
03	CONCRETE	143.8	93.3	119.2	131.3	93.5	112.9	132.7	93.6	113.7	137.7	95.8	117.3	142.1	96.1	119.7	122.6	86.4	105.0
04	MASONRY	179.1	100.4	131.1	158.9	90.6	117.2	159.3	90.6	117.4	165.6	98.8	124.8	168.6	97.9	125.4	159.8	62.3	118.8
05	METALS	125.5	92.3	114.6	105.0	89.9	100.0	105.0	90.0	100.1	107.3	89.9	101.6	108.4	91.9	102.9	115.4	92.1	107.7
06	WOOD, PLASTICS & COMPOSITES	114.5	93.4	102.3	126.1	92.0	106.4	126.1	92.0	106.4	98.3	93.4	95.5	114.1	93.2	102.0	108.7	84.9	94.9
07	THERMAL & MOISTURE PROTECTION	122.5	95.7	111.6	105.4	93.9	100.7	105.4	93.9	100.7	120.7	90.7	108.5	111.3	94.4	104.4	110.8	90.8	102.7
08	OPENINGS	93.4	91.9	93.0	93.4	79.7	89.9	93.4	84.6	91.2	89.1	89.4	89.2	93.4	89.6	92.4	84.5	85.5	84.7
0920	Plaster & Gypsum Board	191.9	93.2	121.4	147.1	91.5	107.4	185.9	91.5	118.5	136.1	92.7	105.1	229.1	93.0	131.9	153.0	84.4	104.0
0950, 0980	Ceilings & Acoustic Treatment	130.9	93.2	106.0	95.8	91.5	93.0	95.8	91.5	93.0	95.8	92.7	93.8	113.0	93.0	99.8	105.8	84.4	91.6
0960	Flooring	134.6	105.3	126.0	132.5	102.5	123.7	132.5	102.5	123.7	130.3	57.3	108.8	132.5	100.3	123.1	128.5	105.3	121.7
0970, 0990	Wall Finishes & Painting/Coating	112.0	95.3	105.2	112.0	95.0	101.5	112.0	95.0	101.5	112.0	87.0	96.6	112.0	88.0	97.2	112.0	91.1	99.1
09	FINISHES	130.9	97.2	112.1	115.2	94.6	103.7	119.7	94.6	105.7	116.6	87.7	100.5	129.5	94.7	110.1	116.9	89.5	101.6
COVERS	DIVS. 10 - 14, 25, 28, 41, 43, 44, 46	140.5	104.0	133.3	140.5	90.7	130.6	140.5	90.7	130.6	140.5	101.7	132.8	140.5	78.8	128.2	140.5	100.9	132.7
21, 22, 23	FIRE SUPPRESSION, PLUMBING & HVAC	100.7	84.0	93.9	103.2	96.1	100.4	103.2	96.1	100.4	103.2	99.7	101.8	103.8	105.9	104.7	99.8	89.1	95.4
26, 27, 3370	ELECTRICAL, COMMUNICATIONS & UTIL.	119.7	87.5	103.1	119.0	75.9	96.7	117.6	75.9	96.1	121.3	85.6	102.9	118.3	93.6	105.6	117.6	85.1	100.8
MF2010	WEIGHTED AVERAGE	121.9	93.7	109.4	114.5	91.1	104.2	114.9	91.3	104.5	116.9	94.4	107.0	118.6	97.3	109.2	113.7	90.4	103.4

CANADA

DIVISION		LAVAL, QUEBEC			LETHBRIDGE, ALBERTA			LLOYDMINSTER, ALBERTA			LONDON, ONTARIO			MEDICINE HAT, ALBERTA			MONCTON, NEW BRUNSWICK		
		MAT.	INST.	TOTAL	MAT.	INST.	TOTAL	MAT.	INST.	TOTAL	MAT.	INST.	TOTAL	MAT.	INST.	TOTAL	MAT.	INST.	TOTAL
015433	CONTRACTOR EQUIPMENT		102.6	102.6		103.7	103.7		103.7	103.7		104.0	104.0		103.7	103.7		102.0	102.0
0241, 31 - 34	SITE & INFRASTRUCTURE, DEMOLITION	93.0	100.9	98.5	110.9	102.8	105.3	110.9	102.2	104.9	115.1	105.3	108.4	109.5	102.3	104.5	94.0	97.8	96.7
0310	Concrete Forming & Accessories	128.2	92.0	96.7	122.4	96.5	99.9	120.7	86.2	90.7	130.9	88.2	93.8	122.3	86.2	90.9	103.1	67.1	71.8
0320	Concrete Reinforcing	157.4	82.4	119.6	161.8	72.5	116.8	161.8	72.5	116.8	142.3	84.7	113.2	161.8	72.5	116.8	148.8	62.3	105.2
0330	Cast-in-Place Concrete	134.5	101.4	121.4	162.9	105.3	140.2	151.1	101.5	131.5	164.1	93.7	136.3	151.1	101.5	131.5	133.1	64.5	106.1
03	CONCRETE	133.5	93.5	114.0	148.1	95.2	122.3	142.1	89.3	116.4	146.2	89.7	118.6	142.2	89.3	116.4	133.2	66.1	100.5
04	MASONRY	159.2	90.6	117.3	176.5	90.7	124.1	158.7	83.8	113.0	183.3	97.7	131.1	158.7	83.8	113.0	157.6	67.6	102.7
05	METALS	105.1	89.9	100.1	125.8	89.1	113.7	107.5	89.0	101.4	124.2	91.6	113.5	107.5	88.9	101.4	104.7	79.4	96.4
06	WOOD, PLASTICS & COMPOSITES	126.3	92.0	106.5	113.0	96.0	103.2	109.8	85.7	95.9	114.5	86.5	98.3	113.0	85.7	97.2	94.7	66.9	78.6
07	THERMAL & MOISTURE PROTECTION	105.9	93.9	101.0	116.4	93.5	107.1	113.6	88.8	103.5	123.3	93.4	111.1	119.6	88.8	107.1	108.9	66.6	91.7
08	OPENINGS	93.4	79.7	89.9	93.4	85.7	91.4	93.4	80.1	90.0	94.4	87.0	92.5	93.4	80.1	90.0	87.4	61.6	80.9
0920	Plaster & Gypsum Board	150.5	91.5	108.4	139.1	95.5	107.9	133.7	84.9	98.8	193.7	86.0	116.8	136.4	84.9	99.6	162.1	65.7	93.3
0950, 0980	Ceilings & Acoustic Treatment	95.8	91.5	93.0	104.9	95.5	98.7	95.8	84.9	88.6	137.2	86.0	103.4	95.8	84.9	88.6	96.7	65.7	76.2
0960	Flooring	132.5	102.5	123.7	132.5	95.4	121.6	132.5	95.4	121.6	134.8	105.3	126.2	132.5	95.4	121.6	115.8	74.3	103.6
0970, 0990	Wall Finishes & Painting/Coating	112.0	95.0	101.5	111.9	108.0	109.5	112.0	84.2	94.8	112.0	97.7	103.2	111.9	84.2	94.8	112.0	53.5	75.9
09	FINISHES	115.6	94.6	103.9	117.8	97.8	106.6	115.3	87.4	99.7	132.1	92.3	109.9	115.4	87.4	99.8	112.6	67.2	87.3
COVERS	DIVS. 10 - 14, 25, 28, 41, 43, 44, 46	140.5	90.7	130.6	140.5	101.3	132.7	140.5	97.8	132.0	140.5	102.6	133.0	140.5	97.8	132.0	140.5	69.5	126.4
21, 22, 23	FIRE SUPPRESSION, PLUMBING & HVAC	99.8	96.1	98.3	103.4	100.8	102.4	103.8	100.9	102.6	100.7	81.2	92.8	103.2	97.3	100.8	103.2	73.8	91.3
26, 27, 3370	ELECTRICAL, COMMUNICATIONS & UTIL.	118.9	75.9	96.7	112.5	89.3	100.5	109.9	89.3	99.3	112.9	85.1	98.6	109.9	89.3	99.3	124.6	64.6	93.7
MF2010	WEIGHTED AVERAGE	114.0	91.1	103.9	120.8	95.3	109.5	116.0	91.8	105.4	121.6	90.2	107.8	116.1	91.1	105.1	114.3	71.7	95.5

CANADA

DIVISION		MONTREAL, QUEBEC			MOOSE JAW, SASKATCHEWAN			NEW GLASGOW, NOVA SCOTIA			NEWCASTLE, NEW BRUNSWICK			NORTH BAY, ONTARIO			OSHAWA, ONTARIO		
		MAT.	INST.	TOTAL	MAT.	INST.	TOTAL	MAT.	INST.	TOTAL	MAT.	INST.	TOTAL	MAT.	INST.	TOTAL	MAT.	INST.	TOTAL
015433	CONTRACTOR EQUIPMENT		104.2	104.2		100.2	100.2		101.7	101.7		102.0	102.0		101.7	101.7		103.9	103.9
0241, 31 - 34	SITE & INFRASTRUCTURE, DEMOLITION	102.6	100.0	100.8	110.6	96.8	101.1	97.2	99.3	98.6	94.6	97.7	96.7	112.7	100.8	104.5	112.6	104.9	107.3
0310	Concrete Forming & Accessories	135.8	90.0	96.0	106.7	63.3	69.0	97.4	75.0	77.9	103.1	66.9	71.6	123.0	91.1	95.2	123.2	93.2	97.1
0320	Concrete Reinforcing	150.6	92.1	121.1	114.2	64.7	89.2	148.8	48.8	98.4	148.8	59.1	103.6	175.1	85.5	129.9	175.1	88.2	131.3
0330	Cast-in-Place Concrete	182.2	98.1	149.1	146.6	73.7	117.9	169.8	74.8	132.4	138.2	64.2	109.1	163.5	90.3	134.7	162.8	91.2	134.6
03	CONCRETE	156.9	93.2	125.9	125.4	67.8	97.4	149.3	70.7	111.0	135.8	65.2	101.4	150.5	89.9	121.0	150.2	91.7	121.7
04	MASONRY	166.1	90.6	120.1	157.3	67.0	102.2	159.0	74.8	107.6	158.0	67.6	102.8	163.3	93.5	120.7	162.9	98.2	123.4
05	METALS	124.2	91.6	113.4	103.9	75.9	94.7	105.8	76.3	96.1	104.7	73.8	94.5	106.7	91.6	101.7	106.8	93.8	102.5
06	WOOD, PLASTICS & COMPOSITES	126.3	90.0	105.3	96.2	61.8	76.3	87.4	74.3	79.8	94.7	66.9	78.6	115.2	91.6	101.6	115.4	92.2	102.0
07	THERMAL & MOISTURE PROTECTION	119.4	93.5	108.9	104.8	67.1	89.5	106.5	73.1	92.9	108.9	65.1	91.1	111.5	89.4	102.5	111.8	91.2	103.4
08	OPENINGS	93.4	79.8	89.9	88.2	59.2	80.9	83.9	67.7	79.9	87.4	58.3	80.1	92.2	87.4	91.0	92.2	91.7	92.1
0920	Plaster & Gypsum Board	171.6	89.3	112.8	131.9	60.5	80.9	155.9	73.4	96.9	162.1	65.7	93.3	155.0	91.3	109.5	156.8	91.9	110.4
0950, 0980	Ceilings & Acoustic Treatment	115.6	89.3	98.2	95.8	60.5	72.5	95.8	73.4	81.0	96.7	65.7	76.2	95.8	91.3	92.8	100.3	91.9	94.8
0960	Flooring	134.1	102.5	124.8	121.1	65.2	104.7	112.1	69.3	99.5	115.8	74.3	103.6	132.5	100.3	123.1	132.5	108.0	125.3
0970, 0990	Wall Finishes & Painting/Coating	112.0	101.1	105.3	112.0	69.2	85.6	112.0	66.5	83.9	112.0	53.5	75.9	112.0	94.5	101.2	112.0	104.9	107.6
09	FINISHES	123.5	93.9	107.0	111.5	64.0	85.0	110.3	73.2	89.6	112.6	67.2	87.3	117.1	93.6	104.0	118.2	97.2	106.5
COVERS	DIVS. 10 - 14, 25, 28, 41, 43, 44, 46	140.5	90.6	130.6	140.5	68.6	126.2	140.5	70.8	126.7	140.5	69.5	126.4	140.5	77.4	128.0	140.5	103.5	133.2
21, 22, 23	FIRE SUPPRESSION, PLUMBING & HVAC	100.7	95.6	98.6	103.8	80.4	94.3	103.2	89.2	97.5	103.2	73.7	91.3	103.2	103.6	103.4	99.8	106.3	102.4
26, 27, 3370	ELECTRICAL, COMMUNICATIONS & UTIL.	123.4	83.0	102.5	120.0	65.6	91.9	119.4	67.3	92.5	119.2	64.6	91.0	119.4	94.9	106.8	118.7	94.5	106.2
MF2010	WEIGHTED AVERAGE	121.9	91.9	108.7	113.4	72.4	95.3	115.3	77.8	98.8	114.1	70.8	95.0	118.0	94.9	107.8	117.1	98.2	108.7

CANADA

DIVISION		OTTAWA, ONTARIO			OWEN SOUND, ONTARIO			PETERBOROUGH, ONTARIO			PORTAGE LA PRAIRIE, MANITOBA			PRINCE ALBERT, SASKATCHEWAN			PRINCE GEORGE, BRITISH COLUMBIA		
		MAT.	INST.	TOTAL	MAT.	INST.	TOTAL	MAT.	INST.	TOTAL	MAT.	INST.	TOTAL	MAT.	INST.	TOTAL	MAT.	INST.	TOTAL
015433	CONTRACTOR EQUIPMENT		103.9	103.9		101.7	101.7		101.7	101.7		103.9	103.9		100.2	100.2		105.7	105.7
0241, 31 - 34	SITE & INFRASTRUCTURE, DEMOLITION	112.3	105.1	107.3	114.9	101.2	105.5	112.7	101.1	104.7	111.2	100.2	103.6	105.5	97.0	99.7	121.2	104.3	109.6
0310	Concrete Forming & Accessories	131.7	90.8	96.1	123.2	89.5	93.9	123.0	92.2	96.2	122.5	75.3	81.5	106.7	63.1	68.8	111.9	89.5	92.4
0320	Concrete Reinforcing	155.4	84.6	119.7	180.2	87.6	133.5	175.1	86.0	130.2	161.8	55.6	108.3	119.4	64.6	91.8	116.8	80.3	98.4
0330	Cast-in-Place Concrete	168.5	92.9	138.7	167.6	86.2	135.5	163.5	92.0	135.4	151.1	78.7	122.6	132.8	73.6	109.5	147.1	105.3	130.6
03	CONCRETE	150.5	90.5	121.3	153.4	88.2	121.6	150.5	91.1	121.6	135.9	73.5	105.5	119.4	67.7	94.2	151.9	93.4	123.4
04	MASONRY	180.1	97.6	129.8	167.0	99.5	125.8	163.3	100.6	125.0	161.7	69.3	105.3	156.3	67.0	101.8	167.9	98.8	125.8
05	METALS	129.3	91.5	116.8	107.3	92.4	102.4	106.7	92.1	101.9	107.5	78.7	98.0	104.0	75.7	94.6	107.3	89.9	101.6
06	WOOD, PLASTICS & COMPOSITES	114.4	90.0	100.3	117.4	88.1	100.5	115.2	90.2	100.8	112.9	76.8	92.0	96.2	61.8	76.3	98.3	85.8	91.1
07	THERMAL & MOISTURE PROTECTION	125.1	92.9	112.0	111.6	90.3	102.9	111.5	95.2	104.9	105.2	74.5	92.8	104.7	65.9	88.9	114.9	89.9	104.8
08	OPENINGS	93.4	88.9	92.2	92.4	86.7	90.9	92.2	89.2	91.5	93.4	67.6	86.9	86.7	59.2	79.8	89.1	85.3	88.2
0920	Plaster & Gypsum Board	230.4	89.7	129.9	155.0	87.7	107.0	155.0	89.9	108.5	130.0	75.6	91.2	131.9	60.5	80.9	136.1	84.9	99.5
0950, 0980	Ceilings & Acoustic Treatment	140.8	89.7	107.0	95.8	87.7	90.5	95.8	89.9	91.9	95.8	75.6	82.5	95.8	60.5	72.5	95.8	84.9	88.6
0960	Flooring	134.6	100.3	124.6	132.5	101.8	123.5	132.5	100.3	123.1	132.5	72.3	114.8	121.1	65.2	104.7	127.3	78.5	113.0
0970, 0990	Wall Finishes & Painting/Coating	112.0	92.6	100.0	112.8	93.1	100.7	112.0	96.9	102.7	112.0	60.8	80.5	112.0	59.0	79.3	112.0	87.6	96.6
09	FINISHES	136.5	92.9	112.2	118.4	92.3	103.8	117.1	94.3	104.4	114.6	73.7	91.8	111.5	62.9	84.4	115.7	86.8	99.6
COVERS	DIVS. 10 - 14, 25, 28, 41, 43, 44, 46	140.5	100.6	132.6	140.0	77.9	127.7	140.5	79.0	128.3	140.5	71.4	126.8	140.5	68.7	126.2	140.5	100.8	132.6
21, 22, 23	FIRE SUPPRESSION, PLUMBING & HVAC	100.7	81.7	93.0	103.0	103.5	103.2	103.2	107.4	104.9	103.2	88.2	97.2	103.8	72.4	91.1	103.2	99.7	101.8
26, 27, 3370	ELECTRICAL, COMMUNICATIONS & UTIL.	112.3	85.9	98.7	120.9	93.6	106.8	117.2	94.5	105.5	119.7	62.8	90.3	120.0	65.6	91.9	117.4	85.6	101.0
MF2010	WEIGHTED AVERAGE	122.9	90.7	108.7	118.8	95.0	108.3	117.7	97.0	108.6	116.0	77.2	98.9	112.3	70.5	93.9	118.1	93.7	107.3

City Cost Indexes

CANADA

DIVISION		QUEBEC, QUEBEC			RED DEER, ALBERTA			REGINA, SASKATCHEWAN			RIMOUSKI, QUEBEC			ROUYN-NORANDA, QUEBEC			SAINT HYACINTHE, QUEBEC		
		MAT.	INST.	TOTAL	MAT.	INST.	TOTAL	MAT.	INST.	TOTAL	MAT.	INST.	TOTAL	MAT.	INST.	TOTAL	MAT.	INST.	TOTAL
015433	CONTRACTOR EQUIPMENT		104.5	104.5		103.7	103.7		100.2	100.2		102.6	102.6		102.6	102.6		102.6	102.6
0241, 31 - 34	SITE & INFRASTRUCTURE, DEMOLITION	100.7	100.1	100.3	109.5	102.3	104.5	115.5	98.5	103.8	93.0	99.7	97.7	92.5	100.9	98.3	93.0	100.9	98.5
0310	Concrete Forming & Accessories	137.2	90.3	96.4	139.2	86.2	93.1	107.0	88.1	90.6	128.1	89.7	94.7	128.1	92.0	96.7	128.1	92.0	96.7
0320	Concrete Reinforcing	141.0	92.2	116.4	161.8	72.5	116.8	128.8	77.8	103.1	112.2	92.1	102.1	157.4	82.4	119.6	157.4	82.4	119.6
0330	Cast-in-Place Concrete	149.3	98.5	129.3	151.1	101.5	131.5	170.8	93.2	140.2	136.9	96.8	121.1	130.1	101.4	118.8	134.5	101.4	121.4
03	CONCRETE	139.0	93.5	116.8	143.2	89.3	116.9	139.9	88.0	114.6	127.7	92.7	110.6	131.3	93.5	112.9	133.5	93.5	114.0
04	MASONRY	173.5	90.7	122.9	158.7	83.8	113.0	167.7	91.1	121.0	158.7	90.6	117.1	158.9	90.6	117.2	159.1	90.6	117.3
05	METALS	124.2	91.8	113.5	107.5	88.9	101.4	110.6	83.7	101.7	104.5	91.1	100.1	105.0	89.9	100.0	105.0	89.9	100.0
06	WOOD, PLASTICS & COMPOSITES	126.9	90.1	105.6	113.0	85.7	97.2	96.3	88.8	92.0	126.1	89.8	105.1	126.1	92.0	106.4	126.1	92.0	106.4
07	THERMAL & MOISTURE PROTECTION	121.4	93.7	110.1	129.1	88.8	112.8	106.7	84.3	97.6	105.4	93.0	100.3	105.4	93.9	100.7	105.6	93.9	100.9
08	OPENINGS	93.4	87.8	92.0	93.4	80.1	90.0	88.2	77.7	85.5	92.9	77.9	89.1	93.4	79.7	89.9	93.4	79.7	89.9
0920	Plaster & Gypsum Board	191.4	89.3	118.4	136.4	84.9	99.6	168.2	88.5	111.2	185.7	89.3	116.8	146.8	91.5	107.3	149.8	91.5	108.2
0950, 0980	Ceilings & Acoustic Treatment	120.7	89.3	99.9	95.8	84.9	88.6	125.7	88.5	101.1	94.9	89.3	91.2	94.9	91.5	92.7	94.9	91.5	92.7
0960	Flooring	134.6	102.5	125.2	134.6	95.4	123.1	121.1	67.4	105.4	132.5	102.5	123.7	132.5	102.5	123.7	132.5	102.5	123.7
0970, 0990	Wall Finishes & Painting/Coating	112.5	101.1	105.4	111.9	84.2	94.8	112.0	87.3	96.8	112.0	101.1	105.3	112.0	95.0	101.5	112.0	95.0	101.5
09	FINISHES	127.5	93.9	108.8	116.0	87.4	100.0	123.0	84.8	101.7	119.5	93.7	105.1	114.9	94.6	103.6	115.3	94.6	103.8
COVERS	DIVS. 10 - 14, 25, 28, 41, 43, 44, 46	140.5	90.7	130.6	140.5	97.8	132.0	140.5	75.1	127.5	140.5	90.1	130.5	140.5	90.7	130.6	140.5	90.7	130.6
21, 22, 23	FIRE SUPPRESSION, PLUMBING & HVAC	100.6	95.7	98.6	103.2	97.3	100.8	99.9	82.3	92.8	103.2	95.6	100.1	103.2	96.1	100.4	98.6	96.1	97.6
26, 27, 3370	ELECTRICAL, COMMUNICATIONS & UTIL.	116.5	83.0	99.2	109.9	89.3	99.3	119.9	87.3	103.1	117.6	83.0	99.7	117.6	75.9	96.1	118.5	75.9	96.5
MF2010	WEIGHTED AVERAGE	119.8	92.3	107.7	116.6	91.1	105.3	116.7	86.2	103.2	114.2	91.6	104.2	114.4	91.1	104.1	113.6	91.1	103.7

CANADA

DIVISION		SAINT JOHN, NEW BRUNSWICK			SARNIA, ONTARIO			SASKATOON, SASKATCHEWAN			SAULT STE MARIE, ONTARIO			SHERBROOKE, QUEBEC			SOREL, QUEBEC		
		MAT.	INST.	TOTAL	MAT.	INST.	TOTAL	MAT.	INST.	TOTAL	MAT.	INST.	TOTAL	MAT.	INST.	TOTAL	MAT.	INST.	TOTAL
015433	CONTRACTOR EQUIPMENT		102.0	102.0		101.7	101.7		100.2	100.2		101.7	101.7		102.6	102.6		102.6	102.6
0241, 31 - 34	SITE & INFRASTRUCTURE, DEMOLITION	94.8	98.6	97.4	111.2	101.2	104.3	106.6	98.5	101.0	101.2	100.8	100.9	93.0	100.9	98.5	93.2	100.9	98.5
0310	Concrete Forming & Accessories	120.5	63.9	71.3	121.9	99.7	102.6	106.9	88.1	90.5	112.0	88.0	91.1	128.1	92.0	96.7	128.1	92.1	96.8
0320	Concrete Reinforcing	148.8	62.5	105.3	124.0	87.4	105.6	119.4	77.8	98.4	112.4	85.6	98.9	157.4	82.4	119.6	148.8	82.4	115.3
0330	Cast-in-Place Concrete	136.2	64.3	107.8	150.9	106.0	133.2	142.0	93.2	122.8	135.5	90.4	117.8	134.5	101.4	121.4	135.6	101.5	122.1
03	CONCRETE	136.0	64.7	101.3	136.1	99.5	118.3	124.0	88.0	106.4	120.0	88.6	104.7	133.5	93.5	114.0	132.7	93.6	113.7
04	MASONRY	176.3	69.8	111.3	174.1	103.5	131.0	164.2	91.1	119.6	159.7	95.1	120.3	159.2	90.6	117.3	159.3	90.6	117.4
05	METALS	104.6	79.9	96.5	106.6	92.5	102.0	104.9	83.7	97.9	105.9	90.8	100.9	105.0	89.9	100.0	105.0	90.0	100.1
06	WOOD, PLASTICS & COMPOSITES	115.9	62.4	85.0	114.3	99.0	105.5	94.6	88.8	91.3	102.6	87.0	93.6	126.1	92.0	106.4	126.1	92.0	106.4
07	THERMAL & MOISTURE PROTECTION	109.2	65.5	91.4	111.5	98.9	106.4	105.4	84.3	96.8	110.3	91.0	102.4	105.4	93.9	100.7	105.4	93.9	100.7
08	OPENINGS	87.3	58.0	79.9	94.4	93.3	94.1	87.2	77.7	84.8	84.9	85.4	85.0	93.4	79.7	89.9	93.4	84.6	91.2
0920	Plaster & Gypsum Board	179.4	61.1	94.9	178.2	98.9	121.6	144.8	88.5	104.5	143.4	86.6	102.8	149.8	91.5	108.2	185.7	91.5	118.4
0950, 0980	Ceilings & Acoustic Treatment	103.1	61.1	75.3	100.3	98.9	99.4	117.5	88.5	98.3	95.8	86.6	89.7	94.9	91.5	92.7	94.9	91.5	92.7
0960	Flooring	128.0	74.3	112.2	132.5	109.7	125.8	121.1	67.4	105.4	124.3	103.8	118.3	132.5	102.5	123.7	132.5	102.5	123.7
0970, 0990	Wall Finishes & Painting/Coating	112.0	71.6	87.1	112.0	109.5	110.5	112.0	87.3	96.8	112.0	101.5	105.5	112.0	95.0	101.5	112.0	95.0	101.5
09	FINISHES	119.7	66.4	90.0	120.8	103.1	110.9	117.8	84.8	99.4	112.4	92.2	101.2	115.3	94.6	103.8	119.5	94.6	105.6
COVERS	DIVS. 10 - 14, 25, 28, 41, 43, 44, 46	140.5	68.9	126.3	140.5	80.5	128.6	140.5	75.1	127.5	140.5	102.1	132.9	140.5	90.7	130.6	140.5	90.7	130.6
21, 22, 23	FIRE SUPPRESSION, PLUMBING & HVAC	103.2	78.9	93.4	103.2	113.8	107.5	99.8	82.3	92.7	103.2	100.0	101.9	103.8	96.1	100.7	103.2	96.1	100.4
26, 27, 3370	ELECTRICAL, COMMUNICATIONS & UTIL.	127.7	84.5	105.4	120.2	96.9	108.2	121.9	87.3	104.1	118.9	94.9	106.5	117.6	75.9	96.1	117.6	75.9	96.1
MF2010	WEIGHTED AVERAGE	116.6	75.3	98.4	117.4	101.6	110.5	113.2	86.2	101.3	112.6	94.5	104.6	114.8	91.1	104.4	114.9	91.3	104.5

CANADA

DIVISION		ST CATHARINES, ONTARIO			ST JEROME, QUEBEC			ST JOHNS, NEWFOUNDLAND			SUDBURY, ONTARIO			SUMMERSIDE, PRINCE EDWARD ISLAND			SYDNEY, NOVA SCOTIA		
		MAT.	INST.	TOTAL	MAT.	INST.	TOTAL	MAT.	INST.	TOTAL	MAT.	INST.	TOTAL	MAT.	INST.	TOTAL	MAT.	INST.	TOTAL
015433	CONTRACTOR EQUIPMENT		101.7	101.7		102.6	102.6		103.3	103.3		101.7	101.7		101.6	101.6		101.7	101.7
0241, 31 - 34	SITE & INFRASTRUCTURE, DEMOLITION	101.1	101.9	101.7	92.5	100.9	98.3	113.3	99.7	103.9	101.2	101.4	101.3	105.2	96.8	99.4	92.7	99.3	97.2
0310	Concrete Forming & Accessories	115.6	94.0	96.8	128.1	92.0	96.7	104.7	72.5	76.7	112.0	88.0	91.1	97.7	61.0	65.8	97.4	75.0	77.9
0320	Concrete Reinforcing	111.5	85.8	98.6	157.4	82.4	119.6	160.7	64.0	112.0	112.4	84.3	98.2	146.6	48.9	97.4	148.8	48.8	98.4
0330	Cast-in-Place Concrete	134.4	95.6	119.1	130.1	101.4	118.8	173.5	85.2	138.7	135.5	92.0	118.4	161.2	63.4	122.6	130.9	74.8	108.8
03	CONCRETE	119.5	93.1	106.6	131.3	93.5	112.9	164.8	75.9	121.5	120.0	89.0	104.9	152.9	60.4	107.8	129.7	70.7	101.0
04	MASONRY	159.3	100.4	123.4	158.9	90.6	117.2	164.8	77.3	111.4	159.4	95.6	120.4	158.1	65.2	101.4	155.9	74.8	106.5
05	METALS	105.9	92.0	101.3	105.0	89.9	100.0	113.8	79.9	102.6	105.9	91.6	101.1	105.8	69.5	93.8	105.8	76.3	96.1
06	WOOD, PLASTICS & COMPOSITES	106.4	93.4	98.9	126.1	92.0	106.4	95.8	71.2	81.6	102.6	87.0	93.6	87.7	60.4	71.9	87.4	74.3	79.8
07	THERMAL & MOISTURE PROTECTION	110.8	96.4	105.0	105.4	93.9	100.7	110.3	75.5	96.2	110.3	90.9	102.4	106.0	63.9	88.9	106.5	73.1	92.9
08	OPENINGS	84.0	90.5	85.6	93.4	79.7	89.9	99.7	66.2	91.2	84.9	85.4	85.0	95.7	52.1	84.7	83.9	67.7	79.9
0920	Plaster & Gypsum Board	137.6	93.2	105.9	146.8	91.5	107.3	172.6	70.1	99.4	143.4	86.6	102.8	156.4	59.1	86.9	155.9	73.4	96.9
0950, 0980	Ceilings & Acoustic Treatment	100.3	93.2	95.6	94.9	91.5	92.7	110.3	70.1	83.8	95.8	86.6	89.7	95.8	59.1	71.5	95.8	73.4	81.0
0960	Flooring	126.9	101.8	119.5	132.5	102.5	123.7	116.5	61.0	100.2	124.3	103.8	118.3	112.2	64.8	98.2	112.1	69.3	99.5
0970, 0990	Wall Finishes & Painting/Coating	112.0	100.9	105.2	112.0	95.0	101.5	112.0	74.7	89.0	112.0	92.1	99.7	112.0	44.5	70.4	112.0	66.5	83.9
09	FINISHES	113.5	96.6	104.1	114.9	94.6	103.6	118.2	70.6	91.6	112.4	91.3	100.6	111.4	60.2	82.9	110.3	73.2	89.6
COVERS	DIVS. 10 - 14, 25, 28, 41, 43, 44, 46	140.5	78.9	128.3	140.5	90.7	130.6	140.5	72.2	126.9	140.5	102.2	132.9	140.5	68.9	126.3	140.5	70.8	126.7
21, 22, 23	FIRE SUPPRESSION, PLUMBING & HVAC	99.8	82.9	92.9	103.2	96.1	100.4	100.1	71.3	88.4	103.2	85.9	96.2	103.2	67.4	88.7	103.2	89.2	97.5
26, 27, 3370	ELECTRICAL, COMMUNICATIONS & UTIL.	119.2	86.0	102.1	118.3	75.9	96.4	116.0	72.3	93.4	116.2	86.0	100.6	118.4	54.7	85.5	119.4	67.3	92.5
MF2010	WEIGHTED AVERAGE	111.7	91.3	102.8	114.4	91.1	104.1	120.3	75.7	100.6	112.3	90.4	102.7	117.0	65.5	94.3	112.8	77.8	97.4

City Cost Indexes

CANADA

DIVISION		THUNDER BAY, ONTARIO			TIMMINS, ONTARIO			TORONTO, ONTARIO			TROIS RIVIERES, QUEBEC			TRURO, NOVA SCOTIA			VANCOUVER, BRITISH COLUMBIA		
		MAT.	INST.	TOTAL	MAT.	INST.	TOTAL	MAT.	INST.	TOTAL	MAT.	INST.	TOTAL	MAT.	INST.	TOTAL	MAT.	INST.	TOTAL
015433	CONTRACTOR EQUIPMENT		101.7	101.7		101.7	101.7		104.0	104.0		102.6	102.6		101.7	101.7		112.1	112.1
0241, 31 - 34	SITE & INFRASTRUCTURE, DEMOLITION	106.5	101.9	103.3	112.7	100.8	104.5	139.0	105.9	116.2	93.2	100.9	98.5	97.2	99.3	98.6	121.1	108.5	112.4
0310	Concrete Forming & Accessories	123.1	91.6	95.7	123.0	91.1	95.2	132.2	99.5	103.7	128.1	92.1	96.8	97.4	75.0	77.9	126.6	87.0	92.2
0320	Concrete Reinforcing	99.7	85.3	92.4	175.1	85.5	129.9	162.1	86.5	124.0	148.8	82.4	115.3	148.8	48.8	98.4	152.8	79.7	116.0
0330	Cast-in-Place Concrete	148.0	94.0	126.7	163.5	90.3	134.7	154.3	114.8	138.8	135.6	101.5	122.1	169.8	74.8	132.4	164.0	101.0	139.2
03	CONCRETE	128.2	91.4	110.3	150.5	89.9	121.0	144.5	102.2	123.9	132.7	93.6	113.7	149.3	70.7	111.0	153.3	90.8	122.9
04	MASONRY	160.1	99.7	123.2	163.3	93.5	120.7	193.0	105.0	139.3	159.3	90.6	117.4	159.0	74.8	107.6	177.9	93.9	126.6
05	METALS	105.7	90.9	100.8	106.7	91.6	101.7	127.1	92.9	115.8	105.0	90.0	100.1	105.8	76.3	96.1	157.9	91.2	135.9
06	WOOD, PLASTICS & COMPOSITES	115.4	90.3	100.9	115.2	91.6	101.6	115.4	98.1	105.4	126.1	92.0	106.4	87.4	74.3	79.8	114.9	85.2	97.7
07	THERMAL & MOISTURE PROTECTION	111.0	93.1	103.8	111.5	89.4	102.5	123.5	100.4	114.1	105.4	93.9	100.7	106.5	73.1	92.2	133.7	89.0	115.5
08	OPENINGS	82.9	88.1	84.2	92.2	87.4	91.0	92.2	96.5	93.3	93.4	84.6	91.2	83.9	67.7	79.9	95.1	84.3	92.4
0920	Plaster & Gypsum Board	170.4	89.9	112.9	155.0	91.3	109.5	176.3	98.0	120.3	185.7	91.5	118.4	155.9	73.4	96.9	162.8	84.1	106.6
0950, 0980	Ceilings & Acoustic Treatment	95.8	89.9	91.9	95.8	91.3	92.8	132.8	98.0	109.8	94.9	91.5	92.7	95.8	73.4	81.0	157.5	84.1	109.0
0960	Flooring	132.5	109.0	125.6	132.5	100.3	123.1	135.8	111.6	128.7	132.5	102.5	123.7	112.1	69.3	99.5	135.0	98.8	124.4
0970, 0990	Wall Finishes & Painting/Coating	112.0	89.5	98.1	112.0	94.5	101.2	114.0	104.9	108.4	112.0	95.0	101.5	112.0	66.5	83.9	111.9	81.4	93.1
09	FINISHES	118.4	94.8	105.3	117.1	93.6	104.0	131.6	102.4	115.4	119.5	94.6	105.6	110.3	73.2	89.6	133.9	88.2	108.4
COVERS	DIVS. 10 - 14, 25, 28, 41, 43, 44, 46	140.5	78.7	128.2	140.5	77.4	128.0	140.5	105.7	133.6	140.5	90.7	130.6	140.5	70.8	126.7	140.5	99.5	132.4
21, 22, 23	FIRE SUPPRESSION, PLUMBING & HVAC	99.8	82.7	92.8	103.2	103.6	103.4	100.6	97.7	99.4	103.2	96.1	100.4	103.2	89.2	97.5	100.6	84.7	94.2
26, 27, 3370	ELECTRICAL, COMMUNICATIONS & UTIL.	117.5	84.5	100.5	118.9	94.9	106.5	114.0	88.1	100.7	119.8	75.9	97.1	117.3	67.3	91.5	115.1	77.7	95.8
MF2010	WEIGHTED AVERAGE	113.1	90.2	103.0	117.9	94.9	107.8	123.0	98.8	112.3	115.1	91.3	104.6	115.1	77.8	98.6	128.1	89.1	110.9

CANADA

DIVISION		VICTORIA, BRITISH COLUMBIA			WHITEHORSE, YUKON			WINDSOR, ONTARIO			WINNIPEG, MANITOBA			YARMOUTH, NOVA SCOTIA			YELLOWKNIFE, NWT		
		MAT.	INST.	TOTAL	MAT.	INST.	TOTAL	MAT.	INST.	TOTAL	MAT.	INST.	TOTAL	MAT.	INST.	TOTAL	MAT.	INST.	TOTAL
015433	CONTRACTOR EQUIPMENT		108.8	108.8		101.8	101.8		101.7	101.7		105.9	105.9		101.7	101.7		101.7	101.7
0241, 31 - 34	SITE & INFRASTRUCTURE, DEMOLITION	121.1	108.0	112.1	108.8	97.7	101.2	96.3	101.8	100.1	121.5	101.6	107.8	97.0	99.3	98.5	128.6	102.3	110.5
0310	Concrete Forming & Accessories	111.9	86.4	89.7	107.0	62.3	68.1	123.1	89.9	94.3	131.6	69.8	77.8	97.4	75.0	77.9	106.9	85.3	88.2
0320	Concrete Reinforcing	116.8	79.6	98.0	115.8	62.7	89.0	109.3	84.6	96.9	143.8	58.2	100.7	148.8	48.8	98.4	118.4	66.6	92.3
0330	Cast-in-Place Concrete	147.1	100.3	128.6	133.5	66.8	101.0	137.7	94.0	120.8	192.3	75.4	146.2	168.1	74.8	131.3	245.0	92.6	185.0
03	CONCRETE	151.9	90.1	121.8	133.5	66.8	101.0	121.3	90.8	106.4	162.7	70.1	117.6	148.4	70.7	110.5	175.5	84.6	131.2
04	MASONRY	167.0	93.9	122.4	170.1	65.3	106.2	159.5	98.5	122.3	188.0	70.9	116.5	158.8	74.8	107.6	177.2	79.2	117.4
05	METALS	107.3	87.2	100.7	110.9	75.6	99.2	105.7	91.4	101.0	154.6	75.5	128.5	105.8	76.3	96.1	109.7	81.6	100.4
06	WOOD, PLASTICS & COMPOSITES	98.3	85.1	90.7	96.2	60.9	75.8	115.4	88.8	100.0	113.4	70.6	88.7	87.4	74.3	79.8	96.2	87.0	90.9
07	THERMAL & MOISTURE PROTECTION	114.9	88.8	104.3	105.3	64.9	88.9	110.8	92.2	103.3	116.3	72.4	98.5	106.5	73.1	92.9	108.2	83.1	98.0
08	OPENINGS	89.1	80.4	86.9	86.7	57.9	79.5	82.6	87.8	84.0	93.4	62.8	85.7	83.9	67.7	79.9	86.7	73.1	83.3
0920	Plaster & Gypsum Board	136.1	84.1	99.0	124.3	59.5	78.0	158.4	88.4	108.4	161.4	69.1	95.5	155.9	73.4	96.9	175.0	86.5	111.8
0950, 0980	Ceilings & Acoustic Treatment	95.8	84.1	88.1	100.3	59.5	73.4	95.8	88.4	90.9	142.8	69.1	94.1	95.8	73.4	81.0	130.2	86.5	101.3
0960	Flooring	127.3	78.5	113.0	121.1	63.2	104.1	132.5	106.1	124.8	133.7	74.8	116.4	112.1	69.3	99.5	121.1	96.9	114.0
0970, 0990	Wall Finishes & Painting/Coating	112.0	81.4	93.1	112.0	58.1	78.7	112.0	93.9	100.8	112.0	56.2	77.6	112.0	66.5	83.9	112.0	84.6	95.1
09	FINISHES	115.7	84.7	98.4	111.7	61.7	83.9	116.6	93.6	103.8	129.7	69.6	96.2	110.3	73.2	89.6	124.3	86.8	103.4
COVERS	DIVS. 10 - 14, 25, 28, 41, 43, 44, 46	140.5	74.5	127.4	140.5	67.9	126.1	140.5	78.1	128.1	140.5	70.5	126.6	140.5	70.8	126.7	140.5	72.5	127.0
21, 22, 23	FIRE SUPPRESSION, PLUMBING & HVAC	103.2	84.6	95.7	103.3	79.0	93.5	99.8	82.3	92.7	100.5	67.9	87.3	103.2	89.2	97.5	103.6	100.9	102.5
26, 27, 3370	ELECTRICAL, COMMUNICATIONS & UTIL.	118.3	77.7	97.4	127.6	65.0	95.3	122.0	86.0	103.4	117.9	68.3	92.3	119.4	67.3	92.5	131.3	90.6	110.3
MF2010	WEIGHTED AVERAGE	118.1	87.2	104.5	116.5	71.3	96.6	112.2	90.0	102.4	128.4	72.4	103.7	115.2	77.8	98.7	123.5	88.9	108.3

Location Factors

Costs shown in RSMeans cost data publications are based on national averages for materials and installation. To adjust these costs to a specific location, simply multiply the base cost by the factor and divide by 100 for that city. The data is arranged alphabetically by state and postal zip code numbers. For a city not listed, use the factor for a nearby city with similar economic characteristics.

STATE/ZIP	CITY	MAT.	INST.	TOTAL
ALABAMA				
350-352	Birmingham	97.6	74.6	87.4
354	Tuscaloosa	96.6	58.6	79.9
355	Jasper	96.8	57.6	79.5
356	Decatur	96.5	60.1	80.5
357-358	Huntsville	96.6	68.1	84.1
359	Gadsden	96.6	57.6	79.4
360-361	Montgomery	97.4	56.6	79.4
362	Anniston	96.0	64.5	82.1
363	Dothan	96.6	52.0	76.9
364	Evergreen	96.2	53.4	77.3
365-366	Mobile	97.7	67.5	84.4
367	Selma	96.3	51.9	76.7
368	Phenix City	97.1	55.9	78.9
369	Butler	96.5	52.0	76.9
ALASKA				
995-996	Anchorage	124.5	114.0	119.9
997	Fairbanks	122.4	115.0	119.1
998	Juneau	120.3	113.9	117.5
999	Ketchikan	134.1	113.9	125.2
ARIZONA				
850,853	Phoenix	97.9	76.6	88.5
851,852	Mesa/Tempe	97.4	72.4	86.3
855	Globe	97.4	62.5	82.0
856-857	Tucson	96.1	72.1	85.5
859	Show Low	97.5	62.9	82.2
860	Flagstaff	100.3	73.9	88.7
863	Prescott	98.2	62.1	82.2
864	Kingman	96.4	73.4	86.3
865	Chambers	96.5	63.0	81.7
ARKANSAS				
716	Pine Bluff	96.7	63.8	82.2
717	Camden	94.1	50.1	74.7
718	Texarkana	95.9	51.3	76.2
719	Hot Springs	93.4	52.9	75.5
720-722	Little Rock	97.0	65.0	82.9
723	West Memphis	95.8	62.8	81.2
724	Jonesboro	96.8	60.7	80.9
725	Batesville	94.2	56.1	77.4
726	Harrison	95.5	52.3	76.5
727	Fayetteville	92.9	54.1	75.8
728	Russellville	94.1	55.4	77.0
729	Fort Smith	97.1	62.5	81.8
CALIFORNIA				
900-902	Los Angeles	100.1	115.5	106.9
903-905	Inglewood	95.3	113.1	103.2
906-908	Long Beach	96.9	113.2	104.1
910-912	Pasadena	95.2	113.6	103.3
913-916	Van Nuys	98.1	113.5	104.9
917-918	Alhambra	97.1	113.5	104.3
919-921	San Diego	99.3	107.8	103.0
922	Palm Springs	96.4	110.9	102.8
923-924	San Bernardino	94.3	110.5	101.4
925	Riverside	98.9	112.4	104.9
926-927	Santa Ana	96.1	110.8	102.6
928	Anaheim	98.9	112.8	105.1
930	Oxnard	100.2	112.3	105.5
931	Santa Barbara	99.5	112.3	105.1
932-933	Bakersfield	100.0	109.5	104.2
934	San Luis Obispo	99.7	109.4	104.0
935	Mojave	97.0	106.9	101.3
936-938	Fresno	100.5	115.7	107.2
939	Salinas	100.5	122.7	110.3
940-941	San Francisco	108.5	142.3	123.4
942,956-958	Sacramento	102.9	118.6	109.9
943	Palo Alto	101.3	129.9	113.9
944	San Mateo	103.8	131.3	115.9
945	Vallejo	102.2	128.2	113.7
946	Oakland	106.3	133.9	118.5
947	Berkeley	105.5	133.2	117.7
948	Richmond	104.7	129.7	115.7
949	San Rafael	106.0	129.5	116.4
950	Santa Cruz	105.0	122.8	112.9

STATE/ZIP	CITY	MAT.	INST.	TOTAL
CALIFORNIA (CONT'D)				
951	San Jose	103.2	135.1	117.3
952	Stockton	100.7	117.2	108.0
953	Modesto	100.7	116.9	107.9
954	Santa Rosa	100.3	135.0	115.6
955	Eureka	101.6	114.2	107.2
959	Marysville	100.9	114.4	106.8
960	Redding	102.8	114.4	107.9
961	Susanville	101.5	114.4	107.2
COLORADO				
800-802	Denver	101.7	84.7	94.2
803	Boulder	98.6	82.9	91.7
804	Golden	100.7	81.7	92.3
805	Fort Collins	102.6	79.1	92.2
806	Greeley	100.0	78.2	90.4
807	Fort Morgan	99.3	82.0	91.7
808-809	Colorado Springs	101.6	81.7	92.8
810	Pueblo	100.4	80.8	91.8
811	Alamosa	101.5	76.7	90.6
812	Salida	101.3	78.3	91.2
813	Durango	101.9	79.6	92.1
814	Montrose	100.8	79.3	91.3
815	Grand Junction	104.2	79.6	93.4
816	Glenwood Springs	101.6	82.2	93.1
CONNECTICUT				
060	New Britain	101.1	118.6	108.8
061	Hartford	102.0	118.7	109.4
062	Willimantic	101.7	118.4	109.1
063	New London	97.6	118.6	106.9
064	Meriden	99.7	118.6	108.0
065	New Haven	102.8	118.6	109.8
066	Bridgeport	102.2	118.7	109.5
067	Waterbury	101.8	118.6	109.2
068	Norwalk	101.8	126.3	112.6
069	Stamford	101.9	126.3	112.7
D.C.				
200-205	Washington	102.4	92.8	98.2
DELAWARE				
197	Newark	99.4	107.9	103.2
198	Wilmington	99.1	108.0	103.0
199	Dover	99.5	107.9	103.2
FLORIDA				
320,322	Jacksonville	98.7	69.4	85.7
321	Daytona Beach	98.8	75.2	88.4
323	Tallahassee	99.4	59.3	81.7
324	Panama City	100.1	60.5	82.7
325	Pensacola	102.6	64.1	85.6
326,344	Gainesville	100.4	69.7	86.9
327-328,347	Orlando	101.8	73.0	89.1
329	Melbourne	101.5	78.2	91.2
330-332,340	Miami	99.6	76.7	89.6
333	Fort Lauderdale	98.1	75.9	88.3
334,349	West Palm Beach	96.7	74.4	86.9
335-336,346	Tampa	99.8	82.1	92.0
337	St. Petersburg	102.5	65.8	86.3
338	Lakeland	99.0	82.0	91.5
339,341	Fort Myers	98.1	75.8	88.3
342	Sarasota	100.0	77.3	90.0
GEORGIA				
300-303,399	Atlanta	98.0	76.5	88.5
304	Statesboro	97.4	54.0	78.3
305	Gainesville	96.1	64.0	82.0
306	Athens	95.5	64.2	81.7
307	Dalton	97.6	60.1	81.1
308-309	Augusta	96.7	64.5	82.5
310-312	Macon	96.4	64.2	82.2
313-314	Savannah	98.1	60.5	81.5
315	Waycross	97.3	61.4	81.5
316	Valdosta	97.6	60.8	81.4
317,398	Albany	97.6	61.5	81.7
318-319	Columbus	97.5	64.6	83.0

Location Factors

STATE/ZIP	CITY	MAT.	INST.	TOTAL
HAWAII				
967	Hilo	110.3	121.2	115.1
968	Honolulu	113.3	121.2	116.8
STATES & POSS.				
969	Guam	131.1	60.8	100.1
IDAHO				
832	Pocatello	100.3	72.2	87.9
833	Twin Falls	101.3	52.5	79.8
834	Idaho Falls	98.9	58.7	81.2
835	Lewiston	107.9	79.7	95.5
836-837	Boise	99.3	72.4	87.4
838	Coeur d'Alene	107.2	76.3	93.6
ILLINOIS				
600-603	North Suburban	98.7	132.4	113.5
604	Joliet	98.6	136.4	115.3
605	South Suburban	98.7	132.4	113.5
606-608	Chicago	99.1	138.9	116.6
609	Kankakee	94.6	127.4	109.1
610-611	Rockford	98.4	124.3	109.8
612	Rock Island	95.9	98.9	97.2
613	La Salle	97.0	120.8	107.5
614	Galesburg	96.9	104.8	100.4
615-616	Peoria	99.5	108.3	103.4
617	Bloomington	96.3	109.3	102.0
618-619	Champaign	99.8	108.1	103.5
620-622	East St. Louis	95.1	108.3	100.9
623	Quincy	96.0	100.1	97.8
624	Effingham	95.3	103.9	99.1
625	Decatur	97.4	106.3	101.3
626-627	Springfield	97.8	107.1	101.9
628	Centralia	93.4	107.0	99.4
629	Carbondale	93.1	102.0	97.1
INDIANA				
460	Anderson	96.1	83.4	90.5
461-462	Indianapolis	99.2	85.6	93.2
463-464	Gary	97.6	108.5	102.4
465-466	South Bend	95.7	82.9	90.1
467-468	Fort Wayne	96.7	78.7	88.8
469	Kokomo	93.7	82.2	88.6
470	Lawrenceburg	92.5	79.6	86.8
471	New Albany	93.6	76.6	86.1
472	Columbus	96.0	81.5	89.6
473	Muncie	97.1	82.6	90.7
474	Bloomington	98.4	82.9	91.6
475	Washington	94.4	83.7	89.7
476-477	Evansville	96.0	86.5	91.8
478	Terre Haute	96.7	86.9	92.4
479	Lafayette	95.7	83.2	90.2
IOWA				
500-503,509	Des Moines	99.1	84.7	92.7
504	Mason City	97.6	67.8	84.5
505	Fort Dodge	97.9	59.7	81.0
506-507	Waterloo	99.7	72.2	87.6
508	Creston	98.1	78.8	89.6
510-511	Sioux City	100.5	70.7	87.3
512	Sibley	98.9	56.5	80.2
513	Spencer	100.6	56.5	81.2
514	Carroll	97.4	72.4	86.4
515	Council Bluffs	101.7	78.8	91.6
516	Shenandoah	98.2	72.3	86.8
520	Dubuque	99.9	79.7	91.0
521	Decorah	98.5	65.7	84.0
522-524	Cedar Rapids	100.7	84.0	93.3
525	Ottumwa	98.4	70.7	86.2
526	Burlington	97.6	77.3	88.6
527-528	Davenport	99.6	90.4	95.5
KANSAS				
660-662	Kansas City	99.5	95.4	97.7
664-666	Topeka	99.2	64.5	83.9
667	Fort Scott	97.7	73.9	87.2
668	Emporia	97.9	69.7	85.4
669	Belleville	99.6	65.5	84.6
670-672	Wichita	98.4	64.1	83.3
673	Independence	99.2	69.5	86.1
674	Salina	100.0	66.9	85.4
675	Hutchinson	94.8	65.4	81.8
676	Hays	98.8	65.8	84.3
677	Colby	99.5	65.8	84.6

STATE/ZIP	CITY	MAT.	INST.	TOTAL
KANSAS (CONT'D)				
678	Dodge City	101.2	66.8	86.0
679	Liberal	98.7	65.8	84.2
KENTUCKY				
400-402	Louisville	97.8	83.5	91.5
403-405	Lexington	96.5	81.8	90.0
406	Frankfort	97.2	83.5	91.2
407-409	Corbin	93.7	72.5	84.4
410	Covington	94.4	95.2	94.8
411-412	Ashland	93.3	98.5	95.6
413-414	Campton	94.5	81.2	88.6
415-416	Pikeville	95.6	91.7	93.9
417-418	Hazard	93.9	81.5	88.4
420	Paducah	92.6	85.7	89.6
421-422	Bowling Green	95.0	83.5	90.0
423	Owensboro	95.2	87.9	92.0
424	Henderson	92.4	87.9	90.4
425-426	Somerset	91.9	79.1	86.2
427	Elizabethtown	91.4	83.9	88.1
LOUISIANA				
700-701	New Orleans	102.3	70.5	88.3
703	Thibodaux	100.2	63.4	84.0
704	Hammond	97.5	56.1	79.3
705	Lafayette	99.9	61.5	83.0
706	Lake Charles	100.1	64.1	84.2
707-708	Baton Rouge	101.2	63.0	84.4
710-711	Shreveport	96.7	56.3	78.9
712	Monroe	97.0	53.6	77.9
713-714	Alexandria	97.1	53.5	77.9
MAINE				
039	Kittery	94.8	80.1	88.3
040-041	Portland	99.5	78.9	90.4
042	Lewiston	98.1	78.9	89.7
043	Augusta	97.6	79.8	89.8
044	Bangor	97.7	78.9	89.4
045	Bath	96.0	79.5	88.7
046	Machias	95.5	79.0	88.2
047	Houlton	95.7	78.9	88.3
048	Rockland	94.7	79.4	88.0
049	Waterville	96.1	79.8	88.9
MARYLAND				
206	Waldorf	98.5	82.4	91.4
207-208	College Park	98.5	84.2	92.2
209	Silver Spring	97.7	82.3	90.9
210-212	Baltimore	100.2	84.7	93.3
214	Annapolis	100.2	83.2	92.7
215	Cumberland	95.8	80.2	89.0
216	Easton	97.5	70.3	85.5
217	Hagerstown	96.8	82.9	90.7
218	Salisbury	97.9	65.3	83.5
219	Elkton	94.7	84.3	90.1
MASSACHUSETTS				
010-011	Springfield	100.0	108.7	103.9
012	Pittsfield	99.6	105.3	102.1
013	Greenfield	97.3	108.4	102.2
014	Fitchburg	96.1	123.3	108.1
015-016	Worcester	100.0	123.5	110.4
017	Framingham	95.5	129.9	110.7
018	Lowell	99.4	128.5	112.3
019	Lawrence	100.5	129.0	113.1
020-022, 024	Boston	102.2	136.8	117.5
023	Brockton	100.8	124.7	111.4
025	Buzzards Bay	94.9	122.1	106.9
026	Hyannis	98.0	122.1	108.6
027	New Bedford	99.9	122.2	109.7
MICHIGAN				
480,483	Royal Oak	93.4	106.2	99.1
481	Ann Arbor	95.8	108.1	101.2
482	Detroit	97.4	110.1	103.0
484-485	Flint	95.5	97.1	96.2
486	Saginaw	95.0	93.5	94.4
487	Bay City	95.2	93.6	94.5
488-489	Lansing	96.4	95.1	95.9
490	Battle Creek	96.0	89.2	93.0
491	Kalamazoo	96.3	87.4	92.4
492	Jackson	94.3	95.9	95.0
493,495	Grand Rapids	97.2	83.0	90.9
494	Muskegon	94.9	84.9	90.5

Location Factors

STATE/ZIP	CITY	MAT.	INST.	TOTAL
MICHIGAN (CONT'D)				
496	Traverse City	93.6	78.7	87.0
497	Gaylord	94.7	72.8	85.1
498-499	Iron mountain	96.7	87.8	92.8
MINNESOTA				
550-551	Saint Paul	101.1	121.5	110.1
553-555	Minneapolis	102.1	123.0	111.3
556-558	Duluth	100.1	112.0	105.4
559	Rochester	100.9	107.6	103.8
560	Mankato	97.7	104.6	100.7
561	Windom	96.4	93.8	95.2
562	Willmar	96.0	105.1	100.0
563	St. Cloud	97.6	120.8	107.9
564	Brainerd	97.6	105.3	101.0
565	Detroit Lakes	99.4	99.3	99.4
566	Bemidji	98.7	101.3	99.9
567	Thief River Falls	98.3	95.8	97.2
MISSISSIPPI				
386	Clarksdale	96.4	55.5	78.4
387	Greenville	100.3	65.1	84.7
388	Tupelo	98.0	57.9	80.4
389	Greenwood	97.6	55.1	78.8
390-392	Jackson	97.4	64.4	82.9
393	Meridian	96.7	66.1	83.2
394	Laurel	97.9	58.8	80.6
395	Biloxi	98.3	60.2	81.5
396	Mccomb	96.3	54.6	77.9
397	Columbus	97.8	56.9	79.8
MISSOURI				
630-631	St. Louis	100.0	105.7	102.5
633	Bowling Green	98.4	90.5	94.9
634	Hannibal	97.3	87.1	92.8
635	Kirksville	100.4	83.8	93.1
636	Flat River	99.4	92.9	96.5
637	Cape Girardeau	99.7	89.6	95.3
638	Sikeston	97.5	85.5	92.2
639	Poplar Bluff	97.0	86.8	92.5
640-641	Kansas City	100.1	108.0	103.6
644-645	St. Joseph	99.3	91.8	96.0
646	Chillicothe	96.1	70.0	84.6
647	Harrisonville	95.7	104.5	99.6
648	Joplin	98.2	70.4	85.9
650-651	Jefferson City	96.2	90.5	93.7
652	Columbia	97.7	92.4	95.4
653	Sedalia	96.6	92.8	94.9
654-655	Rolla	95.2	87.9	92.0
656-658	Springfield	98.5	79.9	90.3
MONTANA				
590-591	Billings	103.5	74.7	90.8
592	Wolf Point	103.1	67.5	87.4
593	Miles City	100.7	69.1	86.8
594	Great Falls	105.0	75.3	91.9
595	Havre	102.0	70.2	88.0
596	Helena	102.6	72.0	89.1
597	Butte	103.5	73.5	90.3
598	Missoula	100.8	72.2	88.2
599	Kalispell	99.6	70.1	86.6
NEBRASKA				
680-681	Omaha	100.0	79.2	90.8
683-685	Lincoln	98.0	73.6	87.2
686	Columbus	96.8	71.7	85.7
687	Norfolk	98.5	75.1	88.2
688	Grand Island	98.7	78.4	89.8
689	Hastings	97.8	79.3	89.6
690	Mccook	97.6	70.7	85.7
691	North Platte	98.1	79.4	89.9
692	Valentine	99.7	67.6	85.6
693	Alliance	99.7	65.8	84.8
NEVADA				
889-891	Las Vegas	99.8	110.7	104.6
893	Ely	98.5	98.1	98.3
894-895	Reno	98.7	93.7	96.5
897	Carson City	97.8	93.5	95.9
898	Elko	97.2	85.3	92.0
NEW HAMPSHIRE				
030	Nashua	100.2	86.7	94.3
031	Manchester	100.7	86.7	94.5

STATE/ZIP	CITY	MAT.	INST.	TOTAL
NEW HAMPSHIRE (CONT'D)				
032-033	Concord	98.6	82.4	91.5
034	Keene	96.9	49.4	76.0
035	Littleton	96.5	56.6	78.9
036	Charleston	96.3	47.1	74.6
037	Claremont	95.3	47.1	74.1
038	Portsmouth	97.8	91.2	94.9
NEW JERSEY				
070-071	Newark	102.7	124.6	112.4
072	Elizabeth	100.6	124.5	111.2
073	Jersey City	99.4	124.4	110.4
074-075	Paterson	101.4	124.5	111.6
076	Hackensack	99.4	124.5	110.5
077	Long Branch	99.0	123.3	109.7
078	Dover	99.7	124.6	110.7
079	Summit	99.6	124.7	110.7
080,083	Vineland	97.6	122.4	108.6
081	Camden	99.4	122.6	109.6
082,084	Atlantic City	98.1	122.1	108.7
085-086	Trenton	99.6	123.6	110.2
087	Point Pleasant	99.7	122.5	109.7
088-089	New Brunswick	100.1	124.3	110.8
NEW MEXICO				
870-872	Albuquerque	98.7	73.6	87.7
873	Gallup	98.8	73.6	87.7
874	Farmington	99.4	73.6	88.0
875	Santa Fe	99.4	73.6	88.0
877	Las Vegas	97.3	73.6	86.8
878	Socorro	96.8	73.6	86.6
879	Truth/Consequences	97.0	70.3	85.3
880	Las Cruces	96.1	69.6	84.4
881	Clovis	97.4	73.5	86.9
882	Roswell	99.3	73.6	88.0
883	Carrizozo	99.4	73.6	88.1
884	Tucumcari	98.1	73.5	87.3
NEW YORK				
100-102	New York	104.8	166.9	132.2
103	Staten Island	100.8	163.6	128.5
104	Bronx	98.9	163.6	127.4
105	Mount Vernon	98.6	133.0	113.8
106	White Plains	98.8	133.0	113.9
107	Yonkers	103.5	133.1	116.5
108	New Rochelle	99.0	133.0	114.0
109	Suffern	98.7	122.0	109.0
110	Queens	101.4	163.8	128.9
111	Long Island City	103.0	163.8	129.8
112	Brooklyn	103.4	163.8	130.0
113	Flushing	102.9	163.8	129.7
114	Jamaica	101.1	163.8	128.7
115,117,118	Hicksville	101.3	145.6	120.8
116	Far Rockaway	103.1	163.8	129.8
119	Riverhead	102.1	146.1	121.5
120-122	Albany	96.7	96.4	96.5
123	Schenectady	97.8	96.1	97.0
124	Kingston	100.9	117.4	108.2
125-126	Poughkeepsie	100.1	128.4	112.6
127	Monticello	99.4	118.0	107.6
128	Glens Falls	92.3	92.6	92.5
129	Plattsburgh	96.9	87.0	92.6
130-132	Syracuse	98.9	94.5	97.0
133-135	Utica	96.8	88.2	93.0
136	Watertown	98.5	93.5	96.3
137-139	Binghamton	98.4	89.2	94.4
140-142	Buffalo	100.3	104.0	101.9
143	Niagara Falls	97.9	106.1	101.5
144-146	Rochester	100.0	94.0	97.3
147	Jamestown	96.9	86.4	92.3
148-149	Elmira	96.7	91.0	94.2
NORTH CAROLINA				
270,272-274	Greensboro	99.7	47.3	76.6
271	Winston-Salem	99.5	46.0	75.9
275-276	Raleigh	99.8	47.6	76.8
277	Durham	100.8	47.9	77.5
278	Rocky Mount	96.8	42.5	72.9
279	Elizabeth City	97.8	40.6	72.6
280	Gastonia	98.6	45.4	75.1
281-282	Charlotte	99.4	47.8	76.7
283	Fayetteville	100.7	49.4	78.1
284	Wilmington	97.4	48.5	75.8
285	Kinston	95.2	43.8	72.6

758

Location Factors

STATE/ZIP	CITY	MAT.	INST.	TOTAL
NORTH CAROLINA (CONT'D)				
286	Hickory	95.5	42.3	72.1
287-288	Asheville	97.8	45.5	74.7
289	Murphy	96.4	33.2	68.5
NORTH DAKOTA				
580-581	Fargo	102.1	65.7	86.0
582	Grand Forks	102.8	53.1	80.9
583	Devils Lake	101.7	57.0	82.0
584	Jamestown	101.9	46.4	77.4
585	Bismarck	100.7	64.4	84.7
586	Dickinson	102.7	58.8	83.3
587	Minot	102.7	70.2	88.4
588	Williston	101.1	58.8	82.4
OHIO				
430-432	Columbus	97.8	90.6	94.6
433	Marion	93.9	85.4	90.2
434-436	Toledo	97.9	98.3	98.1
437-438	Zanesville	94.5	86.1	90.8
439	Steubenville	96.1	93.5	95.0
440	Lorain	97.7	91.3	94.9
441	Cleveland	97.9	99.8	98.7
442-443	Akron	98.7	94.5	96.9
444-445	Youngstown	98.1	90.1	94.6
446-447	Canton	98.2	86.3	93.0
448-449	Mansfield	95.2	87.4	91.7
450	Hamilton	96.6	85.6	91.7
451-452	Cincinnati	96.9	85.8	92.0
453-454	Dayton	96.7	85.2	91.6
455	Springfield	96.6	85.8	91.8
456	Chillicothe	95.2	93.4	94.4
457	Athens	98.0	83.1	91.4
458	Lima	98.4	86.1	93.0
OKLAHOMA				
730-731	Oklahoma City	97.8	60.6	81.4
734	Ardmore	95.0	60.1	79.6
735	Lawton	97.5	61.4	81.6
736	Clinton	96.5	58.8	79.8
737	Enid	97.3	58.7	80.3
738	Woodward	95.1	58.8	79.1
739	Guymon	96.2	31.0	67.5
740-741	Tulsa	97.4	53.5	78.0
743	Miami	93.8	67.9	82.3
744	Muskogee	96.7	39.3	71.4
745	Mcalester	93.6	50.6	74.6
746	Ponca City	94.1	58.7	78.5
747	Durant	94.1	57.7	78.1
748	Shawnee	95.7	56.5	78.4
749	Poteau	93.2	61.4	79.2
OREGON				
970-972	Portland	98.8	100.3	99.5
973	Salem	98.6	98.6	98.6
974	Eugene	98.6	98.5	98.6
975	Medford	100.1	97.0	98.7
976	Klamath Falls	99.8	97.0	98.6
977	Bend	98.8	98.6	98.7
978	Pendleton	94.1	100.0	96.7
979	Vale	91.9	90.5	91.3
PENNSYLVANIA				
150-152	Pittsburgh	99.1	104.6	101.5
153	Washington	95.9	102.1	98.6
154	Uniontown	96.4	100.8	98.3
155	Bedford	97.2	91.7	94.8
156	Greensburg	97.2	100.3	98.5
157	Indiana	96.0	99.1	97.4
158	Dubois	97.6	94.5	96.2
159	Johnstown	97.2	95.8	96.6
160	Butler	92.3	100.5	95.9
161	New Castle	92.4	97.5	94.6
162	Kittanning	92.8	102.4	97.0
163	Oil City	92.4	94.7	93.4
164-165	Erie	94.6	93.6	94.1
166	Altoona	94.8	90.6	93.0
167	Bradford	95.5	93.8	94.7
168	State College	95.0	91.8	93.6
169	Wellsboro	96.1	92.4	94.5
170-171	Harrisburg	97.7	96.1	97.0
172	Chambersburg	95.5	89.5	92.9
173-174	York	96.3	95.2	95.8
175-176	Lancaster	94.3	89.7	92.3

STATE/ZIP	CITY	MAT.	INST.	TOTAL
PENNSYLVANIA (CONT'D)				
177	Williamsport	93.0	79.6	87.1
178	Sunbury	95.1	95.1	95.1
179	Pottsville	94.2	96.8	95.4
180	Lehigh Valley	95.6	113.5	103.5
181	Allentown	97.9	109.2	102.9
182	Hazleton	95.1	97.2	96.0
183	Stroudsburg	94.9	103.0	98.5
184-185	Scranton	98.6	99.5	99.0
186-187	Wilkes-Barre	94.8	98.2	96.3
188	Montrose	94.5	97.0	95.6
189	Doylestown	94.7	120.1	105.9
190-191	Philadelphia	99.9	130.9	113.6
193	Westchester	96.1	122.8	107.9
194	Norristown	95.2	130.2	110.6
195-196	Reading	97.6	100.0	98.6
PUERTO RICO				
009	San Juan	121.3	24.1	78.4
RHODE ISLAND				
028	Newport	99.2	111.1	104.5
029	Providence	100.2	111.1	105.0
SOUTH CAROLINA				
290-292	Columbia	97.8	49.3	76.4
293	Spartanburg	96.8	49.0	75.7
294	Charleston	98.4	57.4	80.3
295	Florence	96.6	49.3	75.8
296	Greenville	96.6	49.0	75.6
297	Rock Hill	96.0	47.2	74.5
298	Aiken	96.8	72.8	86.2
299	Beaufort	97.5	42.5	73.3
SOUTH DAKOTA				
570-571	Sioux Falls	100.7	57.1	81.5
572	Watertown	99.8	51.6	78.5
573	Mitchell	98.6	51.0	77.6
574	Aberdeen	101.6	52.2	79.8
575	Pierre	100.1	53.8	79.7
576	Mobridge	99.3	51.4	78.2
577	Rapid City	101.4	54.6	80.7
TENNESSEE				
370-372	Nashville	96.8	72.2	86.0
373-374	Chattanooga	99.0	66.2	84.6
375,380-381	Memphis	98.2	69.6	85.6
376	Johnson City	98.7	56.7	80.2
377-379	Knoxville	95.4	61.4	80.4
382	Mckenzie	97.1	61.8	81.6
383	Jackson	99.2	63.8	83.6
384	Columbia	95.7	67.1	83.1
385	Cookeville	97.0	60.2	80.8
TEXAS				
750	Mckinney	98.5	50.2	77.2
751	Waxahackie	98.5	57.3	80.3
752-753	Dallas	99.4	66.8	85.0
754	Greenville	98.6	43.3	74.2
755	Texarkana	98.0	51.0	77.2
756	Longview	98.6	40.4	72.9
757	Tyler	99.0	54.4	79.4
758	Palestine	95.2	56.6	78.2
759	Lufkin	95.8	58.3	79.3
760-761	Fort Worth	97.2	63.2	82.2
762	Denton	97.1	47.8	75.4
763	Wichita Falls	98.0	54.8	79.0
764	Eastland	96.5	49.7	75.9
765	Temple	95.1	49.4	75.0
766-767	Waco	97.4	56.4	79.4
768	Brownwood	97.7	49.0	76.2
769	San Angelo	97.3	49.6	76.3
770-772	Houston	99.1	69.8	86.2
773	Huntsville	97.5	55.9	79.2
774	Wharton	98.5	54.0	78.9
775	Galveston	96.5	68.2	84.0
776-777	Beaumont	97.2	62.0	81.7
778	Bryan	94.3	61.7	79.9
779	Victoria	98.6	44.0	74.5
780	Laredo	95.8	52.7	76.8
781-782	San Antonio	96.2	65.0	82.4
783-784	Corpus Christi	98.9	52.3	78.4
785	Mc Allen	98.6	46.6	75.7
786-787	Austin	96.1	58.8	79.6

STATE/ZIP	CITY	MAT.	INST.	TOTAL
TEXAS (CONT'D)				
788	Del Rio	98.0	46.5	75.3
789	Giddings	95.1	56.2	77.9
790-791	Amarillo	98.1	61.0	81.7
792	Childress	97.3	57.2	79.6
793-794	Lubbock	99.6	56.8	80.7
795-796	Abilene	97.9	54.2	78.6
797	Midland	99.4	51.4	78.2
798-799,885	El Paso	96.8	51.2	76.7
UTAH				
840-841	Salt Lake City	101.6	68.4	86.9
842,844	Ogden	97.3	70.0	85.3
843	Logan	99.2	70.0	86.3
845	Price	99.5	65.9	84.7
846-847	Provo	99.7	68.3	85.8
VERMONT				
050	White River Jct.	98.0	54.4	78.8
051	Bellows Falls	96.5	72.7	86.0
052	Bennington	96.8	67.8	84.0
053	Brattleboro	97.2	72.9	86.5
054	Burlington	100.6	65.4	85.1
056	Montpelier	97.4	65.4	83.3
057	Rutland	99.2	65.4	84.3
058	St. Johnsbury	97.9	55.2	79.1
059	Guildhall	96.6	55.0	78.3
VIRGINIA				
220-221	Fairfax	100.1	82.0	92.1
222	Arlington	101.6	79.9	92.0
223	Alexandria	100.6	86.0	94.2
224-225	Fredericksburg	98.8	71.8	86.9
226	Winchester	99.4	62.2	83.0
227	Culpeper	99.3	67.0	85.1
228	Harrisonburg	99.6	63.0	83.4
229	Charlottesville	100.0	62.8	83.6
230-232	Richmond	100.6	66.0	85.3
233-235	Norfolk	101.4	67.6	86.5
236	Newport News	100.4	67.3	85.8
237	Portsmouth	99.9	64.4	84.3
238	Petersburg	99.9	66.0	85.0
239	Farmville	98.9	52.3	78.4
240-241	Roanoke	101.2	59.8	82.9
242	Bristol	98.6	54.3	79.1
243	Pulaski	98.3	50.3	77.1
244	Staunton	99.2	58.2	81.1
245	Lynchburg	99.3	63.6	83.5
246	Grundy	98.6	49.7	77.0
WASHINGTON				
980-981,987	Seattle	102.5	106.3	104.2
982	Everett	102.2	95.6	99.3
983-984	Tacoma	102.2	97.9	100.3
985	Olympia	99.4	97.9	98.7
986	Vancouver	103.5	93.0	98.8
988	Wenatchee	101.8	83.6	93.8
989	Yakima	102.5	92.6	98.1
990-992	Spokane	102.5	82.7	93.7
993	Richland	102.1	86.8	95.4
994	Clarkston	100.4	81.3	92.0
WEST VIRGINIA				
247-248	Bluefield	97.4	82.6	90.9
249	Lewisburg	99.1	82.7	91.9
250-253	Charleston	99.2	89.5	94.9
254	Martinsburg	98.9	79.1	90.2
255-257	Huntington	100.6	91.4	96.6
258-259	Beckley	97.3	86.6	92.6
260	Wheeling	100.7	89.2	95.7
261	Parkersburg	99.6	88.4	94.7
262	Buckhannon	98.8	89.0	94.5
263-264	Clarksburg	99.4	89.1	94.9
265	Morgantown	99.5	88.2	94.5
266	Gassaway	98.7	89.8	94.8
267	Romney	98.7	85.1	92.7
268	Petersburg	98.5	84.9	92.5
WISCONSIN				
530,532	Milwaukee	100.9	108.3	104.2
531	Kenosha	100.8	101.0	100.9
534	Racine	100.1	101.0	100.5
535	Beloit	100.0	96.0	98.3
537	Madison	99.3	97.2	98.4

STATE/ZIP	CITY	MAT.	INST.	TOTAL
WISCONSIN (CONT'D)				
538	Lancaster	97.6	91.5	94.9
539	Portage	96.1	94.5	95.4
540	New Richmond	97.3	94.4	96.0
541-543	Green Bay	101.8	91.6	97.3
544	Wausau	96.5	91.7	94.4
545	Rhinelander	99.8	90.5	95.7
546	La Crosse	98.1	93.2	95.9
547	Eau Claire	99.7	94.3	97.3
548	Superior	97.1	96.9	97.0
549	Oshkosh	97.2	82.6	90.8
WYOMING				
820	Cheyenne	99.1	63.8	83.5
821	Yellowstone Nat'l Park	96.4	59.3	80.1
822	Wheatland	97.5	58.2	80.2
823	Rawlins	99.0	57.0	80.5
824	Worland	97.1	57.0	79.4
825	Riverton	98.1	57.0	80.0
826	Casper	98.9	57.7	80.7
827	Newcastle	96.9	57.0	79.3
828	Sheridan	99.7	60.6	82.4
829-831	Rock Springs	101.0	57.2	81.7
CANADIAN FACTORS (reflect Canadian currency)				
ALBERTA				
	Calgary	129.8	93.1	113.6
	Edmonton	130.0	93.1	113.7
	Fort McMurray	126.5	95.8	113.0
	Lethbridge	120.8	95.3	109.5
	Lloydminster	116.0	91.8	105.4
	Medicine Hat	116.1	91.1	105.1
	Red Deer	116.6	91.1	105.3
BRITISH COLUMBIA				
	Kamloops	116.9	94.4	107.0
	Prince George	118.1	93.7	107.3
	Vancouver	128.1	89.1	110.9
	Victoria	118.1	87.2	104.5
MANITOBA				
	Brandon	116.2	79.0	99.8
	Portage la Prairie	116.0	77.2	98.9
	Winnipeg	128.4	72.4	103.7
NEW BRUNSWICK				
	Bathurst	114.1	70.1	94.7
	Dalhousie	114.1	70.3	94.8
	Fredericton	117.0	75.6	98.8
	Moncton	114.3	71.7	95.5
	Newcastle	114.1	70.8	95.0
	St. John	116.6	75.3	98.4
NEWFOUNDLAND				
	Corner Brook	119.5	70.5	97.9
	St Johns	120.3	75.7	100.6
NORTHWEST TERRITORIES				
	Yellowknife	123.5	88.9	108.3
NOVA SCOTIA				
	Bridgewater	115.6	77.8	98.9
	Dartmouth	117.2	77.8	99.8
	Halifax	119.0	80.3	101.9
	New Glasgow	115.3	77.8	98.8
	Sydney	112.8	77.8	97.4
	Truro	115.1	77.8	98.6
	Yarmouth	115.2	77.8	98.7
ONTARIO				
	Barrie	118.6	96.9	109.1
	Brantford	117.7	101.0	110.4
	Cornwall	117.7	97.2	108.7
	Hamilton	121.9	93.7	109.4
	Kingston	118.6	97.3	109.2
	Kitchener	113.7	90.4	103.4
	London	121.6	90.2	107.8
	North Bay	118.0	94.9	107.8
	Oshawa	117.1	98.2	108.7
	Ottawa	122.9	90.7	108.7
	Owen Sound	118.8	95.0	108.3
	Peterborough	117.7	97.0	108.6
	Sarnia	117.4	101.6	110.5
ONTARIO (CONT'D)	Sault Ste Marie	112.6	94.5	104.6

Location Factors

STATE/ZIP	CITY	MAT.	INST.	TOTAL
	St. Catharines	111.7	91.3	102.8
	Sudbury	112.3	90.4	102.7
	Thunder Bay	113.1	90.2	103.0
	Timmins	117.9	94.9	107.8
	Toronto	123.0	98.8	112.3
	Windsor	112.2	90.0	102.4
PRINCE EDWARD ISLAND				
	Charlottetown	118.2	65.4	95.0
	Summerside	117.0	65.5	94.3
QUEBEC				
	Cap-de-la-Madeleine	114.8	91.3	104.4
	Charlesbourg	114.8	91.3	104.4
	Chicoutimi	113.7	91.6	104.0
	Gatineau	114.4	91.1	104.2
	Granby	114.7	91.0	104.3
	Hull	114.5	91.1	104.2
	Joliette	114.9	91.3	104.5
	Laval	114.0	91.1	103.9
	Montreal	121.9	91.9	108.7
	Quebec	119.8	92.3	107.7
	Rimouski	114.2	91.6	104.2
	Rouyn-Noranda	114.4	91.1	104.1
	Saint Hyacinthe	113.6	91.1	103.7
	Sherbrooke	114.8	91.1	104.4
	Sorel	114.9	91.3	104.5
	St Jerome	114.4	91.1	104.1
	Trois Rivieres	115.1	91.3	104.6
SASKATCHEWAN				
	Moose Jaw	113.4	72.4	95.3
	Prince Albert	112.3	70.5	93.9
	Regina	116.7	86.2	103.2
	Saskatoon	113.2	86.2	101.3
YUKON				
	Whitehorse	116.5	71.3	96.6

R011105-05   Tips for Accurate Estimating

1. Use pre-printed or columnar forms for orderly sequence of dimensions and locations and for recording telephone quotations.

2. Use only the front side of each paper or form except for certain pre-printed summary forms.

3. Be consistent in listing dimensions: For example, length x width x height. This helps in rechecking to ensure that, the total length of partitions is appropriate for the building area.

4. Use printed (rather than measured) dimensions where given.

5. Add up multiple printed dimensions for a single entry where possible.

6. Measure all other dimensions carefully.

7. Use each set of dimensions to calculate multiple related quantities.

8. Convert foot and inch measurements to decimal feet when listing. Memorize decimal equivalents to .01 parts of a foot (1/8″ equals approximately .01′).

9. Do not "round off" quantities until the final summary.

10. Mark drawings with different colors as items are taken off.

11. Keep similar items together, different items separate.

12. Identify location and drawing numbers to aid in future checking for completeness.

13. Measure or list everything on the drawings or mentioned in the specifications.

14. It may be necessary to list items not called for to make the job complete.

15. Be alert for: Notes on plans such as N.T.S. (not to scale); changes in scale throughout the drawings; reduced size drawings; discrepancies between the specifications and the drawings.

16. Develop a consistent pattern of performing an estimate. For example:
    a. Start the quantity takeoff at the lower floor and move to the next higher floor.
    b. Proceed from the main section of the building to the wings.
    c. Proceed from south to north or vice versa, clockwise or counterclockwise.
    d. Take off floor plan quantities first, elevations next, then detail drawings.

17. List all gross dimensions that can be either used again for different quantities, or used as a rough check of other quantities for verification (exterior perimeter, gross floor area, individual floor areas, etc.).

18. Utilize design symmetry or repetition (repetitive floors, repetitive wings, symmetrical design around a center line, similar room layouts, etc.). Note: Extreme caution is needed here so as not to omit or duplicate an area.

19. Do not convert units until the final total is obtained. For instance, when estimating concrete work, keep all units to the nearest cubic foot, then summarize and convert to cubic yards.

20. When figuring alternatives, it is best to total all items involved in the basic system, then total all items involved in the alternates. Therefore you work with positive numbers in all cases. When adds and deducts are used, it is often confusing whether to add or subtract a portion of an item; especially on a complicated or involved alternate.

R011110-10   Architectural Fees

Tabulated below are typical percentage fees by project size, for good professional architectural service. Fees may vary from those listed depending upon degree of design difficulty and economic conditions in any particular area.

Rates can be interpolated horizontally and vertically. Various portions of the same project requiring different rates should be adjusted proportionally. For alterations, add 50% to the fee for the first $500,000 of project cost and add 25% to the fee for project cost over $500,000.

Architectural fees tabulated below include Structural, Mechanical and Electrical Engineering Fees. They do not include the fees for special consultants such as kitchen planning, security, acoustical, interior design, etc.

Civil Engineering fees are included in the Architectural fee for project sites requiring minimal design such as city sites. However, separate Civil Engineering fees must be added when utility connections require design, drainage calculations are needed, stepped foundations are required, or provisions are required to protect adjacent wetlands.

Building Types	Total Project Size in Thousands of Dollars						
	100	250	500	1,000	5,000	10,000	50,000
Factories, garages, warehouses, repetitive housing	9.0%	8.0%	7.0%	6.2%	5.3%	4.9%	4.5%
Apartments, banks, schools, libraries, offices, municipal buildings	12.2	12.3	9.2	8.0	7.0	6.6	6.2
Churches, hospitals, homes, laboratories, museums, research	15.0	13.6	12.7	11.9	9.5	8.8	8.0
Memorials, monumental work, decorative furnishings	—	16.0	14.5	13.1	10.0	9.0	8.3

General Requirements R0111 Summary of Work

R011110-30 Engineering Fees

Typical **Structural Engineering Fees** based on type of construction and total project size. These fees are included in Architectural Fees.

Type of Construction	Total Project Size (in thousands of dollars)			
	$500	$500-$1,000	$1,000-$5,000	Over $5000
Industrial buildings, factories & warehouses	Technical payroll times 2.0 to 2.5	1.60%	1.25%	1.00%
Hotels, apartments, offices, dormitories, hospitals, public buildings, food stores		2.00%	1.70%	1.20%
Museums, banks, churches and cathedrals		2.00%	1.75%	1.25%
Thin shells, prestressed concrete, earthquake resistive		2.00%	1.75%	1.50%
Parking ramps, auditoriums, stadiums, convention halls, hangars & boiler houses		2.50%	2.00%	1.75%
Special buildings, major alterations, underpinning & future expansion		Add to above 0.5%	Add to above 0.5%	Add to above 0.5%

For complex reinforced concrete or unusually complicated structures, add 20% to 50%.

Typical **Mechanical and Electrical Engineering Fees** are based on the size of the subcontract. The fee structure for both are shown below. These fees are included in Architectural Fees.

Type of Construction	Subcontract Size							
	$25,000	$50,000	$100,000	$225,000	$350,000	$500,000	$750,000	$1,000,000
Simple structures	6.4%	5.7%	4.8%	4.5%	4.4%	4.3%	4.2%	4.1%
Intermediate structures	8.0	7.3	6.5	5.6	5.1	5.0	4.9	4.8
Complex structures	10.1	9.0	9.0	8.0	7.5	7.5	7.0	7.0

For renovations, add 15% to 25% to applicable fee.

General Requirements R0121 Allowances

R012157-20 Construction Time Requirements

Table lists average construction time in months for different types, sizes, and values of building projects. Design time runs 25% to 40% of construction time.

Building Type	Size S.F.	Project Value	Construction Duration
Industrial/Warehouse	100,000	$8,000,000	14 Months
	500,000	$32,000,000	19 Months
	1,000,000	$75,000,000	21 Months
Offices/Retail	50,000	$7,000,000	15 Months
	250,000	$28,000,000	23 months
	500,000	$58,000,000	34 months
Institutional/Hospitals/Laboratory	200,000	$45,000,000	31 months
	500,000	$110,000,000	52 months
	750,000	$160,000,000	55 months
	1,000,000	$210,000,000	60 months

R012909-80   Sales Tax by State

State sales tax on materials is tabulated below (5 states have no sales tax). Many states allow local jurisdictions, such as a county or city, to levy additional sales tax.

Some projects may be sales tax exempt, particularly those constructed with public funds.

State	Tax (%)	State	Tax (%)	State	Tax (%)	State	Tax (%)
Alabama	4	Illinois	6.25	Montana	0	Rhode Island	7
Alaska	0	Indiana	7	Nebraska	5.5	South Carolina	6
Arizona	6.6	Iowa	6	Nevada	6.85	South Dakota	4
Arkansas	6	Kansas	5.3	New Hampshire	0	Tennessee	7
California	9.25	Kentucky	6	New Jersey	7	Texas	6.25
Colorado	2.9	Louisiana	4	New Mexico	5.13	Utah	4.75
Connecticut	6	Maine	5	New York	4	Vermont	6
Delaware	0	Maryland	6	North Carolina	5.75	Virginia	4
District of Columbia	6	Massachusetts	6.25	North Dakota	5	Washington	6.5
Florida	6	Michigan	6	Ohio	5.5	West Virginia	6
Georgia	4	Minnesota	6.875	Oklahoma	4.5	Wisconsin	5
Hawaii	4	Mississippi	7	Oregon	0	Wyoming	4
Idaho	6	Missouri	4.225	Pennsylvania	6	Average	5.07 %

Sales Tax by Province (Canada)

GST - a value-added tax, which the government imposes on most goods and services provided in or imported into Canada. PST - a retail sales tax, which three of the provinces impose on the price of most goods and some services. QST - a value-added tax, similar to the federal GST, which Quebec imposes. HST - Five provinces have combined their retail sales tax with the federal GST into one harmonized tax.

Province	PST (%)	QST (%)	GST(%)	HST(%)
Alberta	0	0	5	0
British Columbia	0	0	0	12
Manitoba	7	0	5	0
New Brunswick	0	0	0	13
Newfoundland	0	0	0	13
Northwest Territories	0	0	5	0
Nova Scotia	0	0	0	15
Ontario	0	0	0	13
Prince Edward Island	10	0	5	0
Quebec	0	8.5	5	0
Saskatchewan	5	0	5	0
Yukon	0	0	5	0

R012909-85 Unemployment Taxes and Social Security Taxes

State Unemployment Tax rates vary not only from state to state, but also with the experience rating of the contractor. The Federal Unemployment Tax rate is 6.2% of the first $7,000 of wages. This is reduced by a credit of up to 5.4% for timely payment to the state. The minimum Federal Unemployment Tax is 0.8% after all credits.

Social Security (FICA) for 2012 is estimated at time of publication to be 7.65% of wages up to $110,100.

R012909-90 Overtime

One way to improve the completion date of a project or eliminate negative float from a schedule is to compress activity duration times. This can be achieved by increasing the crew size or working overtime with the proposed crew.

To determine the costs of working overtime to compress activity duration times, consider the following examples. Below is an overtime efficiency and

cost chart based on a five, six, or seven day week with an eight through twelve hour day. Payroll percentage increases for time and one half and double time are shown for the various working days.

Days per Week	Hours per Day	Production Efficiency					Payroll Cost Factors	
		1st Week	2nd Week	3rd Week	4th Week	Average 4 Weeks	@ 1-1/2 Times	@ 2 Times
5	8	100%	100%	100%	100%	100 %	100 %	100 %
	9	100	100	95	90	96.25	105.6	111.1
	10	100	95	90	85	91.25	110.0	120.0
	11	95	90	75	65	81.25	113.6	127.3
	12	90	85	70	60	76.25	116.7	133.3
6	8	100	100	95	90	96.25	108.3	116.7
	9	100	95	90	85	92.50	113.0	125.9
	10	95	90	85	80	87.50	116.7	133.3
	11	95	85	70	65	78.75	119.7	139.4
	12	90	80	65	60	73.75	122.2	144.4
7	8	100	95	85	75	88.75	114.3	128.6
	9	95	90	80	70	83.75	118.3	136.5
	10	90	85	75	65	78.75	121.4	142.9
	11	85	80	65	60	72.50	124.0	148.1
	12	85	75	60	55	68.75	126.2	152.4

R013113-40 Builder's Risk Insurance

Builder's Risk Insurance is insurance on a building during construction. Premiums are paid by the owner or the contractor. Blasting, collapse and underground insurance would raise total insurance costs above those listed. Floater policy for materials delivered to the job runs $.75 to $1.25 per $100 value. Contractor equipment insurance runs $.50 to $1.50 per $100 value. Insurance for miscellaneous tools to $1,500 value runs from $3.00 to $7.50 per $100 value.

Tabulated below are New England Builder's Risk insurance rates in dollars per $100 value for $1,000 deductible. For $25,000 deductible, rates can be reduced 13% to 34%. On contracts over $1,000,000, rates may be lower than those tabulated. Policies are written annually for the total completed value in place. For "all risk" insurance (excluding flood, earthquake and certain other perils) add $.025 to total rates below.

Coverage	Frame Construction (Class 1)			Brick Construction (Class 4)			Fire Resistive (Class 6)		
	Range		Average	Range		Average	Range		Average
Fire Insurance	$.350 to	$ .850	$.600	$.158 to	$.189	$.174	$.052 to	$.080	$.070
Extended Coverage	.115 to	.200	.158	.080 to	.105	.101	.081 to	.105	.100
Vandalism	.012 to	.016	.014	.008 to	.011	.011	.008 to	.011	.010
Total Annual Rate	$.477 to	$1.066	$.772	$.246 to	$.305	$.286	$.141 to	$.196	$.180

Reference Tables

R013113-50   General Contractor's Overhead

There are two distinct types of overhead on a construction project: Project Overhead and Main Office Overhead. Project Overhead includes those costs at a construction site not directly associated with the installation of construction materials. Examples of Project Overhead costs include the following:

1. Superintendent
2. Construction office and storage trailers
3. Temporary sanitary facilities
4. Temporary utilities
5. Security fencing
6. Photographs
7. Clean up
8. Performance and payment bonds

The above Project Overhead items are also referred to as General Requirements and therefore are estimated in Division 1. Division 1 is the first division listed in the CSI MasterFormat but it is usually the last division estimated. The sum of the costs in Divisions 1 through 49 is referred to as the sum of the direct costs.

All construction projects also include indirect costs. The primary components of indirect costs are the contractor's Main Office Overhead and profit. The amount of the Main Office Overhead expense varies depending on the following:

1. Owner's compensation
2. Project managers and estimator's wages
3. Clerical support wages
4. Office rent and utilities
5. Corporate legal and accounting costs
6. Advertising
7. Automobile expenses
8. Association dues
9. Travel and entertainment expenses

These costs are usually calculated as a percentage of annual sales volume. This percentage can range from 35% for a small contractor doing less than $500,000 to 5% for a large contractor with sales in excess of $100 million.

R013113-60 Workers' Compensation Insurance Rates by Trade

The table below tabulates the national averages for Workers' Compensation insurance rates by trade and type of building. The average "Insurance Rate" is multiplied by the "% of Building Cost" for each trade. This produces the "Workers' Compensation Cost" by % of total labor cost, to be added for each trade by building type to determine the weighted average Workers' Compensation rate for the building types analyzed.

Trade	Insurance Rate (% Labor Cost)		% of Building Cost			Workers' Compensation		
	Range	Average	Office Bldgs.	Schools & Apts.	Mfg.	Office Bldgs.	Schools & Apts.	Mfg.
Excavation, Grading, etc.	3.9 % to 18.1%	9.0%	4.8%	4.9%	4.5%	0.43%	0.44%	0.41%
Piles & Foundations	5.1 to 59.5	16.8	7.1	5.2	8.7	1.19	0.87	1.46
Concrete	4.9 to 29.7	12.7	5.0	14.8	3.7	0.64	1.88	0.47
Masonry	4.4 to 28.7	12.1	6.9	7.5	1.9	0.83	0.91	0.23
Structural Steel	5.1 to 124.1	36.9	10.7	3.9	17.6	3.95	1.44	6.49
Miscellaneous & Ornamental Metals	4.2 to 22.6	10.1	2.8	4.0	3.6	0.28	0.40	0.36
Carpentry & Millwork	5.1 to 37	15.0	3.7	4.0	0.5	0.56	0.60	0.08
Metal or Composition Siding	4.9 to 43.3	14.9	2.3	0.3	4.3	0.34	0.04	0.64
Roofing	5.1 to 75.2	28.8	2.3	2.6	3.1	0.66	0.75	0.89
Doors & Hardware	2.8 to 37	10.5	0.9	1.4	0.4	0.09	0.15	0.04
Sash & Glazing	4.9 to 33.1	12.6	3.5	4.0	1.0	0.44	0.50	0.13
Lath & Plaster	2.5 to 30.5	10.8	3.3	6.9	0.8	0.36	0.75	0.09
Tile, Marble & Floors	2.3 to 18.9	8.0	2.6	3.0	0.5	0.21	0.24	0.04
Acoustical Ceilings	3.2 to 30.3	7.8	2.4	0.2	0.3	0.19	0.02	0.02
Painting	4.3 to 28.9	10.7	1.5	1.6	1.6	0.16	0.17	0.17
Interior Partitions	5.1 to 37	15.0	3.9	4.3	4.4	0.59	0.65	0.66
Miscellaneous Items	2.5 to 90.6	14.6	5.2	3.7	9.7	0.76	0.54	1.42
Elevators	1.8 to 12.8	5.9	2.1	1.1	2.2	0.12	0.06	0.13
Sprinklers	1.9 to 17.2	7.0	0.5	—	2.0	0.04	—	0.14
Plumbing	2.7 to 17.1	6.9	4.9	7.2	5.2	0.34	0.50	0.36
Heat., Vent., Air Conditioning	3.8 to 19.3	8.5	13.5	11.0	12.9	1.15	0.94	1.10
Electrical	2.6 to 13.2	5.6	10.1	8.4	11.1	0.57	0.47	0.62
Total	1.8 % to 124.1%	—	100.0%	100.0%	100.0%	13.90%	12.32%	15.95%
		Overall Weighted Average 14.06%						

Workers' Compensation Insurance Rates by States

The table below lists the weighted average Workers' Compensation base rate for each state with a factor comparing this with the national average of 13.7%.

State	Weighted Average	Factor	State	Weighted Average	Factor	State	Weighted Average	Factor
Alabama	19.6%	143	Kentucky	17.0%	124	North Dakota	10.8%	79
Alaska	16.3	119	Louisiana	15.6	114	Ohio	8.7	64
Arizona	9.0	66	Maine	14.0	102	Oklahoma	14.7	107
Arkansas	10.2	74	Maryland	15.6	114	Oregon	12.0	88
California	27.9	204	Massachusetts	12.4	91	Pennsylvania	16.3	119
Colorado	7.1	52	Michigan	20.6	150	Rhode Island	11.9	87
Connecticut	23.1	169	Minnesota	24.1	176	South Carolina	20.1	147
Delaware	9.3	68	Mississippi	13.1	96	South Dakota	16.0	117
District of Columbia	11.2	82	Missouri	17.4	127	Tennessee	13.2	96
Florida	10.8	79	Montana	14.2	104	Texas	10.6	77
Georgia	27.8	203	Nebraska	18.1	132	Utah	7.8	57
Hawaii	9.0	66	Nevada	10.4	76	Vermont	14.6	107
Idaho	10.8	79	New Hampshire	21.1	154	Virginia	8.7	64
Illinois	22.6	165	New Jersey	13.4	98	Washington	11.1	81
Indiana	5.5	40	New Mexico	17.8	130	West Virginia	12.6	92
Iowa	10.8	79	New York	5.9	43	Wisconsin	14.9	109
Kansas	8.0	58	North Carolina	17.6	128	Wyoming	5.7	42
			Weighted Average for U.S. is	14.1% of payroll = 100%				

Rates in the following table are the base or manual costs per $100 of payroll for Workers' Compensation in each state. Rates are usually applied to straight time wages only and not to premium time wages and bonuses.

The weighted average skilled worker rate for 35 trades is 13.7%. For bidding purposes, apply the full value of Workers' Compensation directly to total labor costs, or if labor is 38%, materials 42% and overhead and profit 20% of total cost, carry 38/80 x 13.7% =6.5% of cost (before overhead and profit) into overhead. Rates vary not only from state to state but also with the experience rating of the contractor.

Rates are the most current available at the time of publication.

R013113-60 Workers' Compensation Insurance Rates by Trade and State (cont.)

State	Carpentry – 3 stories or less (5651)	Carpentry – interior cab. work (5437)	Carpentry – general (5403)	Concrete Work – NOC (5213)	Concrete Work – flat (flr., sdwk.) (5221)	Electrical Wiring – inside (5190)	Excavation – earth NOC (6217)	Excavation – rock (6217)	Glaziers (5462)	Insulation Work (5479)	Lathing (5443)	Masonry (5022)	Painting & Decorating (5474)	Pile Driving (6003)	Plastering (5480)	Plumbing (5183)	Roofing (5551)	Sheet Metal Work (HVAC) (5538)	Steel Erection – door & sash (5102)	Steel Erection – inter. ornam. (5102)	Steel Erection – structure (5040)	Steel Erection – NOC (5057)	Tile Work – (interior ceramic) (5348)	Waterproofing (9014)	Wrecking (5701)
AL	29.28	13.11	25.00	12.94	9.22	8.58	11.91	11.91	20.78	14.99	8.43	22.04	16.81	18.81	14.91	8.97	48.49	9.00	10.13	10.13	46.10	24.11	14.15	8.68	46.10
AK	12.44	9.33	13.86	11.33	10.80	6.86	9.06	9.06	32.79	17.54	8.30	16.08	11.68	33.22	13.00	7.53	25.29	7.47	7.41	7.41	25.75	41.66	5.15	5.48	25.75
AZ	11.78	5.44	15.76	6.99	4.24	4.37	4.81	4.81	6.09	10.18	3.64	7.06	7.21	8.53	5.15	4.56	18.34	6.49	9.20	9.20	21.81	10.32	2.89	3.58	21.81
AR	11.58	5.85	10.93	6.39	5.50	4.31	6.73	6.73	7.95	8.55	4.39	7.75	8.88	10.93	12.95	4.70	17.06	5.13	6.99	6.99	28.58	21.04	6.21	2.97	28.58
CA	36.98	36.98	36.98	19.36	19.34	13.20	18.05	18.05	27.35	17.74	16.02	27.71	21.08	23.37	30.46	17.05	64.15	19.29	22.60	22.60	41.64	23.39	2.30	21.08	23.39
CO	7.62	4.14	5.79	6.03	4.35	2.67	4.88	4.88	4.93	7.66	3.52	7.28	5.10	7.13	5.24	3.62	13.42	3.77	5.01	5.01	24.23	9.35	3.77	2.76	21.38
CT	19.80	17.42	27.18	22.71	13.56	7.85	14.62	14.62	19.83	24.28	12.69	27.26	21.51	23.20	18.28	10.85	48.23	13.52	15.52	15.52	64.75	22.28	13.05	6.77	65.75
DE	9.15	9.15	7.05	7.35	6.12	3.18	5.54	5.54	7.90	7.05	7.90	8.28	9.53	11.07	7.90	4.76	17.93	6.13	8.83	8.83	17.30	8.83	5.60	8.28	17.30
DC	8.25	8.16	7.71	12.10	7.65	4.34	9.06	9.06	13.00	6.41	6.70	10.21	5.60	15.02	8.73	7.96	16.80	7.20	15.77	15.77	19.23	10.98	9.00	4.18	19.23
FL	9.33	7.33	10.95	11.45	5.16	5.13	5.96	5.96	8.48	8.55	4.12	8.77	7.91	38.24	11.10	4.82	17.10	6.49	8.06	8.06	19.04	10.13	4.43	4.41	19.04
GA	43.26	19.77	27.19	20.25	15.13	10.28	16.31	16.31	18.41	19.10	12.62	28.72	28.86	23.32	21.44	12.2	75.23	17.15	19.64	19.64	73.28	34.09	12.76	10.18	73.28
HI	7.84	6.87	14.67	7.64	4.33	4.13	5.18	5.18	8.07	5.61	6.55	7.82	6.96	9.14	8.22	4.26	13.77	4.11	6.51	6.51	28.23	9.49	5.22	5.37	20.79
ID	12.77	6.43	10.55	11.76	5.76	3.47	7.06	7.06	9.63	7.67	6.29	7.80	8.43	9.30	13.57	4.27	22.62	7.21	8.55	8.55	32.97	8.02	6.84	4.80	27.50
IL	23.60	15.01	21.72	29.69	12.06	8.24	11.22	11.22	18.09	20.32	12.88	20.85	13.07	21.77	22.24	10.77	36.73	12.98	19.24	19.24	79.82	21.35	16.91	5.78	29.82
IN	8.38	4.15	6.38	4.95	3.41	2.68	4.15	4.15	5.28	6.02	3.24	4.43	4.34	5.34	3.28	2.65	10.05	4.13	4.21	4.21	11.63	5.11	2.89	2.59	11.63
IA	10.34	8.42	12.90	12.72	8.24	4.47	6.88	6.88	9.20	5.82	4.64	8.40	6.76	8.73	10.98	6.40	19.92	4.80	5.70	5.70	36.26	14.65	6.70	4.60	24.10
KS	12.27	5.30	8.09	6.46	5.62	3.67	4.59	4.59	8.35	8.35	3.41	6.73	6.60	7.80	5.04	4.43	13.94	4.61	4.75	4.75	18.81	13.14	5.31	3.99	13.14
KY	12.63	12.30	24.88	12.96	7.51	5.36	11.65	11.65	16.68	13.04	8.83	7.70	11.00	21.98	12.55	5.87	34.41	11.42	10.07	10.07	44.77	17.23	18.88	6.58	44.77
LA	18.20	12.54	20.92	11.92	9.35	6.46	12.02	12.02	14.16	10.50	6.61	13.71	15.78	15.66	10.45	6.08	32.48	12.91	12.19	12.19	38.54	15.25	5.97	5.13	38.54
ME	14.80	10.05	21.30	15.69	9.73	5.57	10.34	10.34	14.40	14.60	5.58	10.56	14.17	12.31	10.83	7.41	24.74	8.22	7.40	7.40	38.54	15.25	5.97	5.13	38.54
MD	17.56	13.32	14.28	19.37	5.51	7.07	7.60	7.60	12.50	14.18	5.58	10.00	7.28	21.24	10.47	7.02	30.67	7.73	10.49	10.49	59.74	22.48	6.69	5.43	22.48
MA	8.68	5.23	9.61	18.85	6.24	2.84	4.35	4.35	9.58	7.78	5.27	10.55	5.09	12.92	4.68	3.50	30.99	5.72	6.89	6.89	54.08	33.00	5.81	2.48	23.75
MI	21.01	11.98	21.01	17.15	10.48	6.35	14.00	14.00	15.12	13.11	13.59	18.32	13.30	59.47	10.34	7.71	42.64	12.58	12.03	12.03	59.47	16.23	11.70	6.95	59.47
MN	19.68	20.00	29.85	10.35	13.68	5.50	10.18	10.18	33.05	22.80	10.63	15.00	14.53	30.90	10.63	8.08	62.73	17.35	9.73	9.73	124.08	8.93	11.70	7.20	8.93
MS	17.32	9.93	13.14	9.84	5.57	4.62	9.84	9.84	11.42	9.11	4.97	12.50	9.60	17.78	10.10	8.54	33.19	6.45	13.01	13.01	30.00	8.32	7.92	4.45	30.00
MO	18.26	12.25	13.57	17.90	11.00	7.12	11.34	11.34	10.74	13.27	9.16	15.43	11.27	18.34	14.38	10.70	36.70	9.28	11.58	11.58	38.70	38.11	13.39	6.50	38.70
MT	13.20	10.71	20.41	12.22	10.37	5.23	12.92	12.92	10.67	19.74	7.41	9.78	8.66	15.08	10.75	7.27	39.34	8.38	9.64	9.64	24.14	9.36	8.36	7.78	9.36
NE	20.53	15.68	20.38	19.08	12.10	7.83	15.30	15.30	12.18	19.13	8.00	18.20	11.23	15.73	13.68	6.22	34.05	11.63	13.15	13.15	44.88	22.70	8.23	6.70	63.83
NV	10.76	7.25	11.35	10.52	7.16	4.80	8.48	8.48	6.75	8.06	3.45	6.62	6.07	8.88	6.41	6.22	17.81	8.84	7.73	7.73	25.26	15.85	13.93	5.16	15.85
NH	20.77	11.63	18.79	25.37	13.01	5.70	13.22	13.22	12.59	13.98	7.47	19.78	24.64	15.55	10.58	10.41	35.43	8.15	11.93	11.93	99.47	28.64	12.59	6.79	99.47
NJ	15.50	9.86	15.50	14.21	9.79	4.45	8.75	8.75	7.99	10.68	12.42	14.48	10.11	14.56	12.42	6.24	38.07	6.71	10.85	10.85	16.29	15.14	7.82	5.82	15.93
NM	19.58	8.17	18.99	13.33	10.11	8.08	8.28	8.28	18.17	11.75	6.68	15.36	11.43	18.20	14.39	8.14	37.47	9.92	12.79	12.79	54.49	40.63	7.66	6.66	54.49
NY	4.93	2.81	6.01	7.46	5.60	2.64	3.94	3.94	6.56	3.28	4.63	6.16	4.85	6.74	2.51	3.31	12.36	4.57	5.46	5.46	10.44	6.29	3.12	3.09	4.57
NC	15.50	11.55	18.63	17.12	8.14	9.96	13.11	13.11	11.87	12.57	9.34	11.05	11.85	16.82	13.63	8.92	33.88	11.87	11.97	11.97	70.40	20.61	8.12	5.41	70.40
ND	10.78	10.78	10.78	4.92	4.92	3.79	4.18	4.18	10.78	10.78	8.88	6.00	6.57	12.68	8.88	6.21	19.53	6.21	12.68	12.68	12.68	12.68	10.78	19.53	8.53
OH	6.82	4.77	6.03	4.98	4.96	3.17	4.83	4.83	9.26	9.83	30.29	7.62	7.64	7.67	7.20	3.94	15.01	4.87	6.17	6.17	12.91	6	6.11	4.90	6.00
OK	13.42	9.17	11.76	11.76	6.34	5.73	10.83	10.83	16.07	19.47	8.45	10.02	8.47	20.25	12.19	7.09	21.13	9.05	13.76	13.76	39.72	23.97	6.50	5.71	39.72
OR	19.79	8.26	13.53	10.34	8.96	5.18	9.03	9.03	10.65	10.89	6.79	11.80	10.85	12.18	9.08	5.38	26.12	6.49	6.82	6.82	20.28	15.61	6.68	5.72	20.28
PA	16.83	16.83	13.22	16.61	11.93	6.69	9.79	9.79	12.75	13.22	12.75	14.24	14.78	17.14	12.75	8.62	33.19	8.74	16.63	16.63	25.25	16.63	9.61	14.24	25.25
RI	9.69	10.13	12.42	13.18	10.51	4.12	8.56	8.56	10.26	13.56	4.93	8.95	9.48	20.95	8.02	5.77	19.37	6.38	6.06	6.06	28.84	20.64	4.60	4.61	28.84
SC	21.43	18.13	22.49	16.89	9.59	11.45	11.78	11.78	16.99	18.58	11.06	13.93	15.50	20.61	15.75	12.36	50.08	13.06	13.91	13.91	41.22	32.53	10.50	6.33	41.22
SD	18.86	13.19	18.65	20.78	8.58	4.84	9.13	9.13	14.61	14.35	6.47	11.88	10.04	15.98	10.23	9.60	31.05	9.72	11.92	11.92	45.18	23.68	6.07	6.02	45.18
TN	16.91	10.01	10.79	11.98	5.04	5.70	11.30	11.30	11.59	8.82	7.16	12.26	9.19	21.55	10.85	5.70	30.21	7.60	8.74	8.74	22.74	23.16	7.39	4.11	22.74
TX	9.68	7.76	9.68	8.61	6.25	5.65	7.33	7.33	7.82	10.32	4.91	8.32	8.04	12.94	8.04	5.47	17.96	13.94	7.16	7.16	29.32	11.95	5.11	5.97	8.11
UT	9.81	5.76	7.93	7.13	5.38	3.09	6.77	6.77	7.13	5.43	4.01	6.86	5.93	7.69	4.67	3.28	21.47	5.41	4.91	4.91	17.18	8.18	3.74	4.02	18.30
VT	13.01	12.79	15.72	11.55	8.77	4.46	10.51	10.51	13.83	14.05	6.01	13.93	9.03	12.33	8.89	8.77	26.46	7.73	10.42	10.42	45.01	23.47	7.90	7.15	45.01
VA	8.91	6.79	7.37	8.34	4.22	3.84	6.23	6.23	6.35	5.77	6.19	6.15	7.43	7.98	6.26	4.34	21.14	5.67	6.19	6.19	25.35	11.51	4.06	2.65	25.35
WA	8.93	8.93	8.93	8.93	8.93	3.27	5.79	5.79	13.83	8.00	8.93	11.74	14.05	20.72	10.68	4.12	19.61	4.58	8.62	8.62	8.62	8.62	7.66	19.61	8.62
WV	13.52	9.63	10.91	11.26	6.25	6.07	9.18	9.18	11.22	9.76	7.21	10.37	10.25	15.47	9.74	6.22	25.10	7.93	8.31	8.31	34.41	16.71	7.86	3.96	32.85
WI	13.04	10.81	16.27	12.16	8.90	5.66	7.72	7.72	12.52	12.83	4.69	15.44	12.07	26.59	10.07	5.82	31.84	8.36	11.02	11.02	21.68	30.03	16.47	5.33	21.68
WY	5.07	5.07	5.07	5.07	5.07	5.07	5.07	5.07	5.07	5.07	5.07	5.07	5.07	5.07	5.07	5.07	5.07	5.07	5.07	5.07	5.07	5.07	5.07	5.07	5.07
AVG.	14.90	10.53	14.96	12.70	8.24	5.58	9.01	9.01	12.57	11.85	7.82	12.10	10.70	16.76	10.78	6.91	28.83	8.47	10.07	10.07	36.86	17.78	8.01	6.41	30.43

R013113-60 Workers' Compensation (cont.) (Canada in Canadian dollars)

Province		Alberta	British Columbia	Manitoba	Ontario	New Brunswick	Newfndld. & Labrador	Northwest Territories	Nova Scotia	Prince Edward Island	Quebec	Saskat- chewan	Yukon
Carpentry—3 stories or less	Rate	6.07	3.93	3.96	4.44	4.62	9.44	3.92	5.70	6.62	12.61	3.74	8.55
	Code	42143	721028	40102	723	238130	4226	4-41	4226	403	80110	B1226	202
Carpentry—interior cab. work	Rate	2.16	2.68	3.96	4.44	4.37	4.24	3.92	4.83	3.69	12.61	2.24	8.55
	Code	42133	721021	40102	723	238350	4279	4-41	4274	402	80110	B1127	202
CARPENTRY—general	Rate	6.07	3.93	3.93	4.44	4.62	4.24	3.92	5.70	6.62	12.61	3.74	8.55
	Code	42143	721028	40102	723	238130	4299	4-41	4226	403	80110	B1202	202
CONCRETE WORK—NOC	Rate	3.86	3.28	6.27	17.86	4.62	9.44	3.92	5.03	3.69	12.86	4.28	8.55
	Code	42104	721010	40110	748	238110	4224	4-41	4224	402	80100	B1314	203
CONCRETE WORK—flat (flr. sidewalk)	Rate	3.86	3.28	6.27	17.86	4.62	9.44	3.92	5.03	3.69	12.86	4.28	8.55
	Code	42104	721010	40110	748	238110	4224	4-41	4224	402	80100	B1314	203
ELECTRICAL Wiring—inside	Rate	1.72	1.71	1.74	3.60	2.09	2.57	2.67	2.68	3.69	4.55	2.24	4.96
	Code	42124	721019	40203	704	238210	4261	4-46	4261	402	80170	B1105	206
EXCAVATION—earth NOC	Rate	2.18	3.62	4.52	5.16	4.62	3.59	3.90	3.16	3.69	8.33	2.21	4.96
	Code	40604	721031	4	711	238190	4214	4-43	4214	402	80030	R1106	207
EXCAVATION—rock	Rate	2.18	3.62	4.19	5.16	4.62	3.59	3.90	3.16	3.69	8.33	2.21	4.96
	Code	40604	721031	40706	711	238190	4214	4-43	4214	402	80030	R1106	207
GLAZIERS	Rate	2.34	3.05	3.96	10.00	4.62	5.25	3.92	5.70	3.69	14.52	4.28	4.96
	Code	42121	715020	40109	751	238150	4233	4-41	4233	402	80150	B1304	212
INSULATION WORK	Rate	1.74	6.88	3.96	10.00	4.37	5.25	3.92	5.70	3.69	12.61	3.74	8.55
	Code	42184	721029	40102	751	238310	4234	4-41	4234	402	80110	B1207	202
LATHING	Rate	3.78	7.08	3.96	4.44	4.37	4.24	3.92	4.83	3.69	12.61	4.28	8.55
	Code	42135	721042	40102	723	238390	4279	4-41	4271	402	80110	B1316	202
MASONRY	Rate	3.84	3.02	3.96	12.39	4.62	5.25	3.92	5.70	3.69	12.86	4.28	8.55
	Code	42102	721037	40102	741	238140	4231	4-41	4231	402	80100	B1318	202
PAINTING & DECORATING	Rate	2.90	3.12	3.44	7.33	4.37	4.24	3.92	4.83	3.69	12.61	3.74	8.55
	Code	42111	721041	40105	719	238320	4275	4-41	4275	402	80110	B1201	202
PILE DRIVING	Rate	3.77	3.92	4.19	6.86	4.62	3.59	3.90	5.03	6.62	8.33	4.28	8.55
	Code	42159	722004	40706	732	238190	4129	4-43	4221	403	80030	B1310	202
PLASTERING	Rate	3.78	7.08	4.48	7.33	4.37	4.24	3.92	4.83	3.69	12.61	3.74	8.55
	Code	42135	721042	40108	719	238390	4271	4-41	4271	402	80110	B1221	202
PLUMBING	Rate	1.72	2.26	2.93	4.06	2.09	2.68	2.67	3.14	2.21	6.44	2.24	4.96
	Code	42122	721043	40204	707	238220	4241	4-46	4241	401	80160	B1101	214
ROOFING	Rate	5.85	6.52	6.74	14.44	4.62	9.44	3.92	7.76	3.69	19.78	4.28	8.55
	Code	42118	721036	40403	728	238160	4236	4-41	4236	402	80130	B1320	202
SHEET METAL WORK (HVAC)	Rate	1.72	2.26	6.74	4.06	2.09	2.68	2.67	3.14	2.21	6.44	2.24	4.96
	Code	42117	721043	40402	707	238220	4244	4-46	4244	401	80160	B1107	208
STEEL ERECTION—door & sash	Rate	1.74	12.03	8.24	17.86	4.62	9.44	3.92	5.70	3.69	20.90	4.28	8.55
	Code	42106	722005	40502	748	238120	4227	4-41	4227	402	80080	B1322	202
STEEL ERECTION—inter., ornam.	Rate	1.74	12.03	8.24	17.86	4.62	9.44	3.92	5.70	3.69	20.90	4.28	8.55
	Code	42106	722005	40502	748	238120	4227	4-41	4227	402	80080	B1322	202
STEEL ERECTION—structure	Rate	1.74	12.03	8.24	17.86	4.62	9.44	3.92	5.70	3.69	20.90	4.28	8.55
	Code	42106	722005	40502	748	238120	4227	4-41	4227	402	80080	B1322	202
STEEL ERECTION—NOC	Rate	1.74	12.03	8.24	17.86	4.62	9.44	3.92	5.70	3.69	20.90	4.28	8.55
	Code	42106	722005	40502	748	238120	4227	4-41	4227	402	80080	B1322	202
TILE WORK—inter. (ceramic)	Rate	2.80	3.34	1.90	7.33	4.37	4.24	3.92	4.83	3.69	12.61	4.28	8.55
	Code	42113	721054	40103	719	238340	4276	4-41	4276	402	80110	B1301	202
WATERPROOFING	Rate	2.90	3.10	3.96	4.44	4.62	4.24	3.92	5.70	3.69	17.81	3.74	8.55
	Code	42139	721016	40102	723	238190	4299	4-41	4239	402	80130	B1217	202
WRECKING	Rate	2.18	4.81	7.34	17.86	4.62	3.59	3.90	3.16	5.61	12.61	2.21	8.55
	Code	40604	721005	40106	748	238190	4211	4-43	4211	402	80110	R1125	202

R013113-80 Performance Bond

This table shows the cost of a Performance Bond for a construction job scheduled to be completed in 12 months. Add 1% of the premium cost per month for jobs requiring more than 12 months to complete. The rates are "standard" rates offered to contractors that the bonding company considers financially sound and capable of doing the work. Preferred rates are offered by some bonding companies based upon financial strength of the contractor. Actual rates vary from contractor to contractor and from bonding company to bonding company. Contractors should prequalify through a bonding agency before submitting a bid on a contract that requires a bond.

Contract Amount		Building Construction Class B Projects			Highways & Bridges					
					Class A New Construction			Class A-1 Highway Resurfacing		
First $ 100,000 bid		$25.00 per M			$15.00 per M			$9.40 per M		
Next 400,000 bid	$ 2,500	plus	$15.00	per M	$ 1,500	plus $10.00	per M	$ 940	plus $7.20	per M
Next 2,000,000 bid	8,500	plus	10.00	per M	5,500	plus 7.00	per M	3,820	plus 5.00	per M
Next 2,500,000 bid	28,500	plus	7.50	per M	19,500	plus 5.50	per M	15,820	plus 4.50	per M
Next 2,500,000 bid	47,250	plus	7.00	per M	33,250	plus 5.00	per M	28,320	plus 4.50	per M
Over 7,500,000 bid	64,750	plus	6.00	per M	45,750	plus 4.50	per M	39,570	plus 4.00	per M

R015113-65 Temporary Power Equipment

Cost data for the temporary equipment was developed utilizing the following information.

1) Re-usable material-services, transformers, equipment and cords are based on new purchase and prorated to three projects.
2) PVC feeder includes trench and backfill.
3) Connections include disconnects and fuses.
4) Labor units include an allowance for removal.
5) No utility company charges or fees are included.
6) Concrete pads or vaults are not included.
7) Utility company conduits not included.

R015423-10　Steel Tubular Scaffolding

On new construction, tubular scaffolding is efficient up to 60′ high or five stories. Above this it is usually better to use a hung scaffolding if construction permits. Swing scaffolding operations may interfere with tenants. In this case, the tubular is more practical at all heights.

In repairing or cleaning the front of an existing building the cost of tubular scaffolding per S.F. of building front increases as the height increases above the first tier. The first tier cost is relatively high due to leveling and alignment.

The minimum efficient crew for erecting and dismantling is three workers. They can set up and remove 18 frame sections per day up to 5 stories high. For 6 to 12 stories high, a crew of four is most efficient. Use two or more on top and two on the bottom for handing up or hoisting. They can

also set up and remove 18 frame sections per day. At 7′ horizontal spacing, this will run about 800 S.F. per day of erecting and dismantling. Time for placing and removing planks must be added to the above. A crew of three can place and remove 72 planks per day up to 5 stories. For over 5 stories, a crew of four can place and remove 80 planks per day.

The table below shows the number of pieces required to erect tubular steel scaffolding for 1000 S.F. of building frontage. This area is made up of a scaffolding system that is 12 frames (11 bays) long by 2 frames high.

For jobs under twenty-five frames, add 50% to rental cost. Rental rates will be lower for jobs over three months duration. Large quantities for long periods can reduce rental rates by 20%.

Description of Component	Number of Pieces for 1000 S.F. of Building Front	Unit
5′ Wide Standard Frame, 6′-4″ High	24	Ea.
Leveling Jack & Plate	24	
Cross Brace	44	
Side Arm Bracket, 21″	12	
Guardrail Post	12	
Guardrail, 7′ section	22	
Stairway Section	2	
Stairway Starter Bar	1	
Stairway Inside Handrail	2	
Stairway Outside Handrail	2	
Walk-Thru Frame Guardrail	2	

Scaffolding is often used as falsework over 15′ high during construction of cast-in-place concrete beams and slabs. Two foot wide scaffolding is generally used for heavy beam construction. The span between frames depends upon the load to be carried with a maximum span of 5′.

Heavy duty shoring frames with a capacity of 10,000#/leg can be spaced up to 10′ O.C. depending upon form support design and loading.

Scaffolding used as horizontal shoring requires less than half the material required with conventional shoring.

On new construction, erection is done by carpenters.

Rolling towers supporting horizontal shores can reduce labor and speed the job. For maintenance work, catwalks with spans up to 70′ can be supported by the rolling towers.

R015423-20　Pump Staging

Pump staging is generally not available for rent. The table below shows the number of pieces required to erect pump staging for 2400 S.F. of building

frontage. This area is made up of a pump jack system that is 3 poles (2 bays) wide by 2 poles high.

Item	Number of Pieces for 2400 S.F. of Building Front	Unit
Aluminum pole section, 24′ long	6	Ea.
Aluminum splice joint, 6′ long	3	
Aluminum foldable brace	3	
Aluminum pump jack	3	
Aluminum support for workbench/back safety rail	3	
Aluminum scaffold plank/workbench, 14″ wide x 24′ long	4	
Safety net, 22′ long	2	
Aluminum plank end safety rail	2	

The cost in place for this 2400 S.F. will depend on how many uses are realized during the life of the equipment.

R015433-10 Contractor Equipment

Rental Rates shown elsewhere in the book pertain to late model high quality machines in excellent working condition, rented from equipment dealers. Rental rates from contractors may be substantially lower than the rental rates from equipment dealers depending upon economic conditions; for older, less productive machines, reduce rates by a maximum of 15%. Any overtime must be added to the base rates. For shift work, rates are lower. Usual rule of thumb is 150% of one shift rate for two shifts; 200% for three shifts.

For periods of less than one week, operated equipment is usually more economical to rent than renting bare equipment and hiring an operator.

Costs to move equipment to a job site (mobilization) or from a job site (demobilization) are not included in rental rates, nor in any Equipment costs on any Unit Price line items or crew listings. These costs can be found elsewhere. If a piece of equipment is already at a job site, it is not appropriate to utilize mob/demob costs in an estimate again.

Rental rates vary throughout the country with larger cities generally having lower rates. Lease plans for new equipment are available for periods in excess of six months with a percentage of payments applying toward purchase.

Rental rates can also be treated as reimbursement costs for contractor-owned equipment. Owned equipment costs include depreciation, loan payments, interest, taxes, insurance, storage, and major repairs.

Monthly rental rates vary from 2% to 5% of the cost of the equipment depending on the anticipated life of the equipment and its wearing parts. Weekly rates are about 1/3 the monthly rates and daily rental rates about 1/3 the weekly rate.

The hourly operating costs for each piece of equipment include costs to the user such as fuel, oil, lubrication, normal expendables for the equipment, and a percentage of mechanic's wages chargeable to maintenance. The hourly operating costs listed do not include the operator's wages.

The daily cost for equipment used in the standard crews is figured by dividing the weekly rate by five, then adding eight times the hourly operating cost to give the total daily equipment cost, not including the operator. This figure is in the right hand column of the Equipment listings under Equipment Cost/Day.

Pile Driving rates shown for pile hammer and extractor do not include leads, crane, boiler or compressor. Vibratory pile driving requires an added field specialist during set-up and pile driving operation for the electric model. The hydraulic model requires a field specialist for set-up only. Up to 125 reuses of sheet piling are possible using vibratory drivers. For normal conditions, crane capacity for hammer type and size are as follows.

Crane Capacity	Hammer Type and Size		
	Air or Steam	Diesel	Vibratory
25 ton	to 8,750 ft.-lb.		70 H.P.
40 ton	15,000 ft.-lb.	to 32,000 ft.-lb.	170 H.P.
60 ton	25,000 ft.-lb.		300 H.P.
100 ton		112,000 ft.-lb.	

Cranes should be specified for the job by size, building and site characteristics, availability, performance characteristics, and duration of time required.

Backhoes & Shovels rent for about the same as equivalent size cranes but maintenance and operating expense is higher. Crane operators rate must be adjusted for high boom heights. Average adjustments: for 150' boom add 2% per hour; over 185', add 4% per hour; over 210', add 6% per hour; over 250', add 8% per hour and over 295', add 12% per hour.

Tower Cranes of the climbing or static type have jibs from 50' to 200' and capacities at maximum reach range from 4,000 to 14,000 pounds. Lifting capacities increase up to maximum load as the hook radius decreases.

Typical rental rates, based on purchase price are about 2% to 3% per month.

Erection and dismantling runs between 500 and 2000 labor hours. Climbing operation takes 10 labor hours per 20' climb. Crane dead time is about 5 hours per 40' climb. If crane is bolted to side of the building add cost of ties and extra mast sections. Climbing cranes have from 80' to 180' of mast while static cranes have 80' to 800' of mast.

Truck Cranes can be converted to tower cranes by using tower attachments. Mast heights over 400' have been used.

A single 100' high material **Hoist and Tower** can be erected and dismantled in about 400 labor hours; a double 100' high hoist and tower in about 600 labor hours. Erection times for additional heights are 3 and 4 labor hours

per vertical foot respectively up to 150', and 4 to 5 labor hours per vertical foot over 150' high. A 40' high portable Buck hoist takes about 160 labor hours to erect and dismantle. Additional heights take 2 labor hours per vertical foot to 80' and 3 labor hours per vertical foot for the next 100'. Most material hoists do not meet local code requirements for carrying personnel.

A 150' high **Personnel Hoist** requires about 500 to 800 labor hours to erect and dismantle. Budget erection time at 5 labor hours per vertical foot for all trades. Local code requirements or labor scarcity requiring overtime can add up to 50% to any of the above erection costs.

Earthmoving Equipment: The selection of earthmoving equipment depends upon the type and quantity of material, moisture content, haul distance, haul road, time available, and equipment available. Short haul cut and fill operations may require dozers only, while another operation may require excavators, a fleet of trucks, and spreading and compaction equipment. Stockpiled material and granular material are easily excavated with front end loaders. Scrapers are most economically used with hauls between 300' and 1-1/2 miles if adequate haul roads can be maintained. Shovels are often used for blasted rock and any material where a vertical face of 8' or more can be excavated. Special conditions may dictate the use of draglines, clamshells, or backhoes. Spreading and compaction equipment must be matched to the soil characteristics, the compaction required and the rate the fill is being supplied.

R015433-15 Heavy Lifting

Hydraulic Climbing Jacks

The use of hydraulic heavy lift systems is an alternative to conventional type crane equipment. The lifting, lowering, pushing, or pulling mechanism is a hydraulic climbing jack moving on a square steel jackrod from 1-5/8" to 4" square, or a steel cable. The jackrod or cable can be vertical or horizontal, stationary or movable, depending on the individual application. When the jackrod is stationary, the climbing jack will climb the rod and push or pull the load along with itself. When the climbing jack is stationary, the jackrod is movable with the load attached to the end and the climbing jack will lift or lower the jackrod with the attached load. The heavy lift system is normally operated by a single control lever located at the hydraulic pump.

The system is flexible in that one or more climbing jacks can be applied wherever a load support point is required, and the rate of lift synchronized.

Economic benefits have been demonstrated on projects such as: erection of ground assembled roofs and floors, complete bridge spans, girders and trusses, towers, chimney liners and steel vessels, storage tanks, and heavy machinery. Other uses are raising and lowering offshore work platforms, caissons, tunnel sections and pipelines.

R024119-10   Demolition Defined

Whole Building Demolition - Demolition of the whole building with no concern for any particular building element, component, or material type being demolished. This type of demolition is accomplished with large pieces of construction equipment that break up the structure, load it into trucks and haul it to a disposal site, but disposal or dump fees are not included. Demolition of below-grade foundation elements, such as footings, foundation walls, grade beams, slabs on grade, etc., is not included. Certain mechanical equipment containing flammable liquids or ozone-depleting refrigerants, electric lighting elements, communication equipment components, and other building elements may contain hazardous waste, and must be removed, either selectively or carefully, as hazardous waste before the building can be demolished.

Foundation Demolition - Demolition of below-grade foundation footings, foundation walls, grade beams, and slabs on grade. This type of demolition is accomplished by hand or pneumatic hand tools, and does not include saw cutting, or handling, loading, hauling, or disposal of the debris.

Gutting - Removal of building interior finishes and electrical/mechanical systems down to the load-bearing and sub-floor elements of the rough building frame, with no concern for any particular building element, component, or material type being demolished. This type of demolition is accomplished by hand or pneumatic hand tools, and includes loading into trucks, but not hauling, disposal or dump fees, scaffolding, or shoring. Certain mechanical equipment containing flammable liquids or ozone-depleting refrigerants, electric lighting elements, communication equipment components, and other building elements may contain hazardous waste, and must be removed, either selectively or carefully, as hazardous waste, before the building is gutted.

Selective Demolition - Demolition of a selected building element, component, or finish, with some concern for surrounding or adjacent elements, components, or finishes (see the first Subdivision (s) at the beginning of appropriate Divisions). This type of demolition is accomplished by hand or pneumatic hand tools, and does not include handling, loading, storing, hauling, or disposal of the debris, scaffolding, or shoring. "Gutting" methods may be used in order to save time, but damage that is caused to surrounding or adjacent elements, components, or finishes may have to be repaired at a later time.

Careful Removal - Removal of a piece of service equipment, building element or component, or material type, with great concern for both the removed item and surrounding or adjacent elements, components or finishes. The purpose of careful removal may be to protect the removed item for later re-use, preserve a higher salvage value of the removed item, or replace an item while taking care to protect surrounding or adjacent elements, components, connections, or finishes from cosmetic and/or structural damage. An approximation of the time required to perform this type of removal is 1/3 to 1/2 the time it would take to install a new item of like kind (see Reference Number R220105-10). This type of removal is accomplished by hand or pneumatic hand tools, and does not include loading, hauling, or storing the removed item, scaffolding, shoring, or lifting equipment.

Cutout Demolition - Demolition of a small quantity of floor, wall, roof, or other assembly, with concern for the appearance and structural integrity of the surrounding materials. This type of demolition is accomplished by hand or pneumatic hand tools, and does not include saw cutting, handling, loading, hauling, or disposal of debris, scaffolding, or shoring.

Rubbish Handling - Work activities that involve handling, loading or hauling of debris. Generally, the cost of rubbish handling must be added to the cost of all types of demolition, with the exception of whole building demolition.

Minor Site Demolition - Demolition of site elements outside the footprint of a building. This type of demolition is accomplished by hand or pneumatic hand tools, or with larger pieces of construction equipment, and may include loading a removed item onto a truck (check the Crew for equipment used). It does not include saw cutting, hauling or disposal of debris, and, sometimes, handling or loading.

R024119-20   Dumpsters

Dumpster rental costs on construction sites are presented in two ways.

The cost per week rental includes the delivery of the dumpster; its pulling or emptying once per week, and its final removal. The assumption is made that the dumpster contractor could choose to empty a dumpster by simply bringing in an empty unit and removing the full one. These costs also include the disposal of the materials in the dumpster.

The Alternate Pricing can be used when actual planned conditions are not approximated by the weekly numbers. For example, these lines can be used when a dumpster is needed for 4 weeks and will need to be emptied 2 or 3 times per week. Conversely the Alternate Pricing lines can be used when a dumpster will be rented for several weeks or months but needs to be emptied only a few times over this period.

Existing Conditions | **R0265 Underground Storage Tank Removal**

R026510-20   Underground Storage Tank Removal

Underground Storage Tank Removal can be divided into two categories: Non-Leaking and Leaking. Prior to removing an underground storage tank, tests should be made, with the proper authorities present, to determine whether a tank has been leaking or the surrounding soil has been contaminated.

To safely remove Liquid Underground Storage Tanks:
1. Excavate to the top of the tank.
2. Disconnect all piping.
3. Open all tank vents and access ports.
4. Remove all liquids and/or sludge.
5. Purge the tank with an inert gas.
6. Provide access to the inside of the tank and clean out the interior using proper personal protective equipment (PPE).
7. Excavate soil surrounding the tank using proper PPE for on-site personnel.
8. Pull and properly dispose of the tank.
9. Clean up the site of all contaminated material.
10. Install new tanks or close the excavation.

Existing Conditions — R0282 Asbestos Remediation

R028213-20   Asbestos Removal Process

Asbestos removal is accomplished by a specialty contractor who understands the federal and state regulations regarding the handling and disposal of the material. The process of asbestos removal is divided into many individual steps. An accurate estimate can be calculated only after all the steps have been priced.

The steps are generally as follows:

1. Obtain an asbestos abatement plan from an industrial hygienist.
2. Monitor the air quality in and around the removal area and along the path of travel between the removal area and transport area. This establishes the background contamination.
3. Construct a two part decontamination chamber at entrance to removal area.
4. Install a HEPA filter to create a negative pressure in the removal area.
5. Install wall, floor and ceiling protection as required by the plan, usually 2 layers of fireproof 6 mil polyethylene.
6. Industrial hygienist visually inspects work area to verify compliance with plan.
7. Provide temporary supports for conduit and piping affected by the removal process.
8. Proceed with asbestos removal and bagging process. Monitor air quality as described in Step #2. Discontinue operations when contaminate levels exceed applicable standards.
9. Document the legal disposal of materials in accordance with EPA standards.
10. Thoroughly clean removal area including all ledges, crevices and surfaces.
11. Post abatement inspection by industrial hygienist to verify plan compliance.
12. Provide a certificate from a licensed industrial hygienist attesting that contaminate levels are within acceptable standards before returning area to regular use.

Existing Conditions — R0283 Lead Remediation

R028319-60   Lead Paint Remediation Methods

Lead paint remediation can be accomplished by the following methods.

1. Abrasive blast
2. Chemical stripping
3. Power tool cleaning with vacuum collection system
4. Encapsulation
5. Remove and replace
6. Enclosure

Each of these methods has strengths and weakness depending on the specific circumstances of the project. The following is an overview of each method.

1. **Abrasive blasting** is usually accomplished with sand or recyclable metallic blast. Before work can begin, the area must be contained to ensure the blast material with lead does not escape to the atmosphere. The use of vacuum blast greatly reduces the containment requirements. Lead abatement equipment that may be associated with this work includes a negative air machine. In addition, it is necessary to have an industrial hygienist monitor the project on a continual basis. When the work is complete, the spent blast sand with lead must be disposed of as a hazardous material. If metallic shot was used, the lead is separated from the shot and disposed of as hazardous material. Worker protection includes disposable clothing and respiratory protection.

2. **Chemical stripping** requires strong chemicals be applied to the surface to remove the lead paint. Before the work can begin, the area under/adjacent to the work area must be covered to catch the chemical and removed lead. After the chemical is applied to the painted surface it is usually covered with paper. The chemical is left in place for the specified period, then the paper with lead paint is pulled or scraped off. The process may require several chemical applications. The paper with chemicals and lead paint adhered to it, plus the containment and loose scrapings collected by a HEPA (High Efficiency Particulate Air Filter) vac, must be disposed of as a hazardous material. The chemical stripping process usually requires a neutralizing agent and several wash downs after the paint is removed. Worker protection includes a neoprene or other compatible protective clothing and respiratory protection with face shield. An industrial hygienist is required intermittently during the process.

3. **Power tool cleaning** is accomplished using shrouded needle blasting guns. The shrouding with different end configurations is held up against the surface to be cleaned. The area is blasted with hardened needles and the shroud captures the lead with a HEPA vac and deposits it in a holding tank. An industrial hygienist monitors the project, protective clothing and a respirator is required until air samples prove otherwise. When the work is complete the lead must be disposed of as a hazardous material.

4. **Encapsulation** is a method that leaves the well bonded lead paint in place after the peeling paint has been removed. Before the work can begin, the area under/adjacent to the work must be covered to catch the scrapings. The scraped surface is then washed with a detergent and rinsed. The prepared surface is covered with approximately 10 mils of paint. A reinforcing fabric can also be embedded in the paint covering. The scraped paint and containment must be disposed of as a hazardous material. Workers must wear protective clothing and respirators.

5. **Remove and replace** is an effective way to remove lead paint from windows, gypsum walls and concrete masonry surfaces. The painted materials are removed and new materials are installed. Workers should wear a respirator and tyvek suit. The demolished materials must be disposed of as hazardous waste if it fails the TCLP (Toxicity Characteristic Leachate Process) test.

6. **Enclosure** is the process that permanently seals lead painted materials in place. This process has many applications such as covering lead painted drywall with new drywall, covering exterior construction with tyvek paper then residing, or covering lead painted structural members with aluminum or plastic. The seams on all enclosing materials must be securely sealed. An industrial hygienist monitors the project, and protective clothing and a respirator is required until air samples prove otherwise.

All the processes require clearance monitoring and wipe testing as required by the hygienist.

R031113-10 Wall Form Materials

Aluminum Forms
Approximate weight is 3 lbs. per S.F.C.A. Standard widths are available from 4" to 36" with 36" most common. Standard lengths of 2', 4', 6' to 8' are available. Forms are lightweight and fewer ties are needed with the wider widths. The form face is either smooth or textured.

Metal Framed Plywood Forms
Manufacturers claim over 75 reuses of plywood and over 300 reuses of steel frames. Many specials such as corners, fillers, pilasters, etc. are available. Monthly rental is generally about 15% of purchase price for first month and 9% per month thereafter with 90% of rental applied to purchase for the first month and decreasing percentages thereafter. Aluminum framed forms cost 25% to 30% more than steel framed.

After the first month, extra days may be prorated from the monthly charge. Rental rates do not include ties, accessories, cleaning, loss of hardware or freight in and out. Approximate weight is 5 lbs. per S.F. for steel; 3 lbs. per S.F. for aluminum.

Forms can be rented with option to buy.

Plywood Forms, Job Fabricated
There are two types of plywood used for concrete forms.
1. Exterior plyform which is completely waterproof. This is face oiled to facilitate stripping. Ten reuses can be expected with this type with 25 reuses possible.
2. An overlaid type consists of a resin fiber fused to exterior plyform. No oiling is required except to facilitate cleaning. This is available in both high density (HDO) and medium density overlaid (MDO). Using HDO, 50 reuses can be expected with 200 possible.

Plyform is available in 5/8" and 3/4" thickness. High density overlaid is available in 3/8", 1/2", 5/8" and 3/4" thickness.

5/8" thick is sufficient for most building forms, while 3/4" is best on heavy construction.

Plywood Forms, Modular, Prefabricated
There are many plywood forming systems without frames. Most of these are manufactured from 1-1/8" (HDO) plywood and have some hardware attached. These are used principally for foundation walls 8' or less high. With care and maintenance, 100 reuses can be attained with decreasing quality of surface finish.

Steel Forms
Approximate weight is 6-1/2 lbs. per S.F.C.A. including accessories. Standard widths are available from 2" to 24", with 24" most common. Standard lengths are from 2' to 8', with 4' the most common. Forms are easily ganged into modular units.

Forms are usually leased for 15% of the purchase price per month prorated daily over 30 days.

Rental may be applied to sale price, and usually rental forms are bought. With careful handling and cleaning 200 to 400 reuses are possible.

Straight wall gang forms up to 12' x 20' or 8' x 30' can be fabricated. These crane handled forms usually lease for approx. 9% per month.

Individual job analysis is available from the manufacturer at no charge.

R031113-30 Slipforms

The slipform method of forming may be used for forming circular silo and multi-celled storage bin type structures over 30' high, and building core shear walls over eight stories high. The shear walls, usually enclose elevator shafts, stairwells, mechanical spaces, and toilet rooms. Reuse of the form on duplicate structures will reduce the height necessary and spread the cost of building the form. Slipform systems can be used to cast chimneys, towers, piers, dams, underground shafts or other structures capable of being extruded.

Slipforms are usually 4' high and are raised semi-continuously by jacks climbing on rods which are embedded in the concrete. The jacks are powered by a hydraulic, pneumatic, or electric source and are available in 3, 6, and 22 ton capacities. Interior work decks and exterior scaffolds must be provided for placing inserts, embedded items, reinforcing steel, and

concrete. Scaffolds below the form for finishers may be required. The interior work decks are often used as roof slab forms on silos and bin work. Form raising rates will range from 6" to 20" per hour for silos; 6" to 30" per hour for buildings; and 6" to 48" per hour for shaft work.

Reinforcing bars and stressing strands are usually hoisted by crane or gin pole, and the concrete material can be hoisted by crane, winch-powered skip, or pumps. The slipform system is operated on a continuous 24-hour day when a monolithic structure is desired. For least cost, the system is operated only during normal working hours.

Placing concrete will range from 0.5 to 1.5 labor-hours per C.Y. Bucks, blockouts, keyways, weldplates, etc. are extra.

R031113-40   Forms for Reinforced Concrete

Design Economy

Avoid many sizes in proportioning beams and columns.

From story to story avoid changing column dimensions. Gain strength by adding steel or using a richer mix. If a change in size of column is necessary, vary one dimension only to minimize form alterations. Keep beams and columns the same width.

From floor to floor in a multi-story building vary beam depth, not width, as that will leave slab panel form unchanged. It is cheaper to vary the strength of a beam from floor to floor by means of steel area than by 2″ changes in either width or depth.

Cost Factors

Material includes the cost of lumber, cost of rent for metal pans or forms if used, nails, form ties, form oil, bolts and accessories.

Labor includes the cost of carpenters to make up, erect, remove and repair, plus common labor to clean and move. Having carpenters remove forms minimizes repairs.

Improper alignment and condition of forms will increase finishing cost. When forms are heavily oiled, concrete surfaces must be neutralized before finishing. Special curing compounds will cause spillages to spall off in first frost. Gang forming methods will reduce costs on large projects.

Materials Used

Boards are seldom used unless their architectural finish is required. Generally, steel, fiberglass and plywood are used for contact surfaces. Labor on plywood is 10% less than with boards. The plywood is backed up with 2 x 4's at 12″ to 32″ O.C. Walers are generally 2 - 2 x 4's. Column forms are held together with steel yokes or bands. Shoring is with adjustable shoring or scaffolding for high ceilings.

Reuse

Floor and column forms can be reused four or possibly five times without excessive repair. Remember to allow for 10% waste on each reuse.

When modular sized wall forms are made, up to twenty uses can be expected with exterior plyform.

When forms are reused, the cost to erect, strip, clean and move will not be affected. 10% replacement of lumber should be included and about one hour of carpenter time for repairs on each reuse per 100 S.F.

The reuse cost for certain accessory items normally rented on a monthly basis will be lower than the cost for the first use.

After fifth use, new material required plus time needed for repair prevent form cost from dropping further and it may go up. Much depends on care in stripping, the number of special bays, changes in beam or column sizes and other factors.

Costs for multiple use of formwork may be developed as follows:

2 Uses	3 Uses	4 Uses
$\dfrac{\text{(1st Use + Reuse)}}{2} = \text{avg. cost/2 uses}$	$\dfrac{\text{(1st Use + 2 Reuse)}}{3} = \text{avg. cost/3 uses}$	$\dfrac{\text{(1st use + 3 Reuse)}}{4} = \text{avg. cost/4 uses}$

R031113-60 Formwork Labor-Hours

Item	Unit	Fabricate	Erect & Strip	Clean & Move	Total Hours 1 Use	2 Use	3 Use	4 Use
Beam and Girder, interior beams, 12" wide	100 S.F.	6.4	8.3	1.3	16.0	13.3	12.4	12.0
Hung from steel beams		5.8	7.7	1.3	14.8	12.4	11.6	11.2
Beam sides only, 36" high		5.8	7.2	1.3	14.3	11.9	11.1	10.7
Beam bottoms only, 24" wide		6.6	13.0	1.3	20.9	18.1	17.2	16.7
Box out for openings		9.9	10.0	1.1	21.0	16.6	15.1	14.3
Buttress forms, to 8' high		6.0	6.5	1.2	13.7	11.2	10.4	10.0
Centering, steel, 3/4" rib lath			1.0		1.0			
3/8" rib lath or slab form			0.9		0.9			
Chamfer strip or keyway	100 L.F.		1.5		1.5	1.5	1.5	1.5
Columns, fiber tube 8" diameter			20.6		20.6			
12"			21.3		21.3			
16"			22.9		22.9			
20"			23.7		23.7			
24"			24.6		24.6			
30"			25.6		25.6			
Columns, round steel, 12" diameter			22.0		22.0	22.0	22.0	22.0
16"			25.6		25.6	25.6	25.6	25.6
20"			30.5		30.5	30.5	30.5	30.5
24"			37.7		37.7	37.7	37.7	37.7
Columns, plywood 8" x 8"	100 S.F.	7.0	11.0	1.2	19.2	16.2	15.2	14.7
12" x 12"		6.0	10.5	1.2	17.7	15.2	14.4	14.0
16" x 16"		5.9	10.0	1.2	17.1	14.7	13.8	13.4
24" x 24"		5.8	9.8	1.2	16.8	14.4	13.6	13.2
Columns, steel framed plywood 8" x 8"			10.0	1.0	11.0	11.0	11.0	11.0
12" x 12"			9.3	1.0	10.3	10.3	10.3	10.3
16" x 16"			8.5	1.0	9.5	9.5	9.5	9.5
24" x 24"			7.8	1.0	8.8	8.8	8.8	8.8
Drop head forms, plywood		9.0	12.5	1.5	23.0	19.0	17.7	17.0
Coping forms		8.5	15.0	1.5	25.0	21.3	20.0	19.4
Culvert, box			14.5	4.3	18.8	18.8	18.8	18.8
Curb forms, 6" to 12" high, on grade		5.0	8.5	1.2	14.7	12.7	12.1	11.7
On elevated slabs		6.0	10.8	1.2	18.0	15.5	14.7	14.3
Edge forms to 6" high, on grade	100 L.F.	2.0	3.5	0.6	6.1	5.6	5.4	5.3
7" to 12" high	100 S.F.	2.5	5.0	1.0	8.5	7.8	7.5	7.4
Equipment foundations		10.0	18.0	2.0	30.0	25.5	24.0	23.3
Flat slabs, including drops		3.5	6.0	1.2	10.7	9.5	9.0	8.8
Hung from steel		3.0	5.5	1.2	9.7	8.7	8.4	8.2
Closed deck for domes		3.0	5.8	1.2	10.0	9.0	8.7	8.5
Open deck for pans		2.2	5.3	1.0	8.5	7.9	7.7	7.6
Footings, continuous, 12" high		3.5	3.5	1.5	8.5	7.3	6.8	6.6
Spread, 12" high		4.7	4.2	1.6	10.5	8.7	8.0	7.7
Pile caps, square or rectangular		4.5	5.0	1.5	11.0	9.3	8.7	8.4
Grade beams, 24" deep		2.5	5.3	1.2	9.0	8.3	8.0	7.9
Lintel or Sill forms		8.0	17.0	2.0	27.0	23.5	22.3	21.8
Spandrel beams, 12" wide		9.0	11.2	1.3	21.5	17.5	16.2	15.5
Stairs			25.0	4.0	29.0	29.0	29.0	29.0
Trench forms in floor		4.5	14.0	1.5	20.0	18.3	17.7	17.4
Walls, Plywood, at grade, to 8' high		5.0	6.5	1.5	13.0	11.0	9.7	9.5
8' to 16'		7.5	8.0	1.5	17.0	13.8	12.7	12.1
16' to 20'		9.0	10.0	1.5	20.5	16.5	15.2	14.5
Foundation walls, to 8' high		4.5	6.5	1.0	12.0	10.3	9.7	9.4
8' to 16' high		5.5	7.5	1.0	14.0	11.8	11.0	10.6
Retaining wall to 12' high, battered		6.0	8.5	1.5	16.0	13.5	12.7	12.3
Radial walls to 12' high, smooth		8.0	9.5	2.0	19.5	16.0	14.8	14.3
2' chords		7.0	8.0	1.5	16.5	13.5	12.5	12.0
Prefabricated modular, to 8' high		—	4.3	1.0	5.3	5.3	5.3	5.3
Steel, to 8' high		—	6.8	1.2	8.0	8.0	8.0	8.0
8' to 16' high		—	9.1	1.5	10.6	10.3	10.2	10.2
Steel framed plywood to 8' high		—	6.8	1.2	8.0	7.5	7.3	7.2
8' to 16' high		—	9.3	1.2	10.5	9.5	9.2	9.0

R032110-10 Reinforcing Steel Weights and Measures

Bar Designation No.**	Nominal Weight Lb./Ft.	U.S. Customary Units			SI Units			
		Nominal Dimensions*			Nominal Weight kg/m	Nominal Dimensions*		
		Diameter in.	Cross Sectional Area, in.^{2}	Perimeter in.		Diameter mm	Cross Sectional Area, cm^2	Perimeter mm
3	.376	.375	.11	1.178	.560	9.52	.71	29.9
4	.668	.500	.20	1.571	.994	12.70	1.29	39.9
5	1.043	.625	.31	1.963	1.552	15.88	2.00	49.9
6	1.502	.750	.44	2.356	2.235	19.05	2.84	59.8
7	2.044	.875	.60	2.749	3.042	22.22	3.87	69.8
8	2.670	1.000	.79	3.142	3.973	25.40	5.10	79.8
9	3.400	1.128	1.00	3.544	5.059	28.65	6.45	90.0
10	4.303	1.270	1.27	3.990	6.403	32.26	8.19	101.4
11	5.313	1.410	1.56	4.430	7.906	35.81	10.06	112.5
14	7.650	1.693	2.25	5.320	11.384	43.00	14.52	135.1
18	13.600	2.257	4.00	7.090	20.238	57.33	25.81	180.1

* The nominal dimensions of a deformed bar are equivalent to those of a plain round bar having the same weight per foot as the deformed bar.

** Bar numbers are based on the number of eighths of an inch included in the nominal diameter of the bars.

R032110-20 Metric Rebar Specification - ASTM A615-81

Grade 300 (300 MPa* = 43,560 psi; +8.7% vs. Grade 40)				
Grade 400 (400 MPa* = 58,000 psi; −3.4% vs. Grade 60)				
Bar No.	Diameter mm	Area mm^2	Equivalent in.^{2}	Comparison with U.S. Customary Bars
10M	11.3	100	.16	Between #3 & #4
15M	16.0	200	.31	#5 (.31 in.^{2})
20M	19.5	300	.47	#6 (.44 in.^{2})
25M	25.2	500	.78	#8 (.79 in.^{2})
30M	29.9	700	1.09	#9 (1.00 in.^{2})
35M	35.7	1000	1.55	#11 (1.56 in.^{2})
45M	43.7	1500	2.33	#14 (2.25 in.^{2})
55M	56.4	2500	3.88	#18 (4.00 in.^{2})

* MPa = megapascals

R032110-40   Weight of Steel Reinforcing Per Square Foot of Wall (PSF)

Reinforced Weights: The table below suggests the weights per square foot for reinforcing steel in walls. Weights are approximate and will be the same for all grades of steel bars. For bars in two directions, add weights for each size and spacing.

C/C Spacing in Inches	Bar Size								
	#3 Wt. (PSF)	#4 Wt. (PSF)	#5 Wt. (PSF)	#6 Wt. (PSF)	#7 Wt. (PSF)	#8 Wt. (PSF)	#9 Wt. (PSF)	#10 Wt. (PSF)	#11 Wt. (PSF)
2"	2.26	4.01	6.26	9.01	12.27				
3"	1.50	2.67	4.17	6.01	8.18	10.68	13.60	17.21	21.25
4"	1.13	2.01	3.13	4.51	6.13	8.10	10.20	12.91	15.94
5"	.90	1.60	2.50	3.60	4.91	6.41	8.16	10.33	12.75
6"	.752	1.34	2.09	3.00	4.09	5.34	6.80	8.61	10.63
8"	.564	1.00	1.57	2.25	3.07	4.01	5.10	6.46	7.97
10"	.451	.802	1.25	1.80	2.45	3.20	4.08	5.16	6.38
12"	.376	.668	1.04	1.50	2.04	2.67	3.40	4.30	5.31
18"	.251	.445	.695	1.00	1.32	1.78	2.27	2.86	3.54
24"	.188	.334	.522	.751	1.02	1.34	1.70	2.15	2.66
30"	.150	.267	.417	.600	.817	1.07	1.36	1.72	2.13
36"	.125	.223	.348	.501	.681	.890	1.13	1.43	1.77
42"	.107	.191	.298	.429	.584	.753	.97	1.17	1.52
48"	.094	.167	.261	.376	.511	.668	.85	1.08	1.33

R032110-50   Minimum Wall Reinforcement Weight (PSF)

This table lists the approximate minimum wall reinforcement weights per S.F. according to the specification of .12% of gross area for vertical bars and .20% of gross area for horizontal bars.

Location	Wall Thickness	Bar Size	Horizontal Steel Spacing C/C	Sq. In. Req'd per S.F.	Total Wt. per S.F.	Bar Size	Vertical Steel Spacing C/C	Sq. In. Req'd per S.F.	Total Wt. per S.F.	Horizontal & Vertical Steel Total Weight per S.F.
Both Faces	10"	#4	18"	.24	.89#	#3	18"	.14	.50#	1.39#
	12"	#4	16"	.29	1.00	#3	16"	.17	.60	1.60
	14"	#4	14"	.34	1.14	#3	13"	.20	.69	1.84
	16"	#4	12"	.38	1.34	#3	11"	.23	.82	2.16
	18"	#5	17"	.43	1.47	#4	18"	.26	.89	2.36
One Face	6"	#3	9"	.15	.50	#3	18"	.09	.25	.75
	8"	#4	12"	.19	.67	#3	11"	.12	.41	1.08
	10"	#5	15"	.24	.83	#4	16"	.14	.50	1.34

R032110-70   Bend, Place and Tie Reinforcing

Placing and tying by rodmen for footings and slabs runs from nine hrs. per ton for heavy bars to fifteen hrs. per ton for light bars. For beams, columns, and walls, production runs from eight hrs. per ton for heavy bars to twenty hrs. per ton for light bars. Overall average for typical reinforced concrete buildings is about fourteen hrs. per ton. These production figures include the time for placing of accessories and usual inserts, but not their material cost (allow 15% of the cost of delivered bent rods). Equipment handling is necessary for the larger-sized bars so that installation costs for the very heavy bars will not decrease proportionately.

Installation costs for splicing reinforcing bars include allowance for equipment to hold the bars in place while splicing as well as necessary scaffolding for iron workers.

R032110-80   Shop-Fabricated Reinforcing Steel

The material prices for reinforcing, shown in the unit cost sections of the book, are for 50 tons or more of shop-fabricated reinforcing steel and include:
1. Mill base price of reinforcing steel
2. Mill grade/size/length extras
3. Mill delivery to the fabrication shop
4. Shop storage and handling
5. Shop drafting/detailing
6. Shop shearing and bending
7. Shop listing
8. Shop delivery to the job site

Both material and installation costs can be considerably higher for small jobs consisting primarily of smaller bars, while material costs may be slightly lower for larger jobs.

R032205-30   Common Stock Styles of Welded Wire Fabric

This table provides some of the basic specifications, sizes, and weights of welded wire fabric used for reinforcing concrete.

	New Designation	Old Designation		Steel Area per Foot				Approximate Weight per 100 S.F.	
	Spacing — Cross Sectional Area (in.) — (Sq. in. 100)	Spacing — Wire Gauge (in.) — (AS & W)		Longitudinal		Transverse			
				in.	cm	in.	cm	lbs	kg
Rolls	6 x 6 — W1.4 x W1.4	6 x 6 — 10 x 10		.028	.071	.028	.071	21	9.53
	6 x 6 — W2.0 x W2.0	6 x 6 — 8 x 8	1	.040	.102	.040	.102	29	13.15
	6 x 6 — W2.9 x W2.9	6 x 6 — 6 x 6		.058	.147	.058	.147	42	19.05
	6 x 6 — W4.0 x W4.0	6 x 6 — 4 x 4		.080	.203	.080	.203	58	26.91
	4 x 4 — W1.4 x W1.4	4 x 4 — 10 x 10		.042	.107	.042	.107	31	14.06
	4 x 4 — W2.0 x W2.0	4 x 4 — 8 x 8	1	.060	.152	.060	.152	43	19.50
	4 x 4 — W2.9 x W2.9	4 x 4 — 6 x 6		.087	.227	.087	.227	62	28.12
	4 x 4 — W4.0 x W4.0	4 x 4 — 4 x 4		.120	.305	.120	.305	85	38.56
Sheets	6 x 6 — W2.9 x W2.9	6 x 6 — 6 x 6		.058	.147	.058	.147	42	19.05
	6 x 6 — W4.0 x W4.0	6 x 6 — 4 x 4		.080	.203	.080	.203	58	26.31
	6 x 6 — W5.5 x W5.5	6 x 6 — 2 x 2	2	.110	.279	.110	.279	80	36.29
	4 x 4 — W1.4 x W1.4	4 x 4 — 4 x 4		.120	.305	.120	.305	85	38.56

NOTES: 1. Exact W—number size for 8 gauge is W2.1
2. Exact W—number size for 2 gauge is W5.4

Concrete | **R0330 Cast-In-Place Concrete**

R033053-50   Industrial Chimneys

Foundation requirements in C.Y. of concrete for various sized chimneys.

Size Chimney	2 Ton Soil	3 Ton Soil	Size Chimney	2 Ton Soil	3 Ton Soil	Size Chimney	2 Ton Soil	3 Ton Soil
75' x 3'-0"	13 C.Y.	11 C.Y.	160' x 6'-6"	86 C.Y.	76 C.Y.	300' x 10'-0"	325 C.Y.	245 C.Y.
85' x 5'-6"	19	16	175' x 7'-0"	108	95	350' x 12'-0"	422	320
100' x 5'-0"	24	20	200' x 6'-0"	125	105	400' x 14'-0"	520	400
125' x 5'-6"	43	36	250' x 8'-0"	230	175	500' x 18'-0"	725	575

R033105-10 Proportionate Quantities

The tables below show both quantities per S.F. of floor areas as well as form and reinforcing quantities per C.Y. Unusual structural requirements would increase the ratios below. High strength reinforcing would reduce the steel weights. Figures are for 3000 psi concrete and 60,000 psi reinforcing unless specified otherwise.

Type of Construction	Live Load	Span	Per S.F. of Floor Area				Per C.Y. of Concrete		
			Concrete	Forms	Reinf.	Pans	Forms	Reinf.	Pans
Flat Plate	50 psf	15 Ft.	.46 C.F.	1.06 S.F.	1.71 lb.		62 S.F.	101 lb.	
		20	.63	1.02	2.40		44	104	
		25	.79	1.02	3.03		35	104	
	100	15	.46	1.04	2.14		61	126	
		20	.71	1.02	2.72		39	104	
		25	.83	1.01	3.47		33	113	
Flat Plate (waffle construction) 20" domes	50	20	.43	1.00	2.10	.84 S.F.	63	135	53 S.F.
		25	.52	1.00	2.90	.89	52	150	46
		30	.64	1.00	3.70	.87	42	155	37
	100	20	.51	1.00	2.30	.84	53	125	45
		25	.64	1.00	3.20	.83	42	135	35
		30	.76	1.00	4.40	.81	36	160	29
Waffle Construction 30" domes	50	25	.69	1.06	1.83	.68	42	72	40
		30	.74	1.06	2.39	.69	39	87	39
		35	.86	1.05	2.71	.69	33	85	39
		40	.78	1.00	4.80	.68	35	165	40
Flat Slab (two way with drop panels)	50	20	.62	1.03	2.34		45	102	
		25	.77	1.03	2.99		36	105	
		30	.95	1.03	4.09		29	116	
	100	20	.64	1.03	2.83		43	119	
		25	.79	1.03	3.88		35	133	
		30	.96	1.03	4.66		29	131	
	200	20	.73	1.03	3.03		38	112	
		25	.86	1.03	4.23		32	133	
		30	1.06	1.03	5.30		26	135	
One Way Joists 20" Pans	50	15	.36	1.04	1.40	.93	78	105	70
		20	.42	1.05	1.80	.94	67	120	60
		25	.47	1.05	2.60	.94	60	150	54
	100	15	.38	1.07	1.90	.93	77	140	66
		20	.44	1.08	2.40	.94	67	150	58
		25	.52	1.07	3.50	.94	55	185	49
One Way Joists 8" x 16" filler blocks	50	15	.34	1.06	1.80	.81 Ea.	84	145	64 Ea.
		20	.40	1.08	2.20	.82	73	145	55
		25	.46	1.07	3.20	.83	63	190	49
	100	15	.39	1.07	1.90	.81	74	130	56
		20	.46	1.09	2.80	.82	64	160	48
		25	.53	1.10	3.60	.83	56	190	42
One Way Beam & Slab	50	15	.42	1.30	1.73		84	111	
		20	.51	1.28	2.61		68	138	
		25	.64	1.25	2.78		53	117	
	100	15	.42	1.30	1.90		84	122	
		20	.54	1.35	2.69		68	154	
		25	.69	1.37	3.93		54	145	
	200	15	.44	1.31	2.24		80	137	
		20	.58	1.40	3.30		65	163	
		25	.69	1.42	4.89		53	183	
Two Way Beam & Slab	100	15	.47	1.20	2.26		69	130	
		20	.63	1.29	3.06		55	131	
		25	.83	1.33	3.79		43	123	
	200	15	.49	1.25	2.70		41	149	
		20	.66	1.32	4.04		54	165	
		25	.88	1.32	6.08		41	187	

R033105-10　Proportionate Quantities (cont.)

4000 psi Concrete and 60,000 psi Reinforcing—Form and Reinforcing Quantities per C.Y.					
Item	**Size**	**Forms**	**Reinforcing**	**Minimum**	**Maximum**
Columns (square tied)	10″ x 10″	130 S.F.C.A.	#5 to #11	220 lbs.	875 lbs.
	12″ x 12″	108	#6 to #14	200	955
	14″ x 14″	92	#7 to #14	190	900
	16″ x 16″	81	#6 to #14	187	1082
	18″ x 18″	72	#6 to #14	170	906
	20″ x 20″	65	#7 to #18	150	1080
	22″ x 22″	59	#8 to #18	153	902
	24″ x 24″	54	#8 to #18	164	884
	26″ x 26″	50	#9 to #18	169	994
	28″ x 28″	46	#9 to #18	147	864
	30″ x 30″	43	#10 to #18	146	983
	32″ x 32″	40	#10 to #18	175	866
	34″ x 34″	38	#10 to #18	157	772
	36″ x 36″	36	#10 to #18	175	852
	38″ x 38″	34	#10 to #18	158	765
	40″ x 40″	32	#10 to #18	143	692

Item	**Size**	**Form**	**Spiral**	**Reinforcing**	**Minimum**	**Maximum**
Columns (spirally reinforced)	12″ diameter	34.5 L.F.	190 lbs.	#4 to #11	165 lbs.	1505 lb.
		34.5	190	#14 & #18	—	1100
	14″	25	170	#4 to #11	150	970
		25	170	#14 & #18	800	1000
	16″	19	160	#4 to #11	160	950
		19	160	#14 & #18	605	1080
	18″	15	150	#4 to #11	160	915
		15	150	#14 & #18	480	1075
	20″	12	130	#4 to #11	155	865
		12	130	#14 & #18	385	1020
	22″	10	125	#4 to #11	165	775
		10	125	#14 & #18	320	995
	24″	9	120	#4 to #11	195	800
		9	120	#14 & #18	290	1150
	26″	7.3	100	#4 to #11	200	729
		7.3	100	#14 & #18	235	1035
	28″	6.3	95	#4 to #11	175	700
		6.3	95	#14 & #18	200	1075
	30″	5.5	90	#4 to #11	180	670
		5.5	90	#14 & #18	175	1015
	32″	4.8	85	#4 to #11	185	615
		4.8	85	#14 & #18	155	955
	34″	4.3	80	#4 to #11	180	600
		4.3	80	#14 & #18	170	855
	36″	3.8	75	#4 to #11	165	570
		3.8	75	#14 & #18	155	865
	40″	3.0	70	#4 to #11	165	500
		3.0	70	#14 & #18	145	765

R033105-10 Proportionate Quantities (cont.)

3000 psi Concrete and 60,000 psi Reinforcing—Form and Reinforcing Quantities per C.Y.

Item	Type	Loading	Height	C.Y./L.F.	Forms/C.Y.	Reinf./C.Y.
Retaining Walls	Cantilever	Level Backfill	4 Ft.	0.2 C.Y.	49 S.F.	35 lbs.
			8	0.5	42	45
			12	0.8	35	70
			16	1.1	32	85
			20	1.6	28	105
		Highway Surcharge	4	0.3	41	35
			8	0.5	36	55
			12	0.8	33	90
			16	1.2	30	120
			20	1.7	27	155
		Railroad Surcharge	4	0.4	28	45
			8	0.8	25	65
			12	1.3	22	90
			16	1.9	20	100
			20	2.6	18	120
	Gravity, with Vertical Face	Level Backfill	4	0.4	37	None
			7	0.6	27	↓
			10	1.2	20	
		Sloping Backfill	4	0.3	31	
			7	0.8	21	
			10	1.6	15	▼

Beams — Live Load in Kips per Linear Foot

Item	Span	Under 1 Kip		2 to 3 Kips		4 to 5 Kips		6 to 7 Kips	
		Forms	Reinf.	Forms	Reinf.	Forms	Reinf.	Forms	Reinf.
Beams	10 Ft.	—	—	90 S.F.	170 #	85 S.F.	175 #	75 S.F.	185 #
	16	130 S.F.	165 #	85	180	75	180	65	225
	20	110	170	75	185	62	200	51	200
	26	90	170	65	215	62	215	—	—
	30	85	175	60	200	—	—	—	—

Item	Size	Type	Forms per C.Y.	Reinforcing per C.Y.
Spread Footings	Under 1 C.Y.	1,000 psf soil	24 S.F.	44 lbs.
		5,000	24	42
		10,000	24	52
	1 C.Y. to 5 C.Y.	1,000	14	49
		5,000	14	50
		10,000	14	50
	Over 5 C.Y.	1,000	9	54
		5,000	9	52
		10,000	9	56
Pile Caps (30 Ton Concrete Piles)	Under 5 C.Y.	shallow caps	20	65
		medium	20	50
		deep	20	40
	5 C.Y. to 10 C.Y.	shallow	14	55
		medium	15	45
		deep	15	40
	10 C.Y. to 20 C.Y.	shallow	11	60
		medium	11	45
		deep	12	35
	Over 20 C.Y.	shallow	9	60
		medium	9	45
		deep	10	40

R033105-10 Proportionate Quantities (cont.)

Item	Size	Pile Spacing	50 T Pile	100 T Pile	50 T Pile	100 T Pile
			3000 psi Concrete and 60,000 psi Reinforcing — Form and Reinforcing Quantities per C.Y.			
Pile Caps (Steel H Piles)	Under 5 C.Y.	24" O.C.	24 S.F.	24 S.F.	75 lbs.	90 lbs.
		30"	25	25	80	100
		36"	24	24	80	110
	5 C.Y. to 10 C.Y.	24"	15	15	80	110
		30"	15	15	85	110
		36"	15	15	75	90
	Over 10 C.Y.	24"	13	13	85	90
		30"	11	11	85	95
		36"	10	10	85	90

Item	Height	8" Thick		10" Thick		12" Thick		15" Thick	
		Forms	Reinf.	Forms	Reinf.	Forms	Reinf.	Forms	Reinf.
Basement Walls	7 Ft.	81 S.F.	44 lbs.	65 S.F.	45 lbs.	54 S.F.	44 lbs.	41 S.F.	43 lbs.
	8		44		45		44		43
	9		46		45		44		43
	10		57		45		44		43
	12		83		50		52		43
	14		116		65		64		51
	16				86		90		65
	18						106		70

R033105-20 Materials for One C.Y. of Concrete

This is an approximate method of figuring quantities of cement, sand and coarse aggregate for a field mix with waste allowance included.

With crushed gravel as coarse aggregate, to determine barrels of cement required, divide 10 by total mix; that is, for 1:2:4 mix, 10 divided by 7 = 1-3/7 barrels.

If the coarse aggregate is crushed stone, use 10-1/2 instead of 10 as given for gravel.

To determine tons of sand required, multiply barrels of cement by parts of sand and then by 0.2; that is, for the 1:2:4 mix, as above, 1-3/7 x 2 x .2 = .57 tons.

Tons of crushed gravel are in the same ratio to tons of sand as parts in the mix, or 4/2 x .57 = 1.14 tons.

1 bag cement = 94#
4 bags = 1 barrel

1 C.Y. sand or crushed gravel = 2700#
1 ton sand or crushed gravel = 20 C.F.

1 C.Y. crushed stone = 2575#
1 ton crushed stone = 21 C.F.

Average carload of cement is 692 bags; of sand or gravel is 56 tons.

Do not stack stored cement over 10 bags high.

R033105-30 Metric Equivalents of Cement Content for Concrete Mixes

94 Pound Bags per Cubic Yard	Kilograms per Cubic Meter	94 Pound Bags per Cubic Yard	Kilograms per Cubic Meter
1.0	55.77	7.0	390.4
1.5	83.65	7.5	418.3
2.0	111.5	8.0	446.2
2.5	139.4	8.5	474.0
3.0	167.3	9.0	501.9
3.5	195.2	9.5	529.8
4.0	223.1	10.0	557.7
4.5	251.0	10.5	585.6
5.0	278.8	11.0	613.5
5.5	306.7	11.5	641.3
6.0	334.6	12.0	669.2
6.5	362.5	12.5	697.1

a. If you know the cement content in pounds per cubic yard,
   multiply by .5933 to obtain kilograms per cubic meter.

b. If you know the cement content in 94 pound bags per cubic yard,
   multiply by 55.77 to obtain kilograms per cubic meter.

R033105-40 Metric Equivalents of Common Concrete Strengths
(to convert other psi values to megapascals, multiply by 0.006895)

U.S. Values psi	SI Value Megapascals	Non-SI Metric Value kgf/cm²*
2000	14	140
2500	17	175
3000	21	210
3500	24	245
4000	28	280
4500	31	315
5000	34	350
6000	41	420
7000	48	490
8000	55	560
9000	62	630
10,000	69	705

* kilograms force per square centimeter

R033105-50 Quantities of Cement, Sand and Stone for One C.Y. of Concrete per Various Mixes

This table can be used to determine the quantities of the ingredients for smaller quantities of site mixed concrete.

Concrete (C.Y.)	Mix = 1:1:1-3/4 Cement (sacks)	Sand (C.Y.)	Stone (C.Y.)	Mix = 1:2:2.25 Cement (sacks)	Sand (C.Y.)	Stone (C.Y.)	Mix = 1:2.25:3 Cement (sacks)	Sand (C.Y.)	Stone (C.Y.)	Mix = 1:3:4 Cement (sacks)	Sand (C.Y.)	Stone (C.Y.)
1	10	.37	.63	7.75	.56	.65	6.25	.52	.70	5	.56	.74
2	20	.74	1.26	15.50	1.12	1.30	12.50	1.04	1.40	10	1.12	1.48
3	30	1.11	1.89	23.25	1.68	1.95	18.75	1.56	2.10	15	1.68	2.22
4	40	1.48	2.52	31.00	2.24	2.60	25.00	2.08	2.80	20	2.24	2.96
5	50	1.85	3.15	38.75	2.80	3.25	31.25	2.60	3.50	25	2.80	3.70
6	60	2.22	3.78	46.50	3.36	3.90	37.50	3.12	4.20	30	3.36	4.44
7	70	2.59	4.41	54.25	3.92	4.55	43.75	3.64	4.90	35	3.92	5.18
8	80	2.96	5.04	62.00	4.48	5.20	50.00	4.16	5.60	40	4.48	5.92
9	90	3.33	5.67	69.75	5.04	5.85	56.25	4.68	6.30	45	5.04	6.66
10	100	3.70	6.30	77.50	5.60	6.50	62.50	5.20	7.00	50	5.60	7.40
11	110	4.07	6.93	85.25	6.16	7.15	68.75	5.72	7.70	55	6.16	8.14
12	120	4.44	7.56	93.00	6.72	7.80	75.00	6.24	8.40	60	6.72	8.88
13	130	4.82	8.20	100.76	7.28	8.46	81.26	6.76	9.10	65	7.28	9.62
14	140	5.18	8.82	108.50	7.84	9.10	87.50	7.28	9.80	70	7.84	10.36
15	150	5.56	9.46	116.26	8.40	9.76	93.76	7.80	10.50	75	8.40	11.10
16	160	5.92	10.08	124.00	8.96	10.40	100.00	8.32	11.20	80	8.96	11.84
17	170	6.30	10.72	131.76	9.52	11.06	106.26	8.84	11.90	85	9.52	12.58
18	180	6.66	11.34	139.50	10.08	11.70	112.50	9.36	12.60	90	10.08	13.32
19	190	7.04	11.98	147.26	10.64	12.36	118.76	9.84	13.30	95	10.64	14.06
20	200	7.40	12.60	155.00	11.20	13.00	125.00	10.40	14.00	100	11.20	14.80
21	210	7.77	13.23	162.75	11.76	13.65	131.25	10.92	14.70	105	11.76	15.54
22	220	8.14	13.86	170.05	12.32	14.30	137.50	11.44	15.40	110	12.32	16.28
23	230	8.51	14.49	178.25	12.88	14.95	143.75	11.96	16.10	115	12.88	17.02
24	240	8.88	15.12	186.00	13.44	15.60	150.00	12.48	16.80	120	13.44	17.76
25	250	9.25	15.75	193.75	14.00	16.25	156.25	13.00	17.50	125	14.00	18.50
26	260	9.64	16.40	201.52	14.56	16.92	162.52	13.52	18.20	130	14.56	19.24
27	270	10.00	17.00	209.26	15.12	17.56	168.76	14.04	18.90	135	15.02	20.00
28	280	10.36	17.64	217.00	15.68	18.20	175.00	14.56	19.60	140	15.68	20.72
29	290	10.74	18.28	224.76	16.24	18.86	181.26	15.08	20.30	145	16.24	21.46

R033105-65   Field-Mix Concrete

Presently most building jobs are built with ready-mixed concrete except at isolated locations and some larger jobs requiring over 10,000 C.Y. where land is readily available for setting up a temporary batch plant.

The most economical mix is a controlled mix using local aggregate proportioned by trial to give the required strength with the least cost of material.

R033105-70   Placing Ready-Mixed Concrete

For ground pours allow for 5% waste when figuring quantities.

Prices in the front of the book assume normal deliveries. If deliveries are made before 8 A.M. or after 5 P.M. or on Saturday afternoons add 30%. Negotiated discounts for large volumes are not included in prices in front of book.

For the lower floors without truck access, concrete may be wheeled in rubber-tired buggies, conveyer handled, crane handled or pumped. Pumping is economical if there is top steel. Conveyers are more efficient for thick slabs.

At higher floors the rubber-tired buggies may be hoisted by a hoisting tower and wheeled to location. Placement by a conveyer is limited to three floors

and is best for high-volume pours. Pumped concrete is best when building has no crane access. Concrete may be pumped directly as high as thirty-six stories using special pumping techniques. Normal maximum height is about fifteen stories.

Best pumping aggregate is screened and graded bank gravel rather than crushed stone.

Pumping downward is more difficult than pumping upward. Horizontal distance from pump to pour may increase preparation time prior to pour. Placing by cranes, either mobile, climbing or tower types, continues as the most efficient method for high-rise concrete buildings.

R033105-85   Lift Slabs

The cost advantage of the lift slab method is due to placing all concrete, reinforcing steel, inserts and electrical conduit at ground level and in reduction of formwork. Minimum economical project size is about 30,000 S.F. Slabs may be tilted for parking garage ramps.

It is now used in all types of buildings and has gone up to 22 stories high in apartment buildings. Current trend is to use post-tensioned flat plate slabs with spans from 22' to 35'. Cylindrical void forms are used when deep slabs are required. One pound of prestressing steel is about equal to seven pounds of conventional reinforcing.

To be considered cured for stressing and lifting, a slab must have attained 75% of design strength. Seven days are usually sufficient with four to five days possible if high early strength cement is used. Slabs can be stacked using two coats of a non-bonding agent to insure that slabs do not stick to each other. Lifting is done by companies specializing in this work. Lift rate is 5' to 15' per hour with an average of 10' per hour. Total areas up to 33,000 S.F. have been lifted at one time. 24 to 36 jacking columns are common. Most economical bay sizes are 24' to 28' with four to fourteen stories most efficient. Continuous design reduces reinforcing steel cost. Use of post-tensioned slabs allows larger bay sizes.

R034105-30   Prestressed Precast Concrete Structural Units

Type	Location	Depth	Span in Ft.		Live Load Lb. per S.F.
Double Tee 8' to 10'	Floor	28" to 34"	60 to 80		50 to 80
	Roof	12" to 24"	30 to 50		40
	Wall	Width 8'	Up to 55' high		Wind
Multiple Tee 8'	Roof	8" to 12"	15 to 40		40
	Floor	8" to 12"	15 to 30		100
Plank or	Roof or Floor		Roof	Floor	40 for Roof
		4"	13	12	
		6"	22	18	
		8"	26	25	
		10"	33	29	100 for Floor
		12"	42	32	
Single Tee 8' to 10'	Roof	28"	40		
		32"	80		
		36"	100		40
		48"	120		
AASHTO Girder	Bridges	Type 4	100		
		5	110		Highway
		6	125		
Box Beam 4'	Bridges	15"	40		
		27"	to		Highway
		33"	100		

The majority of precast projects today utilize double tees rather than single tees because of speed and ease of installation. As a result casting beds at manufacturing plants are normally formed for double tees. Single tee projects will therefore require an initial set up charge to be spread over the individual single tee costs.

For floors, a 2" to 3" topping is field cast over the shapes. For roofs, insulating concrete or rigid insulation is placed over the shapes.

Member lengths up to 40' are standard haul, 40' to 60' require special permits and lengths over 60' must be escorted. Over width and/or over length can add up to 100% on hauling costs.

Large heavy members may require two cranes for lifting which would increase erection costs by about 45%. An eight man crew can install 12 to 20 double tees, or 45 to 70 quad tees or planks per day.

Grouting of connections must also be included.

Several system buildings utilizing precast members are available. Heights can go up to 22 stories for apartment buildings. Optimum design ratio is 3 S.F. of surface to 1 S.F. of floor area.

R034136-90   Prestressed Concrete, Post-Tensioned

In post-tensioned concrete the steel tendons are tensioned after the concrete has reached about 3/4 of its ultimate strength. The cableways are grouted after tensioning to provide bond between the steel and concrete. If bond is to be prevented, the tendons are coated with a corrosion-preventative grease and wrapped with waterproofed paper or plastic. Bonded tendons are usually used when ultimate strength (beams & girders) are controlling factors.

High strength concrete is used to fully utilize the steel, thereby reducing the size and weight of the member. A plasticizing agent may be added to reduce water content. Maximum size aggregate ranges from 1/2" to 1-1/2" depending on the spacing of the tendons.

The types of steel commonly used are bars and strands. Job conditions determine which is best suited. Bars are best for vertical prestresses since they are easy to support. The trend is for steel manufacturers to supply a finished package, cut to length, which reduces field preparation to a minimum.

Bars vary from 3/4" to 1-3/8" diameter. Table below gives time in labor-hours per tendon for placing, tensioning and grouting (if required) a 75' beam. Tendons used in buildings are not usually grouted; tendons for bridges usually are grouted. For strands the table indicates the labor-hours per pound for typical prestressed units 100' long. Simple span beams usually require one-end stressing regardless of lengths. Continuous beams are usually stressed from two ends. Long slabs are poured from the center outward and stressed in 75' increments after the initial 150' center pour.

Length	100' Beam		75' Beam		100' Slab	
Type Steel	**Strand**		**Bars**		**Strand**	
Diameter	0.5"		3/4"	1-3/8"	0.5"	0.6"
Number	4	12	1	1	1	1
Force in Kips	100	300	42	143	25	35
Preparation & Placing Cables	3.6	7.4	0.9	2.9	0.9	1.1
Stressing Cables	2.0	2.4	0.8	1.6	0.5	0.5
Grouting, if required	2.5	3.0	0.6	1.3		
Total Labor Hours	8.1	12.8	2.3	5.8	1.4	1.6
Prestressing Steel Weights (Lbs.)	215	640	115	380	53	74
Labor-hours per Lb. Bonded	0.038	0.020	0.020	0.015		
Non-bonded					0.026	0.022

Labor Hours per Tendon and per Pound of Prestressed Steel

Flat Slab construction — 4000 psi concrete with span-to-depth ratio between 36 and 44. Two way post-tensioned steel averages 1.0 lb. per S.F. for 24' to 28' bays (usually strand) and additional reinforcing steel averages .5 lb. per S.F.

Pan and Joist construction — 4000 psi concrete with span-to-depth ratio 28 to 30. Post-tensioned steel averages .8 lb. per S.F. and reinforcing steel about 1.0 lb. per S.F. Placing and stressing averages 40 hours per ton of total material.

Beam construction — 4000 to 5000 psi concrete. Steel weights vary greatly.

Labor cost per pound goes down as the size and length of the tendon increase. The primary economic consideration is the cost per kip for the member.

Post-tensioning becomes feasible for beams and girders over 30' long; for continuous two-way slabs over 20' clear; also in transferring upper building loads over longer spans at lower levels. Post-tension suppliers will provide engineering services at no cost to the user. Substantial economies are possible by using post-tensioned Lift Slabs.

Concrete | R0345 Precast Architectural Concrete

R034513-10   Precast Concrete Wall Panels

Panels are either solid or insulated with plain, colored or textured finishes. Transportation is an important cost factor. Prices shown in the unit cost section of the book are based on delivery within 50 miles of a plant including fabricators' overhead and profit. Engineering data is available from fabricators to assist with construction details. Usual minimum job size for economical use of panels is about 5000 S.F. Small jobs can double the prices shown. For large, highly repetitive jobs, deduct up to 15% from the prices shown.

2″ thick panels cost about the same as 3″ thick panels, and maximum panel size is less. For building panels faced with granite, marble or stone, add the material prices from those unit cost sections to the plain panel price shown. There is a growing trend toward aggregate facings and broken rib finish rather than plain gray concrete panels.

No allowance has been made in the unit cost section for supporting steel framework. On one story buildings, panels may rest on grade beams and require only wind bracing and fasteners. On multi-story buildings panels can span from column to column and floor to floor. Plastic-designed steel-framed structures may have large deflections which slow down erection and raise costs.

Large panels are more economical than small panels on a S.F. basis. When figuring areas include all protrusions, returns, etc. Overhangs can triple erection costs. Panels over 45′ have been produced. Larger flat units should be prestressed. Vacuum lifting of smooth finish panels eliminates inserts and can speed erection.

Concrete | R0347 Site-Cast Concrete

R034713-20   Tilt Up Concrete Panels

The advantage of tilt up construction is in the low cost of forms and the placing of concrete and reinforcing. Panels up to 75′ high and 5-1/2″ thick have been tilted using strongbacks. Tilt up has been used for one to five story buildings and is well-suited for warehouses, stores, offices, schools and residences.

The panels are cast in forms on the floor slab. Most jobs use 5-1/2″ thick solid reinforced concrete panels. Sandwich panels with a layer of insulating materials are also used. Where dampness is a factor, lightweight aggregate is used. Optimum panel size is 300 to 500 S.F.

Slabs are usually poured with 3000 psi concrete which permits tilting seven days after pouring. Slabs may be stacked on top of each other and are separated from each other by either two coats of bond breaker or a film of polyethylene. Use of high early-strength cement allows tilting two days after a pour. Tilting up is done with a roller outrigger crane with a capacity of at least 1-1/2 times the weight of the panel at the required reach. Exterior precast columns can be set at the same time as the panels; interior precast columns can be set first and the panels clipped directly to them. The use of cast-in-place concrete columns is diminishing due to shrinkage problems. Structural steel columns are sometimes used if crane rails are planned. Panels can be clipped to the columns or lowered between the flanges. Steel channels with anchors may be used as edge forms for the slab. When the panels are lifted the channels form an integral steel column to take structural loads. Roof loads can be carried directly by the panels for wall heights to 14′.

Requirements of local building codes may be a limiting factor and should be checked. Building floor slabs should be poured first and should be a minimum of 5″ thick with 100% compaction of soil or 6″ thick with less than 100% compaction.

Setting times as fast as nine minutes per panel have been observed, but a safer expectation would be four panels per hour with a crane and a four-man setting crew. If crane erects from inside building, some provision must be made to get crane out after walls are erected. Good yarding procedure is important to minimize delays. Equalizing three-point lifting beams and self-releasing pick-up hooks speed erection. If panels must be carried to their final location, setting time per panel will be increased and erection costs may approach the erection cost range of architectural precast wall panels. Placing panels into slots formed in continuous footers will speed erection.

Reinforcing should be with #5 bars with vertical bars on the bottom. If surface is to be sandblasted, stainless steel chairs should be used to prevent rust staining.

Use of a broom finish is popular since the unavoidable surface blemishes are concealed.

Precast columns run from three to five times the C.Y. price of the panels only.

Concrete | R0352 Lightweight Concrete Roof Insulation

R035216-10   Lightweight Concrete

Lightweight aggregate concrete is usually purchased ready mixed, but it can also be field mixed.

Vermiculite or Perlite comes in bags of 4 C.F. under various trade names. Weight is about 8 lbs. per C.F. For insulating roof fill use 1:6 mix. For structural deck use 1:4 mix over gypsum boards, steeltex, steel centering, etc., supported by closely spaced joists or bulb trees. For structural slabs use 1:3:2 vermiculite sand concrete over steeltex, metal lath, steel centering, etc., on joists spaced 2′-0″ O.C. for maximum L.L. of 80 P.S.F. Use same mix

for slab base fill over steel flooring or regular reinforced concrete slab when tile, terrazzo or other finish is to be laid over.

For slabs on grade use 1:3:2 mix when tile, etc., finish is to be laid over. If radiant heating units are installed use a 1:6 mix for a base. After coils are in place, cover with a regular granolithic finish (mix 1:3:2) to a minimum depth of 1-1/2″ over top of units.

Reinforce all slabs with 6 x 6 or 10 x 10 welded wire mesh.

R040130-10 Cleaning Face Brick

On smooth brick a person can clean 70 S.F. an hour; on rough brick 50 S.F. per hour. Use one gallon muriatic acid to 20 gallons of water for 1000

S.F. Do not use acid solution until wall is at least seven days old, but a mild soap solution may be used after two days.

Time has been allowed for clean-up in brick prices.

Masonry | **R0405 Common Work Results for Masonry**

R040513-10 Cement Mortar (material only)

Type N - 1:1:6 mix by volume. Use everywhere above grade except as noted below. - 1:3 mix using conventional masonry cement which saves handling two separate bagged materials.

Type M - 1:1/4:3 mix by volume, or 1 part cement, 1/4 (10% by wt.) lime, 3 parts sand. Use for heavy loads and where earthquakes or hurricanes may occur. Also for reinforced brick, sewers, manholes and everywhere below grade.

Mix Proportions by Volume and Compressive Strength of Mortar

| Where Used | Mortar Type | Allowable Proportions by Volume | | | | Compressive Strength @ 28 days |
		Portland Cement	Masonry Cement	Hydrated Lime	Masonry Sand	
Plain Masonry	M	1	1	—	6	
		1	—	1/4	3	2500 psi
	S	1/2	1	—	4	
		1	—	1/4 to 1/2	4	1800 psi
	N	—	1	—	3	
		1	—	1/2 to 1-1/4	6	750 psi
	O	—	1	—	3	
		1	—	1-1/4 to 2-1/2	9	350 psi
	K	1	—	2-1/2 to 4	12	75 psi
Reinforced Masonry	PM	1	1	—	6	2500 psi
	PL	1	—	1/4 to 1/2	4	2500 psi

Note: The total aggregate should be between 2.25 to 3 times the sum of the cement and lime used.

The labor cost to mix the mortar is included in the productivity and labor cost of unit price lines in unit cost sections for brickwork, blockwork and stonework.

The material cost of mixed mortar is included in the material cost of those same unit price lines and includes the cost of renting and operating a 10 C.F. mixer at the rate of 200 C.F. per day.

There are two types of mortar color used. One type is the inert additive type with about 100 lbs. per M brick as the typical quantity required. These colors are also available in smaller-batch-sized bags (1 lb. to 15 lb.) which can be placed directly into the mixer without measuring. The other type is premixed and replaces the masonry cement. Dark green color has the highest cost.

R040519-50 Masonry Reinforcing

Horizontal joint reinforcing helps prevent wall cracks where wall movement may occur and in many locations is required by code. Horizontal joint reinforcing is generally not considered to be structural reinforcing and an unreinforced wall may still contain joint reinforcing.

Reinforcing strips come in 10' and 12' lengths and in truss and ladder shapes, with and without drips. Field labor runs between 2.7 to 5.3 hours per 1000 L.F. for wall thicknesses up to 12".

The wire meets ASTM A82 for cold drawn steel wire and the typical size is 9 ga. sides and ties with 3/16" diameter also available. Typical finish is mill galvanized with zinc coating at .10 oz. per S.F. Class I (.40 oz. per S.F.) and Class III (.80 oz per S.F.) are also available, as is hot dipped galvanizing at 1.50 oz. per S.F.

R042110-10   Economy in Bricklaying

Have adequate supervision. Be sure bricklayers are always supplied with materials so there is no waiting. Place best bricklayers at corners and openings.

Use only screened sand for mortar. Otherwise, labor time will be wasted picking out pebbles. Use seamless metal tubs for mortar as they do not leak or catch the trowel. Locate stack and mortar for easy wheeling.

Have brick delivered for stacking. This makes for faster handling, reduces chipping and breakage, and requires less storage space. Many dealers will deliver select common in 2' x 3' x 4' pallets or face brick packaged. This affords quick handling with a crane or forklift and easy tonging in units of ten, which reduces waste.

Use wider bricks for one wythe wall construction. Keep scaffolding away from wall to allow mortar to fall clear and not stain wall.

On large jobs develop specialized crews for each type of masonry unit.

Consider designing for prefabricated panel construction on high rise projects.

Avoid excessive corners or openings. Each opening adds about 50% to labor cost for area of opening.

Bolting stone panels and using window frames as stops reduces labor costs and speeds up erection.

R042110-20   Common and Face Brick

Common building brick manufactured according to ASTM C62 and facing brick manufactured according to ASTM C216 are the two standard bricks available for general building use.

Building brick is made in three grades; SW, where high resistance to damage caused by cyclic freezing is required; MW, where moderate resistance to cyclic freezing is needed; and NW, where little resistance to cyclic freezing is needed. Facing brick is made in only the two grades SW and MW. Additionally, facing brick is available in three types; FBS, for general use; FBX, for general use where a higher degree of precision and lower permissible variation in size than FBS is needed; and FBA, for general use to produce characteristic architectural effects resulting from non-uniformity in size and texture of the units.

In figuring the material cost of brickwork, an allowance of 25% mortar waste and 3% brick breakage was included. If bricks are delivered palletized

with 280 to 300 per pallet, or packaged, allow only 1-1/2% for breakage. Packaged or palletized delivery is practical when a job is big enough to have a crane or other equipment available to handle a package of brick. This is so on all industrial work but not always true on small commercial buildings.

The use of buff and gray face is increasing, and there is a continuing trend to the Norman, Roman, Jumbo and SCR brick.

Common red clay brick for backup is not used that often. Concrete block is the most usual backup material with occasional use of sand lime or cement brick. Building brick is commonly used in solid walls for strength and as a fire stop.

Brick panels built on the ground and then crane erected to the upper floors have proven to be economical. This allows the work to be done under cover and without scaffolding.

R042110-50   Brick, Block & Mortar Quantities

Type Brick	Nominal Size (incl. mortar) L	H	W	Modular Coursing	Number of Brick per S.F.	C.F. of Mortar per M Bricks, Waste Included 3/8" Joint	1/2" Joint	Bond Type	Description	Factor
Standard	8 x 2-2/3 x 4			3C=8"	6.75	10.3	12.9	Common	full header every fifth course	+20%
Economy	8 x 4 x 4			1C=4"	4.50	11.4	14.6		full header every sixth course	+16.7%
Engineer	8 x 3-1/5 x 4			5C=16"	5.63	10.6	13.6	English	full header every second course	+50%
Fire	9 x 2-1/2 x 4-1/2			2C=5"	6.40	550 # Fireclay	—	Flemish	alternate headers every course	+33.3%
Jumbo	12 x 4 x 6 or 8			1C=4"	3.00	23.8	30.8		every sixth course	+5.6%
Norman	12 x 2-2/3 x 4			3C=8"	4.50	14.0	17.9	Header = W x H exposed		+100%
Norwegian	12 x 3-1/5 x 4			5C=16"	3.75	14.6	18.6	Rowlock = H x W exposed		+100%
Roman	12 x 2 x 4			2C=4"	6.00	13.4	17.0	Rowlock stretcher = L x W exposed		+33.3%
SCR	12 x 2-2/3 x 6			3C=8"	4.50	21.8	28.0	Soldier = H x L exposed		—
Utility	12 x 4 x 4			1C=4"	3.00	15.4	19.6	Sailor = W x L exposed		-33.3%

Running Bond — Number of Brick per S.F. of Wall - Single Wythe with 3/8" Joints. For Other Bonds Standard Size Add to S.F. Quantities in Table to Left.

Concrete Blocks Nominal Size		Approximate Weight per S.F. Standard	Lightweight	Blocks per 100 S.F.	Mortar per M block, waste included Partitions	Back up
2"	x 8" x 16"	20 PSF	15 PSF	113	27 C.F.	36 C.F.
4"		30	20		41	51
6"		42	30		56	66
8"		55	38		72	82
10"		70	47		87	97
12"		85	55		102	112

R042210-20 Concrete Block

The material cost of special block such as corner, jamb and head block can be figured at the same price as ordinary block of same size. Labor on specials is about the same as equal-sized regular block.

Bond beam and 16″ high lintel blocks are more expensive than regular units of equal size. Lintel blocks are 8″ long and either 8″ or 16″ high.

Use of motorized mortar spreader box will speed construction of continuous walls.

Hollow non-load-bearing units are made according to ASTM C129 and hollow load-bearing units according to ASTM C90.

R050516-30 Coating Structural Steel

On field-welded jobs, the shop-applied primer coat is necessarily omitted. All painting must be done in the field and usually consists of red oxide rust inhibitive paint or an aluminum paint. The table below shows paint coverage and daily production for field painting.

See Division 09 97 13.23 for hot-dipped galvanizing and for field-applied cold galvanizing and other paints and protective coatings.

See Division 05 01 10.51 for steel surface preparation treatments such as wire brushing, pressure washing and sand blasting.

Type Construction	Surface Area per Ton	Coat	One Gallon Covers		In 8 Hrs. Person Covers		Average per Ton Spray	
			Brush	Spray	Brush	Spray	Gallons	Labor-hours
Light Structural	300 S.F. to 500 S.F.	1st	500 S.F.	455 S.F.	640 S.F.	2000 S.F.	0.9 gals.	1.6 L.H.
		2nd	450	410	800	2400	1.0	1.3
		3rd	450	410	960	3200	1.0	1.0
Medium	150 S.F. to 300 S.F.	All	400	365	1600	3200	0.6	0.6
Heavy Structural	50 S.F. to 150 S.F.	1st	400	365	1920	4000	0.2	0.2
		2nd	400	365	2000	4000	0.2	0.2
		3rd	400	365	2000	4000	0.2	0.2
Weighted Average	225 S.F.	All	400	365	1350	3000	0.6	0.6

R050521-20 Welded Structural Steel

Usual weight reductions with welded design run 10% to 20% compared with bolted or riveted connections. This amounts to about the same total cost compared with bolted structures since field welding is more expensive than bolts. For normal spans of 18′ to 24′ figure 6 to 7 connections per ton.

Trusses — For welded trusses add 4% to weight of main members for connections. Up to 15% less steel can be expected in a welded truss compared to one that is shop bolted. Cost of erection is the same whether shop bolted or welded.

General — Typical electrodes for structural steel welding are E6010, E6011, E60T and E70T. Typical buildings vary between 2# to 8# of weld rod per

ton of steel. Buildings utilizing continuous design require about three times as much welding as conventional welded structures. In estimating field erection by welding, it is best to use the average linear feet of weld per ton to arrive at the welding cost per ton. The type, size and position of the weld will have a direct bearing on the cost per linear foot. A typical field welder will deposit 1.8# to 2# of weld rod per hour manually. Using semiautomatic methods can increase production by as much as 50% to 75%.

R050523-10 High Strength Bolts

Common bolts (A307) are usually used in secondary connections (see Division 05 05 23.10).

High strength bolts (A325 and A490) are usually specified for primary connections such as column splices, beam and girder connections to columns, column bracing, connections for supports of operating equipment or of other live loads which produce impact or reversal of stress, and in structures carrying cranes of over 5-ton capacity.

Allow 20 field bolts per ton of steel for a 6 story office building, apartment house or light industrial building. For 6 to 12 stories allow 25 bolts per ton, and above 12 stories, 30 bolts per ton. On power stations, 20 to 25 bolts per ton are needed.

R051223-10   Structural Steel

The bare material prices for structural steel, shown in the unit cost sections of the book, are for 100 tons of shop-fabricated structural steel and include:

1. Mill base price of structural steel
2. Mill scrap/grade/size/length extras
3. Mill delivery to a metals service center (warehouse)
4. Service center storage and handling
5. Service center delivery to a fabrication shop
6. Shop storage and handling
7. Shop drafting/detailing
8. Shop fabrication

9. Shop coat of primer paint
10. Shop listing
11. Shop delivery to the job site

In unit cost sections of the book that contain items for field fabrication of steel components, the bare material cost of steel includes:

1. Mill base price of structural steel
2. Mill scrap/grade/size/length extras
3. Mill delivery to a metals service center (warehouse)
4. Service center storage and handling
5. Service center delivery to the job site

R051223-20   Steel Estimating Quantities

One estimate on erection is that a crane can handle 35 to 60 pieces per day. Say the average is 45. With usual sizes of beams, girders, and columns, this would amount to about 20 tons per day. The type of connection greatly affects the speed of erection. Moment connections for continuous design slow down production and increase erection costs.

Short open web bar joists can be set at the rate of 75 to 80 per day, with 50 per day being the average for setting long span joists.

After main members are calculated, add the following for usual allowances: base plates 2% to 3%; column splices 4% to 5%; and miscellaneous details 4% to 5%, for a total of 10% to 13% in addition to main members.

The ratio of column to beam tonnage varies depending on type of steels used, typical spans, story heights and live loads.

It is more economical to keep the column size constant and to vary the strength of the column by using high strength steels. This also saves floor space. Buildings have recently gone as high as ten stories with 8" high strength columns. For light columns under W8X31 lb. sections, concrete filled steel columns are economical.

High strength steels may be used in columns and beams to save floor space and to meet head room requirements. High strength steels in some sizes sometimes require long lead times.

Round, square and rectangular columns, both plain and concrete filled, are readily available and save floor area, but are higher in cost per pound than rolled columns. For high unbraced columns, tube columns may be less expensive.

Below are average minimum figures for the weights of the structural steel frame for different types of buildings using A36 steel, rolled shapes and simple joints. For economy in domes, rise to span ratio = .13. Open web joist framing systems will reduce weights by 10% to 40%. Composite design can reduce steel weight by up to 25% but additional concrete floor slab thickness may be required. Continuous design can reduce the weights up to 20%. There are many building codes with different live load requirements and different structural requirements, such as hurricane and earthquake loadings which can alter the figures.

					Structural Steel Weights per S.F. of Floor Area				
Type of Building	No. of Stories	Avg. Spans	L.L. #/S.F.	Lbs. Per S.F.	Type of Building	No. of Stories	Avg. Spans	L.L. #/S.F.	Lbs. Per S.F.
Steel Frame Mfg.	1	20'x20'	40	8	Apartments	2-8	20'x20'	40	8
		30'x30'		13		9-25			14
		40'x40'		18	Office	to 10	Various	80	10
Parking garage	4	Various	80	8.5		20			18
Domes (Schwedler)*	1	200'	30	10		30			26
		300'		15		over 50			35

R051223-25 Common Structural Steel Specifications

ASTM A992 (formerly A36, then A572 Grade 50) is the all-purpose carbon grade steel widely used in building and bridge construction.

The other high-strength steels listed below may each have certain advantages over ASTM A992 stuctural carbon steel, depending on the application. They have proven to be economical choices where, due to lighter members, the reduction of dead load and the associated savings in shipping cost can be significant.

ASTM A588 atmospheric weathering, high-strength low-alloy steels can be used in the bare (uncoated) condition, where exposure to normal atmosphere causes a tightly adherant oxide to form on the surface protecting the steel from further oxidation. ASTM A242 corrosion-resistant, high-strength low-alloy steels have enhanced atmospheric corrosion resistance of at least two times that of carbon structural steels with copper, or four times that of carbon structural steels without copper. The reduction or elimination of maintenance resulting from the use of these steels often offsets their higher initial cost.

Steel Type	ASTM Designation	Minimum Yield Stress in KSI	Shapes Available
Carbon	A36	36	All structural shape groups, and plates & bars up thru 8" thick
	A529	50	Structural shape group 1, and plates & bars up thru 2" thick
High-Strength Low-Alloy Quenched & Self-Tempered	A913	50	All structural shape groups
		60	
		65	
		70	
High-Strength Low-Alloy Columbium-Vanadium	A572	42	All structural shape groups, and plates & bars up thru 6" thick
		50	All structural shape groups, and plates & bars up thru 4" thick
		55	Structural shape groups 1 & 2, and plates & bars up thru 2" thick
		60	Structural shape groups 1 & 2, and plates & bars up thru 1-1/4" thick
		65	Structural shape group 1, and plates & bars up thru 1-1/4" thick
High-Strength Low-Alloy Columbium-Vanadium	A992	50	All structural shape groups
Weathering High-Strength Low-Alloy	A242	42	Structural shape groups 4 & 5, and plates & bars over 1-1/2" up thru 4" thick
		46	Structural shape group 3, and plates & bars over 3/4" up thru 1-1/2" thick
		50	Structural shape groups 1 & 2, and plates & bars up thru 3/4" thick
Weathering High-Strength Low-Alloy	A588	42	Plates & bars over 5" up thru 8" thick
		46	Plates & bars over 4" up thru 5" thick
		50	All structural shape groups, and plates & bars up thru 4" thick
Quenched and Tempered Low-Alloy	A852	70	Plates & bars up thru 4" thick
Quenched and Tempered Alloy	A514	90	Plates & bars over 2-1/2" up thru 6" thick
		100	Plates & bars up thru 2-1/2" thick

R051223-30 High Strength Steels

The mill price of high strength steels may be higher than A992 carbon steel but their proper use can achieve overall savings thru total reduced weights. For columns with L/r over 100, A992 steel is best; under 100, high strength steels are economical. For heavy columns, high strength steels are economical when cover plates are eliminated. There is no economy using high strength steels for clip angles or supports or for beams where deflection governs. Thinner members are more economical than thick.

The per ton erection and fabricating costs of the high strength steels will be higher than for A992 since the same number of pieces, but less weight, will be installed.

R051223-35   Common Steel Sections

The upper portion of this table shows the name, shape, common designation and basic characteristics of commonly used steel sections. The lower portion explains how to read the designations used for the above illustrated common sections.

Shape & Designation	Name & Characteristics	Shape & Designation	Name & Characteristics
W	W Shape — Parallel flange surfaces	MC	Miscellaneous Channel — Infrequently rolled by some producers
S	American Standard Beam (I Beam) — Sloped inner flange	L	Angle — Equal or unequal legs, constant thickness
M	Miscellaneous Beams — Cannot be classified as W, HP or S; infrequently rolled by some producers	T	Structural Tee — Cut from W, M or S on center of web
C	American Standard Channel — Sloped inner flange	HP	Bearing Pile — Parallel flanges and equal flange and web thickness

Common drawing designations follow:

W Shape
W 18 x 35
— Weight in Pounds per Foot
— Nominal Depth in Inches (Actual 17-3/4")

American Standard Beam
S 12 x 31.8
— Weight in Pounds per Foot
— Depth in Inches

Miscellaneous Beam
M 8 x 6.5
— Weight in Pounds per Foot
— Depth in Inches

American Standard Channel
C 8 x 11.5
— Weight in Pounds per Foot
— Depth in Inches

Miscellaneous Channel
MC 8 x 22.8
— Weight in Pounds per Foot
— Depth in Inches

Angle
L 6 x 3-1/2 x 3/8
— Length of Long Leg in Inches
— Thickness of Each Leg in Inches
— Length of Other Leg in Inches

Tee Cut from W16 x 100
WT 8 x 50
— Weight in Pounds per Foot
— Nominal Depth in Inches (Actual 8-1/2")

Tee Cut from S12 x 35
ST 6 x 17.5
— Weight in Pounds per Foot
— Depth in Inches (Actual 6-1/4")

Tee Cut from M10 x 9
MT 5 x 4.5
— Weight in Pounds per Foot
— Depth in Inches

Bearing Pile
HP 12 x 84
— Weight in Pounds per Foot
— Nominal Depth in Inches (Actual 12-1/4")

R051223-45   Installation Time for Structural Steel Building Components

The following tables show the expected average installation times for various structural steel shapes. Table A presents installation times for columns, Table B for beams, Table C for light framing and bolts, and Table D for structural steel for various project types.

Table A		
Description	**Labor-Hours**	**Unit**
Columns		
Steel, Concrete Filled		
3-1/2" Diameter	.933	Ea.
6-5/8" Diameter	1.120	Ea.
Steel Pipe		
3" Diameter	.933	Ea.
8" Diameter	1.120	Ea.
12" Diameter	1.244	Ea.
Structural Tubing		
4" x 4"	.966	Ea.
8" x 8"	1.120	Ea.
12" x 8"	1.167	Ea.
W Shape 2 Tier		
W8 x 31	.052	L.F.
W8 x 67	.057	L.F.
W10 x 45	.054	L.F.
W10 x 112	.058	L.F.
W12 x 50	.054	L.F.
W12 x 190	.061	L.F.
W14 x 74	.057	L.F.
W14 x 176	.061	L.F.

Table B				
Description	**Labor-Hours**	**Unit**	**Labor-Hours**	**Unit**
Beams, W Shape				
W6 x 9	.949	Ea.	.093	L.F.
W10 x 22	1.037	Ea.	.085	L.F.
W12 x 26	1.037	Ea.	.064	L.F.
W14 x 34	1.333	Ea.	.069	L.F.
W16 x 31	1.333	Ea.	.062	L.F.
W18 x 50	2.162	Ea.	.088	L.F.
W21 x 62	2.222	Ea.	.077	L.F.
W24 x 76	2.353	Ea.	.072	L.F.
W27 x 94	2.581	Ea.	.067	L.F.
W30 x 108	2.857	Ea.	.067	L.F.
W33 x 130	3.200	Ea.	.071	L.F.
W36 x 300	3.810	Ea.	.077	L.F.

Table C		
Description	**Labor-Hours**	**Unit**
Light Framing		
Angles 4" and Larger	.055	lbs.
Less than 4"	.091	lbs.
Channels 8" and Larger	.048	lbs.
Less than 8"	.072	lbs.
Cross Bracing Angles	.055	lbs.
Rods	.034	lbs.
Hanging Lintels	.069	lbs.
High Strength Bolts in Place		
3/4" Bolts	.070	Ea.
7/8" Bolts	.076	Ea.

Table D				
Description	**Labor-Hours**	**Unit**	**Labor-Hours**	**Unit**
Apartments, Nursing Homes, etc.				
1-2 Stories	4.211	Piece	7.767	Ton
3-6 Stories	4.444	Piece	7.921	Ton
7-15 Stories	4.923	Piece	9.014	Ton
Over 15 Stories	5.333	Piece	9.209	Ton
Offices, Hospitals, etc.				
1-2 Stories	4.211	Piece	7.767	Ton
3-6 Stories	4.741	Piece	8.889	Ton
7-15 Stories	4.923	Piece	9.014	Ton
Over 15 Stories	5.120	Piece	9.209	Ton
Industrial Buildings				
1 Story	3.478	Piece	6.202	Ton

R051223-50   Subpurlins

Bulb tee subpurlins are structural members designed to support and reinforce a variety of roof deck systems such as precast cement fiber roof deck tiles, monolithic roof deck systems, and gypsum or lightweight concrete over formboard. Other uses include interstitial service ceiling systems, wall panel systems, and joist anchoring in bond beams. See Unit Price section for pricing on a square foot basis at 32-5/8" O.C. Maximum span is based on a 3-span condition with a total allowable vertical load of 40 psf.

R051223-80 Dimensions and Weights of Sheet Steel

Gauge No.	Approximate Thickness				Weight		
	Inches (in fractions)	Inches (in decimal parts)		Millimeters	per S.F. in Ounces	per S.F. in Lbs.	per Square Meter in Kg.
	Wrought Iron	Wrought Iron	Steel	Steel			
0000000	1/2"	.5	.4782	12.146	320	20.000	97.650
000000	15/32"	.46875	.4484	11.389	300	18.750	91.550
00000	7/16"	.4375	.4185	10.630	280	17.500	85.440
0000	13/32"	.40625	.3886	9.870	260	16.250	79.330
000	3/8"	.375	.3587	9.111	240	15.000	73.240
00	11/32"	.34375	.3288	8.352	220	13.750	67.130
0	5/16"	.3125	.2989	7.592	200	12.500	61.030
1	9/32"	.28125	.2690	6.833	180	11.250	54.930
2	17/64"	.265625	.2541	6.454	170	10.625	51.880
3	1/4"	.25	.2391	6.073	160	10.000	48.820
4	15/64"	.234375	.2242	5.695	150	9.375	45.770
5	7/32"	.21875	.2092	5.314	140	8.750	42.720
6	13/64"	.203125	.1943	4.935	130	8.125	39.670
7	3/16"	.1875	.1793	4.554	120	7.500	36.320
8	11/64"	.171875	.1644	4.176	110	6.875	33.570
9	5/32"	.15625	.1495	3.797	100	6.250	30.520
10	9/64"	.140625	.1345	3.416	90	5.625	27.460
11	1/8"	.125	.1196	3.038	80	5.000	24.410
12	7/64"	.109375	.1046	2.657	70	4.375	21.360
13	3/32"	.09375	.0897	2.278	60	3.750	18.310
14	5/64"	.078125	.0747	1.897	50	3.125	15.260
15	9/128"	.0713125	.0673	1.709	45	2.813	13.730
16	1/16"	.0625	.0598	1.519	40	2.500	12.210
17	9/160"	.05625	.0538	1.367	36	2.250	10.990
18	1/20"	.05	.0478	1.214	32	2.000	9.765
19	7/160"	.04375	.0418	1.062	28	1.750	8.544
20	3/80"	.0375	.0359	.912	24	1.500	7.324
21	11/320"	.034375	.0329	.836	22	1.375	6.713
22	1/32"	.03125	.0299	.759	20	1.250	6.103
23	9/320"	.028125	.0269	.683	18	1.125	5.490
24	1/40"	.025	.0239	.607	16	1.000	4.882
25	7/320"	.021875	.0209	.531	14	.875	4.272
26	3/160"	.01875	.0179	.455	12	.750	3.662
27	11/640"	.0171875	.0164	.417	11	.688	3.357
28	1/64"	.015625	.0149	.378	10	.625	3.052

Reference Tables

R053100-10 Decking Descriptions

General - All Deck Products

Steel deck is made by cold forming structural grade sheet steel into a repeating pattern of parallel ribs. The strength and stiffness of the panels are the result of the ribs and the material properties of the steel. Deck lengths can be varied to suit job conditions, but because of shipping considerations, are usually less than 40 feet. Standard deck width varies with the product used but full sheets are usually 12″, 18″, 24″, 30″, or 36″. Deck is typically furnished in a standard width with the ends cut square. Any cutting for width, such as at openings or for angular fit, is done at the job site.

Deck is typically attached to the building frame with arc puddle welds, self-drilling screws, or powder or pneumatically driven pins. Sheet to sheet fastening is done with screws, button punching (crimping), or welds.

Composite Floor Deck

After installation and adequate fastening, floor deck serves several purposes. It (a) acts as a working platform, (b) stabilizes the frame, (c) serves as a concrete form for the slab, and (d) reinforces the slab to carry the design loads applied during the life of the building. Composite decks are distinguished by the presence of shear connector devices as part of the deck. These devices are designed to mechanically lock the concrete and deck together so that the concrete and the deck work together to carry subsequent floor loads. These shear connector devices can be rolled-in embossments, lugs, holes, or wires welded to the panels. The deck profile can also be used to interlock concrete and steel.

Composite deck finishes are either galvanized (zinc coated) or phosphatized/painted. Galvanized deck has a zinc coating on both the top and bottom surfaces. The phosphatized/painted deck has a bare (phosphatized) top surface that will come into contact with the concrete. This bare top surface can be expected to develop rust before the concrete is placed. The bottom side of the deck has a primer coat of paint.

Composite floor deck is normally installed so the panel ends do not overlap on the supporting beams. Shear lugs or panel profile shape often prevent a tight metal to metal fit if the panel ends overlap; the air gap caused by overlapping will prevent proper fusion with the structural steel supports when the panel end laps are shear stud welded.

Adequate end bearing of the deck must be obtained as shown on the drawings. If bearing is actually less in the field than shown on the drawings, further investigation is required.

Roof Deck

Roof deck is not designed to act compositely with other materials. Roof deck acts alone in transferring horizontal and vertical loads into the building frame. Roof deck rib openings are usually narrower than floor deck rib openings. This provides adequate support of rigid thermal insulation board.

Roof deck is typically installed to endlap approximately 2″ over supports. However, it can be butted (or lapped more than 2″) to solve field fit problems. Since designers frequently use the installed deck system as part of the horizontal bracing system (the deck as a diaphragm), any fastening substitution or change should be approved by the designer. Continuous perimeter support of the deck is necessary to limit edge deflection in the finished roof and may be required for diaphragm shear transfer.

Standard roof deck finishes are galvanized or primer painted. The standard factory applied paint for roof deck is a primer paint and is not intended to weather for extended periods of time. Field painting or touching up of abrasions and deterioration of the primer coat or other protective finishes is the responsibility of the contractor.

Cellular Deck

Cellular deck is made by attaching a bottom steel sheet to a roof deck or composite floor deck panel. Cellular deck can be used in the same manner as floor deck. Electrical, telephone, and data wires are easily run through the chase created between the deck panel and the bottom sheet.

When used as part of the electrical distribution system, the cellular deck must be installed so that the ribs line up and create a smooth cell transition at abutting ends. The joint that occurs at butting cell ends must be taped or otherwise sealed to prevent wet concrete from seeping into the cell. Cell interiors must be free of welding burrs, or other sharp intrusions, to prevent damage to wires.

When used as a roof deck, the bottom flat plate is usually left exposed to view. Care must be maintained during erection to keep good alignment and prevent damage.

Cellular deck is sometimes used with the flat plate on the top side to provide a flat working surface. Installation of the deck for this purpose requires special methods for attachment to the frame because the flat plate, now on the top, can prevent direct access to the deck material that is bearing on the structural steel. It may be advisable to treat the flat top surface to prevent slipping.

Cellular deck is always furnished galvanized or painted over galvanized.

Form Deck

Form deck can be any floor or roof deck product used as a concrete form. Connections to the frame are by the same methods used to anchor floor and roof deck. Welding washers are recommended when welding deck that is less than 20 gauge thickness.

Form deck is furnished galvanized, prime painted, or uncoated. Galvanized deck must be used for those roof deck systems where form deck is used to carry a lightweight insulating concrete fill.

Wood, Plastics & Comp. — R0611 Wood Framing

R061110-30 Lumber Product Material Prices

The price of forest products fluctuates widely from location to location and from season to season depending upon economic conditions. The bare material prices in the unit cost sections of the book show the National Average material prices in effect Jan. 1 of this book year. It must be noted that lumber prices in general may change significantly during the year.

Availability of certain items depends upon geographic location and must be checked prior to firm-price bidding.

Wood, Plastics & Comp. — R0616 Sheathing

R061636-20 Plywood

There are two types of plywood used in construction: interior, which is moisture-resistant but not waterproofed, and exterior, which is waterproofed.

The grade of the exterior surface of the plywood sheets is designated by the first letter: A, for smooth surface with patches allowed; B, for solid surface with patches and plugs allowed; C, which may be surface plugged or may have knot holes up to 1″ wide; and D, which is used only for interior type plywood and may have knot holes up to 2-1/2″ wide. "Structural Grade" is specifically designed for engineered applications such as box beams. All CC & DD grades have roof and floor spans marked on them.

Underlayment-grade plywood runs from 1/4″ to 1-1/4″ thick. Thicknesses 5/8″ and over have optional tongue and groove joints which eliminate the need for blocking the edges. Underlayment 19/32″ and over may be referred to as Sturd-i-Floor.

The price of plywood can fluctuate widely due to geographic and economic conditions.

Typical uses for various plywood grades are as follows:

AA-AD Interior — cupboards, shelving, paneling, furniture

BB Plyform — concrete form plywood

CDX — wall and roof sheathing

Structural — box beams, girders, stressed skin panels

AA-AC Exterior — fences, signs, siding, soffits, etc.

Underlayment — base for resilient floor coverings

Overlaid HDO — high density for concrete forms & highway signs

Overlaid MDO — medium density for painting, siding, soffits & signs

303 Siding — exterior siding, textured, striated, embossed, etc.

Thermal & Moist. Protec. — R0731 Shingles & Shakes

R073126-20 Roof Slate

16″, 18″ and 20″ are standard lengths, and slate usually comes in random widths. For standard 3/16″ thickness use 1-1/2″ copper nails. Allow for 3% breakage.

Thermal & Moist. Protec. — R0751 Built-Up Bituminous Roofing

R075113-20 Built-Up Roofing

Asphalt is available in kegs of 100 lbs. each; coal tar pitch in 560 lb. kegs. Prepared roofing felts are available in a wide range of sizes, weights and characteristics. However, the most commonly used are #15 (432 S.F. per roll, 13 lbs. per square) and #30 (216 S.F. per roll, 27 lbs. per square).

Inter-ply bitumen varies from 24 lbs. per sq. (asphalt) to 30 lbs. per sq. (coal tar) per ply, MF4@ 25%. Flood coat bitumen also varies from 60 lbs. per sq. (asphalt) to 75 lbs. per sq. (coal tar), MF4@ 25%. Expendable equipment (mops, brooms, screeds, etc.) runs about 16% of the bitumen cost. For new, inexperienced crews this factor may be much higher.

Rigid insulation board is typically applied in two layers. The first is mechanically attached to nailable decks or spot or solid mopped to non-nailable decks; the second layer is then spot or solid mopped to the first layer. Membrane application follows the insulation, except in protected membrane roofs, where the membrane goes down first and the insulation on top, followed with ballast (stone or concrete pavers). Insulation and related labor costs are NOT included in prices for built-up roofing.

Thermal & Moist. Protec.   R0752 Modified Bituminous Membrane Roofing

R075213-30   Modified Bitumen Roofing

The cost of modified bitumen roofing is highly dependent on the type of installation that is planned. Installation is based on the type of modifier used in the bitumen. The two most popular modifiers are atactic polypropylene (APP) and styrene butadiene styrene (SBS). The modifiers are added to heated bitumen during the manufacturing process to change its characteristics. A polyethylene, polyester or fiberglass reinforcing sheet is then sandwiched between layers of this bitumen. When completed, the result is a pre-assembled, built-up roof that has increased elasticity and weatherabllity. Some manufacturers include a surfacing material such as ceramic or mineral granules, metal particles or sand.

The preferred method of adhering SBS-modified bitumen roofing to the substrate is with hot-mopped asphalt (much the same as built-up roofing). This installation method requires a tar kettle/pot to heat the asphalt, as well as the labor, tools and equipment necessary to distribute and spread the hot asphalt.

The alternative method for applying APP and SBS modified bitumen is as follows. A skilled installer uses a torch to melt a small pool of bitumen off the membrane. This pool must form across the entire roll for proper adhesion. The installer must unroll the roofing at a pace slow enough to melt the bitumen, but fast enough to prevent damage to the rest of the membrane.

Modified bitumen roofing provides the advantages of both built-up and single-ply roofing. Labor costs are reduced over those of built-up roofing because only a single ply is necessary. The elasticity of single-ply roofing is attained with the reinforcing sheet and polymer modifiers. Modifieds have some self-healing characteristics and because of their multi-layer construction, they offer the reliability and safety of built-up roofing.

Thermal & Moist. Protec.   R0784 Firestopping

R078413-30   Firestopping

Firestopping is the sealing of structural, mechanical, electrical and other penetrations through fire-rated assemblies. The basic components of firestop systems are safing insulation and firestop sealant on both sides of wall penetrations and the top side of floor penetrations.

Pipe penetrations are assumed to be through concrete, grout, or joint compound and can be sleeved or unsleeved. Costs for the penetrations and sleeves are not included. An annular space of 1" is assumed. Escutcheons are not included.

Metallic pipe is assumed to be copper, aluminum, cast iron or similar metallic material. Insulated metallic pipe is assumed to be covered with a thermal insulating jacket of varying thickness and materials.

Non-metallic pipe is assumed to be PVC, CPVC, FR Polypropylene or similar plastic piping material. Intumescent firestop sealant or wrap strips are included. Collars on both sides of wall penetrations and a sheet metal plate on the underside of floor penetrations are included.

Ductwork is assumed to be sheet metal, stainless steel or similar metallic material. Duct penetrations are assumed to be through concrete, grout or joint compound. Costs for penetrations and sleeves are not included. An annular space of 1/2" is assumed.

Multi-trade openings include costs for sheet metal forms, firestop mortar, wrap strips, collars and sealants as necessary.

Structural penetrations joints are assumed to be 1/2" or less. CMU walls are assumed to be within 1-1/2" of metal deck. Drywall walls are assumed to be tight to the underside of metal decking.

Metal panel, glass or curtain wall systems include a spandrel area of 5' filled with mineral wool foil-faced insulation. Fasteners and stiffeners are included.

Openings — R0813 Metal Doors

R081313-20 Steel Door Selection Guide

Standard steel doors are classified into four levels, as recommended by the Steel Door Institute in the chart below. Each of the four levels offers a range of construction models and designs, to meet architectural requirements for preference and appearance, including full flush, seamless, and stile & rail. Recommended minimum gauge requirements are also included.

For complete standard steel door construction specifications and available sizes, refer to the Steel Door Institute Technical Data Series, ANSI A250.8-98 (SDI-100), and ANSI A250.4-94 Test Procedure and Acceptance Criteria for Physical Endurance of Steel Door and Hardware Reinforcements.

Level		Model	Construction	For Full Flush or Seamless		
				Min. Gauge	Thickness (in)	Thickness (mm)
I	Standard Duty	1	Full Flush	20	0.032	0.8
		2	Seamless			
II	Heavy Duty	1	Full Flush	18	0.042	1.0
		2	Seamless			
III	Extra Heavy Duty	1	Full Flush	16	0.053	1.3
		2	Seamless			
		3	*Stile & Rail			
IV	Maximum Duty	1	Full Flush	14	0.067	1.6
		2	Seamless			

*Stiles & rails are 16 gauge; flush panels, when specified, are 18 gauge

Openings — R0851 Metal Windows

R085123-10 Steel Sash

Ironworker crew will erect 25 S.F. or 1.3 sash unit per hour, whichever is less.

Mechanic will point 30 L.F. per hour.

Painter will paint 90 S.F. per coat per hour.

Glazier production depends on light size.

Allow 1 lb. special steel sash putty per 16" x 20" light.

Openings — R0852 Wood Windows

R085216-10 Window Estimates

To ensure a complete window estimate, be sure to include the material and labor costs for each window, as well as the material and labor costs for an interior wood trim set.

Reference Tables

R087120-10 Hinges

All closer equipped doors should have ball bearing hinges. Lead lined or extremely heavy doors require special strength hinges. Usually 1-1/2 pair of hinges are used per door up to 7'-6" high openings. Table below shows typical hinge requirements.

Use Frequency	Type Hinge Required	Type of Opening	Type of Structure
High	Heavy weight ball bearing	Entrances	Banks, Office buildings, Schools, Stores & Theaters
		Toilet Rooms	Office buildings and Schools
Average	Standard weight ball bearing	Entrances	Dwellings
		Corridors	Office buildings and Schools
		Toilet Rooms	Stores
Low	Plain bearing	Interior	Dwellings

Door Thickness	Weight of Doors in Pounds per Square Foot				
	White Pine	Oak	Hollow Core	Solid Core	Hollow Metal
1-3/8"	3psf	6psf	1-1/2psf	3-1/2 — 4psf	6-1/2psf
1-3/4"	3-1/2	7	2-1/2	4-1/2 — 5-1/4	6-1/2
2-1/4"	4-1/2	9	—	5-1/2 — 6-3/4	6-1/2

R088110-10 Glazing Productivity

Some glass sizes are estimated by the "united inch" (height + width). The table below shows the number of lights glazed in an eight-hour period by the crew size indicated, for glass up to 1/4" thick. Square or nearly square lights are more economical on a S.F. basis. Long slender lights will have a high S.F. installation cost. For insulated glass reduce production by 33%. For 1/2" float glass reduce production by 50%. Production time for glazing with two glaziers per day averages: 1/4" float glass 120 S.F.; 1/2" float glass 55 S.F.; 1/2" insulated glass 95 S.F.; 3/4" insulated glass 75 S.F.

Glazing Method	United Inches per Light							
	40"	60"	80"	100"	135"	165"	200"	240"
Number of Men in Crew	1	1	1	1	2	3	3	4
Industrial sash, putty	60	45	24	15	18	—	—	—
With stops, putty bed	50	36	21	12	16	8	4	3
Wood stops, rubber	40	27	15	9	11	6	3	2
Metal stops, rubber	30	24	14	9	9	6	3	2
Structural glass	10	7	4	3	—	—	—	—
Corrugated glass	12	9	7	4	4	4	3	—
Storefronts	16	15	13	11	7	6	4	4
Skylights, putty glass	60	36	21	12	16	—	—	—
Thiokol set	15	15	11	9	9	6	3	2
Vinyl set, snap on	18	18	13	12	12	7	5	4
Maximum area per light	2.8 S.F.	6.3 S.F.	11.1 S.F.	17.4 S.F.	31.6 S.F.	47 S.F.	69 S.F.	100 S.F.

R092000-50 Lath, Plaster and Gypsum Board

Gypsum board lath is available in 3/8″ thick x 16″ wide x 4′ long sheets as a base material for multi-layer plaster applications. It is also available as a base for either multi-layer or veneer plaster applications in 1/2″ and 5/8″ thick–4′ wide x 8′, 10′ or 12′ long sheets. Fasteners are screws or blued ring shank nails for wood framing and screws for metal framing.

Metal lath is available in diamond mesh pattern with flat or self-furring profiles. Paper backing is available for applications where excessive plaster waste needs to be avoided. A slotted mesh ribbed lath should be used in areas where the span between structural supports is greater than normal. Most metal lath comes in 27″ x 96″ sheets. Diamond mesh weighs 1.75, 2.5 or 3.4 pounds per square yard, slotted mesh lath weighs 2.75 or 3.4 pounds per square yard. Metal lath can be nailed, screwed or tied in place.

Many **accessories** are available. Corner beads, flat reinforcing strips, casing beads, control and expansion joints, furring brackets and channels are some examples. Note that accessories are not included in plaster or stucco line items.

Plaster is defined as a material or combination of materials that when mixed with a suitable amount of water, forms a plastic mass or paste. When applied to a surface, the paste adheres to it and subsequently hardens, preserving in a rigid state the form or texture imposed during the period of elasticity.

Gypsum plaster is made from ground calcined gypsum. It is mixed with aggregates and water for use as a base coat plaster.

Vermiculite plaster is a fire-retardant plaster covering used on steel beams, concrete slabs and other heavy construction materials. Vermiculite is a group name for certain clay minerals, hydrous silicates or aluminum, magnesium and iron that have been expanded by heat.

Perlite plaster is a plaster using perlite as an aggregate instead of sand. Perlite is a volcanic glass that has been expanded by heat.

Gauging plaster is a mix of gypsum plaster and lime putty that when applied produces a quick drying finish coat.

Veneer plaster is a one or two component gypsum plaster used as a thin finish coat over special gypsum board.

Keenes cement is a white cementitious material manufactured from gypsum that has been burned at a high temperature and ground to a fine powder. Alum is added to accelerate the set. The resulting plaster is hard and strong and accepts and maintains a high polish, hence it is used as a finishing plaster.

Stucco is a Portland cement based plaster used primarily as an exterior finish.

Plaster is used on both interior and exterior surfaces. Generally it is applied in multiple-coat systems. A three-coat system uses the terms scratch, brown and finish to identify each coat. A two-coat system uses base and finish to describe each coat. Each type of plaster and application system has attributes that are chosen by the designer to best fit the intended use.

Gypsum Plaster Quantities for 100 S.Y.	2 Coat, 5/8″ Thick		3 Coat, 3/4″ Thick		
	Base	Finish	Scratch	Brown	Finish
	1:3 Mix	2:1 Mix	1:2 Mix	1:3 Mix	2:1 Mix
Gypsum plaster	1,300 lb.		1,350 lb.	650 lb.	
Sand	1.75 C.Y.		1.85 C.Y.	1.35 C.Y.	
Finish hydrated lime		340 lb.			340 lb.
Gauging plaster		170 lb.			170 lb.

Vermiculite or Perlite Plaster Quantities for 100 S.Y.	2 Coat, 5/8″ Thick		3 Coat, 3/4″ Thick		
	Base	Finish	Scratch	Brown	Finish
Gypsum plaster	1,250 lb.		1,450 lb.	800 lb.	
Vermiculite or perlite	7.8 bags		8.0 bags	3.3 bags	
Finish hydrated lime		340 lb.			340 lb.
Gauging plaster		170 lb.			170 lb.

Stucco–Three-Coat System Quantities for 100 S.Y.	On Wood Frame	On Masonry
Portland cement	29 bags	21 bags
Sand	2.6 C.Y.	2.0 C.Y.
Hydrated lime	180 lb.	120 lb.

R092910-10   Levels of Gypsum Drywall Finish

In the past, contract documents often used phrases such as "industry standard" and "workmanlike finish" to specify the expected quality of gypsum board wall and ceiling installations. The vagueness of these descriptions led to unacceptable work and disputes.

In order to resolve this problem, four major trade associations concerned with the manufacture, erection, finish, and decoration of gypsum board wall and ceiling systems have developed an industry-wide *Recommended Levels of Gypsum Board Finish*.

The finish of gypsum board walls and ceilings for specific final decoration is dependent on a number of factors. A primary consideration is the location of the surface and the degree of decorative treatment desired. Painted and unpainted surfaces in warehouses and other areas where appearance is normally not critical may simply require the taping of wallboard joints and 'spotting' of fastener heads. Blemish-free, smooth, monolithic surfaces often intended for painted and decorated walls and ceilings in habitated structures, ranging from single-family dwellings through monumental buildings, require additional finishing prior to the application of the final decoration.

Other factors to be considered in determining the level of finish of the gypsum board surface are (1) the type of angle of surface illumination (both natural and artificial lighting), and (2) the paint and method of application or the type and finish of wallcovering specified as the final decoration. Critical lighting conditions, gloss paints, and thin wall coverings require a higher level of gypsum board finish than do heavily textured surfaces which are subsequently painted or surfaces which are to be decorated with heavy grade wall coverings.

The following descriptions were developed jointly by the Association of the Wall and Ceiling Industries-International (AWCI), Ceiling & Interior Systems Construction Association (CISCA), Gypsum Association (GA), and Painting and Decorating Contractors of America (PDCA) as a guide.

Level 0: Used in temporary construction or wherever the final decoration has not been determined. Unfinished. No taping, finishing or corner beads are required. Also could be used where non-predecorated panels will be used in demountable-type partitions that are to be painted as a final finish.

Level 1: Frequently used in plenum areas above ceilings, in attics, in areas where the assembly would generally be concealed, or in building service corridors and other areas not normally open to public view. Some degree of sound and smoke control is provided; in some geographic areas, this level is referred to as "fire-taping," although this level of finish does not typically meet fire-resistant assembly requirements. Where a fire resistance rating is required for the gypsum board assembly, details of construction should be in accordance with reports of fire tests of assemblies that have met the requirements of the fire rating acceptable.

All joints and interior angles shall have tape embedded in joint compound. Accessories are optional at specifier discretion in corridors and other areas with pedestrian traffic. Tape and fastener heads need not be covered with joint compound. Surface shall be free of excess joint compound. Tool marks and ridges are acceptable.

Level 2: It may be specified for standard gypsum board surfaces in garages, warehouse storage, or other similar areas where surface appearance is not of primary importance.

All joints and interior angles shall have tape embedded in joint compound and shall be immediately wiped with a joint knife or trowel, leaving a thin coating of joint compound over all joints and interior angles. Fastener heads and accessories shall be covered with a coat of joint compound. Surface shall be free of excess joint compound. Tool marks and ridges are acceptable.

Level 3: Typically used in areas that are to receive heavy texture (spray or hand applied) finishes before final painting, or where commercial-grade (heavy duty) wall coverings are to be applied as the final decoration. This level of finish should not be used where smooth painted surfaces or where lighter weight wall coverings are specified. The prepared surface shall be coated with a drywall primer prior to the application of final finishes.

All joints and interior angles shall have tape embedded in joint compound and shall be immediately wiped with a joint knife or trowel, leaving a thin coating of joint compound over all joints and interior angles. One additional coat of joint compound shall be applied over all joints and interior angles. Fastener heads and accessories shall be covered with two separate coats of joint compound. All joint compounds shall be smooth and free of tool marks and ridges. The prepared surface shall be covered with a drywall primer prior to the application of the final decoration.

Level 4: This level should be used where residential grade (light duty) wall coverings, flat paints, or light textures are to be applied. The prepared surface shall be coated with a drywall primer prior to the application of final finishes. Release agents for wall coverings are specifically formulated to minimize damage if coverings are subsequently removed.

The weight, texture, and sheen level of the wall covering material selected should be taken into consideration when specifying wall coverings over this level of drywall treatment. Joints and fasteners must be sufficiently concealed if the wall covering material is lightweight, contains limited pattern, has a glossy finish, or has any combination of these features. In critical lighting areas, flat paints applied over light textures tend to reduce joint photographing. Gloss, semi-gloss, and enamel paints are not recommended over this level of finish.

All joints and interior angles shall have tape embedded in joint compound and shall be immediately wiped with a joint knife or trowel, leaving a thin coating of joint compound over all joints and interior angles. In addition, two separate coats of joint compound shall be applied over all flat joints and one separate coat of joint compound applied over interior angles. Fastener heads and accessories shall be covered with three separate coats of joint compound. All joint compounds shall be smooth and free of tool marks and ridges. The prepared surface shall be covered with a drywall primer like Sheetrock® First Coat prior to the application of the final decoration.

Level 5: The highest quality finish is the most effective method to provide a uniform surface and minimize the possibility of joint photographing and of fasteners showing through the final decoration. This level of finish is required where gloss, semi-gloss, or enamel is specified; when flat joints are specified over an untextured surface; or where critical lighting conditions occur. The prepared surface shall be coated with a drywall primer prior to the application of final decoration.

All joints and interior angles shall have tape embedded in joint compound and be immediately wiped with a joint knife or trowel, leaving a thin coating of joint compound over all joints and interior angles. Two separate coats of joint compound shall be applied over all flat joints and one separate coat of joint compound applied over interior angles. Fastener heads and accessories shall be covered with three separate coats of joint compound.

A thin skim coat of joint compound shall be trowel applied to the entire surface. Excess compound is immediately troweled off, leaving a film or skim coating of compound completely covering the paper. As an alternative to a skim coat, a material manufactured especially for this purpose may be applied such as Sheetrock® Tuff-Hide primer surfacer. The surface must be smooth and free of tool marks and ridges. The prepared surface shall be covered with a drywall primer prior to the application of the final decoration.

Finishes | R0966 Terrazzo Flooring

R096613-10   Terrazzo Floor

The table below lists quantities required for 100 S.F. of 5/8″ terrazzo topping, either bonded or not bonded.

Description	Bonded to Concrete 1-1/8″ Bed, 1:4 Mix	Not Bonded 2-1/8″ Bed and 1/4″ Sand
Portland cement, 94 lb. Bag	6 bags	8 bags
Sand	10 C.F.	20 C.F.
Divider strips, 4′ squares	50 L.F.	50 L.F.
Terrazzo fill, 50 lb. Bag	12 bags	12 bags
15 Lb. tarred felt		1 C.S.F.
Mesh 2 x 2 #14 galvanized		1 C.S.F.
Crew J-3	0.77 days	0.87 days

2′ x 2′ panels require 1.00 L.F. divider strip per S.F.

3′ x 3′ panels require 0.67 L.F. divider strip per S.F.

4′ x 4′ panels require 0.50 L.F. divider strip per S.F.

5′ x 5′ panels require 0.40 L.F. divider strip per S.F.

6′ x 6′ panels require 0.33 L.F. divider strip per S.F.

Finishes | R0972 Wall Coverings

R097223-10   Wall Covering

The table below lists the quantities required for 100 S.F. of wall covering.

Description	Medium-Priced Paper	Expensive Paper
Paper	1.6 dbl. rolls	1.6 dbl. rolls
Wall sizing	0.25 gallon	0.25 gallon
Vinyl wall paste	0.6 gallon	0.6 gallon
Apply sizing	0.3 hour	0.3 hour
Apply paper	1.2 hours	1.5 hours

Most wallpapers now come in double rolls only.
To remove old paper, allow 1.3 hours per 100 S.F.

R099100-10 Painting Estimating Techniques

Proper estimating methodology is needed to obtain an accurate painting estimate. There is no known reliable shortcut or square foot method. The following steps should be followed:

- List all surfaces to be painted, with an accurate quantity (area) of each. Items having similar surface condition, finish, application method and accessibility may be grouped together.
- List all the tasks required for each surface to be painted, including surface preparation, masking, and protection of adjacent surfaces. Surface preparation may include minor repairs, washing, sanding and puttying.
- Select the proper Means line for each task. Review and consider all adjustments to labor and materials for type of paint and location of work. Apply the height adjustment carefully. For instance, when applying the adjustment for work over 8' high to a wall that is 12' high, apply the adjustment only to the area between 8' and 12' high, and not to the entire wall.

When applying more than one percent (%) adjustment, apply each to the base cost of the data, rather than applying one percentage adjustment on top of the other.

When estimating the cost of painting walls and ceilings remember to add the brushwork for all cut-ins at inside corners and around windows and doors as a LF measure. One linear foot of cut-in with brush equals one square foot of painting.

All items for spray painting include the labor for roll-back.

Deduct for openings greater than 100 SF, or openings that extend from floor to ceiling and are greater than 5' wide. Do not deduct small openings.

The cost of brushes, rollers, ladders and spray equipment are considered to be part of a painting contractor's overhead, and should not be added to the estimate. The cost of rented equipment such as scaffolding and swing staging should be added to the estimate.

R099100-20 Painting

Item	Coat	One Gallon Covers			In 8 Hours a Laborer Covers			Labor-Hours per 100 S.F.		
		Brush	Roller	Spray	Brush	Roller	Spray	Brush	Roller	Spray
Paint wood siding	prime	250 S.F.	225 S.F.	290 S.F.	1150 S.F.	1300 S.F.	2275 S.F.	.695	.615	.351
	others	270	250	290	1300	1625	2600	.615	.492	.307
Paint exterior trim	prime	400	—	—	650	—	—	1.230	—	—
	1st	475	—	—	800	—	—	1.000	—	—
	2nd	520	—	—	975	—	—	.820	—	—
Paint shingle siding	prime	270	255	300	650	975	1950	1.230	.820	.410
	others	360	340	380	800	1150	2275	1.000	.695	.351
Stain shingle siding	1st	180	170	200	750	1125	2250	1.068	.711	.355
	2nd	270	250	290	900	1325	2600	.888	.603	.307
Paint brick masonry	prime	180	135	160	750	800	1800	1.066	1.000	.444
	1st	270	225	290	815	975	2275	.981	.820	.351
	2nd	340	305	360	815	1150	2925	.981	.695	.273
Paint interior plaster or drywall	prime	400	380	495	1150	2000	3250	.695	.400	.246
	others	450	425	495	1300	2300	4000	.615	.347	.200
Paint interior doors and windows	prime	400	—	—	650	—	—	1.230	—	—
	1st	425	—	—	800	—	—	1.000	—	—
	2nd	450	—	—	975	—	—	.820	—	—

R131113-20 Swimming Pools

Pool prices given per square foot of surface area include pool structure, filter and chlorination equipment, pumps, related piping, ladders/steps, maintenance kit, skimmer and vacuum system. Decks and electrical service to equipment are not included.

Residential in-ground pool construction can be divided into two categories: vinyl lined and gunite. Vinyl lined pool walls are constructed of different materials including wood, concrete, plastic or metal. The bottom is often graded with sand over which the vinyl liner is installed. Vermiculite or soil cement bottoms may be substituted for an added cost.

Gunite pool construction is used both in residential and municipal installations. These structures are steel reinforced for strength and finished with a white cement limestone plaster.

Municipal pools will have a higher cost because plumbing codes require more expensive materials, chlorination equipment and higher filtration rates.

Municipal pools greater than 1,800 S.F. require gutter systems to control waves. This gutter may be formed into the concrete wall. Often a vinyl/stainless steel gutter or gutter/wall system is specified, which will raise the pool cost.

Competition pools usually require tile bottoms and sides with contrasting lane striping, which will also raise the pool cost.

Special Construction   R1331 Fabric Structures

R133113-10   Air Supported Structures

Air supported structures are made from fabrics that can be classified into two groups: temporary and permanent. Temporary fabrics include nylon, woven polyethylene, vinyl film, and vinyl coated dacron. These have lifespans that range from five to fifteen plus years. The cost per square foot includes a fabric shell, tension cables, primary and back-up inflation systems and doors. The lower cost structures are used for construction shelters, bulk storage and pond covers. The more expensive are used for recreational structures and warehouses.

Permanent fabrics are teflon coated fiberglass. The life of this structure is twenty plus years. The high cost limits its application to architectural designed structures which call for a clear span covered area, such as stadiums and convention centers. Both temporary and permanent structures are available in translucent fabrics which eliminates the need for daytime lighting.

Areas to be covered vary from 10,000 S.F. to any area up to 1000 foot wide by any length. Height restrictions range from a maximum of 1/2 of

width to a minimum of 1/6 of the width. Erection of even the largest of the temporary structures requires no more than a week.

Centrifugal fans provide the inflation necessary to support the structure during application of live loads. Airlocks are usually used at large entrances to prevent loss of static pressure. Some manufacturers employ propeller fans which generate sufficient airflow (30,000 CFM) to eliminate the need for airlocks. These fans may also be automatically controlled to resist high wind conditions, regulate humidity (air changes), and provide cooling and heat.

Insulation can be provided with the addition of a second or even third interior liner, creating a dead air space with an "R" value of four to nine. Some structures allow for the liner to be collapsed into the outer shell to enable the internal heat to melt accumulated snow. For cooling or air conditioning, the exterior face of the liner can be aluminized to reflect the sun's heat.

R133113-90   Seismic Bracing

Sometimes referred to as anti-sway bracing, this support system is required in earthquake areas. The individual components must be assembled to

make a required system.

Example				Additionally, height factors must be taken into account. Add the following percentages to labor for elevated installations:	
	C-Clamp 3/8" rod	2 ea.		15' to 20' high	10%
	Rod, continuous thread 3/8"	10 L.F.		21' to 25' high	20%
	Field, weld 1"	2 ea.		26' to 30' high	30%
				31' to 35' high	40%
				36' to 40' high	50%
				41' to 50' high	60%

Special Construction   R1334 Fabricated Engineered Structures

R133419-10   Pre-Engineered Steel Buildings

These buildings are manufactured by many companies and normally erected by franchised dealers throughout the U.S. The four basic types are: Rigid Frames, Truss type, Post and Beam and the Sloped Beam type. Most popular roof slope is low pitch of 1" in 12". The minimum economical area of these buildings is about 3000 S.F. of floor area. Bay sizes are usually 20' to 24' but can go as high as 30' with heavier girts and purlins. Eave heights are usually 12' to 24' with 18' to 20' most typical.

Material prices shown in the Unit Price section are bare costs for the building shell only and do not include floors, foundations, anchor bolts, interior finishes or utilities. Costs assume at least three bays of 24' each, a 1" in 12" roof slope, and they are based on 30 psf roof load and 20 psf wind load

(wind load is a function of wind speed, building height, and terrain characteristics; this should be determined by a Registered Structural Engineer) and no unusual requirements. Costs include the structural frame, 26 ga. non-insulated colored corrugated or ribbed roofing and siding panels, fasteners, closures, trim and flashing but no allowance for insulation, doors, windows, skylights, gutters or downspouts. Very large projects would generally cost less for materials than the prices shown. For roof panel substitutions and wall panel substitutions, see appropriate Unit Price sections.

Conditions at the site, weather, shape and size of the building, and labor availability will affect the erection cost of the building.

R133423-30   Dome Structures

Steel — The four types are Lamella, Schwedler, Arch and Geodesic. For maximum economy, rise should be about 15 to 20% of diameter. Most common diameters are in the 200' to 300' range. Lamella domes weigh about 5 P.S.F. of floor area less than Schwedler domes. Schwedler dome weight in lbs. per S.F. approaches .046 times the diameter. Domes below 125' diameter weigh .07 times diameter and the cost per ton of steel is higher. See R051223-20 for estimating weight.

Wood — Small domes are of sawn lumber, larger ones are laminated. In larger sizes, triaxial and triangular cost about the same; radial domes cost more. Radial domes are economical in the 60' to 70' diameter range. Most economical range of all types is 80' to 200' diameters. Diameters can

run over 400'. All costs are quoted above the foundation. Prices include 2" decking and a tension tie ring in place.

Plywood — Stock prefab geodesic domes are available with diameters from 24' to 60'.

Fiberglass — Aluminum framed translucent sandwich panels with spans from 5' to 45' are commercially available.

Aluminum — Stressed skin aluminum panels form geodesic domes with spans ranging from 82' to 232'. An aluminum space truss, triangulated or nontriangulated, with aluminum or clear acrylic closure panels can be used for clear spans of 40' to 415'.

R142000-10   Freight Elevators

Capacities run from 2,000 lbs. to over 100,000 lbs. with 3,000 lbs. to 10,000 lbs. most common. Travel speeds are generally lower and control less intricate than on passenger elevators. Costs in the Unit Price sections are for hydraulic and geared elevators.

R142000-20   Elevator Selective Costs   See R142000-40 for cost development.

	Passenger		Freight		Hospital	
A. Base Unit	**Hydraulic**	**Electric**	**Hydraulic**	**Electric**	**Hydraulic**	**Electric**
Capacity	1,500 lb.	2,000 lb.	2,000 lb.	4,000 lb.	4,000 lb.	4,000 lb.
Speed	100 F.P.M.	200 F.P.M.	100 F.P.M.	200 F.P.M.	100 F.P.M.	200 F.P.M.
#Stops/Travel Ft.	2/12	4/40	2/20	4/40	2/20	4/40
Push Button Oper.	Yes	Yes	Yes	Yes	Yes	Yes
Telephone Box & Wire	"	"	"	"	"	"
Emergency Lighting	"	"	No	No	"	"
Cab	Plastic Lam. Walls	Plastic Lam. Walls	Painted Steel	Painted Steel	Plastic Lam. Walls	Plastic Lam. Walls
Cove Lighting	Yes	Yes	No	No	Yes	Yes
Floor	V.C.T.	V.C.T.	Wood w/Safety Treads	Wood w/Safety Treads	V.C.T.	V.C.T.
Doors, & Speedside Slide	Yes	Yes	Yes	Yes	Yes	Yes
Gates, Manual	No	No	No	No	No	No
Signals, Lighted Buttons	Car and Hall	Car and Hall	Car and Hall	Car and Hall	Car and Hall	Car and Hall
O.H. Geared Machine	N.A.	Yes	N.A.	Yes	N.A.	Yes
Variable Voltage Contr.	"	"	N.A.	"	"	"
Emergency Alarm	Yes	"	Yes	"	Yes	"
Class "A" Loading	N.A.	N.A.	"	"	N.A.	N.A.

R142000-30   Passenger Elevators

Electric elevators are used generally but hydraulic elevators can be used for lifts up to 70′ and where large capacities are required. Hydraulic speeds are limited to 200 F.P.M. but cars are self leveling at the stops. On low rises, hydraulic installation runs about 15% less than standard electric types but on higher rises this installation cost advantage is reduced. Maintenance of hydraulic elevators is about the same as electric type but underground portion is not included in the maintenance contract.

In electric elevators there are several control systems available, the choice of which will be based upon elevator use, size, speed and cost criteria. The two types of drives are geared for low speeds and gearless for 450 F.P.M. and over.

The tables on the preceding pages illustrate typical installed costs of the various types of elevators available.

R142000-40   Elevator Cost Development

To price a new car or truck from the factory, you must start with the manufacturer's basic model, then add or exchange optional equipment and features. The same is true for pricing elevators.

Requirement: One-passenger elevator, five-story hydraulic, 2,500 lb. capacity, 12′ floor to floor, speed 150 F.P.M., emergency power switching and maintenance contract.

Example:

Description	Adjustment
A. Base Elevator: Hydraulic Passenger, 1500 lb. Capacity, 100 fpm, 2 Stops, Standard Finish	1 Ea.
B. Capacity Adjustment (2,500 lb.)	1 Ea.
C. Excess Travel Adjustment: 48′ Total Travel (4 x 12′) minus 12′ Base Unit Travel =	36 V.L.F.
D. Stops Adjustment: 5 Total Stops minus 2 Stops (Base Unit) =	3 Stops
E. Speed Adjustment (150 F.P.M.)	1 Ea.
F. Options:	
1. Intercom Service	1 Ea.
2. Emergency Power Switching, Automatic	1 Ea.
3. Stainless Steel Entrance Doors	5 Ea.
4. Maintenance Contract (12 Months)	1 Ea.
5. Position Indicator for main floor level (none indicated in Base Unit)	1 Ea.

Conveying Equipment   R1431 Escalators

R143110-10   Escalators

Moving stairs can be used for buildings where 600 or more people are to be carried to the second floor or beyond. Freight cannot be carried on escalators and at least one elevator must be available for this function.

Carrying capacity is 5,000 to 8,000 people per hour. Power requirement is 2 to 3 KW per hour and incline angle is 30°.

Conveying Equipment   R1432 Moving Walks

R143210-20   Moving Ramps and Walks

These are a specialized form of conveyor 3′ to 6′ wide with capacities of 3,600 to 18,000 persons per hour. Maximum speed is 140 F.P.M. and normal incline is 0° to 15°.

Local codes will determine the maximum angle. Outdoor units would require additional weather protection.

Plumbing   R2201 Operation & Maintenance of Plumbing

R220102-20   Labor Adjustment

Labor Adjustment Factors are provided for Divisions 21, 22, and 23 to assist the mechanical estimator account for the various complexities and special conditions of any particular project. While a single percentage has been entered on each line of Division 22 01 02.20, it should be understood that these are just suggested midpoints of ranges of values commonly used by mechanical estimators. They may be increased or decreased depending on the severity of the special conditions.

The group for "existing occupied buildings", has been the subject of requests for explanation. Actually there are two stages to this group: buildings that are existing and "finished" but unoccupied, and those that also are occupied. Buildings that are "finished" may result in higher labor costs due to the

workers having to be more careful not to damage finished walls, ceilings, floors, etc. and may necessitate special protective coverings and barriers. Also corridor bends and doorways may not accommodate long pieces of pipe or larger pieces of equipment. Work above an already hung ceiling can be very time consuming. The addition of occupants may force the work to be done on premium time (nights and/or weekends), eliminate the possible use of some preferred tools such as pneumatic drivers, powder charged drivers etc. The estimator should evaluate the access to the work area and just how the work is going to be accomplished to arrive at an increase in labor costs over "normal" new construction productivity.

R220105-10   Demolition (Selective vs. Removal for Replacement)

Demolition can be divided into two basic categories.

One type of demolition involves the removal of material with no concern for its replacement. The labor-hours to estimate this work are found under "Selective Demolition" in the Fire Protection, Plumbing and HVAC Divisions. It is selective in that individual items or all the material installed as a system or trade grouping such as plumbing or heating systems are removed. This may be accomplished by the easiest way possible, such as sawing, torch cutting, or sledge hammer as well as simple unbolting.

The second type of demolition is the removal of some item for repair or replacement. This removal may involve careful draining, opening of unions,

disconnecting and tagging of electrical connections, capping of pipes/ducts to prevent entry of debris or leakage of the material contained as well as transport of the item away from its in-place location to a truck/dumpster. An approximation of the time required to accomplish this type of demolition is to use half of the time indicated as necessary to install a new unit. For example; installation of a new pump might be listed as requiring 6 labor-hours so if we had to estimate the removal of the old pump we would allow an additional 3 hours for a total of 9 hours. That is, the complete replacement of a defective pump with a new pump would be estimated to take 9 labor-hours.

Plumbing — R2211 Facility Water Distribution

R221113-50 Pipe Material Considerations

1. Malleable fittings should be used for gas service.
2. Malleable fittings are used where there are stresses/strains due to expansion and vibration.
3. Cast fittings may be broken as an aid to disassembling of heating lines frozen by long use, temperature and minerals.
4. Cast iron pipe is extensively used for underground and submerged service.
5. Type M (light wall) copper tubing is available in hard temper only and is used for nonpressure and less severe applications than K and L.
6. Type L (medium wall) copper tubing, available hard or soft for interior service.
7. Type K (heavy wall) copper tubing, available in hard or soft temper for use where conditions are severe. For underground and interior service.
8. Hard drawn tubing requires fewer hangers or supports but should not be bent. Silver brazed fittings are recommended, however soft solder is normally used.
9. Type DMV (very light wall) copper tubing designed for drainage, waste and vent plus other non-critical pressure services.

Domestic/Imported Pipe and Fittings Cost

The prices shown in this publication for steel/cast iron pipe and steel, cast iron, malleable iron fittings are based on domestic production sold at the normal trade discounts. The above listed items of foreign manufacture may be available at prices of 1/3 to 1/2 those shown. Some imported items after minor machining or finishing operations are being sold as domestic to further complicate the system.

Caution: Most pipe prices in this book also include a coupling and pipe hangers which for the larger sizes can add significantly to the per foot cost and should be taken into account when comparing "book cost" with quoted supplier's cost.

Plumbing — R2240 Plumbing Fixtures

R224000-40 Plumbing Fixture Installation Time

Item	Rough-In	Set	Total Hours	Item	Rough-In	Set	Total Hours
Bathtub	5	5	10	Shower head only	2	1	3
Bathtub and shower, cast iron	6	6	12	Shower drain	3	1	4
Fire hose reel and cabinet	4	2	6	Shower stall, slate		15	15
Floor drain to 4 inch diameter	3	1	4	Slop sink	5	3	8
Grease trap, single, cast iron	5	3	8	Test 6 fixtures			14
Kitchen gas range		4	4	Urinal, wall	6	2	8
Kitchen sink, single	4	4	8	Urinal, pedestal or floor	6	4	10
Kitchen sink, double	6	6	12	Water closet and tank	4	3	7
Laundry tubs	4	2	6	Water closet and tank, wall hung	5	3	8
Lavatory wall hung	5	3	8	Water heater, 45 gals. gas, automatic	5	2	7
Lavatory pedestal	5	3	8	Water heaters, 65 gals. gas, automatic	5	2	7
Shower and stall	6	4	10	Water heaters, electric, plumbing only	4	2	6

Fixture prices in front of book are based on the cost per fixture set in place. The rough-in cost, which must be added for each fixture, includes carrier, if required, some supply, waste and vent pipe connecting fittings and stops. The lengths of rough-in pipe are nominal runs which would connect to the larger runs and stacks. The supply runs and DWV runs and stacks must be accounted for in separate entries. In the eastern half of the United States it is common for the plumber to carry these to a point 5' outside the building.

R224000-50 Water Cooler Application

Type of Service	Requirement
Office, School or Hospital	12 persons per gallon per hour
Office, Lobby or Department Store	4 or 5 gallons per hour per fountain
Light manufacturing	7 persons per gallon per hour
Heavy manufacturing	5 persons per gallon per hour
Hot heavy manufacturing	4 persons per gallon per hour
Hotel	.08 gallons per hour per room
Theatre	1 gallon per hour per 100 seats

Heating, Ventilating & A.C. R2331 HVAC Ducts & Casings

R233100-20 Ductwork

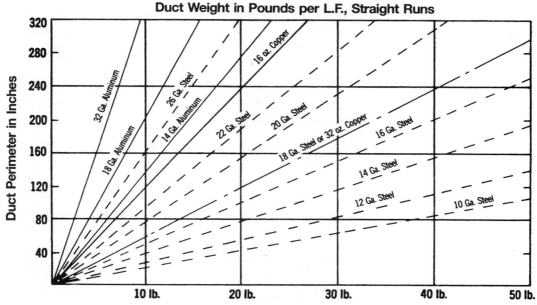

Duct Weight in Pounds per L.F., Straight Runs

Add to the above for fittings; 90° elbow is 3 L.F.; 45° elbow is 2.5 L.F.; offset is 4 L.F.; transition offset is 6 L.F.; square-to-round transition is 4 L.F.; 90° reducing elbow is 5 L.F. For bracing and waste, add 20% to aluminum and copper, 15% to steel.

Heating, Ventilating & A.C. R2334 HVAC Fans

R233400-10 Recommended Ventilation Air Changes

Table below lists range of time in minutes per change for various types of facilities.

Assembly Halls	2-10	Dance Halls	2-10	Laundries	1-3
Auditoriums	2-10	Dining Rooms	3-10	Markets	2-10
Bakeries	2-3	Dry Cleaners	1-5	Offices	2-10
Banks	3-10	Factories	2-5	Pool Rooms	2-5
Bars	2-5	Garages	2-10	Recreation Rooms	2-10
Beauty Parlors	2-5	Generator Rooms	2-5	Sales Rooms	2-10
Boiler Rooms	1-5	Gymnasiums	2-10	Theaters	2-8
Bowling Alleys	2-10	Kitchens-Hospitals	2-5	Toilets	2-5
Churches	5-10	Kitchens-Restaurant	1-3	Transformer Rooms	1-5

CFM air required for changes = Volume of room in cubic feet ÷ Minutes per change.

Heating, Ventilating & A.C. R2350 Central Heating Equipment

R235000-35 Heating (42° degrees latitude)

$$\text{Approximate S.F. radiation} = \frac{\text{S.F. sash}}{2} + \frac{\text{S.F. wall + roof — sash}}{20} + \frac{\text{C.F. building}}{200}$$

R235000-50 Factor for Determining Heat Loss for Various Types of Buildings

General: While the most accurate estimates of heating requirements would naturally be based on detailed information about the building being considered, it is possible to arrive at a reasonable approximation using the following procedure:

1. Calculate the cubic volume of the room or building.
2. Select the appropriate factor from Table 1 below. Note that the factors apply only to inside temperatures listed in the first column and to 0° F outside temperature.

3. If the building has bad north and west exposures, multiply the heat loss factor by 1.1.
4. If the outside design temperature is other than 0° F, multiply the factor from Table 1 by the factor from Table 2.
5. Multiply the cubic volume by the factor selected from Table 1. This will give the estimated BTUH heat loss which must be made up to maintain inside temperature.

Table 1 — Building Type	Conditions	Qualifications	Loss Factor*
Factories & Industrial Plants General Office Areas at 70° F	One Story	Skylight in Roof	6.2
		No Skylight in Roof	5.7
	Multiple Story	Two Story	4.6
		Three Story	4.3
		Four Story	4.1
		Five Story	3.9
		Six Story	3.6
	All Walls Exposed	Flat Roof	6.9
		Heated Space Above	5.2
	One Long Warm Common Wall	Flat Roof	6.3
		Heated Space Above	4.7
	Warm Common Walls on Both Long Sides	Flat Roof	5.8
		Heated Space Above	4.1
Warehouses at 60° F	All Walls Exposed	Skylights in Roof	5.5
		No Skylight in Roof	5.1
		Heated Space Above	4.0
	One Long Warm Common Wall	Skylight in Roof	5.0
		No Skylight in Roof	4.9
		Heated Space Above	3.4
	Warm Common Walls on Both Long Sides	Skylight in Roof	4.7
		No Skylight in Roof	4.4
		Heated Space Above	3.0

*Note: This table tends to be conservative particularly for new buildings designed for minimum energy consumption.

Table 2 — Outside Design Temperature Correction Factor (for Degrees Fahrenheit)									
Outside Design Temperature	50	40	30	20	10	0	-10	-20	-30
Correction Factor	.29	.43	.57	.72	.86	1.00	1.14	1.28	1.43

Reference Tables

R235616-60   Solar Heating (Space and Hot Water)

Collectors should face as close to due South as possible, however, variations of up to 20 degrees on either side of true South are acceptable. Local climate and collector type may influence the choice between east or west deviations. Obviously they should be located so they are not shaded from the sun's rays. Incline collectors at a slope of latitude minus 5 degrees for domestic hot water and latitude plus 15 degrees for space heating.

Flat plate collectors consist of a number of components as follows: Insulation to reduce heat loss through the bottom and sides of the collector. The enclosure which contains all the components in this assembly is usually weatherproof and prevents dust, wind and water from coming in contact with the absorber plate. The cover plate usually consists of one or more layers of a variety of glass or plastic and reduces the reradiation by creating an air space which traps the heat between the cover and the absorber plates.

The absorber plate must have a good thermal bond with the fluid passages. The absorber plate is usually metallic and treated with a surface coating which improves absorptivity. Black or dark paints or selective coatings are used for this purpose, and the design of this passage and plate combination helps determine a solar system's effectiveness.

Heat transfer fluid passage tubes are attached above and below or integral with an absorber plate for the purpose of transferring thermal energy from the absorber plate to a heat transfer medium. The heat exchanger is a device for transferring thermal energy from one fluid to another.

Piping and storage tanks should be well insulated to minimize heat losses.

Size domestic water heating storage tanks to hold 20 gallons of water per user, minimum, plus 10 gallons per dishwasher or washing machine. For domestic water heating an optimum collector size is approximately 3/4 square foot of area per gallon of water storage. For space heating of residences and small commercial applications the collector is commonly sized between 30% and 50% of the internal floor area. For space heating of large commercial applications, collector areas less than 30% of the internal floor area can still provide significant heat reductions.

A supplementary heat source is recommended for Northern states for December through February.

The solar energy transmission per square foot of collector surface varies greatly with the material used. Initial cost, heat transmittance and useful life are obviously interrelated.

R236000-20   Air Conditioning Requirements

BTUs per hour per S.F. of floor area and S.F. per ton of air conditioning.

Type of Building	BTU/Hr per S.F.	S.F. per Ton	Type of Building	BTU/Hr per S.F.	S.F. per Ton	Type of Building	BTU/Hr per S.F.	S.F. per Ton
Apartments, Individual	26	450	Dormitory, Rooms	40	300	Libraries	50	240
Corridors	22	550	Corridors	30	400	Low Rise Office, Exterior	38	320
Auditoriums & Theaters	40	300/18*	Dress Shops	43	280	Interior	33	360
Banks	50	240	Drug Stores	80	150	Medical Centers	28	425
Barber Shops	48	250	Factories	40	300	Motels	28	425
Bars & Taverns	133	90	High Rise Office—Ext. Rms.	46	263	Office (small suite)	43	280
Beauty Parlors	66	180	Interior Rooms	37	325	Post Office, Individual Office	42	285
Bowling Alleys	68	175	Hospitals, Core	43	280	Central Area	46	260
Churches	36	330/20*	Perimeter	46	260	Residences	20	600
Cocktail Lounges	68	175	Hotel, Guest Rooms	44	275	Restaurants	60	200
Computer Rooms	141	85	Corridors	30	400	Schools & Colleges	46	260
Dental Offices	52	230	Public Spaces	55	220	Shoe Stores	55	220
Dept. Stores, Basement	34	350	Industrial Plants, Offices	38	320	Shop'g. Ctrs., Supermarkets	34	350
Main Floor	40	300	General Offices	34	350	Retail Stores	48	250
Upper Floor	30	400	Plant Areas	40	300	Specialty	60	200

*Persons per ton
12,000 BTU = 1 ton of air conditioning

R260519-92 Minimum Copper and Aluminum Wire Size Allowed for Various Types of Insulation

Minimum Wire Sizes

Amperes	Copper THW THWN or XHHW	Copper THHN XHHW *	Aluminum THW XHHW	Aluminum THHN XHHW *	Amperes	Copper THW THWN or XHHW	Copper THHN XHHW *	Aluminum THW XHHW	Aluminum THHN XHHW *
15A	#14	#14	#12	#12	195	3/0	2/0	250kcmil	4/0
20	#12	#12	#10	#10	200	3/0	3/0	250kcmil	4/0
25	#10	#10	#10	#10	205	4/0	3/0	250kcmil	4/0
30	#10	#10	#8	#8	225	4/0	3/0	300kcmil	250kcmil
40	#8	#8	#8	#8	230	4/0	4/0	300kcmil	250kcmil
45	#8	#8	#6	#8	250	250kcmil	4/0	350kcmil	300kcmil
50	#8	#8	#6	#6	255	250kcmil	4/0	400kcmil	300kcmil
55	#6	#8	#4	#6	260	300kcmil	4/0	400kcmil	350kcmil
60	#6	#6	#4	#6	270	300kcmil	250kcmil	400kcmil	350kcmil
65	#6	#6	#4	#4	280	300kcmil	250kcmil	500kcmil	350kcmil
75	#4	#6	#3	#4	285	300kcmil	250kcmil	500kcmil	400kcmil
85	#4	#4	#2	#3	290	350kcmil	250kcmil	500kcmil	400kcmil
90	#3	#4	#2	#2	305	350kcmil	300kcmil	500kcmil	400kcmil
95	#3	#4	#1	#2	310	350kcmil	300kcmil	500kcmil	500kcmil
100	#3	#3	#1	#2	320	400kcmil	300kcmil	600kcmil	500kcmil
110	#2	#3	1/0	#1	335	400kcmil	350kcmil	600kcmil	500kcmil
115	#2	#2	1/0	#1	340	500kcmil	350kcmil	600kcmil	500kcmil
120	#1	#2	1/0	1/0	350	500kcmil	350kcmil	700kcmil	500kcmil
130	#1	#2	2/0	1/0	375	500kcmil	400kcmil	700kcmil	600kcmil
135	1/0	#1	2/0	1/0	380	500kcmil	400kcmil	750kcmil	600kcmil
150	1/0	#1	3/0	2/0	385	600kcmil	500kcmil	750kcmil	600kcmil
155	2/0	1/0	3/0	3/0	420	600kcmil	500kcmil		700kcmil
170	2/0	1/0	4/0	3/0	430		500kcmil		750kcmil
175	2/0	2/0	4/0	3/0	435		600kcmil		750kcmil
180	3/0	2/0	4/0	4/0	475		600kcmil		

*Dry Locations Only

Notes:
1. Size #14 to 4/0 is in AWG units (American Wire Gauge).
2. Size 250 to 750 is in kcmil units (Thousand Circular Mils).
3. Use next higher ampere value if exact value is not listed in table.
4. For loads that operate continuously increase ampere value by 25% to obtain proper wire size.
5. Refer to Table R260519-91 for the maximum circuit length for the various size wires.
6. Table R260519-92 has been written for estimating purpose only, based on ambient temperature of 30°C (86° F); for ambient temperature other than 30°C (86° F), ampacity correction factors will be applied.

R260533-22   Conductors in Conduit

Table below lists maximum number of conductors for various sized conduit using THW, TW or THWN insulations.

Copper Wire Size	1/2" TW	1/2" THW	1/2" THWN	3/4" TW	3/4" THW	3/4" THWN	1" TW	1" THW	1" THWN	1-1/4" TW	1-1/4" THW	1-1/4" THWN	1-1/2" TW	1-1/2" THW	1-1/2" THWN	2" TW	2" THW	2" THWN	2-1/2" TW	2-1/2" THW	2-1/2" THWN	3" THW	3" THWN	3-1/2" THW	3-1/2" THWN	4" THW	4" THWN
#14	9	6	13	15	10	24	25	16	39	44	29	69	60	40	94	99	65	154	142	93		143		192			
#12	7	4	10	12	8	18	19	13	29	35	24	51	47	32	70	78	53	114	111	76	164	117		157			
#10	5	4	6	9	6	11	15	11	18	26	19	32	36	26	44	60	43	73	85	61	104	95	160	127		163	
#8	2	1	3	4	3	5	7	5	9	12	10	16	17	13	22	28	22	36	40	32	51	49	79	66	106	85	136
#6		1	1		2	4		4	6		7	11		10	15		16	26		23	37	36	57	48	76	62	98
#4		1	1		1	2		3	4		5	7		7	9		12	16		17	22	27	35	36	47	47	60
#3		1	1		1	1		2	3		4	6		6	8		10	13		15	19	23	29	31	39	40	51
#2		1	1		1	1		2	3		4	5		5	7		9	11		13	16	20	25	27	33	34	43
#1					1	1		1	1		3	3		4	5		6	8		9	12	14	18	19	25	25	32
1/0					1	1		1	1		2	3		3	4		5	7		8	10	12	15	16	21	21	27
2/0					1	1		1	1		1	2		3	3		5	6		7	8	10	13	14	17	18	22
3/0					1	1		1	1		1	1		2	3		4	5		6	7	9	11	12	14	15	18
4/0						1		1	1		1	1		1	2		3	4		5	6	7	9	10	12	13	15
250 kcmil								1	1		1	1		1	1		2	3		4	4	6	7	8	10	10	12
300								1	1		1	1		1	1		2	3		3	4	5	6	7	8	9	11
350									1		1	1		1	1		1	2		3	3	4	5	6	7	8	9
400											1	1		1	1		1	1		2	3	4	5	5	6	7	8
500											1	1		1	1		1	1		1	2	3	4	4	5	6	7
600												1		1	1		1	1		1	1	3	3	4	4	5	5
700														1	1		1	1		1	1	2	3	3	4	4	5
750														1	1		1	1		1	1	2	2	3	3	4	4

R312316-40 Excavating

The selection of equipment used for structural excavation and bulk excavation or for grading is determined by the following factors.

1. Quantity of material
2. Type of material
3. Depth or height of cut
4. Length of haul
5. Condition of haul road
6. Accessibility of site
7. Moisture content and dewatering requirements
8. Availability of excavating and hauling equipment

Some additional costs must be allowed for hand trimming the sides and bottom of concrete pours and other excavation below the general excavation.

Number of B.C.Y. per truck = 1.5 C.Y. bucket x 8 passes = 12 loose C.Y.

$$= 12 \times \frac{100}{118} = 10.2 \text{ B.C.Y. per truck}$$

Truck Haul Cycle:

Load truck, 8 passes	=	4 minutes
Haul distance, 1 mile	=	9 minutes
Dump time	=	2 minutes
Return, 1 mile	=	7 minutes
Spot under machine	=	1 minute
		23 minute cycle

Add the mobilization and demobilization costs to the total excavation costs. When equipment is rented for more than three days, there is often no mobilization charge by the equipment dealer. On larger jobs outside of urban areas, scrapers can move earth economically provided a dump site or fill area and adequate haul roads are available. Excavation within sheeting bracing or cofferdam bracing is usually done with a clamshell and production

When planning excavation and fill, the following should also be considered.

1. Swell factor
2. Compaction factor
3. Moisture content
4. Density requirements

A typical example for scheduling and estimating the cost of excavation of a 15′ deep basement on a dry site when the material must be hauled off the site is outlined below.

Assumptions:

1. Swell factor, 18%
2. No mobilization or demobilization
3. Allowance included for idle time and moving on job
4. No dewatering, sheeting, or bracing
5. No truck spotter or hand trimming

Fleet Haul Production per day in B.C.Y.

$$4 \text{ trucks} \times \frac{50 \text{ min. hour}}{23 \text{ min. haul cycle}} \times 8 \text{ hrs.} \times 10.2 \text{ B.C.Y.}$$

$$= 4 \times 2.2 \times 8 \times 10.2 = 718 \text{ B.C.Y./day}$$

is low, since the clamshell may have to be guided by hand between the bracing. When excavating or filling an area enclosed with a wellpoint system, add 10% to 15% to the cost to allow for restricted access. When estimating earth excavation quantities for structures, allow work space outside the building footprint for construction of the foundation and a slope of 1:1 unless sheeting is used.

R312316-45   Excavating Equipment

The table below lists THEORETICAL hourly production in C.Y./hr. bank measure for some typical excavation equipment. Figures assume 50 minute hours, 83% job efficiency, 100% operator efficiency, 90° swing and properly sized hauling units, which must be modified for adverse digging and loading conditions. Actual production costs in the front of the book average about 50% of the theoretical values listed here.

Equipment	Soil Type	B.C.Y. Weight	% Swell	1 C.Y.	1-1/2 C.Y.	2 C.Y.	2-1/2 C.Y.	3 C.Y.	3-1/2 C.Y.	4 C.Y.
Hydraulic Excavator	Moist loam, sandy clay	3400 lb.	40%	165	195	200	275	330	385	440
"Backhoe"	Sand and gravel	3100	18	140	170	225	240	285	330	380
15' Deep Cut	Common earth	2800	30	150	180	230	250	300	350	400
	Clay, hard, dense	3000	33	120	140	190	200	240	260	320
	Moist loam, sandy clay	3400	40	170	245	295	335	385	435	475
				(6.0)	(7.0)	(7.8)	(8.4)	(8.8)	(9.1)	(9.4)
	Sand and gravel	3100	18	165	225	275	325	375	420	460
Power Shovel				(6.0)	(7.0)	(7.8)	(8.4)	(8.8)	(9.1)	(9.4)
Optimum Cut (Ft.)	Common earth	2800	30	145	200	250	295	335	375	425
				(7.8)	(9.2)	(10.2)	(11.2)	(12.1)	(13.0)	(13.8)
	Clay, hard, dense	3000	33	120	175	220	255	300	335	375
				(9.0)	(10.7)	(12.2)	(13.3)	(14.2)	(15.1)	(16.0)
	Moist loam, sandy clay	3400	40	130	180	220	250	290	325	385
				(6.6)	(7.4)	(8.0)	(8.5)	(9.0)	(9.5)	(10.0)
	Sand and gravel	3100	18	130	175	210	245	280	315	375
Drag Line				(6.6)	(7.4)	(8.0)	(8.5)	(9.0)	(9.5)	(10.0)
Optimum Cut (Ft.)	Common earth	2800	30	110	160	190	220	250	280	310
				(8.0)	(9.0)	(9.9)	(10.5)	(11.0)	(11.5)	(12.0)
	Clay, hard, dense	3000	33	90	130	160	190	225	250	280
				(9.3)	(10.7)	(11.8)	(12.3)	(12.8)	(13.3)	(12.0)

				Wheel Loaders				Track Loaders		
				3 C.Y.	4 C.Y.	6 C.Y.	8 C.Y.	2-1/4 C.Y.	3 C.Y.	4 C.Y.
	Moist loam, sandy clay	3400	40	260	340	510	690	135	180	250
	Sand and gravel	3100	18	245	320	480	650	130	170	235
Loading Tractors	Common earth	2800	30	230	300	460	620	120	155	220
	Clay, hard, dense	3000	33	200	270	415	560	110	145	200
	Rock, well-blasted	4000	50	180	245	380	520	100	130	180

R312319-90 Wellpoints

A single stage wellpoint system is usually limited to dewatering an average 15' depth below normal ground water level. Multi-stage systems are employed for greater depth with the pumping equipment installed only at the lowest header level. Ejectors with unlimited lift capacity can be economical when two or more stages of wellpoints can be replaced or when horizontal clearance is restricted, such as in deep trenches or tunneling projects, and where low water flows are expected. Wellpoints are usually spaced on 2-1/2' to 10' centers along a header pipe. Wellpoint spacing, header size, and pump size are all determined by the expected flow as dictated by soil conditions.

In almost all soils encountered in wellpoint dewatering, the wellpoints may be jetted into place. Cemented soils and stiff clays may require sand wicks about 12" in diameter around each wellpoint to increase efficiency and eliminate weeping into the excavation. These sand wicks require 1/2 to 3 C.Y. of washed filter sand and are installed by using a 12" diameter steel casing and hole puncher jetted into the ground 2' deeper than the wellpoint. Rock may require predrilled holes.

Labor required for the complete installation and removal of a single stage wellpoint system is in the range of 3/4 to 2 labor-hours per linear foot of header, depending upon jetting conditions, wellpoint spacing, etc.

Continuous pumping is necessary except in some free draining soil where temporary flooding is permissible (as in trenches which are backfilled after each day's work). Good practice requires provision of a stand-by pump during the continuous pumping operation.

Systems for continuous trenching below the water table should be installed three to four times the length of expected daily progress to ensure uninterrupted digging, and header pipe size should not be changed during the job.

For pervious free draining soils, deep wells in place of wellpoints may be economical because of lower installation and maintenance costs. Daily production ranges between two to three wells per day, for 25' to 40' depths, to one well per day for depths over 50'.

Detailed analysis and estimating for any dewatering problem is available at no cost from wellpoint manufacturers. Major firms will quote "sufficient equipment" quotes or their affiliates offer lump sum proposals to cover complete dewatering responsibility.

Description for 200' System with 8" Header		Quantities
Equipment & Material	Wellpoints 25' long, 2" diameter @ 5' O.C.	40 Each
	Header pipe, 8" diameter	200 L.F.
	Discharge pipe, 8" diameter	100 L.F.
	8" valves	3 Each
	Combination jetting & wellpoint pump (standby)	1 Each
	Wellpoint pump, 8" diameter	1 Each
	Transportation to and from site	1 Day
	Fuel for 30 days x 60 gal./day	1800 Gallons
	Lubricants for 30 days x 16 lbs./day	480 Lbs.
	Sand for points	40 C.Y.
Labor	Technician to supervise installation	1 Week
	Labor for installation and removal of system	300 Labor-hours
	4 Operators straight time 40 hrs./wk. for 4.33 wks.	693 Hrs.
	4 Operators overtime 2 hrs./wk. for 4.33 wks.	35 Hrs.

R312323-30 Compacting Backfill

Compaction of fill in embankments, around structures, in trenches, and under slabs is important to control settlement. Factors affecting compaction are:

1. Soil gradation
2. Moisture content
3. Equipment used
4. Depth of fill per lift
5. Density required

Production Rate:

$$\frac{1.75' \text{ plate width x 50 F.P.M. x 50 min./hr. x .67' lift}}{27 \text{ C.F. per C.Y.}} = 108.5 \text{ C.Y./hr.}$$

Production Rate for 4 Passes:

$$\frac{108.5 \text{ C.Y.}}{4 \text{ passes}} = 27.125 \text{ C.Y./hr. x 8 hrs.} = 217 \text{ C.Y./day}$$

Example:

Compact granular fill around a building foundation using a 21" wide x 24" vibratory plate in 8" lifts. Operator moves at 50 F.P.M. working a 50 minute hour to develop 95% Modified Proctor Density with 4 passes.

Earthwork	R3141 Shoring

R314116-40 Wood Sheet Piling

Wood sheet piling may be used for depths to 20' where there is no ground water. If moderate ground water is encountered Tongue & Groove sheeting will help to keep it out. When considerable ground water is present, steel sheeting must be used.

For estimating purposes on trench excavation, sizes are as follows:

Depth	Sheeting	Wales	Braces	B.F. per S.F.
To 8'	3 x 12's	6 x 8's, 2 line	6 x 8's, @ 10'	4.0 @ 8'
8' x 12'	3 x 12's	10 x 10's, 2 line	10 x 10's, @ 9'	5.0 average
12' to 20'	3 x 12's	12 x 12's, 3 line	12 x 12's, @ 8'	7.0 average

Sheeting to be toed in at least 2' depending upon soil conditions. A five person crew with an air compressor and sheeting driver can drive and brace 440 SF/day at 8' deep, 360 SF/day at 12' deep, and 320 SF/day at 16' deep.

For normal soils, piling can be pulled in 1/3 the time to install. Pulling difficulty increases with the time in the ground. Production can be increased by high pressure jetting.

R314116-45 Steel Sheet Piling

Limiting weights are 22 to 38#/S.F. of wall surface with 27#/S.F. average for usual types and sizes. (Weights of piles themselves are from 30.7#/L.F. to 57#/L.F. but they are 15" to 21" wide.) Lightweight sections 12" to 28" wide from 3 ga. to 12 ga. thick are also available for shallow excavations. Piles may be driven two at a time with an impact or vibratory hammer (use vibratory to pull) hung from a crane without leads. A reasonable estimate of the life of steel sheet piling is 10 uses with up to 125 uses possible if a vibratory hammer is used. Used piling costs from 50% to 80% of new piling depending on location and market conditions. Sheet piling and H piles can be rented for about 30% of the delivered mill price for the first month and 5% per month thereafter. Allow 1 labor-hour per pile for cleaning and trimming after driving. These costs increase with depth and hydrostatic head. Vibratory drivers are faster in wet granular soils and are excellent for pile extraction. Pulling difficulty increases with the time in the ground and may cost more than driving. It is often economical to abandon the sheet piling, especially if it can be used as the outer wall form. Allow about 1/3 additional length or more for toeing into ground. Add bracing, waler and strut costs. Waler costs can equal the cost per ton of sheeting.

Earthwork	R3145 Vibroflotation & Densification

R314513-90 Vibroflotation and Vibro Replacement Soil Compaction

Vibroflotation is a proprietary system of compacting sandy soils in place to increase relative density to about 70%. Typical bearing capacities attained will be 6000 psf for saturated sand and 12,000 psf for dry sand. Usual range is 4000 to 8000 psf capacity. Costs in the front of the book are for a vertical foot of compacted cylinder 6' to 10' in diameter.

Vibro replacement is a proprietary system of improving cohesive soils in place to increase bearing capacity. Most silts and clays above or below the water table can be strengthened by installation of stone columns.

The process consists of radial displacement of the soil by vibration. The created hole is then backfilled in stages with coarse granular fill which is thoroughly compacted and displaced into the surrounding soil in the form of a column.

The total project cost would depend on the number and depth of the compacted cylinders. The installing company guarantees relative soil density of the sand cylinders after compaction and the bearing capacity of the soil after the replacement process. Detailed estimating information is available from the installer at no cost.

Earthwork | R3163 Drilled Caissons

R316326-60   Caissons

The three principal types of cassions are:

(1) Belled Caissons, which except for shallow depths and poor soil conditions, are generally recommended. They provide more bearing than shaft area. Because of its conical shape, no horizontal reinforcement of the bell is required.

(2) Straight Shaft Caissons are used where relatively light loads are to be supported by caissons that rest on high value bearing strata. While the shaft is larger in diameter than for belled types this is more than offset by the saving in time and labor.

(3) Keyed Caissons are used when extremely heavy loads are to be carried. A keyed or socketed caisson transfers its load into rock by a combination of end-bearing and shear reinforcing of the shaft. The most economical shaft often consists of a steel casing, a steel wide flange core and concrete. Allowable compressive stresses of .225 f'c for concrete, 16,000 psi for the wide flange core, and 9,000 psi for the steel casing are commonly used. The usual range of shaft diameter is 18″ to 84″. The number of sizes specified for any one project should be limited due to the problems of casing and auger storage. When hand work is to be performed, shaft diameters should not be less than 32″. When inspection of borings is required a minimum shaft diameter of 30″ is recommended. Concrete caissons are intended to be poured against earth excavation so permanent forms which add to cost should not be used if the excavation is clean and the earth sufficiently impervious to prevent excessive loss of concrete.

Soil Conditions for Belling		
Good	Requires Handwork	Not Recommended
Clay	Hard Shale	Silt
Sandy Clay	Limestone	Sand
Silty Clay	Sandstone	Gravel
Clayey Silt	Weathered Mica	Igneous Rock
Hard-pan		
Soft Shale		
Decomposed Rock		

Exterior Improvements | R3292 Turf & Grasses

R329219-50   Seeding

The type of grass is determined by light, shade and moisture content of soil plus intended use. Fertilizer should be disked 4″ before seeding. For steep slopes disk five tons of mulch and lay two tons of hay or straw on surface per acre after seeding. Surface mulch can be staked, lightly disked or tar emulsion sprayed. Material for mulch can be wood chips, peat moss, partially rotted hay or straw, wood fibers and sprayed emulsions. Hemp seed blankets with fertilizer are also available. For spring seeding, watering is necessary. Late fall seeding may have to be reseeded in the spring. Hydraulic seeding, power mulching, and aerial seeding can be used on large areas.

Utilities
R3311 Water Utility Distribution Piping

R331113-80 Piping Designations

There are several systems currently in use to describe pipe and fittings. The following paragraphs will help to identify and clarify classifications of piping systems used for water distribution.

Piping may be classified by schedule. Piping schedules include 5S, 10S, 10, 20, 30, Standard, 40, 60, Extra Strong, 80, 100, 120, 140, 160 and Double Extra Strong. These schedules are dependent upon the pipe wall thickness. The wall thickness of a particular schedule may vary with pipe size.

Ductile iron pipe for water distribution is classified by Pressure Classes such as Class 150, 200, 250, 300 and 350. These classes are actually the rated water working pressure of the pipe in pounds per square inch (psi). The pipe in these pressure classes is designed to withstand the rated water working pressure plus a surge allowance of 100 psi.

The American Water Works Association (AWWA) provides standards for various types of **plastic pipe.** C-900 is the specification for polyvinyl chloride (PVC) piping used for water distribution in sizes ranging from 4″ through 12″. C-901 is the specification for polyethylene (PE) pressure pipe, tubing and fittings used for water distribution in sizes ranging from 1/2″ through 3″. C-905 is the specification for PVC piping sizes 14″ and greater.

PVC pressure-rated pipe is identified using the standard dimensional ratio (SDR) method. This method is defined by the American Society for Testing and Materials (ASTM) Standard D 2241. This pipe is available in SDR numbers 64, 41, 32.5, 26, 21, 17, and 13.5. Pipe with an SDR of 64 will have the thinnest wall while pipe with an SDR of 13.5 will have the thickest wall. When the pressure rating (PR) of a pipe is given in psi, it is based on a line supplying water at 73 degrees F.

The National Sanitation Foundation (NSF) seal of approval is applied to products that can be used with potable water. These products have been tested to ANSI/NSF Standard 14.

Valves and strainers are classified by American National Standards Institute (ANSI) Classes. These Classes are 125, 150, 200, 250, 300, 400, 600, 900, 1500 and 2500. Within each class there is an operating pressure range dependent upon temperature. Design parameters should be compared to the appropriate material dependent, pressure-temperature rating chart for accurate valve selection.

Transportation
R3472 Railway Construction

R347216-10 Single Track R.R. Siding

The costs for a single track RR siding in the Unit Price section include the components shown in the table below.

Description of Component	Qty. per L.F. of Track	Unit
Ballast, 1-1/2″ crushed stone	.667	C.Y.
6″ x 8″ x 8′-6″ Treated timber ties, 22″ O.C.	.545	Ea.
Tie plates, 2 per tie	1.091	Ea.
Track rail	2.000	L.F.
Spikes, 6″, 4 per tie	2.182	Ea.
Splice bars w/ bolts, lock washers & nuts, @ 33′ O.C.	.061	Pair
Crew B-14 @ 57 L.F./Day	.018	Day

R347216-20 Single Track, Steel Ties, Concrete Bed

The costs for a R.R. siding with steel ties and a concrete bed in the Unit Price section include the components shown in the table below.

Description of Component	Qty. per L.F. of Track	Unit
Concrete bed, 9′ wide, 10″ thick	.278	C.Y.
Ties, W6x16 x 6′-6″ long, @ 30″ O.C.	.400	Ea.
Tie plates, 4 per tie	1.600	Ea.
Track rail	2.000	L.F.
Tie plate bolts, 1″, 8 per tie	3.200	Ea.
Splice bars w/bolts, lock washers & nuts, @ 33′ O.C.	.061	Pair
Crew B-14 @ 22 L.F./Day	.045	Day

Change Orders

Change Order Considerations

A change order is a written document, usually prepared by the design professional, and signed by the owner, the architect/engineer, and the contractor. A change order states the agreement of the parties to: an addition, deletion, or revision in the work; an adjustment in the contract sum, if any; or an adjustment in the contract time, if any. Change orders, or "extras" in the construction process occur after execution of the construction contract and impact architects/engineers, contractors, and owners.

Change orders that are properly recognized and managed can ensure orderly, professional, and profitable progress for all who are involved in the project. There are many causes for change orders and change order requests. In all cases, change orders or change order requests should be addressed promptly and in a precise and prescribed manner. The following paragraphs include information regarding change order pricing and procedures.

The Causes of Change Orders

Reasons for issuing change orders include:

- Unforeseen field conditions that require a change in the work
- Correction of design discrepancies, errors, or omissions in the contract documents
- Owner-requested changes, either by design criteria, scope of work, or project objectives
- Completion date changes for reasons unrelated to the construction process
- Changes in building code interpretations, or other public authority requirements that require a change in the work
- Changes in availability of existing or new materials and products

Procedures

Properly written contract documents must include the correct change order procedures for all parties—owners, design professionals and contractors—to follow in order to avoid costly delays and litigation.

Being "in the right" is not always a sufficient or acceptable defense. The contract provisions requiring notification and documentation must be adhered to within a defined or reasonable time frame.

The appropriate method of handling change orders is by a written proposal and acceptance by all parties involved. Prior to starting work on a project, all parties should identify their authorized agents who may sign and accept change orders, as well as any limits placed on their authority.

Time may be a critical factor when the need for a change arises. For such cases, the contractor might be directed to proceed on a "time and materials" basis, rather than wait for all paperwork to be processed—a delay that could impede progress. In this situation, the contractor must still follow the prescribed change order procedures including, but not limited to, notification and documentation.

All forms used for change orders should be dated and signed by the proper authority. Lack of documentation can be very costly, especially if legal judgments are to be made, and if certain field personnel are no longer available. For time and material change orders, the contractor should keep accurate daily records of all labor and material allocated to the change.

Owners or awarding authorities who do considerable and continual building construction (such as the federal government) realize the inevitability of change orders for numerous reasons, both predictable and unpredictable. As a result, the federal government, the American Institute of Architects (AIA), the Engineers Joint Contract Documents Committee (EJCDC) and other contractor, legal, and technical organizations have developed standards and procedures to be followed by all parties to achieve contract continuance and timely completion, while being financially fair to all concerned.

In addition to the change order standards put forth by industry associations, there are also many books available on the subject.

Pricing Change Orders

When pricing change orders, regardless of their cause, the most significant factor is when the change occurs. The need for a change may be perceived in the field or requested by the architect/engineer *before* any of the actual installation has begun, or may evolve or appear *during* construction when the item of work in question is partially installed. In the latter cases, the original sequence of construction is disrupted, along with all contiguous and supporting systems. Change orders cause the greatest impact when they occur *after* the installation has been completed and must be uncovered, or even replaced. Post-completion changes may be caused by necessary design changes, product failure, or changes in the owner's requirements that are not discovered until the building or the systems begin to function.

Specified procedures of notification and record keeping must be adhered to and enforced regardless of the stage of construction: *before, during,* or *after* installation. Some bidding documents anticipate change orders by requiring that unit prices including overhead and profit percentages—for additional as well as deductible changes—be listed. Generally these unit prices do not fully take into account the ripple effect, or impact on other trades, and should be used for general guidance only.

When pricing change orders, it is important to classify the time frame in which the change occurs. There are two basic time frames for change orders: *pre-installation change orders,* which occur before the start of construction, and *post-installation change orders,* which involve reworking after the original installation. Change orders that occur between these stages may be priced according to the extent of work completed using a combination of techniques developed for pricing *pre-* and *post-installation* changes.

The following factors are the basis for a check list to use when preparing a change order estimate.

Factors To Consider When Pricing Change Orders

As an estimator begins to prepare a change order, the following questions should be reviewed to determine their impact on the final price.

General

- *Is the change order work* pre-installation *or* post-installation?

  Change order work costs vary according to how much of the installation has been completed. Once workers have the project scoped in their minds, even though they have not started, it can be difficult to refocus.

Consequently they may spend more than the normal amount of time understanding the change. Also, modifications to work in place, such as trimming or refitting, usually take more time than was initially estimated. The greater the amount of work in place, the more reluctant workers are to change it. Psychologically they may resent the change and as a result the rework takes longer than normal. Post-installation change order estimates must include demolition of existing work as required to accomplish the change. If the work is performed at a later time, additional obstacles, such as building finishes, may be present which must be protected. Regardless of whether the change

occurs pre-installation or post-installation, attempt to isolate the identifiable factors and price them separately. For example, add shipping costs that may be required pre-installation or any demolition required post-installation. Then analyze the potential impact on productivity of psychological and/or learning curve factors and adjust the output rates accordingly. One approach is to break down the typical workday into segments and quantify the impact on each segment. The following chart may be useful as a guide:

Task	Activities (Productivity) Expressed as Percentages of a Workday		
	Means Mechanical Cost Data (for New Construction)	Pre-Installation Change Orders	Post-Installation Change Orders
1. Study plans	3%	6%	6%
2. Material procurement	3%	3%	3%
3. Receiving and storing	3%	3%	3%
4. Mobilization	5%	5%	5%
5. Site movement	5%	5%	8%
6. Layout and marking	8%	10%	12%
7. Actual installation	64%	59%	54%
8. Clean-up	3%	3%	3%
9. Breaks—non-productive	6%	6%	6%
Total	100%	100%	100%

Change Order Installation Efficiency

The labor-hours expressed (for new construction) are based on average installation time, using an efficiency level of approximately 60%–65%. For change order situations, adjustments to this efficiency level should reflect the daily labor-hour allocation for that particular occurrence.

If any of the specific percentages expressed in the above chart do not apply to a particular project situation, then those percentage points should be reallocated to the appropriate task(s). Example: Using data for new construction, assume there is no new material being utilized. The percentages for Tasks 2 and 3 would therefore be reallocated to other tasks. If the time required for Tasks 2 and 3 can now be applied to installation, we can add the time allocated for *Material Procurement* and *Receiving and Storing* to the *Actual Installation* time for new construction, thereby increasing the Actual Installation percentage.

This chart shows that, due to reduced productivity, labor costs will be higher than those for new construction by 5%–15% for pre-installation change orders and by 15%–25% for post-installation change orders. Each job and change order is unique and must be examined individually. Many factors, covered elsewhere in this section, can each have a significant impact on productivity and change order costs. All such factors should be considered in every case.

- *Will the change substantially delay the original completion date?*

  A significant change in the project may cause the original completion date to be extended. The extended schedule may subject the contractor to new wage rates dictated by relevant labor contracts. Project supervision and other project overhead must also be extended beyond the original completion date. The schedule extension may also put installation into a new weather season. For example, underground piping scheduled for October installation was delayed until January. As a result, frost penetrated the trench area, thereby changing the degree of difficulty of the task. Changes and delays may have a ripple effect throughout the project. This effect must be analyzed and negotiated with the owner.

- *What is the net effect of a deduct change order?*

  In most cases, change orders resulting in a deduction or credit reflect only bare costs. The contractor may retain the overhead and profit based on the original bid.

Materials

- *Will you have to pay more or less for the new material, required by the change order, than you paid for the original purchase?*

  The same material prices or discounts will usually apply to materials purchased for change orders as new construction. In some instances, however, the contractor may forfeit the advantages of competitive pricing for change orders. Consider the following example:

  A contractor purchased over $20,000 worth of fan coil units for an installation, and obtained the maximum discount. Some time later it was determined the project required an additional matching unit. The contractor has to purchase this unit from the original supplier to ensure a match. The supplier at this time may not discount the unit because of the small quantity, and the fact that he is no longer in a competitive situation. The impact of quantity on purchase can add between 0% and 25% to material prices and/or subcontractor quotes.

- *If materials have been ordered or delivered to the job site, will they be subject to a cancellation charge or restocking fee?*

  Check with the supplier to determine if ordered materials are subject to a cancellation charge. Delivered materials not used as result of a change order may be subject to a restocking fee if returned to the supplier. Common restocking charges run between 20% and 40%. Also, delivery charges to return the goods to the supplier must be added.

Labor

- *How efficient is the existing crew at the actual installation?*

  Is the same crew that performed the initial work going to do the change order? Possibly the change consists of the installation of a unit identical to one already installed; therefore, the change should take less time. Be sure to consider this potential productivity increase and modify the productivity rates accordingly.

- *If the crew size is increased, what impact will that have on supervision requirements?*

  Under most bargaining agreements or management practices, there is a point at which a working foreman is replaced by a nonworking foreman. This replacement increases project overhead by adding a nonproductive worker. If additional workers are added to accelerate the project or to perform changes while maintaining the schedule, be sure to add additional supervision time if warranted. Calculate the hours involved and the additional cost directly if possible.

- *What are the other impacts of increased crew size?*

  The larger the crew, the greater the potential for productivity to decrease. Some of the factors that cause this productivity loss are: overcrowding (producing restrictive conditions in the working space), and possibly a shortage of any special tools and equipment required. Such factors affect not only the crew working on the elements directly involved in the change order, but other crews whose movement may also be hampered.

As the crew increases, check its basic composition for changes by the addition or deletion of apprentices or nonworking foreman, and quantify the potential effects of equipment shortages or other logistical factors.

- *As new crews, unfamiliar with the project, are brought onto the site, how long will it take them to become oriented to the project requirements?*

  The orientation time for a new crew to become 100% effective varies with the site and type of project. Orientation is easiest at a new construction site, and most difficult at existing, very restrictive renovation sites. The type of work also affects orientation time. When all elements of the work are exposed, such as concrete or masonry work, orientation is decreased. When the work is concealed or less visible, such as existing electrical systems, orientation takes longer. Usually orientation can be accomplished in one day or less. Costs for added orientation should be itemized and added to the total estimated cost.

- *How much actual production can be gained by working overtime?*

  Short term overtime can be used effectively to accomplish more work in a day. However, as overtime is scheduled to run beyond several weeks, studies have shown marked decreases in output. The following chart shows the effect of long term overtime on worker efficiency. If the anticipated change requires extended overtime to keep the job on schedule, these factors can be used as a guide to predict the impact on time and cost. Add project overhead, particularly supervision, that may also be incurred.

Days per Week	Hours per Day	Production Efficiency					Payroll Cost Factors	
		1 Week	2 Weeks	3 Weeks	4 Weeks	Average 4 Weeks	@ 1-1/2 Times	@ 2 Times
5	8	100%	100%	100%	100%	100%	100%	100%
	9	100	100	95	90	96.25	105.6	111.1
	10	100	95	90	85	91.25	110.0	120.0
	11	95	90	75	65	81.25	113.6	127.3
	12	90	85	70	60	76.25	116.7	133.3
6	8	100	100	95	90	96.25	108.3	116.7
	9	100	95	90	85	92.50	113.0	125.9
	10	95	90	85	80	87.50	116.7	133.3
	11	95	85	70	65	78.75	119.7	139.4
	12	90	80	65	60	73.75	122.2	144.4
7	8	100	95	85	75	88.75	114.3	128.6
	9	95	90	80	70	83.75	118.3	136.5
	10	90	85	75	65	78.75	121.4	142.9
	11	85	80	65	60	72.50	124.0	148.1
	12	85	75	60	55	68.75	126.2	152.4

Effects of Overtime

Caution: Under many labor agreements, Sundays and holidays are paid at a higher premium than the normal overtime rate.

The use of long-term overtime is counterproductive on almost any construction job; that is, the longer the period of overtime, the lower the actual production rate. Numerous studies have been conducted, and while they have resulted in slightly different numbers, all reach the same conclusion. The figure above tabulates the effects of overtime work on efficiency.

As illustrated, there can be a difference between the *actual* payroll cost per hour and the *effective* cost per hour for overtime work. This is due to the reduced production efficiency with the increase in weekly hours beyond 40. This difference between actual and effective cost results from overtime work over a prolonged period. Short-term overtime work does not result in as great a reduction in efficiency and, in such cases, effective cost may not vary significantly from the actual payroll cost. As the total hours per week are increased on a regular basis, more time is lost due to fatigue, lowered morale, and an increased accident rate.

As an example, assume a project where workers are working 6 days a week, 10 hours per day. From the figure above (based on productivity studies), the average effective productive hours over a 4-week period are:

$$0.875 \times 60 = 52.5$$

Depending upon the locale and day of week, overtime hours may be paid at time and a half or double time. For time and a half, the overall (average) *actual* payroll cost (including regular and overtime hours) is determined as follows:

$$\frac{40 \text{ reg. hrs.} + (20 \text{ overtime hrs.} \times 1.5)}{60 \text{ hrs.}} = 1.167$$

Based on 60 hours, the payroll cost per hour will be 116.7% of the normal rate at 40 hours per week. However, because the effective production (efficiency) for 60 hours is reduced to the equivalent of 52.5 hours, the effective cost of overtime is calculated as follows:

For time and a half:

$$\frac{40 \text{ reg. hrs.} + (20 \text{ overtime hrs.} \times 1.5)}{52.5 \text{ hrs.}} = 1.33$$

Installed cost will be 133% of the normal rate (for labor).

Thus, when figuring overtime, the actual cost per unit of work will be higher than the apparent overtime payroll dollar increase, due to the reduced productivity of the longer workweek. These efficiency calculations are true only for those cost factors determined by hours worked. Costs that are applied weekly or monthly, such as equipment rentals, will not be similarly affected.

Equipment

- *What equipment is required to complete the change order?*

Change orders may require extending the rental period of equipment already on the job site, or the addition of special equipment brought in to accomplish the change work. In either case, the additional rental charges and operator labor charges must be added.

Summary

The preceding considerations and others you deem appropriate should be analyzed and applied to a change order estimate. The impact of each should be quantified and listed on the estimate to form an audit trail.

Change orders that are properly identified, documented, and managed help to ensure the orderly, professional and profitable progress of the work. They also minimize potential claims or disputes at the end of the project.

Estimating Tips

- The cost figures in this Square Foot Cost section were derived from approximately 11,200 projects contained in the RSMeans database of completed construction projects. They include the contractor's overhead and profit, but do not generally include architectural fees or land costs. The figures have been adjusted to January of the current year. New projects are added to our files each year, and outdated projects are discarded. For this reason, certain costs may not show a uniform annual progression. In no case are all subdivisions of a project listed.

- These projects were located throughout the U.S. and reflect a tremendous variation in square foot (S.F.) and cubic foot (C.F.) costs. This is due to differences, not only in labor and material costs, but also in individual owners' requirements. For instance, a bank in a large city would have different features than one in a rural area. This is true of all the different types of buildings analyzed. Therefore, caution should be exercised when using these Square Foot costs. For example, for court houses, costs in the database are local court house costs and will not apply to the larger, more elaborate federal court houses. As a general rule, the projects in the 1/4 column do not include any site work or equipment, while the projects in the 3/4 column may include both equipment and site work. The median figures do not generally include site work.

- None of the figures "go with" any others. All individual cost items were computed and tabulated separately. Thus, the sum of the median figures for Plumbing, HVAC and Electrical will not normally total up to the total Mechanical and Electrical costs arrived at by separate analysis and tabulation of the projects.

- Each building was analyzed as to total and component costs and percentages. The figures were arranged in ascending order with the results tabulated as shown. The 1/4 column shows that 25% of the projects had lower costs and 75% had higher. The 3/4 column shows that 75% of the projects had lower costs and 25% had higher. The median column shows that 50% of the projects had lower costs and 50% had higher.

- There are two times when square foot costs are useful. The first is in the conceptual stage when no details are available. Then square foot costs make a useful starting point. The second is after the bids are in and the costs can be worked back into their appropriate units for information purposes. As soon as details become available in the project design, the square foot approach should be discontinued and the project priced as to its particular components. When more precision is required, or for estimating the replacement cost of specific buildings, the current edition of *RSMeans Square Foot Costs* should be used.

- In using the figures in this section, it is recommended that the median column be used for preliminary figures if no additional information is available. The median figures, when multiplied by the total city construction cost index figures (see City Cost Indexes) and then multiplied by the project size modifier at the end of this section, should present a fairly accurate base figure, which would then have to be adjusted in view of the estimator's experience, local economic conditions, code requirements, and the owner's particular requirements. There is no need to factor the percentage figures, as these should remain constant from city to city. All tabulations mentioning air conditioning had at least partial air conditioning.

- The editors of this book would greatly appreciate receiving cost figures on one or more of your recent projects, which would then be included in the averages for next year. All cost figures received will be kept confidential, except that they will be averaged with other similar projects to arrive at Square Foot cost figures for next year's book. See the last page of the book for details and the discount available for submitting one or more of your projects.

50 17 00 | S.F. Costs

		UNIT	UNIT COSTS			% OF TOTAL		
			1/4	MEDIAN	3/4	1/4	MEDIAN	3/4
01 0010	**APARTMENTS Low Rise (1 to 3 story)**	S.F.	69.50	87	116				**01**
0020	Total project cost	C.F.	6.25	8.25	10.20			
0100	Site work	S.F.	5.95	8.10	14.30	6.05%	10.55%	14.05%
0500	Masonry		1.37	3.18	5.50	1.54%	3.67%	6.35%
1500	Finishes		7.35	10	12.50	9.05%	10.75%	12.85%
1800	Equipment		2.27	3.45	5.10	2.73%	4.03%	5.95%
2720	Plumbing		5.40	6.95	8.85	6.65%	8.95%	10.05%
2770	Heating, ventilating, air conditioning		3.45	4.25	6.25	4.20%	5.60%	7.60%
2900	Electrical		4.02	5.35	7.20	5.20%	6.65%	8.40%
3100	Total: Mechanical & Electrical	↓	13.95	17.75	22	15.90%	18.05%	23%
9000	Per apartment unit, total cost	Apt.	65,000	99,000	146,000			
9500	Total: Mechanical & Electrical	"	12,300	19,300	25,200			
02 0010	**APARTMENTS Mid Rise (4 to 7 story)**	S.F.	92.50	111	138				**02**
0020	Total project costs	C.F.	7.20	9.95	13.60			
0100	Site work	S.F.	3.69	7.30	13.15	5.25%	6.70%	9.15%
0500	Masonry		6.15	8.45	11.55	5.10%	7.25%	10.50%
1500	Finishes		11.60	15.15	19.10	10.55%	13.45%	17.70%
1800	Equipment		2.68	4.01	5.25	2.54%	3.48%	4.31%
2500	Conveying equipment		1.99	2.53	3.07	1.94%	2.27%	2.69%
2720	Plumbing		5.40	8.65	9.15	5.70%	7.20%	8.95%
2900	Electrical		6.10	8.25	10.05	6.65%	7.20%	8.95%
3100	Total: Mechanical & Electrical	↓	19.45	24.50	29.50	18.50%	21%	23%
9000	Per apartment unit, total cost	Apt.	104,500	123,000	204,000			
9500	Total: Mechanical & Electrical	"	19,700	22,800	28,800			
03 0010	**APARTMENTS High Rise (8 to 24 story)**	S.F.	105	120	145				**03**
0020	Total project costs	C.F.	10.15	11.80	15.10			
0100	Site work	S.F.	3.79	6.15	8.60	2.58%	4.84%	6.15%
0500	Masonry		6.05	11	13.70	4.74%	9.65%	11.05%
1500	Finishes		11.60	14.50	17.15	9.75%	11.80%	13.70%
1800	Equipment		3.36	4.14	5.50	2.78%	3.49%	4.35%
2500	Conveying equipment		2.38	3.61	4.91	2.23%	2.78%	3.37%
2720	Plumbing		7.70	9.10	12.65	6.80%	7.20%	10.45%
2900	Electrical		7.20	9.10	12.25	6.45%	7.65%	8.80%
3100	Total: Mechanical & Electrical	↓	21.50	27.50	33	17.95%	22.50%	24.50%
9000	Per apartment unit, total cost	Apt.	108,500	119,500	166,000			
9500	Total: Mechanical & Electrical	"	23,500	26,800	28,400			
04 0010	**AUDITORIUMS**	S.F.	109	148	212				**04**
0020	Total project costs	C.F.	6.80	9.45	13.60			
2720	Plumbing	S.F.	6.45	9.50	11.40	5.85%	7.20%	8.70%
2900	Electrical		8.40	12.25	20	6.80%	9.05%	11.35%
3100	Total: Mechanical & Electrical	↓	52.50	68.50	94	24.50%	29%	31.50%
05 0010	**AUTOMOTIVE SALES**	S.F.	80.50	109	135				**05**
0020	Total project costs	C.F.	5.30	6.35	8.25			
2720	Plumbing	S.F.	3.67	6.35	6.95	2.89%	6.05%	6.50%
2770	Heating, ventilating, air conditioning		5.65	8.65	9.35	4.61%	10%	10.35%
2900	Electrical		6.35	9.95	13.80	7.25%	8.80%	12.15%
3100	Total: Mechanical & Electrical	↓	20.50	29	35	18.90%	20.50%	22%
06 0010	**BANKS**	S.F.	158	197	250				**06**
0020	Total project costs	C.F.	11.25	15.30	20			
0100	Site work	S.F.	18.05	27.50	40	7.85%	12.95%	16.95%
0500	Masonry		8.20	15.70	28.50	3.36%	6.95%	10.05%
1500	Finishes		14.60	21.50	26.50	5.85%	8.60%	11.55%
1800	Equipment		5.95	13.05	26.50	1%	5.55%	10.50%
2720	Plumbing		4.94	7.05	10.30	2.82%	3.90%	4.93%
2770	Heating, ventilating, air conditioning		9.40	12.55	16.70	4.86%	7.15%	8.50%
2900	Electrical		15	20	26	8.20%	10.20%	12.20%
3100	Total: Mechanical & Electrical	↓	36	48.50	58	16.25%	19.45%	23%
3500	See also division 11020 & 11030 (MF2004 11 16 00 & 11 17 00)							

50 17 00 | S.F. Costs		UNIT	UNIT COSTS			% OF TOTAL		
			1/4	MEDIAN	3/4	1/4	MEDIAN	3/4
13 0010	**CHURCHES**	S.F.	106	135	175				**13**
0020	Total project costs	C.F.	6.60	8.30	10.95			
1800	Equipment	S.F.	1.27	3.04	6.45	.95%	2.11%	4.31%
2720	Plumbing		4.13	5.80	8.50	3.51%	4.96%	6.25%
2770	Heating, ventilating, air conditioning		9.65	12.60	17.85	7.50%	10%	12%
2900	Electrical		8.95	12.30	16.60	7.30%	8.80%	10.95%
3100	Total: Mechanical & Electrical	↓	27.50	37	49	18.30%	22%	25%
3500	See also division 11040 (MF2004 11 91 00)							
15 0010	**CLUBS, COUNTRY**	S.F.	114	137	173				**15**
0020	Total project costs	C.F.	9.20	11.20	15.45			
2720	Plumbing	S.F.	6.90	10.20	23.50	5.60%	7.90%	10%
2900	Electrical		8.95	12.25	16	7%	8.95%	11%
3100	Total: Mechanical & Electrical	↓	47.50	59.50	63	19%	26.50%	29.50%
17 0010	**CLUBS, SOCIAL Fraternal**	S.F.	96	131	175				**17**
0020	Total project costs	C.F.	5.70	8.60	10.30			
2720	Plumbing	S.F.	5.70	7.10	10.75	5.60%	6.90%	8.55%
2770	Heating, ventilating, air conditioning		8.25	9.95	12.80	8.20%	9.25%	14.40%
2900	Electrical		6.85	11.25	12.90	5.95%	9.30%	10.55%
3100	Total: Mechanical & Electrical	↓	35.50	38.50	48.50	21%	23%	23.50%
18 0010	**CLUBS, Y.M.C.A.**	S.F.	119	154	194				**18**
0020	Total project costs	C.F.	5.25	8.80	13.10			
2720	Plumbing	S.F.	7.20	14.35	16.10	5.65%	7.60%	10.85%
2900	Electrical		9.35	11.40	17.05	6.25%	7.95%	9.20%
3100	Total: Mechanical & Electrical	↓	35	42	71	20.50%	21.50%	28.50%
19 0010	**COLLEGES Classrooms & Administration**	S.F.	115	156	207				**19**
0020	Total project costs	C.F.	8.25	12.25	18.90			
0500	Masonry	S.F.	8.45	16	19.55	5.65%	8.05%	10.50%
2720	Plumbing		5.85	12	22.50	5.10%	6.60%	8.95%
2900	Electrical		9.40	14.55	20	7.70%	9.75%	12%
3100	Total: Mechanical & Electrical	↓	39.50	54	72	23.50%	28%	32%
21 0010	**COLLEGES Science, Engineering, Laboratories**	S.F.	214	251	291				**21**
0020	Total project costs	C.F.	12.30	17.95	20.50			
1800	Equipment	S.F.	11.95	27	29.50	2%	6.45%	12.65%
2900	Electrical		17.70	25.50	38.50	7.10%	9.40%	12.10%
3100	Total: Mechanical & Electrical	↓	65.50	78	121	28.50%	31.50%	41%
3500	See also division 11600 (MF2004 11 53 00)							
23 0010	**COLLEGES Student Unions**	S.F.	137	185	225				**23**
0020	Total project costs	C.F.	7.65	10	12.35			
3100	Total: Mechanical & Electrical	S.F.	51.50	55.50	66	23.50%	26%	29%
25 0010	**COMMUNITY CENTERS**	S.F.	113	139	189				**25**
0020	Total project costs	C.F.	7.35	10.55	13.65			
1800	Equipment	S.F.	2.73	4.51	7.35	1.48%	3.01%	5.30%
2720	Plumbing		5.35	9.35	12.75	4.85%	7%	8.95%
2770	Heating, ventilating, air conditioning		8.55	12.45	17.85	6.80%	10.35%	12.90%
2900	Electrical		9.25	12.50	17.95	7.15%	8.95%	10.45%
3100	Total: Mechanical & Electrical	↓	31.50	40	57	19.05%	24%	31%
28 0010	**COURT HOUSES**	S.F.	163	192	265				**28**
0020	Total project costs	C.F.	12.50	14.95	18.85			
2720	Plumbing	S.F.	7.75	10.80	12.25	5.95%	7.45%	8.20%
2900	Electrical		17.30	19.25	28	8.90%	10.45%	11.85%
3100	Total: Mechanical & Electrical	↓	48.50	63.50	69.50	22.50%	27.50%	30%
30 0010	**DEPARTMENT STORES**	S.F.	60	81.50	103				**30**
0020	Total project costs	C.F.	3.23	4.18	5.70			
2720	Plumbing	S.F.	1.87	2.37	3.59	1.82%	4.21%	5.90%
2770	Heating, ventilating, air conditioning		5.45	8.45	12.70	8.20%	9.10%	14.80%

50 17 00 | S.F. Costs

			UNIT	UNIT COSTS			% OF TOTAL			
				1/4	MEDIAN	3/4	1/4	MEDIAN	3/4	
30	2900	Electrical	S.F.	6.90	9.50	11.20	9.05%	12.15%	14.95%	30
	3100	Total: Mechanical & Electrical	↓	12.15	15.55	27	13.20%	21.50%	50%	
31	0010	**DORMITORIES Low Rise (1 to 3 story)**	S.F.	114	158	198				31
	0020	Total project costs	C.F.	6.45	10.45	15.65				
	2720	Plumbing	S.F.	6.85	9.20	11.60	8.05%	9%	9.65%	
	2770	Heating, ventilating, air conditioning		7.25	8.70	11.60	4.61%	8.05%	10%	
	2900	Electrical		7.55	11.50	15.75	6.40%	8.65%	9.50%	
	3100	Total: Mechanical & Electrical	↓	39.50	42	65.50	22%	25%	27%	
	9000	Per bed, total cost	Bed	48,400	53,500	115,500				
32	0010	**DORMITORIES Mid Rise (4 to 8 story)**	S.F.	140	183	226				32
	0020	Total project costs	C.F.	15.45	16.95	20.50				
	2900	Electrical	S.F.	14.90	16.90	23	8.20%	10.20%	11.95%	
	3100	Total: Mechanical & Electrical	"	41.50	83	84.50	25.50%	34.50%	37.50%	
	9000	Per bed, total cost	Bed	20,000	45,600	263,500				
34	0010	**FACTORIES**	S.F.	53	79	122				34
	0020	Total project costs	C.F.	3.40	5.10	8.40				
	0100	Site work	S.F.	6.05	11.05	17.50	6.95%	11.45%	17.95%	
	2720	Plumbing		2.86	5.30	8.80	3.73%	6.05%	8.10%	
	2770	Heating, ventilating, air conditioning		5.55	8	10.80	5.25%	8.45%	11.35%	
	2900	Electrical		6.60	10.45	15.90	8.10%	10.50%	14.20%	
	3100	Total: Mechanical & Electrical	↓	18.85	29	38	21%	28.50%	35.50%	
36	0010	**FIRE STATIONS**	S.F.	106	145	197				36
	0020	Total project costs	C.F.	6.15	8.45	11.25				
	0500	Masonry	S.F.	15.45	28	36.50	8.35%	11.30%	15.70%	
	1140	Roofing		3.43	9.30	10.55	1.90%	4.94%	5.05%	
	1580	Painting		2.66	3.98	4.07	1.37%	1.57%	2.07%	
	1800	Equipment		1.31	2.51	4.63	.62%	1.63%	3.42%	
	2720	Plumbing		5.90	9.40	13.30	5.85%	7.35%	9.45%	
	2770	Heating, ventilating, air conditioning		5.85	9.45	14.60	5.15%	7.40%	9.40%	
	2900	Electrical		7.60	13.20	18	6.85%	8.60%	10.60%	
	3100	Total: Mechanical & Electrical	↓	38.50	49.50	56	18.40%	23%	27%	
37	0010	**FRATERNITY HOUSES & Sorority Houses**	S.F.	105	136	186				37
	0020	Total project costs	C.F.	10.50	10.95	13.15				
	2720	Plumbing	S.F.	7.95	9.10	16.65	6.80%	8%	10.85%	
	2900	Electrical		6.95	15	18.40	6.60%	9.90%	10.65%	
	3100	Total: Mechanical & Electrical	↓	8.55	26.50	32		15.10%	15.90%	
38	0010	**FUNERAL HOMES**	S.F.	111	151	275				38
	0020	Total project costs	C.F.	11.35	12.60	24				
	2900	Electrical	S.F.	4.91	9	9.85	3.58%	4.44%	5.95%	
	3100	Total: Mechanical & Electrical	↓	17.45	25.50	35.50	12.90%	12.90%	12.90%	
39	0010	**GARAGES, COMMERCIAL (Service)**	S.F.	63	97	134				39
	0020	Total project costs	C.F.	4.14	6.10	8.90				
	1800	Equipment	S.F.	3.55	7.95	12.40	2.21%	4.62%	6.80%	
	2720	Plumbing		4.36	6.70	12.25	5.45%	7.85%	10.65%	
	2730	Heating & ventilating		5.70	7.45	9.45	5.25%	6.85%	8.20%	
	2900	Electrical		5.95	9.10	13.15	7.15%	9.25%	10.85%	
	3100	Total: Mechanical & Electrical	↓	13.20	25.50	37.50	12.35%	17.40%	26%	
40	0010	**GARAGES, MUNICIPAL (Repair)**	S.F.	97.50	123	173				40
	0020	Total project costs	C.F.	5.75	7.30	12.60				
	0500	Masonry	S.F.	8.65	16.95	26.50	4.03%	9.15%	12.50%	
	2720	Plumbing		4.14	7.95	15	3.59%	6.70%	7.95%	
	2730	Heating & ventilating		7.10	10.25	19.80	6.15%	7.45%	13.50%	
	2900	Electrical		6.80	10.75	15.60	6.65%	8.15%	11.15%	
	3100	Total: Mechanical & Electrical	↓	31.50	44	63.50	21.50%	23%	28.50%	

50 17 00	S.F. Costs	UNIT	UNIT COSTS			% OF TOTAL		
			1/4	MEDIAN	3/4	1/4	MEDIAN	3/4
41 0010	**GARAGES, PARKING**	S.F.	35.50	52.50	90			**41**
0020	Total project costs	C.F.	3.38	4.59	6.70			
2720	Plumbing	S.F.	1.02	1.58	2.44	1.72%	2.70%	3.85%
2900	Electrical		1.97	2.42	3.80	4.33%	5.20%	6.30%
3100	Total: Mechanical & Electrical	↓	4.04	5.60	7	7%	8.90%	11.05%
3200								
9000	Per car, total cost	Car	15,200	19,100	24,300			
43 0010	**GYMNASIUMS**	S.F.	100	133	176			**43**
0020	Total project costs	C.F.	4.99	6.75	8.30			
1800	Equipment	S.F.	2.38	4.46	8.55	1.98%	3.30%	6.70%
2720	Plumbing		6.35	7.40	9.70	4.65%	6.40%	7.75%
2770	Heating, ventilating, air conditioning		6.80	10.35	21	5.15%	9.05%	11.10%
2900	Electrical		7.70	10.25	13.80	6.60%	8.30%	10.30%
3100	Total: Mechanical & Electrical	↓	28	37	44	19.60%	23.50%	28%
3500	See also division 11480 (MF2004 11 67 00)							
46 0010	**HOSPITALS**	S.F.	192	240	325			**46**
0020	Total project costs	C.F.	14.55	18.10	26			
1800	Equipment	S.F.	4.87	9.35	16.15	.80%	2.53%	4.80%
2720	Plumbing		16.55	23.50	30	7.60%	9.10%	10.85%
2770	Heating, ventilating, air conditioning		24.50	31	43.50	7.80%	12.95%	16.65%
2900	Electrical		21	28.50	40.50	9.90%	11.75%	14.10%
3100	Total: Mechanical & Electrical	↓	62	86	128	28%	33.50%	37%
9000	Per bed or person, total cost	Bed	222,500	306,500	353,000			
9900	See also division 11700 (MF2004 11 71 00)							
48 0010	**HOUSING For the Elderly**	S.F.	94.50	120	147			**48**
0020	Total project costs	C.F.	6.75	9.35	11.95			
0100	Site work	S.F.	6.60	10.20	14.95	5.05%	7.90%	12.10%
0500	Masonry		2.88	10.75	15.70	1.30%	6.05%	11%
1800	Equipment		2.28	3.14	5	1.88%	3.23%	4.43%
2510	Conveying systems		2.30	3.09	4.18	1.78%	2.20%	2.81%
2720	Plumbing		7	8.95	11.25	8.15%	9.55%	10.50%
2730	Heating, ventilating, air conditioning		3.60	5.10	7.60	3.30%	5.60%	7.25%
2900	Electrical		7.05	9.55	12.25	7.30%	8.50%	10.25%
3100	Total: Mechanical & Electrical	↓	26	30	38.50	18.10%	22.50%	29%
9000	Per rental unit, total cost	Unit	88,000	102,500	114,500			
9500	Total: Mechanical & Electrical	"	19,600	22,500	26,300			
50 0010	**HOUSING Public (Low Rise)**	S.F.	79.50	110	143			**50**
0020	Total project costs	C.F.	7.05	8.80	10.95			
0100	Site work	S.F.	10.10	14.55	23.50	8.35%	11.75%	16.50%
1800	Equipment		2.16	3.52	5.35	2.26%	3.03%	4.24%
2720	Plumbing		5.70	7.55	9.55	7.15%	9.05%	11.60%
2730	Heating, ventilating, air conditioning		2.88	5.60	6.10	4.26%	6.05%	6.45%
2900	Electrical		4.80	7.15	9.90	5.10%	6.55%	8.25%
3100	Total: Mechanical & Electrical	↓	23	29.50	33	14.50%	17.55%	26.50%
9000	Per apartment, total cost	Apt.	87,000	99,000	124,500			
9500	Total: Mechanical & Electrical	"	18,600	22,900	25,400			
51 0010	**ICE SKATING RINKS**	S.F.	68	159	174			**51**
0020	Total project costs	C.F.	4.99	5.10	5.90			
2720	Plumbing	S.F.	2.53	4.75	4.86	3.12%	3.23%	5.65%
2900	Electrical		7.25	11.15	11.80	6.30%	10.15%	15.05%
3100	Total: Mechanical & Electrical	↓	12	17	21	18.95%	18.95%	18.95%
52 0010	**JAILS**	S.F.	206	267	345			**52**
0020	Total project costs	C.F.	18.60	26	30.50			
1800	Equipment	S.F.	8.05	24	40.50	2.80%	6.05%	10.35%
2720	Plumbing		19.90	26.50	35	7%	8.90%	13.35%
2770	Heating, ventilating, air conditioning		18.65	25	48	7.50%	9.45%	17.75%
2900	Electrical		21.50	29.50	36.50	8.80%	11.65%	14.95%

50 17 00	S.F. Costs	UNIT	UNIT COSTS			% OF TOTAL		
			1/4	MEDIAN	3/4	1/4	MEDIAN	3/4
52	3100	Total: Mechanical & Electrical	S.F.	56	103	122	27.50%	30%	34%	**52**

			UNIT	1/4	MEDIAN	3/4	1/4	MEDIAN	3/4	
53	0010	**LIBRARIES**	S.F.	132	168	218				**53**
	0020	Total project costs	C.F.	8.95	11.20	14.30				
	0500	Masonry	S.F.	10.45	18.05	30.50	5.80%	7.60%	11.60%	
	1800	Equipment		1.78	4.80	7.25	.34%	1.50%	4.07%	
	2720	Plumbing		4.82	6.85	9.50	3.38%	4.60%	5.70%	
	2770	Heating, ventilating, air conditioning		10.65	18.05	22	7.80%	10.95%	12.80%	
	2900	Electrical		13.40	17.35	22.50	8.35%	10.30%	11.95%	
	3100	Total: Mechanical & Electrical		39.50	51.50	62	20.50%	23%	26.50%	
54	0010	**LIVING, ASSISTED**	S.F.	121	142	168				**54**
	0020	Total project costs	C.F.	10.15	11.85	13.50				
	0500	Masonry	S.F.	3.50	4.23	5.20	2.36%	3.16%	3.90%	
	1800	Equipment		2.68	3.13	4.09	2.06%	2.45%	2.87%	
	2720	Plumbing		10.10	13.55	14.05	6.05%	8.15%	10.60%	
	2770	Heating, ventilating, air conditioning		12	12.55	13.75	7.95%	9.35%	9.70%	
	2900	Electrical		11.80	13.05	15	8.95%	9.95%	10.65%	
	3100	Total: Mechanical & Electrical		33	38.50	44.50	25%	28.50%	31.50%	
55	0010	**MEDICAL CLINICS**	S.F.	123	151	192				**55**
	0020	Total project costs	C.F.	9	11.65	15.50				
	1800	Equipment	S.F.	3.32	7	10.85	1.05%	2.94%	6.35%	
	2720	Plumbing		8.15	11.50	15.40	6.15%	8.40%	10.10%	
	2770	Heating, ventilating, air conditioning		9.75	12.75	18.75	6.65%	8.85%	11.35%	
	2900	Electrical		10.55	15	19.60	8.10%	10%	12.25%	
	3100	Total: Mechanical & Electrical		33.50	46	63	22.50%	27%	33.50%	
	3500	See also division 11700 (MF2004 11 71 00)								
57	0010	**MEDICAL OFFICES**	S.F.	115	143	175				**57**
	0020	Total project costs	C.F.	8.60	11.70	15.80				
	1800	Equipment	S.F.	3.81	7.55	10.70	.68%	4.97%	7.05%	
	2720	Plumbing		6.40	9.85	13.25	5.60%	6.80%	8.50%	
	2770	Heating, ventilating, air conditioning		7.70	11.15	14.70	6.10%	8%	9.70%	
	2900	Electrical		9.25	13.45	18.80	7.60%	9.80%	11.40%	
	3100	Total: Mechanical & Electrical		25	36	52	19.35%	23%	28.50%	
59	0010	**MOTELS**	S.F.	73	106	138				**59**
	0020	Total project costs	C.F.	6.50	8.70	14.20				
	2720	Plumbing	S.F.	7.40	9.40	11.25	9.45%	10.60%	12.55%	
	2770	Heating, ventilating, air conditioning		4.50	6.70	12.05	5.60%	5.60%	10%	
	2900	Electrical		6.90	8.70	10.85	7.45%	9.05%	10.45%	
	3100	Total: Mechanical & Electrical		22	29.50	50.50	18.50%	24%	25.50%	
	5000									
	9000	Per rental unit, total cost	Unit	37,100	70,500	76,500				
	9500	Total: Mechanical & Electrical	"	7,225	10,900	12,700				
60	0010	**NURSING HOMES**	S.F.	114	147	181				**60**
	0020	Total project costs	C.F.	9	11.25	15.35				
	1800	Equipment	S.F.	3.60	4.77	7.95	2.02%	3.62%	4.99%	
	2720	Plumbing		9.80	14.85	17.90	8.75%	10.10%	12.70%	
	2770	Heating, ventilating, air conditioning		10.30	15.65	21	9.70%	11.45%	11.80%	
	2900	Electrical		11.35	14.15	19.25	9.40%	10.55%	12.45%	
	3100	Total: Mechanical & Electrical		27	38	63.50	26%	29.50%	30.50%	
	9000	Per bed or person, total cost	Bed	50,500	63,500	82,000				
61	0010	**OFFICES Low Rise (1 to 4 story)**	S.F.	96.50	125	163				**61**
	0020	Total project costs	C.F.	6.85	9.50	12.50				
	0100	Site work	S.F.	7.75	13.70	20	6.20%	9.70%	13.55%	
	0500	Masonry		3.84	7.65	13.80	2.62%	5.45%	8.50%	
	1800	Equipment		1.02	2.01	5.45	.60%	1.50%	3.50%	
	2720	Plumbing		3.44	5.30	7.80	3.66%	4.50%	6.10%	
	2770	Heating, ventilating, air conditioning		7.60	10.60	15.55	7.20%	10.30%	11.70%	
	2900	Electrical		7.90	11.25	16.05	7.45%	9.65%	11.40%	

		50 17 00 | S.F. Costs		UNIT COSTS			% OF TOTAL			
			UNIT	**1/4**	**MEDIAN**	**3/4**	**1/4**	**MEDIAN**	**3/4**	
61	3100	Total: Mechanical & Electrical	S.F.	21.50	30	45	18.20%	22%	27%	61
62	0010	**OFFICES Mid Rise (5 to 10 story)**	S.F.	102	124	168				62
	0020	Total project costs	C.F.	7.20	9.20	13.10				
	2720	Plumbing	S.F.	3.08	4.78	6.85	2.83%	3.74%	4.50%	
	2770	Heating, ventilating, air conditioning		7.75	11.05	17.65	7.65%	9.40%	11%	
	2900	Electrical		7.55	9.70	14.65	6.35%	7.80%	10%	
	3100	Total: Mechanical & Electrical		19.60	25.50	48	18.95%	21%	27.50%	
63	0010	**OFFICES High Rise (11 to 20 story)**	S.F.	125	158	194				63
	0020	Total project costs	C.F.	8.75	10.95	15.70				
	2900	Electrical	S.F.	7.60	9.30	13.80	5.80%	7.85%	10.50%	
	3100	Total: Mechanical & Electrical		24.50	33	55.50	16.90%	23.50%	34%	
64	0010	**POLICE STATIONS**	S.F.	151	197	249				64
	0020	Total project costs	C.F.	11.95	14.65	20				
	0500	Masonry	S.F.	14.10	25	31.50	6.70%	9.10%	11.35%	
	1800	Equipment		2.39	10.50	16.65	.98%	3.35%	6.70%	
	2720	Plumbing		8.40	16.75	21	5.65%	6.90%	10.75%	
	2770	Heating, ventilating, air conditioning		13.15	17.45	24.50	5.85%	10.55%	11.70%	
	2900	Electrical		16.45	23.50	31	9.80%	11.85%	14.80%	
	3100	Total: Mechanical & Electrical		54	64.50	87.50	25%	31.50%	32.50%	
65	0010	**POST OFFICES**	S.F.	119	146	187				65
	0020	Total project costs	C.F.	7.15	9	10.30				
	2720	Plumbing	S.F.	5.35	6.60	8.35	4.24%	5.30%	5.60%	
	2770	Heating, ventilating, air conditioning		8.35	10.30	11.45	6.65%	7.15%	9.35%	
	2900	Electrical		9.80	13.75	16.35	7.25%	9%	11%	
	3100	Total: Mechanical & Electrical		28.50	37	42	16.25%	18.80%	22%	
66	0010	**POWER PLANTS**	S.F.	825	1,075	2,000				66
	0020	Total project costs	C.F.	22.50	49.50	106				
	2900	Electrical	S.F.	58	123	184	9.30%	12.75%	21.50%	
	8100	Total: Mechanical & Electrical		145	470	1,050	32.50%	32.50%	52.50%	
67	0010	**RELIGIOUS EDUCATION**	S.F.	95.50	125	154				67
	0020	Total project costs	C.F.	5.30	7.60	9.50				
	2720	Plumbing	S.F.	3.99	5.65	7.95	4.40%	5.30%	7.10%	
	2770	Heating, ventilating, air conditioning		10.10	11.45	16.15	10.05%	11.45%	12.35%	
	2900	Electrical		7.60	10.70	14.15	7.60%	9.05%	10.35%	
	3100	Total: Mechanical & Electrical		31	40.50	49.50	22%	23%	27%	
69	0010	**RESEARCH Laboratories & Facilities**	S.F.	146	207	300				69
	0020	Total project costs	C.F.	10.65	21	25				
	1800	Equipment	S.F.	6.35	12.50	30.50	.94%	4.58%	9.10%	
	2720	Plumbing		14.35	18.35	29.50	6.15%	8.30%	10.80%	
	2770	Heating, ventilating, air conditioning		12.85	43	51	7.25%	16.50%	17.50%	
	2900	Electrical		17.20	27.50	45	9.55%	11.50%	15.40%	
	3100	Total: Mechanical & Electrical		55	95.50	137	29.50%	37%	41%	
70	0010	**RESTAURANTS**	S.F.	139	179	233				70
	0020	Total project costs	C.F.	11.65	15.30	20				
	1800	Equipment	S.F.	8.95	22	33.50	6.10%	13%	15.65%	
	2720	Plumbing		11	13.30	17.50	6.10%	8.15%	9%	
	2770	Heating, ventilating, air conditioning		13.95	19.30	23.50	9.20%	12%	12.40%	
	2900	Electrical		14.65	18.10	23.50	8.35%	10.55%	11.55%	
	3100	Total: Mechanical & Electrical		45.50	48.50	63	21%	25%	29.50%	
	9000	Per seat unit, total cost	Seat	5,100	6,775	8,025				
	9500	Total: Mechanical & Electrical	"	1,275	1,700	2,000				
72	0010	**RETAIL STORES**	S.F.	65	87	115				72
	0020	Total project costs	C.F.	4.39	6.25	8.70				
	2720	Plumbing	S.F.	2.35	3.92	6.70	3.26%	4.60%	6.80%	
	2770	Heating, ventilating, air conditioning		5.05	6.95	10.45	6.75%	8.75%	10.15%	
	2900	Electrical		5.85	7.95	11.50	7.25%	9.90%	11.60%	
	3100	Total: Mechanical & Electrical		15.50	19.85	27.50	17.05%	21%	23.50%	

50 17 00	S.F. Costs	UNIT	UNIT COSTS			% OF TOTAL		
			1/4	MEDIAN	3/4	1/4	MEDIAN	3/4
74	0010	**SCHOOLS Elementary**	S.F.	106	132	161				**74**
	0020	Total project costs	C.F.	6.90	8.85	11.45			
	0500	Masonry	S.F.	9.20	16.15	24	5.25%	10.65%	14.05%
	1800	Equipment		2.89	4.88	9	1.83%	3.13%	4.61%
	2720	Plumbing		6.10	8.60	11.50	5.70%	7.15%	9.35%
	2730	Heating, ventilating, air conditioning		9.10	14.50	20.50	8.15%	10.80%	14.90%
	2900	Electrical		10	13.25	16.90	8.45%	10.05%	11.80%
	3100	Total: Mechanical & Electrical	↓	36.50	44.50	55	25%	27.50%	30%
	9000	Per pupil, total cost	Ea.	12,100	18,000	40,400			
	9500	Total: Mechanical & Electrical	"	3,425	4,325	10,900			
76	0010	**SCHOOLS Junior High & Middle**	S.F.	109	134	163				**76**
	0020	Total project costs	C.F.	6.90	8.95	10.05			
	0500	Masonry	S.F.	13.60	17.80	21.50	8%	11.60%	14.35%
	1800	Equipment		3.48	5.60	8.25	1.80%	3.09%	4.86%
	2720	Plumbing		6.35	7.80	9.70	5.30%	6.80%	7.25%
	2770	Heating, ventilating, air conditioning		12.70	15.45	27	8.90%	11.55%	14.20%
	2900	Electrical		10.70	12.90	16.60	7.90%	9.35%	10.60%
	3100	Total: Mechanical & Electrical	↓	35	44.50	55.50	23.50%	26.50%	29.50%
	9000	Per pupil, total cost	Ea.	13,800	18,100	24,400			
78	0010	**SCHOOLS Senior High**	S.F.	113	139	174				**78**
	0020	Total project costs	C.F.	6.85	9.85	16.25			
	1800	Equipment	S.F.	2.98	7	9.75	1.88%	2.91%	4.56%
	2720	Plumbing		6.35	9.55	17.45	5.60%	6.90%	8.30%
	2770	Heating, ventilating, air conditioning		13	14.85	22.50	8.95%	11.60%	15%
	2900	Electrical		11.45	14.85	22.50	8.65%	10.20%	12.25%
	3100	Total: Mechanical & Electrical	↓	37.50	44	74	23.50%	26.50%	28.50%
	9000	Per pupil, total cost	Ea.	10,700	21,700	27,200			
80	0010	**SCHOOLS Vocational**	S.F.	92	133	165				**80**
	0020	Total project costs	C.F.	5.70	8.20	11.35			
	0500	Masonry	S.F.	5.40	13.30	20.50	3.20%	4.61%	10.95%
	1800	Equipment		2.87	7.15	9.85	1.24%	3.10%	4.26%
	2720	Plumbing		5.85	8.75	12.85	5.40%	6.90%	8.55%
	2770	Heating, ventilating, air conditioning		8.20	15.30	25.50	8.60%	11.90%	14.65%
	2900	Electrical		9.55	13.05	17.85	8.45%	11%	13.20%
	3100	Total: Mechanical & Electrical	↓	36	36.50	63	27.50%	29.50%	31%
	9000	Per pupil, total cost	Ea.	12,800	34,200	51,000			
83	0010	**SPORTS ARENAS**	S.F.	80	107	165				**83**
	0020	Total project costs	C.F.	4.35	7.80	10.05			
	2720	Plumbing	S.F.	4.65	7.05	14.90	4.35%	6.35%	9.40%
	2770	Heating, ventilating, air conditioning		10	11.85	16.45	8.80%	10.20%	13.55%
	2900	Electrical		8.35	11.35	14.65	8.60%	9.90%	12.25%
	3100	Total: Mechanical & Electrical	↓	21	37	49.50	21.50%	25%	27.50%
85	0010	**SUPERMARKETS**	S.F.	74	85.50	101				**85**
	0020	Total project costs	C.F.	4.13	4.99	7.55			
	2720	Plumbing	S.F.	4.14	5.20	6.05	5.40%	6%	7.45%
	2770	Heating, ventilating, air conditioning		6.10	8.10	9.85	8.60%	8.65%	9.60%
	2900	Electrical		9.30	10.60	12.60	10.40%	12.45%	13.60%
	3100	Total: Mechanical & Electrical		23.50	25.50	34	23.50%	27.50%	28.50%
86	0010	**SWIMMING POOLS**	S.F.	120	201	430				**86**
	0020	Total project costs	C.F.	9.60	12	13.10			
	2720	Plumbing	S.F.	11.10	12.70	17	4.80%	9.70%	20.50%
	2900	Electrical		9	14.60	21.50	5.75%	6.95%	7.60%
	3100	Total: Mechanical & Electrical	↓	22	56	76.50	11.15%	14.10%	23.50%
87	0010	**TELEPHONE EXCHANGES**	S.F.	159	227	296				**87**
	0020	Total project costs	C.F.	9.90	15.90	22			
	2720	Plumbing	S.F.	6.70	10.40	15.20	4.52%	5.80%	6.90%
	2770	Heating, ventilating, air conditioning		15.60	31.50	39	11.80%	16.05%	18.40%

		50 17 00 | S.F. Costs	UNIT	UNIT COSTS			% OF TOTAL			
				1/4	MEDIAN	3/4	1/4	MEDIAN	3/4	
87	2900	Electrical	S.F.	16.20	25.50	45.50	10.90%	14%	17.85%	87
	3100	Total: Mechanical & Electrical	↓	48	91	128	29.50%	33.50%	44.50%	
91	0010	**THEATERS**	S.F.	100	129	189				91
	0020	Total project costs	C.F.	4.63	6.85	10.05				
	2720	Plumbing	S.F.	3.34	3.62	14.80	2.92%	4.70%	6.80%	
	2770	Heating, ventilating, air conditioning		9.75	11.80	14.60	8%	12.25%	13.40%	
	2900	Electrical		8.75	11.80	24	8.05%	9.35%	12.25%	
	3100	Total: Mechanical & Electrical	↓	22.50	30.50	36	21.50%	25.50%	27.50%	
94	0010	**TOWN HALLS City Halls & Municipal Buildings**	S.F.	112	142	185				94
	0020	Total project costs	C.F.	10.20	11.95	17.20				
	2720	Plumbing	S.F.	4.67	8.75	15.55	4.31%	5.95%	7.95%	
	2770	Heating, ventilating, air conditioning		8.45	16.75	24.50	7.05%	9.05%	13.45%	
	2900	Electrical		10.60	14.75	19.95	8.05%	9.45%	11.65%	
	3100	Total: Mechanical & Electrical	↓	37	46.50	71.50	22%	26.50%	31%	
97	0010	**WAREHOUSES & Storage Buildings**	S.F.	41.50	60.50	88.50				97
	0020	Total project costs	C.F.	2.19	3.42	5.65				
	0100	Site work	S.F.	4.31	8.55	12.95	6.05%	12.95%	19.85%	
	0500	Masonry		2.47	5.95	12.80	3.73%	7.40%	12.30%	
	1800	Equipment		.67	1.45	8.10	.72%	1.69%	5.55%	
	2720	Plumbing		1.39	2.50	4.67	2.90%	4.80%	6.55%	
	2730	Heating, ventilating, air conditioning		1.59	4.48	6	2.41%	5%	8.90%	
	2900	Electrical		2.47	4.65	7.70	5.15%	7.20%	10.10%	
	3100	Total: Mechanical & Electrical	↓	6.90	10.55	21	13.30%	19.10%	26%	
99	0010	**WAREHOUSE & OFFICES Combination**	S.F.	51.50	68.50	93.50				99
	0020	Total project costs	C.F.	2.63	3.81	5.65				
	1800	Equipment	S.F.	.89	1.72	2.56	.52%	1.20%	2.40%	
	2720	Plumbing		1.98	3.45	5.15	3.74%	4.76%	6.30%	
	2770	Heating, ventilating, air conditioning		3.13	4.89	6.85	5%	5.65%	10.05%	
	2900	Electrical		3.44	5.10	8.10	5.75%	8%	10%	
	3100	Total: Mechanical & Electrical	↓	9.55	14.80	23.50	14.40%	19.95%	24.50%	

Square Foot Project Size Modifier

One factor that affects the S.F. cost of a particular building is the size. In general, for buildings built to the same specifications in the same locality, the larger building will have the lower S.F. cost. This is due mainly to the decreasing contribution of the exterior walls plus the economy of scale usually achievable in larger buildings. The Area Conversion Scale shown below will give a factor to convert costs for the typical size building to an adjusted cost for the particular project.

The Square Foot Base Size lists the median costs, most typical project size in our accumulated data, and the range in size of the projects.

The Size Factor for your project is determined by dividing your project area in S.F. by the typical project size for the particular Building Type. With this factor, enter the Area Conversion Scale at the appropriate Size Factor and determine the appropriate cost multiplier for your building size.

Example: Determine the cost per S.F. for a 100,000 S.F. Mid-rise apartment building.

$$\frac{\text{Proposed building area} = 100,000 \text{ S.F.}}{\text{Typical size from below} = 50,000 \text{ S.F.}} = 2.00$$

Enter Area Conversion scale at 2.0, intersect curve, read horizontally the appropriate cost multiplier of .94. Size adjusted cost becomes .94 x $111.00 = $104.00 based on national average costs.

Note: For Size Factors less than .50, the Cost Multiplier is 1.1
For Size Factors greater than 3.5, the Cost Multiplier is .90

Square Foot Base Size							
Building Type	Median Cost per S.F.	Typical Size Gross S.F.	Typical Range Gross S.F.	Building Type	Median Cost per S.F.	Typical Size Gross S.F.	Typical Range Gross S.F.
Apartments, Low Rise	$ 87.00	21,000	9,700 - 37,200	Jails	$ 267.00	40,000	5,500 - 145,000
Apartments, Mid Rise	111.00	50,000	32,000 - 100,000	Libraries	168.00	12,000	7,000 - 31,000
Apartments, High Rise	120.00	145,000	95,000 - 600,000	Living, Assisted	142.00	32,300	23,500 - 50,300
Auditoriums	148.00	25,000	7,600 - 39,000	Medical Clinics	151.00	7,200	4,200 - 15,700
Auto Sales	109.00	20,000	10,800 - 28,600	Medical Offices	143.00	6,000	4,000 - 15,000
Banks	197.00	4,200	2,500 - 7,500	Motels	106.00	40,000	15,800 - 120,000
Churches	135.00	17,000	2,000 - 42,000	Nursing Homes	147.00	23,000	15,000 - 37,000
Clubs, Country	137.00	6,500	4,500 - 15,000	Offices, Low Rise	125.00	20,000	5,000 - 80,000
Clubs, Social	131.00	10,000	6,000 - 13,500	Offices, Mid Rise	124.00	120,000	20,000 - 300,000
Clubs, YMCA	154.00	28,300	12,800 - 39,400	Offices, High Rise	158.00	260,000	120,000 - 800,000
Colleges (Class)	156.00	50,000	15,000 - 150,000	Police Stations	197.00	10,500	4,000 - 19,000
Colleges (Science Lab)	251.00	45,600	16,600 - 80,000	Post Offices	146.00	12,400	6,800 - 30,000
College (Student Union)	185.00	33,400	16,000 - 85,000	Power Plants	1075.00	7,500	1,000 - 20,000
Community Center	139.00	9,400	5,300 - 16,700	Religious Education	125.00	9,000	6,000 - 12,000
Court Houses	192.00	32,400	17,800 - 106,000	Research	207.00	19,000	6,300 - 45,000
Dept. Stores	81.50	90,000	44,000 - 122,000	Restaurants	179.00	4,400	2,800 - 6,000
Dormitories, Low Rise	158.00	25,000	10,000 - 95,000	Retail Stores	87.00	7,200	4,000 - 17,600
Dormitories, Mid Rise	183.00	85,000	20,000 - 200,000	Schools, Elementary	132.00	41,000	24,500 - 55,000
Factories	79.00	26,400	12,900 - 50,000	Schools, Jr. High	134.00	92,000	52,000 - 119,000
Fire Stations	145.00	5,800	4,000 - 8,700	Schools, Sr. High	139.00	101,000	50,500 - 175,000
Fraternity Houses	136.00	12,500	8,200 - 14,800	Schools, Vocational	133.00	37,000	20,500 - 82,000
Funeral Homes	151.00	10,000	4,000 - 20,000	Sports Arenas	107.00	15,000	5,000 - 40,000
Garages, Commercial	97.00	9,300	5,000 - 13,600	Supermarkets	85.50	44,000	12,000 - 60,000
Garages, Municipal	123.00	8,300	4,500 - 12,600	Swimming Pools	201.00	20,000	10,000 - 32,000
Garages, Parking	52.50	163,000	76,400 - 225,300	Telephone Exchange	227.00	4,500	1,200 - 10,600
Gymnasiums	133.00	19,200	11,600 - 41,000	Theaters	129.00	10,500	8,800 - 17,500
Hospitals	240.00	55,000	27,200 - 125,000	Town Halls	142.00	10,800	4,800 - 23,400
House (Elderly)	120.00	37,000	21,000 - 66,000	Warehouses	60.50	25,000	8,000 - 72,000
Housing (Public)	110.00	36,000	14,400 - 74,400	Warehouse & Office	68.50	25,000	8,000 - 72,000
Ice Rinks	159.00	29,000	27,200 - 33,600				

A	Area Square Feet; Ampere	BTUH	BTU per Hour	Cwt.	100 Pounds
ABS	Acrylonitrile Butadiene Stryrene;	B.U.R.	Built-up Roofing	C.W.X.	Cool White Deluxe
	Asbestos Bonded Steel	BX	Interlocked Armored Cable	C.Y.	Cubic Yard (27 cubic feet)
A.C.	Alternating Current;	°C	degree centegrade	C.Y./Hr.	Cubic Yard per Hour
	Air-Conditioning;	c	Conductivity, Copper Sweat	Cyl.	Cylinder
	Asbestos Cement;	C	Hundred; Centigrade	d	Penny (nail size)
	Plywood Grade A & C	C/C	Center to Center, Cedar on Cedar	D	Deep; Depth; Discharge
A.C.I.	American Concrete Institute	C-C	Center to Center	Dis., Disch.	Discharge
AD	Plywood, Grade A & D	Cab.	Cabinet	Db.	Decibel
Addit.	Additional	Cair.	Air Tool Laborer	Dbl.	Double
Adj.	Adjustable	Calc	Calculated	DC	Direct Current
af	Audio-frequency	Cap.	Capacity	DDC	Direct Digital Control
A.G.A.	American Gas Association	Carp.	Carpenter	Demob.	Demobilization
Agg.	Aggregate	C.B.	Circuit Breaker	d.f.u.	Drainage Fixture Units
A.H.	Ampere Hours	C.C.A.	Chromate Copper Arsenate	D.H.	Double Hung
A hr.	Ampere-hour	C.C.F.	Hundred Cubic Feet	DHW	Domestic Hot Water
A.H.U.	Air Handling Unit	cd	Candela	DI	Ductile Iron
A.I.A.	American Institute of Architects	cd/sf	Candela per Square Foot	Diag.	Diagonal
AIC	Ampere Interrupting Capacity	CD	Grade of Plywood Face & Back	Diam., Dia	Diameter
Allow.	Allowance	CDX	Plywood, Grade C & D, exterior	Distrib.	Distribution
alt.	Altitude		glue	Div.	Division
Alum.	Aluminum	Cefi.	Cement Finisher	Dk.	Deck
a.m.	Ante Meridiem	Cem.	Cement	D.L.	Dead Load; Diesel
Amp.	Ampere	CF	Hundred Feet	DLH	Deep Long Span Bar Joist
Anod.	Anodized	C.F.	Cubic Feet	Do.	Ditto
Approx.	Approximate	CFM	Cubic Feet per Minute	Dp.	Depth
Apt.	Apartment	c.g.	Center of Gravity	D.P.S.T.	Double Pole, Single Throw
Asb.	Asbestos	CHW	Chilled Water;	Dr.	Drive
A.S.B.C.	American Standard Building Code		Commercial Hot Water	Drink.	Drinking
Asbe.	Asbestos Worker	C.I.	Cast Iron	D.S.	Double Strength
ASCE.	American Society of Civil Engineers	C.I.P.	Cast in Place	D.S.A.	Double Strength A Grade
A.S.H.R.A.E.	American Society of Heating,	Circ.	Circuit	D.S.B.	Double Strength B Grade
	Refrig. & AC Engineers	C.L.	Carload Lot	Dty.	Duty
A.S.M.E.	American Society of Mechanical	Clab.	Common Laborer	DWV	Drain Waste Vent
	Engineers	Clam	Common maintenance laborer	DX	Deluxe White, Direct Expansion
A.S.T.M.	American Society for Testing and	C.L.F.	Hundred Linear Feet	dyn	Dyne
	Materials	CLF	Current Limiting Fuse	e	Eccentricity
Attchmt.	Attachment	CLP	Cross Linked Polyethylene	E	Equipment Only; East
Avg., Ave.	Average	cm	Centimeter	Ea.	Each
A.W.G.	American Wire Gauge	CMP	Corr. Metal Pipe	E.B.	Encased Burial
AWWA	American Water Works Assoc.	C.M.U.	Concrete Masonry Unit	Econ.	Economy
Bbl.	Barrel	CN	Change Notice	E.C.Y	Embankment Cubic Yards
B&B	Grade B and Better;	Col.	Column	EDP	Electronic Data Processing
	Balled & Burlapped	CO₂	Carbon Dioxide	EIFS	Exterior Insulation Finish System
B.&S.	Bell and Spigot	Comb.	Combination	E.D.R.	Equiv. Direct Radiation
B.&W.	Black and White	Compr.	Compressor	Eq.	Equation
b.c.c.	Body-centered Cubic	Conc.	Concrete	EL	elevation
B.C.Y.	Bank Cubic Yards	Cont.	Continuous; Continued, Container	Elec.	Electrician; Electrical
BE	Bevel End	Corr.	Corrugated	Elev.	Elevator; Elevating
B.F.	Board Feet	Cos	Cosine	EMT	Electrical Metallic Conduit;
Bg. cem.	Bag of Cement	Cot	Cotangent		Thin Wall Conduit
BHP	Boiler Horsepower;	Cov.	Cover	Eng.	Engine, Engineered
	Brake Horsepower	C/P	Cedar on Paneling	EPDM	Ethylene Propylene Diene
B.I.	Black Iron	CPA	Control Point Adjustment		Monomer
Bit., Bitum.	Bituminous	Cplg.	Coupling	EPS	Expanded Polystyrene
Bit., Conc.	Bituminous Concrete	C.P.M.	Critical Path Method	Eqhv.	Equip. Oper., Heavy
Bk.	Backed	CPVC	Chlorinated Polyvinyl Chloride	Eqlt.	Equip. Oper., Light
Bkrs.	Breakers	C.Pr.	Hundred Pair	Eqmd.	Equip. Oper., Medium
Bldg.	Building	CRC	Cold Rolled Channel	Eqmm.	Equip. Oper., Master Mechanic
Blk.	Block	Creos.	Creosote	Eqol.	Equip. Oper., Oilers
Bm.	Beam	Crpt.	Carpet & Linoleum Layer	Equip.	Equipment
Boil.	Boilermaker	CRT	Cathode-ray Tube	ERW	Electric Resistance Welded
B.P.M.	Blows per Minute	CS	Carbon Steel, Constant Shear Bar	E.S.	Energy Saver
BR	Bedroom		Joist	Est.	Estimated
Brg.	Bearing	Csc	Cosecant	esu	Electrostatic Units
Brhe.	Bricklayer Helper	C.S.F.	Hundred Square Feet	E.W.	Each Way
Bric.	Bricklayer	CSI	Construction Specifications	EWT	Entering Water Temperature
Brk.	Brick		Institute	Excav.	Excavation
Brng.	Bearing	C.T.	Current Transformer	Exp.	Expansion, Exposure
Brs.	Brass	CTS	Copper Tube Size	Ext.	Exterior
Brz.	Bronze	Cu	Copper, Cubic	Extru.	Extrusion
Bsn.	Basin	Cu. Ft.	Cubic Foot	f.	Fiber stress
Btr.	Better	cw	Continuous Wave	F	Fahrenheit; Female; Fill
Btu	British Thermal Unit	C.W.	Cool White; Cold Water	Fab.	Fabricated

FBGS	Fiberglass	H.P.	Horsepower; High Pressure	LE	Lead Equivalent
F.C.	Footcandles	H.P.F.	High Power Factor	LED	Light Emitting Diode
f.c.c.	Face-centered Cubic	Hr.	Hour	L.F.	Linear Foot
f'c.	Compressive Stress in Concrete; Extreme Compressive Stress	Hrs./Day	Hours per Day	L.F. Nose	Linear Foot of Stair Nosing
		HSC	High Short Circuit	L.F. Rsr	Linear Foot of Stair Riser
F.E.	Front End	Ht.	Height	Lg.	Long; Length; Large
FEP	Fluorinated Ethylene Propylene (Teflon)	Htg.	Heating	L & H	Light and Heat
		Htrs.	Heaters	LH	Long Span Bar Joist
F.G.	Flat Grain	HVAC	Heating, Ventilation & Air-Conditioning	L.H.	Labor Hours
F.H.A.	Federal Housing Administration			L.L.	Live Load
Fig.	Figure	Hvy.	Heavy	L.L.D.	Lamp Lumen Depreciation
Fin.	Finished	HW	Hot Water	lm	Lumen
Fixt.	Fixture	Hyd.;Hydr.	Hydraulic	lm/sf	Lumen per Square Foot
Fl. Oz.	Fluid Ounces	Hz.	Hertz (cycles)	lm/W	Lumen per Watt
Flr.	Floor	I.	Moment of Inertia	L.O.A.	Length Over All
F.M.	Frequency Modulation; Factory Mutual	IBC	International Building Code	log	Logarithm
		I.C.	Interrupting Capacity	L-O-L	Lateralolet
Fmg.	Framing	ID	Inside Diameter	long.	longitude
Fdn.	Foundation	I.D.	Inside Dimension; Identification	L.P.	Liquefied Petroleum; Low Pressure
Fori.	Foreman, Inside	I.F.	Inside Frosted	L.P.F.	Low Power Factor
Foro.	Foreman, Outside	I.M.C.	Intermediate Metal Conduit	LR	Long Radius
Fount.	Fountain	In.	Inch	L.S.	Lump Sum
fpm	Feet per Minute	Incan.	Incandescent	Lt.	Light
FPT	Female Pipe Thread	Incl.	Included; Including	Lt. Ga.	Light Gauge
Fr.	Frame	Int.	Interior	L.T.L.	Less than Truckload Lot
F.R.	Fire Rating	Inst.	Installation	Lt. Wt.	Lightweight
FRK	Foil Reinforced Kraft	Insul.	Insulation/Insulated	L.V.	Low Voltage
FRP	Fiberglass Reinforced Plastic	I.P.	Iron Pipe	M	Thousand; Material; Male; Light Wall Copper Tubing
FS	Forged Steel	I.P.S.	Iron Pipe Size		
FSC	Cast Body; Cast Switch Box	I.P.T.	Iron Pipe Threaded	M^2CA	Meters Squared Contact Area
Ft.	Foot; Feet	I.W.	Indirect Waste	m/hr.; M.H.	Man-hour
Ftng.	Fitting	J	Joule	mA	Milliampere
Ftg.	Footing	J.I.C.	Joint Industrial Council	Mach.	Machine
Ft lb.	Foot Pound	K	Thousand; Thousand Pounds; Heavy Wall Copper Tubing, Kelvin	Mag. Str.	Magnetic Starter
Furn.	Furniture			Maint.	Maintenance
FVNR	Full Voltage Non-Reversing	K.A.H.	Thousand Amp. Hours	Marb.	Marble Setter
FXM	Female by Male	KCMIL	Thousand Circular Mils	Mat; Mat'l.	Material
Fy.	Minimum Yield Stress of Steel	KD	Knock Down	Max.	Maximum
g	Gram	K.D.A.T.	Kiln Dried After Treatment	MBF	Thousand Board Feet
G	Gauss	kg	Kilogram	MBH	Thousand BTU's per hr.
Ga.	Gauge	kG	Kilogauss	MC	Metal Clad Cable
Gal., gal.	Gallon	kgf	Kilogram Force	M.C.F.	Thousand Cubic Feet
gpm, GPM	Gallon per Minute	kHz	Kilohertz	M.C.F.M.	Thousand Cubic Feet per Minute
Galv.	Galvanized	Kip.	1000 Pounds	M.C.M.	Thousand Circular Mils
Gen.	General	KJ	Kiljoule	M.C.P.	Motor Circuit Protector
G.F.I.	Ground Fault Interrupter	K.L.	Effective Length Factor	MD	Medium Duty
Glaz.	Glazier	K.L.F.	Kips per Linear Foot	M.D.O.	Medium Density Overlaid
GPD	Gallons per Day	Km	Kilometer	Med.	Medium
GPH	Gallons per Hour	K.S.F.	Kips per Square Foot	MF	Thousand Feet
GPM	Gallons per Minute	K.S.I.	Kips per Square Inch	M.F.B.M.	Thousand Feet Board Measure
GR	Grade	kV	Kilovolt	Mfg.	Manufacturing
Gran.	Granular	kVA	Kilovolt Ampere	Mfrs.	Manufacturers
Grnd.	Ground	K.V.A.R.	Kilovar (Reactance)	mg	Milligram
H	High Henry	KW	Kilowatt	MGD	Million Gallons per Day
H.C.	High Capacity	KWh	Kilowatt-hour	MGPH	Thousand Gallons per Hour
H.D.	Heavy Duty; High Density	L	Labor Only; Length; Long; Medium Wall Copper Tubing	MH, M.H.	Manhole; Metal Halide; Man-Hour
H.D.O.	High Density Overlaid			MHz	Megahertz
H.D.P.E.	high density polyethelene	Lab.	Labor	Mi.	Mile
Hdr.	Header	lat	Latitude	MI	Malleable Iron; Mineral Insulated
Hdwe.	Hardware	Lath.	Lather	mm	Millimeter
Help.	Helper Average	Lav.	Lavatory	Mill.	Millwright
HEPA	High Efficiency Particulate Air Filter	lb.; #	Pound	Min., min.	Minimum, minute
		L.B.	Load Bearing; L Conduit Body	Misc.	Miscellaneous
Hg	Mercury	L. & E.	Labor & Equipment	ml	Milliliter, Mainline
HIC	High Interrupting Capacity	lb./hr.	Pounds per Hour	M.L.F.	Thousand Linear Feet
HM	Hollow Metal	lb./L.F.	Pounds per Linear Foot	Mo.	Month
HMWPE	high molecular weight polyethylene	lbf/sq.in.	Pound-force per Square Inch	Mobil.	Mobilization
		L.C.L.	Less than Carload Lot	Mog.	Mogul Base
H.O.	High Output	L.C.Y.	Loose Cubic Yard	MPH	Miles per Hour
Horiz.	Horizontal	Ld.	Load	MPT	Male Pipe Thread

MRT	Mile Round Trip	Pl.	Plate	S.F.C.A.	Square Foot Contact Area
ms	Millisecond	Plah.	Plasterer Helper	S.F. Flr.	Square Foot of Floor
M.S.F.	Thousand Square Feet	Plas.	Plasterer	S.F.G.	Square Foot of Ground
Mstz.	Mosaic & Terrazzo Worker	Pluh.	Plumbers Helper	S.F. Hor.	Square Foot Horizontal
M.S.Y.	Thousand Square Yards	Plum.	Plumber	S.F.R.	Square Feet of Radiation
Mtd., mtd.	Mounted	Ply.	Plywood	S.F. Shlf.	Square Foot of Shelf
Mthe.	Mosaic & Terrazzo Helper	p.m.	Post Meridiem	S4S	Surface 4 Sides
Mtng.	Mounting	Pntd.	Painted	Shee.	Sheet Metal Worker
Mult.	Multi; Multiply	Pord.	Painter, Ordinary	Sin.	Sine
M.V.A.	Million Volt Amperes	pp	Pages	Skwk.	Skilled Worker
M.V.A.R.	Million Volt Amperes Reactance	PP, PPL	Polypropylene	SL	Saran Lined
MV	Megavolt	P.P.M.	Parts per Million	S.L.	Slimline
MW	Megawatt	Pr.	Pair	Sldr.	Solder
MXM	Male by Male	P.E.S.B.	Pre-engineered Steel Building	SLH	Super Long Span Bar Joist
MYD	Thousand Yards	Prefab.	Prefabricated	S.N.	Solid Neutral
N	Natural; North	Prefin.	Prefinished	S-O-L	Socketolet
nA	Nanoampere	Prop.	Propelled	sp	Standpipe
NA	Not Available; Not Applicable	PSF, psf	Pounds per Square Foot	S.P.	Static Pressure; Single Pole; Self-Propelled
N.B.C.	National Building Code	PSI, psi	Pounds per Square Inch		
NC	Normally Closed	PSIG	Pounds per Square Inch Gauge	Spri.	Sprinkler Installer
N.E.M.A.	National Electrical Manufacturers Assoc.	PSP	Plastic Sewer Pipe	spwg	Static Pressure Water Gauge
		Pspr.	Painter, Spray	S.P.D.T.	Single Pole, Double Throw
NEHB	Bolted Circuit Breaker to 600V.	Psst.	Painter, Structural Steel	SPF	Spruce Pine Fir
N.L.B.	Non-Load-Bearing	P.T.	Potential Transformer	S.P.S.T.	Single Pole, Single Throw
NM	Non-Metallic Cable	P. & T.	Pressure & Temperature	SPT	Standard Pipe Thread
nm	Nanometer	Ptd.	Painted	Sq.	Square; 100 Square Feet
No.	Number	Ptns.	Partitions	Sq. Hd.	Square Head
NO	Normally Open	Pu	Ultimate Load	Sq. In.	Square Inch
N.O.C.	Not Otherwise Classified	PVC	Polyvinyl Chloride	S.S.	Single Strength; Stainless Steel
Nose.	Nosing	Pvmt.	Pavement	S.S.B.	Single Strength B Grade
N.P.T.	National Pipe Thread	Pwr.	Power	sst, ss	Stainless Steel
NQOD	Combination Plug-on/Bolt on Circuit Breaker to 240V.	Q	Quantity Heat Flow	Sswk.	Structural Steel Worker
		Qt.	Quart	Sswl.	Structural Steel Welder
N.R.C.	Noise Reduction Coefficient/ Nuclear Regulator Commission	Quan., Qty.	Quantity	St.;Stl.	Steel
		Q.C.	Quick Coupling	S.T.C.	Sound Transmission Coefficient
N.R.S.	Non Rising Stem	r	Radius of Gyration	Std.	Standard
ns	Nanosecond	R	Resistance	Stg.	Staging
nW	Nanowatt	R.C.P.	Reinforced Concrete Pipe	STK	Select Tight Knot
OB	Opposing Blade	Rect.	Rectangle	STP	Standard Temperature & Pressure
OC	On Center	Reg.	Regular	Stpi.	Steamfitter, Pipefitter
OD	Outside Diameter	Reinf.	Reinforced	Str.	Strength; Starter; Straight
O.D.	Outside Dimension	Req'd.	Required	Strd.	Stranded
ODS	Overhead Distribution System	Res.	Resistant	Struct.	Structural
O.G.	Ogee	Resi.	Residential	Sty.	Story
O.H.	Overhead	Rgh.	Rough	Subj.	Subject
O&P	Overhead and Profit	RGS	Rigid Galvanized Steel	Subs.	Subcontractors
Oper.	Operator	R.H.W.	Rubber, Heat & Water Resistant; Residential Hot Water	Surf.	Surface
Opng.	Opening			Sw.	Switch
Orna.	Ornamental	rms	Root Mean Square	Swbd.	Switchboard
OSB	Oriented Strand Board	Rnd.	Round	S.Y.	Square Yard
O.S.&Y.	Outside Screw and Yoke	Rodm.	Rodman	Syn.	Synthetic
Ovhd.	Overhead	Rofc.	Roofer, Composition	S.Y.P.	Southern Yellow Pine
OWG	Oil, Water or Gas	Rofp.	Roofer, Precast	Sys.	System
Oz.	Ounce	Rohe.	Roofer Helpers (Composition)	t.	Thickness
P.	Pole; Applied Load; Projection	Rots.	Roofer, Tile & Slate	T	Temperature; Ton
p.	Page	R.O.W.	Right of Way	Tan	Tangent
Pape.	Paperhanger	RPM	Revolutions per Minute	T.C.	Terra Cotta
P.A.P.R.	Powered Air Purifying Respirator	R.S.	Rapid Start	T & C	Threaded and Coupled
PAR	Parabolic Reflector	Rsr	Riser	T.D.	Temperature Difference
Pc., Pcs.	Piece, Pieces	RT	Round Trip	Tdd	Telecommunications Device for the Deaf
P.C.	Portland Cement; Power Connector	S.	Suction; Single Entrance; South		
P.C.F.	Pounds per Cubic Foot	SC	Screw Cover	T.E.M.	Transmission Electron Microscopy
P.C.M.	Phase Contrast Microscopy	SCFM	Standard Cubic Feet per Minute	TFE	Tetrafluoroethylene (Teflon)
P.E.	Professional Engineer; Porcelain Enamel; Polyethylene; Plain End	Scaf.	Scaffold	T. & G.	Tongue & Groove; Tar & Gravel
		Sch., Sched.	Schedule		
		S.C.R.	Modular Brick	Th.,Thk.	Thick
Perf.	Perforated	S.D.	Sound Deadening	Thn.	Thin
PEX	Cross linked polyethylene	S.D.R.	Standard Dimension Ratio	Thrded	Threaded
Ph.	Phase	S.E.	Surfaced Edge	Tilf.	Tile Layer, Floor
P.I.	Pressure Injected	Sel.	Select	Tilh.	Tile Layer, Helper
Pile.	Pile Driver	S.E.R., S.E.U.	Service Entrance Cable	THHN	Nylon Jacketed Wire
Pkg.	Package	S.F.	Square Foot	THW.	Insulated Strand Wire

Abbreviations

THWN	Nylon Jacketed Wire	USP	United States Primed	Wrck.	Wrecker
T.L.	Truckload	UTP	Unshielded Twisted Pair	W.S.P.	Water, Steam, Petroleum
T.M.	Track Mounted	V	Volt	WT., Wt.	Weight
Tot.	Total	V.A.	Volt Amperes	WWF	Welded Wire Fabric
T-O-L	Threadolet	V.C.T.	Vinyl Composition Tile	XFER	Transfer
T.S.	Trigger Start	VAV	Variable Air Volume	XFMR	Transformer
Tr.	Trade	VC	Veneer Core	XHD	Extra Heavy Duty
Transf.	Transformer	Vent.	Ventilation	XHHW,	Cross-Linked Polyethylene Wire
Trhv.	Truck Driver, Heavy	Vert.	Vertical	XLPE	Insulation
Trlr	Trailer	V.F.	Vinyl Faced	XLP	Cross-linked Polyethylene
Trlt.	Truck Driver, Light	V.G.	Vertical Grain	Y	Wye
TTY	Teletypewriter	V.H.F.	Very High Frequency	yd	Yard
TV	Television	VHO	Very High Output	yr	Year
T.W.	Thermoplastic Water Resistant	Vib.	Vibrating	Δ	Delta
	Wire	V.L.F.	Vertical Linear Foot	%	Percent
UCI	Uniform Construction Index	Vol.	Volume	~	Approximately
UF	Underground Feeder	VRP	Vinyl Reinforced Polyester	∅	Phase; diameter
UGND	Underground Feeder	W	Wire; Watt; Wide; West	@	At
U.H.F.	Ultra High Frequency	w/	With	#	Pound; Number
U.I.	United Inch	W.C.	Water Column; Water Closet	<	Less Than
U.L.	Underwriters Laboratory	W.F.	Wide Flange	>	Greater Than
Uld.	unloading	W.G.	Water Gauge	Z	zone
Unfin.	Unfinished	Wldg.	Welding		
URD	Underground Residential	W. Mile	Wire Mile		
	Distribution	W-O-L	Weldolet		
US	United States	W.R.	Water Resistant		

Index

Index

855

Index

Index

Index

Index

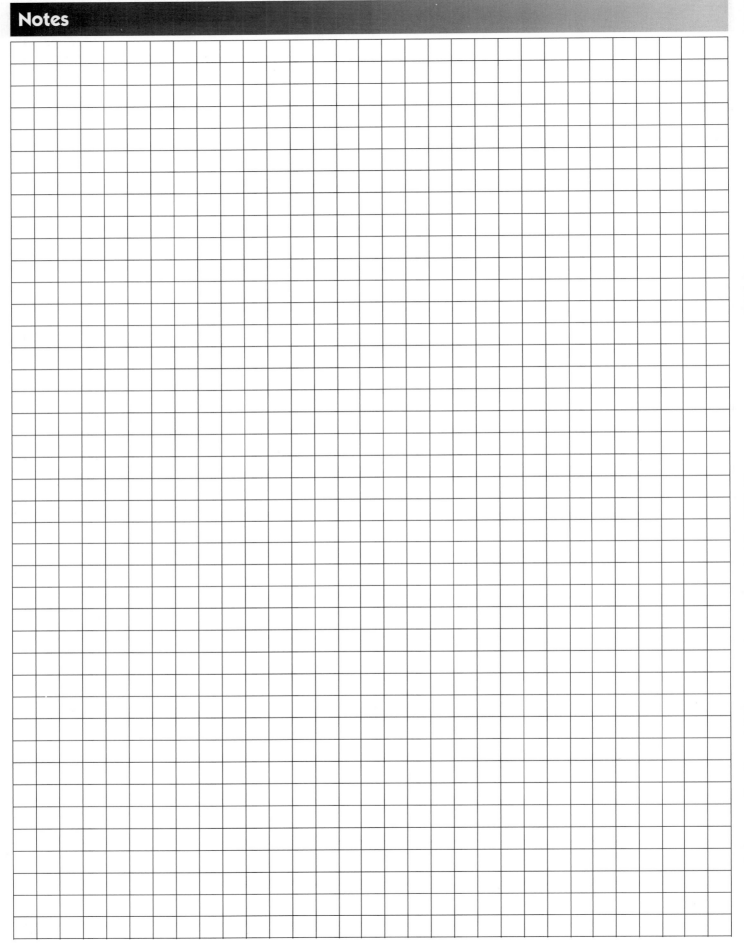

Division Notes

	CREW	DAILY OUTPUT	LABOR-HOURS	UNIT	2012 BARE COSTS				TOTAL INCL O&P
					MAT.	LABOR	EQUIP.	TOTAL	

Division Notes

		CREW	DAILY OUTPUT	LABOR-HOURS	UNIT	2012 BARE COSTS				TOTAL INCL O&P
						MAT.	LABOR	EQUIP.	TOTAL	

Division Notes

	CREW	DAILY OUTPUT	LABOR-HOURS	UNIT	2012 BARE COSTS				TOTAL INCL O&P
					MAT.	LABOR	EQUIP.	TOTAL	

Division Notes

	CREW	DAILY OUTPUT	LABOR-HOURS	UNIT	2012 BARE COSTS				TOTAL INCL O&P
					MAT.	LABOR	EQUIP.	TOTAL	

Division Notes

	CREW	DAILY OUTPUT	LABOR-HOURS	UNIT	MAT.	LABOR	EQUIP.	TOTAL	TOTAL INCL O&P

Table of Contents
Annual Cost Guides
Seminars
Reference Books
New Titles
Order Form

Reed Construction Data/RSMeans— a tradition of excellence in construction cost information and services since 1942

For more information visit the RSMeans website at **www.rsmeans.com**

Unit prices according to the latest MasterFormat!

Book Selection Guide The following table provides definitive information on the content of each cost data publication. The number of lines of data provided in each unit price or assemblies division, as well as the number of reference tables and crews, is listed for each book. The presence of other elements such as equipment rental costs, historical cost indexes, city cost indexes, square foot models, or cross-referenced indexes is also indicated. You can use the table to help select the RSMeans book that has the quantity and type of information you most need in your work.

Unit Cost Divisions	Building Construction	Mechanical	Electrical	Commercial Renovation	Square Foot	Site Work Landsc.	Green Building	Interior	Concrete Masonry	Open Shop	Heavy Construction	Light Commercial	Facilities Construction	Plumbing	Residential
1	571	392	407	495		515	124	319	461	570	505	231	1001	408	174
2	745	264	83	684		965	180	363	219	742	704	439	1174	288	240
3	1539	222	212	906		1342	874	313	1847	1539	1481	386	1567	189	366
4	902	24	0	698		701	183	573	1137	878	593	490	1149	0	405
5	1912	263	168	1022		871	1799	1021	834	1903	1118	893	1905	339	724
6	1986	67	69	1573		401	534	1294	300	1965	380	1667	1602	22	2179
7	1456	208	125	1484		555	778	543	505	1453	26	1161	1516	217	915
8	2081	61	10	2277		296	1155	1765	658	2036	0	1661	2495	0	1412
9	1875	72	26	1670		306	444	1966	386	1817	15	1589	2111	54	1400
10	988	17	10	612		215	27	808	159	988	29	494	1074	231	221
11	1029	208	166	493		126	54	885	28	1014	0	235	1049	169	110
12	562	0	2	318		248	150	1707	12	551	19	364	1757	23	317
13	732	117	111	247		347	128	249	69	731	261	80	760	72	79
14	280	36	0	229		28	0	259	0	279	0	12	299	21	6
21	78	0	16	30		0	0	249	0	78	0	60	430	436	224
22	1144	7289	155	1108		1536	1100	609	20	1106	1660	834	7150	9008	662
23	1201	7048	605	784		157	900	691	38	1121	110	688	5126	1870	411
26	1249	496	9967	1010		763	572	1085	55	1237	539	1095	9903	441	559
27	67	0	261	34		9	0	59	0	67	35	48	263	0	4
28	96	58	124	71		0	21	78	0	99	0	40	139	44	25
31	1477	735	608	798		3186	290	7	1208	1421	3257	600	1540	660	606
32	767	54	0	838		4324	343	359	274	738	1797	380	1645	166	428
33	506	1008	465	205		1964	41	0	228	505	1887	117	1549	1197	115
34	109	0	20	4		144	0	0	31	64	122	0	129	0	0
35	18	0	0	0		166	0	0	0	18	281	0	83	0	0
41	52	0	0	24		7	0	22	0	52	30	0	59	14	0
44	75	83	0	0		0	0	0	0	0	0	0	75	75	0
46	23	16	0	0		162	149	0	0	23	152	0	33	33	0
48	0	0	0	0		0	19	0	0	0	0	0	0	0	0
Totals	23520	18738	13610	17614		19334	9865	15224	8469	22995	15001	13564	47583	15977	11582

Assem Div	Building Construction	Mechanical	Electrical	Commercial Renovation	Square Foot	Site Work Landscape	Assemblies	Green Building	Interior	Concrete Masonry	Heavy Construction	Light Commercial	Facilities Construction	Plumbing	Asm Div	Residential
A		15	0	188	150	577	598	0	0	536	571	154	24	0	1	368
B		0	0	825	2472	0	5572	56	330	1914	368	2032	143	0	2	211
C		0	0	647	864	0	1233	0	1558	146	0	763	245	0	3	588
D		1067	812	712	1798	72	2425	281	769	0	0	1272	1027	1083	4	859
E		0	0	86	259	0	297	0	5	0	0	256	5	0	5	392
F		0	0	0	122	0	122	0	0	0	0	122	3	0	6	357
G		527	294	317	202	3006	668	0	0	535	1372	200	293	685	7	300
															8	760
															9	80
															10	0
															11	0
															12	0
Totals		1609	1106	2775	5867	3655	10915	337	2662	3131	2311	4799	1740	1768		3915

Reference Section	Building Construction Costs	Mechanical	Electrical	Commercial Renovation	Square Foot	Site Work Landscape	Assem.	Green Building	Interior	Concrete Masonry	Open Shop	Heavy Construction	Light Commercial	Facilities Construction	Plumbing	Resi.
Reference Tables	yes	yes	yes	yes	no	yes	yes	yes	yes	yes	yes	yes	yes	yes	yes	yes
Models					107								46			28
Crews	527	527	527	505		527		527	527	527	503	527	503	505	527	503
Equipment Rental Costs	yes	yes	yes	yes		yes		yes	yes	yes	yes	yes	yes	yes	yes	yes
Historical Cost Indexes	yes	yes	yes	yes		yes		yes	yes	yes	yes	yes	yes	yes	yes	no
City Cost Indexes	yes	yes	yes	yes	yes	yes	yes	yes	yes	yes	yes	yes	yes	yes	yes	yes

884

RSMeans Building Construction Cost Data 2012

Offers you unchallenged unit price reliability in an easy-to-use format. Whether used for verifying complete, finished estimates or for periodic checks, it supplies more cost facts better and faster than any comparable source. Over 23,000 unit prices have been updated for 2012. The City Cost Indexes and Location Factors cover over 930 areas, for indexing to any project location in North America. Order and get *RSMeans Quarterly Update Service* FREE. You'll have year-long access to the RSMeans Estimating **HOTLINE** FREE with your subscription. Expert assistance when using RSMeans data is just a phone call away.

$177.95 per copy | Available Sept. 2011 | Catalog no. 60012

RSMeans Mechanical Cost Data 2012

Total unit and systems price guidance for mechanical construction. . . materials, parts, fittings, and complete labor cost information. Includes prices for piping, heating, air conditioning, ventilation, and all related construction.

Plus new 2012 unit costs for:

- Thousands of installed HVAC/controls assemblies components
- "On-site" Location Factors for over 930 cities and towns in the U.S. and Canada
- Crews, labor, and equipment

$174.95 per copy | Available Oct. 2011 | Catalog no. 60022

RSMeans Square Foot Costs 2012

Accurate and Easy To Use

- **Updated price information** based on nationwide figures from suppliers, estimators, labor experts, and contractors
- "How-to-Use" sections, with **clear examples** of commercial, residential, industrial, and institutional structures
- Realistic graphics, offering true-to-life illustrations of building projects
- Extensive information on using square foot cost data, including sample estimates and alternate pricing methods

$187.95 per copy | Available Oct. 2011 | Catalog no. 60052

RSMeans Green Building Cost Data 2012

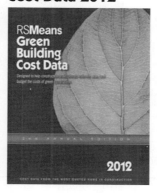

Estimate, plan, and budget the costs of green building for both new commercial construction and renovation work with this first edition of RSMeans Green Building Cost Data. More than 9,000 unit costs for a wide array of green building products plus assemblies costs. Easily identified cross references to LEED and Green Globes building rating systems criteria.

$147.95 per copy | Available Nov. 2011 | Catalog no. 60552

RSMeans Facilities Construction Cost Data 2012

For the maintenance and construction of commercial, industrial, municipal, and institutional properties. Costs are shown for new and remodeling construction and are broken down into materials, labor, equipment, and overhead and profit. Special emphasis is given to sections on mechanical, electrical, furnishings, site work, building maintenance, finish work, and demolition.

More than 47,000 unit costs, plus assemblies costs and a comprehensive Reference Section are included.

$453.95 per copy | Available Nov. 2011 | Catalog no. 60202

RSMeans Commercial Renovation Cost Data 2012

Commercial/Multifamily Residential

Use this valuable tool to estimate commercial and multifamily residential renovation and remodeling.

Includes: New costs for hundreds of unique methods, materials, and conditions that only come up in repair and remodeling, PLUS:

- Unit costs for more than 17,000 construction components
- Installed costs for more than 2,000 assemblies
- More than 930 "on-site" localization factors for the U.S. and Canada

$144.95 per copy | Available Oct. 2011 | Catalog no. 60042

RSMeans Electrical Cost Data 2012

Pricing information for every part of electrical cost planning. More than 13,000 unit and systems costs with design tables; clear specifications and drawings; engineering guides; illustrated estimating procedures; complete labor-hour and materials costs for better scheduling and procurement; and the latest electrical products and construction methods.

- A variety of special electrical systems including cathodic protection
- Costs for maintenance, demolition, HVAC/mechanical, specialties, equipment, and more

$179.95 per copy | Available Oct. 2011 | Catalog no. 60032

RSMeans Electrical Change Order Cost Data 2012

RSMeans Electrical Change Order Cost Data provides you with electrical unit prices exclusively for pricing change orders—based on the recent, direct experience of contractors and suppliers. Analyze and check your own change order estimates against the experience others have had doing the same work. It also covers productivity analysis and change order cost justifications. With useful information for calculating the effects of change orders and dealing with their administration.

$174.95 per copy | Available Dec. 2011 | Catalog no. 60232

RSMeans Assemblies Cost Data 2012

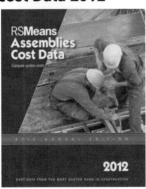

RSMeans Assemblies Cost Data takes the guesswork out of preliminary or conceptual estimates. Now you don't have to try to calculate the assembled cost by working up individual component costs. We've done all the work for you.

Presents detailed illustrations, descriptions, specifications, and costs for every conceivable building assembly—over 350 types in all—arranged in the easy-to-use UNIFORMAT II system. Each illustrated "assembled" cost includes a complete grouping of materials and associated installation costs, including the installing contractor's overhead and profit.

$290.95 per copy | Available Sept. 2011 | Catalog no. 60062

RSMeans Open Shop Building Construction Cost Data 2012

The latest costs for accurate budgeting and estimating of new commercial and residential construction. . . renovation work. . . change orders. . . cost engineering.

RSMeans Open Shop "BCCD" will assist you to:

- Develop benchmark prices for change orders
- Plug gaps in preliminary estimates and budgets
- Estimate complex projects
- Substantiate invoices on contracts
- Price ADA-related renovations

$152.95 per copy | Available Dec. 2011 | Catalog no. 60152

RSMeans Residential Cost Data 2012

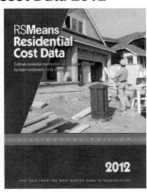

Contains square foot costs for 28 basic home models with the look of today, plus hundreds of custom additions and modifications you can quote right off the page. Includes costs for more than 700 residential systems. Complete with blank estimating forms, sample estimates, and step-by-step instructions.

Now contains line items for cultured stone and brick, PVC trim lumber, and TPO roofing.

$127.95 per copy | Available Oct. 2011 | Catalog no. 60172

RSMeans Site Work & Landscape Cost Data 2012

Includes unit and assemblies costs for earthwork, sewerage, piped utilities, site improvements, drainage, paving, trees and shrubs, street openings/repairs, underground tanks, and more. Contains 78 types of assemblies costs for accurate conceptual estimates.

Includes:

- Estimating for infrastructure improvements
- Environmentally-oriented construction
- ADA-mandated handicapped access
- Hazardous waste line items

$172.95 per copy | Available Dec. 2011 | Catalog no. 60282

RSMeans Facilities Maintenance & Repair Cost Data 2012

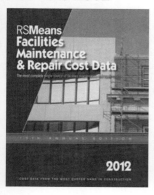

RSMeans Facilities Maintenance & Repair Cost Data gives you a complete system to manage and plan your facility repair and maintenance costs and budget efficiently. Guidelines for auditing a facility and developing an annual maintenance plan. Budgeting is included, along with reference tables on cost and management, and information on frequency and productivity of maintenance operations.

The only nationally recognized source of maintenance and repair costs. Developed in cooperation with the Civil Engineering Research Laboratory (CERL) of the Army Corps of Engineers.

$399.95 per copy | Available Nov. 2011 | Catalog no. 60302

RSMeans Concrete & Masonry Cost Data 2012

Provides you with cost facts for virtually all concrete/masonry estimating needs, from complicated formwork to various sizes and face finishes of brick and block—all in great detail. The comprehensive Unit Price Section contains more than 8,000 selected entries. Also contains an Assemblies [Cost] Section, and a detailed Reference Section that supplements the cost data.

$162.95 per copy | Available Dec. 2011 | Catalog no. 60112

RSMeans Construction Cost Indexes 2012

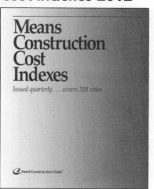

What materials and labor costs will change unexpectedly this year? By how much?

- Breakdowns for 318 major cities
- National averages for 30 key cities
- Expanded five major city indexes
- Historical construction cost indexes

$362.00 per year (subscription) | Catalog no. 50142
$90.50 individual quarters | Catalog no. 60142 A,B,C,D

RSMeans Light Commercial Cost Data 2012

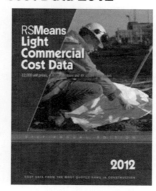

Specifically addresses the light commercial market, which is a specialized niche in the construction industry. Aids you, the owner/designer/contractor, in preparing all types of estimates—from budgets to detailed bids. Includes new advances in methods and materials.

Assemblies Section allows you to evaluate alternatives in the early stages of design/planning.

Over 13,000 unit costs ensure that you have the prices you need. . . when you need them.

$132.95 per copy | Available Nov. 2011 | Catalog no. 60182

RSMeans Labor Rates for the Construction Industry 2012

Complete information for estimating labor costs, making comparisons, and negotiating wage rates by trade for more than 300 U.S. and Canadian cities. With 46 construction trades listed by local union number in each city, and historical wage rates included for comparison. Each city chart lists the county and is alphabetically arranged with handy visual flip tabs for quick reference.

$395.95 per copy | Available Dec. 2011 | Catalog no. 60122

RSMeans Interior Cost Data 2012

Provides you with prices and guidance needed to make accurate interior work estimates. Contains costs on materials, equipment, hardware, custom installations, furnishings, and labor costs. . . for new and remodel commercial and industrial interior construction, including updated information on office furnishings, and reference information.

$182.95 per copy | Available Nov. 2011 | Catalog no. 60092

Annual Cost Guides

For more information visit the RSMeans website at **www.rsmeans.com**

Unit prices according to the latest MasterFormat!

RSMeans Heavy Construction Cost Data 2012

A comprehensive guide to heavy construction costs. Includes costs for highly specialized projects such as tunnels, dams, highways, airports, and waterways. Information on labor rates, equipment, and materials costs is included. Features unit price costs, systems costs, and numerous reference tables for costs and design.

$177.95 per copy | Available Dec. 2011 | Catalog no. 60162

RSMeans Plumbing Cost Data 2012

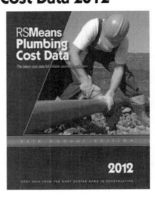

Comprehensive unit prices and assemblies for plumbing, irrigation systems, commercial and residential fire protection, point-of-use water heaters, and the latest approved materials. This publication and its companion, *RSMeans Mechanical Cost Data*, provide full-range cost estimating coverage for all the mechanical trades.

Now contains updated costs for potable water and radiant heat systems and high efficiency fixtures.

$179.95 per copy | Available Oct. 2011 | Catalog no. 60212

2012 RSMeans Seminar Schedule

Note: call for exact dates and details.

Location	Dates	Location	Dates
Las Vegas, NV	March	Las Vegas, NV	October
Washington, DC	April	Orlando, FL	November
San Francisco, CA	June	San Diego, CA	December
Washington, DC	September		
Dallas, TX	September		

1-800-334-3509, ext. 925115

2012 Regional Seminar Schedule

Note: call for exact dates and details.

Location	Dates	Location	Dates
Seattle, WA	January and August	Columbus, GA	June
Dallas/Ft. Worth, TX	January	El Segundo, CA	August
Austin, TX	February	Jacksonville FL	September
Oklahoma City, OK	February	Birmingham, AL	October
Anchorage, AK	March and September	St. Louis, MO	October
Phoenix, AZ	March	Houston, TX	October
San Bernardino, CA	April	Portland, OR	November
Aiea, HI	April	Baltimore, MD	November
Kansas City, MO	April	San Antonio, TX	December
Toronto	May	Raleigh, NC	December
Denver, CO	May		
Bethesda, MD	June		

1-800-334-3509, ext. 925115

Professional Development

Registration Information

Register early... Save up to $100! Register 30 days before the start date of a seminar and save $100 off your total fee. *Note: This discount can be applied only once per order. It cannot be applied to team discount registrations or any other special offer.*

How to register Register by phone today! The RSMeans toll-free number for making reservations is **1-800-334-3509, ext. 925115.**

Individual seminar registration fee - $935. *Means CostWorks®* **training registration fee - $375.** To register by mail, complete the registration form and return, with your full fee, to: RSMeans Seminars, 700 Longwater Drive, Norwell, MA 02061.

Government pricing All federal government employees save off the regular seminar price. Other promotional discounts cannot be combined with the government discount.

Team discount program for two to four seminar registrations. Call for pricing: 1-800-334-3509, ext. 925115

Multiple course discounts When signing up for two or more courses, call for pricing.

Refund policy Cancellations will be accepted up to ten business days prior to the seminar start. There are no refunds for cancellations received later than ten working days prior to the first day of the seminar. A $150 processing fee will be applied for all cancellations. Written notice of cancellation is required. Substitutions can be made at any time before the session starts. **No-shows are subject to the full seminar fee.**

AACE approved courses Many seminars described and offered here have been approved for 14 hours (1.4 recertification credits) of credit by the AACE

International Certification Board toward meeting the continuing education requirements for recertification as a Certified Cost Engineer/Certified Cost Consultant.

AIA Continuing Education We are registered with the AIA Continuing Education System (AIA/CES) and are committed to developing quality learning activities in accordance with the CES criteria. Many seminars meet the AIA/CES criteria for Quality Level 2. AIA members may receive (14) learning units (LUs) for each two-day RSMeans course.

Daily course schedule The first day of each seminar session begins at 8:30 a.m. and ends at 4:30 p.m. The second day begins at 8:00 a.m. and ends at 4:00 p.m. Participants are urged to bring a hand-held calculator, since many actual problems will be worked out in each session.

Continental breakfast Your registration includes the cost of a continental breakfast, and a morning and afternoon refreshment break. These informal segments allow you to discuss topics of mutual interest with other seminar attendees. (You are free to make your own lunch and dinner arrangements.)

Hotel/transportation arrangements RSMeans arranges to hold a block of rooms at most host hotels. To take advantage of special group rates when making your reservation, be sure to mention that you are attending the RSMeans seminar. You are, of course, free to stay at the lodging place of your choice. **(Hotel reservations and transportation arrangements should be made directly by seminar attendees.)**

Important Class sizes are limited, so please register as soon as possible.

Note: Pricing subject to change.

Registration Form

ADDS-1000

Call 1-800-334-3509, ext. 925115 to register or FAX this form 1-800-632-6732. Visit our website: www.rsmeans.com

Please register the following people for the RSMeans construction seminars as shown here. We understand that we must make our own hotel reservations if overnight stays are necessary.

☐ Full payment of $_____enclosed.

☐ Bill me.

Please print name of registrant(s).

(To appear on certificate of completion)

_____

_____

_____

P.O. #:_____
    GOVERNMENT AGENCIES MUST SUPPLY PURCHASE ORDER NUMBER OR TRAINING FORM.

Firm name_____

Address_____

City/State/Zip_____

Telephone no._____ Fax no._____

E-mail address_____

Charge registration(s) to: ☐ MasterCard ☐ VISA ☐ American Express

Account no._____ Exp. date_____

Cardholder's signature_____

Seminar name_____

Seminar City_____

Please mail check to: RSMeans Seminars, 700 Longwater Drive, Norwell, MA 02061 USA

Professional Development

Means CostWorks.com Training

Construction estimating is vital to the decision-making process at each state of every project. Means CostWorks.com works the way you do. It's systematic, flexible and intuitive. In this one day class you will see how you can estimate any phase of any project faster and better.

Some of what you'll learn:

- Customizing Means CostWorks.com
- Making the most of RSMeans "Circle Reference" numbers
- How to integrate your cost data
- Generate reports, exporting estimates to MS Excel, sharing, collaborating and more

Who should attend: architects & engineers, facility managers, contractors, construction estimators, and anyone using this powerful web-based tool.

Facilities Construction Estimating

In this *two-day* course, professionals working in facilities management can get help with their daily challenges to establish budgets for all phases of a project.

Some of what you'll learn:

- Determining the full scope of a project
- Identifying the scope of risks and opportunities
- Creative solutions to estimating issues
- Organizing estimates for presentation and discussion
- Special techniques for repair/remodel and maintenance projects
- Negotiating project change orders

Who should attend: facility managers, engineers, contractors, facility tradespeople, planners, and project managers.

Scheduling with Microsoft Project

Two days of hands-on training gives you a look at the basics to putting it all together with MS Project to create a complete project schedule. Learn to work better and faster, manage changes and updates, enhance tracking, generate actionable reports, and boost control of your time and budget.

Some of what you'll learn:

- Defining task/activity relationships: link tasks, establish predecessor/successor relationships and create logs
- Baseline scheduling: the essential what/when
- Using MS Project to determine the critical path
- Integrate RSMeans data with MS Project
- The dollar-loaded schedule: enhance the reliability of this "must-know" technique for scheduling public projects.

Who should attend: Contractors' project-management teams, architects and engineers, project owners and their representatives, and those interested in improving their project planning and scheduling and management skills.

Site Work & Heavy Construction Estimating

This *two-day* course teaches attendees how to estimate earthwork, site utilities, foundations, and site improvements, using the assemblies and the unit price methods.

Some of what you'll learn:

- Basic site work and heavy construction estimating skills
- Estimating foundations, utilities, earthwork, and site improvements
- Correct equipment usage, quality control, and site investigation for estimating purposes

Who should attend: project managers, design engineers, estimators, and contractors doing site work and heavy construction.

Practical Project Management for Construction Professionals

In this *two-day* course, acquire the essential knowledge and develop the skills to effectively and efficiently execute the day-to-day responsibilities of the construction project manager.

Covers:

- General conditions of the construction contract
- Contract modifications: change orders and construction change directives
- Negotiations with subcontractors and vendors
- Effective writing: notification and communications
- Dispute resolution: claims and liens

Who should attend: architects, engineers, owner's representatives, project managers.

Mechanical & Electrical Estimating

This *two-day* course teaches attendees how to prepare more accurate and complete mechanical/electrical estimates, avoiding the pitfalls of omission and double-counting, while understanding the composition and rationale within the RSMeans mechanical/electrical database.

Some of what you'll learn:

- The unique way mechanical and electrical systems are interrelated
- M&E estimates—conceptual, planning, budgeting, and bidding stages
- Order of magnitude, square foot, assemblies, and unit price estimating
- Comparative cost analysis of equipment and design alternatives

Who should attend: architects, engineers, facilities managers, mechanical and electrical contractors, and others who need a highly reliable method for developing, understanding, and evaluating mechanical and electrical contracts.

Professional Development

Unit Price Estimating

This interactive *two-day* seminar teaches attendees how to interpret project information and process it into final, detailed estimates with the greatest accuracy level.

The most important credential an estimator can take to the job is the ability to visualize construction and estimate accurately.

Some of what you'll learn:
- Interpreting the design in terms of cost
- The most detailed, time-tested methodology for accurate pricing
- Key cost drivers—material, labor, equipment, staging, and subcontracts
- Understanding direct and indirect costs for accurate job cost accounting and change order management

Who should attend: Corporate and government estimators and purchasers, architects, engineers, and others who need to produce accurate project estimates.

RSMeans CostWorks® CD Training

This one-day course helps users become more familiar with the functionality of *RSMeans CostWorks* program. Each menu, icon, screen, and function found in the program is explained in depth. Time is devoted to hands-on estimating exercises.

Some of what you'll learn:
- Searching the database using all navigation methods
- Exporting RSMeans data to your preferred spreadsheet format
- Viewing crews, assembly components, and much more
- Automatically regionalizing the database

You are required to bring your own laptop computer for this course.

When you register for this course you will receive an outline for your laptop requirements.

Also offering web training for CostWorks CD!

Facilities Maintenance & Repair Estimating

This *two-day* course teaches attendees how to plan, budget, and estimate the cost of ongoing and preventive maintenance and repair for existing buildings and grounds.

Some of what you'll learn:
- The most financially favorable maintenance, repair, and replacement scheduling and estimating
- Auditing and value engineering facilities
- Preventive planning and facilities upgrading
- Determining both in-house and contract-out service costs
- Annual, asset-protecting M&R plan

Who should attend: Facility managers, maintenance supervisors, buildings and grounds superintendents, plant managers, planners, estimators, and others involved in facilities planning and budgeting.

Square Foot Estimating using Means CostWorks

This *two day* class uses the leading industry data and a powerful software package to develop highly accurate conceptual estimates for your construction projects. All attendees must bring a laptop computer loaded with the current year *Square Foot Model Costs* and the *Assemblies Cost Data* CostWorks titles.

Some of what you'll learn:
- Introduction to conceptual estimating
- Types of conceptual estimates
- Helpful hints
- Order of magnitude estimating
- Square foot estimating
- Assemblies estimating

Who should attend: architects, engineers, contractors, construction estimators, owner's representatives, and anyone looking for an electronic method for performing square foot estimating.

Assessing Scope of Work for Facility Construction Estimating

This *two-day* practical training program addresses the vital importance of understanding the SCOPE of projects in order to produce accurate cost estimates for facility repair and remodeling.

Some of what you'll learn:
- Discussions of site visits, plans/specs, record drawings of facilities, and site-specific lists
- Review of CSI divisions, including means, methods, materials, and the challenges of scoping each topic
- Exercises in SCOPE identification and SCOPE writing for accurate estimating of projects
- Hands-on exercises that require SCOPE, take-off, and pricing

Who should attend: corporate and government estimators, planners, facility managers, and others who need to produce accurate project estimates.

Reference Books

For more information visit the RSMeans website at **www.rsmeans.com**

Value Engineering: Practical Applications

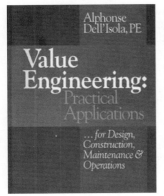

For Design, Construction, Maintenance & Operations
by Alphonse Dell'Isola, PE

A tool for immediate application—for engineers, architects, facility managers, owners, and contractors. Includes making the case for VE—the management briefing; integrating VE into planning, budgeting, and design; conducting life cycle costing; using VE methodology in design review and consultant selection; case studies; VE workbook; and a life cycle costing program on disk.

$79.95 per copy | Over 450 pages, illustrated, softcover | Catalog no. 67319A

The Building Professional's Guide to Contract Documents

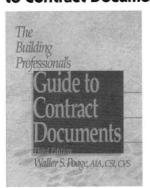

3rd Edition
by Waller S. Poage, AIA, CSI, CVS

A comprehensive reference for owners, design professionals, contractors, and students

- Structure your documents for maximum efficiency.
- Effectively communicate construction requirements.
- Understand the roles and responsibilities of construction professionals.
- Improve methods of project delivery.

$64.95 per copy | 400 pages Diagrams and construction forms, hardcover
Catalog no. 67261A

Cost Planning & Estimating for Facilities Maintenance

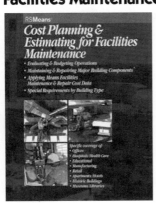

In this unique book, a team of facilities management authorities shares their expertise on:

- Evaluating and budgeting maintenance operations
- Maintaining and repairing key building components
- Applying *RSMeans Facilities Maintenance & Repair Cost Data* to your estimating

Covers special maintenance requirements of the ten major building types

$89.95 per copy | Over 475 pages, hardcover | Catalog no. 67314

Facilities Operations & Engineering Reference

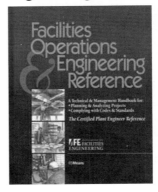

by the Association for Facilities Engineering and RSMeans

An all-in-one technical reference for planning and managing facility projects and solving day-to-day operations problems. Selected as the official certified plant engineer reference, this handbook covers financial analysis, maintenance, HVAC and energy efficiency, and more.

$109.95 per copy | Over 700 pages, illustrated, hardcover | Catalog no. 67318

Complete Book of Framing

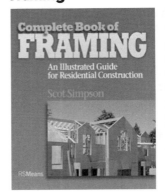

by Scot Simpson

This straightforward, easy-to-learn method will help framers, carpenters, and handy homeowners build their skills in rough carpentry and framing. Shows how to frame all the parts of a house: floors, walls, roofs, door & window openings, and stairs—with hundreds of color photographs and drawings that show every detail.

The book gives beginners all the basics they need to go from zero framing knowledge to a journeyman level.

$29.95 per copy | 336 pages, softcover | Catalog no. 67353

Life Cycle Costing for Facilities

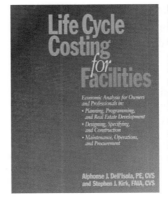

by Alphonse Dell'Isola and Dr. Steven Kirk

Guidance for achieving higher quality design and construction projects at lower costs! Cost-cutting efforts often sacrifice quality to yield the cheapest product. Life cycle costing enables building designers and owners to achieve both. The authors of this book show how LCC can work for a variety of projects — from roads to HVAC upgrades to different types of buildings.

$99.95 per copy | 396 pages, hardcover | Catalog no. 67341

Reference Books

For more information visit the RSMeans website at **www.rsmeans.com**

Interior Home Improvement Costs

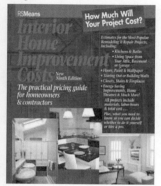

9th Edition

Updated estimates for the most popular remodeling and repair projects—from small, do-it-yourself jobs to major renovations and new construction. Includes: kitchens & baths; new living space from your attic, basement, or garage; new floors, paint, and wallpaper; tearing out or building new walls; closets, stairs, and fireplaces; new energy-saving improvements, home theaters, and more!

$24.95 per copy | 250 pages, illustrated, softcover | Catalog no. 67308E

Exterior Home Improvement Costs

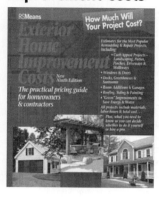

9th Edition

Updated estimates for the most popular remodeling and repair projects—from small, do-it-yourself jobs, to major renovations and new construction. Includes: curb appeal projects—landscaping, patios, porches, driveways, and walkways; new windows and doors; decks, greenhouses, and sunrooms; room additions and garages; roofing, siding, and painting; "green" improvements to save energy & water.

$24.95 per copy | Over 275 pages, illustrated, softcover | Catalog no. 67309E

Builder's Essentials: Plan Reading & Material Takeoff

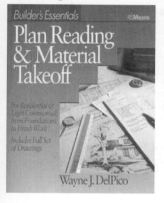

For Residential and Light Commercial Construction
by Wayne J. DelPico

A valuable tool for understanding plans and specs, and accurately calculating material quantities. Step-by-step instructions and takeoff procedures based on a full set of working drawings.

$35.95 per copy | Over 420 pages, softcover | Catalog no. 67307

Means Unit Price Estimating Methods

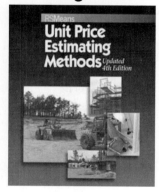

New 4th Edition

This new edition includes up-to-date cost data and estimating examples, updated to reflect changes to the CSI numbering system and new features of RSMeans cost data. It describes the most productive, universally accepted ways to estimate, and uses checklists and forms to illustrate shortcuts and timesavers. A model estimate demonstrates procedures. A new chapter explores computer estimating alternatives.

$59.95 per copy | Over 350 pages, illustrated, softcover | Catalog no. 67303B

Concrete Repair and Maintenance Illustrated

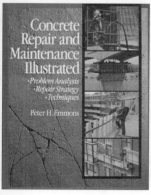

by Peter Emmons

Hundreds of illustrations show users how to analyze, repair, clean, and maintain concrete structures for optimal performance and cost effectiveness. From parking garages to roads and bridges to structural concrete, this comprehensive book describes the causes, effects, and remedies for concrete wear and failure. Invaluable for planning jobs, selecting materials, and training employees, this book is a must-have for concrete specialists, general contractors, facility managers, civil and structural engineers, and architects.

$69.95 per copy | 300 pages, illustrated, softcover | Catalog no. 67146

Total Productive Facilities Management

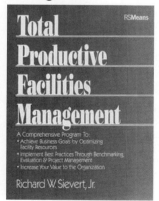

by Richard W. Sievert, Jr.

Today, facilities are viewed as strategic resources. . . elevating the facility manager to the role of asset manager supporting the organization's overall business goals. Now, Richard Sievert Jr., in this well-articulated guidebook, sets forth a new operational standard for the facility manager's emerging role. . . a comprehensive program for managing facilities as a true profit center.

$79.95 per copy | 275 pages, softcover | Catalog no. 67321

Means Illustrated Construction Dictionary

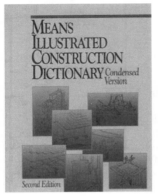

Condensed, 2nd Edition

The best portable dictionary for office or field use—an essential tool for contractors, architects, insurance and real estate personnel, facility managers, homeowners, and anyone who needs quick, clear definitions for construction terms. The second edition has been further enhanced with updates and hundreds of new terms and illustrations . . . in keeping with the most recent developments in the construction industry.

Now with a quick-reference Spanish section. Includes tools and equipment, materials, tasks, and more!

$59.95 per copy | Over 500 pages, softcover | Catalog no. 67282A

Facilities Planning & Relocation

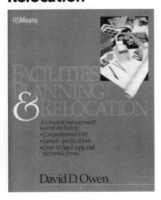

by David D. Owen

A complete system for planning space needs and managing relocations. Includes step-by-step manual, over 50 forms, and extensive reference section on materials and furnishings.

New lower price and user-friendly format.

$89.95 per copy | Over 450 pages, softcover | Catalog no. 67301

Electrical Estimating Methods

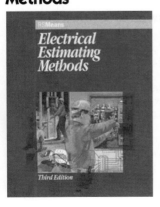

3rd Edition

Expanded edition includes sample estimates and cost information in keeping with the latest version of the CSI MasterFormat and UNIFORMAT II. Complete coverage of fiber optic and uninterruptible power supply electrical systems, broken down by components, and explained in detail. Includes a new chapter on computerized estimating methods. A practical companion to *RSMeans Electrical Cost Data.*

$64.95 per copy | Over 325 pages, hardcover | Catalog no. 67230B

Square Foot & UNIFORMAT Assemblies Estimating Methods

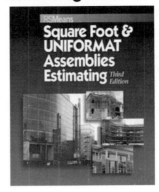

3rd Edition

Develop realistic square foot and assemblies costs for budgeting and construction funding. The new edition features updated guidance on square foot and assemblies estimating using UNIFORMAT II. An essential reference for anyone who performs conceptual estimates.

$69.95 per copy | Over 300 pages, illustrated, softcover | Catalog no. 67145B

Mechanical Estimating Methods

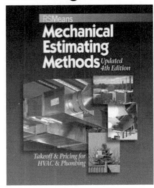

4th Edition

Completely updated, this guide assists you in making a review of plans, specs, and bid packages, with suggestions for takeoff procedures, listings, substitutions, and pre-bid scheduling for all components of HVAC. Includes suggestions for budgeting labor and equipment usage. Compares materials and construction methods to allow you to select the best option.

$64.95 per copy | Over 350 pages, illustrated, softcover | Catalog no. 67294B

Residential & Light Commercial Construction Standards

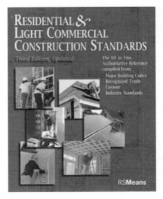

3rd Edition, Updated by RSMeans and contributing authors

This book provides authoritative requirements and recommendations compiled from leading professional associations, industry publications, and building code organizations. This all-in-one reference helps establish a standard for workmanship, quickly resolve disputes, and avoid defect claims. Updated third edition includes new coverage of green building, seismic, hurricane, and mold-resistant construction.

$59.95 | Over 550 pages, illustrated, softcover | Catalog no. 67322B

Construction Business Management

by Nick Ganaway

Only 43% of construction firms stay in business after four years. Make sure your company thrives with valuable guidance from a pro with 25 years of success as a commercial contractor. Find out what it takes to build all aspects of a business that is profitable, enjoyable, and enduring. With a bonus chapter on retail construction.

$49.95 per copy | 200 pages, softcover | Catalog no. 67352

The Practice of Cost Segregation Analysis

by Bruce A. Desrosiers and Wayne J. DelPico

This expert guide walks you through the practice of cost segregation analysis, which enables property owners to defer taxes and benefit from "accelerated cost recovery" through depreciation deductions on assets that are properly identified and classified.

With a glossary of terms, sample cost segregation estimates for various building types, key information resources, and updates via a dedicated website, this book is a critical resource for anyone involved in cost segregation analysis.

$99.95 per copy | Over 225 pages | Catalog no. 67345

How to Estimate with Means Data & CostWorks

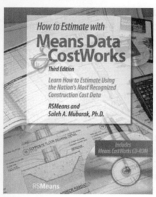

New 3rd Edition
by RSMeans and
Saleh A. Mubarak, Ph.D.

New 3rd Edition—fully updated with new chapters, plus new CD with updated *CostWorks* cost data and MasterFormat organization. Includes all major construction items—with more than 300 exercises and two sets of plans that show how to estimate for a broad range of construction items and systems—including general conditions and equipment costs.

$59.95 per copy | 272 pages, softcover | Includes CostWorks CD
Catalog no. 67324B

Project Scheduling & Management for Construction

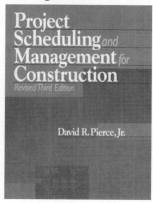

3rd Edition
by David R. Pierce, Jr.

A comprehensive yet easy-to-follow guide to construction project scheduling and control—from vital project management principles through the latest scheduling, tracking, and controlling techniques. The author is a leading authority on scheduling, with years of field and teaching experience at leading academic institutions. Spend a few hours with this book and come away with a solid understanding of this essential management topic.

$64.95 per copy | Over 300 pages, illustrated, hardcover | Catalog no. 67247B

Preventive Maintenance for Multi-Family Housing

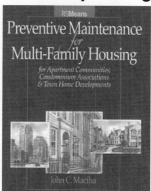

by John C. Maciha

Prepared by one of the nation's leading experts on multi-family housing.

This complete PM system for apartment and condominium communities features expert guidance, checklists for buildings and grounds maintenance tasks and their frequencies, a reusable wall chart to track maintenance, and a dedicated website featuring customizable electronic forms. A must-have for anyone involved with multi-family housing maintenance and upkeep.

$89.95 per copy | 225 pages | Catalog no. 67346

Job Order Contracting Expediting Construction Project Delivery

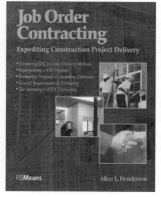

by Allen Henderson

Expert guidance to help you implement JOC—fast becoming the preferred project delivery method for repair and renovation, minor new construction, and maintenance projects in the public sector and in many states and municipalities. The author, a leading JOC expert and practitioner, shows how to:

- Establish a JOC program
- Evaluate proposals and award contracts
- Handle general requirements and estimating
- Partner for maximum benefits

$89.95 per copy | 192 pages, illustrated, hardcover | Catalog no. 67348

Reference Books

For more information visit the RSMeans website at **www.rsmeans.com**

Estimating Building Costs

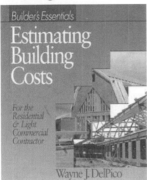

For the Residential & Light Commercial Contractor
by Wayne J. DelPico

Step-by-step estimating methods for residential and light commercial contractors. Includes a detailed look at every construction specialty—explaining all the components, takeoff units, and labor needed for well-organized, complete estimates. Covers correctly interpreting plans and specifications, and developing accurate and complete labor and material costs.

$29.95 per copy | Over 400 pages, illustrated, softcover | Catalog no. 67343

The Homeowner's Guide to Mold

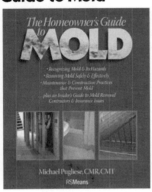

By Michael Pugliese

Expert guidance to protect your health and your home.

Mold, whether caused by leaks, humidity or flooding, is a real health and financial issue—for homeowners and contractors.

This full-color book explains:

- Construction and maintenance practices to prevent mold
- How to inspect for and remove mold
- Mold remediation procedures and costs
- What to do after a flood
- How to deal with insurance companies

$21.95 per copy | 144 pages, softcover | Catalog no. 67344

Landscape Estimating Methods

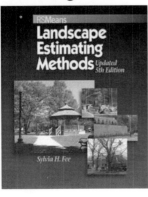

5th Edition

Answers questions about preparing competitive landscape construction estimates, with up-to-date cost estimates and the new MasterFormat classification system. Expanded and revised to address the latest materials and methods, including new coverage on approaches to green building. Includes:

- Step-by-step explanation of the estimating process
- Sample forms and worksheets that save time and prevent errors

$64.95 per copy | Over 350 pages, softcover | Catalog no. 67295C

Building & Renovating Schools

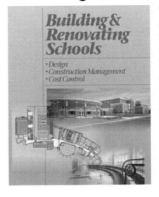

This all-inclusive guide covers every step of the school construction process—from initial planning, needs assessment, and design, right through moving into the new facility. A must-have resource for anyone concerned with new school construction or renovation. With square foot cost models for elementary, middle, and high school facilities, and real-life case studies of recently completed school projects.

The contributors to this book—architects, construction project managers, contractors, and estimators who specialize in school construction—provide start-to-finish, expert guidance on the process.

$99.95 per copy | Over 425 pages, hardcover | Catalog no. 67342

Universal Design Ideas for Style, Comfort & Safety

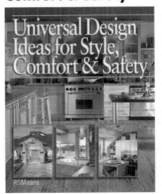

by RSMeans and Lexicon Consulting, Inc.

Incorporating universal design when building or remodeling helps people of any age and physical ability more fully and safely enjoy their living spaces. This book shows how universal design can be artfully blended into the most attractive homes. It discusses specialized products like adjustable countertops and chair lifts, as well as simple ways to enhance a home's safety and comfort. With color photos and expert guidance, every area of the home is covered. Includes budget estimates that give an idea how much projects will cost.

$21.95 | 160 pages, illustrated, softcover | Catalog no. 67354

Plumbing Estimating Methods

3rd Edition
by Joseph Galeno and Sheldon Greene

Updated and revised! This practical guide walks you through a plumbing estimate, from basic materials and installation methods through change order analysis. *Plumbing Estimating Methods* covers residential, commercial, industrial, and medical systems, and features sample takeoff and estimate forms and detailed illustrations of systems and components.

$59.95 per copy | 330+ pages, softcover | Catalog no. 67283B

Understanding & Negotiating Construction Contracts

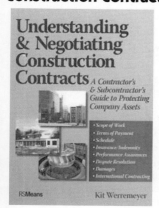

by Kit Werremeyer

Take advantage of the author's 30 years' experience in small-to-large (including international) construction projects. Learn how to identify, understand, and evaluate high risk terms and conditions typically found in all construction contracts—then negotiate to lower or eliminate the risk, improve terms of payment, and reduce exposure to claims and disputes. The author avoids "legalese" and gives real-life examples from actual projects.

$69.95 per copy | 300 pages, softcover | Catalog no. 67350

How Your House Works

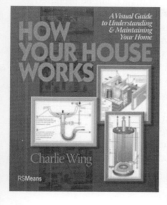

by Charlie Wing

A must-have reference for every homeowner, handyman, and contractor—for repair, remodeling, and new construction. This book uncovers the mysteries behind just about every major appliance and building element in your house—from electrical, heating and AC, to plumbing, framing, foundations and appliances. Clear, "exploded" drawings show exactly how things should be put together and how they function—what to check if they don't work, and what you can do that might save you having to call in a professional.

$21.95 per copy | 160 pages, softcover | Catalog no. 67351

RSMeans Estimating Handbook

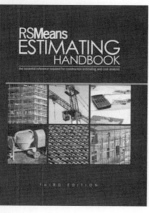

3rd Edition

Widely used in the industry for tasks ranging from routine estimates to special cost analysis projects, this handbook has been completely updated and reorganized with new and expanded technical information.

RSMeans Estimating Handbook will help construction professionals:

- Evaluate architectural plans and specifications
- Prepare accurate quantity takeoffs
- Compare design alternatives and costs
- Perform value engineering
- Double-check estimates and quotes
- Estimate change orders

$99.95 per copy | Over 900 pages, hardcover | Catalog No. 67276B

Spanish/English Construction Dictionary

2nd Edition
by RSMeans and the International Code Council

This expanded edition features thousands of the most common words and useful phrases in the construction industry with easy-to-follow pronunciations in both Spanish and English. Over 800 new terms, phrases, and illustrations have been added. It also features a new stand-alone "Safety & Emergencies" section, with colored pages for quick access.

Unique to this dictionary are the systems illustrations showing the relationship of components in the most common building systems for all major trades.

$23.95 per copy | Over 400 pages | Catalog no. 67327A

The Gypsum Construction Handbook

by USG

An invaluable reference of construction procedures for gypsum drywall, cement board, veneer plaster, and conventional plaster. This new edition includes the newest product developments, installation methods, fire- and sound-rated construction information, illustrated framing-to-finish application instructions, estimating and planning information, and more. Great for architects and engineers; contractors, builders, and dealers; apprentices and training programs; building inspectors and code officials; and anyone interested in gypsum construction. Features information on tools and safety practices, a glossary of construction terms, and a list of important agencies and associations. Also available in Spanish.

$29.95 | Over 575 pages, illustrated, softcover | Catalog no. 67357A

Means Illustrated Construction Dictionary

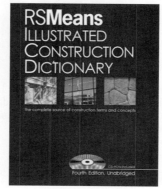

Unabridged 4th Edition, with CD-ROM

Long regarded as the industry's finest, *Means Illustrated Construction Dictionary* is now even better. With nearly 20,000 terms and more than 1,400 illustrations and photos, it is the clear choice for the most comprehensive and current information. The companion CD-ROM that comes with this new edition adds extra features such as: larger graphics and expanded definitions.

$99.95 per copy | Over 790 pages, illust., hardcover | Catalog no. 67292B

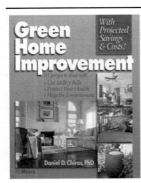

Green
Home Improvement

by Daniel D. Chiras, PhD

With energy costs rising and environmental awareness increasing, people are looking to make their homes greener. This book, with 65 projects and actual costs and projected savings help homeowners prioritize their green improvements.

Projects range from simple water savers that cost only a few dollars, to bigger-ticket items such as HVAC systems. With color photos and cost estimates, each project compares options and describes the work involved, the benefits, and the savings.

$24.95 | 320 pages, illustrated, softcover | Catalog no. 67355

New Titles

Construction Supervision

This brand new title inspires supervisory excellence with proven tactics and techniques applied by thousands of construction supervisors over the past decade. Recognizing the unique and critical role the supervisor plays in project success, the book's leadership guidelines carve out a practical blueprint for motivating work performance and increasing productivity through effective communication.

Features:

• A unique focus on field supervision and crew management

• Coverage of supervision from the foreman to the superintendent level

• An overview of technical skills whose mastery will build confidence and success for the supervisor

• A detailed view of "soft" management and communication skills

$90.00 per copy | Over 450 pages, softcover | Catalog no. 67358

Risk Management for Design and Construction

This book introduces risk as a central pillar of project management and shows how a project manager can be prepared for dealing with uncertainty. Written by experts in the field, *Risk Management for Design and Construction* uses clear, straightforward terminology to demystify the concepts of project uncertainty and risk.

Highlights include:

• Integrated cost and schedule risk analysis

• An introduction to a ready-to-use system of analyzing a project's risks and tools to proactively manage risks

• A methodology that was developed and used by the Washington State Department of Transportation

• Case studies and examples on the proper application of principles

• Information about combining value analysis with risk analysis

$125.00 per copy | Over 250 pages, softcover | Catalog no. 67359

Green Building: Project Planning & Cost Estimating

3rd Edition

Since the widely read first edition of this book, green building has gone from a growing trend to a major force in design and construction.

This new edition has been updated with the latest in green building technologies, design concepts, standards, and costs. Full-color with all new case studies—plus a new chapter on commercial real estate.

$99.95 per copy | Over 450 pages, softcover | Catalog no. 67338B

2012 Order Form

For more information visit the RSMeans website at **www.rsmeans.com**

Qty.	Book no.	COST ESTIMATING BOOKS	Unit Price	Total
	60062	Assemblies Cost Data 2012	$290.95	
	60012	Building Construction Cost Data 2012	177.95	
	60112	Concrete & Masonry Cost Data 2012	162.95	
	50142	Construction Cost Indexes 2012 (subscription)	362.00	
	60142A	Construction Cost Index–January 2012	90.50	
	60142B	Construction Cost Index–April 2012	90.50	
	60142C	Construction Cost Index–July 2012	90.50	
	60142D	Construction Cost Index–October 2012	90.50	
	60342	Contr. Pricing Guide: Resid. R & R Costs 2012	39.95	
	60232	Electrical Change Order Cost Data 2012	174.95	
	60032	Electrical Cost Data 2012	179.95	
	60202	Facilities Construction Cost Data 2012	453.95	
	60302	Facilities Maintenance & Repair Cost Data 2012	399.95	
	60552	Green Building Cost Data 2012	147.95	
	60162	Heavy Construction Cost Data 2012	177.95	
	60092	Interior Cost Data 2012	182.95	
	60122	Labor Rates for the Const. Industry 2012	395.95	
	60182	Light Commercial Cost Data 2012	132.95	
	60022	Mechanical Cost Data 2012	174.95	
	60152	Open Shop Building Const. Cost Data 2012	152.95	
	60212	Plumbing Cost Data 2012	179.95	
	60042	Commercial Renovation Cost Data 2012	144.95	
	60172	Residential Cost Data 2012	127.95	
	60282	Site Work & Landscape Cost Data 2012	172.95	
	60052	Square Foot Costs 2012	187.95	
	62011	Yardsticks for Costing (2011)	174.95	
	62012	Yardsticks for Costing (2012)	176.95	
		REFERENCE BOOKS		
	67329	Bldrs Essentials: Best Bus. Practices for Bldrs	29.95	
	67298A	Bldrs Essentials: Framing/Carpentry 2nd Ed.	24.95	
	67298AS	Bldrs Essentials: Framing/Carpentry Spanish	24.95	
	67307	Bldrs Essentials: Plan Reading & Takeoff	35.95	
	67342	Building & Renovating Schools	99.95	
	67339	Building Security: Strategies & Costs	89.95	
	67353	Complete Book of Framing	29.95	
	67146	Concrete Repair & Maintenance Illustrated	69.95	
	67352	Construction Business Management	49.95	
	67358	Construction Supervision	90.00	
	67314	Cost Planning & Est. for Facil. Maint.	89.95	
	67328A	Designing & Building with the IBC, 2nd Ed.	99.95	
	67230B	Electrical Estimating Methods, 3rd Ed.	64.95	
	67343	Estimating Bldg. Costs for Resi. & Lt. Comm.	29.95	
	67276B	Estimating Handbook, 3rd Ed.	99.95	
	67318	Facilities Operations & Engineering Reference	109.95	
	67301	Facilities Planning & Relocation	89.95	
	67338B	Green Building: Proj. Planning & Cost Est., 3rd Ed.	99.95	
	67355	Green Home Improvement	24.95	
	67357A	The Gypsum Construction Handbook	29.95	
	67357S	The Gypsum Construction Handbook (Spanish)	29.95	
	67148	Heavy Construction Handbook	99.95	
	67323	Historic Preservation: Proj. Planning & Est.	99.95	

Qty.	Book no.	REFERENCE BOOKS (Cont.)	Unit Price	Total
	67308E	Home Improvement Costs–Int. Projects, 9th Ed.	24.95	
	67309E	Home Improvement Costs–Ext. Projects, 9th Ed.	24.95	
	67344	Homeowner's Guide to Mold	21.95	
	67324B	How to Est.w/Means Data & CostWorks, 3rd Ed.	59.95	
	67351	How Your House Works	21.95	
	67282A	Illustrated Const. Dictionary, Condensed, 2nd Ed.	59.95	
	67292B	Illustrated Const. Dictionary, w/CD-ROM, 4th Ed.	99.95	
	67348	Job Order Contracting	89.95	
	67295C	Landscape Estimating Methods, 5th Ed.	64.95	
	67341	Life Cycle Costing for Facilities	99.95	
	67294B	Mechanical Estimating Methods, 4th Ed.	64.95	
	67283B	Plumbing Estimating Methods, 3rd Ed.	59.95	
	67345	Practice of Cost Segregation Analysis	99.95	
	67346	Preventive Maint. for Multi-Family Housing	89.95	
	67326	Preventive Maint. Guidelines for School Facil.	149.95	
	67247B	Project Scheduling & Manag. for Constr. 3rd Ed.	64.95	
	67322B	Resi. & Light Commercial Const. Stds., 3rd Ed.	59.95	
	67359	Risk Management for Design & Construction	125.00	
	67327A	Spanish/English Construction Dictionary, 2nd Ed.	23.95	
	67145B	Sq. Ft. & Assem. Estimating Methods, 3rd Ed.	69.95	
	67321	Total Productive Facilities Management	79.95	
	67350	Understanding and Negotiating Const. Contracts	69.95	
	67303B	Unit Price Estimating Methods, 4th Ed.	59.95	
	67354	Universal Design	21.95	
	67319A	Value Engineering: Practical Applications	79.95	

MA residents add 6.25% state sales tax

Shipping & Handling**

Total (U.S. Funds)*

Prices are subject to change and are for U.S. delivery only. *Canadian customers may call for current prices. **Shipping & handling charges: Add 7% of total order for check and credit card payments. Add 9% of total order for invoiced orders.

Send order to: ADDV-1000

Name (please print) _____

Company _____

☐ Company

☐ Home Address _____

_____

City/State/Zip _____

Phone # _____ P.O. # _____

(Must accompany all orders being billed)

Mail to: RSMeans, 700 Longwater Drive, Norwell, MA 02061

RSMeans Project Cost Report

By filling out this report, your project data will contribute to the database that supports the RSMeans Project Cost Square Foot Data. When you fill out this form, RSMeans will provide a $30 discount off one of the RSMeans products advertised in the preceding pages. Please complete the form including all items where you have cost data, and all the items marked (☑).

$30.00 Discount per product for each report you submit

Project Description (no remodeling projects, please)

☑ Building Use (Office School...)_____

☑ Address (City, State)_____

_____

☑ Total Building Area (SF) _____

☑ Ground Floor (SF) _____

☑ Frame (Wood, Steel...) _____

☑ Exterior Wall (Brick, Tilt-up...) _____

☑ Basement: (check one) ☐ Full   ☐ Partial   ☐ None

☑ Number of Stories _____

☑ Floor-to-Floor Height_____

☑ Volume (C.F.)_____

% Air Conditioned _____ Tons_____

Total Project Cost $ _____

Owner _____

Architect_____

General Contractor _____

☑ Bid Date _____

☑ Typical Bay Size _____

☑ Capacity _____

☑ Labor Force: _____ % Union   _____ % Non-Union

☑ Project Description (Circle one number in each line.)

1. Economy   2. Average   3. Custom   4. Luxury

1. Square   2. Rectangular 3. Irregular 4. Very Irregular

Comments _____

_____

A	☑	General Conditions	$
B	☑	Site Work	$
C	☑	Concrete	$
D	☑	Masonry	$
E	☑	Metals	$
F	☑	Wood & Plastics	$
G	☑	Thermal &Moisture Protection	$
GR		Roofing & Flashing	$
H	☑	Doors and Windows	$
J	☑	Finishes	$
JP		Painting & Wall Covering	$

K	☑	Specialties	$
L	☑	Equipment	$
M	☑	Furnishings	$
N	☑	SpecialConstruction	$
P	☑	ConveyingSystems	$
Q	☑	Mechanical	$
QP		Plumbing	$
QB		HVAC	$
R	☑	Electrical	$
S	☑	Mech./Elec. Combined	$

Please specify the RSMeans product you wish to receive. Complete the address information.

Product Name _____

Product Number _____

Your Name _____

Title _____

Company_____
  ☐ Company
  ☐ Home   Street Address_____

City, State, Zip _____

Return by mail or fax 888-492-6770.

Method of Payment:

Credit Card # _____

Expiration Date _____

Check_____

Purchase Order _____

Phone Number _____
(for those not comfortable mailing in credit card information)

RSMeans
a division of Reed Construction Data
Square Foot Costs Department
700 Longwater Drive
Norwell, MA 02061